Studies in Continental Margin Geology

AAPG Memoir No. 34

Studies in Continental Margin Geology

AAPG Memoir No. 34

Edited by

J.S. Watkins and C.L. Drake

Published by
The American Association of Petroleum Geologists
Tulsa, Oklahoma 74101, U.S.A.

Published July 1983

ISBN: 0-89181-311-X
Library of Congress Catalog Card No. 83-070685

For the Association:
Elected Editor, M.K. Horn
Science Director, E.A. Beaumont
Project Editor, K.M. Stevenson

Dedication

It is appropriate that we dedicate this volume to Hollis Hedberg, one of the world's most influential advocates of the exploration of continental margins. In the 1950s and 1960s, he involved Gulf Oil in the exploration of the West African margin with resultant discoveries in Nigeria and Angola. He initiated the R/V *Gulfrex* program which collected marine seismic, gravity, magnetic, and geochemical data from margins throughout the world. The success of the *Gulfrex* program prompted Gulf to build a newer and larger vessel with greater capabilities. It was appropriately named the R/V *Hollis Hedberg*. Following his retirement from Gulf in 1964, Hollis continued to aggressively support offshore exploration and share his knowledge with students at Princeton.

In addition to his many contributions to Gulf Oil's exploration program, Hollis has served his science well. He has been a member of many committees and panels. He was chairman of the AMSOC Committee, forerunner of the Deep Sea Drilling project, first chairman of the Deep Sea Drilling Project's Safety Panel, and chairman of both U.S. and International Stratigraphic Commissions. Most recently he has aggressively sought rational resolution of problems of Law of the Sea and U.S. offshore leasing. His many publications in the Bulletin of the American Association of Petroleum Geologists and elsewhere have brought him international recognition as a scientist. He has been honored for his many contributions to geology and petroleum exploration. These honors include the Penrose Medal from the Geological Society of America in 1980 (which he served as President in 1959-60), the Sidney Powers Medal in 1963, the Human Needs Award in 1973 from the American Association of Petroleum Geologists, and the Wollaston Medal from the Geological Society of London in 1975, among others. With warm affection and appreciation of his service to geology, asea and ashore, we dedicate this volume to a truly fine gentleman and geologist, Hollis Hedberg.

Table of Contents:

Field Investigations of Margin Structure and Stratigraphy

Model Investigations of Margin Environmental and Tectonic Processes

Foreword

This volume contains papers from a research conference co-sponsored by the American Association of Petroleum Geologists and the University of Texas Marine Science Institute (now the University of Texas Institute for Geophysics) held in Galveston, Texas, in January, 1981.

The volume owes much to two previous volumes with which the editors have been associated and, although it was not planned, the three volumes comprise a trilogy documenting the evolution of scientific thought on continental margin geology over the past decade. The first volume, *Geology of Continental Margins* (Springer-Verlag, 1974), was edited by C. A. Burk and C. L. Drake and resulted from the enthusiasm generated at a 1972 Penrose Conference. This volume brought together new information about continental margins, and in particular newly available multifold seismic reflection data which permitted detailed observations of the deeper structures of many margins.

The second volume, *Geological and Geophysical Investigations of Continental Margins* (American Association of Petroleum Geologists Memoir 29, 1979), edited by J. S. Watkins, L. Montadert, and P. A. Dickerson, resulted primarily from an AAPG-University of Texas Marine Science Institute sponsored research conference in Galveston in 1977. The volume included abundant multifold seismic reflection data which clarified many details of margin structure and stratigraphy. It also anticipated one of the major components of the present volume: the modeling of processes affecting continental margins.

Papers in the present volume utilize markedly improved seismic reflection and refraction data (for example, Aoki et al's remarkable reflection sections showing details of accretion in the Nankai Trough, and Ibrahim and Uchupi's Ocean Bottom Seismograph data from the Gulf of Mexico). Sufficient good quality data are now available to permit the parameterization, modeling, and testing of models of margin mechanisms. Nine papers representing a spectrum of thought on both passive and active margin tectonic processes result from these data. This volume also includes a significant number of papers reporting results of investigations of organic geochemistry and paleoenvironments of continental margins.

The relatively large number of papers reporting investigations of active margins (15 papers, including three grouped with papers on modeling) is indicative of recent advances in understanding of active margin dynamics. For example, in 1974, Seely et al presented multifold seismic reflection evidence for offscraping of oceanic and trench sediments and formation by imbricate thrusting. This work served as a point of departure which led to new data discussed in part in this volume. Investigators now conclude that a diversity of processes, including offscraping, subduction of overpressured sediments, accretionary underplating, and subduction erosion have molded present-day convergent margins.

The bread-and-butter of margin investigation remains the passive or rifted margins. The 22 papers on this subject (including seven grouped with papers on modeling) are indicative of the breadth and depth of investigation of passive margins. These papers cover a wide geographical range, from the Arctic to the North Atlantic, the Gulf of Mexico, the South Atlantic, and the Indian Ocean. Based on multifold seismic reflection data, detailed refraction data, deep well data, and extensive gravity data, they provide significant new insights into margin problems.

Papers on geochemical, depositional, and paleoenvironmental models, while lesser in number than papers on other topics, indicate the direction where important new results are forthcoming. Much remains to be learned about the evolution of organic material deposited on slopes, rises, and abyssal plains. Depositional processes on slopes and rises are also unclear in many areas. Finally, the importance of paleoenvironmental studies is beginning to be widely recognized and we can expect major advances, especially as a result of improved Deep Sea Drilling Program hydraulic-piston coring technology.

We gratefully acknowledge the support of many people and organizations during the research conference and preparation of the volume. The sponsoring organizations, American Association of Petroleum Geologists and the University of Texas Marine Science Institute's Galveston Geophysics Laboratory (now part of the University of Texas Institute of Geophysics) provided moral, logistic, and personnel support. Anne Ginder of UTMSI was particularly helpful. The AAPG Research Activities Subcommittee and the AAPG Research Committee were the immediate sponsors of the research conference.

Gulf Oil Foundation, Texaco Inc., Mobil Oil, and Michael Halbouty provided funds which made it possible for foreign as well as U.S. scientists to attend and present papers, and provided support for attendance by several students. We especially appreciate the efforts of E. E. Sheldon of Gulf Oil Foundation, R. R. Graves of Mobil, and F. A. Seamans of Texaco for assistance in obtaining financial support.

Patty Keyes of Gulf Science and Technology Company (a subsidiary of Gulf Oil Corporation) carried much of the conference organizational load and helped to process the manuscripts. Sally Witt of Gulf Science and Technology Company and Billie Watkins contributed their talents to the illustration of the volume and the organization of the Galveston Conference, respectively. Neither conference nor volume would have been possible without the wholehearted support of Ed Driver and Bob Brodine of Gulf Science and Technology, and Georges Pardo of Gulf Oil Exploration and Production Company.

Joel S. Watkins
Charles L. Drake

Introduction of Hollis Hedberg

by Georges Pardo

Editor's Note: *The following biographical sketch was presented by Georges Pardo as an introduction to Dr. Hollis D. Hedberg during the 1981 Hedberg Research Conference in Galveston, Texas.*

It was almost to the day, 40 years ago when one of my professors at the Institute of Geology in Caracas, Eli Mencher, asked me to accompany him on a field trip to eastern Venezuela. It was then, in San Tome, that I met Hollis Hedberg. At the time, he was one of the prominent geologists concerned with Venezuela in particular and Northern South America in general. It was during that meeting that I noticed a very peculiar characteristic of that man — he took us on a short trip along a section to see some outcrops. He proceeded to walk with enormous strides, his legs appeared to open up from somewhere in the middle of his chest, and without ever stopping he would describe the outcrops, knock samples, bag them, write notes, etc. Little I knew at the time how well I was going to become acquainted with and, in many ways, become a victim of that relentless walking machine. Shortly after that trip I was given a scholarship by Mene Grande, a subsidiary of Gulf, and spent that first summer working in his department. This was the beginning of a remarkable friendship and professional association that has lasted to this day.

Hollis Dow Hedberg was born May 29, 1903, in a Swedish farming community in Kansas on the crowded second floor of a small stone house during one of the worst floods of the region. It happened while nearly everything that the Hedberg family owned — furnishings, cattle, crops — floated away. A midwife was rowed in, but arrived too late. He grew up in the country and by the time he was 8 years old he was plowing the family fields, walking and whistling behind a horse. This background not only begins to explain his unique walk, but also another typical habit: whenever he is absorbed in his thoughts, pondering a problem, he whistles as if the rest of the world ceased to exist. I still remember some of the tunes he whistled while running down a trail or climbing some hill.

In 1920, he graduated from the Falun Rural High School, which had not yet started to suffer from overcrowding, and he and his brother were

part of a class of four.

In the fall of the same year he entered the University of Kansas with the original desire of becoming a journalist. I'm sure this is when he acquired what probably became one of his most useful tools: two powerful index fingers with which he can pound relentlessly and mercilessly any typewriter that comes his way. However, he shifted his interest toward geology and after a one-year interruption to run the farm after his father's death, he graduated with a degree in 1925. He spent 1925-1926 at Cornell University, where he received a master's degree, and in the summer of 1926 went to the Maracaibo Lake, Venezuela, as a sedimentary petrographer for Lago Petroleum Company (at the time, a subsidiary of Standard Oil of New Jersey). In 1928, Lago Petroleum made the kind of mistake that oil company managers, especially personnel managers, have nightmares about. They informed Hollis Hedberg that they had no further use for his services.

He returned to the States and was shortly back in Maracaibo but, this time, working for Gulf Oil Corporation. It was the beginning of 40 years of active participation in Gulf exploration for oil. I proceed with this introduction with a very definite slant toward his accomplishments in Gulf because I feel that this very rich aspect of his life is not generally as well known as others.

He started his career with Gulf as a micropaleontologist and sedimentary petrographer, and then as a stratigrapher and director of the geological laboratory until 1937. It was during those years that he did considerable field and laboratory work in western Venezuela, Colombia, and eastern Venezuela. Most of his effort was dedicated to establishing a solid network of stratigraphic sections for both the Cretaceous and the Tertiary. It was during that time that his very important 1931 paper proposed for the first time that the Cretaceous La Luna limestone was the source of petroleum in northwestern Venezuela. To my knowledge, today, in spite of the great advances in geochemical techniques, this thesis has not been contradicted.

He married Frances Murray in 1932 and their first son was born in 1934. He took a one year leave of absence in 1934-1935 to go to Stanford University to fulfill doctorate degree requirements, which he completed in absentia in 1937. While at Stanford, Dr. Austin Rogers informed him that he would never receive his doctorate unless he took his course. Hedberg told him that his schedule was too crowded and there was no way he could attend. At the orals, Rogers tried his best to flunk Hedberg but did not succeed.

It was during this time that he was almost lost to the world! He frequently went alone to do geological reconnaisance, collect samples, etc., and in those days this type of expedition was not totally without risk. On that particular occasion he went to collect samples along the Misoa River in the Trujillo Mountains and, after filling the bags, he tied them to his belt. By the end of the day he had accumulated a fairly sizable load and, on the way back, he slipped and fell from a ledge into the river which was quite deep at this point. Anybody else might have made an attempt to get rid of the stones, pants, belt, and whatever, but not Hollis Hedberg! He decided that the best thing to do was to crawl along the bottom in the murky water until he reached the steep, slippery rock bank and, only then, did he finally climb out of the water — slowly! And all that with the same gulp of air he took on the way down.

In 1937, the family moved to eastern Venezuela where he was in charge of exploration and geology for Gulf in this new developing area. He first lived near the discovery well in Oficina and, later, in the newly built camp of San Tome. They remained there until 1944. The impact of Hollis Hedberg on the exploration of eastern Venezuela was considerable. The area was expanded from the original Oficina field to one of the major producing areas of the world. Under his direction, hand-run electric logs by Schlumberger were used for the first time in a major way for determination of hydrocarbons, correlation of stratigraphic units, and determination of fault displacements. It was the first time, to my knowledge, that a systematic program was carried on in which seismic, coredrill, and strat-holes were used to explore complex faults and stratigraphically controlled accumulations. It was without a doubt the best planned and carried out exploration program that I have ever witnessed. Hedberg's influence was visible in almost every detail of the operation, whether it was the sample description terminology, the stratigraphic nomenclature, the files, the method of reporting, or the laboratory procedures, including the use of the refractometer to determine the API gravity of the oils. From a petroleum geologist's point of view, this was unquestionably one of the outstanding periods in his life.

During this period he perfected the "Old Fashioned" to the point that he knew exactly the proportions that would make a person of a given age keel over. I could not quite understand why, at a party, his "Old Fashioneds" had labels such as 25 years, 30 years, 40 years, etc. I naturally assumed it was a matter of vintage of whatever was in the glass and, therefore, took a 50 year old one. Well it was a very short party. This is also when I learned the effectiveness of the peculiar walk that I mentioned earlier and his determination to look at reported outcrops that he had not seen before. How many times did I curse myself for having adopted geology as a profession, especially on days like one where we left base camp at 6:30 in the morning and were still going downhill toward some outcrop at 4:30 in the afternoon. I knew that we had to go back up to the camp in order to eat and go to bed. I will never forget Cerro-Azul!

In 1944 the Hedbergs moved to Caracas and Hollis remained there as assistant chief geologist until 1946. They then moved to the New York office where he first became chief geologist and then exploration manager for Gulf's foreign production division. He remained there until 1952, and again, his influence was felt in Gulf's worldwide exploration program. His most remarkable accomplishment of that period was when, after reading numerous publications and re-

ports on Africa, he decided to take an extended tour alone, to take a first-hand look at the reported sections, seeps, tar deposit, etc. He traveled that way for many weeks, reaching spots that 30 years later are impossible to reach. He used every means of transportation that was available to him — bush planes, horses, jeeps, foot, you name it. A latter day Dr. Stanley. The results of his trip and his reporting from the trip were an extremely active exploration campaign by Gulf that extended from the former Spanish Sahara to southwest Africa and produced, over the years, the major discoveries of Nigeria, Cabinda, Zaire, and active exploration still ongoing in Cameroon.

Another anecdote that exemplifies Hedberg's determination is that years later, when he revisited a gas seep near the base of Mt. Cameroon and had only a glass jar with a screw metal top to collect a sample, he made the trip from Cameroon to Harmarville, Pennsylvania, holding that sample jar to be sure that the metal cap did not contact the gas it contained and the water would always wet the seal to avoid contamination. Do you know much time and how many planes it takes to go from Cameroon to Pittsburgh, Pennsylvania?

While in Summit, New Jersey, the Hedbergs had their fifth and last child, a girl, Mary Frances — after four boys.

In 1952, they went to Pittsburgh to corporate headquarters where Hollis was in charge of both domestic and foreign geologic work. They lived in a very beautiful old home on Hulton Road in Oakmont, near Pittsburgh. Their stay there was an interesting kaleidoscope of events reminiscent of James Thurber stories. There was the time when the boys went fishing and decided to keep the fish alive in the bathtub while running water. This resulted in a general flooding of the entire place.

They had a very large lot — several acres — not too far away from one of the famous U.S. golf courses, the Oakmont Country Club. There was dismay and consternation among the neighbors when the three oldest enterprising sons decided to use the large front yard shaded by oak trees as a parking lot during one of the celebrated open championships. This, of course, was nothing compared to the panic that went through that Club when, on a Saturday afternoon, the same three boys decided to start target practice with their rifles in the back yard. They were aiming in the direction of the golf course and golfers began running and hiding for cover as bullets zinged by.

While in Pittsburgh, Hollis Hedberg became chief geologist for Gulf Oil Corporation and later vice-president for exploration. During this time he began to express his belief in the importance of the offshore in the future of energy and sold Gulf management on a long-range commitment in offshore exploration. Thus, in 1967, the ship *Gulfrex* was christened by Mary Frances and started world wide program of reconnaisance of many offshore sedimentary basins around the world. In 1975, when it was decommissioned, it had logged 160,000 miles of surveys. However, the *Gulfrex* had been updated in 1974 by the R/V *Hollis Hedberg*, and today it has logged 120,000 miles of surveying.

Of course, Hollis Hedberg decided that he needed to supplement his work schedule and, in 1959, accepted the chair of professor of geology at Princeton University. The family moved to Princeton, New Jersey, and from that date until 1968, when he officially retired from Gulf, he subjected himself to a frantic commuting schedule between Pittsburgh and Princeton and vice versa. At present, Hollis Hedberg is still associated with Gulf as an advisor to the Corporation, to Gulf Research and Development and to the Exploration and Production Company. He still has an office in Houston, and I think that the only difference between now and before he retired is that he has added Houston to his regular commuting.

So far, I have described a very full and satisfying life. However, this is only part of the story (Hollis Hedberg has managed to pack several lives into that of one normal human being). His extra curricular activities are described in 158 publications and 15 reviews — the first one was an essay in the St. Nicholas magazine in 1916 and the last two, presently in press, are to be published in the proceedings of the meeting on petroleum geology in Beijing, China. The remarkable thing is that the great bulk of the work that went into the preparation, writing and editing of such a volume of paper was done in his spare time. Going through the titles of his papers you can see, to some extent, the continuity and the evolution of Hedberg's professional interests. The early papers are concerned with a subject that has preoccupied him throughout his entire career — diagenesis of sedimentary rocks and origin of petroleum. Later papers, like the very important *Significance of High Wax Oils with Respect to the Genesis of Petroleum* (1968) and more recently *Methane Generation and Petroleum Migration* (1980), describe the evolution of his thinking on the subject.

On the other hand, other subjects show a definite sequence. Many of the early papers, before 1936, are concerned with petrography and micropaleontology. General stratigraphic works, stratigraphic sections, general geology, etc., appear between 1937 and 1950. An increasing concern about stratigraphic procedures appears for the first time in print in *Stage as a stratigraphic unit* (1936). He became strongly involved in stratigraphic nomenclature from the middle 1950's and his work culminated in the publication of *The International Stratigraphic Guide* in 1976. This was unquestionably Hedberg's longest and most arduous work. He spent much time and a great deal of personal income communicating with geologists from all over the world to achieve the consensus that appears in the guide. His motivation to undertake such a task is expressed in the first sentence of the preface: "Stratigraphy is a global matter and international communication and cooperation are necessary if we are to adequately comprehend the picture of the rock strata of the earth as a whole . . ."

This concept of language got Hedberg in hot water many years before. During a discussion with some critics of his concerning having introduced too many new stratigraphic names in eastern Venezuela, he answered by comparing the use of stratigraphic terminology to that of language: an educated western civilized man might have a vocabulary upward of 100,000 words, while a primitive Aborigine from Papua New Guinea might find 1,000 words or so quite adequate to express his thoughts. Well, there went a couple of friends — they were uneducated savages anyway!

During the 1960s and 1970s Hedberg became increasingly interested in marine geology. He was chairman of the AMSOC Mohole Committee NAS in 1962 and 1963 but resigned to protest the pressure to immediately start a Mohole rather than to start with a worldwide shallow depth, dominantly sediment, drilling program. This program eventually became a Joides Deep Sea Drilling Project program. During most of the 1970s he was chairman of the Joides Safety Panel. In 1968 to 1973 he was Chairman of the National Petroleum Council's Subcommittee on Petroleum Resources of the Ocean and became involved in discussion of the law of the sea. In more recent years, he has developed the concept of the geomorphic base of a continent being the most logical guide to political boundaries in the ocean. This concern with territorial limits offshore, combined with his intense interest in the possibility of economic resources in deep water, led to three important happenings: (1) His proposal of a drilling offshore consortium to explore large tracts of federally owned land in the deeper parts of our territory; (2) His critical public evaluation of the U.S./Mexico draft treaty on boundary in the Gulf of Mexico, which led to the Congress requesting the U.S. Geological Survey to evaluate the economic potential of the lower slope before ratifying the treaty; and, (3) His appointment to Reagan's Advisory Energy Task Force during the last presidential campaign. I was very pleasantly surprised in respect to this last item when, during the presidential debate, I heard Reagan make a statement about offshore exploration that sounded extremely familiar. I had heard exactly the same statement a month before while my wife and I were on vacation and spent a few days at the Hedberg's home in New Hampshire. During that time, Hollis Hedberg used me as a sounding board with the draft of his recommendations on offshore exploration. His stand on national energy policy is summarized in a dialogue that took place between Hedberg and a taxi cab driver in Pittsburgh at the height of the energy crisis during the summer of 1978. "Don't you think," asked Hedberg, "that if you had a broken gas gauge in your cab that, instead of blaming me and everybody else for your problem, you should rather get a stick and try to find out how much gas you have in your tank before agreeing to take me to my destination?"

Hollis Hedberg has been interested in, and made contributions to, many other subjects over the years: the International Geologic Congress, the World Petroleum Congress, ECAFE, Associate Editor of the AAPG Bulletin since 1937, and President of the Geological Society of America. He has been interested in the history of geology, space science and much more. For all of these contributions, he has been recognized and rewarded by many institutions. His latest honor was the 1980 Penrose Medal of the GSA.

No introduction of Hollis Hedberg is complete without mentioning travel. If you pick up the phone and call him at his home on any day of the year, you have less than a 50-50 chance to find him there because at that time, he is probably writing some notes or dictating into a tape recorder in an airplane going somewhere. He is a virtuoso of the airline travel guide. He normally makes the reservations himself, certainly never trusts travel agents, secretaries and the like, and finds the most impossible and improbable connections. As a matter of fact, I think that he derives a very special enjoyment in the discovery of these obscure means of air transportation. At any rate, he is the only person I know that was snowbound for 10 days in the Himalayas, witnessed the invasion of Czechoslovakia by the Russians in Prague, and arrived in Rangoon from Beijing without any hotel reservations. You would be amazed at the number of these exciting adventures Frances Hedberg has participated in.

This is a quick and superficial sketch of Hollis Hedberg's life, and I would like to mention his hobbies. Yes, Hollis Hedberg does have time for hobbies! He mixed his pleasures with work of course, and when it came to hiking and camping he was a purist — like never carrying a gas stove on geologic field trips. If you wanted to warm your can of corned beef hash you had to collect wood and start a fire — a very interesting sport in a South American rain forest. All his life he has been quite an excellent athlete — tennis, swimming. He has enjoyed reading all his life, and as a matter of fact, reading aloud to the family is a tradition. But perhaps the one activity which is dearest to him is his vegetable garden. He has always had vegetable gardens. Presently, his garden is in Cape Cod and he grows everything — corn, potatoes, squash, strawberries, you name it and if you go to the Cape you will see how easy it is, if you are not very careful, to get pressed into service.

He and Frances are active square dancers. He collects rare maps, mostly from early explorers of the Caribbean and South America and does have a fine collection of old books from pioneer geologists. He has had a deep interest in the history of geology and has written several articles on the subject. He is profoundly interested in his family background and has actively maintained, with his family, the spirit of tradition and awareness of his roots. He has always been proud of the strength and determination of his ancestors that brought their Swedish customs and traditions to this country. As a matter of fact, there is always a bottle of ice cold Aquavit in his freezer, and, as a good piece of advice, if he invites you for a drink of it, do not attempt to outskaal him.

I have said a lot about what Hollis Hedberg does or has done, but I have not been very explicit about who he is. If you go around and ask his acquaintances what they think of Hollis Hedberg, you will be surprised that, in spite of all his credentials, very few will say that he is a great scientist, that he is an authority, or that he is a famous geologist. You probably will hear that he is a gentleman, that he is generous, that he is kind and that he is a marvelous person. The reason for this is that his professional accomplishments, great as they are, are pale in comparison to the influence he has had on the lives of an untold number of people. All his life he has reached out and tried to bring the best out of anyone he was associated with, whether it was an Indian helper, a professional on his staff, or a student at Princeton. I remember some of his typical advice: "Georges, never become an authority — it would mean only that you are stuffed and can't learn any more, not that you know all there is to know." "No, don't worry about conventions, express your opinion as you see it but be prepared to be responsible for it." "Don't ever judge people from what you hear, judge them by what they are to you."

The most admirable feature of his personality is the persistence and determination that will never let him allow a good idea, from him or anybody else, to die because of lack of effort and sacrifice.

Section 1: Field Investigations of Margin Structure and Stratigraphy

The investigation of the earth differs fundamentally from the investigations of many other scientific phenomena in one key aspect: in physics, chemistry, or biology, for example, the investigator designs and controls an experiment, while geologists or geophysicists see only the end product of an experiment. They must reconstruct the experiment and discover its nature. They do this by studying selected field areas where they believe that the earth has conducted an experiment which reveals the nature of some geological process.

Insight gained from field studies provides the basis for development of conceptual and mathematical models, which can be tested using mathematical models or by further field studies. It is not surprising, then, that over half the papers in this volume report field investigations and, in addition, a number of papers devoted mainly to models rely heavily on result of field investigation (In some cases, attribution of papers to "Model Investigations . . ." or "Field Investigations . . ." was rather arbitrary).

Field investigations of rifted margins reported herein deal mainly with the offshore; locations range from the Arctic to Southern Africa, from the United States to Japan, to Australia and to India. Topics include the evolution of continental margins during prerift, synrift, and post rift stages (for example, Brice et al), the anomalous breaking away and subsequent drifting of a continental crustal fragment (Nunns), the development of Tertiary deltas (Willumsen and Cote), the crustal structure of the Gulf of Mexico (Hall et al, Ibrahim and Uchupi), and others. Data quality and quantity continues to improve (the East Coast of the U.S. is a notable example) and quality data are now available from areas of limited accessibility such as the Arctic, or where thick sedimentary sections have impeded studies of deeper structure (for example, U.S. East Coast and Gulf of Mexico). Finally, it is important, as Pratsch reminds us, that in spite of many advances in technique and in tectonic ideology, we do not lose sight of the lessons of the past.

Field investigations of convergent margins document a growing understanding of structural diversity. Aoki et al beautifully show details of offscraping; Biju-Duval et al report an uplifted and exposed margin section in Hispaniola, including relatively undisturbed ocean crust and sediments from the seaward side of the extinct trench, while White and Louden explore the effects of subduction of a thickly sedimented ocean plate. Four papers report results from studies of ancient margins, the best places to study the deep structure of convergent margins. Although ancient margin structure is complex, it is here that we must look to complete our understanding of margin tectonics and stratigraphy. These papers thus sample an important trend in margin research.

Rifted Margins

Tectonics and Sedimentation of the South Atlantic Rift Sequence: Cabinda, Angola

Suzan E. Brice
Michael D. Cochran
Georges Pardo
Alan D. Edwards
Gulf Exploration and Production Company
Houston, Texas

Mesozoic rifting of the South Atlantic produced a succession of five distinct tectonic regimes: Prerift, Synrift (I and II), Postrift, and Regional Subsidence. These tectonic episodes are recognized in the sedimentary section of Cabinda, Angola. Each has a characteristic structural style and stratigraphy and is bounded by a major unconformity. The total sequence reaches a maximum cumulative thickness of about 12,000 m and ranges in age from Jurassic to Holocene.

The tectonics and sedimentation of the West African continental margin provide a model for the geologic processes associated with the South Atlantic rift. Cabinda, Angola lies in a median position along this rift zone (Figures 1 and 2). It will be used as a type section for describing the tectonics and sedimentation associated with the opening of the South Atlantic. While some local variations are expected along the South American-African rift zone, it appears that the major tectono-stratigraphic units described in this paper are regionally significant. The authors have not had access to seismic data or well logs from the conjugate margin of Brazil but the review literature (Asmus and Ponti, 1973; Campos, Ponte, and Miura, 1974) indicates that the tectonics and the stratigraphy of the Southern Brazil marginal basins are similar to those of the Cabinda margin. The tectono-stratigraphic framework of Cabinda, Angola, developed in this paper appears to be applicable to the Southern Brazil margin.

The separation of Africa and South America, which began in the early Mesozoic, dominated the geologic

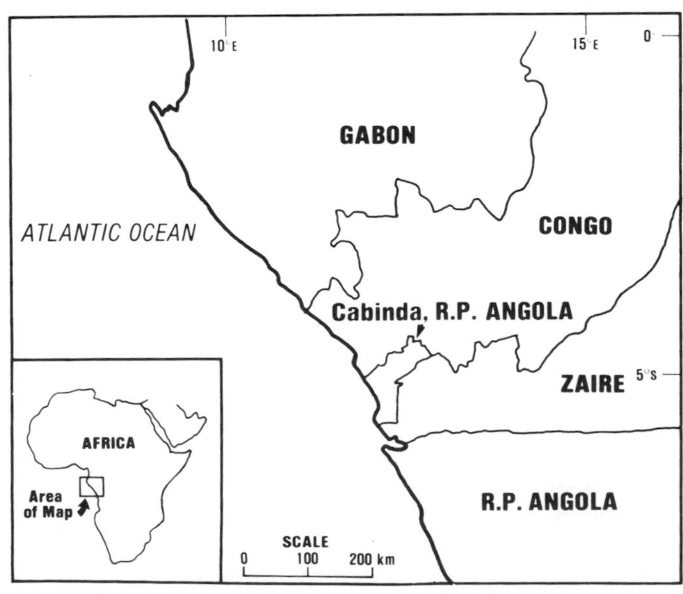

Figure 1 — Location map, Cabinda, R. P. Angola.

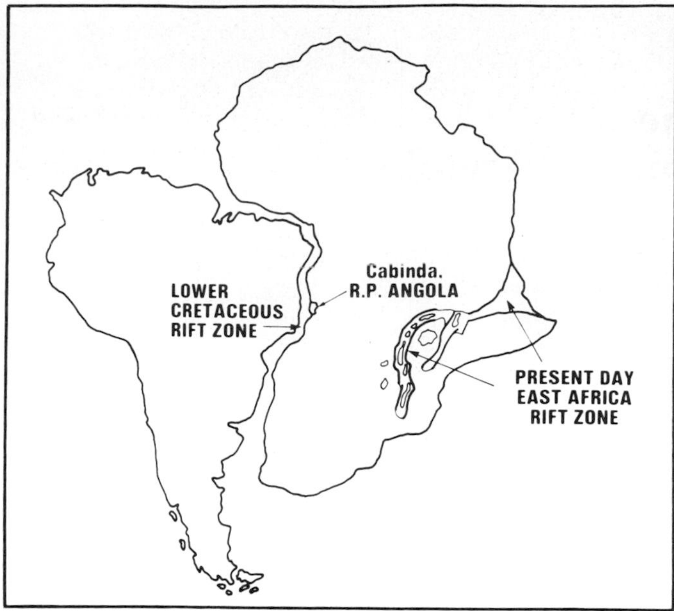

Figure 2 — Pre-separation reconstruction of South America and Africa; East African rift system is shown for scale.

history of Cabinda from its initial onset to the present day. Varying tectonic regimes associated with the separation have greatly influenced the stratigraphy. The sequence thickness is over 12,000 m and the sequence ranges from Jurassic to Holocene in age.

This paper summarizes the geologic history of Cabinda and shows how it was affected by various tectonic regimes as the area changes from an intracratonic basin to an open-marine shelf. Our emphasis is on how the tectonics influenced the stratigraphy.

We depart from the traditional format in that few references are given in the text because the bulk of information is derived from Gulf Oil Corporation files and has not been previously published. Where applicable, available published literature was studied and taken into consideration (Caflisch, 1975; Vidal, Joyes, and Vanveen, 1975; Lehner and DeRuiter, 1977; Caflisch, 1979; Brice and Pardo, 1980). Formation names are not given, as they have not been previously published. Their definition is not within the scope of this paper.

Data collected during 20 years of oil exploration and production in Cabinda were used to make this study. Seismic, gravity, magnetic, well log, biostratigraphic, and lithostratigraphic data are combined to make an integrated regional study.

The discussion of the pre-salt section is from a report in preparation, which discusses the results in more detail and defines the lithostratigraphic nomenclature. The discussion of the post-salt sequence has been strongly influenced by results of studies being done by co-workers within Gulf. Credit should be given to: M. D. Bush and D. W. Lewis, who have been working on the stratigraphy of the salt and post-salt sequence of Cabinda; G. A. Seiglie, who has been working on the biostratigraphy of the marine Cretaceous of Cabinda; C. A. Rachwal, who studied the salt tectonics of this area; and R. Spaw, who studied the lithostratigraphy of the post-salt carbonates in the Cabinda-Zaire area. Although the authors' independent work forms the core of this work, data were supplied by the above studies and by past studies done by numerous Gulf geologists and geophysicists.

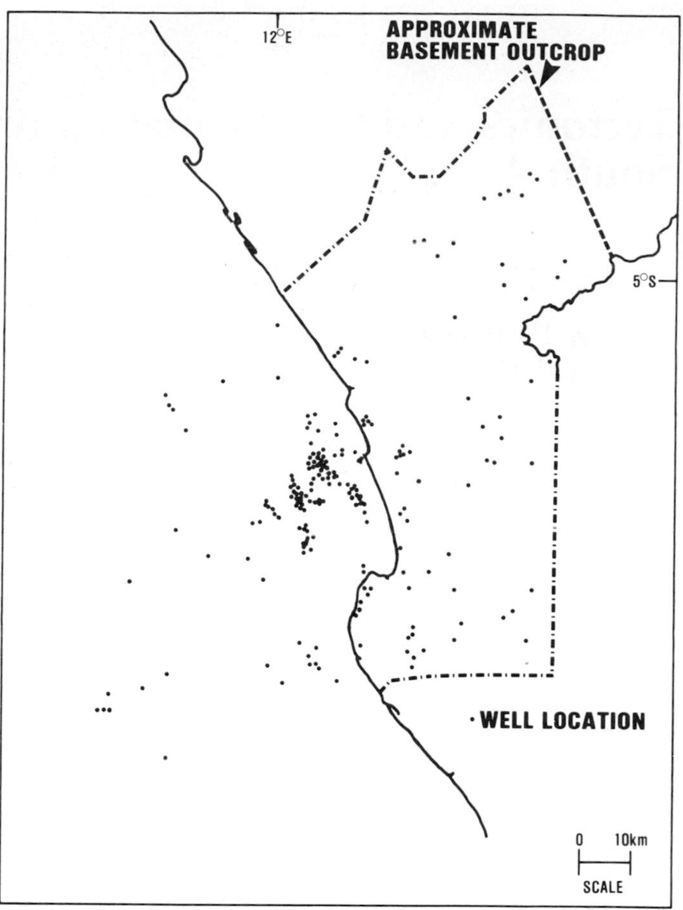

Figure 3 — Well location map.

Many vintages of seismic data were used, but only the recent data allow a seismic-stratigraphic interpretation. A salt layer acts as a seismic-distorting blanket, making conventional seismic interpretation difficult. Modern vintage migrated seismic data tied carefully to the wells permit a confident interpretation beneath the salt. Above the salt, seismic data quality is good. However, a detailed biostratigraphic framework tied to true amplitude migrated seismic data was required before a seismic-stratigraphic interpretation was made. Seismic sections used in this paper are perpendicular to the present coastline. The precise location of these data is not given due to ongoing exploration in the area.

The well control is not uniformly distributed across the area (Figure 3). It is concentrated on highs; hence, in many cases it penetrates only a thin section of sediments above the basement. This is especially true for the pre-salt section. Some wells which were drilled for pre-salt objectives are off-structure in the post-salt section, thus providing important post-salt downdip control.

TECTONO-STRATIGRAPHIC UNITS

Tectonic regimes associated with the continental separation greatly influenced stratigraphy. In Cabinda, the tectonic movements are divided into five episodes (Table 1), from the oldest to youngest:

1) Prerift with gentle tectonism;
2) Synrift I with strong tectonism;
3) Synrift II with moderate tectonism;
4) Postrift with gentle tectonism;
5) Regional Subsidence with major tilting.

Each of these episodes is marked by a characteristic tectonic and stratigraphic style. They are (with one exception, Synrift II and Postrift), separated from each other by a major unconformity. Because of their distinctive structural and stratigraphic character, each has a seismic signature permitting mapping outside of well control. Each stratigraphic unit has an associated lithology: 1)Prerift-sandy clastics; 2)Synrift I-shaly clastics and carbonates; 3)Synrift II-evaporites, carbonates, and sandy clastics; 4)Postrift-sandy clastics and carbonates; and 5)Regional Subsidence-shaly clastics.

Prerift

Prerift sediments were deposited unconformably on faulted metamorphic basement prior to major continental rifting. With the onset of active rift faulting, the Prerift sequence underwent extensive block faulting and erosion. Fault blocks with dips over 15 to 20° and bounded by a major unconformity characterize the unit. The Prerift lithostratigraphy consists of a sandy fluvio-lacustrine sequence. Its total thickness is unknown, but over 1,000 m of Prerift clastics were penetrated in some wells. The sequence consists of massive, clean, well-sorted sands and massive siltstones which were deposited in and around a broad, shallow lake system in a gently subsiding intracratonic basin. A layer of volcanics up to 30 m thick is often encountered at the top of this sequence. Gulf has dated these volcanics radiometrically at 140 m.y. ± 5 m.y. old, placing them at the Jurassic-Cretaceous boundary. This age gives an indication of the time of active rifting. The age of the oldest depositional unit

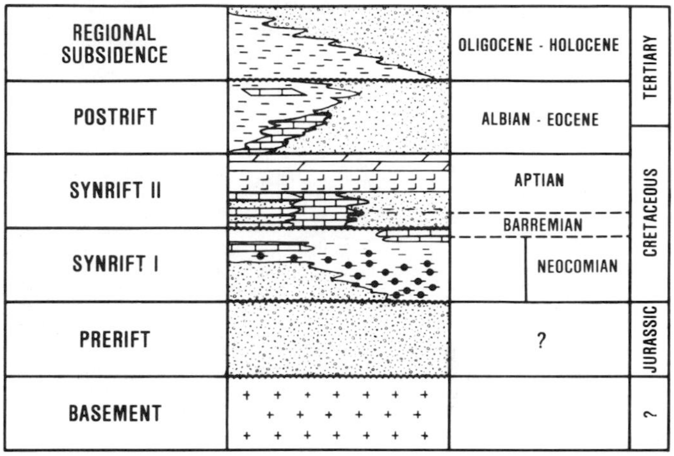

Figure 4 — Stratigraphic column showing tectono-stratigraphic units.

Table 1. Tectono — Stratigraphic Summary

Tectono-Stratigraphic Unit	Structural Style	Lithologic Character	Seismic Signature
Regional Subsidence	Shale domes and diapirs Discordance Channel cuts	Shaly clastics High pressure shale Turbidites Regressive sequence	Weak discontinuous reflections Chaotic patterns Cut-and-fill patterns Progradational offlap patterns Low seismic velocity
Postrift	Gentle conformable folds at top Complex halokinesis and faulting at base	Marine clastics and carbonates Non marine red beds Transgressive sequence	weak conformable reflections at top Complex patterns with strong reflections at base Variable seismic velocity
Synrift II	Halokinesis Complex faulting	Evaporites Halite with high potash content Dolomites at top	Transparent zone bounded by strong reflections Complicated patterns High seismic velocity
	Internally concordant Gentle dips Few faults	Carbonates Sandy clastics Shallow lacustrine	Concordant events Strong, continuous reflections grading into transparent zones
Synrift I	Internally concordant Gentle dips Few faults	Shaly clastics High organic-matter concentrations Deep lacustrine sequence	Concordant events Strong, low frequency events Truncated sequences Low seismic velocity
Prerift	Block faulted Angular unconformity Steeply dipping beds	Sandy clastics Fluvio-lacustrine Occasional volcanics	Weak, discontinuous reflections Angular reflection patterns Truncated sequences High seismic velocity

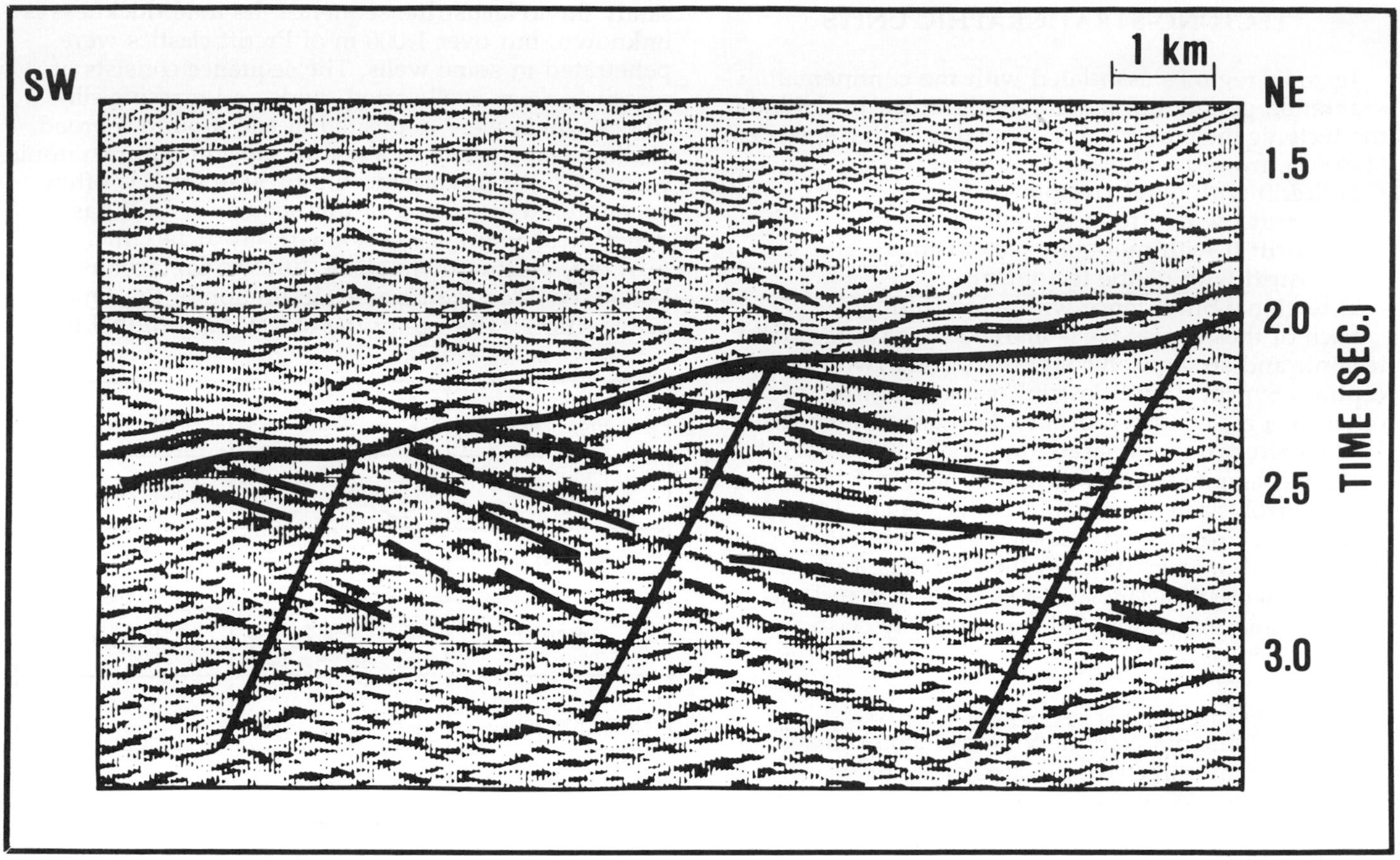

Figure 5 — True amplitude, migrated seismic section showing typical Prerift character.

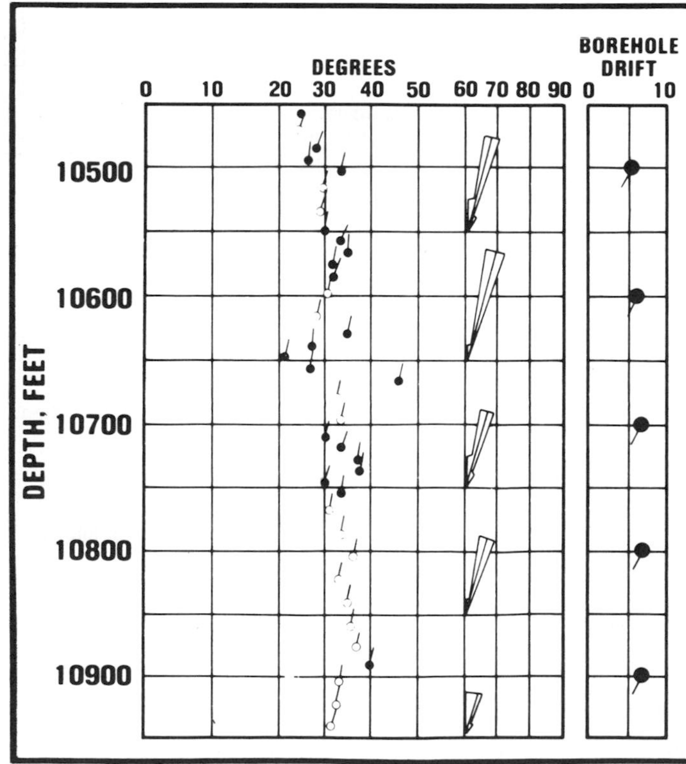

Figure 6 — Dipmeter log from typical Prerift sequence.

in the Prerift section is not known.

The combination of structural style and lithologic character gives the Prerift sequence its unique seismic character. Although seismic data quality is fair to poor, seismic identification is often possible. The Prerift seismic sequence is a truncated package of weak, steeply dipping reflections (Figure 5). Seismic velocities are about 5,400 m/sec. The interpretation of steeply dipping beds is confirmed by dipmeter surveys (Figure 6) which show dips averaging 20°.

Synrift

Active rift faulting beginning in Early Cretaceous formed a deep graben lake basin subparallel to the present coastline. This rifting phase, which continued through the Neocomian, is divided into two subphases: Synrift I and Synrift II, each initiated by a major tectonic event.

Synrift I

The initial phase of rift faulting consisted of uplift and basement block faulting; the tilted, eroded fault blocks formed a graben lake system which underwent rapid subsidence and was infilled with lacustrine sediments. The facies relationships are complicated, but the structural style is simple with internally concordant bedding and gentle dips. Fewer faults cut this

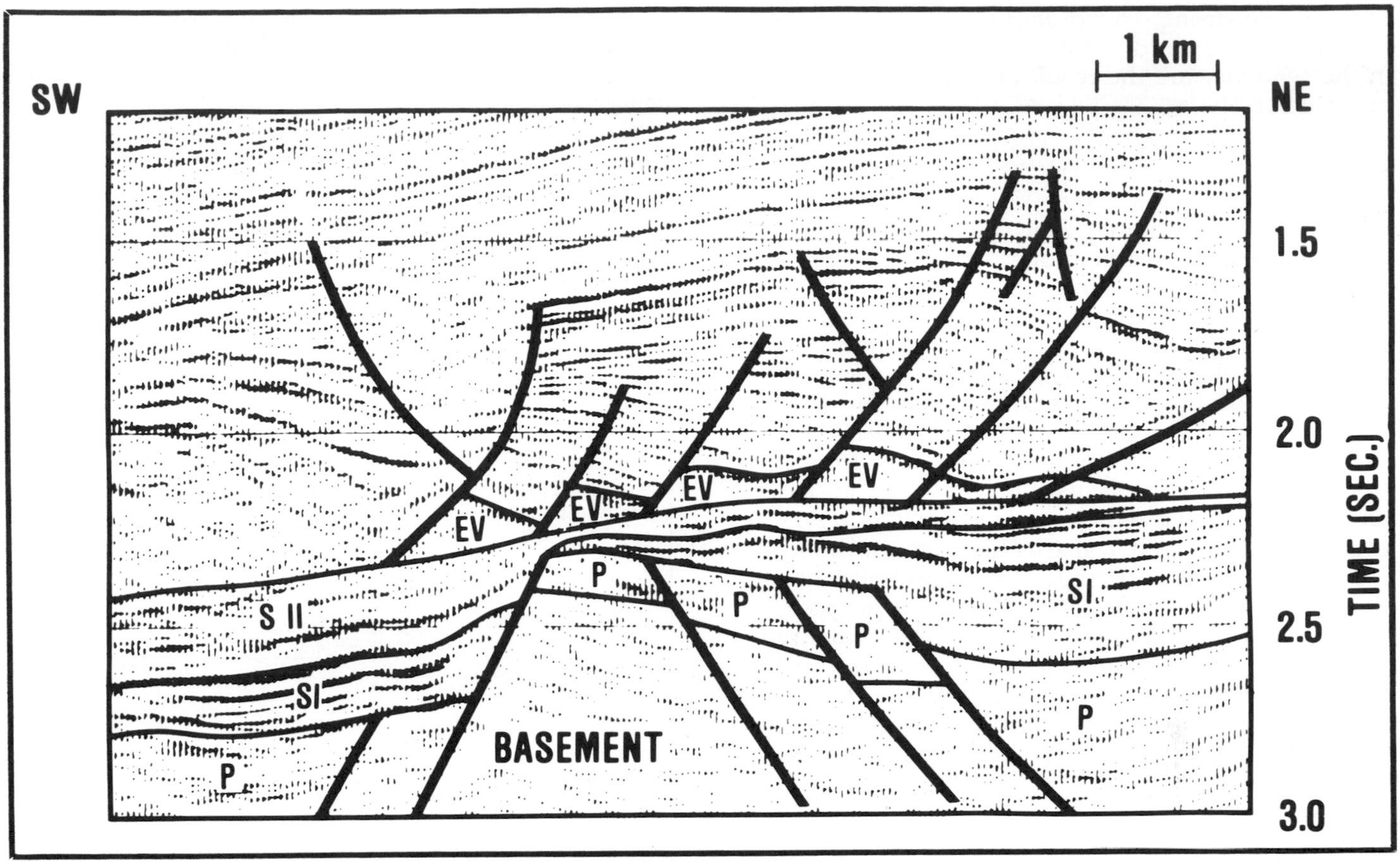

Figure 7 — Seismic section showing typical Synrift I and Synrift II seismic signature and the unconformity separating the two sequences. Ev = evaporites; SII = Synrift II; SI = Synrift I; and, P = Prerift.

sequence than the Prerift. In general, Synrift I sediments onlap the Prerift unit but they are occasionally interrupted laterally by major late faulting (Figure 7). An erosional unconformity marked the end of Synrift I deposition.

The Synrift I lithology is characterized by shaly, organic-rich lacustrine sediments. The graben lake system was filled by lacustrine turbidites which grade laterally and upward into an organically rich, deep lacustrine dolomitic shale. This organically rich dolomitic shale was deposited in an anoxic lake basin, which is approximately 80 km wide and extends from Angola to northern Gabon (as evidenced in Gulf Oil Co. proprietary data).

The lateral continuity of the unique physical properties of the organic-rich dolomite shale make these beds particularly useful as time-stratigraphic markers. On well logs, this shale is characterized by intervals of extremely low densities and low sonic velocities and correspondingly high resistivities (Figure 8). Densities as low as 1.8 gm/cc and velocities as low as 2,200 m/sec are common. Lab analyses show up to 20% organic matter in these zones (Brice, Kelts, & Arthur, 1980). These low density-high resistivity zones were laterally correlated over 80 km distances. They are useful in dating basement uplifts and subbasin development.

As the lake shallowed the organic-rich shale graded upward into shallow lacustrine shales and carbonates.

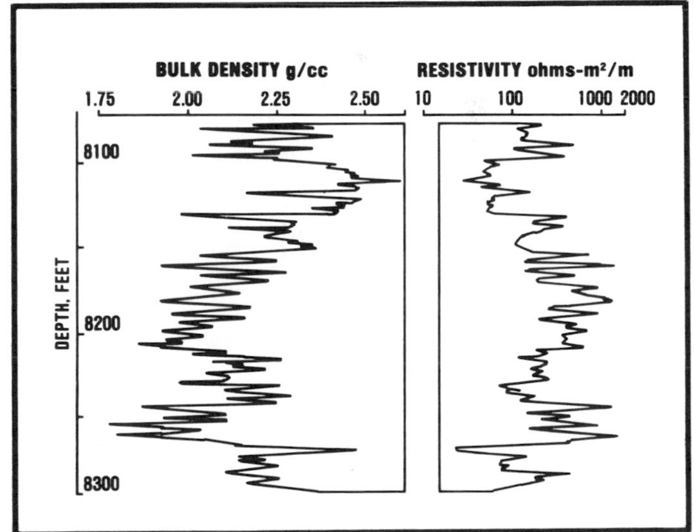

Figure 8 — Density and resistivity logs from a typical Synrift I sequence.

Algal limestone banks formed on the high standing blocks. A period of major tectonics and erosion terminated this shallowing upward lacustrine sequence. Synrift I sediments maximum thickness is 1,500 m.

The seismic character of the Synrift I section differs greatly from that of the Prerift sction. Synrift I reflections are strong, concordant (Figure 7), and correlatable over great distances. Low seismic velocities in the sequence are the result of the thick section of organic-rich shales.

Synrift II

Another period of major basement movements, in which some of the earlier zones of faulting were reactivated, marked the beginning of Synrift II. Westward subsidence and erosion of high-standing blocks followed. Throughout the Synrift II period, and into the early part of the Postrift period, minor tectonic movement locally affected sedimentation.

The Synrift II sediments are a transitional sequence marking the change from non-marine to marine conditions. They consist of lacustrine carbonates and sands and alluvial clastics which grade upward into an evaporite sequence that marks the onset of the marine incursion. The section ranges from Barremian to Aptian in age and is divided for descriptive purposes into two subunits showing distinctly different structural styles. The structural style of the lower part of the Synrift II sequence is similar to that of the Synrift I sequence. It has generally concordant bedding with gentle dips and is cut by few faults, expect around young basement highs. A major unconformity lies at the base of the Synrift II sequence (Figure 7). The upper evaporite sequence of the Synrift II unit has a complicated structural style.

Synrift II sedimentation began with the deposition of up to 500 m of shallow lacustrine carbonates and sands in sub-basins formed by movement along ex-

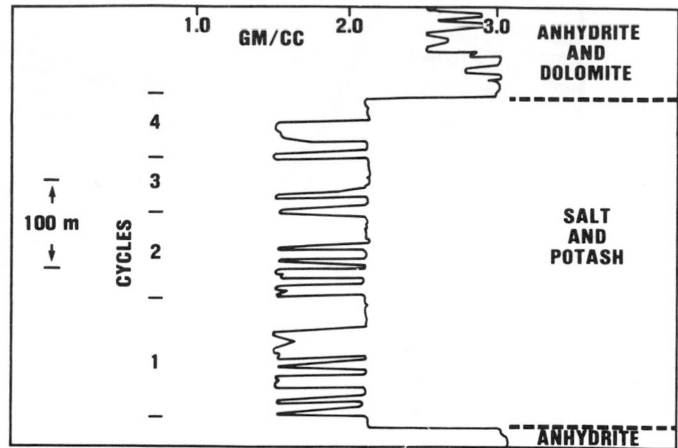

Figure 9 — Density log from a typical Synrift II evaporite sequence.

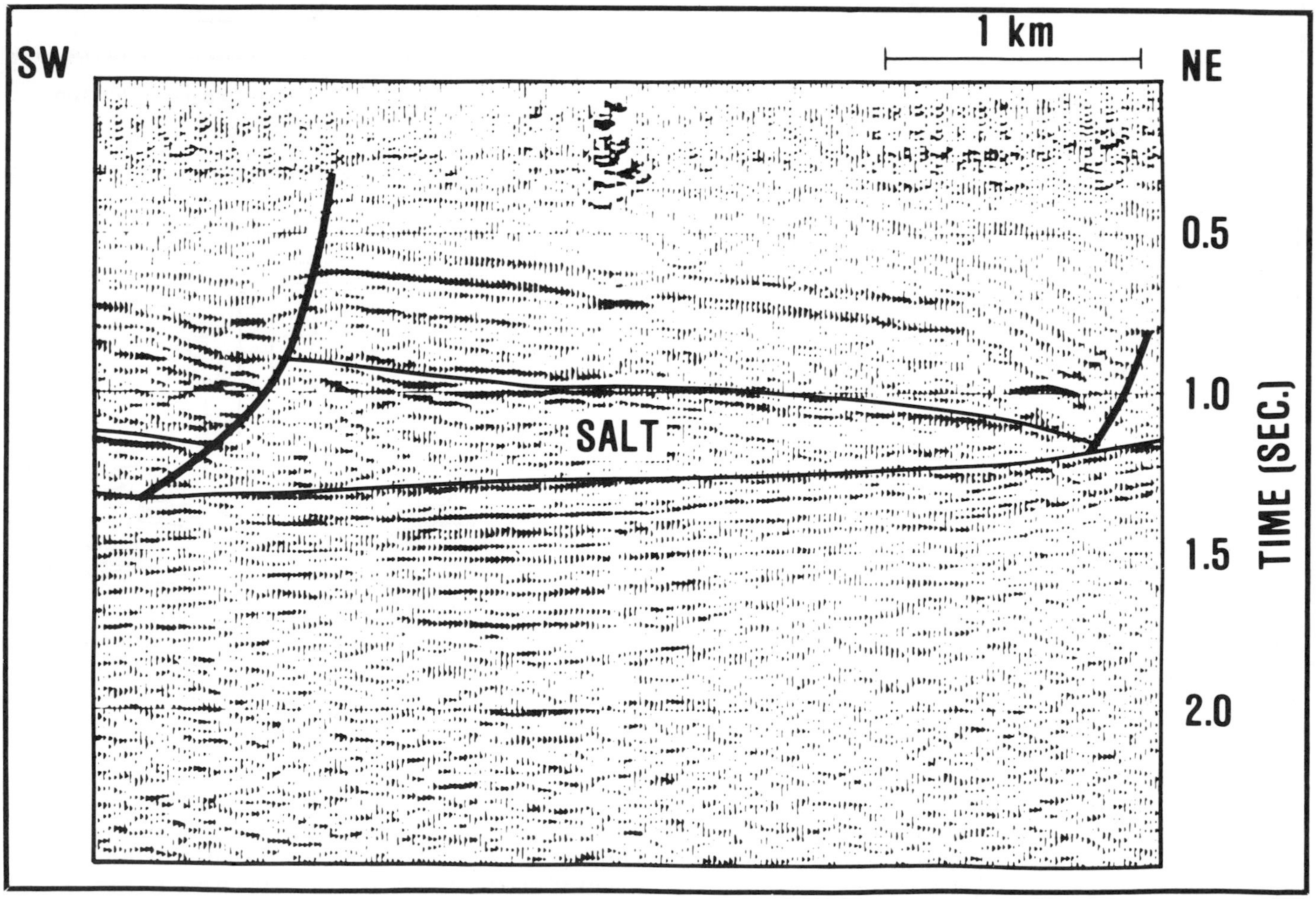

Figure 10 — True amplitude, migrated seismic section showing the character of the Synrift II evaporites.

isting faults. The unit thins to the east and over basement highs and grade onshore into a uniform 50 m sand and shale alluvial-plain deposit.

The first marine incursion of the basin occurred toward the end of the Synrift II period. Saline waters entered the basin from the south. This incursion was initially confined to the sub-basins, but towards the end of the Aptian it covered most of the area with a thick evaporite sequence (up to 1000 m in the Cabinda area). These evaporites are characterized by high potash content and low anhydrite content, suggesting that the saline waters were super-saturated and the conditions highly evaporitic. At least four desiccation cycles are recognized in Cabinda, and they appear similar to those reported in Congo by Belmonte, Hirtz, and Wenger (1965). These cycles are laterally consistent and can be correlated over much of eastern Cabinda. Markers within these cycles can be used as time-stratigraphic markers to date salt movements and determine the nature of the movements (Figure 9).

The deposition of a 50 m thick dolomitic unit over all the area marked the end of the evaporite sequence. This unit appears conformable with the overlying Postrift sequence.

The Synrift II sequence has two characteristic seismic signatures; one for the lower part and a different one for the upper part. The lower part is similar to that of the Synrift I sequence. Where carbonates are more massive, it is marked by strong, concordant reflections; where carbonates are thinly interbedded with sand, it varies from strong reflection zones to transparent zones. The reflection at the basal unconformity is generally strong, and the Synrift II seismic velocity is significantly higher than that of the Synrift I unit (Figure 7).

The upper part, the evaporite sequence, is often seismically transparent, although some reflections from the potash beds have been recognized. A strong reflector marks the top of the sequence (Figure 10). Basal reflection character is variable because of facies changes in the lower part of the Synrift II unit. The seismic velocity tends to be high, but not as high as that of pure halite because of the high potash content of the sequences. Reflection patterns at the top of the evaporites may show complex structure as a result of salt tectonics (Figure 11).

Postrift

The Postrift period was marked by gentle regional subsidence. Local movement along some Synrift II fault systems as well as salt movement continued dur-

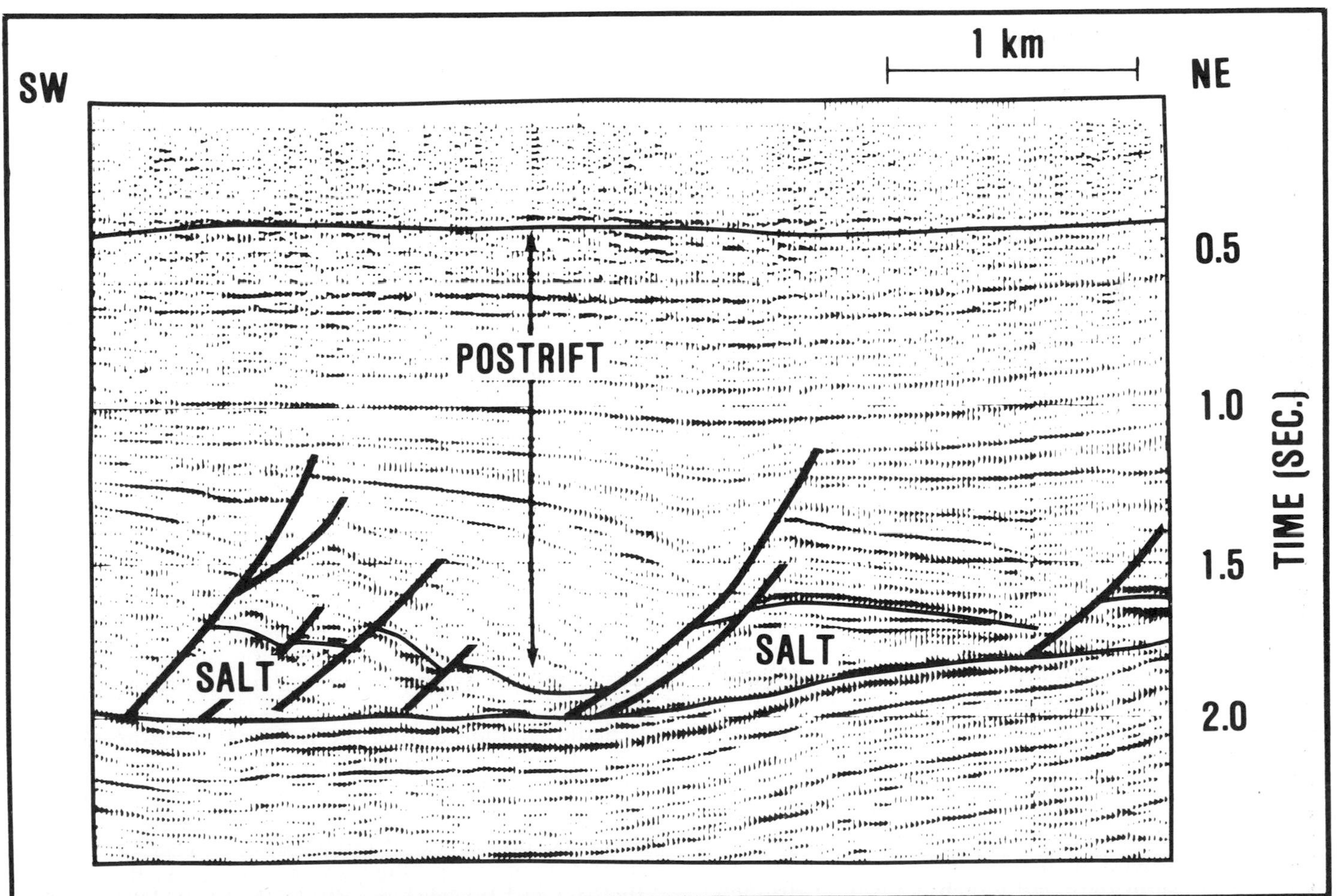

Figure 11 — True amplitude, migrated seismic section showing the character of the Postrift and the carbonate growth faults.

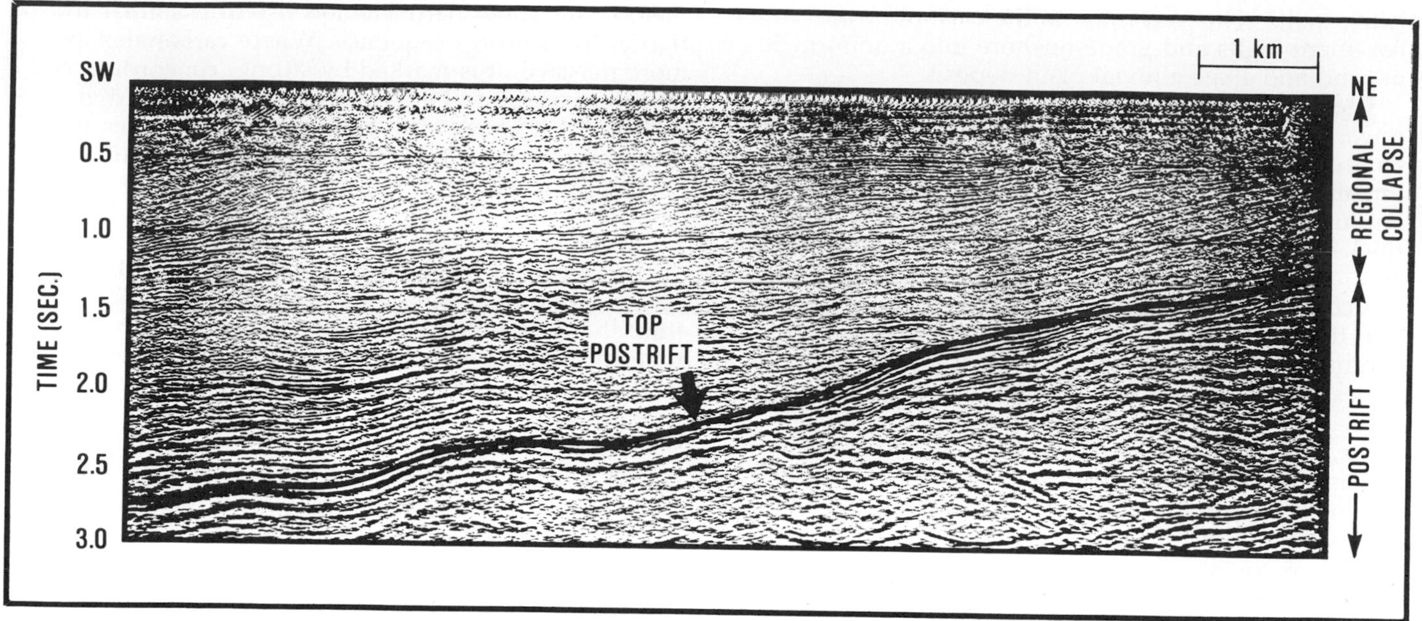

Figure 12 — True amplitude, migrated seismic section showing the boundary at the top of Postrift.

ing the early Postrift period. However, the tectonics and sedimentation were dominated by a regional crustal oscillation and accompanied by a major marine transgression, followed by a regression. Up to 2,000 m of sediments were deposited.

The structural style of the basal Postrift section is characterized by salt movement and associated faulting. Regional subsidence triggered ductile flow and brittle fracture of the Synrift II evaporite sequence. These movements created growth faults which in turn modified local sedimentation. Many of these structures formed during the deposition of carbonates producing "carbonate growth faults" (Figure 11) analogous to the typical clastic growth faults. Bacoccoli, Morales, and Campos (1980) and Carozzi et al (1979) discussed similar "carbonate growth faults" in the Postrift section of offshore Brazil.

Broad gentle structures are common in the upper portion of the Postrift unit. They are cut by few faults and show internal concordance (Figure 11). The top of the Postrift unit is marked by high geomorphic relief, which appears to correspond to the paleoshelf and slope break. It represents a depositional surface which is, in part, erosional as well as nondepositional (Figure 12).

Postrift sedimentation was initially transgressive, beginning with a nonmarine red bed sequence which grades upward into a nearshore carbonate-clastic unit. As subsidence continued, this unit graded into deep water shales and marls. The transgression, which was from the southwest, continued through the Campanian. At its maximum extent the sea covered most of the present offshore and part of onshore Cabinda. Subsidence slowed at the end of the Campanian; a regional regression followed which continued through the Paleogene during which, a sequence of nearshore-marginal marine clastics and carbonates were deposited.

The Postrift seismic character grades from complex reflection patterns at its base, upward into gentle concordant reflection patterns (Figure 11). The reflection strength decreases up the section and to the west, probably because of a transition from mixed carbonate-clastic sediments to more clastic sediments. Seismic velocity is proportional to the carbonate content of the sequence.

Regional Subsidence

At the end of Paleogene, the seaward edge of the continent foundered producing a strong westward tilt. During the Oligocene and the Miocene, a thick regressive clastic sequence was unconformably deposited across the older shelf sequence. This clastic sequence built the continental shelf to its present position.

The structural style of the Regional Subsidence sequence is typical of a rapidly deposited regressive sequence. It consists mainly of structures related to the tectonics of gravity sliding and differential loading. It is marked by high pressure shale movement and related growth faulting. Because of eustatic changes, the sequence was occasionally subjected to extensive channelling (probably related to the proto-Zaire River drainage), producing numerous cut-and-fill structures (Figure 13).

Sedimentation during the Regional Subsidence period resulted in a regressive, shaly, clastic sequence. Deep water shales and turbidites grade upward to nearshore sands and shales, followed by an alluvial sand section. This unit ranges in thickness up to 6,000 m and in age from Oligocene to Holocene. It is thin in the east and thickens rapidly to the west. Deep grabens near the present shelf edge may contain up to 6,000 m of this unit.

Discontinuous and weak seismic reflections are

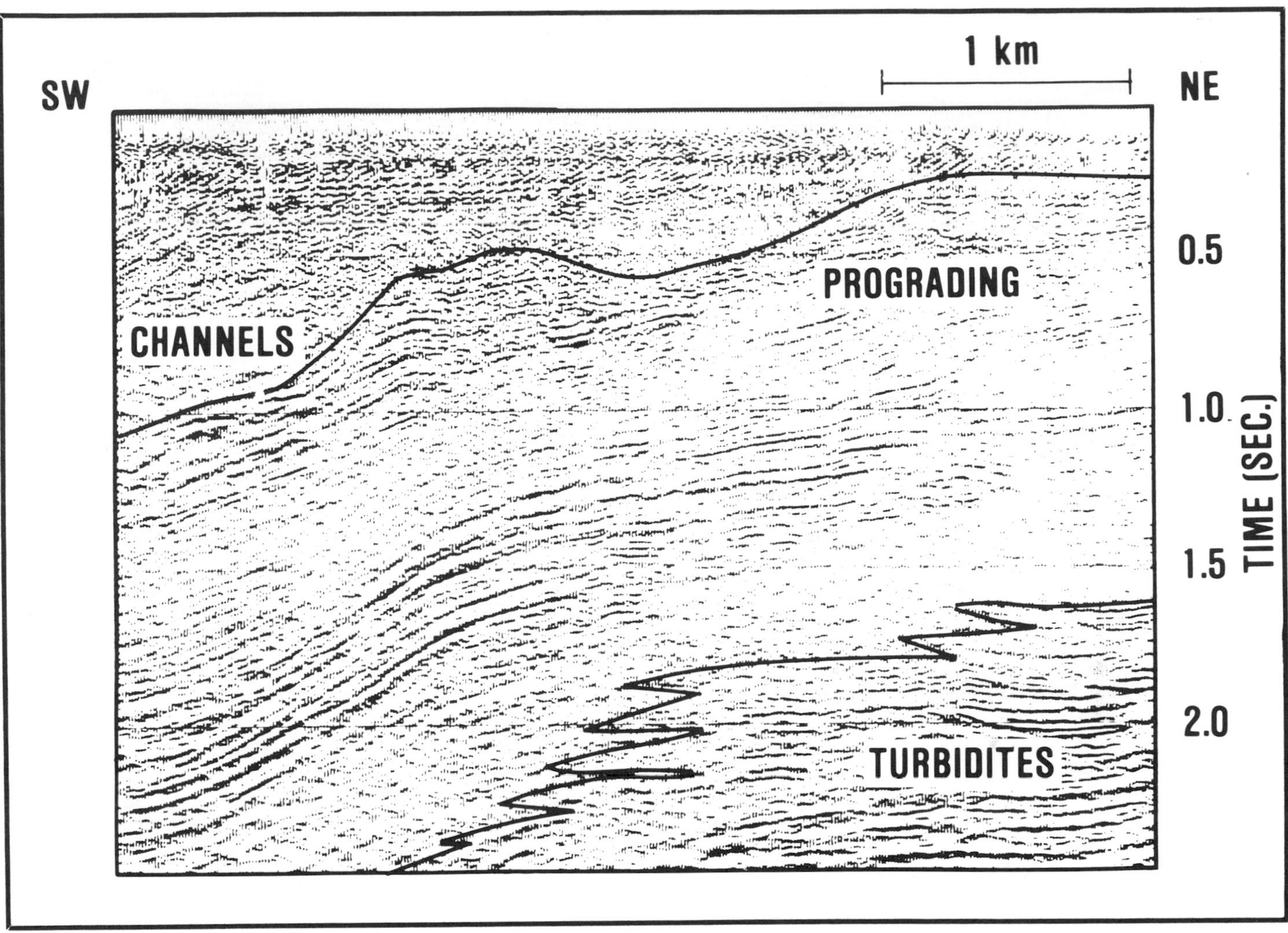

Figure 13 — True amplitude, migrated seismic section showing a typical Regional Subsidence sequence.

common in the Regional Subsidence unit. The typical seismic sequence shows chaotic patterns, cut-and-fill patterns, progradational offlap patterns, and domal patterns (Figure 13). Seismic velocities are low and dependent on depth of burial.

GEOLOGIC HISTORY

The geologic history of Cabinda is the result of the superposition of five tectonic regimes; each affected the structural style and stratigraphy of the area. Table 1 summarizes tectono-stratigraphic units and their structural style, lithologic character, and seismic signature.

The Prerift unit was deposited in a slowly subsiding intracratonic basin. This basin was filled with fluvio-lacustrine clastics (Figure 14a). A major tectonic pulse related to the initial separation of Africa and South America occurred at the end of the Jurassic. This pulse resulted in strong basement deformation: uplift and block faulting (Figure 14b). Resultant rift faulting formed a deep graben lake system. As this graben lake system subsided, intralake highs were submerged, and a deep anoxic lake basin formed, extending from Angola to northern Gabon. The lake basin was filled with the Synrift I lacustrine sequence (Figure 14c). Deep lacustrine sediments (turbidites and organic-rich dolomitic shales) graded upward into green shales and carbonates as the lake basin was infilled during the Neocomian. Continued sporadic basement movement resulted in additional structural adjustment (Figure 14d) during this time. This depositional cycle was brought to a close by a pulse of basement movement which reactivated many fault systems (Figure 14e).

Continued movement along the basement faults formed sub-basins, and subsequent erosion of the high blocks filled these sub-basins with the Synrift II sequence of nonmarine carbonates and sands. This shallow lacustrine unit thins to a veneer of alluvial sands and shales in the onshore portion of Cabinda. As subsidence continued through the Aptian, the region was invaded by saline waters from the south. The subbasins were filled with a potash-rich evaporite which grades upward into a dolomitic unit (Figure 14f). The Synrift II episode ended with an increase in the subsidence rate at the beginning of the Albian.

Concomitant with the increase in subsidence rate

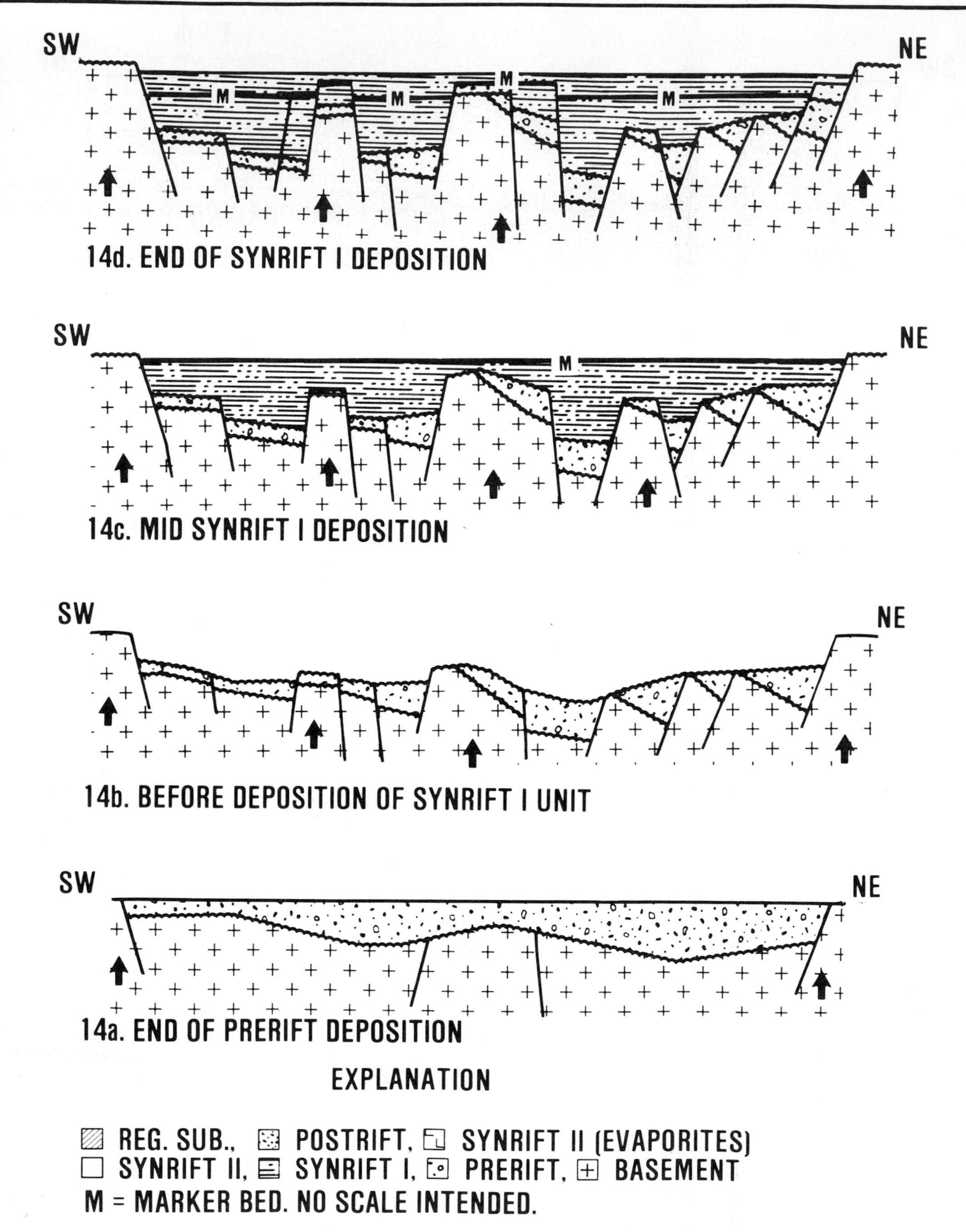

Figure 14 (a-d) — Generalized reconstruction of the geologic history: Cabinda, Angola.

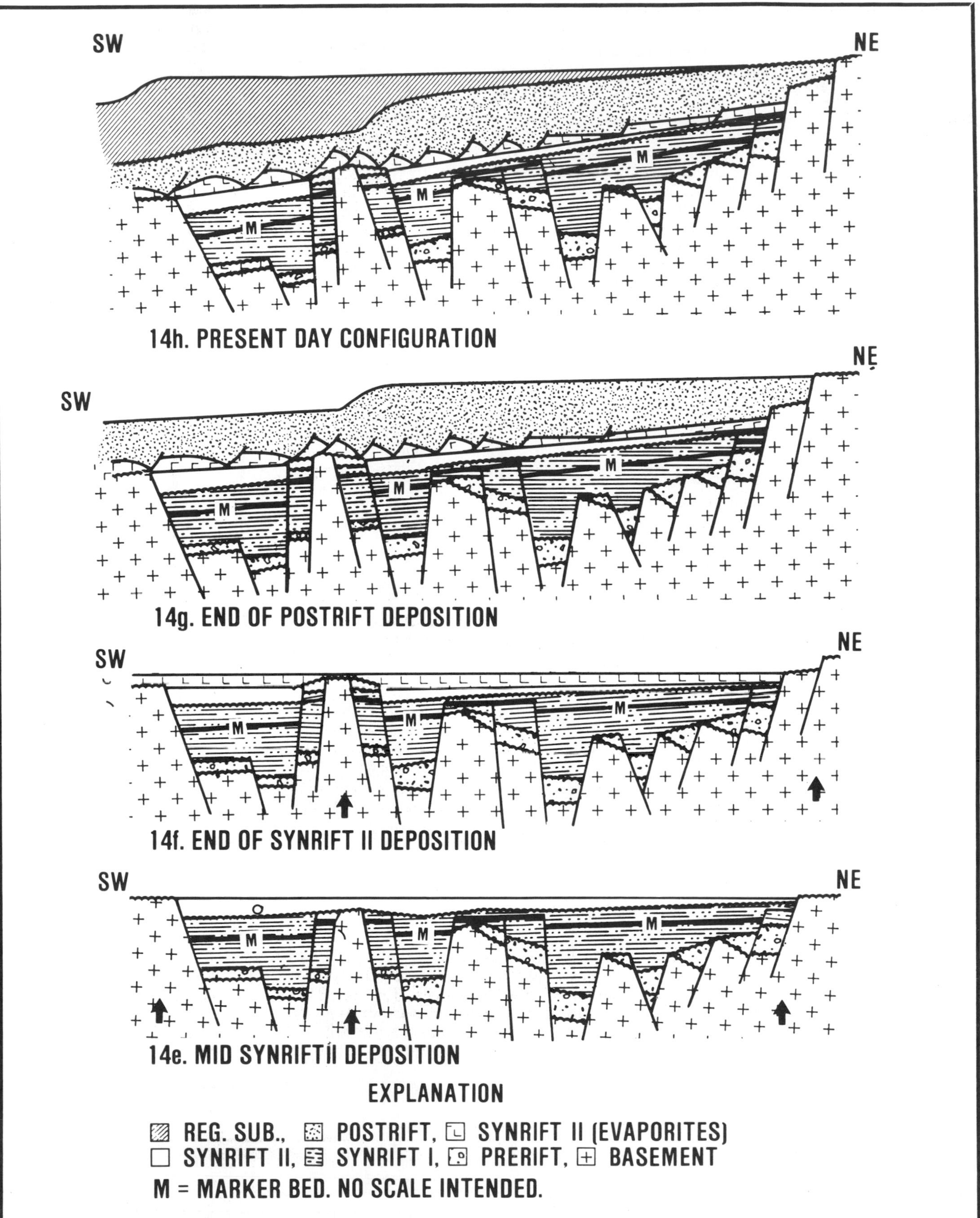

Figure 14 (e-h) — Generalized reconstruction of the geologic history: Cabinda, Angola.

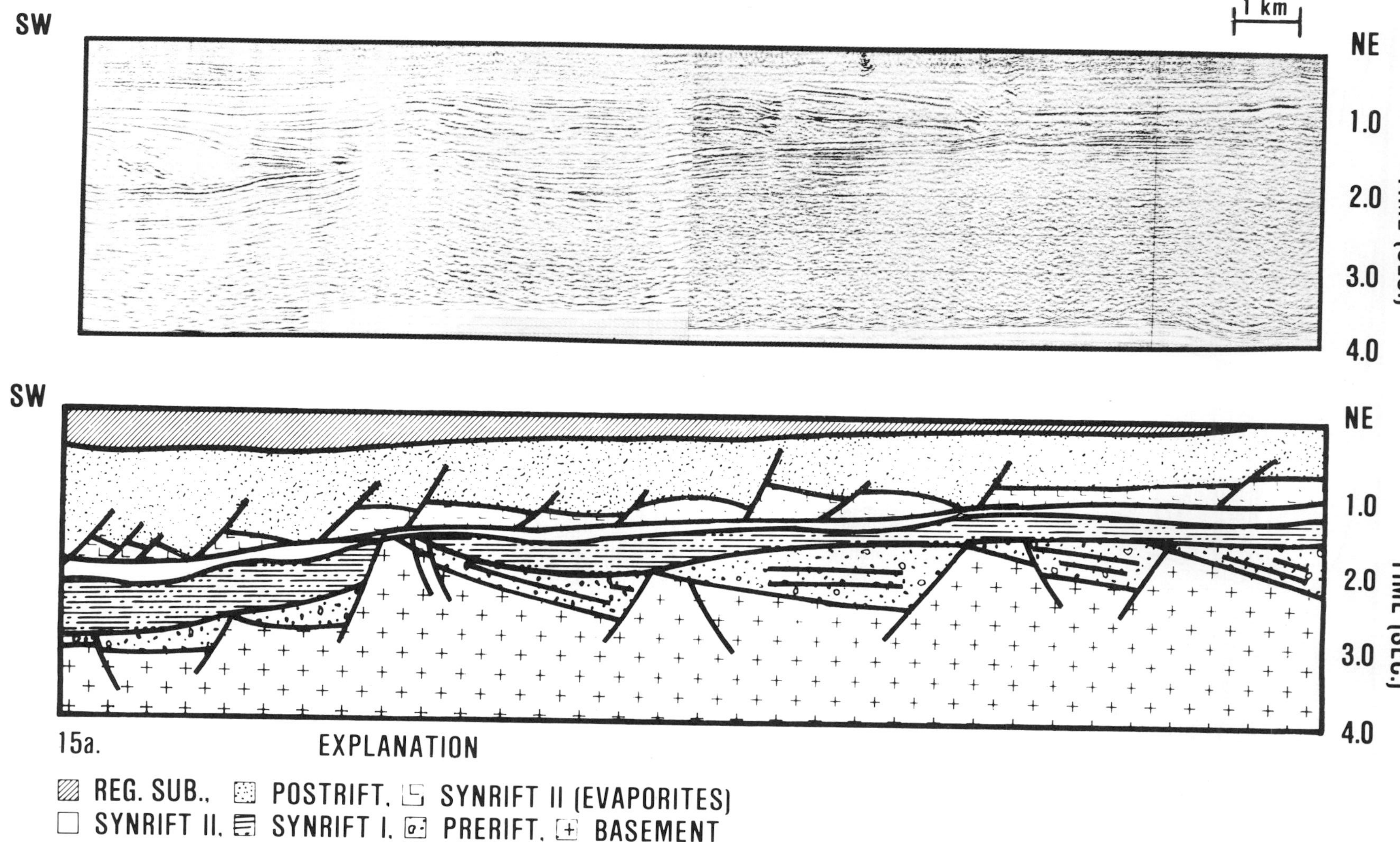

Figure 15a — Regional, true amplitude, migrated seismic line and interpretation; northeast.

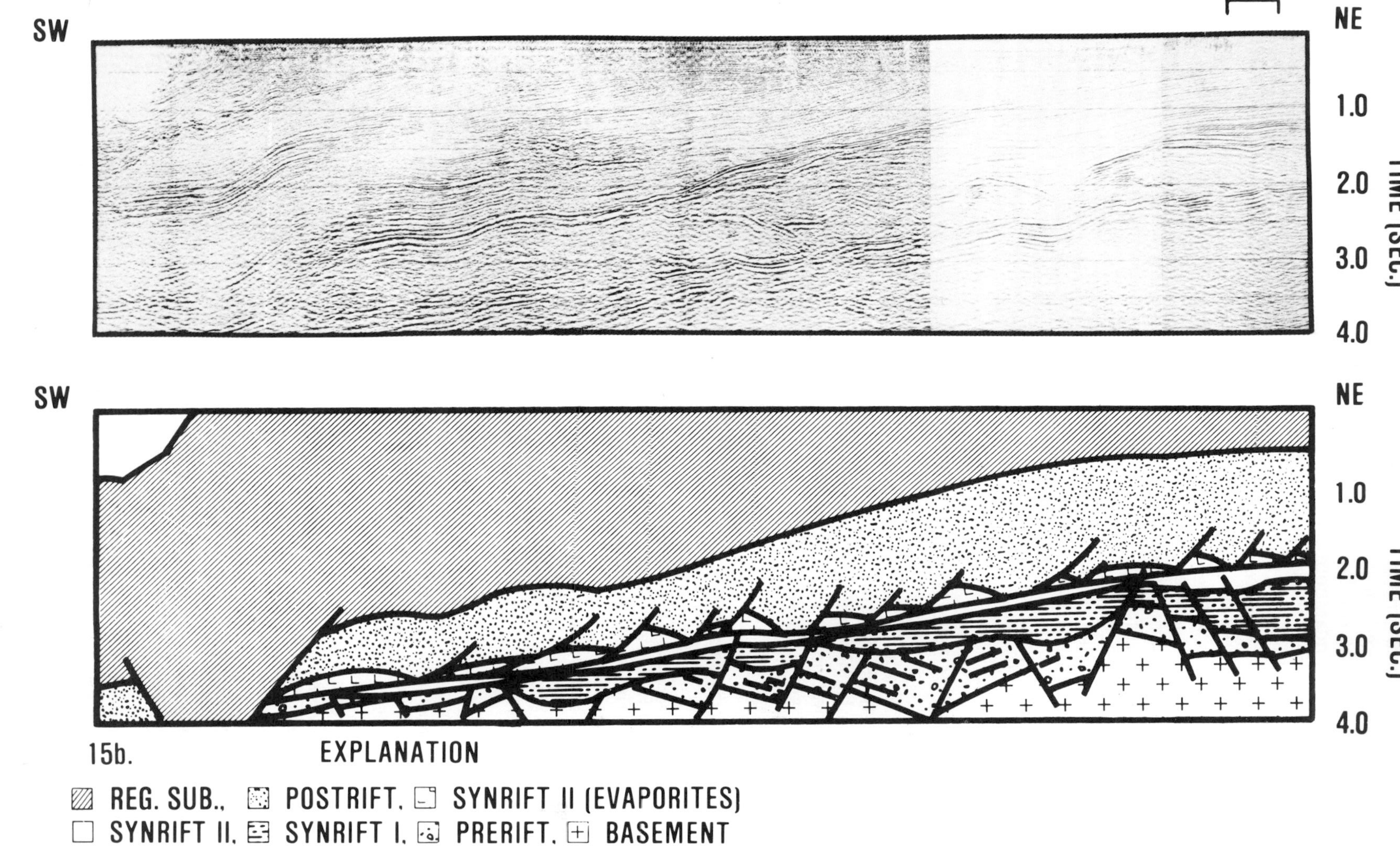

Figure 15b — Regional, true amplitude, migrated seismic line and interpretation; southwest.

was a transgression of marine waters into the Cabinda area from the southwest. This transgressive cycle continued throughout the Campanian and at its maximum extent marine waters covered most of offshore Cabinda and part of the onshore. The Postrift sequence of nonmarine red beds, nearshore carbonates and sands, and deep water marine shales and marls was deposited. Salt movement resulted in growth faults and minor salt domes. Subsidence ceased at the end of the Campanian and a regressive cycle of nearshore carbonates and sands was deposited (Figure 14g). As the region was tilted to the west, a Neogene regressive sequence of shaly clastics was unconformably deposited over the Postrift sequence. During deposition, the shales became overpressured and moved and the salt was reactivated (Figure 14h).

The regional history and tectonics are summarized in Figure 15a and Figure 15b. This section crosses Cabinda from northeast to southwest and is tied to Gulf proprietary well data.

ACKNOWLEDGMENTS

We thank Cabinda Gulf Oil and SONANGOL (Sociedade Nacional de Combustiveis de Angola) for the release of this material. Gulf Oil Exploration and Production Co., Central Exploration Group-International supported this work as a part of its West Africa regional study.

REFERENCES CITED

Asmus, H. E., and F. C. Ponte, 1973, The Brazilian marginal basins, *in* A. E. Nairn and F. G. Stehli, eds., The ocean basins and margins, V. 1, The South Atlantic: New York, Plenum, p. 87-133.

Bacoccoli G., R. G. Morales, and O.A.J. Campos, 1980, The Namorado oil field — A major oil discovery in the Campos Basin, Brazil, *in* M. T. Halbouty, ed., Giant oil and gas fields of the decade 1968-1978: AAPG Memoir 30, p. 329-338.

Belmonte, Y., P. Hirtz, and R. Wenger, 1965, The salt basins of the Gabon and the Congo (Brazzaville) — Tentative paleographic interpretation, *in* Salt basins around Africa: London, The Institute of Petroleum, p. 55-74.

Brice, S. E., and G. Pardo, 1980, Hydrocarbon occurrences in nonmarine, pre-salt sequence of Cabinda, Angola (Abs.): AAPG Bulletin, v. 64, p. 681.

———, K. R. Kelts, and M. A. Arthur, 1980, Lower Cretaceous lacustrine source beds from early rifting phases of South Atlantic (Abs.): AAPG Bulletin, v. 64, p. 680-681.

Caflisch, L., 1975, Comments to the paper L'exploration petroliee au Gabon et au Congo, *in* Exploration and Transportation, Tokyo, proceedings, 9th World Petroleum Congress, v. 3: London, Applied Science Publishers Ltd., p. 199-200.

———, 1979, Oil exploration in pre-salt and post-salt sequence of West Africa — results in Cabinda (Abs.): AAPG Bulletin, v. 63, p. 427.

Campos, C.W.M., F.C. Ponte, and K. Miura, 1974, Geology of the Brazilian continental margin, *in* C. A. Burk and C. L. Drake, eds., The geology of continental margins: New York, Springer-Verlag, p. 447-461.

Carozzi, A. V., et al, 1979, Depositional-diagenetic history of Macae carbonate reservoirs (Albian-Cenomanian), Campos Basin, offshore Rio de Janeiro, Brazil (abs): AAPG Bulletin, v. 63, p. 429.

Lehner, P., and P.A.C., DeRuiter, 1977, Structural history of Atlantic margin of Africa: AAPG Bulletin, v. 61, p. 961-981.

Vidal, J., R. Joyes, and J. Van Veen, 1975, L'exploration petroliere au Gabon et au Congo, *in* Exploration and Transportation, Tokyo, proceedings, 9th World Petroleum Congress, v. 3: London, Applied Science Publishers Ltd., p. 149-165.

Growth Faulting and Salt Diapirism: Their Relationship and Control in the Carolina Trough, Eastern North America

William P. Dillon
Peter Popenoe
John A. Grow
Kim D. Klitgord
B. Ann Swift
Charles K. Paull
Katharine V. Cashman

U.S. Geological Survey
Woods Hole, Massachusetts

The Carolina Trough is a long, linear, continental margin basin off eastern North America. Salt domes along the trough's seaward side show evidence of active diapirism and a normal growth fault along its landward side has been continually active at least since the end of the Jurassic. This steep fault extends to a strong reflection event at about 11 km depth that may represent the top of a salt layer. We infer that faulting is caused by seaward flow of salt from the deep part of the trough into domes, thereby removing support for the overlying block of sedimentary rock. Diapirs off eastern North America seem to be concentrated in the Carolina Trough and Scotian Basin, where basement seems to be thinner than in other basins off eastern North America, south of Newfoundland. Thinner basement, probably due to greater stretching during rifing, may have resulted in earlier subsidence below sea level, a longer life for the salt evaporating pans in these basins, and thus a thicker salt layer, which would be more conducive to diapirism.

The Carolina Trough is one of the four major basins off the United States' east coast, and it is unusual in its configuration and its history of postrift faulting and diapirism. This paper considers the causes of this faulting, the relationship of faulting to salt diapirism, and the reasons for localization of these processes on the eastern North American continental margin.

The Carolina Trough (Figure 1) is long, narrow, and linear, unlike the other east coast basins. It is about 450 km long and 40 km wide and also unlike the other basins, it does not seem to be segmented along the extensions of oceanic fracture zones, although it is terminated by such features. This effect is most notable at its southern end (Figure 1), where the extension of the Blake Spur Fracture Zone abruptly separates the Carolina Trough from the Blake Plateau Basin to the south.

A major system of normal faults extends for more than 300 km along the northwestern (landward) side of the basin and a linear group of diapirs is located on the basin's southeastern side (Figure 1). The diapirs are considered to be cored by salt because the chlorinity values distinctly increase downward in short sediment cores taken on top of them (F.T. Manheim, unpublished data, 1980). Distribution of the diapirs indicated in Figure 1 is based on the grid of multichannel seismic profiles shown, plus a much denser grid of single-channel seismic lines and a long-range sidescan-sonar survey. Sidescan-sonar was useful because many diapirs disrupt the sea floor.

EVIDENCE FOR MAJOR GROWTH FAULTING AND DIAPIRISM

To examine the structure of the Carolina Trough, we consider three adjacent multichannel seismic lines: BT1, 32, and TD6 (locations shown in Figure 1). These three profiles cross the central part of the trough and are about 60 km apart.

Seismic profile BT1

A part of profile BT1, showing a complete crossing of the trough, is presented in Figure 2 and its interpretation is shown in Figure 3. We interpret that an unconformity, giving rise to diffractions, dips to the southeast at the left side of the profile segment and extends beneath a set of very strong subhorizontal reflectors at 6 to 7 seconds. The unconformity is considered to be the postrift unconformity or breakup unconformity (Falvey, 1974), and the strong subhorizontal reflections are considered to arise from salt

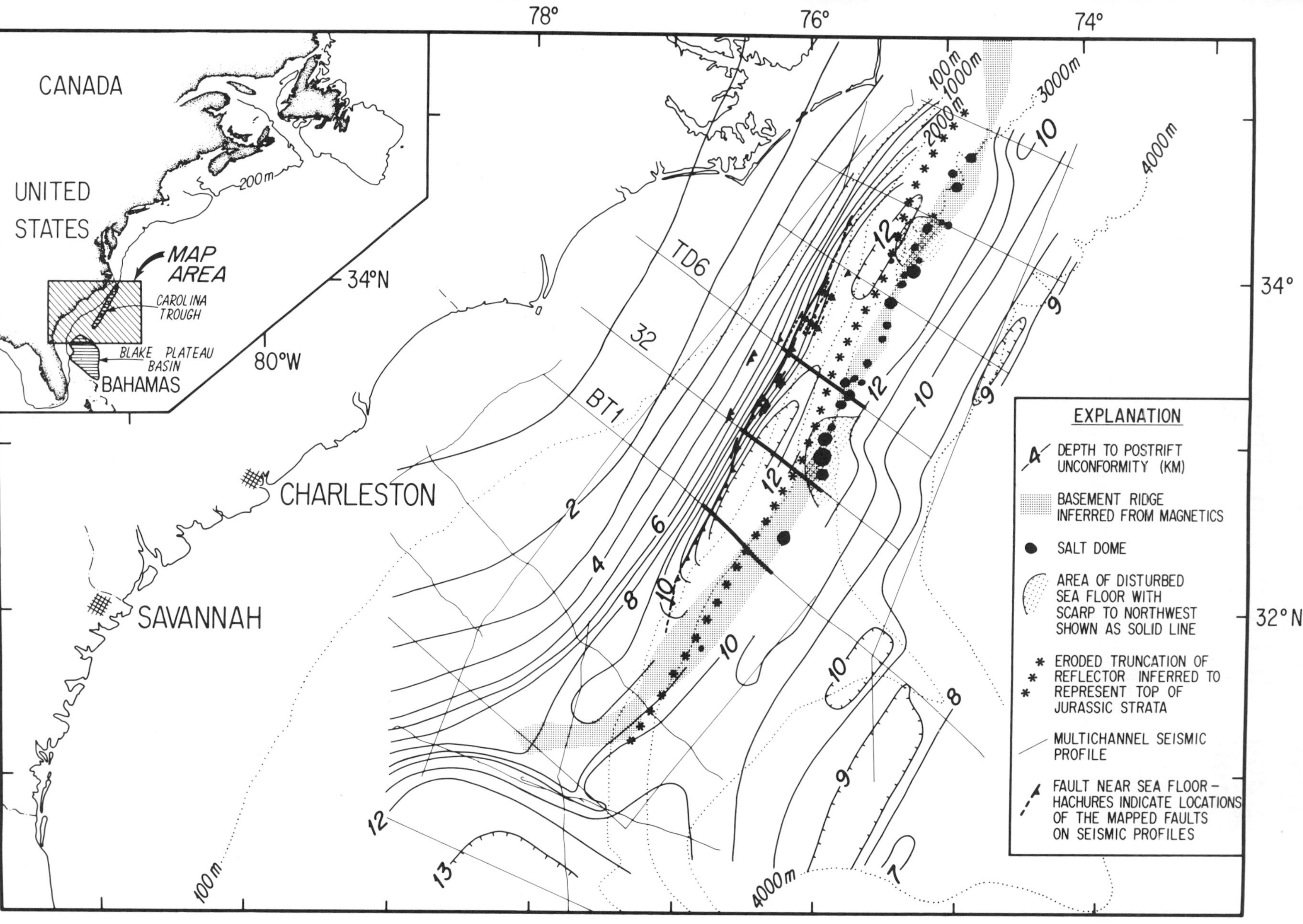

Figure 1 — The Carolina Trough. Contours showing depth in kilometers to the postrift unconformity are based on multichannel seismic-reflection profiles. The difficulty in defining this horizon is discussed in the text. Near-surface locations of faults are mapped. Locations of faults and diapirs are based on multichannel seismic profiles (locations shown), a much more dense coverage of single-channel seismic profiles and, for the diapirs, a single *Gloria* sidescan swath run along the axis of the group of diapirs. Hachures on faults show downthrown side and also indicate location of a profile crossing the fault. Lightly dotted lines are bathymetric contours. The parts of the three seismic lines shown in Figures 2, 7, and 11 are thickened.

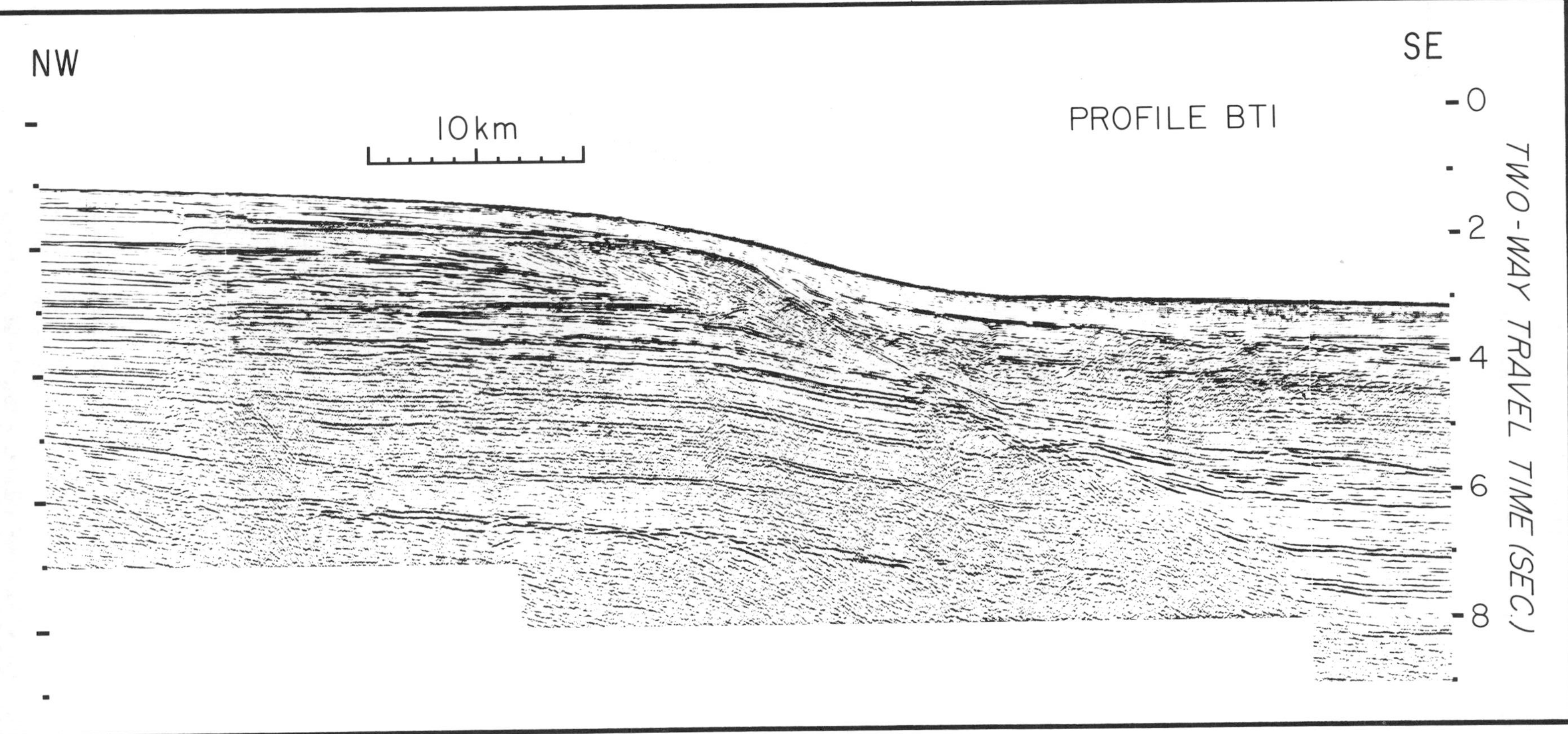

Figure 2 — Part of seismic profile BT1 that crosses the Carolina Trough (northwest is to left). Profile location is shown in Figure 1. Data were collected by Geophysical Service Incorporated (GSI) using a 1310-cu in (21,467 cu cm) airgun array, and a 48-trace, 3600-m streamer. Data were processed by GSI for a 36-fold stack. The strong set of reflections at 6 to 7 seconds to the left of center may arise from a salt layer. The bottom simulating reflector (BSR) at 0.3 to 0.5 seconds subbottom depth in the middle and right side of the figure probably represents the base of gas hydrate cementation of sediment (Dillon, Grow, and Paull, 1980; Paull and Dillon, 1981). No reflections indicative of major reef development are apparent in this profile.

(although they are marked in the interpretations with reversed L's for salt and rhombs to signify the possibility of a dolomite layer, which also could produce a strong reflection).

The postrift or breakup unconformity originally was defined as forming by erosion during a postulated "final uplift pulse associated with pre-breakup upwelling in the mantle," representing "the youngest cycle of subaerial erosion in a marginal basin." It was thought to be "very nearly the same age as the oldest oceanic crust in the adjacent deep ocean basin" (Falvey, 1974). The emphasis on subaerial erosion was Falvey's.

The reflection event that we called the postrift unconformity is mapped in Figure 1 to define the trough. The postrift unconformity is definable without controversy northwest of the inferred salt pinchout (Figure 2), where it is considered an eroded surface on Paleozoic basement and early Mesozoic continental deposits. To best define the Carolina Trough, we contoured the extension of this surface, atop Paleozoic and early Mesozoic rocks, where it continues southeastward beneath the inferred salt in the deepest part of the trough. It is unclear whether this horizon is truly the postrift unconformity because we do not know the relative age of the salt. Did seawater find its way into the trough during the rifting stage, resulting in the salt forming as a synrift deposit? In that case, the post-rift unconformity should be extended across the top of the salt. Alternatively, did the salt form just after the rift to drift change? At that time newly formed oceanic crust might have floated deep isostatically and formed a channelway, allowing oceanic water to finally enter the trough. In the latter case, the post-rift unconformity would occur beneath the inferred salt, if it existed at all. A subaerially eroded post-rift unconformity would not exist in the deepest part of the trough if deposition had been continuous there after initial subsidence in the rifting stage. The problem of the nature and location of the postrift unconformity in the Carolina Trough is unsolvable at present.

The unconformity is identified with difficulty seaward of the pinchout of the strong "salt" reflector, and contours are dropped where identification becomes too extremely imaginative. Much of the obscuring of basement seems due to diffractions arising at the eroded paleoslope formed of truncated Jurassic and Cretaceous strata. Some interpreters of magnetic data place a magnetic basement ridge in this region of obscured seismic returns (Klitgord and Behrendt, 1979), whereas others model no such ridge (Hutchinson and others, this volume). If no ridge is modeled, the magnetic anomaly is considered to be due to the contact of basements of different magnetic susceptibilities. We avoided the controversy by simply indicating the location of the proposed ridge (Klitgord and Behrendt, 1979) with a dot pattern (Figure 1). The point is significant because the diapirs are aligned on the magnetic anomaly that can be modeled as a ridge, and the possibility of control of diapir location by basement structure should be considered.

Several faults are indicated in Figure 3. The dominant fault, shown on the left of the figure, is observ-

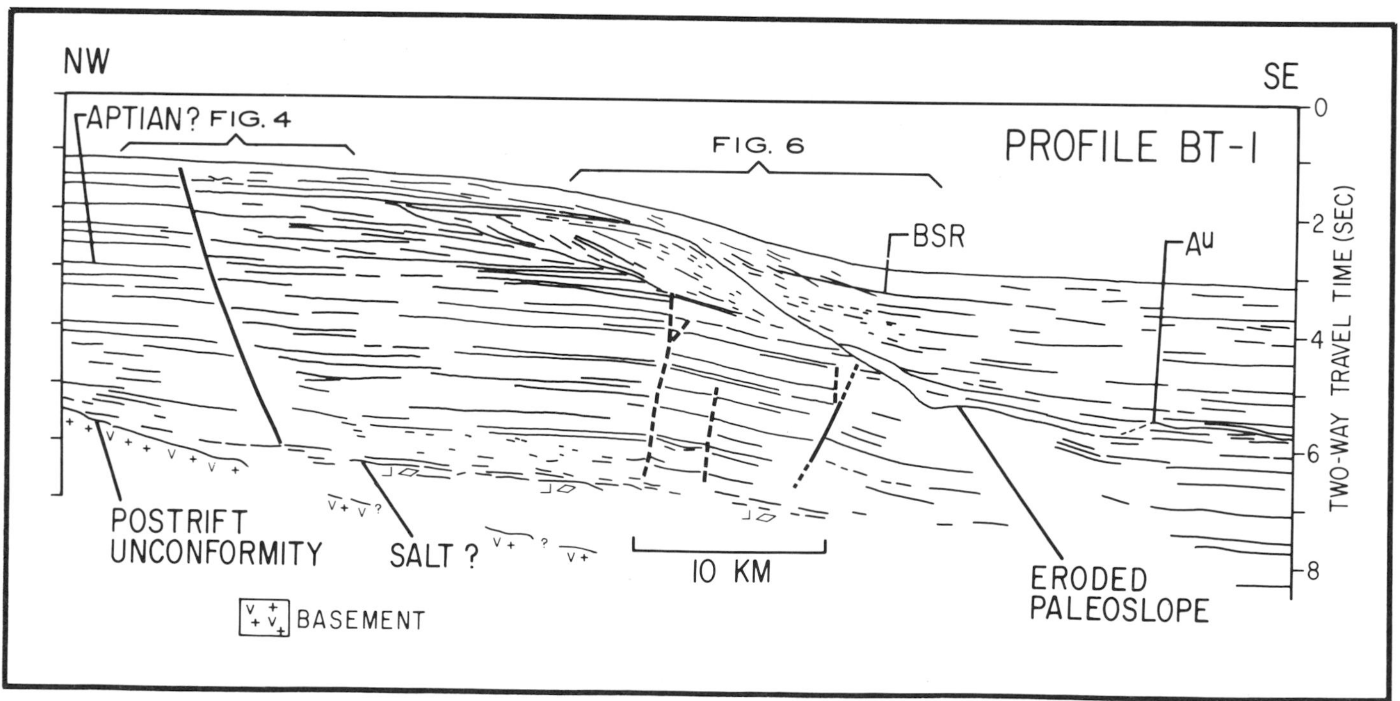

Figure 3 — Interpretation of the section of seismic profile BT1 shown in Figure 2. Light lines indicate reflections, heavy lines show interpreted faults. Brackets show location of detailed record photos shown in Figures 4 and 6. Stratigraphic identifications in this and other profile interpretations (Figures 8 and 12) are based on long-distance extrapolation from drilled horizons and should be considered preliminary. BSR, bottom simulating reflector; Au, horizon A unconformity.

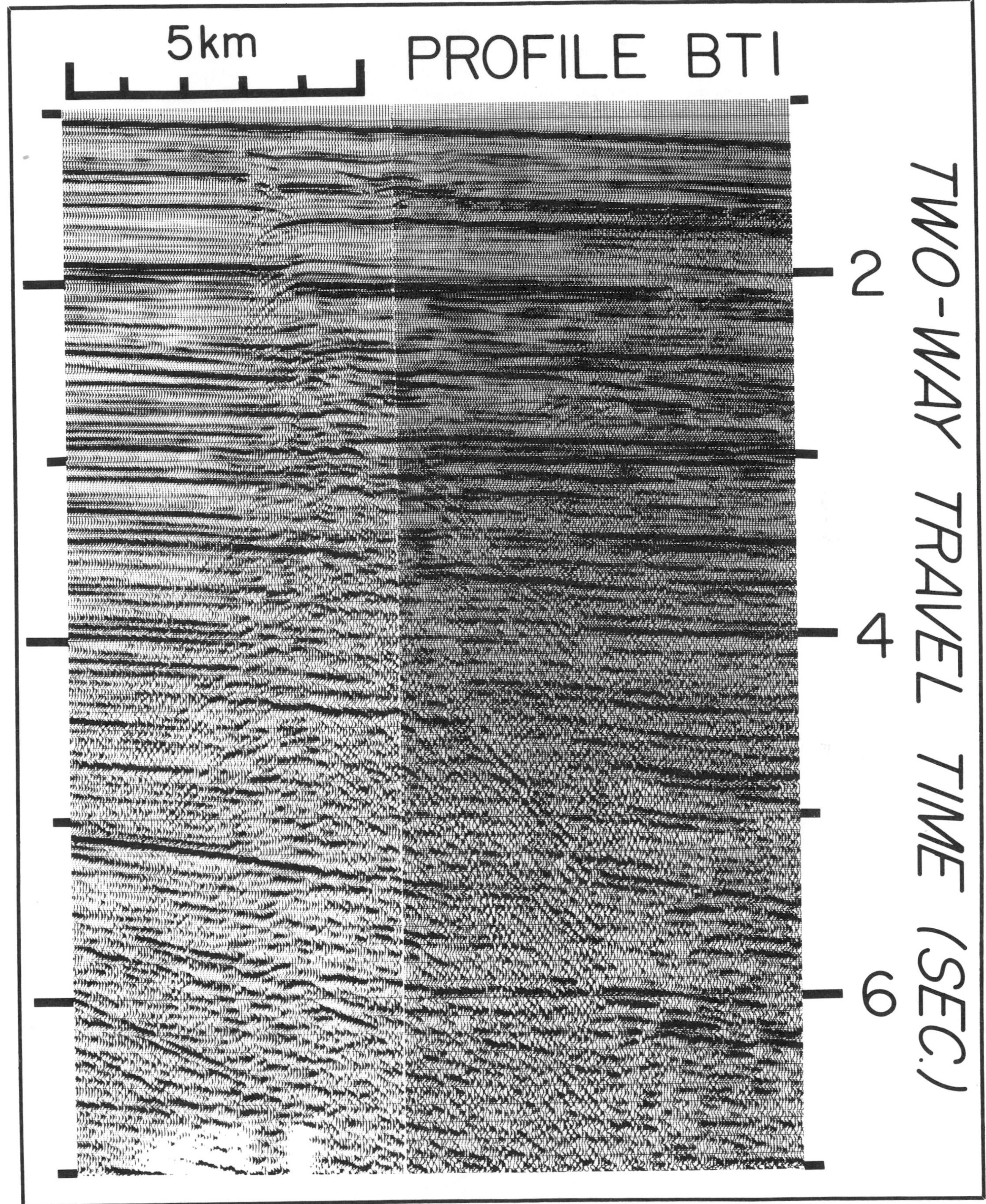

Figure 4 — Detail of seismic profile BT1, showing main growth fault. Location is shown in Figure 3.

able in many profiles. Its near-surface location is mapped in Figure 1 on the basis of both multichannel profiles and a much denser grid of single channel profiles. Hachures on the faults (Figure 1) show the downthrown side and the locations of control, where our profiles cross the fault. The main fault, shown at the left sides of Figures 2 and 3, is presented enlarged in Figure 4. We believe that certain distinctive packages of reflection events can be matched across the fault on profile BT1, allowing us to calculate throws at various depths. A plot of these data (Figure 5) shows that throw increases fairly smoothly as depth increases, indicating that the fault was active during sediment deposition (i.e., a growth fault). We believe that the fault should be termed a growth fault (Ocamb, 1961) because it shows evidence of movement during the deposition. An equivalent term is "contemporaneous fault," defined by Hardin and Hardin (1961). Our stratigraphic estimates are not sufficiently developed in this area to make a throw versus age plot. However, assuming that the long-term sedimentation rate did not vary greatly, Figure 5 suggests that movement on the growth fault at the three locations graphed was at an approximately constant average rate. Throw is observed to increase downward at least as deep as a horizon inferred to be the top of the Jurassic (the salt is inferred to be of Early Jurassic age). Thus, the fault has been active at least since the end of the Jurassic and probably earlier. Below the inferred top of Jurassic, reflections cannot be matched across the fault.

The fault seems to continue steeply to the interpreted salt layer. On profile BT1, the fault is located well landward of the paleoshelf-slope break; it does not appear to curve and flatten into bedding, and it does not seem to have associated antithetic faults expected to exist at a fault having a curved fault plane. Thus, it is not characterized by features associated with ordinary slump-type growth faults of the continental margin. The lack of curvature on this fault is, of course, best documented on a depth converted section. One is shown in Figure 21 for a nearby crossing of the fault.

Three other faults are interpreted to exist seaward of the main growth fault (Figures 3, 6). No major movement seems to have taken place across these faults, as distinct major reflector packets are identified on both sides of the fault and show little offset.

Major erosion of outer shelf strata occurred during the cutting of the horizon A unconformity in the

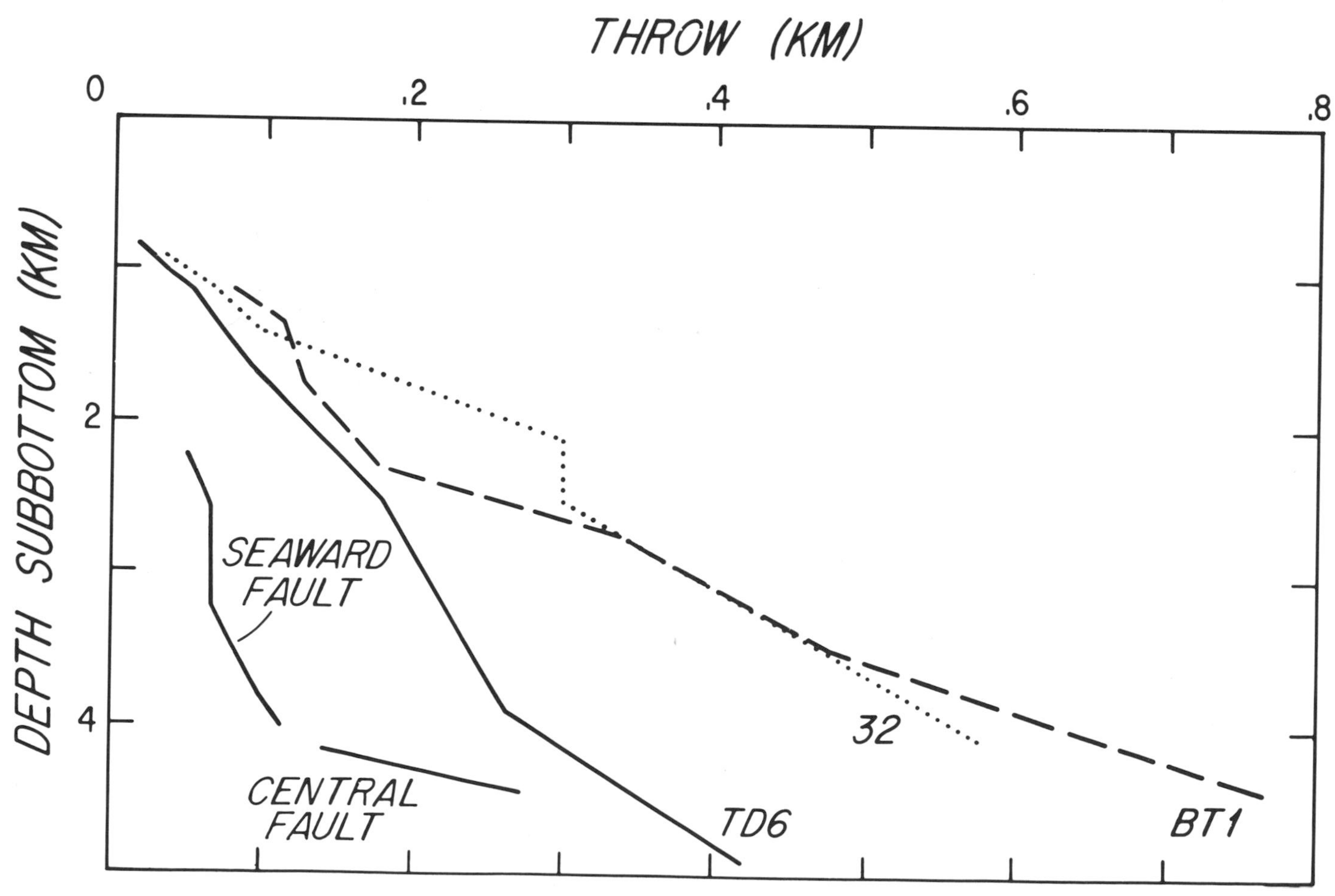

Figure 5 — Plot of throw versus depth for the faults of the three profiles discussed in the text. Solid lines identify the three fault patterns plotted for profile TD6.

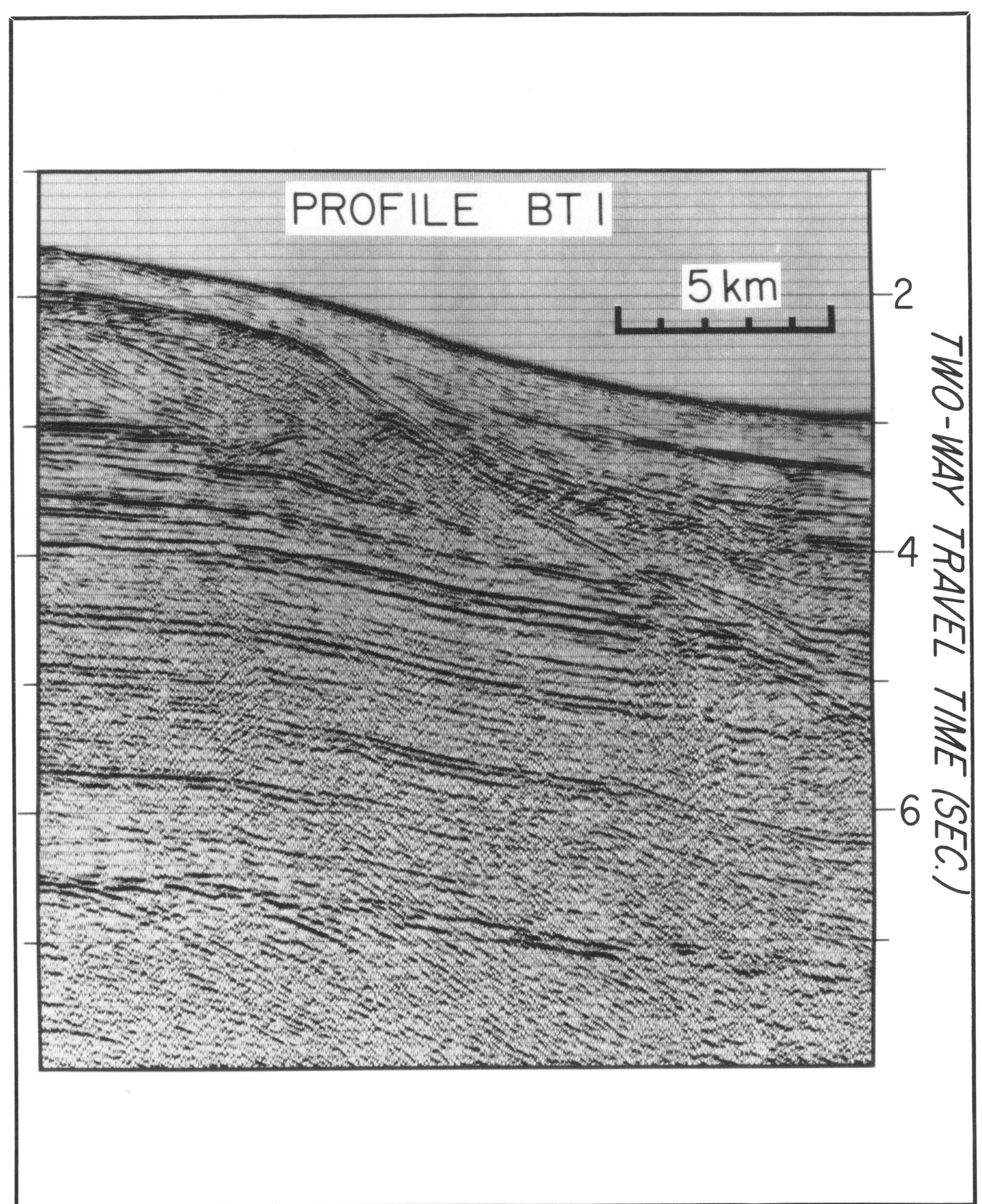

Figure 6 — Detail of seismic profile BT1 showing eroded shelf edge and several faults. Location and interpretation are shown in Figure 3.

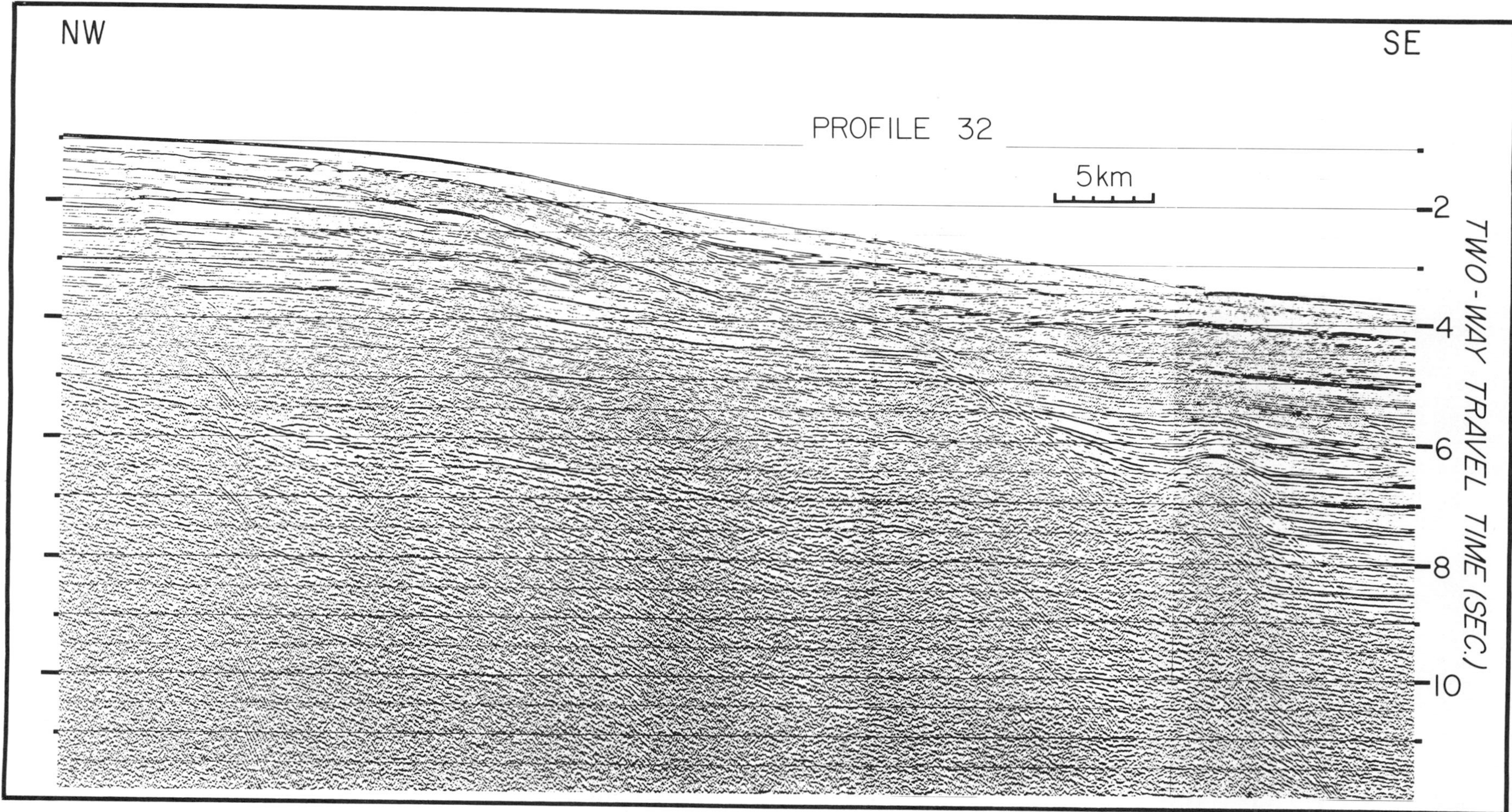

Figure 7 — Part of seismic profile 32 that crosses the Carolina Trough (northwest is to left). Profile location is shown in Figure 1. Data were collected by Geophysical Service Incorporated using a 2000-cu in (32,774 cu cum) airgun source and 48-trace, 3600-m streamer. It was processed for 48-fold stack by GSI.

western North Atlantic (Tucholke, 1979), and the results of this erosion are apparent in Figures 3 and 6. At least one previous episode of erosional retreat and progradational advance of the shelf edge also is apparent in this profile.

Seismic profile 32

Profile 32 (Figures 7, 8) also shows the major fault at the landward side of the Carolina Trough and the strong reflections inferred to come from salt. The fault (Figure 9) shows increased throw as depth increases (Figure 5) in a pattern very similar to that of the fault on profile BT1. A small diapiric upwarp is present just seaward of the eroded Jurassic and Cretaceous strata of the paleoslope (Figures 8, 10). Several subhorizontal reflection events are present beneath the paleoslope in the time section (left side of Figure 10, 4 to 8 seconds). These subhorizontal reflections, when depth corrected, show a landward dip due to the seaward-thickening water layer. This reverse dip is considered post-depositional and not associated with structure of a shelf-edge reef or bank; rather, the rotation of beds down toward the trough axis probably was caused by subsidence of the strata in the main part of the trough to the west, due to differential thermal subsidence of basement and salt withdrawal. The reduced coherence of reflection events beneath the steepest part of the eroded paleoslope probably is caused by interference of diffraction patterns generated at the rough interface of the paleoslope.

Reflections from strata just below the sea floor are partially obscured on line 32 by the blanking effect presumably produced by gas hydrates formed in the sediments. This is particularly noticeable on the left and right sides of Figure 10. The base of the gas hydrate cemented zone forms a bottom-simulating reflector (BSR) at 0.4 to 0.5 seconds below the sea floor at the pressure-temperature limit for gas hydrate stability. A similar BSR at 0.3 to 0.5 seconds subbottom is apparent in profile BT1 (Figure 2). The BSR is very well developed in the southern part of the Carolina Trough region (Shipley et al, 1979; Dillon, Grow, and Paull, 1980; Paull and Dillon, 1981).

Seismic profile TD6

Like the other profiles shown, profile TD6 displays a main growth fault, which is part of the system mapped in Figure 1, and also some faults to the southeast that dip landward and cannot be mapped with our line spacing (Figures 11, 12; detail of profile showing faults, Figure 13). As in the previous cases shown, the landward-dipping faults are thought to terminate at depth in a salt layer, and a well-developed salt dome is present (Figure 14). The landward-dipping faults also show growth fault characteristics on this profile (Figure 5). Unlike the other profiles, TD6 displays some apparently antithetic faults that join the main growth fault near inflections of the fault plane (Figure 13). The deeper sections of the fault, between the inflections, become progressively steeper with increasing depth.

Downward steepening of the fault plane of the main growth fault (left fault, Figure 13) cannot be an artifact of the seismic profile because increasing velocities of deeper rocks has the effect of apparently bending *up* the fault plane in a time section, rather

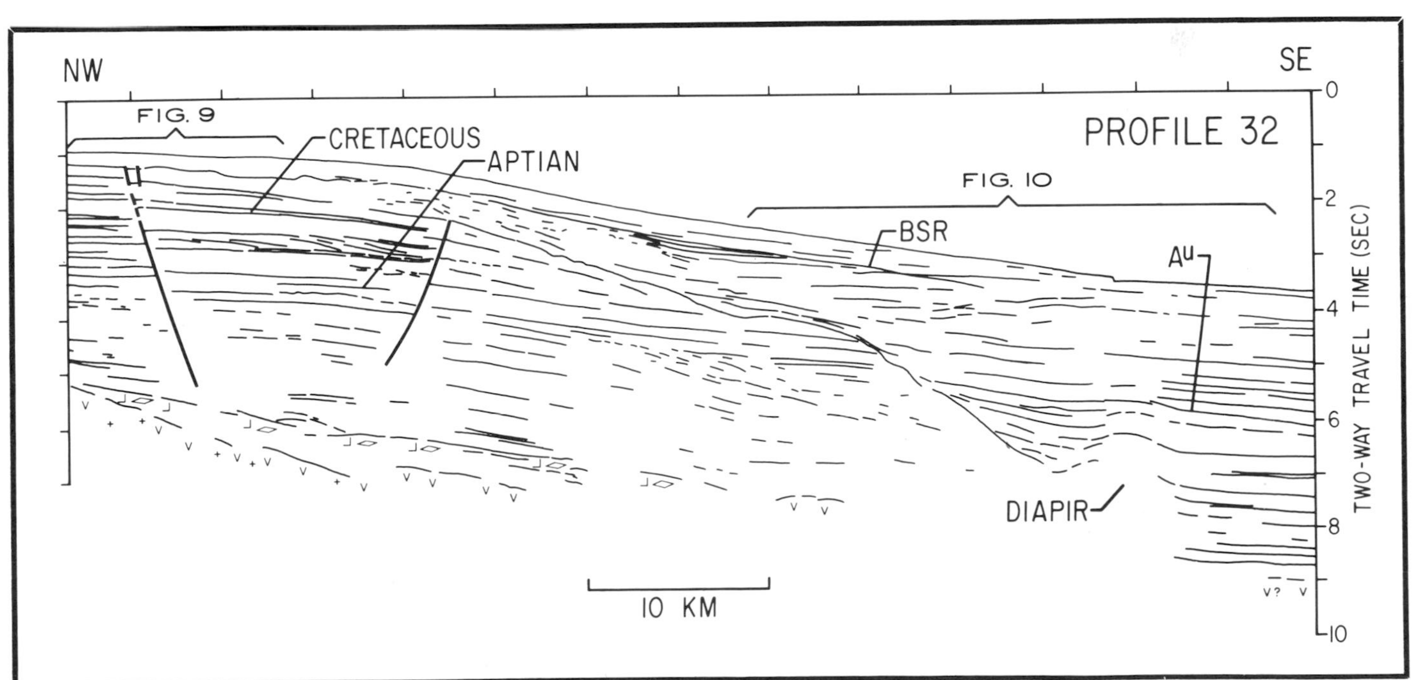

Figure 8 — Interpretation of the section of seismic profile 32 shown in Figure 7. Symbols are the same as those used in Figure 3.

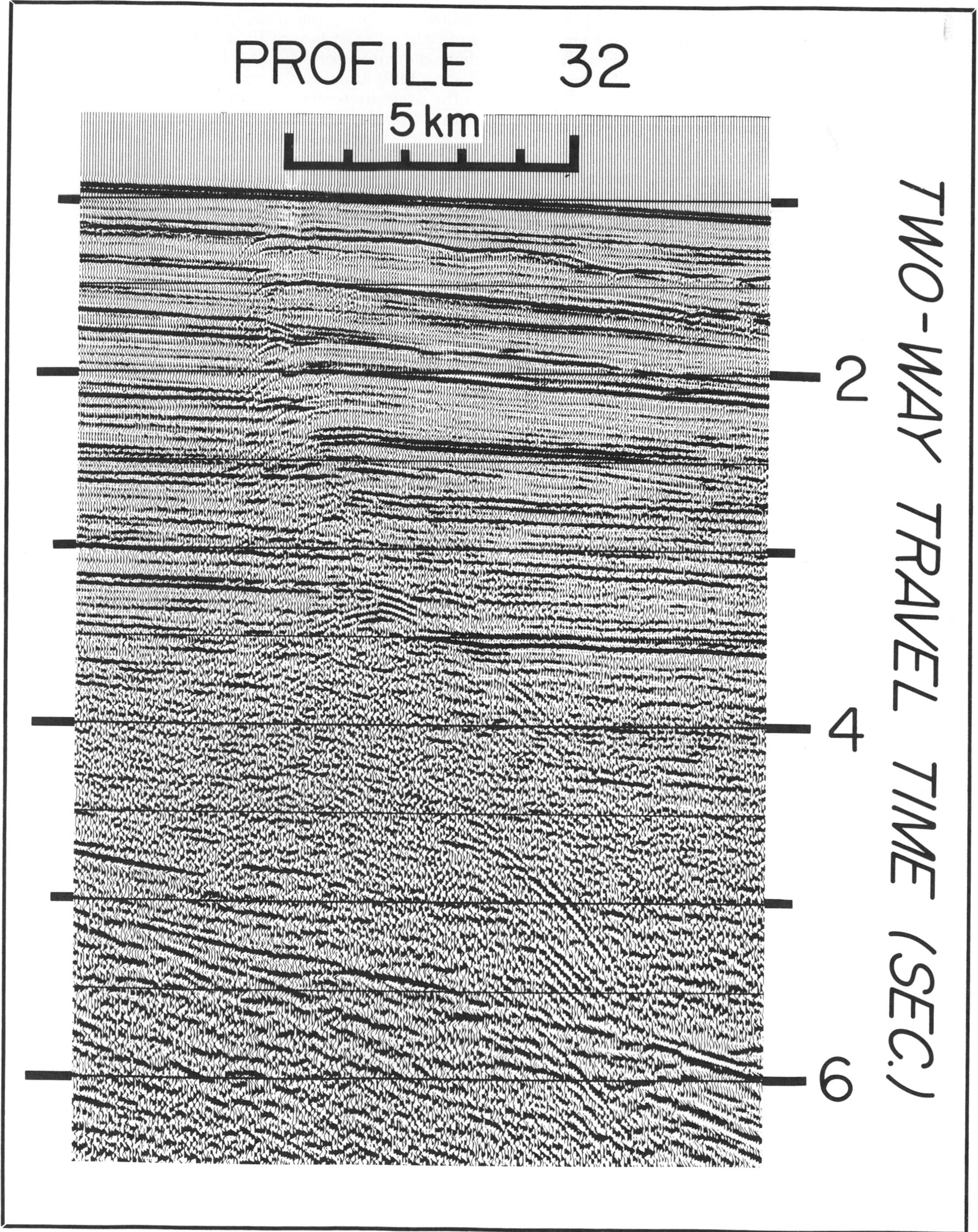

Figure 9 — Detail of seismic profile 32 showing main growth fault. Location is shown in Figure 8.

than bending it down. When these faults are replotted to true scale with no vertical exaggeration, the increasing dip with depth on the main growth fault remains. The curve of the two landward-dipping faults is nearly removed in a depth-corrected plot, indicating that those faults are almost planar.

We acknowledge that the geometry of these faults is radically different from the downward-flattening growth faults with antithetic normal faults and rollover structures that are ordinarily found on continental margins where high deposition rates prevail (Hardin and Hardin, 1961; Ocamb, 1961; Short and Stäuble, 1967; Lehner, 1969; Bishop, 1973; Bruce, 1973; Edwards, 1976, 1981; Weimer and Davis, 1977; Rider, 1978; Harding and Lowell, 1979). In the ordinary growth faults, not only do the faults flatten seaward, but the main faults all dip seaward except for the antithetic faults terminating in the main faults. Conversely, in the Carolina Trough we consistently observed landward-dipping faults that do not appear to terminate at depth in a seaward-dipping fault. Ordinary, concave-up growth faults show extensional features near the fault because the seaward movement of the upper block by gravity gliding tends to result in opening of a gap at the shallow, more steeply dipping part of the fault. Such a gap does not actually open, of course, because it is filled by collapse of shallow strata, either by fracture (antithetic normal faults) or by plastic subsidence (rollover anticlines). Such structures are not observed in the Carolina Trough faults. Indeed, the deeper strata, in some cases (Figure 13), actually seem to have undergone shortening and been folded against the main fault. The ordinary growth fault pattern is consistent with a model in which a block slides seaward by gravity gliding (Crans and Mandl, 1981a, b), so perhaps such a model is not applicable in the Carolina Trough.

RECENCY OF MOTION OF FAULTS AND DIAPIRS

The multichannel profiles indicate that the main fault breaking the strata on the west side of the deep Carolina Trough is a growth fault, as demonstrated by the plot of throw versus depth (Figure 5). By our

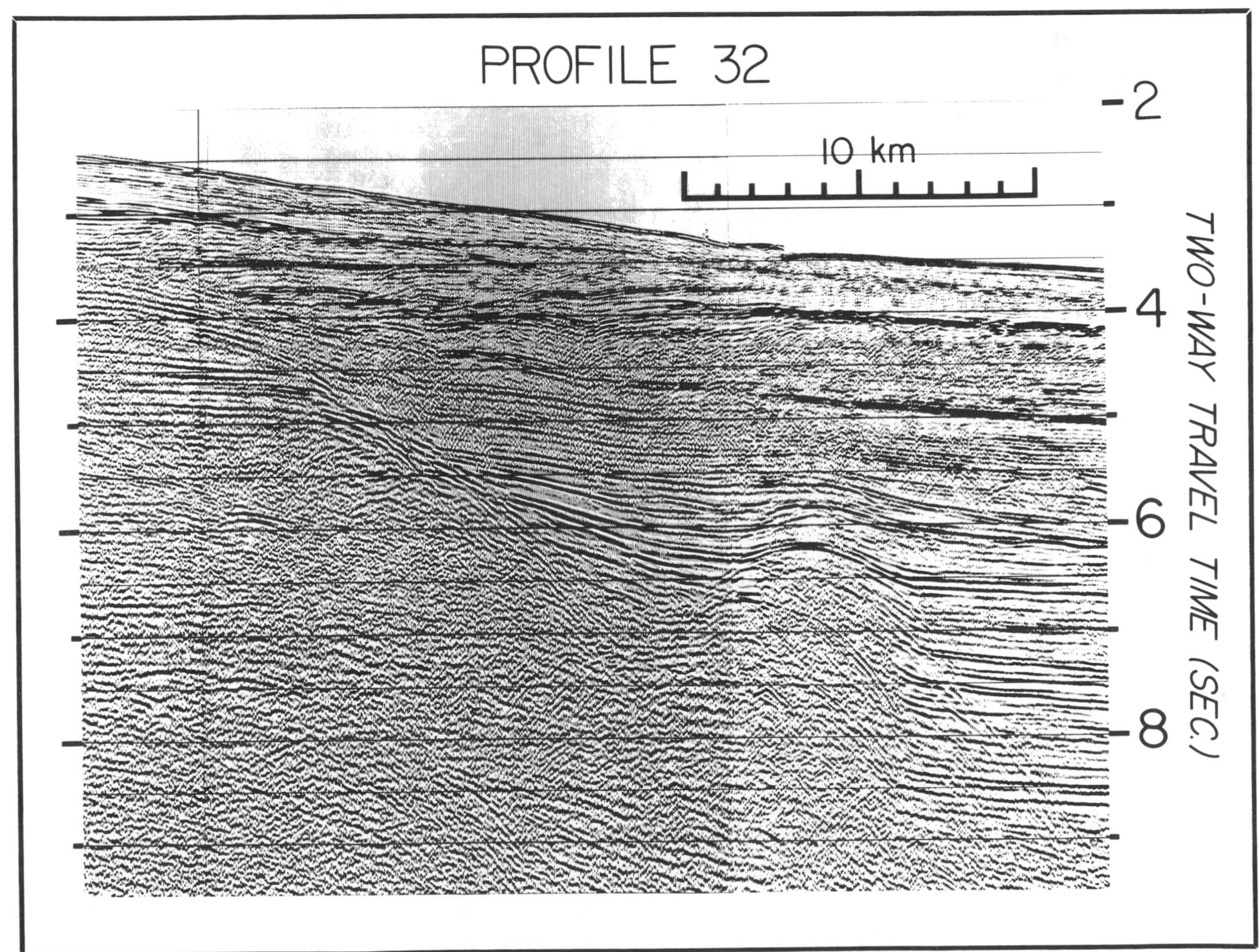

Figure 10 — Detail of seismic profile 32, showing eroded paleoslope and diapir. Location is shown in Figure 8.

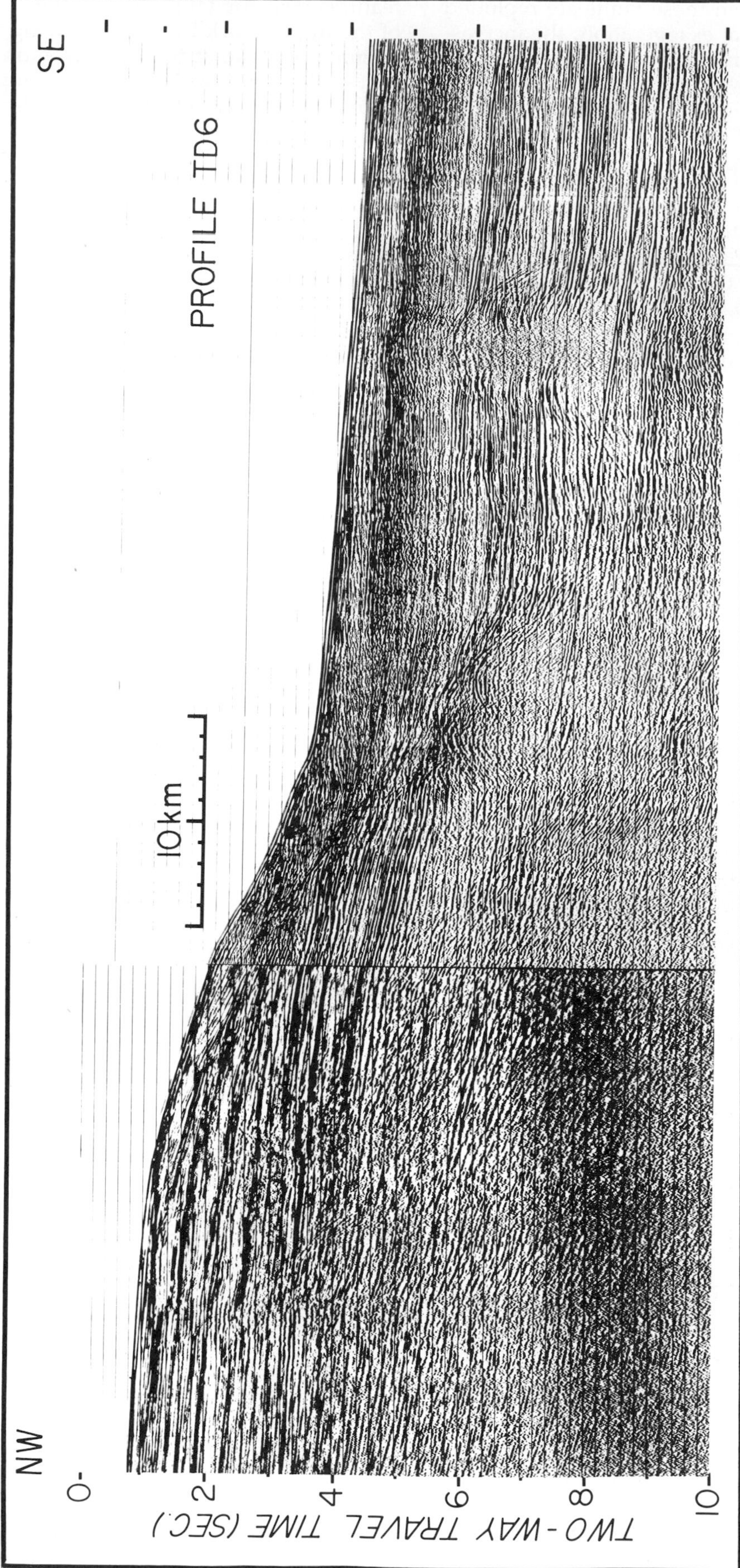

Figure 11 — Part of seismic profile TD6 that crosses the Carolina Trough (northwest is to left). Profile location is shown in Figure 1. Data were collected by Teledyne using a 2160-cu in (35,396 cu cum) array of airguns and a 48-trace, 3600-m streamer. Data were processed by the U.S. Geological Survey for a 36-fold stack.

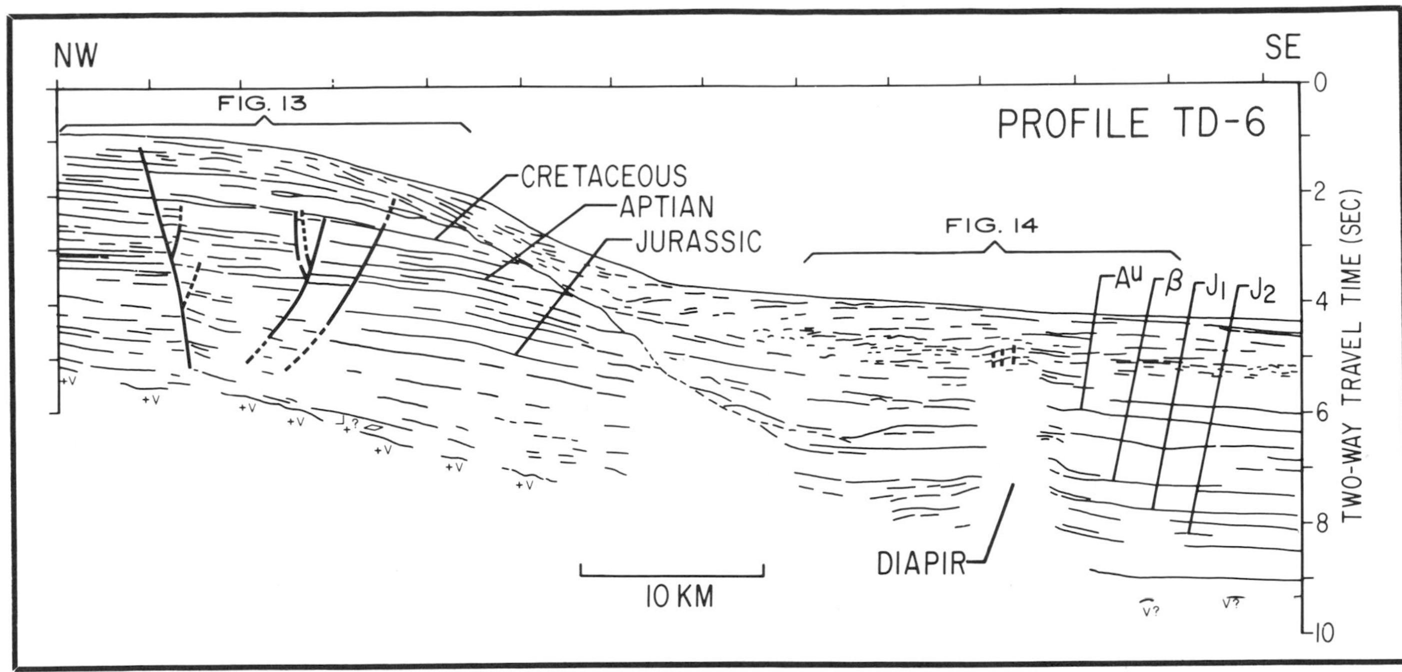

Figure 12 — Interpretation of the section of seismic profile TD6 shown in Figure 11. Symbols are the same as those used in Figure 3.

extrapolated estimates of stratigraphic horizons, we see increasing offsets back to at least the end of the Jurassic. Resolution of multichannel profiles does not allow us to define the most recent movement on the fault, so we have collected high-resolution single-channel profiles for that purpose (Sylvester, Dillon, and Grow, 1979). A high-resolution profile across the fault, 20 km northeast of profile 32 (Figure 15), shows that the effect of the fault extends to within 0.04 seconds of the sea floor (about 30 m). Breaks are not evident at that level, however, and we may be observing draping at very shallow subbottom depths. By matching packets of reflectors we conclude that a throw of about 35 m (slightly less than 0.05 seconds) exists at 200 m subbottom (0.27 seconds subbottom, or about 1.3 seconds below sea surface). Any fault showing effects so close to the sea floor in an area of deposition must be considered active.

Our detailed surveys show that diapirism is presently active because salt diapirs deform the sea floor in an area of active sedimentation. For example, Figure 16 shows a strike line through diapirs that offset the sea floor just northeast of profile 32; Figure 17 shows a dip line through the diapir that is shown in Figure 16 to have the greatest sea-floor relief. The small scarp about 13 km upslope from the diapir in Figure 17 also appears on line 32 (Figure 10). *Gloria* sidescan- sonar records demonstrate that the scarp is arcuate and about 50 km long (Figure 18). The location of this scarp and a second one to the north, as well as areas of hummocky topography adjacent to them, are mapped in Figure 1. Two high-resolution profiles across the scarp are presented in Figures 19 and 20. Strata seaward of the scarp do not appear extensively disrupted on the multichannel profile (Figure 10), but Figures 17 and 19 show disturbance, and the sidescan record shows that the sea floor is hummocky within the arc of the scarp. Subsidence of the sea floor east of the scarp has removed support for slope sediments, resulting in a series of small slump faults (Figure 20). We conclude that the sea floor east of the scarp collapsed due to salt solution and the uplifting of the sea floor by diapiric salt flow. This collapse is continuing to create the hummocky topography.

STRUCTURAL MODEL — RELATIONSHIP BETWEEN GROWTH FAULT AND DIAPIRS

The locations of the main growth fault and the salt diapirs clearly are related to the morphology of the Carolina Trough. The growth fault is located on the landward side of the deep basin of the trough (Figure 1). The diapirs are found on the seaward side of the trough and are located on a magnetic anomaly that may arise from a basement ridge (Figure 1). The diapirs also are seaward of, and trend parallel with, the subcrop of the top of Jurassic(?) strata where it is truncated at the eroded paleoslope (line of asterisks, Figure 1). Certainly the Jurassic shelf edge was seaward of this eroded-back position, but analysis of seismic data suggests that the Jurassic shelf edge was not far seaward of the position and, therefore, the diapirs probably rose beneath the Jurassic continental slope.

We suggest that salt was deposited in the deepest part of the Carolina Trough, and that it was loaded by sediments during the Jurassic and began to flow seaward and migrate into rising domes (Parker and McDowell, 1955; Humphris, 1979). The location of the domes probably was controlled by a shallowing of

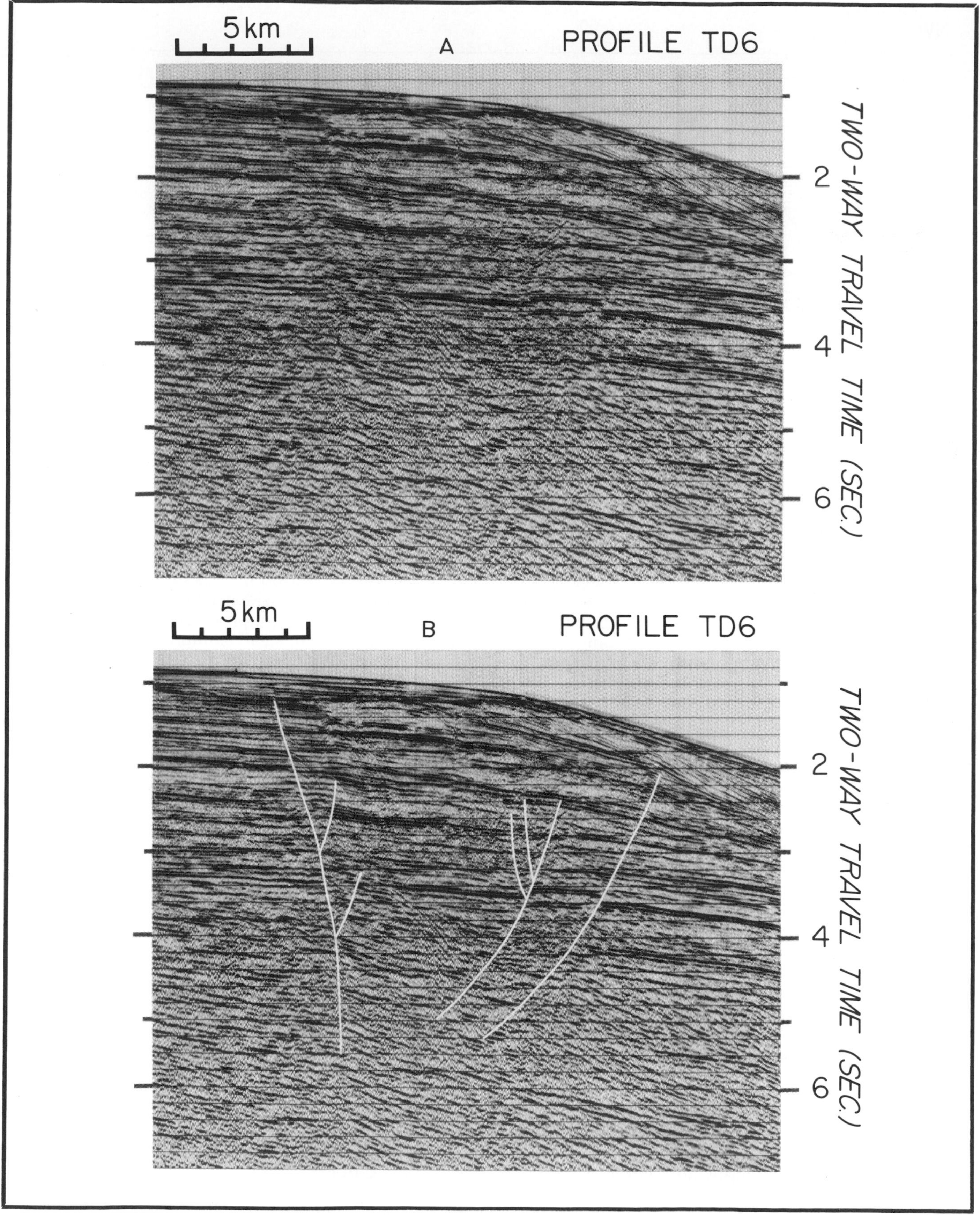

Figure 13 — Detail of seismic profile TD6 showing faults. Location is shown in Figure 12. Part A shows unmarked record; part B shows interpreted faults.

basement at the seaward side of the trough, which caused the salt to begin to flow upward and by the position of the shelf edge, seaward of which overburden pressure on the salt was reduced. Removal of a volume of salt resulted in subsidence of the block of sedimentary rock above the area of the original salt-depositing pan. At the landward side of the pan, subsidence of that block caused a fracture in the sedimentary strata and, because the flow of salt continued for a long period (and still continues), the fault was active throughout this period. Thus, the growth fault formed because of continual removal of support from a major block of the continental margin strata by salt flow, and the location of the fault marks the landward limit of significant salt deposition. This volume-transfer model is indicated graphically on a depth-converted seismic profile in Figure 21. Narrow, half-barbed arrows indicate subsidence along the fault and broad arrows show proposed salt flow.

Such a volume-transfer model requires that the volume lost in subsidence of the block of strata in the Carolina Trough must be equal to the volume of salt removed, which is represented mainly by the volume in the domes. Neither of these volumes can be calculated accurately, but by measuring the area of the subsiding block and estimating an average throw that we infer from profiles, and by measuring the area of observed domes and estimating an average height, we

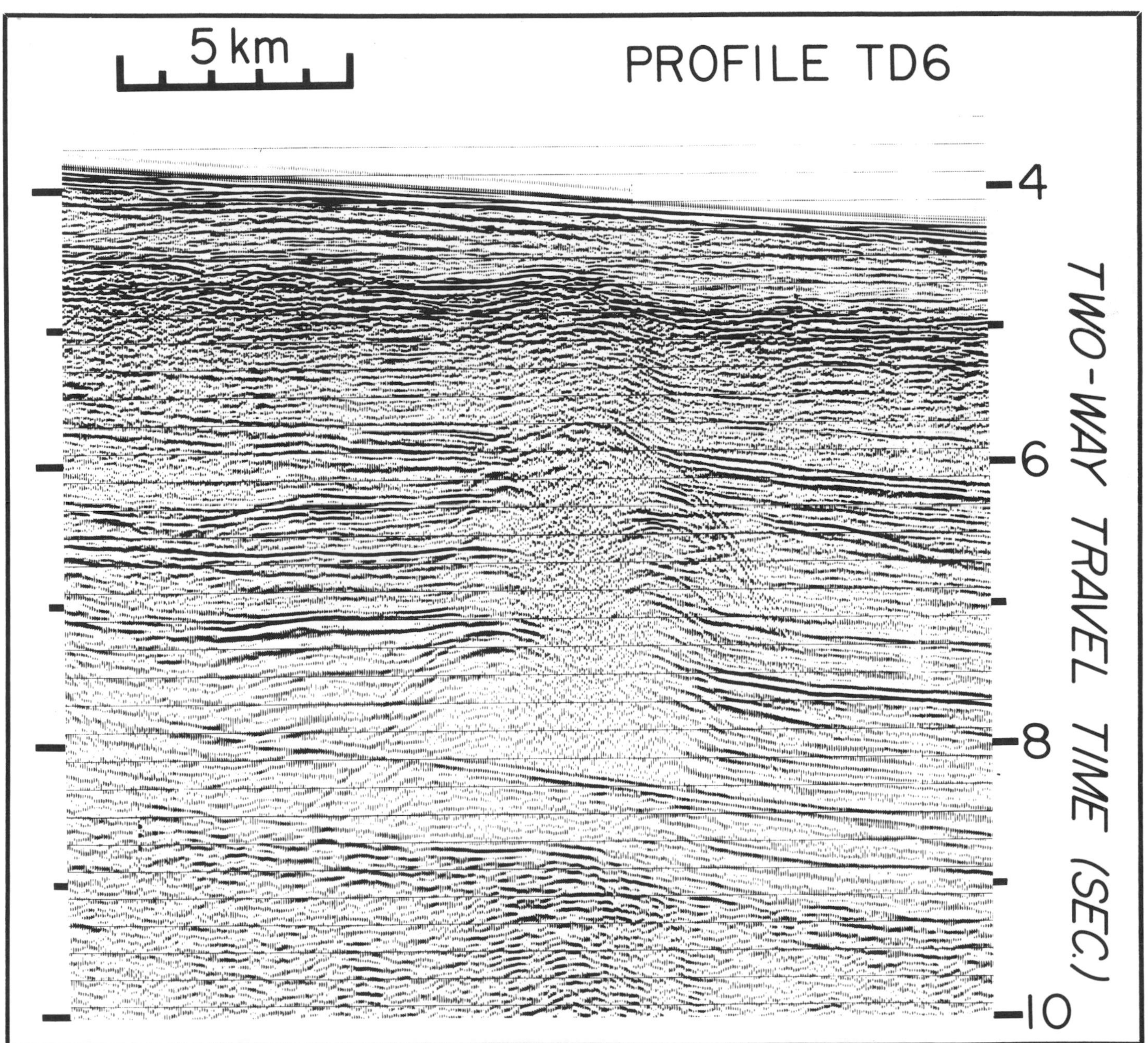

Figure 14 — Detail of seismic profile TD6 showing diapir. Location is shown in Figure 12. Reflections that apparently pass through the diapir at about 8.2 seconds depth and deeper are multiple reflections.

can make approximations of salt volume lost and gained. Both will be minimum values. The subsidence calculation will be a minimum because our estimated throw is measured no deeper than the top of the Jurassic owing to the difficulty in matching reflector packages at greater depth in the seismic record, and so throw probably is greater than we determined. Salt-dome volume represents a minimum because we must have missed some domes. Calculated values are 4,400 cu km lost in subsidence and 4,100 cu km of salt added to the slope in the salt domes. The agreement probably is fortuitous, but the general correspondence is encouraging for our structural model.

Faulting due to salt withdrawal has been identified by various authors (Lehner, 1969; Seglund, 1974; Hospers and Holtke, 1980), but generally it is associated with an extensive layer of salt. In the Carolina Trough a narrow, linear salt-depositing basin resulted in formation of a linear fault when salt flow and subsidence occurred. Seglund (1974) defined faults due to salt withdrawal as "collapse faults," but since this term requires the inference of salt flow, rather than being simply descriptive of fault geometry, we prefer the term growth fault. Apparently the block of strata within the Carolina Trough subsided nearly vertically. A more common situation seems to entail generation of a series of seaward-dipping growth faults by seaward-directed gravity gliding of blocks of strata,

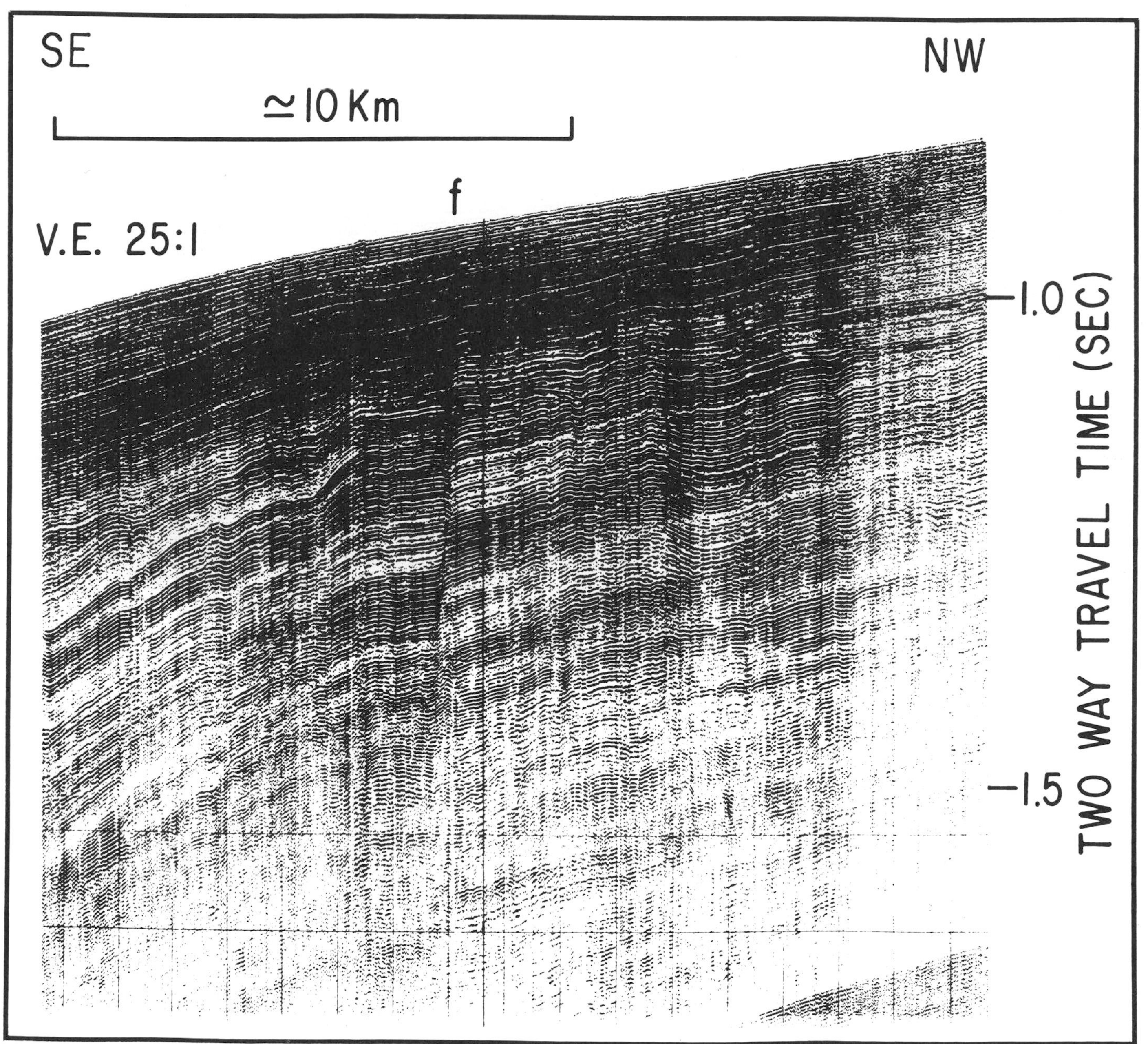

Figure 15 — Single-channel, high-resolution airgun profile across the growth fault. Fault is marked by f. Profile is located 20 km northeast of profile 32.

perhaps on a lubricating layer of salt or shale (Cloos, 1968; Bishop, 1973; Bruce, 1973; Edwards, 1976; Weimer and Davis, 1977; MacPherson, 1978). Such seaward movement commonly generates extensional structures (rollover anticlines and antithetic normal faults) at the steeper shallow part of the concave-up fault (Short and Stäuble, 1967; Bruce 1973; Edwards, 1976, 1981; MacPherson, 1978; Harding and Lowell, 1979). As noted in the discussion of profile TD6 (Figures 12, 13), we have a totally different structural style in the faults of the Carolina Trough. The main seaward-dipping growth fault is essentially planar, as shown in the depth converted profile of Figure 21, or, in one crossing, even concave-down (Figures 12, 13); faults at the outer part of the Carolina Trough dip landward rather than seaward, and we have some evidence of shortening of strata by folding against the growth fault in the deeper part of the trough (Figure 13).

Why does the faulting of the Carolina Trough take

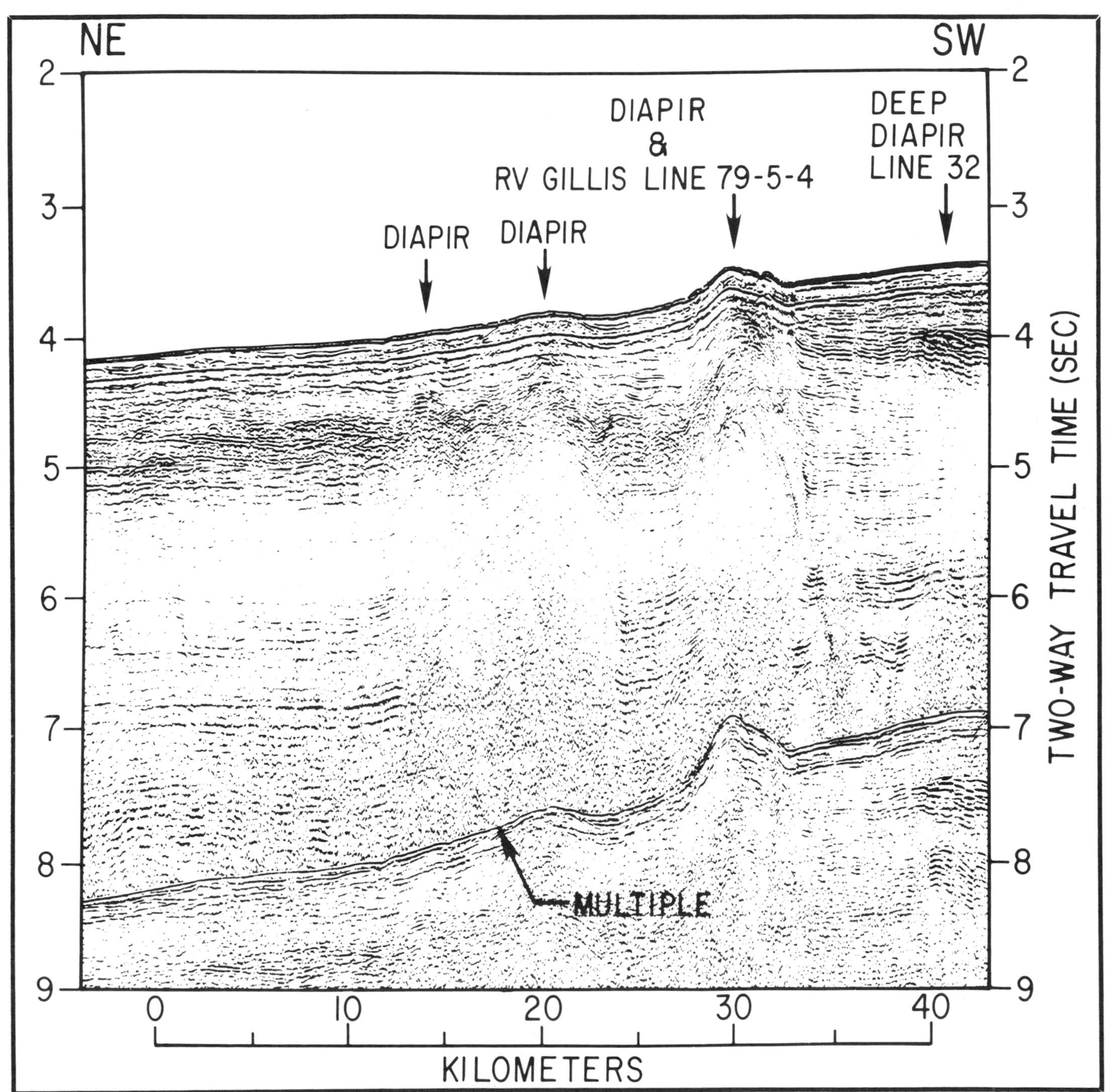

Figure 16 — Single-channel airgun profile along diapir trend near profile 32.

this unusual pattern? One reason may be the presumed narrow zone of salt deposition (perhaps 30 to 40 km wide) which did not provide an extensive lubricating layer on which the old slope and rise deposits could slide seaward; thus they acted as a buttress. The basement ridge inferred from magnetics at the seaward side of the trough may have added to this buttressing effect. Furthermore, the rising salt domes themselves may have created a discontinuous uplifted ridge at the seaward side of the trough. This ridge could have inhibited seaward gliding of the sediment block within the trough by creating a tendency for sediment to slide landward into the trough. The landward-dipping faults may result partly from relative subsidence off that ridge, in a manner similar to that suggested by Quarles (1953) or Bruce (1973).

You would expect vertical subsidence of the block of sediment between two troughward-dipping fault sets to produce some compression, and we have evidence of that in the folding against the fault plane. You

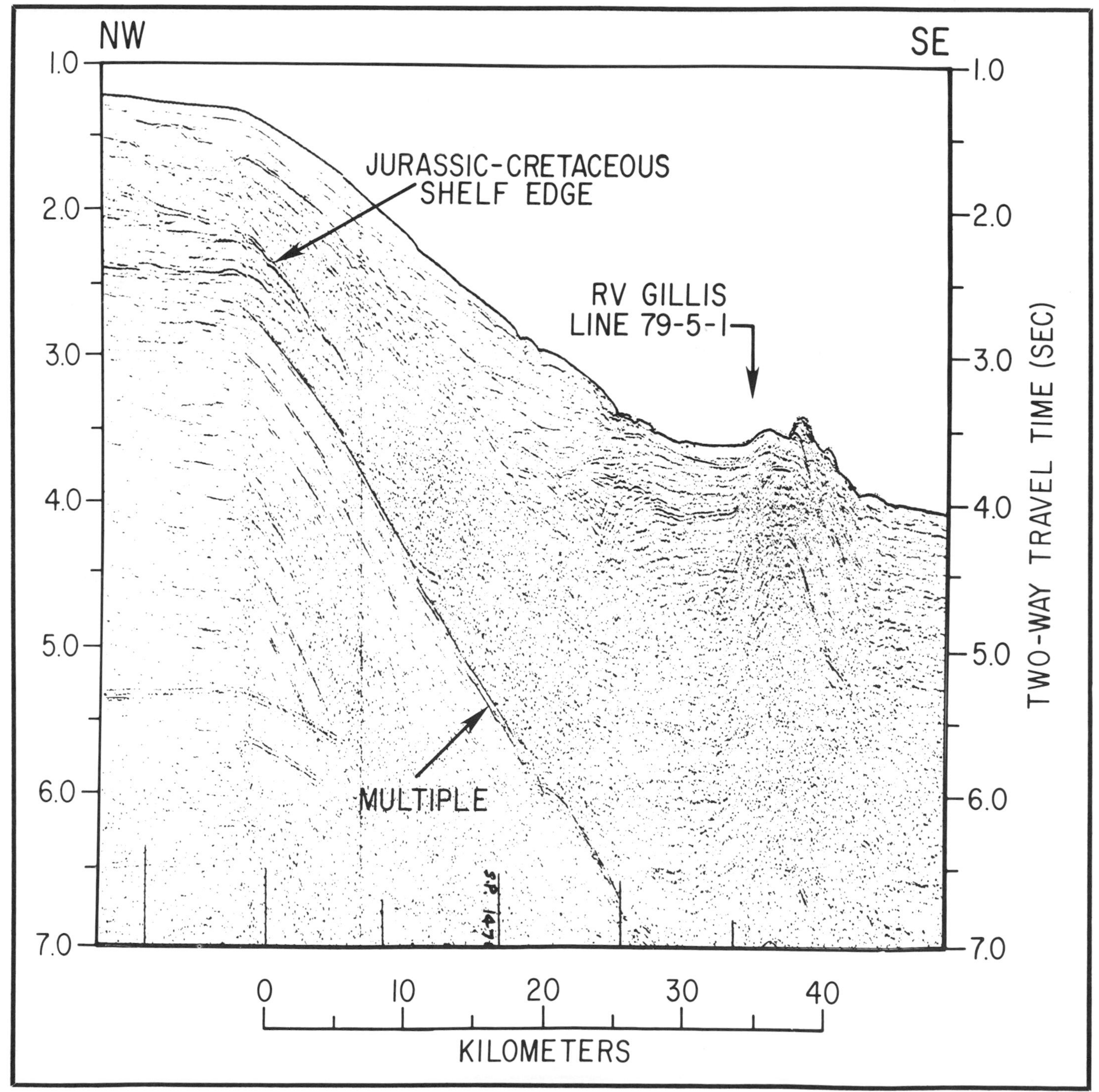

Figure 17 — Single-channel airgun profile across diapir that has disrupted sea floor. The scarp is less obvious here than in Figures 19 and 20; it occurs 13 km northwest (left) of the highest peak of the diapir.

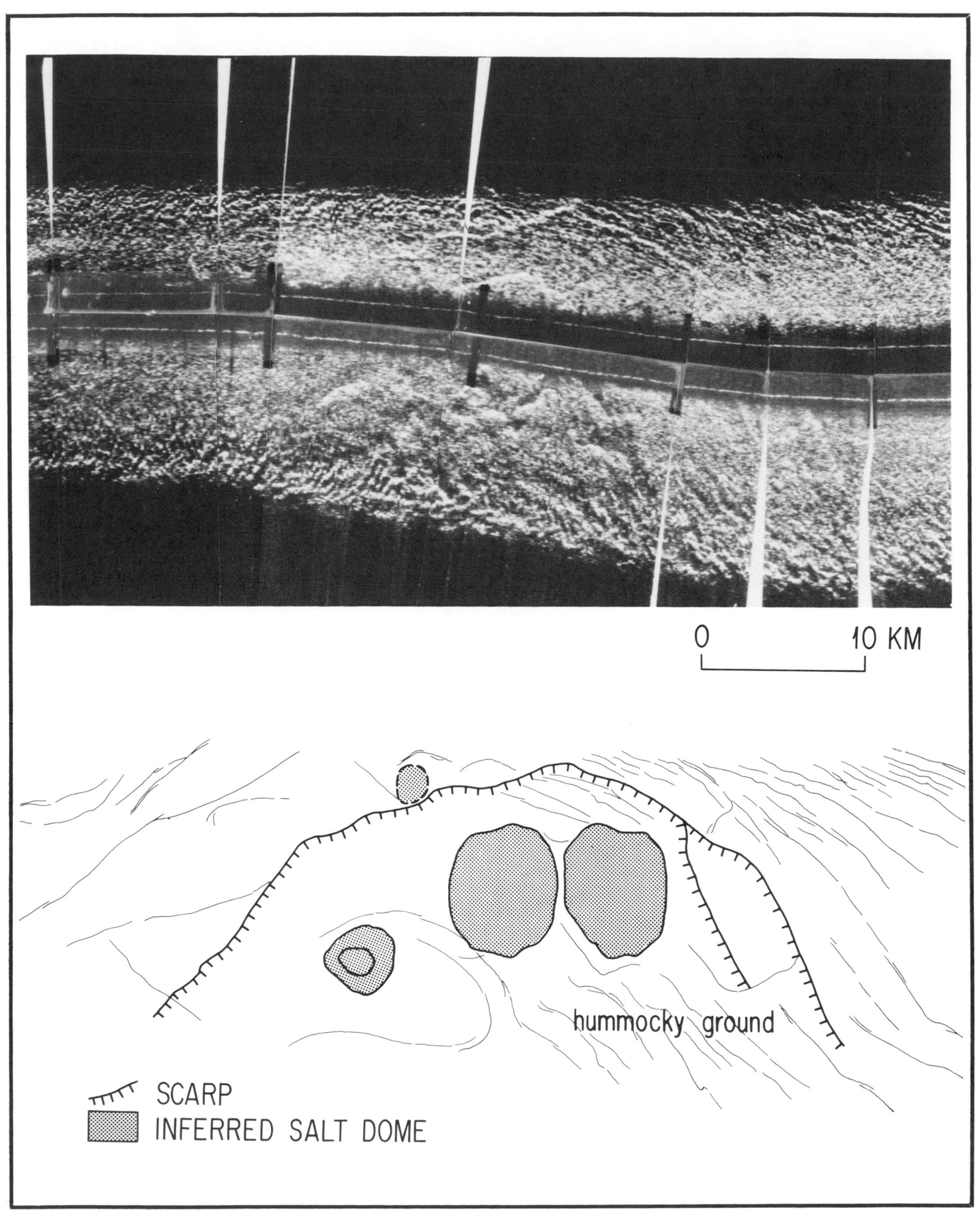

Figure 18 — Long-range (*Gloria*) sidescan-sonar record of part of the diapir trend with interpretation. This shows the southern of the two areas of hummocky sea floor, bounded by scarps to the northwest, that are indicated in Figure 1.

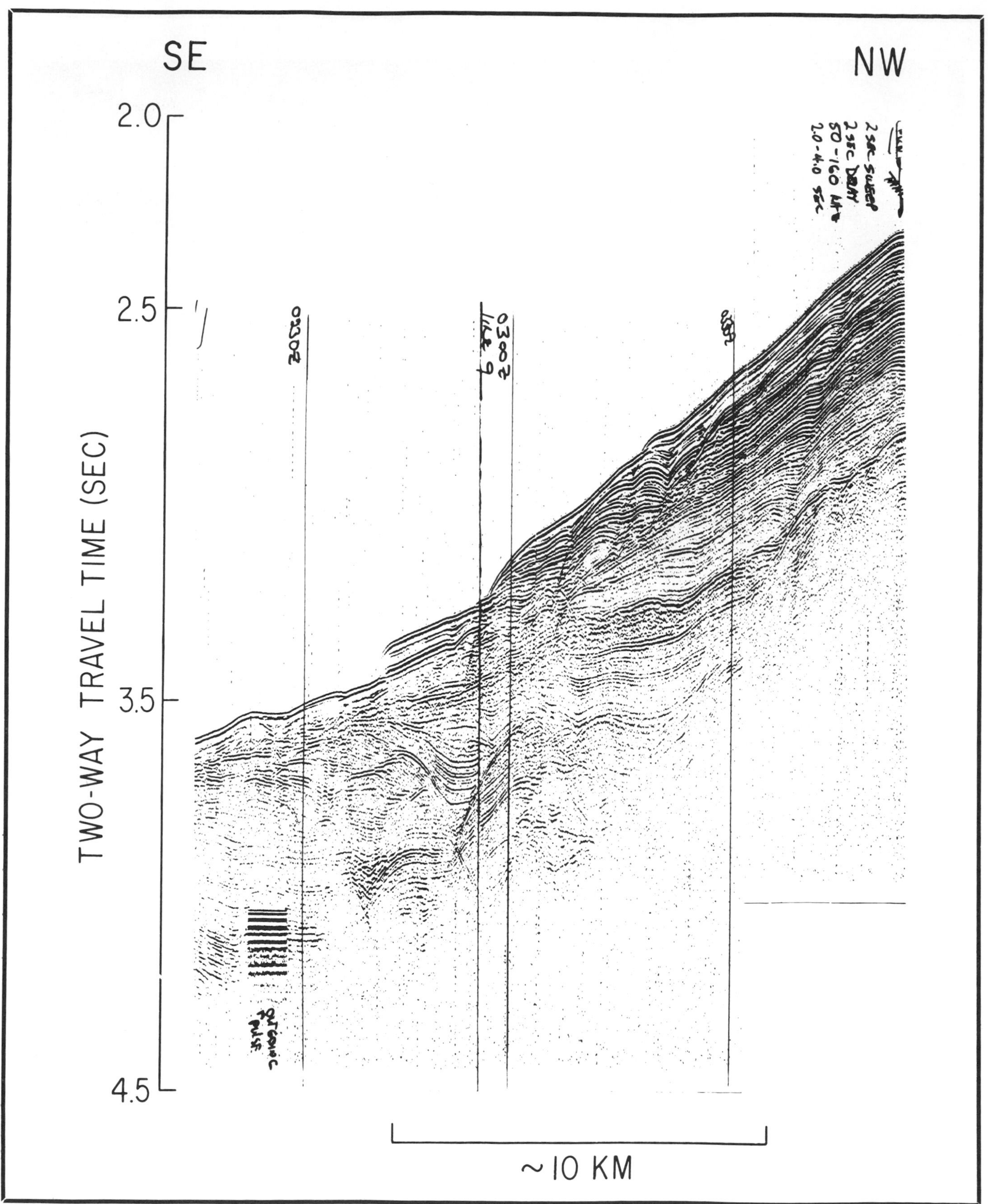

Figure 19 — Single-channel, high-resolution airgun profile across the southern part of the scarp that is shown in the *Gloria* sidescan sonar-record of Figure 18. Note disturbed strata and irregular sea floor downslope from the scarp. Vertical exaggeration is 14:1.

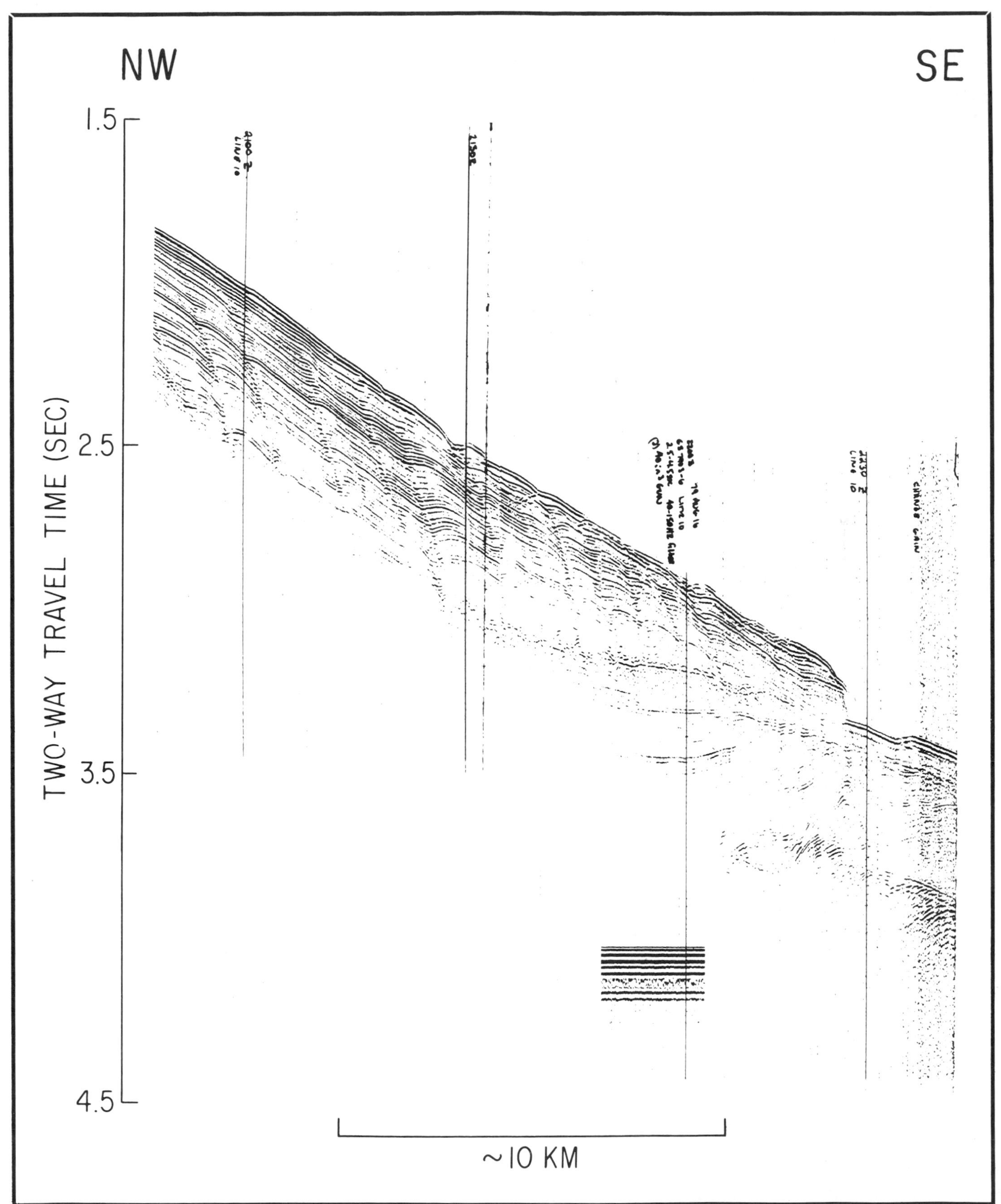

Figure 20 — Single-channel, high-resolution airgun profile across the southern part of the scarp that is shown in the *Gloria* sidescan-sonar record of Figure 18. Note fairly evenly spaced small slump faults, apparently resulting from removal of support at the scarp. Vertical exaggeration is 11:1.

should also expect downfaulting, with a downward steepening fault plane (concave down), to produce compression against the fault that would be greater on the shallower part of the fault plane than at depth. That is, there would be a tendency to lift up the shallower part of the subsiding seaward block, the part above the flatter, shallower part of the fault. Such a situation is interpreted at the main growth fault (left fault) in Figures 12 and 13. We propose that the apparently antithetic faults that join the main growth fault at its inflections (left fault, Figure 13) are agents to release such compression. Because the strata to the right of the apparently antithetic faults are subsiding and the faults dip to the left, they are by definition reverse faults. In such a case, where downfaulting occurs at a steep angle against a normal fault that steepens with increasing depth, compression is generated against the fault plane. That compression, which is greater at the shallow part of the fault, can be released along reverse faults that terminate in the main normal fault. We informally have termed such reverse faults "wedge faults."

REGIONAL CONTROLS ON SALT DEPOSITION AND DIAPIR FORMATION

Diapirs are common on the seaward sides of the Carolina Trough (Figure 1) and also the Scotian Basin (Jansa and Wade, 1975), yet, although profiling coverage is intensive, they appear to be relatively scarce elsewhere on the eastern North American continental margin (Figure 22). Diapirs are present to a very minor extent east of Florida in the Bahamas (Ball, Gaudet, and Leist, 1968), in the Baltimore Canyon Trough (Grow, 1980), and on eastern Georges Bank. The domes on eastern Georges Bank are an extension of the Scotian Basin diapirs (Uchupi and Austin, 1979). Salt has been drilled in the Bahamas (Tator and Hatfield, 1975), Baltimore Canyon Trough (Grow, 1980), and Georges Bank (Amato and Simonis, 1980). Consequently, we conclude that salt deposition probably was nearly universal during early stages of continental margin formation off eastern North America (also suggested by Evans, 1978), but that domes formed extensively only in two areas, the Scotian Basin and the Carolina Trough. We question the armor-plate theory — that thick salt may exist, yet not flow into domes because of a strong, rigid, rock layer above it. Some zones of relative weakness should be expected, allowing passage of salt. We prefer the argument that presence of diapirs simply indicates the former presence of a thick layer of salt. Therefore, we suggest that a thickness of salt sufficient to allow generation of extensive groups of domes accumulated only in two regions.

Why are thick deposits of salt concentrated only in these two basins? Some earlier conclusions (Burke, 1975) based on less data do not seem to apply. Thick

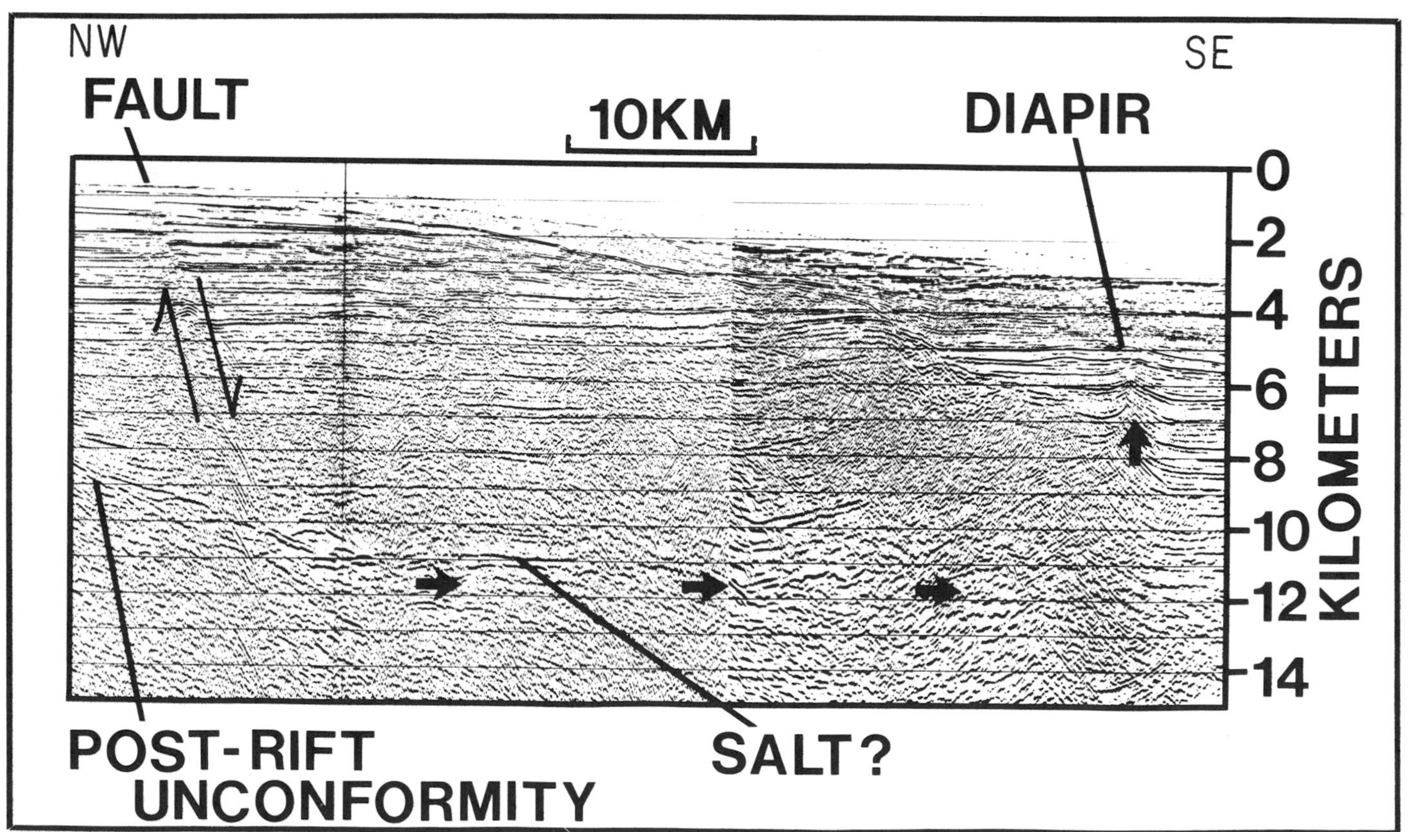

Figure 21 — Depth-converted part of seismic profile 32 across the Carolina Trough. Compare this to Figure 7 that shows the same profile in a time section. Single-barbed arrows show relative motion on the growth fault that we believe has resulted from salt flow into diapirs. Direction of inferred salt flow is shown by broad arrows. Vertical exaggeration is only 2:1.

salt deposits are not restricted to the vicinity of an ocean spillway (Figure 22); they do not seem to be restricted by obvious high tectonic barriers or former hot-spot locations and they probably are not controlled by latitudinal climatic zones. Indeed, a continent as large as Pangea probably always had a dry zone near its center, which would mean that the entire newly-formed Atlantic margin was located in a dry area.

We propose that thickness of salt in continental margin basins and, ultimately, the presence or absence of extensive salt domes were controlled by timing of early basin subsidence. The timing was determined by the amount of stretching and thinning of basement during the rifting stage. Basins having greater thinning of basement would have subsided below sea level sooner and, therefore, received oceanic waters for a longer time before the opening ocean developed a circulation connected to the world ocean that terminated salt deposition. Because they underwent a longer period of salt deposition, these early-subsiding basins probably accumulated a thicker salt layer. The few available crustal sections across the margin, based on gravity and refraction, support this conclusion by

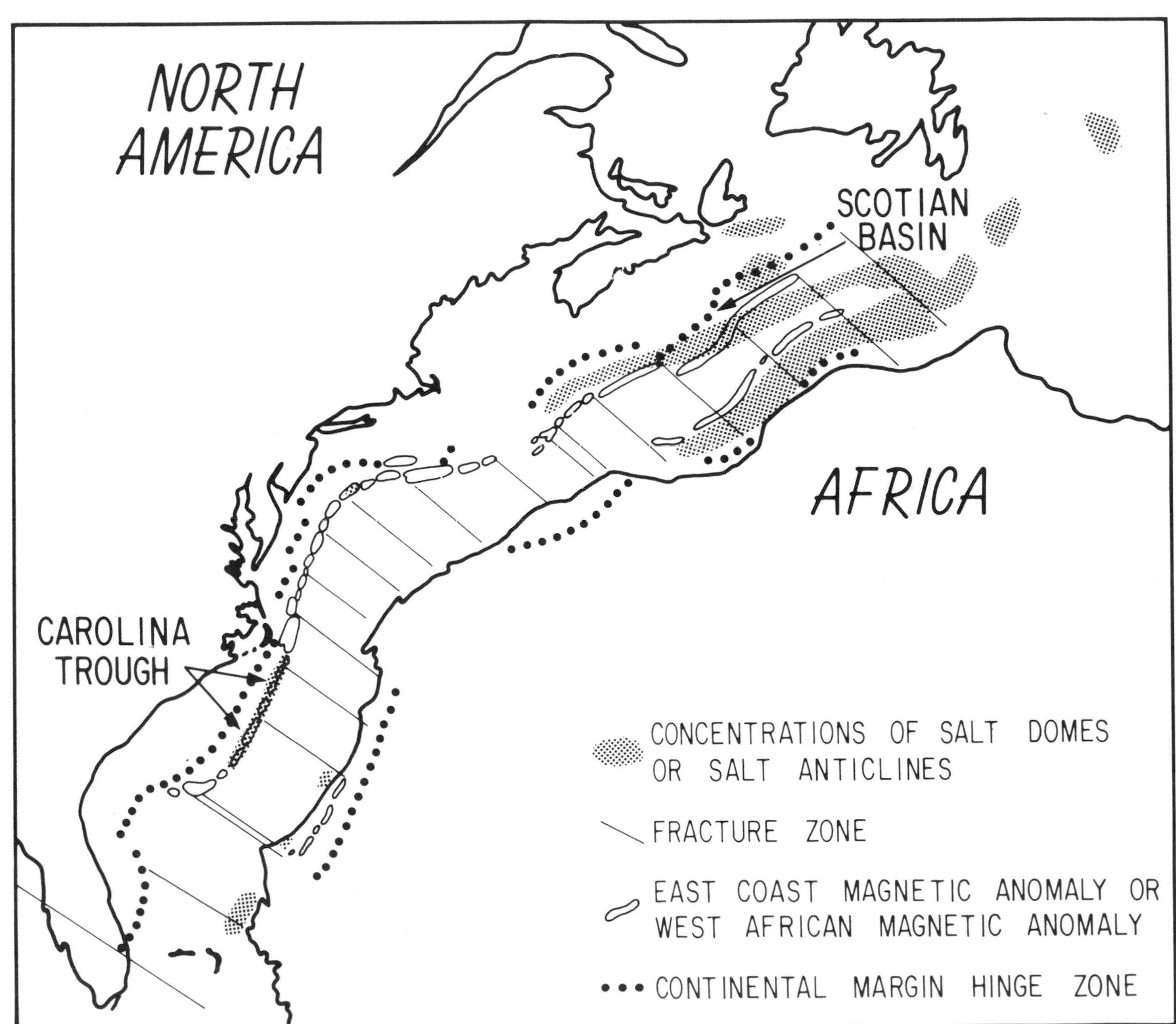

Figure 22 — Reconstructed locations of major continental blocks during early ocean opening in Middle Jurassic time. Locations are shown for known diapirs or diapir groups that presumably were forming by this time (Aymé 1965; Ball, Gaudet, and Leist, 1968; Templeton, 1970; Grunau et al, 1975; Jansa and Wade, 1975; Hinz, 1977; Uchupi and Austin, 1979; Grow, 1980; Jansa, Bujak, and Williams, 1980).

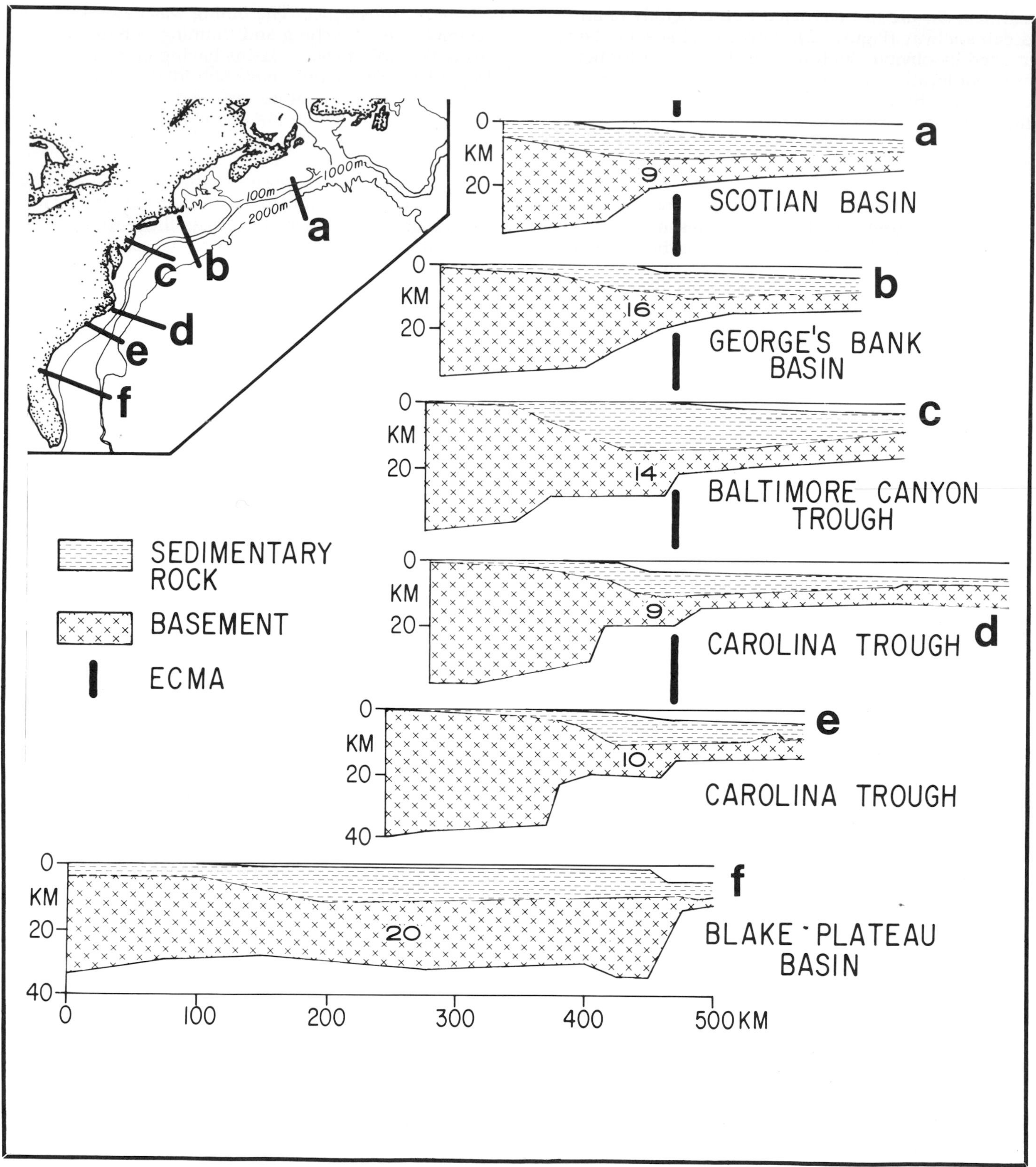

Figure 23 — Simplified crustal sections across the basins of eastern North America based on gravity and refraction (Keen et al 1975; Grow, Mattick and Schlee, 1979; Kent, Grow and Dillon, 1979; Grow, 1980; Hutchinson and others, this volume). Numbers in section indicate basement thickness at the locations of the numbers, which are placed approximately at the centers of basins. More recent interpretation of the Scotian Basin data suggest that the basin basement may be somewhat thicker than indicated, perhaps 11 km. (C.E. Keen, C. Beaumont and R. Boutilier, this volume). The profiles are aligned along the East Coast Magnetic anomaly (ECMA) except for the Blake Plateau Basin profile, where the ECMA is not present. Section "b" actually is to the west of the main part of the Georges Bank Basin and crosses a sub-basin of the Georges Bank sytem; it is well away from the diapirs that extend down from the Scotian Basin.

showing that basins associated with many diapirs (Carolina Trough and Scotian Basin) have basement thicknesses at the basin axis of 9 to 10, perhaps 11, km (Figure 23). Conversely, other basins, those without abundant diapirs, show basement thicknesses of 14 to 20 km.

SUMMARY

1. The Carolina Trough is a long (450 km), narrow (40 km), linear basin off eastern North America.

2. A group of salt diapirs is aligned along the seaward side of the trough. The diapirs are located at a magnetic anomaly on the seaward side of the basin that may mark a basement ridge or step that could have controlled upward flow of salt.

3. A normal fault follows the landward side of the Carolina Trough. It is a growth fault, as is shown by a pattern of throws increasing as depths increase. Unlike ordinary continental margin slump-type faults, the growth fault does not flatten into bedding nor does it have antithetic normal faults. The fault generally continues steeply down to a strong reflector that may represent the salt horizon in the trough. At one location, the fault steepens at depth and may have reverse faults associated with it that were created by this steepening of the fault plane.

4. The growth fault probably resulted from removal of support as salt flowed seaward into the domes from the deep part of the trough. This transfer of volume from the deep trough resulted in slow subsidence of a large block of strata as the domes rose.

5. The growth faulting and related flow of salt probably began in Jurassic time, as is indicated by increasing offsets seen in deep strata. Both faulting and diapirism seem to still be active, because strata near the sea bottom are offset at the fault and disturbed sea floor is associated with salt flow.

6. Only the Carolina Trough and Scotian Basin have widespread salt domes off eastern North America, probably because they were the only basins to accumulate large thicknesses of salt. Basements beneath those two troughs also seem to be thinner than those beneath other basins, probably due to greater thinning by stretching during the rifting phase. This may have resulted in greater early subsidence of the Carolina Trough and Scotian Basin and, thus, in more time to accumulate salt.

ACKNOWLEDGMENTS

Mahlon M. Ball and John S. Schlee reviewed this manuscript and offered valuable suggestions. We thank Margaret Clare Mons-Wengler for preparing the manuscript and Patricia Forrestel and Jeffrey Zwinakis for their drafting assistance.

REFERENCES CITED

Amato, R. V., and E. K. Simonis, 1980, Geologic and operational summary, cost No. G-2 well, Georges Bank area, North Atlantic OCS: U.S. Geological Survey Open-File Report 80-269, 116 pp., 3 plates.

Aymé, J. M., 1965, The Senegal salt basin *in* Salt Basins around Africa: Elsevier, Amsterdam, The Institute of Petroleum, p. 83-90.

Ball, M. M., R. M. Gaudet and G. Leist, 1968, Sparker reflection seismic measurements in Exuma Sound, Bahamas (Abs); American Geophysical Union, Transcripts v. 49, p. 196-197.

Bishop, W. F., 1973, Late Jurassic contemporaneous faults in north Louisiana and south Arkansas: AAPG Bulletin, v. 57, p. 858-877.

Bruce, C. H., 1973, Pressured shale and related sediment deformation — mechanism for development of regional contemporaneous faults: AAPG Bulletin, v. 57, p. 878-886.

Burke, Kevin, 1975, Atlantic evaporites formed by evaporation of water spilled from Pacific Tethyan and southern oceans: Geology, v. 3, p. 613-616.

Cloos, Ernst, 1968, Experimental analysis of Gulf Coast fracture patterns: AAPG Bulletin, v. 52, p. 420-444.

Crans, W., and G. Mandl, 1981a, On the theory of growth faulting, part II (b) — genesis of the "unit": Journal of Petroleum Geology, v. 3, p. 333-355.

———, and ———, 1981b, On the theory of growth faulting, Part II (c) — genesis of the "unit": Journal of Petroleum Geology, v. 3, p. 455-576.

Dillon, W. P., J. A. Grow, and C. K. Paull, 1980, Unconventional gas hydrate seals may trap gas off southeast U.S.: Oil and Gas Journal, v. 78, no. 1, p. 124, 126, 129, 130.

Edwards, M. B., 1976, Growth faults in Upper Triassic deltaic sediments, Svalbard: AAPG Bulletin, v. 60, p. 341-355.

———, 1981, Upper Wilcox Rosita delta system of south Texas — Growth-faulted shelf-edge deltas: AAPG Bulletin, v. 65, p. 54-73.

Evans, Robert, 1978, Origin and significance of evaporites in basins around Atlantic margin: AAPG Bulletin, v. 62, p. 223-234.

Falvey, D. A., 1974, The development of continental margins in plate tectonic theory: Australian Petroleum Exploration Association Journal, v. 14, p. 95-106.

Grow, J. A., 1980, Deep structure and evolution of the Baltimore Canyon Trough in the vicinity of the COST No. B-3 well, *in* P.A. Scholle, ed., Geological studies of the COST No. B-3 well, United States mid-Atlantic continental slope area U.S. Geological Survey Circular 833, p. 117-132.

———, R. E. Mattick, and J. S. Schlee, 1979, Multichannel seismic depth sections and interval velocities over outer continental shelf and upper continental slope between Cape Hatteras and Cape Cod, *in* J. S. Watkins, Lucien Montadert, and P. W. Dickerson, eds., Geological and geophysical investigations of continental margins: AAPG Memoir 29, p. 65-83.

Grunau, H. R., et al, 1975, New radiometric ages and seismic data from Fuerteventura (Canary Islands), Maio (Cape Verde Islands), and São Tomé (Gulf of Guinea), *in* Progress in geodynamics: North-Holland, Amsterdam, Geodynamics Scientific Report #13, p. 90-110.

Hardin, F. R., and G. C. Hardin Jr., 1961, Contemporaneous normal faults of Gulf Coast and their relation to flexures: AAPG Bulletin, v. 45, p. 238-248.

Harding, T. P., and J. D. Lowell, 1979, Structural styles, their plate tectonic habitats, and hydrocarbon traps in petroleum provinces: AAPG Bulletin, v. 63, p. 1016-1058.

Hinz, Karl, 1977, Bericht über den Fahrtabschnitt M 46/1, Hamburg-Casablanca, 8.10.77-5.11.77, der METEOR — Westafrikafahrt 1977: Bundesanstalt für Geowissenschaften und Rohstoffe, Archiv — Nn. 78 906, 40 p.

Hospers, J., and J. Holtke, 1980, Salt tectonics in block 8/8 of the Norwegian sector of the North Sea: Tectonophysics, v. 68, p. 257-282.

Humphris, C. C. Jr., 1979, Salt movement on continental slope, northern Gulf of Mexico: AAPG Bulletin, v. 63, p. 782-798.

Jansa, L. F., J. P. Bujak, and G. L. Williams, 1980, Upper Triassic salt deposits of the western North Atlantic: Canadian Journal of Earth Sciences, v. 17, p. 547-559.

Jansa, L. F., and J. A. Wade, 1975, Geology of the continental margin off Nova Scotia and Newfoundland, *in* W. J. M. van der Linden, and J. A. Wade, eds., Offshore geology of eastern Canada, v. 2, Regional geology: Geological Survey of Canada, Paper 74-30, 51-105.

Keen, C. K., et al, 1975, Some aspects of the ocean-continent transition at the continental margin of eastern North America, *in* W. J. M. van der Linden, and J. M. Wade, eds., Offshore geology of eastern Canada, v. 2, Regional geology: Geological Survey of Canada, Paper 74-30, p. 189-197.

Kent, K. M., J. A. Grow, and W. P. Dillon, 1979, Gravity studies of the continental margin off northern Florida: Geological Society of America, Abstracts with Programs, v. 11, no. 4, p. 184.

Klitgord, K. D., and J. C. Behrendt, 1979, Basin structure of the U.S. Atlantic Continental Margin, *in* J. S. Watkins, L. Montadert, and P. W. Dickerson, eds., Geological and geophysical investigations of continental margins: AAPG Memoir 29, p. 85-112.

Lehner, Peter, 1969, Salt tectonics and Pleistocene stratigraphy on continental slope of northern Gulf of Mexico: AAPG Bulletin, v. 53, p. 2431-2479.

MacPherson, B. A., 1978, Sedimentation and trapping mechanism in Upper Miocene Stevens and older turbidite fans of southeastern San Joaquin Valley, California: AAPG Bulletin, v. 62, p. 2243-2274.

Ocamb, R. D., 1961, Growth faults of south Louisiana: Transactions of the Gulf Coast Association of Geological Societies, v. 11, p. 139-175.

Parker, T. J., and A. N. McDowell, 1955, Model studies of salt-dome tectonics: AAPG Bulletin, v. 39, p. 2384-2470.

Paull, C. K., and W. P. Dillon, 1981, Appearance and distribution of the gas hydrate reflection in the Blake Ridge region, offshore southeastern United States: U.S. Geological Survey Miscellaneous Field Studies Map, MF 1252.

Quarles, Miller, Jr., 1953, Salt-ridge hypothesis on origin of Texas Gulf Coast type of faulting: AAPG Bulletin, v. 37, p. 489-508.

Rider, M. H., 1978, Growth faults in Carboniferous of western Ireland: AAPG Bulletin, v. 62, p. 2191-2213.

Seglund, J. A., 1974, Collapse-fault systems of Louisiana Gulf Coast: AAPG Bulletin, v. 58, p. 2389-2397.

Shipley, T. H., et al, 1979, Seismic evidence for widespread possible gas hydrate horizons on continental slopes and rises: AAPG Bulletin, v. 63, p. 2204-2213.

Short, K. C., and A. J. Stäuble, 1967, Outline of geology of Niger delta: AAPG Bulletin, v. 51, p. 761-779.

Sylwester, R. E., W. P. Dillon, and J. A. Grow, 1979, Active growth fault on seaward edge of Blake Plateau, *in* Gill, Dan and D. F. Merriam, eds., Geomathematical and Petrophysical Studies in Sedimentology: Oxford, Pergamon Press, p. 197-209.

Tator, B. A., and L. E. Hatfield, 1975, Bahamas present complex geology: Oil and Gas Journal, Part 1, v. 73, No. 43, p. 172-176; Part 2, v. 73, N. 44, p. 120-122.

Templeton, R. S. M., 1970, The geology of the continental margin between Dakar and Cape Palmas, *in* The geology of the east Atlantic continental margin, 4, Africa: United Kingdom, Institute of Geological Sciences Report No. 70/16, p. 47-60.

Tucholke, B. E., 1979, Relationship between acoustic stra tigraphy and lithostratigraphy in the western North Atlantic Basin *in* Tucholke et al, eds., Initial reports of the deep sea drilling project, v. 43, Washington, D.C., U.S. Government Printing Office, p. 827-846.

Uchupi, Elazar, and J. A. Austin, 1979, The geologic history of the passive margin off New England and the Canadian Maritime Provinces: Tectonophysics, v. 59, p. 53-69.

Weimer, R. J., and T. L. Davis, 1977, Stratigraphic and seismic evidence for Late Cretaceous growth faulting, Denver Basin, *in* C. E. Payton, ed., AAPG Memoir 26, p. 277-299.

Post-Paleozoic Succession and Structure of the Southwestern African Continental Margin

I. Gerrard*
G. C. Smith*
Johannesburg, South Africa

Four seismic horizons — T, R, P and L from bottom to top — form major seismic sequence boundaries that range in age from Mesozoic to Cenozoic. Horizon T is regarded as the rift-onset unconformity and horizon R as the drift-onset unconformity, with the post-R succession being drift-stage deposits. Three periods of Mesozoic to Cenozoic igneous activity are represented, but, owing to extensive alteration, the few age determinations are suspect. The drift-onset unconformity R, with dipping reflections beneath it, can be traced westward to where it is juxtaposed with the top of typical reflection-free oceanic crust. We believe that this marks the position of the continent-ocean boundary and that the distinctive magnetic and gravity anomaly G, much nearer the coast, is produced by the pinchout of the T-to-R wedge with its substantial basaltic component.

The continental margin under discussion, over 2000 km long, extends from Cape Fria at about 18°S in the north to the southern tip of the Agulhas arch at about 37°S. It is located on the western side of southwest Africa (Namibia) and the Republic of South Africa (Figure 1). Although most of the data presented are from the continental shelf and upper slope, the area of investigation also encompasses the easternmost part of the deep-sea Cape Basin and the easternmost part of the Walvis Ridge.

Emery et al (1975) provide a summary of the data published up to 1974 on the geology of the southern African, west-coast, offshore region. Subsequent published work which discusses the Mesozoic-Cenozoic deposits and the continental-oceanic boundary of the west coast are Dingle and Scrutton (1974), Scrutton and Dingle (1974), Bolli et al (1975), Dingle (1976), Rabinowitz (1976), Scrutton (1976), Bolli et al (1978), Dingle (1978), Rabinowitz (1978), Scrutton (1978), Dingle (1979), du Plessis (1979), McLachlan and McMillan (1979), and Van der Linden (1980).

The geological and geophysical interpretations presented here are based on reflection seismic and well data and on many unpublished reports in the possession of the state-financed Southern Oil Exploration Corporation (SOEKOR). G. C. Smith did the geophysical interpretation and prepared the maps; I. Gerrard did the geological interpretation and wrote the text.

The data available include over 30,000 km of commercial seismic surveys of varying standard acquired initially by different contractors and later by SOEKOR (from 1969 on). Ten wells (A-A1, A-C1, A-C2, Ba-A1, Ba-A2, K-A1, K-A2, K-B1, K-D1, and Kudu 9a-1) have been drilled on the continental shelf in the area stretching from west of the Orange River mouth to near Saldanha Bay to the south (Figure 1). DSDP holes 360 to 363 lie in deeper waters. We have used these data to decide which of the conflicting hypotheses regarding the continent-ocean boundary appears more realistic. We also briefly mention the origin of the Walvis Ridge.

THE MESOZOIC SUCCESSION

Regional setting

The continental margin of southwestern Africa is a typical, fully developed Atlantic type, having formed as described by Dewey and Bird (1970a), Dickinson

*Presently with: Southern Oil Exploration Corporation.

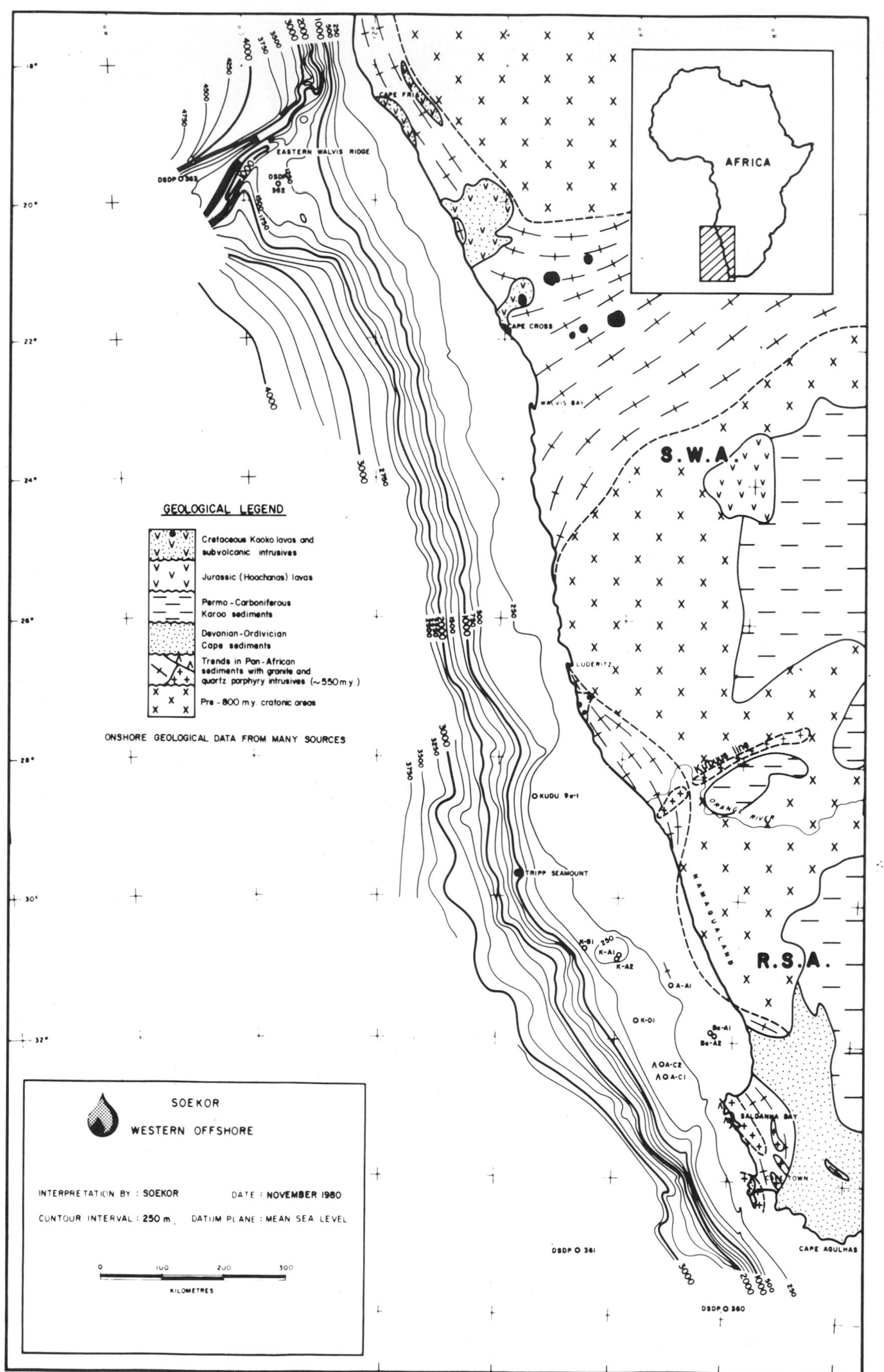

Figure 1 — Map of bathymetry and simplified onshore geology.

(1974) and Falvey (1974). The margin is described by Dingle (1976) as a tensional, clean-break crust, locally displaced by transverse marginal fracture zones. This type of margin develops at the divergent edge of lithosperic plates (Dickinson, 1974). Distinctive lithotectonic units, as described by Dickinson (1974), Falvey (1974) and Schneider (1972), have been deposited along this margin. The stratigraphy consists of prerift, rift-valley, narrow ocean or early drift, and drift or open ocean lithotectonic sequences (Figure 2).

Regional seismic reflection coverage has allowed the mapping of four main seismic horizons — T, R, P, and L — over much of the area. Other seismic reflectors have been mapped locally. Figures 3 and 4 show diagrammatic sections through the boreholes to which the main seismic reflectors have been tied by well velocity surveys. Figure 5 is a generalized profile showing features typical of the continental margin of southwestern Africa. Many of the features in this profile are apparent on the composite seismic section shown in Figure 6. Seismic horizons T, R, and P are sequence boundaries as described by Vail et al (1977). The P-to-L interval consists of a thick sedimentary succession and encompasses a number of seismic sequences. The isopach map of the R-to-sea floor interval, i.e. the sediments deposited since the onset of drifting, indicates the presence of a major sedimentary wedge with a depocentre off the Namaqualand coast (Figure 7). This has long been termed the Orange Basin by SOEKOR and others. Similar smaller basins defined by this interval are present off Lüderitz and Walvis Bay to the north. The term 'basin' is used here for the area of greatest accumulation of sediment and not basin *sensu stricto.* When the total sediment accumulation is split into lower Cretaceous early drift-stage sediments (R-to-P, Figure 8), the upper Cretaceous late drift-stage sediments (P-to-L, Figure 9), and the Cenozoic sediments (L-to-sea floor, Figure 10), it becomes apparent that the Orange Basin is essentially an area of accumulation of Cretaceous sediments.

The sparse well coverage and the variable depths to which the wells penetrated the sedimentary pile, with most not extending down to basement, allows only a generalized picture of the regional lithological framework and of depositional environments.

Basement Formations

Basement formations present on the adjoining coastal area vary considerably in age, lithology, and tectonic history (Figure 1). The greater part of the continental margin is likely to be underlain by high-grade metamorphites belonging to the pre-800 million years cratonic areas and by the generally lower-grade metamorphites of the Pan-African belt (approximately 550 million years). The latter are intruded by the Cape Granite plutons in the south. Associated with these are quartz porphyries at Saldanha Bay. Further north near the Orange River, granite plutons and alkaline intrusives of similar age to the Cape Granites form the Kuboos-Bremen line and in the far north numerous syntectonic granites (not shown in Figure 1) are present in the wide Pan-African belt stretching inland from the Walvis Bay area. The Devonian to Ordovician Cape sediments would be present only in the south. Though the Permo-Carboniferous Karoo sediments occupy significant areas inland their presence offshore south of Walvis Bay is believed to be unlikely. North of Walvis Bay they may be present offshore, but would be regarded as intracratonic rather than basement.

Only three wells intersected basement below their Cretaceous sediments. Well A-A1 recovered green phyllites assigned to the low-grade metaphorphites of the Pan-African belt.

The basement underlying the basalt of the T-to-R wedge at well A-C1 apparently consists of the same type of acid lava as that of well A-C2, drilled 20 km north of A-C1 and east of the T-to-R hingeline. Thus, acid lava appears to form basement in the A-C area, but its regional distribution is still unknown. Its stratigraphic position is uncertain, and although it could be related to the approximately 550 m.y. old quartz porphyry, forming extensive outcrops at Saldanha Bay 120 km southeast, it could represent a Jurassic acid igneous episode not present onshore in southwestern Africa but well-documented on the South American conjugate margin.

Data on the densities and seismic velocities of the basement acid lava were obtained from geophysical well logs. Seismic velocities range from 4,800 to 5,100 m/sec, with an average of 5,000 m/sec, over the 70 m penetrated in A-C1. They range from 4,400 to 4,900 m/sec, with an average of 4,600 m/sec, over the 140 m penetrated in A-C2. The density data indicate that a slightly weathered zone could be present below the unconformities T at A-C1 and R (=T) at A-C2.

Densities are tabulated in Table 1.

Table 1. Density data below unconformities T at A-C1 and R(=T) at A-C2.

Well	A-C1	A-C2
Range in weathered zone	2.48-2.56	2.48-2.56
Average in weathered zone	2.54 (N=9)	2.52 (N=27)
Range below weathered zone	2.57-2.62	2.53-2.60
Average below weathered zone	2.59 (N=20)	2.56 (N=17)

N=the number of samples measured

Lithotectonic sequences

An intracratonic basin stage (Falvey, 1974), or prerift sequence, present for instance in the Reconcarvo basin of Brazil (Asmus and Ponte, 1973; Ponte and Asmus, 1976) and the west Australian basins (Jones and Pearson, 1972; Veevers, 1972) could be present off the west coast. However, as wells seldom penetrate the full succession, and as it is unlikely that such intracratonic fill could be distinguishable on seismic sections from what is regarded as the continental basement, its presence or absence over much of the area offshore is not likely to be resolved in the near future. There is some onshore indication that such deposits occur locally on

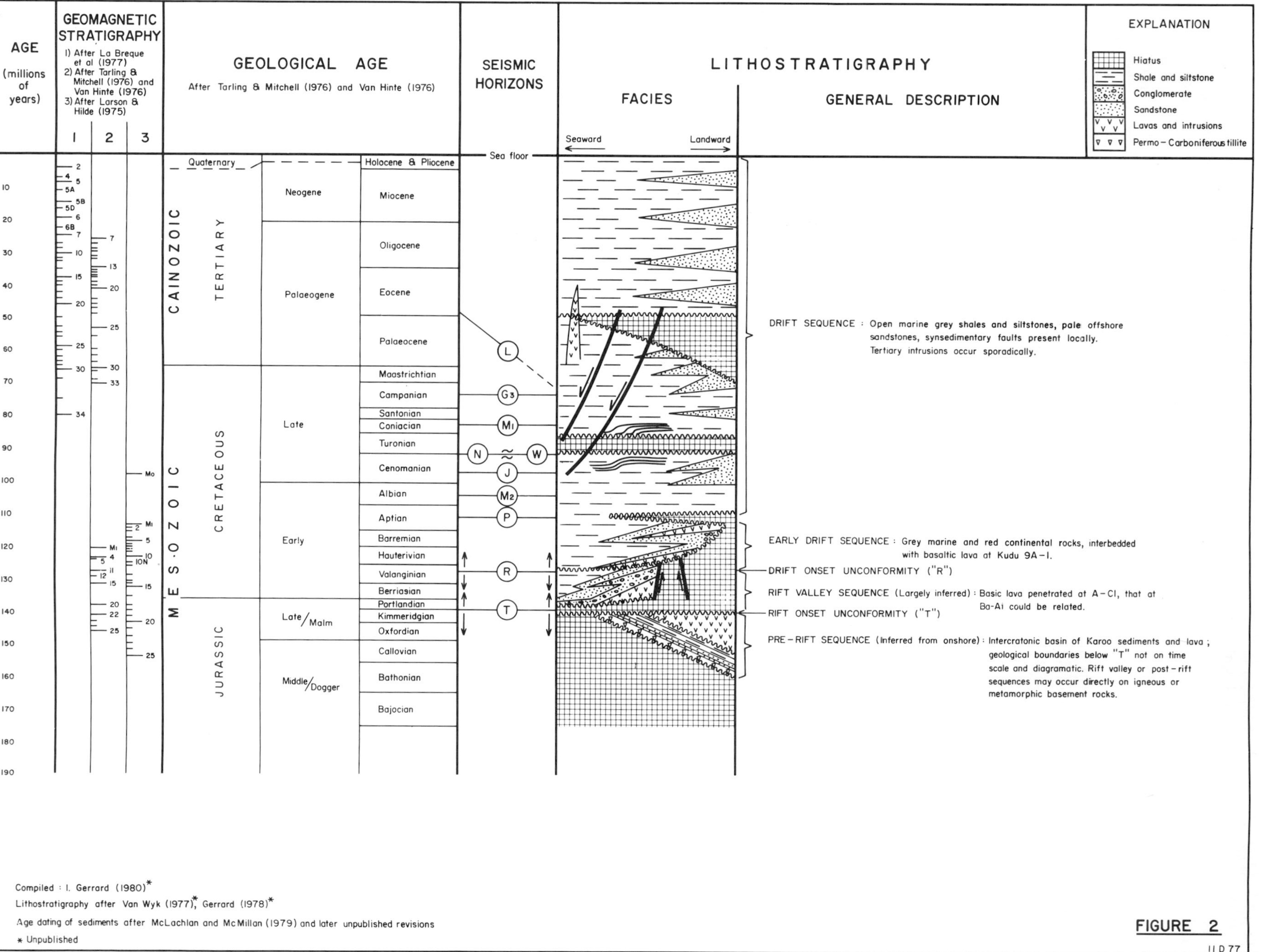

Figure 2 — Provisional chronostratigraphic and lithotectonic framework of southwestern Africa.

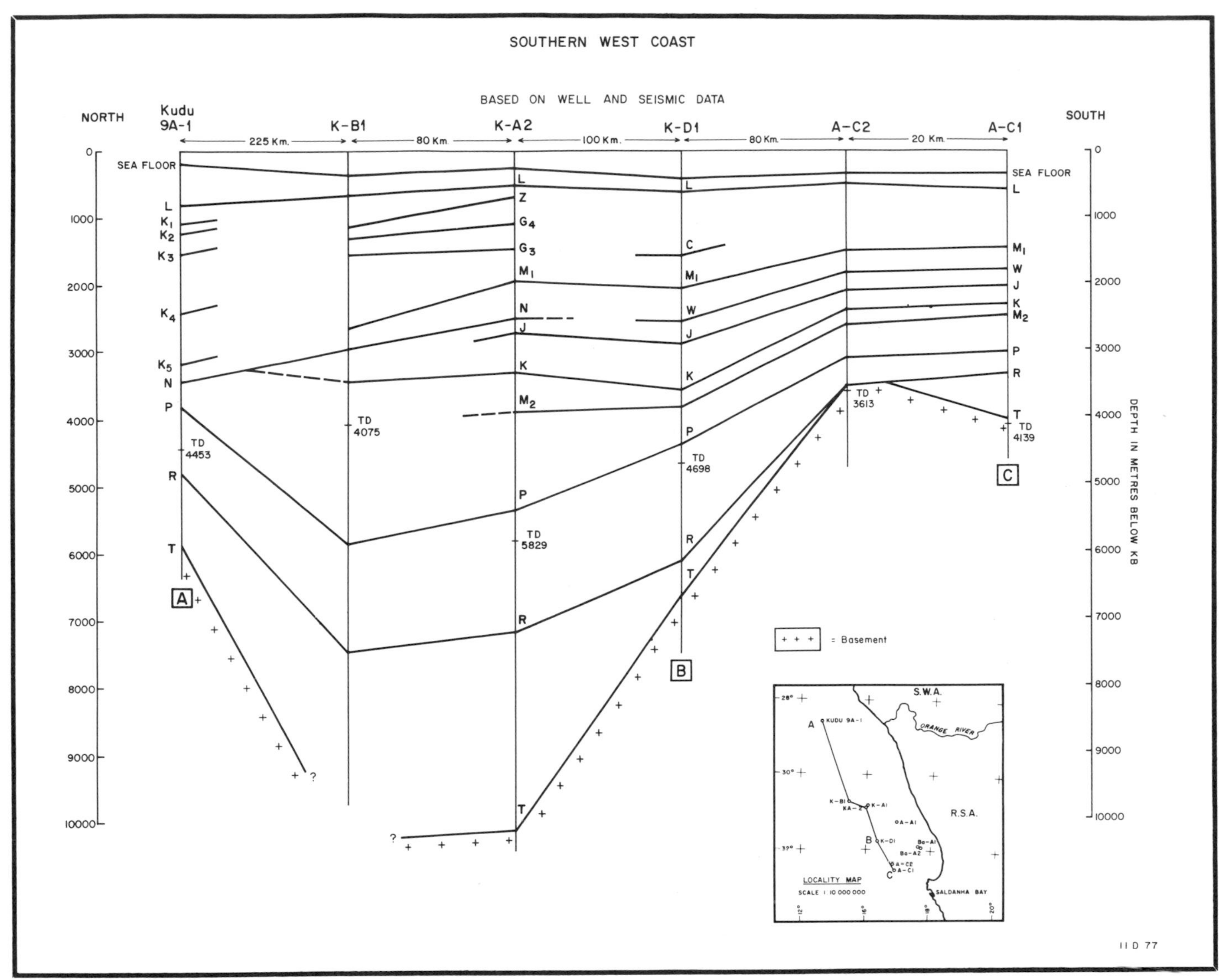

Figure 3 — Southern West Coast regional marker correlation, based on well and seismic data.

the immediately adjacent coastal area between Cape Cross and Cape Fria, where a stratigraphic test well intersected deposits which could be of this type beneath a blanket of Cretaceous lavas. The well passed through a considerable thickness of Karoo clastics (Permo-Carboniferous), indicating that downfaulted or downwarped remnants of Karoo sediments could be present offshore.

The rift-valley stage is an essential ingredient in Atlantic-type margin development (Falvey, 1974; Schneider, 1972). Graben and half-graben related to this stage have been detected on the south-western African continental margin from near Walvis Bay in the north to south of Cape Town (figs. 6, 11). However, their recognition and delimitation is dependent on the distribution of the seismic surveys and on the quality of the seismic records available.

In the area extending south from the Orange River to about the latitude of the K-A wells, the graben and half-graben which have been mapped in greatest detail occur about 40 km from the coast and vary in width from 2 to about 35 km. They are aligned north-northwest parallel or subparallel with the coast. The wider graben seem to have irregular floors with distinct antiformal and synformal configurations, but it is uncertain whether this reflects paleotopography or minor faulting of the graben floor. In the troughs, the reflectors within the faulted fill generally appear conformable to horizon T, which forms the floor. Minor compaction effects are apparent when tracing seismic horizons immediately above the graben across their rims to the adjoining areas.

South from Saldanha Bay, the graben are narrow (3 to 10 km), are aligned north to south and north-northwest and occur about 35 km offshore.

About 3 to 40 km west of the graben, and 50 to 130 km from the coast, a wedge-shaped and generally westward-thickening seismic sequence, the T-to-R interval, has been mapped on a sub-continental scale from near the Walvis Ridge in the north to south of Cape Town (Figures 6 and 11). Horizon T, the base of the sequence, is regarded as marking basement wherever the wedge is present. T is regarded as the rift-onset unconformity (Falvey, 1974).

Generally, the eastern pinch-out of the T-to-R wedge is due to erosion. Horizon R, the drift onset unconformity, truncates the sequence. However, locally the eastern boundary of this wedge is faulted.

Horizon T can be followed west 20 to 30 km from the edge of the wedge before becoming too indistinct to be traced on seismic sections, owing to the decrease in signal-to-noise ratio at great depths. In some areas off the Namaqualand coast, horizon T appears to be synformal and begins rising again in a westerly direction. The T-to-R wedge could be part of a half-graben similar to the wider ones present in the Orange River to Saldanha shelf segment. Consequently, faulting affecting the T-to-R sequence could be present to the west. As horizon T has not yet been identified on the seismic sections toward the ocean basin, its relationship to the typical oceanic crust present farther west is unclear. To the west of the T-to-R pinch-out, reflections below horizon R can be seen dipping generally west, perhaps faulted down to the east in places, and truncated by a generally smooth horizon R. Continuing west, horizon R is juxtaposed with the top of typical reflection-free oceanic crust with a rugose character. On some seismic sections a distinct faulted boundary is apparent between the two types of crust.

The horizon at the base of the graben sequences to the east is regarded as the equivalent of horizon T, described above as the base of the wedge. The correlation is not certain because the graben fill is nowhere continuous with the T-to-R wedge. However, the eastern graben and the more western T-to-R wedge are both likely to be related to the breakup of the continents because they have been truncated by horizon R, the drift-onset unconformity, and are subparallel to the present coast line and continental shelf. Where horizon T is truncated by horizon R, then horizon R forms the base of the Cretaceous succession.

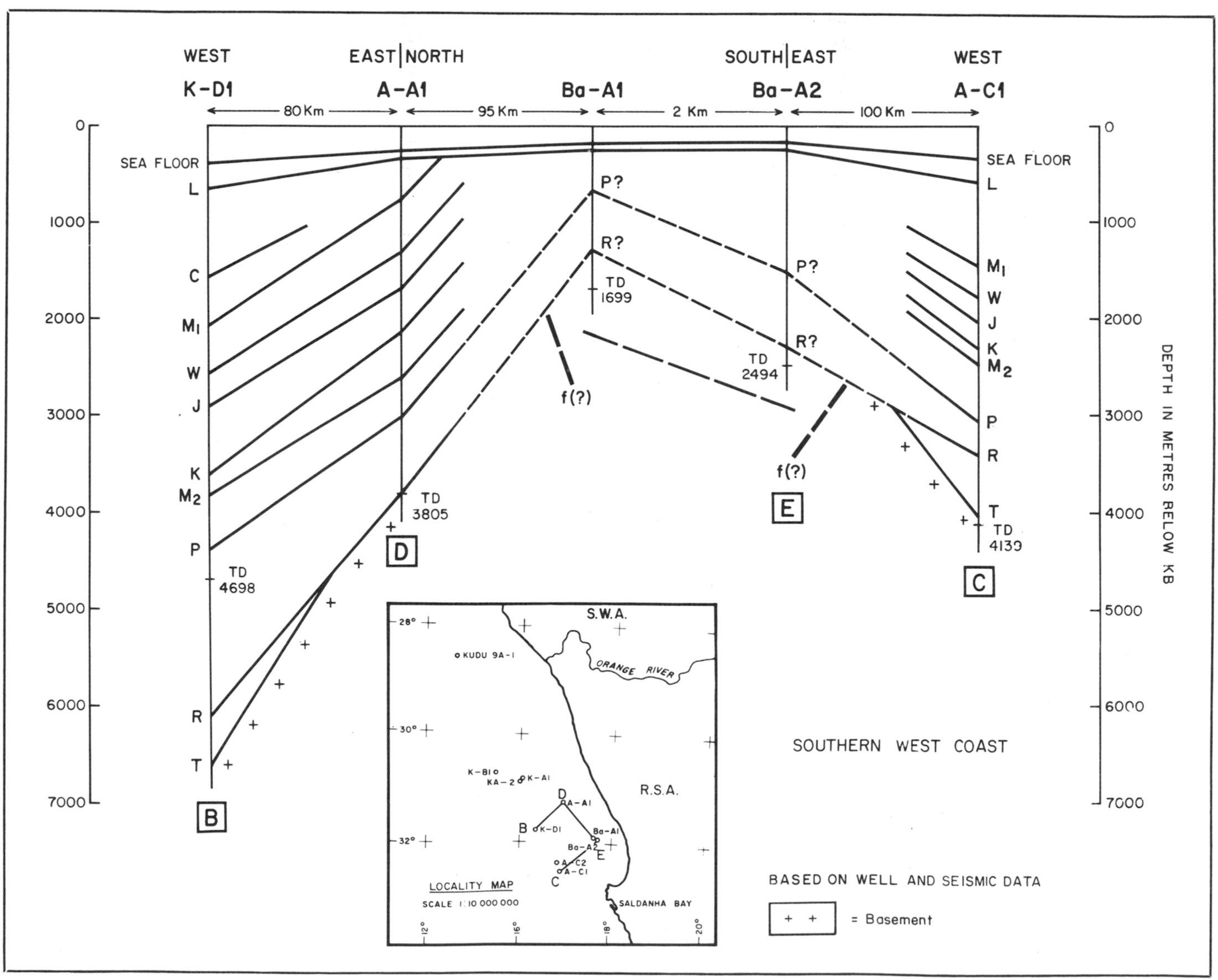

Figure 4 — Southern West Coast regional marker correlation, based on well and seismic data.

Within the wedge sequence, some reflections are roughly conformable to T with R truncating them at a distinct angle. But there is also some onlap onto T with a low angle erosion surface at horizon R. Both situations appear on different seismic sections. Locally, a discontinuous but sometimes quite strong reflector, at an angle intermediate between that of the underlying horizon T and overlying horizon R, can be observed. We suspect that it could mark a major lithological change, possibly from mainly lava in the lower part to mainly sediment in the upper portion.

Thickness contours of the T-to-R wedge are essentially parallel with the pinch-out or hinge-line trend, but as horizon R, the upper bounding surface, is an unconformity, that trend may differ slightly from the original one. The parallelism of the deeply buried T-to-R hinge line with that of the rift valleys and the present coast (Figure 11) suggests that from as early as horizon R times, major marginal subsidence and warping occured along the same trends, except locally, where ancient lines of weakness were the controlling factors. Thus, local but distinct changes in direction of the T-to-R hinge-line trend are present south of Cape Town (Gerrard, 1979b), north of Saldanha Bay, opposite the Orange River, and at two localities north of Walvis Bay. That presence opposite the Orange River could be related to an ancient line of weakness indicated by intrusives of the Kubuus-Bremen trend, which range in age from 490 to 550 million years (Allsop et al, 1979). The two changes in pinch-out trend north of Walvis Bay are regarded as relating to the two subparallel lines of weakness marked by early Cretaceous (130 million years) subvolcanic intrusions. (Miller, 1980). The latter Cretaceous alignments are parallel with the Pan-African (550 million years) trend in that area.

No drilling has been undertaken in the graben or half-graben east of the T-to-R hinge line, but northwest of Saldanha Bay the T-to-R wedge was penetrated by well A-C1, sited just down-dip from the pinch-out. At this locality, the T-to-R interval occurs at a more shallow depth than elsewhere along the coast. It is 690 m thick and consists mainly of basic lavas, with lesser amounts of acid alkaline lava and possibly subordinate sediments or pyroclastics. The presence of basic and acidic alkaline volcanic rocks within a rift-valley fill is typical of the rift valley stage of continental breakup (Dewey and Bird, 1970b; Schneider 1972). The mainly basic lavas of the T-to-R interval are tentatively regarded as mid-Jurassic to late-Jurassic age. The nearest onshore volcanics that may be correlative are the mid-Jurassic Hoachanas basalts, which have an average age of 168 million years (Gidskehaug, Creer, and Mitchell, 1975; Siedner and Mitchell, 1976) and outcrop about 1,000 km to the north-northwest (Figure 1), and lavas (162 million years) of the Suurberg

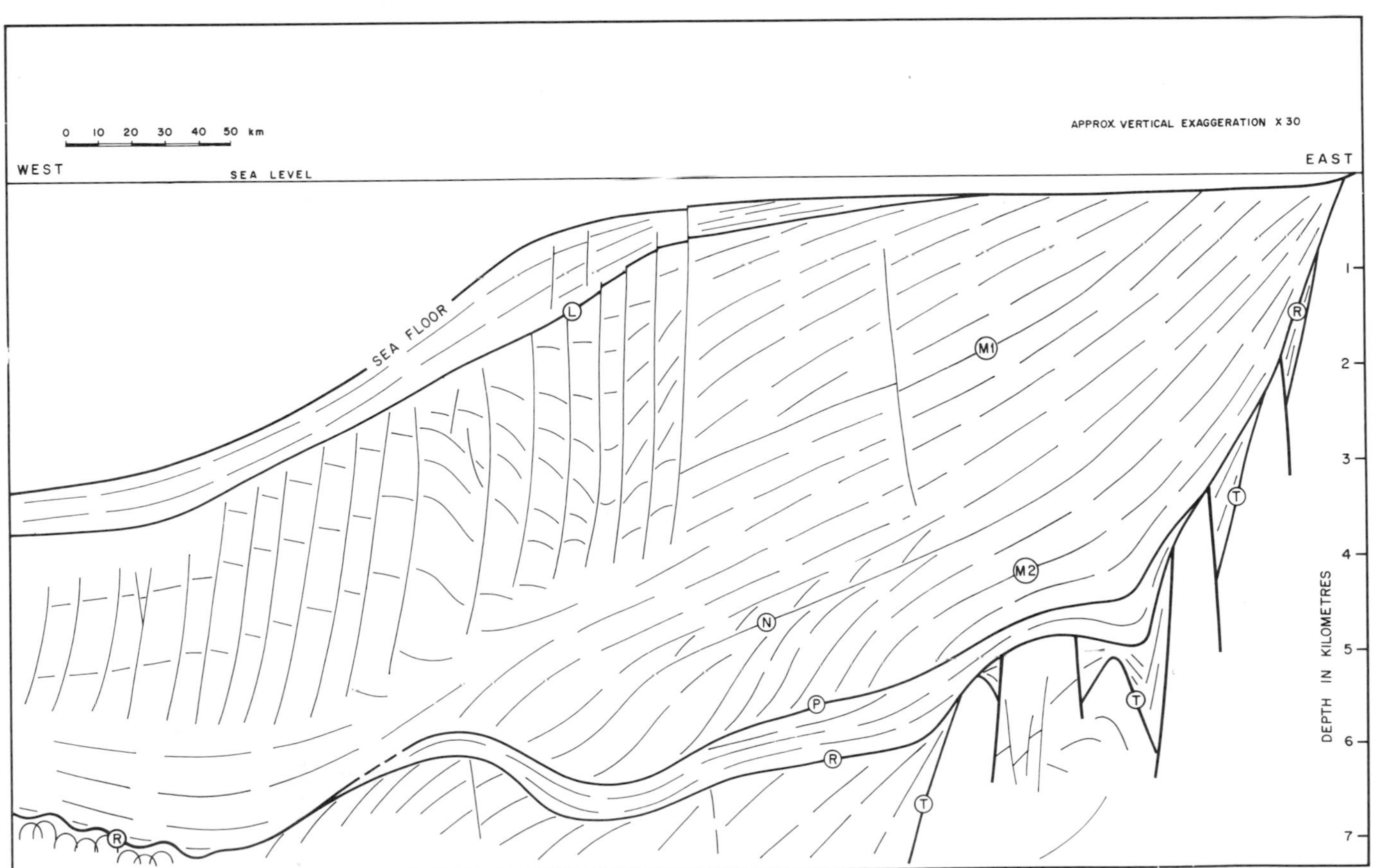

Figure 5 — Diagrammatic section across the Orange Basin.

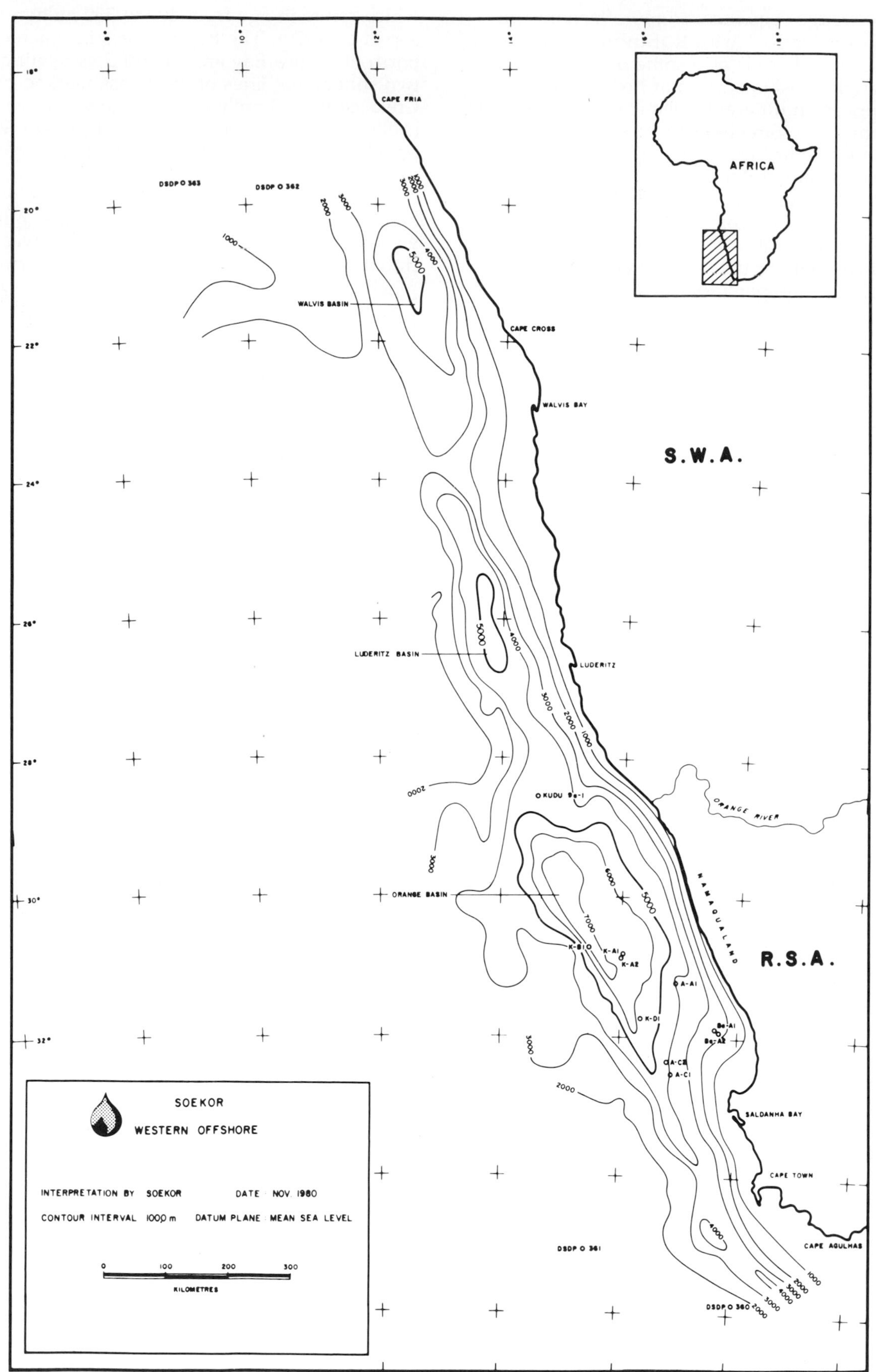

Figure 7 — Isopach map between sea floor and horizon R.

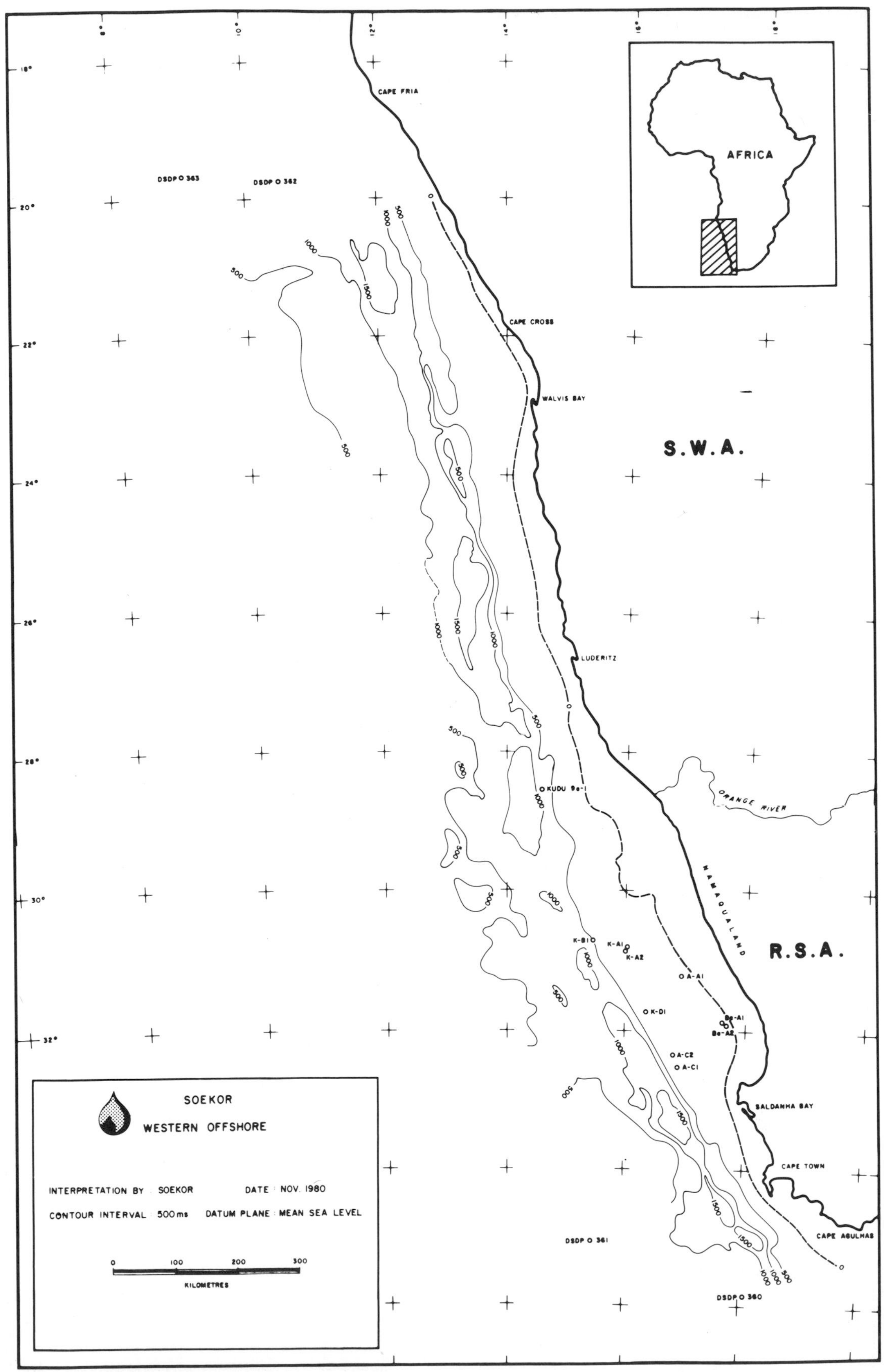

Figure 10 — Two-way time thickness map between sea floor and Horizon L.

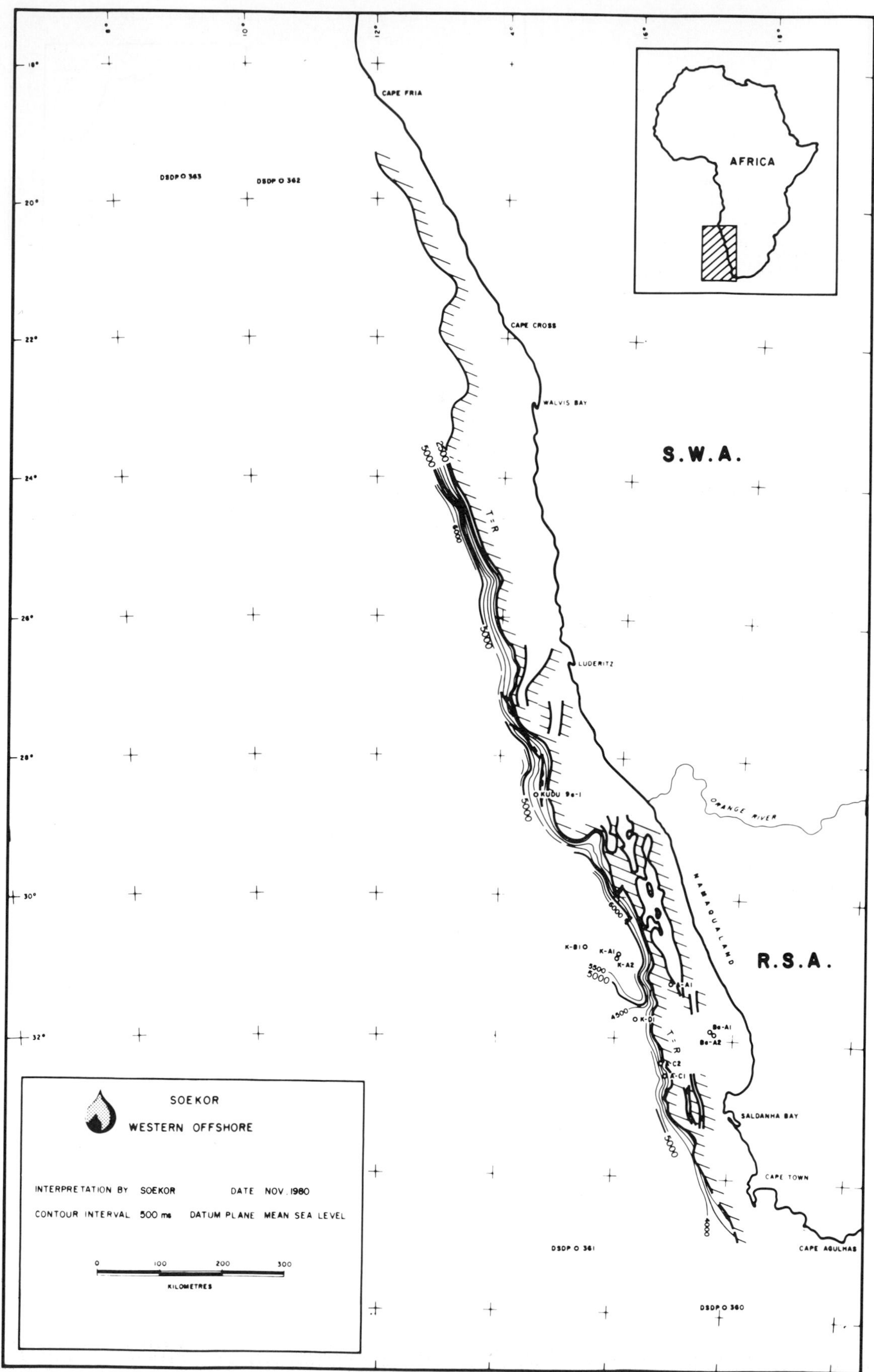

Figure 11 — Two-way time contour map of Horizon T.

Group (Hill, 1975; Winter, 1973) underlying the late Jurassic to early Cretaceous Enon conglomerate of the Algoa basin about 800 km to the east-southeast. No chemical analyses are available to indicate whether the basic lavas have continental or oceanic affinities.

Well logs indicate that the densities of the basic lava range from 2.53 to 2.89, averaging 2.67 gm/cc. This can be interpreted as indicating a large proportion of amygdaloidal basalt, with the lava with the lowest density values being highly amygdaloidal. The alkaline acid lavas intercalated with these basic lavas have densities ranging from 2.52 to 2.60. The seismic velocities of the basic lava, derived from sonic logs, range from 4,100 to 6,100 m/sec, averaging 5,000 m/sec. This is the same as that from the underlying acid lava assigned to the basement at A-C1, but there is a density difference of 0.08 gm/cu cm between them.

The stratigraphic position of the lava and volcanoclastics penetrated at the Ba-A1 well is uncertain, but they may belong to the T-to-R sequence.

Radioactive sandstones and conglomerates containing a high percentage of acid lava cobbles and pebbles are present above the T-to-R sequence. It is therefore possible that some of the intervals within the T-to-R succession interpreted as acid alkaline lava could consist of conglomerates or sandstones, or both. The A-C1 well was drilled close to the pinch-out position of the T-to-R interval, in a region where some truncation of reflections by horizon R is apparent. It is therefore likely that to the west of the A-C1 well, a younger succession of T-to-R rocks is present overlying the volcanics. This succession could consist predominantly of sediments. The fairly regular trending pinch-out of the T-to-R wedge could give rise to a linear magnetic anomaly, due to the magnetic susceptibility contrast between the basic lava of the wedge and the other lithologies present below R farther to the east. For the same reason a gravity anomaly could be expected. This topic is discussed in a later section.

The drift-onset unconformity is considered to be represented by horizon R. This is the base of the Mesozoic succession over much of the eastern part of the Orange Basin where it is not underlain by the T-to-R wedge or the graben (Figure 12). The R-to-P interval could thus represent a proto-oceanic sequence (post breakup stage of Falvey, 1974), formed both adjacent to and in an incipient new oceanic basin or Red Sea stage of margin development. Such basins are commonly filled with sediments typical of restricted environments, as well as lavas. The basaltic lavas intersected in the Kudu 9A-1 well are considered part of such a sequence.

As mentioned in the previous section, horizon R truncates the underlying T-to-R successions of both the major wedge and the graben fills underlying the Orange and Lüderitz basins. Shoreward, horizon R is truncated by horizon P, also an unconformity, of mid-Aptian age. Eastward of the R-to-P pinch-out, horizon P forms the base of the Mesozoic succession. In a seaward direction, under the present lower continental slope, the R-to-P interval thins until horizons R and P appear to merge at the limits of resolution of the seismic sections. The thinning is due in part to offlapping within the R-to-P sequence. North of the Lüderitz basin, horizons R and P appear to merge on seismic sections.

The two-way time structure map (Figure 12) of horizon R indicates a fairly uniform westerly dip for this horizon though gradients are less under part of the Orange basin. The contours are essentially parallel with the coast, though the shallower portion of horizon R north of Saldanha Bay and opposite the Orange River has major changes in trend which results in an embayment off the Namaqualand coast. The contour spacing is too wide to indicate the steepening of the gradient of horizon R to the west of the underlying T-to-R hinge line. This superimposed hinge line on horizon R is regional and is conspicuous on many seismic sections. It could be due to compactional effects in the underlying T-to-R interval, or due to greater subsidence near the T-to-R wedge caused by basic lavas in the lower part of it. On conversion of the time-structure map to a depth map (Figure 13), a low discontinuous ridge at the western part of the R unconformity becomes apparent. The ridge is not seen on the time maps because of the distorting effects of the varying thickness of the water layer. The depth map emphasizes the existence of the distinct structural low area off the Namaqualand coast.

Note that the structure contours of horizon R east of the low angle R-to-P erosional truncation are those of the compound surface R = P. Although an unknown amount of sediment from the R-to-P interval has been removed by erosion to the east of the R-to-P pinch-out, once the pre-Mesozoic basement was reached the effectiveness of erosion in lowering the low gradient R surface could have been reduced so that the R = P surface probably does not stand far below the original R surface.

Although it has not been mapped along the entire continental margin, the two-way time thickness map (Figure 8) of the R-to-P interval indicates that this interval forms a broad, deep basin off Namaqualand. By contrast, the interval occurs as a narrower and shallower trough north of the Orange River. Thus, the Orange Basin came into existence at an early stage in the evolution of the continental margin. It is unlikely that extensive deltas could form along much of the west coast as only short drainages probably developed on the uplifted coastal arch, as was the case for the east coast of South Africa (de Swardt and Bennet, 1974). However, an ancestral Orange River may have dumped a considerable load onto the young continental shelf. Accelerated subsidence due to loading could account for the substantial thickness of the R-to-P interval preserved off Namaqualand. The R-to-P interval is rarely affected by faulting, and not even by rejuvenation of rift faults, and though its thickness has locally been partly controlled by compaction of the underlying graben fills, it disconformably overlies the graben of the rifted sequence. The regional hinge line on horizon R seems to coincide with a facies boundary in the R-to-P interval that separates generally thinner continental red beds in the east from the thicker, gray,

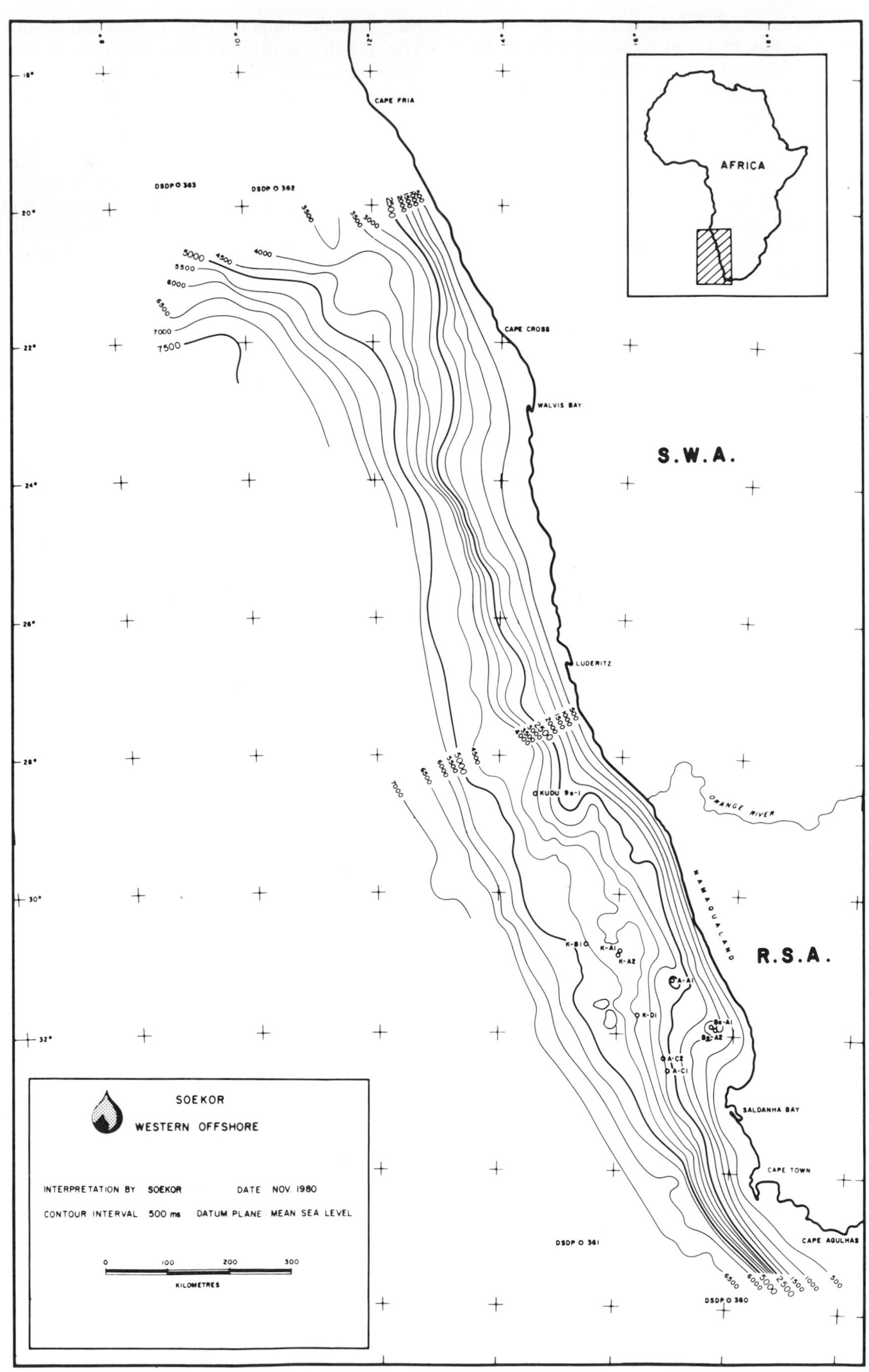

Figure 12 — Two-way time contour map of Horizon R.

restricted-marine or lacustrine sediments in the west. Tertiary intrusions occur on the hinge line in the A-C area.

There appears to be a significant variation in lithology of the R-to-P interval from north to south. At the Kudu 9a-1 well in the north, that part of the interval penetrated consists mainly of marine shales resting on gas-bearing sandstone overlying and intercalated with basaltic lava. South of the Orange River, wells A-C1, A-C2, K-A1, and K-D1 encountered mainly sandstones and, locally, conglomerates of shallow-water origin. Closer inshore, in the south, wells A-A1, Ba-A1, and Ba-A2 intersected continental red bed sequences with minor lavas or igneous intrusions.

Well Ba-A1 also intersected some conglomerate below horizon P. The sedimentary succession is underlain by lavas and volcano-clastics, but these could be part of the T-to-R sequence of the rift valley fill. A core from the R-to-P interval at A-C2 contained cobbles and pebbles of quartz-feldspar porphyry, which could be derived from that exposed at Saldanha Bay, and a few of an orthoquartzite, probably derived from the Ordovician Table Mountain Group (Figure 1). The nearest position to this well where basement would have been exposed at P times (the R-to-P pinch-out position) is about 36 km east. But as the R-to-P sequence onlaps onto horizon R, basement outcrops could have been much closer during the deposition of the earlier stages of the R-to-P sequence.

Based on biostratigraphic data from the wells on the continental shelf (McLachlan and McMillan, 1979 and later unpublished amendment), horizon P is dated as mid-Aptian and the upper portion of the R-to-P interval as early Aptian and older. Hence, the lower part of this sequence could be Barremian or older. Horizon P underlies the present shelf and slope, and is regarded as more or less equivalent to horizon AII of the basinal and rise areas of the eastern Cape Basin and the eastern Walvis Ridge (Emery et al, 1975). Horizon AII was dated by DSDP wells 361 and 363 (Bolli et al, 1978) as upper Aptian. The presence of the equivalent of the R-to-P interval on the eastern Walvis Ridge is indicated by Emery et al (1975; Figure 12) and by Dingle and Simpson (1976; Figure 3), where sediment occurs between acoustic basement and horizon AII. Hence, early drift sediments are also present on this portion of the Walvis Ridge. In DSDP well 363, drilled on the eastern part of the Walvis Ridge, the equivalent to the upper part of the R-to-P interval is formed by the lower part of lithologic units 2 and 3. The lowermost Aptian unit 3 is a shallow-water calcarenite (Bolli et al, 1978), with the overlying sediments indicating a progressive deepening of the depositional environment. Emery et al (1975, Figures 5 to 10, 12, and 36) show the distribution of sediments between acoustic basement and horizon AII over the eastern Cape Basin. We interpret these sediments as the deep water equivalents of our R-to-P interval, which, closer to the shore, generally consists of shallow-water sediments. Toward the eastern ends of some of these figures dipping reflectors are truncated by AII and these sequences abut sharply against acoustic (oceanic) basement. We interpret the AII horizon at these localities as the equivalent of our compound R=P horizon and the sequence below as the equivalent of our T-R wedge. On other figures the dipping sequence is truncated by a deeper reflector than AII, and here AII is probably equal to P and the deeper reflector to R.

Horizon AII pinches out to the west against oceanic crust near magnetic anomaly MO, which is dated as Aptian (Bolli et al, 1978).

At DSDP well 361, drilled in the eastern Cape Basin, the Aptian lithologic unit 7 (equivalent to the upper part of the R-to-P sequence) consists of sapropelic shales interbedded with siltstones and massive silty sandstones. We interpret these as turbidites deposited over oceanic crust. Though not penetrated, the oceanic crust was dated as Barremian on the basis of magnetic anomaly correlation (Bolli et al, 1978).

Although the entire R-to-P sequence was not penetrated in Kudu 9a-1, a total of just over 130 m of lava was intersected in the lower part of the interval actually drilled; this is believed to be of Barremian age and is correlated with the approximately 120 m.y. old Kaoko lavas (Gidskehaug, Creer, and Mitchell, 1975; McLachlan and McMillan, 1979, Figure 2 and p. 171; Siedner and Miller, 1968; Siedner and Mitchell, 1976) present onshore in southwest Africa, about 900 km north, where the lavas overlie the Cretaceous Etjo or Plateau sandstone (Figure 1). Depending on their thickness and degree of alteration, these lavas in the R-to-P interval are expected to give rise to magnetic and gravity anomalies that would be superimposed on those due to the lavas of the underlying T-to-R interval.

The important lavas present within this interval at the Kudu 9a-1 well were not encountered in the A-A1, A-C1, A-C2, Ba-A1, or Ba-A2 wells which were drilled through the R-to-P sequences. Only minor flows or intrusions were intersected in these wells. As only part of the R-to-P succession was intersected at the K-A2 and K-D1 wells, it is not known whether the lavas are present in the area between the A-C wells and the Kudu well.

Once drifting had proceeded beyond a certain stage, deltaic, interdeltaic, and open-marine prograding depositional environments prevailed. The prograding sequences are indicative of considerable relative sea level changes.

As the underlying R-to-P interval is so thin over some of the offshore area, horizon P could only be mapped with confidence under the Lüderitz and Orange basins (Figure 14).

Due to the contour spacing used, only the major change in trend of the structure contours of horizon P near the Orange River is conspicuous; this coincides with changes in trend of structure contours on the underlying horizons T and R. Horizon L, which forms the upper boundary of the interval, is an unconformity and truncates Paleocene, Maastrichtian, and Campanian sediments on the middle to outer shelf. Closer inshore, it is cut across uplifted Cenomanian and Albian sediments (Figure 15).

South of the Orange basin, north of the Lüderitz basin, and seaward, the two-way time thickness map

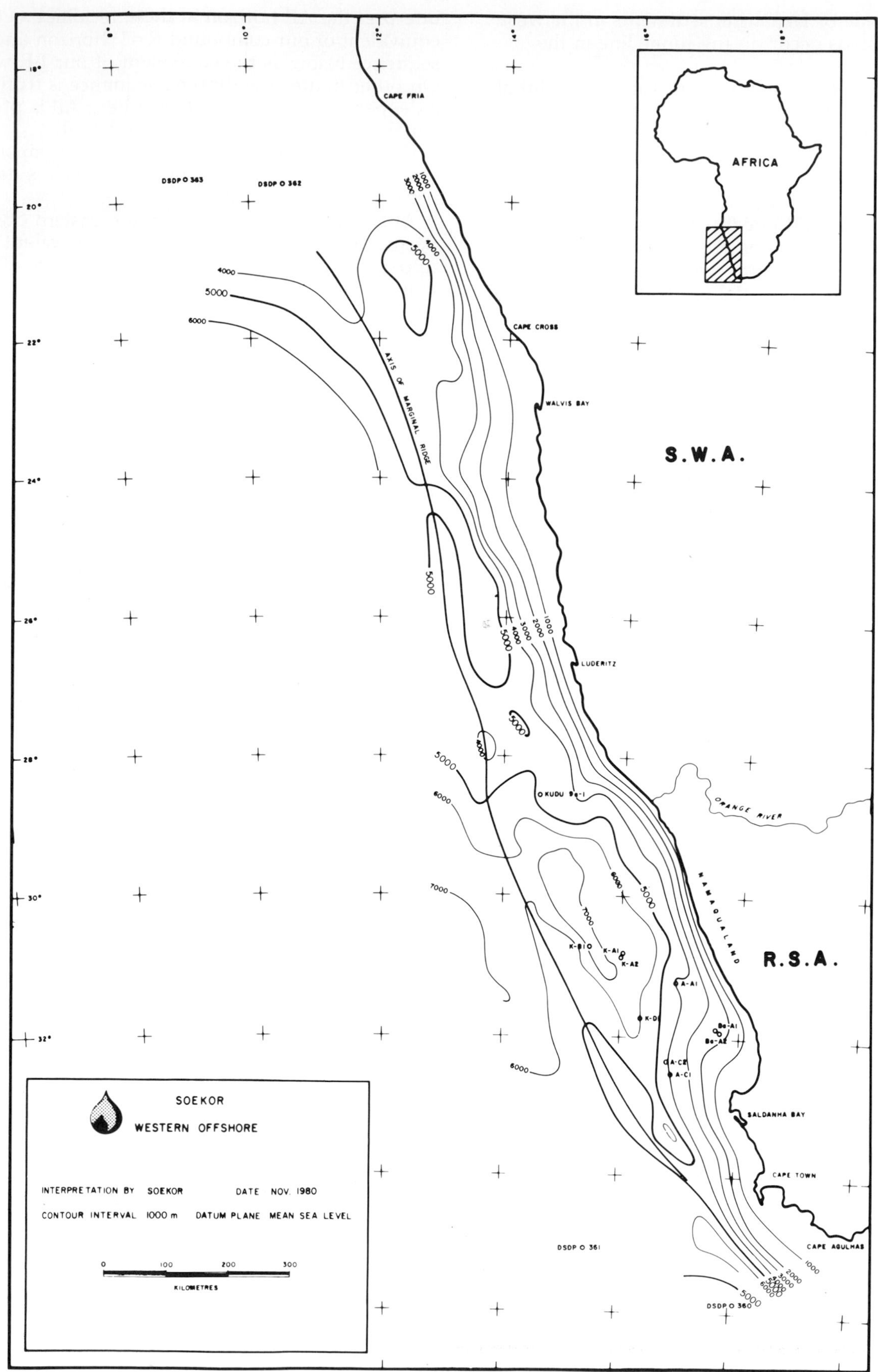

Figure 13 — Depth map of Horizon R.

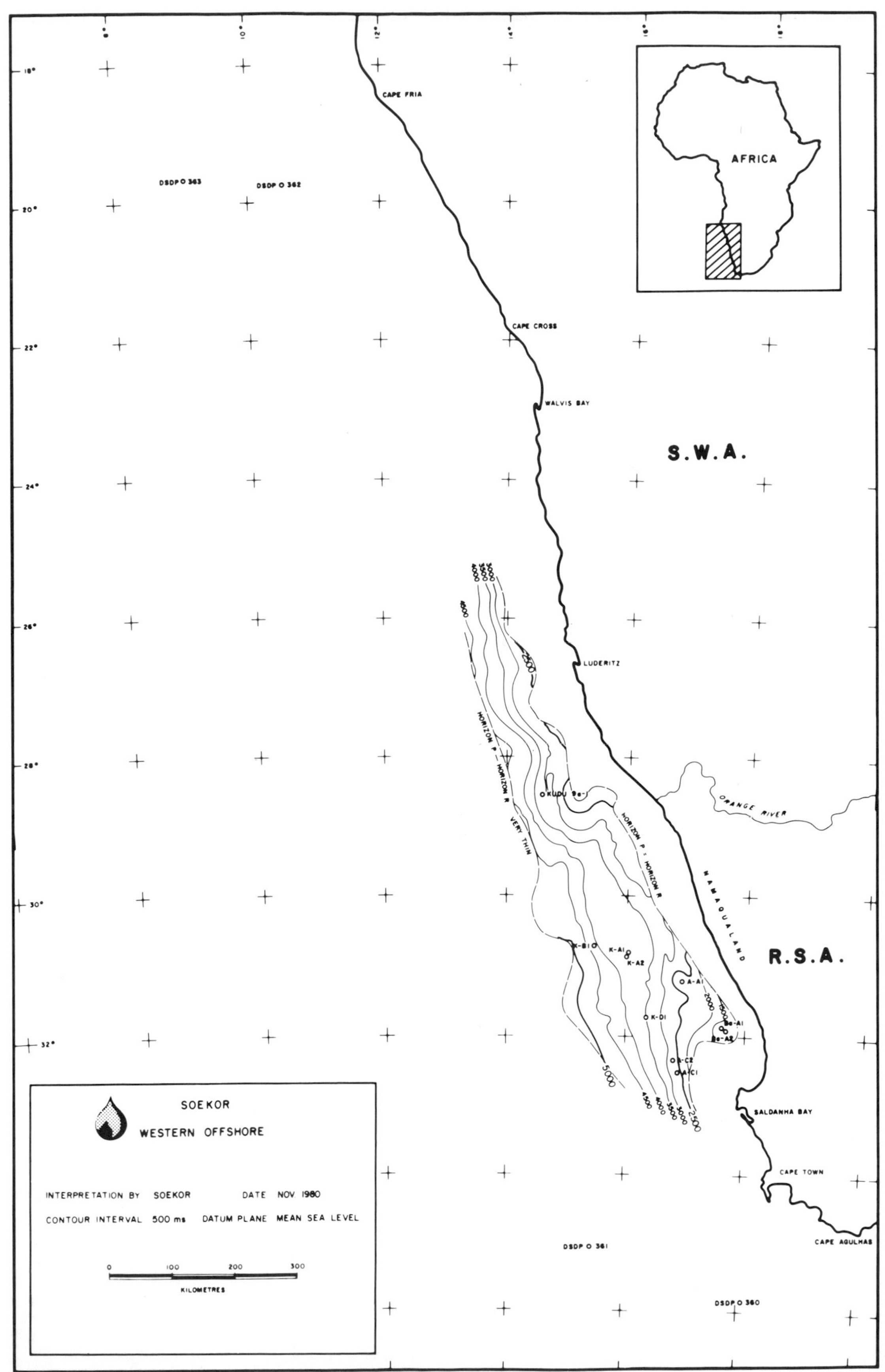

Figure 14 — Two-way time contour map of Horizon P.

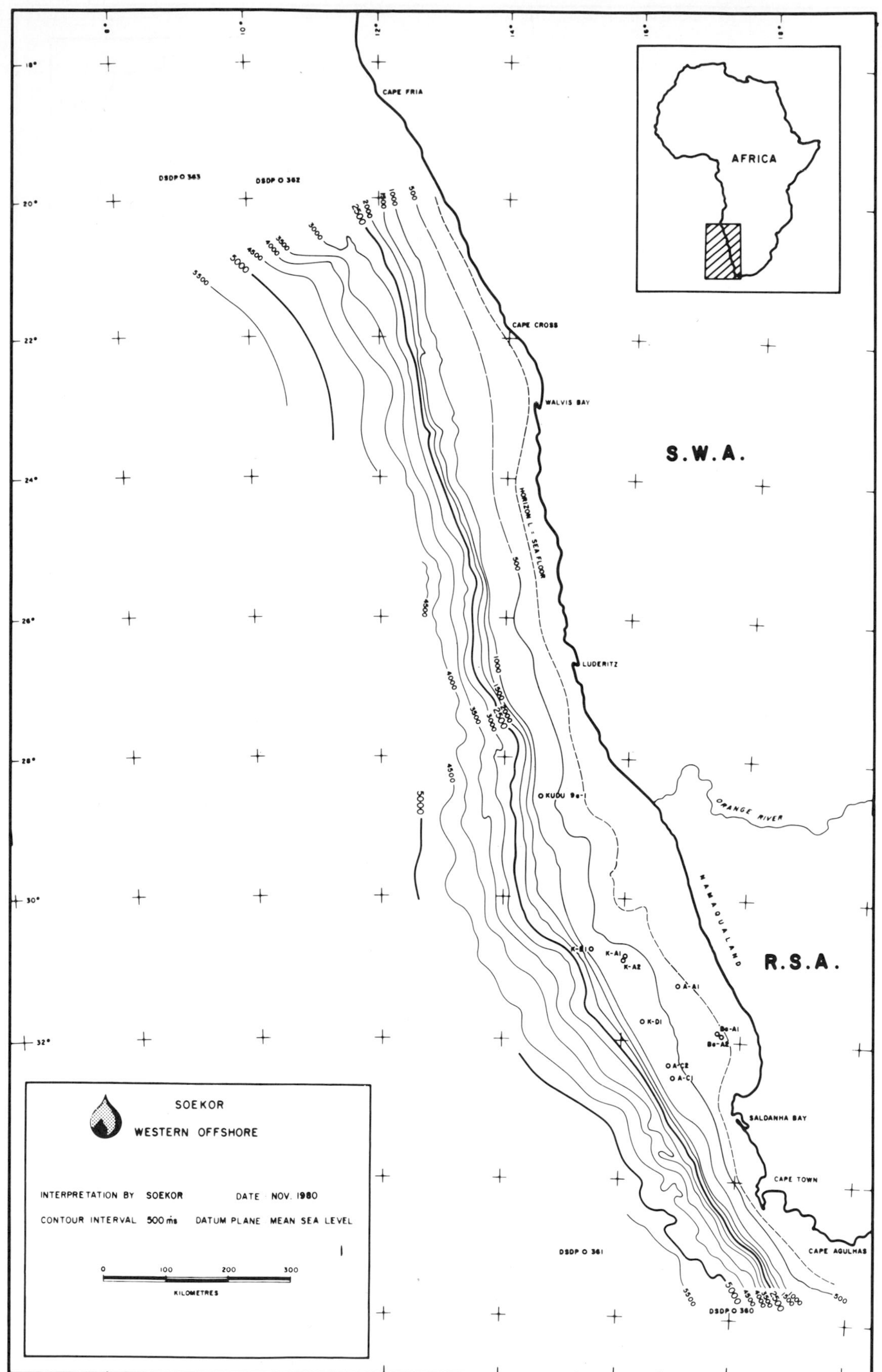

Figure 15 — Two-way time contour map of Horizon L.

of the P-to-L interval (Figure 9) probably includes an unresolvable slice of the R-to-P interval.

The P-to-L interval, where drilled, consists mainly of shales and sandstones, in varying proportions. In the Orange basin, these sediments range in age from late Aptian to Maastrichtian and Paleocene. In the southeastern portion of the Orange basin, upper Cretaceous sediments have been eroded away and horizon L truncates Cenomanian to Albian sediments. Two major superimposed prograded sequences separated by a distinct seismic horizon, mappable on a subregional scale, have been recognized on seismic sections over the southern half of the Orange basin. Well data indicate that the top of the lower prograded sequence is Cenomanian and the base of the upper is Coniacian. The sequence boundary represents a considerable hiatus, as the entire Turonian succession is missing (I. K. McMillan, personal communication, 1980). In both prograding sequences there is a progressive up- and out-building of the ancient shelf edges. Where mapped, these parallel the present coast. The axes of the thickest preserved sediment accumulations of the P-to-L interval and the R-to-P interval both parallel the coast.

A roughly north to northwest trending linear zone of synsedimentary faults, subparallel with the present coast, occurs within these Cretaceous sediments. The zone of faulting coincides with palaeoshelf breaks in the underlying prograded sequences. The amount of rotation within the fault blocks generally decreases eastward and the faults die out downward before horizon P is reached. The upper ends of many of the faults displace the base of the Tertiary sediments at horizon L.

CENOZOIC SUCCESSION

Regional setting

The Cenozoic sediments accumulated in elongated depocentres generally seaward of the Mesozoic depocentres (Figure 10). The main source of Tertiary sediments was probably the uplifted Cretaceous accumulations at the basin margin to the east.

The L-to-sea floor sequence is regarded as the continuation of open-ocean sedimentation started in post-horizon P times. When total post-drift sediment accumulation (sea floor to horizon R) is considered, the Tertiary component is relatively insignificant and has little influence on the configuration of the main Orange, Lüderitz, or Walvis depocentres (compare Figures 16 and 10).

Although Tertiary intrusions are sporadically distributed in the offshore basins, volcanic activity has been locally more intensive (for instance, in the Ba area). In the A-C area, some of these intrusions are present near the T-to-P pinch-out and have intruded on the hinge line in the overlying R-to-P interval.

Drift stage: L-to-sea floor

Although open ocean sedimentation continued during much of the Tertiary, a lowering of relative sea level occurred near the Paleocene-Eocene boundary. Close to the coast, Horizon L marks the base of the Tertiary with thin Eocene or younger sediments resting on a Cretaceous succession that ranges in age from Coniacian to Albian. Toward the outer shelf, horizon L marks the base of the Eocene which rests on Maastrichtian or the lowermost Tertiary (Palaeocene) sediments. The horizon probably becomes a disconformity, and finally a conformity, oceanwards.

Structure contours on horizon L (Figure 15) show the shape of the early Tertiary continental shelf, shelf break, and continental slope. Tertiary onlaps on L are present near the ancient shelf break at the level of L. This ancient shelf break is seaward of the highest breaks observed in the underlying P-to-L sequence. Over much of the middle and inner shelf, the Tertiary sediments are thin and they reach a considerable thickness only at the outer shelf. The linear Tertiary depocentre (Figure 10) is more seaward of the P-to-L depocentres as would be expected under conditions of out-building of a shelf.

The faulting that commonly affects horizon L is due to rejuvenation along the earlier synsedimentary faults in the P-to-L succession. High-resolution seismic profiling (Gerrard, 1979a) has indicated that some of these faults extend through thin Tertiary sediments to, or very close to, the sea floor. Details of the structure within the Tertiary succession are found in Dingle (1973) and Van Andel and Calvert (1971).

Very little is known about the lithology of west coast Tertiary rocks since, when drilling through these shallow and thin sediments, the cutting returns are displaced up the hole onto the sea floor and are not sampled. Samples become available only when the lower part of the Tertiary succession extends below the surface casing. On the basis of occasional bit samples and gamma-ray log interpretation, the Upper Tertiary sediments are believed to be sandy and shelly limestone and calcareous sandstones. Claystones and sandstones appear to form the bulk of the Lower Tertiary sediments.

Alignment of some Tertiary intrusions along the T-to-R pinch-out and along the superimposed hingeline in the R-to-P interval in the A-C area, lends credence to Moore's (1976) suggestion that the onshore Tertiary lines of uplift and intrusion are essentially parallel with the coast.

CONTINENT-OCEAN BOUNDARY AND AGE OF CONTINENTAL BREAK-UP

Positions of the continent-ocean boundary, (COB) shown in Figure 17 are based on Scrutton (1973b), Emery et al (1975), Rabinowitz (1976), Scrutton (1976), Dingle (1979), Rabinowitz and La Breque (1979), and Smith (1980). These interpreters have assessed the COB in terms of magnetics, gravity and seismic reflection and refraction data or combinations of these.

Emery et al (1975) and Scrutton (1973b), using variable seismic refraction control, interpreted their gravity data as indicating a broad continent-ocean transition zone. Emery et al (1975) believe that the shoreward

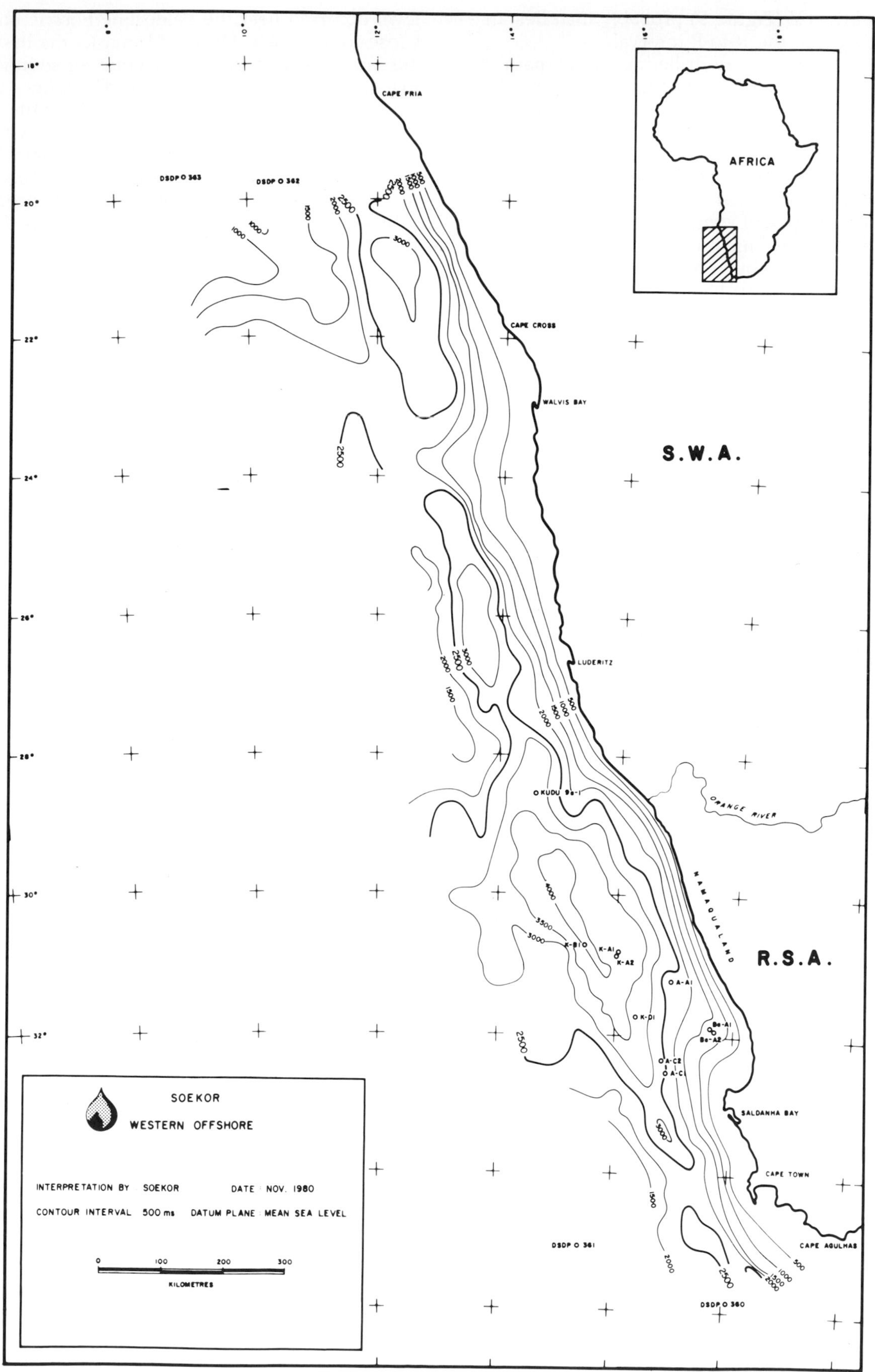

Figure 16 — Two-way time thickness map between sea floor and Horizon R.

side of the 50 mgal contour is underlain by continental crust and the seaward side of the 300 mgal contour by oceanic crust (see Figure 17). They have two buried basement ridges, interpreted from the magnetic and gravity data within the zone of transitional crust. Dingle (1979), who draws on data from Scrutton (1973b, 1976, 1978), placed the COB near the lower continental slope.

Smith (1980) places the COB the farthest offshore (Figure 17). He bases his location on a change in the reflection character of horizon R and the reflections beneath it. On the western end of Figure 6, diffraction patterns mark the base of the reflecting sequence. Below this event, seismic sections are totally structureless except for diffractions typical of oceanic crust. Following the diffractions eastward, we see character change to a smoother reflection (horizon R,) beneath which we see dipping reflections indicating a structured sequence. This change occurs on some seismic lines at a down-to-the-west fault, which is interpreted as the continent-ocean boundary with rift and prerift sequences abutting against oceanic crust.

Larson and Ladd (1973), Rabinowitz (1976), and Rabinowitz and LaBreque (1979) place the COB closest to the coast on the basis of their interpretation of magnetic anomalies (anomaly M of Larson and Ladd, the same as anomaly G(=M13) of Rabinowitz). Scrutton (1978) interprets the gravity anomaly coincident with magnetic anomaly G as a basement ridge and places the COB considerably further offshore. Emery et al (1975) agree with Scrutton regarding the placement of the COB and infer an inner basement ridge largely concident with magnetic anomaly G.

Data presented in this paper favor the interpretations of both Emery et al (1975) and Scrutton (1976, 1978), with the magnetic and gravity anomalies due not to the COB, but due to a westward-thickening wedge of mainly basic lava of the T-to-R interval. In the Walvis Bay to Orange River sector of the coast, there is good coincidence of the suggested buried basement ridge (Emery et al, 1975), magnetic anomaly G (Rabinowitz, 1976) and the T-to-R pinch-out. South of the Orange River the suggested buried ridge, anomaly G and the T-to-R pinch-out are still close together but show a varying degree of divergence (Figure 17). Basic lavas within the long graben paralleling the coast to the east of the R-to-P wedge would give rise to the observed gravity and magnetic anomalies.

In the offshore between Walvis Bay and Saldanha Bay, both Dingle's (1979) and Smith's (1980) COB's are essentially parallel with the Mesozoic Cape Basin magnetic anomaly lineations. South, the COB's transgress from younger to older magnetic lineations (Figure 18). This suggests a difference in age with the continent-ocean boundary being older in the south. However, before discussing this possibility, the differences in dating and correlation of the magnetic anomalies should be examined.

Du Plessis (1979) and Van der Linden (1980) draw attention to the discrepancies in dating the Mesozoic magnetic anomalies in the Cape Basin, as well as to

Table 2. Breakup age of continental margin (million years).

Magnetic anomaly Correlator	**Magnetic reversal scale** Larson and Hilde (1975)	Van Hinte (1976a and 1976b)
Rabinowitz M13 = G	130	129.5
Du Plessis M22	147.5	141.5⁺

⁺Du Plessis (1979) actually suggests 142 million years for the breakup.

differences in correlations of these anomalies. For example, Du Plessis (1979, Figure 4.2), correlates Rabinowitz's (1976, 1979) anomaly G = M13 with his anomaly, M22. Although Du Plessis (1979) accepts Rabinowitz's (1976, 1979) COB, which is labelled M22 by the former and G = M13 by the latter, it results in a rather different set of ages for the separation of the South American and the African continents, as indicated in Table 2.

The oldest postulated age for this breakup is 165 million years (Emery et al, 1975), but data presented below using the seismic COB suggest that the breakup age varies from south (oldest) to north (youngest).

SOEKOR's seismic sections covering a large area of the continental margin (Smith, 1980), as well as the profiles of Emery et al (1975, Figures 6, 7, 9, 10, and 12) and Lehner and De Ruiter (1976, 1977), indicate that a considerable portion of the magnetic anomaly sequence west of anomaly G is underlain by continental, rather than oceanic crust. This has been proven by two offshore wells in the southeastern part of the Orange basin. Van der Linden's (1980) suggestion, based on profiles of Lehner and De Ruiter (1976, 1977), that the eastern Walvis Ridge is a continental fragment covered by basalt coincides with this interpretation.

The COB of both Smith (1980) and Dingle (1979), based on data other than magnetics, trends across the magnetic anomaly directons (Figures 17 and 18) with some of the correlated magnetic lineations occurring over continental crust and some over oceanic crust. This suggests that either unrelated magnetic anomalies are being correlated, or that some of the lavas formed during oceanic crust formation were extruded over the adjacent continental crust. Both situations can exist; well data indicate that lavas occur at two main stratigraphic levels at some localities and that their magnetisation can affect magnetic correlations.

The transgression of the COB, based on seismic data, across the correlated magnetic anomalies indicates that the marginal rift is oldest in the south and becomes slightly younger to the north. Depending on whether the magnetic anomaly correlation of Du Plessis (1979) or of Rabinowitz (1976, 1979) is used, and on whether Larson and Hilde's (1975) or Van Hinte's (1976a, 1976b) datings of magnetic reversals are accepted, various ages for the continental breakup of west Gondwanaland, using the seismic COB (Smith, 1980), can be obtained (tabulated in tables 3 and 4).

Table 4 shows that the age span of the continental breakup, from south to north, can vary from 4 to 13

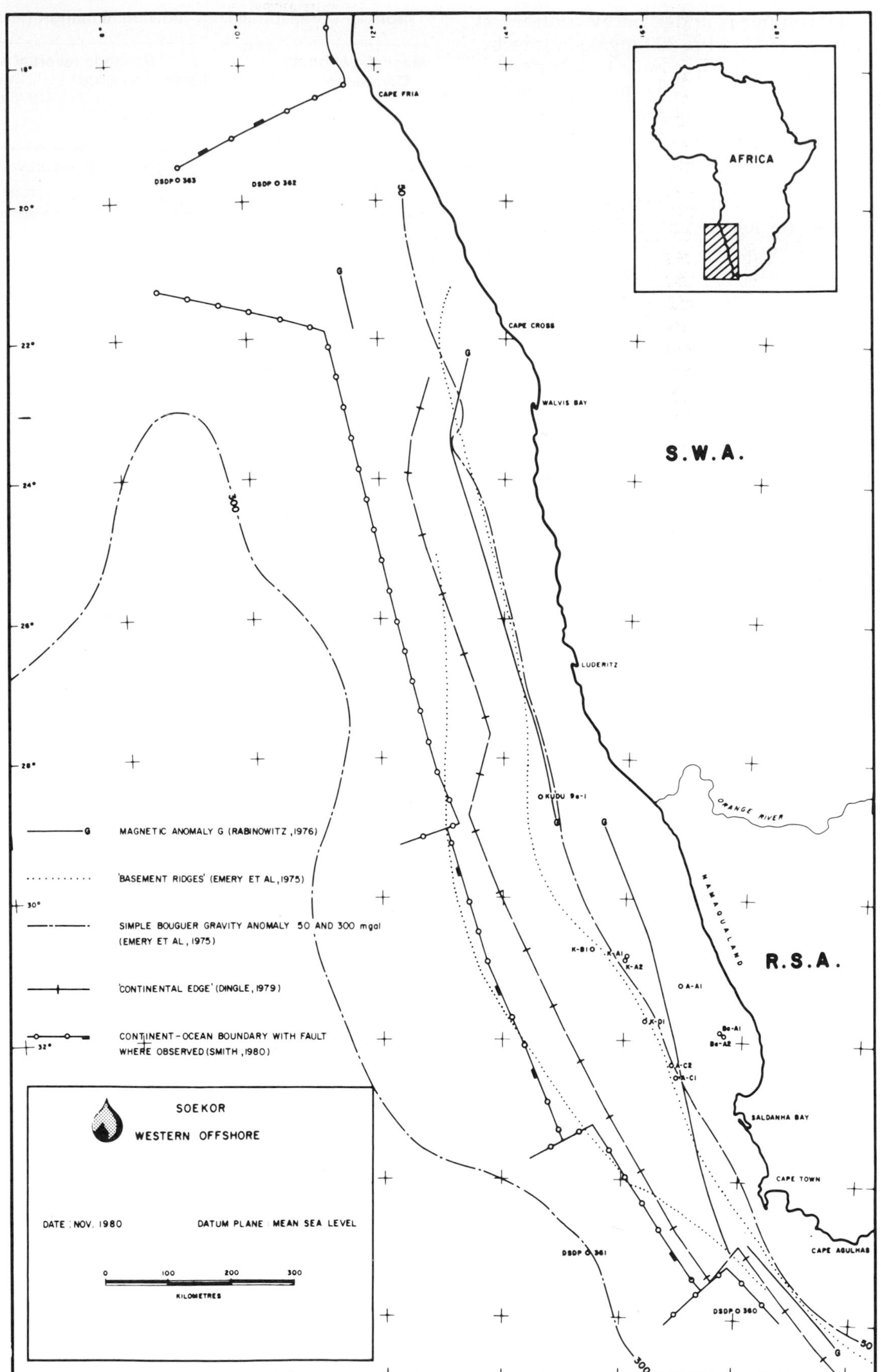

Figure 17 — Interpreted positions of continent-ocean boundary and basement ridges.

Table 3. Position of continental margin relative to magnetic anomalies.

Magnetic anomaly correlator	Anomalies	
	South	North
Rabinowitz	M12	M3
Du Plessis	M21	between M17-M18

million years, depending on the magnetic correlations used and the dating of these anomalies.

It is possible that the horsts and graben mapped off the west coast represent aborted marginal rifts, and that continental separation occurred more to the west. This is consistent with the westward movement of the COB when followed from opposite Cape Agulhas in the south, to opposite the Orange River mouth in the north. This suggestion agrees with the mechanisms proposed by Du Plessis (1979) and Van der Linden (1980) for the various ages of oceanic crust established north and south of the Walvis Ridge. The latter author regards the eastern Walvis Ridge as a continental fragment dislocated by offsets.

The COB, according to seismic evidence, lies well west of anomaly G (Rabinowitz, 1976, 1979) in the area north of Saldanha Bay. This means that magnetic lineations east of Rabinowitz's anomaly M3 could be caused by basic lava overlying thin or transitional continental crust. These particular anomalies have thus been correlated with many that are developed over actual oceanic crust south of Saldanha Bay, where this COB is most markedly discordant to the magnetic lineations. Not unexpectedly, on the eastern side of the COB (seismic criteria) oceanic lavas would overrun continental crust and there the associated magnetic anomalies would be part of the easternmost Cape Basin Mesozoic oceanic sequence.

Magnetic anomaly G (Rabinowitz, 1976, 1979) can be followed on a sub-continental scale. It provides the easternmost magnetic correlation when the magnetic anomalies of the lavas on continental crust are correlated consecutively eastward from the last true oceanic correlation in the west. These ocean sequence correlations can then be tied south, with the anomaly sequence over true oceanic crust present south of Saldanha Bay. Two situations are then possible: First the magnetic anomalies on the continental crust are unrelated to the magnetic anomalies of the ocean series. The apparent correlation with magnetic anomalies of the Mesozoic oceanic sequence is fortuitous. This would be the situation if the lavas in the T-to-R wedge are part of the rift sequence preceeding drifting. Because of the parallelism of the rifting and fracturing with extrusion and accumulation of these lavas on the continental crust, and the parallelism of these trends with the COB, the magnetic lineations of the early lavas on the continental crust would be roughly parallel with magnetic lineations of the later developed oceanic sequence. These linear magnetic anomalies could easily be mistaken for oceanic anomalies. A variation to this particular situation is that the magnetic anomalies related to the lavas of the R-to-P interval would be superimposed on those of the underlying T-to-R sequence. Second, the magnetic anomalies are due mainly to the lavas present in the early drift R-to-P interval. These have been extruded along fractures extending vertically through both the continental crust and the intervening T-to-R wedge. The fractures having propagated laterally north into the continental crust from the major rifts separating oceanic and continental crust, as suggested by the stepped COB west of Cape Agulhas and Saldanha Bay (Figure 18). Since the COB is more westward in position and progressively younger northward, it is envisaged that the extruded lavas become younger in a westerly direction. Those lavas present on the continental crust north of Saldanha Bay would be coeval with those of the oceanic crust to the south. Thus, the correlation of their magnetic anomalies would be legitimate, remembering that north of Saldanha Bay these magnetic lineations cannot be used for dating the crust but only for dating the accumulation of the particular lavas. However, the easternmost anomaly G, attributable to the lavas of the T-to-R interval, would thus not be part of the oceanic sequence.

Table 4. Breakup age of continental margin (millions of years).

Correlator	Date of magnetic reversals (million years)			
	Larson and Hilde (1975)		Van Hinte (1976a and 1976b)	
	South	North	South	North
Rabinowitz	128	115	129	121
Du Plessis	145	138	138	134

CONCLUSIONS

Regionally developed seismic sequences, T-to-R, R-to-P, P-to-L, and L-to-sea floor, have been mapped off the southwestern African coast and are comparable with lithotectonic units recognized elsewhere along Atlantic-type continental margins.

Well A-C1, at the southern margin of the Mesozoic Orange basin, indicated a substantial thickness of basic lava within the lower part of the T-to-R wedge. Acid lava occurs beneath the basalts and below horizon T and appears to be similar to that occurring as basement east of the T-to-R pinch-out and directly below horizon R at well A-C2 (about 20 km north). A distinctive gravity and magnetic anomaly (G) is located over the T-to-R pinch-out in the A-C1 area. The pinch-out can be traced on seismic sections north and south of the A-C area. A correlatable gravity and magnetic anomaly occurs over the pinch-out along most of the coast and there appears to be a causal relationship between the two. This particular anomaly has previously been interpreted by some researchers as indicative of the COB, but seismic reflection data indicate that the COB lies considerably farther west. Thus, those linear magnetic anomalies, correlated in

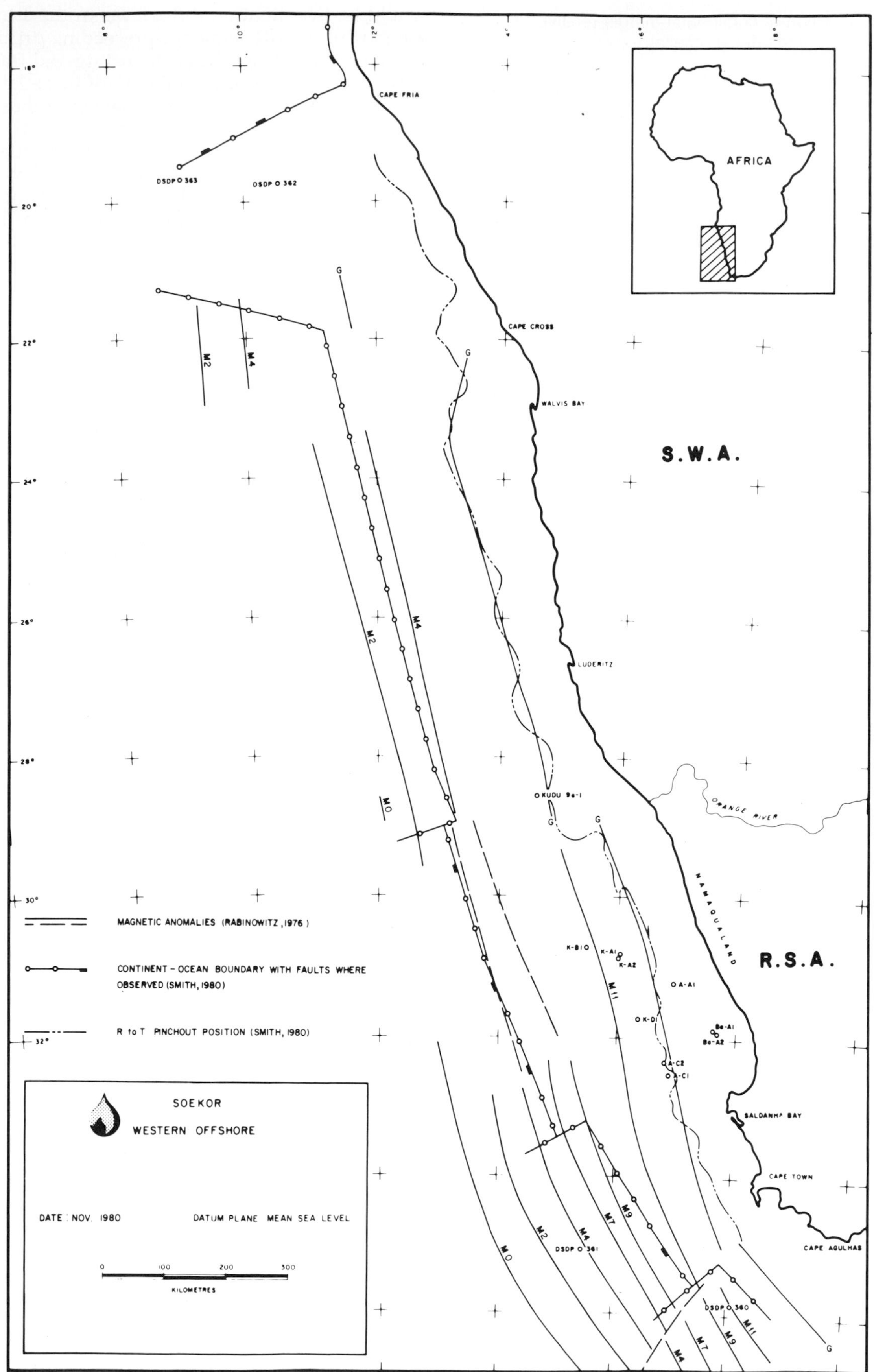

Figure 18 — Interpreted magnetic anomalies and observed seismic features.

the literature with the oceanic M series but occurring between the seismic COB and anomaly G, are regarded as being located over continental crust where a varying amount of basic material has been extruded over, or intruded into, continental coastal rocks.

North-to-northwest trends are apparent on the structure and isopach maps for various stratigraphic levels. These trends date from the earliest rifting phase to the various phases of later subsidence following drifting and into the Holocene, with the present coastline having the same alignment. Moore (1976) discussed this trend in terms of onshore Tertiary warping and intrusion alignments. The persistence of this trend could be because of the dense basic lavas within the T-to-R interval, causing additional loading of the crust as compared to sediment loading only. The T-to-R interval would exert a strong influence on marginal subsidence over time, with the consequent shoreward uplifts perhaps accounting for the lack of a coastal plain underlain by Mesozoic and Cenozoic sediments onshore.

The structural and isopach maps also indicate that local trends, interpreted as being related to onshore northeast structural alignments, are superimposed on the regional trend. These structural cross-trends occur north of Saldanha Bay, opposite the Orange River mouth, and in the Walvis Bay area. Opposite the Orange River mouth the cross-trends appear to have influenced sediment accumulation since the start of drifting until about the end of the Cenomanian.

ACKNOWLEDGMENTS

We thank the Management of SOEKOR for permission to publish. We are indebted to colleagues, some no longer with SOEKOR, for useful discussions over many years and for the contributions to our geological and geophysical thinking which appear in many unpublished company reports. The authors accept responsibility for the views expressed and any acts of omission. We thank Dr. A. M. J. de Swardt, Dr. C. L. Drake, and Dr. J. S. Watkins for their editing and valuable suggestions for improving the manuscript.

REFERENCES CITED

Allsop, H. L., et al, 1979, Rb-Sr and U-Pb geochronology of late Pre-Cambrian-early Proterozoic igneous activity in the Richtersveld (South Africa) and southern southwest Africa: Transactions of the Geological Society of South Africa, v. 82, p. 185-204.

Asmus, H. E., and F. C. Ponte, 1973, The Brazilian marginal basins, *in* A. E. M. Nairn and F. G. Stehli, eds., The ocean basins and margins, v. 1: The South Atlantic, New York-London, Plenum Press, p. 87-133.

Bolli, H. M., et al, 1975, Basins and margins of the eastern south Atlantic: Geotimes, v. 20, p. 22-24.

———, ———, 1978, Initial reports of the deep sea drilling project, v. 40: Washington, D.C., U.S. Government Printing Office.

De Swardt, A. M. J., and G. Bennet, 1974, Structural and physiographic development of Natal since the Late Jurassic: Transactions of the Geological Society of South Africa, v. 77, p. 309-322.

Dewey, J. F., and J. M. Bird, 1970a, Mountain belts and the new global tectonics: Journal of Geophysical Research, v. 75, p. 2625-2647.

———, and ———, 1970b, Plate tectonics and Geosynclines: Tectonophysics, v. 10, p. 625-638.

Dickinson, W. R., 1974, Plate tectonics and sedimentation: Society of Economic Paleontologists and Mineralogists, Special Publication, n. 22, p. 1-27.

Dingle, R. V., 1973, The geology of the continental shelf between Lüdertiz and Cape Town (southwest Africa), with special reference to Tertiary strata: Geological Society of London Journal, v. 129, p. 338-363.

———, 1976, A review of the sedimentary history of some post-Permian continental margins of Atlantic-type: Annales of the Brazil Academy of Science, supplement, v. 48, p. 67-80.

———, 1978, South Africa, *in* M. Moullade and A. E. M. Nairn, eds., The Phanerozoic geology of the world, II, The Mesozoic, A: Amsterdam-Oxford-New York, Elsevier, p. 410-434.

———, 1979, Sedimentary basins and basement structures on the continental margin of southern Africa, *in* A. du Plessis and R. V. Dingle, eds., Papers on marine geoscience: Geological Survey of South Africa, Department of Mines, Bulletin, n. 63.

———, and R. A. Scrutton, 1974, Continental breakup and the development of post-Paleozoic sedimentary basins around southern Africa: Geological Society of America Bulletin, v. 85, p. 1467-1474.

———, and E. S. W. Simpson, 1976, The Walvis Ridge — a review, *in* C. L. Drake, ed., Geodynamics: Progress and Prospects: American Geophysical Union, p. 160-176.

du Plessis, A., 1979, The evolution of the southeastern Atlantic ocean — a review, *in* A. du Plessis and R. V. Dingle, eds., Papers on marine geoscience: Geological Survey Department of Mines, Bulletin, n. 63.

Emery, K. O., et al, 1975, Continental margin of western Africa — Cape St. Francis (South Africa) to Walvis Ridge (southwest Africa): AAPG Bulletin, v. 59, p. 3-59.

Falvey, D. A., 1974, The development of continental margins in plate tectonic theory: The Australian Petroleum Exploration Association Journal, v. 14, p. 95-106.

Gerrard, I., 1979a, Preliminary report of cruise of R. V. *Thomas B. Davie* to the K-B1 drillsite: Unpublished, Southern Oil Exploration Corporation report.

———, 1979b, Preliminary report on west coast offshore geology between Cape Columbine and Cape Agulhas: Unpublished SOEKOR report.

Gidskehaug, A., K. M. Creer, and J. G. Mitchell, 1975, Paleomagnetism and K-Ar ages of the southwest African basalts and their bearing on the time of initial rifting of the south Atlantic ocean: Geophysical Journal of the Royal Astronomical Society, v. 42, p. 1-20.

Hill, R. S., 1975, The Geology of the northern Algoa basin, Port Elizabeth: Annales of the University of Stellenbosch, Series Al (geology), v. 1, p. 105-192.

Jones, D. K., and G. R. Pearson, 1972, The tectonic elements of the Perth basin: The Australian Petroleum Exploration Association Journal, v. 12, p. 17-22.

La Breque, J. L., D. V. Kent, and S. C. Cande, 1977, Revised magnetic polarity time-scale for Late Cretaceous and Cenozoic time: Geology, v. 5, p. 330-335.

Larson, R. L., and T. W. C. Hilde, 1975, A revised time-scale of magnetic reversals for the early Cretaceous and late Jurassic: Journal of Geophysical Research, v. 80, p. 2586-2594.

———, and J. W. Ladd, 1973, Evidence for the opening of

the south Atlantic in the Early Cretaceous: Nature, v. 246, p. 209-212.

Lehner, P., and P. A. D. de Ruiter, 1976, Africa's Atlantic margin typified by string of basins: Oil and Gas Journal, v. 74, n. 45, p. 252-266.

———, and ———, 1977; Structural history of Atlantic Margin of Africa: AAPG Bulletin, v. 61, p. 961-981.

McLachlan, I. R., and I. K. McMillan, 1979, Microfaunal biostratigraphy, chronostratigraphy and history of Mesozoic and Cenozoic deposits on the coastal margin of South Africa, *in* A. M. Anderson and W. J. van Biljon, eds., Some sedimentary basins and associated ore deposits of South Africa: Geological Society of South Africa, Special Publication 6, p. 161-181.

Miller, R. McG., 1980, Phanerozoic intrusives in southwest Africa/Namibia, *in* L. E. Kent, compiler, Stratigraphy of South Africa: Handbook 8, Geological Survey, Department of Mineral and Energy Affairs, p. 632-638.

Moore, A. E., 1976, Controls of post-Gondwana alkaline volcanism in southern Africa: Earth and Planetary Science Letters, v. 31, p. 291-296.

Ponte, F. C., and H. E. Asmus, 1976, The Brazilian marginal basins; current state of knowledge: Annales of the Brazilian Academy of Science, supplement, v. 48, p. 215-239.

Rabinowitz, P. D., 1976, Geophysical study of the continental margin of southern Africa: Geological Society of America Bulletin. v. 87, p. 1643-1653.

———, 1978, Geophysical study of the continental margin of southern Africa: Discussion and reply: Geological Society of America Bulletin, v. 89, p. 791-796.

———, and J. L. La Breque, 1979, The Mesozoic South Atlantic Ocean and evolution of its continental margin: Journal of Geophysical Research, v. 84b, p. 5973-6002.

Schneider, E. D., 1972, Sedimentary evolution of rifted continental margins, *in* R. Shagam et al, eds., Studies in Earth and Space Sciences: Geological Society of America, Memoir 132, p. 109-118.

Scrutton, R. A., 1973 , Gravity results from the continental margin of south-western Africa: Geophysical Researches, v. 2, p. 11-21.

———, 1976, Crustal structure at the continental margin of South Africa: Geophysical Journal of the Royal Astronomical Society, v. 44, p. 601-623.

———, 1978, Geophysical study of the continental margin of southern Africa: Discussion and reply: Geological Society of America Bulletin, v. 89, p. 791-796.

———, and R. V. Dingle, 1974, Basement control over sedimentation on the continental margin west of southern Africa: Transactions of the Geological Society of South Africa, v. 77, p. 253-260.

Siedner, G., and J. A. Miller, 1968, K-Ar age determinations on basaltic rocks from southwest Africa and their bearing on continental drift: Earth and Planetary Science Letters, v. 4, p. 451-458.

———, and J. G. Mitchell, 1976, Episodic Mesozoic volcanism in Namibia and Brazil, a K-Ar Isochron study bearing on the opening of the South Atlantic: Earth and Planetary Science Letters, v. 30, p. 292-302.

Smith, G. C., 1980, 1:1 000 000 maps showing westerly limit of seismic reflections below horizon R and interpreted continent-ocean boundary: Southern Oil Exploration Corporation, unpublished information.

Tarling, D. H., and J. G. Mitchell, 1976, Revised Cenozoic polarity time-scale: Geology, v. 4, p. 133-136.

Vail, P. R., et al, 1977, Seismic stratigraphy and global changes of sea level, *in* C. E. Payton, ed., Seismic Stratigraphy — applications to hydrocarbon exploration: AAPG Memoir 26.

Van Andel, T. H., and S. E. Calvert, 1971: Evolution of sediment wedge, Walvis Shelf, South West Africa: Journal of Geology, v. 79, p. 585-602.

Van Hinte, J. E., 1976a, A Jurassic time scale: AAPG Bulletin, v. 60, p. 489-497.

———, 1976b, A Cretaceous time scale: AAPG Bulletin, v. 60, p. 498-516.

Van der Linden, W. J. M., 1980, Walvis Ridge, a piece of Africa?: Geology, v. 8, p. 417-421.

Veevers, J. J., 1972, Evolution of the Perth and Carnarvon basins: The Australian Petroleum Exploration Association Journal, v. 12, p. 52-54.

Winter, H. de la R., 1973, Geology of the Algoa Basin, South Africa, *in* G. Glant, ed., Sedimentary basins of the African Coasts, part II: Association of African Geological Surveys, p. 17-48.

Rifting History and Structural Development of the Continental Margin North of Alaska

Arthur Grantz
Steven D. May
U.S. Geological Survey
Menlo Park, California

Seismic-reflection profiles in the Alaskan Beaufort Sea and onshore geology indicate that the continental margin north of Alaska is of Atlantic type. Rifting appears to have begun in earliest Jurassic time, about 190 to 185 m.y. ago, when crustal extension created a rift-valley system beneath the Beaufort shelf and part of the adjacent coastal plain. Subsequent crustal warming caused rift-margin uplift and erosion, created a breakup unconformity, and initiated breakup and seafloor spreading in the Canada Basin about 125 m.y. ago. Subsequent cooling caused rapid subsidence of the margin, which was followed by vigorous progradation of the present continental terrace of the Beaufort Sea beginning in Albian time.

GENERAL FEATURES OF THE CONTINENTAL MARGIN

The continental margin north of Alaska faces the Canada basin of the Arctic Ocean across a gently arcuate continental-shelf break and slope (Figure 1). The present morphology of this shelf and slope is the result of continental-terrace progradation and gravity-driven slope processes in a region of general tectonic stability. However, this morphologically simple continental terrace, the Alaskan Beaufort shelf, also conceals large structures related to both continental rifting and postrifting tectonism.

As pointed out by Rickwood (1970), stratigraphic and structural evidence in northern Alaska suggests that the continental margin north of Alaska is Atlantic type (passive) and was formed by rifting during Early Cretaceous (Neocomian) time. This report presents an overview of the rift and postrift structures and of the geologic history of this margin, based on a study of the general configuration of basement and major suprabasement structures beneath the Beaufort shelf and slope. Figure 2 shows the bathymetry, major geographic features, and oilfields of the region, as well as the locations of the 24-channel common-depth-point (CDP) seismic-reflection profiles which constitute our principal data base. These profiles were shot by the U.S. Geological Survey in 1977 from the research vessel *S. P. Lee* and are being processed at the Survey's office in Menlo Park, California. Note that the outer edge of the continental shelf ("structural shelf break" on Figures 5, 10, and 15) lies 60 to 110 km offshore. The shelf break is defined by a combination of geomorphic and structural criteria; in general, it separates structurally and stratigraphically more coherent rocks beneath flat or gently sloping seabed from structurally and stratigraphically more complex rocks beneath steeper gradients of the continental slope. In the text and figures, depths to features observed on seismic-reflection records are given in seconds of two-way seismic-reflection time and (or) kilometers below sea level; a rough conversion of reflection time to depth is given in Figure 3. Numerical ages of geologic units or events are from the time scale of Van Eysinga (1978).

Several features, and the absence of others, indicate that the northern Alaska continental margin is extensional and Atlantic type. The margin truncates the cratonic Arctic platform, which extends from northern Alaska and the continental shelf in the Chukchi Sea to a hinge zone beneath the Beaufort shelf (Figure 5) without folding or other evidence of compression. Early Mississippian to earliest Cretaceous stable-shelf

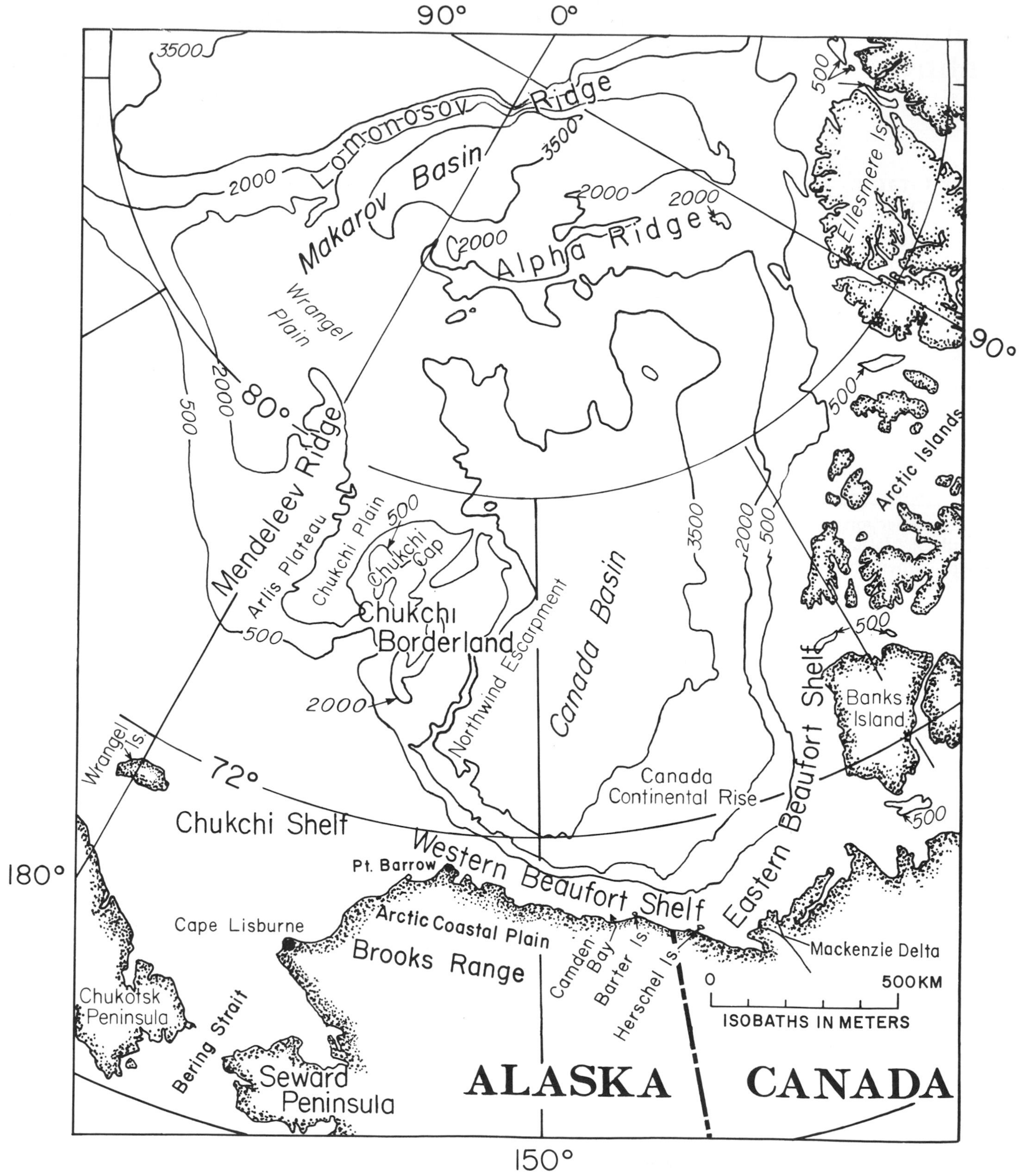

Figure 1 — Continental margin north of Alaska and major physiographic features of Arctic Ocean.

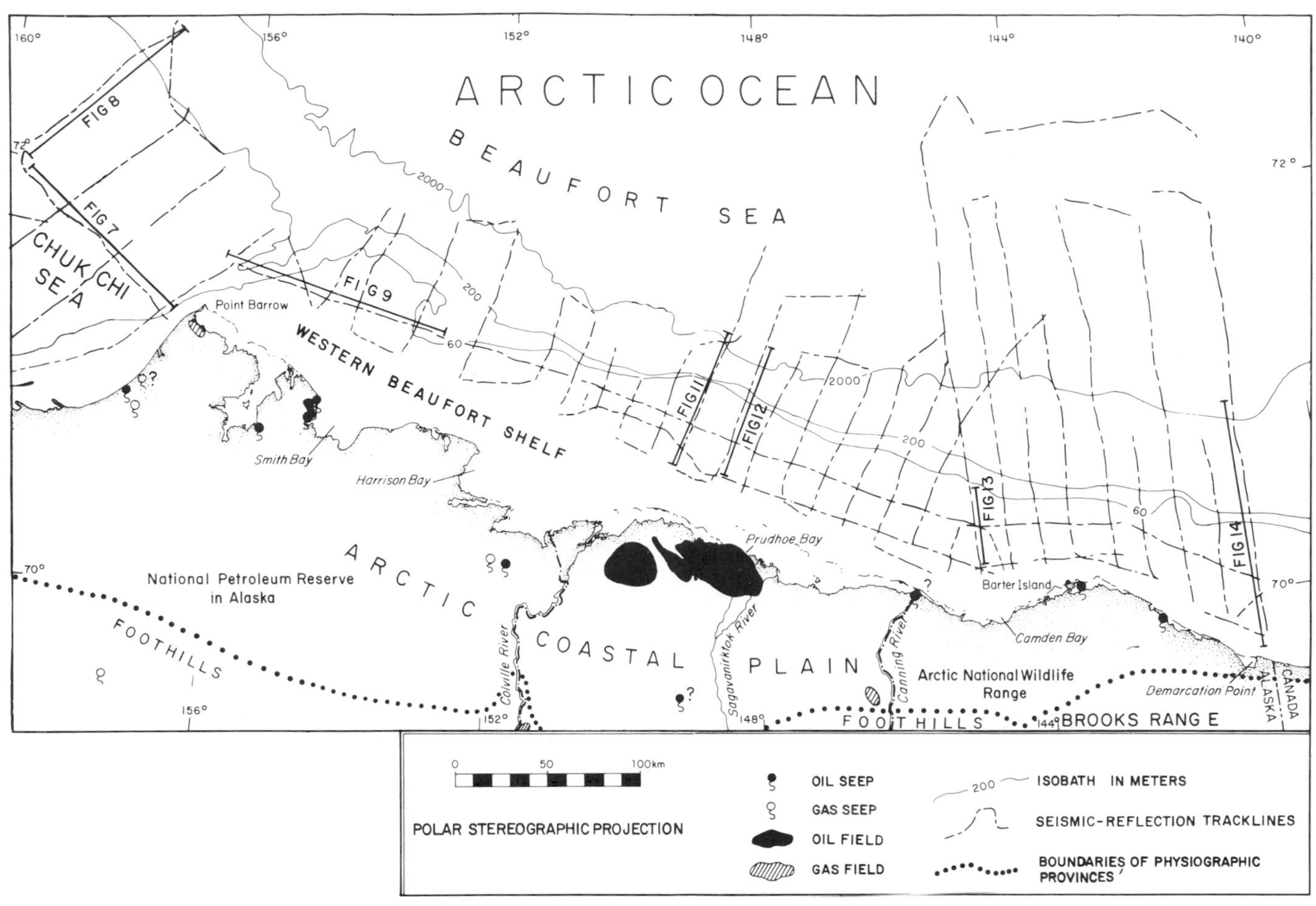

Figure 2 — Major geographic and bathymetric features of Alaskan (western) Beaufort shelf and vicinity, showing locations of oilfields and gas fields and multichannel CDP seismic-reflection profiles.

sediment deposited on the platform is now severed from a sourceland (Barrovia of Tailleur, 1973) that lay to the north, in the present southern Canada basin area. A graben and a subsidence hinge zone beneath the Alaskan Beaufort shelf parallel the shelf break and resemble the extensional structures of Atlantic-type margins, as discussed by Falvey (1974). As in typical passive margins, the continental terrace in the Alaskan Beaufort Sea is structurally simple and was deposited on a subsiding shelf.

East of Alaska, the gently arcuate Beaufort shelf and slope trend into an area of structural complexity and thickened Jurassic to Cenozoic sedimentary rocks in the Mackenzie delta. The trend of the shelf and slope then swings sharply north at the east side of the delta. To the west, the arcuate shelf loses morphologic definition where it intersects the Chukchi continental borderland at the steep east-facing Northwind Escarpment (Figure 1).

Several characteristics indicate that the Canada basin north of Alaska is oceanic. Its abyssal plain attains depths of about 4,000 m off Point Barrow, and it is underlain by more than 2 km (Hall, 1973), probably by about 4.0 to 4.5 km (Grantz, Eittreim, and Dinter, 1979, p. 267), of sediment. Thus, basement beneath the basin is 6 to 8.5 km deep, which is typical for old ocean basins. More directly, Lg-phase earthquake surface waves, which require continental paths for their propagation, are blocked along teleseismic paths that cross the Canada basin (Oliver, Ewing, and Press, 1955). Surface-wave dispersion across the Arctic Ocean, including the Canada basin (Hunkins, 1963), indicates that where water depth in the Arctic Ocean exceeds 2 km, the crust is between 6 and 15 km thick and resembles crust beneath other ocean basins of similar depth. Seismic-refraction measurements in the central Canada basin (Mair and Lyons, 1981) reveal an oceanic layer with $V_P = 4.3$ km/sec overlain by a sedimentary layer 4 to 5 km thick.

TECTONIC FRAMEWORK

Reconnaissance seismic-reflection studies in the Beaufort and northern Chukchi seas, regional stratigraphic and structural relations in northern Alaska, and inferences from bathymetry indicate that the continental margin off northern Alaska consists of three areal segments of quite different geologic development — a segmentation thought to be inherited from the geometry of rifting. The characteristics of these segments and a geometric model to account for their origin were discussed by Grantz, Eittreim, and Dinter (1979) and Grantz, Eittreim, and Whitney (1981). Figure 4 illustrates the model, somewhat revised as a result of the present study.

The model illustrated in Figure 4 elaborates the suggestion of Carey (1958), Rickwood (1970), and Tailleur (1969, 1973) that the Canada basin opened by rifting and rotational spreading about a pole to the south or southeast. In Figure 4, this pole is placed in the Mackenzie delta region at latitude 69.1° N, longitude 130.5° W. This location is based on matching of the 1,000-m isobaths on opposite sides of the basin between the Mackenzie delta and the Northwind Escarpment. The fit is a rough approximation because the specific position of the 1,000-m isobath on the Alaskan Beaufort shelf is determined by postrift continental shelf progradation, which ranges from less than 5 to 100 km.

The Chukchi sector of the segmented margin lies west of the postulated Chukchi-Northwind fracture zone (Figure 4), which is inferred from the alignment of the steep Northwind Escarpment and the offset, or jog, in the north edge of the cratonic Arctic platform of northern Alaska at the east end of the North Chukchi basin. The high-standing flat-topped blocks that constitute the Chukchi continental borderland may be fragments of the Arctic platform that moved relatively northward from a minor spreading axis beneath the

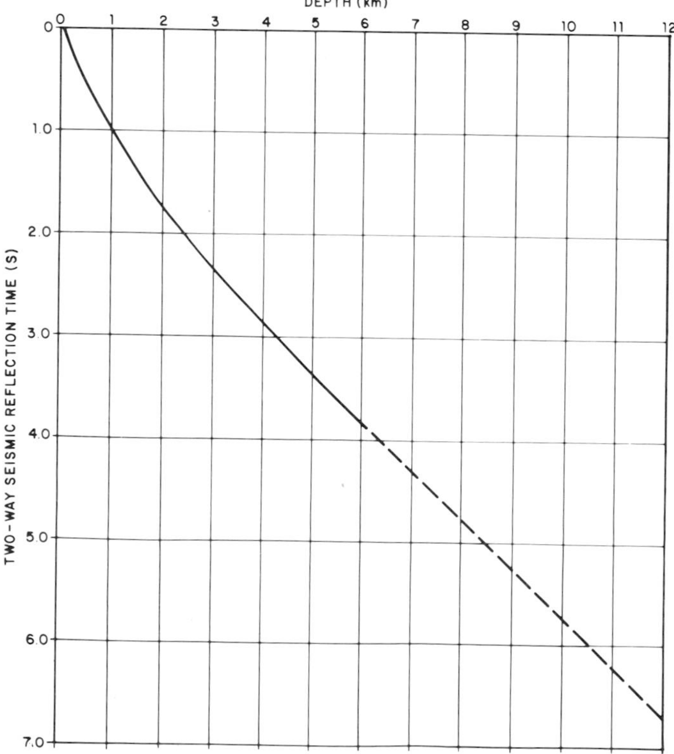

Figure 3 — Generalized average seismic-reflection time as a function of depth for Beaufort shelf, derived from seismic-stacking-velocity measurements.

North Chukchi basin. This basin is postulated to have formed on thinned continental crust of the Arctic platform, rather than on new oceanic crust. Tracing of seismic-stratigraphic units from the North Slope indicates that the North Chukchi basin contains a thick section of Cretaceous and Tertiary clastic strata. The south margin of the basin is a hinge zone north of which the Cretaceous and Tertiary beds thicken rapidly.

The Barrow sector of the segmented margin (Figures 4 and 5) lies between the Chukchi-Northwind fracture zone and a pronounced shoreward bend in the subsidence hinge line between Prudhoe Bay and the Canning River (Figures 5 and 10). The rift in this sector is thought to traverse the south flank of Barrovia, the sourceland for the Mississippian to Neocomian (earliest Cretaceous) stable-shelf deposits of the Arctic platform. Rifting also occurred locally within the northern part of the platform, where it created the large Dinkum graben beneath the shelf in the east half of the Barrow sector (Figures 5 and 10). Rifting in the Barrow sector was accompanied by Jurassic and Early Cretaceous graben-filling sedimentation. Vigorous Cretaceous and less vigorous Tertiary sedimentation completed filling of the Dinkum graben and prograded the present continental terrace. Deformation of the prograded sediment is mainly the result of gravitational forces, and most of the shelf in the Barrow sector is underlain by flat or broadly warped strata. Progradation seaward of the hinge line created the thick sedimentary prism of the deep Nuwuk basin (Figure 5) in the western part of the Barrow sector.

East of the bend in the subsidence hinge line, and extending to the Makenzie delta, lies the Barter Island sector of the continental margin (Figures 4 and 5). The hinge line and the edge of the Arctic platform lie near the coast in this sector, and the entire shelf is underlain by the thick sedimentary prism of the Kaktovik basin (Figure 5). In contrast to the Barrow sector, the strata beneath the shelf in the Barter Island sector are deformed by both diapiric and compressional folds, and probably lie on transitional or oceanic crust. Earthquakes and both Neogene and Quarternary faulting and folding indicate that the area is still active tectonically. It has been postulated (Grantz, Eittreim, and Dinter, 1979, p. 184; Grantz, Dinter, and Biswas, 1982) that late Cenozoic tectonism in the Barter Island sector is related to uplift and 25 to 50 km of northward translation of the northeastern Brooks Range with respect to the coastal plain to the west, which lies in the Barrow sector. The northward swing in the outcrop of Paleozoic rocks in the eastern Brooks Range (Figure 4) is due mainly to uplift but in part to northward translation.

STRATIGRAPHIC FRAMEWORK

Lerand (1973) conveniently divided the bedded rocks of western Arctic North America into three stratigraphic sequences — Franklinian, Ellesmerian, and Brookian — of contrasting tectonic style, provenance, and lithology. These sequences, in the modified form adopted by Grantz, Eittreim, and Whitney (1981), are used here to discuss the stratigraphy of the Alaskan Beaufort shelf (Figure 6).

The Franklinian sequence of northern Alaska consists of strongly deformed and mildly metamorphosed to undeformed and unmetamorphosed Middle Cambrian to Upper Devonian marine and subordinate nonmarine miogeoclinal and eugeoclinal sedimentary rocks. Argillite and graywacke of the Franklinian sequence, containing Ordovician and Silurian graptolites and chitinozoans, were encountered in wells near Prudhoe Bay and Barrow (Carter and Laufeld, 1975). These beds dip 20 to 70° in many cores and constitute acoustic basement beneath the Arctic coastal plain of northern Alaska; they are thought to extend beneath the Chukchi shelf, where they are thick and return coherent seismic reflections.

Structural deformation and erosion during the Late Devonian and Early Mississippian Ellesmerian orogeny consolidated the Franklinian rocks into a craton and eroded the extensive Arctic platform of northern Alaska and adjacent offshore. Deposition of stable-shelf clastic rocks and carbonate rocks of northern provenance on the platform created the Ellesmerian sequence of Early Mississippian to earliest Cretaceous age in northern Alaska. These strata are only slightly deformed over most of the platform. Formations in the Ellesmerian sequence are typically tens or hun-

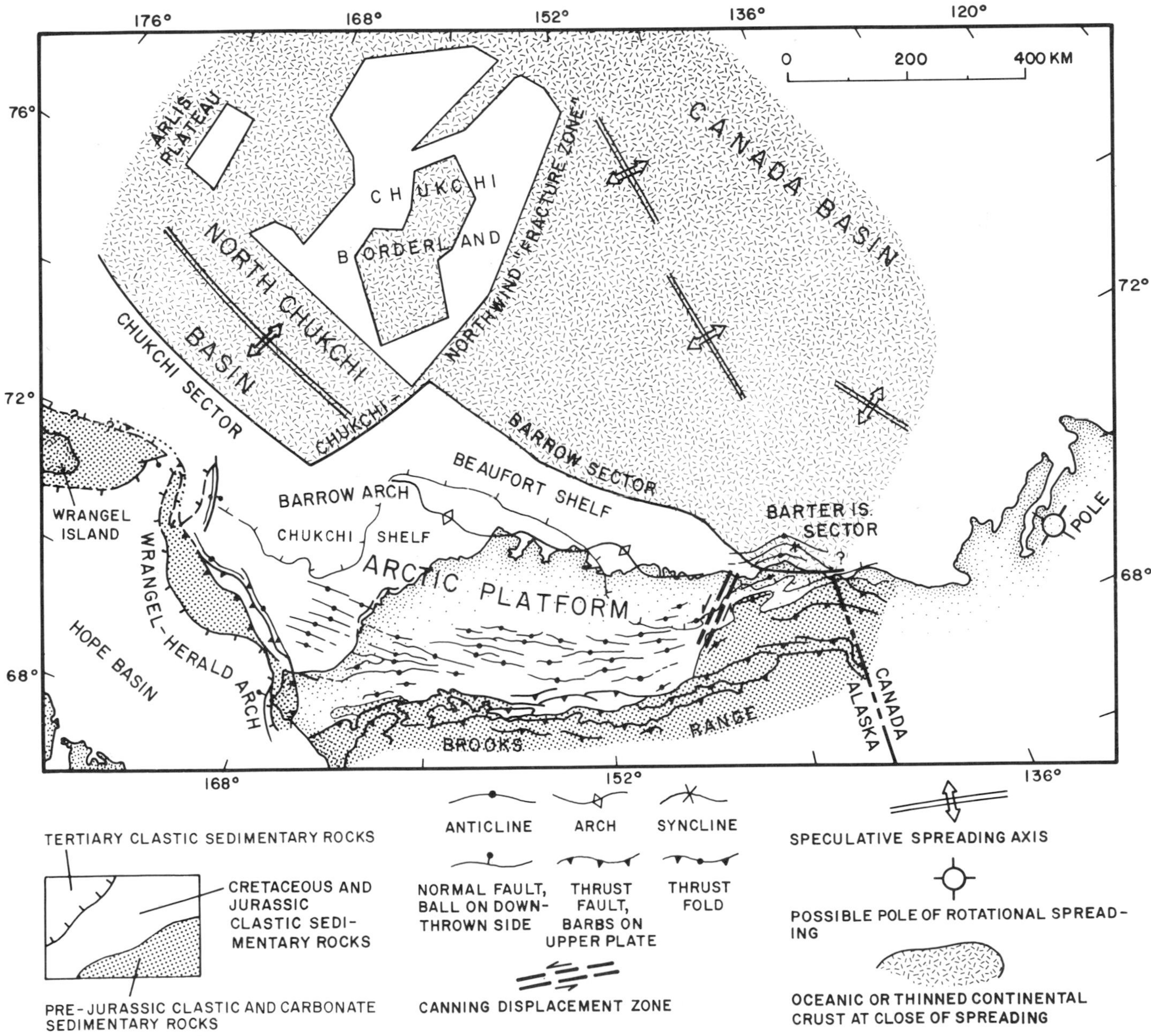

Figure 4 — Tectonic model of northern Alaska continental margin and generalized geologic map of northern Alaska (revised from Grantz, Eittreim, and Dinter, 1979, Figures 4 and 5).

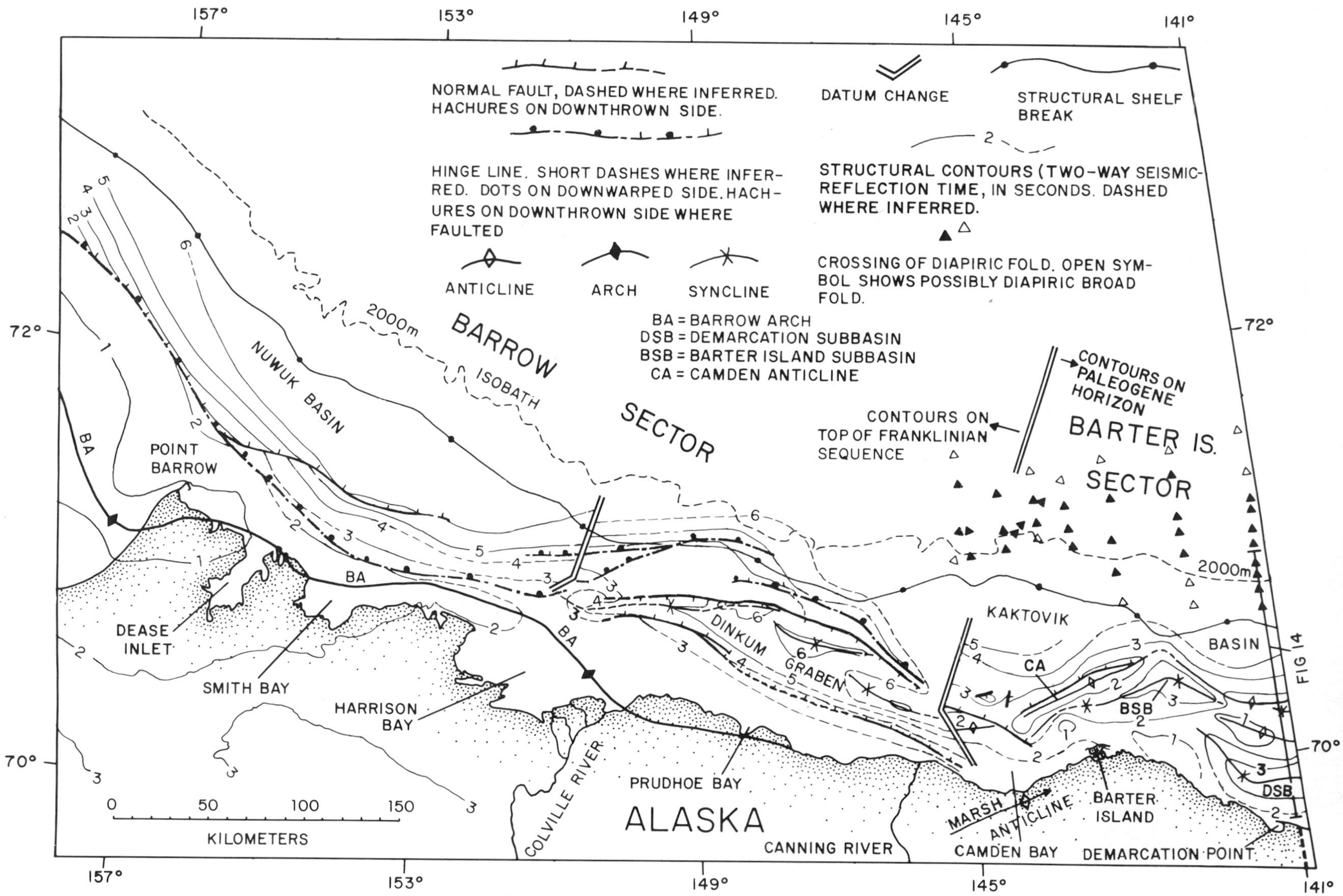

Figure 5 — Structural geologic map of Alaskan Beaufort shelf, showing generalized isochrons on top of Franklinian sequence west of Camden Bay and on a Paleogene (middle Eocene?) horizon east of Camden Bay. Onshore and nearshore data between Point Barrow and Colville River from Miller et al (1979).

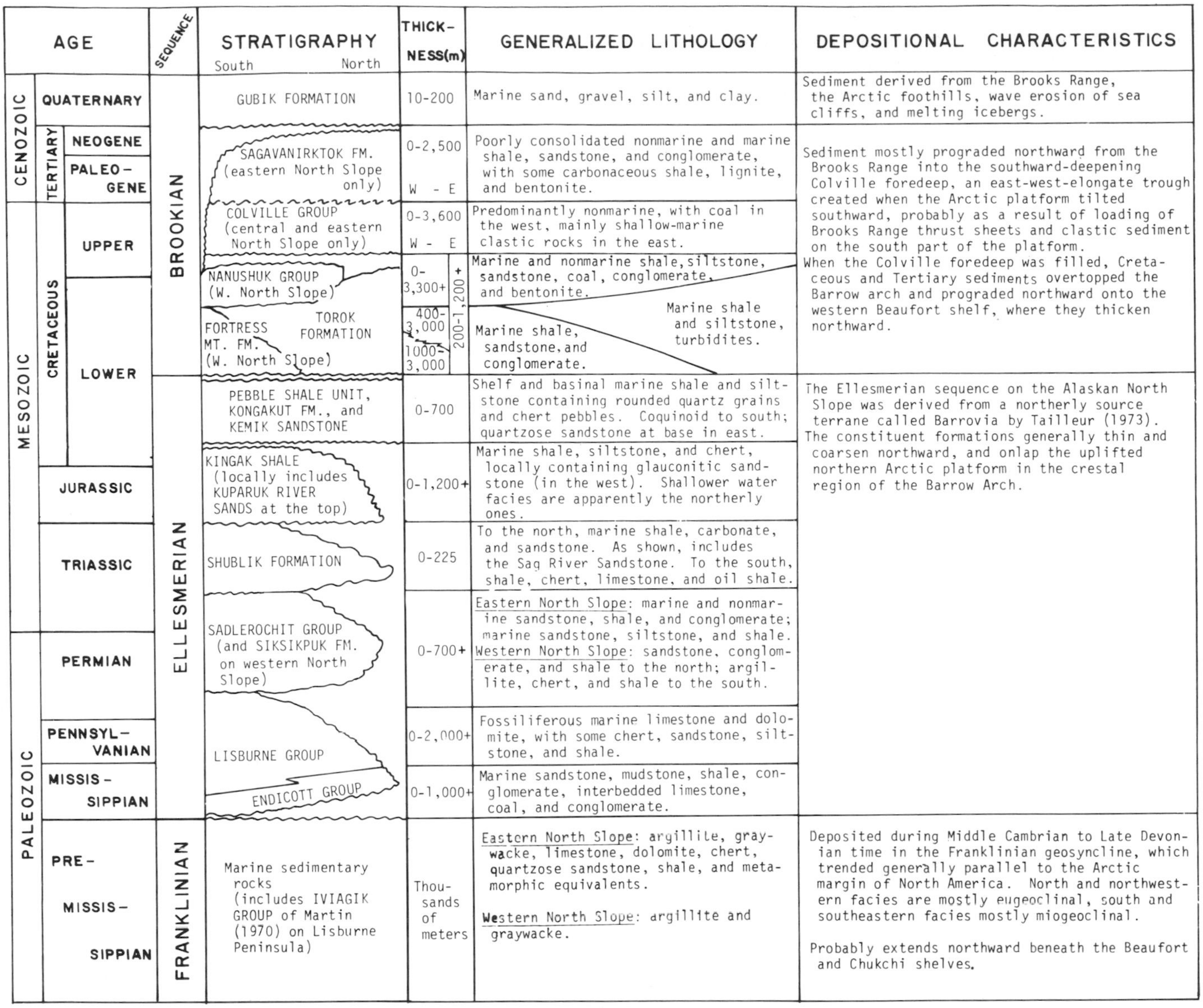

AGE			SEQUENCE	STRATIGRAPHY (South — North)	THICKNESS(m)	GENERALIZED LITHOLOGY	DEPOSITIONAL CHARACTERISTICS
CENOZOIC	QUATERNARY		BROOKIAN	GUBIK FORMATION	10-200	Marine sand, gravel, silt, and clay.	Sediment derived from the Brooks Range, the Arctic foothills, wave erosion of sea cliffs, and melting icebergs.
CENOZOIC	TERTIARY	NEOGENE; PALEO-GENE	BROOKIAN	SAGAVANIRKTOK FM. (eastern North Slope only)	0-2,500 W - E	Poorly consolidated nonmarine and marine shale, sandstone, and conglomerate, with some carbonaceous shale, lignite, and bentonite.	Sediment mostly prograded northward from the Brooks Range into the southward-deepening Colville foredeep, an east-west-elongate trough created when the Arctic platform tilted southward, probably as a result of loading of Brooks Range thrust sheets and clastic sediment on the south part of the platform. When the Colville foredeep was filled, Cretaceous and Tertiary sediments overtopped the Barrow arch and prograded northward onto the western Beaufort shelf, where they thicken northward.
MESOZOIC	CRETACEOUS	UPPER	BROOKIAN	COLVILLE GROUP (central and eastern North Slope only)	0-3,600 W - E	Predominantly nonmarine, with coal in the west, mainly shallow-marine clastic rocks in the east.	
MESOZOIC	CRETACEOUS	UPPER / LOWER	BROOKIAN	NANUSHUK GROUP (W. North Slope)	0-3,300+ (200-1,200+)	Marine and nonmarine shale, siltstone, sandstone, coal, conglomerate, and bentonite.	
MESOZOIC	CRETACEOUS	LOWER	BROOKIAN	FORTRESS MT. FM. (W. North Slope); TOROK FORMATION	400-3,000; 1000-3,000 (200-1,200+)	Marine shale, sandstone, and conglomerate. Marine shale and siltstone, turbidites.	
MESOZOIC	CRETACEOUS	LOWER	ELLESMERIAN	PEBBLE SHALE UNIT, KONGAKUT FM., and KEMIK SANDSTONE	0-700	Shelf and basinal marine shale and siltstone containing rounded quartz grains and chert pebbles. Coquinoid to south; quartzose sandstone at base in east.	The Ellesmerian sequence on the Alaskan North Slope was derived from a northerly source terrane called Barrovia by Tailleur (1973). The constituent formations generally thin and coarsen northward, and onlap the uplifted northern Arctic platform in the crestal region of the Barrow Arch.
MESOZOIC	JURASSIC		ELLESMERIAN	KINGAK SHALE (locally includes KUPARUK RIVER SANDS at the top)	0-1,200+	Marine shale, siltstone, and chert, locally containing glauconitic sandstone (in the west). Shallower water facies are apparently the northerly ones.	
MESOZOIC	TRIASSIC		ELLESMERIAN	SHUBLIK FORMATION	0-225	To the north, marine shale, carbonate, and sandstone. As shown, includes the Sag River Sandstone. To the south, shale, chert, limestone, and oil shale.	
PALEOZOIC	PERMIAN		ELLESMERIAN	SADLEROCHIT GROUP (and SIKSIKPUK FM. on western North Slope)	0-700+	Eastern North Slope: marine and nonmarine sandstone, shale, and conglomerate; marine sandstone, siltstone, and shale. Western North Slope: sandstone, conglomerate, and shale to the north; argillite, chert, and shale to the south.	
PALEOZOIC	PENNSYL-VANIAN		ELLESMERIAN	LISBURNE GROUP	0-2,000+	Fossiliferous marine limestone and dolomite, with some chert, sandstone, siltstone, and shale.	
PALEOZOIC	MISSIS-SIPPIAN		ELLESMERIAN	ENDICOTT GROUP	0-1,000+	Marine sandstone, mudstone, shale, conglomerate, interbedded limestone, coal, and conglomerate.	
PALEOZOIC	PRE-MISSIS-SIPPIAN		FRANKLINIAN	Marine sedimentary rocks (includes IVIAGIK GROUP of Martin (1970) on Lisburne Peninsula)	Thousands of meters	Eastern North Slope: argillite, graywacke, limestone, dolomite, chert, quartzose sandstone, shale, and metamorphic equivalents. Western North Slope: argillite and graywacke.	Deposited during Middle Cambrian to Late Devonian time in the Franklinian geosyncline, which trended generally parallel to the Arctic margin of North America. North and northwestern facies are mostly eugeoclinal, south and southeastern facies mostly miogeoclinal. Probably extends northward beneath the Beaufort and Chukchi shelves.

Figure 6 — Generalized stratigraphy of northern Alaska and adjacent Beaufort shelf.

dreds of meters thick, and texturally mature sandstone and conglomerate are common.

Uplift and deformation in the region of the present central and southern Brooks Range, beginning in Jurassic time, created a southern sourceland for the thick synorogenic and postorogenic clastic deposits that flooded the Arctic platform during late Mesozoic and Tertiary time. The southerly dipping depositional paleoslope of the platform, which had been established during Mississippian time, was now augmented by the load of Jurassic and Cretaceous sediment and nappes from the new tectonically active sourceland to the south. The tilting created the Colville foredeep, which received thousands of meters of Cretaceous prodelta and intradelta deposits (Figure 6). As the northern part of the Arctic platform subsided after rifting, Brookian clastic materials filled the foredeep. By middle Albian time, this sediment overtopped the structurally highest northern part of the platform and began to prograde the present Beaufort shelf. The structurally high part of the platform, modified by postrift subsidence, created the regional Barrow arch of central-coastal and west-coastal northern Alaska (Figure 5). The Brookian sequence thins over the arch, and west of Point Barrow it is locally less than .3 km thick on its broad crest. Near the Brooks Range, these beds exceed 8 km in original thickness, and seaward of the subsidence hinge line beneath the Beaufort shelf they are more than 10 km thick.

Temporal and areal variations in the distribution of uplifts and clastic-sediment source terranes in the Brooks Range resulted in an inhomogeneous distribution of the Brookian clastic formations. Thus, the Albian Torok formation and Nanushuk group, which are thick beneath the southern and central parts of the western North Slope, are thin or absent over the north-central part of the North Slope and the adjacent offshore area east of Harrison Bay. Where the Torok and Nanushuk thin, the Turonian and Senonion Colville group and Tertiary clastic rocks thicken; these latter rocks are probably more than 10 km thick beneath the eastern Alaskan Beaufort shelf. Locally, as at the Exxon Alaska State No. 1 well on Flaxman Island near Camden Bay, Cretaceous beds are only about 100 m thick, and the post-Franklinian section is mostly Tertiary. To the east of Camden Bay, however, both Lower and Upper Cretaceous bedded rocks and, in places, Jurassic rocks crop out on the Arctic coastal plain. Albian beds, which are absent on the central North Slope, are represented by as much as 3,000 m of flysch in the mountains and coastal plain of the northern Yukon Territory (Young, Myhr, and Yorath, 1976).

PRINCIPAL STRUCTURAL FEATURES OF THE ALASKAN BEAUFORT SHELF

Franklinian sequence basement

Cretaceous, locally Jurassic, bedded rocks are the oldest that return coherent seismic reflections of regional extent on the Beaufort shelf east of Dease Inlet and north of the wedgeout of the main body of Ellesmerian rocks (Figure 10). West of Dease Inlet, a thick sequence of well-bedded rocks underlies both the Cretaceous beds north of the wedgeout and the Ellesmerian beds south of it; these older well-bedded rocks are interpreted to belong to the Franklinian sequence. The basal beds that rest on the Franklinian beds north of the wedgeout, as mapped on seismic-reflection profile 2783 (Figure 7) and other profiles in the Chukchi Sea, project onshore into the basal beds above a diachronous angular unconformity (discussed in the next section) that is encountered in many wells near Barrow. These basal beds consist of very gently dipping Neocomian rocks that overlie strongly deformed, mildly metamorphosed argillite and graywacke of Ordovician and Silurian (Franklinian) age (Carter and Laufeld, 1975). These Franklinian beds are overlain by strongly folded sub-Ellesmerian Lower and Middle Devonian chert pebble conglomerate and plant-bearing carbonaceous lutite and sandstone in the Topagoruk No. 1 and South Meade No. 1 wells in the northern part of the coastal plain south of Barrow (Brosgé and Tailleur, 1971; Brosgé and Tailleur, oral communication, 1981). The unit of well-bedded rocks (Ds.), 0.8 sec (2 km) or more thick, that constitutes the upper part of the Franklinian section in Figure 7 may consist of these beds.

The Franklinian beds beneath the shelf east of Dease Inlet return only sporadic packets of coherent seismic reflections of limited horizontal and vertical extent. This behavior is consistent with the observation that the argillite and graywacke encountered in wells on the adjacent North Slope commonly are severely deformed and dip steeply. Strong deformation is also compatible with the occurrence of granitic rocks that are probably more than 332 ± 10 m.y. old (Bird et al, 1977) at the bottom of the East Teshekpuk No. 1 well, on the northern part of the coastal plain between Smith and Harrison Bays. These rocks appear to postdate and, presumably, intrude the Ordovician and Silurian argillite and graywacke.

Rocks that return coherent seismic reflections and that we correlate on structural grounds with Franklinian basement occur on the west end of reflection profile 778 (Figure 9) in the Beaufort Sea, as well as on several profiles, including 2783 and 1783 (Figures 7 and 8), in the northeastern Chukchi Sea. These coherent reflections underlie the midshelf north of Dease Inlet, where they occur to depths of about 3.5 sec beneath the lowermost Cretaceous unconformity. They fade out eastward into noncoherent reflections near the hinge line on profile 778. The gradational boundary or fadeout zone between Franklinian rocks with coherent reflections and those with noncoherent reflections extends southwestward into the northeastern Chukchi Sea (southeast end of profile 2783, (Figure 7), where it is 10 to 35 km off the Alaskan coast. This fadeout zone is thought to mark the transition between strongly deformed argillite and graywacke to the east, and gently deformed beds of the same lithology to the west. The extent of the coherent terrane is at least 180 km north to south by 220 km east to west.

The coherent Franklinian reflectors in the northeastern Chukchi Sea extend downward at least 4 sec (9.2 km) beneath the unconformably overlying Cretaceous beds (Figures 7 and 8). The upper part of this Franklinian section has seismic-interval and sonobuoy-refraction velocities that average 4.75 (range, 4.15 to 5.35 or 4.15 to 6.00) km/sec. The upper part also contains a thick clastic wedge that thickens and, presumably, coarsens to the west or northwest. We interpret the main body of these beds as deep-marine-basin and submarine-fan deposits of early Paleozoic age (probably Ordovician and Silurian). The locally occurring overlying unit of well-bedded Devonian(?) rocks (unit Ds, Figure 7) may consist of nonmarine Devonian deposits. The lower part of the Franklinian section, which is characterized by strong parallel reflectors with reflection and sonobuoy-refraction interval velocities of 5.35 to 7.30 km/sec, may consist of upper Proterozoic or Cambrian carbonate or metamorphic rocks.

Morphology of the basement surface

A strong reflector that marks the top of the Franklinian sequence can be followed to depths of between 6 and 7 sec below the sea floor on most of the seismic profiles. This reflector is a diachronous unconformity; its oldest parts were cut during the Late Devonian and Early Mississippian Ellesmerian orogeny, and its youngest parts by local erosion during the final stages of uplift associated with late Neocomian rifting and breakup. The youngest part of the unconformity is labelled "BU" on Figures 7, 8, 9, 11, and 12. Erosion at this angular unconformity apparently removed 6 km or more of upper Franklinian beds and an un-

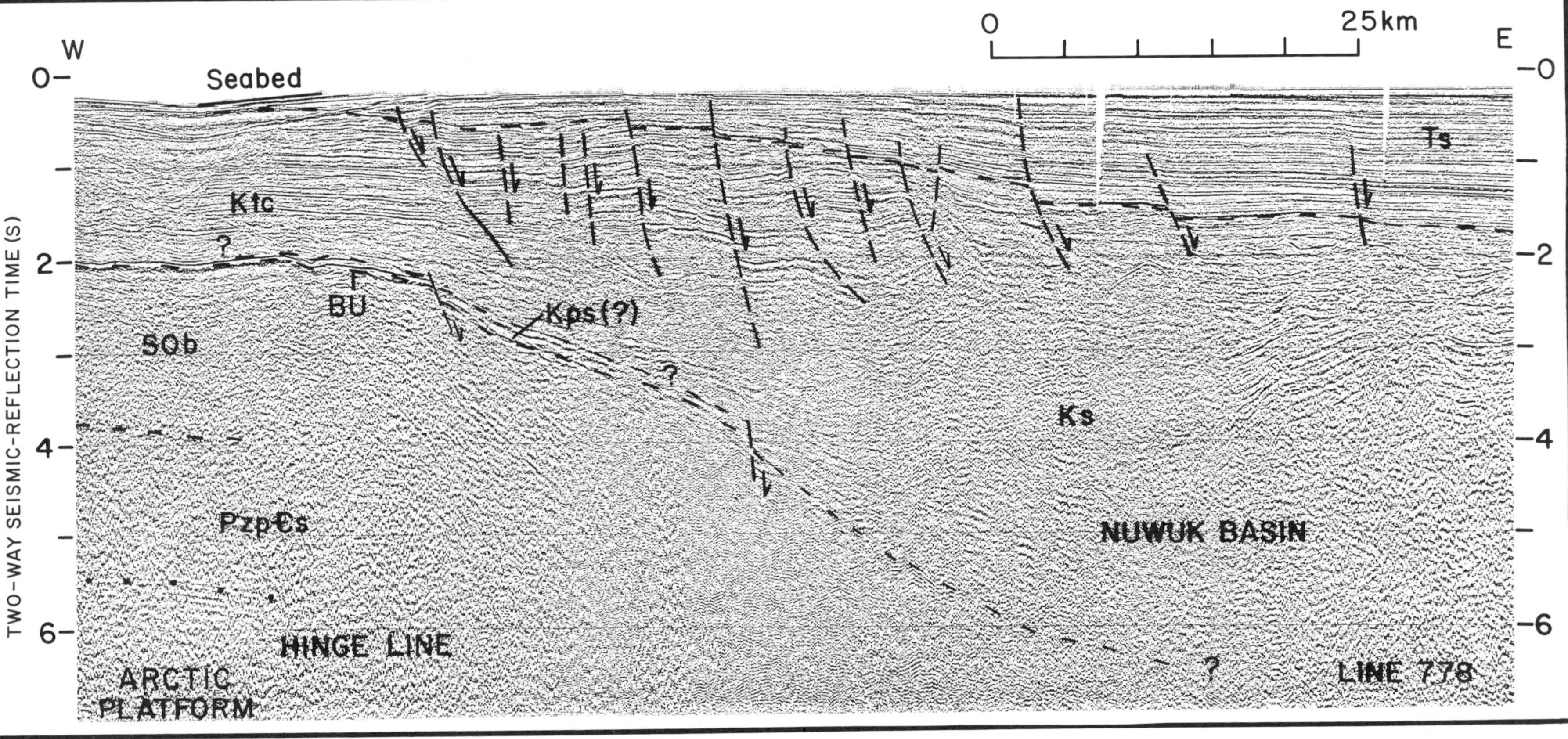

Figure 9 — Oblique crossing of rifted continental margin along CDP seismic-reflection profile 778 northeast of Point Barrow, showing Arctic platform, breakup unconformity (BU), subsidence hinge line, and Nuwuk basin. Inferred acoustic units: Ts, Tertiary and possibly Upper Cretaceous clastic sedimentary rocks; Ktc, Torok formation (Albian) and Colville group (Upper Cretaceous) marine and possibly some paralic sedimentary rocks; Kps(?), Pebble shale unit (Hauterivian to Barremian condensed shale) and (or) bottom-set beds of basal part of Torok formation (Albian); Ks, Lower and Upper Cretaceous clastic sedimentary rocks of Nuwuk basin; SOb, inferred basinal deposits, tentatively correlated with Orodivician and Silurian marine argillite and graywacke of subsurface northern Alaska; PzpCs, Upper Precambrian(?) and lower Paleozoic bedded, presumable sedimentary rocks. Note concentration of growth faults over steep slope north of subsidence hinge line, and extensive slumping and channeling in Cretaceous beds of Nuwuk basin. See Figures 2 and 10 for location of profile.

known thickness of Ellesmerian strata from the Arctic platform in the northern Chukchi Sea (Figure 7). The unconformity can be mapped (Figures 5 and 10) from the western part of the study area to the vicinity of the Canning River. East of the Canning River it is generally beyond the reach of our seismic-reflection system. Conversion of our isochrons to depth can be approximated from Figure 3.

The top-of-Franklinian-basement surface beneath the Beaufort shelf increases in both depth and morphologic complexity from west to east (Figures 5 and 10). In the Barrow sector the surface slopes gently and is only locally faulted beneath the inner shelf. Along the Barrow arch the surface is as little as 0.25 sec deep west of Point Barrow and almost 3 sec deep near the Canning River. From the arch the surface slopes gently seaward to a subsidence hinge line or hinge zone that is less than 10 to almost 100 km seaward of the crest of the Barrow arch. Basement at the hinge line is 1.5 to 2 sec deep northwest of Point Barrow and more than 4 sec deep off the Canning River.

The hinge line is an inflection at which the seaward slope of the Franklinian surface increases from about 1 to 4°, to as much as 12 to 16°. In places, the hinge zone is broken by down-to-the-basin normal faults (Figures 8 and 10). West of Point Barrow, a fault swings into the hinge zone from the steepened slope to the north and displaces it (down to the north) as much as 1.7 sec (3.4 km).

Where the hinge line underlies the inner shelf, it is a morphologically simple inflection of the basement surface; where it underlies the outer shelf, as it does opposite the western part of the Dinkum graben (Figures 5 and 10), the hinge line is a compound structure. The hinge line swings seaward from the inner shelf west of the Colville River to the outer shelf off the west end of the Dinkum graben by means of an echelon offset beneath the inner shelf and a bifurcation near the shelf break. It swings back toward the inner shelf between Prudhoe Bay and the Canning River by means of another, larger echelon step (shown in Figures 10 and 12). Near the Canning River, the trend of the hinge line also swings to the southeast, striking toward Camden Bay at about a 50° angle to the shelf break. This swing in trend is an important contributor to the structural contrast between the Barrow and Barter Island sectors.

The steepened Franklinian-basement surface seaward of the hinge line descends to observed depths of more than 6 or 7 sec (10.5 to 12.6 km) subsea, where we lose it on the seismic reflection profiles. The slope of this surface is generally smooth, but in a few places is broken by normal faults with down-to-the-north displacements as large as 0.5 to 1.7 sec (1-3.4 km).

In the Barter Island sector, Franklinian basement is, in most places, beyond the reach of our seismic-reflection system. In two areas, however, western Camden Bay and the southern limb of the Demarcation subbasin (Figure 5), the basement surface appears to dip north on the reflection profiles. Although the data are too sparse for a firm conclusion, they suggest that the subsidence hinge line may trend southeast to the Camden Bay area, and from there easterly beneath the inner Beaufort shelf and the northern coastal plain near Barter Island (Figure 5).

Principal sedimentary prisms overlying basement

Ellesmerian sequence

The age of the sedimentary rocks that cover Franklinian basement beneath the Beaufort shelf is not well known. Between Point Barrow and a point between Prudhoe and Camden bays, basement in the crestal region of the Barrow arch is overlain by a northward-thinning wedge of Mississippian to lowermost Cretaceous (Neocomian) stable-shelf clastic and carbonate deposits of the Ellesmerian sequence. The Ellesmerian section ranges from 0 to 1,200 m thick near the coast but generally wedges out northward. This is due to thinning and erosional truncation south of our seismic-reflection lines, which, in most places, are 20 km or more offshore. Seaward of the Prudhoe Bay area, however, moderately reflective beds that may belong to the Ellesmerian sequence extend northward into the Dinkum graben (Figures 11 and 12). Figure 10 shows a highly generalized and approximate position of the featheredge of these rocks.

The uppermost formation of the Ellesmerian sequence is the informally-named Pebble shale unit, a transgressive organic marine deposit containing rare floating polished sand grains and some pebbles of quartz and chert. The unit rests on both Ellesmerian and Franklinian rocks of the Beaufort shelf and North Slope. It has a low seismic velocity and typically returns strong seismic reflections. Offshore it appears to be about 0 to 200 msec (0 to 400 m(?)) thick. Onshore, beneath the northern part of the Arctic coastal plain from the Colville delta west, the Pebble shale unit is 70 to 125 m thick and contains microfossils that suggest a Hauterivian to Barremian age, although the highest beds may, in places, range into the lower Albian (Molenaar, 1981, p.9-11).

Brookian sequence

Landward of the subsidence hinge line in the Barrow sector (Figures 5 and 10) the Ellesmerian and Franklinian rocks of the Beaufort shelf are overlain by about 0.25 to more than 4 sec (0.25 to 6.3 km) of Albian through Tertiary northward-prograded clastic strata of the Brookian sequence (Figure 6). The Brookian rocks are thinnest on the Barrow arch near Point Barrow and thicken eastward and northward toward Camden Bay and the Canada basin. Seaward of the hinge line are two large and deep sedimentary prisms that merit designation as sedimentary basins. The prism that underlies the outer shelf and slope from Harrison Bay west is here named the "Nuwuk basin," after the Eskimo name for nearby Point Barrow and an abandoned old village at the point. The prism that underlies the shelf and slope from Camden Bay east is here named the "Kaktovik basin," after the Eskimo

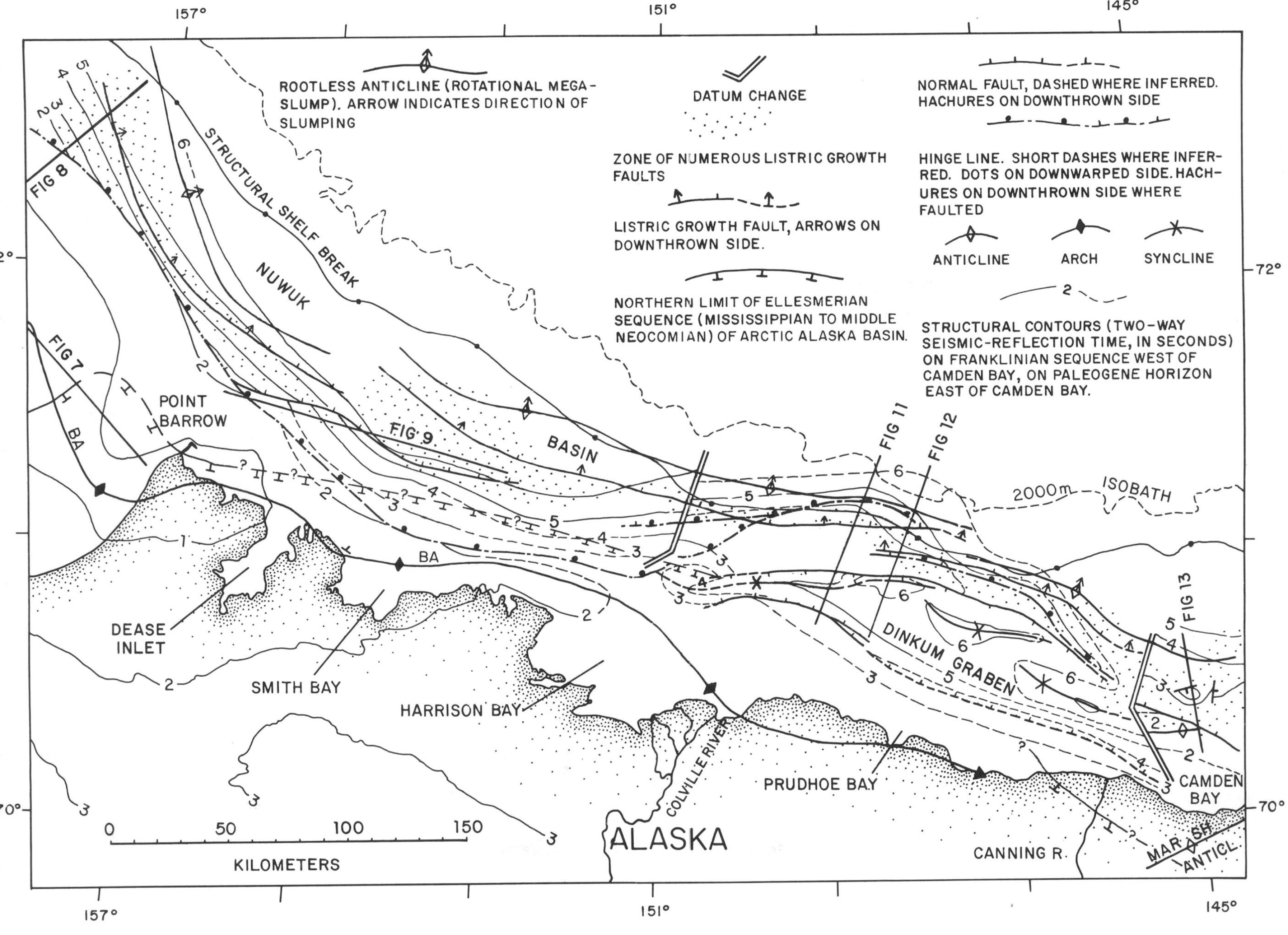

Figure 10 — Structural geologic map of Alaskan Beaufort shelf west of Camden Bay, showing listric normal and growth faults and regionally extensive rootless anticlines (rotational megaslumps). Onshore and nearshore data between Point Barrow and Colville River from Miller et al (1979) and Guldenzopf et al (1980).

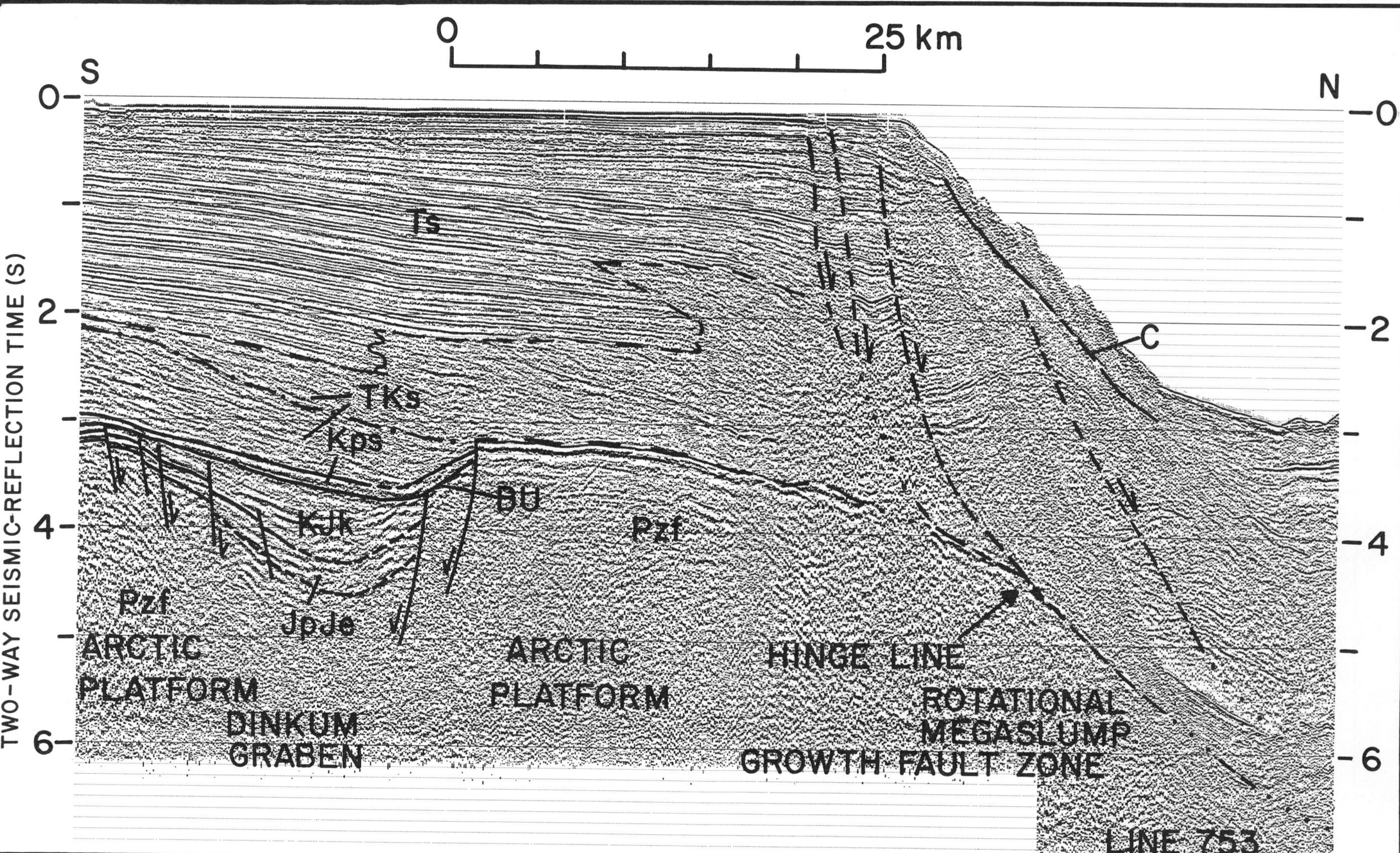

Figure 11 — CDP seismic-reflection profile 753, a dip line across continental margin northwest of Prudhoe Bay, showing rifted Arctic platform, Dinkum graben, subsidence hinge line, breakup unconformity (BU), and growth faults and associated rotational megaslump near shelf break. Interpreted acoustic units: Ts, topset beds in Tertiary regressive sequence in upper part of postbreakup progradational sedimentary prism; TKs, Cretaceous and lower Tertiary foreset and some bottom-set beds in mainly transgressive lower part of postbreakup progradational sedimentary prism. Dot-dashed line within TKs unit is a rough estimate for base of Tertiary, projected from onshore; Kps, Pebble shale unit (Hauterivian to Barremian condensed shale) and possibly bottom-set beds of basal part of Brookian sequence; KJk, weakly reflective beds, Kingak Shale(?) in Dinkum graben; JpJe, moderately reflective beds (sandstone ? and possibly carbonate ?) of Ellesmerian sequence in Dinkum graben (Jurassic or older); Pzf, Franklinian sequence (Cambrian to Devonian); C, basal reflector of zone of gas hydrate. Dinkum graben is interpreted as a rift-valley basin. According to Falvey (1974, p. 103), passive margins with such basins underwent intermediate rates of development. See Figures 2 and 10 for location of profile.

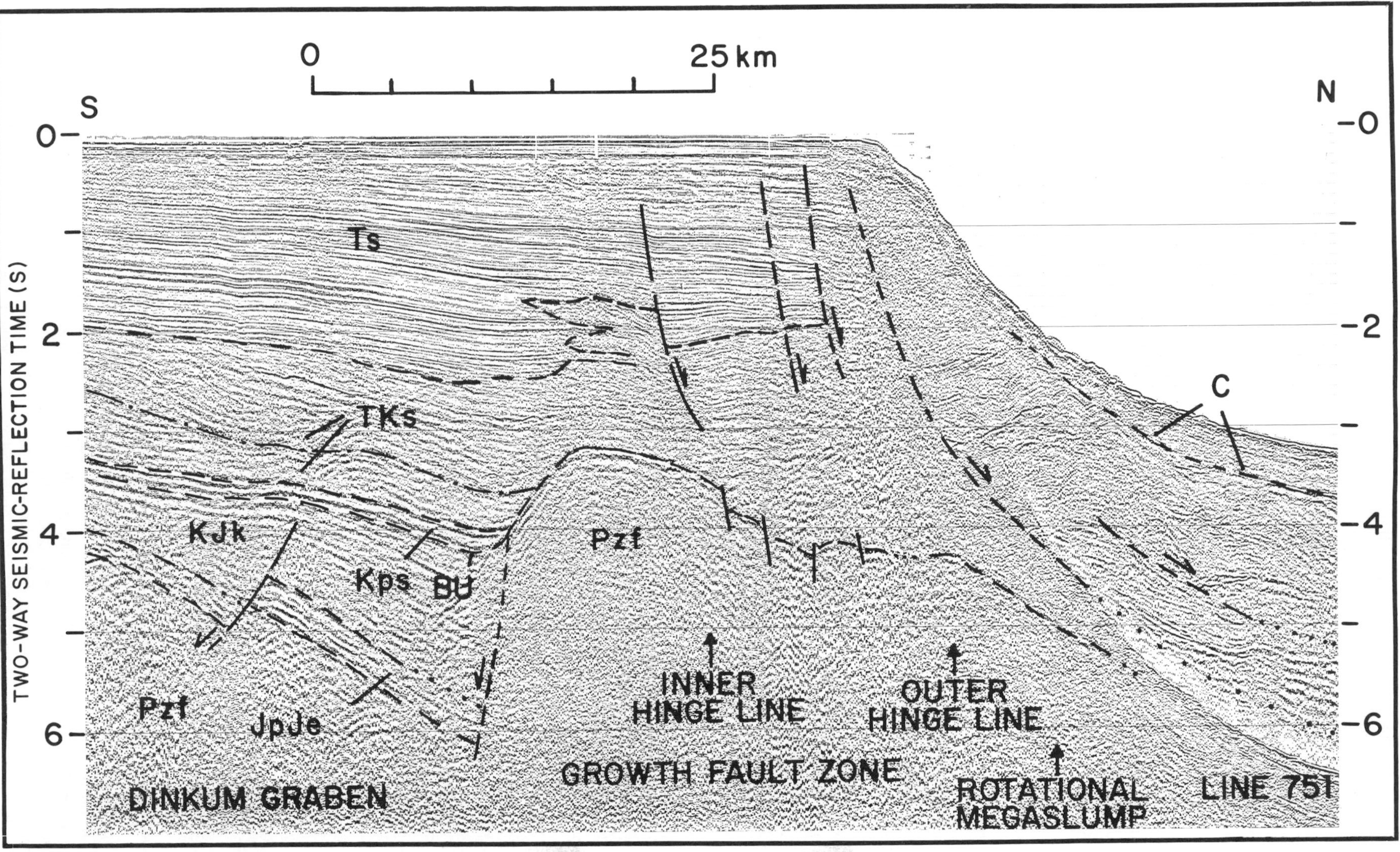

Figure 12 — CDP seismic-reflection profile 751, a dip line across continental margin north of Prudhoe Bay, showing Dinkum graben, compound subsidence hinge line, breakup unconformity (BU), and zone of growth faults and associated rotational megaslump overlying hinge zone. Interpreted acoustic units same as in Figure 11. See Figures 2 and 10 for location of profile.

name for Barter Island and the modern Eskimo village of Kaktovik.

Nuwuk basin

The sedimentary section in the Nuwuk basin (Figures 5, 9, and 10) appears to be as much as 7 sec (about 12.6 km) thick beneath the midshelf off Smith Bay and may be even thicker beneath the outer shelf. Extrapolation from subsurface data onshore and reflector geometry offshore suggest that Albian and Late Cretaceous marine beds form the bulk of the basin fill. The upper part of the section, which consists of almost undeformed gently seaward dipping beds with low seismic-reflection interval velocities (2.2 to 2.4 km/sec), is interpreted to be Tertiary. The lowest beds may be Neocomian on the basis of regional relations.

Kaktovik basin

An exceptionally thick sedimentary section and a moderately complex stratigraphy and structure characterize the Kaktovik basin (Figures 5, 13, and 14). The thickness of the basin fill, which probably rests on Franklinian basement or transitional crust beneath the inner shelf and on transitional or oceanic crust beneath the outer shelf, exceeds our record length of 7 sec (12.5 km). Although the oldest post-Franklinian beds are inferred from regional relations to consist of Jurassic and Cretaceous marine sedimentary rocks, projection of onshore data (Reiser et al, 1980; Norris, 1977a, 1977b) suggests that a northward-thinning wedge of Ellesmerian beds may underlie the south side of the Demarcation subbasin. If such beds do occur there, however, they are not evident on our

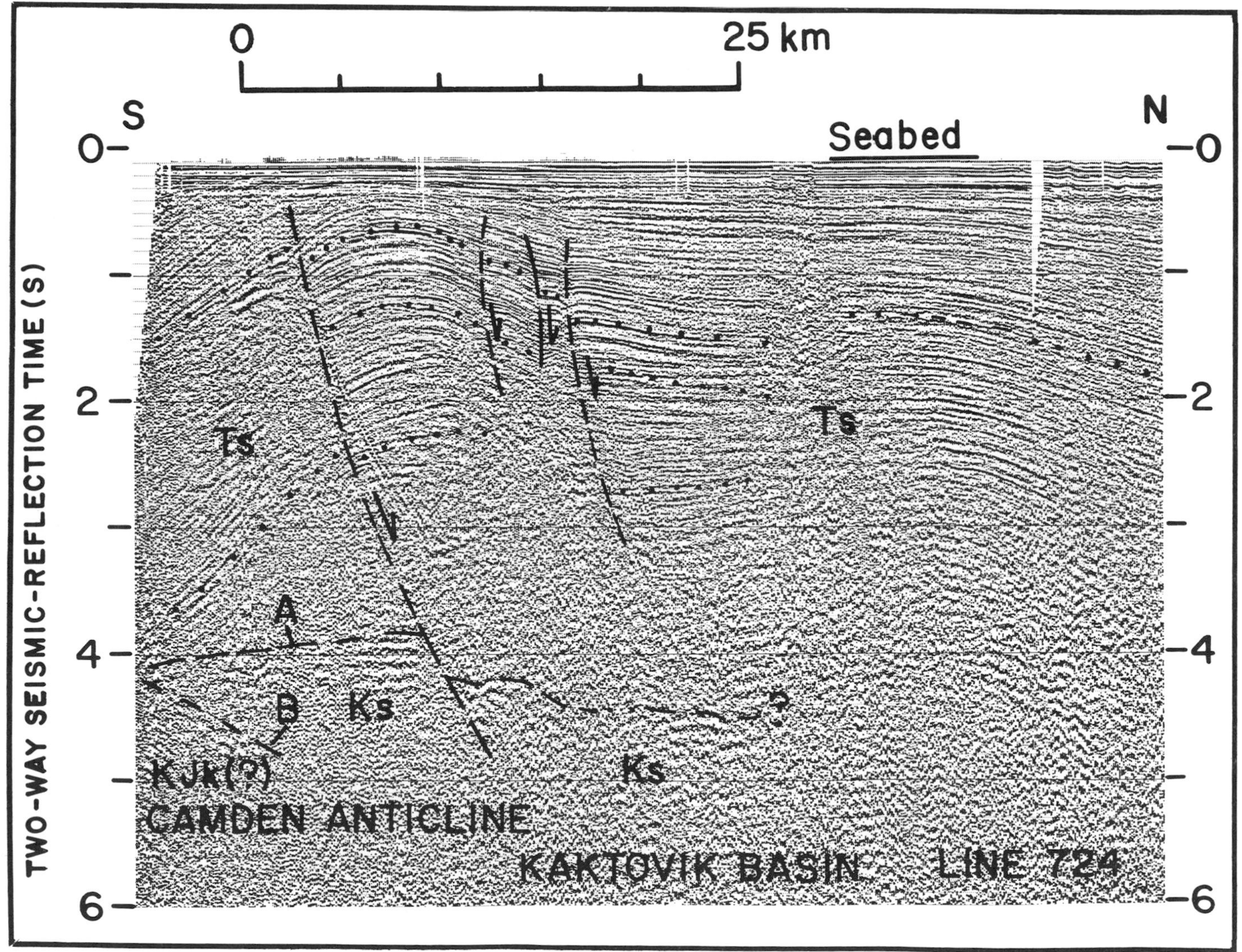

Figure 13 — CDP seismic-reflection profile 724, a dip line across Camden anticline, a large east-northeast-striking detachment fold off Camden Bay in the western part of Barter Island sector. Fold is inferred, from discordance between horizon A and higher beds, to be detached near 4 sec isochron. Detachment surface or zone does not itself return a reflection on our records. Horizon A is estimated by rough projection from onshore to be near base of Tertiary sedimentary rock unit Ts. Horizon B may be at top of a thin sequence of Lower Cretaceous bottomset shale beds that possibly overlies the Jurassic and Lower Cretaceous Kingak Shale (KJk). Unit Ks may be Cretaceous marine sedimentary rocks. Detachment-fault surface at base of fold is cut by seaward-dipping listric normal faults.

seismic-reflection profiles. West of Barter Island, Jurassic and Cretaceous beds of the Kaktovik basin are at least locally thin or absent, and the Tertiary beds lie close to basement, beneath the inner shelf. In the Exxon Alaska State No. 1 well on Flaxman Island, near the mouth of the Canning River, lower Tertiary beds are about 100 m above Franklinian basement, and only Cretaceous beds occur in between.

East of Barter Island, the Kaktovik basin includes two subbasins: the Barter Island and Demarcation (Figures 5 and 14). The basin contains a section of presubbasin strata of varying, and, in many places, considerable thickness. These two subbasins contain bedded rocks, units Ts and Ns on Figure 14, with sonobuoy velocities of 2.1 to 3.5 km/sec. These rocks are inferred, by extrapolation from the Canadian Beaufort shelf (Hea et al, 1980), to consist of middle and upper Tertiary clastic sedimentary deposits. In the Demarcation subbasin, beneath the south end of profile 714 (Figure 14), these beds are 4.4 sec (7.2 km) thick. The structural contours shown in the Barter Island sector of the Beaufort shelf, which is underlain by the Kaktovik basin (Figure 5), are drawn on the base of the middle and upper Tertiary beds. The section that underlies these beds consists of irregularly bedded strata of Jurassic, Cretaceous, and possible early Paleogene age (unit TJs, Figure 14) that range in thickness from less than 0.5 sec (1 km) to 5 sec (9 km) or more. These beds appear to thin beneath the axial region of the Demarcation subbasin and to thicken to the north and south. Thus, on seismic profile 714, the Jurassic to early Paleogene rocks (unit TJs) are 2.2 sec (4.5 km) thick beneath the southern limb of the subbasin, about 1.2 sec (2.6 km) thick beneath its axis, and more than 4.5 sec (8.5 km) thick beneath the diapiric anticline or shale ridge north of the subbasin.

The thickest sections of unit TJs are in places structurally disordered, apparently by diapirism. Diapirism also appears to account for the structural depressions occupied by the Barter Island and Demarcation subbasins, the thinning of unit TJs beneath the thickest part of the Demarcation subbasin, and the observed strong variations in the thickness of these rocks from place to place. The unit can be recognized beneath the south ends of all of our seismic reflection profiles east of Barter Island, and it also probably extends beneath the coastal plain to the south.

Extrapolation of data from outcrops in the coastal northern Yukon Territory and from an interpreted seismic-reflection profile across the Canadian Beaufort shelf (Norris, 1977a, 1977b), and of outcrop data from northeastern Alaska (Reiser et al, 1980) onto the eastern Alaskan Beaufort shelf, suggests that unit TJs consist of the Kingak Shale (Jurassic and lower Neocomian) and of Cretaceous and lower Tertiary clastic sedimentary rocks. Sonobuoy refraction velocities in these rocks of 3.0 to 4.0 km/sec (3.0 to 3.8 km/sec on profile 714, Figure 14) are compatible with this interpretation. If the interpretation is correct, the apparently diapiric Kingak Shale that crops out on Niguanak Ridge 30 km southeast of Barter Island, and the overlying beds of the Neocomian Pebble shale unit and the Upper Cretaceous Colville group (Reiser et al, 1980), belong to this sequence.

The prominent angular unconformity between 1 and 2 sec subbottom over the diapiric-shale ridge in profile 714 separates structurally orderly Oligocene(?) and Eocene rocks (unit Ts, Figure 13) above from the diapirically disrupted Jurassic to early Paleogene rocks below. This angular unconformity is correlated with an unconformity between strongly and weakly diapirically disrupted strata that is widespread on the Canadian Beaufort shelf (Hea et al, 1980), where it is dated as middle Eocene. A fairly complete Tertiary section lies above and below this unconformity on the Canadian shelf, and a thick section of Jurassic and Cretaceous marine strata can be inferred to underlie the unconformity on the basis of test wells, seismic-reflection profiles, and extrapolation from the outcrop. The structural contours on the "Paleogene horizon" in the Kaktovik basin (Figure 5) are probably not strictly isochronous, due to line spacing and the absence of subsurface control. These contours follow the "middle Eocene" unconformity (U-1, Figure 14), where it can be mapped; elsewhere they follow reflectors that correlate with the unconformity but do not separate beds that are differentially deformed.

Dinkum graben

The deep structure of the Beaufort shelf between the Colville delta and Camden Bay is dominated by a large asymmetric graben, in places only a half-graben, that is 10 to perhaps 40 km wide (Figures 5 and 10). This structure dies out on the west near where the hinge line steps from the inner to the outer shelf off Harrison Bay; on the east, it disappears beneath the thick sedimentary section in the Kaktovik basin. The trend of this graben suggests that it may extend eastward from Camden Bay beneath the inner shelf and coastal plain of the Barter Island sector and include the area underlain by the thick diapiric sections of Jurassic to early Paleogene rocks (unit TJs, Figure 14) east of Barter Island. The presently mapped length of the graben exceeds 170 km. For convenience in discussion, the graben is here named for the Dinkum Sands, which lie nearby off the mouth of the Sagavanirktok River.

A large south-dipping normal fault on the north side of the Dinkum graben (Figures 5, 10, 11, and 12) is the principal graben-forming structure. Stratigraphic displacement on this fault ranges from 1 to at least 2.3 sec (2 to 4.7 km) on the seismic-reflection profiles. The north-dipping normal fault or faulted flexure on the south side of the graben (Figure 11) has a much smaller displacement. Bedding of sedimentary rocks within the graben typically diverges to the north, and dips are 2.5 to 12° northerly; southward dips, however, are found near the north-bounding fault on some seismic-reflection profiles. Because of the northerly slope of basement and the relative displacement on the bounding faults, the synclinal axis of the fill generally lies on the north side of the graben. Deepest Franklinian basement beneath the presently mapped parts of the graben is 7 sec (12.6 km) below sea level.

The sedimentary section in the Dinkum graben appears to be at least 2.3 sec (about 4.7 km) thick, although the age of the beds cannot be directly determined. The graben fill is overlain by a thick sequence of beds that, from their position and seismic-reflection character, are identified as the marine Seabee formation of Turonian age in the lower part of the Colville group and overlying Cretaceous and Tertiary beds (see Figures 6, 11, and 12). Between these beds and the graben fill lies a thin sequence of beds interpreted to represent the Hauterivian to Barremian Pebble shale unit and possibly bottom-set beds at the base of the overlying Brookian sequence. If these correlations are correct, the main graben fill is pre-Hauterivian.

The existence of as much as 1.4 km of relief along the northbounding fault of the Dinkum graben during Turonian and perhaps earlier Cretaceous time (Figures 11 and 12), when the fault was overstepped by sedimentary deposits, suggests that the fault, fault scarp, and graben fill were also young features at that time. It is a reasonable inference from these relations that the graben contains Jurassic and Lower Cretaceous fill. This inference is somewhat supported by a clastic wedge present in the graben (Figure 12) that thickens northward, toward the north-bounding fault, in a manner encountered in syndepositional grabens. These relations do not, however, preclude the possibility that the deepest beds in the graben are pre-Jurassic Ellesmerian, perhaps as old as the Endicott group (Mississippian).

Other evidence for the age of the graben fill is found on seismic-reflection profiles off the Canning River. There, beds inferred to be late Neocomian by extrapolation from onshore are underlain by beds of the graben fill that are at least 1.4 sec (2.3 km) thick. The flat-lying inferred pre-late-Neocomian beds rest on north-sloping basement at a buttress unconformity, whereas bedding in the inferred late Neocomian rocks parallels both bedding in the graben fill and, where it oversteps the fill, the top of basement. These relations suggest that the graben fill is not much older than the overlying late Neocomian beds and that the fill was deposited on and against a newly tilted basement surface which is part of the graben structure.

Less direct evidence for the age of the graben fill comes from the Canadian Beaufort coast, where a sedimentary prism of the marine Jurassic and earliest Cretaceous Kingak Shale (Figure 16) rests on pre-Ellesmerian rocks (Norris, 1977a; Hea et al, 1980). The Jurassic and earliest Cetaceous shale prism and the Dinkum graben may be the same or analogous structures because they strike toward each other and because an outcrop of apparently diapiric Kingak Shale lies between them on the Arctic coastal plain 30 km southeast of Barter Island (Figure 16).

Graben faulting ceased by Seabee (Turonian) time because beds interpreted as Seabee overstep the faults (profiles 751 and 753, Figures 11 and 12). The profiles also show that at least 0.7 sec (1.4 km) of morphologic relief remained at the graben walls when progradation of the Beaufort shelf began in middle Cretaceous time. Final filling of the graben by postrift-faulting progradation was not completed until late Late Cretaceous or early Tertiary time.

Suprabasement structures

Growth faults and megaslumps

Three large anticlines, or anticlinal trends, as long as 250 km, underlie the outer shelf and slope of the Barrow sector seaward of the hinge line (Figures 5, 10, 11, and 12). Where these structures are shallow enough to be studied, they consist, at the core, of large backrotated blocks that moved downslope along systems of seaward-dipping listric growth faults. These faults descend to or near the seaward-sloping top of Franklinian basement, which they then follow to depths below the deepest seismic reflections at about 6 sec. Thus, the folds are rootless and akin to rotational megaslumps. The systems of listric faults that bound the megaslumps shoreward are generalized as single normal faults on Figures 5 and 10.

These listric faults and rotated blocks are best developed in Albian and Upper Cretaceous beds in the lower and middle parts of the suprabasement section. Although displacement dies out rapidly in the upper part of the section (inferred Tertiary), many of the faults displace Quaternary beds (see Figure 15). A few of the listric faults affect Holocene beds and even the seabed and are, therefore, still active (Grantz and Dinter, 1980; Grantz, Dinter, and Biswas, 1982). Continued growth and drape of Upper Cretaceous(?) and Tertiary beds over the rotated blocks created broad anticlines in the overlying shallow section.

The seismic-reflection profiles are too widely spaced to determine whether each of the three long anticlines consists of a single regionally continuous slump block or of a series of aligned structurally related blocks. The draped structure in the higher beds is, however, continuous or semicontinuous. From Figures 11 and 12 it is evident that gravitational failures in Brookian beds at or just above the seaward-sloping basement surface north of the subsidence hinge line created these elongate folds.

Detachment folds

Northeast-striking linear folds, 40 to 90 km long and about 20 km wide, underlie the inner shelf and coastal plain in the western part of the Barter Island sector. These folds occur on both sides of the subsidence hinge line, as projected into the area from the Barrow sector (Figure 5). At the Marsh anticline, which underlies the coastal plain near Camden Bay, Pliocene beds near the core of the fold dip more than 55°, and middle Pleistocene beds on one flank dip 18° (Morris, 1954; Reiser et al, 1980). The dips in the Pliocene and Pleistocene beds, as well as several morphologic features (Grantz, Dinter, and Biswas, 1982), indicate that folding was a Quaternary event and may be continuing. Offshore, the large Camden anticline (CA, Figures 5, 13, and 15) has dips of 10° in Tertiary beds and 0.5 to 1.5° in upper Quaternary beds. The folded Quaternary beds, spatially associated shallow earthquakes, and some late Quaternary and Holocene faulting (Figure 15) indicate that the offshore folds are also young and probably still active.

Several features indicate that the anticlines are rootless detachment folds. The anticlines are long and straight, have large ratios of length to width and, at the Marsh anticline, steep dips in the core. In these respects, they resemble the numerous Laramide (broadly defined) detachment folds of the foothills province on the north side of the Brooks Range. Seismic-reflection profile 724 (Figure 13) shows a structural discontinuity at a depth of 3.8 to 4.2 sec (5.9 to 6.8 km), which is interpreted to be the detachment surface at the base of the folds. Detachment is interpreted to be in the lowest Tertiary shale.

On the west, at the boundary between the Barrow and Barter Island sectors and along the projection of the Canning displacement zone (Grantz, Dinter, and Biswas, 1982), the detachment folds change in trend from southwestward to west-northwestward, and die out. The west to northwest trending segments parallel the local trend of the hinge line. On the east, the detachment folds die out or merge with mainly older structures in the subprovince of irregular diapiric folds east of Barter Island. Shallow earthquakes in northeastern Alaska are concentrated in the zone of compressional folds and in the adjacent western part of the northeastern Brooks Range; they die out in the Canning displacement zone (Figure 15). The fact that these earthquakes do not extend northward of the northern limb of the detachment fold (Figure 15) indicates that this fold marks the structural front of the postulated thinskinned deformation. Spatial relations suggest that the detachment folds, the earthquakes, the displacement zone, and uplift of the northeastern Brooks Range are genetically related late Cenozoic features.

Diapiric folds on the shelf

East of Barter Island, the inner and central Beaufort shelf is underlain by a large complex shale ridge, or diapiric anticline, with multiple structural cul-

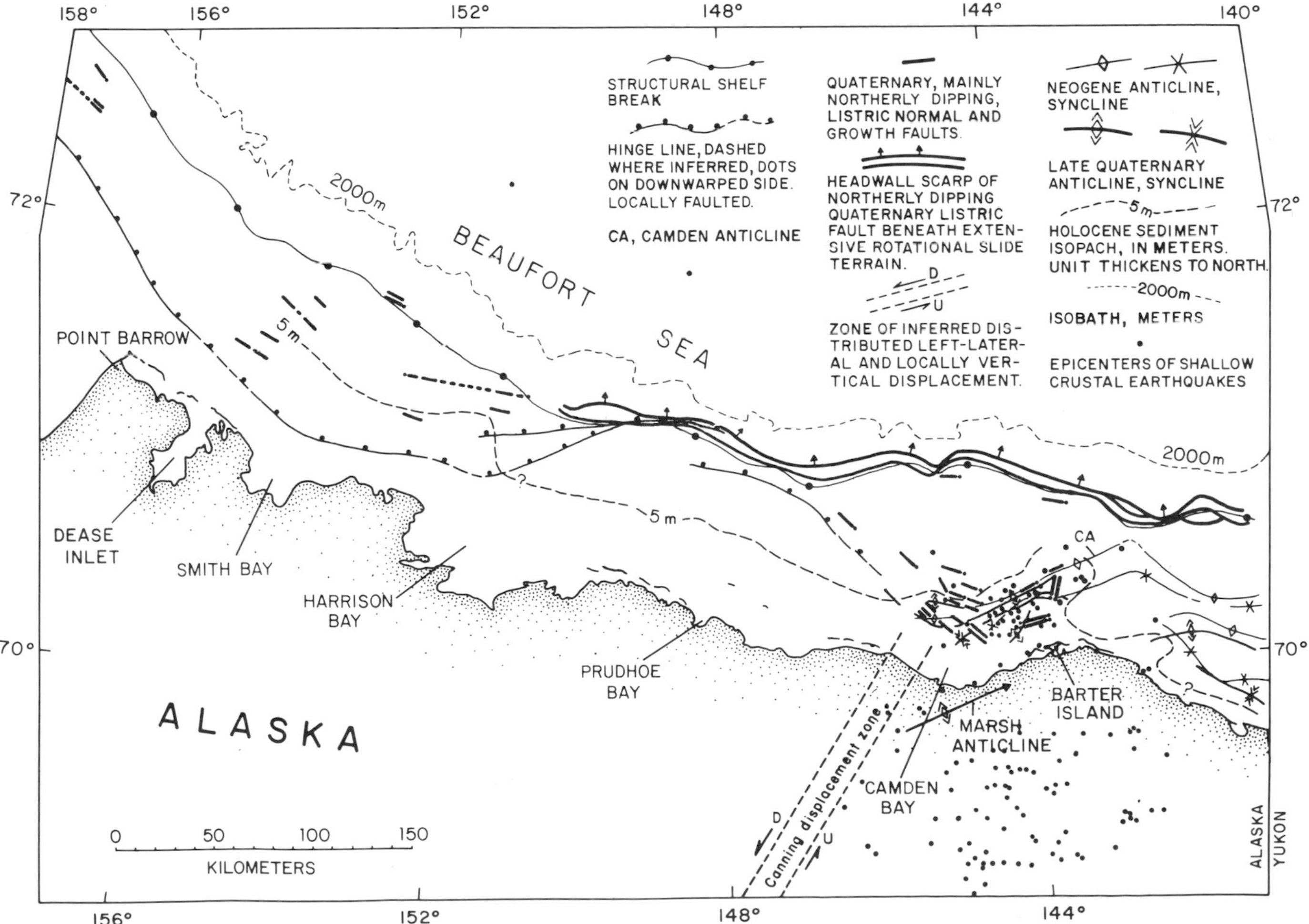

Figure 15 — Quaternary geologic structures of Alaskan Beaufort shelf. CA, large northeast-striking Camden anticline, a detachment fold that is locus of northernmost earthquakes in western part of the Barter Island sector.

minations and complex internal structure (Figure 5). This ridge is flanked on the south by the penecontemporaneous and genetically related Barter Island and Demarcation synclinal subbasins. An additional similar anticlinal uplift may underlie the inner shelf east of Barter Island. The diapiric folds are 40 to more than 75 km long, commonly 5 to 10 km wide, and have as much as 3.5 sec (6.5 km) of structural relief. Figure 14 shows a typical reflection profile across the diapiric fold belt. Geographic restriction of these diapiric structures to the region east of Barter Island indicates the presence of a section of weak sedimentary rocks that is thicker or more deeply buried there than to the west.

Structurally disturbed bedded rocks in the cores of the diapiric anticlines extend to depths that exceed 6.5 sec (11 km) subsea. The core zones contain upraised domains in which coherent seismic reflectors are absent, flanked by sections with poorly-developed to well-developed bedding. The flanking bedded sections generally dip away from the central upraised domains. Sonobuoy-refraction velocities in the core rocks range from about 3.0 to 3.8 km/sec (Figure 14), values appropriate for upper Mesozoic and lower Tertiary clastic rocks buried to these depths.

We interpret the structure on profile 714 (Figure 14) to be a multistage shale-rich diapiric intrusive that moved out of an extension or analog of the Dinkum graben into the overlying Tertiary sedimentary beds. The first stage of activity is recorded in an unconformity (U-1, Figure 14) that overlies the uplifted core of the diapiric anticline at subsurface depths of 1 to 2 sec; the structurally high region of this unconformity is a rough erosional surface. Between 1.8 sec and a little below 3 sec on the north flank of the Demarcation subbasin the unconformity is offset by normal faults between the core rocks and unrumpled but south-dipping well-bedded rocks in the basin. These normal faults are not extensional or gravity features, because the footwalls moved up. The normal faults, and thinning of beds from thicks near the basin axis toward the diapiric anticline, demonstrate that sedimentation was penecontemporaneous with this early stage of diapirism. Onlap of beds midway (in thickness) in the Demarcation subbasin section onto the highest parts of the main, erosionally rough unconformity U-1 at the top of the diapiric-shale ridge demonstrates that this early phase of diapirism generated about half the structural relief between the diapiric anticline and the Demarcation subbasin. In several places, high spots on unconformity U-1 are the sources of packets of foreset beds that indicate local and, probably, rapid sedimentation in the section immediately above the unconformity.

A second stage of diapirism is represented by angular unconformity U-2 (Figure 14), which lies 0.7 to 0.8 sec subsea over the crest of the diapiric anticline. A third stage is represented by angular unconformity U-3 and the southward dip of beds in the uplifted north flank of the Demarcation subbasin (Figure 14). The young (late Tertiary) folding associated with the formation of unconformity U-3 affected virtually the entire section in the Demarcation subbasin and generated about half the structural relief (in thickness) between the basin and the diapiric anticline. Thinning of beds in the upper part of the section from the axis of the subbasin toward the diapiric anticline indicates that sedimentation was also penecontemporaneous with the latest stage of diapirism. Seismic-reflection interval velocities in bedded rocks in the lower part of the subbasin section, which were intruded (pierced) by the diapiric rocks, are about 3.4 to 3.7 km/sec; interval velocities in postintrusion beds overlying the crest of the diapiric anticline are 2.1 to 2.3 km/sec. These interval velocities are appropriate for lower and upper Tertiary clastic strata at these depths.

Residual diapiric activity may possibly be continuing in the diapiric-fold subprovince. Slightly thickened late Quarternary sediment and a shallow depression of the seabed (Dinter, 1981) overlie the Demarcation subbasin, and a slight elevation of the seabed (Dinter, 1981) overlies the diapiric anticline to the north. These gentle differential movements, which may be due to differential compaction and/or residual diapiric motion, appear to be aseismic. The few earthquakes epicenters reported from the subprovince of diapiric folds (Biswas and Gedney, 1978) were located near its margins.

The initiation of diapiric movement can be tentatively dated from relations on the Canadian Beaufort shelf (Norris, 1977a; Hea et al, 1980), where the combined results of drilling and seismic-reflection surveys have served to date the widespread diapirism that occurs there. Canadian workers have observed a widespread middle Eocene unconformity separating older beds that are strongly affected by diapirism from younger beds that are much less affected. We provisionally correlate erosional unconformity U-1 with the middle Eocene unconformity of the Canadian Beaufort shelf.

Diapiric structures of the slope and rise

Numerous diapiric structures underlie the continental slope and adjacent rise in the Barter Island sector (Figure 5). As observed at 42 reflection-profile crossings, these structures are much smaller and have more regular cross-sectional geometries than do the large diapiric anticlines of the shelf to the south. The structures of the slope and rise are symmetrical diapiric folds or domes, rather than piercement structures or irregular shale ridges. Although the reflection profiles are too widely spaced to allow us to map these structures, they suggest that the structures are subcircular or moderately elongate and that individual folds are not of regional extent.

The diapiric folds are typically 2 to 10 km wide on seismic crossings, and the largest folds have at least 1.5 sec (1.25 km) of structural relief. The most northerly structures (100 km north of the shelf break) die out and are overlain by undisturbed beds about 2 sec (2 km) below the seabed. The tops of the diapirs are progressively higher in the section landward, and beginning about 50 to 60 km from the shelf break they buckle the seabed and act as dams for clastic sediment and landslide debris moving downslope.

The diapiric anticlines appear to constitute a westward extension of the shale-diapir province of the western Canadian Beaufort shelf. Seismic-reflection interval velocities of about 1.8 to 3.5 km/sec in the intruded beds (Eittreim and Grantz, 1979), and lateral tracing of reflectors, suggest that these diapiric structures penetrate Tertiary and Quarternary beds. Lateral tracing of reflectors also indicates that those structures which are closest to the shelf, and which are strongly developed in the shallow part of the section and disrupt the seabed, originated in lower Tertiary and possibly deeper beds. The broad folds farthest from the shelf, which die out upward at about 2 sec subbottom and are no more than broad warps at 3.5 sec subbottom, are less obviously diapiric. We assign these broad folds to the diapiric province because they are contiguous with the strong folds and weaken progressively away from them. If correctly interpreted as diapiric, these folds must have originated at a considerable depth below our deepest records, possibly in early postrift Lower Cretaceous sedimentary deposits of the Canada Basin or in prebreakup Jurassic rift-valley sedimentary deposits.

DISCUSSION

Comparison of northern Alaska with Atlantic-type margins

We compare here the Beaufort shelf with typical rifted margins, as described and, in part, modeled by Falvey (1974). Falvey's model postulates both thermal expansion and thermal metamorphic events that are out of phase and create two cycles of uplift (doming), erosion, and subsidence during the formation of a passive margin. The first cycle of uplift is held to accompany formation of a rift-valley system, and the second cycle to accompany breakup and the beginning of sea-floor spreading. Recent work in the Bay of Biscay by Montadert et al (1979, p. 1048-1051), however, suggests that rift valleys may be created by brittle fracturing in the upper crust following thinning in the underlying ductile lower crust. This thinning is held to be caused by extension without domal uplift and erosion or the creation of a rift-onset unconformity. In Montadert et al's model, crustal warming is caused by an isostatic rise in the mantle beneath the thinned crust and is an effect, not a cause, of rift-valley formation.

In the absence of well control on the Alaskan Beaufort shelf, we have compared the geometry of seismic reflections beneath the shelf, tentatively dated by extrapolation and by analogy with the subsurface of northern Alaska, with the features of passive margins as described by Falvey (1974) and modified after Montadert et al (1979). Such margins contain four lithotectonic elements and a breakup unconformity. Each of these four lithotectonic elements — continental basement, a precursor intracratonic basin, one or more rift valleys, and a postbreakup progradational sedimentary prism — and the breakup unconformity appear to be present in more or less typical form and relative timing beneath the Alaskan Beaufort shelf. One feature of Falvey's model that has not yet been identified on the Beaufort shelf is the presence of volcanic rocks in the rift-valley sedimentary fill. Volcanic rocks, however, appear to be inconspicuous or absent on many passive margins (e.g., the Bay of Biscay, Montadert et al, 1979).

The first lithotectonic element, continental basement, clearly exists beneath the northern Alaska margin. Rifting sundered the Arctic platform (Figures 4, 7, 8 and 9), a cratonic area that has been little deformed since the Devonian. In the northeastern Chukchi Sea (Figures 7 and 8), the Arctic platform rests on more than 9 km of virtually undeformed lower Paleozoic and possible Proterozoic strata that have seismic-reflection geometries characteristic of sedimentary prisms. The platform is in turn overlain by the Arctic Alaska shelf basin of late Paleozoic and early Mesozoic age. The sourceland for this basin, Barrovia in Tailleur (1973), lay beneath the present outer Beaufort shelf and southern Canada Basin. The Arctic Alaska basin appears to be the second lithotectonic element — a prerift intracratonic basin. Falvey's model provides for a simple subsidence basin containing fluviodeltaic and shallow-marine sedimentary rocks that formed 50 to perhaps 150 million years before breakup. The Arctic Alaska basin received Lower Mississippian to Valanginian fluviodeltaic and marine sediment, and its north margin was adjacent to the rift beneath the present Beaufort shelf. According to one reconstruction of prerift northern Alaska (Grantz, Eittreim, and Dinter, 1979, Figure 13), the Sverdrup basin of the Canadian Arctic Islands was contiguous with the Arctic Alaska basin before rifting. The section in the Sverdrup basin is chronostratigraphically and lithostratigraphically similar to that in the Arctic Alaska basin, and the two may be parts of a single prerift sedimentary basin. If this reconstruction is correct, then the rift traversed the combined Arctic Alaska-Sverdrup basin somewhere northwest of Point Barrow. The interval between the formation of these basins 340 to 335 m.y. ago and the breakup about 125 m.y. ago is about 210 to 215 million years, somewhat longer than the interval of 50 to 150 million years in Falvey's model.

The Dinkum graben, well-developed on the Alaskan Beaufort shelf (Figures 5, 10, 11, and 12) resembles the third element in Falvey's model — a rift valley. The graben contains a northward-thickening clastic-sediment wedge, interpreted to be Jurassic and Neocomian, that is at least 3.3 km thick. These Jurassic and Neocomian beds are considered to have been syntectonic with graben-boundary faulting. An additional 1 km of section beneath these beds may also be Jurassic graben fill, although a prerift Ellesmerian age appears more likely. If the main graben fill is Jurassic and Neocomian, then the graben is a rift valley related to subsequent late Neocomian breakup and sea-floor spreading.

The earliest structural manifestation of rifting off northern Alaska is the unconformity at the base of inferred Jurassic fill in the Dinkum graben (base of weakly reflective beds, unit KJk, or of moderately reflective beds, unit JpJe, in Figures 11 and 12). Although we cannot date this unconformity directly, if the Dinkum graben and the Jurassic sedimentary prism along the Canadian Beaufort coast (Figure 16) are the same or analogous and correlative structures, then the age of the basal unconformity in Canada may date the onset of rifting off Alaska. In Canada, strata in the Jurassic prism consist of shale of southeasterly provenance that thickens rapidly seaward. Lithofacies maps of Jurassic to Berriasian strata in the Beaufort-Mackenzie basin by Young, Myhr, and Yorath (1976, Figures 4, 5, and 7) show that these beds coarsen toward presumed clastic-sediment sources near the Vittrekwa embayment, including the Eagle arch, the Rat high, and the Keele-Old Crow landmass, in the general area of the southern Richardson Mountains. Their source is thus in the area of the apex, or pole of the postulated rotational rift that formed the Canada Basin. The Jurassic shale oversteps Upper Triassic marine beds of nearshore facies and northerly provenance in the British Mountains of the northern Yukon Territory, and rests on beds as old as Proterozoic along the adjacent Beaufort seacoast and offshore (Norris, 1977a, 1977b). The timing of the reversal in provenance and of overstepping of the Triassic beds in early Early Jurassic, because beds of latest Triassic (Rhaetian) and earliest Jurassic (Hettangian and early Sinemurian) age have not been found in northeastern Alaska (Detterman et al, 1975, p. 7) or the coastal northern Yukon Territory (Poulton, 1978, p. 449). If this hiatus dates the rift-onset unconformity, it is about 185 to 190 m.y. old and formed about 60 to 65 million years before breakup.

A strong erosional unconformity that morphologically resembles, and is interpreted to be, a breakup unconformity overlies the Kingak Shale and older Ellesmerian units in the Dinkum graben and mainly Franklinian rocks beneath the shelf west and

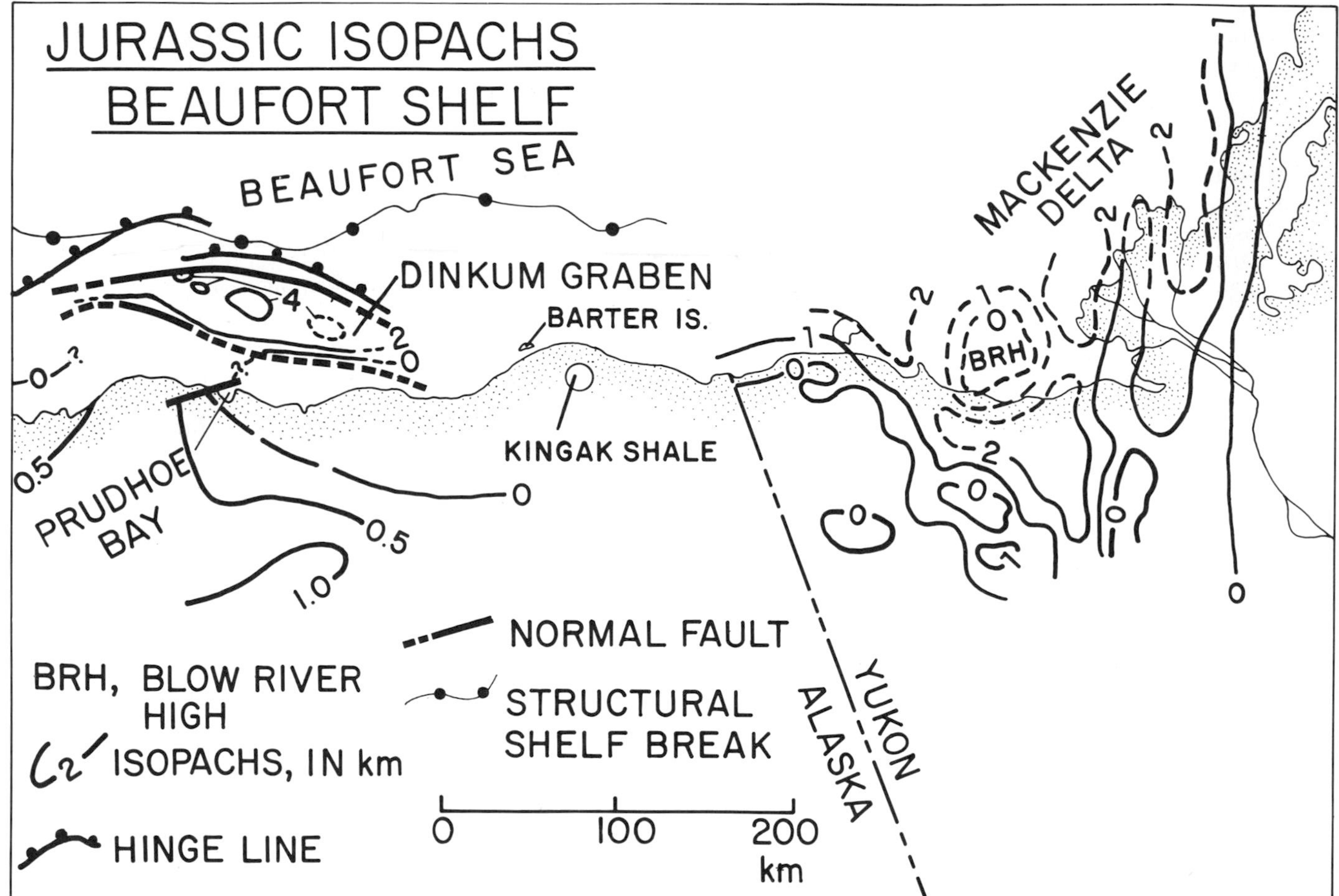

Figure 16 — Thickness of Jurassic and lowest Cretaceous sedimentary rocks in Mackenzie delta region (Hea et al, 1980, Fig. 16), central northern Alaska (Pessel et al, 1978), and Alaskan Beaufort shelf. Sedimentary wedge in Dinkum graben may include pre-Jurassic beds of Ellesmerian sequence as well as Jurassic clastic rocks. Zero isopachs in Prudhoe Bay area are truncation edges at breakup unconformity at base of the Pebble shale unit, and not original limit of sediment. According to Hea et al (1980), Blow River high (BRH) is an uplifted shale mass probably floored by Jurassic shale. In test wells and outcrops in northern Alaska, and in both onshore and offshore Mackenzie delta region, Jurassic and lowest Cretaceous rocks consist predominantly of shale and siltstone belonging to the Kingak Shale.

north of the graben. The unconformity is overlain in turn by a thin, strongly reflective seismic interval that resembles the reflections returned by the condensed, highly organic, Hauterivian to Barremian Pebble shale unit of the North Slope (Kps in Figures 7, 8, 9. 11, and 12). The high organic content and thinness (about 100 m) of the Pebble shale unit in coastal northern Alaska indicate that by Hauterivian time, northerly clastic-sediment sources were feeble in northern Alaska. The Pebble shale unit may, in places, be difficult to distinguish on reflection profiles from basal or distal bottom-set beds at the base of the overlying Brookian sequence. These beds may have been mapped with the Pebble shale on some of the seismic profiles.

The gap in the stratigraphic record at the breakup unconformity increases northward, and beneath the Beaufort shelf and northeast Chukchi Sea the unconformity merges with the diachronous regional angular unconformity that separates the Franklinian and Ellesmerian sequences. Southward, the unconformity merges into a conformable contact between the Pebble shale unit and the Jurassic-Lower Cretaceous Kingak Shale in northern Alaska (Molenaar, 1981). In the northeast Chukchi Sea, and in places in the Beaufort shelf, the north-sourced Pebble shale unit appears to wedge out northward and to be overstepped by the Torok formation (Albian) and probably in places by younger beds at the base of the south-sourced Brookian sequence. The region of the breakup unconformity north of the wedge-out is thought to have been the feeble sourceland of detritus for the synbreakup Pebble shale unit. Cessation of Pebble shale unit deposition marked the beginning of postbreakup subsidence; the overstepping of the Pebble shale by the southerly-sourced Torok formation of the Brookian sequence initiated deposition of the postbreakup progradational sedimentary prism of the Beaufort shelf. If the breakup unconformity is correctly identified, then sea-floor spreading in the Canada basin began near the beginning of Hauterivian time, about 125 m.y. ago. The increase in seaward slope of the breakup unconformity and the top of basement at the subsidence hinge line is also typical of passive margins. This increase is thought to reflect crustal thinning caused by heating, uplift, and erosion of the crust adjacent to the rift, and by rapid early postrift subsidence due to marked cooling and subsequent sedimentary loading.

The postrift progradational sedimentary prism of the Beaufort shelf (Figures 8, 9, 11 and 12) is typical of passive margins and is the fourth element in Falvey's (1974) model. Above the commonly occurring strong reflector interpreted to represent the Pebble shale unit and possibly a thin section of basal Brookian shale (Kps, Figures 8, 9, 11 and 12) are thick sections of Torok, Late Cretaceous and Paleogene foreset beds interpreted to be prodelta shale with sandstone or siltstone interbeds (Ktc or TKs, Figures 8, 9, 11 and 12). These beds, as much as 2.5 to 3.0 km thick, are regressive deposits that followed the strong marine transgression that resulted from rapid early postbreakup cooling and subsidence of the middle and outer Beaufort shelf. These deposits are succeeded by about 3 km of Senonian and Tertiary topset beds, interpreted to be intradelta shale, sandstone, and conglomerate (upper part of unit TKs and unit Ts, Figures 11 and 12). These beds mark accelerated progradation of the Beaufort shelf in response to the exponential decline in the rate of postbreakup subsidence along rifted margins, as predicted by Sleep (1971).

Correlation of rifting history with events in adjacent areas

The rifted continental margin north of Alaska forms the present north boundary of the Arctic Alaska plate of Tailleur (Oil and Gas Journal, 1976) and Newman, Mull, and Watkins (1977), and some workers have estimated the age of sea-floor spreading in the Canada basin from the age of collision events along the present south and southwest margins of the plate. Fujita and Newberry (1982) interpret the southwest boundary of the plate to be the South Anyui suture, a continent-to-continent collision zone in northeastern Siberia. This suture separates Chukotka, part of the Arctic Alaska plate, from the Omolon massif, which they consider to have been welded to the Siberian plate during Late Jurassic time. Fujita and Newberry conclude that Chukotka and the Omolon massif were separated by oceanic crust that began to be consumed by subduction, possibly along both margins, during Middle Jurassic time, and that collision occurred during Hauterivian time.

In Alaska, the south boundary of the Arctic Alaska plate is generally placed at the root zone of the far-traveled nappes of the Brooks Range; this root zone lies along the south side of the range. Most of these nappes consist of continental, stable-shelf, and shelf-basin or oceanic clastic rocks, carbonate rocks, and chert of Devonian to Neocomian age. The topmost nappes consist of ultramafic and mafic rocks (Snelson and Tailleur, 1968; Martin, 1970; Roeder and Mull, 1978). The age of initiation of thrusting has been estimated from the emplacement age of the ultramafic and mafic nappes and from the age of a flysch basin that formed in front of the nappes during the early phases of thrusting. Patton et al (1977) reported that K-Ar dates on mafic rocks in the upper nappes suggest emplacement during Middle or Late Jurassic time. The deposits of the flysch basin, the Okpikruak formation, are reported by Mull et al (1976) to contain olistostromes and polymict conglomerate derived from advancing nappes in the earliest Cretaceous beds, and by Martin (1970) to contain detritus from Brooks Range sources in the Berriasian beds and abundant ultramafic boulders in the Valanginian beds. These relations indicate that thrusting began during and, possibly, before the deposition of the Okpikruak. Thus, thrusting in the Brooks Range may have begun during Middle or Late Jurassic time and was apparently well underway by early Okpikruak (Berriasian or possibly latest Jurassic) time.

If we accept Fujita and Newberry's (1982) timing for the closing of the Chukotka-Omolon gap (Middle Jurassic to Hauterivian), then the initiation of sea-floor spreading in the Canada basin during the Hauterivian was contemporaneous with only the last phase of closing, or with collision. Brooks Range thrusting, which is considered to be the result of a collisional event, also began much earlier than spreading in the Canada basin if the Middle and Late Jurassic K-Ar ages on mafic rocks in the upper nappes date emplacement. On the other hand, if the Okpikruak formation dates collision and the initiation of thrusting, then these events began some 15 million years or more before spreading. Although there is some uncertainty in the dating, we conclude that the closing of the seaway between Chukotka and the Omolon massif was largely completed, and Brooks Range thrusting well underway, when sea-floor spreading began in the Canada basin. This suggests that: 1) opening of the Canada basin was not genetically linked to formation of the Brooks Range nappes by underthrusting at a collision zone along the south side of the Arctic Alaska plate, and 2) a large transform fault extended from the Arctic to the Pacific basin and separated the accreting Canada basin and the Arctic Alaska plate from the Eurasian Arctic and Siberia.

SUMMARY

The known and inferred structural history of the Atlantic-type continental margin north of Alaska consists of the following principal events:

(1) Consolidation of an early Paleozoic eugeoclinal terrane into a craton, and cutting of a low relief surface across the newly consolidated terrane by Late Devonian or Early Mississippian time, forming the stable Arctic platform of northern Alaska.

(2) Deposition of stable-shelf clastic and carbonate rocks of northerly provenance within the intracratonic Arctic Alaska basin between Early Mississippian and Neocomian time. Sedimentation in the basin, which rests on the Arctic platform, began about 210 million years before breakup along the present passive margin. Off northern Alaska, the rift zone of the passive margin probably cut across marginal facies of the Arctic Alaska basin.

(3) Crustal extension and subsidence, and formation of a rift-valley system, beneath the Beaufort shelf and, in places, the northern part of the Arctic coastal plain during Early Jurassic time, about 60 to 65 million years before breakup. On the eastern Beaufort shelf this subsidence caused a reversal of provenance from northerly during Late Triassic to southeasterly during Early Jurassic time. The rift-valley system received thick sections of Jurassic and earliest Cretaceous marine and possibly some nonmarine strata.

(4) Late prebreakup uplift and erosion adjacent to the rift, and formation of a breakup unconformity, by Hauterivian time, about 125 m.y. ago. Breakup and initiation of sea-floor spreading in the Canada basin began at about that time.

(5) Cooling and subsidence of the northern part of the Arctic platform, and rapid subsidence of the erosionally thinned region north of the subsidence hinge line, creating the Nuwuk and Kaktovik basins and initiating transgressive sedimentation. By middle Albian time, about 105 m.y. ago, subsidence of the northern part of the Arctic platform and filling of the Colville foredeep permitted Albian clastic materials from southern (Brookian) sources to overtop the Barrow arch and to begin prograding post-rift-valley, postbreakup sedimentary prisms of the Beaufort shelf and the Canada basin.

(6) Beginning in middle Late Cretaceous time, exponential reduction in the rate of cooling and subsidence, and continued influx of clastic materials from southerly sources, resulting in accelerated progradation of the Beaufort shelf sedimentary prism.

(7) Beginning in Late Cretaceous or Paleogene time, gravity failures in the sedimentary prisms seaward from the subsidence hinge line, generating normal and growth faults and related rotational megaslumps or slope anticlines of regional extent. Some of these growth faults are still active.

(8) Middle Eocene and perhaps earlier flowage of Jurassic, Cretaceous, and possible early Paleogene beds, creating large diapiric shale ridges or anticlines and flanking synclinal subbasins beneath the Beaufort shelf east of Barter Island. Some of these ridges were exposed to submarine and, perhaps, subaerial erosion. Renewed uplift during the Neogene significantly increased structural relief on the diapiric shale ridges and flanking subbasins.

(9) Neogene and Quarternary uplift and relative northeastward translation of the northeastern Brooks Range along the Canning left-lateral displacement zone. This uplift and translation is postulated to have formed a shallow low-angle detachment surface that generated the detachment folds and faults and caused the earthquakes observed in Neogene and Quarternary strata of the coastal plain and shelf in the western part of the Barter Island sector (Grantz, Dinter, and Biswas, 1982).

ACKNOWLEDGMENTS

John S. Schlee, Michael S. Marlow, Paul R. Gucwa, and Charles G. Mull critically reviewed the manuscript and offered many helpful suggestions for its improvement. Lyn Smith drafted the illustrations and Winifred Trollman typed the manuscript.

REFERENCES CITED

Bird, K. J., et al, 1977, Granite on the Barrow arch, northeast NPRA, *in* K. M. Johnson, ed., The United States Geological Survey in Alaska: Accomplishments during 1977: U.S. Geological Survey, Circular 772-B, p. B24-B25.

Biswas, N. N., and L. Gedney, 1978, Seismotectonic studies of northeast and western Alaska: Fairbanks, University of Alaska Geophysics Institute Administrative Report, 45 p.

Brosgé, W. P., and I. L. Tailleur, 1971, Northern Alaska petroleum province, *in* I. H. Cram, ed., Future petroleum provinces of the United States — their geology and potential: AAPG Memoir 15, v. 1, p. 68-99.

Carey, S. W., 1958, The tectonic approach to continental

drift, *in* Carey, S. W., ed., Continental drift, a symposium: Hobart, Tasmania University, p. 177-355.

Carter, Claire, and Sven Laufeld, 1975, Ordovician and Silurian fossils in well cores from North Slope of Alaska: AAPG Bulletin, v. 59, p. 457-464.

Detterman, R. L., et al, 1975, Post-Carboniferous stratigraphy, northeastern Alaska: U.S. Geological Survey Professional Paper, 886, 46 p.

Dinter, D. A., 1981, Holocene sediments on the middle and outer continental shelf of the Beaufort Sea north of Alaska: U.S. Geological Survey Miscellaneous Investigation Series Map I-1182-B, scale 1:500,000.

Eittreim, S., and Arthur Grantz, 1979, CDP seismic sections of the western Beaufort continental margin: Tectonophysics, v. 59, p. 251-262.

Falvey, D. A., 1974, The development of continental margins in plate tectonic theory: Australian Petroleum Exploration Association Journal, v. 14, p. 95-106.

Fujita, Kazuya, and J. T. Newberry, 1982, Tectonic evolution of northeastern Siberia and adjacent regions: Tectonophysics, v. 89, p. 337-357.

Grantz, Arthur, and D. A. Dinter, 1980, Constraints of geologic processes on western Beaufort Sea oil developments: Oil & Gas Journal, v. 78, n. 18, p. 304-319.

———, S. Eittreim, and D. A. Dinter, 1979, Geology and tectonic development of the continental margin north of Alaska: Tectonophysics, v. 59, p. 263-291.

———, ———, and O. T. Whitney, 1981, Geology and physiography of the continental margin north of Alaska and implications for origin of the Canada Basin, *in* A. E. M. Nairn, Michael Churkin, Jr., and F. G. Stehli, eds., The ocean basins and margins: Plenum Publishing Corporation, New York, v. 5, p. 439-492.

———, D. A. Dinter, and N. N. Biswas, in press, Map, cross sections,and chart showing late Quaternary faults, folds, and earthquake epicenters on the Alaskan Beaufort shelf: U.S. Geological Survey, Miscellaneous Investigations Series Map I-1182-C, scale 1:500,000.

Guldenzopf, E. C., et al, 1980, National Petroleum Reserve, Alaska, Summary geological report fiscal year 1980: Tetra Tech, Inc., prepared for Husky Oil National Petroleum Resources Operations, Inc., under contract to U.S. Geological Survey, 168 p.

Hall, J. K., 1973, Geophysical evidence for ancient sea-floor spreading from Alpha Cordillera and Mendeleyev Ridge *in* Max G. Pitcher, ed., Arctic geology: AAPG Memoir 19, p. 542-561.

Hea, J. P., et al, 1980, Post-Ellesmerian basins of Arctic Canada; their depocentres, rates of sedimentation, and petroleum potential, *in* A. D. Miall, ed., Facts and principles of world petroleum occurrence: Canadian Society of Petroleum Geology, Memoir 6, p. 447-488.

Hunkins, Kenneth, 1963, Submarine structure of the Arctic Ocean from earthquake surface waves, *in* Proceedings of the Arctic Basin Symposium, 1962: Washington, D.C., Arctic Institute of North America, p. 3-8.

Lerand, Monti, 1973, Beaufort Sea, *in* R. G. McCrossam, ed., The future petroleum provinces of Canada — their geology and potential: Canadian Society of Petroleum Geology, Memoir 1, p. 315-386.

Mair, J. A., and J. A. Lyons, 1981, Crustal structure and velocity anistrophy beneath the Beaufort Sea: Canadian Journal of Earth Science, v. 18, p. 724-741.

Martin, A. J., 1970, Structure and tectonic history of the western Brooks Range, De Long Mountains and Lisburne Hills, northern Alaska: Geological Society of America Bulletin, v. 81, p. 3605-3622.

Miller, C. C., et al, 1979, National Petroleum Reserve, Alaska, Summary geophysical report fiscal year 1979: Tetra Tech, Inc., prepared for Husky Oil National Petroleum Resources Operations, Inc., under contract to U.S. Geological Survey, 47 p.

Molenaar, C. M., 1981, Depositional history and seismic stratigraphy of Lower Cretaceous rocks, National Petroleum Resources: U.S. Geological Survey Open-File Report. 81-1084.

Montadert, L., et al, 1979, Rifting and subsidence of the northern continental margin of the Bay of Biscay, *in* Initial reports of the deep sea drilling project, v. 48: Washington, D.C., U.S. Government Printing Office, p. 1025-1060.

Morris, R. H., 1954, Reconnaissance study of the Marsh anticline, northern Alaska: U.S. Geological Survey Open-File Report, 6 p.

Mull, C. G., et al, 1976, New structural and stratigraphic interpretations, central and western Brooks Range and Arctic Slope, *in* E. H. Cobb, ed., The United States Geological Survey in Alaska: Accomplishments during 1975: U.S. Geological Survey Circular 733, p. 24-26.

Newman, G. W., C. G. Mull, and N. D. Watkins, 1977, Northern Alaska paleomagnetism, plate rotation, and tectonics, *in* A. Sisson, ed., The relationship of plate tectonics to Alaskan geology and resources: Alaska Geological Society Symposium Proceedings, p. C1-C7.

Norris, D. K., 1977a, Geologic map of parts of Yukon Territory, District of Mackenzie, and District of Franklin, *with* Structure sections, by D. K. Norris and C. J. Yorath: Geological Survey of Canada Open-File Report 399, 4. p., scale 1:1,000,000.

———, 1977b, Geologic maps of Yukon territory and northwest territories: Geological Survey of Canada Open-File Report 499, 7 p., 4 sheets, scale 1:250,000.

Oliver, Jack, M. Ewing, and F. Press, 1955, Crustal structure of the Arctic regions from the Lg phase: Geological Society of America Bulletin, v. 66, p. 1063-1074.

Oil & Gas Journal, 1976, Geologist calls Prudhoe one of a kind: v. 74, n. 44, p. 32.

Patton, W. W., Jr., et al, 1977, Preliminary report on the ophiolites of northern and western Alaska, *in* R. G. Coleman and W. P. Irwin, eds., North American ophiolites: Oregon Department of Geology and Mineral Industries, Bulletin 95, p. 51-58.

Pessel, G. H., J. A. Leversen, and I. L. Tailleur, 1978, Generalized iospach and net sand maps of Lower Cretaceous sands, eastern North Slope petroleum province, Alaska: U.S. Geological Survey Miscellaneous Field Studies Map MF-928-J, scale 1:500,000.

Poulton, T. P., 1978, Pre-late Oxfordian Jurassic biostratigraphy of northern Yukon and adjacent Northwest territories, *in* C. R. Stelck and B. D. E. Chatterton, eds., Western and arctic Canadian biostratigraphy: Geological Association of Canada, Special Paper 18, p. 445-471.

Reiser, H. N., et al, 1980, Geologic map of the Demarcation Point quadrangle, Alaska: U. S. Geological Survey Miscellaneous Investigations Series, Map I-1133, scale 1:250,000.

Rickwood, F. K., 1970, The Prudhoe Bay field, *in* W. L. Adkison and M. M. Brosgé, eds., Proceedings of the geological seminar on the North Slope of Alaska: Los Angeles, AAPG, Pacific Section Meeting, p. L1-L11.

Roeder, Dietrich, and C. G. Mull, 1978, Tectonics of Brooks Range ophiolites, Alaska: AAPG Bulletin, v. 62, n. 9, p. 1696-1702.

Sleep, N. H., 1971, Thermal effects of the formation of Atlantic continental margins by continental breakup: Geophysical Journal of the Royal Astronomical Society, v. 24,

p. 325-351.

Snelson, Sigmund, and I. L. Tailleur, 1968, Large-scale thrusting and migrating Cretaceous foredeeps in western Brooks Range and adjacent regions of northwestern Alaska (Abs.): AAPG Bulletin, v. 52, n. 3, p. 567.

Tailleur, I. L., 1969, Rifting speculation on the geology of Alaska's North Slope, part 2: Oil & Gas Journal, v. 67, n. 39, p. 128-130.

———, 1973, Probable rift origin of Canada basin, *in* Max G. Pitcher, ed., Arctic geology: AAPG Memoir 19, p. 526-535.

Van Eysinga, F. W. B., 1978, Geological time table, third edition: Amsterdam, Elsevier.

Young, F. G., D. W. Myhr, and C. J. Yorath, 1976, Geology of the Beaufort-Mackenzie basin: Geological Survey of Canada, Paper 76-11, 65 p.

Geodynamics of the Transitional Zone from the Moma Rift to the Gakkel Ridge

A. F. Grachev
Institute of Physics of the Earth
Academy of Sciences
Moscow, USSR

The Moma Rift, the Laptev Sea rift zone, and the mid-oceanic Gakkel Ridge form the Arctic rift system. It took shape about 50 m.y. ago due to opening of the Eurasian Basin with respect to a rotation pole in the southern Moma rift (Northeastern Asia). A thinned crust, anomalous mantle, local seismicity, and linear grabens characterize the present structure of the Laptev sea floor. These features were caused by the invasion of the mid-oceanic Gakkel Ridge onto a shelf area. The southernmost Arctic rift system (i.e. the Moma Rift) developed as a superimposed tension structure within a Northeastern Asia orogenic zone.

The Chersky mountain system where the Moma Rift is situated was discovered in 1926 by S. V. Obruchev. In the 1930s, D. I. Mushketov noted a seismic belt in the Arctic Ocean, and 30 years later it became evident that this belt was related to a mid-oceanic ridge (now the Gakkel Ridge). Discovering the Gakkel Ridge led to the question about its onshore continuation like some other known mid-oceanic ridges.

Heezen and Ewing (1961) thought that Pri-Verkhoyansk foredeep was a continental extension of the same kind. The author (Grachev et al, 1967) concluded that the Moma trough is the most probable continuation of the Gakkel Ridge deep structure. Special field studies (Grachev, Karasik, and Puminov, 1970; Grachev, 1973, 1977) confirmed the rift nature of the Moma trough and allowed the Moma rift zone to be distinguished. The latter, together with the Gakkel Ridge, form the Arctic rift system (Grachev, 1975, 1977). Later, rifting in the Chersky mountain system in northeastern Asia was widely appraised in geological and geophysical literature (Churkin, 1972; Patyk-Kara and Grishin, 1972; Conant, 1973; Minster et al, 1974).

The Moma Rift problem provoked great interest in the search for the boundary between the Eurasian and North American lithosphere plates. The determination of the poles of relative movements of the Eurasian and North American plates showed that the Moma Rift and the Eurasian Basin of the Arctic Ocean form a single zone of the lithosphere extension (Grachev and Karasik, 1974; Minster et al, 1974; Karasik and Grachev, 1977). Considering this, discussing the transitional zone from the Moma Rift to the Gakkel Ridge is particularly interesting. Thus, it is necessary to consider the major features of recent structure of the Moma Rift, the Laptev Sea floor, the adjacent continental rise, and the southeastern edge of the Gakkel Ridge.

THE MOMA RIFT

The Moma Rift is situated within the Mesozoic fold belt in the northeastern USSR. It stretches northwest from the Seimchan-Byiandyn depression in the south to the Laptev Sea in the north, and it is 1200 km long and 100 km wide. The Moma Rift is represented by a system of troughs which consist of two major linear depressions with surrounding uplifts, such as the Chersky mountain system in the southwest, and the Moma, Andrei-Tag, and Selennyakh ridges in the northeast. The highest block uplift does not exceed 1600 m and subsidence in a rift valley is more than

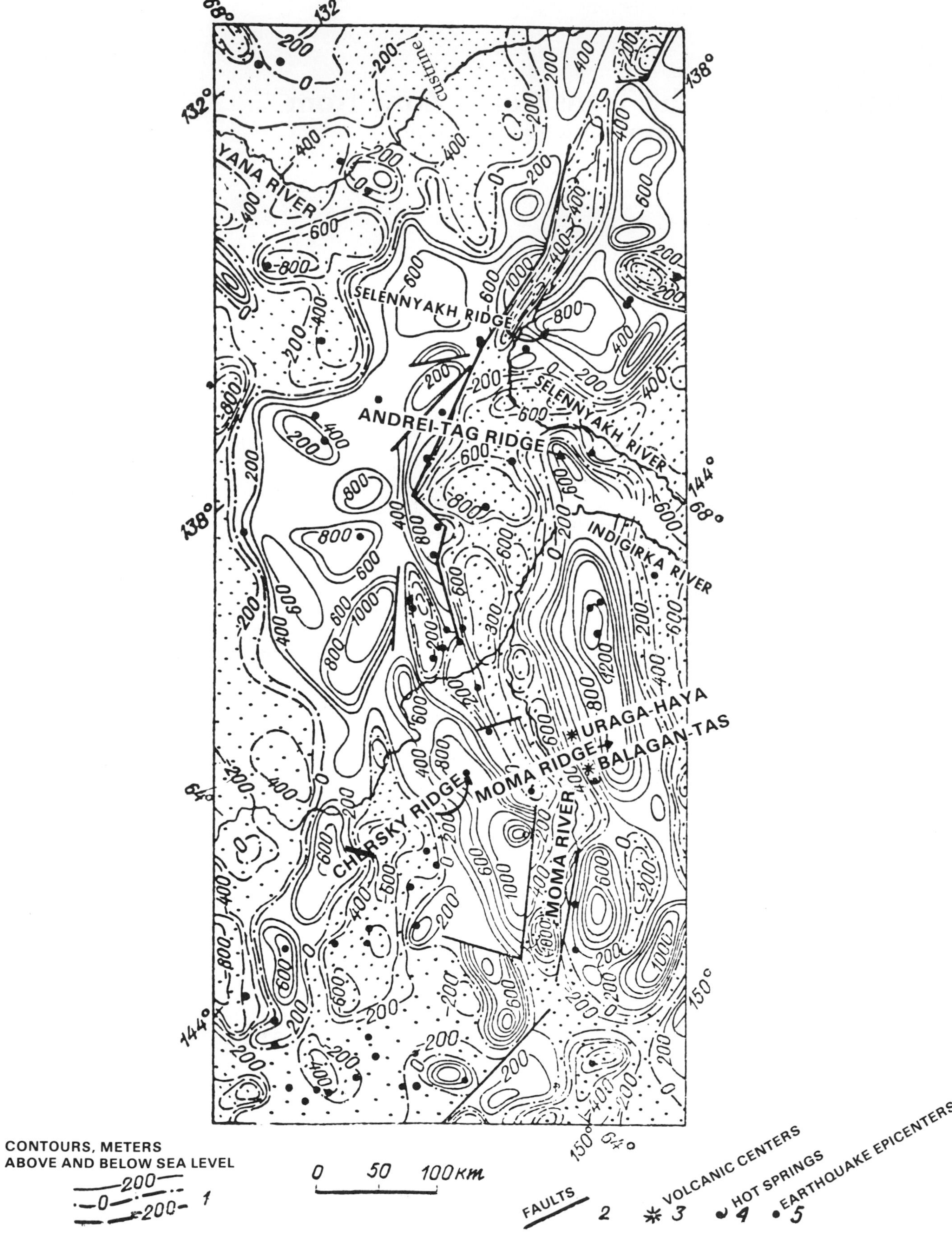

Figure 1 — Map of a recent structure of the Moma continental rift (Grachev, 1973).

800 m (Figure 1).

Two stages are recognized in the history of the Moma Rift. The first (Pliocene to Early Quaternary) is characterized by the accumulation of thick, well-sorted alluvial-lacustrine deposits. Coarse clastics (molasse) resting unconformably on the lower unit were deposited owing to the growth of mountain ridges during the second stage.

The boundary between the rift valley and Chersky Range is associated with high-angle normal faults. Vertical displacement is about 2 km while a gentle flexure of strongly faulted Mesozoic folded basement is observed in the junction of the Moma Ridge and rift valley. Detailed field studies of rocks adjoining this zone revealed a regular system of northwestern striking open fractures (to 1 m wide) accompanied by hot springs.

Along this tension zone a recent volcanism has developed, marked by two Late Quaternary volcanoes: Balagan-Tas cinder cone and Uraga-Chaya rhyolitic dome.

Chemical composition of the Balagan-Tas basalts shows a weak dispersion of all oxides and the presence of normative olivine and nepheline. These basalts are alkaline and similar to volcanics of continental rifts such as Baikal and East African rifts (Grachev, 1977).

The Uraga-Chaya rhyolitic dome is 30 km north of the Balagan-Tas Volcano. The dome is the first proved manifestation of recent acid volcanism in northeastern Eurasia.

The Moma Rift is a zone of high seismic activity. Usually, seismic stations record shallow earthquakes (less than 20 km deep) with magnitudes of 4 to 4.5 (Kozmin, 1975). Evidence for ancient catastrophic earthquakes has been found in the region.

The recent structure of the Moma Rift in the southeast is highly inherited from Mesozoic basement structure, while the Mesozoic framework changes in the northwestern part. A superimposed pattern of recent structures is easily established in the Polousny Range (Patyk-Kara and Grishin, 1972) where the mid-Quaternary West Polousny fault system stretches into a spreading structure of the Laptev Sea floor (Figure 2).

STRUCTURE OF THE TRANSITIONAL ZONE

In this paper the transitional zone is the entire Laptev Sea continental margin to the northwest end of the Moma Rift.

The earthquake focal zone, which joined the Moma Rift at the latitude of the upper Yana River, is subject to sharp transverse displacement towards the Lena River delta, where it merges with a seismically active zone passing from the Eurasian Basin through the Laptev Sea (Figure 2). Avetisov and Golubkov (1971) first pointed out this transverse zone, which has no particular features except for earthquake epicentres. We confirm that this zone is an early transform fault (Grachev, 1973).

Trend analysis has been applied to studies of particular features of the Laptev Sea floor (Grachev and Mishin, 1975). We draw attention to the trend map (Figure 3a) where the isoline pattern of depths (in meters) shows two distinct directions: northeastern isolines passing from the Ust'-Chatanga flexure and northwestern isolines passing through the Ust'-Lena flexure. These seem to divide the Laptev Sea basin into four segments. These two structural lines are confirmed by the rose diagram of the Siberian platform fault strikes (Mezhvilk and Murzina, 1972).

The trend map enables us to recognize the main regular features of the Laptev Sea floor structure; moreover, the map of trend bias from a real surface, showing a local component, may be used as a base for structural subdivisions (Grachev and Mishin, 1975; Figures 3b). East-Laptev, Ust'-Yana, and South-Laptev uplifts are examples of positive deviations; Ust'-Lena, West-Laptev, and Belkovsky flexures are examples of negative ones.

Figure 4 shows the recent structure of the Laptev Sea floor (Grachev, 1973) with values for uplifts and subsidences (shown without correction for thickness of Neogene-Quaternary sediments), major faults, local structural features, and earthquake epicentres. Here, a zero isoline separates areas of uplift and subsidence.

The northernmost isoline within the Laptev Seafloor outlines the continental slope and directly joins it. To the south it goes into a subsidence area, where periclines of the Taimyr mountain-building area and Novosibyrskaya uplift are situated to the east and west. Southward there are undifferentiated South-Laptev and strongly differentiated Ust'-Yana uplifts.

A zone of northwest movements, represented by a system of linear depressions assigned to the Ust'-Lena subsidence zone, occurs against this background. The northwest end can be traced toward the continental slope, and at 76°N it joins the northeast striking subsidence zone (a part of the continental slope structure) where the buried extension of the Gakkel Ridge is determined by geophysical data and trend analysis. Thus, the zone of the most distinct tectonic movements of the Laptev Sea is an extension of the Eurasian Basin structure. These movements are characterized by the same axis of symmetry. We can obtain approximate outlines of recent distinct movements in this zone if we project southward the boundary trend of the Eurasian Basin continental slope.

Valid evidence supporting the structural position of this zone is obtained by defining a pole of rotation with respect to the opening of the Eurasian Basin. The coordinates of this pole allow us to locate the above zone in the sector.

Seismicity is the last factor which points to a solitary position exposed in the shelf structure. The distribution of earthquakes (Figure 4) implies their complete location within this zone. Note that accuracy of epicentre definition does not exceed 50 to 70 km.

At present a correct determination of focal mechanism has been obtained for only a single earthquake with an epicentre near the continental slope. This earthquake (γ = 78.2° N, λ = 126.7° E) occurred on August 25, 1964; it had a magnitude of 6.25 and a depth of hypocenter of about 30 km. The axis of ten-

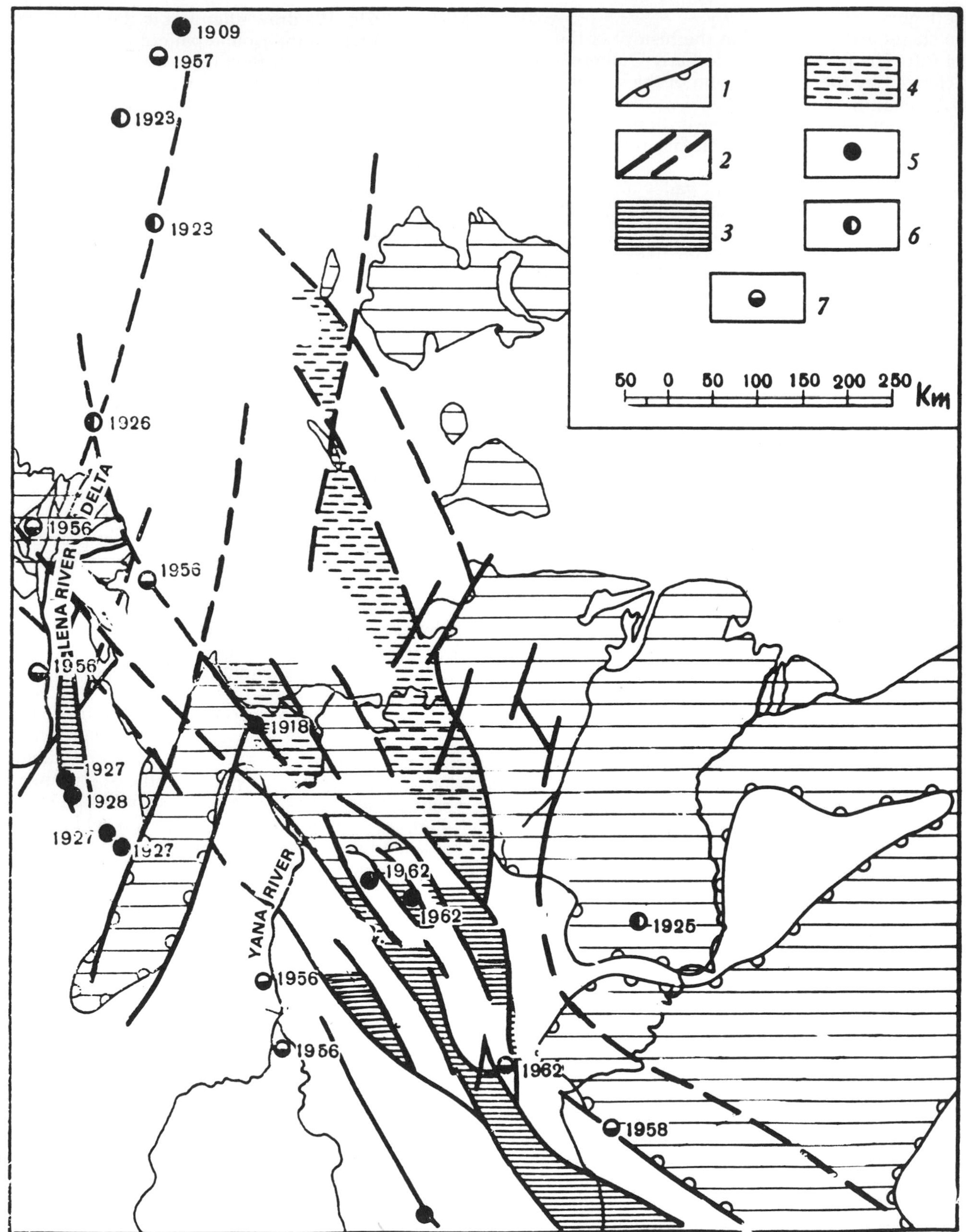

Figure 2 — Diagram showing West-Poluosny fault system. 1 — Primoria and Kolyma lowlands; 2 — fractures manifested during the recent tectonic stage, determined and inferred; 3 — basins; 4 — submerged blocks within Primoria lowland and shelf determined by geophysical data; earthquake epicentres and the year of earthquake; 5 — M=6; 6 — M=5 to 6; 7 — M=4 to 5 (Patyk-Kara and Grishin, 1972).

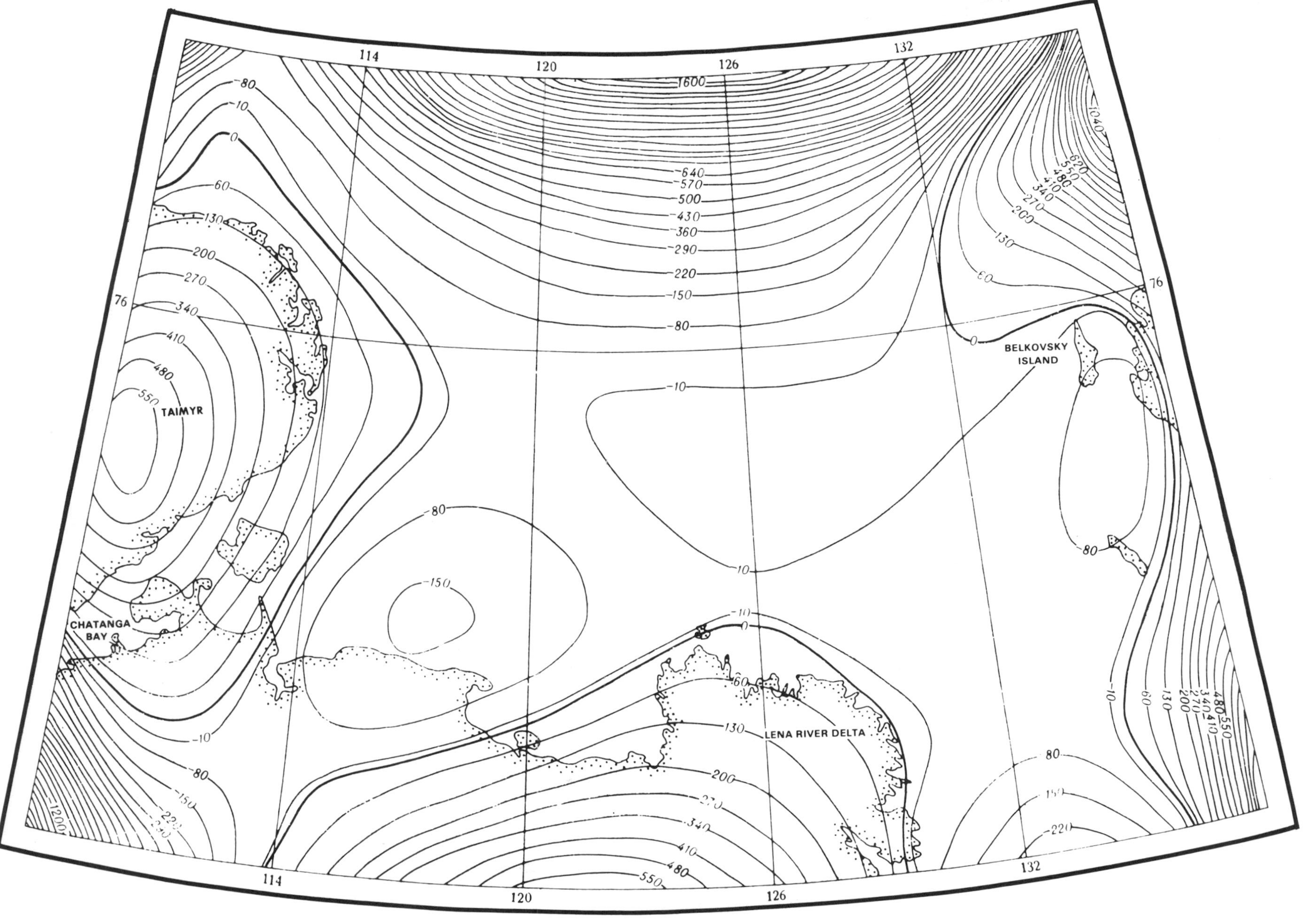

Figure 3a — Map of a trend surface for depths of the Laptev Sea floor (Grachev and Mishin, 1975).

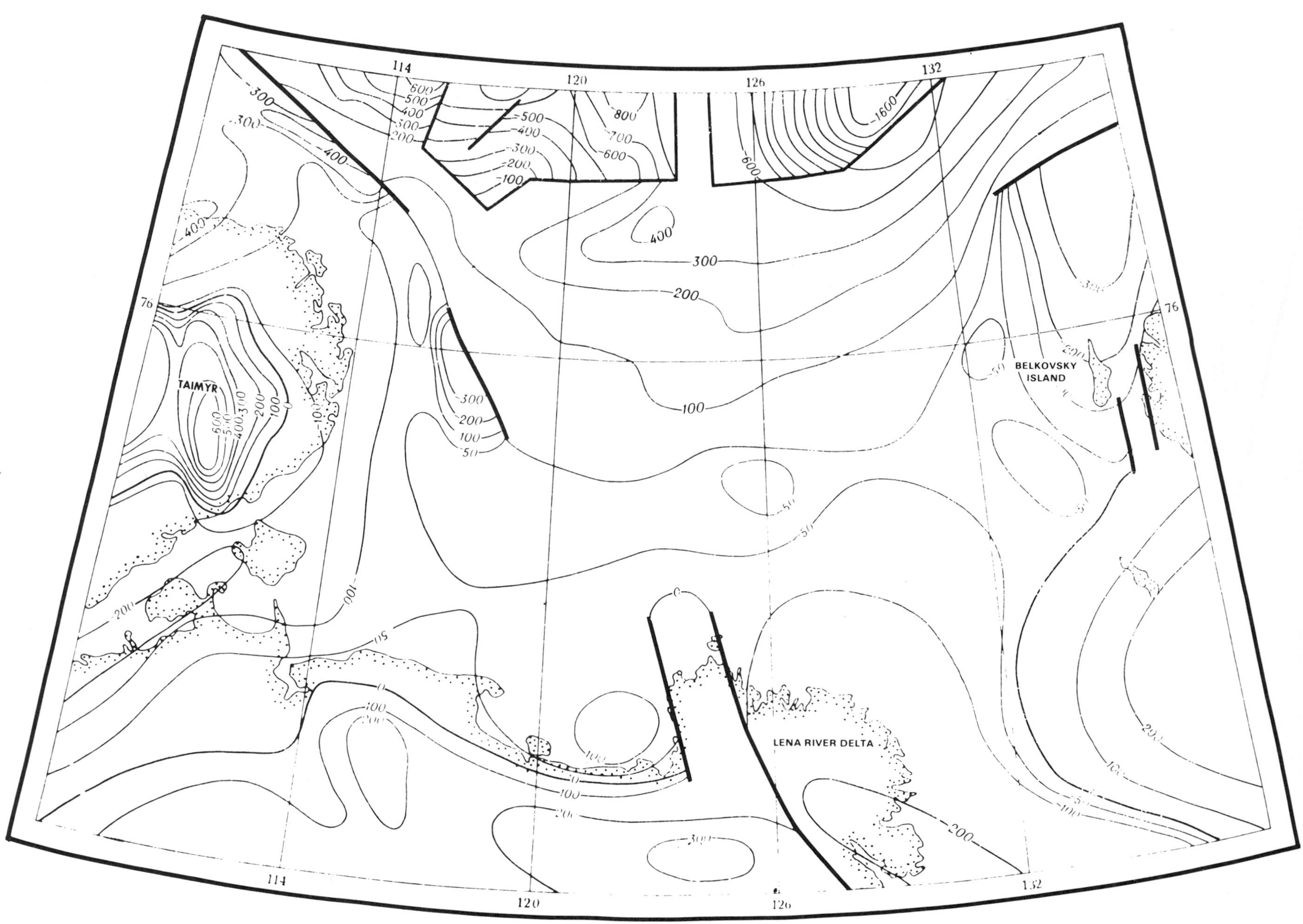

Figure 3b — Map of trend surface deviations from a real surface for the Laptev Sea (Grachev and Mishin, 1975).

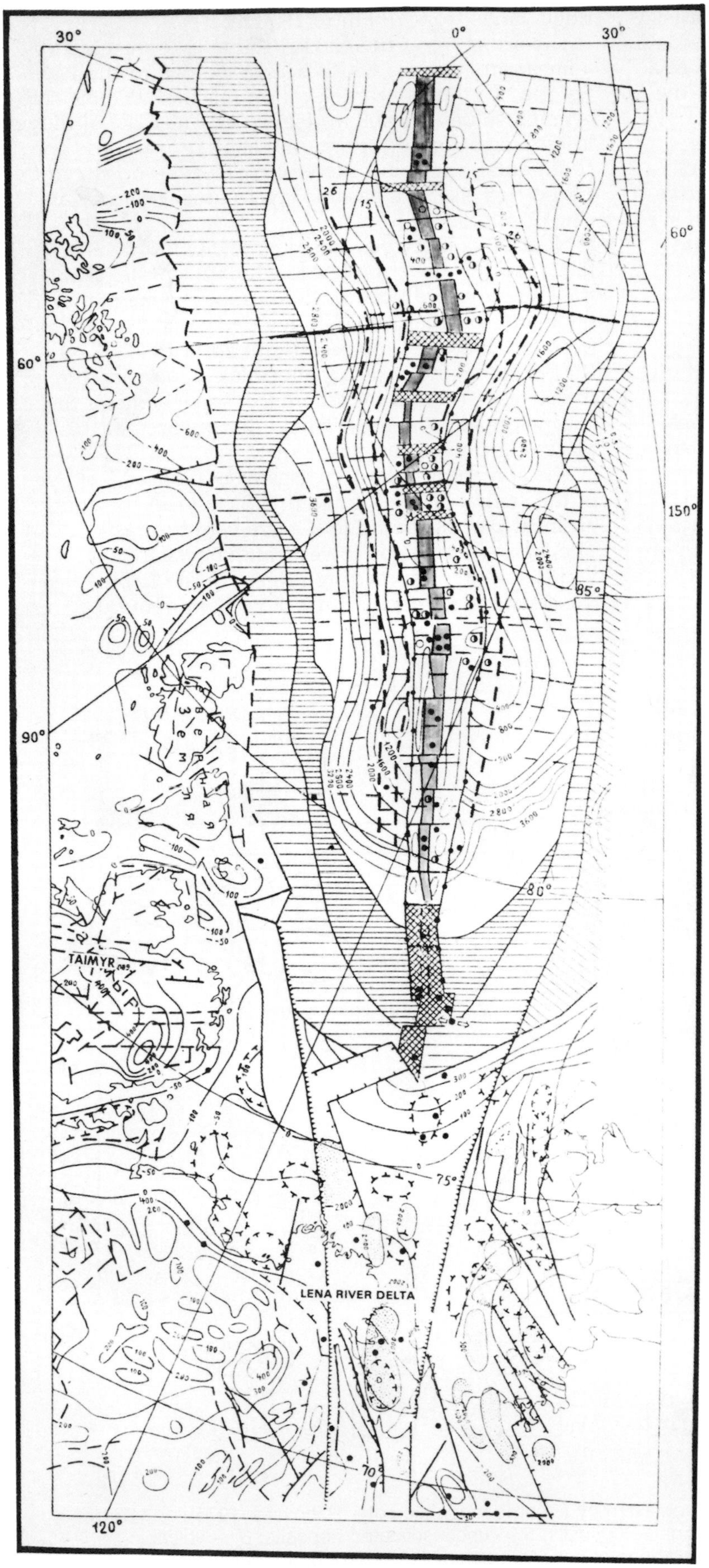

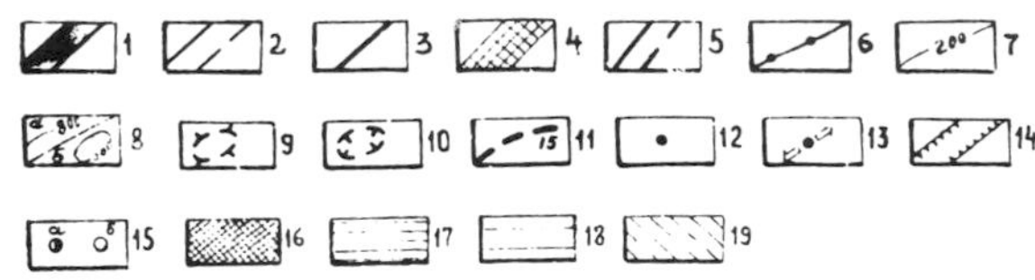

Figure 4 — Map of recent tectonics of the Eurasian basin and its margin (Grachev and Karasik, 1974). 1 — rift valley of the Gakkel Ridge; 2 — transform fault on oceanic crust; 3 — transform fault separating areas of a single magnetic province; 4 — zones of transform faults; 5 — faults on the continental crust; 6 — boundary of the Gakkel Ridge mid-oceanic crust; 7 — isobaths of recent tectonic movements; 8a — isopachs of sedimentary cover for oceanic plates; 8b — isopachs of sedimentary cover for the Laptev Sea shelf; 9 — local recent uplifts; 10 — Local recent subsidence; 11 — isochrons of oceanic crust (m.y.); 12 — earthquake epicentres; 13 — stress pattern in earthquake foci; 14 — area of embryonic tension of the Earth's crust; 15a — concordant submarine volcanoes; 15b — discordant submarine volcanoes; 16 — extension of the Gakkel Ridge spreading structure not reflected in topography; 17 — foredeep compensated by sedimentation; 18 — transition zone; 19 — displaced and submerged parts of the Barents-Kara Sea shelf (Lomonosov Ridge).

sion is horizontal, the direction is approximately east to west (azimuth 80°), and the axis of compression is subvertical. So seismic data confirm the existence of a stress field which is characteristic of rift zones on the continental slope (Sykes, 1965).

If we consider that the discussed zone of movements passes into an area of active young tectonic movements and seismicity of the Primoria lowland and northern part of the Chersky Ridge mountain system, then a connection between the Moma Rift and the Gakkel Ridge rift zone becomes quite evident.

MID-OCEANIC GAKKEL RIDGE

The Gakkel Ridge is the northernmost fragment of the global system of oceanic ridges and it is a direct extension (through the Mona and Knipovich Ridges) of Mid-North-Atlantic Ridge. It divides the Eurasian Basin into two equal parts, and the coincidence with the median line is better than for the other known ridges (Grachev, Karasik, and Puminov, 1970; Grachev and Naryshkin, 1978). The Gakkel Ridge extends more than 1700 km up to 79° N and in the southwest is slightly reflected in topography. The width of the ridge varies from 80 km to 160 km.

The structure of the Gakkel Ridge is represented by a linear arch whose width decreases toward the Laptev Sea. The maximum elevation reaches 1000 m, and average values are 400 m (Figure 4). The Gakkel Ridge loses expression in topography as it approaches the continental slope, but the structure of the ridge can be traced as a rise of a basaltic layer (Grachev and Karasik, 1974).

Near the Laptev Sea the magnetic field intensity falls, and the structure becomes more simple. Due to the anomaly conflucence, the number of magnetic onomalies on the Gakkel Ridge decreases from 47 to 6 (Karasik, 1980). At the southern end of the Gakkel Ridge the magnetic field pattern is characterized by the absence of polarity difference; so all signs of the ocean-spreading magnetic field disappeared at this place (Figure 5; Rzhevsky, 1975).

The spreading rate decreases from 0.6 to 0.7 cm/yr in the central Gakkel Ridge to 0.3 cm/yr in the southernmost part of the ridge (Grachev and Karasik, 1974). The latter value is the lowest rate known for the world system of spreading ridges. In the Laptev Sea the separation velocity must have been about 0.1 cm/

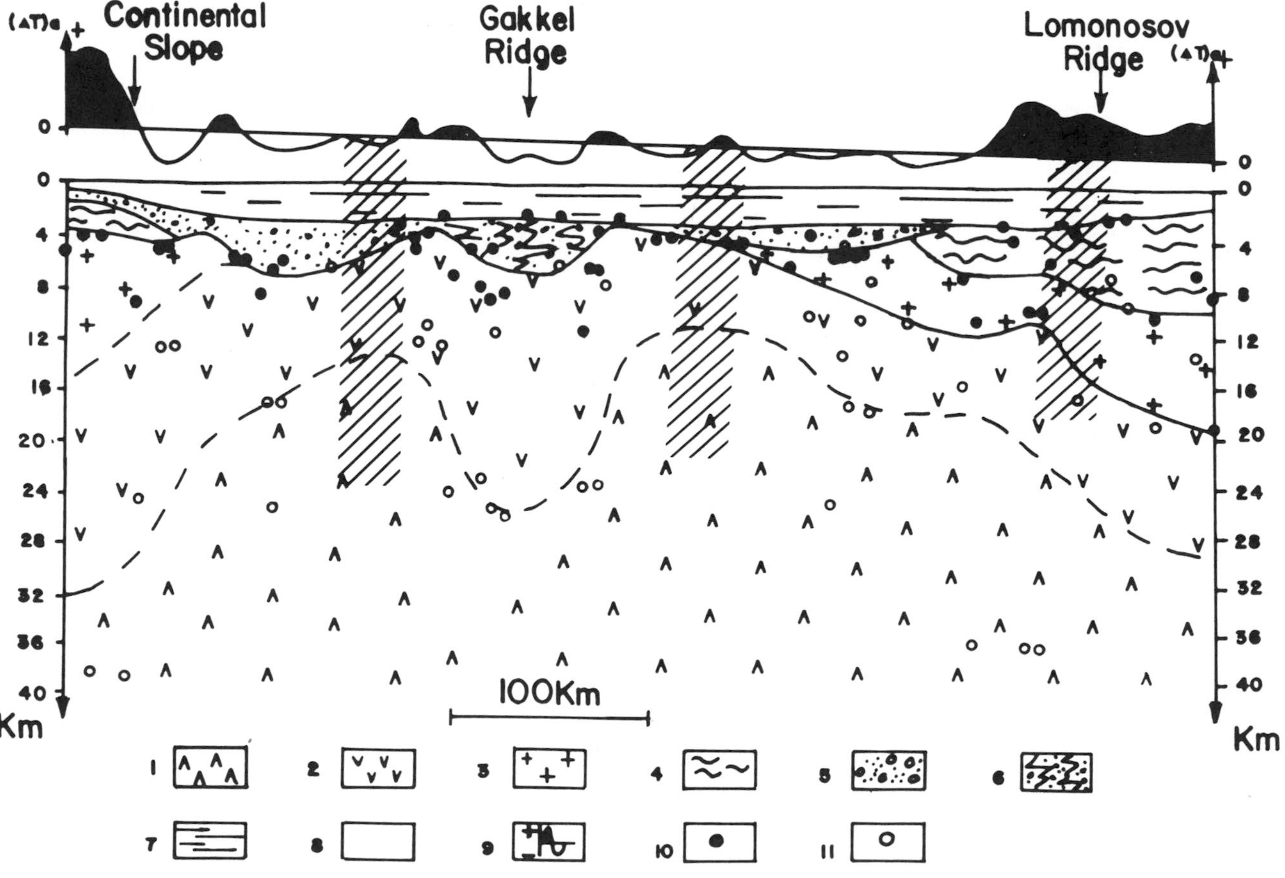

Figure 5 — Crustal section of the southern Gakkel Ridge obtained by airborne magnetic data (Rzhevsky, 1975). 1 — upper mantle; 2 — basaltic layer; 3 — granitic layer; 4 — consolidated deposits; 5 — unconsolidated deposits; 6 — volcanic-sedimentary strata of a rift zone; 7 — water; 8 — weakened zone of uncertain genesis; 9 — plot of anomalous magnetic field Δ T_a; 10 — depth of upper edges of magnetic bodies; 11 — depth of lower edges of magnetic bodies.

yr. Thus, sea-floor spreading on the southern Gakkel Ridge is followed by continental lithospheric tension in the Laptev Sea, resulting in crustal thinning and neck formation (Grachev and Karasik, 1974).

Deep seismic sounding along the profile from the Lena River mouth to the Yana mouth through the northern tip of Buor-Haya cape proves the existence of a neck. (Kogan, 1974) (Figure 6). This study showed the presence of a suboceanic crust far from the continental slope. The peculiar feature of the crustal section is that the granite layer is practically absent (1.0 to 2.5 km) while the thickness of a basaltic layer underlain by low velocity mantle (Vp — 7.5 km/s) is high (19 to 20 km). It is significant that a series of great faults, with a high amplitude of vertical displacements, was observed in a deep seismic sounding zone (Kogan, 1974).

Results of deep seismic sounding and the above mentioned data on the Laptev Sea imply the existence of a zone with deep structural changes in the continental lithosphere, caused by tension processes due to extension of the Gakkel Ridge structure into this area. Nevertheless, changes in the crustal structure of the Laptev Sea floor have not yet achieved continental lithosphere breakup. Considering that the Eurasian Basin, the Laptev Sea rift zone, and the Moma Rift are developing relative to the same pole of rotation (Figure 7) we can propose that the velocities of horizontal movements within each of three segments are different. The difference between 0.3 cm/yr in the southern Eurasian Basin and 0.1 cm/yr in the Moma Rift is not obvious. We may attribute this to proximity of the pole of separation.

CONCLUSION

At present we recognize three types of junction areas between the mid-oceanic ridges and continental structures. The first, Afar type, is characterized by a direct transition of the Sheba Ridge through a system of the transform faults into the Afar triangle. The second type, Californian, is where the northern end of the East Pacific Rise through the San-Andreas fault joins with the system of mid-oceanic ridges (Gorda, Juan de Fuca, and Explorer). In the third type, Arctic

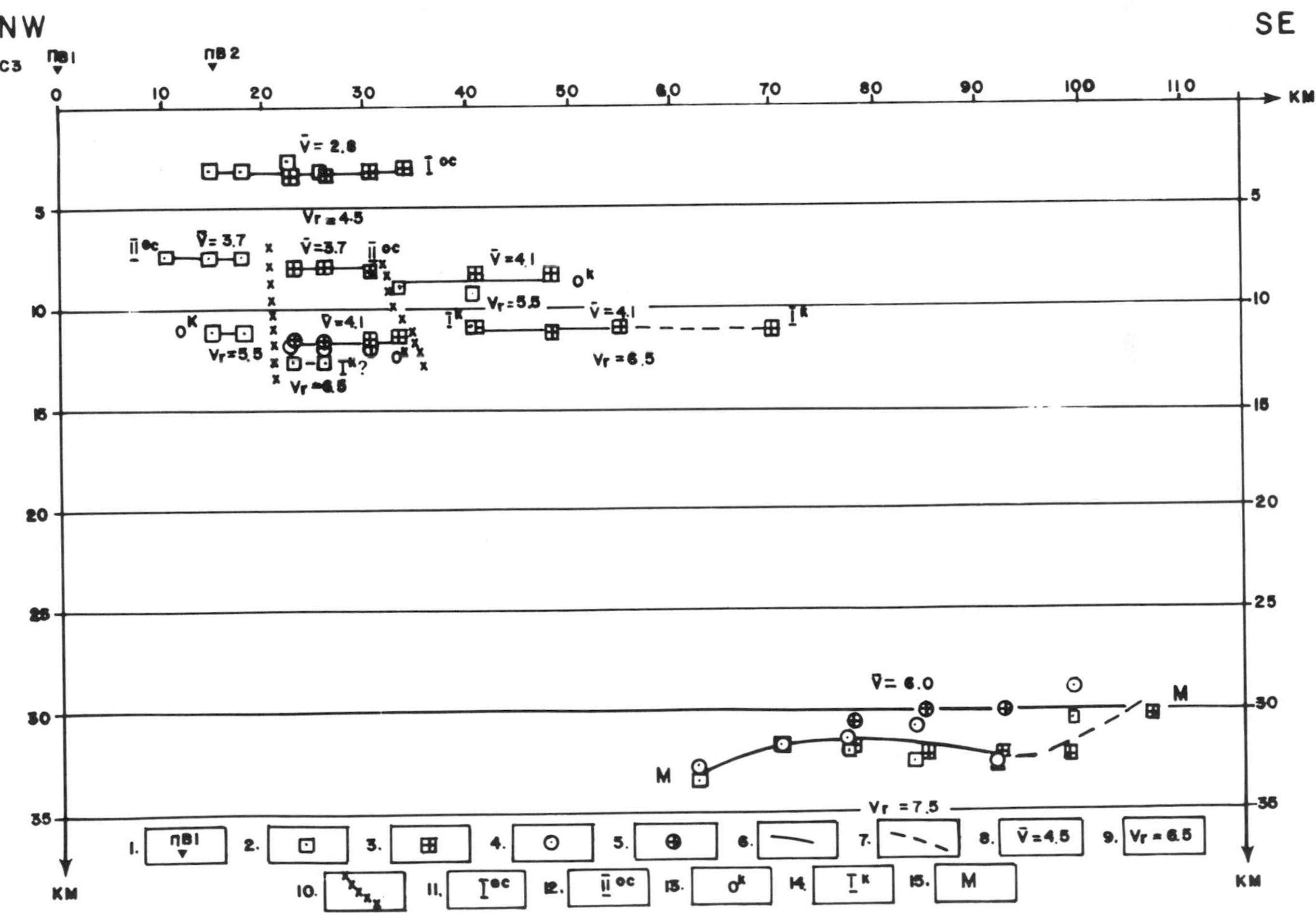

Figure 6 — Seismic profile across the southern Laptev Sea from the Lena mouth to the Yana mouth (Kogan, 1974). 1 — explosion points, 2 and 3 — depths determined by refraction; 4 and 5 — depths determined by reflections; 6 — seismic boundaries plotted by means of uncertain data; 8 and 9 — average (effective) and boundary velocities, km/s; 10 — fracture zones; 11 — seismic boundaries top of crystalline basement; 12 — seismic boundaries situated within a folded basement section; 13 — top of folded basement; 14 — approximately corresponding to the top of a basaltic layer; 15 — M-discontinuity.

Rift system, the junction is quite different.

The junction area of the Gakkel Ridge and the Laptev Sea is a transverse 'butt-end' type and it provides a unique example of direct transition of sea-floor spreading to continental rifting. In this case, there is no transform fault but wrench faulting begins to develop.

The uniqueness of this behavior lies in the decrease of separation rate along the axis of opening. This phenomenon can be attributed to the proximity of the pole of rotation between Eurasian and North American lithosphere plates. It is a single region in the Earth where the transitional zone reflects an intermediate stage of kinematics of moving lithosphere plates.

It is important to note that the pole of rotation of the Eurasian Basin changed its position from the Laptev sea area in Paleogene, to the modern site behind the Moma continental rift.

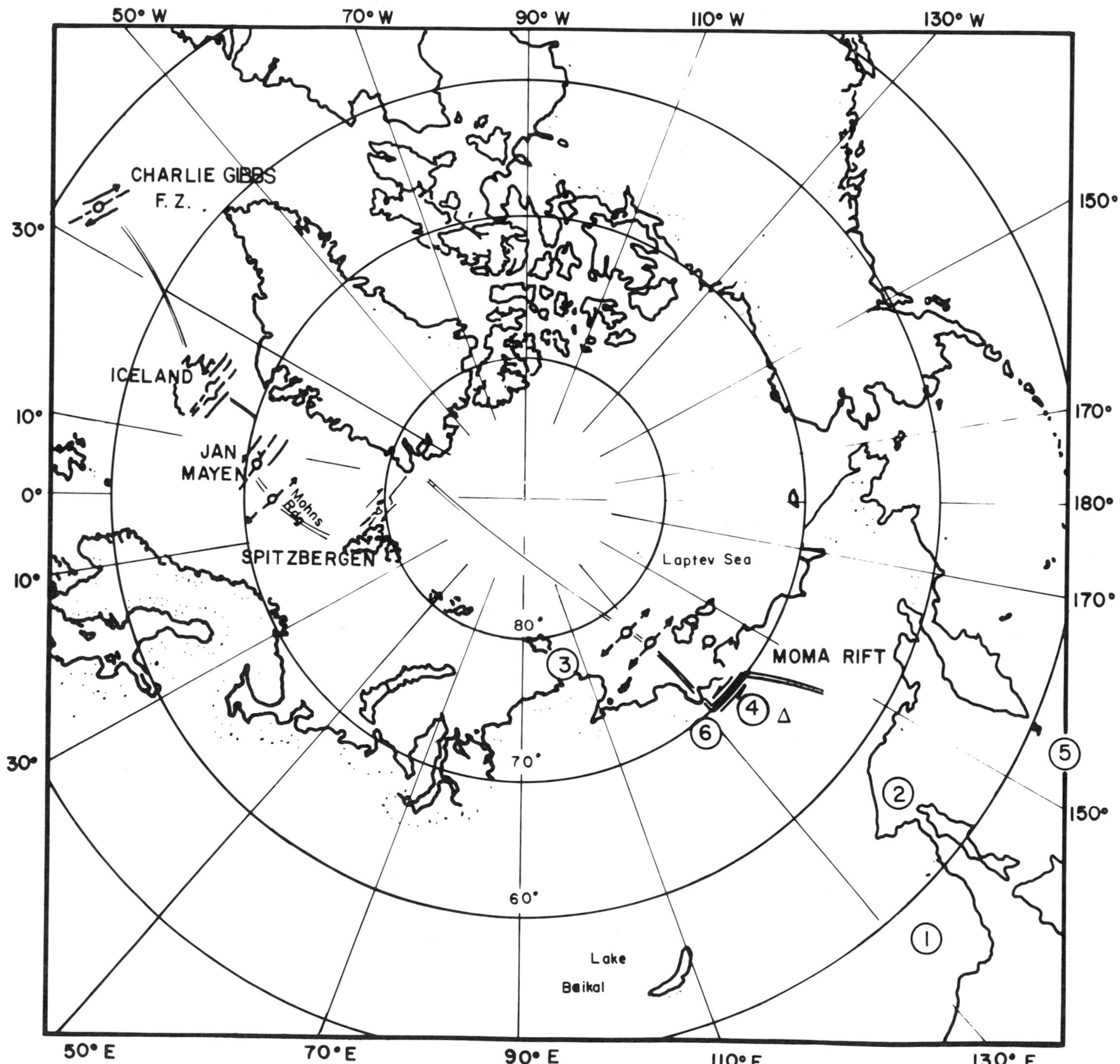

Figure 7 — Position of opening poles of the Eurasian Basin. 1 — McKenzie and Parker (1967); 2 — Pitman and Talwani (1972); 3 — LePichon (1968); 4 — Pitman and Talwani (1972); 5 — Chase (1972); 6 — Minster et al (1974); Δ — Karasik (1977).

REFERENCES CITED

Avetisov, G. P., and V. S. Golubkov, 1971, Tectono-seismic zonation of the Arctic basin and adjacent regions, *in* Geology and mineral resources of the northern Siberian platform, n. 8: Leningrad, Nauchno-Issledovatel'skiy Institut Geologii Arktiki, Trudy, p. 25-30 (in Russian).

Chase, C. G., 1972, The N plate problem of plate tectonics: Oxford, Geophysical Journal of the Royal Astronomical Society, v. 29, p. 117-122.

Churkin, M., Jr., 1972, Western boundary of the North American continental plate in Asia: Geological Society of America, Bulletin, v. 83, n. 4, p 1027-1036.

Conant, D. A., 1973, Six new focal mechanism solutions for the Arctic and a center of rotation for plate movements: Seismological Society of America Bulletin, v. 63, p. 1120-1131.

Grachev, A. F., 1973, The Moma continental rift, *in* Geophysical surveys in the Arctic, n. 8: Leningrad, Nauchno-Issledovatel'skiy Institut Geologii Arktiki, Trudy, p. 56-75 (in Russian).

———, 1975, Evolutionary sequence of rift zones of the Earth, in Rifting problems: Irkutsk, p. 15-16.

———, 1977, Rift zones of the Earth: Leningrad, Nedra Publishers, 247 p. (in Russian).

———, and A. M. Karasik, 1974, Sea floor spreading and tectonics of the Eurasian basin, *in* Geotectonic implications to the prospects of mineral resources on the Arctic shelf: Leningrad, Naucho-Issledovatel'skiy Institut Geologii Arktiki, Trudy, p. 19-33 (in Russian).

———, and V. I. Mishin, 1975, Construction of recent tectonics maps on the basis of trend analysis: Geomorfologia, n. 2, p. 63-70 (in Russian).

———, and G. D. Naryshkin, 1978, Main features of sea floor topography of the Eurasian basin of the Arctic Ocean: Vestnik Leningradskogo Universiteta, Geologiya, Geografiya, n. 12, p. 94-102 (in Russian).

———, A. M. Karasik, and A. P. Puminov, 1970, Explanatory notes to the map of the recent tectonics of Arctic and Subarctic: Moscow, Nedra Publishers, 39 p. (in Russian), scale 1:5,000,000.

———, et al, 1967, On the rift system in the Arctic: Uchenye zapiski Niiga, Series regionalnaya geologia, n. 10, p. 65-70 (in Russian).

Heezen, B. C., and M. Ewing, 1961, The mid-oceanic ridge and its extension through Arctic basin, *in* G. K. Raasch, ed., Geology of the Arctic: Toronto, Toronto University Press, p. 374-398.

Karasik, A. M., 1980, Main peculiar features of the history and structure of the sea floor of the Arctic basin inferred from airborne magnetic data, *in* Marine Geology, sedimentology, sedimentary petrography and geology of the ocean: Leningrad, Nedra Publishers, p. 178-193 (in Russian).

———, and A. F. Grachev, 1977, Structures on the crust of oceanic type, *in* Explanatory notes to the Tectonic Map of the North Polar region of the Earth: Leningrad, Nauchno-Issledovatel'skiy Institut Geologii Arktiki, Trudy, p. 178-193 (in Russian).

Kogan, A. L., 1974, Seismic studies using CMRW and DSS from marine ice on the Arctic sea shelf (examplified by the Laptev Sea), *in* Geophysical surveys in the Arctic, n. 9: Leningrad, Nauchno-Issledovatel'skiy Institut Geologii Arktiki, Trudy, p. 33-38 (in Russian).

Koz'min, B. M., 1975, Main stresses in earthquake foci of the northeast USSR, *in* New data on geology of Yakutiya: Yakutsk, p. 117-125 (in Russian).

LePichon, X., 1968, Sea-floor spreading and continental drift: Journal of Geophysical Research, v. 73, p. 3661-3698.

McKenzie, D., and R. L. Parker, 1967, The north Pacific — An example of tectonics on a sphere: Nature, v. 216, p. 1276.

Mezhvilk, A. A., and G. A. Murzina, 1972, Faults on Siberian platform, *in* Geology and oil and gas possibilities of the Soviet Arctic: Leningrad, Nauchno-Issledovatel'skiy Institut Geologii Arktiki, Trudy, p. 82-86.

Minster, I. B., et al, 1974, Numerical modelling of instantaneous plate tectonics: Oxford, Geophysical Journal of the Royal Astronomical Society, v. 36, p. 541-576.

Nikolaev, N. I., and A. A. Naimark, 1973, Recent tectonics of the northeast USSR and Kamchatka, *in* Recent tectonics, recent deposits and a man: Moscow, Moscow University, n. 5, p. 161-192 (in Russian).

Patyk-Kara, N. G., and M. A. Grishin, 1972, A location of the Polyusny Ridge in the structure of the northeast USSR and its recent tectonics: Geotektonika, n. 4, p. 90-98 (in Russian).

Pitman, W. C., III, and M. Talwani, 1972, Sea-floor spreading in the North Atlantic: Geological Society of America Bulletin, v. 83, p. 619-646.

Rzhevsky, N.N., 1975, On the question of anomalous magnetic field on the southern part of the Gakkel Ridge, *in* Tectonics of the Arctic, n. 1: Leningrad, Nauchno-Issledovatel'skiy Institut Geologii Arktiki, Trudy, p. 18-20 (in Russian).

Sykes, R. L., 1965, The seismicity of the Arctic: Seismological Society of America, Bulletin, v. 55, p. 501-517.

The Rotational Origin of the Gulf of Mexico Based on Regional Gravity Data

D. J. Hall*
T. D. Cavanaugh**
J. S. Watkins***
K. J. McMillen****
*Gulf Science and Technology Company
Pittsburgh, Pennsylvania*

Regional free-air gravity data from the Gulf of Mexico define deep-seated linear features which we interpret as outer marginal basement highs. Highs are arranged symmetrically around the deep water Gulf. Abrupt changes in trend occur along three well-defined zones on both sides of the central Gulf. The marginal high off Galveston parallels the Cretaceous Edwards Reef trend 230 km to the northwest. We interpret the seaward limits of these outer marginal highs as close to the landward edges of oceanic crust. The crust in the area from the continental hinge zone (near which the Edwards Reef developed) to the outer high is thinned, faulted, and intruded by mafic dikes, but probably has a nearly continental overall composition.

We infer that the Gulf opening followed a pattern of early rifting and subsequent sea-floor spreading. Our model implies that the thick salt deposits underlying the modern Texas slope were deposited on sediments overlying oceanic crust. Thinner salt deposits overlie pre-salt sediments on rift-stage crust both northwest of the Texas outer marginal high and in the Sigsbee knolls. A change in the location of the sea-floor spreading center led to separation of the two main salt depo-centers beneath the Sigsbee Knolls and along the Texas-Louisiana shelf and slope. Recent paleomagnetic evidence (L. Sanchez Barreda, personal communication, 1981) indicates a post-Permian, 24° clockwise rotation of Chiapas relative to Oaxaca (this is consistent with our rotational model). The Salina Cruz fault, crossing the Isthmus of Tehuantepec between the Permian outcrops in Chiapas and Oaxaca, was probably a major transform fault active during the Gulf opening.

"The problem of the origin of the Gulf thus reduces primarily to the question of how the Yucatan block moved away from Texas and mainland Mexico," Dickinson and Coney (1980).

". . . the most striking thing about the proceedings is what is left out rather than what is included. There is no discussion of the nature of rifting prior to continental decoupling, and only in passing is there any discussion of the distribution of oceanic crust. Information on both of these subjects is needed to reconstruct the geologic history of the Gulf of Mexico," Uchupi (1980).

The long-standing problem of the Gulf of Mexico's origin depends critically on present distribution of oceanic crust and the history of rifting and subsequent drifting of the adjacent continental land masses. Buffler et al (1980) adapt the general outlines of the passive margin model of Scheupbach and Vail (1980) to the Gulf, but only schematically indicated the present distribution of oceanic crust. Scheupbach and Vail (1980) suggest that outer marginal basement highs can occur near the edges of oceanic crust, and that free-air gravity data can be used to map them.

To test formation models and early development of the Gulf of Mexico, the continental margins group at Gulf Science and Technology Company has studied and integrated data from the R/V GULFREX and R/V HOLLIS HEDBERG, gravity and aeromagnetic surveys, published literature and contractor data. Free-air gravity data discussed in this paper are important regionally. When interpreted in the light of current

*Presently with: Zenith Petroleum, Houston, Texas.
**Presently with: Champlin Petroleum, Oklahoma City, Oklahoma.
***Presently with: Gulf Oil Exploration and Production, Houston, Texas.
****Presently with: Gulf Oil Exploration and Production, Bakersfield, California.

models of passive continental margin formation and development (e.g. Scheupbach and Vail, 1980), these data suggest a rotational origin for the deep Gulf of Mexico basin and surrounding continental margins.

GEOLOGICAL BACKGROUND

Distribution of Oceanic Crust in the Gulf of Mexico

The 1980 Symposium at Louisiana State University (Pilger, 1980) achieved general consensus that oceanic crust underlies at least part of the central Gulf of Mexico basin. However, the exact distribution of oceanic crust remained a critical element in the problem of the Gulf's origin (Uchupi, 1980). Four recent interpretations are diagrammatically shown in Figure 1.

Salvador and Green (1980) developed a Kimmeridgian reconstruction (Figure 1a) from analysis of stratigraphic data surrounding the Gulf. The transform fault along the west Florida shelf edge and the relatively narrow strip of oceanic crust in the central Gulf basin are key features. The Yucatan block is high-standing and underlain by continental crust. The true boundary of the block extends far north of the Campeche Scarp, in agreement with the results of OBS refraction surveys carried out at the University of Texas Marine Science Institute (Buffler et al, 1980; Ibrahim et al, 1981; and Ibrahim and Uchupi, this volume). In this reconstruction (Figure 1a), oceanic crust is less than about 200 km wide.

White (1980) interpreted the limits of Pre-Mesozoic continental crust as more than 1000 km apart across the central Gulf (Figure 1c). Oceanic or greatly attenuated continental crust underlies the intervening areas, with the exception of the Wiggins Arch. The crustal margins are drawn based on the location of

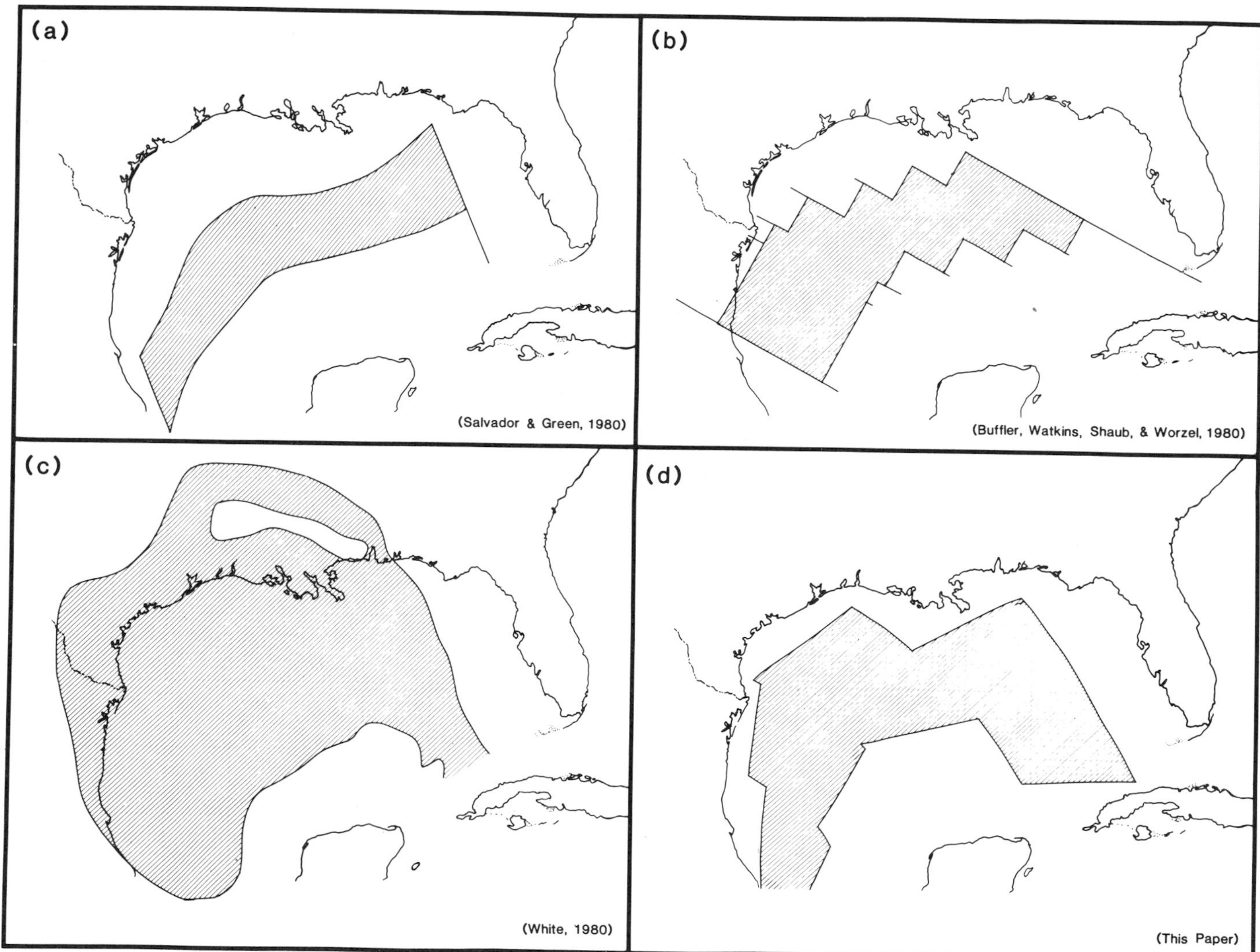

Figure 1 — Distribution of crustal types around the Gulf of Mexico according to four different recent interpretations.
a. Kimmeridgian reconstruction after Salvador and Green (1980). Ruled area is oceanic crust.
b. Diagrammatic Late Jurassic oceanic spreading pattern after Buffler et al (1980). Ruled area is oceanic crust. Surrounding area is largely transitional crust.
c. Limits of pre-Mesozoic continental crust after White (1980). Ruled area is oceanic or greatly attenuated continental crust. White area in Louisiana is the Wiggins Arch.
d. Outline of oceanic crust based on interpretation of free-air gravity data in this paper (Figure 4).

Jurassic evaporites. Two main centers of evaporite deposition, the Sigsbee-Campeche Knolls and the Gulf Coast Salt Dome Province, overlie oceanic or attenuated continental crust.

Buffler et al (1980) inferred plate boundaries and spreading directions in the Late Jurassic along transforms roughly parallel with fracture zones in the North Atlantic and reported shear zones in Mexico (Figure 1b). Although their diagram is only schematic, on a regional basis it explains the symmetry in salt distribution and crustal type across the deep Gulf.

Here, we apply regional free-air gravity data to delineate the boundaries of oceanic crust (Figure 1d). In our interpretation, basement features, which are the sources for linear free-air gravity highs, arranged symmetrically around the Gulf mark the edges of oceanic crust. The distribution of these basement features leads to a rotational model for the Gulf's origin.

Outer Marginal Basement Highs

Outer marginal basement highs characterize the edge of oceanic crust on many rifted continental margins (Scheupbach and Vail, 1980). Typically, the highs are relatively narrow (10 to 50 km) features persisting for hundreds of kilometers, roughly parallel with modern shorelines and often offset by zones that may mark changes in trend. The contact with oceanic crust is abrupt and can be mapped with regional geophysical data. The highs form during the late rift phase of divergent margin development, possibly as the result of a poorly understood thermal-mechanical event immediately preceding the formation of oceanic crust (Scheupbach and Vail, 1980).

The highs themselves are rift-stage or transitional crust located immediately landward of oceanic crust. True continental crust, characterized by thicknesses of about 35 kilometers and compositions roughly equivalent to average granodiorite, typically lies well inland (200 to 300 km) of outer marginal highs. A reasonably sharp contact between continental and rift-stage crust often occurs as a major hinge zone in the basement (Watts and Steckler, 1981). Under normal conditions, the two contacts (continental with rift-stage crust and rift-stage with oceanic crust) are roughly parallel and mappable. We interpret the intervening zone as faulted, thinned, metamorphosed, and dike-injected transitional basement with the imprint of the rifting stage well preserved.

Outer highs can be located with regional geophysical data. Their location provides information useful in the reconstruction of former plate positions and the history of sea-floor spreading. Because these features are linear and extend over long distances, their identification on seismic dip sections and their extension between seismic lines using free-air gravity are straightforward. For example, Scheupbach and Vail (1980) used isostatic gravity anomalies to show the signature of the outer marginal high offshore part of West Africa (Figure 2). In this case, the anomaly extends over 15° of latitude and has an amplitude of about 20 milligals. We used free-air gravity anomalies supplemented by magnetic anomalies, but the results are essentially the same.

DATA

Gravity Map

The regional gravity map (Figure 3) was largely compiled from R/V GULFREX free-air anomaly data. There are almost 20,000 km of marine data included on the map. The coverage is good for the deep Gulf of Mexico and extremely good on the northern shelf and the southern two-thirds of the Florida shelf. The biggest problem was the number of misties which are generally of the order of ±5 milligals, but can be as high as 40 milligals. Some data were discarded, but where it seemed appropriate, linear corrections were made by using the gravity from intersecting lines as control values. The contour interval on the map

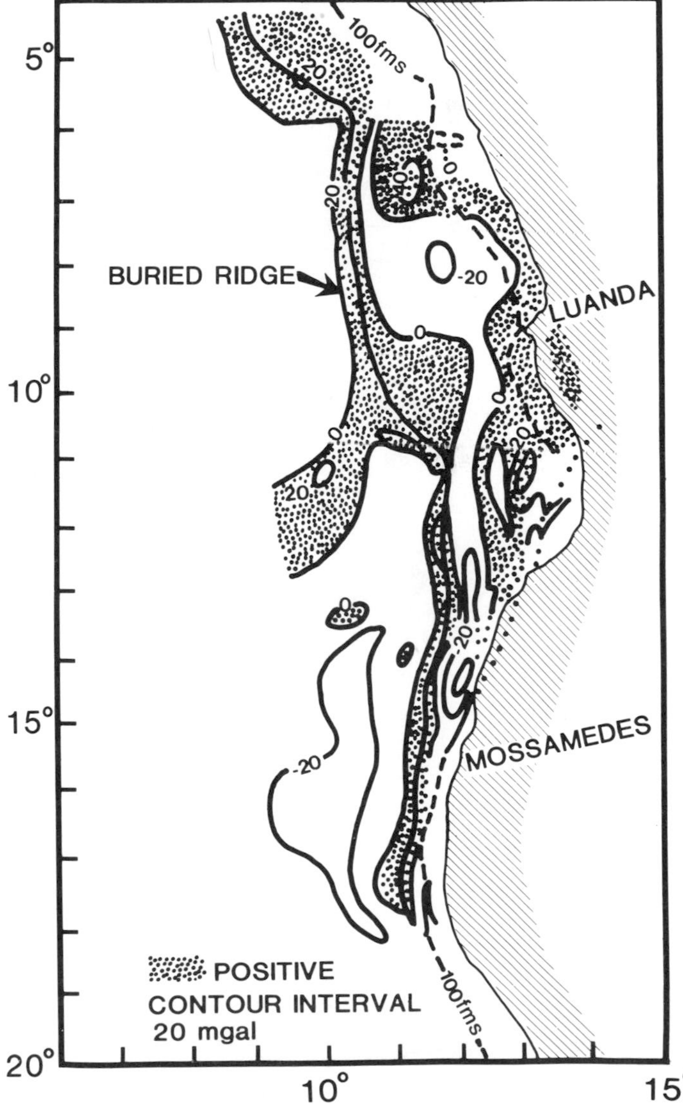

Figure 2 — Isostatic gravity anomaly map of part of offshore West Africa with interpreted outer marginal basement high (buried ridge), after Scheupbach and Vail (1980).

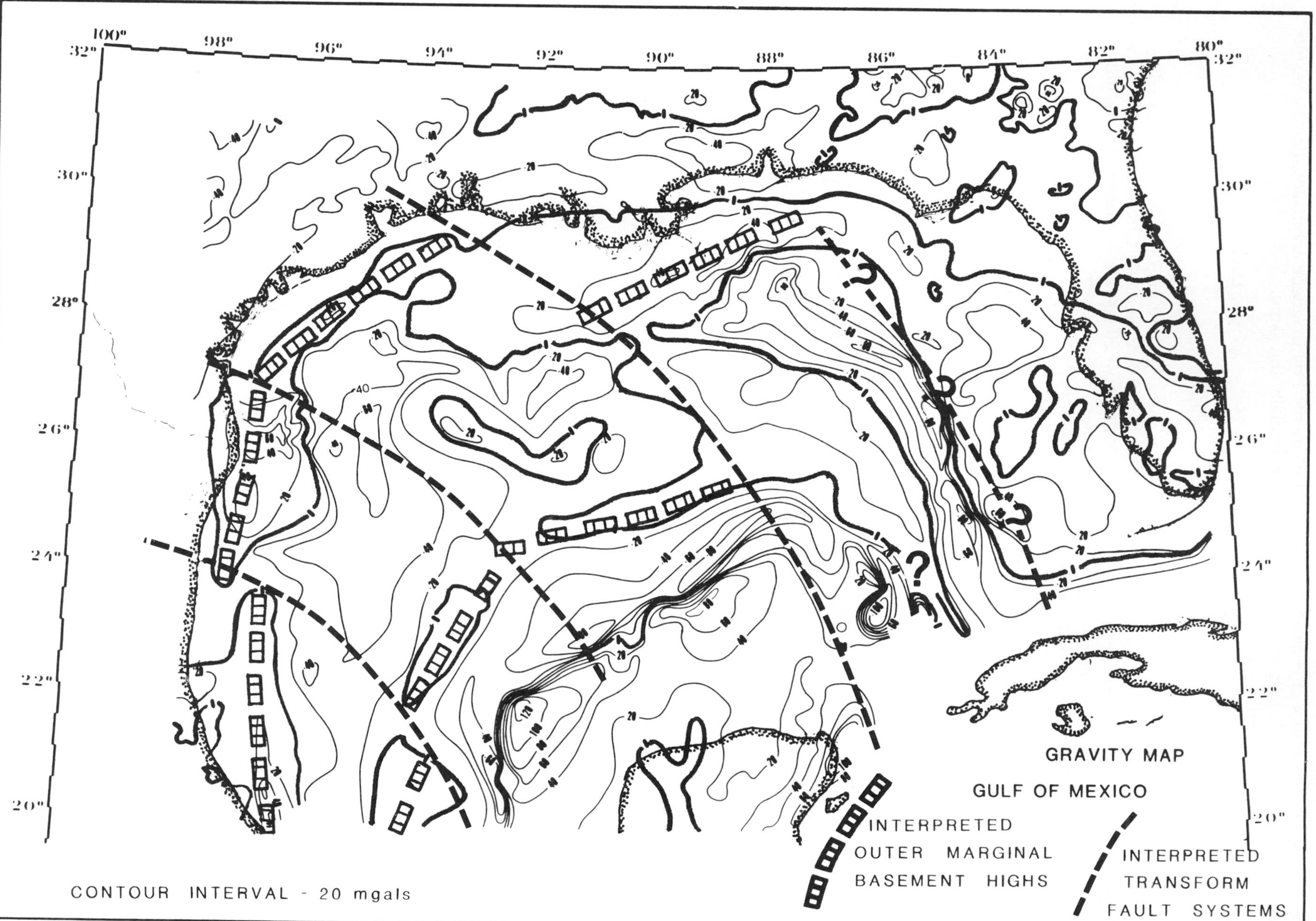

Figure 3 — Regional gravity map (Free-air at sea; Bouguer on land) compiled largely from R/V GULFREX data and some published data from Krivoy, Eppert, and Pyle (1976). Interpreted outer marginal basement highs are outlined by linear gravity highs on this map but are locally confirmed by other regional geophysical data. Transform fault systems are much more approximate and are drawn concentric to a pole located 90 km northeast of Acapulco in Mexico (latitude 17° 22′N, longitude 99° 15′W). Rigorous pole location takes the complications of spherical projection into account and this location should still be considered tentative.

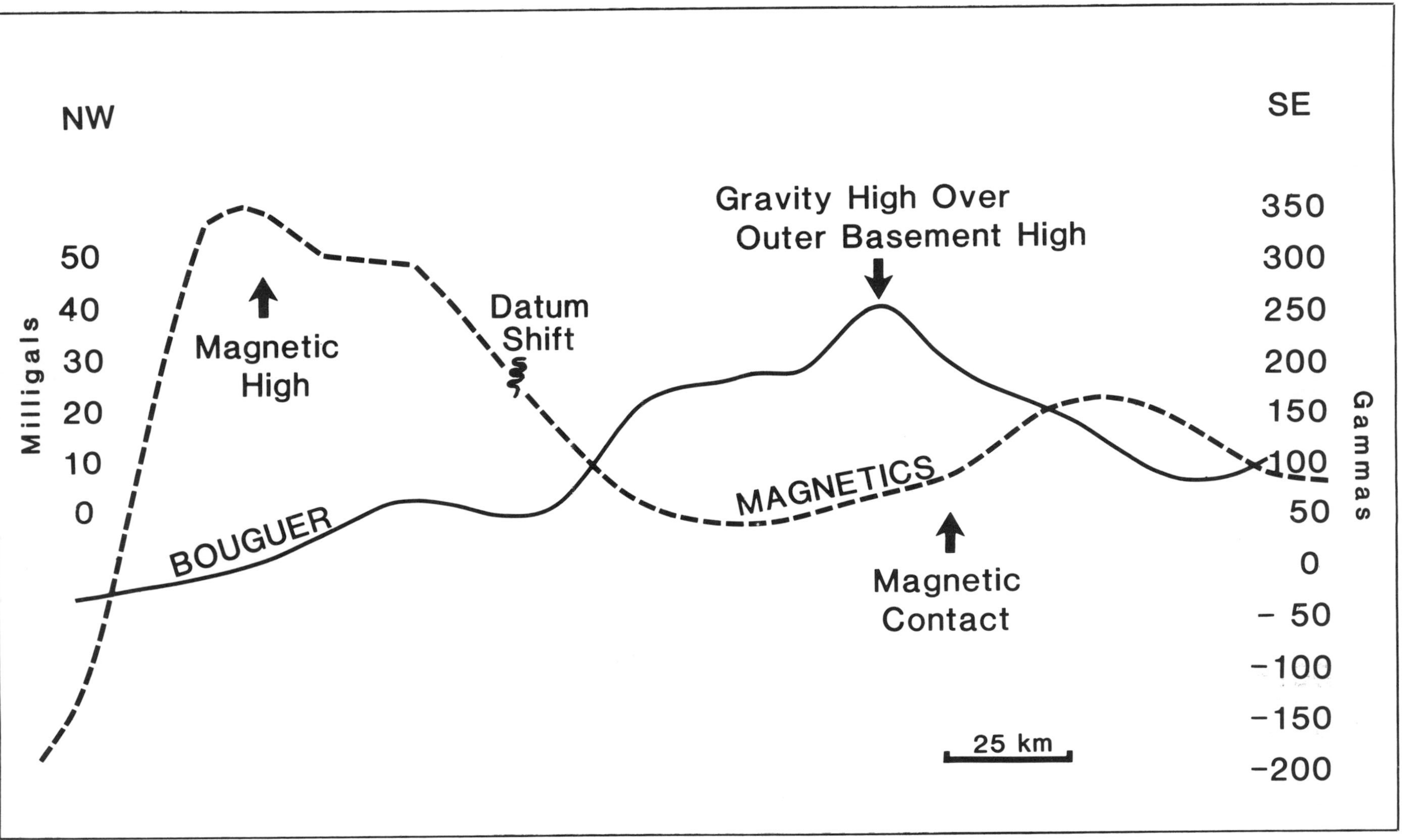

Figure 4 — Generalized regional Bouguer gravity and aeromagnetic total field profiles taken perpendicular to the coast near Galveston, Texas. Note the gravity low southeast of our interpreted outer basement high. (Bouguer anomalies are roughly equivalent to free-air anomalies in this area because of shallow water and flat sea floor.)

is 20 milligals.

The R/V GULFREX did not collect data on the Yucatan shelf, on large portions of the east Mexico shelf, or in the extreme northeast Gulf of Mexico. In these areas, free air gravity was calculated at quarter-degree grid points from bathymetry and Bouguer values obtained from the USGS Bouguer gravity map of the Gulf of Mexico (Krivoy, Eppert, and Pyle, 1976). For evaluations near sea level the Bouguer and free-air gravity values are essentially the same, so the Bouguer contours were used to extend the free-air map over the coastal regions.

Gravity and Magnetic Profile in the Northwest Gulf

Regional Bouguer gravity and magnetic total field profiles were generalized from maps of the northwest Gulf of Mexico near Galveston (Figure 4). The Bouguer gravity includes land data and seabottom gravity and is repeatable within 0.5 milligals. The northeast trend of the gravity anomaly in Figure 4 is shown in Figures 3 and 5. Bouguer anomalies and free-air anomalies are comparable in land areas with low relief and in marine areas with shallow water and flat bottom topography.

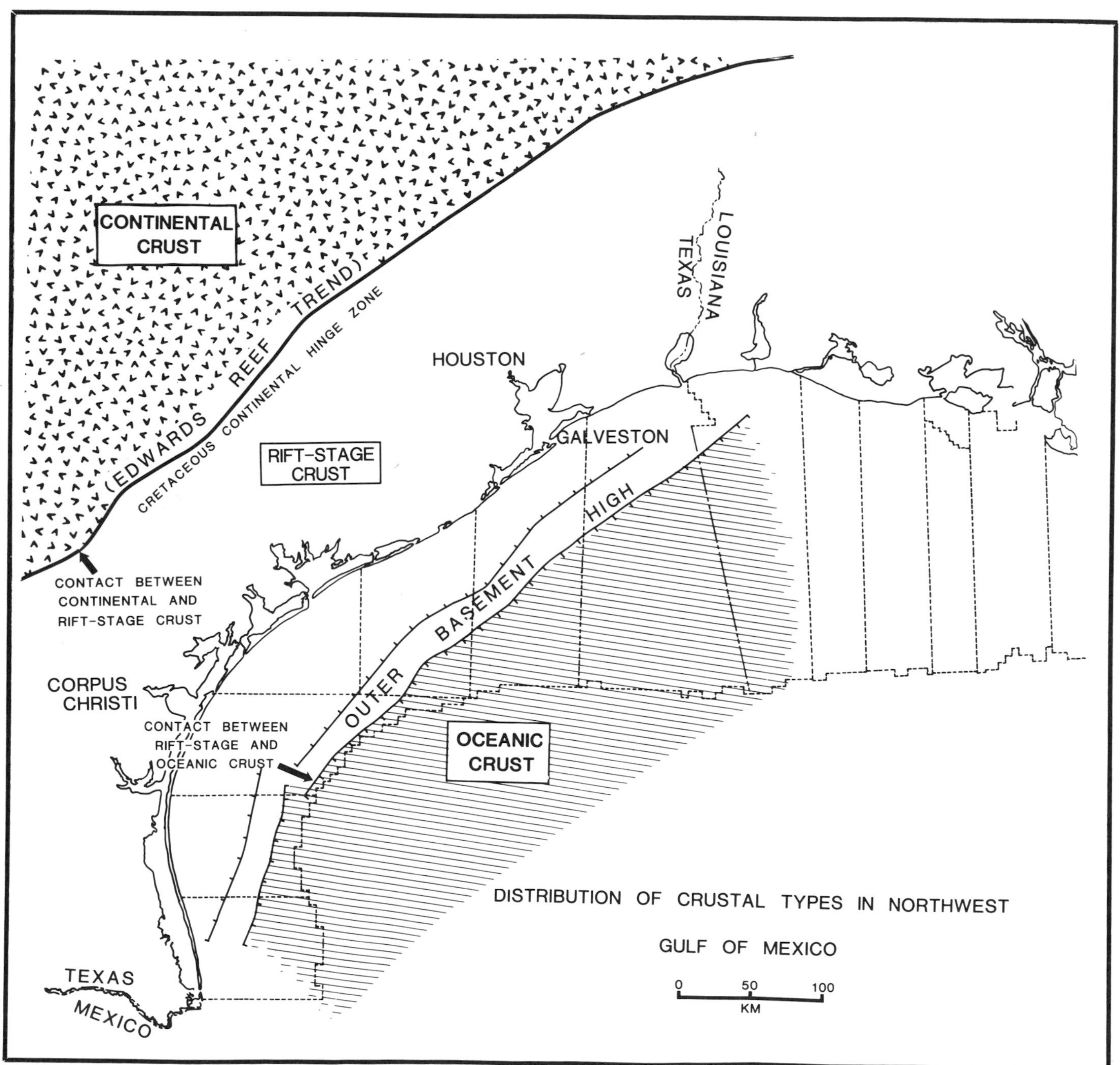

Figure 5 — Distribution of crustal types in northwest Gulf of Mexico, according to the interpretation in this paper. Mesozoic oceanic crust formed southeast of the outer basement high by normal sea-floor spreading processes.

INTERPRETATION

Outer Marginal Basement Highs in the Northwest Gulf of Mexico

In the northwest Gulf of Mexico, linear regional free-air gravity anomalies outline the interpreted outer basement highs. In Figure 4, magnetic and Bouguer gravity profiles perpendicular to the coast near Galveston demonstrate the correspondence between the two independent sets of data and their interpretation. The 20-milligal gravity high over the outer basement feature follows the Texas shelf from Mexico to the Louisiana border (Figures 3 and 5). The magnetic anomaly just southeast of the gravity high (Figure 4) is typical of this latitude and this profile direction from a contact between more magnetic basement rock to the southeast and less magnetic basement rock to the northwest. The consistent geographic association of the gravity with the magnetic anomaly is best explained by a basement source. The parallelism of the Edwards Reef trend (Figure 5), which is close to the Cretaceous hinge zone in the continental crust, and the outer basement high in the northwest Gulf provides additional evidence that both features have regional tectonic significance.

The 400-gamma magnetic high located near the northwest side of the profile (Figure 4) may be analogous to the U.S. East Coast and other continental-margin magnetic anomalies. These anomalies have been interpreted to mark the contact between oceanic and continental crust (e.g. Grow, Mattick and Schlee, 1979). Our interpretation differs: we place oceanic crust about 100 km southeast and infer that the anomaly overlies rift-stage or transitional crust. The 400-gamma anomaly in Figure 4 may have its source in a mafic dike complex injected into faulted rift-stage crust. Alternatively, the anomaly may be due to mafic or ultramafic rocks emplaced in the crust during late Paleozoic Appalachian-Quachita orogenic events.

The zone of rift-stage crust in south Texas extends 230 km from the Edwards Trend to the outer margin high (Figure 5). The crust in this region may be expected to show considerable evidence of rifting, dike intrusion, volcanism, and terrestrial rift-basin sedimentation. Associated with the rifting stage, some type and degree of deeper crust metamorphism is probable (Falvey and Middleton, 1981).

The regional gravity profile (Figure 4) shows values gradually increasing to the southeast. We believe this is due to the density contrast between rift-stage crust and oceanic crust. The outer basement high is the source for the high-frequency 20-milligal anomaly located 75 km from the southeast edge of the profile. Steep gradients, 20 km apart, overlie the edges of the high. The 30 milligal, moderate-frequency gravity low near the southeast end of the profile is thought to be due to a relatively thick, low density sedimentary section overlying oceanic crust southeast of the outer basement high.

In Figure 6, a two-dimensional, slab-like model was used to calculate a gravity anomaly matching the ob-

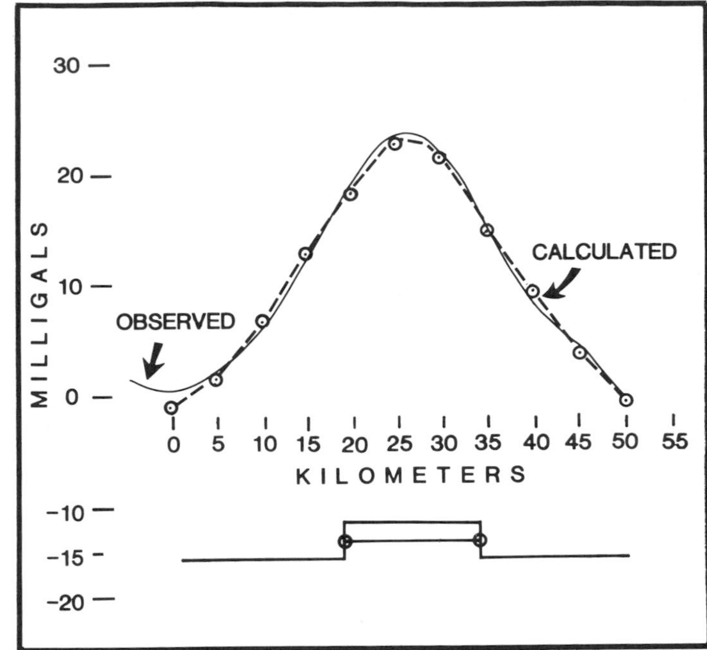

Figure 6 — Observed and calculated gravity profiles over outer basement ridge. The high is probably regionally isostatically compensated.

served field over the basement high. The modeling restricts the product of density contrast (g/cu cm) and slab thickness (km) to 2.22, but the inherent ambiguity does not allow you to uniquely determine either value. Using 2.80 as the density for rift-stage crust and assuming the basement high is surrounded by salt (2.2 g/cu cm), we calculate that the top of the high would be 12.2 km below sea level and the relief on the high would be 3.7 km. If the basement high were surrounded by sediments (2.4/ cu cm), the top of the high would be 11.2 km below sea level and the relief would be 5.6 km. In either case, the midpoints of the basement high are 14 km below sea level and 15 km apart. The basement high in our model is not isostatically compensated locally.

We suggest that the thick salt present in the modern Texas slope was deposited in a basin on oceanic crust southeast of the outer marginal high. The 30-milligal low at the southeast end of our profile (Figure 4) is consistent with a thick salt basin. Salt deposits overlying oceanic crust have been reported for the Red Sea (Humphris, 1978; Girdler and Styles, 1974; Blank, 1979; S. Hall, personal communication, 1979) and the Mediterranean (Hsu et al, 1978).

In Figure 7, we suggest a possible sequential development of the outer margin high as reconstructed in Airy isostatic models. The development is consistent with a gradually cooling and subsiding rift-stage and oceanic crust and mantle.

In the Jurassic isostatic reconstruction (Figure 7) thin salt lies on rift-stage crust over a pre-salt section. Hot crust and mantle have correspondingly lower densities. Deeper water deposition, suggested by the wide lateral continuity, and seismic character of the Cretaceous Challenger formation (Addy, Buffler, and

Worzel, 1979), was due to the cooling and resulting subsidence of more dense crust and mantle (Figure 7). In our model, sea-floor spreading had ceased by this stage because Challenger reflectors, undisturbed except by later differential subsidence, diapiric activity, and growth faulting, can be correlated from the deep central Gulf to the shelf overlying rift-stage crust. Later development of the northwest Gulf followed a pattern of slow subsidence and sedimentary accumulation gradually progressing toward the Tertiary (Figure 7).

Diapiric structures, soft-sediment deformation, and growth-faulting produced the complicated modern geometry. Salt movement in the Sigsbee Scarp (Amery, 1969, Humphris, 1978, and Buffler et al, 1980) further modified the oceanward leading edge of the thick salt mass under the influence of accumulating Tertiary sedimentary units.

Distribution of Outer Marginal Basement Highs Around the Gulf Basin

Regional free-air gravity anomalies (Figure 3) outline the outer marginal basement high in the northwest Gulf of Mexico. Similar linear gravity highs are arranged symmetrically around the western Gulf basin as indicated in the map. The contact between oceanic and transitional crust described by Buffler et al (1980) roughly coincides with the two linear gravity highs close to the Sigsbee Knolls, northwest of the Campeche Scarp in deep water. Both of these linear highs appear to have rotated about 24° clockwise, away from corresponding linear highs of the same length and amplitude in the Texas shelf. In our interpretation, these features were formerly continuous and are now separated by Mesozoic oceanic crust as in the Scheupbach and Vail (1980) model. The correspondence of these features across the deep Gulf basin is good considering the variations in bathymetry and sediment cover and the regional nature of these marine gravity data.

Rigid plate rotation of 24°, with a pole located at 17° 22′N, 99° 15′W (90 km northeast of Acapulco), brings the three pairs of ridges in the western Gulf together. Offsets in the linear trends of the anomalies are obvious locations for concentric transform systems, as indicated in Figure 3. The gradually wider separation of basement ridges, and consequently wider oceanic crust in the northern Gulf, is expected of spreading plate motion on a sphere relatively near a rotational pole.

In the eastern and southeastern Gulf, the continent-ocean boundary is not as well defined (Figure 3). Bathymetric effects obscure the linear free-air gravity anomaly southeast of the Mississippi delta. However, we suspect deep transform faulting close to and along the border of the west Florida shelf (Salvador and Green, 1980). Refraction data indicate normal oceanic crust present west of the shelf edge (Ibrahim et al, this volume). Rigid rotation of 24° is consistent with these observations and places the southern analog of the Louisiana outer basement high to the south of our

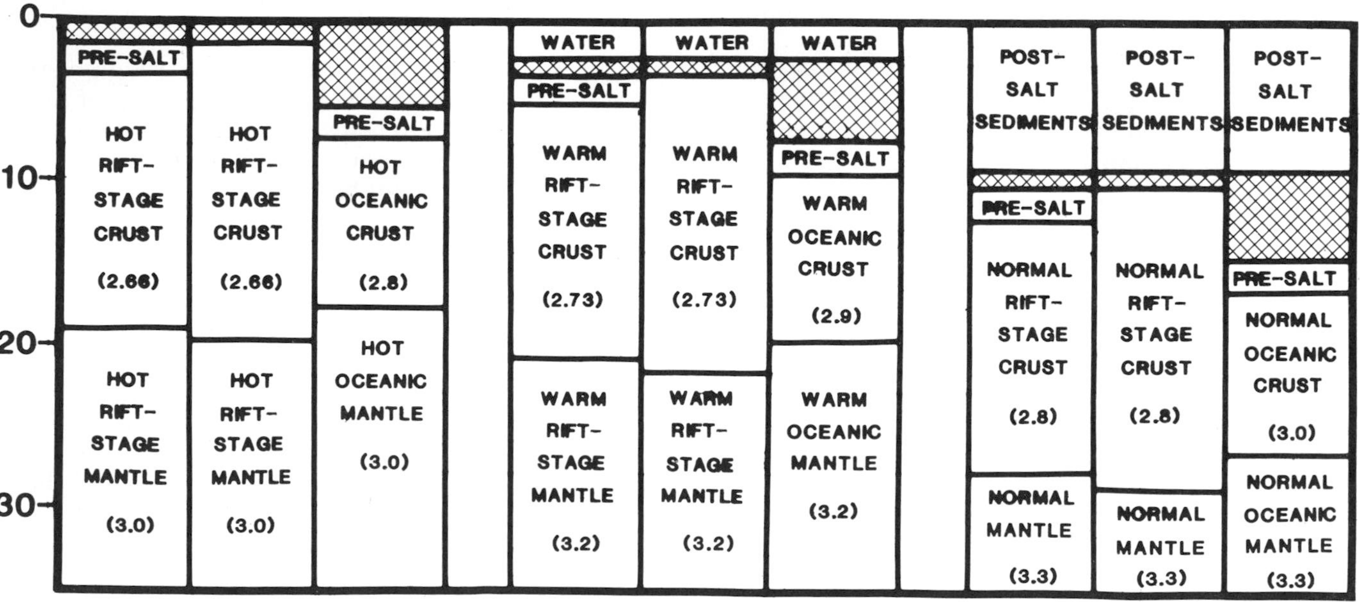

Figure 7 — Schematic isostatic models of region around outer marginal high suggesting a possible history of development and subsidence. Figures in parentheses are assumed densities, vertical scale is in kilometers. Jurassic salt was deposited over pre-salt and basement of lower density due to thermal effects. Continuous Challenger seismic event suggests deep water deposition over somewhat cooler and subsided crust and upper mantle in the Cretaceous. Tertiary post-salt deposits overlie previous section on normal density oceanic and rift-stage crust and mantle.

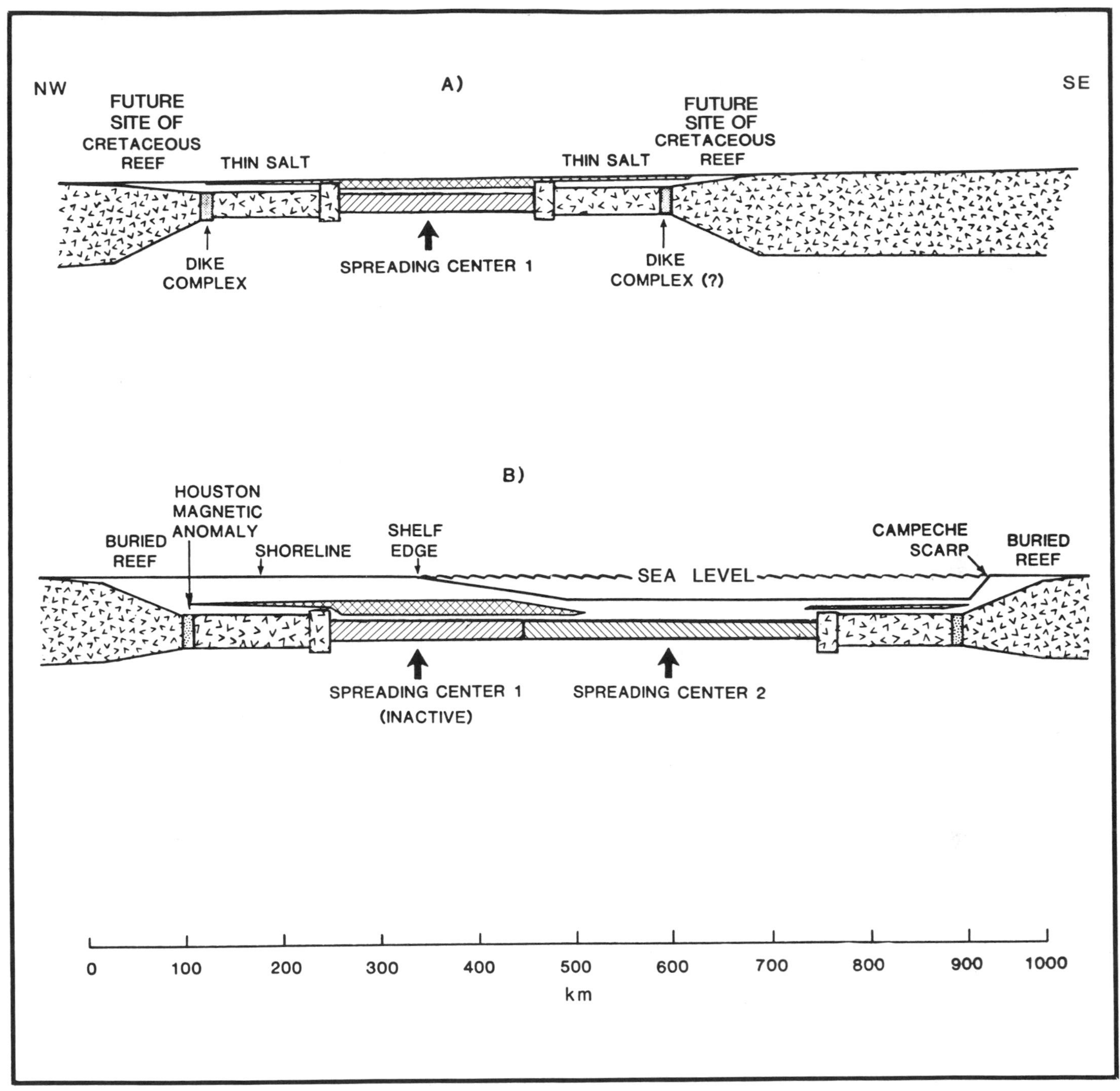

Figure 8 — Diagrammatic cross sections from Texas across the Gulf to Yucatan to show our preferred model of the origin of the Gulf. **A.** Symmetrical spreading characterizes the Jurassic cross section. Salt deposits, while thicker over oceanic crust, spread out of the central basin onto rift-stage crust. A change in spreading-center location nearly coincides with the cessation of main salt deposition. Further subsidence leads to the deposition of the deep water Challenger after spreading is curtailed. Cretaceous reef complexes develop on both continental crustal hinges. **B.** Subsidence continues after deposition of the Challenger following thermal decay laws. Originally contiguous salt deposits are separated due to spreading from center 2. Note that spreading must have continued postsalt in order to separate the northwestern from the southeastern deposits (Buffler et al, 1980), and spreading must have effectively ceased pre-Challenger to preserve Challenger continuity from the deep Gulf to the shelf. Extremely rapid spreading is not necessarily implied; at rates of 10 cm/yr the production of 300 km of center 2 crust would take less than 6% of Jurassic time. Since cessation of spreading, subsidence, and deposition, diapirism and growth-fault tectonics have characterized the Gulf Basin. Carbonate deposits predominate on the Yucatan side and thick Tertiary clastic deposits on the Texas side of the Gulf, at least partly due to the much larger drainage area of the North American Continent and its epeirogenic and tectonic activity. The inferred dike complex associated with the Houston magnetic anomaly (also shown at the left of Figure 5) may have been injected at any time; at least one possibility is during the Cretaceous as a result of extensional fracturing related to the subsidence of rift-stage crust. A similar high-amplitude magnetic anomaly is also present along at least part of the Yucatan side of the Gulf (Moore and Castillo, 1974).

data in the complex structural zone northwest of Cuba. The basement in this region is characterized by rotated and faulted basement blocks (Buffler and Shaub, personal communication, 1981) that might be expected to form in the tectonically complicated interplay of rift and transform systems with rift-stage and oceanic crust. Further deformation may have accompanied the partial subduction of the Bahamian platform under Cuba in the Eocene (Dickinson and Coney, 1980).

More data are needed in the Straits of Tehuantepec of Mexico, south of our map (Figure 3), to evaluate the implications of our regional reconstruction. Although complicated by Tertiary volcanics and the long-range effects of Pacific subduction, the Isthmian salt basin has long been considered a major tectonic zone with the through-going and possibly reactivated Salina Cruz fault system (Martin and Case, 1975).

Late Rift-Stage and Drift-Stage Evolution

Our model suggests that the Late Jurassic configuration across the Gulf resembled Figure 8. If spreading was symmetrical, the thin salt deposited over rift-stage crust southeast of the spreading center could only have been offset from the main thickness of salt (deposited on oceanic crust) by a second spreading center developing close to the Jurassic edge of Campeche rift-stage crust (Figure 8). Salt deposition largely ceased by full development of this second spreading center. Relatively rapid Jurassic spreading followed the main salt deposition and ended prior to deposition of the mid-Cretaceous Challenger unit, which appears to have been continuous across nearly all the Gulf on both rift-stage and oceanic crustal types (before diapirs disrupted it in the northern Gulf). An alternative to the ridge-jump hypothesis is

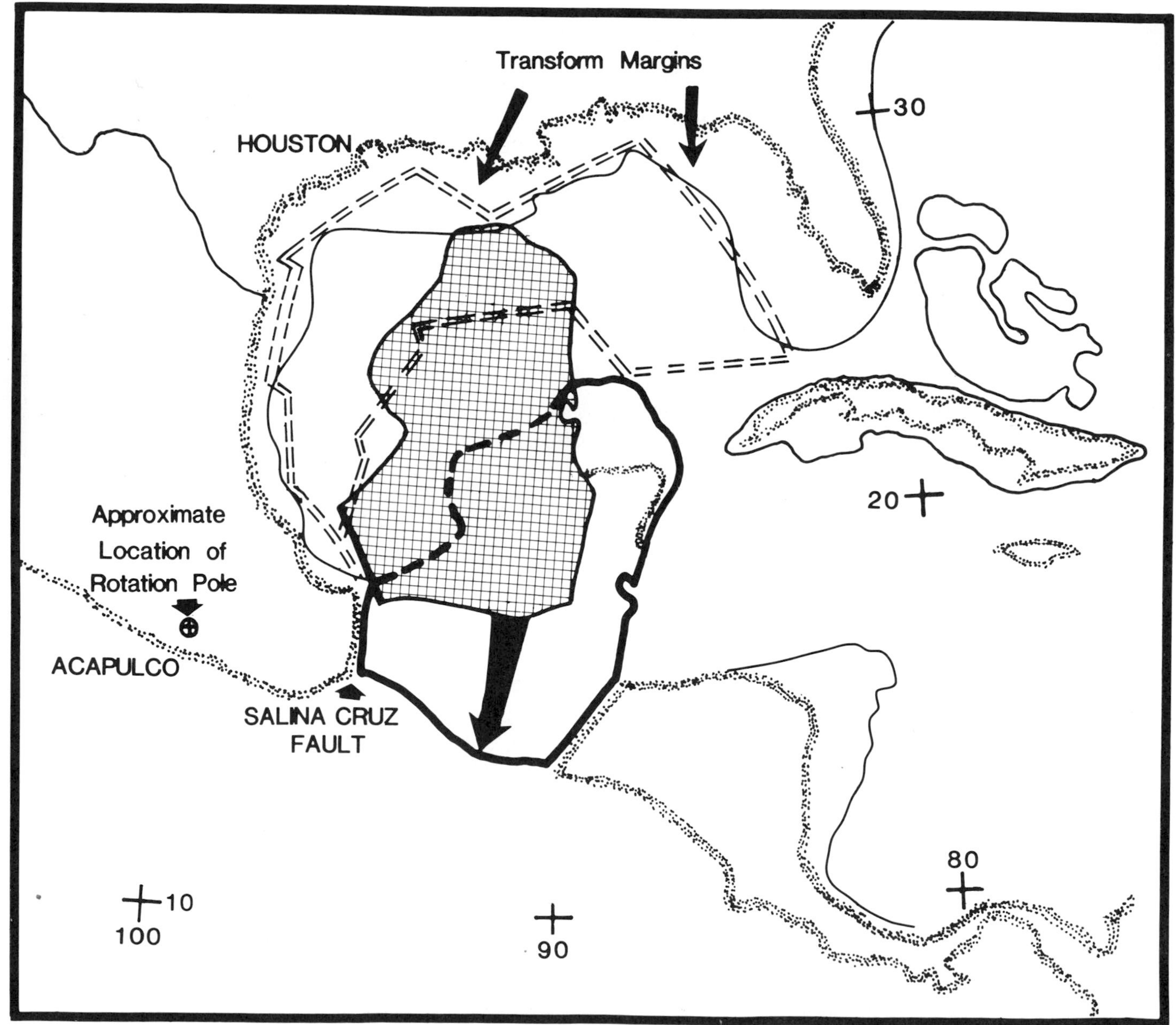

Figure 9 — Sketch map showing pre-drifting and present location of Yucatan block.

asymmetric spreading with new ocean crust forming close to the Campeche outer marginal basement high. In this hypothesis, main salt deposition ceased due to circulation changes associated with an ever-widening Jurassic ocean floor.

In either case, symmetrical reef systems close to the continental hinge zones on both sides of the Gulf were separated in the Cretaceous by relatively deep water and continuously subsiding rift-stage and oceanic crust. The much larger volume of younger sediments on the northern side of the Gulf is due to the correspondingly larger drainage area and sediment flux of central North America relative to the micro-continental dimensions of the Yucatan block.

Figure 9 is a sketch map showing our reconstruction of the motion of the Yucatan block away from Texas and mainland Mexico. Transform margins concentric to the rotational pole northeast of Acapulco are located in the west Florida shelf, offshore Louisiana, and the southernmost Gulf of Mexico. The Salina Cruz fault is interpreted as a major transform fault separating mainland Mexico from the Yucatan block.

We did not deal specifically with the eastern Gulf of Mexico but realize some of the regional implications of our interpretation. The eastern edge of the Yucatan block, when rotated back to its original position, and the west edge of the Florida platform define a roughly triangular gap in the continental crust of Triassic North America. This gap is similar in size and shape to the Chortis block (White, 1980) composed of Honduras, Nicaragua, and the offshore Nicaragua rise.

The transform fault along the west Florida shelf strikes toward the complex structural regime and offsets the Ouachitas from the Appalachians. It could have been localized by a reactivation of fundamental basement discontinuities.

SUMMARY

In summary, we make the following statements:

Thick salt in the Texas-Louisiana slope of the Gulf of Mexico rests on pre-salt sediments overlying oceanic crust.

Normal ocean-floor spreading processes can account for the distribution of the crust and subsequent sedimentary history.

Both transform fault and rifted passive margins exist around the Gulf of Mexico. The differences between these two types of margins deserve considerable future study.

The outer marginal high model of Scheupbach and Vail (1980) seems to fit the Gulf of Mexico well, although the formation mechanism of the outer highs remains poorly understood.

Finally, the western Gulf of Mexico originated by the Yucatan block rotating 24° clockwise away from Texas and mainland Mexico.

REFERENCES CITED

Addy, S. K., R. T. Buffler, and J. L. Worzel, 1979, Correlation of seismic reflectors from the shelf, slope, and abyssal regions in the northeastern Gulf of Mexico: Geological Society of America, Abstracts with Programs, p. 378.

Amery, G. B., 1969, Structure of the Sigsbee Scarp, Gulf of Mexico: AAPG Bulletin, v. 53, p. 2480-2482.

Blank, H. R., Jr., 1979, Tertiary continental margin in southwest Saudi Arabia: Status of current investigations (Abs.): Eos, American Geophysical Union Transactions, v. 60, n. 18, p. 375.

Buffler, R. T., et al, 1980, Structure and early geologic history of the deep central Gulf of Mexico basin, *in* Rex H. Pilger, ed., The origin of the Gulf of Mexico and the early opening of the central North Atlantic Ocean: Baton Rouge, Proceedings, Symposium at Louisiana State University, p. 3-16.

Dickinson, W. R. and P. J. Coney, 1980, Plate tectonic constraints on the origin of the Gulf of Mexico, *in* Rex H. Pilger, ed., The origin of the Gulf of Mexico and the early opening of the Central North Atlantic Ocean: Baton Rouge, Proceedings, Symposium at Louisiana State University, p. 27-36.

Falvey, D. A., and M. F. Middleton, 1981, Passive continental margins: evidence for a prebreakup deep crustal metamorphic subsidence mechanism, *in* Proceedings, 26th International Geological Congress, Geology of Continental Margins Symposium: Oceanologica Acta, p. 103-114.

Girdler, R. W., and P. Styles, 1974, Two stage Red Sea floor spreading: Nature, v. 247, n. 5435, p. 7-11.

Grow, J. A., R. E. Mattick, and J. S. Schlee, 1979, Multichannel seismic depth sections and interval velocities over outer continental shelf and upper continental slope between Cape Hatteras and Cape Cod, *in* J. S. Watkins, L. Montadert, and P. W. Dickerson, eds., Geological and geophysical investigations of continental margins: AAPG Memoir 29, p. 65-82.

Hsu, K. H., et al, 1978, History of the Mediterranean salinity crises, *in* Initial reports of the deep sea drilling project; Washington, D.C., U.S. Government Printing Office, v. 42, part 1, p. 1053-1078.

———, et al, 1981, Crustal structure in Gulf of Mexico from OBS and multichannel reflection data: AAPG Bulletin, v. 67, p. 1207-1229.

Humphris, C. C., Jr., 1978, Salt Movement on continental slope, northern Gulf of Mexico, *in* A. H. Bouma et al, eds., Framework, facies and oil-trapping characteristics of the upper continental margin: AAPG Studies in Geology No. 7, p. 69-85.

Ibrahim, A. K., and E. Uchupi, 1982, Continental/oceanic crustal transition in the Gulf Coast geosyncline: this volume.

———, et al, 1981, Crustal structure in Gulf of Mexico from OBS and multichannel reflection data: AAPG Bulletin, v. 67, p. 1207-1229.

Krivoy, H. L., H. C. Eppert, Jr., and T. E. Pyle, 1976, Simple Bouguer anomaly map of the Gulf of Mexico and adjacent land areas: U.S. Geological Survey Geophysical Investigations Map GP-912.

Martin, R. G., and J. E. Case, 1975, Geophysical studies in the Gulf of Mexico, *in* A. E. M. Nairn and F. G. Stehli, eds., The ocean basins and margins, v. 3: New York, Plenum Press, p. 65-106.

Moore, G. W., and L. D. Castillo, 1974, Tectonic evolution of the Southern Gulf of Mexico: Geological Society of America, Bulletin, v. 85, p. 607-618.

Pilger, R. H., Jr., 1980, The origin of the Gulf of Mexico and the early opening of the central north Atlantic Ocean: Baton Rouge, Proceedings, Symposium at Louisiana State University, 103 p.

Salvador, A., and A. R. Green, 1980, Opening of the Caribbean Tethys (origin and development of the Caribbean and the Gulf of Mexico), *in* 26th International Geological Congress, Geology of the Alpine chains born of the Tethys: B.R.G.M. Memoire n. 115, p. 224-229.

Scheupbach, M. A., and P. R. Vail, 1980, Evolution of outer highs on divergent continental margins, National Research Council, Geophysics Study Committee, Continental Tectonics: National Academy of Sciences, Studies in Geophysics, p. 50-60.

Uchupi, E., 1980, Book review, of R. H. Pilger, Jr., ed., The origin of the Gulf of Mexico and the early opening of the Central North Atlantic Ocean: Science, v. 209, p. 798.

Watts, A. B., and M. S. Steckler, 1981, Subsidence and tectonics of Atlantic-type continental margins, *in* Proceedings, 26th International Geological Congress, Geology of Continental Margins Symposium: Oceanologica Acta, p. 143-153.

White, G. W., 1980, Permian-Triassic continental reconstruction of the Gulf of Mexico-Caribbean area: Nature, v. 283, p. 823-826.

Deep Structure and Evolution of the Carolina Trough

D. R. Hutchinson
J. A. Grow
K. D. Klitgord
B. A. Swift
U.S. Geological Survey
Woods Hole, Massachusetts

Multichannel seismic-reflection data together with two-dimensional gravity and magnetic models suggest that the crustal structure off North Carolina consists of normal continental crust landward of the Brunswick magnetic anomaly (BMA), rift-stage crust in the 80-km-wide zone between the BMA and the East Coast magnetic anomaly (ECMA), and normal oceanic crust seaward of the ECMA.

The deep structure along passive continental margins has been studied in recent years from magnetic, seismic-refraction, seismic-reflection, and free-air and isostatic gravity data to determine the nature of the boundary between oceanic and continental crust (Worzel and Shurbet, 1955; Drake and Nafe, 1968; Drake, Ewing, and Stockard, 1968; Taylor, Zietz, and Dennis, 1968; Emery et al, 1970; Talwani and Eldholm, 1972, 1973; Keen and Keen, 1974; Rabinowitz, 1974; Keen, et al, 1974; Rabinowitz, Cande, and LaBreque, 1976; Schlee et al, 1976; Konig and Talwani, 1977; Rabinowitz and LaBrecque, 1977; Talwani et al, 1978; Grow, Bowin, and Hutchinson, 1979; Ludwig et al, 1979; Talwani et al, 1979; Sheridan et al, 1979). These studies suggest that the ocean-continent boundary does not coincide with a specific bathymetric feature such as the shelf-slope break, and that it probably is not a single boundary but a broad zone. Some researchers suggest that the edge of continental crust or oceanic crust can be correlated with specific magnetic anomalies, isostatic gravity anomalies, or crustal compressional-wave velocities. We have examined magnetic and gravity models calculated along two lines (line IPOD and line 32) that cross the U.S. Atlantic continental margin off North Carolina (Figure 1). In this report, we describe the structure and evolution of the Carolina continental margin, based on the interpretation of the seismic-reflection profiles and gravity and magnetic models.

The Carolina platform and trough, which are the focus of this study, are a segment of the system of platforms and troughs that underlie the continental shelf along the United States Atlantic margin (Klitgord and Behrendt, 1979; Grow, Bowin, and Hutchinson, 1979; Folger, 1979). The Carolina platform, a gently eastward dipping, flat-topped platform, forms the basement surface beneath the coastal plain and shelf from Virginia to Georgia (Figure 1). The Carolina trough is located offshore of the Carolina platform. The trough is the narrowest (60 to 80 km) and most linear (450 km) of the Mesozoic basins of the margin. It is bounded to the north by the Baltimore Canyon trough, to the south by the Blake Plateau basin, and to the west by the Carolina platform. The boundary to the west is approximated by the position of the Brunswick magnetic anomaly (BMA) and to the east by the position of the East Coast magnetic anomaly (ECMA, Figure 1). The trough underlies the outer shelf, slope, and upper rise (Figure 2). Analyses of seismic profiles and magnetic depth-to-basement estimates indicate that basement depths on the Carolina platform range from 1 to 5 km, whereas within the Carolina trough, they reach 11 km (Klitgord and Behrendt, 1979). The Blake Outer Ridge (Figures 1 and 2) covers the southern part of the trough and represents a post-Oligocene sedimentary wedge formed by current-controlled deposition (Markl, Bryan, and Ewing, 1970; Bryan, 1970) rather than a major basement feature.

Geophysical data indicate that the basement structure seaward of the Carolina trough is comparably linear at least as far east as the Blake Spur magnetic anomaly. Seismic basement is a series of ridges, troughs, and scraps that are parallel with the ECMA (Klitgord and Grow, 1980), as is the general pattern of magnetic anomalies (Klitgord and Behrendt, 1979). The linearity and simplicity of the Carolina trough make it well-suited for two-dimensional gravity and magnetic modeling. In addition, seismic data on line 32 show the simplest and clearest view of basement structure near the ECMA anywhere along the U.S. Atlantic margin.

Our objectives in modeling the free-air gravity and magnetic anomalies are: 1) to combine the well-defined shallow structure with some simple assumptions about crustal densities in order to determine the deep crustal and upper mantle structure; 2) to determine whether reasonable magnetizations assumed for the upper crustal structure can explain the major magnetic anomalies along this part of the margin (the BMA and the ECMA); 3) to infer the width of the zone that separates continental from oceanic crust; and, 4) to compare the results of these models with those of two previous models (Grow, Bowin,and Hutchinson,) for the Baltimore Canyon trough and the Long Island platform regions to the north.

MULTICHANNEL DATA

The multichannel data were collected by using a 3600-m-long, 48-channel streamer and a tuned airgun array totaling 1700 cu in for line IPOD and 2,000 cu in for line 32. The IPOD line was shot in 1974 by Digicon, Inc., and was processed in 1975 by Geophysical Services, Inc. (GSI). Shipboard procedures, processing methods, and a detailed interpretation were described by Grow and Markl (1977). Line 32 was shot and processed by GSI in 1978 using similar techniques (for example, deconvolution, time-variant filtering, time variant scaling, velocity analysis every 3 km). Parts of the IPOD line (Shotpoint-SP, SP 700 to 1,200) and line 32 (SP 3,900 to 4,600) at the ECMA (Figure 2) were migrated prior to depth conversion. Navigation for the IPOD line was controlled by LORAN C, Doppler sonar, and satellite systems. For line 32, navigation

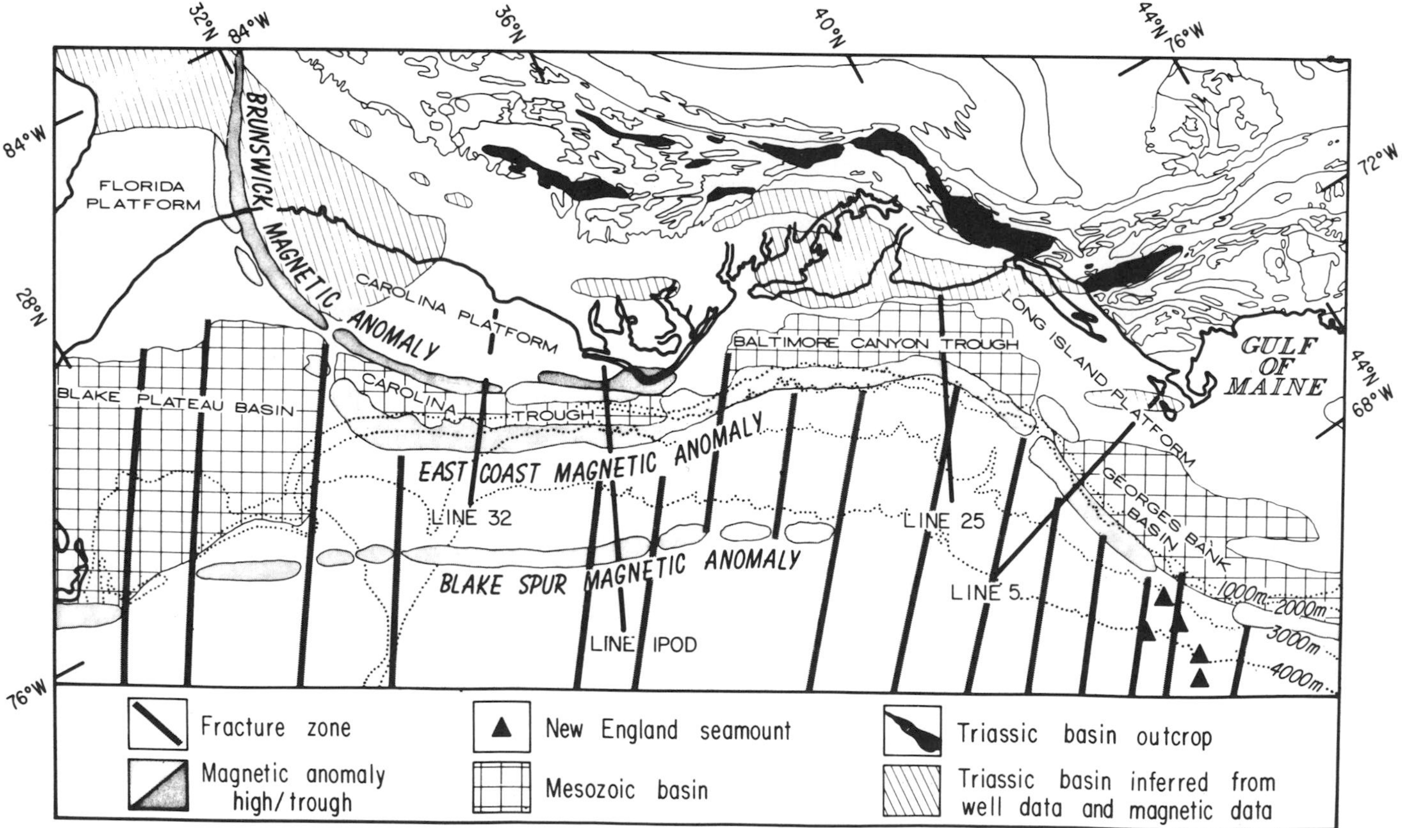

Figure 1 — Index map showing the major structural elements and magnetic anomalies of the U.S. Atlantic continental margin and the locations of lines IPOD, 32, 5, and 25 (modified from Klitgord and Behrendt, 1979). Bathymetry in meters is shown by dotted lines.

was controlled by ARGO, LORAN C, satellite, and Doppler sonar.

Depth sections and line-drawing interpretations, including interval velocities, averaged from the 3-km scans are shown in Figures 3 (line IPOD) and 4 (line 32). The sediment ages are assigned on the basis of correlations with the Hatteras Light No. 1 well (Skeels, 1950; Rona and Clay, 1967; Grow and Markl, 1977), DSDP (Deep Sea Drilling Project) wells (University of Caifornia, Scripps Institute of Oceanography, LaJolla, 1972; Benson et al, 1976; Tucholke and Mountain, 1979; Klitgord and Grow, 1980), and USGS (U.S. Geological Survey) coreholes (Hathaway et al, 1976, 1979). Mesozoic deposits are considerably thicker than Cenozoic deposits except along the seaward end of line 32 near the Blake Outer Ridge. High-amplitude reflectors beneath the lower slope are correlated with the Jurassic shelf-edge sediments, possibly carbonate bank or reef deposits (Grow and Markl, 1977; Klitgord and Grow, 1980). On both lines, these paleoshelf deposits are approximately 10 km landward of the axis of the ECMA. On line IPOD, the present shelf edge has retreated 35 km from the Jurassic shelf edge; on line 32, it has retreated approximately 100 km. Most major erosional events in the area appear to have taken place during the Tertiary. The Tertiary slope and upper rise section on line IPOD has been cut back, leaving Upper Cretaceous sedimentary rocks exposed on the lower slope. On line 32, a thicker Tertiary wedge covers the Jurassic-Lower Cretaceous shelf edge, because of the construction of the Blake Outer Ridge by Tertiary currents. Much erosion accompanied by landward retreat of the shelf edge has taken place off the Carolina margin during Tertiary time.

A diapir complex, appearing to originate in the deep Jurassic sediments, occurs at the ECMA on both lines. The absence of short-wavelength gravity or magnetic anomalies over these features suggests that they are salt or sediment, rather than igneous bodies. Recently acquired single-channel and multichannel

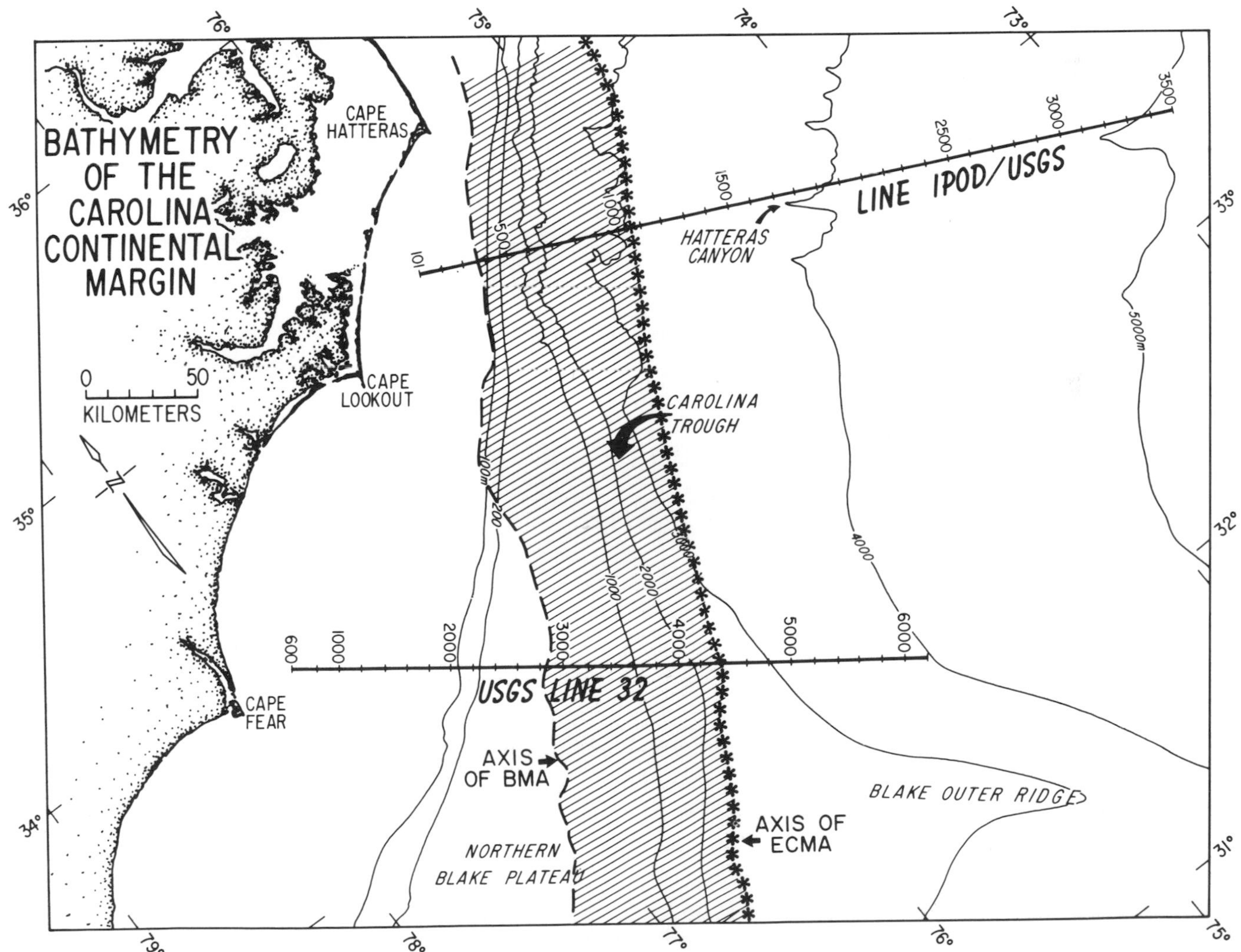

Figure 2 — Bathymetric map of the Carolina margin showing shotpoint locations for lines IPOD and 32. The axis of the ECMA runs along the maximum value of the anomaly; the axis of the BMA follows the steepest gradient between the positive and negative peaks of the anomaly. Note the proximity of the Blake Outer Ridge to the outer end of line 32. Isobaths in meters.

seismic data (Dillon et al, this volume) indicate that at least 26 diapirs occur in a narrow band that corresponds to the ECMA for the length of the Carolina trough. Although Jurassic salt horizons have been drilled on the Georges Bank shelf (Amato and Simmonis, 1980) and Upper Triassic to Lower Jurassic salt horizons drilled on the Nova Scotia shelf are inferred to be connected with a broad diapir or sedimentary ridge along the slope (Jansa and Wade, 1975; Jansa, Bujak, and Williams, 1980), few salt diapirs have been found on the remaining U.S. Atlantic margin. Three inferred salt structures have been found, however, along the ECMA in the Baltimore Canyon trough (Grow, 1980).

A major growth fault occurs on line 32 from SP 3,300 to 3,400 (Figure 4). Offset across the fault decreases from 600 m at the inferred top of the Jurassic to 250 m near the inferred top of the lower Cretaceous. Therefore, the fault appears to have been a growth fault active in the Jurassic and Createaceous. Offset in the Cenozoic section can be detected by high-resolution seismic-reflection profiles (Dillon et al, this volume).

The unconformity at the base of the sediments is called the breakup unconformity, according to the terminology of Falvey (1974). This unconformity marks a period of erosion associated with thermal and tectonic uplift during the rifting stage and continued after the onset of true sea-floor spreading in the Early Jurassic. Rapid subsidence caused by thermal cooling of the lithosphere and sediment loading (Watts and Steckler, 1979) resulted in a thick wedge of Lower Jurassic through Cretaceous sediments being deposited within the Carolina trough and an onlapping wedge transgressing onto the Carolina platform. The sediments above the breakup unconformity are inferred to be postrift. The beveled character of the breakup unconformity on line 32 indicates that the rifted margin over the Carolina platform, and possibly within the California trough itself, was probably above sea level for a considerable period of time during rifting and after the initiation of sea-floor spreading. In contrast, rifting within the Bay of Biscay and at other margins is thought to have taken place in a submarine environment without a significant amount of uplift during the rifting stage (Montadert, Roberts, and DeCharpal, 1979; Kent, 1980).

Subbasement reflectors occur on line 32 from SP 2,300 to SP 2,800 within the uppermost basement (Brunswick graben, Figure 4). Because these reflectors dip steeply landward, do not roll over, and contrast with the gently seaward dipping reflections above the

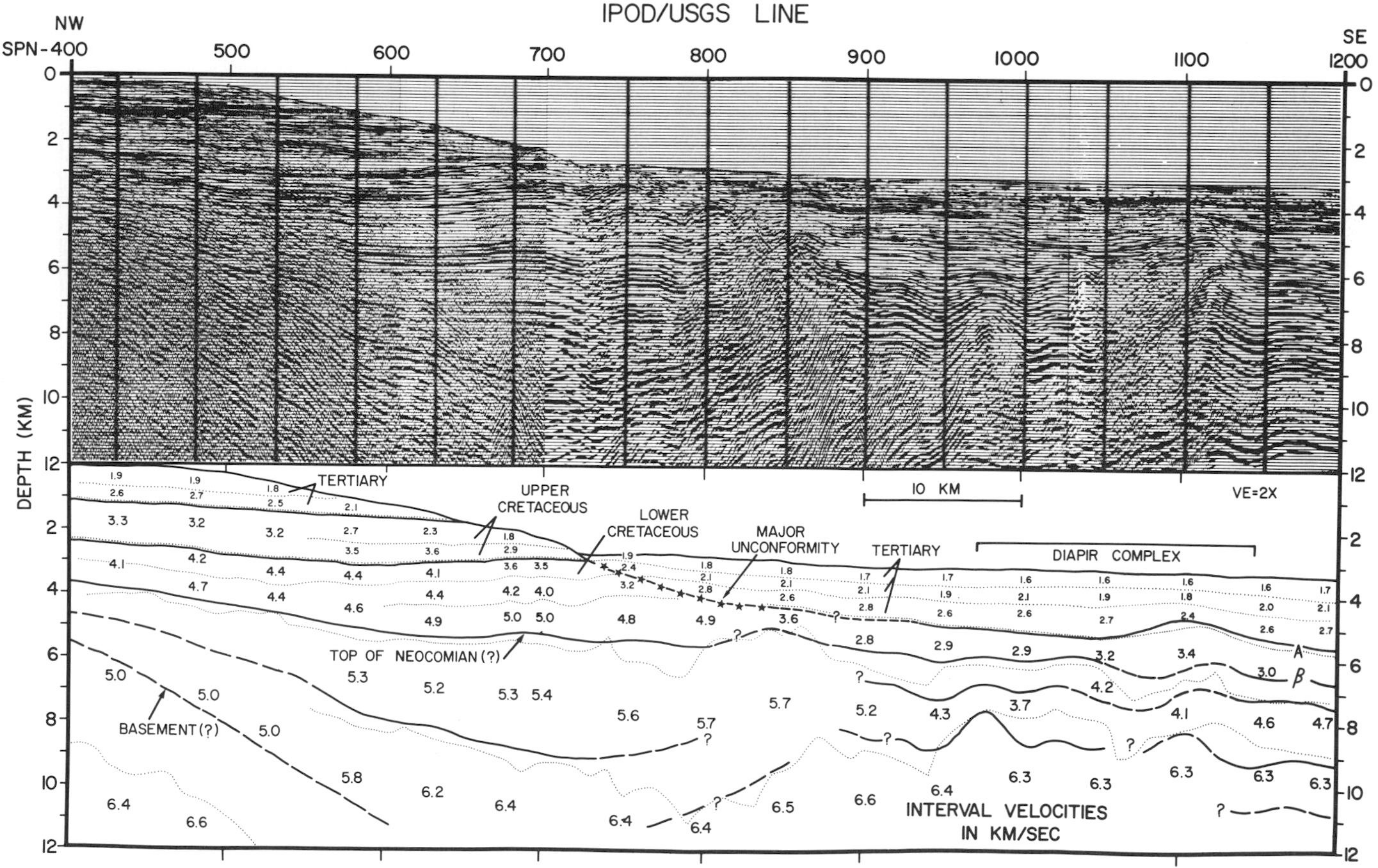

Figure 3 — Depth section and line drawing interpretation for Line IPOD/USGS between shotpoints 400 and 1,200. Interval velocities used for depth conversion were averaged every 5 km along horizontal bands separated by dotted lines. After Grow and Markl (1977).

basement, they are probably not multiples or diffractions. The subparallel aspect of these reflectors suggests sedimentary layering, although the outlines of the inferred basin are not clearly defined acoustically. We call this structure the Brunswick graben because of its location beneath the BMA. The position of the graben beneath the breakup unconformity suggests it contains prerift sediments as described by Falvey (1974). The lack of reflector convergence toward the top of the graben suggests that the sediments are not synrift (deposited at the time of rifting) as described by Montadert, Roberts, and DeCharpal (1979). Whether the sediments are Triassic or Paleozoic cannot be determined from these data alone. Other real subbasement events may exist on line 32, but none are clear or strong enough that an origin attributable to multiples, diffractions, or processing artifacts cannot be discounted. Our interpretation of the subbasement structure of line 32 differs considerably from the thrust fault and decollement interpretation of Harris and Bayer (1979).

Line 32 reflection data near the ECMA suggest that if a basement ridge exists, it is not of major dimension. Undisturbed, coherent, sedimentary reflectors can be traced seaward beneath the shelf to SP 4,000 at depths of more than 11 km and landward beneath the rise to SP 4,400 at depths near 11 km (Figure 4). Acoustic basement in the intervening 20-km-wide zone (SP 4,000 to SP 4,400) is more shallow (5 to 6 km). The narrowness (20 km) of this zone of limited penetration, as well as the undisturbed nature and similar depths of the reflectors on either side of this zone, suggest that a major basement feature does not exist. Although these data are not conclusive, this interpretation is significantly different from earlier hypotheses that a basement high underlies the ECMA (Burke, 1968; Drake and Nafe, 1968; Taylor, Zietz, and Dennis, 1968; Klitgord and Behrendt, 1979; Grow, Bowin, and Hutchinson, 1979; Grow, Mattick, and Schlee, 1979). On line IPOD, the width of this zone of poor resolution is about 40 km; therefore, line 32 represents a major improvement in our understanding of the deep sedimentary and basement structure near the ECMA.

Acoustic reflectors within the crust are occasionally found seaward of the ECMA. A deep crustal reflector, at 11 seconds two-way travel time, occurs on the IPOD line seaward of the Blake Spur magnetic anomaly (SP 2,600; Grow and Markl, 1977). Depth conversion (performed by using interval velocities and refraction velocities of standard oceanic crust) indicates that this reflector occurs at 12 to 14 km depth (Grow and Markl, 1977). This reflector is interpreted as the Moho discontinuity and limits the Moho depth at the seaward end of line IPOD in our gravity model.

The interval velocities for the sedimentary units, presented in Figures 3 and 4, generally increase with depth and across the shelf and slope. Velocities decrease seaward across the rise. A complete discussion of the velocity data on the IPOD line is presented in Grow and Markl (1977). Because the data on line 32 and the IPOD line are similar, the reader is referred to Grow and Markl (1977) and Grow, Mattick, and Schlee, (1979) for a discussion of the velocity determinations, distributions, and implications.

MAGNETIC DATA

Three well-defined magnetic lineations parallel the continental margin south of Cape Hatteras. The major magnetic feature is the East Coast magnetic anomaly (ECMA; Figure 5), a positive high-amplitude anomaly which extends from Canada to Georgia (Drake, Heirtzler, and Hirschman, 1963; Taylor, Zietz, and Dennis, 1968; Klitgord and Behrendt, 1979). It is interpreted primarily as an edge effect between highly magnetized oceanic crust and less magnetized sedimentary units overlying transitional or continental crust (Keen, 1969; Klitgord and Behrendt, 1979). Thus, the ECMA is a possible marker for the edge of oceanic crust. The Brunswick magnetic anomaly (BMA) is landward of and parallel with the ECMA off the Carolina margin (Figure 5). The BMA consists of a coupled positive and negative anomaly, which extends south from Cape Hatteras parallel with the ECMA before curving onshore at Brunswick, Georgia (Taylor, Zietz, and Dennis, 1968; Pickering, Higgins, and Zietz, 1977; Klitgord and Behrendt, 1979). The magnetic high of the BMA has been called the inner branch of the ECMA (Taylor, Zietz, and Dennis, 1968; Pickering, Higgins, and Zietz, 1977). The Blake Spur magnetic anomaly (BSMA) lies parallel with and seaward of the ECMA, stretching from the Bahamas to 36°N latitude (Taylor, Zietz, and Dennis, 1968; Vogt, 1973; Kiltgord and Behrendt, 1979). A landward-facing basement scarp is beneath the BSMA along the entire length of the Carolina margin (Klitgord and Grow, 1980). The magnetic anomalies and basement features between the ECMA and BSMA have a similar linear character. In general, all these magnetic features are indicative of a simple crustal structure that is oriented parallel with the ECMA.

Magnetic depth-to-basement solutions from the Carolina margin (Figure 6) show that the basement structure is linear and two dimensional (Klitgord and Behrendt, 1979; Behrendt and Klitgord, 1980). The Carolina platform is characterized by basement depths of less than 6 km and the absence of major basement features. A small trough occurs beneath the BMA and we call this feature the Brunswick graben (Figure 6). Magnetic depth estimates within the Carolina trough exceed 11 km (Figure 6). The ridge that marks the seaward edge of the trough (depths shallow from 9 to 11 km in the trough, to 7 to 8 km along the ridge; Figure 6) underlies the ECMA and probably marks the edge of oceanic crust (Klitgord and Behrendt, 1979). The acoustic data discussed in the previous section indicate that this ridge may not be as large or continuous as the magnetic depth estimates show.

GRAVITY DATA

Gravity data over lines IPOD and 32 were collected aboard the R/V *Fay* in 1976 using one of the Woods

Hole Oceanographic Institution's vibrating-string-accelerometers (Bowin, Aldrich, and Folinsbee, 1972). Navigation included range-range LORAN C, hyperbolic LORAN C, and satellite. The accuracy of the gravity data is estimated at 2 to 3 mgal. Additional details of the processing and reduction were given by Grow, Bowin, and Hutchinson, (1979).

Free-air gravity anomalies over the Carolina margin (Figure 7) are dominated by the shelf-edge positive anomaly (0 to +60 mgal) and slope minimum (−80 to 0 mgal), which are attributed to the edge effect caused by changing water and mantle depths (Worzel, 1965, 1966). The Carolina trough is not clearly expressed in the gravity map because of the masking effect of the shelf-edge positive anomaly, but the Carolina platform is characterized by many short-wavelength anomalies whose source is within the shallow basement. Line 32 crosses an unusually large, local negative anomaly at its landward end. A regional long-wavelength negative anomaly characterizes the continental rise and has been attributed to mantle structure (Grow, Bowin, and Hutchinson, 1979). The relative high over the Blake Outer Ridge results from the sedimentary deposits, as no obvious basement ridge is present (Markl, Bryan, and Ewing, 1970; Hollister, et al, 1972).

GRAVITY MODELS

The two-dimensional gravity models were constructed by converting interval-velocity information from the multichannel lines into densities using the Nafe/Drake relationship between compressional velocities and densities (Ludwig, Nafe, and Drake, 1971). We used interval velocity-density relationships only for the sedimentary section. We assume that beneath the sediments, a two-layer crustal structure exists in which a homogeneous upper crust (with a density of 2.7 gm/cu cm) overlies a homogeneous lower crust (with a density of 3.0 gm/cu cm). We infer that the shape of the boundary between the upper and lower crust is similar to that of the 7.1 km/second refractor indentified off New Jersey (Sheridan et al, 1979) and northwest Africa (Weigel and Wissman, 1977). It approximates the boundary between oceanic layers 2 and 3 at the seaward side of the margin and dips gently landward beneath the rise and slope, in

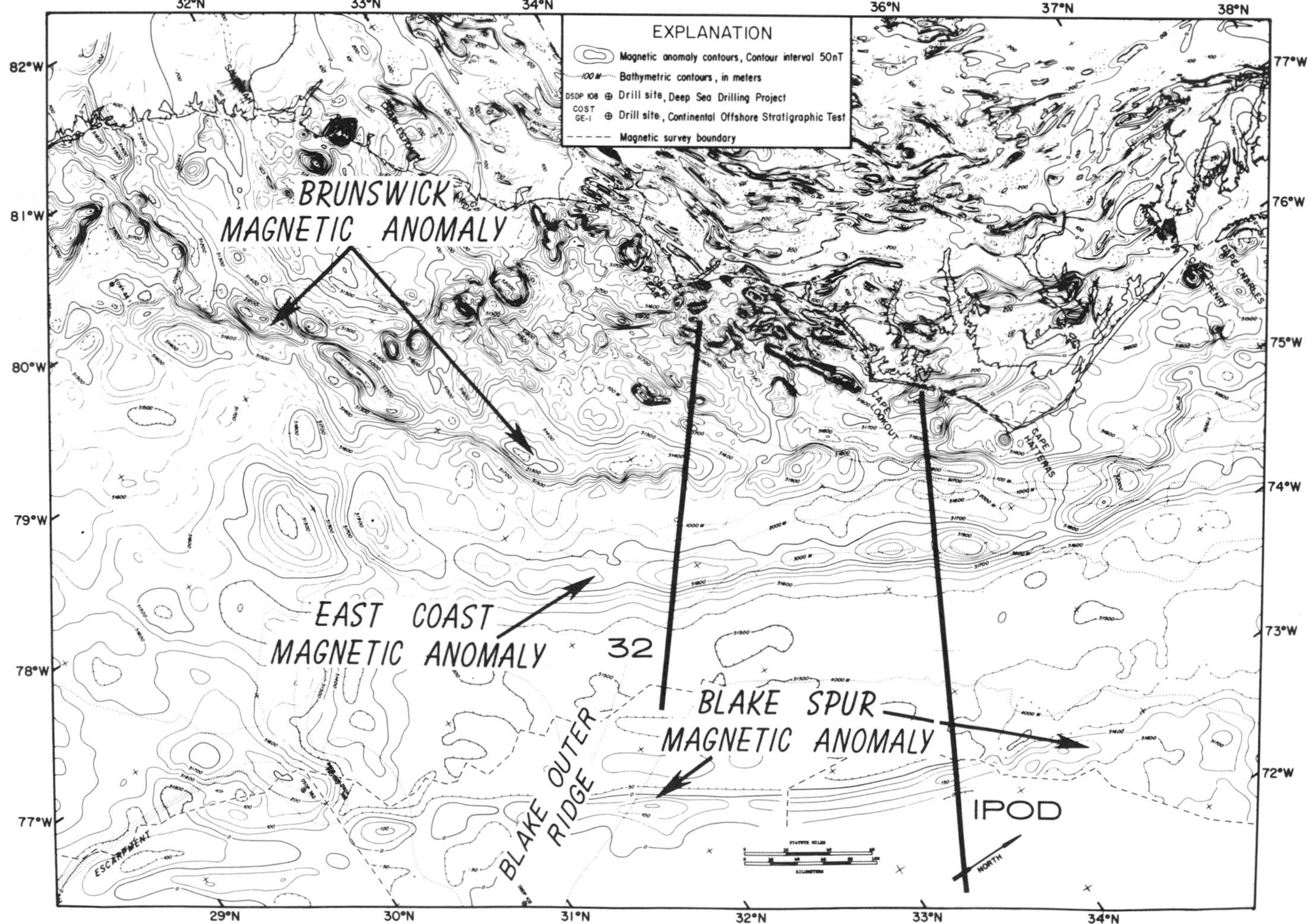

Figure 5 — Magnetic anomaly map of the Carolina continental margin. From Klitgord and Behrendt (1979).

agreement with the average crustal model presented by Sheridan et al (1979) for the New Jersey margin. We assume this boundary is flat and horizontal as it extends landward.

The depth to Moho on the seaward end of the IPOD line is indicated by the depth of the deep reflector (12 to 14 km deep) mentioned previously. We assume a slightly greater depth to Moho of 15 km along the seaward end of line 32 because the sedimentary section there is thicker than that at the end of line IPOD. Because the sedimentary layers were defined by the reflection data, and because the upper-lower crust boundary was assumed, the depth to Moho represents the free variable in the model.

The gravity models for lines 32 and IPOD are shown in Figures 8 and 9. The onshore gravity data are Bouguer values taken from Mann (1962). The shape of the sedimentary wedge onshore was determined from the depth-to-basement determined from wells 1, 2, 3, 5, 6, and 7 (Table 1), structure contours (Denison, Ravelling, and Rouse, 1967; King, 1969; Murphy, 1972), and the location of basement outcrops (King and Beikman, 1974). Figures 8 and 9 show the locations of magnetic depth solutions and of acoustic basement.

In the model of line 32 (Figure 8), the density of the sediments increases as depth increases to a maximum of 2.6 gm/cu cm beneath the shelf and upper rise. The sediments thin beneath the outer rise, and sedimentary densities at the seaward end of the line range from 1.8 to 2.4 gm/cu cm. A pronounced hinge line at 5.5-km depth near SP 3,000 coincides with the BMA marking the edge of the Carolina trough. We modeled small density changes across the growth fault at SP 3,300.

Moho is inferred to be the boundary separating the 3.0 and the 3.3 gm/cu cm bodies. It gradually shallows from 40 km at the landward edge of the model to 37 km at SP 2,500. From there, it jumps abruptly, in 30 km, to 22 km at SP 3,100. It is slightly inclined beneath the trough, then rises again from 22.5 km (SP 4,300) to 16 km (SP 4500) in a distance of 10 km. The Moho shows minor relief from SP 4,500 seaward. The two abrupt changes in Moho depth coincide with the

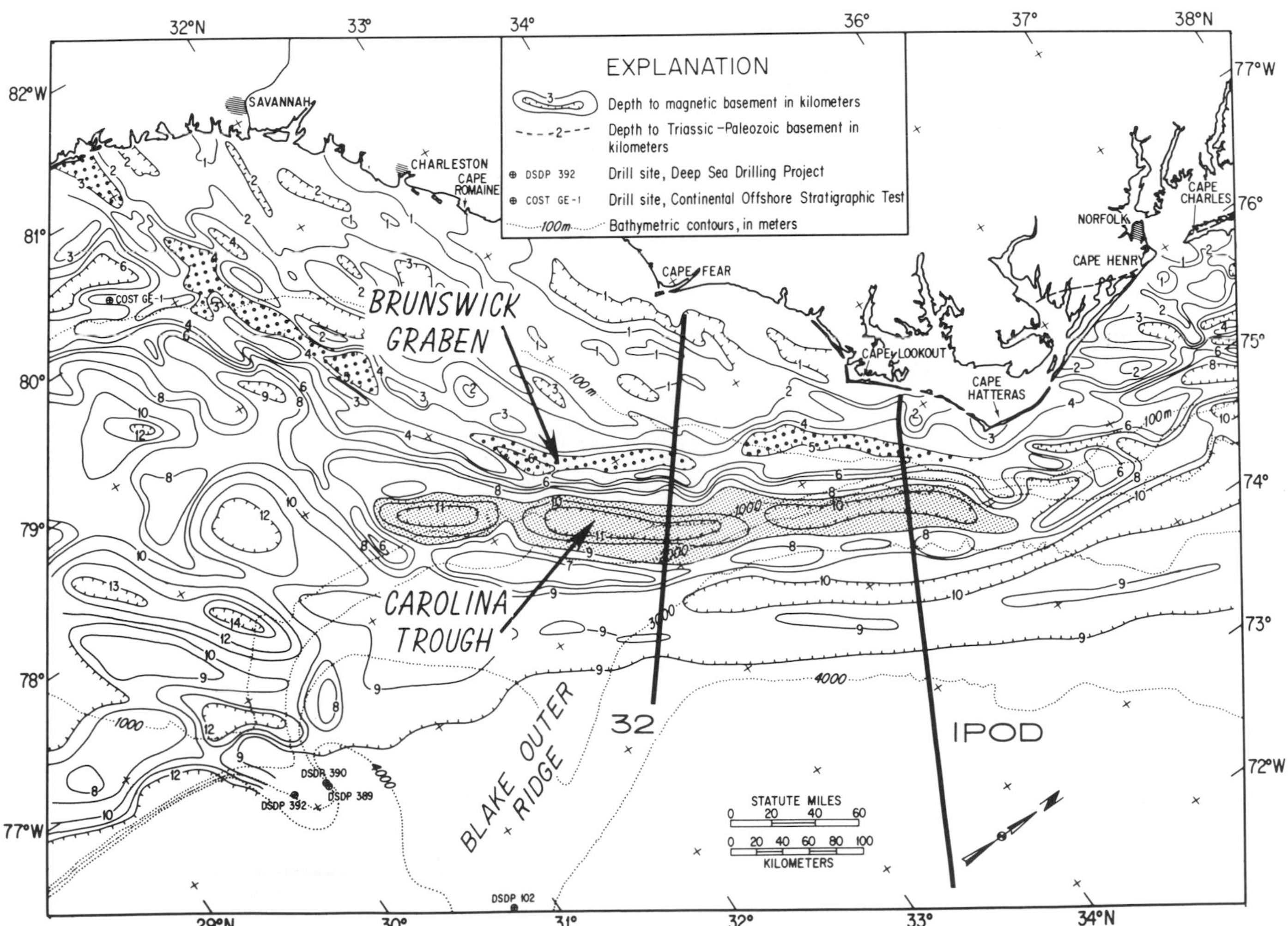

Figure 6 — Depth-to-basement contours for the Carolina continental margin based on magnetic-depth estimates. The Brunswick graben and Carolina trough are shaded within the 5 km and 9 km contours respectively. From Klitgord and Behrendt (1979).

locations of the BMA and the ECMA, along the edges of the Carolina trough.

No reliable velocity (and therefore, no density) information exists in the multichannel data for the Brunswick graben. Available density data from drilled and exposed Triassic sediments on land indicate that the densities are probably less than the surrounding bedrock by 0.1 to 0.2 gm/cu cm (Mann and Zablocki, 1961; Sumner, 1977). We modeled this body with a density of 2.5 gm/cu cm to approximate well-compacted sediments. This keeps its size within the limits suggested by the acoustic and magnetic data and fits the gravity curve. The Brunswick graben cannot be omitted from the model without introducing an unrealistic short-wavelength bump on the crust-mantle boundary.

The major negative gravity anomaly that exists beneath the inner shelf on line 32 (Figure 7 and 8) requires a large low-density body to fit the observed data. We modeled this anomaly by a body within the uppermost crust of 2.5 gm/cu cm density, although we have no evidence other than the gravity anomaly to substantiate its existence. A similarly shaped gravity low north of Cape Hatteras has been modeled as a low-density igneous intrusion (having a density of 2.65 gm/cu cm) within a crust of density equal to 2.8 gm/cu cm (Murphy, 1972). The Hilton Park well (well 2; table 1; Figures 7 and 8) at Cape Fear bottomed in granite (Maher, 1971). We suspect that part of this anomaly could be explained by a low-density granitic body.

Continental and oceanic crusts are labeled on the model (Figure 8), as well as the intervening zone which we have called rift-stage crust (Falvey, 1974). A complete discussion of the crustal structure follows the section on magnetic models.

The model for line IPOD (Figure 9) is similar to that of line 32 because the sedimentary section thickens seaward of a hinge line and the densities beneath the shelf and rise increase as depth increases. Differences exist which do not significantly alter the final results of this model from the results of line 32: the slope is steeper; maximum depth to acoustic basement within the trough is 1 km less; total sediment thickness is only 9 km; the unconformity separating a significant lateral density break is steeper and is exposed on the slope; and, no growth fault exists. The Moho has a similar shape; it shallows from 30 to 20 km beneath

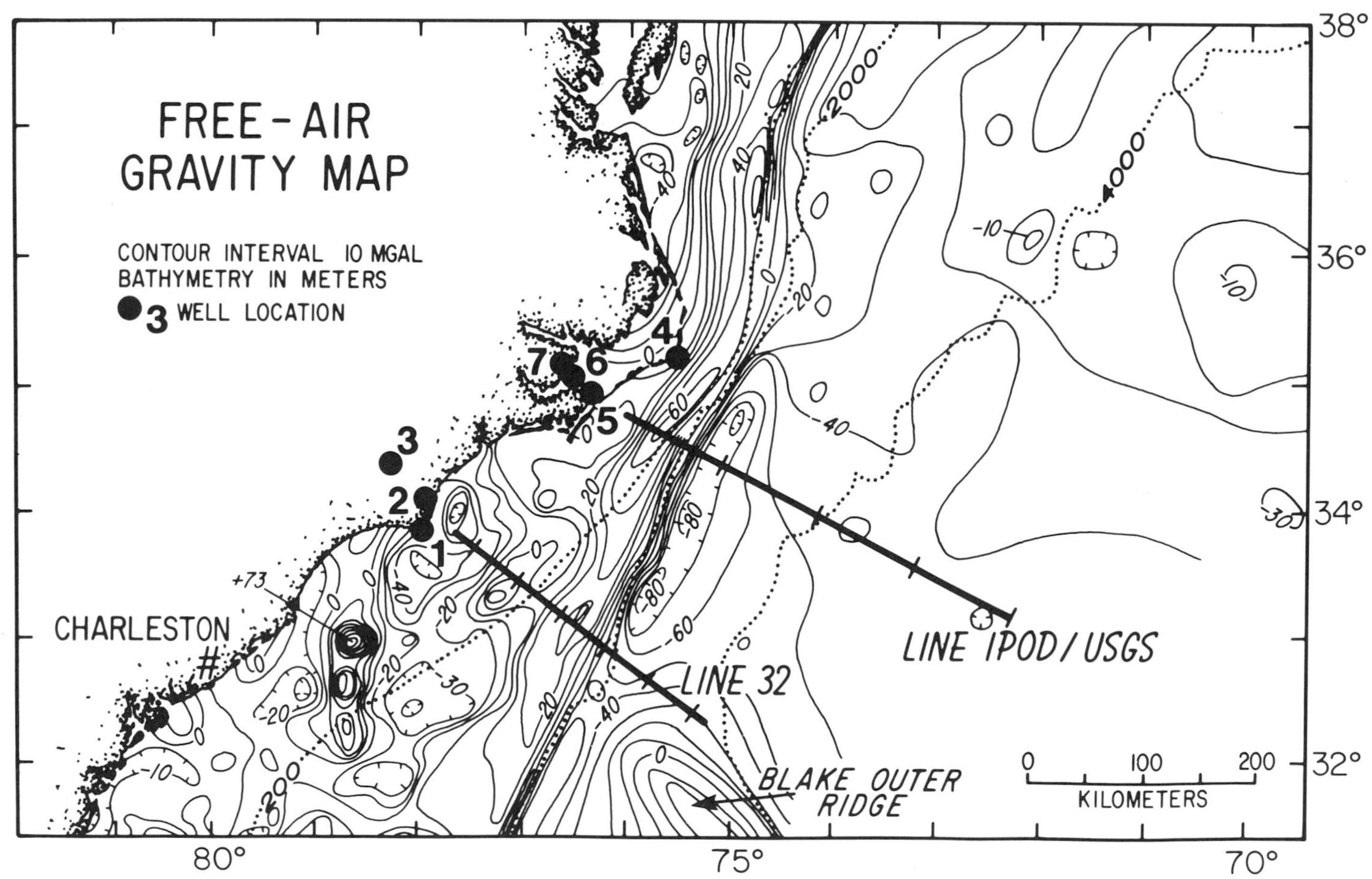

Figure 7 — Free-air gravity anomaly map for the continental margin off North Carolina, showing the locations of lines IPOD and 32 (modified from Grow, Bowin, and Hutchinson, 1979). Tick marks on lines IPOD and 32 correspond to every one-thousand shotpoints. Well locations shown in Figure 7 are described in Table 1.

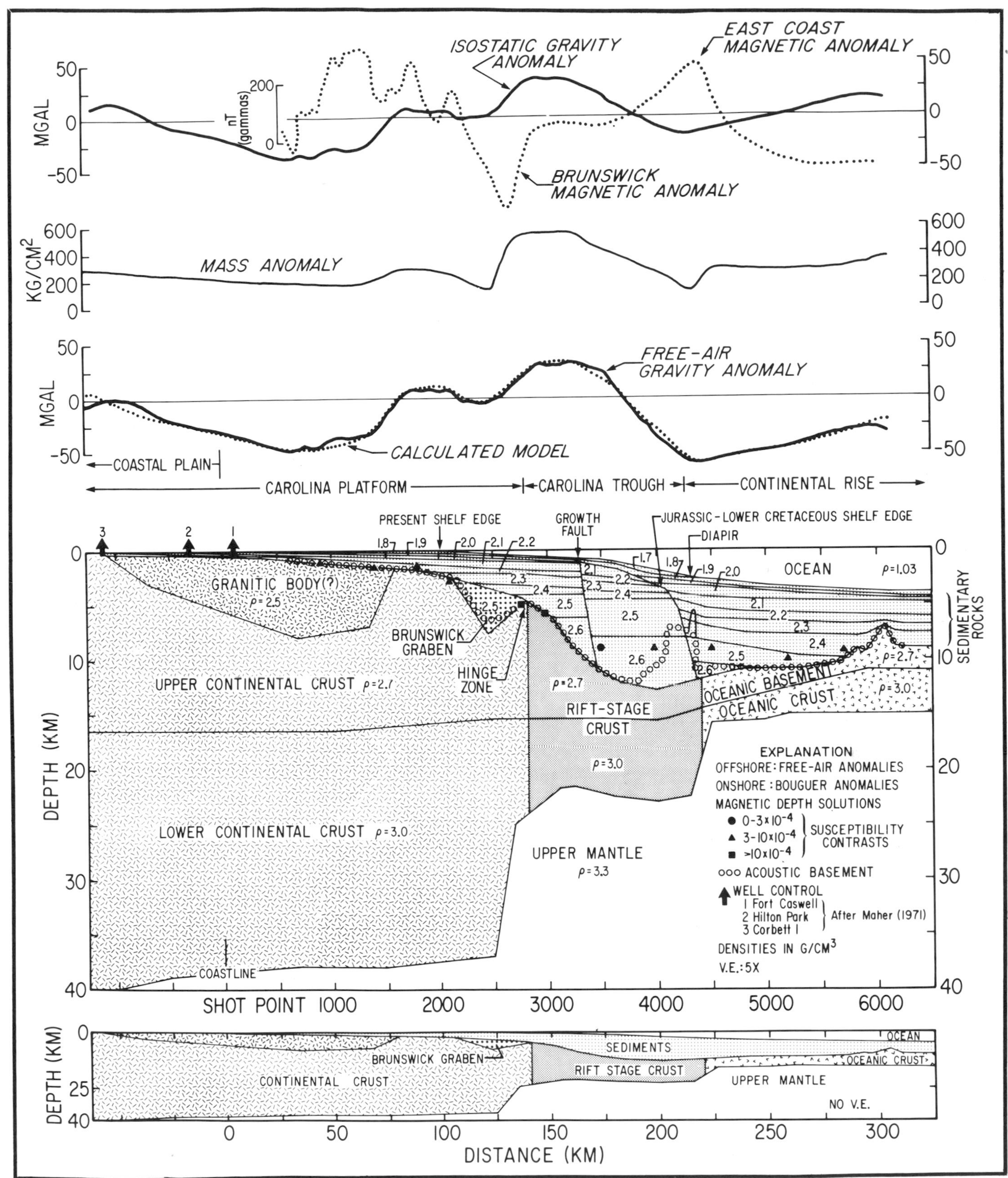

Figure 8 — Gravity model for line 32 off Cape Fear. The base of the sedimentary section is defined by multichannel reflection data, except as indicated by the location of acoustic basement. Acoustic evidence along line 32, unlike that along line IPOD, defines the body modeled as the Brunswick graben beneath the Brunswick magnetic anomaly. The model shown in this figure requires a second low-density body beneath the inner shelf, which may be a low-density granitic body. The transition from continental to oceanic crust is abrupt, and the fairly narrow transition zone is labeled here as rift-stage crust.

the BMA and from 20 to 14 km beneath the ECMA. It has little relief seaward of the ECMA and dips gradually from 30 to 38 km landward of the BMA before becoming level under the coastal plain.

A gravity model computed off Cape Hatteras by Worzel and Shurbet (1955) shows that the Moho rises from continental to oceanic depths in a distance of about 80 km. They placed the edge of the continent at the 1000-fm (1.8 km) isobath. On line IPOD, which is about 80 km south of this gravity line, Moho rises from continental to oceanic depths in a distance of approximately 70 km. The edge of the continent is be-

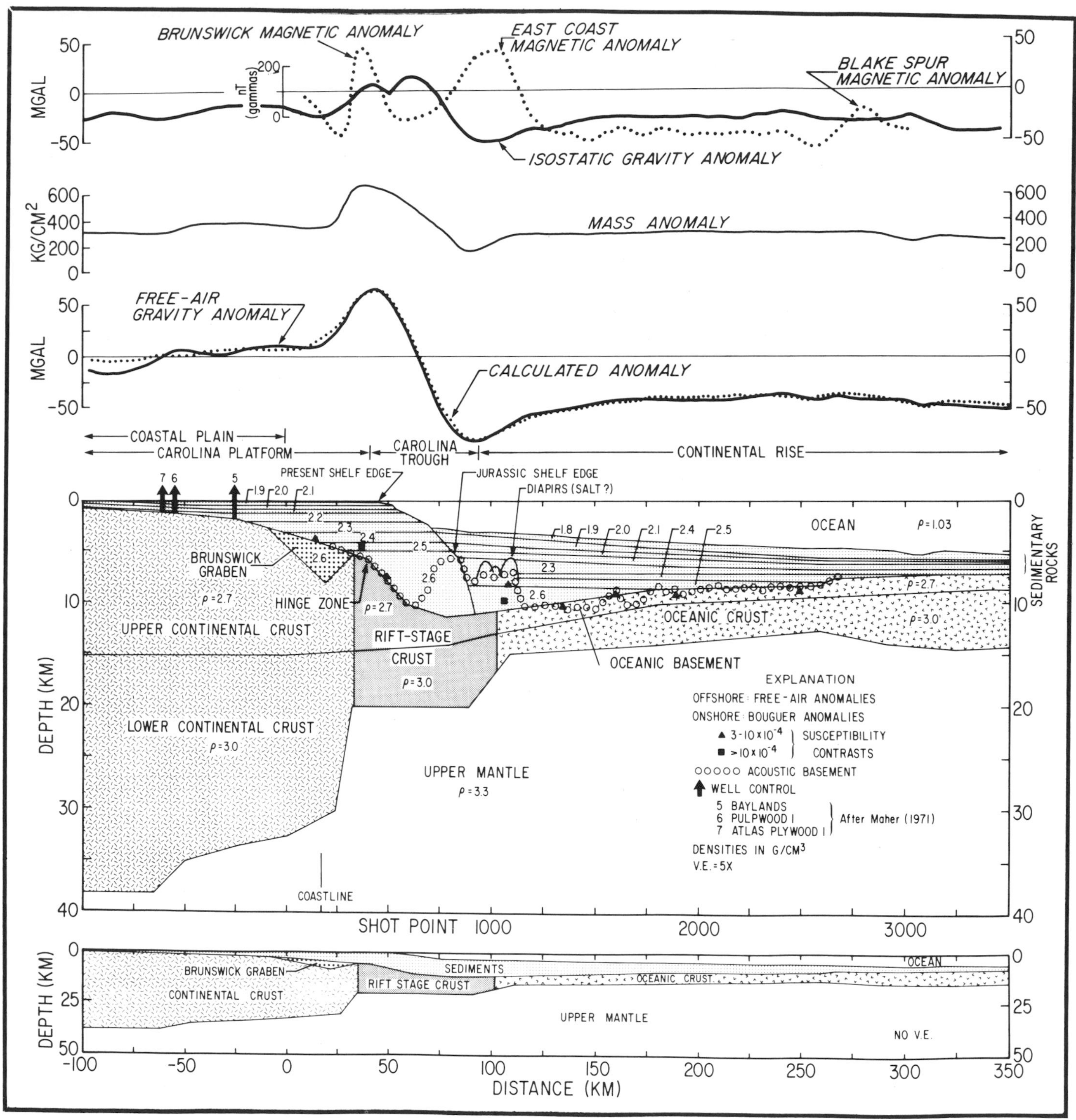

Figure 9 — Gravity model for line IPOD off Cape Hatteras. Multichannel reflection data define the base of the sedimentary section except beneath the ECMA as indicated by symbols for acoustic basement. The Moho is constrained at 14 km depth seaward of SP 2,600 (Grow and Markl, 1977). The model requires a low-density body beneath the BMA. We have no acoustic evidence for this body but interpret it as a prerift Triassic(?) graben called the Brunswick graben. Note the similarity between this gravity model and that for line 32 (Figure 8).

neath the outer shelf, or about 45 km landward of the Worzel and Shurbet (1955) estimate. This model, computed 25 years ago with little sediment thickness or crustal thickness control, is remarkably consistent with our data.

Some of our assumptions need clarification. The assumption of two-dimensionality is appropriate to the linear shape of the Carolina trough. However, the presence of diapiric (salt?) structures beneath the ECMA on both lines present a non-two-dimensional problem. We chose not to include diapiric bodies in our model, because of their small dimension compared to the scale of the model. Line 32 has two additional complications at the seaward end of the line: the presence of a basement high (SP 6,000) and its oblique crossing of the edge of the Blake Outer Ridge gravity high (SP 5,000 to SP 6,200). The Blake Outer Ridge is a recent sedimentary feature (Markl, Bryan, and Ewing, 1970; Bryan, 1970; Hollister et al, 1972) rather than the result of a major crustal structure. The observed and calculated gravity are not in complete agreement in this area, and we emphasize that our model focuses on the structure beneath the BMA and the ECMA, where the assumption of two-dimensionality does hold.

We do not specifically model lateral density variations within the sediments or crust — variations that are geologically reasonable and likely. Lateral density changes in the sedimentary section are likely to be significant, especially in the vicinity of the high-amplitude, high-velocity reflectors associated with the Jurassic-Cretaceous shelf edge (Figures 3 and 4) and in regions of reef deposits (Sheridan, 1974; Schlee et al, 1976; Grow, Mattick, and Schlee, 1979; Sheridan et al, 1979). We tried to minimize these uncertainties within the sedimentary section by constructing isodensity contours from the interval-velocity/density conversion rather than from reflector geometry, but better velocity control on each profile would improve placement of these isodensity contours.

We modeled oceanic and continental crust as an upper, 2.7-gm/cu cm layer and a lower, 3.0-gm/cu cm layer overlying a homogeneous upper mantle of density 3.3 gm/cu cum. Although lateral density variations in the crust and mantle probably exist, we suspect that these variations are small enough that changes in the relative depth of the 2.7/3.0 cu cm and 3.0/3.3 gm/cu cm density boundaries can account for them.

The two-layer crustal configuration agrees with the average oceanic crustal structure derived from a review of refraction data along the U.S. margin (Sheridan et al, 1979). We used the Sheridan et al (1979) model because seismic-refraction data from the East Coast Onshore-Offshore Experiment (ECOOE) yielded conflicting interpretations of the velocity structure of the crust beneath the Carolina trough. These interpretations are superimposed on our gravity model in Figure 10. The velocity model of oceanic crust from

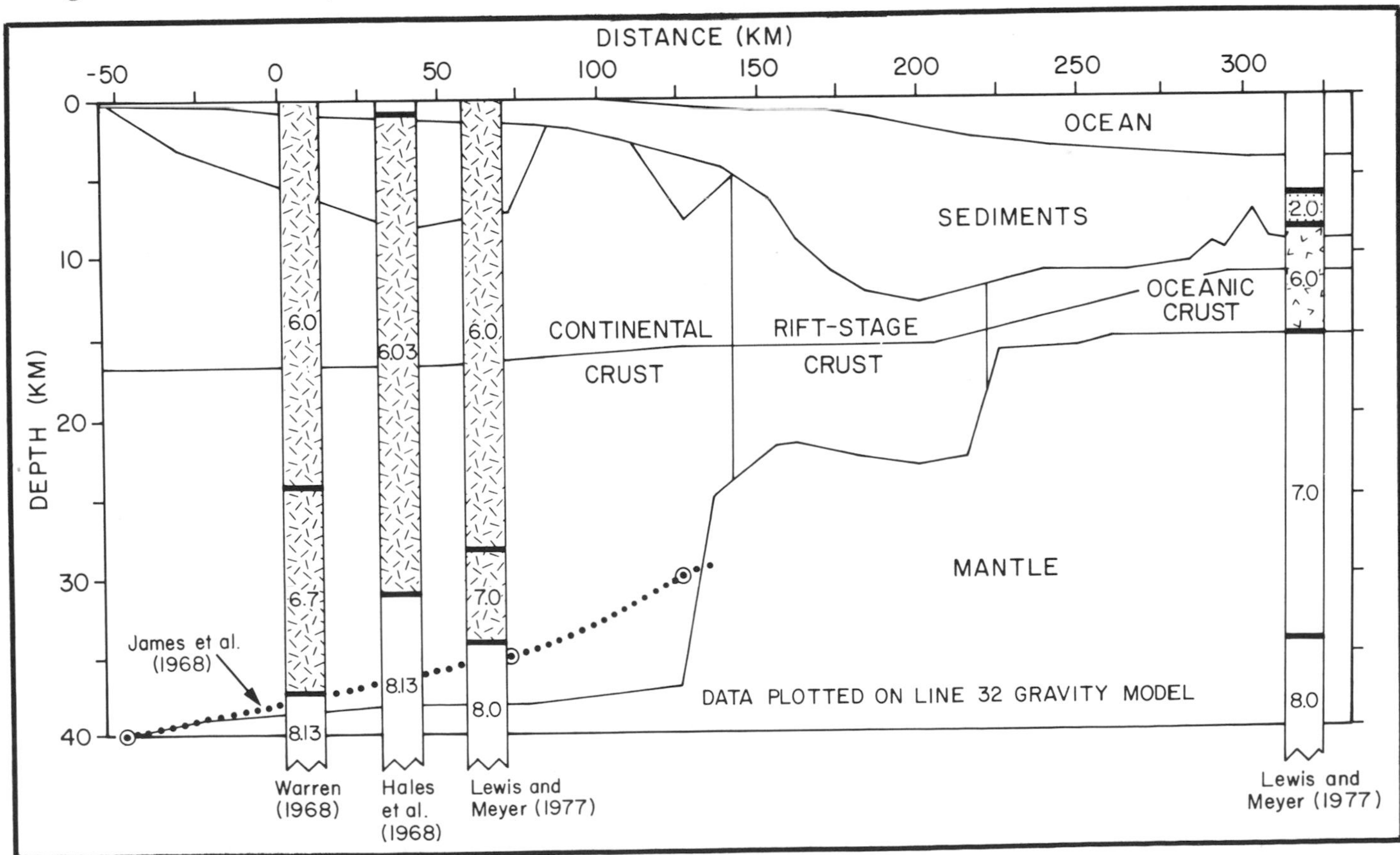

Figure 10 — Summary of published velocity structures of the crust beneath the Carolina trough superimposed on the line 32 gravity model (from Figure 8). The refraction data are from the unreversed southern line of the East Coast Onshore-Offshore Experiment (ECOOE). References are shown in the figure.

ECOOE shows a thin sedimentary layer above a 6.0 km/second velocity crust that overlies a low-velocity mantle at 15 km depth (Lewis and Meyer, 1977). This structure is inconsistent with existing models of normal oceanic crust (Spudich and Orcutt, 1980) and with the model of average oceanic crust along the margin (Sheridan et al, 1979). Continental crust is interpreted from travel-time analysis of the ECOOE data as a single layer of constant velocity (Hales et al, 1968), a double layer of separate but constant velocities (Warren, 1968), and a single layer having a basal-velocity gradient (Lewis and Meyer, 1977). Our model of continental crust uses two layers separated by the landward extension of the Sheridan et al (1979) oceanic boundary at about 15 km depth. The shape and position of this boundary beneath continental crust are not well constrained. Depths to mantle from the ECOOE are estimated at 30 km (Hales et al, 1968), 34 km (Lewis and Meyer, 1977), and 37 km (Warren, 1968). Time-term analysis of the ECOOE data shows a mantle shallowing from 40 km beneath the coastal plain to 30 km beneath the continental slope (James, Smith, and Steinhant, 1968). Given the conflicting interpretations, the data uncertainties, and the unreversed character of the ECOOE line in the Carolina trough, our estimate of the crust/mantle surface is not inconsistent with the refraction results. Until more velocity and density data become available for the Carolina trough region, this simple two-layered density structure of the crust represents a reasonable working model.

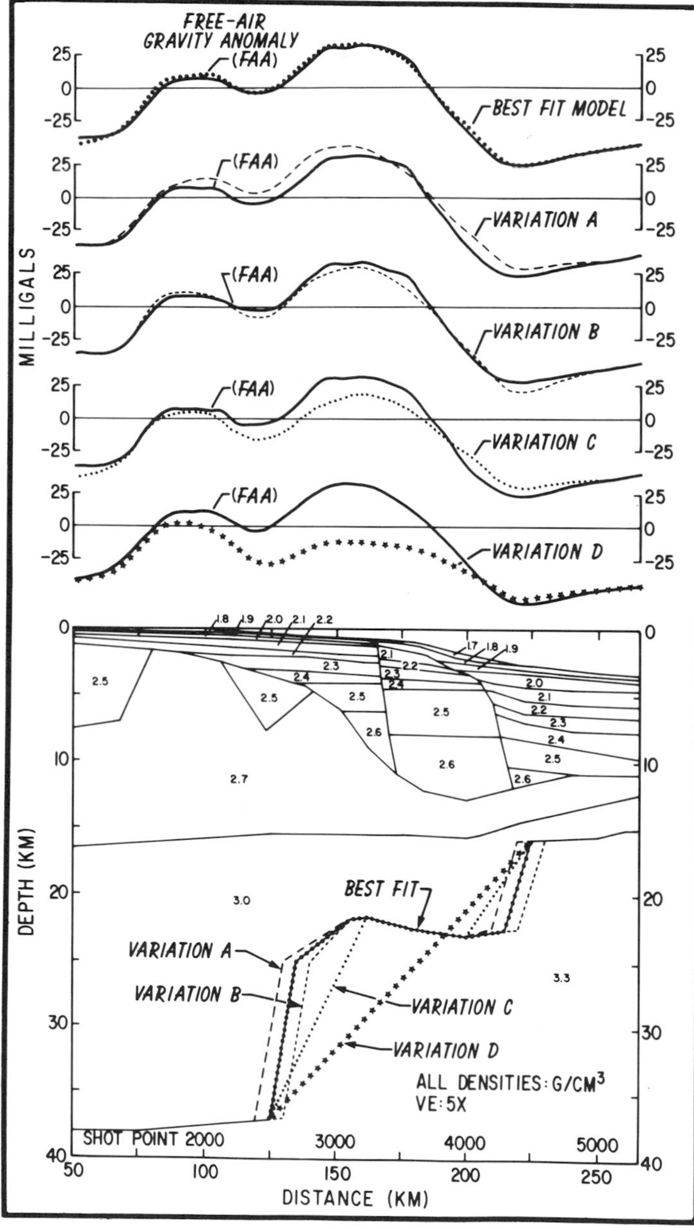

Figure 11 — Limits on the crust/mantle boundary for line 32. All bodies in the model were held constant except the lower crust/upper mantle interface which was varied 5 km landward (A), 5 km seaward (B), partially smoothed (C), and completely smoothed (D) from the best-fit estimate. This figure illustrates the sensitivity of the model to changes in the location of the crust/mantle boundary.

Our gravity models shown in Figures 8 and 9 combine the shallow structure, determined from the multichannel seismic and magnetic-depth-solution data, with an assumed two-layered crustal structure. This leaves the Moho as the free variable. Figure 11 shows the effect that changing the shape of the Moho has on the accuracy of the model for line 32. Four variations are shown: Variation A, in which the steep slopes of the Moho have been shifted 5 km landward; Variation B, in which the steep slopes have been shifted 5 km seaward; Variation C, in which the steep slopes have been partially smoothed; and Variation D, in which the steep slopes have been replaced by a uniform smoothed boundary. Each model shows misfit with the observed data ranging from 7 mgal (Variation B) to 45 mgal (Variation D). This compares with the best-fit model, in which the difference between the observed and computed curves is less than 4 mgal. This misfit demonstrates how sensitive the model is to the slope and position of the crust-mantle interface.

MAGNETIC MODEL ALONG LINE 32

The essential features of the magnetic field along line 32 are a series of short wavelength but relatively low amplitude anomalies landward of the BMA, the large-amplitude negative anomaly of the BMA, and the pronounced positive anomaly of the ECMA at the seaward end of the line. We constructed a magnetic model for line 32 on the basis of two-dimensional source bodies to determine whether the BMA and the ECMA could be explained by structures consistent with the gravity model (Figure 12). The magnetic model assumes that the anomalies caused by the bodies landward of the ECMA are induced from the Earth's present field, which, at the latitude of line 32, is approximately 53,650 nanoteslas (nT; 1 nT = 1 gamma). It has an inclination of 66° and a declination of 10°W. The bodies seaward of the ECMA have a remanent magnetization similar to that expected from oceanic crust. The base of the sediments was used as the upper surface of a 4 to 5-km-thick source layer, and assigned magnetizations were the best fitting, geologically reasonable estimates. The observed magnetic data are from the 1975 U.S. Geological Survey — LKB Resources, Inc. aeromagnetic survey flown at 450 m above sea level (Klitgord and Behrendt, 1977; 1979; Behrendt and Klitgord, 1980).

The computed fit of the magnetic model is in good agreement with the observed magnetic data (Figure 12) despite the simplifications of the model and the absence of information on the magnetic properties of rocks beneath the margin.

The body inferred to be the Brunswick graben in the gravity model is specified as a non-magnetic unit in the magnetic model, which is consistent with our interpretation of low-density sediments filling the graben. If the sediments in this unit are intercalated with basaltic or diabase dikes, flows, or sills, such as those exposed in Triassic grabens along the coastal plain (Klein, 1969; Van Houten, 1977), then a zero magnetization is incorrect. If flows, sills or dikes having a remanent component in a reverse direction are present within the graben, they would tend to improve the fit by increasing the amplitude of the negative anomaly. Alternatively, if the graben is filled

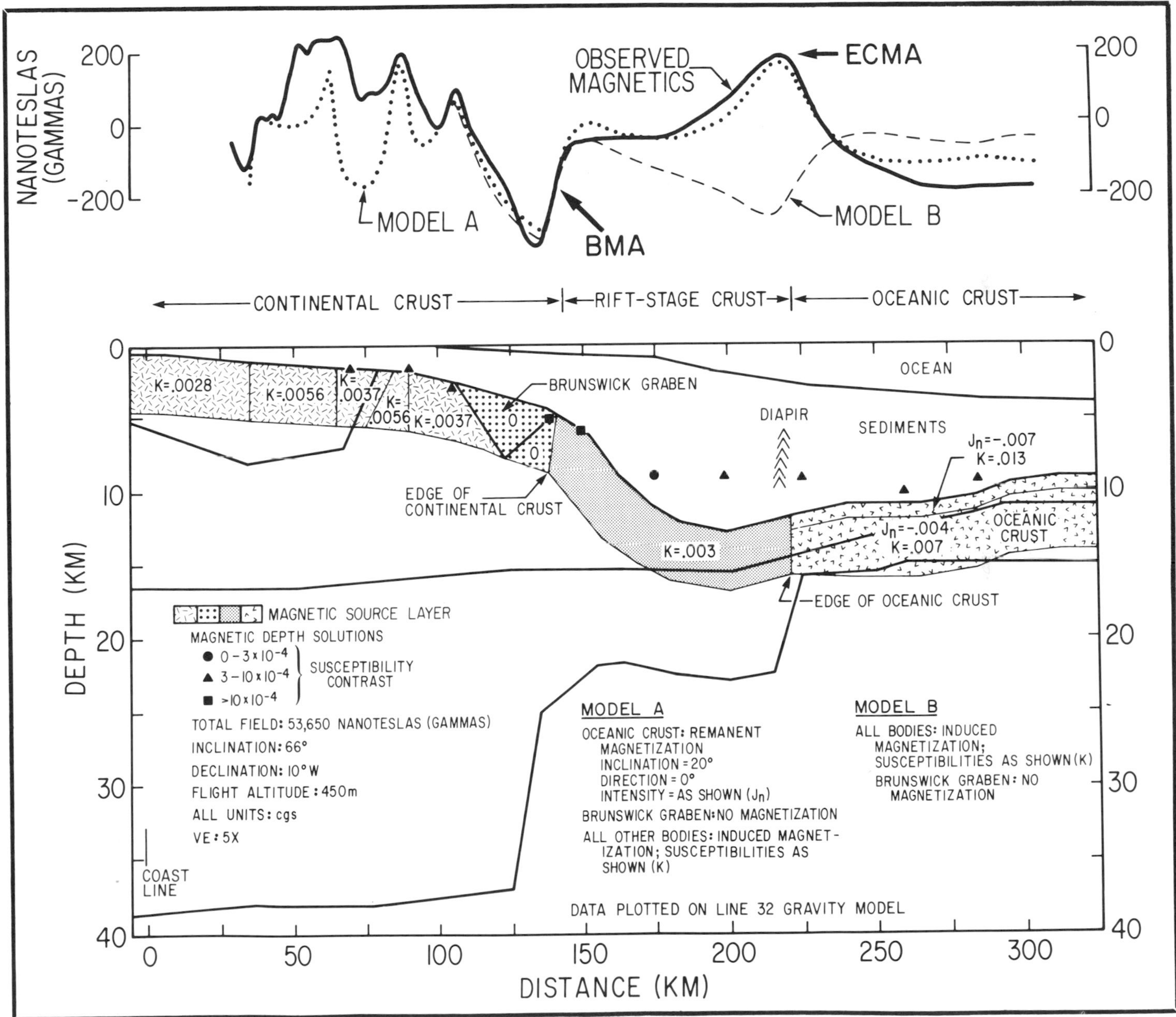

Figure 12 — Magnetic model of line 32. The model assumes a source layer about 4 km thick consisting of bodies defined by the gravity model. Model A assumes induced magnetization for continental and rift-stage crust, zero magnetization for the Brunswick graben, and remanent magnetization for oceanic crust. Model B assumes all induced magnetizations except zero magnetization for the Brunswick graben. The Brunswick and East Coast magnetic anomalies are easily explained by Model A, suggesting that the Brunswick magnetic anomaly is caused by a non-magnetic body at the edge of continental and rift-stage crust, and that the East Coast anomaly originates as the edge effect of magnetic oceanic crust.

by magnetic source rocks having a component of normal magnetization, then the strong negative anomaly could be produced either by increasing the postulated susceptibilities of the surrounding rocks, or by increasing the postulated thickness of the magnetic source layer. Until more data on the age and composition of the rocks within this graben are collected, we prefer the simplest model in which the graben represents a non-magnetic body.

The Brunswick graben in the magnetic model is larger than that in the gravity model in order to make a more exact fit. Because the size of the body is not adequately defined by the seismic data nor uniquely defined by the gravity data, its larger size is still consistent with the available information. This body immediately underlies the BMA and is its prime source in this model.

The computed magnetic model shows a larger edge effect at the eastern margin of the Brunswick graben than is observed in the data along line 32. We have not tuned this slight misfit because the shape of the body more than adequately explains the shape of the anomaly, and we have no conclusive evidence to justify adjusting the postulated boundaries or postulated magnetizations of the body. Although such a large positive edge effect is not present in the observed field of line 32, a large peak is associated with the BMA on the IPOD line (Figure 9).

The body inferred to be granitic in the gravity model (Figure 8) underlies the short-wavelength, but relatively low-amplitude, anomalies beneath the inner shelf. Since granites show large variability in magnetic properties (Telford et al, 1976, p. 121), we divided this body into several smaller bodies to simulate magnetization contrasts that might, in a general way, explain the frequency content of the anomaly. The contrasts used in the model are greater than those typically found in granites, although they are within the range found in metamorphic rocks. Continental basement beneath the shelf is magnetically similar to that of the slate belt and coastal plain (Taylor, Zietz, and Dennis, 1968; Zietz, 1970; Klitgord and Behrendt, 1979; Hatcher and Zietz, 1980), suggesting that Paleozoic plutonic and metamorphic rocks of the Piedmont extend offshore. The magnetic bodies within our model are consistent with an Appalachian-type basement composed of different lithologies and contrasting magnetizations. We put no geologic interpretations on these bodies and emphasize that the model is non-unique.

Because recent studies have described the ECMA as the edge effect between the more highly magnetized oceanic crust and the relatively nonmagnetic continental or transitional crust (Keen, 1969; Keen et al, 1974; Klitgord and Behrendt, 1979; Behrendt and Klitgord, 1980), we specified that the seaward end of line 32 consists of a 1-km-thick layer of highly magnetized material interpreted as oceanic basalt. This overlies a thicker, less magnetized layer interpreted as oceanic crust. These bodies terminate along a vertical edge beneath the ECMA, which abuts considerably less magnetic material. The magnetizations of the oceanic material are reversed in agreement with earlier models of the East Coast margin (Emery et al, 1970, Figure 33; Rabinowitz, 1974; Steiner and Helsey, 1975). Recent studies of Triassic-Jurassic polarity intervals show that magnetization may have been reversed during the late Triassic-early Jurassic (Opdyke and McElhinny, 1965; Burek, 1970; McElhinny and Burek, 1971; Steiner and Helsley, 1975; Steiner, 1980), although a relatively low field strength (de Boer, 1968; Taylor, Zietz, and Dennis, 1968) and a rapidly changing magnetic pole position (de Boer, 1968; de Boer and Snider, 1979; Steiner, 1980) introduced scatter into the directional data. Our model, like that of Keen (1969), shows that a reversely magnetized oceanic edge next to a layer having an induced positive magnetization can easily explain the ECMA. As in other parts of the model, the fit is not perfect. We attribute this to details of the structure and magnetization that are presently unresolvable.

Table 1. Summary of well data.

Well No.[1]	Well name[2]	Altitude (ft)	Total depth (ft)	Altitude of pre-Cretaceous rocks (ft)	(km)	Basement lithology
1	Fort Caswell	11	1,543	−1,529	−0.466	Metamorphic rocks
2	Hilton Park	9	1,330	−1,100	−0.335	Granite
3	Corbett 1	23	765	− 667	−0.203	Unknown
4	Hatteras Light 1	24	10,054	−9,854	−3.003	Granite
5	Baylands	0	5,607	−5,561	−1.695	Unknown
6	Pulpwood 1	11	3,667	−3,646	−1.111	Granite
7	Atlas Plywood 1	11	3,425	−3,403	−1.037	Granite

[1]this paper. Well locations are shown in Figure 7.
[2]from Maher (1971)

The body between the BMA and the ECMA is modeled as a uniformly magnetized body of susceptibility 0.0030 (cgs units). A wide variety of igneous and some metamorphic rocks have susceptibilities near this value (Telford et al, 1976, p. 121), making a precise estimate of the body's lithology impossible. The model shows that the deepening topography is sufficient to explain the relatively flat magnetic field between the BMA and the ECMA. Alternatively, the large depth-to-basement in this region would obscure any short-wavelength anomalies associated with small susceptibility contrasts within the basement rocks. The source could be composed of rocks of slightly different susceptibilities that together average about 0.0030. We call this transitional body rift-stage crust and discuss it in greater detail in the following section.

The magnetic anomalies of the IPOD line differ from those of line 32 primarily in the larger amplitude and broader wavelength of the BMA. Preliminary models from line 32 indicated that the larger amplitude could be generated by making the susceptibility contrast between the Brunswick graben and rift-stage crust larger by .001 to .002 emu, or well within existing uncertainties. Models also indicated that the broader wavelength could be explained by a combination of this larger amplitude and flattening the seaward boundary of the graben source layer. We did not model the magnetics of the IPOD line since we did not feel the additional information would modify the general results of the line 32 model.

The major assumptions of this model are that the source layer is about 4 km thick, that the structures are two dimensional, that the computed anomalies landward of the ECMA are caused primarily by induced magnetization, and that anomalies seaward of the ECMA result from remanent magnetization. The 4-km-thick layer is a simple approximation of the source that corresponds to seismic layers 2 and 3 for oceanic crust. It is arbitrary for the continental and rift-stage crust. A thinner or thicker layer would not cause unrealistic adjustments to the susceptibility values, although thickening a uniformly magnetized layer would make it more difficult to match the short wavelength anomalies on the shelf. The two-dimensionality argument is similar to that presented for the gravity model: non-two-dimensional features occur at the landward end of line 32 over the Carolina platform, but in the area of the BMA and the ECMA, the geophysical anomalies are linear in the direction perpendicular to the profile.

The assumption of induced magnetization is valid for continental crust because most granitic and metamorphic rocks have low values of the ratio of remanent to induced magnetization (the Q value; for a discussion of the Q values, see Nagata, 1969, p. 397). The assumption of remanent magnetization for oceanic rocks is valid because oceanic rocks tend to have high Q values (Vacquier, 1972, p. 23). Figure 12, model B shows the curve computed if the entire model has induced magnetization. This alternate curve shows major misfit on the oceanic side, indicating that an induced field does not explain the ECMA. The important features of this model are that the anomalies landward of the ECMA are reasonably explained by induced magnetization, whereas the anomalies at and seaward of the ECMA are more readily explained by a negative thermoremanent magnetization (TRM). Since the anomalies are caused by relative magnetization variations, a positive magnetization could be assumed for the oceanic crust. This assumption would require considerably higher magnetizations for the continental and rift-stage crusts.

The magnetic depth-to-source estimates of Klitgord and Behrendt (1979) coincide closely with the depth to acoustic basement landward of the Carolina trough. They are as much as 3 km more shallow than basement estimates determined from the multichannel analysis at the seaward edge of the Carolina trough and over oceanic crust (Figure 6). The two shallow solutions at the EMCA are at the edges of the opaque acoustic zone where basement might be at shallower depths. These depth estimates may also be too shallow if the model assumptions about sources are incorrect. The two source estimates over oceanic crust are within 1 km of acoustic basement, well within the expected accuracy of either the magnetic depth estimates or of the velocity information at that depth. Our model uses basement depths determined from the multichannel data rather than depths determined from magnetic data. The use of a slightly more shallow body at the EMCA would have negligible effect on our model's gross features. Clearly, more data are needed to define the geometry, magnetization, and existence of the source features.

DISCUSSION

A geologic cross section of the Carolina trough, based on the multichannel data and on gravity and magnetic models of line 32, is shown in Figure 13. Age relations and locations of carbonate complexes, diapirs, subbasement bodies, continental, rift-stage crusts, oceanic crusts, and the inferred Moho boundary are outlined. The boundaries of the sedimentary units differ from the boundaries of the density polygons (Figure 8) because the sedimentary units are defined by reflector geometry: the polygons are defined from isodensity contours.

The edge of oceanic crust is placed beneath the axis of the ECMA because major changes in crustal structure occur in its vicinity. These include changes in the depth to Moho, crustal thickness, magnetic properties and a zone of weakness along which diapirs have formed.

The edge of continental crust is placed beneath the BMA, also because of major changes in crustal structure. These are changes which effect the depth to Moho, the crustal thickness, the basement slope (the hinge zone), the basement structure (the Brunswick graben), and the magnetic properties of the crust. The fact that the BMA swings westward onto the continent suggests that the anomaly and its associated graben are not uniquely related to the edge of continental crust, and that the correlation between the BMA and the edge of continental crust is probably more coincidental than causal. The best criterion for defining the edge of continental crust is the change in crustal thickness. However, in areas where information on the deep structure is not available, the hinge-zone represents the next best criterion. We use the BMA in the Carolina trough because it is a continuous geophysical anomaly which coincides with each of these criteria.

The gravity and magnetic models of the rift-stage crust indicate that simple density and uniform magnetization assumptions can explain the observed gravity and magnetic fields. They also indicate that the lithologic and structural properties may be reasonably simple at this scale, although the inherent non-uniqueness of these models and the lack of resolution at depth makes this conclusion tentative. In the absence of drilling, modern and exposed analogues of this crust (for example, the east African rift system, the Rio Grande rift system, or the Bay of Biscay) are probably the best indicators of its composition. The rift-stage crust underlying the Carolina trough must have formed before the initiation of normal sea-floor spreading, which is postulated to have begun in the earliest Jurassic time (186 m.y. to 195 m.y. ago; Cornet, 1977; Vogt, 1973; de Boer and Snider, 1979; Grow, 1980) and after the initiation of rifting, which could be Late Triassic (Cornet, 1977) or older.

Basement Structure Near the ECMA

It has been postulated that a basement ridge beneath the outer shelf and rise is part of the North American passive margin (Drake, Ewing, and Sutton, 1959; Burk, 1968; Drake and Nafe, 1968; Drake, Ewing, and Stockard, 1968; Emery et al, 1970; Keen and Keen 1974; Klitgord and Behrendt, 1979; Grow, Bowin, and Hutchinson, 1979; Grow, Mattick, and Schlee, 1979; Schuepbach and Vail, 1980). Poor acoustic penetration makes the multichannel data from the Carolina trough inadequate to resolve whether a basement ridge exists beneath the ECMA. However, the gravity and magnetic models show that the structure of the margin can be explained without a significant basement ridge. The lack of agreement between the magnetic depth estimates and acoustic (oceanic) basement on the seaward end of line 32 suggests that the assumptions behind the magnetic depth estimate method and upon which the basement ridge is based (Klitgord and Behrendt, 1979) may require modification near the ECMA.

Multichannel seismic-reflection data from USGS line 25 in the Baltimore Canyon trough have been interpreted as showing no basement ridge based on the existence of reflectors projected through the zone of limited acoustic penetration beneath the ECMA (Grow, 1980). Clearly, gravity and magnetic modeling information and additional deep geophysical information are needed from other parts of the margin to test whether this elusive basement ridge exists.

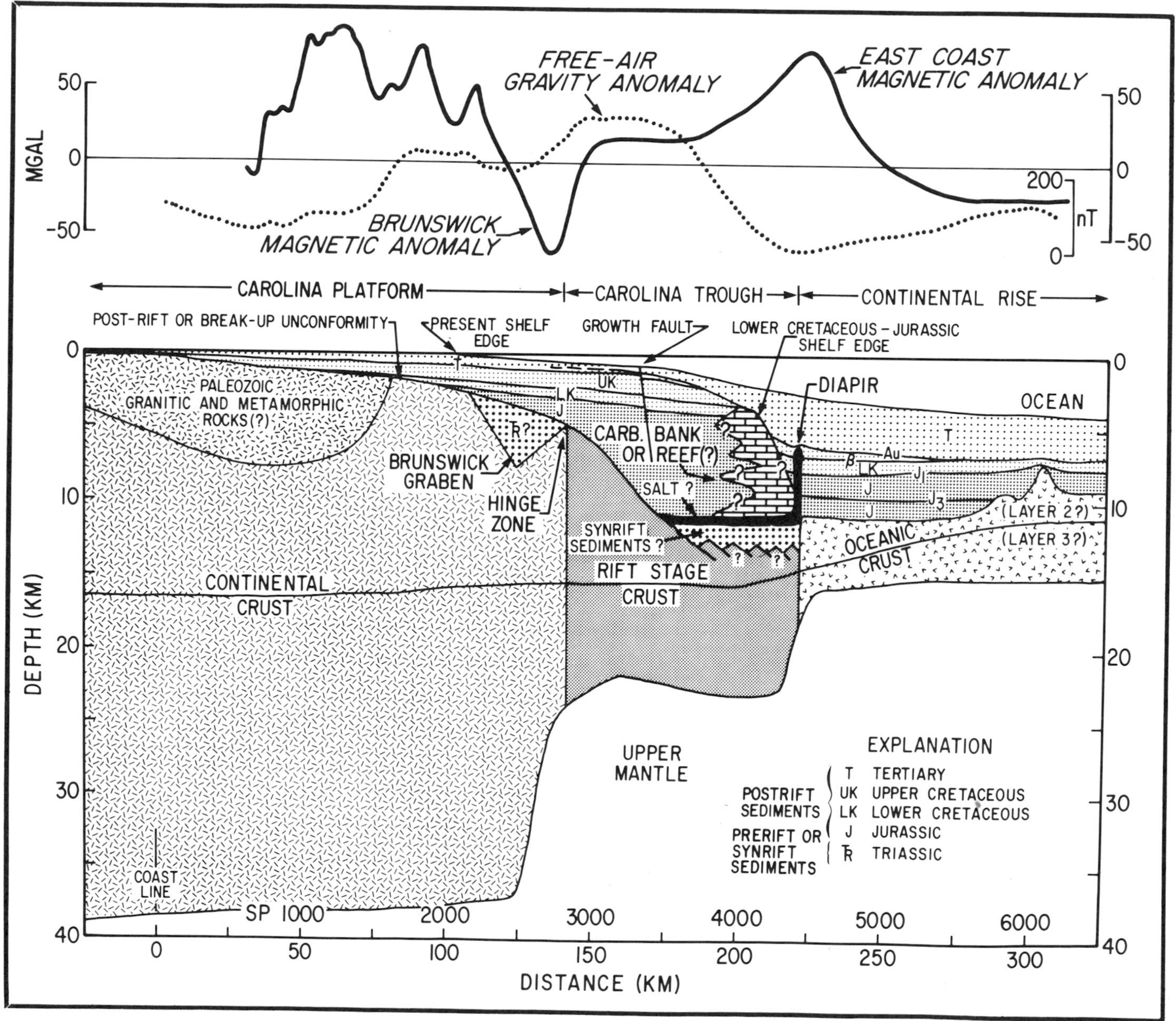

Figure 13 — Schematic cross section of the crustal structure along line 32, based on the multichannel reflection data and the gravity and magnetic models.

Brunswick Graben

The magnetic model indicates that the Brunswick graben underlies and is the source of the Brunswick magnetic anomaly. The density and magnetization of the material filling the trough is lower than that of the surrounding basement rocks. This, together with the parallel subbasement reflectors on line 32, are the criteria for postulating a sediment-filled graben. Magnetic depth estimates (Figure 6) indicate that this graben can be traced more or less continuously from Cape Hatteras to Brunswick, Georgia, and therefore represents a significant basement feature along the Carolina-Georgia margin (Klitgord and Behrendt, 1979).

Laterial variations in the composition of the rocks filling the graben probably occur. On the IPOD line, the density of the graben (2.6 gm/cu cm) and amplitude of the BMA (360 nT) exceed those of line 32 (2.5 gm/cu cum and 300 nT, respectively). This suggests higher density magnetic source rocks, such as diabase sills, dikes, or volcanic rocks, might exist near the IPOD line. Our data suggest that higher density and highly magnetic igneous rocks are probably not present in large quantities near line 32.

Models of the BMA on land, where it extends from Brunswick, Georgia, to the Alabama line, link the magnetic low with a Triassic graben (Popenoe, 1977; Gohn et al, 1978; Nishenko and Sykes, 1979) and the magnetic high with a rift-related mass of mafic rock or a crystallized magma chamber of an old (Paleozoic?) island arc (Pickering, Higgins, and Zietz, 1977). Land wells in the BMA low area near the coast contain Paleozoic rhyolitic tuff (Applin, 1951; Chowns and Williams, in preparation), indicating that Triassic rocks may actually be north of the BMA onshore, and that Paleozoic volcanic rocks are the source of the BMA. The Paleozoic rhyolites in the subsurface of Georgia have very low susceptibilities and are nearly flat-lying (Popenoe, personal communication, 1980). Thus, they would be identical in geophysical expression to Triassic sediments.

The age of the sediments postulated to fill the Brunswick graben offshore is uncertain. The exposed rift-related grabens in the coastal plain and Piedmont are Late Triassic and Early Jurassic in age (Cornet, 1977). However, data from the wells mentioned in the previous paragraph suggest that the rocks causing the BMA onshore are Paleozoic. Either of these ages is prerift, which is compatible with our observations that the layered units within the Brunswick graben are truncated unconformably by the breakup unconformity (Figure 4).

The Brunswick graben's formation age is also uncertain. We favor the interpretation that the graben was formed during the rifting event as a downdropped trough. Because the sediments within this downdropped block can be older than the downdropping, this interpretation allows the graben to be synrift, whereas the infilling sedimentary rocks could be prerift.

Our magnetic model shows that the source of the BMA (the Brunswick graben) is very different from the source of the ECMA (the boundary between oceanic and rift-stage crust). Therefore, the BMA is different from, and not an inner branch of, the ECMA, as originally postulated (Taylor, Zietz, and Dennis, 1968; Zietz, 1970). Our offshore and onshore data are insufficient to determine whether the BMA onshore (oriented east to west) has the same source as the BMA offshore (oriented north to south). Any interpretation of the BMA onshore source needs to include its offshore association with the Brunswick graben.

Rift-Stage Crust

The boundary between the oceanic and continental crusts of passive Atlantic-type margins has been described based on refraction results (Keen et al, 1974; Ludwig et al, 1979), isostatic gravity anomalies (Talwani and Eldholm, 1972, 1973; Rabinowitz, 1974; Rabinowitz and LaBrecque, 1977), magnetic anomalies (Keen, 1969; Keen and Keen, 1974), the location of a basement ridge (Burk, 1968; Emery et al, 1970; Schuepbach and Vail, 1980), isobaths (Worzel and Shurbet, 1955), and specific structural or topographic features (Talwani and Eldholm 1972; Rabinowitz, 1976; Rabinowitz, Cande, and LaBrecque, 1976). Recent theories on crustal structure across rifted continents have called for a transition zone that separates oceanic from continental crust and is considered to be either thinned continental crust (Sheridan, 1974; Sheridan et al, 1979; Montadert, Roberts, and DeCharpal, 1979; Bott, 1979) or thickened oceanic crust (Rabinowitz and LaBrecque, 1977). We used the information from our gravity and magnetic models of the Carolina trough to define the edge of continental crust at the BMA and the edge of oceanic crust at the ECMA. We prefer to call the intervening crust rift-stage crust, according to Falvey (1974), because such a term encompasses a crust having both oceanic and continental properties and relates that crust to the processes that created it.

Rift-stage crust includes crust that has been heated, thinned, eroded, metamorphosed, subsided, faulted, and extensively intruded for as long as 50 million years prior to actual breakup (Falvey, 1974). These processes are commonly proposed for the rifting process (Mohr, 1972; McConnell, 1972; Hutchinson and Engels, 1972; Cordell, 1978; Montadert, Roberts, and DeCharpal, 1979) and are probably sufficient to create a crust that is different from either continental or oceanic crust. Since the Carolina trough lies between the BMA and the ECMA (Klitgord and Behrendt, 1979), or between the edges of continental and oceanic crust, the entire trough is underlain by rift-stage crust.

The thickness and width of the zone of rift-stage crust for lines 32 and IPOD are presented in Figure 14, together with those of USGS line 5 from the Long Island platform (Grow, Bowin, and Hutchinson, 1979) and USGS line 25 off New Jersey (Grow, 1980; Grow and Hutchinson, unpublished data). On the northern lines, the edge of oceanic crust is placed beneath the axis of the ECMA but because no equivalent of the BMA exists this far north, the edge of continental

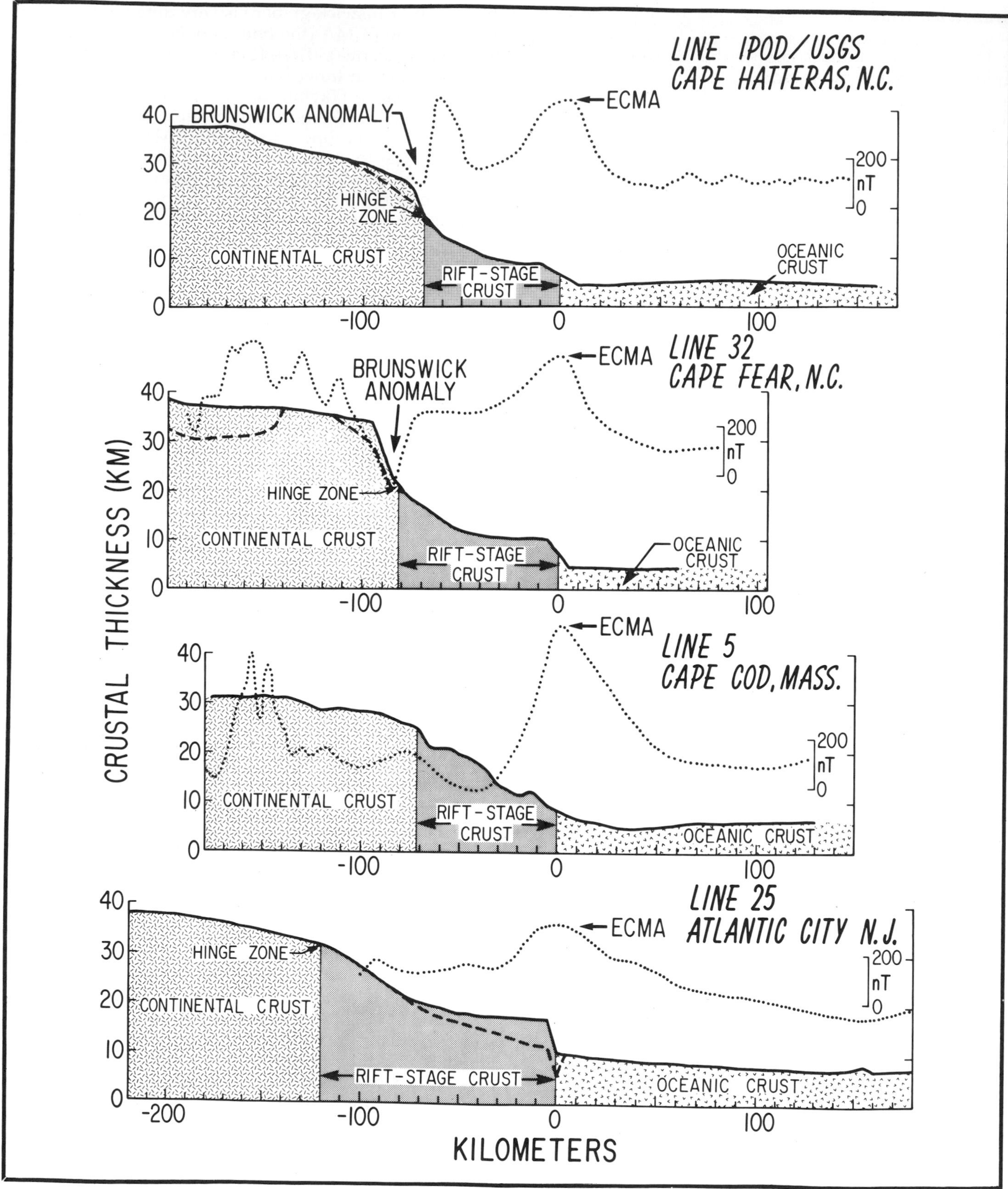

Figure 14 — Thickness of the crust for lines IPOD (Cape Hatteras), line 32 (Cape Fear), line 5 (Long Island) and line 25 (Cape May) shown on Figure 1. Thickness is the sum of the layers having densities of 2.7 and 3.0 gm/cu cm. For lines 5 and 25, the edge of continental crust corresponds to the hinge zone; the edge of oceanic crust is inferred to coincide with the axis of the ECMA. The data are plotted with respect to the ECMA axis. Dashed lines indicate crustal thickness excluding the subbasement bodies. Line 5 is modified from Grow, Bowin, and Hutchinson (1979). Line 25 is from Grow and Hutchinson (unpublished data, 1981).

crust is placed at the change in crustal thickness. On line 25, a hinge zone coincides with the edge of continental crust. No pronounced hinge zone exists on line 5. These boundary criteria are modifications of the 25 and 8-km-thickness criteria used by Grow, Bowin, and Hutchinson (1979) to define the edges of normal-thickness continental and oceanic crust. Crustal thickness of the rift-stage crust ranges from 7 to 20 km for line 32, and from 6.5 to 18 km for line IPOD. Its width is 80 km for line 32, and 70 km for line IPOD. Rift-stage crust on line 5 is approximately 70 km wide and is from 8.5 to 28 km thick (Figure 14). On line 25, it is 120 km wide and is from 10.5 to 32 km thick (Figure 14).

Results from lines IPOD, 32, and 5 indicate that the zone of rift-stage crust is fairly narrow (70 to 80 km). Our results agree with estimates of the ocean-continent transition-zone width from other passive margins in the Atlantic Ocean. The width off Nova Scotia is estimated at 70 km (Keen et al, 1974), and that off Argentina is estimated at 60 km (Ludwig et al, 1979). Specific structural or topographic features that have been equated with the ocean-continent boundary off South Africa (Agulhas fracture zone; Rabinowitz, 1976; Rabinowitz, Cande, and LaBrecque, 1976), South America (Falkland Escarpment; Rabinowitz, Cande and LaBrecque, 1976), and Norway (Vøring plateau; Talwani and Eldholm, 1972) imply fairly narrow transition zones at these margins. The zone of rift-stage crust on line 25 is approximately 120 km wide, significantly wider than the 70 to 80-km-wide zone of the other lines. Crustal thickness is also slightly greater. The Baltimore Canyon trough is a much wider and deeper basin than the Carolina trough (Grow, 1980), which may explain its wider and thicker rift-stage crust. Additional models from other parts of the margin are necessary to investigate the range of variations of this rift-stage crust.

Crustal Configuration at the Initiation of Sea-Floor Spreading

Recent analyses of passive-margin subsidence rates indicate that three processes can account for most of the observed subsidence: 1) an initial rapid subsidence due to extensional rifting which precedes sea-floor spreading, and depends on the original crustal thickness and the amount of stretching (McKenzie, 1978); 2) sediment loading during the rifting and postrift evolution (Watts and Ryan, 1976); and 3) lithospheric cooling as the ridge migrates away from the margin (Parsons and Sclater, 1977).

By isostatically unloading the postrift sediments from line 32 and adding a thermal expansion factor, we can account for the second and third processes of the subsidence curve. Then, we can approximate conditions in the Carolina trough prior to postrift sediment loading and lithospheric cooling (just before the initiation of sea-floor spreading; Figure 15). We make no assumptions about the initial rapid subsidence part of the curve.

The structure of the line 32 gravity model was used to construct theoretical, 40-km-thick mass columns of the crust and mantle, located from the present coastline to the initial position of the Mid-Atlantic Ridge, which we assume is the ECMA. Postrift sediments were removed in one step and each column was isostatically balanced locally against a reference column, located landward of the coastline. This mathematical unloading caused the basement to rebound to the level shown by the dashed line in Figure 15.

We then applied, a smooth, sinusoidal thermal expansion curve, which produces maximum uplift of 3 km at the ECMA and decreases to zero uplift at the present coastline. This accounts for the increased heating during rifting (dotted line, Figure 15) and assumes that a passive margin, like normal oceanic crust, subsides approximately 3 km as the ridge crest migrates away from the continent (Parsons and Sclater, 1977).

This exercise, despite its simplification, relates to the crustal structure of the Carolina trough in three ways: 1) the model indicates that the rift-related thermal expansion of the sediment-free crust would result in an erosional surface, which we see in the data as the breakup unconformity. Elevating the breakup unconformity to 1 km above sea level landward of the BMA suggests that heating effects assumed for that part of the margin were overestimated in the model. 2) the model suggests that sediments within the Brunswick graben could have been tilted landward by thermal uplift near the ECMA. This mechanism, together with the listric faulting postulated to occur during breakup (Montadert, Roberts, and DeCharpal, 1979), could account for the observed subbasement reflectors of the Brunswick graben. 3) the model shows that the Carolina trough was well below sea level at the onset of sea-floor spreading. Determining the age and method of salt formation at the ECMA and/or within the Carolina trough remains a problem (Dillon et al, this volume). If the salt was deposited in a shallow-water environment, it must predate the onset of sea-floor spreading (prerift or synrift). Alternatively, if salt deposition occurred by evaporation at depths well below mean sea level, such as that postulated to have taken place in the Mediterranean during the Miocene (Hsu, Cita, and Ryan, 1973), then the age of the salt could be late synrift or early postrift. Upper Triassic salt deposits from the Grand Banks, Newfoundland, are interpreted as shallow-water evaporites formed within an extensive and complex graben system intermittently fed by Tethyan waters (Jansa, Bujak, and Williams, 1980). If the Carolina trough evaporites are part of this larger system, then an Upper Triassic, late synrift age is consistent with this model's results.

SUMMARY

We modeled the deep crustal and upper mantle structure of two lines over the Carolina trough where we have excellent multichannel seismic-reflection data to define the shallow structure. Our conclusions from these models are:

(1) The entire Carolina trough is underlain by crust

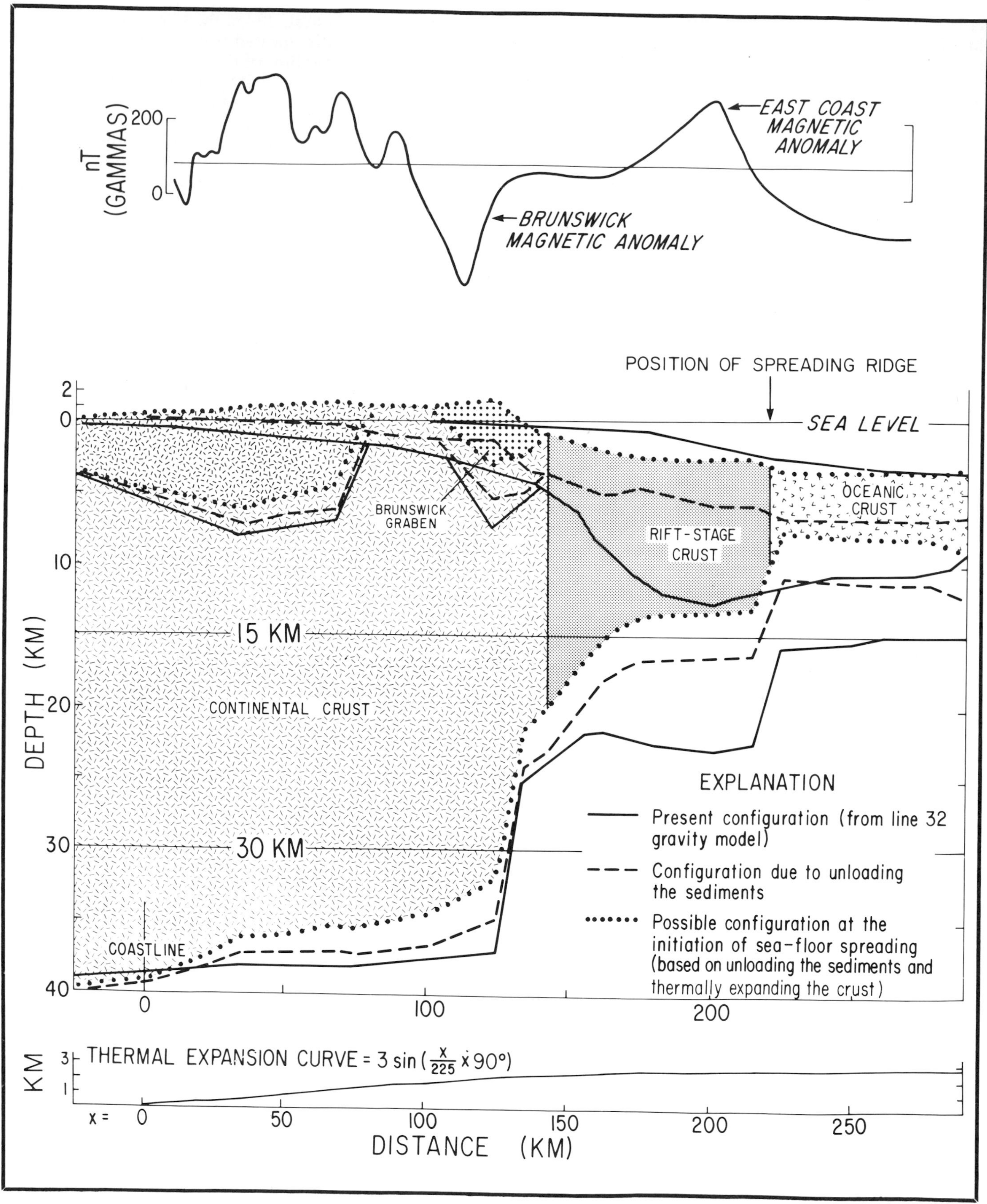

Figure 15 — Approximation of the geometry of the crust beneath line 32 resulting from mathematically unloading the sediments and thermally expanding the lithosphere. This figure is a simplified approximation of what the Carolina margin could have looked like at the onset of true sea-floor spreading. Note the uplift of the Brunswick graben above sea level.

that we distinguish from normal oceanic or continental crust and call rift-stage crust, according to Falvey (1974). The width of this crust, and therefore the trough, is 70 km on line IPOD and 80 km on line 32. These widths are similar to the 70-km-wide zone found by gravity modeling of line 5 on the Long Island Platform, but narrower than that found for line 25 on the Baltimore Canyon trough. In our models, this crust is characterized by a simple density structure and sources having uniform magnetic properties.

(2) The seaward edge of the Carolina trough coincides with the edge of oceanic crust and the ECMA. The magnetic model confirms the hypothesis of Keen (1969) that the ECMA is the edge effect caused by more magnetic oceanic crust with reversed polarity abutting less magnetic continental or rift-stage crust.

(3) The landward edge of the Carolina trough coincides with the seaward edge of continental crust, with a hinge zone in the breakup unconformity, and with the Brunswick magnetic anomaly. We use the BMA to mark the edge of continental crust, but the hinge zone may be a better indicator in parts of the margin where the BMA does not exist.

(4) Gravity and magnetic models require a low-density nonmagnetic body beneath the Brunswick magnetic anomaly; we call this feature the Brunswick graben. Acoustic evidence suggests that a subbasement layered body exists at this location on line 32, and we suspect that the Brunswick graben formed during the Triassic by downdropping of prerift sediments. The graben formation is probably related to rifting between North America and Africa. However, our data do not rule out the trough could be a Paleozoic or older feature. The graben structure is the primary cause of the BMA.

(5) The gravity model for line 32 requires a large, low-density body beneath the inner shelf and coastal plain. We infer that this body represents low-density granitic material in the upper crust, similar to that modeled for the Albemarle Sound gravity low north of Cape Hatteras by Murphy (1972). However, the magnetic model suggests that the area of this low-density body contains more mafic components similar to igneous and metamorphic rocks of the Piedmont. Thus, these magnetic data suggest that metamorphic rocks may be present at the basement surface near the gravity low and that the source of the gravity deficiency may be deeper than our model shows. A deeper low-density body would not change the gross features of the model.

(6) The acoustic evidence for a basement ridge associated with the ECMA within the Carolina trough is ambiguous, but the gravity and magnetic models do not require such a ridge. Thus, the postulated basement ridge beneath the continental margin of North America either may not exist or may be less continuous and smaller than originally hypothesized.

ACKNOWLEDGMENTS

We thank Peter Popenoe, Charlie Paull, Carl Bowin, and Hugh Rowlett for critically reviewing the manuscript and offering valuable suggestions for its improvement. We also thank Patty Forrestel, Jeff Zwinakis, and Dann Blackwood for drafting and photographing all figures, and Claire Daly, who typed the many preliminary and final versions of the manuscript. The gravity and magnetic models were prepared by use of programs written by Evelyn Wright and Gerry Evenden. Doug Peeler assisted in the data reduction.

REFERENCES CITED

Amato, R. V., and E. K. Simonis, 1980, Geologic and operational summary, COST No. G-2 well, Georges Bank area, North Atlantic OCS: U.S. Geological Survey Open-File Report 80-269, 116 p.

Applin, P. L., 1951, Preliminary report on buried pre-Mesozoic rocks in Florida and adjacent states: U.S. Geological Survey Circular 91, 28 p.

Behrendt, J. C., and K. D. Klitgord, 1980, High-sensitivity aeromagnetic survey of the U.S. Atlantic continental margin: Geophysics, v. 45, p. 1813-1846.

Benson, W. E., et al, 1976, Deep-sea drilling in the North Atlantic: Geotimes, v. 21, p. 23-26.

Bott, M. H. P., 1979, Subsidence mechanisms at passive continental margins, *in* J. S. Watkins, L. Montadert, and P. W. Dickerson, eds., Geological and geophysical investigations of continental margins: AAPG Memoir 29, p. 3-9.

Bowin, C. O., T. C. Aldrich, and R. A. Folinsbee, 1972, VSA gravity meter system — Tests and recent developments: Journal of Geophysical Research, v. 77, p. 2018-2033.

Bryan, G. M., 1970, Hydrodynamic model of the Blake Outer Ridge: Journal of Geophysical Research, v. 75, p. 4530-4537.

Burek, P. J., 1970, Magnetic reversals: their application to stratigraphic problems: AAPG Bulletin, v. 54, p. 1120-1139.

Burk, C. A., 1968, Buried ridges within continental margins: Transactions of the New York Academy of Sciences, v. 20, p. 397-409.

Cordell, L., 1978, Regional geophysical setting of the Rio Grande rift: Geological Society of America Bulletin, v. 89, p. 1073-1090.

Cornet, B., 1977, Palynostratigraphy and age of the Newark Supergroup: University Park, Pennsylvania, Pennsylvania State University, unpublished Ph.D. dissertation, 505 p.

de Boer, J., 1965, Paleomagnetic indications of megatectonic movements in the Tethys: Journal of Geophysical Research, v. 70, p. 931-944.

———, 1968, Paleomagnetic differentiation and correlation of Late Triassic rocks in the central Appalachians (with special reference to the Connecticut Valley): Geological Society of America Bulletin, v. 79, p. 600-626.

——— and F. G. Snider, 1979, Magnetic and chemical variations of Mesozoic diabase dikes from eastern North America: evidence for a hot spot in the Carolinas?: Geological Society of America Bulletin, v. 90, p. 185-198.

Denison, R. E., H. P. Ravelling, and J. T. Rouse, 1967, Age and descriptions of subsurface basement rocks, Pamlico and Albemarle Sound areas, North Carolina: AAPG Bulletin, v. 51, p. 268-272.

Dillon, W. P., et al, 1982, Growth faulting and salt diaperism. Their relationship and control in the Carolina

Trough, eastern North America: this volume.

Drake, C. L., and J. E. Nafe, 1968, Transition from ocean to continent from seismic refraction data: Upper Mantle Symposium, 11th Pacific Scientific Congress, American Geophysical Union Monograph 12, p. 174-186.

———, M. Ewing, and G. H. Sutton, 1959, Continental margins and geosynclines: The east coast of North America north of Cape Hatteras, *in* L. H. Ahrens et al, eds., Physics and chemistry of the earth: London, Pergamon Press, v. 3, p. 110-198.

———, J. Heirtzler, and J. Hirschman, 1963, Magnetic anomalies off eastern North America: Journal of Geophysical Research, v. 68, p. 5259-5275.

———, J. I. Ewing, and H. Stockard, 1968, The continental margin of the Eastern United States: Canadian Journal of Earth Sciences, v. 5, p. 993-1010.

Emery, K. O., et al, 1970, Continental Rise off eastern North America: AAPG Bulletin, v. 54, p. 44-108.

Falvey, D. A., 1974, The development of continental margins in Plate Tectonic theory: Australian Petroleum Exploration Association Journal, v. 14, p. 95-106.

Folger, D. W., 1979, Evolution of the Atlantic continental margin of the United States, *in* M. Talwani, W. Hay, and W. B. F. Ryan, eds., Deep drilling results in the Atlantic Ocean: Continental margins and paleoenvironment: American Geophysical Union, Maurice Ewing Series 3, p. 87-108.

Gohn, G.S., et al 1978, Regional implications of Triassic or Jurassic age for basalt and sedimentary red-beds in the South Carolina coastal plain: Science, v. 202, p. 887-890.

Grow, J. A., 1980, Deep structure and evolution of the Baltimore Canyon trough in the vicinity of the COST No. B-3 well, *in* P. A. Scholle, ed., Geological studies of the COST No. B-3 well, United States Mid-Atlantic continental slope area: U.S. Geological Survey Circular 833, p. 117-125.

———, and R. G. Markl, 1977, IPOD-USGS multichannel seismic reflection profile from Cape Hatteras to the Mid-Atlantic Ridge: Geology, v. 5, p. 625-630.

———, C. O. Bowin, and D. R. Hutchinson, 1979, The gravity field of the U.S. Atlantic continental margin: Tectonophysics, v. 59, p. 27-52.

———, R. E. Mattick, and J. S. Schlee, 1979, Multichannel seismic depth sections and interval velocities over outer continental shelf and upper continental slope between Cape Hatteras and Cape Cod, *in* J. S. Watkins, L. Montadert, and P. W. Dickerson, eds., Geological and geophysical investigations of continental margins: AAPG Memoir 29, p. 65-83.

Hales, A. L., et al, 1968, The East Coast onshore — offshore experiment, I. The first arrival phases: Seismological Society of America Bulletin, v. 58, p. 757-819.

Harris, L. D., and K. C. Bayer, 1979, Sequential development of the Appalachian orogen above a master decollement — a hypothesis: Geology, v. 7, p. 568-572.

Hatcher, R. D., Jr., and I. Zietz, 1980, Tectonic implications of regional aeromagnetic and gravity data from the southern Appalachians, *in* D. R. Wones, ed., Blacksburg, Virginia, Proceedings, "The Caledonides in the USA," IGCP Project 27-Caledonide Orogen, 1979 meeting: Virginia Polytechnic Institute and State University, Memoir 2, p. 235-244.

Hathaway, J.C., et al, 1976, Preliminary summary of the 1976 Atlantic margin coring project of the U.S. Geological Survey: U.S. Geological Survey Open-File Report 76-844, 217 p.

———, ———, et al, 1979, U.S. Geological Survey core drilling on the Atlantic shelf: Science, v. 206, p. 515-527.

Hollister, C. D., et al, 1972, Sites 102-103-104-Blake Bahama Outer Ridge (northern end), *in* LaJolla University of California Scripps Institute of Oceanography, Initial reports of the deep sea drilling project, v. 11: Washington, D.C., National Science Foundation, p. 135-218.

Hsu, K. J., M. B. Cita, and W. B. F. Ryan, 1973, The origin of the Mediterranean evaporites, *in* LaJolla, University of California, Scripps Institute of Oceanography, 1969-1972, Initial reports of the deep sea drilling project, v. 13: Washington, D.C., National Science Foundation, p. 1203-1231.

Hutchinson, R. W., and G. G. Engels, 1972, Tectonic evolution in the southern Red Sea and its possible significance to older rifted continental margins: Geological Society of America Bulletin, v. 83, p. 2989-3002.

Irving, E., and M. R. Banks, 1961, Paleomagnetic results from the Upper Triassic lavas of Massachusetts: Journal of Geophysical Research, v. 66, p. 1935-1939.

James D. E., T. J. Smith, and J. S. Steinhart, 1968, Crustal structure of the Middle Atlantic States: Journal of Geophysical Research, v. 73, p. 1983-2007.

Jansa, L. F., and J. A. Wade, 1975, Geology of the continental margin off Nova Scotia and Newfoundland, *in* W. J. M. Van der Linden, and J. A. Wade, Offshore geology of eastern Canada, v. 2, Regional geology: Geological Survey of Canada Paper 74-30, v. 2, p. 51-105.

———, J. P. Bujak, and G. L. Williams, 1980, Upper Triassic salt deposits of the western North Atlantic: Canadian Journal of Earth Sciences, v. 17, p. 547-559.

Keen, C. E., and M. J. Keen, 1974, The continental margins of Eastern Canada and Baffin Bay, *in* C. A. Burke, and C. L. Drake, eds., The geology of continental margins: New York, Springer, p. 381-389.

———, et al, 1974, Some aspects of the ocean-continent transition at the continental margin of eastern North America, *in* W. J. M. Van der Linden, and J. A. Wade, eds., Offshore geology of eastern Canada, v. 2, Regional geology: Geological Survey of Canada Paper 74-30, p. 189-197.

Keen, M. J., 1969, Possible edge effect to explain magnetic anomalies off the eastern seaboard of the U.S.: Nature, v. 222, p. 72-74.

Kent, Sir P., 1980, Vertical tectonics associated with rifting and spreading, *in* Sir P. Kent et al, eds., The evolution of passive continental margins in light of recent deep drilling results: London, The Royal Society, p. 125-132.

King, P. B., compiler, 1969, Tectonic map of North America: Washington, D.C., U.S. Geological Survey, scale 1:5,000,000.

———, and H. E. Beikman, compilers, 1974, Geologic map of the United States: U.S. Geological Survey, scale 1:2,500,000.

Klein, G. deV., 1969, Deposition of Triassic sedimentary rocks in separate basins, eastern North America: Geological Society of America Bulletin, v. 80, p. 1825-1832.

Klitgord, K. D., and J. C. Behrendt, 1977, Aeromagnetic survey of the U. S. Atlantic continental margin: U.S. Geological Survey, Miscellaneous Field Studies map MF-913, scale 1:250,000.

———, and ———, 1979, Basin structure of the U.S. Atlantic margin, *in* J. S. Watkins, L. Montadert, and P. W. Dickerson, eds., Geological and geophysical investigations of continental margins: AAPG Memoir 29, p. 85-112.

———, and J. A. Grow, 1980, Jurassic seismic stratigraphy and basement structure of western Atlantic magnetic quiet zone: AAPG Bulletin, v. 64, p. 1658-1680.

Konig, M., and M. Talwani, 1977, A geophysical study of

the southern continental margin of Australia: Great Australian bight and western sections: Geological Society of America Bulletin, v. 88, p. 1000-1014.

Lewis, B. T. R., and R. P. Meyer, 1977, Upper mantle velocities under the east coast margin of the U.S.: Geophysical Research Letters, v. 4, p. 341-344.

Ludwig, W. J., J. E. Nafe, and C. L. Drake, 1971, Seismic refraction, *in* The sea: New York, Wiley-Interscience, v. 4, part 1, p. 53-84.

———, et al, 1979, Structure of Colorado Basin and continent-ocean crust boundary off Bahia Blanca, Argentina, *in* J. S. Watkins, L. Montadert, and P. W. Dickerson, eds., Geological and geophysical investigations of continental margins: AAPG Memoir 29, p. 113-124.

Maher, J. C., 1971, Geologic framework and petroleum potential of the Atlantic coastal plain and continental shelf: with a section on stratigraphy by J. C. Maher and E. R. Applin: U.S. Geological Survey Professional Paper 659, 98 p.

Mann, V. I., Bouguer gravity map of North Carolina: Southeastern Geology, v. 3, p. 207-220.

———, and F. S. Zablocki, 1961, Gravity features of the Deep River Wadesboro Triassic basin of North Carolina: Southeastern Geology, v. 2, p. 191-215.

Marine, I. W., and G. E. Siple, 1974, Buried Triassic basin in the central Savannah River area, South Carolina and Georgia: Geological Society of Americ Bulletin, v. 85, p. 311-320.

Markl, R. G., G. M. Bryan, and J. I. Ewing, 1970, Structure of the Blake — Bahama outer ridge: Journal of Geophysical Research, v. 75, p. 4539-4555.

McConnell, R. B., 1972, Geological development of the rift system of eastern Africa: Geological Society of America Bulletin, v. 83, p. 2549-2572.

McElhinny, M. W., and P. J. Burek, 1971, Mesozoic paleomagnetic stratigraphy: Nature, v. 232, p. 98-102.

McKenzie, D., 1978, Some remarks on the development of sedimentary basins: Earth and Planetary Sicence Letters, v. 40, p. 25-32.

Mohr, P. A., 1972, Regional significance of volcanic geochemistry in the Afar Triple Junction, Ethiopia: Geological Society of America Bulletin, v 83, p. 213-222.

Montadert, L., D. G. Roberts, and O. DeCharpal, 1979, Rifting and subsidence of the northern continental margin of the Bay of Biscay, *in* LaJolla, University of California Scripps Institution of Oceanography, Initial report of the deep sea drilling project, v. 47: Washington, D.C., National Science Foundation, p. 1025-1060.

Murphy, C. J., III, 1972, Modeling the northeastern North Carolina gravity low: Chapel Hill, North Carolina, University of North Carolina, unpublished Master's thesis.

Nagata, T., 1969, Reduction of geomagnetic data and interpretation of anomalies, *in* P. J. Hart, ed., The Earth's crust and upper mantle: American Geophysical Union Geophysical, Monograph 13, p. 391-398.

Nishenko, S. P., and L. R. Sykes, 1979, Fracture zones, Mesozoic rifts and the tectonic setting of the Charleston, South Carolina earthquake of 1886 (Abs.): Eos, transactions, American Geophysical Union, v. 60, p. 310.

Opdyke, N. W., and M. W. McElhinny, 1965, The reversal at the Triassic — Triassic boundary and its bearing on the correlation of Karro igneous activity in southern Africa: Eos, Transactions, American Geophysical Union, v. 46, p. 65.

Parsons, B., and J. G. Sclater, 1977, An analysis of the variation of ocean floor bathymetry and heat flow with age: Journal of Geophysical Research, v. 82, p. 803-825.

Pickering, S. M., Jr., M. W. Higgins, and I. Zietz, 1977, Relation between the Southeast Georgia Embayment and the onshore extent of the Brunswick magnetic anomaly (Abs.): Eos, Transactions, American Geophysical Union, v. 58, p. 432.

Popenoe, P., 1977, A probable major Mesozoic rift system in South Carolina and Georgia (Abs.): Eos, Transactions, American Geophysical Union, v. 58, p. 432.

Rabinowitz, P. D., 1974, The boundary between oceanic and continental crust in the western North Atlantic, *in* C. A. Burk, and C. L. Drake, eds., The geology of continental margins: New York, Springer, p. 67-84.

———, 1976, Geophysical study of the continental margin of southern Africa: Geological Society of America Bulletin, v. 87, p. 1643-1653.

———, and J. L. LaBrecque, 1977, The isostatic gravity anomaly: key to the evolution of the ocean — continent boundary at passive continental margins: Earth and Planetary Science Letters, v. 35, p. 145-150.

———, S. C. Cande, and J. L. LaBrecque, 1976, The Falkland Escarpment and Agulhas Fracture Zone: the boundary between oceanic and continental basement at conjugate continental margins: An. Acad. bras. Ciênc, v. 48, p. 241-251.

Rona, P. A., and C. S. Clay, 1967, Stratigraphy and structure along a continuous seismic reflection profile from Cape Hatteras, North Carolina to the Bermuda Rise: Journal of Geophysical Research, v. 72, p. 2107-2130.

Schlee, J., et al, 1976, Regional geologic framework off northeastern United States: AAPG Bulletin, v. 60, p. 926-951.

Schuepbach, M. A., and P. R. Vail, 1980, Evolution of outer highs on divergent continental margins, *in* Studies in geophysics, continental margins: Washington, D.C., National Academy of Sciences, p. 50-61.

Sheridan, R. E., 1974, Atlantic continental margin of North America, *in* C. A. Burk, and C. L. Drake, eds., The geology of continental margins: New York, Springer, p. 391-407.

———, et al, 1979, Seismic refraction study of the continental edge off the eastern United States, *in* C. E. Keen, ed., Crustal properties across passive margins: Tectonophysics, v. 59, p. 1-26.

Skeels, D. C., 1950, Geophysical data on the North Carolina coastal plain: Geophysics, v. 15, p. 409-425.

Spudich, P., and J. Orcutt, 1980, A new look at the seismic velocity structure of the oceanic crust: Reviews of Geophysics and Space Physics, v. 18, p. 627-645.

Steiner, M. B., 1980, Investigation of the geomagnetic field polarity during the Jurassic: Journal of Geophysical Research, v. 85, p. 3572-3586.

———, and C. E. Helsley, 1975, Reveral pattern and apparent polar wander for the Late Jurassic: Geological Society of America Bulletin, v. 86, p. 1537-1543.

Sumner, J. R., 1977, Geophysical investigation of the structural framework of the Newark-Gettysburg Triassic basin, Pennsylvania: Geological Society of America Bulletin, v. 88, p. 935-942.

Talwani, M., and O. Eldholm, 1972, Continental margin off Norway: a geophysical study: Geological Society of America Bulletin, v. 83, p. 3575-3606.

———, and ———, 1973, Boundary between continental and oceanic crust at the margin of rifted continents: Nature, v. 241, p. 325-330.

———, J. L. Worzel, and M. Landisman, 1959, Rapid gravity computations for two dimensional bodies with application to the Mendocino submarine fracture zone: Journal of Geophysical Research, v. 64, p. 49-59.

———, et al, 1978, The margin south of Australia — a continental margin paleorift, *in* I. E. Ramberg and E. R. Neumann, eds., Tectonics and geophysics of continental rifts: Dordrecht, Holland D. Reidel Publishing Company, p. 203-219.

———, et al, 1979, The crustal structure and evolution of the area underlying the magnetic quiet zone on the margin south of Australia, *in* J. S. Watkins, L. Montadert, and P. W. Dickerson, eds., Geological and geophysical investigations of continental margins: AAPG Memoir 29, p. 151-175.

Taylor, P. T., I. Zietz, and L. S. Dennis, 1968, Geologic implications of aeromagnetic data for the eastern continental margin of the United States: Geophysics, v. 33, p. 755-780.

Telford, W. M., et al, 1976, Applied geophysics: New York, Cambridge University Press, 860 p.

Tucholke, B. E., and G. S. Mountain, 1979, Seismic stratigraphy, lithostratigrahy and paleosedimentation patterns in the North American Basin, *in* M. Talwani, W. Hay, and W. B. F. Ryan, eds., Deep drilling results in the Atlantic Ocean: continental margins and paleoenvironment: American Geophysical Union, Maurice Ewing Series 3, p. 58-86.

University of California, Scripps Institution of Oceanography, LaJolla, 1972, Initial reports of the deep sea drilling project, v. 11: Washington, D.C., National Science Foundation, 1077 p.

Vacquier, V., 1972, Geomagnetism in marine geology: New York, Elsevier, 185 p.

Van Houten, F. B., 1977, Triassic — Liassic deposits of Morocco and eastern North America: comparison: AAPG Bulletin, v. 61, p. 79-99.

Vogt, P. R., 1973, Early events in the opening of the North Atlantic, *in* D. H. Tarling, and S. K. Runcorn, eds., Implications of continental drift to the earth sciences, v. 2: London, Academic Press, p. 693-712.

Warren, D. H., 1968, Transcontinental geophysical survey (35° - 39°N) seismic refraction profiles of the crust and upper mantle from 74° to 87°W longitude: U.S. Geological Survey, Miscellaneous Geologic Investigations Map I-535-D, scale 1:1,000,000.

Watts, A. B., and W. B. F. Ryan, 1976, Flexure of the lithosphere and continental margin basins: Tectonophysics, v. 36, p. 25-44.

———, and M. S. Steckler, 1979, Subsidence and eustacy at the continental margin of eastern North America, *in* M. Talwani, W. Hay, and W. B. F. Ryan, eds., Deep drilling results in the Atlantic Ocean: continental margins and paleoenvironment: American Geophysical Union, Maurice Ewing Series 3, p. 218-234.

———, and ———, in press, Subsidence and tectonics of Atlantic-type continental margins: Oceanologica Acta.

Weigel, W., and G. Wissman, 1977, A first crustal section from seismic observations of Mauretania: Publication of the Polish Academy of Sciences, Institute of Geophysics A-4, v. 115, p. 369-380.

Worzel, J. L., 1965, Pendulum gravity measurements at sea, 1936-1959: New York, Wiley, 422 p.

———, 1966, Structure of continental margins and development of ocean trenches, *in* W. H. Poole, ed., Continental margins and island arcs: Geological Survey of Canada Paper 66-15, p. 357-397.

———, and G. L. Shurbet, 1955, Gravity anomalies at continental margins: Proceedings, National Academy of Sciences (U.S.A.), v. 41, p. 458-469.

Zietz, 1970, Eastern continental margin of the United States, part I, a magnetic study, *in* A. E. Maxwell, ed., The Sea: New York, Wiley — Interscience, v. 4, part 2, p. 311-320.

Continental Oceanic Crustal Transition in the Gulf Coast Geosyncline

Abou-Bakr K. Ibrahim *
U.S. Nuclear Regulatory Commission
Washington, D.C.

Elazar Uchupi
Woods Hole Oceanographic Institution
Woods Hole, Massachusetts

Seismic refraction measurements indicate that the transition from rifted continental crust to oceanic crust takes place at a water depth of over 3000 m northwest of Cuba between the Florida and Campeche escarpments. Along the eastern flank of the Mississippi Embayment, the transition occurs at about 2500 m deep and on the embayment itself inboard of the coast. Off south Texas the boundary between the rifted continental crust and oceanic crust is near the shelf's edge, off Mexico the boundary is on the continental slope, and off Campeche Bank the boundary is about 100 km northwest of Campeche Escarpment. In the northern Gulf an oceanic crustal high may lie beneath the upper continental slope. This high served as a foundation for a Mesozoic reef. Maximum sediment accumulation took place along the contact between the rifted continental crust and oceanic crust.

The Gulf of Mexico, with its tectonic fabric of carbonate platforms and clastic wedges which were deformed by mud and salt piercement structures and massive gravitational slides, has been the site of numerous scientific publications. The present study uses geophysical data from previous studies, supplemented by four reversed refraction measurements on the eastern Texas shelf, to construct two cross sections of the Gulf Coast geosyncline (Figure 1). From the nature of these crustal profiles we determine the location of the boundary between the continental and oceanic crust in the northern Gulf of Mexico off the Mississippi Embayment and Texas.

DATA BASE

Geophysical measurements used to compile the two cross sections include those of Antoine and Ewing (1962), Cram (1961), Warren, Healy, and Johnson, (1966), Hales, Helsley, and Nation, (1970), and Dorman et al (1972) on the coastal plain and adjacent shelf, and Ewing, Antoine, and Ewing (1960), Antoine and Ewing (1962), Swolfs (1967), Ibrahim and Latham (1978), and Ibrahim et al (1981) on the continental margin and adjacent deep sea. Additional information on the crustal structure of the region was obtained from four reversed refraction stations recorded on the eastern Texas shelf during the present investigation (Figure 1). These refraction stations recorded on the shelf consisted of four instrument set-ups spaced 25 km apart. Total profile length was 75 km. Bottom hydrophone arrays, with leads either to an instrument boat or a buoy with a radio receiver, were used as receiving packages. Shot points were 1.5 km apart with a 500-pound charge for each shot. Figures 2 and 3 show examples of the record sections, and Figs. 4 and 5 display travel-time curves constructed from these records. Due to the large shots, only the first ground arrivals were used in the analyses. Data from the travel time curves were fitted by the least squares method, and the inverse slope of the lines provided the apparent velocities. These velocities and the intercept times were used to calculate the thickness of the layers. From these data the crustal sections displayed in Figures 6 and 7 were constructed (see Tables 1 and 2).

*Formerly with: The University of Texas, Marine Science Institute Geophysics Laboratory, Galveston, Texas.

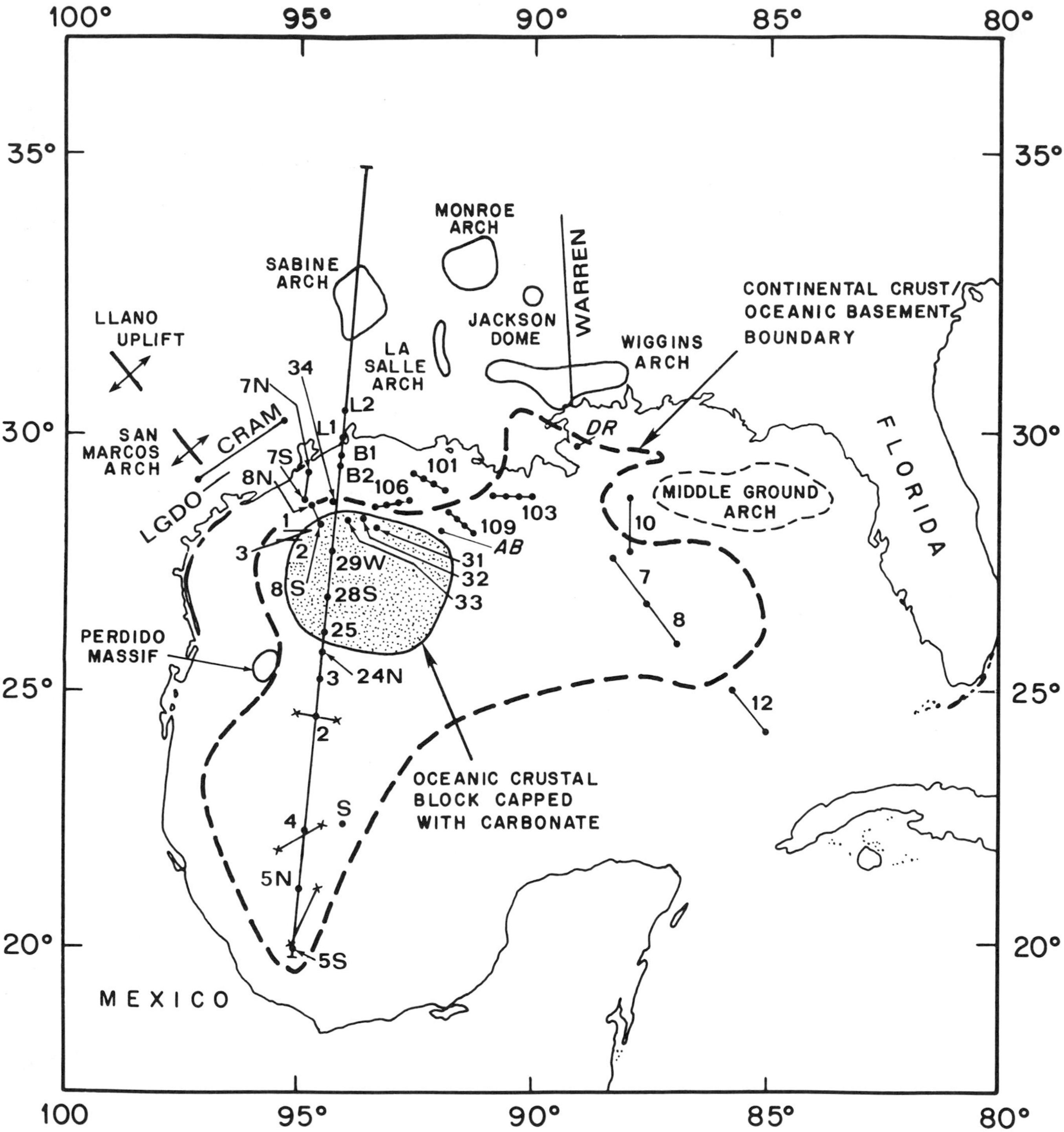

Figure 1 — Chart showing locations of seismic refraction stations and crustal cross sections. The heavy dashed line indicates the boundary between rifted continental crust and oceanic crust and was constructed from data described in this report, supplemented by data from Buffler et al, (1980; Figure 10). Sources of refraction stations are as follows: 5S, 5N, 4, 3, and 2 (Ibrahim et al, 1981); S (Swolfs, 1967); 24N, 25, 28S, 29W, 31, 32, 33, and 34 (Ewing, Antoine, and Ewing, 1960); 1, 2 and 3 (Ibrahim and Latham, 1978); 8 and 7 (Antoine and Ewing, 1962); B2, B1, L1, and L2 (Hales, Helsey, and Nation, 1970); Cram (Cram, 1961); LDGO (Dorman et al 1972); Warren (Warren, Healy, and Jackson, 1966); 101, 103, 106, and 109 (present investigation). Basement tectonic elements of the coastal plain and margin from King (1969), Martin (1978) and Watkins et al, (1978). AB = Alderdice Bank; DP = Alderdice Bank; DP = Door Point. Stations 24N, 25, 28S, and 29W are projected on the western cross section.

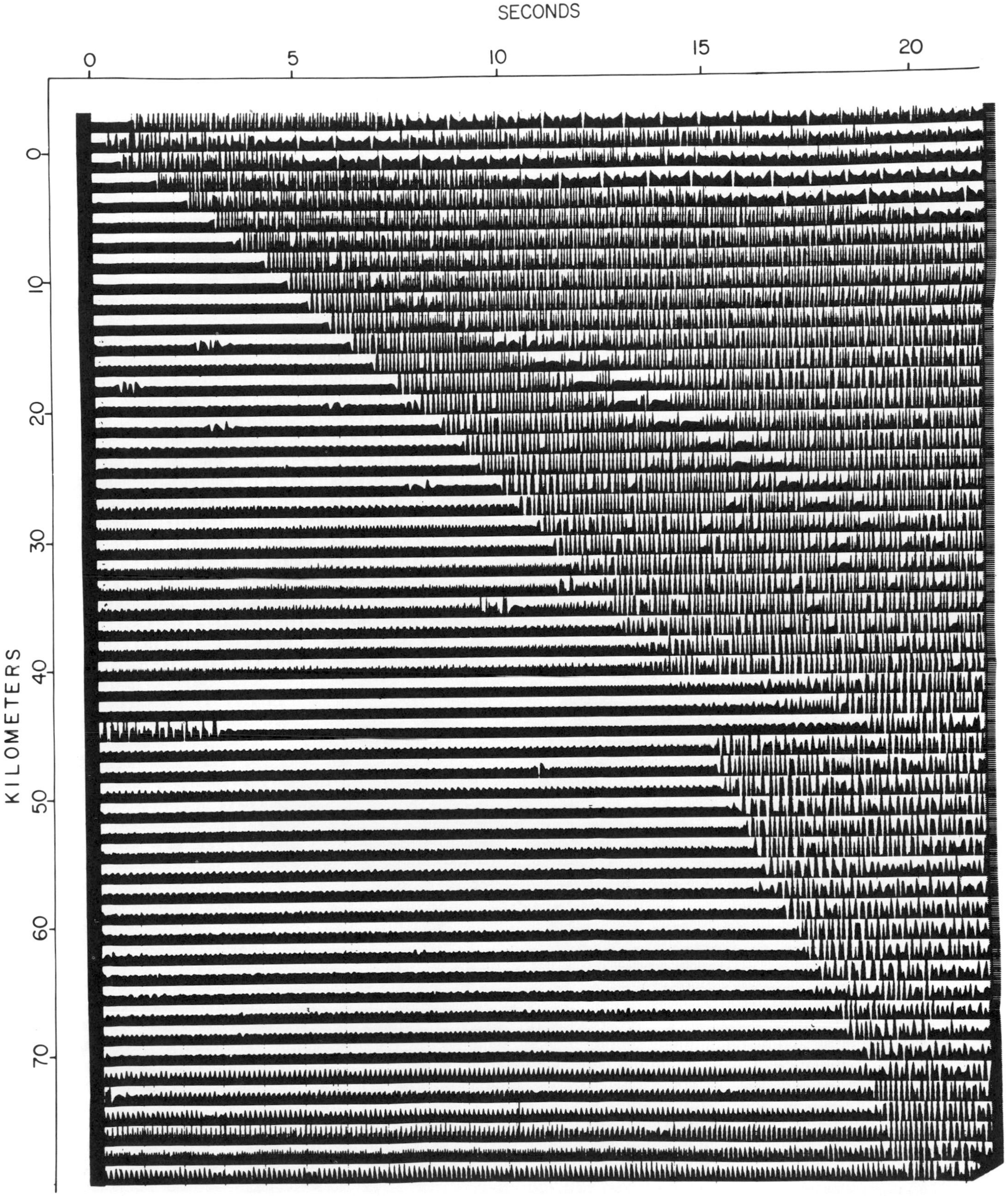

Figure 2 — Record section for Station 106. See Figure 1 for location of station.

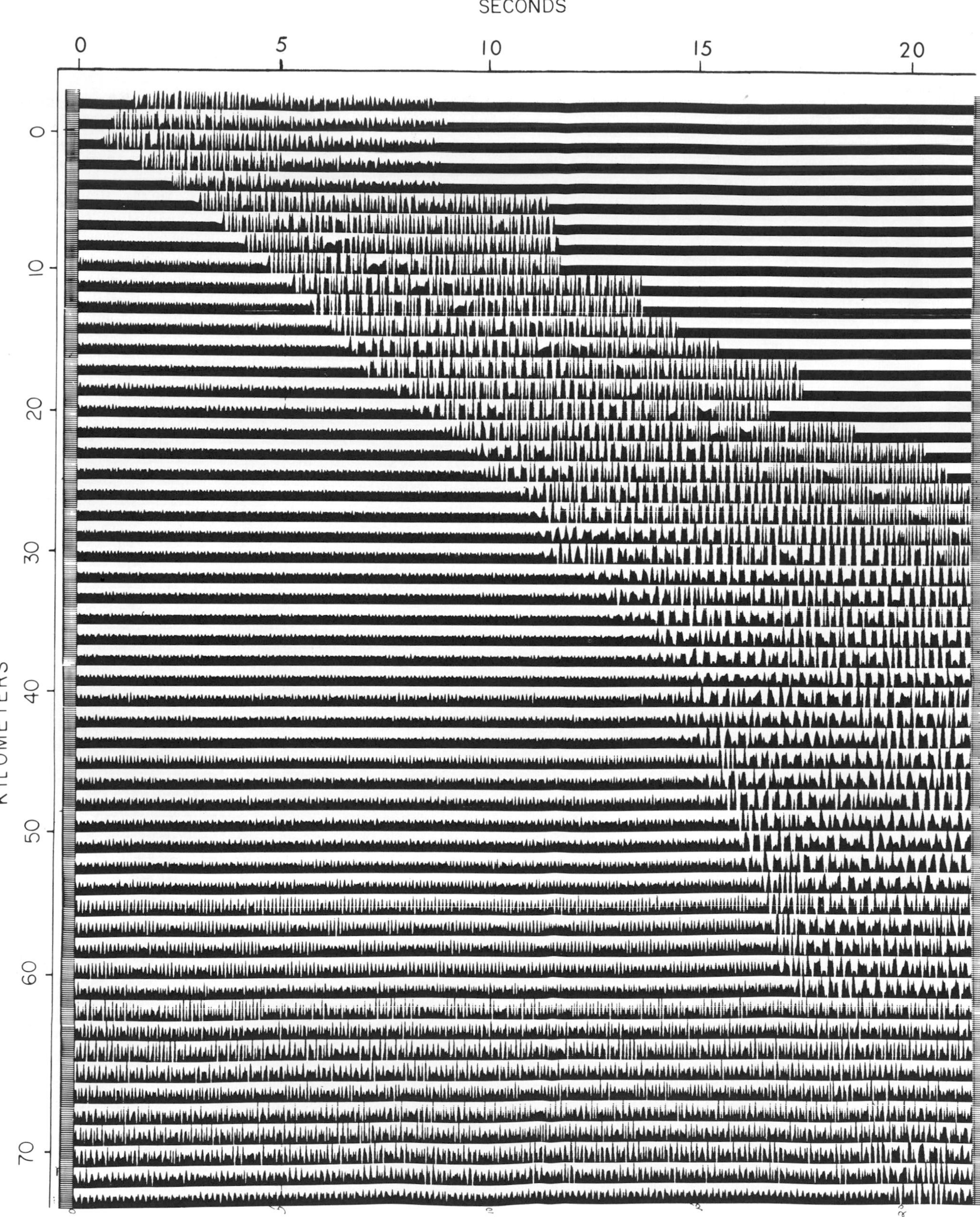

Figure 3 — Record section for Station 109. See Figure 1 for station location.

CROSS SECTIONS

The cross section at the eastern end of the Gulf Coast geosyncline extends from west of Cuba to the Mississippi fan and from the fan up the axis of the Mississippi Embayment (Figure 8). On the western end of the geosyncline the cross section extends from the Gulf of Campeche to the northern limit of the coastal plain (Figure 9). In the cross section, we interpreted velocities of 7.8 km/sec (7.6 km/sec on profile 10 in the eastern section; Figure 8) or greater as mantle. In the eastern cross section mantle can be traced the length of the whole section reaching a maximum depth of 40 km below sea level near the southern end of the Mississippi Embayment. From here, mantle shoals gradually to less than 27 km deep at the profile's northern end. Along the western cross section the mantle surface undulates seaward of the Sigsbee Escarpment, displaying a prominent high near profile 4 and a smaller high near the Sigsbee Escarpment (Figure 9). Northward of this smaller high, mantle plunges rapidly and reaches a depth of nearly 35 km below sea level landward of the coast. Refraction velocities of 6.3 to 6.9 km/sec are interpreted as low continental crust north of Cuba, along the eastern flank of the Mississippi Embayment, and beneath the coastal plain (Figures 8 and 9). Velocities ranging from 6.5 to 6.8 km/sec along profiles 103 and 109 along the shelf (Figures 6 and 7) and 6.3 to 7.2 km/sec in the deep Gulf and in the high landward of the Sigsbee Escarpment (Figures 8 and 9) are interpreted as oceanic crust. Velocities from 5.8 to 6.0 km/sec in the Mississippi Embayment and Texas coast plain were interpreted by Worzel and Watkins (1973) and Warren, Healy, and Jackson (1966) as Paleozoic igneous and metamorphic rocks. As Worzel and Watkins point out, this crustal unit may contain some sedimentary rocks since some of the velocities obtained from this unit fall in this range. Along the eastern profile the surface of this crustal layer, which can be traced to near the coast, is broadly undulating with a depression corresponding to the Mississippi Embayment and a high corresponding to Wiggins arch. The 5.8 to 6.0 km/sec unit appears to be missing south of the Mississippi delta (Figures 6 and 7; Profiles 103 and 109) with the Mesozoic-Cenozoic strata resting directly on the 6.5 to

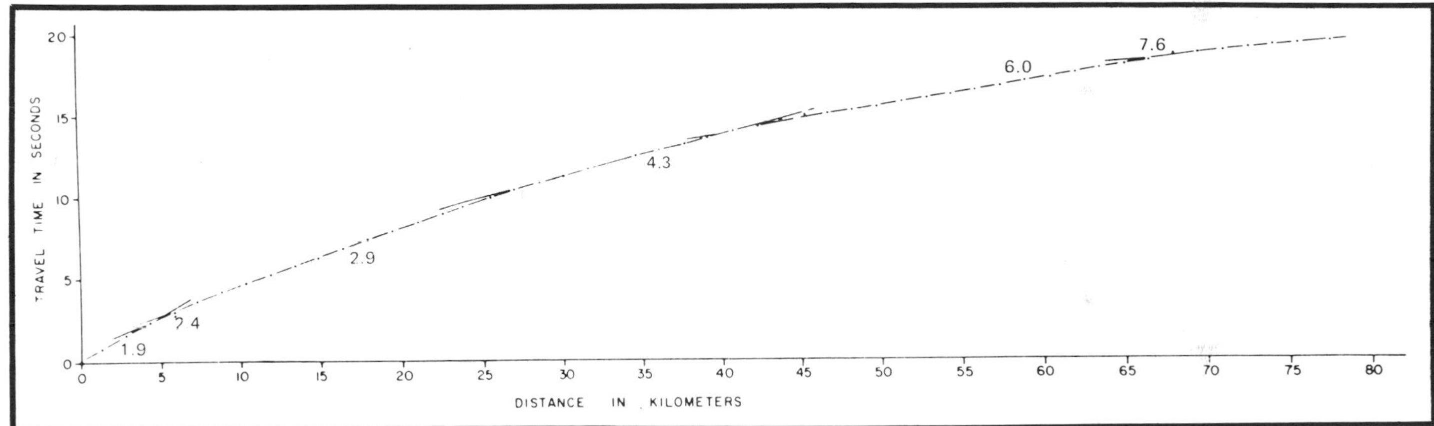

Figure 4 — Travel-time curve for Station 106.

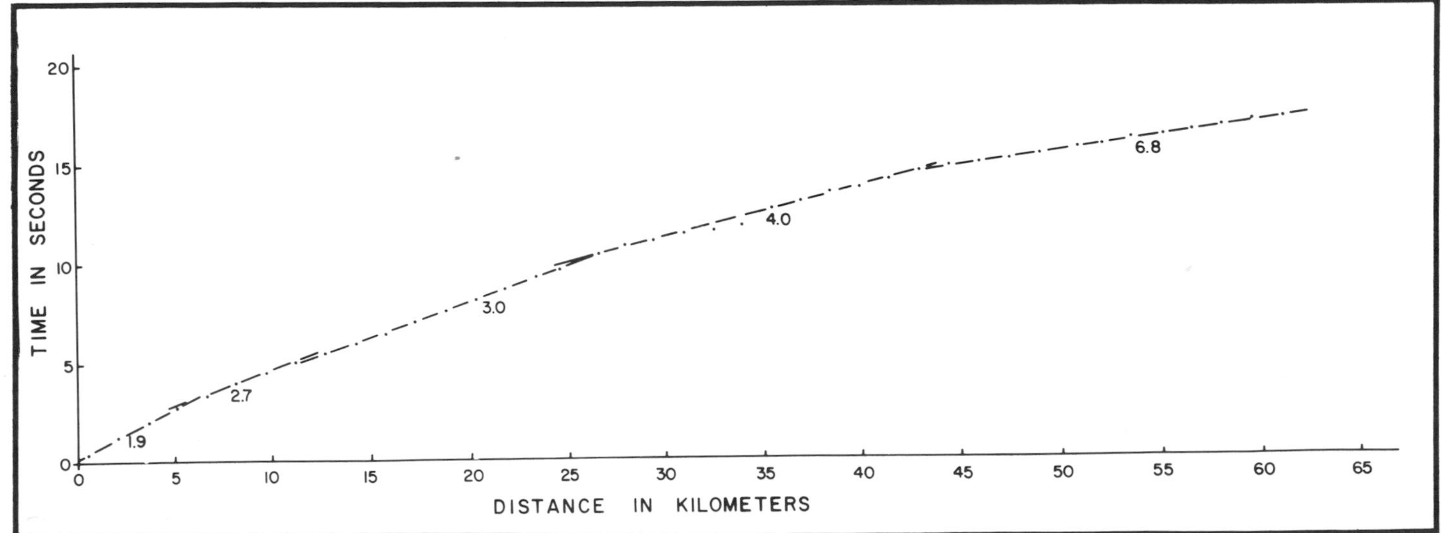

Figure 5 — Travel-time curve for station 109.

6.8 km/sec layer which we believe represents oceanic crust. Along the western profile the surface of the 5.8 to 6.0 km/sec is also broadly undulating, pinching out near the shelf's edge. In the deep Gulf Coast, the layer with a velocity of 4.5 to 5.0 km/sec, which varies in thickness from 2 to 5 km, is made up of oceanic basement (layer 2) and the Challenger sedimentary unit of pre-middle Cretaceous age (Buffler et al, 1980; Figures 8 and 9). The unconformity which forms the top of this unit has been dated as middle Cenomanian.

Following Bally's (1975) suggestion, we interpreted the material beneath the upper continental slope with a velocity of 5.3 km/sec (station 28, Figure 9) and the material beneath the shelf's edge with a velocity of 5.7 km/sec (stations 3 and 8, Figure 9) as platform carbonates resting on either oceanic basement or oceanic crust. The 5.1 km/sec unit recorded on station 34 on the outer shelf may be carbonate debris derived from this platform. We point out, however, that station 28 was recorded in a region highly tectonized by salt. Thus, the refraction velocity obtained at this station is of questionable quality. Within the rest of the sedimentary apron velocities of 4 km/sec and higher are Jurassic and Early Cretaceous evaporites and carbonates (Figures 6, 7, 8, and 9). Velocities lower than 4 km/sec represent Late Cretaceous and Cenozoic clastics intruded by salt and mud.

CONTINENTAL OCEANIC CRUST BOUNDARY

On both cross sections (Figures 8 and 9), the surface of the Paleozoic terrane in the Gulf Coast geosyncline is broadly undulating. We believe this texture is due to crustal extensions, and that structures such as Middle Ground arch (Figure 1) are the products of this rifting episode. These highs appear to be aligned in a northwest to southeast direction (Figure 1) and may be located along the northwest extension of the southern edge of the North American plate. A similar rifting also appears to have affected the pre-Mesozoic rocks northwest of Cuba and northwest of Campeche Bank (Worzel and Burk, 1978; Buffler et al, 1980). Between the Campeche and Florida escarpments the rifted terrain can be traced to near the coast of Cuba

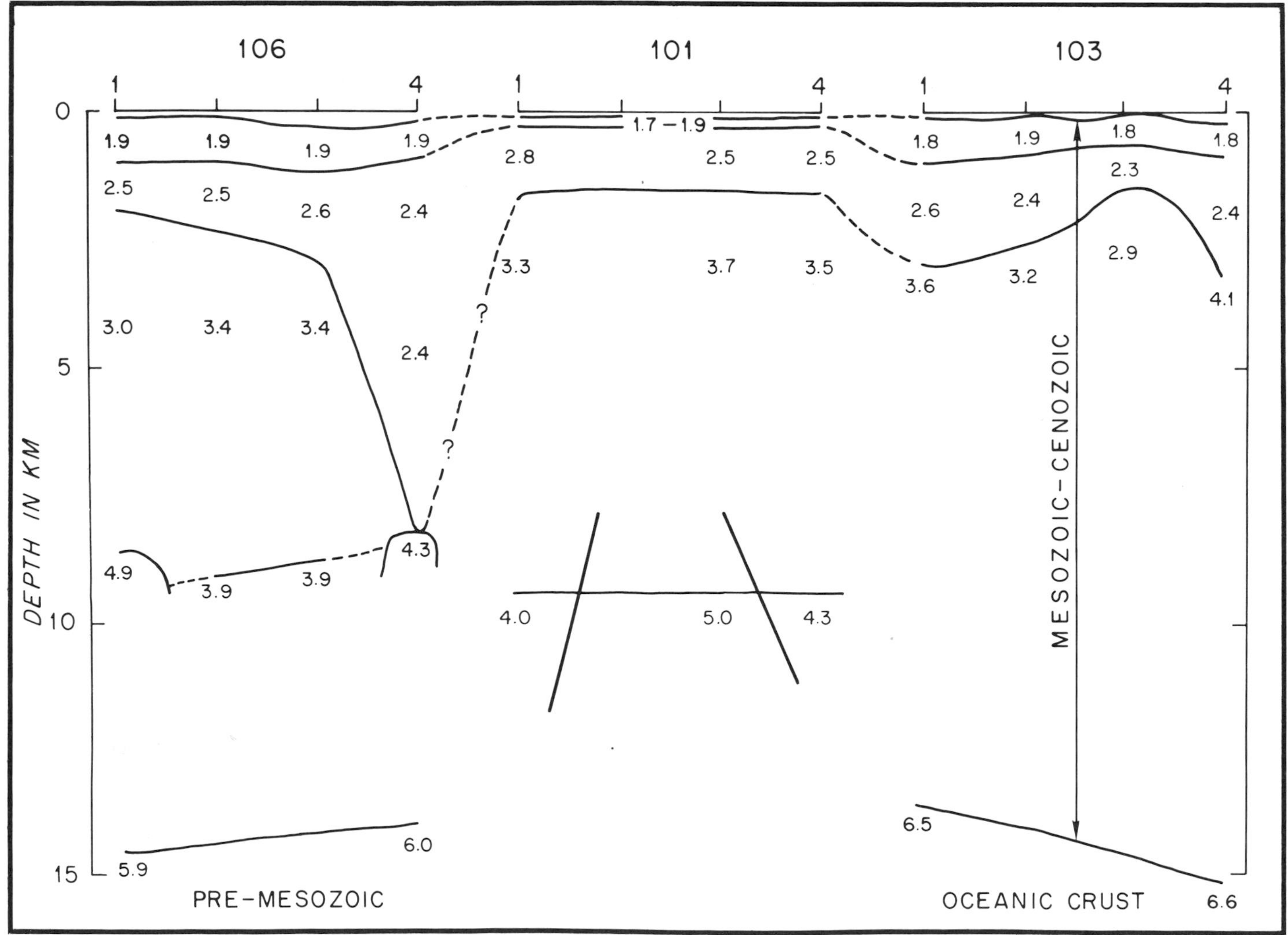

Figure 6 — Cross section of the East Texas Shelf showing the velocities of the layers in km/sec. Along profile 103 sediments rest on a crustal layer with a velocity of 6.5 to 6.6 km/sec, whereas on 106 they rest on a layer with a velocity of 5.9 to 6.0 km/sec. See Figure 1 for section location.

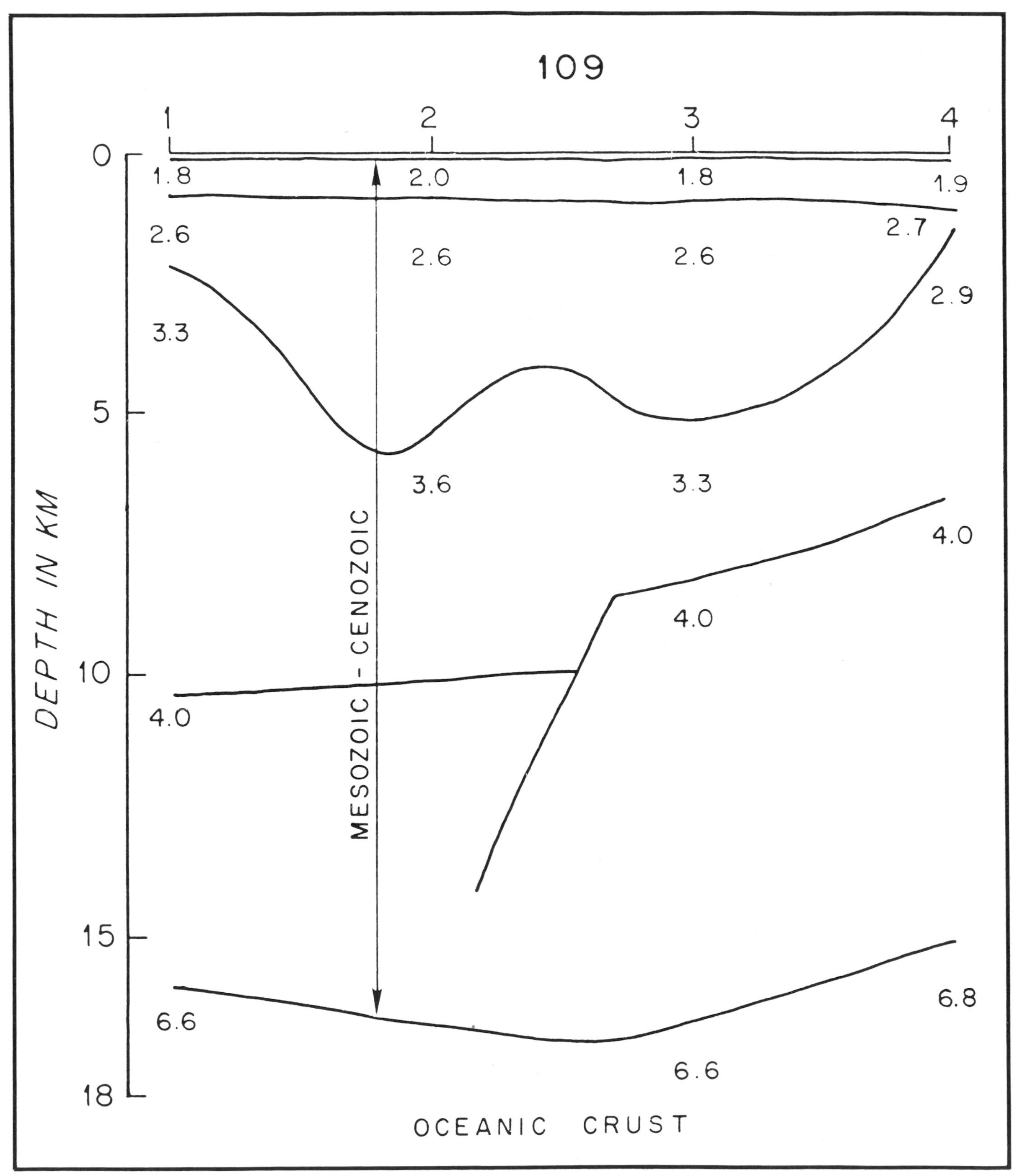

Figure 7 — Cross section near the shelf's edge off eastern Texas showing apparent velocities of the various layers in km/sec. The crustal layer with a velocity of 5.7 to 6.0 km/sec layer is missing from 109, similar to profile 103 (Figure 6), and the sediments rest directly on the 6.6 to 6.8 km/sec which may represent oceanic crust. See Figure 1 for section location.

Table 1. Crustal layer velocities and thicknesses[1].

Station 101
1 — 1.5 (.1)[2]; 1.7-1.9 (.15); 2.8 (1.3); 3.3 (7.8); 4.0
3 — 1.5 (.1); 1.7-1.9 (.2); 2.5 (1.2); 3.7 (7.8); 5.0
4 — 1.5 (.1); 1.7-1.9 (.15); 2.5 (1.3); 3.5 (7.7); 4.3

Station 103
1 — 1.5 (.1); 1.8 (.8); 2.6 (1.9); 3.6 (10.6); 6.5
2 — 1.5 (<.1); 1.9 (.7); 2.4 (1.7); 3.2
3 — 1.5 (<.1); 1.8 (.6); 2.3 (.8); 2.9
4 — 1.5 (.2); 1.8 (.6); 2.4 (2.3); 4.1 (11.8); 6.6

Station 106
1 — 1.5 (.15); 1.9 (.8); 2.5 (.9); 3.0 (6.6); 4.9 (6.0); 5.9
2 — 1.5 (.1); 1.9 (.9); 2.5 (1.4); 3.4 (6.7); 3.9
3 — 1.5 (.3); 1.9 (.9); 2.6 (1.8); 3.4 (5.9); 3.9
4 — 1.5 (.2); 1.9 (.7); 2.4 (7.4); 4.3 (5.7); 6.0

Station 109
1 — 1.5 (.2); 1.8 (.7); 2.6 (1.3); 3.3 (7.5); 4.0 (5.3); 6.6
2 — 1.5 (.2); 2.0 (.7); 2.6 (2.2); 3.6
3 — 1.5 (.2); 1.8 (.8); 2.6 (4.0); 3.3 (3.0); 4.0 (8.0); 6.6
4 — 1.5 (.2); 1.9 (1.0); 2.7 (.6); 2.9 (4.4); 4.0 (8.[illegible]); 6.8

1 For locations of stations see Figure 1 and Table 2.
2 First number is velocity of the layer in km/sec; number in parenthesis is the thickness of the layer in km.

Table 2. Locations of refraction stations.

Station	Location
Station 101	
1	92°40′W, 29°18′N
2	92°28′W, 29°12′N
3	92°09′W, 29°04′N
4	91°54′W, 28°58′N
Station 103	
1	90°58′W, 28°48′N
2	90°41′W, 28°48′N
3	90°24′W, 28°48′N
4	90°08′W, 28°48′N
Station 106	
1	93°30′W, 28°28′N
2	93°13′W, 28°30′N
3	93° 0′W, 28°33′N
4	92°45′W, 28°38′N
Station 109	
1	91°59′W, 28°32′N
2	91°48′W, 28°25′N
3	91°34′W, 28°17′N
4	91°25′W, 28°08′N

where it appears to disappear beneath the Mesozoic-Paleogene island arc accretionary wedge.

Northwest of Cuba, Buffler et al (1980) place the change from rifted continental crust to oceanic crust in water depths of about 3000 m (Figures 1 and 8). Along the eastern flank of the Mississippi fan the boundary is at the proximal end of the fan at, about 2500 m deep (Figure 8). Farther west our data indicate that the continental/oceanic crust boundary along the western side of the Mississippi Embayment may be landward of the present coast and that the Mississippi delta may rest on oceanic crust (Figures 1, 6, and 7). This re-entrant of the oceanic crust supports Burke and Dewey's (1973) contention that the Mississippi Embayment is a Mesozoic aulocogen. It is interesting to note that Alderdice Bank at 28°04.4′N; 92°0.78′W (Figure 1), a site of a large outcrop of Late Cretaceous basalt (76.8 ± 3 m.y.; Rezak and Tieh, 1980), and the Cretaceous basalt intrusive (age 82 ± 8 m.y.) described by Braunstein and McMichael (1976) from Door Point are along the flanks of the structural low. Farther west our data indicates that the transition from rifted continental to oceanic crust occurs near the shelf's edge where the 5.8 to 6.0 km/sec layer pinches out (Figure 1 and 9). Farther west off Mexico, Buffler et al, (1980) place the boundary on the continental slope at or landward of the 3000 m water contour. Off Yucantan these same authors place the boundary 100 km northwest of the Campeche Escarpment (Figure 1).

EVOLUTION OF THE GULF OF MEXICO

The Atlantic Ocean and adjacent marginal seas, such as the Gulf of Mexico, are recent geologic features formed by the breakup of Pangea in the Mesozoic. Incipient rifting in the North Atlantic appears to have been initiated in the late Triassic (Cousminer and Manspeizer, 1976). The extensional tectonic fabric present in crustal cross sections of the Gulf Coast geosyncline was formed during this rifting episode. It is into this intra-continental rift system that intermittent seawater flooding under hot and arid climatic conditions resulted in the deposition of evaporites. Off eastern North America the evaporites were deposited in latest Triassic to earliest Jurassic (Jansa, Bujck, and Williams, 1980). In the Gulf of Mexico evaporite deposition atop the rifted terrain took place later in the Middle Jurassic (Buffler et al, 1980). In the North Atlantic, actual continental de-coupling and emplacement of oceanic crust by sea-floor spreading took place at the peak of igneous activity along the eastern seaboard of North America 180 m.y. ago. According to Salvador and Green (in press), the Gulf of Mexico continental separation took place somewhat later 143 m.y. ago. During the drifting episode the evaporite deposits were separated into their eastern and western masses in the North Atlantic and northern and southern masses in the Gulf of Mexico (Buffler et al, 1980).

As the continental fragments began drifting to their present positions, open water circulation was established and a massive system of reefs and carbonate platforms flourished along the Atlantic and Gulf of Mexico margins. They also flourished atop isolated highs such as the one beneath the continental slope off Texas and Jordan Knoll in the Straits of Florida (Bryant et al, 1969). The increased thickness of the 4.5 to 5.0 km/sec layer at the base of the Sigsbee Escarpment and the presence of 5.1 km/sec material on the outer shelf off south Texas inciated that the carbonate platform beneath the upper continental slope was an important sediment source both landward and seaward. The major influx of clastic material, which began in Late Cretaceous, coupled with a major transgression at about the same time resulted in either the burial or extinction of the reef complex bordering the Gulf of Mexico and North Atlantic. The progradation of this massive clastic wedge from the Late Cretaceous to early Holocene reduced the Gulf of Mexico to its present dimension (Wilhelm and Ewing, 1972). Basinward, flow of the Middle Jurassic evaporites, due to sediment loading, in the northwestern Gulf further reduced the size of the basin. Other modifications of the Gulf of Mexico depocenter resulted from massive gravity sliding or compression of the Cenozoic deposits (Mexican Ridges in the southwestern Gulf of

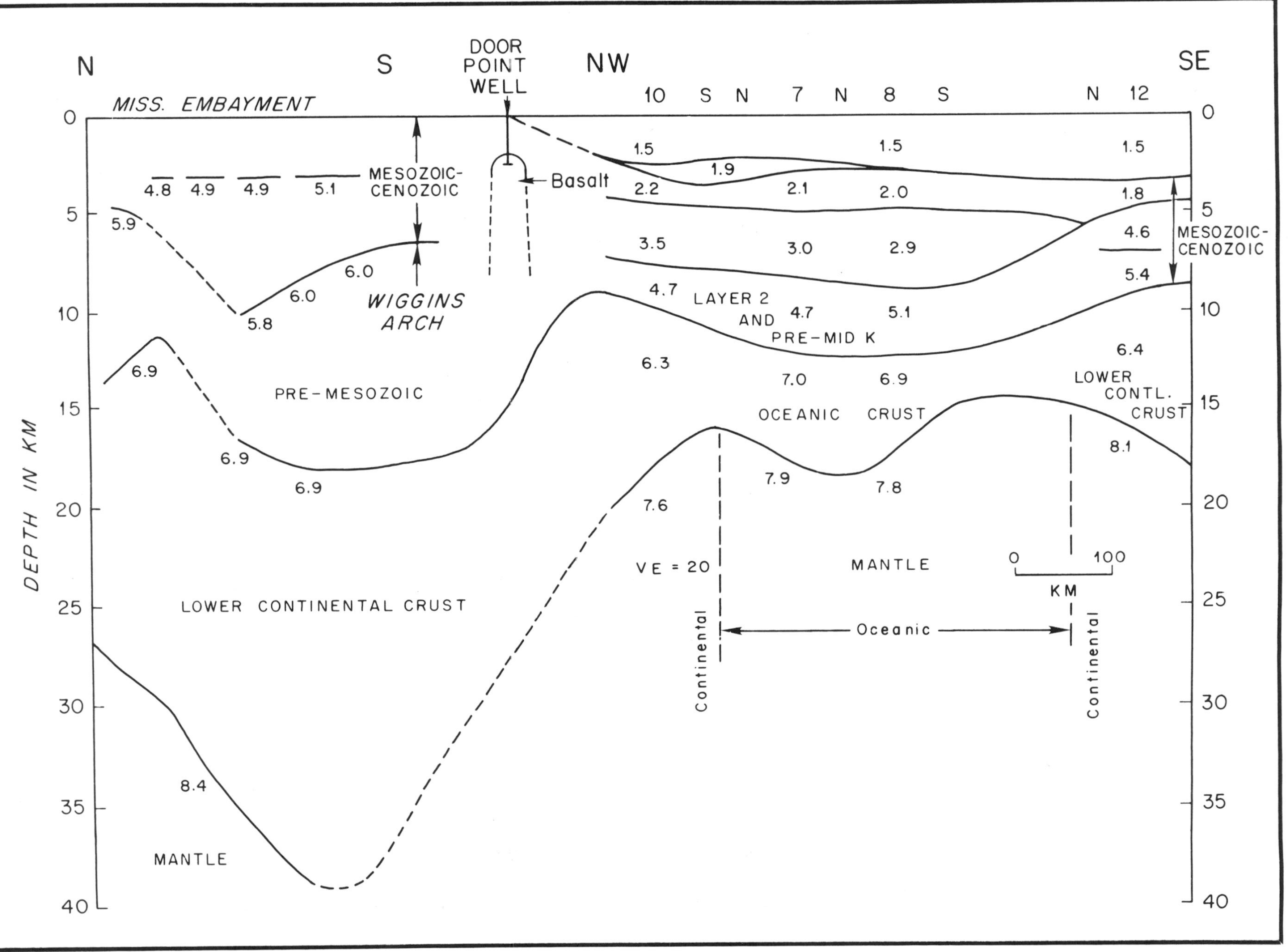

Figure 8 — Crustal section at the eastern end of the Gulf Coast geosyncline. Section of the Mississippi Embayment is from Warren, Healy, and Jackson (1966), and the section from the Mississippi Fan to northwest of Cuba from Ibrahim et al (1981). Door Point well is from Braunstein and McMichael (1976). For section location see Figure 1.

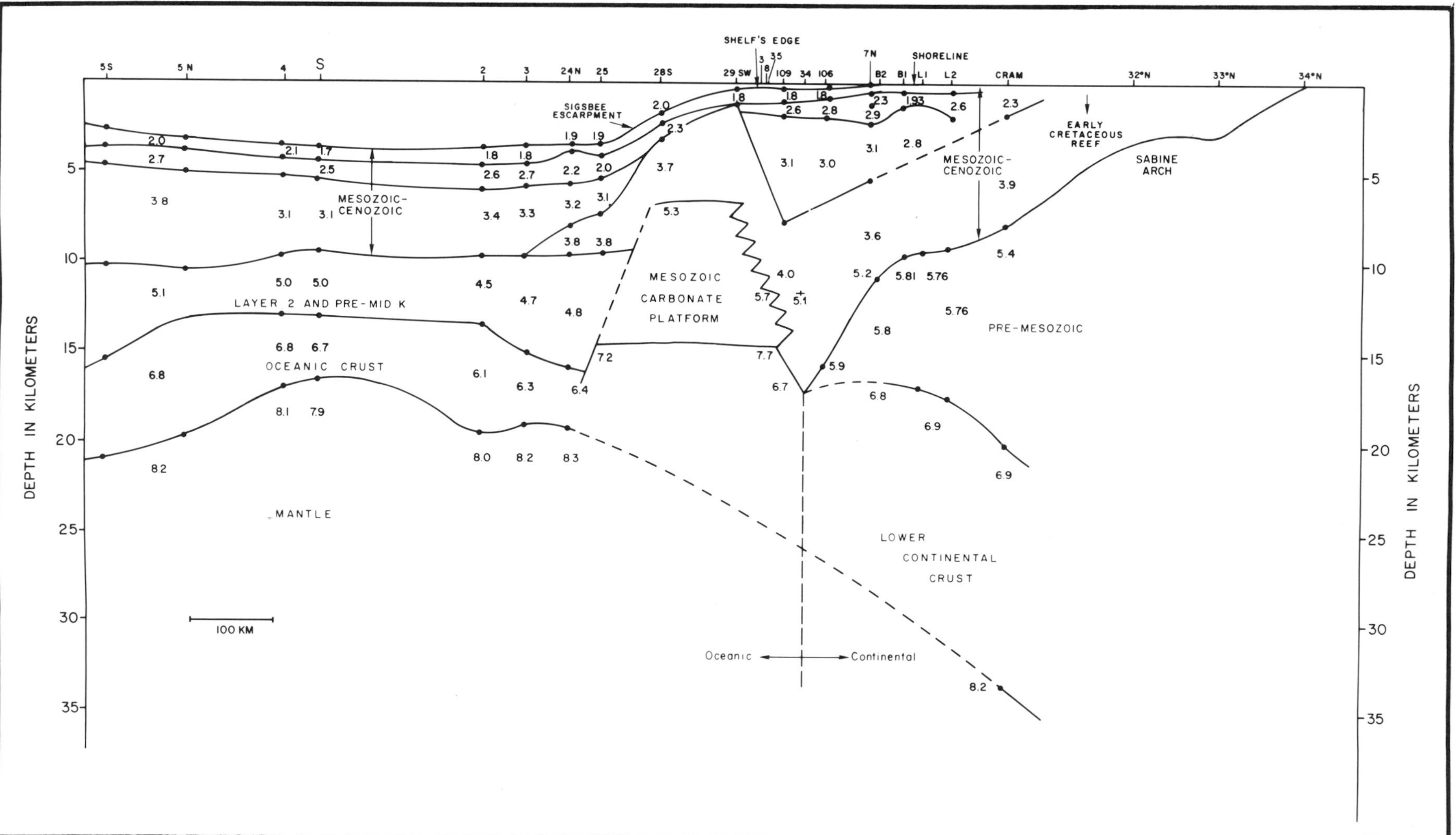

Figure 9 — Cross section of the western end of the Gulf Coast geosyncline. Profile was constructed from data recorded on profiles indicated on Figure 1 and depth to basement (Pre-Mesozoic rocks) contours on the tectonic map of North America (King, 1969).

Mexico; Buffler et al, 1979). Northward thrusting of the Mesozoic-Paleogene island arc sequence at the western end of the Greater Antilles Ridge (Cuba) over the rifted continental crust, between the Campeche and Florida escarpments and the southern edge of the Bahama carbonate platform, further restricted the depocenter.

ACKNOWLEDGMENTS

We wish to thank Shell Oil Company for providing the geophysical data on the shelf off East Texas, and R. A. Stephen and D. G. Aubrey of Woods Hole Oceanographic Institution for their comments during the preparation of this report. Funding for E. Uchupi was provided by the Office of Sea Grant (Grant NA80AA-D-00077). Contribution No. 4831 of the Woods Hole Oceanographic Institution, Woods Hole, Massachusetts. Contribution number 462 of the Institute of Geophysics, University of Texas, Austin, Texas.

REFERENCES CITED

Antoine, J. W., and J. Ewing, 1962, Seismic refraction measurements on the margins of the Gulf of Mexico: Journal of Geophysical Research, v. 68, p. 1975-1996.

Bally, A. W., 1975, A geodynamic scenario for hydrocarbon occurrences: panel discussion, Global tectonics and petroleum occurrence: Transcripts of the World Petroleum Congress, 11 p.

Braunstein, J., and C. E. McMichael, 1976, Door Point a buried volcano in southeast Louisiana: Gulf Coast Association of Geological Societies, Transactions, v. 26, p. 79-80.

Bryant, W. R., et al, 1969, Escarpments, reef trends, and diapiric structures, eastern Gulf of Mexico: AAPG Bulletin, v. 53, p. 2506-2542.

Buffler, R. T., et al, 1979, Anatomy of the Mexican Ridges, southwestern Gulf of Mexico, *in* J. S. Watkins, I. Montadert, and P. W. Dickerson, eds., Geological and geophysical investigations of continental margins: AAPG Memoir 29, p. 319-325.

———, ———, 1980, Structure and early geologic history of the deep central Gulf of Mexico basin, *in* R. H. Pilger, Jr., ed., The origin of the Gulf of Mexico and the early opening of the central North Atlantic Ocean: Baton Rouge, Louisiana, proceedings, Louisiana State University, School of Geoscience Symposium, p. 3-16.

Burke, K., and J. F. Dewey, 1973, Plume-generated triple junctions: key indicators in applying plate tectonics to old rocks: Journal of Geology, v. 81, p. 406-433.

Cousminer, H. L., and W. Manspeizer, 1976, Triassic pollen date Morocan High Atlas and the incipient rifting of Pangea as middle Carnian: Science, v. 191, p. 943-945.

Cram, I. H., 1961, A crustal structure survey in south Texas: Geophysics, v. 26, p. 560-573.

Dorman, J., et al, 1972, Crustal section from seismic refraction measurements near Vitoria, Texas: Geophysics, v. 37, p. 325-336.

Ewing, J., J. Antoine, and M. Ewing, 1960, Geophysical measurements in the western Caribbean Sea and in the Gulf of Mexico: Journal of Geophysical Research, v. 65, p. 4087-4126.

Hales, A. L., C. E. Helsley, and J. B. Nation, 1970, Crustal structure study on Gulf Coast of Texas: AAPG Bulletin, v. 54, p. 2040-2057.

Ibrahim, A. K., and G. V. Latham, 1978, A comparison between sonobuoy and ocean bottom seismograph data and crustal structure of the Texas shelf zone: Geophysics, v. 43, p. 514-527.

———, et al, in press, Crustal structure in the Gulf of Mexico from OBS and multi-channel reflection data: AAPG Bulletin, v. 65, p. 1207-1229.

Jansa, L. F., J. P. Bujck, and C. L. Williams, 1980, Upper Jurassic salt deposits of the western North Atlantic: Canadian Journal of Earth Science, v. 17, p. 547-559.

King, P. B., 1969, Tectonic map of North America: U.S. Geological Survey, scale 1:5,000,000, 2 sheets.

Martin, R. G., 1978, Northern and eastern Gulf of Mexico continental margin: stratigraphic and structural framework, *in* H. Bouma, G. T. Moore, and J. M. Coleman, eds., Framework, facies, and oil-trapping characteristics of upper continental margin: AAPG Studies in Geology No. 7, p. 21-42.

Rezak, R., and T. T. Tieh, 1980, Basalt on Louisiana outer shelf (Abs.): Transactions of the American Geophysical Union, v. 61, p. 989.

Salvador, A., and A. R. Green, in press, Opening of the Caribbean Tethys: Paris, Proceedings, International Geological Congress, 1980.

Swolfs, H. S., 1967, Seismic refraction studies in the southwestern Gulf of Mexico: Texas A&M Master's thesis, 42 p.

Warren, D. H., J. H. Healy, and W. H. Jackson, 1966, Crustal seismic measurements in southern Mississippi: Journal of Geophysical Research, v. 71, p. 3437-3458.

Watkins, J. S., et al, 1978, Occurrence and evolution of salt in deep Gulf of Mexico, *in* H. Bauma, G. T. Moore, and J. M. Coleman, eds, Framework, facies and oil-trapping characteristics of the upper continental margin: AAPG Studies in Geology No. 7, p. 43-65.

Wilhelm, O., and M. Ewing, 1972, Geology and history of the Gulf of Mexico: Geological Society of America, Bulletin, v. 83, p. 575-600.

Worzel, J. L., and J. S. Watkins, 1973, Evolution of the northern Gulf Coast deduced from geophysical data: Gulf Coast Association of Geological Societies, Transactions, v. 23, p. 84-91.

———, and C. A. Burk, 1978, Margins of Gulf of Mexico: AAPG Bulletin, v. 62, p. 2290-2307.

Structural Framework and the Evolutionary History of the Continental Margin of Western India

Bhoopal R. Naini*
Lamont-Doherty Geological Observatory and Department of Geological Sciences Palisades, New York

Manik Talwani
Gulf Research and Development Company Pittsburgh, Pennsylvania

Sediment and upper crustal structure derived from sonobuoys, together with other geophysical evidence, divides the continental margin of western India into two provinces: the western basin and the eastern basin. The western basin exhibits well-developed sea-floor spreading type magnetic anomalies, whereas the eastern basin has no significant correlatable anomalies. The two basins are divided by the Chagos-Laccadive and the Laxmi ridges. The Laxmi ridge has a velocity structure similar to that of the eastern basin; however, the total crustal thickness of the ridge (more than 20 km) is much greater than that of the eastern basin. We propose that the western margin of India has a two phase evolutionary history: a phase of rifting followed by a phase of sea-floor spreading in early Tertiary time.

The sea-floor spreading history of western India and the adjoining Arabian Sea has been known in a general sense (McKenzie and Sclater, 1971; Whitmarsh, 1974; Norton and Sclater, 1979). However, the details relating to its evolution are tentative, as the data are rather sparse. Most present knowledge comes from reconnaissance cruises of the International Indian Ocean Expedition (Udintsev, 1975). The prerift position of western India in Gondwanaland, and the process leading to its separation are a matter of conjecture. The answer to these problems lies to a considerable extent along the present day western margin of India. Figures 1 and 2 show a generalized framework of the Arabian Sea and the western margin of India.

Lamont-Doherty Geological Observatory research vessels *Robert D. Conrad* and *Vema,* each spent two months carrying out detailed field work in the Arabian Sea and the western continental margin of India during 1974 and 1977, respectively (Figure 3). Topography, gravity, magnetic, and seismic reflection data were gathered on a continuous basis along the ships' tracks. In addition, 67 refraction profiles were shot during these cruises. The refraction profiles were recorded by 53 short-range and 14 long-range sonobuoys (Figure 4). The data from these different cruises, with emphasis on seismic refraction data, together with data from reconnaissance cruises, other cruises in the Arabian Sea, and published refraction results of Closs, Bungenstock, and Hinz (1969), form the basis of this study.

Numerous regional structures and sedimentary basins are located along the continental margin of western India and the contiguous Arabian Sea (Naini, 1980). Important among these are the Chagos-Laccadive ridge, Laxmi ridge, Prathap ridge, Murray ridge, Owen fracture zone, Carlsberg Ridge, and the largest sedimentary body of the Arabian Sea, the Indus cone (Figure 1). This study concentrates on the Chagos-Laccadive and Laxmi ridges, which parallel the western margin of India and the deep sea region east of this ridge complex (Figure 2).

DATA REDUCTION

Lamont research vessels *Vema* and *Conrad* were both equipped with 3.5 kHz and 12 kHz precision depth recording devices. In the present work, all depths are in corrected meters. The bathymetric data are presented as a contour map (Figure 2). Revised bathymetric charts for the Arabian Sea have been contoured at a

*Presently with: Sohio Petroleum Company, Dallas, Texas.

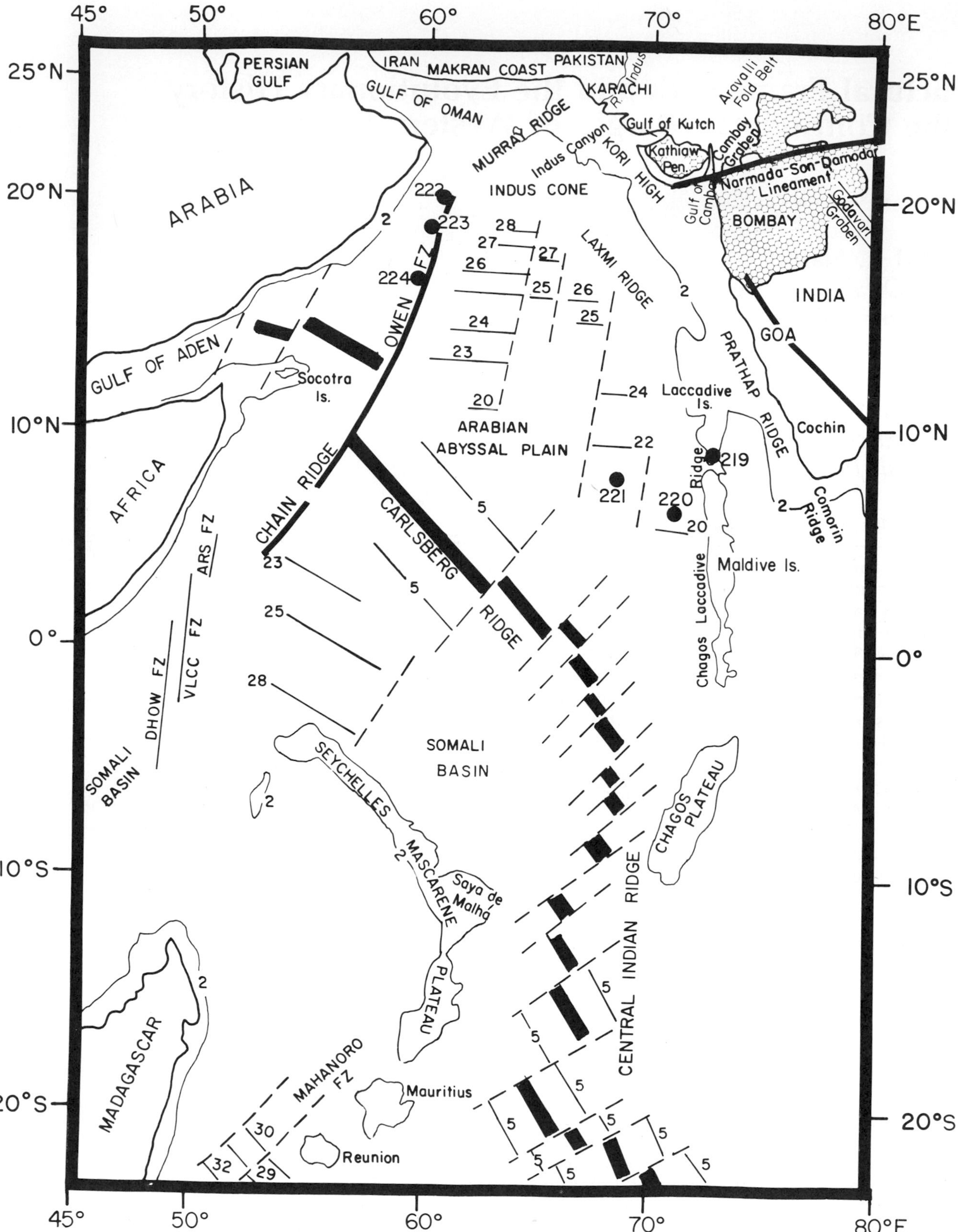

Figure 1 — Generalized map of the Arabian Sea showing various physiographic features. For the purpose of clarity only the 2 km isobath is shown around continents, islands, plateaus and ridges. Thick black bars, offset by dashed lines, represent segments of mid-ocean ridge crests, and fracture zones (FZ), respectively. Thin lines with numbers correspond to magnetic anomalies (after Naini, 1980; Norton and Sclater, 1979; Bunce and Molnar, 1977; Schlich, 1975; Whitmarsh, 1974; and McKenzie and Sclater, 1971). DSDP Leg 23 sites (Whitmarsh et al, 1974) are shown by solid circles. Also shown are important lineaments (thick black lines) and extent of Deccan lava flows or traps (patterned) in India (Eremenko and Negi, 1968).

primary contour interval of 500 m using the new Lamont data and data from all cruises in the International Indian Ocean expedition atlas (editor, Udintsev, 1975; Table 1). Additional contours were used where data permitted. The maps were originally contoured at a mercator scale of 1:2,000,000, latitude 33°. A new U.S. Navy bathymetric chart, based on detailed soundings, shows some other interesting features (V. Kolla, personal communications, 1981). These include seamount chains which run from the slope to the shelf in a manner reminiscent of the New England seamount chain in the western North Atlantic.

Gravity data were, in general, collected on a continuous basis along the Lamont-Doherty ships' tracks. Both research vessels were equipped with Graf-Askania GSS-2 surface ship gravimeters mounted on stable platforms (Graf and Schulze, 1961). Errors in gravity measurements at sea (off leveling, cross coupling, and Eötvös) are reviewed in detail by Talwani, Early, and Hayes (1966).

Revised free-air anomaly charts for the Arabian Sea have been contoured at a primary contour interval of 25 mgal (Figure 5). The charts were originally constructed at the same scale as the bathymetric chart. All the data from previous cruises in the International Indian Ocean Expedition Atlas (Udintsev, 1975) were used in conjunction with the new data (Table 1).

A "two-dimensional" isostatic gravity anomaly map has also been compiled for the Arabian Sea. Simple Airy isostasy was assumed for isostatic compensation. A depth of compensation of T = 30 km is used together with density contrasts of 1.67 and 0.5 gm/cc across the crust-water and crust-mantle interfaces, respectively. The 30 km depth of compensation corresponds to the average sea level crustal thickness in India, calculated by Aravamudhu, Verma, and Quereshy (1970). Thus, the Moho depth under sea is given by the following formula:

$$H = 30 - \frac{d(\rho_c - \rho_w)}{(\rho_m - \rho_c)} ,$$

where H is depth to the Moho-discontinuity; d is the water depth; ρ_w is density of water; ρ_c is density of crust; and ρ_m is density of upper mantle.

Table 1. Cruises and types of data used to study the Northeastern Arabian Sea.

Vessel	Cruise ID	Gravity	Magnetics	Seismics	Topo
Vema	V-19	X	X	X	X
Vema	V-33	X	-	X	X
Vema	V-34	X	X	X	X
Conrad	C-9	X	X	X	X
Conrad	C-17	X	X	X	X
Meteor	MTR	X	-	-	X
Owen	OW	X	X	-	-
Oceanographer	OSS	X	-	-	-
Lusiad	LU	X	-	-	X
Glomar	GL	-	X	X	X

The gravity computations all assume two-dismensionality and the "two-dimensional" isostatic gravity map has been contoured at 25 mgal intervals (Figure 6). "Two-dimensional isostatic anomalies" are also presented as selected projected profiles across the continental margin of India and over the Laxmi ridge (Figure 7).

The total intensity of the magnetic field (H) along the Lamont-Doherty ships' tracks was measured continuously by towing a Proton Precession Magnetometer astern of the ships. The magnetic anomaly at any point is deduced by subtracting the International Geomagnetic Reference Field (Cain et al, 1968) from the observed field. The magnetic data are presented as profiles along the ships' tracks on a mercator projection, originally plotted at a scale of 1:2,000,000 at 33° latitude (Figure 8). The observed anomalies were correlated with the marine magnetic time scale of Heirtzler et al (1968) and La Brecque, Kent, and Cande (1977) in the context of sea-floor spreading (Figures 8 and 9).

Seismic data in this study were acquired utilizing conventional (short-range Navy) sonobuoys, as well as commercial (long-range) sonobuoys. The sonobuoys yield two types of information: 1) wide-angle seismic reflections through which interval velocities of sediment are obtained; and, 2) refractions through which interface velocities are obtained. Techniques for acquiring sonobuoy data and procedures for data reduction are described by LePichon, Ewing, and Houtz (1968) and Houtz, Ewing, and LePichon (1968).

An air gun sound source was used at source-receiver distances of about 25 km. Crustal refraction work was carried out using long-range sonobuoys built commercially by Fairfield Industries (79 MHz configuration, channels A to D; SB76). Data at short ranges (typically about 10 to 25 km) were first recorded as variable density profiles using the Lamont 21-cubic-inch air gun. These air gun profiles were extended until the reflection hyperbolae from sediment and basement were well defined (Figure 10). At longer source-receiver distances (typically beyond 10 to 25 km) where the returning refracted waves were weak, the air gun sound source was replaced by explosives (Figure 10). At the same time, data were simultaneously recorded as traces on an SIE recorder. Explosive charges started at two pounds and were gradually increased to 130 pounds. The traces were read to the nearest one-hundredth of a second. A time-distance graph interpreted from the data recorded by a typical long-range sonobuoy is shown in Figure 11.

Sink rates of the shots were computed by tying long threads with flags at one meter intervals for all sizes of charges. As the charges started sinking, times were recorded through the descent of each flag and an average sinking rate was derived. Data were corrected for the shot and receiver depths and the burn time of the fuse. Proper topographic corrections were then applied to the arrival data, and the data were interpreted according to the conventional techniques as described by Ewing (1963).

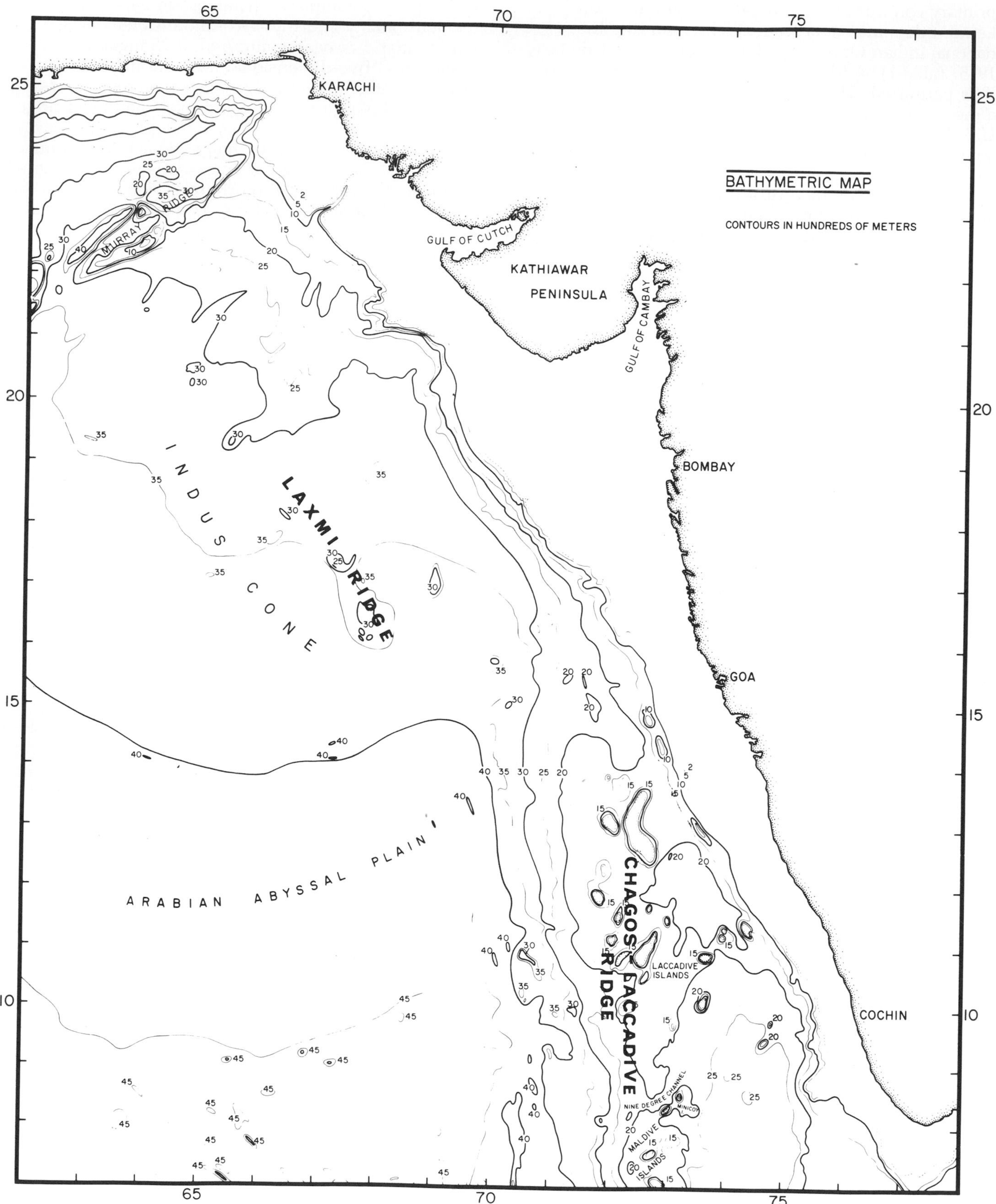

Figure 2 — Detailed bathymetric map of the northeastern Arabian Sea and adjoining continental margin of India showing various physiographic features discussed in the text. Contours are in hundreds of corrected meters, at 500 m intervals. Also shown is the 200 m isobath (dotted line).

RESULTS AND DISCUSSION

Physiography

Important bathymetric features of the Arabian Sea, as outlined in the bathymetric map (Figure 2), are discussed below. The average shelf break along the western margin of India occurs at about 200 m depth. The shelf width is variable, being the widest off Bombay (~ 350 km), and narrower to the south (~ 60 km off Cochin) and north (~ 150 km between Karachi and the Gulf of Kutch). Between Goa and the coast of Kathiawar, a normal continental slope and rise is observed. From Cochin to Goa, isolated topographic highs between 2,500 and 3,000 m deep are present on the continental slope and rise. These highs are continuous with the subbottom highs and form a continuous structural high (Prathap ridge) between 7° and 15° north (Naini, 1980). North of the Kathiawar coast the margin morphology is modified by the Indus cone.

Southeast of Karachi, the Indus River empties into the Arabian Sea and forms a deep submarine canyon known as the Indus canyon. The Indus canyon and the numerous channels formed by the sediment-laden Indus waters form a terrain of rough topography known as the Indus cone. This topography extends southward from the mouth of the river to about 20°N and modifies the continental margin morphology.

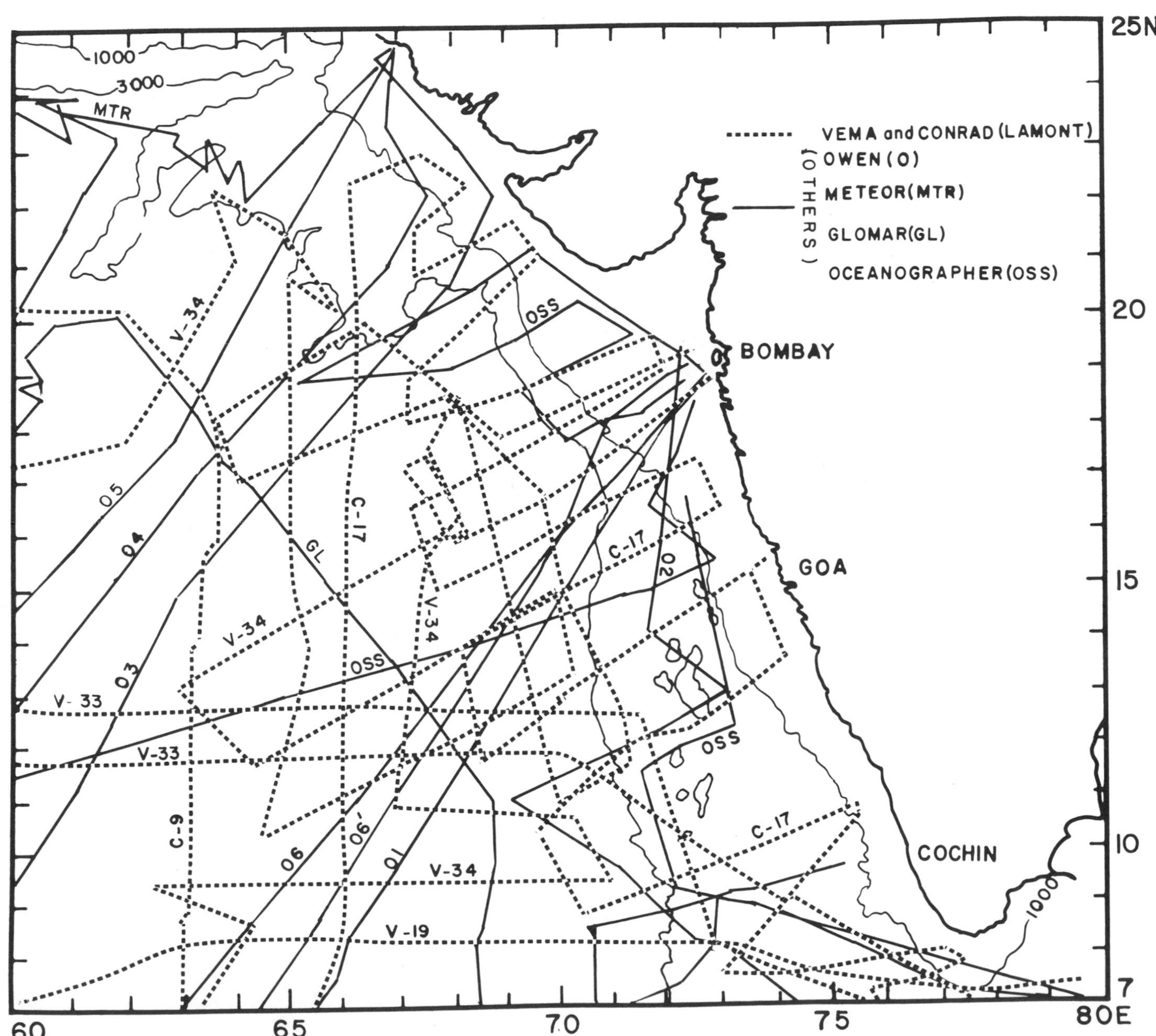

Figure 3 — Track chart showing coverage of the northeastern Arabian Sea and continental margins of India and Pakistan, by ships belonging to various nationalities and sources, including R/V *Vema* and R/V *Robert D. Conrad* of Lamont. See Table 1 for the types of data collected during each cruise.

The Arabian abyssal plain extends between about 10°N and 15°N and lies between 4,000 and 4,500 m depth. This abyssal plain is terminated to the east by the Chagos-Laccadive ridge. South of 10°N, average depth of the abyssal plain is greater than 4,500 m, and consists of numerous small topographic highs and lows associated with the northern flank of the Carlsberg Ridge.

The Chagos-Laccadive ridge extends between 7°N and 15°N in the region contoured here. It extends farther southward (off the bounds of this map), to about 10°S, in an arcuate fashion (Heezen and Tharp, 1964). The ridge is asymmetrical in appearance, with a steeper eastern flank. The apex of the ridge lies, on the average, less than 1,000 m deep with occasional coral atolls and volcanic islands, such as the Lac-

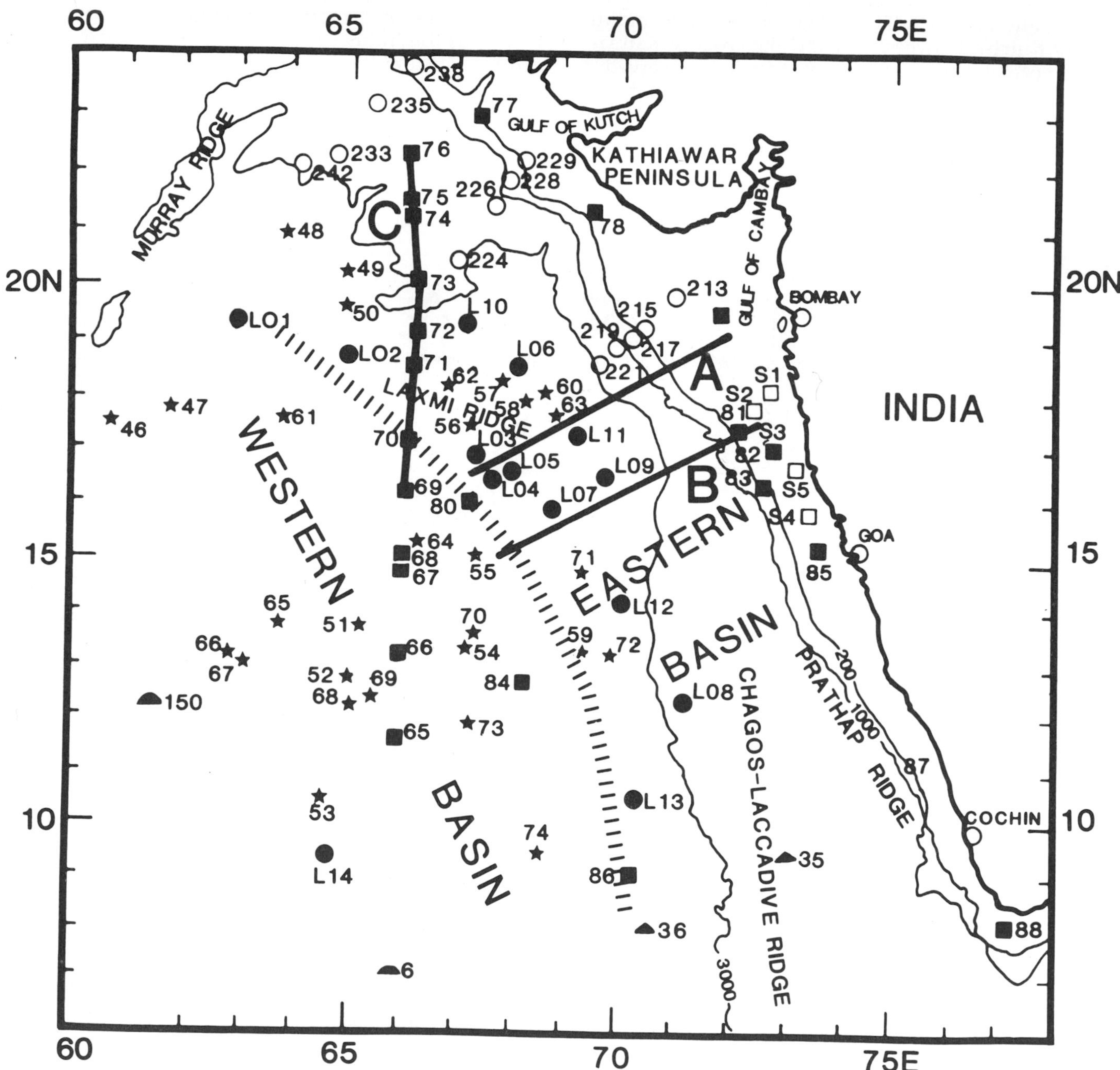

Figure 4 — Map showing location of seismic reflection and gravity profiles shown in Figure 7 of the text. Also shown are the location of the wide-angle seismic reflection and refraction (sonobuoys — both short and long range), as well as two-ship refraction stations. Stars denote (R/V *Vema*, solid squares are R/V *Conrad*, and numbers with prefix-L (solid circles) are long range sonobuoy stations. Refraction stations indicated by open circles are from Closs, Bungenstock, and Hinz (1969); numbers with prefix-S (open squares) are from Rao (1970); solid triangles are from Francis and Shor (1966); and, solid half circles are from Neprochnov (1961), as reported in the IIOE atlas (Udintsev, 1975). The boundary between the eastern and western basins discussed in the text is shown by the thick dashed line.

cadives, topping the ridge crest.

The Murray ridge, believed to be the northward extension of the Owen fracture zone (Matthews, 1966; Barker, 1966), is oriented northeast to southwest. The northeast trend merges into the continental margin topography north of Karachi. The Murray ridge, at the shallowest portion, is less than 500 m deep. Northwest of this ridge and parallel with it, is a trough with depths in excess of 4,000 m.

In addition to the features described above, there are numerous isolated topographic highs in the region. Prominent among these is an approximate

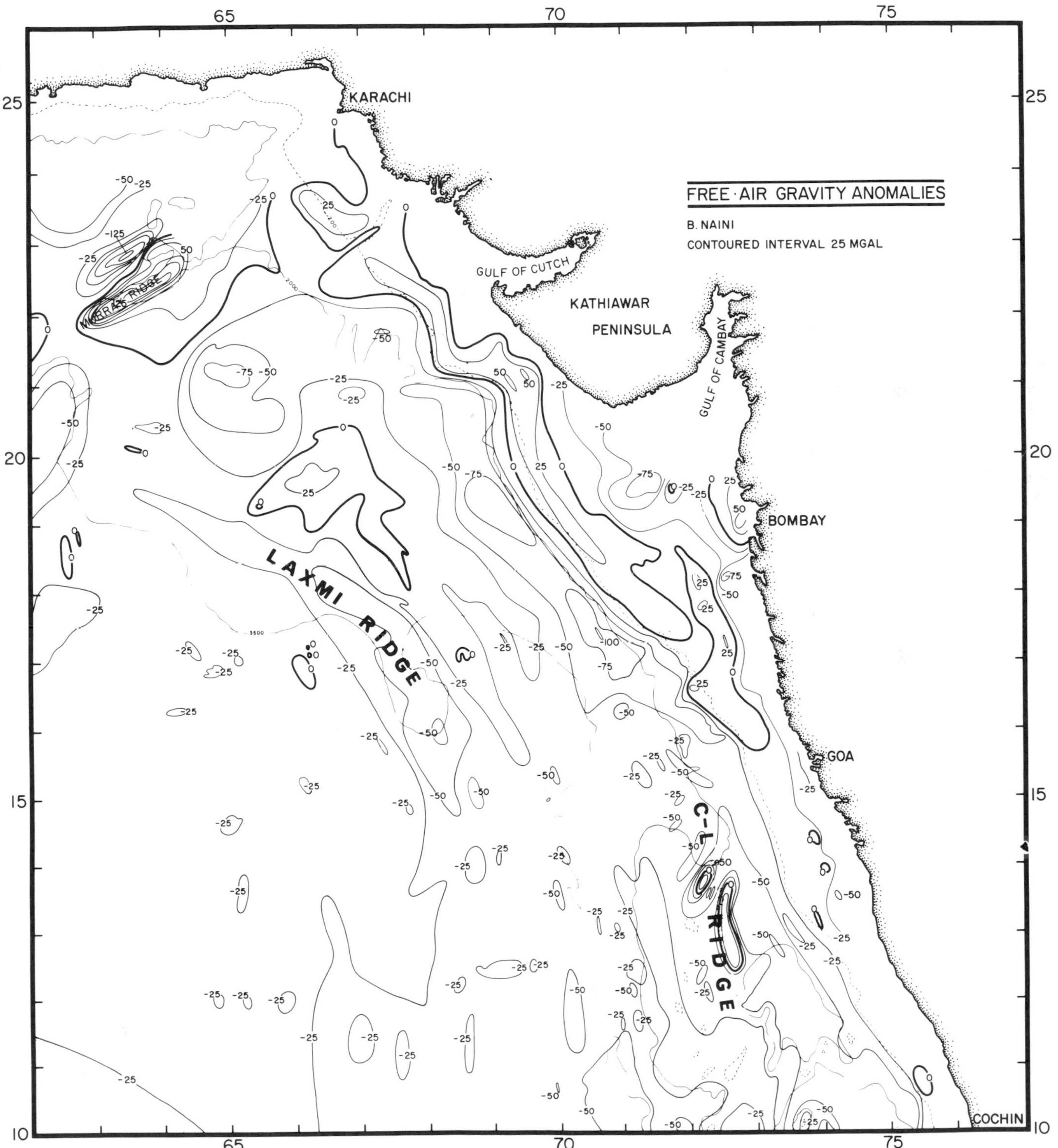

Figure 5 — Free-air gravity anomaly map for the western margin of India and the adjoining Arabian Sea. Contours are in milligals at 25 mgal intervals. Selected bathymetric contours are shown by thin lines. 200 m isobath is shown by thin dashed line.

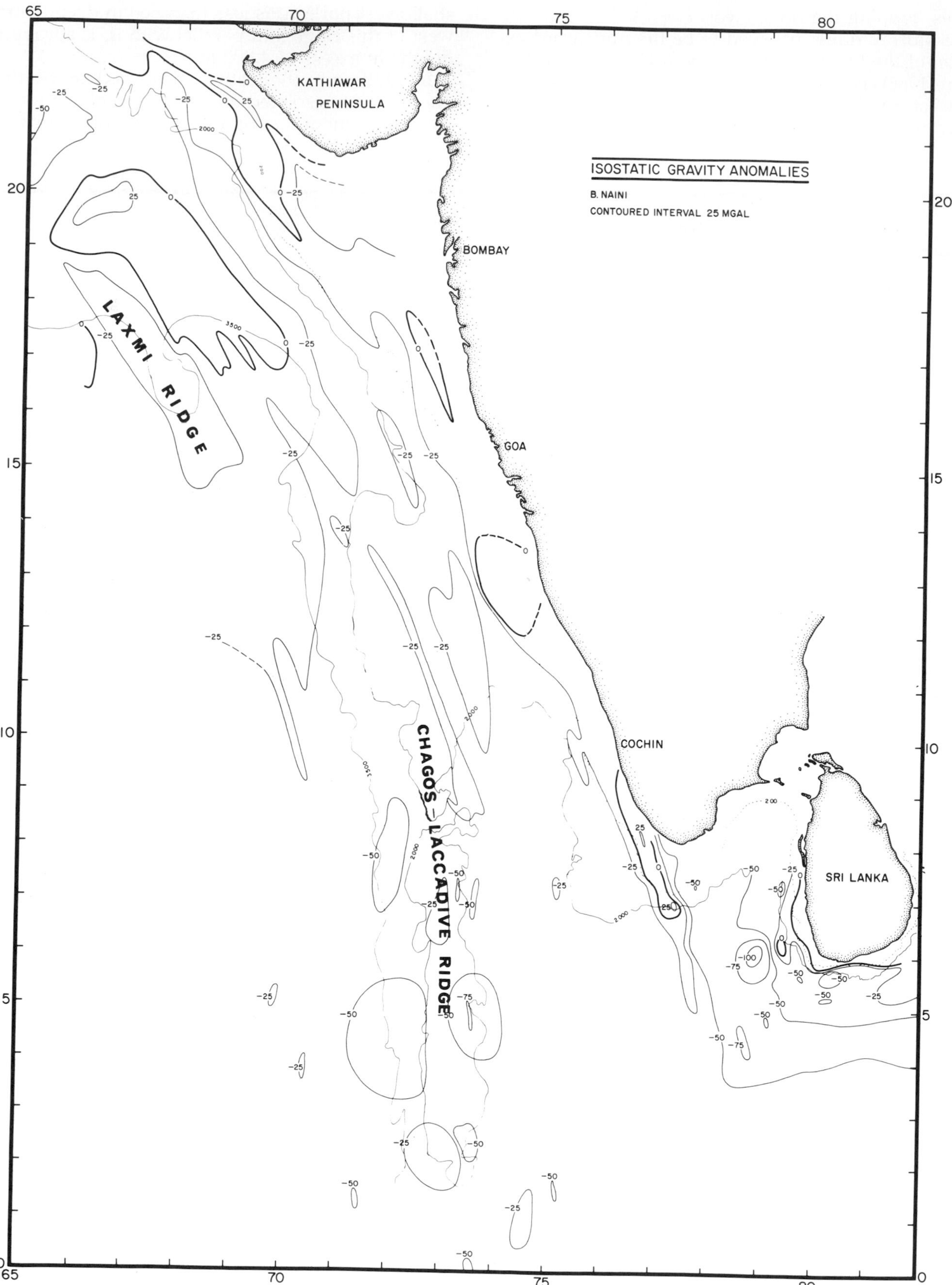

Figure 6 — Computed two-dimensional, Airy (T = 30 km) isostatic anomalies for the western margin of India and the adjoining Arabian Sea. Contours are in milligals at 25 mgal intervals. Selected bathymetric contours are shown by thin lines. Thin dashed line corresponds to 200 m isobath.

northwest to southeast trending high, with depths less than 3,000 m, located between the 3,000 and 3,500 m isobaths in the Indus cone-Arabian abyssal plain region. Naini and Talwani (1977) named this the Laxmi ridge.

Free-Air Gravity Anomalies

Figure 5 shows the free-air gravity anomaly map for the study area. The regional free-air gravity field is negative over the entire Arabian Sea. The Indian Ocean is, in general, marked by negative regional free-air anomalies (Talwani and Kahle, 1975). Superimposed on this regional field are relative negative and positive anomalies.

The largest free-air gravity values (> + 100 and < −100 mgal) are distributed over the Murray ridge and form contiguous positive and negative belts of linear anomalies. Both of these belts trend approximately northeast to southwest. The positive belt, with free-air values greater than +125 mgal, is to the south and coincides with the topographic high associated with the Murray ridge. The negative belt, with free-air values less than −125 mgal, is coincident with the trough located north of the Murray ridge.

The free-air gravity anomalies along the continental margin off western India and the adjoining deeper Arabian Sea have values ranging between about −75 and +50 mgal. These anomalies form contiguous sublinear belts of positive and negative values and trend approximately northwest to southeast, parallel with the trend of the present day western India shelf edge. A linear belt of positive anomalies with values reaching 50 mgal extends along the entire length of the margin, centered approximately over the shelf edge. South of Goa, the shelf edge high is not as pronounced as it is to the north. Seaward of this positive belt, a negative belt of anomalies with values less than −75 mgal is located over the continental slope and rise between the latitudes of Goa and the Kathiawar Peninsula. However, the negative anomalies located over the continental slope and rise of the entire Indian margin do define a linear trend. A belt of relative positive anomalies with values greater than −25 mgal is located southwest of Goa. These relative positive anomalies are coincident with the approximate crestal region of the Chagos-Laccadive ridge.

Between the latitudes of Goa and the Kathiawar Peninsula, contiguous linear belts of positive and negative anomalies are found in the deep sea. The westernmost of these, defining a negative belt with free-air values less than −25 mgal, is centered over the Laxmi ridge. The positive belt of anomalies is located to the northeast, between the belt of negative anomalies located over the continental slope and rise and the Laxmi ridge. The location of this positive belt coincides with the broad structural high located north of the Laxmi ridge (Naini, 1980).

Free-air gravity anomalies near the shelf edge inherently have the "edge effect." It is difficult to decide which part of the anomaly is caused by topography and its compensation, and which part is due to mass or density inhomogeneities within, or below, the crust. The edge effect is more pronounced over steeper slopes. Since the slopes along the western margin of India are steep, to avoid edge effects we computed the isostatic anomalies. However, isostatic anomalies may be in error because of the lack of knowledge of type and amount of compensation. For stable margins, a considerable change in isostatic parameters causes only a minor change in isostatic anomalies. Therefore, even though the isostatic model may be wrong in detail, the isostatic anomalies provide useful information.

"Two-Dimensional" Isostatic Gravity Anomalies

Isostatic anomaly values in the area range from < −75 mgal to > +25 mgal (Figure 6), and the regional field is negative. Superposed on this regional field are relative positive and negative anomalies. Southwest of India, between 71°E to 74°E and 2°N to about 9°N, a linear north-south trend is defined by alternate relative negative and positive anomalies. An approximately north-south belt of relative positive and negative isostatic anomalies, south of 10°N, are associated with the crest and flanks of the Chagos-Laccadive ridge. At about the latitude of Cochin, the isostatic anomaly trends change to northwest. The belts of relative positive and negative anomalies are approximately parallel with the trend of the present day shelf edge of western India. Isolated relative positive anomalies are found along the shelf edge and they are especially prominent close to the southern tip of India. A negative isostatic anomaly belt, located between 14°N to 19°N and 65°E to 69°E, is coincident with the Laxmi ridge. Pronounced negative gravity anomalies (values less than −100 mgal) are observed south of the southern tip of India and have been discussed earlier by Kahle et al (1976 and 1981).

Isostatic gravity anomalies are more subdued than free-air gravity anomalies. The isostatic correction reduces the shelf edge free-air high in the isostatic anomaly. In many places (e.g. between Goa and the Kathiawar Peninsula) it nearly eliminates the free-air anomaly. An isostatic high observed between the Laxmi ridge and the continental margin coincides with the structural high located northeast of the Laxmi ridge (Figure 7). The Laxmi Ridge exhibits negative isostatic gravity anomalies, amounting to about 30 mgal (Figure 7).

Near elimination of the shelf edge free-air anomalies by isostatic corrections indicates that the observed free-air gravity anomalies over the shelf edge off western India are principally caused by the "edge effect" of margin topography and its compensation.

Gravity lows are present over relatively slow spreading centers (< 2.5 cm/yr), both active and extinct, and have been cited in the literature (e.g. Kristoffersen and Talwani, 1977; Cochran, 1979; Weissel and Watts, 1979). The Laxmi ridge is associated with a gravity low of 30 to 40 mgal with a wavelength of about 200 km. The magnitude of the gravity low associated with this ridge is of the right order, as compared with those of the Mid-Atlantic Ridge and

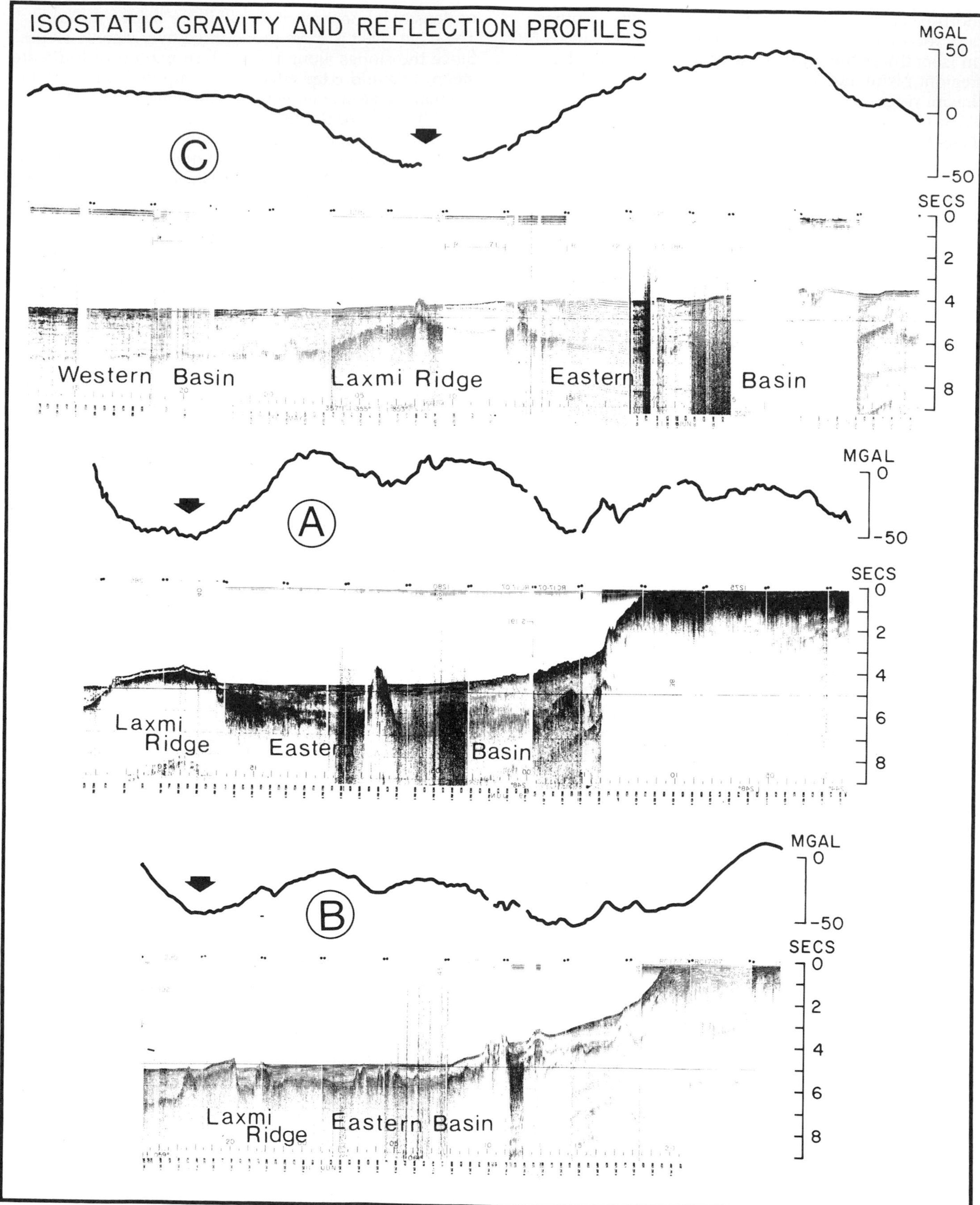

Figure 7 — Selected seismic reflection and two-dimensional isostatic gravity anomaly profiles across the Laxmi Ridge. Note the negative isostatic anomaly (shown with arrow), centered exactly on top of the Laxmi ridge. See text for explanation of the negative anomaly and Figure 4 for location of the profiles.

the extinct spreading centers such as in the Coral Sea and the Labrador Sea (Kristoffersen and Talwani, 1977; Cochran, 1979; Weissel and Watts, 1979), the wavelength is quite large. The Mid-Atlantic Ridge, as well as the extinct spreading centers mentioned above, have a definite magnetic signature associated with them. In contrast, the Laxmi ridge has no definite magnetic signature of significant extent, as is shown later. Thus, we rule out the possibility that the Laxmi ridge is an active or extinct spreading center.

Magnetic Anomalies

Total intensity residual magnetic anomalies are plotted as profiles along ships' tracks (Figure 8). A tentative correlation between tracks was made to determine the possible strike of the magnetic anomalies. In correlating, we chose the conspicuous anomalies, such as those located between 12°N to 17°N and 60°E to 64°E. We found that the anomalies here strike approximately east to west. We assigned numbers to these anomalies with the help of the marine magnetic time scale of Heirtzler et al (1968) and La Brecque, Kent, and Cande (1977) and with the help of magnetic anomaly identifications in the Arabian Sea by earlier workers, such as McKenzie and Sclater (1971), Whitmarsh (1974), Schlich (1975), and Norton and Sclater (1979). The observed anomalies matched well with synthetic anomalies generated by a ridge (spreading at variable rates) located at 10°S with a paleostrike of N75°E, and now observed at 15°N with a near east to west strike (Figure 9).

We identified sea-floor spreading type magnetic anomalies ranging between anomaly 20 (45 million years) and 28 (64 million years) of the marine magnetic time scale (Heirtzler et al, 1968; La Brecque, Kent, and Cande, 1977). Where the anomalies were not consistent for long distances and where they are not part of a continuous sequence, we did not assign numbers to anomalies. However, we traced such anomalies from one track to another. We also identified with dots anomalies that appeared to be sea-floor spreading type anomalies on isolated tracks, but could not be definitely correlated from one track to another. The final magnetic anomaly identification is shown in Figure 8. It is clear from this map that a well-developed, marine magnetic anomaly sequence, with anomalies 20 to 28, is present only in the western part of the area between about 10°N to 20°N and 60°E to 65°E. East of 65°E, anomalies 20 through 24 are found south of 15°N and west of 70°E. This sequence of anomalies is offset along numerous fracture zones (Figure 8). Similar fracture zones were postulated by Whitmarsh (1974). However, our identification of anomalies and fracture zones is better constrained by more data.

To the north and east of anomaly sequence 25 to 28 (Figure 8), the magnetic lineations strike more nearly northwest to southeast. We did not assign anomaly numbers to these lineations, since they did not obviously match any part of the marine magnetic time sequence. We believe that these anomalies could have a non-sea-floor spreading origin. It is interesting that these lineations are approximately parallel with the present day shelf edge off western India.

Earlier studies in this area, including those of McKenzie and Sclater (1971), Whitmarsh (1974), Schlich (1975), and Norton and Sclater (1979), show magnetic lineations to be east to west, east of the 25 to 28 anomaly sequence. On the contrary, we show that some of these lineations swing to the southeast. The earlier works, cited above, assume the presence of fracture zones extending through some of these anomalies, thus maintaining the east to west strike of the remaining anomalies. The anomaly identification in these works was based on a single track, and the anomaly azimuths were assumed parallel with the sequence in the west. However, our identification is based on more data and closer track spacing. We arrived at the northwest to southeast anomaly strike based on correlations made at very closely spaced track crossings (e.g. at about 65°E and 17°N, where three cruise tracks nearly intersect).

Thus, the magnetic anomalies in the Arabian Sea and the adjoining continental margin of India seem to be of two kinds. First, the east to west striking lineations (anomalies 20 to 28) in the west are generated by sea-floor spreading. Second, lineations striking approximately northwest to southeast in the east might have a non-sea-floor spreading origin. The Laxmi and Chagos-Laccadive ridges, which have structural expression for large distances, fail to exhibit any consistent correlatable magnetic signatures of significant extent. The synthetic magnetic model (Figure 9) suggests that the oceanic crust underlying the western part of the area was generated in the southern hemisphere by a ridge located at 10°S and oriented approximately east to west, at variable spreading rates. This oceanic crust was generated between about 45 and 64 m.y. ago, according to the magnetic time scale of La Brecque, Kent, and Cande (1977). Since the spreading ridge was located in the southern hemisphere, the crust generated at this ridge later crossed the equator and reached its present location, and hence, the observed anomalies are negative. The lineations striking northwest to southeast, with limited extent having no obvious correlation to the marine magnetic time scale, are located over the Laxmi and Chagos-Laccadive ridges proper and the region east of them. Since this region is underlain by anomalous crustal structure, as discussed later, it may lend further support to our suggestion that the origin of these anomalies is not due to normal sea-floor spreading.

Crustal Structure

Data warrants dividing the Arabian Sea into two broad regions. For the purposes of this study, we call these the "western basin" and the "eastern basin" (Figure 4). These two basins are approximately divided by the Chagos-Laccadive and Laximi ridge complex. The Laxmi ridge and the Chagos-Laccadive ridges are discussed separately.

Combined results from both the short range and

long range sonobuoys, together with the published refraction results of Closs, Bungenstock, and Hinz (1969; Figure 3 shows locations) for the western and eastern basins are presented in Tables 2 and 3, respectively. Velocity information and velocity-depth relations from the basins are shown as histograms in Figure 12, and as crustal columns in Figure 13, respectively.

The Western Basin

In this basin, water depth ranges between about 3.4 and 4.3 km, with an average of 3.8 km, sediment thickness ranges between 1.3 km and 4.2 km, with an average of 2.58 km, and sediment velocity range between 1.7 and 3.8 km/sec, with an average of 2.49 km/

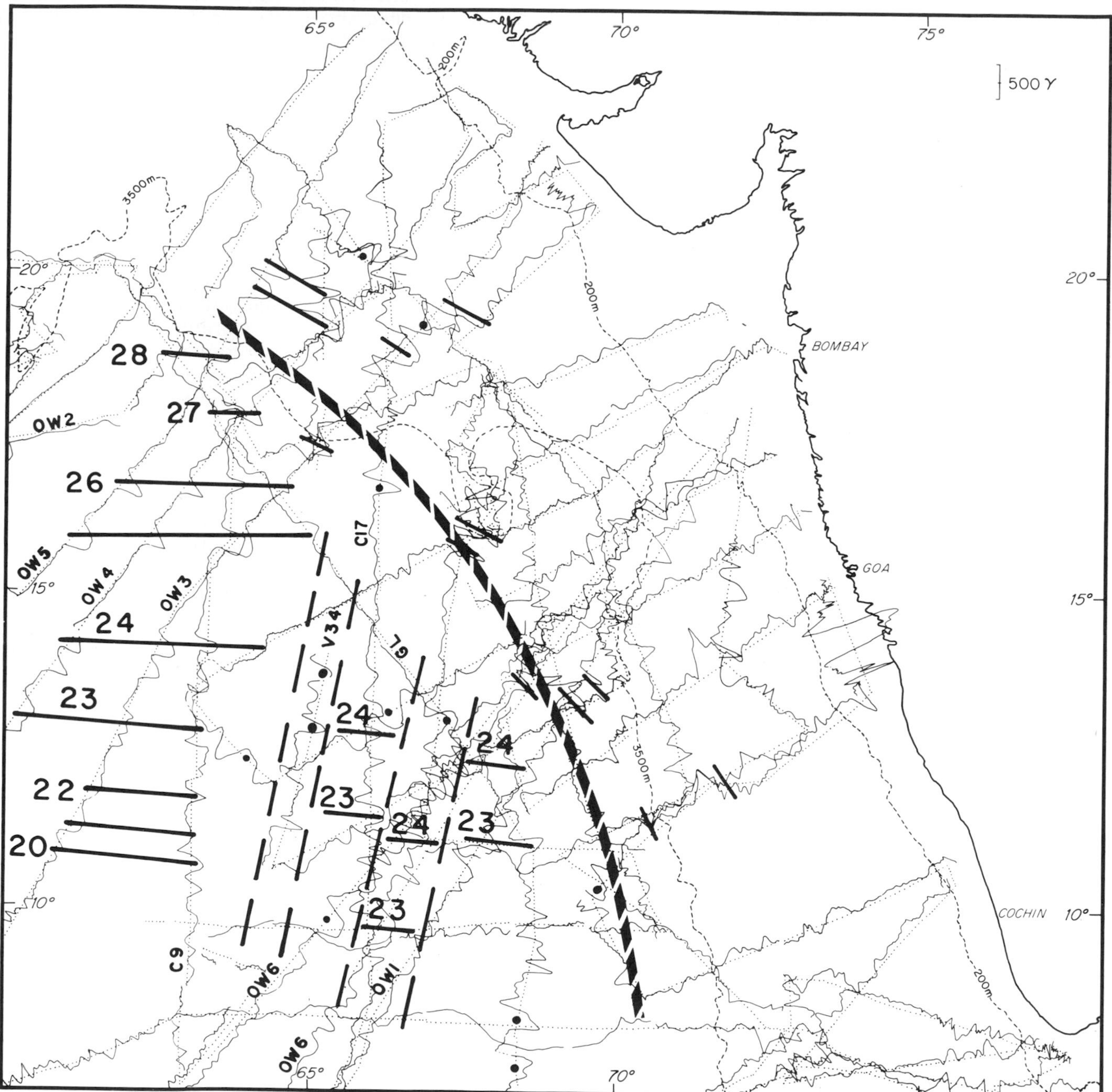

Figure 8 — Residual total intensity magnetic anomalies plotted along ships' tracks. Thick lines with numbers are magnetic anomalies matched with marine magnetic time scale of Heirtzler et al (1968) and La Brecque, Kent, and Cande (1977) (cf. Figure 9). Magnetic lineations that do not correlate with the time scale are shown as lines without numbers. Isolated, prominent anomalies are shown with dots. Long dashed lines represent fracture zones. Thick lines diagonally dashed correspond to the boundary between the eastern and western basins, discussed in the text. See Figure 3 for the track identification. Only those tracks that are used for the projection (Figure 9) are identified here. The prefixes C, GL, OW and V stand for *Conrad, Glomar, Owen* and *Vema*, respectively.

sec. The velocity histogram (Figure 12a) indicates that the sediment velocities are distributed over the whole spectrum of values between 1.5 and 4.0 km/sec. Values between 4.0 and 5.0 km/sec are not observed. This break in velocity structure represents a possible change from sediment to crustal rocks (basalt).

The crust underlying the sediment is represented by velocities ranging between 5.0 and 7.0 km/sec. The crustal velocity distribution shows two peaks (Figure 12a): The first has a velocity range of 5.5 to 6.0 km/sec; the second one ranges between 6.5 to 7.0 km/sec. The average velocity and thickness of the upper crustal layer is 5.51 ± 0.2 km/sec and 1.69 ± 0.38 km, respectively. The upper bound of this layer coincides

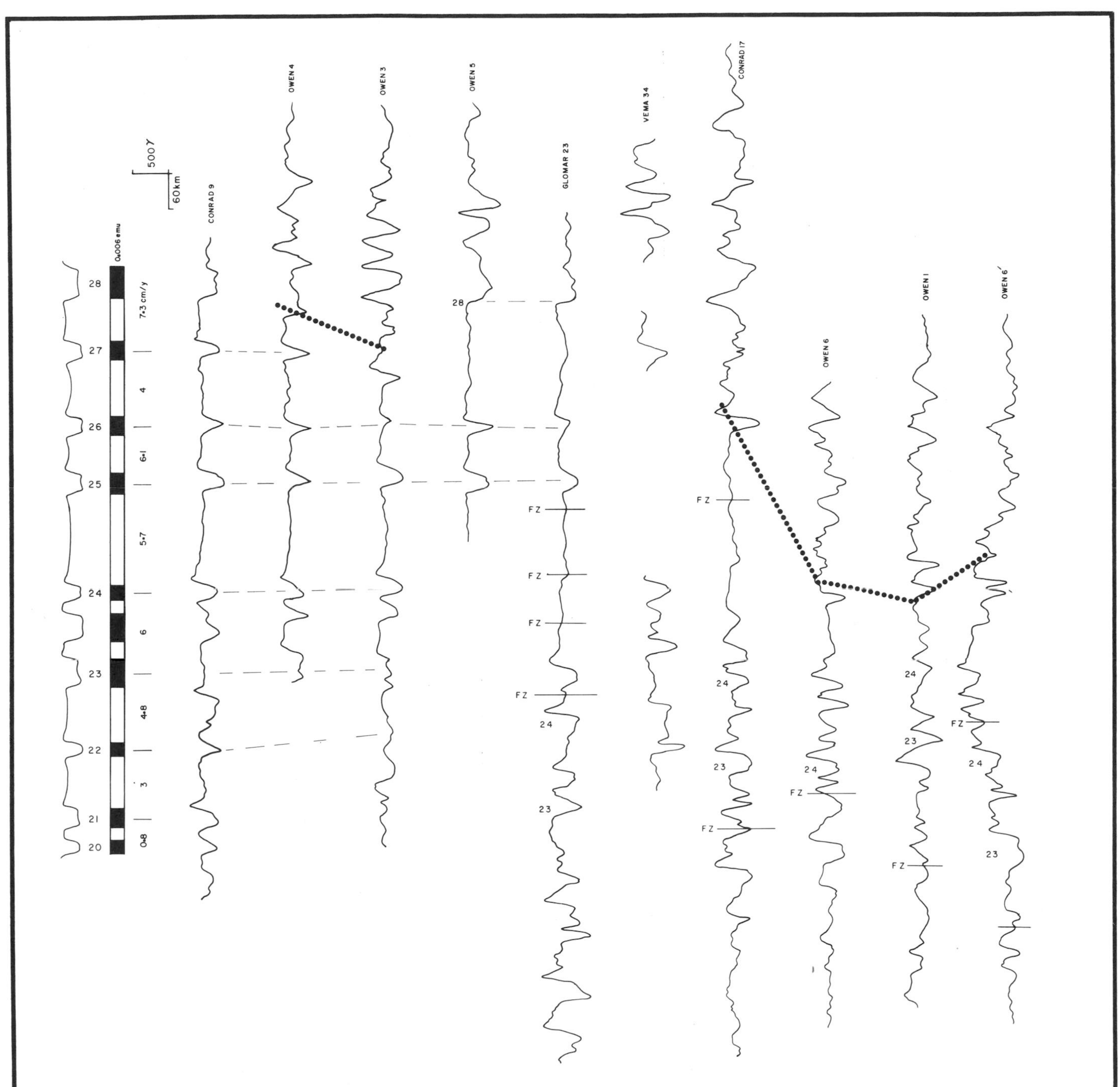

Figure 9 — Selected magnetic anomalies in the Arabian Sea projected onto N0°E. For location of profiles see Figures 8 and 3. Synthetic profile shown on top (black/white; normally/reversely magnetized blocks) was generated by the following parameters: paleo-ridge strike, N105°E; present location of anomalies, 15°N on a line striking due north; spreading rates per limb, variable as shown under the synthetic blocks; remanent inclination, -24°; remanant declination, -15°; θ = 185°; intensity of magnetization, 0.006 EMU.; depth of magnetization, 500 m; plot azimuth, 0°; present inclination, 20°; and present declination, -1°. Dotted line corresponds to the boundary between eastern and western basins discussed in the text.

Table 2. The western basin, showing velocity (V) in km/sec and thickness (H) in km.

Buoy ID	Latitude deg-min	Longitude deg-min	Water depth km	V_1	H_1	V_2	H_2	V_3	H_3	V_4	H_4	V_6	H_6	V_7	H_7	V_8
67V34	13 – 06.4	63 – 13.5	4.039	1.89±.07	0.53	2.50±.15	1.62									
66V34	13 – 48	63 – 33	3.992	2.23±.13	0.65	2.50±.14	0.69	2.77±.17	0.70	2.99±.25	0.71	5.1	1.52	6.5	2.84	8.1
65V34	13 – 51.4	63 – 38.2	3.928	2.01±.06	0.77	2.45±.14	0.60	3.29±.08	0.89	3.54±.36	0.63					
61V34	17 – 35.2	63 – 54	3.505	1.87±.13	0.66	2.51±.20	0.56	2.85±.14	1.38	3.76±.21	1.04	5.4	2.31	6.8	3.63	8.0
L01V34	19 – 06.4	62 – 55.5	3.493	1.89±.12	0.42	2.28±.20	0.51			3.5	1.84	6.2	1.50	6.9	3.6	8.0
65C17	11 – 20.0	66 – 02.0	4.220	1.98±.07	0.46	2.18±.08	0.44	2.48±.16	0.40			5.4	1.41	6.65		
68V34	12 – 29.2	65 – 08.2	4.008	2.29±.09	1.53			2.76±.24	0.49							
52V34	12 – 47.2	65 – 02.2	3.996	2.08±.07	0.52	2.24±.06	0.93	2.94±.11	0.60			5.6	1.12	6.6	4.66	7.9
69V34	12 – 34.5	65 – 16.5	4.043	2.08±.18	0.68	2.20±.12	0.67	2.82±.13	0.66			5.8	2.06	6.8	2.82	8.3
66C17	12 – 54.3	66 – 01.3	4.030	1.73±.04	0.30	2.59±.08	0.72	3.0*	0.93			5.75	1.51	6.6		
67C17	14 – 32.6	66 – 04.3	3.780	1.91±.04	0.35	2.30±.07	0.59	2.85±.13	2.55			5.6				
68C17	14 – 41.9	66 – 04.7	3.840	2.11±.03	0.94	2.48±.11	1.02			3.6*	1.59	5.5	2.08	6.8		
64V34	15 – 14.2	66 – 12.4	3.703	2.05±.21	0.74	2.35±.18	1.00	3.05±.19	0.62	3.48±.12	1.02	5.1	1.78	6.7	2.65	8.2
69C17	15 – 59.6	66 – 06.4	3.730	1.98±.02	0.94	2.63±.09	1.12			3.61±.05	1.01	5.5	1.40	6.7		
70C17	16 – 53.8	66 – 08.9	3.580	1.83±.05	0.53	2.24±.09	0.94	3.20±.13	1.54	3.81±.18	0.76	5.6				
73V34	11 – 53.2	66 – 57.4	4.128	2.23±.09	0.68					3.5*	0.70	5.7	2.43	6.6		
84C17	12 – 34.8	68 – 19.6	4.120	1.76±.05	0.42			2.88±.58	0.67	3.7	0.34	5.5	1.70	6.6		
53V34	11 – 12.0	65 – 28.2	4.198	1.80±.11	0.50	2.12±.09	0.48	2.71±.14	0.51			5.2	1.69	6.7		
54V34	13 – 10.1	67 – 07.1	3.983	2.23±.14	1.29			2.94±.22	0.57			5.4	1.28	6.6		
70V34	13 – 41.2	67 – 14.2	3.814	1.82±.07	0.60	2.29±.11	0.65	3.17±.21	0.79	3.5*	0.62	5.7	1.30	6.6	2.41	8.3
55V34	14 – 54.4	67 – 21	3.769	1.79±.07	0.88			3.00±.08	1.41			5.3	1.84	6.6	2.42	8.1
74V34	09 – 51	68 – 20.2	4.340	1.97±.09	0.70	2.20±.15	0.34	3.29±.19	1.16							
L02V34	18 – 38.2	65 – 10.5	3.376	2.18±.12	1.03	2.59±.13	0.97	3.10±.13	0.90	3.70±.28	1.30	5.6	1.53	6.6	2.85	8.0
Mean			3.88	2.00	0.71	2.35	0.79	2.96	0.96	3.57	0.96	5.51	1.69	6.67	3.03	8.10
Standard Deviation			0.25	0.17	0.30	0.15	0.31	0.22	0.55	0.24	0.43	0.21	0.39	0.11	0.76	0.14

*Assumed velocity. Numbers following the sign '±' refer to standard deviation of velocity (interval velocity only). Velocity values without standard deviation refer to refraction (interface) velocity.

Table 3. The eastern basin, showing velocity (V) in km/sec and thickness (H) in km.

Station ID	Latitude deg-min	Longitude deg-min	Water Depth	V_1	H_1	V_2	H_2	V_3	H_3	V_4	H_4	V_5	H_5	V_6	H_6	V_7	H_7	V_8	H_8†
86C17	09 – 07.4	70 – 13.2	4.340	2.31±.31	0.32							4.8	0.74			6.2	0.94	7.2	
L13V34	10 – 33.4	70 – 23.2	3.779	1.74±.14	0.52									5.4	1.80	6.1	2.88	7.2	8.30
L08V34	11 – 54.4	71 – 10.5	2.135	1.65±.04	0.41	2.12±.14	0.42					4.4	0.66	5.6	1.49	6.3	4.69	7.2	8.40
59V23	13 – 01.5	69 – 14.2	4.210	1.89±.11	0.33	2.08±19	0.34	3.0	1.40					5.3					
72V34	13 – 16.1	69 – 27	4.017	1.70±.08	0.77					3.8	0.67			5.5	1.18	6.4			
L12V34	14 – 16.5	70 – 00	3.628	2.0*	0.43							4.2	1.7	5.7	1.93	6.3	1.84	7.3	7.99
71V34	14 – 48.4	69 – 12	3.839	1.76±.12	0.52			3.0*	0.50					5.4	1.22	6.2	3.04	7.0	
80C17	15 – 59.3	67 – 20.9	3.720	1.75±.19	0.47	2.46±.34	0.50							5.6					
L05V34	16 – 27.4	67 – 59.3	3.136	1.82±.17	0.28	2.27±.20	0.49					4.3	0.73	5.4	3.51	6.2	4.13	7.2	10.70
L04V34	16 – 25.5	67 – 42	2.811	1.80±.10	0.26			3.03	0.30					5.4	2.10	6.2	5.18	7.1	7.30
L03V34	16 – 40.1	67 – 38.2	3.048	1.75±.14	0.40	2.28±.20	0.34					4.2	0.65	5.2	1.39	6.2	2.80†		
56V34	17 – 08.2	67 – 17.2	3.472	2.17±.17	0.25					3.58±.30	0.72			5.3					
62V34	18 – 18.4	66 – 55.5	3.279	1.81±.07	0.79			3.04±.16	1.05	3.5	1.01			5.3	2.01	6.4			
71C17	18 – 22.5	66 – 13.7	3.240	1.66±.04	0.42	2.31±.04	0.52	3.0*	0.58			4.75							
72C17	18 – 51.7	66 – 12.9	3.040	2.07±.05	0.52	2.17±.05	0.98	3.15±.06	0.92			4.9	1.28	5.85					
50V34	16 – 48.4	65 – 03.4	3.538	2.18±.07	1.60			3.28±.18	1.98			4.2	1.18	5.3	0.81	6.3	0.70	7.2	
49V34	20 – 14.2	65 – 10.1	3.091	1.69±.16	0.33	2.37±.09	0.70	2.83±.17	0.97	3.1	0.26	3.6							
48V34	20 – 57.4	64 – 05.2	3.293	2.27±.05	1.47			3.17±.15	1.09			4.0							
L07V34	15 – 51	68 – 37.5	3.779	2.04±.09	0.70							4.6	1.71	5.2	1.07	6.3	3.24	7.4	5.79
L09V34	16 – 10.1	69 – 43.5	3.685	2.23±.16	0.84									5.3	1.2	6.1	2.12	7.1	7.26
L11V34	17 – 22.1	69 – 03.4	3.519	1.78±.08	0.63	2.3*	0.71					4.6	0.41	5.4	2.73	6.4	0.7	7.3	6.47
63V34	17 – 55.5	68 – 30.4	3.357	2.05±.09	0.67	2.35±.07	0.77	3.0*	1.22					5.4	2.13	6.4			
57V34	18 – 20.2	67 – 55.1	3.294	2.00±.03	1.06	2.64±.18	0.80			3.47±.24	0.79	4.1	0.40	5.1	1.80	6.3	2.58	7.1	
60V34	17 – 51	68 – 45.4	3.328	1.78±.10	0.75	2.42±.11	0.79	3.24±.31	0.59	3.5	0.42			5.3	1.40	6.3			
L06V34	18 – 34.5	68 – 07.1	3.369	1.94±.11	0.47	2.30±.10	0.84	3.01±.11	1.13			4.5	1.38	5.6	1.32	6.4	2.09	7.3	6.65
L10V34	19 – 01.1	67 – 18	3.161	2.20±.07	0.93	2.43±.10	0.61	3.10±.21	1.07			4.5	1.03	5.6	1.30	6.3	2.78	7.0	9.00
73C17	19 – 25.2	66 – 16.6	2.846	2.03±.05	0.76	2.48±.03	0.71	2.89±.12	0.50	3.26±.13	0.97								
74C17	20 – 54.3	66 – 11.2	2.360	2.07±.03	1.02	2.72±.04	0.80			3.44±.11	1.40	4.8	1.03	5.8					
75C17	21 – 02.1	66 – 10.1	2.506	1.75±.09	0.50	2.64±.08	0.67	3.18±.07	1.54										
76C17	22 – 04.8	66 – 07.5	2.460	1.70±.04	0.52	2.65±.05	1.07	3.16±.19	0.77	3.48±.39	0.96	4.2	0.34	5.5					
224	20 – 19	67 – 09	2.985	v(z)	1.70			3.17	1.10			4.42	2.0			6.6			
221	18 – 25	69 – 41	2.495	v(z)	0.65					3.26	0.85	4.4	4.0			6.5			
226	21 – 13	67 – 60	1.850	v(z)	1.95					3.5	2.0	4.3	1.8			6.5			
219	18 – 42	70 – 09	0.850†	1.7 + 1.03	1.23	2.27	1.67			3.47	4.4					6.0			
Mean			3.23	1.91	0.72	2.39	0.72	3.11	1.04	3.49	1.28	4.46	1.29	5.43	1.63	6.3	2.79	7.19	
Standard Deviation			0.58	1.21	0.43	0.19	0.31	0.16	0.42	0.15	1.17	0.27	0.89	0.19	0.62	0.14	1.32	0.12	

* Assumed velocity. Numbers following the sign '±' refer to standard deviation of velocity (interval velocity only). Values without standard deviation refer to refraction (interface) velocity.
† Minimum Thickness.

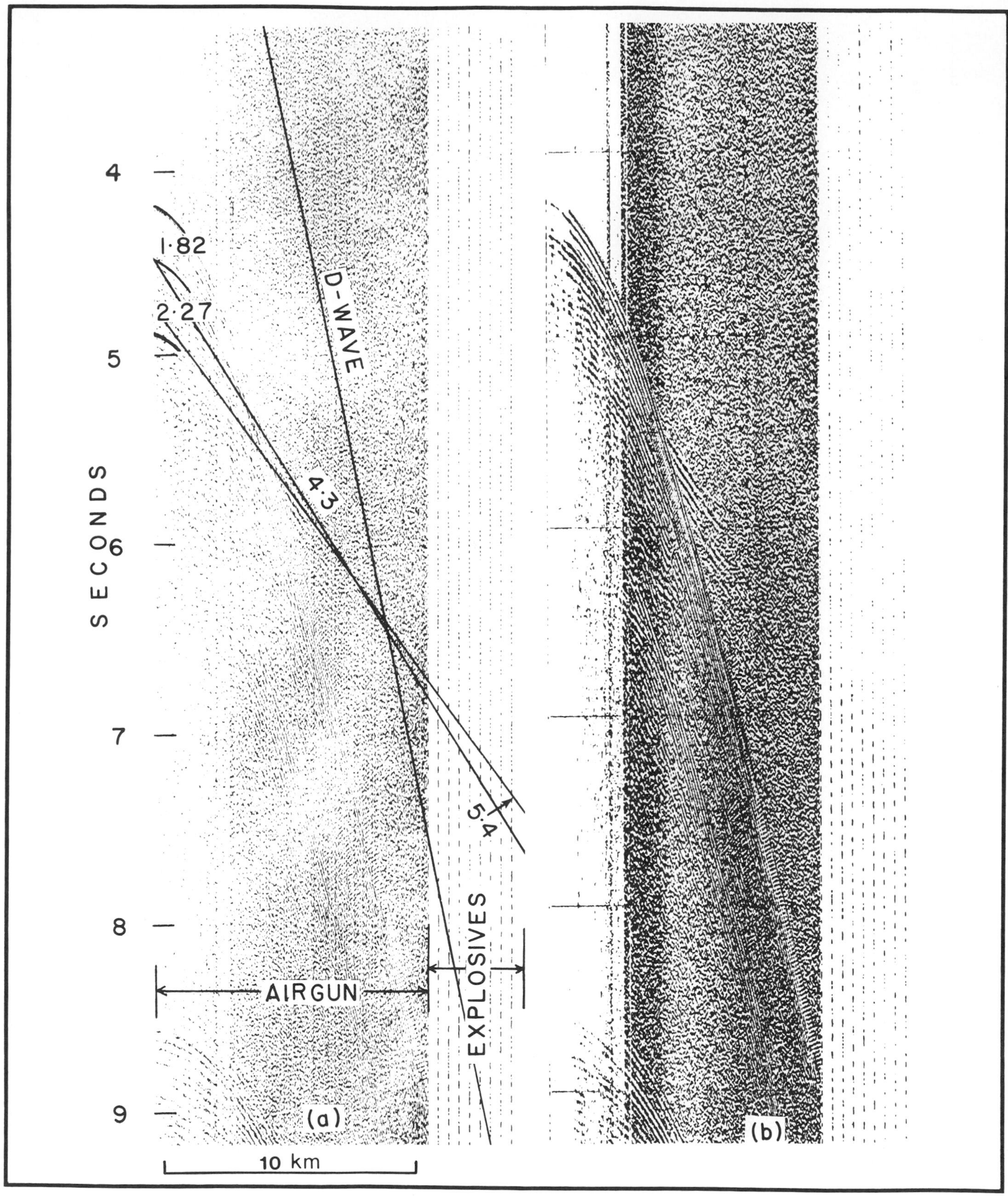

Figure 10 — A typical sonobuoy (corresponding to station L05; Figure 4), wide-angle reflection and refraction record. Note the record was made initially using air-gun sound source, but as returning seismic signals got weaker, the air gun was replaced by explosives (b). The interpretation of the sonobuoy is shown in (a). Numbers wiithin hyperbolic traces are interval velocities; those on straight line segments are refraction velocities. D-Wave is the direct water wave. Scale on left is vertical two-way travel time.

with the acoustic basement of the reflection profiles. The second crustal layer has a mean velocity of 6.67 ± 0.11 km/sec and has an average thickness of 3.02 ± 0.4 km. Moho was observed at nine stations (Figure 13a) with velocities ranging between 7.9 and 8.3 km/sec. The average depth to Moho is about 11.5 km. The average crustal column representing the western basin is shown in Figure 14.

The velocity structure of the western basin is similar to that of the oceanic basin crust observed elsewhere (e.g. Raitt, 1963; Ewing, 1969; Christensen and Salisbury, 1975). Therefore, the western basin is considered to be typically oceanic. Evidence for oceanic crust under the western basin also comes from sea-floor spreading type magnetic anomalies present over the basin, as discussed earlier. The latter data also suggests that the oldest basement under the basin is about 64 m.y. old.

Laxmi Ridge

The detailed survey of the western margin of India, by R/V *Conrad* in 1974, indicated the presence of isolated submarine structures with negative gravity anomalies that were located approximately along a line extending northwest to southeast between 15° and 18°N (Figure 2). Naini and Talwani (1977) interpreted these isolated structures to be part of a continuous structural feature. This submarine structure was subsequently named the Laxmi ridge. The results of recent R/V *Vema* cruises confirmed the continuity of the Laxmi ridge.

To minimize the corrections involved with changes in topography, a small grid survey along the crest of the ridge (Figure 3) was carried out to choose the best site for the deployment of sonobuoys. The profiles were run approximately parallel with the strike of the ridge and over its crest. Three long range sonobuoys were deployed. The first sonobuoy (Station L03, Figure 4) gave only limited results (Figure 13b) as it was defective. The second and third profiles (Stations LO4 and LO5, Figure 4) were recorded within half a degree of each other and in opposite directions. The long range sonobuoy profile for station L05 and a travel-time versus range and shot size plot are shown in Figures 10 and 11, respectively. These records are typical of data from all of the long range sonobuoys.

The average water depth over the Laxmi ridge is about 2.8 km. The average sediment thickness is about 500 m, with pockets of thicker ponded sedi-

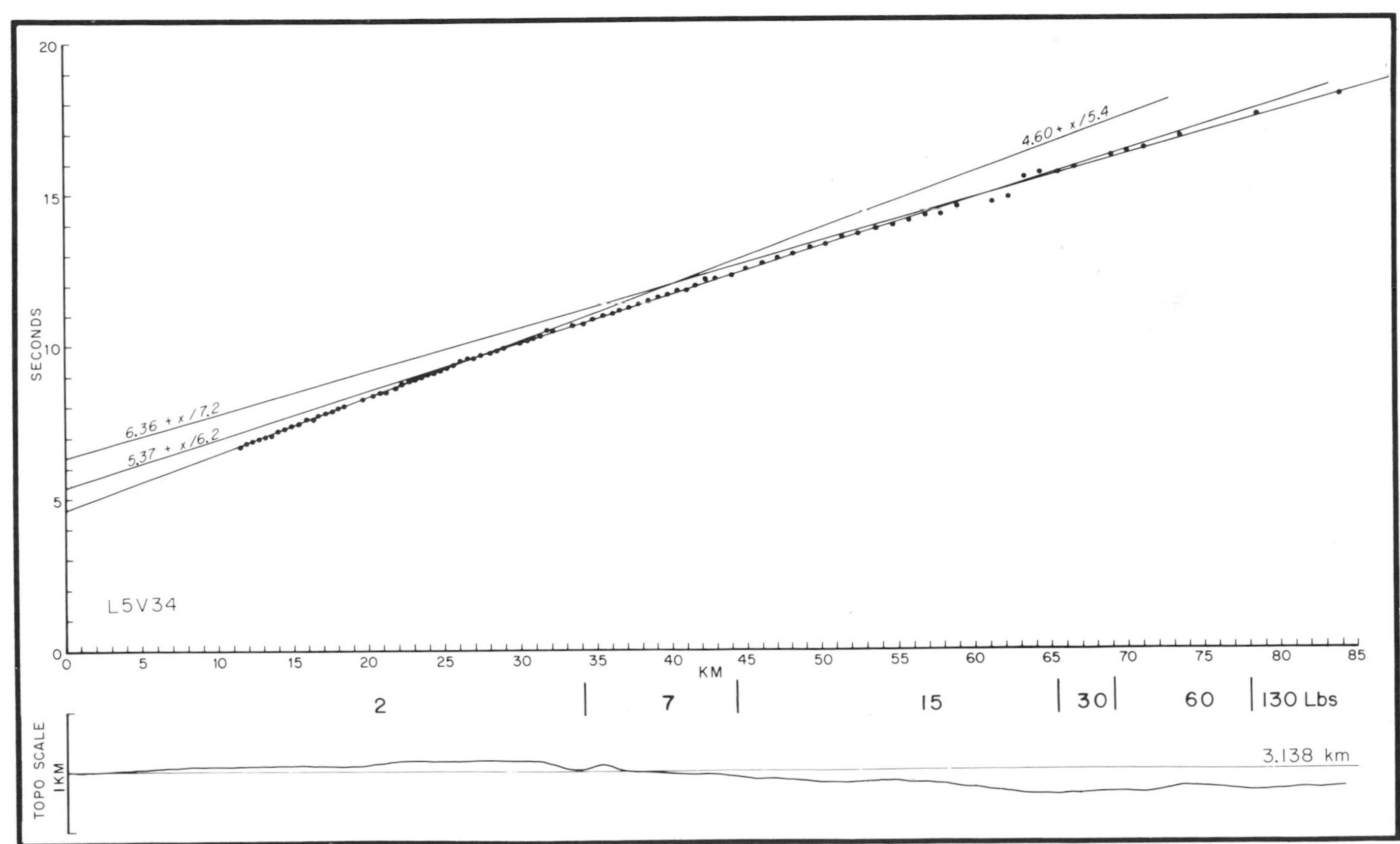

Figure 11 — A typical time-distance graph and its interpretation, corresponding to long range sonobuoy station L05 (see Figure 4 for location). The sonobuoy was originally recorded with airgun sound source but later switched to explosives (cf. Figure 10). Horizontal scale is range in kilometers derived from direct-water wave time through the surface sound channel. Shot size is shown at the bottom of the range scale. The topographic profile along the sonobuoy run is shown at the bottom. The arrival data is corrected for topography, assuming a velocity of 4.5 km/sec. The average of topography (shown as straight line with 3.138 km) is taken as the base line for the corrections.

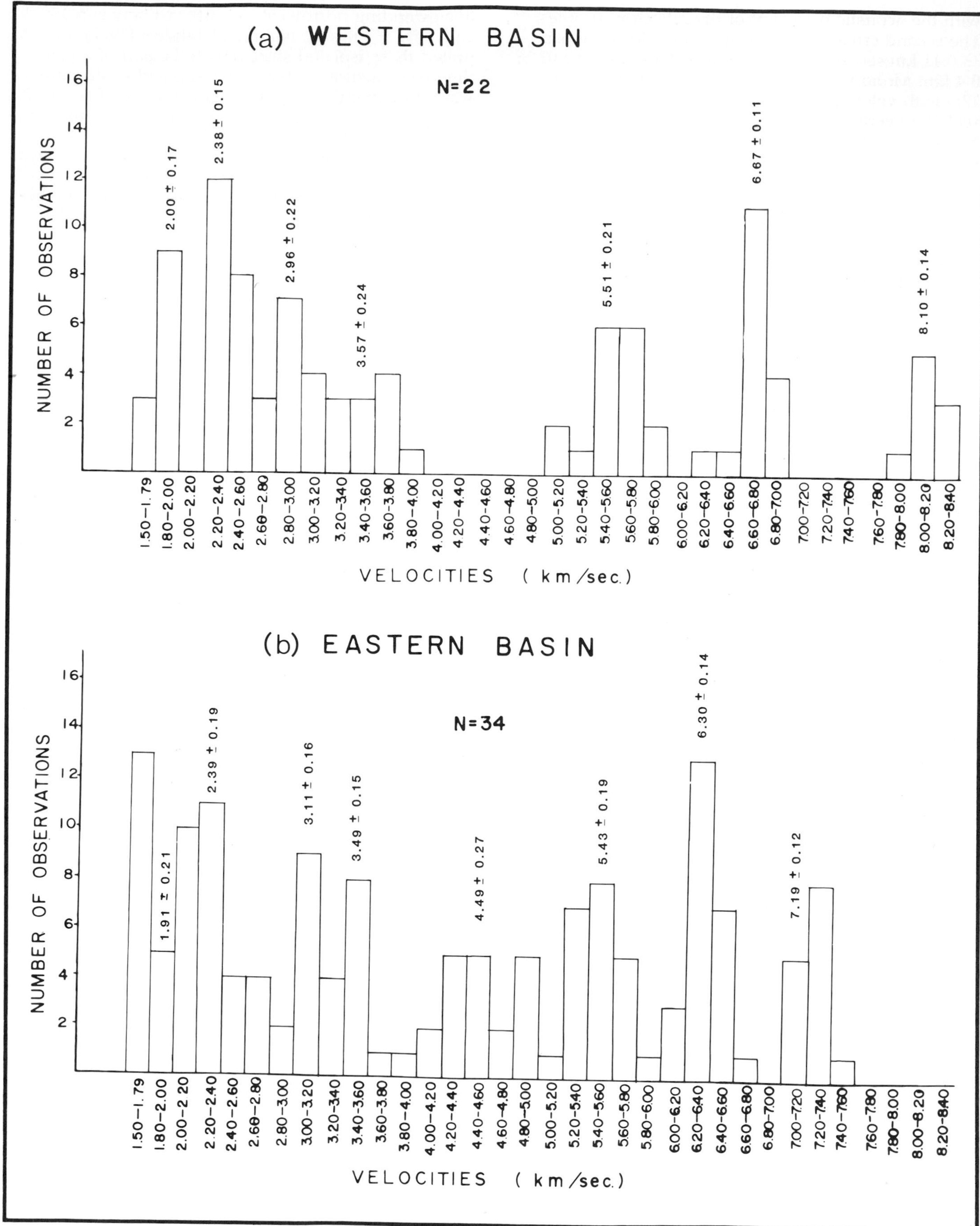

Figure 12 — Velocity histograms for the western basin (a) and eastern basin (b). See Figure 4 for location of the basins.

ments. The velocity of the sediment ranges between 1.7 and 3.9 km/sec. At stations L03 and L05 (Figure 13b), a basal sediment (or altered igneous) layer with average velocity and thickness of 4.3 km/sec and 0.7 km, respectively, was found. Underlying the sediment were three crustal layers: a 2.4 km thick layer with a velocity of 5.3 km/sec, overlying a 4.5 km thick layer with a velocity of about 6.2 km/sec, overlying a 7.2 km/sec layer. The thickness of the bottom layer was not established. Moho was not observed at any of the stations even though one of the profiles (L05) was extended to a range of 84 km (Figure 11). Assuming a velocity of 8.2 km/sec for the upper mantle, the average minimum depth to Moho under the Laxmi ridge is greater than 21 km (Figure 14).

Before discussing the various possible origins of the Laxmi ridge, we briefly summarize the available geophysical information. Structurally, the ridge is a positive feature. It is flat-topped and largley buried, as revealed by the seismic reflection records in Figure 7. The orientation of the ridge is northwest and it is covered by only a thin veneer of sediments. It has magnetic anomalies that are rather irregular in character and which contrast with the sea-floor spreading type anomalies to the southwest in the following ways: a) the anomalies over the Laxmi ridge are somewhat smaller in amplitude; b) they are not lineated very strongly; and, c) many lineations are northwest, contrasting to the mainly east to west orientation of the identified seafloor spreading type anomalies. The isostatic gravity anomalies are negative over the Laxmi ridge, which is most unusual for a positive structural feature. The isostatic gravity anomalies cannot be ascribed to a large thickness of low density sediments.

Seismically, the structure of the Laxmi ridge is very unusual for an oceanic feature. Below the thin cover of sediments, the top crustal layer with a velocity of 5.4 km/sec and 2.4 km thickness is comparable with the average oceanic crustal layer 2 of the global oceans. However, the layer with velocity 6.2 km/sec is seldom found in oceanic areas and if the 7.15 km/sec layer is considered a part of the crust, the minimum crustal thickness of 21 km is unusually large for oceanic structures.

Aseismic ridges such as the Shatsky rise (Den et al, 1969), the Nazca ridge (Cutler, 1977), and the Ontong — Java Plateau (Murauchi et al, 1973; Hussong, Wipperman, and Kroenke, 1979) tend to have large crustal thicknesses (up to 20 km), but unlike the Laxmi ridge they have normal crustal velocities and positive gravity anomalies (Kienle, 1971; Watts and Talwani, 1975; Watts, Bodine, and Bowin, 1979).

The crustal structure observed under the Laxmi ridge is comparable to that of Iceland (Palmason and Saemundson, 1974) as well as Hawaii (Furumoto, Campbell, and Hussong, 1971). However, these areas differ from the Laxmi ridge in that they are areas of active volcanism. The 7.2 km/sec layer under Iceland and Hawaii is associated with a hotter than normal mantle rather than with the crust (for the Laxmi ridge we have considered the 7.2 km/sec layer a crustal layer, although we are not absolutely certain about its origin). Furthermore, the large positive free-air gravity anomalies over Iceland and Hawaii suggest basic differences between them and this ridge. The presence of large negative anomalies in the absence of surficial low density material suggests that the structure of the Laxmi ridge differs to considerable depths from that of the surrounding areas.

Oceanic features, such as the Agulhas Plateau (Tucholke, Houtz, and Barrett, 1981) and the Rockall Plateau (Scrutton, 1971), with velocities and thicknesses similar to that observed under Laxmi ridge have been classified as continental in origin or as microcontinents. Other structures of the global oceans with similar velocity structures, such as Galicia Bank (Black et al, 1964), Flemish Cap, and Porcupine Bank (Gray and Stacey, 1970), are also considered continental in origin. Another example is the Seychelles-Mascarene Plateau (Shor and Pollard, 1963; Davies and Francis, 1964; and Matthews and Davies, 1966), where crustal layers of velocities 5.6, 6.3, and 6.8 to 7.0 km/sec and a crustal thickness of over 20 km have been found.

Crustal studies of India, utilizing earthquake body waves by Dube, Bhayana, and Chaudhury, (1973) and Kaila, Reddy, and Narain, (1968), have yielded velocities for P_g, P^*, and P_n phases similar to those observed under the Laxmi ridge. Surface wave studies by Tandon (1973) and refraction studies by Varadarajan and Behl (1971), in the Bombay area, yielded a velocity for the Deccan trap basalts of about 5.3 km/sec, which is similar to the upper crustal layer under the Laxmi ridge. Therefore, we believe that it is likely that the Laxmi ridge is of continental origin and that its crustal structure is comparable to that of Indian subcontinent. Use of seismic velocities to determine the nature of origin of the crust is always a hazardous exercise. This suggests that the foregoing result should be considered with caution.

Chagos-Laccadive Ridge

The Chagos-Laccadive ridge is a linear north to south ridge that extends over 2,200 km between the Chagos archipelago, at about 10°S, and the Laccadive Islands, at about 12°N (Figure 1). Numerous hypotheses have been proposed for its origin. Narain, Kaila, and Verma (1968), based on the seismic refraction work of Francis and Shor (1966), suggested that the ridge forms the transition between the oceanic crust to the west and continental crust to the east. The ridge was explained as a former transform fault by Fisher, Sclater, and McKenzie (1971), Sclater and Fisher (1974), and Norton and Sclater (1979). A hot spot origin (Wilson, 1963) for the ridge was proposed by Francis and Shor (1966), Dietz and Holden (1970), Morgan (1972), and Whitmarsh (1974). Avraham and Bunce (1977) also postulated such an origin for parts of the ridge. According to this hypothesis, the ridge is the manifestation of the motion of the Indian plate over a fixed hot spot in the asthenosphere. A microcontinent origin for this ridge was proposed by, among others, Davis (1928), Krishnan (1968), and Heezen and Tharp (1964). Avraham and Bunce (1977) suggested that the Maldives part of the ridge is a microcontinent.

Earlier in the paper, we proposed that the Laxmi ridge was of continental origin. Morphologically speaking, the Laxmi ridge seems to form the northwestward extension of the Chagos-Laccadive ridge. Thus, the northern part of the Chagos-Laccadive ridge and the Laxmi ridge could have a common origin. A single long-range sonobuoy refraction profile shot over the flanks of the Chagos-Laccadive ridge (Station L08; Figure 4 and Figure 13b) yielded results identical to those on the Laxmi ridge. Further evidence for the continental origin of the northern part of the Chagos-Laccadive ridge comes from the results of the DSDP Site 219 located on the crest of the ridge. The sedimentation history at Site 219 (Whitmarsh, 1974) suggested that the site was once closer to the Indian landmass, or perhaps was a part of the main Indian landmass. Petrologic studies of Site 219 samples by Siddiquie and Sukheswala (1976) indicated acidic or rhyolitic tuffs in the basal units, suggestive of possible continental affinities. It was further suggested that the tuffs found at Site 219 are similar to the rhyolitic tuffs found in the Deccan traps on land. Even though the age of the Deccan traps is widely debated (e.g. Sukheswala and Poldervart, 1958; Rama, 1964; Krishnan, 1966; Eremenko and Negi, 1968; Wadia, 1970) it is generally agreed that they were emplaced about 65 m.y. ago (Krishnan, 1968; Wellman and McElhinny, 1970; Kaneoka and Haramura, 1973; Molnar and Francheteau, 1975). Drilling over the crest of the Chagos-Laccadive ridge by D/V *Glomar Challenger* at DSDP Site 219, located about 1,000 km south of the nearest Deccan traps exposure on the west coast of India, gave a basement minimum age of 62 million years (Whitmarsh et al, 1974). The true basement age of the ridge may therefore be greater than 62 m.y. old and could be comparable to the age of the Deccan traps. Therefore, a hot spot origin for this ridge is not favored.

This study does not critically examine the different hypotheses for the origin of the Chagos-Laccadive ridge. However, the new siesmic refraction data that we mention above do favor a continental fragment hypothesis for at least the northern part of the Chagos-Laccadive ridge.

The Eastern Basin

The eastern basin is more shallow than the western basin, and the water depths range between 1.85 and 4.3 km, with an average of 3.23 km. Eastern basin sediment, with velocities less than 5.0 km/sec, is thinner than western basin sediment and ranges from 0.3 km to over 4 km thick. The velocity histogram (Figure 12b) indicates peak sediment velocity values of 1.5 to 2.0 km/sec, 2.0 to 2.4 km/sec, 3.0 to 3.2 km/sec, 3.4 to 3.5 km/sec, and 4.2 to 4.6 km/sec.

For velocities greater than 5.0 km/sec (crustal layer?), the maximum number of occurrences (as seen in the histogram; Figure 12b) are found for the group having 5.4 to 5.6 km/sec, 6.2 to 6.4 km/sec, and 7.2 to 7.4 km/sec. The average values of velocities and thicknesses of the individual crustal layers are: 5.42 ± 0.19 km/sec; 6.28 ± 0.14 km/sec and 7.19 ± .12 km/sec; and, 1.63 ± 0.62 km and 2.79 ± 1.32 km, respectively. The average crustal column representing the eastern basin is shown in Figure 14. Since 7.19 km/sec is the last refracting horizon in the eastern basin, its lower bound could not be estimated. However, assuming the upper mantle with 8.2 km/sec velocity to lie underneath this layer, the minimum depth to Moho could be estimated. All the long-range sonobuoy profiles in the eastern basin extended to an average range of 65 km. At this range, the minimum depth to Moho is about 17 km. Gravity measurements discussed earlier suggest a crustal thickness of over 16 km, if one assumes local Airy type isostatic compensation, which agrees with that determined by seismic refraction. Thus, the seismic refraction and gravity observations suggest that the crust under the eastern basin is thicker than average oceanic crust (11 km) by about 40%. However, the water depth in the average ocean basins is greater than 5 km, whereas in the easten basin it is only 3.23 km. The greater crustal thickness observed under the eastern basin is not anomalous if one assumes the basin to be isostatically compensated on a local scale.

The eastern basin has high-velocity basal sediments. The velocity range of 4.0 to 5.0 km/sec may be associated with the sediment that lies just over the basement. However, in some cases this high velocity may be partly associated with the uppermost part of the basement, or material derived from altered basement. The average thickness of this high velocity layer is 1.28 km, whereas the sediment overlying this layer is 1.84 km thick.

Crustal structure under the eastern basin can be compared with that of the Indian shield. The upper crustal layer can be compared with that of the Deccan traps, or a sedimentary layer such as limestone; and the underlying crustal velocities of 6.2 and 7.2 km/sec can then be compared with the P_g and P^* phases under the Indian peninsula, as well as with that under the Indo-gangetic plain. However, the thicknesses of these layers are very different. The total crustal thickness under the eastern basin is nearly half that under the northern Indian plain. Even though the generalized velocity structure under the Indian shield and eastern basin is similar, the difference in total crustal thickness, clearly, is important.

Rifting as a possible mechanism in the evolution of the western margin of India is suggested by the geologic history of the west coast of India. The Gulf of Cambay (Figure 1) is a graben filled with Tertiary sediment (Raju, 1968). The Cambay basin is marked by step faults, which extend in a north to south direction into the Gulf of Cambay, both on the western and eastern sides (Raju, 1968). It is believed that the graben formed by extensional tectonics (Raju, 1968; and Chowdhary, 1975). The Deccan volcanics, together with the eruption of alkaline igneous rocks on land (Sukheswala and Udas, 1964), took place dur-

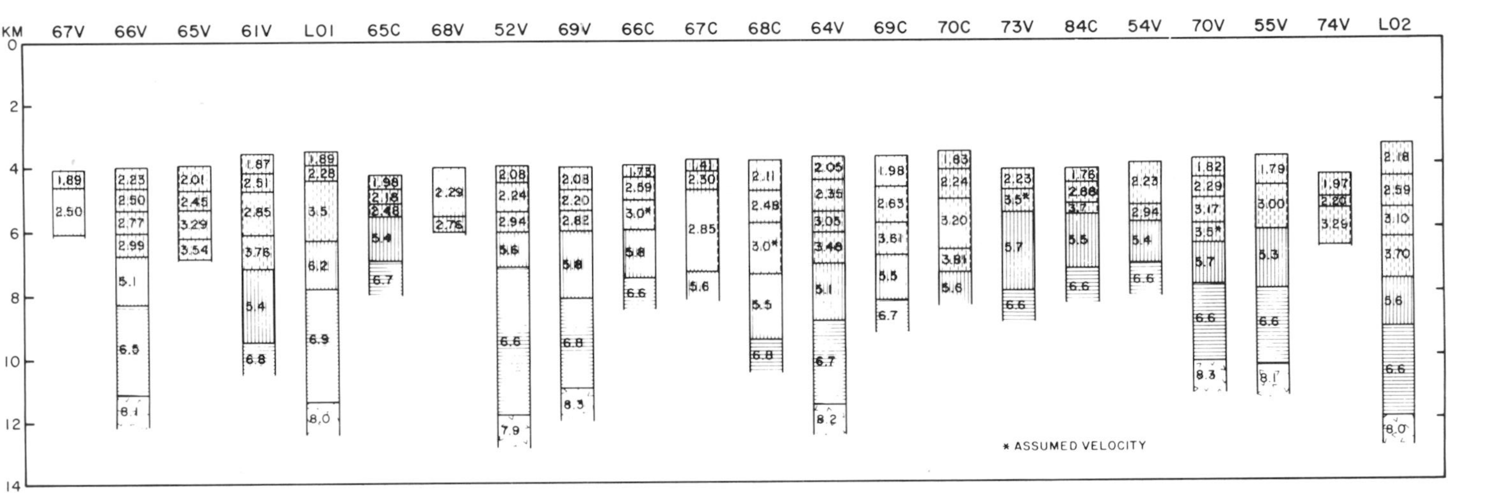

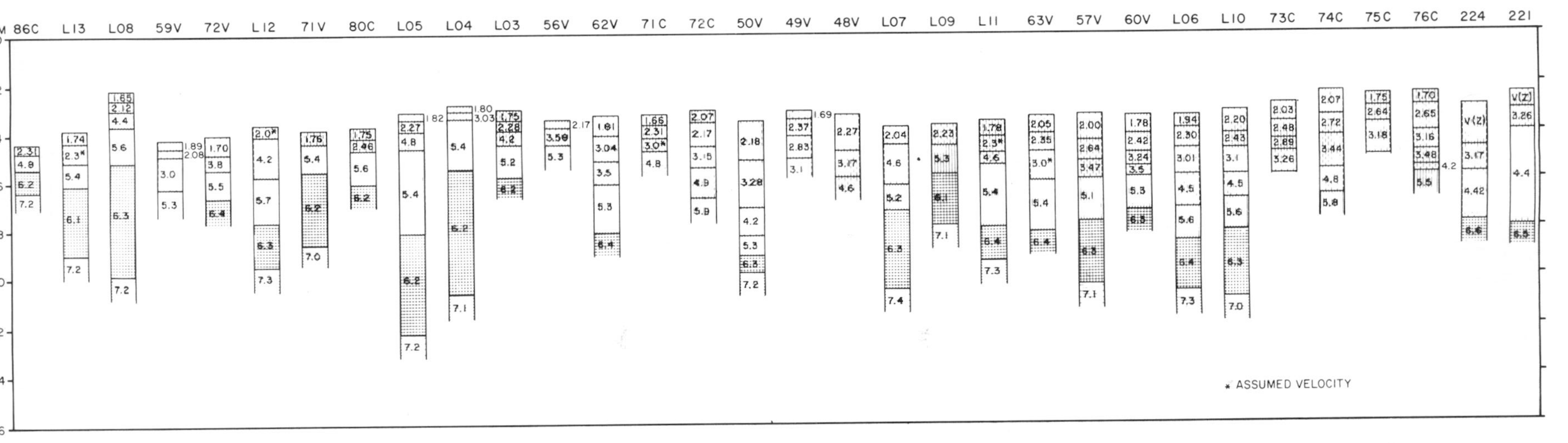

Figure 13 — Crustal structure of the western basin (a) and the eastern basin (b), depicted as crustal columns at each station. See Figure 4 for location of stations.

ing the latest Cretaceous to earliest Tertiary time. Chowdhury (1975) suggested that following this volcanism, unsuccessful rifting began in the region of the Cambay basin. The distribution of various structures, the structural grain along the western margin of India, the patterns of sediment distribution, and the geologic history of the western shelf of India (Ramaswamy and Rao, 1980) point to a rifting origin for the western margin of India (Naini, 1980).

We suggest that the eastern basin may have evolved by a thermally induced process, through which the entire region along the western margin of India was thermally updomed and subject to extensional forces, followed by thinning and rupturing of continental crust (Falvey and Middleton, 1981). Continental crust rupturing was followed by dike injection and volcanic outpourings along the entire region of the eastern basin, rather than along a single spreading center. The setting was perhaps similar to that of the present Afar, as suggested by Barberi and Varet (1976). Active sea-floor spreading took place only in the western basin, commencing about anomaly 28 time. According to this scheme, the Laxmi ridge and the northern part of Chagos-Laccadive ridge are continental slivers that were left behind between initial rifting along the margin and the later phase of active sea-floor spreading in the western basin.

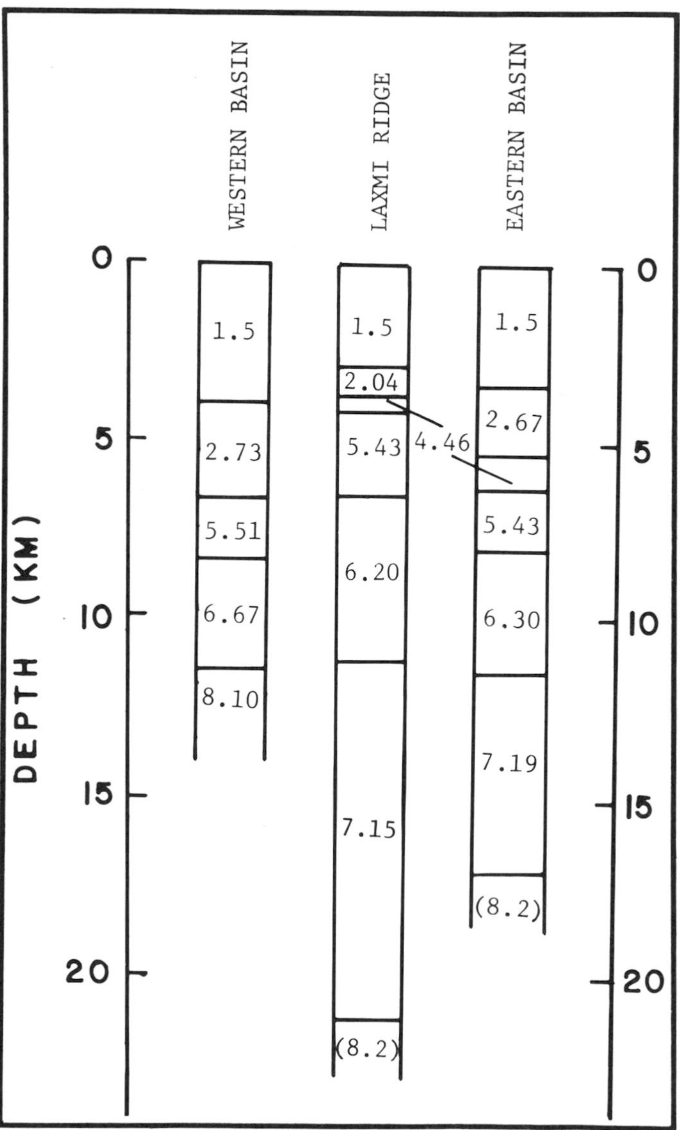

Figure 14 — Average crustal columns representing western basin, eastern basin and Laxmi ridge.

There are many points of similarity in structure between the eastern basin and the area of the magnetic quiet zone off southern Australia, as reported by König and Talwani (1977) and Talwani et al (1979). It is difficult to prove, in either case, that the area was not created by sea-floor spreading. However, we favor a process of alteration and subsidence of continental crust.

CONCLUSIONS

The continental margin of western India is marked by a number of regional structures that approximately parallel the present shelf edge. Important among these are the Chagos-Laccadive ridge and the newly discovered aseismic Laxmi ridge, which is located at the northern end of the Chagos-Laccadive ridge and continues northwest toward the Murray ridge. The Laxmi ridge has a basement relief of up to 2 km, is underlain by over 20 km thick crust, and is characterized by negative isostatic gravity anomalies. Crustal structure under this ridge, together with negative gravity anomalies and the evolutionary history of the adjacent margin, led us to believe that it might be of continental origin.

The Chagos-Laccadive ridge has a crust that is thicker-than-normal oceanic crust and is located in a setting similar to that of the Laxmi ridge. DSDP results and seismic refraction data at a single site over the northern part of the ridge suggest, through similarity with the Laxmi ridge, a possible continental origin for the Chagos-Laccadive ridge. Thus, we believe that the northern part of the Chagos-Laccadive ridge, located west of the continental margin of India, and the Laxmi ridge might be continental slivers that resulted from rifting along India's western margin.

The continental margin of western India and the adjoining Arabian Sea can be divided into two regions, based on the velocity of the sediment and underlying crust: the western basin and the eastern basin. The basins are divided approximately along a line parallel with the present day shelf edge of western India, and coincident with the western limit of the Laxmi ridge and Chagos-Laccadive ridge complex. The western basin is underlain by oceanic crust and was generated by the process of sea-floor spreading, beginning about 64 m.y. ago (Anomaly 28 of marine magnetic time scale).

The eastern basin is underlain by crust that is thicker than normal oceanic crust, but thinner than normal continental crust. This basin evolved by rifting and lacks magnetic anomalies associated with sea-floor spreading. The crust under the eastern basin is, therefore, transitional or rift stage crust.

We suggest that the northeastern Arabian Sea has a two-phase evolutionary history. The western margin of India and the adjoining eastern basin evolved by a process of rifting during the latest Cretaceous. During the rifting phase, the Cambay graben, the eastern basin, and the Chagos-Laccadive ridge-Laxmi ridge complex came into existence. Beginning in the Early Paleocene, active sea-floor spreading began along centers of spreading (oriented east to west) during which the western basin evolved.

ACKNOWLEDGMENTS

This work would not have been possible without the active efforts and cooperation of many people at the observatory and at sea. We thank the master, officers, and crew of research vessels *Vema* and *Conrad*. We are grateful to Robert Leyden for actively participating in the long-range crustal refraction experiments at sea, and Venkatarathnam Kolla for assisting through many phases of this work. Thanks also to our colleagues R. Houtz, V. Kolla, M. König, and H. Rowelette for critical reviews of the manuscript. Naini thanks John Ewing and Philip Rabinowitz for discussions, advice, and guidance. He also thanks Joel Watkins and David Hall for providing the opportunity to present this work at the Hedberg Symposium.

Financial support came through numerous contracts and grants under National Science Foundation GA-27281 and OCE-7683382, an Office of Naval Research N00014-67-A-0108-004 and N00014-75-C-0210. Lamont-Doherty Geological Observatory Contribution 3250.

REFERENCES CITED

Arayamadhu, P. S., B. K. Verma, and M. N. Quereshy, 1970, Isostatic studies in south India; Bulletin of National Geophysical Research Institute of India, v. 8, n. 3-4, p. 41-50.

Avraham, Z. B., and E. T. Bunce, 1977, Geophysical study of the Chagos-Laccadive ridge, Indian Ocean: Journal of Geophysical Research, v. 82, n. 8, p. 1295-1305.

Barberi, F., and J. Varet, 1976, Nature of the Afar crust — a discussion, *in* A. Pilger and A. Rosler, eds., Afar depression of Ethopia, International Communications on Geodynamic Science, Report n. 14: E.S. Verlagsbuchland-Stuttgart-Federal Republic of Germany, p. 375-378.

Barker, P. F., 1966, A reconnaissance survey of the Murray Ridge; Royal Society of London, Philosophical Transcripts, Series A, v. 259, p. 187-197.

Black, M., et al, 1964, Three nonmagnetic seamounts off the Iberian coast: Quarterly Journal, Geological Society of London, v. 120, p. 447-517.

Bunce, E. T., and P. Molnar, 1977, Seismic reflection profiling and basement topography in the Somali Basin: possible fracture zones between Madagascar and Africa: Journal of Geophysical Research, v. 82, p. 5305-5331.

Cain, J. C., et al, 1968, Computation of the main geomagnetic field from spherical harmonic expansion: Greenbelt, Maryland, Data User's Note NGSDC 68-11.

Chowdhary, L. R., 1975, Reversal of basement-block motions in Cambay basin, India, and its importance in petroleum exploration: AAPG Bulletin, v. 59, p. 85-96.

Christensen, N. I., and M. H. Salisbury, 1975, Structure and constitution of the lower oceanic crust: Review of Geophysics and Space Physics, v. 13, p. 57-86.

Closs, H., H. Bungenstock, and K. Hinz, 1969, Results of seismic refraction measurements in the northern Arabian Sea, a contribution to the International Indian Ocean Expedition: Meteor Research Results, C, n. 3, p. 1-28.

Cochran, J. R., 1979, An analysis of isostasy in the world's oceans, 2, Mid ocean ridge crests: Journal of Geophysical Research, v. 84, p. 4713-4729.

Cutler, S. T., 1977, Geophysical investigations of the Nazca Ridge: American Geophysical Union, Transcripts, v. 58, p. 1230.

Davies, D., and T. J. G. Francis, 1964, The crustal structure of the Seychelles Bank: Deep-Sea Research, v. 11, p. 921-927.

Davis, M. D., 1928, The coral reef problem: New York, American Geographical Society, Special Publication 9, p. 524-534.

Den, N., et al, 1969, Seismic-refraction measurements in the northwest Pacific basin: Journal of Geophysical Research, v. 74, p. 1421-1434.

Dietz, R. D., and J. C. Holden, 1970, Reconstruction of Pangaea breakup and dispersion of continents, Permian to present: Journal of Geophysical Research, v. 75, p. 4939-4963.

Dube, R. K., J. C. Bhayana, and H. M. Chaudhury, 1973, Crustal structure of the Peninsular India: Pure and Applied Geophysics, v. 109, p. 1718-1727.

Eremenko, N. A., and B. S. Negi, 1968, Tectonic map of India, with accompanying guide, N. A. Eremenko and B. S. Negi, eds.: Dehra Dun, India, Oil and Natural Gas Commission of India, p. 15.

Ewing, J. I., 1963, Seismic refraction and reflection measurements,. M. N. Hill, ed., The sea, v. 3: New York, Interscience Publishers.

———, 1969, Seismic model of the Atlantic, *in* P. J. Hart, ed., The Earth's crust and upper mantle: Washington, D. C., American Geophysical Union, Geophysical Monogram Series, v. 13, p. 220-225.

Falvey, D. A., and M. F. Middleton, 1981, Passive continental margins: evidence for a prebreakup deep crustal metamorphic subsidence mechanism, *in* Geologie des Marges Continentales Revue Europeenne d'oceanologie: Cedex, France, Oceanologica Acta, p. 103-114.

Fisher, R. L., J. G. Sclater, and D. P. McKenzie, 1971, The evolution of the central Indian Ridge, western Indian Ocean: Geological Society of America, Bulletin, v. 82, p. 553.

Francis, T. J. G., and G. G. Shor, 1966, Seismic refraction measurements in the northwest Indian Ocean: Journal of Geophysical Research, v. 71, p. 427-449.

Furumoto, A. S., J. F. Campbell, and D. M. Hussong, 1971, Seismic refraction surveys along the Hawaiian Ridge, Kanai to Midway Islands: Seismological Society of America, Bulletin, v. 61, p. 147-166.

Graf, A., and R. Schulze, 1961, Improvements on the sea gravimeter GSS2: Journal of Geophysical Research, v. 66, p. 1813-1821.

Gray, F., and A. P. Stacey, 1970, Gravity and magnetic interpretation of Porcupine Bank and Porcupine Bight: Deep Sea Research, v. 17, p. 467-475.

Heezen, B. C., and M. Tharp, 1964, Physiographic diagram of the Indian Ocean, the Red Sea, the South China Sea, the Sulu Sea and the Celebes Sea (with descriptive sheet): New York, Geological Society of America Inc.

Heirtzler, J. R., et al, 1968, Marine magnetic anomalies, geomagnetic field reversals, and motions of the ocean floor and continents: Journal of Geophysical Research, v. 73, p. 2119-2136.

Houtz, R., J. Ewing, and X. Le Pichon, 1968, Velocity of deep sea sediments from sonobuoy data: Journal of Geophysical Research, v. 73, p. 2615-2641.

Hussong, D. M., L. K. Wipperman, and L. W. Kroenke, 1979, The crustal structure of the Ontong Java and Manihiki Oceanic Plateaus: Journal of Geophysical Research, v. 84, p. 6003-6010.

Kahle, H. G., et al, 1976, Geophysical study on the continental margin South of India and west of Sri Lanka: Eos, Transactions, American Geophysical Union, v. 57, p. 933.

———, ———, 1981, A marine geophysical study of the "Common Ridge", north central Indian Basin: Journal of Geophysical Research, v. 86, p. 3807-3814.

Kaila, K. L., P. R. Reddy, and H. Narain, 1968, Crustal structure in the Himalayan foothills area of north India, from 'P' wave data of shallow earthquakes: Seismological Society of America, Bulletin, v. 58, p. 597-612.

Kaneoka, I., and H. Haramura, 1973, K/Ar age of successive lava flows from the Deccan Traps, India: Earth and Planetary Science Letters, v. 18, p. 229-236.

Kienle, J., 1971, Gravity and magnetics measurements over Bowers ridge and Shirshov ridge, Bering Sea: Journal of Geophysical Research, v. 76, n. 29, p. 7138-7153.

König, M., and M. Talwani, 1977, A geophysical study of the southern continental margin of Australia, Great Australian Bight and western sections: Geological Society of America, Bulletin, v. 88, p. 1000-1014.

Krishnan, M. S., 1966, Tectonics of India, Symposium on tectonics: India, National Institute of Science, Bulletin, n. 32, p. 1-36.

———, 1968, Geology of India and Burma, 5th edition: Higginbothams Madras, 536 pp.

Kristoffersen, Y., and M. Talwani, 1977, Extinct triple function south of Greenland and the Tertiary motion of Greenland relative to North America: Geological Society of America, Bulletin, v. 88, p. 1037-1049.

La Brecque, J. L., D. V. Kent, and S. C. Cande, 1977, Revised magnetic polarity time scale for Late Cretaceous and Cenozoic time: Geology, v. 5, p. 330-335.

Le Pichon, X., J. Ewing, and R. E. Houtz, 1968, Deep sea sediment velocity determination made while reflection profiling: Journal of Geophysical Research, v. 73, p. 2597-2614.

Matthews, D. H., 1966, The Owen Fracture Zone and the northern end of the Carlsberg Ridge: Royal Society of London, Philosophical Transcripts, Series A, v. 259, p. 172-186.

———, and D. Davies, 1966, Geophysical studies of the Seychelles Bank: Royal Society of London, Philosophical Transcripts, Series A, v. 259, p. 227-239.

McKenzie, D. P., and J. G. Sclater, 1971, The evolution of the Indian Ocean since the Late Cretaceous: Geophysical Journal, Royal Astronomical Society, v. 25, p. 437-528.

Molnar, P., and J. Francheteau, 1975, Plate tectonic and paleomagnetic implications for the age of the Deccan Traps and the magnetic anomaly time scale: Nature, v. 255, p. 128-130.

Morgan, W. J., 1972, Deep mantle convection plumes and plate motions: AAPG Bulletin, v. 56, p. 203-213.

Murauchi, S., et al, 1973, Seismic refraction measurements on the Ontong Java Plateau northeast of New Ireland: Journal of Geophysical Research, v. 78, p. 8653-8663.

Naini, B. R., 1980, A geological and geophysical study of the continental margin of western India, and the adjoining Arabian Sea including the Indus Cone: New York, Columbia University, Ph.D. thesis, unpublished, p. 173.

———, and M. Talwani, 1977, Sediment distribution and structures in the Indus Cone and the western continental margin of India (Arabian Sea): Eos, Transcripts, American Geophysical Union, v. 58, p. 405.

Narain, H., K. L. Kaila and R. K. Verma, 1968, Continental margins of India: Canadian Journal of Earth Sciences, v. 5, p. 1051-1065.

Neprochnov, Yu. P., 1961, Sediment thickness of the Arabian sea basin (in Russian), Dokl. Akad. Nauk., v. 139, p. 177-179.

Norton, I. O., and J. G. Sclater, 1979, A model for the evolution of the Indian Ocean and the breakup of Gondwanaland: Journal of Geophysical Research, v. 84, p. 6803-6830.

Palmason, G., and K. Saemundson, 1974, Iceland in relation to the Mid-Atlantic Ridge: Annual Review of Earth and Planetary Science Letters, v. 2, p. 25-50.

Raitt, R. W., 1963, The crustal rocks, *in* M. N. Hill, ed., The sea, v. 3: New York, Interscience Publishers, p. 85-109.

Raju, A. T. R., 1968, Geologic evolution of Assam and Cambay Tertiary basins of India: AAPG Bulletin, v. 52, p. 2422-2437.

Rama, S. N. I., 1964, Potassium/argon dates of some samples from Deccan Traps: New Delhi, 22nd International Geological Congress, part 7, p. 139-140.

Ramaswamy, G., and K. L. N. Rai, 1980, Geology of the continental shelf of the west coast of India, *in* A. D. Miall, ed., Facts and principles of world petroleum occurrence, Memoir 6: Canadian Society of Petroleum Geologists, p. 801-821.

Rao, T. C. S., 1970, Seismic and magnetic surveys over the continental shelf off Konkan coast, *in* Hyderabad, India, Proceedings, Second Symposium on Upper Mantle Project.

Schlich, R., 1975, Structure et age de l'Ocean Indien occidental: Societe Geologique de France, Memoire 6, 103 p.

Sclater, J. G., and R. L. Fisher, 1974, Evolution of the east central Indian Ocean, with emphasis on the tectonic setting of the Ninetyeast Ridge: Geological Society of America, Bulletin, v. 85, p. 683-702.

Scrutton, R. A., 1971, The crustal structure of Rockall Plateau microcontinent: Geophysical Journal, Royal Astronomical Society, v. 27, p. 259-275.

Shor, G. G., Jr., and D. D. Pollard, 1963, Seismic investigations of Seychelles and Saya de Malha banks, northwest Indian Ocean: Science, v. 142, p. 48-49.

Siddiquie, H. N., and R. N. Sukheswala, 1976, Occurrence of rhyolitic tuffs at deep sea drilling project site 219 on the Laccadive Ridge: Eos, Transcripts, American Geophysical Union, v. 57, p. 410.

Sukheswala, R. N., and A. Poldervart, 1958, Deccan basalts of the Bombay area, India: Geological Society of America, Bulletin, v. 69, p. 1475.

Sukheswala, R. N., and G. R. Udas, 1964, The carbonatities of Amba Dongar, India: Some structural considerations: 22nd International Geological Congress of India, Report, p. 109.

Talwani, M., and H. G. Kahle, 1975, Map of free-air gravity anomalies in the Indian Ocean, *in* G. Udintsev, ed., The international Indian Ocean expedition: Moscow, Geological Geophysical Atlas of the Indian Ocean.

———, W. P. Early, and D. E. Hayes, 1966, Continuous analog computation and recording of cross-coupling and off-levelling errors: Journal of Geophysical Research, v. 71, p. 2079-2090.

———, et al, 1979, The crustal structure and evolution of the area underlying the magnetic quiet zone on the margin south of Australia, *in* J. S. Watkins, L. Montadert, and P. W. Dickerson, eds., Geological and geophysical investigations of continental margins: AAPG Memoir 29, p. 151-175.

Tandon, A. N., 1973, Average thickness of the Deccan Traps between Bombay and Koyna from dispersion of short-period Love waves: Pure and Applied Geophysics, v. 109, p. 1693-1699.

Tucholke, B. E., R. E. Houtz, and D. M. Barrett, 1981, Continental crust beneath the Agulhas Plateau, southwest Indian Ocean: Journal of Geophysical Research, v. 86, p. 3791-3806.

Udintsev, G. B., 1975, Geological and geophysical Atlas of the Indian Ocean: Moscow, Academy of Sciences, 151 p.

Varadarajan, S., and G. N. Behl, 1971, Refraction velocities of trap basalts: International Association of Volcanology and Chemistry of the Earth's Interior, Bulletin Volcanologique, v. 35, p. 790-798:

Wadia, D. N., 1970, Geology of India: London, Macmillan and Company Ltd., 536 p.

Watts, A. B. and M. Talwani, 1975, Gravity field of the northwest Pacific Ocean Basin and its margin: Geological Sociey of America, Special Map Series, MC-9.

———, J. Bodine, and C. Bowin, 1979, Free-air gravity field, *in* D. E. Hayes, ed., A geophysical atlas of east and southeast Asian Seas: Geological Society of America, Map and Chart Series, MC-25.

Weissel, J. K., and A. B. Watts, 1979, Tectonic evolution of the Coral Sea Basin; Journal of Geophysical Research, v. 84, p. 4572-4582.

Wellman, P., and M. W. McElhinny, 1970, K-Ar age of the Deccan Traps, India: Nature, v. 227, p. 595-596.

Whitmarsh, R. B., 1974, Some aspects of plate tectonics in the Arabian Sea, *in* Initial reports of the deep sea drilling project, v. 23: Washington, D.C., U.S. Government Printing Office, p. 527-535.

———, et al, 1974, Site 219, *in* Initial reports of the deep sea drilling project, v. 23: Washington, D.C., U.S. Government Printing Office, p. 35-115.

Wilson, J. T., 1963, Hypothesis of earth's behavior: Nature, v. 198, p. 925-929.

The Structure and Evolution of the Jan Mayen Ridge and Surrounding Regions

Alan Nunns*
Department of Geological Sciences
University of Durham
Durham, United Kingdom

A new tectonic scheme is proposed for the evolution of the southern Norwegian-Greenland Sea. The Jan Mayen fragment began to split from the East Greenland margin at the end of anomaly 20 times (43 Ma). Until anomaly 7 time (27 Ma), complementary "fan-shaped" spreading occurred about the newly formed Kolbeinsey axis to the west of the Jan Mayen block and about the Aegir axis in the Norway Basin to the east. The Jan Mayen block rotated 28° anticlockwise before spreading ceased in the Norway Basin and became regular in the Iceland Plateau. The primary cause of the shift of axis may have been the anticlockwise reorientation of spreading direction which followed the cessation of spreading in the Labrador Sea.

As defined by the presence of the relatively transparent basement reflector, "Horizon 0," the Jan Mayen block is about 90 km wide and extends from 70°N to at least as far south as 67°N. A reconstruction to the beginning of anomaly 24 time suggests that Horizon 0 is basaltic and represents either early phase oceanic crust formed by rapid subaerial accretion or plateau basalt overlying continental crust. During its separation from Greenland, the southern part of the block was sheared sinistrally, resulting in volcanism and *en echelon* graben formation. The Oligocene unconformity, "Horizon A" postdates the initial separation and was probably caused by the major mid-Oligocene drop in sea level.

One of the processes which sometimes occurs at a passive continental margin is separation of a fragment from the margin because of a shift of spreading axis. This paper deals with a ridge jump in the southern Norwegian-Greenland Sea, which led to the detachment of the Jan Mayen block from the East Greenland margin.

According to Johnson and Heezen (1967) early spreading in the southern Norwegian-Greenland Sea took place about the now extinct Aegir axis (Figure 1). This was followed by a westward shift of axis which isolated the Jan Mayen block from the Greenland margin. Spreading about the presently active Kolbeinsey axis then led to the formation of the Iceland Plateau. In contrast, spreading in the north and south has taken place symmetrically about the Mohns and Reykjanes axes respectively.

This model is confirmed and extended on the basis of extensive geophysical investigations. Major recent contributions include an evolutionary synthesis for the entire Norwegian-Greenland Sea (Talwani and Eldholm, 1977) which was based, in part, on deep sea drilling (Talwani, et al, 1976), multichannel seismic surveys of the Jan Mayen block and East Greenland margin (Gairaud et al, 1978; Hinz and Schlüter, 1979a, 1979b), and a detailed aeromagnetic survey of the Iceland Plateau (Vogt, Johnson, and Kristjansson, 1980).

This paper aims to elucidate the kinematics and also the dynamics of the Jan Mayen block separation. It is mainly based on a study of sea floor spreading magnetic anomalies in the southern Norwegian-Greenland Sea. I show that the Jan Mayen block did not split from Greenland in a straightforward manner. Instead, there was simultaneous spreading about the Aegir axis and the newly formed Kolbeinsey axis for a considerable length of time. The plate tectonic reconstructions which are presented allow me to draw certain inferences about the structure of the Jan Mayen block and the present East Greenland margin. Some of the conclusions which are presented in this paper are similar to those of Mirlin (1979), Mirlin, Popov, and Finger (1980a), and Mirlin, Kostoglodov, and Suzyumov (1980b) who also analyzed the magnetic anomaly pattern north of Iceland. The present study was conducted independently of Mirlin's studies.

Magnetic Anomalies over the Southern Norwegian-Greenland Sea

Figure 2 shows a compilation of marine magnetic anomaly data obtained from the National Geophysical

*Presently with: Gulf Research and Development Company, Pittsburgh, Pennsylvania.

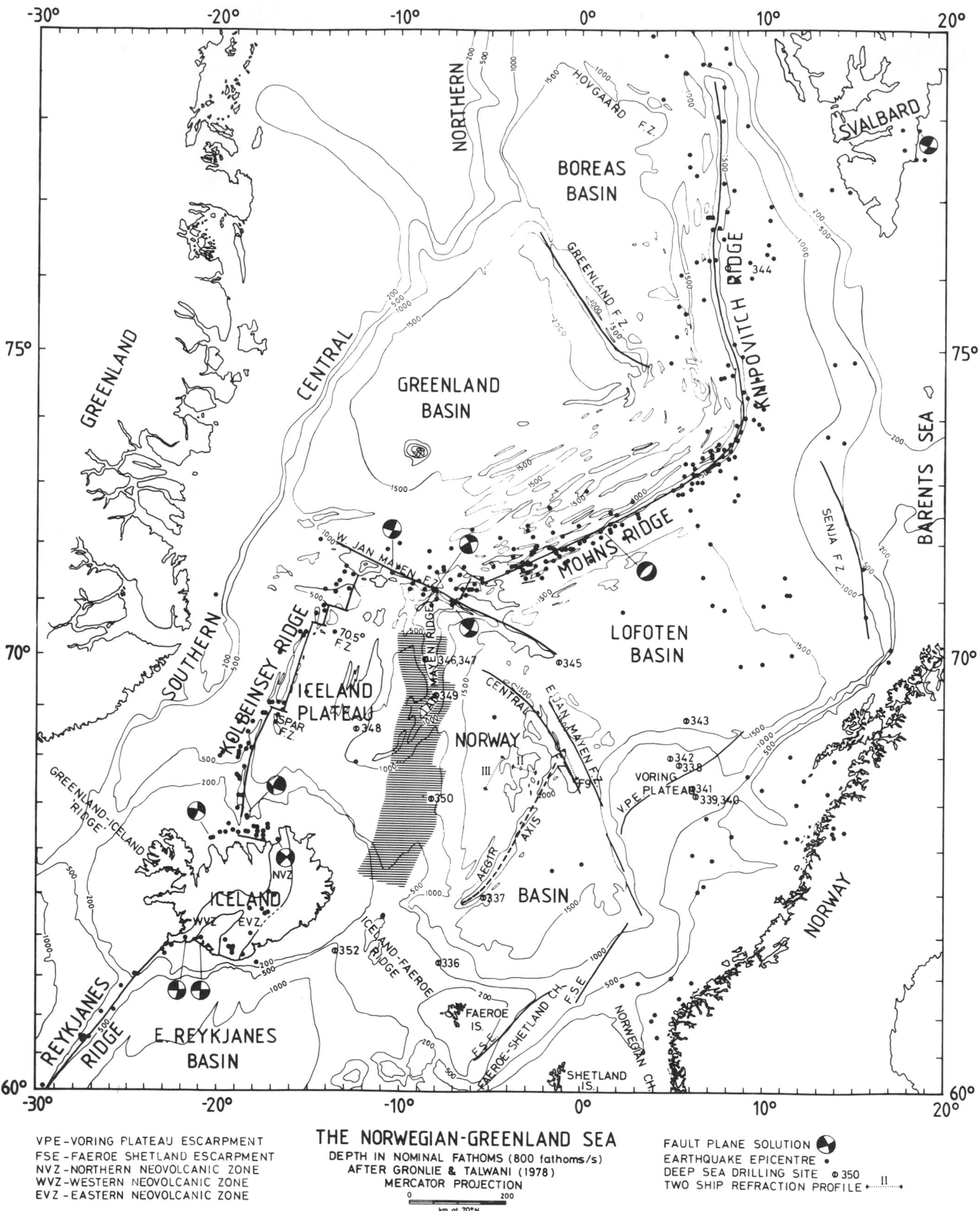

Figure 1 — The Norwegian-Greenland Sea, redrawn with slight modifications from various charts in the Geophysical Atlas of the Norwegian-Greenland Sea (Grønlie and Talwani, 1978). The probable extent of the Jan Mayen block is shaded.

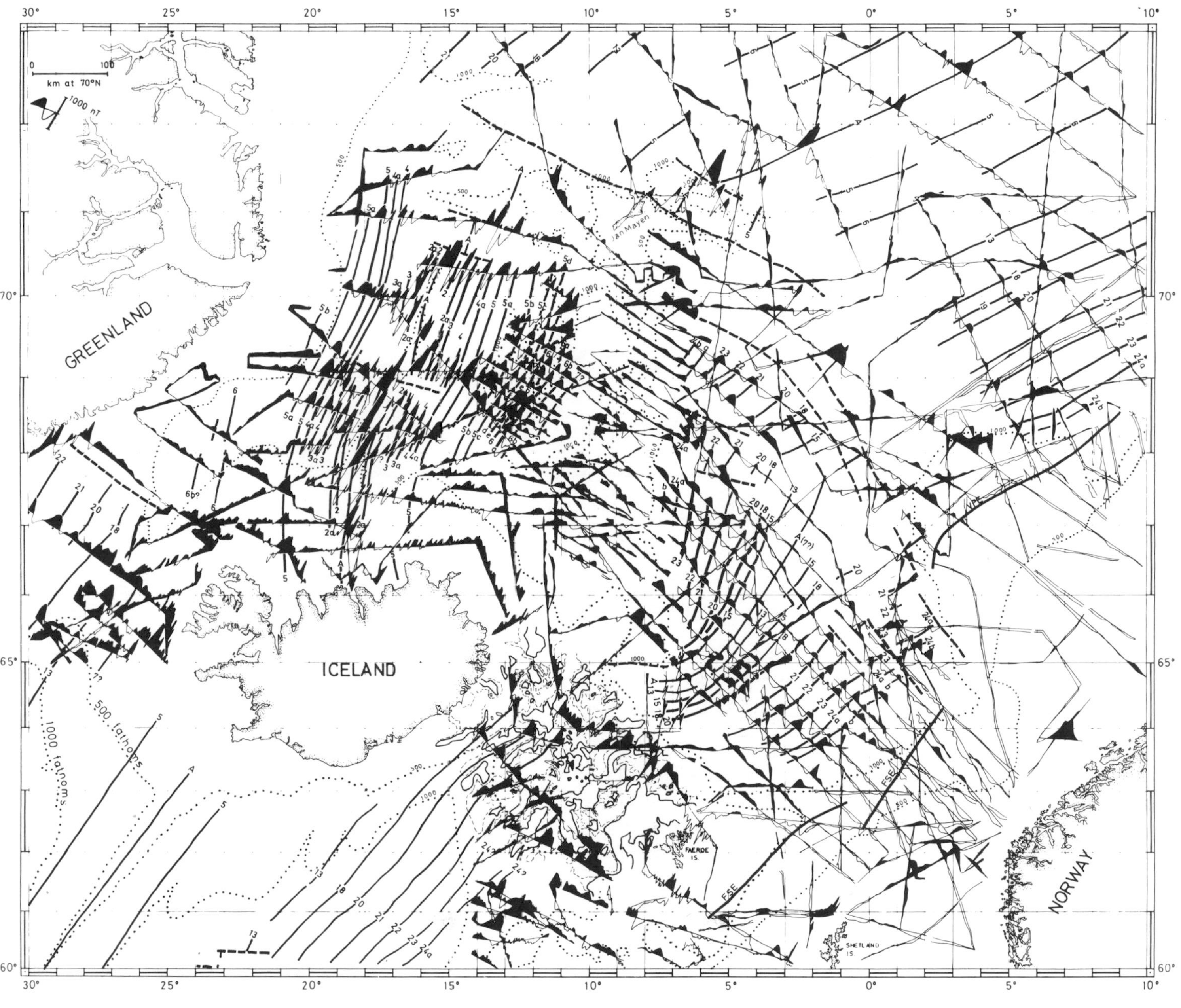

Figure 2 — Residual total field magnetic anomalies over the southern Norwegian-Greenland Sea and surrounding areas (from marine surveys). The anomalies are projected N 30°E with the ship's track as a base. Also shown for the Iceland-Faeroe Ridge are the zero (bold) and +200 nT (dashed) contours from the magnetic survey of Fleischer et al (1974).

and Solar-Terrestrial Data Center. It includes additional profiles from University of Durham surveys (Nunns and Peacock, in press) and from Talwani et al (1979). Also shown are contours of total field anomaly over the Iceland-Faeroe Ridge after Fleischer et al (1974).

The correlations and identifications over the Iceland Plateau are based on the detailed aeromagnetic survey of Vogt, Johnson, and Kristjansson (1980). The oldest anomaly identified on both sides of the Kolbeinsey axis is 6B, which has an age of 24 Ma according to the revised Cenozoic timescale of Hailwood et al (1979) adopted in this study. Over the Jan Mayen block the magnetic field is generally quiet, or at least does not exhibit recognizable sea floor spreading lineations (Grønlie, Chapman, and Talwani, 1979; Navrestad and Jorgensen, 1979; Vogt, Johnson, and Kristjansson, (1980). The correlations and identifications over the Norway Basin (Nunns and Peacock, in press) differ significantly from those previously published (Talwani and Eldholm, 1977). The oldest anomaly on both sides of the Aegir axis is 24B which formed at about 52 Ma, shortly after Greenland and Europe began to separate.

In the southern part of the central Norwegian-Greenland Sea, the anomaly lineations about the Mohns Ridge are based on the detailed aeromagnetic survey data reported by Vogt, et al (1978) and Phillips et al (in preparation), and agree very well with those of Talwani and Eldholm (1977).

Lineations about the Reykjanes Ridge are from Voppel, Srivastava, and Fleischer (1979). Possible continuations of anomalies 21, 20 and 18 across the Iceland Faeroe Ridge are shown, in accordance with the suggestion that the Icelandic transverse ridge (Bott, 1974) formed parallel with the area farther south (Voppel, Srivastava, and Fleischer, 1979). The anomaly identifications across the Greenland-Iceland Ridge (Vogt, Johnson, and Kristjansson, 1980) support this hypothesis. The fracture zone immediately north of the Denmark Straits (Figure 2) was postulated by Vogt, Johnson, and Kristjansson (1980). It marks a clear discontinuity in the magnetic anomaly pattern but is apparently not evident in basement topography (H. C. Larsen, personal communication).

Boundaries of the Jan Mayen Block

Physiographically, the Jan Mayen block includes the flat topped Jan Mayen Ridge, which extends southward from the Jan Mayen Island to about 68.25°, and the more southerly complex of narrow discontinuous ridges which is separated from the main ridge by a northeast trending trough. To make a meaningful reconstruction of the southern Norwegian-Greenland Sea, it is desirable to locate the line which separates the oceanic crust of the Iceland Plateau from the older material associated with the Jan Mayen block.

This study defines the Jan Mayen block from seismic reflection records (mainly those presented by Gairaud et al, 1978) by the presence of Horizon 0, which is a distinct, smooth, locally faulted basement reflector of relatively transparent character (Gairaud et al, 1978). Horizon 0 is different from the flat-lying basaltic basement characteristic of much of the Iceland Plateau (the opaque layer of Talwani and Eldholm, 1977), and also from the more typical hummocky oceanic basement of the Norway Basin. It is therefore possible to define western and eastern boundaries of the Jan Mayen block. It is clear from the profiles presented by Gairaud et al (1978) that Horizon 0 is older than the typical oceanic crust at the western edge of the Norway Basin; therefore, it is certain that the Jan Mayen block existed prior to the westward axis shift.

Figure 3 shows a conservative western boundary of

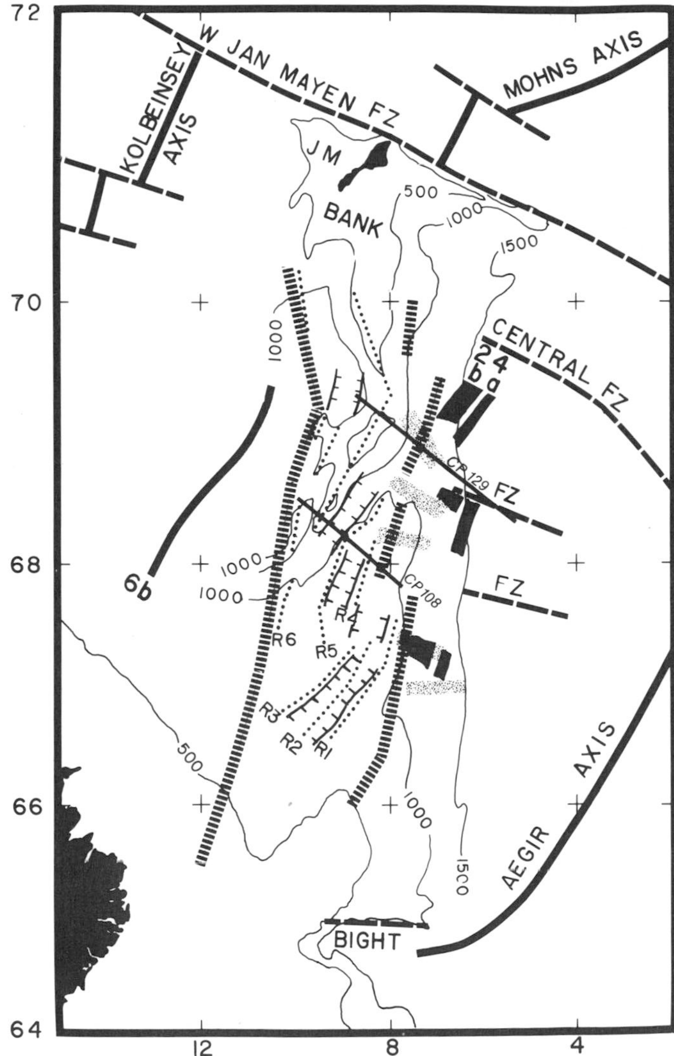

Figure 3 — The Jan Mayen block, showing the eastern and western boundaries inferred from seismic reflection records (stripes). Within the block, basement ridges (dotted) are delineated on the basis of linear free air gravity anomaly maxima (Grønlie, Chapman, and Talwani, 1979) and are numbered R1 to R6 in the southern region, following Talwani and Eldholm (1977). The locations of anomalies 6B and 24 are shown west and east of the block, respectively. Also shown (stippled) are the transition zones across which the Moho apparently deepens about 4 km, passing from the Norway Basin to the Jan Mayen block (determined by free air gravity modeling along a number of University of Durham survey lines).

the Jan Mayen block lying just outside the westernmost discontinuous line of acoustically transparent basement peaks. This line diverges to the south from anomaly 6B. The region between the western boundary and anomaly 6B is characterized by a flat acoustic basement, presumed to represent oceanic flood basalt. It is not certain that the Jan Mayen fragment does not extend westward beneath this region. However, the free air gravity field over this region (Grønlie, Chapman, and Talwani, 1979) is quite smooth, in contrast to that over the Jan Mayen block which is characterized by linear anomalies reflecting basement peaks and troughs. It is reasonable to assume that there are no hidden structures in this region.

The eastern boundary of the Jan Mayen block shown in Figure 3 is roughly parallel with the western boundary. Between the eastern and western boundaries, particularly within the basement rifts, Horizon 0 is not always evident but is inferred to lie at depth beneath a layer of flood basalt (Gairaud et al, 1978). The eastern boundary has a fairly consistent relationship with the position of anomaly 24 in the Norway Basin and also with the zone of westward crustal thickening, as inferred from the analysis of free air gravity anomalies. It is often suggested that the Jan Mayen block is underlain by continental crust. This paper considers the possibility that the block may be underlain by oceanic crust formed at an anomalous elevation during the earliest spreading phase. I consider the nature of Horizon 0 in more detail below.

No seismic profiles are available to resolve the question of the extension of the Jan Mayen block south of 67°N, but it is likely that the western boundary extends to the edge of the Iceland shelf as indicated. The northern extension of the block beyond 70° is also doubtful. Grønlie, Chapman, and Talwani (1979) and Larsen (1980) suggest that the northernmost shallow part of the Jan Mayen Ridge (the Jan Mayen bank) may be considerably younger than the main part of the ridge.

Fan-Shaped Spreading in the Norway Basin

A key to understanding the evolution of the southern Norwegian-Greenland Sea lies in the fan-shaped magnetic anomaly pattern of the Norway Basin (Talwani and Eldholm, 1977; Nunns and Peacock, in press) which is shown in zebra stripe form in Figure 4 on a polar stereographic base chart (Grønlie and Talwani, 1978). From the Aegir axis outward, the fan-shaped pattern is defined by the following predominantly positive bands: 7?-12, 13 (discontinuous), and 15-18. Outside this zone anomalies 20 to 24B are regularly developed, although they are affected by a number of offsets.

Talwani and Eldholm (1977) suggest that spreading about a near pole may have been responsible for the convergent pattern, but do not locate a pole except to specify that it needs to be near the southern end of the Aegir axis. In the present study a single rotation pole was found (Figure 4), accounting for the pattern of the central anomalies and for a number of bathymetric features in the Norway Basin.

For the purpose of studying near pole rotations, the projection of Figure 4 is effectively distortion free. The pole position in Figure 4 (64.87°N, 12.3°W) was found by matching the western edge of the anomaly 15 to 18 band on the northwestern side of the axis with corresponding lineation on the southeastern side. The curvature of the lineations constrains pole position quite severely; rotation is 22.2°. The fit is good, as indicated by the match between southeastern lineation and the square symbols. With respect to the pole, these symbols are uniformly 22.2° from their counterparts which are aligned along the northwestern lineation.

Using the same pole, rotation angles were determined to match the inner edges of the anomaly 15 to 18 band and the outer edges of the central band (13.0° and 4.2°). The goodness of fit is indicated with triangular and circular symbols. The same pole can satisfactorily account for the entire fan-shaped spreading phase.

Figure 5 is a plot of angular separation versus absolute age. The angular separations obtained from Figure 4 (adding a value for the end of anomaly 20 time, see below) fall on a straight line of slope 2.3° Ma. Extrapolated to zero, this curve implies that spreading ceased at anomaly 10 time (32 Ma), but the spreading rate may have slowed down at the end. Accretion beyond anomaly 7 time appears to be ruled out by the absence of the following negative anomaly, even in the northern basin where spreading rates are similar to those about the Mohns Ridge.

A number of bathymetric features provide evidence consistent with the pole position shown in Figure 4. First, the ridges and troughs of the central Jan Mayen Fracture Zone are parallel with small circles about the pole. The central fracture zone is primarily defined by a ridge of about 1 km relief which runs from 69.7°N, 5°W to 67.8°N, 0.6°W (Grønlie and Talwani, 1978). Indications of a buried ridge are present beyond both these limits. The ridge is paralleled to the southwest by a topographic depression and a more pronounced linear gravity low (Grønlie and Talwani 1978). It is suggested that the curved fracture zone came into existence during the beginning of rotation about a near pole, and that it replaced the eastern Jan Mayen Fracture Zone as the locus of transform motion between the Aegir axis and the Mohns Ridge.

Within the central zone of the Norway Basin there are a number of smaller scarps which are indicated schematically in Figure 4 and which are also concentric about the proposed pole. Closure back to about anomaly 18 time also causes almost complete closure of east to west trending bathymetric bight at the southern end of the Aegir axis. According to the present interpretation (see below), the bight represents a pre-existing fracture zone which opened when fan-shaped spreading began.

Various lines of bathymetric evidence provide control on the position of the pole independent of magnetic anomaly lineations. The central Jan Mayen Fracture Zone provides a strong constraint on the

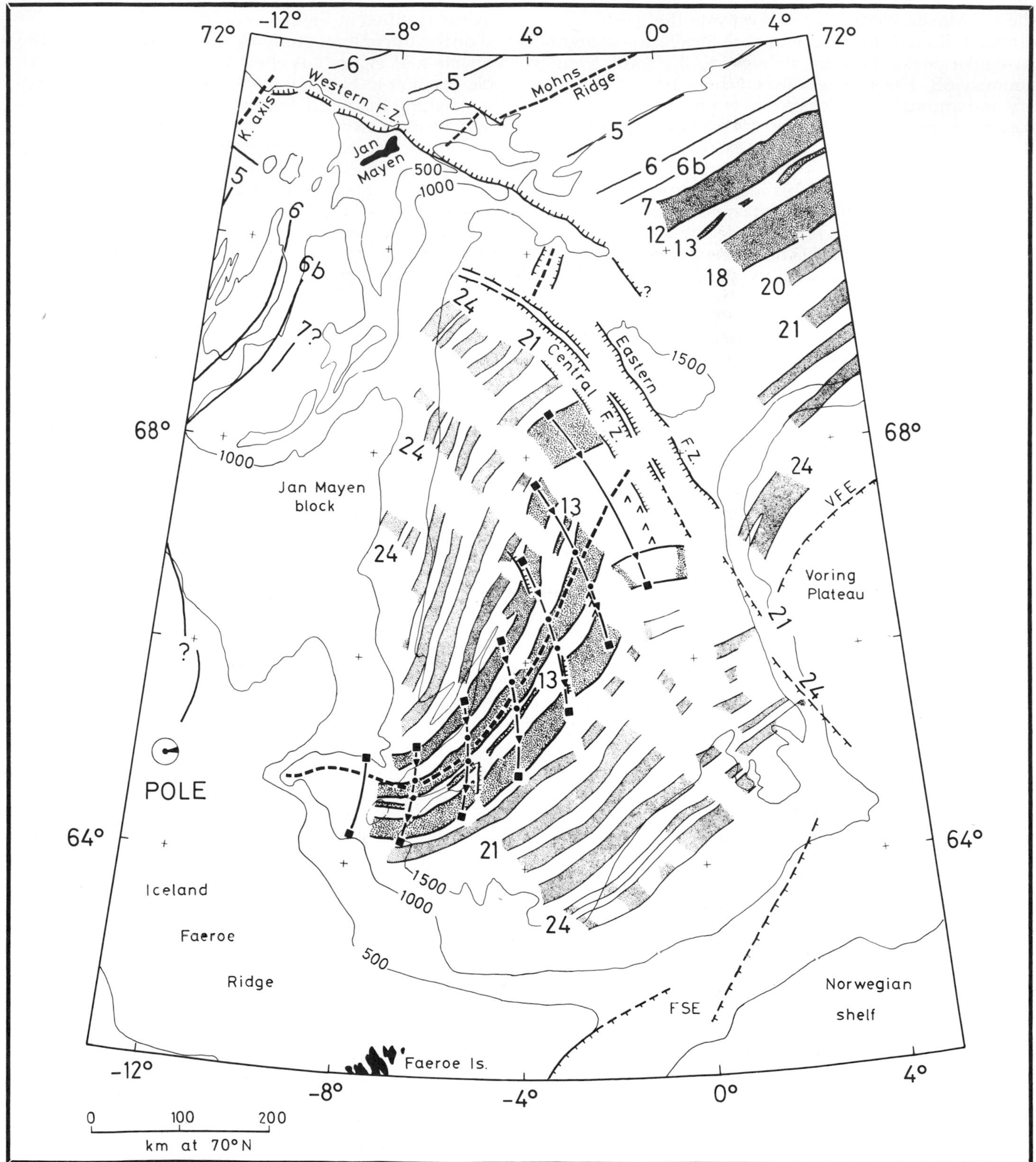

Figure 4 — Polar stereographic projection of the Norway Basin and environs, showing magnetic anomaly lineations, physiographic features, and flow lines for fan-shaped spreading. From the centre of the basin outward, the heavily stippled stripes correspond to the anomaly 7? to 12, 13, and 15 to 18 bands; the lightly stippled stripes correspond to anomalies 20, 21, 22, 23, 24A, and 24B. Dashed lines delineate spreading axes (extinct or active) while unbroken and broken hachured lines mark bathymetric scarps and buried basement scarps respectively. In the Norway Basin, flow lines for fan-shaped spreading are shown, with the solid square, triangular and circular symbols corresponding to 41.0, 36.6, and 33.1 Ma isochrons respectively. With respect to the pole, the symbols on the eastern side of the Aegir axis are at a constant angular distance from those on the western side.

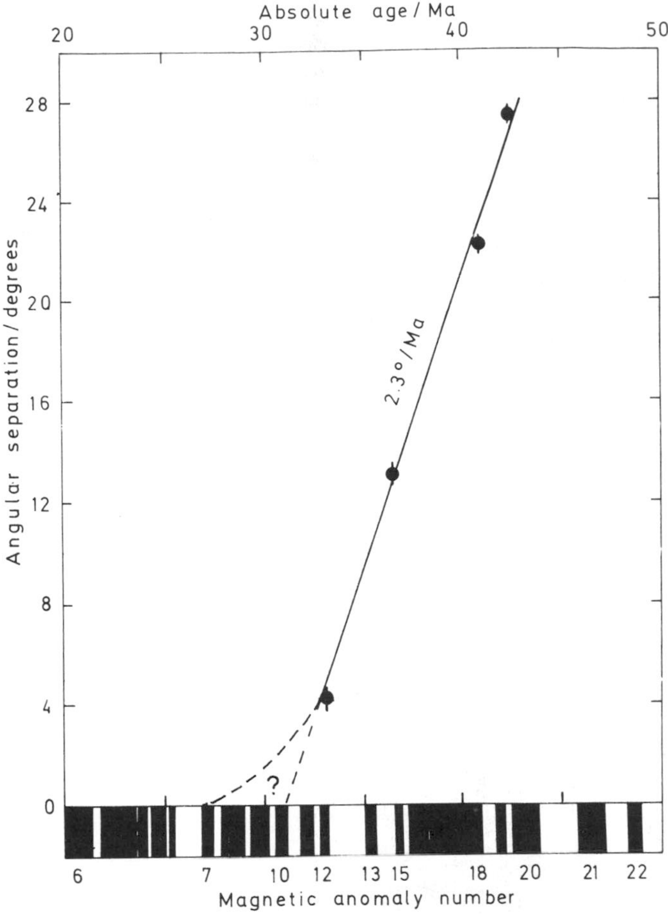

Figure 5 — Spreading curve for the Norway Basin during fan-shaped spreading: angular separation versus absolute age according to the Hailwood et al (1979) time scale.

position of the pole with respect to variation in a northwest to southeast direction, but a less strong constraint in the perpendicular "radial" direction. The pole must be farther southwest than about 66.5°N, 8.5°W, otherwise compression during opening is implied across the southern Aegir axis. The pole cannot lie much farther southwest than its proposed position and still give a realistic closure of the bathymetric bight. The lineations in the central zone of the basin constrain the pole to lie within a narrow west-southwest to east-northeast trending region of limited extent. Numerous trial poles were considered, all gave roughly equivalent overall solutions, with varying emphasis on the relative importance of magnetic lineations and bathymetric features. Nearly all these poles lie within an oval area about 40 km long and 30 km wide, stretching in a southwesterly direction away from the pole shown in Figure 4.

Reconstruction of the Norway Basin at the end of anomaly 20 time (42.7 Ma), shown in Figure 6, was achieved by closing 27.4° about the pole in Figure 4. Anomalies 20 to 24 are parallel on either side of the axis and the anomaly offsets match fairly well. In the south, the bathymetric bight is completely closed. The inferred position of anomaly 20 age crust over the Iceland-Faeroe Ridge is shown. It is suggested that a 200 km transform fault existed in the vicinity of the closed bathymetric bight. The eastern Jan Mayen Fracture Zone offsets the axis sinistrally by a similar amount at the basin's northern end.

Upon reconstruction beyond anomaly 20 time, the Jan Mayen bank cannot be accommodated south of the eastern Jan Mayen Fracture Zone; this supports the contention (Grønlie, Chapman, and Talwani, 1979; Larsen, 1980) that the bank is underlain by oceanic crust. A straightforward interpretation of Figure 6 suggests that the bank is underlain by crust of greater age than anomaly 20.

It is evident from the reconstruction that the Norway Basin was enclosed prior to about anomaly 21 time, when the Iceland-Faeroe Ridge, the Voring Plateau, and parts of the Jan Mayen Ridge were probably emergent (Grønlie, 1979).

The Complementary Axis: Fan-Shaped Spreading about the Kolbeinsey Axis

As inferred by Talwani and Eldholm (1977) simultaneous fan-shaped spreading, convergent to the north, must have occurred in some region west of the Aegir axis during the period that fan-shaped spreading took place in the Norway Basin. However, their suggestion that this spreading took place south of the Jan Mayen Ridge, in an area encompassing basement ridges 1, 2, and 3 (Figure 3), is not accepted because it is believed that this region is similar at depth to the rest of the Jan Mayen block (Gairaud et al, 1978). Moreover, this region is not large enough to account for all the fan-shaped spreading which must have taken place.

The simplest hypothesis is that complementary spreading took place about the newly formed Kolbeinsey axis. Figure 7 shows a reconstruction of the Iceland Plateau to anomaly 6B time. This was achieved by rotating the Jan Mayen block by -5.5° clockwise about a pole situated at 69°N, 130°E. This pole position represents a rounded average of total opening poles determined for anomaly 5 and 13 times by Talwani and Eldholm (1977) and for anomaly 6 time by Phillips et al (in preparation). The amount of finite rotation was determined concordantly from the results of Talwani and Eldholm (1977), and of Phillips et al (in preparation), by interpolating the curves of total angular separation versus time.

Figure 7 shows a satisfactory correspondence between the position of anomaly 6B on the Greenland margin and the reconstructed position of the eastern anomaly 6B (Vogt, Johnson and Kristjansson, 1980). Assuming that spreading in the Norway Basin ceased at about anomaly 7 time (and neglecting the slight age difference between anomaly 7 and anomaly 6B), Figure 7 is believed to represent the configuration about the Kolbeinsey axis when fan-shaped spreading ceased.

Because the Jan Mayen block diverges south from the Greenland coastline, Figure 7 clearly shows that complementary fan-shaped spreading may have taken

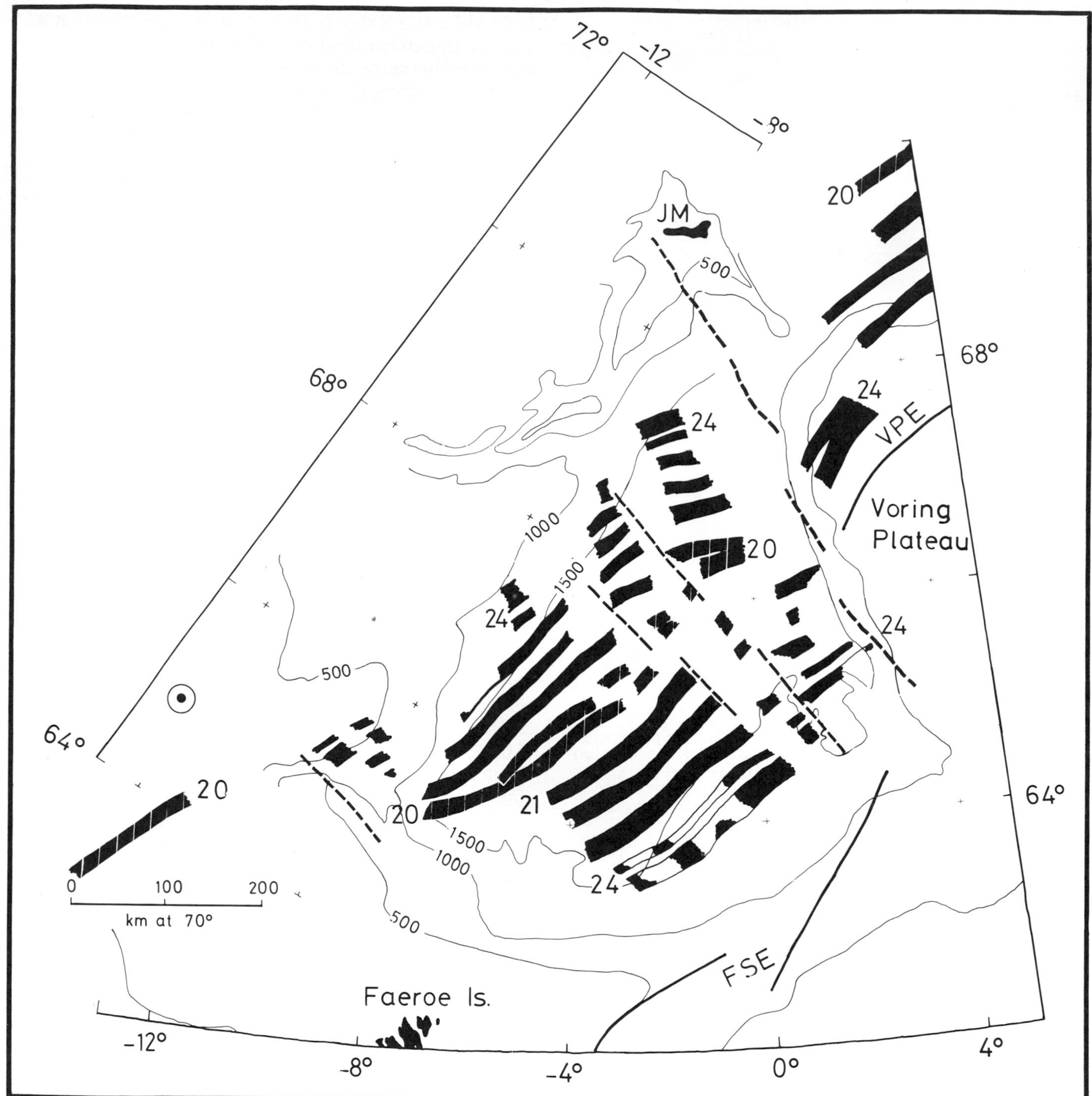

Figure 6 — Reconstruction of the Norway Basin to the end of anomaly 20 time (42.7 Ma).

place about the Kolbeinsey axis prior to anomaly 7 time. In the absence of well-defined magnetic lineations the finite closure pole is not well constrained. However, closure should bring the north scarp of the Norway Basin bathymetric bight into line with the Denmark Straits fracture zone, and the western margin of the Jan Mayen block to a reasonable position seaward of the Greenland coast. Also, the closure should be consistent with the relative closure for Eurasia-Jan Mayen and Eurasia-Greenland.

To align the ends of the Denmark Strait facture zone and the bathymetric bight, the closure pole should lie along the line XX′ (Figure 7), which is the perpendicular bisector of the line joining these features. The pole cannot lie far southward along this line or the northern part of the Jan Mayen block moves away from Greenland during closure. Alternatively, if the pole lies farther than about 72°N the amount of clockwise rotation that can be accommodated, without causing an unrealistic overlap of the

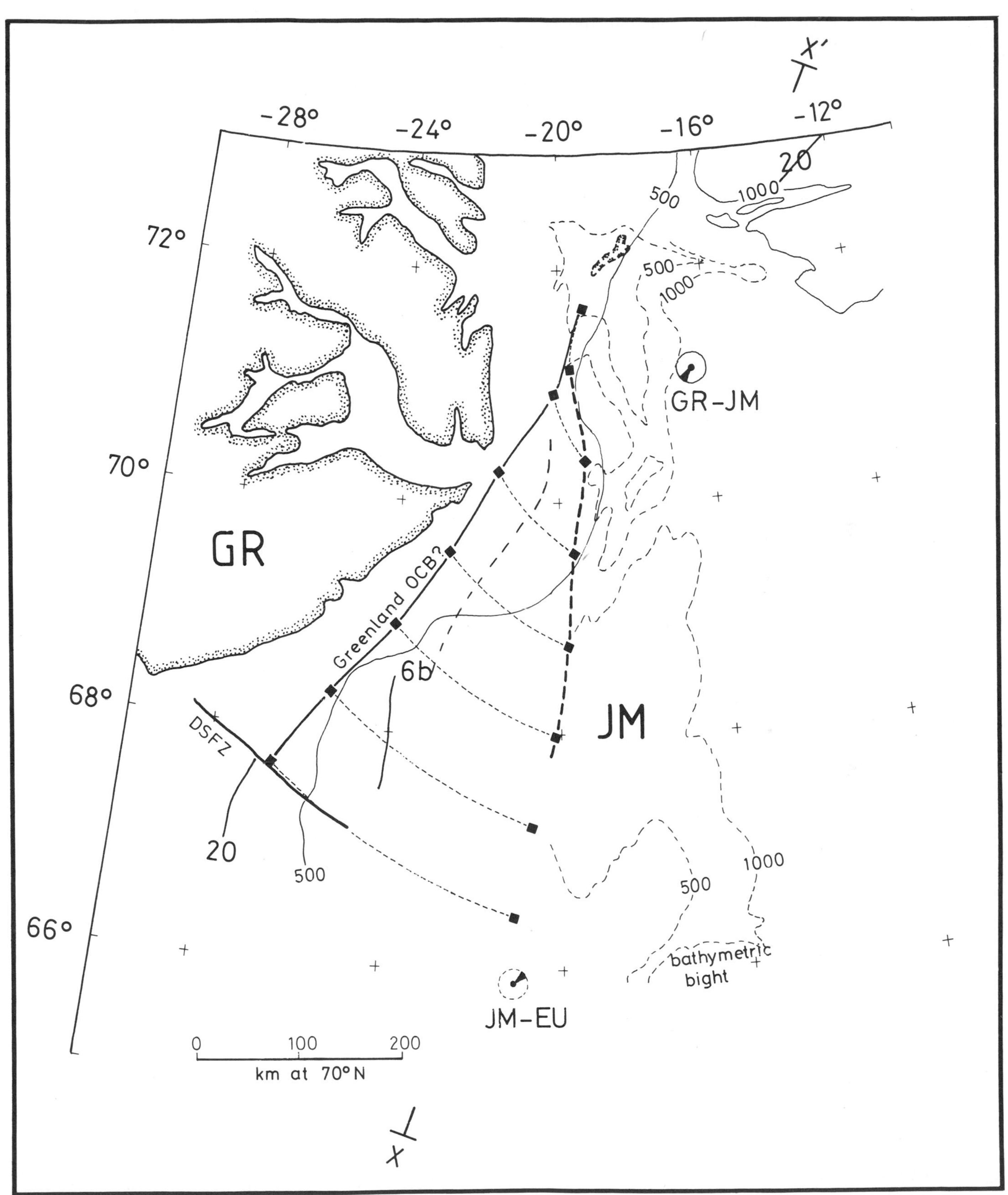

Figure 7 — Reconstruction of the Jan Mayen block relative to Greenland, at anomaly 6B time (24 Ma). Lineations and contours are shown dashed for the Jan Mayen block, solid for Greenland. The proposed finite rotation pole (GR-JM) closes JM against GR at 43 Ma. XX' is the line along which the pole must lie to bring the bathymetric bight in line with the Denmark Straits fracture zone (DSFZ).

Jan Mayen bank, is significantly smaller than the amount of rotation which took place in the Norway Basin. In this case, the composition of the Greenland-Jan Mayen rotation, with the Jan Mayen-Eurasia rotation, gives rise to a total spreading azimuth which is too far south of east, and to a total opening pole which lies south of the area instead of north.

The chosen pole position is at 71.12°N and 16.36°W in the reference frame of Greenland (Figure 7). In Figure 7 the western margin of the Jan Mayen block is indicated by the line of square symbols. If the Jan Mayen block is rotated by −29.0° about the chosen pole, this line, shifted to a satisfactory position off the Greenland coast, lies just seaward of the anomaly 20 lineation south of the Denmark Strait fracture zone. This is consistent with the formation of the Kolbeinsey axis at the end of anomaly 20 time by direct northward propagation from the Reykjanes axis.

Composition of the Jan Mayen-Greenland opening about this pole and the Eurasia Jan-Mayen opening gives a total opening of 3.12° about a pole situated at 48.1°N, 118.3°E. This pole position lies within the scatter area of determined finite difference poles for anomaly 7 to anomaly 20 time. For example, Phillips et al (in preparation) position the anomaly 13 to 18 pole at 39.6°N, 116.2°E, the anomaly 12 to 13 pole at 55.0°N, 130.4°E, and the anomaly 6 to 12 pole at 55.4°N, 157.9°E. The amount of rotation (3.1°) is also consistent with the results of Phillips et al. These results support the proposed complementary spreading regime, particularly as the composite pole position is sensitive to the position of the Greenland-Jan Mayen pole.

Near Scoresby Sund, the Greenland Shelf forms a conspicuous semi-circular salient (Vogt and Perry, 1978) which is delineated in Figure 7 by the 500 fathom isobath. Hinz and Schlüter (1979b) suggested that on the basis of multichannel seismic profiles the ocean-continent boundary lies near the shelf break, and the deepest coherent reflector beneath the salient is underlain by pre-drift sediments possibly containing hydrocarbons. Conversely, Vogt, Johnson, and Kristjansson (1980) suggest that the ocean-continent boundary lies much closer to the coast, contending that published sonobuoy and multichannel seismic data do not preclude such an interpretation. Like Brooks (1979) they believe that rapid deltaic progradation in late Tertiary times formed the Scoresby Salient. The reconstruction in Figure 7 supports this interpretation as the proposed ocean continent boundary lies over 100 km landward of the shelf edge. Note that Hinz and Schlüter (1979b) delineate a rapid basement depth increase just east of the Scoresby Sund mouth, corresponding in position to the boundary marked in Figure 7.

Closure of the Southern Norwegian-Greenland Sea

Figure 8 shows a closure of the southern Norwegian-Greenland Sea to the beginning of anomaly 24B time (52 Ma). I assume that spreading prior to 43 Ma was governed by the early spreading pole of Talwani and Eldholm (1977), 5.74°S, 124.9°E, in the reference frame of Eurasia. Figure 6 shows that 3.3° of rotation is required to close the Norway Basin from 43 to 52 Ma. (This compares with 3.5° extrapolated from the total opening rotations of Phillips et al, in preparation, for anomalies 20, 23, and 24).

Figure 8 shows good correspondence between the lineations, except in the north where there is a 20 km overlap. The Jan Mayen block (south of the Jan Mayen bank) fits well between the eastern Jan Mayen Fracture Zone and the Faeroe Islands. Figure 8 does not show the bathymetric bight because it formed within ocean crust and hence is reconstructed farther east than the 52 Ma isochron.

Structure and Evolution of the Jan Mayen Block

In this section some aspects of the geology of the Jan Mayen block are considered in light of the reconstructions presented. Figure 9 shows line drawings of two representative multichannel seismic profiles across the Jan Mayen block, redrawn from Gairaud et al (1978). CP129 crosses the Jan Mayen Ridge. CP108 crosses the southern tip of the Jan Mayen Ridge and the northern part of the disrupted zone to the south.

Nature of Horizon 0

Following Gairaud et al (1978) I consider the rocks underlying Horizon 0 to predate anomaly 24, but the nature of the rocks remains uncertain. Detailed comparison of sonobuoy refraction results with seismic reflection profiles shows no evidence of a continuous sedimentary basin beneath Horizon 0. The mean refraction velocity associated with Horizon 0 is greater than 5 km/sec. Analysis of free air gravity anomalies does not indicate low density sediment beneath Horizon 0, although overall crustal thickness appears to increase to at least 18 km, passing from the Norway Basin to the Jan Mayen block (Nunns, 1980).

A number of possibilities exist which are consistent with the constraint that relatively dense, high velocity rocks underlie Horizon 0. The first one I consider is that part or all of the Jan Mayen block was formed by spreading prior to anomaly 24 time.

On the basis of seismic reflection studies of the Voring Plateau, Mutter (1981) and Mutter, Talwani, and Stoffa (in press) infer that there were two phases in the early spreading history of the Norwegian-Greenland Sea. Immediately following the initiation of drift, rapid crustal accretion took place about a subaerial axis. Lava was extruded at high rate and flowed for a considerable distance from the spreading center. During this phase, a smooth elevated basement surface was produced, characterized by seaward dipping interval reflectors which correspond to lava units. After a few million years, the rate of crustal accretion decreased and the spreading axis subsided below sea level. Lava flow lengths diminished, producing a typical oceanic rough basement with a chaotic internal structure.

In the Norway Basin, seismic data indicate that much of the crust between the Faeroe-Shetland Es-

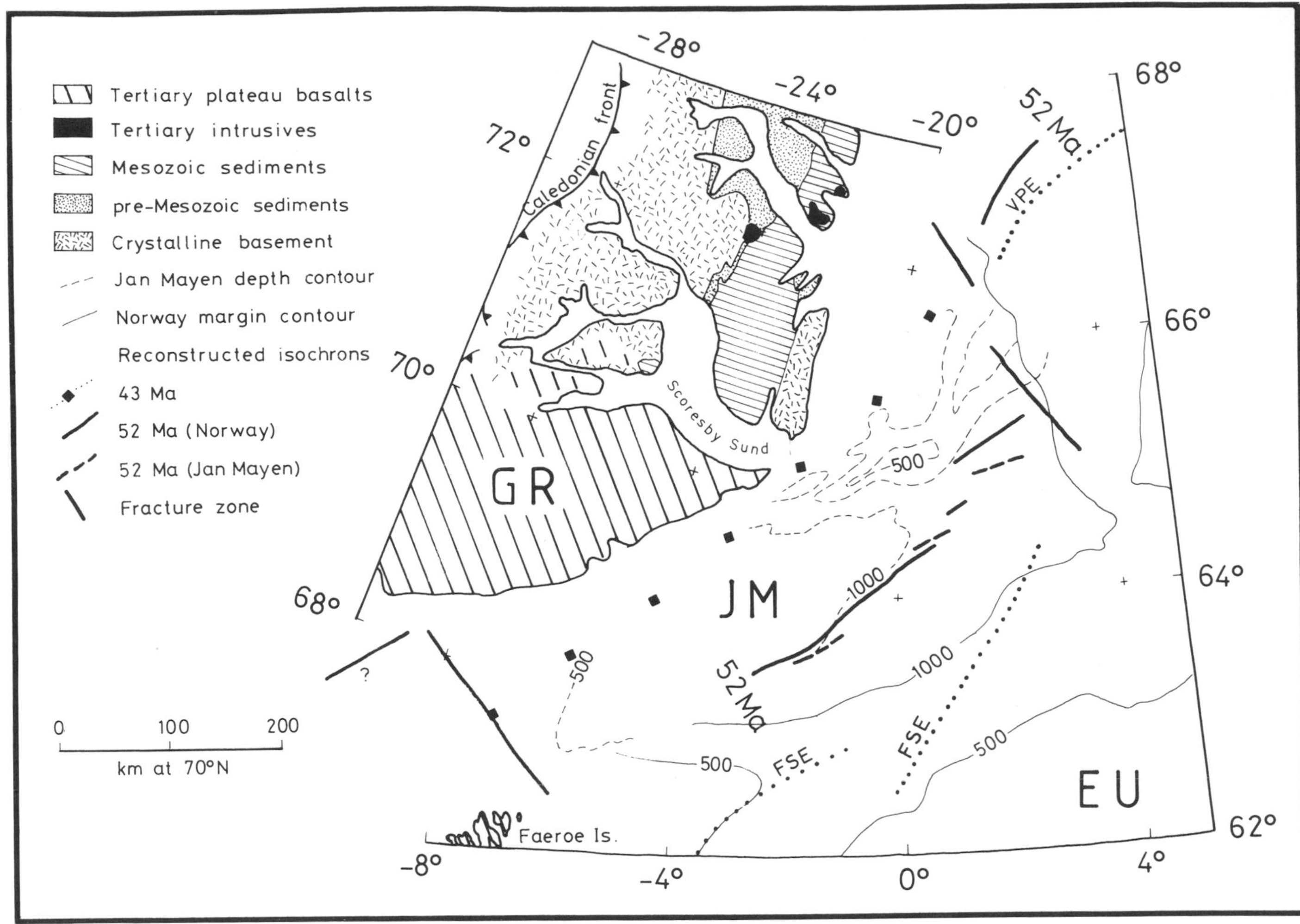

Figure 8 — Closure of the southern Norwegian-Greenland Sea to the beginning of anomaly 24 time.

carpment and anomaly 24 was produced by early phase spreading (Talwani and Eldholm, 1977; M. Talwani and J. C. Mutter, personal communication). The smooth-rough basement transition is approximately coincident with the 52 Ma isochron (Figure 8) and lies considerably seaward of the ocean-continent transition zone. Unless spreading prior to anomaly 24 time was very asymmetric, it is likely that part or all of the Jan Mayen block was also produced by early phase spreading. Note that Horizon 0 is smooth and characterized by seaward dipping internal reflectors (Figure 9). If this model is correct, the eastern boundary of the Jan Mayen block (Figure 3) corresponds to a smooth-rough basement transition (Figure 9) rather than to the boundary between a microcontinental fragment and oceanic crust. The fact that the magnetic field over the Jan Mayen block is quiet does not preclude the possibility that basaltic rocks underlie Horizon 0, as they may be uniformly magnetized.

Even if spreading did not produce the Jan Mayen block, it is possible that Horizon 0 is underlain by plateau basalt. Figure 8 shows the geology of East Greenland schematically, simplified and extrapolated over areas of ice cover (Escher and Watt, 1976). Reconstruction suggests two possible sources for the East Greenland flood basalts, which extend inland from the Blosseville Coast: first, they may have been erupted from the rift south of the Denmark Straits fracture zone and from the fracture zone itself (i.e., from the "hot spot" of the incipient Icelandic transverse ridge); or, second, at least some of the lavas may have spread laterally from the site of the Aegir axis. In either case, it is difficult to avoid the conclusion that most, if not all, of the southern Jan Mayen block must have also been covered by great thicknesses of flood basalt. It is possible that the adjacent land area north of Scoresby Sund was once covered by basalts (Soper et al, 1976) and if this is the case, remnants of the lava pile are also expected under the Jan Mayen Ridge. Conversely, Hinz and Schlüter (1979b) suggest that under the Jan Mayen Ridge, Horizon 0 is comprised of crystalline basement overlain on its dip slope by a relatively thin Mesozoic sequence. This is a plausible model considering the geology of Liverpool and Jameson Landes. It is also consistent with the occurrence of dipping reflectors within Horizon 0.

Nature of Horizon A

The post-drift sediments of the Jan Mayen block are divided into two sequences by the important reflector, Horizon A (Gairaud et al, 1978). Horizon A is a clear erosional unconformity which is flat under the top of the Jan Mayen Ridge, but cuts deeply into the lower sequence under the eastern ridge flank. Sediments of the upper sequence, where drilled near the ridge crest, are unconsolidated sandy muds while those of the lower sequence are lithified sandy mudstones (Talwani et al, 1976).

Johnson and Heezen (1967) originally suggested that the Horizon A unconformity was caused by uplift and erosion concomitant with the spreading axis shift. Drilling on the ridge crest shows, however, that the unconformity was formed not before anomaly 17 time and possibly as late as anomaly 7 time (correlating the biostratigraphic ages given by Talwani et al, 1976, with the revised timescale of Hailwood et al, 1979). Therefore, it is suggested that Horizon A is unrelated to the separation event which began at the end of anomaly 20 time. Instead, Horizon A may be due to the major mid-Oligocene drop in sea level at about 30 Ma (Vail, Mitchum, and Thompson, 1977). Recognition of an unconformity similar to Horizon A on the East Greenland margin south of the Denmark Straits, where no ridge shift occurred (Larsen, 1980), supports this suggestion.

Faulting and Volcanism

According to the proposed scheme, the initial separation stage of the Jan Mayen block from Greenland lasted from 43 to 27 Ma. During this period the southern Jan Mayen block moved at right angles away from Greenland, while the northern part may have moved slightly down the margin.

During separation, the Jan Mayen block was affected by faulting and volcanism more intensely in the southern region. Seismic profiles across the block (Gairaud et al, 1978) show that faulting postdated the deposition of the lower post-drift sediments and predated Horizon A. On each seismic profile crossing the southern Jan Mayen block (Gairaud et al, 1978) there is a graben bounded by outward dipping fault blocks. The tectonic pattern is characterized by an *en echelon* displacement of graben toward the eastern edge of the block as you move south, as shown in Figure 3. Any particular ridge which lies east of a graben in the north, and is eastward dipping, rotates to be westward dipping farther south where it lies west of a graben (illustrated by the tilt change of Jan Mayen Ridge from CP129 to CP108 in Figure 9). This configuration of *en echelon* rifts may have resulted from the shear imparted to the Jan Mayen block as it separated rotationally from Greenland.

Within the southern Jan Mayen block there was basaltic volcanism before and after faulting and tilting. Gairaud et al (1978) identify flow basalt overlying Horizon 0 on the eastward dipping slope of ridge 1 (Figure 3). This basalt was drilled at DSDP Site 350 (Talwani, 1976) and was dated at 41 Ma, which agrees with the 43 Ma age proposed for the initiation of the split. After faulting and tilting ceased basalt flooded the low-lying regions of the Jan Mayen block, presumably from the west (Gairaud et al, 1978). The final stage of effusion may have taken place after fan-shaped spreading ceased. If extensive lateral flow of lava was involved, as seems likely in view of the continuity of the opaque layer in the eastern Iceland Plateau, it may have been partially responsible for the obliteration of magnetic anomalies immediately west of the Jan Mayen block.

Evolution of the Northeast Atlantic Since Anomaly 20 Time

This section proposes an explanation for the evolution of the southern Norwegian-Greenland Sea in terms of the overall northeast Atlantic evolution, which was elucidated by Vogt and Avery (1974), Laughton (1975), Kristoffersen and Talwani (1977), and Phillips et al (in preparation), among others. Figure 10 shows reconstructions of the North Atlantic and Arctic oceans, utilizing the polar stereographic base chart of Phillips et al (in preparation).

At the end of anomaly 20 time (Figure 10a) the Aegir axis was offset to the east relative to the main trend of the mid-Atlantic ridge. Prior to this, an active triple junction existed at the mouth of the Labrador Sea. While spreading took place in the Labrador Sea, Greenland had a northerly component of motion relative to Eurasia and North America, which separated east to west. As the arrows in Figure 10a indicate, the direction of spreading between Greenland and Eurasia trended roughly northwest to southeast, parallel with the transform faults bounding the Norway Basin.

At the end of anomaly 20 time, spreading in the Labrador Sea ceased and Greenland became fixed to the North American plate, which continued to move as it had before, relative to Eurasia. Consequently, the spreading direction in the northeast Atlantic reoriented anticlockwise (Vogt and Avery 1974) and became parallel with the arrows shown in Figure 10b.

The new spreading direction was not parallel with the transform faults bounding the Norway Basin. If only sinistral offsets in the axis had been present, the corresponding transform faults would have become "leaky" and spreading would have continued much as before, except in the oblique direction. However, the dextral offset in the axis south of the Norway Basin spreading prevented this from taking place, because the new direction caused compression across the transform fault.

It is suggested that the resulting stresses caused propagation of a new rift northward along the line of the Kolbeinsey axis, and the detachment of the Jan Mayen block. Spreading in the new direction continued, with simultaneous fan-shaped spreading on both sides of the block, until the Jan Mayen microplate reoriented sufficiently and free movement was able to take place. Spreading in the Norway Basin then ceased and all of the spreading took place about

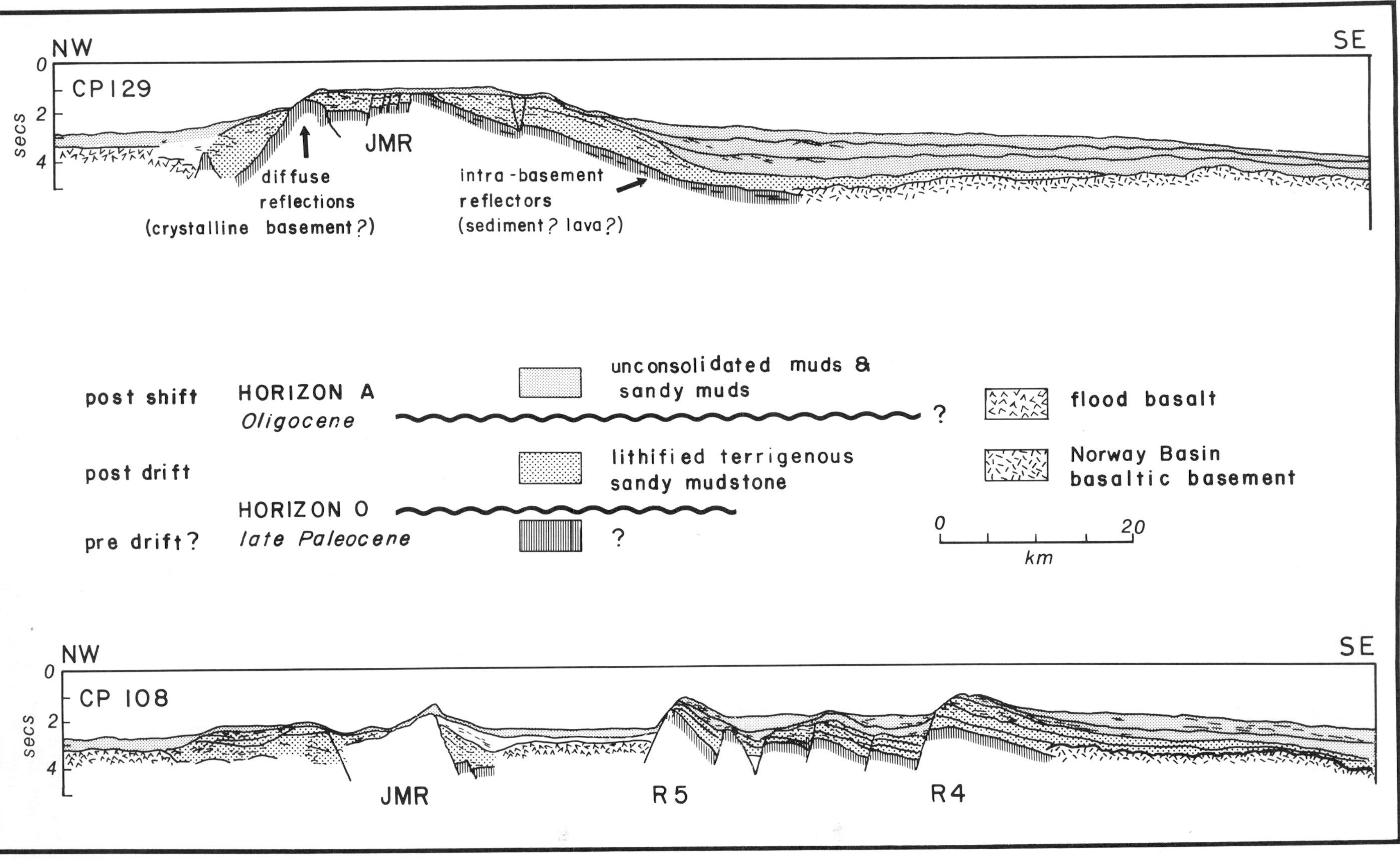

Figure 9 — Line drawing of 24 channel seismic profiles CP 129 and CP 108, crossing the Jan Mayen Ridge (JMR) and southern ridge complex, after Gairaud et al (1978). Figure 3 shows profile locations. The sediment lithologies and ages are based on drilling at DSDP Sites 346, 347, and 349 (Figure 1) on the Jan Mayen Ridge crest. The age of the flood basalt with respect to Horizon A is uncertain.

the Kolbeinsey axis.

Reconstruction in Figure 10b illustrates the situation at anomaly 5 time, after the Aegir axis became inactive. The figure shows the closely spaced fracture zones which formed about the Reykjanes Ridge after the end of anomaly 20 time and disappeared prior to anomaly 5 time (Vogt and Avery, 1974; Voppel and Rudloff, 1980). Vogt and Avery (1974) suggest an explanation for the appearance and disappearance of these fractures. A different explanation may be plausible in light of the present evolutionary model. It is suggested that compressive stresses originating at the transform fault propagated southward, causing fracture zones to form. Because of decoupling across the Jan Mayen fracture zone, stresses were not transmitted northward, so fractures were not formed about the Mohns Ridge. When the stresses opposing the new spreading direction were dissipated by the reorientations within the southern Norwegian-Greenland Sea, the fractures south of Iceland disappeared and an oblique spreading regime was established.

According to the model presented above, the separation of the Jan Mayen block resulted from plate interactions across transform zones, following a change in spreading direction and not from a spontaneous shift of axis. The width of the detached fragment may have been determined by the position of the Reykjanes axis when the spreading direction

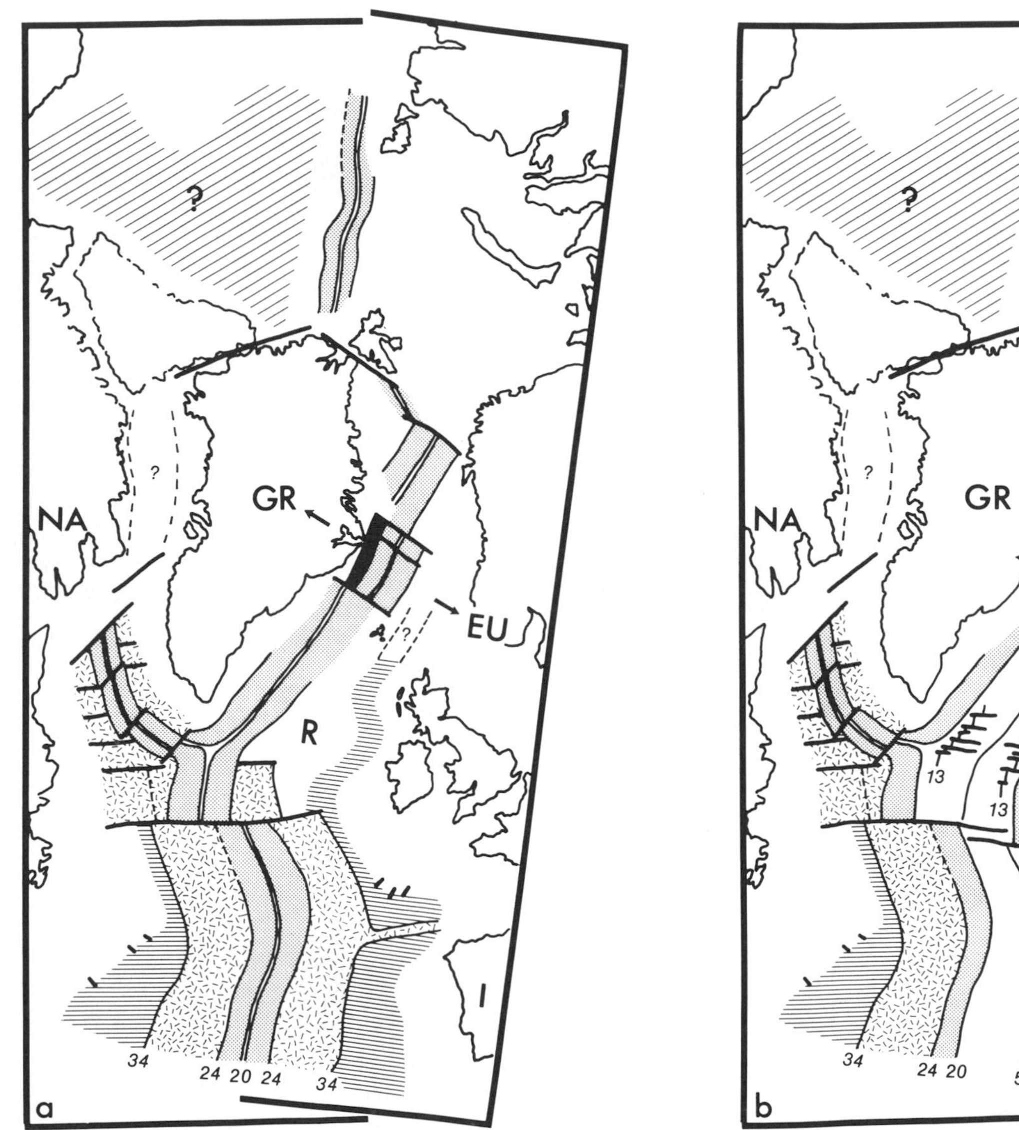

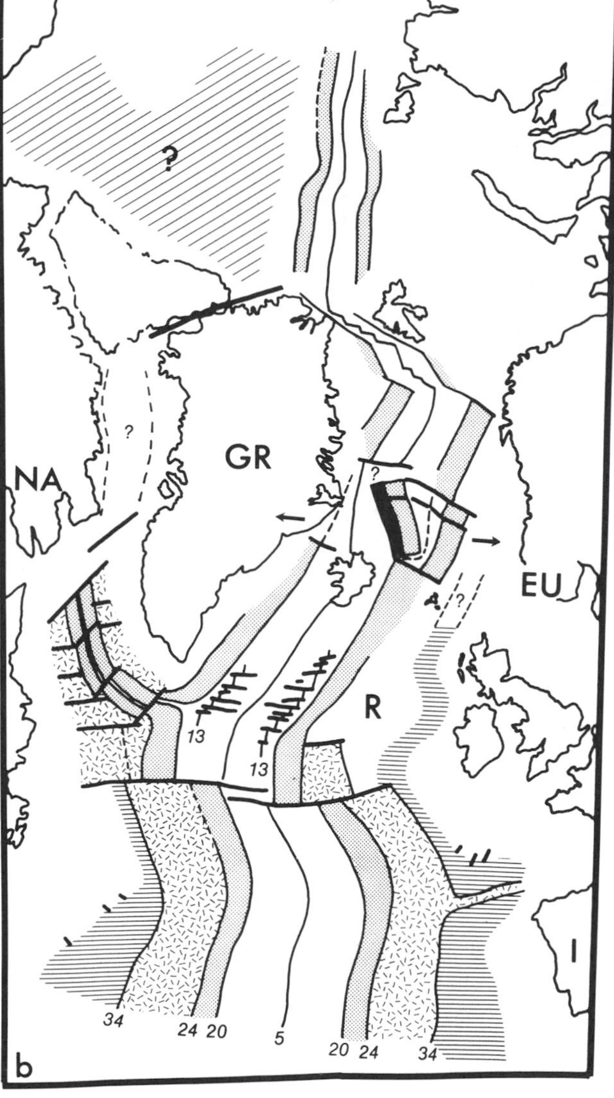

Figure 10 — Plate tectonic evolution of the northeast Atlantic from the end of anomaly 20 time (43 Ma) to anomaly 5 time (10 Ma). Based on a polar stereographic chart of Phillips et al (in preparation). The anomaly lineations given by Phillips et al are augmented by data from: Vogt and Avery (1974); Kristoffersen and Talwani (1977); Kristoffersen (1978); Srivastava (1978); Sibuet et al (1979); Voppel, Srivastava, and Fleischer (1979); Voppel and Rudloff (1980); Vogt, Johnson, and Kristjansson (1980); the present study. NA — North America, GR — Greenland, EU — Eurasia, R — Rockall microcontinent, I — Iberia. The Jan Mayen block is shown black.

change took place, rather than any pre-existing margin feature. However, the fact that the northward propagating line of separation remained parallel with the Aegir axis may have been determined by a preferential weakness in this direction.

ACKNOWLEDGMENTS

This study was made at the University of Durham, where the author was sponsored by the Association of Commonwealth Universities. It is a pleasure to thank Professor M. H. P. Bott and J. H. Peacock for stimulating advice and supervision. I also gratefully acknowledge useful discussions with T. L. Armstrong, K. Gunnarsson, and A. M. Rogan, and the use of data provided by Dr. C. Tapscott and Dr. J. D. Phillips.

The manuscript was typed by L. Hayes at Lamont-Doherty Geological Observatory, where Professor M. Talwani and Dr. P. D. Rabinowitz kindly provided facilities. The paper was improved following thoughtful reviews by Dr. B. Collins and Professor J. E. van Hinte, and following consideration of information provided by Professor M. Talwani and J. C. Mutter.

REFERENCES CITED

Bott, M. H. P., 1974, Deep structure, evolution and origin of the Icelandic Transverse Ridge, *in* L. Kristjansson ed., Geodynamics of Iceland and the North Atlantic Area: Holland, D. Reidel Publishing Company, p. 33-47.

Brooks, C. K., 1979, Geomorphological observations at Kangerdlugssuaq, East Greenland: Meddelelser om Grønland, Geoscience, n. 1, p. 3-21.

Escher, A. and S. Watt eds., 1976, Geology of Greenland: Copenhagen, The Geological Survey of Greenland, 603 pages.

Fleischer, U., et al, 1974, Die Struktur des Island-Färöer-Rückens aus geophysikalischen Messungen: Deutsche Hydrographische Zeitschrift, v. 27, p. 97-113.

Gairaud, H., et al, 1978, The Jan Mayen Ridge, synthesis of geological knowledge and new data: Oceanologica Acta, v. 1, p. 335-358.

Grønlie, G., 1979, Tertiary paleogeography of the Norwegian-Greenland Sea: Norsk Polarinstitutt Skrifter, n. 170, p. 49-61.

———, and M. Talwani, 1978, Geophysical Atlas of the Norwegian-Greenland Sea. Palisades, New York, Lamont-Doherty Geological Observatory, Vema Research Series, n. 4.

———, M. Chapman, and M. Talwani, 1979, Jan Mayen Ridge and Iceland Plateau: origin and evolution: Norsk Polarinstitutt, Skrifter, n. 170, p. 27-48.

Hailwood, E. A., et al, 1979, Chronology and biostratigraphy of northeast Atlantic sediments, DSDP leg 48, *in* L. Montadert, et al, Initial reports of the Deep sea drilling project, v. 48: Washington, D. C., U. S. Government Printing Office, p. 1119-1141.

Hinz, K., and H. U. Schlüter, 1979a, The North Atlantic — results of geophysical investigations by the Federal Institute for Geosciences and Natural Resources on North Atlantic Continental Margins: Oil Gas — European Magazine, v. 3, p. 31-38.

———, and ———, 1979b, Continental margin off East Greenland: 10th World Petroleum Congress, Heyden and Sons Ltd., Special Paper 7, 14 pages.

Johnson, G. L., and B. C. Heezen, 1967, Morphology and evolution of the Norwegian-Greenland Sea: Deep Sea Research v. 14, p. 755-771.

Kristoffersen, Y., 1978, Seafloor spreading and the early opening of the North Atlantic: Earth and Planetary Science Letters, v. 38, p. 273-290.

———, and M. Talwani, 1977, Extinct triple junction south of Greenland and the Tertiary motion of Greenland relative to North America: Geological Society of America; Bulletin, v. 88, p. 1037-1049.

Larsen, H. C., 1980, Geological perspectives of the East Greenland continental margin: Geological Society of Denmark, Bulletin, v. 29, p. 77-101.

Laughton, A. S., 1975, Tectonic evolution of the North-East Atlantic; a review: Norges Geologiske Undersoekelse Skrifter, n. 316, p. 169-193.

Mirlin, Ye.G., 1979, Movements and deformations of lithosphere plates in the Icelandic region of the North Atlantic and the character of the Faeroes-Greenland threshold: Washington, D. C., American Geophysical Union, Geotectonics (English edition), v. 12, p. 455-465.

———, K. V. Popov, and D. L. Finger, 1980a, Age of the ocean floor in the Icelandic region: Washington, D. C., American Geophysical Union, Oceanology (English edition), v. 19, p. 696-700.

———, V. V. Kostoglodov, and A.Ye. Suzyumov, 1980b, Cenozoic plate tectonics in the Iceland region of the North Atlantic: Washington, D. C., American Geophysical Union, Oceanology (English edition), v. 20, p. 168-172.

Mutter, J. C., 1981, Layered oceanic basement complex of the Norwegian margin: a heuristic model: Washington, D. C., EOS (American Geophysical Union Transactions), v. 62, p. 407.

———, M. Talwani, and P. L. Stoffa, in press, Origin of finely layered igneous stratigraphy in oceanic crust off the Norwegian margin: "subaerial seafloor spreading": Geology.

Navrestad, T., and F. Jorgensen, 1979, Aeromagnetic investigations on the Jan Mayen Ridge: Oslo Norwegian Petroleum Society, Norwegian Sea Symposium.

Nunns, A. G., 1980, Marine geophysical investigations in the Norwegian-Greenland Sea between the latitudes of 62°N and 74°N: University of Durham, Ph.D. thesis.

———, and J. H. Peacock, in press, Correlation, identification and inversion of magnetic anomalies in the Norway Basin, *in* A. Vogel, ed., The Structure, Evolution and Dynamics of the Norwegian-Greenland Sea: Wiesbaden West Germany, Vieweg, Earth Evolution Sciences, v. 2.

Sibuet, J-C, et al, 1979, Initial reports of the deep sea drilling project, v. 47, part II; Washington, D. C., U. S. Government Printing Office, 787 p.

Soper, N. J., et al, 1976, Biostratigraphic ages of Tertiary basalts on the East Greenland continental margin and their relationship to plate separation in the Northeast Atlantic: Earth and Planetary Science Letters, v. 32, p. 149-157.

Srivastava, S. P., 1978, Evolution of the Labrador Sea and its bearing on the early evolution of the North Atlantic: Geophysical Journal of the Royal Astronomical Society, v. 52, p. 313-357.

Talwani, M., and O. Eldholm, 1977, Evolution of the Norwegian-Greenland Sea: Geological Society of American, Bulletin, v. 88, p. 969-999.

———, et al, 1976, Initial reports of the Deep sea drilling project, v. 38: Washington, D. C., U. S. Government Printing Office, 1256 p.

———, ———, 1979, Survey at sites 346, 347, 348 and 350;

the area of the Jan Mayen Ridge and the Icelandic Plateau, *in* M. Talwani. et al, eds., Initial reports of the deep sea drilling project, supplement to v. 38, 39, 40 and 41: Washington, D. C., U. S. Government Printing Office, p. 465-488.

Vail, P. R., R. M. Mitchum, and S. Thompson, 1977, Seismic stratigraphy and global changes of sea level, part 4: global cycles of relative changes of sea level, *in* C. E. Payton, ed., Seismic stratigraphy — applications to hydrocarbon exploration: AAPG Memoir 26, p. 83-97.

Vogt, P. R., and O. E. Avery, 1974, Detailed magnetic surveys in the northeast Atlantic and Labrador sea: Journal of Geophysical Research, v. 79, p. 363-389.

———, and R. Perry, 1978, Post-rifting accretion of continental margins in the Norwegian-Greenland and Labrador Seas: Morphological evidence: Washington, D. C., EOS (American Geophysical Union Transactions), v. 59, p. 1204.

———, G. L. Johnson, and L. Kristjansson, 1980, Morphology and magnetic anomalies north of Iceland: Journal of Geophysics, v. 47, p. 67-80.

———, et al, 1978, The ocean crust west and north of the Svalbard Archipelago: synthesis and review of new results: Polarforschung, v. 48, p. 1-19.

Voppel, D., and R. Rudloff, 1980, On the evolution of the Reykjanes Ridge south of 60°N between 40 and 12 million years before present: Journal of Geophysics, v. 47, p. 61-66.

———, S. P. Srivastava, and U. Fleischer, 1979, Detailed magnetic measurements south of the Iceland-Faeroe Ridge: Deutsche Hydrographische Zietschrift, v. 32, p. 154-172.

Wedge Tectonics Along Continental Margins

J. C. Pratsch
Mobil Exploration and Producing Services Inc.
Dallas, Texas

The geologic concept of "wedge tectonics" allows us to interpret tectonic deformation results in terms of interactions of compression, tension, and shear. The emphasis lies on contemporaneous deformation in intimately related stress fields of different styles, not on intensity of deformation or on quantification of stress. Wedge tectonics occur at all times, in all dimensions, and in all geologic settings. On continental margins, wedge tectonic interpretations explain many observable tectonic features such as arc fragmentation, wrench faulting along mobile belt fronts, land and oceanward thrusting along arc flanks, and contemporaneous tension tectonics in backarc and forearc basins during arc thrusting. In orogenic belts, foreland tension appears to be contemporaneous with orogenic compression and shear. Examples from European Paleozoic passive and active margins (or margin phases) show the applicability of the concept.

The tectonic style, the geometry of deformed geologic rocks and bodies, and the development of a basin are not results of statistical or unpredictable forces. A few common natural laws govern the geologic processes of the past and of the present. To logically explain data and observations obtained during area evaluations, several geologic concepts have been proposed and utilized. "Wedge tectonics," presented here, cannot be called new as portions have been described by many before. It has remarkable power for explanation and prediction, and is valuable for hydrocarbon or mining exploration, for seismicity and paleomagnetic studies, and for many other fields. This paper is purposely limited to geologic continental margin processes, but wedge tectonic concepts can be applied in any other environment, basin class or geologic study.

DEFINITIONS AND PRINCIPLES

The principles of wedge tectonics lie in the interaction of all three main types of deformation: compression, tension, and shear. Numerous definitions for single phases of rock deformation have been proposed and utilized, yet the basic principles are quite simple (Dubey 1980; Kupfer, 1968; Figure 1). Material under compression reacts in two ways: 1) It becomes com-

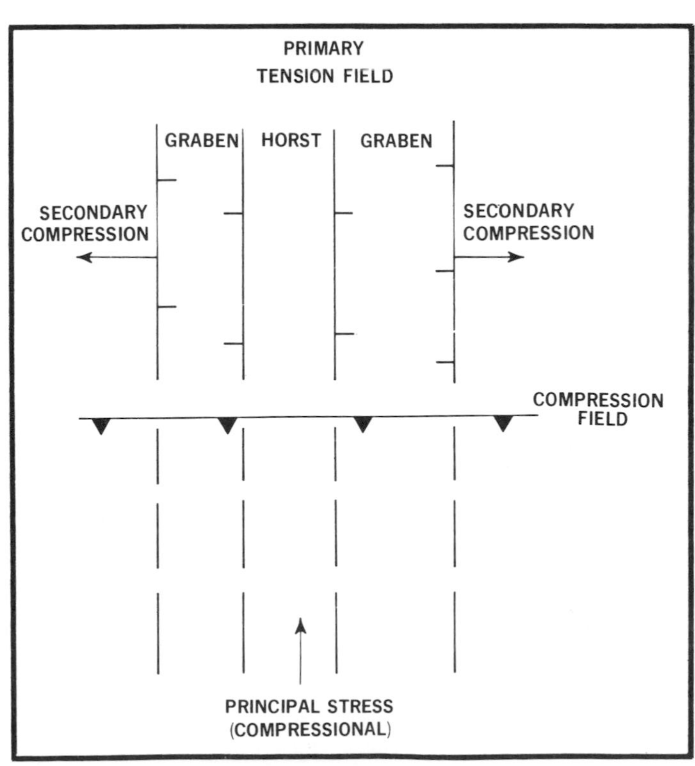

Figure 1 — Composite stress distribution, A.

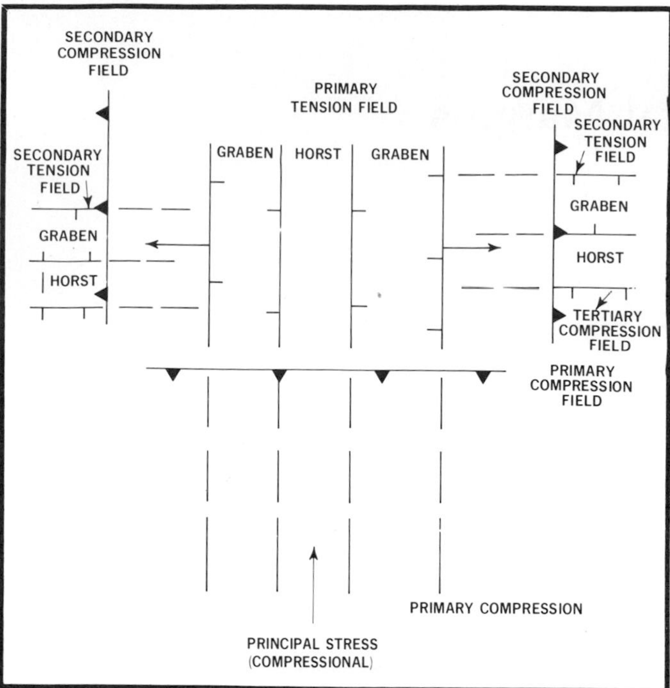

Figure 2 — Composite stress distribution, B.

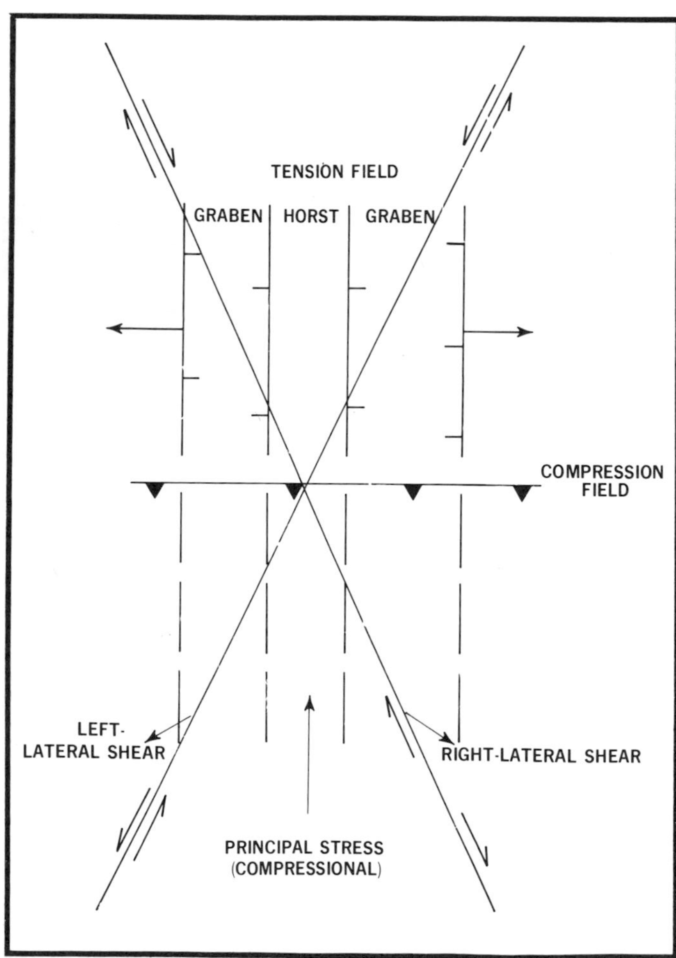

Figure 3 — Composite stress distribution, C.

pressed and shortened in the direction of compression; 2) it becomes lengthened and extended in a direction normal to the compressional direction. The primary extension results in a secondary compression component that acts perpendicular to the primary principal stress (Figure 2). Similarly, secondary lengthening and extension occurs parallel with and normal to the secondary compression direction. The theoretical third "generation" of compression and tension is parallel with the primary direction and cannot be separated from it.

First and second generation compression and tension would be the only tectonic deformation styles responding to unilateral compressional stress if the materials involved were completely uniform; tension would be the preferred deformation style with brittle material, compression with plastic material. However, geologic rock material is not homogeneous and this will affect the type of final deformation.

Because of these material differences, rocks respond to unilateral stress by tension, as described, and also by shearing. Two sets of "conjugate" shear planes, together with first and second order compression and tension directions, thus form the basic frame of wedge tectonics (Figure 3).

First-order shearing leads to second-order compression and tension in response to shear forces involved (Figure 4). The theoretical geometric relations between these features are well known (Wilcox, Harding and Seely, 1973). The actual angular relations again depend on specific material properties.

Common complications arise where two "sets" of tectonic wedges interact (Figure 5). Here, tectonic directions can be utilized by deformation components of either wedge unit; the trend directions are equal or

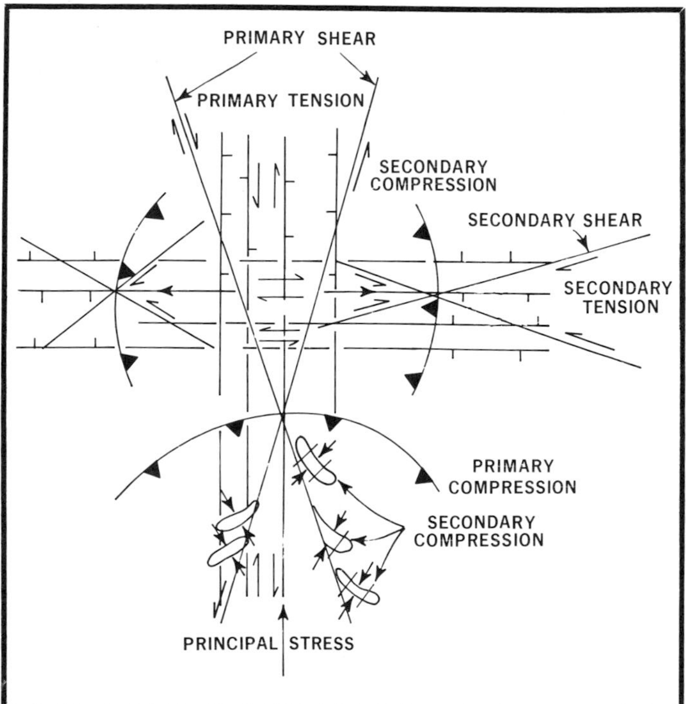

Figure 4 — Composite stress distribution, D.

close, but the relative deformation direction will be opposed. The "final" direction depends on the relative deformation intensity of the specific tectonic components. In practice, shear directions may be similar but shear motions may vary with time along the same trend.

Geometric relations between shear planes and higher order compression and tension are interesting to consider when following examples. Where shear planes are offset in en echelon arrangement, or where shear planes undergo a change in their direction, local zones of compression or tension develop (Figure 6). These higher order zones are common in all geologic dimensions and have high practical importance from developing regional tectonic trends and local basins or anticlinal features, to developing fracture and influencing fluid migration (Moody, 1973; Moody and Hill, 1956).

In summary, wedge tectonics describe the intimately interrelated occurrence of compression, tension, and shear deformation in all cases of structural deformation (Figure 6). Where one deformation style is known, the other two will have left some indication of their activity and will be predictable. Not always are all three components and all possible orders equally developed. Pre-existing material differences and pre-existing older deformation results lead to preference of one component or one deformation type over others. Still, the potential for all components and orders are present in any given area.

There is also no way to relate the intensity or visibility of any one deformation style or component (or order) to a certain quantitatively determinable intensity of overall deformation. In other words, second-order shears may be well-developed where first-order components are only vaguely manifested. And there is no relation between measurable offset along certain shears and related higher-order zones. Examples of such qualitatively satisfying, but quantitatively unresolved, deformations can be found in many textbooks

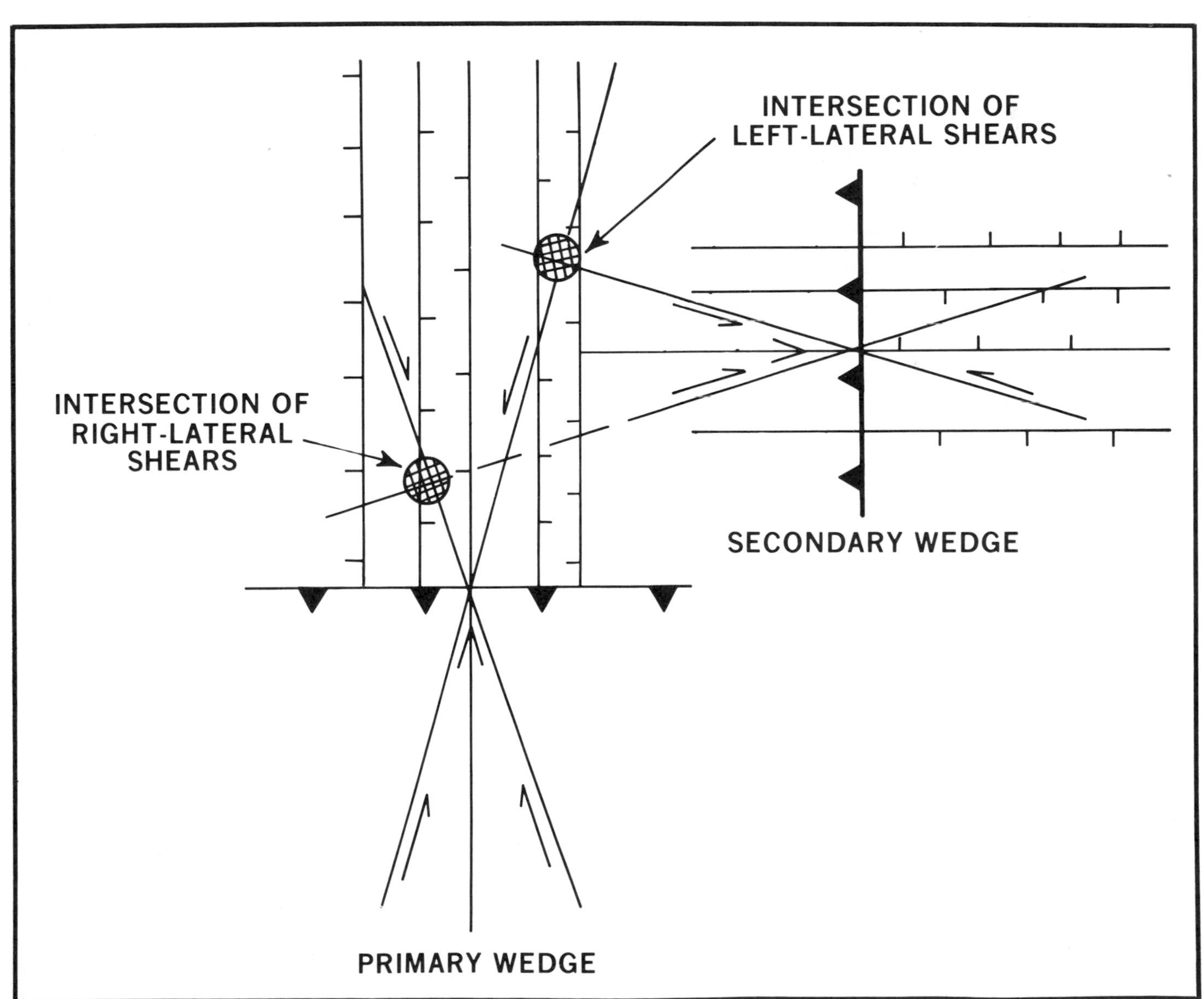

Figure 5 — En-echelon shear faults.

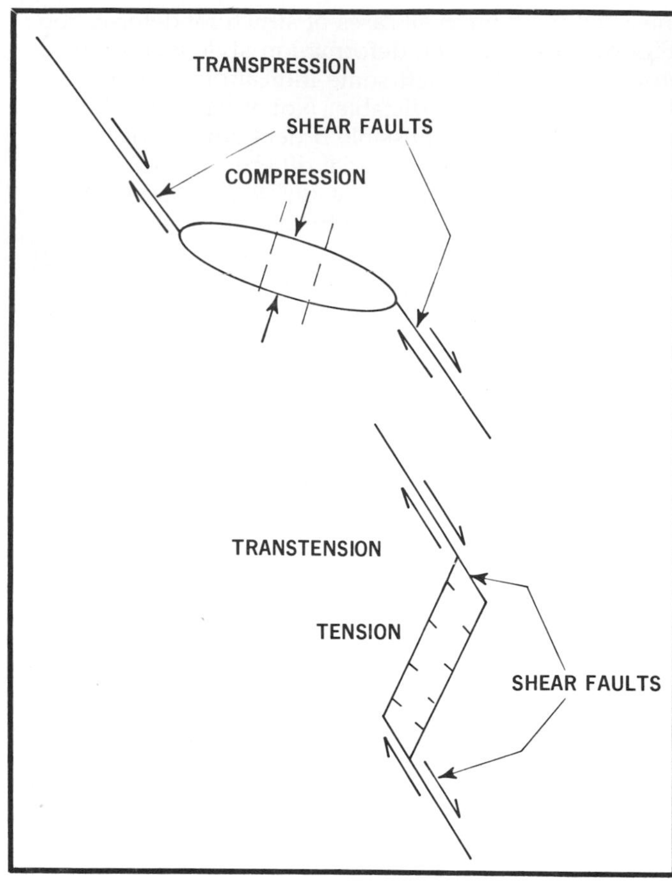

Figure 6 — Composite stress distribution, E.

and field case descriptions (Fitch, 1972; Molnar and Tapponnier, 1975; Caire, 1979; Brunn and Burollet, 1979; Dubey, 1980; Jacob, 1980; Seagall and Pollard, 1980). This is where the predictability of tectonic processes finds its limits.

PALEOZOIC CONTINENTAL MARGIN, NORTHWEST GERMANY

Wedge tectonic concepts were recently applied to the northwest German basin during studies of applied regional geology (Pratsch, 1979; 1980; in press) to explain present distributions of major gas accumulations and to predict additional exploration targets.

The northwest German basin (Figure 7) is the southeastern extension of the greater northwest Europe or North Sea basin, and as such, part of the Paleozoic continental margin along the Fenno-Scandian shield. Limited in the north by the east to southeast trending Ringkobing Fyn High, and in the south by the northeast trending Variscan mobile belt, the northeast German basin has a triangular shape open to the west. This configuration is commonly used when regional thickness and facies distributions of individual Paleozoic units are discussed (Ziegler, 1978; Plein, 1979). However, internal basement regional structure indicates basement highs and lows trending north, northeast, and northwest (Figure 8). Available regional stratigraphic maps clearly show the synsedimentary nature of these features (Pratsch, 1979) since at least late Paleozoic time, and possibly since early Paleozoic time. Basement horsts and grabens were active throughout observable geologic time, affecting sedimentation, regional structure, and distribution of hydrocarbons. Age and type of Paleozoic and early Mesozoic rocks in the basin are shown on Figure 9, a regional cross section is shown on Figure 10, and the distribution of Pre-Permian geologic units is shown on Figure 11. Clearly, the influence of basement features on sediment distribution is evident. Pre-Permian erosion and subsequent continued deposition show a major break in the region's geologic history and the impact of basement block motions.

Paleogeographic and present tectonic trend directions in the basin (Figure 12) are especially interesting here: The basement horsts and grabens trend north to north-northeast in the southern basin (north of about N53); northeast in the northern portion (north of about N54); northwest in the east (east of E10); and northwest and northeast south of about N53.

These trends are part of a common tectonic deformation system when, on a regional scale, wedge tectonic principles are utilized: North, the basin is bordered by the early Paleozoic Caledonian continental margin as part of the Russian-Fennoscandian shield (during continental margin development compression, normal to the margin, resulted in north-northeast trending tensional features along which elongate basement blocks developed). South, compression related to the middle to late Paleozoic Variscan mobile belt, itself trending east to northeast, led to the development of north-northwest to north trending tensional basement structure.

The two tensional systems meet about in the middle of the basin, which explains the change in regional strike long the river Elbe ("Elbe lineament, lineament A" in the literature). At the same time, northwest trending right-lateral shear and northeast trending left-lateral shear develop. Northwest trending right-lateral shear system is more dominant and more active because it is parallel with the western continental margin trend of the Russian shield cratonic block of truly continental dimension. There is evidence for both shear directions and shear systems in surface and subsurface geologic data in northwest Germany, as well as for north to south trending tensional features. Shear system interactions, belonging to the northern and southern semi-regional compressional systems (Caledonide in the north; Variscan in the south), explain the observable variation in shear trends. The influence of basement shear on the present location and geometry of salt structures in the basin, originating from Permian, Triassic, and Jurassic evaporite formations (Pratsch, 1979; Figure 4) is especially clear. The well-known principle of inversion tectonics (Voigt, 1963), where previous sedimentary troughs become compressed, folded, and inverted, finds a logical explanation in shear motions of specific basement faults.

Wedge tectonic processes of the dimensions of the

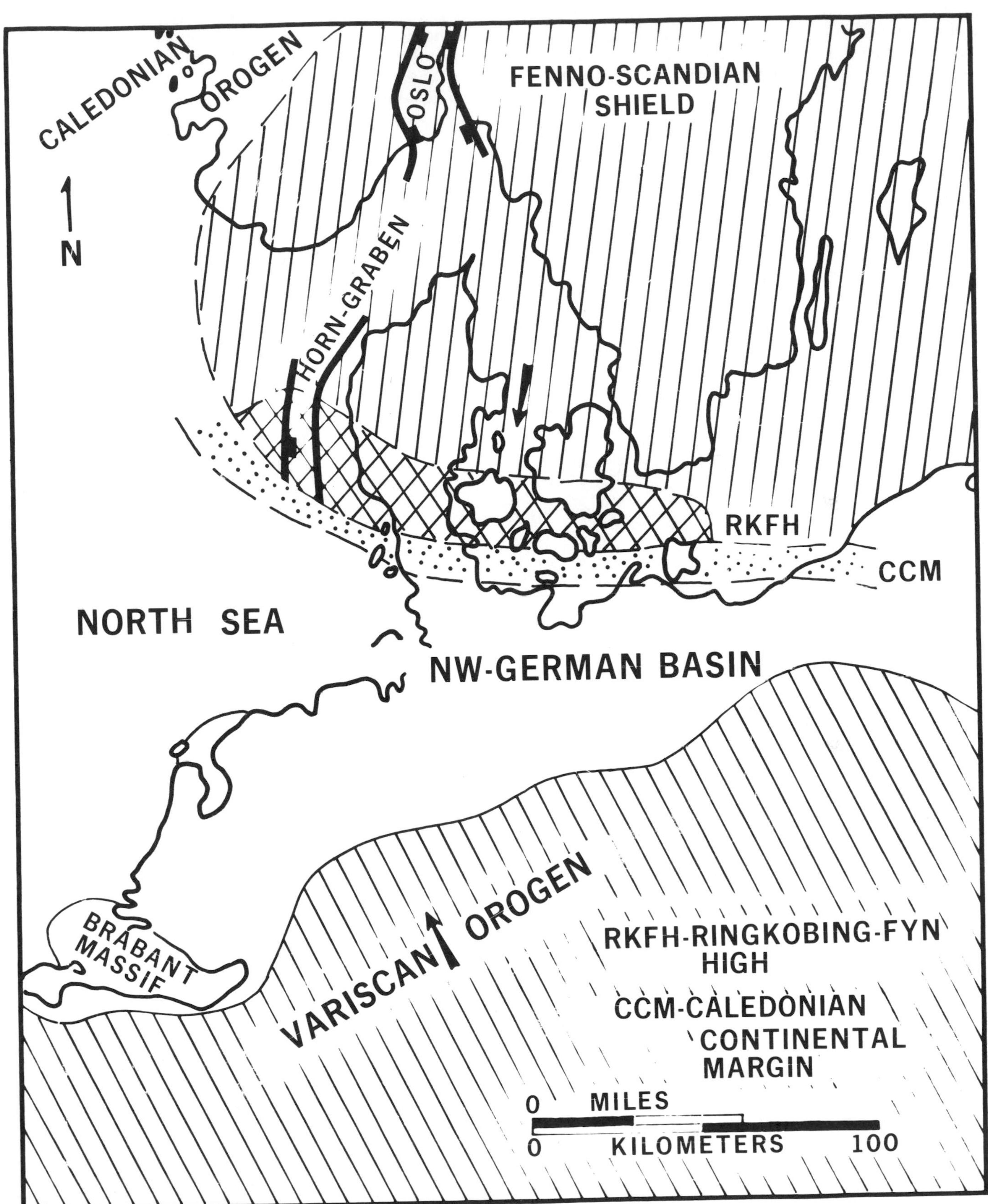

Figure 7 — The northwest German basin location map.

northwest German basin cannot stand isolated. They must be connected with and part of larger features, as well as related to local phenomena. For the northwest European basin as a whole, Ziegler (1975) shows continental dimensions of deformation and deformation history (p. 165): "The earth's crust under most of Europe and the North Sea appears to be formed by a complex checkerboard of cratonic fragments which lies along the southwest margin of the Fenno-Scandian Russia-Asia craton and extends eastward to Iran. The Tornquist Line and its various *en echelon* successors mark the boundary between this 'Fragmented Europe' and the more coherent craton to the northeast."

The complicated regional deformation picture of the northwest German basin thus is resolved when wedge tectonic principles are applied. The result is a complex

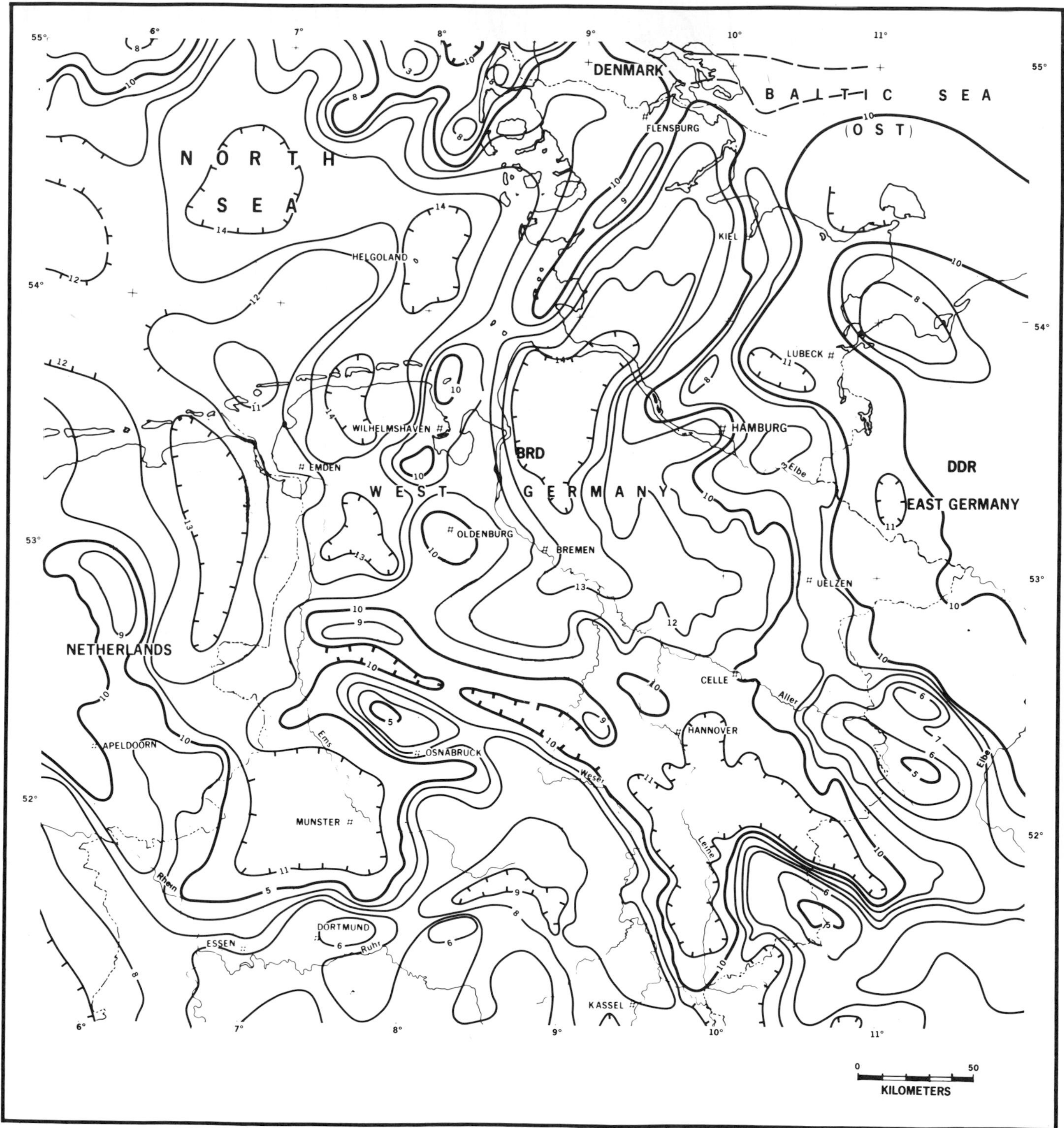

Figure 8 — Regional basement structure of northwest Germany. Taken from gravity-magnetic data.

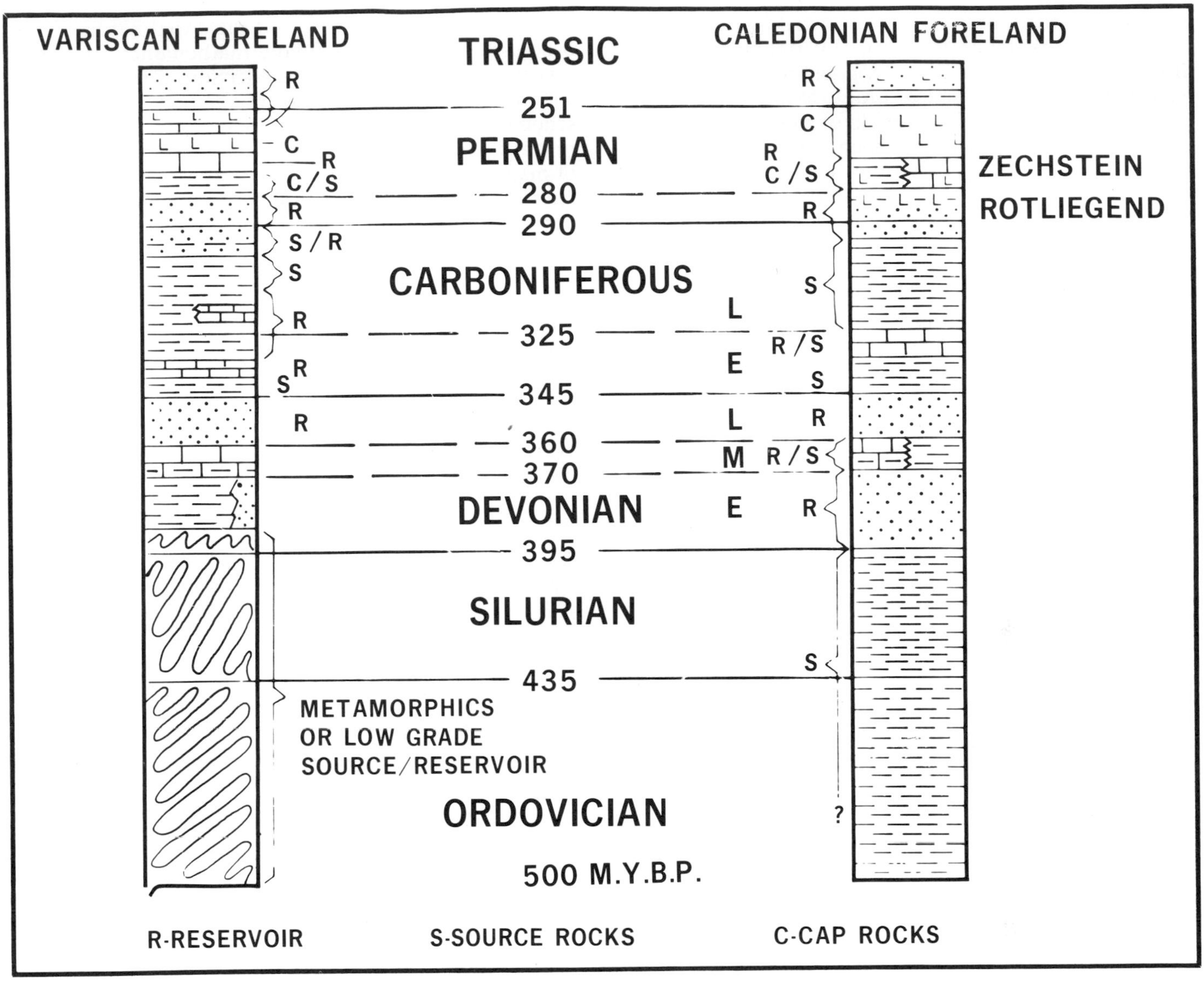

VARISCAN FORELAND
TRIASSIC
CALEDONIAN FORELAND
R
251
C
PERMIAN
R
C/S
ZECHSTEIN
280
ROTLIEGEND
290
S/R
S
CARBONIFEROUS
L
R
325
R/S
E
R
S
345
L
R
360
M R/S
370
DEVONIAN
E R
395
SILURIAN
435
METAMORPHICS
OR LOW GRADE
SOURCE/RESERVOIR
ORDOVICIAN
?
500 M.Y.B.P.
R-RESERVOIR
S-SOURCE ROCKS
C-CAP ROCKS

Figure 9 — Prospective section of northwest Germany.

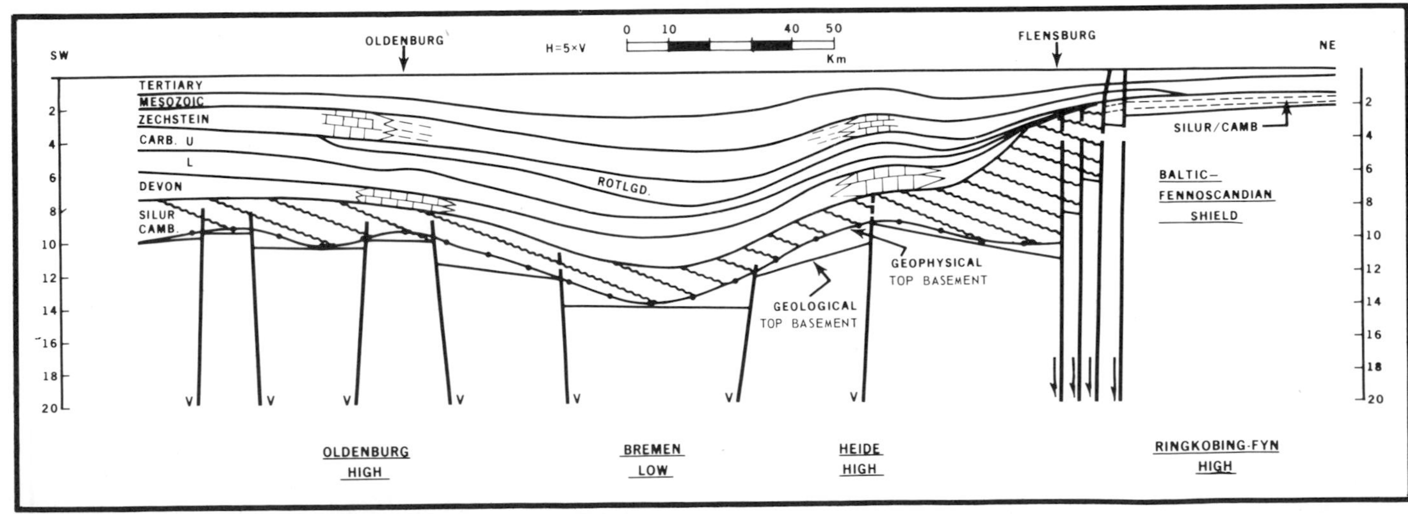

Figure 10 — Regional cross section of northwest Germany.

of tectonic directions that appear to follow established theoretical laws, given the normal variation in trend directions. Of interest is the long-lasting deformation history since early Paleozoic times, following even older trends. The motions inside of this complex deformation system must have been continuous, if one considers the system as a whole, while motions on individual faults may have been discontinuous.

Commonly, a certain variation in tectonic trend directions in a specific area indicates that natural processes do not rigidly follow mathematical or geometric rules. This may cause rejection of the wedge tectonic concept or may lead to complicated tectonic schemes including block rotations, tectonic phases, reversal of motions along the same fault, or alterations of main stress directions. For example, there is a difference of about 20° between north-northwest trending Variscan cross faults and north trending Tertiary

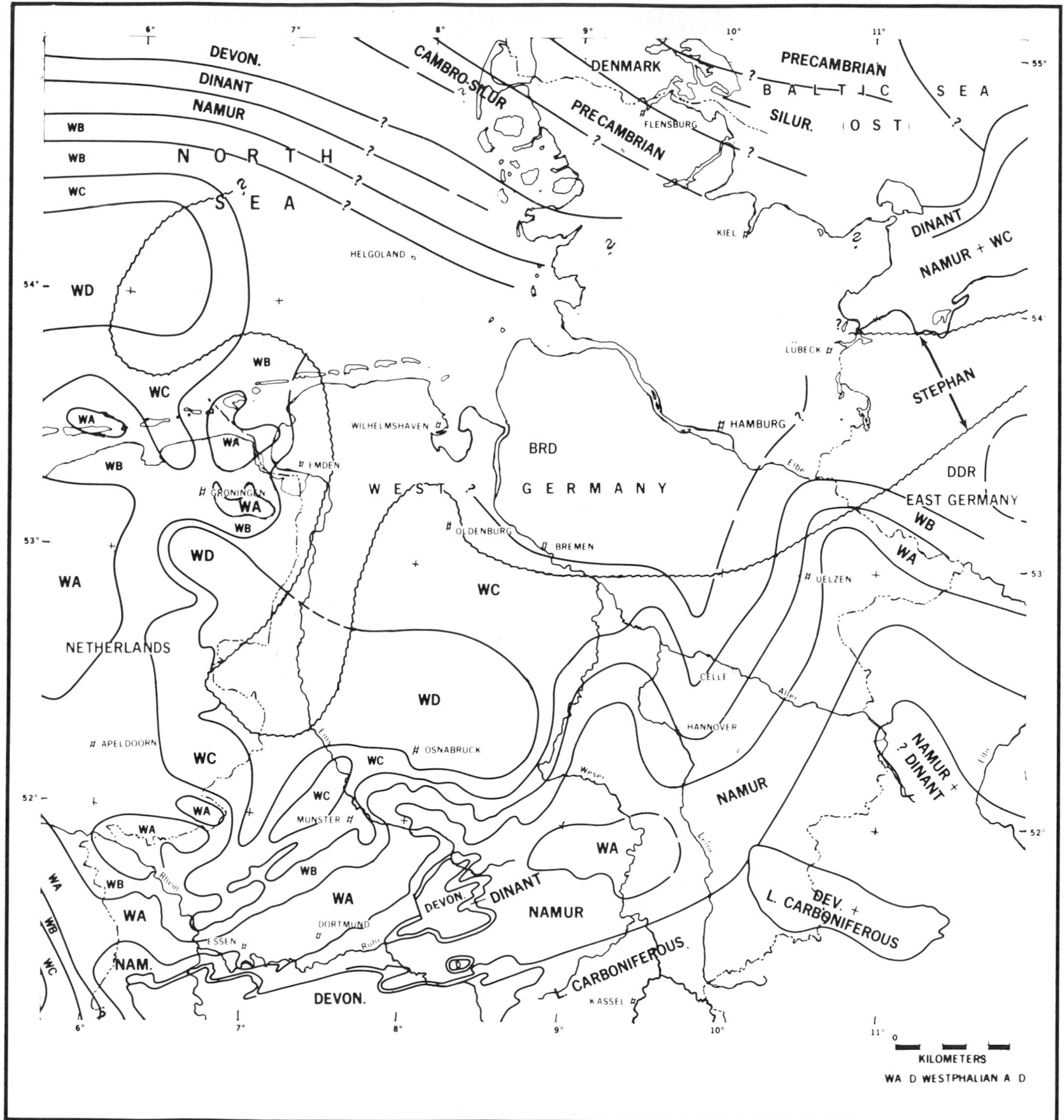

Figure 11 — Pre-Permian geology of northwest Germany.

tensional faults in southern Hannover/Westphalia, Germany. There is, however, also a geographic exclusion in that Tertiary and Variscan faults do not occur in exactly the same area. Block rotations or stress differences between late Paleozoic and Tertiary times cannot be ruled out, nor can Tertiary normal faults developed over northerly pre-existing zones of weakness. Averaging tectonic directions is a major concern in any regional-tectonic systhesis.

It appears possible to expand these findings in the northwest German basin east to the Russian craton and west to the British Islands, indicating the truly

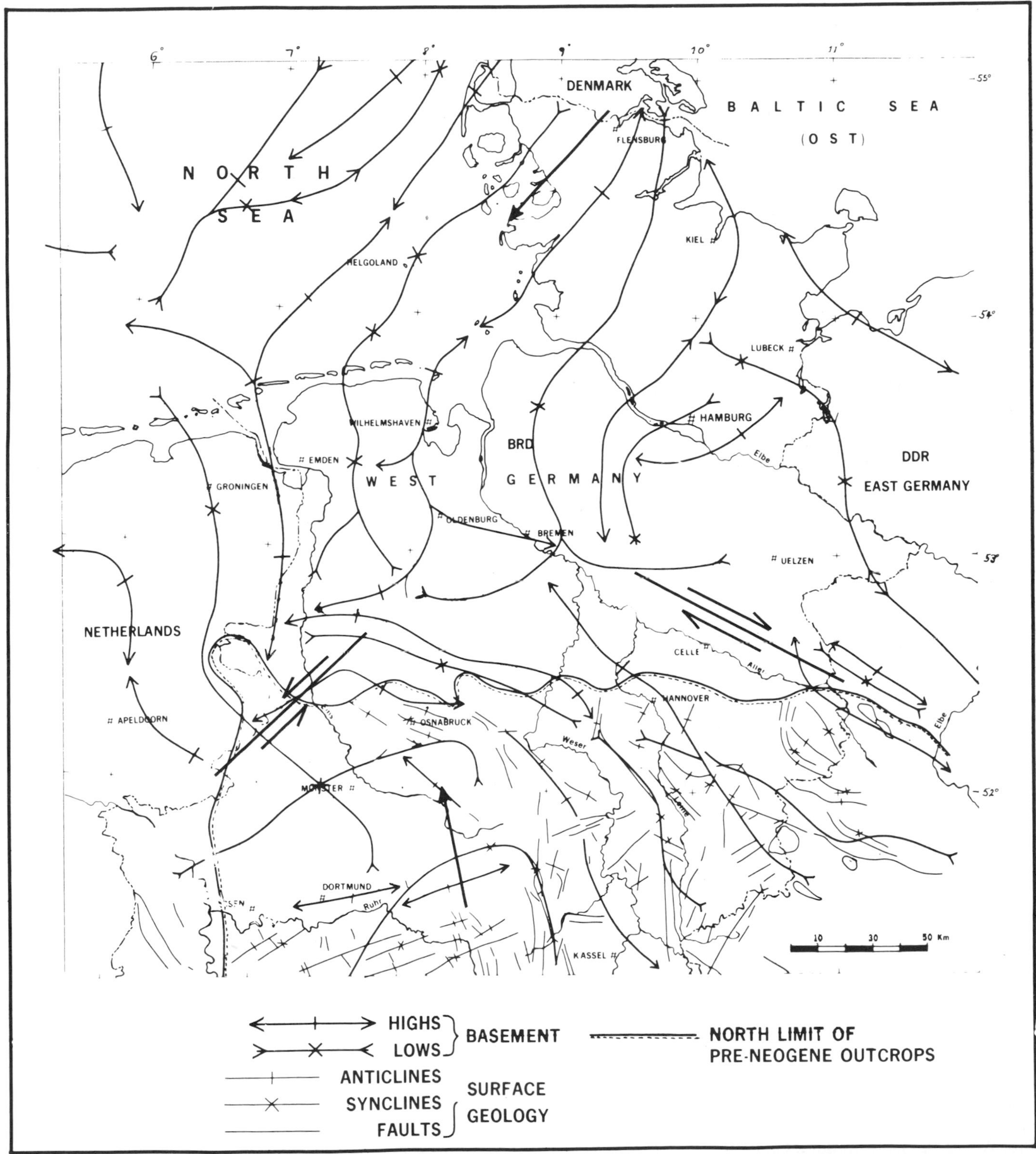

Figure 12 — Regional structural trends of northwest Germany.

regional character of the wedge-tectonic geologic-historical processes here as indicated in the adjacent Netherlands (Van Wijhe, Lutz, Kaasschieter, 1980), in northwest-France (Gapais and Le Corre, 1980), and in northern Europe (Bergerat and Geyssant, 1980).

The initial goal of the study-explanation of type, timing, and geometry of tectonic deformations is reached with application of wedge tectonic concepts. The second goal, explanation of presently known gas accumulations in the basin, is a by-product of this analysis: Synsedimentary basement horsts act as preferred sites for reservoir development and as focal points of regional gas migration from adjoining synsedimentary lows (Pratsch, 1980). The final goal, prediction of future exploration targets, is a logical step from that point.

CONCLUSIONS

In the concept of wedge tectonics, the interrelations of the three actual or potential stress components (compression, tension, and shear) are utilized. They leave testimony in the form of rock deformation of variable intensity but of predictable style and geometry. The interrelationship of stress components exists in space and in time, in any dimension, and in any orientation.

The wedge tectonic concept also has predictive power where one or two of the stress components are not directly observable. Predictions will include type of deformation, geometry of resultant deformation, and its timing. Although such predictions will be mainly qualitative, they can be important in applied geology, such as oil or gas exploration and production, mining, water movements, seismicity, and construction. Quantitative tectonic predictions, like site of specific deformation and intensity of deformation (for example, distance or length of fractures), will be possible only in especially favorable cases.

ACKNOWLEDGMENTS

We appreciate the permission of Mobil Oil Management to publish this paper as a contribution to the January, 1981, AAPG Research Conference on Continental Margins. My special thanks to Dr. J. Halsey, Mobil Oil Research and Development Company, Dallas, Texas, who introduced me to wedge tectonics and who clarified many facets of the concept through his immense knowledge in this field.

REFERENCES CITED

Bergerat, F., and J. Geyssant, 1980, La fracturation tertiaire de l'Europe du Nord: resultat de la collision Afrique — Europe: Paris, Compte Rendus Academie Science, D, v. 290, p. 1521-1524.

Brunn, J. H., and P. F. Burollet, 1979, Island Arcs and the origin of folded ranges: Geologie en Mijnbouw, v. 58, p. 117-126.

Caire, A., 1979, Géotectonique Giratoîre: Geologie en Mijnbouw, v. 58, p. 241-252.

Dubey, A. K., 1980, Model experiments showing simultaneous development of folds and transcurrent faults: Tectonophysics, v. 65, p. 69-84.

Fitch, T. J., 1972, Plate convergence, transcurrent faults and internal deformation adjacent to southeastern Asia and the western Pacific: Journal of Geophysical Research, v. 77, p. 4432-4460.

Gapais, D., and C. Le Corre, 1980, Is the Hercynian belt of Brittany a major shear zone?: Nature, v. 288, p. 574-576.

Jacob, K. H. B., 1980, Oblique subduction and rifting along ocean-continent transform boundaries: Eos, Transactions, American Geophysical Union, v. 61, p. 358.

Kupfer, D. H., 1968, A proposed deformation diagram for the analysis of fractures and folds in orogenic belts: Proceedings, 13th International Geological Congress, p. 219-232.

Molnar, P., and P. Tapponnier, 1975, Cenozoic tectonism of Asia: effect of continental collision: Science, v. 189, p. 419-426.

Moody, J. D., 1973, Petroleum exploration aspects of wrench-fault tectonics: AAPG Bulletin, v. 57, p. 449-476.

———, and M. J. Hill, 1956, Wrench-fault tectonics: Geological Society of America Bulletin, v. 67, p. 1207-1246.

Plein, E., 1979, Das deutsche erdöl und erdgas: Jahreshefte Gesellschaft fuer Naturkunde Württemberg, v. 134, p. 5-33.

Pratsch, J. C., 1979, Regional structural elements in northwest Germany: Journal of Petroleum Geology, v. 2, p. 159-180.

———, 1980, Basement deformation and basement structure in the northwest German basin: Geologische Rundschau, v. 69, p. 609-621.

———, in press, Regional structural elements and major gas accumulations in northwest Europe: Hannover, Seminar on Exploration for Gas, European Economic Community.

Seagall, P., and D. D. Pollard, 1980, Mechanics of discontinuous faults: Journal of Geophysical Research, v. 85, p. 4337-4350.

Van Wijhe, D. H., M. Lutz, and J. P. H. Kaasschieter, 1980, The Rotliegend in the Netherlands and its gas accumulations: Geologie en Minjbouw, v. 59, p. 3-24.

Voigt, E., 1963, Über Randtröge vor Schollenrändern and ihre Bedeutung im Gebiet der mitteleuropäischen Senke und angrenzender Gebiete: Zeltschrift für Deutsche Geologische, v. 114, p. 378-422.

Wilcox, R. E., T. P. Harding, and D. R. Seely, 1973, Basic wrench tectonics: AAPG Bulletin, v. 57, p. 74-96.

Ziegler, P. A., 1978, Northwestern Europe: tectonics and basin development: Geologie en Mijnbouw, v. 57, p. 627-654.

Ziegler, W. H., 1975, Outline of the geological history of the North Sea, *in* A. W. Woodland, ed., Petroleum and the continental shelf of northwest Europe: New York, J. Wiley and Sons.

Seismic Stratigraphy of the Georges Bank Basin Complex, Offshore New England

J. S. Schlee
U.S. Geological Survey
Woods Hole, Massachusetts

J. Fritsch
Bundesanstalt für Geowissenschaften und Rohstoffe
Hannover, Federal Republic of Germany

A regional synthesis of 3,350 km of multichannel seismic-reflection profiles over the Georges Bank area reveals that subsidence concentrated in several areally restricted rift basins of Early Jurassic age and older. Beneath the northwestern half of Georges Bank area, these small sub-basins form narrow grabens within a structurally shallow platform (4 km deep). Beneath the southern half of Georges Bank and Nantucket Shoal area are three deep sub-basins containing more than 10 km of sediments. The sedimentary fill is divided into six seismic units that reveal a change from an older, thick sequence of rift-filling (synrift) continental and evaporitic deposits (0 to 8 km thick) of Late Triassic to Early Jurassic age, to younger, more widespread open-marine shelf sequences of carbonate and clastic sedimentary rocks of Jurassic and younger ages.

Between 1973 and 1979 the U.S. Geological Survey (USGS) and the Bundesanstalt für Geowissenschaften und Rohstoffe (BGR) of the Federal Republic of Germany collected approximately 3,350 km of multichannel seismic reflection profiles over the Georges Bank area (Figure 1) in a continuing effort to help assess the resource potential of the area. In 1976 and 1977, two Continental Offshore Stratigraphic Test (COST) wells were drilled into the basin. These data, plus those from shallow holes, submersible dives, and dredge hauls, provide a fairly complete picture of the geologic history of Georges Bank area. Our purpose is to describe the complex of basins under Georges Bank — their shape, sedimentary fill, and history — and to compare them with nearby basins.

All USGS profiles were collected by Digicon Inc., Teledyne Exploration, and Geophysical Services, Inc., between 1973 and 1978. Each company used a slightly different airgun configuration. Arrays of 4 to 23 guns (total volume of 1,200 to 2,160 cu in) were operated at pressures of 1800 to 2000 psi. Hydrophone streamers were 2.4 km long (1973 data) and 3.6 km long (1974, 1975, 1977, and 1978). Incoming signals were recorded on a Texas Instruments DFS 3, 4, or 5 system and the shotpoint interval was 50 m. Tapes were later processed to include a common-depth-point gather, velocity analysis, normal moveout corrections, vertical summations of 2 on 1 horizontal stacks, and time-variant filtering. Twelve-fold processing was done on the shelf part of line 1, and 24-fold processing was done on the outer shelf-slope part of the same line. Other lines were processed at either 36 or 48 fold (see Grow, Mattick, and Schlee, 1979). Digicon Inc. processed line 1; Geophysical Services Inc. processed lines 5, 18, 19, 20, 21, 22, and 33; and Teledyne Exploration processed line 16. Lines 7, 8, and 12 were processed by the USGS seismic group on the Phoenix I system in Denver.

In 1979, BGR ran a 4,700 km grid of multichannel seismic reflection profiles across the seaward part of Georges Bank and the outer shelf, slope and rise off New Jersey and Delaware using the Prakla-Seismos vessel *Explora.* Seismic acquisition system consisted of: 1) two arrays of Bolt airguns, total of 24 guns (total capacity 23.45 liters; 1,430 cu in) operated at 2000 psi and towed 15 m below sea level; 2) a 2,400 m streamer with a 50-m group interval; 3) a Texas Instruments DFS-5 recording system (recording length 10 sec with 4 ms sampling rate); and, 4) an integrated navigation and data system consisting of a satellite receiver, doppler sonar, and Loran C. Tapes were processed in Hannover, West Germany by Prakla-Seismos to yield 24-fold time sections. Velocity scans were computed every 3.6 km.

GEOLOGIC SETTING

The complex of sub-basins known as the Georges Bank basin is situated between the Long Island platform to the southwest and the La Have platform to the northeast (Figure 1). The general outline of the basin is given by Drake, Ewing, and Sutton (1959), Maher (1971), Emery and Uchupi (1972), Schultz and Grover (1974), Ballard and Uchupi (1975), Schlee (1978), Austin et al (1980), and Schlee and Jansa (1981). Earlier studies showed an elongate trough containing 6 to 10 km of sediment under Georges Bank. Schultz and Grover (1974) illustrated some of the structure complexities along the northeast side of the trough; Mattick et al (1974) indicated that the central thickest part of the basin was built across block-faulted crystalline rocks which shallowed to the southeast and northwest. Largely on the basis of proprietary oil company multichannel seismic reflection profiles, Ballard and Uchupi (1975) described an elongate trough containing more than 4 seconds of sediment, underlying the southern edge of the bank and built over a series of northeast trending Triassic and Jurassic rift basins. Uchupi, Ballard, and Ellis (1977) inferred that a basement ridge borders the southside of the Georges Bank, and separates the shelf sedimentary wedge beneath Georges Bank basin from the eugeoclinal wedge underlying the continental rise. Austin et al (1980) used the oil company profiles (Ballard and Uchupi, 1975), six-channel profiles collected by Woods Hole Oceanographic Institution, and three USGS multichannel lines (Schlee et al, 1976) to outline the structurally complex Georges Bank basin whose geometry is related to subsidence controlled by an intricate pattern of deep-seated faults. Their data showed over 13 km of sediments southeast of the Yarmouth Arch and 11 km in narrow basin under the central part of the bank.

The general outline of the basin (Figure 2) shows that it is really one main basin and several sub-basins. Two sub-basins (Nantucket and Atlantis) are in the transition zone with the Long Island platform, and two are associated with the Yarmouth Sag adjacent to the La Have platform. The sub-basins appear to be oriented northeast, subparallel to the tectonic trend of the Appalachians (Williams, 1978) and the Triassic age rift basins in the Gulf of Maine (Ballard and Uchupi, 1975). The south central part of the basin is also built

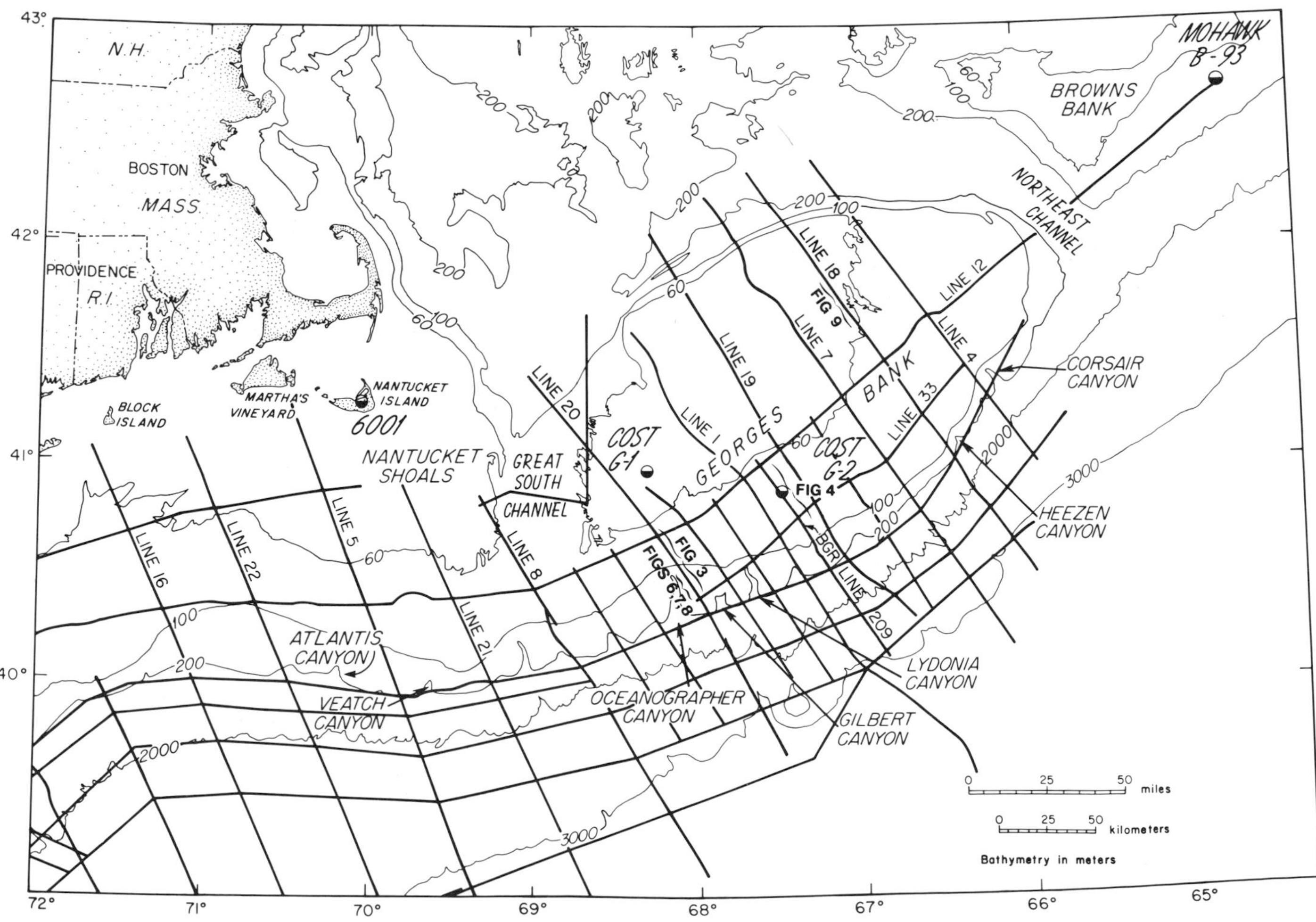

Figure 1 — Georges Bank basin area showing locations of Continental Offshore Stratigraphic Test (COST) wells, the Nantucket hole (USGS 6001), the Shell Mohawk B-93 well on Browns Bank, and multichannel seismic-reflection profiles used in this study.

over a deeper rift basin. Beneath the outer edge of the shelf acoustic basement is not clearly defined, so ideas about the transition to oceanic crust and the arrangement of deeply buried deposits are largely conjecture. The overall aspect of the basin is a broadly subsided feature underlain by an intricately rifted basement.

SEISMIC STRATIGRAPHY

Objectives and Techniques

Analysis of multichannel seismic-reflection profiles had three purposes: 1) delineation of chronostratigraphic units on the profiles; 2) inference of the paleoenvironment under which the units were deposited; and, 3) comparison of seismic stratigraphy with the stratigraphy in the COST wells and with other stratigraphic data. Examination of two profiles (Line 20 and BGR Line 209; Figure 1) gives the lateral and vertical change in the character of seismic reflections, and the inferences of sedimentary rock associations. Figure 3 spans 40 km of Line 20, transecting the central and seaward part of the main Georges Bank basin. Figure 4 spans 35 km of BGR Line 209 over the site of the COST G-2 well (Figure 1). Towards the bottom of both profiles, a group of high continuity, high amplitude reflections characterize the base of the sedimentary section. They show up on the Line 20 profile between shotpoints 3,200 (off the left side of Figure 3) to 2,400, and vertically, between acoustic basement and 3 seconds (two-way travel time). The reflections probably represent the sequence of interbedded carbonate rocks and evaporites as drilled in the COST G-2 well (Figure 4). Strata represented by these characteristic reflections occupy the central part of the basin. To the southeast (right side of Figure 3), the reflections at equivalent depth change character and become discontinuous and distinctly clinoform. This change of acoustic character is interpreted to represent a progradation of the platform-reef complex to the southeast during Jurassic. The clinoform reflectors climb in the section as part of a shelf-edge complex that built seaward over earlier formed slope deposits. On Figure 4, acoustic basement is not apparent and continuous reflections occupy the section between 2.25 and 4 seconds. They continue horizontally beyond shotpoint 800 to the southeast, where they fade out in a conspicuous structureless mass buried beneath the continental slope; the mass is thought to be a car-

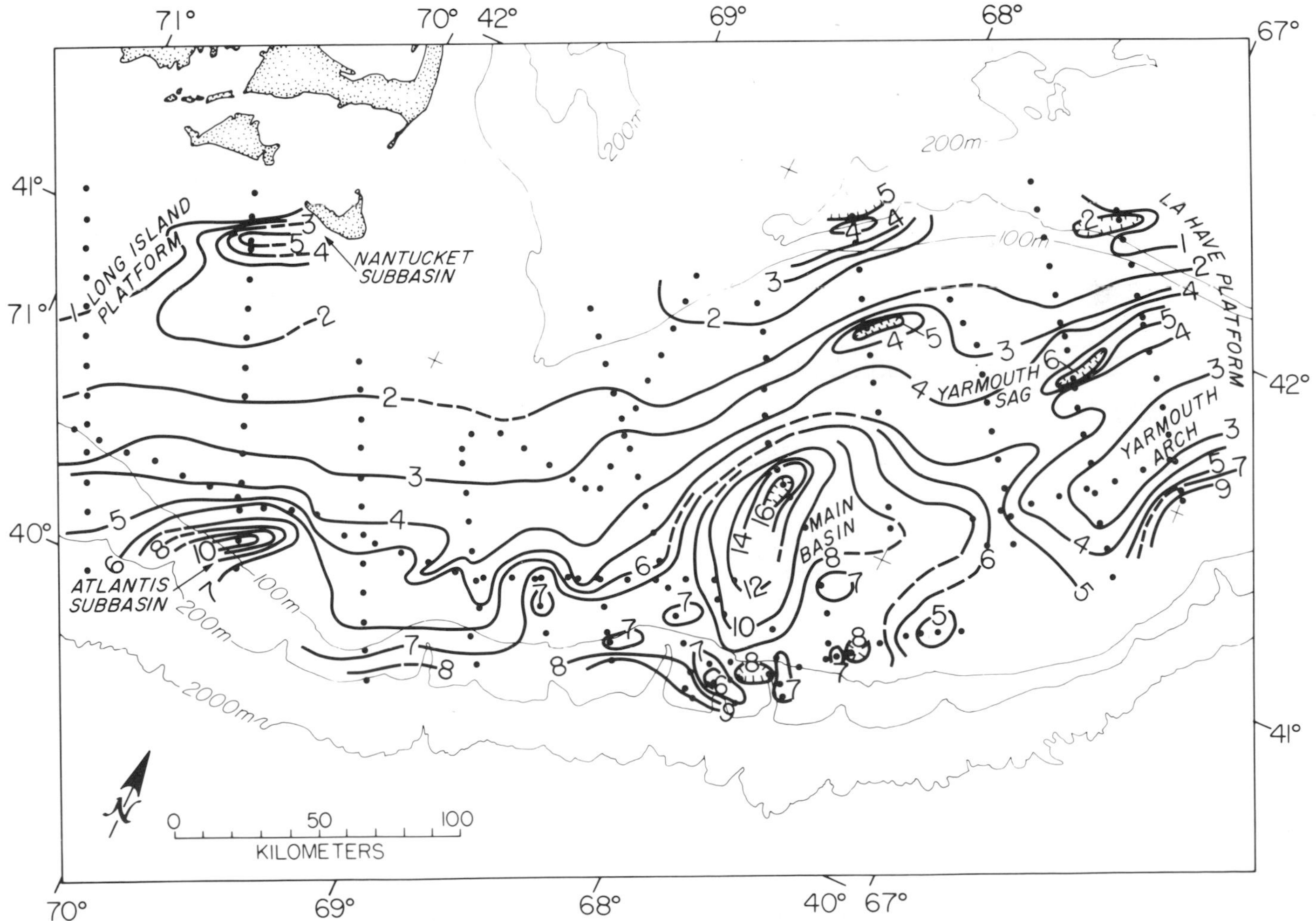

Figure 2 — Isopach map of the Georges Bank area showing sedimentary rock thickness, inferred to be Late Triassic (?) and younger (in kilometers). Dots indicate control for the isopachs.

bonate platform (Schlee, Dillon, and Grow, 1979). The broad warps at the profile's base in Figure 4 are inferred to be caused by reef-like carbonate buildups that probably have foundations on less deeply buried basement blocks.

To the northwest on Line 20, basal reflections within the carbonate-evaporite section onlap a shoaling acoustic basement; inshore of Figure 3, upper reflections in the carbonate-evaporite sequence to become less continuous and lower amplitude. The change probably represents a facies transition to a mixed marine carbonate and nonmarine section. The COST G-1 well (Amato and Bebout, 1980) 60 km to the northwest encountered a sequence of interbedded dolomite, limestone, red-brown shale, and sandstone in the probable equivalent age section (Figure 5).

Above the inferred carbonate-evaporite section is a group of reflections with strongly variable character. These reflections appear to represent interfingering of nonmarine and marine shelf sediments during the Late Jurassic and Early Cretaceous (Figures 3, 4, and 7). This group of reflections is approximately 1 to 1.5 seconds thick (within interval 1.3 to 2.8 seconds on Figure 3, and 1.3 to 2.4 seconds on Figure 4) and shows a lateral change from discontinuous variable amplitude reflections alternating with fairly continuous reflections in the northwest, to moderately continuous reflections to the southeast (Figure 7). As shown by Figures 4B and 7B, the zones of discontinuous reflections are inferred to be tongues of nonmarine deposits that built out into the basin. The tongues of discontinuous reflections appear to extend farther to the southeast in the younger part of the interval, where they change laterally in several kilometers to more continuous reflections. Still farther southeast, the interval changes as reflections lose their continuity and amplitude beneath the outer continental shelf in the vicinity of the carbonate bank-reef complex. To the northwest, reflections within the interval become more discontinuous and variable in amplitude and probably are indicative of a sequence of alluvial-coastal deposits.

Upper Jurassic and Lower Cretaceous strata drilled in the COST G-1 are red-brown shale interbedded

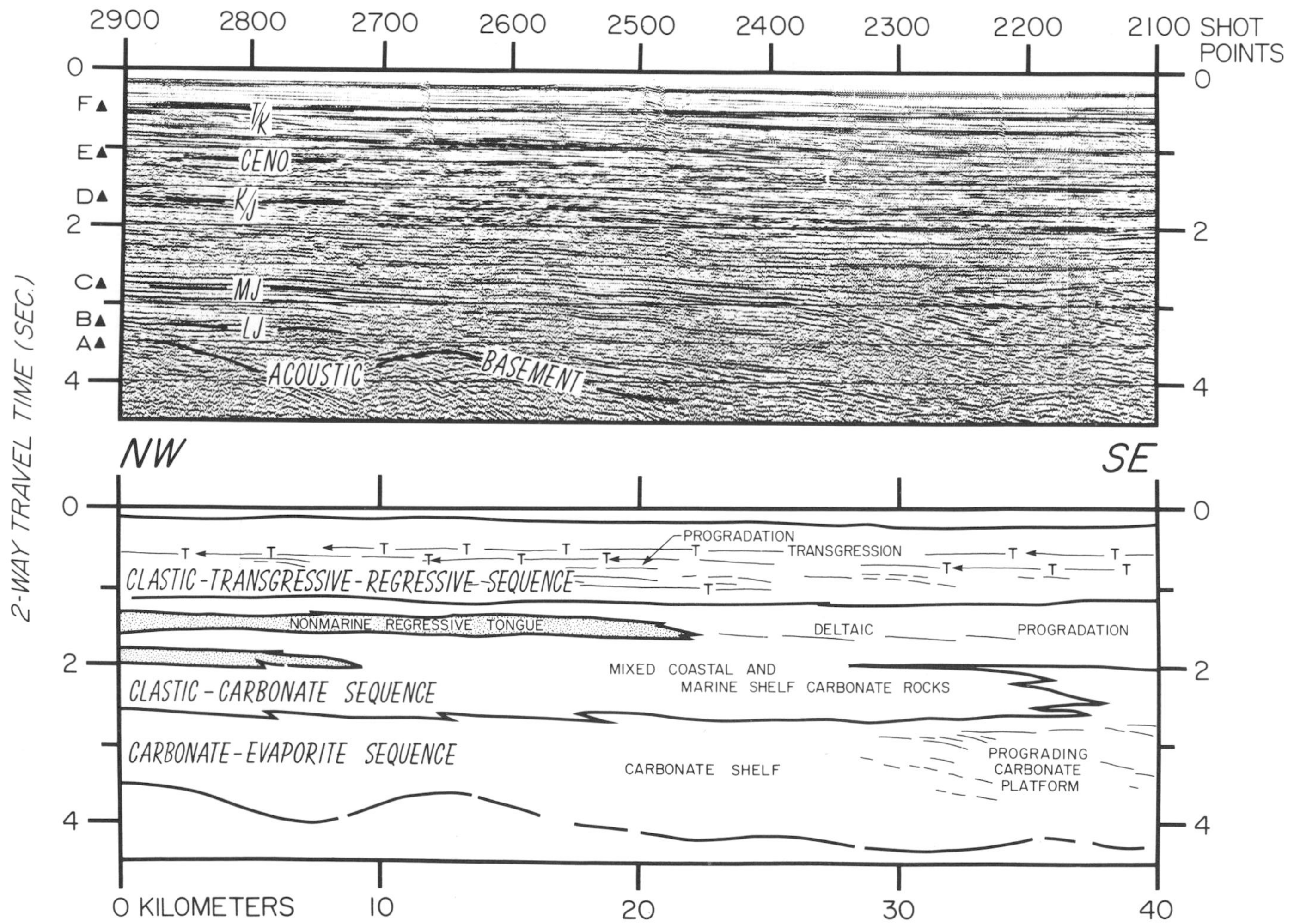

Figure 3 — (A) Multichannel seismic reflection profile 20, between shotpoints 2100 and 2900. Locations shown in Figure 1. T/K, Tertiary-Cretaceous boundary; CENO., Cenomanian; K/J, Cretaceous-Jurassic boundary; MJ, Middle Jurassic; LJ, Lower Jurassic. Letters at the left edge designate seismic sequences discussed in the next section. **(B)** Line drawing showing interpretation of major constructional phases along the same line. T-T marks a thin transgressive unit.

with thick beds of fine-to-medium grained sandstone (Figure 5), whereas strata of the same age in cost G-2 are interbedded red-gray, very fine-to-medium grained sandstone, shale, and limestone (Amato and Simonis 1980; Amato and Bebout, 1980); further, the interval is 400 m thicker in the G-2 well than in the G-1.

On the Scotian margin, the stratigraphic equivalent of this seaward part of the middle interval is probably the shelf-carbonate platform facies of the Abenaki Formation (Eliuk, 1978; Given, 1977). For the inshore part of the interval (a mixed carbonate coastal facies) possible equivalent units are the Mic-Mac and Mohawk Formations.

In the uppermost interval (less than 1.2 seconds, two-way travel time; Figures 3, 4, and 8), zones of faintly continuous reflections alternate with zones (1 to 2 wavelets thick) of reflections showing high continuity and moderate amplitude. In some of the faintly continuous zones, reflections can show a broad progradational pattern of reflections dipping gently toward the seaward part of the line. The broad pattern of outbuilding, as shown by faint reflections, is interlayered with the thin continuous zones of reflections (arrows, Figures 4, 7) — an arrangement we interpret to represent slow regressive progradation of the shelf followed by fairly rapid broad marine transgression (thin continuous reflections would represent sequence of an extensive limestone). Between the progradation and the transgression is an erosional break, as indicated by an inconspicuous unconformity at the top of the progradational sequences; the unconformity is detected by the low-angle truncation of the broadly foreset reflections in a progradation sequence. The evidence for rapid transgressions following relative sea level drops (T's, Figure 8) is most apparent in strata of Late Jurassic and younger age. Outbuilding apparently took place in the latest Jurassic, in the Early Cretaceous, in the latest Cretaceous, and in the Tertiary.

Equivalents to the uppermost interval within COST G-1 and G-2 holes (Late Cretaceous and younger) are interbedded gray calcareous shale and sandstone. The strata become coarser grained, more lignitic, and less calcareous toward the northwestern part of the bank

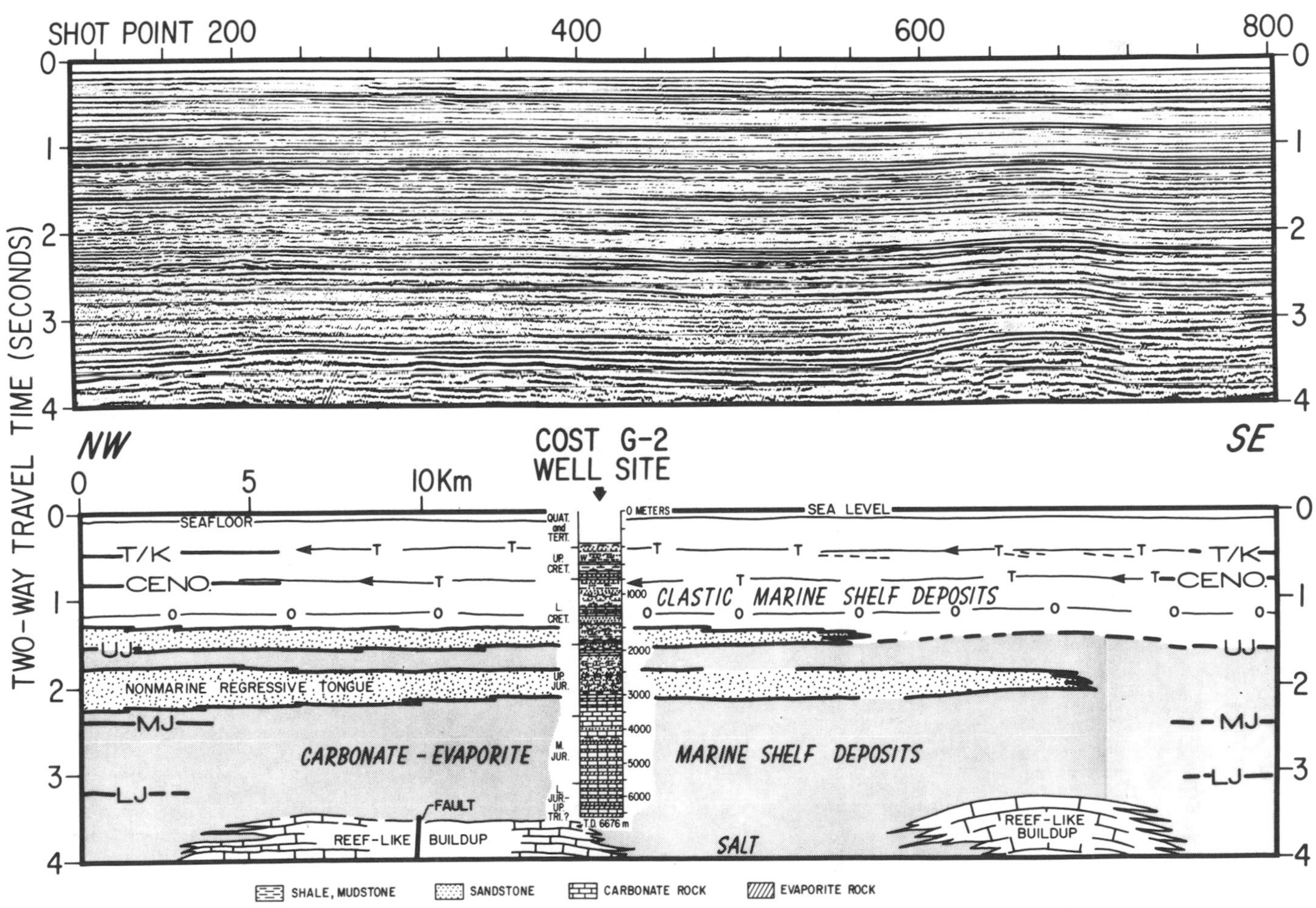

Figure 4 — **(A)** A part of BGR line 209 across the main Georges Bank basin and through the site of the COST No. G-2 well. Location of section shown on Figure 1. **(B)** Line drawing showing interpretation of profile — major sedimentary associations, COST G-2 site lithology, and inferred ages of sequences. T/K, Tertiary-Cretaceous boundary; CENO., Cenomanian; UJ, near the top of the Upper Jurassic section; MJ, within the Middle Jurassic section; LJ, near the base of the Lower Jurassic section; T-T, a thin transgressive unit. O- O, is the correlation with the "O marker," a zone of limestone within the Missisauga Formation beneath the Scotian margin (Jansa and Wade, 1975).

(G-1). The G-2 well has thin limestone, dolomite, and chalk beds interstratified with the shale and sandstone.

Sequence Delineation

The pattern of postrift margin development as shown on Line 20 and BGR Line 209 (Figures 3, 4, 6, 7, and 8) gives a general view of how Georges Bank basin formed during the past 200 million years. Major seismo-stratigraphic units were mapped, integrating all the profiles into a basin-wide analysis. The approach is modified from Vail, Mitchum, and Thompson (1977a,b) and consists of: 1) subdivision of the seismic profiles into intervals that represent depositional sequences ("a stratigraphic unit composed of a relatively conformable succession of genetically related strata and bounded at its top and base by unconformities or their correlative conformities." Mitchum, Vail, and Thompson, 1977, p. 53); 2) analysis of the seismic facies to show their thickness, interval velocity, and inferred environment of deposition; and, 3) correlation of the seismic stratigraphy with the COST well logs and sea floor samples.

Six major seismic units are delineated on Georges Bank in Table 1. The boundaries are basin-wide unconformities at the base of the Tertiary, in the middle Cretaceous (Cenomanian), at the base of the Cretaceous, in the Middle Jurassic, and in the Lower Jurassic. These inferred ages can be compared with the ages of unconformities detected in the COST G-1 and G-2 wells (Scholle, Krivoy, and Hennessy, 1980; Scholle, Schwab, and Krivoy, 1980). Hiatuses occur in the late Kimmeridgian and between the Barremian and Hauterivian, the Aptian and Cenomanian, the

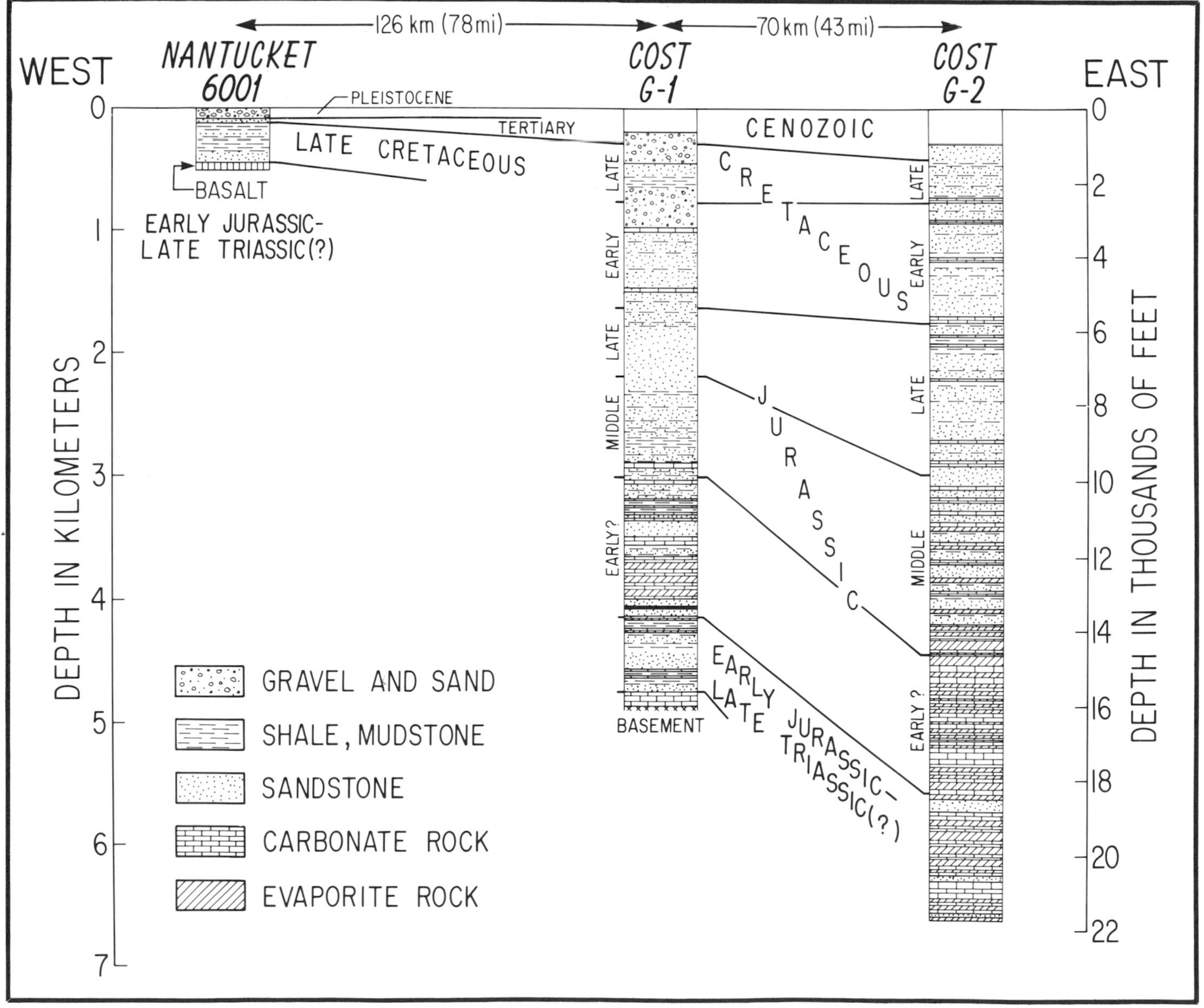

Figure 5 — Lithologic logs of the Nantucket hole (USGS 6001) and Continental Offshore Stratigraphic Test (COST G-1 and G-2 holes. Data taken from Folger, et al (1978), Scholle, Krivoy, and Hennessy (1980), and Scholle, Schwab, and Krivoy (1980). Locations of the holes are shown on Figure 1.

Table 1. Seismic units of the Georges Bank basin.

Unit designation	Age of COST well locations	Lower boundary	Internal properties			Range of averaged interval veolcities	Paleoenvironment interpretation
			Configuration	Continuity	Amplitude		
F	Quaternary and Tertiary.	Conformable over truncated edge of unit E under east-central Georges Bank and shelf south of Cape Cod.	Parallel	High to moderate.	Moderate to high.	1.60-2.25 km/s	Marine shelf with outbuilding along southeast side.
E	Cenomanian to base of Tertiary.	Mainly conformable over southern half of basin; onlap over truncated edge of Unit D.	Parallel	Moderate to low.	Moderate to low.	1.74-2.90 km/s	Mainly a marine shelf; limited shelf-edge progradation; some interbedded nonmarine beds.
D	Lower Cretaceous.	Conformable to the south and onlapping to the north.	Parallel	Moderate to low.	Moderate variable.	1.8-3.45 km/s	Marine shelf with the remnant of a carbonate buildup along the southside and a nonmarine plain to the north.
C	Upper Jurassic.	Mainly conformable.	Parallel	Low to moderate; high under outer shelf.	High to moderate; variable.	2.55-4.71 km/s	A carbonate-rich marine shelf giving way to nonmarine deposits to the north.
B	Middle and Lower Jurassic.	Channeling and onlap over truncated and nontruncated older units.	Parallel and divergent at a low angle.	High to low.	High to moderate.	2.18-6.19 km/s	Restricted and non-restricted prograding marine shelf.
A	Lower-most Jurassic and pre-Jurassic.	Onlap; fault boundary.	Divergent and parallel.	Low to high.	High to low.	2.48-6.40 km/s	Non-marine and restricted marine shelf.

Cenomanian and Turonian, the Turonian and Coniacian, the Santonian and the Eocene, the Eocene and the Miocene-Pliocene, and the Miocene-Pliocene and the Pleistocene (C. W. Poag, personal communication, Sept. 1980).

We examined unconformities in Scotian margin wells (Ascoli, 1976) and in the COST B-3 well (Poag, 1980) in the northern Baltimore Canyon trough to see which hiatuses could be interregional. Ascoli (1976) lists the main hiatuses on the Scotian margin to be in rocks of the Berriasian, Aptian-early Albian, Cenomanian-Turonian, Maestrichtian, and late Eocene age, although not all the time-gaps were represented in the 15 wells Ascoli examined. At the COST B-3 well (Poag, 1980), unconformities were detected between rocks of Hauterivian and Barremian Age, late Cenomanian and early Turonian Age, early Turonian and Coniacian age, early Maestrichtian to early Eocene age, late Eocene and late Oligocene Age, and early Miocene to middle Miocene age. These drilling results indicate that interregional unconformities are present between the Hautervian to Barremian, Cenomanian to Turonian, Turonian to Coniacian, latest Cretaceous to Early Tertiary, and Eocene to Late Oligocene age. For the section of Cretaceous and Cenozoic age, seismic records appear to detect two of these (Cenomanian to Turonian; base of Tertiary); the others (Late Eocene to Late Oligocene, Turonian to Coniacian) are seen clearly on a few of the profiles.

For seismic facies analysis, the interpretive approach of Sangree and Widmier (1977, 1979) is used together with the stratigraphy from the two COST holes to identify the types of seismic facies present (Table 2). Table 2 shows that the arrangement and character of seismic reflections (continuity and amplitude) are most important in delineating different seismic facies. Sangree and Widmier based their environmental inferences on drill hole data keyed to multichannel seismic reflection profiles collected in the same area.

The arrangement of reflectors within depositional sequences can be important in distinguishing different types of environment. Those sequences deposited on an ancient shelf usually show parallel to low-angle — divergent arrangement of reflectors. Sediment se-

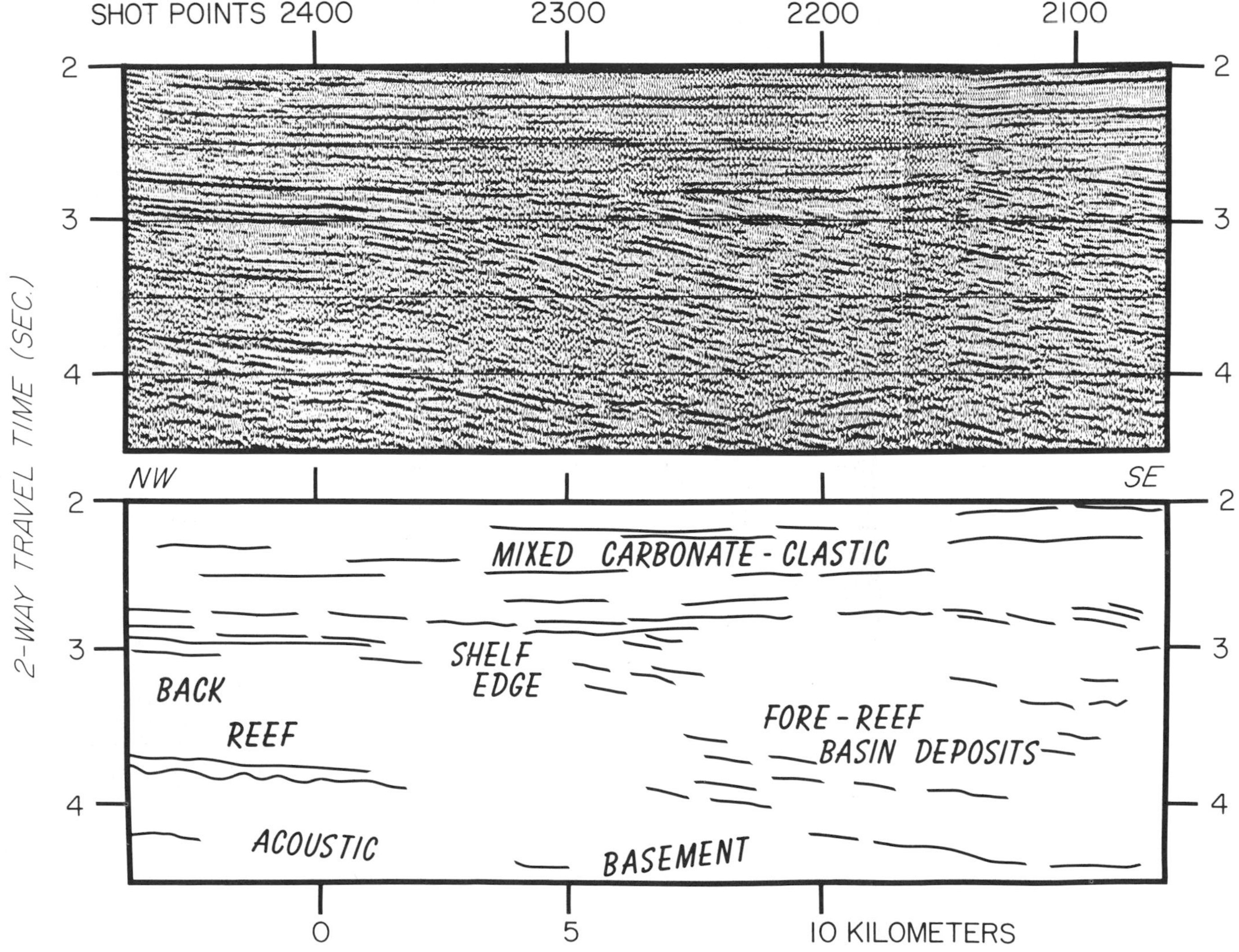

Figure 6 — (A) Enlargement of the bottom part of seismic reflection profile 20, between shotpoints 2080 and 2475. **(B)** Interpretative line drawing of the same part profile 20 identifies major sedimentary association.

Table 2. Seismic facies present in the Georges Bank basin.

Seismic facies unit	Environmental facies interpretation	External form	Interval configuration	Reflection geometry at boundaries
Low continuity and variable amplitude.	Nonmarine — river and associated marginal marine transport processes.	Sheet or wedge	Parallel to divergent	Concordant at the top and concordant to gentle onlap and downlap at base.
High continuity and high amplitude.	Shallow open marine shelf with wave transport; high and low energy conditions; possible fluvial deposits interbedded with swamp and coastal deposits.	Sheet or wedge	Parallel to divergent	Concordant at top and concordant to gentle onlap and downlap at base.
High continuity and high amplitude changing laterally to low continuity and spotty diffraction events.	Shallow restricted to open marine shelf changing laterally to a carbonate bank.	Sheet changing to a bulbous mass.	Parallel to hummocky.	Concordant or onlap at top; concordant to unknown at base.
Low amplitude	Sand: Fluvial and nearshore littoral processes. Shale: Marine shelf sediments deposited under low-energy regime from turbidity currents and by wave transport.	Sheet or wedge	Parallel to divergent	Discordant at top and concordant to gentle downlap or onlap at base.
Moderate continuity and low to moderate amplitude.	Mixed inshore shelf deposits formed in low to modest energy conditions and nonmarine coastal deposiits.	Sheet or wedge	Parallel to divergent	Concordant or low angle truncation at top; concordant or onlap at the base.
Sigmoid — low relief mound.	Shelf delta complex.	Low, broad mound	Sigmoid along depositional dip; onlap and lens shaped parallel to depositional strike.	Concordant at top and downlap and onlap at base.
Oblique — progradational	Shelf-slope boundary; variable energy conditions from high in toplap truncated area to low energy in clinoform and fondoform (prodelta zones).	Fan-shaped	Parallel to upper part; oblique along depositional dip; and parallel or gently oblique to sigmoid parallel to depositional strike.	Toplap truncation to downlap at base.

quences forming shelf-edge and deltaic deposits result in a sigmoid progradational pattern of reflectors. If high energy conditions affected the deltaic progradation and if sediment bypassed the delta, then the reflectors might show an oblique progradational pattern. These deposits might be expected to grade laterally into other shelf or upper slope deposits along the ancient shelf edge.

The high continuity of reflections indicates a contrast in depositional energy conditions operating over a wide area. These conditions are expected on a broad marine shelf where fluctuating high and low energy states result in the deposition of interbedded sandstone, limestone, or shale; agents of active sediment transport would be waves and bottom currents. Sangree and Widmier (1977, 1979) also noted that reflections with high continuity can also be returned from fluvial sands interbedded with widespread coal and marsh clays that were deposited in a swampy coastal plain. Conversely, low-continuity reflections indicate the presence of lithologic units of limited areal extent, such as discontinuous channel sands interbedded with flood plain clays in an alluvial plain. To recognize this type of an environment the seismic amplitude must be coupled with the continuity.

Seismic amplitude provides a qualitative guide to the density contrast of adjacent reflections. Combined with low reflection continuity, variable amplitude implies that nonmarine sequences are present: nonmarine environments commonly contain small channel sands whose densities contrast strongly with those of adjacent clays, resulting in discontinuous high amplitude reflections. Interbedded silts may provide less contrast to the clay, and therefore may yield lower amplitude reflections. Thus, the characteristic pattern for nonmarine sequences tends to be low-continuity, variable-amplitude reflections (Table 2).

Marine shelf sequences (characterized by a parallel arrangement of reflections having high amplitude and high continuity) change laterally toward the southern side of Georges Bank to broad carbonate bank-platform complexes that fringe the seaward side of the basin (Schlee, Dillon and Grow, 1979; Figures 4, 5, and 6). The change is marked by a loss of reflection continuity and amplitude, and by the presence of spotty diffraction events that outline the seaward edge of the carbonate buildup. Commonly, the interval velocities in the bank-platform complex (Figure 11) range from 5 to 6 km/second whereas interval velocities in the marine-shelf sequence behind the complex are 3 to 5 km/second. Also characteristic is the shape of the carbonate mass — a hummocky seaward facing front and

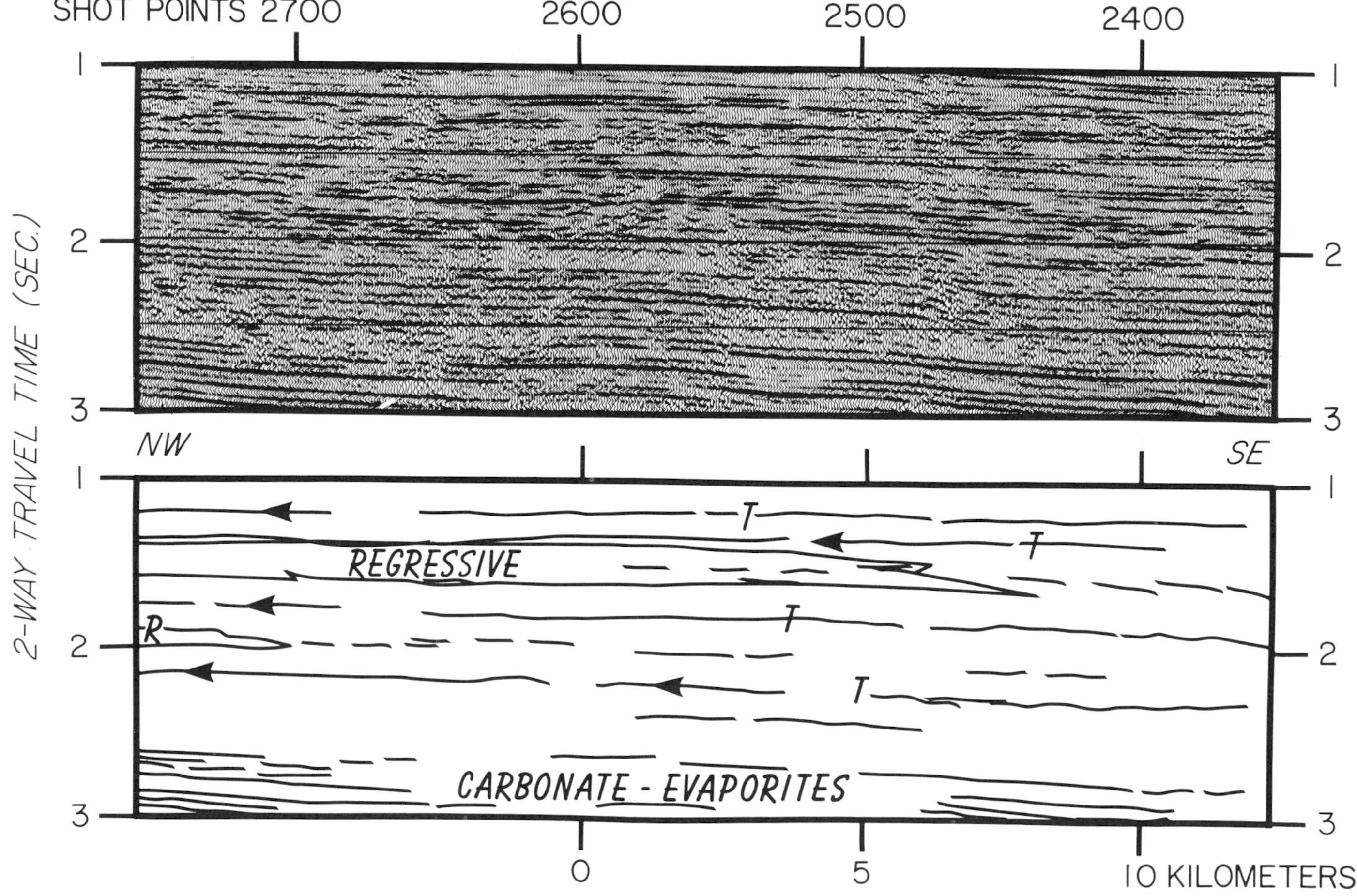

Figure 7 — (A) Enlarged part of seismic reflection profile 20, between shotpoints 2350 and 2760. **(B)** Interpretative line drawing identifies major sedimentary associations and inferred transgressive (T) and regressive (R) phases of these associations.

a flattened to slightly bulging top. Bubb and Hatlelid (1977) give a more detailed discussion of carbonate facies, as seen on seismic reflection profiles.

In other facies, reflection amplitudes may be weak or may not appear; this pattern suggests sedimentary rocks of a uniform lithology, usually sandstone or shale. A thick section of sandstone can form in a littoral zone where beach and barrier bar sands build a complex association with inner shelf deposits. A thick section of shale might be expected across a marine shelf in deeper parts of the basin where depositional conditions are more uniform. Thus, both shale- and sand-prone sequences could be expected to change laterally to a sequence in which parallel reflections have high continuity and high amplitude (shallow-marine shelf paleoenvironment): Toward the landward edge of a depositional basin, the sand-prone section should interfinger with a nonmarine facies (represented by reflectors having low continuity and variable amplitude). The shale-prone section might change to one of the many seismic facies indicative of a slope paleoenvironment (Sangree and Widmier, 1979).

Paleoenvironments described in Table 2 are mainly shelf or coastal plain types. Two facies (sigmoidal-mound seismic facies and oblique-progradational seismic facies) indicate a shelf or upper slope setting. The main characteristic of most "shelf" seismic facies is a parallelism of reflections. The reflections diverge in some areas because of differing rates of subsidence and they can become discontinuous because of a change in the types of sediment and in the depositional environment.

SEISMIC SEQUENCES

Unit A

The oldest unit (A) overlies an irregular acoustic basement of probable block-faulted continental crust. The age of unit is open to question because some Canadian geologists (J. A. Wade, unpublished data, 1981) infer the graben fill to be late Paleozoic (Pennsylvanian and Permian?). On the other hand, Given (1977) inferred a Permo-Triassic(?) age for prerift red beds situated in a half-graben beneath the inner Scotian Shelf and a Trassic-Early Jurassic age for the rift deposits shown on a profile across the outer Scotian Shelf, 350 km northeast of Georges Bank.

A Late Triassic-Early Jurassic age is inferred for unit A because terrestrial sediments in grabens and half-grabens exposed in nearby New England are that age (Cornet 1977). Van Houten (1977) inferred two stages

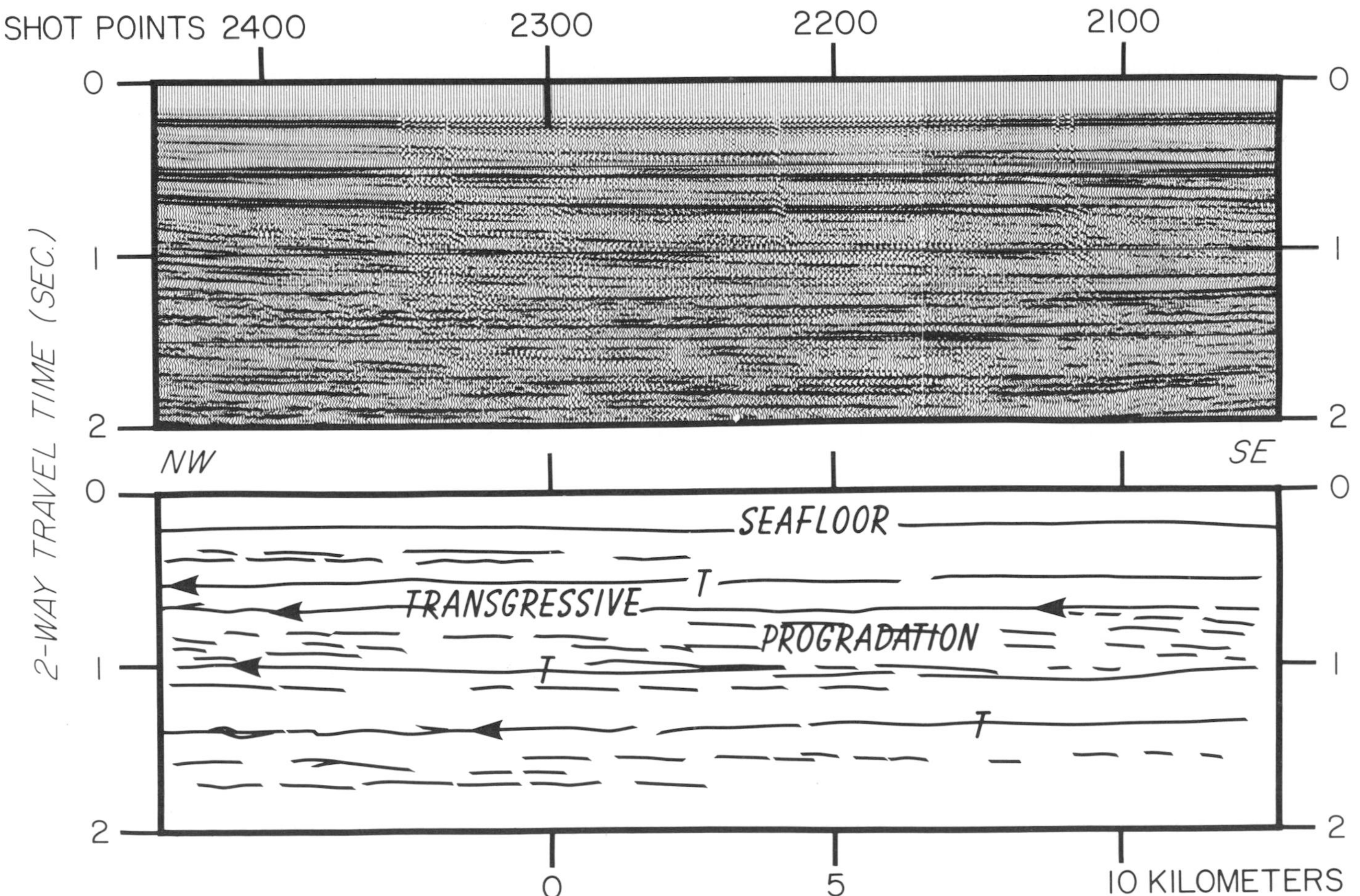

Figure 8 — (A) Enlarged part of the top two seconds of seismic-reflection profile 20 between shotpoints 2050 and 2440. **(B)** Interpretative line drawing of the same part of the profile, showing inferred transgressive (T) and regressive (prograding) phases of shelf buildup.

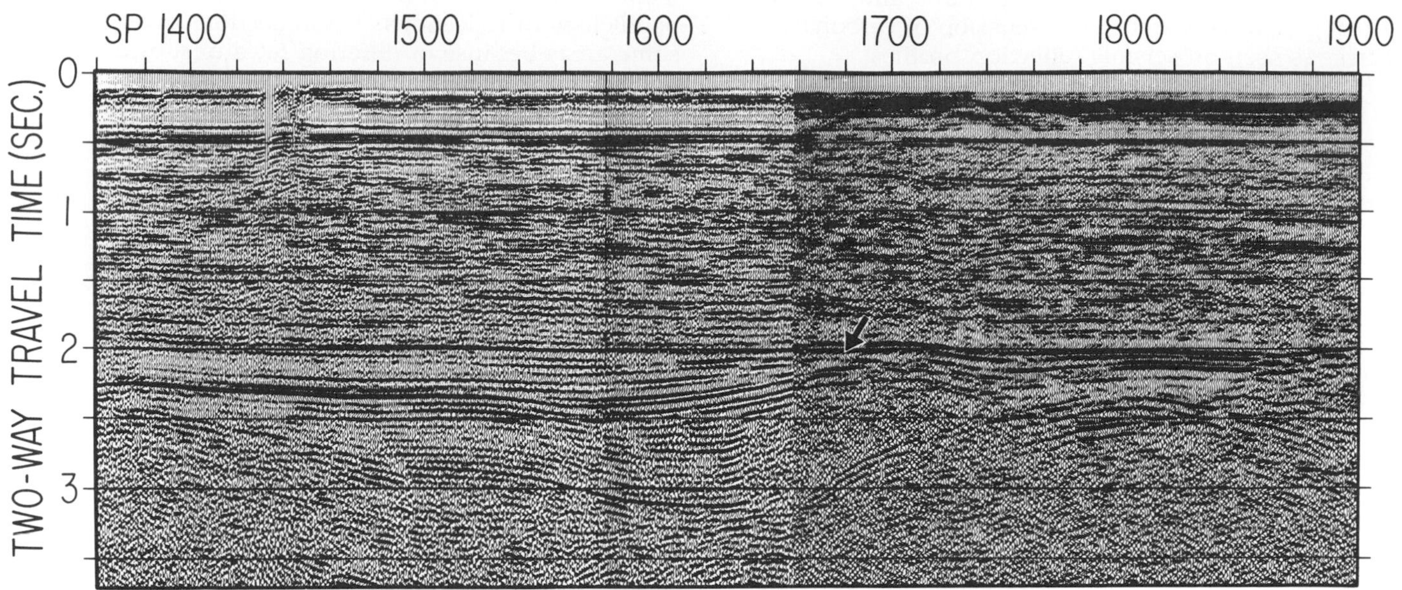

Figure 9 — Part of seismic profile 18, between shotpoints 1360 and 1900 (see Figure 1 for location). Notice conspicuous unconformity (arrow) forming the upper boundary of inferred rift deposits within a graben. High amplitude reflectors at 2 seconds (two-way travel time) probably mark carbonate rocks in this part of the basin. Every 100 shotpoints is 5 km.

in the development of these rift deposits: 1) rifting to the west giving rise to a string of fault basins and sags stretching from Georgia to Nova Scotia; and later 2) basin pull-apart under the present shelf. Most likely in platform areas (La Have and Long Island) where continental crust may be thicker than under basins, discrete systems of grabens created subbasins (Nantucket, Atlantis, Figure 2). But beneath basins and sub-basins where the crust may be thin, stretching resulted in a broad sag in which a thick sequence of red beds and salt accumulated (Argo and Eurydice Formations in the Scotian basin). Then, in a correlation of conspicuous reflections from the Scotian margin to eastern Georges Bank, Wade (1977; Figure 5B) showed that the Triassic and Jurassic rift basins on the flank of the La Have platform (marked by obvious unconformities) disappear southeastward where basinal sag exists; within the areas of basinal sag the unconformities are not obvious, but warped high amplitude, high continuity reflections suggestive of an evaporite-carbonate sequence are present.

In the COST G-1 and G-2 wells, the paucity of fossils does not allow exact dating in the oldest parts of the sedimentary sections; however, Scholle, Krivoy, and Hennessy (1980) and Scholle, Schwab, and Krivoy (1980) inferred an Early Jurassic (?) age for a sequence of dolomites and red beds in the bottom of the G-1 well and "not older than Upper Triassic" age for a sequence of evaporitic and carbonate rocks at the bottom of the G-2 well.

In this paper, the upper limit of seismic unit A is an unconformity most conspicuous above the many subbasins that flank the adjacent platforms (Figure 9). The unconformity is shown by an onlap of basal reflections in unit B over the truncated terminations of reflections in unit A. Because unit A is confined mainly to grabens and half-grabens along the northern basin floor, its reflections are divergent rather than like those reflections in the younger units. In the Falvey (1974) model this erosional gap would be the breakup unconformity, signifying a change from the rifting phase to the continental separation ("drift") phase of margin formation.

Isopachs of unit A are irregular (Figure 10) and trend northeast in the central and northeastern parts of the bank and below the shelf south of Cape Cod. Thickest accumulation is in the main basin (almost 9 km), and toward the Northeast Channel beneath the outer part of Georges Bank (Figure 1). Throughout much of the basin, unit A forms a thin layer (less than 0.5 km).

Average interval velocities determined from stacking of seismic reflection data range from 2.48 km/second under inner shelf south of Martha's Vineyard, to 6.40 km/s in the area of thickest sediment accumulation. Highest values (5 to 6 km/second from the south central part of the basin) compare with 5.0 to 7.3 km/second measured at the bottom of the COST G-1 and G-2 wells. A few of the lines that cross the main basin show higher interval velocities for unit B than for unit A; this velocity reversal is interpreted to signify possible thick accumulations of salt in the bottom of the main basin as part of unit A. Averaged interval velocities of unit A along the western end of Line 33 (Figures 1 and 10) are consistently lower than equivalent velocities in unit B (Figure 11); unit A velocities are mainly 5 to 6 km/second, a range somewhat higher than the 4 to 5 km/second given by Gardner, Gardner, and Gregory (1974) for salt.

The inferred lithologies suggest that nonmarine and restricted marine conditions prevailed during deposition of unit A. The central main basin contains marine

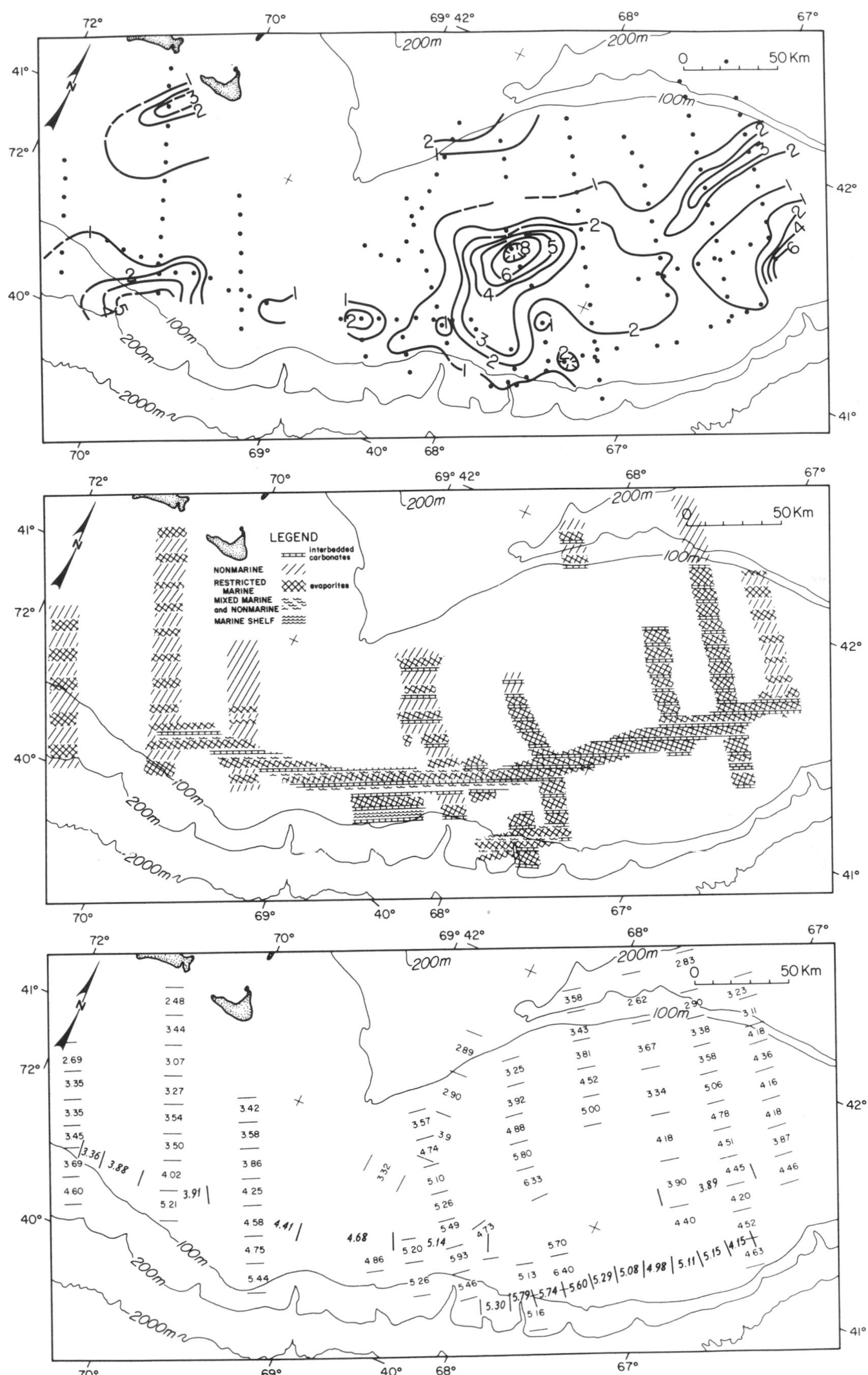

Figure 10 — Unit A thickness in kilometers (top); inferred depositional environment (middle); and averaged interval velocity in kilometers per second (bottom). Dots indicate control for thickness. Velocity values are the arithmetic mean of 3 to 5 shotpoints between the bars. Italicized values are averages for profiles that parallel the coastline.

limestones associated with evaporitic deposits, which indicates that a restricted seaway existed here and to the northeast. Salt flowage is obvious on line 12 where it crosses the Northeast Channel, and on the 6-channel profiles interpreted by Austin et al (1980), and Uchupi, Ballard, and Ellis (1977). A BGR profile near Heezen Canyon, along the outer continental shelf east of Line 4, clearly shows a shallow diapir that originates at least 4 seconds below sea level, and the top of which is about 0.8 second below sea level. On Figure 4, the reef-like buildup shown between shotpoints 620 and 720 could have its foundation on a broad salt swell. To the west and northwest of the main basin, conditions appear to have been more continental during deposition of unit A and a few strong continuous reflections suggest the presence of interbedded evaporite deposits. The inference of environmental patterns find support in the lithologies drilled in COST holes (Figure 5).

To the southwest in the Baltimore Canyon trough, the rocks of probable equivalent age show similar depositional facies (mainly restricted-marine grading to nonmarine inshore to the northwest) and a similar irregular isopach pattern (Schlee 1981; Figure 7); maximum thickness is 9 km toward the outer part of the trough. Averaged interval velocities range from 3.8 km/second to 6.5 km/second. In both the Baltimore Canyon trough and the Georges Bank area marine incursions took place during the rifting phase of margin formation.

Unit B

Unit B is similar to the underlying unit in seismic character, except that it is thinner and more widespread. It onlaps the truncated edges of unit A above the breakup unconformity, particularly toward the rifted northwest and northeast parts of the area. Unit C conformably overlies it although some channeling and low-angle truncation is evident at the upper boundary. Near the COST G-2 well, the upper and lower boundaries of unit B are conformable. Toward the southern edge of the main basin, the boundaries are lost in the massive carbonate buildup that borders the southern flank of the area (Schlee, Dillon, and Grow, 1979).

The isopach map (Figure 11) shows that unit B is distinguished in the main basin, but not toward the inner Long Island platform and La Have platform (south of Cape Cod and east end of Georges Bank) where units B and C combine mainly because unit C pinches out. Unit B is more than 2.5 km thick over the main basin and thins to less than 0.5 km to the northwest. It is thickest in the same general area as unit A, but shows more widely spaced contours than the older unit.

Through correlation with the COST G-2, the age of unit B is inferred (Figure 5) to be Early-Middle Jurassic; unit B is found at approximately 2.5 to 3.5 seconds and it correlates with a section of interbedded limestone, anhydrite, and sandstone. Unit B presumably encompasses several globally recognized hiatuses (Vail, Mitchum, and Thompson, 1977b; Figure 2). The boundaries are inferred to coincide with two moderately conspicuous relative changes in sea level shown by the Vail curve; the Sinemurian-Hettangian shift for the lower boundary, and the Callovian-Bathonian shift for the upper boundary. Sparse faunal remains obtained in the COST G-2 well do not permit paleontological documentation of these hiatuses.

Seismic reflections within unit B are parallel and show low to high continuity and variable amplitude. In the main basin, reflections are most continuous and even; toward the northwest as unit B thins, continuous reflectors are fewer and low-continuity, variable-amplitude reflectors comprise more of the unit. This trend probably indicates a more clastic, noncarbonate section of arkosic red beds (G-1 well). A similar change was noted by Given (1977; Figure 5) for the Scotian margin where several holes documented the change from dolomite and shale under the southeastern part of the basin to evaporite and red beds under the northwestern basin edge (part of the Iroquois and Mohican formations).

Interval velocity averages show a progression from low values (2.5 to 3.5 km/second) in the northwest where the unit is thin, more nonmarine, and less deeply buried, to high values in the carbonate-evaporitic sequences under the main basin to the southeast, where values are in the 5 to 6 km/second range. They are similar to the range of interval velocities in unit A and support the inference that dolomite, limestone, and evaporitic deposits are widespread under the southern part of Georges Bank.

Unit C

Rocks of Late Jurassic age constitute a broadly distributed unit that attains a maximum thickness of 4 km along the bank's southern side. (Figure 12). Averaged interval velocities increase in the same direction, and the unit becomes more marine to the south, continuing the trend present in units A and B. In the COST holes, a section of interbedded shale and sandstone (G-1) changes to the east into interbedded sandstone and limestone (G-2) deposited under shelf conditions (Scholle, Krivoy, and Hennessy, 1980; Schwab, and Krivoy, 1980).

Unit C thickens from slightly less than 1 km along the northern side of the basin to more than 3 km on the southern side (Figure 12). As in the older units, a subcircular depocenter underlies the central part of the main basin. The lower boundary is conformable with unit B in most profiles, except those where channeling and truncation of the unit B are evident or where onlap by basal reflectors of unit C is apparent. The upper boundary of unit C is also conformable on many profiles, but on some the boundary is marked by erosion, truncation, and channeling. To the southwest and northeast, unit C is not distinguishable from unit B. Toward the southeast part of the bank all continuity of reflectors is lost in the carbonate platform (Figure 3, Seismic Stratigraphy section).

Seismic reflectors in unit C are somewhat similar to those in unit A and B, but unit C reflectors are less continuous. Within the basin, reflectors tend to be

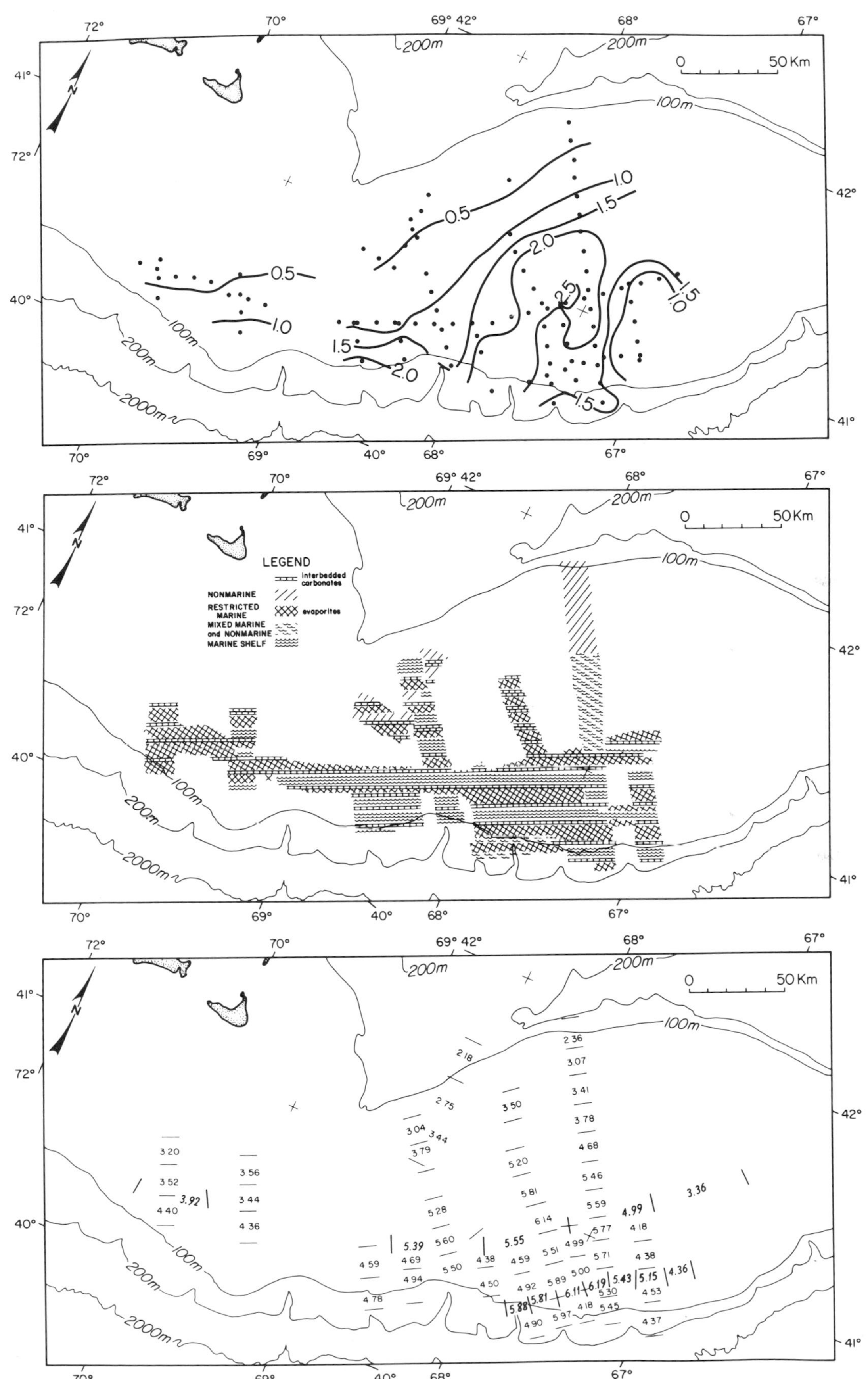

Figure 11 — Unit B thickness in kilometers (top); inferred depositional environment (middle); and averaged interval velocity in kilometers per second (bottom). Dots indicate control for thickness. Velocity values are the arithmetic mean of 3 to 5 shotpoints between the bars. Italicized values are averages for profiles that parallel the coastline.

parallel and of variable amplitude (mainly moderate to high). As line 20 (Figure 3) shows, the unit is part of a mixed nonmarine and marine shelf sequence (more carbonate rich toward the southeast). Some of the inferred nonmarine sequences appear as zones of variable-amplitude, low-continuity reflectors that become more continuous to the southeast.

The lithologic section, penetrated in the COST wells (Scholle, Krivoy, and Hennessy, 1980; Scholle, Schwab, and Krivoy, 1980), and the interval velocities support the seismic inferences. The COST G-2 well (Figure 5) drilled an 1800-m section (Upper Jurassic) of interbedded micritic-to-oolitic limestone and very fine-to-medium grained silty red and gray sandstone containing some beds of varicolored shale, probably deposited at inner shelf depths (0 to 30 m). The COST G-1 well (Figure 5) penetrated 1450 m of Upper Jurassic interbedded medium-to-coarse grained sandstone and red-brown shale. Thin beds of limestone and coal are also part of an alluvial marine shelf sequence (upper part). The intermediate interval velocities (4 to 4.5 km/second) characterizing unit C beneath the seaward part of the basin (Figure 12) could indicate a sequence of carbonate and noncarbonate sedimentary rocks. Interval velocities are low (2.5 to 3 km/second) to the northwest (Figure 12), an area characterized by nonmarine-transitional sandstone and shale. Interval velocity ranges for unit C rocks are from 3.0 to 4.3 km/second (G-1) and from 3.4 to 5.0 km/second (G-2).

At the seaward ends of two profiles, a pronounced irregular hummocky pattern of reflections suggests a reef buildup (Bubb and Hatlilid, 1977). During deposition of unit C, the whole basin was probably a broad shelf characterized by carbonate deposition along the southwest side (Schlee, Dillon and Grow, 1979) and bordered to the northwest by a low alluvial coastal plain of fluctuating width.

The most complete lithologic seismic study of an equivalent stratigraphic unit was made by L. S. Eliuk (1978) for the Abenaki Formation (Scotian margin). He distinguished a major mid-Jurassic transgression and found that the thick Baccaro Limestone member was deposited along the ancient shelf edge as a shallow-water carbonate platform; a clayey neritic moat separated sandy nearshore sediments from the platform. Eliuk (1978) mapped sedimentary facies (oolitic bar, mud shoal, shelf-edge stromatoporoid-coral-algal reef cyclothems) within the carbonate mass related these facies to shifts in sea level, and correlated them with the curve of Vail, Mitchum and Thompson (1977b; Figure 2).

In the Baltimore Canyon trough, equivalent age sediment was sampled in the COST B-3 well. It consists of oolitic limestone of Tithonian age overlying a clastic wedge of terrestrial interbedded sandstone, micaceous shale, and coal beds (Amato and Simonis, 1979). The oldest rocks are early Kimmeridgian age; and they were penetrated at 4691 to 4822 m deep. Poag (1980) thought these rocks represented "alternating shallow-marine and coastal marsh deposits," part of a back-reef sequence. The B-3 well never penetrated the main carbonate mass, which is either deeper or at an equivalent depth 10 km east of the well (Schlee and Grow 1980; Grow 1980).

Unit D

Unit D of Early Cretaceous age marks the transition from a dominantly carbonate shelf to a noncarbonate shelf in a seaward thickening sedimentary wedge (Figure 13) whose thickness exceeds 1 km in the southeastern part of the Georges Bank area. In the COST G-2 well, unit D consists of interbedded chalky limestone, sandstone, and shale; the same age sequence is mainly sandstone and interbedded shale deposited under inner shelf conditions at the COST G-1 well.

An isopach map (Figure 13) shows that unit D is widespread although it is much thinner than units A through C. The unit ranges in thickness from less than 0.25 km along the northern side of the basin to 1.25 km south of Cape Cod. Towards the northeast, unit D thins (0.5 km) and much of it is missing. Tertiary age rocks directly overlie Neocomian limestones in Heezen and Corsair Canyons (suggested by Ryan et al, 1978). Correlation of seismic profiles with the COST wells indicates that the upper boundary is a Cenomanian unconformity and the lower boundary is the base of the Cretaceous. Both boundaries are close to fluctuations in sea level (Vail, Mitchum, and Thompson, Figure 2) and the lower boundary is generally conformable beneath the southern part of the basin. Toward the northern edge of the basin and the platforms, basal reflectors onlap older units along a contact that shows channeling and low angle truncation of older reflectors. At the upper boundary, reflectors are similar and are conformable over the south-central and northeast part of the basin. Along the northern flank of the basin, channeling and low angle truncation of unit D is evident.

The averaged interval velocity ranges from 1.88 to 3.45 km/second (Figure 13). Low values are associated with the thinner, less deeply buried part of the unit while the high values (more than 3 km/second) are found toward the outer part of the basin, usually in areas with thickest sediment accumulation.

Reflectors are mainly parallel and of low-to-moderate continuity and moderate amplitude. As shown in Figure 13, the facies trend from sequences of nonmarine rocks (probably sandstone and shale) under the northern part of the basin and adjacent platform, to mixed marine shelf and nonmarine sequences under the main part of the basin. The nonmarine rocks extend as tongues into the marine south of the basin (Figures 3, 4). Along the basin's southern edge, and south of the Long Island Shelf, the high intensity and continuity of some reflectors indicate that carbonate rocks are interbedded with other marine shelf sequences. On the seaward ends of Lines 20 and 1 (southern Georges Bank, Figure 1), unit D changes to a hummocky pattern of reflectors suggestive of the carbonate buildup seen in units B and C; thus the carbonate platform persisted into the lower Cretaceous just as it did on the seaward edge of the Baltimore Canyon

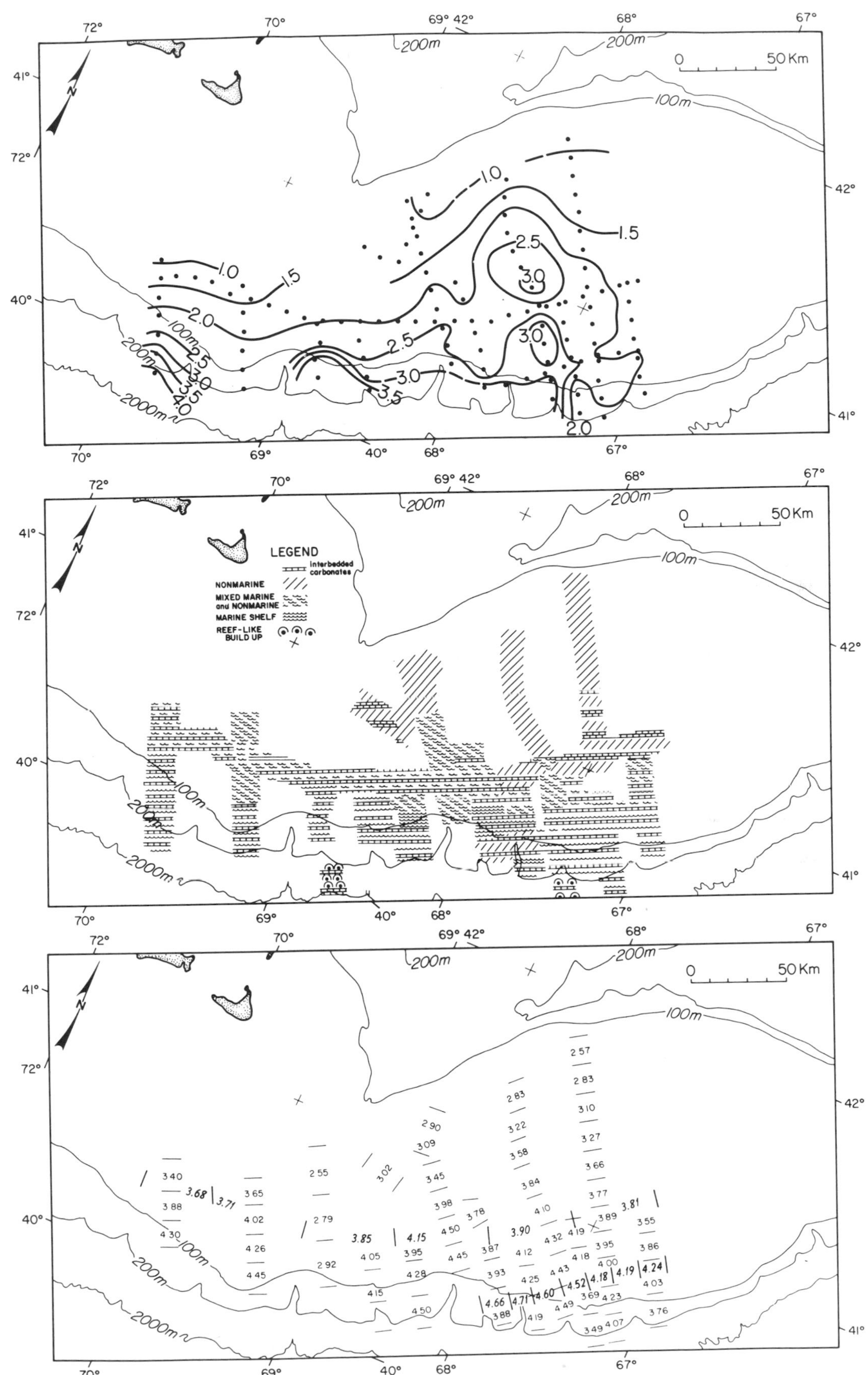

Figure 12 — Unit C thickness in kilometers (top); inferred depositional environment (middle); and averaged interval velocity in kilometers per second (bottom). Dots indicate control for thickness. Velocity values are the arithmetic mean of 3 to 5 shotpoints between the bars. Italicized values are averages for profiles that parallel the coastline.

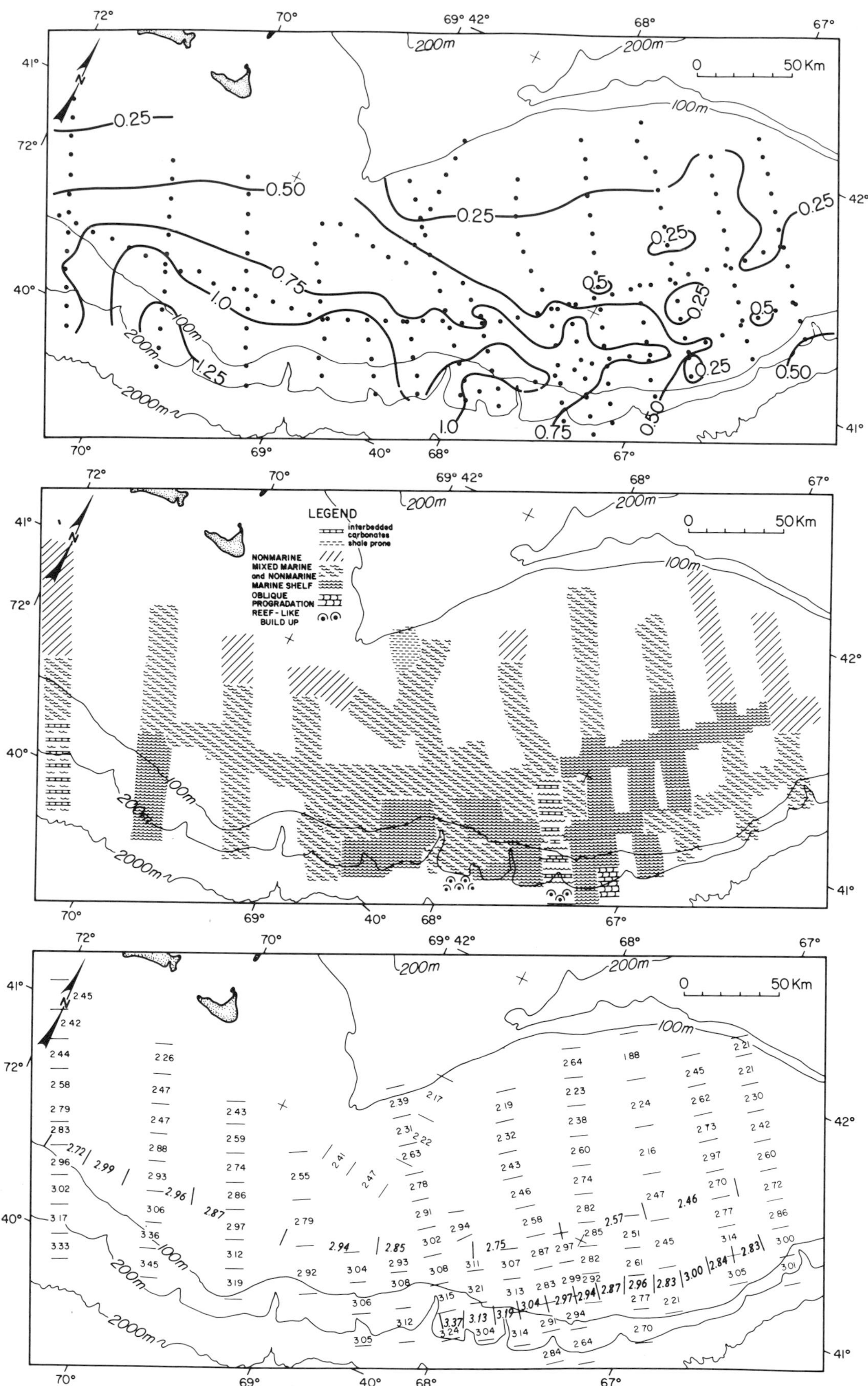

Figure 13 — Unit D thickness in kilometers (top); inferred depositional environment (middle); and averaged interval velocity in kilometers per second (bottom). Dots indicate control for thickness. Velocity values are the arithmetic mean of 3 to 5 shotpoints between the bars. Italicized values are averages for profiles that parallel the coastline.

trough (Schlee, Dillon and Grow, 1979; line 2).

The COST wells show a similar range of paleoenvironmental bathymetry, mainly littoral to inner shelf although deeper water shelf conditions prevailed during the Aptian (COST G-2; Scholle, Krivoy, and Hennessy, 1980; Scholle, Schwab, and Krivoy, 1980). COST G-2 contains more limestone in unit D than the G-1 well, possibly indicating a more open marine shelf. Farther out in the basin, Ryan et al (1978) sampled an outcrop of bioclastic limestone (biopelsparite) of Neocomian age (based on calpionellids) in Heezen Canyon. Fossils suggest that the limestone accumulated as part of a reef tract facing open ocean.

Unit E

The upper limit of Unit E is an unconformity that marks the Tertiary-Cretaceous boundary. This unconformity is characterized by conformity of the basal reflectors from the unit above and by low-angle truncation of the uppermost reflectors in unit E in the eastern and central parts of the basin and the outer shelf south of Cape Cod. Some downlap of basal reflections in the overlying unit (F) is evident under the outer shelf area. Over much of the rest of the basin, the upper boundary is conformable and evidence of erosion is lacking.

Reflecting a basinlike pattern of subsidence, unit E forms a broad sheet that increases in thickness from less than 0.25 km under northwestern Georges Bank, to slightly more than 1 km south of Nantucket Shoals (Figure 14). It has been drilled in four holes: 1) In the COST G-1 hole it is 520 m thick (Scholle, Krivoy, and Hennessy, 1980); 2) On Nantucket it is 310 m thick (Folger et al, 1978); On Martha's Vineyard it is 176 m thick (Hall, Poppe, and Ferrebee, 1980); 4) At the Shell Mohawk B-93 well on Browns Bank (Figure 1), equivalent age strata are 700 m of interbedded shale and sandstone (Ascoli, 1976).

Inferred paleoenvironmental patterns (Figure 14) are chiefly of a marine shelf; a nonmarine paleoenvironmental is inferred for parts of the unit deposited on the La Have and Long Island platforms. Reflections indicating the presence of shale are distinguished on the northwest (inner shelf) part of Line 5 (Figure 1) and mark the seaward continuation of a nonmarine deltaic shale sequence drilled in both the Nantucket and Martha's Vineyard holes (Folger et al, 1978; Hall, Poppe, and Ferrebee, 1980). More prevalent nonmarine and mixed marine-nonmarine facies on the northern side of the basin continue a pattern seen in units C and D.

Throughout the main part of the basin, units E and F are characterized by alternating zones of continuous reflectors (Figure 8) that probably represent transgressive marine and regressive deltaic phases of shelf deposition (see discussion of Line 20 and BGR Line 209 in "Seismic Stratigraphy"). Toward the seaward edge of the basin, several seismic lines show progradation of unit E into deep water and the building of a constructional slope (Schlee, Dillon, and Grow, 1979).

Unit E was sampled in the COST G-1 and G-2 wells; it is interbedded gray shale, sand, and gravel (and some lignitic stringers) deposited under transitional- to outer-shelf conditions (G-1 well), and light gray glauconitic calcaerous shales interbedded with limestone and sandstone (G-2) deposited mainly under an outer shelf paleoenvironment (Scholle, Krivoy, and Hennessy, 1980; Scholle, Schwab, and Krivoy, 1980).

Unit E was sampled in several submarine canyons indenting the shelf south of Cape Cod and Georges Bank. In Atlantis and Veatch canyons (Figure 1), bathyal siltstones of Maestrichtian and Santonian Age were sampled from the submersible ALVIN (Valentine, 1978) and 400 m of Coniacian (?) through Maestrichtian rocks were drilled in 1967 (Weed et al, 1974; Poag, 1978). In Lydonia, Corsair, and Oceanographer canyons (Figure 1) dredge hauls and samples obtained in ALVIN dives (Gibson, Hazel, and Mello, 1968; Ryan et al, 1978) of sandstone, clay, and calcareous mudstone of Late Campanian to Maestrichtian age indicate depositional depths ranging from sublittoral to bathyal (probably canyon fill). Valentine, Uzmann, and Cooper (1980) described 300 m of Santonian and younger sandstones that exist in Oceanographer Canyon.

Averaged interval velocities of unit E range from 1.74 to 2.90 km/second and follow the pattern distribution seen in the lower units. The values in the vicinity of the COST wells (2 to 2.2 km/second) compare with a range of 1.7 km/second to 2.6 km/second obtained by sonic logs from the same interval in the holes. The interval velocities (lower by 0.5 km/second from the velocities of similar aged units in the Baltimore Canyon trough) are in the range of velocities that characterize partly consolidated sandstone and shale (Gardner, Gardner, and Gregory, 1974).

Unit F

Unit F is a thin widespread blanket of marine and glacial sediment (averaged interval velocity 1.60 to 2.25 km/second). It onlaps older units and downlaps toward the outer shelf edge, and accumulated at a much slower rate than the older units accumulated. As shown by the isopach map (Figure 15), unit F is only 0.1 to 0.4 km thick; the thickest part is along the southern edge of the shelf. In a structure contour map, Emery and Uchupi (1965) showed that depth to the top of the Cretaceous ranges from less than 300 m at the northern part of the bank to more than 900 m at the southern edge. Austin et al (1980) showed thickness from less than 250 m to more than 1,500 m in the vicinity of Oceanographer Canyon. Valentine, Uzmann, and Cooper (1980) located the Cretaceous-Tertiary boundary at 915 m deep in Oceanographer Canyon.

Unit F appears to have formed on a marine shelf that prograded to the southeast. On most of the profiles, reflectors show good continuity and moderate to high amplitude (see discussion of Line 20 and BGR Line 209 in "Seismic Stratigraphy"). South of Cape

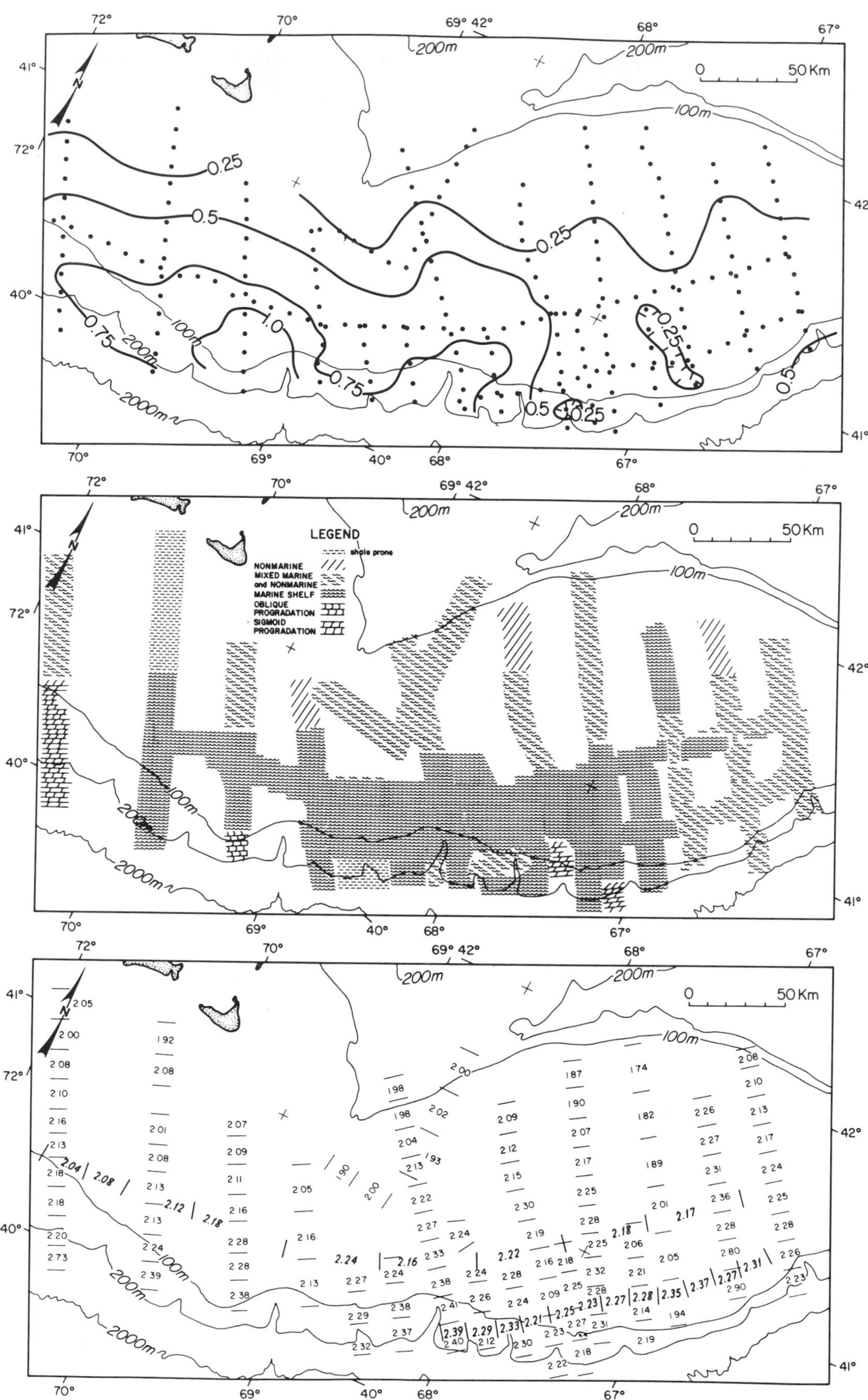

Figure 14 — Unit E thickness in kilometers (top); inferred depositional environment (middle); and averaged interval velocity in kilometers per second (bottom). Dots indicate control for thickness. Velocity values are the arithmetic mean of 3 to 5 shotpoints between the bars. Italicized values are averages for profiles that parallel the coastline.

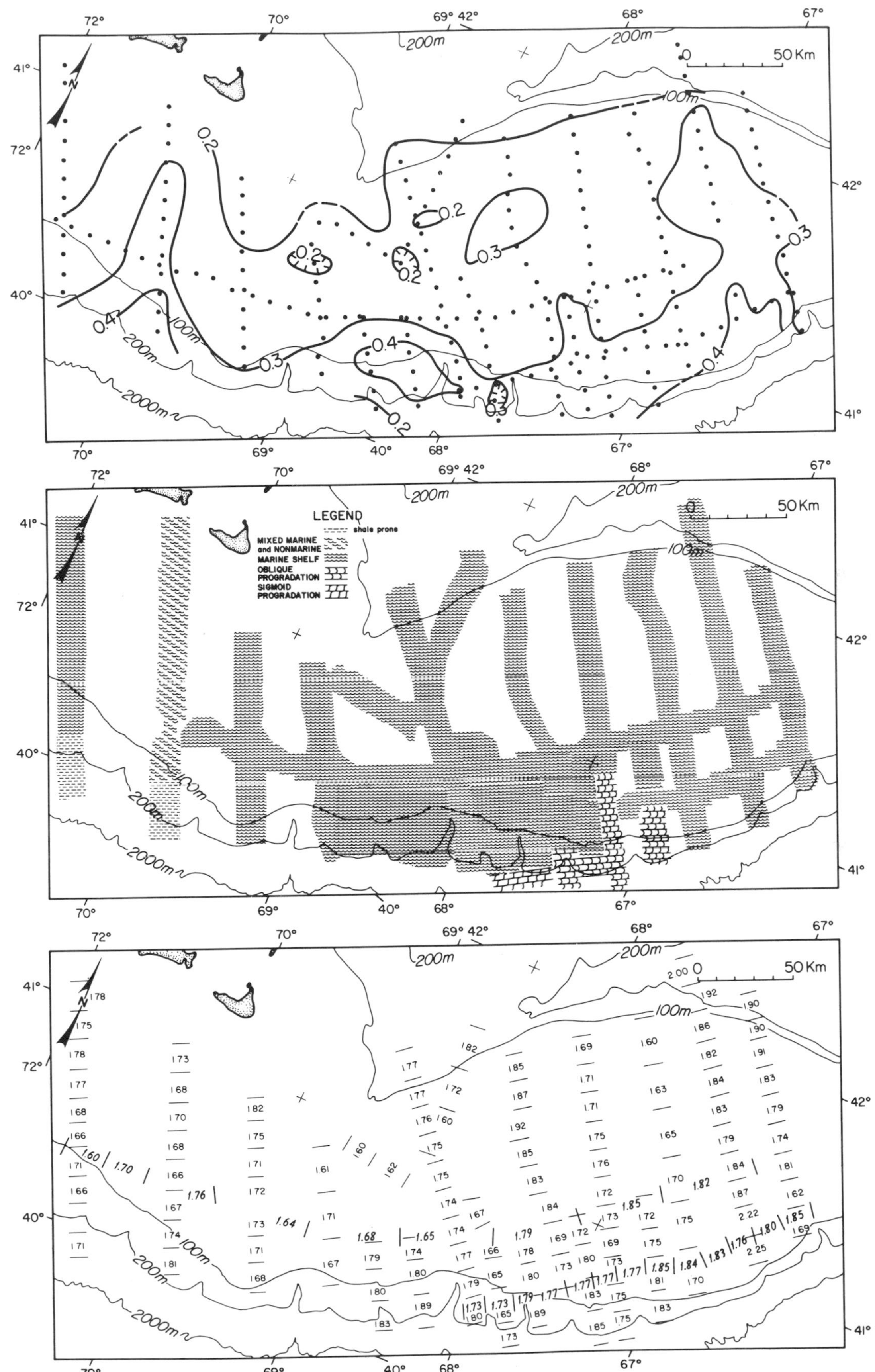

Figure 15 — Unit F thickness in kilometers (top); inferred depositional environment (middle); and averaged interval velocity in kilometers per second (bottom). Dots indicate control for thickness. Velocity values are the arithmetic mean of 3 to 5 shotpoints between the bars. Italicized values are averages for profiles that parallel the coastline.

Cod (Line 5), the character of the reflectors suggests that of a mixed marine shelf and nonmarine paleoenvironment. Poag (1978) reviewed the results from shallowing drilling and dredging on Georges Bank and found the Tertiary rocks were mainly of shelf origin. Valentine, Uzmann, and Cooper (1980) sampled some chalks and silty claystones deposited in outer shelf-upper slope environments from Oceanographer Canyon. Ryan and others (1978) recovered Eocene brown silty mudstones, calcarenites, and chalk of canyon fill and pelagic origin in Oceanographer, Corsair, and Heezen canyons. Separated from older beds in the unit by a conspicuous unconformity (Lewis and Sylwester, 1976) is an extensive blanket of Pleistocene drift over Georges Bank. The blanket consists of hemipelagic silts and clay that discontinuously mantle the continental slope and submarine canyons (Ryan et al, 1978; Valentine, Uzmann, and Cooper, 1980).

Averaged interval velocities of unit F (Figure 15) are low and indicate poorly consolidated sediment. Unit F was drilled at seven sites during the Atlantic Margin Coring (AMCOR) Project (Hathaway et al, 1979) and these holes revealed loose sand and punky clay in the upper 100 m.

Additional hiatuses appear to exist within unit F and at least one can be traced locally in Line 5 (Figure 1). On high resolution profiles over Georges Bank, Lewis and Sylwester (1976) recognized a hiatus between the Pleistocene drift and gently dipping Tertiary coastal plain strata. Single-channel seismic reflection profiles over the northern part of the bank revealed a hiatus, inferred to be caused by fluvial erosion during the Pliocene (Oldale et al, 1974). Thus, additional unconformities are probably present within unit F. To the southwest in the Baltimore Canyon trough, hiatuses have been inferred (Schlee, 1981) in the Oligocene and late Miocene because the Cenozoic section is two to three times thicker than it is in the Georges Bank basin. The thinness of the unit over the bank makes its subdivision difficult.

DISCUSSION

Rift-drift Sequence Geometry

The complex geometry of the Georges Bank basins are seen in an isopach map (Figure 2). Rather than being a single broad open feature like the Baltimore Canyon trough, they began as a group of scattered small sub-basins associated with a slightly bigger basin under the south central part of Georges Bank. The pattern appears to result from the rifting of continental crust adjacent to two broad platforms. Rifting is manifest as grabens and half-grabens of limited area extent. Areal magnetic anomaly patterns (Klitgord and Behrendt, 1979) indicate that the volcanic rocks associated with these fault basins extend well beyond the basins themselves and have a northeast trend. Depth to acoustic basement in these fault troughs ranges from 4 to 10 km. Under south Central Georges Bank, the basin broadens to an area of at least 4,800 sq.km (1800 sq. mi); here, depths of 10 to 16 km are similar to those recorded for the Baltimore Canyon trough and the Scotian basin. The depth to acoustic basement beneath the Outer Continental Shelf and Slope is unknown because the carbonate buildup masks it. Schultz and Grover (1974) suggested that basement rises in this critical area to around 6.7 km (22,000 ft), based on the Yarmouth Arch extending along the south side of the basin. Austin et al (1980) inferred a depth of 11 to 13 km to acoustic basement in areally restricted parts of the basin, and a depth of 7 km to basement beneath the outer shelf and upper slope. Klitgord and Behrendt (1979) inferred that magnetic basement is 8 to 10 km below sea level under the outer shelf, based on an aeromagnetic survey of the area. Schlee et al (1976) show that acoustic basement is covered by 5 to 7 km of sediment under the lower continental slope-upper continental rise south of Georges Bank and the Long Island Shelf. Thus, the critical transition zone between oceanic and continental crust appears to shoal though it is extensively rifted.

Comparing unit A's thickness (Figure 10) with the thickness of units B through F (Jurassic through Cenozoic) shows that subsidence was rapid and in several scattered depocenters at the beginning of basin formation. It became more concentrated in one depocenter and less rapid through time (Figure 16). Horsts and grabens are most obvious on the platform flanks. Following continental separation of Africa and North America in the Early Jurassic, the area of basin subsidence expanded to cover most of the bank. As noted above, evidence of a breakup unconformity (Falvey, 1974) is most apparent on the platforms where older deposits filling the grabens are unconformably overlain by the sedimentary wedge deposited during the subsidence phase of margin formation. As in the Scotian basin, the unconformity is not easily delineated within the main central basin, probably because the rifting phase was not separated from the subsidence phase.

Sedimentary Associations

The change in tectonic setting from narrow rifts to a broadly subsiding basin was accompanied by a change in lithology. The basal sequence of evaporite and carbonate deposits is overlain by the mixed sequence of sandstones, shales, and limestones; in turn the mixed sequence is overlain by a sequence of shale, siltstone, and thin limestone beds. This pattern of margin sedimentation extends to the continental slope (Schlee, Dillon, and Grow, 1979) where probable reef-flank, deepwater fan, and slump deposits underlie the upper continental rise. But within the shelf area along much of the Atlantic continental margin, sedimentary facies appear remarkably similar from basin to basin (see Given, 1977; Wade, 1977; Eliuk, 1978; Schlee, 1981).

As shown in Figure 17, sedimentary associations for a particular age are somewhat similar. For rocks within a system, lateral changes take place (Figure 5)

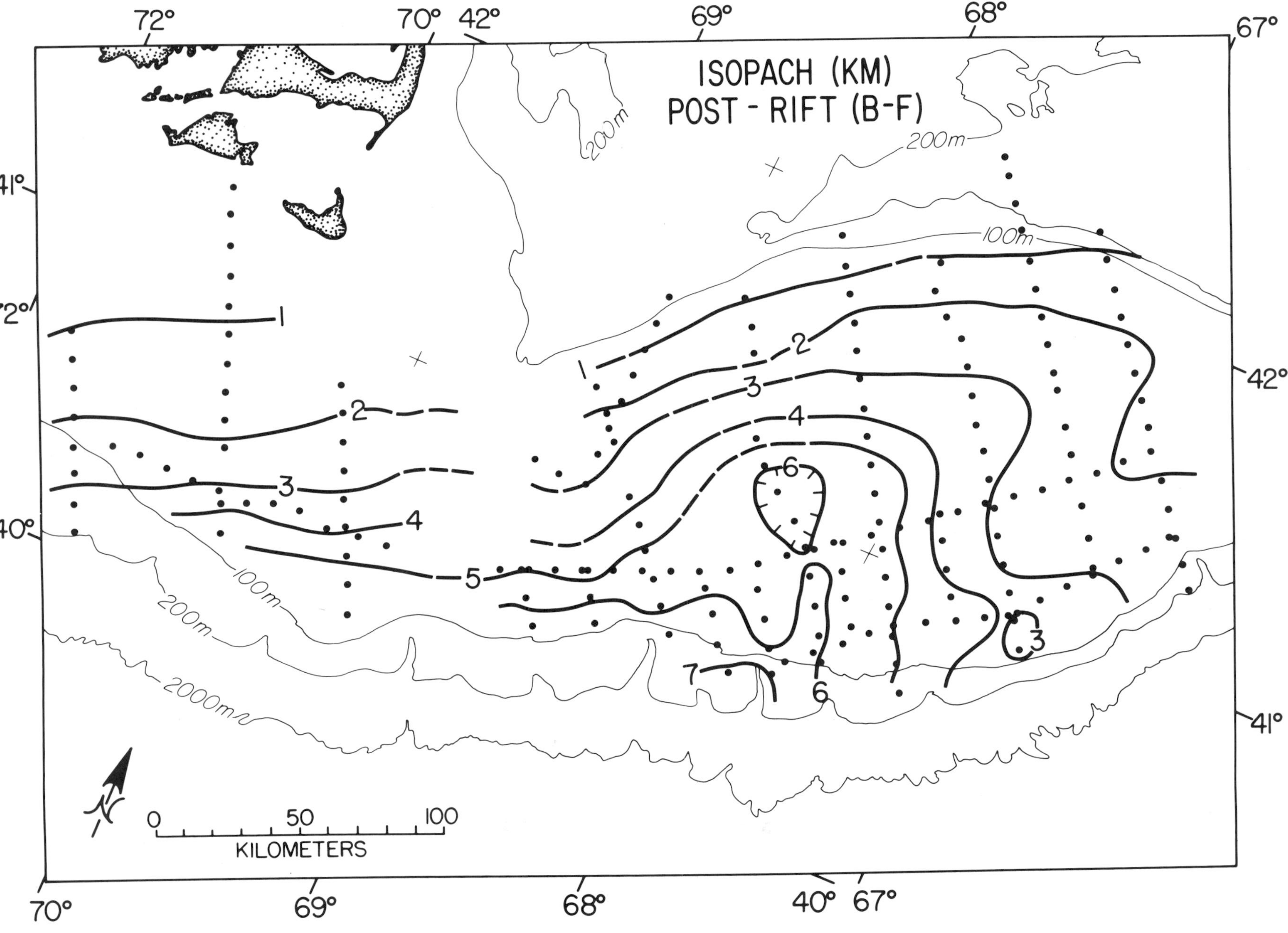

Figure 16 — Isopach map (in kilometers) of sedimentary rocks inferred to be of Jurassic age and younger (units B through F). Dots indicate control.

where nonmarine clastic rocks dominate toward the presumed landward part of the basin complex. This is obvious on the Scotian margin, where the Mic-Mac H-86 well reveals an older rift sequence of red beds that grades laterally into interbedded salt, anhydrite, and carbonate rocks of Early Jurassic and Late Triassic(?) age in the Mohican I-100 well (Jansa and Wade, 1975; Ascoli, 1976; Given, 1977). The same lateral change persisted throughout the Jurassic as a series of carbonate banks was built along the seaward part of the basin.

The main difference between Georges Bank basin and basins of adjacent areas is thickness. The Cretaceous system on Georges Bank (1,250 to 1,400 km) is thinner than that in the northern Scotian basin (2,750 to 3,050 m). Drilled equivalent age units in the Jurassic are thicker in the Georges Bank area (4,875 m from the top of the Jurassic to the top of the Argo Formation) than on the Scotian margin (1,675 m). This apparent difference in thickness reflects in part drilling on structural highs within the basin. In addition, the locus of maximum sedimentation for the Late Jurassic was seaward of the Scotian shelf, where more than 2 km of Late Jurassic sediment is inferred to exist (Jansa and Wade, 1975). The post-Cretaceous section is thinner beneath Georges Bank than that in the areas to the northeast or southwest (650 m for the COST G-2, 1,750 m for the COST B-3 and Mohican I-100 well). In all three areas, the wells show that sediment accumulation was rapid during the Jurassic and that rates diminished thereafter. For Georges Bank, at least 4,875 m of rock accumulated during the first 54 million years of basin development in the vicinity of the COST G-2 well. Only 1,750 m of sediment accumulated in the last 141 million years and most of this, 1,380 m or 79% was deposited during the Cretaceous (141 to 64 m.y. ago).

Carbonate Buildup

The nature of the carbonate platform or bank has been studied in most detail on the Scotian margin where it is a highly varied, subsiding feature "that rimmed the Jurassic western North Atlantic continental shelf, was influenced by sea level changes and was locally exposed and subjected to fresh water leaching and dolomitization" (Eliuk, 1978, p. 425). He distinguished several facies also seen in Unit C. On the Scotian margin, the platform was separated from the nearshore sandy basement ridge zone by a neritic shelf moat in which carbonate and noncarbonate muds accumulated; the carbonate content of the mud depended on shifts of sea level during the Late Jurassic. Areally, the carbonate buildup was separated into two banks by a delta that built out in the Sable Island vicinity. The delta influenced shelf-edge morphology; a ramplike shelf break formed in which clastic and carbonate sediments are interbedded. To the west (La Have platform) in the area where a rimmed-carbonate platform, similar to the modern Great Bahama platform, prevailed, the shelf break was sharp and the ancient slope dropped away at an angle of 20 to 30°.

On Georges Bank the same progradational arrangement of reflectors (Figure 3) attests to migration of the shelf edge to the southeast, probably in response to the outbuilding of tongues of deltaic sediment. Migration took place during the Late Jurassic and is most evident beneath southwestern Georges Bank (Figure 18). Along much of the rest of the bank, a rimmed platform marked by upward growth and a relatively fixed shelf break (Schlee, Dillon, and Grow, 1979; Figures 4, 5, 6) is present. Toward Long Island, the platform is buried by 1 to 2 km of younger sediment; seaward of the eastern end of Georges Bank, on the flank of the La Have platform, the ancient shelf-edge platform appears to be exposed through cutback of slope (Figure 18) in the area where Ryan et al (1978) collected specimens of reef-tract Neocomian limestone (see discussion, unit D).

An important aspect of post-depositional history that influences hydrocarbon potential of the carbonate platform is the diagenesis of the limestones. For the Baccaro Limestone member, Eliuk (1978) proposed a diagenetic model based on detailed examination of 22 wells. He found several porous zones in the Baccaro member that he equated to drops of sea level, subaerial exposure, and leaching of limestone in the vadose zone. Dolomitization was secondary and Eliuk thought it formed at the base of a freshwater lens. During high stands of sea level, the offshore bank was covered by muddy sediments (partly as mounds) containing radial ooids and grapestone. It will be interesting to see if similar zones are found beneath Georges Bank in equivalent age rocks.

African Counterpart

A reconstruction of Africa and North America (Klitgord and Behrendt, 1979) shows that the Georges Bank area was opposite the Aaiun basin of Western Sahara and Morocco. It and its offshore extension (Cape Bojador marginal basin) are similar to the Georges Bank area in the thickness, age, and type of sedimentary section (von Rad and Einsele, 1980; von Rad and Arthur, 1979). An isopach map (von Rad and Arthur, 1979; Figure 1) of the Mesozoic and younger sedimentary rocks shows that the basin is a seaward thickening wedge more than 14 km beneath the outer shelf-upper slope. Although the data do not show deep structure, the wedge does not appear to be broken into isolated sub-basins and ridges like its North American counterpart.

Several onshore and offshore holes drilled in the Aaiun basin reveal a sequence of units similar to those in the Georges Bank. A few holes drilled inland penetrated 1 to 2 km of Middle to Upper Jurassic shallow-water carbonate deposits and marlstones, part of a broad carbonate platform thought to be more than 8 km thick under the main part of the basin. Lower Cretaceous Wealden-type deltaic deposits built over the Late Jurassic limestones and dolomites give rise to a seaward progression of alluvial delta-plain sands, lagoonal clays, intertidal muds, and deltaic sands-prodeltiac clays. After major mid-Cretaceous erosional break, late Cenomanian to Tur-

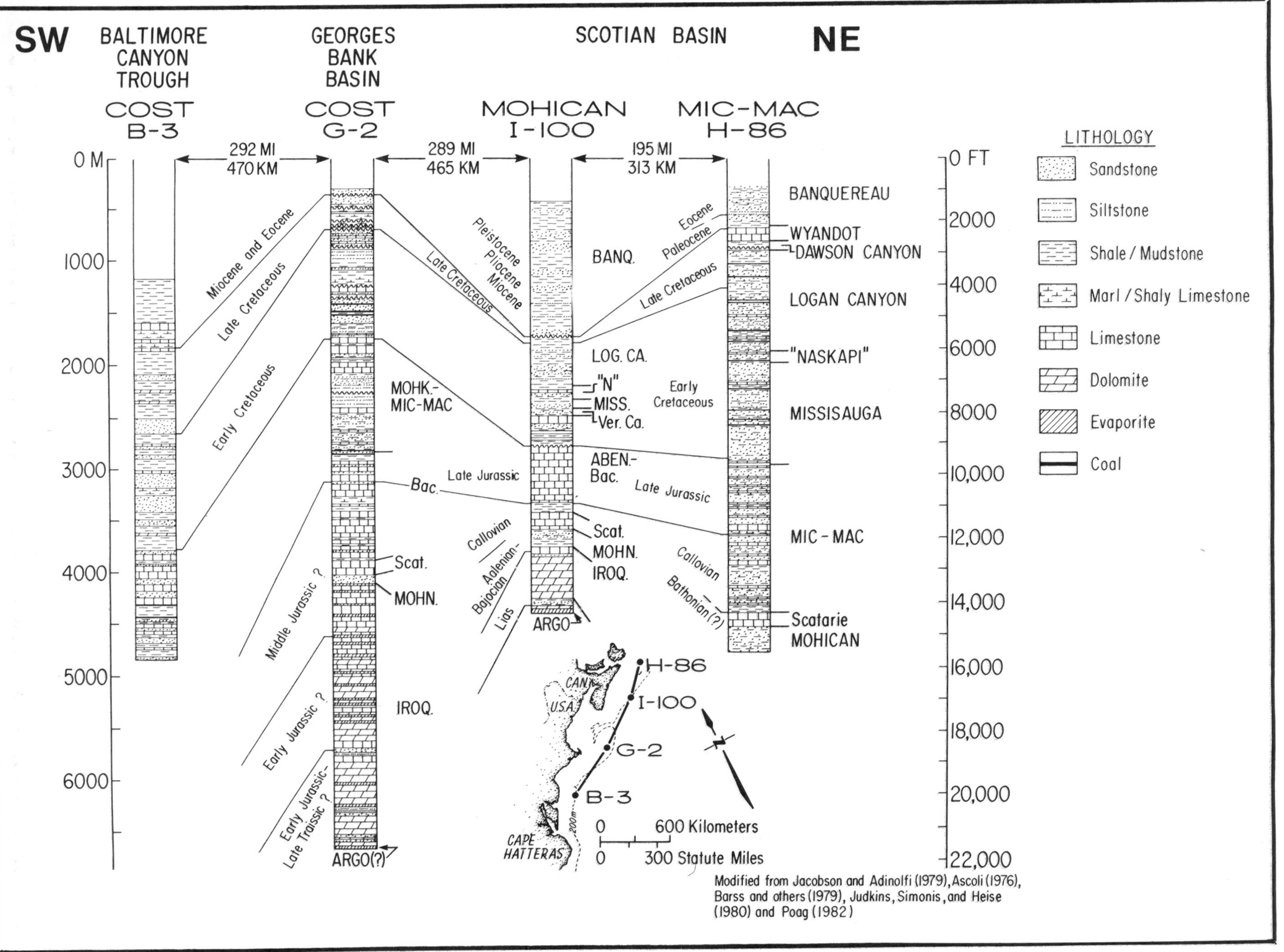

Figure 17 — Lithological sections from four wells drilled on the Atlantic margin. Datum is mean sea level. See Figure 5 for explanation of the lithologic symbols.

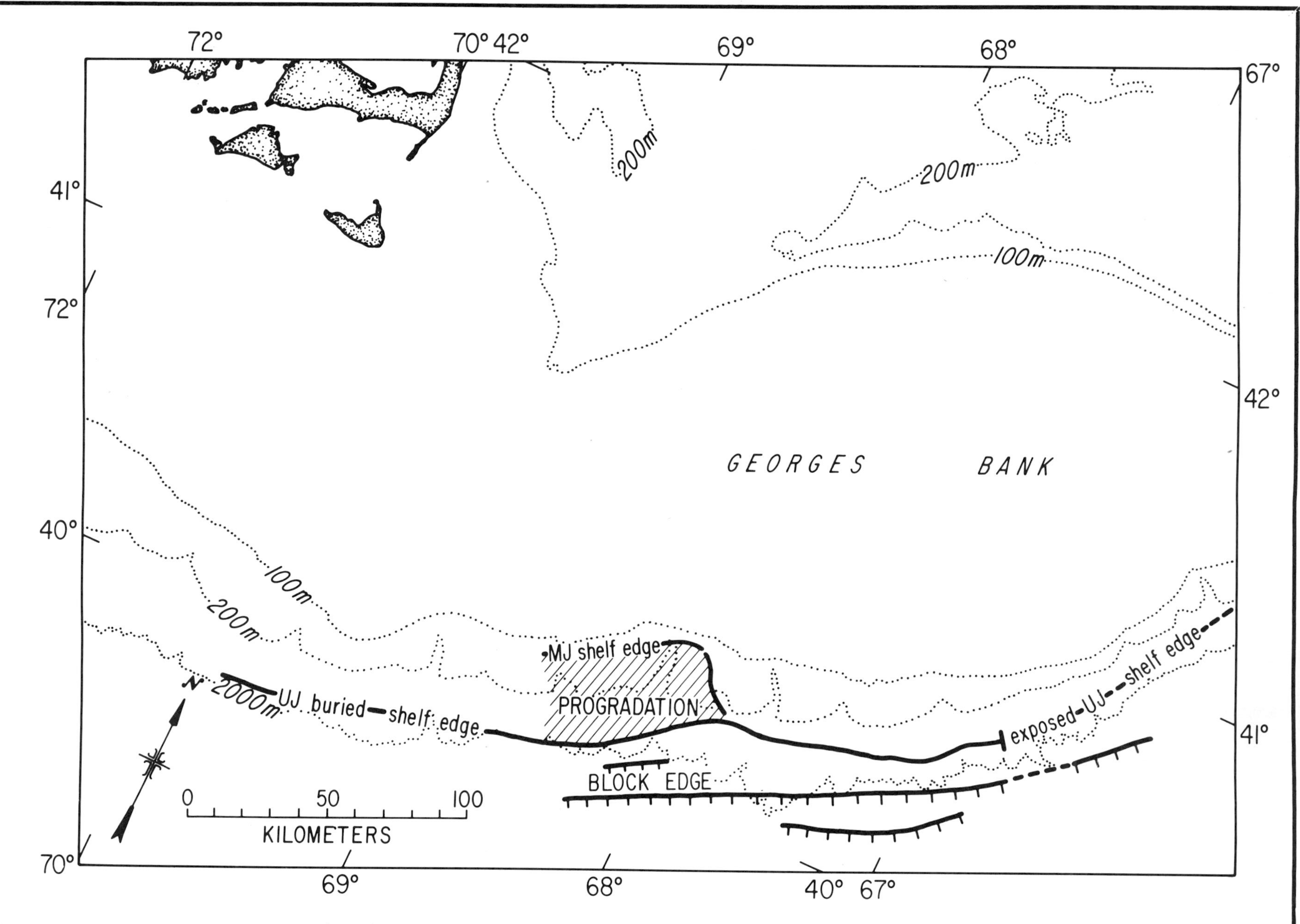

Figure 18 — Map showing the location of the Middle to Late Jurassic shelf edges. Cross-hatch area shows where a 40-km migration of the shelf edge is inferred. Dashed line indicates where the Late Jurassic-Early Cretaceous carbonate platform may be exposed because of continental slope cutback. UJ, Upper Jurassic; MJ, Middle Jurassic.

onian anaerobic marls were deposited over the outer Aaiun basin in a major expansion of the mid-water oxygen minimum zone into surface waters of the shelf (von Rad and Arthur, 1979). In the last 100 million years, marine shelf chalks, marls, and silty shale accumulated on the shelf at slower rates than those in the Early Cretaceous; Lower Cretaceous deposits are 0.3 to 1.1 km thick. Like the Georges Bank area and the Baltimore Canyon trough, the Aaiun basin has its thickest Cenozoic deposits (2 km) beneath the upper continental rise and its Cenozoic deposits form a blanket 1 km thick covering the shelf (von Rad and Einsele, 1980; Figure 2).

Certain tectonic features and the basin geometry are more reminiscent of the Baltimore Canyon trough (Grow and Sheridan, 1981) than they are of Georges Bank. Beneath the West African continental slope is a broad "slope anticline," best shown by reflections from deltaic strata thought to be Early Cretaceous age (Hinz, 1979; Figure 2). The anticline is cut by antithetic faults, and parallels the strike of the slope in a broad arc. The faults probably formed through a combination of differential subsidence (basin subsiding faster than slope area) and isostatic rebound due to substantial slope cutback during the middle Tertiary. The profile used by Hinz (1979; Figure 2) lacked a pronounced buried carbonate shelf-edge; instead reflections from presumed Lower Cretaceous strata made a flattened arc to the pronounced middle Tertiary unconformity where they were cut out. A similar arrangement of reflections was shown by Schlee, Dillon, and Grow (1979; Figure 9) for the Baltimore Canyon trough, where erosion (probably Oligocene) cut back so far that the front of the Jurassic carbonate platform was removed. The seaward edge of the buried carbonate platform is preserved along most of the slope seaward of Georges Bank, except for the eastern part. Here, a diminished amount of post-Jurassic shelf subsidence, active slope erosion, and building of a thin continental rise wedge combined to expose rather than to bury limestones of Early Cretaceous age.

The isopach map of the Aaiun basin (von Rad and Arthur, 1979; Figure 1) looks more like the isopach map of the Baltimore Canyon trough (Schlee, 1981) than the map of Georges Bank (Figure 2). The fragmentation of the West African margin does not appear to have resulted in the complex pattern of basement blocks seen on Georges Bank. The thickness contours outlining the Aaiun basin are fairly evenly-spaced and not too convoluted. Their pattern is more indicative of a divergent marginal basin than of a basin possibly formed by transform motion as the two continents separated.

The different behavior in basin subsidence between Georges Bank and the West African margin may relate to crustal geology beneath the two areas. The Aaiun — Cape Bojador basin is built adjacent to the Reguibat uplift, a Precambrian crystalline terrane part of the West African Shield. Georges Bank is built over the faulted and folded roots of the Appalachians adjacent to a broad structural reentrant (Williams, 1978) in the mountain chain. Relaxation of the compressional fabric in the deformed Paleozoic rocks underlying much of Georges Bank may have resulted in a broad zone of rifting, just as relaxation resulted in rifting onshore in New England and the Canadian Maritime provinces.

SUMMARY

One main basin and several smaller sub-basins exist beneath Georges Bank. Some aspects of their tectonic setting and sedimentary fill are similar to those of other basins along eastern North America (Schlee and Jansa, 1981; Grow and Sheridan, 1981). Like the Scotian margin, Georges Bank is built over a complexly faulted basement whose continued movement during the early stages of basin formation probably influenced sedimentary facies and thickness. The deepest, most areally restricted part of the main basin is inferred to contain as much as 8 km of Lower Jurassic and older evaporitic and carbonate rocks; the COST G-2 well appears to have penetrated part of this sequence. Middle and Upper Jurassic nonmarine terrigenous clastic rocks and marine carbonate rocks (0.4 km thick) were the initial deposits in the subsidence phase of basin formation and signify a transition to an open-shelf environment. Buildup of the carbonate rocks probably started on elevated basement blocks, eventually forming a massive platform that covered much of the southern half of the area. In the western part, the platform appears to have prograded seaward 20 km, perhaps in response to deltaic outbuilding. During the Cretaceous, transgressive and regressive marine and nonmarine clastic sedimentary rocks and thin limestones buried the earlier platform as the broad pattern of basin subsidence continued at a diminished rate. During Cenozoic, the continental slope was cutback periodically and sediment slowly accumulated on the bank. Similar to other eastern North American basins, the interval of basin development spans the Late Triassic(?) to the present. During this interval, more than 10 m of sediment accumulated in several basins and sub-basins.

ACKNOWLEDGMENTS

The paper benefited from the comments of many colleagues, among them John Grow, Kim Klitgord, and C. Wylie Poag. Wylie Poag, Robert Mattick, Elizabeth Winget, and Elizabeth Goode read the manuscript and suggested many improvements. The USGS-Bundesanstalt für Geowissenschaften and Rohstoffe Marine Atlantic cooperative was initiated by Karl Hinz (BGR) and John Behrendt (USGS). As a result of the BGR EXPLORA cruise over the Georges Bank area in 1979, additional profiles were collected over the outer shelf and slope; they enabled us to obtain a much tighter grid of isopachs, a better tie to the COST G-2 well, and a clearer view of the arrangement of deep sedimentary reflectors. Drafting was done by Patty Forrestel and Jeffrey Zwinakis. Numerous versions of the manuscript were typed by Peggy Mons-Wengler.

REFERENCES CITED

Amato, R.V., and E.K. Simonis, 1979, Geological and operational summary, COST No. B-3 well, Baltimore Canyon trough area, Mid-Atlantic OCS: U.S Geological Survey Open-File Report 79-1159, 118 p.

———, and J.W. Bebout, 1980, Geologic and operational summary, COST No. G-1 well, Georges Bank area, North Atlantic OCS: U.S. Geological Survey Open-File Report 80-268, 112 p.

———, and E.K. Simonis, 1980, Geologic and operational summary, COST No. G-2 well, Georges Bank area, North Atlantic OCS: U.S. Geological Survey Open-File Report 80-269, 116 p.

Ascoli, P., 1976, Foraminiferal and ostracod biostratigraphy of the Mesozoic-Cenozoic, Scotian shelf, Atlantic Canada: Maritime Sediments Special Publication 1, Part B, Paleoecology and biostratigraphy, p. 653-771.

Austin, J.A., et al, 1980, Geology of New England passive margin: AAPG Bulletin, v. 64, n. 4, p. 501-526.

Ballard, R. D., and Elazar Uchupi, 1975, Triassic rift structure in Gulf of Maine: AAPG Bulletin, v. 59, n. 7, p. 1041-1072.

Barss, M.S., J. P. Bujak, and G. L. Williams, 1979, Palynological zonation and correlation of 67 wells, eastern Canada: Canada Geological Survey Paper 78-24, 118 p.

Bubb, J.N., and W. G. Hatlelid, 1977, Seismic stratigraphy and global changes of sea level, part 10: seismic recognition of carbonate buildups, *in* C. E. Payton, ed., Seismic stratigraphy - applications to hydrocarbon exploration: AAPG Memoir 26, p. 185-204.

Cornet, Bruce, 1977, The palynostratigraphy and age of the Newark super group: Pennsylvania State University Unpublished Ph.D. Dissertation, 505 p.

Drake, C. L., Maurice Ewing, and G. H. Sutton, 1959, Continental margins and geosynclines: the east coast of North America north of Cape Hatteras, *in* L. H. Ahrens et al, eds., Physics and Chemistry of the Earth, v. 3: London, Pergamon Press, p. 110-198.

Eliuk, L. S., 1978, The Abenaki formation, Nova Scotia shelf, Canada — a depositional and diagenetic model for a Mesozoic carbonate platform: Bulletin of Canadian Petroleum Geology, v. 26, n. 4, p. 424-514.

Emery, K. O., and Elazar Uchupi, 1965, Structure of Georges Bank: Marine Geology, v. 3, p. 349-358.

———, and ———, 1972, Western north Atlantic Ocean: topography, rocks, structure, water, life and sediments: AAPG Memoir 17, 532 p.

Falvey, D.A., 1974, The development of continental margins in place tectonic theory: Australian Petroleum Exploration Association Journal, v. 14, p. 95-106.

Folger, D. W., et al, 1978, Stratigraphic test well, Nantucket Island Massachusetts: U.S. Geological Survey Circular 773, 28 p.

Gardner, G. H. F., L. W. Gardner, and A. R. Gregory, 1974, Formation velocity and density — the diagnostic basics for stratigraphic traps: Geophysics, v. 39, n. 6, p. 770-780.

Gibson, T. G., J. E. Hazel, and J. F. Mello, 1968, Fossiliferous rocks from submarine canyons off the northeastern United States: U.S. Geological Survey, Professional Paper 600-D, p. D222-D230.

Given, M. M., 1977, Mesozoic and early Cenozoic geology of offshore Nova Scotia: Bulletin of Canadian Petroleum Geology, v. 25, n. 1, p. 63-91.

Grow, J. A., 1980, Deep structure and evolution of the Baltimore Canyon trough in the vicinity of the COST No. B-3 well, *in* P. A. Scholle, ed., Geological studies of the COST No. B-3 well, United States Mid-Atlantic continental slope area: U.S. Geological Survey Circular 833, p. 117-126.

———, and R. E. Sheridan, 1981, Deep structure and evolution of the continental margin off eastern United States: Paris, Oceanologica Acta, No. SP, Actes 26 Congres International de Geologie, Colloque Geologie des Margeo Continentales, 7-17, p. 11-19.

———, R. E. Mattick, and J. S. Schlee, 1979, Multichannel depth sections over outer continental shelf and upper continental slope between Cape Hatteras and Cape Cod, *in* J. C. Watkins, L. Montadert, and P. W. Dickerson, eds., Geological and geophysical investigations of the continental margins: AAPG Memoir 29, p. 65-83.

Hall, R. E., L. J. Poppe, and W. M. Ferrebee, 1980, A stratigraphic test well, Martha's Vineyard, Massachusetts: U. S. Geological Survey Bulletin 1488, 19 p.

Hathaway, J. C., et al, 1979, U.S. Geological Survey core drilling on the Atlantic Shelf: Science, v. 206, n. 4418, p. 515-527.

Hinz, K., 1979, Seismic sequences of Cape Bojador, northwest Africa, *in* U. von Rad, et al, (eds.), Initial reports of the deep sea drilling project, v. 47, part 1, p. 485-489.

Jacobson, S. and F. Adinolfi, 1979, Plate 3, Geological cross section of Mesozoic and Cenozoic sediments on the Atlantic continental margin from the southeast Georgia embayment to the Scotian shelf, *in* R. V. Amato and E. K. Simonis eds., Geologic and operational summary, COST No. B-3 well, Baltimore Canyon trough area, Mid-Atlantic OCS: U.S. Geological Survey Open-File Report 79-1159.

Jansa, L. F., and J. A. Wade, 1975, Paleogeography and sedimentation in the Mesozoic and Cenozoic of southeastern Canada, *in* C. J. Yorath, E. R. Parker, and D. J. Glass, eds., Canada's continental margins and offshore petroleum exploration: Canadian Society of Petroleum Geologists Memoir 4, p. 79-102.

Judkins, T. W., E. K. Simonis, and B. A. Heise, 1980, Correlation with other wells, *in* R. V. Amato, and E. K. Simonis, eds., Geological and operational summary, COST No. G-2 well, Georges Bank area, North Atlantic OCS U.S. Geological Survey Open-File Report 80-269, p. 33-36.

Klitgord, K. D., and J. C. Behrendt, 1979, Basin structure of the U.S. Atlantic Margin, *in* J. S. Watkins, L. Montadert and P. W. Dickerson, eds., Geological and geophysical investigations of continental margins: AAPG Memoir 29, p. 85-112.

Lewis, R. S., and R. E. Sylwester, 1976, Shallow sedimentary framework of Georges Bank: U.S. Geological Survey Open-File Report 76-874, 14 p.

Maher, J. C., 1971, Geologic framework and petroleum potential of the Atlantic Coastal Plain and Continental Shelf: U.S. Geological Survey Professional Paper 659, 98 p.

Mattick, R. E., et al, 1974, Structural framework of the United States Atlantic outer continental shelf north of Cape Hatteras: AAPG Bulletin, v. 58, n. 6, part 2, p. 1179-1190.

Mitchum, R. M., P. R. Vail, and S. Thompson III, 1977, Seismic stratigraphy and global changes of sea level, part 2: The depositional sequence as a basic unit for stratigraphic analysis, *in* C. E. Payton, ed., Seismic stratigraphy--applications to hydrocarbon exploration: AAPG Memoir 26, p. 53-62.

Oldale, R. N., et al, 1974, Geophysical observations on northern part of Georges Bank and adjacent basin of Gulf of Maine: AAPG Bulletin, v. 58, n. 12, p. 2411-2427.

Poag, C. W., 1978, Stratigraphy of the Atlantic continental shelf and slope of the United States: Annual Review of

Earth and Planetary Sciences, v. 6, p. 251-280.
———, 1980, Foraminiferal stratigraphy, paleoenvironments, and depositional cycles in the outer Baltimore Canyon trough, *in* P. A. Scholle, ed., Geological studies of the COST No. B-3 well, United States Mid-Atlantic continental slope area: U.S. Geological Survey Circular 833, p. 44-65.
———, 1982, Foraminiferal and seismic stratigraphy, paleoenvironments, and depositional cycles in the Georges Bank basin, *in* P. A. Scholle and C. R. Wenkam, eds., Geological studies of the COST G-1 and G-2 wells, United States North Atlantic outer continental shelf: U.S. Geological Survey Circular, n. 861, p. 43-91.
Ryan, W. B. F., et al, 1978, Bedrock geology in New England submarine canyons: Oceanologica Acta, v. 1, n. 2, p. 233-254.
Sangree, J. B., and J. M. Widmier, 1977, Seismic stratigraphy and global changes of sea level, part 9: Seismic interpretation of clastic depositional facies *in* C. E. Payton, ed., Seismic stratigraphy-applications to hydrocarbon exploration: AAPG Memoir 26, p. 165-184.
———, and ———, 1979, Interpretation of depositional facies from seismic data: Geophysics, v. 44, n. 2, p. 131-160.
Schlee, J. S., 1978, Geology of Georges Bank *in* J. J. Fishers, ed., New England marine geology, new concepts in research and teaching and bibliography of New England marine geology, 1870-1970: Proceedings, 26th Annual Meeting of the New England Section, National Association of Geology Teachers, p. 88-92.
———, 1981, Seismic stratigraphy of the Baltimore Canyon trough: AAPG Bulletin, v. 65, n. 1, p. 25-53.
———, and J. A. Grow, 1980, Seismic stratigraphy in the vicinity of the COST No. B-3 well *in* P. A. Scholle, ed., Geological studies of the COST No. B-3 well, United States Mid-Atlantic continental slope area: U.S. Geological Survey Circular 833, p. 111-116.
———, and L. F. Jansa, in 1981, The paleoenvironment and development of the eastern North American continental margin: Paris, Oceanologic Acta, No. SP, Actes 26 Congres International de Geologie, Colloque Geologie des marges continentales, 7-17, p. 71-80.
———, W. P. Dillon, and J. A. Grow, 1979, A structure of the continental slope off the eastern United States, *in* L. J. Doyle, and O. H. Pilkey, eds., Geology of continental slopes: Society of Economic Paleontologists and Mineralogists, Special Publication 27, p. 95-118.
———, et al, 1976, Regional framework off northeastern United States: AAPG Bulletin, v. 60, n. 6, p. 926-951.
Scholle, P. A., H. L. Krivoy, and J. L. Hennessy, 1980, Summary chart of geological data from the COST No. G-1 well, U.S. North-Atlantic outer continental shelf: U.S. Geological Survey Oil and Gas Investigation Chart, OC 104, 1 sheet.
———, K. A. Schwab, and H. L. Kirvoy, 1980, Summary chart of geological data from the COST No. G-2 well, U.S. North-Atlantic outer continental shelf: U.S. Geological Survey Oil and Gas Investigation Chart, OC 105, 1 sheet.
Schultz, L. K., and R. L. Grover, 1974, Geology of Georges Bank basin: AAPG Bulletin, v. 58, n. 6, part 2, p. 1159-1168.
Uchupi, E., R. D. Ballard, and J. P. Ellis, 1977, Continental slope and upper rise off western Nova Scotia and Georges Bank: AAPG Bulletin, v. 61, p. 1483-1492.
Vail, P. R., R. M. Mitchum, Jr., and S. Thompson III, 1977a, Seismic stratigraphy and global changes of sea level changes, part 3: Relative changes of sea level from coastal onlap, *in* C. E. Payton, ed., Seismic stratigraphy — applications to hydrocarbon exploration: AAPG Memoir 26, p. 63-81.
———, ———, and ———, 1977b, Seismic stratigraphy and global changes of sea level, part 4: Global cycles of relative changes of sea level, *in* C. E. Payton, ed., Seismic stratigraphy — applications to hydrocarbon exploration: Memoir 26, p. 83-97.
Valentine, P. C., 1978, Shallow subsurface stratigraphy of the continental margin off southeastern Massachusetts (Abs.): Geological Society of America Abstracts with Programs, v. 10 (2), p. 90.
———, J. R. Uzmann, and R. A. Cooper, 1980, Geology and biology of Oceanographer Canyon: Marine Geology, v. 38, p. 283-312.
Van Houten, F. B., 1977, Triassic-Liassic deposits of Morocco and eastern North America — comparison: AAPG Bulletin, v. 61, n. 1, p. 79-99.
von Rad, U., and M. A. Arthur, 1979, Geodynamic, sedimentary and volcanic evolution of the Cape Bojador continental margin (northwest Africa) *in* M. Talwani, W. Hay, and W. B. F. Ryan, eds., Deep drilling results in the Atlantic Ocean: Continental margins and paleoenvironment: American Geophysical Union, Maurice Ewing Series 3, p. 187-203.
———, and G. Einsele, 1980, Mesozoic — Cainozoic subsidence history and paleobathymetry of the northwest Africa continental margin: Philosophical Transactions of the Royal Society of London, Series A, v. 294, n. 1409, p. 37-50.
Wade, J. A., 1977, Stratigraphy of Georges Bank basin — Interpretation from seismic correlation to the western Scotian shelf: Canadian Journal of Earth Science, v. 14, n. 10, p. 2274-2283.
Weed, E. G. A., et al, 1974, Generalized pre-Pleistocene geologic map of the northern United States Atlantic continental margin: U.S. Geological Survey Miscellaneous Geological Investigations Series Map I-861, scale 1:1,000,000.
Williams, Harold, 1978, Tectonic lithofacies map of the Appalachian Orogene: Memorial University of Newfoundland, Map No. 1, 2 sheets, scale 1:1,000,000.

Mesozoic-Cenozoic Sedimentary and Volcanic Evolution of the Starved Passive Continental Margin off Northwest Australia

Ulrich von Rad
Bundesanstalt für Geowissenschaften und Rohstoffe
F.R. Germany

Neville F. Exon
Bureau of Mineral Resources, Geology & Geophysics
Canberra City, Australia

The 125 to 160 m.y. old, sediment-starved passive margin off northwest Australia is ideally suited for a study of early rift, breakup, juvenile and mature ocean paleoenvironmental and geodynamic evolution. Triassic to Tertiary rocks, dredged and cored at 50 stations along the flanks of the Wallaby, Exmouth and Scott Plateaus, allow the dating of seismic sequences and correlation with coeval formations drilled in Northwest Shelf petroleum exploration wells.

The northwest Australian rifted margin is a good example of a sediment-starved passive continental margin. Because sedimentation rates were comparatively low after the Oxfordian to Neocomian breakup of Gondwanaland between western Australia and India, seismic profiling and drilling can penetrate very old parts of the sequence and resolve the early stages of margin evolution. These stages cannot be studied as easily in most mature Atlantic-type passive margins, which are usually blanketed by sediment wedges 10 to 15 km thick.

Similar starved passive margins exist in the northeast Atlantic (Biscay, Rockall), where recent deep-sea drilling has considerably improved our understanding of early rifting, crustal attenuation, and subsidence (Montadert et al, 1979a; Montadert et al, 1979b). Ancient counterparts now exposed on land are the Tethyan margins in the southern Alps and the external zones of the Apennines (Bernoulli, Kälin, and Patacca, 1979). Some of the most urgent problems of the development and structural history of passive margins, including the environment of early rifting, the onset of spreading after breakup, and post-rift evolution and subsidence, can be best studied by deep-sea drilling selected transects across starved passive margins.

We used a less expensive, but nevertheless efficient approach. During two Bundesanstalt für Geowissenschaften und Rohstoffe (BGR) cruises — the *Valdivia* 16 (VA16), 1977 cruise (Hinz et al, 1978) and the *Sonne* 8 (SO8), 1979, cruise (von Stackelberg et al, 1980) — we attempted to dredge the well-established seismic sequences, where they crop out along the steep rifted flanks of the sunken marginal plateaus, typical of the northwest Australian margin. Fifty stations recovered a great variety of Triassic to Tertiary pre-, syn- and post-breakup sediments, volcanic rocks, and volcanogenic sediments by dredging and coring the outer plateau margins between 12° and 26°S. This improved our understanding of the geological evolution of the Wallaby, Exmouth, and Scott Plateaus, which was previously inferred from seismic profiles calibrated by commercial wells on the Northwest Shelf. The preliminary results of both cruises are discussed in Hinz et al (1978) and von Stackelberg et al (1980), and the structural re-interpretation of the Exmouth Plateau margins in a paper by Exon et al (in press).

The objective of this paper is to describe and to interpret the lithology of the Triassic to Tertiary rocks dredged and cored during both cruises, and to correlate their facies with coeval sediments sampled from six representative petroleum exploration wells on the adjacent Northwest Shelf. Our synthesis presents the

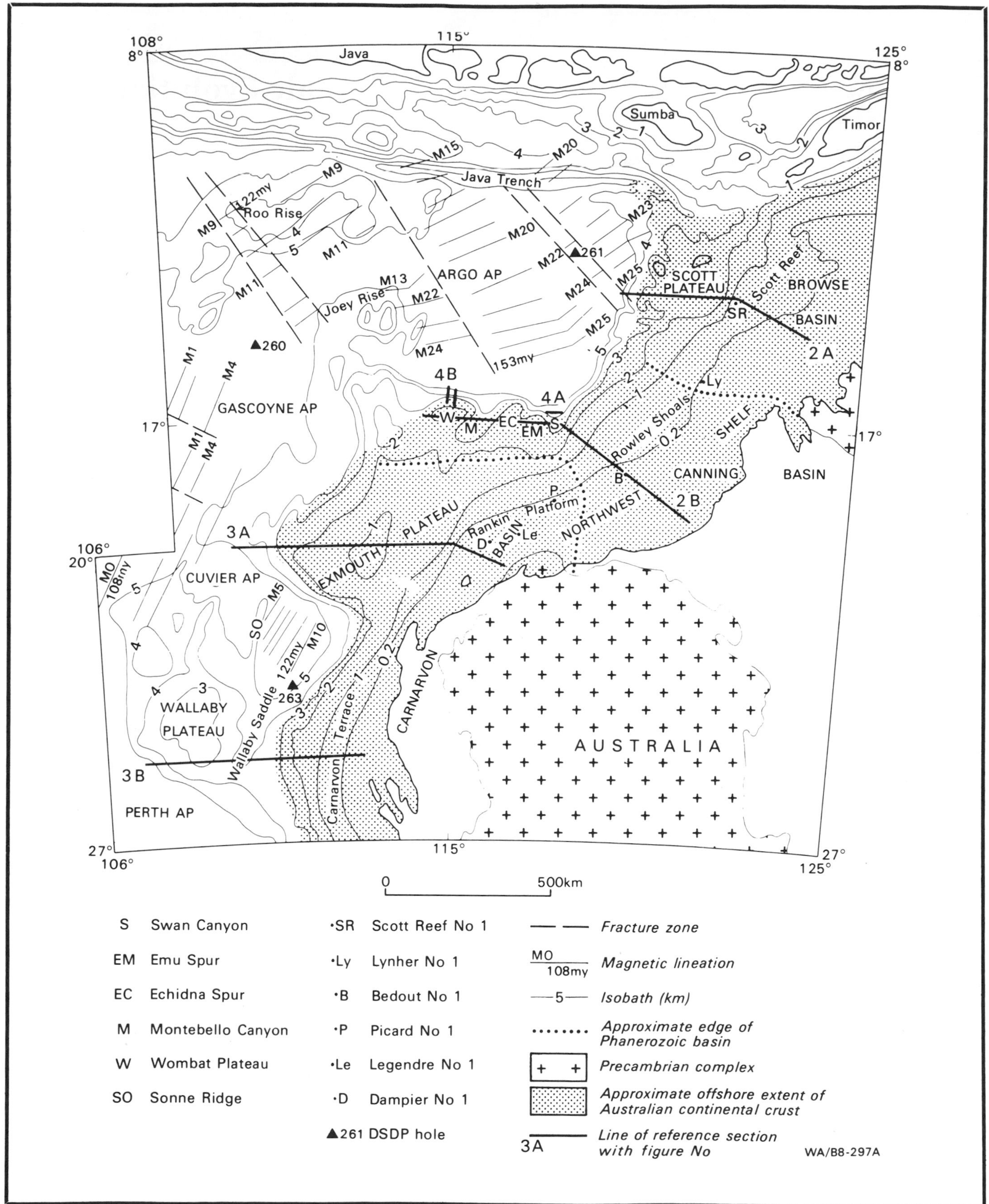

Figure 1 — Regional and tectonic setting of the Scott, Exmouth and Wallaby Plateaus off northwest Australia, showing general *Valdivia* 16 and *Sonne* 8 sampling areas and selected commercial shelf wells. Bathymetry after Falvey and Veevers (1974) and Veevers and Cotterill (1978); magnetic lineations after Heirtzler et al (1978) and Larson et al (1979). Heavy dashed lines = reference sections (Figures 2A, 2B, 3A, 3B).

paleoenvironment and the history of sedimentation, volcanism, and subsidence, especially before, but also during and after, the Oxfordian to Neocomian breakup of Gondwanaland and formation of a small proto-Indian Ocean.

STRUCTURE AND SEISMIC STRATIGRAPHY OF THE CONTINENTAL MARGIN

Structure

The Northwest Australian continental margin between 12° and 26°S consists of three major Paleozoic to Tertiary sedimentary basins located offshore from or between the Precambrian massifs of the Kimberley and Pilbara Blocks (Powell, 1976; Veevers et al, 1974). Our studies concentrate on the evolution of three submerged marginal plateaus: the Scott Plateau in the north, which is the offshore extension of the Browse Basin; the Exmouth Plateau, off the Canning and Carnarvon Basins; and the Wallaby Plateau in the south (Figure 1). A great deal of seismic data is available from the Exmouth Plateau (Exon and Willcox, 1978, 1980; Wright and Wheatley, 1979); a considerable amount from the Scott Plateau (Stagg, 1978; Hinz et al, 1978); and some from the Wallaby Plateau (Symonds and Cameron, 1977). Figure 2 shows schematic cross sections extending from the continental shelf towards the marginal plateaus. The approximate edge of continental crust shown in Figure 1 is defined by seismic and magnetic evidence.

The origins and geology of the Scott, Exmouth and Wallaby Plateaus are rather different. We regard the Scott Plateau as a continental block (Stagg and Exon, 1982), although Veevers and Cotterill (1978) suggested that it is a volcanic buildup related to sea-floor spreading — an epilith. The Exmouth Plateau is widely accepted as a continental block. According to Veevers and Cotterill (1978) and von Stackelberg et al (1980), the Wallaby Plateau is an epilith and not a continental block, as postulated by Symonds and Cameron (1977).

Magnetic lineations (Figure 1) suggest that the Scott Plateau and northern Exmouth Plateau separated from Gondwanaland in the Callovian (Heirtzler et al, 1978); this interpretation is supported by the Oxfordian sediments overlying oceanic crust in DSDP hole 261 (Veevers et al, 1974). Magnetic lineations suggest that the southern margin, and most if not all of the western margin, of the Exmouth Plateau began forming in the Neocomian (Larson et al, 1979). The age of the oldest (Albian) sediments from DSDP hole 260 (Veevers et al, 1974) is compatible with the magnetic information. Magnetic lineations from the Cuvier and Perth Abyssal Plains (Larson et al, 1979; Markl, 1974) indicate that the southwestern margin of Australia separated from "Greater India" in the Neocomian. So, if the Wallaby Plateau is an epilith, it must have formed as separation proceeded.

The Scott Plateau can be regarded as the subsided outer margin of the Browse Basin (Allan, Pearce, and Gardner, 1978). The northern Exmouth Plateau is continuous with, and similar to, the Rowley Sub-basin of the offshore Canning Basin (Warris, 1976). We regard the central and southern parts of the Exmouth Plateau as part of the Carnarvon Basin, described by Thomas and Smith (1974). The Wallaby Plateau is an oceanic feature unrelated to the nearby Carnarvon Basin.

Seismic Stratigraphy

Seismic stratigraphy and reflector nomenclature were first established on the Exmouth Plateau by Willcox and Exon (1976) and Exon and Willcox (1978), and were extended to the Wallaby Plateau by Symonds and Cameron (1977), and to the Scott Plateau by Stagg (1978). The earlier interpretation of the northern Exmouth Plateau was modified due to the results of the *Sonne* 8 (1979) cruise (Exon et al, in press). Figure 5 illustrates the relationship between reflector nomenclature, Northwest Shelf stratigraphy, and the samples described here.

The *F unconformity* (Figures 2 and 3) is the oldest normally seen on seismic records. It can be observed in the Browse Basin (Scott Reef No. 1) but is not present on the Scott Plateau. On the Exmouth Plateau it is a rift-onset unconformity marking the top of tilted fault blocks, and is probably of latest Triassic age. It is not present on the Wallaby Plateau. The sequence underlying the F unconformity is of Triassic and older age. In the Browse Basin (Figure 2A) there is probably a thick Permian and Triassic sequence, deposited in a northeast-trending depocenter which persisted through the Mesozoic, and was apparently a rift-valley (Powell, 1976). Basement rocks were apparently being eroded on the Scott Plateau at the time. In the Canning Basin (Figure 2B), as in the Browse Basin, there is a thick Permian and Triassic sequence below the F reflector. On the Exmouth Plateau (Figure 3A) the Triassic sequence is variably reflecting and more than 1,000 m thick. Seismic and sampling evidence suggests that it generally consists of fluviodeltaic sediments; a sequence of intermediate volcanics (Late Triassic and earliest Jurassic in age) preceded or accompanied rifting on the northern and western margins of the Wombat Plateau (Figure 2B).

The *E unconformity* is the breakup unconformity of the Scott Plateau, Browse Basin, Rowley Terrace, Dampier Sub-basin of the Carnarvon Basin, and the northern Exmouth Plateau. It is not identifiable on the Exmouth Plateau proper, and is absent from the Wallaby Plateau. It is an angular unconformity which was completely planed off in many areas; fault displacements are generally minor, except in marginal areas. It is basically of Callovian age. The F-E sequence consists typically of paralic detrital sediments of Early and Middle Jurassic age (Figure 5). Coal measures are widely present, and shelf carbonates are widespread on the northern Exmouth Plateau. The sequence is thin and patchy on the Scott Plateau and the Exmouth Plateau proper, and absent from the Wallaby Plateau. On the northern Exmouth Plateau it is variably reflecting and up to 3,000 m thick; it is also thick in the Browse Basin, the Rowley Sub-basin, and the deeper parts of the Dampier Sub-basin (Figures 2 and 5).

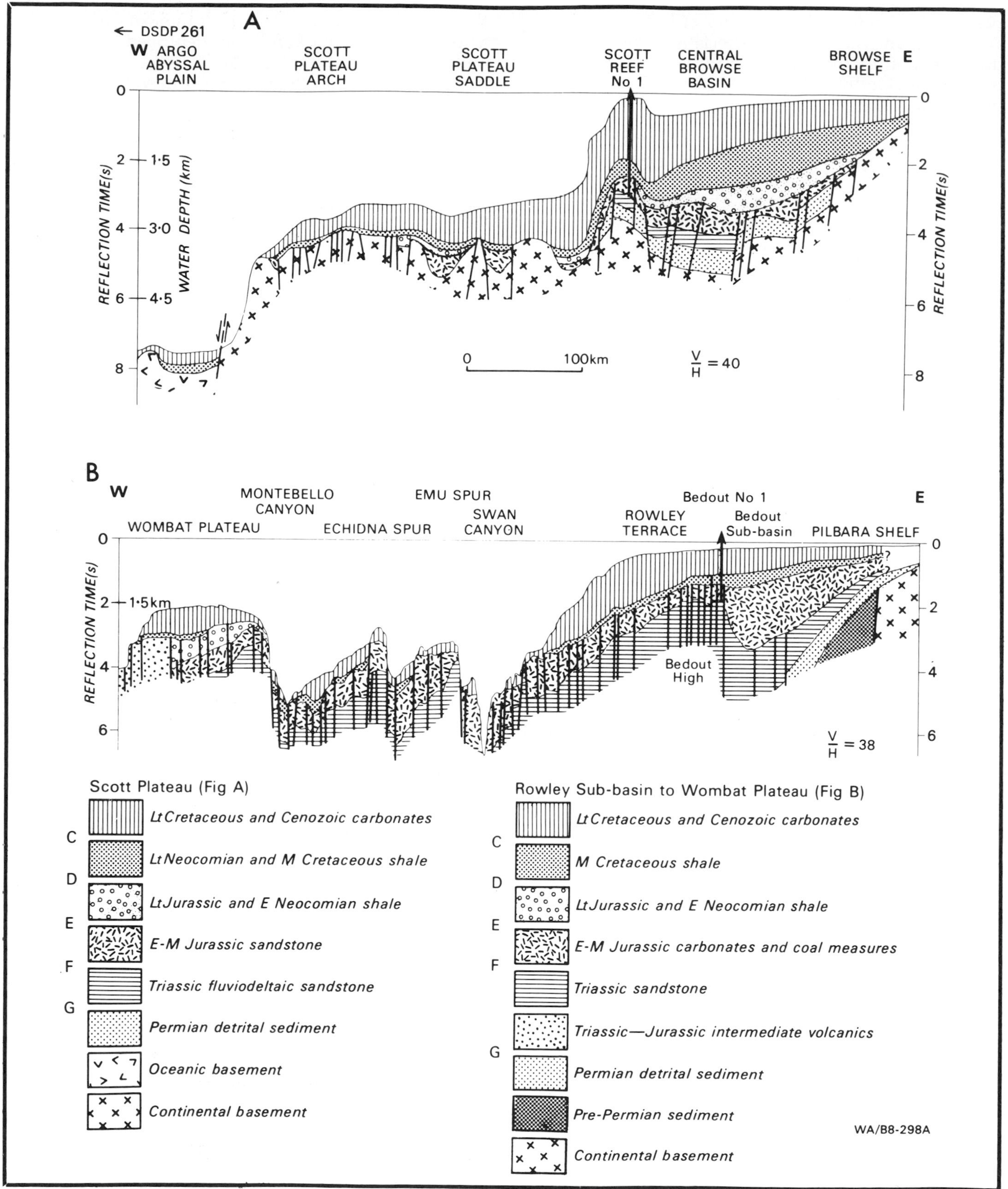

Figure 2 — Generalized tectonic cross-sections across Scott Plateau and Browse Basin (A), and across northern Exmouth Plateau and Bedout Sub-basin of Canning Basin (B). Scott-Browse cross-section based on BMR 18/061 west of Scott Reef, and on Powell (1976) and Allen et al (1978) east of Scott Reef. Exmouth-Bedout cross-section based on BMR 17/079 west of Rowley Terrace, Gulf Au8 between Rowley Terrace and Bedout No. 1, and processed Burmah Oil Company line (Veevers and Johnstone, 1974) east of Bedout No. 1. Approximate position of cross-sections shown in Figure 1.

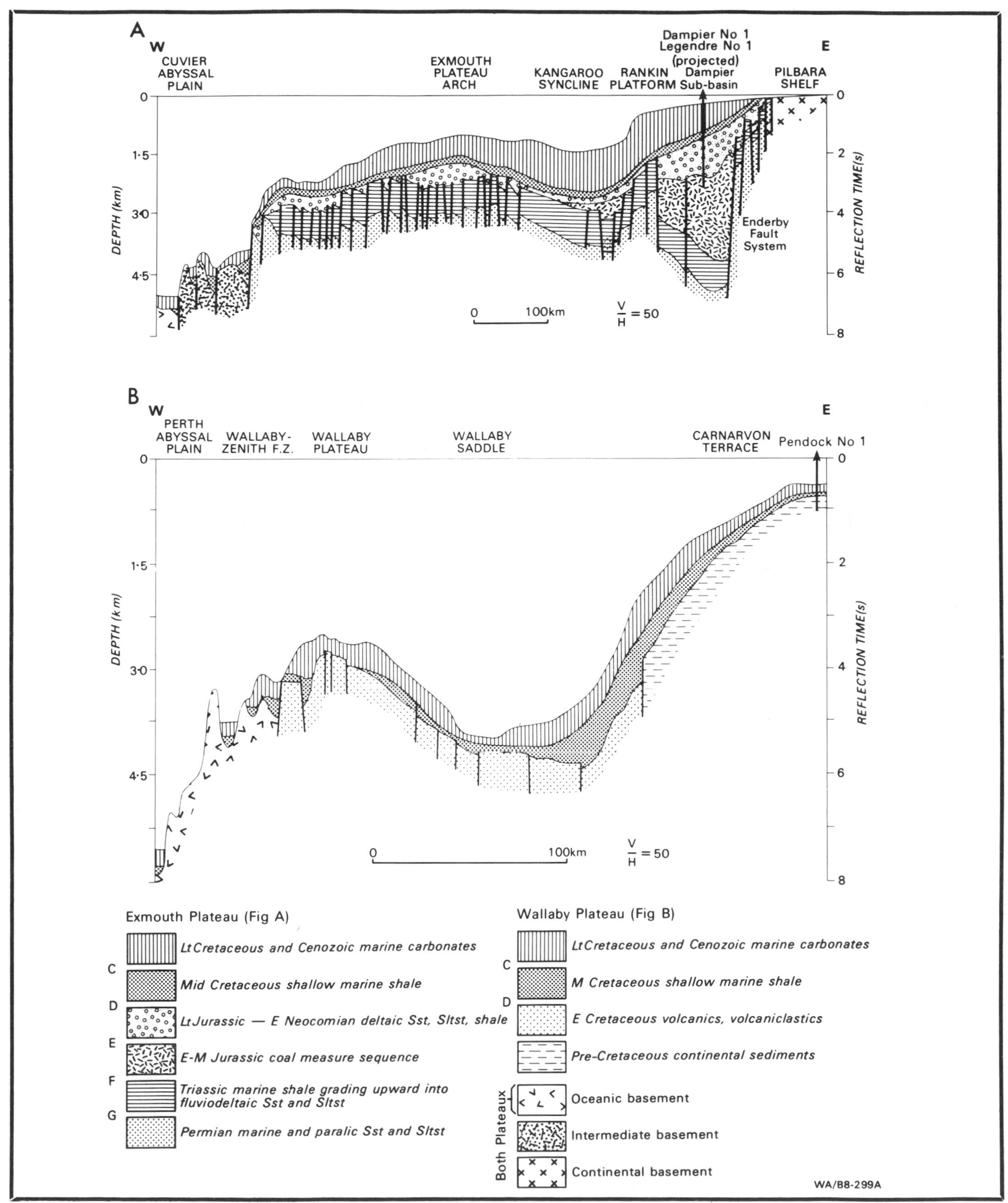

Figure 3 — Generalized tectonic cross-sections across central Exmouth Plateau and Dampier Sub-basin of the Carnarvon Basin (A), and across Wallaby Plateau and Carnarvon Terrace (B). Exmouth-Dampier cross-section based on BMR 17/068, and published company cross-sections (Thomas and Smith, 1974; Powell, 1976). Wallaby-Carnarvon cross-section based on BMR 17/046 (after Symonds and Cameron, 1977). Approximate position of cross-sections shown in Figure 1.

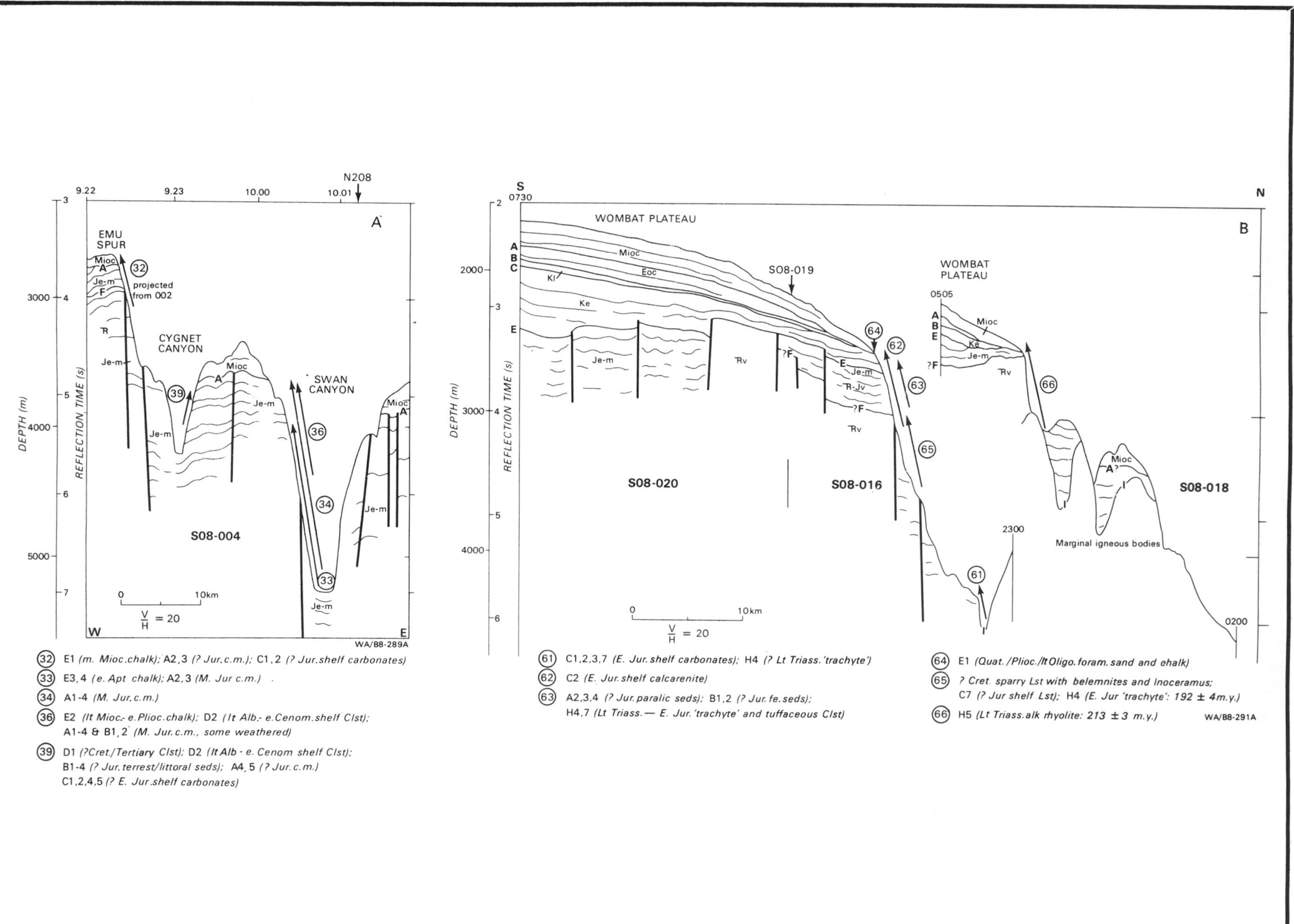

Figure 4 — Line drawing of seismic profiles across Swan Canyon-Emu Spur(A) and northern flank of northern Wombat Plateau (B), with seismic stratigraphy and sampling results. Profiles on northern Exmouth Plateau (see von Stackelberg et al, 1980, their Figure 5). Reflector nomenclature in Figure 6. I is top of igneous basement.

The *D reflector* is not easily recognizable in the Browse Basin, the Rowley Sub-basin, and much of the Dampier Sub-basin. Although it is the Neocomian breakup unconformity of the Exmouth Plateau proper, it seldom displays any angularity, but marks the top of a deltaic sequence. Only on the Wallaby Plateau (Figure 2D) is it a marked angular unconformity, separating dipping older beds from onlapping younger beds. The D reflector is generally little displaced by later faulting. The E-D sequence is thin or absent on the Scott Plateau, in the Browse Basin, and in the Rowley Sub-basin (Figure 5). It consists of deltaic sandstone and siltstone, and prodelta mudstone on the Exmouth Plateau and in the Barrow Sub-basin of the Carnarvon Basin. At least two deltas existed; both were as much as 2,000 m thick, one building north across Exmouth Plateau, the other north into the Barrow Sub-basin.

The Exmouth Plateau delta is weakly to moderately reflecting, and shows strong progradation. Seismic evidence suggests that it is of similar age to the Barrow Formation of the Barrow Sub-basin, i.e. latest Jurassic and early Neocomian. The muddy pro-deltaic sequence is acoustically almost transparent. The E-D sequence appears to be present in the Wombat Plateau area, where it is up to 1,000 m thick, but not elsewhere along the northern margin (Figure 2B). On the Wallaby Plateau the sequence below the D uncon-

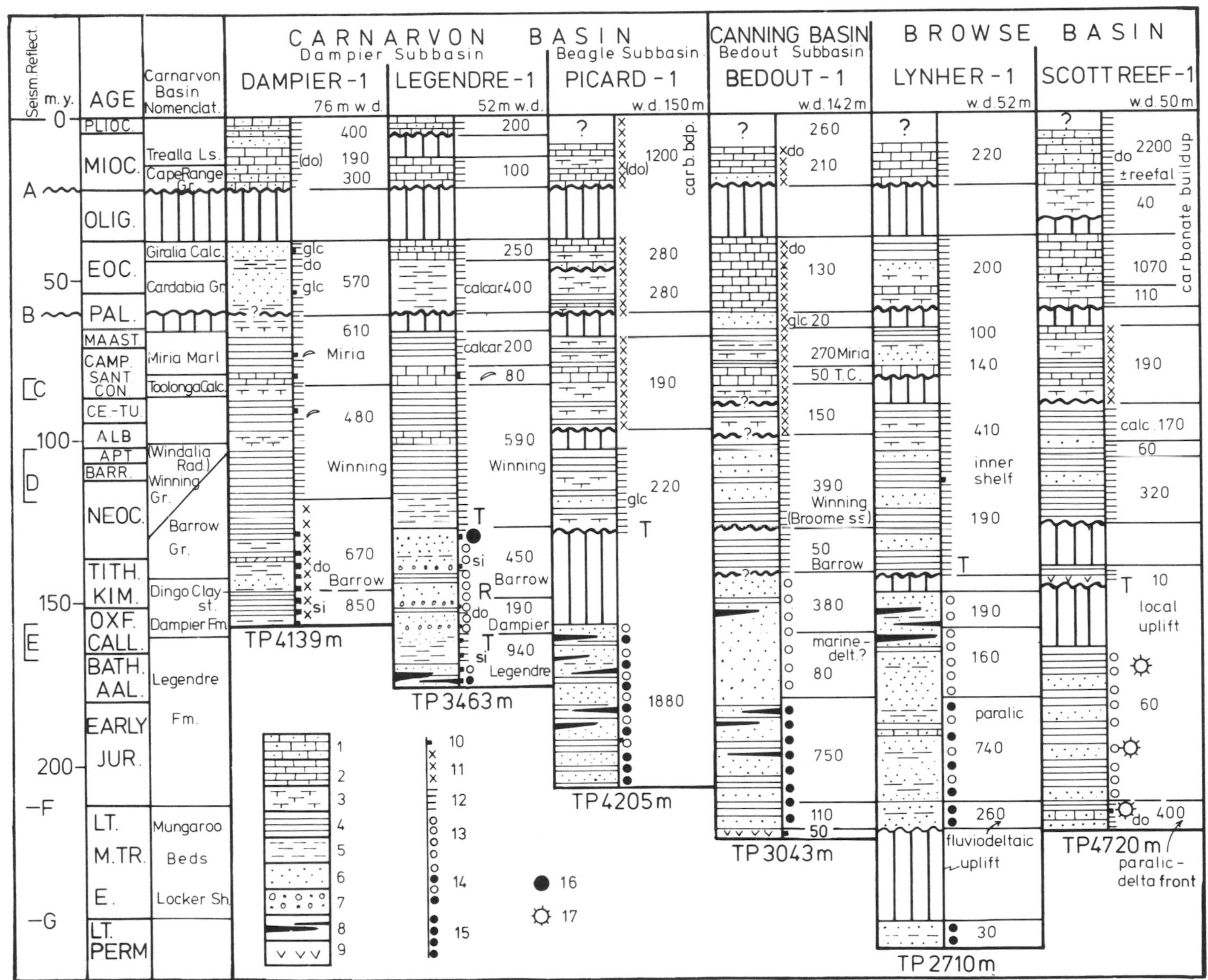

Figure 5 — Generalized lithostratigraphy of six selected Northwest Shelf petroleum exploration wells, extending from Scott Reef No. 1 in the northeast, to Dampier No. 1 in the southwest. Reflector nomenclature as in Figure 6. Wells located in Figure 1. Symbols: 1 = calcarenite; 2 = calcilutite, calcisiltite, chalk; 3 = marl (stone); 4 = claystone, shale; 5 = siltstone; 6 = fine to medium sandstone; 7 = coarse sandstone to conglomerate; 8 = lignite, coal; 9 = volcanics & volcaniclastics; 10 = investigated core samples (Table 2). Environments: 11 = outer shelf to bathyal; 12 = inner shelf; 13 = paralic to marine deltaic; 14 = paralic to fluviodeltaic; 15 = terrestrial (fluvial); 16 = oil shows; 17 = gas shows; T = transgression, R = regression; do = dolomite; glc = glauconite; si = siderite; calcar. = calcarenite; w.d. = water depth; bdp = buildup; shell symbol = mollusks; calc = calcilutite; TP = total penetration. All thicknesses in meters.

formity appears to consist of dipping volcanic and volcaniclastic rocks (von Stackelberg et al, 1980). It is well-bedded and more than 2,000 m thick.

The C *reflector* marks the base of the carbonates which characterize the Late Cretaceous and Cenozoic. Carbonate deposition replaced detrital deposition along the western Australian margin in the Santonian (Veevers & Johnstone, 1974). The reflector is a gentle angular unconformity in places, and is seldom cut by faults. The D-C sequence is generally present throughout the region, and from 100 to 400 m thick (Figure 6). It is the Winning Group of the Carnarvon Basin, and consists largely of shallow-marine claystone with occasional beds of siltstone, sandstone, or marl. The sequence is everywhere acoustically transparent to weakly reflecting. It is thin or missing in areas which were structurally high at the time; on some horst blocks on the Scott Plateau and northern Exmouth Plateau (Figure 2), and on local highs on the Wallaby Plateau (Figure 3B).

The *carbonate (post-C) sequence* is several thousand meters thick in places on the Northwest Shelf, where it is often dominated by prograded Miocene shelf calcarenite. It is much thinner on the plateaus, where it consists largely of marl and pelagic carbonate (Figures 2 and 5). The average thickness of the sequence on the Scott, Wombat, and Wallaby Plateaus is about 500 m, and on the Exmouth Plateau about 700 m (Figure 2). The carbonate sequence is cut by two major unconformities (*reflectors B and A*) of Paleocene and Oligocene age, respectively. The unconformities were probably a result of the strong flow of cold, oxygen-rich water around Australia at those times, causing non-deposition, erosion, or carbonate solution (Exon and Willcox, 1978).

On the plateaus the Upper Cretaceous (C-B) sequence is generally strongly reflecting and less than 200 m thick, although there are some bands 400 m thick on the Exmouth Plateau. Numerous small diffractions suggest some gravitational movement on the Scott and Wombat Plateaus. The Paleocene-Eocene (B-A) sequence is moderately reflecting, and probably lithologically similar to the C-B sequence (mudstone, marl, pelagic carbonate). It is up to 600 m thick on the Exmouth Plateau, where small diffractions suggest gravitational movement, but it is much thinner on the other plateaus. The Miocene to Recent (A-seabed) sequence is generally acoustically transparent, consisting largely of homogeneous pelagic carbonates. It is several hundred meters thick on the Wombat Plateau, where its lower half is well-bedded, and may consist of interbedded claystone, marl, and pelagic carbonate.

LITHOFACIES AND STRATIGRAPHY

The most important data of the dredged and cored pre-Quaternary VA16 and SO8 samples are tabulated in Hinz et al (1978, Table 2) and von Stackelberg et al (1980, Table 1). Altogether, 111 thin sections were studied from a great variety of rocks to ascertain their texture, fabric, composition, biogenic content, and paleoenvironmental significance (Plates 1 to 4). For comparison we also investigated 46 thin sections from the Triassic to Eocene cores of six petroleum exploration shelf wells (Table 2; Figure 6). Table 1 explains the lithofacies types distinguished, which are related to the regional stratigraphy in Figure 6. Selected samples were studied by XRD analysis. Volcanic rocks were investigated by optical and geochemical methods (VA 16: M. Mohr, Hannover, H.-U. Schmincke, Bochum; SO 8: J. Emmermann, Karlsruhe; A.L. Jaques, Canberra) and dated by the K/Ar method (H. Kreuzer, BGR; Table 3). The biostratigraphy is based mainly on the determination of calcareous nannoplankton, foraminifera, and palynomorphs (von Stackelberg et al, 1980; Quilty, in press). The following description covers mainly the newly discovered lithostratigraphic units from the outer plateau margins.

Mid-Triassic to Upper Triassic Sediments

Mid-Triassic rocks were dredged only from the southern Wombat Plateau (northern Exmouth Plateau) below seismic reflector F. They consist of gray, very poorly-sorted, fine-grained graywacke (lithofacies A3, Table 1) with mica and plant remains grading into quartz- and mica-rich, carbonaceous muddy siltstone to silty claystone (lithofacies A2/3). The palynological assemblages indicate deposition in a brackish environment (D. Burger, personal communication).

Upper Triassic to lowermost Liassic rhyolitic and intermediate volcanics (lithofacies H)

During a major rift phase at the Triassic/Jurassic boundary, a very distinctive silica-rich to intermediate volcanic rock suite erupted, which is chemically and petrographically different from the basaltic suite recovered along the Wallaby and Scott Plateau margins. One rock type (lithofacies H4, Table 1), which contains feldspar laths in a fluidal (trachytic to pilotaxitic) orientation and a few plagioclase microphenocrysts and irregular amygdules (Plate 4:1), can be chemically described as a mugearite or silica-undersaturated trachyte (sensu lato; Table 1; R. Emmermann, personal communication). The K/Ar age is 192 $\pm$ 8 million years (Hettangian), which correlates well with the occurrence of these rocks below the mid-Jurassic reflector E (Table 3).

The other volcanic rock type (H5) is a very silica-rich (68%) and K_2O-rich alkali rhyolite with porphyritic texture and large euhedral sanidine phenocrysts in a fine-grained matrix of alkali feldspar, opaques and titanite pseudomorphs (Plate 4:2). The fresh sanidine phenocrysts were separated and gave a very reliable K/Ar age of 213 $\pm$ 3 million years whereas the whole rock K/Ar age is 206 $\pm$ 6 million years (latest Triassic to earliest Liassic; Table 3).

Lower Jurassic shallow-water carbonates (lithofacies C)

This very heterogeneous group of shallow-water marine carbonates is restricted to the northern margin of the Exmouth Plateau and to parts of the Dampier Sub-

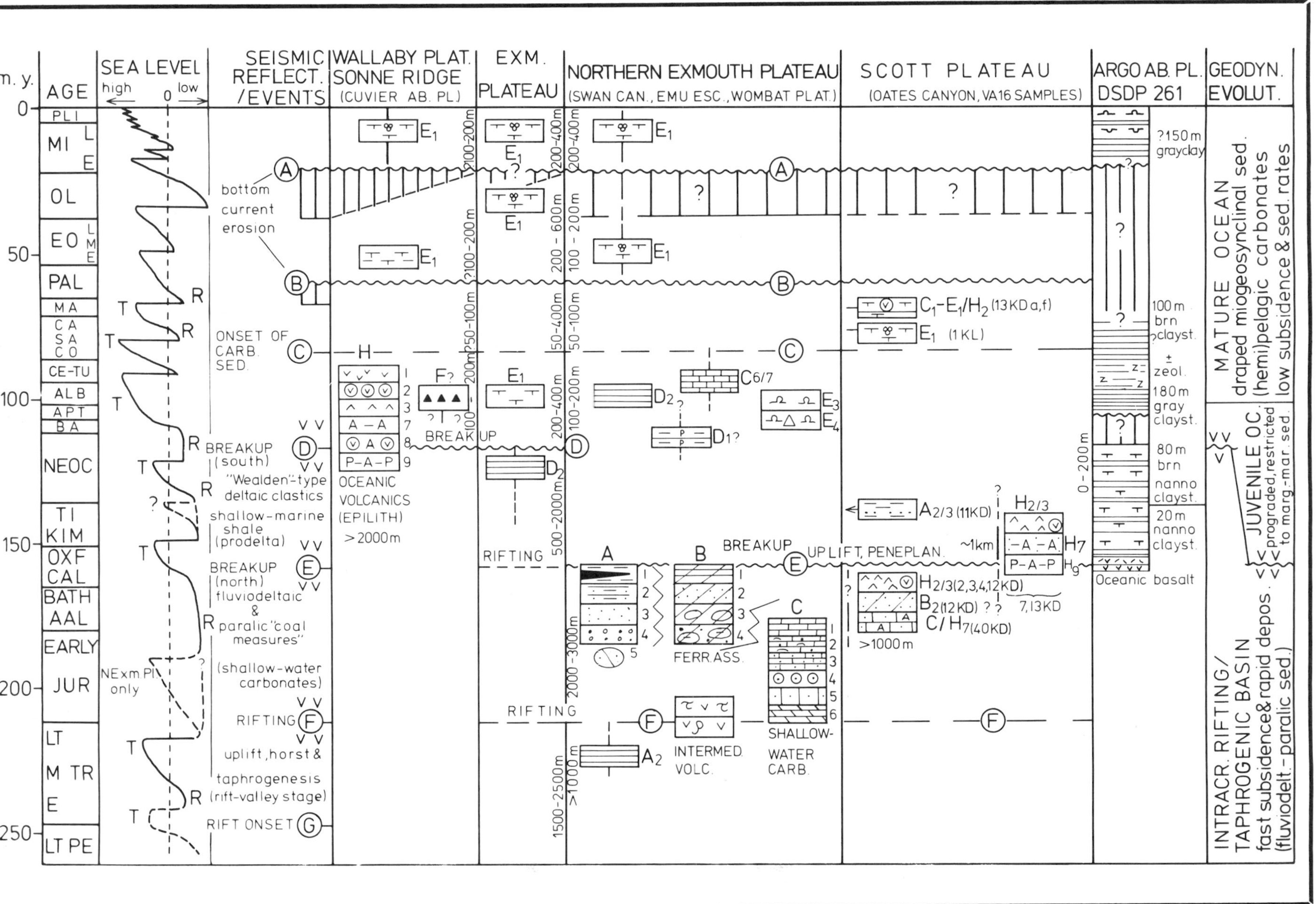

Figure 6 — Stratigraphic distribution of dredged and cored pre-Quaternary lithotypes from the Wallaby, Exmouth, and Scott Plateaus with seismic stratigraphy, sea level fluctuations and generalized geodynamic evolution. DSDP Site 261 for comparison (Veevers, Heirtzler et al, 1974). Sea level fluctuations generalized from Vail, Mitchum et al (1978), adapted to the west Australian regional regression-transgression history from various sources (T = peak transgressions, R = peak regressions). E 1, D 2, etc., designate lithofacies types (Table 1). Lithological symbols are self-explanatory (Table 1).

basin and its northeastern extension, the Beagle Trough.

Although many limestones were too highly recrystallized for biostratigraphic determination, a few biocalcarenites (especially SO8-61KD/1-3; lithofacies C2/3) contained benthic foraminifera (Nodosariidae), probably indicating deposition in an inner shelf environment. The foraminiferal fauna is of late Sinemurian to Pliensbachian age (Quilty, in press; B. Zobel, personal communication). Common macrofossils in lithofacies C2/3 include crinoids (*Isocrinus* and *Pentacrinites*), thin-shelled *Inoceramus* fragments, belemnites (? *Belemnopis* sp), brachiopods, ostracods and calcareous algae.

At the time of the SO8 cruise this was the first record of marine Liassic limestones from western Australia. However, recently Crostella and Barter (1980) reported a thin (?early) Sinemurian "basal limestone" unit underlying the Dingo Claystone in the Dampier Sub-basin (Dampier No. 1, Figure 5). This limestone is oolitic and contains skeletal fragments of crinoids and mollusks very similar to our lithofacies C2.

Although some of the C lithofacies types might be younger than Liassic, they are briefly described together below (Table 1):

Micritic limestones (calcilutite to calcisiltite; C1) usually contain quartz and sometimes recrystallized foraminifera and mollusks. The matrix is a homogeneous micrite, partly recrystallized to microsparite. A biomicrite to biopelmicrite from the Scott Plateau (VA16-13KD/a) contains foraminifera, pelecypods, and oncoidally coated grains. According to Wilson's (1975) definition of Standard Micro Facies, these sediments should have been deposited under quiet conditions on the outer shelf.

We also recovered biocalcarenites (C2) with abundant foraminifera, mollusks (partly including *Inoceramus* sp.) and echinoderms (mainly crinoids), and some quartz (Plate 2:1). Many bioclasts are oncoidally coated, and undifferentiated carbonate intraclasts and (?algal) lumps are common. The matrix is a dense micrite, partly recrystallized to microsparite. One sample (SO8-62KD/1) is a biopelsparite containing micritic peloids or (?fecal) pellets and calcareous algae, in addition to foraminiferal and molluscan remains.

Quartzose biocalcarenites (C3) show transitions to facies C2 and C5 and contain 20% quartz and 10% feldspar, as well as foraminiferal and crinoid fragments. They reflect an open-marine environment. Traces of volcanic rock fragments indicate reworking of the Late Triassic to earliest Jurassic volcanics of the nearby northern Wombat Plateau (Plate 2:5).

A very coarse crinoid biosparite is a rudstone, normally observed in the fore-reef area. Two types are distinguished: a) a poorly-sorted algal-mollusk-echinoderm-biocalcarenite to/-rudite with abundant foraminifera, mollusks, colonial corals, echinoid spines, crinoids, bryozoans, brachiopods, fish teeth, phylloid (?) calcareous algae, and many algally coated bioclasts (Plate 2:6); and b) a well-sorted monomict crinoid biosparite with 95% crinoid fragments, welded together by eutaxial calcite overgrowth (Plate 2:3). A very similar, but dolomitized crinoid breccia of Late Triassic (?) age occurs in the Scott-Reef No. 1 well (Plate 2:4; Table 2).

A well- to moderately-sorted calcite-cemented quartz arenite or highly quartzose calcarenite (C5) consists of 40% quartz, some feldspar, reworked clayey ironstone fragments, crinoid, and mollusk fragments, cemented by sparry calcite (Plate 3:1).

Dolomitized, recrystallized sparry limestone (C6) contains *Inoceramus* prisms, Rhynchonellidae, ostracods, bryozoans, and crinoids. Yellowish-brown dolomite (dolosparite; C7) contains traces of detrital quartz, feldspar, chert, and volcanic rock and glass fragments, as well as not recrystallized echinoderms and bivalves. The dolomite rhombs are markedly zoned and contain dark (organic?) nuclei (Plate 2:2). The low nucleation rate suggests late-diagenetic dolomitization of a bioclastic limestone similar to facies C2, C4, or C6 (H. Zankl, personal communication).

(Middle) Jurassic coal measure sequence (lithofacies A): Legendre Formation — Dampier Formation/Dingo Claystone

Rocks belonging to this lithofacies association, which is common along the northern margin of the Exmouth Plateau but rare on the Scott Plateau (Figure 6), are present in most Northwest Shelf wells (Figure 7; Table 2; Plate 1:5, 1:6, 1:7).

This heterogeneous group of carbonaceous clastic sediments closely resembles the characteristic cyclic coal measures of the Late Carboniferous of North America, or the Permian and Triassic of Australia. In general, these sediments are carbonate-free and barren, except for pollen and spores in A1 and A2/3, which generally indicate a middle Jurassic (pre-Callovian, J 4-5) age, and a non-marine to brackish environment (D. Burger, personal communication).

A black, vitreous, immature subbituminous coal (60% C) with comparatively low reflectivity (A1) occurs in thin stringers (less than 5 cm thick) between A2 mudstones (Plate 1:1). This coal was found only along Swan Canyon (northern Exmouth Plateau) and dated late Middle to early Late Jurassic (in one case ? Toarcian).

A gray, carbonaceous silty claystone or fissile shale (A2) is very rich in quartz, mica, kaolinite, and carbonaceous material (tiny plant fragments, etc., up to 10%), oriented parallel with the bedding plane. The massive shales alternate with thin, well-laminated siltstone beds (Plate 1:2 and 1:3). Although the sediment is highly bioturbated, no fossils other than palynomorphs are present.

The A2 facies grades into carbonaceous quartz siltstones or very fine sandstones (A3). In addition to abundant quartz, they contain common feldspar (mainly plagioclase), mica, traces of heavy minerals, and rock fragments (chert, shale, crystalline rocks). The clayey matrix is predominantly kaolinite. Some very poorly-sorted clayey sand-stones are subgraywackes rich in rock fragments. These rocks are often parallel-bedded and small-scale cross-laminated.

The A3 siltstones and very fine sandstones alternate

Table 1 — Lithofacies associations and types of investigated pre-Quaternary *Valdivia* 16, *Sonne* 8 and, Northwest Shelf samples. The lithofacies designations are also used in Table 2 and throughout the text. For detailed station, core and dredge data of *Valdivia* 16 cruise (see Table 2 in Hinz et al, 1978); for those of *Sonne* 8 cruise see von Stackelberg et al (1980; their Table 1).

Lithofacies Associations and Types	Age	Paleoenvironment	Areas
A: Coal measure association	(mid-) Jurassic	fluviodeltaic-paralic	N Exm. Plat. (sub-lithotypes 1-6)
(1) black vitreous coal, (2) carbonaceous, quartzose & micaic silty claystone, (3) ± carbonaceous qtz siltst to v.f. sst, (4) fine to med qtz sst., (5) pyrite concretions, (6) pink siltst./claystone.		(terrestrial flood plain, coal swamp, delta, ?lagoonal-estuarine, ? prodeltaic)	(Exm. Plat.: 2) (?Scott Plat.: 3) NW Shelf (1-5)
B: Ferruginous association	? Jurassic	? fluviodeltaic, later subaerially exposed to arid conditions reworking of B1 (littoral)	N. Exm. Plat. (1-4)
(1) brown clayey ironstone, (2) ferruginous qtz sdst. (sdy ironstone), (3) ferruginous concretions & boxstones, (4) brown ironstone breccia ("Trümmereisenerz")	(? late Jurassic)		(Scott Plat.: 2)
C: Shallow-water carbonates (mostly pre-Tertiary)	? Jurassic	shelf:	N. Exm. Plat. (1-7)
(1) micritic limest., (2) foram.-moll.-echin. biocalcarenite & biopelmicrosparite, (3) qtzose biocalcarenite, (4) v. coarse crinoid biosparite and algal-mollusk-echin. biocalcarenite, (5) calcite-cem. qtz. arenite, (6) recryst. sparry lst., (7) dolomicrite to -sparite (± siderite)	C2/6: middle Liassic (lt. Sinemurian-Pliensbachian)	quiet water (1, 2, 6), perireefal/bank tops (4), inner shelf-littoral (5);	? Scott Plat. (1, 2) (?NW Shelf: part of Dampier Subbasin)
D: Marine claystone/siltstone	Lt. Jur. -E. Cretac.	restricted marine (2)	N Exm. Plat. (1, 2)
(1) phosphatized claystone, (2) micac. clayey siltst. ± glauc., (3) glauc. -qtz sst, ± phosphatic, calcar. to dolomitic (to biospar.)		poorly oxygenated, lagoon., estuar.,? prodelta; ? outer shelf (1, 3)	Exm. Plat. (2) Scott Plat. (1) NW Shelf (2, 3)
E: Pelagic, ± siliceous chalk	mid-Cret. to Tert.	hemipelagic (2, 5)	(NW Shelf +) all plateaus (1) N Exm. Plat. (1-4) Scott Plat. (5)
(1) semiconsol. foram. nanno chalk, (2) qtzose & radiol. chalk, (3) weakly silicified rad. chalk, (4) radiol. porcellanite (5) pel. foram. (-mollusk) biomicrite to marlstone	Tertiary E. Cret. (E.Apt.) Lt. Cret. (Camp. - Maastr.)	eupelagic (1, 3, 4), ± bathyal; ? upwelling (3, 4)	
F: Quartz chert/orthoquartzite	?early Mesozoic (1)	? pelagic (1);	Carnarvon Ter. (1)
(1) ferrugin. qtz chert, (2) recryst. orthoquartzite	late Triassic (2)	diagen. matured qtz-aren. (2)	NW Shelf (2)
G: Fe/Mn nodules (1) and crusts (2)	subrecent	slow in-situ growth	all plateaus
H: Volcanic and volcaniclastic rocks (± altered)	"mid-Cretaceous" (pre-89 m.y.)	? shallow-water extrusion (post-breakup)	Wallaby Plat./ Sonne Ridge (1-2) Scott Plateau
(1) tholeiitic basalt (olivine tholeiite), (2) ?tholeiitic basalt breccia (3) differentiated alkali basalt (?hawaiite) (4) silica-undersat. "trachyte" (s.l.) or mugearite (feldsp.-rich) (5) alkali rhyolite, (6) rhyolitic pumice, (7) altered tuff to lapillistone, (8) volcaniclastic sst/breccia, (9) phosphatized white pyroclastic claystone (10) tuff, altered to zeolite (phill.) or palygorskite	Lt. Jur./E. Cret. (pre-132 m.y.) early Liass. (pre-192 m.y.) (5): Lt. Triass. (213 m.y.) Lt. Jur./E. Cret.; "mid-Cretac." ?Lt. Jur./E. Cret.	subaerial or shallow-water tuffs, reworked in high-energy envir. and diagen. altered	N. Exm. Plat. (Wombat Pl.) Scott Plat., N. Exm., Sonne R., Wall. Plat. Scott Plat.

with moderately- to well-sorted fine to medium quartz sandstones (A4) with abundant, well-rounded quartz, common feldspar, and rock fragments (crystalline rocks, chert), all embedded in a clayey (mainly kaolinitic) matrix (Plate 1:4). The sandstones grade from impure quartz arenites to subarkoses. The fabric is mostly massive, sometimes flaggy, and rarely parallel to cross-bedded.

Dark gray ellipsoidal "pyrite concretions" (A5) actually consist of a pyritized moderately-sorted, medium quartz sandstone.

Color (and possibly lack of carbonaceous material) differentiates the pink to dusky red siltstones/claystones (A6) from the A2 silty claystones. These sediments suggest an oxidizing nonmarine (?flood plain) environment.

Mesozoic (? Jurassic) ferruginous facies association (lithofacies B)

These reddish-brown, unfossiliferous ferruginous sediments may be subaerially weathered analogs of the coal measure rock types and, therefore, also of Jurassic age. They are restricted to the northern Ex-

Table 2 — Location, age, rock description (XRD- and thin-section analysis), lithofacies, and paleoenvironment of studied samples from six Burmah Oil Company Northwest Shelf wells. Lithofacies associations (A-H), see Table 1. Wells: Scott Reef No. 1: 14°58′S, 117°37.35′E; Lynher No. 1: 15°56.4′S, 121°05′E; Bedout No. 1: 18°14.7′S, 119°23.4′E; Picard No. 1: 18°58′S,

Well	Core	Depth Below Rotary Table (feet)	Subbottom Depth (m)	Location Water Depth (w.d.in m)	Age
Scott Reef No. 1	1	14362′10-12.5″		offshore	
	1	14363′ 9-10.5″	4318	Browse	? Late
	1	14370′10-11″		Basin	Triassic
	1	14372′ 6- 8″	4320	w.d.:50	
Lynher No. 1	1	4088′ 0- 2″	1179	Browse	Tithon.-
	1	4091′ 9-11″	1180	Basin	Neocom.
Bedout No. 1	1	9961′ 9-12″	2857	offshore	(? pre-)
	1	9973′ 3- 6″	2861	Canning B.	mid
	1	9976′ 7-12″	2862	w.d.:170	Triassic
Picard No.1	1	7630′ 8- 9″	2175	Beagle	middle
	1	7232′ 6- 8″	2176	Subbasin	Jurassic
	1	7636′ 6- 9″	2178	of offsh.	
	3	13810′10-12″	4059	Canning	early
	3	13820′ 8-10″	4062	Basin	Jurassic
	3	13822′ 0- 2″	4063	w.d.:141	
Legendre No.1 (19°40.3′S,116°43.9′E)	4	4161′10-12″	1207		? Santon.
	10	6343′11-12″	1872		early Neoc.
	11	6979′ 4- 6″	2066		?Tith.-e.Neoc.
	12	7725′10-12″	2293	Dampier	
	12	7727′ 6- 9″	2294	Sub-	Late
	13	8155′ 6- 9″	2424	basin	Jurassic
	13	8157′ 9-12″	2425	of	(Kimmer.
	13	8163′ 8-11″	2427	offshore	to Oxford.)
	14	8318′ 3- 6″	2474	Car-	
	14	8327′ 6- 8″	2477	narvon	
	16	9758′ 3- 7″	2913	Basin	M.Jur.
	16	9758′ 0- 3″	2913		(Bathon.)
	17	10446′ 1- 4″	3123	w.d.:52	
	18	11108′10-12″	3324		Bath.-Bajoc
Dampier No.1 (19°24′S,116°O′E)	1	3563′ 9-11″	1001		Eocene
	1	3580′ 8-10″	1007		
	2	4336′ 0- 2″	1236		Paleocene
	2	4338′ 2- 4″	1237	Dampier	
	5	6472′10.5-12″	1887	Sub-	Maastr.-
	6	7121′ 3- 5″	2085	basin	Santon.
	7	7773′ 3- 5″	2284	of	Alb-Sant.
	8	8603′ 6- 8″	2537	offshore	? mid-Cret.
	9	9303′ 7- 8.5″	2750	Car-	? early
	10	9512′ 0- 2″	2814	narvon	Neocomian
	11	10059′11-12″	2981	Basin	? Tithon.
	11	10067′ 8-10″	2983		(J 1)
	12	10780′ 6- 8″	3200	w.d.: 76	
	13	11443′ 2- 4″	3403		late
	13	11456′ 9-11″	3407		Jur.(J 2)
	14	12184′ 8-10″	3628		? Oxford.
	16	13569′ 1- 3″	4051		(J 3)
	16	13582′10-12″	4055		

mouth Plateau and the Scott Plateau (only sample VA16-12KD; Figure 6).

The reddish-brown clayey ironstone ("Toneisenstein", B1) consists of clay minerals, silt-sized quartz, mica-chlorite, and abundant secondary goethite cement.

The reddish-brown sandy ironstones (B2) consist of massive, well- to moderately-sorted, medium to coarse, goethite-cemented quartz sandstones (Plate 3:2). Calcite cement is rare, and biogenic remains are absent. An oolitic sandy ironstone was dredged from the Scott Plateau (VA16-12KD/c).

Brown ferruginous concretions, crusts and bands (B3) are common. Sometimes they include semi- or unlithified clayey sediment ("boxstones"), into which goethite "stalactites" precipitated. The concretions have the composition of clayey to sandy ironstones with up to 50% quartz and abundant goethite cement.

A dark-brown ironstone brecia ("Trümmereisenerz", B4) has fragments of dense clayey ironstone, up to 3 cm long, embedded in a matrix of very coarse goethite-cemented quartz sandstone. Besides Fe-oxides, silica also mobilized during early diagenesis; quartz and goethite filled the veins.

117°37.3′E. Legendre No. 1 and Dampier No. 1, see Table. Abbreviations: M.M. = Miria Marl, T.C. = Toolonga Calcilutite, W.G. = Winning Group; sh. = shelf; v.f. = very fine.

Rock Description (thin-section, XRD analysis)	Lithofacies	Remarks	Environment	Formation
orthoquartzite,well-sort.,homogen.,mature	? F	compacted	? beach	?
dolomicrosparite.recryst.,dol.,ech.& mol. frag.,qtz	C	pellets,?algae	v.shallow-marine	
quartzose dolosparite w.abund.crinoid fragm.	C	crin.not recryst.	v.shallow-marine	
dolomitized crinoid breccia,quartzose	C	syntax.overgr.	v.shallow-marine	
clayey glauconite sdst.,quartzose,dolomitic	A/D	friable	outer shelf (low.sed.rate)	Barrow
kaolinitic silty claystone qtz,mica,sid.,pyr.	A	aggrpolarization		
coarse volcanic conglom.w.alt.basalt pebbles	H	smect.matrix	?	2
highly alt."basalt",partly brecciated	H			
porphyritic volc.rock w.abund.feldsp.phenocr.	H			
qtzose silty clayst.w.well sort.v.f.qtz sdst.	A	well lam./bedd.	deltaic to littoral	Legendre
med.to coarse qtz-arenite,min.mature,	A	massive		
well sorted,overgrowth,silica cem.,grain contacts	A	massive		
kaolinitic,qtzose,micaceous silty claystone	A	sider.concret.	paralic-fluviodeltaic (? prodelta)	
clayey-v.f.sdy qtz siltst w.black carbonac.laminae	A	lamin.,graded		
qtz-arenite,well sort.,homogen.,sideritic	A	massive		
foramin.-mollusk biomicrite,silty-sdy,bedded	E	Inocer.prisms	hemipel.(out.sh)	T.C.
med.-coarse qtz-glauc.sdst,siderite-cemented	D	mod.sort.,massive	? middle-outer shelf	Barrow
clayey quartz graywacke,pyr.concret.,graded	A	compacted		
v.coarse qtz arenite-cgl.,sideritic	A	Precambr.	shallow-mar. delta-topset; high-energy tidal channel deposits	Dampier
massive,qtz overgrowth,stylolites;polymodal	A	band.chert frag.		
coarse quartz arenite,diagen.mature, ? graded	A	pyrite concret.		
polymodal quartz-chert conglom.,calcite-	A	massive		
cemented,some second.sid.,pyr.& silica cem.	A	massive		
v.poorly sorted sandy siltst.=qtz graywacke	A	bedd.-lamin.	delta foreset slope below wave action: prodelta (or lagoon ?)	Legendre
v.poorly sort. quartz graywacke	A	massive		
well sort.v.f.qtz sdst.(subarkose),kaol.,sider.	A	laminated		
well sort.coarse qtz siltst.(subarkose)	A	lensy lamination		
v.well sort.coarse qtz siltst,compacted,calc.veins	A	laminated		
mod.sort.micac.-carbonac.qtz siltst.(shale)	A	lamin.,bioturb.	? coal swamp	
poorly sort.calcar.qtz-glauc.-foram.sdst	C/D	Nummulites ?	? outer shelf	Cardabia
mod.sort.qtzose,glauc.-phosph.rich biosparite	C/D	peloids,algae	? carb.platform	
well sort.calcite-cemented f.qtz sdst(qtz-aren.)	C	macrosparite	middle-outer shelf	
mod.sort.clayey qtz(glauc.)arenite(v.f.sdst.)	A/D			
silty shale,qtzose,glauc.,micac.,carbonac.	A	pyritization	? outer shelf	M.M.
foramin.nanno chalk(calcilut.,packst.)	E	lensy lamination	pelag.-epicont.	T.C.
lamin.foram.-mollusk marlst.w.Inocer.prisms	E	bioturbation	hemipel.-shelf	W.G.
black silty shale,qtzose,carbonac.,micac.	A	pyritic	restr.-mar.	Barrow
clayey f-v.f.sdst w.better sort.sdst lenses	A	mottling	marine delta	
clayey f.-m.sdst.,qtzose,chlor.,kaol.,mic.,pyr.	A	silty clayst.l.		
calcilut.w.grad.,mod.sort.lam.v.f.qtz sdst	A/C	dol.;mollusks	neritic-paralic(lagoon delta plain)	Dampier/Dingo
v.poorly sort.clayey-silty v.f.qtz sdst.	A	v.f.sdst.lenses		
quartzose silty calcilutite,slightly dolomitic	C	pyr.concret.		
v.poorly sort.muddy sdst.w.thin siderite layers	A	laminated	outer neritic (prodeltaic)	
black silty shale,qtzose,micac.,carbonaceous	A	laminated		
silty shale,qtzose,micac.,carbonaceous	A	pyrite concr.		
silty to v.f.sandy qtartzose shale,pyrit	A	laminated		
clayey siltst.-silty clayst.,qtzose,moll.fragm.	A	pyrite concr.		

Late Jurassic-Neocomian shallow-water marine clastic sediments (Barrow Group) = lithofacies D

In contrast to the terrestrial to paralic carbonaceous silty claystones of Triassic to mid-Jurassic age, the few upper Jurassic (?Tithonian) to Neocomian claystones dredged from the margins of the Exmouth and Scott Plateaus (Figure 6) are fully marine. These rocks consist of noncalcareous or slightly calcareous, clastic, restricted shallow-marine sediments. Two types were observed:

(1) Phosphatized clayey quartz sandstone to silty claystone (D1), with 50 to 80% cryptocrystalline collophane cement (Plate 3:3) and of unknown age, occurs on the northern Exmouth Plateau (SO8-39KD); and,

(2) Micaceous clayey siltstone with some glauconite (D2) was recovered from the southwest and south Exmouth Plateau. A similar massive clayey siltstone with phosphatic grains, wood and leaf remains, pelecypods, and ammonites was dredged from the Scott Plateau (VA16-11KD; Figure 6; Hinz et al, 1978). These clayey siltstones contain a marine dinoflagellate flora and are probably of uppermost Jurassic (Tithonian) age. Another slightly calcareous (?prodeltaic) claystone (SO8-51KD) contains Early Cretaceous radiolarians.

Mid-Jurassic to Early Cretaceous volcanic and volcaniclastic rocks (lithofacies H)

Middle to late Jurassic volcanism is well-known from the Scott Plateau vicinity (Scott Reef No. 1, Ashmore Reef No. 1; Powell, 1976), so it is not surprising that *R/V Valdivia* dredged a great variety of volcanic and volcaniclastic rocks from the steep walls of Oates Canyon at the plateau margin (Hinz et al, 1978). The least weathered of these basalts were dated by the K/Ar method and gave an apparent minimum age of 128 to 132 million years (Table 3), or an Upper Jurassic to lowermost Cretaceous age for the Scott Plateau volcanics.

All volcanic rocks dredged from the Scott Plateau belong to a suite of normal to differentiated basalts with 1% phenocrysts (mostly plagioclase; altered olivine, rarely pyroxene). Most plagioclases and some pyroxenes are fresh; olivine is always replaced by layer silicates and some samples are highly vesicular and oxidized. Mineralogical and chemical composition indicate differentiated alkali basalts.

Most sedimentary rocks dredged from the Scott Plateau margins contain volcanogenic components and have hence to be classified as tuffs to tuffaceous sandstones or siltstones (H7). These pyroclastic sediments are commonly parallel, cross- and convolute laminated, and full of phosphatic ooids (Plate 4:6).

Completely phosphatized pyroclastic claystones containing cryptocrystalline collophane (apatite), X-ray amorphous matter, feldspar, barite, and smectite are also present.

A very heterogeneous suite of altered volcanic rocks and volcaniclastic sediments was recovered from the Sonne Ridge in the Cuvier Abyssal Plain, along the northern, eastern, and southern margins of the Wallaby Plateau, and from the western extremity of the Canarvon Terrace (von Stackelberg et al, 1980). In fact, the Wallaby Plateau yielded only volcanic or volcani-

Table 3 — K-Ar data from volcanic rocks dredged from the Scott, Exmouth, and Wallaby Plateaus (chemistry by W. Harre and H. Raschka, K/Ar dating by H. Kreuzer, all BGR, Hannover). 1000 to 500 /μm (first value) and 500 to 135 /μm (second value) fractions of the ground whole rock samples were analyzed. The higher loss on ignition (LOI) values of the fine fractions indicate a higher degree of alteration and consequently more reduced K/Ar dates. All whole rock dates are regarded as minimum ages. Decay constants according to the Sydney convention (Steiger and Jäger, 1977). Age interpretation based on time scale of Armstrong (1978) and BGR results of northwest German Cretaceous glauconites. Errors calculated to 95% confidence limits of analytical precision. + slightly, + + moderately, + + + highly altered.

Sample No.	Rock Type	Alteration	LOI (wt.%)	Potassium (wt.%)	Atmospheric Argon (10^{-7}ccSTP/g)	Radiogenic Argon (10^{-7}ccSTP/g)	K-Ar Date (m.y.)	Interpretation Towards a Stratigraphic Age
Scott Plateau, Southwest Oates Canyon								
VA16-13KD/h	alkali basalt	+ +	1.78	0.955 ± .013	19.5 ± .5	51.0 ± .8	132 ± 3	at least
		+ +	2.44	0.905 ± .013	22.6 ± .5	48.6 ± .8	133 ± 3	Valanginian
								to Berriasian
VA16-13KD/i	alkali basalt	+ +	2.82	1.071 ± .015	17.6 ± .5	57.0 ± .9	132 ± 3	(earliest Cretac.)
	(? hawaiite)	+ +	3.14	1.096 ± .015	17.05 ± .35	56.5 ± .8	128 ± 3	
Northern Exmouth Plateau (Wombat Plateau-N)								
	SiO_2 undersatur.							
S08-65KD	"trachyte" s.l.	+	1.78	2.09 ± .03	8.3 ± .6	165.4 ± 2.2	193 ± 4	at least
	(feldspar-rich)	+	1.78	2.12 ± .03	7.8 ± .5	165.0 ± 2.2	190 ± 4	Pliensbachian
	alkali rhyolite		.80	6.23 ± .08	3.2 ± 1.6	532 ± 7	207 ± 4	minimum estimated
S08-66KD	with large sanidine phenocrysts		.90	6.07 ± .07	2.6 ± 1.6	515 ± 7	206 ± 4	age of extrusion
	Sanidine, hand-picked			5.12 ± .04	7.3 ± 3.0	450 ± 5	213 ± 3	Rhaetian/Hettangian
Wallaby Plateau (South)								
S08-170KD/1	olivine	+ + +	3.41	0.772 ± .011	(!) 37.0 ± .5	27.25 ± .45	88.6 ± 1.9	discordant !
	tholeiite	+ + +	3.69	0.797 ± .012	(!) 44.3 ± .6	26.35 ± .5	83.1 ± 2.0	older than Turonian

Table 4 — Correlation of lithostratigraphic units from Scott, Exmouth, and Wallaby Plateaus with equivalent formations from the adjacent basins of the Northwest Shelf (after various sources). Ex = Exmouth Plateau; N.Ex. = Northern Exmouth Plateau; Sc = Scott Plateau; W/SR = Wallaby Plateau/Sonne Ridge. Scott Plateau correlated with Browse Basin (e.g., Scott Reef No. 1, Lynher No. 1); northern Exmouth Plateau with offshore Canning Basin (Bedout Sub-basin, Bedout No. 1); and western and southwestern Exmouth Plateau with offshore Carnarvon Basin (Beagle and Dampier Sub-basins with Picard No. 1, Dampier No. 1, Legendre No. 1).

Age	Outer Plateau Margins	Northwest Shelf Equivalent Formations	Paleoenvironment & Remarks to Northwest Shelf Formations
Tertiary Maastr. to Coniac.	pelagic foraminiferal nanno	Cardabia Group - Trealla 1s (carbonate buildup/progradation)	shelf carbonates (calcaren., marls, sst., ± dolomitic), much glc + ph. Scott Reef No. 1: 3 km reefal carbonates
	chalk (facies E, bathyal) W, Ex, N. Ex, Sc.	Miria Marl Toolonga Calcilutite — highly condensed	± calcar. claystone, marl, dolomite (Inoceramus prisms) foram.-mollusk biomicrite-marlstone (Inoc. prisms)
Turon. to Albian		± calcareous shales & marls — section	
late Early Cretac.	W/S.R. breakup oceanic basalt & volcaniclastics	Winning Group	transgressive shallow-marine: radiolarites
Apt. Barr. Neoc.	(rad.) chalk/porcellan. (E1, 3, 4) N. Ex, Ex	Windalia Radiolarite Muderong Shale	marlst., ± calc. shale, black shale (w. for. & mol.)
E. Neoc. to late Jur.	marine clayst. (D2) ?phosphat. clayst. (D1) (Ex) ?Tithonian carbonac. silty marg.-mar. clayst. (A2, Sc.)	Barrow Formation	shallow-marine (?prodelta) with regressive sand cycles (Bedout No.1: Broome Sst) black shale, calcar. shale, marl, sst. Legendre-1: tidal flats, delta platform, partly emergent
Lt.- M. Jur.	Sc. (?N. Ex) volcanics: ± different. alkali basalt + tuffs etc	m.-lt. Jur. post-breakup volc. (Scott Reef No. 1)	similar to coal measure sequence, except more marine (always glauc., mol); also restricted (much py.; siderite, black shale). Dolomicrite layers. Well-sort. qtz-arenite. coal swamps near large deltas?
M. Jur.	ferruginous association (B-facies)	Dingo Claystone Dampier Form. (more distal, basinal prodelta)	deltaic sst., siltstone, silty clayst., frequent coal progradation of fluvial complex:delta foreset slope (?prodelta) to delta platform (channel deposits, tidal sand flats) in Legendre No.1
(-E. Jur.)	paralic coal measure sequence (A)	Legendre (Learmonth Form.) fluviodeltaic w. some marine influence	
E. Jur. (?Sin./ Pliensb.)	shallow-water carbonate association (N. Ex.)	"Basal Limestone" (Sinemurian) in Dampier Sub-basin (e.g. Brigadier No. 1)	marine transgression limestone also rich in echinoderms & crinoids (Crostella & Barter, 1980)
Earliest Jur. - latest Triass.	intermed.-acid synrift volcanics Wombat Pl. (N.Ex.)	local (?pre) mid-Triassic basaltic volcanism (Bedout No. 1)	altered ?basalt, volcaniclastic congl. etc
Late to mid-Triassic	paralic silty shale (A2; N.Ex.)	Mungaroo Beds	fluviodeltaic (rift valley)

clastic rocks and the eastern Wallaby Plateau only volcaniclastic rocks, apart from Cenozoic pelagic chalks and oozes.

Altered tholeiitic basalts (H1) were recovered from the Sonne Ridge and the Wallaby Plateau (Plate 4:3). Only one somewhat questionable K/Ar age was obtained from an olivine tholeiite (Table 3; southern Wallaby Plateau): the apparent age is 89 m.y. old ("older than Turonian"), but this is a minimum age. Thus the Barremian (120 m.y. old) age of the onset of spreading in the Cuvier Abyssal Plain and of volcanism on the Wallaby Plateau, (Johnson et al, 1980; Larson et al, 1979) from magnetic evidence, is not contradicted by our data.

A volcanic breccia (H2), probably of tholeiitic composition, occurs along the southern margin of the Wallaby Plateau. Highly altered, amygdaloidal basalt fragments (diameter of 2 to more than 30 mm) were cemented by secondary smectite and phillipsite cement.

Altered, slightly differentiated alkali basalts (possibly hawaiites, i.e. andesine-bearing alkali olivine basalt; H3) were recovered from Sonne Ridge and the northern margin of the Wallaby Plateau. Because of their high degree of alteration, classification is difficult (very low MgO, but very high secondary K_2O and Fe_2O_3 contents).

Altered tuff to lapillistone and poorly-sorted tuffaceous sand-to siltstone (H7), with various admixtures of non-volcanic (epiclastic) components, are very common on the Wallaby Plateau. Originally most of these rocks were vitric tuffs, although crystal (sanidine-rich) tuffs and lithic tuffs were also recognized. Later, some tuffs were probably reworked under shallow-water conditions and mixed with terrigenous (up to 20% quartz) and authigenic components (glauconite).

A multicolored, poorly-sorted volcaniclastic sandstone or breccia (H8) occurs on Sonne Ridge and the Wallaby Plateau (Plate 4:4,5). One sample is a silicified volcanogenic breccia with 50% secondary quartz cement. Another sample from Sonne Ridge is a volcanoclastic breccia with subangular to rounded fragments (1 mm to 10 cm) of a porphyritic basalt. A few of these basalt pebbles are amygdaloidal, highly angular, and impregnated by Fe-oxides. They probably formed at comparatively shallow water depths and have no original cement. The pore space was later filled by authigenic smectite and by blocky phillipsite (-chabazite) cement. This marked cementation is due to the submarine diagenesis of the basaltic clasts which were themselves altered to smectite.

White phospatized pyroclastic claystones (H9) are also common along Sonne Ridge and the northern margin of Wallaby Plateau. The main constituent is cryptocrystalline collophane ("apatite"), which partially replaced a tuffaceous claystone.

Mid-Cretaceous shales and radiolarian chalks (?Winning Group) = lithofacies D, E, F

Offshore equivalents of the transgressive Winning Group were only rarely recovered from the margins of the Exmouth Plateau. The wide range of lithofacies types points to rapid horizontal and vertical facies differentiation.

Table 5 — Source rock potential of Jurassic sediments (coal measure sequence), dredged from the northern Exmouth Plateau. Localities and water depths in von Stackelberg et al (1980, their Table 1). Analyses by Australian Mineral Development Laboratories (Adelaide) and J. Koch & H. Wehner (BGR = *). TOC = total organic carbon; R_o = vitrinite reflectance; vitr. = vitrinite; exin. = exinite; asph. = asphaltenes. (...) = R_o—values considered atypical (too low for coal, too high for sample 32

Sonne 8 sample	Sediment type	Facies	TOC (%)	Domin. org. matter	R_0 (%)	EOM ppm	SAT ppm
28 KD/1	siltst./clst.	A2	0.09	vitr.	(.45)	107	14.8
29 KD/2	clst./sst.	A2/3	1.05	vitr.	.30	309	30.9
32 KD/5	siltst./sst.	A2/3	<0.01	vitr.	(.62)	62	10.8
33 KD/1B	clst./sst.	A2/3	2.35	exin.	.31	1446	66.5
34 KD/4	} subbitumin.	A1	48.5	vitr.	(.2)	22715	795
*34 KD/4	} coal/lignite	A1	59.9	vitr.	.36 .39	14900	871
36 KD/2	silty clst.	A2	0.79	vitr.	.26	604	39.3
*36 KD/5	xylite	A1/2	49.4	vitr.	.27 .34	1500	nd

A Lower Aptian radiolarian nanno chalk (E3) from the northern Exmouth Plateau contains up to 30% radiolarians, traces of sponge spicules, and fish remains, but no terrigenous debris. It can be correlated with the Aptian Windalia Radiolarite of the Carnarvon Basin (Table 4). These sediments are commonly transformed into weakly silicified radiolarian chalks with about 15% of the siliceous skeletons replaced by opal-CT. Further diagenesis and silica mobilization produced radiolarian porcellanites (E4).

A vitreous ferruginous chert (lithofacies F) which was dredged only from the slope west of the Carnarvon Terrace (SO8-173KD) has a different composition. Although circular fossil remains suggest the original presence of radiolarians, this completely recrystallized rock now contains 95% micro- to macrocrystalline quartz cement and 5% Fe oxides (Plate 3:6). Since pre-Cretaceous radiolarian-rich sediments are known from only one locality in the Carnarvon Basin, the quartz chert may be of mid-Cretaceous age, as are the E4 porcellanites.

A fissile shale (D2) was dated by pollen as late Albian to earliest Cenomanian, an age which is generally supported by thick-shelled *Inoceramus* prisms in that dredge haul.

Upper Cretaceous to Tertiary pelagic foraminiferal nanno-chalks (lithofacies E)

On the Northwest Shelf carbonate sedimentation started in the Coniacian/Santonian (Toolonga Calcilutite), whereas older radiolarian chalks on the Exmouth Plateau date back to the early Aptian.

The oldest foraminiferal nanno chalks (E1) recovered from the Scott Plateau are of early Campanian age (VA16-1KL). A middle Maastrichtian chalk (VA16-13KD/b,c,) forms the matrix of an intrabasalt limestone or volcanic breccia with altered basalt and tuff fragments (Figure 6; Plate 3:5).

Semiconsolidated foraminiferal nanno chalks recovered from the Wallaby and Exmouth Plateaus are of late Paleocene to early Eocene, middle Oligocene, late Oligocene to early Miocene, and middle Miocene to Pliocene ages (von Stackelberg et al, 1980). These sediments are probably hemipelagic in origin.

CORRELATION OF THE OUTER PLATEAU LITHOFACIES WITH FORMATIONS FROM ADJACENT SHELF WELLS.

Because of the inherent deficiencies of the dredging technique, stratigraphic records from the outer margins of the Wallaby, Exmouth, and Scott Plateaus are fragmentary. However, a very dense coverage of this continental margin by seismic profiles and their control by more than 60 offshore wells allows comparatively good correlations with the stratigraphy of the adjacent Northwest Shelf and coastal basins. Table 4 shows our attempt to correlate the lithofacies and stratigraphy of the outer plateau margins (Figure 6) with formations from the Northwest Shelf wells; especially those six wells which we sampled and studied in detail (Table 2; Figure 5). From the Triassic to the time of breakup (Late Jurassic or Neocomian in different areas) most plateau formations correlate very well to the adjacent shelf wells. However, only the Aptian

KD/5 with extremely low TOC!); EOM = extractable organic matter; SAT, AROM = saturated, aromatic hydro-carbons; POLAR = polar compounds; ASPH = asphaltenes; Pr = pristan; Ph = phytan. All samples are immature (source rating is potential, if sediments reached organic maturity elsewhere in the section), nd = no data.

AROM ppm	POLAR ppm	ASPH ppm	$Pr/n.C_{17}$	$Ph/n.C_{18}$	Pr/Ph	Source Rock Potential	Remarks
7.4	16.6	53.5	0.96	0.38	1.55	poor	
10.2	51.6	204.2	1.22	0.40	1.58	fair	
nd	nd	15.5	0.89	0.48	1.31	poor	?potential gas source
28.9	235.7	620.3	0.67	0.31	1.83	good	potential oil source
363.4	5020	6564.6	nd	nd	nd	good	algal mat. + terr. plants
3084	3486	3607	nd	nd	nd	good	herbac. plants + woody roots
2.4	258.5	105.1	0.71	0.26	2.43	fair	terrestr. org. matter
nd	nd	nd	nd	nd	nd	?	gymnosp. wood

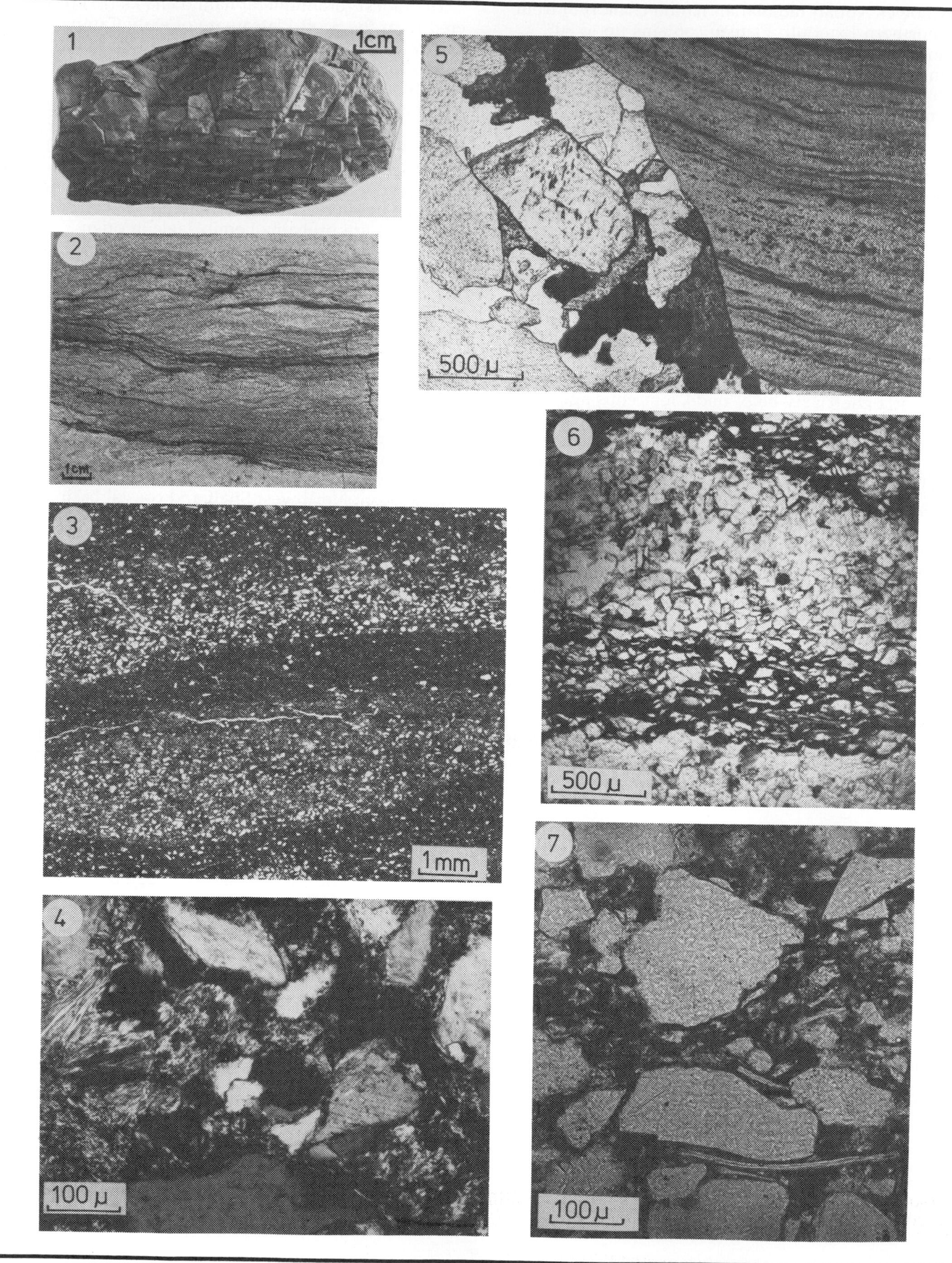
1
1cm
2
1cm
3
1mm
4
100 µ
5
500 µ
6
500 µ
7
100 µ

Windalia Radiolarite and the Santonian Toolonga Calcilutite of the post-Neocomian facies of the outer plateau margins compare with the Northwest Shelf.

GEODYNAMIC AND PALEOENVIRONMENTAL EVOLUTION OF THE NORTHWEST AUSTRALIAN CONTINENTAL MARGIN

Structural evolution during the rift stage

The intracratonic basins of the northwest Australian continent were subject to tensional rifting intermittently from Permian times to the separation of the Scott and northern Exmouth Plateau from the adjacent northerly parts of Gondwanaland in the mid-Jurassic. Hence, the Scott and Exmouth Plateaus are characterized by extensive normal faulting.

The margin of the Scott Plateau developed by rifting parallel with the northeasterly-trending spreading axes, documented by magnetic lineations on the Argo Abyssal Plain (Figure 1), and by transform faulting normal to the spreading axes (Stagg and Exon, 1982). Northeasterly-trending normal faults are not well-documented, and major displacements are confined to the outermost margin. Several large horsts and grabens on the outer margin of the plateau (Figure 7 in Stagg, 1978) formed by northwest to southeast faults with vertical displacements of 1,000 m or more parallel with the transform direction. One such fault, the North Wilson Transform (Hinz et al, 1978), continues onto the abyssal plain, and appears to form the western edge of Australian continental crust (Figure 1). Evidence from the Browse Basin suggests that tensional faulting began in the Triassic. Faulting seldom extends above the E unconformity in the Browse Basin or on the Scott Plateau (Figure 2A), except on the outermost margin. This indicates that movement generally terminated in the Callovian, when this part of Gondwanaland broke up.

The northern Exmouth Plateau is separated from the remainder of the Exmouth Plateau by an east-west trending hinge-line, which effectively marks the southern limit of the Canning Basin in that region (Figure 1; Exon et al, in press). The structurally complex northern plateau is cut by northeasterly-trending normal faults, developed in a tensional regime. Apparently, rifting started in the Triassic, forming large horsts and grabens with displacements of thousands of meters (Figure 2B). Most faults extend only to the E unconformity, indicating that movement terminated in the Callovian, when the margin was formed by separation from a continental fragment to the north. However, some faults cut the C unconformity (Figure 2B), suggesting that they were rejuvenated in the Cenozoic.

The western and southern Exmouth Plateau is cut by numerous northeast-trending normal faults which strongly displace the F unconformity, but seldom extend to the D horizon (Figure 3A). Away from the plateau margins, displacements seldom exceed 200 m. These faults presumably began forming in response to tension in the Triassic, and ceased to move after the northern margin formed in the Callovian. An abortive rift graben, related to the Callovian breakup, extends southwest from the Swan Canyon to the Kangaroo Syncline (Figure 1 in von Stackelberg et al, 1980).

Triassic early rift paleoenvironment and volcanism

During Late Permian, Triassic, and Early Jurassic rifting, a multiple rift-valley arch system of fault-bounded troughs formed along the present northwest border of the Australian continent and rapidly filled with extremely thick, mainly terrestrial taphrogenic sediments (Powell, 1976). This rift graben system was periodically flooded from the north or northwest by a shallow Paleotethys, the western extension of Panthalassa between Gondwanaland and eastern Eurasia.

Mid-Triassic shales and sandstones, dredged from the northern Exmouth Plateau, reflect a regressive, brackish to terrestrial environment similar to the thick fluviodeltaic Mungaroo Beds of the Carnarvon Basin (Powell, 1976; Thomas and Smith, 1974). A different marine facies characterizes sediments from the Scott Reef No. 1 well in the Browse Basin further north. There, an Upper Triassic orthoquartzite probably represents a littoral sand, which was later recrystallized.

During a latest Triassic/earliest Liassic intracratonic rift phase, a suite of a highly differentiated rhyolitic to trachytic rocks erupted along the northern Exmouth (Wombat) Plateau, probably under subaerial to very

---

Plate 1 — Jurassic "Coal measure sequence" (A)

(1) Black, vitreous, laminated, immature subbituminous coal (A1), 5 cm thick seam alternating with silty claystone. SO8-34KD/4 (Swan Canyon, northern Exmouth Plateau, w.d. 5180-3700 m).
(2) Silty claystone (A2) with layers and lenses of wavy laminated silty very fine quartz sandstone (subgraywacke; A3). Note flaser bedding. SO8-32 KD/2 (Emu Escarpment, northern Exmouth Plateau; polished section).
(3) Quartzose, ferruginous silty claystone (A2) alternating with thin indistinctly graded and cross-laminated layers of quartz siltstone (A3). SO8-33 KD/1A (Swan Canyon, northern Exmouth Plateau; thin-section).
(4) Clayey medium quartz sandstone (quartz graywacke; A4) with abundant quartz, common feldspar, mica-chlorite, crystalline rock fragments, and a kaoliniticsilica matrix. SO8-32 KD/9 (see 2). Crossed nicols.
(5) Late Jurassic polymodal, very coarse quartz arenite to breccia (A4) with abundant quartz, common glauconite, pyrite, and feldspar, cemented by calcite-siderite(?). Note large pebble of banded ferruginous chert, probably derived from Precambrian Pilbara Massive. Legendre No. 1, core 12, (7725′; Dampier Sub-basin, Northwest Shelf).
(6) Mid-Jurassic laminated clayey micaceous siltstone (A2) with abundant dark (? carbonaceous and pyritic) matrix, alternating with compacted very fine quartz sandstone. Picard No. 1, core 1 (7630′), Beagle Sub-basin (Northwest Shelf).
(7) Mid-Jurrassic, bedded silty sandstone with abundant angular quartz, common muscovite and feldspar, and kaolinitic matrix. Legendre No. 1, core 14 (8318′). Dampier Sub-basin (Northwest Shelf).

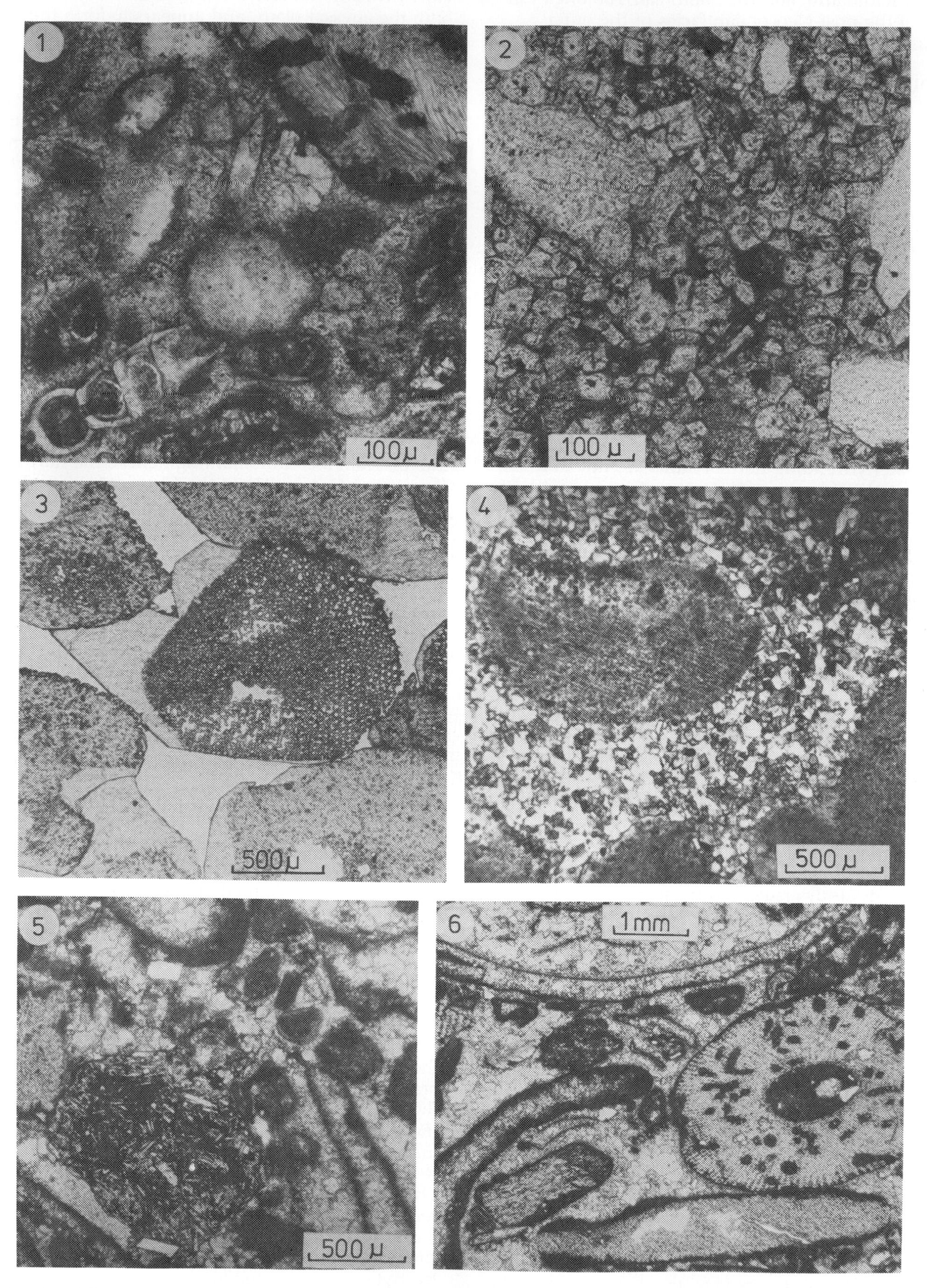
1
100µ
2
100 µ
3
500µ
4
500 µ
5
500 µ
6
1mm

shallow-marine conditions. Similar early-rift rhyolitic volcanism is known from the Lord Howe Rise off southeast Australia, and preceded the opening of the Tasman Sea by about 15 million years (McDougall and Van der Lingen, 1974). Triassic volcanism is known also from the Northwest Shelf (Bedout No. 1), where (?pre)mid-Triassic, porphyric basaltic rocks and volcaniclastic conglomerates were penetrated (Figure 5; Table 2).

Liassic transgression

The shallow-marine carbonate association, recorded from the northern Exmouth Plateau (von Stackelberg et al, 1980) and parts of the Dampier Sub-basin (Crostella and Barton, 1980), documents a mid-Liassic (Late Sinemurian to Pliensbachian) transgression of the warm Tethys sea onto a subsiding margin, covered with fluviodeltaic sediments and volcanics. At the same time, the central Exmouth Plateau was eroded, whereas in most parts of the coastal basins paralic to terrestrial sedimentation continued.

The fact that some biocalcarenites from the northern Exmouth (Wombat) Plateau contain oncoidally coated "trachyte" fragments (Plate 2:5), proves that the Liassic sea eroded and reworked the early-rift volcanics on the Northern Exmouth Plateau. In general, the depositional environment of the biomicrites (C1, Table 1), and most biocalcarenites and microbioclastic calcisiltites (C2), was the open shelf with comparatively quiet water below wave base.

For other limestones, however, algal coatings (Plate 2:5,6) and calcareous algae (Dasycladacea and Codiacea) suggest a very shallow, subtidal environment (Flügel, 1978). For the quartz-rich biocalcarenites (C3) we envisage a shallow-marine (?inner shelf) environment with intermediate to high-energy conditions. It is possible that the biosparite (C4) was reworked by tidal and other currents and deposited as sand waves along an outer reef or carbonate bank flank (Jenkyns, 1971). Also some polymict biocalcarenites reflect a high-energy, inner-shelf environment of deposition.

The calcite-cemented quartz arenites (C5) associated with these limestones can possibly represent transgressive, winnowed sediment from a high-energy nearshore or beach environment.

The crystallized limestones (C6) and dolomites (C7) were originallly deposited as bioclastic carbonates in an inner-shelf environment with weak to intermediate bottom currents. The recrystallization and dolomitization occurred during later (?subaerial) diagenesis, similar to that of the Upper Triassic dolomites of Scott No. 1 well.

Mid-Jurassic regression and deltaic sedimentation (coal measure sequence)

In the Browse Basin and the Barrow-Dampier-Beagle Sub-basins east of the Rankin Platform, 1 to 2 km of Lower to Middle Jurassic clastic sediments (Legendre or Learmonth Formation) were deposited on large northwest prograding fluvial delta plains (Figure 6; Table 4). These deltaic sediments indicate a relatively humid climate and the proximity of a moderately high-relief hinterland: the Precambrian Pilbara Massif. Beds or stringers of black immature coal indicate a "delta abandonment facies" with a subaqueous peat bog stage following the progradational facies of the delta plain (Reading, 1978).

A similar "coal measure sequence" (A, Table 1) was deposited on the northern Exmouth Plateau during the mid-Jurassic regression. Most of the typical facies of upward coarsening cyclothems are recognized: immature, subbituminous coals (A1) are interbedded with carbonate-free, carbonaceous kaolinitic, and micaceous silty claystones (A2) with thin layers and lenses of well-sorted silt. The quartz siltstones and very fine sandstones (A3) are often parallel and cross-laminated on a small scale. The sedimentary structures and the lack of marine indicators suggest a fluviodeltaic environment. The abundance of pyrite concretions indicates an anoxic to mildly reducing depositional environment. The fine to medium sandstones (A4) along the northern margin of the Exmouth Plateau reflect high-energy fluviatile to paralic (barrier-lagoon?) conditions. Coarse to conglomeratic sandstones (with abundant reworked quartz and chert pebbles) occur in the Dampier Sub-basin (Legendre Formation), proximal to the Pibara Massif source area. They can be interpreted as channel sands from the subtidal part of the delta platform, whereas the siltstones and claystones of the Dampier Formation or Dingo Claystone represents a more distal, prodeltaic environment (Tables 2, 4; Figure 5).

Plate 2 — Thin-sections of early Jurassic shallow-water carbonate association (C)

(1) Poorly-sorted echinoderm-mollusk-foraminiferal biocalcarenite (C2). Note microsparitic cement, large mollusk fragments, and benthonic foraminifera (miliolids, nodosariids). SO8-61KD/3 (Wombat Plateau, northern Exmouth Plateau).
(2) Dolomitized biocalcarenite (C8) with terrigenous quartz and not recrystallized echinoderm fragments. Note zoned large dolomite rhombs with dark (? carbonaceous) nuclei. The low nucleation rate suggests late diagenetic dolomitization. SO8-49 KD/1 (Wombat Plateau, northern Exmouth Plateau).
(3) Very coarse crinoid breccia (biosparite, C4). Note abundant free pore space and syntaxial calcite overgrowth on crinoid fragments. SO8-39KD/3B. Cygnet Canyon (northern Exmouth Plateau).
(4) Latest Triassic quartzose dolosparite with large (not recrystallized) echinoderm (crinoid) fragments (C 4/7). Scott Reef No. 1, core 1 (14363'). Browse Basin (Northwest Shelf).
(5) Poorly-sorted algal-mollusk-echinoderm biocalcarenite (C4) with reworked volcanic rock (? trachyte) fragment, and oncoidally (?) coated bioclasts (micritized ?). SO8-43KD/2 (Wombat Plateau, northern Exmouth Plateau).
(6) Same as 5 with macrosparitic calcite cement and mollusk, algal and echinoderm fragments, all with micritic envelopes (boring micro-organisms). SO8-43 KD/2 (Wombat Plateau, northern Exmouth Plateau).

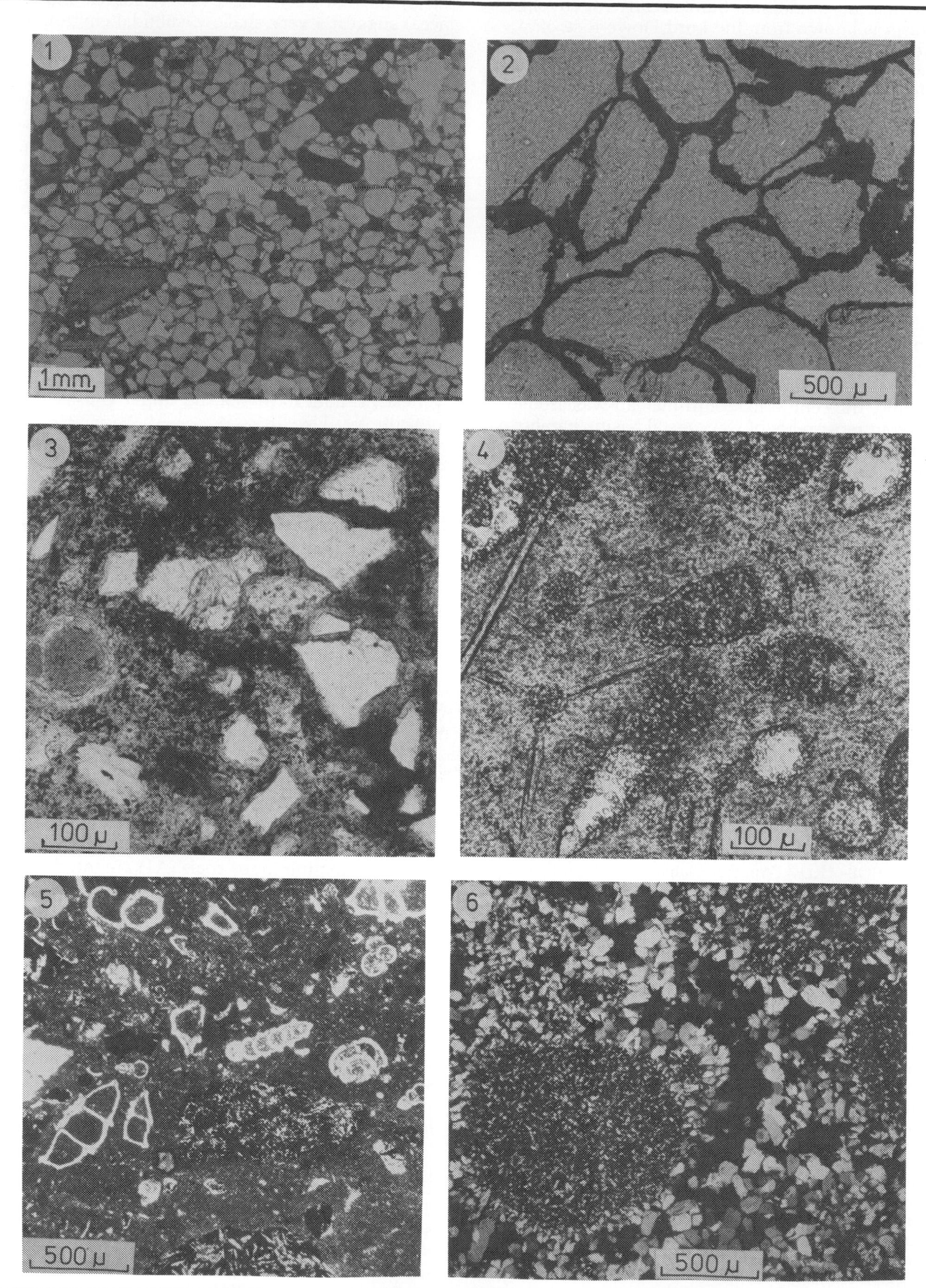
1
1mm
2
500 μ
3
100 μ
4
100 μ
5
500 μ
6
500 μ

For the lithotypes of the ferruginous association (B, Table 1), found only on the northern Exmouth Plateau and on Scott Plateau (one sample), we envisage an age and paralic environment similar to that for the coal measure sequence. The ferruginization was probably due to a later (?Upper Jurassic) emergence and impregnation by Fe-rich weathering solutions under arid conditions. Locally, ironstone breccias (B4) were deposited, probably in a littoral environment similar to that of the well-known "Trümmereisenerz" of Salzgitter in northwest Germany.

Structural evolution after the mid-Jurassic breakup of Gondwanaland

In the Scott Plateau/Browse Basin and northern Exmouth Plateau/Canning Basin regions the breakup of Gondwanaland started in Callovian times. This major rifting event led to the creation of Upper Jurassic oceanic crust in the Argo Abyssal Plain along an ENE-trending spreading center (Figure 6; Heirtzler et al, 1978). The ocean that developed between northwest Australia and a continental fragment to the north widened and carried the fragment northward into a subduction zone where it disappeared. The ocean-continent boundary generally developed along the axis of the ENE-trending early Mesozoic rift valley; except for several transform faults that cut the margin normal to the spreading direction (east of the Echidna and Emu Spurs; Figure 1). Extensive normal faulting accompanied the breakup of the northern Exmouth Plateau margin in the Callovian. Peneplanation of the horst and graben relief followed the breakup.

The western and southern Exmouth Plateau margins formed in the Neocomian; the western margin by breakup along a spreading center, the southern margin by transform faulting (Larson et al, 1979; Johnson et al, 1980).

The northeast-trending landward-tilted fault blocks of the western Exmouth Plateau margin step down to the Gascoyne Abyssal Plain (Figure 3A), indicating that this margin was the eastern flank of a rift valley prior to breakup. South of the Cape Range Fracture Zone, the southern margin of the Exmouth Plateau, a fragment of Greater India moved northwest during the Gondwanaland breakup. Faults with throws of hundreds of meters parallel the margin. Total displacement exceeds 2000 m in the west, and 3000 m in the south. New oceanic crust formed in the Cuvier Abyssal Plain about 118 to 121 m.y. ago and subsided steadily from the late Neocomian onward, while the southern Exmouth Plateau was thermally uplifted (Exon et al, in press). This event cut off the southerly source of the Neocomian deltaic sequence on the Exmouth Plateau.

Since faulting on the Wallaby Plateau generally cuts the Neocomian breakup unconformity (D), and seldom cuts the C unconformity (Figure 2D), most vertical movement occurred soon after breakup in the Early Cretaceous. Apparently the northern margin of the plateau was not displaced relative to the Cuvier Abyssal Plain because the D unconformity can be traced gently downward to join the top of the plain's oceanic basement. The complex western margin of the Wallaby Plateau may represent the original edge of the epilith, or it may have experienced some rifting. The steep, straight southern margin, the Wallaby-Zenith Fracture Zone, is an old transform fault with horizontal and vertical displacement between the Wallaby Plateau and the oceanic crust of the Perth Abyssal Plain to the south.

The Upper Jurassic-Lower Cretaceous juvenile ocean stage and post-breakup volcanism

An Upper Jurassic (?Tithonian) transgressive sea levelled the block-faulted lower Mesozoic rocks and deposited shallow-marine, fine-grained clastic sediments in parts of the Browse, Canning, and Carnarvon Basins (Upper Jurassic to Neocomian Dingo Claystone and Barrow Formation). The Tithonian mudstone of the Scott Plateau might have been deposited in an estuarine or prodeltaic environment under restricted conditions, as were the siltstones (D2) and the quartz sandstones (D1) of the southern Exmouth Plateau (Figure 6), where the depocenter of a huge Tithonian-Neocomian delta was located. Paralic to terrestrial clastics accumulated on prograding deltas extending landward from this shallow sea. This latest Jurassic to early Neocomian sedimentary cycle coincided with the formation of Wealden-type deltas around the margins of the North Atlantic during a major global regression (von Rad and Arthur, 1979).

The Callovian to Oxfordian fragmentation of the Scott and Exmouth Plateau regions of Gondwanaland was associated with intense volcanism (extrusion of differentiated alkali basalts). Predominantly explosive

---

Plate 3 — Thin-sections of varied sedimentary rocks

(1) Moderately-sorted, calcite-cemented crinoid-rich medium quartz arenite (C5). SO8-39 KD/3A (Jurassic; Cygnet Canyon, northern Exmouth Plateau).
(2) Highly porous, moderately-sorted, ferruginous medium quartz sandstone (sandy ironstone, B2). Goethite rims around (iron-stained) quartz grains. SO8-36 KD/4A (Jurassic; northern Exmouth Plateau).
(3) Phosphatized, poorly-sorted sandy-silty quartzose claystone (D1) with 50% phosphate (collophane) cement, phosphatized foraminiferid, quartz, feldspar, and chert fragments. SO8-39 KD/10 (Cretaceous/Tertiary; Cygnet Canyon, northern Exmouth Plateau).
(4) Early Aptian silicified radiolarian nanno chalk (porcellanite, E4). Sponge spicules and opal-CT replaced radiolarians in Opal-CT/nannomicrite matrix. SO 8-33 KD/2C (Swan Canyon, northern Exmouth Plateau).
(5) Mid-Maastrichtian foraminiferal (*Globotruncana* sp.) nannomicrite (chalk) with angular basalt and tuff pebbles. VA16-13 KD (c) (Oates Canyon southwest, Scott Plateau).
(6) Slightly ferruginous vitreous quartz chert with numerous micro- to cryptocrystalline circular relict bodies (? former siliceous fossils), cemented by macrocrystalline quartz. SO-173 KD/1 (late Mesozoic; slope of Carnarvon Terrace, southeast of Wallaby Plateau).

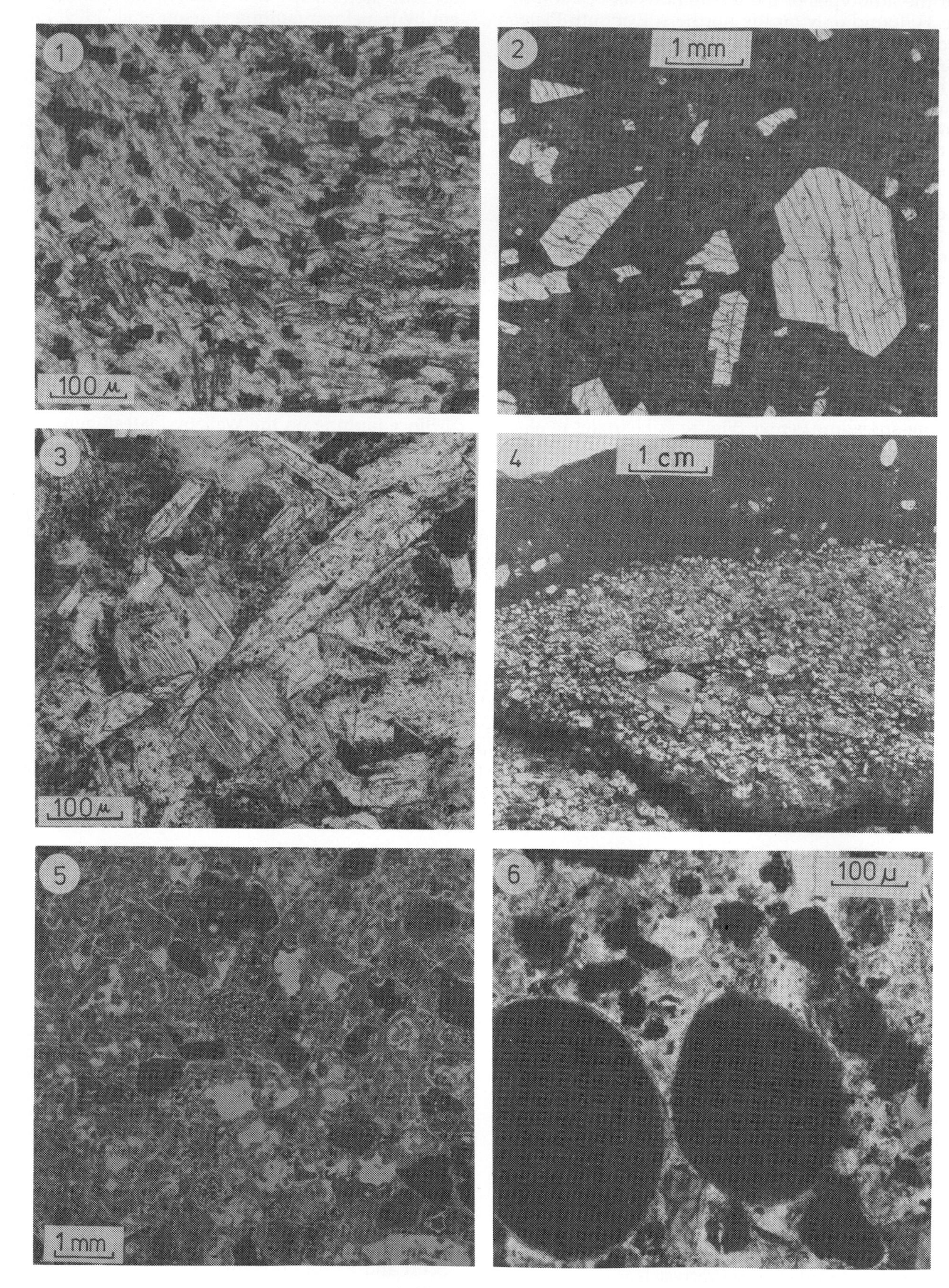
1
100 μ
2
1 mm
3
100 μ
4
1 cm
5
1 mm
6
100 μ

volcanism produced a variety of parallel- and cross-laminated tuffs and tuffaceous sandstones or siltstones. Subsequently these ashes were reworked in a high-energy marine environment.

A third volcanic climax came in late Early Cretaceous times (about 125 to 118 m.y. ago), when the abnormally thick oceanic crust (epilith) of the Wallaby Plateau was created. This brief, but spectacular, volcanic event followed the Upper Neocomian breakup of this part of Gondwanaland, between the present day Gascoyne and Perth Abyssal Plains (Veevers and Cotterill, 1978; Larson et al, 1979; Johnson et al, 1980). More than 1 km of layered volcanic flows and volcaniclastic rocks are present on the flanks of the Wallaby Plateau (von Stackelberg et al, 1980). According to A.L. Jacques (unpublished BMR Report), the suite contains *olivine tholeiites* with trace element characteristics resembling "transitional tholeiites" or oceanic floor basalts, as well as slightly differentiated alkaline varieties (hawaiites). Such "transitional" tholeiitic basalts are found in anomalous ridge segments, oceanic islands (such as Iceland), oceanic fracture zones, aseismic ridges, and seamounts (Bryan et al, 1976). They are generally interpreted as reflecting lower degrees of partial melting than normal ocean-floor basalts, or derivation from a large-ion-lithophile (LIL) element-enriched mantle source, or both. Hence the geochemistry of the basalts supports an oceanic (epilith) rather than continental nature. We therefore envisage that the Wallaby Plateau had a similar origin to Iceland, but with abrupt cessation of volcanism and subsidence below sea level after the oceanic ridge jumped to its new position further northwest.

Plate 4 — Volcanic and volcaniclastic rocks

(1) Early Liassic feldspar-rich volcanic rock (silica-undersaturated trachyte s.l. H3) with lath-shaped microlites of plagioclase aligned along flow lines ("fluidal", trachytic texture). K/Ar age: 190-193 million years. (Table 3). S08-65 KD (northern Wombat Platau, northern Exmouth Plateau).
(2) Late Triassic alkali rhyolite with porphyritic texture (large sanidine phenocrysts in alkali feldspar groundmass with opaques and titanite pseudomorphs. K/Ar age: 213 ± 3 million years. S08-66 KD (northern Wombat Plateau, northern Exmouth Plateau).
(3) Vescular tholeiitic basalt with intersertal texture and large plagioclase laths and clinopyroxene phenocrysts and ore in a greenish phyllosilicate matrix. SO8-168 KD/1B (?mid-Cretaceous; southern Wallaby Plateau).
(4) Volcaniclastic breccia/conglomerate (debris flow deposit?) covered by thick ferromanganese crust. SO8-165 KD/1 (eastern Wallaby Plateau).
(5) Very coarse tuffaceous sandstone/conglomerate with more or less altered basalt pebbles, clay-replaced tuff and glass fragments and rare quartz in a brown smectite matrix. Most fragments are rimmed by a phyllosilicate cement. SO8-165 KD/1 (mid-Cretaceous; eastern Wallaby Plateau).
(6) Highly altered phosphatized tuffaceous sandstone (H9) with two large phosphatic ooids (?), apatite, quartz, feldspar, tuff fragments and a collophane matrix. VA16-7 KD (e) (Jurassic; northern wall of Oates Canyon, Scott Plateau).

Upper Cretaceous to Cenozoic mature ocean stage

On the Northwest Shelf Coniacian/Santonian carbonate sedimentation indicates the mature ocean stage, characterized by unrestricted thermohaline oceanic circulation and high plankton productivity (Veevers and Johnstone, 1974). When Greater India began moving northwest, about 125 m.y. ago, the removal of southerly and westerly sources of sediment supply caused a sharp decline of the terrigenous influx. The isolation from downslope allochthonous continent-derived sediments resulted in waning sedimentation rates, and a thin cover of pelagic sediments began accumulating on the plateaus (compare with Dingle and Scrutton, 1979). Erosion or non-deposition prevailed during early Paleocene, mid-Eocene, and Oligocene times. These hiatuses were probably caused by intensified bottom current activity and helped keep average Cenozoic sedimentation rates very low (Figure 7). The shelf also experienced buildup and progradation of up to 3 km of shallow-water marls and calcareous clastics on a tilted surface, a "carbonate ramp" (Figures 5 and 7).

Paleobathymetry and subsidence history

Figure 7 compares the Mesozoic-Cenozoic subsidence history and paleobathymetry of the Dampier Sub-basin (Dampier No. 1 well) with that of the northern Exmouth Plateau (based on seismic and dredging evidence). These curves are not corrected for postdepositional compaction (von Rad and Einsele, 1980) or sea level fluctuations (Watts and Steckler, 1979), allowing only a rough estimate of the margin's total subsidence rates, since the biostratigraphic boundaries, hiatus durations, paleodepths, and seismic correlations are often ambiguous. After the removal of the effects of sediment loading (and of the water load at the Exmouth Plateau during the past 110 m.y.), "tectonic" or "thermal" subsidence rates are also calculated and shown as boxed values in Figure 7.

The subsidence history of the 5-km-thick Paleozoic sediments of the stable intracratonic basins is obscure. The Triassic to Jurassic rates indicate a similar uniform subsidence for both areas. Intermediate tectonic subsidence rates (up to 45m/m.y.) were compensated by relatively rapid rates of terrestrial to shallow-water rift valley sedimentation (total subsidence rates up to 90m/m.y.). After the Callovian breakup (reflector E) the tectonic subsidence rates decreased considerably in both areas (10 to 20 m/million years). In the Dampier Sub-basin, the tilting of the shelf was compensated by the progradation of clastic sediments or by a "carbonate ramp," so the water depth remained very shallow. At the outer plateaus (the northern Exmouth Plateau) the tectonic subsidence rates were slightly smaller (10 m/million years) than in the Dampier Sub-basin. However, because only a very thin layer of pelagic chalks draped this isolated part of the margin, the sea floor gradually deepened to its present depth of 2 km over the northern Exmouth Plateau.

As was the case with the Rockall Plateau margin,

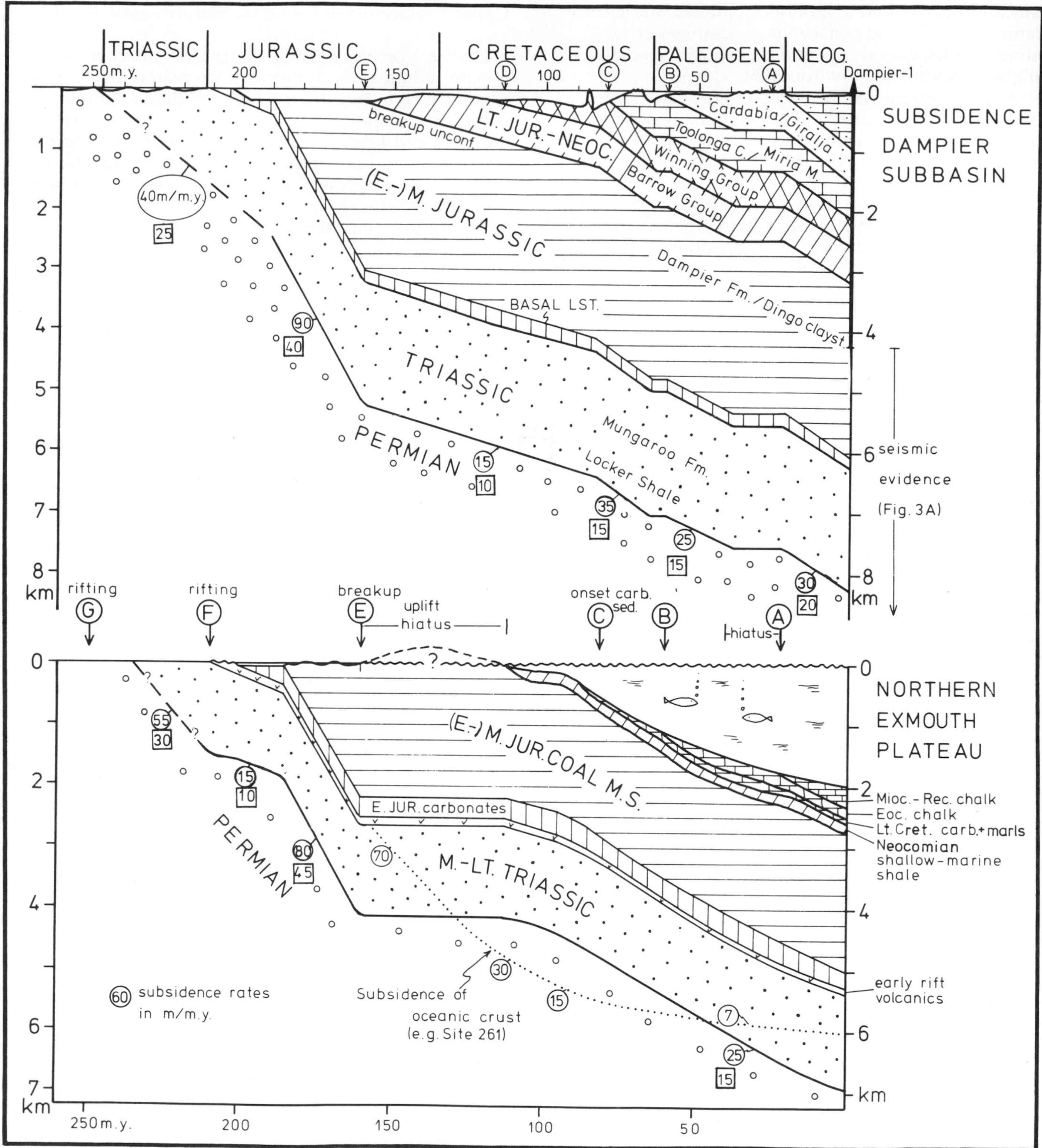

Figure 7 — Mesozoic-Cenozoic subsidence history and changes in water depth in Dampier Sub-basin (well Dampier No. 1, extrapolated by seismic evidence; Jurassic stratigraphy after Crostella and Barter, 1980, their Figure 17) and northern Exmouth Plateau (Figure 3A). Late Cretaceous paleodepths after Apthorpe (1979); northern Exmouth Plateau thicknesses seismically determined (see von Stackelberg et al, 1980; their Figure 7). The right-hand margin of each diagram shows the present thickness and depth below sea level of the different stratigraphic units which can be back-tracked through time. All encircled numbers designate total subsidence rates in m/m.y., whereas numbers in square boxes designate backstripped "tectonic" (thermal) subsidence rates due to thinning crust without the influence of sediment (and water) loading. Subsidence curves are not corrected for compaction or seal level fluctuations. Subsidence curve for oceanic crust of Site 261 for comparison (after Sclater et al, 1977).

the Precambrian-Paleozoic crust of the Exmouth Plateau may have experienced heating, causing uplift and volcanism, before rifting by brittle fracture of the upper crust and ductile flow in the lower crust could take place (Roberts and Montadert, 1980). The tectonic component of subsidence probably resulted from thinning of the continental crust (to about 20 km) under the Exmouth Plateau by early Mesozoic crustal stretching; or from the addition of heat into the crust from a low-velocity, low-density mantle, resulting in phase changes and followed by cooling (S.V. Sobolev, personal communication). After the onset of spreading, when the Exmouth Plateau was near sea level, tectonic subsidence rates were roughly comparable to those of the cooling and sinking crust of the adjacent Indian Ocean (Figure 7; Sclater, Abbott, and Thiede, 1977).

Hydrocarbon Potential

The *Sonne* 8 and *Valdivia* 16 cruises provided new information on the potential for hydrocarbon accumulations beneath the Exmouth, Scott and Wallaby Plateaus (von Stackelberg et al 1980).

The Exmouth Plateau contains as much as 10 km of Phanerozoic strata with suitable source rocks, reservoir rocks, and structures for the generation and trapping of hydrocarbons (Exon and Willcox, 1978). Numerous Triassic-Jurassic fault blocks, and stratigraphic traps in a large Neocomian delta in the south are obvious exploration targets. According to Wright and Wheatley (1979), the fault blocks consist of Middle to Upper Triassic sediments, probably sandstone and shale; a probable gas chimney is associated with the highest fault block, and fault blocks have a higher potential for oil than the Neocomian delta. Results of the *Sonne* 8 cruise confirmed the previously established picture, but indicated that the northern Exmouth Plateau was more akin to the Canning Basin than to the Carnarvon Basin. On the northern Exmouth Plateau more than 1,000 m of Triassic paralic sandstone and shale is covered by 2,500 m of Lower Jurassic shelf carbonates and Middle Jurassic coal measures, which are overlain in turn by 500 m of younger carbonates. Possible source rocks, in the form of early Mesozoic siltstone, shale and coal, and potential reservoir rocks, in the form of porous Jurassic quartzose sandstone (coal measure sequence) and calcarenite, are present.

Source rock analyses on eight samples of mid-Jurassic siltstone, shale, and coal (A-facies) are given in Table 5 and interpreted here, following comments by K. Jackson (BMR) and H. Wehner (BGR). The clastic rocks contain up to 2.35% total organic carbon. In most cases the extracted organic matter consists of dominant asphaltenes, subdominant resins, and subordinate saturated and aromatic hydrocarbons. The distribution of n-alkanes in the saturates reaches a peak either in the range C_{25} to C_{29} (with a marked tendency for a predominance of odd-numbered alkanes), or in the range C_{18} to C_{20} (without that tendency). Vitrinite reflectance of the coal varies from 0.20% to 0.39%. Hence the coalification of the Jurassic sediments, which were probably never buried by more than 1 to 2 km of sediment, did not exceed a stage beyond the boundary between lignite and subbituminous coal (von Stackelberg et al, 1980). Vitrinite is normally the dominant constituent of the organic matter. The high exinite content (86%) in sample 33/1B marks it a potential oil source sediment. Overall the results show that the source rock potential for gas of the clastic rocks is poor to fair (good), and that of the coal is fair to good. Based on the vitrinite reflectance data and the saturated hydrocarbon gas chromatograms, however, all samples of Table 5 are immature; the rocks are submature at the present burial depth and temperature level.

Twenty-five cores from four areas high on the Exmouth Plateau were isotopically analyzed by E. Faber and W. Stahl (BGR) for gaseous hydrocarbons adsorbed to surface sediments (von Stackelberg et al, 1980). Most stations yielded too little methane for reliable isotope determinations. However, the sample from one station above the gas chimney, reported by Wright and Wheatley (1979), yielded approximately 18 parts/billion methane with a $\delta^{13}C$ of -40%. This methane was probably generated by sapropelic marine source material within the oil window.

By late 1980, eight petroleum exploration wells were drilled, plugged, and abandoned on the Exmouth Plateau. Publicly available information indicates that no significant oil shows were found in the wells, but that gas shows were fairly common. The largest gas accumulations found in these wells were those at Scarborough No. 1 and Eendracht No. 1 wells on the crest of the plateau. Since gas is not presently a commercial proposition in the water depths prevailing on the plateau, the exploration results announced so far are not encouraging.

For an average geothermal gradient of 20°C/km on the Exmouth Plateau (Kangaroo Syncline; Wright and Wheatley, 1979) and an effective heating time of 100 m.y., the upper boundary of oil generation (75°C; $R_o = 0.7$; "oil window") will be at least 3.5 km below the sea floor. However, not even in the Neocomian delta do the post-Jurassic sediments exceed a thickness of 2 km. This downgrades the chances for the maturation of organic matter in the Upper Triassic to mid-Jurassic sediments to commercial hydrocarbon accumulations — a problem typical of starved continental margins.

Due to the thinness of its Mesozoic section, the petroleum potential of the Scott Plateau is also low (Stagg, 1978). The pre-breakup section, however, includes Paleozoic sediments which might have some prospectivity. Veevers and Cotterill (1978) suggest that Stagg's pre-breakup section consists of a Jurassic buildup of oceanic volcanics, an epilith, which rules out the possibility of Paleozoic sediments, and further downgrades the plateau's prospects. The *Valdivia* sampling did yield a variety of basic extrusive rocks from near the breakup unconformity (E), but Hinz et al (1978) and Stagg and Exon (1979) associated them with normal volcanism on continental crust, rather than regarding them as part of an epilith. If the Scott Plateau is a continental block its petroleum prospects can be regarded as poor to moderate; but, if it is an epilith, they are very poor indeed. In either case, the structural lows to the east and south of the plateau have a much thicker section of Mesozoic sediments, and can be regarded as fairly prospective. A technological problem is the great water depth (generally more than 2000 m) over much of the region.

The petroleum potential of the Wallaby Plateau is negligible, because it appears to consist of a very thick pile of late Lower Cretaceous volcanic and volcaniclastic rocks, overlain by a veneer of mid-Cretaceous shale and Upper Cretaceous and Cenozoic carbonates.

ACKNOWLEDGMENTS

The German Federal Ministry for Research and Technology (BMFT) sponsored and financed the BGR cruises with R/V *Valdivia* and *Sonne*. We are grateful to the chief scientists of both cruises, K. Hinz (BGR) and U. von Stackelberg (BGR), and to several shore-based scientists. We thank especially R. Emmermann (Karlsruhe), H. U. Schmincke (Bochum), A. L. Jaques (BMR), H. Raschka (BGR), M. Mohr (BGR), and P. Müller (BGR) for geochemical and petrographical analyses of volcanic rocks. H. Kreuzer and W. Harre (both BGR) for the K/Ar age dates, H. Rösch (BGR) for XRD analyses, K. Jackson (BMR) and H. Wehner (BGR) for the interpretation of hydrocarbon source rock analyses, B. Zobel (BGR) for micropaleontological information and advice in the microfacial interpretation, and D. Burger (BMR) for palynological information. H. Jones (BMR), Chuck Drake, and Joel Watkins, kindly reviewed the manuscript. N. F. Exon publishes with the permission of the Director, Bureau of Mineral Resources.

REFERENCES CITED

Allan, G.A., L.G. Pearce, and W.E. Gardner, 1978, A regional interpretation of the Browse Basin: Australian Petroleum Exploration Association Journal, v. 18, p. 23-33.

Apthorpe, M.C., 1979, Depositional history of the upper Cretaceous of the Northwest Shelf, based upon foraminifera: Australian Petroleum Exploration Association Journal, v. 79, n. 1, p. 74-89.

Armstrong, R.L., 1978, Pre-Cenozoic Phanerozoic time scale, *in* computer file of critical dates and consequences of new and in-progress decay-constant revisions: AAPG, Studies in Geology, v. 6, p. 73-90.

Bernoulli, D., O. Kälin, and E. Patacca, 1979, A sunken continental margin of the Mesozoic Tethys-the northern and central Apennines, *in* Symposium "Sedimentation jurassique W — européen ": A.S.F. publ. spec., n. 1, p. 197-210.

Bryan, W.B., et al, 1976, Inferred geologic settings and differentiation in basalts from the deep sea drilling project: Journal of Geophysical Research, v. 81, n. 23, p. 4285-4304.

Crostella, A., and T. Barter, 1980, Triassic-Jurassic depositional history of the Dampier and Beagle Sub-basins, Northwest Shelf of Australia: Australian Petroleum Exploration Association Journal, v. . 20, n. 1, p. 25-33.

De Charpal, O., et al, 1978, Rifting, crustal attentuation and subsidence in the Bay of Biscay: Nature, v. 275, n. 5682, p. 706-711.

Dingle, R.V., and R.A. Scrutton, 1979, Sedimentary succession and tectonic history of a marginal platform (Goban-Spur, southwest of Ireland): Marine Geology, v. 33, p. 45-69.

Exon, N.F., and J.B. Willcox, 1978, Geology and petroleum potential of Exmouth Plateau area off Western Australia: AAPG Bulletin, v. 62, p. 40-72.

———, and ———, 1980, The Exmouth Plateau; Stratigraphy, structure, and petroleum potential: Bureau of Mineral Resources Bulletin, v. 199, 52 p.

———, U. Von Rad, and U. Von Stackelberg, in press. The geological development of the margins of the Exmouth Plateau off Northwest Australia: Marine Geology.

Falvey, D.A., and J.J. Veevers, 1974, Physiography of the Exmouth and Scott Plateaus, western Australia, and adjacent northeast Wharton Basin: Marine Geology,v. 17, p. 21-59.

Flügel, E., 1978, Mikrofazielle Untersuchungsmethoden von Kalken: Springer, 454 p.

Heirtzler, J.R., et al, 1978, The Argo Abyssal Plain: Earth and Planetary Science Letters, v. 41, p. 21-31.

Hinz, K., et al, 1978, Geoscientific investigations from the Scott Plateau off Northwest Australia to the Java trench: Bureau of Mineral Resources Journal of Australian Geology and Geophysics, v. 3, p. 319-340.

Jenkyns, H.C., 1971, Speculations on the genesis of crinoidal limestones in the Indian Ocean between India and Australia: Earth and Planetary Science Letters, v. 47, p. 131-143.

Johnson, B. D., C. M. Powell, and J. J. Veevers, 1980, Early spreading history of the Indian Ocean between India and Australia: Earth and Planetary Science Letters, v. 47, p. 131-143.

Larson, R.L., et al, 1979, Cuvier Basin-a product of ocean crust formation by Early Cretaceous rifting off western Australia: Earth and Planetary Science Letters, v. 45, p. 105-114.

Markl, R.G., 1974, Evidence for the breakup of eastern Gondwanaland by the Early Cretaceous: Nature, v. 251, p. 196-200.

McDougall, I., and G.J. Van der Lingen, 1974, Age of rhyolites of the Lord Howe Rise and the evolution of the southwest Pacific Ocean: Earth and Planetary Science Letters, v. 21, p. 117-126.

Montadert, L., et al, 1979a, *in* M. Talwani et al, eds., Deep drilling results in the Atlantic Ocean; continental margins and paleoenvironment: Washington, D.C., American Geophysical Union, Maurice Ewing Series, v. 3, p. 154-186.

———, ———, 1979b, Initial reports of the deep sea drilling project: Washington, D.C., U.S. Government Printing Office, v. 48, 1183 p.

Powell, D.E., 1976, The geological evolution of the continental margin off Northwest Australia: Australian Petroleum Exploration Association Journal, v. 16, n. 1, p. 13-23.

Quilty, P.G., in press, Early Jurassic foraminifera from the Exmouth Plateau, western Australia: Journal of Paleontology.

Reading, H.G., 1978, Sedimentary environments and facies: Blackwell Scientific Publishers, 557 p.

Riech, V., and U. von Rad, 1979, Silica diagenesis in the Atlantic Ocean — diagenetic potential and transformations, *in* M. Talwani, et al (eds.), Deep drilling results in the Atlantic Ocean; continental margins and paleoenvironment: American Geophysical Union, Maurice Ewing Series, v. 3, p. 315-340.

Roberts, D.G., and L. Montadert, 1980, Contrasts in the structure of the passive margins of the Bay of Biscay and Rockall Plateau: Royal Society of London, Philosophical Transactions, Series A, v. 294, p. 97-103.

Sclater, J.G., D. Abbott, and J. Thiede, 1977, Paleobathymetry and sediments of the Indian Ocean, *in* J. R. Heirtzler et al, eds., Indian Ocean geology and biostratigraphy: American Geophysical Union, Washington, D.C., p. 25-59.

Stagg, H.M.J., 1978, The geology and evolution of the Scott Plateau: Australia Petroleum Exploration Association Journal, v. 18, n. 1, p. 34-43.

———, and N. F. Exon, 1979, Western margin of Australia — Evolution of a rifted arch system, discussion: Geological Society of American Bulletin, v. 90, m. 1, p. 795-797.

———, and ———, 1982, Geology of the Scott Plateau and Rowley Terrace: Bureau of Mineral Resources Bulletin, v. 213, p. 1-47.

Steiger, R.H., and E. Jäger, 1977, Subcommission on Geochronology — convention on the use of decay constants in geo- and cosmochronology: Earth and Planetary Science Letters, v. 36, p. 359-362.

Symonds, P.A., and P. J. Cameron, 1977, The structure and stratigraphy of the Carnarvon Terrace and Wallaby Plateau: Australian Petroleum Exploration Association Journal, v. 17, n. 1, p. 30-41.

Thomas, B. M., and D. M. Smith, 1974, A summary of the petroleum geology of the Carnarvon Basin: Australian Petroleum Exploration Association Journal, v. 14, n. 1, p. 66-76.

Veevers, J.J., and D. Cotterill, 1978, Western margin of Australia — evolution of a rifted arch system: Geological Society of America Bulletin, v. 98, p. 337-355.

———, and M. H. Johnstone, 1974, Comparative stratigraphy and structure of the western Australian margin and the adjacent deep ocean floor, in J. J. Veevers et al, eds., Reports of the deep sea drilling project: U.S. Government Printing Office, Washington, D.C., v. 27, p. 571-585.

———, et al, 1974, Initial reports of the deep sea drilling project: Washington, D. C., U.S. Government Printing Office, v. 27, 1060 p.

Von Rad, U., and M. A. Arthur, 1979, Geodynamic, sedimentary and volcanic evolution of the Cape Bojador continental margin (northwest Africa), *in* M. Talwani et al, eds., Deep drilling results in the Atlantic Ocean — continental margins and paleoenvironment: Washington, D.C., American Geophysical Union, Maurice Ewing Series, v. 3, p. 187-203.

———, and G. Einsele, 1980, Mesozoic-Cainozoic subsidence history and paleobathymetry of the northwest African continental margin (Aaiun Basin to DSDP site 397): Royal Society of London, Philosophical Transactions, Series A, v. 294, p. 37-50.

Von Stackelberg, U., et al, 1980, Geology of the Exmouth and Wallaby Plateaus off Northwest Australia; sampling of seismic sequences: Bureau of Mineral Resources Journal of Australian Geology and Geophysics, v. 5, p. 113-140.

Warris, B.J., 1976, Canning Basin, off-shore, *in* C. L. Knight, ed., The economic geology of Australia and Papua-New Guinea, v. 3, Metals: Australasian Institute of Mining and Metallurgy, Monograph n. 7, p. 185-188.

Watts, A.B., and M.S. Steckler, 1979, Subsidence and eustacy at the continental margin of eastern North America, *in* M. Talwani et al, eds., Deep drilling results in the Atlantic Ocean; continental margins and paleoenvironment: Washington, D.C., American Geophysical Union, Maurice Ewing Series, v. 3, p. 218-234.

Willcox, J.B., and N.F. Exon, 1976, The regional geology of the Exmouth Plateau: Australian Petroleum Exploration Association Journal, v. 16, n. 1, p. 1-11.

Wilson, J.L., 1975, Carbonate Facies in geologic history: Springer, 471 p.

Wright, A.J., and T.J. Wheatley, 1979, Trapping mechanisms and the hydrocarbon potential of the Exmouth Plateau, western Australia: Australia Petroleum Exploration Association Journal, v. 19, n. 1, p. 19-29.

Tertiary Sedimentation in the Southern Beaufort Sea, Canada

P. S. Willumsen
R. P. Cote
Gulf Canada Resources Inc.
Alberta, Canada

Five Tertiary deltaic cycles are identified in the Mackenzie Basin. These cycles added approximately 35,000 sq. mi. to the continental shelf, and deposited more than 100,000 cu mi of sediment. The five deltaic cycles followed a distinct counter-clockwise progradational pattern into the Mackenzie Basin with sedimentation beginning in the southwestern part of the basin and shifting northeast. The area distribution of the three major depositional facies for each of the five deltaic cycles, namely the delta plain, the delta front, and the prodelta facies, is outlined. The recognition of a turbiditic subfacies within the prodelta sediments is of major importance and provides new and deeper prospects for petroleum exploration in the Mackenzie Basin.

This paper presents an explorationist's view of the geological evolution of the Mackenzie Basin during the Tertiary.

Gulf Canada Resources, Inc. has been involved in petroleum exploration in the Mackenzie delta/Beaufort Sea area since its inception in the 1960s. During this period there has been a need to incorporate all available geological and geophysical data in order to guide exploration strategy. This paper represents the latest update of this continual effort, which was performed primarily by geologists and geophysicists in Gulf Canada's exploration and research departments.

REGIONAL SETTING

The Beaufort Sea is in the southern part of the Arctic Ocean, extending from the western edge of the Canadian Arctic Islands to Point Barrow in Alaska. Figure 1 outlines the physiography of the southern Beaufort Sea between Banks Island and Prudhoe Bay. Within the Canadian portion of the Beaufort Sea the shaded area highlights the Mackenzie Basin, which is filled with great thicknesses of Tertiary and Mesozoic sediments. The Mackenzie Basin covers about 35,000 sq mi and is a sedimentary depocenter bounded by the Cordillera to the southwest and by a major fault system to the southeast. Seaward, the Mackenzie Basin extends to approximately the present-day shelf edge. The Mackenzie delta, the focal point of this basin, can be subdivided into two parts; the larger, offshore portion underlying the continental shelf and a smaller, onshore part occupied by the Arctic coastal plain including the present-day Mackenzie delta. The areas on the coastal plain underlain by Tertiary sediments are indicted in a coarse stipple on Figure 1.

METHODOLOGY

The maps and cross sections in this paper result from integration of well information and seismic data. Approximately 150 wells have been drilled to date in the region; 20 were drilled offshore and the remainder on land. Seismic data cover the area in a grid averaging 1 to 2 mi between lines.

Paleontological age determinations are the primary tool in establishing the stratigraphic subdivision of the study area, and are useful in determining the depositional environment of any given time slice. These data, combined with lithologic and seismic data, develop a basic geologic model for the Mackenzie Basin.

POST CRETACEOUS ISOPACH

This paper deals with sedimentation in the Mackenzie Basin only during the Tertiary. References to the basin's formation are in Young, Myhr, and Yorath

(1976) and Jones (1980).

Figure 2 shows the total thickness of post-Cretaceous, mainly Tertiary sediments, contoured at 5,000 ft intervals. A maximum thickness of over 30,000 ft is centered directly north of the present Mackenzie delta. Areas of relatively thin Tertiary sediments to the north and west are attributed to a series of diapiric shale ridges which are discussed later. The map of the study area (Figure 2) falls within the black frame shown on the regional map (Figure 1). More notable fields, such as Parsons Lake and Taglu, and wells such as Ukalerk and Kopanoar, are shown on this and succeeding figures. The present-day Mackenzie Delta is highlighted as a hachured area on Figure 2.

Figure 3 shows a generalized north-south cross section through the central part of the Mackenzie Basin. Paleozoic and Mesozoic rocks, which define the basin floor, are shown in heavy hachured pattern near the base of the section. Within the Tertiary basin five deltaic cycles are identified and the depositional facies are coded for each cycle. The vertical exaggeration of the cross section is about 15:1.

STRATIGRAPHY/DEPOSITIONAL HISTORY

The columnar section of Figure 4 displays the informal names and ages of the various deltaic sequences. The ages derive from paleontological and palynological age determinations by Gulf, which are not always in accordance with the age determination cited in the literature (compare Young, Myhr, and Yorath, 1976; Lerand, 1973; Staplin, 1976).

The first Tertiary deltaic cycle deposited in the basin is the Paleocene Moose Channel Formation. Following a transgression indicated by wide-spread thick deposits of shale, the late Paleocene to Eocene "Taglu Delta" sequence (Reindeer Fm., Mountjoy, 1967) built out into the basin. This sequence today lies mainly beneath the Mackenzie Delta while the three younger deltas lie beneath the present Beaufort Sea shelf. The

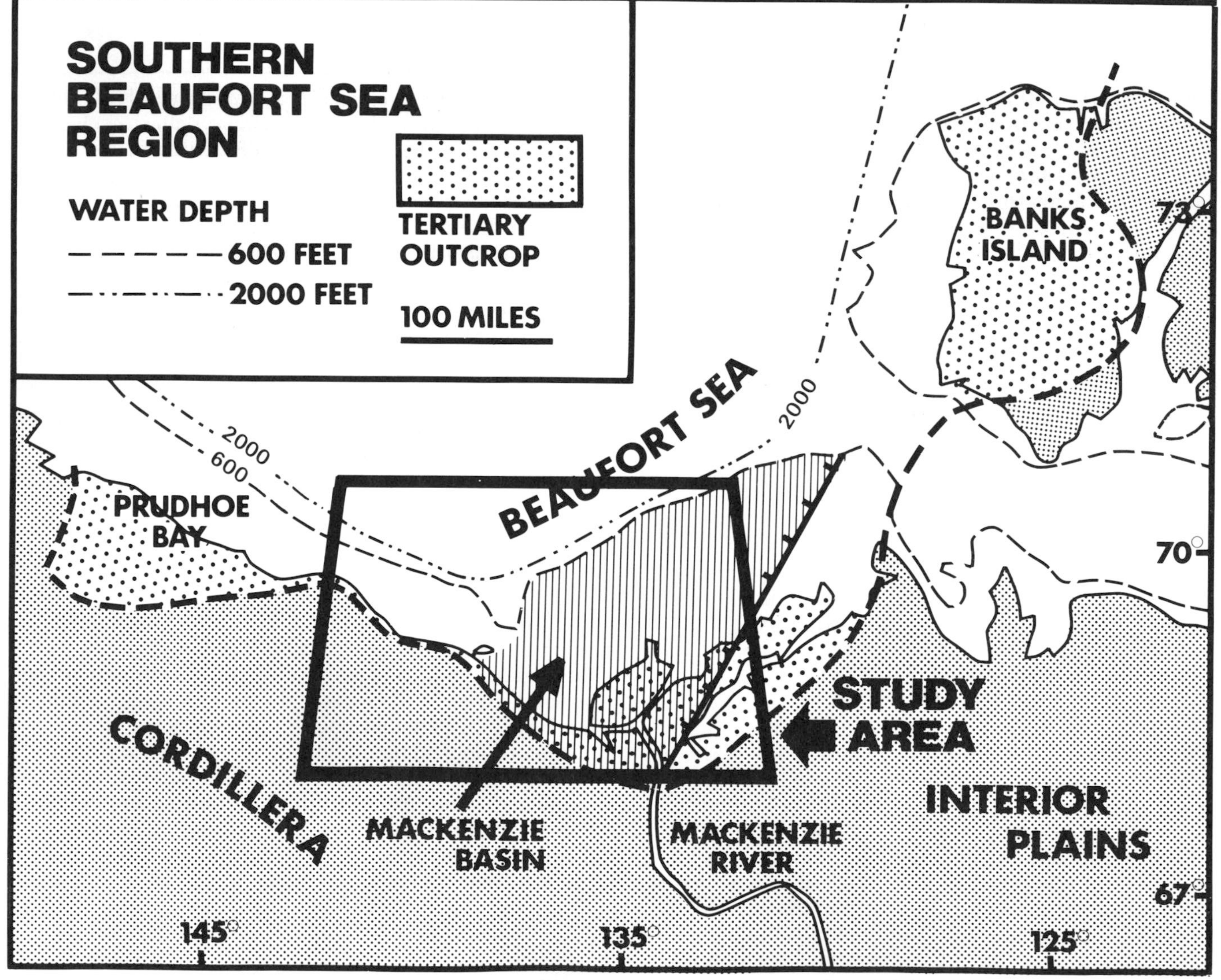

Figure 1 — Index map. The coarsely stipled area indicates the portion of the coastal plain underlain by Tertiary sediments. Heavy frame indicates area of study.

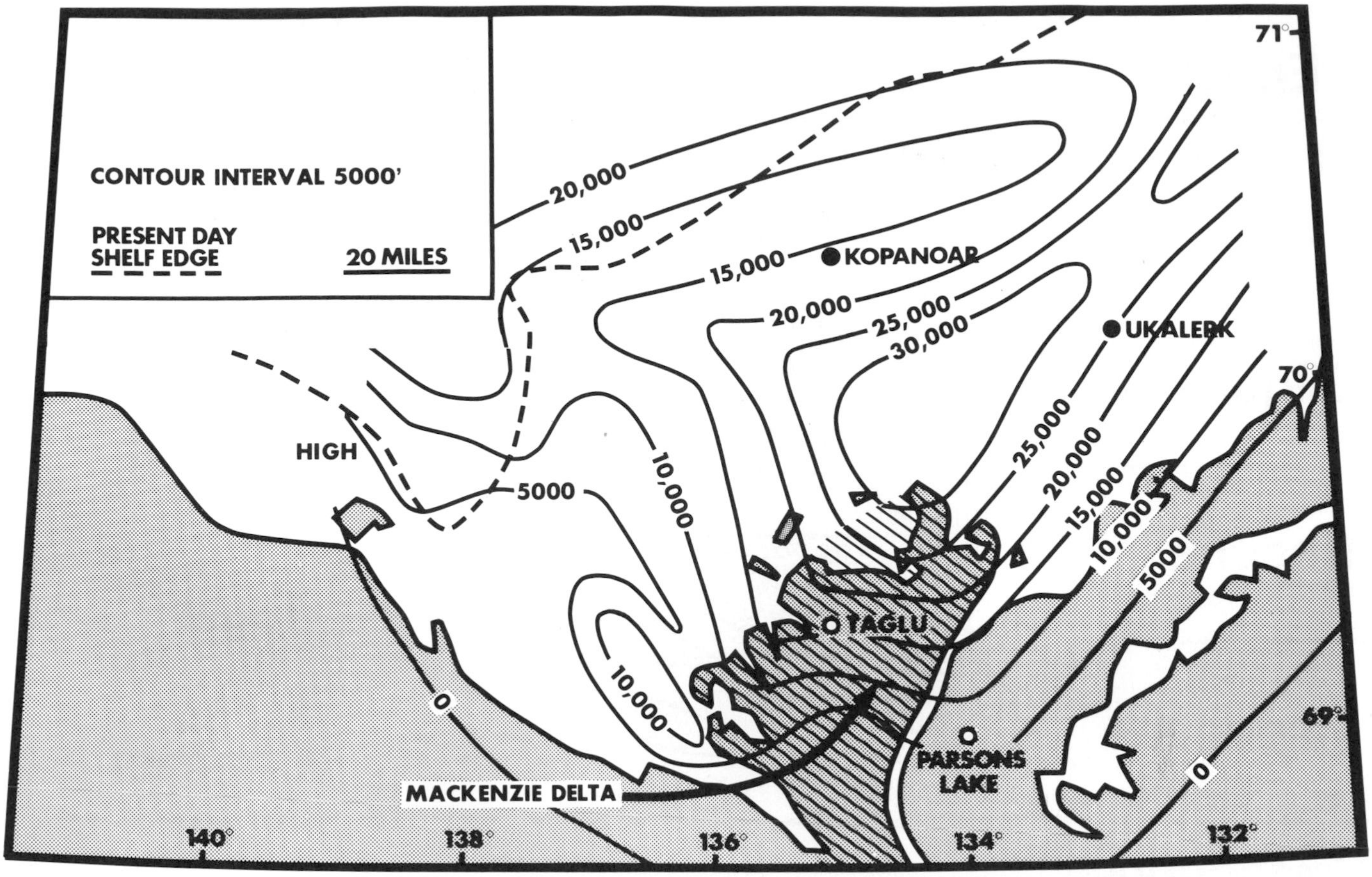

Figure 2 — Post-Cretaceous sediment thicknesses in the Mackenzie delta region. Present-day delta is indicated by cross-lining.

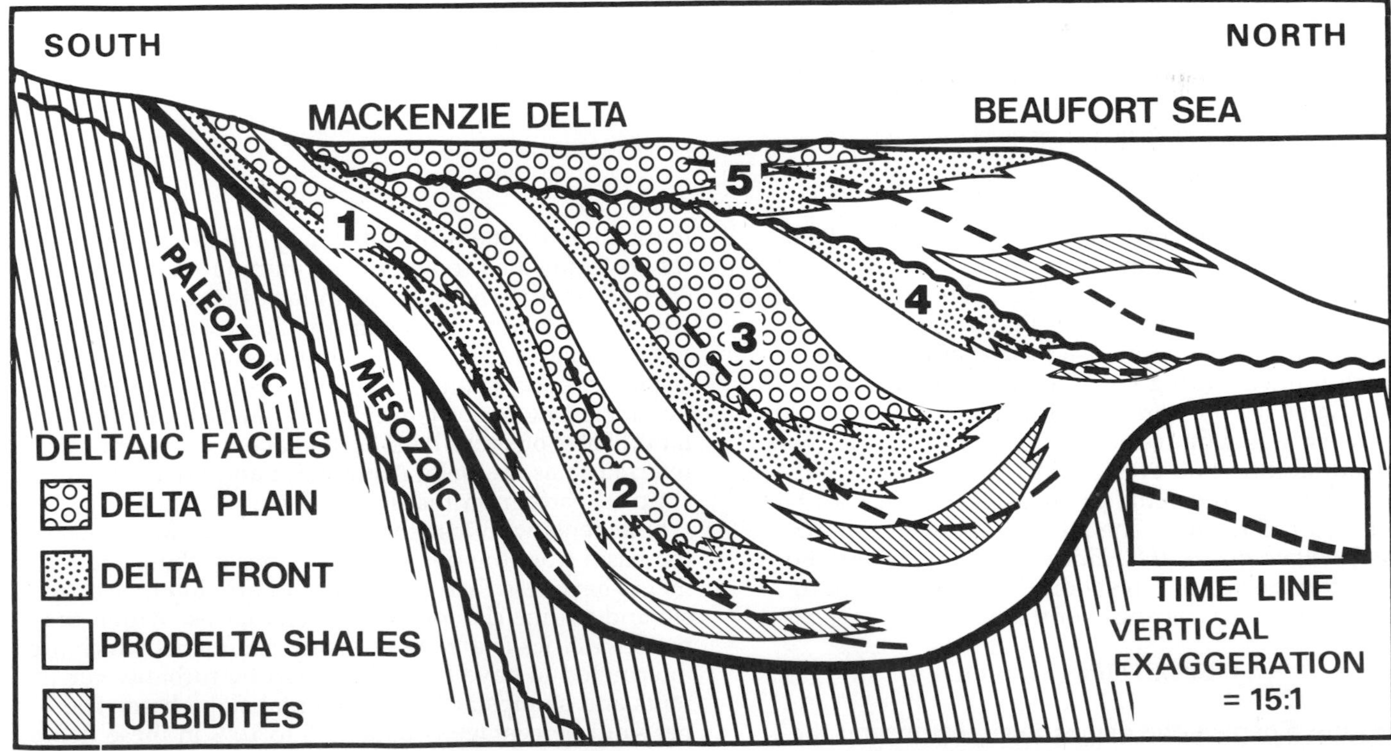

Figure 3 — Generalized cross section of the Mackenzie delta. 1-Paleocene Moose Channel delta; 2-Eocene Taglu delta; 3-Oligocene Pullen delta; 4-Akpak delta and; 5-Beaufort delta.

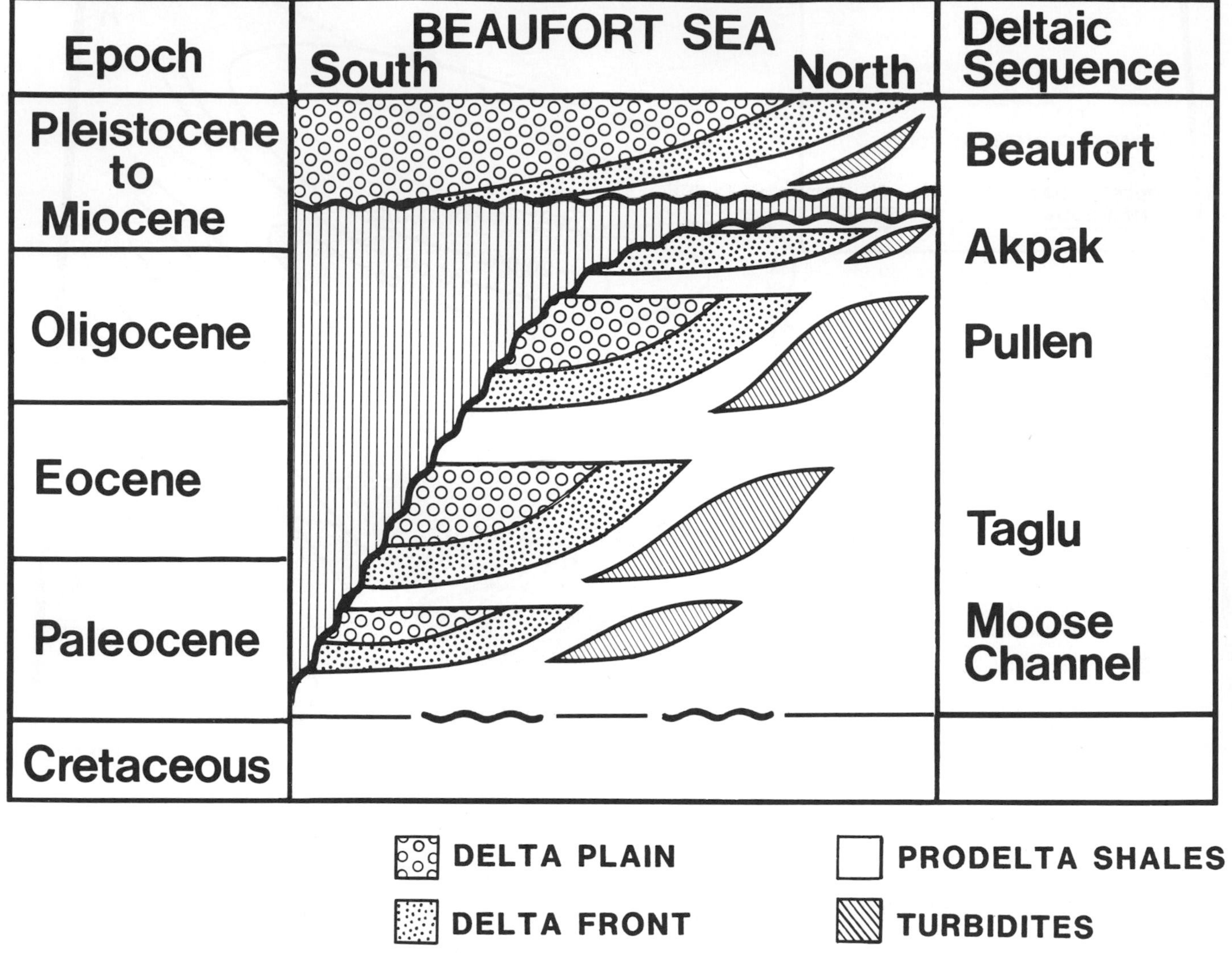

Figure 4 — Ages and informal names of delta sequences.

third deltaic sequence, the Oligocene "Pullen delta" (informal name used in the oil industry; Hea et al, 1980) is separated from the "Taglu delta" by thick widespread shale and is overlain by the "Akpak Sequence" (informal name used internally in Gulf) which formed during late Oligocene and early Miocene times. In the early to middle Miocene most of the basin was tectonically unstable, resulting in widespread faulting and erosion. A major unconformity formed at this time cuts progressively older rocks in a southward direction. The last (compare Hea et al, 1980) deltaic pulse, the Miocene to Plio/Pleistocene "Beaufort sequence," rests on the truncated older deltaic sequences as shown on Figure 3.

DEPOSITIONAL MODEL

Figure 5 depicts the depositional features on the shelf and slope in an area influenced by deltaic sedimentation. Sediments carried to the depocenter are deposited on the delta plain and in the nearshore marine environment (in this study the delta plain facies are defined as sediments containing more than 35% net sand). The delta plain facies is coded as open circles on the illustrations and is dominated by channel pointbar sands and lake deposits with abundant coals. Sediments containing 15 to 35% net sand are dominated by deposits originating in distributary mouthbars, offshore bars, and shallow bays. They are referred to as the delta front facies and are coded in a stippled pattern on the illustrations.

Sediments with less than 15% net sand form toward the basin from the delta front. Typically these sediments have less than 5% sand and are dominated by silt and marine clay of the prodelta facies. Within the prodelta facies is a zone of sandy sediments which are interpreted to have been deposited by turbidity currents and slumps at the continental slope base. The net sand percentage is usually 5 to 10% in these deep water sediments, which are coded in a hachured pattern on the illustrations. As the delta progrades into the basin the sedimentary facies recording these de-

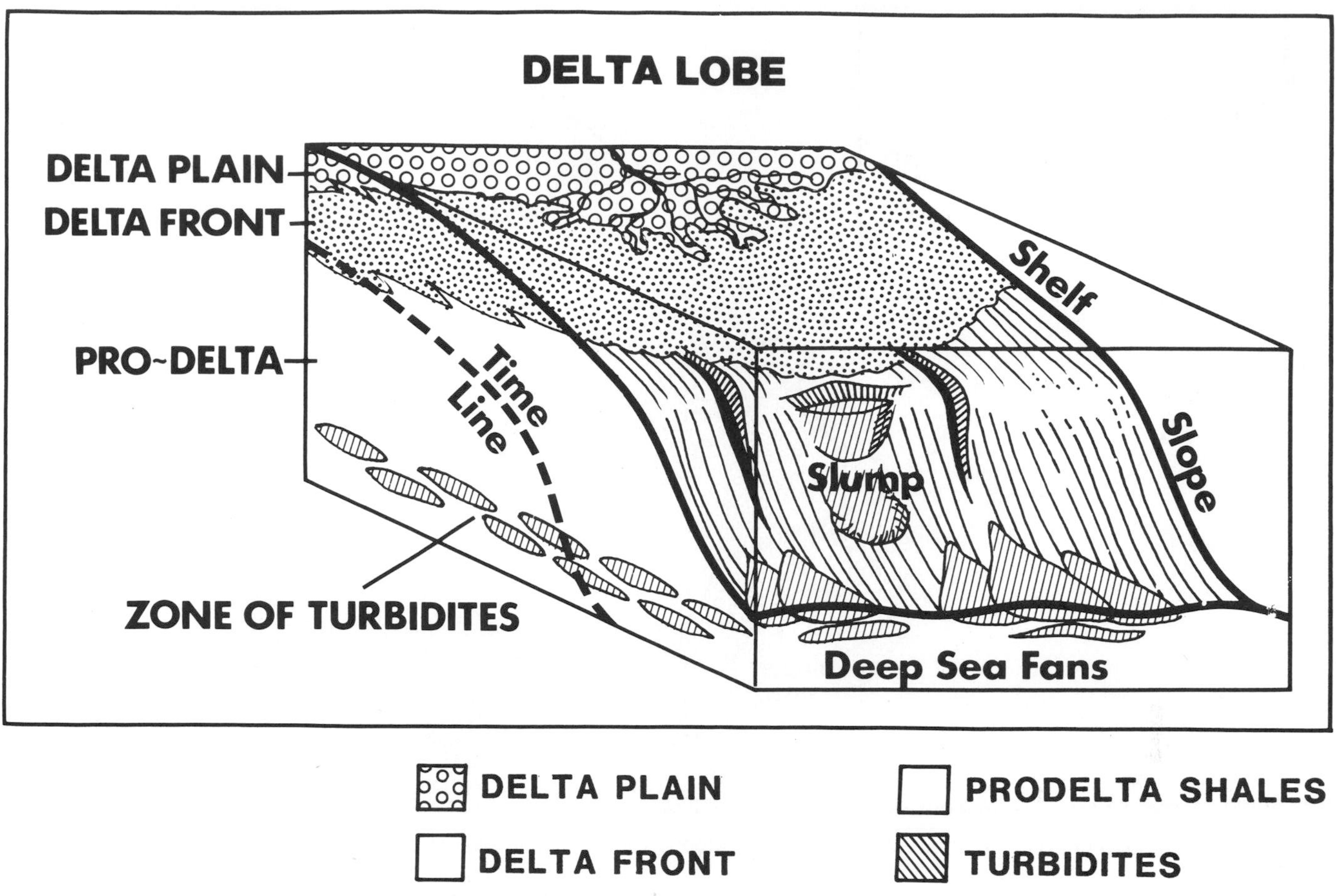

Figure 5 — Schematic drawing showing inferred depositional environments.,

positional environments form a continuous vertical sequence. In a typical vertical column the facies succession begins with prodelta shales, including turbidites, progressing upward into delta front then delta plain sediments. The only sandy sediments found beyond the shelf edge are turbidites.

PALEOGEOGRAPHY; PALEOCENE MOOSE CHANNEL FORMATION

The first deltaic cycle deposited in the Mackenzie Basin during the Tertiary is the Moose Channel Formation. This unit overlies shales and silts of late Cretaceous and early Paleocene age, and is restricted to the southwestern part of the basin. Figure 6 and succeeding maps of the five delta sequences are isopachs of the gross sand interval, contoured in feet. The map includes the lowest to the uppermost sand in the delta sequence and excludes the thick shale section between the individual deltaic cycles. Within this interval the net sand percentage varies according to the depositional environment. Note that the gross sand intervals for the shelf and slope derived sediments are contoured independently.

Two large indentations in the erosional zero edge to the south are caused by erosion over structural highs (Figure 6). It is believed that the Blow High to the west is partly syndepositional, since it is the result of a series of diapiric ridges extending in a northwest to southeast direction. The Kipnik High in the lower centre of Figure 6 is most likely a postdepositional tectonic feature.

In the middle to late Paleocene, marine shales were deposited over the entire area either representing an intermission in clastic material influx from the source areas or a rapid subsidence of the basin. These shales are overlain by the late Paleocene to Eocene Taglu delta.

EOCENE TAGLU DELTA SEQUENCE

The Taglu delta prograded into the basin late in the Paleocene and at about the end of the Eocene had reached its maximum areal extent along the shelf edge at the heavy dashed line marking the transition between the shelf and slope (Figure 7). To date this deltaic sequence is the major exploration target in the southwestern portion of the basin. The depocenter of this sequence is located near the Taglu field, for which the unit is named. Note that the depocenter is displaced basinward from the Moose Channel depocenter, and that the gross sand interval reaches approximately 6000 ft thick, about twice the thickness of the previous unit. The depositional strike trends approximately northwest to southeast and parallels a series of syndepositional diapiric ridges located under the outer shelf and slope areas (compare to Figure 12)

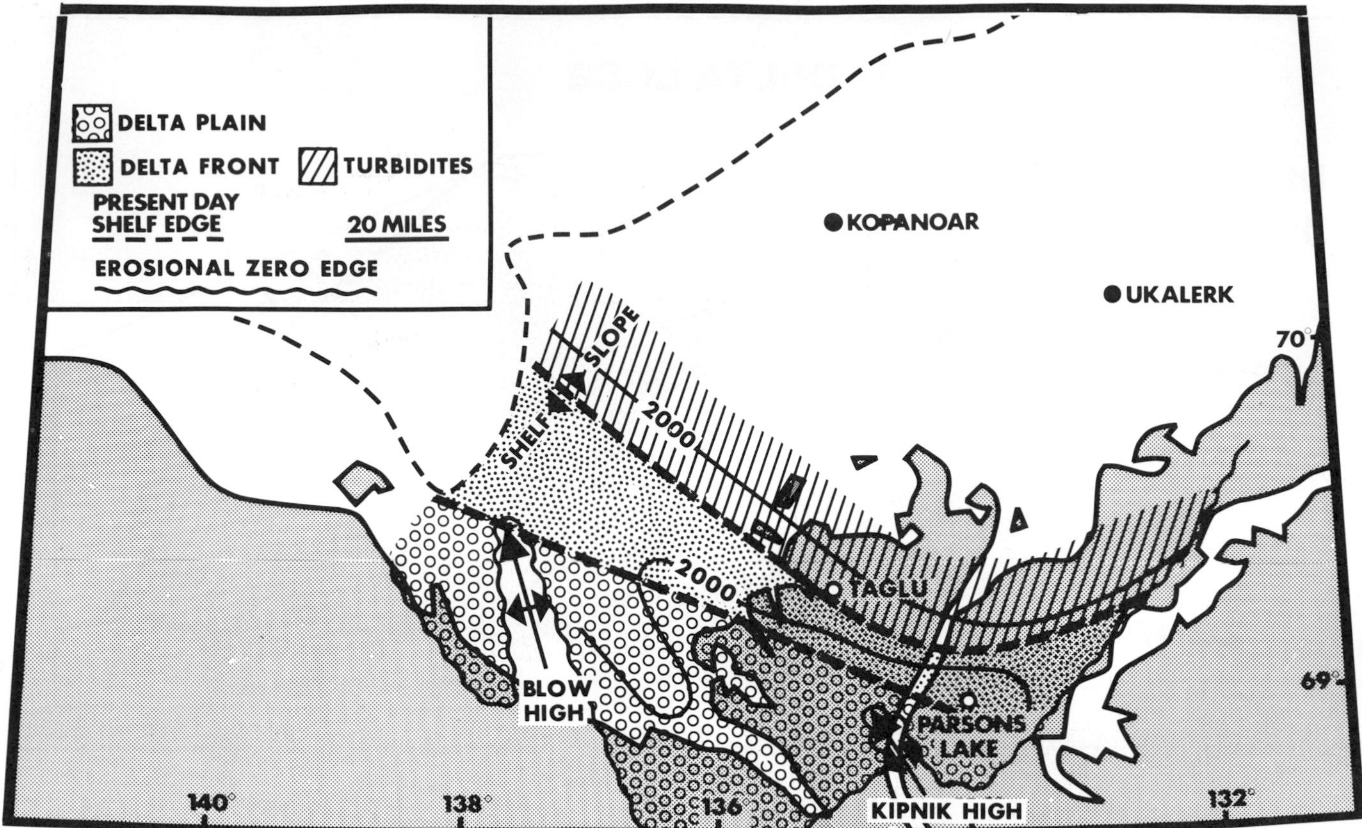

Figure 6 — Gross sand intervals in shelf and slope facies, Moose Channel delta sequence.

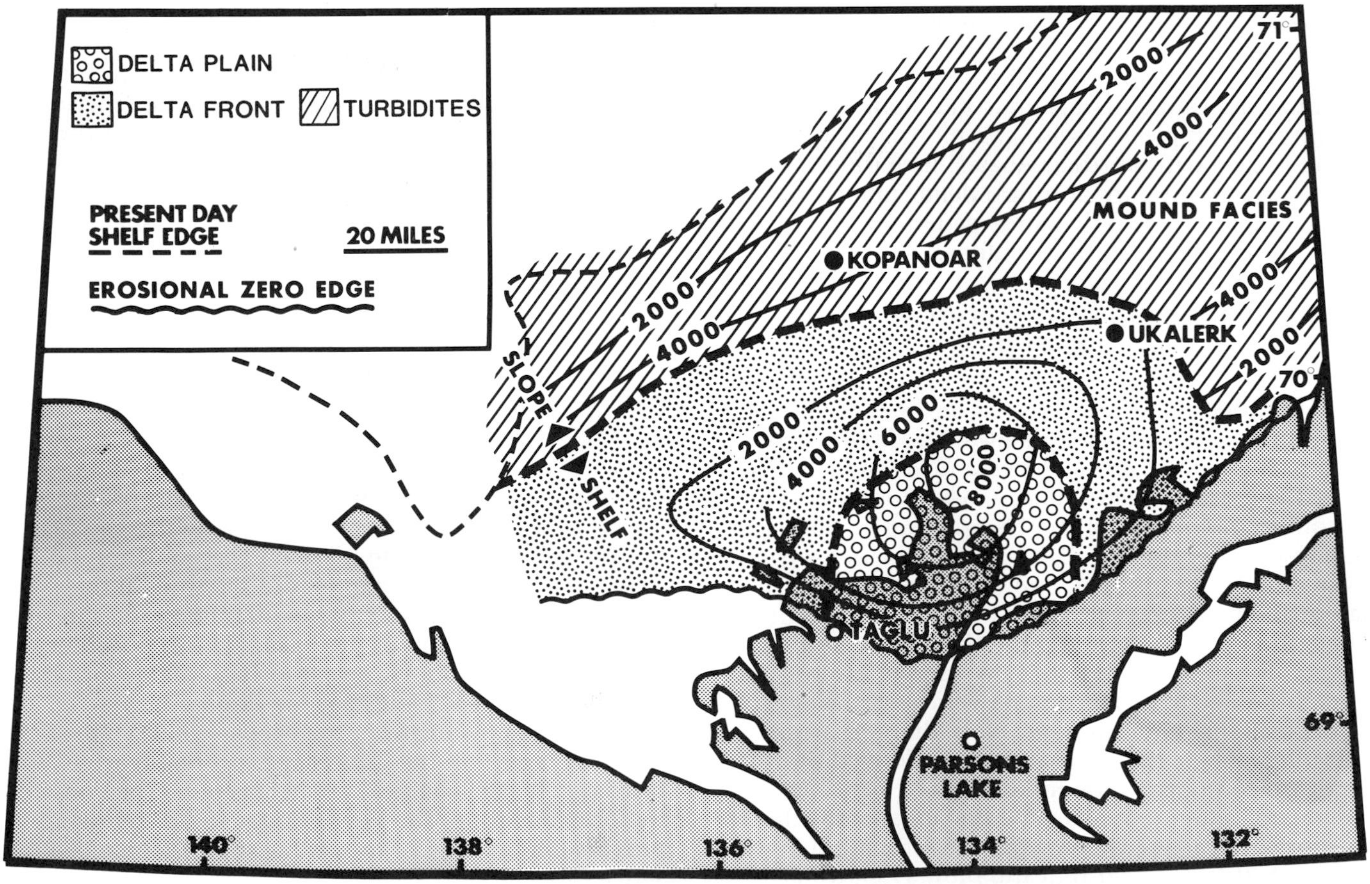

Figure 7 — Gross sand intervals in shelf and slope facies, Taglu delta sequence.

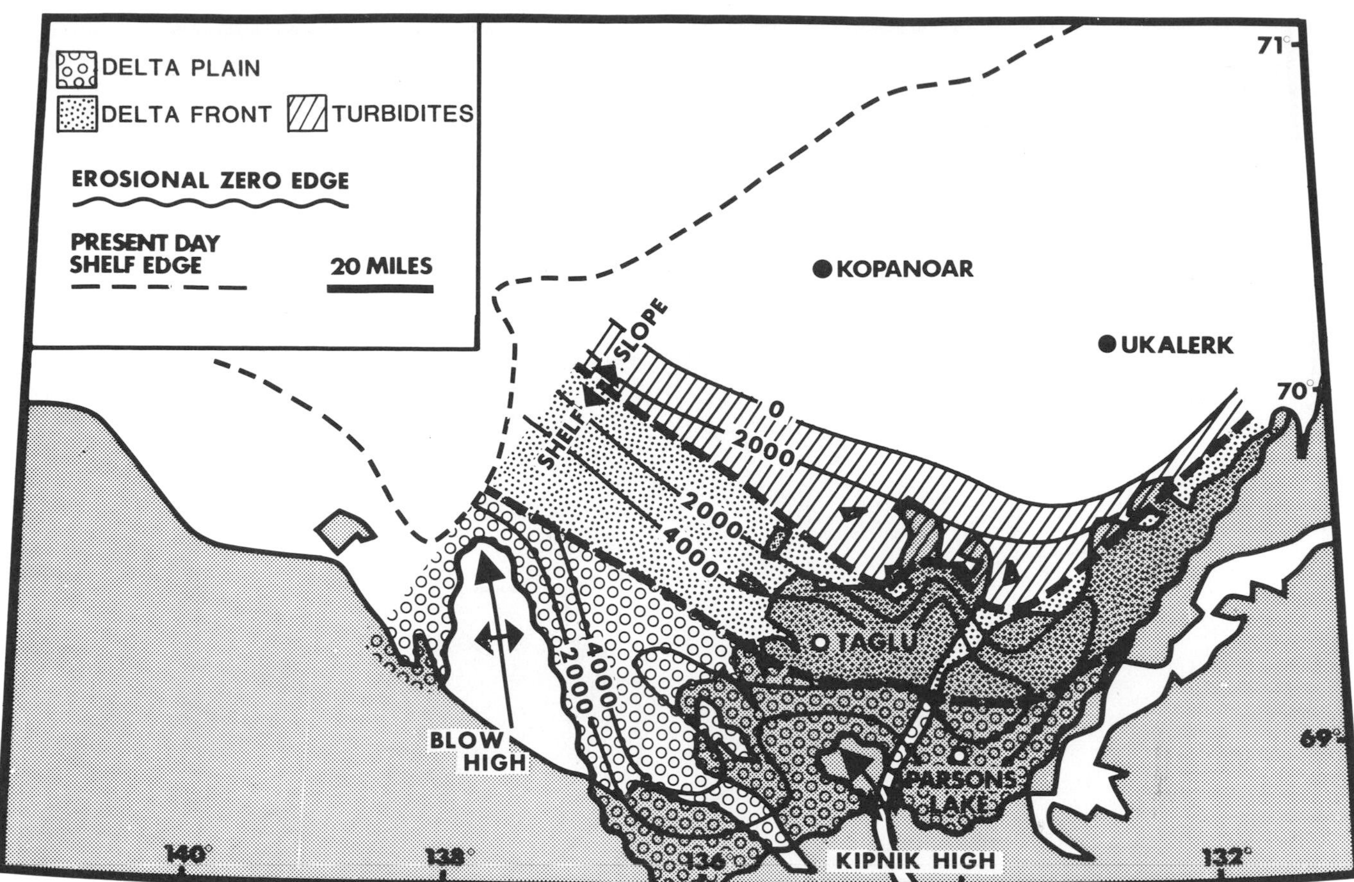

Figure 8 — Gross sand intervals in shelf and slope facies, Pullen delta sequence.

and changes to parallel the northeast trending coastline east of Taglu.

In the turbidite component of the Taglu delta sequence, approximately 3000 ft of gross sand interval have been penetrated under the Taglu shelf edge. The Kipnik High is postdepositional with respect to this deltaic sequence. The Taglu delta sequence terminated toward the end of the Eocene by a thick sequence of marine shales deposited in late Eocene and early Oligocene time.

OLIGOCENE PULLEN DELTA SEQUENCE

Following transgression, renewed deltaic deposition began early in the Oligocene. Informally this deltaic sequence is called the Pullen delta, since its depocenter is located near Pullen Island off the northern tip of the Mackenzie delta. In this area the gross sand interval reaches a maximum 8400 ft thick, making the Pullen delta sequence the thickest deltaic unit in the Mackenzie basin. This delta is almost exclusively confined to the modern Beaufort Sea shelf and the northernmost part of the Mackenzie delta.

During the Oligocene, the Pullen delta prograded into the basin in a north to northeasterly direction, and at its climax it pushed the shelf edge about 50 mi offshore from the previous Eocene shelf edge. During the Paleocene and Eocene the shelf edge had a northwest to southeast orientation, but during the Oligocene the shelf edge migrated counterclockwise to an east to west orientation. This is indicated by the trend of the contours on Figure 8 and is especially evident in a series of syndepositional diapiric ridges trending roughly east to west through the Kopanoar area (Figure 12).

Delta plain sediments in the area near the tip of the modern Mackenzie delta dominate the deltaic deposition. The net sand content and the thickness decrease towards the Pullen shelf edge in a concentric pattern away from the depocenter.

The turbiditic sands found within the prodelta shales have been penetrated in several wells. In the Kopanoar discovery well these turbidite sands yielded oil at a calculated rate of 12,000 barrels/day. In the Ukalerk well, which is close to the Pullen shelf edge, sands of turbiditic origin have been encountered within the prodelta shales below the relatively thin shelf sediments.

The turbiditic sediments are more than 4000 ft thick just north of Ukalerk, and generally trend parallel with the shelf edge and the coastline to the southeast.

Turbidites in the Kopanoar area show thinning over east to west orientated diapiric ridges, indicating that diapirism was syndepositional. The area northeast of Ukalerk lacks diapirs, but has a system of compound mounds which presumably represent slumped material associated with deep sea fans. None of these mounds has been drilled as yet; hence, their origin

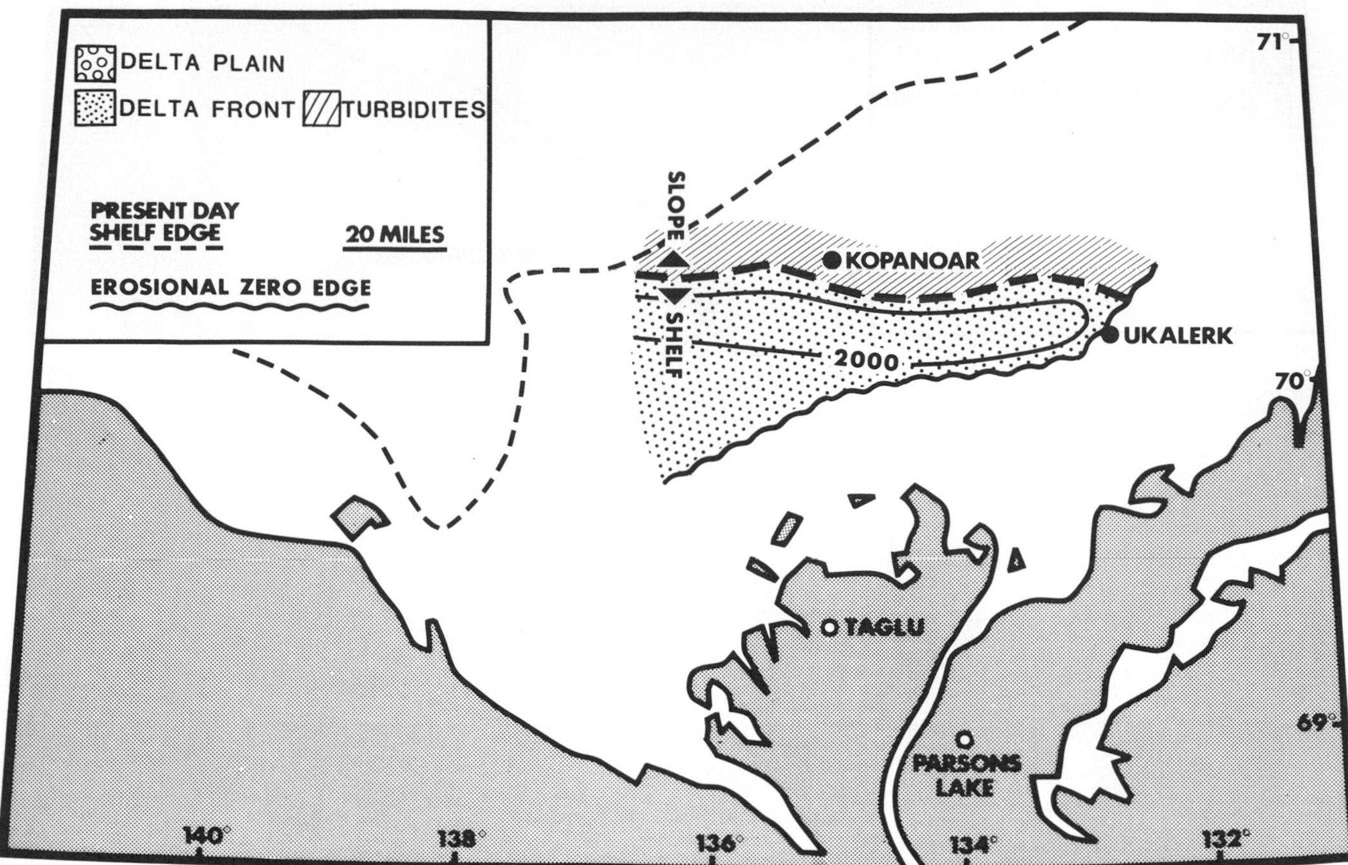

Figure 9 — Gross sand intervals in shelf and slope facies, Akpak delta sequence.

remains uncertain. The Pullen delta sequence is the major offshore exploration target and hosts all the significant hydrocarbons discovered to date in that area.

LATE OLIGOCENE/EARLY MIOCENE AKPAK DELTA SEQUENCE

The cross section in Figure 3 illustrates that another transgression terminated the Pullen delta, causing marine shales to be deposited on top of the delta sequence.

In some wells sands are present at the top of this transgressive shale, indicating yet another delta cycle. Paleontological information indicated a late Oligocene to early Miocene age for this sequence, which we informally call the Akpak delta (Figure 9). The map of this unit is largely based on seismic information, indicating the late Oligocene to early Miocene deltaic cycle reaches a maximum gross sand interval of about 3000 ft. The position of the shelf edge is interpreted from very pronounced, large-scale foresets on seismic data (compare Figure 13).

The Akpak sequence is a minor deltaic cycle extending the shelf edge a short distance beyond the Pullen shelf edge. The limited size of the sequence seems to result from its short lifespan near the Oligocene/Miocene boundary.

In the early Miocene the whole Beaufort region experienced widespread normal faulting and erosion.

Figures 3 and 4 show a regional unconformity, indicated in a heavy black line, cutting progressively older units in a southern direction. This unconformity can be traced over the entire southern Beaufort Sea region and represents one of the better seismic markers in the area. The unconformity is smooth and only disturbed by subsequent deposition. Since it erodes the tops of tilted fault blocks to the south, it is clear that extensive faulting and uplift occured prior to or during erosion.

MIOCENE — PLEISTOCENE BEAUFORT SEQUENCE

After a period of tectonic instability and erosion, active deposition began again in mid-Miocene times and continued through to the Plio/Pleistocene. This time span is represented by what is proposed as the Beaufort formation (Hea et al, 1980).

The Beaufort deltaic sequence (Figure 10) apparently prograded in a northwesterly direction. The sedimentary package reaches thicknesses of some 10,000 ft, but a gross sand interval greater than 5000 ft has not been observed. The depocenter for the unit is located 40 to 50 miles shoreward of the Plio/Pleistocene shelf edge. Turbidites are not prominent in this unit, but they are found in some wells in the outer shelf area.

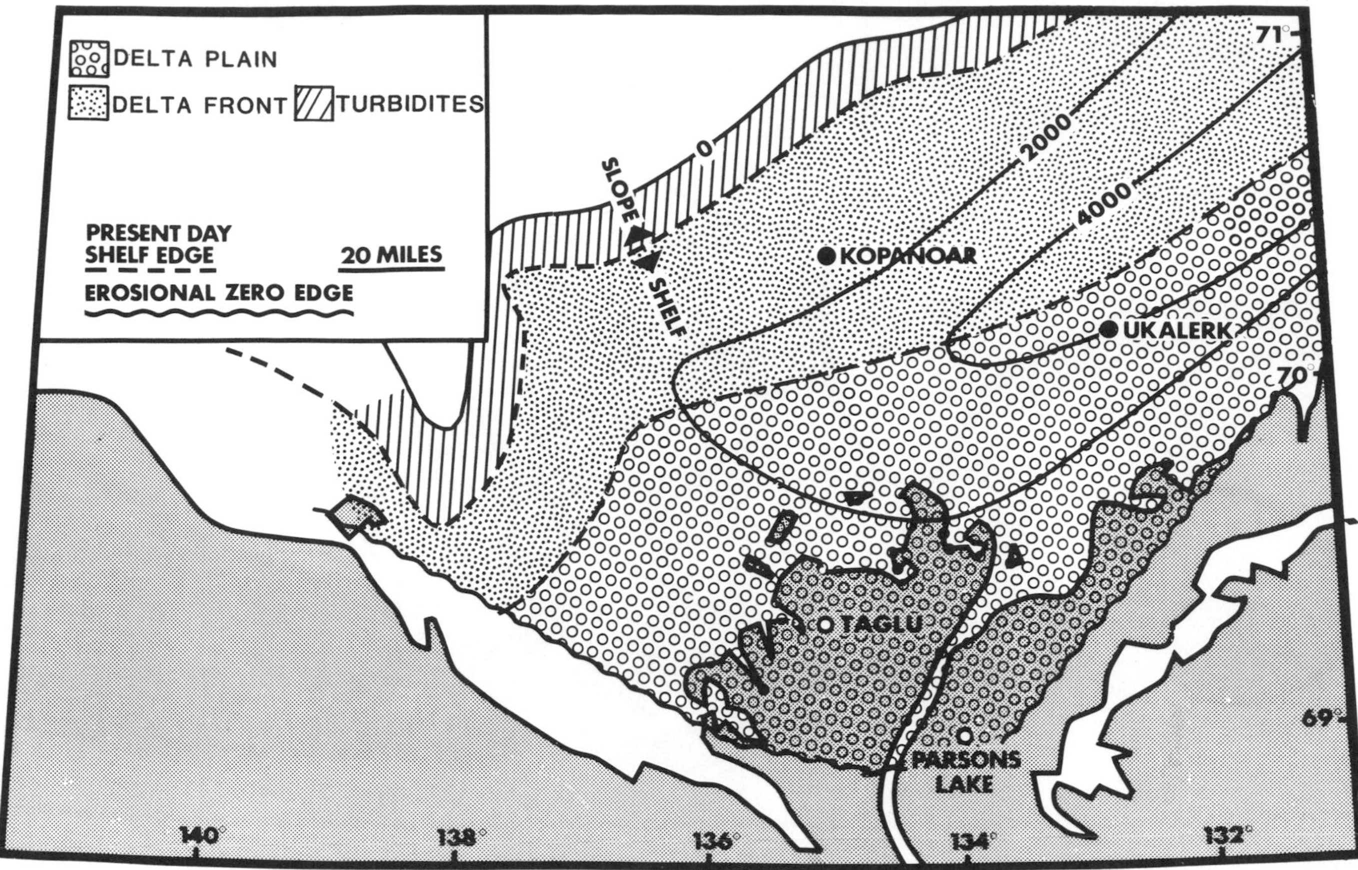

Figure 10 — Gross sand intervals in shelf and slope facies, Beaufort delta sequence.

SHELF EDGE PROGRESSION

The map on Figure 11 summarizes the shelf edge position at various times during the Tertiary. At the end of the Cretaceous, the shelf edge was located along the recent shorelines to the southwest and the southeast. During the Tertiary, the shelf edge was displaced into the Mackenzie basin through deltaic deposition in five cycles. Its position at the termination of each cycle is shown on the figure. The arrow follows the general direction of transport and displays a distinct counterclockwise rotation with time. Deposition during the Tertiary added an area of about 35,000 sq mi to the continental shelf in the Mackenzie basin area with sediment thicknesses reaching a maximum of about 30,000 ft.

TIMING OF TECTONIC FEATURES

Figure 12 shows the major tectonic features of the study area, where main tectonic elements consist of faults and diapirs. The diapirs form long ridges parallel with the depositional strike at the time of their formation. The are thought to be cored by Cretaceous shales, and were formed in the continental slope area as the deltaic cycles prograded into the basin. As the deltaic deposition moved further into the basin, ridges formed roughly parallel with the shelf edge. In some cases the shelf sediments prograde over diapirs formed at an earlier stage, which is the case for the Paleocene and Eocene deltaic cycles in the southwestern part of the study area.

Comparing Figure 11 and 12 shows a strong similarity in the trends of diapiric ridges and shelf edges for the various stages. This supports the interpretation that sedimentation began in the southwestern part of the basin and migrated counterclockwise with time.

The only major break in the Tertiary sequence occurred in the early Miocene. The first phase of this event was normal faulting, which is present in most of the basin. The faults indicated on Figure 12 are only the most prominent of the multitude of faults dating back to this time. Down-to-the-basin throw dominates, but some area display faults with throw in the opposite direction. The overall impression is one of extensional tectonics accompanied by uplift and erosion.

TERTIARY STRATIGRAPHY/STRUCTURE

The seismic line on Figure 13 is about 60 mi long, runs north to south in the western part of the study area, and is distal to the main depocenters. This particular line was chosen because it crosses two sets of diapiric ridges of Eocene and Oligocene age respectively, and three wells are located on or near it.

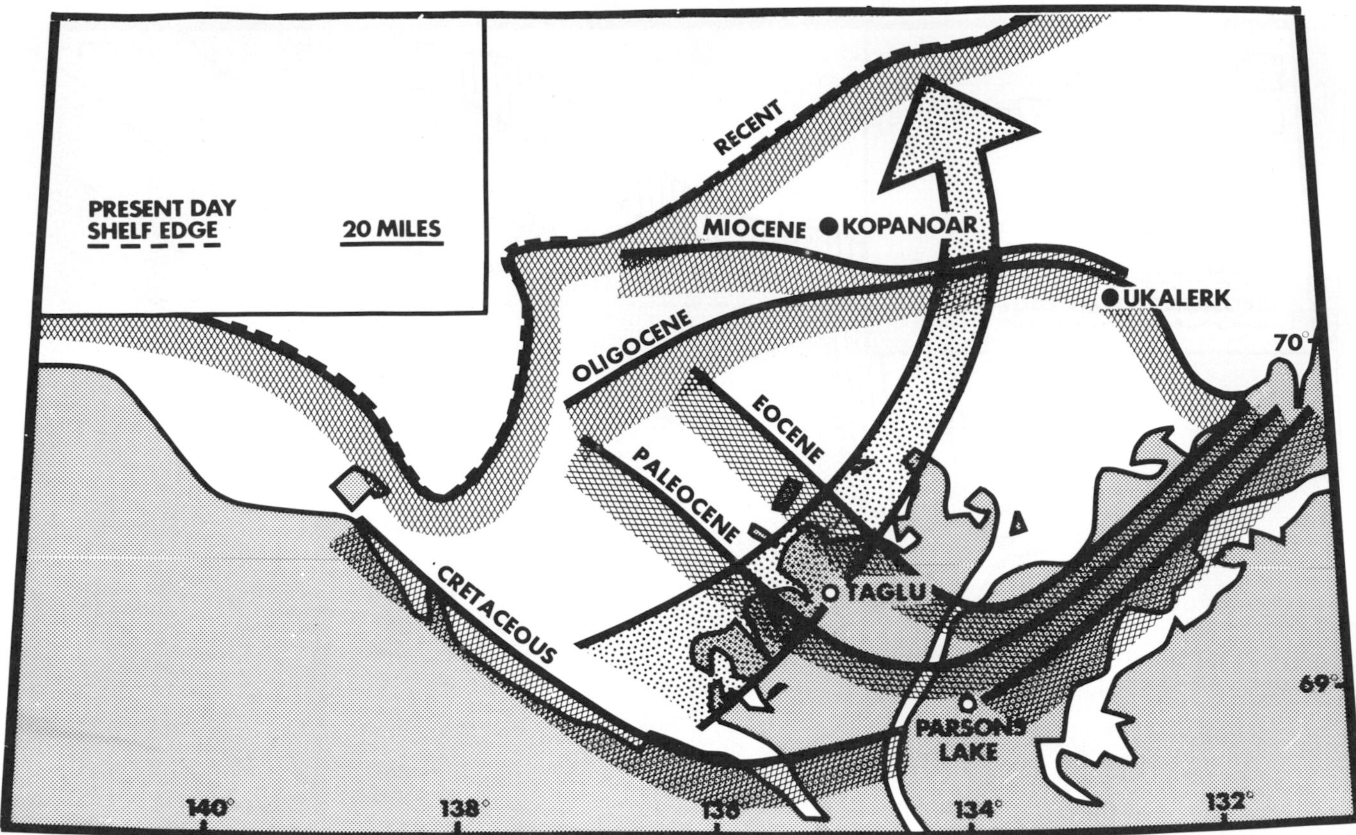

Figure 11 — Location of present-day and paleo-shelf edges.

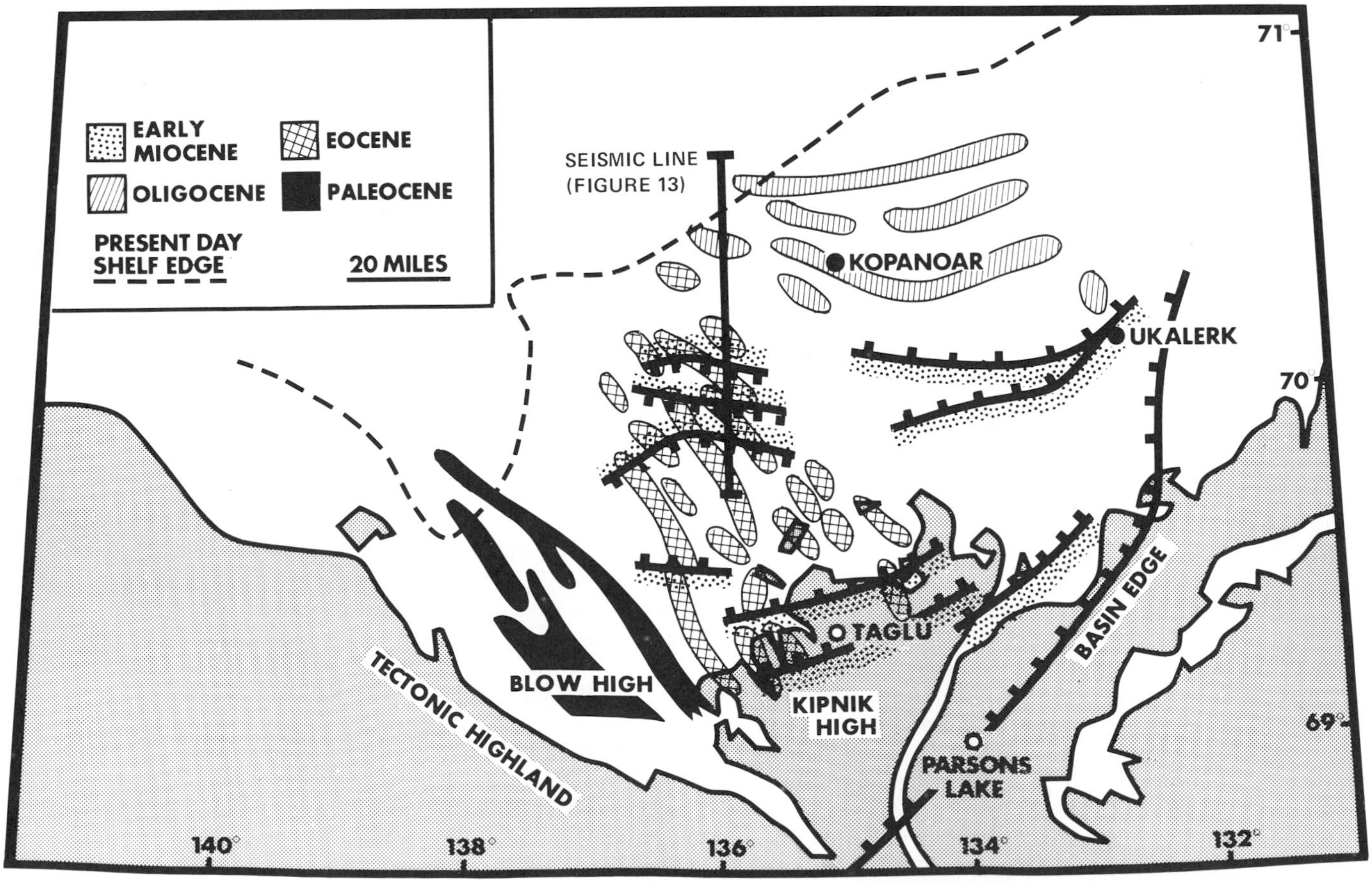

Figure 12 — Timing of diapiric activity and faulting. Diapirism coincided with seaward progression of deposition.

The Nektoralik well is located on the present-day shelf edge on the extreme left; two other wells, Tarsiut and North Netserk, are at the southern end of the section.

The slanted hachure represents mainly shaly, supposedly Cretaceous, sediments, that form cores of the diapirs present in the section. The vertically hachured portions represent the gross sand interval for the individual deltaic cycles discussed previously. The unhachurred part of the seismic line consists of Tertiary shales, which lie between deltaic sequences.

The Eocene Taglu delta has a thick package of distal delta front shelf sediments of North Netserk. Further north at the Tarsiut well the unit consists of fairly dirty sand which have been interpreted as turbidites. The Taglu gross sand interval gradually wedges out to the north, indicating a southerly source. The Oligocene Pullen delta sequence is quite thin in the area because it is distal to its depocenter located some 60 mi east. At Tarsiut and North Netserk the sequence develops in an outer shelf facies, but it is turbiditic in the Nektoralik well to the north. (The age determination used here is different from the one published by Jones, Brache, and Lentin, 1980). The transition from shelf to turbiditic sediments occurs in about the middle of the seismic profile, showing that the shelf sediments respond only very mildly, if at all, to the underlying diapirs. This suggests that the diapirs to the south were not active in the Oligocene. To the north the Pullen sediments thin over the crests of diapirs, which accordingly are interpreted as syndepositional. This line of evidence is used to show the diachronous nature of diapirism, which becomes progressively younger in a northward direction, although the diapirs are all derived from the same, possibly Cretaceous, material.

The Akpak sequence is quite heavily eroded in the section's southern end. The heavy black line on top of the Akpak sequence is the regional early Miocene unconformity. Erosion occurred after the extensional faulting as the unconformity cuts cleanly across the faults, and is fairly smooth and undisturbed. In the left center, the unconformity suddenly changes slope which represents the shelf edge position at the termination of the Akpak sequence.

Sediments above the unconformity range in age from mid-Miocene to Pleistocene, and, apart from a relatively narrow band at the top of the section, consist of mainly prodelta shales.

On the left side of the seismic section the recent shelf edge stands out quite sharply, and it is seen that the present-day bathymetry is mimicked in sediments below. The uppermost reflectors tend to be horizontal, while the deeper reflectors dip north, parallel with the present-day continental slope. The junction of these two differently inclined segments defines the position of the shelf edge. The shelf edge migrates progressively deeper and shifts to the south as you follow the section to the right from the present-day shelf edge. The seismic line illustrates the progradation of the shelf edge in the post Oligocene time, and similar evidence is found in older parts of the section.

ACKNOWLEDGMENTS

The authors extend their thanks to Gulf Canada Resources Inc. for permission to publish this paper. Thanks also to some of Gulf's exploration partners in the Beaufort Sea, especially Dome Petroleum, for permitting the release of information of confidential nature.

REFERENCES CITED

Hea, J. P., et al, 1980, Post-Ellesmerian basins of Arctic Canada: Their depocentres, rates of sedimentation and petroleum potential, *in* Facts and principles of world petroleum occurrence: Canadian Society of Petroleum Geologists, Memoir 6, p. 447-488.

Jones, P. B., 1980, Evidence from Canada and Alaska on plate tectonic evolution of the Arctic Ocean Basin: Nature, v. 285, p. 215-217.

———, J. Brache, and J. K. Lentin, 1980, The geology of the 1977 offshore hydrocarbon discoveries in the Beaufort-Mackenzie basin: Bulletin of Canadian Petroleum Geologists, v. 28, n. 1, v. 28, n. 1, p. 81-102.

Lerand, M., 1973, Beaufort Sea in the future petroleum provinces of Canada: Canadian Society of Petroleum Geologists, Memoir 1, p. 315-386.

Mountjoy, E. W., 1967, Upper Cretaceous and Tertiary stratigraphy, northern Yukon territory and northwestern district of Mackenzie: Geological Survey of Canada, Paper 66-16, p. 1-70.

Staplin, F. L., 1976, Tertiary biostratigraphy, Mackenzie delta region, Canada: Bulletin of Canadian Petroleum Geology, v. 24, n. 1, p. 117-136.

Young, F. G., D. W. Myhr, and C. J. Yorath, 1976, Geology of the Beaufort-Mackenzie basin: Geological Survey of Canada, Paper 76-11, p. 1-65.

Neogene and Quaternary Development of the Lower Continental Rise off the Central U.S. East Coast

Brian E. Tucholke
Woods Hole Oceanographic Institution
Woods Hole, Massachusetts

Edward P. Laine
Graduate School of Oceanography
University of Rhode Island
Narragansett, Rhode Island

Two major phases occurred in the Neogene and Quaternary construction of the lower continental rise north of Cape Hatteras. Intense circulation of abyssal boundary currents during the late Paleogene eroded a regional unconformity (Horizon A^u) along the lower continental rise, but by early Miocene time a first phase of very rapid sedimentation replaced this erosional regime. Sediments injected into deep water formed large abyssal fans at the mouths of the Hudson Canyon and the Norfolk-Washington Canyon. At the eastern perimeter of these fans the abyssal Western Boundary Undercurrent interacted with the edge of the Gulf Stream, redistributed fine-grained sediments, and formed the half-kilometer-high Hatteras Outer Ridge. A second phase of development occurred near the onset of northern hemisphere glaciation 2.5 to 3.0 m.y. ago when abyssal currents severed the connection between the Hatteras and Gulf Stream Outer Ridges and formed an erosional basin in its place. Shortly thereafter, this basin and the basins behind the Hatteras Outer Ridge were flooded by turbidites; these eventually buried the western flank of the Hatteras Outer Ridge and ponded to form the present lower continental rise terrace. Turbidity currents have modified the shape and surficial structure of many of the sediment waves on the outer ridge, but the Western Boundary Undercurrent continues to be important in sculpting the lower continental rise.

The abyssal North Atlantic Ocean is unique in having numerous, prominently developed "sedimentary outer ridges." These ridges have formed primarily as depositional loci beneath abyssal current systems. In the northern North Atlantic, for example, current-controlled deposition has formed the Feni, Hatton, Gardar, and Eirik ridges along basin margins. Geological and geophysical studies indicate that these ridges have developed in the Cenozoic, possibly as recently as Neogene time (Jones et al, 1970; Johnson, Vogt, and Schneider, 1971; Ruddiman, 1972; Roberts, 1975). Prominent sedimentary ridges in the western North Atlantic basin include the Blake and Bahama Outer Ridges (Markl, Bryan, and Ewing, 1970; Ewing and Hollister, 1972), the Caicos and Greater Antilles Outer Ridges north of the Bahama Banks and Puerto Rico Trench, respectively (Tucholke and Ewing, 1974), and the Hatteras Outer Ridge (Figure 1). The Blake, Bahama, Caicos, and Greater Antilles Outer Ridges developed by current-controlled deposition beginning in Late Paleogene or Early Neogene time.

In the western North Atlantic, the stage was set for development of these ridges in the early to middle Oligocene, when the continental rise of the eastern United States was deeply eroded by a southward-flowing abyssal current system, probably the precursor of the present Western Boundary Undercurrent (Tucholke and Mountain, 1979; Miller and Tucholke, in press). The erosion truncated beds ranging in age from Lower Cretaceous to middle Eocene, and it formed an angular unconformity (Horizon A^u) that slopes gently eastward along the deep-water continental margin. This unconformity has been penetrated at DSDP Sites 4, 5, 99, 100, and 101 near the western end of the Bahama Banks, at Site 391 in the Blake-Bahama Basin, and at Sites 105 and 106 on the lower continental rise off New York (Tucholke, 1979).

By early Miocene time, rapid deposition replaced the erosional conditions, but, in contrast to the pre-Horizon A^u depositional environment (pelagic and downslope accumulation), sedimentation now was strongly influenced by abyssal currents. Major depositional loci above Horizon A^u developed into the Blake and Bahama Outer Ridges (Ewing and Hollister, 1972; Tucholke and Mountain, 1979), and the Caicos and the Greater Antilles Outer Ridges formed above flat-lying, time-equivalent beds (Tucholke and Ewing, 1974).

The Hatteras Outer Ridge also is developed above Horizon A^u along the lower continental rise off the eastern United States, between Cape Hatteras and Long Island. However, it is morphologically defined only near its southwestern end where it lies just east of the Hatteras Transverse Canyon (Figure 1); it was there that the name Hatteras Outer Ridge first was applied (Rona, Schneider, and Heezen, 1967). Farther north, the western flank of the outer ridge has been buried by flat-lying turbidites to form the lower continental rise terrace (Laine, 1977; Asquith, 1979).

The origin of the Hatteras Outer Ridge, and more

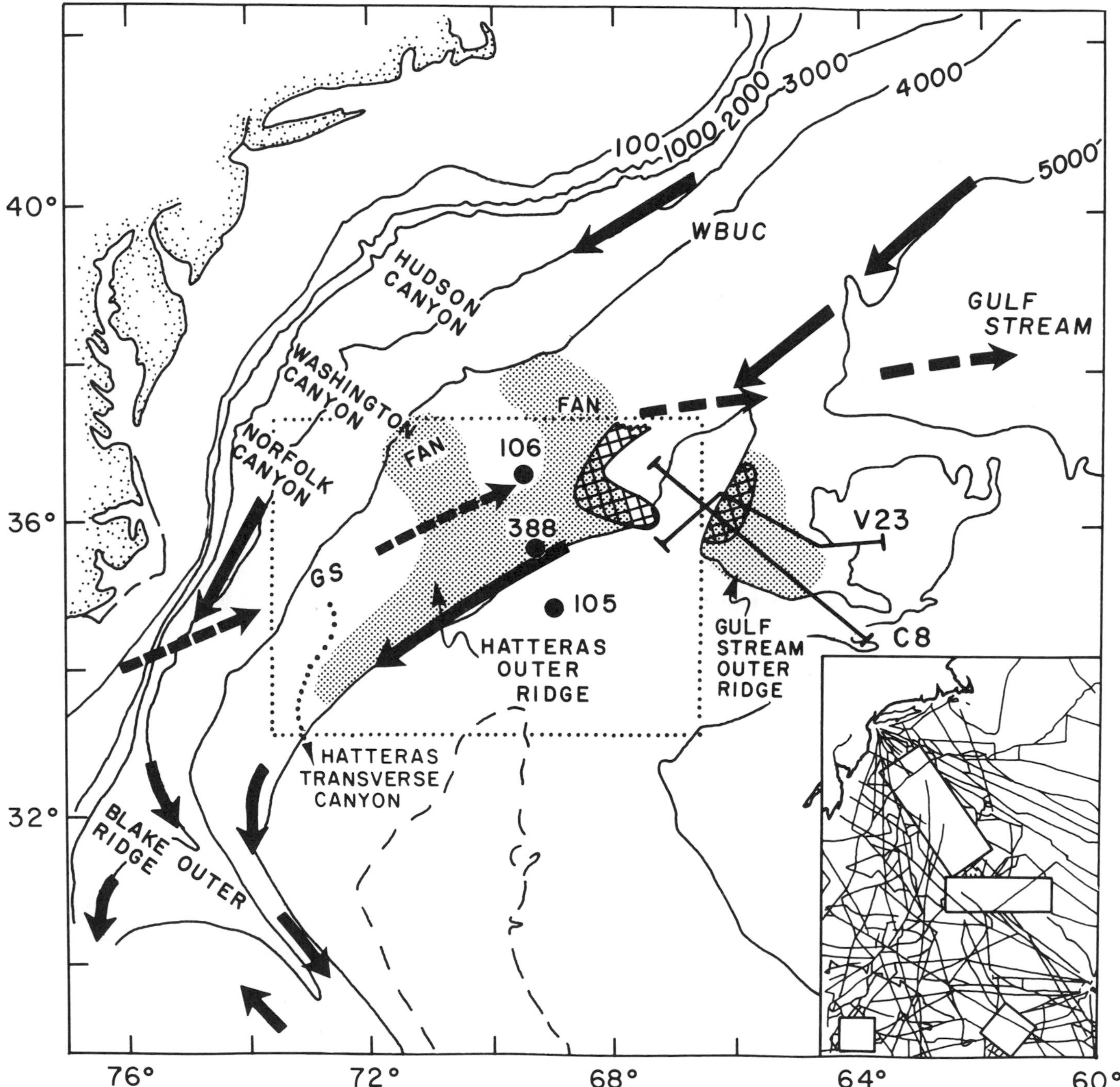

Figure 1 — Index map of the study region. Present Hatteras and Gulf Stream Outer Ridges are shaded, with erosional outcrops crosshatched. Modern axes of Gulf Stream (dashed arrows) and shallow and deep parts of Western Boundary Undercurrent (solid arrows) are shown. Dots show DSDP sites; dotted box shows location of Figure 4. Seismic track control insert at lower right (boxes have dense data coverage).

specifically the "lower continental rise hills" that form its exposed eastern flank, has been widely contested in the literature. Ballard (1966) and Emery et al (1970) suggested a gravity-slide origin, but Rona (1969) noted that the lower continental rise hills are oriented generally east-west to northwest-southeast at a large angle to the regional and sedimentary-ridge contours. This does not support a gravity slide origin. Rona (1969) and Fox, Heezen, and Harian (1968) favored the hypothesis that the "hills" were abyssal sediment waves whose deposition was controlled by the southerly flowing Western Boundary Undercurrent (Heezen, Hollister, and Ruddiman 1966). Asquith (1979) accepted the general concept that the Hatteras Outer Ridge was formed by current-controlled deposition, but he explained the "hills" as erosional remnants lying between turbidity current pathways. He also inferred that these remnants subsequently were modified by the Western Boundary Undercurrent. Benson et al (1978) also concluded that the lower continental rise and Hatteras Outer Ridge were of composite origin with abyssal currents and turbidity currents of primary importance.

Unfortunately, none of the above studies addressed a reinterpretation of the large quantity of seismic reflection data in the region of the Hatteras Outer Ridge. In this report, we discuss the evolution of the Hatteras Outer Ridge and the lower continental rise hills on the basis of data from DSDP borehole results and low-frequency (<150Hz) seismic stratigraphy. We conclude that the genesis of both the Hatteras Outer Ridge and the "lower continental rise hills" is consistent with the timing and nature of models of current-controlled deposition previously invoked to explain the development of the Blake-Bahama, Caicos, and Greater Antilles Outer Ridges. However, in the case of the Hatteras Outer Ridge, Quaternary turbidites filled the basins west of the outer ridge. By late Pleistocene time, turbidity currents also breached the outer-ridge crest; they substantially channeled and modified the shallow structure of the exposed lower continental rise hills on the eastern flank of the Hatteras Outer Ridge.

METHODS

As already noted, the morphologic expression of the Hatteras Outer Ridge is largely masked by a cover of Pleistocene turbidites which form the lower continental rise terrace (Figure 2). Therefore, the time-stratigraphic interval approximately including these turbidites must be removed in order to determine the paleomorphology of the outer ridge. This interval is best known at DSDP Site 106, which was drilled through the ponded terrace turbidites over the west flank of the Hatteras Outer Ridge (Figures 1 and 3). The drill site penetrated about 350 m of hemipelagic mud that contains silt and sand beds and correlates with the seismically stratified turbidites. The underlying outer-ridge sediments are more uniform, dark green-gray hemipelagic muds that have less pronounced seismic reflectors (Hollister et al, 1972). Although the contact between turbidites and the outer ridge sediments dates to about 1.8 million years old at Site 106, stratigraphically lower turbidites are present in the deeper part of the basin to the west (Figures 2 and 4). Comparisons with observed turbidites thickness and accumulation rates at Site 106 indicate that the base of the turbidites in the deepest part of the basin is about 2.8 m.y. old. This age closely agrees with that determined for the initiation of significant turbidite influx elsewhere in the North Atlantic, for example near Site 382 along the Sohm Abyssal Plain (Laine, 1980) and at Site 113 in the northern Atlantic (Laughton et al, 1972). The age also correlates with the beginning of northern hemisphere glaciation about 2.5 to 3.0 m.y. ago (Berggren, 1972; Shackleton and Kennett, 1975; Kennett, 1977).

We studied the paleomorphology of the Hatteras Outer Ridge in seismic profiles collected by Lamont-Doherty Geological Observatory, Woods Hole Oceanographic Institution, and the U.S. Geological Survey, Woods Hole. The map in Figure 4 represents the approximate morphology of the Hatteras Outer Ridge 2.8 m.y. ago, prior to the initial influx of turbidites. The map was prepared by removing the full thickness of turbidites in the terrace basins plus 0.1

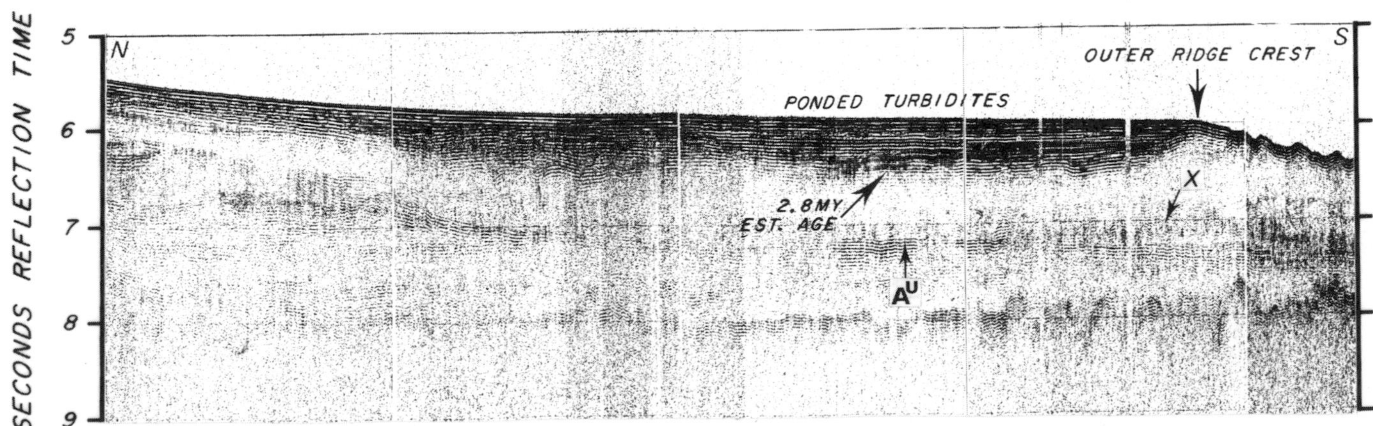

Figure 2 — *Conrad*-16 single-channel seismic profile across the Hatteras Outer Ridge and across the adjacent turbidite basin that forms the lower continental rise terrace. Location in Figure 4.

sec (about 80 m) of surface sediment in areas where the turbidites are absent. The turbidites are easily differentiated by their continuous, flat-lying and parallel acoustic stratification. Outside the turbidite areas, the 0.1 sec of sediment removed approximates the past 2.8 million years of accumulation at DSDP Sites 105 and 388 (Hollister, et al, 1972; Benson et al, 1978). Because the features mapped were simple and not likely to be significantly distorted by the contrast in compressional wave velocity between sediments and sea water, we chose to present the map in two-way travel time rather than true depth. This follows standard practice (Emery et al, 1970; Emery and Uchupi, 1972) and allows ready comparison with other reflection profiles recorded in time rather than depth.

RESULTS

Neogene Sedimentary Framework

Morphology

With Quaternary turbidites and hemipelagic sediments removed, the late Pliocene morphology of the Hatteras Outer Ridge is clearly expressed (Figure 4). The axis of the outer ridge is slightly sinuous and parallels the present regional depth contours, extending northeast to southwest. The outer ridge terminates rather abruptly at its southern end at the position of the Hatteras/Pamlico Canyon system (33°N), and at its northern end just south of the Hudson Canyon system (36.5°N). Its total length is about 550 km, and minimum depths along the ridge crest are about 6.0 sec (4500 m) below sea level. The general morphology and contour-parallel development of this outer ridge are characteristic of current-controlled sedimentation. At its southwest end, the outer ridge tails off in a manner like that observed at the downstream ends of the Blake and Caicos Outer Ridges. However, at the northeast end seismic profiles show that the outer ridge has experienced significant late Pliocene to Pleistocene erosion, and it therefore is difficult to reconstruct the late Pliocene morphology (Figures 3 and 4). Just to the east, the Gulf Stream Outer Ridge (Laine, 1977; Laine and Hollister, 1981) has undergone similar erosion along its western face (Figures 1 and 5). Except in this erosional zone, the late Pliocene morphology of the Gulf Stream Outer Ridge probably was very similar to its present morphology. Thus, the shape of these two outer ridges and the fact that their erosional zones face one another suggest that they formed a continuous sedimentary ridge in late Pliocene time. We discuss details of the evolution of this ridge system later in the text.

In the late Pliocene, the Hatteras Outer Ridge was joined to the continental margin by two broad sedimentary saddles; these were oriented perpendicular to the regional contours and were situated seaward of Hudson Canyon and the confluence of the Norfolk and Washington Canyons (Figure 4). The surfaces of these saddles are slightly deeper (~100 m) than the crest of the Hatteras Outer Ridge, and together with the outer ridge they bound two prominent sedimentary basins. The southwestern basin is deepest (600 m below the outer-ridge crest) just east of the Hatteras Transverse Canyon axis and it opens southward into the Atlantic basin. The larger basin to the northeast is up to 450 m deeper than the adjacent

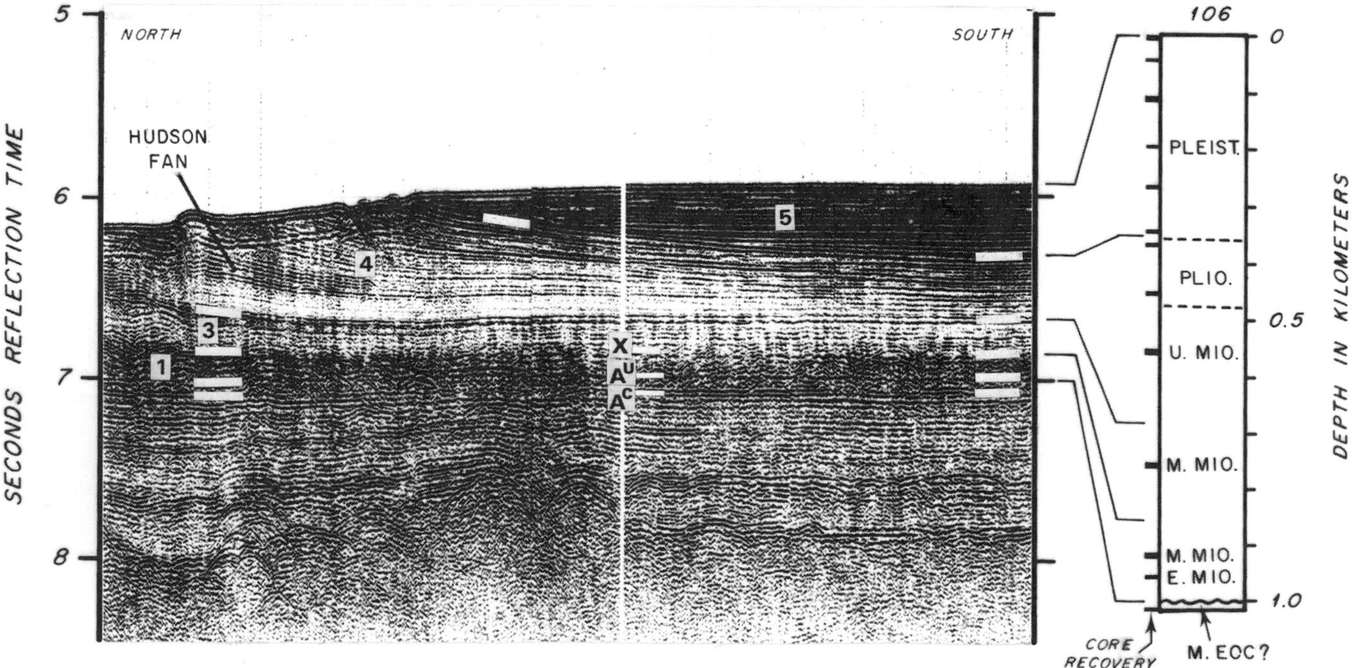

Figure 3 — Monitor record of *Conrad*-21 multichannel seismic line across southeastern edge of Hudson fan, with correlation to DSDP Site 106. Note that accretionary sequence 2 (Figure 6) either was removed by erosion at Horizon X or was not deposited in this area. Also note truncation of beds at sea floor on left side of profile. Location in Figure 4.

outer-ridge crest. Because the saddles rimming these basins both occur seaward of major submarine canyon/channel systems, it is probable that they are submarine fans. Their internal structure also supports this conclusion, as discussed below.

Stratigraphy

The seismic stratigraphy of the lower continental rise in this region is poorly defined in most existing reflection profiles. However, recently acquired USGS profiles (R/V *Fay*) and multichannel seismic profiles obtained on R/V *Conrad* Cruise 21 show much better definition of the internal structure of the Hatteras Outer Ridge and the adjacent continental rise than was available in previous studies (Figures 3 and 6). The limited coverage by these profiles precludes detailed analysis of spatial relationships in the sedimentary evolution of the area, but it does provide enough information to analyze gross temporal changes in sedimentary patterns.

The erosional unconformity, Horizon A^u, forms a flat-lying reflector that truncates deeper, nearly horizontal beds above acoustic basement (Figure 6). At DSDP Site 105, just 50 km southeast of the profile in Figure 6, Horizon A^u separates upper Miocene from probably Upper Cretaceous sediments (Hollister et al, 1972). Just above Horizon A^u, two features are readily

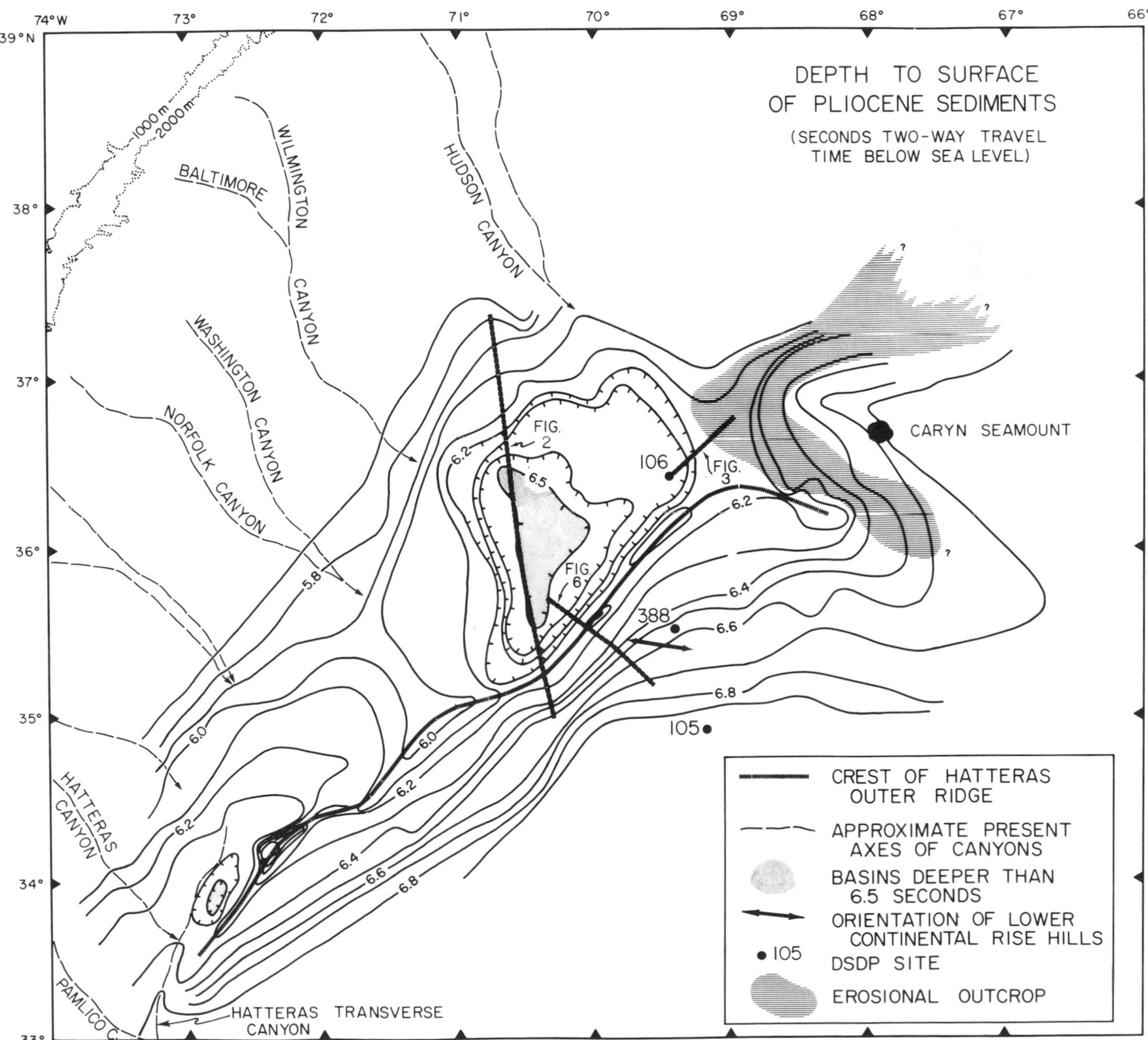

Figure 4 — Late Pliocene morphology of lower continental rise region (depths in seconds, two-way travel time below sea level). Bold lines give profile locations in Figures 2, 3, and 6.

apparent: 1) bedding planes are much more uneven than in deeper strata, and 2) migrating sediment waves are clearly developed beneath both flanks of the outer ridge. The co-occurrence of these features leaves little doubt that abyssal currents played a major role in forming the Hatteras Outer Ridge, beginning immediately above Horizon A^u.

Detailed evolution of the Hatteras Outer Ridge is complex. For example, the sediments are interrupted by at least two significant seismic discontinuities and by several minor discontinuities. For purposes of discussion, the section can be divided into a number of accretionary sequences (Figure 6). Initial deposition above Horizon A^u occurred west of the present outer ridge crest (accretionary sequence 1). Internal structure of these sediments suggests that they are sediment waves with wave-crest migration toward the east. The corresponding stratigraphic interval farther north and west contains a thinner sequence of flat-lying, well-bedded sediments. These features suggest that the initial accretion was influenced by currents in the area just west of the present ridge crest and that more evenly-bedded, possibly fan-lobe sediments were deposited still farther to the west.

Accretionary sequence 2 (Figure 6) shows marked progradation eastward onto Horizon A^u and beneath the present crest of the Hatteras Outer Ridge. This sequence is capped by Horizon X (Markl, Bryan, and Ewing, 1970), which in most single-channel seismic profiles appears as a weak and diffuse reflecting zone rather than as a distinct reflector (Figure 2). The *Conrad*-21 profile shows that Horizon X is an irregular, diffracting interface that, in places, truncates deeper reflectors (Figures 3 and 6). The diffracting character is reminiscent of current-eroded furrows presently observed on the Bahama Outer Ridge (Hollister et al, 1974). Horizon X therefore may be an erosional unconformity with ubiquitous small-scale relief. Traced north along the *Conrad*-21 profile to DSDP Site 106, Horizon X dates approximately to the middle of the middle Miocene, although no unconformity was observed at the drill site because of widely-spaced coring intervals (Figure 3). Eastward, Horizon X merges with Horizon A^u near the right side of the profile in Figure 6 and just short of DSDP Site 105, where upper Miocene sediments probably overlie the unconformity (Hollister et al, 1972).

Sediments below Horizon X and within the second accretionary sequence are characterized by several lens-like subsequences, each with contorted and discontinuous reflectors (Figure 6). However, the degree to which diffractions from Horizon X contribute to the apparent discontinuity of these underlying reflectors is uncertin. In the area of sediment waves at the northwest end of the profile in Figure 6, the sediment waves show continued development and Horizon X is atypically free of diffractions. Currents, therefore, must have continued to control sedimentation in this specific zone without significant erosion.

It is difficult to interpret the mechanisms responsible for deposition of sequence 2 farther to the east. The eastward onlap onto Horizon A^u suggests prograding fan deposits, and the total thickness of sequences 1 and 2 increases markedly toward the southwest in the R/V *Fay* profiles (K. Klitgord, personal communication, 1980; Klitgord and Grow, 1980).

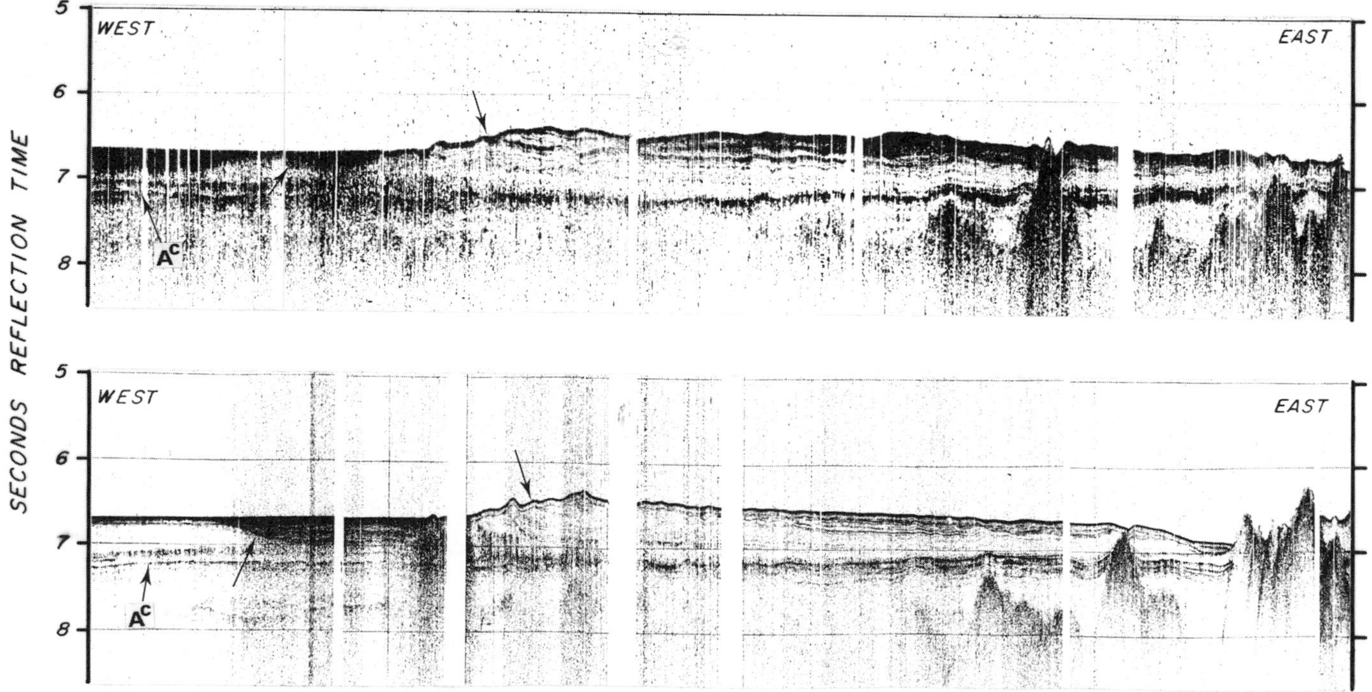

Figure 5 — *Conrad*-8 (top) and *Vema*-23 (bottom) seismic profiles across Gulf Stream Outer Ridge. Arrows show erosional outcrops on west face of outer ridge, and erosional surface in basin between the Hatteras and Gulf Stream Outer Ridges. This basin is now partly filled by Pleistocene turbidites. Locations in Figure 1.

Therefore, it appears that the Norfolk-Washington fan is the source of the sediments. The contorted and discontinuous reflectors suggest some slumping and debris flows within sequence 2, and there may also be molding and erosion by currents. However, current effects do not dominate the depositional record as they do farther west.

Accretionary sequence 3 (Figure 6) contains better defined and smooth reflecting interfaces as well as several lens-shaped subsequences near the center of the profile. We interpret these as fan-lobe deposits from the Norfolk-Washington Canyon system to the southwest. Abyssal currents probably modified sedimentation patterns in this central zone because sediment waves are developed northwest and southeast of the zone. The sequence is thicker beneath the Hatteras Outer Ridge than in the basin to the northwest. It also thickens toward the Norfolk-Washington fan. These observations suggest sediment sources both from current-controlled deposition and from the canyon-fan system. Sequence 3 is capped by a reflector that, in places, unconformably cuts subjacent bedding planes. At DSDP Site 106 this reflector falls within a 192 m coring gap, but it probably dates to the upper middle or basal upper Miocene.

Accretionary sequence 4 contains upper Miocene through upper Pliocene sediments (Figures 3 and 6). It is in this stratigraphic interval that the Hatteras Outer Ridge shows its principal morphologic development. Compared to underlying sequences, there is a clear eastward shift in the axis of maximum current-controlled deposition and sediment waves become widely and coherently developed (Figure 6). The sediment waves are best developed on the eastern flank of the ridge where they migrate upslope toward the ridge crest. The orientation of the waves is east-west (lower continental rise "hills" arrow, Figure 4), indicating that migration is to the north. The less well-developed sediment waves on the western flank of the ridge also migrate toward the ridge crest, but their exact orientation is unknown. Within this sedimentary sequence, sediment waves also are common along the perimeter of the adjacent fans, but they are absent within the intervening basins. Deposition of this sequence clearly was dominated by abyssal currents.

Sequence 4 (Figure 6) is thinner in the basins west of the outer ridge, but it thickens beneath both the Norfolk-Washington and Hudson fans. The thickening is most marked in the Hudson fan (Figure 3). Therefore, the principal development of the outer part of this fan must have occurred in the upper Miocene and Pliocene. These sediments are well-stratified and have evenly-spaced, coherent reflectors, characteristic of outer-fan deposition.

Model of Deposition

Seismic reflection data show that the Neogene development of the lower continental rise apparently was controlled by downslope sediment movement on abyssal fans and by abyssal currents. However, the relative importance of each sedimentation mechanism through time, and in particular the change in zonation of bottom currents, is difficult to assess. We present a depositional model that is admittedly speculative, but that explains the observed depositional sequences.

Following the intense current-erosion that excavated Horizon A^u, abyssal currents appear to have become weaker in the early Miocene. Sediment was deposited above the unconformity by fallout of suspended par-

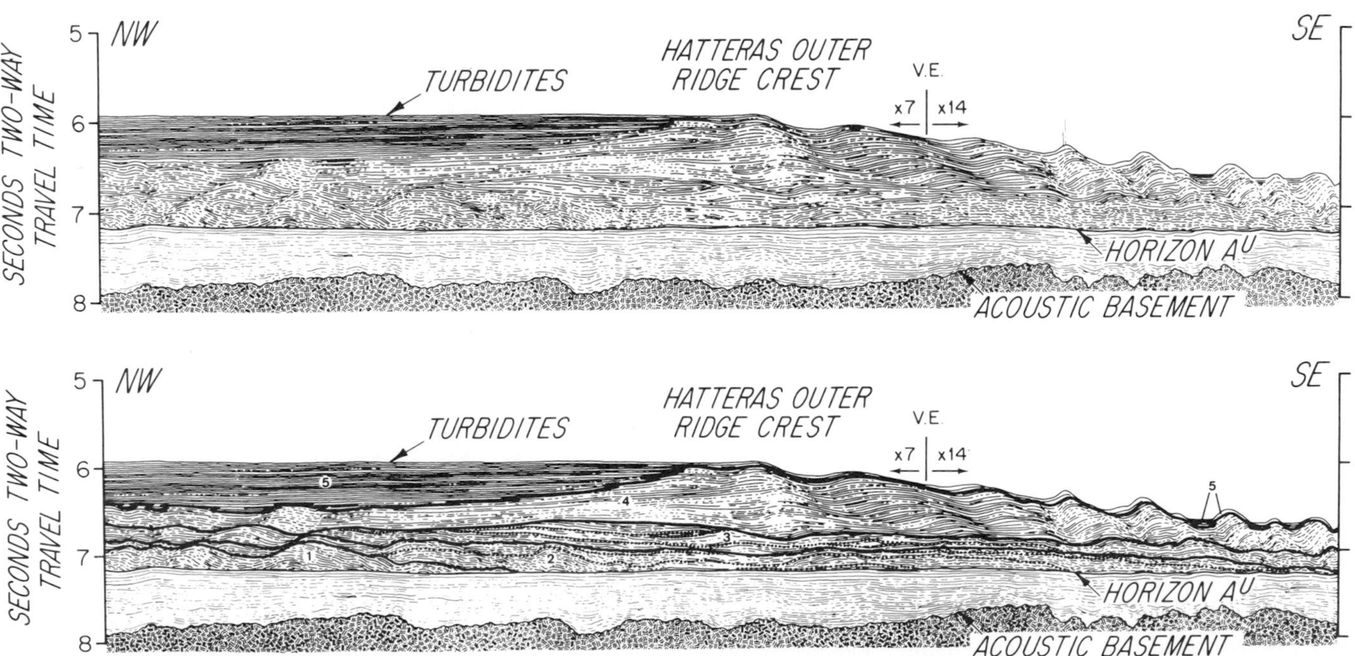

Figure 6 — Tracings of single-channel monitor record from *Conrad*-21 multichannel seismic line across Hatteras Outer Ridge. Lower tracing shows the accretionary sequences and reflecting horizons discussed in text. Location in Figure 4.

ticulate matter from the currents, and by downslope sediment movement that formed abyssal fans (Figure 7, A). The principal abyssal current system is presumed to be the percursor of the modern Western Boundary Undercurrent (WBUC), which flows southwest parallel to the bathymetric contours of the continental rise (Heezen, Hollister, and Ruddiman, 1966). The Gulf Stream flowed in the opposite direction and probably seaward of the main part of the WBUC, and it may have touched bottom at least intermittently on the continental rise. The position of the WBUC principally landward of the Gulf Stream is suggested in a hydrodynamic model of WBUC/Gulf Stream interaction, used by Bryan (1970) to explain the development of the Blake Outer Ridge (Figure 7, A). Shear zones between the oppositely-directed Gulf Stream and WBUC were developed on the western and eastern edges of the Gulf Stream. These zones probably were the loci of preferential sediment accumulation (for example, Tucholke and Ewing, 1974). Beneath the western shear zone, development of the upper continental rise was accentuated. Beneath the eastern shear zone, initial nucleation of the linear Hatteras Outer Ridge occurred (Figure 7, B). Sediment deposited from the currents was probably transported from upstream sources and locally entrained from turbidity currents traversing the submarine fans.

Following current-controlled deposition of accretionary sequence 1 (Figure 6), sequences 2 and 3 suggest increased dominance of fan-type deposition

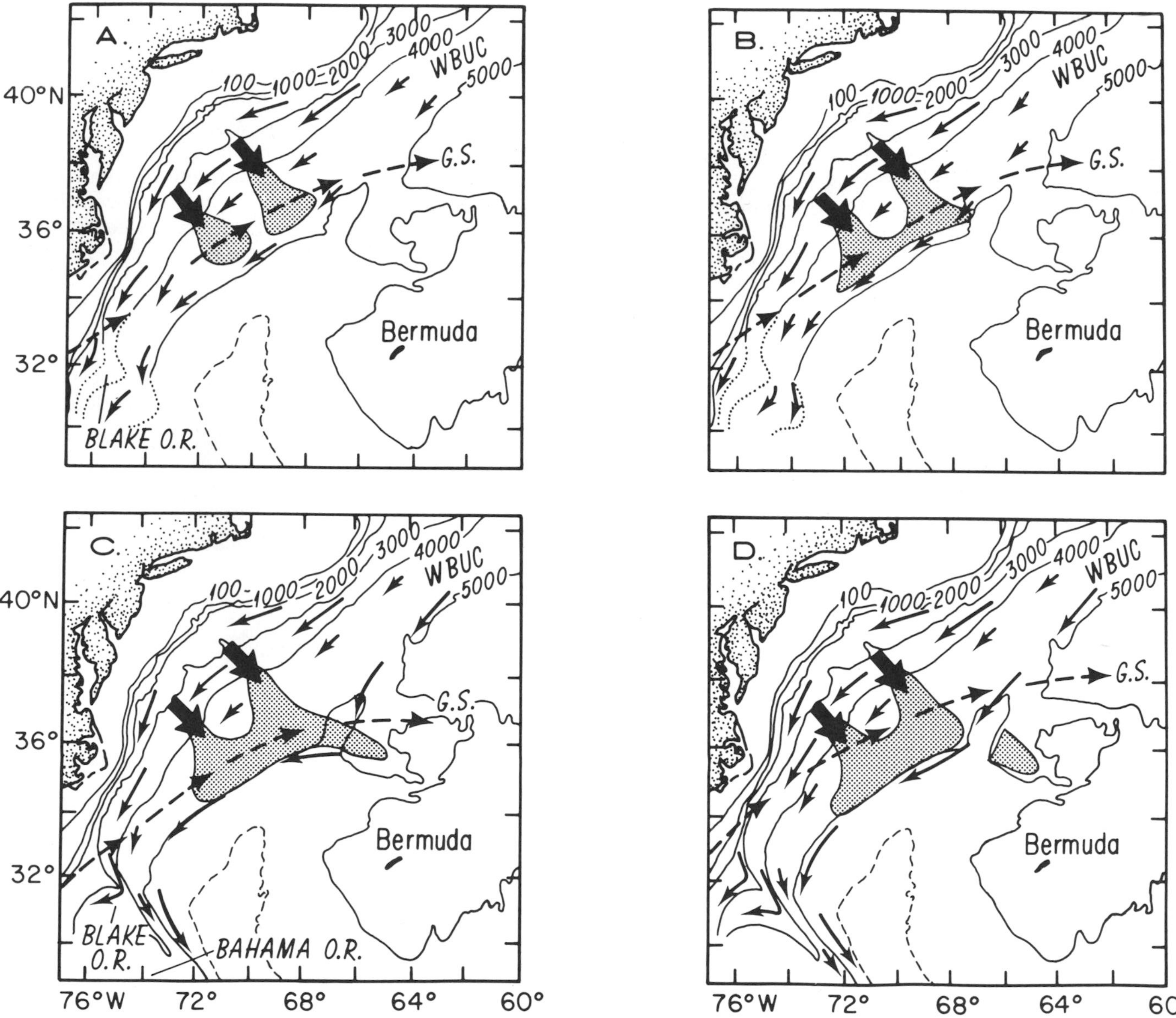

Figure 7 — Schematic early Miocene to late Pliocene development of the Hatteras and Gulf Stream Outer Ridges and the Norfolk-Washington and Hudson fans. Present bathymetric contours in meters (solid lines) and generalized paleo-contours (dotted lines) are shown. In Western Boundry Undercurrent (WBUC), length of arrow increases with inferred strength of flow. G. S. is Gulf Stream axis.

with lesser current influence. This phenomenon probably is locally accentuated in the profile of Figure 6, where seaward progradation of the fan reached into deeper water and into the shear zone between the WBUC and the eastern edge of the Gulf Stream. Continued formation of sediment waves, and erosional episodes at Horizon X and at the boundary between accretionary sequences 3 and 4, indicates that bottom currents continued to be active in these middle Miocene sequences.

An eastward shift in the axis of most rapid deposition occurred at the beginning of sequence 4 (middle to upper Miocene, 10 to 12 m.y. ago). The exact cause of this shift is uncertain, but it may relate to eastward movement of the axis of the Gulf Stream. Pinet, Popenoe, and Nelligan (1981) suggested that sea floor topography on the Blake Plateau forced such eastward shifts during times of lowered sea level. Thus, the significant and sustained sea level lowering documented for the early late Miocene by Vail, Mitchum, and Thompson (1977) may have diverted the Gulf Stream to a more seaward position (Figure 7,C).

It also appears that the deepest part of the WBUC became better developed. This is suggested by the coherent deposition and northward migration of sediment waves along the eastern flank of the outer ridge; such waves, trending at 30° to 35° to the regional contours, characteristically migrate "upstream" and upslope (Embley and Langseth, 1977). The increased circulation of the deep part of the WBUC may relate to entry of significant volumes of Antarctic bottom water into the basin, following a major middle to late Miocene increase in Antarctic ice cover accompanied by probably cooling climate (e.g. Kennett, 1977). Note that the Bahama Outer Ridge and northwestern Greater Antilles Outer Ridge also began to form about this time, apparently in response to the same deep circulation pattern (Tucholke and Ewing, 1974; Sheridan, Golovchenko, and Ewing, 1974). The required shallow and deep southerly flows of most strongly circulating bottom water (Figure 7,C) seem to agree with modern observations of the WBUC. The current presently is well-documented at depths of 1,000 to 3,500 m (e.g. Barrett, 1965; Richardson, 1977). It appears to have diminished flow near 4,000 m, and it includes strong southerly flow below 4,800 m (Richardson and Wimbush, 1980; Tucholke, Wright, and Hollister, 1973). Unfortunately, only widely-spaced measurements now define this pattern, and the pattern therefore could be regionally variable.

Throughout the late Miocene and Pliocene (sequence 4), the Hudson fan developed rapidly at its juncture with the Hatteras Outer Ridge. The rapid sediment input, together with the interaction of the WBUC and the Gulf Stream at a probable crossover point, greatly facilitated construction of the Hatteras Outer Ridge. As noted earlier, we believe that this construction also extended eastward to include the present Gulf Stream Outer Ridge (Figures 1 and 7,C). The eastward construction was probably similar to that of the Blake Outer Ridge, which was deposited somewhat earlier in a region of WBUC/Gulf Stream crossover (Figure 7,A; Bryan, 1970). By late Pliocene time, the Hatteras Outer Ridge was constructed nearly to its present height. It extended from Hatteras Transverse Canyon northeastward for more than 900 km into the Gulf Stream Outer Ridge.

Late Pliocene and Quaternary Processes

Following the major constructional phase of the Hatteras/Gulf Stream Outer Ridge system (accretionary sequence 4), pronounced erosion occurred at the northern end of the Hatteras Outer Ridge (Figures 1 and 4). In this region, the highly stratified beds of the Hudson fan and the less stratified sediments of the outer ridge have both been truncated by bottom-current erosion (Figure 3). Traced to DSDP Site 106, the youngest eroded beds date to the late Pliocene, near the boundary between accretionary sequences 4 and 5. The western flank of the Gulf Stream Outer Ridge exhibits comparable erosion (Figure 5). It is separated from the Hatteras Outer Ridge by a broad, shallow, apparently erosional trough that is now filled with Pleistocene turbidites (Figure 1). We infer that the Western Boundary Undercurrent breached the Hatteras — Gulf Stream Outer Ridge system and eroded this trough in late Pliocene to early Pleistocene time (Figure 7,D). The turbidites filling the trough were most likely derived from the Hudson Canyon system, which now debouches into a broad swale at the northern end of the Hatteras Outer Ridge and at the western margin of the trough. Turbidites slope gently southward in the trough and are continuous with those in the northern Hatteras Abyssal Plain.

As inferred from sedimentation rates at DSDP Site 106, turbidites also began filling the basins landward of the Hatteras Outer Ridge in late Pliocene time, about 2.8 m.y. ago. Continued ponding formed accretionary sequence 5 and the present lower continental rise terrace (Figures 2 and 6). As the basin filled, turbidity currents eventually breached the crest of the Hatteras Outer Ridge, flowed down its eastern flank, and formed small turbidite ponds in swales on the ridge flank (Asquith, 1979). DSDP Sites 105 and 388 were drilled in or near such small ponds (Hollister et al, 1972; Benson et al, 1978). The oldest sandy and presumably graded turbidite beds at these sites date to the *Pseudoemiliania lacunosa* Zone (NN19), indicating that the ridge crest was breached no later than 0.5 m.y. ago.

Asquith (1979) proposed that the lower continental rise hills on the eastern flank of the outer ridge are erosional remnants that lie between turbidity current pathways. The surfaces of these remnants were presumably molded by the Western Boundary Undercurrent to form shallow (~100 m), dipping beds like those in sediment waves. However, the presence of migrating sediment waves is now well documented throughout the entire stratigraphic section comprising the Hatteras Outer Ridge (Figure 6). Thus, it is apparent that the sediment waves were modified by turbidity currents, but they controlled, rather than resulted from, turbidite dispersal onto the adjacent Hatteras Abyssal Plain.

CONCLUSIONS

Analysis of the seismic stratigraphy of the lower continental rise and its correlation with DSDP borehole data suggests the following Neogene and Quaternary sedimentary evolution: Following an early to middle Oligocene episode of intense erosion by the abyssal Western Boundary Undercurrent, current velocities decreased and current-controlled deposits began accumulating above Horizon A^u no later than early Miocene time. At the same time, abyssal fan deposits prograded eastward across the unconformity. Along the eastern perimeter of the Norfolk-Washington and Hudson fans, bottom currents reworked and deposited sediments to form the nucleus of the southwest-northeast elongated Hatteras Outer Ridge. The locus where sediments initially accreted most rapidly was the low-velocity shear zone between the parallel, but oppositely directed, Western Boundary Undercurrent and Gulf Stream. Sediments deposited on the ridge were probably pirated from local turbidity currents and transported from upstream sources in the two flows.

The two abyssal fans continued to develop during the middle Miocene and they locally dominated the depositional framework near the Hatteras Outer Ridge. Contorted middle Miocene sediments with discontinuous reflectors also suggest local slumping and debris flows on the Norfolk-Washington fan. At its juncture with the Hatteras Outer Ridge, the most prominent development of the Norfolk-Washington fan appears to have occurred during the middle Miocene. The most prominent development of the Hudson fan, at its juncture with the Hatteras Outer Ridge, occurred in the late Miocene.

Middle Miocene bottom currents continued to rework the fan sediments and form local sediment waves. A pulse of accelerated current flow near the middle of the middle Miocene caused strong dissection of the sea floor and formed local unconformities. Both of these features are now observed in the strongly diffracting Horizon X.

An eastward shift in the axis of maximum current-controlled deposition occured near the beginning of the late Miocene and continued at least through the Pliocene. The bulk of the present Hatteras Outer Ridge was formed in a shear zone between the eastern edge of the Gulf Stream and the deepest limb of the Western Boundary Undercurrent during this accretionary episode. The northern end of the Hatteras Outer Ridge was extended eastward, forming a sedimentary prism continuous with the present Gulf Stream Outer Ridge. The extension of the outer ridge may relate to an eastward veering of the Gulf Stream and a resultant crossover with the deep-flow component of the Western Boundary Undercurrent. The interaction between these flows probably controlled ridge development in a manner similar to that modeled for the Blake Outer Ridge (Bryan, 1970). In addition, deposition in this region was accentuated by an abundant supply of sediments from the adjacent Hudson fan.

In late Pliocene to early Pleistocene time, the Hatteras/Gulf Stream Outer Ridge connection was breached by the Western Boundary Undercurrent and an erosional basin formed in its place. It is possible that this event correlates with the culmination about 3.8 m.y. ago of a steady intensification in the Gulf Stream circulation (Kaneps, 1979). This intensification could have shifted the crossover site between the Gulf Stream and Western Boundary Undercurrent away (presumably northward) from the outer ridge, allowing the unimpeded Western Boundary Undercurrent to erode the ridge system.

Shortly following this event and coincident with the onset of strong northern hemisphere glaciation, the erosional trough and the basins west of the Hatteras Outer Ridge were flooded with turbidites. The basins west of the outer ridge were filled by about 0.5 m.y. ago and turbidity currents spilled over the ridge crest. These flows were channeled by sediment waves that had developed along the eastern flank of the outer ridge, but they locally modified the sediment waves and formed small sediment ponds before dispersing onto the northern Hatteras Abyssal Plain.

ACKNOWLEDGMENTS

Tucholke's research support was provided by National Science Foundation Grant OCE-76-02038, by Office of Naval Research Contract N00014-75-C-0210 at Lamont-Doherty Geological Observatory, and by ONR Contract N00014-79-C-0071 at Woods Hole Oceanographic Institution. Laine was supported by ONR Contracts N00014-76-C-0226 and N00014-76-C-0262 NR 083-004. We thank C. C. Windisch (chief scientist, *Conrad* Cruise 21-01) for obtaining, and allowing us to use, the profiles in Figures 3 and 6. We also thank K. Klitgord (USGS, Woods Hole) for providing access to the R/V *Fay* seismic profiles. The manuscript was reviewed by E. Uchupi and J. I. Ewing. Contribution No. 4826 of Woods Hole Oceanographic Institution.

REFERENCES CITED

Asquith, S. M., 1979, Nature and origin of the lower continental rise hills off the east coast of the United States: Marine Geology, v. 32, p. 165-190.

Ballard, J. A., 1966, Structure of the lower continental rise hills of the western North Atlantic: Geophysics, v. 31, p. 506-523.

Barrett, J. R., Jr., 1965, Subsurface currents off Cape Hatteras: Deep-Sea Research, v. 12, p. 173-184.

Benson, W., et al, 1978, Initial reports of the Deep sea drilling project, v. 44: Washington, D.C., U.S. Government Printing Office, 1005 p.

Berggren, W. A., 1972, Late Pliocene-Pleistocene glaciation, *in* A. S. Laughton et al, Initial reports of the Deep sea drilling project, v. 12: Washington D.C., U.S. Government Printing Office, p. 953-963.

Bryan, G. M., 1970, Hydrodynamic model of the Blake Outer Ridge: Journal of Geophysical Research, v. 75, p. 4530-4537.

Embley, R. W., and M. G. Langseth, 1977, Sedimentation processes on the continental rise of northeastern South

America: Marine Geology, v. 25, p. 279-297.
Emery, K. O., and E. Uchupi, 1972, Western North Atlantic Ocean: topography, rocks, structure, water, life, and sediments: AAPG Memoir 27, 532 p.
———, et al, 1970, Continental rise off eastern North America: AAPG Bulletin, v. 54, p. 44-108.
Ewing, J. I., and C. D. Hollister, 1972, Regional aspects of deep-sea drilling in the western North Atlantic, *in* C. D. Hollister, et al, Initial reports of the Deep sea drilling project, v. 11: Washington, D.C., U.S. Government Printing Office, p. 951-973.
Fox, P. J., B. C. Heezen, and A. M. Harian, 1968, Abyssal anti-dunes: Nature, v. 220, p. 470-472.
Heezen, B. C., C. D. Hollister, and W. F. Ruddiman, 1966, Shaping of the continental rise by deep geostrophic contour currents: Science, v. 152, p. 502-508.
Hollister, C. D., et al, 1974, Abyssal furrows and hyperbolic echo traces on the Bahama Outer Ridge: Geology, v. 5, p. 395-400.
———, et al, 1972, Initial reports of the deep sea drilling project, v. 11: Washington, D.C., U.S. Government Printing Office, 1077 p.
Johnson, G. L., P. R. Vogt, and E. D. Schneider, 1971, Morphology of the northeastern Atlantic and Labrador Sea: Deutsche Hydrographische Zietschrift, v. 24, p. 49-74.
Jones, E. J. W., et al, 1970, Influences of Norwegian Sea overflow water on sedimentation in the northern North Atlantic and Labrador Sea: Journal of Geophysical Research, v. 75, p. 1655-1680.
Kaneps, A. G., 1979, Gulf Stream: velocity fluctuations during the Late Cenozoic: Science, v. 204, p. 297-301.
Kennett, J. P., 1977, Cenozoic evolution of Antarctic glaciation, the circum-Antarctic Ocean, and their impact on global paleoceanography: Journal of Geophysical Research, v. 82, p. 3843-3860.
Klitgord, K. D., and J. A. Grow, 1980, Jurassic seismic stratigraphy and basement structure of the western North Atlantic magnetic quiet zone: AAPG Bulletin, v. 64, p. 1658-1680.
Laine, E. P., 1977, Geological effects of the Gulf Stream system in the North American Basin: Ph.D. thesis, Massachusetts Institute of Technology-Woods Hole Oceanographic Institution Joint Program in Oceanography, 164 p.
———, 1980, New evidence from beneath the western North Atlantic for the depth of glacial erosion in Greenland and North America: Quaternary Research, v. 14, p. 188-198.
———, and C. D. Hollister, 1981, Geological effects of the Gulf Stream system on the northern Bermuda Rise: Marine Geology, v. 39, p. 277-310.
Laughton, A. S., et al, 1972, Site 113, *in* A. S. Laughton et al, Initial reports of the Deep sea drilling project, v. 12: Washington, D.C., U.S. Government Printing Office, p. 255-312.
Markl, R. G., G. M. Bryan, and J. I. Ewing, 1970, Structure of the Blake-Bahama Outer Ridge: Journal of Geophysical Research, v. 75, p. 4539-4555.
Miller, K., and B. E. Tucholke, in press, Development of Cenozoic abyssal circulation south of the Greenland-Scotland Ridge, *in* M. H. P. Bott et al, eds., Structure and development of the Greenland-Scotland Ridge — new methods and concepts, NATO conference series 4: Marine Sciences, Plenum Press.
Pinet, P. A., P. Popenoe, and D. F. Nelligan, 1981, Gulf Stream: reconstruction of Cenozoic flow patterns over the Blake Plateau: Geology, v. 9, p. 266-270.
Richardson, P. L., 1977, On the crossover between the Gulf Stream and the Western Boundary Undercurrent: Deep-Sea Research, v. 24, p. 139-159.
Richardson, M. J., and M. Wimbush, 1980, An exceptionally strong near-bottom current on the continental rise of Nova Scotia: Eos, American Geophysical Union, Transactions, v. 61, p. 1014.
Roberts, D. G., 1975, Marine geology of the Rockall Plateau and Trough: Philosophical Transactions of the Royal Society of London, A, v. 278, p. 447-509.
Rona, P. A., 1969, Linear "lower continental rise hills" off Cape Hatteras, Journal of Sedimentary Petrology, v. 39, p. 1132-1141.
———, E. D. Schneider, and B. C. Heezen, 1967, Bathymetry of the continental rise off Cape Hatteras: Deep-Sea Research, v. 14, p. 625-633.
Ruddiman, W. F., 1972, Sediment redistribution on the Reykjanes Ridge: seismic evidence: Geological Society of America Bulletin, v. 83, p. 2039-2062.
Shackleton, N. J., and J. P. Kennett, 1975, Late Cenozoic oxygen and carbon isotopic changes at DSDP Site 284: Implications for glacial history of the Northern Hemisphere and Antarctica, *in* J. P. Kennett et al, Initial reports of the Deep sea drilling project, v. 29: Washington, D.C., U.S. Government Printing Office, p. 801-807.
Sheridan, R. E., X. Golovchenko, and J. I. Ewing, 1974, Late Miocene turbidite horizon in Blake-Bahama Basin: AAPG Bulletin, v. 58, p. 1797-1805.
Tucholke, B. E., 1979, Relationships between acoustic stratigrahy and lithostratigraphy in the western North Atlantic basin, *in* B. E. Tucholke, et al, Initial reports of the Deep sea drilling project, v. 43: Washington, D.C., U.S. Government Printing Office, p. 827-846.
———, and J. I. Ewing, 1974, Bathymetry and sediment geometry of the Greater Antilles Outer Ridge and vicinity: Geological Society of America Bulletin, v. 85, p. 1789-1802.
———, and G. S. Mountain, 1979, Seismic stratigraphy, lithostratigraphy, and paleosedimentation patterns in the western North Atlantic: American Geophysical Union, Maurice Ewing Series, v. 3, p. 58-86.
———, W. R. Wright, and C. D. Hollister, 1973, Abyssal circulation over the Greater Antilles Outer Ridge: Deep-Sea Research, v. 20, p. 973-995.
Vail, P. R., R. M. Mitchum, Jr., and S. Thompson, III, 1977, Seismic stratigraphy and global changes of sea level, part 4: Global cycles of relative changes of sea level, *in* C. E. Payton, ed., Seismic stratigraphy — applications to hydrocarbon exploration: AAPG Memoir 26, p. 83-97.

Convergent Margins

Detailed Structure of the Nankai Trough from Migrated Seismic Sections

Yutaka Aoki
Toshiro Tamano
Susumu Kato
Japan Petroleum Exploration Co., Ltd.
Ohtemachi, Tokyo, Japan

The structure of the inner trench slope of the Nankai trough is studied using migrated seismic records to avoid diffraction effects. In the western Nankai trough, the migrated seismic sections show systematic deformation of trench fill turbidites by thrust faults at the foot of the inner slope, forming linear ridges parallel with the trench. The underlying sediments of the Shikoku basin, on the other hand, are subducted and not deformed. The angle of subduction is steeper in the eastern Nankai trough than in the western part. Alternatively, accreted sediment of turbidite origin is subjected to deformation mostly by folding and large thrust faults are localized in the lower section of accreted sediments. No appreciable undeformed layer exists in the eastern Nankai trough.

The Nankai trough is the northern boundary of the Shikoku basin, which is a part of the Philippine plate. Here, we consider the Philippine plate as subducting under the Eurasian plate. So in plate tectonic terms, the Nankai trough is a trench and is sometimes alternatively called the southwest Japan trench or Shikoku trench (terminology for trenches is used in the following discussion). The trench is rather shallow, generally between 4,000 to 4,800 m and only several hundred meters deeper than the adjacent Shikoku basin.

The four major differences between the Nankai trough and the Japan trench, Izu-Mariana trench, and Nanseishoto (Ryukyu) trench in the northwest Pacific are outlined below. The first difference is in the topography of the sea floor. The 1,000 m high Zenisu ridge extends from Zenisu in the southern Izu peninsula parallel with the Nankai trough. Dips of the outer and inner slope of the trough become almost equal south of Omaezaki. The axis of the Zenisu ridge is only 30 km from the axis of the Nankai trough and the ridge differs in height and length from a marginal swell, thus it is referred to as a spur or an outer ridge. The ridge is about 400 km long and declines in height toward the west until it disappears south of the Kii peninsula. From south of the Kii peninsula to the Kyushu — Palau ridge the Nankai trough is broad without an outer ridge, in contrast to the eastern part. In this article, the part with no outer ridge is called the western Nankai trough and the part with the ridge is called the eastern Nankai trough.

The second difference between Nankai trough and the other trenches is that, in the outer trench slope of the Nankai trough, no normal faults appear that suggest tensional stress features in the subducting plate. This is seen from the sea floor topography and internal structure of the oceanic sediment.

The third difference regards seismicity and focal mechanisms. The Benioff zone associated with subducting Philippine plate at the trough is difficult to define. This is because the distribution of deep focus earthquakes is not concentrated in a narrow zone. According to recent data however, the distribution of upper mantle earthquakes indicates a gently inclined Benioff Zone in the western Nankai trough and a steep zone in the eastern trough (Yamazaki Oida, and Aoki, 1980). The maximum depth of the Benioff Zone is considered to be about 80 km in the western trough and about 50 km in the eastern trough (Kanamori, 1972; Shiono, 1974) and is very shallow in comparison with other trench systems in the northwest Pacific. Focal mechanisms indicate that earthquakes along the Nankai trough are of the thrust type (Kanamori, 1972), which is consistent with subduction. But sub-crustal earthquakes have a tensional axis parallel with the Nankai trough (Shiono, 1977).

The fourth difference is that relatively high heat flow (compared with other trenches) is observed in the Nankai trough. Even around the trench, Watanabe, Langseth and Anderson (1977), obtained a value of 2 to 3 HFU.

Despite these differences, single channel seismic reflection records show many features characteristic of other trenches (Hilde, Wageman and Hammond, 1969; Ludwig, Den, and Murauchi, 1973; Moore and Karig, 1976; Murauchi and Asanuma, 1977; Inoue, 1978), such as oceanic basement extending from the trench beneath the landward slope. However, very little detail is resolved in these records due to lack of acoustic penetration and high vertical exaggeration. At the base of inner trench slope off Shikoku, site 298 was drilled 611 m deep during Leg 31 of the Deep Sea Drilling Project (DSDP). This site was about 5 km landward of the trench axis. The upper part of the sedimentary layer consisted of trench fill turbidites that were interpreted as being isoclinally folded by the compressive stress associated with subduction (Ingle et al, 1975a; Moore and Karig, 1976).

The structure of the outer slope, the trench fill turbidite, and inner slope are described here using several multichannel seismic reflection records across the Nankai trough which have much greater detail than previous records. Figure 1 shows the tracks lines and the parameters of data acquisition and data processing are listed in Table 1. Lines K54-1-2, S-4, and S-8 were acquired by Japan Petroleum Exploration (JAPEX) and lines H-1, I-1 by Japan National Oil Corporation (JNOC). Kaiyo Maru of JAPEX obtained all data.

MIGRATED REFLECTION SEISMIC RECORDS

Even on unmigrated multichannel seismic records of the Nankai trough, the subduction of oceanic basement is well defined a few tens of kilometers landward of the trough. However, interpretation of the inner slope structure from unmigrated seismic profiles is usually very difficult due to diffractions which can be reduced or eliminated by the migration process (Figure 2). The migrated data show the internal structure much more clearly, so the following discussion is based on the migrated reflection records.

Line K54-1-2

This line is located about 70 km southwest of DSDP hole 298 and trends roughly perpendicular to the Nankai trough axis (Figure 1). The Southeast end is on the outer slope and about 80 km from DSDP hole 297 on the outer trench swell. Lower record in Figure 2 shows a part of migrated time section; the Nankai trough is at the right end of the section. At the base of the inner slope, well developed imbricated sediment wedges or thrust-faulted blocks are shown clearly with the thrust faults which produced them. Ridges produced by the thrust faults are distributed systematically at a spacing of 1.5 to 2.0 km and have relative heights of 100 to 300 m above the small intervening troughs. The topographic ridge runs parallel with the Nankai trough (Ingle et al., 1975 a; Moore and Karig, 1976; JAPEX, unpublished data). These ridges trap sediment from the land and the troughs between the ridges are eventually filled with nonaccreted sediment. Troughs have progressively less fill toward the trench. In fact, the trough nearest the trench has essentially no fill, whereas landward troughs are completely filled. The fill dips slightly landward near the sea floor and has increasingly steeper dips with increasing depth, indicating continued rotation from uplift of the ridge. Unfortunately the record becomes more obscure landward and it is difficult to clearly discriminate between trough fill and tilted trench deposits. Heights of all the ridges are about equal and it appears that an incipient ridge is just beginning to form in the trench, based on gentle warping of these sediments. These observations indicate that a large initial amount of deformation occurs at the foot of the trench slope, but that deformation continues within the accreted sediment landward at a much lesser rate.

Table 1. Data acquisition parameters and processing sequence.

	K54-1-2	I-1, H-1	S-8, S-4
Air gun volume (c.i)	2090	1570	2000
Shot interval (m)	50	50	50
Group interval (m)	50	50	50
Number of groups	48	48	48
Sample interval (ms)	4	4	4
Recorded in	1980	1978	1978
Data Process	1. Demultiplex 2. CDP sorting 3. Gain recovery 4. Deconvolution 5. Filter 6. Velocity analysis 7. NMO correction 8. Stack 9. Datum correction 10. F-K migration	1. Demultiplex 2. CDP sorting 3. Gain recovery 4. Deconvolution 5. Velocity analysis 6. NMO correction 7. Stack 8. Deconvolution 9. Filter 10. Datum correction 11. Finite difference migration	1. Demultiplex 2. CDP sorting 3. Gain recovery 4. Deconvolution 5. Velocity analysis 6. NMO correction 7. Stack 8. Deconvolution 9. Filter 10. Datum correction 11. F-K migration (S-8) 12. Kirchoff migration (S-4)

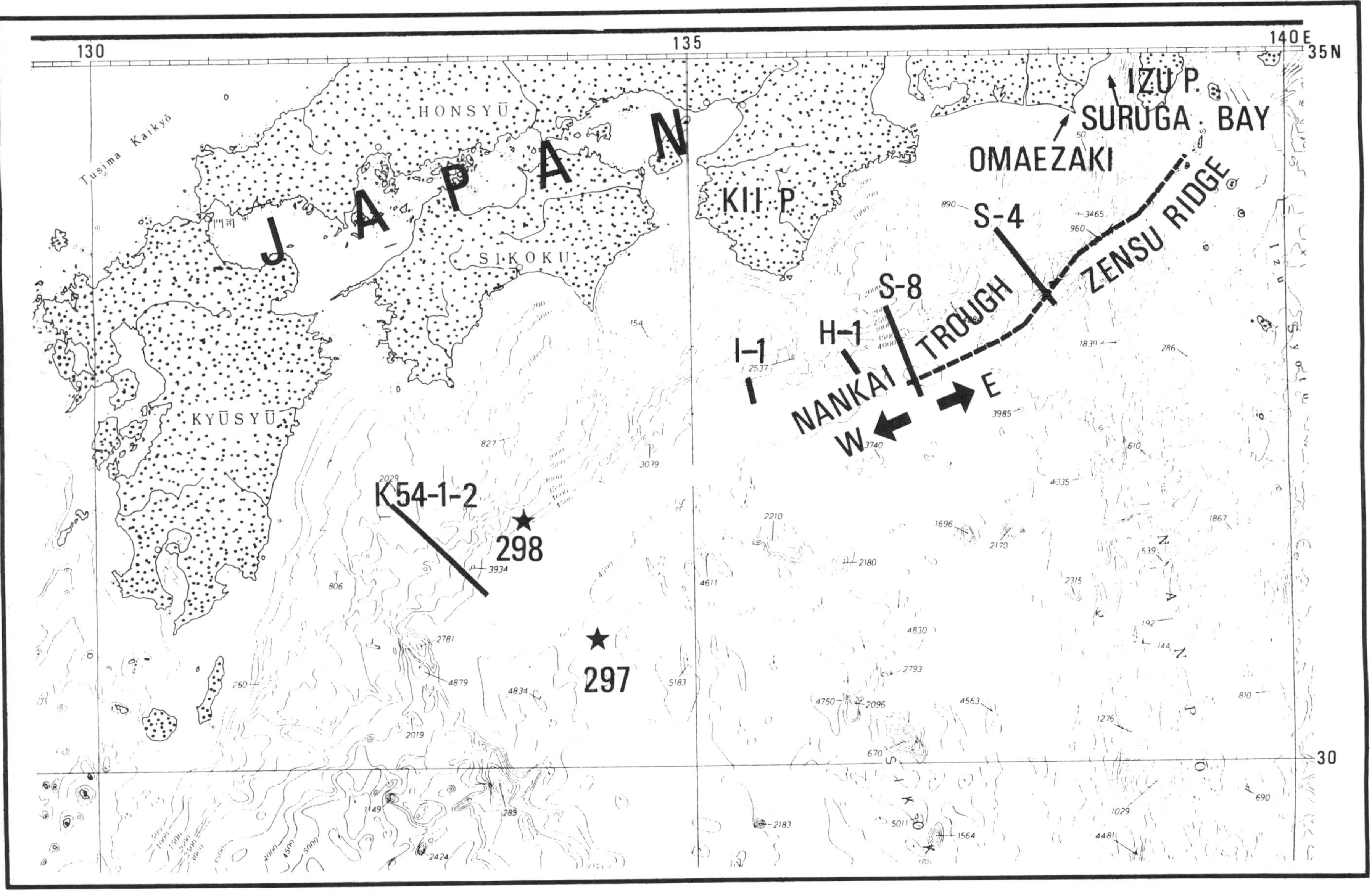

Figure 1 — Tracks of multichannel reflection seismic lines across the Nankai trough. Star indicates the location of DSDP hole 298 and 297.

According to the results of DSDP hole 298 (Ingel et al., 1975a) and seismic surveys (Ludwig, Den, and Murauchi, 1973; Moore and Karig, 1976; Karig and Sharman, 1975; Inoue, 1978), the trench fill and therefore the accretionary wedge is probably a mud and sand turbidite. Shikoku basin sediment, on the other hand, is a pelagic sediment principally composed of clay (Ingle et al., 1975b) and the boundary between the trench fill and Shikoku basin sediment beneath it appears to be marked by a distinct reflector. Thus the trench wedge fill is probably a unit with distinct, well-defined reflectors on reflection records. In the western Nankai trough, the trench fill is 100 to 1000 m thick and 10 to 20 km wide. Note that in Figure 2 the Shikoku basin sediment, where subducted, is almost undeformed despite the imbrication of the overlying trench deposits. The decollement between these two units is a characteristic feature of the western Nankai trough and is especially well defined on line K54-1-2 (Figure 2). In Figure 2, the thickness of the undeformed layer is consistently about 0.3 sec (two-way travel time) near the trench and landward. Minor undulations are present with amplitudes of about 50 ms (two-way travel time), but the undulations correspond to changes of the sea floor topography and have resulted from velocity effects due to the topography. Below the undeformed layer is the acoustic basement. DSDP drilling and refraction studies (Murauchi et al., 1968) show that it corresponds to the igneous oceanic crust. The boundary between the basin sediment (oceanic layer 1) and the igneous oceanic crust (Oceanic layer 2) constitutes a distinctive reflector and serves as a good marker on the subducting oceanic crust. This distinct reflector, commonly seen on reflection seismic records from trench systems all over the world, often appears in a time section like a seaward dipping reflector due to velocity effects. In the eastern Nankai trough the velocity pull-up effect is also commonly observed, but in the western part, the inclination of this boundary remains almost constant or the velocity pull-up effect is not pronounced as is shown on the entire seismic record of line K54-1-2 (Figure 3). The lack of a pronounced velocity pull-up may relate to the fact that in the western Nankai trough, the initial subduction angle, which corresponds to the inclination of the outer slope sea floor topography, is small compared to other trench systems (see Figure 4, Karig and Sharman, 1975) and accreted sediment thickness increases gradually so that accreted section velocity increase by compaction or dehydration is also gradual. Below the oceanic basalt layer, possibly primary reflections are observed in several places that may come from oceanic layers 2 and 3. Though these reflectors cannot be established with confidence, they may indicate layering in layers 2 and 3.

Beginning at point B in Figure 3, systematically distributed thrust faults, as observed at the foot of the inner slope, disappear. In Figure 3, the sea floor between B and C is much smoother than the topography between A and B or C and D. The ridges are spaced farther apart and covered by slope sediment. Between B and C, there is a bottom simulating reflection that may be from the base of a gas hydrate (Tucholke, Bryan, and Ewing, 1977; Shipley, 1979). This reflection obscures the structure of the slope sediment and it is difficult to discern thrust faults corresponding to the ridges. The change in structural character between segment T and B and B and C may correspond to a stripping of the subducting sediment layer from the igneous ocean crust. Undulation of the subducting layer and the oceanic basement between B and C cannot be totally attributed to velocity effects. This is evident from the amplitude of reflections and their correspondence to sea floor topography. Between B and C, reflections dip gently just above the basement and more steeply in the upper portion. These events presumably indicate thrust faults which are developing in the previously undeformed subducting sediment layer. This indicates that basin sediment is scraped from the oceanic basement and accumulates in the overlying accretionary complex.

The subducting igneous oceanic crust can be traced for at least 60 km landward from the trench axis (T). If the landward event of low frequency shown by the arrow in Figure 3 really corresponds to the oceanic basement, the distance increases up to 70 km. At point C, more distinct ridges nearly 500 m high form the edge of the forearc basin where the sediment is about 3000 m thick. This basin corresponds to a deep sea terrace, defined by Hoshino (1969), and forms a part of the Hiuga Basin. Therefore the trench slope break (Dickinson, 1973; Karig and Sharman, 1975) on the line K54-1-2 corresponds to point C. Dipping reflection events suggesting thrust or shear are not recognizable landward of point C.

Line I-1

Line I-1 also trends perpendicular to the Nankai trough and is located south of the Kii peninsula (Figure 1). The migrated time section (Figure 4) shows the foot of the landward slope of the Nankai trough where the structure is basically the same as in line K54-1-2 (Figure 2 and Figure 3). The trench fill is thrust-faulted, but the underlying Shikoku basin sediment is subducted as a coherent layer on the igneous oceanic crust. A fault plane reflection from a major thrust fault appears clearly and is indicated by an arrow in the figure, reflection events from other thrust faults are generally not as clear as in K54-1-2. However, thrust fault positions are identifiable from the pattern of the bedding plane reflections in the thrust slices. Thrust fault spacing, assumed from the coherent reflection packets within the folded sediment, is about 2.5 km and the lengths are nearly equivalent to, or slightly larger than, those on K54-1-2.

Ridges corresponding to thrust slices are low and have not formed significant troughs. Accordingly, the slope's base has the topography of a terrace 8 to 9 km wide. At the landward end of this terrace, the sea floor suddenly rises several hundred meters forming a big ridge or a scarp. However, no thrust fault related to this ridge is recognized in the seismic record. Perhaps many small thrusts, instead of one dominant re-

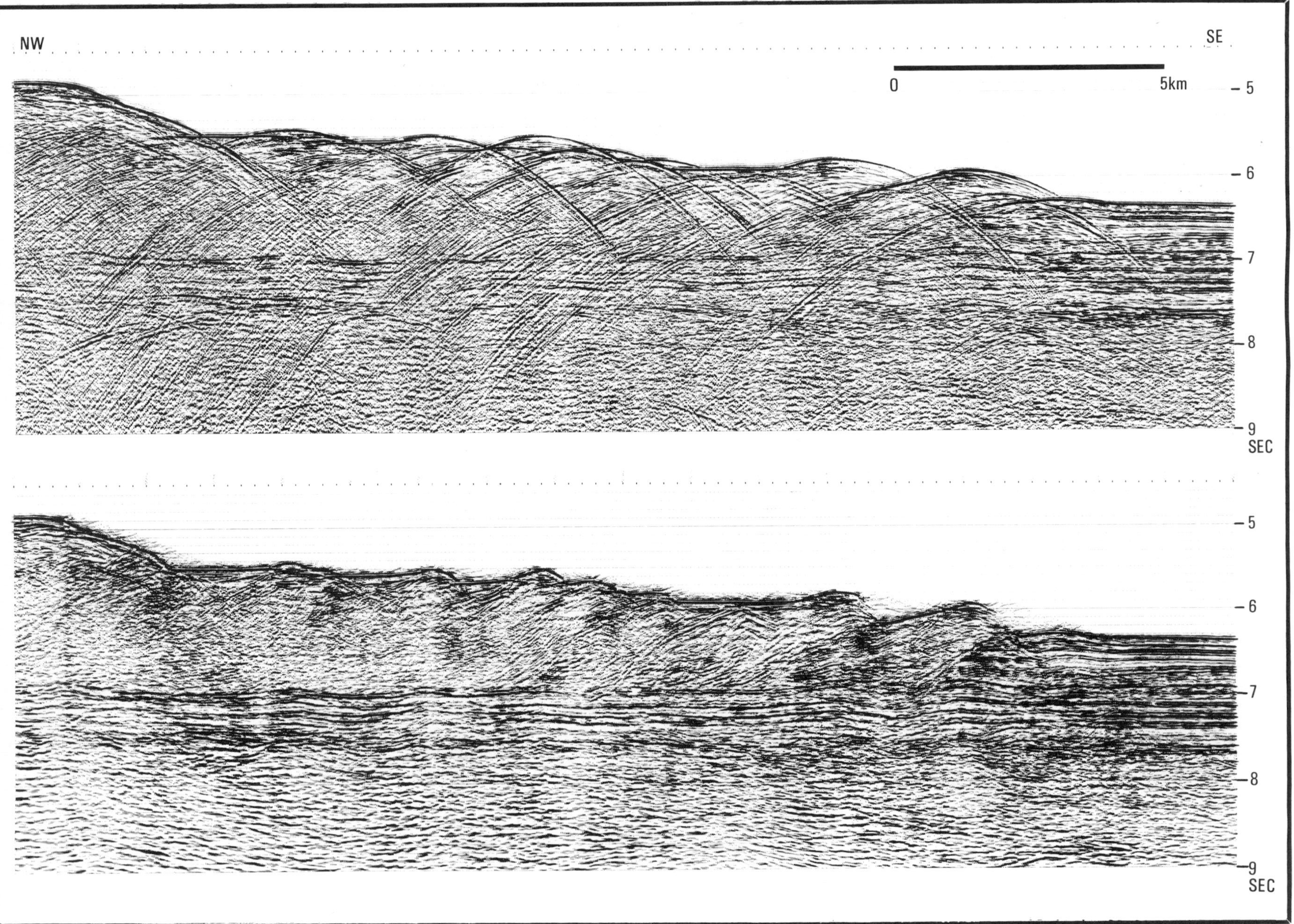

Figure 2 — Unmigrated (upper) and migrated (lower) time sections across the foot of the inner slope portion of line K54-1-2. Resolution of faults and structure of the youngest thrust slices is much clearer as a result of the migration.

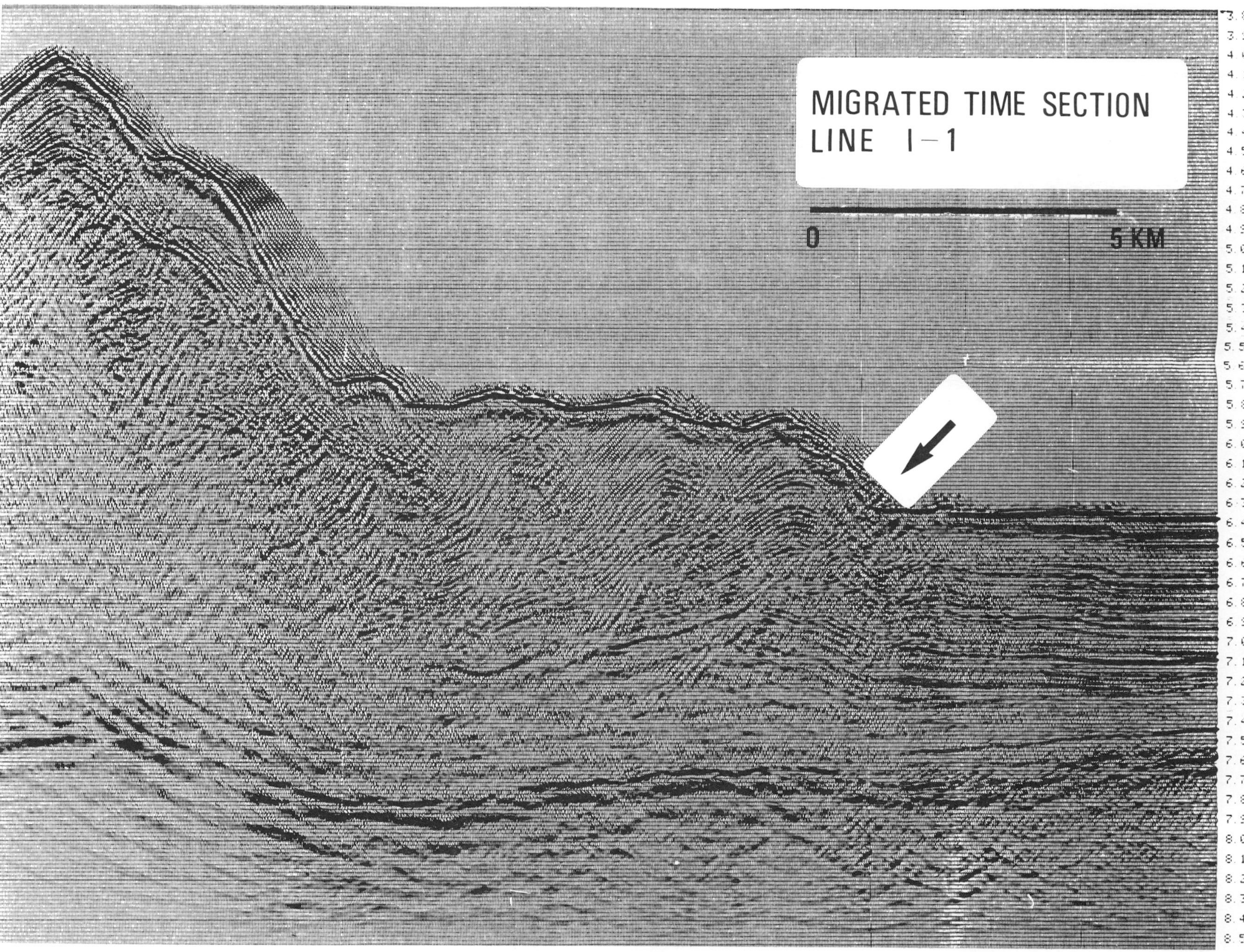

Figure 4 — Migrated time section of the line I-1, showing a series of thrust slices at the foot of the slope. Overlying the oceanic basement is undeformed basin sediment. Arrow indicates a probable thrust fault.

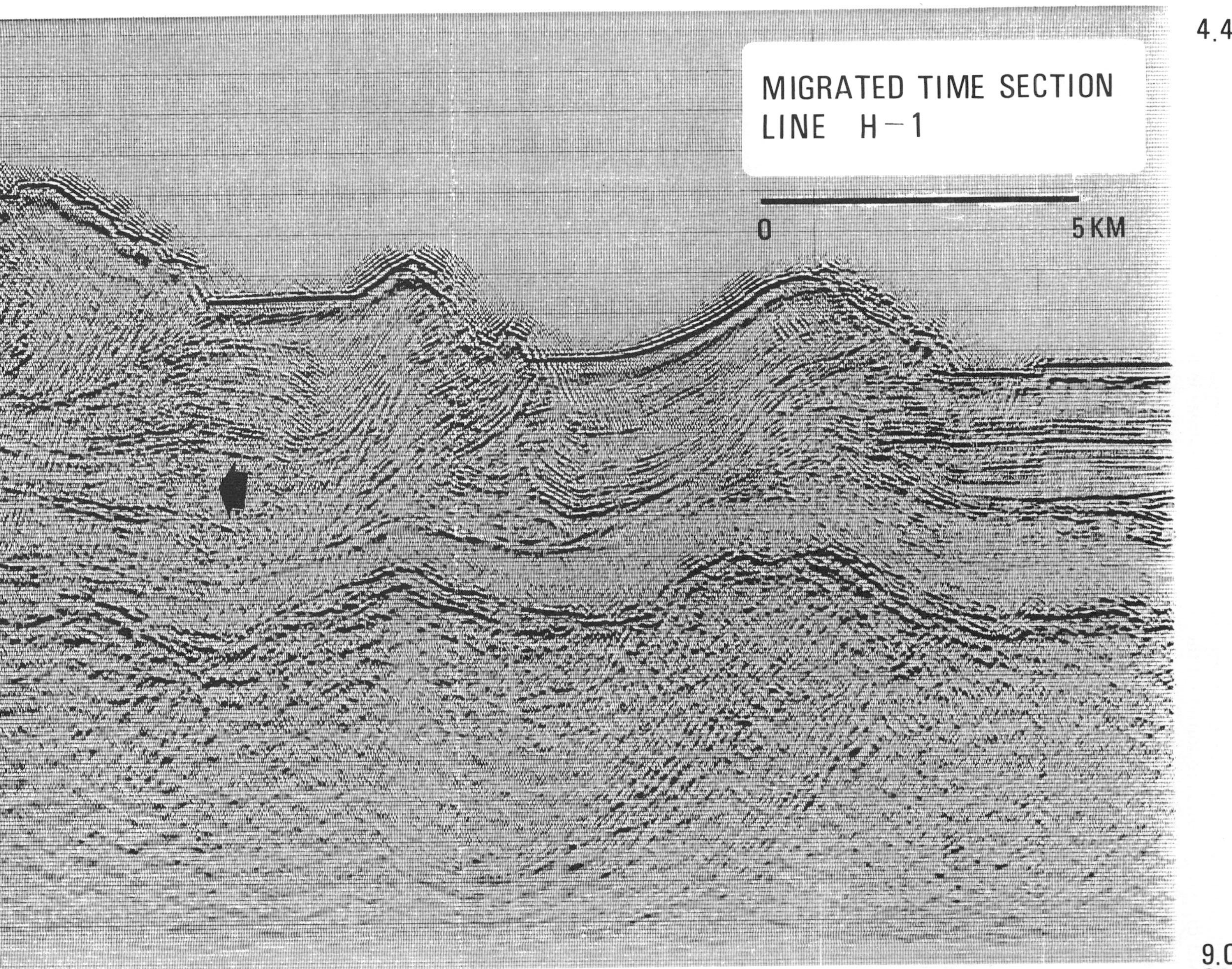

Figure 5 — Migrated time section of the line H-1, showing systematic development of thrust slices under the lower slope.

verse fault, formed this big ridge or scarp, because landward dipping events are found on the seaward slope of the ridge. A bottom simulating reflection is also observed in this record at a water depth of 3500 to 4500 m. Seismic reflectors diminish in clarity landward of the distinct fault (arrow in Figure 4); seaward, the trench fill shows initial deformation by minor thrust faults. The tilting of beds generally increases towards the land, thus the observations in record I-1 are similar to those in record K54-1-2 and the deformation features include subduction of sediment beneath an imbricated trench fill section. Here, the subducted layer is characterized by a faint zone due to weak reflection amplitude. The subducting oceanic basement appears to dip seaward on the left side of the figure, but converted from time to depth, the record shows the dip of the oceanic basement is nearly constant.

Line H-1

Line H-1 is also located south of Kii peninsula of Honshu (Figure 1) just east of I-1. The accretionary process shown in this seismic record (Figure 5) is basically the same as that shown in K54-1-2 (Figure 2,3) and I-1 (Figure 4). The trench fill is thrust-faulted and is separated by a decollement from unfaulted Shikoku basin sediment with low reflection amplitude, which remains coherent during the first stage of accretion. Note again that deformation in the seawardmost thrust slice does not significantly effect bedding. Even in the second thrust slice, bedding is fairly well-preserved. The structure of these two thrust slices indicates that the seaward one is younger than the landward one because bedding is better preserved and less steeply tilted in the seaward thrust slice.

The landward thrust slice was apparently tilted by horizontal compressional stress during development of the seaward thrust slice. The slight landward dip of the slope basin sediment surface suggests that continued rotational movement took place. These observations indicate that the landward thrust slice already existed when the seaward thrust slice formed, and that the landward slice is affected by continued compressional stress resulting in shortening and block rotation. As thickening of the landware thrust slice indicates, thickening of the adjacent landward block may have also occurred during accretion. Perhaps the accreted Shikoku basin sediment played an important role. A particular stratum, indicated by arrows in Figure 5, is very similar to the basin sediment in appearance and it is probably a detached segment of coherent Shikoku basin sediment. Currently a coherent layer is being underthrust beneath it. In contrast to the almost constant thickness of the subducted sediment layer on line K54-1-2, the subducted layer in H-1 thins toward land. Thinning by consolidation or dehydration is quite possible beneath the accreted sediment, but this is not clear because of insufficient accuracy in interval velocity data.

In the line H-1, the trench fill is broken into only two imbricate slices, namely the paired ridges and slope basins. In the line K54-1-2 (Figure 2), the width of the systematic zone of imbrication is about 20 km, whereas in line H-1, this zone is only about 10 km wide and similar to the configuration of line I-1 (Figure 5). The interval between ridges in the zone of imbrication is 2.5 km or less in K54-1-2 and I-1, but is about 5 km in line H-1. A characteristic of the highly distorted zone, located landward of systematic imbrication zone, is the dipping landward reflection events. In addition to the thrust or shear represented by these reflections, folded structure has been well developed and it apparently resembles the structure of the landward slope of the eastern Nankai trough.

S-8

This line is located southeast of the Kii peninsula at the west end of the eastern Nanakai trough (as this paper defines) and crosses the Nankai trough at nearly right angles (Figure 1). Figure 6 shows the migrated time section of the entire line with a reduced horizontal scale. Here, the wedge of trench fill is seen very well. The Nankai trough appears much like a typical trench in water about 4,000 m deep. Sediment thickness at the trench axis, including both trench fill and Shikoku basin sediment, is around 1,300 m of which the upper 800 m is fill. In this record the trench sediment sequence has an upper wedge-shaped fill underlain by a lower wedge-shaped packet, which is tilted and pinches out at point B (Figure 6). Both packets rest on the Shikoku basin sediment. It is unclear whether the lower sediment packet is a previous trench fill tilted by uplift of the ridge at point C (Fig. 6), or a hemipelagic sequence similar to one found seaward of the middle America trench (Moore et al, 1979) and the Japan trench (Langseth et al, 1978).

Point C in Figure 7, shows a topographic high associated with an acoustic basement high. The basement high is an extension of the Zenisu ridge and differs from an outer swell or marginal swell. A fault occurs between B and C (Figure 6), and the displacement of the oceanic basement is greater than the displacement of Shikoku basin sediment overlying it. This reverse fault probably originated from stress in the oceanic crust. Landward of the trench axis, structure indicates that the inner slope is mostly accreted trench sediment. The systematic ridges and slope basins associated with thrusts in the western Nankai trough are not found here and ridges several hundred meters high are distributed at intervals of about 10 km (Figure 6, ridge D and ridge F). Between them is ridge E, which was buried by terrigeneous slope basin sediment. This is evident in Figure 7 which shows a part of the slope at less vertical exaggeration and an enlarged scale. A thrust fault that probably contributed to formation of the buried ridge is identified by the displacement of a folded layer and is drawn on the record (Figure 7). Other thrust faults can be inferred from truncated reflections in this record, thus characterizing the foot of the slope as an accretionary complex.

It is not clear that Shikoku basin sediment is subducted in a continuous unbroken layer, as in the

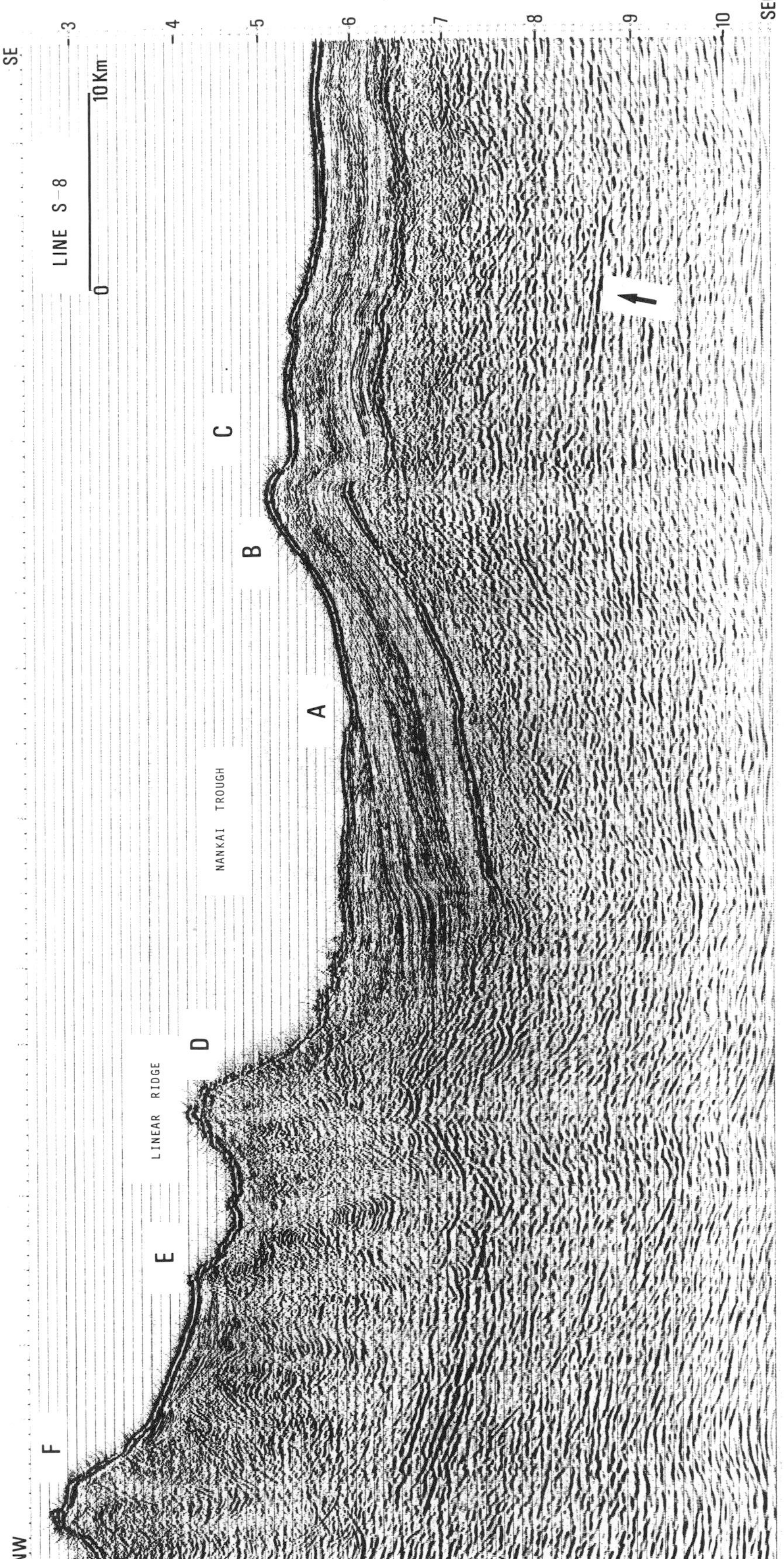

Figure 6 — Migrated time section of the line S-8. The topographic high at point C corresponds to the southwest end of Zenisu ridge. Shikoku basin sediment is interrupted below ridge A, but possible undeformed basin sediment is seen in the left portion of the section.

western Nankai trough, because a high in the oceanic basement under the ridge at D (Figure 6) breaks the continuity of the accreted sedimentary layers. Some of the Shikoku basin sediment continues landward of the basement high (Figure 7). Although the basement high is certainly accentuated by a velocity pull-up beneath the ridge (point D, Figure 6), it disrupts sediment on the ocean crust and is thus likely to be an irregularity on the igneous crust. Perhaps ridge development at D was triggered as this high was first subducted. One additional feature of record S-8 is that coherent reflections are recognizable in the igneous oceanic crust. Interval seismic velocity (P-wave) of the layer above the horizon indicated by an arrow in Figure 6, is about 6,000 m/sec, which compares well with refraction velocities in this area (Murauchi et al, 1968). This deeper reflection may be from the Moho discontinuity (Aoki et al, 1982).

Line S-4

Line S-4 is about 150 km southwest of the mouth of Suruga bay and 120 km northeast of line S-8 (Figure 1). The bathymetry of the Nankai trough is nearly symmetrical (Figure 8) and the Zenisu ridge, which extends to the topographic high located in the outer slope of Line S-8 (Hydrographic Department, 1976), is on the outer slope. Magnetic anomalies and high velocities from reflection seismic surveys indicate that this topographic high consists of igneous rocks (JAPEX, unpublished date). Igneous rocks were dredged from the southwest flank of this ridge near the location of this line (Inoue, 1978).

The structure in the migrated seismic reflection record is surprisingly asymmetrical (Figure 8). Trench fill laps monoclinally onto Zenisu ridge, whereas it is folded and thrust-faulted at the foot of the trench inner slope, indicating compressional deformation as is shown in the other seismic sections which traverse the Nankai trough. The acoustic basement extends under the inner slope and can be traced to the landward end of the record (Figure 8). These observations imply that the accretionary process occurred in association with the subduction of the Philippine plate. Accretion here seems somewhat different from that seen in other seismic records of the western Nankai trough. Landward dipping reflections which may correspond to thrust faults do not reach the sea floor and as a result, there are no conspicuous ridges on the lower slope. It is evident from the structure shown by bedding reflections that trench turbidite has been subjected to folding. Although artifacts of the migration process mask the fine structure of the landward portion, folding is dominant in shallower parts whereas thrust-faulting prevails at depth in the accretionary complex. This may mean that sediment rigidity is greater in the lower portion of accreted sediment so that brittle fracture can occur. Consequently, considerable compressional stress is accommodated by folding and thrusting in the eastern Nankai trough.

The reflections in the Nankai trough show that sedimentary strata thicken toward land (northwestward). If sediment is deposited nearly horizontally as is shown by the present sea floor, then either the landward side has always been subsiding or the seaward side has been elevated. From other evidence, the beginning of the subduction which formed the present Nankai trough is assumed to be about 3 million years (Ingle et al, 1975b). Before that time, the area of the Nankai trough was probably flat and the present trough topography formed later. Nevertheless, landward thickening of the Nankai trough sediment as shown in S-4 is considered to be unrelated directly to the formation of the present trough. This landward thickening is a peculiarity of the eastern Nankai trough which possesses the seaward topographic high. Consequently, the present landward thickening of trench fill is considered to be closely related to the Zenisu ridge formation.

The Zenisu ridge structure is not clear from the seismic data, because almost no coherent events are recognizable except within the uppermost sedimentary layer. If the event inclined landward (C, Figure 8) is considered to be a thrust, then it is consistent with reverse fault affecting the oceanic crust on line S-8.

DISCUSSION

The differences between the eastern and western Nankai troughs suggest that sea floor morphology or sediment character may influence the tectonics in an accretionary zone. Along the entire trough, the trench filled with a sediment sequence 10 to 20 km wide which is being accreted into the lower slope of the trench. Along the western Nankai trough, the front part of the accretionary complex is a series of upthrust ridges and intervening troughs (some 10 to 20 km wide) that form a relatively shallow dipping sea floor along the lower slope; along the eastern part, there is a pronounced ridge in the basement that flanks the seaward side of the trench. Among other differences are the seismology (Seno, 1977; Yamazaki, Oida, and Aoki, 1980; Hirahara, 1980; Kanamori 1972; Shiono, 1974) and the thickness of trench fill at the front of the subduction zone. Essentially, the differences combine so that the accretionary complex thickens more rapidly along the eastern Nankai trough than along the western trough. We propose that faulting in the thicker eastern accretionary complex requires greater stress. Thus, the upper, more plastic sediment is folded, while strata below these sediments are broken by thrust faults spaced at large (10 km wide) intervals rather than the smaller 2.5 to -5 km intervals common at the front of the western accretionary complex. Along the eastern Nankai trough the more rapid thickening of the accretionary complex is achieved by the combination of a steep trench slope above and a more rapidly plunging subduction zone below. Because greater stress is needed to deform the accretionary complex, perhaps part of the stress associated with convergence is accommodated at a zone of weakness at the Zenisu ridge. Thrust or reverse faults appear to have accompanied uplift of the ridge, rather than being entirely taken up within the

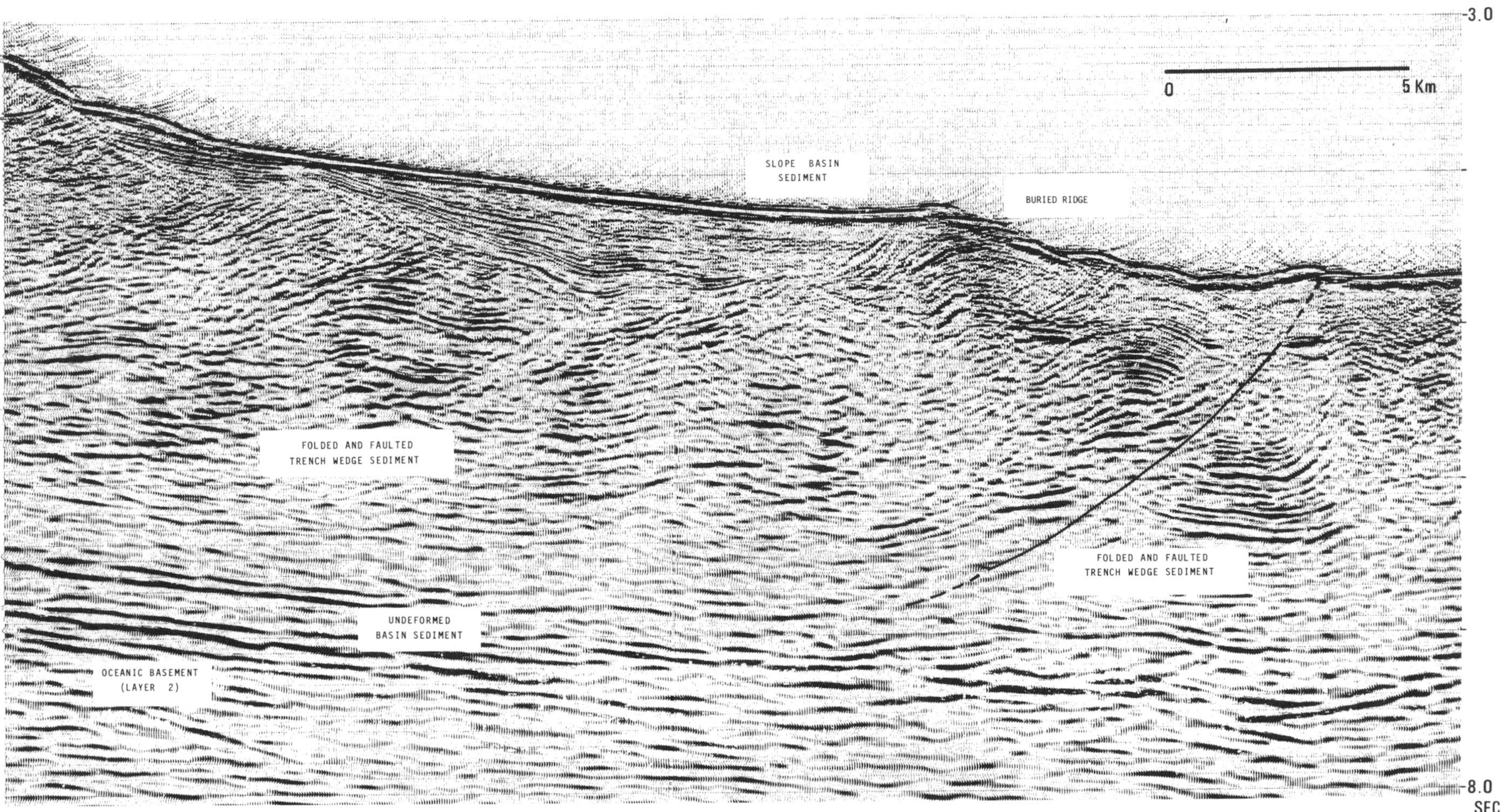

Figure 7 — Enlarged portion of line S-8. Folded and faulted structure of the accreted trench sediment is clearly shown. Displaced, but correlatable reflections at the right of the section indicate a thrust fault (heavy line).

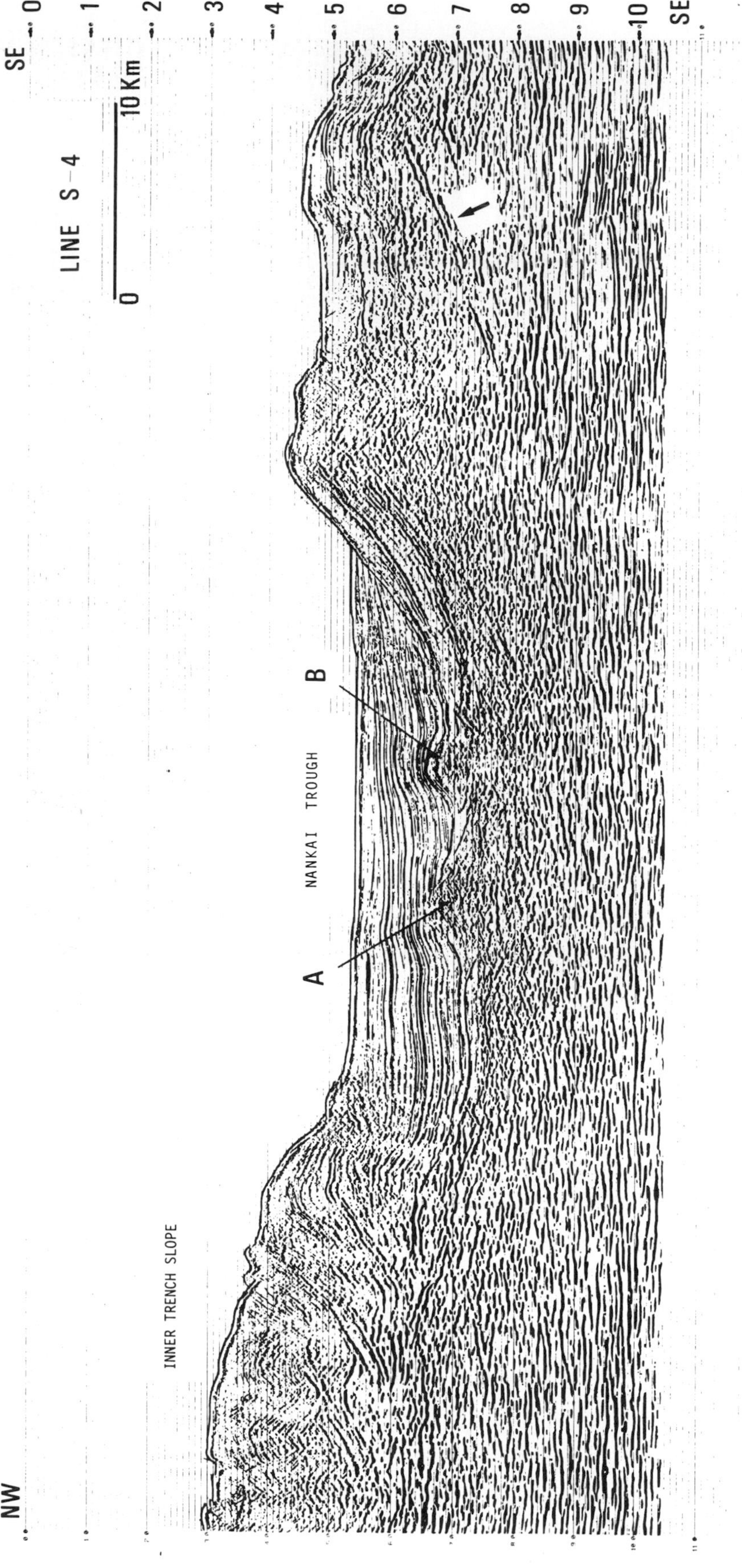

Figure 8 — Migrated section of S-4. The topographic high right of the Nankai trough is the Zenisu ridge, which has a velocity characteristic of the oceanic crust. The structure of the inner slope is essentially the same as in other lines, but folding and faulting are highly developed. A and B are possible basement highs.

accretionary complex.

In the western Nankai trough the accretion can continue as long as the sediment is thin and has not dewatered. Once dewatered, the competence and thickness of the sediment increases and it begins reacting much like the accretionary complex along the eastern Nankai trough. Therefore, the frontal ridge and trough accretionary complex in the western Nankai trough gives way landward to an accretionary structure similar to that along the eastern trough.

SUMMARY

Along the Nankai trough, the Philippine plate is subducted beneath the Asian plate and despite the nearly constant rate of convergence, the eastern and western parts of the trough have pronounced differences in structure. Along the western part of the trough sediment that fills the trench axis is scraped off and accreted to the slope, resulting in a sequence of thrust faults at 2.5 to 5 km spacing that is clearly observed in migrated seismic records. Each thrust produces a pronounced ridge with associated slope basins at the sea floor. Below the accreted fill, and separated from it by a decollement, is a sequence of coherent reflections from relatively undeformed, subducted Shikoku basin sediment. These reflections can be followed up to 60 km landward from the trench. The sediments overlay and are carried along by the subducting igneous ocean crust, which dips only 2 to 3 km. About 5 to 15 km landward from the trench the accreted sediment overlying the decollement displays an abrupt change in tectonic style and structural character that includes only widely-spaced thrust faults and, near the surface, broad folds. The structural change is also accompanied by a breakup of the subducted, previously undeformed sediment layer and accretion of detached segments of this layer underneath the steeply dipping reflections of the accretionary complex; in short, a process of accretion known as underplating (Watkins and Moore, 1981).

The angle of subduction is steeper along the eastern Nankai trough than along the west, and the thickness of accreted sediment increases suddenly from the trench axis. Large scale folds develop even at the foot of the slope with associated thrust faults. In both eastern and western Nankai troughs, the dip of the thrust faults generally becomes steeper landward which indicates continuing uplift and rotation of the accreted sediment.

The reflection profiles (time sections) show highs in the oceanic basement, at least in part caused by a velocity pull-up effect under the linear ridges generated by thrusting. Simple velocity effects cannot explain some of these highs, and it is considered likely that some of the linear ridges are formed or triggered by the subduction of basement highs which existed on the incoming oceanic basement.

ACKNOWLEDGMENT

The authors are grateful to JNOC Japan and JAPEX who kindly gave permission to use their data. We also acknowledge the assistance by our many colleagues, including the party of R/V Kaiyomaru, in collecting data and the staffs of data processing center. The discussion with R. Von Huene, who critically reviewed the manuscript, was very valuable.

REFERENCES CITED

Aoki, Y., et al, 1982, Compressional velocity analysis for suboceanic basement reflectors in the Japan trench and Nankai trough, based on multichannel seismic reflection profiles: Geodynamic Project Final Report, n. 1.

Dickinson, W.R., 1973, Widths of modern arc-trench gaps proportional to past duration of igneous activity in associated magmatic arcs: Journal of Geophysical Research, v. 78, p. 3376-3389.

Hilde, T. C., J. M. Wageman, and W. T. Hammond, 1969, The structure of Tosa terrace and Nankai trough off southeastern Japan: Deep-Sea Research, v. 16, p. 67-76.

Hirahara, K., 1980, Three dimensional P-wave velocity distribution in southwestern Japan: Annual Meeting of the Seismological Society of Japan, n. 2 (in Japanese).

Hoshino, M., 1969, On the deep sea terrace: La Mer, v. 3, p. 222-224.

Hydrographic Department, Maritime Safety Agency of Japan, 1976, Bathymetric chart of Iro saki to Muroto saki.

Ingle, J. C., Jr., et al, 1975a, Site 298, Initial reports of the deep sea drilling project: Washington, D. C., U.S. Government Printing Office, v. 31, p. 37-350.

———, ———, 1975b, Site 297, Initial reports of the deep sea drilling project: Washington, D.C., U.S. Government Printing Office, v. 31, p. 275-316.

Inoue, E., 1978, Investigation of the continental margin of southwest Japan: Geological Survey of Japan, Cruise Report, n. 9.

Kanamori, H., 1972, Tectonic implications of the 1944 Tonankai and the 1946 Nankaido earthquakes: Physics of the Earth and Planetary Interiors, v. 5, p. 129-139.

Karig, D. E., and G. F. Sharman, III, 1975, Subduction and accretion in trenches: Geological Society of America Bulletin, v. 86, p. 377-389.

Langseth, M., et al, 1978, Near the Japan trench transects begin: Geotimes, v. 23, p. 22-26.

Ludwig, W. J., S. Den, and S. Murauchi, 1973, Seismic reflection measurements of southwest Japan margin: Journal of Geophysical Research, v. 78, p. 2508-2516.

Moore, G. F., and D. E. Karig, 1976, Development of sedimentary basins on the lower trench slope: Geology, v. 4, p. 693-697.

Moore, J. C., and D. E. Karig, 1976, Sedimentology, structural geology and tectonics of the Shikoku subduction zone: Geological Society of America Bulletin, v. 87, p. 1259-1268.

———, et al, 1979, Progressive accretion in the Middle America trench, southern Mexico: Nature, v. 281, p. 638-642.

Murauchi, S., and T. Asanuma, 1977, Seismic reflection profiles in the western Pacific, 1965-1974: Tokyo, University of Tokyo Press.

———, et al, 1968, Crustal structure of the Philippine sea: Journal of Geophysical Research, v. 73, p. 3143-3171.

Seno, T., 1977, The instantaneous rotation vector of the Philippine Sea plate relative to the Eurasian plate: Tectonophysics, v. 42, p. 209-226.

Shiono, K., 1974, Travel time analysis of relatively deep earthquakes in southwest Japan with special reference to underthrusting of the Philippine Sea plate: Osaka City

University, Journal of Geosciences, v. 18, p. 37-59.
———, 1977, Focal mechanisms of major earthquakes in southwest Japan and their tectonic significance: Journal of the Physics of the Earth, v. 25, p. 1-26.
Shipley, T. H., et al, 1979, Seismic-reflection evidence for the widespread occurrence of possible gas hydrate horizons on continental slopes and rises: AAPG Bulletin, v. 63, p. 2204-2213.
Tucholke, B. E., G. M. Bryan, and J. I. Ewing, 1977, Gas-hydrate horizons detected in seismic-profiler data from the western North Atlantic: AAPG Bulletin, v. 61, p. 698-707.
Walanabe, T., M. G. Langseth, and R. N. Anderson, 1977, Heat flow in back-arc basins of the western Pacific: American Geophysical Union, Maurice Ewing Series 1, p. 137-161.
Watkins, J. S., J. C. Moore, and Leg 66 Shipboard Party, 1981, Accretion underplating, subduction and tectonic evolution, middle America trench, southern Mexico — Results from DSDP leg 66, *in* Geology of continental margins: Oceanological Acta, v. 4 supplement, p. 214-224.
Yamazaki, F., T. Oida, and H. Aoki, 1980, Subduction of the Philippine plate in Tokai district: Seismological Society of Japan Annual Meeting, n. 2 (in Japanese).

Active Margin Processes: Field Observations in Southern Hispaniola

B. Biju-Duval
G. Bizon
A. Mascle
C. Muller
Institut Francais du Petrole
Rueil-Malmaison, France

Geophysical and geological investigations suggest processes of sedimentary accretion and evolution of a forearc basin in the Southern Hispaniola region. Several different sedimentological and structural domains are defined on multichannel seismic reflection lines offshore. Other domains, including pelagic abyssal plain, trench, accretionary wedge, and forearc basin, are identified in the onshore outcrops in Southern Hispaniola. Onland deposits of the pelagic and trench domains (Southern Peninsula and Enriquillo trough), located north of Beata Ridge, are considered slightly different from the offshore ones because of collision between the thick crusts of Beata Ridge and the Central Cordillera. Accretionary wedge and forearc basin domains (Sierra de Neiba, Sierra El Numero, San Cristobal basin) provide sedimentological and stratigraphical data concerning the evolution of the margin. Widespread distribution of Oligocene heterometric and polymicritic conglomerates indicate an "erosional crisis," which could date the beginning of collision processes to the west.

The Los Muertos Trench (Figure 1; Garrison et al, 1972; Matthews and Holcombe, 1976; Ladd and Watkins, 1979) marks the boundary between the obducting Greater Antilles island arc and the subducting Venezuela Basin crust to the south. Seismic data from offshore Puerto Rico and Hispaniola show that an accretionary wedge of offscraped oceanic sediments and possibly infolded and faulted slope apron sediments (Ladd and Watkins, 1979) lies north of the trench. A well-developed forearc basin lies north of the accretionary wedge. Overall, the trench-wedge-forearc basin sequence closely resembles island arc sequences observed around in the Caribbean (Chase and Bunce, 1969; Marlow et al, 1974; Westbrook, 1975, 1981; Biju-Duval et al, 1978, Mascle et al, 1979; Mascle, Lajat, and Nely, in press) and elsewhere (Seely, 1979; Moore et al, 1980).

The Los Muertos Trench terminates at about 71.5°W longitude where it abuts the Beata Ridge and onshore the Bahoruco Range (Figure 2). The tectonic elements (forearc basin accretionary wedge, and oceanic crust), extend onshore and outcrop in southwestern Hispaniola, thereby providing an unusual opportunity to study a relatively young tectonic sequence. Outcrops are excellent, better than those along the more typical Lesser Antilles active margin (a few square kilometers of exposure in Barbados and a few good exposures in southern Trinidad) and better than those in largely-covered northern Colombia, where the deformed offshore margin merges and passes to the western Andes.

Structural deformation, folding, reverse faulting, and overthrusting were described for many years in the southern part of the Hispaniola (Butterlin, 1960; Llinas, 1972; Arick, 1941, *in* Bowin, 1975; Michael and Millar, 1977; discussion in Lewis, 1980.) Deformation affects rocks from Cretaceous to Neogene, but tectonic interpretation of these data remained unclear until now: relationships between possible superimposed paleotectonics and application of plate tectonic concepts are still poorly understood.

New concepts of nappes and superimposed tectonics with a southward vergence in southern Hispaniola, recently proposed by Bourgois et al (1979a and 1979b), partly clarify the question. But such concepts need to be integrated into the Caribbean geological history and to be documented with more detailed stratigraphical and sedimentological observations because of the structural complexity of this area. Moreover, because of southern Hispaniola's location between a strike-slip extensional system to the west (Burke, Grippi, and Sengor, 1980) and a subduction-like sys-

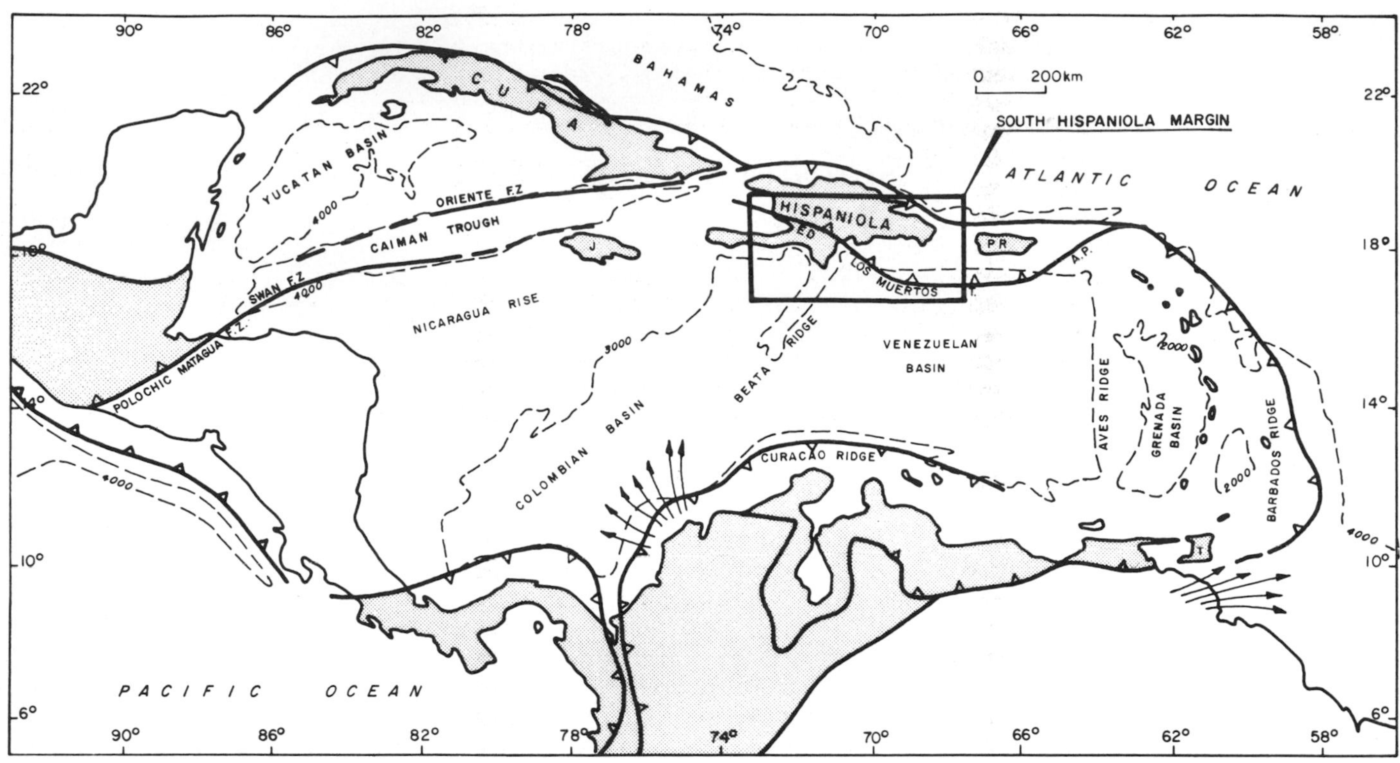

Figure 1 — Location map. J = Jamaïca; P.R. = Puerto-Rico; E.D. = Enriquillo Depression; A.P. = Anegada Passage; T. = Trinidad.

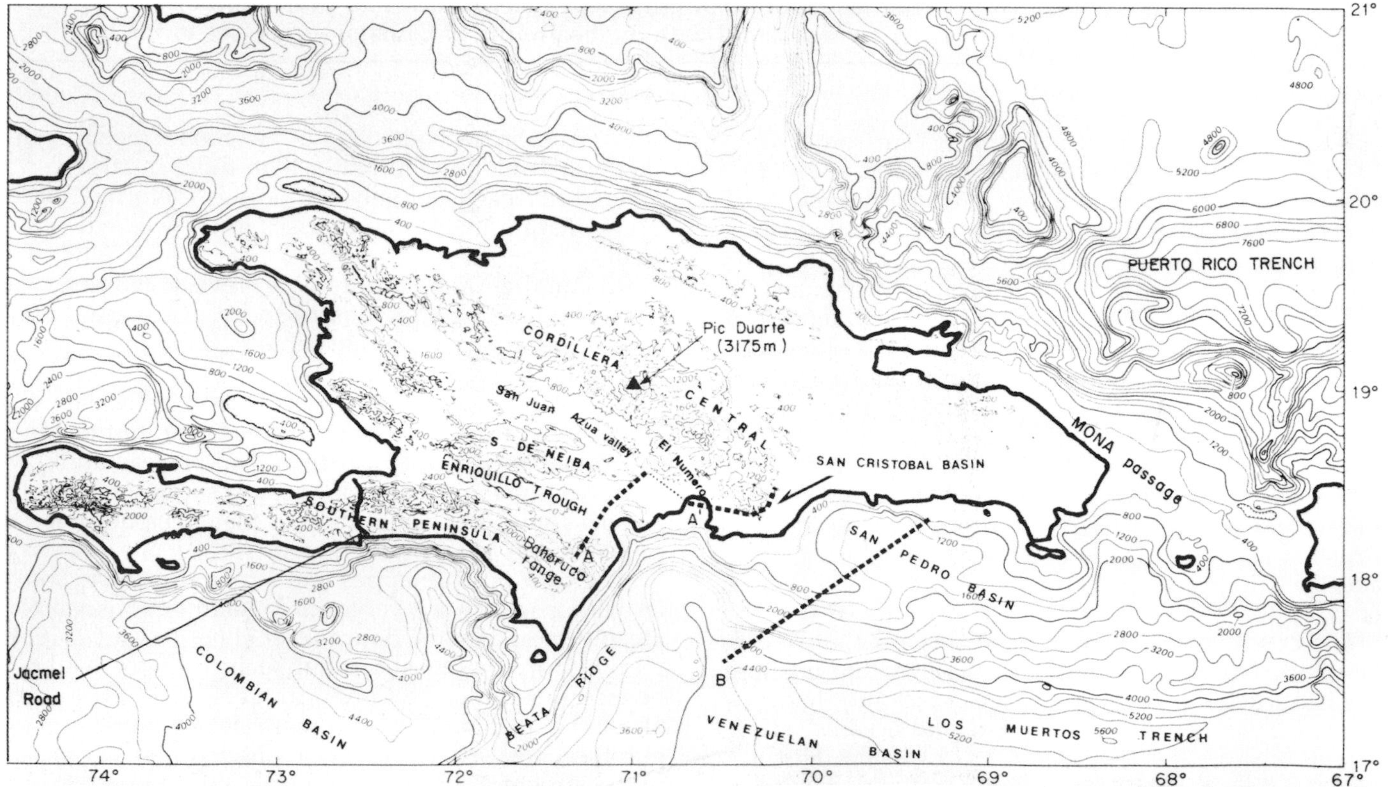

Figure 2 — Physiographic and bathymetric map of Hispaniolan area (modified from Bowin, 1975). Dashed lines: location of A and B geological cross sections in Figure 4.

tem to the east and immediately north of the composite Caribbean crust (Colombian Basin, Beata Ridge, Venezuelan Basin), various mechanisms (subduction, collision, large transcurrent horizontal motions) may have been dominant during parts of the geological history. Exposures of a complete stratigraphic sequence from Upper Cretaceous to Quaternary in southern Hispaniola permit investigation of the complete recorded geological history of this margin.

This paper presents results of field studies conducted in 1978, 1979, and 1980 south of Hispaniola, and proposes comparisons between land and marine data along the southern Hispaniolan margin.

FIELD STUDIES

Structure of the offshore margin is inferred from a depth-converted section of a CEPM multichannel seismic reflection line. Existing Geological maps of onshore southern Hispaniola are small scale maps (Butterlin, 1960; Zoppis de Serra, 1969). With the exception of recent, but geographically limited, works by Maurasse et al (1977), Maurasse (1979), and Mercier de Lepinay et al (1979) in Haiti, and by Llinas (1972) and Michael (1978) in the Dominican Republic, few detailed stratigraphic data or maps were available until recently (see the Ninth Caribbean Geological Conference, 1980). This paper reports results of sampling for stratigraphic and paleoenvironmental data, surveys of sediment thicknesses, and accumulation rate evaluations from the Southern Peninsula and Bahoruco Range to the San Pedro basin (Figure 2, 3). It also reports sampled sections and photogeologic interpretation done to determine lateral variations and general structural trends. The structural complexity of this area and the lack of detailed mapping limits us to a preliminary sketch of the onshore structure and comparison with the different offshore structural provinces. Nannofossils and foraminiferal assemblages provide a first set of age stratigraphic data which clarify some previous uncertainties in this area.

GEOLOGIC SETTING

Seismic reflection data suggest that the offshore margin south of Dominican Republic is an active margin with oceanic crust underthrusting the island arc. Figures 3 and 4 show the main elements from south to north. The Los Muertos Trench is a narrow and arcuate depression in which the reflecting horizons of the Venezuelan Basin (pelagic sediments and oceanic crust) deepen toward the north. In some places a flat

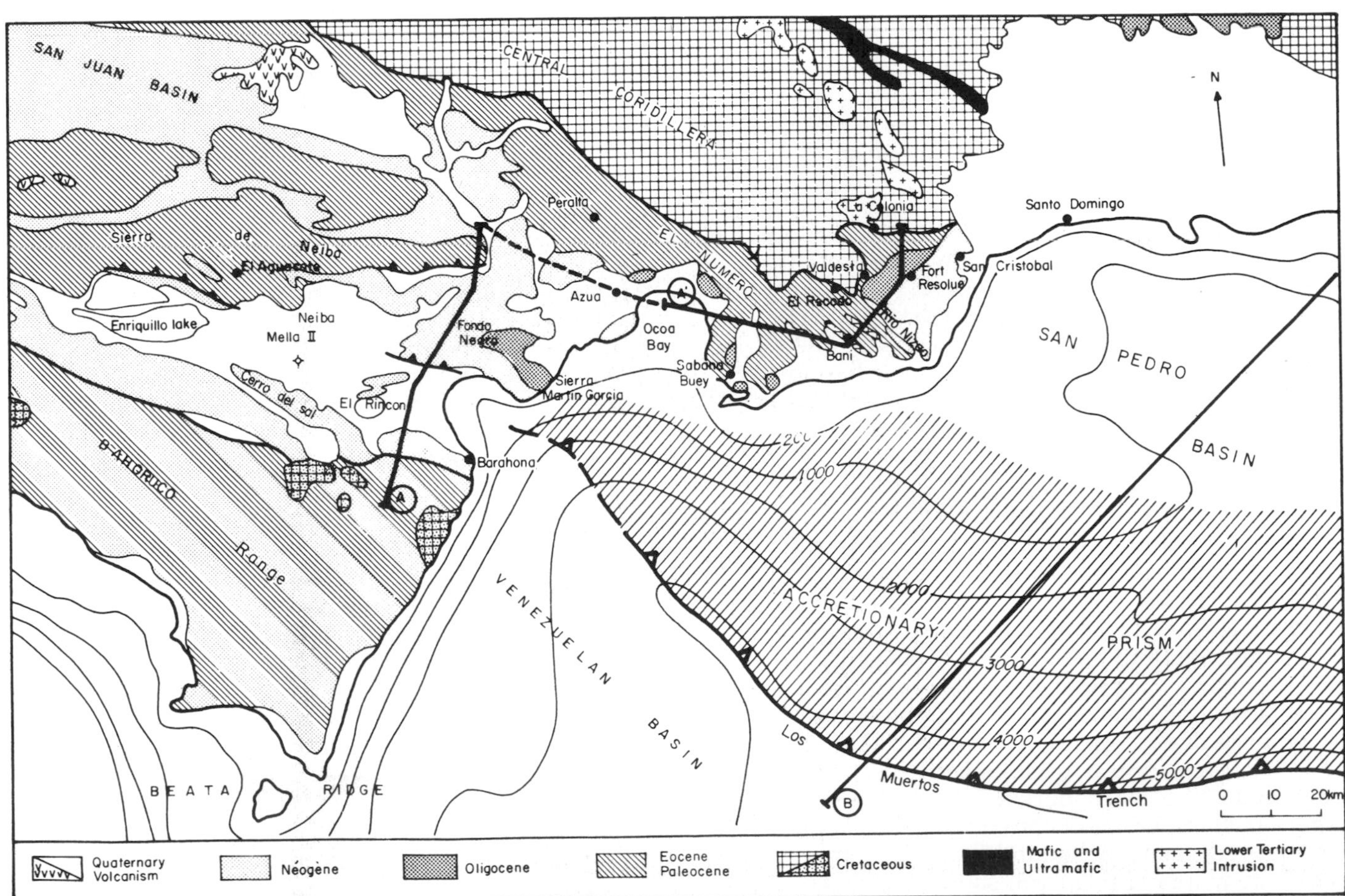

Figure 3 — Geological sketch map of Southern Hispaniolan margin.

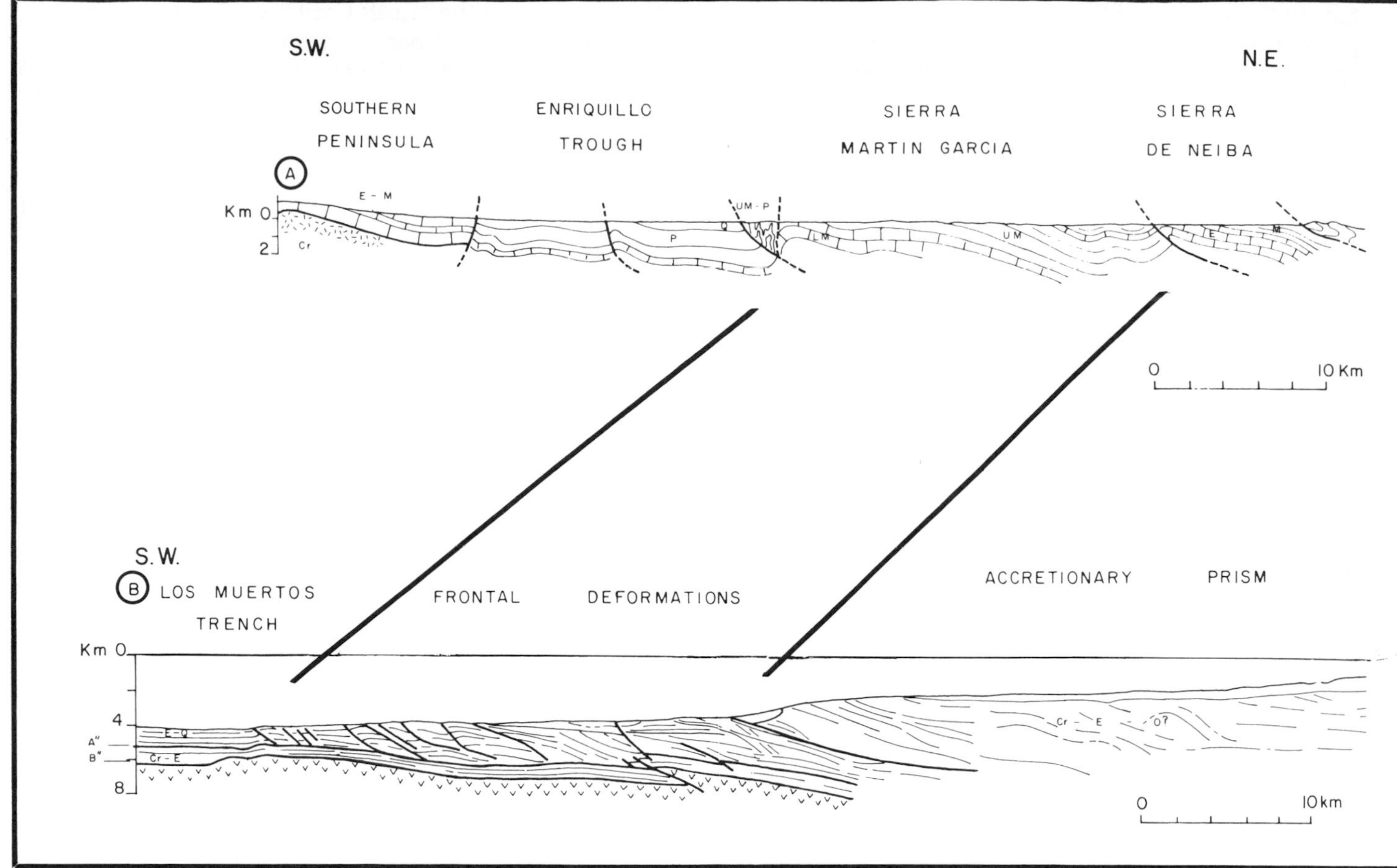

Figure 4 — Comparative geological cross-sections through onshore (A-A') and offshore (B) Southern Hispaniolan margin (location on Figure 2 and 3). Cr = Cretaceous; Pal = Paleocene; E = Eocene; O = Oligocene; M = Miocene; LM = Lower Miocene; UM = Upper Miocene; P = Pliocene; an, Q = Quaternary.

seabottom topography is due to local turbiditic infilling of Neogene to Recent age (Ladd and Watkins, 1979). The lower slope shows progressive deformation of the post lower Middle Eocene sediments. Shortening of the sedimentary cover is probably due to northward underthrusting of the oceanic basement and offscraping of abyssal plain sediments. The section is less clear upslope, where numerous landward dipping reflectors are observed below a thin undeformed hemipelagic sequence.

The San Pedro basin is evident on the upper slope (water depth: 1500 m). It rests on the tectonized wedge to the south, and is gently deformed. To the north, sedimentary layers progressively onlap Mesozoic basement. Field data indicate that the age of this sedimentary infilling is Oligocene to Recent.

This sequence (north of the Venezuelan Basin) of a trench, a tectonized wedge, and an upslope basin is that of a classical active margin assemblages (Seely, 1979). This structural pattern extends westward until terminated along the Bahoruco Range, north of Beata Ridge which appears to be an uplifted block of oceanic crust between Venezuelan and Colombian Basins.

Onshore mapping (Figure 3) shows a sequence similar to that observed offshore. The Enriquillo depression is filled by very thick, predominantly clastic sediments of Late Miocene to Quarternary age (Llinas, 1972; Bowin, 1975). Young folds and salt diapirs occur, especially on the southern edge of the depression. The west-east trend of this narrow trough is a recent feature oblique to the older tectonic trends, as shown on the geological map (Figure 3). Pelagic facies lie northwest of the present boundary. The northern edge of Enriquillo depression is a reverse fault system demonstrating recent shortening, which is well-documented in the Sierra de Neiba (Bourgois et al, 1979a). The Azua-Ocoa Basin has several small, elongated Neogene depressions extending from the Venezuelan Basin to the northwestern part of Haiti (San Juan Basin, Artibonite Basin). Lithologic and stratigraphic sequences from Late Cretaceous to Oligocene age, which outcrop between Azua-Ocoa and San Cristobal Basins, resemble the tectonized wedge of a former Greater Antilles active margin.

The thrust belt seen between Ocoa Bay and the Bani (Figure 3) underlies the Oligocene-Neogene San Cristobal Basin, which is the extreme end of the offshore San Pedro Basin. To the northeast, the older strata rest unconformably on the metamorphosed Cretaceous rocks (Bowin, 1966) whereas, to the southwest, they are partly incorporated in the deformed belt (described later).

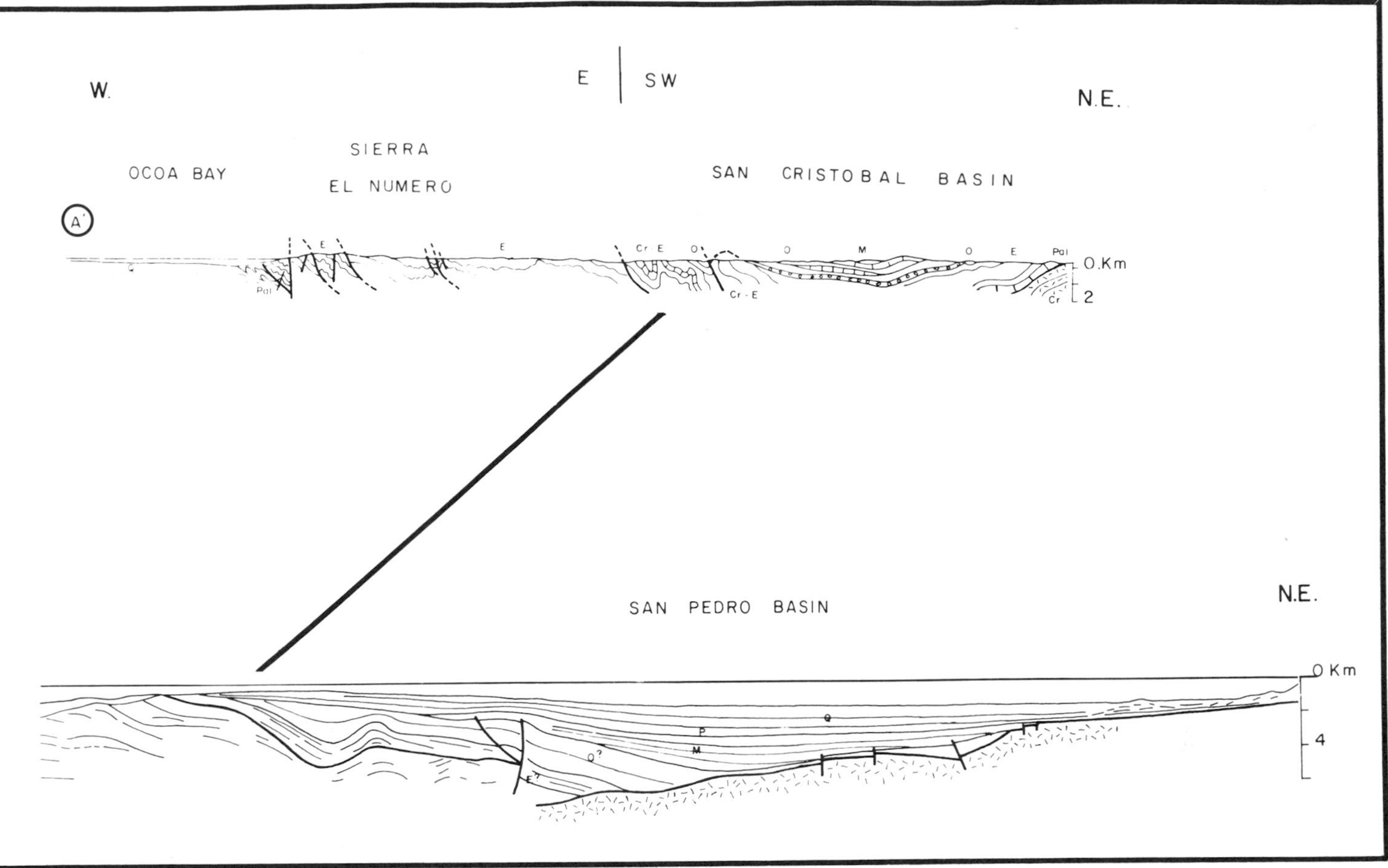

Structural and lithologic relationships suggest that onshore rocks to the west are equivalent to those observed offshore to the east. For example, we see from south to north a pelagic domain consisting of the Venezuelan Basin, Beata Ridge, and Bahoruco Range; a narrow elongated trough showing evidence of northward underthrusting (Los Muertos trench and Enriquillo depression); then a tectonized wedge with southward overthrusts (the continental slope of the Dominican Republic and southern folded edge of the Cordillera Central, including the Sierra de Neiba, and Sierra el Numero); and finally the San Pedro-San Cristobal forearc basin.

Below, we discuss in more detail the structural domains outlined above.

RESULTS

Pelagic Domain

The pelagic domain in the Bahoruco Range of the southern Dominican Republic consists of pelagic limestone of Late Cretaceous to Pliocene age, overlying volcanics and pyroclastics. Tectonic complexities have so far prevented measurement of a single, intact stratigraphic section, but a composite section can be constructed aided by recent cuts made for the Port-au-Prince to Jacmel Road in the Southern Peninsula of Haiti (Figure 5). These data suggest the following sequence.

The deepest stratigraphic levels consist of massive basalt flows, pillow lavas, dolerites, diabases, and diorites (Butterlin, 1960; Llinas, 1972; Maurasse, 1979). They are commonly associated with pelagic sedimentary rocks; this led to the comparison with the Caribbean oceanic crust, which may have been uplifted to form these land exposures (Butterlin, 1977; Maurasse et al, 1977).

The exact relationships between basalts and sedimentary rocks, the age of the older strata, and the origin of the flows are still under dicussion (see the review in Maurasse, 1980). However, along the road from Port-au-Prince to Jacmel, basement rocks consist of a complex assemblage of volcanic flows and tuffs, interbedded siliceous limestones, and chaotic melanges of siliceous limestone blocks in a volcaniclastic matrix (Figure 6). From foraminiferal determinations, their age is Senonian. Because of the scarcity of other good outcrops, the paleoenvironment is difficult to identify with precision but appears to be a relatively deep marine area with pelagic sedimentation and with volcanic seamounts providing lavas and volcaniclastic slope and deep basin deposits. Turbidites and debris-flow deposits have been observed in several localities. To explain the "chaotic" character of some outcrops, Mercier de Lepinay et al, (1979) proposed Late Cretaceous nappe emplacement. Despite possible older deformation, most of the reverse fault contacts affect Late Miocene strata and perhaps Pliocene sedimentary strata (Butterlin, 1960). A large part of the Senonian chaotic facies may be due to synsedimentary faulting, erosion, or transport by gravity sliding along slopes during the Late Cretaceous. Recent faulting may have locally accentuated the chaotic character. Butterlin (1960), Maurasse et al (1977) and Mercier de Lepinay et al, (1979) described older sedimentary rocks interbedded with the volcanic rocks, which demonstrate relative permanency of a marine volcanic environment since the Middle Cretaceous (Reeside, 1947). In our investigations, Early and Middle Cretaceous ages were only found in pebbles from a Maestrichtian conglomerate. New field investigations are necessary to clarify the initial age of this sequence.

Petrologic and geochemical studies of the volcanic rocks led Maurasse (1980) to propose the emplacement of the magmas in an environment "involving spreading related to a mantle plume activity behind the initial Greater Antilles Arc." Recent geochemical analyses of basaltic rocks we collected in Barahona area (Dominican Republic) show that they are anorogenic and non-alkaline tholeiites of possible abyssal origin (Girard, 1981).

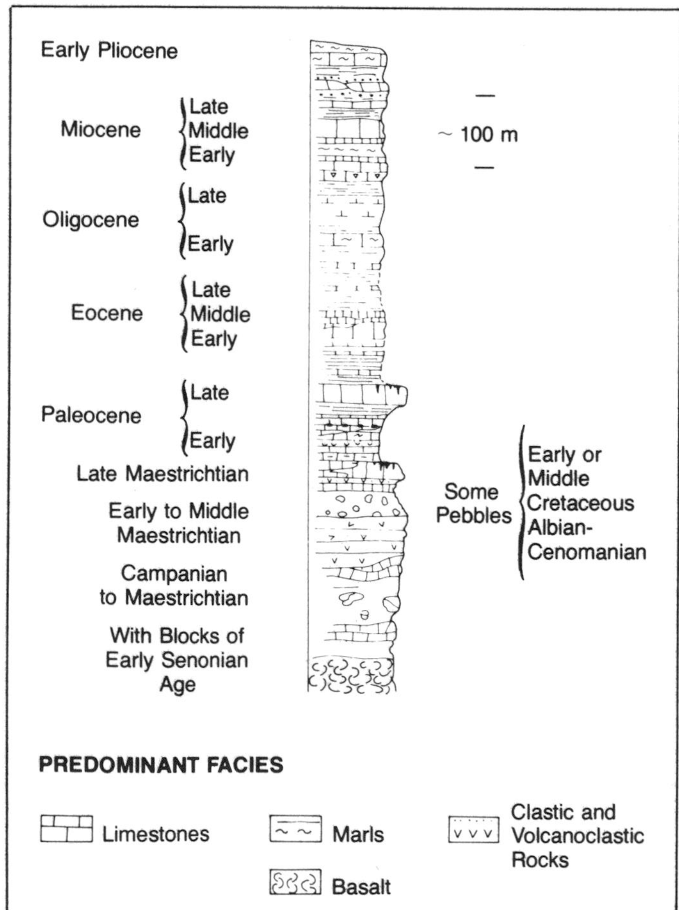

Figure 5 — Synthetic sedimentary column reconstructed from different sections along Jacmel Road, Southern Peninsula, Haiti.

No sharp boundary between the basement complex and the overlying sedimentary column was found because several volcanic post-Senonian events occurred in the Southern Peninsula. In the extreme western part of the peninsula, the Middle Maestrichtian Macaya limestones are overlain by thick basalt flows (see Butterlin, 1960). Near St-Louis, volcaniclastic influxes into the pelagic environment document active magmatism during Early Paleocene time; and in the Sierra de Neiba, thick volcanic flows associated with volcaniclastic and pelagic deposits, melanges associated with turbiditic deposits, and breccias indicate that similar environments persisted at least until middle Eocene time.

Several reconnaissance sections along rivers or roads provide additional data to the Jacmel road section presented on Figure 5. Data from these sections show that Maestrichtian to Pliocene deposits are generally pelagic carbonates: cherty limestones, siliceous limestones, radiolarian-rich limestones, marls, and chalks are encountered at different stratigraphic levels. Definition of depositional depths remains difficult. Most of these pelagic limestones accumulated above the carbonate compensation depth. Occurrence of calcareous turbidites with large reworked forams, terrigenous turbidites, slumping, chanelling, and erosional or nondepositional surfaces, favors a complex paleobathymetry with locally synsedimentary listric faults (Figure 7). Shallow environments are attested by algal or bioclastic limestones, which are observed at different stratigraphic levels (Paleocene, Oligocene, late Miocene) and reinforce the notion of highly variable sea-bottom topography through the Cenozoic (or Tertiary).

Sedimentation rates varied from place to place. High organic productivity and high rate of subsidence are necessary to explain the thick Maestrichtian radiolarian limestones of the Macaya area. The morphology of the sea bottom probably largely controlled hiatuses, compaction, or thick accumulations.

The Beata Ridge, immediately south of the Bahoruco Range, shows a complex morphology with scattered volcanic seamounts rising several hundred meters above the surrounding sea-bottom. The normal pelagic sedimentation, as in the surrounding Venezuelan and Colombian deep-sea basins, is locally influenced by slope deposition, erosion, and local influxes from the seamount sources (Figure 16 in Case, 1975). We thus interpret the paleoenvironment of the Southern Peninsula in a similar way, with occurrence of shallow seamount and ridges, slope aprons, and discontinuous deep sea depressions during the whole Tertiary depositional history. Emergence of the whole peninsula could be very recent, at least post-early Pliocene; the present high culminations (up to 2,600 m) and strong erosional processes should be Pliocene and Quaternary in age.

Figure 6 — Example of chaotic facies in Senonian series. Note dislocated pelagic limestone blocks in a tuffaceous volcanic matrix. A = general view; B = detail.

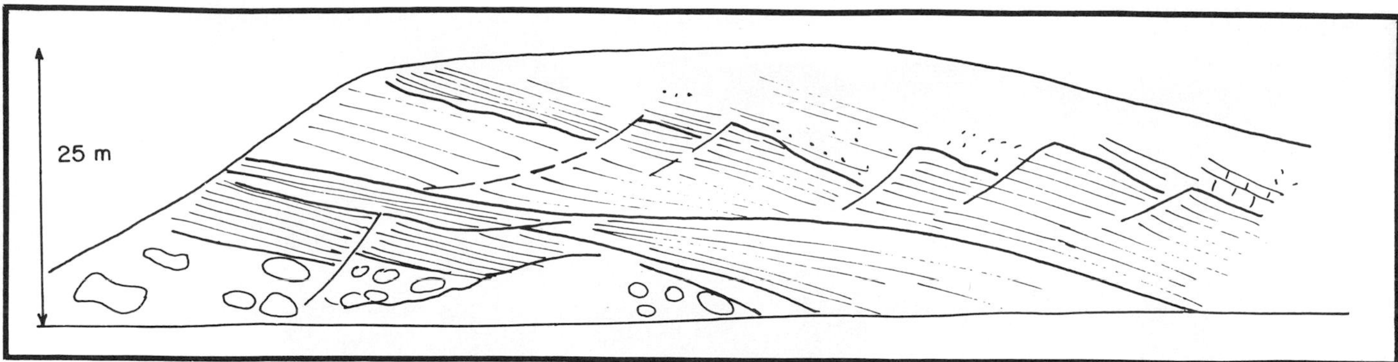

Figure 7 — Listric synsedimentary faults in late Miocene chalk and calcarenites. Jacmel Road (Haiti).

NEOGENE-QUATERNARY TROUGH

The Enriquillo trough is a flat-surfaced depression crossing Hispaniola from east to west. Its average altitude is close to sea level, but its lowest point was nearly 44 meters below sea level in 1950 (Lake Enriquillo). Eastward, elongated and narrow hills on the sides at Lakes Enriquillo and Rincon are anticlinal features with an elevation of a few tens of meters.

Oil-well data show high sedimentation and subsidence rates during late Neogene-Quaternary time. At the southern edge of the depression, a narrow anticline at Cerro del Sal has a core of massive salt and gypsum, overlain by a clastic shallow-water sequence which was dated as late Miocene by nannoplankton (NN 11 zone). It also contains a considerable amount of reworked Eocene to middle Miocene fauna. Ostracode assemblages (Van den Bold, 1975) suggest that the sequence ranges from upper Miocene to Pliocene. These observations indicate a restricted depositional environment immediately north of the pelagic domain exposed in the mountains to the south, and suggest a major paleogeographic change at the end of Miocene in this area.

To the northeast and near the seashore, evaporites are in tectonic contact with the Sierra Martin Garcia anticline. The core of the anticline was not explored, but along the road from Barahona to Azua (Fondo Negro Section, Figure 8) one can observe a lower sequence of bioclastic limestones, calcareous siltstones and sandstones, and marls dated as middle to late Miocene (NN 9 to NN 11 nannozones), which we consider outershelf or upperslope deposits. Also present is a thick turbiditic Late Miocene sequence (2,800 meters) including marls, calcareous siltstones, and sandstones deposited in open-marine conditions. Faunal assemblages, commonly poorly-preserved, and distal turbidites indicate deposition in a relatively deep basin, near the carbonate compensation depth. The upper part of this sequence, Late Miocene as determined from a poor pelagic microfauna, is progressively marked northward by infilling and shallowing; indicated by more-and-more proximal turbidites, coarser material, reworked shallow faunas (oysters, corals, gastropods), channelling, and conglomeratic foresets. Uppermost in the section is a shallow-marine to brackish-deltaic sequence with the sandy bioclastic and coral limestones, shell beds, marls, calcareous sands, and, at the top, thick conglomerates. Late Miocene nannofossils (NN 11) are found in a few marl layers. These sequences are folded (Figure 4) and overthrust by Eocene-Miocene pelagic strata of the Sierra de Neiba. Thus, the northern part of the Enriquillo depression was successively marked by: initiation to the north of a deep and narrow trough during middle-late Miocene time; high rates of sedimentation and subsidence, followed by infilling of the trough with reworking of shallow material along slopes, and establishment of shallow-marine to subcontinental conditions during late Miocene (early Pliocene?); and, folding and overthrusting linked with crustal shortening simultaneously with rapid subsidence of the Enriquillo depression. Deformation extended over a large area including the southern edge of the trough and the Southern Peninsula.

ACCRETIONARY COMPLEX

As mentioned above, the area between the Enriquillo depression and the high mountains of the Cordillera Central is characterized by numerous folded and highly tectonized faulted hills separated by elongated Neogene-Quaternary depressions.

The Sierra de Neiba anticlinal trend either overthrusts the folded late Miocene clastic sequence (Fondo Negro section) or is directly in tectonic contact with the Pliocene-Quaternary fill of the Enriquillo depression. There, along the new road from Neiba to El Aguacate, one finds exposures of Eocene to middle-late Miocene pelagic facies: marls, cherty and chalky limestones (including thin turbidites with large forams), and sporadic shallow-water facies. The sequence resembles that described in the Southern Peninsula, and it probably belonged to the same depositional area from Eocene to Middle Miocene time, before the initiation of the Enriquillo trough. Intense folding (Figure 11), low angle reverse faults, and southward vergence of complicated slices of allochtonous material are observed (Bourgois et al 1979a). This tectonism is at least of Late Miocene age because pelagic marls of this age (NN 10 nannozone) are involved in the deformation. Regional observations suggest that the major deformation is early Pliocene or younger.

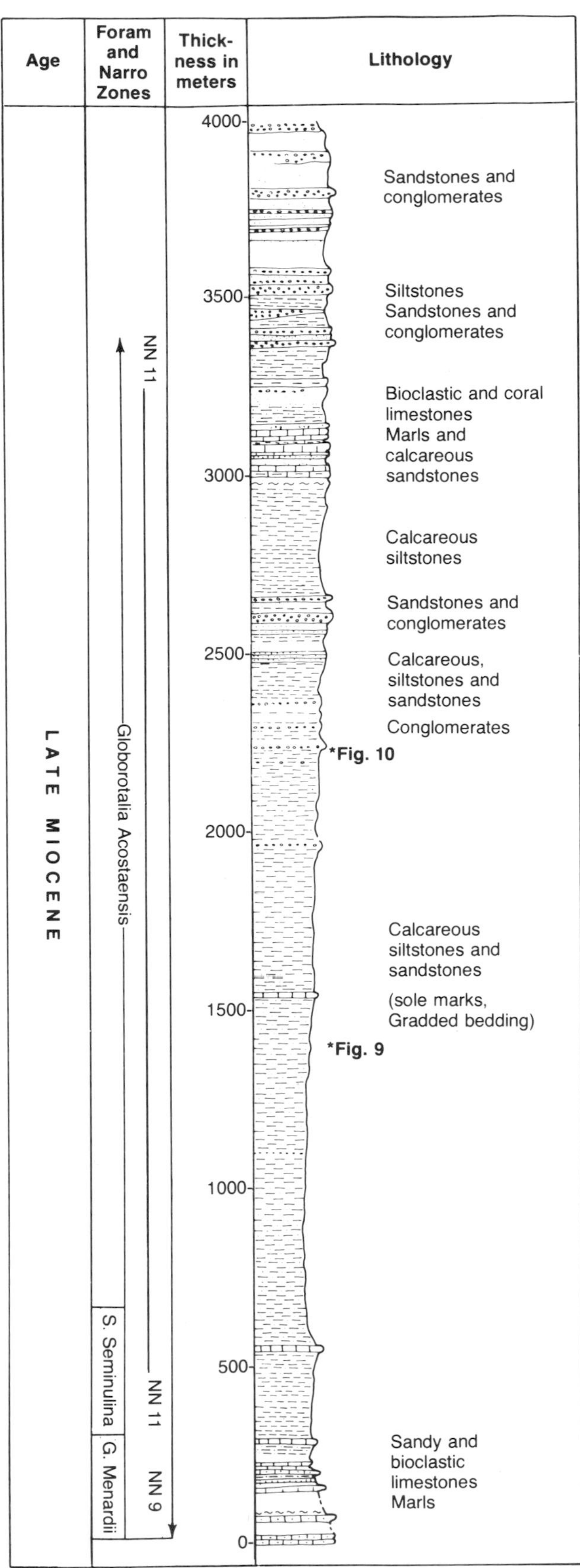

Figure 8 — Fondo Negro section. Thickness calculated from topographic and geological surveys along the road and the Yaque del Sur river; ages are from more 100 samples.

Figure 9 — Thin calcareous sandy turbidites of Late Miocene age along the Fondo Negro road (Figure 8).

Figure 10 — Coarse-grained turbidites of Late Miocene age along the Fondo Negro road (Figure 8).

This Sierra de Neiba belt consists of several overthrust structures overlain towards the north by the apparently isoclinally folded Neogene fill of the San Juan-Azua-Ocoa Basin, where geophysical exploration shows synsedimentary deformation and Miocene reverse faulting (Nemec, 1980).

Eastward in the Azua-Ocoa Basin, sedimentary and structural relationships between the limited outcrops (large anticlines near the city of Azua, northern edge of the basin) were not studied. The different facies (bioclastic, sandy, reefoidal limestones, pelagic limestones, and conglomerates), with ages from Paleocene to lower Miocene, along with different tectonic features (faulted contacts, overthrusts) need to be explored to define the evolution and tectonic style of this mainly Quaternary infilled depression.

In the northeastern part of the study area, a tectonized wedge of Late Cretaceous-Paleogene strata forms the southern flank of the Central Cordillera. It is particularly well-exposed along the new road from Azua to Bani (Figure 3). Deformation is intense with numerous abnormal contacts. We mapped the major elements on photogeological interpretation (Figure 12) and supplemented this with detailed stratigraphic sampling. Stratigraphic inversions and northwest to southeast trending thrust faults were documented. Large olistostromes and numerous transverse and secondary normal faults complicate the general pattern.

In the Sierra El Numero, one observes the following lithologic sequences (Figure 13): lower Paleocene to upper Eocene terrigenous turbiditic facies which constitute the dominant facies; and, a second facies consisting of a condensed association of pelagic siliceous limestones, radiolarites and laminated marls. Thick Paleocene is constituted by marls and fine to coarse grained sandstones, with abundant volcanic material. Slumping and turbidites suggest deposition along a paleoslope. Pelagic deposits of late Paleocene to Middle Eocene ages are marked by occurrence of thin calcareous and sandy turbidites; they outcrop either as continuous beds in anticlinal cores, generally close to thrust faults, or as highly deformed tectonic sheets (Figure 14) imbricated in the clastic series. The thick Middle-Late Eocene clastic turbiditic sequence includes a chaotic zone interpreted as an olistostrome with exotic rocks of different facies and ages and sizes up to tens of meters (Figure 15). Pebbles of metamorphic origin indicate erosion of the Mesozoic central cordillera. Reworking of shallow-water Eocene facies in the matrix indicates erosion of the upper slope edge and accumulation of hemipelagic sediments on an unstable slope. Eocene pelagic limestones and radiolarites imply rapid uplift and erosion of former pelagic sediments. Sections along the main road and along rivers also contributed data used in Figure 12. There, we found Late Cretaceous sediments in the tectonized wedge. These sediments contain terrigenous turbiditic pelagic facies of Late Maestrichtian age (road from Bani to Valdesia); breccias, volcaniclastics and pyroclastics of Campanian to Maestrichtian age along the Bani River (El Recodo Road); and, boulders and conglomerates in a Maestrichtian matrix west of Bani.

Figure 11 — Highly-folded, finely-bedded pelagic Eocene limestones in the Sierra de Neiba.

Lower Paleogene sequences in tectonic units differ in that thick volcaniclastics of Sierra El Numero are unknown along Valdesia Road where lower-middle Paleocene consists of thin pelagic limestones on top of Maestrichtian sequences. The lower Eocene consists of thick siliceous limestones east of Bani, but is missing in the Valdesia Road section where middle Eocene coarse sands rest directly on the middle Paleocene.

The mid-late Eocene turbiditic sequence varies in thickness. Included chaotic facies, interpreted as olistostromes, occur in several units outcropping between Ocoa River and the city of Bani. The "Eocene a blocs" from Bourgois et al (1979b) can correspond to our chaotic facies, but probably includes tectonized slices of different ages.

Lower to middle Oligocene turbiditic sequences are observed in several localities: in places in stratigraphic continuity with upper Eocene turbidites (Valdesia Road), or disconformably on the Cretaceous (Valdesia Road) or Eocene rocks. South of Sierra El Numero such mid-Oligocene turbidites include conglomerates and exotic blocks (Sabana Bay section, Figure 13).

From this area which is linked to the deformed wedge observed offshore on the seismic records, we infer that pelagic sequences from a relatively deep and quiet environment were offscraped as slices or perhaps nappes at the southern part of the wedge (Sierra de Neiba, parts of the inner belt). Further, slopes and trench-derived sediments accumulated between Late Cretaceous and Oligocene time were accreted in tectonic units of the inner part of the prism. Middle to Late Eocene shortening is documented by melanges, including deep and shallow facies and material derived from the Central Cordillera. Erosion of inland mountains, slope disposition, rapid lateral variations,

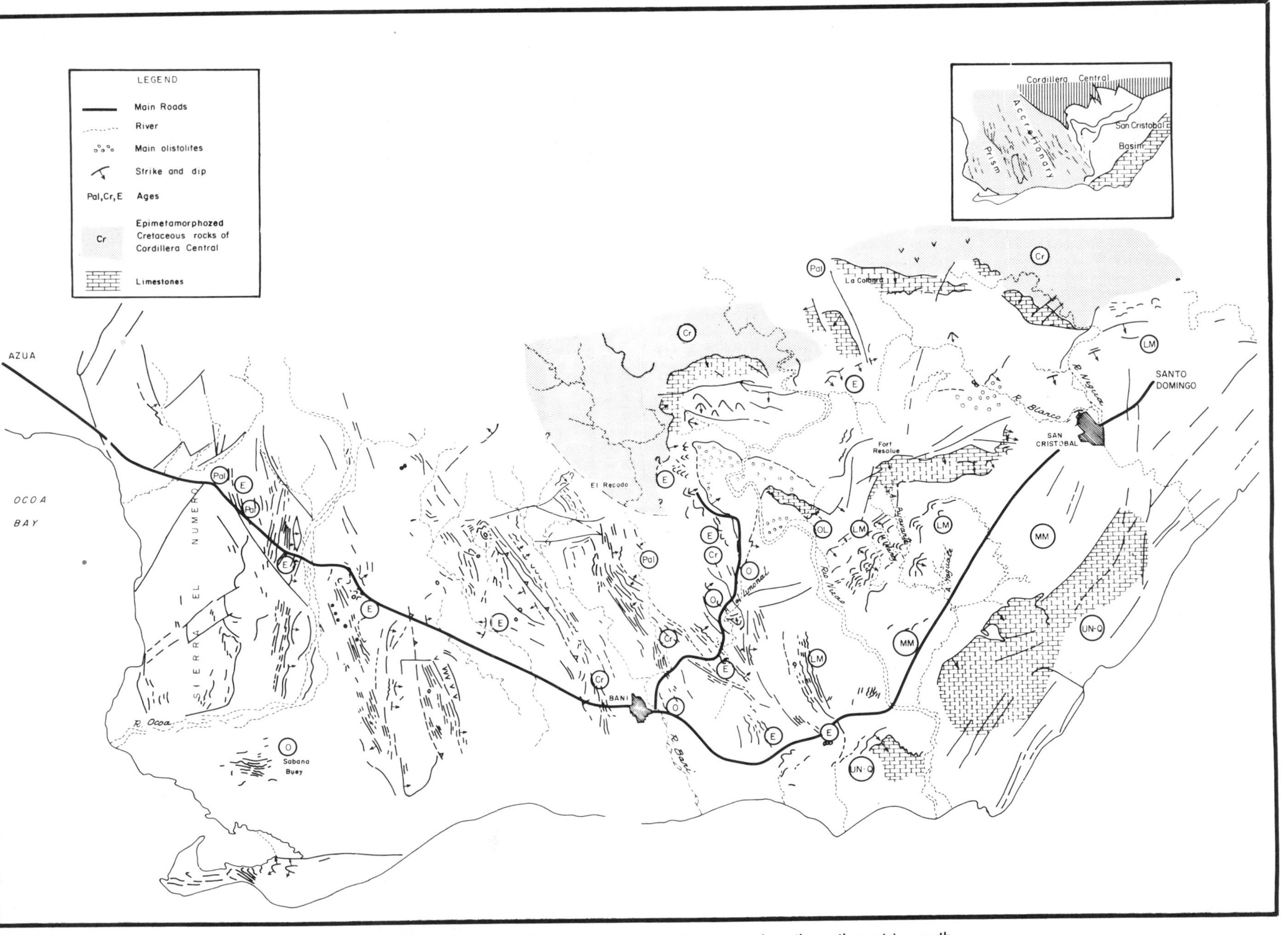

Figure 12 — Photogeological sketchmap of the Sierra El Numero/San Cristobal area. Note the general north-northwest to south-southeast trend of isoclinaly deformed Paleogene series and the tectonic repetition of olistostromes in the accretionary prism, on the left. The right side shows the changes of facies in the Oligo-Miocene gently deformed San Cristobal Basin. Cr = Cretaceous; Pal = Paleocene; E = Eocene; O = Oligocene; LM = Lower Miocene; MM = Middle Miocene; UN-Q = Upper Neogene-Quaternary.

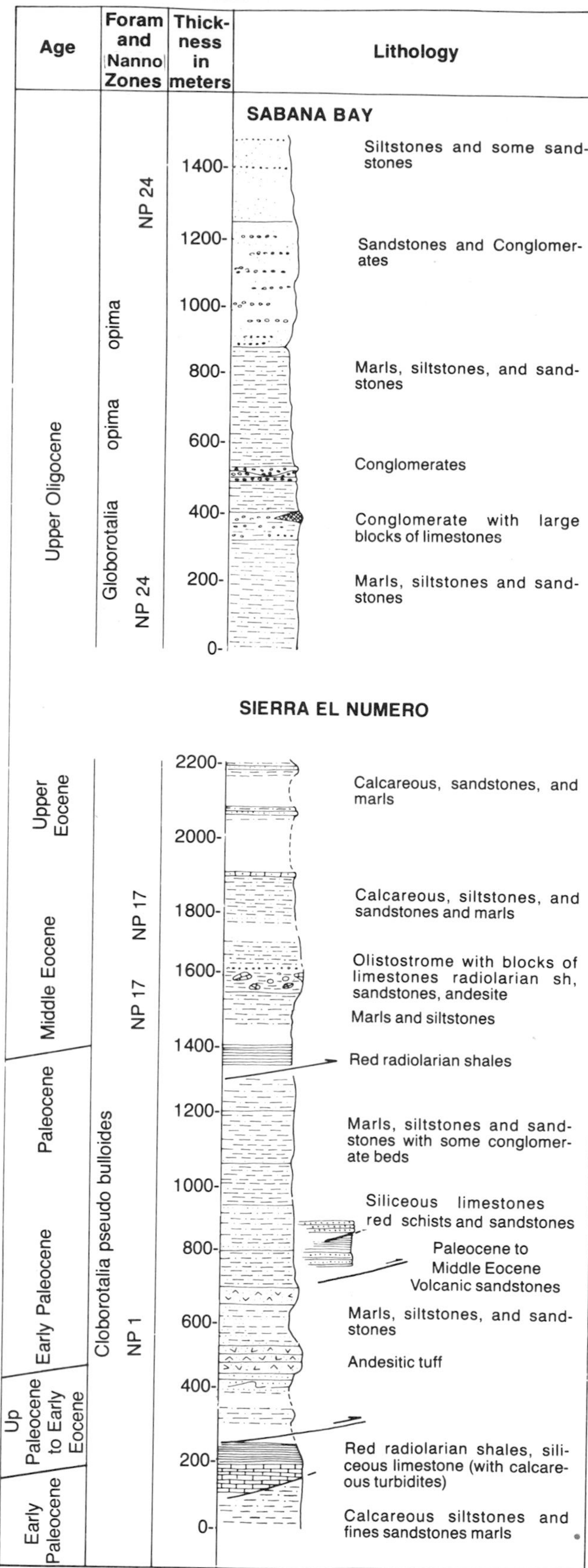

Figure 13 — Sierra El Numero-Sabana Bay sections. Thickness uncertain because of faults.

Figure 14 — Highly-folded pelagic limestones of Late Paleocene-Early Eocene age in the Sierra El Numero. Limestones consist of alternating pelagic foram-rich and radiolarian-rich limestones, siliceous limestones, fine calcarenitic, or sandy turbidites.

Figure 15 — Olistolites in a fine-grained clastic Late Middle Eocene sequence. Pebbles and blocks range in age from Cretaceous to Middle Eocene and include metamorphics, volcanics, radiolarian shales, shallow-water limestones, and pelagic siliceous limestones.

and unconformities indicate that offscraping and accretionary processes took place during this time. Conglomerates, coarse sandstones, and marls of Late to Middle Eocene age (observed in fault contact with Late Paleocene-Early Eocene pelagic limestones, Peralta Road section) and thick Late Oligocene conglomerates south of Sierra El Numero (Sabana Bay section) are interpreted as deposits in slope basins. They formed at the same time as the formation of the accretionary prism while subduction was still active.

To the south, Eocene-Middle Miocene pelagic rocks of Sierra de Neiba and Late Neogene slope-trench deposits of Fondo Negro section were subsequently deformed and overthrusted the Enriquillo depression.

FOREARC SAN CRISTOBAL BASIN

Tertiary sediments of the San Cristobal Basin form a broad syncline whose structure and infilling is asymmetrical (Figure 12, 16, 17). This basin appears to be

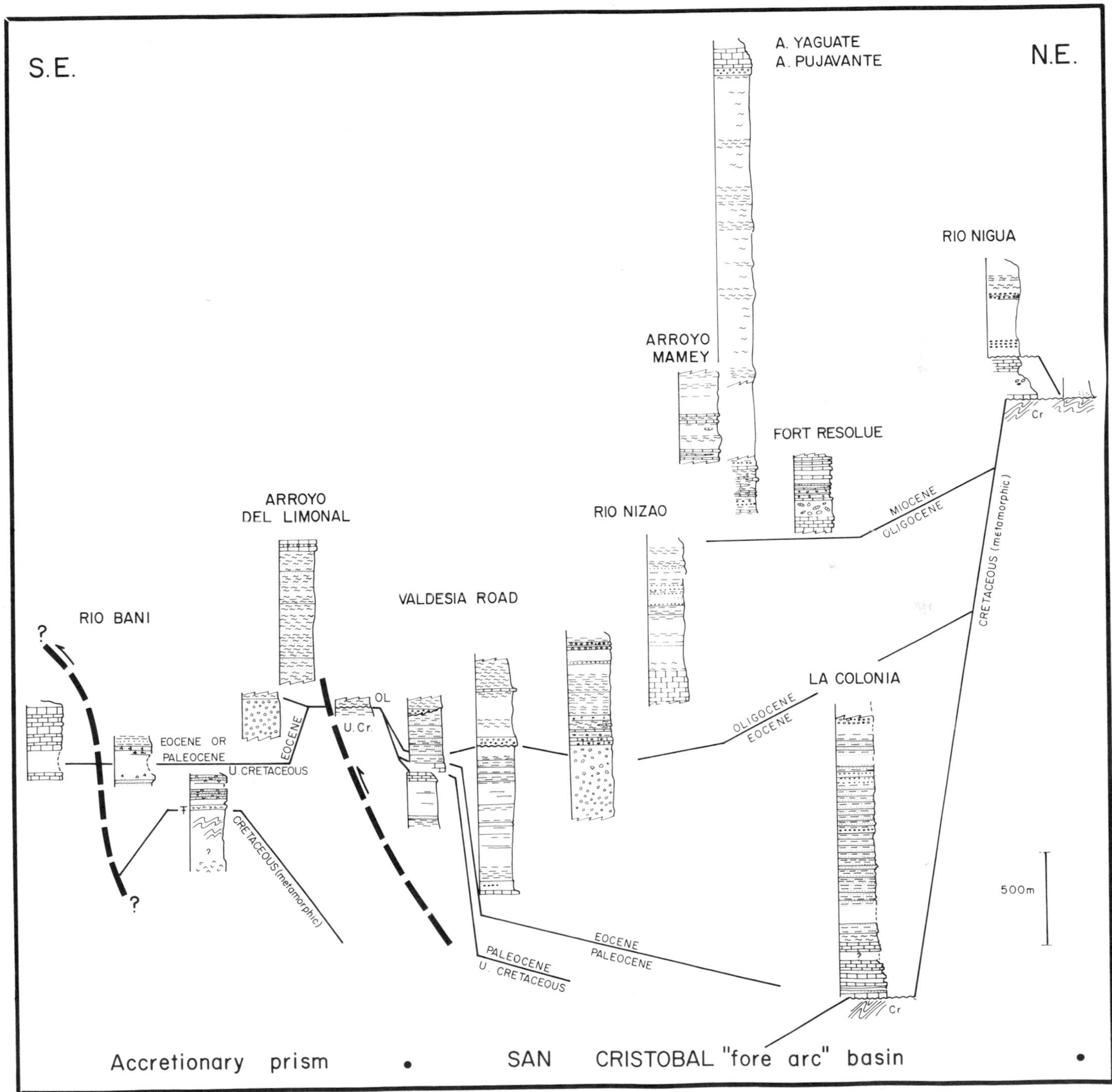

Figure 16 — Stratigraphic sections of the San Cristobal Basin. (location on Figure 12). Note differences between the northeastern and southwestern edges of the basin: to the northeast, Tertiary sequences onlap the metamorphosed Cretaceous basement; to the southwest, lateral changes of facies, unconformities, and thickness variations indicate the permanency of tectonic unstability (diagrammatic presentation on Figure 17).

the landward extension of the marine San Pedro Basin. In the northeast (Figure 16), a complete sequence of Paleocene to Middle Miocene overlies metamorphic rocks of Cordillera Central. To the west (Nizao river; Valdesia road sections, Figure 16), Oligocene rocks unconformably overly or are included within deformed sediments of the inner part of the prism. Unconformities and rapid changes of facies throughout the sedimentary sequence indicate progressive deformation of the basin.

Massive algal limestones unconformably overly metavolcaniclastic Cretaceous sediments (Tireo formation; Bowin, 1966). Bioclasts of algae, corals, bryozoans, gastropods, echinoids, and large forams (discocyclinidae, miliolidae), with benthic and planktic small forams, collectively indicate an outer shelf environment in which volcanic debris accumulated during upper Paleocene or Lower Eocene. Early to Middle Eocene rocks consist to several meters of sandy bioclastic limestones, followed by laminated pelagic limestones and marls with only planktic forams and then nanno-foram assemblages.

Clastic sediments first appear in basal upper Eocene. A thick sequence of turbidites with poorly preserved forams, radiolaria, and few nannofossils indicate deepening of the basin with terrigenous influx. Isolated pebbles, lenses of conglomerates, and interbedded coarse conglomeratic layers show reworking of Paleocene and Eocene rocks rapidly redeposited along slope or in the basin. The turbidite sequence is overlain by thick conglomerates as yet undated, but probably of Late Eocene-Lower Oligocene, as seen in the Nizao section.

To the east, Miocene rocks onlap the conglomerate sequence and north of San Cristobal, the Cretaceous meta-volcanics. The Miocene facies vary from algal limestones, to bioclastic and sandy limestones, marls, siltstones, and conglomerates, indicating generally a shallow-marine environment (San Cristobal section, Figure 16). These rocks contain Early to Middle Miocene (NN 1 to NN 6 nannozones) fossils with abundant Oligocene reworked fossil assemblages.

The uppermost part of the sequence consists of conglomerates and reefoidal limestones with ages ranging from Late Miocene to Quaternary.

In the southwest, immediately west of Nizao River, the Paleocene-Eocene sequence becomes highly tectonized with numerous lateral changes. In the same area we observe changes of facies in the Oligocene series. Near the Valdesia dam, the lower Oligocene (uppermost part of Eocene?) consists of several hundred meters of polymictitic coarse conglomerates. Boulders and pebbles of metamorphic rocks and Eocene limestones accumulated rapidly. Thicknesses vary

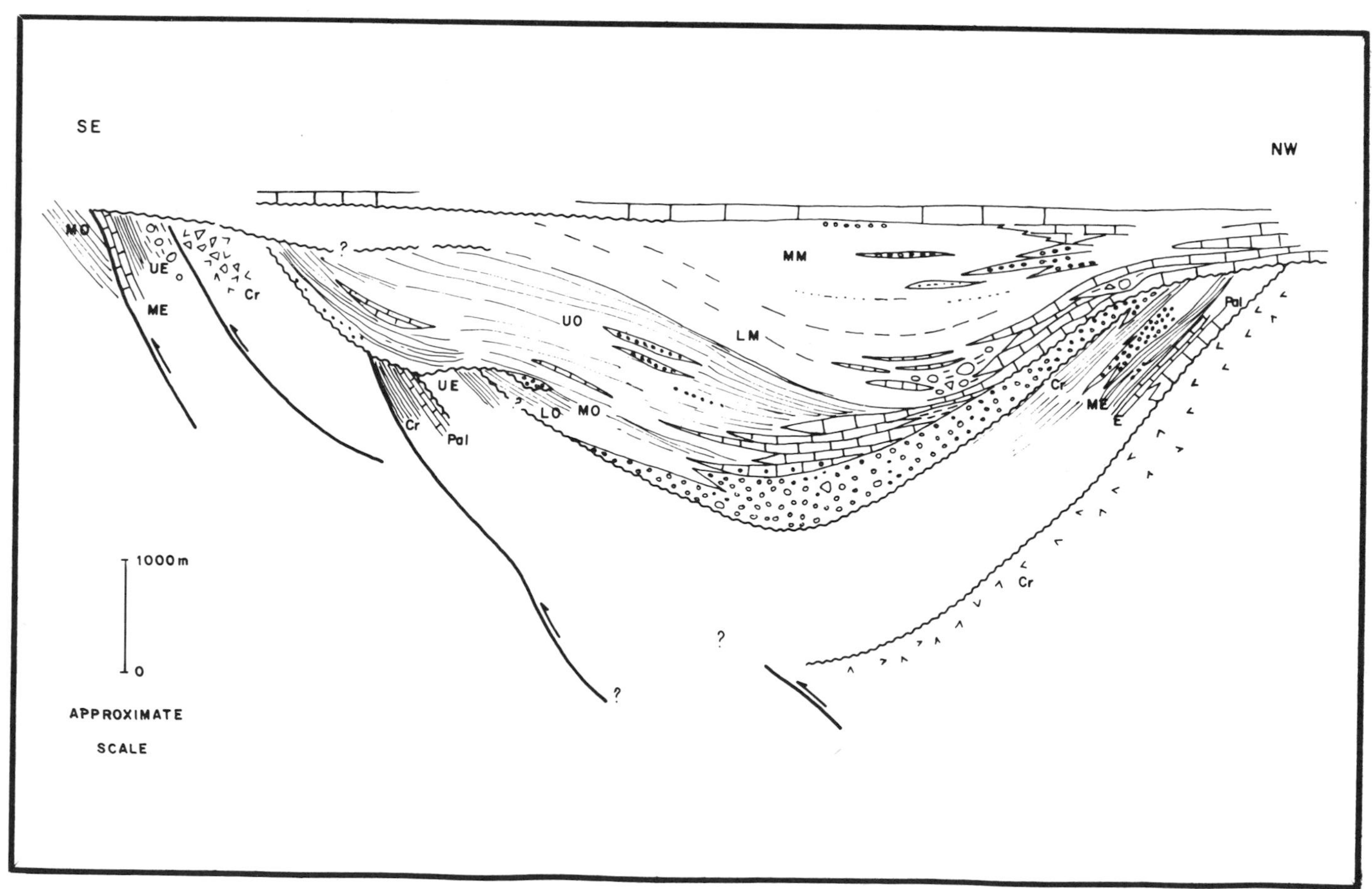

Figure 17 — Diagram illustrating repartition of facies and unconformities in the San Cristobal basin. Note the permanency of the northeastern hinge line where shallow-water or shelf facies predominates; the rapid change to basinal facies to the southwest, especially during Oligocene; and, the migration to the northeast of the Miocene depocenter.

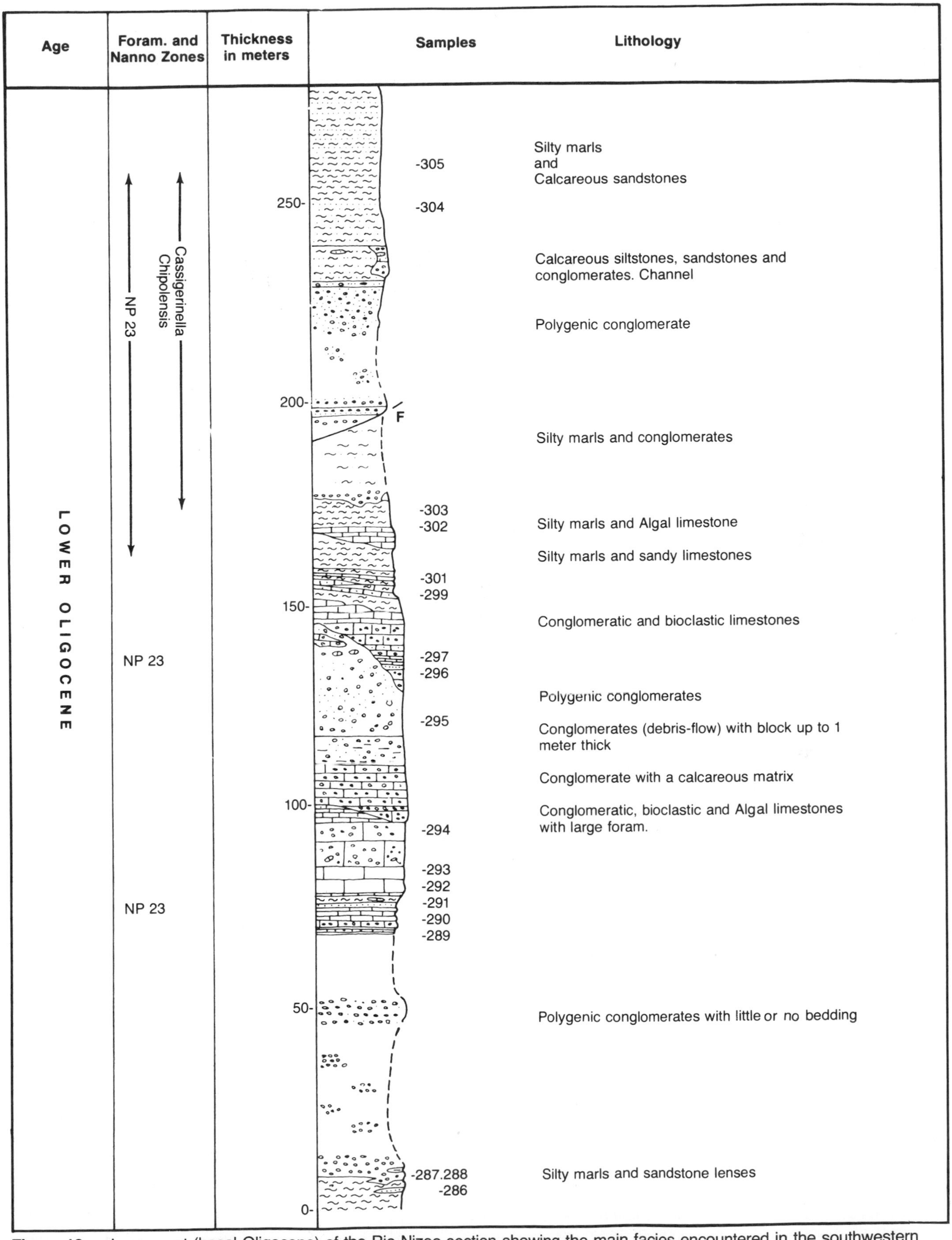

Age	Foram. and Nanno Zones	Thickness in meters	Samples	Lithology
LOWER OLIGOCENE	NP 23; Cassigerinella Chipolensis		-305	Silty marls and Calcareous sandstones
		250-	-304	
				Calcareous siltstones, sandstones and conglomerates. Channel
				Polygenic conglomerate
		200-	F	Silty marls and conglomerates
			-303 -302	Silty marls and Algal limestone
				Silty marls and sandy limestones
			-301 -299	
		150-		Conglomeratic and bioclastic limestones
	NP 23		-297 -296	
				Polygenic conglomerates
			-295	Conglomerates (debris-flow) with block up to 1 meter thick
				Conglomerate with a calcareous matrix
		100-		Conglomeratic, bioclastic and Algal limestones with large foram.
			-294	
	NP 23		-293 -292 -291 -290 -289	
		50-		Polygenic conglomerates with little or no bedding
			-287.288 -286	Silty marls and sandstone lenses
		0-		

Figure 18 — Lower part (basal Oligocene) of the Rio Nizao section showing the main facies encountered in the southwestern edge of the basin. Poorly-bedded and unsorted conglomerates and bioclastic-pebbly limestones, numerous occurrences of channelling, and erosional surfaces are interpreted as evidences of slope deposition.

greatly and the conglomerates pinch out a few kilometers west of Nizao River. Along the river they are overlain by clastic sequences (Figure 18) ranging from shallow-water bioclastic limestones to fine-grained, marly siltstones. Most facies are fine-grained and terrigenous with lenses of conglomerates. Graded bedding, channeling, large foreset systems, debris flows (Figure 19) indicate deposition along a slope or an unstable area. Conglomeratic, bioclastic, and algal limestones with reworked corals and large forams (Figure 12) appear locally. These sediments are interpreted as being deposited as part of an upper slope apron. Great thicknesses of these middle to upper Oligocene sediments implies high rates of sedimentation and subsidence.

Synsedimentary deformation is found in the west. Along the Valdesia Road, we observed continuous turbidite deposition from Late Eocene to Early Oligocene, followed by an internal disconformity in the Middle Oligocene. Locally, we observed microbreccia with large forams (Figure 20). The upper Oligocene overlies older, folded sediments (Cretaceous to Eocene) in angular unconformity. Slumping, channelling with coarse infilling and reworking of material (Figure 21) are common. Clastic middle Oligocene sediments near Bani are overthrust by Eocene pelagic limestone. Southward, near the seashore, Eocene and Oligocene sediments rest unconformably on accretive Cretaceous, Eocene and Oligocene sediments.

Figure 19 — Two thick beds of heterometric polymicritic conglomerates inside a clastic sequence, interpreted as debris flow deposits downslope of the Late Oligocene basin (Rio Nizao section).

Lower Miocene rocks are well-exposed between San Cristobal and the Nizao River. In the Fort Resolue area, thick coral, algal, and bioclastic limestones of Aquitanian age seem to conformably overly older deposits. As schematically shown in the photogeological sketchmap (Figure 12) and on the diagram of Figure 17, these rocks pass in to breccias with large boulders and blocks derived from erosion of calcareous banks and resedimentation of calcareous beds and coral debris along with terrigenous influx along channels and slopes (Figure 22). Limestone was deposited until the end of Early Miocene when about 2000 m of sandy siltstone and marl accumulated in an open-marine environment.

High sedimentation rates followed by diminution of subsidence rates may explain the emergence of the basin after Late Miocene. Conglomerates and uplifted coral limestones, which fringe the present coast line, prelude the emergence of the San Cristobal basin. Marine sedimentation continued in the marine San Pedro basin.

In summary, sediments of the San Cristobal basin landward of the accretionary prism document the Oligocene and later evolution of the inner part of the prism. Widening and deepening of the basin is deduced from facies distribution. Quieter conditions in Early and Middle Miocene may relate either to migration of tectonic deformation to the southwest, or to a temporary cessation of the subduction and blanketing of the accretionary prism. Emergence in Late Miocene may result from processes occurring in the west when Beata Ridge collided the prism.

CONCLUSIONS

Geophysical investigations of active margins provided information about their geometry and structure (Seely, 1979). However, geophysical data do not define the sedimentary environments or the tectonic details. Moreover, results of attempts to drill and core tectonized wedges are limited (Hussong, Uyeda et al, 1980; Von Huene, et al, 1980; Von Huene, Aubouin et al, 1980; Watkins et al, 1981). The tectonic style of deformation and sedimentary facies distribution within prisms remains unclear. In several places, active margins are uplifted and emerge on islands or continents where more direct field investigation is undertaken (Moore et al, 1980). That is the case in southern Hispaniola where geological studies shed new light on the sedimentary and tectonic processes along a convergent margin.

Emergence of deep offshore structures results from unusual tectonic events. For example, Ravenne et al (1977) and Mascle et al (1977) describe how heterogeneities in the subducting plate explain segmentation of active margins. Failed subsubduction of thick crust (not necessarily of continental type) may provoke collision-like structures and uplift. We propose that collision and partial subduction of thick Beata Ridge crust, beneath thin Central Cordillera crust, uplifted the margin of southwestern Hispaniola. To the east, subduction of thin Venezuelan Basin crust beneath thinner island crust continued. If this is true, studies of south Hispaniola can provide useful data about collision processes and their relationship to mountain building.

Our preliminary investigations show two main tectonic events: A Middle to Late Eocene event followed by emergence and erosion of the Central Cordillera, as attested by the widespread distribution of polymicritic coarse breccias and conglomerates on both sides of the range (Palmer, 1979); and the second event, Pliocene to Recent in age, uplifted the southern tectonic features.

Field studies between Enriquillo trough and San Cristobal basin provide useful sedimentological, stratigraphical, and structural data which we interpret to indicate offscraping of pelagic sediments, initially accumulated in a relatively deep environment (abyssal plain and related seamounts). These sediments are now observed in the inner prism (Sierra El Numero, Bani area) and on its southern edge (Sierra de Neiba).

The data also document overthrusting of slope and outershelf sediments tectonically incorporated and mixed with accreted oceanic deposits (subduction kneading of Scholl et al, 1980) observed from Sierra El Numero to Nizao River. Thus, the southern Central Cordillera folded belt appears to be an accretionary prism containing sediments offscraped from oceanic crust and infolded and faulted slope-apron sediments (Figure 23).

Tectonic style varies from place to place, but clear similarities are evident between land and marine sections (Figure 14). In particular, landward dipping reflectors on the seismic profiles resemble isoclinally northeastward dipping series and thrusts observed on the field (Figure 12). Large and relatively simple overthrusts associated with mainly pelagic sequences occur along the southern edge. Complicated and discontinuous slices of pelagic abyssal plain and slope-forearc material indicate imbrication and secondary stacking in the inner part where erosional processes were also very active. Abrupt, numerous changes of facies from Paleocene to Middle Miocene, distribution of Middle Eocene to Oligocene olistostromes, and internal unconformities show continuous instability.

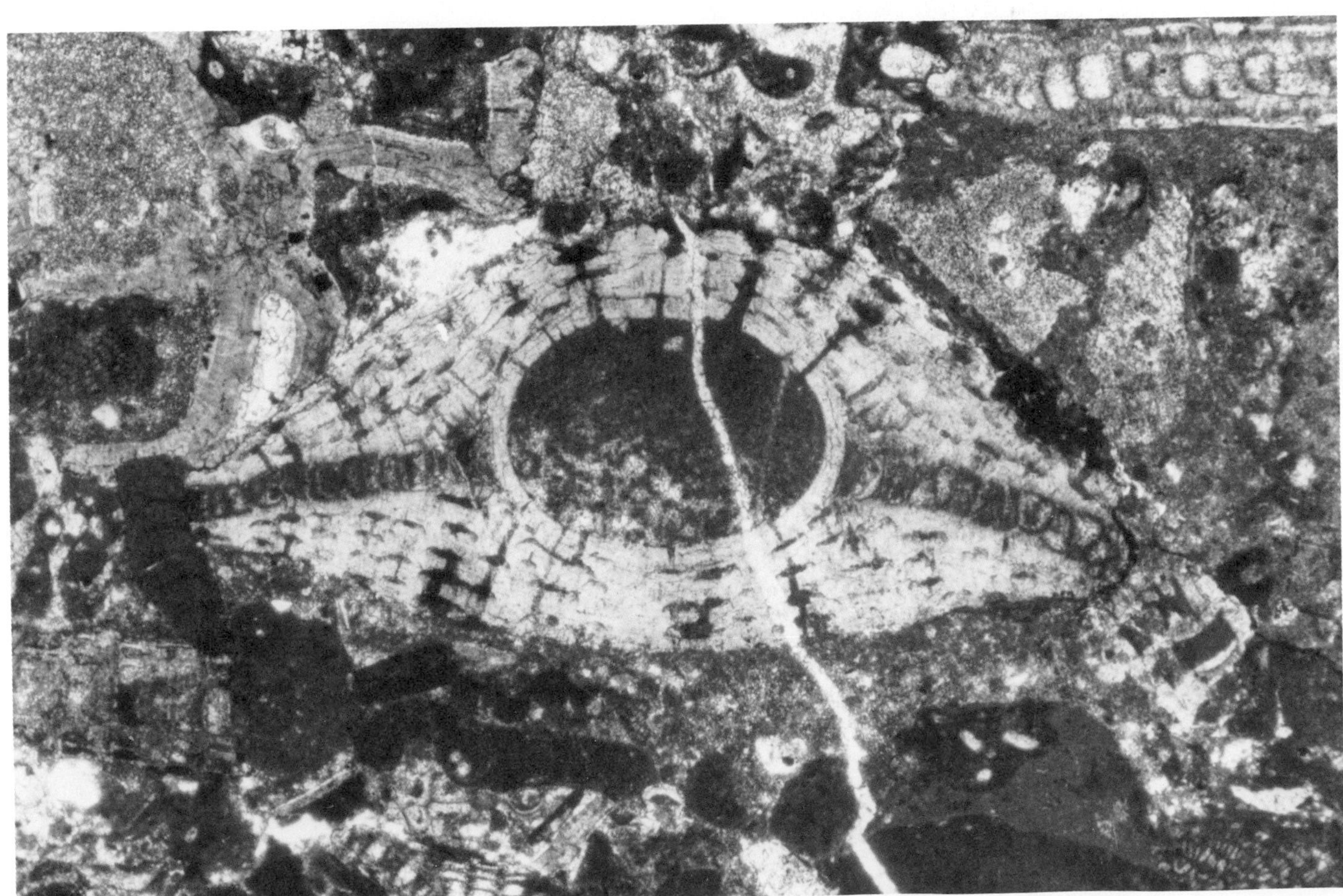

Figure 20 — Thin section of a biomicrite with large Lepidocyclina and association of benthonic organisms, diversified and relatively poorly-sorted bioclastic. Age = Oligocene (above G. opima zone).

Figure 21 — Oligocene heterometric and polymicritic conglomerates. A: Upper Eocene (?)-Lower Oligocene poorly-sorted and massive conglomerates (in Colonia Section). B: Very coarse clastic sequence with angular unsorted boulders infilling small channel. They rest above a mid-Oligocene unconformity (Valdesia road).

Figure 22 — Megaclasts observed in the lowermost Miocene limestones (Fort Resolue road). They are interpreted as *in situ* breccias along paleocliffs or paleochannels, on the edge of a bank of shallow-water reefoidal limestones.

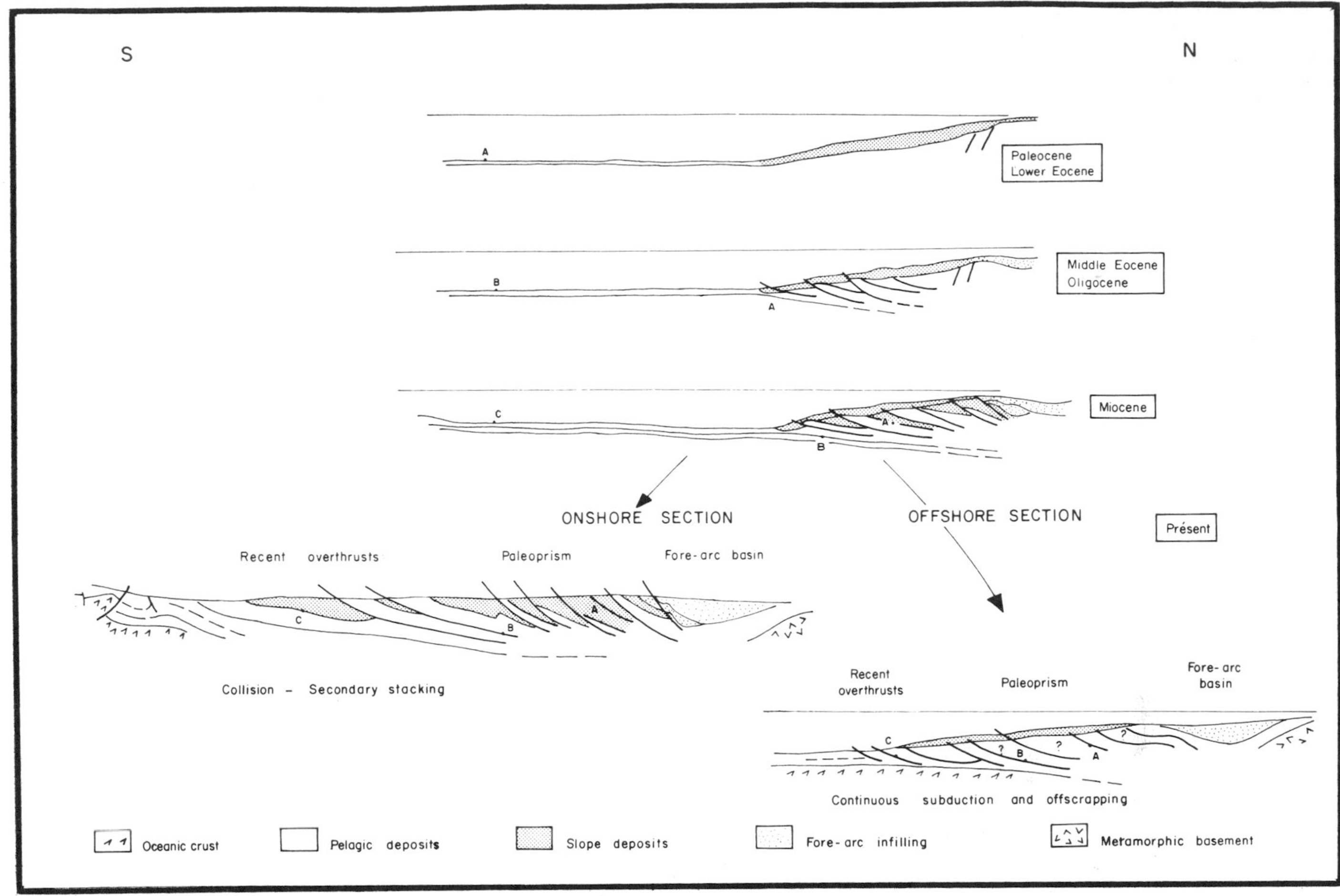

Figure 23 — Tentative reconstruction of the evolution of an active margin with both offscrapping of pelagic sediments and stacking of slope deposits from the Southern Hispaniola margin.

Analyses of sedimentary structures, faunal assemblages, sedimentation rates indicate Eocene widening and deepening of the fore-arc basin; followed by Oligocene slope accumulation linked with deformation and migration of the axis of subsidence in Miocene times, with partial infilling in late Neogene.

ACKNOWLEDGMENT

We are specially indebted to R. Llinas and I. Tavares for useful discussions and a short field-trip on the Southern Dominican Republic outcrops; to J. Saunders, A. Eva, P. Jung, J. Geister, and L. Montadert, with whom some preliminary investigations were done; and, to A. Chartier, who surveyed all the sections.

We thank also Fritz Pierre-Louis, who provided facilities for Haiti field-work. Finally, we are grateful to the French Committee of Petroleum Marine Studies, which gave permission to use the seismic line.

REFERENCES CITED

Biju-Duval, B., et al., 1978, Seismic investigations in the Colombia, Venezuela and Grenada basins and on the Barbados Ridge for future IPOD drilling: Geologie in Mijnbouw, v. 57, p. 105-116.

Bourgois, J., et al, 1979a, Découverte d'une tectonique tangentielle récente a vergence sud dans la Sierra de Neiba (ile d'Hispaniola, republique Dominicaine, Grandes Antilles): Paris, Academie des Sciences, Serie D, v. 289, p. 257-260.

———, ———, 1979b, L'Eocene a blocs d'Ocoa (Republique Dominicaine, Grandes Antilles); temoin d'une tectonique tangentielle a vergence sud dans l'ile d'Hispaniola: Societe Geologique de France, Bulletin, v. 21, p. 759-764.

Bowin, C., 1966, Geology of central Dominican Republic; a case history of part of an island arc: Geological Society of America, Memoir 98, p. 11-84.

———, 1975, The geology of Hispaniola, *in* A.E.M. Nairn and F.G. Stehli, eds. The ocean basins and margins, v.3, The Gulf of Mexico and the Caribbean: New York, Plenum Press, p. 501-552.

Burke, K., J. Grippi, and A.M. Celal Sengor, 1980, Neogene structure in Jamaica and the tectonic style of the northern Caribbean plate boundary zone: Journal of Geology, v. 88, n. 4.

Butterlin, J., 1960, Geologie générale et régionale de la Republique d'Haiti: Travaux et Memorie Mst Htes Etudes de l'Amerique Latine, Universite de Paris, v. 6, 194 p.

———, 1977, Geologie structurale de la Region des Caraibes: Paris, Editions Masson, 259 p.

Case, J. E., 1975, Geophysical studies in the Caribbean Sea, *in* A.E.M. Nairm and F.G. Stehli, eds., The ocean basins

and margins; v. 3, The Gulf of Mexico and the Caribbean: New York, Plenum Press, p. 107-180.
Chase, R.L., and E.T. Bunce, 1969, Underthrusting of the eastern margin of the Antilles by the floor of the western North Atlantic Ocean, and origin of the Barbados Ridge: Journal of Geophysical Research, v. 74, p. 1413-1420.
Edgar, N.T., et al, 1973, Initial reports of the deep sea drilling project: Washington, D.C., U.S. Government Printing Office, v. 15, 1137 p.
Garrison, L.E., et al, 1972, U.S. Geological Survey, international decade of ocean exploration, leg 3: Geotimes, v. 17, n. 3, p. 14-15.
Girard, D., in press, These 3e cycle: University Bretagne Occ.
Hussong, D., et al, in press, Initial reports of the deep sea drilling project: Washington, D.C., U.S. Government Printing Office, v. 60.
Ladd, J.W., and J.S. Watkins, 1979, Tectonic development of trench-arc complexes on the northern and southern margins of the Venezuela basin, *in* Geological and geophysical investigations of continental margins: AAPG, Memoir 29, p. 363-371.
Lewis, J.F., 1980, Resumen de la Geologia de la Hispaniola: Santo Domingo, Field Guide, 9th Caribbean Geological Conference.
Llinas, R.A., 1972, Geologia del area Polo-Duverge, Cuenca de Enriquillo, Codia: Colegio Domin. de Lugen, Arquitec y Agrim., part 1, n. 31, p.55-65, and part 2, n. 32, p. 40-53.
Marlow, M.S., et al, 1974, Tectonic transition zone in the northeastern Caribbean: U.S. Geological Survey, Journal of Research, v. 2, p. 289-302.
Mascle, A., D. Lajat, and G. Nely, in press, Sediments deformation linked to subduction and to argilokines in the Southern Barbados Ridge from multichannel seismic surveys: Trinidad, Proceedings, 4th Latino-American Congress.
———, et al, 1977, Sediments and their deformations *in* active margins of different geological settings, *in* Proceedings, International symposium on the geodynamics of the Southwest Pacific, Noumea 1976: Paris, Technip, p. 327-344.
———, ———, 1979, Estructura y evolucion de los margenes este y sur del Caribe, *in* Analisis de los problemas del Caribe: Bulletin B.R.G.M., section 4, n. 3-4, p. 171-184.
Matthews, J., and T. Holcombe, 1976, Possible Caribbean underthrusting of the Greater Antilles along the Muertos Trough: Guadeloupe, Proceedings, 12th Caribbean Geological Conference, p. 235-242.
Maurasse, F., et al, 1977, Ophiolite complex of the southern peninsula of Haiti, *in* A view of the Caribbean crust (Abs.): Proceedings, 8th Caribbean Geological Conference.
———, 1979, Upraised Caribbean sea floor below acoustic reflector B at the southern Peninsula of Haiti: Geologie en Mijnbouw, v. 58, n. 1.
———, 1980, Relations between the geologic setting of Hispaniola and the origin and evolution of the Caribbean: Proceedings, 1st Collegium sur la Geologie d'Haiti, p. 263-277.
Mercier de Lepinay, B., et al, 1979, Sedimentation chaotique et tectonique tangentielle maestrichtienne dans la presqu'ile du Sud d'Haiti (ile d'Hispaniola, Grandes Antilles): Academie des Sciences, Comptes Rendus Hebdormadaires des Seances, v. 289, n. 13.
Michael R., 1978, Geology of the southwestern flank of the cordillera central, Dominican Republic: George Washington University, M.S. thesis.
———, and J. Millar, 1977, Tertiary geology of the southwestern flank of the cordillera central, Dominican Republic (Abs.): Curacoa, Proceedings, 13th Caribbean Geological Conference, p. 123-124.
Moore, G.P., et al, 1980, Sedimentology and paleobathymetry of neogene trench slope deposits, Nias Island, Indonesia: Chicago, Journal of Geology, v. 88.
Nemec, M.C., 1980, A two phase model for the tectonic evolution of the Caribbean: Santo Domingo, 9th Caribbean Geological Conference (Abs).
Palmer, H. C., 1979, Geology of the Moncion-Jarabacoa area, Dominican Republic, *in* B. Lidz and F. Nagle, eds., Hispaniola; tectonic focal point of the northern Caribbean: Miami Geological Society, p. 29-68.
Ravenne, C., et al, 1977, Model of a young intra-oceanic arc; the New Hebrides island arc, *in* Geodynamics in south west Pacific: Paris, Editions Technip, 413 p.
Reeside, J.B., Jr., 1947, Upper Cretaceous ammonites from Haiti: U.S. Geological Survey, Professional Paper, n. 214-A, p. 1-11.
Scholl, D.W., et al, 1980, Sedimentary masses and concepts about tectonic processes at underthrust ocean margins: Geology, v.8, p. 564-568.
Seely, D.R., 1979, The evolution of structural high bordering major forearc basins, *in* Geological and geophysical investigations of continental margins: AAPG Mem. 29.
Van den Bold, W.A., 1975, Neogene biostratigraphy (ostracoda) of southern Hispaniola: Bulletin of American Paleontologists, v. 66, n. 286.
Von Huene, R., and J. Aubouin, et al, 1980, Leg 67, Deep sea drilling project, Mid-America transect off Guatemala: Geological Society of America Bulletin, v. 91, p. 421-432.
———, ———, 1980, Summary of Japan trench transect, *in* Scientific party, initial reports of the deep sea drilling project, 56, 57: Washington, D.C., U.S. Government Printing Office.
Watkins, J.S., et al, 1981, Initial reports of the deep sea drilling project: Washington, D.C., U.S. Government Printing Office, v. 66.
Westbrook, G.K., 1975, The structure of the crust and upper mantle in the region of Barbados and the Lesser Antilles: Geophysical Journal of the Royal Astronomical Society, v. 43, p. 201-242.
———, 1981, The Barbados Ridge complex; tectonics of a mature forearc system, *in* J. K. Legget, eds., Trench and forearc sedimentation and tectonics: London, Blackwells.
Zoppis de Serra, R., 1969, Atlas Geologico y Mineralogico de la Republica Dominicana: Scale 1:250,000.

Episutural Oligo-Miocene Basins along the North Venezuelan Margin

B. Biju-Duval
A. Mascle
Institut Francais Du Petrole
Rueil-Malmaison, France

H. Rosales
G. Young
Petroleos De Venezuela S.A.
Caracas, Venezuela

Multichannel seismic reflection surveys across the North Venezuelan margin between 64 and 70° W provide new data on the three major structural elements of the area; the Venezuelan Basin margin along the Curacao Ridge, the episutural Falcon and Bonaire Basins, and the southern Guyana Shield continental crust. The Curacao Ridge consists of a thick, elongated belt of deformed sediments; it resulted from Neogene underthrusting of Venezuelan Basin crust. The formation of the onshore Falcon and offshore Bonaire Basins appears to have begun in Middle or Late Eocene. Structures observed in these basins suggest regional compression with attendant folding, faulting, and differential vertical movement. Northern elements of the Guyana Shield, including the Merida Andes to the west, appear to have been involved to some extent in tectonic movements of the margin in Cretaceous to Pliocene times.

The North Venezuelan margin area extends roughly 500 km from the Guajira Peninsula to Trinidad-Tobago and the Atlantic margin to the east (Figure 1). Its present structure developed in early Mesozoic times when North American and African-South American plates diverged, creating a proto-Caribbean seaway and associated passive margins connected with the Tethys Ocean to the northeast. There is some evidence (Bellizzia et al, 1980) of additional tectonic activity as early as Jurassic, but this activity remains poorly understood. The Venezuelan Basin, as we know it today, probably formed in Middle Cretaceous when South America shifted away from Africa with the opening of the central and southern Atlantic Ocean. Basin formation was probably complete by the Turonian, as no younger basement has been discovered in DSDP holes in the Venezuelan Basin proper (Saunders et al, 1973).

From Late Cretaceous to Late Eocene, northern South American continental margins were again subjected to strong tectonism as island arcs and collision complexes formed, overthrusting the Guyana Shield and its sedimentary cover to the south (Bellizzia, 1972; Maresh, 1974; Stephan et al, 1980). Jurassic to Eocene series now exposed in the Cordillera de la Costa and in the Araya-Paria peninsula along the northern coast of Venezuela provide evidence of these events. These series outcrop in almost all islands to the north from Los Monjes to Tobago (Figure 1). These rocks are for the most part metamorphosed and form the basement beneath the Falcon and Bonaire basins, described below.

Marine data discussed in this paper come from a multichannel seismic reflection survey carried out three years ago by Institut Francais du Petrole and Instituto Technologico Venezolano del Petroleo. Seismic lines, previously shot by the Comite d'Etudes Petrolieres Marines in the deep Venezuelan Basin, are also used. Dating of prominent reflectors of the Bonaire Basin is based on data from wells drilled in the La Vela Bay, Golfo Triste and on the eastern Tortuga shelf. Land data were gathered from several publications (Hedberg, 1950; Anonymous 1977; Stephan et al, 1977; Wozniak and Wozniak, 1979; Gonzales de Juana, Inturralde De Arozena, and Picard, 1980; Stephan et al, 1980). Microstructural studies done a few years ago in the Falcon area provide additional data for the discussion of tectonic processes (Wozniak and Wozniak, 1979; Mascle et al, in press).

This paper shows three cross-sections (Figure 1, 2) illustrating major structural elements along the

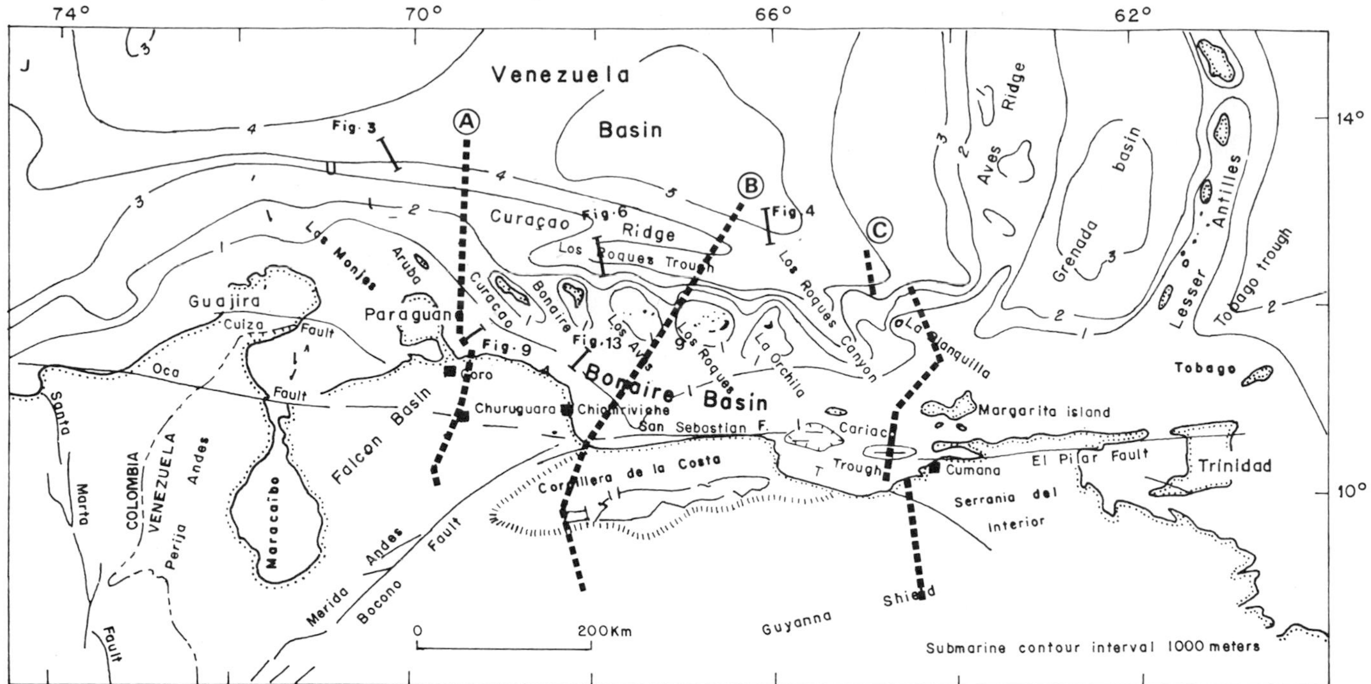

Figure 1 — North Venezuelan margin (modified from Silver, 1975). Dashed lines are the location of sections on Figure 2.

Northern Venezuelan margin and, in addition, illustrate in some detail post-Eocene tectonism seen in the seismic sections.

VENEZUELAN BASIN AND LOWER CONTINENTAL SLOPE

The Venezuelan Basin and the lower continental slope are the object of several recent geophysical investigations (Edgar, Ewing, and Hennion, 1970; Silver, 1975; Talwani et al, 1977; Biju-Duval et al, 1978; Ladd and Watkins, 1979). It was also studied by DSDP drilling holes (146-149, 150, and 153; Saunders et al, 1973). The basin is generally considered oceanic with an anomalously thick and heterogenous crust (Figure 2). The top of the crust generally coincides with the B″ reflector, which is Late Cretaceous in age. In some places sub-B″ reflectors can represent volcanic or volcano-sedimentary layers above the true oceanic crust; in a small area in the southeastern corner of the basin, B″ lies above a rough basement more nearly typical of oceanic crust in other ocean basins.

The overlying upper Cretaceous to Recent sedimentary cover shows rapid changes in thickness and facies along the margin. Sedimentary hiatuses associated with paleocurrents were demonstrated at DSPD hole 150 (Holcombe and Moore, 1977). To the west and at the foot of the margin, sediments are less than 2,000 m thick (Figure 3) but they reach thicknesses of 5,000 m at 66°W (Figure 4). This is partly due to a 2,000-m-thick Neogene turbidite sequence which seems to have accumulated in response to a southward or southeastward tilting of the basement in early Miocene.

The sedimentary cover is highly deformed at the foot of the Venezuelan margin, where it forms the Curacao Ridge. Figure 3 illustrates where sediments were probably scraped off the basement along low-angle thrust faults. The tectonic style is less obvious along other lines (for instance, Figure 4) but high angle reverse or transcurrent faults are inferred.

Silver (1975) interpreted the Curacao Ridge as a small accretionary wedge, due to the southward underthrusting of the Venezuelan Basin crust below the South America borderland. Mascle et al (in press) pointed out, however, that underthrusting is probably a Neogene phenomenon because: 1) the turbidites filling the "trench" at the foot of the Curacao Ridge are no older than Early Miocene; 2) the volume of the Curacao Ridge is small, relative to basin sedimentary thicknesses; and, 3) no Benioff Zone is present, as would be the case if an extensive slab of oceanic crust was subducted. These lines of evidence, although debatable, favor a relatively young origin for the Curacao Ridge.

The deformed belt extends farther west along the Colombian margin where the Magdalena deep sea fan sediments are involved in deformation (Krause, 1971). To the east, it seems to end at 65° 30°W (north of Los Roques Canyon). Further east (Figure 5), the structure at the foot of the margin is not clear; the Venezuelan Basin crust seems to extend below the lower continental slope, but overlying sediments do not show evidence of compressive deformation related to present underthrusting.

The northern edge of the South American continental crust is evident north of the islands of Aves, Los Roques, and La Orchila where it forms a steep

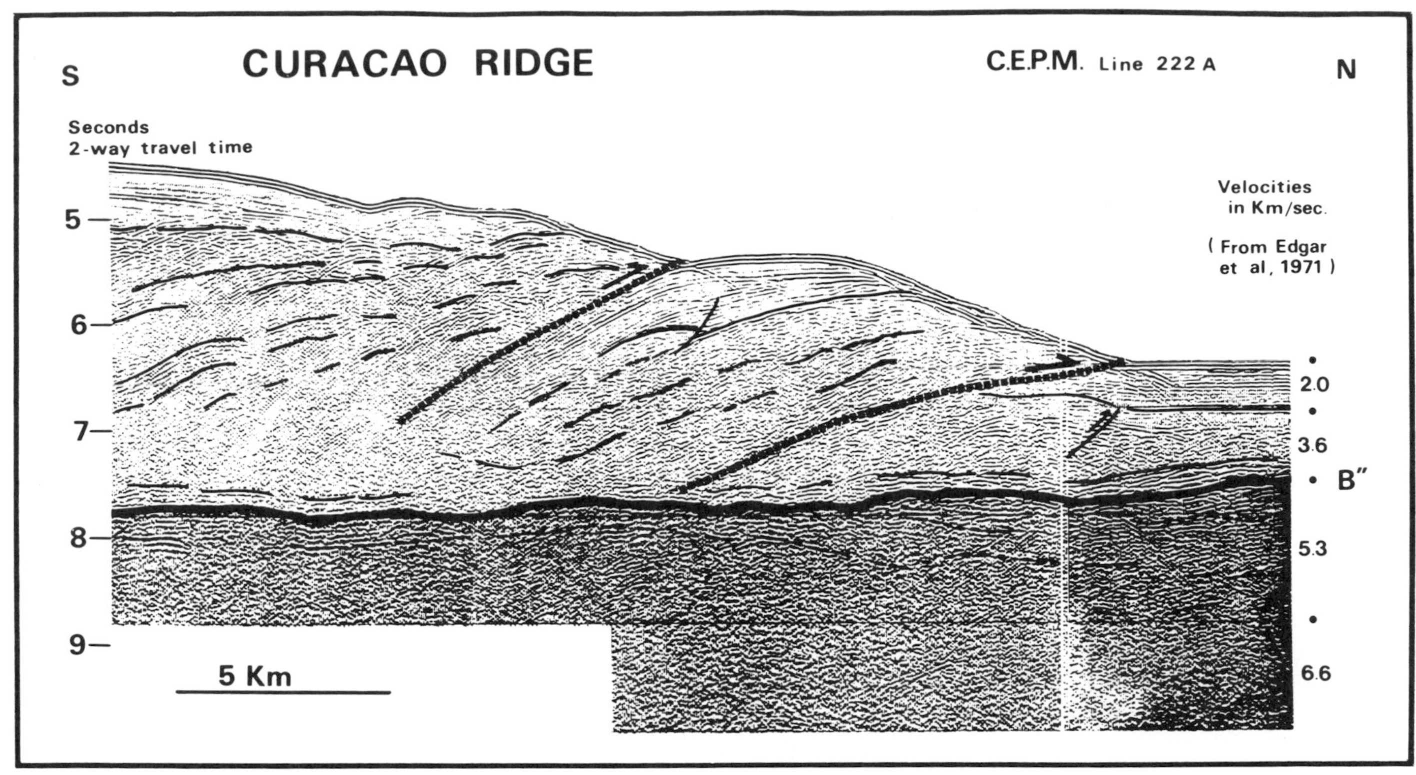

Figure 3 — Frontal overthrust of the Curacao Ridge north of Aruba Island. Location on Figure 1. The B" reflector represents the top of volcanic layers or the top of an anomalously thick oceanic crust. Dated as Late Turonian to Early Senonian age in nearby DSDP holes.

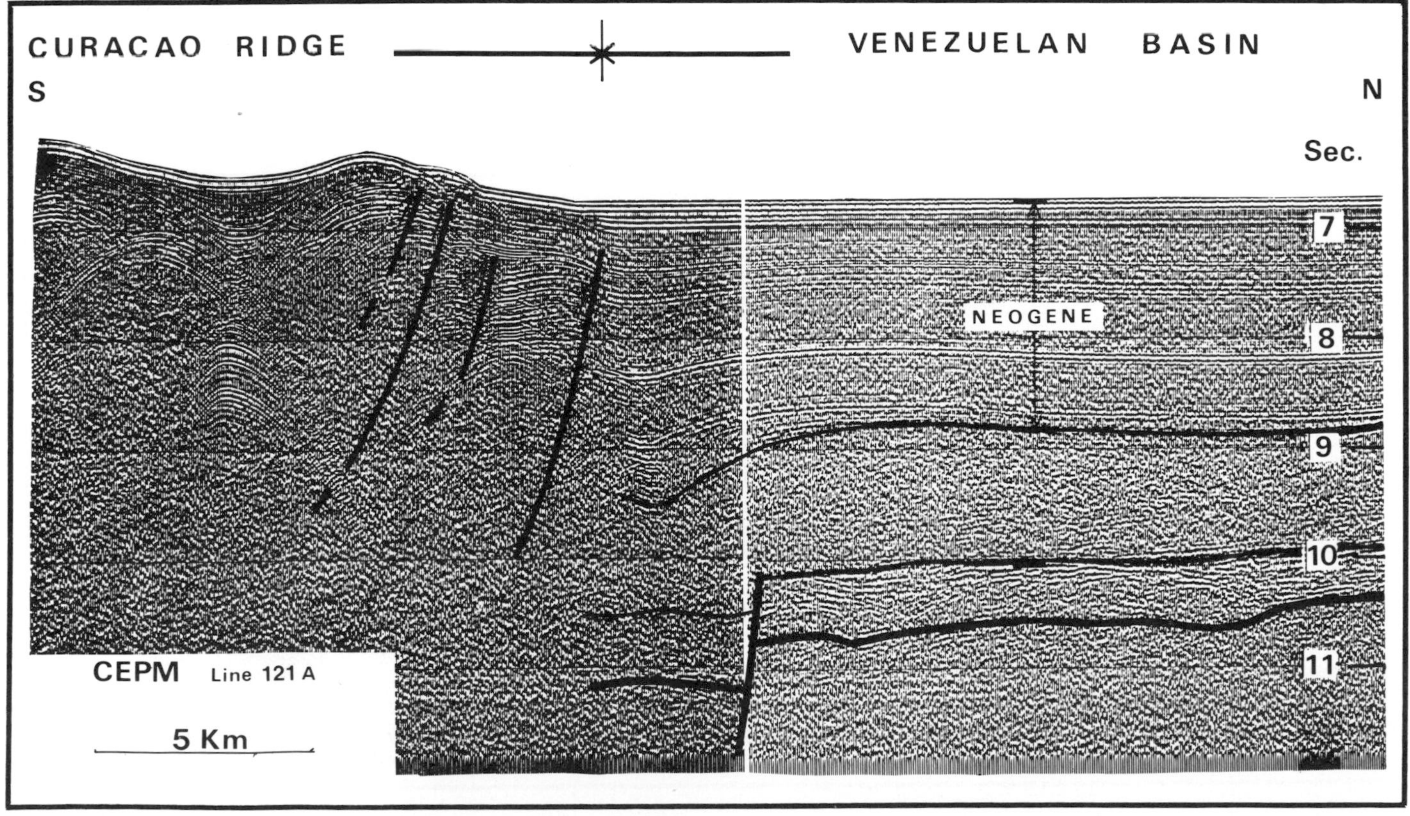

Figure 4 — Frontal deformations of the Curacao Ridge north of La Orchila Island. Location on Figure 1. The lower reflector on the right represents the top of the oceanic crust. Sediments above are of Late Cretaceous to Recent age, according to DSDP holes.

slope locally devoid of sedimentary cover. At its foot, the elongated Los Roques Trough forms a small "fore-arc basin" containing sediments behind the Curacao Ridge. Los Roques Trough sediments are also being progressively incorporated at the back of the accretionary prism (Figure 6).

FALCON AND BONAIRE BASINS

Although the Falcon Basin lies onshore and the Bonaire Basin lies offshore, they occupy similar structural positions along the margins and have similar geologic histories.

Upper Middle Eocene rocks on Curacao and Bonaire consist of conglomerates, sandstones, and bioclastic limestones. These rest unconformably on folded, faulted, and metamorhosed sedimentary and volcanic rocks (MacGillavry, 1977). Clay, sandstone, and some limestone beds of the same age also outcrop in southeastern Falcon (Hunter, 1972), but their contact with the underlying unit is not observed. Outcrops of Upper Eocene shales, sandstones, and limestones are

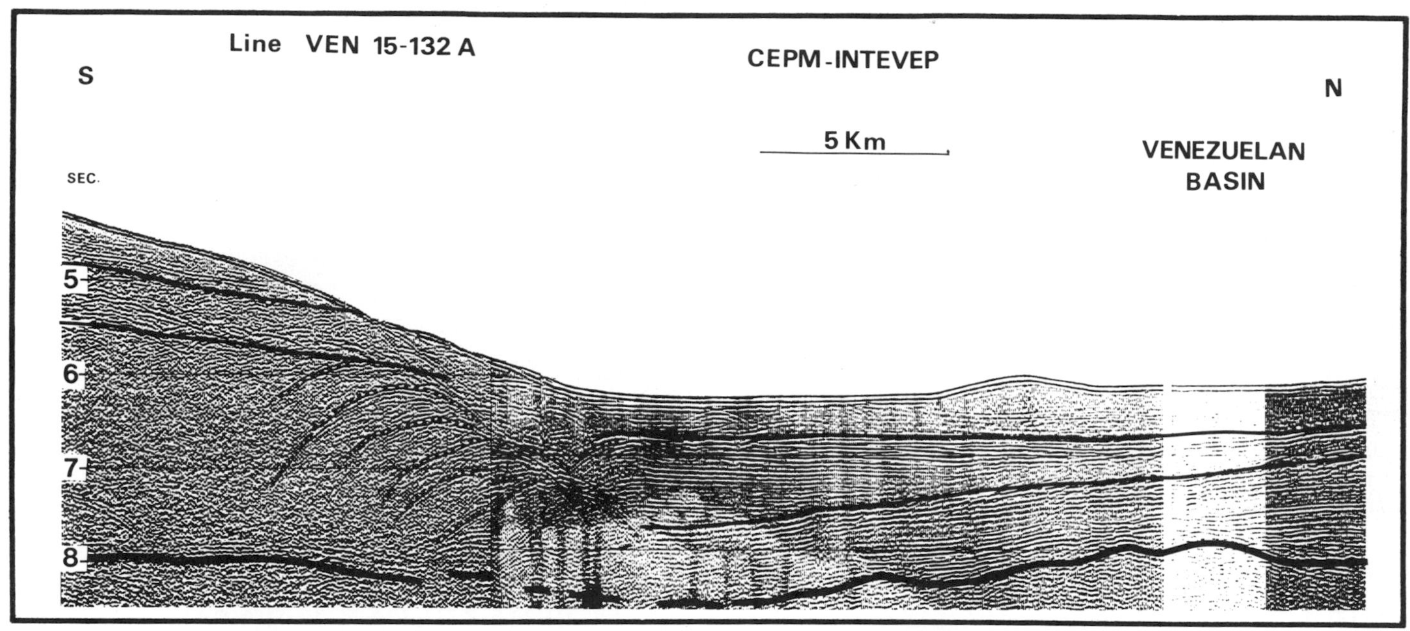

Figure 5 — Structure of the lower continental slope north of La Blanquilla Island. Location on Figures 1 and 2-C. The lower reflector on the right is the top of the oceanic crust. It seems to extend below the continental slope to the left.

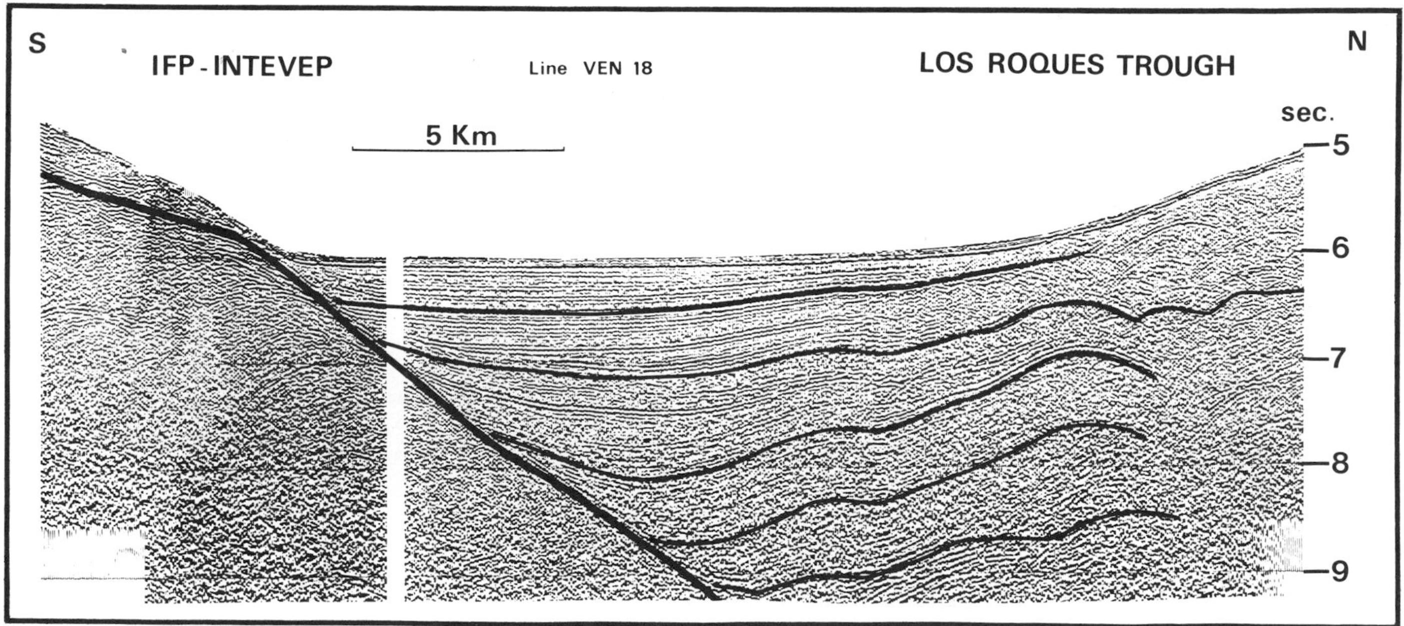

Figure 6 — Los Roques Trough north of Los Aves Islands. Location on Figure 1. To the left, sediments progressively onlap the metamorphic basement of the north Venezuelan islands. To the right, they are progressively deformed and incorporated to the Curacao Ridge accretionary wedge.

known in the Falcon area (Cerro Mission; Hunter, 1972). They were also reported from Golfo Triste wells where they unconformably overly Lower Eocene metamorphic series (Figure 7). Middle Eocene shales, sandstones, and conglomerates are also known on Margarita island at the eastern end of the Bonaire Basin (Punta Carnero Group; Hunter 1972; Munoz, 1973). These clastic sediments indicate erosion of growing reliefs probably to the south (present Cordillera de la Costa).

The Falcon Basin probably began subsiding largely in the Oligocene (Coffiniere et al, 1970). A diversified paleogeography developed in late Oligocene where the relatively deep water Peccaya clays (depositional water depth between 100 and 1,000 m according to foraminiferal association; Wozniak and Wozniak, 1979) pass laterally into shallower water clays, sandstones, and conglomerates of the San Juan de la Vega and Patiocitos formations. This paleogeography persisted into the lower Middle Miocene. For instance, the Lower Miocene San Luis and Canderalito limestones are bioclastic and reefal sediments which were deposited at the subsiding southern edge of the Paraguana High. At the same time deeper, or at least open-marine, conditions prevailed south and east for the deposition of upper Peccaya and Agua Clara clays.

Oligocene to Middle Miocene paleoenvironments were characterized by subsiding troughs, growth faults and surrounding remnant highs that can be recognized at the southern edge of the present Aves — Los Roques Island platform and in the former Bonaire Basin in seismic sections. Figure 8 shows seismic sequences below the Middle Miocene unconformity where differential movement between basement highs and foundered, faulted narrow depressions controlled distribution of the sedimentary sequences.

Beginning in Middle Miocene, the depositional environment changed as the Bonaire Basin deepened and acquired its present shape. This regional subsidence is demonstrated in La Vela Bay (Figure 9) where Middle Miocene to Pleistocene sediments show characteristic sigmoidal seismic sequences over a prograding shelf. It is also shown by the homogeneous turbiditic and hemipelagic Middle Miocene to Recent cover which drapes previous structures in the north central Bonaire Basin (Figure 8). This subsidence created the present day basin, now 2,000 m below sea level at its deepest point.

Other parts of the margin were not affected by subsidence. To the north, the previously emerged Bonaire, Curacao and Aruba islands sank below sea level in Middle Miocene. They then began a slow, discontinuous emergence as indicated by elevated coastal calcareous terraces (Herweijer, Buison, and Herweijer, 1977). To the east, a shallow-water environment continued in the Falcon into Middle and Upper Miocene, with development of bioclastic or reefal limestone (Caujarao and Cumarebo formations, for instance). Pelagic fauna (Wozniak and Wozniak, 1979) indicate an open-marine environment to the north and east, toward the Bonaire Basin.

Following tectonic deformation, regressive shallow-water facies developed in the Falcon Basin during the Pliocene-Pleistocene period. Its central part was also

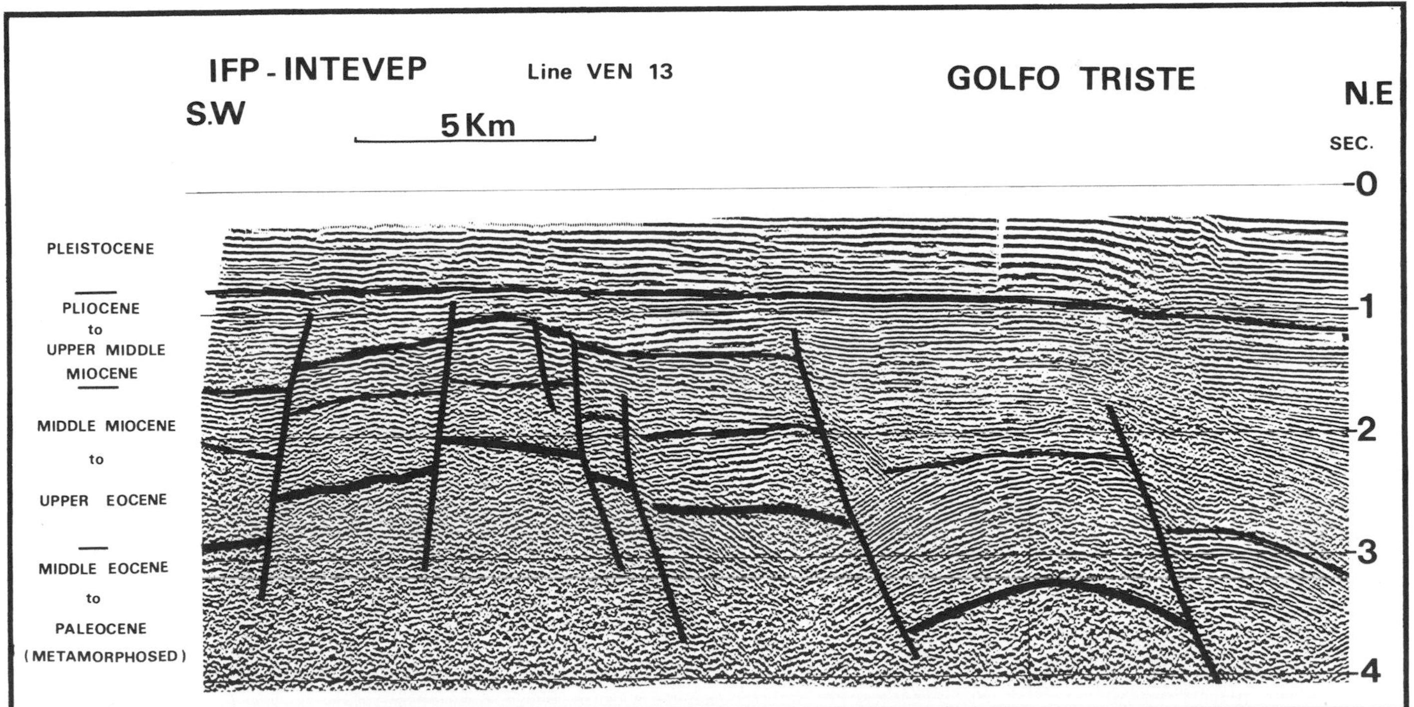

Figure 7 — The Golfo Triste east of Chichiriviche. Location on Figure 1 and 2-B. Ages on the left are from well data. The lowest reflector represents Middle Eocene to Paleocene metamorphic basement. Part of the large vertical displacement along faults appears to be a post-Middle Miocene.

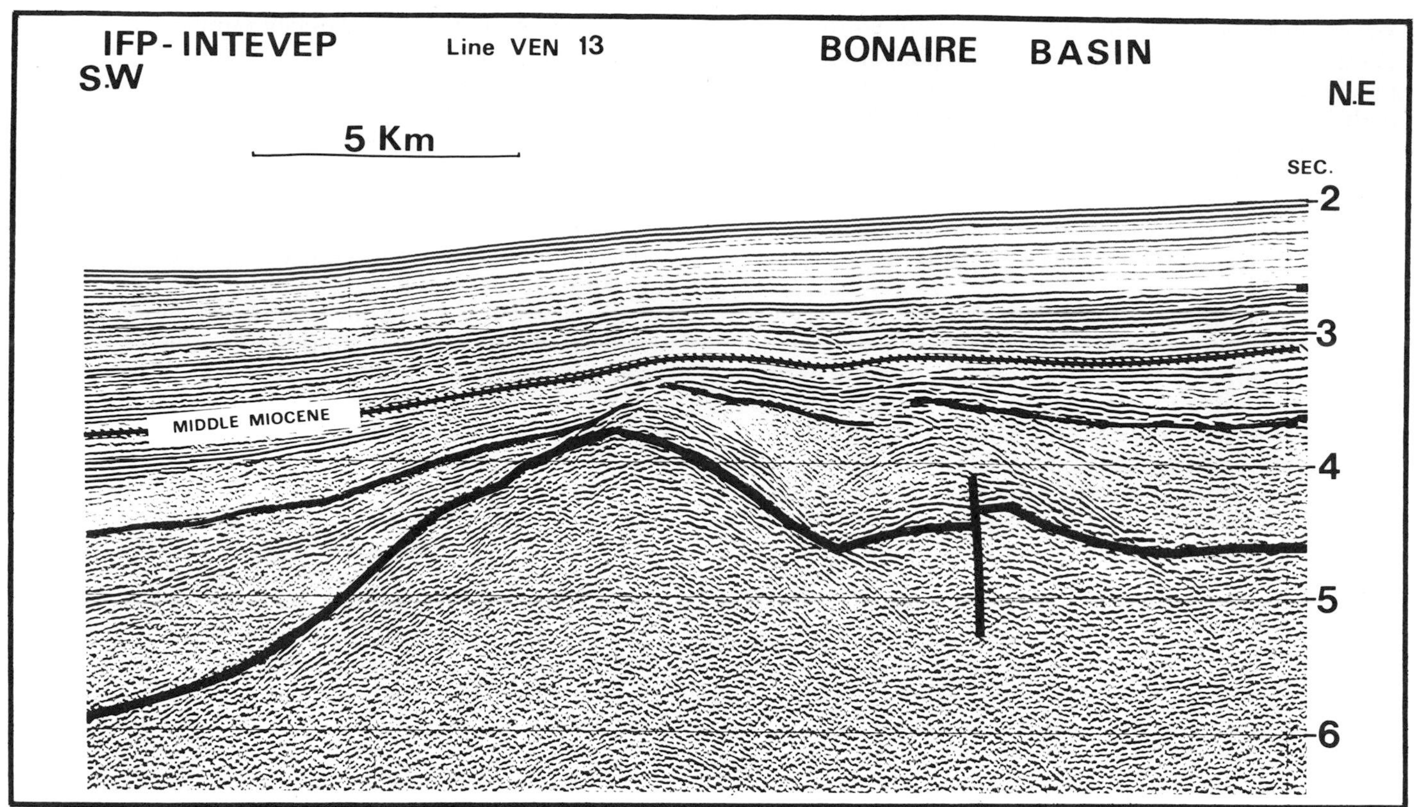

Figure 8 — Northern Bonaire Basin south of Los Aves Islands. The lowest reflector is the top of the Mesozoic to Early Tertiary metamorphic basement. The differences in the seismic facies of the overlying sediments on each part of the central basement high may be related to different paleoenvironments: shelf on the right and slopes to basinal on the left. From Middle Miocene time, sedimentation seems to be homogenous over the entire area.

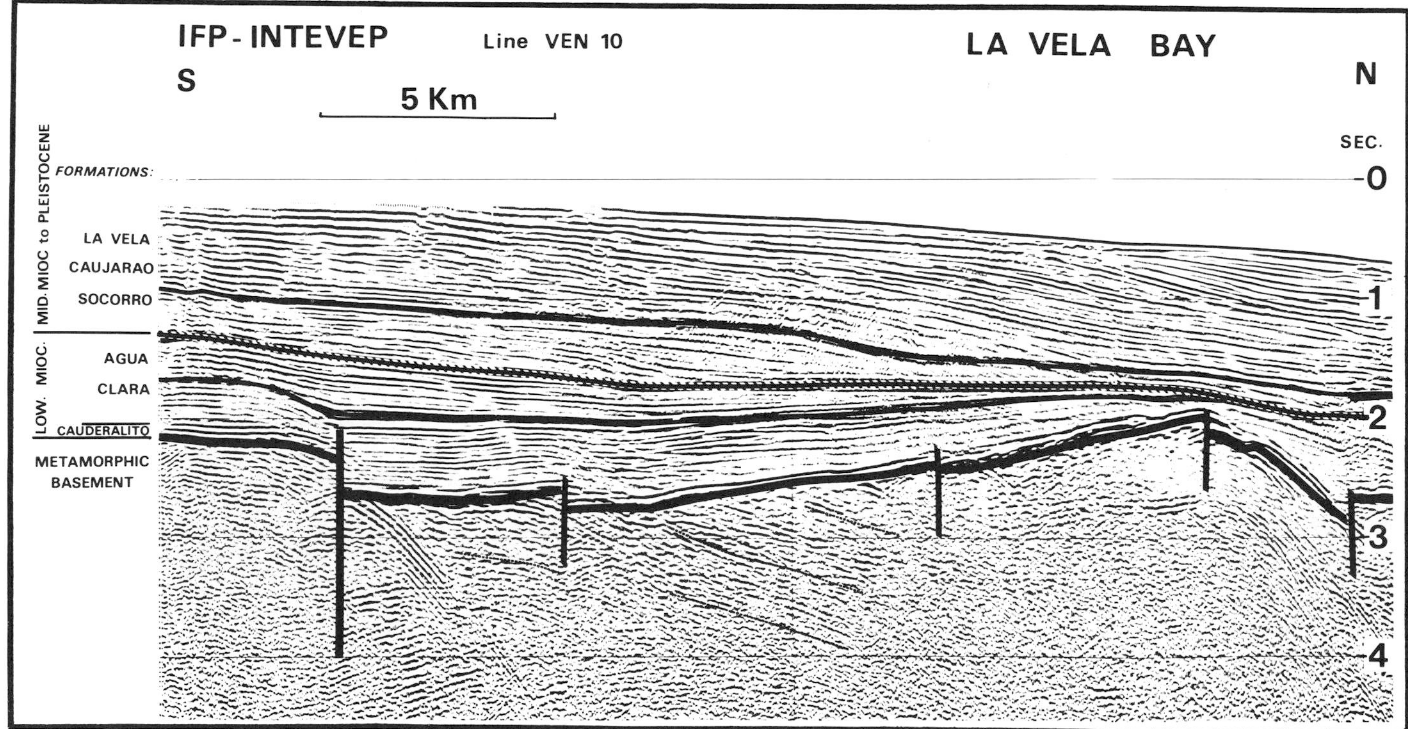

Figure 9 — La Vela Bay. Location on Figure 1. Ages on the left are from well data. La Vela, Caujarao, and Socorro formations are mainly claystones and sandstones with some calcareous beds. Agua Clara formation is the monotonous sequence of gray claystones with some thin calcareous levels. The Canderalito formation is a bioclastic limestone. The metamorphic basement is composed of schists, gneiss, and mafic rocks.

eroded as Oligocene and Miocene pebbles are found in the upper Pliocene-lower Pleistocene Coro conglomerate.

The tectonic history of the Falcon and Bonaire Basins is marked by vertical movement, crustal shortening, and sediment folding. Large vertical movements were recorded for the first time in the Upper Oligocene when subsiding troughs began forming in the Falcon Basin, and probably in the Bonaire Basin. Basaltic igneous rocks intruded at that time now outcrop as interbedded sills (Coronel, 1970). The tectonic regime consisted of more or less east-west compressive stresses (Figure 10), consistent with the pull-apart model Muessig (1978) proposed to explain the beginning of subsidence in the Falcon and Bonaire Basins. The lower Miocene paleogeography was probably the same but compressive stresses were along a N 040°E strike.

However, from Middle Miocene to Recent, the Caribbean crust is being thrust under the Venezuelan margin to the south. Vertical movement

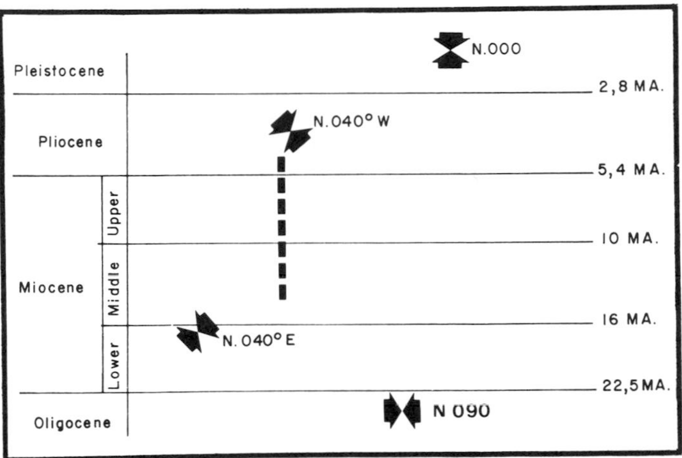

Figure 10 — Direction of regional shortening in the central and northern Falcon Basin (from Wozniak and Wozniak, 1979, Mascle et al, in press). The N 040°W appears to be the more important stress direction in the field; it was active from Middle Miocene to Pliocene.

remained an important factor controlling sedimentation. The present Bonaire Basin has subsided continuously except for the north, where the present island probably stayed close to sea level.

Folding and reverse or transcurrent faulting is observed along the southern border of the Bonaire Basin. The direction of shortening was N 40°W from the Middle Miocene to the Pliocene and more or less north to south in the Pleistocene (Figure 10). Main structures of the Falcon Basin are due to these tectonic shortenings (Figure 11). Seismic profiles in the Bonaire Basin show either paleostructures, as on the east Tortuga shelf (Figure 13); or, active structures such as the high angle reverse faults at the origin of the lower slope to the south (Figure 13). There is also evidence of strike-slip motion (Figure 14).

THE GUYANA SHIELD

The Guyana Shield is a stable platform where sedimentation has taken place from the Proterozoic to the Pleistocene. It is not within the scope of this paper to discuss this long geological history, but we will briefly recall the main Tertiary tectonic events and the way they are related to the Venezuelan margin.

Southern terminations of regional sections (Figure 2) in the Guyana Shield and the Merida Andes exhibit three different structural patterns. In the west, the Merida Andes form a northeast-southwest trending range where Paleozoic to Neogene rocks outcrop. These rocks were uplifted (up to 5,000 m at the Pic Simon Bolivar) in Pliocene and Pleistocene time (Stephan et al, 1977). At the same time, overturned folds and reverse faults with a small amount of overthrusting developed toward the edge of the range, to the northwest and southeast. To the northeast, Upper Cretaceous to Eocene sediments of the Cordillera de la Costa overthrust the Merida Andes.

The central section (Figure 2) illustrates the southward overthrusting of the Cordillera de la Costa onto Guyana Shield. Major tectonism with stacking of

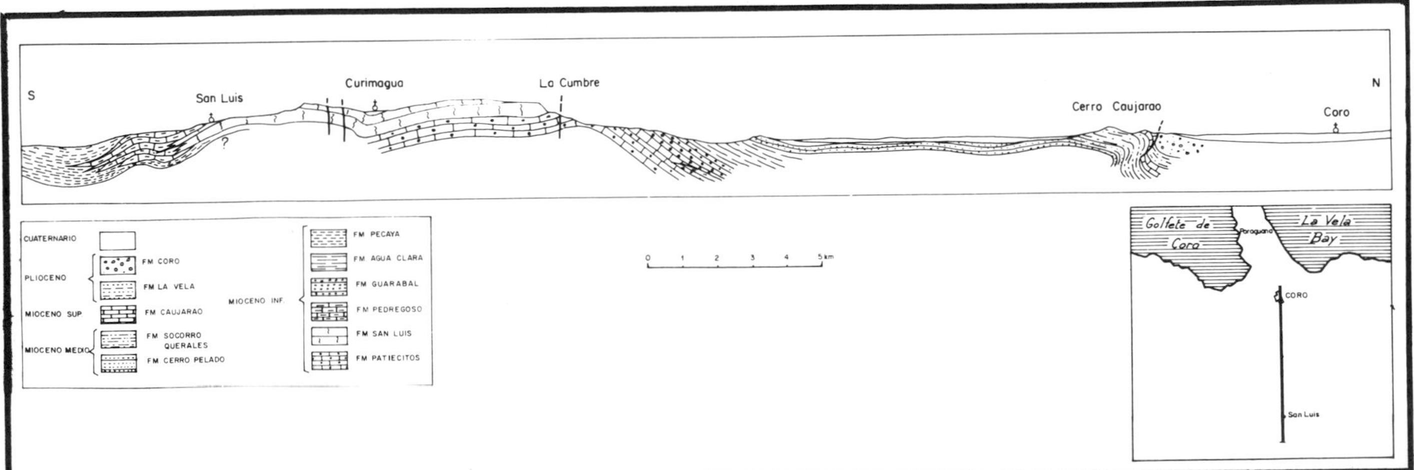

Figure 11 — Geological section in the north-central Falcon Basin. Folding is related to the N 040°W and north-south shortening (Figure 10). The previous N 040°E shortening is expressed on the field by potential shears, tension gaps, and tectonic stylolites mainly.

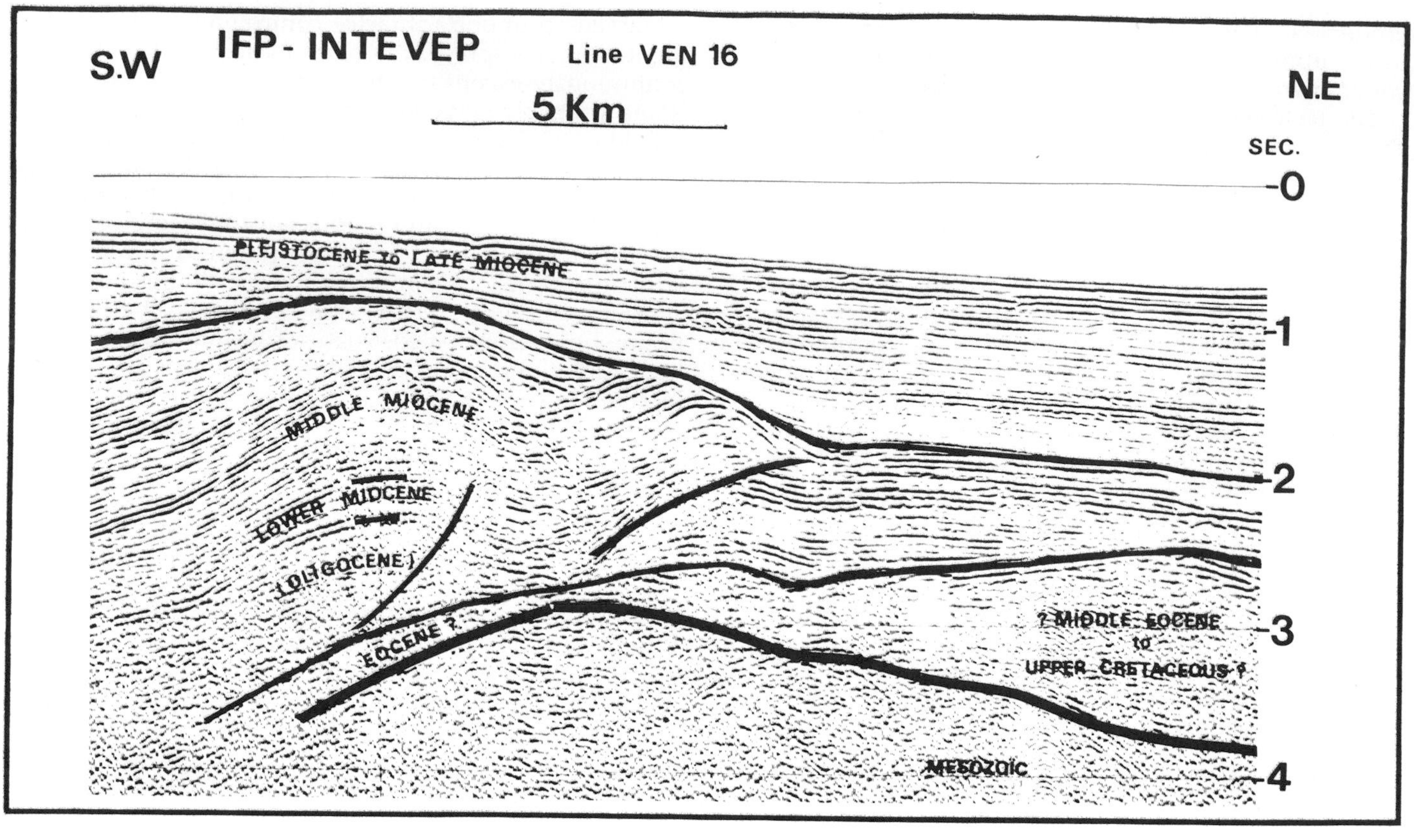

Figure 12 — Paleostructures of late Middle Miocene age on the upper continental slope north of Margarita Island. Location on Figure 1 and 2-C. Ages on the left are from well data. Lowest reflectors are the top of the Mesozoic metamorphic basement.

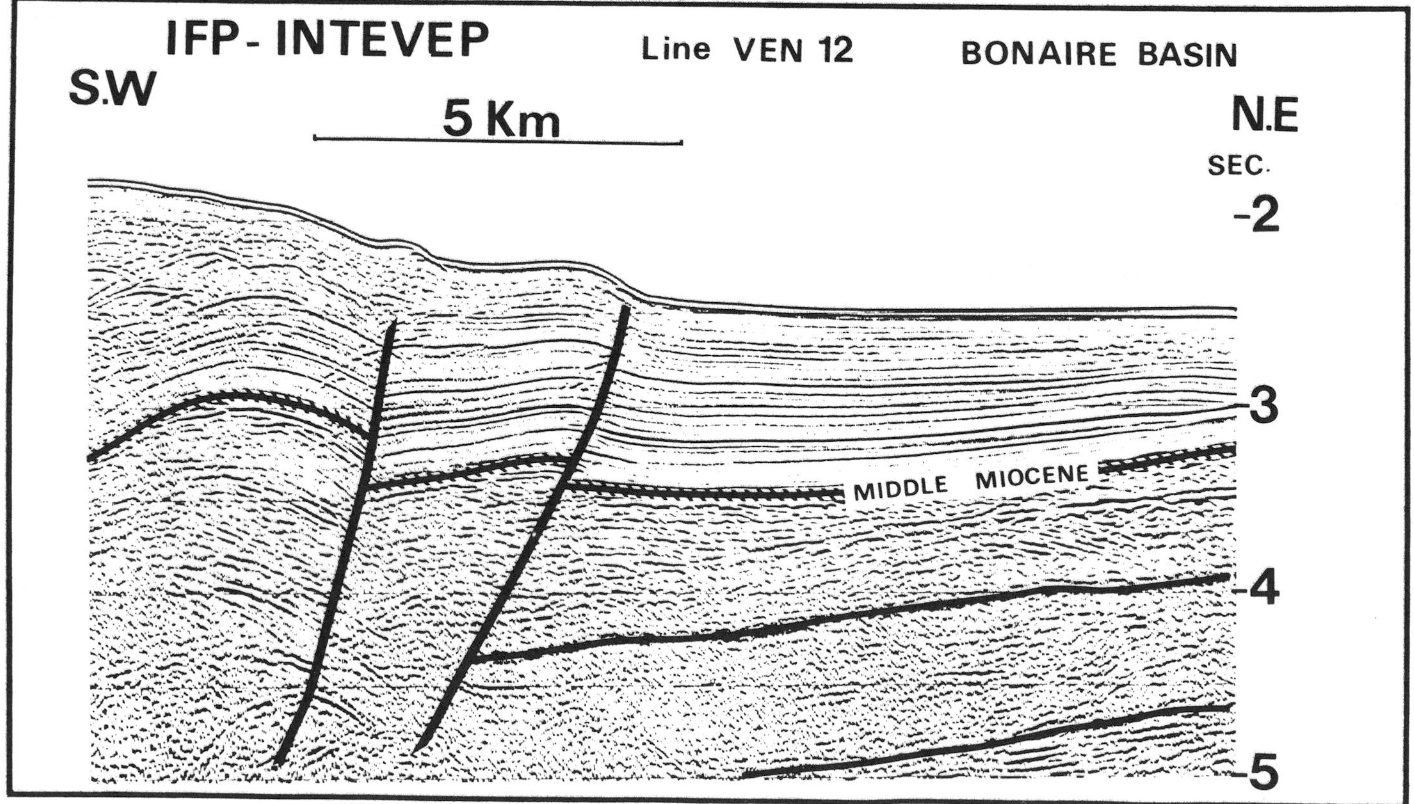

Figure 13 — Recent reverse faults and related folds along the lower slope of the southwestern Bonaire Basin. Location on Figure 1.

nappes occurred in Late Cretaceous to Eocene time, and renewed overthrusting occurred in Late Miocene along low angle to flat thrust faults (Beck, 1978).

In the Serrania del Interior to the east, east-northeast to west-southwest trending folds expose Middle Cretaceous to Neogene sediments. To the south, southward overturned folds developed into the Late Miocene (Gonzales de Juana, Iturralde de Arozena, and Picard, 1980). Large transcurrent faults are mapped in this area and in some places recent displacement can be demonstrated. For instance, close to the City of Cumana where east-west and northwest-southeast trending faults intersect, the Pleistocene Cumana and Caiguire formations are deformed and even locally overturned.

These three examples show that in Neogene time, regional shortening occurred far to the south of the present continental edge. Parts of the Lower Tertiary deformed belts of the Cordillera de la Costa have been rejuvenated but deformation also affect wide areas of the quiet Guyana Shield sedimentary cover.

DISCUSSION

As documented by geophysical data, shortening occurred in Neogene time along the North Venezuelan margin. Direction of shortening ranged from N 040°W in Middle Miocene-Pleiocene, to roughly north-south in Pleistocene time in the Falcon area. This shortening led to sediment deformation in two areas (Figure 15, 16); (1) at the foot of the continental slope where the Caribbean oceanic crust underthrusts the South American plate forming the Curacao Ridge; and, (2) on the South American continent between the southern part of Bonaire Basin and the northern edge of the Guyana Shield 300 km south of Caribbean subduction.

During the same period, differential subsidence occurred along the northern edge of the South American continent (Bonaire Basin and Island Platform) between normally faulted blocks.

A relatively simple mathematical model explains the distribution of tectonic stresses over the North Venezuelan margin (N'Gokwey, 1982). The subducting Caribbean crust appears mechanically linked with the South American plate and is transmitting part of the stress due to plate motion. Distention observed along the northern edge of the continent appears to be due to bending of the top of the subduction zone. Consumption of oceanic crust is now partly inhibited by the thickness of the Caribbean Crust, as well as by the oblique nature of the plate collision with largely strike slip motion.

From a regional point of view, the north-south component of shortening along the South Caribbean

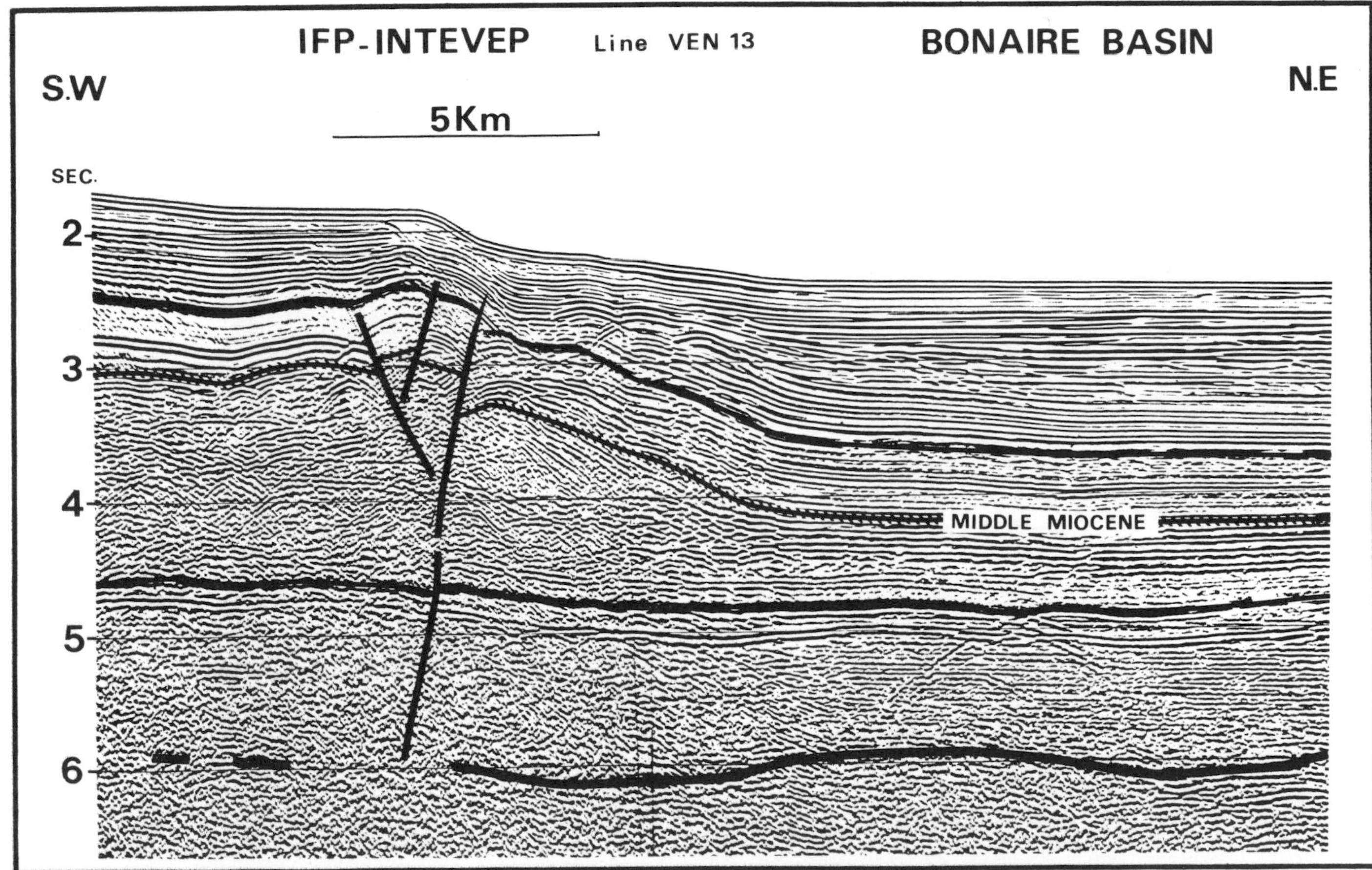

Figure 14 — Recent deformations along the lower slope of the southern Bonaire Basin. Location on Figure 1 and 2-B. The structural style is not clear but strike-slip motion is inferred for the main fault and disharmonic folding is expected below the Middle Miocene reflector. The lowest reflector would be the top of the metamorphic basement.

margin, (discussed in this paper), as well as along the North Caribbean margin (Biju-Duval et al, 1982, this volume) appears due to the small amount of convergence between North and South America. The Caribbean oceanic crust transmits part of the stresses without large internal deformation, as do other small oceanic basins, e.g., the Mediterranean (Letouzey and Tremolieres, 1980).

In conclusion, according to the Bally and Snelson (1980) classification, the North Venezuelan margin represents a megasuture with a complex Mesozoic-Cenozoic geological history. The present configuration of the margin (Figure 16) developed since the lower or middle Miocene; that is, underthrusting of the Caribbean oceanic crust below the South American Borderland (B-subduction zone) is probably not older than Miocene. The tectonic events recorded in the Neogene to the south (Guyana Shield) are due to crustal shortening related to the northward underthrusting of the Guyana Shield below the Merida Andes and Cordillera de la Costa system (A-subduction zone). This global pattern was somewhat different in the Oligocene when transcurrent movement along the margin may have been the main process. Partly metamorphosed and highly deformed Jurassic to Eocene sediments exposed in the Cordillera de la Costa and in the northern island represent remnant structures from older margins and oceanic basins which have largely disappeared through subduction and collision processes. The Falcon and Bonaire Basins are episutural basins which developed since upper middle Eocene unconformably above these deformed series.

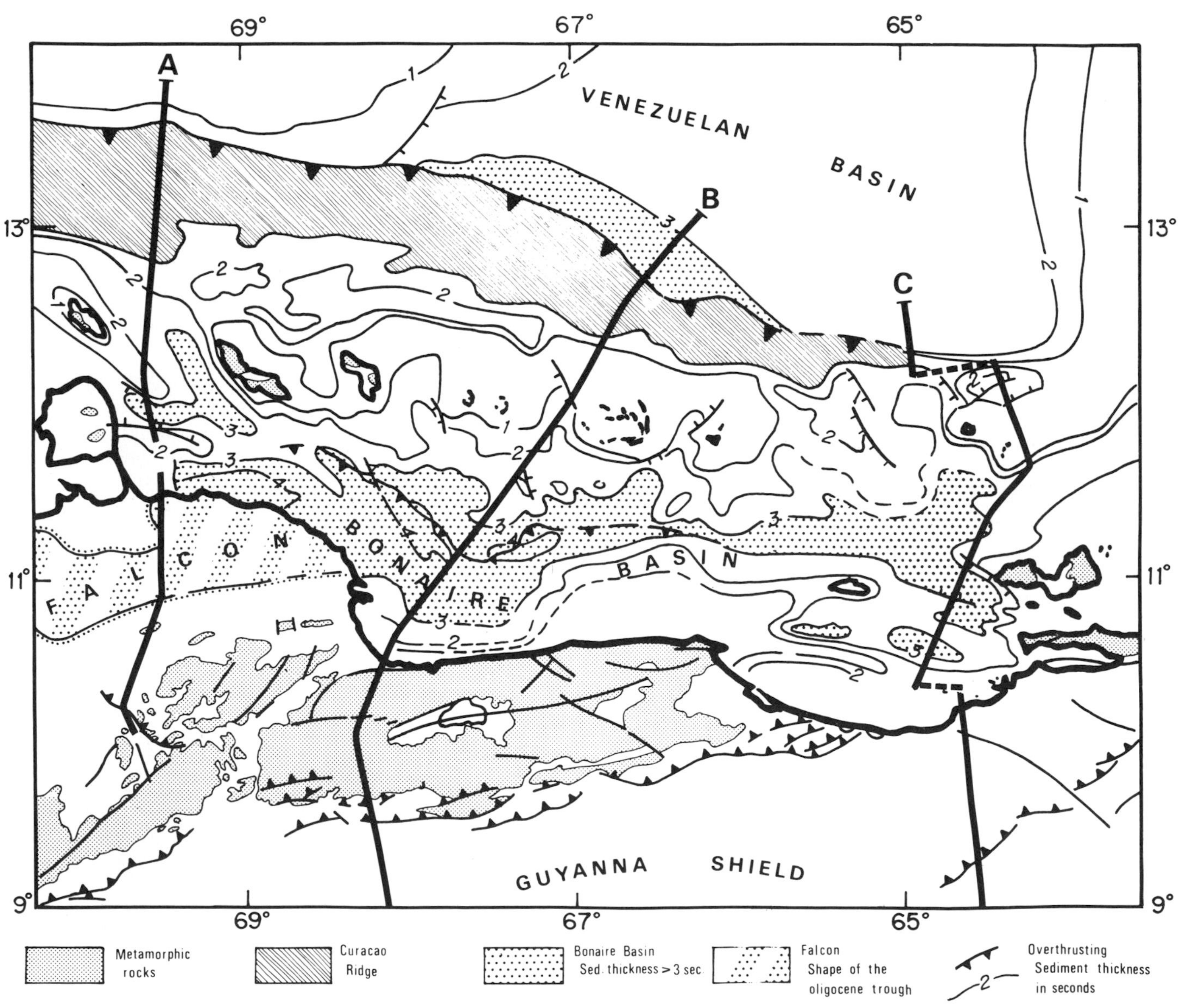

Figure 15 — Simplified structural map of the North Venezuela margin with the sediment thickness in seconds two-way travel time in the Bonaire and Venezuelan Basins. The Falcon Basin is drawn in the shape of the Oligocene subsiding trough, from data in Coffiniére et al (1970). Other land data come from the Venezuelan Geologic Map (Anon., 1977). Lines A, B, and C are the location of the sections on Figure 2.

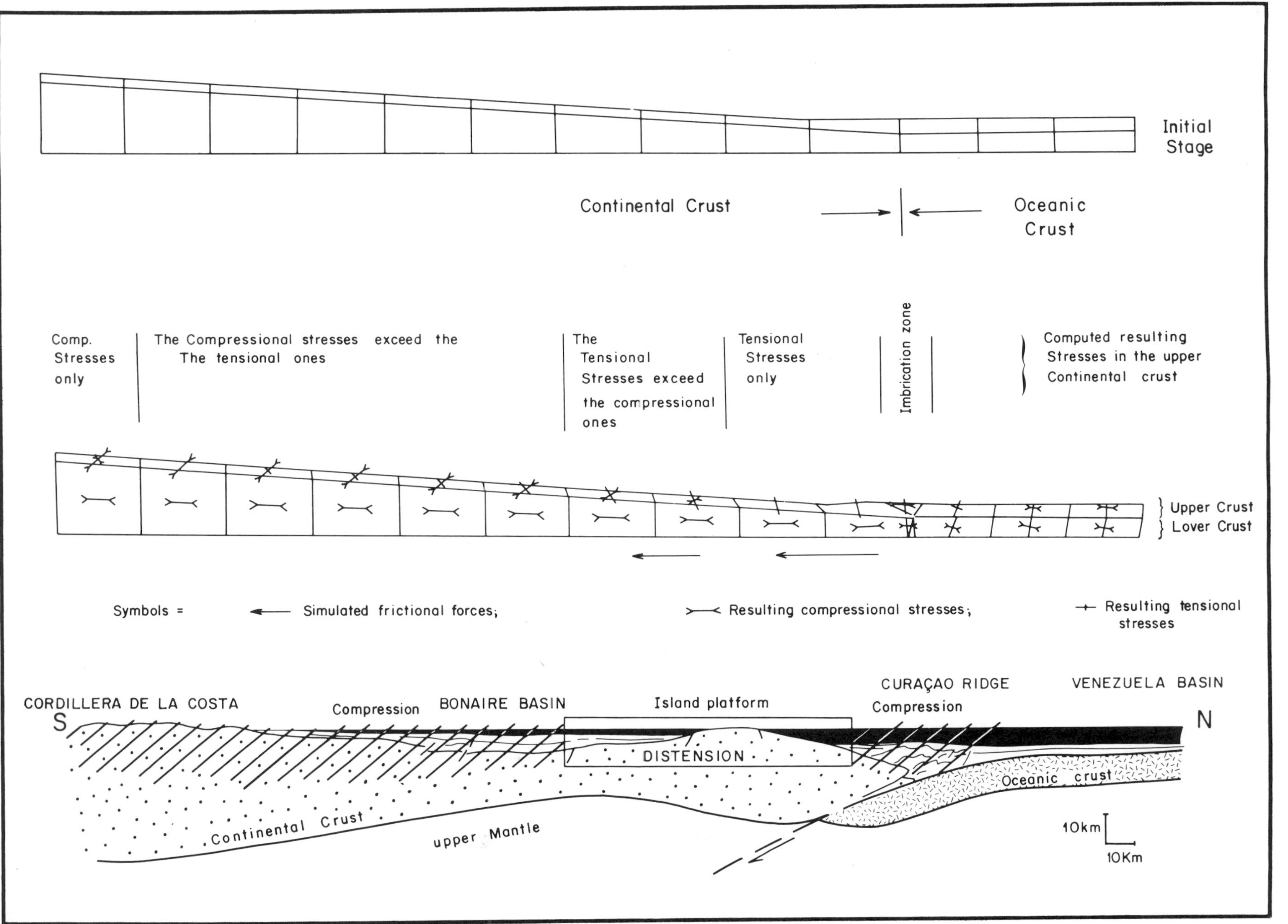

Figure 16 — Observed deformations across the North Venezuelan margin; comparison with mathematical model using the finite element method of a young subduction zone where continental and oceanic crust are in mechanical contact. This was simulated using frictional forces at the lower part of the continental crust (arrows; simplified from N'Gokwey, 1981).

REFERENCES CITED

Anonymous, 1977, Geologic Map of Venezuela: Caracas Scale, Ministerio de Energia y Minas, Direccion de Geologia, 1:500,000.

Bally, A.W., and S. Snelson, 1980, Realms of subsidence in facts and principles of world petroleum occurrence: Canadian Society of Petroleum Geologists, Memoir 6, p. 9-94.

Beck, C., 1978, Polyphasic tertiary tectonics of the interior range in the central part of the western Caribbean Chain, Guarico State, northern Venezuela: Geologie in Mijnbouw, v. 57, p. 99-104.

Bellizzia, A., 1972, Sistema montanosso del Caribe, borde sur de la placa Caribe, es una cordillera allochtona ? : Margarita, Proceedings, 6th Caribbean Geological Conference, p. 247-258.

———, et al, 1980, La chaine caraibes, du Mesozoique a l'Actuel; tectogenese et modele d'evolution geodynamique (Abs. with figs.): Societe Geologique de France, p. 34.

Biji-Duval, B., et al, 1978, Seismic investigations in the Colombia, Venezuela, and Grenada Basins, and on the Barbados Ridge for future IPOD drilling: Geologie in Mijnbouw, v. 57, p. 105-116.

———, ———, 1982, Active margin processes; field observations in southern Hispaniola: AAPG Memoir 34.

Coffiniere, P., et al, 1970, Synthese paleogeographique et petroliere du Venezuela occidental (lere partie): Revue de l'Institut Francais du Petrole, v. 25, n. 12, p. 1449-1492.

Coronel, G., 1970, Igneous rocks of central Falcon: Caracas, Boletin Informativo, v. 13, n. 5, p. 155-161.

Edgar, N.T., J. Ewing, and J. Hennion, 1971, Seismic refraction and reflection in the Carribean Sea: AAPG Bulletin, v. 55, p. 833-870.

Gonzales de Juana, C., J. M. Iturralde, and X. Picard, 1980, Geologia de Venezuela y de sus cuencas petroliferas: Caracas, Tomo I y II, Ediciones Foninves, 1031 p.

Hedberg, H.H., 1950, Geology of the eastern Venezuela Basin (Anzoategui-Monagas-Sucre-Eastern Guarico portion): Geological Society of America Bulletin, v. 61, p. 1173-1216.

Herweijer, J.P., P.H. de Buison, and J.P. Herweijer, 1977, Neogene and Quarternary geology and geomorphology, *in* Guide to the field excusions on Curacao, Bonaire, and Aruba, Netherland Antilles: Curacao, Proceedings, 8th Caribbean Geology Conference, p. 39-55.

Holcombe, T.L., and W.S. Moore, 1977, Paleocurrents in the eastern Caribbean; geologic evidence and implications: Marine Geology, n. 2, p. 35-56.

Hunter, V.F., 1972, A Middle Eocene flysche from East Falcon, Venezuela: Margarita, 6th Caribbean Geological Conference, Memoir, p. 126-130.

Krause, D.C., 1971, Bathmetry, geomagnetism, and tectonics of the Caribbean Sea north of Colombia, *in* Caribbean geophysical, tectonic, and petrologic studies: Geological Society of America, Memoir 130, p. 35-54.

Ladd, J., and J. Watkins, 1979, Tectonic development of trench arc complexes on the northern and southern margins of the Venezuelan Basin, *in* Geological and geophysical investigations of continental margins: AAPG Memoir 29, p. 363-371.

Letouzey, J., and P. Tremolieres, 1980, Paleo-stress fields around the Mediterranean since the mesozoic derived from microtectonics; comparison with plate tectonic data Geologie des chaines Alpines issues de la Tethys, p. 261-273.

MacGillavry, H.J., 1977, Tertiary formations, *in* Guide to the field excursions on Curacao, Bonaire, and Aruba, Netherland Antilles: Curacao, Proceedings, 8th Caribbean Geological Conference, p. 36-38.

Maresh, W., 1974, Plate tectonic origin of the Caribbean mountain system of northern South America, discussions and proposal: Geological Society of America Bulletin, v. 85, p. 669-682.

Mascle, A., et al, in press, Neogene compressional events on the north Venezuelan margin: Trinidad and Tobago, Proceedings, 4th Latin American Geological Congress.

Muessig, K.W., 1978, The central Falcon igneous suite, Venezuela; alkaline basaltic intrusions of Oligocene-Miocene age, Geologie in Mijnbouw, v. 57, p. 261-266.

Munoz, N.G., 1973, Geologia sedimentaria del flysch Eoceno de la isla de Margarita, Venezuela: Escuela Geologia Minas, Universidad Central de Venezuela, Geos, n. 20, p. 5-64.

N'Gokwey, K., 1982, Application de la methode des elements finis a la connaissance des contraintes et deformations au front d'une marge active: Rueil-Malmaison, France. These de Docteur Ingenieur, Institut Francais du Petrole.

Saunders, J., et al, 1973, Cruise synthesis, *in* Initial reports of the deep sea drilling project: Washington, D.C., U.S. Government Printing Office, v. 15, p. 1077-1111.

Silver, E.A., 1975, Geophysical study of the Venezuelan borderland: Geological Society of America Bulletin, v. 86, p. 213-226.

Stephan, J.E., et al, 1977, Guia de la excursion n. 6 Cordillera de los Andes y Surco de Barquisimeto: Caracas, Proceedings, 5th Congreso Geologico Venezolano.

———, ———, 1980, La chaine Caraibe du Pacifique a l'Atlantique; Geologie des chaines Alpines issues de la Tethys, p. 38-59.

Talwani, M., et al, 1977, Multichannel seismic study in the Venezuelan Basin and the Curacao Ridge, *in* Island arcs, deep sea trenches, and back arc basins: American Geophysical Union, Maurice Ewing Series 1, p. 93-98.

Wozniak, J., and M.H. Wozniak, 1979, Geologia de la parte Nor-Central de la Sierra de Falcon, Hogas Coro y Cabure: Caracas, Ministerio de Energia y Minas, Direccion de Geologia.

Banda Arc Tectonics: The Significance of the Sumba Island (Indonesia)

F. H. Chamalaun
A. E. Grady
C. C. von der Borch
Flinders University of South Australia
Bedford Park, South Australia

H. M. S. Hartono
Geological Research and Development Center
Bandung, Indonesia

Current geological and geophysical knowledge of the Sumba (Indonesia) Island is reviewed with consideration of recent fieldwork. Sumba does not exhibit features of the subduction tectonics of the Sunda Arc system (to the west of Sumba), nor the collision tectonics of the Banda Arc system to the east. It is supposed that Sumba is a continental fragment from Australia to the south or from Sundaland to the north, that became trapped behind the eastern Java Trench. The data does not provide convincing support for either hypothesis, but appears to favor an Australian origin. We propose critical studies that aid in resolving the origin of Sumba.

Sumba (Indonesia) Island has long been regarded as an enigma. In reviewing the structural history of the Indonesian Archipelago, Umbgrove (1949) noted that Sumba did not fit in with other major structural units with regard to its position or geological history. He pointed out that a synthesis of the region's tectonics must clear up "the problem of Sumba Island." Despite the development since Umgrove wrote about new concepts in global tectonics, we appear to be no nearer to an acceptable explanation for the origin of Sumba.

Geographically, Sumba Island is situated at the eastern extremity of the Sunda Arc (Figure 1). The Sunda Arc comprises the volcanic arc and trench (subduction) system from at least northwest Sumatra to Sumba and is characterized by the subduction of the Indian Ocean floor at the Java Trench, below Sundaland. To the east of Sumba lies the double chain of volcanic and nonvolcanic islands called the Banda Arc system, which extends from Flores, through Timor and Wetar, to Seram and Buru (Figure 1). It is commonly accepted (Audley-Charles and Milsom, 1974; Chamalaun and Grady, 1978; Von der Borch, 1979; Bowin et al, 1980) as a continent–island arc collision zone, involving the Australia-New Guinea continental mass and a subduction zone at its northwesterly margin.

Despite its geographical position, Sumba appears to share few characteristics with either the Sunda Arc or Banda Arc system, as noted by Umbgrove (1949), Katili (1971), Audley-Charles (1975) and Nishimura et al (1980).

With regard to the Sunda Arc, Sumba lies just east of and outside the negative isostatic anomaly belt of Vening Meinesz (Kuenen, Umbgrove, and Meinesz, 1934), the free air anomaly belts (Watts, Bondine, and Bowin, 1978), and the strong Bouguer gravity gradient (Green et al, 1979), that characterize the Java Trench subduction zone.

Similarly, the Outer Arc Ridge (Sedimentary Arc) is exposed at islands such as Nias, Southwest of Sumatra, and continues as a topographic expression (Java Ridge) south of Java, pinches out west off Sumba, and is separated from Sumba by the 4-km-deep Lombok Basin. The Java Trench, typically 6 to 7 km deep south of Java, shallows to 4 km deep southeast of Sumba while changing trend from easterly to southeasterly.

Hinz et al (1978) noted that the nature of the ocean floor is different between two VALDIVIA traverses across the eastern extremity of the Java Trench south of Sumba. West of latitude 119°E the topography is typically oceanic. However, to the east it resembles the crust of the southern flank of Timor. Hinz et al (1978) consider the crust between the Java Trench and Scott

Plateau to be continental in origin, and they introduce the North Wilson transform fault (Figure 1) to delineate the two areas.

It appears that some of the distinctive characteristics of the Java Trench system become ill-defined near Sumba. However, the volcanic arc north of Sumba is more or less continuous from the Sunda Arc to the Banda Arc in its essential calcalkaline characteristics and SR^{87}/SR^{86} distribution (Foden and Varne, in press; Abbott, personal communication, 1981). Centers of recent volcanism are offset north of Sumba; slightly more to the north in the islands west of Flores than in west or middle Flores (Audley-Charles, 1975). Another such offset appears between Alor and Wetar. On Lombok and Sumbawa there are isolated centers of high potassium calcalkaline volcanism, which might be indicative of deep, across-arc fracture zones (Foden and Varne, in press).

Seismicity transition from the Sunda to the Banda Arc was recently examined by Cardwell and Isacks (1978), and they concluded that the subducted slab is continuous across the Sumba sector. Below we discuss some observations which we believe suggest that the continuation of the seismicity pattern is perhaps less simple.

Geologically, Sumba does not appear to be readily correlated with other Sunda Arc islands. It lacks signs of imbrication like those found at Nias (Karig et al, 1979) and suspected to be present along the Sunda Arc outer arc ridge. Sumba also does not show any definitive geological characteristics that could be correlated with Sundaland (e.g. Kalimantan, Sulawesi). Turning to the east and the Banda Arc system, one finds that Sumba lies on a submarine ridge, which also contains the small islands of Savu and Roti. The geology of Savu, which is about halfway between Sumba and Timor, shows several elements in common with Timor and Roti. This suggests continuity along at least that part of the submarine ridge. However, these similarities are absent on Sumba. The ridge on Sumba strikes southeast while the main trend of the Banda Arc system, as exemplified by Timor and the islands immediately to the east of it, is southwest. The strong Bouguer gradient that marks Timor's north coast and extends along the Banda Arc (Chamalaun, Lockwood, and White, 1976) swings sharply south just west of Sumba. The most striking difference between Sumba and Timor is their tectonic characteristics. While Sumba shows very little deformation, Timor (<400 km away) shows very intense and extensive deformation, involving the pre-Permian basement and much of the younger formations. The deformation is almost certainly due to a collision between the Australian continental margin and a subduction zone, but the details of the collision process and the ensuing deformation are still a matter of debate (Fitch and Hamilton, 1974; Carter, Audley-Charles, and Barker, 1976; Chamalaun and Grady, 1978; Hamilton, 1979; Bowin et al, 1980).

Sumba, situated at the junction of the Banda and Sunda Arc systems, does not possess the typical collision tectonics of the former, nor the subduction tectonics of the latter. This article summarizes the geological and geophysical data for Sumba, in particular considering some observations during recent (1980) field work program. We then review the various hypotheses explaining the origin and subsequent geological history of Sumba.

STRATIGRAPHY AND SEDIMENTATION

The stratigraphy of Sumba is discussed by several workers, most significantly the summary by Van Bemmelen (1970). More recent studies are presented by Meiser et al (1965). Caudri (1934) published data on Sumba's Tertiary biostratigraphy. In the following section, the stratigraphy of Sumba is briefly summarized in light of a recent field trip to the island. We emphasize that the summary is preliminary, based largely on field observations. Paleontological examinations of specimens have not yet been made.

Generalized Stratigraphy

Figure 2 illustrates generalized stratigraphic columns of Sumba, modified from Van Bemmelen (1970). Elements such as the Eocene sands and limestones are omitted from these diagrams due to lack of data and of continuous vertical sequences.

A carbonaceous Mesozoic strata sequence of unknown thickness with no observable basement forms the substrate of the island. These rocks were tilted and intruded by numerous volcanic dikes and plutons, as discussed in the next section. Unconformities separate these Mesozoic strata from considerably less deformed Tertiary and Quarternary deposits totaling more than 1 km in thickness (Van Bemmelen, 1970). The Paleogene components of the Tertiary are generally not well exposed, at least in readily accessible areas. However, in western Sumba the Paleogene dominantly comprises volcanic diamictites and marine limestones which are separated from the significantly thick Miocene marine section by an unconformity (Van Bemmelen, 1970). Overlying this break the 1-km-thick marine Miocene Sumba Formation (Geological Survey of Indonesia), also known as the Kananggar Formation of Kinser and Deiperink (Van Bemmelen, 1970), dominates the stratigraphic sequence and much of the island's scenery.

A spectacular series of raised Quaternary coral reefs and reef talus cap the stratigraphic section in coastal areas. The stratigraphic units exposed on Sumba are discussed in detail below.

Pre-Tertiary Strata

In the present study, only Mesozoic rocks cropping out along the coast immediately south of Waikabubak and southeast of Pegunungan Tanadaro (Figure 3) were sampled. These sediments are typified by their relatively high carbon content (which often gives them a light to dark gray color), and fine grainsize (dominantly siltstones with starved ripples of fine sand and occasional units of interbedded flaggy medium sandstones). Tuffaceous components, particularly crystal

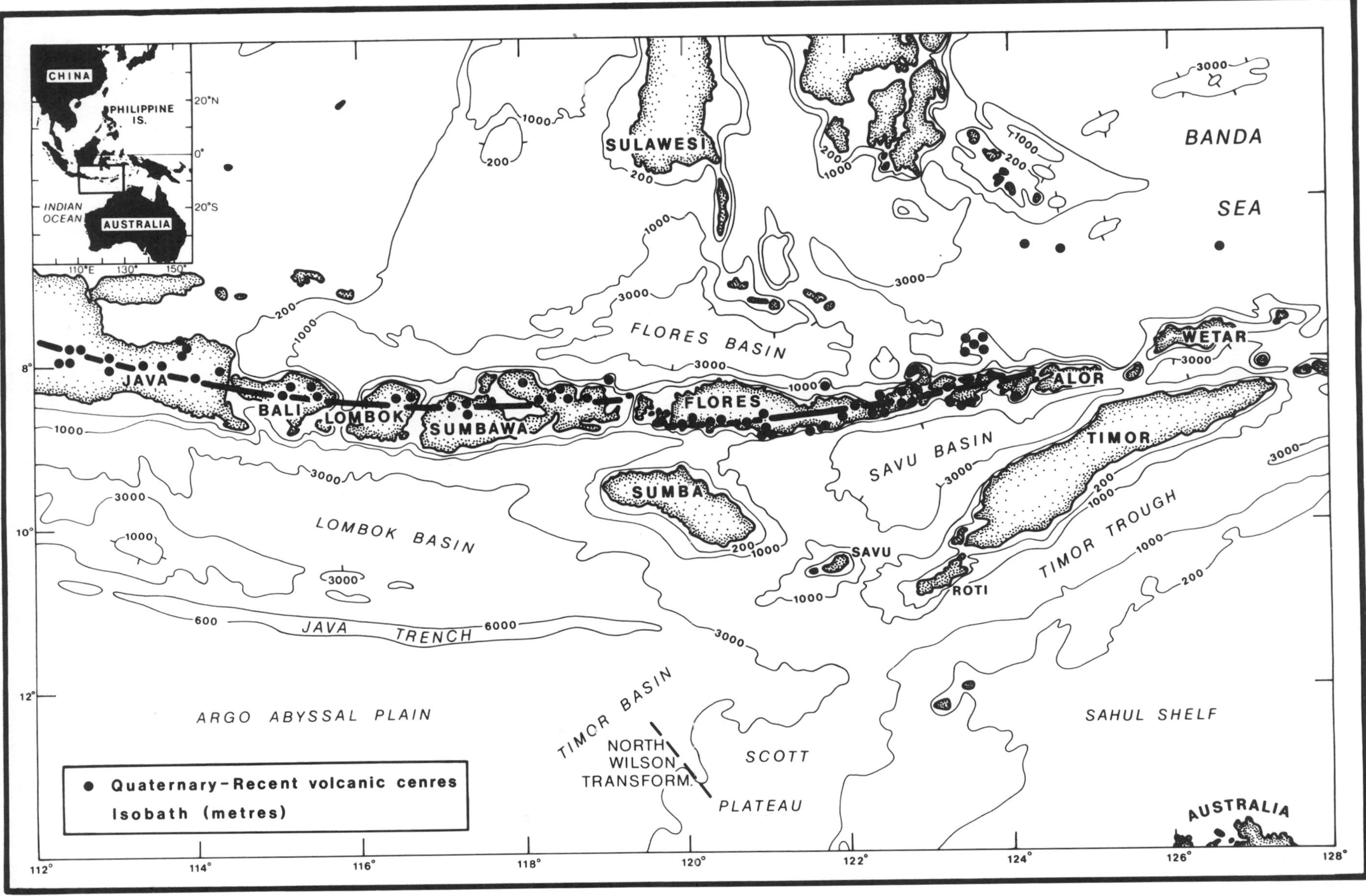

Figure 1 — Map of Sumba region with principal features. The trend of recent volcanism is shown by the heavy dashed line.

tuffs (feldspar), are also widespread. Isolated large-scale slump structures appear in some pre-Tertiary sections. Carbonized wood and leaves are rare to common components of some of the gray siltstones. Unfortunately, all attempts to isolate polynomorphs so far have failed, possibly due to destruction by relatively high temperatures to which the rocks were submitted during their pre-Tertiary history.

Rare large molluscs occur in some of the carbonaceous siltstone units. These appear to be a species of Inoceramus, suggesting a Jurassic to Cretaceous age for the pre-Tertiary strata associated with these occurrences. The only other faunal elements isolated so far in this study are rare calcareous concretions from some of the carbonaceous siltstones. These contain abundant, poorly-preserved and, as yet, unidentified microfossils and possible wood fragments.

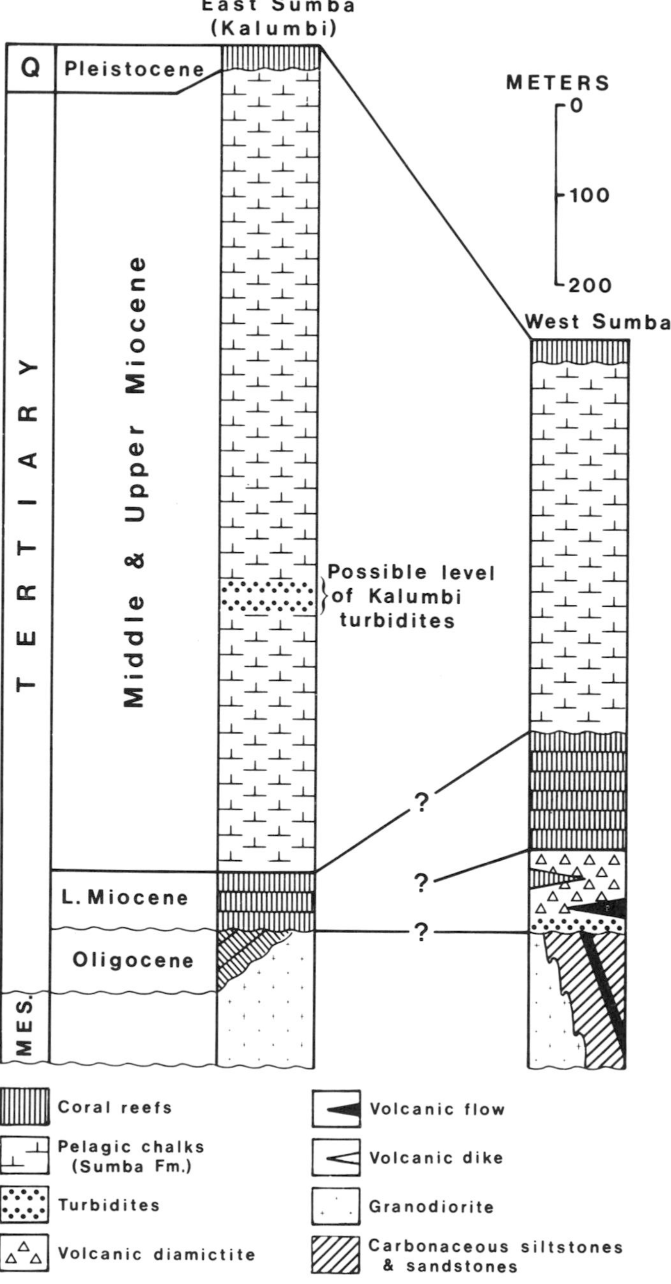

Figure 2 — Generalized stratigraphic sections of east and west Sumba, modified after Van Bemmelen (1970).

Paleogene Strata

Outcropping Paleogene rocks between Waikabubak and the adjacent south coast include Eocene and Oligocene marine limestones and sandstones, not shown in Figure 3. Volcanic agglomerates occur near the Bondokapu Complex (Figure 3) to the west. However, Paleogene sedimentation might be relatively insignificant compared with that of Mesozoic and Neogene times. This suggests that Sumba stood mostly above or just below sea level during much of the Paleogene.

Neogene Strata

Volcanic diamictites of possible Miocene age form thick sequences which are well exposed along the road and coastline south of Waikabubak. These diamictites contain variegated volcanic clasts, mainly of acid composition, in a fine-grained glassy matrix. Channelled sands are common within the diamictite sequence, with tuffaceous siltstones, lignites, and pumiceous tuffs forming associated facies. Bedded lavas are found in at least one locality, and localized coral reef buildups occur within the diamictite sequence. Silicified logs are a common association, particularly in the coastal exposures of diamictites south of Waikabubak where the volcanogenic sediments directly overlie pre-Tertiary basement.

Above the volcanic diamictite unit near Waikabubak lies the Sumba (Kananggar) Formation, a flat-lying, undeformed sequence of pelagic chalks and reefal limestones containing significant tuffaceous units rich in pumiceous clasts. Reef buildups and related back-reef calcareous grainstones form significant geomorphic elements, mainly as low hills, in the western Sumba region. This is due to enhanced cementation of reef facies carbonates by sparry calcite, which renders the reefs and associated calcareous grainstones considerably more resistant to erosion than the intercalated lagoonal wackestones and packstones.

A vertical sequence measured through part of the Sumba Formation contains, within the pelagic chalks, a well developed turbidite sequence at least 100 m thick. These turbidites crop out at river level, thus their relationships with underlying strata are uncertain. They may be of middle to upper Miocene age, as shown on the stratigraphic columns in Figure 2. Intercalated thin and thick bedded turbidites occur, comprising arkosic gravels and sands, large formainifera, and abundant resedimented pumice. The sand and gravel units have sharp bases and are crudely graded, often with convolute bedding. Individual sand and gravel beds vary from 30 cm to 1 m thick. Occasional channel-fill units, possibly representing former fan valleys, occur up to 8 m thick. Pelagic layers between the sands comprise white chalks similar to those characterising the Sumba Formation elsewhere. Some sections of the turbidites are dominated by multi-

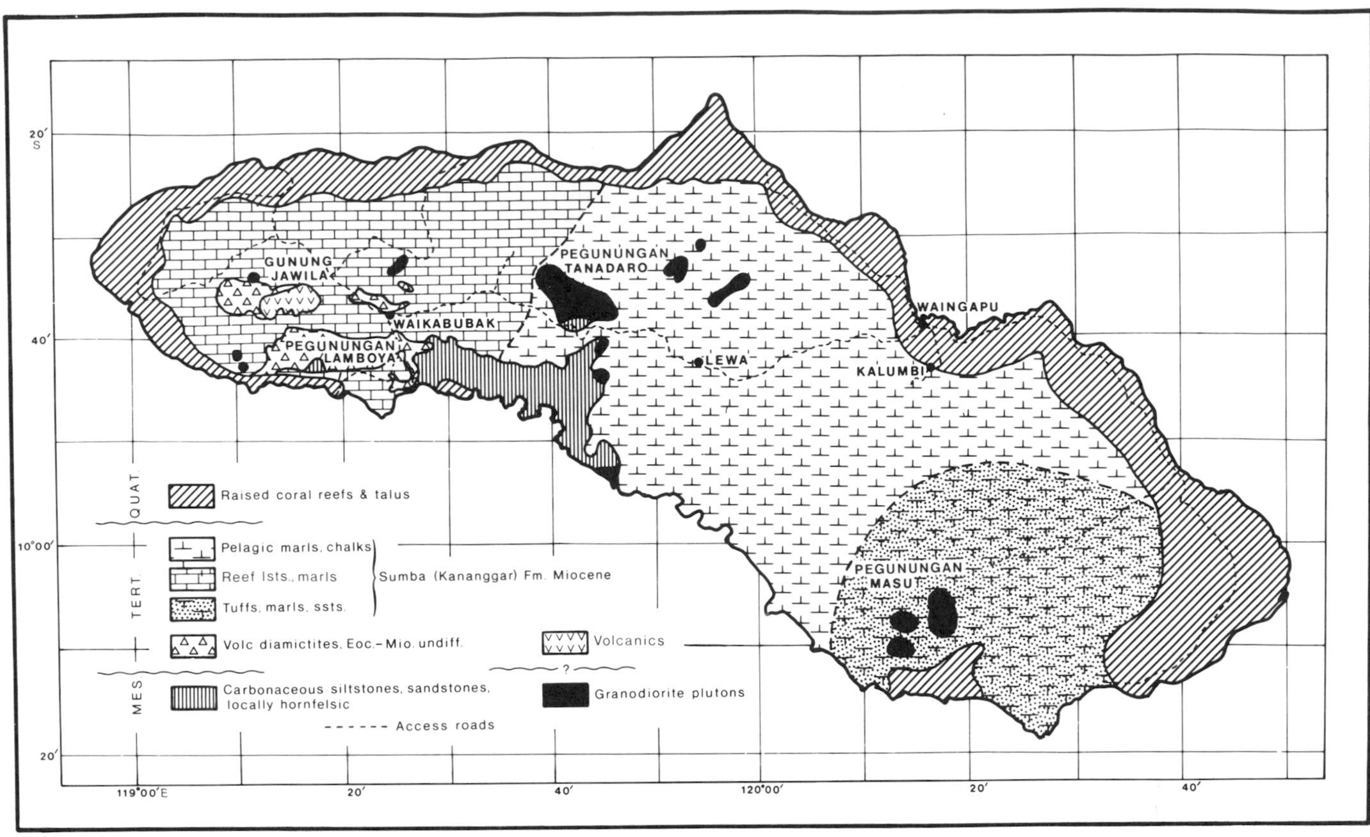

Figure 3 — Generalized geological map of Sumba after Meisser et al (1965).

storied lithologies of sandy turbidites or channel fills, often containing distorted pelagic limestone clasts measuring meters in length, as well as abundant pumice fragments.

In vertical succession, the turbidite sequence near Kalumbi appears to pass up section to the massive channel sands described above. Above these sands tuffaceous marls, intercalated with increasingly tuff-free pelagic chalk layers, appear to dominate another 40 meters up section. Ultimately, the tuff component decreases significantly and is replaced by clean, white pelagic chalks which may constitute the remainder of the 1 km thick Sumba Formation in this area.

Quaternary Raised Coral Reefs

Quaternary raised coral reef terraces, containing reef and associated back-reef and fore-reef elements, overlie the Sumba Formation along the island's north coast. Quaternary fore-reef talus with localized conglomerate-filled submarine channels overlie pre-Tertiary strata along portions of the southwest coast near Waikabubak (Figure 3). In all areas, evidence of considerable Quaternary uplift of Sumba is indicated by the reefs and associated submarine deposits which are now elevated tens to hundreds of meters above present sea level. Although no studies of these raised reefs are published, it is likely that a significant rate of uplift is continuing at the present day.

Sedimentary History

The interpretation of the stratigraphic sequences described in this study generally supports the description of the geological evolution of Sumba described in Van Bemmelen (1970). This is outlined in Figure 4, which is a generalized diagram summarizing the key stratigraphic and igneous events of the island in terms of the Sumba block's vertical movements.

The pre-Tertiary sequence of Jurassic or Cretaceous carbonaceous sediments may represent a former Tethyan passive continental margin sequence, deposited in a variety of subareal, lacustrine, and submarine sedimentary environments in a subsiding rifted continental setting. The relatively high carbon content and the presence of carbonized plant debris point to deposition within an oxygen minimum zone, or on the floor of an anoxic basin which received an abundant supply of continental sediments and terrigenous organic matter. The existence of carbonaceous sediments may relate to one or more of the widely documented world wide Cretaceous anoxic events.

Although such correlations are not proven, there remains little doubt that rapid terrigenous deposition occurred at least during mid to late Mesozoic times in a relatively restricted and at least partly marine basin, which in turn may have been a continental rift developed prior to a phase of sea-floor spreading during the breakup of Gondwanaland. Bedded, possibly coeval volcanics and ubiquitous tuffs form an association consistent with such proposed early rift tectonics.

Following Mesozoic sedimentation a magmatic and volcanic episode occurred, possibly related to the significant phase of uplift and widespread subareal erosion that followed and unroofed the granodiorite plutons and developed the widespread pre-Tertiary unconformity.

It is evident from Figure 3 that much of what is now Sumba was subareal to shallow-marine during the Paleogene and early Neogene. The Eocene and Miocene volcanic diamictites, the latter with abundant silicified logs and some lava flows, attest catastrophic volcanic mudflows relating to contemporaneous volcanism. These mudflows occasionally flowed into shallow-marine and logoonal regions, as evidenced by intercalated coral reef buildups and lignites of possibly lagoonal origin.

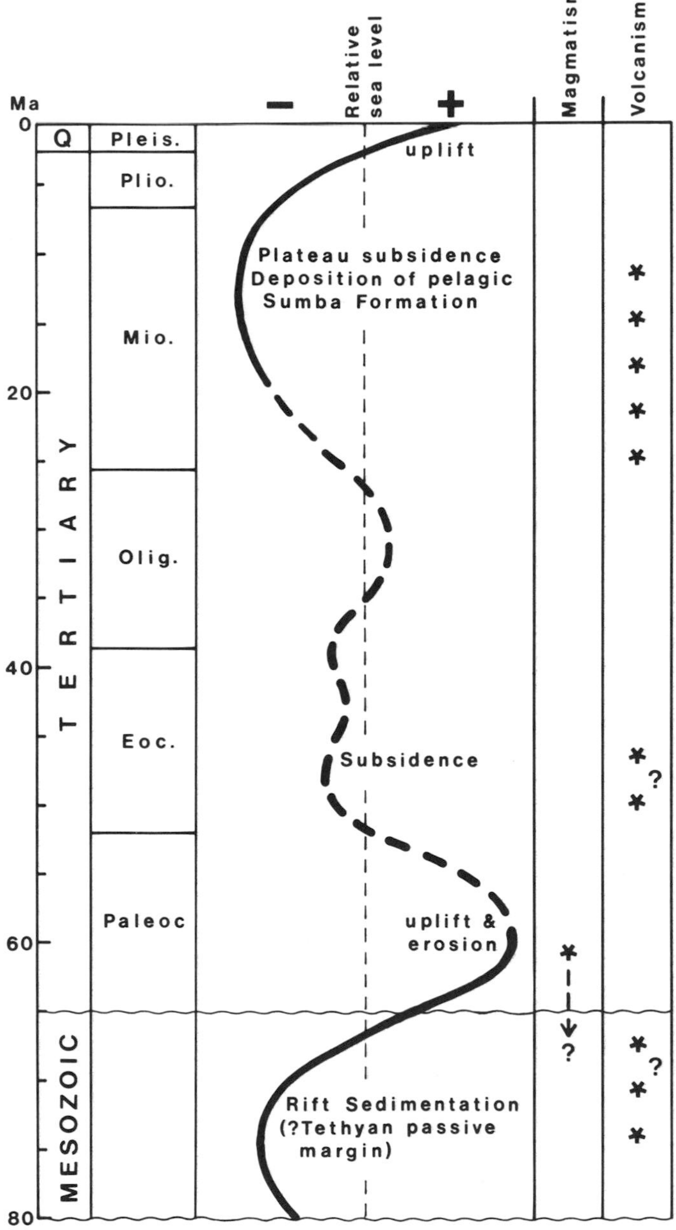

Figure 4 — Interpretation of vertical tectonic components of Sumba based on stratigraphic information. Possible periods of magmatism and volcanism are also indicated.

During the Miocene, much of Sumba subsided to depths well below sea level. A widespread shallow platform with abundant coral reefs initially dominated the region. This eventually subsided below the maximum depth of coral growth, at which time pelagic chalks and marls dominated the sedimentation particularly in eastern Sumba. Areas of east Sumba apparently subsided more deeply than western regions, but the sediment interface remained always above the local Miocene carbonate compensation depth. Deeper regions, at least in the east, received abundant sedimentation from the activity of turbidity currents followed by deposition of pelagic chalks. The turbidites originated from unknown sources, but are postulated to have come from what may have been high-standing areas near Pegunungan Masu to the southeast (Figure 3). Upward coarsening and thickening sand and gravel units described in the river gorge section near Kalumbi suggest deep sea fan or suprafan progradation, with the thick gravelly sand units representing fan valley fill emplacement near a prograding fan apex.

Abundant pumiceous volcanic debris entrained in the turbidites and interbedded as tuffs in the pelagic chalks of the Sumba Formation suggest active volcanism occurring nearby, possibly waning during later stages of the Miocene. Considered overall, the pelagic chalk section (overlying basinal turbidites which in turn may overlie reef limestones) is consistent with subsidence of the Sumba block, possibly coupled with adjacent cooling and subsiding oceanic crust during a sea floor spreading episode.

The final phase of vertical tectonism recorded by the sediment column is the regional uplift, which began in the Pliocene or Quaternary and continues to the present. Raised Quaternary reefs and talus deposits, which occur in northern and southern Sumba, may imply uplift with no significant tilting; detailed neotectonic movements have not yet been established. It appears certain that this latest phase of Sumba uplift correlates with its relative recent involvement with the Banda Arc region of convergent tectonics.

IGNEOUS ACTIVITY

Igneous rocks, while not predominant rock types in Sumba, form a very important component of its geology and contribute significantly to its distinctive character. A problem which is still not resolved for many occurrences is determining the emplacement age. Therefore, the chronologically based groupings used below are not absolutely established to date due to insufficient detailed data. Figure 4 includes a summary of the time variation in igneous activity for the island.

Pre-Tertiary

The Cretaceous -? Jurassic sedimentary sequence contains rare, scattered basaltic to andesitic flows (the degree of alteration evident in these rocks makes exact classification difficult without recourse to detailed geochemical data). Early reports, Van Bemmelen (1970), on this stratigraphic unit place considerable emphasis on igneous content which, from our field experience, is probably due mainly to pervasive younger dikes and intrusions referred to later in this section.

The only other evidence for contemporary (or preceeding) igneous activity found in Pre-Tertiary rocks is from the previously mentioned igneous-derived material occurring as a clastic component within some beds. Throughout the observed sequence occasional very thin beds, usually comprising coarse sandy sediments, contain angular plagioclase and epidote grains and rare fragments of feldspathic volcanic rock. On the southern slopes of Pegunungan Tanadaro, some relatively massive beds of coarse volcaniclastic breccia within the pre-Tertiary sequence suggest derivation from an adjacent eruptive centre. The exact age of these beds is unknown; they appear to conformably overlie a northerly dipping sequence of the pre-Tertiary (presumed Cretaceous) carbonaceous dark shales, siltstones, and silty sandstones. The whole sequence appears to be uncomformably overlain by the Miocene limestones.

Tertiary

Dikes and small intrusions

One of the most striking features of the pre-Tertiary sequence is the almost pervasive occurrence of generally thin (< 5 m) dikes and small intrusions of dacitic to andesitic composition. Again the intensity of alteration precludes more specific classification of rock types. In traverse mapping, the observed spacing of these intrusions is from several meters to less than 100 m. They seem to be remarkably constant in composition, which is similar to that of the larger dioritic-granodioritic intrusions (described below) and may be comagmatic and coeval with the larger intrusions. They show no evidence of internal deformation except along rarely observed younger faults. These intrusions are not known to intersect any of the units in the Tertiary succession. On this basis their age is presumed to be late Cretaceous to earliest Tertiary.

Granodiorite-diorite-gabbro intrusions

Scattered across the island, mainly in the centre and southern half, are exposures of granodiorite, diorite, and (rarely) gabbro which occur as inliers within the Tertiary sequence or as intrusions within the pre-Tertiary sequence. Intrusion size ranges from extremely small bodies to bosses or stocks; the largest intrusions occur at Pegunungan Tanadaro in the centre of the island. They are all medium to coarse grained and non-foliated, and show no signs of pervasive post-crystallisation deformation. All exhibit, in variable intensity, post-crystallization alteration.

These intrusions, like the apparently related dykes, clearly post-date the Cretaceous to ? Jurassic sediments. There is still some uncertainty about the upper limits on their intrusion age because, although they appear to be overlain by most if not all of the Tertiary sediments, uncertainties still remain concerning detailed contact relationships with some basal Tertiary units. Very meagre preliminary radiometric age determinations (Geological Research and Development Center of Indonesia, written communication, 1980) suggest a minimum age of early Paleocene.

Paleogene lavas, ignimbrites and volcaniclastic breccias

Paleogene volcanic rocks occur in three main masses: in and around Pegunungan Masu (eastern south coastal); north and east of Pegunungan Tanadaro (central); and, in and east of Pegunungan Lamboya (western south coastal). Other small isolated outcrops occur along the south coastal region. The rock types represented in this group are basaltic-andesitic-dacitic lavas, small intrusions, ignimbrites (particularly at Gunung Bondokapu west of Waikabubak and about 10 km north of Lewa, and other as yet unrecognized occurrences may exist), and volcaniclastic breccias containing clasts of acid to basic composition. All occurrences observed so far show evidence of significant post-crystallization alteration, but no evidence of pervasive deformation. In most localities, strata of this group have moderate dips suggesting at least local tilting.

The exact age, or age range, is difficult to establish. Some members intrude, and others appear to unconformably overlie the Cretaceous to ? Jurassic sequence; others are interbedded with or closely related to sedimentary strata of Paleogene or Eocene age. On local and regional scales, rocks of this group are unconformably overlain by Miocene sediments. Very sparse, preliminary radiometric age dating suggests a minimum age of late Eocene for some members (Geological Survey of Indonesia, written communication, 1980).

Miocene lavas and volcaniclastic breccias

On Gunung Jawila (western Sumba), in the vicinity of Waikabubak (particularly south of the town) and on the eastern side of Pegunungan Masu, are exposures of basaltic-andesitic lavas and related volcaniclastic breccias mentioned in an earlier section. Rocks of this group uncomformably overlie the pre-Tertiary sediments and overlie with apparent unconformity various units of Paleogene sediments and volcanic rocks. They are interbedded with and overlain by lower Miocene reefal limestones and clastic sediments. Representatives of this group show some post-crystallization alteration effects which are generally less intense than those for all older igneous rocks. They show no evidence of pervasive deformation, and bedding dips are gentle.

Middle and Upper Miocene Tuffs

Within the Miocene pelagic chalk-reef limestone sequence (see stratigraphy and sedimentation) are many thin units characterized by pumiceous tuff content. The origin of this volcanic component is unknown but it may have been derived from the volcanic inner Banda Arc, which is nearby to the north.

Metamorphism

Regional

Published accounts of the geology of Sumba place little emphasis on the effects of regional metamorphism. Roggeveen (1932) recorded the occurrence of epidote, chlorite, and calcite as secondary minerals within igenous rocks.

Our investigations of material from past expeditions to the island, supplied by the Geological Research and Development Center, Bandung, and from our own recent field work show that low-grade metamorphism has significantly affected most pre-Miocene rocks. Neocrystalline sericite, epidote, chlorite, calcite, actinolite, and serpentine minerals are the dominant indicators of metamorphism in both igneous and sedimentary rocks of appropriate compositions. Less commonly, prehnite and/or pumpellyite and/or zeolite minerals are associated products in igneous or igneous-derived rocks. Alteration varies from intense (rocks in which the original texture is almost destroyed) to incipient (rocks showing only partial sericite-epidote-carbonate alteration of feldspar, and chlorite-epidote-carbonate alteration of mafic minerals). Secondary veining is often pervasive, with carbonate, quartz, epidote, chlorite, and plagioclase (in decreasing order of abundance) as the common vein minerals.

The assemblages developed indicate that lower greenschist or sub-greenschist facies (probably prehnite-pumpellyite metagreywacke facies) metamorphism was widespread. Its effects are most noteworthy in the pre-Tertiary rocks, but are also evident in the Tertiary dikes and intrusions (all sizes), and the Paleogene lavas, ignimbrites, and volcaniclastic breccias. Miocene igneous and igneous-derived rocks show incipient alteration along grain boundaries and internal fractures, possibly due to weathering.

Contact

Some intrusions show localized and subdued contact metamorphic effects, but these are mainly restricted to slight recrystallization or baking of the contact rocks. Only in the Pegunungan Tanadaro granodiorite are the effects marked. On its southern margin, the zone of baking is perhaps 1 km wide and pelitic beds in this zone show pervasive neocrystallization of actinolite within the groundmass.

Structural Geology

A striking feature in the geology of Sumba is the general lack of evidence for intense deformation on both mesoscopic and macroscopic scales. Available data are insufficient for detailed discussion, but the following general comments can be made.

Pre-Tertiary

The Cretaceous to Jurassic sequence is characterized in general by lack of cleavage; locally a very weak, steeply dipping cleavage develops. Broad open folds in the region south of Pegunungan Tanadaro appear to have generally sub-horizontal easterly trending fold axes. Internal strain was apparently very low, evidenced by undeformed microfossils and other originally spherical microstructures. Observed bedding has low to moderate dips and is not overturned except in minor-slump structures. Macro-fold wavelengths seem to be near kilometers, while no significant shear zones, mylonite zones, or thrust faults have been observed.

Tertiary

All Tertiary rocks show no signs of penetrative mesoscopic deformation and bedding dips are low to moderate. In Paleogene sequences, bedding orientation is fairly consistent in small areas but variable regionally. Available data are too meagre for a systematic analysis. Broad folding, block tilting, and variable initial repose orientation are all possible models, not necessarily mutually exclusive, to explain this effect.

Neogene strata are strongly characterized by low dips, and open, gentle folds are developed but difficult to distinguish from variations in initial repose orientation. On a regional scale, dips are either away from pre-Neogene basement highs, or generally to the north.

Unconformities

The basal Tertiary unconformity shows evidence of original moderate relief, however the effects of later faulting have not been adequately analysed. Similar comments can be made about the basal Neogene unconformity.

A low angle unconformity was reported (Laufer, 1950) within the Miocene pelagic chalk-tuff sequence about 25 km west of Waingapu, beside the road from Waingapu to Lewa. This feature was not reported elsewhere and its extent and significance are unknown.

Faulting

All pre-Quaternary rocks are intersected by steeply dipping faults. Vertical movements of up to 400 m are known but no accurate picture of fault-related strain is available. The orientation of fault strikes is quite variable on the regional scale, and east to east-southeast striking faults are common, as are those striking north to north-northeast and others striking generally northwest. No data are available on age relationships between these groups, and all pre-Quaternary units are affected by members of each of these orientation groups. At the present state of mapping there is no indication of zones of significant faulting.

GEOPHYSICS

The geophysical data base for Sumba is even more scanty than the geological one, and much of it is of a preliminary or reconnaissance nature. The key areas of interest are the nature of the crust near Sumba, and any evidence that might suggest the presence of transform faults. Green et al (1979) published a Bouguer gravity map which incorporates the results of a detailed gravity survey carried out by the Geological Research and Development Center of Indonesia under the supervision of Untung. The Bouger gravity field is virtually featureless and gives an anomaly of + 160 to + 200 Mgal. No other geophysical surveys (seismic or magnetic) that might help to delineate the deep crustal structure of Sumba are reported. Although several shipborne surveys were carried out near the island, none of the data are published or readily accessible. In 1977 we participated in a VALDIVIA Cruise (Hinz, 1977) which collected gravity, magnetic, and reflection data between Timor and Sumba and the Savu Sea. In addition, an unreversed refraction profile was shot along the northeast coast of Sumba. The data is still not completely processed but preliminary indications (J. Sunderland, written communication, 1980) suggest that Sumba is continental with a 24 km thick crust. The crust appears to thicken toward Timor, where it is estimated at 40 km (Chamalaun, Lockwood, and White, 1976; Bowin et al, 1980). This crustal thickening appears to be confirmed by the gravity gradient east of Sumba. During the VALDIVIA survey no conspicuous geophysical features were discovered north of Sumba, but there was an indication of some energy loss in the refraction shooting between Savu and Sumba.

Cardwell and Isacks (1978) interpreted the seismicity pattern to show that the Benioff zone from the Sunda Arc to the Banda Arc is continuous across Sumba. They examined the seismicity using various "windows" oriented perpendicular to the volcanic arc. They observed that in a window 4.5° wide (west of Sumba), the Benioff zone dipped steeply down; in a similar sized window across Timor, the dip was less steep and the Benioff zone showed a pronounced flattening. Between these two windows they combined all the data over a 6° wide window (which includes Sumba) to show that the Benioff zone here is intermediate between the two. Since their intepretation of continuity across Sumba depends strongly on their choice of window, we re-examined their Sumba window using the same data set with the following result. Figure 5 shows the shallow (<100 km) and intermediate to deep (300 to 700 km) earthquake centers. Note that west of longitude 121°E the shallow

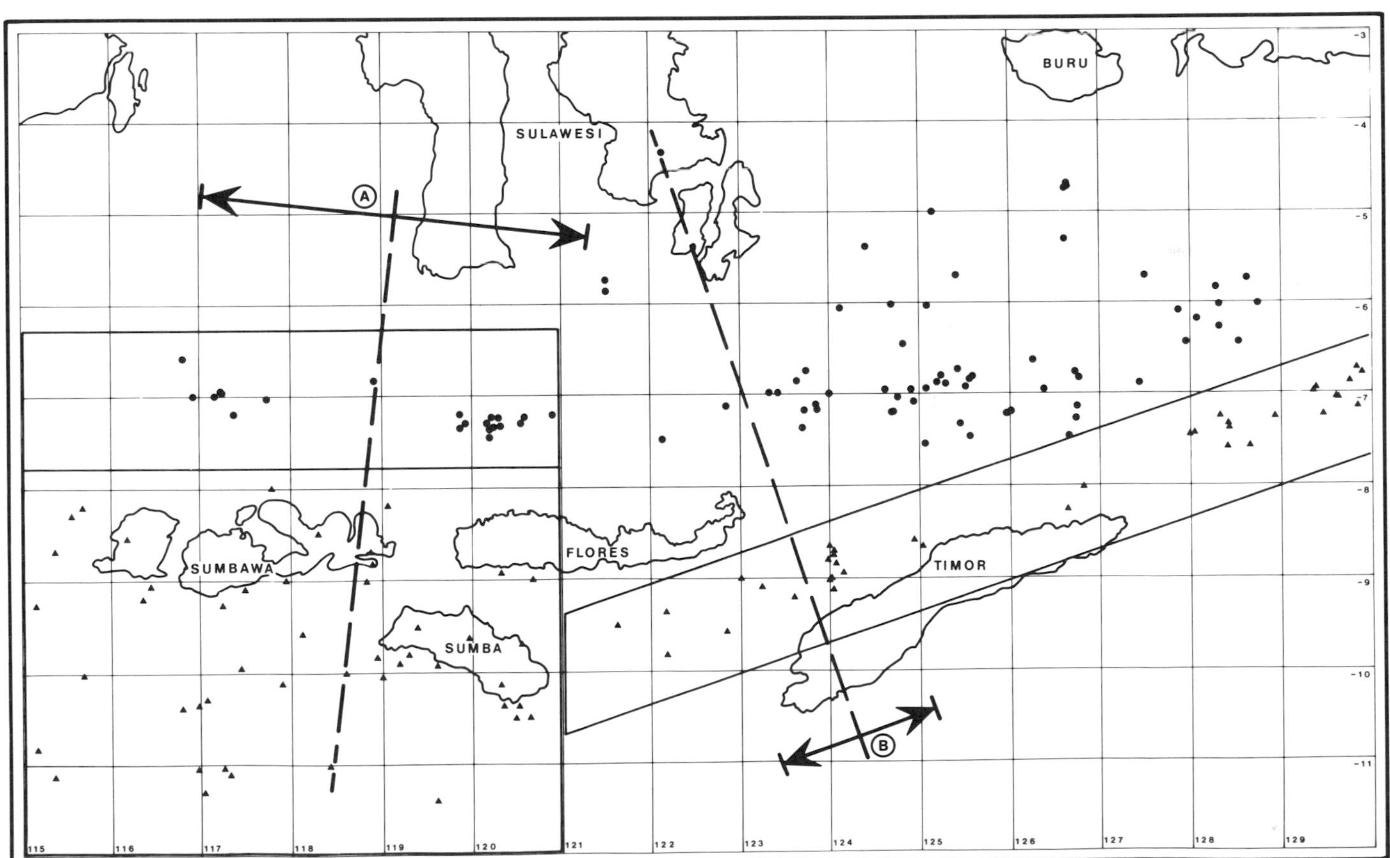

Figure 5 — Seismicity map based on the data of Cardwell and Isacks (1978). Triangles = epicenters with depth less than 100 km. Solid circles = epicenters with depth greater than 100 km. Regions delineated by solid lines are discussed in the text. A and B = windows and central plane of projection of seismic foci used in Figure 6.

earthquakes lie in a broad, 4.5° wide zone, whereas to the east they are confined to a narrow 1.5° zone. Intermediate depth earthquakes show a reverse pattern, with a narrow zone to the west and broad zone to the east. The change in the two patterns appears sharp and distinctive. The shapes of the corresponding Benioff zones are shown in Figures 6a and 6b. West of longitude 121°, the Benioff zone appears to be typical of an oceanic subduction zone, with a broad area of shallow earthquakes and a steeply dipping zone. To the east, shallow earthquakes are almost absent and the Benioff zone less steep, eventually flattening at depth. Further east, the depth and extent of the flattening diminishes. We submit that these observations admit an alternative interpretation, in which it is supposed that the Benioff zone is not necessarily continuous, but changes character quite sharply at about 121° longitude. Furthermore, we suggest that this discontinuity in the seismicity is probably the best evidence so far to locate an East Sumba transform fault. We also note that on this interpretation, as well as on the Cardwell and Isacks (1978) interpretation, Sumba is now situated directly behind (north of) the active Java Trench subduction zone.

Paleomagnetism provides direct evidence for the paleogeographical position of Sumba relative to other land masses by comparing paleomagnetic poles obtained from formations of comparable age. The igneous Tertiary and pre-Tertiary rocks, as well as the tuffaceous members of the Tertiary limestone cover, provide prime paleomagnetic targets. Igneous activity shown in Figure 4 suggests that an upper Cretaceous to Miocene polarwander curve may be constructed, provided the target rocks exhibit a stable and reliable paleomagnetism. However, accessibility to sample sites is poor, weathering is severe, and at this stage many of the critical stratigraphical relationships are uncertain. It may be some time before a clear paleomagnetic picture emerges.

The first paleomagnetic study was reported by Otofuji et al (1980) and Nishimura et al (1980); their poles are shown in Figure 7a. In 1979 and 1980, we collected additional paleomagnetic samples. The 1979 collection was designed primarily to test the paleomagnetic suitability of various rock types. The 1980 collection, comprising some 100 samples, is now being processed. Figure 7a shows our preliminary 1979 results. The poles were obtained after extensive thermal and alternating field demagnetization.

Comparing the two Sumba data sets, we note that there is reasonable agreement between the two investigations in the distribution of poles, particularly with regard to the main groups. However, there are uncertainties with regard to the age relationships between the two collections. This is partly due to the general lack of age control, but also because the information published by Nishimura et al (1980) is insufficient to relate to our information with any confidence. A critical discrepancy is between our results PS_2 and their IS143, IS144, and IS145. In our case, the result is obtained from rhyolites in the southwest and microgranites in the central east, to which we assigned an upper Cretaceous or lower Eocene age. The samples from Nishimura et al (1980) are mudstones from the western north coast and described as Jurassic.

A second difficulty arises with respect to our pole

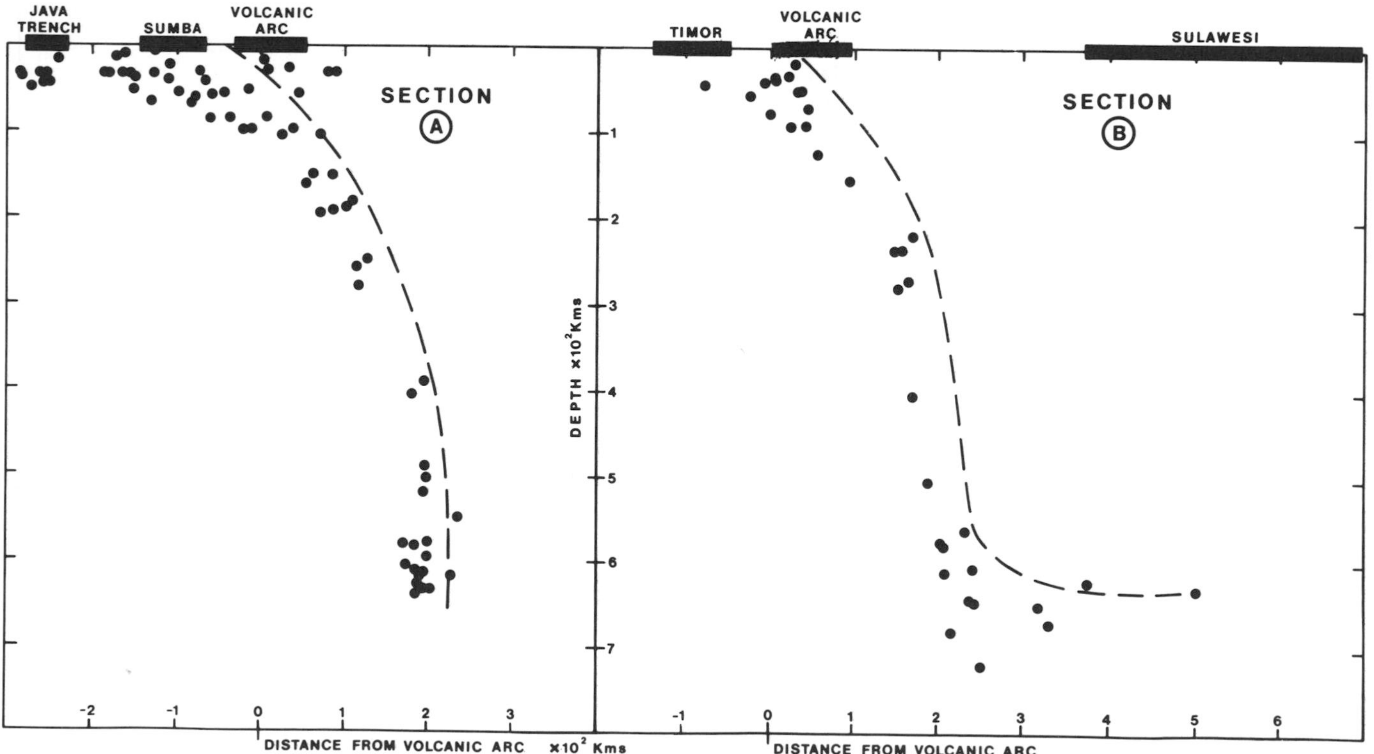

Figure 6 — Benioff zones corresponding to sections A and B of Figure 5.

PS1 and their pole IS137. Ours is based on three samples of the central granodiorite and three associated dikes in central Sumba; theirs is based on what is described as a Miocene mudstone. Again, the granodiorite age is Cretaceous rather than Miocene (see above). Direction and polarity seem to be in superficial agreement.

In view of these difficulties and the early stage of paleomagnetic investigations, it would be unwarranted to attempt a detailed interpretation of Sumba's paleogeography. However, as a first attempt you could compare the poles in Figure 7a with the poles obtained for Australia in Figure 7b and those from Sundaland (Figure 7c). No obvious correlation appears. For Australia, Otofuji et al (1980) suggested that a counterclockwise rotation of Sumba, relative to Timor, would coincide the Jurassic (?) pole with the Mesozoic or Permian of Timor (Chamalaun, 1977a, 1977b). Furthermore, such a rotation would bring the axis of Sumba in line with Timor. They suggested a rotation over 70°, with the pole of rotation just west of Roti. If Timor was part of the Australian continental margin since the Permian (Chamalaun 1977b; Chamalaun and Grady 1978), then this rotation would cause a considerable mismatch for poles younger than the Jurassic. Furthermore the rotation through 70° would cause Sumba to overlie the Scott Plateau, which is believed to have been part of the Australian margin since the pre-Permian (Hinz et al, 1978). On the other hand, if Timor was never part of the Australian margin, or at least not since the Jurassic, then such a rotation can be checked by obtaining Mesozoic paleomagnetic data for Timor or comparable late Mesozoic and early Tertiary formations.

For Sundaland, the main poles for comparison are the Cretaceous of Malaya (McElhinny, Haile, and Crawford, 1974), Borneo (Haile, McElhinny, and McDougall, 1977), and the Jurassic of Sulawesi (Haile, 1978). The agreement between these poles was interpreted by Haile, McElhinny, and McDougall (1977) to mean that all three areas formed a unified landmass in the Cretaceous. To effect a fit between these and Sumba, a relative rotation of 180° is required. (The Cretaceous pole obtained by Sasajima et al, 1980b, appears to be anomalous and awaits further work.)

Although present data from Sumba are not particularly reliable, they do show significant departures from either Australia or Sundaland. Since the island is not markedly deformed, this observation may be significant by suggesting that Sumba may have undergone a drift episode, independently from Australia and Sundaland.

TECTONIC HYPOTHESES

The previous section makes the point, as others have made before, that the geological characteristics of Sumba do not readily fit its tectonic position either as part of the Sunda Arc or Banda Arc Systems.

In delineating the tectonic history of Sumba, the first and most critical question is that of Sumba's origin. Assuming that in pre-Jurassic times Australia and Sundaland were separated by the Tethys ocean, you may consider three classes of hypotheses:

(1) Sumba was a microcontinent by itself or part of a larger, now fragmented, continent within Tethys, de-

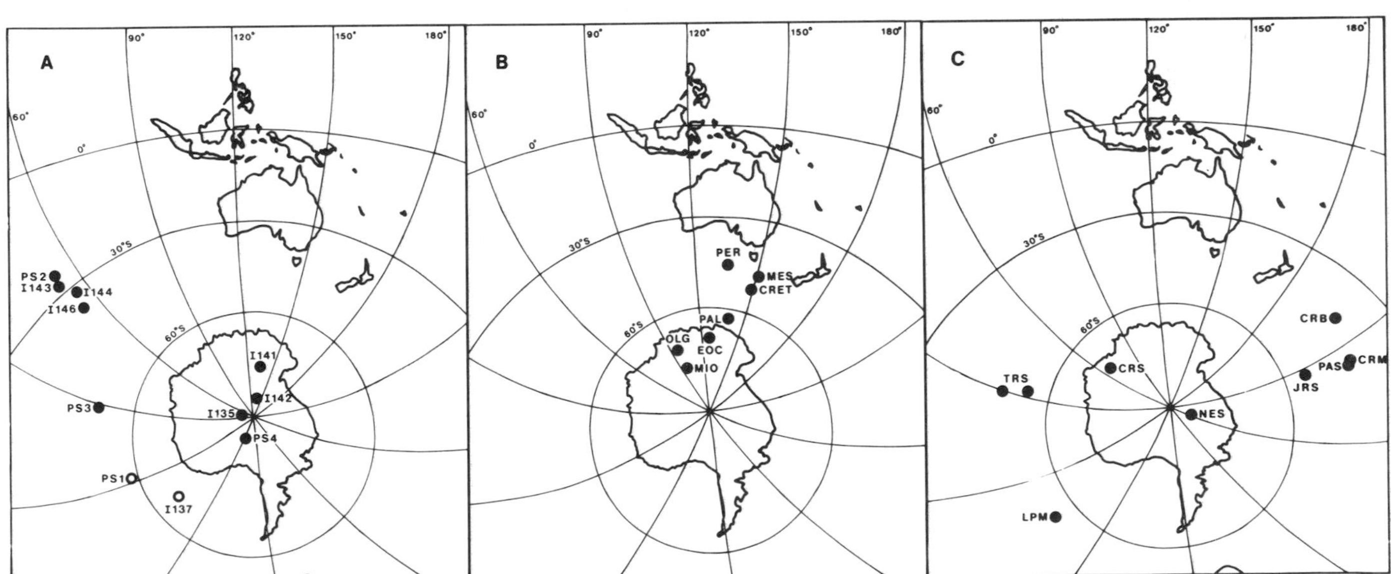

Figure 7 — Paleomagnetic poles. Solid circles, south poles. Open circles, north poles. **(A)** Paleomagnetic poles for Sumba. Poles identified by *I* from Otofuji et al, (1980). Poles with *P* recent 1979 collection (see text). **(B)** Paleomagnetic poles from Australia (McElhinny, 1973). PER = Permian, MES = Mesozoic, CRET = Cretaceous, PAL = Paleogene, EOC = Eocene, OLG = Oligocene, and MIO = Miocene. **(C)** Paleomagnetic poles from Sundaland. TRS = Triassic from Sumatra (Haile, 1979; Sasajima et al, 1980a), CRM = Cretaceous Malaya, LPM = Paleozoic Malaya (McElhinny et al, 1974), NES = Neogene Sulawesi, PAS = Paleogene Sulawesi, CRS = Cretaceous Sulawesi (Sasajima et al, 1980b), JRS = Jurassic Sulawesi (Haile, 1978), and CRB = Cretaceous Kalimantan (Haile et al, 1977).

scribed as the "Within Tethys" origin.

(2) Sumba originated as part of the Australian continental margin, described as the "South of Tethys" origin.

(3) Sumba may have been part of Sundaland described as the "North of Tethys" origin.

We will consider these hypotheses in light of our current knowledge. While we are aware that Sumba should not be treated in isolation from the remainder of the archipelago, we cannot examine any but the most obvious consequences on a regional scale.

"Within Tethys"

The "Within Tethys" hypothesis assumes that Sumba was a separate continental fragment (or perhaps part of a former larger mass) within the Tethys ocean. You might invoke scenarios similar to those of Sasajima et al (1980a) for Sumatra, or Coney, Jones, and Monger (1980) for allochtonous western North America terranes. The principal difficulty with this approach is that it introduces too much freedom, making it nearly impossible to devise adequate tests.

Pre-Tertiary sediments suggest that Sumba was not isolated at that time because of the need to postulate a significant sediment source. This means that Sumba is a fragment of a pre-existing major continental mass, which is now difficult to identify. You could consider that other islands of the Banda Arc system (possibly including the Sula Spur) are also fragments of such a pre-existing continental mass, but then you would need to propose a mechanism for its fragmentation. This brings us outside the realm of permissible speculations. Furthermore, although admittedly the geology of most of the other islands is poorly understood, we have found no features that interlink their geology with Sumba. Paleomagnetism is unlikely to provide an adequate test, as it requires formations of preferably Permian but at least undoubted Jurassic age. These have not been found. We therefore consider this hypothesis as a 'default' solution.

"South of Tethys"

The simplest of the "South of Tethys" hypotheses supposes that Sumba is derived from the northern Australian continental margin. Audley-Charles (1975) discussed this type of hypothesis in detail and proposed that Sumba rotated clockwise relative to Australia along a Sumba fracture zone, which he suggested had southwest to northeast orientation along the eastern edge of the Wharton Basin.

Sumba's exact position on the Australian continental margin is problematical, but a reasonable choice is a position near the Scott Plateau. This position involves a minimum rotation, and it is consistent with Warris's (1973) interpretation of the sedimentary regime of Australia's northwest and Sahul shelves. Warris (1973) summarized data obtained from oil exploration in the region and suggested that a western land mass must have existed just west of the Scott Plateau. He also suggested that the tectonics of the Sahul shelf were indicative of a rift-drift regime which began in the Jurassic and terminated in the early Cretaceous.

It is tempting to identify Sumba as part of such a western land mass. The proposal implies that Jurassic strata on Sumba should be correlated with the Jurassic strata on the Sahul shelf. At this stage, Jurassic age strata on Sumba is conjectural and no paleontological or paleoenvironmental data are available to attempt such a correlation. Similarly, the suggested correlation of the Jurassic basalts in the Ashmore Reef and Cartier Reef wells and the volcanics in the pre-Tertiary strata of Sumba (Audley-Charles, 1975) is somewhat speculative. However, it should be possible to test this hypothesis through geochemical and paleoenvironmental studies.

Rifting of the western land mass and hence Sumba would probably be related to the opening of the Wharton Basin. Falvey (1972), Larson (1975), and Heirtzler et al, (1978) discussed the presence of Jurassic linear magnetic anomalies in the Wharton Basin and Argo abyssal plain. DSDP 261 (Veevers et al, 1974) confirmed an Oxfordian age for the oldest sediments in the plain. Heirtzler et al (1978) showed that anomaly M25 (153 m.y.old) follows the Australian continental margin and presumably indicates the age of initial drifting. Spreading in the Argo abyssal plain continued at least until anomaly M5 (119 m.y. old) in the western part. However, there is a significant discontinuity between the eastern and western parts, which either implies a spreading cessation in the western part between M22 (148 m.y.old) and M13 (131 m.y.old), or a ridge jump at M13. Neither the anomaly trends nor the postulated fracture zones are sufficiently mapped to determine the position of the spreading pole. A spreading pole near Roti, as proposed by Otofuji et al (1980), is not consistent with the anomaly trends. The opening of the Wharton Basin is presumably a direct consequence of the opening of the Indian Ocean in the wake of India's northward drift from Gondwanaland. The scenario implies that the Wharton Basin spreading center rifted Sumba from the Australian continental margin, and Sumba drifted north ahead of the spreading center (see Figure 8). Since there is no present trace of the spreading center, we must make a further assumption that the spreading center, and most of the paleo-Wharton Basin ocean floor, was subducted along a subduction zone that formed south of Sumba.

The important point is that the formation of the subduction zone ceased the northward drift of Sumba and left Sumba stranded north of it. Thus, according to the model, Sumba did not collide with an island arc and was not involved in significant imbrication, hence its lack of tectonic deformation. In effect, the trench protected Sumba from compressive stresses.

The subduction zone is now most readily identified as the present eastern extension of the Java Trench. If this is true, it follows that the eastern Java Trench is a relatively recent feature that did not exist prior to Sumba's arrival at its present position. You could speculate that the Java Trench originally terminated south of Sumbawa and was connected through transform faults with active features to the north (Hilde,

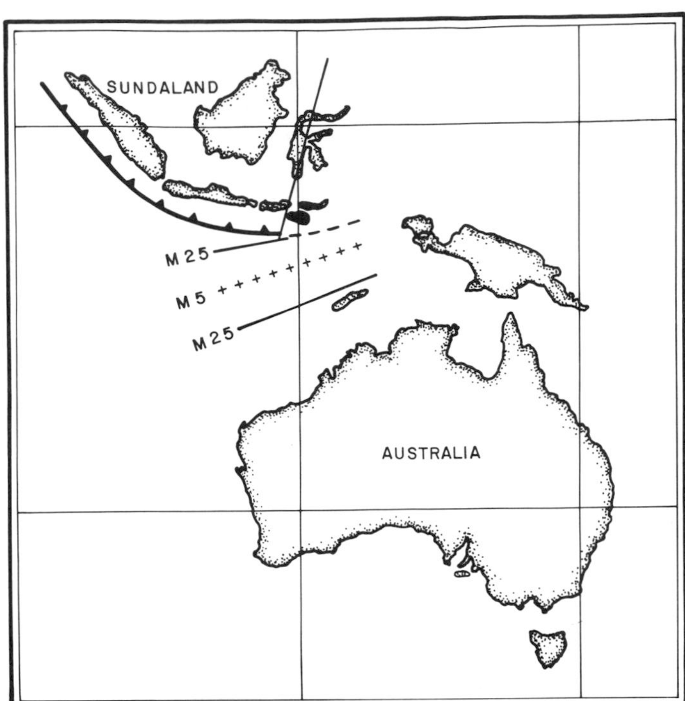

Figure 8 — "South of Tethys" model. Cartoon depicts the arrival of Sumba at its present position before the formation of the East Java Trench (shown dashed) M5, the M25 magnetic anomalies associated with the spreading of the Argo abyssal plain, and the rifting of Sumba from Australia. Figure not to scale.

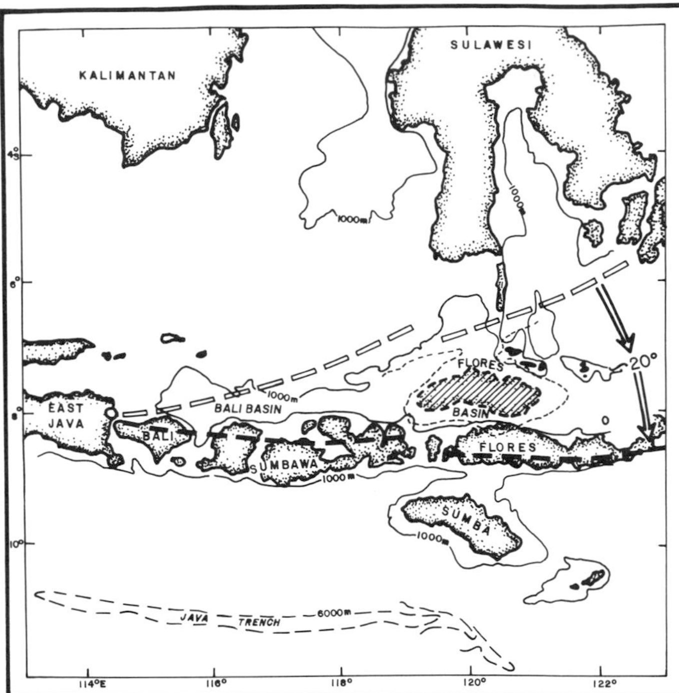

Figure 9 — "North of Tethys" model. Diagram showing the positions of Sumba (diagonal shading) as part of Sundaland before its clockwise rotation about a pole near Bali. The volcanic arc (heavy dashed lines) is included in the rotation. Selected isobaths are shown, with depths indicated in meters.

Uyeda, and Kroenke, 1977). After Sumba's arrival near Sundaland, the Java Trench grew eastward. Alternatively, you might suppose that the original Java Trench subduction system skirted the southeastern corner of Sundaland (north of Sumba) and then jumped to its present position (Audley-Charles, 1975). Considering the tectonics of Sumba, both explanations are nearly equivalent.

Although the timing of the events is still ill-defined, we suspect that Sumba arrived at its current position between the Eocene and Miocene. The paleomagnetic test of the hypothesis would consist of showing that the pre-Eocene, and particularly Jurassic strata, have poles closely related to those from Australia; whereas from the Eocene onward the poles should correspond to those from Sundaland. The above scenario does not provide a ready explanation for the Paleogene volcanism on Sumba. However, you might suppose that when the subduction zone south of Sumba initially formed it had a relatively shallow dip. This, combined with the presence of the original rifting fractures on Sumba, could have lead to minor amounts of continental type volcanism.

In Figure 8, we show the paleogeography at the time that the subduction zone was about to form south of Sumba. However, since relative rotations between Australia and Sundaland and between Australia and Sumba are not determined, actual distances are hypothetical. The figure implies that the rift formed between Sumba and Timor while both were part of the Australian margin. Thus, Timor did not rift from Australia, but eventually collided with the newly formed subduction zone. For reasons given previously (Chamalaun and Grady, 1978), we prefer this explanation for the tectonics of Timor. However, you might argue that if you postulate Sumba to have rifted from the Australian margin, then you could also postulate the same for Timor, as suggested by Bowin et al (1980). Timor, like Sumba, would then be situated north of the newly formed subduction zone and its tectonic deformation would be explained in a way which does not apply to Sumba. One possibility may be that it simply collided with the Australian margin earlier than Sumba. Clearly, the scenario outlined for Sumba does not provide a significant clue regarding the tectonics of Timor.

"North of Tethys"

Hamilton (1979) favoured the "North of Tethys" model. The prime feature of the model is that Sumba is part of a microcontinental fragment derived by rifting from the southeastern edge of the Sunda continental mass to the north. The most likely origin of Sumba would then be from the present Flores Basin (Hamilton, 1979; Figure 9).

The model implies that, as a result of a significant change in plate dynamics (probably the early Eocene Australia-Antarctic rifting), a new segment of the Sunda subduction zone was initiated along or near the southern margin of the Sunda continental mass. This change caused a new zone of rifting which cut partly through the continental mass and by progressive back-arc spreading rotated the separated microcontinental mass (including Sumba), together with its developing magmatic arc, to the south (about 20° clockwise, for a rotation pole chosen near western Bali; Figure 9). Such spreading and rotation probably began in the Eocene and might have been operative into the Neogene. From Eocene to present, Sumba should have acted as the forward continental crust buttress against which any subduction effects might have occurred (e.g. accretion, erosion, kneading, etc.; Scholl et al, 1980). The Cretaceous-Paleocene magmatism on Sumba might reflect early rifting processes, although it seems reasonable to expect much larger volumes of basaltic volcanic rocks than are evident. The Miocene volcanics on Sumba might relate to changes of subduction zone dip as it evolved, or they might reflect local changes in subduction dynamics in response to the Miocene changes in Pacific Plate dynamics (Hide, Uyeda, and Kroenke, 1977).

The lack of Tertiary-Quaternary imbrication-melange in Sumba requires all such processes to have occurred south of Sumba between the island and the eastern end of the Java Trench. The geological characteristics of this area are not known, as detailed data are not available. However, unless subduction erosion was a dominant process in the Sumba sector, a southward protrusion of the arc-trench gap, directly associated with Sumba, should be expected. The reverse appears to be true (between longitude 118 and 120°E). Clearly, there is a need for detailed marine geological and geophysical information between Sumba and the eastern end of the Java Trench.

The model implies that there should be continuous continental crust between Sumba and Flores — Sumbawa to the north because the magmatic arc supposedly developed in the continental crust. The sedimentary basins developed in this region should have the characteristics of forearc basins and be relatively undeformed. This feature of the model cannot be assessed at present because we lack suitable data. Such data could be gathered concurrently with any marine survey carried out near Sumba.

An important initial test concerns the degree of correlation between the prerifting (Eocene) geology of Sumba and that of the area immediately adjacent to its proposed prerifted position. Most of that area is below sea level and covered by younger Tertiary sediments, so correlation with the exposed geology of the southwestern part of the south arm of Sulawesi must be attempted. There, inliers of high pressure-low temperature pre-Cretaceous basement are overlain unconformably by a Createaceous-Tertiary sequence (Hamilton, 1979). The lower part of that sequence is red radiolarian chert interbedded with schistpebble conglomerate and overlain by Albian siliceous shale. These are overlain by a thick, dominantly clastic sequence of shale, turbidite siltstone, greywacke, and minor limestone, part of which is possibly Paleocene (Hamilton, 1979). In Sumba there are no known equivalents of the pre-Cretaceous metamorphic rocks, nor of the radiolarian chert. However, Sumba's pre-Tertiary sedimentary rocks might correlate with the shale, turbiditic sillstone, greywacke sequence of Sulawesi. Neither sequence has been studied in detail so no firm conclusions can be drawn.

Another prime test of the model concerns paleomagnetic data. If Sumba originated in the present Flores Basin and began rotating southward (clockwise) in the Eocene, then Sumba's pre-Eocene rocks should have paleomagnetic poles which relate, by appropriate anticlockwise rotation, to equivalent age paleomagnetic poles from southern Sulawesi and Kalimantan. The post-Eocene paleolatitudes should correlate well with those for Sundaland, but as shown above that is not a critical test of this particular model. The paleomagnetic data presented earlier in this article show that the Cretaceous paleomagnetic poles for Sumba cannot be related to those for Sundaland (Sulawesi, Kalimantan, and Malaysia) by rotations implicit in this model.

SUMMARY

Although Sumba is situated at the junction of the Sunda Arc and Banda Arc systems, its geology shows very little involvement with the geodynamics of either. It appears that Sumba became trapped between the eastern extremity of the Java Trench and the volcanic arc, either by a southward migration from Sundaland, or a northward migration from Australia. Available data are insufficient to convincingly distinguish between these two hypotheses. We feel, however, that in view of the spreading history of the Argo abyssal plain and the evidence for rifting at the northwestern Australian margin, an Australian origin for Sumba is the more likely hypothesis.

REFERENCES CITED

Audley-Charles, M. G., 1975, The Sumba fracture: A major discontinuity between eastern and western Indonesia: Tectonophysics, v. 26, p. 213-218.

——— and J. S. Milson, 1974, Comment (on) plate convergence, transcurrent faults, and internal deformation adjacent to southeast Asia and the western Pacific: Journal of Geophysical Research, v. 79, p. 4980-4981.

Bowin, C., et al, 1980, Arc continent collision in Banda Sea region: American Association of Petroleum Geologists Bulletin, v. 64, p. 868-915.

Cardwell, R. K., and B. L. Isacks, 1978, Geometry of the subducted lithosphere beneath the Banda Sea in eastern Indonesia from siesmicity and fault plane solutions: Journal of Geophysical Research, v. 83, p. 2825-2838.

Carter, D. J., M. G. Audley-Charles, and A. J. Barber, 1976, Stratigraphical analysis of island arc-continental margin collision in eastern Indonesia: Geological Society of London Journal, v. 132, p. 179-198.

Caudri, C. B. M., 1934, Tertiary deposits of Soemba: Amsterdam, H. J. Paris.

Chamalaun, G. H., 1977a, Paleomagnetic reconnaissance result from the Maubisse formation east Timor and its tectonic implications: Tectonophysics, v. 42, p. 17-26.

——— 1977b, Paleomagnetic evidence for the relative positions of Timor and Australia in the Permian: Earth and Planetary Science Letters, v. 34, p. 107-112.

——— and A. E. Grady, 1978, The tectonic development of Timor: A new model and its implications for petroleum exploration: Australian Petroleum Exploration Association Journal, v. 18, p. 102-108.

——— K. Lockwood, and A. White, 1976, The Bouguer gravity field and crustal structure of eastern Timor: Tectonophysics, v. 30, p. 241-259.

Coney, P. J., D. L. Jones, and J. W. Monger, 1980, Cordilleran suspect terranes: Nature, v. 288, p. 329-331.

Falvey, D. A., 1972, Spreading in the Wharton Basin (northeast Indian Ocean) and the breakup of eastern Gondwanaland: Australian Petroleum Association Journal, v. 12, p. 86-88.

Fitch, T. J., and W. Hamilton, 1974, Reply to comments by M. G. Audley-Charles and T. S. Milsom on paper: Plate convergence transcurrent faults and internal deformation adjacent to southeast Asia and the Western Pacific: Journal of Geophysical Research, v. 79, p. 1982-1985.

Foden, J. D., and R. Varne, in press, Petrogenetic and tectonic implications of near coeval calcalkaline to highly alkaline volcanism on Lombok and Sumbawa Island in the Eastern Sunda Arc, *in* The geology and tectonics of eastern Indonesia: Bandung, Indonesia, Geological Research and Development Centre, Special Publication n. 2.

Green, R., et al, 1979, Bouguer gravity anomaly map of Indonesia, with marginal text: Armidale, University of New England Press.

Haile, N. S., 1978, Reconnaissance paleomagnetic results from Sulawesi, Indonesia, and their bearing on paleogeographic reconstructions: Tectonophysics, v. 46, p. 77-85.

——— 1979, Paleomagnetic evidence for the rotation and northward drift of Sumatra: Geological Society of London Journal, v. 136, p. 541-545.

——— M. W. McElhinny, and I. McDougall, 1977, Paleomagnetic data and radiometric ages from the Cretaceous of west Kalimantan (Borneo), and their significance in interpreting regional structure: Geological Society of London Journal, v. 133, p. 133-144.

Hamilton, W., 1979, Tectonics of the Indonesian region: U.S. Geological Survey, Professional Paper 1078.

Heirtzler, J. R., et al, 1978, The argo abyssal plain: Earth and Planetary Science Letters, v. 41, p. 21-31.

Hilde, T. W. C., S. Uyeda and L. Kroenke, 1977, Evolution of the western Pacific and its margin: Tectonophysics, v. 38, p. 145-165.

Hinz, K., 1977, Bericht über den Fahrtabschnitt VA-16-2C, darwin udjung pandang 11.3.77-23.3.77 der Valdivia — southeast Asian Fahrt: Hanover, Bundesanstalt fur geowissenshaften und Rohstoffe.

——— et al, 1978, Geoscientific investigations from the Scott Plateau off northwest Australia to the Java Trench: Bureau of Mineral Resources, Journal of Australian Geology and Geophysics, v. 3, p. 319-340.

Karig, D. E., et al, 1979, Structure and Cenozoic evolution of the Sunda Arc in the central Sumatra region, *in* J. S. Watkins, L. Montadert, and P. W. Dickinson, eds., Geological and Geophysical Investigations of Continental Margins: AAPG Memoir 29, p. 223-237.

Katili, J. A., 1971, A review of the geotectonic theories and tectonic maps of Indonesia: Earth Science Review, v. 7, p. 143-163.

Kuenen, P. H., J. H. F. Umbgrove, and F. A. Vening Meinesz, 1934, Gravity geology and morphology of the east Indian Archipelago, *in* Gravity expeditions at sea, 1923-1932: Netherlands Geodetic Commission, p. 107-194.

Larson, R. L., 1975, Late Jurassic sea floor spreading in the eastern Indian Ocean: Geology, v. 3, p. 69-71.

Laufer, R., 1950, Geology and morphology of west and central Sumba: Organization for Scientific Research, News, v. 12, p. 161-166.

McElhinny, M. W., 1973, Paleomagnetism and plate tectonics: Cambridge, Cambridge University Press, 357 p.

——— N. S. Haile, and A. R. Crawford, 1974, Paleomagnetic evidence shows Malay Peninsula was not part of Gondwanaland: Nature, v. 252, p. 641-643.

Meiser, P., et al, 1965, Hydrogeological map of the isle of Sumba. Scale 1:250,000: Indonesia Geological Survey Bandung.

Nishimura, S., et al, 1980, Physical geology on the Sumba, Sumbawa, and Flores Islands, *in* S. Nishimura, ed., Physical geology of Indonesian island arcs: Kyoto, Kyoto University Press, p. 47-50.

Otofuji, Y., et al, 1980, Paleomagnetic evidence for the paleoposition of Sumba Island, Indonesia, *in* Nishimura, ed., Physical geology of Indonesian island arcs: Kyoto, Kyoto University Press, p. 59-66.

Roggeveen, P. M., 1932, Abyssische und hypabyssische eruptiv gesteine der insel Soemba, Niederlandish Ost Indien: v. 35, p. 878-890.

Sasajima, S. M., et al, 1980a, Paleomagnetic studies on Sumatra Island: on the possibility of Sumatra being part of Gondwanaland, *in* S. Nishimura, ed., Physical geology of Indonesian island arcs: Kyoto, Kyoto University Press, p. 13-22.

——— et al, 1980b, Paleomagnetic studies combined with fission track dating on the western arc of Sulawesi east Indonesia, *in* S. Nishimura, ed., Physical geology of Indonesian island arcs: Kyoto, Kyoto University Press, p. 13-22.

Scholl, O. W., et al, 1980, Sedimentary masses and concepts about tectonic processes at underthrust ocean margins: Geology, v. 5, p. 564-568.

Umgrove, J. H. F., 1949, The structural history of the East Indies: Cambridge, Cambridge University Press, 63 p.

Veevers, J. J., et al, 1974, Initial reports of the deep sea drilling project, Leg. 27: Washington, D.C., U.S. Government Printing Office, 1060 p.

Van Bemmelen, R. W., 1970, The geology of Indonesia (second edition): Government Printing Office, The Hague, v. 1A, 732 p.

Von der Borch, C. C., 1979, Continent-island arc collision in the Banda Arc: Tectonophysics, v. 54, p. 169-193.

Warris, B. J., 1973, Plate tectonics and the evolution of the Timor Sea, northwest Australia: Australian Petroleum Exploration Association Journal, v. 13, p. 13-18.

Watts, A. B., J. H. Bondine, and C. O. Bowin, 1978, Free air gravity field, *in* D. E. Hayes, ed., A geophysical atlas of east and southeast Asian Seas: Geological Society of American Map and Chart Series, MC-25.

The Southern Uplands Accretionary Prism: Implications for Controls on Structural Development of Subduction Complexes

J. K. Leggett
Department of Geology
Imperial College of Science & Technology
London

D. M. Casey
Department of Geology and Mineralogy
University of Oxford
Oxford

Subduction complexes preserving extensive coherent strata beneath lower slope sediments can be more useful than those dominated by melange for studying the sequence of deformation caused by accretion. In the Southern Uplands, a 70 to 80-km-wide coherent accretionary complex comprising imbricated thin ocean-floor (basalt, metalliferous sediment, radiolarian chert, black graptolitic shale) and thick trench (volcaniclastic greywacke) deposits records 50 to 60 million years of accretion during northward subduction of Iapetus oceanic crust under Ordovician and Silurian southern Scotland, then part of the southern margin of ancient North America (Laurentia). Like many more recent coherent accretionary terranes, the Southern Uplands complex appears to have developed about a slow convergence-high sediment input subduction zone. Competency contrasts in the subducting sequence controlled the mesoscopic and macroscopic structural style and determined the type of strata accreted. Detailed mapping of three new map areas shows a generalized tectonic history in which compressive stress is taken up by: 1) initial decollement (offscraping) commonly at or above a black shale unit (the Moffat Shale Group) underlying the greywackes; 2) local fold development, during and after isolation of discrete linear packets of offscraped strata; 3) variable cleavage development, commonly transecting folds; 4) strain-hardening, tightening, and eventual locking of each packet so that deformation switches to a new packet at the base of the inner trench slope; 5) subsequent more-or-less bedding-parallel (commonly intense) imbrication within packets as they are uplifted and rotated.

Recent IPOD drilling on active continental margins has been directed toward understanding the tectonic processes operating at the leading edge of the overriding plate during subduction. The interesting and variable results achieved pose new problems which emphasize the need for studying uplifted emergent forearcs and supposed ancient examples in orogenic belts. This paper analyzes the structural history of a Lower Paleozoic subduction complex in Scotland's Southern Uplands, and compares it with other ancient or emergent accretionary forearcs. We review the detailed stratigrahy and evidence for interpreting the Southern Uplands as an accretionary prism in a companion paper (Leggett, McKerrow, and Casey, 1982).

The Southern Uplands subduction complex is characterized by lack of melange and by division of the accreted ocean floor and trench sediments into stratigraphically distinct fault-bounded tracts. In this paper, data from three new map areas emphasize the importance of imbrication in deformation during accretion of otherwise coherent strata. Convergence rate, convergence angle, sediment input, and stratigraphy of the subducting sediments are important in our comparison of the Southern Uplands structural history with those of other accretionary forearcs. Concepts of tectonic processes on active margins are as defined in Moore, Watkins and Shipley (1981), Watkins et al (1981), and Scholl et al (1981).

THE SOUTHERN UPLANDS ACCRETIONARY COMPLEX

Regional Setting

During the Early Paleozoic, Scotland was part of the southern continental margin of the North American (Laurentian) plate along the Iapetus Ocean's northern border. Northward subduction shaped orogenic events from at least Late Cambrian through Early Devonian time (summaries in Leggett, McKerrow, and Casey, 1982; McKerrow, this volume). The Southern Uplands lie immediately north of the suture along which the Iapetus ocean finally closed (Figure 1), and south of a contemporaneous magmatic arc (Phillips, Stillman, and Murphy, 1976).

Deformed Lower Ordovician through Middle Silurian rocks form the accretionary complex, and are exposed over some 12,000 sq km south of the Southern Upland Fault in Scotland. They are similarly exposed along strike in Ireland. Generally, Ordovician strata comprise *in* upward succession, basalt, metalliferous sediment, radiolarian chert, black shale and greywacke; Silurian rocks comprise black shale and greywacke. Greywackes occupy most of the outcrop area and are interpreted as both trench deposits and ocean floor clastics (Leggett, 1980). Leggett (1979) interprets the underlying strata as pelagic-hemipelagic ocean floor deposits, including local slivers of uppermost oceanic crust (layer 2). Figure 2 summarizes the stratigraphy of the Southern Uplands. Interpretating the area as an accretionary complex is based on its paleo-forearc setting, on the presence of strata of ocean floor affinities, and on the structural configuration. Major reverse strike faults separate tracts in which structural younging is to the northwest (toward the ancient continent) and stratigraphic sequences become progressively younger southeast (toward the ancient ocean), as should be the case in accretionary complexes according to the Seely, Vail, and Walton (1974) model. Figure 3 summarizes the interpreted history of the Southern Uplands accretionary prism.

Most of the Lower Paleozoic strata north of the Southern Uplands are hidden under younger cover. On the west coast immediately north of the Southern Upland Fault, the Lower Ordovician Ballantrae ophiolite (Figure 1) may represent an accreted portion of Iapetus Ocean crust (Leggett, McKerrow, and Casey, 1982). It is overlain unconformably by Ordovician through Silurian slope-shelf sediments, suggesting that the Southern Uplands accretionary complex does not extend beneath the Midland Valley very far north of the Southern Upland Fault, if at all.

The supra-ophiolitic succession at Ballantrae, and Silurian successions in other Midland Valley inliers (where the base of the succession is not seen) record the history of a forearc (upper slope) basin. Although entirely marine in the Ballantrae area, the succession in inliers to the northeast contains Middle Silurian (Wenlock) terrestrial deposits overlying Lower Silurian (Llandovery) turbidites. Much, if not all, of the sediment was derived from the accretionary complex to the south, where a trench slope break emerged in the middle Silurian (Figure 3b; Leggett, 1980). Geophysical evidence suggests that this forearc was underlain by continental crust (Figure 3b; discussion in Leggett, McKerrow, and Casey 1982). In the terminology of Dickinson and Seely (1979), the Early Paleozoic Midland Valley basin is a constructed forearc basin and the Scottish portion of the Southern Laurentian continental margin evolved from a simple sloped forearc (Figure 3a) to a terrestrial ridged forearc (Figure 3b) during about 55 million years (latest Llandeilo to late Wenlock) of accretionary history (Figure 2).

Direct evidence for the contemporaneous magmatic arc is the early and mid-Ordovician calc-alkaline volcanicity of the South Mayo Trough and the Tyrone inlier in northern Ireland (Phillips et al, 1976). That andesitic volcanoes were once abundant north of the Southern Uplands, both in Silurian and Ordovician times, is suggested by the volcaniclastic nature of much of the Northern and Central Belt (Figure 2) flysch (Walton, 1965), and numerous metabentonites intercalated in the condensed oceanic pelagic sections (Leggett, 1979). Some early Ordovician granitoids in the Grampian Highlands and Northwest Highlands are probably subduction-related, and may have fed volcanoes since eroded (Simpson et al, 1979; Brown, 1979). Lower Devonian andesites are exposed across the Midland Valley and Grampain Highlands, and show across-arc geochemical trends similar to those of modern continental margin arcs (Thirlwall, 1981). They indicate that subduction may have continued into the early Devonian, although biostratigraphic uncertainties and the lack of marine Devonian strata in the accretionary complex suggest that these andesites may be of latest Silurian age (Thirlwall, 1981).

There is evidence that subduction occurred prior to the accretionary phase. Subduction-related processes presumably caused the latest Cambrian/earliest Ordovician Grampian Orogeny, but this problem is much disputed and is beyond the scope of this paper (see McKerrow, this volume). Radiometric ages of granite boulders from supra-ophiolitic conglomerates near Ballantrae (Figure 1) suggest the presence of a latest Cambrian-early Ordovician arc in the southwestern part of the Midland Valley, according to Longman et al (1979). If so, the arc-trench gap is remarkably narrow (<100 km between the Southern Upland Fault and Highland Boundary Fault on the north border of the Midland Valley). On this basis we suggest that subduction erosion may have occurred along the margin prior to onset of the accretionary phase. Another possibility, difficult to assess, is that strike-slip movement parallel or sub-parallel with the continental margin might have removed a late Cambrian/early Ordovician forearc terrane (compare to pre-late Miocene Middle America margin off Mexico, Moore et al, 1979). The first accretion in the Southern Uplands (mid-late Ordovician followed the first appearance of thick turbidites along the northern Iapetus margin (Figure 2, sequence 1).

Final closure of the Iapetus Ocean along the suture line south of the Southern Uplands in Early Devonian

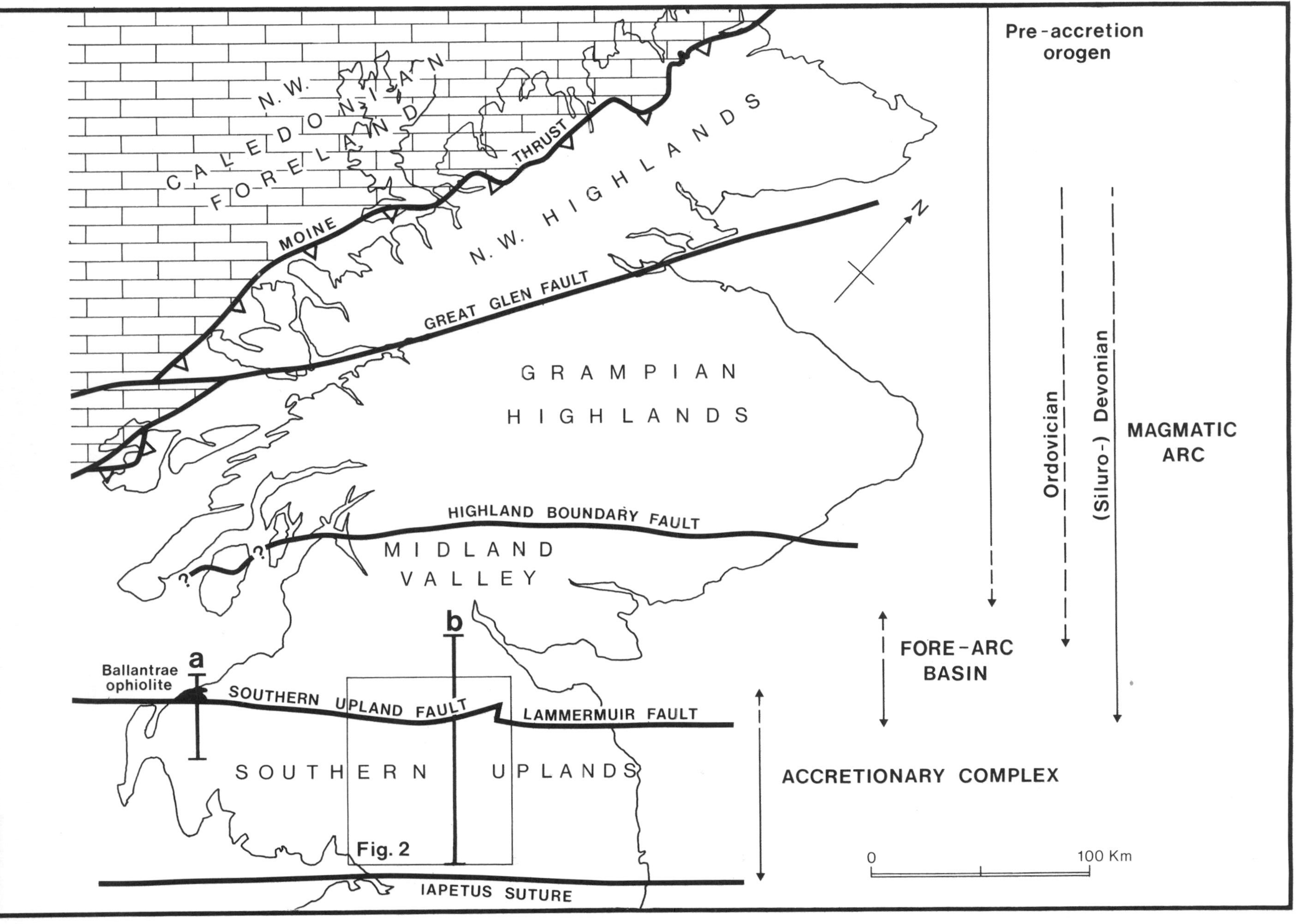

Figure 1 — Geologic provinces of the northern British Caledonides in Scotland, and extent of mid-Ordovician through Silurian geotectonic zones relating to northward subduction from the Iapetus suture line. Known (outcropping) extent of each zone shown by solid line; dashed lines indicate possible extensions of unknown distance. Area of Figure 2 shown and a, b are reconstructed profiles shown in Figure 3.

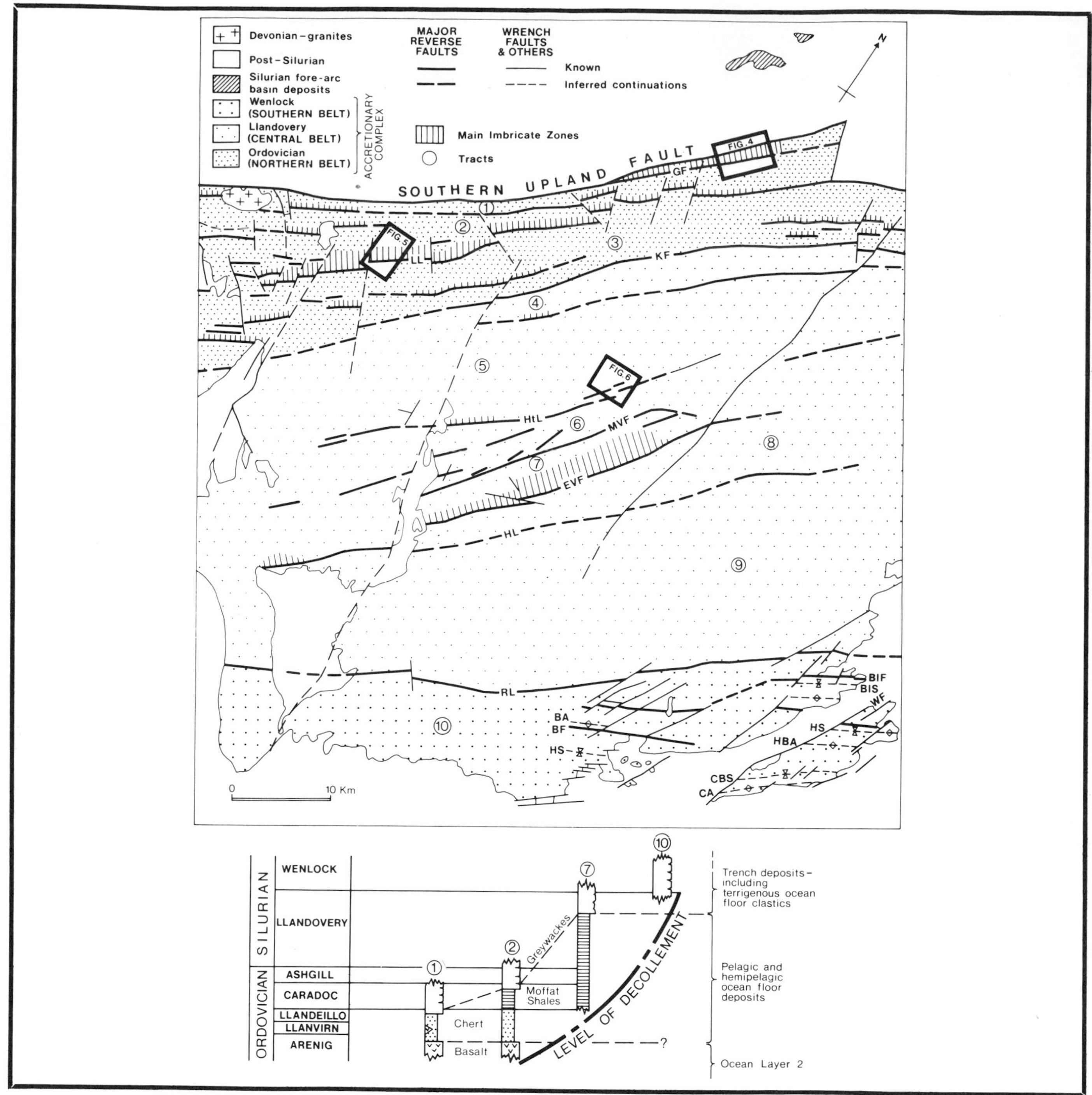

Figure 2 — Geologic map of central Southern Uplands and part of Midland Valley (after Leggett, McKerrow, and Casey, 1982). Major reverse faults divide the Southern Uplands into tracts (numbered 1 to 10 as in Leggett et al, 1979, and Leggett McKerrow, and Casey, 1982). The tracts contain distinctive sequences of rocks interpreted as ocean floor and trench deposits. Predominant structural younging of strata within tracts is northwestward (toward the ancient continent), but the sequences become younger from tract to tract toward the southeast (oceanward). We interpret tracts as original discrete packets of off-scraped strata. However, we anticipate that some tracts will be further subdivided by future mapping; unrecognized tract-bounding faults probably occur, especially in tracts 5 and 9. The inset of four representative sequences illustrates how the level of basal decollement rises in younger (Silurian) tracts as the incompetent black shale unit thickens. Biostratigraphic resolution is good in the pelagic-hemipelagic deposits, but is poor in the greywackes, except for sequence 10; greywackes in sequences 1, 2, and 7 may be confined to just one graptolite zone (Leggett, McKerrow and Casey, 1982). The sequences in tracts numbered on the map are traceable laterally where bounded to both northwest and southeast by a mapped major reverse fault. Structures: BA — Bush Anticline; BF — Bush Fault; BLF — Berryfell Fault; BS — Berryfell Syncline; CA — Caddroun Anticline; CBS — Caddroun Burn Syncline; EVF — Ettrick Valley Fault; GF — Grassfield Fault; HBA — Hope Burn Anticline; HS — Hope Syncline; HtL — Hartfell Line; KF — Kingledores Fault; LL — Leadhills Line; MVF — Moffat Valley Fault; RL — Riccarton Line; WF — Wolfhopelee Fault.

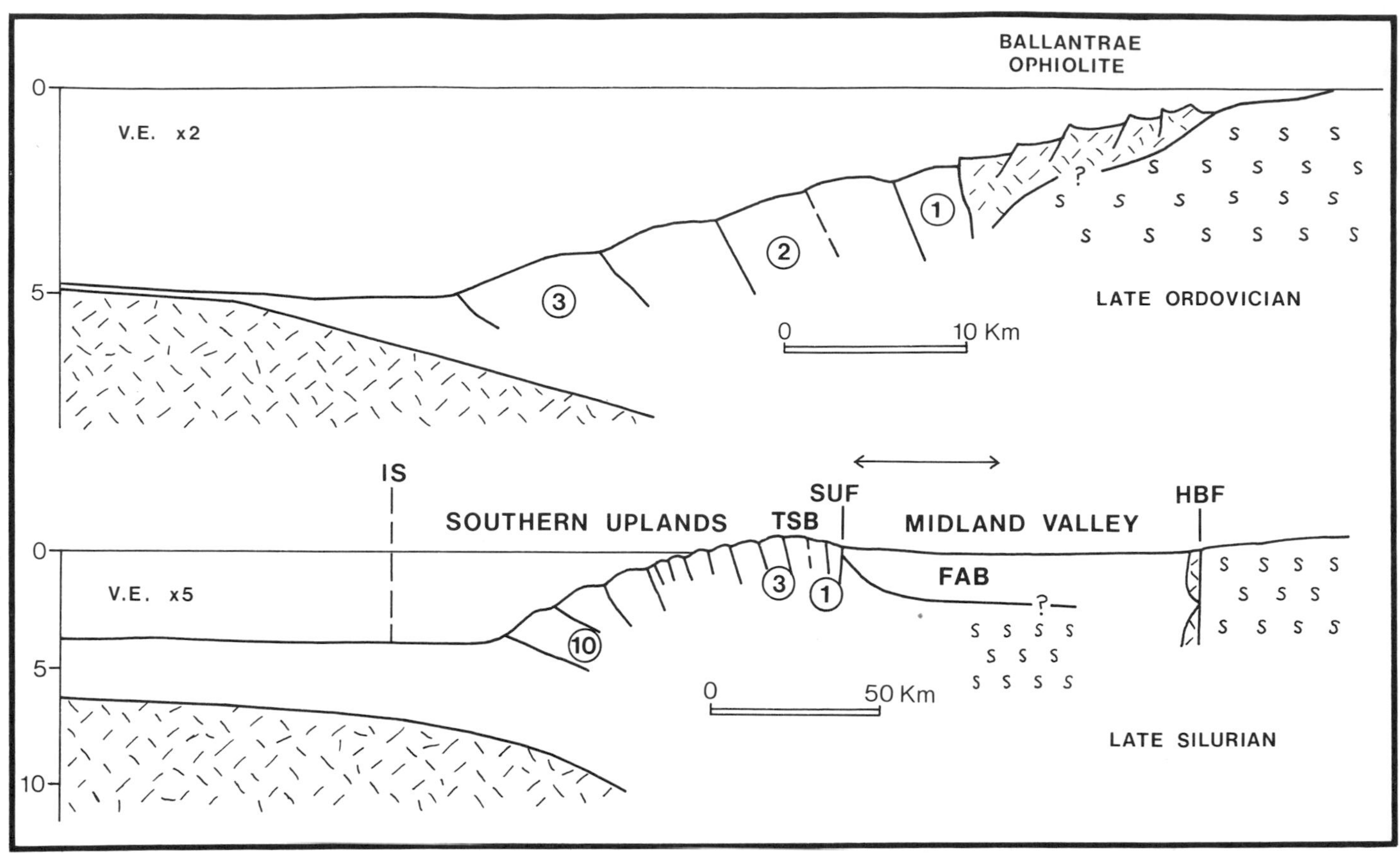

Figure 3 — Reconstructed profiles of Southern Uplands accretionary prism in late Ordovician (based on section a, Figure 1) and late Silurian (based on section b, Figure 1; Leggett, McKerrow, and Casey, 1982). Positions of tracts 1, 2, 3, 10 (as in Figure 2) shown for reference. Continental basement is *'s'* ornament, ophiolitic rocks dashed. FAB — Forearc basin; HBF — Highland Boundary Fault; IS — Future position of Iapetus suture; SUF — Southern Upland Fault; TSB — Trench slope break.

times did not result in the intense, nappe-producing deformation found in Norway. Closure-related tectonism in the Southern Uplands accretionary complex seems to be limited to mesoscopic brittle deformation, localized within about 50 km from the suture (Phillips, Flegg, and Anderson, 1979). Phillips, Stillman and Murphy (1976) relate the absence of nappe structures to oblique collision, with earlier closure in the northeast generating dextral strike-slip along the suture. This hypothesis, based on the conclusion that volcanism ceased in mid-Ordovician time on the southeast Iapetus margin, is questioned by Johnson, Sanderson, and Soper (1979). Thirlwall (1981) uses geochemical trends in Silurian or Lower Devonian andesites to propose oblique convergence with east-west relative plate motion between the Scottish-American (Laurentian) and European plates. This explains nappe formation during collision in Scandinavia and the Shetlands, while allowing preservation of the primary structures of the Southern Uplands accretionary prism.

Structural History

Weir (1979) and Leggett, McKerrow, and Casey (1982) summarize the structural history of the Southern Uplands, and Anderson and Cameron (1979) describe the best-exposed individual transect of the belt on the Ards and Lecale peninsulas of the Irish continuation.

Major east northeast trending strike faults, with indeterminable though clearly substantial downthrows to the south, dominate the Southern Uplands (Figure 2). These faults are commonly manifested by zones of imbricated basalt, chert, and/or black shale (basal lithologies) up to about 2 km wide (e.g. Ettrick Valley Imbricate Zone, Figure 2). In the next section we discuss the geology of two imbricate zones in the Northern Belt. Where well-developed, the imbricate zones mark the main strike faults clearly, suggesting in some areas complex "braiding" of the fault planes. The major faults in greywacke terrane are more difficult to map, but some are marked by intense shearing (e.g. Hawick Line on Geological Survey one-inch sheet 16, Figure 2). In the case of the Orlock Bridge Fault (the continuation of the Kingledores Fault in northeast Ireland Figure 2), a 50 m wide breccia zone shows evidence of repeated reactivation (Anderson and Cameron, 1979). However, in other cases, such as the Riccarton Line (Figure 2), a fault's existence is difficult to prove other than on stratigraphic grounds (Warren, 1964).

D_1 folds in the Southern Uplands are open-to-isoclinal, predominantly dextral when viewed Northeast, with steeply dipping or slightly overturned long limbs (summary in Stringer and Treagus, 1981). Almost all axial traces are parallel with the strike faults (i.e. northeast-southwest to east northeast-west southwest) and formed in the same compressive stress regime (Lumsden et al, 1967; Warren, 1964; Anderson and Cameron, 1979). Even in well-exposed areas, individual major folds rarely map out for any distance along strike because of common shallow plunges and along-axis variations in style and attitude (Stringer and Treagus 1980, 1981; Walton, 1965). Consequently, Figure 2 shows major folds only in an intensively studied portion of the Wenlock greywackes (Southern Belt). Folds are commonly in discrete packets with flat-lying or northward-dipping fold envelopes, separated by thicker sections of homoclinal, vertical or steeply-dipping, predominantly northwest younging strata. Major folds show north-younging limbs of up to 4 km and shorter south-younging limbs of up to 1.7 km (see sections in Anderson and Cameron, 1979; Leggett, McKerrow and Casey, 1982; Stringer and Treagus, 1981). Large-scale northeast-verging compound monoclines, comprising "flat limbs" of intense buckle folds with horizontal fold envelopes and "vertical limbs" of northwest-younging homoclinal strata, are recognized in some transects (summary in Weir, 1979). However, axial zones are faulted where exposed and there is no clear differentiation of "vertical" and "flat" belts in the best-exposed transects (Anderson and Cameron, 1979; Stringer and Treagus, 1981).

All large-scale folds and most mesoscopic folds in the Southern Uplands formed at about the same time as the strike faults, although the deformational sequences vary from area to area. For example, Eales (1979) argues that major faulting preceded folding in the Ettrick Valley Imbricate Zone (Figure 2), whereas Cook and Weir (1979) argue that folding preceded reverse faulting in an area along strike to the southwest. These differences are consistent with accretionary tectonics (McKerrow, Leggett and Eales, 1977; Leggett, McKerrow, and Eales, 1979; Leggett et al, 1979; Weir, 1979; Anderson and Cameron, 1979; Phillips, Flegg, and Anderson, 1979). By analogy with modern accretionary margins we can interpret the major structures of the Southern Upland as follows. Initial decollement for each accreted packet occurred along incompetent levels in the subducting pelagic-hemipelagic sequence, forming an intially low-angle thrust, approximately parallel with bedding. Later, underthrusting rotated the packet and steepened the initial faults. In this interpretation, folding can occur before, during, and after decollement, and faults and fold axes have similar orientations since they formed in response to the same stress vectors (Leggett, McKerrow, and Casey, 1982). After initial accretion (incorporating the new packet into the lower trench slope) deformation during rotation can involve further faulting, resulting in complex zones of imbrication, further folding, or both. The next section shows that imbrication seems to be the more important process.

Cleavage imposed during D1 deformation commonly transects the folds, having been imposed after early folding and initial rotation of bedding (Leggett et al, 1979; Stringer and Treagus 1980, 1981; Leggett, McKerrow, and Casey, 1982). Within a zone extending about 50 km north of the Iapetus suture this S1 cleavage is rotated anticlockwise as much as 20° with respect to F_1 axial planes. Phillips, Flegg, and Anderson (1979) relate this rotation to post-closure dextral shear along the suture line. Stringer and Treagus (1981) explain the rotation by cleavage imposing on already-inclined bedding non-orthogonal with respect to bulk strain axes, late in the deformation history of each accreted slice. In this explanation the cleavage is still technically a D_1 (accretion-related) feature, and the non-orthogonal relationship is a manifestation of oblique convergence (Stringer and Treagus, 1981).

Post D_1 features are limited to mesoscopic folds, with local associated crenulaton cleavage, which are more common near the suture (Phillips, Flegg, and Anderson, 1979). In the Ards-Lecale transect F_2 folds verge southeast and F_3 folds verge northwest (Cameron and Anderson, 1979). Both trend northeast to southwest, reflecting an important phase of shortening near the suture probably related to final closure of the Iapetus, but do not affect the broad structural pattern of the Southern Uplands (Phillips, Flegg, and Anderson, 1979).

Another important structural feature of the Southern Uplands is wrench faulting. Most of the wrench faults are oriented northwest-southwest to north northeast-south southwest, and have sinistral displacements which sometimes exceed 1 km (Figure 2). North-south and north northwest-south southeast trending wrench faults also occur. The faults represent late-stage extension along the belt axis, resulting from continuing north northwest-south southeast principal stress vectors (e.g. Anderson and Cameron, 1979; Eales, 1979). The exact chronology is uncertain since continuing movement has sheared lamprophyre dykes of probable Early Devonian age emplaced along some of the wrench faults in Ireland and Scotland (Anderson and Cameron, 1979; Weir, 1979). Some of these dykes pre-date D_2 deformation (Stringer and Treagus, 1981). Weir (1979) argues for two discrete phases of wrench faulting, and it seems likely that wrench movements might be related to deformation prior to and during the closure of the Iapetus. Wrench faults reactivated by extension in Permo-Triassic times formed basin-bounding normal faults (Figure 2).

CONTROLS ON STRUCTURAL STYLE IN ACCRETIONARY FOREARCS

The structural configuration of modern accretionary complexes comparable with the Southern Uplands encourages us to describe them as coherent, as opposed to disrupted (melange-dominated), accretionary complexes. These terms refer only to the accreted material; that is, the material below the slope sequence.

On the basis of scaled models, calculated strain rates, and modern examples, Cowan and Silling (1978), Moore (1979), and Moore and Karig (1980) conclude that slow convergence rates and rapid sediment input to the trench facilitate accretion of discrete intact packets of strata. Prolonged accretion under such conditions therefore generates a coherent accretionary complex. We argue that the evolutionary history of the Southern Uplands, a mature coherent accretionary complex, supports these conclusions. Graptolite faunas, giving excellent biostratigraphic control in most of the ocean floor sediments preserved in the Southern Uplands, indicate that accretion took place for a minimum of 45 to 50 million years from mid-Ordovician (latest Llandeilo or early Caradoc) to late Silurian (late Wenlock or early Ludlow). During most of this time accretion was "efficient:" ocean floor lithologies were scraped off the descending plate along with turbidites in 7 of the 10 tracts mapped in Figure 2. There is no evidence for appreciable sediment subduction accompanying accretion, as seems to be the case along the Middle America Trench off Mexico (Watkins et al, 1981). During the first 15 to 20 million years of accretion (inception to end of the Ordovician), a belt of accreted sediment some 10 to 25 km wide (tracts 1, 2, and 3, Figure 2) was added to the southern Laurentian forearc in the Southern Uplands. An equivalent width of accreted trench sediment in the IPOD Leg 66 area of the Middle America Trench, notwithstanding the inefficient accretion already mentioned, took only 10 million years to be emplaced (Moore et al, 1979). After 50 to 60 million years, further accretion in the Southern Uplands led to the addition of a belt of accreted strata in total 70 to 80 km wide. In the well-studied Sumatra forearc an accretionary complex some 40 to 50 km wider than that of the Southern Uplands, and largely comprising melange, was built up in only 20 million years or so (Karig et al, 1980). Clearly the Southern Uplands accretionary complex was constructed at a much slower rate than two of the best studied modern subduction complexes.

Critically, both the Mexican and Sumatran examples occur above moderate to fast subducting oceanic plates (Moore et al, 1979; Karig et al, 1980). Slow convergence between the Laurentian and European plates, and hence slow subduction below southern Scotland, is indicated by the relatively long period during which the Southern Uplands accretionary complex was constructed. Phillips, Stillman, and Murphy (1976) also argue for a slow convergence rate (less than 2 cm/yr) but base their conclusions on what we consider to be tenuous grounds: the cessation of arc volcanism as an indicator of the end of subduction along the margin.

Sedimentation rates in the Southern Uplands turbidites were moderately high (about 150 to 300 m/million years excluding compaction effects; Leggett, 1980). With this moderate to high sedimentation rate (S) and associated slow convergence rate (C) the Southern Uplands accretionary complex compares well with several modern examples for which data are available. These include Washington-Oregon (C = 2 to 3 cm/yr and S = 140 to 940 m/million years since Pliocene; Kulm and Fowler, 1974); Makran (C nearly 5 cm/yr and S unknown but high in Neogene; White, 1982); Shikoku (C about 2 cm/yr and S locally around 850 m/million years in Pleistocene; Moore and Karig, 1976); and the southern Lesser Antilles (C about 2 cm/yr and S unknown but high; Westbrook, 1982). Critically, seismic reflection and some drilling data indicate that all these complexes offscraped coherent strata in the recent past (Moore and Karig, 1976; Farhoudi and Karig, 1977; Snavely, Wagner, and Lender, 1982; Westbrook, 1982; White, 1982). The Southern Uplands therefore provides an example from the ancient record supporting the tenet that prolonged accretion along slow convergence margins with high sediment input will generate coherent subduction complexes.

Imbrication

Three map areas illustrate the important effects of imbrication in the Southern Uplands accretionary complex. Two are in imbricate zones in the Northern Belt, where basal lithologies crop out, and one is in greywacke terrane of the Central Belt (Figure 2).

1) The Noblehouse area. Along the northern border of the Southern Uplands the oldest tectonic slice (Coulter-Noblehouse tract, tract 1 in Figure 2) contains a sequence of basalt, metalliferous sediment, chert, and greywacke rich in ophiolitic detritus (Marchburn greywackes of Floyd, 1975). The most informative exposures are in the Noblehouse area (NT 184501, Figure 4), where unnamed north-northwest flowing streams, several marking wrench faults, drain a gentle compound scarp topography immediately south of the Southern Upland Fault. The Coulter-Noblehouse tract is as little as 700 m wide in this area, being juxtaposed by a major strike fault against a different sequence to the south (Figure 4).

Peach and Horne (1899) and Lamont (1975) ascribed repetition of Coulter-Noblehouse lithologies across the tract to large-scale folding. However, by using criteria such as pillow lava shapes and grading in greywacke and siltstone, it is possible to show that the strata young northwesterly throughout the area. Basalts are demonstrably basal in the sequence in the Ironstone Cottages Burn, where they are overlain on the northwest by metalliferous sediments and bedded cherts, and fault bounded at the base (Figure 4). The basalts form linear ridges, locally up to 200 m wide and traceable for several kilometers along strike.

Imbrication intensity is particularly evident in the Ironstone Cottages Burn where six faults occur within 130 m, all but one parallel or subparallel with bedding. Here, as elsewhere in this tract, the Marchburn greywackes overlie the cherts with minor interdigitation. However, the intense imbrication found in the Noblehouse map area makes the relative stratigraphic position of a locally important grey cherty mudstone/siltstone unit uncertain (Figure 4).

2) The Abington area. The sequence in tract 2 of Figure 2 is well-exposed in hills flanking the Clyde valley near Abington (Figure 5). The sequence in tract 2 differs from that in tract 1, having black shale between the chert and the greywacke. Also, the basalt may be slightly older than that in tract 1, (see discussion in Leggett, McKerrow, and Casey, 1982) and the cherts are of a notably different facies (Leggett,

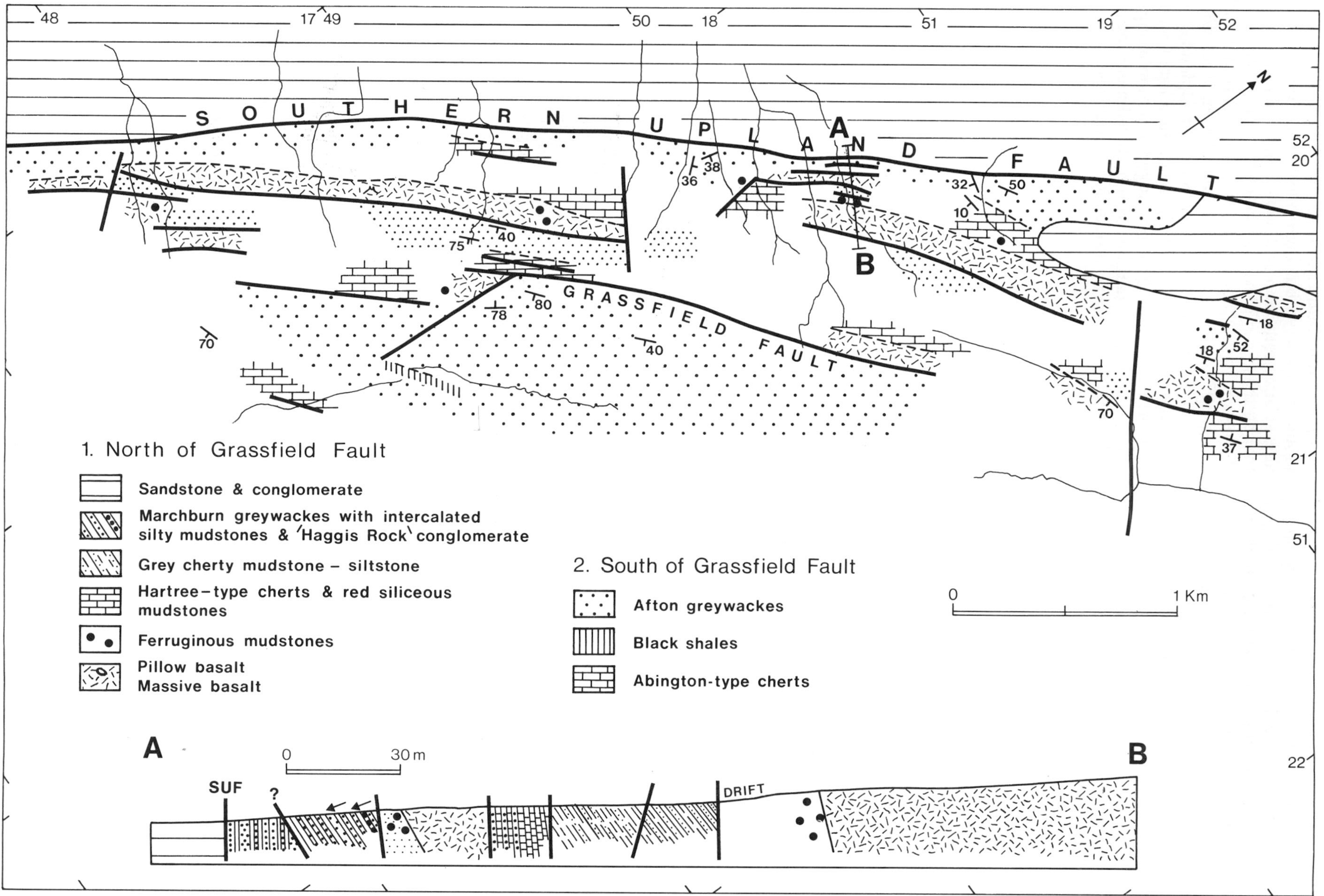

Figure 4 — Geologic map of Noblehouse — Grassfield area, Borders Region, and cross-section in Ironstone Cottages Burn (A to B). On either side of Grassfield Fault the stratigraphy differs markedly (details of chert facies and greywacke facies are in Leggett, 1978, and Floyd, 1975, respectively).

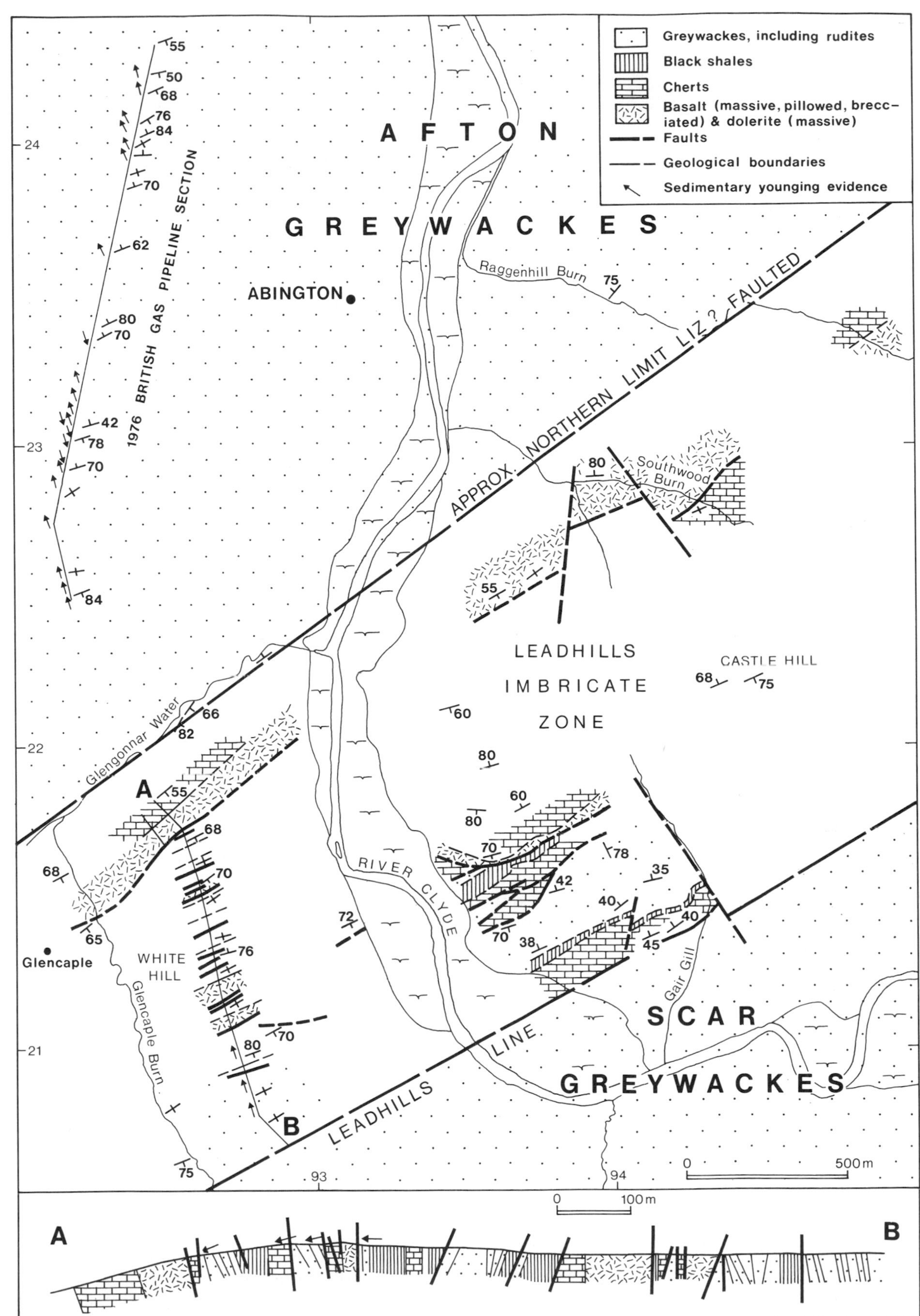

Figure 5 — Geologic map of Abington area, Clyde Valley, and section across White Hill along 1976 British Gas pipeline section (from Leggett et al, 1979).

1979). Additionally, Floyd (personal communication) shows that the greywacke petrography differs considerably between tracts 1, 2, and 3. West of Abington, Hepworth, McMurtry, and Oliver (1982) identify two additional fault-bounded tracts containing different stratigraphic sequences, but the extension of these east of the Clyde valley is uncertain (Figure 2).

Basal lithologies of tract 2 crop out in the Leadhills Imbricate Zone, north of the southern boundary (the Leadhills Line) of the tract. The 1976 British Gas pipeline excavation, crossing the regional strike almost at right angles, exposed a continuous section across the Leadhills Imbricate Zone on White Hill (Figure 5). In less than 1 km of section there are 11 major reverse strike faults which are parallel or subparallel with bedding and form a crude fan arrangement across the section. Four basalt strips, ranging from 20 to 90 m thick, illustrate best the nature of the faulting. All are fault-bounded at the base and overlain (in two cases with a clearly concordant contact) by bedded gray chert. The basal faults are thin (up to 30 cm across) crush zones with irregular quartz veins, or thicker more diffuse zones of shearing cutting the underlying lithology. The black shales are particularly affected, in most cases showing evidence of intense shearing which has destroyed original lamination and locally produced scaly argillite.

The intensity of imbrication revealed in the pipeline section is not evident in outcrop. Laterally tracing individual imbricate slices is very difficult. One exception is the northernmost basaltic body, which is a distinctive uniform coarse dolerite, traceable more than 3 km on both sides of the Clyde Valley (Figure 5). One particularly well-exposed area on the southwest flank of Castle Hill shows that the reverse faults braid over relatively short distances. For this reason we do not attempt to draw a detailed map of the imbricate zone where it is less completely exposed (Figure 5).

3) The Megget Valley area. A 4 m diameter tunnel drilled during 1978 to 1980 nearly 8 km across strike in part of the Central Belt provides an excellent opportunity to study faulting in greywacke terranes (Figure 2). The area, shown in Figure 6, straddles tracts 5 and 6 of Figure 2 in a strip where the boundary fault (Hartfell Line) is difficult to map due to sparse black shale marker bands. The greywackes consist of alternations of massive and thick-bedded, largely medium-to-coarse grained and locally pebbly sandstone, and thin-bedded (centimeter to decimeter scale) fine sandstone and siltstone. For the most part, sedimentary contacts between these two facies are abrupt. Thick (up to 2 km) units dominated by either facies are recognizable, but alterations also occur on a scale of less than 10 m (Figure 7). Bedding dips steeply northwest, younging for the most part the same way. Minor folds, localized to thin-bedded units, disturb the homoclinal sequence in places; most plunge north to northeast at shallow angles. Thinly bedded rocks have a locally developed cleavage which strikes at an angle 15° clockwise from bedding and dips steeply northwest. One major fold, plunging steeply west, swings the strata parallel with the tunnel trend between km 2 to 3 (see plan of tunnel,Figure 7); the significance of this structure is presently unclear.

The tunnel section is dominated by strike faults, in most cases bedding-parallel or nearly so (Figures 7, 8). Faulting concentrates in the bedded greywacke and siltstone units, and is common along lithological boundaries (Figure 6, 7). The faults are very numerous, although many may have only minor displacements. Over 200 occur between km 0 and 2.

Many of the faults (all shown on Figure 7, section X-Y) are marked by decimeter to meter-wide gouges of scaly clay. This is an intensely sheared rock which fractures along innumerable an anastomosing, dark polished slip planes so that it superficially resembles faulted black shale. However, lenses of less-altered protolith (some original lamination still visible) occur in thicker gouges in siltstone units. Intense shearing in these has locally altered the rock still further to a gray plastic clay. The scaly clay is greenish and less common in the massive greywackes.

The one 10 m section of black shale exposed in the tunnel at 1.4 km is faulted on both sides. The sediment is intensely sheared, but some graptolite fragments are preserved. At the surface this black shale band maps out as a lens along a major fault between massive greywacke and siltstone units. A similar black shale southwest of the dam site lenses out along a major strike fault, and is represented at 0.7 km in the tunnel by a thick zone of scaly clay along the fault. Later brittle faults, at high angles to bedding and characterized by brecciated gouge, are only a minor component in the tunnel section.

Summary

During about 55 million years of Ordovician-Silurian subduction under the British portion of the southern margin of Laurentia, a belt of ocean floor and trench sediment at least 70 km wide was accreted to the continental margin. The accreted material in Scotland is preserved in discrete strike-fault-bounded and stratigraphically distinct tracts. We recognize 10 of these tracts and future mapping will certainly subdivide them. They range in thickness (i.e. width in present outcrop) from less than 500 m to about 10 km (Figure 2). The strata within each tract are intact (i.e. free from melange) though much deformed by folding and imbrication.

The distinctive stratigraphic sequences recognizable in individual tracts, and the major strike faults (initially thrusts) which bound them, are in some cases mappable across the Southern Uplands for more than 200 km (Leggett, McKerrow, and Eales, 1979; Leggett et al, 1979; Leggett, McKerrow, and Casey, 1982). Therefore, we conclude that for each tract the fundamental decollement during accretion was the major reverse fault bounding the south side. Much of the internal deformation within packets must have resulted from continued underthrusting and consequent rotation after the rocks were incorporated in the lower trench slope. Evidence of multiple reactivation along

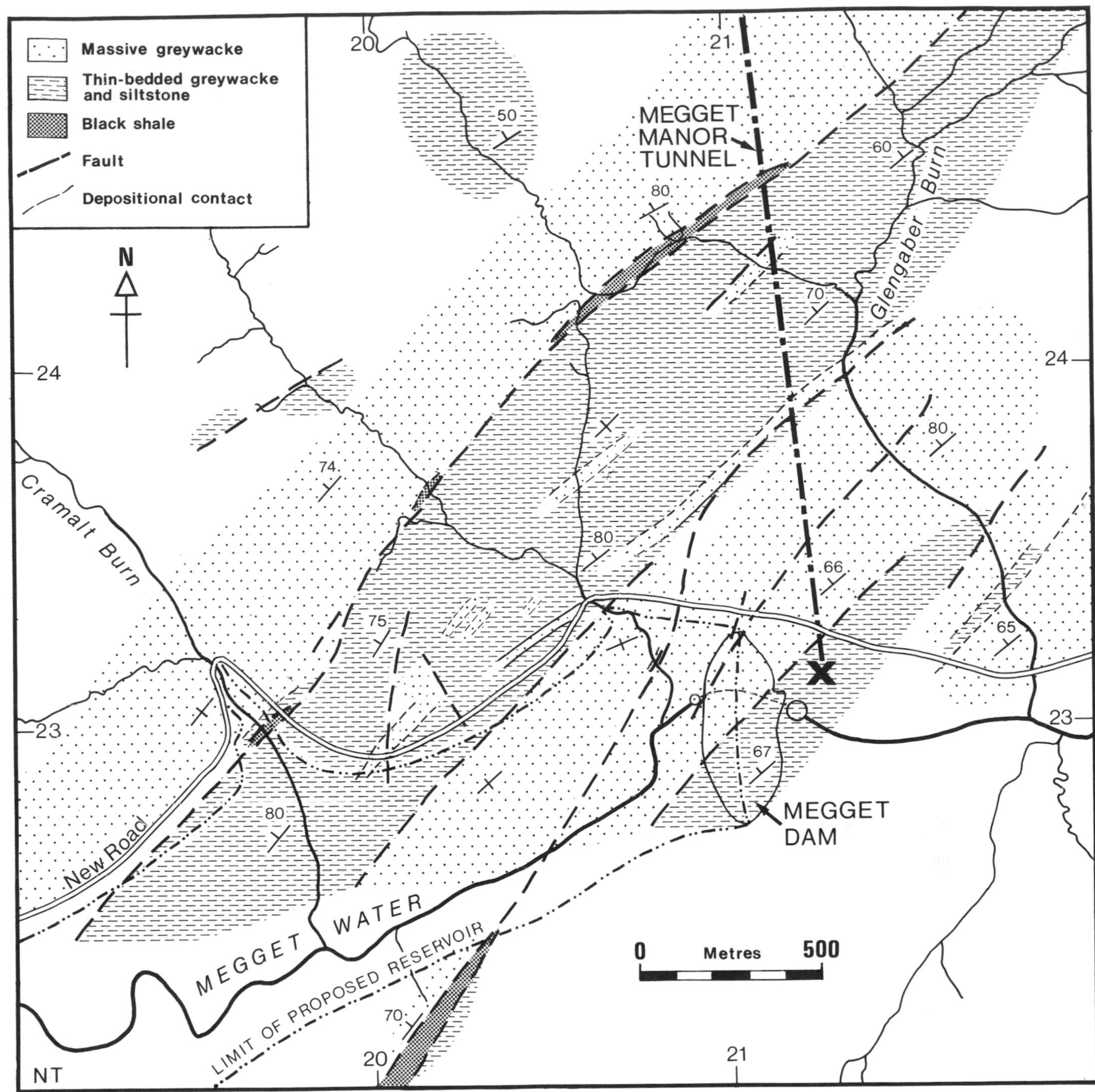

Figure 6 — Geologic map of Megget Dam area, Megget Valley. Sedimentary younging consistently to the northwest.

strike faults supports this conclusion. The map areas described show the importance of imbrication in this post-decollement deformation.

COMPARISON WITH OTHER ACCRETIONARY FOREARCS

The most striking features of the Southern Uplands accretionary complex are the coherent strata present throughout the outcrop, and the relative ease with which stratigraphically distinct tracts (individual accretionary packets) can be recognized, at least where oceanic pelagic-hemipelagic lithologies crop out. Structural complications disguise the boundaries of individual accretionary packets in melange-dominated accretionary complexes. For example, it is very difficult to map out boundaries between discrete packets in the Franciscan Complex of California (Bachman, 1982) and the Shimanto Belt of Japan (Taira et al, 1982). Some workers suggest that the incoherence of seismic records, below certain supposedly accretionary lower trench slopes, indicates that trench sediment may be continually "bulldozed" into the accretionary complex rather than uplifted in discrete thrust-bounded packages (Shipley et al, 1982). For this reason our discussion concentrates on modern and ancient accretionary complexes comprising coherent strata. We omit the evolution of lower slope sedimentary sequences because, with minor possible local exceptions, there is no evidence that successions originally mantling accreted ocean floor and trench sediments have been preserved in the Southern Uplands (Leggett, McKerrow, and Casey, 1982).

Sequence of deformation

Accreted packets of trench and ocean floor material in the Southern Uplands average 2 km or more in thickness (Figure 2); some are traceable more than 200 km. This moderately regular spacing and lateral continuity indicates that the fundamental decollement plane (sole thrust) for each packet propogated when the package was still relatively undeformed. In the early stages of deformation of packets containing thick black shale, we envisage a main thrust in or at the base of the incompetent black shale horizon (Figure 2) and strain-hardening by simple contemporaneous folding in the overlying greywackes. If generated at an early stage of accretion, the intensity of imbrication now seen in the accreted tracts in the areas described would not allow the stress transmission uniformity necessary to create such uniform packets.

Initial stages of accretion involving similar simple folding and fault generation occur at the foot of the lower trench slope in the Shikoku (Moore and Karig, 1976), Washington-Oregon (Carson, Yuan, and Myers, 1974), and Makran (White, 1982) forearcs. Off Shikoku, in the Nankai Trough, seaward-verging folds within the lowermost accretionary packet in the inner trench slope are similar to those of the Southern Uplands, and show near-parallel relationship between bedding and bounding thrust faults (a feature common in the Southern Uplands). Each of the accreted packets at the foot of the Shikoku inner trench slope is topographically expressed by a ridge traceable up to 10 km parallel with the trench (Moore and Karig, 1976). Similar ridges at the foot of the Washington-Oregon slope are uplifted open anticlines of Astoria Fan sediment. These anticlines verge landward due to an overpressured mud layer at depth in the subducting sequence (Seely, 1977). At the base of the Makran inner trench slope off Pakistan, spectacular frontal folds 3 to 4 km wide verge toward the ocean and are traceable for several hundred kilometers. Progressive tilting of ponded lower slope sediments suggests that the folds are uplifted and rotated between bounding thrusts (White, 1982).

In the exposed part of the Makran accretionary prism in the coast ranges of Pakistan and Iran, accretionary basement composed of folded flysch underlies slope sediments preserved in lenticular basins (Farhoudi and Karig, 1977). The deformation in the accretionary complex is much greater than deformation at the base of the lower slope. Accretionary material comprising the Oligo-Miocene Panjgur Formation (Hunting, 1961) is mapped in the region of Turbat, western Pakistan (Ahmed, 1969). Here, reverse faults dominate the outcrop pattern, strike east-west perpendicular to the northward subducting ocean crust below the Gulf of Oman (Jacob and Quittmeyer, 1979), and dip steeply northward. Ahmed (1969) reports that "most of the outcrops are in severely faulted zones" and "complete undisturbed sequences are rare . . . ." Nonetheless, the formation comprises well-cemented orthoquartzitic sandstone interbedded with shale and is fundamentally intact (Hunting 1961). Parallel to the strike-faults are anticlines, often with faulted limbs (Ahmed, 1969). Both reverse faults and folds are cut, mostly with minimal displacement, by numerous northeast to southwest trending conjugate wrench faults (Ahmed, 1969, Figure 3). Structural parallels with the Southern Uplands, at least on the megascopic level, are remarkable (Figure 2). The main differences are in the preservation of recognizable slope sediment in the Makran.

Similar increases in structural complexity upslope and inboard across subduction complexes are recorded elsewhere (e.g. von Huene, Moore, and Moore, 1979 for the Aleutians; Kulm and Fowler, 1974, and Snavely, Wagner, and Landes, 1980, for Washington-Oregon). Von Huene (1979) states the problem succinctly for the Aleutians: "Seismic reflection data show the beginning of subduction and the subduction complex, which is the end product, can be seen in exposures on land. The intermediate processes that change trench and slope sediment into a subduction complex are largely inferential because indirect geophysical methods cannot yet resolve such complex structure."

What might have transformed initially simple thrust-bounded fold packets in the Southern Uplands accretionary complex into the complex imbricated terranes seen today? Based on our mapping, we suggest that strain hardening (Moore and Karig, 1976) transferred stress from rotated and tightened folds to shear zones along both pre-existing strike faults and zones of competency contrast. Our mapping indicates this happened predominantly in the incompetent pelagic — hemipelagic basal lithologies, but also within heterogeneous turbidite sequences. Wrench faults may have also been important. Dubey (1980) shows experimentally that compression in one direction (such as might be expected above a subduction zone) can cause early generation of transcurrent faults (Figure 9). They initially form at an acute angle (bisected by the principal compressive stress vector) of 35 to 45° which widens with continuing compression. Folds formed at the same time are commonly confined to wrench fault-bounded areas. The folds have varying geometries and interlimb angles along hinge, and are deflected through interference with faults. Such early formation of wrench faults in the Southern Uplands could explain lack of major fold axes traceable along strike for any distance. Detection of early-formed faults in the Southern Uplands is difficult because limited movement along them persists with continuing compression, so that they appear to be the latest deformational stage. Similar comments apply to the

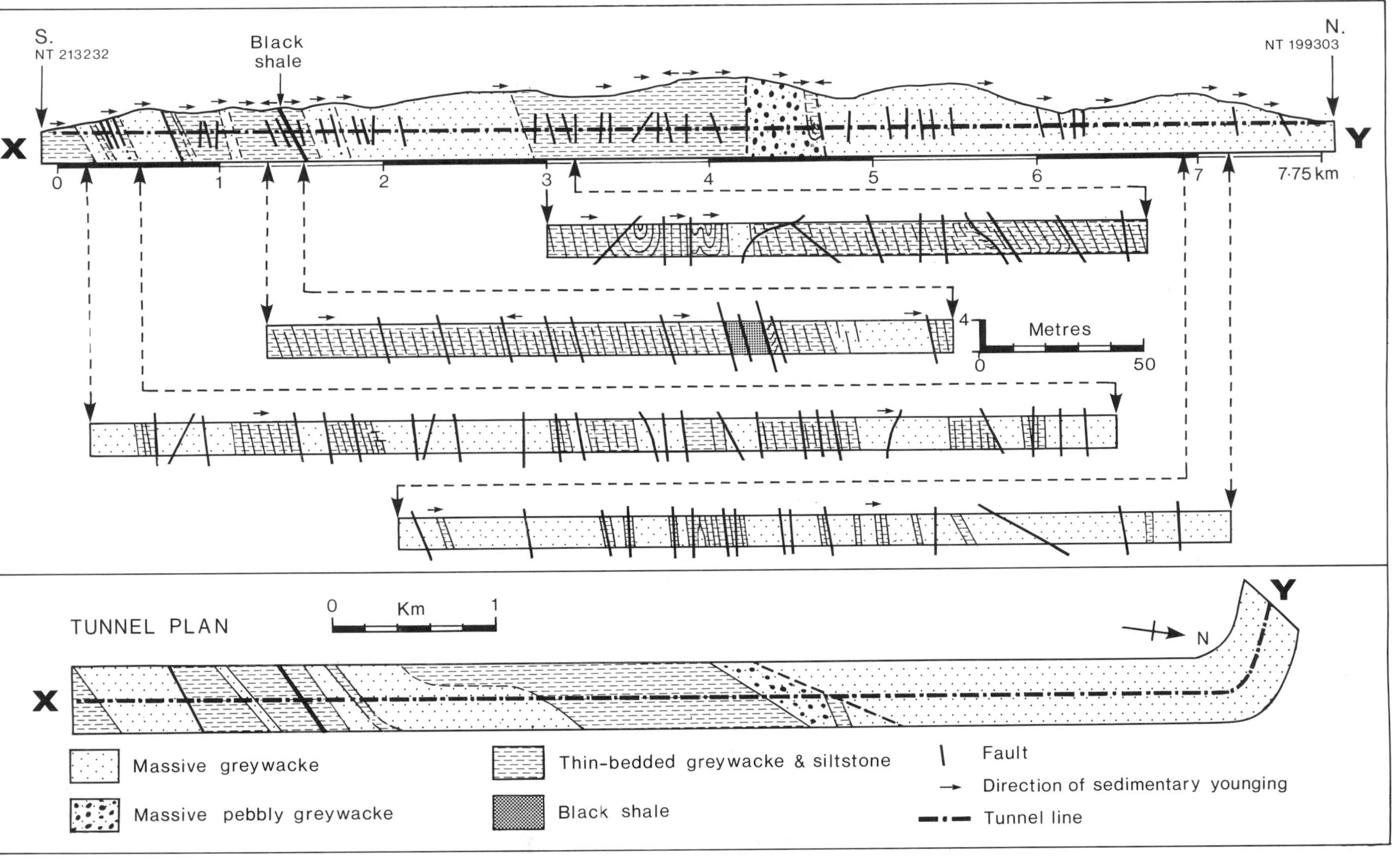

Figure 7 — Geology of the Megget-Manor tunnel. For location see Figure 6. The larger scale sections show details of representative portions of the tunnel section.

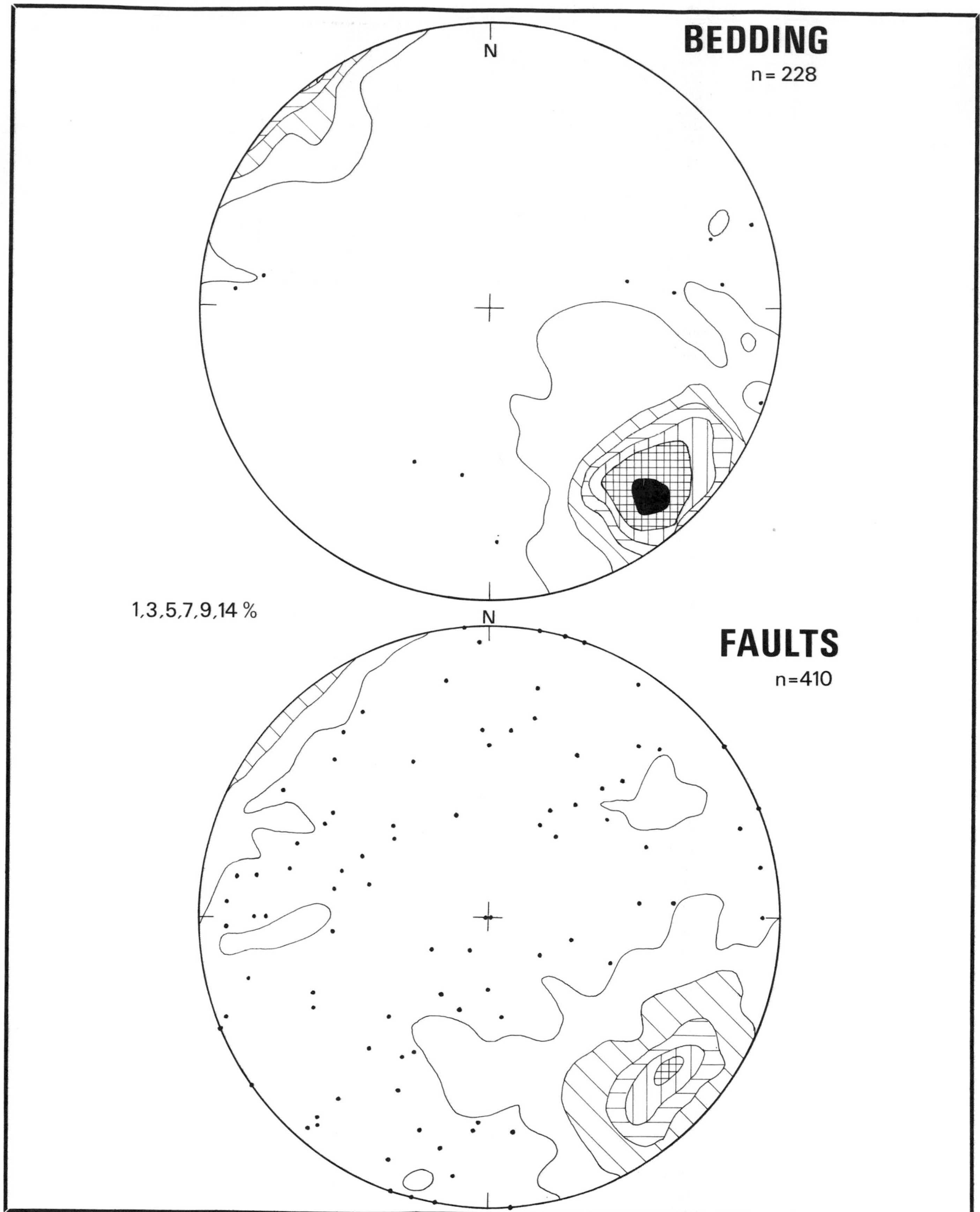

Figure 8 — Lower hemisphere projections of poles to bedding and faults for the Megget-Manor tunnel section. Faults are predominantly bedding-parallel, or nearly so. The faults scattered away from the maximum are mainly recorded from homogenous massive greywacke sections.

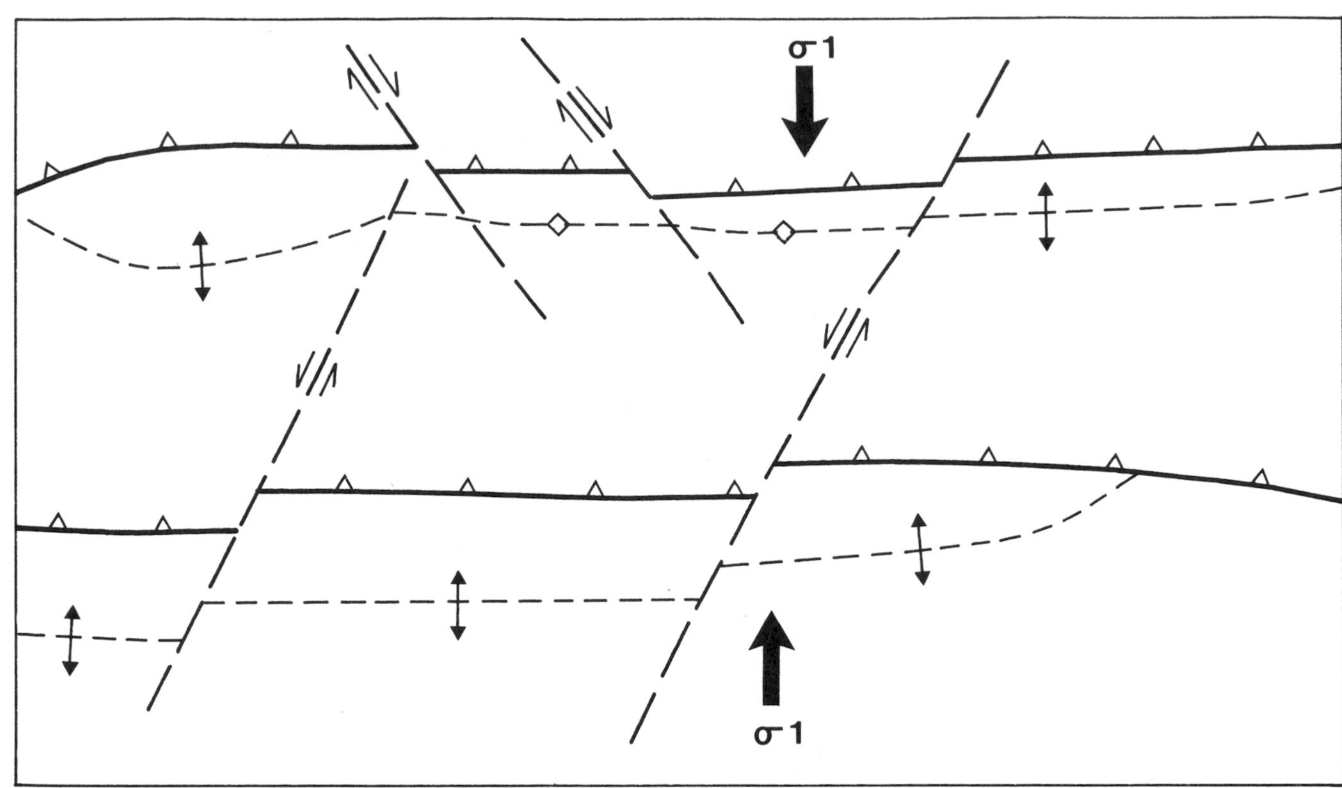

Figure 9 — Idealized structural history of the Makran accretionary complex (adapted from Hunting, 1961), considered equally applicable to the Southern Uplands. Dominant compression is from bottom of diagram upward. Early-formed major folds (e.g. dashed anticlines) are separated into packets by essentially coeval thrust faults (solid lines with teeth on overthrusting side). Wrench faults, forming in response to the same maximum principal stress (σ^1) may form early, dividing accretionary terrane into structural blocks with independent subsequent evolution; minor movement along these wrench faults continues after folding and strike fault movements end.

Makran (Figure 9). Fold generation becomes significantly inhibited by the wrench faults in the later stages of shortening in Dubey's experimental models. In an accretionary complex below which underthrusting was continuous, subsequent strain must be taken up in imbricate zones or by slip along already existing strike faults.

Method of offscraping

Stratigraphic sequences in the Southern Uplands accreted packets suggest that in the early stages of accretion (tracts 1, 2, and 3; Figure 2) the whole ocean floor pelagic section was accreted along with the trench turbidite section. Local slivers of uppermost oceanic Layer 2 basalt were also scraped off the descending plate. In later accreted packets (tracts 4, 5, 6, and 7; Figure 2) decollement occurred stratigraphically higher; that is, within black shales or at the base of turbidites. We attribute this progressive loss of the basal lithologies to selective subduction (Moore, 1975). In the Southern Uplands, progressive thickening of a highly incompetent hemipelagic shale unit (the Moffat Shales), from 0 to about 100 m in successive accreted slices (Figure 2), apparently controlled the level of decollement in most slices (Leggett, McKerrow, and Eales, 1979). Basal chert and basalt sequencs are preserved where the shale unit is absent or thin (less than about 10 m) in tracts 1, 2, and 3 of Fig. 2.

The basalts, nowhere more than 200 m thick, may in some cases be sills or flows within the chert sequence, but the widespread occurrence of metalliferous sediments within and above the basalts suggests that many are slivers of true oceanic basement (Leggett, 1979). They were perhaps accreted in similar fashion to slivers of Nazca Plate basalt incorporated in parts of the lower slope of the Peru-Chile trench. Kulm et al (1982) suggest that these are bounded by thrusts which are propagated in weak levels of the ocean crust such as basalt breccia layers, and refract to the surface by exploiting pre-existing normal faults inherited from tensional stress as the ocean crust bent down into the trench. Basaltic breccia is a common, though not abundant, component of the Southern Uplands basalt.

Of course, our loosely-defined coherent and disrupted classes are conceptual end-members between which there is considerable gradation. We anticipate that individual accretionary complexes may commonly change along or across strike from one categorization to the other, as along the Sunda Arc (Moore et al, 1980), if the requisite controlling parameters of subduction change. Additionally, convergence rate and sediment input are not the only factors governing structural style in accretionary complexes, though they may be the most important. Other possible parameters include stratigraphy of the subducting sediment pile and obliquity of convergence.

Seely (1977) describes the importance of stratigraphy of the subducting sedimentary sequence, stressing the importance of overpressured mudstone sequences in taking up shear stress and allowing development of landward-verging folds in overlying turbidites. In the Southern Uplands, the Moffat Shale Group undoubtedly provided an over-pressured zone at the base of the subducting sequence; its major effect was to localize decollement and concentrated subsequent imbrication.

We already mentioned evidence for oblique convergence under the Southern Uplands (Thirlwall, 1981; Stringer and Treagus, 1981). Moore et al (1980) relate increasingly coherent accretion in the Sunda Arc, northwest of Sumatra, in part to an increasingly oblique component of convergence. The same effect may have operated along the Laurentian margin in northern Britain, preserving coherent strata in the Southern Uplands accretionary complex. Additionally, Stringer and Treagus (1981) argue that steep plunges on certain D_1 folds in the Southern Uplands are a manifestation of oblique compression.

Finally, Moore, Watkins, and Shipley (1981) suggest that in certain subduction complexes relatively coherent offscraped strata may be preserved in a thick "rind" at high tectonic levels while disrupted rocks are underplated, to be preserved at deeper levels of the same complex. This interesting model awaits further testing. Moore et al propose that in forearcs with the same geothermal gradients as modern Mexico (IPOD Leg 66 area, on which they base their arguments) zeolite facies metamorphism should characterize the offscraped rind and prehnite-pumpellyite facies the underplated deposits. The model was applied successfully to a Tertiary subduction complex exposed on Kodiak Island, Alaska (Moore and Allwardt, 1980). Prehnite-pumpellyite facies metamorphism dominates the Southern Uplands rocks (Oliver and Leggett, 1981). Perhaps original offscraped packets in the Southern Uplands accretionary complex extended to deeper levels than those recognized by Moore, Watkins, and Shipley (1981) off Mexico.

ACKNOWLEDGMENTS

We are indebted to T. A. Anderson, N. Lundberg, and J. Platt for improving the manuscript with their reviews, and to W. S. McKerrow, D. Karig, and R. von Huene for valuable discussions and advice during the preparation of the paper. We also thank the Natural Environment Research Council for fieldwork support.

REFERENCES CITED

Ahmed, S. S., 1969, Tertiary geology of part of South Makran, Baluchistan, West Pakistan: AAPG Bulletin, v. 53, p. 1480-1499.

Anderson, T. B., and T. D. J. Cameron, 1979, A structural profile of Caledonian deformation in Down, *in* The Caledonides of the British Isles — reviewed: Geological Society of London, Special Publication 8, p. 263-267.

Bacchman, S. B., 1982, The Coastal Belt of the Franciscan: youngest phase of northern California subduction, *in* Trench-Forearc Geology: Geological Society of London Special Publication 10, p. 401-407.

Brown, G. C., 1979, Geochemical and geophysical constraints on the origin and evolution of Caledonian granites, *in* The Caledonides of the British Isles — reviewed: Geological Society of London Special Publication 8, p. 645-651.

Carson, B., J. Yuan, and P. B. Myers, Jr., 1974, Initial deep-sea sediment deformation at the base of the Washington continental slope; a response to subduction: Geology, v. 2, p. 561-564.

Cook, D. R., and J. A. Weir, 1979, Structure of the Lower Palaeozoic rocks around Cairnsmore of Fleet, Galloway: Scottish Journal of Geology, v. 15, p. 187-202.

Cowan, D. S., and R. M. Silling, 1978, A dynamic, scaled model of accretion at trenches and its implications for the tectonic evolution of subduction complexes: Journal of Geophysical Research, v. 83, p. 5389-5396.

Dickinson, W. R., and D. R. Seely, 1979, Structure and stratigraphy of forearc regions: AAPG Bulletin, v. 63, p. 2-31.

Dubey, A. K., 1980, Model experiments showing simultaneous development of folds and transcurrent faults: Tectonophysics, v. 65, p. 69-84.

Eales, M. H., 1979, Structure of the Southern Uplands of Scotland, *in* The Caledonides of the British Isles — reviewed: Geological Society of London Special Publication 8, p. 269-273.

Farhoudi, G., and D. E. Karig, 1977, Makran of Iran and Pakistan as an active arc system: Geology, v. 5, p. 664-668.

Floyd, J. D., 1975, The Ordovician rocks of west Nithsdale: doctoral thesis, University of St. Andrews (unpublished).

Hepworth, B. C., M. J. McMurtry, and G. J. H. Oliver, 1982, Sedimentology, volcanism, structure and metamorphism of the Northern Belt of the Southern Uplands of Scotland — A Lower Palaeozoic fore-arc complex, *in* Trench-forearc geology: Geological Society of London Special Publication 10, p. 521-534.

Hunting Survey Corporation, 1961, Reconnaissance geology of part of West Pakistan: Miracle Press, Ontario.

Jacob, K. H., and R. L. Quittmeyer, 1979, The Makran region of Pakistan and Iran: trench-arc system with active plate subduction, *in* A. Farah and K. A. De Jong, eds., Geodynamics of Pakistan: Geological Survey of Pakistan, Quetta, p. 305-317.

Johnson, M. R. W., D. J. Sanderson, and N. J. Soper, 1979, Deformation in the Caledonides of England, Ireland and Scotland, *in* The Caledonides of the British Isles — a review: Geological Society of London Special Publication 8, p. 165-186.

Karig, D. E., et al, 1980, Structural framework of the fore-arc basin, N. W. Sumatra: Journal of Geological Society of London, v. 137, p. 77-91.

Kulm, L. D., and G. A. Fowler, 1974, Cenozoic sedimentary framework of the Gorda-Juan de Fuca plate and adjacent continental margin — reviewed: *in* Society of Economic Paleontologists and Mineralogists, Special Publication 19, p. 212-229.

Lamont, A., 1975, Noblehouse, Lamancha, *in* G. Y. Craig and P. McL. D. Duff, eds., The geology of the Lothians and south east Scotland — An excursion guide: Scottish Academic Press, p. 158-166.

Leggett, J. K., 1979, Oceanic sediments from the Ordovician of the Southern Uplands, *in* The Caledonides of the British Isles — reviewed: Geological Society of London

Special Publication 8, p. 495-498.
———, 1980, The sedimentological evolution of a Lower Palaeozoic accretionary fore-arc in the Southern Uplands of Scotland: Sedimentology, v. 27, p. 401-417.
———, W. S. McKerrow, and M. H. Eales, 1979, The Southern Uplands of Scotland: A Lower Palaeozoic accretionary prism: Journal of Geological Society of London, v. 136, p. 755-770.
———, ———, and D. M. Casey, 1982, Anatomy of a Lower Palaeozoic accretionary complex: the Southern Uplands of Scotland, *in* Trench-forearc geology: Geological Society of London Special Publication 10, p. 495-520.
———, 1979, The northwestern margin of the Iapetus Ocean, *in* The Caledonides of the British Isles — reviewed: Geological Society of London Special Publication 8, p. 499-512.
Longman, C. D., B. J. Bluck, and O. van Breemen, 1979, Ordovician conglomerates and the evolution of the Midland Valley: Nature, v. 280, p. 578-581.
Lumsden, G. I., et al, 1967, The geology of the neighborhood of Langholm: Memoirs of the Geological Survey of Scotland, sheet 11, 255 pp.
McKerrow, W. S., J. K. Leggett, and M. H. Eales, 1977, Imbricate thrust model of the Southern Uplands of Scotland: Nature, v. 297, p. 237-239.
Moore, G. F., and D. E. Karig, 1980, Structural geology of Nias Island, Indonesia: implications for subduction zone tectonics: American Journal of Science, v. 280, p. 193-223.
———, et al, 1980, Variations in geologic structure along the Sunda fore-arc, northeastern Indian Ocean, *in* The tectonic and geologic evolution of southeast Asian Seas and islands: American Geophysical Union, Geophysical Monograph 23, p. 145-160.
Moore, J. C., 1975, Selective subduction: Geology, v. 3, p. 530-532.
———, 1979, Variation in strain and strain rate during underthrusting of trench deposits: Geology, v. 7, p. 185-188.
———, and D. E. Karig, 1976, Sedimentology, structural geology and tectonics of the Shikoku subduction zone, southwestern Japan: Geological Society of America Bulletin, v. 87, p. 1259-1268.
———, and A. Allwardt, 1980, Progressive deformation of a Tertiary trench slope, Kodiak Islands, Alaska: Journal of Geophysical Research, v. 85, p. 4741-4756.
———, J. S. Watkins, and T. H. Shipley, 1981, Summary of accretionary processes, deep sea drilling project leg 66: offscraping, underplating, and deformation of the slope apron, *in* Initial reports of the deep sea drilling project, v. 66: Washington D.C., U.S. Government Printing Office.
———, et al, 1979, Progressive accretion in the Middle America Trench, Southern Mexico: Nature, v. 281, p. 638-642.
Oliver, G. J. H., and J. K. Leggett, 1980, Metamorphism in an accretionary prism: prehnite-pumpellyite facies metamorphism of the Southern Uplands of Scotland *in* Transactions of the Royal Society of Edinburgh: Earth Sciences, v. 71, p. 235-246.
Peach, B. N., and J. Horne, 1899, The Silurian rocks of Britain, v. 1, Scotland: Memoir of the Geological Survey of Scotland, 749 pp.
Phillips. W. E. A., C. J. Stillman, and T. Murphy, 1976, A Caledonian plate tectonic model: Journal of the Geological Society of London, v. 132, p. 579-609.
———, A. M. Flegg, and T. B. Anderson, 1979, Strain adjacent to the Iapetus suture in Ireland, *in* The Caledonides of the British Isles — reviewed: Geological Society of London Special Publication 8, p. 257-262.
Scholl, D. W., et al, 1981, Sedimentary masses and concepts about tectonic processes at underthrust ocean margins: Geology, v. 8, p. 564-568.
Seely, D. R., 1977, The significance of landward vergence and oblique structural trends on trench inner slopes, *in* M. Talwani and W. C. Pitman III, eds., Island arcs, deep sea trenches and back arc basins: Washington D.C., American Geophysical Union, p. 187-198.
———, P. R. Vail, and G. G. Walton, 1974, Trench slope model, *in* C. A. Burk, and C. L. Drake, eds., The geology of continental margins: New York, Springer-Verlag, p. 249-260.
Shipley, T. H., et al, 1982, Tectonic processes along the Middle America Trench inner slope, *in* Trench-forearc geology: Geological Society of London Special Publication.
Simpson, P. R., et al, 1979, Uranium mineralization and granite magmatism in the British Isles: Philosophical Transactions of the Royal Society of London, v. A291, p. 385-412.
Snavely, P. D., Jr., H. C. Wagner, and D. L. Lander, 1980, Interpretation of the Cenozoic geologic history, central Oregon continental margin: Cross section summary: Geological Society of America Bulletin, v. 91, p. 143-146.
Stringer, P., and J. E. Treagus, 1980, Non-axial planar S_1 cleavage in the Hawick Rocks of the Galloway area, Southern Uplands, Scotland: Journal of Structural Geology, v. 2, p. 317-331.
———, and ———, 1981, Asymmetrical folding in the Hawick Rocks of the Galloway area, Southern Uplands: Scottish Journal of Geology, v. 17, p. 129-148.
Taira, A., et al, 1982, The Shimanto Belt of Japan: Cretaceous-Lower Miocene sedimentation in fore-arc basin to deep-sea trench environments, *in* Trench-forearc geology: Geological Society of London Special Publication 10, p. 5-26.
Thirlwall, M., 1981, Implications for Caledonian plate tectonic models of chemical data from volcanic rocks of the British Old Red Sandstone: Journal of Geological Society of London, v. 138, p. 123-138.
von Huene, R., 1979, Structure of the outer convergent margin off Kodiak Island, Alaska, from multichannel seismic records: AAPG Memoir 29, p. 261-272.
———, G. W. Moore, and J. C. Moore, 1979, Cross section, Alaska peninsula — Kodiak island — Aleutian trench: Summary: Geological Society of American Bulletin, v. 90, p. 427-430.
Walton, E. K., 1965, Lower Paleozoic rocks *in* G. Y. Craig, ed., The geology of Scotland: Edinburgh, Oliver & Boyd, p. 161-227.
Warren, P. T., 1964, The stratigraphy and structure of the Silurian rocks southeast of Hawick, Roxburghshire: Quarterly Journal of the Geological Society of London, v. 120, p. 192-222.
Watkins. J. S., et al, 1981, Accretion, underplating, subduction and tectonic evolution — Middle America Trench, southern Mexico: Results from leg 66 DSDP, *in* International Geological Congress: Paris, Special Publication, v. c3, p. 214-224.
Weir, J. A., 1979, Tectonic contrasts in the Southern Uplands: Scottish Journal of Geology, v. 15, p. 169-186.
Westbrook, G. K., 1982, The Barbados ridge complex: tectonics of a mature forearc system, *in* Trench-forearc geology: Geological Society of London Special Publication 10, p. 275-290.
White, R. S., 1982, Deformation of the offshore Makran accretionary sediment prism, *in* Trench-forearc geology: Geological Society of London Special Publication 10, p. 357-372.

Multichannel Seismic Survey of the Colombia Basin and Adjacent Margins

Richard S. Lu
Institute for Geophysics
University of Texas at Austin
Austin, Texas

Kenneth J. McMillen
Gulf Science and Technology Company
Pittsburgh, Pennsylvania

Twelve-fold multichannel seismic data reveal two types of seismic stratigraphy in the Colombia Basin: basin-floor-type and fan-type. Basin-floor-type stratigraphy occurs in areas with thin sediment cover in the central basin and parts of the lower Nicaragua Rise; whereas the fan-type stratigraphy with thick sediment cover occupies the western and eastern regions of the Colombia Basin. The reflection zone above acoustic basement in the basin-floor-type province and in each region of the fan-type province is divided into four seismic intervals separated by three laterally persistent reflectors of unknown ages.

The acoustic basement shoals toward the Costa Rica coast and dips down toward the Panama and Colombia margins. The Panama and Colombia margins at the southern boundary of the Colombia Basin consist of a convergence zone where basin sediments have been underthrust and deformed into active-margin-type structures. Structural detail of the deformed belt indicates that the directions of convergence at the northern Panama and Colombia coasts have been different from those of the present day relative motions between the Nazca, South American, and Caribbean plates.

The Colombia Basin is one of the major basins in the Caribbean plate. It is bounded on the northwest by the Nicaragua Rise. To the east, the Beata Ridge separates the Colombia Basin from the Venezuela Basin. These two basins are connected by the Aruba Gap between the southern end of Beata Ridge and the South American continental shelf. The Colombia Basin is bounded on the south by the Neogene volcanic belt of Central America, the Panama Isthmus, and the northwestern Colombia accretionary zone (Figure 1; Case and Holcombe, 1980).

Uplifted oceanic crust occurs near many of the Colombia Basin margins. Seismic refraction measurements suggest that the crust of northern lower Nicaragua Rise is similar to oceanic crust except for a thicker third layer (16 km; Edgar, Ewing, and Hennion, 1971). Basalt, diabase, and sedimentary rocks dredged and cored from the steep western slope of the Beata Ridge, and refraction data suggest that the ridge is composed of Colombia Basin sediments and oceanic crust (with thicker third layer similar to that of the Nicaragua Rise) uplifted in Late Cretaceous to Early Tertiary time (Fox et al, 1970). The ridge may represent a recently active compressional zone (Fox and Heezen, 1975).

Costa Rica contains clastic sedimentary rocks outcropping at the Caribbean coast with middle Tertiary volcanics in the central part. Tectonically emplaced marine sediments and oceanic crust occur along the Pacific coast. Panama can be divided into eastern and western regions at roughly the Canal Zone. Western Panama has oceanic basalts and extensive middle to late Tertiary and Quaternary volcanics (Terry, 1957). Eastern Panama consists mainly of uplifted oceanic basalts and pelagic sediments (Terry, 1957; Bandy and Casey, 1973) with only a few scattered Tertiary and Quaternary volcanics (Case, 1974) bounding the Chucunaque basin. This basin and the basaltic uplifts bounding it are a structural extension of the Atrato basin and Baudo uplift of Colombia (Case, 1974).

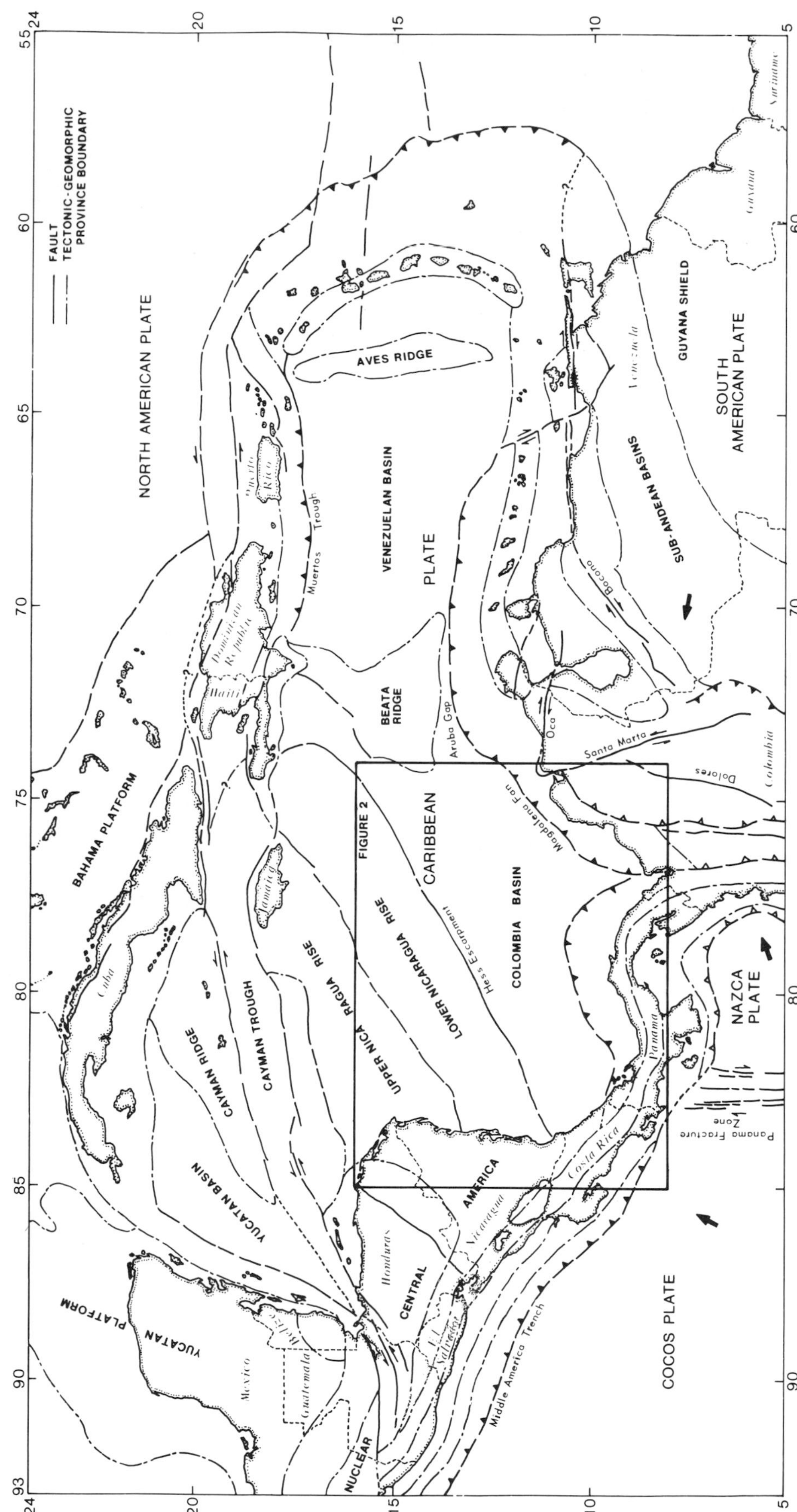

Figure 1 — Principal geological features of the Caribbean region. Adopted from Case and Holcombe (1980). Solid arrows indicate the directions of relative motions of the Cocos, the Nazca, and the South American plates with respect to the Caribbean plate. The rectangle in the lower left indicates the area discussed in this report.

The geology of northwestern Colombia consists mainly of older massifs and a wide belt of oceanic rocks accreted during the Tertiary (Duque-Caro, 1979), indicating the past existence of a subduction zone along the northwest Colombia margin. Malfait and Dinkelman (1972) presumed that this zone persisted throughout Late Cretaceous until Early or Middle Oligocene time. The igneous rocks of northwest Colombia give general clues to the subduction history. Mesozoic plutonic rocks, mostly of Triassic to Jurassic age, occur in the Santa Marta massif. Tertiary plutonic rocks consist mainly of Middle Eocene age batholiths of the Guajira-Santa Marta region (Irving, 1975). Thus, Mesozoic and Cenozoic subduction appears episodic or slow, although much accreted terrane is associated with it.

Previous seismic refraction measurements indicate that Colombia Basin crust is oceanic, although its thickness, 12 to 18 km, is greater than typical oceanic crust (Ewing, Antoine, and Ewing, 1960; Edgar, Ewing and Hennion, 1971; Ludwig, Houtz, and Ewing, 1975; Houtz and Ludwig, 1977). Crustal ages can be inferred from magnetic and heat flow data. Christofferson (1976) correlated east-west anomalies from south to north with anomalies 27 to 37 of the paleomagnetic time scale (62 to 80 m.y.; LaBrecque, Kent, and Cande, 1977). Average heat flow from 37 measurements in the Colombia Basin is 1.36 ± 0.50 μcal/cm^2 sec, giving an age of 65 to 80 m.y. (Late Cretaceous) according to either the plate or the boundary layer model (Sclater, Jaupart, and Galson, 1980).

Earlier studies of reflection profiles acquired in the Colombia Basin suggested that the A″ and B″ reflectors underlie upper sedimentary units similar to those of the Venezuela Basin (Edgar, Ewing, and Hennion, 1971; Ludwig, Houtz, and Ewing, 1975). Later seismic studies showed the identifications of A″ and B″ to be doubtful (Houtz and Ludwig, 1977). Analyses of new multichannel seismic data obtained in the Colombia Basin enable us to clarify the seismic stratigraphy of that basin. Structural analysis of different seismic intervals and comparison to those of the neighboring areas also shed light on the tectonic history of the basin and its surrounding margins.

The northern margins of Panama and Colombia consist of an extensive fold belt (Case, 1974; Bowin, 1976) which is part of a deformed zone extending from Costa Rica to eastern Venezuela. Several origins have been proposed for this belt, ranging from compressive and strike-slip tectonics to gravity sliding (Case, 1974) and mud diapirism (Duque-Caro, 1979), or combinations of all of them. Convergence seems to be at least partly responsible for formation of the Venezuelan belt based partly on relative plate-motion solutions for North and South America (Jordan, 1975; Ladd, 1976) and partly on seismic evidence from the deformed belt itself (Silver, Case, and MacGillvary, 1975; Ladd and Watkins, 1979). The multichannel seismic data we present here helps resolve the question of the origin of the deformed belt offshore of northern Panama and Colombia. Greater penetration of the sound source allows resolution of deeper structure, and processing techniques, such as migration, eliminate confusing diffraction patterns.

DATA COLLECTION AND PROCESSING

The University of Texas Institute for Geophysics is currently engaged in regional studies of Caribbean Sea tectonics, with emphasis directed toward understanding the structure and origin of the margins. Preliminary results are now available from the Colombia Basin where 12-fold multichannel seismic reflection data were collected (Figure 2). All the CT1 lines were shot in the summer of 1978, and the rest were shot in 1975 and 1977. The sound source used was 3 or 4 1,500-cu in Bolt® airguns operated at 350 to 500 psi or a Maxipulse® system. Signals were received by a 2,400-m-long streamer and recorded digitally on a Texas Instruments DFS 10,000 system in 1975 and a DFS III system in 1977 and 1978. Other geophysical measurements included high-resolution profiling (sparker and 3.5 kHz), sonobuoy refraction, and magnetic observations. The navigation was conducted with a satellite-positioning system of high accuracy.

Multichannel data were demultiplexed and processed on a TEMPUS® system using Petty-Ray Geophysical software. The normal processing sequence consisted of: (1) static correction and bad-shot editing, (2) sorting, (3) velocity analysis for normal moveout, (4) stacking based on rms velocity profiles, (5) bandpass filtering, and (6) gain function application. Lines CT1-21, CT1-22, and CT1-25 were migrated with a wave-equation program to remove diffraction interferences. The velocity data for stacking came from the velocity analyses of normal moveout for shallow penetration (generally 2.5 seconds sub-bottom) and from refraction velocities published by Ewing, Antoine, and Ewing (1960) for deeper velocities where reflection returns were weak.

Magnetic and bathymetric data were merged with navigation data. Bathymetry was corrected with Matthews' table (Matthews, 1939).

INTERPRETATION

We present the interpretation in two sections: (1) seismic stratigraphy of the Colombia Basin, and (2) structural style of the marginal deformed belt.

Seismic Stratigraphy of the Colombia Basin

Examination of the stacked CDP seismic sections reveal two types of seismic stratigraphy in the Colombia Basin, here termed basin-floor-type and fan-type.

Basin-floor-type stratigraphy mainly occupies the central Colombia Basin (Figure 3). Two typical sections of this type of stratigraphy, selected from line CT1-12, are presented in Figure 4. The reflection zone above the acoustic basement (KI_4) is divided into four seismic intervals termed CBI1, CBI2, CBI3, and CBI4, which are separated by three laterally persistent reflectors KI, KI_2, and KI_3. CBI1 and CBI4 contain laterally continuous reflectors, whereas CBI2 and CBI3 are characterized by chaotic zones and abundant diffractions of unknown source. The interval velocities shown in Figure 4 were calculated from the rms stacking velocities; interval boundaries for interval velocities do not all fall on stratigraphic boundaries. In both Figure 4A and 4B, the three shallow velocities are comparable to those of sedimentary rocks. KI_4 of rough topography resembles the top of layer 2 in other oceans, and is interpreted here to be the top of oceanic crust. The laterally extensive reflectors beneath CBI4 in the western part of Figure 4A are not parallel with reflectors in the upper seismic intervals and might be due to side echos. They could also represent tilted basin formations bounded at northeast by a fault near shot point 4,800. KI_4 drops to a deeper horizon at the fault and, to the west, continues to be beneath these inclined reflectors.

Fan-type stratigraphy occurs in the eastern and western parts of the Colombia Basin (Figure 3). Two seismic sections of this stratigraphy, one selected from line CT1-12 for the western region and one selected from line CT1-27 for the eastern region, are presented in Figure 5. Figure 5A shows that the reflection zone above the acoustic basement KII_4 in the western region can be divided into four seismic intervals (CBII1, CBII2, CBII3, and CBII4) separated by three laterally

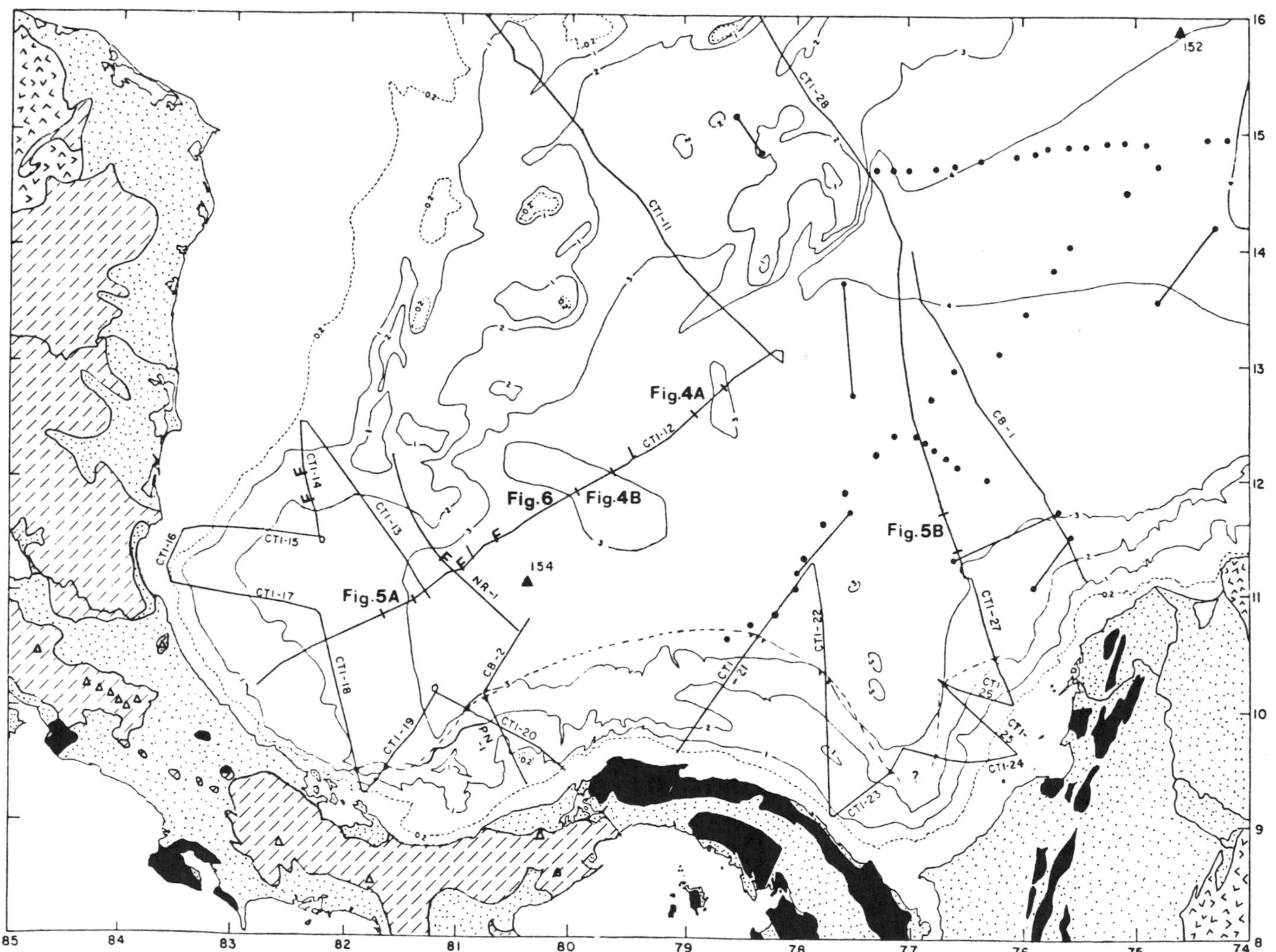

Figure 2 — Trackline map of R/V Ida Green in the Colombia Basin. Multichannel seismic and magnetic data were collected along the tracks of projects CT1 (Caribbean Tectonics Phase 1), CB (Colombia Basin), PN (Panama) and NR (Nicaragua Rise). F indicates fault found on the seismic profiles. Solid triangles are locations of DSDP sites. Solid circles are station locations of sonobuoy refraction measurements of Ludwig, Houtz, and Ewing (1975) and Houtz and Ludwig (1977). Solid circles connected by solid lines are locations of two-ship refraction lines of Ewing, Antoine, and Ewing (1960). Dashed line is the seaward edge of the deformed belt offshore Panama and Colombia. Bathymetric contours are simplified from USGS Open File Map 75-146. The contour interval is 1 km with 200 m contour added along the margin. The land geologic map is simplified from the tectonic map of North America of King (1969) and from Case and Holcombe (1980). Dotted areas are sedimentary rocks of Tertiary and Quaternary ages. Areas marked by slants are volcanic rocks of Tertiary and Quaternary ages. Open triangles in these areas are sites of historical active volcanoes. Black areas are basalts and ultramafic rocks, mainly of the Cretaceous age. Areas marked by random angles are metamorphic and plutonic rocks, mainly of the Paleozoic and Precambrian ages.

persistent reflectors (KII_1, KII_2, and KII_3). The seismic interval $CBII_4$, whose velocities suggest carbonates, lithified clastics, or volcanic rocks, is thin compared to upper intervals. In the eastern region, the reflection zone above the acoustic basement $KIII_4$ can also be divided into four seismic intervals (CBIII1, CBIII2, CBIII3, and CBIII4) separated by three laterally persistent reflectors ($KIII_1$, $KIII_2$, and $KIII_3$), as seen in Figure 5B. Reflector and seismic interval characteristics of this region are different from those of the western region. The interval boundaries for interval velocities calculated from rms velocities (Figure 5) again do not fall exactly on the stratigraphic boundaries. The deepest calculated velocity, 4.56 km/sec, of the western region in Figure 5A is higher than that of the eastern region shown in Figure 5B, indicating different lithologies. Acoustic basements KII_4 and $KIII_4$ both resemble the top of oceanic crust in other ocean basins.

Figure 6 illustrates how basin-floor stratigraphy changes to fan-type stratigraphy in the western Colombia Basin. The reflectors and seismic intervals of the basin-floor stratigraphy lose their identities near shot point 6,600.

Basin-floor stratigraphy is also observed in areas of the lower Nicaragua Rise, lapping against the westernmost Hess Escarpment, and offshore western Panama (Figure 3). On the Hess Escarpment and offshore western Panama, the observed basin-floor stratigraphy is interpreted to be preserved due to uplifting and is not affected by the surrounding sedimentation. To the north, on the lower Nicaragua Rise, we interpret basin-floor stratigraphy as due to tectonic quiescence.

In the rise province (Figure 3), sediments are disturbed or very thin. No clear stratigraphic pattern is evident.

Total sediment thickness above KI_4 in the basin-floor province is 1 to 1.5 seconds two-way travel time (Figure 4, 6, and 7), much thinner than the average of 3 seconds in fan-type areas (Figures 5, 6, and 7). Sediments in basin-floor areas are mainly pelagic and hemipelagic (Saunders et al, 1973). Thickness variations and pinch-outs related to bottom currents and basement structure occur in seismic intervals below KI_2, in contrast to the relative uniformity in thicknesses of seismic intervals above KI_2 (Figures 4 and 6). Recent tectonic activity is evident locally (for example, near shot point 5,800 shown in Figure 6), although there has been little tectonic activity in the Colombia Basin since KI_2 time. In the western part of the Colombia Basin, deposition of terrigeneous sediments

Figure 3 — Stratigraphic province map of Colombia Basin and lower Nicaragua Rise. Multichannel seismic profiles along dotted-line tracks (Figure 2) were used for the construction of the map. Boundaries between provinces are not well-defined due to the lack of data.

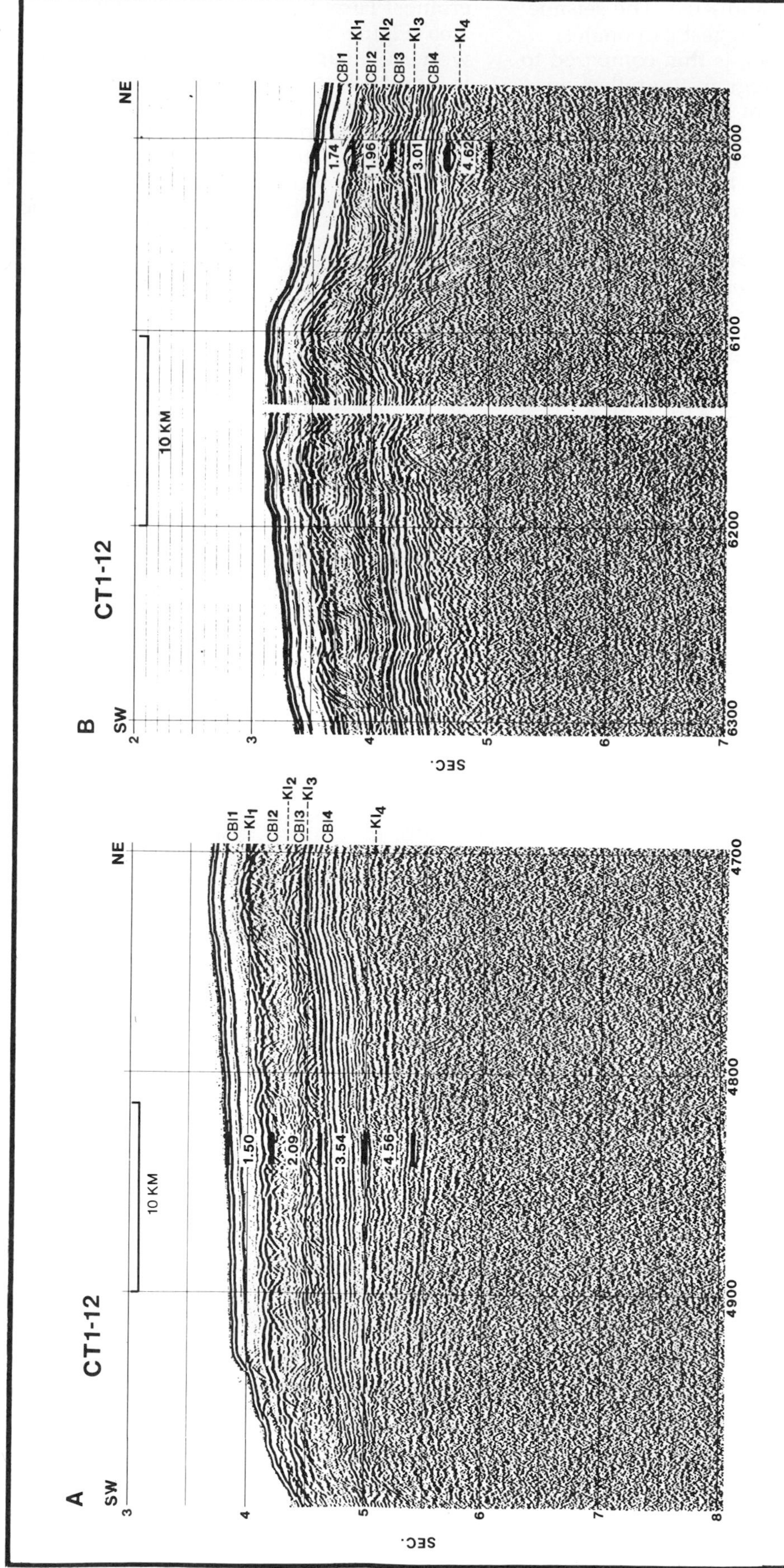

Figure 4 — Seismic sections of line CT1-12 (refer to Figure 2 for locations) illustrating the basin-floor-type stratigraphy in the Colombia Basin. Vertical scale is the two-way reflection time. Interval velocities (km/sec) between solid bars were calculated from rms velocities obtained from velocity analyses. Seismic intervals and reflectors are annotated at the right of the seismic section. Numbers at bottom are shot point numbers.

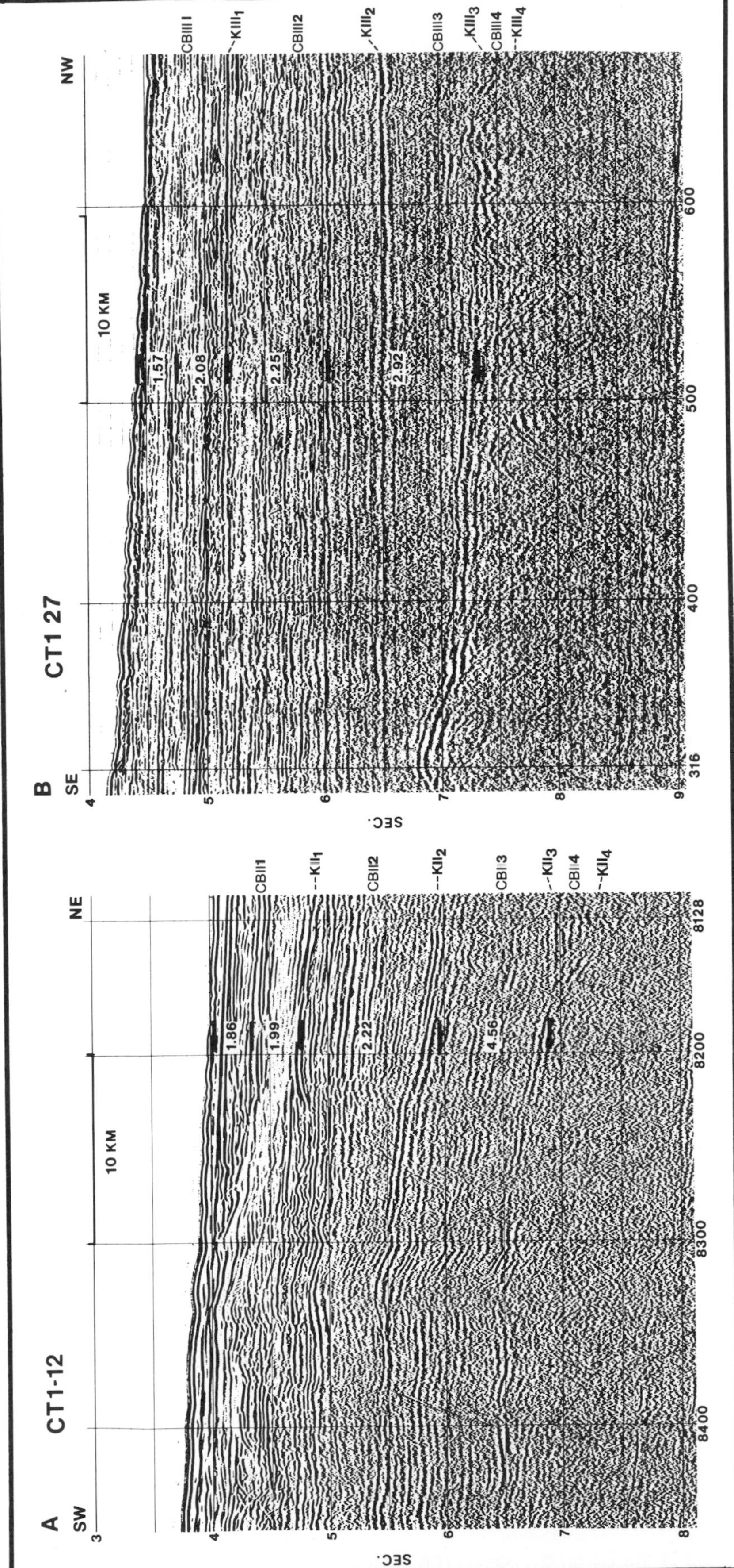

Figure 5 — Seismic sections of lines CT1-12 and CT1-27 (refer to Figure 2 for locations) illustrating the fan-type stratigraphy in the Colombia Basin. Vertical scale is the two-way reflection time. Interval velocities (km/sec) between solid bars were calculated from rms velocities obtained from velocity analyses. Seismic intervals and reflectors are annotated at the right of the seismic section. Numbers at bottom are shot point numbers.

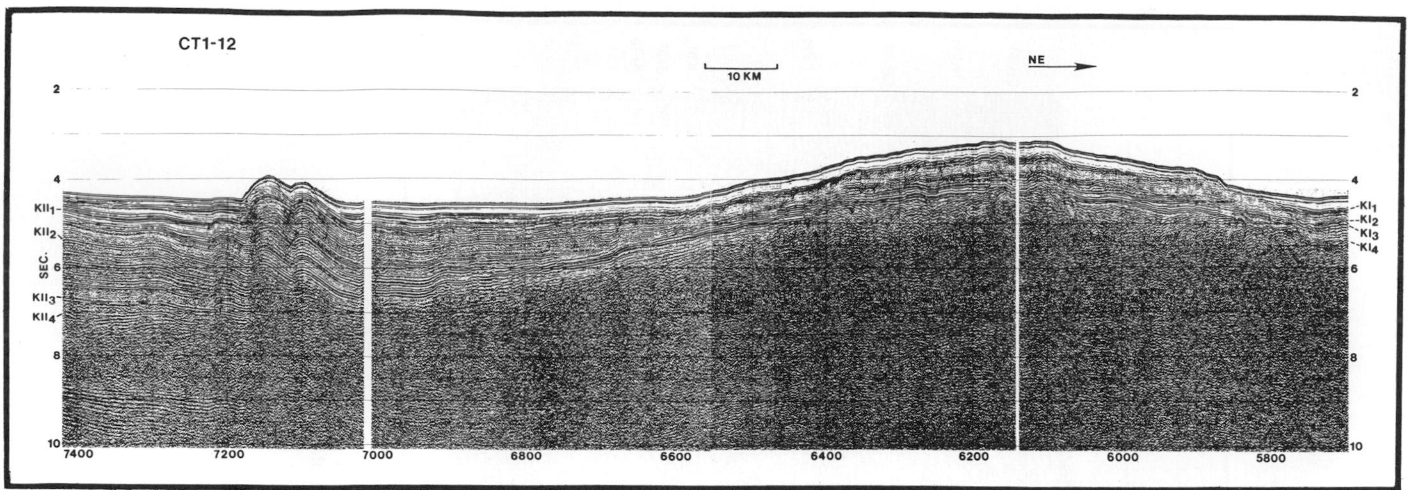

Figure 6 — Seismic section of line CT1-12 (refer to Figure 2 for location) illustrating how basin-floor stratigraphy changes to outer fan-type stratigraphy southwestward. Vertical scale is the two-way reflection time. Seismic reflectors are annotated at both sides of the seismic section. Numbers at bottom are shot point numbers.

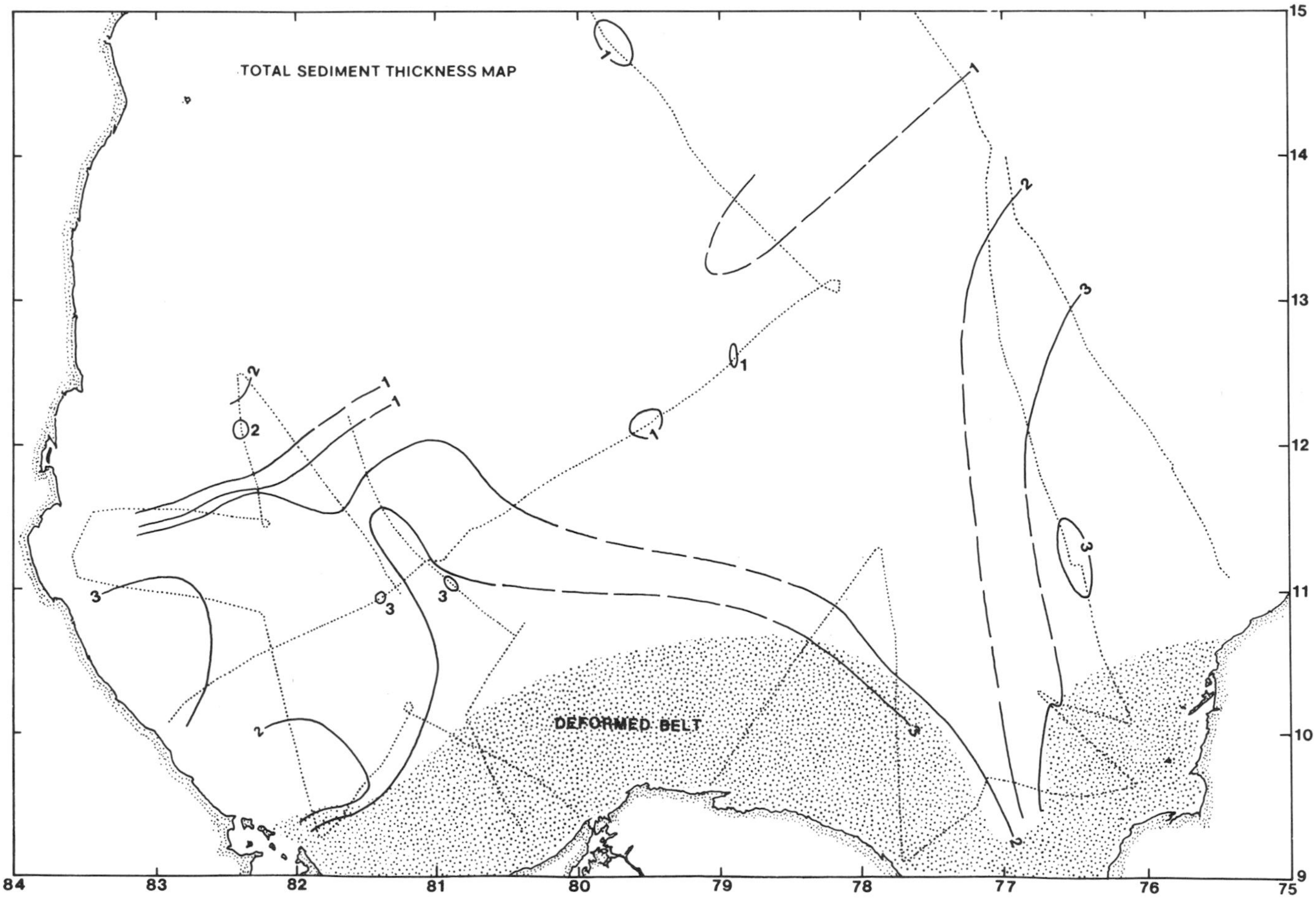

TOTAL SEDIMENT THICKNESS MAP

DEFORMED BELT

Figure 7 — Total sediment thickness map of the Colombia Basin. The contour interval is 1 second in two-way reflection time. Dotted lines indicate the tracks of seismic reflection data utilized for this map. Dotted area is the deformed belt province (Figure 3).

from Central America has formed a submarine fan (Figure 3) evident in the upper part of seismic section in Figures 5A and 6. Sources for older sediments in the deeper part of the seismic section are uncertain. The Magdalena Fan occupies in the eastern part of the Colombia Basin (Figure 3) where rivers in South America have discharged large amounts of sediments. Erosional features can be observed in the shallower part of the seismic section in Figure 5B. Sediment facies shown in Figures 5A and 5B are complicated and divisions of seismic intervals described earlier are somewhat arbitrary. Total sediment thickness increases and the basement deepens toward the margins of Panama and Colombia (Figures 7 and 8), where convergence of the Colombia Basin crust with the Panama and Colombia results in a deformed belt.

Structural Style of the Marginal Deformed Belt

A zone of deformed sediments occurs along the northern Panama and northwest Colombia margins. General features of this belt include (Figures 9, 10, 11 and 12): (1) a seaward belt of folded and faulted sediments; (2) a middle-slope or shelf basin with variable amounts of sedimentary fill; (3) small lower-slope basins; (4) landward dipping reflectors that pass beneath the deformed lower slope; and, (5) a prominent, continuous, bottom-simulating reflector (BSR) produced by the acoustic contrast between gas-hydrated sediment and underlying sediment with free gas (Shipley et al, 1980). The deformed sediment belt extends eastward along the Panama and Colombia coasts from profile CT1-18 (Figures 2 and 3). The deformed belt is arcuate offshore Panama and widest east of the canal along profiles CT1-21 and CT1-22. The Colombia portion of the deformed belt originates in the Gulf of Uruba where it faces the Panama belt, then continues north-northeastward until it disappears beneath the Magdalena Fan. The belt reappears and continues eastward along the northern Colombia and Venezuela margins (Case, 1974).

Colombia Basin oceanic crust dips toward the margin offshore Panama and Colombia (Figures 8, 9, and 10). Greater sediment thickness offshore Colombia makes the top of oceanic crust difficult to recognize, but deeper sedimentary reflectors dip toward Colombia along profiles CT1-24, CT1-25, CT1-26, and CT1-27 (Figures 2, 8, and 10). Profiles CT1-19, CT1-12, and CT1-17 off western Panama and eastern Costa Rica show oceanic crust rising landward. The crustal

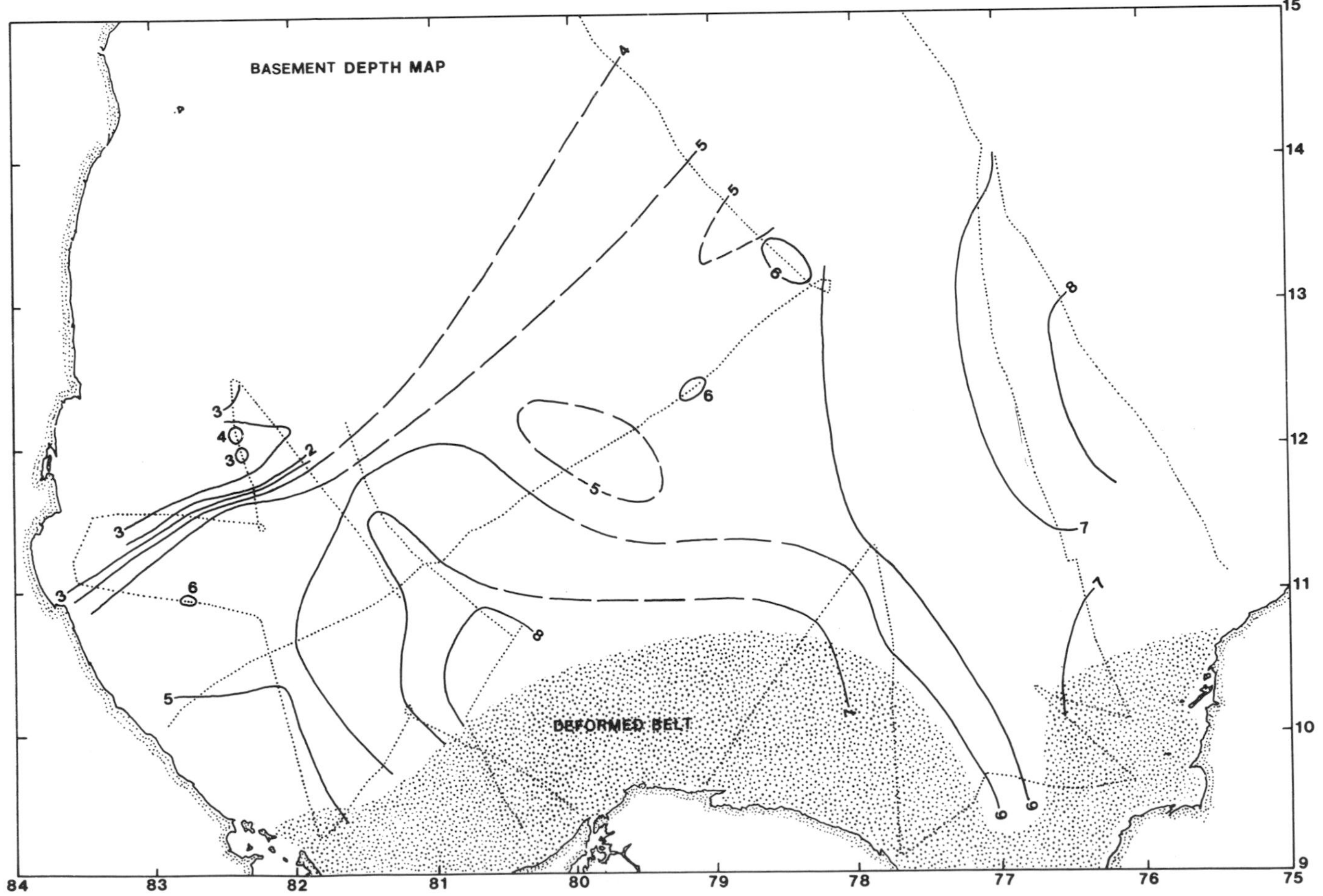

Figure 8 — Basement depth map of the Colombia Basin. The contour interval is 1 second in two-way reflection time. Dotted lines indicate the tracks of seismic reflection data utilized for this map. Dotted area is the deformed belt province (Figure 3).

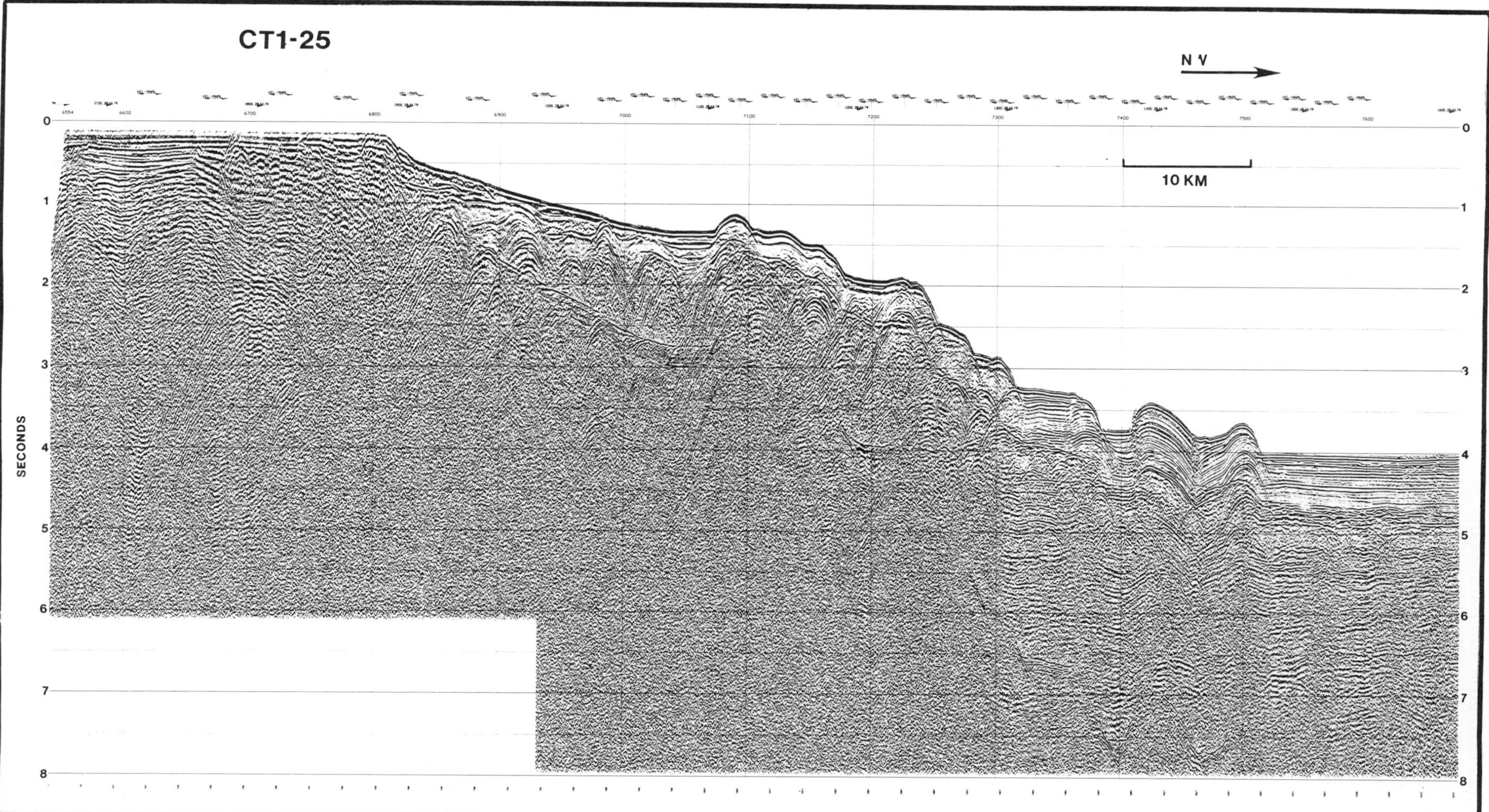

Figure 10 — Twelve-fold stacked seismic section of line CT1-25. Fan-type stratigraphic sequence underthrusts the margins in this area.

configuration pattern suggests that depression of the crust offshore eastern Panama and Colombia is due to sediment loading and tectonic underthrusting. In areas where oceanic crust dips toward the margins, Colombia Basin stratigraphic sequences thicken toward the margins (Figures 7 and 8).

Figures 11 and 12 are line-drawings of inferred structural details of the deformed belt. Migrated stacked sections show that many steeply-dipping reflectors are real and not caused by diffractions. Several fault-bounded blocks can be seen on profiles CT1-21 and CT1-22 offshore Panama and each block is defined by a set of landward or seaward-dipping reflectors and a drag fold on the upper part of the block (Figures 9 and 11). The fault-plane orientation and location of the drag fold show that all faults in the deformed slope sequence are thrust faults (Figure 9). These thrust faults occur in sets, dipping to the south in the southern part of the deformed belt, northward in the middle of the belt, and southward again in the northern, or seaward, part of the belt (Figure 11). Deformation on profiles CT1-19 and CT1-20 (Figure 2) consists of southward-dipping faults only (profiles not shown in this paper). On profile CT1-21, normal faulting is indicated in the landward-dipping reflector sequence beneath the thrust-faulted zone, which further suggests decollément above the normal-faulted zone (Figure 11).

Lower slope structures of Colombian lines CT1-24, 25, and 26 consist of two parts (Figures 10 and 12): (1) mid-slope landward-tilted blocks separated by thrust faults; and, (2) base-of-slope deformation. Distinct parallel reflectors present in the base-of-slope structures suggest deformation of Colombia Basin turbidites. Two structures occur at the base of the lower slope: (1) a block uplifted without tilting and apparently bounded on the seaward side by a high-angle reverse fault; and, (2) a block of uplifted and tilted turbidites bounded by landward- and seaward-dipping thrust faults (Figures 10 and 12). This latter structure superficially resembles a "flower" structure (Harding and Lowell, 1979) created by strike-slip faulting.

No apparent deformation occurs on the slope of the Magdalena Fan traversed on line CB-1 (Figures 2 and 3, profile not shown). High sedimentation may mask deformation; alternatively, a hypothetical seaward extension of the right-lateral Oca Fault could offset the deformed belt beneath the Magdalena Fan. Middle- and upper-slope details are similar for the Panama and Colombia margins and differ only in the amount of sedimentary cover present. Both margins contain a mid-slope structural high underlain by deformed sediments (Figures 9, 10, 11, and 12). Landward of the high, a mid-slope terrace with small sediment ponds occurs offshore eastern Panama (Figures 9 and 11) with a steep inner slope and narrow shelf. Indistinct, landward-dipping reflectors occur beneath the upper slope of line CT1-23 (Figure 2). In western Panama, a thicker shelf and slope sequence fills the mid-slope terrace. Offshore Colombia, indistinct landward-dipping reflectors (Figures 10 and 12), landward of the structural high, underlie a wide shelf and shelf basin.

DISCUSSION

Seismic Stratigraphy and Evolution of the Colombia Basin

We are unable to obtain ages for basin-floor reflectors KI_1, KI_2, KI_3, and KI_4 as no DSDP drill site is located in the province with basin-floor stratigraphy (Figures 2 and 3).

Sites 154 and 502 (collected with the hydraulic piston cores) were located on a small rise of folded sediments in the fan-type stratigraphy province (Saunders et al, 1973; Prell and Gardner, 1980), which is the southeastern extension of the folded and faulted structure between shot points 7,000 and 7,200 in Figure 6. DSDP Site 154 (Figure 2) was reached only to Miocene sediments (volcanic ash, silt, clay) with 278 m of penetration (Saunders et al, 1973). Penetration was less at Site 502. The estimated thickness of CBII1, assuming its P-wave velocity to be 1.6 km/sec, is 400 m near Site 154 (between shot points 7,000 and 7,200; Figure 6). Thus, we infer that the Pliocene unconformity between the nannoplankton calcareous clay and volcanic sands, silts, and clay, discovered at about 160 m beneath water bottom at site 154, is a minor unconformity within CBII1.

At site 153, drilled in the Aruba Gap, Coniacian basalt was reached at 776 m. Previous studies suggest that this horizon is probably not true oceanic basement (top of layer 2): velocities of 3.7 to 3.9 km/sec were obtained from multichannel seismic data processing beneath this horizon (Ludwig, Houtz, and Ewing, 1975). We are unable to correlate either Site 153 units or reflectors reported by Hopkins (1973) with reflectors in the eastern Colombia Basin. However, the possible real oceanic basement at Site 153, as interpreted by Hopkins (1973) from a seismic section to lie beneath another series of deeper strong reflectors, resembles KI_4, KII_4, and $KIII_4$.

Diebold et al (1981) concluded that the oceanic basement lies below the smooth B″ in the western Venezuela Basin. This smooth B″ was drilled at Site 146/149 to be composed of the Coniacian basalt (Saunders et al, 1973). Thus, the age of oceanic crust in the western Venezuela Basin could be older (e.g. Early Cretaceous or even Jurassic). The oceanic basement interpreted by Diebold et al (1981) from one seismic section in the western Venezuela Basin resembles KI_4, KII_4, and $KIII_4$. Basin-floor reflectors above KI_4 in the Colombia Basin also resemble those of the western Venezuela Basin (compare Figure 4 of this paper with Figures 4 and 5 of Ladd and Watkins, 1980). Moreover, seismic interval characteristics and interval velocities in basin-floor areas in the Colombia Basin are very similar to those recognized in the southeastern Venezuela Basin (Talwani et al, 1977) with $KI_2 = A''$, $KI_3 = X$, and $KI_4 = B''$. This B″ is rough and interpreted to be the oceanic crust by Talwani et al (1977) and Diebold et al (1981). Thus, we speculate that the Colombia Basin and Venezuela Basin oceanic crust is one piece with $KI_4 = KII_4 = KIII_4 =$ oceanic crust of the Venezuela Basin. Later, different sources of sedimentation

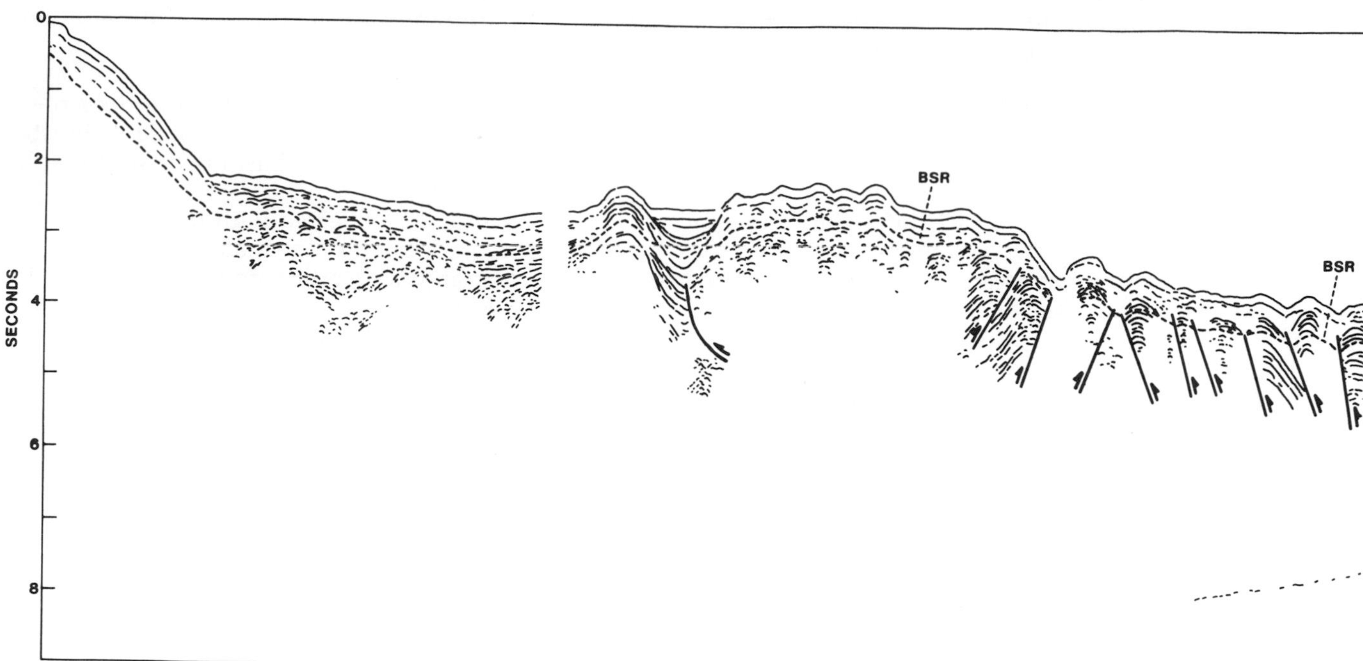

Figure 11 — Interpretative seismic reflection profile of line CT1-21 (Figure 9). See text for explanations.

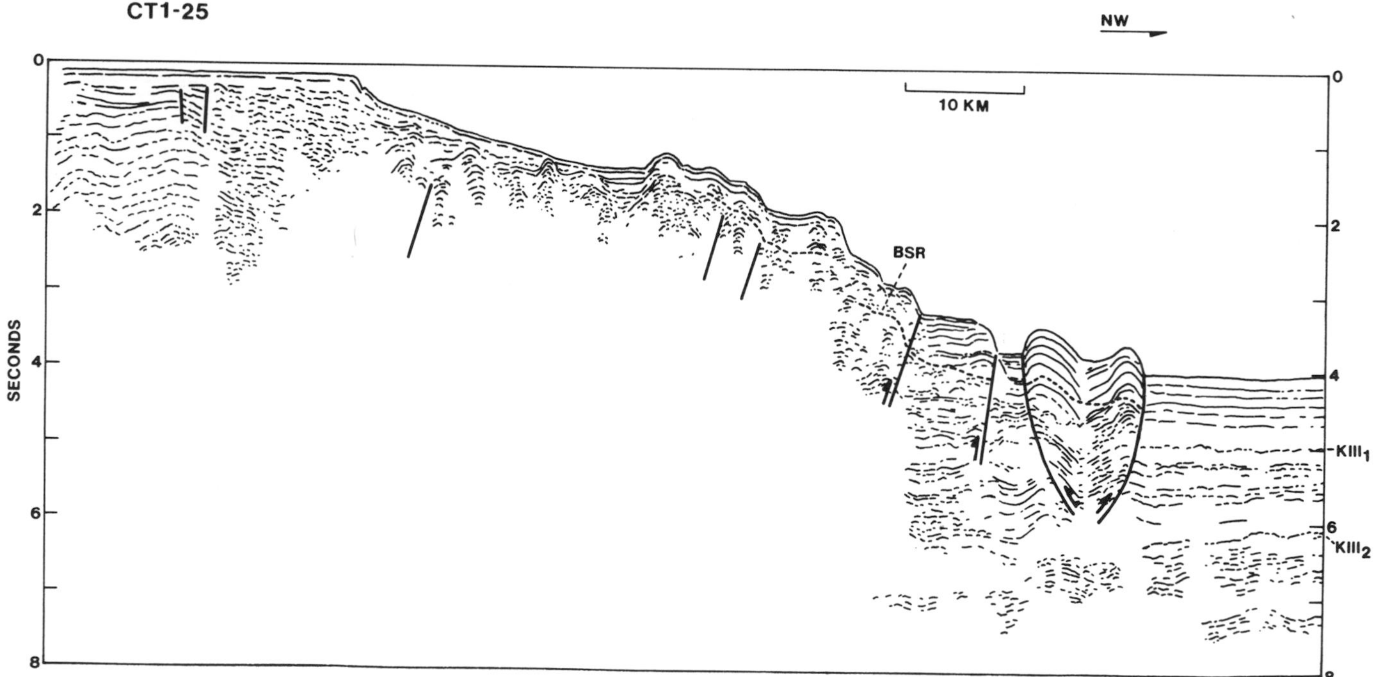

Figure 12 — Interpretative seismic reflection profile of line CT1-25 (Figure 10). See text for explanations.

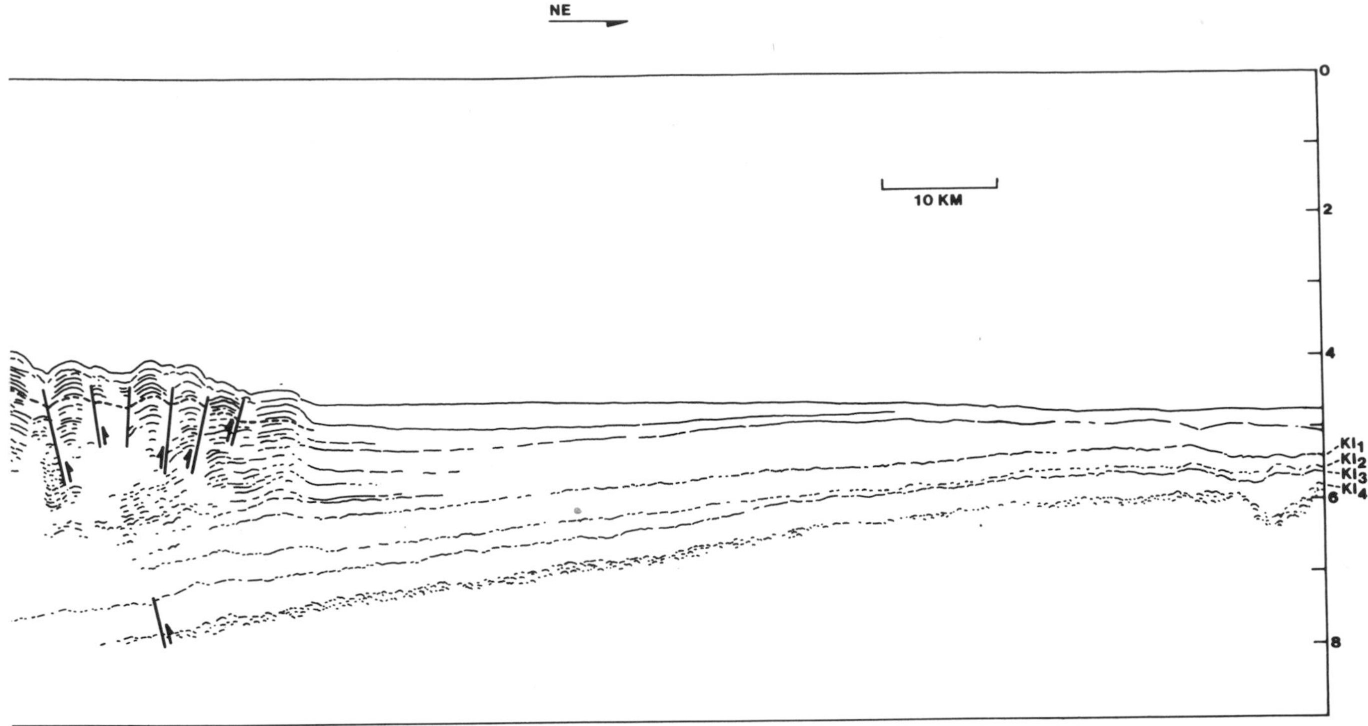

and different tectonic histories for different parts of the Caribbean have caused the variations in stratigraphy and structure presently observed. In this sense, KI_2 time is Eocene and the weakening of bottom currents, since KI_2-reflector time, could be associated with the developments of Central American isthmus and Lesser Antilles island arc in Eocene time.

Marginal Deformed Belt: Model and Evolution

The deformed belts of Panama and northwest Colombia resemble active margins elsewhere: (1) oceanic crust dips landward with little deformation beneath clear compressed sediments above; (2) a chaotic wedge of presumably offscraped sediments is located immediately landward; and, (3) a forearc basin developed between the accretionary wedge and the shore. Offshore eastern Panama, the basin has little sediment fill whereas the western Panama Basin is partially filled and the basin offshore Colombia is completely filled.

Offshore Panama, the forearc structure is best developed along profiles CT1-21 (Figures 9 and 11) and CT1-22. Quaternary volcanics found in the northern part of the eastern Panama (Case and Holcombe, 1980) suggest that subduction has actually occurred. If past relative motion between the Nazca and Caribbean plates were the same as the present day plate motion (5.4 cm/yr at an azimuth of N71°E; Minster and Jordan, 1978), we would expect strike-slip motion in western Panama instead of the observed thrust faulting and compression. Thus, either the past relative plate motion is different from that of the present day motion, or the lithosphere here behaves in a way that relative motion along the northern Panama coast is different from that calculated by Minster and Jordan (1978) for the Nazca and Caribbean plates. Plastic deformation or the existence of microplates in Panama could account for the difference.

In the Colombia offshore area, convergence and subduction have also occurred, as shown by the similarity of its margin and active margin structures. However, forearc structures are less developed (Figures 10 and 12), indicating a different convergence regime with slower convergence rate than that offshore Panama. This is supported by the predicted present-day relative plate motion between the South American and Caribbean plates of 2.3 cm/yr at an azimuth of N77°W. It is not known whether an actual subduction process is taking place now. Also, large terrigeneous sediment input along with continental shelf outgrowth (Duque-Caro, 1979) may have affected the development of the deformed belt structure. Fault and shear zones in northwestern Colombia further complicates the structural style.

Synthetic cross sections of the margins have been constructed by combining results from the present geophysical survey with onshore geologic models. The Panama model (Figure 13) utilizes Case's cross-sectional interpretation of Panama based mainly on magnetic and gravity measurements (Case, 1974). In this model the central Chucunaque basin is flanked by uplifts of oceanic crust. Basalts and cherts (Bandy and Casey, 1973) of Late Cretaceous age confirm the oceanic nature of these uplifts. Case projected a normal fault at the shoreline to separate the deformed

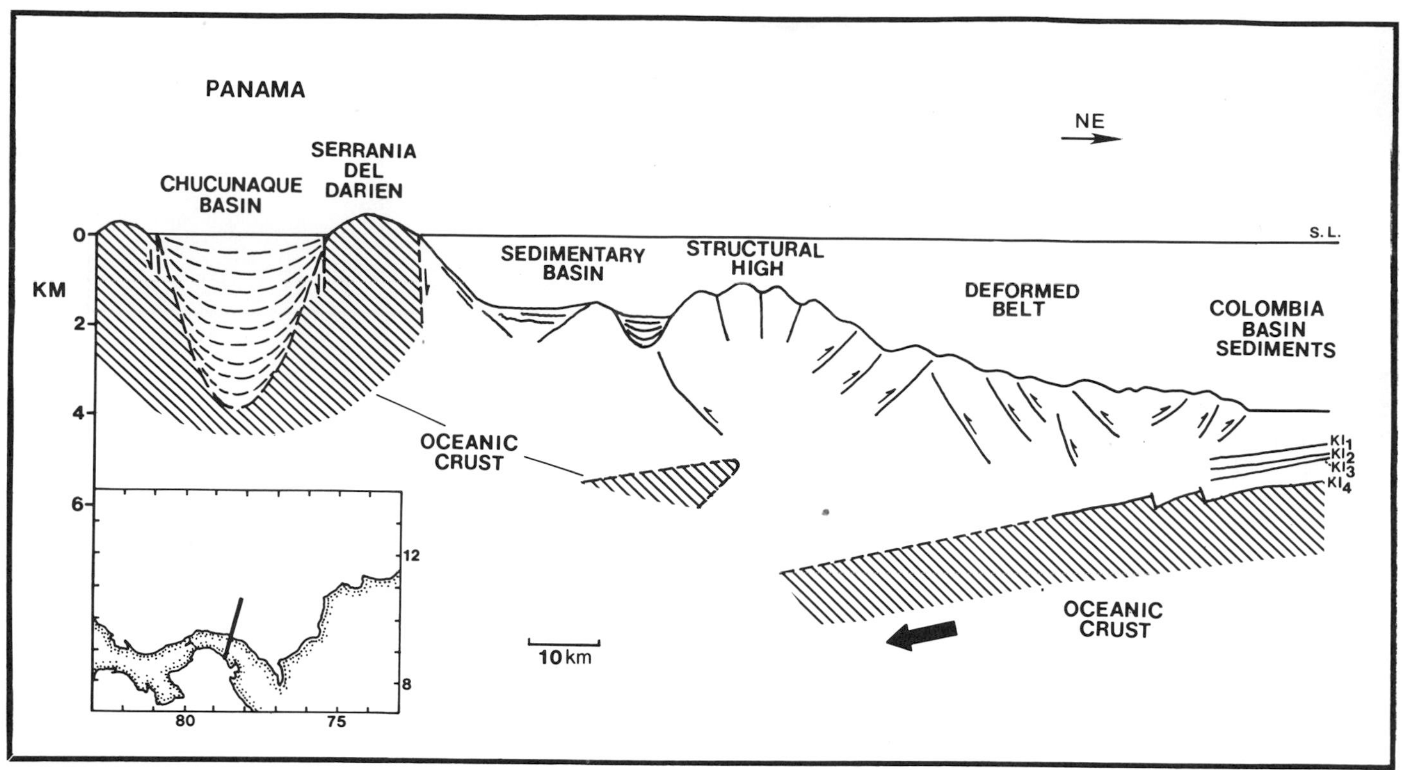

Figure 13 — Synthetic cross section of the Panama margin. Index map indicates the location of the cross section.

belt from the oceanic uplifts. We retain this fault because we have no evidence to disprove it. Magnetic anomalies on line CT1-22 above the mid-slope terrace suggest the presence of oceanic crust at moderate depths. Therefore, we project a sliver of emplaced oceanic crust within the inner part of the margin. The lower deformed slope is underlain by the landward-dipping oceanic crust and accreted fan and basin-floor sequences. This margin differs from most active margins in having two sedimentary basins behind the outer structural high. The Chucunaque basin rests on trapped oceanic crust, whereas the northern basin may rest on accreted terrane (Seely, 1979).

The Colombia margin (Figure 14) consists of a series of accreted terranes, beginning with the Middle Eocene age San Jacinto Belt (Duque-Caro, 1979). Accreted terrane also extends under the shelf basin and the lower slope. This corresponds to the Sinu Belt of Duque-Caro (1979). Deformation of Colombia Basin turbidites at the base of lower slope on line CT1-25 shows that tectonic activity continues to present (Figures 10 and 12). The younger deformed belt, near the foot of the lower slope with steep faults, uplifted and untilted turbidite sequences, and apparent "flower structures," suggests oblique convergence and a component of strike-slip motion. The landward deformed belt consisting of fault blocks bounded by steeply-dipping reflectors resembles accretionary zones of active margins with convergence nearly perpendicular to the margin.

Paleobathymetric data of Bandy and Casey (1973) suggest that the eastern Panama isthmus of uplifted oceanic crust and overlying sediment sequence started to evolve in Late Oligocene-Early Miocene time. Landward thickening of the sediment sequence (seismic intervals above KI_4; Figure 9), observed on the multichannel seismic profiles, indicates that a basin was present in eastern Panama by Late Oligocene time. Thereafter, eastern Panama has been uplifted and decoupled from the Colombia Basin crust, and subduction and island arc structures have developed along with the entrapment of some slivers of oceanic crust beneath the accretionary wedge. This whole process may be caused by the convergence of North America and South America since Oligocene time (Ladd, 1976; Minster and Jordan, 1978). Final closure of the isthmus may have occurred as recently as Late Pliocene time (Keigwin, 1978). Western Panama began forming as a volcanic arc in Oligocene time (Malfait and Dinkelman, 1972) and was preceded by uplift occurring in the Middle and Late Eocene similar to that in Costa Rica and its offshore area. Decoupling and the formation of a deformed belt north of western Panama probably began after Oligocene as its forearc type strucfture is not as well-developed as eastern Panama's.

Based on land geologic studies, Duque-Caro (1979) proposed intermittent convergence at the northwestern Colombia margin from Late Cretaceous to present, resulting in the outgrowth of the margin with younger accreted terranes. Older, onshore accreted terranes of northwestern Colombia are associated with the convergence of Farallon and South American plates in a nearly East to West direction perpendicular to the margin from Late Cretaceous to Eocene. The deformed belt offshore northwestern Colombia has

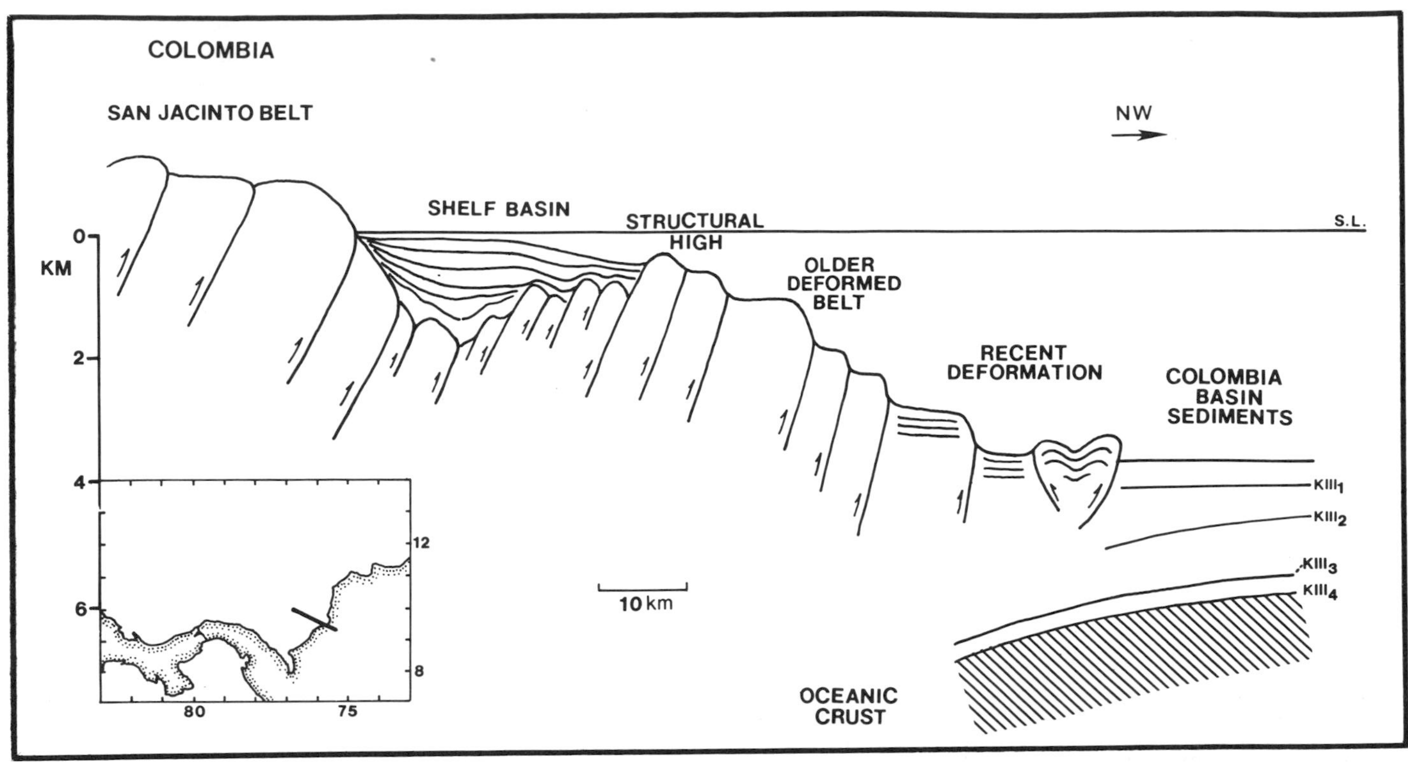

Figure 14 — Synthetic cross section of the Colombia margin. Index map indicates the location of the cross section.

been developing since Oligocene when North America and South America started to converge. Since then, the relative motion between the South American and Caribbean plates has been at about the same direction (i.e. N77°W). Convergence in the offshore northwestern Colombia area should have been nearly perpendicular to the margin. Recent oblique convergence and strike-slip motion occurring here may be associated with the initiation of a strike-slip fault zone in northwestern South America consisting of the Dolores magashear, Santa Marta Fault, Oca Fault, and Bocono Fault. Northwestern South America may not be an integral part of the South American rigid plate. Relative motion between the Caribbean plate and this broken zone is no longer the same as that between the South American and Caribbean plates.

ACKNOWLEDGMENTS

We thank the crew and scientific staff aboard R/V Ida Green in 1975, 1977, and 1978, on cruises IG15-6, IG24-8, IG29-1, and IG29-4, for their help in collecting the geophysical data utilized in this paper. During the preparation of the manuscript, R. T. Buffler, W. Gose, T. L. Holcombe, J. W. Ladd, and J. D. Phillips provided many hours of helpful discussion. Carl Bowin provided magnetic data collected by the Woods Hole Oceanographic Institution offshore Panama and Colombia. John W. Ladd provided some original multichannel seismic stacked sections of the Lamont-Doherty Geological Observatory in the southeastern Venezuela Basin. Critical reviews of the manuscript by J. E. Case, T. L. Holcombe, J. W. Ladd, and an unknown reviewer were most valuable. Special thanks are due to Linda Lu for typing and editing the manuscript. Financial support of this research is provided by the industrial associates of the University of Texas Institute for Geophysics.

REFERENCES CITED

Bandy, O. L., and R. E. Casey, 1973, Reflection horizons and paleobathymetric history, eastern Panama: Geological Society of America Bulletin, v. 84, p. 3081-3086.

Bowin, C. O., 1976, Caribbean gravity field and plate tectonics: Geological Society of America, Special Paper 169, 79 p.

Case, J. E., 1974, Oceanic crust forms basement of eastern Panama: Geological Society of America Bulletin, v. 85, p. 645-62.

———, and Holcombe, T. L., 1980, Geologic-tectonic map of the Caribbean region: U.S. Geological Survey, Miscellaneous Investigations Series Map 1-1100.

Christofferson, E., 1976, Colombia Basin magnetism and Caribbean plate tectonics: Geological Society of America Bulletin, v. 87, p. 1255-1258.

Diebold, J. B., et al, 1981, Venezuela Basin crustal structure: Journal of Geophysical Research, v. 86, p. 7901-7923.

Duque-Caro, H., 1979, Major structural elements and evolution of northwestern Colombia, *in* J.S. Watkins, L. Montadert, and P. W. Dickerson, eds., Geological and geophysical investigations of continental margins: AAPG Memoir 29, p. 329-351.

Edgar, N. T., J. I. Ewing, and J. Hennion, 1971, Seismic refraction and reflection in Caribbean Sea: AAPG Bulletin, v. 55, n. 6, p. 833-870.

Ewing, J., J. Antoine, and M. Ewing, 1960, Geophysical measurements in the western Caribbean Sea and in the Gulf of Mexico: Journal of Geophysical Research, v. 65, n.

12, p. 4087-4126.
Fox, P. J., and B. C. Heezen, 1975, Geology of the Caribbean crust, *in* A.E.M. Nairn and F. G. Stehli, eds., The ocean basins and margins, v. 3: New York, Plenum Press, p. 421-466.
———, et al, 1970, The geology of the Caribbean crust, I: Beata Ridge: Tectonophysics, v. 10, p. 495-513.
Harding, T. P., and J. D. Lowell, 1979, Structural styles, their plate-tectonic habitats and hydrocarbon traps in petroleum provinces: AAPG Bulletin, v. 63, n. 7, p. 1016-1058.
Hopkins, H. R., 1973, Geology of the Aruba Gap abyssal plain near Deep sea drilling project site 153, *in* N. T. Edgar, et al, eds, Initial reports of the deep sea drilling project, v. 15: Washington, D.C., U.S. Government Printing Office, p. 1039-1050.
Houtz, R. E., and W. J. Ludwig, 1977, Structure of Colombia Basin, Caribbean Sea, from profiler-sonobuoy measurements: Journal of Geophysical Research, v. 82, n. 30, p. 4861-4867.
Irving, E. M., 1975, Structural evolution of the northernmost Andes, Colombia: U.S. Geological Survey, Professional paper 846, 47 p.
Jordan, T. H., 1975, The present-day motions of the Caribbean plate: Journal of Geophysical Research, v. 80, n. 32, p. 4433-4439.
Keigwin, L. D., Jr., 1978, Pliocene closing of the isthmus of Panama, based on biostratigraphic evidence from nearby Pacific Ocean and Caribbean Sea cores: Geology, v. 6, p. 630-634.
King, P. B., 1969, Tectonic map of North America: Washington, D.C., United States Geological Survey.
LaBrecque, J. L., D. V. Kent, and S. C. Cande, 1977, Revised magnetic polarity time scale for late Cretaceous and Cenozoic time: Geology, v. 5, p. 330-335.
Ladd, J. W., 1976, Relative motion of South America with respect to North America and Caribbean tectonics: Geological Society of America Bulletin, v. 87, p. 969-976.
———, and J. S. Watkins, 1979, Tectonic development of trench-arc complexes on the northern and southern margins of the Venezuela Basin, *in* J. S. Watkins, L. Montadert, and P. W. Dickerson, eds., Geological and geophysical investigations of continental margins: AAPG Memoir 29, p. 363-371.
———, and ———, 1980, Seismic stratigraphy of the western Venezuela Basin: Marine Geology, v. 35, p. 21-41.
Ludwig, W. J., R. Houtz, and J. Ewing, 1975, Profiler-sonobuoy measurements in Colombia and Venezuela Basins, Caribbean Sea: AAPG Bulletin, v. 59, n. 1, p. 115-123.
Malfait, B. T., and M. G. Dinkelman, 1972, Circum-Caribbean tectonic and igneous activity and the evolution of the Caribbean plate: Geological Society of America Bulletin, v. 83, p. 251-272.
Matthews, D. J., 1939, Tables of the velocity of sound in pure water and sea water for use in echo-sounding and sound ranging: London, Hydrographic Department, Admiralty, 52 p.
Minster, J. B., and T. Jordan, 1978, Present-day plate motions: Journal of Geophysical Research, v. 83, n. B11, p. 5331-5354.
Prell, W. L., and J. V. Gardner, 1980, Hydraulic piston coring of late Neogene and Quaternary sections in the Caribbean and equatorial Pacific: Preliminary results of Deep sea drilling project leg 68: Geological Society of America Bulletin, part 1, v. 91, p. 433-444.
Saunders, J. B., et al, 1973, Cruise synthesis, *in* N. T. Edgar et al, eds., Initial reports of the Deep sea drilling project, v. 15: Washington, D.C., U.S. Government Printing Office, p. 407-414.
Sclater, J. G., C. Jaupart, and D. Galson, 1980, The heat flow through oceanic and continental crust and the heat loss of the earth: Review of Geophysics and Space Physics, v. 18, n. 1, p. 269-311.
icSeely, D. R., 1979, The evolution of structural highs bordering major forearc basins, *in* J. S. Watkins, L. Montadert, and P. W. Dickerson, eds., Geological and geophysical investigations of continental margins: AAPG Memoir 29, p. 245-260.
Shipley, T. H., et al, 1980, Seismic evidence for widespread possible gas hydrate horizons on continental slopes and rises: AAPG Bulletin, v. 63, n. 12, p. 2204-2213.
Silver, E. A., J. E. Case, and H. J. MacGillvary, 1975, Geophysical study of the Venezuelan borderland: Geological Society of America Bulletin, v. 86, p. 213-226.
Talwani, M., et al, 1977, Multichannel seismic study in the Venezuela Basin and the Curacao Ridge, *in* M. Talwani and W. C. Pitman III, eds., Island arcs, deep sea trenches and back arc basins: Washington, D.C., American Geophysical Union, p. 41-56.
Terry, R. A., 1957, A geological reconnaissance of Panama: California Academy of Science, Occasional Paper 23, 91 p.

Arc, Forearc, and Trench Sedimentation and Tectonics; Amlia Corridor of the Aleutian Ridge

D. W. Scholl
T. L. Vallier
A. J. Stevenson
U.S. Geological Survey
Menlo Park
California 94025

A broad spectrum of geological and geophysical information has recently been collected within the Amlia corridor (173°W longitude) of the Aleutian Ridge. The ridge's upper crustal rocks can be divided into three rock series: lower, middle, and upper. The Aleutian Ridge is fundamentally a massive, little deformed antiform of lower series rocks produced by voluminous submarine volcanism in Eocene and perhaps earlier Tertiary time. Erosional debris from the dying arc accumulated over its flanks as the middle and upper series deposits of Oligocene through Holocene age; these deposits are as much as 4 to 5 km thick. A slightly deformed mass of tectonically thickened trench deposits underlies the lower part of the trench's landward slope. This accretionary wedge was added in post-middle Miocene time to the ridge's igneous framework of lower series rocks.

INTRODUCTION

The Aleutian Ridge, which forms the northern margin of the Pacific Basin, is a classic example of an ensimatic magmatic arc (Figure 1; Coats, 1962; Shor, 1964). Prior to our study no systematic geologic examination of a transverse sector of the ridge had been initiated, although many of the islands of the 2,200-km-long ridge had been examined geologically (for example, Schmidt, Serova, and Dolmatova, 1973, and the numerous U.S. Geological Survey Bulletins of the 1028 series), and reconnaissance offshore geological and geophysical surveys had been carried out since the early 1950s (Gibson and Nichols, 1953; Gates and Gibson, 1956). Investigations began in 1978 to resolve more clearly the geologic evolution of the Aleutian Ridge by examining its rock record in a 200-km-wide sector. This sector, the Amlia corridor, crosses the ridge at 173°W longitude and includes Amlia and Atka Islands, two of the Andreanof Islands of the east-central Aleutians (Figure 1).

The north-trending Amlia corridor traverses four physiographic provinces: (1) the abyssal floor of the Aleutian Basin to the north; (2) the Aleutian island arc, which is the summit cordillera of the Aleutian Ridge; (3) the Aleutian Terrace; and (4) the Aleutian Trench to the south (Figure 1). In geomorphic terms, the Aleutian Basin is a backarc basin, the island arc is a volcanic or magmatic arc, the Aleutian Terrace is a forearc basin, and the seaward and landward trench slopes and separating flat trench floor define a deep-sea or oceanic trench. In this paper the expression "island arc" or "arc" refers only to the geomorphic form and underlying rocks of the ridge's island-crested summit cordillera and the immediately flanking slopes of this large geanticline, and not the forearc area of the Aleutian Terrace or landward slope of the adjacent Aleutian Trench (Figure 2). We recognize that the ridge's igneous framework or massif, which is thickest beneath the arc, in parts extends beneath the forearc area. Bathymetric detail of the Amlia corridor is shown on the charts of Nichols and Perry (1966) and the NOS Seamap Series of the Department of Commerce (1973).

Following a north to south traverse through the physiographic provinces, we here present a description of the initial and seemingly most important geological and geophysical results from Amlia corridor

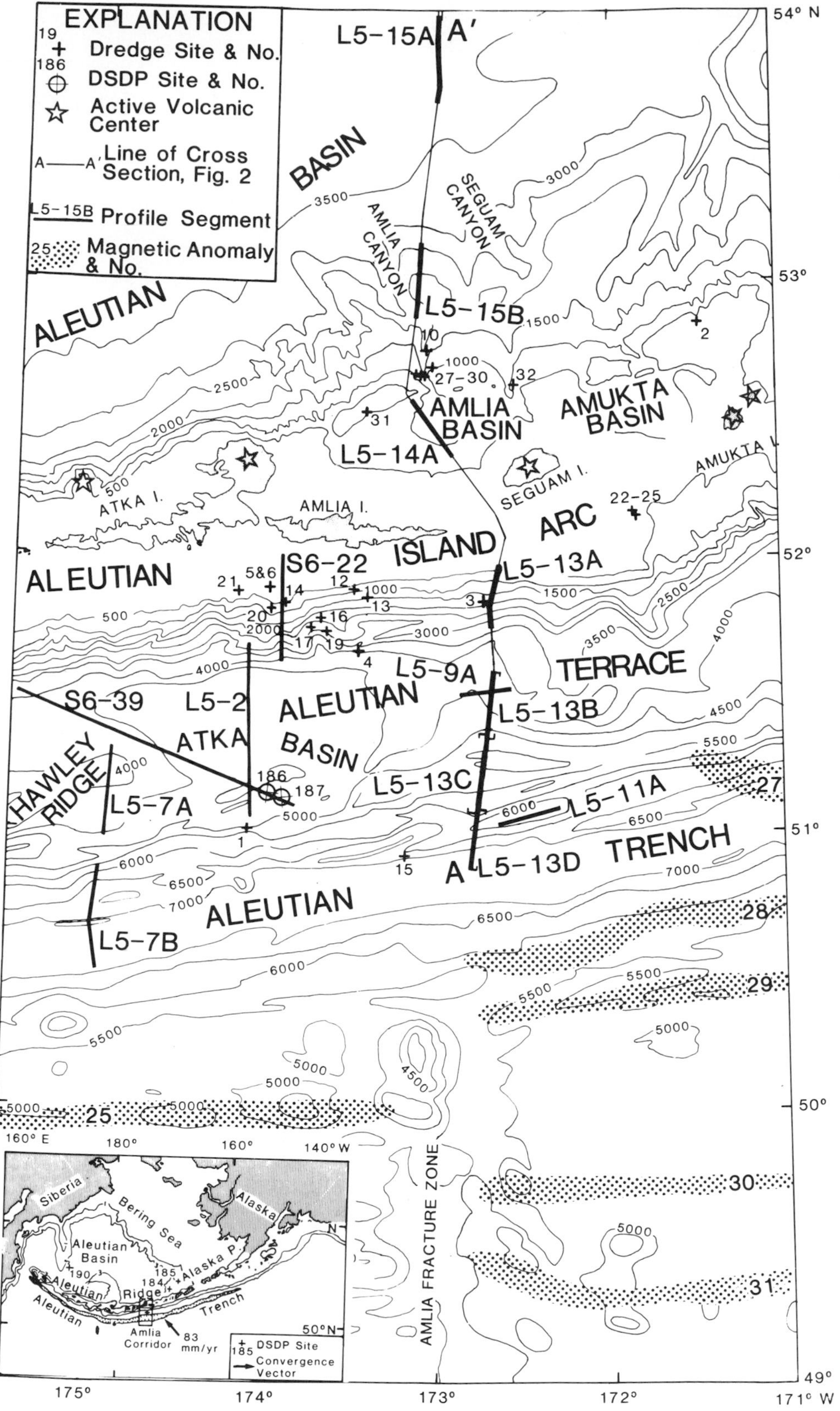

Figure 1 — Index map of the Amlia corridor region of the Aleutian Ridge, which comprises the Aleutian island arc, the Aleutian Terrace, and the inner or landward slope of the Aleutian Trench. Depth contours are in velocity-corrected meters. Dredge numbers correspond to the last number of the dredge sites listed on Table 1. The vector of Pacific-North America movement of 83 mm/yr is taken from Minster and Jordan (1978). The Lambert conformal base is from NOS Seamap Series (Department of Commerce, 1973).

investigations and preliminary interpretations of these results and their implications for the geological evolution of the Pacific's northern rim.

REGIONAL FRAMEWORK

The Aleutian Ridge separates the north Pacific Basin from the Bering Sea (Figure 1). Although lofty strato-volcanoes of basaltic to andesitic composition form the ridge's mountainous edifices (Coats, 1950), they constitute only a small fraction of its submerged bulk. The great mass of the ridge, which is centered beneath its island arc, is a 20- to 24-km thick crustal framework of volcanic, plutonic and sedimentary rocks (Coats, 1956; Shor, 1964; Carder et al, 1967; Marlow et al, 1973; Scholl, Buffington, and Marlow, 1975a; and Scholl; Marlow, and Buffington, 1975b). The arcuate chain of active and dormant volcanoes rises above a prominent and partly buried erosional unconformity that has been cut across Tertiary rocks of the island arc (Gates, Fraser, and Snyder, 1954). The Aleutian island arc is therefore a mostly submerged flat-topped mountain range — a southward curving guyot-like cordillera that rises approximately 7,000 m above its southern or Pacific base at the Aleutian Trench and 3,600 m above its northern or Bering Sea base (Figure 1).

Oceanic crust of probable Cretaceous age underlies the Aleutian Basin north of the ridge (Cooper, Marlow, and Scholl, 1976a). Immediately south of the ridge oceanic lithosphere of the northward-dipping Pacific plate is of Late Cretaceous or early Tertiary age. Pacific lithosphere obliquely underthrusts (320°T, or 30° off normal; Figure 1) the Amlia corridor at 80 to 85 mm/yr (Minster and Jordan, 1978). Magnetic anomalies associated with the crustal rocks of this lithosphere generally strike westward (Figure 1). They are offset in a left-lateral sense by the north-trending Amlia Fracture Zone (173°05′W longitude), thus slightly younger (10 million years) and shallower igneous crust underlies the western half of the corridor's Pacific area (Hayes and Heirtzler 1968; Grim and Erickson 1969; Erickson and Grim, 1969).

INFORMATION BASE

Subbottom geophysical information was initially collected within the Amlia corridor by Shor (1964), Malahoff and Erickson (1969), Hayes and Ewing (1970), and Marlow et al (1973). Regional geopotential maps that include the corridor have been issued by the U.S. Department of Commerce (1973) and by Watts (1975). In 1971 two DSDP (Deep Sea Drilling Project) holes, sites 186 and 187, were drilled within the corridor along the southern edge of the Aleutian Terrace (Figure 1; Creager et al, 1973). Geological and geophysical interpretations in part based on the drilling results have been presented by Scholl and Creager (1973), Grow (1973b), and Stewart (1978). These data and additional offshore geophysical information collected in 1970 by the Scripps Institution of Oceanography were synthesized and interpreted by Grow (1973a,b). His work represents the first attempt to combine marine geological and geophysical data for the purpose of deducing the tectonic evolution of the Aleutian Ridge within the Amlia corridor. More recently Scholl, Marlow, and Buffington (1975b) reported on the structure and origin of the corridor's two summit depressions, Amlia and Amukta basins. The only studies of the corridor's active volcanic centers have been carried out on the large vent complex of Atka Island (Marsh, 1980).

New studies specifically concentrated in the Amlia corridor include the completion of reconnaissance geologic mapping of Amlia and neighboring Atka Islands (Hein and McLean, 1980; McLean et al, 1980; Hein, McLean, and Vallier, 1981) and related laboratory and petrographic studies of their exposed rocks. Recently completed offshore studies include a detailed single-channel seismic reflection examination of the trench floor in 1978 (U.S.G.S. Cruise L8-78), single-channel reflection profiling and refraction measurements combined with bottom dredging in 1979 (U.S.G.S. Cruise S6-79), and 24-channel reflection profiling, bottom sampling, and sonobuoy refraction studies in 1980 and 1981 (U.S.G.S. Cruises L5-80 and L9-81). Gravity and magnetic data and bathymetric information were gathered during each of these investigations. Navigation was controlled by satellite fixes augmented by LORAN-C (rho-rho method) positions. Some of the results of these latest investigations are reported by Scholl, Vallier, and Stevenson (1982a).

DESCRIPTION OF RESULTS

General

Acoustic reflection profiles that traverse the Aleutian Ridge commonly reveal three stratigraphically and structurally distinguishable rock units. The descriptive notation *upper, middle,* and *lower* series has been adopted to designate them (Figure 2). In many areas the series are vertically superimposed, and, in a general way, each can be temporally correlated to exposed onshore rock sequences (Scholl, Buffington, and Marlow, 1975a; Scholl, Marlow, and Buffington, 1975b). In offshore areas the upper Cenozoic beds of the upper series are characteristically either slope-conforming or basin-filling deposits. Except between the landward trench slope, upper series beds are typically little deformed. Over the crest of the arc they unconformably overlie the wave-planed surface or summit platform cut across middle and lower series rocks, and beneath the Aleutian Terrace upper series beds typically thin updip and unconformably against middle series strata. Beds of the upper series typically have good internal reflectivity and laterally coherent, prominent reflection horizons. Onshore, massive rocks of the young volcanic centers are included in the upper series.

Middle series deposits are more deformed and, beneath the summit area of the arc, dips as steep as 10° or higher occur (Figure 2). Dipping beds of the middle series are truncated by the ridge's summit platform, but below the sloping flanks of the arc the attitude of these deposits is parallel or subparallel to that of the sea floor. Beneath these slopes, and also the adjacent Aleutian Terrace, middle series beds unconformably overlie rocks of the lower series. Along the arc's crestal region, thick (greater than 500 m) sections of middle Tertiary volcanic and sedimentary rocks are exposed in sequences of middle series deposits. Middle series beds exhibit acoustically strong to weak internal reflectivity, and acoustic stratification can be laterally persistent or discontinuous and irregular.

In contrast, rocks of the underlying lower series are internally poorly reflective. Beneath the lower flanks of the arc and adjacent Aleutian Terrace and Aleutian Basin (Bering Sea), the upper surface of the lower series is a persistent, irregular reflection interface (designated below as an Strong, Subbottom, Irregular Reflector-type reflection horizon). In certain areas weak but laterally coherent acoustic horizons occur within the lower series. Rocks of the lower series are acoustically massive accumulations, and are primarily slope-forming rather than basin-filling sequences.

Regional information implies that lower series rocks of the ridge's crestal area are dominantly submarine volcanic flows and related hypabyssal intrusive masses and coarse volcaniclastic debris. Seaward, beneath the submerged flanks of the ridge, the middle series is dominantly marine sedimentary deposits. Except near late Cenozoic volcanic centers, upper series beds are chiefly marine sedimentary sequences of volcaniclastic detritus and pelagic debris.

The acoustically identifiable offshore rock series and their onshore counterparts record important phases of the ridge's geologic evolution (Scholl, Buffington, and Marlow, 1975a; Scholl, Marlow, and Buffington, 1975b). Because regional information implies that the evolutionary phases of the ridge are time and space dependent, the three rock series do not constitute exactly coeval chronostratigraphic sequences along the length of the ridge. Nonetheless, a regional chronostratigraphic scheme can be loosely fitted to the ridge's exposed and geophysically observed submarine rock sequences. Exposed and submarine sedimentary and igneous rocks of the upper series, which along the ridge crest are younger than the oldest part of the truncation surface of the summit platform, are thought to be deposits of primarily late Miocene but possibly middle Miocene through Holocene age (roughly the past 10 to 13 million years), those of the middle series formed chiefly during Oligocene through early or middle middle Miocene time (35 to 13 m.y. ago), and lower series rocks are mostly upper Eocene and probably older units. Rocks older than middle Eocene are unknown from the Aleutian Islands, but they may constitute the hypothetical initial (or basement) series of Scholl, Buffington, and Marlow (1975a) and Scholl, Buffington, and Marlow (1975b).

A columnar section included on Figure 2 diagrammatically portrays the ridge's "ideal" three-part chronostratigraphic framework of middle Eocene and younger rocks — a framework that with time is increasingly dominated by the formation of sedimentary rather than igneous rock sequences. Schmidt (1978) has suggested a similar scheme that is most applicable to the insular or exposed rocks of the island arc. He defines four rock "horizons," each is assigned a formal name that implies regional temporal equivalents.

Implicit and essential to our three-part chronostratigraphic framework, which attempts to include offshore rock sequences, is the notion that although each series is assigned a general time interval, similar lithologic units or facies can appear in any of them, no series is exclusively one rock or depositional type, and the time frame or span of a series is somewhat variable along the length of the ridge. The time span of the series indicated by the heavy vertical lines along the left side of the columnar section on Figure 2 indicate our present estimate of the age range of the three series in the Amlia corridor; the dashed or extended line is our estimate of the age range elsewhere along the ridge.

Aleutian Basin

The abyssal floor of the Aleutian Basin (3,600 m) of the Bering Sea rises gently southward (0.2°) toward the base of the Amlia sector of the Aleutian island arc (Figures 2 and 3). This area of the basin floor is underlain by a 2 to 3 km-thick blanket of internally reflective or acoustically stratified deposits of the upper and middle series overlying a roughly equally thick sequence of lower series rocks characterized by indistinct and laterally irregular and discontinuous acoustic stratification. Although middle series beds are not as internally reflective as those of the upper series, an acoustically distinctive reflection horizon does not separate the two series. A prominent reflection horizon, however, designated SIR-B (Strong, Subbottom Irregular Reflector-Bering Sea) on Figures 2 and 3, marks the top of the lower series beds.

Interval velocities (based on sonobuoy and multichannel velocity data) within the coherently stratified upper and middle series increase downsection from 1.6 km/sec near the sea floor to about 3.5 km/sec immediately above the SIR-B reflector. The underlying lower series comprises rocks with interval velocities between about 4.0 and 5.5 km/sec. Velocities typical of oceanic crust occur at a subbottom depth of 4.5 to 5.0 km (Ben-Avraham and Cooper, 1982). Acoustic scattering by the lower series typically inhibits clear resolution of the surface of the oceanic crust in this area (Figures 3). The acoustically distinct lower series may only exist near the island arc. Across the central area of the Aleutian Basin, reflection profiles typically do not reveal an SIR-B horizon, but rather a smooth, terminal reflection surface that is either the surface of the basin's igneous oceanic crust or a lithified sedimentary section immediately overlying it (Cooper, Marlow, and Ben-Avraham, 1982).

Based on the stratigraphic sequences penetrated at relatively nearby DSDP sites 184 and 185, and the more distant site 190 (Figure 1; Creager, Scholl et al, 1973), the upper series is predominantly diatomaceous clayey silt, turbidite deposits, and rare limestone beds

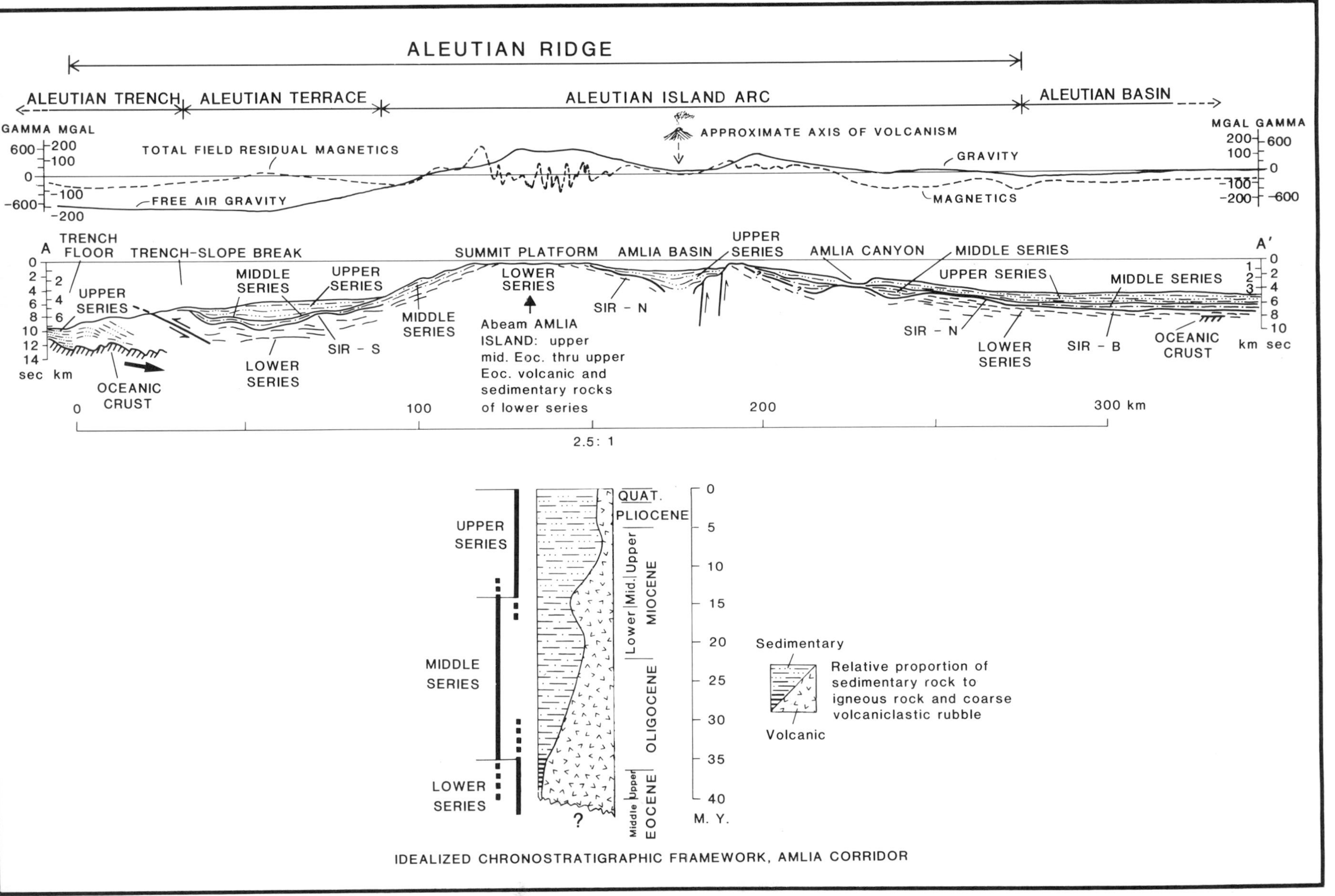

Figure 2 — Major geomorphic elements and generalized structural section, based on reflection events and horizons, of upper crustal rocks of the Aleutian Ridge, Amlia corridor. Subbottom information is based chiefly on initially-processed multifold reflection profiles along line A-A', Figure 1. SIR-B, -N, -S horizons are regionally persistent strong, irregular reflection horizons that typically separate lower from middle or upper series deposits. (See Figures 3, 5, 7, 8, 9, and 10.). Lower figure is an idealized chronostratigraphic scheme adaptable to the onshore and offshore rock sequences of the Amlia corridor and to the Aleutian Ridge in general. The age range of these rock series are thought to vary along the ridge, a likelihood implied by the extended and dashed range bars left of the column.

Table 1. Aleutian Ridge dredge haul locations, water depths, rock ages, and major lithologies from USGS cruises S6-79, L5-80, and L9-81.

Dredge No.	General Location (and series)	Latitude and Longitude	Water Depth (m)	Ages	Major Lithologies
L5-80-1	Top of landward slope of Aleutian trench (late series)	50°59.8′N 174°06.3′W	5000	Late Pliocene-early Quaternary	Limestone and mudstone
L5-8-2	Northern edge of summit platform north of Amukta Island (lower or middle series)	52°52.1′N 171°28.2′W	500-400	Unknown	Volcaniclastic breccia
L9-81-3	Upper slope south of Seguam Pass (upper series with included clasts of middle series)	51°48.9′N 172°49.5′W	2250	Late Pliocene - early Quaternary with clasts of late Oligocene	Sandstone and conglomerate
L9-81-4	Lowermost slope south of Amlia Island. (middle series)	51°37.6′N 173°27.5′W	3680-3700	Late Pliocene (silt) Early middle Miocene(ss)	Soft siltstone and hard sandstone
L9-81-5 & 6	Summit platform south of Amlia Pass (lower series)	51°53.3′N 174°00.3′W	80	Unknown	Flow rocks and volcanic sandstone
L9-81-10	East wall, upper reaches of Amlia Canyon (upper series)	52°43.8′N 173°03.8′W	1500-1850	Late middle Miocene and middle Pliocene	Sandstone and mudstone
S6-79-12	Upper slope south of Amlia Island (middle series)	51°51.6′N 173°32.2′W	1000-900	Early Oligocene	Sandstone
S6-79-13	Upper slope south of Amlia Island (middle series)	51°50.0′N 173°28.1′W	1700-1650	Early Oligocene	Sandstone and tuff
S6-79-14	Uppermost slope south of Amlia Island (middle series)	51°48.9′N 173°56.3′W	500-400	Late Oligocene-early Miocene	Sandstone
S6-79-15	Inner wall of the Aleutian Trench (upper series)	50°53.1′N 173°12.9′W	7050-6600	Early Quaternary	Mud
S6-79-16	Middle slope south of Amlia Island (middle series)	51°45.8′N 173°43.4′W	2650-2450	Oligocene	Sandstone and siltstone
S6-79-17	Middle slope south of Amlia Island (middle and upper (?) series)	51°44.4′N 173°45.2′W	3100-2950	Oligocene and middle Miocene	Sandstone
S6-79-19	Middle slope south of Amlia Island (middle series)	51°43.5′N 173°41.3′W	2900-2750	Early to middle Oligocene	Sandstone
S6-79-20	Uppermost slope south of Amlia Island (upper series)	51°48.7′N 174°00.4′W	93-700	Late Miocene-early Pliocene	Sandstone
S6-79-21	Summit platform south of Atka Island (lower series)	51°52.1′N 174°12.9′W	120	Unknown	Flow rocks and volcanic sandstone
S6-79-22	Summit platform south of Amukta Pass and Seguam Island (lower series)	52°08.9′N 171°51.9′W	450-300	Middle Eocene-early Oligocene	Tuff, sandstone and siltstone
S6-79-25	Summit platform south of Amukta Pass and Seguam Island (lower series)	52°09.6′N 171°52.2′W	260	Unknown	Flow rocks and tuff
S6-79-27	West wall of Amlia Canyon, north side Aleutian Ridge (upper series)	52°39.2′N 173°07.8′W	1550-1500	Early Pliocene-Holocene	Sandstone, siltstone and tuff
S6-79-28	West wall of Amlia Canyon (upper series)	52°39.7′N 173°09.9′W	1350-1225	Middle late Miocene early Pliocene	Breccia and conglomerate
S6-79-29	West wall of Amlia Canyon (upper series)	52°39.8′N 173°09.9′W	1200-1100	Late Miocene-early Pliocene(?)	Sandstone and breccia
S6-79-30	East wall of Amlia Canyon (upper series)	52°41.1′N 173°04.3′W	1550-1400	Middle middle Miocene	Sandstone, breccia, and conglomerate
S6-79-31	Fault scarp on northwest side of Amlia Basin (lower or middle series)	52°30.8′N 173°27.4′W	580-300	Unknown	Volcaniclastic breccia and flow rocks
S6-79-32	East wall of Seguam Canyon, North side of Aleutian Ridge (upper series)	52°37.5′N 172°35.2′W	975-900	Late Miocene-early Pliocene	Sandstone and breccia

of at least late Miocene through Quaternary age. The lithology of the underlying middle and lower series beds is unknown.

Island Arc

Northern Flank

Flat-lying upper and middle series beds of the Aleutian Basin can be traced southward to the base of the northern flank of the Aleutian Island arc (Figures 1 and 2). At this geomorphic junction the attitude of the beds swings upward to conform to the slope angle (about 3.5°) of the arc's descending flank. Beds of the two series therefore constitute a dip-slope sequence of prominently stratified deposits. As is the case in the Aleutian Basin, no distinctive acoustic horizon separates middle and upper series beds. Over the lower part of the northern flank their combined thickness is 1 to 2 km. This section of stratified deposits is actually the southwestern extension of the even thicker pile of probable Oligocene and younger sedimentary deposits of neighboring Umnak Plateau (Scholl and Creager, 1973; Ben-Avraham and Cooper, 1982; Figure 1).

A persistent and typically strong, irregular reflecting horizon, SIR-N (Strong, Subbottom, Irregular Reflector-North), underlies the lower reaches of the northern ridge flank and delineates the surface of the lower series (Figure 3). The indistinctly or irregularly layered rocks of this series thicken beneath the base of the northern flank where the series is more than 2 to 3 km thick. In general, the acoustic velocity of lower series beds exceeds 4 km/sec. Although layering within the series is only locally coherent and distinct, stratification is roughly parallel to the inclination of the lower part of the ridge flank (Figure 3). The acoustic interfaces SIR-N and SIR-B, although different in character, are presumably roughly age-equivalent horizons.

Traced upslope, toward the crest of the arc, middle series beds thin and thicken, respectively, over structural highs and depressions in the underlying framework rocks of the lower series (Figure 2). Near the crest of the ridge northward inclined deposits of the middle series either wedge out (upslope) against lower series beds or extend southward beneath the summit platform. Here lithified and inclined beds of the middle series are presumed to be truncated by the regional erosion surface of the summit platform, which, in many areas, is unconformably overlain by semi-lithified deposits of the upper series.

Beneath the upper (southern) reaches of the arc's northern flank, beds of the upper series form a dip-slope sequence. Traced upslope these beds either pinch out against underlying middle and lower series rocks, or drape over the northern edge of the summit platform; they either shallowly bury truncated strata of older rocks or appreciably thicken in the arc's summit basins (Figures 2 and 5). A prominent reflection horizon typically separates upper series beds from underlying and more lithified and deformed middle and lower series deposits. In contrast, beneath the crestal region of the ridge the acoustic facies of middle and lower series beds are both characterized by weak and laterally discontinuous reflection horizons. Except locally, the two older series cannot be acoustically separated on most reflection records.

Rocks dredged from the arc's northern flank, especially from outcrops along the upper reaches of Amlia and Seguam Canyons (Stations 10, 27, 28, 29, 30, and 32, Figure 1), document that the upper series includes fine-grained marine beds of middle Miocene through Pleistocene age (Table 1). Middle Miocene sandstone beds probably form the basal unit of the upper series, but strata of this age have not been confidently traced above the erosional surface of the summit platform. Most of the recovered rocks of the upper series are incompletely indurated terrigenous deposits that are rich in diatomaceous debris and, prior to middle Pliocene time, a pollen and spore assemblage derived from a nearby conifer-dominated forest.

Samples of the underlying rocks of either the middle or lower series, typically red and brown volcaniclastic sandstone and breccia, were recovered with certainty from the north side of the Aleutian Ridge only at dredge sites 2 and 31 (Table 1). These dredge stations are located at or near the top of the arc's northern flank (Figure 1). Recovered rocks of the older two series are unfossiliferous and, therefore, their deposition age is unknown. The red and brown oxidized colors of these rocks, the lack of marine fossils, and the poorly sorted clayey and silty matrices suggest that the recovered fragmental breccia beds may be non-marine deposits.

Summit Platform, Islands, and Basins

In the vicinity of Amlia Island the submerged and geomorphically flat crestal area of the Aleutian Ridge — the summit platform — is 50 to 60 km wide (Figures 1 and 2). The platform is less than 200 m deep, but large islands (Amlia, Amukta, Seguam and Atka) rise as high as 1,500 m above it, and geomorphic basins (Amlia and Amukta basins) descend to depths of 1,000 m (Figure 2).

The summit platform owes its geomorphic form to wave-base erosion across rocks of mostly the lower and middle series, although the platform also cuts into upper series rocks of the corridor's late Cenozoic volcanic centers (Figure 6). Flat-lying or gently deformed marine deposits as old as late Miocene in age, and possibly as old as middle Miocene, unconformably overlie the submerged erosional surface. Truncated beds believed to be the lower series include lithified sandstone; siltstone and eruptive rocks have been dredged from submerged outcrops at stations 5, 6, and 21 through 25 (Figure 1, Table 1). Closely spaced high-amplitude magnetic anomalies attest that magmatic rocks probably dominate the rocks underlying the summit platform (Figure 2). Calcareous nannofossils from associated sedimentary rocks established an extrusion time of middle Eocene or possibly early Oligocene age (Bukry, 1980, written communications) for much or perhaps the bulk of the truncated igneous rocks.

Amukta (and smaller, adjacent Chagulak Island) and Seguam islands are young stratovolcanoes of basic and intermediate composition. These edifices are aligned northeasterly along the center of the ridge crest; they rise above and bury the erosional surface of the summit platform. A cluster of volcanic centers issuing rocks of dominantly andesitic basalt composition form the northeastern part of Atka Island (Marsh, 1980). This vent complex, the largest along the central Aleutians, is part of an alignment trend of eruptive centers positioned along the northern side of the ridge crest and therefore offset from the Seguam-Amukta trend (Scholl, Marlow, and Buffington, 1975b). The volcanoes of the Seguam-Amukta trend probably began to form in Quaternary time, an estimated age based on regional rather than specific information (Coats, 1950, 1962; Scholl, Marlow, and Buffington, 1975b, Scholl et al, 1976). More northerly volcanic edifices of Atka Island are known to have begun to form in latest Miocene time (Marsh, 1980;

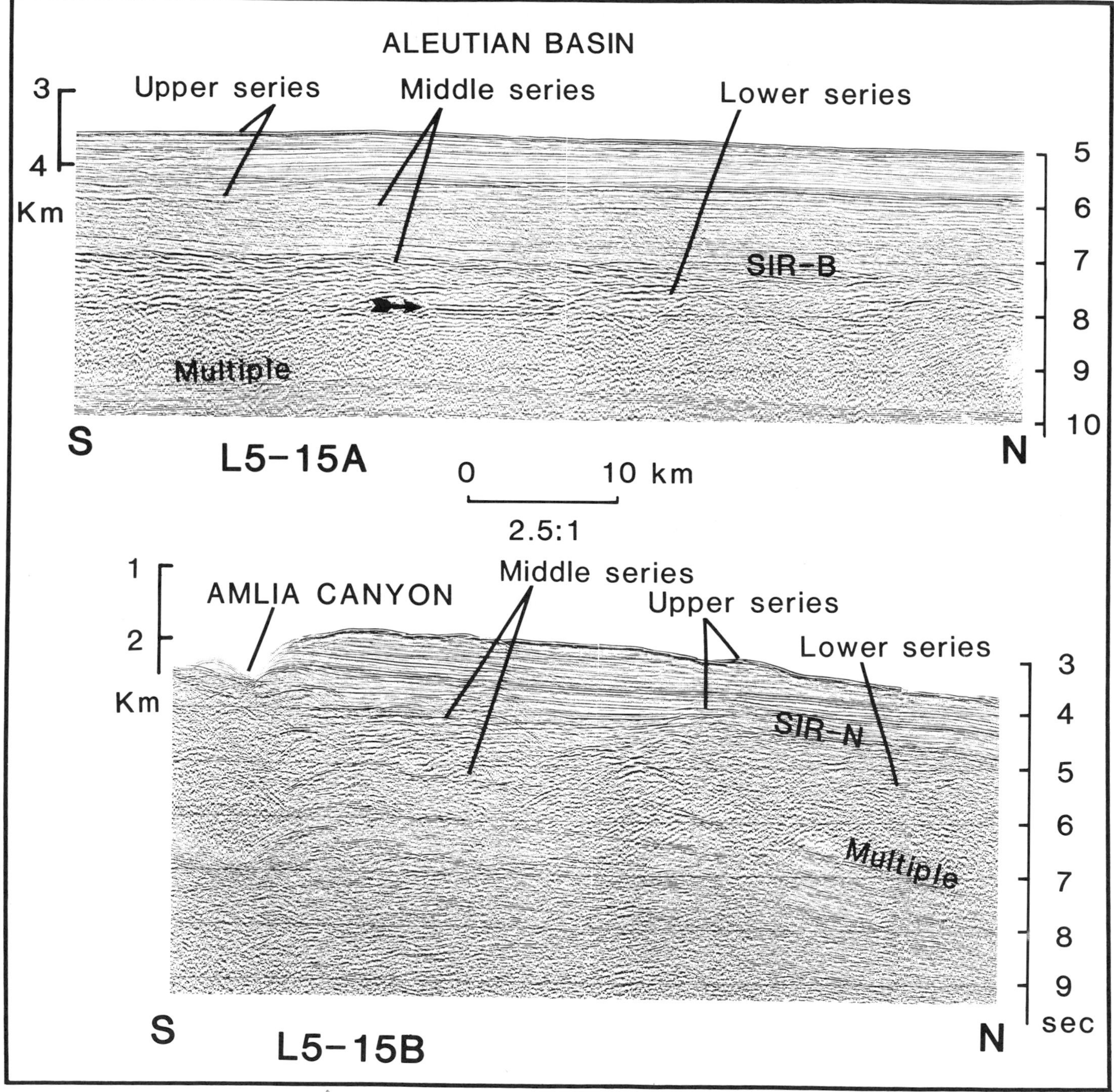

Figure 3 — Segments of 24-fold seismic reflection profiles (first-stack display) that cross the southern edge of the Aleutian Basin (upper) and lower part of adjoining northern slope of the Aleutian arc (lower). Arrow on upper profiles indicates approximate position of igneous oceanic crust. See Figure 5 for continuation of 24-fold record across the Aleutian Ridge.

Hein, McLean, and Vallier, 1981). The corridor's active volcanoes and immediately underlying but genetically associated upper Tertiary effusives are therefore eruptive rocks of the upper series.

Amlia Island, which is positioned toward the southern side of the summit platform (Figure 1), is constructed solely of little deformed volcanic, volcaniclastic, and sparsely fossiliferous sedimentary rocks containing a late middle Eocene to possibly early Oligocene nannoflora (Figure 4; McLean et al, 1980, Hein, McLean, and Vallier, 1981). The whole-rock K-Ar ages of one flow and an associated dike are somewhat younger (32 to 33 m.y. old), but a small gabbro body is at least of late middle Eocene age (40 m.y. old; Figure 4). Alteration of the radiometrically dated rocks, as shown by their abundant secondary minerals and high water content, suggests that argon has been lost; the radiometric dates must therefore be regarded as minimal ages. Considering all factors, we presume that the exposed rocks of Amlia Island are chiefly of late Eocene age (35 to 40 m.y. old). Similar but more deformed Paleogene rocks occur on neighboring Atka Island, but the section exposed here includes intrusive granodioritic plutons and probably related eruptive rocks of middle Miocene age (Hein and McLean, 1980; Hein, McLean, and Vallier, 1981). Near the plutons, deformation and thermal alteration of lower series rocks may be intense.

We presently regard the upper Eocene eruptive and subordinate sedimentary rocks of Amlia and Atka islands as exposed sections of the ridge's lower series. Middle series rocks are represented on Atka Island by the intrusive and associated eruptive masses of early middle Miocene age (15.1 m.y. old; Pb/U date). Upper series rocks are those eruptive and volcaniclastic deposits that have formed at its still-active northeastern vent complex.

On Amlia Island the lower series rocks are at least 500 m thick (Figure 4). The exposed strata define a gentle, south-dipping (5 to 15°) monocline that strikes eastward parallel to the length of the island and the axis of the arc. Because epizonal plutons of the middle series do not intrude the Amlia section, as they do on adjacent Atka Island, thermal alteration is limited to low temperature effects (up to prehnite-pumpellyite facies). Although subaerial deposits are probably present, particularly at some of the higher elevations, the bulk of the lower series exposed on Atka and Amlia islands accumulated in relatively shallow water along the crestal or upper flanks of an upper Eocene volcanic arc.

The volcanic rocks of the lower series of Amlia Island range compositionally from basalt to rhyolite. Both thoeliitic and calc-alkaline differentiation trends are exhibited on AMF geochemical plots (Figure 4). The more mafic samples, typically selected from pillow outcrops, show a slightly elevated yet rather flat distribution of rare-earth elements. More differentiated andesitic and rhyolitic rocks have large enrichments in the light rare-earth elements, a relation typical of island arc rocks in general. Significantly, the upper Eocene effusive rocks of Amlia Island differ little, geochemically, from the upper series eruptive rocks of the active volcanic edifices of Atka Island (Marsh, 1980).

Amlia and Amukta basins, two 60-km-long, west-trending structural depressions, lie along the northern edge of the summit platform (Figure 1; Perry and Nichols, 1966). These geomorphic basins, each approximately 1000 m deep, are associated with the general axis of late Cenozoic volcanism and are possibly genetically related to the offset alignment trends of corridor volcanoes (Figure 2; Scholl, Marlow, and Buffington, 1975b; Spence, 1977). Amlia and Amukta basins are grabens or half-grabens that attest to late Cenozoic extensional rifting of the arc's crestal region. Both basins are partially filled with gently deformed beds of the upper series (Figure 5). Folds increase down-section in structural relief and are commonly flanked by high-angle growth faults. Lower and probably middle series rocks form the basins' structural basement. Preliminary analyses of sonobuoy refraction data and depth to magnetic anomalies imply that the young sedimentary sections are several kilometers thick. Seguam volcano appears to have erupted into these beds and thereby geomorphically divided Amlia and Amukta basins, which are probably structurally contiguous.

Upper Miocene and younger beds of poorly consolidated sandstone and siltstone that underlie the arc's northern flank can be traced upslope into the sedimentary section of Amlia and Amukta basins (Table 1; Figure 1). Although marine deposits as old as middle middle Miocene age have been recovered from the base of the acoustically recognized upper series along the upper reaches of Amlia Canyon (Station 30, Figure 1), which heads in Amlia Basin, only beds as old as late Miocene have been definitely traced into Amlia and Amukta basins. We can only speculate that their sedimentary sections might include basinal deposits 12 to 13 m.y. old.

Southern Flank

Within the Amlia corridor the southern or Pacific flank of the island arc is underlain by a thick (more than 1,000 m) sequence of undulatory but acoustically stratified slope deposits that dip southward roughly parallel to the inclination of the sea floor (Figure 2). The bulk of this upper dip-slope sequence is constructed of middle series beds. Although a number of prominent reflection horizons occur within the middle series, except toward the base of the southern slope, a distinct acoustic basement or SIR-type horizon does not delineate the top of the underlying lower series, which is characterized acoustically by reflectively faint, and more irregularly layered rocks (Figures 6 and 7). South of Amlia Island, middle and lower series beds can be traced downslope beneath younger and virtually horizontal basinal deposits of the upper series underlying the floor of the Aleutian Terrace (Figure 1). Traced updip, toward the island, lower and middle series beds are truncated by the planar surface of the southern extension of the arc's summit platform (Figures 6 and 7). The seaward inclination of both series is

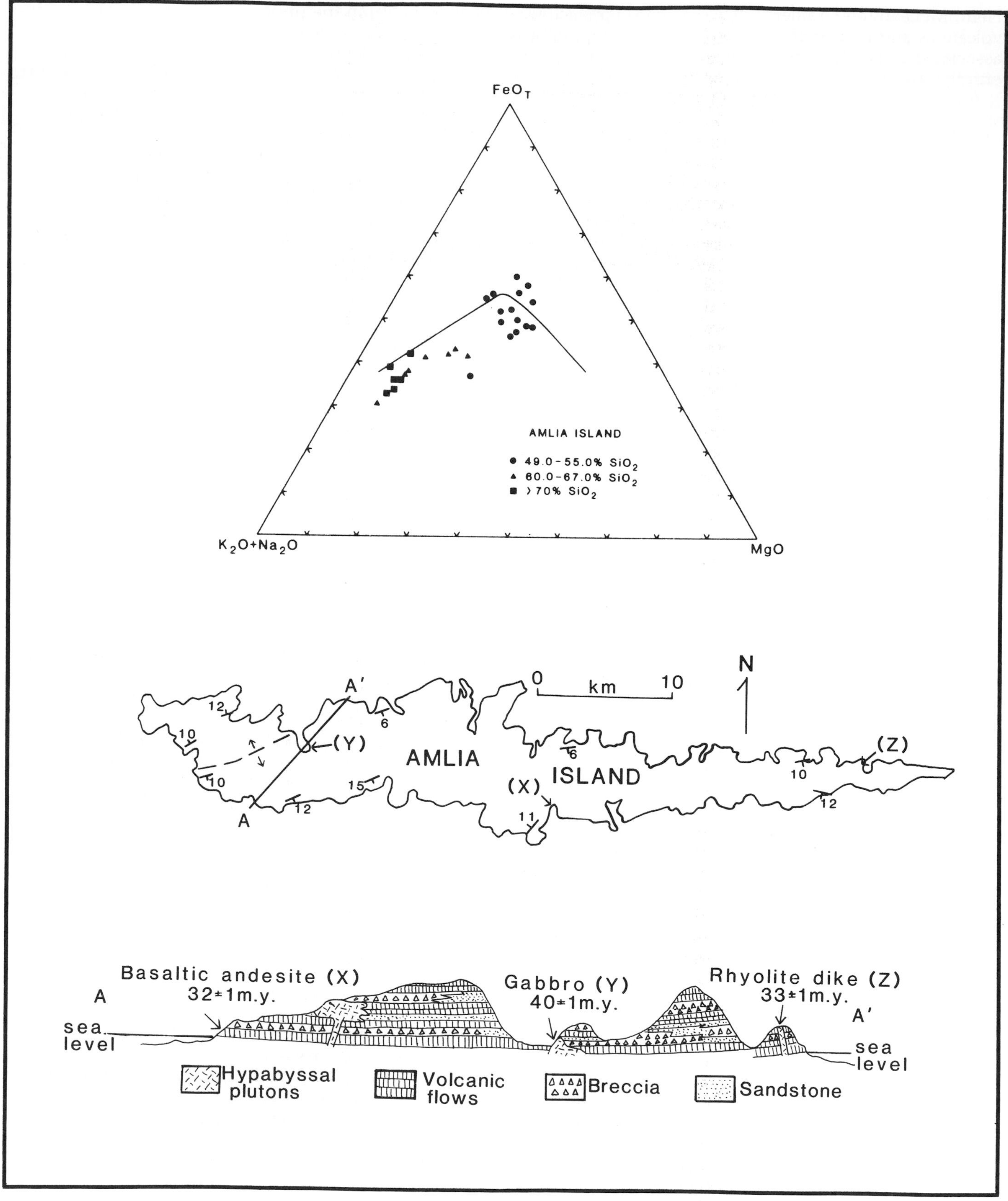

Figure 4 — Upper drawing is AMF diagram (total iron as FeO) of igneous rocks exposed on Amlia Island. Line separates theoleiitic (above) from calc-alkaline (below) fields (after Kuno, 1968). Lower two drawings are sketch geologic map and idealized cross section, respectively, of Amlia Island based on the work of McLean et al (1980) and Hein, McLean, and Vallier (1981). Ages of whole-rock dated (K/Ar) rocks are projected onto the cross section. The nannoflora of interbedded sedimentary deposits indicate an age of late middle Eocene to possibly earliest Oligocene.

about 10°, virtually equal to that of the concordantly dipping volcanic and subordinate sedimentary rocks of the lower series that crop out on adjacent Amlia Island (Figure 1). However, beneath the slope incompletely processed reflection records appear to reveal a slight angular discordance between the two series.

Dredging of submarine outcrops from the upper 500 to 700 m of the slope sequence recovered semi-consolidated to weakly consolidated middle series beds of diatomaceous sandstone, siltstone and mud stone of early Oligocene through early middle or possibly middle middle Miocene age (Table 1). Beds of mostly late Miocene or younger age were sampled near the top of the dip-slope section (Stations 2, 3, and 20, Figure 1, Table 1); but it is evident that in many areas upper series deposits are thin or absent (Figures 6 and 11). Thicker sections of the late series may have been present prior to the late Cenozoic episode of canyon cutting that extensively eroded the arc's southern and northern flanks (Scholl, Marlow, and Buffington, 1975b). Sandstone beds of the middle series are fine-grained and moderately well sorted; only a small amount of clay-sized material is present. In comparison to the insular exposures of slightly older but firmly lithified sedimentary beds of the lower series, lower Oligocene and younger beds of the middle series are virtually unlithified and unaltered beneath the southern flank of the arc. Significantly, these slope beds are rich in essentially unaltered detritus shed from the arc in late Paleogene and early Neogene time. Presumably, in the Amlia area, exposure and erosion of significantly altered volcanic source rocks did not occur until after early Miocene time.

Aleutian Terrace

The Aleutian Terrace is a forearc basin (Figure 1). Its gently sloping or basinal profile extends smoothly southward for 50 to 60 km to a series of west-trending antiformal ridges that define it seaward limit (Figure 2; Grow, 1973a, 1973b). Within the Amlia corridor the terrace's central or basinal area, Atka Basin, descends to a depth of about 4,700 m (Figure 1). The outer highs, cresting at a depth between 3000 and 5000 m, also define the trench-slope break (Figures 2, 8, and 10); DSDP holes 186 and 187 were drilled in the vicinity of the trench-slope break (Figures 1, and 11).

Multichannel profiles document that the terrace is underlain by an upper, basinal section of internally reflective and coherently stratified upper series beds, a medial section of middle series strata characterized by somewhat less planar stratification, and a basal section of relatively weakly reflecting and broadly undulatory lower series strata (Figure 2). A commonly strong, irregular reflector, the SIR-S (Strong, Subbottom, Irregular Reflector-South) horizon (Figures 7, 8, and 9), separates middle and lower series deposits. The subsurface SIR-S horizon can be traced southward and downward from the lower reaches of the arc's southern flank, deep (4 to 5 km) beneath the terrace floor, and upward to shallowly underlie the outer terrace highs (Figure 2). The top of the "older terrace sediment" defined by Grow (1973a, Figure 6) is roughly equivalent to the top of the middle series as defined here.

Beneath the central area of Atka Basin the combined thickness of the stratified deposits of the upper and middle series is as much as 5 km (reflection time = 4.0 sec), a section only slightly thinner than that deduced by Grow (1973a) based on the geophysical data then available. The younger, mostly late middle Miocene through Quaternary beds of the upper series constitute about two-thirds of the terrace's stratified section. Interval velocities in the upper series beds increase downward to about 3.0 to 3.2 km/sec at the top of the middle series, which is typically discernable as a prominent, undulating reflection horizon (Figures 9 and 10). The upper series is constructed of large packets of lens-shaped depositional units. These units include both well-stratified sequences and more massively bedded deposits that lack coherent internal reflectivity (Figures 7, 8, and 11; Marlow et al, 1973; Grow, 1973a, 1973b). Deposition of the packets geomorphically formed the terrace by infilling the differentially warped surface of the SIR-S horizon and overlying middle series section. Changes in the location of the maximum thickness of the packets attest to both landward and seaward shifts in depocenters caused by basement warping (Figure 11). The principal direction of shifting has been landward in concert with relative uplift of the outer-terrace antiformed highs, however.

At DSDP site 186, which is located near the crest of a low outer-terrace high (Figures 1 and 11; Grow, 1973a), a seaward-thinning sequence of upper series deposits was penetrated to a subbottom depth of 926 m (Creager et al, 1973). The sampled section is chiefly diatomaceous clay and sand rich in volcaniclastic detritus (Stewart, 1978). Pleistocene beds constitute the upper one-third of the drilled sequence; deposits at the base of the hole are no older than about middle lower Pliocene (4 to 4.2 m.y. old, John Barron, written communication, 1981). A displaced mass of middle Miocene beds was intersected within the lower beds of the upper Pliocene section; and upper Miocene deposits were evidently reached at the base (370 m) of nearby Site 187, which is located over a second and more seaward antiformal high (Figure 11). We deduct from these findings that upper Miocene beds form part of the older deposits of the acoustically defineable upper series of the terrace section. The displaced middle Miocene deposits are either samples of the underlying middle series which can be traced on multifold records to the structural cores of the antiforms (Figure 10), or representative of the oldest beds of the upper series. Perhaps of considerable importance is the fact that the benthic foraminiferal fauna of the displaced middle Miocene block imply that its original depositional depth (as shallow as 2,000 m) may have been significantly shallower than the block's present depth (5,300 m).

Multichannel reflection and sonobuoy refraction data attest that the velocities within middle series strata vary from about 3.5 to 4.5 km/sec. Beneath the floor of Atka Basin these beds are approximately 2 km thick; but unlike the overlying upper series, middle series beds do

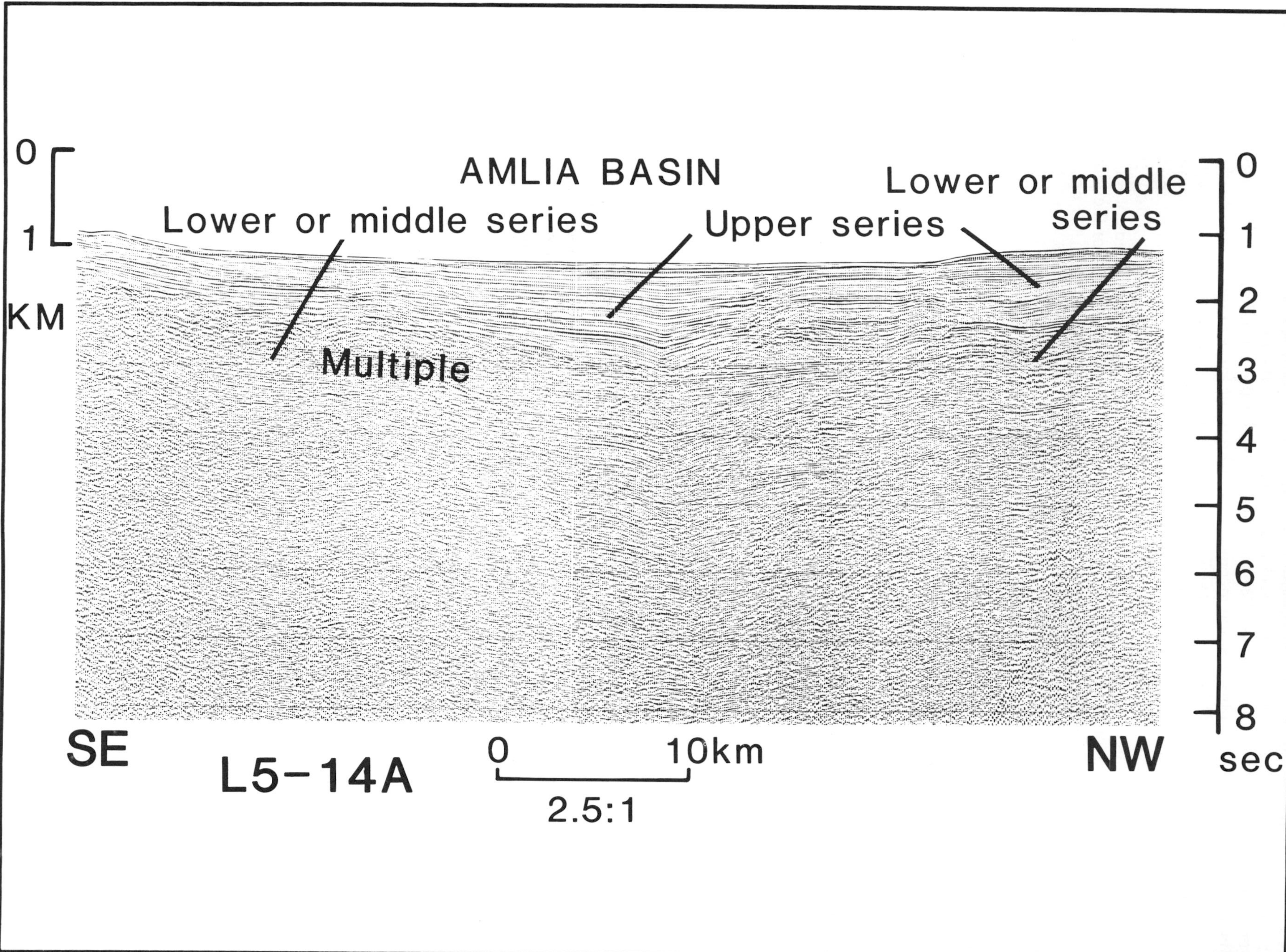

Figure 5 — Segment of 24-fold seismic reflection profile (first-stack display) that crosses the western part of Amlia Basin (Figure 1). Upper series deposits are probably of late Miocene and younger ages.

not appear to significantly thicken here as a basinal sequence. Limited information, however, implies that the series thins southward against the basement cores of the terrace's seaward antiformal highs.

Refraction velocities recorded along the SIR-S horizon, which tops the underlying lower series, are typically near 5.3 km/sesc. Compressional velocities of this magnitude are typical of the second layer of igneous oceanic crust and also the main or upper crustal layer of the Aleutian arc (Shor, 1964). Below the terrace the thickness of the lower series is unknown, but a strong refractor at 6.2 to 7.3 km/sec is commonly recorded 1 to 2 km beneath the SIR-S surface. This refraction horizon is evidently equivalent to the 6.4 km/sec crustal velocity reported by Grow (1973a). Weekly coherent and undulating reflection horizons (on multichannel records) can be traced laterally within the lower series for tens of kilometers (Figures 7 and 9). These horizons occur as deep as 5 km (2 sec) below the SIR-S interface, which itself is a broadly undulating surface (relief of 1 to 2 km in 5 to 10 km). The SIR-S horizon appears to unconformably underlie middle series beds, but the attitude of this series is generally parallel or subparallel to that of the SIR-S interface.

Aleutian Trench

Landward Slope

The trench's inner or landward slope descends in a series of ridges and basins from the trench-slope break at a depth near 5,000 m to the trench floor at about

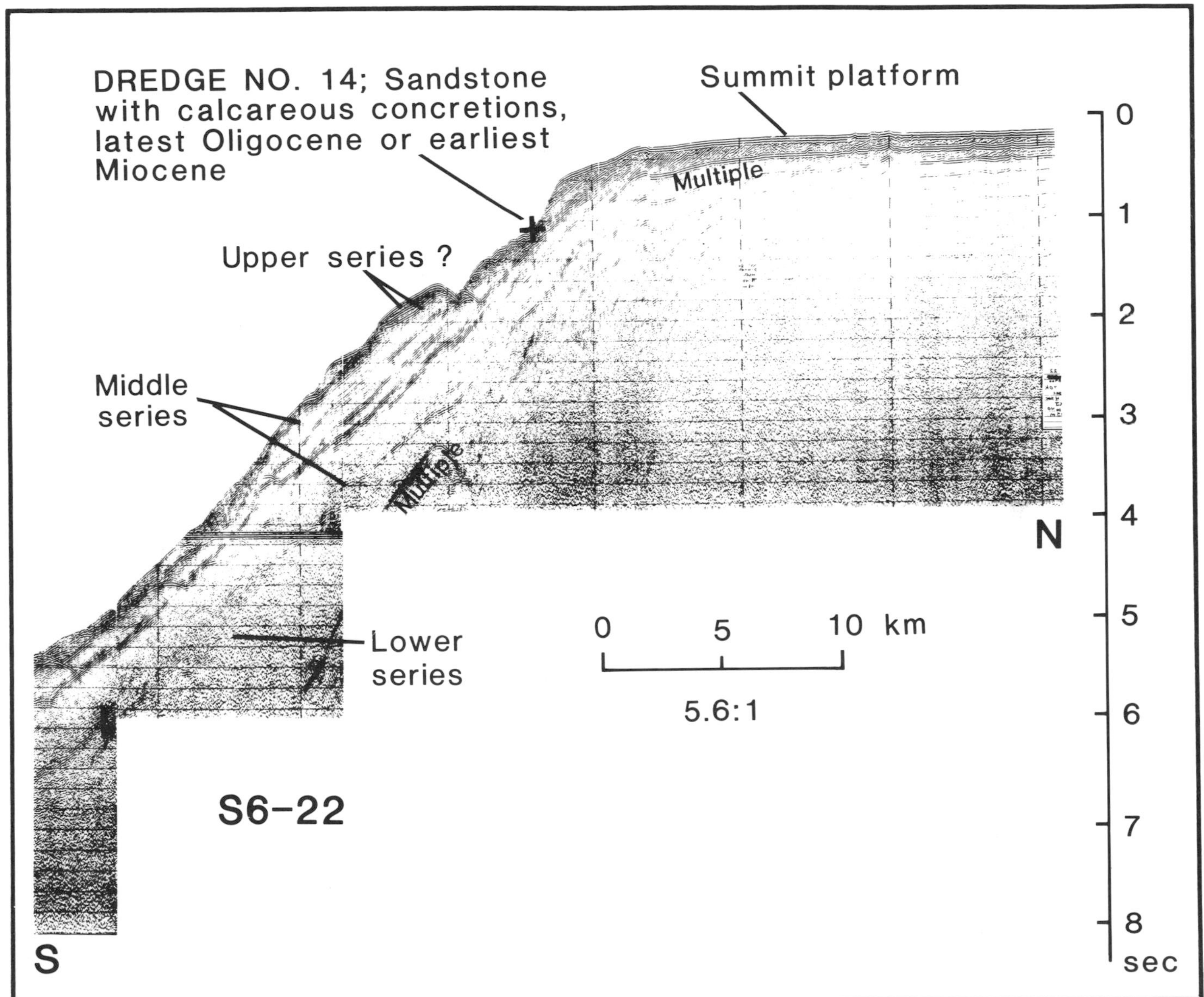

Figure 6 — Single-channel reflection profile that crosses the Pacific slope of the arc south of Amlia Island (Figure 1). Summit platform, a wave-planed erosion surface, is cut across south dipping (10°) beds as old as late Eocene (northern side) and possibly as young as early Miocene (near shelf edge). An SIR-type reflector does not separate middle and lower series beds. Upper series strata are absent or only thinly mantle the slope.

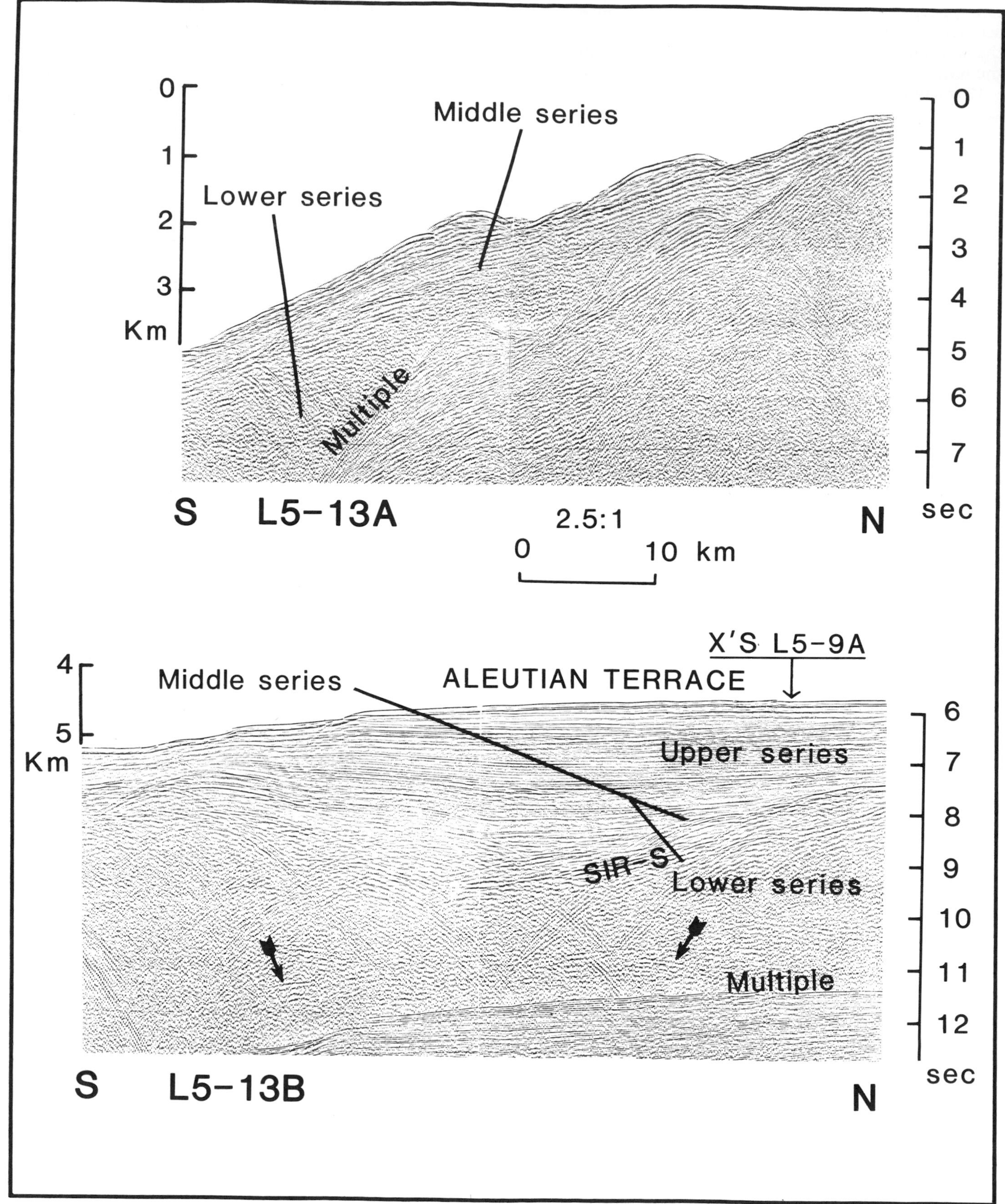

Figure 7 — Segments of 24-fold seismic reflection profiles (first-stack display) that cross the southern slope of the island arc (upper; compare with Figure 6) and central area of the adjacent Aleutian Terrace (lower). Arrows locate weak but laterally persistent layering within the lower series (see Figure 9).

7,300 m (Figures 1 and 8) (Perry and Malahoff, 1981). Grow (1973a) reports that the trench-facing slopes of some of these ridges are as steep as 30 to 35°. Although the internal structure of the inner slope is only partially resolved on initially processed multifold records (Figures 8 and 10), the north-dipping surface of the underlying oceanic basement can be acoustically tracked beneath its entire 30 to 40 km width (Figure 2). Beneath the trench-slope break, oceanic crust lies at a subbottom reflection time of 6 to 7 sec, which roughly corresponds to a subsurface depth of 8 to 10 km (13 to 15 km below sea level; Figures 8 and 10).

The basement surface rises southward toward the trench floor. The northward inclination of the oceanic crust is roughly 7 to 8°, and its attitude and subsurface position is similar to that shown on Model B of Grow's (1973a) structural interpretations of the Amlia forearc area. The angle of descent beneath the inner slope is approximately equal to the crust's northward plunge (6 to 7°) beneath the turbidite wedge of the adjacent trench floor. The subsurface topography of the descending crust is either relatively smooth (profile segment L5-7A, Figure 10, lower) or rough (profile segment L5-13D, Figure 8 lower).

The oceanic basement beneath the lower or southern half of the inner slope is overlain by a trenchward-thinning mass of relatively low-density, low-velocity deposits (Grow, 1973a). These sedimentary beds exhibit acoustic stratification that typically dips gently northward (landward) at angles equal to or somewhat steeper than the underlying oceanic crust (Figures 8 and 10). Interval velocities increase downward to about 3.5 km/sec at a subbottom depth near 4 km. This velocity structure is similar to that of the upper series of the Aleutian Terrace. Profile segments L5-13D (Figure 8), L5-7A (Figure 10), and initial studies of migrated records reveal that the trenchward-thinning mass is not complexly deformed. Profile segment L5-11A (Figure 9), which trends parallel to the slope, further shows that acoustic layering, which is caused by low-angle depositional and tectonic surfaces (McCarthy et al, 1982), is laterally coherent for tens of kilometers. The mass of sedimentary deposits underlying the toe or lower reaches of the inner trench slope is a thrust-thickened sequence of trench-floor deposits that can be assigned to the ridge's upper series.

Trench Floor

A landward thickening, wedge-shaped mass of upper series deposits underlies the flat floor of the Amlia sector of the Aleutian Trench (Figure 2). The west-southwest-trending (260°T) sector is approximately 20 km wide and lies at a depth near 7,300 m. Although virtually horizontal, the trench floor slopes gently landward (0.2° or 3.3 m/km) toward a thalweg or trench low along the base of the inner wall. Pacific oceanic crust of latest Cretaceous and earliest Tertiary age dips northward 4 to 10° beneath the sector's sedimentary wedge (Figure 2). The wedge is thickest (3.7 to 4.0 km) along the base of the trench's inner wall near and east of its crossing of the Amlia Fracture Zone (Figures 1 and 8). West of the fracture zone, over slightly younger oceanic crust, the thickness of the turbidite wedge is only 1 to 2 km. DSDP drilling in the trench south of the Alaskan mainland and surface cores collected in the Amlia Sector imply that the wedge is chiefly constructed of silty and possibly sandy turbidite beds interbedded with diatomaceous hemipelagic mud (Horn, Horn, and Delach, 1970; Piper, von Huene, and Duncan, 1973; von Huene, 1974). The wedge's oldest bed is estimated to be about 0.5 m.y. old (Scholl, Vallier, and Stevenson, 1982).

Subsurface strata of the wedge dip gently northward toward the base of the Aleutian Ridge. The angle of dip increases downsection to about 5 to 6° immediately above oceanic basement. Slight undulations on the trench floor reflect subsurface antiforms and synforms that strike generally westward parallel or slightly oblique to the axial trend of the trench. The amplitude of folding typically increases downsection, and these structures are superimposed above ridges and swales, respectively, in the underlying oceanic basement. West-striking, high-angle faults and abrupt flexures border the southern flanks of some of the antiforms; adjacent to these structures the northern side of the trench floor is typically displaced upward (Figure 10). A similar pattern of up-to-arc displacement disrupts the northward descending oceanic crust seaward of the trench floor (Scholl, Vallier, and Stevenson, 1982a). Although high-angle north-dipping reverse faults offset turbidite beds and the trench floor near the base of the landward wall, extensional-mechanism earthquakes are associated with those faults that also offset the underlying oceanic crust (Stauder, 1968a,b; Frohlich et al, 1980).

Seaward Slope

The southern or seaward slope of the Aleutian Trench descends toward its flat axial floor in a series of west-trending ridges and troughs (Perry and Malahoff, 1981). The average inclination of this slope increases trenchward from about 2° between the 5,500 and 6,000 m isobaths, to 5° between the 6,500 and 7,000 m bathymetric contours (Figure 1). The surface of this descending surface is sawtoothed in profile, outlining asymmetric ridges with steep, south-facing slopes and gentler northfacing or trenchward inclined flanks. The strike of the asymmetric ridges is generally westward, parallel to the regional trend of magnetic anomalies (Malahoff and Erickson, 1969; Grim and Erickson, 1969).

Igneous crust seaward of the trench is overlain by a relatively thin (200 to 300 m) blanket of pelagic and hemipelagic deposits that can be traced northward beneath the beds of the trench fill. With increased depth of burial beneath the trench fill, acoustic resolution of the compacted sequence of pelagic beds is typically lost (Figures 8 and 10). Based on magnetic anomalies, and the sedimentary sections drilled at north Pacific DSDP sites (Kulm et al, 1973; Creager, 1973) and recovered in piston cores south of the trench (Horn, Delach, and Horn, 1969, Horn, Horn, and Delach, 1970; Opdyke and Foster, 1970), the pelagic sequence probably consists of hemipelagic mud

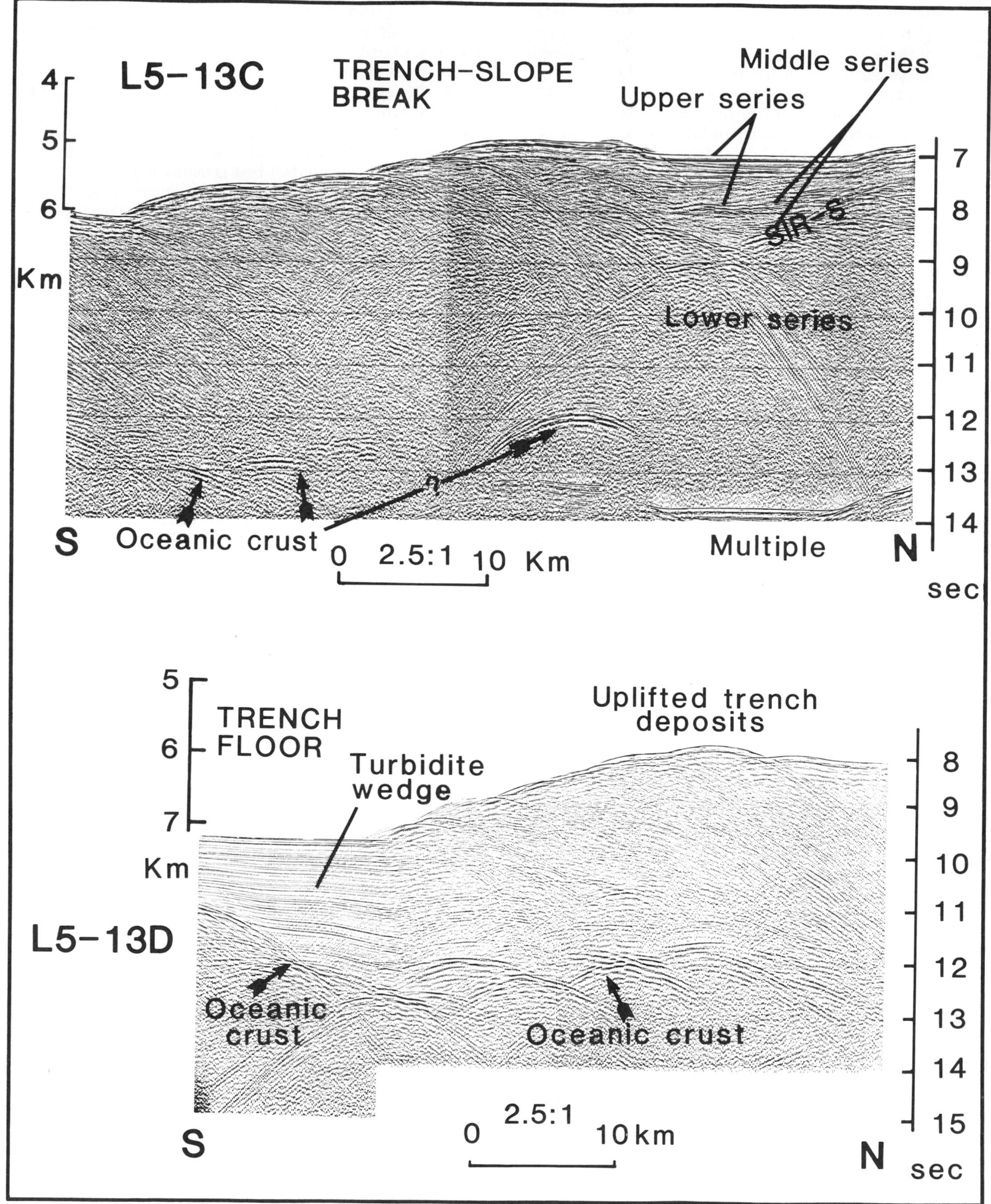

Figure 8 — Segments of 24-fold seismic reflection profiles (first-stack display) that cross the seaward edge of the Aleutian Terrace (upper) and lower part of the landward slope of the Aleutian Trench and adjacent trench floor (lower; Figure 1).

and diatom ooze of Neogene and Quaternary age, and abyssal clay and chalk beds of Paleogene age. The strata are temporally equivalent to the ridge's lower, middle, and upper series, but because the oceanic sequence was not deposited against or over the rock framework of the ridge, they are not considered to be part of its chronostratigraphic rock series.

PRELIMINARY INTERPRETATIONS

Arc Crest

Except for its distal forearc region (trench-slope break and landward trench slope) the structural fabric of the upper crustal rocks of the Aleutian Ridge is that of a massive antiform. Igneous processes have been dominant in constructing the antiform's bulk, which crests and is thickest beneath the axial line of the island arc. The igneous framework or massif of the antiform is roughly 200 to 250 km wide at its base, extending from at least the trench-slope break northward to the base of the arc's northern or Bering Sea flank (Figure 2). Sedimentary, mass movement, and to a lesser extent tectonic processes have shaped the upper several kilometers of the arc's Bering Sea and Pacific flanks, which are, respectively, the gently dipping northern and upper part of the southern flanks of the antiform. Although its broad geanticlinal form has been shaped more by constructional rather than deformational processes, the geomorphic profile of the arc's crestal area has been significantly reshaped by mid and late Cenozoic tectonic and erosional processes.

Three field relations displayed within the Amlia corridor record a two-stage evolutionary history for the crestal area of the island arc. First, the bulk of the rocks underlying the summit of the arc are volcanic with lesser volumes of sedimentary units of the lower series that accumulated along or near the crestal region of an upper Eocene arc. Second, Oligocene and early Miocene eruptive masses have not been identified, although they may have been removed by late Cenozoic erosion, but lower series rocks are intruded and thermally altered by early middle Miocene granodioritic plutons and are possibly overlain by coeval volcanic deposits. And third, the arc's broad summit platform reflects a surface of wave-base erosion that was cut chiefly during the past 10 to 13 million years across a subsiding and extensionally fractured framework of broadly warped Paleogene and lower Neogene rocks. Coincident with cutting and subsidence, the crestal area was extensionally rifted and thick upper Miocene and younger marine deposits infilled structurally deepening summit grabens.

The first relation documents that constructional magmatic activity had built the arc above sea level at least by 40 m.y. ago. Emplacement of volcanic and intrusive rocks had probably supported rapid growth of the arc in earlier Eocene or Paleocene time. Because there is no direct information about these older rocks (named the initial series by Scholl, Buffington, and Marlow, 1975a; see also Schmidt, 1978), the role of tectonism, for example, the stacking of obducted slabs of oceanic crust, in initiating or sustaining the natal growth of the arc cannot be evaluated. Also, the age when the ridge began to form has not been determined, although regional relations suggest latest Cretaceous or earliest Tertiary time (Scholl, Buffington, and Marlow, 1975a; Cooper, Scholl, and Marlow, 1976b).

The second field relation implies that rapid volcanic growth of the ridge diminished, or virtually ended, in late Paleogene and early Neogene time, and significant space-dependent thermal alteration of Paleogene volcanic rocks was caused by the early middle Miocene plutonic episode, and possibly by an older Oligocene plutonic event (Citron et al, 1980). The third field observation vouches that at least by post-middle Miocene time geomorphic destructional processes of erosion and extensional tectonism greatly dominated growth by constructional volcanism.

Despite many uncertainties about the exact timing of events, we infer from the results of corridor and broader regional studies that rapid growth of the arc by constructional magmatism greatly slowed, but did not end, in late Eocene or early Oligocene time (30 to 35 m.y. ago). A transition history of diminished or diminishing volcanism continued until about 15 m.y. ago, when an important igneous event resulted in the upper crustal emplacement of middle Miocene felsic plutons. The subsequent history of the ridge crest is marked by the regional cutting of the summit platform, which testifies to the effective end of the arc's magmatic growth. Final destruction of the ridge crest has been slowed only by the construction of the corridor's existing but volumetrically small volcanic centers beginning in latest Miocene time (Marsh, 1980). Extensional rifting of the summit platform to form Amlia and Amukta basins, which contain a thick fill of mostly upper Miocene and younger beds of the upper series, may be linked to the initiation of this late Cenozoic volcanic episode (Scholl, Marlow, and Buffington, 1975b).

Reduced to its fundamental evolutionary phases, the history of the corridor's crestal area can therefore be viewed as involving (1) an initial, pre-Oligocene construction phase characterized by arc volcanism, and (2) a subsequent and mostly destructional phase during which the relative (to volcanic growth) rate of crestal denudation and tectonic collapse progressively increased and, at least by 10 to 13 m.y. ago, ultimately exceeded volcanic growth. The chronostratigraphic column included on Figure 2 diagrammatically portrays this two-part history.

Northern or Bering Sea Flank

The northern flank of the arc or of the ridge's antiform appears to record a two-phase history of volcanic growth and subsequent sedimentary burial of its framework rocks. For example, the SIR-N horizon presumably delimits the top of lower series igneous rocks that were emplaced beneath the flanks of the arc in mostly pre-Oligocene time. Accumulation of the overlying sedimentary deposits of the middle series presumably attests to significant slowing of flank igne-

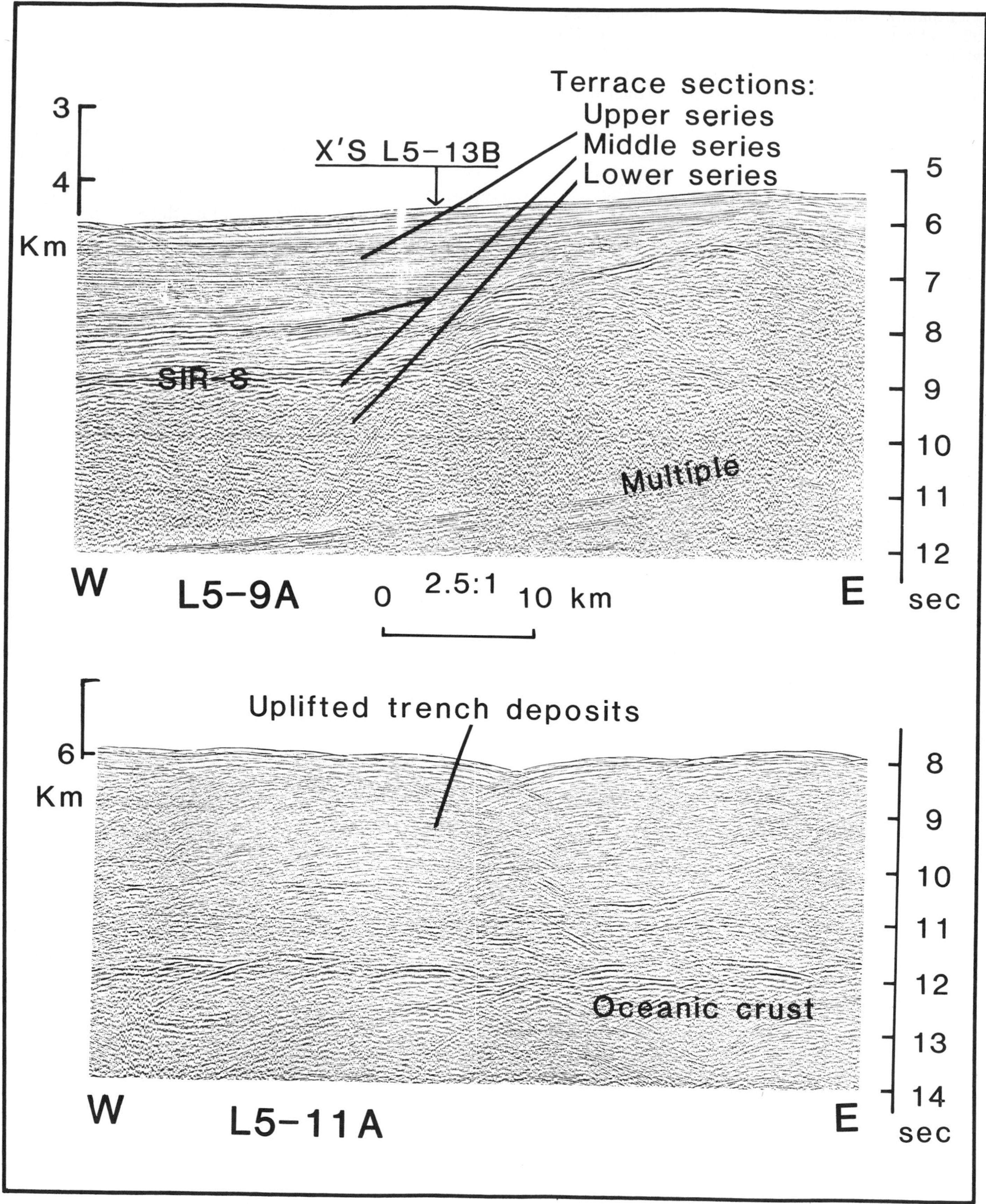

Figure 9 — Segments of 24-fold seismic reflection profiles (first-stack display) that trend parallel with the axis of the Aleutian Terrace (upper), and along the west-trending crest of a large bathymetric ridge that parallels the trench floor near the base of the landward trench slope (lower; Figure 1). Middle series beds appear to thin eastward over a thickened mass of lower series beds.

ous activity. Deposition of the middle series beds probably also signifies that by early Oligocene time large areas of the crest had built above sea level and processes of subaerial erosion, mass wasting, and explosive volcanism could thereby create large supplies of clastic debris. Alternatively, because middle series beds appear to thin over relatively flat-topped basement highs of lower series rocks, much of the width of the present northern flank may have subsided from a late Eocene position at or above sea level. If the prominent SIR-N horizon is in part a subaerial or shallow-water erosional unconformity, then it is likely that the structurally contiguous Umnak Plateau was also a shallowly submerged or subaerial terrane in Paleogene time.

Deposition of the stratified upper series, probably beginning in middle Miocene time, marks the end of the transition along the ridge crest from the constructional buildup of summit volcanism to the destructional dominance of subaerial erosion and wave-base flattening. In late Cenozoic time, owing to the effects of glaciation and oscillating sea levels, erosion of the ridge was accelerated. As a consequence large submarine canyons were cut into the northern flank and deep sea fans of arc detritus were deposited at its base (Scholl, Marlow, and Buffington, 1975b).

Pacific Flank and Aleutian Terrace

South of Amlia Island, Oligocene and lower Neogene deposits record the southward shedding of sandy and silty erosional detritus from a subaerially exposed and forested ridge crest. These stratified marine deposits of the middle series accumulated above massive volcanic and volcaniclastic rocks emplaced during the arc's pre-Oligocene phase of rapid magmatic growth.

The relatively thin (in comparison to the arc's northern flank) section of upper series deposits can be attributed to either the lack of deposition on the arc's steep (10°) southern flank, or to erosional stripping during the intense episode of late Cenozoic canyon cutting (Scholl, Marlow, and Buffington, 1975b). The arc's Pacific flank is nearly three times steeper than the gentler northern or Bering slope (3.5°). Considering that canyon cutting rather equally affected both flanks of the arc, we suppose that slope steepness may have been the dominant factor inhibiting the accumulation of a thick upper series on the arc's Pacific flank. Because this slope is underlain by a thick and prominently stratified section of middle series beds, the south-facing slope must have been tectonically steepened after about middle Miocene time. Evidence for this steepening in the southward tilt of the surface of the summit platform to a depth of 1000 m off Atka and nearby Adak Island.

Beneath the adjacent Aleutian Terrace the two-phase evolutionary history of the ridge's antiformal mass is recorded by the terrace's two-tiered structural fabric, a lower basement sequence of generally massive rock (lower series) and an overlying section of stratified sedimentary deposits (middle and upper series). The origin and geologic implications of the bowl-shaped SIR-S horizon, which separates lower and middle series deposits (Figure 10), are only vaguely understood. Gravity, magnetic and preliminary processed multifold reflection lines and refraction velocities imply that the SIR-S horizon overlies a complex but crudely layered mass of igneous rocks that extends southward from the crest of the arc to the upper part of the landward slope of the Aleutian Trench. Formation of the basinal profile of the SIR-S horizon trapped upper and possibly middle series deposits; and, as they thickened, especially during the past 10 to 13 m.y., the geomorphic form of the terrace evolved.

The original geometry of the surface of the lower series (SIR-S) cannot be confidently reconstructed from existing data. Therefore, it is unknown whether the terrace's basinal structure existed in early Tertiary time. But preliminary interpretations suggest that middle series beds of Oligocene age thin southward against the basement cores of presumed lower series rocks that underlie the terrace's seaward antiformal highs. We provisionally suppose that the middle series beds constitute the initial sedimentary deposits of the Aleutian Terrace basin, but these beds are fundamentally slope rather than basinal accumulations. They are presumably the transitional deposits laid down along the flanks of the arc during its change from rapid magmatic growth to extensional collapse and erosional leveling.

Upper series beds of the terrace's stratified sequence probably began to accumulate about 10 to 13 m.y. ago in middle or late Miocene time. These deposits, which are as thick as 3 km, are known to include pelagic and hemipelagic sediment rich in diatomaceous debris, turbidite beds, and massive debris flows (Figure 11). Erosion of the ridge crest and mass wasting of its sloping flank supplied the terrigenous beds of the upper series. Elevation of the outer antiformal highs relative to the terrace floor was rapid and coeval with the accumulation of the upper series. Evidence justifying this conclusion includes: 1) the deeply swaled profile of time-depositional surface of upper series beds, and the fact that these surfaces are most elevated along the seaward side of the terrace (Figures 2, and 11); 2) observations that at DSDP 186, which is located near the crest of an outer-terrace antiform, turbidite beds are common in lower and upper Pliocene beds whereas pelagic and hemipelagic beds dominate the overlying Pleistocene section; and 3) displaced blocks of middle Miocene terrace sediment occur in the upper Pliocene section at Site 186. Additionally, the late Cenozoic arching of Hawley Ridge (Figures 2 and 10), an outer-terrace antiform of exceptional size (200 km long, 25 km wide, and 1500 km high; Grow, 1973a), relative to the terrace floor effectively divides part of Atka Basin into inner and outer depocenters for Pleistocene beds (Figure 11).

The great size of Hawley Ridge impressively documents that tectonic processes in Quaternary time rapidly and regionally elevated the seaward edge of the Aleutian Terrace or caused substantial subsidence of its depositional floor.

Reflection profiles reveal both geomorphic and subsurface evidence that faulting is associated with the relative uplift of the outer antiformal highs. Steeply dipping subsurface faults have also been resolved along the base of the northern or landward flank of Hawley Ridge (Grow, 1973a; Figure 11). These west-striking breaks appear to be high-angle, down-to-basin normal faults, but initially migrated multifold records imply that the offset may reflect deep-seated reverse faulting. Grow (1973a) speculates that normal faulting adjacent to Hawley Ridge, and possibly other outer terrace antiforms, may be the surface expression of uplift caused by deep-seated thrusting. The hypothetical north-dipping thrusts, the likely deformational consequence of crustal underthrusting or subduction, are revealed on preliminary processed multifold records. Although few earthquakes occur within the forearc section, seismic activity is associated with the outer structural highs (LaForge and Engdahl, 1979; Chen, Frohlich, and Lathram, 1982).

Alternatively, because a prominent reflection horizon underlying the Pacific margin of northern Japan has been determined to be a subaerial unconformity of early and middle Tertiary age (von Huene et al, 1980), the similar appearing SIR-S horizon of the Amlia corridor is conceivably in part a deeply-subsided shallow water or even subaerial landscape of Eocene age. This possibility, which is supported by the relative shallow water middle Miocene fauna recovered at DSDP site 186 (Creager, Scholl and Creager, 1973), implies that the basin of the Aleutian Terrace may reflect a combination of original constructional relief produced by arc-type volcanism, and subsequent relief produced by differential subsidence presumably caused by deep-seated processes beneath the present forearc region.

Aleutian Trench

The inner or landward slope of the Aleutian Trench is underlain chiefly by rotated masses or blocks of gently deformed trench deposits. In agreement with the interpretations of Grow (1973a), migrated multifold records reveal that this accretionary wedge has been thickened by internal thrusting (McCarthy et al, 1982). The steep fronts (30 to 35°) of the step-like ridges that descend the trench's landward slope rise above the low-angle thrust planes as they approach the sea floor. However, beneath and seaward of the flat trench floor reflection profiles and extensional earthquakes document that the igneous oceanic crust is ruptured by high-angle, south-dipping normal faults. These antithetic (relative to the north dip of the oceanic crust) or up-to-arc basement faults may therefore assist in elevating trench deposits into the accretionary wedge. If extensional faulting disrupts the oceanic basement beneath the wedge, then some of the scarps of the trench's landward slope may reflect near-vertical offsets of the thrust slices by basement-controlled fracturing.

The nature of the contact of the accreted wedge with lower series rocks that underlie the vicinity of the trench-slope break has not been fully resolved. It seems likely, however, that the wedge has been thrust beneath uplifted lower series beds (Grow, 1973a; Figure 2). Considering that relative to the terrace floor the antiformal highs of the trench-slope break rose rapidly in post-middle Miocene time, it is likely that the accretionary wedge may only be a late Cenozoic structure.

The volumetric mass of the accreted wedge is a clue to its age. For example, within the 200-km-wide Amlia Corridor the volume of the accreted mass is roughly 20,000 cu km (200 x 40 x 2.5 km), the rate of underthrusting measured normal to the trench is 73 km/million years, and the typical thickness of the virtually undeformed turbidite beds underlying the trench floor at the base of the slope is 1.5 to 2.0 km. Accordingly, for a trench wedge of constant or steady-state thickness, the accreted mass of the adjacent inner trench slope is at least 1 m.y. old. This age, which must be regarded as a minimum one, is supported by the recovery of lower Quaternary slope deposits from the trench's landward wall (Station 15, Figure 1; Table 1).

Because the velocity structure of the accreted wedge is only slightly higher than that of the virtually undeformed trench deposits, a volume correction for tectonic dewatering of the accreted wedge is not considered to be an important age-increasing factor. More significantly, because it is unlikely that the trench wedge, which is presently dominated by rapidly deposited glacial age deposits (Scholl, 1974; Scholl, Vallier, and Stevenson, 1982a), has maintained a steady-state thickness (Helwig and Hall, 1974), the age of the accreted mass could easily be an order of magnitude older. Also, because sediment subduction has likely occurred, especially at times when the trench wedge was relatively thin (Scholl, Marlow, and Cooper, 1977), the volume of the preserved accreted prism is likely to be only a partial measure of the age of its oldest beds.

Based on the uplift history of the outer highs of the Aleutian Terrace, we estimate the oldest sediment in the accreted wedge is about 10 m.y. old. Accordingly, the turbidite deposits of older Tertiary trench wedges have been either subducted beneath the Aleutian Terrace or never deposited in the trench's corridor sector. This latter circumstance is considered the more probable because the trench's chief sediment-contributing area is thought to be the glaciated drainages of the eastern Gulf of Alaska (von Huene and Kulm, 1973; Scholl, 1974; Scholl, Vallier, and Stevenson, 1982a). Rapid delivery of terrigenous debris from these glaciated drainages began over the floor of the northeastern Pacific in middle Miocene time (Piper, 1973; von Huene and Kulm, 1973). It is likely, therefore, that at about the same time terrigenous debris was also shunted westward or downtrench to nourish a thickening wedge in the distant (2,000 km) Amlia sector. However, we are mindful of the possibility that turbidite deposition on the deep-sea floor of the Gulf of Alaska may have been widespread even in Eocene and Oligocene time (Hamilton, 1967; 1973; Scholl and Creager, 1973; Stevenson, Scholl; and Vallier, in press).

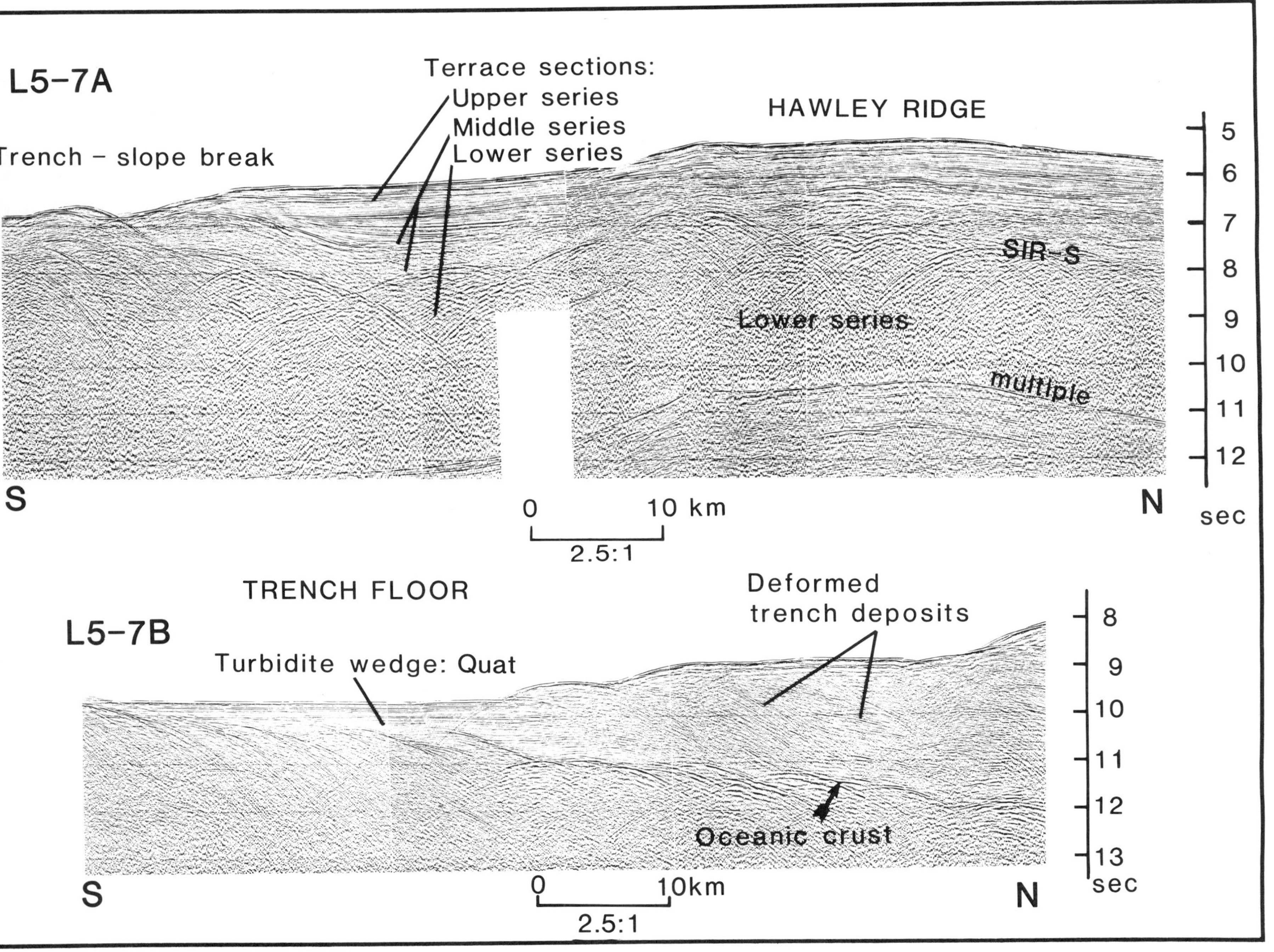

Figure 10 — Segments of 24-fold seismic reflection profiles (first-stack display) that cross the eastern end of Hawley Ridge, seaward side of the Aleutian Terrace (upper), and the floor and landward slope of the adjacent Aleutian Trench (lower; Figure 1).

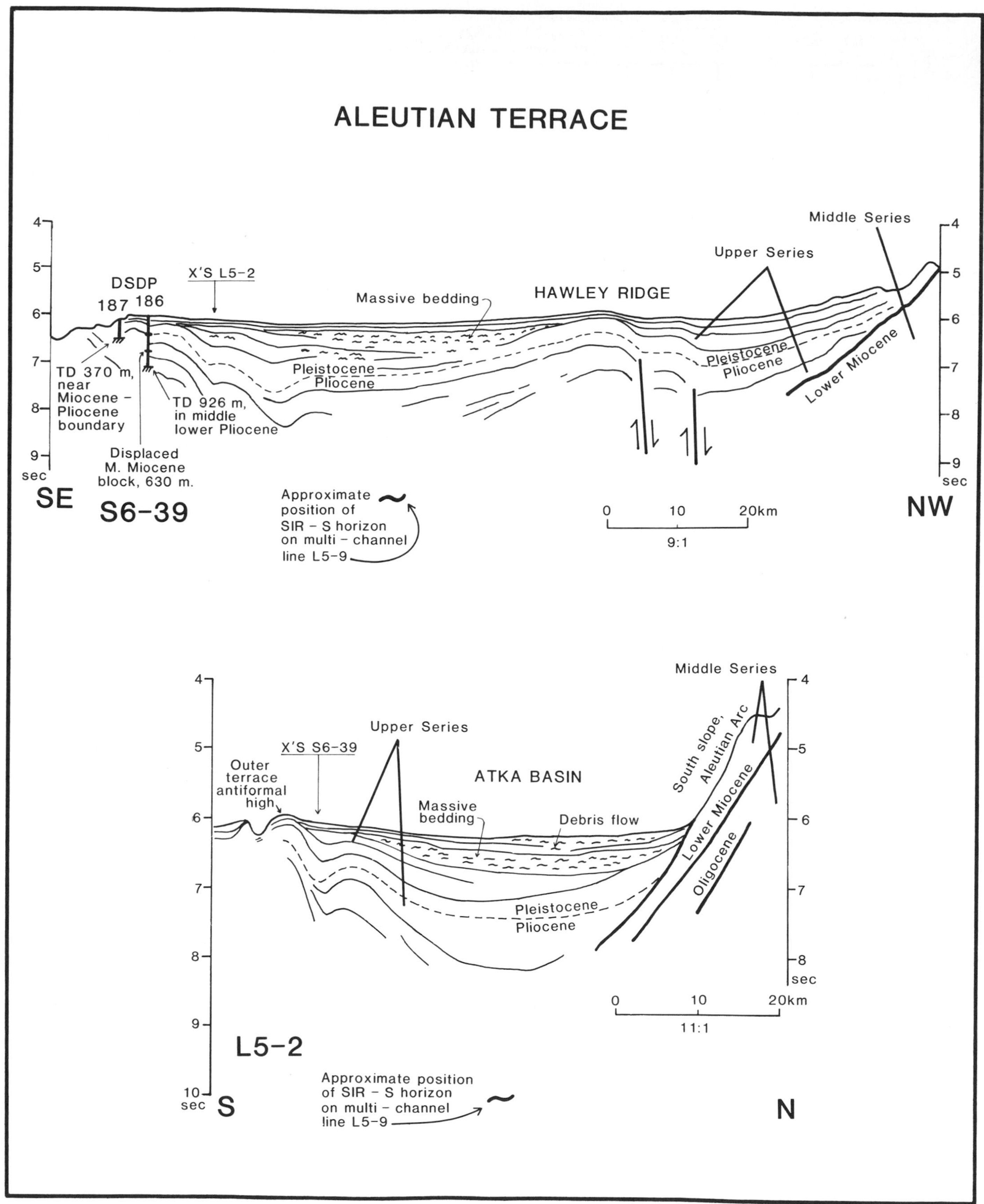

Figure 11 — Interpreted line drawings of single-channel reflection profiles that cross the Aleutian Terrace. Growth of outer antiformal highs and of Hawley Ridge in late Cenozoic time shifted and split depocenters for upper series deposits of Pliocene and Quaternary age (see also, Marlow et al, 1973, Fig. 4; and Grow, 1973a, Fig. 4).

Much of the undulatory flexuring of the trench's turbidite wedge of upper series beds can be ascribed to differential compaction over buried basement relief (Scholl, Vallier, and Stevenson, 1982a). However, high-angle ruptures and abrupt flexuring also attest to tectonic deformation. Upper crustal earthquakes imply that normal faulting has caused the progressive elevation of more northerly or ridgeward blocks of basement and overlying turbidite deposits. Except near the base of the landward trench slope, the dominance of extensional faulting of the trench's turbidite wedge is different, therefore, than the effects of compressional deformation described for the Peru-Chile Trench (Prince et al, 1974; Hussong, Odegard, and Wipperman, 1975, Hussong et al, 1976; Prince and Kulm, 1975; Prince and Schweller, 1978; Schweller, Kulm and Prince, 1981), and the middle America Trench (Heezen and Rawson, 1977; Shipley et al, 1980; Watkins et al, 1981).

Normal faulting in the Amlia corridor presumably records stress relief of the Pacific plate as it bends northward and downward toward the Aleutian subduction zone (Abe, 1972; Caldwell et al, 1976; Spence, 1977; Jones et al, 1978; Chapple and Forsyth, 1979; Frohlich et al, 1980). Moreover, except along the base of the landward wall, west-trending folds in the Amlia trench fill, whether formed by differential compaction or basement warping, do not record a component of compression attributable to the convergence of the Pacific and North America plates. Instead, these subtrench-floor structures are either directly or indirectly caused by extensional faulting along preexisting zones of crustal weakness that parallel magnetic anomalies. These anomalies are fortunately parallel or subparallel to the strike of the trench within the Amlia corridor (Figure 1). They were originally formed south of the Pacific-Kula spreading center in latest Cretaceous and earliest Tertiary time.

SUMMARY AND CONCLUDING REMARKS

Except for the area seaward of the trench-slope break, the fundamental structure of the Aleutian Ridge within the Amlia corridor is that of a massive antiform. Constructional magmatic processes created the bulk of its architecture in early and middle Paleogene time, especially that great body of igneous rocks that forms the geanticlinal mass of the island arc. These rocks, the lower series, built the ridge above Pacific oceanic crust of probable Cretaceous age. It is not known if tectonic or magmatic processes formed the ridge's deep-seated root or natal structures. But at least by 40 m.y. ago (latest middle Eocene) the accumulation of lower series rocks had built the island-arc geanticline above the ridge's 200 to 250-km-wide foundation to reach sea level, where a tree-covered crown of eruptive centers issued mafic to felsic products similar to those erupted today. Beneath the submerged flanks of this emergent ridge lower series framework rocks were subsequently buried by Oligocene and younger sedimentary deposits of the middle and upper series. Except near intrusive bodies, the lower and middle series rocks of the present island arc (i.e., the ridge exclusive of its forearc basin and landward trench slope) have not been significantly thermally altered or tectonized, and upper series beds (middle Miocene to present) are virtually undeformed sequences of basinal and slope mantling deposits.

Magmatic growth of the ridge along the Amlia corridor appears to have effectively ceased over its off-axis flanks and significantly waned along the crest of the arc in late Eocene or early Oligocene time. By the end of early middle Miocene time (14 to 15 m.y. ago), at the conclusion of a plutonic episode, the rate of crestal volcanism had diminished to the point where the arc's summit area was being rapidly leveled by erosional processes. Despite renewed arc construction beginning 5 to 6 m.y. ago along the existing chain of Aleutian volcanoes, destructional forces during the past 10 to 13 million years have been the dominant geologic processes shaping the arc's geomorphic form. These processes include the subaerial and glacial denudation of insular areas, the cutting of the summit platform by wave-base erosion, and the canyon-scarring of its sloping flanks by sluicing sediment masses.

Because the summit platform was cut across elevated and lithified marine deposits of Paleogene and possible early Neogene age, waning of voluminous magmatic growth was accompanied and probably in part preceeded by crestal uplift and broad warping. Erosional flattening of the summit area presumably attests to subsequent submergence — much as the summit platforms of guyots are evidence for their subsidence below wave-base erosion. Within the Amlia corridor subsidence was differential and marked by extensional faulting, block rotation, and the foundering of grabens or half-grabens along the northern side of the summit platform. These summit basins, which are filled with upper series deposits supplied by crestal denudation, are aligned with the trends of the arc's late Cenozoic volcanic centers. The unconformity that separates upper series deposits from broadly deformed and thermally altered lower series rocks is the most prominent structural discordance recorded along the crestal region of the arc.

Throughout its early destructional phase (roughly Oligocene through early middle Miocene time) the sloping Bering Sea and Pacific flanks of the island arc received a 1 to 3-km-thick blanket of stratified terrigenous and pelagic debris. These middle series deposits buried the irregular and acoustically prominent (SIR-S and -N reflection horizons) upper surface of the lower series. Subsurface relief exhibited by the lower series is as much as 3 km, this relief is in part tectonic in origin, but some may be original constructional relief between eruptive centers. Considering the implications of the DSDP findings off northern Japan, it is possible that the deeply submerged SIR-S and -N horizons were originally at shallower depths. Accordingly, post-Eocene submergence and differential subsidence of the forearc and backarc flanks of the ridge may have created much of the geometry of its sediment-buried basement surface.

Beneath the ridge's forearc area the surface of lower

series rocks was prominently warped after about middle Miocene time to form most of the basement relief of the structural basin of the Aleutian Terrace. Concommitantly, the crestal area of the ridge subsided regionally and the resultant cutting of the summit platform supplied large volumes of sediment to the terrace area. It filled rapidly with as much as 3 km of upper Neogene and Quaternary beds of the ridge's upper series. As the terrace structure continued to form and fill, the crestal area was extensionally fractured and volcanic activity rekindled along the line of the existing eruptive centers. Although tectonic activity obviously affected the ridge crest, the principal focus of tectonism during the past 10 to 13 million years has nonetheless been along the seaward flank of the Aleutian Terrace and adjacent landward slope of the Aleutian Trench. Thrust faulting is associated with the relative elevation of the terrace's outer highs, but beneath the terrace floor the surfaces of the deeply downwarped SIR-S horizon and overlying strata of the middle and upper series are not significantly offset by faults.

It is important to emphasize that many forearc basins geomorphically similar to the Aleutian Terrace are believed to be underlain by both a basement of offscraped or accreted oceanic deposits and igneous crust, and a basin that formed by the subsidence of the deep underlying surface of the oceanic lithosphere (for example, Karig and Sharman, 1975; and Seely, 1979). Although this model may yet prove its applicability to the Amlia corridor, existing geophysical and geological data indicate that the terrace's basement framework is constructed of arc-type igneous rocks that can be traced northward from south of the trench-slope break to outcrops of volcanic rocks of Eocene age on Amlia and Atka islands. This interpretation supports the expansive basement framework of arc volcanic rocks modeled by Grow (1973a, Model A). If, in fact, the terrace and its seaward rampart of antiformal highs are underlain by igneous rocks of the arc's lower series, then the origin of the terrace's structural basement cannot be linked to tectonic accretion of oceanic rocks at the base of the early Tertiary Aleutian Ridge. However, as noted, the basinal structure of the terrace may in part reflect differential subsidence of the ridge's forearc areas, and this subsidence could be effected by changes in the position or geometry of the subducting oceanic crust.

Tectonic accretion of trench deposits has formed the lower part of the landward slope of the Aleutian Trench. Accretion probably began no earlier than late Miocene time when west-flowing turbidity currents transporting terrigenous detritus from glaciated drainages of the Gulf of Alaska began to reach the trench's Amlia sector. Prior to this time relatively thin trench sections of pelagic and terrigenous deposits were probably subducted rather than offscraped — which is the case off northern Japan (von Huene, Langseth, and Nasu, 1980) and Guatemala (von Huene, Aubouin et al, 1980) during the past 10 to 20 million years. In early and middle Tertiary time the landward slope of the Aleutian Trench (in the Amlia corridor) may have consisted chiefly of arc-related lower series rocks that were being tectonically foreshortened and thinned in place by processes of subduction erosion (Scholl et al, 1980).

Turbidite deposits underlying the existing trench floor are exceptionally thick (2.0 to 4.0 km) owing to the effects of oscillating Pleistocene sea levels and deep erosion caused by extensive glaciation of Gulf of Alaska drainages. Beneath the flat trench floor of the Amlia corridor the turbidite wedge has been tectonically deformed chiefly by basement faulting associated with extensional earthquakes. Bending of the Pacific plate into the Aleutian subduction zone evidently causes extensional rupturing across normal fault planes that dip south or antithetic to the regional northward dip of the oceanic crust. The consequent up-to-arc offsets may thereby assist in elevating trench deposits into the toe of the accretional wedge, where they are thickened or stacked but not significantly deformed by north-directed thrusting. It is likely that basement extension causes normal faults to locally cut the thrust-thickened accretionary wedge.

The horizontal or north-to-south structural fabrication of the Aleutian Ridge in the Amlia corridor can be viewed as involving two separate but tectonically linked processes: 1) the construction of the ridge's antiformal massif of mostly igneous rocks, and 2) the formation of the ridge's outer forearc region of the trench-slope break and adjacent landward trench slope. The evolution of the ridge's antiform, fundamentally a vertical process, was itself completed in two steps, a mostly Eocene and possibly Paleocene phase of voluminous growth by submarine volcanism across a 200 to 250-km-wide base, and a subsequent history of greatly diminished magmatic growth rate and the increasing dominance of erosional reduction attended by summit and possibly extensive flank subsidence. Complex deformation played no part in the vertical evolution of the ridge's island arc geanticline. However, tectonism was important to the structural evolution and horizontal lengthening of the ridge's forearc region.

The forearc basin of the Aleutian Terrace also evolved in a two stage manner: 1) an initial, early Tertiary formation of its basement framework colaterally with the more rapid buildup of the igneous core of the adjacent island arc, and 2) the subsequent differential warping of the terrace's framework rocks to form its basin chiefly in post middle Miocene time. The adjacent landward slope of the Aleutian Trench also exhibits a compound evolution, the formation of a landward framework of lower series rocks and a younger, seaward mass of trench-floor deposits that have been thrust against and, presumably, partially beneath this framework. It seems likely that the tectonic emplacement of the accretionary wedge may have contributed to the inplace deformation of lower series rocks and the consequent formation of much of the terrace's basinal structure.

The two-part horizontal and vertical formative rhythm of the Aleutian Ridge is probably covariant with fundamental changes in the tectonic setting of the Pacific's northern rim. This presumption has been the subject of lengthy speculations by Grow and Atwater (1970), Marlow et al (1973), Scholl, Buffington, and Marlow (1975a), DeLong and McDowell (1975), and DeLong, Fox and McDowell, (1978), and Scholl, Vallier, and Stevenson (1982b). Schmidt (1978), however, views the evolution of the ridge as unrelated to the presumed general consequence of plate tectonics.

Most likely, the double faceted, or on-off, evolution of the ridge's magmatic framework is linked to either the middle Eocene change in relative (to hotspots) motion of the Pacific plate recorded by the bend in the Emperor-Hawaii Seamount chain (Scholl, Vallier, and Stevenson, 1982b), or perhaps the earlier or coeval reorganization of the Kula-Pacific-Farallon spreading ridges (Byrne, 1979; Butler and Coney, 1981). Reduction of igneous growth could have been caused by the subduction of the Kula-Pacific spreading center beneath the Aleutian Ridge in either late Paleocene or Oligocene time (Marlow et al, 1973; DeLong, Fox, and McDowell, 1978). We note here that within the Amlia corridor the transition 30 to 35 m.y. ago from rapid volcanic growth to the increasing domination of erosional destruction appears to have occurred at least 10 million years after the time of major plate motion changes and reorganizations and the earliest time of ridge subduction. If the Kula-Pacific ridge underthrust the Aleutian Ridge in Oligocene rather than Paleocene time, it had ceased being a spreading center for at least 20 million years, and a melt-throttling thermal effect would not be expected to attend its subduction as part of the Pacific plate.

The second, or deformational, stage of the evolution of the ridge's outer forearc region may be related to accelerated motion between the Pacific and North American plates beginning in early Neogene time. This circumstance could speculatively cause colateral increases in sediment supply to the Aleutian Trench from elevated Alaskan drainages, extensional collapse and heightened erosional leveling of the arc's summit area in association with minor volcanic growth, and, in conjunction with the accretion of thick trench deposits to its landward slope, the transmission of compressional stresses to the rock fabric of the ridge's forearc area and the consequent formation of the basin of the Aleutian Terrace.

Crucial to relating the geologic evolution of the Aleutian Ridge to its Pacific tectonic setting is the accurate dating of the cessation time of rapid magmatic growth and the beginning of progressive destruction of its summit area; a process that has been slowed only by two brief periods of crestal magmatic growth. Equally crucial to determining the structural origin of the ridge's magnificent forearc basin is the resolution of the age and geologic origin of the SIR-S horizon, and the dating of the onset of tectonic accretion of trench deposits to the ridge's seaward edge.

ACKNOWLEDGMENTS

We wish to acknowledge the help of many of our colleagues, but especially Roland von Huene, U.S. Geological Survey, and J. Casey Morre, University of California, Santa Cruz, who critically reviewed our interpretations, despite their provisional and obviously speculative nature. Paleontologic age determinations of sedimentary samples were provided by John Barron (diatoms), David Bukry (nannos), and Norman Frederiksen (spore and pollen). M. L. (Bill) Silberman supervised K/Ar age determinations on igneous samples from Amlia and Atka Islands. Reduction of multichannel seismic reflection data was supervised by Don Tompkins and toiled over by Dennis Mann. Reduction of most of the gravity and magnetic data was executed by John Childs. Computation of sonobuoy refraction and multichannel reflection velocities were supervised by Jill McCarthy; and Merid Dates carried out most of the sedimentological and X-ray analyses on rock and sediment samples. Dorothy Sicard handled all word-processing phases of manuscript preparation and editing. We extend our genuine thanks to these associates and other colleagues who generously devoted their time and talents to assist us.

REFERENCES CITED

Abe, K. M., 1972, Lithospheric normal faulting beneath the Aleutian Trench: Physics of the Earth and Planetary Interiors, v. 3, p. 190-198.

Ben-Avraham, Z., and A. K. Cooper, 1982, The early evolution of the Bering Sea by collision of ocean rises and north Pacific subduction zones: Geological Society of America Bulletin, v. 92, p. 485-495.

Butler, R. F., and P. J. Coney, 1981, A revised magnetic polarity time scale for the Paleocene and early Eocene and implications for Pacific plate motion: Geophysical Research Letters, v. 8, p. 301-304.

Byrne, T., 1979, Late Paleocene demise of the Kula-Pacific spreading center: Geology, v. 7, p. 341-344.

Caldwell, J. G., et al, 1976, On the applicability of a universal elastic trench profile: Earth and Planetary Science Letters, v. 31, p. 239-246.

Carder, D. S., et al, 1967, Seismic wave arrivals from Longshot, 0° to 27°: Seismological Society America Bulletin, v. 57, p. 573-590.

Chapple, W. M., and D. W. Forsyth, 1979, Earthquakes and bending of plates at trenches: Journal of Geophysical Research, v. 84, p. 6729-6747.

Chen, A. T., C. Frohlich, and G. V. Lathram, 1982, Seismicity of the forearc marginal wedge (accretionary prism): Journal of Geophysical Research, v. 87, p. 3679-3690.

Citron, G.P., et al, 1980, Tectonic significance of early Oligocene plutonism on Adak Island, central Aleutian Islands: Geology, v.8, p. 375-379.

Coats, R. R., 1950, Volcanic activity in the Aleutian Arc: U.S. Geological Survey Bulletin 974-B, p. 35-47.

———, 1956, Reconnaissance geology of some western Aleutian Islands, Alaska: U.S. Geological Survey Bulletin, 1028-E, p. 83-100.

———, 1962, Magma type and crustal structure in the Aleutian arc, *in* G. A. MacDonald, and H. Kuno, eds., The crust of the Pacific Basin: American Geophysical Union, Monograph 6, p. 92-109.

Cooper, A. K., M. S. Marlow, and A. W. Scholl, 1976a, Mesozoic magnetic lineations in the Bering Sea marginal basin: Journal of Geophysical Research, v. 81, p. 1916-1934.

———, ———, and Z. Ben-Avraham, 1982, Multichannel seismic evidence bearing on the origin of Bowers Ridge,

Bering Sea: Geological Society of America Bulletin, v. 92, p. 474-484.

———, D. W. Scholl, and M. S. Marlow, 1976b, Plate Tectonic model for the evolution of the Bering Sea Basin: Geological Society of America Bulletin, v. 87, p. 1119-1126.

Creager, J. S., et al, 1973, Initial reports of the deep sea drilling project, v. 19: Washington, D.C., U.S. Government Printing Office, 913 p.

DeLong, S. E., P. J. Fox, and F. W. McDowell, 1978, Subduction of the Kula Ridge at the Aleutian Trench: Geological Society of America Bulletin, v. 89, p. 83-95.

———, and F. W. McDowell, 1975, K-Ar ages from the Near Islands, western Aleutian Islands, Alaska; indication of a mid-Oligocene thermal event: Geology, v. 3, p. 691-694.

Department of Commerce, 1973, NOS Seamap Series, No. 16648-14B, and 16648-14M, Pacific Ocean (Scale 1:1,000,000). Washington, D.C., National Oceanographic and Atmospheric Administration.

Erickson, B. H., and P. J. Grim, 1969, Profiles of magnetic anomalies south of the Aleutian Island arc: Geological Society of America Bulletin, v. 80, p. 1387-1390.

Frohlich, C., et al, 1980, Ocean bottom seismograph measurements in the central Aleutians: Nature, v. 286, p. 144-145.

Gates, O., G. D. Fraser, and G. L. Snyder, 1954, Preliminary report on the geology of the Aleutian Islands: Science, v. 119, p. 446-447.

———, and W. Gibson, 1956, Interpretation of the configuration of the Aleutian Ridge: Geological Society of America Bulletin, v. 67, p. 127-146.

Gibson, W. M., and H. Nichols, 1953, Configuration of the Aleutian Ridge, Rat Islands-Semisopochnoi Island to west of Buldir Island, Alaska: Geological Society of America Bulletin, v. 64, p. 1173-1187.

Grim, P. J., and B. H. Erickson, 1969, Fracture zones and magnetic anomalies south of the Aleutian Trench: Journal of Geophysical Research, v. 76, p. 1488-1494.

Grow, J. A., 1973a, Crustal and upper mantle structure of the central Aleutian arc: Geological Society of America Bulletin, v. 84, p.-2169-2192.

———, 1973b, Implications of deep sea drilling, sites 186 and 187 on island arc structure, *in* J. S. Creager, D. W. Scholl, et al, eds., Initial reports of the deep sea drilling project, v. 19: Washington, D.C., U.S. Government Printing Office, p. 799-803.

Grow, J., and T. Atwater, 1970, Mid-Tertiary tectonic transition in the Aleutian Arc: Geological Society of America Bulletin, v. 81, p. 3715-3722.

Hamilton, E. L., 1967, Marine geology of abyssal plains in the Gulf of Alaska: Journal of Geophysical Research, v. 72, p. 4189-4213.

———, 1973, Marine Geology of the Aleutian Abyssal Plain: Marine Geology, v. 14, p. 295-325.

Hayes, D. E., and M. Ewing, 1970, Pacific boundary structure, *in* Maxwell, ed., The sea, v. 4; the earth beneath the sea-concepts, part 2, regional observations: New York, Wiley-Interscience, p. 29-72.

———, and J. R. Heirtzler, 1968, Magnetic anomalies and their relation to the Aleutian Island arc: Journal of Geophysical Research, v. 73, p. 4637-4646.

Heezen, B. C., and M. Rawson, 1977, Visual observations of the sea floor subduction line in the Middle America Trench: Science, v. 196, p. 423-426.

Hein, J. R., and H. McLean, 1980, Reconnaissance geology of Atka Island, Central Aleutian Islands, Alaska: Cordilleran Meeting, Geological Society of America, Abstract with Programs, v. 12, n. 3, p. 110.

———, ———, and T. L. Vallier, 1981, Reconnaissance geologic map of Atka and Amlia islands, Alaska: U.S. Geological Survey, Open-File Report 81-159.

Helwig, J., and G. A. Hall, 1974, Steady-state trenches?: Geology, v. 2, p. 309-316.

Holmes, M. L., R. E. von Huene, and D. A. McManus, 1972, Seismic reflection evidence supporting underthrusting beneath the Aleutian Arc: Journal of Geophysical Research, v. 77, p. 959-964.

Horn, D. R., M. N. Delach, and B. M. Horn, 1969, Distribution of volcanic ash layers and turbidites in the north Pacific: Geological Society of America Bulletin, v. 80, p. 1715-1724.

———, B. M. Horn, and M. N. Delach, 1970, Sedimentary provinces of the north Pacific, *in* J. D. Hays, ed., Geological investigations of the north Pacific: Geological Society of America, Memoir 126, p. 1-21.

Hussong, D. M., 1976, Crustal structure of the Peru-Chile Trench; 8°12'S latitude, *in* G. H. Sutton, M. H. Manghnani, and R. Moberly, eds., The geophysics of the Pacific Ocean basin and its margin (Woollard volume): American Geophysical Union, Geophysical Monograph 19.

Hussong, D. M., M. E. Odegard, and L. K. Wipperman, 1975, Compressional faulting of the oceanic crust prior to subduction in the Peru-Chile Trench: Geology, v. 3, p. 601-604.

Jones, G. M., et al, 1978, Fault patterns in outer trench walls and their tectonic significance: Journal of Physics of the Earth, v. 26, p. 85-101.

Karig, D. E., and G. F. Sharman, 1975, Subduction and accretion in trenches: Geological Society of America Bulletin, v. 86, p. 377-389.

Kulm, L. D., et al, 1973, Initial report of the deep sea drilling project, v. 18: Washington, D. C., U.S. Government Printing Office, 1077 p.

Kuno, H., 1968, Differentiation of basalt magmas, *in* H. H. Hess, and A. Poldervaart, eds., Basalts: New York, Interscience, John Wiley and Sons, v. 2, p. 623-688.

LaForge, R. L., and E. R. Engdahl, 1979, Tectonic implications of seismicity in the Adak Canyon region, central Aleutians: Bulletin of the Seismological Society of America, v. 69, p. 1515-1532.

Malahoff, A., B. H. Erickson, 1969, Gravity anomalies over the Aleutian Trench: Transactions, American Geophysical Union, v. 50, p. 552-555.

Marlow, M. S., et al, 1973, Tectonic history of the central Aleutian arc: Geological Society of America Bulletin, v. 84, p. 1555-1574.

Marsh, B. D., 1980, Geology and petrology of northern Atka, Aleutian Islands, Alaska: Geological Society of America, Abstracts with Programs, v. 12, n. 7, p. 476.

McCarthy, J., et al, 1982, Mechanisms of accretion along the Aleutian Trench: Eos, Transactions, American Geohysical Union, v. 63, p. 1115.

McLean, H., et al, 1980, Geology of Amlia Island, Aleutian Islands, Alaska: Cordilleran Meeting, Geological Society of America, Abstracts with Programs, v. 12, n. 32, p. 119.

Minister, B. J., and T. H. Jordan, 1978, Present-day plate motions: Journal of Geophysical Research, v. 83, p. 5332-5354.

Nichols, H., and R. B. Perry, 1966, Bathymetry of the Aleutian Arc, Alaska, scale 1:4000,000: U.S. Coast and Geodetic Survey Monogram v. 3, maps.

Opdyke, N. D., and J. H. Foster, 1970, Paleomagnetism of cores from the north Pacific, *in* J. D. Hays, ed., Geological investigations of the north Pacific: Geological Society of America, Memoir 126, p. 83-119.

Perry, R. B., and A. Malahoff, 1981, Micromorphology and microtectonics of the Aleutian Trench to the south of Adak, Alaska, as interpreted from Seabeam data: Eos, American Geophysical Union, Transactions, v. 62, p. 305.

———, and H. Nichols, 1966, Geomorphology of Amka Basin, Aleutian Arc: Geographic Reviews, v. 56, p. 570-576.

Piper, D. J., R. von Huene, and J. R. Duncan, 1973, Late Quarternary sedimentation in the active eastern Aleutian Trench: Geology, v. 1, p. 19-22.

Prince, R. A., and L. D. Kulm, 1975, Crustal rupture and the initiation of imbricate thrusting in the Peru-Chile Trench: Geological Society of America Bulletin, v. 87, p. 1639-1653.

Prince, R. A., et al, 1974, Significance of uplifted turbidite basins on the seaward wall of the Peru Trench: Geology, v. 2, p. 601-611.

———, and W. J. Schweller, 1978, Dates, rates and angles of faulting in the Peru-Chile Trench: Nature, v. 271, p. 743-745.

Schmidt, O. A., 1978, Akademiya Tektonika Geologicheskiy Komandorskikh Ostrovoi i struktura Aleutskoi Griady: Moscow Akademiya Nauk SSSR Geological Institut, Trudy, v. 320, Izdatel'stvo Nauka, 100 p. (in Russian).

———, M. Ya. Serova, and L. M. Dolmatova, 1973 stratigraphy and paleontological features of the volcanic rock series on the Komandorski Islands: Akademiya Nauk SSSR Izyestiya, Seriya Geologicheskaya n. 11, p. 77-78.

Scholl, D. W., 1974, Sedimentary sequences in north Pacific trenches, *in* C. A. Burk, and C. L. Drake, eds., The geology of continental margins: New York, Springer-Verlag, p. 493-504.

———, and J. S. Creager, 1973, Geologic synthesis of Leg 19 (DSDP) results; far north Pacific, Aleutian Ridge, and Bering Sea, *in* J. S. Creager, D. W. Scholl, et al, eds., Initial reports of the deep sea drilling project, v. 19: Washington, D. C., U.S. Government Printing Office, p. 897-913.

———, H. G. Greene, and M. S. Marlow, 1970, Eocene age of the Adak Paleozoic(?) rocks, Aleutian Islands, Alaska: Geological Society America Bulletin, v. 81, p. 3583-3592.

———, E. D., Buffington, and M. S. Marlow, 1975a, Plate tectonics and the structural evolution of the Aleutian-Bering Sea region, *in* R. B. Forbes, ed., Contributions to the geology of the Bering Sea Basin and adjacent regions: Geological Society of America Special Paper 131, p. 1-31.

———, M. S. Marlow and E. C. Buffington, 1975b, Summit basins of the Aleutian Ridge, North Pacific: Bulletin, v. 59, p. 799-816.

———, ———, and A. K. Cooper, 1977, Sediment subduction and offscraping at Pacific margins, *in* M. Talwani, and W. C. Pitman, III, eds., Island arcs, deep sea trenches, and back-arc basins: Washington, D. C., Maurice Ewing Series, 1, American Geophysical Union, p. 199-210.

———, T. L. Vallier, and A. J. Stevenson, 1982a, Sedimentation and deformation in the Amlia sector of the Aleutian Trench: Marine Geology, v. 48, p. 105-134.

———, ———, and ———, 1982b, First-order effects of Tertiary interactions between the Pacific and North American plates — evidence from the Aleutian Ridge: Eos, Transaction of the American Geophysical Unions, v. 63, p. 913.

———, et al, 1976, Episodic Aleutian Ridge igneous activity, implications of Miocene and younger submarine volcanism west of Buldir Island: Geological Society of America Bulletin, v. 87, p. 547-554.

———, et al, 1980, Sedimentary masses and concepts about tectonic processes at underthrust ocean margins: Geology, v. 8, p. 564-568.

Schweller, W. J., L. D. Kulm, and R. A. Prince, 1981, Tectonics, structures and sedimentary framework of the Peru-Chile Trench, *in* L. D. Kulm, et al, eds., Studies of the Nazca plate and adjacent convergence zone: Geological Society of America, Memoir 154, p. 323-350.

Seely, D. R., 1979, The evolution of structural highs bordering major forearc basins, *in* J. S. Watkins, eds., Geological and geophysical investigations of continental margins: AAPG Memoir 29, p. 245-260.

Shipley, T. H., et al, 1980, Continental margin and lower slope structures of the Middle America Trench near Acapulco (Mexico): Marine Geology, v. 35, p. 68-82.

Shor, G. G., Jr., 1964, Structure of the Bering Sea and the Aleutian Ridge: Marine Geology, v. 1, p. 213-219.

Spence, W., 1977, The Aleutian arc; tectonic blocks, episodic subduction, strain diffusion, and magma generation: Journal of Geophysical Research, v. 82, p. 213-230.

Stauder, W., 1968a, Tensional character of earthquake foci beneath the Aleutian trench with relation to sea floor spreading: Journal of Geophysical Research, v. 73, p. 7693-7701.

———, 1968b, Mechanism of the Rat Island earthquake sequence of February 4, 1965, with relation to island arcs and sea-floor spreading: Journal of Geophysical Research, v. 73, p. 3847-3858.

Stevenson, A. J., D. W. Scholl, and T. L. Vallier, in press, Tectonic and geologic implications of the Zodiac fan, Aleutian Abyssal Plain, northeast Pacific: Geological Society of America Bulletin.

Stewart, R. J., 1978, Neogene volcaniclastic sediments from Atka Basin, Aleutian Ridge: AAPG Bulletin, v. 62, p. 87-97.

von Huene, R., 1974, Modern trench sediments, *in* C. A. Burk, and C. L. Drake, eds., The geology of continental margins: New York, Springer-Verlag, p. 207-211.

———, et al, 1980, Leg 67; The deep sea drilling project mid-America trench transect off Guatemala: Geological Society of America Bulletin, part I, v. 91, p. 421-432.

———, and L. D. Kulm, 1973, Tectonic summary of Leg 18, *in* L. D. Kulm, et al, eds., Initial reports of the deep sea drilling project, v. 18: Washington, D. C., U.S. Government Printing Office, p. 961-976.

———, M. Langseth, and N. Nasu, 1980, Summary, Japan Trench transect, *in* Scientific party: Initial reports of the deep sea drilling project, v. 56, part 1: Washington, D.C., U.S. Government Printing Office, p. 473-488.

Watkins, J. S., et al, 1981, Accretion, underplating, subduction and tectonic evolution, middle America trench, southern Mexico — results from Leg 66 DSDP: *in* Paris, Proceedings, 26th International Geological Congress, Geology of Continental Margins Symposium: Oceanologica Acta, May, p. 213-294.

Watts, A. B., 1975, Gravity field of the northwest Pacific Ocean basin and its margin; Aleutian Island arc-trench system: Geological Society of America Map and Chart Series, MC-10.

Emplacement of the Zambales Ophiolite into the West Luzon Margin

W. J. Schweller*
D. E. Karig
Department of Geological Sciences
Cornell University
Ithaca, New York

The Zambales Ophiolite in western Luzon is a large fragment of oceanic crust that was uplifted several kilometers without being obducted onto a continental margin. Deposition of pelagic limestone on the ophiolite during the late Eocene through Oligocene gave way to deposition of ophiolite-derived clastics during the early Miocene. The uplift of the western edge of the ophiolite probably was related to the initiation of subduction along the Manila Trench in the late Oligocene, but the Zambales crust predates the oldest crust in the adjacent South China Basin by about 8 million years. A new sandstone petrology method traces the uplift and erosional history of the ophiolite through the changing compositions of the ophiolite-derived clastics.

Emplacement of a large ophiolite into the overriding plate of a subduction system is a rare and poorly understood geologic event, but one with far-reaching consequences for the development of convergent plate boundaries. Large, intact ophiolites such as the Semail (Oman) and Bay of Islands (Newfoundland) complexes appear to have been emplaced by obduction of a slab of oceanic crust and upper mantle onto a continental margin (Coleman, 1977); smaller but more numerous thrust-slice ophiolites are often incorporated into complexly deformed belts and accretionary prisms (Irwin, 1977; Karig, 1981). Geologic evidence for the sequence of events that leads to the emplacement of an ophiolite is rarely preserved. Erosion and post-emplacement orogenies usually obscure or destroy records of timing and causes of initial detachment of the fragment of ocean crust, its uplift and obduction onto a continental edge.

The Zambales ophiolite avoids many of these difficulties. This large, young, and relatively intact ophiolite is still part of an active arc system and has not yet suffered the complexities of obduction. Sediments around the ophiolite record almost all of the uplift history of the Zambales complex. This paper introduces a new method for evaluating the provenance of sediments with a significant ophiolite source component, then applies this method, in combination with stratigraphy, regional geology, and geophysical data, to constrain the emplacement history of the Zambales ophiolite in western Luzon.

REGIONAL TECTONICS AND LUZON GEOLOGY

The western Pacific region around the Philippines is an area of complex present plate tectonics and controversial Tertiary evolution (Figure 1). A combination of east-dipping and west-dipping subduction zones brackets the Philippine archipelago, with the west-dipping Philippine Trench in the south and the east-dipping Manila Trench bordering the western margin of Luzon. Earthquake seismicity indicates that a new subduction system may be forming at the eastern margin of Luzon as a northward extension of the Philippine Trench (Fitch, 1972; Cardwell, Isacks, and Karig, 1980). Elsewhere in the Philippines, zones of deep seismicity not connected to presently active trenches suggest that subduction zones have changed their configuration in the recent past (Cardwell, Isacks, and Karig, 1980). These changes in subduction-

*Presently with: Gulf Science and Technology Company, Pittsburgh, Pennsylvania.

zone polarity may be conducive to ophiolite formation.

The collision of island arcs and continental fragments with trenches in the Philippines region has caused numerous changes in the position and polarity of subduction zones around Luzon during the Tertiary. Several models have been proposed for the tectonic evolution of Luzon and its surrounding marginal basins. Murphy (1973) postulated a flip of subduction zone polarity around northern Luzon from west-dipping to east-dipping in the Pliocene. Karig (1973) proposed a change around Luzon from westward to eastward subduction in the late Miocene. Hilde, Uyeda, and Kroenke, (1977) showed two west-dipping subduction zones in the west Philippine Basin area during the late Oligocene, but their position relative to Luzon is not clear. Ben-Avraham (1978) suggested a late Pliocene or later initiation of the Manila Trench west of Luzon while the Philippine Trench remained active from pre-middle Miocene to the present. Bowin et al's (1978) model includes a subduction reversal from westward to eastward around Luzon in the late Oligocene, and northward propogation of the Philippine Trench in the late Pleistocene. De Boer et al (1980) envisioned westward subduction related to the Sierra Madre arc in the Eocene and early Oligocene. This changed to eastward subduction at the Manila Trench in the late Oligocene with reactivation of

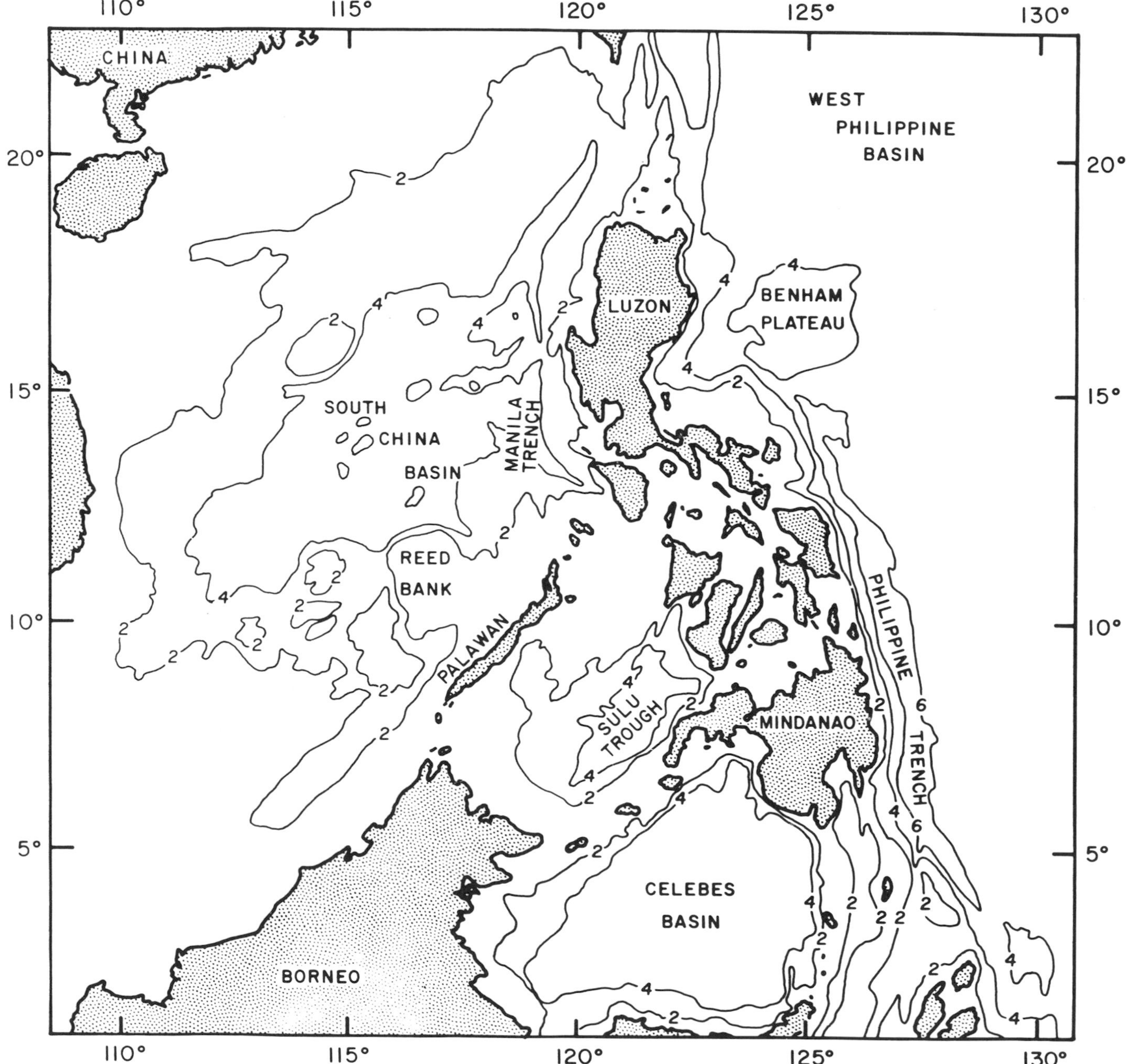

Figure 1 — Regional physiography and tectonic features of the Philippines and surrounding areas of the western Pacific.

westward subduction at the Philippine Trench in the Pliocene. The disagreement among these models reflects the paucity of reliable geologic constraints on subduction from both marine geophysical surveys and subduction-related rocks exposed on the islands.

Most of the major geologic features of Luzon can be attributed to various episodes of subduction-related volcanism, plutonism, basin filling, and deformation. Three north-south trending mountain belts frame northern central Luzon: The Sierra Madre, Cordillera Central, and the Zambales (Figure 2). The Sierra Madre range is only vaguely understood and is probably more complex than now known. At present, the range is thought to represent parts of an east-facing arc system of Cretaceous and early Tertiary age. Ultramafic thrust sheets are reported along the east coast in the northern Sierra Madre, flanked to the west by a chain of calc-alkaline plutons that intrude a thick volcanic and volcaniclastic sequence (Philippine Bureau of Mines, 1963). Our field work in early 1981 discovered a large Cretaceous ophiolite along the west flank of the southern Sierra Madre, capped by a volcanic turbidite sequence and then by pelagic limestone. This Cretaceous ophiolite probably formed in a marginal basin behind the east-facing arc. Mid-Tertiary plutonic-volcanic centers are superimposed on this ophiolite terrane (Anonymous, 1977). The Cordillera Central has been the site of arc magmatism for much of the Tertiary (Balce et al, 1980), and its history may extend back to the late Cretaceous. These igneous rocks have obscured the earlier history of the range, but there may be local blocks of even older metamorphic basement (Anonymous, 1977).

In contrast to the Sierra Madre and Central Cordillera, the Zambales range consists of a massive block of gabbro and peridotite over 100 km long and 30 km wide with diabase and basaltic volcanics along its eastern and northern flanks (Figure 3). A chain of late Tertiary to Quaternary calc-alkaline volcanoes covers the southern part of this range and continues south toward Mindoro, while a line of dacite plugs extends northward along the eastern edge of the range (De Boer et al, 1980). Tertiary strata overlie the east and west flanks of the Zambales and are described below in detail. The Central Valley basin, filled by more than 10 km of sediments derived from the Sierra Madre arc and the Zambales ophiolite and volcanic centers, separates the Zambales range from the rest of Luzon. The northern end of the Zambales complex plunges gently northward beneath the South China Sea, while the western edge of the range is truncated along a linear scarp that drops to a low, narrow coastal platform.

Despite its great size and accessibility from a major city (Manila is within 100 km along paved highways) the Zambales range only recently received attention as one of the world's major ophiolites. Although relief across the range is over 1,500 m, the wet tropical climate and dense vegetation limits exposures and increases the difficulties of access. Since 1975, the Geological Survey Division of the Philippine Bureau of Mines has undertaken the arduous task of mapping the Zambales range at 1:50,000 scale. All data in this study concerning the interior of the range (Figure 3) are taken from unpublished maps and reports of the Bureau of Mines. The Bureau of Mines also provided logistical support for the authors' field work on the basalts and sediments along the east flank and the sediments and structural relationships along the west flank.

GEOLOGY OF THE ZAMBALES OPHIOLITE

Igneous Sequence

The Zambales ophiolite extends more than 120 km along western Luzon, and widens to almost 40 km in its central portion (Figure 2). The interior of the Zambales range consists of peridotite, dunite, and gabbro, the latter as layered (cumulate) and massive types (Philippine Bureau of Mines, unpublished reports). Mapping, by Darwin Rossman of the U.S. Geological Survey in the early 1960s (open file report) and by the Philippine Bureau of Mines since 1975, shows a nearly continuous band of gabbro roughly 10 km wide along the eastern side. Large areas of gabbro and periodotite intermix over the western half of the range (Figure 3). Dunite occurs as elongate masses generally less than 200 m thick near the contact between gabbro and peridotite, and is often associated with chromite ore bodies (Stoll, 1958; Rossman, 1964).

Diabase and basalt form a band 2 to 5 km wide along the eastern edge of the range (Philippine Bureau of Mines, unpublished mapping; Figure 3). Small intrusive bodies of leucocratic rocks often mark the transition from the gabbro section to the diabase sills. We traversed a sheeted sill complex with units 0.5 to 5 m thick that dip 25 to 40° eastward toward the Central Valley at several locales along the east flank. Within the eastern parts of this sill complex, layers of altered pillow basalt are sandwiched between sills and become increasingly common upward (eastward) in the section. The attitudes of the pillows indicate that the sills were nearly horizontal; both the sill and the pillows appear to have been tilted eastward 30 to 40°.

The uppermost volcanic sections of the ophiolite, exposed along the eastern edge of the range (Figure 3), are predominantly basaltic rubble several hundred meters thick, with minor amounts of recognizable pillow basalts and thin sills or flows. No zones of interbedded sediment, analogous to those drilled in the upper oceanic crust at DSDP sites (Site 485; Lewis et al, 1979) were observed within this section. However, rare clasts of fine-grained limestone were found within volcanic breccias at the top of the volcanic section.

This sequence of lithologic units is similar to that of other ophiolites (Coleman, 1977) but there is no agreement as to whether the deeper igneous layers are of normal thickness or have been disturbed by faulting. Geochemical data for the igneous suite suggests both island arc and ocean (or back-arc) basin affinites (Hawkins, Evans, and Bacuta, 1980; De Boer et al, 1980). However, the absence of shallow-water debris or coarse volcanoclastics in the capping sediments,

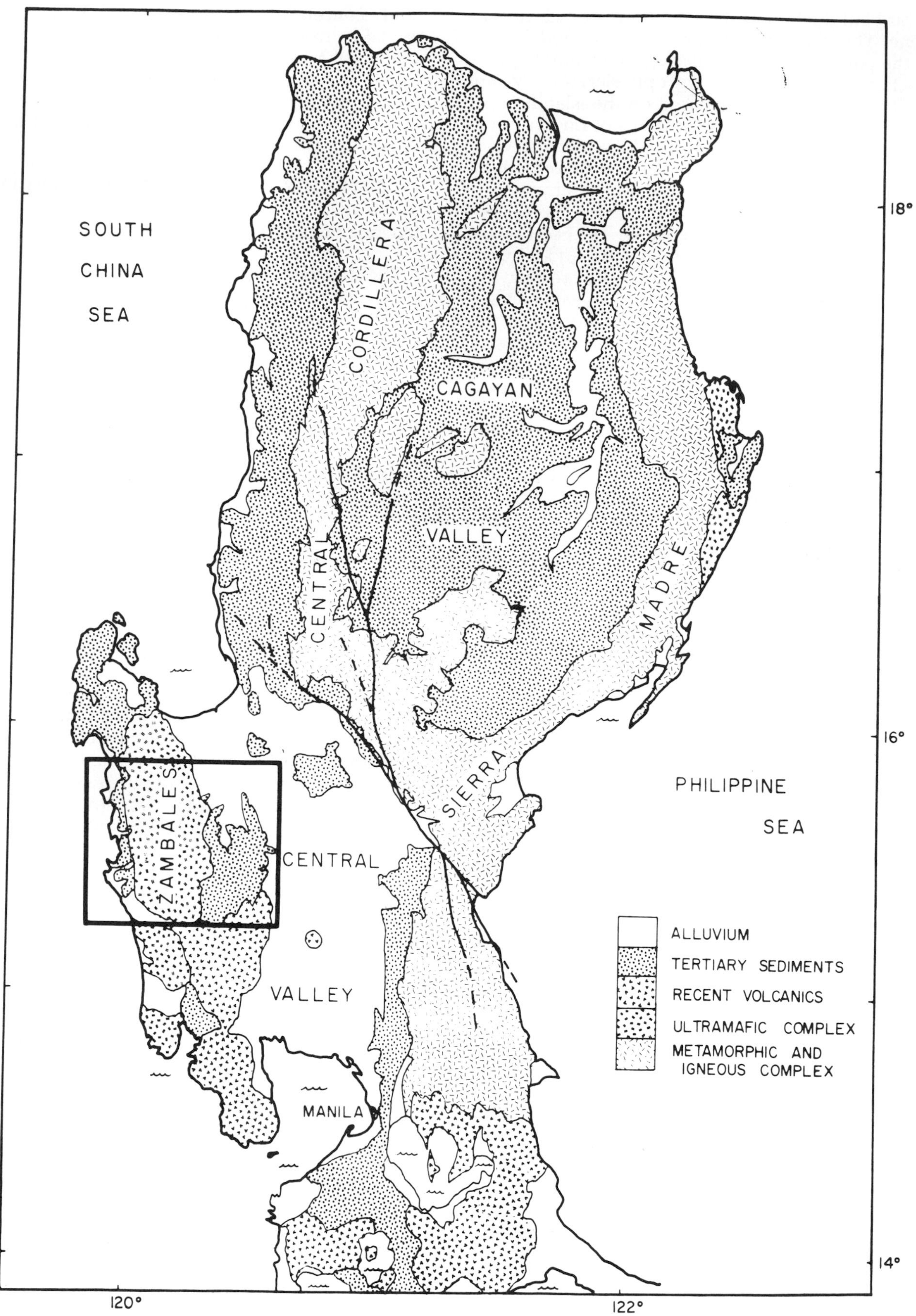

Figure 2 — Generalized geology of Luzon, adapted from Geologic Map of the Philippines (Bureau of Mines, 1963).

as would be expected near an arc system, argues strongly against an arc origin. Attitudes of sills, pillow basalts, and overlying sediments, as well as the overall distribution of lithologies (Figure 3) suggest that the eastern part of the Zambales ophiolite dips eastward beneath the Central Valley at approximately 30°. A northeast or east-northeast strike of the sill-dike complex (Violette, 1981, written communication) and of linear magnetic anomalies mapped over the eastern flank of the range and the Central Valley (San Jose Oil Company, proprietary report; Sano, written communication, 1981) may represent the relict trend of the spreading center that formed the ophiolite. However, this present trend may have undergone significant tectonic rotation from its original orientation in the Paleogene (De Boer et al, 1980).

East Flank Sedimentary Section

A Tertiary sediment sequence exposed along the east flank of the Zambales ophiolite records generally shallowing paleoenvironments and an increasing input of coarser-grained clastic material from the late Eocene through the Pliocene (Figure 5). The basal sediment over the Zambales volcanics is a white, fine-grained pelagic limestone with thin tuff layers 2 to 20 cm thick. This unit, the Bigbiga limestone member of the Aksitero formation (Amato, 1965) is approximately 40 m thick at the type section along the Aksitero River (Figures 3 and 6); although the thickness varies along the east flank to as little as 7.5 m at the Moriones River. Amato (1965) assigned the Aksitero formation to the upper Eocene through lower Oligocene with a bathyl (deep marine) paleoenvironment, probably deeper than 1 km. Garrison et al (1979) corroborated Amato's late Eocene age assignment and concluded that the limestone represents deep-water sedimentation with no input of terrigenous detritus.

Previous studies (Corby et al, 1951; Divino-Santiago, 1963; Bandy, 1963; Amato, 1965; Garrison et al, 1979) described the basal contact of the Aksitero limestone as unconformable over the Zambales complex, giving the ophiolite a pre-Eocene formation age. However, our field work discovered clear outcrops, demonstrating that limestone deposition began virtually simultaneous with the last stages of submarine volcanism that formed the Zambales volcanics (Schweller and Karig, 1979). In some cases, limestone fills voids and

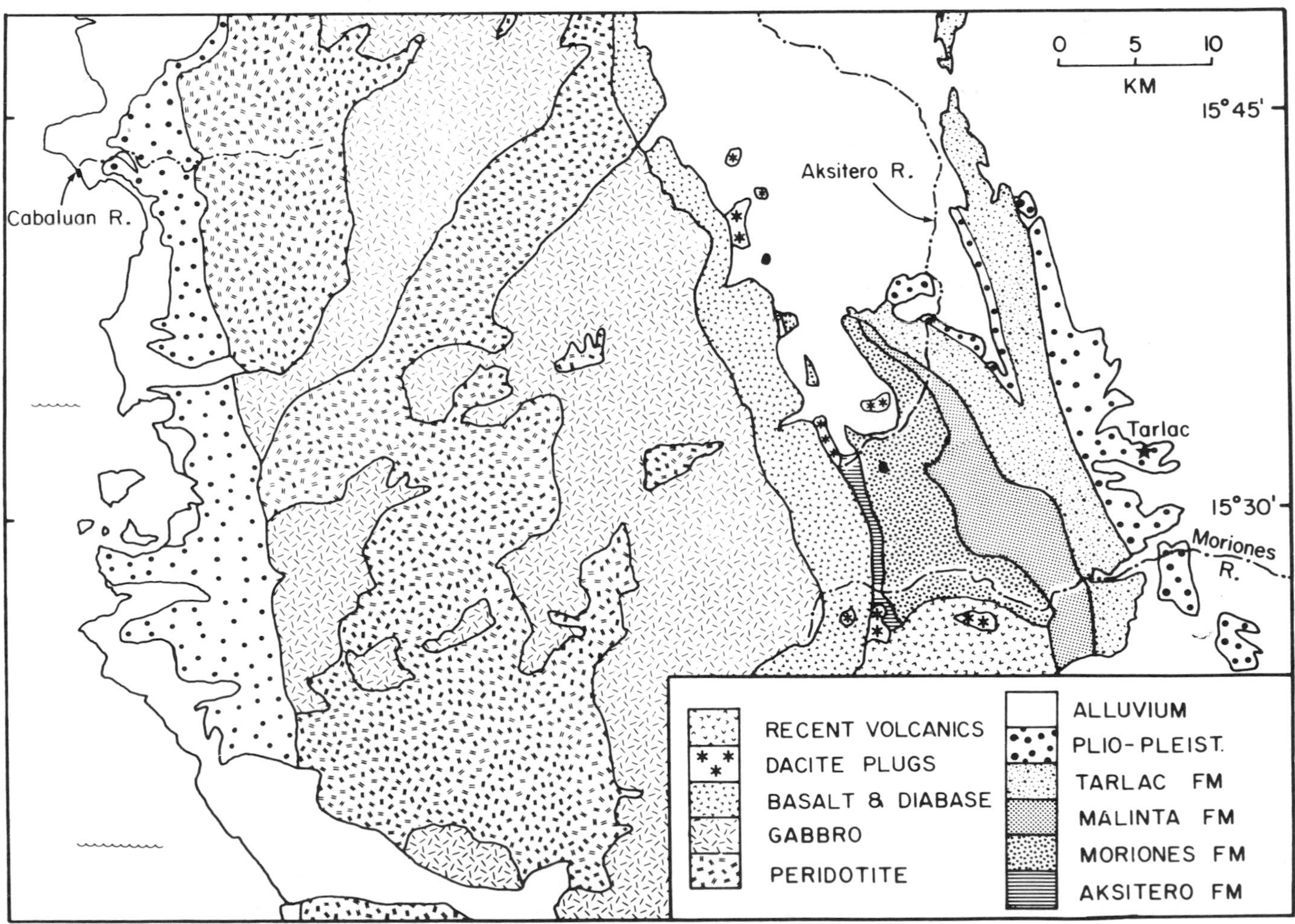

Figure 3 — Geology of the central Zambales range, after unpublished maps of the Philippine Bureau of Mines, with Aksitero formation outcrops revised by field work of the authors. Dashed line marks approximate section line of Figure 9.

cracks within angular volcanic breccia and pillow basalt, and is continuous upward into the bedded limestones. In other outcrops, angular volcanic clasts are supported in a limestone matrix, with this breccia resting directly on massive volcanics and overlain by bedded limestone. In at least two localities, a lens of reddish brown mudstone up to 2 m thick separates the volcanics from the Aksitero limestone section. This mudstone may be related to hydrothermal exhalations from newly-formed ocean crust. All of the types of basement contacts are analogous to situations observed by submersible studies at mid-ocean spreading centers (Ballard and van Andel, 1977). This ophiolitic crust along the eastern flank of the Zambales range is thus assigned a late Eocene age, equivalent to the oldest Aksitero formation sediment.

Outcrops of limestone on ophiolitic volcanics become more tuffaceous north of the type section along the east flank. At the Barlo Mine area near the northern end of the ophiolite (Figure 2), over 100 m of white, fine-grained bedded tuffs with rare radiolaria lie depositionally on the pillow basalts (Bryner, 1967; our field mapping). No terrigenous or coarse volcanic components have been recognized in these tuffs. Provisional dating of the sparce radiolarian fauna indicates a late Eocene age (A. Sanfillipo, written communication, 1979), making these tuffs correlative with the lower Askitero formation. If this tentative age is valid, it suggests a strong northward increase in ash deposition in the late Eocene, perhaps indicating that a volcanic arc lay not far north of the present northern outcrop limit of the ophiolite. The high proportion of ash also favors a back-arc basin setting over a mid-ocean ridge origin for the Zambales crust, although a narrow ocean basin with a nearby arc cannot be ruled out.

An abrupt change in lithologic character without an apparent hiatus marks the boundary between the lower and upper members of the Aksitero formation. Above about 45 m in the type section, silty and sandy volcaniclastic turbidites dominate the sediment section, although thin interbeds of limestone continue through most of the upper member. Brownish-gray graded sand and silt beds 10 to 50 cm thick alternate with thinner beds of fine-grained bluish-gray limestone; the proportion of limestone decreases upward from about 50% at the base of the upper member to less than 10% near the top. This change in sedimentation character probably represents the encroachment of a channelized turbidite depositional system over an ocean floor area that was previously isolated from coarse-grained detritus.

A depositional hiatus between the upper Aksitero formation and the lower Miocene Moriones formation (Figure 5) was defined on the basis of a paleontologic break (Amato, 1965). However, field evidence for this hiatus is weak at the type sections along the Aksitero River. There is no significant change in the attitude of bedding nor general lithologic trends across the supposed hiatus, and certainly no evidence for subaerial exposure and erosion within this section. The detrital mineralogy shows no break between the upper Aksitero formation and the overlying strata (Schweller and Bachman, in press). Paleontologic studies by Hashimoto (1980) and Schweller et al (in press) indicate a break of, at most, 1 or 2 million years between the type sections along the Aksitero River. Along the Moriones River, about 10 km south of the type sections, the entire Aksitero formation consists of 7.5 m of upper Eocene pelagic limestone overlain by volcaniclastic turbidites of the Moriones formation. The formational contact is sharp and approximately parallel with bedding. There is no evidence for subaerial exposure and erosion of the top of the limestone, but a hiatus of several million years is apparently represented. Thus, the time span represented by the pre-Miocene hiatus over the Aksitero formation seems highly variable.

The lower Miocene Moriones formation is a thick sequence of sandstone, mudstone, and conglomerate overlying the upper Aksitero clastic section (Figures 3 and 5). The thickness of the Moriones formation is approximately 1 km at the type section along the Aksitero River (Roque, Reyes, and Gonzales, 1972). Thin-bedded sandstone and mudstone predominate in the lower part of this formation, although a few coarse sandstone beds up to 1 m thick occur within the lowest 100 m of the section. The maximum grain size of sandstones increases gradually up through the section, with pebble and cobble conglomerates entering the sequence above about 300 m. Thin parallel bedding in the lower section changes to thicker beds and channel structures with as much as 2 m of relief on lens-shaped beds across a 20 m wide outcrop. Channel facies begin at about 200 m above the base of the formation in medium to coarse sandstone with occasional mudstone clasts. The first conglomerates to appear contain the entire igneous suite of the ophiolite: serpentenized ultramafics, gabbro, diabase, and fine-grained Aksitero-type limestone. The well-rounded cobbles, together with fragments of carbonized wood, document subaerial exposure and deep erosion of part of the ophiolite during the early Miocene. Paleocurrents and depositional facies indicate turbidite sedimentation filling the Central Valley basin from a western source.

West Flank Sedimentary Section

The strata that flank the ophiolite on the west are only locally well-exposed and are quite different from the sediment cover of the east flank. The sedimentary sequence, which appears to underlie much of the narrow coastal plain, begins with a transgressive series in depositional contact with the ultramafic and gabbroic units of the ophiolite. Above a thin basal sequence of ultramafic-clast conglomerate and shallow-marine (back-reef?) clastics, a limestone reef facies extends along much of the western edge of the ophiolite as a prominent ridge. A limestone unit along the western flank was originally named the Santa Cruz Limestone and was dated as late Oligocene (N3) (Hashimoto, 1980) on the basis of large foraminifera. This limestone is overlain by a poorly understood sequence of silty mudstones and calcarenites deposited in somewhat deeper water conditions. Both the supposedly Oligocene limestone and the overlying strata are being studied in more detail to clarify their ages and paleoenvironments.

This transgressive sequence indicates that the deeper levels of the ophiolite were above sea level by the end of the Oligocene and that the source was close to the present western edge of the ophiolite. The apparent deepening of paleoenvironments upward from the basal conglomerates through the reef facies to the fine-grained clastics overlying the Santa Cruz Limestone further implies an episode of submergence and a possible reduction in the elevation of the ophiolite. Although further data are needed to constrain the magnitude and timing of uplift and subsidence along the western edge of the ophiolite, much of the present relief may be much younger than the initial phase of uplift and may be due to a second uplift episode.

Enigmatic boulders of radiolarian chert are prominent in the Cabaluan River area (Figure 3). Our mapping shows that chert occurs as large tectonic lenses up to 100 m across within a band of serpentinite, mylonitized ultramafics, and low-grade schists that extend several kilometers southward from the Cabaluan River along the western edge of the ophiolitic massif. The radiolaria appear to be too recrystallized to permit identification, but they must predate the late Oligocene sediments at the base of the west flank sediment sequence. These cherts are totally unlike any lithology found along the east flank of the ophiolite, but smaller amounts of chert are present in a highly sheared and disrupted assemblage of ophiolitic lithologies about 20 km northwest of Barlo (Figure 2). Large amounts of similar chert are found associated with serpentinite melange units along the western coast of northern Luzon (Smith, 1907; G. Haeck, unpublished field mapping, 1980).

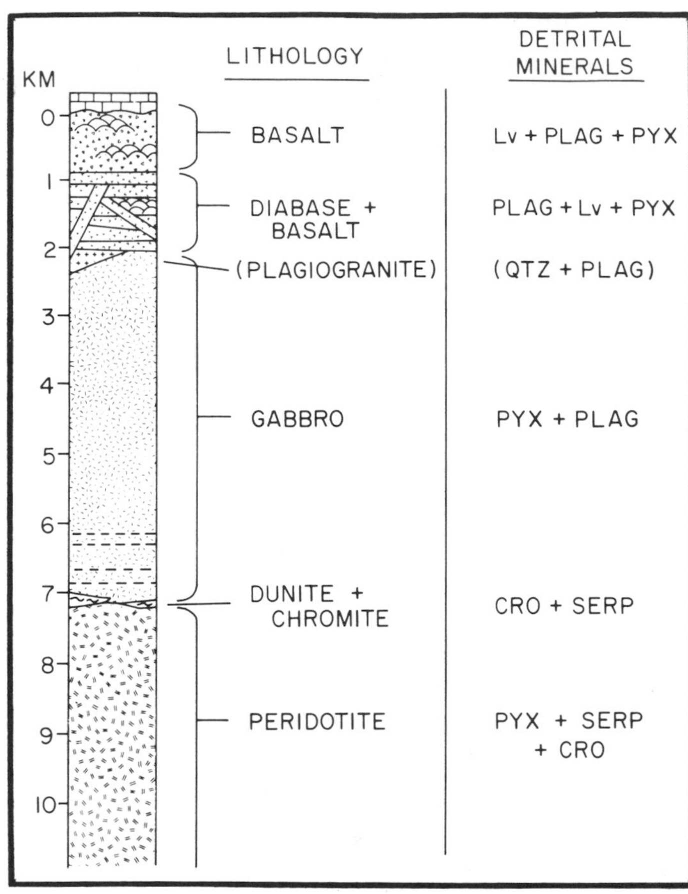

Figure 4 — Igneous sequence of the Zambales ophiolite. Thicknesses of layers calculated from maps and traverses of the east flank (Philippine Bureau of Mines, unpublished reports). Detrital mineralogy predicted from average igneous petrology.

SANDSTONE COMPOSITION TRENDS

The conglomerates in the Moriones formation provide strong evidence for uplift and exposure of the deeper levels of the Zambales Ophiolite in the early Miocene. Conglomerate clast compositions can be related directly to the vertical segregation of igneous rock types in an ophiolite (Figure 4) to infer the level of erosion and exposure. Unfortunately, conglomerates are not common in the Moriones formation and are entirely absent in the lowest Moriones and upper Aksitero sections (Figure 5), so this method gives only a crude history of uplift for the Zambales ophiolite. Sandstone is a much more abundant component of the sedimentary section, and would give a more detailed uplift history if the different ophiolite layers could be related to distinctive sandstone compositions.

Sandstone composition, as determined by point counts of thin sections, is a powerful and widely used tool for reconstructing the paleogeography of source regions in a variety of tectonic settings (Dickinson, 1970; Dickinson and Suczek, 1979). Detrital mineral components reflect the petrology of source terranes and can discriminate among cratonic, magmatic arc, and recycled orogen provenances (Dickinson and Suczek, 1979) as well as provide information about the evolution of clastic sources around a basin (Ingersoll, 1978). However, to our knowledge, no previous study has used these techniques to evaluate the emplacement of a large ophiolite. A new method for this purpose is briefly introduced here; a further discussion of the technique and its applications is given in Schweller and Bachman (1982).

The correlation between sandstone composition and different provenances or terranes is usually established empirically through an analysis of modern and ancient sandstones with known sources. Because of a lack of published data on ophiolite-derived sandstone, the method used here relies on sandstone compositions inferred directly from the well-defined and consistent mineralogy of the layered ophiolite sequence (Coleman, 1977). Figure 4 shows the correspondence of detrital minerals to rock types within a typical ophiolite sequence.

Triangular diagrams are used here to evaluate two aspects of ophiolite-derived sandstones. The commonly used triangular diagrams (QFL, Dickinson and

Suczek, 1979) ignore the mafic minerals that constitute a large percentage of the lower sections of ophiolites, and are of little use for evaluating ophiolitic sandstone compositions. A new mineralogic component, *Mu*, is defined here as the sum of pyroxene, olivine, serpentine, and chromite grains. This component accounts for the mafic minerals that make up most of the ultramafic layer, as well as about half of the gabbro section. Most of the remainder is plagioclase (Figure 4).

A triangle diagram with apices representing mafics (Mu), stable quartzose grains (Q), and volcanic rock fragments (Lv) can be used to discriminate among ophiolitic, magmatic arc, and continental provenances (Figure 7). Sand derived from continental or cratonic sources generally contains a high percentage of quartz, while sand from magmatic arcs is rich in volcanic rock fragments (Dickinson and Suczek, 1979). The fields predicted for sands from magmatic arcs and those from the upper layers of ophiolites overlap on

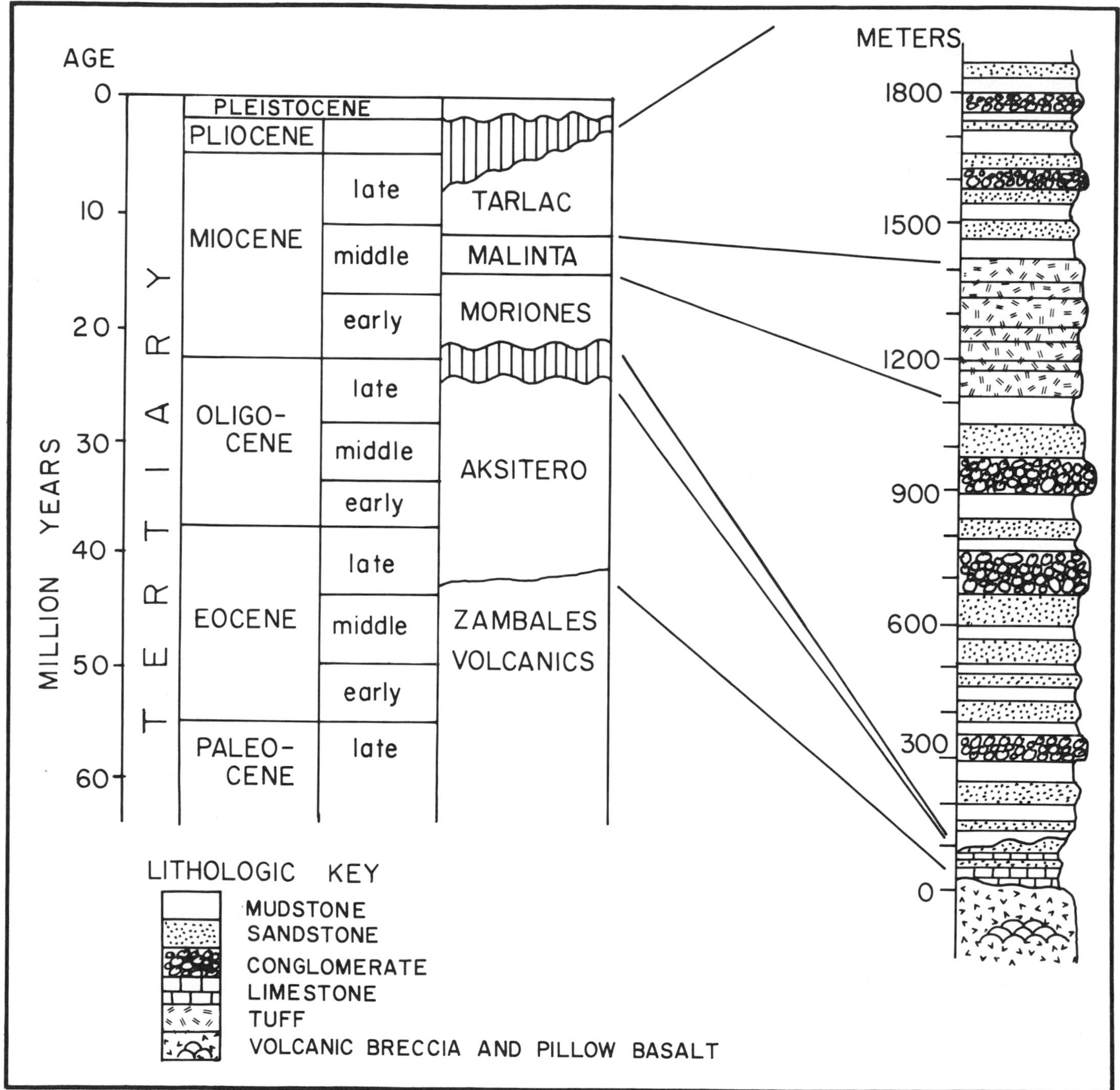

Figure 5 — Stratigraphic section across the Zambales east flank. Neogene sections from Roque, Reyes, and Gonzales (1972) and unpublished reports of the Philippine Bureau of Mines; Aksitero formation from the authors' field work. Lithologic symbols are schematic averages of lithologies.

the QLvMu diagram, since both terranes are predominantly basic volcanic rocks. Geochemical analyses are generally necessary to distinguish between juvenile island arc and ocean floor volcanics (Garcia, 1978).

A second triangle diagram uses the distinct petrologic layering of large ophiolites to evaluate the erosion level into an ophiolite (Figure 8). The upper levels of an ophiolite contain predominantly volcanic rocks, the gabbro layer consists mainly of plagioclase and pyroxene, and the ultramafic section contains olivine, pyroxene, and minor chromite, with olivine commonly weathering to serpentine (Figure 4). Substituting a plagioclase apex (P) for stable quartose grains (Q) expands the linear ophiolite trend on the QLvMu diagram (Figure 8). Plagioclase increases to a maximum proportion within the upper gabbro layer, then dwindles to nearly zero in the ultramafics (Figure 4). Erosion down into a large, nearly flat-lying ophiolite should produce sandstone compositions that follow this trend from predominantly volcanic rock fragments to mixed mafics and plagioclase, and finally into predominantly mafic grains.

Results

Sandstone compositions in the Aksitero and Moriones formations show three distinct mineralogic suites. Only general trends are presented here; a more complete analysis of this sedimentary sequence in terms of ophiolite-derived sandstone compositions is given in Schweller and Bachman (1982).

Sandstone from midway through the Moriones formation near the first conglomerates is nearly 100% serpentine, with minor amounts of pyroxene and chromite. This sandstone is virtually identical in character to modern sands in the Cabaluan River

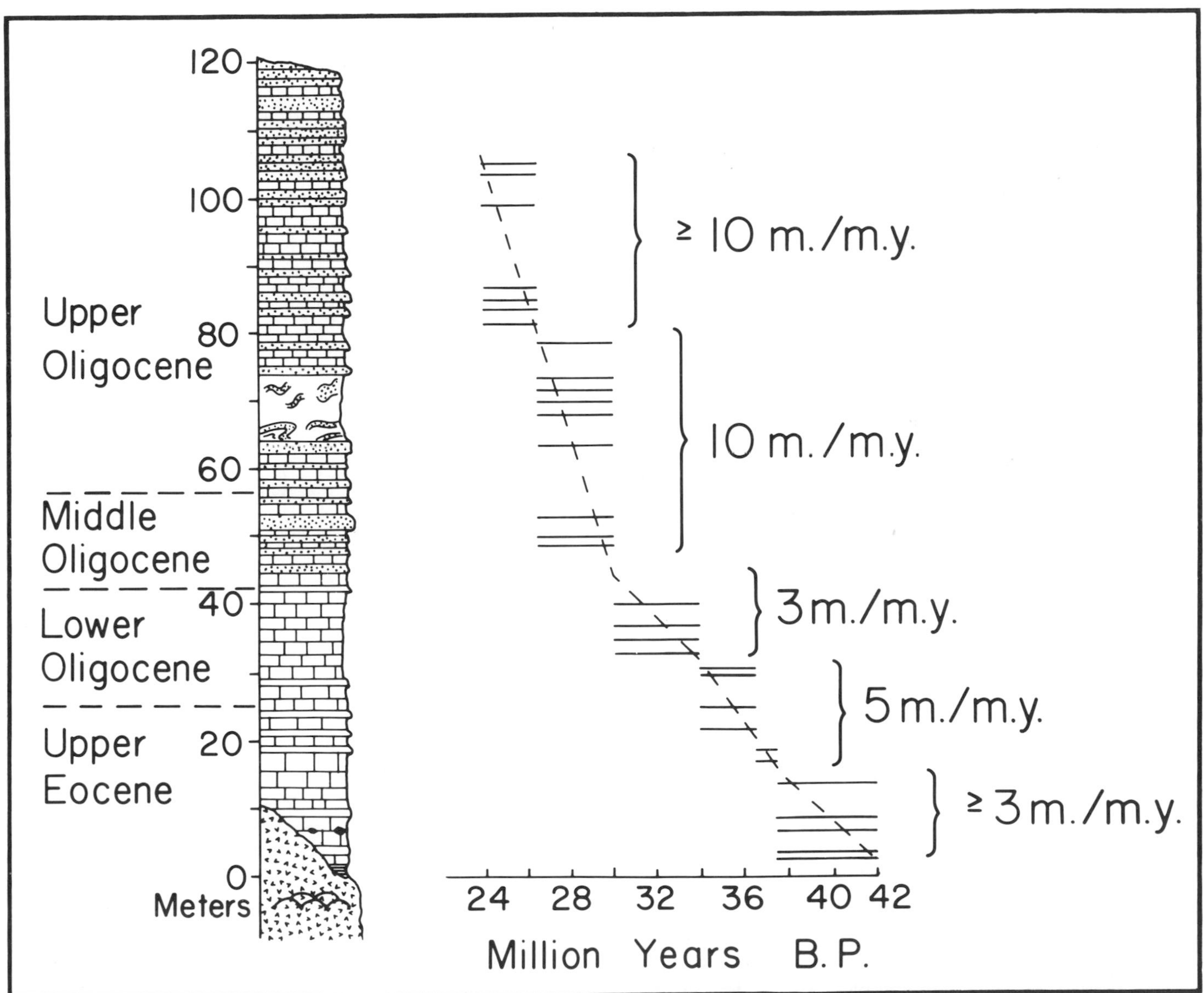

Figure 6 — Stratigraphic section of the Aksitero formation at the type section, based on field measurements by the authors with age control from calcareous nannofossils (Schweller et al, 1982). See Figure 5 for key to symbols.

(Schweller and Bachman, 1982) whose drainage area is entirely within ultramafics and gabbros (Figure 3). These serpentine sandstones are interbedded with sandstones having a more diverse mineralogy: in addition to abundant serpentine and minor pyroxene and chromite, they contain volcanic lithics, plagioclase, and sedimentary lithic grains. This latter type of sandstone represents a mixed source from the lower and upper layers of the ophiolite, comparable to large modern rivers draining the eastern flank of the Zambales range (Schweller and Bachman, 1982).

Near the base of the Moriones formation, sandstone units contain abundant plagioclase with smaller amounts of volcanic lithics and pyroxene. The large percentage of monocrystalline feldspar grains suggests a gabbroic source with some contribution from a volcanic terrane. The lack of chromite and serpentine indicates that erosion of the ophiolite had progressed down into the gabbro layer but not into the ultramafics at this time.

Sandstone in the upper Aksitero formation is typically severely weathered and altered. The only consistently identifiable constituent is volcanic lithic grains, which make up most of the less-altered beds. These early sediments cannot unequivocally be proven to derive from the upper layer of the ophiolite. However, their position in the stratigraphic sequence at a time slightly older than beds with a gabbro provenance suggests such an origin.

The sandy base of some of the graded tuff beds in the lower Aksitero limestone section have a completely different composition than the upper Aksitero sandstones. These tuff beds contain angular plagioclase crystals with a few pumice fragments in a fine-grained glass-shard matrix. No volcanic lithic grains are found in these beds. Garrison et al (1979) attribute these ash beds to pyroclastic debris from underwater volcanic eruptions. While they may also represent primary ash falls from subaerial eruptions, they do not appear to be derived from previously erupted volcanic rocks.

STRUCTURAL DATA

East Flank

Strata along the east side of the Zambales Range dip eastward beneath the Central Valley at moderate angles. Dips in the Aksitero formation, as well as gross layering in volcanic breccias within the uppermost ophiolitic volcanic layer, range from 20 to 30°. Bedding in the limestone locally bends around basement irregularities, and minor faults with offsets of a few decimeters are common in the lowest 20 m above volcanic basement. Dips in the lower Moriones formation immediately above the Aksitero section begin at 20 to 25° and flatten gradually to an average of about 15° approximately 1 km east of the ophiolite (Figure 9). Open folds form anticlinal ridges in the Neogene strata of the Central Valley (Figure 3), but appear to be relatively recent structures not related to any emplacement mechanisms.

Zones of disrupted bedding up to 5 m thick are interlayered within thick sections of undisturbed eastward dip in the lower Moriones Formation along the Aksitero River. The earliest such zone appears midway through the upper Aksitero Formation (Figure 6). These zones are interpreted as mass-flow deposits on the basis of highly contorted internal structures and sharp depositional contacts with the underlying and overlying beds. The disturbed-bedding zones in the lower Moriones Formation consist of a

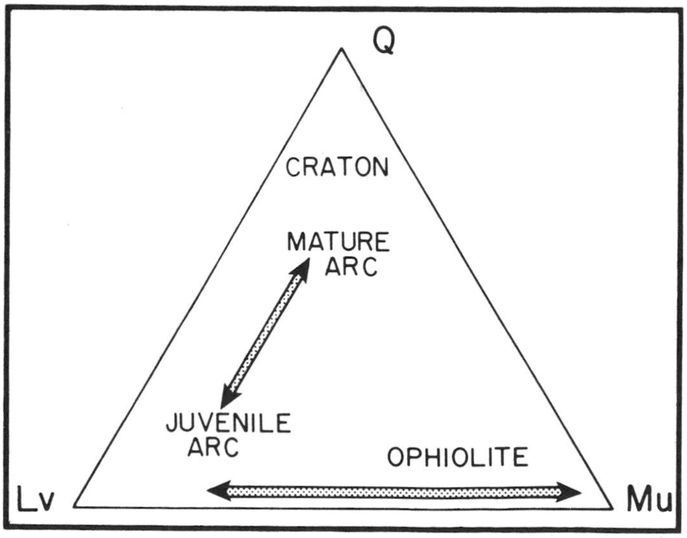

Figure 7 — QLvMu diagram for determining provenance from sandstone composition. Compositions of craton and magmatic arc sands from Dickinson and Suczek (1979). Definitions of Q and Lv components from Dickinson (1970); Mu is defined here as the sum of pyroxene, olivine, serpentine, and chrormite grains.

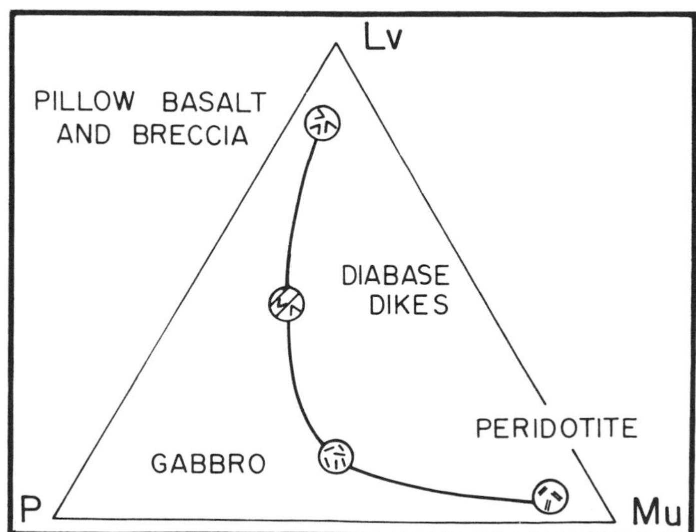

Figure 8 — PLvMu diagram for determining the level of erosion into an ophiolite suite from sandstone composition. P represents monocrystalline plagioclase grains; other components as in Figure 7.

foraminifera-rich silty mudstone with well-rounded pebbles and cobbles of serpentinized periodtite and gabbro, while the disturbed zone in the upper Aksitero Formation contains only mudstone and sandy siltstone. The cobble-bearing deposits imply proximity to a moderately high paleoslope and a source of stream- or surf-rounded ophiolite fragments. Their initial occurrence in the lower Moriones Formation indicates considerable relief and deep erosion of the ophiolite in the early Miocene.

West Flank

The steep western flank of the Zambales range is defined by a sharp linear structure that can be traced on satellite imagery for over 50 km along its north-south trend (Figure 3). On the ground, this zone is marked by sharply upturned strata that include conglomerates and sandstones derived mainly from ultramafics. The flexure is probably controlled by a very high angle fault at depth (Figure 9). Poor exposures preclude a detailed description of the boundary zone in most areas, but better exposures (with possibly anomalous geological conditions) near the Cabaluan River (Figure 3) provide some insight into the nature of the ophidite's western edge.

Dips of the west flank strata generally reach 45 to 75° W along the flexure zone, but become overturned to 55° E in the Cabaluan River section (Figure 9). This overturning may be related to a slide block of ultramafic rocks which overlies older stream gravels in an outcrop a few hundred meters distant. At the Cabaluan River and elsewhere, dips of the flanking strata flatten to less than 20° W within 100 to 200 m west of the edge of the ultra-mafics. The age of the flexure, which presumably is associated with uplift that created some of the present relief of the range, is not yet closely constrained.

Tectonic lenses of pink chert and of low grade schistose metamorphic occur in a highly sheared zone, with mylonitized ultramafics and serpentine, along the western edge of the ultramafic massif. The metamorphics have nearly vertical foliation that trends generally north-northeasterly. These metamorphics were apparently formed during an early phase of deformation, then juxtaposed against unmetamorphosed rocks by later shearing movement. These phases of deformation may be related to the Oligocene uplift of the ophiolite, while the upwarping of sedimentary strata probably dates from a younger phase of deformation.

EMPLACEMENT MODEL AND TECTONIC HISTORY

Using the sandstone composition trends and biostratigraphic data, together with regional geology and geophysics, a preliminary model for the Tertiary evolution of the Zambales ophiolite can be constructed. The Zambales ophiolite began as a piece of ocean floor at a site isolated from significant amounts of land detritus or coarse volcanic debris. Volcaniclastic sedimentation dominates many marginal basins either as coarser-grained turbidites, associated with large back-

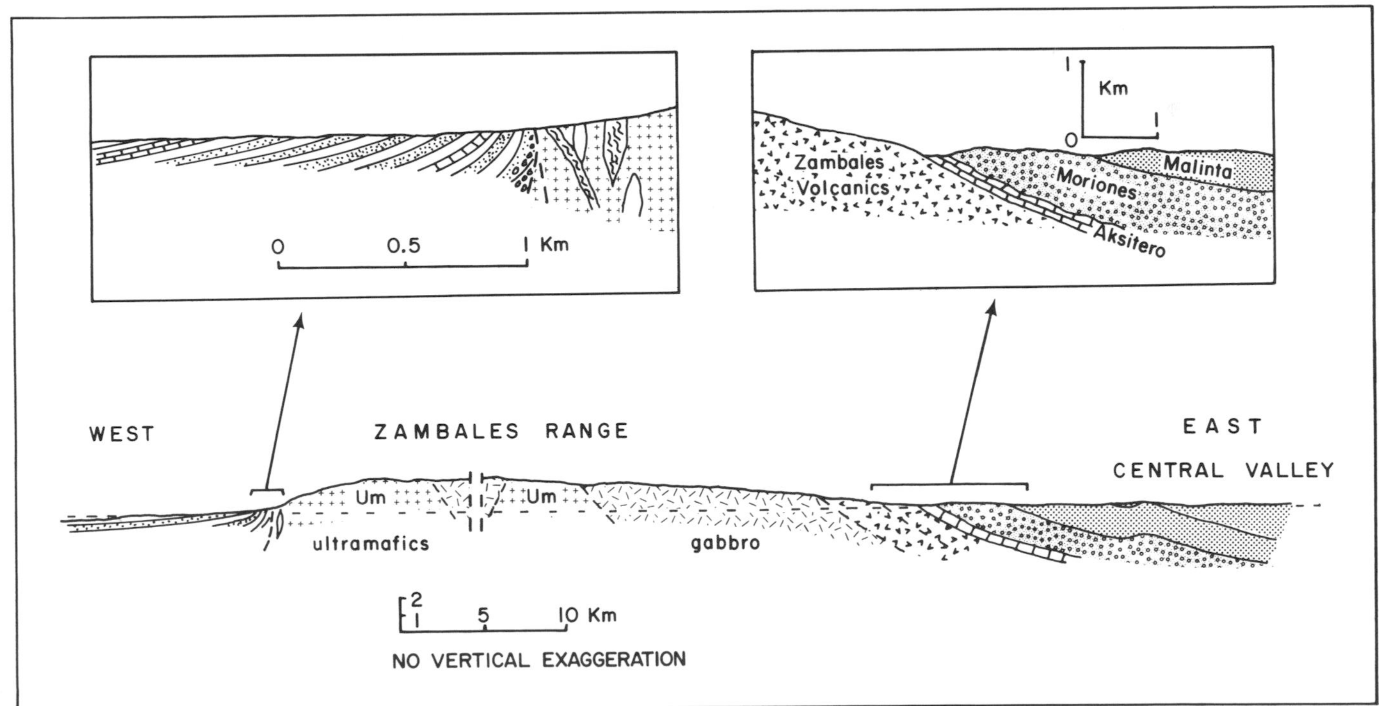

Figure 9 — Schematic structural section across the Zambales ophiolite and Central Valley (section line indicated on Figure 3). West Flank inset represents the Cabaluan River section; east flank inset shows area near the Aksitero River. Central Valley structures based on seismic reflection profiles and surface trends.

arc sediment aprons, or as finer-grained hemipelagic muds (Karig and Moore, 1975). Marginal basins adjacent to a continental margin, such as the south China basin (Figure 1), are dominated by terrigenous mud and turbidites (Damuth, 1980). The Zambales crust received neither terrigenous material or abundant volcaniclastic debris for 6 to 8 million years after its formation, but did receive persistent small amounts of fine-grained ash of intermediate composition over this period (Schweller et al, in press). These factors imply an original setting far from a continental margin and a considerable distance, perhaps a few hundred kilometers, from an active volcanic arc. A setting in the middle of a large marginal basin satisfies these requirements, as does a segment of a mid-ocean ridge with a nearby island arc system. Toward the north, the proportion of fine-grained ash in the basal sediments overlying the ophiolite increases, which suggests that the arc may have been in that direction relative to the Zambales crust, although subsequent tectonic rotation may have significantly changed the orientation of the Zambales with respect to Luzon and its surroundings.

Perhaps the simplest model for the evolution of the Zambales ophiolite is that the Zambales crust formed as part of a wide interarc basin behind an active magmatic arc in the late Eocene (Figure 10A). Such a configuration requires considerable tectonic rotation of the Zambales crust relative to the Sierra Madre in order to position the tuffaceous sediments of the northern end of the ophiolite nearest to the arc. Ash beds at DSDP Site 292, just east of Luzon (Figure 1), record continuous arc activity at the Sierra Madre from the late Eocene through the Miocene (Donnelly, 1975), but the earlier development of the Sierra Madre is largely unknown.

Uplift of the western side of the Zambales began in the Oligocene (Figure 10B), while the detrital evidence for erosion of the ophiolite along the eastern flank appears somewhat later, in the lower Miocene strata. The west flank uplift may coincide with the initiation of eastward subduction at the Manila trench, although uplift of limited areas along a major transform fault could also explain the structures and early sediments along the west flank. Such a strike-slip margin could be subsequently changed to a trench by a change in relative plate motions. A pulse of igneous activity near the end of the Oligocene in the southern Central Cordillera (Balce et al, 1980) and a maximum of ash layers at DSDP Site 292 just east of Luzon (Donnelly, 1975) may signal the onset of Manila Trench subduction.

By the early Miocene, large amounts of ophiolite-derived detritus began filling the Central Valley basin (Figure 10C). The abundance of ultramafic debris in the Moriones formation implies that a sizable area of the Zambales ophiolite had been uplifted several kilometers and eroded down through the basalt, diabase, and gabbro sections by this time. However, the area at the present eastern edge of the ophiolite had not yet tilted significantly from its original attitude, implying a relatively sharp upward bending of the western part of the ophiolite in the early Miocene.

Continued erosion and uplift of the eastern flank of the ophiolite had nearly filled the Central Valley basin with several kilometers of clastic sediments by the end of the Miocene (Figure 10D), giving a configuration quite similar to the present. This uplift was probably caused by continued subduction and growth of an accretionary prism at the Manila trench through the Miocene. The calc-alkaline valcanoes along the southern part of the Zambales reflect the later stages of this subduction system (De Boer et al, 1980).

At present, the Zambales ophiolite has been uplifted several kilometers and exposed down to its ultramafic roots. However, although the ophiolite has been emplaced into an apparently stable position with respect to Luzon, it has not yet been obducted in the classic sense of being thrust onto a continental margin (Coleman, 1971). If subduction at the Manila trench continues long enough to close the South China Basin,

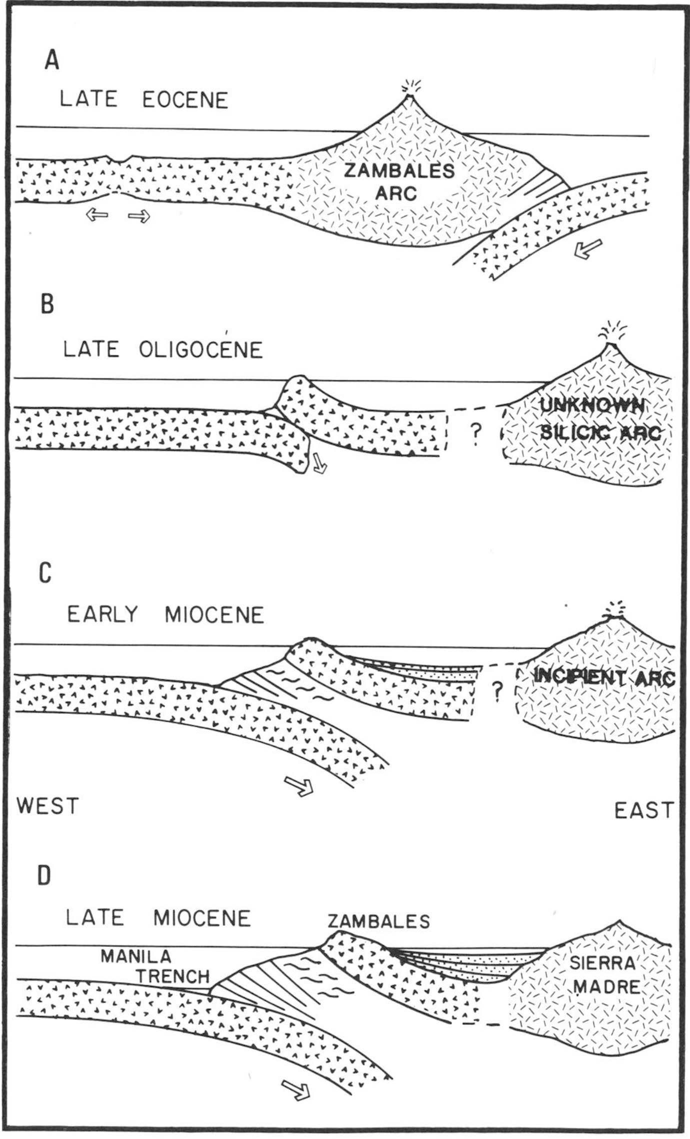

Figure 10 — Model for the tectonic evolution of the Zambales ophiolite as discussed in the text.

Luzon would collide with the Asia continental margin in a manner similar to Taiwan at present. Such a collision would be favorable for the obduction of the Zambales ophiolite onto the Asia margin, with the Sierra Madre forming the new edge of the continent.

CONCLUSIONS

The Zambales ophiolite formed in the late Eocene in a deep marine environment and remained isolated from land-derived sediments for at least 6 million years, although it received thin layers of fine-grained volcanic ash throughout this period. By the late Oligocene, volcaniclastic turbidites began to supercede pelagic limestone deposition at the present east flank of the Zambales. At the same time, ultramafic-clast conglomerates and shallow-marine facies were deposited along the west flank, implying that a narrow area of the ophiolite was uplifted several kilometers near the present western edge of the range. By the early Miocene, abundant ultramafic detritus was shed eastward into the Central Valley basin. Sediments derived from the Zambales and the Sierra Madra ranges nearly filled the Central Valley basin to sea level by the end of the Miocene.

Sandstone compositions show the progressively deeper erosion into the ophiolite's igneous layers during the late Oligocene and early Miocene. The mafic-rich ophiolite detritus requires introducing a new component to the standard QFL triangle diagrams in order to trace the evolution of ophiolite provenances. Sandstone composition can be an effective tool for tracing the emplacement history of a large ophiolite.

The Zambales ophiolite is not simply a piece of the South China Basin because the ophiolite's paleontologically-determined late Eocene age (approximately 40 million years) is almost 10 million years older than the oldest crust in the South China Basin (Taylor and Hayes, 1980). Spreading trends in the Zambales defined by magnetic anomalies and structures in the deeper igneous layers strike northwesterly. Paleomagnetic data suggest strong rotation of the Zambales block (De Boer et al, 1980; M. Fuller, written communication).

The earliest uplift of the western edge of the ophiolite, probably in the late Oligocene, cannot be firmly associated with the initiation of subduction at the Manila trench. The linear western edge represents several kilometers of vertical offset, but strike-slip motion may have played an important role in its early history. The tectonic lenses of chert and metamorphic rocks along the western edge of the ophiolite are exotic lithologies found nowhere else around the Zambales, and may have been emplaced by transcurrent faulting.

This study roughly outlines the development of the Zambales ophiolite, especially in the vertical dimension. However, it is apparent that the plate tectonic framework for the emplacement of the ophiolite against Luzon is much more complex than was originally envisioned. This paleotectonic history may have included strike-slip faulting and rotations as well as back-arc spreading and the initiation of a subduction zone. We must better understand the Luzon Paleogene tectonic configuration to apprehend the emplacement sequence for the Zambales Ophiolite.

ACKNOWLEDGMENTS

We thank S. B. Bachman for assistance in the field, assistance with the sandstone analysis, and numerous discussions. We also thank G. Bacuta of the Philippine Bureau of Mines (PMB) for aid with chromite evaluations; G. Haeck for assistance with analysis of metamorphic rocks; and O. Crispin, G. Casteneda, G. Balce, R. Villiones and T. Apostol of the PBM for assistance with field studies. Supported by National Science Foundation grant OCE-79-19164.

REFERENCES CITED

Amato, F. L., 1965, Stratigraphic paleontology in the Philippines: Philippine Geologist, v. 20, p. 121-140.

Anonymous, 1977, Report on geological survey of northeastern Luzon: Japan International Cooperation Agency and Metal Mining Agency of Japan, 45 p.

Balce, G. R., et al, 1980, Geology of the Baguio district and its implication on the tectonic development of the Luzon Central Cordillera: Geology and Paleontology of Southeast Asia, v. 21, p. 265-287.

Ballard, R. D., and T. H. van Andel, 1977, Morphology and tectonics of the inner rift valley at latitude 36°50'N on the Mid-Atlantic ridge: Geological Society of America Bulletin, v. 88, p. 507-530.

Bandy, O. L., 1963, Cenozoic planktonic foraminiferal zonation and basinal development in the Philippines: AAPG Bulletin, v. 47, p. 1733-1745.

Ben-Avraham, Z., 1978, The evolution of marginal basins and adjacent shelves in east and southeast Asia: Tectonophysics, v. 45, p. 269-288.

Bowin, C., et al, 1978, Plate convergence and accretion in Taiwan-Luzon region: AAPG Bulletin, v. 62, p. 1645-1672.

Bryner, L., 1967, Geology of the Barlo Mine and vicinity, Dasol, Pangasinan province, Luzon, Philippines: Philippine Bureau of Mines Report of Investigation n. 60, 55 p.

Cardwell, R. K., B. L. Isacks, and D. E. Karig, 1980, The spatial distribution of earthquakes, focal mechanism solutions and subducted lithosphere in the Philippine and northeastern Indonesian islands: American Geophysical Union, Monograph 23, p. 1-36.

Coleman, R. G., 1971, Plate tectonic emplacement of upper mantle peridotites among continental edges: Journal of Geophysical Research, v. 76, p. 1212-1222.

———, 1977, Ophiolites — ancient oceanic lithosphere?: New York, Springer-Verlag, 229.

Corby, W. G., et al, 1951, Geology and oil possibilities of the Philippines: Philippine Bureau of Mines Technical Bulletin n. 21, 363 p.

Damuth, J. E., 1980, Quaternary sedimentation processes in the south China basin as revealed by echo-character mapping and piston-core studies, *in* D. E. Hayes, ed., The tectonic and geologic evolution of southeast Asian seas and islands: American Geophysical Union, Monograph 23, p. 105-125.

De Boer, J., et al, 1980, The Bataan orogene: eastward subduction, tectonic rotations, and volcanism in the western Pacific (Philippines): Tectonophysics, v. 67, p. 251-282.

Dickinson, W. R., 1970, Interpreting detrital modes of gray-

wacke and arkose: Journal of Sedimentary Petrology, v. 40, p. 695-707.
———, and C. A. Suczek, 1979, Plate tectonics and sandstone compositions: AAPG Bulletin, v. 63, p. 2164-2182.
Divino-Santiago, P., 1963, Planktonic foraminiferal species from west side of Tarlac province, Luzon central valley: Philippine Geologist, v. 17, p. 69-99.
Donnelly, T. W., 1975, Neogene explosive volcanic activity of the western Pacific: Sites 292 and 296, Deep sea drilling project Leg 31, *in* D. E. Karig et al, eds., Initial reports of the deep sea drilling project, v. 31: Washington, D.C. U.S. Government Printing office, p. 577-598.
Fitch, T. J., 1972, Plate convergence, transcurrent faults and internal deformation adjacent to southeast Asia and the western Pacific: Journal of Geophysical Research, v. 77, p. 4432-4460.
Garcia, M. O., 1978, Criteria for the identification of ancient volcanic arcs: Earth Science Reviews, v. 14, p. 147-165.
Garrison, R. E., et al, 1979, Petrology, sedimentology, and diagenesis of hemipelagic limestone and tuffaceous turbidites in the Aksitero formation, Central Luzon, Philippines: U.S. Geological Survey, Professional Paper 1112, 16 p.
Hashimoto, W., 1980, Geologic development at the Philippines: Contributions to the Geology and Paleontology of Southeast Asia, v. 217, p. 83-190.
Hawkins, J. W., C. Evans, and G. Bacuta, 1980, Petrology of the Zambales range ophiolite, Luzon, Philippine Islands: gabbro, diabase and basalt: Eos, Transactions, American Geophysical Union, v. 61, p. 1154.
Hilde, T. W., S. Uyeda, and L. Kroenke, 1977, Evolution of the western Pacific and its margin: Tectonophysics, v. 38, p. 145-165.
Ingersoll, R. V., 1978, Petrofacies and petrologic evolution of the Late Cretaceous fore-arc basin, northern and central California: Journal of Geology, v. 86, p. 335-352.
Irwin, W. P., 1977, Ophiolitic terranes of California, Oregon, and Nevada, *in* North American ophiolites: Portland, Oregon, Oregon Department of Geology and Mineral Industries Bulletin 95, p. 75-92.
Karig, D. E., 1973, Plate convergence between the Philippines and the Ryukyu Islands: Marine Geology, v. 14, p. 153-168.
———, 1981, Initiation of subduction zones, *in* J. K. Legget, ed., Trench and fore-arc sedimentation and tectonics: Geological Society of London, Special Publication.
———, and G. F. Moore, 1975, Tectonically controlled sedimentation in marginal basins: Earth and Planetary Science Letters, v. 26, p. 233-238.
Lewis, B. T. R., et al, 1979, Leg 65 drills into young ocean crust: Geotimes, v. 24, n. 8, p. 16-18.
Murphy, R. W., 1973, The Manila trench-West Taiwan foldbelt: a flipped subduction zone: Geological Society of Malaysia, Bulletin G, July 1973, p. 27-42.
Philippine Bureau of Mines, 1963, Geological map of the Philippines, Sheet ND 51, City of Manila.
Roque, V. P., Jr., B. P. Reyes, and B. A. Gonzales, 1972, Report on the comparative stratigraphy of the east and west sides of the mid-Luzon central valley, Philippines: Mineral Engineering Magazine, September 1972, p. 11-62.
Rossman, D. L., 1964, Chromite deposits of the northcentral Zambales range, Luzon, Philippines: U.S. Geological Survey, Open File Report, 65 p.
Schweller, W. J., and D. E. Karig, 1979, Constraints on the origin and emplacement of the Zambales ophiolite, Luzon, Philippines: Geological Society of America, Abstracts with programs, v. 11, p. 512-513.
———, and S. B. Bachman, in press, Uplift and erosional history of the Zambales ophiolite, Luzon, from sandstone compositions: Journal of Sedimentary Petrology.
———, et al, in press, Sedimentation history and biostratigraphy of ophiolite-related tertiary sediments, Luzon, Philippines: Geological Society of America Bulletin.
Smith, W. D., 1907, The Asbestos and manganese deposits of Ilocos Norte, with notes on the geology of the region: Philippine Journal of Science, v. 2, p. 145-175.
Stoll, W. C., 1958, Geology and petrology of the Masinloc chromite deposits, Zambales, Luzon, Philippine Islands: Geological Society of America Bulletin, v. 69, p. 419-440.
Taylor, B., and D. E. Hayes, 1980, The tectonic evolution of the south China basin: American Geophysical Union, Monograph 23, p. 89-104.

Evolution of the Sialic Margin in the Central Western United States

R. C. Speed
Department of Geological Sciences
Northwestern University
Evanston, Illinois

The cryptic perimeter of Precambrian sialic North America lies within the cordillera of the western United States. Between this perimeter and the Pacific basin is an assembly of displaced terranes. Ages and origins of segments of the sialic margin, orogenic features of the continental foreland, and times of attachment of displaced terranes provide a partial record of the plate tectonic evolution of western North America. The same area records Mississippian and Early Triassic tectonic events, probably passive margin-arc collisions. An active margin developed late in the Triassic, accompanied by a continental arc and foreland deformation in a zone that spanned most of what is now the Great Basin. The shortened foreland cover may have been underlain by an extensive decollement. The sialic margin in California and Idaho is Mesozoic and the probable product of tectonic removal of tracts of North America and earlier accreted terranes.

A complete and resolvable record of late Precambrian and Phanerozoic continental margin tectonism in the United States cordillera seems to exist only in central Nevada. A comparable record has not been recognized elsewhere in the cordillera for various reasons: fragments containing an earlier margin were removed, the record of earlier events was obscured by younger tectonism, or, conceivably, earlier events left no imprint on such regions. The focus of this paper is on the interpretation of tectonic processes that affected the edge of ancient sialic North America in Paleozoic and Mesozoic time in the central United States cordillera and especially in the central Nevada region. Other problems addressed are: 1) identification of the cryptic sialic margin, 2) age and origin of various reaches of the sialic margin, and 3) reasons why the record of Paleozoic and older margin tectonism is preserved in central Nevada.

TECTONIC PROVINCES

Western North America can be divided into four provinces based on Phanerozoic motions and structures; approximate boundaries as shown on Figure 1. Province 1, the innermost, remained relatively undeformed by continental margin tectonics. It is the craton or sialic nucleus characterized by thick ancient crystalline crust, platform cover strata, and a history of heterogeneous vertical motions that may have been coupled to global rather than local plate processes (Sloss and Speed, 1974). Province 2 (Figure 1) is also underlain by ancient sialic crust but has undergone Phanerozoic deformation (shortening, extension, large uplift, translation, or rotation) probably related to margin tectonics. Such crust probably consists of fragments that are parautochthonous with respect to one another and to North America. The cover of province 2, however, probably includes exotic terranes that were obducted across the sialic margin.

Province 3 comprises displaced terranes (Coney, Jones, and Monger, 1980) which may or may not be exotic with respect to North America and which have accreted to one another and/or the outer margin of sialic North America. Such terranes include microplates which represent lithospheric fragments of magmatic arc, continental, or oceanic origin and packets of deformed basinal sedimentary rocks and megafragments that were assembled at least partly as accretionary prisms. Province 3 also contains autochthonous sedimentary and magmatic rocks that are contemporary with or younger than times of accretion (examples: Idaho batholith, Tertiary cover shown in

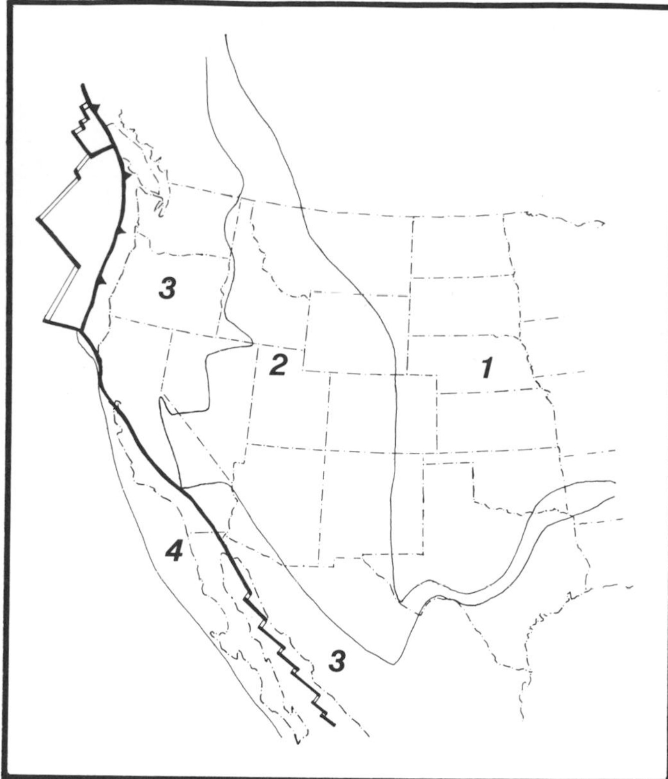

Figure 1 — Tectonic provinces of western continental North America. Province boundaries and welded continent-ocean boundaries are light lines; plate boundaries are heavy lines. Province 1: sialic Precambrian continent undeformed in Phanerozoic; 2: sialic Precambrian continent deformed in Phanerozoic; 3: region of displaced terranes now attached to North America; 4: displaced terranes now moving with respect to North America.

Figure 2). Province 4 consists of terranes that are currently moving with respect to North America and, for the moment, lie on the Pacific plate west of the San Andreas-Gulf of California system. It is not clear how much displacement between the Pacific and North American plates is being taken up by slip and/or strain east of the San Andreas fault. Estimates of relative velocity are as great as 2 cm/yr for intraplate western North America, but loci of preferred displacement are unknown (Minster and Jordan, 1978). Thus, the boundary between provinces 3 and 4 is poorly established, and some regions assumed to be fixed in province 3 may be in motion with respect to the sialic nucleus.

MARGIN OF ANCIENT SIALIC CRUST IN WESTERN NORTH AMERICA

The outer boundary of province 2 is the present margin of contiguous sialic Precambrian North America in the western United States. The ages and origins of different intervals of this boundary, therefore, should provide a partial record of plate margin tectonics that affected western North America. The sialic margin is a basement feature and is generally cryptic due to burial by younger rocks or nappes or obscuration by magmatism and metamorphism. Thus, the location of its surface trace (Figure 2) is approximate and requires explanation, given below.

The traditional stratigraphic approach has been to establish loci of outermost outcrops of autochthonous Lower Paleozoic and latest Precambrian strata of inferred continental platform or shelf facies (Roberts et al, 1958; Kay, 1960, Kay and Crawford, 1964; Stewart and Poole, 1974). The sialic margin presumably lies below or slightly outboard of these outcrops. In central Nevada (Figure 2), Paleozoic carbonate strata seem to be mainly autochthonous, at least at a local scale, and their outermost exposures may correctly mark the innermost possible limit of the sialic margin. Outside central Nevada (Figure 2), shelf and platform facies have been affected by intracontinental Mesozoic deformation (Figure 3), and their present distribution is a poor indication of their original extent (Dover, 1980; Speed; Speed and Sleep, 1982).

Another indicator of the surface trace of the cryptic sialic margin is the initial isotopic strontium ratio contour ($(^{87}Sr/^{86}Sr)_o$ in Mesozoic plutonic rocks thought to be autochthonous with respect to host rocks. The 0.706 contour, taken from Kistler and Peterman (1973), Armstrong, Taubeneck, and Hales, (1977), and R. W. Kistler (1981, written communication), is approximate because of limited sampling density, but the strontium data are better resolved than those of any other discriminator. The rationale for interpretation of this isotopic ratio is that continental lithospheres are older and rubidium-enriched relative to oceanic or island arc lithospheres; hence, magmas generated at the base of or within continental lithospheres should contain more radiogenic strontium. Empirically, magmas that have emerged from ancient continental lithospheres have initial strontium isotopic ratios ≥ 0.706 whereas those from oceanic or island arc lithospheres are largely between 0.702 and 0.705.

Lead isotopic ratios form spatial patterns in the western United States that support the delineation of the sialic margin by strontium isotopes (Doe, 1973; Zartman, 1974). Similarly, muscovite-bearing peraluminous Mesozoic plutons, thought to have derived from continental lithospheric sources, lie chiefly continentward of the 0.706 line (Miller and Bradfish, 1980). The 0.706 line is congruent with outermost Paleozoic shelf facies outcrops in Nevada (Figure 2).

Geophysical data in Nevada permit the assignment of the sialic margin to the vicinity of the 0.706 strontium isotopic contour. A strong regional Bouguer gravity gradient (average 0.5 mgal/km over 100 km; Cogbill, 1979) lies close to, or includes, the 0.706 line. The gradient belt may reflect thickening toward the continent and/or a decrease in density of the crust from the edge of Precambrian North America. Seismic station delays and poorly resolved refraction depths to the Moho in Nevada suggest thicker crust east and south of the 0.706 line (Eaton, 1963; Hill and Pakiser, 1966; Prohdehl, 1979; Cogbill, 1979; Stauber, 1980; K. F. Priestly, written communication, 1981). The apparently thinner and denser crust in the region of less

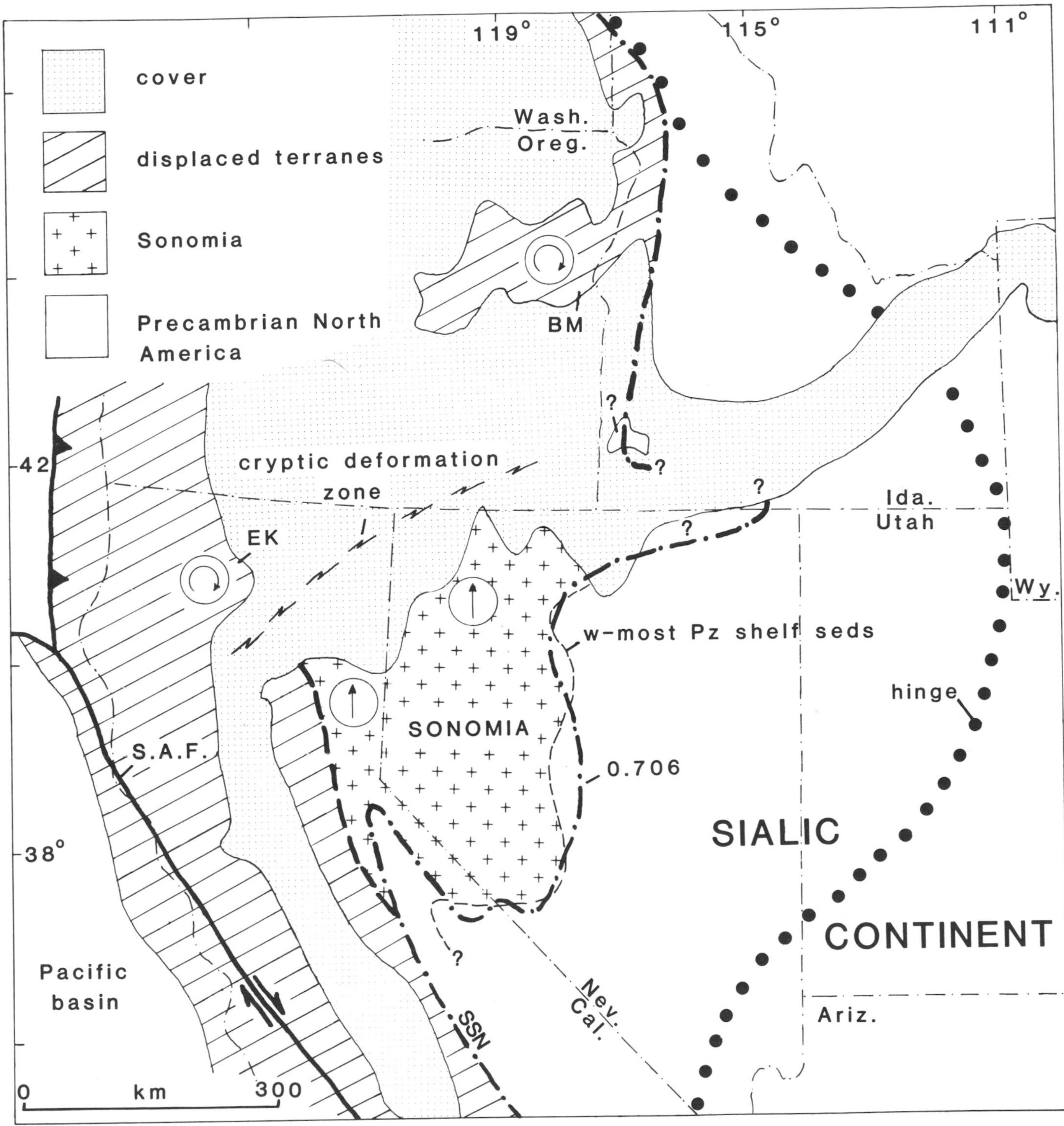

Figure 2 — Map of part of central United States cordillera. Dots = late Precambrian-Paleozoic platform-shelf hinge; heavy dash-dot line = $(^{87}Sr/^{86}Sr)_0$ = 0.706 contour which is taken to approximate present margin of ancient sialic continent; light dashed line = outermost exposures of Paleozoic shelf strata; heavy dashed line = suture between Sonomia and younger displaced teranes; circled straight arrows = Cretaceous magnetizations with North American orientations; circled curved arrows = Cretaceous and early Cenozoic magnetizations rotated clockwise with respect to North America; cover consists of autochthonous Cenozoic sedimentary and volcanic rocks. SSN = southern Sierra Nevada; BM = Blue Mountains; SAF = San Andreas Fault.

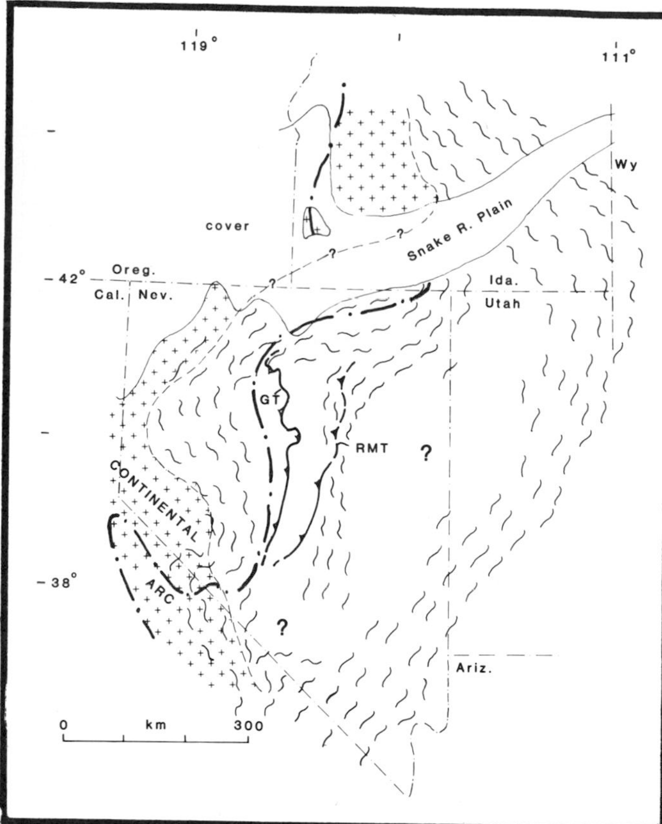

Figure 3 — Map of some tectonic features in part of United States cordillera. Short curved lines = Mesozoic cordilleran thrust and fold belt; RMT = continentward limit of exposures of Golconda allochthon (Early Triassic); heavy dash-dot line = $^{87}Sr/^{86}Sr)_0$ = 0.706 which is taken to approximate present margin of ancient sialic continent.

radiogenic initial strontium in Nevada is in agreement with the inference that an exotic noncontinental terrane (Sonomia, Figure 2) exists outboard of the 0.706 line.

In westernmost Nevada and eastern California, gravity and refraction data (Eaton, 1963; Cogbill, 1979) indicate thicker and/or less dense crust in a northerly-trending zone. This zone is interpreted as the root of a Mesozoic continental arc athwart the ancient sialic margin.

Although the geophysical crustal structure of Nevada admits interpretation by the distribution of ancient lithospheric units, the alternative of a modern tectonic origin is equally plausible. Crustal thicknesses could be entirely late Cenozoic and a product of laterally different rates of extensional thinning by basin-range tectonics. Lateral differences in crustal density could have been created by basalt diking proportional to extension rates (Lachenbruch and Sass, 1978). No independent evidence indicates, however, that differential extension rates exist, or that extension rates are correlative with crustal or lithospheric thickness.

To conclude, present data indicate that the edge of Precambrian North America lies near the 0.706 strontium isotopic contour in the area of Figure 2. Hereafter, the sialic margin and the 0.706 contour are assumed to be coincident.

DISPLACED TERRANES

The division of province 3 (Figure 1) into discrete tectonic units and the interpretation of ages of final motion relative to North America, the degree of pre-collision assembly, and the transport directions of these units are subjects of much current effort (e.g. Coney, Jones, and Monger, 1980; Beck, 1980; Monger, Price, and Tempelman-Kluit, 1982). I focus here on the relation of displaced terranes to the evolution of the sialic margin of the western United States.

The oldest displaced terrane lodged permanently against the western perimeter of sialic North America is Sonomia (Figure 2), thought to consist of sequential Paleozoic arc terranes that accreted in northwestern Nevada in Early Triassic time (Speed, 1977, 1979). Sonomia added significant local girth to Triassic North America and may have originally been more extensive than it is now. The boundary of North America where Sonomia sutured was probably a passive continental margin of Paleozoic and late Precambrian age.

Lodging of younger displaced terranes seems to have occurred during the existence of an active margin along western North America. Displaced terranes of younger attachment ages that contact sialic North America and Sonomia consist mainly of arc-related rocks and melange of varied protolith ages (Schweikert and Cowan, 1975; Irwin, 1977; Vallier, Brooks, and Thayer, 1977; Saleeby et al, 1978; Brooks and Vallier, 1978; Dickinson and Thayer, 1978; Saleeby, 1982). In the southern Sierra Nevada (SSN, Figure 2), melange accreted in Late Triassic or Early Jurassic time along a regional suture that is approximately coincident with the 0.706 line (Saleeby, 1982). Isotopic data imply that melange is sutured to sialic North America, although intense post-collision metamorphism, magmatism, and deformation obscure the relationship between the rocks east of the 0.706 line in the southern Sierra Nevada and those of the sialic continent. Northward in the Sierra Nevada, the same suture apparently cuts the older boundary between Sonomia and the sialic continent (Figure 2). In the northern Sierra Nevada, accretion of outboard terranes to Sonomia was probably contemporaneous with suturing to the south.

Sialic North America contacts displaced terranes in western Idaho (Figure 2) where east to northeast trending belts of Paleozoic and Triassic igneous rocks and melange in the Blue Mountains (BM, Figure 2) discordantly intersect the .706 line. Paleomagnetic data indicate these terranes may have been relatively south of their current contact with North America in late Triassic time (Jones, Silberling, and Hillhouse, 1977; Hillhouse, Grommé, and Vallier, 1982), but by the Late Jurassic or Early Cretaceous, they were at or near present latitudes (Wilson and Cox, 1980). Thus, the displaced arc rocks of the Blue Mountains may have migrated north to their current suture after other terranes accreted to the sialic margin and to Sonomia in California. The age of attachment in Idaho is within the Cretaceous according to dates of plutons that stitch the contact (85 to 90 million years) and those that are cut by the suture (130 to 140 million years) (Hamilton, 1963; Brooks and Vallier, 1973; T. L. Vallier, 1981, oral communication).

It can also be inferred that relative motion occurred between Triassic North America (which includes Sonomia) and displaced terranes of the Klamath (EK, Figure 2) and Blue Mountains (BM, Figure 2) regions in both Late Mesozoic and Early Tertiary time. Mesozoic and Eocene rocks in both outboard regions have substantial clockwise paleomagnetic declination anomalies (Simpson and Cox, 1977; Beck et al, 1978; Wilson and Cox, 1980) with respect to the North American magnetic declinations of Cretaceous age in Sonomia, as indicated in Figure 2. In contrast, magnetizations in rocks of Sonomia and its Mesozoic cover, determined in two regions (Figure 2) (Hannah and Verosub, 1980; Russell et al, 1982) were reset during Cretaceous magmatism but indicate autochthoneity of Sonomia with respect to North America since at least the Cretaceous. The loci of such motions, probably below Cenozoic cover (Figure 2), and their kinematics are uncertain.

Paleozoic and Mesozoic rocks in the eastern Klamath Mountains of northern California (EK, Figure 2) were provisionally included (Speed, 1979) in the Sonomia microplate. The declination anomalies now indicate the Klamath tract is either exotic to Sonomia or was detached from Sonomia.

The foregoing discussion demonstrates the large lateral variation in age of final attachment (Early Triassic to Cretaceous) of displaced terranes to the margin of Precambrian North America in the central U.S. cordillera.

REGIONAL TECTONIC EVENTS IN PROVINCE 2

The Phanerozoic succession of regional tectonic events recorded by structures and rocks in province 2 of the sialic continent began with a passive continental margin forming in late Precambrian time. Autochthonous and parautochthonous late Precambrian and early Paleozoic shelf strata are widely exposed in the area of Figure 2 between the platform-shelf hinge and the sialic margin (.706 line). Lower strata of this sequence compose a marginward-thickening prism. Stewart (1972, 1976) and Stewart and Suczek (1977) consider this prism to have been deposited, starting perhaps at 850 m.y. ago, during the arching, rifting, and early drifting phases of the development of a western North America passive margin. These strata are thought to lie above older Precambrian beds and crystalline rocks with structural trends differing from strikes of isopachs of the overlying clastic prism. Development of the late Precambrian passive margin may have greatly reshaped western North America from its prior configuration (Stewart, 1976). Early Paleozoic strata that succeed the clastic prism are mostly carbonate rocks indicative of shallow marine subsiding shelf environments.

The Antler orogeny defined in central Nevada (Roberts et al, 1958) began in Early Mississippian time and was the first widespread disturbance of the continental shelf after creation of the passive margin. Antler events were: 1) emplacement of the Roberts Mountains allochthon of early Paleozoic deformed oceanic, sedimentary, and volcanic rocks at least 130 km inboard of the sialic margin (Figure 3); 2) subsidence of a foreland basin in the Mississipian just continentward of and colinear with the allochthon (Poole, 1974); 3) filling the foreland basin with sediment from the allochthon; and 4) uplift of the allochthon and nearby shelf in Pennsylvanian and Permian time. No contemporaneous magmatism, metamorphism, or pervasive deformation of the overridden shelf rocks accompanied the orogeny.

The next event was the Sonoma orogeny (Silberling and Roberts, 1962) in Early Triassic time when the regional Golconda allochthon (Figure 3) was emplaced above the Roberts Mountains allochthon and subjacent continental shelf. The Golconda allochthon (Silberling, 1973; Speed, 1977, 1979) contains oceanic sedimentary and volcanic rocks of late Paleozoic age that were obducted as much as 100 km inboard of the sialic margin. Like the Antler orogeny, the Sonoma imposed no thermal effects and little deformation on rocks below the Golconda allochthon. In contrast to the Antler, however, the Sonoma generated a minor foreland basin that is recognized only in the southernmost reach of the Sonoma belt. The correlative timing between emplacement of the Golconda allochthon and collision of Sonomia implies the two events were related, as discussed later.

Following the Sonoma orogeny and attachment of Sonomia, a Triassic carbonate platform (Nichols and Silberling, 1977; Speed, 1978) developed along much of the sialic margin and a short distance inland in Nevada. West of the platform, a deep sedimentary basin developed in the region underlain by Sonomia in part of Triassic time (Speed, 1978). The absence of Triassic platform and basin deposits elsewhere along the sialic margin suggests either that Triassic paleogeography differed markedly from that of Nevada or, more probable, that such deposits were tectonically removed later in Mesozoic time.

The emergence of siliceous magmatic rocks in a probable belt, shown as a continental arc in Figure 3, signaled the onset of active margin tectonism in western North America (Burchfiel and Davis, 1972). The belt includes scattered remnants of generally poorly-dated volcanogenic rocks and plutons whose ages range from Late Triassic to Late Cretaceous (Burchfiel, Pelton, and Sutter, 1970; Smith et al, 1971; Armstrong and Suppe, 1973; Kistler and Peterman, 1973; Speed, 1978; Dunne, Gulliver, and Sylvester, 1978; Russell, 1981). The belt is within ancient sialic crust in California (Burchfiel and Davis, 1972) and apparently crosses into Sonomia in Nevada and continues north as far as northwestern Nevada. There, Russell (1981) found Upper Triassic, Jurassic, and possibly Cretaceous arc rocks in a volcanic pile built on the floor of a deep marine basin.

Although layered rocks of the magmatic belt are generally deformed, they are probably not greatly displaced. This is because: 1) they occur in a long belt that is partly within the sialic continent; 2) their strontium isotopic composition changes appropriately at the Sonomia boundary; and, 3) there seem to be no anomalies with respect to North America in Jurassic and Cretaceous magnetization of rocks of the magmatic belt (Gromme and Merrill, 1965; Gromme, Merrill, and Verhoogan, 1967; Hannah and Verosub, 1980; Russell et al, 1982).

The trend of the magmatic belt in western Nevada suggests that plutonic rocks of central Idaho (Idaho batholith) are a northward continuation below the Snake River Plain (Figure 3). It is noteworthy that in contrast to the long duration of Mesozoic magmatism in Nevada and California, Idaho batholith plutonic rocks give radiometric dates that are mainly Late Cretaceous (Armstrong et al, 1977).

Another effect of Mesozoic active margin tectonism in the central western United States was deformation of the foreland between the continental arc and the Paleozoic platform-shelf hinge during Jurassic and Cretaceous time. (Figures 2 and 3; Silberling and Roberts, 1962; Armstrong, 1968; Burchfiel, Pelton, and Sutter, 1970; Royse, Warner, and Reese, 1975; Speed, 1978; Oldow, 1981; Russell, 1981). In the area of Figure 3, foreland structures of Mesozoic age consist mainly of thrust nappes and folds in Phanerozoic and late Proterozoic strata. Such structures represent shortening under nonmetamorphic conditions and without evident involvement of the crystalline basement. The foreland deformation zone seems to occur in single belts in Idaho and southeastern California. In what is now the Great Basin, however, the zone branches into three apparently discrete belts that enclose enclaves of relatively little Mesozoic deformation in at least the upper few kilometers of Phanerozoic cover.

The westernmost of the three Mesozoic deformation belts in the Great Basin (Figure 3) includes the continental arc rocks, the ancient sialic margin and suprajacent Triassic platform edge strata, and the Triassic basin fill that lay between. Shortening directions in this belt are mainly easterly but locally vary from northeasterly to southeasternly, perhaps due to later rotations. Movements occurred in Early and Late Jurassic and Early and medial Cretaceous times (Speed, 1978, Speed and Kistler, 1979; Russell, 1981). Dating of displacements in the western belt, however, is too sparse to suggest whether deformation was temporally continuous or spatially progressive.

A central belt of deformation at about 117° (Figure 3) includes thrusts and folds that were generated at least partly in the Early Cretaceous, and probably before and after that time (Nolan, Merriam, and Blake, 1974; Smith and Ketner, 1977). Displacements include easterly older-over-younger transport and throw greater than several kilometers. The eastern or Sevier belt (Armstrong, 1968) contains continent-verging thrusts that caused 100 to 150 km of contraction in the cover. Sevier thrusting began in the Late Jurassic or probably earlier (Allmendinger and Jordan, 1981), and frontal breakthroughs migrated progressively east until cessation in Paleocene time.

The western of the two enclaves between thrust belts is defined by the lack of deformation of Triassic shelf strata and autochthonous Late Paleozoic beds. The eastern enclave is identified by the regional concordance of Paleozoic and Cenozoic strata and lack of Mesozoic tectonic structures other than gentle folds in the Paleozoic cover (Armstrong, 1968; Hose and Blake, 1974). Substantial Mesozoic deformation, however, occurred in the eastern enclave at depths of 10 km or more. This is indicated by strongly folded metamorphic rocks with dated synkinematic recrystallization and post-kinematic intrusions (Howard, 1980); Miller, 1980; Snoke, 1980) that were uplifted during Cenozoic extension (Davis and Coney, 1979; Wernicke, 1981).

The similarity of shortening directions and the at least partial contemporaneity of displacements in the three Mesozoic thrust belts in the Great Basin imply they were kinematically related during foreland deformation. Displacements in each belt may have transferred locally down into a ductile substratum such that the intervening enclaves are rooted. Alternatively, the displacements that surfaced in the thrust belts may have emerged from a regional decollement below the Mesozoic Great Basin. In this case, the two enclaves are meganappes that underwent quasi-rigid translations during foreland shortening. The existence of Mesozoic deformation at deep levels in the eastern enclave supports the second hypothesis.

Reconstructions of the Sevier belt indicate about 50% shortening (Armstrong, 1968; Royce, Warner, and Reese, 1975). Applying this figure to the other two belts, the foreland cover of the Great Basin probably contracted at least 300 km in Jurassic and Cretaceous time.

A consequence of Mesozoic thrusting in the western belt was the redistribution of Paleozoic and Mesozoic rocks and obfuscation of much of earlier tectonic record. Stratigraphy and tectonostratigraphy of central Nevada underwent major reshuffling in the western belt (Figure 3) with uncertain amounts of transport on individual thrusts. Dover (1980) contended that in central Idaho, rocks such as those of the Roberts Mountains allochthon in Nevada may have been displaced by Mesozoic thrusts 100 km east of their late Paleozoic position. The Paleozoic tectonic record seems to be preserved in the tectonic enclaves of Nevada, but it, too, may have been thrust eastward by 150 km or more.

Cenozoic events in the area shown in Figure 2 include Laramide high-angle thrusting east of the earlier thrust and fold belt, local supracrustal extension, infrastructural uplift (Armstrong, 1972; Davis and Coney, 1979), and basin-range uplift and extension. Cumulative Cenozoic extensions in the Great Basin have been mainly east-west as a principal direction, but the magnitude and distribution of displacements are poorly known. The eastern half of the region has probably extended more (50 to 100%; Wernicke, 1981) than the western half (10 to 30%; Stewart, 1978). Such estimates suggest that the western edge of the cover of the Mesozoic foreland in the Great Basin could have at least regained its pre-shortening Triassic position by Cenozoic extension.

AGES AND ORIGINS OF THE SIALIC MARGIN

Next, the ages and tectonic origins of segments of the margin of Precambrian North America in central United States cordillera (Figure 2) are considered. I propose that the late Precambrian passive margin may be preserved in the western tectonic enclave of central Nevada but that elsewhere, the sialic margin is younger and a result of Mesozoic tectonic removal.

Facies differences among autochthonous and parautochthonous Lower Paleozoic shelf strata in central Nevada permit interpretation of inner and outer shelf environments (Kay, 1960; Kay and Crawford, 1964; Stewart and Poole, 1974; Matti and McKee, 1977; Rowell, Rees, and Suczek, 1979). Moreover, shoal-water carbonate rocks that locally form outermost outcrops of the shelf succession are considered as deposits on a ridged outer shelf edge (Kay and Crawford, 1964; Matti and McKee, 1977). Because such rocks are approximately suprajacent to the sialic margin, it can inferred that the margin in central Nevada is a preserved segment of the late Precambrian — Paleozoic passive margin or is not far inboard of the passive margin. Thus, the width of the subsiding shelf in central Nevada was about 600 km, assuming Mesozoic shortening and Cenozoic extension canceled one another.

The Early Triassic suture between Sonomia and the sialic continent in central Nevada can therefore be inferred to be the locus of collision between an island

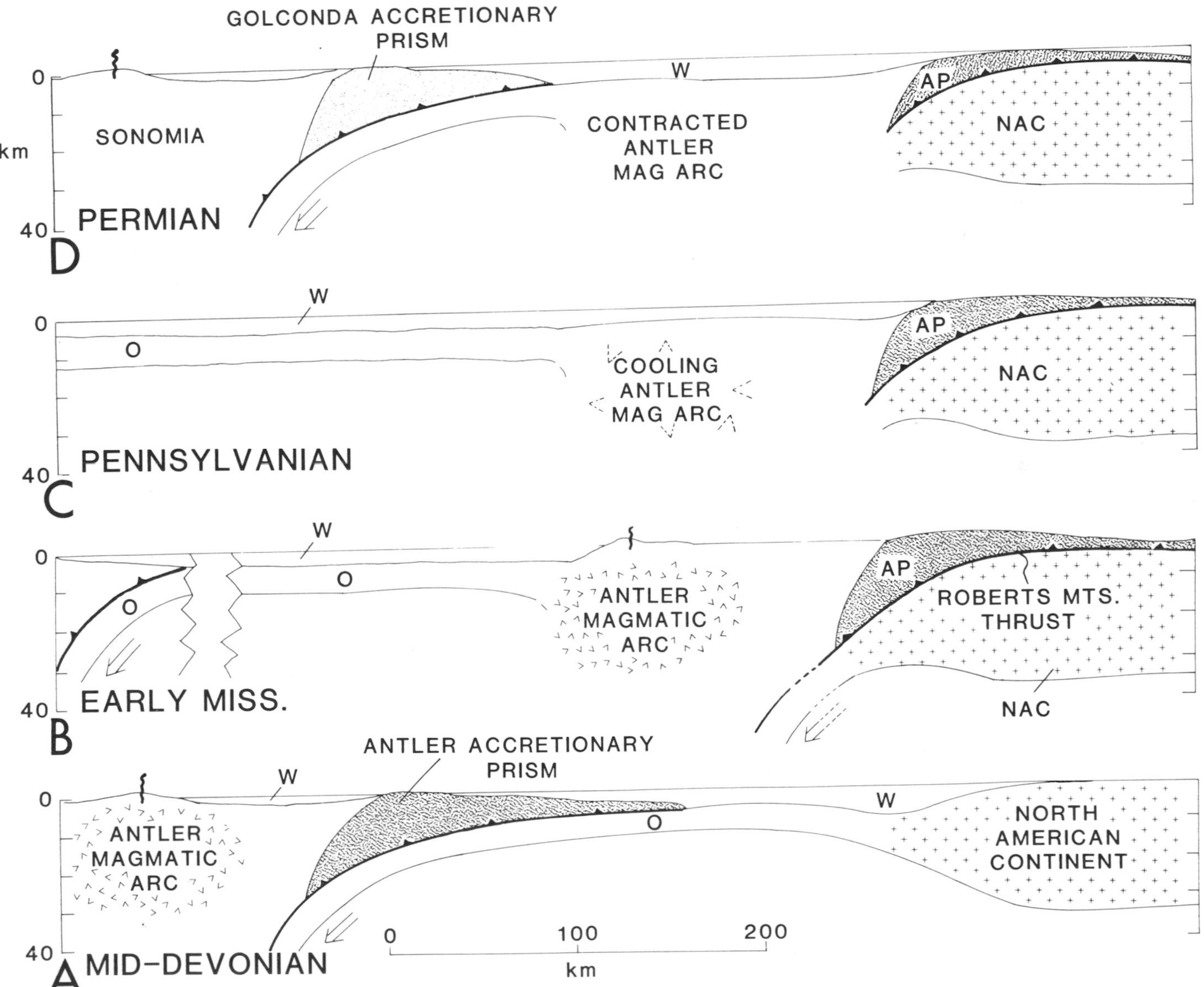

Figure 4 — Sequential model sections showing Paleozoic tectonics at passive margin of central United States cordillera. Continent held fixed. O = oceanic crust; W = water; AP = accretionary prism; NAC = North American Continent. See text for discussion.

arc terrane and the long-standing passive margin of western North America. This implies a westward-dipping subduction zone during Sonomia's closure, a configuration supported by the absence of volcanism inboard of the sialic margin. If the present sialic and ancient passive margins are nearly coincident, the collision probably stopped at the onset of subduction of sialic North America, as shown in Figure 4B.

The geology of the sialic margin in Idaho differs markedly from that in central Nevada (Figure 2). If Triassic strata and subjacent tectonostratigraphic sequences of western and central Nevada originally existed to the north in eastern Oregon and western Idaho (Figure 2), they have largely been removed. Moreover, the present sialic margin in Idaho must be well inland of the Precambrian passive margin, implying that part of the sialic continent was also removed, as noted by Hamilton (1976). Other evidence for tectonic removal west of the 0.706 line in Idaho and Oregon is: 1) northward convergence and intersection of the sialic margin and Paleozoic hinge (Figure 2); 2) absence of Sonomia and other accretants of pre-Cretaceous ages of attachment; and, 3) absence of continental arc rocks in Idaho with radiometric ages older than Late Cretaceous.

It is inferred that a large chunk of sialic continent, together with early accreted terranes and continental arc rocks, was removed in mid-Mesozoic time from eastern Oregon and western Idaho. The removal was conceivably by subduction, but more likely by rafting away with right slip and/or rifting motions with respect to the remnant sialic margin. The microplate so created may now reside in Washington and British Columbia as parts of composite Terrane I (Monger, Price, and Templeman-Kluit, 1982) which shows early and mid-Mesozoic magnetic inclinations about equivalent to latitudes of Nevada and Idaho (Irving, Monger, and Yole, 1980). Following removal of the continental fragment from Idaho and Oregon and probable generation of oceanic lithosphere in its wake, the displaced Blue Mountain terranes (Figure 2) moved from the west and into place in the Cretaceous. They perhaps rotated as they entered the tectonic embayment north of Sonomia (Figure 2). The Cretaceous collision may account for the apparently young ages of continental magmatic rocks in Idaho.

Models that do not employ tectonic removal in Idaho must evoke: 1) fundamentally different Triassic and Paleozoic tectonic histories of the sialic margin in Nevada and Idaho; 2) unusual passive margin tectonics in which the sialic margin cuts the shelf hinge; and, 3) tectonic means of keeping the sialic margin in Idaho free of lodged displaced terranes until the Cretaceous.

The north to northwest trending sialic margin in the southern Sierra Nevada of California (SSN, Figure 2) is probably a truncation (Hamilton and Myers, 1966). The main evidence is the apparent absence of Sonomia rocks in the southern Sierra Nevada and the colinearity of the west edges of Sonomia and the sialic continent in the Sierra Nevada. The age of truncation is evidently Triassic, between times of Sonomia's collision and the accretion of younger displaced terranes. Burchfiel and Davis (1972) interpreted the truncation as a Permian or Triassic transform margin, and Speed (1978) related it to the emplacement of Sonomia.

PASSIVE MARGIN TECTONICS IN CENTRAL NEVADA

Late Precambrian and Paleozoic passive margin and shelf features of the western United States were spared strong Jurassic and Cretaceous deformation, redistribution, and tectonic erosion only in the tectonic enclaves of Nevada. It is noteworthy that structures clearly arising from the Mississippian Antler and Early Triassic Sonoma orogenies have not been identified beyond central Nevada. Thus, the record of Paleozoic and Triassic continental margin tectonics of the western United States must be interpreted from this small region, as considered below.

The late Precambrian passive margin was apparently unaffected by plate boundary tectonics until the Early Mississippian Antler orogeny. The characteristics of this orogeny in comparison with modern tectonic analogs seem best explained by passive margin collision with an island arc system, (Moores, 1970; Dickinson, 1977; Speed 1977). The following model is from Speed and Sleep (1982). An exotic Antler island arc (Figure 4a) fronted by a large accretionary prism moved continentward in the Devonian, subducting oceanic lithosphere attached to the passive margin of western North America. Closure continued until the sialic continent attempted to underthrust the arc in Early Mississippian time (Figure 4b). Buoyancy of the sialic crust arrested closure and the accretionary prism became the Roberts Mountains allochthon, and convergence was taken up at unknown oceanward site.

The magmatic arc, now sutured to the continent (Figure 4b) and without subduction-related heating, cooled and contracted. Some 40 million years later, the continent was once again adjoined by a deep oceanic basin (Figure 4c).

The obducted accretionary prism created an enormous subaerial mountain belt on the outer shelf. The vertical load of the prism caused elastic or elastico-viscous subsidence of the shelf with a basin-arch inflection 50 to 150 km east of the prism (allochthon) toe. A linear foreland basin arose from the ocean-facing slope of the loaded shelf and the continent-facing slope of the beached accretionary prism. Sediment eroded from the prism was transported mainly to the foreland basin in the Early Mississippian, but later to both flanks as the sutured magmatic are contracted and sank. Erosion of the prism caused outer shelf uplift by unloading and isostacy, as recorded in the late Paleozoic uplift of the Antler orogenic belt.

This model accounts for: 1) emplacement of deformed oceanic rocks over nearly undeformed shelf strata without accompanying magmatism or metamorphism; 2) the foreland basin-highland evolution as a consequence of vertical loading of the shelf by the prism; and, 3) the continental margin regaining its former configuration after collision except for the existence of erosional remnants of the obducted accretionary prism on the outer shelf and foreland basin fill inboard.

The Early Triassic Sonoma orogeny followed a similar sequence of events. Late in the Permian another terrane, the magmatic arc of Sonomia and its accretionary prism (Figure 4d), converged on the passive margin of western North America (Speed, 1977, 1979). By this time the Antler magmatic arc had almost completely contracted and was probably subductible. The collision of Sonomia and the continent repeated the scene in Figure 4b, except that a new accretionary prism (the Golconda allochthon) was emplaced across the continental slope and outer shelf and the external parts of the older allochthon.

As in the Antler orogeny, Sonomia's collision involved no significant deformation, metamorphism, or magmatism within the sialic continent. Moreover, like the postulated Antler magmatic arc, Sonomia contracted after collision and formed a deep water basin external to the sialic margin in northwestern Nevada (Speed, 1978). Unlike the Antler orogeny, however, the Aonoman event did not apparently create a foreland basin along the entire length of the belt. This might be explained by the Golconda's having a smaller width (and probable lesser thickness) than the Roberts Mountains allochthon. The smaller size would have yielded smaller deflection of the shelf and may have precluded significant subaerial exposure of the prism (Figure 4b), therefore rendering the foreland basin inconspicuous because of unavailable abundant coarse sediment. Alternatively, erosion may have removed the evidence of a Sonoman foreland basin.

To conclude, the main features of the two passive margin collisions are the obducted oceanic rocks and synorogenic sediments. If later erosion had completely stripped the allochthons and foreland basin fills, the occurence of such collisional events would be hard to detect.

ACTIVE MARGIN TECTONICS

In the Triassic, the margin of western North America changed from passive to active in the central United States cordillera south of Idaho. (Burchfiel and Davis, 1972). The truncation of the sialic continent and Sonomia in California was presumably related. The position of the Mesozoic continental arc (Figure 3) indicates that the preserved Precambrian passive margin in central Nevada was well inland from the Mesozoic subduction trace and protected by the bulwark of Sonomia. This raises questions concerning why the passive margin features were not obliterated by Mesozoic foreland deformation and why the deformation belt is so wide and contains enclaves of rocks with less Mesozoic deformation in Nevada.

The Mesozoic thrust and fold belt seems to occur mainly between the continental arc and Paleozoic hinge which are separated by highly varied distances along the North American foreland. If displacement was laterally constant in the foreland deformation zone, mean shortening strain in the supracrustal rocks would vary inversely with the belt width and be minimum at central Nevada latitudes. Furthermore, strain might not be uniformly distributed in the wide zone of the belt but rather be concentrated at loci, such as ramps in crustal layering, where stress would focus during lateral loading.

The discrete thrust belts in Nevada may each be related to layering deflections: the margin of the sialic continent in the western branch and the Paleozoic hinge in the eastern, and the crustal depression due to loading by the Roberts Mountains allochthon in the central. Each may have localized breakthroughs of displacement to the surface from a decollement at depth. Thus, the passive margin in Nevada may have caused nappes to pile against it but was not itself subject to severe disruption.

CONCLUSIONS

The present margin of ancient sialic North America in the central western United States varies greatly in age and origin. In central Nevada, the sialic margin may be a preserved segment of the late Precambrian passive margin of western North America. Elsewhere in the central cordillera, the sialic margin is Mesozoic and is the inferred product of rafting away of fragments of sialic continent together with exotic masses that had accreted earlier.

Ages of final attachment of displaced terranes to the present sialic margin also vary with position. The oldest seems to be Early Triassic in Nevada, and the youngest, Cretaceous in Idaho.

Mid-Mesozoic tectonism in the foreland deformation belts affected the marginal zone of the sialic continent and widely obscured earlier tectonostratigraphic and structural relationships. Important exceptions are large tectonic enclaves of lesser Mesozoic deformation in central Nevada, around which the thrust belt branches and rejoins. The enclave contains the only well-preserved record of earlier margin tectonics in the western United States.

Data from the central Nevada enclaves show that major tectonic events affected the passive margin of North America in Early Mississippian and Early Triassic times before the development of an active margin in California in the Triassic. In both events, magmatic arc systems collided with the sialic margin following the underriding of large accretionary prisms by the slope and outer shelf of the continent. The major orogenic effect on the sialic continent of such encounters was the emplacement of the accretionary prisms and, in one case, the generation of an extensive foreland basin.

The Mesozoic foreland thrust and fold belt that developed during active margin tectonism apparently spans the full width of the Mesozoic continent between continental arc and Paleozoic platform-shelf hinge. Branches in the thrust belt in central Nevada may follow older decelvities in crustal layering.

ACKNOWLEDGMENTS

I am grateful to L. L. Sloss, E. L. Miller, and B. J. Russell for manuscript review and to T. L. Vallier for permission to mention his unpublished dating of a suture in Idaho. This work was supported by National Science Foundation Grant EA-7911150.

REFERENCES CITED

Allmendinger, R. W., and T. E. Jordan, 1981, Mesozoic evolution, hinterland of the Sevier orogenic belt: Geology, v. 9, p. 309-314.

Armstrong, R. L., 1968, Sevier orogenic belt in Nevada and Utah: Geological Society of America, Bulletin, v. 79, p. 429-458.

———, 1972, Low-angle (denudation) faults, hinterland of the Sevier orogenic belt, eastern Nevada and western Utah: Geological Society of America, Bulletin, v. 83, p. 1729-1754.

———, and J. Suppe, 1973, Potassium-argon geochronometry of Mesozoic igneous rocks in Nevada, Utah, and southern California: Geological Society of America, Bulletin, v. 84, p. 1375-1392.

———, W. H. Taubeneck, and P. O. Hales, 1977, Rb-Sr and K-Ar geochrometry of Mesozoic granitic rocks and their Sr isotopic composition, Oregon, Washington and Idaho: Geological Society of America, Bulletin, v. 88, p. 397-441.

Beck, M. E., Jr., 1980, Paleomagnetic record of late-margin tectonic proceses along the western edge of North America: Journal of Geophysical Research, v. 85, p. 7115-7131.

———, et al, 1978, Paleomagnetism of the middle Tertiary Clarno Formation, north-central Oregon: Constraint on models for tectonic rotation (Abs.): Washington, D.C., Eos, v. 59, p. 1058.

Brooks, H. C. and T. L. Vallier, 1978, Mesozoic rocks and tectonic evolution of eastern Oregon and western Idaho: Pacific Coast Paleogeography Symposium 2, Mesozoic Paleogeography of the western U.S., Society of Economic Paleontologists and Mineralogists, p. 133-146.

Burchfiel, B. C., and G. A. Davis, 1972, Structural framework and evolution of the southern part of the Cordilleran orogen, western United States: American Journal of Science, v. 272, p. 97-118.

———, P. J. Pelton, and J. Sutter, 1970, An early Mesozoic deformation belt in south-central Nevada — southeastern California: Geological Society of America Bulletin, v. 81, p. 211-215.

Cogbill, A. H., 1979, Relationships of crustal structure and seismicity, western Great Basin: Ph.D. thesis, Northwestern University, 289 p.

Coney, P. J., D. L. Jones and J. W. H. Monger, 1980, Cordilleran suspect terranes: Nature, v. 288, p. 329-333.

Davis, G. H., and P. J. Coney, 1979, Geological development of the cordilleran metamorphic core complexes: Geology, v. 7, p. 120-124.

Dickinson, W. R., 1977, Paleozoic plate tectonics and the evolution of the Cordilleran continental margin: Pacific Coast Paleogeography Symposium 1, Paleozoic Paleogeography of the Western U.S., Society of Economic Paleontologists and Mineraologists, p. 137-157.

———, and T. P. Thayer, 1978, Paleogeographic and paleotectonic implications of Mesozoic stratigraphy and structure in the John Day inlier of central Oregon: Pacific Coast Paleogeography Symposium 2, Mesozoic Paleogeography of the Western U.S., Society of Economic Paleontologist and Minerologists, p. 395-408.

Doe, B. R., 1973, Variations in lead-isotopic compositions in Mesozoic granitic rocks of California — A preliminary investigation: Geological Society of America, Bulletin, v. 84, p. 3513-3526.

Dover, J. H., 1980, Status of Antler orogeny in central Idaho — clarifications and constraints from the Pioneer Mountains: Rocky Mountain Paleogeography Symposium 1, Paleozoic Paleogeography of the West-Central U.S., Society of Economic Paleontologist and Mineralogists, p. 371-386.

Dunne, G. C., R. M. Gulliver, and A. G. Sylvester, 1978, Mesozoic evolution of rocks of the White, Inyo, Argus, and State Ranges, eastern California: Pacific Coast Paleogeography Symposium 2, Mesozoic Paleogeography of the Western U.S., Society of Economic Paleontologists and mineralogists, p. 189-209.

Eaton, J. P., 1963, Crustal structure from San Francisco, California, to Eureka, Nevada, from seismic refraction measurements: Journal of Geophysical Research, v. 68, p. 5789-5806.

Gromme, C. S., and R. T. Merrill, 1965, Paleomagnetism of Late Cretaceous granitic plutons in the Sierra Nevada, California: further results: Journal of Geophysical Research, v. 70, p. 3407-3420.

———, R. T. Merrill, and J. Verhoogen, 1967, Paleomagnetism of Jurassic and Cretaceous plutonic rocks in the Sierra Nevada, California, and its significance for polar wandering and continental drift: Journal of Geophysical Research, v. 72, p. 5661-5684.

Hamilton, W., 1963, Metamorphism in the Riggins region, western Idaho: U.S. Geological Survey Professional Paper, 436, 95 p.

———, 1976, Tectonic history of west-central Idaho: Geological Society of America Abstracts with Programs, v. 8, p. 378.

———, and W. B. Myers, 1966, Cenozoic tectonics of the western United States: Review of Geophysics, v. 4, p. 509-549.

Hannah, J. L., and K. L. Verosub, 1980, Tectonic implications of remagnetized upper Paleozoic strata of the northern Sierra Nevada: Geology, v. 8, p. 520-524.

Hill, D. P., and L. C. Pakiser, 1966, Crustal structure between NTS and Boise, Idaho, from seismic refraction measurements, *in* The Earth beneath the Continents: American Geological Union Monograph Series 10, p. 391-419.

Hillhouse, J. W., C. S. Grommé, and T. L. Vallier, 1982, Paleomagnetism and Mesozoic tectonics of the Seven Devils volcanic arc, northeastern Oregon: Journal of Geophysical Research, v. 87, p. 3777-3794.

Hose, R. K., and M. C. Blake, Jr., 1976, Geology and mineral resources of White Pine County, Nevada: Nevada Bureau of Mines and Geology Bulletin, n. 85, 32 p.

Howard, K. A., 1980, Metamorphic infrastructure in the northern Ruby Mountains, Nevada: Geological Society of America, Memoir 153, p. 335-349.

Irving, E., J. W. H. Monger, and R. W. Yole, 1980, New paleomagnetic evidence for displaced terranes in British Columbia: Geological Association of Canada Special Paper 20, p. 441-456.

Irwin, W. P., 1977, Review of Paleozoic in the Klamath Mountains: Pacific Coast Paleogeography Symposium I, Paleozoic Paleogeography of the Western U.S., Society of Economic Paleontologists and Mineralogists, p. 454-491.

Jones, D. L., N. J. Silberling, and J. W. Hillhouse, 1977, Wrangellia, a displaced terrane in northwestern North America Canadian Journal of Earth Sciences, v. 14, p. 2565-2577.

Kay, M., 1960, Paleozoic continental margin in central Nevada, western: Proceedings, 21st International Geological Congress, 12, p. 93-103.

———, and J. P. Crawford, 1964, Paleozoic facies from the miogeosynclinal to eugeosynclinal belt in thrust slices, central Nevada: Geological Society of America, Bulletin, v. 75, p. 425-454.

Kistler, R. W., and Z. E. Peterman, 1973, Variations in Sr, Rb, K, Na and initial $^{87}Sr/^{86}Sr$ in Mesozoic granitic rocks and intruded wall rocks in Central California: Geological Society of American Bulletin, v. 84, p. 3489-3512.

Lachenbruch, A. H., and J. H. Sass, 1978, Models of an extending lithosphere and heat flow in the Basin and Range province: Geological Society of America, Memoir 152, p. 209-251.

Matti, J. C., and E. H. McKee, 1977, Silurian and Lower Devonian paleogeography of the outer continental shelf of the cordilleran miogeocline, central Nevada: Pacific Coast Paleogeography Symposium 1, Society of Economic Paleontologists and Mineralogists, p. 181-217.

Miller, C. F., and L. J. Bradfish, 1980, An inner Cordilleran belt of muscovite-bearing plutons: Geology, v. 8, p. 412-416.

Miller, D. M., 1980, Structural geology of the northern Albion Mountains, south-central Idaho: Geological Society of America, Memoir 153, p. 399-426.

Minster, J. B., and T. H. Jordan, 1978, Present-day plate motions: Journal of Geophysical Research, v. 83, p. 5331-5354.

Monger, J. W. H., R. A. Price and D. J. Tempelman-Kluit, 1982, Tectonic accretion and the origin of the two major metamorphic and plutonic welts in the Canadian Cordillera: Geology, v. 10, p. 70-75.

Moores, E., 1970, Ultramafics and orogeny, with models of the U.S. Cordilleran and the Tethys: Nature, v. 228, p. 837-842.

Nichols, K. M., and N. J. Silberling, 1977, Stratigraphy and depositional history of the Star Peak Group (Triassic), northwestern Nevada: Geological Society of America, Special Paper 178, 73 p.

Nolan, T. B., C. W. Merriam, and M. C. Blake, Jr., 1974, Geologic map of the Pinto Summit quadrangle, Nevada: U.S. Geological Survey Miscellaneous Geological Investigations Map I-79B, scale 1:316.

Oldow, J. S., 1981, Kinematics of late Mesozoic thrusting, Pilot Mountains, Nevada: Journal of Structural Geology, v. 3, p. 39-51.

Poole, F. G., 1974, Flysch deposits of the Antler foreland basin: Society of Economic Paleontologists and Mineralogists, Special Publication 22, p. 58-82.

Prohdehl, C., 1979, Crustal structure of the western United States: U.S. Geological Survey, Professional Paper 1034, 74 p.

Roberts, R. J. et al, 1958, Paleozoic rocks of north-central Nevada: AAPG Bulletin, v. 42, p. 2813-2857.

Rowell, A. J., M. N. Rees, and C. A. Suczek, 1979, Margin of the North American continent in Nevada during late Cambrian time: American Journal of Science, v. 279, p. 1-18.

Royse, F., Jr., M. A. Warner, and D. L. Reise, 1975, Thrust belt structure geometry and related stratigraphic problems, Wyoming, Idaho — northern Utah: Rocky Mountain Association Geologists Symposium, p. 41-54.

Russell, B. J., 1981, Pre-Tertiary paleogeography and tectonic history of the Jackson Mountains, northwestern Nevada: Evanston, Illinois, Ph.D. thesis, Northwestern University, 205 p.

———, et al, 1982, Cretaceous magnetizations in northwestern Nevada and tectonic implications: Geology, v. 10, p. 423-428.

Saleeby, J. B., 1982, Polygenetic ophiolite belt of the California Sierra Nevada — Geochronological and tectonostratigraphic development: Journal of Geophysical Research, v. 87, p. 1803-1824.

———, et al, 1978, Early Mesozoic paleotectonic-paleogeographic reconstruction of the southern Sierra Nevada region: Pacific Coast Paleogeography Symposium 2, Mesozoic Paleogeography of the Western U.S., Society of Economic Paleontologists and Mineralogists, p. 311-336.

Schultz, K. L., and S. Levi, 1981, Paleomagnetism of the Upper Cretaceous Hornbrook formation — implications for tectonic rotation of the Klamath Mountains providence: Washington, D.C., Eos, v. 62, p. 854.

Schweikert, R. W., and D. S. Cowan, 1975, Early Mesozoic evolution of the western Sierra Nevada, California: Geological Society of America Bulletin, v. 86, p. 1329-1336.

Silberling, N. J., 1973, Geologic events during Permian-Triassic time along the Pacific margin of the U.S.: Alberta Society of Petroleum Geologists Memoir 2, p. 345-362.

———, and R. J. Roberts, 1962, Pretertiary stratigraphy and structure of northwestern Nevada: Geological Society of American Special Paper 72, 58 p.

Simpson, R. W., and A. Cox, 1977, Paleomagnetic evidence for tectonic rotation of the Oregon Coast Range: Geology, v. 5, p. 585-589.

Sloss, L. L., and R. C. Speed, 1974, Relationships of cratonic and continental margin tectonic episodes: Society of Economic Paleontologists and Mineralogists Special Publication 22, p. 98-120.

Smith, J. G., et al, 1971, Mesozoic granitic rocks of northwestern Nevada: a link between the Sierra Nevada and Idaho batholiths: Geological Society of America, Bulletin, v. 82, p. 2933-2944.

Smith, J. F., Jr., and K. B. Ketner, 1977, Tectonic events since early Paleozoic in the Carlin-Pinon Range area, Nevada: U.S. Geological Survey Professional Paper 867C, 18 p.

Snoke, A. W., 1980, Transition from infrastructure to suprastructure in the northern Ruby Mountains, Nevada: Geological Society of America, Memoir 153, p. 257-334.

Speed, R. C., 1977, Island arc and other paleogeographic terranes of late Paleozoic age in the western Great Basin: Pacific Coast Paleogeography Symposium 1, Paleozoic Paleogeography of the Western U.S., Society of Economic Paleontologists and Mineralogists, p. 349-362.

———, 1978, Paleogeographic and plate tectonic evolution of the early Mesozoic marine providence of the western Great Basin: Society of Economic Paleontologists and Mineralogists Pacific Coast Paleogeography Symposium 2, Mesozoic Paleogeography of the Western U.S., p. 253-270.

———, 1979, Collided Paleozoic microplate in the western United States: Journal of Geology, v. 87, p. 279-292.

———, and R. W. Kistler, 1979, Cretaceous volcanism, Excelsior Mountains, Nevada: Geological Society of America, Bulletin, v. 91, p. 392-398.

———, and N. H. Sleep, 1982, Antler orogeny and foreland basin: A model: Geological Society of America, Bulletin, v. 93, p. 815-828.

Stauber, D. A., 1980, Crustal structure in the Bottle Mountain heat flow high in northern Nevada from seismic refraction profiles and Rayleigh wave phase: Ph.D. thesis, Stanford University, 316 p.

Stewart, J. H., 1972, Initial deposits of the Cordilleran geosyncline: evidence of a late Precambrian (850 m.y.) continental separation: Geological Society of America, Bulletin, v. 83, p. 1345-1360.

———, 1976, Late Precambrian evolution of North America: plate tectonics implication: Geology, v. 4, p. 11-15.

———, 1978, Basin and range structure in western North America — a review: Geological Society of America, Memoir 152, p. 1-31.
———, and F. G. Poole, 1974, Lower Paleozoic and uppermost Precambrian Cordilleran miogeocline, Great Basin: Society of Economic Paleontologists and Mineralogists, Special Publication 22, p. 27-57.
———, and C. A. Suczek, 1977, Cambrian and latest Precambrian paleogeography and tectonics in the western United States: Pacific Coast Paleogeography Symposium 1, Society of Economic Paleontologist and Mineralogists, p. 1-19.
Vallier, T. C., H. C. Brooks, and T. P. Thayer, 1977, Paleozoic rocks of eastern Oregon and western Idaho: Pacific Coast Paleogeography Symposium 1, Paleozoic Paleogeography of the Western U.S., Society of Economic Paleontologist and Mineralogists, p. 455-467.
Wernicke, B., 1981, Low-angle normal faults in the Basin and Range Province: nappe tectonics in an extending orogen: Nature, v. 291, p. 645-647.
Wilson, D., and A. Cox, 1980, Paleomagnetic evidence for tectonic rotation of Jurassic plutons in the Blue Mountains, eastern Oregon: Journal of Geophysical Research, v. 85, p. 3681-3689.
Zartman, R. E., 1974, Lead isotopic provinces in the Cordillera of the western United States and their geologic significance: Economic Geology, v. 69, p. 792-805.

Stratigraphy, Sedimentation, and Tectonic Accretion of Exotic Terranes, Southern Coast Ranges, California

J. G. Vedder
D. G. Howell
Hugh McLean
U. S. Geological Survey
Menlo Park, California

The southern Coast Ranges west of the San Andreas fault consist of two composite tectono-stratigraphic terranes. Paleomagnetic and geologic relations indicate that pre-early Eocene rocks in both terranes are allochthonous to the California region, and Late Cretaceous to early Eocene northward drift of 1,500 km is suspected. Basement rocks of the Salinian composite terrane are composed of middle Cretaceous granite plutons and older high-temperature metasedimentary rocks. The Salinian composite terrane is bordered on the west by the Sur-Obispo composite terrane, which includes a melange of the Franciscan assemblage (San Simeon terrane) that is structurally overlain by a Middle Jurassic ophiolite and Upper Jurassic and Cretaceous forearc-basin strata (Stanley Mountain terrane). Along the east edge of the Sur-Obispo composite terrane, an upper Campanian to lower Maestrichtian fluvio-deltaic sequence containing 102-m.y.-old to 126-m.y.-old granitic boulders lends credence to the postulated suturing of these two terranes in Late Cretaceous time. A regional unconformity that developed sometime in the span of middle(?) Paleocene to early (?) Eocene time apparently crosses both terranes. We suggest that an allochthon composed of the Sur-Obispo and Salinian composite terranes, as well as other terranes, was accreted to the southern California region in latest Paleocene or earliest Eocene time. The hypothetical Santa Lucia-Orocopia allochthon (new name) represents the amalgamated composite of these mutually exotic terranes.

Evidence for exceptionally large amounts of translation and subsequent accretion along the western margin of North America is beginning to elicit new interpretations of the origin of the California Coast Ranges. Lateral offsets of hundreds of kilometers that resulted from Neogene plate interactions along the West Coast are supplemented by newly proposed movement involving thousands of kilometers of pre-Neogene transport. These novel concepts allow more flexible constraints and broader perspectives on the magnitude of late Mesozoic and early Tertiary tectonic events. We expand upon some of these ideas and add speculative reasoning concerning the development of the southern Coast Ranges and adjoining areas (Figure 1).

In a thought-provoking paper on allochthonous terranes in the North American Cordillera, Coney, Jones, and Monger (1980) stated that more than 70% of the region is composed of geologic provinces which seem to have been transported and accreted to the North American craton largely during Mesozoic and early Cenozoic time. All of the Coast Ranges are included in their "suspect" terranes. Champion, Gromme, and Howell's (1980) preliminary interpretation of paleomagnetic data from the southern Coast Ranges proposed as much as 2500 km of northward transport for one tract of sedimentary rocks 65 to 75 m.y. old. Petrologic study (Ross, 1978) indicated that the Cretaceous and older crystalline rocks in the southern Coast Ranges have no known correlatives in the Sierra Nevada and Mojave Desert and may be exotic to California. Similarly, the suite of Precambrian to middle Mesozoic crystalline rocks in the San Gabriel, Chocolate, and Orocopia Mountains seems out of place when compared with crystalline rocks of the cratonic areas in the southwestern United States and northwestern Mexico (Silver, 1971). Haxel and Dillon (1978) interpreted the Precambrian to Mesozoic gneissic and granitic rocks of the southwestern Mojave Desert as remnants of a hypothetical allochthon that overrode oceanic sediments in Late Cretaceous or Paleocene time. Further conjecture is aroused by the resemblance of basement rocks in southern Mexico (Ortega, Anderson, and Silver, 1977) to rocks in the composite terranes of southwestern California, both in protolith and metamorphic grade. The exotic nature of the terranes of the San Gabriel and Chocolate-Orocopia Mountains, however, is not universally accepted.

Some workers envision an initially west-verging thrust that separates crystalline rocks of cratonic North America in the hanging wall from oceanic rocks in the foot wall (Burchfiel and Davis, 1981; Crowell, 1981; Yeats, 1981).

Southern Coast Ranges west of the San Andreas fault consist of two distinct, composite terranes (Figure 2) that may have been allochthonous to the California region before middle Tertiary time. The Sur-Obispo composite terrane is the westernmost of the two and is composed principally of structurally superposed rock units: one, a lower thrust plate made up of melange of the Franciscan assemblage; the other, an upper thrust plate composed of the Middle Jurassic Coast Range ophiolite and an overlying Upper Jurassic through Cretaceous sedimentary cover (Fig-

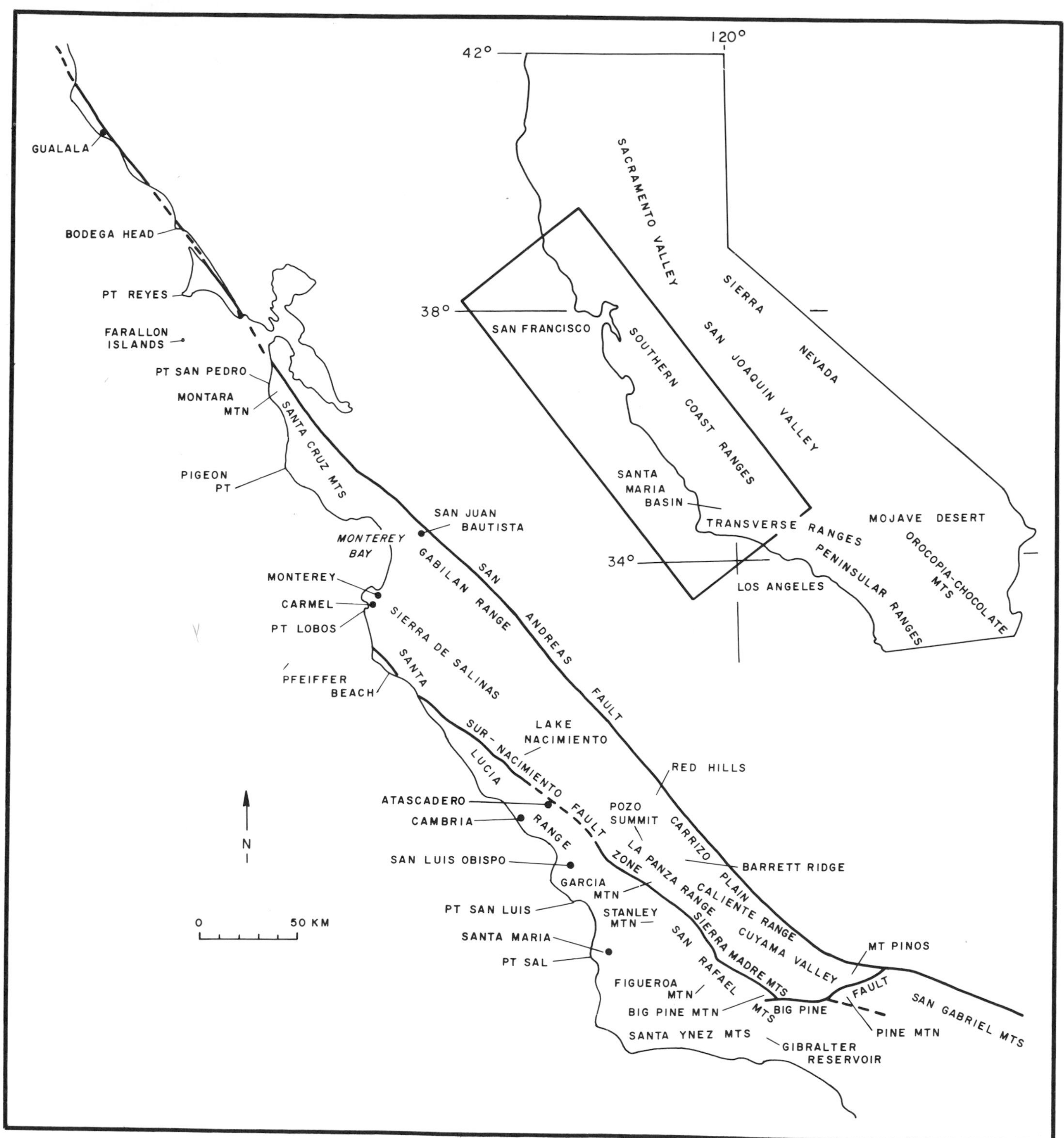

Figure 1 — Map showing locations of places referred to in the text.

ures 3 and 4). Paleocene and early Eocene strata are conspicuously absent from all but the southern part of this terrane. Compositionally and stratigraphically, rocks of the upper plate are comparable to the Great Valley sequence east of the San Andreas fault even though their sites of deposition may have been far apart.

The Salinian composite terrane is bordered by the San Andreas fault zone on the east and the Sur-Nacimiento fault zone on the west (Figures 2 and 3). Throughout much of the Coast Ranges sector, middle Cretaceous granitic plutons (Ross, 1972a; Mattinson, 1978) intrude a complex suite of metasedimentary rocks of unknown age (Compton, 1966), the so-called Sur Series of Trask (1926). Depositionally overlying these crystalline basement rocks is a Campanian to

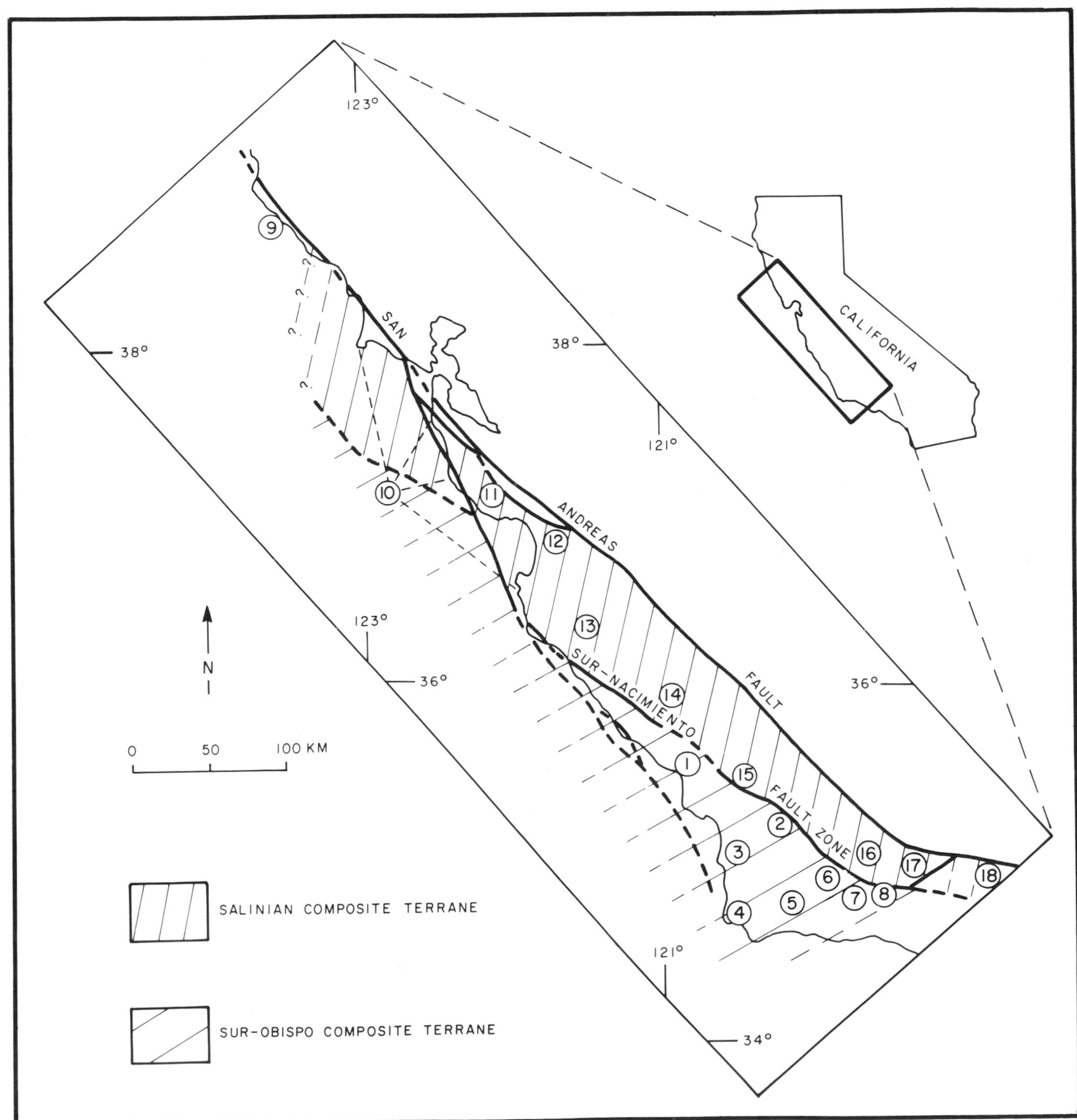

Figure 2 — Salinian and Sur-Obispo composite terranes; northern and southern boundaries for each terrane are not defined. Circled numbers show location of stratigraphic columns in correlation chart, Figure 5.

Paleocene epiclastic suite of deep-marine to fluvio-deltaic lithofacies (Figures 4 and 5). A latest Paleocene and earliest Eocene regional unconformity (about 60 to 55 m.y. ago) is now recognized. The oldest known strata that have direct counterparts on the east side of the San Andreas fault are early Eocene in age (about 55 m.y. ago; Nilsen and Clarke, 1975).

Delineation of the southern boundaries of the Coast Ranges terranes is equivocal. The Sur-Obispo composite terrane is obscured by late Cenozoic structures and strata in the western Transverse Ranges. A pre-Cretaceous plutono-metamorphic subterrane in the southeasternmost Salinian composite terrane seems to extend southward as far as the San Gabriel Mountains (Ross, 1978). It has been suggested that these mountains are a late Cenozoic offset equivalent of the Chocolate-Orocopia Mountains (Crowell, 1975). Pre-Tertiary crystalline rocks in both of these mountain areas are rootless and constitute the upper plate of the Vincent-Orocopia-Chocolate Mountains thrust system (Haxel and Dillon, 1978; Crowell, 1981).

Although paleomagnetic data are not yet available from the Sur-Obispo composite terrane, geologic relations indicate that it was separate from its eastern neighbor until latest Cretaceous time. Suturing about 75 m.y. ago is inferred from a Campanian-Maestrichtian fluvio-deltaic sequence that presumably crossed the southeastern margin of the Sur-Obispo composite terrane (Vedder, Howell, and McLean, 1980b). Included in this fluvio-deltaic sequence are alluvial-fan deposits composed chiefly of granitic boulders of probable Salinian provenance. Thus, the Sur-Obispo basement rocks also may have been sundered from a source as far south as southern Mexico. However, the parent terrane remains unidentified. Using other criteria for timing, Page (in press) concluded that the two terranes welded later, perhaps in late Paleocene to early Eocene time.

The combination of stratigraphic-structural relations and paleomagnetic data suggest that at least 1,500 km of northward drift occurred in some segments of coastal California during the interval 75 to 55 m.y. ago, and that a variety of exotic terranes were emplaced in southwestern California by lateral accretion by late Paleocene or early Eocene time (60 to 55 m.y. ago). If these concepts of drift and accretion are accepted, a collage of amalgamated exotic terranes, including the Sur-Obispo, Salinian, and Tujunga terranes of southern California (Blake, Jones, and Howell, 1982), formed a single displaced piece of crust. We propose the name Santa Lucia-Orocopia allochthon for this displaced crust. Additional northward transport of the Salinian and Sur-Obispo composite terranes resulted from right slip on the Neogene San Andreas fault system.

Emplacement of these alien terranes in California occurred during, and may have been linked to, other events in the southwestern United States. These events include: the Laramide orogeny (Coney, 1979), a Late Cretaceous to Eocene magmatic hiatus (Cross and Pilger, 1978), early Tertiary peneplanation across the Peninsular Ranges (Minch, 1979), contrasting high-relief dissection resulting from wrenching in the Coast Ranges, movement on the Vincent-Orocopia- Chocolate Mountains thrust system, and metamorphism of the Pelona-Orocopia schist (Ehlig, 1975, 1981; Haxel and Dillon, 1978). We consider only the events that seem to be directly tied to our reconstructions.

Because they provide some of the best clues for deciphering paleogeography, details of stratigraphy are emphasized for selected rock sequences in specific areas (especially the southern Coast Ranges). For brevity, we describe these rock sequences without using stratigraphic names; instead, nomenclature, thickness, and correlation are shown in chart form (Figure 5). Our knowledge of the rocks and structures of the San Gabriel Mountains and southwestern Mojave Desert is limited, and we rely on descriptions by workers familiar with those areas, particularly Ehlig (1981), Crowell (1981), Haxel and Dillon (1978), and Burchfiel and Davis (1981). Nevertheless, speculation about the origins of these areas seems warranted.

SALINIAN COMPOSITE TERRANE

Basement rocks

Basement rocks of the Salinian composite terrane form a complex suite of Precambrian and younger high-temperature metamorphic rocks and late Mesozoic plutonic rocks (Compton, 1966; Ross, 1972a, 1978; Mattinson, 1978) that are unrelated to basement rocks in the contiguous terranes.

According to Ross (1977), the main body of the pre-intrusive metasedimentary rocks consists predominantly of gneiss, granofels, and impure quartzite, with smaller amounts of schist and marble. All of these suggest a protolith of thin-bedded, quartz-rich clastic strata admixed with subordinate calcareous rocks. Wiebe (1966) inferred a shallow-marine (shelf) depositional environment for these rocks. Ross (1977) recognized two other distinct bodies of metasedimentary rocks: the Red Hills-Barrett Ridge slice of Precambrian(?) gneiss, and the metagraywacke of the Sierra de Salinas. These distinct parts imply subterranes and an earlier history of tectonic assembly. None of the subterranes has known counterparts in the cratonic rocks of the western North American Cordillera. In nature of protolith, however, the metagraywacke of the Sierra de Salinas may be equivalent to the Pelona and Orocopia Schists (Haxel and Dillon, 1978).

The granitic rocks of the Salinian composite terrane have been divided into four subterranes (blocks) by Ross (1978) based on petrographic similarity and structural coherence. The largest of these subterranes is the "central block," which extends from the Monterey Bay area to the Carrizo Plain. Four rock units are included in the "central block": (1) granodiorite-quartz monzonite of the La Panza Range, which underlies an area of about 1,000 sq km, (2) quartz diorite-granodiorite in the Gabilan Range, (3) hornblende-biotite quartz diorite south of Carmel in the Santa Lucia Mountains, and (4) the porphyritic granodiorite of Monterey. Ross's (1977) "western block" consists predominantly

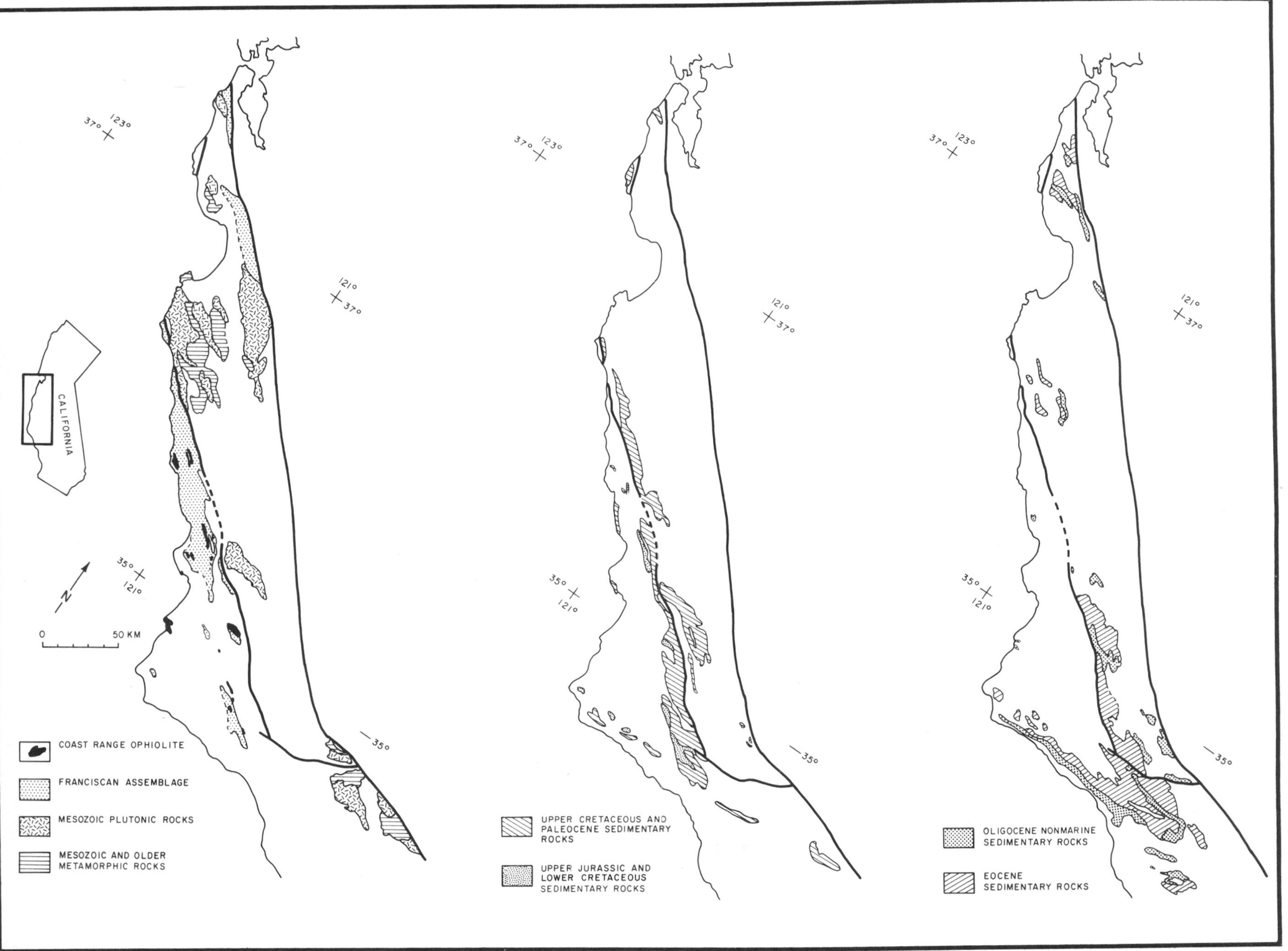

Figure 3 — Generalized distribution of outcrops of major rock units discussed in text. Gualala and Point Reyes areas are north of map edge.

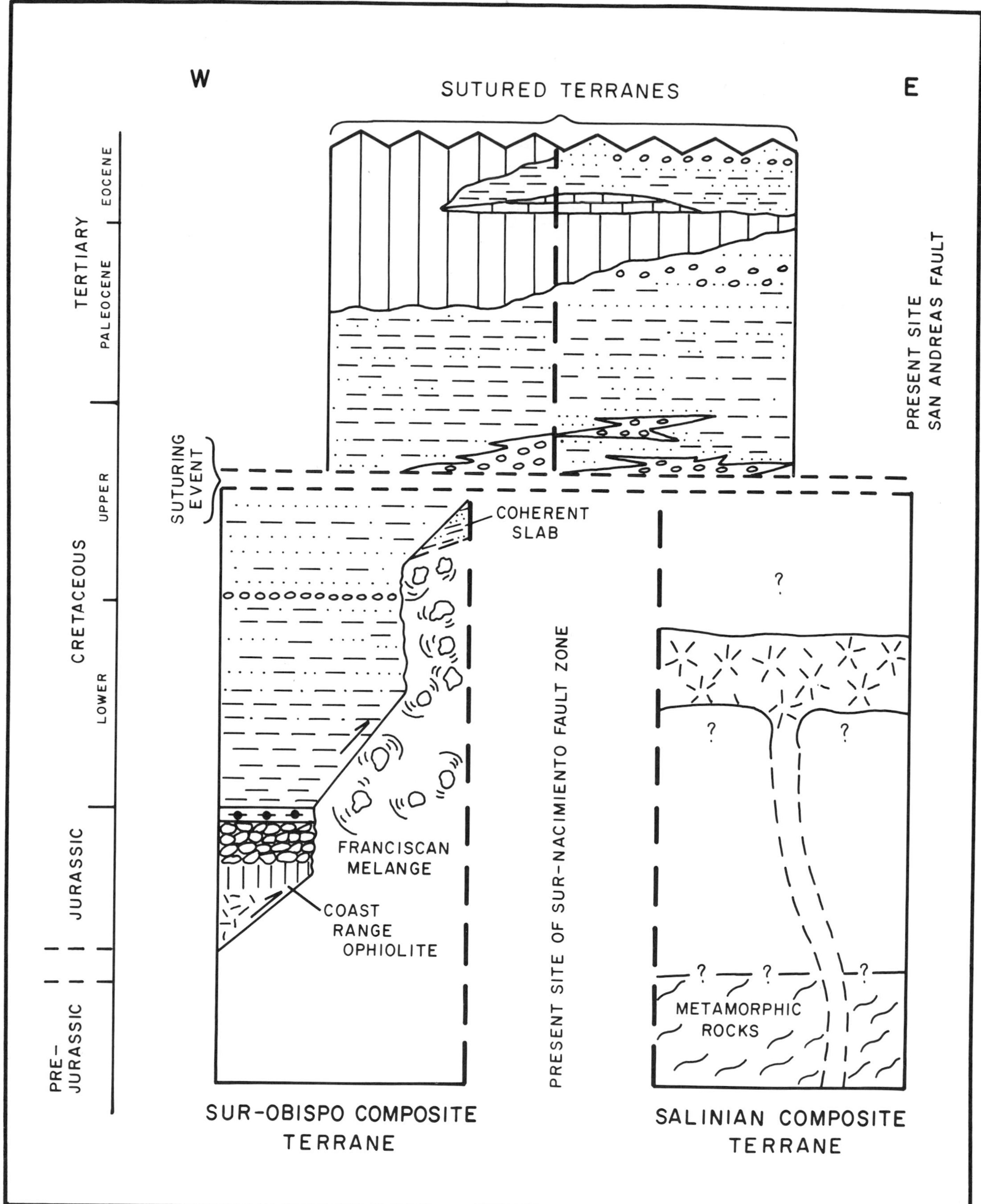

Figure 4 — Generalized stratigraphic columns for the Salinian and Sur-Obispo composite terranes, which presumably were separate until Campanian time when they were sutured to form a new amalgamated terrane.

of hypersthene quartz diorite together with small bodies of aplite-alaskite in the northwestern Santa Lucia Range. The "southeastern block" is characterized by felsic granitic rocks in the Red Hills-Barrett Ridge and Mount Pinos area, whereas the "northern block" consists largely of quartz diorite in the Montara Mountain, Point Reyes-Bodega Head, and Farallon Islands areas.

Published radiometric apparent ages for the granitic rocks are late Mesozoic and most are Late Cretaceous. Samples from the Santa Lucia, Gabilan, and La Panza Ranges dated by K/Ar methods gave apparent ages of 82 to 92 million years (Curtis, Evernden, and Lipson, 1958), which were later refined to 78 to 88 million years (Evernden and Kistler, 1970). Hart (1976) gave anomalously young K/Ar apparent ages of 55 to 72 million years on partly choloritized biotite from the western La Panza Range. Because the eroded granite surface is overlain by late Campanian strata, some of these apparent ages probably reflect a post-emplacement event. Rb/Sr dates (Ross; 1972b) on samples from the Gabilan Range yielded a 110 ± 5 m.y. isochron. U/Pb dates from sphene, feldspar, and apatite samples in the Santa Lucia Range gave a 79 ± 2 m.y. isochron, and zircon separates from this area, as well as Bodega Head and Point Reyes, suggest magmatic crystallization about 104 m.y. ago (Mattinson, Hopson, and Davis, 1972; Mattinson, 1978). Ross (1977) concluded that the Rb/Sr and U/Pb data suggest emplacement ages of 100 to 110 million years for much of the granitic suite of rocks in the Salinian composite terrane. Different thermal histories postulated by Mattinson (1978) imply that the northern plutons had a simple thermal history, in contrast to those in the central region where more complex and prolonged high-temperature conditions prevailed. Additional sampling and analyses may reveal a broader range in plutonism ages.

Initial $^{87}S_r/^{86}S_r$ ratios from 18 granitic rock samples in the Salinian composite terrane are all greater than 0.7060 (Kistler and Peterman, 1978). Ehlig and Joseph (1977), reported an initial $^{87}S_r/^{86}S_r = 0.7084 \pm 0.0002$ from quartz monzonite in the La Panza Range. Although available data imply homogeneity and intrusion through sialic crust, further work may show more variability for the intrusions within the terrane as a whole.

Sedimentary sequences

Sequences of clastic strata (Figures 4 and 5), no older than Campanian (Howell and Vedder, 1978), depositionally overlie the eroded crystalline basement rocks of the Salinian composite terrane. These Upper Cretaceous and Paleocene strata are sporadically exposed along the coastal area between Gualala and Point Lobos, throughout the Santa Lucia Range, and in the La Panza-Caliente-Sierra Madre Mountains region. At some places, such as Point Reyes, Point San Pedro, and Point Lobos, the oldest exposed beds resting on basement are Paleocene. If Cretaceous strata are present, they have been overlapped. At other places, such as the Sierra Madre Mountains and eastern Caliente Range, the basement rocks as well as the Upper Cretaceous beds presumably are buried beneath Paleocene or younger strata. In the San Gabriel Mountains, Maestrichtian and Paleocene rocks depositionally abut crystalline basement. The Upper Cretaceous through Paleocene sedimentary rocks in the Salinian composite terrane represent various rapidly changing depositional environments and shifts in sedimentation loci that may indicate sinuous shorelines and borderland-like topographic configurations. Furthermore, progradational and retrogradational cycles within these depositional sequences suggest a wrench-tectonic regime (Howell and Vedder, 1978). A widespread late Paleocene hiatus seems to have disrupted deposition, but it is poorly documented because of sparse paleontologic evidence.

Upper Cretaceous and Paleocene sedimentary rocks — The stratigraphy and depositional history of the predominantly coarse-clastic Upper Cretaceous sedimentary rocks have been reviewed area-by-area in Howell et al (1977) and Howell and Vedder (1978). In general, environments include outer to inner submarine-fan facies that are locally interrupted by upper slope, shallow marine, and nonmarine facies. Exposed thicknesses are variable; they range from less than 400 m in the northern Santa Lucia Range to as much as 3400 m in the La Panza Range (Figure 5). Dominant paleocurrent directions are west-northwest, west, southwest, and south. Although faunal data are sparse in the Upper Cretaceous rocks, an age span of upper(?) Campanian through lower (?) Maestrichtian is provisionally assigned on the basis of benthic foraminifers and mollusks. The most carefully studied Upper Cretaceous section in the terrane is at Pigeon Point, where paleomagnetic data from submarine-fan facies imply as much as 2500 km of northward translation since deposition (Champion, Gromme, and Howell, 1980).

Scattered throughout the Salinian composite terrane are patches of early Paleocene strata that lie conformably on Upper Cretaceous strata or noncomformably on crystalline basement. With the exception of the Gualala section (Figure 5), the Paleocene strata apparently are separated from early Eocene rocks by an unconformity. The Gualala section is also anomalous in that the underlying Upper Cretaceous strata structurally overlie spilite (Wentworth, 1968) and are not known to be depositional on granitic rocks. The relationship of the Gualala strata to those of the rest of the terrane is conjectural; Graham and Berry (1979) speculated that the early Eocene part of the section is a western continuation of a submarine-fan complex that once extended eastward into the Sierran province.

In the northwest part of the Salinian block at Point Reyes, Point San Pedro, and Point Lobos, Paleocene strata lie unconformably on granitic basement. Because these strata crop out in sea cliffs and the underlying basement surface dips seaward, it is probable that downdip to the west the Paleocene strata conceal Upper Cretaceous rocks that abut the granitic base-

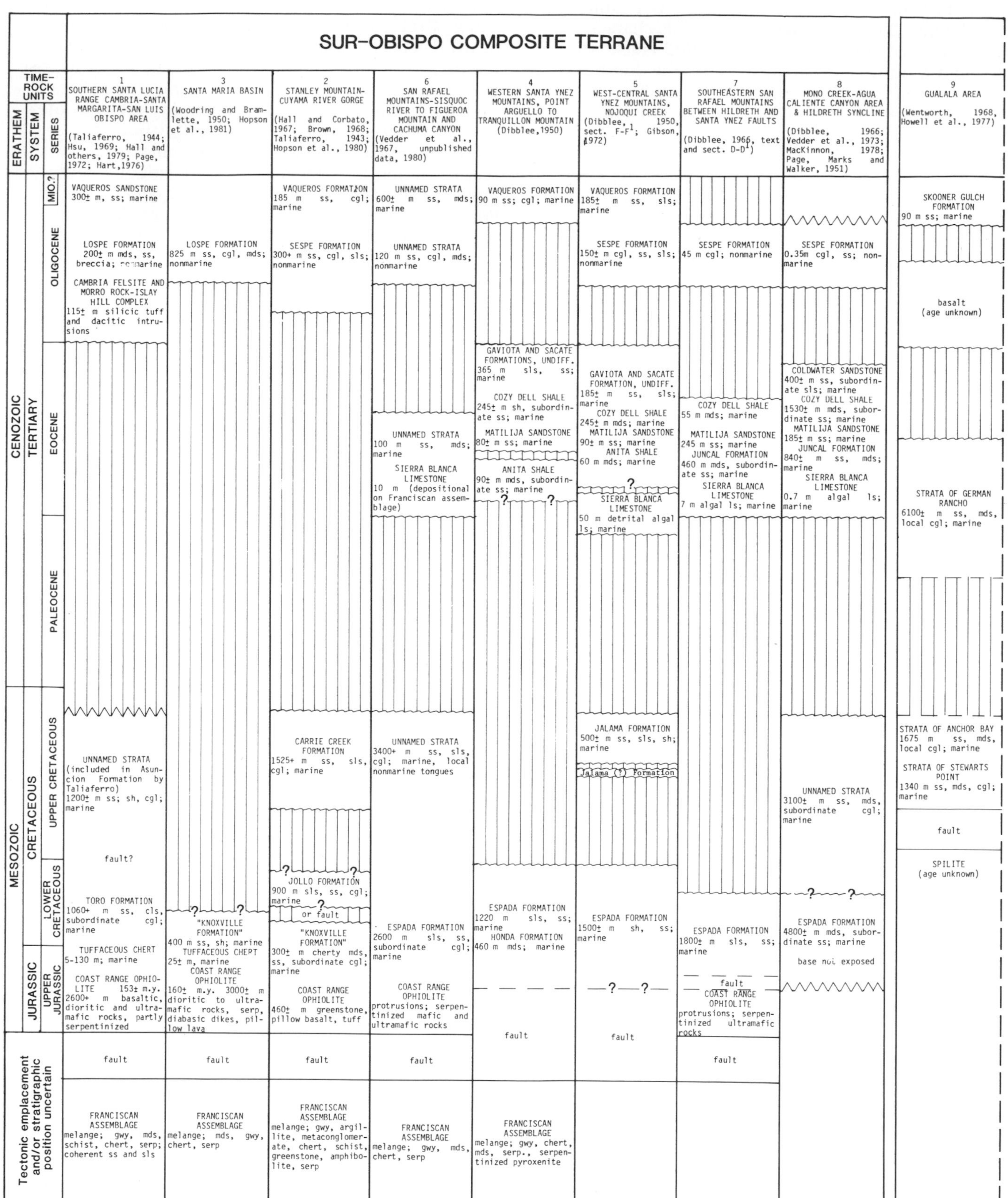

Figure 5 — Correlation chart illustrating known and inferred stratigraphic and structural discontinuities in the Sur-Obispo and Salinian composite terranes. Stratigraphic units as well as maximum or aggregated thicknesses and predominant rock types are shown for places specified. Principal references are given at the head of each column; locations are indicated on Figure 2.

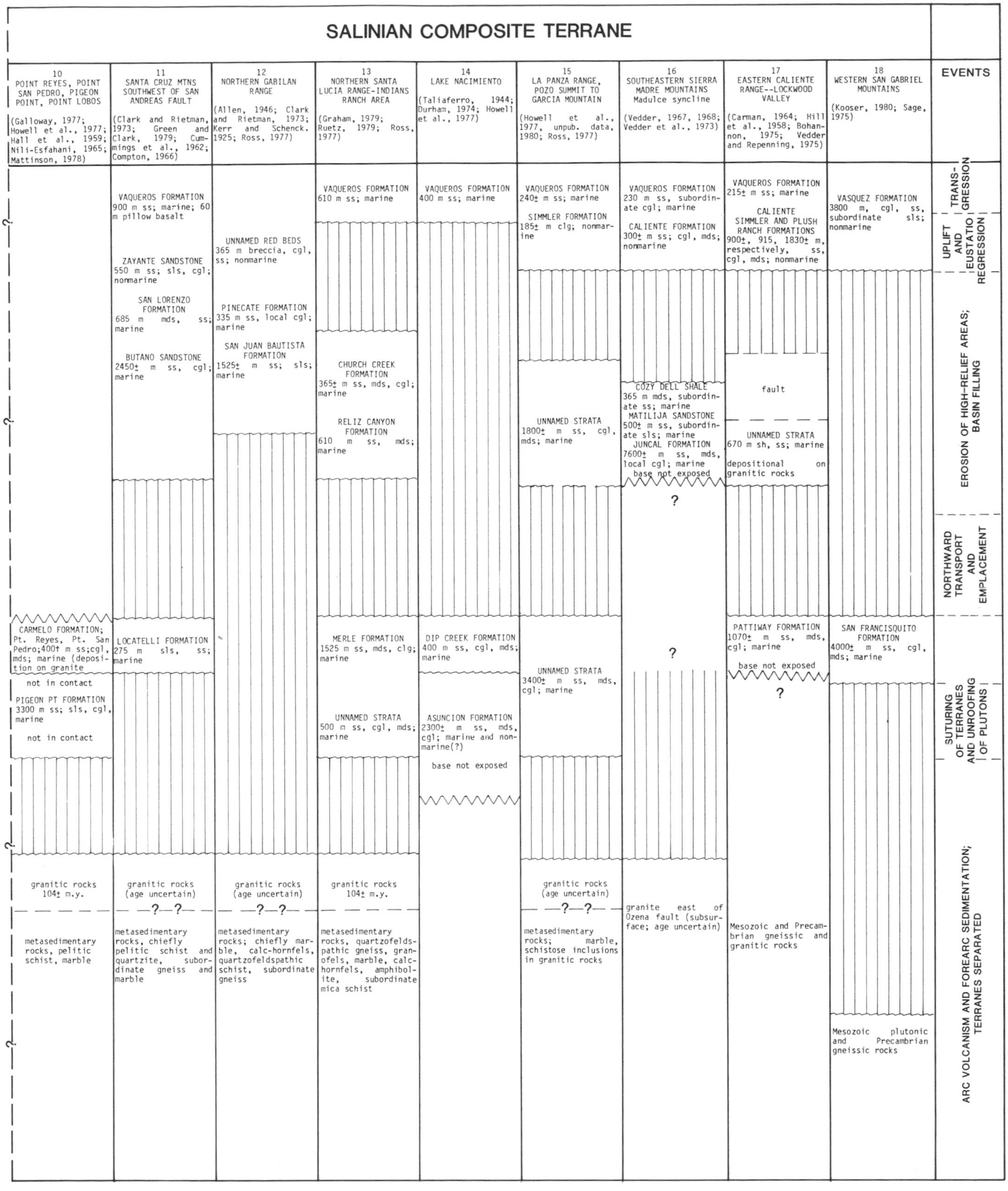

SALINIAN COMPOSITE TERRANE

No.	Locality	References	Units (top to bottom)
10	POINT REYES, POINT SAN PEDRO, PIGEON POINT, POINT LOBOS	(Galloway, 1977; Howell et al., 1977; Hall et al., 1959; Nili-Esfahani, 1965; Mattinson, 1978)	CARMELO FORMATION; Pt. Reyes, Pt. San Pedro;400± m ss;cgl, mds; marine (deposition on granite not in contact PIGEON PT FORMATION 3300 m ss; sls, cgl, marine not in contact granitic rocks 104± m.y. metasedimentary rocks, pelitic schist, marble
11	SANTA CRUZ MTNS SOUTHWEST OF SAN ANDREAS FAULT	(Clark and Rietman, 1973; Green and Clark, 1979; Cummings et al., 1962; Compton, 1966)	VAQUEROS FORMATION 900 m ss; marine; 60 m pillow basalt ZAYANTE SANDSTONE 550 m ss; sls, cgl; nonmarine SAN LORENZO FORMATION 685 m mds, ss; marine BUTANO SANDSTONE 2450± m ss, cgl; marine LOCATELLI FORMATION 275 m sls, ss; marine granitic rocks (age uncertain) ?—? metasedimentary rocks, chiefly pelitic schist and quartzite, subordinate gneiss and marble
12	NORTHERN GABILAN RANGE	(Allen, 1946; Clark and Rietman, 1973; Kerr and Schenck. 1925; Ross, 1977)	UNNAMED RED BEDS 365 m breccia, cgl, ss; nonmarine PINECATE FORMATION 335 m ss, local cgl; marine SAN JUAN BAUTISTA FORMATION 1525± m ss; sls; marine granitic rocks (age uncertain) ?—? metasedimentary rocks; chiefly marble, calc-hornfels, quartzofeldspathic schist, subordinate gneiss
13	NORTHERN SANTA LUCIA RANGE-INDIANS RANCH AREA	(Graham, 1979; Ruetz, 1979; Ross, 1977)	VAQUEROS FORMATION 610 m ss; marine CHURCH CREEK FORMATION 365± m ss, mds, cgl; marine RELIZ CANYON FORMATION 610 m ss, mds; marine MERLE FORMATION 1525 m ss, mds, clg; marine UNNAMED STRATA 500 m ss, cgl, mds; marine granitic rocks 104± m.y. metasedimentary rocks, quartzofeldspathic gneiss, granofels, marble, calc-hornfels, amphibolite, subordinate mica schist
14	LAKE NACIMIENTO	(Taliaferro, 1944; Durham, 1974; Howell et al., 1977)	VAQUEROS FORMATION 400 m ss; marine DIP CREEK FORMATION 400 m ss, cgl, mds; marine ASUNCION FORMATION 2300± m ss, mds, cgl; marine and nonmarine(?) base not exposed
15	LA PANZA RANGE, POZO SUMMIT TO GARCIA MOUNTAIN	(Howell et al., 1977, unpub. data, 1980; Ross, 1977)	VAQUEROS FORMATION 240± m ss; marine SIMMLER FORMATION 185± m clg; nonmarine UNNAMED STRATA 1800± m ss, cgl, mds; marine UNNAMED STRATA 3400± m ss, mds, cgl; marine granitic rocks (age uncertain) ?—? metasedimentary rocks; marble, schistose inclusions in granitic rocks
16	SOUTHEASTERN SIERRA MADRE MOUNTAINS Madulce syncline	(Vedder, 1967, 1968; Vedder et al., 1973)	VAQUEROS FORMATION 230 m ss, subordinate cgl; marine CALIENTE FORMATION 300± m ss; cgl, mds; nonmarine COZY DELL SHALE 365 m mds, subordinate ss; marine MATILIJA SANDSTONE 500± m ss, subordinate sls; marine JUNCAL FORMATION 7600± m ss, mds, local cgl; marine base not exposed ? ? granite east of Ozena fault (subsurface; age uncertain)
17	EASTERN CALIENTE RANGE--LOCKWOOD VALLEY	(Carman, 1964; Hill et al., 1958; Bohannon, 1975; Vedder and Repenning, 1975)	VAQUEROS FORMATION 215± m ss; marine CALIENTE SIMMLER AND PLUSH RANCH FORMATIONS 900±, 915, 1830± m, respectively, ss, cgl, mds; nonmarine fault UNNAMED STRATA 670 m sh, ss; marine depositional on granitic rocks PATTIWAY FORMATION 1070± m ss, mds, cgl; marine base not exposed ? Mesozoic and Precambrian gneissic and granitic rocks
18	WESTERN SAN GABRIEL MOUNTAINS	(Kooser, 1980; Sage, 1975)	VASQUEZ FORMATION 3800 m, cgl, ss, subordinate sls; nonmarine SAN FRANCISQUITO FORMATION 4000± m ss, cgl, mds; marine Mesozoic plutonic and Precambrian gneissic rocks

EVENTS (top to bottom):
- TRANSGRESSION
- UPLIFT AND EUSTATIC REGRESSION
- EROSION OF HIGH-RELIEF AREAS; BASIN FILLING
- NORTHWARD TRANSPORT AND EMPLACEMENT
- SUTURING OF TERRANES AND UNROOFING OF PLUTONS
- ARC VOLCANISM AND FOREARC SEDIMENTATION; TERRANES SEPARATED

Biostratigraphic subdivision of the Paleocene sequences is uncertain, as the strata generally are coarse-grained and sparsely fossiliferous. U-Pb isotopic ages for granitic and ophiolitic rocks are from Mattinson (1978) and Hopson, Mattison, and Pessagno (1981).

ment. No Eocene strata are known to overlie these rocks. Characteristically, these Paleocene sequences are faulted and exposed thicknesses are variable. Most beds are turbidites or other associated mass-flow depositional units. Conglomerate at Point San Pedro is mostly locally derived granitic debris, whereas at Point Reyes and Point Lobos, exotic siliceous volcanic clasts predominate. Environments of deposition of all three localities are typically submarine-canyon to middle-fan settings of relatively small, submarine-fan systems.

East of Point San Pedro and Pigeon Point in the Santa Cruz Mountains, a thick sequence of clastic strata constitutes the early Tertiary section (Cummings, Touring, and Brabb, 1962). The Paleocene sequence consists principally of mudstone and fine-grained sandstone that are interbedded with local lenses of conglomerate. These rocks and their associated fauna suggest low-energy traction and hemipelagic deposition in shelf and upper slope environments. This setting represents a facies marginal to the aforementioned small submarine-fan settings at Point Reyes and Point San Pedro.

In the northern Santa Lucia Range, beds of Paleocene flysch are part of a retrogradational sequence that begins with Upper Cretaceous rocks depositionally abutting crystalline basement (Ruetz, 1979). This onlap steps eastward while moving upsection, as inferred for the Point Lobos to Point Reyes sections. In the central part of the San Lucia Mountains around Lake Nacimiento, a shallow-marine Paleocene clastic sequence conformably overlies nonmarine and shallow-marine Upper Cretaceous strata (Taliaferro, 1944). These rocks are unconformably overlain by nonmarine Oligocene sedimentary rocks (Durham, 1974).

In the La Panza-Garcia Mountain area, extremely sparse and indefinite faunal data result in poor resolution of boundaries between Upper Cretaceous, Paleocene, and Eocene strata. However, regional relations suggest a gradual overlap and a gap in deposition that possibly represents late Paleocene and/or early Eocene time (Figure 5). The thick Upper Cretaceous and Paleocene sequence grades upward into coarser grained beds that include large conglomerate lenses at the top of the section on Garcia Mountain. These lithofacies define a progradational cycle that changes upsection from outer-fan fringe to middle-fan to inner-fan channel facies (Howell and Vedder, 1978). Coarse clastic strata near the base of the exposed section in the northwesternmost part of the Sierra Madre Mountains presumably represent the waning stages of this cycle. Measured current indicators show a diverse flow pattern, generally southeast, south, and west, with west dominating in the upper part (Chipping, 1972). Interbedded mudstone, sandstone, and conglomerate of Paleocene age in the eastern Caliente Range probably represent inner-fan channel facies (Vedder, 1975). Ten to twenty kilometers east of Cuyama Valley, a correlative sandstone and siltstone section may be shallow marine and nonmarine.

Exceptionally thick Upper Cretaceous and Paleocene strata in the western and north-central San Gabriel Mountains consist principally of sandstone and conglomerate turbidites (Kooser, 1980). Shallow-marine, crossbedded sandstone lenses occur in the section exposed in the north-central part of the mountains where Upper Cretaceous strata are absent. Paleocurrent directions and overlap relations on basement rocks indicate north and northeast source areas; deposition was chiefly on restricted submarine fans.

Eocene and Oligocene sedimentary rocks — A thick Eocene section (Figure 5), composed principally of coarse clastic material (Nilsen, 1971) unconformably overlies Paleocene rocks in the Santa Cruz Mountains. These rocks are offset correlatives of rocks on the east side of the San Andreas fault in the San Joaquin Valley (Nilsen and Clarke, 1975). Together they make up the relatively large Butano-Point of Rocks submarine-fan system. Fine-grained Eocene strata east of Monterey Bay near San Juan Bautista possibly are shelf deposits marginal to this submarine-fan system.

Eocene strata unconformably overlie the Paleocene section in the northern part of the Santa Lucia Mountains (Graham, 1979). Locally, algal limestone beds resting on Salinian basement attest to an orogenic episode between Paleocene and Eocene depositional sequences (Figure 6). The clastic part of the Eocene section represents deep-water, submarine-fan deposition that changed upsection to shallow-marine conditions by early Oligocene time (Graham, 1979; Link and Nilsen, 1980).

In the northwestern Sierra Madre Mountains, the known Eocene section typically is coarse-grained and conglomeratic, particularly in the upper part. Chipping (1972) concluded that much of this sequence was deposited in inner fan environments near a basin margin and that paleocurrents were west and southwest. The extraordinarily thick sequence of early and middle Eocene strata in the southeastern Sierra Madre Mountains consists chiefly of turbidites that may represent overall progradation. Chipping (1972; Figure 4)

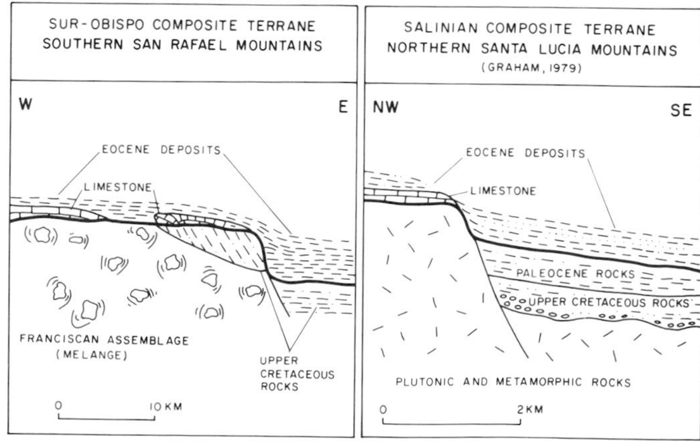

Figure 6 — Schematic diagrams depicting the high depositional relief and overlapping of Eocene strata on both the Sur-Obispo and Salinian composite terranes. The heavy line represents the surface where Eocene rocks were deposited. Vertical scale is exaggerated.

shows random paleocurrent directions that range from north and northwest to west and southwest. Isolated patches of Eocene marine shale occur on the south blank of Mount Pinos (Carman, 1964). Southeast of the Big Pine fault and east of Pine Mountain, the Eocene section coarsens and nearshore depositional environments are reflected in the strata that lap onto basement (Howell, 1975).

Nonmarine beds of late Eocene(?) and Oligocene(?) age are sparsely distributed throughout the Salinian composite terrane from the Santa Cruz Mountains southward into the Cuyama Valley-Mount Pinos area. In the Santa Cruz Mountains and northern Santa Lucia Range, the nonmarine beds intertongue with, and overlie, marine beds of Oligocene age. In the La Panza Range and all other areas to the southeast, they are everywhere unconformable on older rocks including basement (Carman, 1964; Vedder and Brown, 1968; Dibblee, 1973, 1976; Bartow, 1978). Most of these terrestrial sediments were deposited in restricted basins in alluvial-fan, flood-plain, and lacustrine systems, and much of the detritus was locally derived. For example, sandstone boulders as large as 3 m in diameter eroded from underlying strata are the dominant clasts in conglomerate beds in the southeastern La Panza Range and northwestern Sierra Madre Mountains. In this area, imbrication and diminution of clast size indicate northward and northeastward transport into a closed basin (Vedder and Brown, 1968; Bartow, 1978), in which as much as 1,200 m of sediment accumulated in local pockets. A probably correlative section in the Mount Pinos area was derived from underlying Eocene strata, as well as nearby basement rock sources (Carman, 1964).

Throughout the southern part of the Salinian composite terrane, transgressive shallow-marine facies of late Oligocene to early Miocene age overlie the nonmarine beds. In the Santa Cruz Mountains, however, similar beds rest on late Eocene to early Oligocene deep-marine facies and intertongue with Oligocene shallow-marine facies rather than red beds (Cummings, Touring, and Brabb, 1962; Clark and Reitman, 1973). In the eastern Caliente Range, marine facies grade laterally to nonmarine.

Sandstone and conglomerate compositions

Upper Cretaceous sandstone and conglomerate beds crop out in the coastal areas of Gualala and Pigeon Point, northern Santa Lucia Range, La Panza Range, and San Rafael Mountains. The sandstone is mainly arkosic or feldspathic arenite (Figure 7). Total quartz ranges from 45 to 60%, total feldspar from 36 to 49%, total unstable rock fragments from 5 to 9%, potassium feldspar from 18 to 24%, and biotite from 5 to 8%. Ratios of quartz, plagioclase, and potassium feldspar in sandstone are simlar to ratios in underlying granitic rocks, especially biotite quartz monzonite. Unstable rock fragments include porphyritic-aphanitic andesite and dacite, quartz-mica schist, quartz-epidote granofels, and rare quartzite. Ratios of volcanic rock to total unstable rock fragments range from 0.16 to 0.45 (Table 1).

Upper Cretaceous conglomerates of the Salinian composite terrane consist mainly of locally derived granitic and gneissic basement rocks and presumably exotic intermediate and porphyritic siliceous volcanic rocks. An exception is in the Gualala area, where gabbroic clasts predominate in the upper of two Upper

Table 1. Summary of Sur-Obispo and Salinian sandstone petrography from Lee-Wong and Howell (1977), and McLean (unpublished data, 1981).

Locality	Age	Q	F	L	%K	Q_T	P	K	Mica	P/F	V/L
Areas east of Sur-Nacimiento fault zone											
Northern Santa Lucia Range	Campanian and/or Maestrichtian	45	49	5	20	48	29	23	6	.55	.16
Lake Nacimiento	Campanian and/or Maestrichtian	46	46	9	23	50	21	29	8	.40	.38
La Panza Range	Campanian and/or Maestrichtian	60	36	5	18	62	18	19	5	.48	.23
Do.	Campanian and/or Maestrichtian	45	49	5	24	48	24	27	5	.46	.45
Northern Santa Lucia Range	Paleocene	64	35	1	16	64	21	15	7	.60	.00
Santa Madre Mts.	Eocene	42	44	14	13	48	32	20	4	.63	.13
Areas west of Sur-Nacimiento fault zone											
Northern San Rafael Mts.	Albian to Campanian	49	45	7	15	50	30	17	6	.65	.28
Central San Rafael Mts.	Albian to Campanian	38	60	1	11	39	47	14	17	.77	.00
Do.	Albian to Campanian	38	45	16	13	46	35	19	8	.65	.31
Do.	Albian to Campanian	49	41	10	6	55	34	11	6	.76	.36
Northern Santa Lucia Range	Late Cretaceous	38	57	5	10	41	48	12	5	.81	.65
Pfeiffer Beach	Late Cretaceous	39	48	14	21						
Cambria	Late Cretaceous	39	49	11	7	44	39	16	7	.69	.08
Point San Luis	Late Cretaceous	39	52	9	11	43	43	13	8	.76	.12

Q — Quartz
F — Feldspar
L — Lithic fragments
K — Potassium feldspar
Q_T — Total quartzose grains
P — Plagioclase
V — Volcanic rock fragments

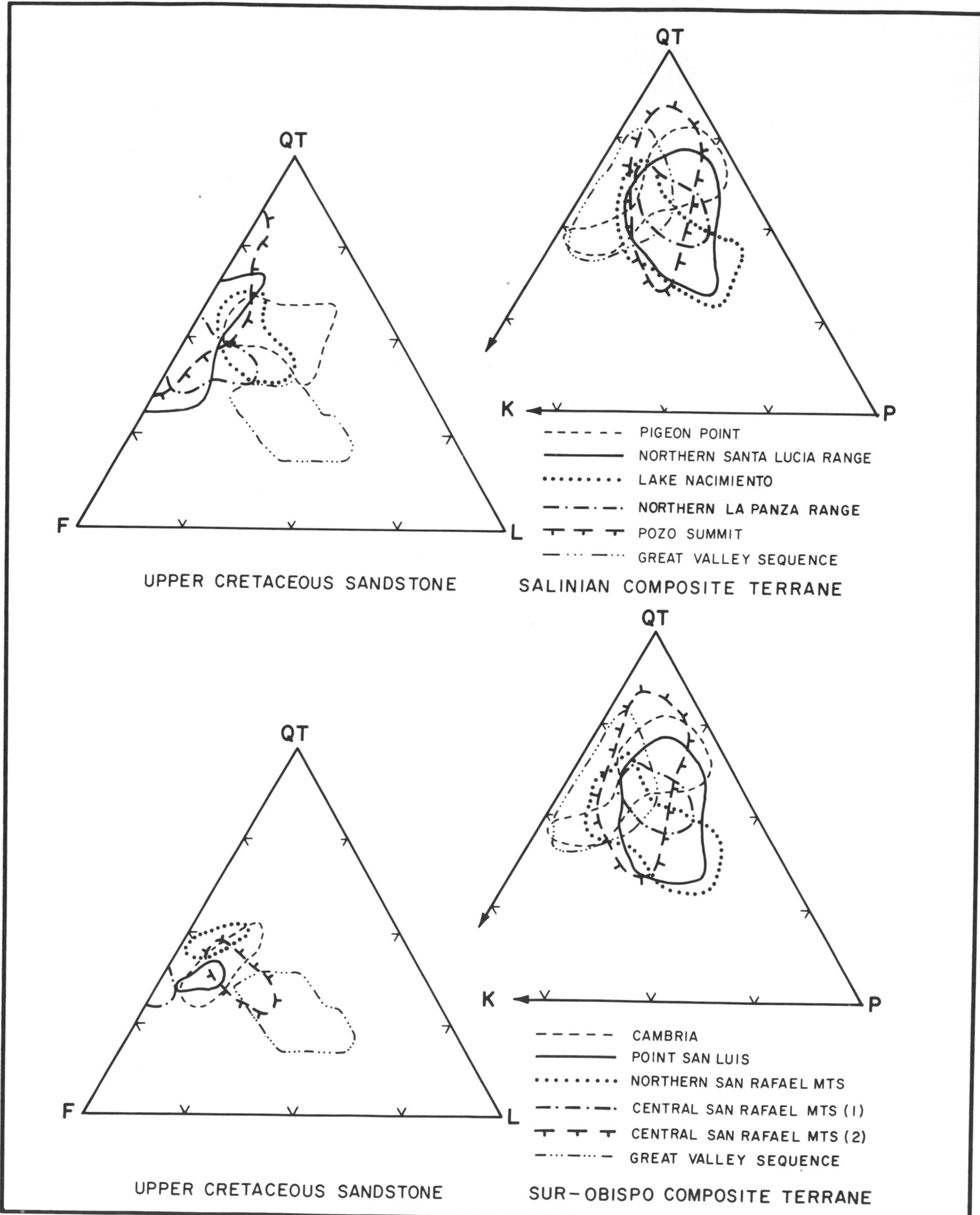

Figure 7 — QFL and QPK diagrams for Upper Cretaceous sandstone from Sur-Obispo and Salinian composite terranes. Modified from Lee-Wong and Howell (1977) and Bourgeois (1980).

Cretaceous units (Wentworth, 1968). Local clast populations vary widely depending on sedimentary environment (submarine fan-channel facies, and alluvial-fan and paralic conglomerate) and proximity to basement. Clast counts by Ruetz (1979) from submarine-fan facies in the northern Santa Lucia Range indicate that locally derived granitic, gneissic, and schistose clasts predominate over sandstone, quartzite, and porphyritic volcanic rocks. At Lake Nacimiento, however, exotic porphyritic rocks make up about 90% of the conglomerate clasts.

Conglomerate exposed along the west side of Pozo Summit in the La Panza Range contains high percentages of granitic, gneissic, and quartzite clasts in beds that were deposited in paralic environments and that probably abut crystalline basement rocks. Beds upsection contain well-rounded and well-sorted pebbles and cobbles of porphyritic silicic volcanic rocks, intermediate volcanic rocks, and gray quartzite. This suggests mixing of transported clasts with locally derived clasts by a far-reaching fluvial system. About 20 km west of Pozo Summit, submarine-fan conglomerate low in the section contains granite, porphyritic volcanic rocks, quartzite, and gneiss (in decreasing order of abundance). In summary, the Upper Cretaceous conglomerates of the Salinian terrane contain clasts that were derived from local crystalline basement together with clasts of well-rounded, abrasion-resistant siliceous volcanic rocks that were transported into the area from an as yet unidentified source terrane (Table 2). Clasts of chert, graywacke, argillite, mafic and ultramafic rock are either rare or absent.

Conglomerate clasts in the Upper Cretaceous and Paleocene beds in the western San Gabriel Mountains are chiefly cobbles and boulders of locally derived granitic and gneissic rocks and exotic porphyritic siliceous volcanic rocks. The most abundant clasts in the Paleocene conglomerate beds in the eastern Caliente Range have a similar composition.

Paleocene strata at Point Reyes consist of coarse, sandy conglomerate containing well-rounded pebbles of chert, and porphyritic volcanic rocks along with subangular fragments of granitic rock up to 1.0 m in long dimension. Submarine-canyon deposits at Point Lobos are principally sandstone and conglomerate; the clasts are predominantly volcanic rocks that accompany subordinate granitic clasts and rare chert, quartz, jasper and rhyolite tuff (Table 2). A local source is inferred for the large granitic clasts, but the source of volcanic clasts is unknown. Paleocene strata at Point San Pedro are composed of thin-bedded sandstone and mudstone turbidites, interbeds of pebble-cobble conglomerate composed of locally derived granite and schist, and subordinate dark-gray hornfels or argillite, limestone, and gneiss. Exotic siliceous volcanic rocks are also present.

In the northern Santa Lucia Mountains, Paleocene sandstone, mudstone, and conglomerate beds, locally called the Merle Formation (Ruetz, 1979), unconformably overlie crystalline basement rocks. At places the Paleocene strata overlie Upper Cretaceous strata without apparent discordance, and the sandstone and conglomerate beds are similar in composition to their Upper Cretaceous counterparts (Ruetz, 1979). Sandstone samples (Figure 8; Table 1) are chiefly feldspathic arenite; total quartz ranges from 36 to 82%, total feldspar from 17 to 62%, and total unstable rock fragments from 0 to 5%. Granitic rock fragments and abundant orthoclase and microcline indicate a granitic source terrane. Volcanic rock fragments are rare.

Eocene sandstone samples (Figure 8; Table 1) from the Sierra Madre Mountains also are similar in composition to Upper Cretaceous rocks farther north, although the Eocene sandstone contains slightly higher percentages of granitic and metamorphic (mostly quartz-mica schist) rock fragments and potassium feldspar. In contrast to Eocene conglomerate beds that typically contain abundant silicic and intermediate volcanic clasts, volcanic rock fragments are rare in the sandstone.

Basin analysis summary

Facies studies of Upper Cretaceous and Paleocene sedimentary sequences suggest that borderland-like realms of deposition prevailed throughout the Salinian composite terrane. Although only fragments of these depositional systems are preserved, enough remain to demonstrate progradational and retrogradational cycles (Howell and Vedder, 1978). These cycles, together with the basin configurations and sediment compositions, imply that active wrench tectonism accompanied deposition of volcanic detritus derived from somewhere in the western Cordillera. During Eocene time, borderland-like topography persisted but the regions of high relief were probably inherited from the earlier episodes of wrenching, which possibly included collision and suturing together of the hypothetical Santa Lucia-Orocopia allochthon (Figures 4 and 5). Gradual basin filling resulted in broad depositional aprons and shoaling. Renewed tectonism and sea level lowering, beginning in Oligocene time, combined to create restricted internal-drainage basins that accumulated nonmarine sediments. The shoaling and renewed tectonism, together with the prevailing terrestrial conditions, possibly signify the initial interaction of the Pacific and North American plates. Subsequently, in late Oligocene and early Miocene time, shallow-marine deposits migrated across the terrestrial beds and signaled widespread development of shelf-and-slope sedimentary sequences in an evolving borderland setting that may have resulted from early activity on the San Andreas fault system (Blake et al, 1978).

SUR-OBISPO COMPOSITE TERRANE

Basement rocks

West of the Sur-Nacimiento fault zone, basement rocks consist of two main types: Franciscan assemblage and Coast Range ophiolite. The Franciscan assemblage (San Simeon terrane of Blake, Howell, and Jones, 1982) is chiefly melange that is exposed along the southwestern flanks of the Santa Lucia Range and San Rafael Mountains, and in the western Santa Ynez

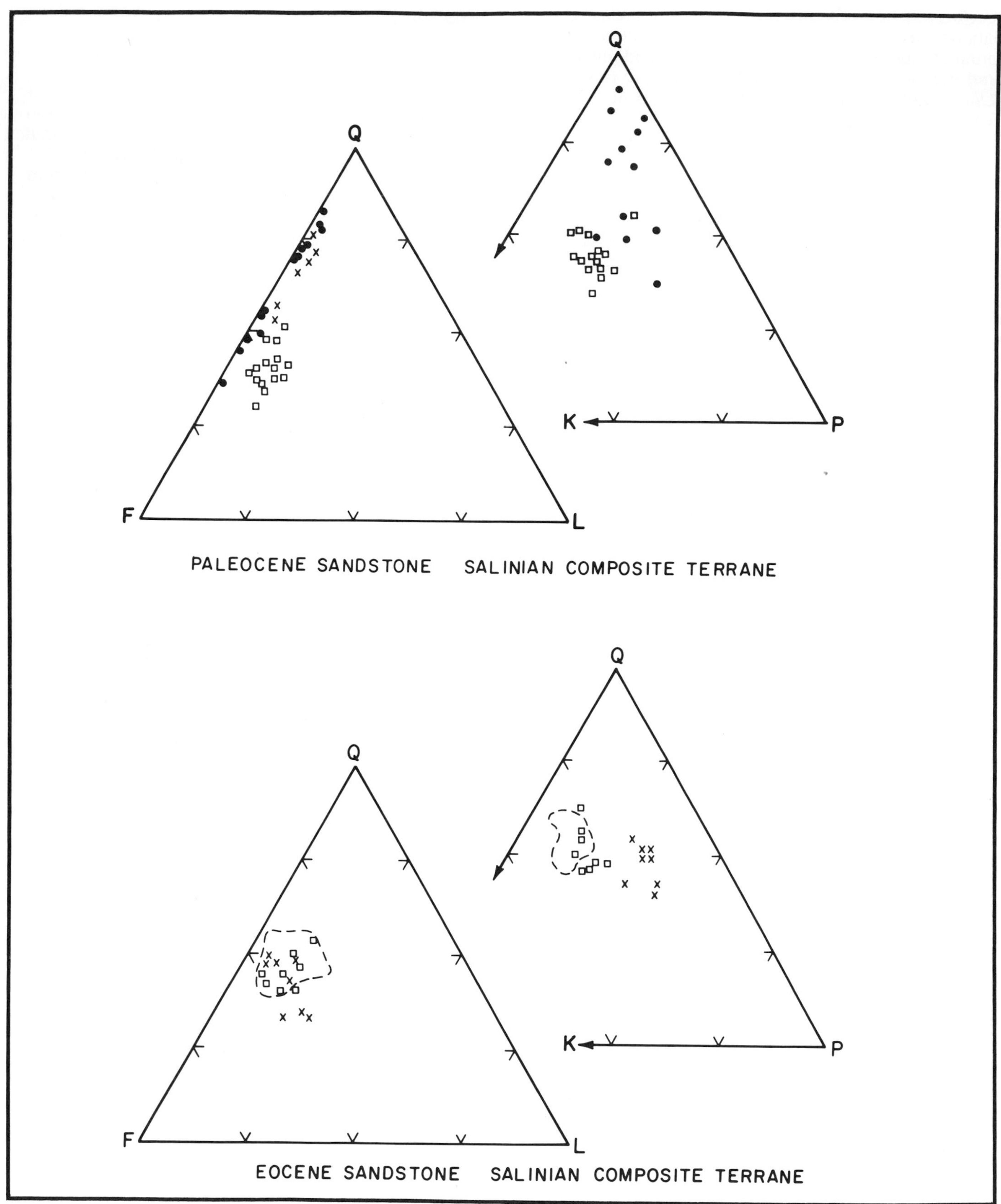

Figure 8 — QFL and QPK diagrams for Paleocene and Eocene sandstone in the Salinian composite terrane. Paleocene sandstone: Gualala — squares (Graham and Berry, 1979); and northern Santa Lucia Range — dots (McLean, unpublished data, 1981) and crosses (Ruetz, 1979). Eocene sandstone: Santa Cruz Mountains — squares (Clarke, 1973); northern Santa Lucia Range — dashes (Link and Nilsen, 1980); and, northern Sierra Madre Mountains — crosses (McLean, unpublished data, 1981).

Mountains. Three small isolated belts of structurally coherent rocks assignable to the Franciscan occur along the coastal edge of the Santa Lucia Range. The ophiolite extends discontinuously from the southern Santa Lucia Range into the San Rafael Mountains and underlies parts of the Santa Maria basin.

Franciscan assemblage — Blocks of chert, graywacke, metavolcanic rocks, serpentine, and blueschist embedded in pervasively sheared mudstone typically constitute the melanges in the Franciscan assemblage. Fossils are extremely rare in the melanges; Page (1970) reported Cretaceous palynomorphs at several places, and Seiders, Pessagno, and Harris (1979) noted Triassic radiolarians in chert pebbles from conglomerate blocks.

Isoclinally folded, structurally coherent strata assigned to the Franciscan assemblage occur in three small subterranes at Pfeiffer Beach (Underwood, 1977), Cambria, and Point San Luis (Hsu, 1969; Howell et al, 1977; Hall et al, 1979). In general, rocks in these areas are composed largely of sandstone and minor amounts of siltstone and conglomerate. The sandstones in all of these belts have Q-F-L ratios that approximate those in both the Salinian composite terrane and the Great Valley sequence. All three of these sedimentary subterranes may represent fragments of slope basins developed on Franciscan basement into which ponded sediments were later tectonically kneaded (Smith, Howell, and Ingersoll, 1979). Greenstone, volcanic breccia, pillow basalt, and chert are present within these coherent "slabs." A composite thickness of nearly 4,100 m is estimated for the stratal sequence at Cambria (Howell et al, 1977).

Coast Range ophiolite — Remnants of dismembered oceanic crust are exposed in the Santa Lucia Range, San Rafael Mountains, and west of Santa Maria (Figure 3). A nearly complete sequence about 3 km thick crops out in the vicinity of Point Sal (Figure 5), where Hopson, Mattinson, and Pessagno (1981) divided the sequence into several units which include from bottom to top (1) gabbro cumulates, (2) dioritic rocks and plagiogranite, (3) a mafic dike-and-sill complex, and (4) massive and pillow lavas, chiefly basalt. These igneous units are overlain by Upper Jurassic chert, sandstone and shale. Together, these constitute the Stanley Mountain terrane of Blake, Howell, and Jones (1982). In the Santa Lucia Range north of San Luis Obispo, remnants of the ophiolite consist of serpentinized ultramafic rocks, pillow basalt, altered diabasic rocks, and dike-and-sill complexes (Page, 1972). Northwest of Cambria the ophiolitic rocks are largely dioritic and diabasic (Hall and others, 1979). At Stanley Mountain in the southeasternmost Santa Lucia Range, interbedded andesite and basalt flows, pillow lavas, and tuffs were mapped by Brown (1968) in the upper plate of a thrust that overrides melange of the Franciscan assemblage. Wedge-like protrusions of serpentinized harzburgite, pyroxenite, cumulate gabbro and other ultramafic rocks disrupt the Upper Jurassic and Lower Cretaceous sedimentary section near Figueroa Mountain in the southwestern San Rafael Mountains (Vedder, Howell, and McLean, 1980a).

Sedimentary sequences

Remnants of a very thick sequence of Upper Jurassic through Upper Cretaceous forearc-basin deposits extend from the central Santa Lucia Range southward into the San Rafael Mountains and western Santa Ynez Mountains (Figures 3 and 5). These rocks closely resemble, and are coeval with, the Great Valley sequence of Bailey, Blake, and Jones (1970). This sequence is typified by exposures along the west sides of the Sacramento and San Joaquin Valleys where the section is as much as 12 km thick. Truncation by faults and unconformities disrupts the section everywhere within the Sur-Obispo terrane; the most nearly complete succession of beds probably is in the southeastern San Rafael Mountains (Figure 5), where Upper Jurassic, Lower Cretaceous, and Upper Cretaceous strata are exposed (MacKinnon, 1978). In the subsequent descriptions, emphasis is placed on newly studied sequences in the southern Santa Lucia Range and San Rafael Mountains.

Upper Jurassic and Lower Cretaceous sedimentary rocks — At Stanley Mountain in the southeastern Santa Lucia Range, breccia lenses composed chiefly of greenstone fragments directly overlie altered pillow basalt of an ophiolite sequence and are succeeded by thin-bedded tuffaceous chert and cherty mudstone in a section 50 to 70 m thick. These siliceous strata, which at places include thin limy beds, contain Late Jurassic radiolarian assemblages (D. L. Jones, oral communication, 1980). Partly equivalent beds are present in a thrust-enclosed tract near Stanley Mountain (Brown, 1967; Vedder, Howell, and McLean, 1980a), where the structural and stratigraphic relations were misinterpreted by earlier workers. Unusual among these fault-bounded strata is a lens of pebble-boulder conglomerate that includes abundant angular limestone and andesite clasts as large as 1 m (Table 2). The limestone fragments contain Tithonian mollusks, and the andesite and metamorphic clasts suggest a continental source. Overlying the cherty sequence is a poorly-exposed section of interbedded mudstone and sandstone, above which are sparse lenses of chert-pebble conglomerate that contain Valanginian mollusks.

Lower Cretaceous sequences probably are most nearly complete in the San Rafael Mountains, where Valanginian, Hauterivian or Barremian, and Albian fossils have recently been reported (McLean, Howell, and Vedder, 1977; Vedder, unpublished data, 1980) in a sequence of thin-bedded graded strata northeast of Figueroa Mountain. In this area, the Upper Jurassic section is incomplete or absent, presumably as a result of fault truncation and protrusion by ultramafic rocks along zones of weakness. The Valanginian to Albian rocks are largely alternating thin beds of mudstone and sandstone turbidites with minor amounts of chert-pebble conglomerate. Together, these turbidites and conglomerate closely resemble other Lower Cretaceous basin-plain and submarine fan sequences in the Santa Lucia Range and Santa Ynez Mountains. In the San Rafael Mountains, a lenticular conglomerate unit 650 m thick persists for 16 km along the north flank of

Table 2. Conglomerate assemblages (abundance in percent except as noted⁺).

	Granitic	Porphyritic volcanic	Gneiss	Schist	Quartzite	Argillite	Sand-stone	Silt-stone	Vein quartz	Banded tuff	Volcanic breccia	Green-stone	Chert	Lime-stone
Cretaceous Conglomerate Assemblages (Salinian Terrane)														
Upper Cretaceous														
Northern Santa Lucia Range (Ruetz, 1979)	30-65	3-7	10-40	10-20	2-5		2-5	1-2				0-3		
Lake Nacimiento	15	90			T				T	T	T			
Northern La Panza Range	40	24	11	2	15		7		1					
Do.	5	90			3				T	1				
Do.														
Fanglomerate	xxx	x	x	x	xxx									
Fluvial conglomerate		40			60					T				
Fluvial conglomerate		80	T	T	20									
Central La Panza Range														
Fanglomerate	35	3	35	2	25									
Inner Fan Channel	14	8	18	2	58									
Cretaceous Conglomerate Assemblages (Sur-Obispo Terrane)														
Upper Cretaceous														
Sierra Madre Mts.*	x	x			x	x			x		x			
Do.	xxx	x	xx		xx	x				x		x		
Southern Santa Lucia Range	x	x			x		x			x				
Stanley Mtn. area	x	x			x								x	
Do.	xx	xxx		x	xx	x							x	
San Rafael Mts.	10	70				15	R	R					R	
Do.	25	70			5	3								
Do.	25	65				3	3						4	
Lower Cretaceous and Upper Jurassic(?)														
Stanley Mtn. area		xxx					x						x	xx
Do. (Brown, 1968)	R	30		R		R	10						50	R
San Rafael Mts.	R				R	x	x		R				xxx	
Southern Santa Lucia Range (Seiders, Pessagno, and Harris, 1979)				x	x		x		x				xxx	
*Southwest of Sur-Nacimiento fault zone														

Paleocene Conglomerate Assemblages (Salinian Terrane)

	Granitic	Porphyritic volcanic	Gneiss	Schist	Quartzite	Amphibolite	Hornfels	Sandstone	Mudstone	Limestone or Marble	Chert
Gualala (Wentworth, 1968)	xx	xx	x	xx	x	x				x	
Point Reyes	xx	xxx									
Point San Pedro	xxx	x		x			x			x	
Point Lobos (Nili-Esfahani, 1965)	5+	90+									4+
Santa Cruz Mts. (Cummings, Touring, and Brabb, 1962)	xx	xxx	x	x	xxx						
Northern Santa Lucia Range (Ruetz, 1979)											
P1	35	5	35	15				5	5		
P2	55	3	10	15				3	5		
P3	50	5	15	25				5	2		

Eocene Conglomerate Assemblages (Salinian Terrane)

	Granitic	Porphyritic volcanic	Gneiss	Schist	Quartzite	Marble	Sandstone	Chert	Pegmatite
Santa Cruz Mts. (Clarke, 1973)	xx	x	x	x	x	x	x	x	
Northern Santa Lucia Range (Link and Nilsen, 1980)	41	40	5 (plus schist)		7	2			2
Sierra Madre and Garcia Mountains	xx	xxx	x	x	xx		x		

⁺Rock types listed in visual estimates of abundance
xxx Abundant
xx Common
x Present
R Rare
T Trace

Figueroa Mountain and separates the Lower Cretaceous turbidites from those in the Upper Cretaceous sequence. This conglomerate is unusual because it contains intermediate to mafic volcanic and hypabyssal detritus in varying amounts and directional features that directly oppose those in younger conglomerates (McLean, Howell, and Vedder, 1977; Nelson, 1979). Cenomanian fossils were reported by Vedder et al (1967) from the top of this conglomerate unit and from directly overlying beds. Page, Marks, and Walker (1951) and Dibblee (1966) recognized neither a lithologic nor a faunal discontinuity between Lower and Upper Cretaceous strata farther southeast in the San Rafael Mountains. The chronostratigraphic boundary is uncertain because of fossil sparsity, but it is now known to be lower in the section than indicated by Dibblee (1966).

Upper Cretaceous and Paleocene sedimentary rocks, San Rafael, Sierra Madre, and Santa Ynez Mountains — Because of the diverse nature of Upper Cretaceous rocks in the Sur-Obispo composite terrane, stratigraphic descriptions focus on the sequences exposed in the San Rafael Mountains and on the southwest flank of the Sierra Madre Mountains west of the Sur-Nacimiento fault zone. Other sequences are referenced and depicted in Figure 5. In general, the lower beds in the section exposed in the central San Rafael Mountains consist of thin, alternating graded sandstone and mudstone which contain Cenomanian mollusks northeast of Figueroa Mountain (Vedder et al 1967) and Turonian foraminifers east and northeast of Gibralter Reservoir (Howell et al, 1977). Upsection in the same areas, these thin-bedded strata grade into thicker-bedded, coarser-grained facies which include lenticular thin-bedded to thick-bedded sandstone and mudstone-siltstone beds, and thin to thick lenses of pebble-boulder conglomerate. Turonian to lower Maestrichtian foraminifers and mollusks are reported from these sequences (Vedder et al, 1967; Vedder, Dibblee, and Brown, 1973; Howell et al, 1977; MacKinnon, 1978). Most of the Upper Cretaceous strata in the San Rafael Mountains probably represent deposition on a subsea-fan system that included basin-plain, mid-fan, and possibly inner-fan channels (Howell et al 1977). Sediment-transport directions generally range from northwest to southwest.

One significant exception to the prevailing Late Cretaceous depositional pattern is a sequence of beds that indicates paralic environments along a narrow, 35-km-long belt at the northeast edge of the southern Sur-Obispo composite terrane (Vedder and Brown, 1968; Vedder, Howell, and McLean, 1977, 1980b). Because of its importance for tectonic reconstructions (Figure 4), this paralic sequence requires a brief description: it consists of a marine-nonmarine-marine regressional and transgressional cycle that is recognized only in a small segment of the composite terrane. Along the southwest flank of the Sierra Madre Mountains, on the southwest side of the Sur-Nacimiento fault zone, submarine-fan facies grade upward into crossbedded sandstone and lenses of organized conglomerate that probably represent shallow-marine depositional environments. These shallow-marine beds are overlain by massive, pebble-boulder conglomerate and subordinate amounts of variegated mudstone and crossbedded sandstone, all of which have a nonmarine aspect. The conglomerate occurs in lenticular, internally disorganized beds as thick as 30 m; the clasts are mainly granitic and as large as 2 m, and the matrix is composed of very coarse-grained angular quartz, feldspar, and biotite grains. An alluvial-fan and braided-stream depositional environment is inferred. Flow directions are generally south and west. Upward gradation and westward tonguing into carbonaceous crossbedded sandstone and fossiliferous mudstone indicate a reversion to shallow-marine conditions. Farther upsection, a recurrence of submarine-fan deposits consists chiefly of interbedded siltstone and graded sandstone. Where they are cut by the Sur-Nacimiento fault zone, the shallow-marine-nonmarine strata are as much as 1,500 m thick, but are not known to extend more than 5 km from the fault at the surface. Fossil mollusks from below the nonmarine section suggest assignment to the Campanian and possibly lower Maestrichtian Stages. Mollusks, including ammonites, from above it have lower Maestrichtian affinities. Three zircon dates from granitic boulders range from 102 to 126 million years (J. M. Mattinson, written communication 1980), and initial $^{87}Sr/^{86}Sr=0.7040$ (R. W. Kistler, written communication, 1980). Hence, late Mesozoic granitic plutons were eroded on the northeast side of the present site of the Sur-Nacimiento fault zone, and a Salinian provenance is inferred. Late Cretaceous (Campanian) suturing of the Salinian and Sur-Obispo terranes is further implied by other stratigraphic and sedimentologic relations (Figure 4).

In the west-central Santa Ynez Mountains, clastic marine strata constitute the Upper Cretaceous section (Dibblee, 1950; Almgren, 1973), the lower part of which consists of sandstone turbidites that show northward paleocurrent trends, a direction opposite the prevailing trend in the San Rafael Mountains. The upper part of the section is composed largely of sandstone and mudstone that contain foraminiferal and molluscan assemblages indicative of upper slope and nearshore environments, respectively (Dailey and Popenoe, 1966).

Strata of Paleocene age are known only in the southernmost part of the terrane. Gibson (1972, 1976) described an unconformity between Paleocene algal limestone containing Franciscan detritus and Upper Cretaceous beds at places in the western Santa Ynez Mountains. Comstock (1975) reported intertonguing upper Paleocene limestone and siltstone beds in a conformable Upper Cretaceous to lower Eocene sequence along the Big Pine fault in the southern San Rafael Mountains, but the biostratigraphic relations are questionable and the faunal data should be reexamined.

Eocene and Oligocene sedimentary rocks — Eocene strata have a limited distribution in the Sur-Obispo composite terrane and apparently are restricted to the southern part. In the southeastern San Rafael Mountains, lower Eocene bathyal mudstone beds rest without discordance on Maestrichtian beds south of the

Big Pine fault. In places, the mudstone beds grade laterally into lenticular algal limestone-breccia beds. Within 20 km to the west and southwest, Eocene limestone, mudstone, and sandstone beds are unconformable on the Franciscan assemblage (Figure 6). Franciscan detritus characterizes the basal beds in these areas. Directly north of the Big Pine fault at Big Pine Mountain, about 900 m of lower (?) Eocene sandstone is conformable on Maestrichtian(?) flysch; south of the fault, the entire Eocene section is nearly 3,000 m thick and consists of submarine-fan deposits that grade upsection into slope and shelf deposits. Similar environments and an overall progradational cycle persist in the Santa Ynez Mountains. In combination, the widespread absence of Eocene strata in the Sur-Obispo composite terrane north of the Transverse Ranges, the generally conformable contact between Upper Cretaceous marine beds and Oligocene nonmarine beds northeast of Santa Maria, and the emergent San Rafael high (Reed and Hollister, 1936; Nilsen and Clarke, 1975) suggest a region of high relief early in Eocene time.

Nonmarine strata that are unconformable on late Eocene and older rocks occur at a few places in the Sur-Obispo composite terrane south of the Cambria area (Figure 3). These nomarine beds are locally present in the southern Santa Lucia Range, western Santa Maria basin, and San Rafael Mountains, where they are overlain by, or intertongue with, upper Oligocene and lower Miocene marine deposits or pyroclastic rocks. In the Santa Ynez Mountains they are broadly distributed and grade westward into marine beds. Typically, the nonmarine sequences are conglomeratic, lenticular, and restricted in thickness and distribution (Figure 5). Together, the nonmarine character, limited distribution, and variable thickness imply vertical tectonics including graben development and possible beginnings of wrench tectonics.

Sandstone and conglomerate compositions, Sur-Obispo composite terrane

In contrast to the nearly homogenous composition of the Upper Cretaceous sandstones in the Salinian composite terrane, those in the Sur-Obispo composite terrane show stratigraphically controlled petrofacies. Apparently, the petrofacies change occurred during early Late Cretaceous time in the Stanley Mountain-San Rafael Mountains area. Because this change is an indicator of the timing of major tectonic events and the unroofing of plutons in the source area, it is briefly described.

Thin sandy layers that are interbedded with cherty mudstone and tuffaceous chert directly above the ophiolite near Stanley Mountain consist exclusively of grains of volcanic plagioclase and altered tuff. Upper Jurassic sandstone beds upsection from the cherty beds contain quartz, plagioclase, and mafic volcanic rock fragments; potassium feldspar is absent. Brown (1967) noted that Valanginian (?) sandstone beds have abundant quartz (30 to 50%), common feldspar (15 to 25%), and smaller amounts of lithic fragments (5 to 10%). In the Albian to Cenomanian section northeast of Figueroa Mountain, Nelson (1979) reported abundant mafic rock fragments and little or no potassium feldspar. Farther north in the Sur-Obispo composite terrane, Tithonian to Valanginian sandstones were described by Gilbert and Dickinson (1970) as "feldspathic" and devoid of potassium feldspar in the lower part of the section. Uppermost Cretaceous sandstones throughout the terrane have compositions (Lee-Wong and Howell, 1977) very much like those in correlative strata in both the Salinian composite terrane (Hart, 1976) and the Great Valley sequence of the Sacramento Valley (Dickinson and Rich, 1972). In general, quartz and feldspar, including potassium feldspar, are abundant constituents and lithic fragments are sparse (Table 1). Presumably, the regional change in sandstone composition resulted from the first stage of unroofing of batholithic rocks in the western Cordillera during Cenomanian and later Cretaceous time.

In the Sur-Obispo composite terrane, differences in clast composition are evident between Upper Jurassic to Lower Cretaceous conglomerates and Upper Cretaceous conglomerates (Table 2). Taliaferro (1943) and most subsequent workers in the Santa Lucia Range recognized such differences. Moreoever, a similar change in clast composition occurs in the Great Valley sequence east of the San Andreas fault. In general, the oldest conglomerates (Tithonian and Valanginian) are dominated by a chert-graywacke-mafic volcanic assemblage of clasts. At some places, locally derived limestone fragments are abundant, particularly low in the section. Middle Cretaceous conglomerates (Albian to Cenomanian) are characterized by a heterogeneous mixture of clasts that includes pebbles and cobbles similar to those in the older conglomerates as well as volcanic rocks of intermediate composition and granitic rocks. Latest Cretaceous conglomerates (Campanian and Maestrichtian) contain abundant potassic granitic rocks, quartzite, porphyritic siliceous volcanic rocks and, less commonly, gneissic and schistose rocks (Table 2).

Basin analysis summary

Although only small fragments of the depositional system remain, preserved sedimentary facies indicate that a forearc basin persisted throughout the Sur-Obispo composite terrane during Late Jurassic and Cretaceous time. The original dimensions and configuration of this forearc basin are conjectural. However, the present distribution of the basin remnants suggests that it extended far beyond the confines of the terrane itself. Near the close of Cretaceous time, this vast basin began disrupting with an episode of tectonic wrenching that included removal of parts of the opposing edges and subsequent suturing of the Salinian and Sur-Obispo terranes (Page, 1981; Page, in press). Upper Cretaceous and Paleocene deposition seems to reflect the development of restricted, borderland-like basins. The regional middle(?) Paleocene to early(?) Eocene unconformity indicates an orogenic event that resulted in high relief. Basin filling in Eocene time culminated with nonmarine conditions in localized troughs during the Oligocene or early Miocene and was followed by shallow-marine transgression and subsequent formation of deep-marine basins in Miocene time.

Significant among these phases of basin evolution are: 1) widespread Upper Jurassic and Lower Cretaceous basin-plain and outer submarine-fan facies; 2) progradation and localization of Upper Cretaceous paralic facies along a segment of the southeastern edge of the Sur-Obispo terrane boundary; 3) regional discordance at the base of the Eocene together with the high-relief topography; 4) general absence of marine basins in Oligocene time; and, 5) regeneration of Neogene marine basins in a borderland setting. These sequential changes in basin characteristics indicate that the late Mesozoic forearc basin gradually was overwhelmed by an influx of sediment derived from unroofed plutons in the western Cordillera, and that the basin subsequently was fragmented and converted to a basin and ridge system by a shift from subduction to transform tectonics.

TECTONICS

Major Fault Systems

The large horizontal dislocations inferred from paleomagnetic data require major fault systems along the edges of the displaced teranes. Possibly contributing to the Mesozoic and Cenozoic displacements of the Salinian and Sur-Obispo composite terranes are the so-called proto-San Andreas, the Neogene San Andreas fault system, and the Sur-Nacimiento fault zone. Each has a specific kinematic history. Major thrust systems such as the Vincent-Orocopia-Chocolate Mountains thrust are related more to accretionary tectonics than to long distance transport of allochthanous terranes.

Proto-San Andreas fault system — To explain apparently discrepant amounts of estimated cumulative offset on the northern and southern segments of the San Andreas fault and to accomodate the total postulated movement on this transform boundary, an ancestral counterpart has been proposed (Suppe, 1970; Nilsen and Clarke, 1975). Although there is no direct evidence for throughgoing pre-Miocene dextral faults, regional paleogeographic and tectono-stratigraphic reconstructions imply that a regionally extensive right-lateral system was active during latest Cretaceous and Paleogene time. Nilsen (1978) inferred proto-San Andreas activity including borderland-like topography in western California, uplifted land masses west of the San Joaquin Valley, diminution of magmatism in southwestern California, basin-and-range structure in eastern California and adjacent areas, and a possible rifted-margin basin north of the Gulf of California. Johnson and Normark (1974) accounted for much of the supposed proto-San Andreas offset by Neogene movements not only on the San Andreas, but also on subsidiary strike-slip faults that presumably sliced and lengthened the Salinian terrane. Cumulative slip on the Neogene San Andreas fault system is therefore the sum of slip on the family of northwest-oriented strike-slip faults that track through the broad zone of the California Coast Ranges and that join, or are interwoven with, the San Andreas (Graham and Dickinson, 1978).

Sur-Nacimiento fault zone — Antiquity, subsequent modification, and complexity typify the fault zone that separates the Salinian and Sur-Obispo terranes (Figure 3) (Vedder and Brown, 1968; Page, 1970, 1981). Northwestward from San Luis Obispo, the Sur-Nacimiento fault zone probably marks the boundary between rocks of the Franciscan assemblage on the west and granitic plutons and associated wallrocks on the east. Southeastward, the zone includes the Rinconada fault of Dibblee (1976), which presumably is the surface expression of the Sur-Nacimiento fault zone and which juxtaposes Upper Cretaceous and Paleogene sedimentary rocks along much of its trace.

One of the enigmatic aspects of the Sur-Nacimiento fault zone is that reliable pre-Oligocene offset piercing points have not been identified, perhaps because of profound lateral displacements. Other unmatched features include: 1) the contrast in Mesozoic stratigraphic sequences on opposite sides; 2) neither granitic rocks on the southwest side nor evidence of Franciscan detritus in pre-Miocene sedimentary rocks on the northeast side; 3) absence of contact metamorphism on the southwest side; and, 4) presence of post-Campanian and pre-Oligocene thrust faults on the southwest side (Page, 1970; 1981; in press). An important implication of these mismatches is that large pieces of the original Salinian terrane were sliced off or subducted before or during emplacement of the Sur-Obispo composite terrane (Page, 1981; in press).

Vincent-Orocopia — Chocolate Mountains thrust system — Segments of a large thrust fault system have been mapped through the San Gabriel Mountains and southern Mojave Desert as far as western Arizona (Haxel and Dillon, 1978; Burchfiel and Davis, 1981; Crowell, 1981; Ehlig, 1981). This thrust system, which has been severely deformed and later disrupted by strike-slip faults, generally is called the Vincent-Orocopia-Chocolate Mountains thrust, even though individual traces are exposed only in isolated windows and may not represent the same fault. Mylonitic rocks generally separate Precambrian to Mesozoic crystalline rocks in the upper plate from greenschists in the lower plate. The schists probably were metamorphosed near the end of Cretaceous or early in Tertiary time (Ehlig, 1975), but the age of the protolith is unknown. Rocks have not been found beneath the greenschists and relative displacement directions have not been firmly established (Jacobson, 1980), although local structures suggest northeast vergence (Ehlig, 1975; Haxel and Dillon, 1978; Haxel, 1981). Investigators agree that parts of the region are allochthonous.

Inferences from paleomagnetic and stratigraphic data

The sequence of events proposed by the advocates of plate-tectonics involve three general episodes on

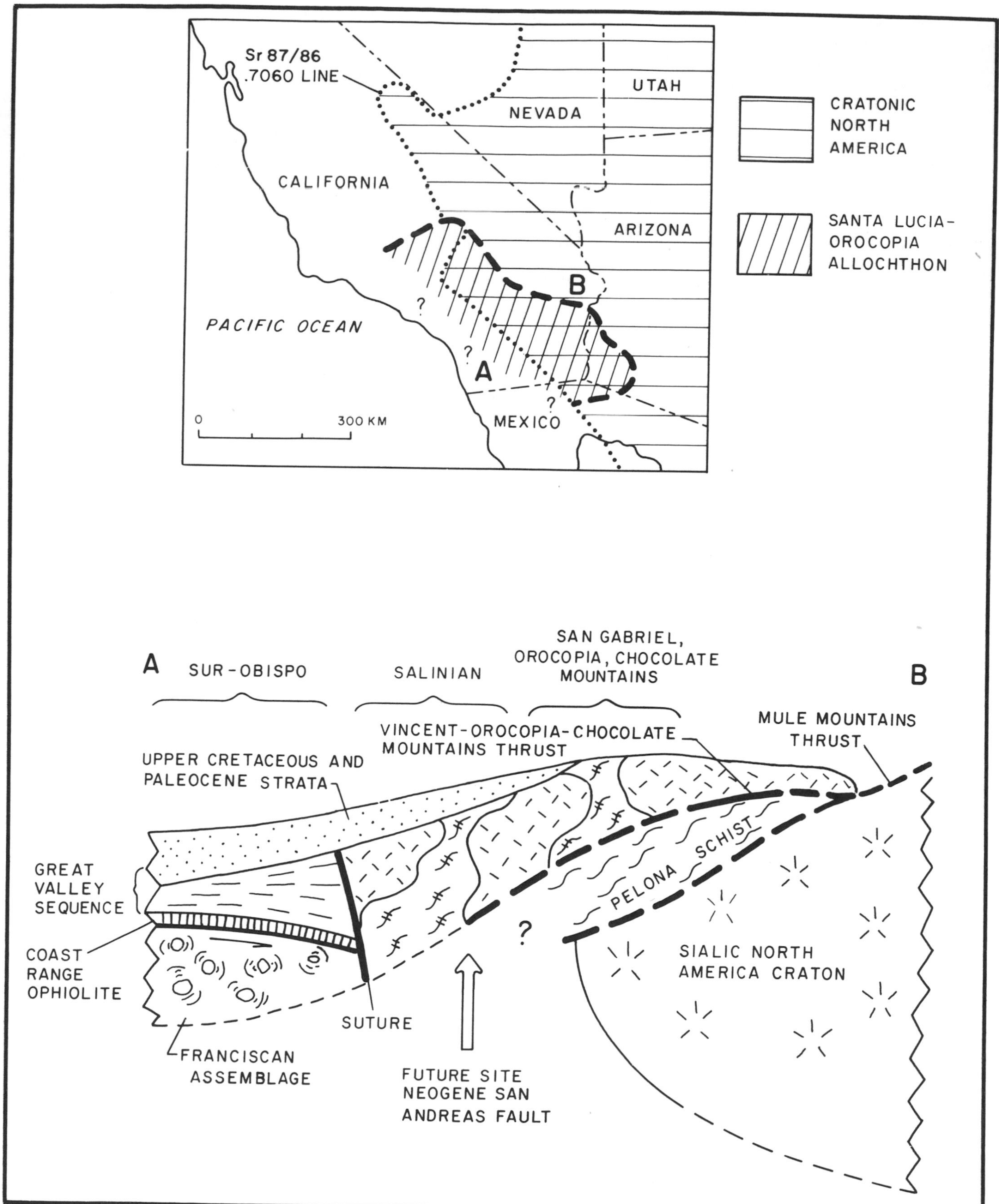

Figure 9 — Diagrammatic map and section showing the possible early Eocene location and crustal structure of the hypothetical Santa Lucia-Orocopia allochthon and its terrane elements. Line of initial $^{87}Sr/^{86}Sr = 0.7060$ generalized from Kistler and Peterman (1978).

the West Coast: 1) a Late Jurassic to early Tertiary subduction regime in an arc-trench gap configuration (Dickinson, 1970, 1972); 2) proto-San Andreas faulting and a borderland setting during an early to middle Tertiary episode of oblique subduction along a flattened Benioff zone (Coney, 1979); and, 3) continental-margin transform faulting resulting in tectonic splintering and additional borderland development (Atwater, 1970; Johnson and Normark, 1974). If, however, the recent paleomagnetic studies of Champion, Gromme, and Howell (1980) are accepted and basin analyses by Howell and Vedder (1978) are correct, then the timing and nature of the continental margin origin must be reinterpreted. The paleomagnetic studies include data from Upper Cretaceous sedimentary strata at Pigeon Point, where an initial remanent magnetization of anomolously low inclinations indicates deposition about 2,500 km farther south than the present site. Basement rock is not exposed at Pigeon Point, but offshore geophysical data (Hoskins and Griffiths, 1971; McCulloch et al, 1980) suggest the presence of granitic rocks. This basement presumably underlies the southern part of the offshore Farallon-Pigeon Point structural high, along which granitic rocks are exposed in places. Thus, the Pigeon Point strata probably are part of the Salinian composite terrane, and this terrane, therefore, is allochthonous. Compositional differences led Ross (1977, 1978) to suspect that the Salinian rocks might be exotic. (Ross (1977, 1978) was unable to match any of the metamorphic basement rocks with supposed offset counterparts south of the Sierra Nevada. If the evidence for a far-traveled exotic terrane is valid, the need for a proto-San Andreas fault, which had 100 to 200 km of rightslip, is eliminated. Instead, long-distance translation, accompanied by amalgamation of allochthonous terranes, is required.

Sedimentary rock sequences of the Salinian and Sur-Obispo composite terranes also indicate that the previously proposed three-phase tectonic history for the California region is untenable. Progradational and retrogradational cycles through the Late Cretaceous and Paleocene Salinian composite terrane suggest wrench tectonics during this time (Howell and Vedder, 1978). Additionally, the clast content and transport direction of a 1,500-m-thick fluvio-deltaic sequence along the southwest flank of the Sierra Madre Mountains seem to substantiate the premise that the Sur-Obispo and Salinian composite terranes were attached by Campanian time (about 75 m.y. ago). Thus, both composite terranes, which came from different parent terranes, must have traveled northward side by side as an allochthon. Despite the remarkable paralleism between the Cretaceous sedimentary environments of the Sur-Obispo composite terrane and those in the Great Valley sequence, the original deposition sites must have been latitudinally far apart.

The hypothetical allochthon in the southern Coast Ranges may extend far southward and possibly include the basement rocks of the northwestern San Gabriel Mountains, which are inferred to be continuations of the Barrett Ridge slice of Salinian basement (Ross, 1977). The San Gabriel Mountains, in turn, are believed to be a detached basement segment of the Chocolate and Orocopia Mountains (Crowell, 1962). Although its leading edge has not been recognized, the hypothetical amalgamated composite that links the terranes of the southern Coast Ranges and southwestern Mojave Desert is informally called the Santa Lucia-Orocopia allochthon. It includes the "single allochthon" of Haxel and Dillon (1978) and corresponds to the upper plate of the Vincent-Orocopia-Chocolate Mountains thrust system (Figure 9). Possibly incoprorated in it are the Pelona, Orocopia, and Rand Schists, whose origins are uncertain and whose underpinnings are unknown. They, too, may have come from low latitudes, or they may represent California ocean-margin deposits that were metamorphosed during collision of the hypothetical Santa Lucia-Orocopia allochthon with North America. Alternatively, Burchfiel and Davis (1981), Crowell (1981), and Yeats (1981) suggested that these schist bodies may be correlative with the Franciscan assemblage, and instead of being structurally superimposed on sialic North America, they are structurally beneath it.

The hypothetical allochthon must have been emplaced before early Eocene time (pre-55 m.y. ago), because rocks of this age lap across it onto autochthonous California terranes, as indicated by reconstruction of the Butano-Point of Rocks fan system of Nilsen and Clarke (1975). This collision event probably occurred sometime between 65 and 55 m.y. ago. This time interval partly corresponds to: 1) volcanic quiescence along segments of the west coast (Armstrong et al, 1969; Armstrong, 1978; Cross and Pilger, 1978); 2) regional coastal uplift as reflected by K/Ar apparent ages (Miller and Morton, 1980); 3) the age of the Vincent thrust and metamorphism of the Pelona and Orocopia Schists (Haxel and Dillon, 1978); and, 4) the age span of the regional unconformity recognized on the Sur-Obispo and Salinian composite terranes and the San Gabriel and Chocolate-Orocopia Mountains terranes.

CONCLUSIONS

Because speculative reasoning is used to arrive at many of our conclusions, their full acceptance must await testing. Until then, we risk their negation with the belief that some will be proven sound. Before Campanian time, the Sur-Obispo composite terrane presumably was part of an arc-trench system analogous to that of the Sierra Nevada-Great Valley pair. The Salinian composite terrane is a fragmented part of some continental margin, and its plutonic rocks suggest a middle Cretaceous continental arc. By the end of Campanian time, these two terranes were completely sutured as part of a single allochthon (Figure 9). But the event must have been preceded by, or involved with, removal of a western segment of the Salinian composite terrane and an eastern segment of the Sur-Obispo composite terrane (Page, 1981; in press). Paleomagnetic data from part of the hypothetical

Santa Lucia-Orocopia allochthon imply approximately 2,500 km of northward drift between Late Cretaceous and middle Eocene time (about 75 to 55 m.y. ago).

During the proposed episode of northward drift, the two composite terranes mutually shared a phase of wrench faulting and continental-margin sedimentation in a borderland configuration that probably developed along a transform-fault system (Howell, McLean, and Vedder, 1980). The abundance of clasts of porphyritic siliceous volcanic rocks in many of the Upper Cretaceous and lower Tertiary sequences of both terranes indicates contiguity with the western Cordillera at some place yet to be identified.

Judging from reconstructed relative plate motions (Atwater, 1970), Neogene faulting may have generated 1,000 km of cumulative right slip between the continental margin and eastern California. If newly acquired paleomagnetic data are valid, then at least 1,500 km of pre-55-m.y.-old northward drift must also be accommodated. The sedimentary rocks used for the paleomagnetic measurements are late Campanian and/or early Maestrichtian. This Late Cretaceous age suggests that during the interval about 75 to 65 m.y. ago, a continental transcurrent fault, perhaps in the present vicinity of southern Mexico, contributed to the northward motion. Sometime within the span about 65 to 55 m.y. ago, which envelops the age of the erosional unconformity throughout the hypothetical allochthon, northward transport may have transpired without physical connection to the Cordillera to the east. Inferred large northward motions are not predicted by currently accepted plate reconstructions involving the North American and Farallon plates (Atwater and Molnar, 1973; Coney, 1979; Engebretson, Gordon, and Cox, 1980). However, the Kula or some other vanished plate may have interacted with North America in a way that induced long range transport along the continental margin. Approximately 10 cm/yr of northward motion is required for this lost plate if the estimated 1,500 km of minimum slip occurred during the proposed 15 to 20 million year interval.

Emplacement of the hypothetical Santa Lucia-Orocopia allochthon between 60 and 55 m.y. ago created a region of high relief. Throughout Eocene and early Oligocene time, the high-relief areas were eroded and the basins largely filled. By late Oligocee time, relief was subdued and the entire California margin was affected by an eustatically controlled marine regression and possibly by regional uplift. Approximately 30 m.y. ago, the west margin of California began interacting with the Pacific plate, and fault-bounded troughs and thick, areally restricted breccia beds developed. Regional right-hand shear and right slip along the Sur-Nacimiento fault zone probably resulted in the formation of small basins and local zones of thrusting. Certainly by middle Miocene time (about 15 m.y. ago), major continental transform faults were the dominant tectonic factor.

ACKNOWLEDGMENTS

Suggestions and comments by B.M. Page, D. S. McCulloch, C. A. Hall, Jr., and J. Bourgeois helped to improve the final report. We benefited from discussions with many colleagues, who are also involved with the problems of Cordilleran geology. We are grateful to D. E. Champion for sharing his paleomagnetic data and to D. L. Jones for providing biostratigraphic data on Mesozoic fossils.

REFERENCES CITED

Allen, J. A., 1946, Geology of the San Juan Bautista quadrangle, California: California Division of Mines and Geology Bulletin, n. 133, p. 5-76.

Almgren, A. A., 1973, Upper Cretaceous Foraminifera in southern California, *in* Cretaceous stratigraphy of the Santa Monica Mountains and Simi Hills: Society of Economic Paleontologists and Mineralogists, Pacific Section, Field Trip Guidebook, p. 31-44.

Armstrong, R. L., 1978, Cenozoic igneous history of the U.S. Cordillera from latitude 42° to 49° north, *in* R. B. Smith, and G. P. Eaton, eds., Cenozoic tectonics and regional geophysics of the western Cordillera: Geological Society of America, Memoir n. 152, p. 265-282.

———, et al, 1969, Space-time relations of Cenozoic silicic volcanism in the Great Basin of the western United States: American Journal of Science, v. 267, p. 478-490.

Atwater, T., 1970, Implications of plate tectonics for the Cenozoic tectonic evolution of western North America: Geological Society of America Bulletin, v. 81, p. 3513-3536.

———, and P. Molnar, 1973, Relative motion of the Pacific and North American plates deduced from sea-floor spreading in the Atlantic, Indian, and South Pacific Oceans, *in* R. L. Kovach, and A. Nur, eds., Proceedings, Conference on Tectonic Problems of the San Andreas fault system: Stanford, California, Stanford University Publications, v. 13, p. 136-148.

Bailey, E. H., M. C. Blake, Jr., and D. L. Jones, 1970, On-land Mesozoic oceanic crust in California Coast Ranges: U. S. Geological Survey Professional Paper 700-C, p. C70-C81.

Bartow, J. A., 1978, Oligocene continental sedimentation in the Caliente Range area, California: Journal of Sedimentary Petrology, v. 48, n. 1, p. 75-78.

Blake, M. C., Jr., et al, 1978, Neogene basin formation in relation to plate-tectonic evolution of San Andreas fault system, California: AAPG Bulletin, v. 62, n. 3, p. 344-372.

———, D. G. Howell, and D. L. Jones, 1982, Map of geologic terranes of California: U.S. Geological Survey Open-file Report 82-593, scale 1:500,000.

Bohannon, R. G., 1975, Mid-Tertiary conglomerates and their bearing on Transverse Range tectonics, southern California, *in* San Andreas fault in southern California: California Division of Mines and Geology, Special Report 118, p. 75-82.

Bourgeois, J., 1980, Sedimentology and tectonics of Upper Cretaceous rocks, southwest Oregon: Madison, Wisconsin, University of Wisconsin, Ph.D. thesis.

Brown, J. A., 1967, Probable thrust contact between Franciscan formation and Great Valley sequence northeast of Santa Maria, California (Abs.): Geological Society of America, Cordilleran Section, Abstracts with Program, p. 23.

———, 1968, Thrust contact between Franciscan Group and Great Valley sequence northeast of Santa Maria, California: Los Angeles, California, University of Southern California, Ph.D. thesis, 234 p.

Burchfiel, B. C., and G. A. Davis, 1981, Mojave Desert and

environs, *in* W. G. Ernst, ed., The geotectonic development of California: Englewood Cliffs, New Jersey, Prentice-Hall, Inc., p. 217-252.

Carman, M. F., Jr., 1964, Geology of the Lockwood Valley area, Kern and Ventura Counties, California: California Division of Mines and Geology, Special Report 81, 62 p.

Champion, Duane, S. Gromme, and D. G. Howell, 1980, Paleomagnetism of the Cretaceous Pigeon Point Formation and the inferred northward displacement of 2,500 km for the Salinian Block, California: Washington, D.C., EOS, American Geophysical Union Transactions, v. 61, n. 46, p. 948.

Chipping, D. H., 1972, Early Tertiary paleogeography of Central California: AAPG Bulletin, v. 52, n. 3, p. 480-493.

Clark, J. C., and J. D. Reitman, 1973, Oligocene stratigraphy, tectonics, and paleogeography southwest of the San Andreas fault, Santa Cruz Mountains and Gabilan Range, California Coast Ranges: U.S. Geological Survey, Professional Paper 783, 18 p.

Clarke, S. H., 1973, The Eocene Point of Rocks Sandstone: provenance, mode of deposition and implications for the history of offset along the San Andreas fault in central California: Berkeley, California, California University, Ph.D. thesis, 302 p.

Compton, R. R., 1966, Granitic and metamorphic rocks of Salinian block, California Coast Ranges, *in* E. H. Bailey, ed., Geology of northern California: California Division of Mines and Geology Bulletin, n. 190, p. 277-287.

Comstock, S.C., 1975, Upper Cretaceous and Paleogene stratigraphy along the western Big Pine fault, Santa Barbara County, California *in* Future energy horizons of the Pacific Coast, Paleogene symposium and selected technical papers: Long Beach, California, AAPG, Society of Economic Paleontologists and Mineralogists, and Society of Exploration Geophysicists, Pacific Sections Annual Meeting, 629 pp.

Coney, P. J., 1971, Cordilleran tectonic transition and motion of the North American plate: Nature, v. 233, p. 462-465.

———, 1978, Mesozoic-Cenozoic Cordilleran plate tectonics, *in* R. B. Smith, and G. P. Eaton, eds., Cenozoic tectonics and regional geophysics of the western Cordillera: Geological Society of America, Memoir n. 152, p. 33-50.

———, 1979, Tertiary evolution of Cordilleran metamorphic core complexes, *in* Cenozoic paleogeography of the western United States: Society of Economic Paleontologists and Mineralogists, Pacific Section, Pacific Coast Paleogeography Symposium 3, p. 14-28.

———, and S. J. Reynolds, 1977, Cordilleran Benioff zones: Nature, v. 270, p. 403-406.

———, D. L. Jones, and J. W. H. Monger, 1980, Cordilleran suspect terranes: Nature, v. 288, n. 27, p. 329-333.

Cross, T. A., and R. H. Pilger, Jr., 1978, Constraints on absolute motion and plate interaction inferred from Cenozoic igneous activity in the western United States: American Journal of Science, v. 278, n. 7, p. 865-902.

Crowell, J. C., 1962, Displacement along the San Andreas fault, California: Geological Society of America, Special Paper 71, 61 p.

———, 1975, The San Andreas fault in southern California, *in* J. C. Crowell, ed., San Andreas fault in southern California, a guide to the San Andreas fault from Mexico to Carrizo Plain: California Division of Mines and Geology, Special Report 118, p. 7-27.

———, 1981, An outline of the tectonic history of southeastern California, *in* W. G. Ernst, ed., The geotectonic development of California: Englewood Cliffs, New Jersey, Prentice-Hall, Inc., Rubey, v. 1, p. 583-600.

Cummings, J. C., R. M. Touring, and E. E. Brabb, 1962, Geology of the northern Santa Cruz Mountains, California, *in* Geologic guide to the gas and oil fields of northern California: California Division of Mines and Geology and Bulletin, n. 181, p. 179-220.

Curtis, G. H., J. F. Evernden, and J. L. Lipson, 1958, Age determinations of some granitic rocks in California by the potassium-argon method: California Department of Natural Resources, Division of Mines and Geology, Special Report 54, 16 p.

Dailey, D. D., and W. P. Popenoe, 1966, Mollusca from the Upper Cretaceous Jalama Formation, Santa Barbara County, California: University of California Publications in Geological Sciences, v. 65, 27 p.

Dibblee, T. W., Jr., 1950, Geology of Southwestern Santa Barbara County, California: California Division of Mines and Geology Bulletin, n. 150, 95 p.

———, 1966, Geology of the central Santa Ynez Mountains, Santa Barbara County, California: California Division of Mines and Geology Bulletin, n. 186, 99 p.

———, 1973, Stratigraphy of the southern Coast Ranges near the San Andreas fault from Cholame to Maricopa, California: U.S. Geological Survey, Professional Paper 764, 45 p.

———, 1976, The Rinconada and related faults in the southern Coast Ranges, California, and their tectonic significance: U.S. Geological Survey, Professional Paper 981, 55 p.

Dickinson, W. R., 1970, Relations of andesites, granites, and derivative sandstones to arc-trench tectonics: Reviews of Geophysics and Space Physics, v. 8, p. 813-860.

———, 1972, Evidence for plate-tectonic regimes in the rock record: American Journal of Science, v. 272, p. 551-576.

———, and E. I. Rich, 1972, Petrologic intervals and petrofacies in the Great Valley sequence, Sacramento Valley, California: Geological Society of America Bulletin, v. 83, p. 3007-3024.

Durham, D. L., 1974, Geology of the southern Salinas Valley area, California: U.S. Geological Survey, Professional Paper 819, 111 p.

Ehlig, P. L., 1975, Basement rocks of the San Gabriel Mountains, south of the San Andreas fault, southern California, *in* San Andreas fault in southern California: California Division of Mines and Geology, Special Report 118, p. 177-186.

———,. and S. E. Joseph, 1977, Polka Dot Granite and correlation of La Panza quartz monzonite with Cretaceous batholithic rocks north of Salton Trough, *in* D. G. Howell, J. G. Vedder, and K. McDougall, eds., Cretaceous geology of the California Coast Ranges, west of the San Andreas fault: Society of Economic Paleontologists and Mineralogists, Pacific Section, Pacific Coast Paleogeography Field Guide 2, p. 91-96.

———, ———, 1981, Origin and tectonic history of the basement terrane of the San Gabriel Mountains, central Transverse Ranges, *in* W. G. Ernst, ed., The tectonic development of California: Englewood Cliffs, New Jersey, Prentice-Hall, Inc., Rubey Volume 1, p. 253-283.

Engebretson, D., R. Gordon, and A. Cox, 1980, Relative motions between the North America, Pacific, Kula, and Farallon plates from 130 m.y. to present: Washington, D.C., Eos, Transactions, American Geophysical Union, v. 61, n. 4, p. 947.

Evernden, J. F., and R. W. Kistler, 1970, Chronology of emplacement of Mesozoic batholithic complexes in California and western Nevada: U.S. Geological Survey, Professional Paper 623, 42 p.

Galloway, A., 1977, Geology of the Point Reyes peninsula, Marin County, California: California Division of Mines and Geology Bulletin, n. 202, 72 p.

Gibson, J. M., 1972, The Anita Formation and "Sierra Blanca Limestone" (Paleocene and early Eocene), western Santa Ynez Mountains, Santa Barbara County, California, *in* D. W. Weaver, ed., Guidebook, Central Santa Ynez Mountains, Santa Barbara County, California: AAPG and Society of Economic Paleontologists and Mineralogists, Pacific Sections, p. 16-19.

———, 1976, Distribution of planktonic Foraminifera and calcareous nannoplankton, Late Cretaceous and early Paleogene, Santa Ynez Mountains, California: Journal of Foraminiferal Research, v. 6, n. 2, p. 87-106.

Gilbert, W. C., and W. R. Dickinson, 1970, Stratigraphic variations in sandstone petrology, Great Valley sequence, central California coast: Geological Society of America Bulletin, v. 81, p. 949-954.

Graham, S. A., 1979, Tertiary paleotectonics and paleogeography of the Salinian block, *in* J. M. Armentrout, M. R. Cole, and H. TerBest, Jr., eds., Cenozoic paleogeography of western United States: Society of Economic Paleontologists and Mineralogists, Pacific Section Symposium 3, p. 45-52.

———, and W. R. Dickinson, 1978, Evidence for 115 kilometers of right slip on the San Gregorio-Hosgri fault trend: Science, v. 199, p. 179-181.

———, and K. D. Berry, 1979, Early Eocene paleogeography of the central San Joaquin Valley: Origin of the Cantua Sandstone, *in* J. M. Armentrout, M. R. Cole, and H. Terbest, Jr., eds., Cenozoic paleogeography of western United States: Pacific Section Symposium 3, Society of Economic Paleontologists and Mineralogists, p. 119-128.

Greene, H. G., and J. C. Clark, 1979, Neogene paleogeography of the Monterey Bay area, California, *in* J. M. Armentrout, M. R. Cole, and H. TerBest, Jr., eds., Cenozoic paleogeography of western United States: Society of Economic Paleontologists and Mineralogists, Pacific Section Symposium 3, p. 277-296.

Hall, C. A., D. L. Jones, and S. A. Brooks, 1959, Pigeon Point formation of Late Cretaceous age, San Mateo County: AAPG Bulletin, v. 43, p. 2855-2859.

Hall, C. A., Jr., and C. E. Corbato, 1967, Stratigraphy and structure of Mesozoic and Cenozoic rocks, Nipomo quadrangle, southern Coast Ranges, California: Geological Society of America Bulletin, v. 78, p. 559-582.

———, et al, 1979, Geologic map of the San Luis Obispo-San Simeon region, California: U.S. Geological Survey, Miscellaneous Investigations Series Map I-1097, scale 1:48,000.

Hart, E. W., 1976, Basic geology of the Santa Margarita area, San Luis Obispo County, California: California Division of Mines and Geology Bulletin, n. 199, 45 p.

Haxel, Gordon, 1981, Late Cretaceous and early Tertiary orogenesis, south central Arizona: Geological Society of America, Abstracts with Programs, v. 13, n. 2, p. 60.

———, and John Dillon, 1978, The Pelona-Orocopia Schist and Vincent-Chocolate Mountain thrust system, southern California, *in* D. G. Howell, and K. A. McDougall, eds., Mesozoic paleogeography of the western United States: Society of Economic Paleontologists and Mineralogists, Pacific Section, Pacific Paleogeography Symposium 2, p. 453-470.

Hill, M. L., S. A. Carlson, and T. W. Dibblee, Jr., 1958, Stratigraphy of Cuyama Valley-Caliente Range area, California: AAPG Bulletin, v. 42, n. 12, p. 2973-3000.

Hopson, C. A., J. M. Mattinson, and E. A. Pessagno, Jr., 1981, Coast Ranges Ophiolite, western California, *in* W. G. Ernst, ed., The geotectonic development of California: Englewood, New Jersey, Prentice-Hall, Rubey Volume 1, p. 418-510.

Hoskins, E. G., and S. R. Griffiths, 1971, Hydrocarbon potential of northern and central California offshore, *in* I. H. Cram, ed., Future petroleum provinces of the United States — Their geology and potential: AAPG Memoir 15, v. 1, p. 212-228.

Howell, D. G., 1975, Early and middle Eocene shoreline offset by the San Andreas fault, southern California, *in* J. C. Crowell, ed., San Andreas fault in southern California: California Division of Mines and Geology, Special Report 118, p. 69-74.

———, and J. G. Vedder, 1978, Late Cretaceous paleogeography of the Salinian block, California, *in* D. G. Howell, and K. A. McDougall, eds., Mesozoic paleogeography of the western United States: Society of Economic Paleontologists and Mineralogists, Pacific Section, Pacific Coast Paleogeography Symposium 2, p. 107-116.

———, Hugh McLean, and J. G. Vedder, 1980, Late Cretaceous suturing and translation of the Salinian and Nacimiento blocks, California: Washington, D.C., Eos, Transactions, American Geophysical Union, v. 61, n. 46, p. 948.

———, et al, 1977, Review of Cretaceous geology, Salinian and Nacimiento blocks, Coast Ranges of central California, *in* D. G. Howell, J. G. Vedder, and K. McDougall, eds., Cretaceous geology of the California Coast Ranges, west of the San Andreas fault: Society of Economic Paleontologists and Mineralogists, Pacific Section, Pacific Coast Paleogeography Field Guide 2, p. 1-46.

Hsu, K. J., 1969, Preliminary report and geologic guide to Franciscan melanges of the Morro Bay-San Simeon area, California: California Division of Mines and Geology, Special Publication n. 35, 46 p.

Jacobson, C. E., 1980, Deformation and metamorphism of the Pelona Schist beneath the Vincent thrust, San Gabriel Mountains, California, Los Angeles, California; California University, Ph.D. thesis, 231 p.

Johnson, J. D., and W. R. Normark, 1974, Neogene tectonic evolution of the Salinian block, west-central California: Geology, v. 2, n. 1, p. 11-14.

Kerr, P. F., and H. G. Schenck, 1925, Active thrust faults in San Benito County, California: Geological Society of America Bulletin, v. 36, n. 3, p. 465-494.

Kistler, R. W., and Z. E. Peterman, 1978, Reconstruction of crustal blocks of California on the basis of initial strontium isotopic compositions of Mesozoic granitic rocks: U.S. Geological Survey, Professional Paper 1071, 17 p.

Kooser, M. A., 1980, Stratigraphy and sedimentology of the San Francisquito Formation, Transverse Ranges, California: Riverside, California, California University, Ph.D. thesis, 201 p.

Lee-Wong, Florence, and D. G. Howell, 1977, Petrography of Upper Cretaceous sandstone in the Coast Ranges of central California, *in* D. G. Howell, J. G. Vedder, and K. McDougall, eds., Cretaceous geology of the California Coast Ranges, west of the San Andreas fault: Society of Economic Paleontologists and Mineralogists, Pacific Section, Pacific Coast Paleogeography Guide 2, p. 47-56.

Link, M. H., and T. H. Nilsen, 1980, The Rocks Sandstone, an Eocene sand-rich deep-sea fan deposit, northern Santa Lucia Range, California: Journal of Sedimentary Petrology, v. 50, n. 2, p. 583-601.

MacKinnon, T. C., 1978, The Great Valley sequence near Santa Barbara, California, *in* D. G. Howell, and K. A. McDougall, eds., Mesozoic paleogeography of the western United States: Society of Economic Paleontologists and Mineralogists, Pacific Section, Pacific Coast Paleogeography Symposium 2, p. 483-507.

Mattinson, J. M., 1978, Age, origin, and thermal histories of some plutonic rocks from the Salinian block of California: Contributions to Mineralogy and Petrology, v. 67, p. 233-245.

———, C. A. Hopson, and T. E. Davis, 1972, U-Pb studies of plutonic rocks of the Salinian block, California: Washington, D.C., Carnegie Institute, Yearbook n. 71, p. 571-576.

McCulloch, D.S., et al, 1980, Regional geology, petroleum potential, and environmental geology in the northern part of proposed lease sale 73, offshore central and northern California: U.S. Geological Survey, Open-File Report 80-2007, p. B2-B68.

McLean, Hugh, D. G. Howell, and J. G. Vedder, 1977, An unusual Upper Cretaceous conglomerate in the central San Rafael Mountains, Santa Barbara County, California, *in* D. G. Howell, J. G. Vedder, and K. A. McDougall, eds., Cretaceous geology of the California Coast Ranges, west of the San Andreas fault: Society of Economic Paleontologists and Mineralogists, Pacific Section, Pacific Coast Paleogeography Field Guide 2, p. 79-84.

Miller, F. K., and D. M. Morton, 1980, Potassium-argon geochronology of the eastern Transverse Ranges and southern Mojave Desert, southern California: U.S. Geological Survey, Professional Paper 1152, 30 p.

Minch, J. A., 1979, The Late Mesozoic-early Tertiary framework of continental sedimentation, northern Peninsular Ranges, Baja California, Mexico, *in* Eocene depositional systems San Diego, California: Society of Economic Paleontologists and Mineralogists, Pacific Section, p. 43-68.

Nelson, A. S., 1979, Upper Cretaceous depositional environments and provenance indicators in the central San Rafael Mountains, Santa Barbara County, California: Santa Barbara, California, California University, Master's thesis, 153 p.

Nili-Esfahani, Alireza, 1965, Investigation of Paleocene strata, Point Lobos, Monterey County, California: Los Angeles, California, University of California, Master's thesis, 159 p.

Nilsen, T. H., 1971, Sedimentology of the Eocene Butano Sandstone, a continental borderland submarine-fan deposit, Santa Cruz Mountains, California: Washington, D.C., Geological Society of America Annual Meeting, Abstracts with programs, p. 659-660.

———, 1978, Late Cretaceous geology of California and the problem of the proto-San Andreas fault, *in* Howell, D. G., and McDougall, K. A., eds., Mesozoic paleogeography of the western United States: Society of Economic Paleontologists and Mineralogists, Pacific Section, Pacific Coast Paleogeography Symposium 2, p. 559-573.

———, and S. H. Clarke, Jr., 1975, Sedimentation and tectonics in the early Tertiary continental borderland of central California: U.S. Geological Survey, Professional Paper 925, 64 p.

Ortega, G. F., T. H. Anderson, and L. T. Silver, 1977, Lithologies and geochronology of the Precambrian craton of southern Mexico: Geological Society of America, Abstracts with Programs, v. 9, n. 7, p. 1121-1122.

Page, B. M., 1970, Sur-Nacimiento fault zone of California: Continental margin tectonics: Geological Society of America Bulletin, v. 81, n. 3, p. 667-690.

———, 1972, Oceanic crust and mantle fragment in subduction complex near San Luis Obispo, California: Geological Society of America Bulletin, v. 83, n. 4, p. 957-972.

———, 1981, The Southern Coast Ranges, *in* W. G. Ernst, ed., The geotectonic development of California: Englewood Cliffs, New Jersey, Prentice-Hall, Inc., Rubey Volume 1, p. 329-415.

———, in press, Migration of Salinian composite block, California, and disappearance of fragments: American Journal of Science.

———, J. G. Marks, and G. W. Walker, 1951, Stratigraphy and structure of mountains northeast of Santa Barbara, California: AAPG Bulletin, v. 35, p. 1727-1780.

Reed, R. D., and J. S. Hollister, 1936, Structural evolution of southern California: AAPG, 157 p.

Ross, D. C., 1972a, Petrographic and chemical reconnaissance of some granitic and gneissic rocks near the San Andreas fault from Bodega Head to Cajon Pass, California: U.S. Geological Survey, Professional Paper 698, 92 p.

———, 1972b, Geologic map of the pre-Cenozoic basement rocks, Gabilan Range, Monterey and San Benito Counties, California: U.S. Geological Survey, Miscellaneous Field Studies Map MF-357, scale 1:125,000.

———, 1973, Are the granitic rocks of the Salinian block trondhjemitic?: U.S. Geological Survey Journal of Research, v. 1, n. 3, p. 251-254.

———, 1976, Map showing distribution of metamorphic rocks and occurrences of garnet, coarse graphite, sillimanite, orthopyroxene, clinopyroxene, and plagioclase amphibolite, Santa Lucia Range, Salinian block, California: U.S. Geological Survey, Miscellaneous Field Investigations Map MF-791, scale 1:500,000.

———, 1977, Pre-intrusive metasedimentary rocks of the Salinian block, California — A paleotectonic dilemma, *in* J. H. Stewart, C. H. Stevens, and A. E. Fritsche, eds., Paleozoic paleogeography of the western United States: Society of Economic Paleontologists and Mineralogists, Pacific Section, p. 371-380.

———, 1978, The Salinian block — A Mesozoic granitic orphan in the California Coast Ranges, *in* D. G. Howell, and K. A. McDougall, eds., Mesozoic paleogeography of the western United States: Society of Economic Paleontologists and Mineralogists, Pacific Section, Pacific Coast Paleogeography Symposium 2, p. 509-522.

Ruetz, J. W., 1979, Paleocene submarine fan deposits of the Indians Ranch area, Monterey County, California, *in* Tertiary and Quaternary geology of the Salinas Valley and Santa Lucia Range, Monterey County, California: Society of Economic Paleontologists and Mineralogists, Pacific Coast Paleogeography Field Guide 4, p. 13-24.

Sage, Orrin, Jr., 1975, Sedimentological and tectonic implications of the Paleocene San Francisquito Formation, Los Angeles County, California, *in* J. C. Crowell, ed., San Andreas fault in southern California: California Division of Mines and Geology, Special Report 118, p. 162-169.

Seiders, V. M., E. A. Pessagno, Jr., A. G. Harris, 1979, Radiolarians and conodonts from pebbles in the Franciscan assemblage and the Great Valley sequence of the California Coast Ranges: Geology, v. 7, p. 37-40.

Silver, L. T., 1971, Problems of crystalline rocks of the Transverse Ranges: Geological Society of America, Abstracts with Programs, v. 3, n. 2, p. 193-194.

Smith, G. W., D. G. Howell, and R. V. Ingersoll, 1979, Late Cretaceous trench-slope basins of central California: Geology, v. 7, p. 303-306.

Suppe, J., 1970, Offset of late Mesozoic basement terrains by the San Andreas fault system: Geological Society of America Bulletin, v. 81, p. 3253-3258.

Taliaferro, N. L., 1943, Geologic history and structure of the central Coast Ranges of California: California Division of Mines and Geology Bulletin, n. 188, p. 119-163.

———, 1944, Cretaceous and Paleocene of Santa Lucia Range, California: AAPG Bulletin, v. 28, n. 4, p. 449-521.

Trask, P. D., 1926, Geology of the Point Sur quadrangle, California: California University, Department of Geological Sciences Bulletin, v. 16, n. 6, p. 118-186.

Underwood, M. B., 1977, The Pfeiffer Beach slab deposits, Monterey County, California: Possible trench-slope basins, *in* D. G. Howell, J. G. Vedder, and K. McDougall, eds., Cretaceous geology of the California Coast Ranges west of the San Andreas fault: Society of Economic Paleontologists and Mineralogists, Pacific Section, Pacific Coast Paleogeography Field Guide 2, p. 57-70.

Vedder, J. G., 1968, Geologic map of Fox Mountain quadrangle, Santa Barbara County, California: U.S. Geological Survey, Miscellaneous Geologic Investigations Map I-547, scale 1:24,000.

———, 1975, Juxtaposed Tertiary strata along the San Andreas fault in the Temblor and Caliente Ranges, California, *in* J. C. Crowell, ed., San Andreas fault in southern California: California Division of Mines and Geology, Special Report 118, p. 234-240.

———, and R. D. Brown, Jr., 1968, Structural and stratigraphic relations along the Nacimiento fault in the southern Santa Lucia Range and San Rafael Mountains, California, *in* W. R. Dickinson, and A. Grantz, eds., Proceedings, Conference on Geologic Problems of San Andreas Fault System: Stanford, California, Stanford University Publications, Geological Sciences, School of Earth Sciences, v. 11, p. 242-259.

———, and C. A. Repenning, 1975, Geologic map of the Cuyama and New Cuyama quadrangles, San Luis Obispo and Santa Barbara Counties, California: U.S. Geological Survey, Miscellaneous Investigations Map I-876, scale 1:24,000.

———, T. W. Dibblee Jr., and R. D. Brown Jr., 1973, Geologic map of the upper Mono Creek-Pine Mountain area, California: U.S. Geological Survey, Miscellaneous Geologic Investigations Map I-752, scale 1:48,000.

Vedder, J. G., D. G. Howell, and Hugh McLean, 1977, Upper Cretaceous redbeds in the Sierra Madre-San Rafael Mountains, California, *in* D. G. Howell, J. G. Vedder, and K. A. McDougall, eds., Cretaceous geology of the California Coast Ranges, west of the San Andreas fault: Society of Economic Paleontologists and Mineralogists, Pacific Coast Paleogeography Field Guide 2, p. 71-78.

———, ———, and ———, 1980a, Structural and stratigraphic relations of pre-Tertiary rocks on the perimeter of the Santa Maria basin (Abs.): AAPG Bulletin, v. 64, n. 3, p. 450.

———, ———, and ———, 1980b, Upper Cretaceous redbeds: evidence for early suturing of the Nacimiento and Salinian Blocks, California (abs.): Geological Society of America, Abstracts with Programs, v. 12, p. 157-158.

———, et al, 1967, Reconnaissance geologic map of the central San Rafael Mountains and vicinity, Santa Barbara County, California: U.S. Geological Survey, Miscellaneous Geologic Investigations Map I-487, scale 1:48,000.

Wentworth, C. M., 1968, Upper Cretaceous and lower Tertiary strata near Gualala, California, and inferred large right slip on the San Andreas fault, *in* W. R. Dickinson, and Arthur Grantz, eds., Proceedings, Conference on Geologic Problems of the San Andreas fault system: Stanford, California, Stanford University Publications, Geological Sciences, School of Earth Sciences, v. 11, p. 130-143.

Wiebe, R. A., 1966, Structure and petrology of the Ventana Cones area, California: Stanford California, Stanford University, Ph.D. dissertation, 95 p.

Woodring, W. P., and M. N. Bramlette, 1950, Geology and paleontology of the Santa Maria district, California: U.S. Geological Survey, Professional Paper 222, 185 p.

Yeats, R. S., 1981, Quaternary flake tectonics of the California Transverse Ranges: Geology, v. 9, n. 1, p. 16-20.

The Makran Continental Margin: Structure of a Thickly Sedimented Convergent Plate Boundary

Robert S. White
Keith E. Louden
Bullard Laboratories
Cambridge University
Cambridge, United Kingdom

Two long refraction lines parallel with the structural strike show that the Moho of the subducting oceanic Arabian plate dips north beneath the Makran accretionary prism at an angle of less than 2°. The downgoing plate carries a thickness of about 7 km of sediment into the accretionary wedge. The accreted sediment rapidly consolidates with a concommitant increase in seismic velocity. Slope sediments are trapped in narrow, well-lineated slope basins formed between uplifted fold ridges, and the dip of sediment layers within these slope basins records the progressive tilting of the margin towards the coast. We document the shapes of channels along which downslope sediment movement occurs. Decoupling layers within the sediments, caused by high pore pressures generated by the non-expulsion of pore water from overpressured shale sections, control the deformation in the frontal 70 km of the accretionary prism.

The Makran subduction zone in the Gulf of Oman, northwest Indian Ocean, is a good example of one end-member of the different convergent margins in which the sediment pile on the downgoing plate is thick. This leads to an accretionary sediment prism up to several hundred kilometers wide, beneath which the subducting plate dips shallowly. The Makran subduction zone is about 900 km long, stretching from near the Strait of Hormuz in the west to Karachi in the east (Figure 1a). The oceanic part of the Arabian plate descends beneath the continental Eurasian plate to the north about 5 cm/yr. West of the Strait of Hormuz in the Persian Gulf region, the convergence changes to a continent-continent type boundary with the continental part of the Arabian plate in collision with the Eurasian plate. The eastern termination of the Makran subduction zone is marked by a triple junction between the Eurasian, Arabian, and Indian plates near Karachi. The seismically active left-lateral Chaman and Ornach-Nal fault systems mark the eastern edge of the Makran accretionary wedge.

This article discusses the structure of the frontal 150 km of the accretionary system, where it lies offshore beneath the Gulf of Oman; the accreted wedge continues some 400 to 500 km further north across the Makran of Iran and Pakistan, where it is presently exposed above sea level. The results presented here continue the description of the structure of the offscraped sediment on the Makran continental margin by White (1981) based mainly on seismic reflection profiles. You are referred to that paper for details of the structure, which we summarize here. We discuss the structure at three levels: first, the deep configuration of the Makran continental margin based on preliminary results from two long seismic refraction lines; second the tectonics of the higher structural levels of the accretionary prism deduced from seismic reflection profiles; and third, the interaction of slope sedimentation with the tectonics.

DEEP STRUCTURE

The northward dip of the Moho beneath the accretionary prism was measured using two seismic refraction lines shot parallel with the Makran margin. An offshore line (marked 'OFF' on Figure 1b) was positioned over the Gulf of Oman abyssal plain above undeformed sediments, and a parallel inshore line (marked 'IN' on Figure 1b) was run 90 km to the north on the continental margin. Each refraction line was 180 km long with one ocean bottom hydrophone (OBH), and either four (inshore line) or five (offshore line) free-floating, internally-recording radio sonobuoys (SB). The sonobuoys were deployed at the ends

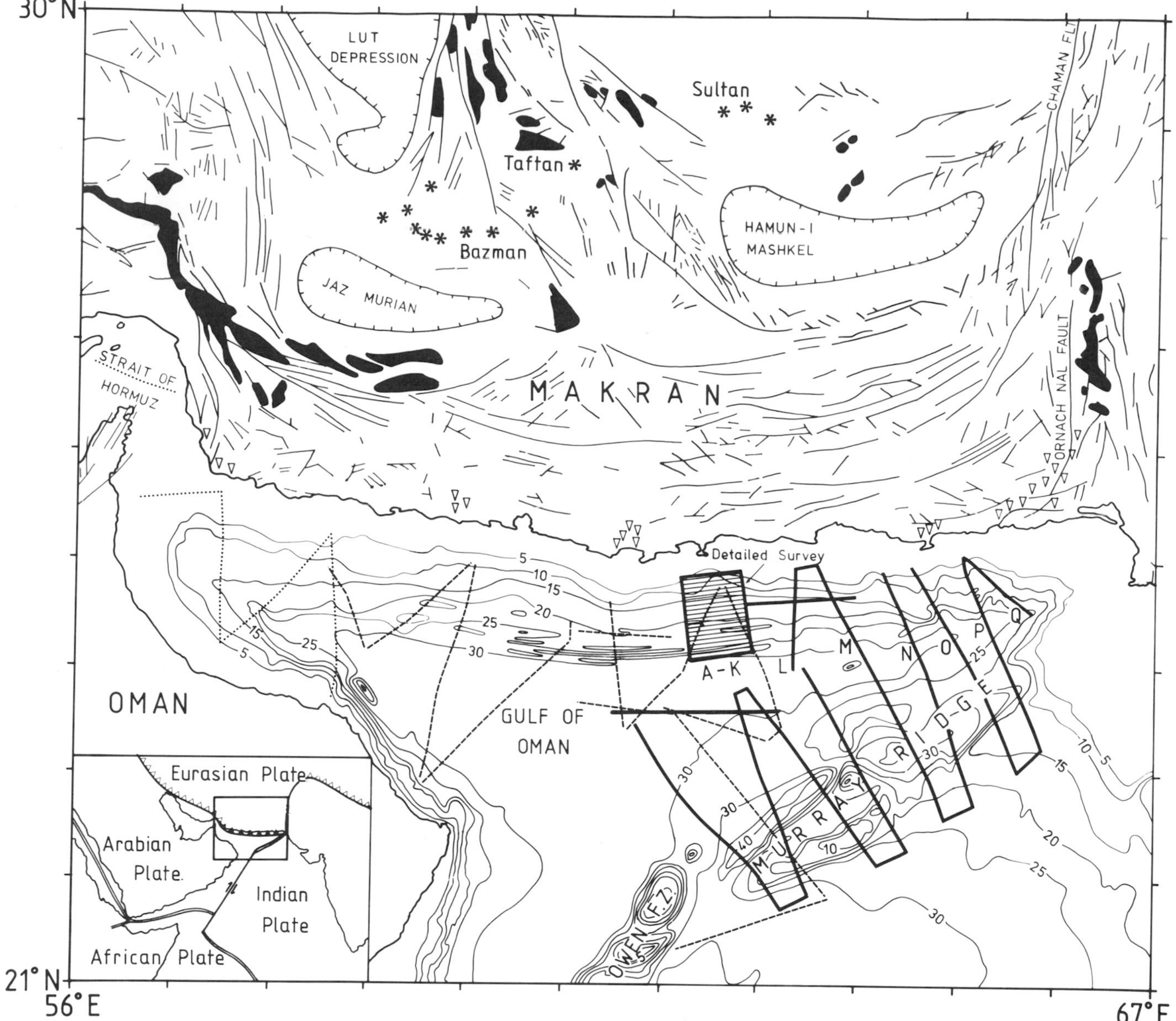

Figure 1a — Map showing general tectonic setting of the Makran subduction zone in the Gulf of Oman, northwest Indian Ocean, and the position of survey tracks at sea, all with continuous seismic profiles, gravity, magnetic, and bathymetric measurements. Dotted tracks are form RV ATLANTIS II Leg 96 (13); dashed tracks from RRS SHACKLETON Leg 3/75; solid tracks from RRS SHACKLETON Leg 1/80. Stars show locations of volcanic centers, triangles the positions of recently active mud vents, solid areas mark ophiolitic assemblages and "coloured melanges," and thin lines show the major faults on land. Diagram adapted from White (1981), using information compiled from Ahmed (1969), Barker (1966), Berberian (1976), Hudson et al (1954), Jacob & Quittmeyer (1979), Kazmi (1979), Nowroozi (1976), and Snead (1964). Inset shows general setting of the area with the plate boundaries.

and the midpoint of each profile, giving reversed refraction velocity determinations on the upper and lower crust. A continuous seismic reflection profile was made along each line to enable us to correct the travel-times for variations in the depth of the basement. Free-sinking, chemically-fused explosives were fired approximately every 1.5 km along the refraction lines, using alternately small (25 kg) and large (87.5 to 200 kg) charges to give good arrivals at near and distant receivers (see Figure 2 for shot and receiver positions). Higher resolution velocity determinations of the uppermost sediment were made on a previous cruise (SHACKLETON 3/75) by deploying disposable sonobuoys close to the location of the offshore seismic reflection line.

Topographic corrections for the sea floor relief were made to the mean water depth along each line, which was 3.3 and 1.8 km for the offshore and inshore lines respectively. These travel time corrections assumed that the deep refractors paralleled the sea floor and used least-squares fit refraction velocities found iteratively from the corrected first arrival travel times. Corrections were negligible for the offshore line over the abyssal plain (Figure 3a), unlike the inshore line

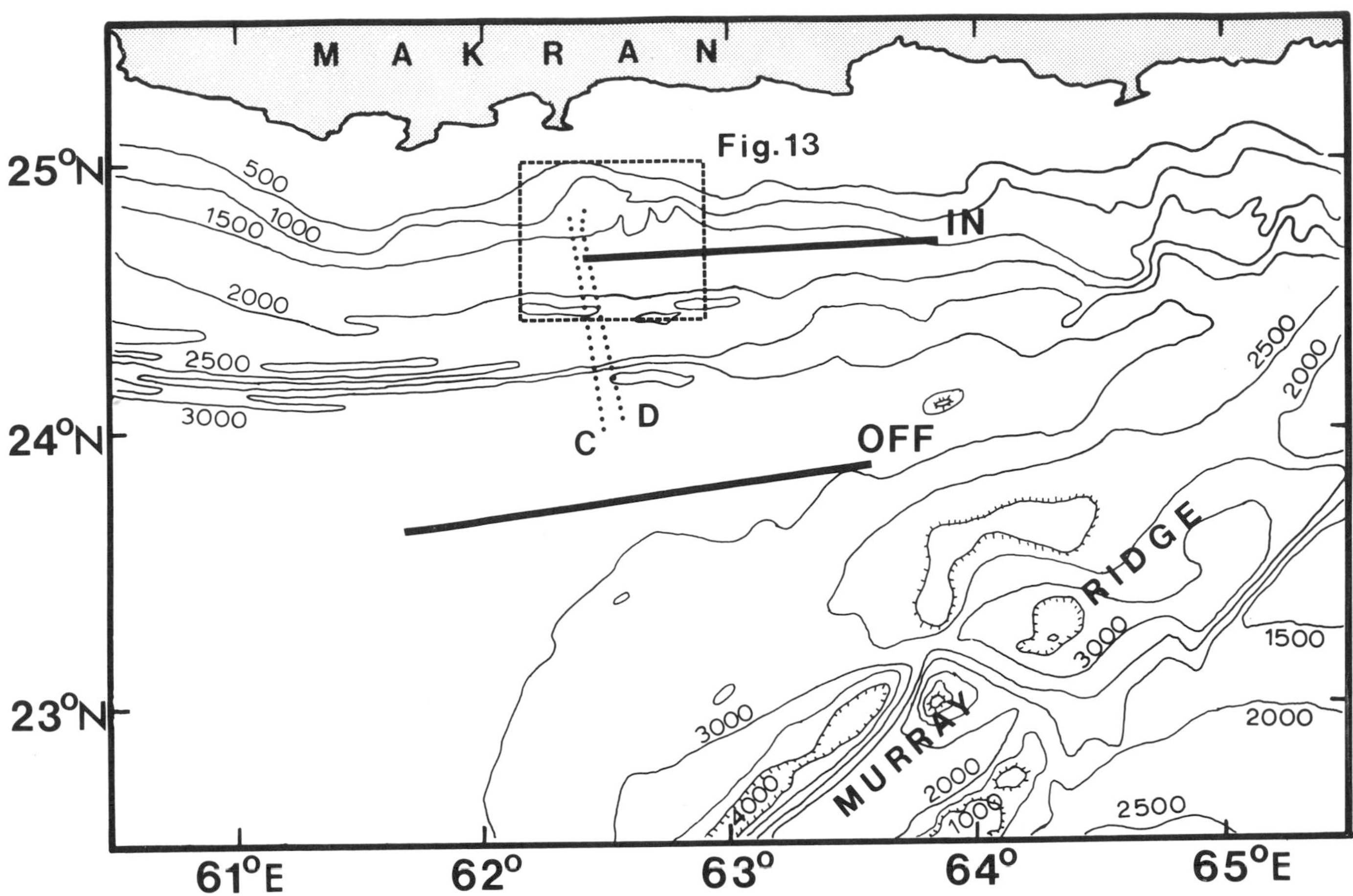

Figure 1b — Enlargement of part of Figure 1a, showing location of survey lines discussed in this paper. Box is area of detailed survey shown enlarged in Figure 13. Heavy lines mark locations of offshore (OFF) and inshore (IN) refraction lines; maps of the areas containing these lines in Figure 2. Seismic reflection profiles along lines C and D (dotted lines) are illustrated in Figures 7 and 8.

where sea floor relief is considerable (Figure 3b). However, strong currents caused the sonobuoys on the offshore line to drift considerable distances (Figure 2a), but except for the two westernmost sonobuoys, SB1 and SB2, they remained over the abyssal plain. In analyzing the refraction data reported here, we did not use arrivals from these two westernmost sonobuoys. On the inshore line the sonobuoys remained close to the shooting line (Figure 2b). Similar travel time corrections were made for the offshore line, where the irregular basement reflector is seen at 7 to 8.5 seconds two-way time (Figure 3a). No such coherent basement reflector is visible on the reflection profile along the inshore line (Figure 3b), where folding and possible hydrate layers complicate the deeper reflected arrivals. Therefore, it was not possible to correct the refraction travel times for variations in sediment thickness on the inshore line.

First arrivals from the upper crust and mantle are seen on all the sonobuoy record sections. In Figure 4a, we illustrate the split profile at sonobuoy 3 on the offshore line. The trace amplitudes are corrected for shot size, and we made an allowance for geometric spreading by multiplying the amplitude of each trace by its range. Examples of the record sections at two of the sonobuoys on the inshore line are shown in Figure 4b.

First arrival travel times from all instruments where they are clearly determined are combined onto a reduced time-distance graph for each profile in Figures 5a and 5b. This approach is feasible because the individual record sections show no consistent evidence for large variations in slopes or intercepts due to dipping layers. Scatter in arrival times is caused primarily by uncertainty in basement depths under the sonobuoys, particularly for SB5 and SB6 on the offshore line which drifted north of the profile and over thicker sediments. The arrivals on both profiles can be di-

Table 1. Least squares fit velocities and intercepts from first arrivals of explosives.*

	Velocity (km/sec)	Intercept Time (sec)	Number of points
OFFSHORE LINE	4.48 ± 0.08	6.48 ± 0.08	9
	7.94 ± 0.10	9.46 ± 0.13	30
INSHORE LINE	3.93 ± 0.07	3.70 ± 0.12	22
	7.92 ± 0.12	9.38 ± 0.12	24

*Errors quoted are one standard deviation.

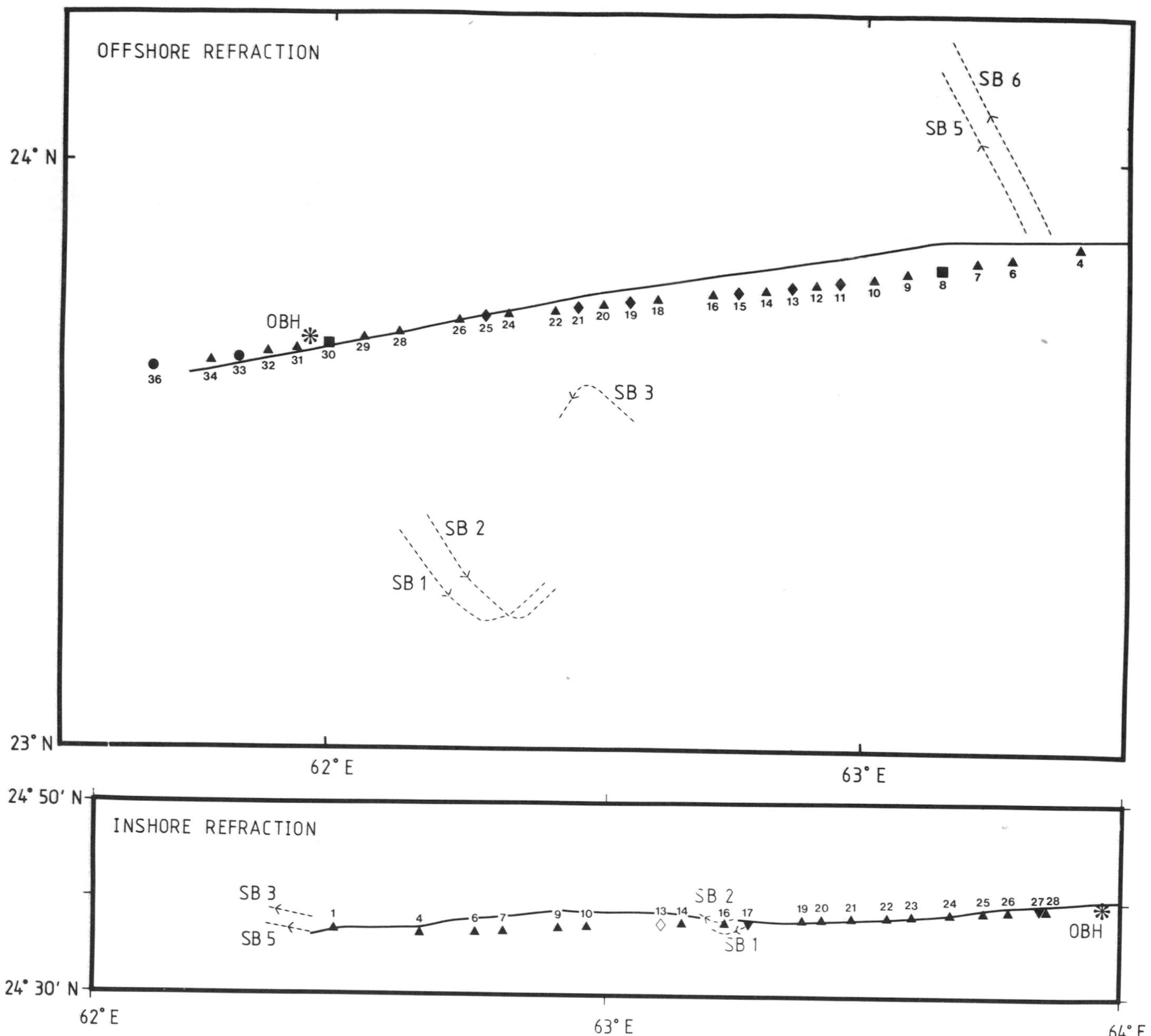

Figure 2 — Detailed maps of (a) offshore and (b) inshore refraction profiles. Broken lines show sonobuoy drift paths estimated from intersections of successive ranging arcs. Solid line denotes location of continuous seismic reflection profile (illustrated in Figure 3). Numbered symbols show shot positions: triangles 25 kg, inverted triangles 75 kg, filled diamonds 87.5 kg, open diamonds 100 kg, circles 150 kg, squares 200 kg.

vided into those from two refractors: an upper crustal refractor with a straight line least-squares fit velocity of 3.93 km/sec (inshore), or 4.48 km/sec (offshore); and, a mantle arrival with a velocity of 7.92 km/sec (inshore), or 7.94 km/sec (offshore). The least squares fit velocities and intercepts, together with their standard deviations, are in Table 1. We do not include the OBH travel times because of their relatively poor signal-to-noise ratio compared with that of the sonobuoys.

The most striking feature of the travel time plots is that the relatively low velocity crustal arrivals break directly to mantle velocities without any lower crustal returns appearing as first arrivals. This is a consequence of the thick sediment pile in this region masking the returns from the lower crust. Although the refraction velocities on the inshore and offshore lines are similar, the break from upper crustal to mantle arrivals occurs at a much greater range on the inshore than on the offshore line. This results from a thicker sediment layer on the inshore line.

It is unlikely that the crust consists of entirely uniform velocity material of 3.9 to 4.5 km/sec directly overlying mantle. But if we make this simple assumption, we calculate the crustal structure shown in Figure 6A. The refraction lines did not sample the up-

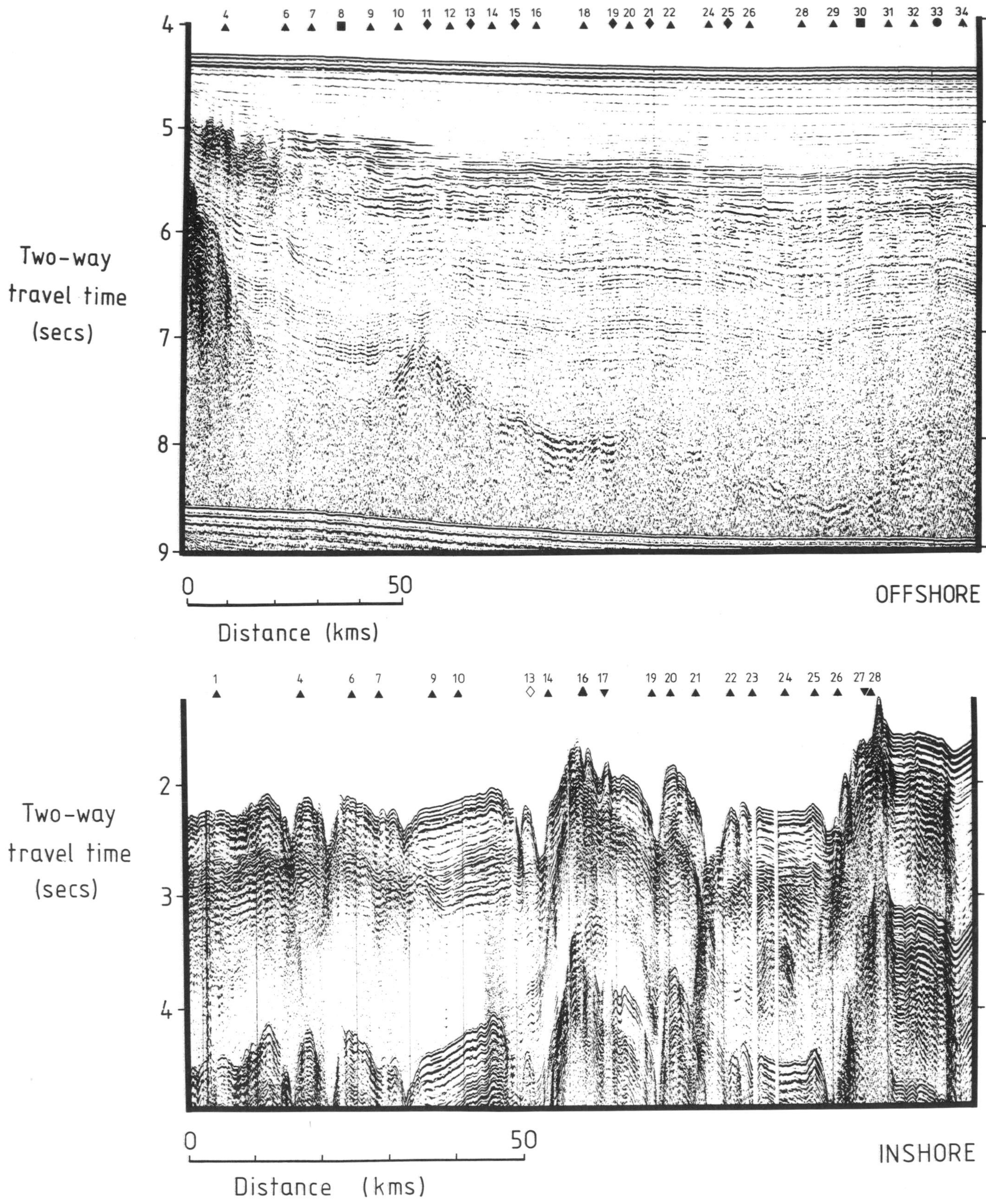

Figure 3 — Single-channel continuous seismic reflection profile along (a) offshore and (b) inshore refraction lines (see Figure 2 for location of profiles). Numbered symbols show shot locations (see Figure 2 caption for key to symbols). Profiling system comprised a Geoméchanique streamer with a single airgun fired every 12 seconds (a) offshore profile used a 2.6 litre (160 cu in) airgun fired at 13.8 MN/sq m (2000 psi); (b) inshore line used a 4.9 litre (300 cu in) airgun at 11.7 MN/sq m (1700 psi).

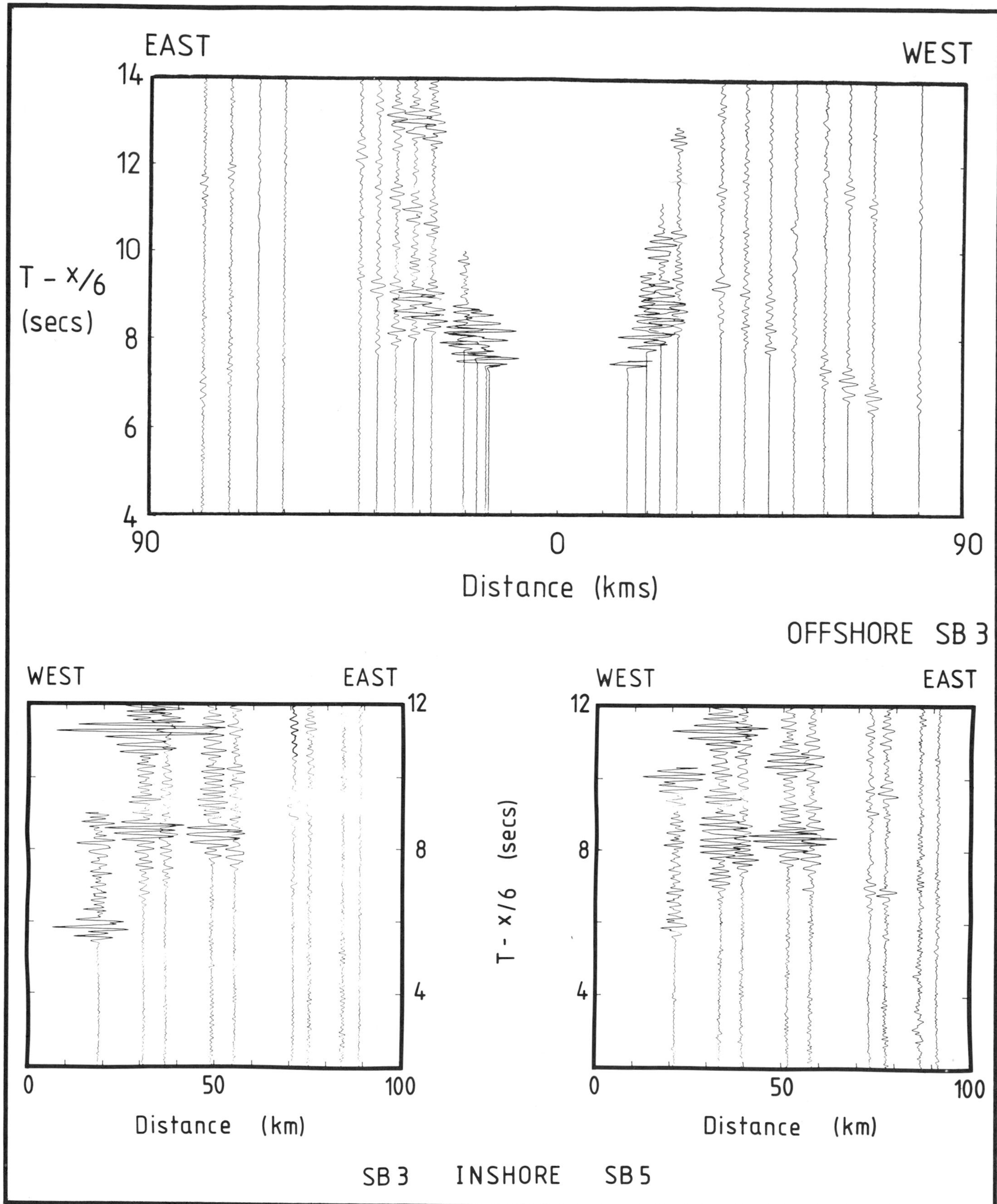

Figure 4 — Typical sonobuoy record sections from (a) offshore and (b) inshore refraction lines. The trace amplitudes are corrected for shot size and for geometric spreading. Band pass filters from 4 to 15 Hz are applied. Travel times are corrected for sea floor topography and are reduced with a reduction velocity of 6 km/sec.

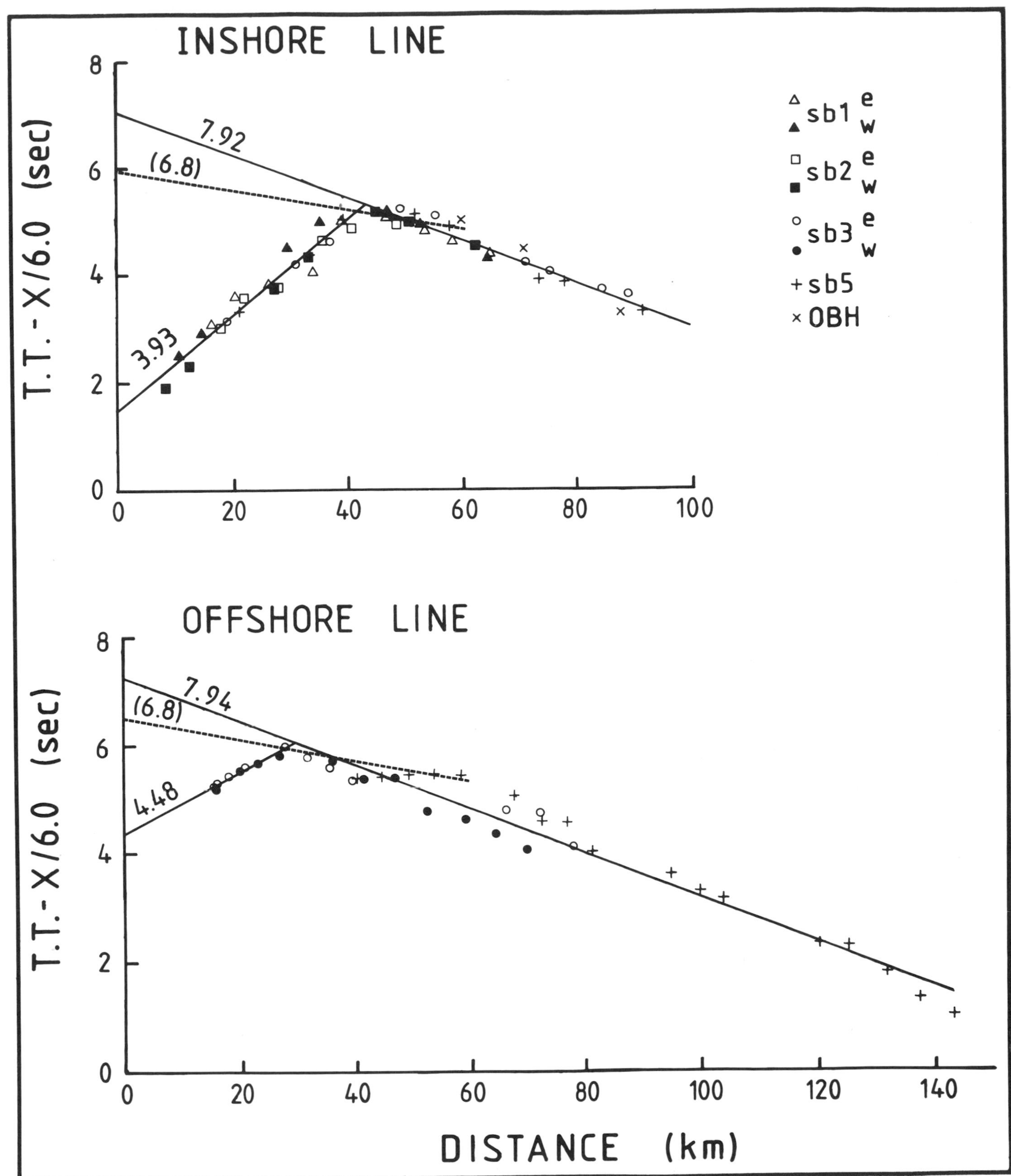

Figure 5 — Combined first arrival travel times from the sonobuoys on (a) inshore and (b) offshore lines, corrected for the sea floor and, on the offshore line, the basement topography. Solid lines show the least squares fit lines with velocities and intercepts as listed in Table 1. The broken lines show the maximum thickness of a hidden layer of assumed velocity 6.8 km/sec, which yields the velocity structure depicted in Figure 6b. Figures on the lines show the least squares fit velocities in km/sec. Note the abrupt break from first arrivals with velocities typical of consolidated sediment to those typical of the mantle.

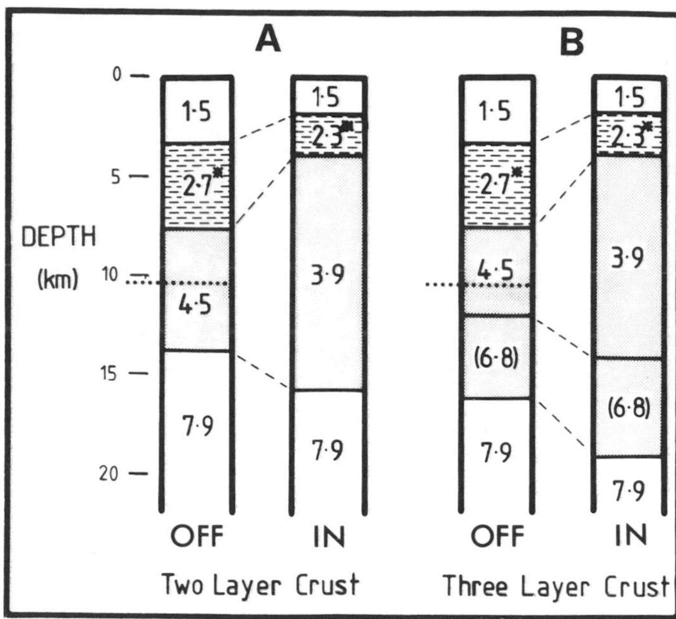

Figure 6 — (A) Velocity-depth sections derived from offshore and inshore refraction profiles. These assume that the crust consists of two uniform velocity layers with velocities and intercepts calculated from the first arrivals (see solid lines in Figure 5), which are overlain by unconsolidated sediment with a velocity gradient measured from several variable angle reflection-refraction profiles (see Figure 3 of White and Klitgord, 1976). Interval velocity of unconsolidated sediment indicated by a star. **(B)** Velocity-depth sections constructed as in **A**, but with the addition of the maximum permissible thickness of a hidden basal layer of assumed velocity 6.8 km/sec (see broken lines in Figure 5). Note the increased thickness of sediment and the greater depth to the Moho on the inshore line compared to the offshore line. Dotted lines show depth of basement reflector observed on offshore line.

permost unconsolidated sediment, so in the top of the crustal column we adopted the velocity structure reported by White and Klitgord (1976; Figure 3). They analyzed variable angle reflection — refraction profiles from the Gulf of Oman and incorporated other measurements by Closs, Bugenstock, and Hinz, (1969), and found that velocity increases smoothly with depth in a closely similar way to that reported from other areas of thick terrigenous accumulations. The resultant crustal structure (Figure 6A) yields a minimum depth to the Moho of 14 km beneath the offshore line and 16 km beneath the inshore line.

A more realistic model allows higher velocity material at the base of the crust. Since we did not observe high velocity first arrivals, we calculated the maximum thickness of a "hidden layer" of assumed velocity 6.8 km/sec by constructing that refractor which does not quite appear as a first arrival (shown by the broken lines on Figure 5). This yields the crustal structures shown in Figure 6B. The maximum thickness of the hidden layer is 4 to 5 km. A lower assumed velocity for the hidden layer would yield a correspondingly decreased thickness. Other than a deepening of the Moho by 2 to 3 km, the general characteristics of the three-layered model, which includes the hidden layer, are the same as those of the two-layered model. The velocities and crustal thickness of the inshore line are in remarkably close agreement with Niazi, Shimamura, and Matsu'ura (1980), whose results were from three 180-kg-charges fired into an array of high gain land seismographs and ocean bottom seismometers at the western end of the Makran accretionary prism.

We draw several useful conclusions from these preliminary crustal models. First, we conclude that the crust beneath the Gulf of Oman is not of continental type, because the high velocity basal crustal layers can only be thin. It seems probable that the thick sediment pile is underlain by oceanic crust because if we insert a hidden layer of velocity 6.8 km/sec. (typical of oceanic layer 3), its maximum thickness of 4 to 5 km agrees with that generally found in the oceans (e.g. Raitt, 1963; Christensen and Salisbury, 1975).

Second, the Moho increases in depth on the inshore line by 2 to 3 km, representing a northward dip between 1.3 and 1.9° over the 93 km distance between the two refraction lines. Such shallow dips to the subducting plate have also been observed on other convergent plate boundaries where there are equally large trench to volcanic chain separations: for example, in the eastern Aleutians (von Huene, 1972; Shor & von Huene, 1972) and Barbados (Westbrook, 1975). Karig, Caldwell, and Parmentier (1976) show that this is probably a consequence of the plate's elastic bending under the extensive accumulation of accretionary sediments between the volcanic arc and the zone of initial sediment deformation. Models based on other data for the Makran margin, such as gravity anomalies (White, 1979b), seismic reflection records (White & Klitgord, 1976; White, 1981), the overall configuration and width of the sediment prism (Farhoudi & Karig, 1977), and the earthquake seismicity (Jacob & Quittmeyer, 1979), also are consistent with a shallow dip to the subducting plate under the outer accretionary prism.

Third, the sedimentary section thickens by almost 4 km between the offshore and the inshore lines, representing about a 40% increase in total thickness (Figure 6). This is accommodated partly by the northward dip of the Moho and partly by the 1.5 km decrease in sea floor depth. The deformation within the accretionary prism which results in a thickening of the sediment section is discussed in the following section. Between the offshore and inshore refraction lines, the ratio of unconsolidated to consolidated sediment changes considerably. The uppermost unconsolidated layer is actually about 2 km thinner on the inshore line, decreasing from 4 km thick on the offshore line; the lower consolidated layer with a velocity of 3.9 to 4.5 km/sec increases in thickness by almost 6 km, a proportionate increase of over 100%. The resolution of our first arrival least-squares fits is not very good, and it is probable that the lowermost 1 or 2 km of the layer we have designated as consolidated sediment is actually volcanic basement (layer 2 of the oceanic crust) with a velocity higher than 5 km/sec. If so, this means that the proportionate increase of consolidated sediment on the inshore line is even greater than estimated above.

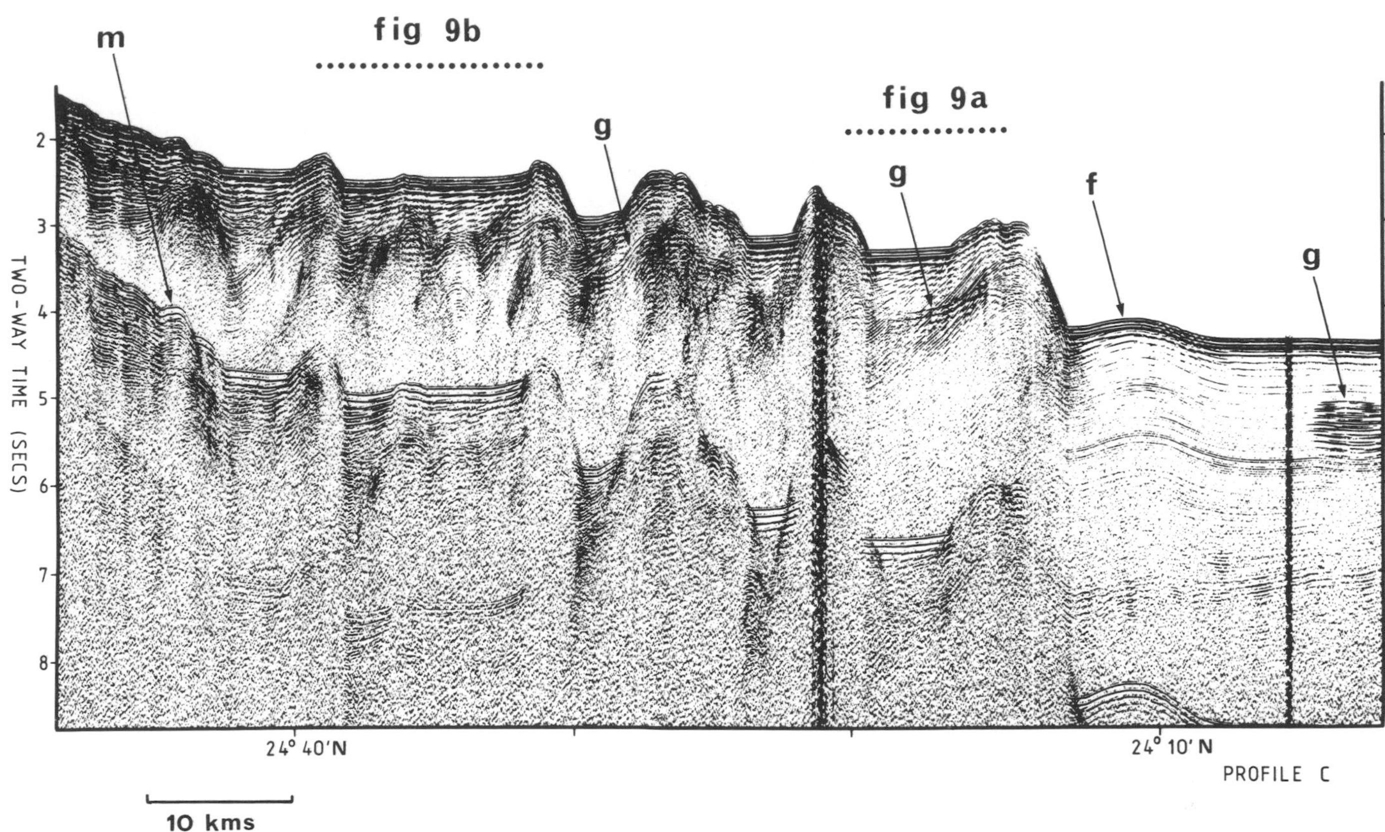

Figure 7 — Continuous seismic reflection profile along line C across the Makran continental margin (see Figure 1 for location). Profiling system comprised a single channel Geomécanique streamer with a 2.6 litre (160 cu in) airgun fired once every 12 seconds at a pressure of 13.8 MN/sq m (2000 psi). No processing other than band pass filtering between 5 to 70 Hz has been applied. Vertical exaggeration at sea floor is 7:1. Details of the interfold basins marked "a" and "b" are shown enlarged in Figures 9a and 9b. Frontal fold marked by "f," reflector at base of hydrated sediments by "g," and water layer multiple by "m."

We suggest that as the large pile of unconsolidated sediment beneath the abyssal plain is carried into the accretionary prism of the Makran subduction zone, it becomes rapidly dewatered and consolidated by tectonic processes. We see evidence for this process in the structure of the upper accretionary wedge, which is revealed by the reflection profiles discussed later in this article. The unconsolidated sediment at the top of the section on the inshore line, which was shot along a slope basin on the upper continental margin, represents slope sediments deposited on top of the accreted material.

DEFORMATION IN THE ACCRETIONARY PRISM

Our seismic reflection profiles reveal structural details of the uppermost sedimentary layers, indicating the way in which the sediment section becomes thickened within the accretionary wedge. Reflectors can be discerned up to 4 seconds two-way travel time (typically 5 to 6 km) beneath the sea floor of the abyssal plain, and these can often be traced into the frontal, or southernmost, fold of the continental margin. Within the main part of the accretionary belt the dips of the uplifted and faulted ridges are so large that little of their internal structure can be observed.

Nevertheless, we can draw a number of conclusions regarding the style of deformation inside the sedimentary prism from clues provided by the structure of sediments deposited in interfold basins on top of the offscraped material.

The overall structure of the sediment prism, comprising a series of open folds and intervening basins, is illustrated in profile C (Figure 7) which runs from north to south perpendicular to the continental margin. Initially gently dipping sediments beneath the abyssal plain become folded at the frontal fold ("f" on Figure 7) and subsequently uplifted by about 1.25 km as they are incorporated into the accretionary wedge. Northward toward the coast, the pattern of open ridges and basins continues until the ridge tops become buried by subsequently deposited sediment.

Here are a few words of warning concerning the illustrated reflection profiles. The vertical exaggeration is approximately 7:1, so dips of layers in Figure 7 are gentle and the structure less condensed than is normal in many other subduction zones. This may be a consequence of the thick sediment pile brought into the accretionary system by the subducting plate. Also, the profile is an unprocessed time section, so the water layer multiple (marked "m" in Figure 7) is prominent and the single airgun source produces a long wavetrain of bubble pulses. Finally, there are

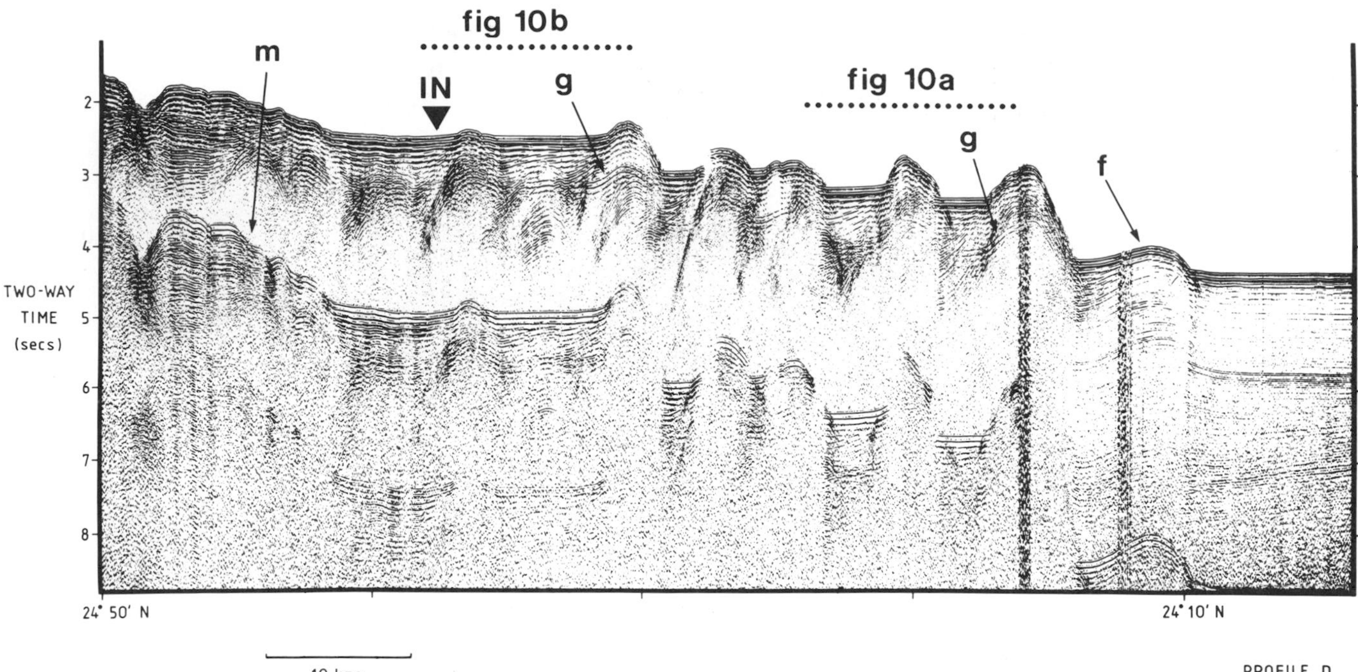

Figure 8 — Continuous seismic reflection profile along line D parallel with, and approximately 7 km away from, line C. Details of profiling system are discussed in caption to Figure 7. Segments of the interfold basins are shown enlarged in Figures 10a and 10b. Intersection with inshore refraction line is marked by "IN." Vertical exaggeration at sea floor approximately 7:1.

numerous strong reflections generated at the base of a hydrated gas layer ("g" on Figure 7); these reflectors mimic the shape of the sea floor and are artifacts dependent on the pressure-temperature conditions of the free gas to gas hydrate phase change. They are not reflectors defining the internal structures. Further details about these prominent gas reflectors are in White (1977b, 1979a, 1981) and Hutchison et al (1981).

The kinematics of the initial deformation in the frontal fold are reported in detail by White (1977a, 1981), and are not repeated here other than to point out that the initial buckling is confined to the uppermost 2.5 to 3 km of unconsolidated sediment. Beneath this, we observe a décollement within the frontal fold which separates the buckled layers from underlying, flat lying sediments. The décollement probably occurs in shale layers weakened by overpressured pore waters; the depth at which the décollement occurs (2.5 to 3 km) is typical of the amount of burial needed for overpressuring in a rapidly deposited sedimentary sequence (Chapman, 1974).

The frontal fold's amplitude changes along strike, reaching a maximum height of 400 m above the abyssal plain before decreasing again. On profile D (Figure 8), which is parallel with profile C but about 10 km further east, the frontal fold increases in amplitude and the seaward limb becomes much steeper. There is some evidence of localized slumping off the front of the new fold, forming the herringbone pattern seen on Figure 7.

The next stage in incorporating the sediment on the subducting plate into the accretionary wedge is for the frontal fold to be compressed above the décollement zone and uplifted until the fold's crest is about 1.25 km above the abyssal plain. The maximum observed amplitude attained by the frontal fold along the entire Makran continental margin is only 400 m, so a delicate balance appears to exist in the frontal fold between continued folding and upthrusting. This balance is probably governed by the degree of compression necessary before the originally weak, unconsolidated sediments of the abyssal plain are sufficiently strong to support a major fault. Once the frontal fold is added onto the front of the accretionary prism, a new fold forms to the south of it. Through this, the accretionary wedge grows seaward at around 10 km/Ma (White, 1981).

The sediments within the slope basins immediately behind the most seaward uplifted fold ridge dip gently landward and the dip increases with depth (see enlargements of profile C in Figure 9a and of profile D in Figure 10a). The prominent hydrate-free gas reflector is marked "g". The increase in dip with depth may be caused by progressive basin tilting. One possible mechanism for producing such tilting is continued motion on imbricate thrust faults, such as was postulated in other accretionary prisms by Seely, Vail, and Walton (1974), Karig & Sharman (1975), and Moore & Karig (1976). The cartoon in Figure 11 shows a possible configuration of the imbricate thrust faults. Such thrust faults are notoriously difficult to detect on seismic reflection profiles, and although the structural complexity prevents observation of any faults in our records, we believe that they provide the best explanation for the generation of dipping layers in the uppermost sediments.

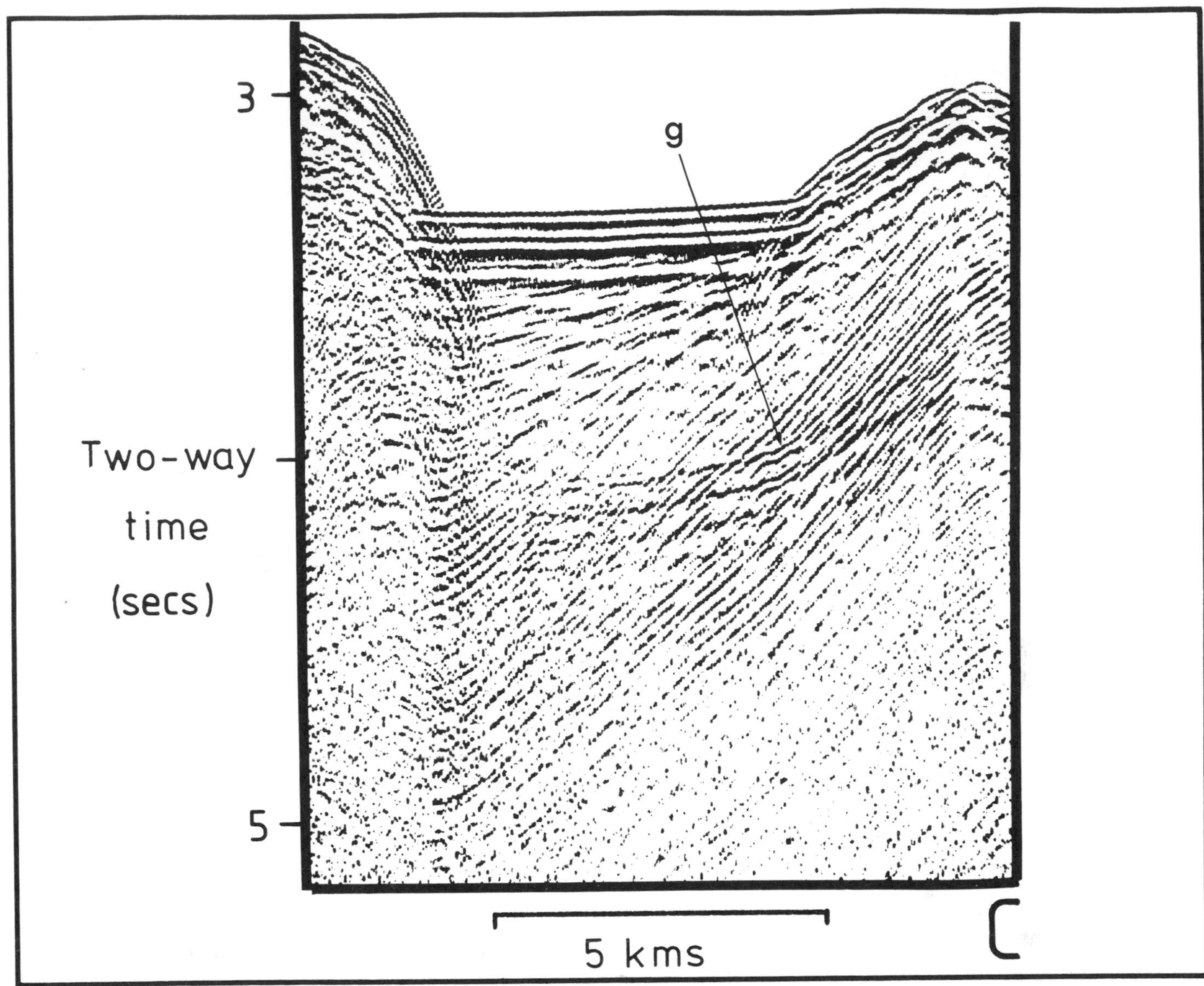

Figure 9a — Enlargement of parts of the continuous seismic reflection profile along line C showing details of the structure within the interfold basin near the front of the accretionary fold belt. Location of the enlarged section shown on Figure 7. Note the increase of dip with depth in the interfold basin sediments. Vertical exaggeration at sea floor is approximately 7:1. Reflectors marked "g" are from the base of the gas hydrate.

There is another possible explanation for tilting in the most seaward slope basin, although it cannot be generalized (as can the imbricate thrust model) to account for tilting in subsequently deposited layers further toward the coast. This possibility arises because sediments in the seaward basin may have accumulated in the region between the crest of the frontal fold and the seaward facing scarp marking the edge of the adjacent uplifted fold ridge. If, as seems to be the case, the frontal fold increases in height slowly over a considerable period of time, the sediments concurrently accumulating above the landward limb of the fold will become progressively tilted with the most recently deposited (most shallow) sediments experiencing the least tilting. The asymmetry in the shape of the frontal fold produced by these ponded sediments is clear on both profiles C and D (Figures 7 and 8). Thus, in the southernmost slope basin, we cannot distinguish between dipping sediment layers generated on the landward side of the frontal fold during the course of its growth, and those resulting from tilting accompanying imbricate thrusting. In the more landward slope basins this ambiguity does not exist. In those, even the most recently deposited slope sediments exhibit increasing dips with depth. This could only be produced by tectonic tilting subsequent to the emplacement of the slope basins.

Toward the coast, the slope basins become progressively filled with sediment and the ridge tops are eventually buried (Figures 9b, 10b). Once the ridges are completely buried, the sediment moves freely down the margin and attains a relatively uniform slope (northern ends of Figures 7 & 8). Prior to this stage, deposition is mainly from currents fed through

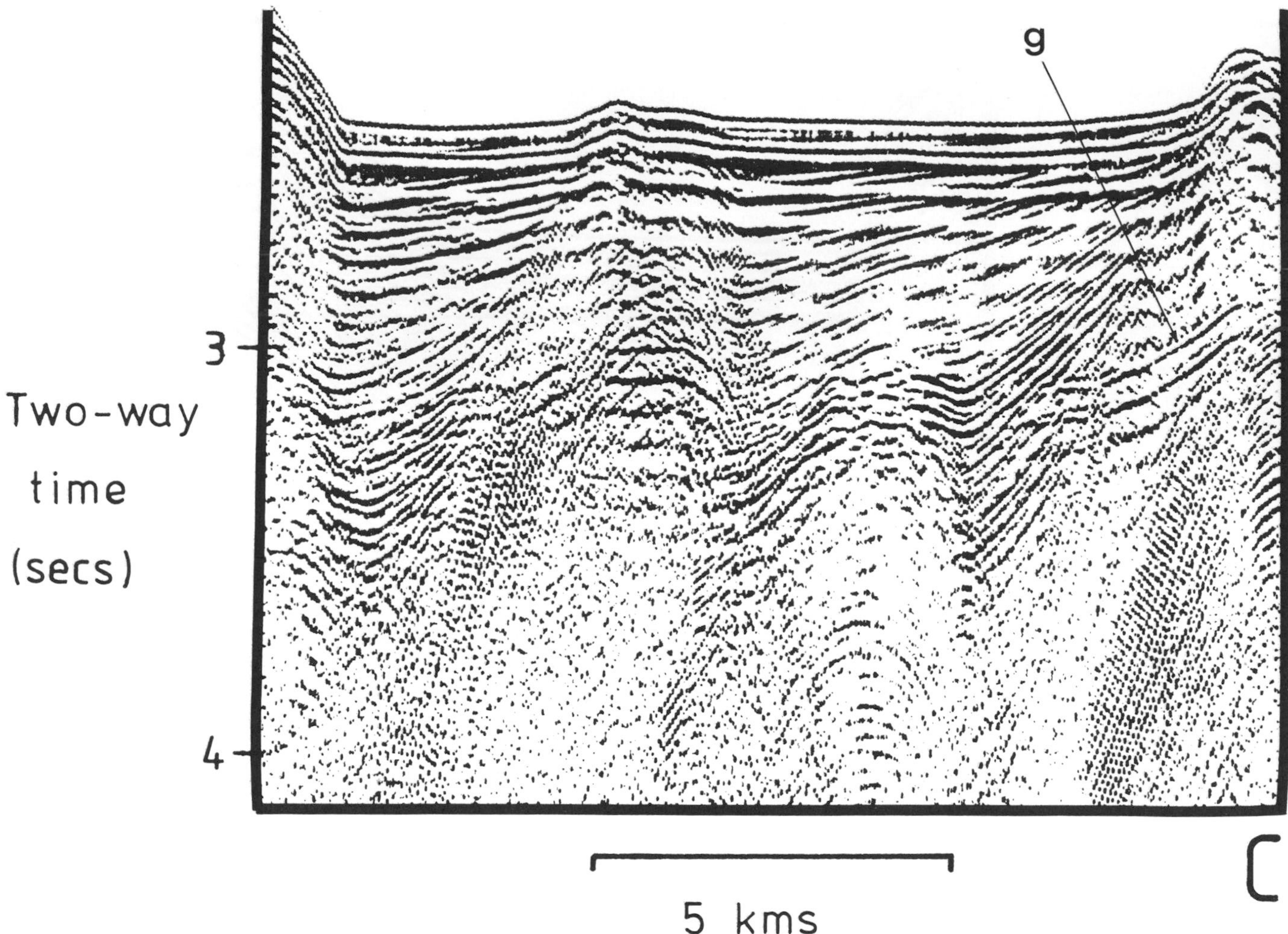

Figure 9b — Enlargement of part of the continuous seismic reflection profile along line C showing details of the structure within the interfold basins. Location is to the north where the infilling sediment begins to overtop the fold ridges. Location of enlarged section shown on Figure 7. Note the increase of dip with depth in the interfold basin sediments. Vertical exaggeration at sea floor is approximately 7:1. Reflectors marked "g" are from the base of the gas hydrate.

channels cut across the ridges then flowing along the lineated slope basins. The consistent increase of landward dip with depth in the originally flat-lying slope basin sediments attests to continued tectonic tilting across the offshore Makran margin. The low frequency bubble pulse of the profiles in Figures 9b and 10b limits resolution to about 150m. High frequency reflection profiles across the same basins show that layers within only a few meters of the sea floor exhibit increasing landward dips with depth. This suggests that imbricate faulting which causes tilting of the surface is active at least 70 km north of the frontal fold. There is no evidence of internal buckling of the slope basin sediments; the separation between adjacent fold ridges remains constant across the offshore accreted sediments, from the front of deformation to the region near the coast where the ridges become completely buried by slope sediments. We conclude that, other than at the frontal fold, the shortening and thickening at least across the offshore part of the accretionary belt is accommodated by imbricate thrusting rather than by folding. If we had accurate knowledge of the sedimentation rates in the slope basins, we could reconstruct the history and distribution of the tectonic tilting across the margin. Unfortunately, as we show in the next section, the rate of sediment deposition depends on the configuration of feeder channels and this develops and changes with time. Deposition rates are likely to be low while the slope basin is isolated by the ridges on either side, and then suddenly increase as a feeder channel cuts through to the basin. All we can say is that tectonic tilting continues across the entire margin.

Although the décollement zone observed in the middle of the sediment pile in the frontal fold cannot be traced further into the accretionary wedge (because on our profiles the steep dips obscure deep reflectors), White (1981) argues that a zone of decoupling within the sedimentary section continues 60 to 70 km shoreward from the frontal fold. This argument is based on the very low average bathymetric slope of only 0.6°. Slippage probably occurs along sedimentary horizons weakened by high pore pressures (Chapple, 1978), producing a décollement above deeper flat-lying hor-

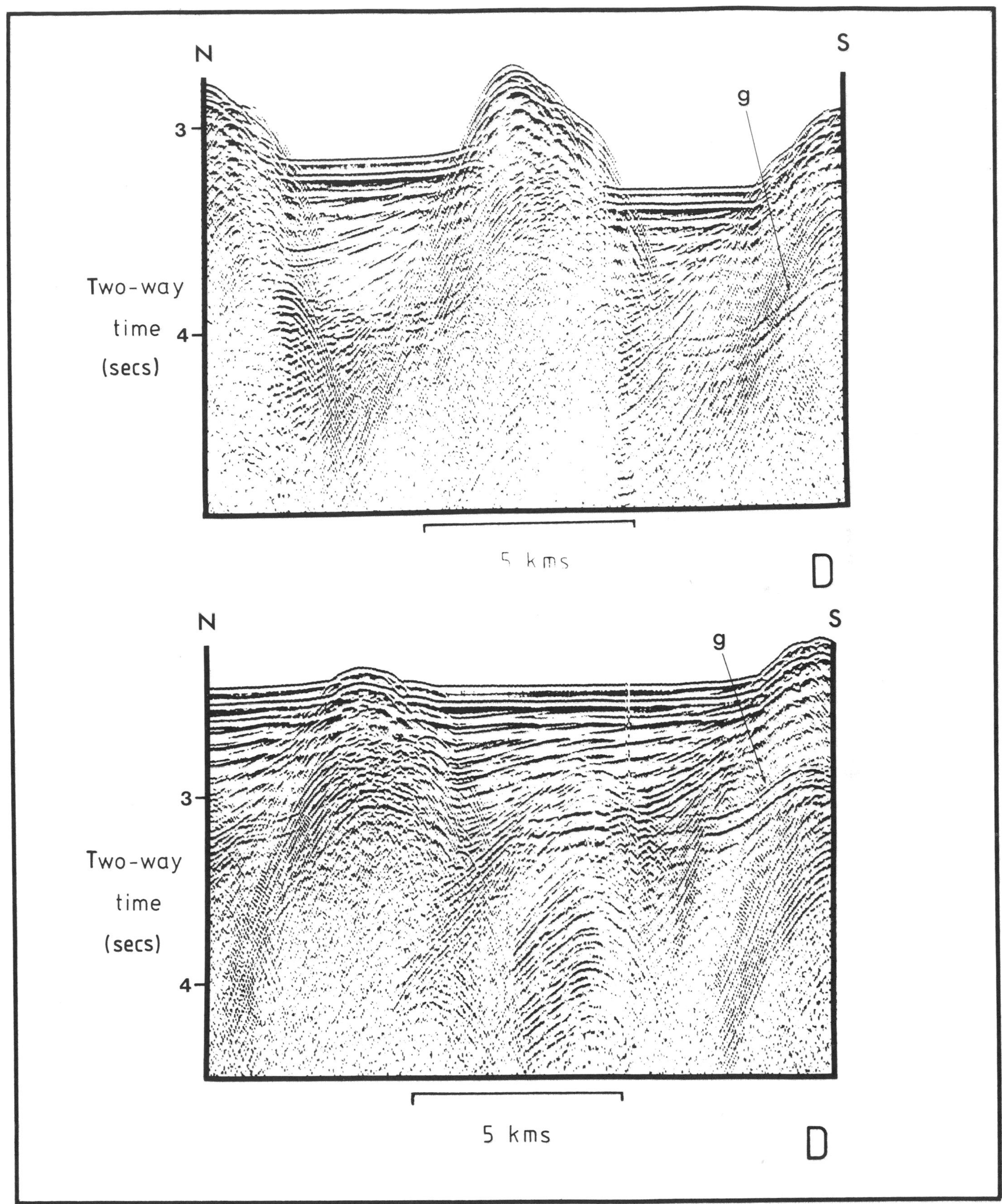

Figure 10 — Enlargements of parts of the continuous seismic reflection profile along line D at locations marked on Fig. 8. Note the increase of dip with depth in the interfold basinal sediments. Vertical exaggeration at sea floor is approximately 7:1. Reflectors marked "g" are from the base of the gas hydrate.

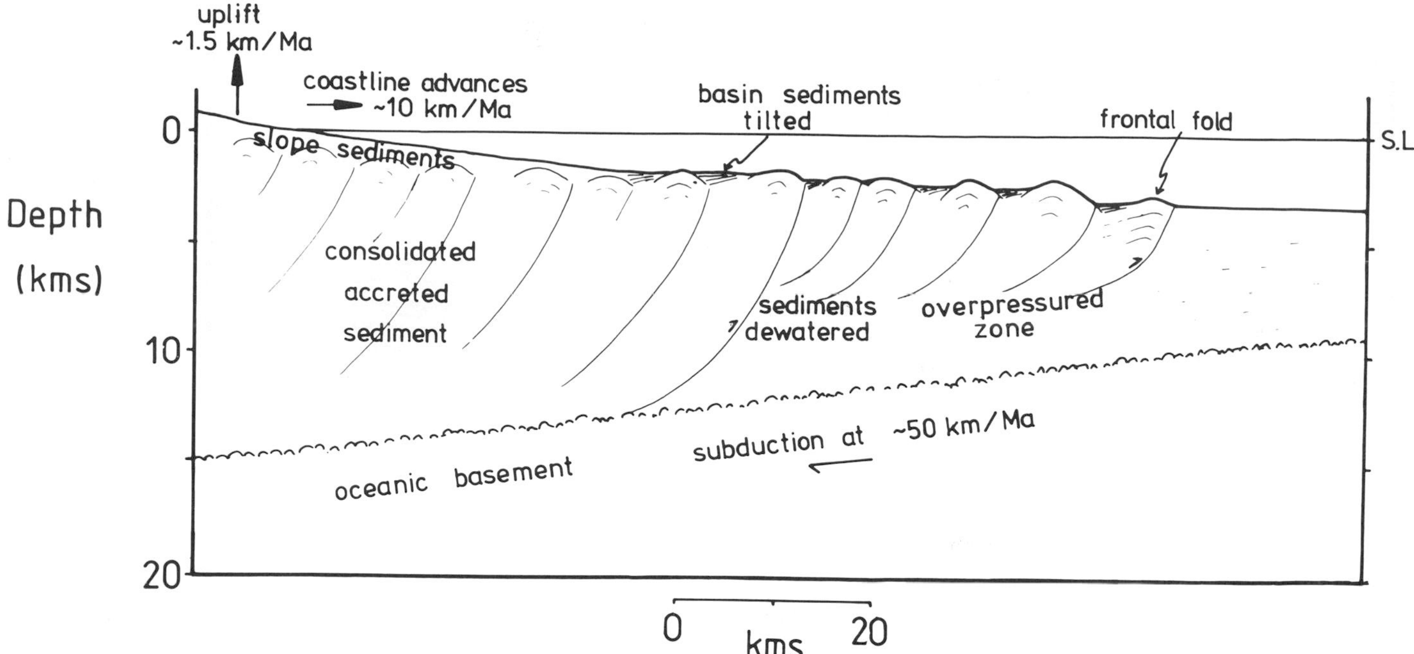

Figure 11 — Schematic diagram of possible configuration of imbricate thrust faults on a north to south transect across the Makran accretionary prism. Rates of uplift and seaward movement of the front of the accretionary prism are discussed in the text. Vertical exaggeration is approximately 7:1.

izons similar to that reported by Westbrook (1975) and sampled by DSDP holes 541 and 542 (Scientific Party, 1981) in the frontal 50 km of the Barbados accretionary wedge. Emplacement of the frontal fold increases the pore pressures in the underlying sediments, and with clay diagenesis releasing additional water the high pore pressures are likely to exist within the accretionary prism for several milion years before they are released through the sedimentary pile. The decoupling zone level may change with time within the frontal part of the accretionary belt as the abnormally high pore pressures migrate.

Some 70 km landward from the front of deformation, the bathymetric slope becomes much greater and we postulate that in this region, the decoupling zone in the sediments is no longer present and the basement becomes involved in imbricate thrusting (Figure 11). Rapid uplift of the coastal area, typically at rates of 1.5 to 2.0 mm/yr (Vita-Finzi, 1975, 1979), is associated with the basement involvement and accompanied by extensive seismicity (Page et al, 1979) and mud diapirism (Sondhi, 1947; Snead, 1964; Ahmed, 1969).

The generation of open, well-lineated ridge and slope basin topography underlain by décollement zones within the lower part of the accretionary wedge is dependent on the input of thick, relatively uniform sediments on the subducting plate. In the eastern part of the Makran continental margin, a large basement ridge lies within the sediments feeding into the subduction zone. As it enters the accretionary system the ridge and basin structure becomes immediately chaotic (Figures 12 and 13, White, 1981), with a much steeper overall bathymetric slope than in the simpler part of the margin that we illustrated in Figures 7 through 10.

INTERACTION OF TECTONICS AND SEDIMENTATION

Sediment incorporated into the accretionary wedge has two distinctive sources. First, there is the sediment scraped off the subducting plate, which rapidly becomes consolidated by tectonic processes. This material may be reworked by erosion and slumping off the uplifted ridges and redeposition in slope basins. Second, there is the sediment carried across the margin from the adjacent coastal regions or deposited on top of the accreted material from windborne detritus. In the case of the Makran, much of the sediment eroded from the coastal regions may come from previously accreted sediment now exposed above sea level because the accretionary prism extends such a long way onshore. Sedimentation rates over the abyssal plain are high, about 250 to 400 m/Ma (Hutchison et al, 1981). The material is terrigenous, often with a large percentage of carbonates, and up to half of it may be aeolian in origin (Stewart et al, 1965; Stoffers and Ross, 1979).

Slumping as a means of redepositing sediment from the ridges into the slope basins is surprisingly rare. One of the few examples we observed is shown in Figure 12. The lower part of Figure 12 is a 10 kHz echo sounder record, orientated north to south across a narrow slope basin about 30 km landward of the frontal fold. On the right hand edge is a ridge whose crest lies about 1,900 m deep with a very steep slope down to the basin at 2,360 m. The slope is so steep that we see no direct echoes from it. At the foot of the scarp is a hummocky surface (indicated by an arrow) caused by slumped sediments. In the top of Figure 12

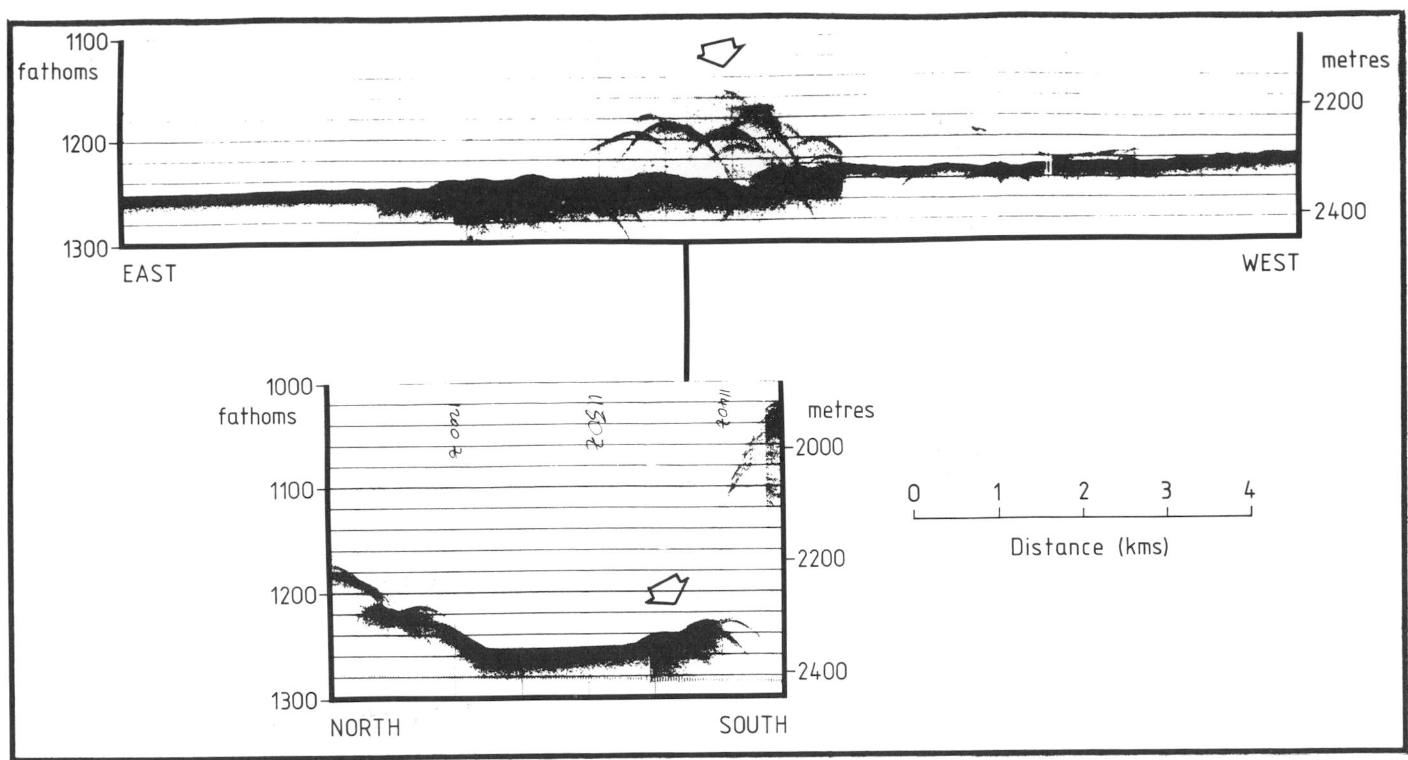

Figure 12 — Copies of 10 kHz echo-sounder records across a small slump at the base of an uplifted fold ridge on the Makran continental margin (see Figure 1b for location). The lower record crosses a small basin perpendicular to the strike; note the slump structures marked by an arrow at the foot of a 400 m scarp and the northward tilting surface of the basin. The upper record is along the interfold basin. The arrow marks the position of side reflections from the adjacent scarp; a hummocky surface extending about 8 km laterally marks the slumped sediment. Also note the general dip to the east, which suggests that bottom currents have fed material from the west along this particular slope basin.

we show the bathymetry recorded along an east to west profile down the slope basin. The track runs so close to the ridge base that we see side echoes (marked by arrow). It is also clear that slumped material extends several kilometers along the foot of the scarp.

Small-scale slumping also occurs off the seaward limb of the frontal fold, as was already mentioned. However, there is no sign of large scale slumping off the considerably steeper scarp on the seaward limb of the adjacent uplifted fold ridge. We conclude that the sediment in the accretionary wedge is much more consolidated than the sediment beneath the abyssal plain, in agreement with the increased sedimentary seismic velocities observed on the inshore refraction line.

The series of well-lineated fold ridges in the accretionary prism control the sedimentary processes in the intervening basins by forming barriers to downslope sediment movement. The bathymetry of the upper part of the continental margin is shown in Figure 13. This was taken from the northern part of a detailed survey with a dense track coverage (see Figure 6 of White, 1981). To the south of the area shown in Figure 13 the survey clearly shows the well-lineated fold ridge and basin morphology, as typified by our profiles across the margin (Figures 7 & 8). In Figure 13, a long scarp about 400 m high runs along strike at 24° 30′N, marking the approximate boundary between the unfilled slope basins to the south and the region, to the north, where subsequently deposited material completely fills the basins. In this northern region two separate channel systems cut across the upper slope sediments; the major channels are marked by broken lines on Figure 13.

Sediment channels running south from the coast are deeply incised into the upper slope sediments and cut across the buried structural lineations. As soon as the flow reaches a fold ridge which is not buried, it is deflected through 90° and runs along the long, narrow slope basins. The flow direction along the slope basins is either east or west. The two channel systems we have marked are deflected in opposite directions. Within the remainder of the detailed survey the gentle bathymetric slopes along the basins indicate that the turbidity currents, which have deposited the infilling basinal sediments, have come from both east and west in different basins. Consequently, the provenance of the slope basin sediments could be quite different even in adjacent basins.

Echo-sounder records from selected crossings of the eastern distribution system record the changing morphology of the channels as they cross the active margin (Figure 14). The cross section at the bottom of Figure 14 shows the steep slope where the channels are incised in the northern part of the area; here the average gradient over 5 km is about 8°. As the gradi-

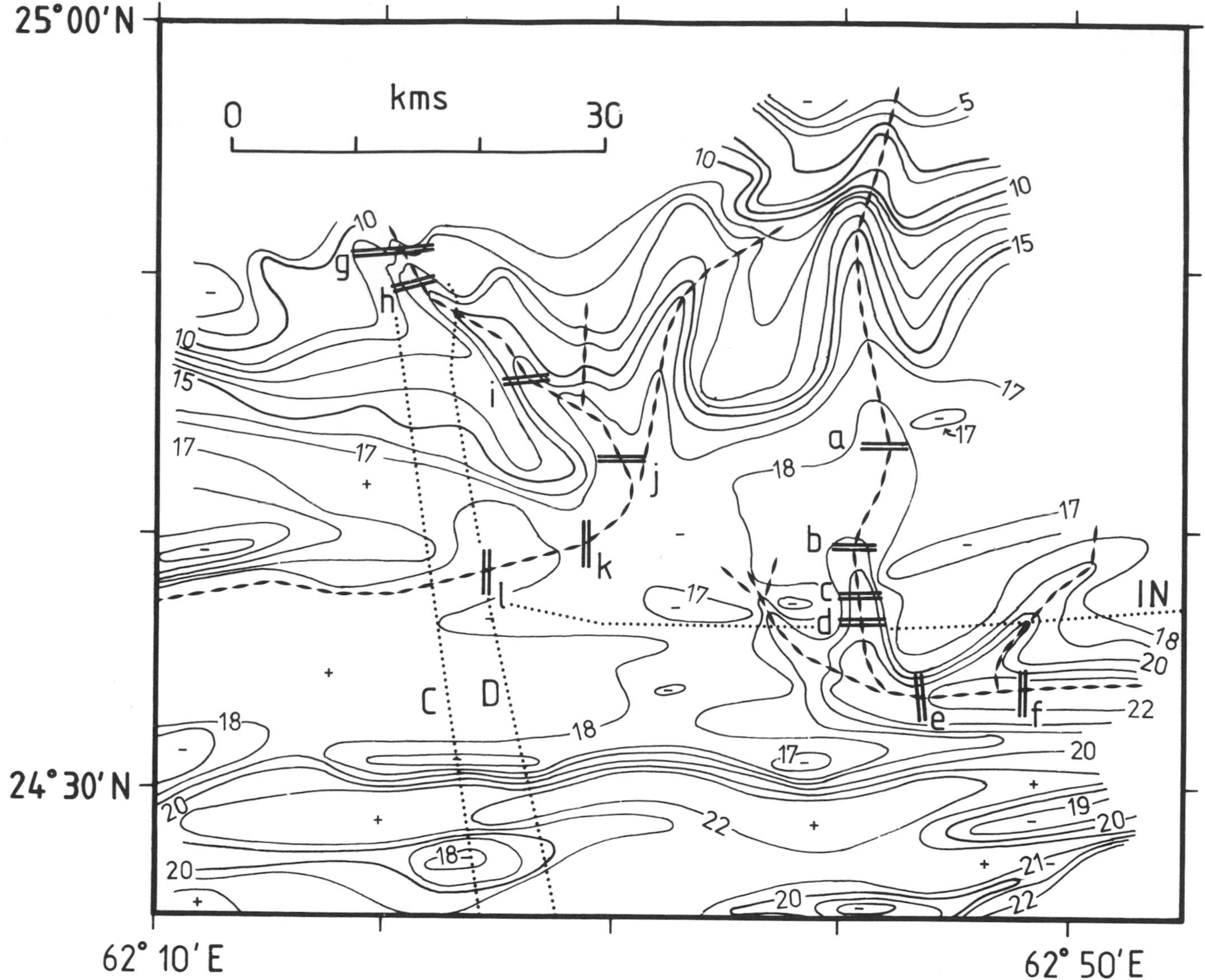

Figure 13 — Bathymetry in the northern part of the detailed survey in area delineated by box in Figure 1b. Contours in hundreds of meters. Broken lines show the courses of two major sea floor channel systems which feed sediment from the coastal region into the interfold basins. Echo-sounder records of the crossings of these channels at points marked by letters are illustrated in Figures 14 and 15. Dotted lines show positions of profile C (Figure 7), profile D (Figure 8), and profile along inshore refraction line (Figure 3b).

ent flattens over a sediment bench at about 1,800 m (980 fm) deep, the channel becomes much more shallow ("a", Figure 14), producing a series of small terraces. Although in previous times the flow was dammed by the ridge at the basin's southern edge (Figure 13), it has now raised the sediment level sufficiently to overflow into the next slope basin to the south which is some 400 m deeper. Continued tectonic uplift of the margin may have caused the channel to become deeply incised where it cuts through the ridge ("c", Figure 14). Not only is the channel over 300 m deep at this point, but a small inflection has been produced in the cross section where the channel cuts back into the sediments. As the flow is directed eastward by the next ridge, the gradient again becomes more gentle and the channels open out forming small terraces as they deposit material in the narrow, empty slope basin ("e" and "f", Figure 14).

The western channel system shown in Figure 15 begins with deeply incised channels on the upper slope ("g"), which coalesce as they are traced southward ("h" and "i"). The channels here ignore the structural trends which are buried beneath a steep sediment apron. As the flow reaches the wide infilled basin, the channels rapidly open out and become more shallow ("j"), turning westward along the slope basin. Initially a series of small terraces are formed ("k"), but as the sediment load is progressively dropped the channel becomes nothing more than a shallow depression ("l") running along the infilled slope basin with a gradient of less than 0.1°. We do not see any overflow into the deeper basin to the south within the area of the detailed survey, although in places the bounding southern ridge is almost overtopped.

Active tectonism on the margin affects the sediment distribution pattern by first producing long ridges which form barriers to downslope movement. Continued uplift then causes sediment channels to become incised in the ridges they cross. On the Makran margin we see no evidence of slope basin bypassing because the basins are so long that they capture any material moving across the margin. Deposition rates within any particular slope basin will remain low while the basin is isolated by the adjacent ridges, increasing abruptly as a new feeder channel cuts through from the next basin upslope. The thick sediments in the Gulf of Oman probably entered via the ends of the basin, skirting the ends of the subduction zone rather than moving across the efficient sediment traps of the accretionary wedge.

MASS BALANCE

Our seismic refraction and reflection determinations of the offshore accretionary prism shape allow us to make an approximate mass balance between the sediment entering the subduction zone on the Arabian plate, and the amount accumulated in the accretionary pile. At best, this is only an order of magnitude calculation because we must assume, as others have (e.g. Karig et al., 1981), that the configuration, rates of convergence, and sediment input have remained the same up to the present. We therefore ignore the volumetric changes caused by metamorphism at the base of the sediment pile, since this is a second order effect. We also leave out of our mass balances the erosion of sediment from that part of the accretionary wedge now exposed on land in the Makran of Iran and Pakistan, and its deposition in slope basins on the continental margin. This omission is justified because the movement of sediment, though substantial, remains within the accretionary system and the efficient sediment traps on the offshore part of the accretionary wedge prevent most of the eroded material from being lost by moving out onto the abyssal plain beyond the deformational front. We assume that the accretionary wedge is two-dimensional, so that significant amounts of sediment do not move laterally. Our figures are for a 1-km-wide strip along strike over 1 million years, using the geometry and rates indicated on the cartoon in Figure 11.

The volume of sediment per kilometer strip carried in to the accretionary system is the thickness (7 km) derived from the offshore refraction line, times the convergence rate (50 km/Ma), which gives 350 cu km/Ma. Add the extra amount from southward migration of the deformational front at 10 km/Ma (White, 1981), and the total input is 420 cu km/Ma. This volume, however, contains more than 4 km thickness of unconsolidated sediment (Figure 6), which rapidly dewaters within the accretionary wedge. If we adopt the porosity versus compressional wave velocity relationships for silt-clays and turbidites suggested by Hamilton (1978), and the observed sediment compressional

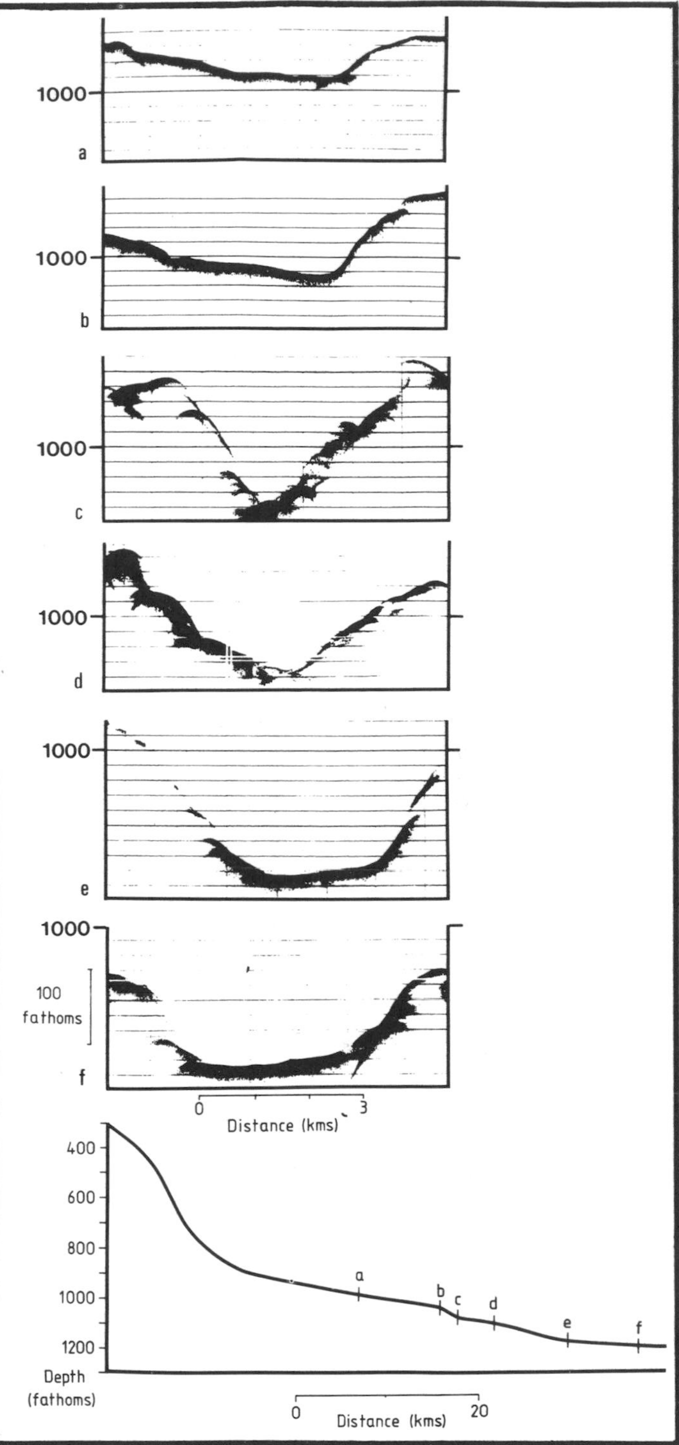

Figure 14 — Cross section at bottom shows profile along the channel marked by the broken line in Figure 13 in the eastern part of the detailed survey area. Note the local increase in slope between crossings "b" and "c" where the channel cuts through an uplifted ridge which can be seen on Figure 13. Copies of the 10 kHz echo-sounder records are shown at crossings "a" through "f." At "a" and "b," the channel is shallow where it crosses the broad upper slope. At "c" and "d," it is deeply incised where it cuts through an uplifted ridge. At "e" and "f," it is turned eastward along the structural strike and follows a narrow interfold basin. Tick marks on the echo sounder records are at 1,000 fathoms depth, scale lines are 20 fathoms apart. 100 fathoms equal 183 m.

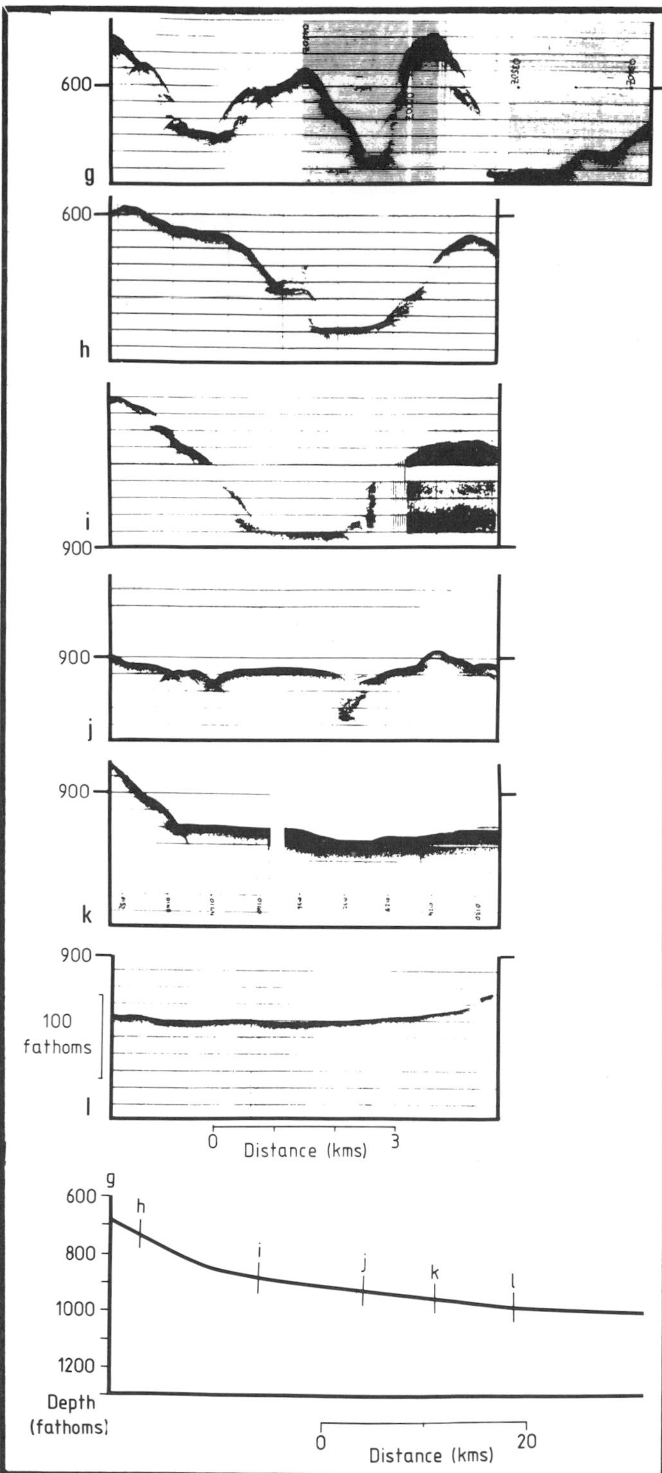

Figure 15 — Cross section at bottom shows profile along the channel marked by the broken line in Figure 13 in the western part of the detailed survey area. Copies of the 10 kHz echo-sounder records show deeply incised channels at crossing "g" in the north, which coalesce into channel "h" and continue southward down a 1 to 2 km wide channel "i" cutting across the structural trends. By crossing "j," the channels have flattened out on the upper slope basin and rapidly decrease in depth as the flow turns westward along the slope basin at "k" and "l" (with a gradient of less than 0.1°). Tick marks on the echo-sounder records are at 600 or 900 fathoms, scale lines are 20 fathoms apart. 100 fathoms equal 183 m.

wave velocities in the Gulf of Oman abyssal plain from White and Klitgord (1976), we find that if all the pore water is expelled then the 4.2 km of unconsolidated sediment compacts to under 3 km of rock. The total sediment input of 420 cu km/Ma compacts to about 330 cu km/Ma when the deeper consolidated part is included.

To balance the consolidated sediment input against the rate of growth of the accretionary wedge, we must first assume the position of some vertical boundary within the wedge behind which no further deformation or thickening occurs. As a first estimate, we assume that the coastline marks the approximate position of this boundary. At this point, the depth to the base of the sediment is 14 km, derived from the northward dip defined by the two refraction lines (Figure 6). The rate of regression of the coastline is about 10 km/Ma (White, 1981), so the volume of material added to the accretionary prism across this boundary is about 140 cu km/Ma. This is less than one-half the sediment input. Our conclusion is that deformation and thickening of the accreted sediments continues up to 350 km north of the present coastline in order to account for the sediment input. This agrees with our earlier comments based on the structures observed in the offshore margin. Active faulting occurs across the entire Makran and these rough mass balances suggest that deformation continues within both the offshore and onshore portions of this unusually wide accretionary zone.

SUMMARY

The open, well-lineated morphology of the offshore Makran accretionary wedge results from the input of a very thick, unconsolidated sediment pile on the shallowly subducting plate. The deformational processes, described in detail in previous sections, probably depend strongly on the movement of pore waters within the accreting sediments. For instance, in the frontal fold the stratal shortening is less than 1%, but this is sufficient to consolidate the terrigenous sediments enough to support the major faulting necessary to uplift the fold to 1.25 km above the abyssal plain. Slumping is seen off the seaward limb of the unconsolidated frontal fold sediments, yet is rare on the much steeper slopes of the tectonically consolidated, accreted material on the continental margin. The consolidation of sediments within the accretionary prism is probably caused by water expulsion and results in the marked increase in seismic velocities observed on the inshore refraction line.

The décollement zone observed within the sediment pile in the frontal fold is attributed to overpressured shale. Beneath the frontal 60 to 70 km of the accretionary wedge, the open fold and basin structure and the low average bathymetric slope suggests that decoupling layers continue to be present in the underlying sediments. The rapidly accumulated terrigenous sediment pile contains a great deal of water, both as free pore water and as water bound up in clay structures. The restricted expulsion of this water is probably responsible for the production of high pore pressures in layers along which slippage can occur (Powers, 1967; Perry and Hower, 1972). The accretionary prism becomes more brittle and allows imbricate faulting to extend down to the basement (Delany and Helgeson, 1978; Wang, 1980) only after the excess water and associated abnormally high pore pressure escapes from the sediment pile. Where the incoming sediment is thinner, or more variable, the simple pattern of open folding seen in the front 70 km is not developed and the overall bathymetric slope is much steeper as faulting initially extends deeper through the sediments. Excess water escapes through the sediments and may be assisted in its passage by the dipping fault planes. Near the coast the frequent mud diapirs also contribute to water removal (Snead, 1964).

ACKNOWLEDGMENTS

The data discussed in this article were collected on Leg 1/80 of RRS SHACKLETON. We are grateful to the officers and crew of the ship and to the scientific party for their assistance. The work was funded by the Natural Environment Research Council (NERC) under grant GR3/1651 to Dr. D. H. Matthews. Robert S. White acknowledges personal support from the NERC and Emmanuel College, Cambridge. Department of Earth Sciences, Cambridge contribution number 152.

REFERENCES CITED

Ahmed, S. S., 1969, Tertiary geology of part of south Makran Baluchistan, West Pakistan: AAPG Bulletin, v. 53, p. 1480-1499.

Barker, P. F., 1966, A reconnaisance survey of the Murray ridge: Philosophical Transactions of the Royal Society of London, Series A, v. 259, p. 187-197.

Berberian, M., 1967, Contribution to the seismotectonics of Iran, part 2 : Geological Survey of Iran, Report n. 39, 516 p.

Chapman, R. E., 1974, Clay diapirism and overthrust faulting: Geological Society of America Bulletin, v. 85, p. 1597-1602.

Chapple, W. M., 1978, Thin-skinned fold-and-thrust belts: Geological Society of America Bulletin, v. 89, p. 1189-1198.

Christensen, N. I., and M. H. Salisbury, 1975, Structure and composition of the lower oceanic crust: Reviews of Geophysics and Space Physics, v. 13, p. 57-86.

Closs, H., H. Bugenstock, and K. Hinz, 1969, Ergebnisse seismischer Untersuchungen im nordlichen Arabischem Meer ein Beitrag zur Internationalen Indischen Ozean Expedition: Meteor ofrschungsergebnisse, Reihe C, Heft 2, 28 pp.

Delany,. J. M., and H. C. Helgeson, 1978, Calculation of the thermodynamic consequences of dehydration in subducting oceanic crust to 100 kb and 800°C: American Journal of Sciences, v. 278, p. 638-686.

Farhoudi, G., and D. E. Karig, 1977, Makran of Iran and Pakistan as an active arc system: Geology, v. 5, p. 664-668.

Hamilton, E. L., 1978, Sound velocity-density relations in sea floor sediments and rocks: Acoustical Society of America Journal, v. 63, p. 366-377.

Hudson, R. G. S., A. McGugan, and D. M. Morton, 1954, The structure of the Jebel Hagab area, Trucial Oman: Geological Society of London, Quarterly Journal, v. 60, p. 121-157.

Hutchison, I., et al, 1981, Heat flow and age of the Gulf of Oman: Earth and Planetary Science Letters, v. 56, p. 252-262.

Jacob, K. H., and R. C. Quittmeyer, 1979, The Makran region of Pakistan and Iran: trench-arc gap with active plate subduction, *in* A. Farah and K. A. DeJong, eds., Geodynamics of Pakistan: Quetta, Geological Survey of Pakistan, p. 305-318.

Karig, D. E., and G. F. Sharman, 1975, Subduction and accretion in trenches: Geological Society of America Bulletin, v. 86, p. 377-389.

———, J. G. Caldwell, and E. M. Parmentier, 1976, Effects of accretion on the geometry of the descending lithosphere: Journal of Geophysical Research, 81, p. 6281-6291.

———, et al, 1981, Morphology and shallow structure of the lower trench slope off Nias island, Sunda arc *in* D. E. Hayes, ed., The tectonic/geologic evolution of southeast Asia: American Geophysical Union, Monograph.

Kazmi, A. H., 1979, Preliminary seismotectonic map of Pakistan, scale 1: 2,000,000: Quetta, Geological Survey of Pakistan, Map series.

Moore, G. F., and D. E. Karig, 1976, Development of sedimentary basins on the lower trench slope: Geology, v. 4, p. 693-697.

Niazi, M., H. Shimamura, and M. Matsu'ura, 1980, Microearthquakes and crustal structure off the Makran coast of Iran: Geological Research Letters, v. 7, p. 297-300.

Nowroozi, A. A., 1976, Seismotectonics provinces of Iran: Seismological Society of America Bulletin, v. 66 p. 1249-1276.

Page, W. D., et al, 1979, Evidence for the recurrence of large-magnitude earthquakes along the Makran coast of Iran and Pakistan: Tectonophysics, v. 52, p. 533-547.

Perry, E., and J. Hower, 1972, Late-stage dehydration in deeply buried pelitic sediments: AAPG Bulletin, v. 56, p. 2013-2021.

Powers, M. C., 1967, Fluid-release mechanisms in compacting marine mudrocks and their importance in oil exploration: AAPG Bulletin, v. 51, p. 1240-1254.

Raitt, R. W., 1963, The crustal rocks, *in* M. N. Hill, ed, The sea, c. 3: New York, Wiley-Interscience, p. 85-102.

Scientific Party, 1981, Near Barbados ridge: scraping off, subduction scrutinized: Geotimes v. 26, n. 10, p. 24-26.

Seely, D. R., P. R. Vail, and G. G. Walton, 1974, Trench slope model, *in* Burk and Drake, eds., The geology of continental margins: New York, Springer-Verlag, p. 249-260.

Shor, G. G., Jr., and R. von Huene, 1972, Marine seismic refraction studies near Kodiak, Alaska: Geophysics, v. 37, p. 697-700.

Snead, R., 1964, Active mud volcanoes of Baluchistan, West Pakistan: Geographical Review, v. 54, p. 545-560.

Sondhi, V. P., 1947, The Makran earthquake, 28th November 1945, the birth of new islands: Indian Minerals, v. 1, n. 3, p. 146-154.

Stewart, R. A., O. H. Pilkey, and B. W. Nelson, 1965, Sediments of the northern Arabian Sea: Marine Geology, v. 3, p. 411-427.

Stoffers, P., and D. A. Ross, 1979, Late Pleistocene and Holocene sedimentation in the Persian Gulf — Gulf of Oman: Sedimentary Geology, v. 23, p. 181-208.

Vita-Finzi, C., 1975, Quaternary deposits in the Iranian Makran: Geographical Journal, v. 141, p. 415-420.

———, C., 1979, Contributions to the quaternary geology of

southern Iran: Geological and Mining Survey of Iran, Report No. 47.

von Huene, R., 1972, Structure of the continental margin and tectonics of the eastern Aleutian trench: Geological Society of America Bulletin, v. 83, p. 3613-3626.

Wang, C., 1980, Sediment subduction and frictional sliding in a dubduction zone: Geology, v. 8, p. 530-533.

Westbrook, G. K., 1975, The structure of the crust and upper mantle in the region of Barbados and the Lesser Antilles: Royal Astronomical Society, Geophysical Journal, v. 43, p. 201-242.

White, R. S., 1977a, Recent fold development in the Gulf of Oman: Earth and Planetary Science Letters, v. 36, p. 85-91.

———, 1977b, Seismic bright spots in the Gulf of Oman: Earth and Planetary Science Letters, v. 37, p. 29-37.

kr0———, 1979a, Gas hydrate layers trapping free gas in the Gulf of Oman: Earth and Planetary Science Letters, v. 42, p. 114-120.

———, 1979b, Deformation of the Makran continental margin *in* A. Farah and K. A. DeJong, eds., Geodynamics of Pakistan: Geological Survey of Pakistan, p. 295-304.

———, 1981, Deformation of the Makran accretionary sediment prism in the Gulf of Oman (northwest Indian Ocean) *in* J. K. Leggett, ed., Trench and fore-arc sedimentation and tectonics: Geological Society of London, p. 69-84.

———, and K. D. Klitgord, 1976, Sediment deformation and plate tectonics in the Gulf of Oman: Earth and Planetary Science Letters, v. 32, p. 199-209.

The Northwest Margin of the Iapetus Ocean During the Early Paleozoic

W. S. McKerrow
Department of Geology and Mineralogy
Oxford University
Oxford, England

The locations, in space and time, of the collisions between island arcs and the eastern margin of North America suggest that the Iapetus Ocean plate moved north relative to the continent during the Ordovician. But the final collisions of North America with Scandinavia and Avalon suggest relative westward movement of the oceanic plate in the Late Silurian and Early Devonian.

The theory of Plate Tectonics permits many geologists to develop analogies between their own field areas and other parts of the world; the recognition of these analogies often leads to a greater understanding of the areas being compared. The appreciation of these tectonic processes has led to a multitude of "models" to explain the geological development of particular regions. The areas bordering the Iapetus Ocean perhaps have more than their share of such speculative accounts, most of them are variations on Dewey (1969), who gave us the model for later "models." My justification for writing yet another speculative paper is two-fold: first, I want to summarize the spatial and time relationships of events which took place along 5,000 km of the margin of a single continent over 130 million years; and second, if Applachian and Caledonian field geologists make observations in their field areas to prove or disprove my speculations, then this synthesis will make a small contribution to our science.

It is now well-established that, during the early Paleozoic, a wide ocean was present to the east of North America. This ocean was first postulated, on the basis of faunal differences (Wilson, 1966), before plate tectonic processes were fully understood. It was named Iapetus (Harland and Gayer, 1972) after the father of Atlas, from whom the Atlantic Ocean takes its name. The precise suture line where the ocean finally closed can, in most regions, be determined by faunal differences in the Cambrian and Ordovician rocks on either side (e.g. McKerrow and Cocks, 1977). Until the Ashgill, the different shelf faunas in North America and Scandinavia show that the ocean was wide enough (perhaps 1,000 km) to separate animals (like brachiopods and trilobites) which had pelagic larval stages, but it was not until the Ludlow that benthic ostracodes and fresh-water fish became established on both sides (McKerrow and Cocks, 1976). In the northern Appalachians, Neuman (1972) recognized Early Ordovician brachiopod faunas which were different from contemporary forms in both North America and Europe, and he concluded they were living on oceanic islands. The significance of Neuman's observations is discussed in detail below. The paleontological indications of an ocean and oceanic islands are confirmed by numerous structures typical of modern active margins, by the widespread occurrence of calc-alkaline igneous rocks, by the presence of ophiolites, and by differences in paleomagnetic fields in contemporary rocks on either side of the suture where the ocean finally closed.

Recent syntheses of northern Appalachian and Caledonian regional geology suggest that the history of the Early Paleozoic Iapetus Ocean is represented by a series of orogenic events spread over 130 million years from Early Ordovician to Early Devonian. These orogenic events appear to include: (1) prolonged uplift and igneous activity associated with subduction; (2) shorter episodes of deformation, metamorphism, and thrusting related to the collision of oceanic islands (island arcs, oceanic volcanoes, and perhaps microcontinents) with the eastern margin of the North American continent; (3) major dextral faults associated with strike-slip along plate margins; and (4) the closure of the last remnants of the ocean when different parts of northern Europe and the Avalon Platform collided at different times with North America.

In this contribution, the structures on the northwest margin of the Iapetus Ocean, from Long Island Sound to Greenland, are summarized with particular reference to their chronology. Although I have personally carried out field investigations in Greenland, the British Isles, and eastern Canada, I am relying heavily on field data and conclusions contained in several symposium volumes, notably those edited by Zen et al (1968), Kay (1969), Bassett (1976), Bowes and Leake (1978), Tozer and Schenk (1978), Harris, Holland, and Leake (1979), and Wones (1980). Many of my conclusions depend on a precise correlation between American and British stratigraphic zones (Berry, 1976; Williams, 1976). In a review like this, even more critical is the correlation between stratigraphic and radiometric time scales (Figure 1). Although there are still many uncertainties, the adoption of internationally agreed decay-constants (Steiger and Jager, 1977) enables us to make some better estimates; in this paper we use the scale of McKerrow, Lambert, and Chamberlain (1980) throughout. This review relies heavily on these recent advances in chronostratigraphy.

The eastern margin of North America, south of Long Island Sound, is essentially a history of the Paleozoic ocean between Africa and North America (McKerrow and Ziegler, 1972). The only former part of Africa occurring in the northern Appalachians is Nova Scotia, south of the Chedabucto Fault (Schenk, 1971). The southern Appalachians form a different, though analagous, development. They are not considered further here.

ORDOVICIAN HISTORY OF THE MARGIN IN NEW ENGLAND

Rodgers (1968) published a map showing the transition from Early Ordovician platform carbonates to deeper water clastic sediments, and concluded that this line marked the margin of the North American continent (Figure 2). East of this line, the geology of the northern Appalachians is highly complex, and numerous tectonic and metamorphic events and the widespread occurrence of calc-alkaline igneous rocks indicate that many areas were situated above contemporary subduction zones. In contrast, west of Rodgers' line there is little evidence of any Early Paleozoic igneous activity, and the principal tectonic structures are nappes thrust westward over the old continental margin. The absence of any calc-alkaline igneous rocks west of Rodgers' line leads to the conclusion that during the Early Paleozoic, a passive margin persisted from Long Island Sound to Newfoundland. It follows that the areas with Ordovician and earlier calc-alkaline rocks must have collided with North America after the Arenig.

Robinson and Hall (1980) recently produced a tectonic synthesis of southern New England. They conclude that the remnants of a distinct "Bronson Hill Plate" are present east of Rodgers' line, and that this oceanic island moved toward North America during the Early Ordovician above subducting oceanic crust. (The Bronson Hill terrain is shown in its present position in Figure 3). Many of the rocks making up the Bronson Hill terrain are now metamorphosed, but Robinson and Hall conclude that the terrain is based on Late Precambrian, Cambrian, and Ordovician rocks which were perhaps part of an Ordovician arc (Robinson and Hall, 1980, p. 80). Similar interpretations for the Magog area of Quebec are also postulated (St. Julien and Hubert, 1975). The rocks and structures of the Bronson Hill terrain appear to continue northeast (Hussey, 1968; St. Julien and Hubert, 1975; Osberg, 1978) to southern Quebec and northern Maine, including the region where the Early Ordovician Shin Brook Formation (Neuman, 1968) is exposed. This formation contains brachiopods which Neuman (1972) suggests may have been isolated from North America by their position on islands far out in the Iapetus Ocean.

The westward movement of the Bronson Hill island arc eventually resulted in its collision with North America; the Taconic Orogeny is marked by continental margin sediments thrusting over the platform edge. The climax of this orogeny is dated (Bird and Dewey, 1971) as occurring in the *D. clingani* and *P. linearis* zones (Figure 1).

We do not know the rates of subduction which were present on the west side of the Bronson Hill island arc, nor do we know how much (if any) later strike slip faulting shifted it along the continental margin after collision. However, low subduction rates are inferred from studies in the Southern Uplands of Scotland (see below), and there are no records of large strike-slip movements in the Taconic Mountains (Zen, 1968).

The end of the *D. clingani* zone was around 450 m.y. ago, so if it moved 2 cm/year, the Bronson Hill island arc would have been 1,000 km from the continent 500 m.y. ago; this is the position shown in Figure 2. The direction of relative movement between the Iapetus Ocean and North America is discussed below in light of all the known indications from along the whole plate boundary.

ORDOVICIAN HISTORY OF THE MARGIN IN QUEBEC AND NEW BRUNSWICK

Structures of the Bronson Hill terrain do not appear to reach the Maine/New Brunswick border. Between Houlton, Maine, and Matapedia, Quebec, Late Ordo-

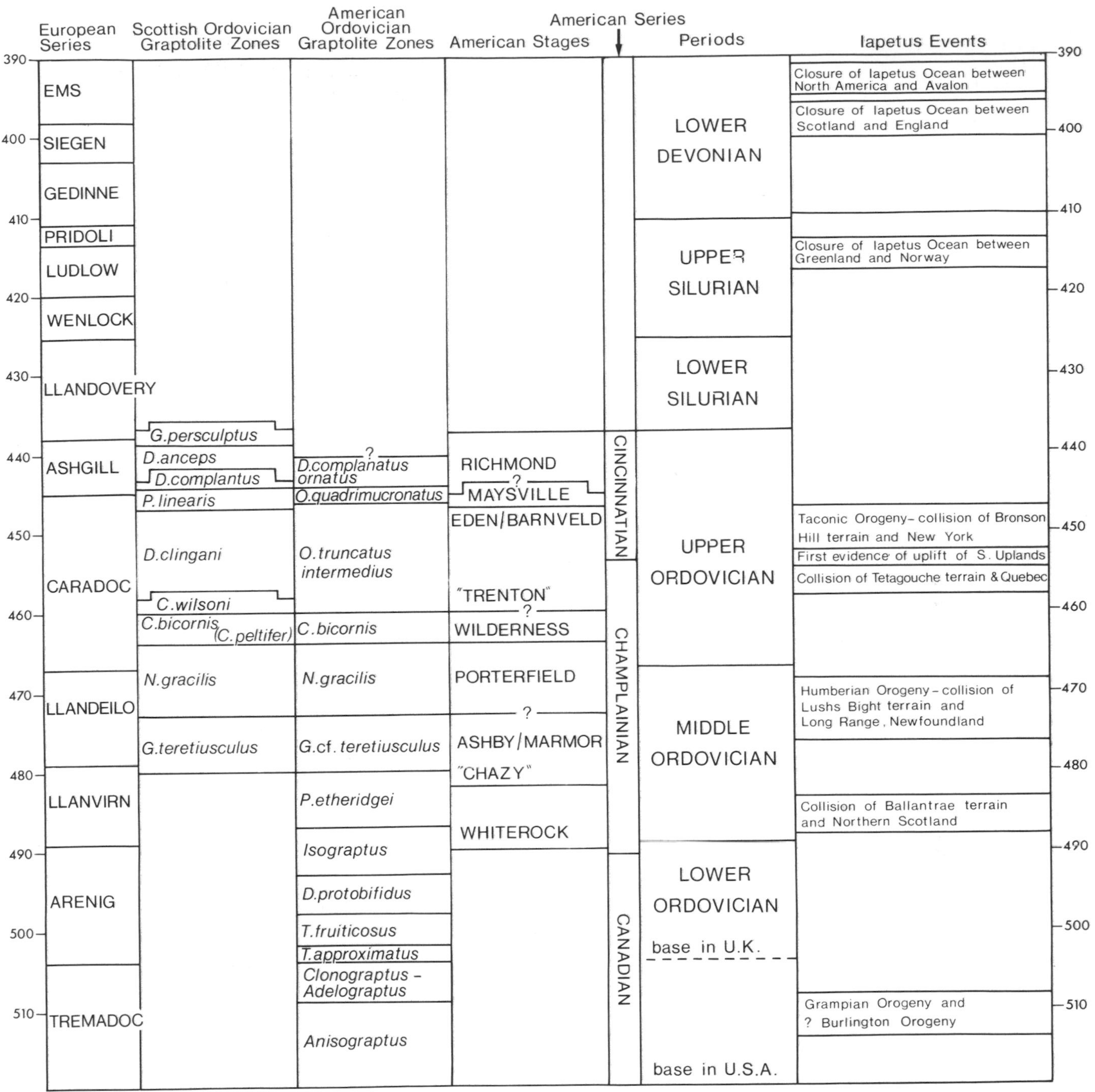

Figure 1 — Stratigraphic correlation chart showing relations between the British series and Scottish graptolite zones (Williams, 1976, p. 61), the American series and graptolite zones (Williams, 1976; Berry, 1976, p. 160; Ross, 1976, p. 78), the events discussed in the paper, and the time scale from isotopic dates (McKerrow, Lambert, and Chamberlain, 1980).

vician and Early Silurian turbidites in the Matapedia Basin show current directions toward the south (Ayrton et al, 1969; McKerrow and Ziegler, 1971) and I believe that this basin deepened to the south to link up with the Iapetus Ocean (Figure 3).

The Matapedia Basin separates rocks assigned to the Bronson Hill terrain (to the west) from Ordovician volcanic rocks of the Tetagouche Group (to the east). The Tetagouche Group includes deformed early Caradoc rocks (Skinner, 1974) which are overlain unconformably by Late Llandovery beds. Although the precise age of deformation cannot be determined directly, several formations (e.g. Carys Mills Formation) in and around the Matapedia Basin show a continuous sequence from the *Orthograptus truncatus* var *intermedius* zone upward into the Lower Llandovery (Ayrton et al,

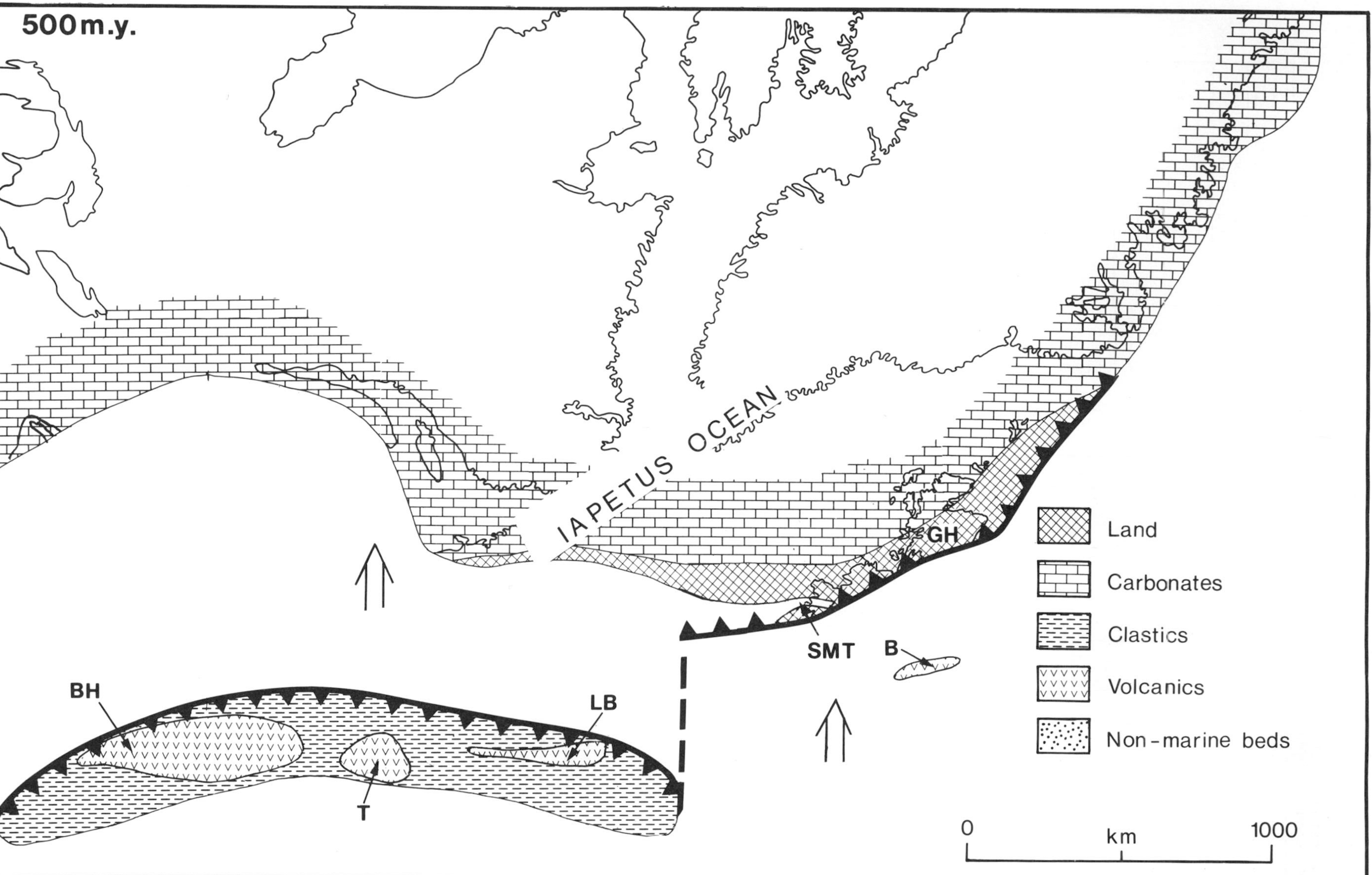

Figure 2 — Possible configuration of the western margin of North America 500 m.y. ago (Arenig) showing the estimated positions of active margins (thick lines with ticks on the over-riding plate edge) and oceanic islands. Large arrows indicate movement of parts of the Iapetus Ocean relative to North America. The thick dashed line is a postulated transform fault separating those parts of the ocean moving north (Iapetus plate) from those parts on the North American plate. Three island arc terrains (BH = Bronson Hill; T = Tetagouche; LB = Lushs Bight) were possibly part of the same arc as indicated. SMT = South Mayo Trough of western Ireland; GH = Grampian Highlands; B = Ballantrae oceanic edifice. BH, T, LB and B are shown in positions assuming subduction rates of 2 cm/yr. The Grampian Highlands are part of the highland area resulting from the Grampian Orogeny. Outlines of present day land (from conical projections to reduce distortion) are shown for geographical reference, and for the same reason subsequent strike faults are ignored.

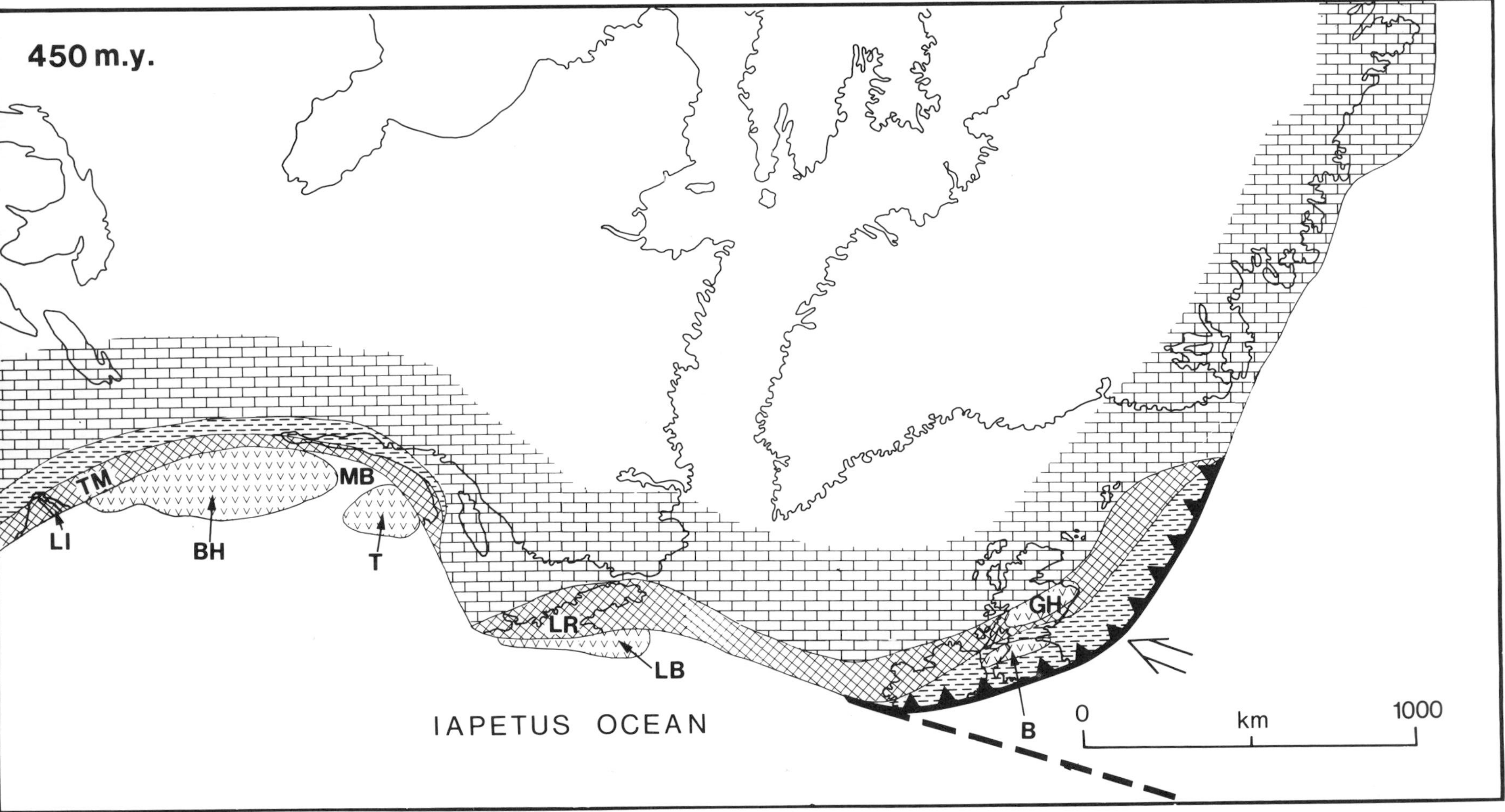

Figure 3 — The possible configuration of the western margin of North America 450 m.y. ago (late Caradoc) showing the estimated position of the active margin (thick line). By this time the island arcs (BH = Bronson Hill, T = Tetagouche, and LB = Lushs Bight) have collided with the continent to form the Taconic Mountains (TM) and high land in the Long Range (LR) of western Newfoundland. LI = Long Island, New York; MB = Matapedia Basin, New Brunswick. Volcanoes are present in the Grampian Highlands (GH), Scotland; to the south the Ballantrae edifice (B) has impinged on the continental margin. South of Ballantrae the Southern Uplands accretionary prism develops. The thick dashed line represents an inferred transform fault separating the parts of the Iapetus Ocean being subducted from the parts now attached to the North American plate (it is drawn parallel with the similar fault in Figure 4). Shading as in Figure 2.

1969; McKerrow and Ziegler, 1971), it seems probable that the Tetagouche Group deformed prior to the deposition of this sequence, in the earlier part of the lengthy (McKerrow, Lambert, and Chamberlain, 1980) *D. clingani* zone, which is equivalent to the *O. truncatus* var *intermedius* zone (Figure 1). It is possible that the Tetagouche terrain also includes *D. clingani* zone or earlier rocks in southern Gaspe; if so, it would be demarcated from the Ordovician North American continent by a belt containing ophiolites and shale-melange, the internal domain of St. Julien and Hubert (1975). Like the Bronson Hill terrain, the Tetagouche rocks include Ordovician volcanics which are ascribed (Rast, Kennedy, and Blackwood, 1976) to emplacement above a southeastward dipping subduction zone. The Tetagouche terrain is also similar to the Bronson Hill terrain in containing a site (in York County, New Brunswick) where Neuman (1972) recorded an Early Ordovician fauna similar to the oceanic island fauna of Maine.

Figure 3 shows the Tetagouche terrain in its position after collision (in the *D. clingani* zone) with North America. Figure 2 shows it in a position 950 km south of the continental margin; this is where it would be situated if it moved north at 2 cm/year during the 45 million years between the early Arenig and the late Caradoc. It would, of course, be further south if subduction rates were greater than 2 cm/year.

ORDOVICIAN HISTORY OF THE MARGIN IN NEWFOUNDLAND

Since the work of Bird and Dewey (1971), there have been numerous structural interpretations of the Ordovician rocks of western and central Newfoundland (Williams and Stevens, 1974; Kennedy, 1975; Rast, Kennedy, and Blackwood, 1976; Williams and Max, 1980; Currie, Pickerill, and Pajari, 1980), and I do not intend to summarize all the diverse views which have been presented. All authors agree that Rodgers and Neale (1963) were correct when they postulated that the ophiolites and deep water clastic sediments of western Newfoundland had been thrust over platform carbonates, in an event now known as the Humberian Orogeny. In the Bay of Islands region, the Humberian Orogeny reached its climax in the *N. gracilis* zone (Bird and Dewey, 1971), but around Hare Bay, north of the Long Range, the thrusting may have been slightly earlier (Dallmeyer, 1977).

The Burlington Peninsula, east of the Long Range, consists of deformed Late Precambrian and Paleozoic rocks (of the Fleur de Lys Supergroup), which are divided into western and eastern areas by the ophiolites of the Baie Verte lineament. The rocks in the eastern area are intruded by pre-Acadian granodiorites (Kennedy, 1975) which have been notoriously difficult to date (Pringle, 1978; Mattinson, 1977). The sediments of the western Fleur de Lys appear to have been deformed in a pre-Arenig Burlington Orogeny (Kennedy, 1975), but Rb-Sr and U-Pb ages from volcanic rocks east of the Burlington Peninsula suggest that the eastern sequence includes rocks of Silurian and Early Devonian age (Pringle, 1978; Mattinson, 1977). This suggests that the deformation here may be later (i.e. Humberian or Acadian). The nature of the deformation of the Fleur de Lys Supergroup is still a matter of debate (Kennedy, 1975; Williams, 1977), and so is its relation to the ophiolites of th Baie Verte lineament. Kennedy (1975) interprets the lineament as the collision line between an oceanic terrain and the margin of North America during the Burlington Orogeny; Williams and St. Julien (1978) view the lineament as the continuation of the Taconic suture in Quebec.

Along the eastern shores of the Burlington Peninsula, a further belt of ophiolites (the Snooks Arm Group) was emplaced over the eastern margin of the Fleur de Lys. These ophiolites lie west of the Lushs Bight Group, which are interpreted as island arc volcanics and related sediments (Dean, 1978). The Lushs Bight terrain may consist both of volcanics and older deformed rocks (Williams and Payne, 1975). The collision of this island arc with the Burlington Peninsula appears to cause the Humberian Orogeny (Bird and Dewey, 1971; Nelson and Casey, 1979), when the ophiolites of the Snooks Arm Group were obducted over the eastern Burlington Peninsula. The Bay of Islands ophiolites were also emplaced at this time, but it is still not clear whether they were derived from the Baie Verte lineament (Williams, 1977) or from further east (Kennedy, 1975).

The rocks of the Lushs Bight Group extend eastward along the south side of Notre Dame Bay as far as New World Island. They are now exposed north of the Lobster Cove and Lukes Arm faults (Dean and Strong, 1977), which appear to have been active during the Late Ordovician and Early Silurian when the Lushs Bight terrain north of the fault provided a source for the thick sedimentary sequences of New World Island (McKerrow and Cocks, 1978, 1981; M. Watson, written communication). The Llandovery Big Muddy Cove Group contains large olistoliths of volcanic and pyroclastic rocks identical to parts of the Lushs Bight Group. From this area, Neuman (1972, 1976) describes an Arenig oceanic island fauna, very similar to those of New Brunswick and Maine. Younger beds in the same sequence yield Llandeilo and Caradoc shelly faunas which are of undoubted North American affinities (McKerrow and Cocks, 1978). The paleontological information thus suggests that this Ordovician sequence was distant from North America in the Arenig and yet close to it by the end of the Llandeilo. The faunal changes reinforce the tectonic models, suggesting movement of the Lushs Bight island arc from a mid-ocean situation in the Arenig to collision (the Humberian Orogeny) with North America in the late Llandeilo.

Figure 3 shows the Lushs Bight terrain in its post-collision position, and Figure 2 shows it 600 km south of the continental margin, where it would have been situated assuming subduction rate of 2 cm/year. It is perhaps significant that the three oceanic islands of Bronson Hill, Tetagouche, and Lushs Bight lie along a continuously curved arc when put into their possible Arenig positions. The differences in timing (Figure 1) of the Humberian Orogeny of Newfoundland and the Taconic Orogeny of New England, New Brunswick, and Quebec may well be due to the irregular shape of the North American margin during the Ordovician.

ORDOVICIAN HISTORY OF THE MARGIN IN THE BRITISH ISLES

Three major faults trend northeastward across Scotland and northwestern Ireland: the Moine Thrust, the Great Glen Fault, and the Highland Boundary Fault. These faults separate regions with different geological developments. Northwest of the Moine Thrust, Early Ordovician and Early Cambrian carbonates with basal sandstones rest unconformably on Precambrian sediments and gneisses. The faunas in these carbonates are comparable with contemporary faunas on the North American carbonate platform, and there is little doubt that northwest Scotland was part of North America (i.e. west of the Iapetus Ocean) in the Lower Paleozoic.

Between the Moine Thrust and the Great Glen, Precambrian Moine metasediments were metamorphosed in several events prior to 730 million years and again around 455 to 430 million years (van Breemen, Aftalion, and Johnson, 1979), i.e. in the latest Ordovician and early Silurian (McKerrow, Lambert, and Chamberlain, 1980). In this area, the Carn Chuinneag granite has an age of 560 ± 10 million years (Pidgeon and Johnson, 1974) and many other plutons range in age from 456 to 400 million years (van Breeman, Aftalion, and Johnson, 1979; Pidgeon and Aftalion, 1978).

In Scotland, south of the Great Glen Fault, a very thick sequence of late Precambrian and Cambrian sediments is present in the Grampian Highlands north of the Highland Boundary Fault: the Dalradian Supergroup (Harris et al, 1978). The Dalradian rocks are also present in northwest Ireland, but the southwestward continuation of the Great Glen and Highland Boundary faults is debateable. The Dalradian Supergroup is dominantly a shelf sequence in its lower part and has a greater development of turbidites and mafic pillow lavas in its upper part (Harris et al, 1978). The youngest beds in Scotland contain Lower Cambrian trilobites with American affinities (Cowie, Rushton, and Stubblefield, 1972) while a probable Middle Cambrian sponge is present in western Ireland (Rushton and Phillips, 1973).

Dalradian rocks were deformed and metamorphosed in the Grampian Orogeny (Lambert and McKerrow, 1977), which may have occurred in the Late Cambrian or Early Ordovician before (or after?) the deposition of the late Tremadoc Lough Nafooey Group of western Ireland (Ryan, Floyd and Archer, 1980). Although I postulate (Lambert and McKerrow, 1977) that the Grampian Orogeny may have been related to the impingement of a spreading ridge on a trench, there are many other possibilities. The region was probably on continental crust above an active margin: age dates on granitoid intrusions in the Grampian Highlands and in western Ireland range from 514 ± 7 (Pankhurst and Pidgeon, 1976) to 389 ± 4 (J.A. Clayburn, written communication, 1980), though there is a scarcity of reliable dates earlier than 489 million years (Pankhurst, 1970). Other possible causes of the Grampian Orogeny include: collision of an oceanic island with the continental margin along the southern edge of the Grampian Highlands, but there is no evidence for this; or, the Grampian Highlands could have once been detached from North America (like the Bronson Hill, Tetagouche and Lushs Bight islands). The Dalradian rocks suffered polyphase deformation and metamorphism like the rocks on some of these old island arcs, but they do not contain the same quantity of pre-collision volcanic rocks.

Ordovician fossils are reported in rocks near the Highland Boundary Fault in Scotland. Although these rocks are deformed, they do not appear metamorphosed to the same degree as the adjacent Dalradian, and I suspect that they were deposited after the climax of the Grampian Orogeny. In western Ireland, the South Mayo Trough (Figure 2) was developed after the Grampian Orogeny (Dewey, 1971) and contains over 12 km of Early Ordovician sediments resting on pillow basalts. It appears to have been a small marginal basin on the edge of the North American continent (Ryan and Archer, 1977).

In Scotland, south of the Grampian Highlands, much of the Midland Valley is covered by Upper Paleozoic rocks, but in the southwest (around Girvan) the Early Ordovician Ballantrae Volcanic Group is exposed. Anomalously thick basaltic pillow lavas and interbedded volcanoclastics appear to represent a volcanic edifice (or aseismic ridge) rather than a normal section of oceanic crust (Leggett, McKerrow, and Casey, 1981); they have also been interpreted as part of an island arc (Longman, Bluck, and van Breeman, 1979; Bluck et al, 1980). Trace element analyses suggest hot spot basalts, island arc tholeiites, and ocean floor basalts (Wilkinson and Cann, 1974). It seems probable that subduction northward, beneath the Grampian Highlands, carried the Ballantrae edifice toward the trench. The date of impingement with the trench is probably Llanvirn, as the youngest Ballantrae rocks and the oldest overlying sediments are both assigned to this series (Ingham, 1978).

Upper Llanvirn to Wenlock sediments which lie above the Ballantrae Volcanic Group show a large variety of facies and thickness changes. Middle and Late Ordovician sedimentation appears to have been influenced by contemporary faulting with major downthrows to the southeast (Williams, 1962). The paleoslope was also southeast until the middle Caradoc (Ingham, 1978). Up to this time, much of the detritus in these beds appears to have derived from erosion of the Ballantrae Volcanic Group, but clasts of granite, vein quartz, and metamorphic debris are present (Williams, 1962; Ingham, 1978). This suggests that by Llandeilo times the Ballantrae rocks were adjacent to a metamorphic source region, like the Grampian Highlands.

From the middle Caradoc on, the Girvan paleoslope was more variable. By the *D. clingani* zone there are indications that there was a slope down to the northwest (Ingham, 1978), perhaps reflecting the creation of a trench slope break in the growing accretionary prism to the south (Leggett et al, 1979; Leggett, McKerrow, and Casey, 1981). The Southern Uplands accretionary prism developed over a long period (455 to 420 million years) and is devoid of melange and contains only a few olistostromes and slump beds; these facts suggest very slow subduction rates (Leggett, 1980). For this reason, slow subduction rates were employed in constructing Figure 3, where the Ballantrae edifice is shown in its present position relative to northern Scotland, and Figure 2, where it is shown in its possible Arenig position assuming a subduction rate of 2 cm/yr.

The area affected by the Grampian Orogeny extends from the Shetland Islands (north of Scotland) to western Ireland (north and south of the South Mayo Trough). It did not extend north to Greenland. No subduction appears to have occurred along the east Greenland coast at any time during Early Paleozoic (Haller, 1971), and the only indications of a Tremadoc orogeny are in a Middle and Upper Cambrian stratigraphic break in the platform successions (Cowie, 1971). It was suggested (Phillips, Kennedy, and Dunlop, 1969) that the pre-Arenig deformation of the western Fleur de Lys in Newfoundland may represent a westward continuation of the Grampian Orogeny; this area is shown as a continuation of the highland area of Scotland in Figure 2.

SILURIAN AND EARLY DEVONIAN HISTORY OF THE MARGIN IN THE CALEDONIDES

In east North Greenland, Early Paleozoic carbonates continue until the Upper Llandovery. They are followed by latest Llandovery (C_6) and Wenlock turbidites derived from the east. These are the youngest rocks involved in the deformation related to the east Greenland Caledonide thrust belt (Dawes, 1971; Hurst and McKerrow, 1981). After the Middle Wenlock, the Caledonian Orogeny is marked by the development of large nappes which travelled from the east or southeast.

At the same time, in Norway, nappes were derived from the west or northwest (Gee and Wilson, 1974; Gee, 1978). It appears that Greenland collided with Scandinavia in the Late Silurian following subduction eastward below Norway (Roberts and Gale, 1978; Mykkeltveit, Husebye, and Oftedahl, 1980). The collision resulted in a region uplifting along the continental suture, which first provided a source for clastic sediments and then the source from which the Greenland and Scandinavian nappes were derived (Hurst and McKerrow, 1981; Gee, 1975). This uplifted area is now represented by exposures of Precambrian rocks along the Greenland and Norwegian coasts.

The geology of margins of the northern Iapetus Ocean suggests strongly that Norway and Greenland were adjacent to each other in the Late Silurian. Nowhere else along the whole margin is there evidence of Late Silurian thrusting. Thus, at present I doubt the paleomagnetic evidence (Kent and Opdyke, 1978; Van der Voo, French, and French, 1979) which suggests considerable offset at this time. At least it appears that Greenland and Scandinavia were opposite each other, though Avalon may have been a long way south (see below).

In Britain, subduction continued southeastward under England until the Early Silurian, but abundant calc-alkaline rocks show that it persisted northwestward below Scotland through much of the Early Devonian (Thirwall, 1981). So in the Wenlock, the northern parts of the Iapetus Ocean were simultaneously subducted eastward below Scandinavia and northwestward below Scotland. This discrepancy suggests the presence of a transform fault (Figure 4).

This fault separates different parts of the ocean being subducted in different directions; they were perhaps also being subducted at different rates. Collision was distinctly earlier (Ludlow or Pridoli) between Greenland and Norway than between Scotland and England (Early Devonian). The exact time of collision in Britain is not known, but it is unlikely that it occurred during the continuous deposition of Silurian (from Llandovery to Pridoli) in the English Lake District or during the deposition of the Lower Old Red Sandstone (Pridoli to Siegen) of the Scottish Midland Valley (House et al, 1977). The Ems is the earliest time when suitable stratigraphic break is present on both sides of the Iapetus suture in Britain; this is very likely the time of continental collision.

SILURIAN AND EARLY DEVONIAN HISTORY IN THE NORTHERN APPALACHIANS

In Newfoundland, the Iapetus suture is marked by the Reach Fault on the north coast and by the Cape Ray Fault in the southwest (McKerrow and Cocks, 1977). This line was recognized by paleontological differences; it is not marked by ophiolites nor by outwardly directed thrusts (as in Norway and Greenland), nor is it adjacent to Silurian and Lower Devonian calc-alkaline igneous rocks as in Scotland.

In southern Notre Dame Bay and New World Island, large (probably dextral) transcurrent faults separate terrains with very distinct Late Ordovician and Early Silurian sedimentary sequences (Horne and Helwig, 1969; McKerrow and Cocks, 1981). Evidence suggests considerable contemporary faulting during this time, but no subduction.

The Iapetus suture in New Brunswick lies along the Fredericton Trough (McKerrow and Ziegler, 1971). Southern New Brunswick, northern Nova Scotia, and coastal New England are generally considered part of the Avalon Platform (Poole, 1967; Schenk, 1978; Williams, 1979; Rast, 1980; Williams and Max, 1980), and it is probable that this platform continued south to Rhode Island (Robinson and Hall, 1980).

The rocks of the Avalon Platform (from eastern Newfoundland to Rhode Island) appear to be founded on a late Precambrian island arc. The Cambrian and Ordovician faunas show a strong resemblance to those of England and Wales, where similar late Precambrian volcanics form a base to the Lower Paleozoic sedi-

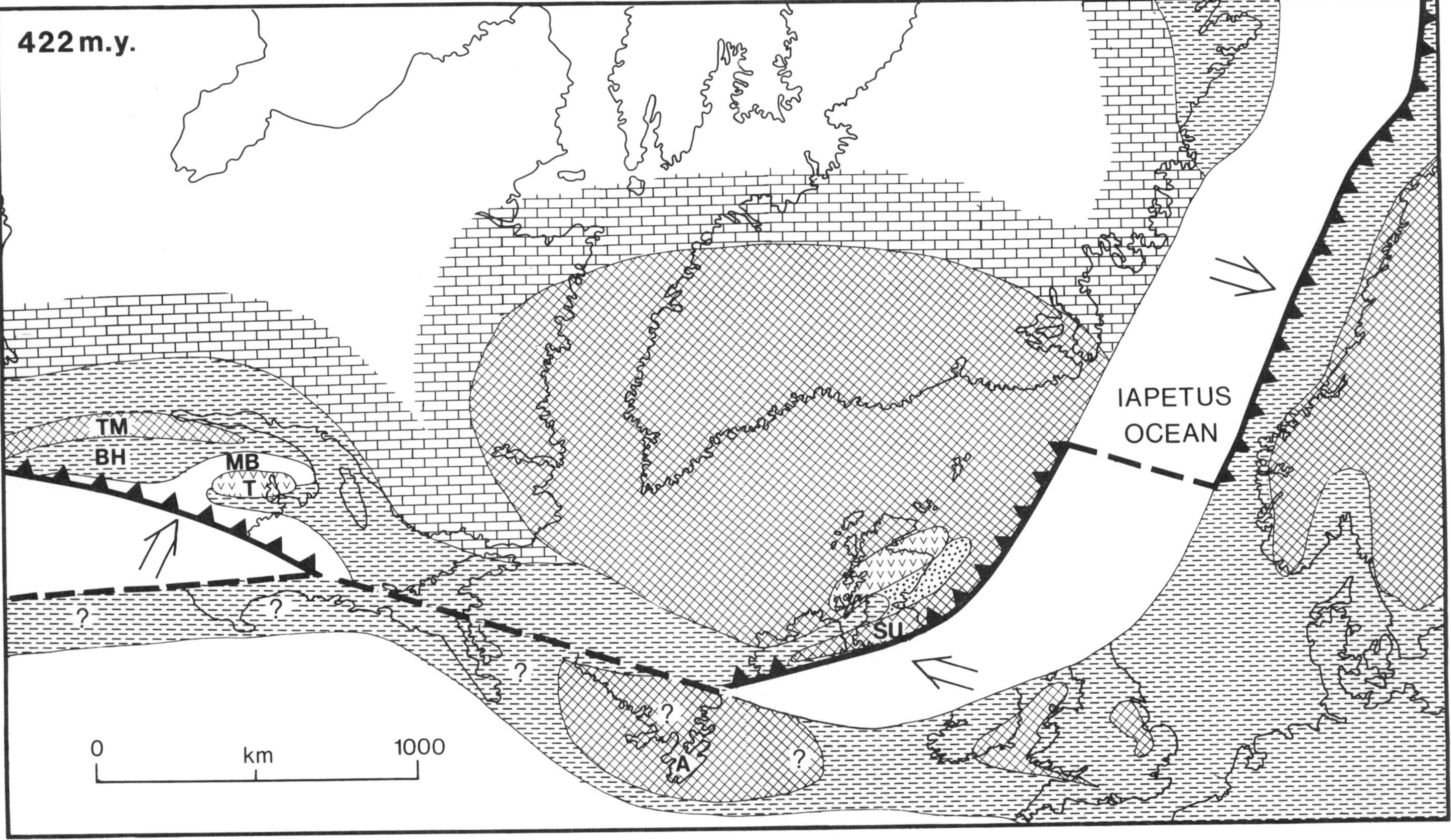

Figure 4 — Possible configuration of the Iapetus Ocean 422 m.y. ago (Wenlock). The orientation of the transform fault in Newfoundland indicates the direction of movement of the ocean subducting under the accretionary prism in the Southern Uplands (SU) of Scotland. A parallel transform is postulated to separate the ocean subducting under Norway. There may have been ocean floor south of the Newfoundland transform: Avalon (A) could have lain far south. The Matapedia Basin (MB) still persisted between the Bronson Hill (BH) and Tetagouche (T) terrains. TM = Taconic Mountains. Shading as in Figure 2.

ments. There is a probability that the whole region east of the Iapetus was situated as shown in Figure 4. However, similarity of geology does not necessarily mean that there was geographic proximity (similar events are occurring today on many widely separated continental margins); it could be that the paleomagnetic conclusions of Van der Voo, French, and French (1979) are correct and that the Avalon Platform lay far south of New England until the Late Devonian (hence the queries on Figure 4).

If transcurrent movement was present during the Late Ordovician and Silurian along the margin of North America in Newfoundland, the relative motion of the Iapetus plate to North America must be parallel with the suture. This motion fits with that surmised for the movement in Scotland and northwestern Ireland, where Phillips, Stillman and Murphy (1976) report strong dextral shear along the suture zone associated with final collision. In addition, we have evidence from the Southern Uplands accretionary prism (Leggett, 1980, Leggett and Casey, this volume) of very slow rates of subduction, so the probable positions of England and southeastern Ireland (Figure 4) relative to North America can be fixed from Late Ordovician times until the final collision in Early Devonian.

The situation in the northern Appalachians is much more uncertain. Not only is the position of Avalon doubtful (and its extension to Rhode Island), but there appears to have been a distinct small oceanic plate off New Brunswick. The presence of Silurian andesites in northern New Brunswick and southern Quebec (Poole, 1967; McKerrow and Ziegler, 1971) suggests northward subduction below this region, which is incompatible with the relative plate movements inferred for Britain and Newfoundland (Figure 4).

The southern part of Nova Scotia was perhaps a part of Africa throughout the Early Paleozoic (Schenk, 1978); it is omitted from Figures 2, 3, and 4.

CONCLUSIONS

Field geologists often plot data on 1 : 10,000 scale maps to learn more about the space distributions of rocks. From this they make deductions about structure and geological history. Similarly, there are several conclusions to be made from Figures 2, 3, and 4, even though the data are plotted on a very different scale. The maps are taken from conical projections so minimum distortion is involved.

As explained above, the island arcs in Figure 2 are plotted with reference to their post-collision position, their time of collision, and an assumed subduction rate of 2 cm/yr (the rate of subduction estimated for the Southern Uplands). Subduction was below the Bronson Hill, Tetagouche, and Lushs Bight island arc (the arrow south of the island arc in Figure 2 shows movement of the overriding plate with the island arc on its leading edge). Movement was probably within 20° of the direction shown. A movement toward the northwest is prohibited by the St. Lawrence Promontory; while northeastward movement results in three preserved parts of the island arc being situated *en echelon*, an improbable, though possible, arrangement.

In Britain, the Ordovician oceanic plate subducted northward below Scotland. This entails the presence of a transform fault between the ocean to the west (which was subducted southward) and the ocean to the east. The fault is shown on Figure 2, orientated parallel with the assumed direction of movement of the island arc in the west. In Arenig time, the South Mayo Trough was filled by a thick turbidite sequence, resting on pillowed basalts; it may have been a Gulf of California type marginal basin (Ryan and Archer, 1977), but its origin is not yet proved. In the Grampian Highlands (Figure 2) and their western extension through Ireland and possibly Newfoundland, the Dalradian rocks were uplifted and intruded by calc-alkaline rocks.

The accretionary prism of the Southern Uplands of Scotland commenced to develop 450 m.y. ago (late Caradoc). It shows no change in its development from this time until the Late Silurian, so the direction of Caradoc subduction (Figure 3) is shown as it was in the Wenlock (Figure 4). This implies a shift from the direction of the Arenig (Figure 2), perhaps associated with the collision of the island arc (Bronson Hill, Tetagouche and Lushs Bight). Again the evidence suggests that with subduction below Scotland but no subduction in the northern Appalachians, a transform fault must be present between Ireland and Newfoundland. This fault is shown in a possible position, with its orientation parallel with the Silurian orientation (Figure 4).

It is likely that Greenland collided with Norway in the Ludlow or Pridoli (420 to 411 million years) so these two continental margins are shown parallel (Figure 4). During the Silurian, subduction was eastward under Norway but northward under Scotland, so a transform fault is shown on Figure 4 parallel with the transform in Newfoundland. Rate of subduction under Scotland is known to have been very slow, but the rate below Norway may have been normal as collision took place between Greenland and Norway earlier (Late Silurian) than in Britain (Early Devonian).

In western Newfoundland there is no evidence of subduction, but many large contemporary faults are present. It is likely that the North American margin in this region was a transform fault. It is this fault which is used to determine the relative direction of movement between the northern part of the Iapetus Ocean and North America, as the directions fit with the dextral shear associated with collision in Britain and with the directions of thrusting in Greenland and Norway.

Although Avalon (and its extension along coastal New England) is shown on Figure 4, its exact position is very uncertain. Paleomagnetic evidence suggests that it may have been situated much further to the south.

The Newfoundland transform fault is shown (Figure 4) to extend southwestward offshore from New Brunswick. The presence of Late Silurian andesites in northern New Brunswick and Quebec indicate subduction below this area which is incompatible with motion along the Newfoundland transform, so perhaps a separate oceanic plate was involved.

Evidence from field mapping in the northern Appalachians and the European and Greenland Caledonides suggests that continental accretion occurred by parts of an island arc colliding in the Appalachians and by an oceanic island colliding in Scotland, as well as by the sedimentary accretionary prism of the Scottish Southern Uplands.

Sutures between these former oceanic islands and the North American continental margin are often marked by outcrops of ophiolites. By contrast, ophiolites are rare along the suture where the Iapetus Ocean finally closed in the Late Silurian and Early Devonian.

The varied nature of accretion, along the 5,000 km between Long Island Sound and northern Greenland, results in a variety of structures. Although the northern Appalachian-Caledonian orogen extends throughout this long distance, most of the collision and accretionary events are restricted to much shorter distances, each event being normally confined to only a few hundred kilometers of the continental margin. Thus, it is neither possible nor logical to recognize individual structural zones along the entire length of the orogen (Fyffe, 1977).

ACKNOWLEDGMENTS

I am grateful to R. J. Arnott, J. A. P. Clayburn, L. R. M. Cocks, R. A. Fortey, J. K. Leggett, R. J. Norris, C. Pudsey, S. W. Richardson, and M. Watson for discussing the various topics included in this paper, and to Gillian M. Collins for drafting the diagrams. Part of my field work was funded by a research grant from Natural Environment Research Council of the United Kingdom. The Geological Survey of Greenland financed and organized my field work in east North Greenland in 1980.

REFERENCES CITED

Ayrton, W. G., et al, 1969, Lower Llandovery of the northern Appalachians and adjacent regions: Geological Society of America Bulletin, v. 80, p. 459-484.

Bassett, M. G., ed., 1976, The Ordovician System: Cardiff, University of Wales Press and National Museum of Wales, 696 p.

Berry, W. B. N., 1976, Aspects of correlation of North American shelly and graptolitic faunas, *in* M. G. Bassett ed., The Ordovician System: Cardiff, University of Wales Press, p. 153-169.

Bird, J. M., and J. F. Dewey, 1971, Lithosphere plate-continental margin tectonics and the evolution of the Appalachian Orogen: Geological Society of America Bulletin, v. 81, p. 1031-1060.

Bowes, D. R., and B. E. Leake, ed., 1978, Crustal evolution in northwestern Britain and adjacent regions: Geological Journal Special Issue 10, 492 p.

Bluck, B. J., et al, 1980, Age and origin of Ballantrae ophiolite and its significance to the Caledonian orogeny and Ordovician time scale: Geology, v. 8, p. 492-495.

Cowie, J. W., 1971, The Cambrian of the North American Arctic regions, *in* C. H. Holland ed., The Cambrian of the new world: London, Interscience, p. 325-383.

———, A. W. A. Rushton, and C. J. Stubblefield, 1972, A correlation of the Cambrian rocks of the British Isles: Geological Society of London Special Report 2, 42p.

Currie, K. L., R. K. Pickerill, and G. E. Pajari, 1980, An early Paleozoic plate tectonic model of Newfoundland: Earth and Planetary Science Letters, v. 48, p. 8-14.

Dallmeyer, R. D., 1977, Diachronous ophiolite obduction in western Newfoundland, evidence from $^{40}Ar/^{39}Ar$ ages of the Hare Bay metamorphic auriole: American Journal of Science, v. 277, p. 61-72.

Dawes, P. R., 1971, Precambrian to Tertiary of northern Greenland, *in* A. Escher, and W. S. Watts, eds., Geology of Greenland: Grønlands Geologiske Undersøgelse, p. 248-303.

Dean, P. L., 1978, The volcanic stratigraphy and metallogeny of Notre Dame Bay, Newfoundland: Memorial University of Newfoundland, Geology Report 7, 205p.

———, and D. F. Strong, 1977, Folded thrust faults in Notre Dame Bay, Newfoundland: American Journal of Science, v. 277, p. 97-108.

Dewey, J. F., 1969, Evolution of the Appalachian/Caledonian orogen: Nature, v. 22, p. 124-9.

———, 1971, A model for the Lower Palaeozoic evolution of the southern margin of the early Caledonides of Scotland and Ireland: Scottish Journal of Geology, v. 7, p. 219-240.

Fyffe, L. R., 1977, Comparison of some tectonostratigraphic zones in the Appalachians of Newfoundland and New Brunswick: Discussion: Canadian Journal of Earth Science, v. 14, p. 1468-1469.

Gee, D. G., 1975, A tectonic model for the central part of the Scandinavian Caledonides: American Journal of Science, v. 275-A, p. 468-515.

———, 1978, The Swedish Caledonides — a short synthesis, *in* E. T. Tozer and P. E. Schenk, eds., IGCP Project 27, Caledonian-Appalachian Orogen of the North Atlantic region: Geological Survey of Canada Paper, 78-13, p. 63-72.

———, and M. R. Wilson, 1974, The age of orogenic deformation in the Swedish Caledonides: American Journal of Science, v. 274, p. 1-9.

Haller, J., 1971, Geology of the East Greenland Caledonides: Interscience, London, 413p.

Harland, W. B., and R. A. Gayer, 1972, The Arctic Caledonides and earlier oceans: Geological Magazine, v. 109, p. 289-314.

Harris, A. L., C. H. Holland, and B. E. Leake, eds., 1979, The Caledonides of the British Isles — reviewed: Edinburgh, Scottish Academic Press, 768 p.

———, et al, 1978, Ensialic basin sedimentation: the Dalradian Supergroup, *in* D. R. Bowes, and B. E. Leake, eds., Crustal evolution in northwestern Britain and adjacent regions: Geological Journal Special Issue 10, p. 115-138.

Horne, G. S. and J. Helwig, 1969, Ordovician stratigraphy of Notre Dame Bay, Newfoundland, *in* M. Kay, ed., North Atlantic — Geology and continental drift: AAPG Memoir 12, p. 388-407.

House, M. R., et al, 1977, A correlation of Devonian rocks of the British Isles: Geological Society of London, Special Report 7, 110 p.

Hurst, J. M., and W. S. McKerrow, 1981, The Caledonian nappes of eastern North Greenland: Nature, v. 290, n. 5809, p. 772-774.

Hussey, A. M., 1968, Stratigraphy and structure of southwestern Maine, *in* E-An Zen et al., eds., Studies of Appalachian geology — northern and maritime: New York,

Interscience Publications, p. 291-301.
Rast, N., 1980, The Avalonian plate in the northern Appalachians and Caledonides, *in* Wones, D. R., ed., The Caledonides in the U.S., Virginia Polytechnic Institute and State University, Department of Geological Sciences, Memoir 2, p. 63-66.
———, M. J. Kennedy, and R. F. Blackwood, 1976, Comparison of some tectonostratigraphic zones in the Appalachians of Newfoundland and New Brunswick: Canadian Journal of Earth Sciences, v. 13, p. 868-875.
Roberts, D., and G. H. Gale, 1978, The Caledonian-Appalachian Iapetus Ocean, *in* D. H. Tarling, ed., Evolution of the earth's crust: London, Academic Press, p. 255-342.
Robinson, P., and L. M. Hall, 1980, Tectonic synthesis of southern New England, *in* D. R. Wones, ed., The Caledonides in the USA: Virginia Polytechnic Institute and State University, Department of Geological Sciences, Memoir 2, p. 73-82.
Rodgers, J., 1968, The eastern edge of the North American continent during the Cambrian and Early Ordovician, *in* E-An Zen et al, eds., Studies of Appalachian geology: northern and maritime: New York Interscience Publications, p. 141-149.
———, and E. R. W. Neale, 1963, Possible "Taconic" klippen in western Newfoundland: Geological Association of Canada Proceedings, v. 16, p. 83-94.
Ross, R. J., 1976, Ordovician sedimentation in the western United States, *in* M. G. Bassett, ed., The Ordovician System: Cardiff, University of Wales Press and National Museum of Wales, p. 73-105.
Rushton, A. W. A., and W. E. A. Phillips, 1973, A *Protospongia* from the Dalradian of Clare Island, County Mayo, Ireland: Paleontology, v. 16, p. 231-237.
Ryan, P .D., and J. B. Archer, 1977, The south Mayo trough: a possible Ordovician Gulf of California-type marginal basin in the west of Ireland: Canadian Journal of Earth Sciences, v. 14, p. 2453-2461.
———, P. A. Floyd, and J. B. Archer, 1980, The stratigraphy and petrochemistry of the Lough Nafooey Group (Tremadocian), western Ireland: Journal of the Geological Society of London, v. 137, p. 443-458.
St. Julien, P., and C. Hubert, 1975, Evolution of the Taconic Orogen in the Quebec Appalachians: American Journal of Science, v. 275A, p. 337-362.
Schenk, P. E., 1971, Southeastern Atlantic Canada, northwestern Africa and continental drift: Canadian Journal of Earth Sciences, v. 8, p. 1218-1251.
———, 1978, Synthesis of the Canadian Appalachians, *in* E. T. Tozer and P. E. Schenk, eds., IGCP Project 27, Caledonian-Appalachian Orogen of the North Atlantic region: Geological Survey of Canada Paper 78-13, p. 111-136.
Skinner, R., 1974, Geology of Tetagouche Lakes, Bathhurst and Nepisiguit Falls map areas, New Brunswick: Geological Survey of Canada Memoir 371, 133p.
Steiger, R. H., and E. Jäger, 1977, Subcommission on geochronology: convention on the use of decay-constants in geo- and cosmochronology: Earth and Planetary Science Letters, v. 36, p. 359-362.
Thirwall, M. F., 1981, Plate tectonic implications of chemical data from volcanic rocks of the British Old Red Sandstone: Journal of the Geological Society of London, v. 138, p. 123-138.
Tozer, E. T., and P. E. Schenk, eds., 1978, IGCP Project 27, Caledonian-Appalachian Orogen of the North Atlantic region: Geological Survey of Canada, Paper 78-13, 242 p.
van Breemen, O., M. Aftalion, and M. R. W. Johnson, 1979, Age of the Loch Borrolan complex, Assynt, and late movements along the Moine Thrust Zone: Journal of the Geological Society of London, v. 136, p. 489-495.
van der Voo, R., A. N. French, and R. B. French, 1979, A paleomagnetic pole position from the folded Upper Devonian Catskill redbeds, and its tectonic implications: Geology, v. 7, p. 345-348.
Wilkinson, J. M., and J. R. Cann, 1974, Trace elements and tectonic relationships of basaltic rocks in the Ballantrae igneous complex, Ayrshire: Geological Magazine, v. 111, p. 35-41.
Williams, A., 1962, The Barr and Lower Ardmillan Series (Caradoc) of the Girvan district, southwest Ayrshire, with descriptions of the brachiopods: Geological Society of London Memoir 3, 267 p.
———, 1976, Plate tectonics and biofacies evolution as factors in Ordovician correlation, *in* M. G. Bassett, ed., The Ordovician System: Cardiff, University of Wales Press and National Museum of Wales, p. 29-66.
Williams, H., 1977, Ophiolitic melange and its significance in the Fleur de Lys supergroup, northern Appalachians: Canadian Journal of Earth Sciences, v. 14, p. 987-1003.
———, 1979, The Appalachian Orogen in Canada: Canadian Journal of Earth Sciences, v. 16, p. 792-807.
———, and R. K. Stevens, 1974, The ancient continental margin of eastern North America, *in* C. A. Burke and C. L. Drake, eds., The geology of continental margins: New York, Springer-Verlag, p. 781-796.
———, and J. G. Payne, 1975, The Twillingate Granite and nearby volcanic groups: an island arc complex in northeastern Newfoundland: Canadian Journal of Earth Sciences, v. 12, p. 982-995.
———, and P. St. Julien, 1978, The Baie Verte-Brompton line in Newfoundland and regional correlations in the Canadian Appalachians: Geological Survey of Canada, Paper 78-1A, p. 225-229.
———, and M. D. Max, 1980, Zonal subdivision and regional correlation in the Appalachian Orogen, *in* D. R. Wones, ed., The Caledonides in the USA; Virginia Polytechnic Institute and State University, Department of Geological Sciences, Memoir 2, p. 57-62.
Wilson, J. T., 1966, Did the Atlantic close and then re-open?: Nature, v. 211, p. 676-681.
Wones, D. R., ed., 1980, The Caledonides in the USA: Virginia Polytechnic Institute and State University, Department of Geological Sciences, Memoir 2, 329 p.
Zen, E-An, 1968, Nature of the Ordovician orogeny in the Taconic area, *in* E-An Zen et al, eds., Studies in Appalachian geology: northern and maritime: New York, Interscience Publications, p. 129-139.
———, et al, eds., 1968, Studies of Appalachian geology: northern and maritime: New York, Interscience Publications, 475 p.
Ingham, J. K., 1978, Geology of a continental margin 2: middle and late Ordovician transgression, Girvan, *in* D. R. Bowes and B. E. Leake, eds., Crustal evolution in northwestern Britain and adjacent regions: Geological Journal Special Issue 10, p. 163-176.
Kay, M., ed., 1969, North Atlantic — geology and continental drift, a symposium: AAPG Memoir 12, 1082 p.
Kennedy, M. J., 1975, Repetitive orogeny in the northeastern Appalachians — new plate models based on Newfoundland examples: Tectonophysics, v. 28, p. 39-87.
Kent, D. V., and N. D. Opdyke, 1978, Paleomagnetism of the Devonian Catskill Redbeds: evidence for motion of the coastal New England-Canadian Maritime region relative to cratonic North America: Journal of Geophysical Research, v. 83, p. 4441-4450.
Lambert, R. St. J. and W. S. McKerrow, 1977, The Grampian

Orogeny: Scottish Journal of Geology, v. 12, p. 271-292.

Leggett, J. K., 1980, The sedimentological evolution of a Lower Palaeozoic accretionary forearc in the Southern Uplands of Scotland: Sedimentology, v. 27, p. 401-417.

———, W. S. McKerrow, and R. H. Eales, 1979, The Southern Uplands of Scotland: a lower Paleozoic accretionary prism, Journal of the Geological Society of London, v. 136, p. 755-770.

———, ———, and D. M. Casey, 1981, The anatomy of Lower Paleozoic accretionary forearc complex — the Southern Uplands of Scotland, *in* J. K. Leggett, ed., Trench Forearc Geology: Geological Society of London, Special Publication, 10, p. 495-520.

———, et al, 1979, The northwestern margin of the Iapetus Ocean, *in* A. L. Harris, C. H. Holland, and B. E. Leake, eds., The Caledonides of the British Isles — reviewed: Edinburgh, Scottish Academy Press, p. 499-512.

Longman, C. D., B. J. Bluck, and O. van Breeman, 1979, Ordovician conglomerates and the evolution of the Midland Valley: Nature, v. 280, p. 578-581.

Mattinson, J. M., 1977, U-Pb ages of some crystalline rocks from the Burlington Peninsula, Newfoundland, and implications for the age of the Fleur de Lys metamorphism: Canadian Journal of Earth Sciences v. 14, p. 2316-2324.

McKerrow, W. S., and A. M. Ziegler, 1971, The Lower Silurian paleogeography of New Brunswick and adjacent areas: Journal of Geology, v. 79, p. 635-646.

———, and ———, 1972, Paleozoic oceans: Nature — Physical Science, v. 240, p. 92-94.

———, and L. R. M. Cocks, 1976, Progressive faunal migration across the Iapetus Ocean: Nature, v. 263, p. 304-6.

———, and ———, 1977, The location of the Iapetus suture in Newfoundland: Canadian Journal of Earth Sciences, v. 14, p. 488-495.

———, and ———, 1978, A lower Paleozoic trench-fill sequence, New World Island, Newfoundland: Geological Society of America Bulletin, v. 89, p. 1121-1132.

———, and ———, 1981, Stratigraphy of eastern Bay of Exploits, Newfoundland: Canadian Journal of Earth Sciences, v. 18, p. 751-764.

———, R. St. J. Lambert, and V. E. Chamberlain, 1980, The Ordovician, Silurian and Devonian time scales: Earth and Planetary Science Letters, v. 51, p. 1-8.

Mykkeltveit, S., E. S. Husebye, and C. Oftedahl, 1980, Subduction of the Iapetus Ocean crust beneath the Møre Gneiss Region, southern Norway: Nature, v. 288, p. 473-475.

Nelson, K. D., and J. F. Casey, 1979, Ophiolitic detritus in the Upper Ordovician flysch of Notre Dame Bay and its bearing on the tectonic evolution of western Newfoundland: Geology, v. 7, p. 27-31.

Neuman, R. B., 1968, Paleogeographic implications of Ordovician shelly fossils in the Magog Belt of the northern Appalachian region, *in* E-An Zen et al, ed., Studies of Appalachian geology: northern and maritime: New York, Interscience Publications, p. 35-48.

———, 1972, Brachiopods of Early Ordovician volcanic islands: Montreal, Proceedings, 24th International Geological Congress, v. 7, p. 297-303.

———, 1976, Early Ordovician (late Arenig) brachiopods from Virgin Arm, New World Island, Newfoundland: Bulletin of the Geological Survey of Canada, v. 261, p. 11-61.

Osberg, P. H., 1978, Synthesis of the geology of the northeastern Appalachians, USA, *in* E. T. Tozer and P. E. Schenk, eds., IGCP Project 27, Caledonian-Appalachian Orogen of the North Atlantic region: Geological Survey of Canada Paper 78-13, p. 137-147.

Pankhurst, R. J., 1970, The geochronology of the basic igneous complexes: Scottish Journal of Geology, v. 6, p. 83-107.

———, and R. T. Pidgeon, 1976, Inherited isotope systems and the source region pre-history of early Caledonian Granites in the Dalradian Series of Scotland: Earth and Planetary Science Letters, v. 31, p. 55-68.

Phillips, W. E. A., M. J. Kennedy, and G. M. Dunlop, 1969, Geologic comparison of western Ireland and northeastern Newfoundland, *in* M. Kay, ed., North Atlantic — geology and continental drift: AAPG Memoir 12, p. 194-211.

Phillips, W. E. A., C. J. Stillman, and T. Murphy, 1976, A Caledonian plate tectonic model: Journal of the Geological Society of London, v. 132, p. 579-609.

Pidgeon, R. T., and M. R. W. Johnson, 1974, A comparison of zircon U-Pb and whole rock Rb-Sr systems in three phases of the Carn Chuinneag granite, northern Scotland: Earth and Planetary Science Letters, v. 24, p. 105-112.

———, and M. Aftalion, 1978, Cogenetic and inherited zircon U-Pb systems in granites: Paleozoic granites of Scotland and England, *in* D. R. Bowes and B. E. Leake, eds., Crustal evolution in northwestern Britain and adjacent regions: Geological Journal Special Issue 10, p. 183-248.

Poole, W. H., 1967, Tectonic evolution of the Appalachian region of Canada: Geological Association of Canada Special Paper 4, p. 9-51.

Pringle, I. R., 1978, Rb-Sr ages of silicic igneous rocks and deformation Burlington Peninsula, Newfoundland: Canadian Journal of Earth Sciences, v. 15, p. 293-300.

Section 2: Model Investigations of Margin Environmental and Tectonic processes

Modeling goes hand-in-hand with data collection. Data constrain the models while models identify gaps in the data and suggest lines of investigation. In the earth sciences, as in other sciences, we base conclusions upon models whether we are trying to spot an oil well or determine the tectonics of a large region.

In this section, we deal with two types of models: environmental and tectonic. The first of these categories is concerned with environmental factors; the nature of transport and deposition of the sedimentary bodies on continental margins and characteristics of their geochemistry. These models are process-oriented and, in the tradition of Lyell, lead us to a better understanding of the effects of these processes in the past and their relevance to hydrocarbon accumulation. The second category of models is also process-oriented, but concerns the thermal and tectonic processes that have brought the margins to their present state. The idea that passive margins are rifted and that accretionary margins are compressional has been around a long time and is not new. What is new is a body of data sufficient to expand the idea into more sophisticated models that more closely approximate the real earth. By the same token, these models indicate to us that the data base in the transition zone from continent to ocean is not yet sufficient to be certain of the exact nature of the processes shaping the margins.

While the papers in this section are too diverse to mention individually, we would like to call attention to the group of papers about tectonic processes. The last few years have seen significant advances in our understanding of the tectonics of margins, largely due to a widespread effort at mathematically modeling thermal history and subsidence following rifting. Not surprisingly, the majority of the papers in the tectonic processes group report results of such modeling. We are pleased that so many investigators in this field submitted papers to the volume and that we can offer this up-to-date summary of recent thinking about passive (and not-so-passive) margins.

Environmental Processes

Offshore Brazil — Twelve Years of Oil Exploration

Giuseppe Bacoccoli
Basins Interpretation Division, Exploration Department
Petrobras
Brazil

By the end of 1979, a total of 307,929 km of seismic lines, 309,709 km of aeromagnetic profiles, and 94,067 km of gravity profiles had been surveyed offshore Brazil for oil exploration purposes. Drilling 452 exploratory wells resulted in a success ratio of 1:5.6. Recoverable reserves of 116.2 x 10^6 cu m of oil and 20.9 × 10^9 cu m of gas were discovered.

At present, the following aspects related to oil in offshore Brazil can be highlighted: oil fields have been discovered in all basins from Ceara to Campos; important source rocks occur in lacustrine shales of the rift stage; despite the presence of deep Tertiary depocenters, significant discoveries related to deltaic sediments are lacking; primary porosity in Cretaceous sandstones decreases sharply with depth; nevertheless, in some basins, development of secondary porosity is inferred at depths of about 4,000 m; a combination of structural, stratigraphic and paleogeomorphic factors control the accumulation in several oil fields; and, oil exploration offshore Brazil has proved economically attractive, despite the relatively small size of the oil fields as compared to the ones found in highly prolific oil provinces.

THE REGIONAL SETTING

The Brazilian continental shelf is divided into six tracts (Figure 1): (1) The Amazonas tract comprises the Cretaceous Cassipore Graben (a Tertiary limestone shelf) and the terrigenous Amazonas cone, as well as the Para-Maranhao and offshore Barreirinhas basins. (2) The northern tract includes the two important producing basins, Ceara and Potiguar, plus Piaui basin. (3) The northeastern tract contains the apparently shallow Paraiba-Pernambuco basin and the Sergipe-Alagoas and North Bahia offshore basins. (4) The eastern tract comprises several small basins (Camamu, Almada, Jequitinhonha, and Cumuruxatiba) together with the large Espirito Santo basin and the shallow volcanic rocks of the Abrolhos Bank. (5) The Campos tract, which is bounded to the north by the Vitoria arch and to the south by the Cabo Frio arch, is the most important offshore oil province yet encountered. (6) The Southern tract includes Santos and Pelotas basins. Despite extensive exploration in these two basins, no discoveries have been made either by PETROBRAS or international exploration companies under risk clause contract.

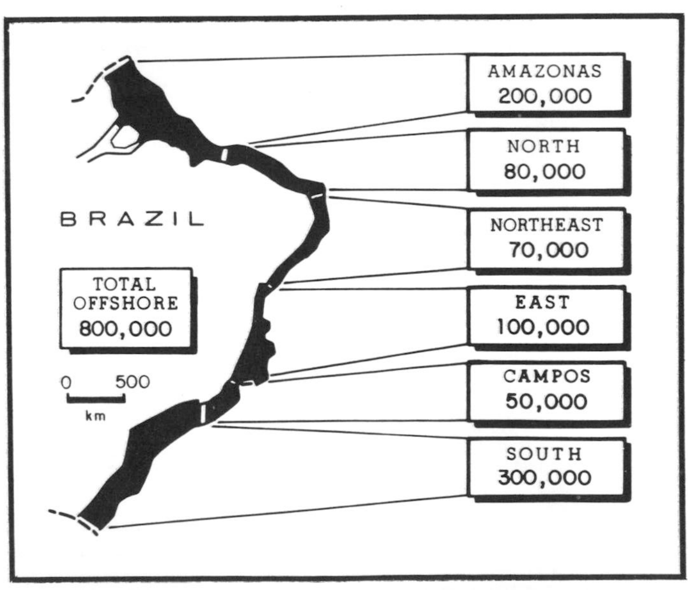

Figure 1 — The Brazilian offshore can be divided into six tracts with relatively distinct geological settings. Numbers indicate offshore areas out to the 200 m isobath. Area in square kilometers.

EXPLORATION ACTIVITY

By the end of 1979 PETROBRAS and other exploration companies had surveyed a total of 307,929 km of seismic profiles, 309,709 km of aeromagnetic profiles, and 94,067 km of gravity profiles offshore Brazil. Table 1 shows the distribution of PETROBRAS seismic surveys for oil exploration during the last twelve years per each offshore tract.

In addition, outside exploration companies have collected 43,842 Km of seismic data and drilled 26 wells.

Seismic data have been acquired using various seismic sources such as aquapulse, air-gun, vapor-choc, and various system designs. Record quality generally ranges from fair to good; local poor-quality record is mainly related to subcrops of thick, high velocity carbonate and basalt layers or complex structural configuration within the faulted rift stage sequence. High resolution seismic was shot for relatively shallow targets, and whenever possible, the acoustic impedance technique was applied to help defining oil accumulation boundaries in stratigraphic sandstone and carbonate traps. Three-dimensional seismic surveys were also shot for the same purpose.

Table 1. Distribution of PETROBRAS seismic surveys for oil exploration during the last 12 years (per each offshore tract).

Offshore Tract	Seismic Survey (km)	Exploratory Wells
Amazonas	69,429	47
North	45,015	70
Northeast	25,067	104
East	49,121	92
Campos	34,525	127
South	40,930	12
Total	264,087	452

From 1968 to the end of 1979, 618 wells (all categories) were drilled and from 1976 to 1979, international companies drilled 26 wildcats in the Amazonas and south tracts. Concentration of wells relates closely to oil discoveries. The producing Campos and northeastern basins contain more than 100 exploratory wells each. Fewer wells were drilled in the north coast, which is also productive. Despite the small oil accumulations discovered to date in the east coast, 92 wells were drilled in the area. The south coast (no discoveries) and the Amazonas (one small gas field discovery) are the less drilled tracts, with 12 and 47 exploratory wells respectively.

EXPLORATION RESULTS

Of the total 452 exploratory wells drilled by PETROBRAS, 81 are classified as oil producers and 9 as gas producers. The calculated oil success ratio is 1:5.6. Of the 354 wildcats, 44 are classified as oil producers and 8 as gas producers with an oil success ratio of about 1:8.0.

Most of the recoverable reserves of oil are in Campos basin (78.3%). The north and the northeast coast each contain about 10% of the total reserves, while the east coast contributes with less than 1%. No oil discoveries have been made in the Amazonas and south coasts, as shown in Figure 2. The situation is similar with respect to gas discoveries; 54.6% of the gas reserves are in Campos basin and the northeastern coast contributes with 28.1%. At the end of 1979, cumulative offshore Brazil oil production amounted to 10.1 $\times 10^6$ cu m with proven reserves of 106.1 $\times 10^6$ cu m. Geologists estimate the ultimate volume of oil already discovered offshore Brazil may be twice the present proved reserves or about 1.5 billion barrels.

Figure 3 shows average recoverable reserves of oil

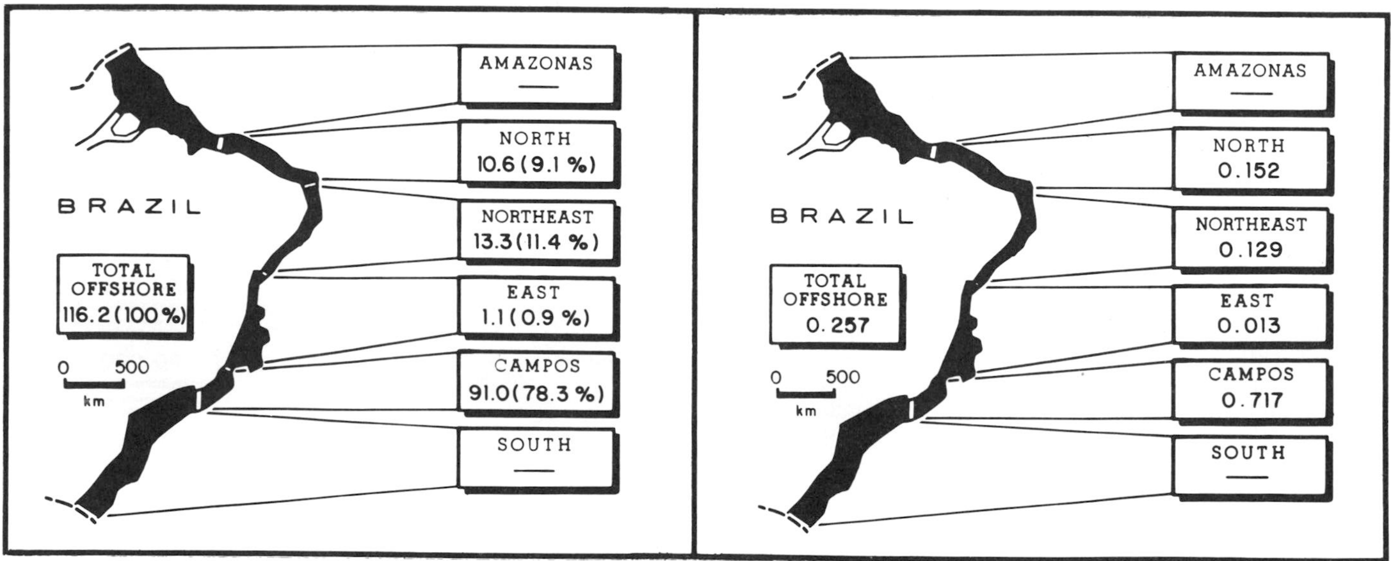

Figure 2 — Distribution of proven reserves of oil in millions of cubic meters per tract offshore Brazil, as of December 1979.

Figure 3 — Average proven reserves of oil discovered by each PETROBRAS exploratory well offshore Brazil, as of December 1979. Numbers are in millions of cubic meters.

discovered by each PETROBRAS exploratory well. The average of the total offshore Brazil recoverable reserves is 0.257 × 10^6 cu m (1.6 million barrels) of oil per well. Offshore Brazil averages 83 cu m (522 barrels) per meter drilled while the average in Campos basin is 213.6 cu m (1,343 barrels) per meter.

Thus, despite the relatively small volumes of oil found offshore Brazil, exploration is economically attractive given present oil prices in the international market. Including geophysical surveys and other indirect costs, PETROBRAS estimates a cost of less than $5.00 (U.S.) to locate a new recoverable barrel of oil in the continental shelf.

TWELVE YEARS OF OIL EXPLORATION

In the early 1960s it was considered that the continental shelf could offer adequate conditions for oil prospecting. At that time, PETROBRAS was intensively involved in onshore exploration of a country with more than 3 × 10^6 sq. km of land basin area. In addition unsolved technological problems inhibited exploration in Brazil's ocean areas. Nevertheless, seismic surveys were carried out offshore Sergipe/Alagoas, Bahia, Espirito Santo, and Campos in early 1960s. Data quality was very poor, but information was obtained concerning the discovery of salt domes offshore Espirito Santo. In the late 1960s, all the northeast, east, and Campos basins were surveyed by gravity out to the 50 m isobath. In 1968, after obtaining good quality seismic data, the first well was drilled offshore Espirito Santo. The next year, the first oil field, Guaricema, was discovered offshore Sergipe.

During the last 12 years, exploration activity moved continuously toward deeper waters and deeper geological targets. The first wildcat deeper than 5,000 m was drilled in 1975 (Figure 4). Although the curve of average final depth is influenced by the great number of shallow wildcats drilled nearshore in some basins, from 1976 on the tendency is one of increasing average depths.

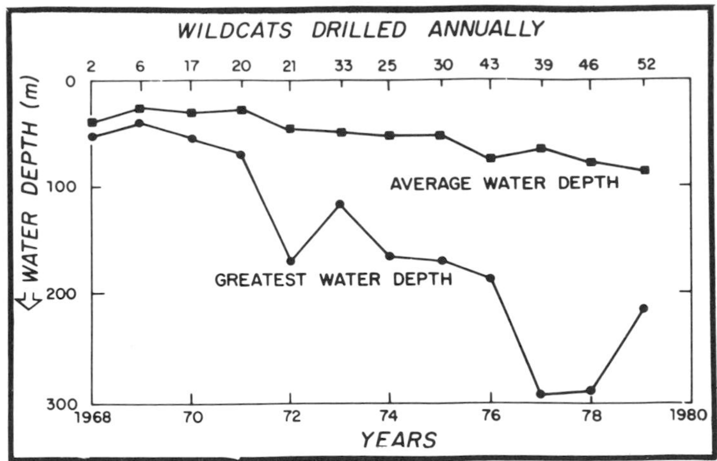

Figure 5 — Water depths for wildcats drilled by PETROBRAS offshore Brazil.

The average wildcat water depth remained below 50 m until 1975, but it is now near 100 m. Discoveries made since 1974 in the Campos basin in water depths greater than 100 m contributed significantly to the increased average depth. A few wildcats have been drilled in nearly 300 m of water, and at least three have discovered oil pools in these depths (Campos basin). Practically all important oil fields discovered so far in Campos basin, including Pampo, Enchova, Garoupa, Cherne, and Namorado fields, are at greater than 100 m deep. Some of these fields (East-Enchova and South-Namorado) are in 200 m water depths.

During 1980, PETROBRAS operated 27 mobile drilling rigs offshore Brazil; including 14 jackups and 13 semi-

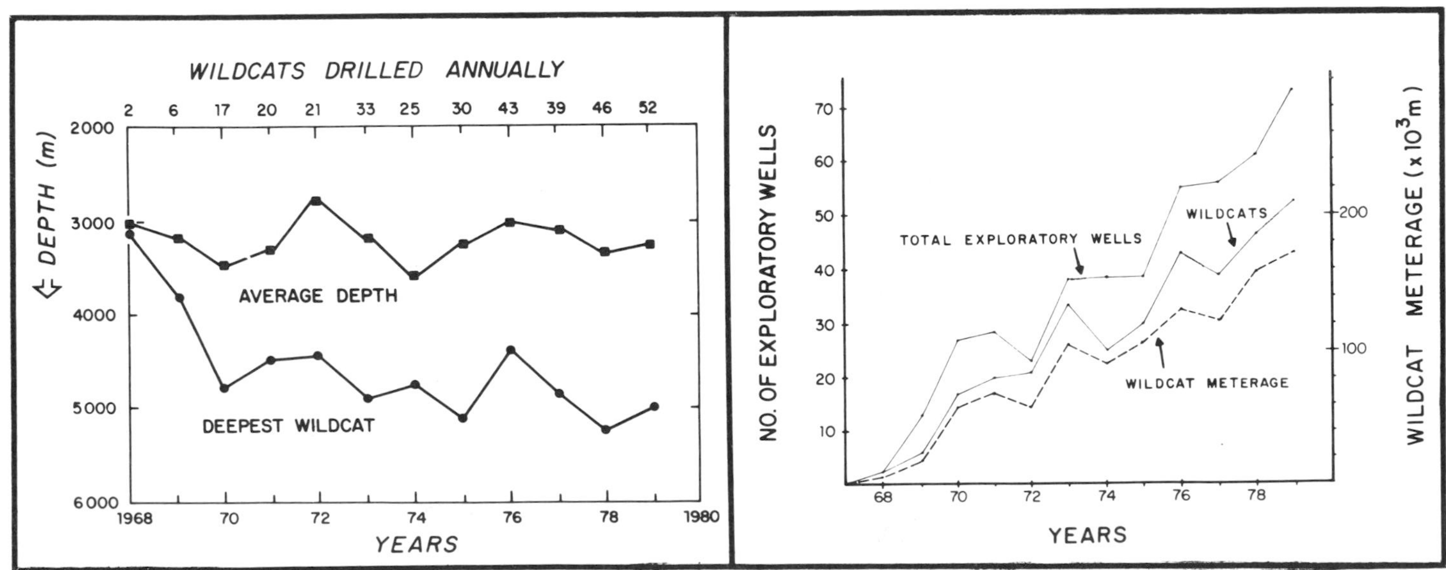

Figure 4 — Depth of wildcats drilled by PETROBRAS offshore Brazil.

Figure 6 — Yearly exploratory well activity by PETROBRAS offshore Brazil.

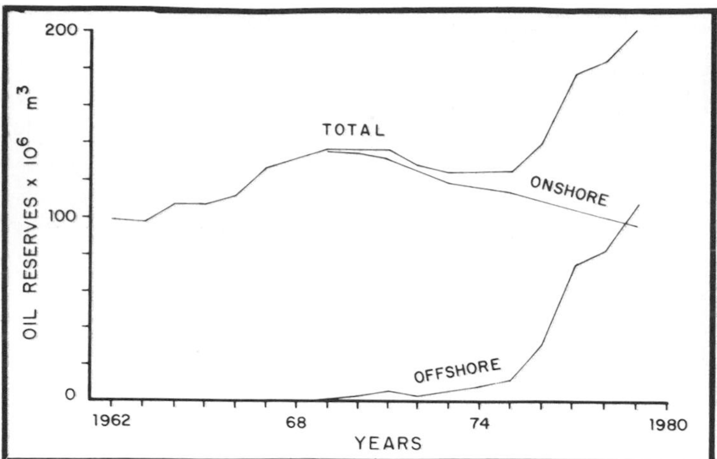

Figure 7 — Brazilian recoverable oil reserves from 1962-1979.

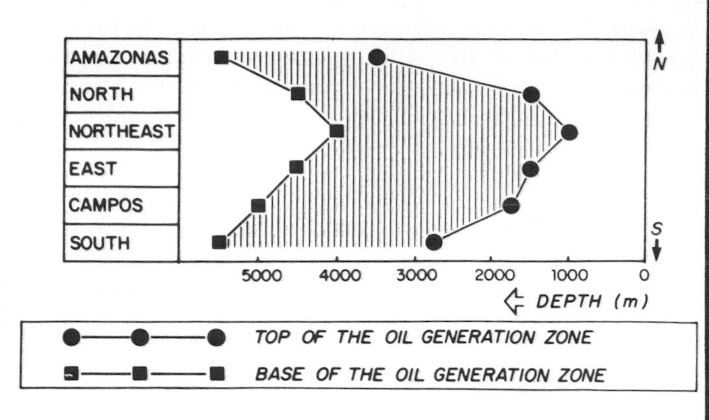

Figure 8 — Schematic representation of the oil generation window along the continental shelf of Brazil.

submersibles and/or drilling vessels. Exploration activity offshore Brazil has continuously increased over the last 12 years (Figure 6). Important discoveries include:

(1) 1969 — the discovery of Guaricema oil field offshore Sergipe;

(2) 1973 — the discovery of Ubarana oil field, offshore Rio Grande do Norte, in the north coast;

(3) 1975 — the discoveries of Garoupa and Namorado oil fields in the Campos basin; and,

(4) 1976 — the discovery of Xareu, offshore Ceara, in the north coast.

Brazilian offshore oil reserves have continuously increased, as shown on Figure 7. For the first time, after substantial oil reserves in the Campos basin had been proved, oil reserves offshore Brazil were larger than onshore reserves.

THE GEOLOGIC SETTING

Exploration offshore Brazil was first concentrated near the mouths of important rivers and in Tertiary depocenters, searching for ancient deltas and major oil provinces like the Niger. However, the energy of the depositional environment offshore Brazil appears inadequate for the development of important deltas.

At the end of 1979, the distribution of the proven reserves of oil with respect to the age of sediments was as follows:

Tertiary and Late Cretaceous — 17.06×10^6 cu m (14.6%); Cenomanian/Albian — 80.515×10^6 cu m (69.2%); Aptian/pre-Aptian — 18.701×10^6 cu m (16.0%).

Concentration in the Cenomanian/Albian is due to Campos basin reservoirs of Garoupa, Pampo, Cherne, and Namorado. If this distribution is representative for offshore Brazil, younger sequences have a low oil potential with less than 15% of the total reserves. Gas reserves were as follows:

Tertiary and Late Cretaceous — $4,588.7 \times 10^6$ m^3 (21.9%); Cenomanian/Albian — $8,879.2 \times 10^6$ m^3 (42.4%); Aptian/pre-Aptian — $7,458.0 \times 10^6$ m^3 (35.6%); total — $20,926.0 \times 10^6$ m^3 (100%).

Considering both oil and gas, and considering that practically all exploratory wells have drilled the pre-mid Cretaceous sequence, and that only a few have reached older rocks, Early Cretaceous rocks have a better potential than Tertiary-Late Cretaceous rocks.

Figure 8 compares the depth of the oil generation window with location along the Brazilian coast. The oil window is deep-seated in the Amazonas and south coast, but becomes more shallow to the northeast. Early Cretaceous rocks are generally mature and are presently found in the north and northeast tracts. The deep top of the oil generation window may be due to the thick Late Tertiary sequence, but in the south coast the Tertiary is thinner and the oil generation zone is still deep. We believe that the thermal history of the Brazilian continental shelf is related to the separation of the African and South American plates with stable, "cold" conditions of passive margin first reached in the Amazonas and south coast.

The distribution of proven oil reserves with depth is asymetric (Figure 9). Largest reserves (more than 40×10^6 cu m, 250 million barrels) are located in the 3,000 to 3,500 m depth range. Observed depth ranges are due in part to the Namorado oil field in Campos basin, and in part to several other factors. First, the top of the oil generation zone is relatively deep. Second, younger sequences have low potential because of unfavorable geological settings (geochemical, stratigraphic, and structural). Third, older sequences, with the exception of the well-structured Tertiary sequence in the Amazonas Cone, exhibit better structure. And fourth, the older rift-stage sequence coincides with a worldwide maxima in source rock deposition.

PETROBRAS has researched the diagenesis of terrigenous rocks to better understand porosity development in deeply buried sandstones. Lower Cretaceous sandstone reservoirs are subject to almost complete obliteration of porosity when buried beneath more than 3,000 m of cover. At greater depths, diagenesis can create secondary porosity which significantly improves reservoir properties. Figure 10 shows a porosity versus depth graph of a typical offshore Brazil well in an Early Cretaceous continental terrigenous sequence. Here, porosity drops below 10% at about

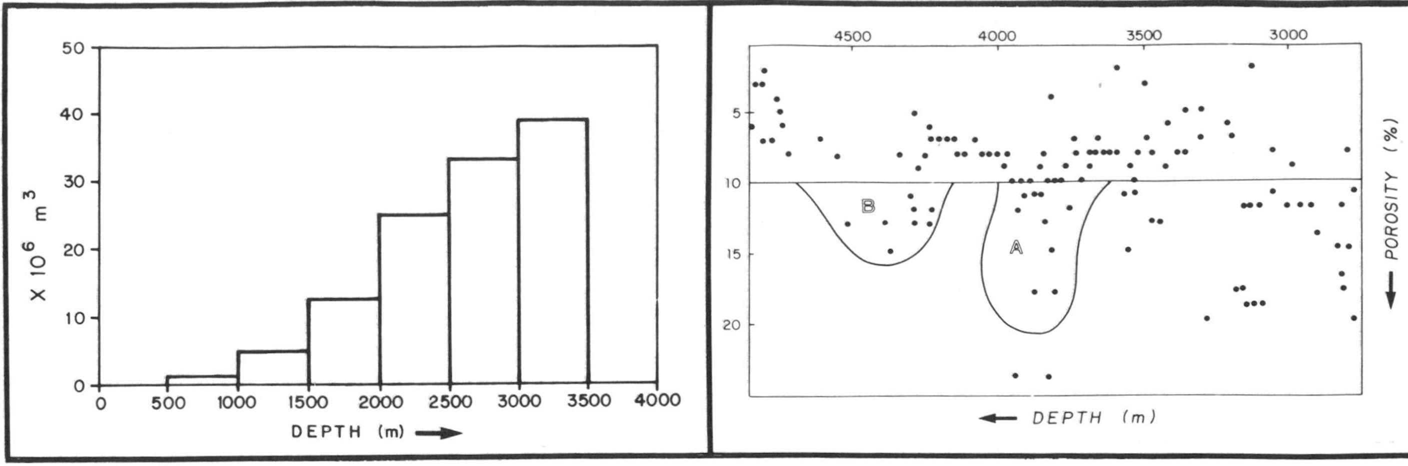

Figure 9 — Depth range of the recoverable reserves of oil discovered offshore Brazil from 1968-1979.

Figure 10 — Sandstone porosity versus depth in a well offshore Brazil. Based on density log data.

3,200 m, but two deeper zones (A and B) show better porosity because of secondary porosity due to diagenesis. However, we lack predictive models establishing the location and depth of secondary porosity zones. Early reservoir oil migration, before the diagenetic porosity loss, may prevent reservoir obliteration due to oil which would avoid circulation of subsurface waters.

Carbonate rocks are the other important potential reservoir rock-type offshore Brazil. Generally speaking, carbonate reservoirs are found in several environments. They occur in Tertiary, open-marine, limestone shelves rich in organic matter and with reefs. They also occur in the Cenomanian/Albian marine carbonate shelf, poor in organic content but with oolitic-pisolitic banks and/or secondary porosity. Finally, they occur as pre-Aptian continental lacustrine carbonates developed near paleo-highs of the rift stage. Oil fields in onshore and offshore carbonate rocks have been found in all of these environments except in the Tertiary.

Several authors, Asmus and Ponte (1973), Asmus and Porto (1972), Ponte, Dauzacker, and Porto (1978), have published papers describing the main sedimentary sequences offshore Brazil and comparing the west coast of Africa with the origin of these sequences regarding plate tectonics. For this reason, we present (Figure 11) a schematic geologic section of all facies offshore Brazil based on a typical configuration of the east coast or Campos basin. The three main sequences are a pre-Aptian continental rift-stage sequence, an Aptian to Albian transitional environment sequence, and a Late Cretaceous to Tertiary open-marine sequence.

Schematic oil traps are shown in black in the section. In the lower sequence, faulted blocks may trap oil in sandstone when contacting organically rich lacustrine shales. The Caioba field, offshore Sergipe in the northeastern coast, is an example of such a trap. Sometimes, pre-Aptian oil is found in lacustrine limestones (as in the Badejo oil field, Campos basin), in structural traps, beneath unconformities, or in porosity-controlled stratigraphic traps in limestones.

Reservoirs in the Aptian-Albian sequence may occur as basal conglomerates overlain by evaporites such as in the Camorim oil field, offshore Sergipe. The Albian carbonates may be folded and faulted by salt movement in such a way that they are adjacent to Late Cretaceous

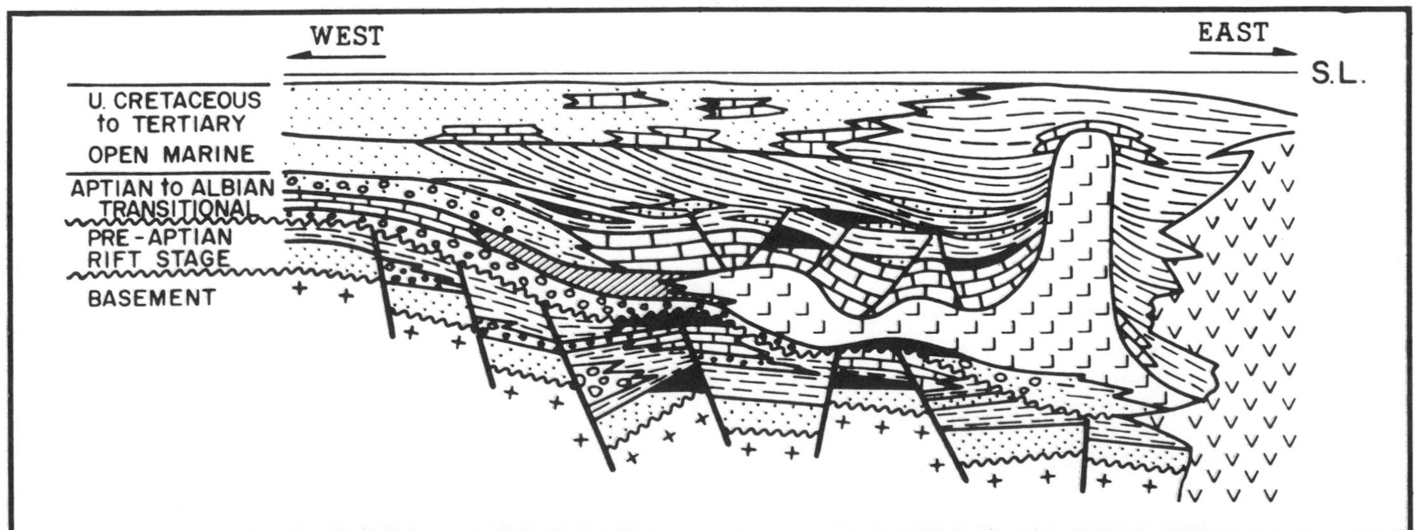

Figure 11 — Schematic geologic dip-section offshore Brazil.

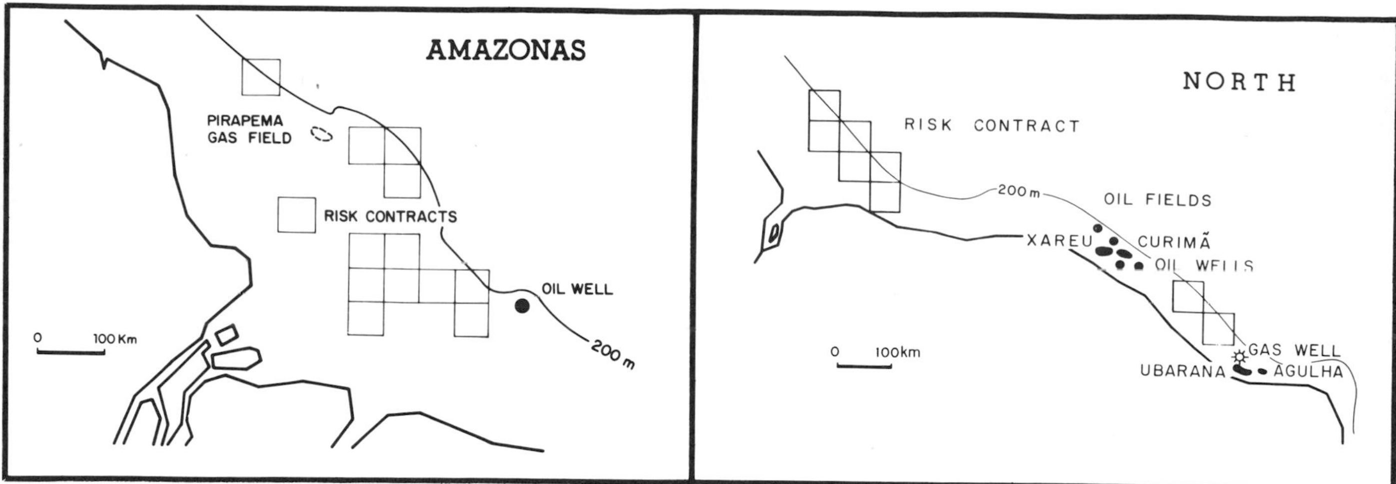

Figure 12 — Amazonas tract. Squares indicate blocks leased to international oil companies.

Figure 13 — North tract, offshore Brazil. Squares indicate blocks leased to international oil companies.

marine shale source rocks. If porosity is present, oil fields such as Garoupa or Pampo, in Campos Basin, can be found.

In the northeast and in the north coast, the Aptian-Albian sequence may be very thin or absent. In these cases, an unconformity between the older continental faulted sequence and the younger open marine sequence has resulted in several oil fields. Curima, offshore Ceara in the north coast, is an example of such an accumulation.

The Late Cretaceous-Tertiary sequence lacks large structures (with exception to the Amazonas) but contains stratigraphic traps, mostly turbiditic and paleogeomorphic traps related to paleocanyons; for example, the Guaricema oil field offshore Sergipe, and the Enchova oil field in the Campos basin.

Salt domes are present in several basins but we have not discovered any oil associated to these structures.

AMAZONAS COAST

The structural similarity of the Amazonas Cone with that of other very prolific deltaic areas (Figure 12) led to intense exploratory activity during the last 12 years. By the end of 1979, PETROBRAS had drilled 47 exploratory wells and international contractors had drilled 12 wildcats. Discoveries in this area consist of a small accumulation of biogenic gas (Pirapema field) found by PETROBRAS while drilling a "bright spot" anomaly, and a 1980 oil discovery in well 1-PAS-9 (oil well in Figure 12). Production in this well comes from fractured Tertiary limestone. PETROBRAS is still evaluating the discovery and other exploratory wells near the 1-PAS-9 are scheduled for drilling. Several other obvious shallow prospects in this area have been unsuccessfully tested, but many prospects remain to be tested in the Cretaceous Cassipore basin.

THE NORTHERN COAST

The northern coast (Figure 13) seems to be the second most promising area, following the Campos Basin. By the end of the 1979 PETROBRAS had drilled 70 exploratory wells and discovered 10.6 x 10^6 cu m of proven reserves of oil and about 3 x 10^9 cu m of gas.

PETROBRAS discoveries are located in two basins: Ceara on the northwest, and Rio Grande do Norte on the southeast. In the Ceara basin, oil accumulations are controlled by an unconformity where Late Cretaceous, open-marine shales directly overlie the Early Cretaceous continental sequence. Normal faulting also controls some oil fields. Several accumulations are related to an elongate horst near the present shelf edge, the "External High." After the first discoveries (Xareu and Curima), commercial flows of oil and gas have been obtained from other wildcats, as shown in Figure 13. The geologic setting of Rio Grande do Norte basin is similar to that of Ceara. The main oil field, Ubarana, is trapped by an unconformity at the top of the continental sequence or by normal faults. This field, still under delimitation, now covers an area in excess of 50 sq km. The Agulha field is a small stratigraphic trap in turbidites of the marine sequence.

In both Ceara and Rio Grande do Norte basins, high geothermal gradients create a shallow oil generation zone. The main exploratory problem is the low porosity of the sandstone reservoirs in the continental sequence, due to diagenesis. Some structures on the Ceara basin's western flank may be related to compression. PETROBRAS intends to drill several wildcats in this area which, according to geochemical data, seems to be gas prone.

All accumulations discovered in the northern tract lie in shallow water (less than 60 m). Production facilities are simple and the new fields can come on stream quickly.

THE NORTHEASTERN COAST

The first oil field offshore Brazil was discovered in the northeastern tract (Figure 14) in 1969. Due to the physical continuity with the onshore producing basin of Sergipe/Alagoas, exploration activity in this area remained high throughout the last 12 years. PETRO-

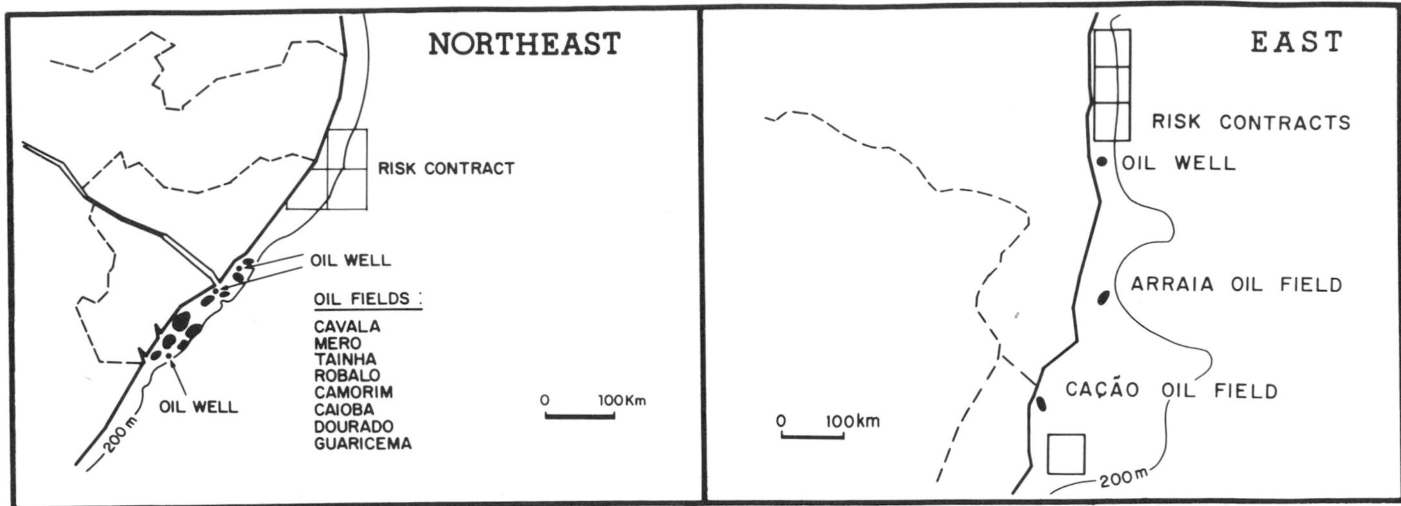

Figure 14 — Northeast tract, offshore Brazil.

Figure 15 — East tract, offshore Brazil.

BRAS has drilled 104 exploratory wells, and discovered 13.3 $\times$ 10^6 cu m of proven reserves of oil and almost 6 $\times$ 10^9 cu m of gas. Some of the oil fields, Mero, Cavala, and perhaps Tainha (Figure 14), are small accumulations of no economic interest. Others (e.g., Guaricema) have produced for several years and are coming to the secondary recovery phase. Several isolated oil producing wells are being reanalyzed in order to locate extension wells.

The Caioba oil field is on a fault block and production comes from sandstone in the basal continental section. Robalo and Camorim are related to fault blocks in which an unconformity atop the continental sequence traps hydrocarbons. Production in these fields is mostly from Aptian conglomerates. Many accumulations (Guaricema, Dourado, Mero, and Tainha) are in turbidite stratigraphic traps. Although offshore Sergipe/Alagoas is entering a mature stage of exploration, several new prospects will soon be drilled.

The top of the oil generation zone is, in comparison with other areas offshore Brazil, relatively shallow in the northeast. The lower continental sequence contains abundant structures and there are some problems regarding loss of porosity with depth in sandstones and conglomerates. There is good potential for gas prospecting between 3,000 and 4,000 m.

The northern sector has few structures, a shallow basement and volcanics. No wells have been drilled there but PETROBRAS plans to perform a stratigraphic test. In the south, one wildcat yielded unfavorable stratigraphic results and no oil shows. International contractors (for example, Union Oil) have signed contracts for some blocks offshore Alagoas.

THE EASTERN COAST

The eastern coast (Figure 15) represents one of the main offshore Brazil exploration challenges. From the south of Bahia to offshore Espirito Sauto states, there are well-defined structures, favorable stratigraphic sequences, good reservoir rocks, good oil source rocks, and direct evidence of hydrocarbons. Despite these, 92 exploratory wells have produced modest proven reserves of 1.1 $\times$ 10^6 cu m of oil and 270 $\times$ 10^6 cu m of gas.

In the northern part of this sector, basement arches separate several small basins from one another. Although the stratigraphic setting resembles that of the prolific onshore Reconcavo basin, no important offshore discoveries have been made.

In 1979, PETROBRAS discovered oil in the 1-BAS-37 wildcat (oil well in Figure 15). Production comes from Aptian carbonates and basement rocks in a fault block overlain by evaporites. Extension wells drilled near the wildcat were dry or uncommercial.

The Arraia oil field consists of a small stratigraphic accumulation in marine turbidites. Extension wells were drilled with no commercial results and the field goes into production in the near future.

Offshore Espirito Santo offers an adequate geological setting for oil and gas exploration, but as yet the only producing field is Cacao, a small paleogeomorphic trap covered by Late Cretaceous marine

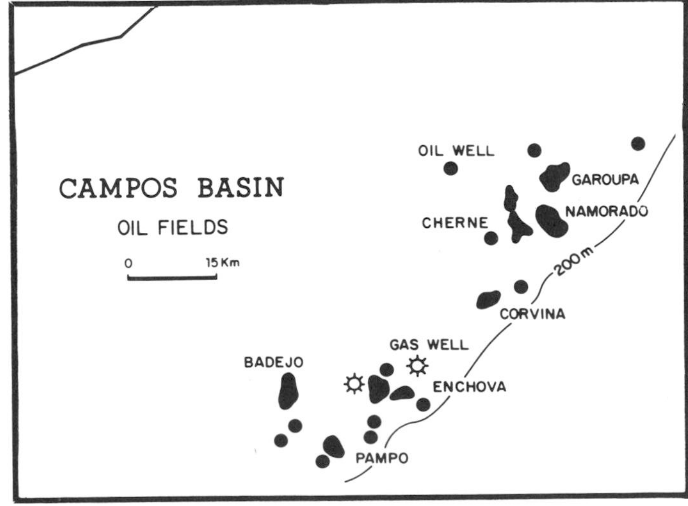

Figure 16 — Campos tract, offshore Brazil, showing oil and gas discoveries.

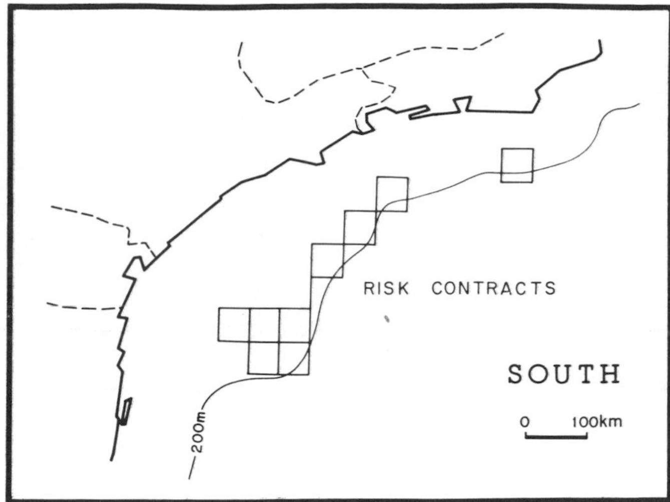

Figure 17 — Santos area, in the south tract of the Brazilian offshore.

shales. The area contains interesting deep prospects due to a thick marine sequence.

In the Abrolhos Bank, east of Espirito Santo Basin, Tertiary limestones overlie a volcanic complex. A reef would render the area attractive if source rocks were present.

CAMPOS

The Campos basin (Figure 16) is presently the most important oil province offshore Brazil. After drilling 127 exploratory wells, proven reserves are 91×10^6 cu m of oil and 11×10^9 cu m of gas. Most of the oil fields, such as Garoupa, Namorado, Cherne, and Pampo, are related to salt-controlled structures. Reservoirs are Albian limestones, as in Garoupa and Pampo, or Middle Cretaceous turbidites, as in Cherne and Namorado. There are source rocks in the Late Cretaceous marine shales, which are spatially related to reservoirs through the listric salt-movement-induced-faults. Some accumulations, such as Corvina and Enchova, are related to stratigraphic and/or paleogeomorphic traps in Tertiary turbidites. Badejo and other recent discoveries produce from lacustrine, normally faulted limestones, from beneath the Aptian-pre-Aptian unconformity, or from porous zones in the limestone. Extension wells will evaluate several new discoveries.

All discoveries in Campos are in waters deeper than 100 m. Some fields are already producing through provisional facilities and a production platform is now being installed in Garoupa oil field.

THE SOUTHERN COAST

No discoveries have been made on the southern coast (Figure 17), following drilling of 26 wildcats (12 by PETROBRAS and 14 by international contractors). This tract can be divided into two basins; Santos in the northeast, and Pelotas in the south. Deep-seated salt in the Santos basin produces remarkable structures. Pelotas is mostly a monotonous homoclinal basin. Source rocks are rare in both basins and in spite of unfavorable results, several prospects are scheduled for new wildcats.

CONCLUSIONS

Although oil and gas discoveries offshore Brazil are insufficient to meet domestic consumption needs, they are significant considering the complexity of exploration and the geological setting of an Atlantic-type passive margin. Exploration is economically attractive, especially in view of the present oil prices in the international market. After 12 years of experience, PETROBRAS and international contractors should now be able to increase the success ratio by looking for more specific prospects.

REFERENCES CITED

Asmus, H. E., and F. C. Ponte, 1973, The Brazilian marginal basins, *in* The ocean basins and margins: New York, Plenum Press, v. 1, p. 87-133.

Asmus, H. E. and R. Porto, 1972, Classificacao das bacias sedimentares brasileiras segundo a tectonica de placas: Belem, 26th Congresso de Geologia, v. 2, p. 67-90.

Ponte, F C., M. V. Dauzacker, and R. Porto, 1978, Origem e acumulacao de petroleo nas bacias sedimentares Brasileiras: Rio de Janeiro, Proceedings of 1st Brazilian Petroleum Congress, p. 121-147.

Oil and Gas on Passive Continental Margins

Colin Barker
Department of Geosciences
The University of Tulsa
Tulsa, Oklahoma

A simple classification of organic matter types is combined with models of sedimentary environments to predict generalized trends for the distribution of normal crudes, waxy crudes, and gas on passive margins. The sequence from continental rift to open-marine conditions produces an idealized vertical organic matter type profile that has abundant terrestrial organic matter deep and aquatic organic matter shallow. In contrast, the vertical organic matter type profile that develops in deltas prograding across continental margins is from aquatic organic matter deep to terrestrial organic matter shallow. These distributions of organic matter types exercise initial control on the location of gas, waxy crudes, and normal crudes. There may be subsequent modifications and redistributions due to migration, maturation and alteration.

It is now widely accepted that the earth's surface is made up of a set of mobile plates. Interactions between moving plates produce major near-surface features, such as mountain ranges and sedimentary basins. Any concept that provides insight into the development of sedimentary basins is important to petroleum geology because much of the world's petroleum resources are located in sediments laid down in a basin setting. Basins have been classified within the plate tectonic framework using criteria such as the nature of the underlying crust, relationship to stable cratonic areas, rate of sediment fill, heat flow, and geothermal gradient. Data for oil and gas in these basins have been incorporated as an empirical afterthought and have not been tied genetically to the basin classification. To classify basins purely in terms of their physical and inorganic characteristics, when petroleum is the major consideration, is an exercise comparable to classifying wines by the shapes of the bottles they are in. Regularities in the occurrence of oil and gas in basins of different types (Klemme, 1975, 1977, 1980; Bally, 1975; Bally and Snelson, 1980) suggest that oil and gas distribution in different basins is not random, but is related to basin setting and development.

In the last decade there have been considerable advances in understanding and documenting the conditions required to generate oil and gas, and the influence of organic matter type in controlling the nature of the petroleum generated (Tissot and Welte, 1978; Hunt, 1979). These developments occurred at the same time that the ideas of plate tectonics were developed into a unifying hypothesis for many aspects of the earth sciences. This paper is a preliminary attempt to integrate some of the ideas of organic geochemistry into the framework provided by plate tectonics. In this broad brush treatment, a simplified view of tectonic setting and organic matter type is used to show that there are apparent regularities in the distribution of petroleum which would merit further study and documentation.

"Petroleum" includes materials with a wide range of chemical compositions and physical properties, ranging from low density natural gas to heavy crude oils and carbonaceous solids that have API gravities less than 10 (specific gravity ⟩1.00). Modern analytical techniques have been used to characterize these materials and information is available about the relative amounts of various compound types. Improving analytical capabilities, particularly the combination of gas chromatography with mass spectrometry, has stimulated the investigation of compounds with characteristic structures ("biological markers") even though these may account for only trival percentages of the total crude. These studies are justified because they help in understanding geochemical processes and because the information is useful in correlating crude oils to their source rocks, grouping oils into families (especially when some members of the group are degraded), and establishing the past thermal history of potential source intervals. However, the average refiner does not much care about details such as the ratio of (17α, 21β) H-hopane to (17β, 21α) H-hopane. For most practical purposes only gross chemical characteristics, like percent sulfur or wax content, are important.

There is, however, a significant economic distinction between oil and gas. The major consumers of natural gas are in the industrialized areas of North America and west Europe and gas finds outside of these regions are much less desirable than oil. Although the rising cost of gas (relative to oil) is making gas more attractive, it is still difficult and expensive to transport. A gas find in a hostile or economically unfavorable area has to be very large to justify exploitation. Thus, it is important to include in an exploration philosophy some predictive capability in establishing gas-prone versus oil-prone areas.

Oil and gas are generated from organic matter in sediments influenced by rising temperature that accompanies increasing depth of burial. Time (i.e. geologic age) has a significant, but secondary, role. The generation process modifies the chemical nature of the insoluble "kerogen" and the extractable "bitumens" by converting kerogen to bitumen, with the former becoming more carbon-rich and the latter more hydrogen-rich. Ultimately, high temperatures convert all types of organic matter to graphite and methane (Figure 1) but in the early stages of generation the nature of the petroleum products is controlled by the type of organic matter.

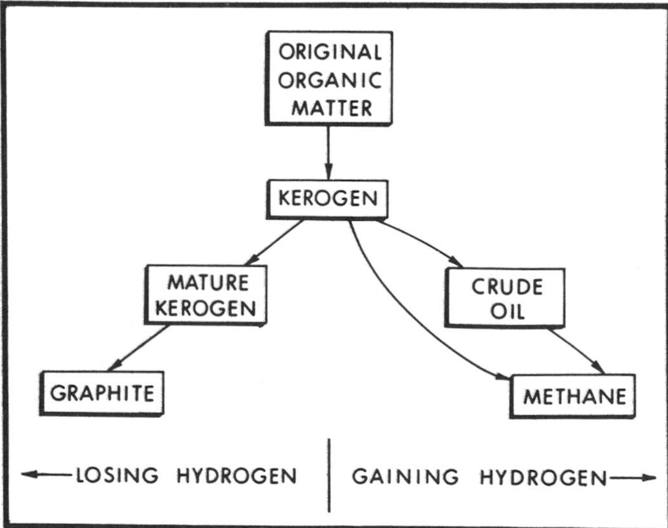

Figure 1 — Schematic illustration of the redistribution of hydrogen during the generation of oil and gas from organic matter (Barker, 1979a).

ORGANIC MATTER TYPES

The amount and type of organic matter incorporated in sediments varies widely and depends on depositional environment. The environment may also control the survival of organic materials by influencing the oxidizing or reducing nature of the water and sediments.

One of the most direct ways of establishing the nature of the organic matter in a sediment, or older rock, is to remove the mineral matrix by solution in mineral acids (HF/HC1) and examine the kerogen microscopically. Remains of organisms are frequently recognized by their characteristic morphologies and spores, pollen, cuticle, and wood remnants identified. However, much material obtained in this way has no discernable structure and is called amorphous. In addition, sediments may contain organic materials that have been recycled from older sediments. These different types of organic matter have different roles in the generation of oil and gas and are discussed individually.

Aquatic Organic Matter

Lower unicellular aquatic plants, such as phytoplankton, are a major source of the particulate and dissolved organic matter in sea water. Although much of this is oxidized and ultimately recycled to the biosphere, a small percentage can become incorporated into sediments and buried. Organic matter derived in this way generally retains no recognizable morphology and is frequently called "amorphous." It is important to note that while most algal debris is amorphous, there are other types of amorphous organic matter that are derived from different organisms and have very different chemical compositions.

In the water column and in sediments, bacteria actively degrade organic matter while building their own body structures. Their remains can also become part of the biogenic materials in sediments and may account for a significant fraction of it (Lijmbach, 1975).

Terrestrial Organic Matter

Terrestrial organic matter is derived from organisms that grew on the land surface and they show some important differences from organisms that grow in water. For an aquatic organism, the water provided support whereas land surface organisms must develop systems to provide their own structural support. This support usually involves woody-type materials made up of roughly equal amounts of cellulose and lignin. Both of these are biopolymers and are built up from sugar and phenylpropane monomers respectively. On burial, cellulose is fairly rapidly degraded by hydrolysis to give its constituent sugars, which are not only water soluable but also readily metabolized by bacteria. Lignin, in contrast, survives better and is more resistant to bacterial degradation because of its dominant phenolic character (which is bacteriocidal). Organisms growing in water are supported by the water and do not need structural support, and so lignin-derived organic matter is characteristic of terrestrial plants but not aquatic organisms.

A second feature of land plants is their development of protective coatings to prevent dehydration and loss of essential body fluids. These external coatings also help to minimize mechanical damage and protect plants from micro-organisms (Hadley, 1980). Biochemically resistant compounds have evolved to fulfill these roles and for most plants surface waxes, cutin and suberin are quantitatively important components. The surface waxes are esters of long chain acids and alcohols with chain lengths up to about 35 carbon atoms. On burial, they generate paraffin chains in which those with odd numbers of carbon atoms are particularly abundant, especially for carbon numbers of 27, 29, and 31. The surface waxes provide the only documented source of very long chain normal paraffins (more than about 25 carbon atoms). Cutin is also a very paraffinic material and is made up mainly of chains containing about seventeen carbon atoms that are linked through oxygen bridges and contain hydroxyl groups (Holloway, 1977). Eglinton (personal communication to Tissot and Welte, 1976) reported that hydroxyl acids typical of cutin and suberin seem to have been found only in higher terrestrial plants. Since protection against dehydration is not required by an organism growing in water, surface waxes do not develop. As a result, long chain compounds are not found in aquatic organisms.

Recycled Organic Matter

Clastic sediments are formed predominantly from the weathered and transported remains of pre-existing rocks and it is quite common for some remnants of the older, once-buried organic matter to be remobilized and redeposited. Recycled material is usually present as angular, compact particles that generally have higher densities and darker color than the rest of the organic matter. Fusain (fossil charcoal) is also dark and angular and even though it is being incorporated into sediments for the first time, it will have no active role in petroleum generation and can be classed with the recycled material which it often resembles morphologically and chemically.

Controls on Organic Matter Distribution

The distribution of various organic matter over the earth's surface is non-uniform and there are major regional differences in their productivities. The highest productivities for aquatic organic matter are in those oceanic areas where upwelling brings deep, nutrient-rich waters into the sunlight and stimulates the growth of phytoplankton, which form the base of the food chain. Oceans at high latitudes, particularly those around Antarctica, are especially productive. Upwelling also occurs along the west coasts of southern-hemisphere continent so that some areas of high productivity occur in this setting. The Humboldt Current off the coast of South America provides one example.

In contrast, high productivity for terrestrial organic matter is restricted to near-equatorial settings and currently includes the Amazon basin, central Africa and parts of the Far East and northeast Australia. Climate exercises the major influence on the distribution of land-surface organic matter with high productivities being favored by hot, humid conditions while the polar regions have no terrestrial organic matter. Throughout geologic time, the positions of the continents have changed relative to one another and also relative to the poles. While the polar regions have not always had ice caps they have always been cold, so high concentrations of terrestrial organic matter have never developed at very high latitudes. Also, terrestrial organic matter will be absent from Ordovician and older rocks because these predate the evolution of land-surface organisms.

The vast majority of sediments preserved in the geologic record were laid down in aquatic environments, most of which were marine. If terrestrial organic matter is to be preserved, it must be transported into such a depositional setting. Rivers have a dominent role in carrying the organic matter derived from land-surface organisms to the coasts, and in depositing it in nearshore sediments such as those in deltas. River systems are controlled by climate and topography with barriers like mountain ranges, or conduits like rift valleys, having an important influence. This is well-illustrated by the Amazon River, which used to flow westward into the Pacific. As the west-moving South American plate overrode the east Pacific Nazca plate and developed a subduction zone, the Andes were raised and formed a barrier in the path of the Amazon. The river flow was reversed and the Amazon now drains to the Atlantic coast (deOliveira, 1956; James, 1971). Because of this reversal, all the terrestrial organic matter from the Amazon basin is now being transported towards the Atlantic coast of South America, rather than the Pacific coast. Thus, sediments deposited in the east Pacific will have only minor input from the terrestrially-derived organic material (Figure 2).

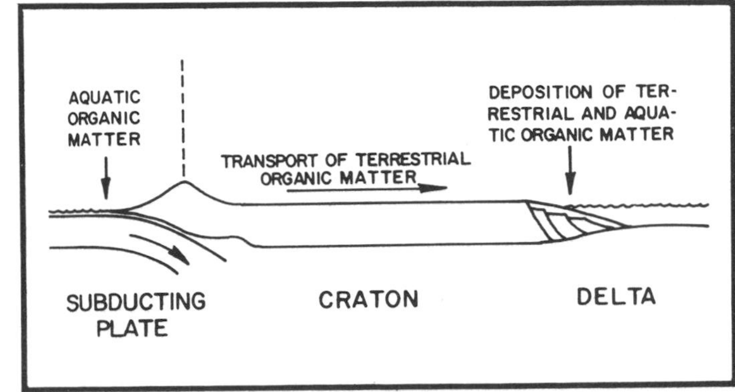

Figure 2 — Section through South America showing how the Andes restrict drainage to the Pacific so that most of the terrestrial organic matter is now being transported to the passive margin of the plate.

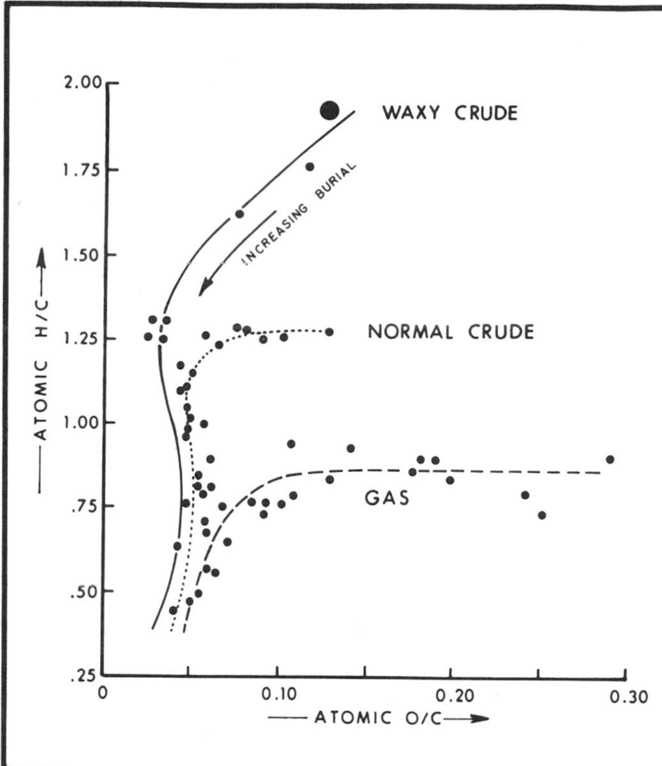

Figure 3 — Van Krevelen diagram showing kerogen evolution paths. The data points are from Tissot et al (1974) who used them to define the main evolution paths for different types of kerogen. The large black dot shows the elemental composition of a mixture of cutin and surface wax and its position on the Type I evolution path.

Terrestrially-derived organic matter may make a quantitively important contribution to sediments deposited in deltaic or lacustrine environments. For example, Tissot and Welte (1978) suggest that the organic matter in the Eocene lake sediments of the Uinta and Green River basins represented large quantities of bacterial biomass produced by degradation of higher plant materials. This is consistent with data for elemental composition since an equal mixture of long chain wax and cutin plots on the Type 1 track of a van Krevelen diagram (Figure 3), as does the Green River shale kerogen. Very paraffinic crude oils often seem to be a major product from lacustrine sediments (as discussed below) and the Altamont and Bluebell crudes from the Uinta basin have pour points of over 100°F (38°) (Lucas and Drexler, 1975).

Survival of the organic matter depends on physical conditions in the sediments. In oxygenated environments, the organic matter is converted to carbon dioxide and water in a reaction that is, in effect, the opposite of photosynthesis, although aerobic bacteria generally play a major role in the oxidation. If the organic materials are protected from contact with oxygen they are more likely to be preserved, but degradation by sulfate-reducing bacteria and fermenters is still possible. When a large amount of organic matter is introduced into an oxygenated environment, the oxygen may be used completely in oxidizing only part of the organic matter so that the environment becomes reducing and the rest of the organic matter survives. On the average, only about 0.1% of the original, living organic matter survives into sediments in oxygenated environments. In reducing systems it can average as much as 4%. Demaison and Moore (1980) discuss some of the geological aspects of the development of oxic and anoxic zones.

Survival of organic matter also depends on climate because micro-organisms act much more effectively at higher temperatures. Much of the terrestrial organic matter developed in humid tropical areas, like the tropical rain forests, is rapidly recycled and does not survive long enough to be transported to a depositional environment that will be preserved into the geologic record. Since the rate of microbial attack decreases more rapidly with temperature than does the growth of land surface organic matter, there is a peak in temperature latitudes for the amount of terrestrial material that can be incorporated into marine sediments. Although few data are available, it seems likely that the various types of organic materials have different tolerances to degradation. Biochemically inert materials, such as surface coastings, should survive best. Lignins, which contain the bacteriocidal phenol group, are known to survive degradation much better than the cellulosic materials that form the other major component of wood.

Sedimentation rate is probably also a factor in survival because low rates leave the organic matter exposed for longer periods, leading to more severe microbial degradation. At the other extreme, high sedimentation rates dilute the organic matter with clastics and produce relatively low organic matter contents. Intermediate sedimentation rates probably provide optimum conditions for high organic matter contents (Dow, 1979).

Burial

The various types of organic matter in sediments have different chemical compositions and molecular structures and respond differently to the rising temperature that accompanies increasing depth of burial. Elemental composition is one useful way of summarizing organic matter composition and can be expressed in terms of atomic ratios of oxygen-to-carbon and hydrogen-to-carbon for display on a van Krevelen diagram (Tissot et al, 1974). These diagrams can be used to show how compositions change with increasing thermal maturity (Figure 3).

Woody, organic matter starts with a high oxygen-to-carbon ratio because both cellusose and lignin have abundant oxygen-containing functional groups (mainly hydroxyl and methoxyl). The hydrogen-to-carbon ratio is generally low because the lignin biopolymer is built up from aromatic monomers, which have lower hydrogen contents than saturated molecules. Much of the cellulose is lost rather early and reduces the oxygen content. Subsequent rising temperature due to burial produces a further loss of

Table 1. Summary of the main organic matter categories and their dominant petroleum products.

SOURCE OF ORGANIC MATTER	TYPE OF ORGANIC MATTER	ROLE	CHEMICAL STRUCTURE AND DEGRADATION PRODUCTS	PETROLEUM TYPE
TERRESTRIAL	LIGNIN (monomer)	STRUCTURAL SUPPORT	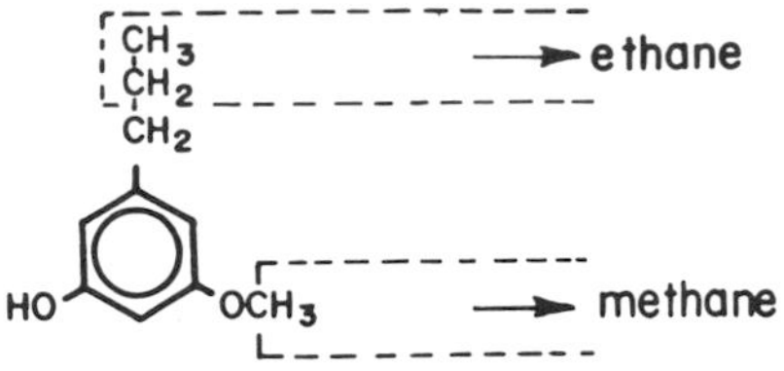 CH_3 CH_2 → ethane; CH_2; HO, OCH_3 → methane	NATURAL GAS
	SURFACE WAXES CUTIN	PREVENT EVAPORATION	WAX: 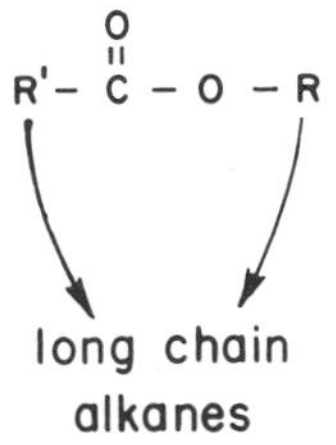 R'–C(=O)–O–R → long chain alkanes; CUTIN: C–R–C–R'–C → n-alkanes	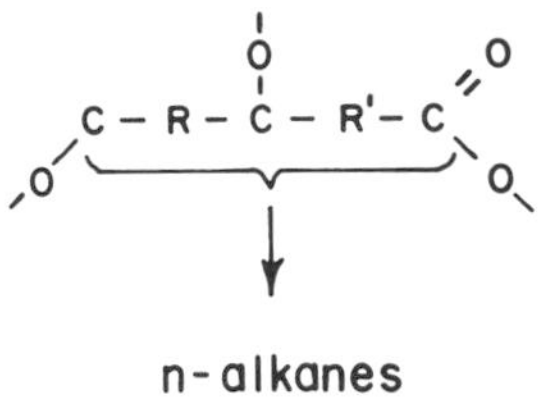 WAXY CRUDE OILS
AQUATIC	LIPIDS		isoprenoids, naphthenes, n-alkanes	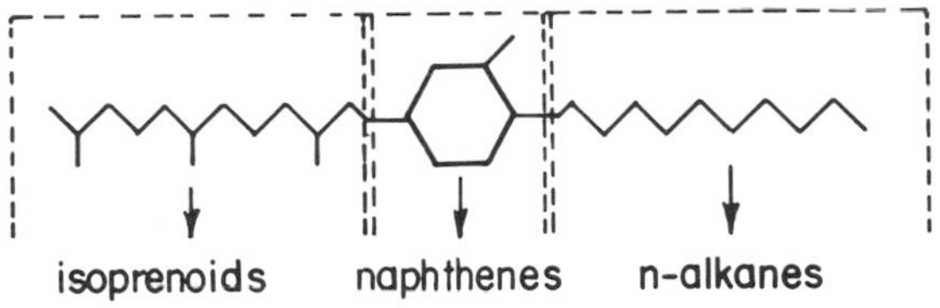 NORMAL CRUDE OILS

oxygen as carbon dioxide and water, and at higher temperatures low molecular weight hydrocarbons, are produced. These come from cleavage of the short side chains, probably producing ethane first and then methane at higher temperatures (Table 1). This oxygen-rich, hydrogen-poor kerogen was classed as Type III by Tissot et al (1974) and is generally recognized as a gas-generating type. An extreme example of organic matter derived mainly from woody recursors is provided by coal, which generates substantial amounts of methane as rank increases during coalification. Most of the gas in the southern North Sea was formed in this way (Lutz, Kaasschieter, and van Wijhe, 1975).

Cutin and esters in the surface coating of land plants both lose carbon dioxide on burial and generate paraffin-rich materials. In the case of the wax esters, the paraffins have a strong odd-even preference in carbon chain length with the 27, 29, and 31 carbon chains being particularly abundant. The cutin produces a smoother distribution of normal paraffins and these are often of shorter chain length, frequently in the range from C_{17} to C_{20}.

Thus, both of the major components in surface coatings on terrestrial plants generate very paraffinic crude oil on burial. These oils can have more than 20% paraffins and have very high pour points (many are solids at room temperature). For example, in the South Java Sea, oils have pour points of greater than 70°F (21°C) and show marked odd-even periodicity in the normal paraffins with a maximum at nC_{29}. The source rock for these oils contains 95% terrestrially-derived organic matter and of this, 74% is cuticle (i.e. surface coating) (Sutton, 1977). Other paraffinic oils with high pour points occur in marginal marine environments (Hedberg, 1968); in continental settings (Uinta basin; Daqing, China); in deltas (Niger, Mackenzie, Gippsland basin); and on continental margins (Brazil, Gabon).

Much of the amorphous organic matter in sedimentary rocks is of presumed algal origin and is quite rich in hydrogen but poor in oxygen. It corresponds to the Type II of Tissot et al (1974) (Figure 3). On burial, this material generates the compounds found in non-waxy crude oils and the normal paraffins which are present show no predominance of odd carbon number chains. The paraffins are generally most abundant around nC_{20}. A major amount of naphthenes (cycloparaffins) is generated along with isoprenoids, aromatics, and naphtheno-aromatics which produces crudes that have low pour points and normal API gravities.

The petroleum products produced by the various types of organic matter are summarized in Table 1 which shows that there is an economically important distinction between the products from land-derived organic matter (which makes gas and waxy crudes) and aquatic organic matter (which produces the normal range of crude oils). Because petroleum is a relatively hydrogen-rich product, it is generated in smaller amounts by the hydrogen-poor Type III kerogen and so this type of material is not a very prolific generator.

This simple three-fold classification of organic matter into 1) "aquatic," which is largely amorphous and of presumed algal origin; 2) woody; and, 3) surface coatings is used in the ensuing discussions. Wood and surface coatings are often discussed together and called "terrestrial." While much more comprehensive classifications could be developed, even though organic matter has been characterized in considerable detail, the proposed groupings appear to account for the major differences among petroleums, particularly the economically important distinction between oil and gas.

RIFTS AND ABORTED RIFTS

Moving lithospheric plates can interact in several ways. The limiting possibilities for two plates are: a) direct collision leading to overriding of one plate by another, or to suturing; b) slide-by involving major transform faulting; or, c) mutual movement apart with the introduction of new organic crust in the area between the diverging fragments of a previous single continent. This paper is concerned with oil and gas in rifts and passive margins that develop during continental breakup.

The processes that cause breakup of a continent through rifting are not fully understood. In many cases the initial stage, at least locally, appears to involve a doming of the crust, probably over the site of a mantle plume or "hot spot" (Morgan, 1971). Kinsman (1975) pointed out that for long rifted margins, such as those characteristics of the Atlantic, the rift is probably made up of sections associated with plumes together with interplume segments that show less, if any, doming.

The East African rift system provides a modern example of rifts developing in an uplifted area. Both reservoir rocks and source rocks can develop in this setting. Erosion of the elevated margins of normal fault blocks and scarps by short active streams may deposit reservoir quality fluvial-deltaic sediments in the rift. In the area of the southern Red Sea rift, for example, the Dogali Series of Ethiopia, the Trap Series of Yemen, and the Shumazi Group of Saudi Arabia developed in this way (Lowell et al, 1975). Petroleum source rocks can form in the lakes that frequently develop in the fault-bounded valleys of the rift. The lacustrine sediments receive a considerable amount of land surface-derived organic matter, along with the clastic input being derived from uplift erosion. They also incorporate organic matter from organisms growing in the lake. However, the terrestrially-derived organic matter survives degradation better than the algal-derived material which is largely oxidized (Prahl, Bennett, and Carpenter, 1980). In many areas, continentally-derived source rocks are the source of waxy crudes or gas. The lacustrine sediments of the Uinta basin, which have sourced the very high pour point paraffinic oils of fields like Altamont and Bluebell, provide a well-documented example, while Chinese continentally-sourced oils are paraffinic and waxy (Wanli, Yongkang, and Ruigi, 1981). Other examples from rifted environments are discussed later.

The early stage of rift development often occurs well above sea level and no marine sedimentation is possible. With time, the domed-up areas associated with the rift subside to sea level and below, and the rift stage is terminated by the insertion of new oceanic crust into the rift axis and the separation of the original continent fragments along new plate boundaries. Some rifts have not evolved in this way and have remained at an early stage of development. These aborted rifts, or "aulacogens," are still important because they control local drainage and therefore the transport of land-derived organic matter to the marine (or marginal marine) depositional environment. A good example is the Benue rift, which is the aborted arm of a rift-rift-rift triple junction in which the other two arms developed into oceans as part of the evolution of the South Atlantic. The Benue rift has controlled the drainage pattern and the Niger River now flows through this aulacogen and brings with it the land-derived organic matter that survives transport from the area north of the rift. This is dumped (along with the transported clastic material) in the Niger delta, where the dominant petroleum types reflect the organic matter type so that waxy crude oils and gas are common (Evamy et al, 1978). Much of the gas is not being developed for economic reasons and many of the non-waxy crudes are bacterially-degraded waxy crudes. The Nile (Burollet, 1980) and the Mississippi (Burke and Dewey, 1973) provide other examples of major river systems (and hence delta locations) controlled by aborted rifts.

PASSIVE MARGINS

Early stages of rifting are dominated by fluviolacustrine sedimentation because the regional doming places the rift floor well above sea level. Restricted conditions often develop in rift lakes and stratification in the water column leads to deep anoxic layers where preservation of organic matter is favored (Demaison and Moore, 1980). Later, subsidence of the uplifted rift flanks may be caused by decreasing temperature, sediment loading, or extensional stress. These may act alone or, more likely, in combination. As the base of the grabens subside below sea level, marine sedimentation becomes possible. The form it takes depends on local conditions.

If ocean water flows into an environment in which circulation is restricted (Figure 4) and climatic condi-

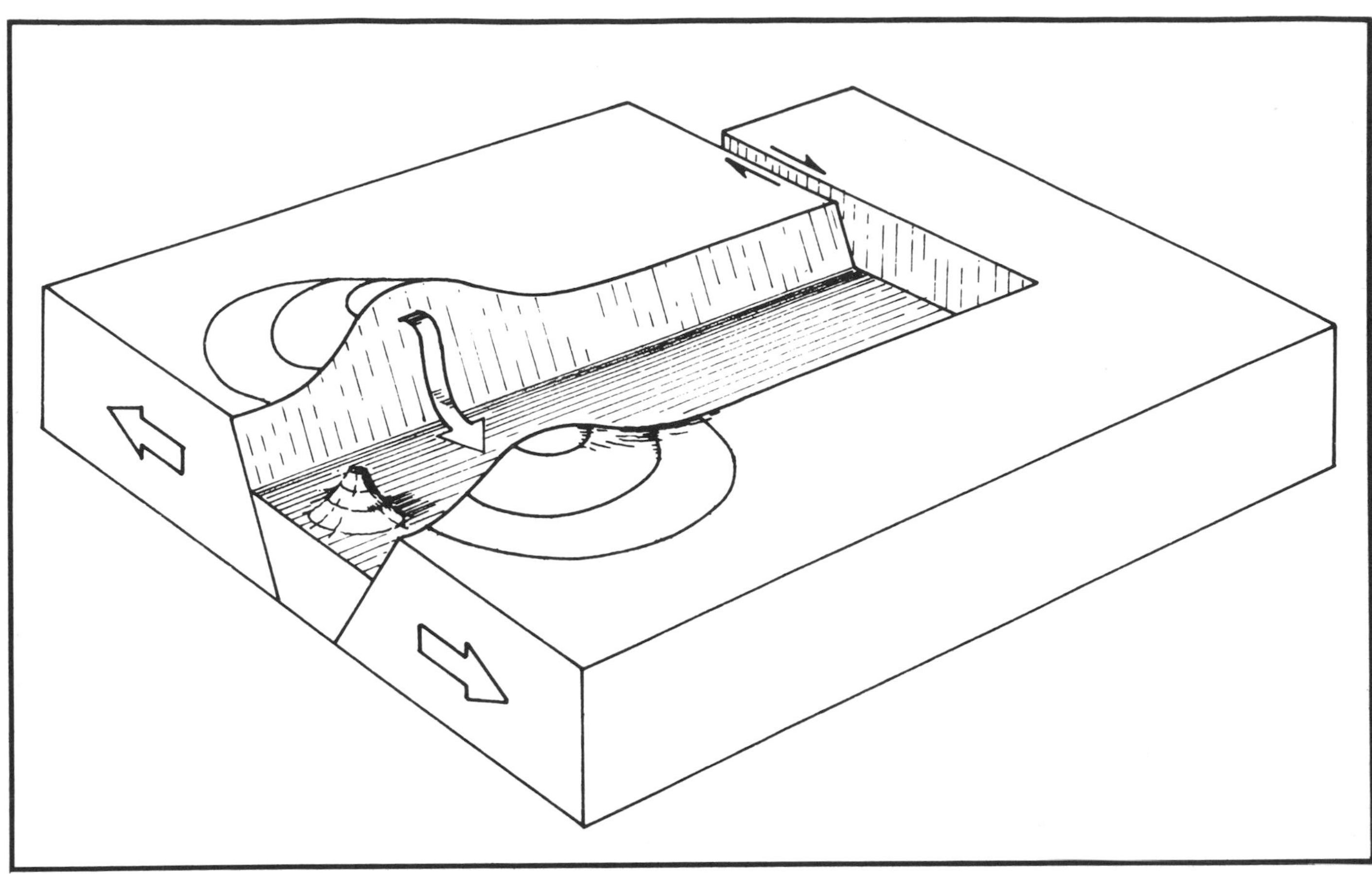

Figure 4 — Development of a restricted environment during rifting. This could represent the proto-South Atlantic with the transform stretching from Nigeria westward. The southern restriction could be provided by the Walvis-Rio Grande ridge.

tions favor evaporation, then extensive evaporite deposits may form. The Aptian salt basins along the margins of the South Atlantic provide a good example (Evans, 1978). The original rifts occurred in the middle of a continent and such a location tends to be arid, particularly when it is located at low latitudes. In this particular case, circulation was restricted by a transform margin at the north and by the Walvis-Rio Grande volcanic ridge at the south. When the North and South Atlantic were finally linked, conditions for restricted circulation were eliminated and evaporites did not continue to develop, thus explaining their absence along the north coast of Brazil and the corresponding African coasts of Ghana and Nigeria (Evans, 1978).

There are three possibilities for the development of evaporites as rifts evolve into continental margins: 1) The incursion of oceanic waters can produce evaporites overlying the previously deposited lacustrine and continental sediments; 2) The evaporites may be deposited more or less directly on the new oceanic crust being produced in the rift axis in which case there is only thin sediment sequence underlying the salt; and 3) No evaporite sequence develops because the necessary climatic and oceanic circulation conditions do not develop. This situation is probably common where the rift opens to a large ocean. Examples include the southern end of the proto-South Atlantic rift, the northern end of the Viking graben in the North Sea, and the Labrador shelf.

When evaporites are present on continental margins they generally separate the underlying terrestrially-dominated sediments from the marginal marine or open marine sediments above. Since salt is an excellent seal for petroleum, it provides an ideal trap for petroleum generated in the subsalt section. Later salt movements, such as diapirism, may lead to structures in the overlying sediments that provide suitable locations for petroleum accumulation. Salt also has a role in petroleum generation in some areas by disturbing the local geothermal gradient. The high thermal conductivity of the evaporites can lead to higher temperatures at shallower depths than in areas lacking salt.

The sequence of events on a developing passive margin is shown schematically in Figure 5. The initial continental sediments with their terrestrially-derived organic matter are overlain by increasingly marine sediments (with or without evaporites) so that the idealized organic matter profile shows increasing percentages of aquatic organic matter with decreasing depth. This distribution influences the location of gas, waxy (paraffinic) crudes, and normal crudes on passive margins. Examples from several continental margins are discussed below.

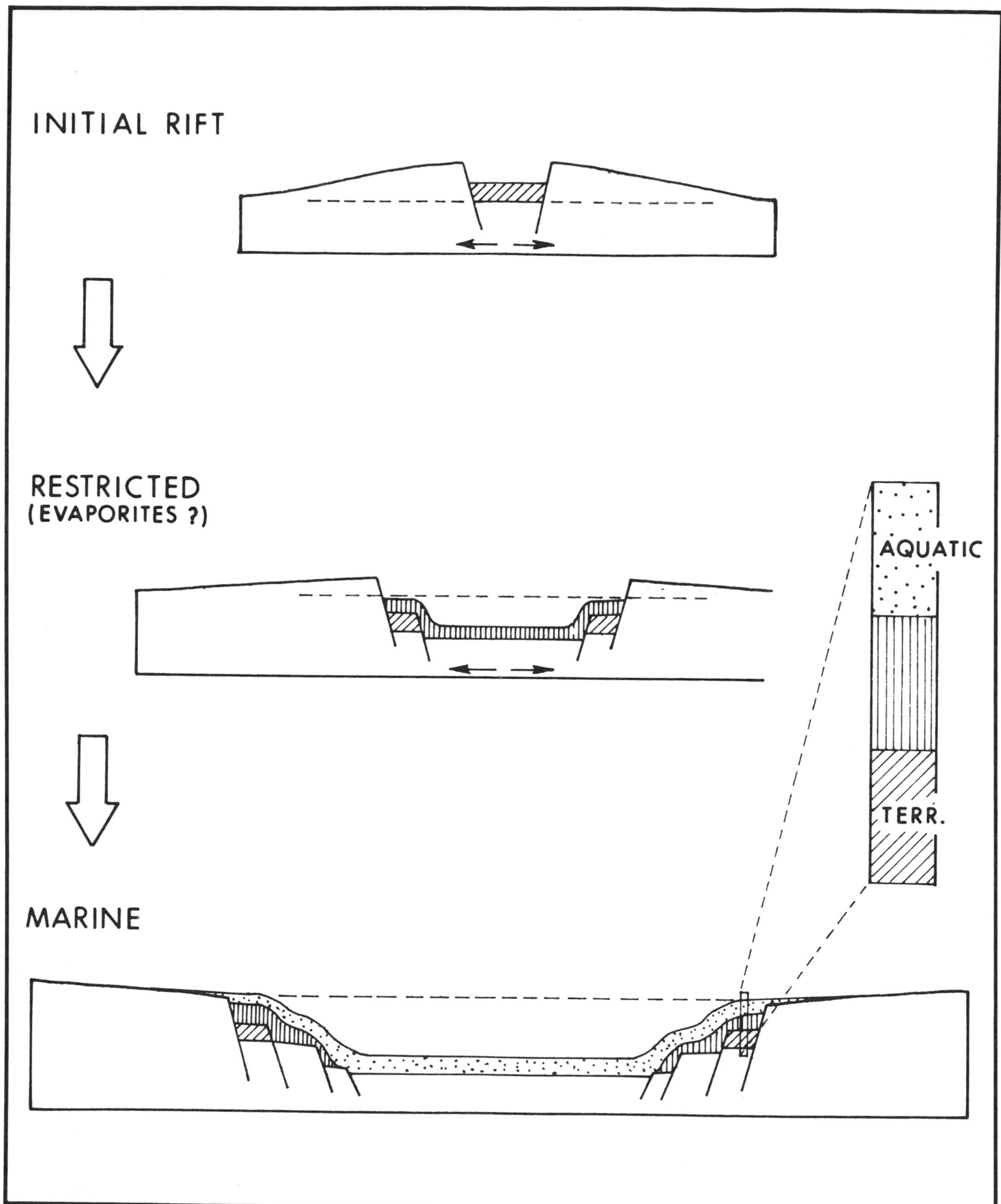

Figure 5 — Schematic illustration of the development of an aquatic-over-terrestrial organic matter type profile on a passive margin. This also shows diminishing terrestrial organic matter in a seaward direction. The dotted line gives the relative position of sea level.

PETROLEUM ON PASSIVE MARGINS

Brazil

The remarkable similarity in the shapes of the west Africa and eastern South America coasts has long been advanced as evidence that they were once contiguous. Computer fits (Bullard, Everett, and Smith, 1965), continuity of structures and geology (Allard and Hurst, 1969; Torquato and Cordani, 1981), and continuity of geologic ages (Hurley et al, 1967) support this view. The separation of South America from Africa appears to have started in the Late Jurassic-Early Cretaceous and since then sedimentary basins have developed on both sides of the Atlantic.

In the western South Atlantic, 8000 km of Brazilian coast is sub-divided into about a dozen basins (Asmus and Ponte, 1973) but knowledge of these basins varies considerably and depends largely on the extent of petroleum exploration. However, the general sequence of events seems to have involved 1) a pre-rift stage with major regional uplift, 2) an Afro-Brazilian depression preceeding rifting, 3) an intracratonic rift-valley stage 4) a proto-oceanic gulf stage, and 5) an open ocean stage (Ponte, Fonseca, and Morales 1971). This produced 1) a lower sequence composed of lacustrine clastic sediments; 2) an intermediate sequence of evaporites separating the lower continental sequence from the upper marine sediments; and 3) an upper sequence of sediments deposited in an open, circulating marine environment (Asmus and Ponte, 1973). Ponte, Fonseca, and Carozzi (1980) have described these sequences using a classification based on the depositional environments of the petroleum source rocks. Source rocks formed in the intracratonic-rift valley stage are dominated by nonmarine organic matter and have produced very high pour point waxy crude oils.

In the Sergipe-Alagoas basin, oils from the deep continental section are waxy with high paraffic contents and high pour points. In the oils sourced from the overlying marine section, the normal alkanes are quantitatively less important, shorter chain lengths are more important (Figure 6), and pour points are low (Ferreira and Gaglianone, 1980). A similar distribution of waxy crudes occurs in the other basins along the Brazilian continental margin (Ferreira, personal communication, 1980).

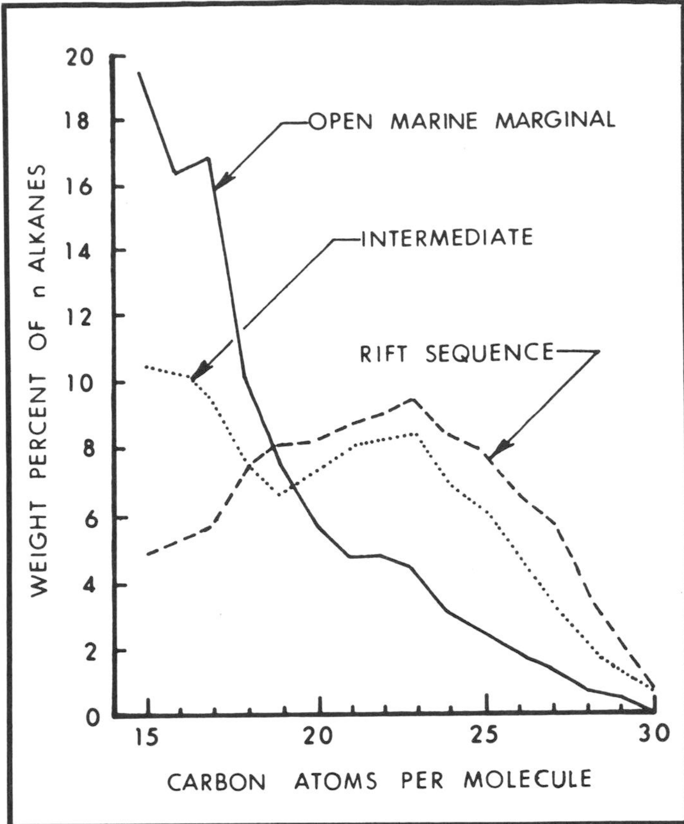

Figure 6 — Relative abundances of straight chain paraffins (alkanes) as a function of chain length for oils from various depositional environments on the evolving Brazilian margin (from Ferreira and Gaglianone, 1980).

West Africa

Between the present Niger delta and the Walvis ridge there are four major sedimentary basins which closely resemble the corresonding basins on the Brazilian margins (Franks and Nairn, 1973). The Mossamedes basin in the south has no commercial petroleum production (St. John, 1980) but the Gabon, Bas Congo-Cabinda, and Cuanza basins are all productive.

Stratigraphic succession in the Gabon basin can be divided into three broad units with an intermediate evaporite sequence separating underlying continental sediments from the overlying marine sequence (Brice and Pardo, 1980). Although in the Cuanza basin to the south the continental section is of limited extent (Brognan and Verrier, 1966), it is a major feature in the stratigraphic sequence of the Gabon basin (Brink, 1974) where the combined thickness of presalt units may exceed 25,000 ft (7650 m). In the Congo basin, subsalt fluvial and lacustrine sediments range up to 11,500 ft (3,500 m) thick. They are overlain by black bituminous shales which immediately underlie the Aptian salt (Hourq, 1966). Source rocks in the subsalt section of these basins are organic-rich, lacustrine dolomitic shales which attain thicknesses up to 900 m (Brice, Kelts, and Arthur, 1980). They were laid down in deep lakes where anoxic bottom conditions developed, and total organic carbon contents of these sediments can be as high as 20% percent (Brice, Kelts, and Arthur, 1980). The organic matter is dominantly Type 1 (Claret, Jardine, and Robert, 1981).

Production from the continentally-derived sediments below the salt in these West African basins is dominated by high pour point waxy crudes, though there is some condensate. The general character of these oils is shown by the nature of the refinery streams quoted by Aalund (1976). The Gamba stream (Gabon) has an API gravity of 31.8°, a pour point of 73.4°F (23° C), and 0.11% sulfur. In contrast, the Emeraude oil from the post-salt section of the Congo

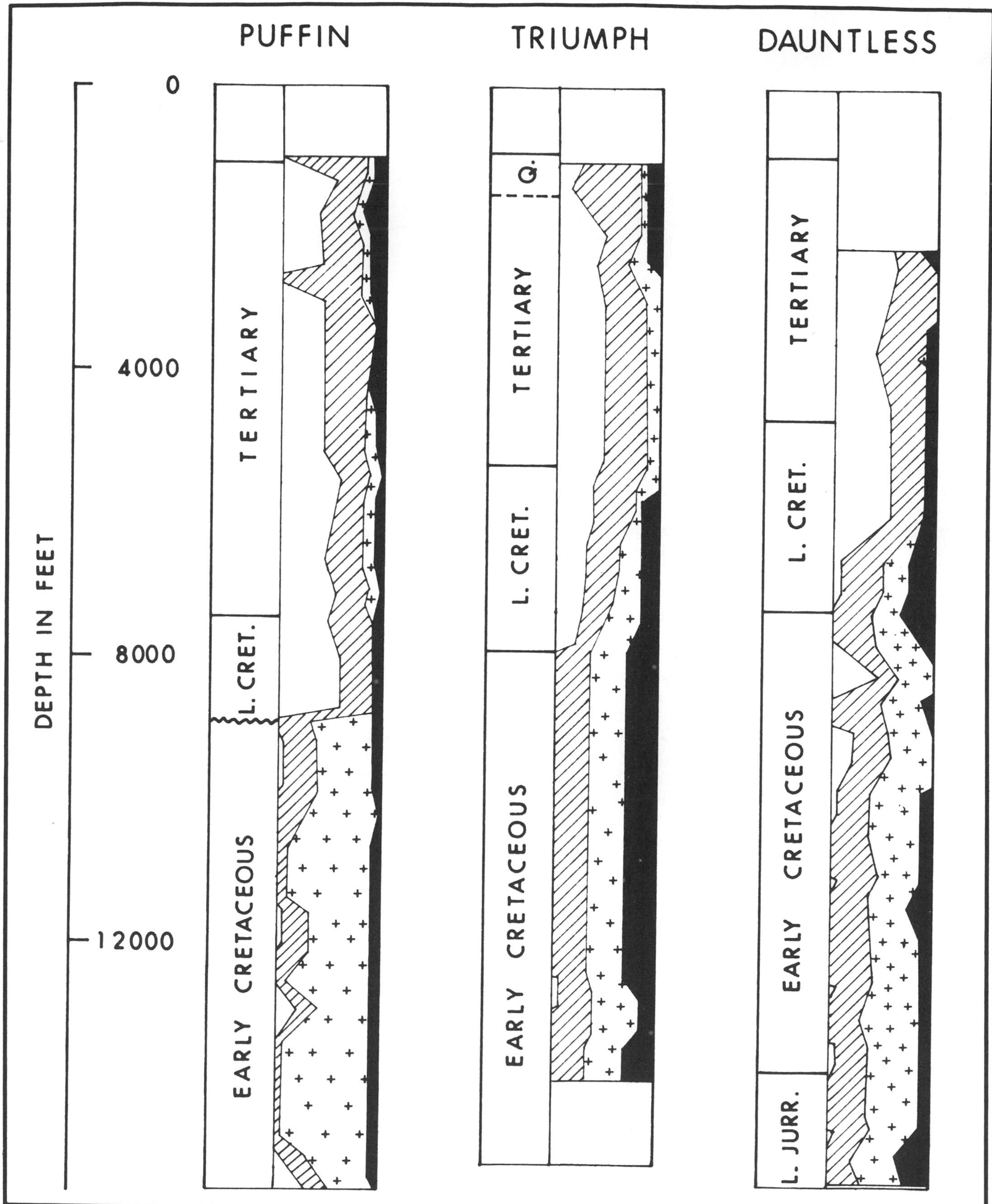

Figure 7 — Organic matter type profiles in wells drilled on the Scotian shelf (from Bujak, Barss, and Williams, 1977a,b). The white area on the left corresponds to amorphous (presumed aquatic) organic matter and the black area on the right represents coaly and recycled material. The shaded and strippled areas indicate different types of terrestrially-derived organic matter.

basin has an API gravity of 23.6°, a pour point of −36°C, and 0.5% sulfur (it appears to have been bacterially degraded; Claret et al, 1977). Cabinda produces oils from the shallow Iabe formation ("Lago crude") and a deeper "Lacula crude" which is often lighter and of higher pour point (Anon, 1981a). The Cabinda refinery stream has an API gravity of 32.9°, a pour point of 65°F (18.3°C), and 0.15% sulfur (Aalund, 1976).

Eastern Canada

The continental margin of eastern Canada is a passive, divergent margin produced by the separation of North Africa and Eurasia from Canada in the Mesozoic. Three principle stages of Mesozoic-early Cenozoic rifting led to somewhat different tectonic styles along the margin (Jansa and Wade, 1975; Umpleby, 1979; Purcell, Umpleby, and Wade, 1980). East of the Nova Scotia uplift, rifting and subsequent subsidence produced a series of grabens with a variety of salt structures. North of this region of the Grand Banks and East Newfoundland, basins are defined by two major transform faults on the southwest and northeast. Basement uplifts and large block-faulted basins were developed along with strike-slip faults, growth faults, and salt tectonism. North of the bounding transform fault is the Labrador shelf. Here, there are basement uplifts and block-faulted basins, often offset by strike-slip movement, but there is no salt north of the Cartwright arch.

Petroleum exploration has been active and fairly successful along the eastern Canadian margin. Profiles of organic matter type have been published for many of the wells drilled offshore and selected profiles from the Scotian shelf (Bujak, Barss, and Williams, 1977a,b) are given in Figure 7. These follow the generalized trends established for passive margins (Figure 5) and show that amorphous organic matter has its highest concentration in the shallower parts of the section, where it can account for more than 50% of the organic matter. Only the deeper sections, which are dominated by various types of terrestrially-derived organic matter, have had enough temperature-time exposure to reach maturity. Consequently, only the terrestrially-dominated organic matter has generated petroleum, and in this case it is predominantly gas. Thus, the Scotian Basin and Labrador shelf areas are both gas-dominated areas. A few occurrences of oil do occur and are associated with salt piercement structures, anomalously high geothermal gradients, or, in the case of the Sable 4H-58 well, the presence of high percentages of aquatic organic matter in the marine Verrill Canyon Formation.

Further offshore there is major oil accumulation at Hibernia. This is 150 km east of Newfoundland and probably contains in excess of 2 billion bbl. Figure 5 clearly shows the trend of decreasing terrestrial organic matter with increasing distance from shore. The pour point of the Hibernia crude is about 50°F (Anon, 1981b) and it appears to be transitional, suggesting that postrift sources may have made an important contribution. Rifts that do not become restricted (and develop evaporites) may have marine incursions which develop high organic matter contents and provide good oil source rocks.

Baltimore Canyon

The continental shelf of the northern part of eastern United States is analagous to that of Nova Scotia and developed as North America and Eurasia/Africa separated. Petroleum exploration along this continental margin is at an early stage and, in the most extensively drilled area of the Baltimore Canyon, is generally disappointing. Here, only marginally commercial gas has been reported. In additon to oil company wells, two COST wells have been drilled in the Baltimore Canyon area. The COST No. B-2 well (Scholle, 1977) is nearest shore and penetrated more nonmarine section than the COST No. B-3 well (Scholle, 1980) which is further offshore. In both wells, organic matter was generally immature except in the deepest section. There was also a general trend towards increasing terresterial organic matter with depth. The bottom section of both wells is more nonmarine. For example, in the COST No. B-3 well coals occur in the Lower Cretaceous and Upper Jurassic. Neither well penetrated the salt, which is presumed to overlie Triassic nonmarine lake and swamp deposits. Interestingly, gas has recently been reparted from the prerift sediments on the corresponding coast of Africa about 20 miles east of Essaouria in Morocco (Anon, 1982).

Western Australia

The structural development of the western margin of Australia followed the classical sequence of events for rifted margins formation. Veevers (1974) and Lofting, Grostella, and Halse (1975) described these as follows: 1) rifting to form a complex of grabens in the Permian or late Jurassic to Early Cretaceous; 2) rupture and formation of juvenile ocean in the late Jurassic to Early Cretaceous; 3) subsidence of the rifted margin and deposition of marine shale; and, 4) development of an ocean with carbonates on the shelf and upper slope. The major sedimentary basins along the coast now contain Permian to Tertiary sediments and are being actively explored. Major gas production has been established in the Rankin trend with total reserves of over 12 Tcf of gas and 320 million bbl of condensate in Triassic and Jurassic reservoirs (Weaver, Houde, and Smitherman, 1980). The oil accumulations in this area appear to be younger and shallower than the gas, but the Barrow Island oil has a pour point of 30 to 60°F and a sulfur content of 0.04%. Alexander, Kagi and Woodhouse (1981) found that in this area the shallow Windalia oil was predominantly naphthenes and aromatics with less than 5% normal paraffins. In contrast, the deeper Jurassic sediments produced shows of paraffinic oil.

Failed Rifts

Some of the rift systems that failed to open into oceans remained as major sedimentary basins and developed into important petroleum provinces. The Sirte

Table 2. Properties of oils from the Reconcavo basin, Brazil (data from Ghignone and Andrade, 1970).

Field	°API	Pour Point °C(°F)	Paraffins %	Sulfur %
Agua Grande	40-41	32(90)	24	0.04
Dom Joao	36-38	29(84)	15	0.05
Miranga	38-41	32(90)	—	<0.1
Candeis	29-32	30(86)	—	0.09
Buracia	33-35	36(97)	—	—

basin in Libya and the Cambay basin in India provide examples, and in both cases the dominant production is waxy crude. Waxy crudes are also well-developed in the Reconcavo-Tucono rift basin of eastern Brazil and its pour points often exceed 90° F (32°C; Table 2). There is also some minor gas production in this basin.

The Sirte basin indents the Mediterranean coast of Libya and trends roughly south to southeast. It is mainly an upper Mesozoic and Tertiary feature developed on an old basement and eroded Paleozoic surface. Regional horst and graben trends began to develop in mid-Cretaceous. The Sarir oil field is producing from sands close to the basement at 9000 ft (2745 m) and the 37° API oil is highly paraffinic with a wax content of 19%. Pour points range from 55 to 75°F (13 to 24°C) and sulfur is less than 0.25% (Sanford, 1970). The oil in the Amal field described by Roberts (1970) is also highly paraffinic with a pour point of 65°F (18.3°C) and an API gravity of 35°). Pour points for oils from other fields in the Sirte basin are listed by Hedberg (1968) and generally range from 30 to 75°F (−1 to 24°C).

The Cambay basin is located in the northwest corner of India and is part of a rift system that extends offshore to include the Bombay High field (Chowdhary, 1975). Oils in the carbonate reservoirs of the Bombay High are highly paraffinic. Other paraffinic oils also occur in the landward part of the rift and are described by Nerouchev, Paskov, and Bhattachanya (1968). Pour points range up to 90°F (32°C, Hedberg, 1968), though there are some low pour point oils that appear to be bacterially degraded. There is also a limited amount of gas production.

Waxy Crudes or Gas?

The preceeding examples show that organic matter associated with continentally-derived sediments can generate gas from woody organic matter or very paraffinic, waxy crude oils from surface coatings. From an exploration viewpoint, it is important to predict whether gas or waxy crude will be the dominant product in any particular area. This involves consideration of the conditions controlling the relative amounts of wood and surface coatings. Here, two factors are important: first, the relative amount of the biological materials that are produced by the local flora or transported to the area of sedimentation, and second, the ability of these materials to survive degradation during transport, deposition, and accumulation. Both the composition of the plants and the degree of degradation depend on climate. We noted an extreme example earlier in that terrestrial organic matter is absent from polar regions while algal organic matter may be abundant in polar seas. Many factors influence climate, but proximity to the equator is a rough general guide.

The basins bordering the South Atlantic that now produce waxy crudes from the rift and prerift sediments were all within 20° of the equator at the time of rifting (based on the reconstructions of Habicht, 1979). The Reconcavo-Tucano embayment, the Sirte basin, and the Cambay basin were also within 20° of the equator during active rifting. Other occurrences of waxy crude oils extend the range much further north and south. In contrast, the major gas-prone areas on passive margins appear to have been more than 20° north or south at the time of rifting. The most plausible explanation is not that the organic matter types growing in the various zones were different (since there is no biochemical evidence to support this) but rather that the different types of organic matter survived degradation to different extents, with the more resistant outer coating of terrigenous organisms resisting degradation better than lignin. Lignin is degraded by fungi, partricularly under aerobic conditions. This would lead to a preferential accumulation of waxy-crude prone organic matter in near-equatorial regions. Since coal is essentially concentrated Type III (woody) organic matter, the statistics for coal distribution can be used to clarify the factors controlling lignin distribution. Blackett (1961) lists no important occurrences of coals between 25°N and 35°S, with the dominant deposits being in temperate regions at the time of coal formation. These preliminary views on the factors controlling the distribution of gas as opposed to waxy crudes must be interpreted with caution until source rock character and petroleum production from passive margins are better documented.

DELTAS

As continental blocks rift, separate, and develop into oceans, clastic material can be transported by rivers and deposited on the subsiding margin to build major delta sequences. Aborted rifts also exercise control over drainage and have an important role in establishing delta position as discussed above.

Along with the clastic sediments, rivers also transport organic matter from the land surface to the continental margin and bring nutrients which stimulate the growth of aquatic organisms offshore. Thus, the deltaic environment has a variety of organic matter types. These different materials are not uniformly distributed and the terrestrial organic matter has its highest concentration near the point of input. Algal-derived aquatic organic matter is most abundant somewhat further offshore. This leads to a generallized trend within a "time-slice" of terrestrial organic matter nearshore and aquatic organic matter offshore (Barker, 1979a; Figure 8). When these sediments are buried deeply enough to generate petroleum, the nearshore products are gas and waxy crude while fur-

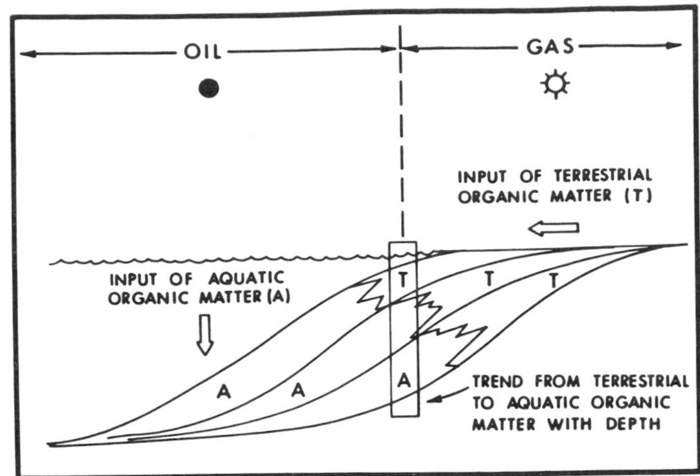

Figure 8 — Schematic representation of the distribution of terrestrial and aquatic organic matter in a prograding delta (Barker, 1979a). (T-terrestrial organic matter; A-aquatic organic matter).

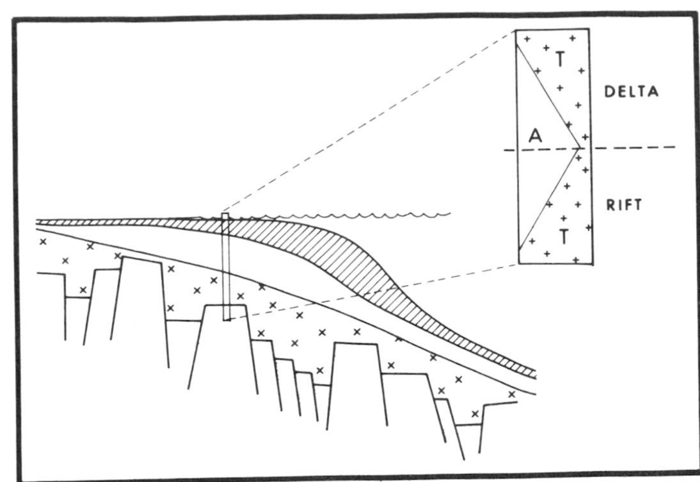

Figure 9 — Schematic illustration of a combined organic matter type profile that develops when a deltaic sequence progrades over a passive margin. (T-terrestrial organic matter; A-aquatic organic matter).

ther from the paleoshoreline the petroleum products are normal crude oil. This trend is observed in the Orinoco delta (Michaelson, 1976), the Gippsland basin (Griffith and Hodgson, 1971), and the Mahakam delta (Magnier, Oki, and Kartaadiputra, 1975). Waxy crudes are quite common in deltas and it is speculated that the full sequence from nearshore is probably gas to waxy crude to normal crude.

This treatment of deltas suggests that normal (i.e. non-paraffinic) crude oils should be more abundant in highly constructive delta systems and that waxy crudes and gas should dominate in highly destructive deltas. Latitude does not seem to be a controlling factor because waxy crudes are common from the Niger delta (almost on the equator) to the Mackenzie delta and the Latrobe formation of the Gippsland basin, which both formed at very high latitudes. Waxy crudes are not common in the Mississippi delta and more work needs to be done to clarify the relative importance of the type of organic matter being deposited and the highly constructive nature of this delta.

As the delta progrades into its receiving basin, sediments enriched in terrestrial organic matter are deposited over more aquatic rich sediments so that at any given location a vertical profile of terrestrial over aquatic organic matter develops (Barker, 1979b; Figure 8). Note that this profile has exactly the opposite sequence to the one that develops on passive margins, and leads to a different vertical distribution of oil and gas. The Tom O'Connor field in the prograding delta sequence of South Texas provides an excellent example of a vertical sequence from shallower gas to deeper oil (Mills, 1970).

LIMITATION OF SIMPLE MODELS

Multiple Environments

At any geographic location the style of sedimentation may change considerably through geologic time and the factors controlling the distribution of different organic matter types will also change. As rift valleys widen and evolve into passive margins, for example, a deltaic sequence may prograde over them. Since in the underlying sediments the type of organic matter and its distribution remain unchanged, organic matter type profiles can be stacked one on top of the other. In the example of a delta sequence deposited on a passive margin, the terrestrial-over-aquatic profile of the deltas will overlie the aquatic-over-terrestrial profile of the rift/passive margin stage of evolution. This idealized sequence is shown in Figure 9.

Growth of deltas onto downwarped rifted margins is not uncommon, though, in most cases, the full sequence has not yet been drilled. Off the Labrador shelf many wells have penetrated the postrift sequence developed in the Tertiary and some of the organic matter type profiles are given in Figure 10. The terrestrial-over-aquatic profile that has developed is clear.

Thermal Maturity

Organic matter type exercises initial control on the nature of the petroleum generated (normal crude versus waxy crude versus gas), but the composition of petroleum may undergo major changes after it has accumulated in the reservoir. These changes can be induced by rising temperature or by external influences such as water washing and bacterial degradation (Milner, Rogers, and Evans, 1977). In the case of thermal maturation, the normal trend is for oils to get lighter and evolve through high API gravity oils into condensates, wet gas, and, finally, dry gas.

Significant changes in the areal distribution of oil and gas can result from the thermal cracking of crude oil to gas. This is well-illustrated in the deltaic sequence of the Lower Wilcox of South Texas. Data from Fisher and McGowen (1967) are plotted in Figure 11 to show how the gas-to-oil ratio changes with increasing

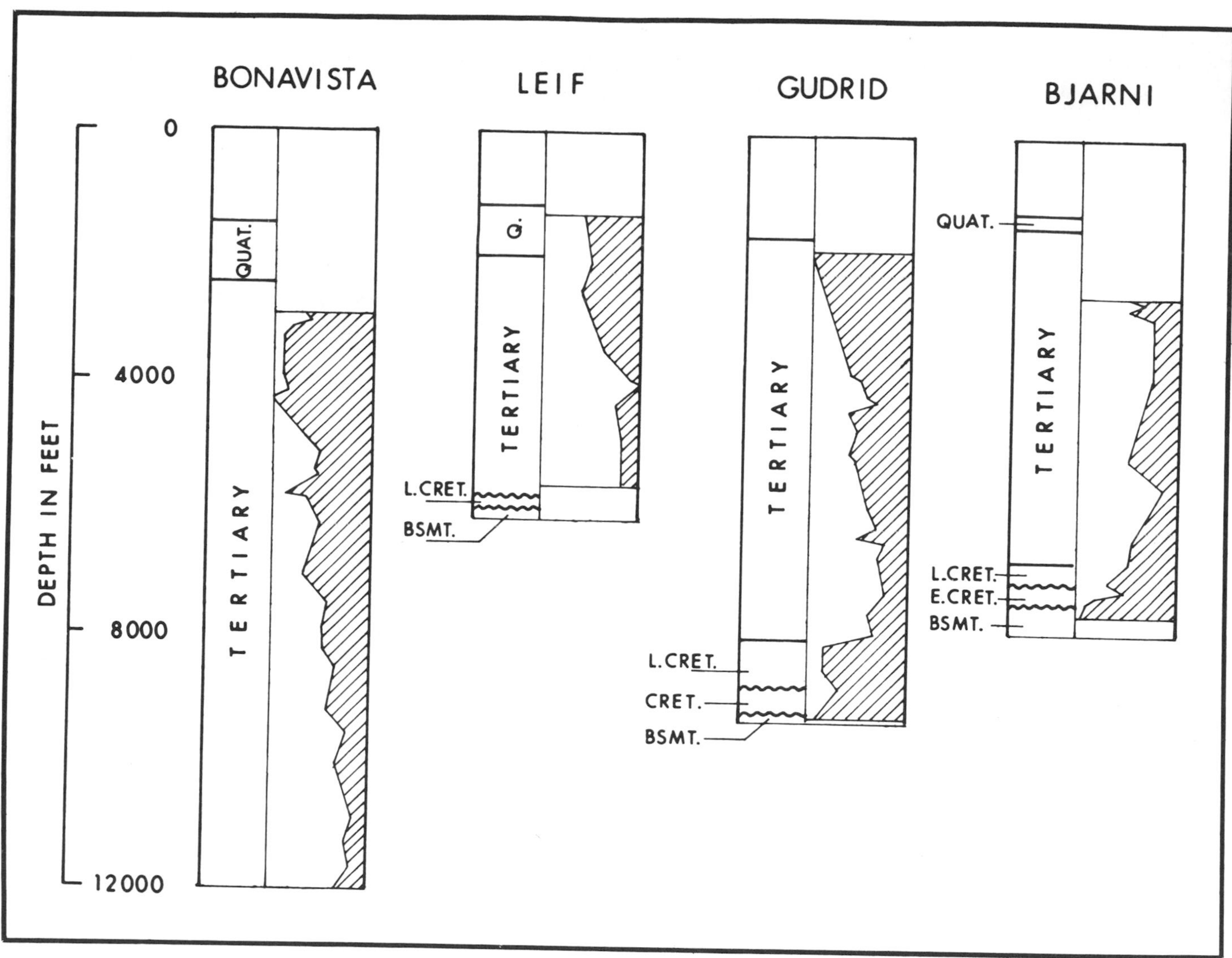

Figure 10 — Organic matter type profiles in wells drilled on the Labrador shelf (from Bujak, Barss, and Williams, 1977a,b). The white area on the left corresponds to amorphous (presumed aquatic organic matter) while the shaded area on the right represents various types of terrestrially-derived organic matter.

distance from the paleoshoreline. Nearshore the ratio is about 2 but decreases seawards to 1 as oil becomes relatively more abundant. However, with continuing increase in distance from shore to shore, gas-oil ratio climbs to 2, then 11 and is over 77 in the environment furthest from shore. The simple model for a delta developed above predicted oil as the dominant petroleum type far from shore. In the case of the L. Wilcox the offshore sediments are now buried at 17-19,000 feet and the geothermal gradient is above average so that crude oil is being thermally cracked to give gas. Thermally cracked oil is a much more prolific generator of gas than the normal process of gas generation from a woody type of kerogen.

Bacterial degradation

Oils derived from the surface coating type of terrestrial organic matter have paraffin contents which lead to high pour points. Since gas chromatograms, or even percent paraffins, are often not available in the literature high pour point can be taken as an indication of terrestrial source. In some situations this interpretation must be used with care because bacterially-degraded, terrestrially-sourced oils can have low pour points. Both field studies and laboratory simulations show that bacteria which degrade crude oils in the subsurface consume the straight chain ("normal") paraffins in preference to any other compound type (Milner, Rogers, and Evans, 1977; Bailey et al, 1973). This preferential removal of the paraffins reduces the wax content of the oil and can produce a dramatic decrease in pour point, for example from 75°F to < − 35°F (24°C to < −19°C) in some Mackenzie delta oils (Burns, Hogarth, and Milner, 1975). A similar reduction in pour point is found in Niger delta crudes (Evamy et al, 1978) where the degraded, low pour point oils are shallow and the waxy, high pour point crudes are deeper (even in the same field).

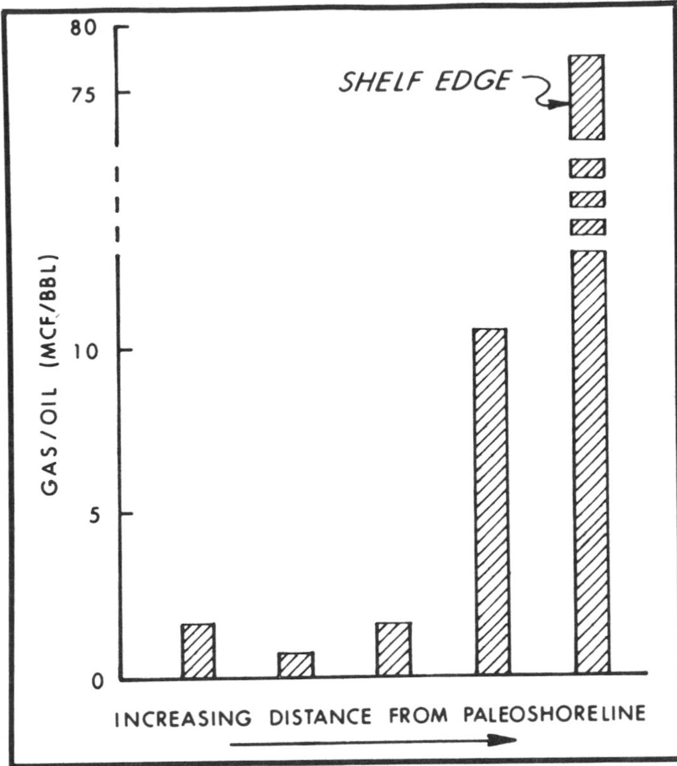

Figure 11 — Variation of gas/oil ratio with increasing distance from paleoshoreline for the Lower Wilcox, Texas (data from Fisher and McGowen, 1967).

Migration

Since petroleum is a fluid, it is mobile in the subsurface and commercial accumulations may be located over 100 km from the source rocks where generation occurred. Geochemical correlation techniques have been used to document long distance migration in intracratonic basins such as the Williston basin (Williams, 1974), the Denver basin (Clayton and Swetland, or 1980), and the Powder River basin (Momper and Williams, 1979). In this structurally simple setting, petroleum can move considerable distances updip by buoyancy because there is good continuity in the permeable carrier beds. The permeable beds in deltas on continental margins are broken up by growth faulting and lack long distance continuity, while on rifted margins the block faulting that leads to horst and graben development also breaks up the pearmeable carrier beds. In both deltas and rifted margins short range migration leads to the development of a larger number of separate petroleum accumulations. These can be quite large if the local source rock is prolific, as, for example, in the North Sea.

Additional complications arise when oil and gas are migrating together, and the original distributions that are controlled by organic matter type may be modified considerably. An example from the Niger delta illustrates this point. In the Niger delta, much of the oil and gas is reservoired in rollover anticlines up against growth faults. When one of these traps is filled to the spill point any additional petroleum it receives causes spilling off the bottom and into the next updip trap through the growth fault (assuming that this is acting as a conduit; Figure 12). Even if the reservoir contains a substantial gas cap over the oil, oil will be spilled in a process of differential entrapment (Gussow, 1954). In a series of reservoirs the net effect is to develop a sequence from gas downdip to oil updip. In a delta setting, updip is generally nearshore so that a trend from oil nearshore to gas offshore is produced (Barker, 1979b). This is, of course, exactly the opposite of the distribution of oil and gas deduced from the simple delta model discussed earlier. Thus, prediction of gas distribution in a deltaic setting requires an evaluation of the relative importance of differential entrapment.

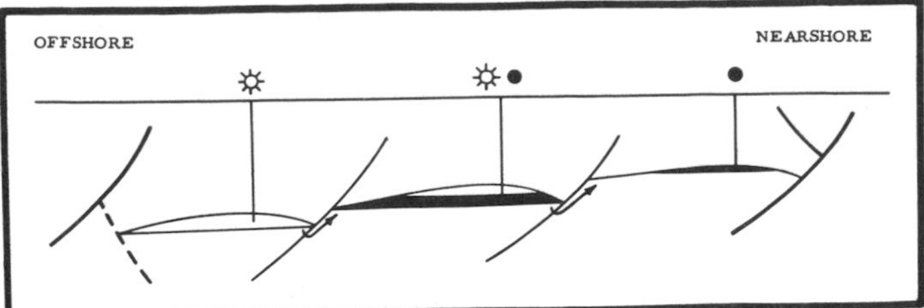

Figure 12 — Illustration of the way in which petroleum migration and differential entrapment can lead to gas downdip (offshore) and oil updip (nearshore) in a deltaic setting such as the Niger delta.

CONCLUSIONS

This paper combines a simple classification of organic matter types with simplified models of sedimentary environments to predict generalized trends for the distribution of normal crudes, waxy crudes, and gas on passive margins. The sequence from continental rift to open-marine conditions produces an idealized vertical organic matter type profile with abundant terrestrial organic matter deep and aquatic organic matter shallow. In contrast, the vertical organic matter type profile that develops in prograding deltas is from aquatic organic matter deep to terrestrial organic matter shallow. Deltas also show a trend from more terrestrial organic matter nearshore to aquatic offshore. These distributions of organic matter types exercise initial control on the location of gas, waxy crudes, and normal crudes. There may be subsequent modifications and redistributions due to migration, maturation, and alteration.

The treatment of organic matter types and their relationships to depositional setting has been necessarily general, but the models can be refined as more data become available. At present, only a limited number of passive margins have been thoroughly explored and very few wells have penetrated the complete sedimentary sequences. Data (such as shows) from wells that are subsequently not commercial rarely appear in the literature, and even for productive areas the geochemical data are very limited. I hope that this paper encourages publication of pertinent data and stimulates the appropriate extension, modification, or revision of the generalized trends presented.

REFERENCES CITED

Aalund, L. R., 1976, Wide variety of world crudes gives refiners range of charge stocks: Oil and Gas Journal, v. 74, n. 13, p. 87-89.

Alexander, R., R. I. Kagi, and G. W. Woodhouse, 1981, Geochemical correlation of Windalia oil and extracts of Winning Group (Cretaceous) potential source rock, Barrow Subbasin, Western Australia: AAPG Bulletin, v. 65, p. 235-250.

Allard, G. E., and V. J. Hurst, 1969, Brazil-Gabon geologic link supports continental drift: Science, v. 163, p. 528-532.

Anonymous, 1981a, Oil production hike due off Cabinda: Oil and Gas Journal, v. 79, n. 46, p. 42-45.

———, 1981b, St. John's symposium provides priceless peek at Hibernia discovery: Oilweek, v. 32, n. 4, p. 4.

———, 1982. Morocco yields pre-Jurassic strike: Oil and Gas Journal, v. 80, n. 3, p. 69.

Asmus, H. E., and F. C. Ponte, 1973, The Brazilian marginal basins, *in* A. E. Nairn and F. G. Stehli, eds., The Ocean Basins and Margins, The South Atlantic: Plenum Publishing Company, v. 1, p. 87-133.

Bally, A. W., 1975, A geodynamic scenario for hydrocarbon occurrences: Tokyo, Proceedings, 9th World Petroleum Congress; Essex, England, Applied Science Publishing, v. 2, p. 33-44.

———, and S. Snelson, 1980, Realms of subsidence *in* A. D. Miall, ed., Facts and Principles of World Petroleum Occurrence: Canadian Society of Petroleum Geology, Memoir 6, p. 9-94.

Barker, C., 1979a, Organic geochemistry in petroleum exploration: AAPG Continuing Education Course Note Series, n. 10, 159 p.

———, 1979b, Generation and accumulation of oil and gas in deltas, in N. H. Hyne, ed., Pennsylvanian Sandstones of the Mid-Continent: Tulsa Geological Society Special Publication, n. 1., p. 83-96.

Blackett, P. M. S., 1961, Comparison of ancient climates with the ancient latitudes deduced from rock magnetic measurements: Royal Society of London Proceedings, Series A., v. 263, p. 1-30.

Brice, S.E., and G. Pardo, 1980, Hydrocarbon occurrences in nonmarine, pre-salt sequence of Cabinda, Angola (Abs.): AAPG Bulletin, v. 64, p. 681.

———, K. R. Kelts, and M. A. Arthur, 1980, Lower Cretaceous lacustrine source beds from early rifting phases of South Atlantic (Abs.): AAPG Bulletin, v. 64, p. 680-681.

Brink, A. H., 1974, Petroleum Geology of Gabon basin: AAPG Bulletin, v. 58, p. 216-235.

Brognon, G. P. and Verrier, G. R., 1966. Oil and geology in Cuanza basin of Angola: AAPG Bulletin, v. 50, p. 108-158.

Bujak, J. P., M. S. Barss, and G. L. Williams, 1977a, Offshore East Canada's organic type and color and hydrocarbon potential, part 1: Oil and Gas Journal, v. 75, n. 14, p. 198-202.

———, ———, and ———, 1977b, Offshore East Canada's organic type and color and hydrocarbon potential, part 2: Oil and Gas Journal, v. 75, n. 15, p. 96-100.

Bullard, E., J. E. Everett, and A. G. Smith, 1965, The fit of the continents around the Atlantic: Royal Society of London, Philosophical Transactions, Series A, 258, p. 41-51.

Burke, K., and J. F. Dewey, 1973, Plume-generated triple junctions: key indicators in applying plate tectonics to old rocks: Journal of Geology, v. 81, p. 406-433.

Burns, B. J., J. T. C. Hogarth, and C. W. D. Milner, 1975, Properties of Beaufort Basin liquid hydrocarbons: Bulletin of Canadian Petroleum Geology, v. 23, p. 295-303.

Burollet, P. F., 1980, Petroleum potential of the Mediterranean basins: Canadian Society of Petroleum Geology, Memoir 6, p. 707-221.

Chowdhary, L. R., 1975, Reversal of basement-block motions in Cambay Bay, India, and its importance in petroleum exploration: AAPG Bulletin, v. 59, p. 85-96.

Clayton, J. L., and P. J. Swetland, 1980, Petroleum generation and migration in Denver basin: AAPG Bulletin, v. 64, p. 1613-1633.

Claret, J., S. Jardine, and P. Robert, 1981, The diversity of oil source rocks: Geological context and economic implications as suggested by four examples: Pau, Bulletin des Centres de Recherches, v. 5, p. 383-412.

———, et al, 1977, Un exemple d'huile biodegradee a basse teneur en soufre: le gisement d'Emeraude (Congo) *in* R. Campos and J. Goni, eds., Advances in Organic Geochemistry, 1975: Madrid, Enadimsa, p. 507-522.

Demaison, G. J., and G. T. Moore, 1980, Anoxic environments and oil source bed genesis: AAPG Bulletin, v. 64, p. 1179-1209.

de Oliveira, A. I., 1956, Brazil, in W. F. Jenks, ed., Handbook of South American Geology: Geological Society of America, Memoir 65.

Dow, W. G., 1979, Petroleum source beds on continental slopes and rises: AAPG Memoir 29, p. 423-442.

Eglinton, G., and R. J. Hamilton, 1967, Leaf epicuticular waxes: Science, v. 156, p. 1322-1324.

Evamy, B. D., et al, 1978, Hydrocarbon habitat of tertiary Niger delta: AAPG Bulletin, v. 62, p. 1-39.

Evans, R., 1978, Origin and significance of evaporites in basins around Atlantic margin: AAPG Bulletin, v. 62, p. 223-234.

Ferreira, J. C., and P. C. Gaglianone, 1980, Sergipe-Alagoas basin, Barzil. Source rock characterization and evaluation (Abs): AAPG Bulletin, v. 64, p. 705.

Fisher, A. G., and J. H. McGowen, 1967, Depositional systems of the Wilcox Group of Texas and their relationship to occurrence of oil and gas: Gulf Coast Association of Geological Societies, Transactions, v. 17, p. 105-125.

Franks, S., and A. E. M. Nairn, 1973, The equatorial marginal basins of west Africa, in A. E. Nairn and F. G. Stehli, eds., The Ocean Basins and Martins: New York, Pluenum Publishing Company, The South Atlantic, v. 1, p. 301-350.

Garb, F. A., 1981, Oil and gas in China: World Oil, v. 192, n .2 , p. 35-41.

Ghignone, J. I., and G. Andrade, 1970, General geology and major oil fields of Reconcavo basin, Brazil, in M. T. Halbouty, ed., Geology of Giant Petroleum Fields: AAPG Memoir 14, p. 337-358.

Griffith, B. R., and E. A. Hodgson, 1971, Offshore Gipsland basin fields: Australian Petroleum Exploration Association Journal, v. 11, p. 85-89.

Gussow, W. C., 1954, Differential entrapment of oil and gas: a fundamental principle: AAPG Bulletin, v. 38, p. 816-853.

Habich, J. K. A., 1979, Paleoclimate, paleomagnetism, and continental drift: AAPG Studies in Geology No. 9, 31 p.

Hadley, N. F. 1980, Surface waxes and integumentary permeability: American Scientist, v. 68, p. 546-553.

Hedberg, H. D., 1968, Significance of high wax oils with respect to genesis of petroleum: AAPG Bulletin, v. 52, p. 736-750.

Holloway, P. J., 1977, Aspects of cutin structure and formation: Biochemical Society Transcripts, v. 5, p. 1263-1266.

Hourq, V., 1966, Le bassin cotier congolais, in D. Reyre, ed., Sedimentary basins of the African coasts: New Delhi,

1964 Symposium, p. 197-206.
Hunt, J. M., 1979, Petroleum geochemistry and geology: San Francisco, W. H. Freeman, 617 p.
Hurley, P. N., et al, 1967, Test of continental drift by comparison of radiometric ages: Science, v. 157, p. 495-500.
James, D. E., 1971, Plate tectonic model for the evolution of the central Andes: Geological Society of American Bulletin, v. 82, p. 3325-3346.
Jansa, L. F., and Wade, A., 1975, Geology of the continental margin off Nova Scotia and Newfoundland, in W. J. M. Van der Linden and J. A. Wade, eds., Offshore Geology of Eastern Candada: Geological Survey of Canada, Paper 74-30, v.2.
Kinsman, D. J. J., 1975, Rift valley basins and sedimentary history of trailing continental margins, in A. G. Fischer and S. Judson, eds., Petroleum and Global Tectonics: Princeton University Press, p. 83-126.
Klemme, H. D., 1975, Giant oil fields related to their geologic setting: a possible guide to exploration: Bulletin of the Canadian Petroleum Geologists, v. 23, p. 30-66.
———, 1977 , One-fifth of reserves lie offshore: Oil and Gas Journal, Petroleum 2000, v. 75, n. 35, p. 108-128.
Klemme, H. D., 1980, Petroleum basins — classification and characteristics: Journal of Petroleum Geologists, v. 3, p. 187-207.
Lijmbach, G. W. M., 1975, On the origin of petroleum: London, Proceedings, 9th World Petroleum Conference, Applied Science Publishers, v. 2, p. 357-369.
Lofting, M. J. W., A. Grostella, and J. W. Halse, 1975, Exploration results and future prospects in the northern Australian region: London, Proceedings, 9th World Petroleum Congress, v. 2.
Lowell, J. D., et al, 1975, Petroleum and late tectonics of the southern Red Sea, in A. G. Fischer and S. Judson, eds., Petroleum and Global Tectonics: Princeton University press, p. 129-153.
Lucas, P. T., and J. M. Drexler, 1975, Altamont-Bluebell — A major naturally fractured stratigraphic trap, Uinta Basin, Utah, in J. Braunstein, ed., North American Oil and Gas Fields, AAPG Memoir 24, p. 121-135.
Lutz, M., J. P. H. Kaasschieter, and D. H. van Wijhe, 1975, Geological factors controlling Rotliegend gas accumulations in the Mid-European basin: Proceedings, 9th World Petroleum Congress, v. 2, p. 93-103.
Magnier, P., T. Oki, and L. W. Kartadiputra, 1975, The Mahakam delta, Kalimantan, Indonesia: Proceedings, 9th World Petroleum Congress, v. 2, p. 239-250.
Michelson, J. E., 1976, Miocene deltaic oil habitat, Trinidad: AAPG Bulletin, v. 60, p. 1502-1519.
Mills, H. G., 1970, Geology of Tom O'Connor Field, Rufugio County, Texas: AAPG Memoir 14, p. 292-300.
Milner, C. S. D., M. A. Rogers, and C. R. Evans, 1977, Petroleum transformations in reservoirs: Journal of Geochemical Exploration, v. 7, p. 101-153.
Momper, J. A., and J. A. Williams, 1979, Geochemical exploration in the Powder River basin: Oil and Gas Journal, v. 77, n. 50, p. 129-134.
Morgan, W. J., 1971, Convection plumes in the lower mantle: Nature, v. 230, p. 42-43.
Nerouchev, S. G., Y. V. Paskov, and S. N. Bhattacharya, 1968, Crudes of the Cambay basin and their migration: India Oil and Natural Gas Commission Bulletin, v. 5, p. 31-37.
Ponte, F. C., J. R. Fonseca, and R. G. Morales, 1977, Petroleum geology of the eastern Brazilian continental margin: AAPG Bulletin, v. 61, p. 1470-1482.
———, ———, and A. V. Carozzi, 1980, Petroleum habitats in the Mesozoic-Cenozoic of the continental margin of Brazil: Canadian Society of Petroleum Geology, Memoir 6, p. 857-886.
Prahl, F. G., J. T. Bennett, and R. Carpenter, 1980, The early diagenesis of aliphatic hydrocarbons and organic matter in sedimentary partiulates from Dabob Bay, Washington: Geochimica et Cosmochimica Acta, v. 44, p. 1967-1976.
Purcell, L. P., D. C. Umpleby, and J. A. Wade, 1980, Regional geology and hydrocarbon occurrences off the east coast of Canada, in A. D. Miall, ed., Facts and principles of world petroleum occurence: Canadian Society of Petroleum Geology, Memoir 6, p. 551-566.
Roberts, J. M., 1970, Amal field, Libya, in M. T. Halbouty, ed., Geology of Giant Petroleum Fields, AAPG Memoir 14, p. 438-448.
Sanford, R. M., 1970, Sarir oil field, Libya-Desert surprise, in M. T. Halbouty, ed., Geology of Giant Petroleum Fields: AAPG Memoir 14, p. 449-476.
Scholle, P. A., 1977. Geological studies on the COST No. B-2 well, United States Mid-Atlantic Outer Continental Shelf area: U.S. Geological Survey, Circular 750.
———, 1980, Geological studies of the COST No. B-3 well, United States Mid-Atlantic continental slope area: U.S. Geological Survey Circular 833, 132 p.
St. John, B., 1980, Sedimentary basins of the world: AAPG Map and accompanying text.
Sutton, C., 1977, Depositional environments and their relation to chemical composition of Java Sea crude oils: Manila, Association of Southeast Asian Nations, Committee for Coordination of Joint Prospecting for Mineral Resources in Asian offshore areas, Seminar on Generation and Maturation of Hydrocarbons in Sedimentary Basins.
Thomas, B. M., 1979, Geochemical analysis of hydrocarbon occurrences in northern Perty basin, Australia: AAPG Bulletin, v. 63, p. 1092-1107.
Tissot, B., et al, 1974, Influence of nature and diagenesis of organic matter in formation of petroleum: AAPG Bulletin, v. 58, p. 499-506.
Tissot, B. P., and D. H. Welte, 1978, Petroleum formation and occurrence: New York, Springer-Verlag, 538 p.
Toquato, J. R., and U. G. Cordani, 1981, Brazil-Africa geological links: Earth-Science Reviews, v. 17, p. 155-176.
Umplebey, D. C., 1979, Geology of the Labrador Shelf: Geological Survey of Canada, Paper 79-13, 34 p.
Veevers, J. J., 1974, Western continental margin of Australia, in C. A. Burke and C. L. Drake, eds., The ecology of Continental Margins: New York, Springer-Verlag, p. 605-616.
Wanli, Y., L. Yongkang, and G. Ruigi, 1981, Formation and evolution of nonmarine petroleum in the Songliao basin, China: Scientific Research and Design Institute of Daging Oil Field, p. 1-22.
Weaver, O. D., Y. Houde, and J. Smitherman, 1980, Western Australia: oil and gas potential — part 1: Oil and Gas Journal, v. 78, n. 2, p. 139-142.
Williams, J.A., 1974, Application of oil-correlation and source rock data to exploration in the Williston basin: AAPG Bulletin, v. 58, p. 1243-1252.

Intraslope Basins in Northwest Gulf of Mexico: A Key to Ancient Submarine Canyons and Fans

Arnold H. Bouma*
U.S. Geological Survey
Corpus Christi, Texas

The hummocky, diapirically deformed Texas-Louisiana continental slope includes three major types of intraslope basins: blocked-canyon intraslope basin, interdomal basin, and collapse basin. Major sand bodies present in the blocked-canyon intraslope basins are used to determine the sedimentary history of the Tertiary and Quaternary of this area. During relative lowering of sea level, coarse silt and sand temporarily stored near the shelf break were transported by gravity mechanisms and deposited within submarine canyons tending to nullify diapiric movement. Seismic records show these sediments as transparent to semitransparent onlapping seismic reflections. During low stands and rises in sea level, large amounts of mud were transported mainly as mud turbidites (recorded as indistinct, parallel, onlapping seismic reflections), and it is during this stage that diapiric activity begins. Pelagic and hemipelagic sediments dominate during sea level high stands (recorded as distinct parallel seismic reflections) and drape the canyons and surrounding sea floor. Extensive uplift breaks up the depositional units and eventually disrupts the continuity of canyon systems. Reconstruction of ancient submarine canyons can be achieved by correlating identified blocked-canyon basins; such a correlation is based on the assumption that only a few canyon systems presently exist.

The continental slope off Texas and Louisiana in the Gulf of Mexico is characterized by very hummocky topography (Figure 1; Martin and Bouma, 1978), caused primarily by underlying diapirs. According to several investigators (Martin, 1978), most of these diapirs formed by the vertical movement of Middle and Upper Jurassic Louann salt. Most of these salt bodies are mantled by Tertiary shale, and some of the diapirs may consist entirely of shale. Locally the shale crops out; in a few places, salt or caprock is probably exposed (Figure 12). However, most of the diapiric bodies are covered by younger Tertiary and Quaternary sediments that range in thickness from a few meters to several hundred (Bouma, Martin, and Bryant, 1980). Between the diapirs on the continental slope, there are intraslope basins; some are nearly filled to capacity with sediment.

The irregular topography of this part of the Gulf of Mexico continental slope is maintained by the continuing upward motion of the diapirs. This upward diapiric movement is generally considered variable geographically and through time, being dependent upon deposition and removal of sediments (Bouma, 1981). On the outer continental shelf, examples of slow or near-zero action are found where erosion took place during the last low sea level stand, and onlapping of younger sediments is seen on seismic records. Reefs cap several diapirs, mainly in a band above the shelf break. On the upper continental slope, a few examples of near-zero upward motion of diapirs are found where rather thick overlying unconsolidated sediments do not show upward bulging.

Typically, no two adjacent diapirs are the same size (Martin, 1980) or have the same characteristics regarding the overlying sediments. Generalizations must be handled carefully, but certain similarities do exist. The presently known bathymetry is based on available cruise tracks and not on a dense regular grid; thus, the true rugged character of the slope cannot be shown. Various charts (Uchupi, 1967, 1975; Holland, 1970; Bergantino, 1971; Sorensen et al, 1975; and Martin and Bouma, 1978) present good overall impressions of the morphology and the location of most hillocks and depressions. However, they cannot be used for bottom navigation or as a base for detailed studies.

*Presently with: Gulf Research and Development Company, Pittsburgh, Pennsylvania.

MORPHOLOGIC AND REFLECTION SEISMIC CHARACTERISTICS

On the basis of present knowledge, one can approach the problem of submarine canyons and deep-sea fans on and adjacent to this physiographic province by studying the depressions rather than attempting to reconstruct direct canyon routes. The application of seismic stratigraphy and the collection of long piston cores helps to identify different types of depressions; this information allows attempted correlation studies. Most of the depressions on this continental slope are intraslope basins. Three major types have been distinguished so far: 1) blocked-canyon intraslope basin, 2) interdomal basin, and 3) collapse basin (Bouma and Garrison, 1979; Bouma, 1981; Martin and Bouma, 1981).

A blocked-canyon intraslope basin forms when the thalweg of a submarine canyon is blocked by an upward-moving diapir. The sediments in the axis of such a basin contain sand layers overlain by hemipelagic deposits. Bottom-transport characteristics are seen on seismic reflection profiles as onlaps of reflections onto the flanks of the diapirs. Hemipelagic deposits interbedded in the canyon fill show up as draping units on the seismic records. The Gyre Basin (Figure 2) is the best described example in this category (Bouma et al, 1978).

An interdomal basin forms when a group of coalescing, upward-moving diapirs surrounds a section of sea floor that remains more or less at its original depositional depth. Direct bottom-transported sediment cannot enter such depressions, and the only deposition that takes place comes from hemipelagic and pelagic sediments that drape the sea floor, or from local slumping. The Orca Basin is the only known example of an interdomal basin. The Orca Basin is exceptional in that it contains a hypersaline anoxic bottom layer of water about 200 m thick (Trabant and Presley, 1978).

The collpase basin is only found on crests of diapirs, and as a rule, is the smallest of the three types. Collapse basins are caused either by tensional collapse of the sediments covering the diapir or by solution of salt from the top of a diapir and collapse of the overburden (Martin and Bouma, 1978; Bouma, Martin, and Bryant, 1980). The normally flat or rounded bottom of the basin is surrounded by higher rims. The seismic reflections are parallel, but draping sediment is highly disturbed by both growth and normal faults (Figure 3). The sediments are supposedly pelagic and hemipelagic. Most areas like the Carancahaua Basin (Figure 4) show blanketing reflections, but present studies that define these types of depressions have not been completed.

Submarine Canyon Systems

As a first step toward identifying intraslope basins, one can survey the areas, especially using sparker, minisparker, and small airgun, to identify the seismic patterns of each depression. The next step is to select all blocked-canyon intraslope basins and to divide that group based on the ratio of onlapping to draping of seismic zones and the percentage of fill. This step should enable the investigator to make reconstructions of submarine canyon systems. I believe that only a limited number of canyons or canyon systems exist on

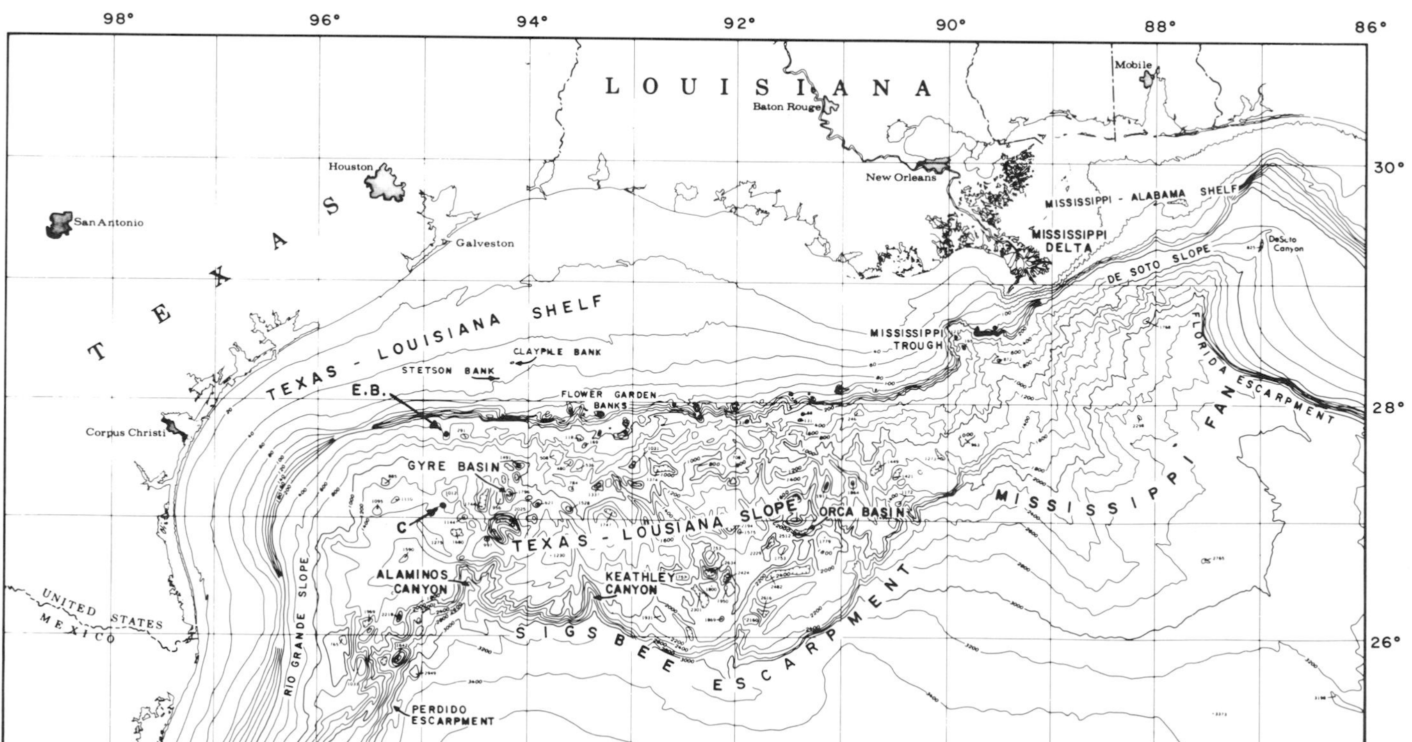

Figure 1 — General bathymetry of the northwest Gulf of Mexico. Contour intervals 20 m for the zone 0 to 200 m, 200 m thereafter. EB is East Breaks Basin; C is Carancahua Basin. From Martin and Bouma, 1978.

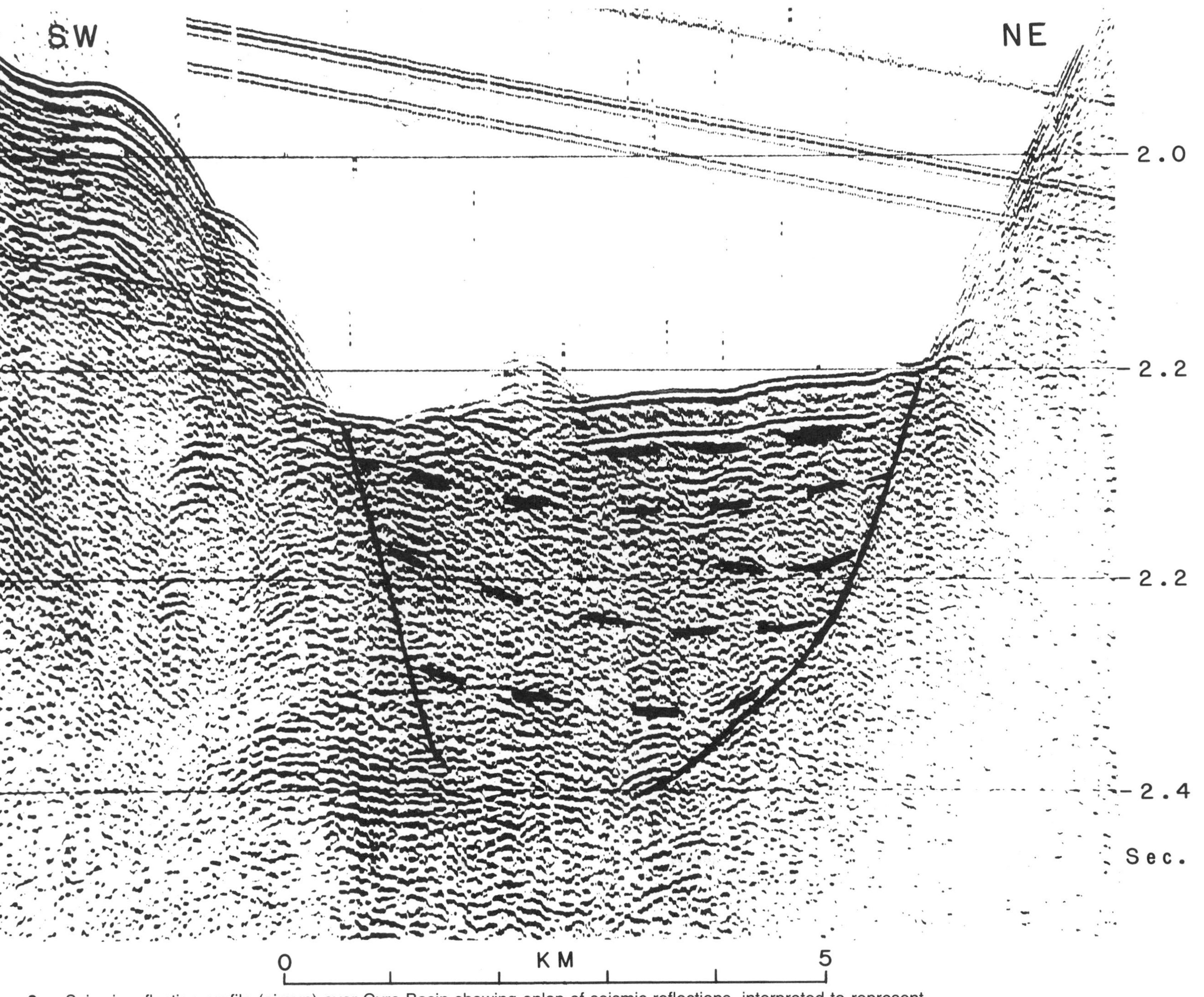

Figure 2 — Seismic-reflection profile (airgun) over Gyre Basin showing onlap of seismic reflections, interpreted to represent deposits of bottom-hugging gravity flows.

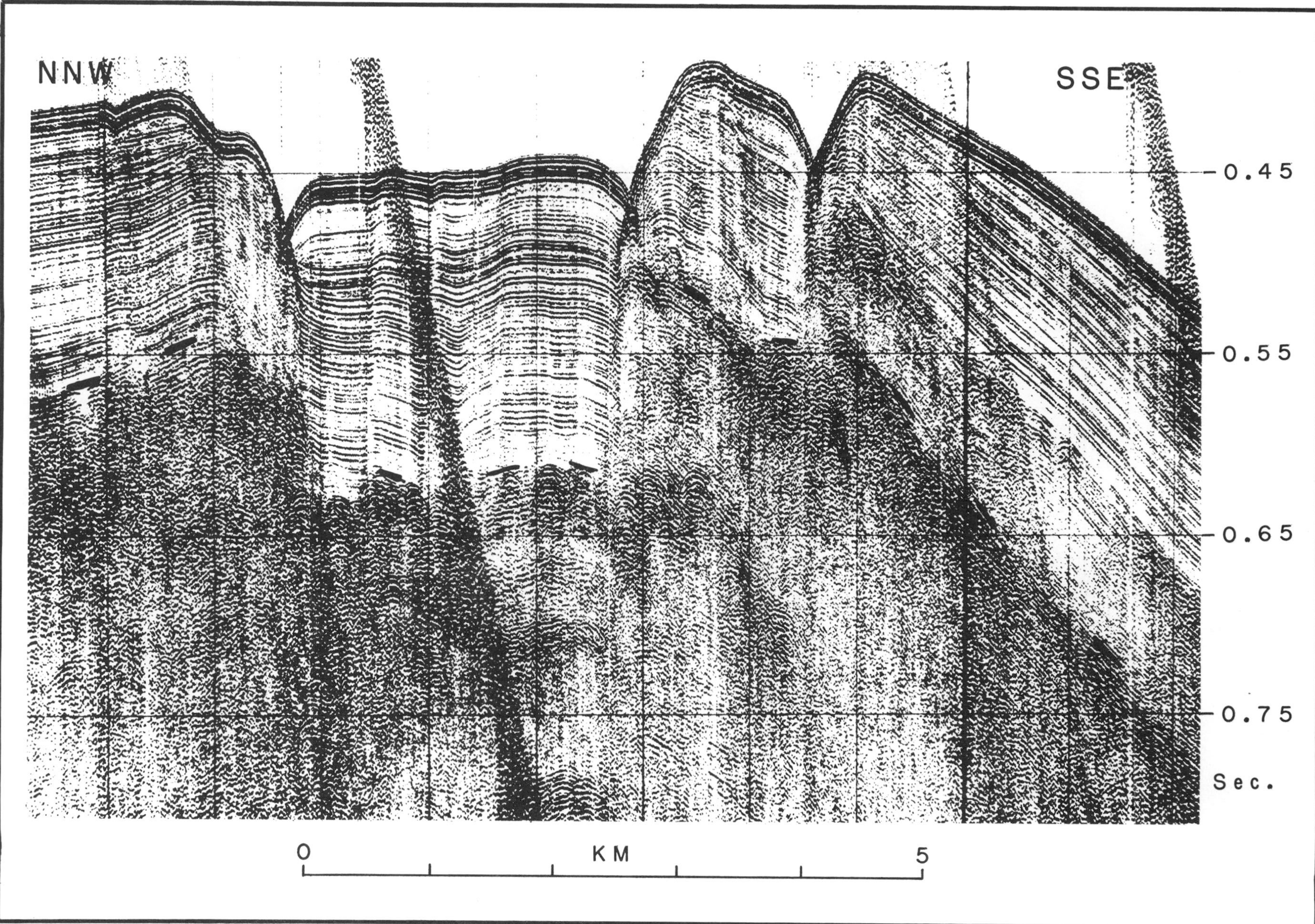

Figure 3 — Seismic-reflection profile (minisparker) over a collapse basin (East Breaks Basin at about 27°45′N and 94°50′W) showing a graben, faulting, tilting of blocks, and draping sediments.

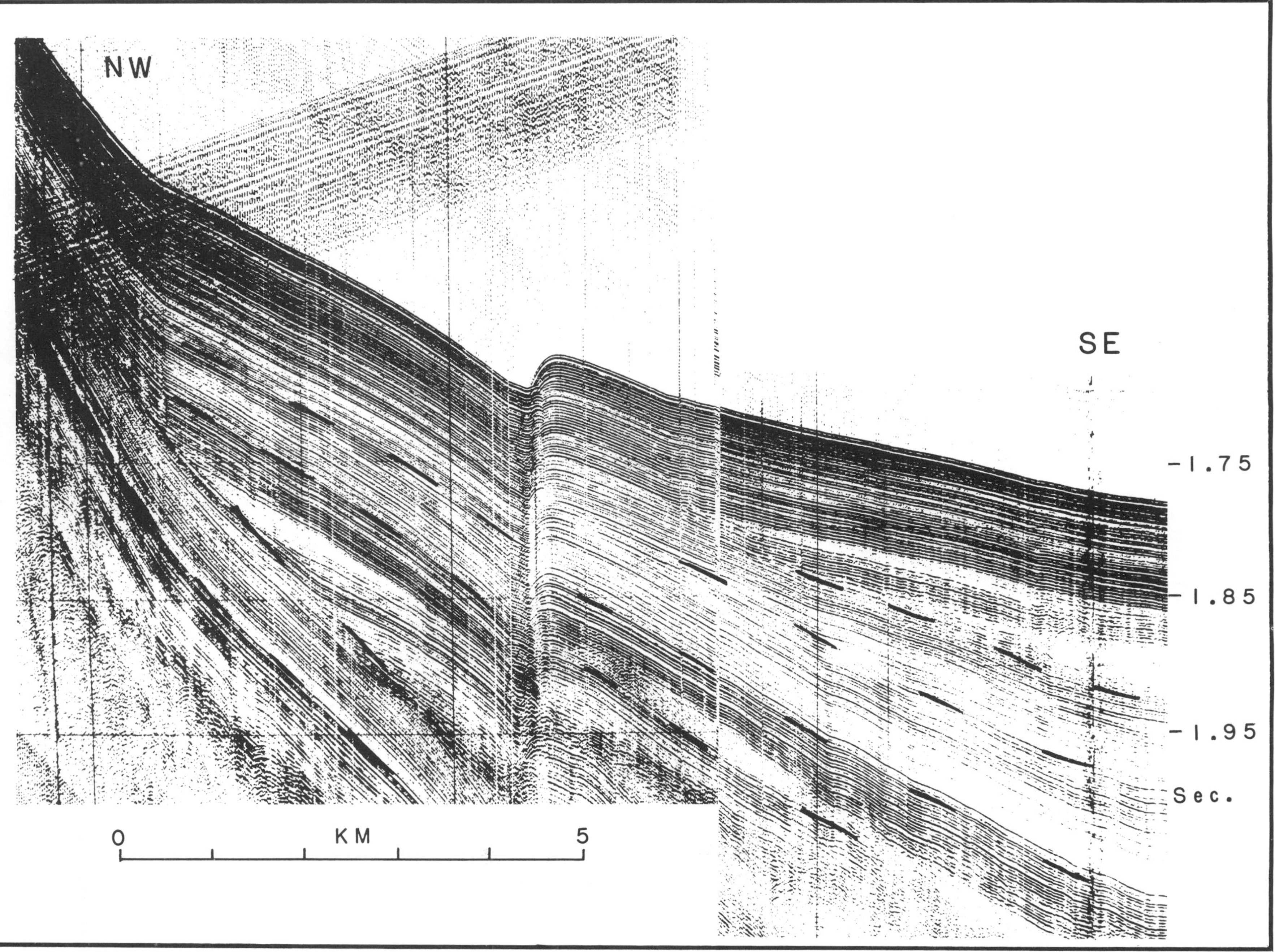

Figure 4 — Seismic-reflection profile (minisparker) of the sediment cover over a diapir forming one side of Carancahua Basin (location of record about 27°08′N, 94°45′W). The record shows seismic wedges and sediment draping.

the Texas-Louisiana Continental Slope, and that once formed, they maintained their overall position throughout that period of the geologic record during which sandy material was intermittently supplied. Because of the major change of river-discharge activity from west to east during the Tertiary, we are likely to find the oldest canyon fills in the west and the younger ones further to the east.

The origin and growth of submarine canyons and the transport and deposition of sediment are much debated (Bouma, 1981). Any of the early interdiapiric depressions can only become part of a submarine canyon system if available source material can be transported to and through it. A topographic depression may have to be filled by either mass transport (slumps, debris flow, turbidity current), or by hemipelagic sediments to the level of a spill point before a downcurrent section of the continental slope becomes part of such a canyon; finally, a continuous operative canyon system is formed that terminates at the Sigsbee Escarpment and forms the small coalescing fans on the continental rise.

Several complications should be considered. Assuming the northern Gulf of Mexico has always had a rather low and broad coastal area as well as a wide shelf during the Tertiary and Quaternary, any relative change in sea level would have an enormous effect on the availability of sandy and muddy sediment. Transport of sandy material (see below) into the head or heads of a submarine canyon (Shepard and Dill, 1966) can establish a continuous system across the continental slope if sufficient material is available. Slope sediments are probably transported via a canyon rather than deposited to form a fan; fans form on the rise and on the abyssal plain seaward of canyon mouths (Figure 5A). Sediment loading affects diapiric growth, and parts of the sea bottom underlain by diapirs will rise upward. Continued bottom transport causes ponding and scouring, and the thalweg continues to be active. Once such bottom transport ceases, upward movement of diapiric ridges (Figure 5B) and isolated bodies (Figure 5C) will cause blockage. Remaining or renewed deposition may move into interridge depressions, forming elongated fans (Figure 5B), or may start filling the last basin to which it has access. Continued diapirism may even break up existing sediment bodies into smaller ones, which in time become incorporated in the stratigraphic column as isolated sand bodies (Figure 5D). Diapiric motion steepens slopes (Bouma and Garrison, 1979; Martin and Bouma, 1981) and may cause slumping, which in retrogressive steps may move large amounts of material locally. The variation in diapiric activity, the influence of sea level changes, the effect of slumping, and the types of transport and deposition are reflected in the sediments that fill the intraslope basins and

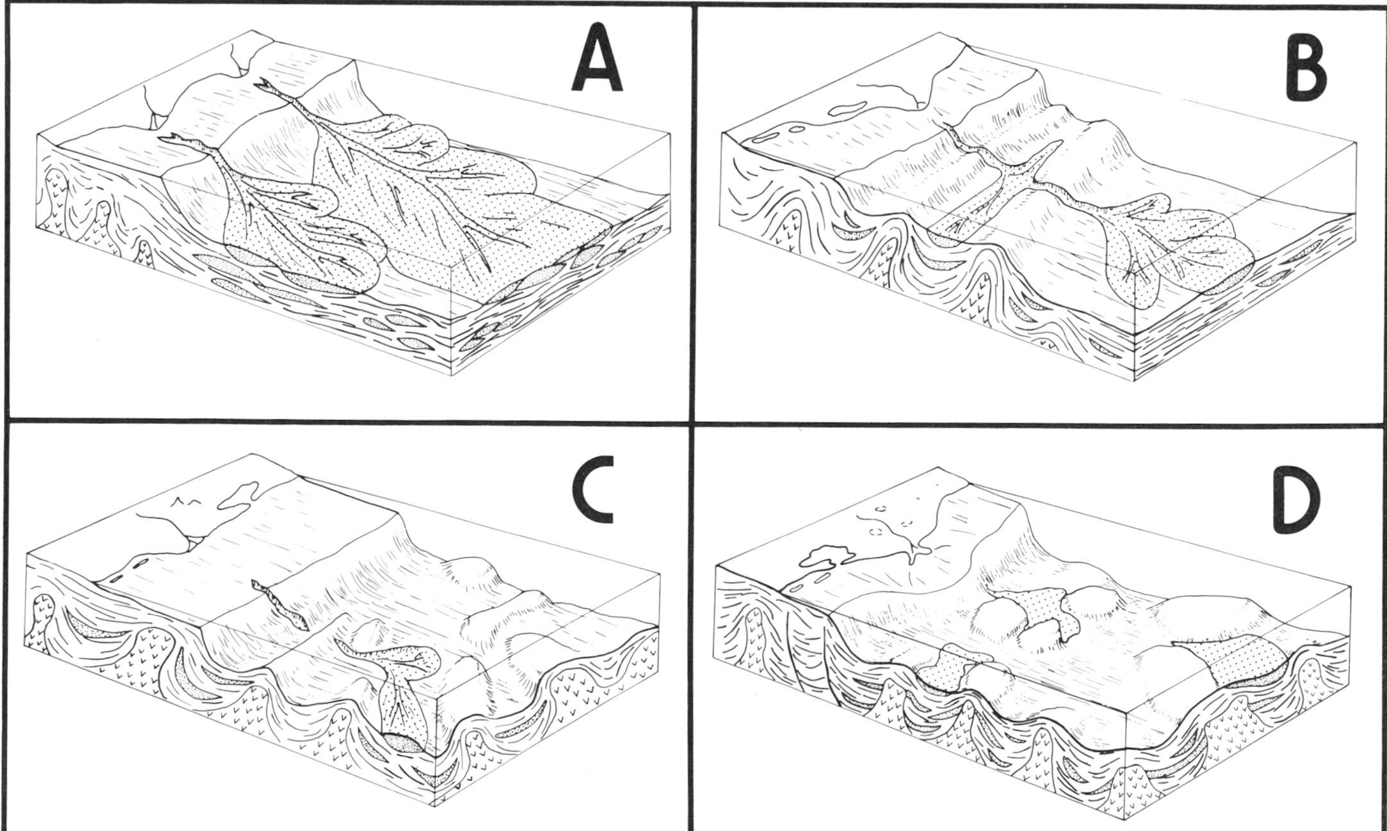

Figure 5 — Schematics showing: **A**, classical coalescing fans on lower slope and continental rise; **B**, diapiric ridge blocking a canyon which resulted in deposition in the trough to a spillpoint level, after which the canyon becomes active; **C**, the breaking apart of canyon and fan by a diapir; **D**, the breaking apart of a fan is segments by diapirs.

cover the diapirs. The records show a cyclic pattern in seismic-reflection characteristics that can be explained by the above-mentioned mechanisms (Bouma, 1981).

SEISMIC-REFLECTION PATTERNS

On airgun (5 and 40 cu in) and minisparker records, a small number of seismic-reflection patterns are observed in the younger sediments. The most important patterns are distinct sets of parallel reflections, sets of indistinct parallel reflections, and acoustically semitransparent and chaotic zones. A set having a certain reflection pattern can be rather uniform in thickness, wedge-shaped, or irregular in shape (see also Mitchum, Vail, and Sangree, 1977). Discontinuities are rather common, as are reflections that onlap the flanks of diapirs or drape over entire hillocks (Vail, Mitchum, and Thompson, 1977; Vail and Hardenbol, 1979).

In many undisturbed strata, a vertical sequence of seismic-reflection patterns consists of a zone of acoustically semitransparent reflections, overlain by a set of indistinct and discontinuous parallel reflections. These, in turn, are capped by a set of distinct parallel reflections (Figure 6). Some records show a distinct contact at the bottom of the set of semitransparent reflections; others display a better sequence boundary at the bottom of the set that has distinct parallel reflections (Figures 6 and 7). The distinctness of the boundary between two reflection patterns reflects local depositional changes.

During a relative lowering of sea level, large quantities of material are removed from the continent and the inner shelf by river transport to maintain or obtain an adjusted equilibrium gradient in the downriver section. The large amount of sandy and clayey material that must be moved is transported across the newly exposed shelf to the water edge. If a depression is available, a delta system may be constructed, building out rapidly because of the excess sediment available. If such a location is not available, the sediment moves by longshore transport to a more favorable site.

Continued lowering of sea level, which seems to be a fast process (Vail, Mitchum, and Thompson, 1977), may move the detritus to the outer shelf or upper slope. Once large amounts of materials are stored temporarily in reach of slope transport, they may become unstable. If so, slumping probably results and, in retorgressive steps, large amounts of material are set in motion. If the slumps take up small amounts of water or lose their internal strength, they change into debris flows; continued dilution results in turbidity currents (Middleton and Hampton, 1976; Hampton, 1979). If a submarine canyon is available across the continental slope and major debris flows or turbidity currents form, the resulting deposits may cause a set of transparent or semitransparent onlapping reflections. If the amounts deposited are small, the beginning of this set of reflections probably is indistinct.

If the thalweg of the submarine canyon is dammed near the slumping area, there are probably fewer slumps transformed into debris flows, and turbidity currents may not form at all. This situation may also be recorded as chaotic reflections. Damming of the thalweg forces ponding of the depression to the level of a spill point before continued transport takes place. Because some small chaotic reflections are seen in this semitransparent zone, many slump deposits are assumed to be either small or thin.

The thickness of the semitransparent reflection set varies from depression to depression and probably depicts either the amount of material transported and deposited or the duration of this period of lowered sea level.

Once the relative lowering of sea level slows down or stops and the rivers no longer bring large amounts of material to the coastline, the sand-size sediments are replaced by mud. The succeeding rise in sea level and the high stand make it impossible to move sand to the continental slope when wide shelves are involved. Initially, bottom-hugging mud currents may move through canyons and create rather indistinct parallel seismic onlapping reflectors. Once pelagic and hemipelagic sedimentation becomes dominant, the deposits show up as draping sets of distinct parallel reflections. The possible thick accumulation of mud on top of sand deposits (as at Shell Hole 19 near the Gyre Basin; Lehner, 1969; Bouma et al, 1978; Martin and Bouma, 1981) may cause instabilities that result in slumping again; especially when slopes are steepened because of consolidation of the deposited mud, and when the upward motion of the diapirs results from sediment loading on their lower flanks. Consolidation of the sandy deposits and the initial mud results in a lowering of the sediment surface, allowing deposition. Because the sandy sediment is transported by gravity mechanisms along the sea bottom, the sediment burden varies from place to place and is much thicker in the basins and lower flanks than higher up the side of the diapirs. The sediment pressure either starts or increases the speed of the upward diapiric motion. Such increased diapiric movement probably become effective after sea level reaches a low position. Besides slumping that results from the diapir movement and related slope steepening, the only deposition taking place is the hemipelagic and pelagic types. The resulting thickness is a function of time and amount of suspended matter available for deposition.

Comparison of airgun and minisparker records (which have different frequency ranges) shows similar sedimentary sequences but at different scales (Figures 3 and 8). The thinner sequences visible on the minisparker records must refer to smaller sea level variations. Nevertheless, these sequences represent sufficient relative lowering of sea level to transport enough sand or mud to the continental slope to affect the seismic reflection pattern (Figures 3 and 8).

SELECTED INTRASLOPE BASINS

The Gyre Basin, near latitude 27°15′N and 94°10′W, is the only well-described example of a blocked-canyon intraslope basin on the Texas-Louisiana Continental Slope (Bouma et al, 1978). The basin is about 33 km long from north to south, and about 15 km from south-southwest to north-northeast. The rim of the basin is 900 to 1,100 m deep, and the deepest part is slightly deeper than 1,650 m (Figure 9). The east and west slopes are relatively steep, from 11° to 12°. Slopes along the length axis are 1° to 2°.

The surficial sediments consist of a slightly silty clay containing abundant pelagic foraminifera and some peteropods near the surface, and silty-sand laminae at depths greater than 2 m below the mudline. High-resolution reflection profiles show that most of the slope deposits are disrupted by slumping (Figure 10). Accumulation rates based on foraminifera range from 9 to 18 cm/1,000 years in the central part of the basin to about 5 cm/1,000 years on the rims (Sidner, Gartner, and Bryant, 1978). A corehole drilled in 1966 (Hole 10-15) in an intraslope basin 23 km northeast of Gyre Basin, in a water depth of 1,327 m, contains 14 to 30 m of sandstone, according to electrical logs (Woodbury, Spotts, and Akers, 1978). Lehner (1969) described the geology of a hole (Shell Oil Company, corehole 19E) drilled on the topographic high forming the western flank of Gyre Basin, about 42 km southwest of the center of the basin. Salt was penetrated at 170 m; this salt is overlain by 18 m of turbidite sand. This, in turn, is capped by 148 m of shale, to which a late Pleistocene age is given. Assuming the sand in the axis

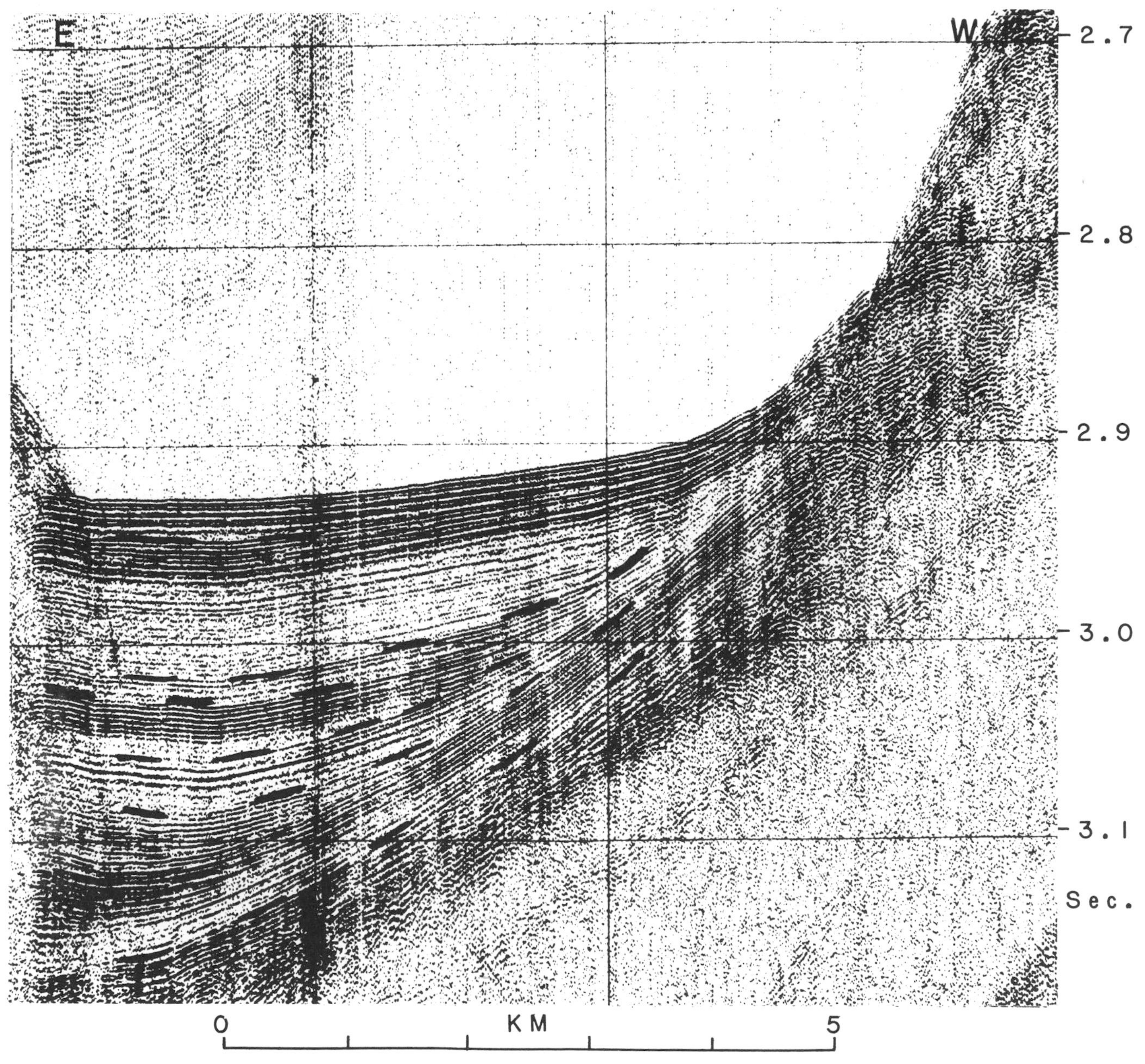

Figure 6 — Seismic reflection profile (minisparker) of a canyon fill near 27°12′N and 91°24′W. Seismic onlaps are slightly masked because of the width of the sound pulse and the small-scale irregular contours of the diapirs. Several low-angle unconformities can be distinguished. Sequence boundaries are placed at the bottom of each set of semitransparent or indistinct parallel reflectors.

of Gyre Basin is equivalent to the sand in core 19E, which has an age of 10,000 to 18,000 years and a depositional depth similar to the Gyre Basin sand (400 m deeper), then the average upward velocity of this diapir ranges from 2 to 4 cm/yr. Such a fast upward motion causes slope steepening and possible slope instability. This theoretical approach shows that the processes mentioned earlier may be extremely fast in a geological sense, especially because the calculated numbers are averages (Bouma and Garrison, 1979; Bouma, Martin, and Bryant, 1980; Martin and Bouma, 1981).

The only studied interdomal basin is the Orca Basin, centered about latitude 26°55′N and longitude 91°20′ W (Shokes et al, 1977; Trabant and Presley, 1978; McKee et al, 1978; Bouma, Martin, and Bryant, 1980). It is elbow-shaped, about 33 km long, and 15 km wide. The basin rim is 1,700 to 1,900 m deep, whereas depths greater than 2,400 m are present in both the northern and southern parts of the basin (Figure 11). This intraslope basin is exceptional in that at its bottom is a 200-m-thick, anoxic, high-salinity brine which has a temperature of 5.6°C and a salinity of 303‰. The values for the overlying water are 4.2°C and 36‰, respectively. The interface between the normal sea water and the brine is sharp and shows up on minisparker, airgun, and CDP records (Figure 12). Salinity gradients in the black mud underlying the brine decrease down core, suggesting that the salt does not come from diapirs below the basin floor, but either from the side or from higher on the surrounding diapirs. Penrose and Kennett (1979) showed that the black, well-laminated, anaerobic nonbioturbated sediments that are rich in organic matter contain well-preserved fossils. Radiolarians, peteropods, and delicate spinose foraminifera dominate in the completely pelagic sediment. Seismic-reflection records suggest that the salt core in the diapir on the

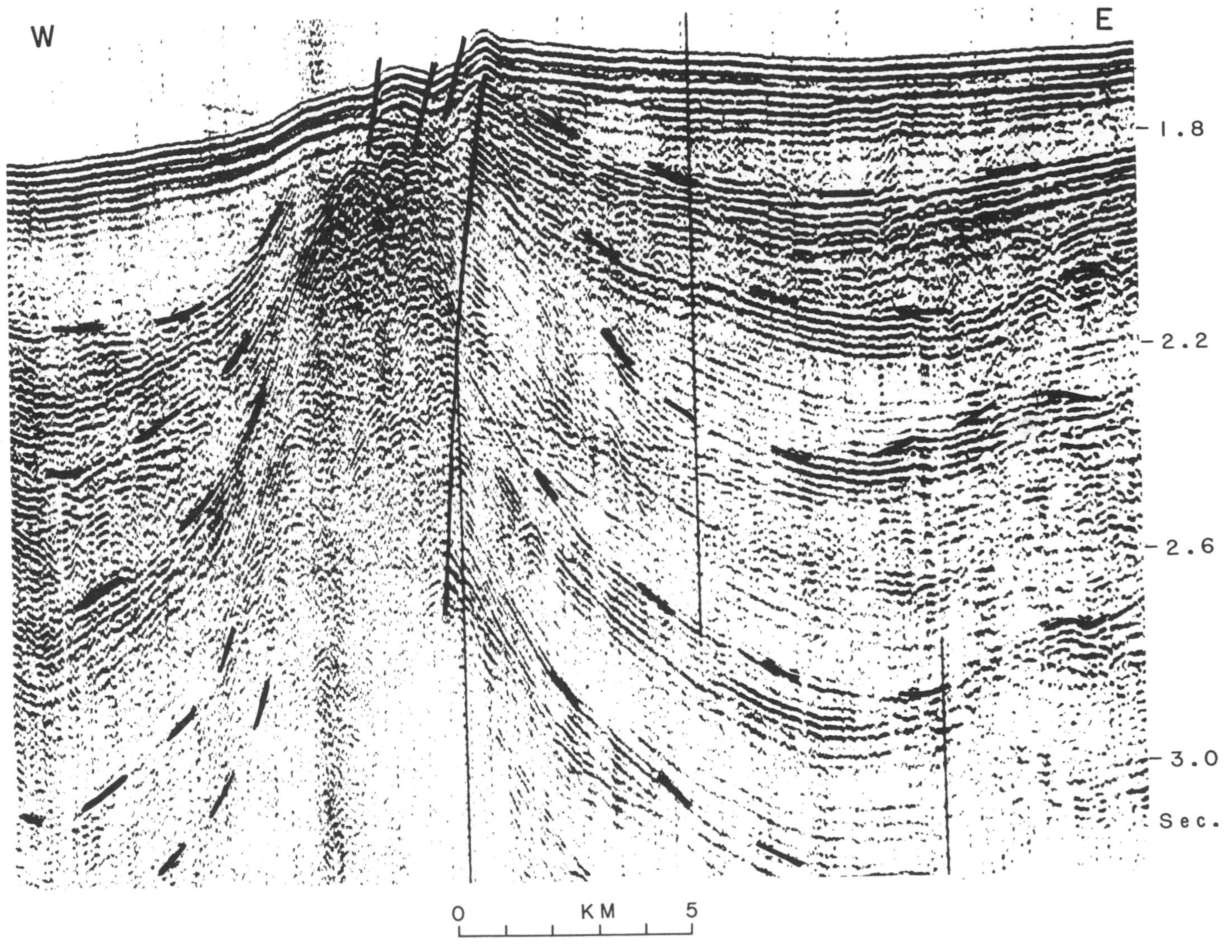

Figure 7 — Seismic reflection profile (airgun) in Green Canyon Area near 27°27′N and 90°23′W, showing five sedimentary sequences. Note the differences in thickness of the sequences on both sides of the diapiric ridge. All reflectors show onlap onto the diapiric flanks.

northeast side of the basin comes very close to, or reaches, the surface (R. G. Martin, personal communication, 1979). Minisparker and airgun profiles do not show any distinct sedimentary fill in the basin, which probably indicates a very thin fill rather than attenuation of sound at the sea water-brine interface (Figure 12). Additional complications are due to seismic hyperbolae from the many slumps present. Addy and Behrens (1980) estimated that on the basis of diffusion rates, the brine pool started forming about 7,900 years ago.

Presently available data suggest that the upward moving diapirs caused slope steepening, which led to massive slumping and the exposure of part of the salt core. Dissolution of salt formed a brine which, because of its density, maintained identity and ran down the slope into the basin. Similar brine flows are reported by Texas A&M University oceanographers from dives in a submersible on the Flower Garden Banks on the outer continental shelf south of Galveston, Texas (R. Rezak, personal communication, 1980). Brine formation will continue until the solution cavity collapses. Whether such a process is repeated several times, completely stopped, or is still ongoing is not known.

Collapse basins seem to be more common than was thought, according to data from our own cruises and from seismic records. They range from very small,

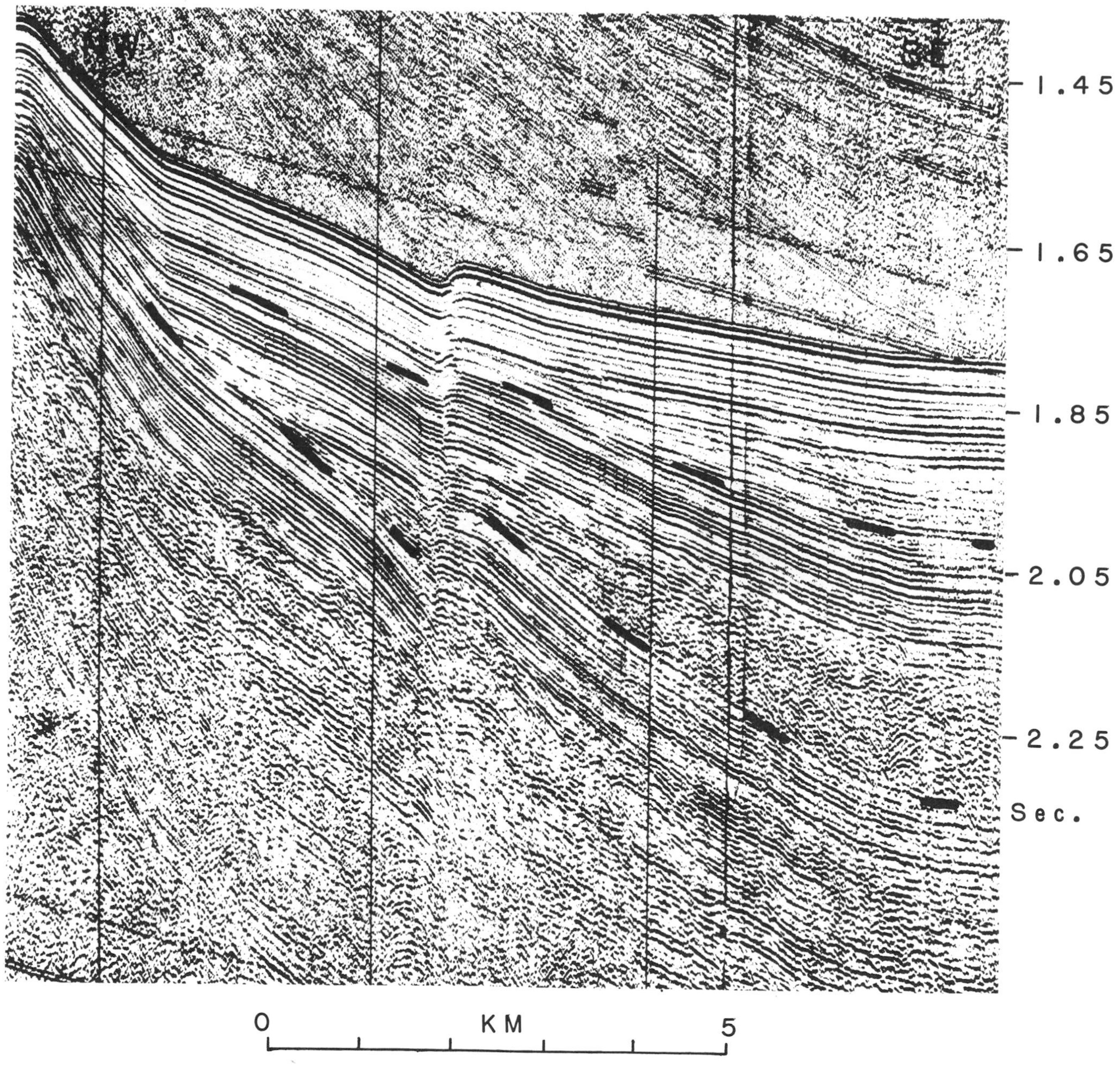

Figure 8 — Seismic reflection profile (airgun) of the section presented in Figure 3. Comparison of both records shows that similar sedimentary sequences are present at different scales. Note that the zone of semitransparent reflections near 2.05 seconds shows chaotic reflections that probably represent slumps. Upflank, the reflections become more and more distinct.

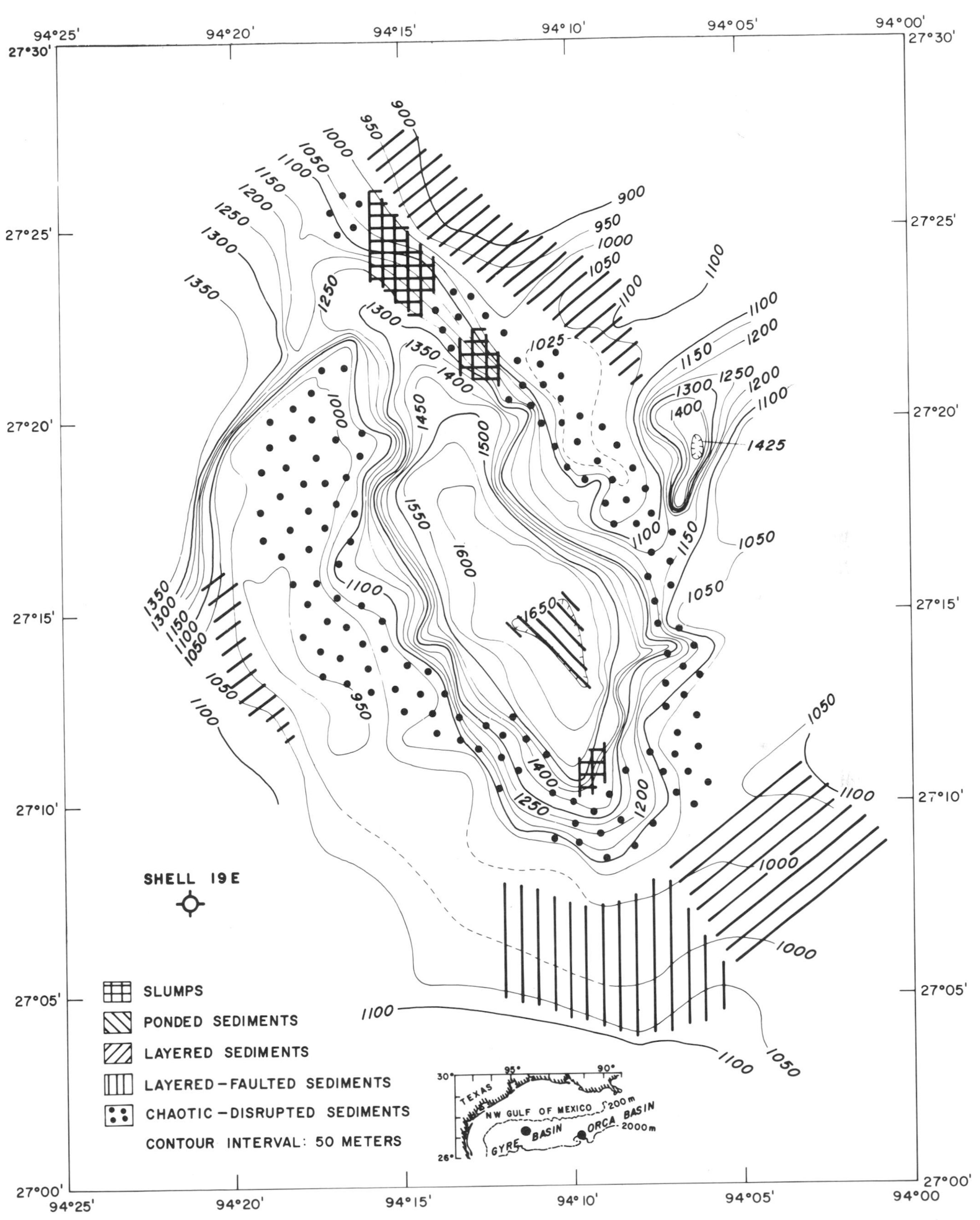

Figure 9 — Bathymetric chart of Gyre Basin with an overprint of seismic facies based on 3.5 kHz records (after Bouma et al, 1978).

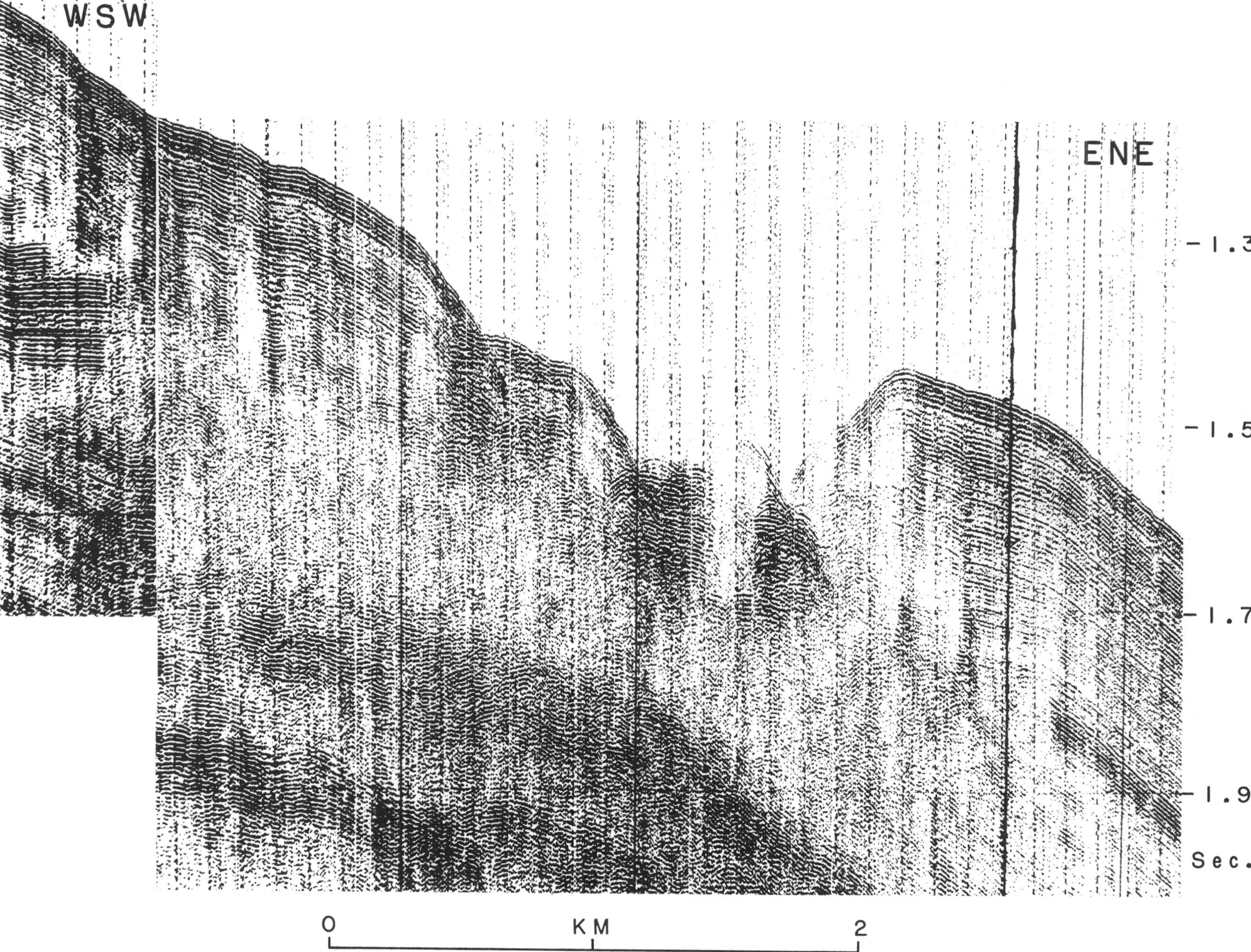

Figure 10 — Seismic reflection profile (minisparker) over northern side of Gyre Basin, showing large-scale and small-scale slumping and faults (pull aparts) at the surface as well as inside the sediment cover (e.g. left-hand side of figure below 1.35 seconds).

more or less oval to lenticular simple depressions bounded by faults, to very complex-shaped features, where correlation between survey lines as close as 3 mi apart becomes difficult (see also Sidner, Gartner, and Bryant, 1978). Two of these basins are partly surveyed thus far, East Breaks Basin and Carancahua Basin, and little information is available. Figure 4 shows a seismic reflection profile over one of those graben-like features over the crest of a diapiric uplift. It is bounded by growth faults and normal faults. The bottom of the depression is higher than the surrounding sea floor. The seismic reflection patterns suggest that sediments are hemipelagic and pelagic, which is supported by a few piston cores.

CONCLUSIONS

Three types of intraslope basins are recognized on the Texas-Louisiana Continental Slope: 1) blocked-canyon intraslope basins, 2) interdomal basins, and 3) collapse basins. Seismic stratigraphic studies show that only the blocked canyon type basin shows onlapping of seismic reflections onto diapiric flanks, suggesting deposition from bottom-hugging gravity flows. This mechanism is verified by a limited number of gravity and piston cores. The presence of "turbidite sands" (Lehner, 1969) on top of at least one salt diapir supports the idea that upward motion of salt bodies still exists at a rate that may average as high as 2 to 4 cm/yr. The presence of 148 m of shale of late Pleistocene age, overlying sand (in Shell hole 19E southwest of Gyre Basin), also indicates that mud deposition can be very high and presumably took place in a depression rather than on top of the topographic high.

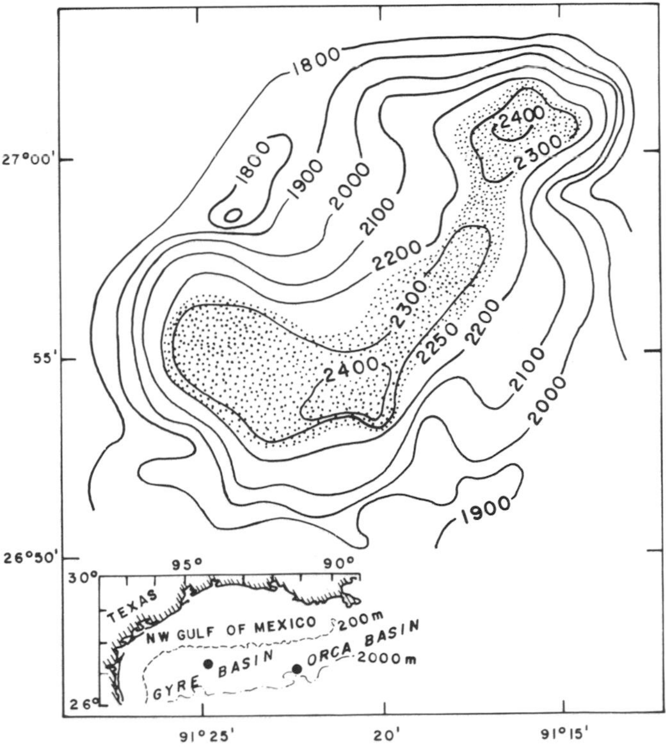

Figure 11 — Bathymetric chart of Orca Basin, in meters, showing the approximate location of the brine (dotted pattern). Figure 12 crosses the southwestern area. (after Trabant and Presley, 1978).

I assume that only a limited number of canyons or canyon systems (e.g. Alaminos Canyon: Bouma, Bryant, and Antoine, 1968; Bouma, Chancey, and Merkel, 1972) were present on this continental slope. Once a canyon forms, it may maintain its character as long as sediment is supplied from shallow areas. Once such supply ceases, the diapiric activity can destroy the continuous thalweg by damming, and a subsequent sediment supply may follow a nearby alternate route, making a single canyon into a canyon system.

I have suggested (Bouma, 1981) the following working hypothesis: During a relative lowering of sea level, large amounts of sand and coarse silt are carried onto the continental slope, either directly or in stages involving temporary depocenters on the shelf. Continued lowering of sea level may cause high overburden pressures, perhaps causing sediment to slump in retrogressive steps. Most of these slumps lose their internal strength or take up water and change into debris flows and turbidity currents. During low level sea stand, a rise in sea level and a high stand, the transport of sand and silt ceases and is followed for some time by the transport of large amounts of mud. Finally, the quantity of fine-grained detrital material diminishes and a mixture of hemipelgic and pelagic deposits slowly blankets the sea floor.

My hypothesis is that only a limited number of sand bodies exist on the Texas-Louisiana Continental Slope, as far as horizontal coverage is concerned. Reoccupation of canyons results in the formation of sand units in a vertical sense, separated by hemipelagic and pelagic sediments. Therefore, the first search for sand bodies (reservoir rocks) should be carried out in present canyon systems that can be reconstructed by a search for blocked-canyon intraslope basins.

ACKNOWLEDGMENTS

Figure 5 is based on three figures found in the August 1978 edition of *Essobron*. Exxon Production Research Company kindly gave permission to publish the redrafted block diagrams. Discussion with several colleagues from the oil industry changed my original interpretations of the lithologic characteristics of the different reflection zones. Their comments are highly appreciated. I want to thank Ray Martin (U.S. Geological Survey), Joel Watkins (Gulf Science and Technology Company), and Bill Behrens (University of Texas, Marine Science Institute, Galveston) for their constructive remarks while reviewing this manuscript.

REFERENCES CITED

Addy, S. K., and E. W. Behrens, 1980, Time of accumulation of hypersaline anoxic brine in Orca Basin (Gulf of Mexico): Marine Geology, v. 37, p. 214-252.

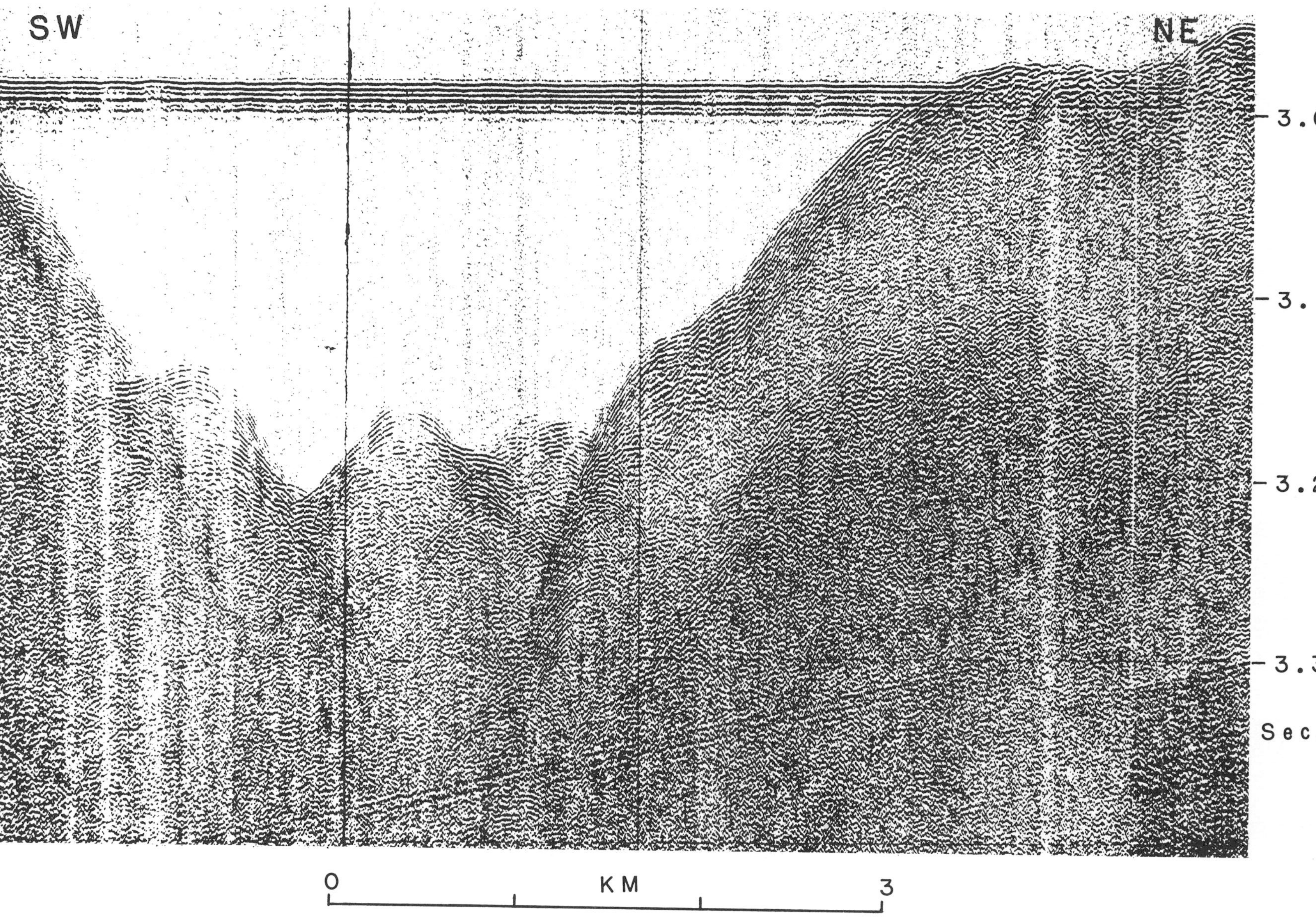

Figure 12 — Seismic reflection profile (minisparker) of a line crossing Orca Basin. Note the distinct reflector between brine and overlying sea water. Slumping is apparent in the basin, as is the lack of any significant younger Tertiary to Holocene sediment cover.

Bergantino, R. M., 1971, Submarine regional geomorphology of the Gulf of Mexico: Geological Society of America Bulletin, v. 82, p. 741-752.

Bouma, A. H., 1981, Depositional sequences in clastic continental slope deposits, Gulf of Mexico: Geo-Marine Letters, v. 1, n. 2.

———, and L. E. Garrison, 1979, Intraslope basins, Gulf of Mexico: Geological Society of America Bulletin (Abs. with programs), 1979 Annual Meeting, San Diego, p. 329.

———, W. R. Bryant, and J. W. Antoine, 1968, Origin and configuration of Alaminos Canyon, northwestern Gulf of Mexico: Gulf Coast Association Geological Society Transcripts, v. 18, p. 290-296.

———, O. Chancey, and G. Merkel, 1972, Alaminos Canyon area, *in* R. Rezak and V. J. Henry, eds., Contributions on the geological oceanography of the Gulf of Mexico: Texas A&M University Oceanography Studies, v. 3, Houston, Gulf Publishing Company, p. 153-179.

———, R. G. Martin, and W. R. Bryant, 1980, Shallow structure of upper continental slope, central Gulf of Mexico: Houston, Texas, Proceedings, Offshore Technology Conference, Offshore Technology Conference 3913, p. 583-592.

———, et al, 1978, Intraslope basins in northwest Gulf of Mexico, *in* A. H. Bouma, G. T. Moore, and J. M. Coleman, eds., Framework, facies, and oil-trapping characteristics of the upper continental margin: AAPG Studies in Geology 7, p. 289-302.

Hampton, M. A., 1979, Buoyancy in debris flow: Journal of Sedimentary Petrology, v. 49, p. 753-758.

Holland, W. C., 1970, Bathymetry maps, eastern continental margin, U.S.A., Sheet 3 of 3, Northern Gulf of Mexico: published by the AAPG, scale 1:1,000,000.

Lehner, P., 1969, Salt tectonics and Plesistocene stratigraphy on continental slope of northern Gulf of Mexico: AAPG Bulletin, v. 53, p. 2431-2479.

Martin, R. G., 1978, Northern and eastern Gulf of Mexico continental margin: stratigraphic and structural framework, *in* A. H. Bouma, G. T. Moore, and J. M. Coleman eds., Framework, facies, and oil-trapping characteristics of the upper continental margin: AAPG Studies in Geology 7, p. 21-42.

———, 1980, Distribution of salt structures in Gulf of Mexico region: map and descriptive text: U.S. Geological Survey Miscellaneous Field Studies Map MF-1213, 8 p., 2 sheets.

———, and A. H. Bouma, 1978, Physiography of the Gulf of Mexico, *in* A. H. Bouma, G. T. Moore, and J. M. Coleman, eds., Framework, facies, and oil-trapping characteristics of the upper continental margin: AAPG Studies in Geology 7, p. 3-19.

———, and ———, 1981, Evidence of active diapirism and engineering constraints, Texas-Louisiana slope, northwest Gulf of Mexico: Marine Geotechnology, v. 5, n. 1.

McKee, T. R., et al, 1978, Holocene sediment geochemistry of continental slope and intraslope basin areas, northwest Gulf of Mexico, *in* A. H. Bouma, G. T. Moore, and J. M. Coleman, eds., Framework, facies, and oil-trapping characteristics of the upper continental margin: AAPG Studies in Geology 7, p. 313-326.

Middleton, G. V., and M. A. Hampton, 1976, Subaqueous sediment transport and deposition by sediment gravity flows, *in* D. J. Stanley and D. J. P. Swift, eds., Marine Sediment Transport and Environmental Management: New York, John Wiley & Sons, p. 197-218.

Mitchum, R. M., Jr., P. R. Vail, and J. B. Sangree, 1977, Seismic stratigraphy and global changes of sea level, Part 6: Stratigraphic interpretation of seismic reflection patterns in depositional sequences, *in* C. E. Payton, ed., Seismic stratigraphy-application to hydrocarbon exploration: AAPG Memoir 26, p. 117-133.

Penrose, N. L., and J. P. Kennett, 1979, Anoxic and aerobic basins in the northern Gulf of Mexico: comparison of microfossil preservation: Geological Society of America Bulletin (Abs. with programs), p. 493.

Shepard, F. P., and R. F. Dill, 1966, Submarine canyons and other sea valleys: New York, Rand McNally, 381 p.

Shokes, R. F., et al, 1977, Anoxic, hypersaline basin in the northern Gulf of Mexico: Science, v. 196, p. 1443-1446.

Sidner, B. R., S. Gartner, and W. R. Bryant, 1978, Late Pleistocene geologic history of outer Texas continental shelf and upper continental slope, *in* A. H. Bouma, G. T. Moore, and J. M. Coleman, eds., Framework, facies, and oil-trapping characteristics of the upper continental margin: AAPG Studies in Geology 7, p. 243-266.

Sorensen, F. H., et al, 1975, Preliminary bathymetric map of the Gulf of Mexico region: U.S. Geological Survey Open-File Map 75-140, scale 1:2,500,000.

Trabant, P. K., and B. J. Presley, 1978, Orca Basin, anoxic depression on the continental slope, northwest Gulf of Mexico, *in* A. H. Bouma, G. T. Moore, and J. M. Coleman, eds., Framework, facies, and oil-trapping characteristics of the upper continental margin: AAPG Studies in Geology 7, p. 303-311.

Uchupi, E., 1967, Bathymetry of the Gulf of Mexico: Gulf Coast Association of Geology Socs. Transcripts, v. 17, p. 161-172.

———, 1975, Physiography of the Gulf of Mexico and Caribbean Sea, *in* A. E. M. Nairn and F. G. Stehli, eds., Ocean basins and margins: the Gulf of Mexico and the Caribbean: New York, Plenum Press, v. 3, p. 1-64.

Vail P. R., and J. Hardenbol, 1979, Sea-level changes during the Tetiary: Oceanus, v. 22, p. 71-79.

———, R. M. Mitchum, Jr., and S. Thompson, III, 1977, Seismic stratigraphy and global changes of sea level, part 3: relative changes of sea level from coastal onlap, *in* C. E. Payton, ed., Seismic stratigraphy-application to hydrocarbon exploration: AAPG Memoir 26, p. 63-81.

Woodbury, H. O., J. H. Spotts, and W. H. Akers, 1978, Gulf of Mexico continental-slope sediments and sedimentation, *in* A. H. Bouma, G. T. Moore, and J. M. Col;eman, eds., Framework, facies, and oil-trapping characteristics of upper continental margin: AAPG Studies in Geology, 7, p. 117-137.

Cap Ferret Deep Sea Fan (Bay of Biscay)

F. Coumes
J. Delteil
H. Gairaud
Society Nationale Elf-Aquitaine
Pau, France

C. Ravenne
Institut Francais du Petrole
Rueil, France

M. Cremer
Institut de Geologie du Bassin d'Aquitaine
Talence, France

The geological history of the Cap Ferret deep sea fan in the Bay of Biscay has been reconstructed using multichannel seismic reflection data, piston coring, echo sounding, and seabeam bathymetric data. Cap Ferret canyon is flat bottomed and several adjacent channels feed into it from the north. The canyon is tectonically controlled by a fault bordering its southern flank. Its continuation into deeper water is characterized by the buildup of sedimentary deposits which form a levee bordered on the south by a flat and complex channel. This linguoid morphology is of purely sedimentary construction. The southern area is further complicated by extensions of the Santander and Torrelavega canyons. Seismic stratigraphic interpretation of an extensive grid of multichannel seismic data (Miniflexichoc) provides the basis for interpretation of the geological history of the fan, a history that begins in the Upper Miocene following faulting of the continental platform.

In 1978 Elf-Aquitaine, associated with l'Institut Francais du Petrole and Total Oil Company, conducted a survey of the Cap Ferret canyon and deep sea fan which extends across the continental shelf of southwestern France into the Bay of Biscay. The purpose of the study was to increase knowledge of the mechanisms involved in constructing the canyon and the fan, of their internal structure and geological history and of the size and distribution of sand bodies with an eye toward reservoir properties.

The Bay of Biscay extends east from France in the north to Spain in the south. Cap Ferret canyon lies between the American passive continental shelf to the northeast and the Landais plateau to the south. The plateau is the offshore extension of the Parentis basin (Figure 1). Cap Breton canyon is located at the northern limit of the Iberian plate, just in front of the subduction zone. Santander and Torrelavega canyons extend north across the Iberian margin in the study area and join the Cap Ferret channel in its lower part.

Previous studies showed a complex morphology in front of Cap Ferret canyon. This was assumed to be a deep sea fan and was chosen as the target. The region was studied through numerous different techniques: a seabeam bathymetric survey; direct observation of the bottom in the shallow part of the fan using the submersible, CYANA; piston coring of distinct subenvironments of the fan; 3.5 kHz echo sounding of the deepest part of the fan; and, high resolution seismic profiling using Miniflexichoc.

RESULTS

Bathymetry and Geomorphology

The bathymetry of the area was determined by a seabeam survey carried out aboard the research vessel, JEAN CHARCOT. In the eastern part of the study area, the initial point of the canyon is characterized by a dense network of converging narrow channels (Figure 1). The canyon proper is flat bottomed and is bordered on the north by a thick accumulation of sediments with a flat surface in contrast to the steep relief to the south. This flat zone is affected by adjacent channels coming in from the north. The steep, straight southern margin of the canyon is most striking, and corresponds to a major fault that controls channelling and sediment transport across the continental margin.

Continuing to the west, there is a zone of coalescent channels and sharp and disturbed relief resulting from the interaction of the sedimentary inputs from the Cap Ferret, Santander, and Torrelavega canyons. Farther west, and on the north side of the main channel, sediments piled up to form an elongate and asymetric morphology, or levee. The southern side is more complicated, not only because of the Santander and Torrelavega canyons coming in from the south, but by the meandering Cap Breton canyon which joins from the east.

Figure 2 presents a simplified morphological sketch of the area. The principal features are: a canyon 4 to 5 km in width with a straight, flat bottom; V-shaped channels coming in from the north across a blanket of sediment; the hook-shaped channel of Santander canyon entering from the south; and, the main northern levee, reaching 200 m above the bottom of the main channel.

CYANA Observations

The submersible, CYANA, permitted observation of the upper part of Cap Ferret canyon. Observations made during one diving operation is summarized schematically in Figure 3. The northern flank of the canyon has a main slope of 33°, which has been affected somewhat by slumping and sliding as indicated by resulting scars. The flat bottom of the canyon has ridges and furrows parallel with the canyon axis. These features suggest that erosional processes are taking place in the canyon. However, there is a blanket of Holocene mud on the walls and the floor of the canyon, suggesting that the present transport of sediments in the canyon is at a low rate.

Piston Coring

Figure 4 summarizes some biostratigraphic results from studies made on cores recovered from various sub-environments of the fan. Cores 11 and 12 indicate a hiatus between the Würm 2 to 3 glacial stages and the Holocene, indicating active erosion in this interval near the mouth of Cap Ferret canyon. Cores 4 to 8 point to significant Holocene sedimentation in Cap Breton canyon while only small amounts of Holocene sediment were deposited in Cap Ferret canyon. On the main levee, north of the Cap Ferret channel (Core 10), sedimentation continued without interruption throughout the Würm glacial stage, in contrast to the Torrelavega channel in which erosion dominates during the Würm (Core 1).

From the piston core studies we can summarize the sedimentological history of the study area during Würm 2 to 3, Würm 3, and the Holocene. During Würm 2 to 3, transport and deposition are dominant in every part of the fan. During Würm 3, erosion is

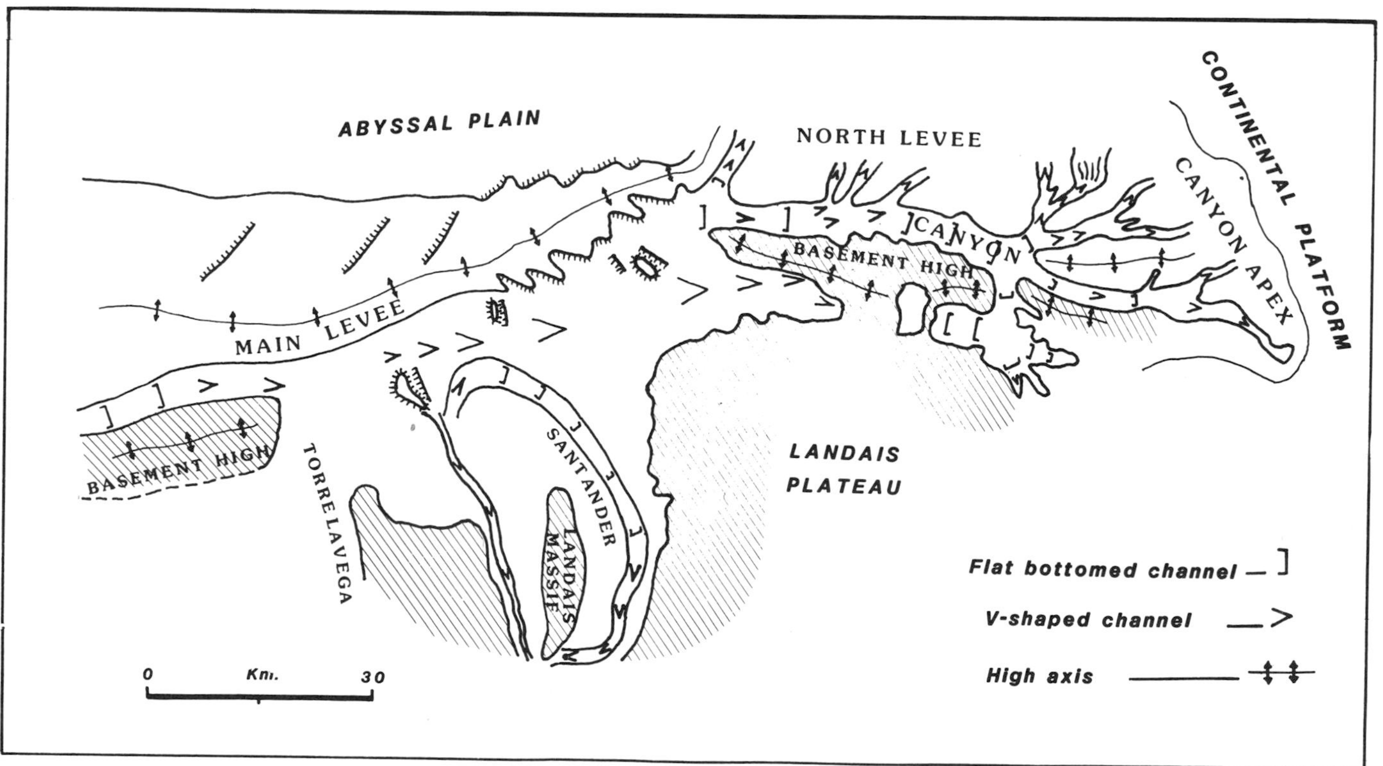

Figure 2 — Simplified morphological sketch map of Cap Ferret canyon. The canyon separates the continental platform of the Amorican margin from the Landais plateau to the south.

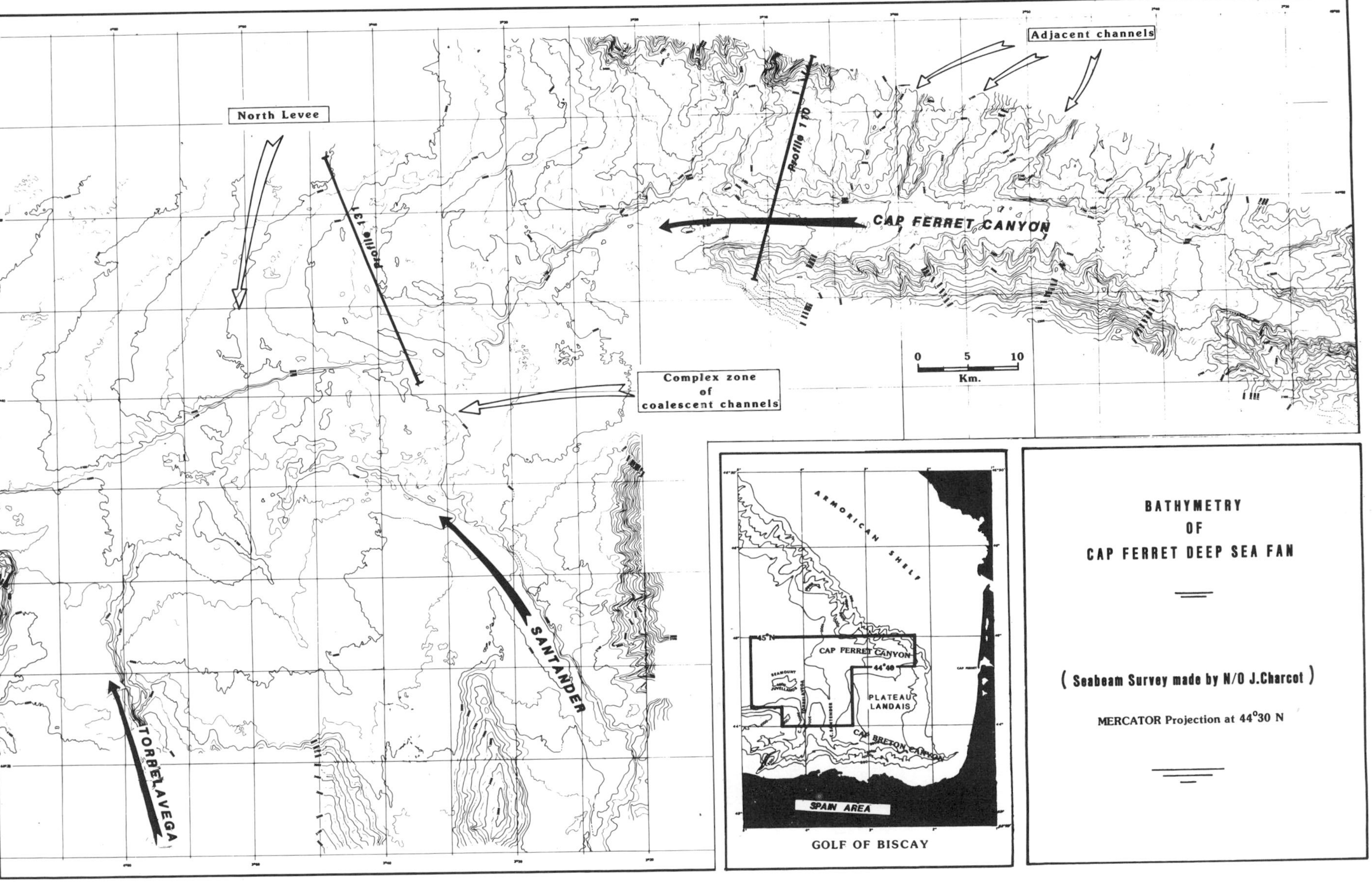

Figure 1 — Location map and bathymetry of Cap Ferret canyon and deep sea fan and associated features.

active in the main channel, but sedimentation dominates on the main levee. In the Holocene, deposition ceases on the levee and sediments run down the main channel from the south.

All of the sediments recovered were composed of sand, silt, and clay organized in thin-bedded turbidites. The distribution of these sediments with regard to their sand and silt content is shown in Figure 5. It appears that the highest percentages of sand and silt are located in the channels where there are more than 20 cm of sand per meter. The sand and silt percentage is much lower on the main levee.

Echo Sounding

The soundings obtained from 3.5 kHz echo-sounding surveys were classified following the system developed by J. Damuth of Lamont-Doherty Geological Observatory (1977). The most important types found were: distinct echoes with continuous sub-bottom reflectors (D.A.R. or I B type of Damuth), seemingly restricted to the levees; indistinct, prolonged, fuzzy echoes with no sub-bottom reflectors (P. U. type or II B of Damuth), located only in the main channel; and, indistinct hyperbolic echoes with varying vertex elevations above the sea floor (H. I. or type III of Damuth), found only to the south of the main channel. There appears to be good correlation of these echo types with differences in the nature and the internal structure of the piston cores as suggested by sedimentary analysis (Figure 6).

High Resolution Seismic Survey

Two areas were covered by a network of high resolution Miniflexichoc seismic reflection surveys. Profile 110 (Figure 7), located at the entrance of Cap Ferret canyon, shows clearly a flat, U-shaped channel. The northern part is covered with a blanket of sediments, deeply eroded by adjacent, southward-flowing channels. This sedimentary blanket seems to be made up of turbidites flowing laterally from the main channel, indicating that the canyon of Cap Ferret is, in reality, a channel trapped in a depression of the basement.

Seismic stratigraphic interpretation (Payton, 1977) was carried out on the data from the upper part of this canyon. Although the study of the acoustic basement is not one of the purposes of the survey, it is worth noting that there is a difference in tectonic style between the northern and the southern borders of the canyon. The northern border, or American margin, has a smooth tectonic style, while the southern contains a rather complex ridge with very complicated internal structure and strong deformation of strata.

Four distinct chronological sequences fill the canyon and lie on an irregular basement without acoustic reflectors (Figure 7).

Sequence A.

These are early deposits in the structural depression. The lower part of these sediments seems to be contemporaneous with the latest basement structural movements. The origin of the sediments is likely from turbidity currents.

Sequence B.

These sediments are relatively continuous with the underlying sequence. They are characterized by a high level of sediment transport energy with the occurrence of channels and with thick lateral accretion. They seem to represent a very active period of sediment

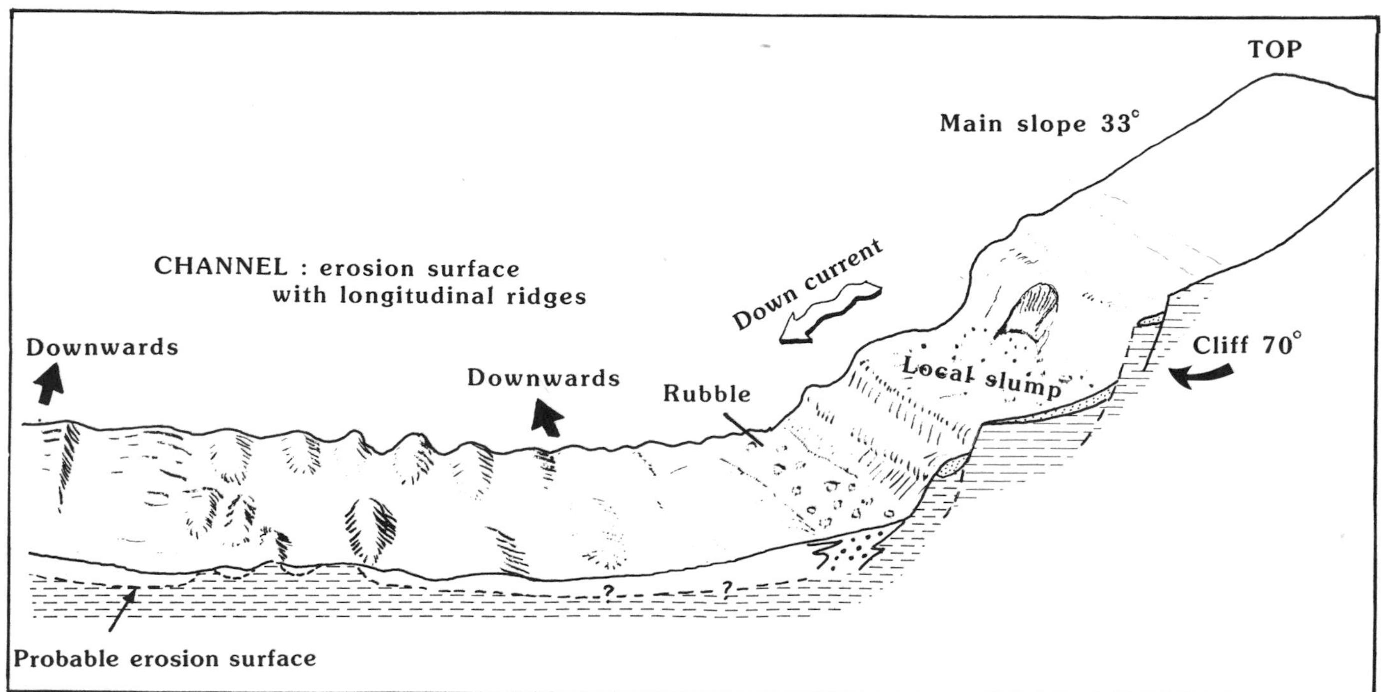

Figure 3 — Simplified morphology of Cap Ferret canyon as determined by observations from the submersible, CYANA.

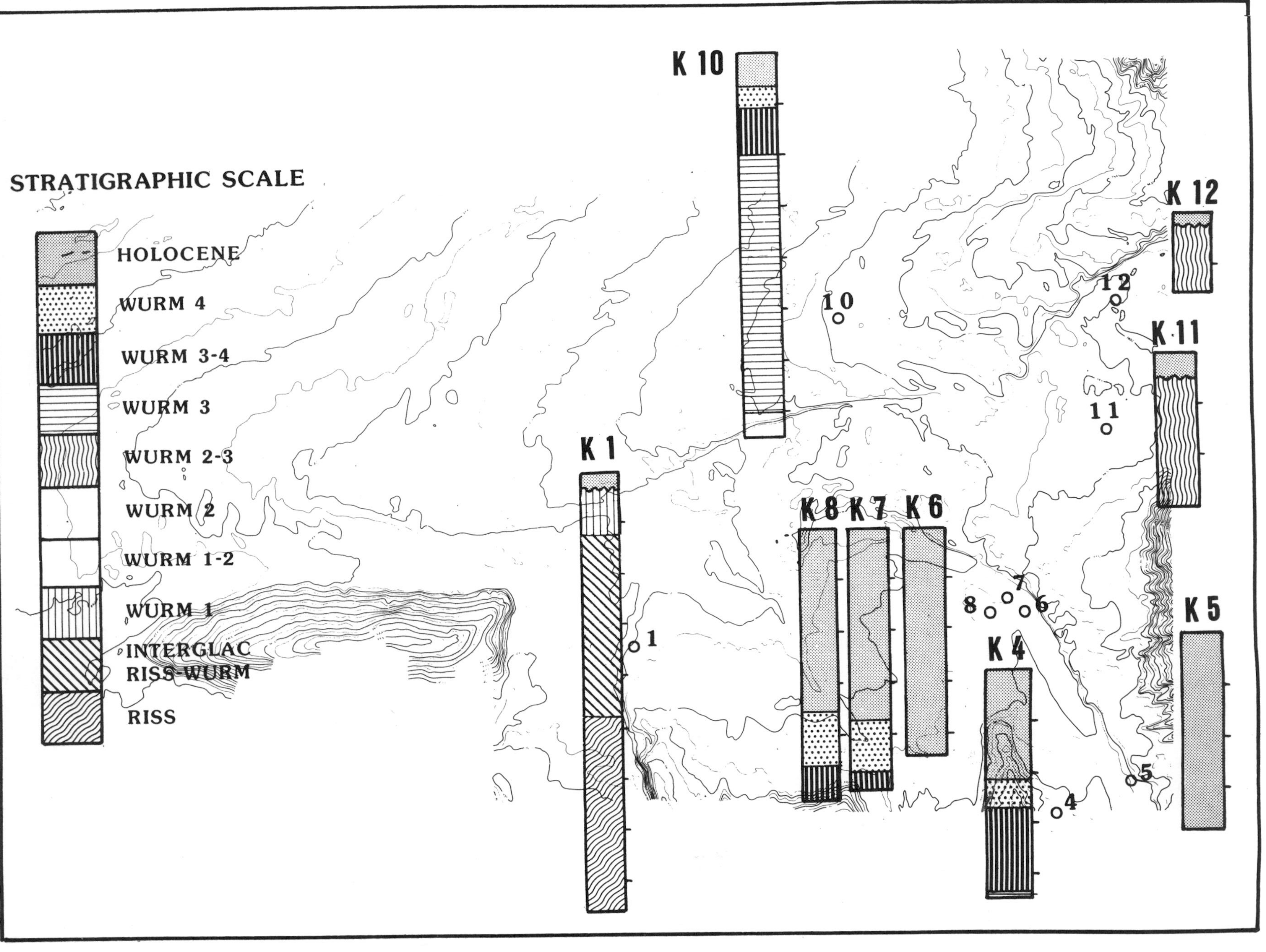

Figure 4 — Late Pleistocene and Holocene stratigraphy in piston cores from Cap Ferret canyon and adjacent areas.

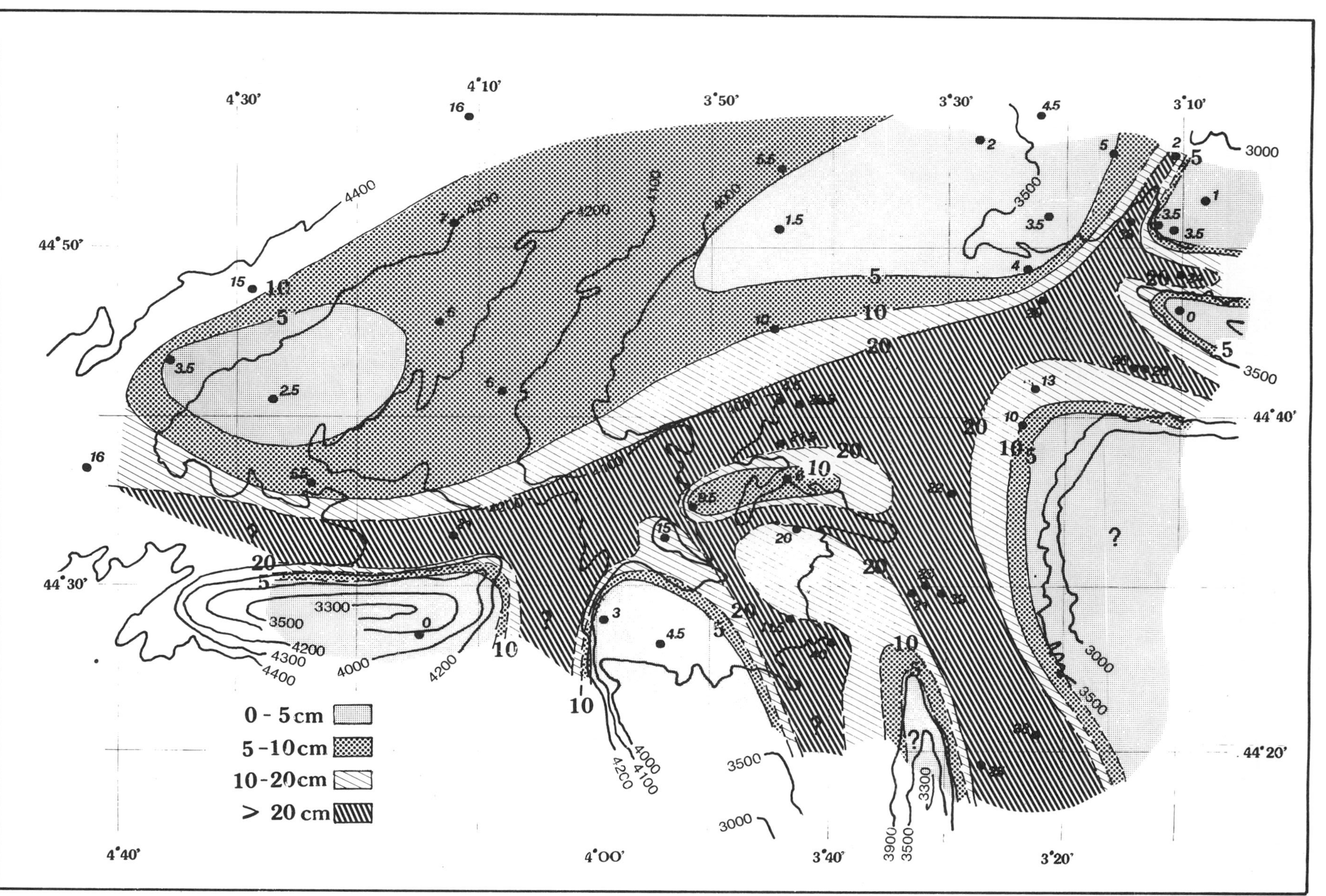

Figure 5 — Silt and sand, per meter of piston core in the Cap Ferret canyon area.

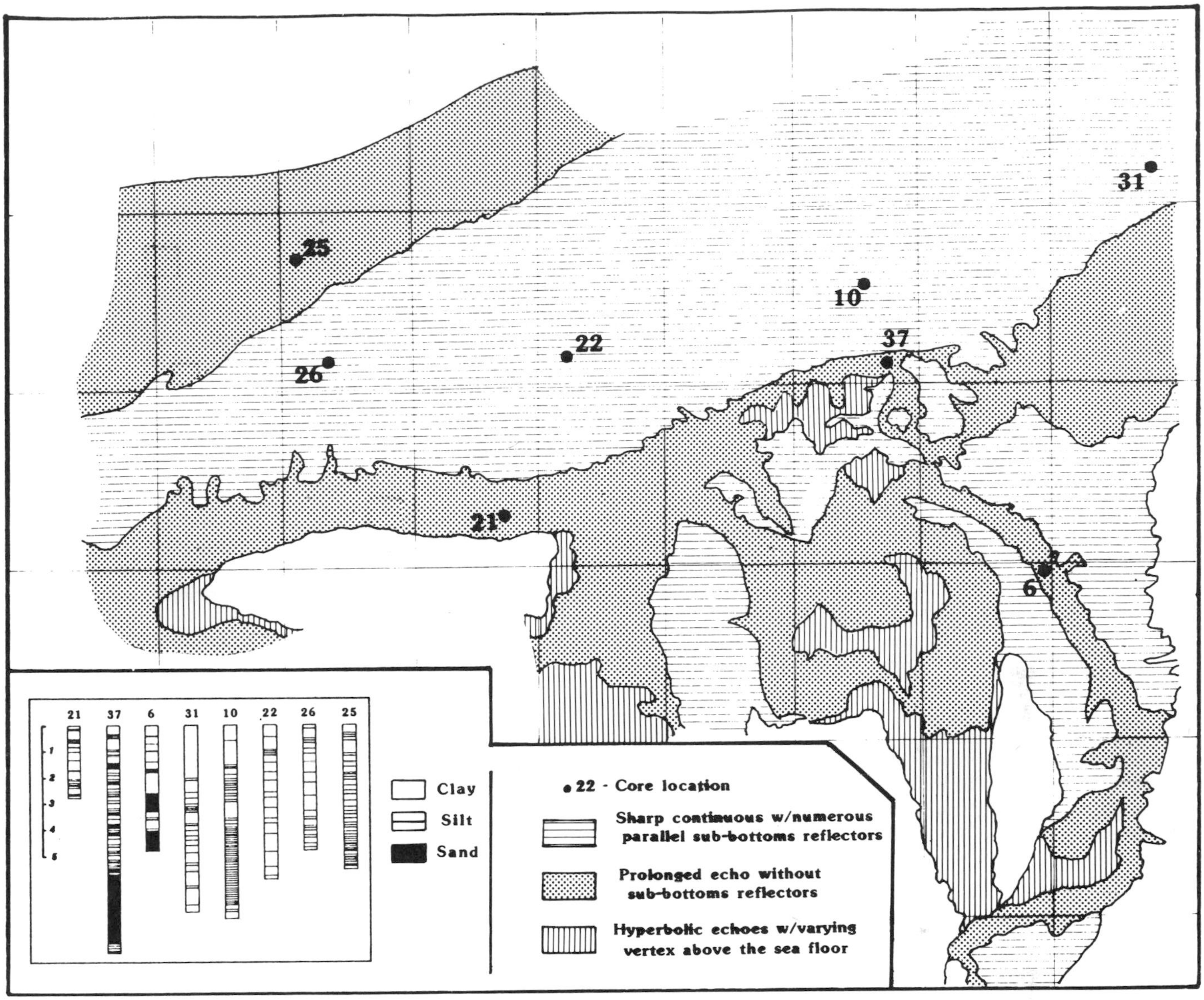

31
10
37
22
25
26
21
6
21 37 6 31 10 22 26 25
Clay
Silt
Sand
• 22 - Core location
Sharp continuous w/numerous parallel sub-bottoms reflectors
Prolonged echo without sub-bottoms reflectors
Hyperbolic echoes w/varying vertex above the sea floor

Figure 6 — Echo character from 3.5 kHz sounding survey of the Cap Ferret canyon area. Horizontal lines correspond to type IB of Damuth (1975), dotted pattern to type IIB, and vertical lines to type III.

SSW
NNE
LATERAL ACCRETION
MAIN CHANNEL
Unit D
Unit C
Unit B
Unit A
BASEMENT
I SEC
TD

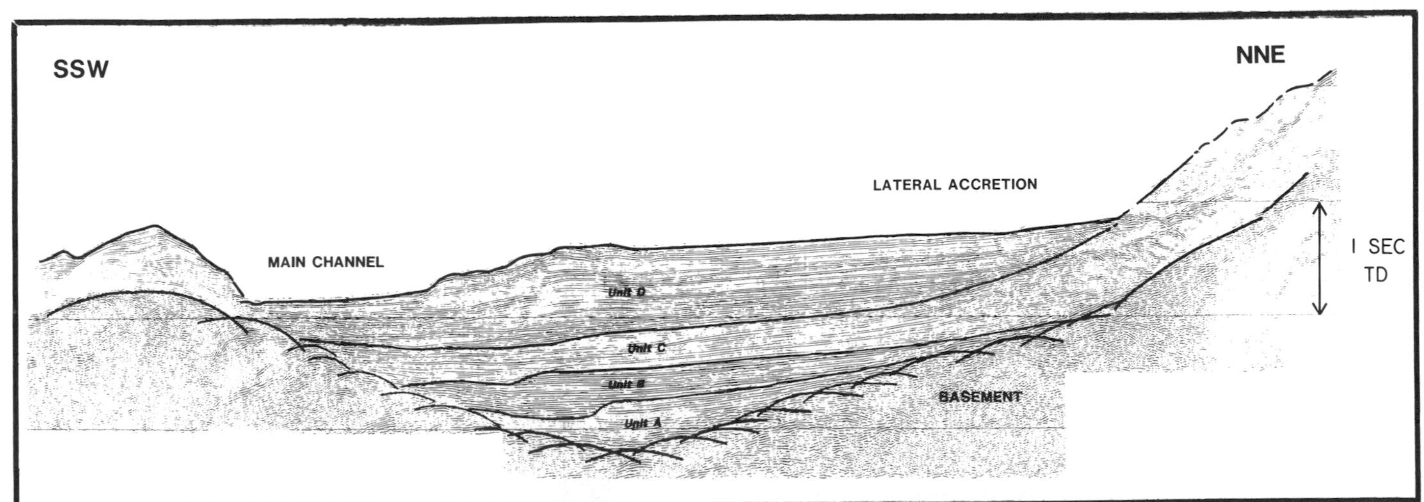

Figure 7 — Miniflexichoc seismic reflection profile across Cap Ferret canyon. Profile is approximately 20 km long (Profile 110; see location map Figure 1).

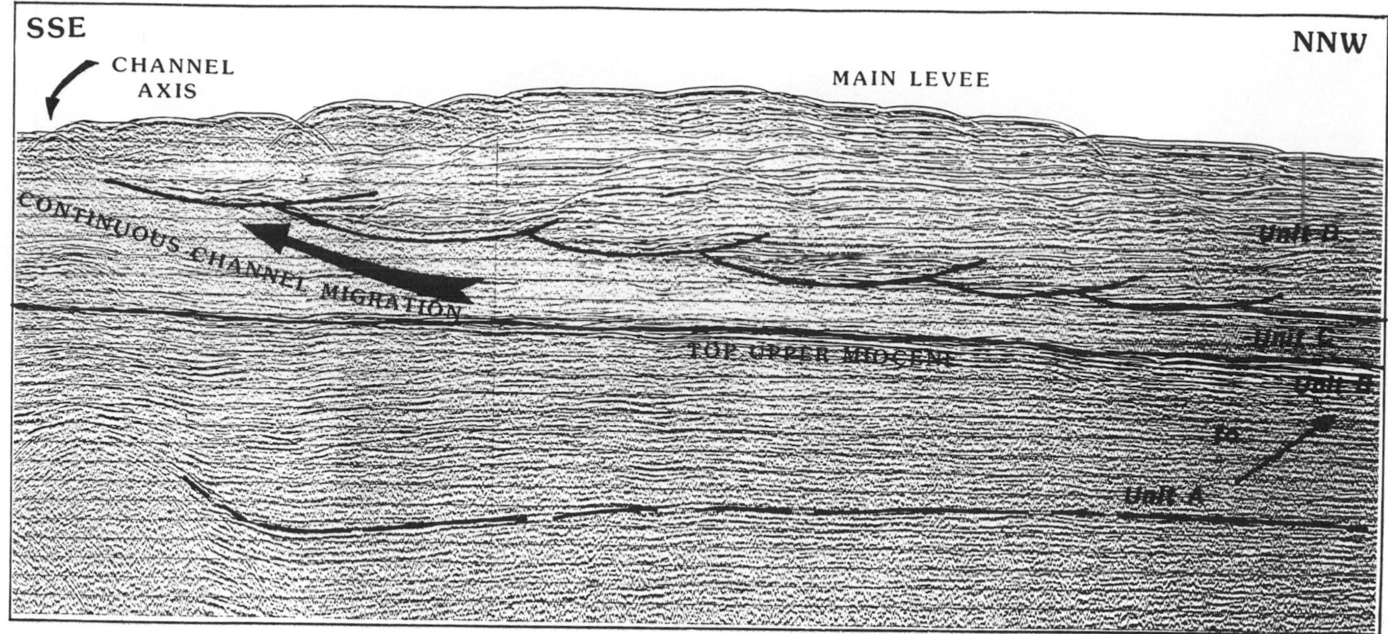

Figure 8 — Miniflexichoc seismic reflection profile across the main levee on the north side of Cap Ferret channel. Profile is approximately 22 km long (Profile 131; see location map Figure 1).

transport along the canyon axis between the shelf and the abyssal plain.

Sequence C.

The boundary between this sequence and the underlying sequence is usually of erosional origin. The main seismic feature is an acoustically transparent facies that seemingly corresponds to a period of low sediment transport energy smoothing the previous topography. The sequence characteristically has a flat top and occurs also in the deepest part of the fan.

Sequence D.

This sequence is marked by buildup of thick sedimentary levees on the northern side of the main channel. Several building episodes with several construction cycles were found.

In the deep part of the fan, the distributary area, the most striking feature is the lateral migration of the main channel (Figure 8). This continuous migration may reach as much as 40 km from north to south.

A strong marker (Unit B), taken to be the top of the Miocene, marks the first stage of construction of the fan. This flat reflector leads us to conclude that the main levee is of pure sedimentary construction without tectonic control. The maximum thickness of this unit reaches 1,500 m. The migrating levee is built on top of this and constructed by turbidity currents which travelled down the canyon and then spilled over the northern flank of the deeper channel.

CONCLUSIONS

Three exploration wells, located on the platform very close to the head of the canyon, allow us to date the formations and identify the significant events in the geological history.

The first Eocene-Oligocene sediments were deposited on a structured Cretaceous erosional surface. At the end of the Oligocene, erosion by channelling occurred, followed by infilling of the channels. During the Miocene, progradation of the platform is the main process responsible for the sedimentation. During the last stage of the Miocene, the Cap Ferret canyon appeared and cut deeply into the previous Miocene deposits. This canyon is tectonically controlled by a major east to west fault which breaks the continental platform into two parts. The southern part, the Landais plateau, is tilted slightly to the south.

In the depression resulting from this fault, a channel appeared and found its way through the remains of prograding formations previously deposited and residue left by lateral sliding. In the abyssal plain, a main levee was constructed with the adjoining channel migrating progressively southward. Any levee produced on the southern flank of the channel was destroyed by the east to west migration of the Santander channel.

REFERENCES CITED

Coumes, F., et al, 1979, Etude des eventails detzitiques Mofonds du Golfe de Gascogne analyse ge'omorjhologique de la carte bathymetzique de canyon de Cap Ferret et de ses abords: Bulletin de la Socete Geologique de France, v. 21, n. 5, p. 563-568.

Damuth, J.E., 1975, Echo character of the western equatorial Atlantic floor and its relationship to the dispersal and distribution of terrigenous sediments: Marine Geology, v. 18, p. 17-45.

Payton, C.E., ed., 1977, Seismic stratigraphy - applications to hydrocarbon exploration: AAPG Memoir 26, 515 p.

Slope Readjustment During Sedimentation on Continental Margins

G. C. Dailly
Societe Nationale Elf-Aquitaine
Pau, France

The physical characteristics of successive layers of sediments at the time of deposition vary widely; in particular, the equilibrium angle of the foreset slope of sedimentation. When a layer of soft sediments, like shale, with a low slope of deposition (approximately 1°) succeeds a layer of harder sediments with steeper slope, like limestone and marl, a period of slope readjustment takes place. During this period, a dumping of sediments at the base of the limestone "cliff" accompanied by synchronous regressive erosion of the top of the "cliff" by submarine canyon occurs. The Gabon sedimentary basin and the Aquitaine Basin provide examples of the slope readjustment theory. The importance and possible generalization of this theory are discussed in this article. This phenomenon is concomitant with more general effects such as tectonic phases, isostatic readjustment, eustatic changes, and crust movements.

The infilling of many sedimentary basins results from the accumulation of successive prograding slices of sediments which accumulate in front of and above each other (Figure 1 Gabon; Figure 2, France). Many other examples of progradation are found in the literature, particularly in the deltaic sequences of the Mississippi and Niger rivers (Dailly, 1975). In these deltaic sequences, the nature of the sediments is relatively homogeneous (sands and shales) and successive slices of sediments pile up with relatively uniform foreset dips. In the case of shaly sedimentation, the dip is quite gentle, sometimes as low as one-half a degree (Ireton shale of Alberta; Stoakes, 1980). In most basins, sediments are more diversified; for instance, carbonates, marls, shales, silicified shales, and prograde with steeper foreset dips. The continental margins of Africa and the Americas (north and south) show the passage from Jurassic to Cretaceous carbonate sedimentation to Cretaceous to present clastic sedimentation. This paper addresses the problem of various successive depositional foreset slopes in heterogeneous sediment types, and describes in par-

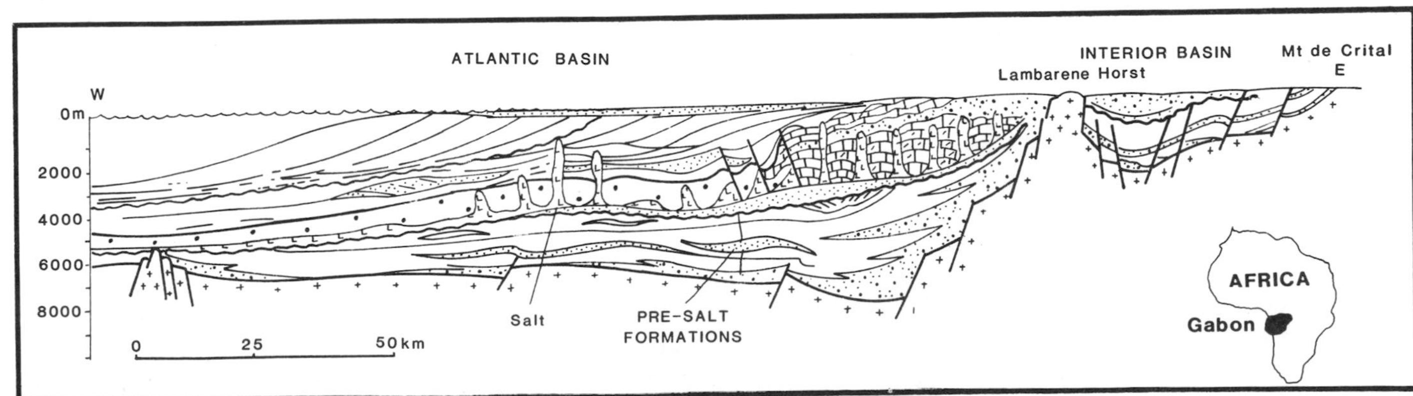

Figure 1 — Schematic cross section of Gabon sedimentary basin (Vidal, Joyes, and Van Veen, 1975).

ticular what happens at the base of a steep carbonate slope until a new low angle depositional slope for clastics is established.

SLOPE READJUSTMENT

A previous publication (Dailly, 1975) examined progradation of a deltaic sequence with successive layers of sediments of uniform foreset slope. To show the relative importance of subsidence and progradation, a diagrammatic illustration (Figure 3) was presented. This schematic view does not take into account many of the factors which affect the growth of sedimentary lenses; for example, isostatic subsidence and flexure of the lithosphere. Slumping of sands and shales tends to correct any anomalously steep slopes, resulting in roughly the same dip to slopes of deposition, provided the supply of sediments is homogeneous. This schematic approach shows the angle of the megafacies lines with the time lines.

In Figure 4, slices 6 to 8 with a low angle of deposition correspond to clastic infilling at the base of the cliff. Starting at slice 9, the slope is readjusted and clastic sedimentation proceeds normally until slice 12. From slices 12 to 15, a new carbonate progradation occurs followed by a period of infilling (slope readjustment) and so on. Slices 1 to 32 are time slices; 1 is the oldest and 32 the youngest.

We assume that dumping of sediments at the base of a cliff is accompanied by simultaneous erosion at the top of the cliff. The letter E marks this erosion which will be concentrated preferentially in canyons. From period 4 to 9, 15 to 21, and 25 to 32, the top of the three cliffs of Figure 4 will be submitted to regressive erosion.

Duration of a Period of Slope Readjustment

The process of slope readjustment described above is nothing new to most geologists. It encompasses several phenomena like turbiditic sedimentation, onlap, offlap; and stratigraphic gaps, all of which are well known to explorationists. However, the magnitude of the time span that a slope takes to readjust, as estimated in Gabon and Brazil, is surprisingly high, near several tens of millions of years.

Figure 5 is a schematic representation of a slope adjustment; the horizontal scale is exaggerated. A period of prograding limestone sedimentation with a foreset slope arbitrarily taken as 4° to a water depth of 5,000 m is followed by a period of shale sedimentation with a foreset slope of 1°. The rate of progradation, which in a sedimentary basin can be judged by the seaward regression of the shore line, was taken at 1 km/million years and is the same for the carbonate as for the clastics. These figures are gross averages taken from the Gabon sedimentary basin, as I detail later in this article. The time necessary to adjust the slope from 4 to 1° is the time necessary to infill the base of the cliff (gray triangular patches ABC called turbidity infilling in Figure 5). This time span depends on the size of the turbiditic infilling and the rate of supply of sediments (rate of progradation). The size of the turbiditic infilling is determined by the value of the slopes (1° and 4°) and the water depth. A slope of 1° drops 17.6 m every kilometer. Therefore, to reach a depth of 5 km, a distance of 5000/17.6 = 285 km is necessary. A 4° slope only needs 71 km to drop from sea level to sea bottom.

Triangle ABC in Figure 5, marked "Turbidite Infilling," represents the sediments necessary to infill the

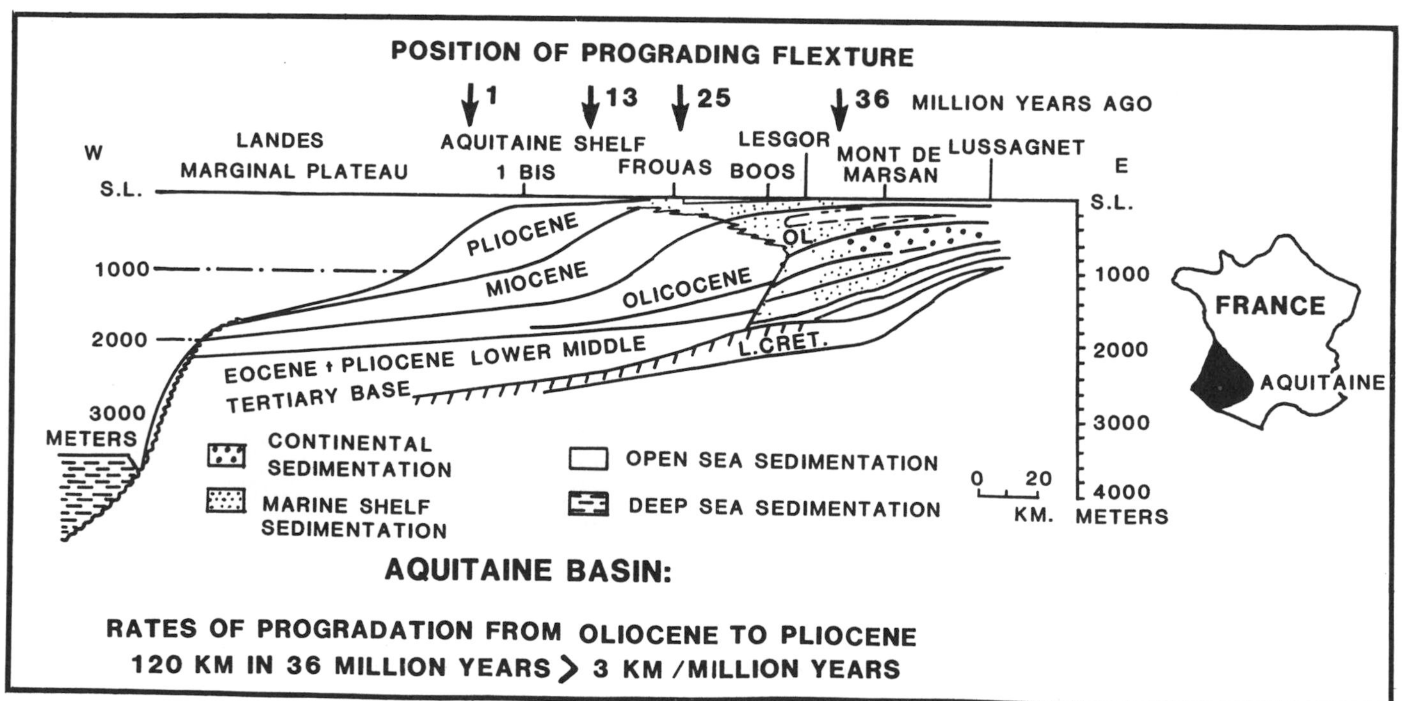

Figure 2 — Schematic cross section showing progradation of Cenozoic and recent deposits in Aquitaine Basin (Montadert et al, 1973).

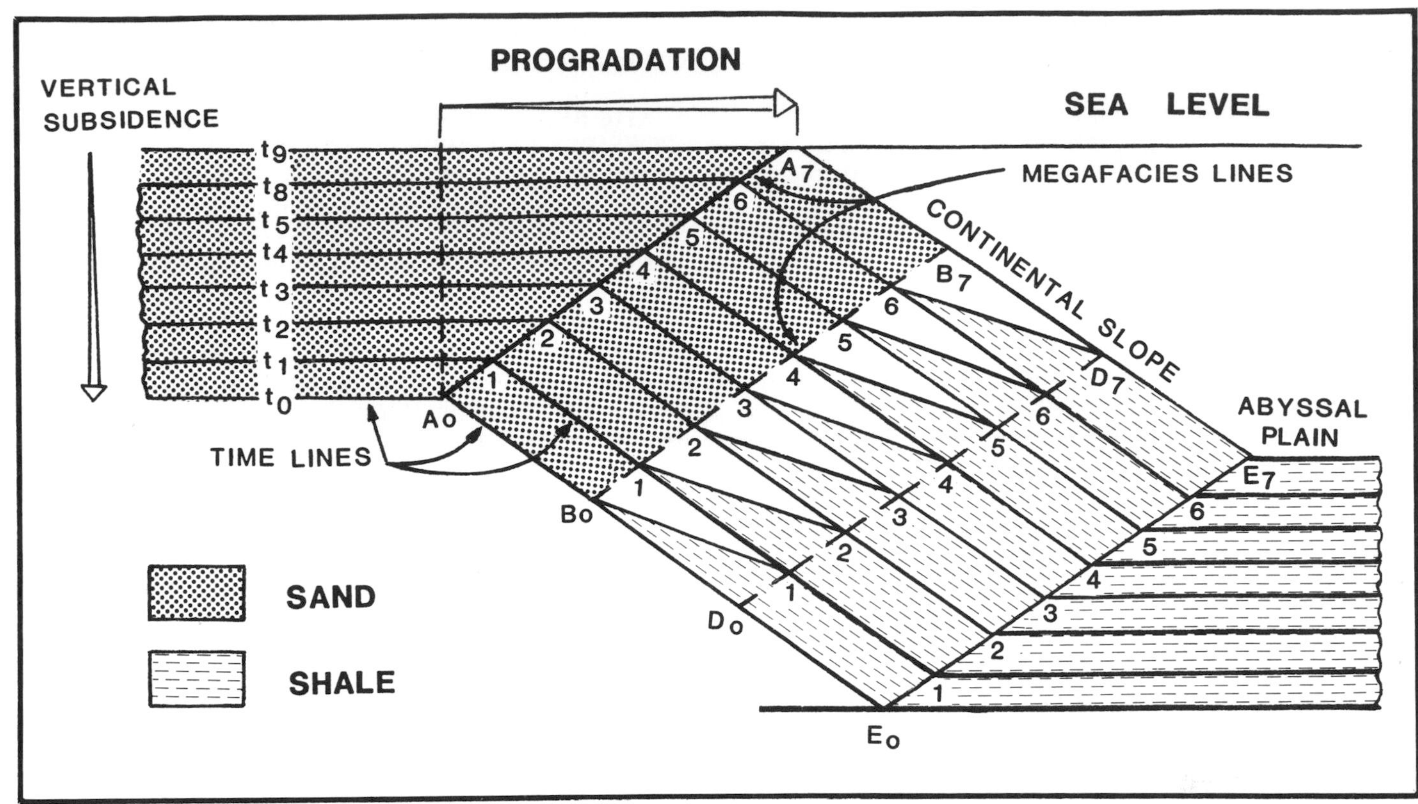

Figure 3 — Regressive Megasequence; progradation with vertical subsidence (Dailly, 1975).

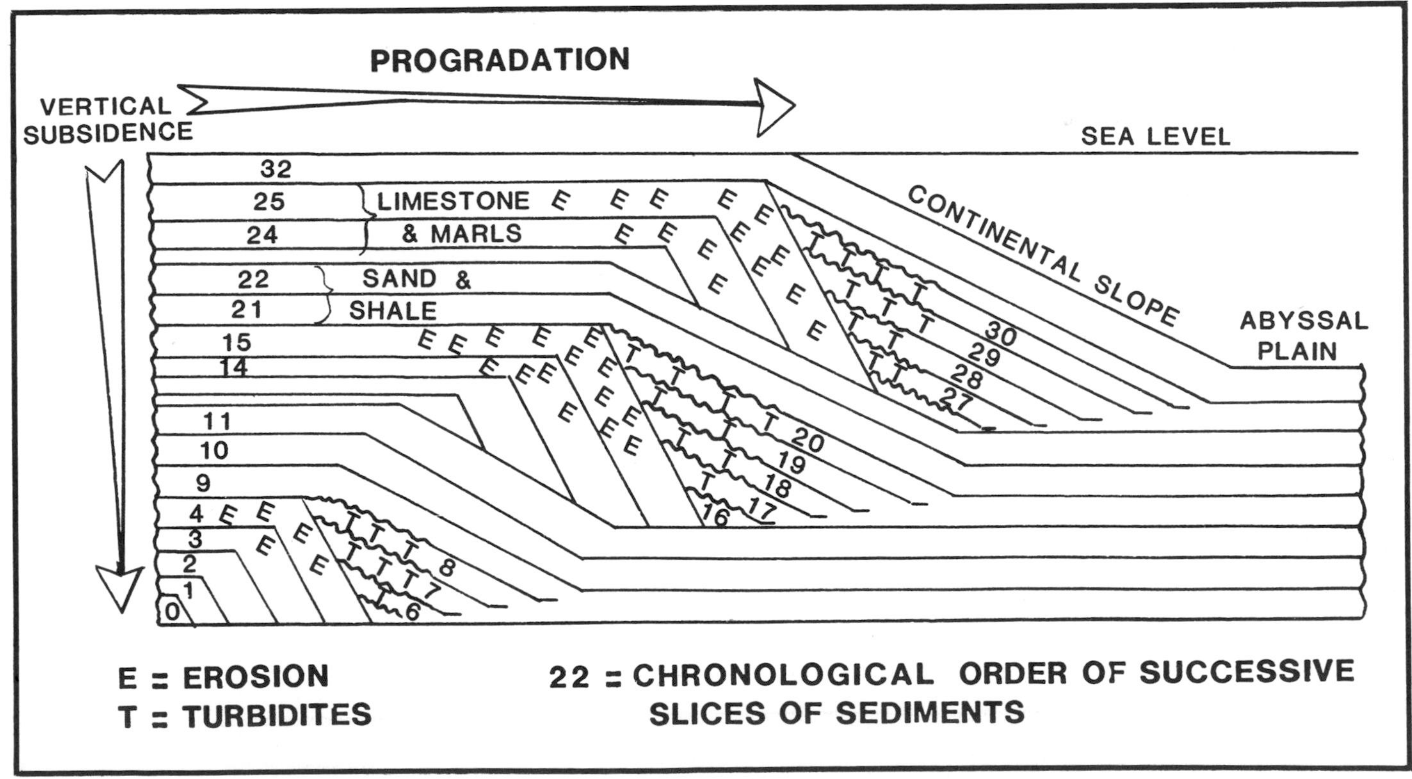

Figure 4 — Effect of successive changes in equilibrium slopes in prograding sediments.

base of the cliff, assuming there is no contemporaneous erosion on top of the cliff. This turbidite triangle has a base of 285-71 = 214 km. It requires the same amount of time to infill the triangle as to infill half of rectangle ABCD. If a source of sediments were available and the shoreline prograded at an average speed of 1 km/million years, in this idealized case it would take 214 million years to infill the rectangle or 107 million years to infill triangle ABC.

Take the value of 107 million years with care, as the water depth of 5 km is quite exaggerated. A water depth of 2,000 m reduces this slope readjustment time to 43 million years. However, the main factor governing the size of the turbiditic infilling is the 1° slope of the shale and sandstone. This value is a normal value for these types of sediments. Sometimes the depositional slope for shale sedimentation, as the Ireton shale in Alberta, is of 7 m/km (less than .5°), according to Stoakes (1980). The 4° slope for limestone and marl, or silicified shale, is probably underestimated.

The stratigraphic gap on top of the limestone cliff will be as large as the time of infilling, and probably larger due to regressive erosion. My conclusion is that if the idea of necessary slope adjustment is true, then this phenomenon is a major geologic event characterized by a period of several millions (sometimes tens-of-millions) of years during which clastic sediments onlap on the limestone, or hard sediments.

What is the average reaction of a geologist who finds an erosional contact between a shallow limestone covered by an onlapping shale facies with a stratigraphic gap of several million years? Most probably he thinks of a transgression related to a tectonic phase. Will he even consider that this could be explained by a regressive process, and represents the end of a period of slope readjustment?

THE BRAZILIAN CONTINENTAL MARGIN: A POSSIBLE EXAMPLE OF SLOPE READJUSTMENT

The Brazilian continental margin is the western Atlantic equivalent of the Gabon continental margin. The idealized geologic cross section (Figure 6) shows all the aspects of the theoretical model (Figure 5).

Following the Cretaceous carbonate, two periods of sand and shale deposition took place, starting with a turbiditic infilling which is called "transgressive." I am convinced that the term "transgressive" means that the clastic infilling is progressively onlapping on the base of the cliff, giving a "transgressive" aspect. When the slope is readjusted, the transgressive infilling phase and normal progradation take place on this new readjusted slope.

Figure 6 clearly shows the erosion and stratigraphic gap on top of the carbonate cliff. The analogies between the theoretical model and this schematic cross section are too striking for pure coincidence, and I am convinced that a major slope readjustment took place in Brazil after the limestone deposition.

THE GABON BASIN: A POSSIBLE EXAMPLE OF SLOPE ADJUSTMENT

Summary of the Geology of the Gabon Sedimentary Basin

This summary stresses only points which are important to demonstrate the phenomenon of slope adjustment. Figure 1 shows the overall architecture of

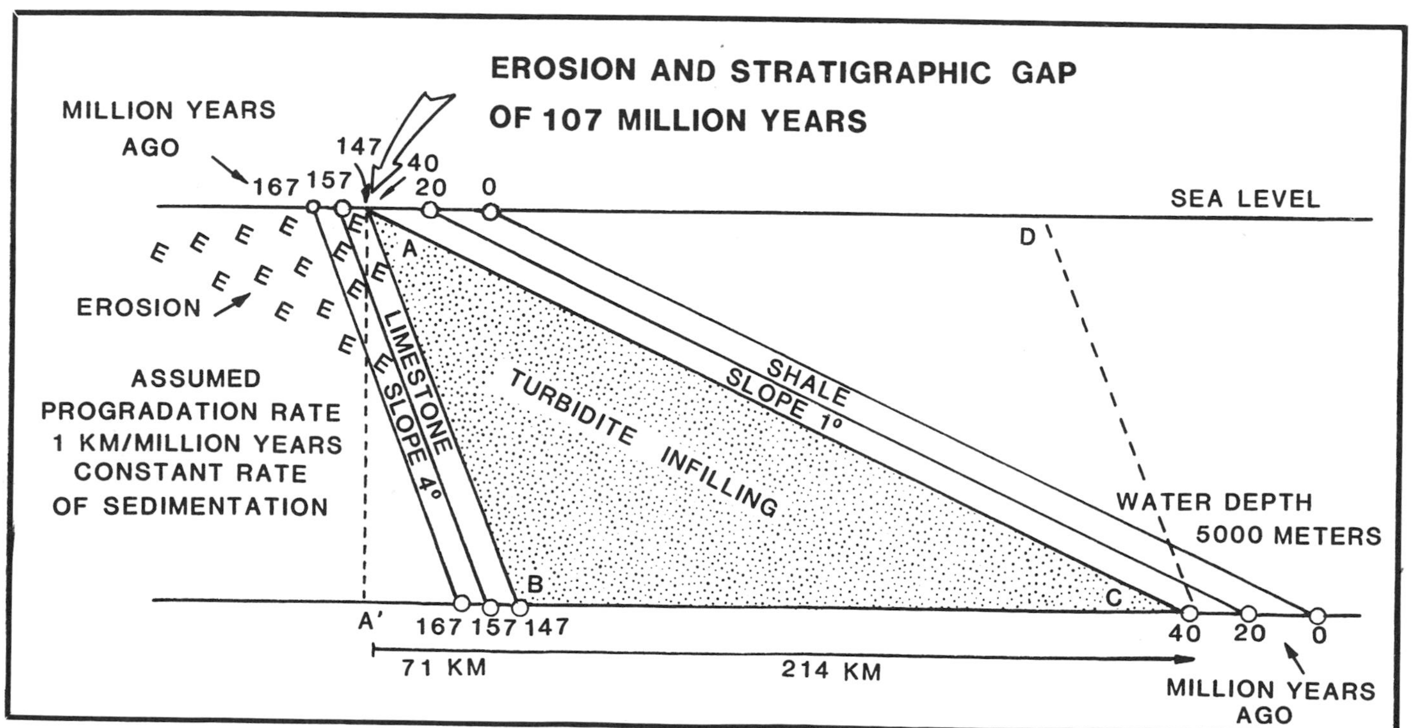

Figure 5 — Effect of change in the equilibrium slope in prograding sediments.

the Gabon sedimentary basin (Vidal, Joyes, and Van Veen, 1975). Above the salt, the prograding character of successive layers of sediments is clearly visible.

The stratigraphic scale is schematized in Figure 7 (R. Wenger, personal communication, 1973).

As in Brazil, the lithologic succession of sediments above the Aptian salt begins with a thick carbonate sequence: the Madiela limestone 2,000 to 1,000 ft thick, which is followed by a succession consisting mainly of sand and shale series. However, in the sand and shale series some "hard" layers, with an assumed higher dip than the foreset, exist; for instance, limestone beds in the Turonian and silicified shale and carbonate of the Ozouri formation of Eocene age.

A detailed facies zonation has been established by Elf GABON geologists (R. Wenger, personal communication, 1973) that distinguishes between continental, littoral and marine facies. A regional map with thickness and facies limit is given in Figure 7.

Progradation of the Continental to Littoral Facies Boundary

Figure 8 shows successive boundaries between continental and littoral facies. The overall regression of facies is visible even though some minor anomalies occur; for instance, the facies boundary 4 is sometimes eastward of line 3. This may indicate minor transgression or poor control in the facies limit definition.

The rate of regression of these facies boundaries averages 1 km/million years, which is quite low. This explains why in the schematic model of slope readjustment (Figure 5), the infilling of the base of a limestone cliff could be evaluated in several millions or tens of millions of years. This rate is lower than the rate of regression of the Niger and Mississippi deltas, which average 1 to 5 km/million years. For these more homogeneous deltaic sequences with a much higher rate of sedimentation, the slope adjustment periods will be reduced.

Figure 9 shows the facies boundaries between littoral and marine sediments for successive layers of sediments. Overall regression is also clear here, and the rate is of 1 km/million years.

Figure 10 is a composite map deduced from Figures 3 and 4. The area between lines C-1 and C-2 represents the overall regression of the continental to littoral facies limit, and the area between lines M-1 and M-2 represents the overall regression of the littoral to marine facies.

Figure 11 is a typical cross section showing migration of marine, littoral, and continental facies boundaries.The area between lines M-2 and C-1 exhibits progradation of sediments from Cenomanian to Paleocene at a rate of 1 km/million years.

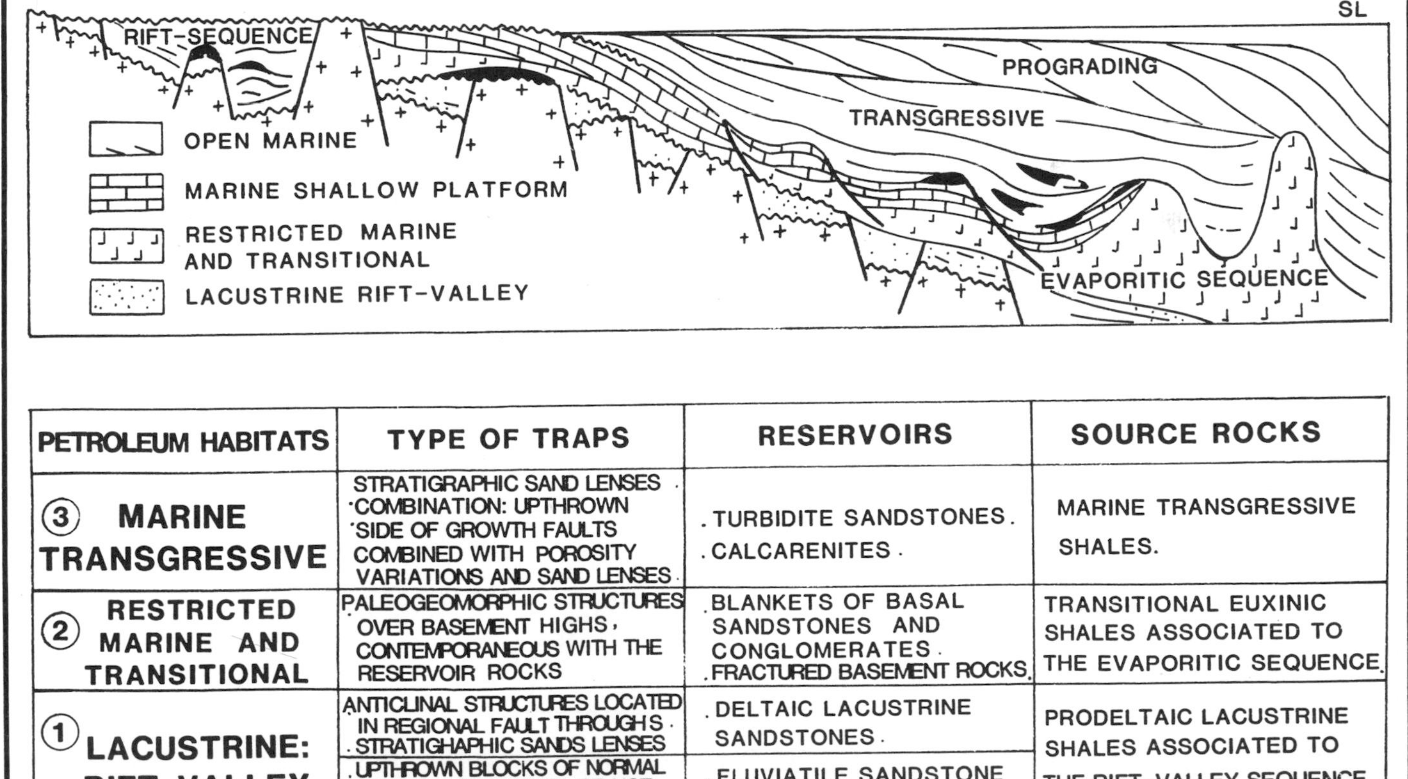

PETROLEUM HABITATS	TYPE OF TRAPS	RESERVOIRS	SOURCE ROCKS
(3) MARINE TRANSGRESSIVE	STRATIGRAPHIC SAND LENSES . ·COMBINATION: UPTHROWN SIDE OF GROWTH FAULTS COMBINED WITH POROSITY VARIATIONS AND SAND LENSES.	. TURBIDITE SANDSTONES. . CALCARENITES.	MARINE TRANSGRESSIVE SHALES.
(2) RESTRICTED MARINE AND TRANSITIONAL	PALEOGEOMORPHIC STRUCTURES OVER BASEMENT HIGHS, CONTEMPORANEOUS WITH THE RESERVOIR ROCKS	. BLANKETS OF BASAL SANDSTONES AND CONGLOMERATES. . FRACTURED BASEMENT ROCKS.	TRANSITIONAL EUXINIC SHALES ASSOCIATED TO THE EVAPORITIC SEQUENCE.
(1) LACUSTRINE: RIFT-VALLEY	ANTICLINAL STRUCTURES LOCATED IN REGIONAL FAULT THROUGHS. . STRATIGHAPHIC SANDS LENSES	. DELTAIC LACUSTRINE SANDSTONES.	PRODELTAIC LACUSTRINE SHALES ASSOCIATED TO THE RIFT-VALLEY SEQUENCE.
	. UPTHROWN BLOCKS OF NORMAL FAULTS, ASSOCIATED OR NOT TO LOCAL UNCONFORMITIES.	. FLUVIATILE SANDSTONE BLANKET.	

Figure 6 — Idealized geologic cross section of the Brazilian continental margin, with the main characteristics of the petroleum habitats (from Celso Ponte, Fonseca, and Carozzi, 1980).

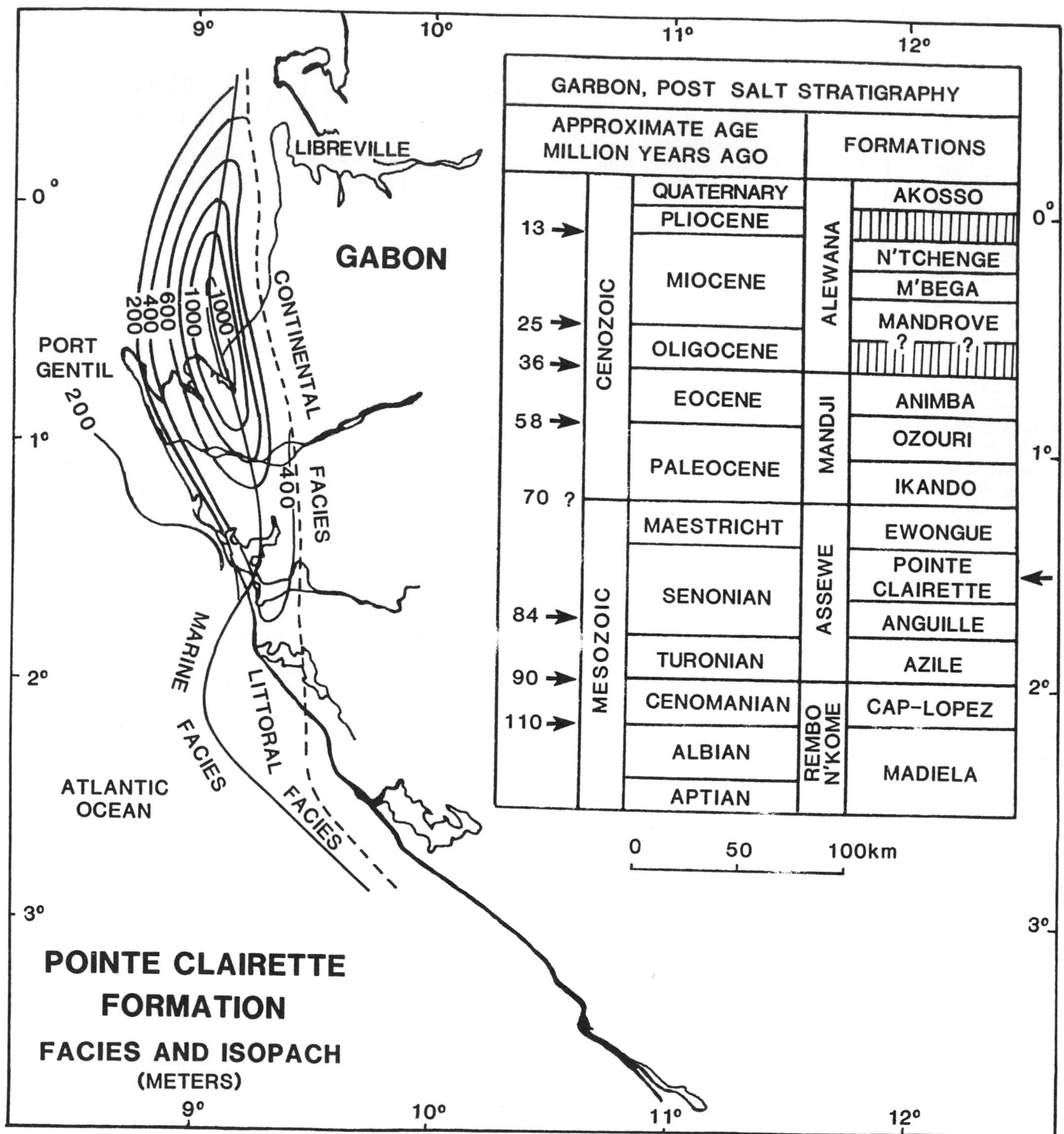

Figure 7 — An example of synthetic maps prepared by ELF GABON geologists (R. Wenger, personal communication, 1974).

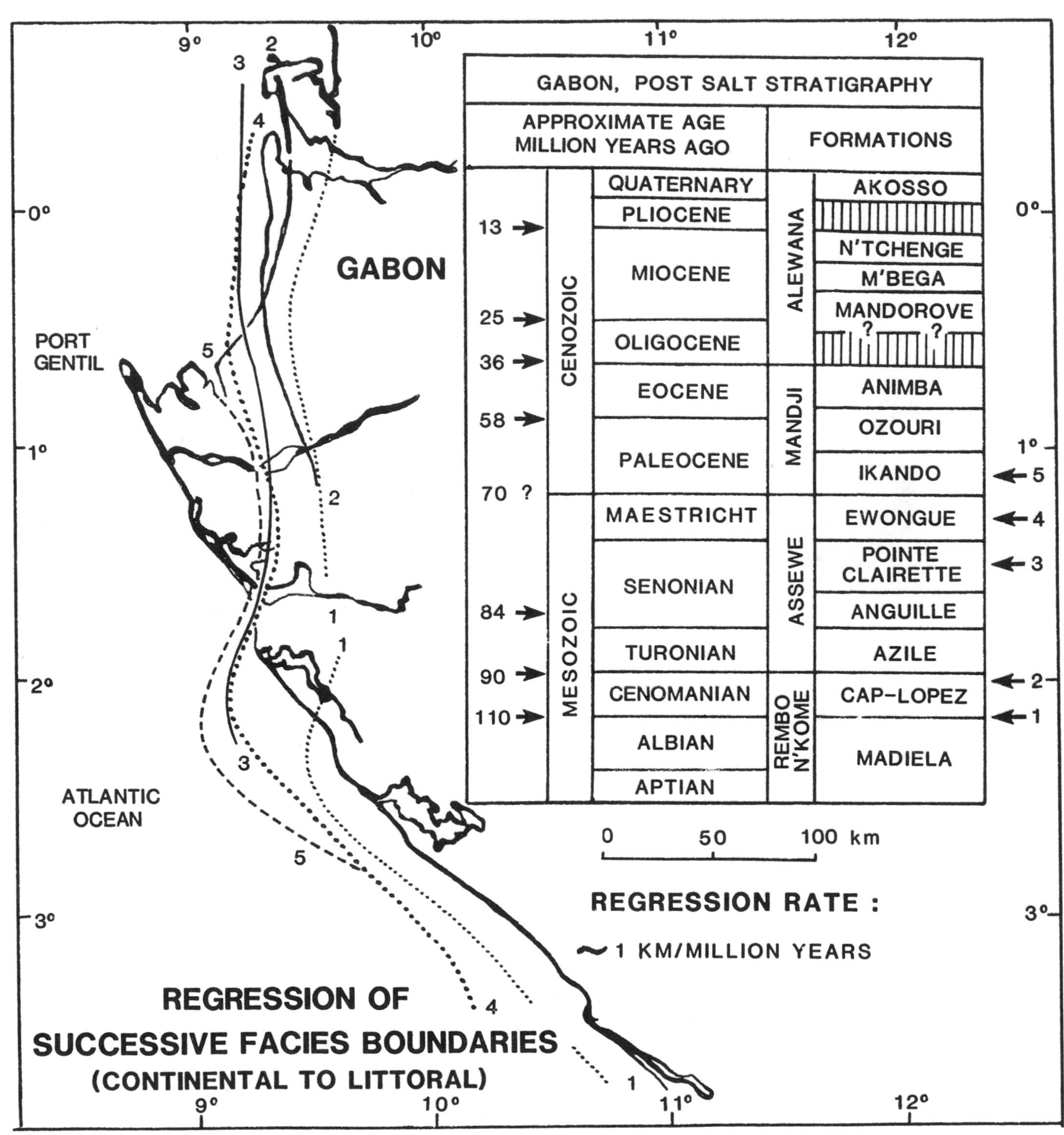

Figure 8 — Composite map of successive facies boundaries (continental to littoral). Figures on the boundaries refer to the same figures on stratigraphic scale (adapted from R. Wenger, personal communication, 1974).

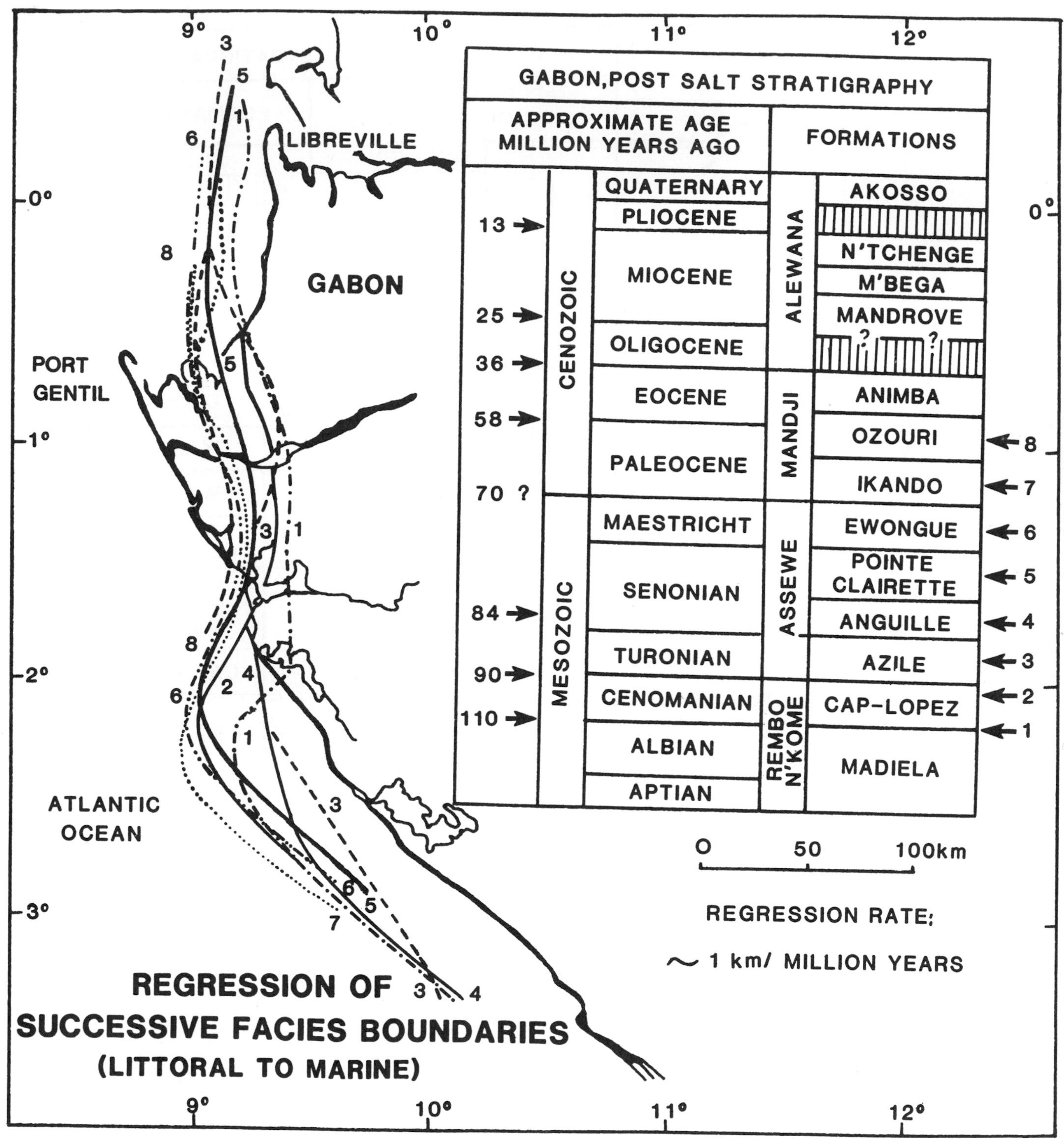

Figure 9 — Composite map of facies boundaries (littoral to marine). (Adapted from R. Wenger, personal communication, 1974).

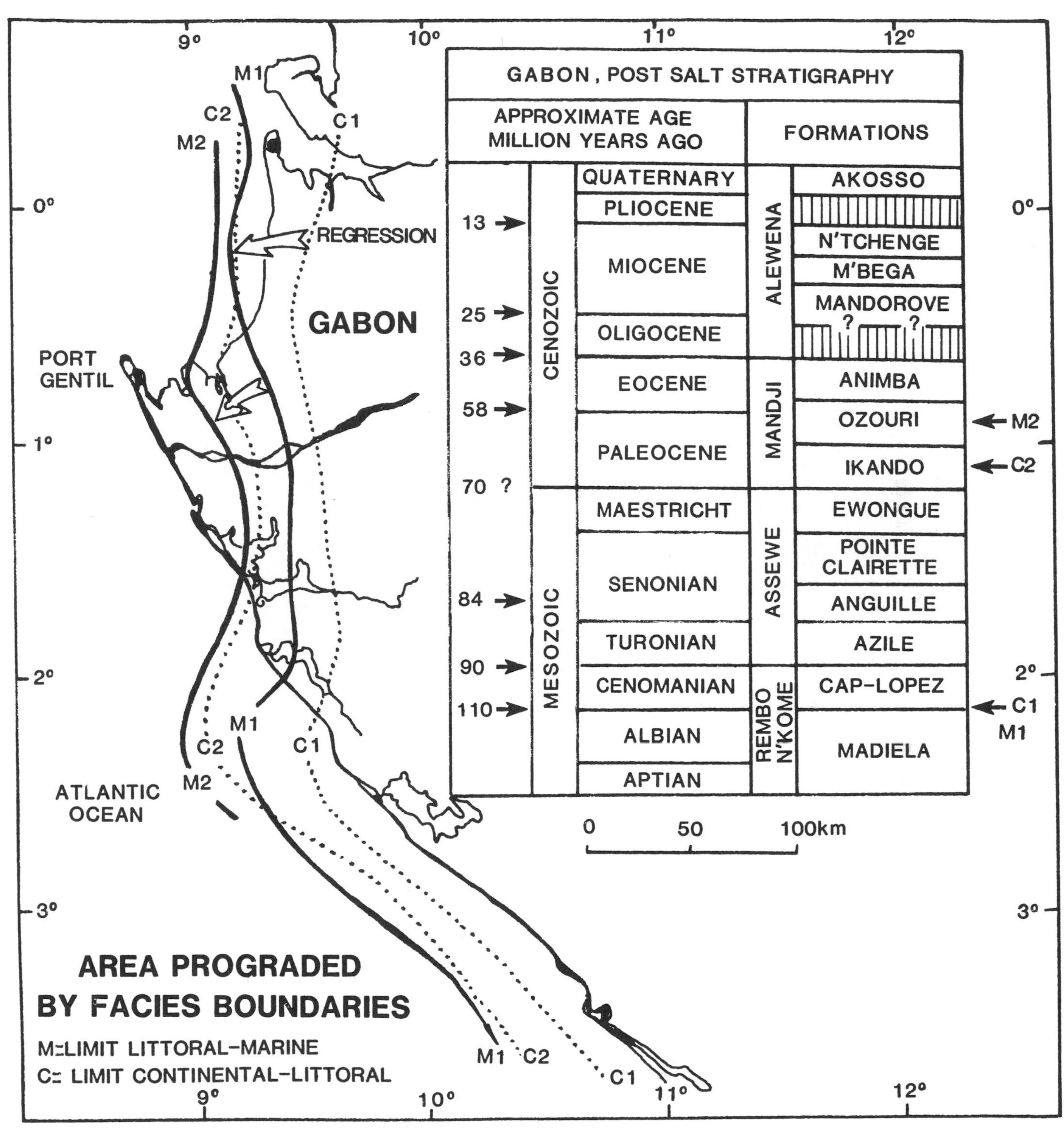

Figure 10 — Progradation of facies boundaries (adapted from R. Wenger, personal communication, 1974).

Submarine Erosion

I briefly mentioned in the lithological succession of sand and shale sediments the existence of harder rock made up of silicified shales, porcellanite chert, and littoral carbonate of the Ozouri Formation. This oil producing formation is well-known for its abrupt disappearance by erosion which is abundantly proven in the subsurface (Animba oil fields, etc.). The erosion is illustrated on Figure 12, which is based on the more than 100 wells drilled through this formation in the Mandji Peninsula. The wells are generally located on top of salt domes. The regional synchronism of the erosion has not been attributed to a tectonic phase, and remains unexplained. My explanation is that after deposition of the hard silicified shale, a period of slope adjustment took place.

Sand Distribution at the Base of Madiela "Cliff"

The Madiela limestone was deposited as a prograding limestone as its general facies zonation shows. The facies grades progressively from continental to deep water in the following order: continental, lagunal neritic, neritic barrier, and bathyal. After the Madiela limestone, the sediments are mainly shale and sandstone with some minor carbonate phases such as the Turonian carbonates. If a period of slope adjustment took place after each carbonate or silicified shale episode, a schematic cross section of Gabon would look more like Figure 4 than like Figure 5 (Brazil). The tentative proof that there was a slope readjustment period from Madiela deposition is given by the general sand distribution above it. Three maps of the sand distribution are given here (Figures 13, 14,

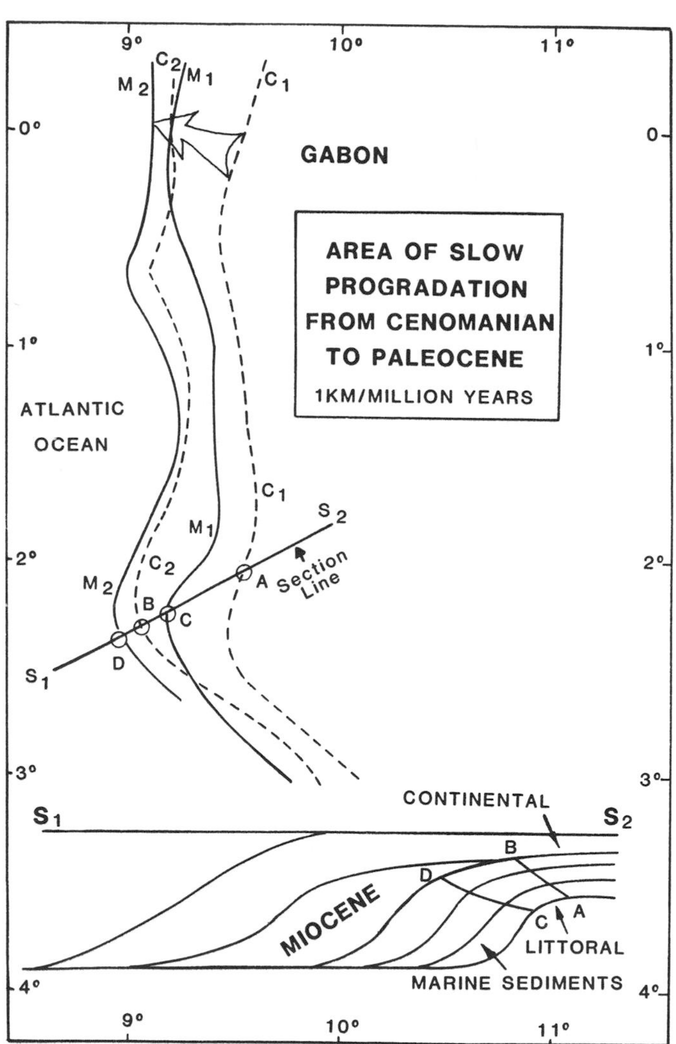

Figure 11 — Cross section of the prograding facies boundaries.

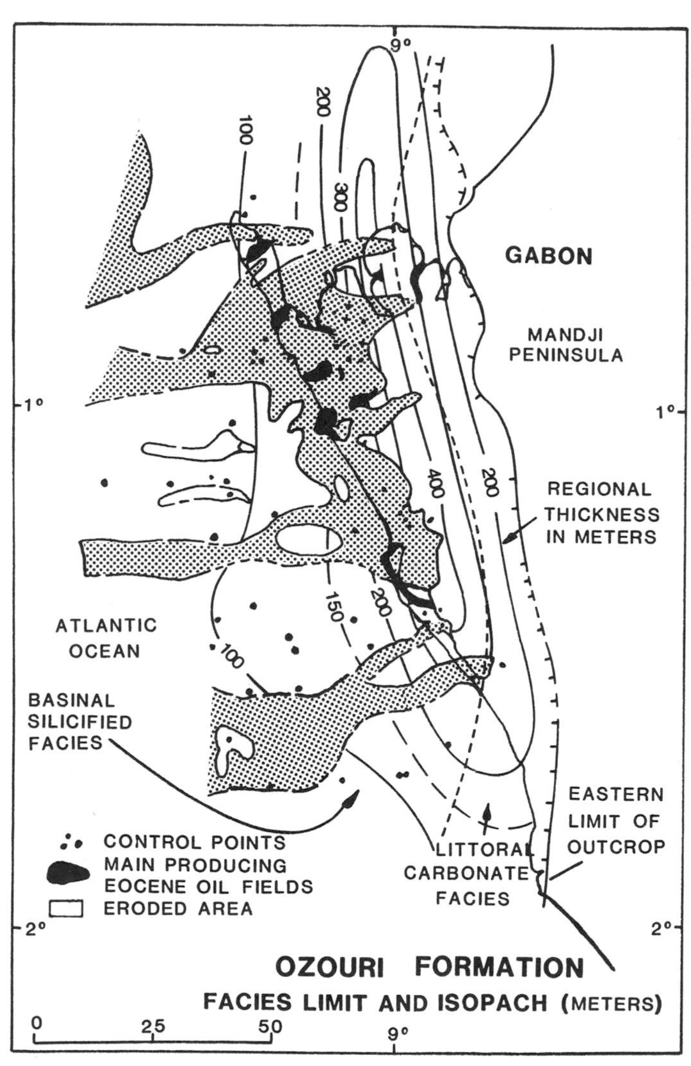

Figure 12 — The tremendous erosion of the marine silicified facies of the Ozouri Formation (adapted from R. Wenger, personal communication, 1974).

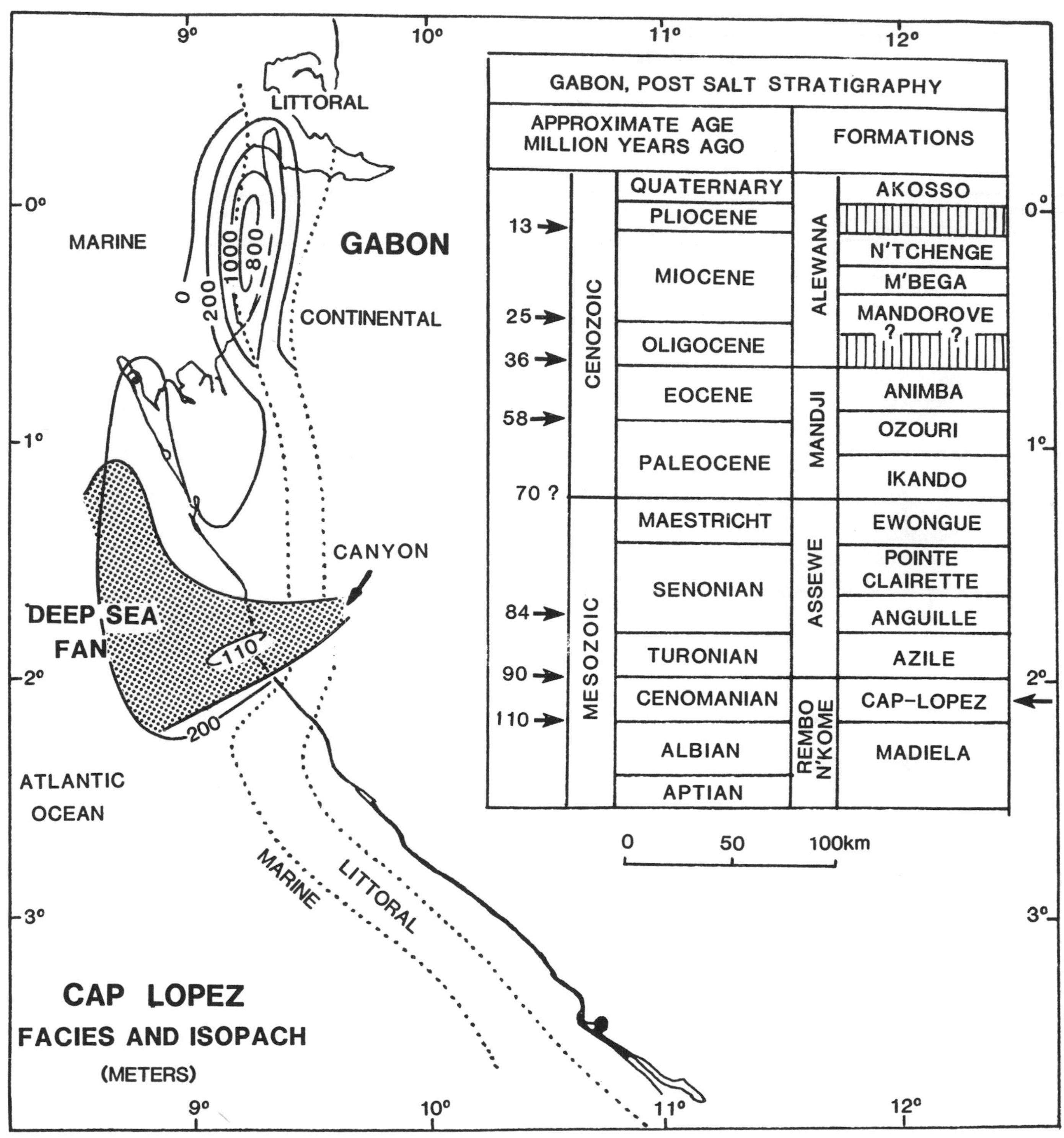

Figure 13 — Distribution of sand facies in the first formation above the Madiela limestone (adapted from R. Wenger, personal communication, 1974).

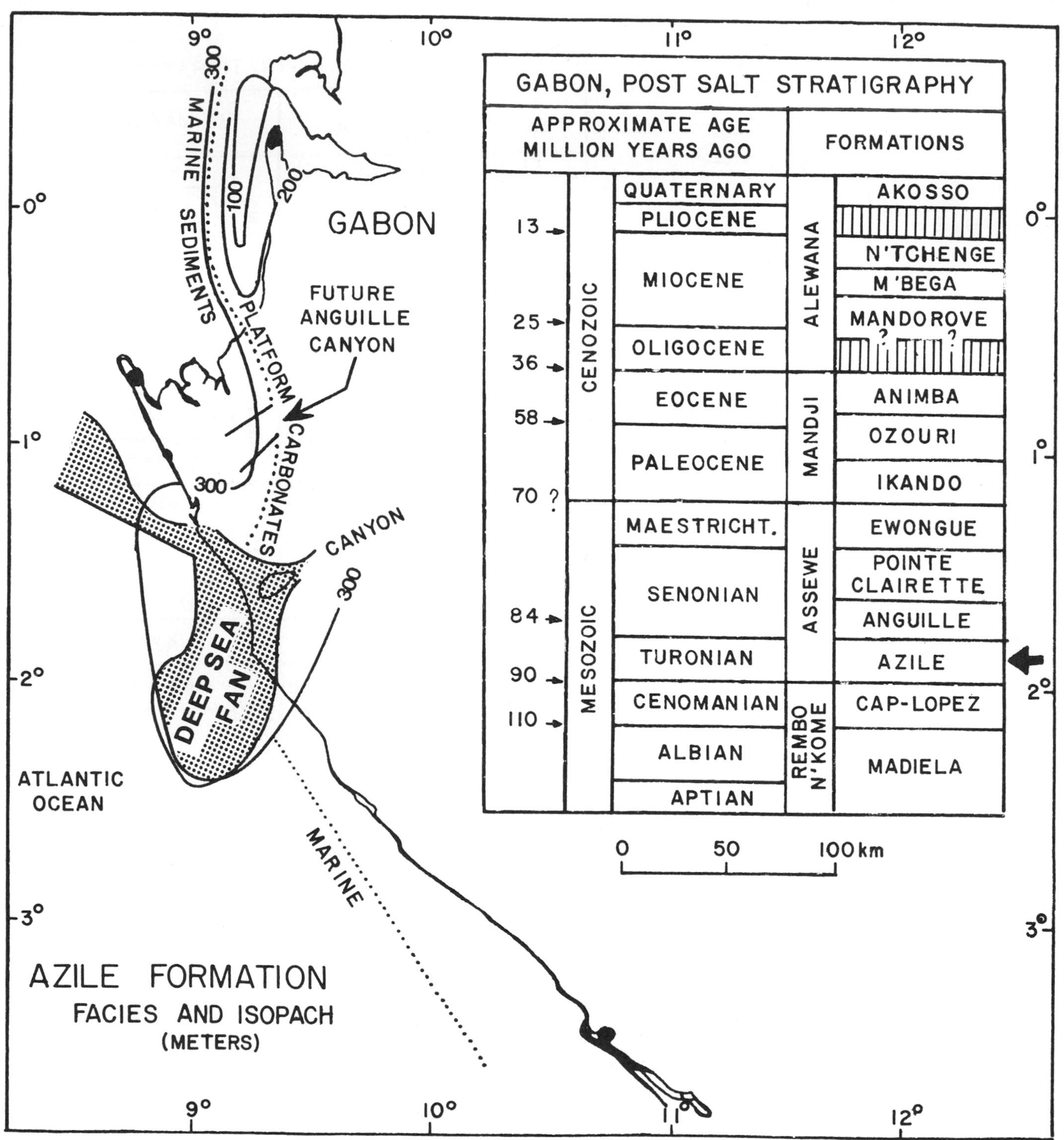

Figure 14 — Distribution of sand facies in the second formation above the Madiela limestone (adapted from R. Wenger, personal communication, 1974).

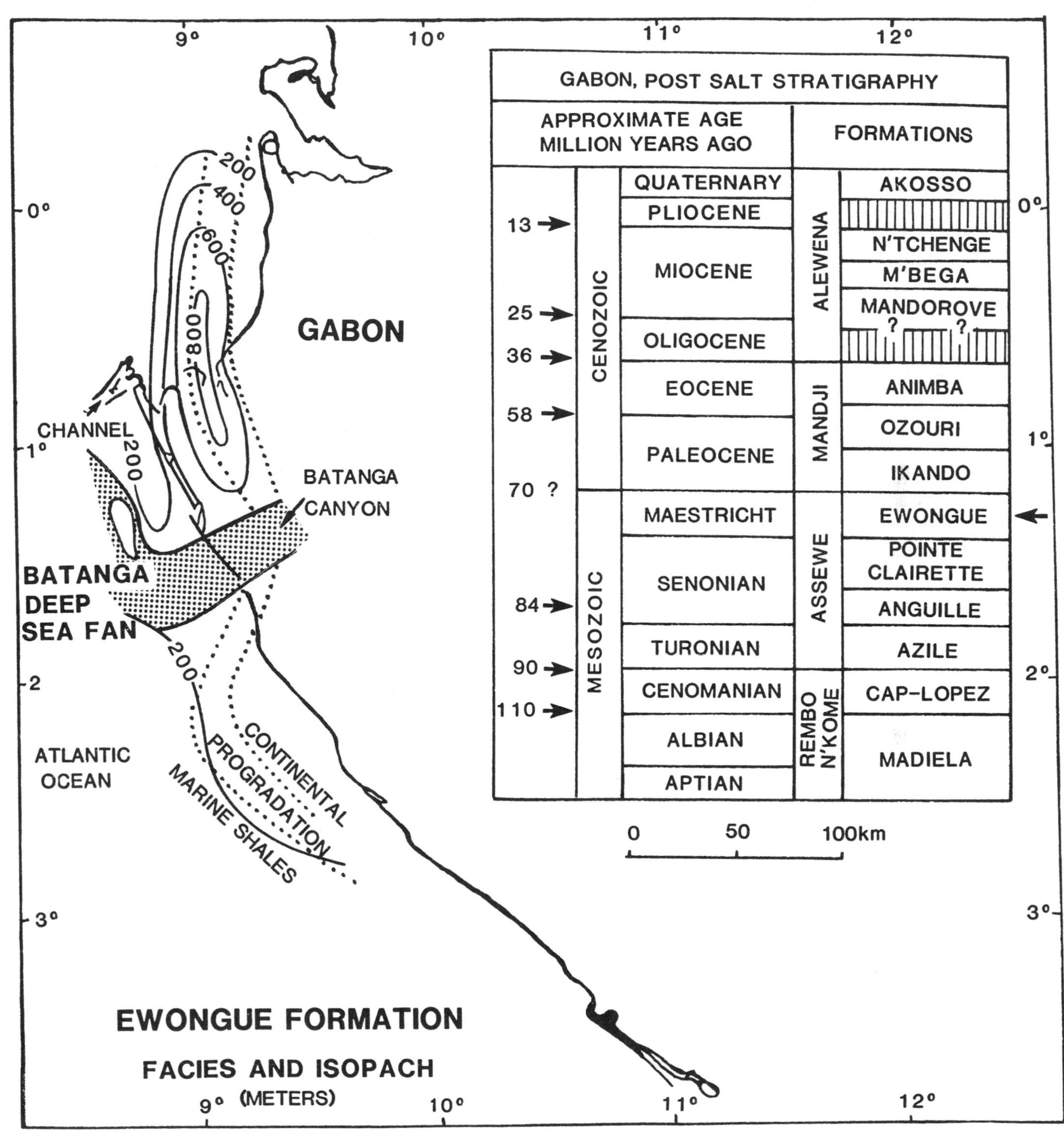

Figure 15 — Distribution of sand facies in Maestrichtian time (adapted from R. Wenger, personal communication, 1974).

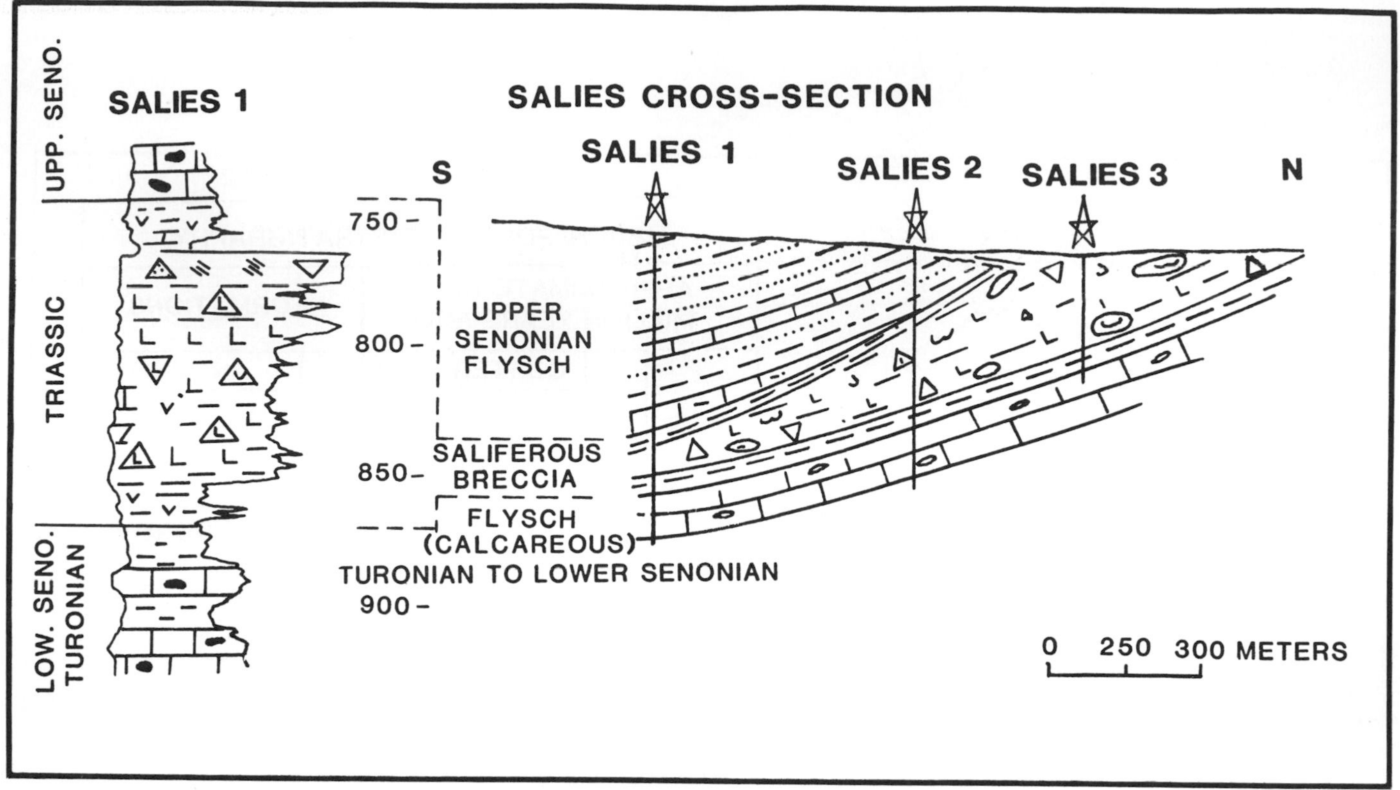

Figure 16 — Triassic continental facies (salt) interbedded in marine cretaceous sediments (Henry and Zolnai, 1971).

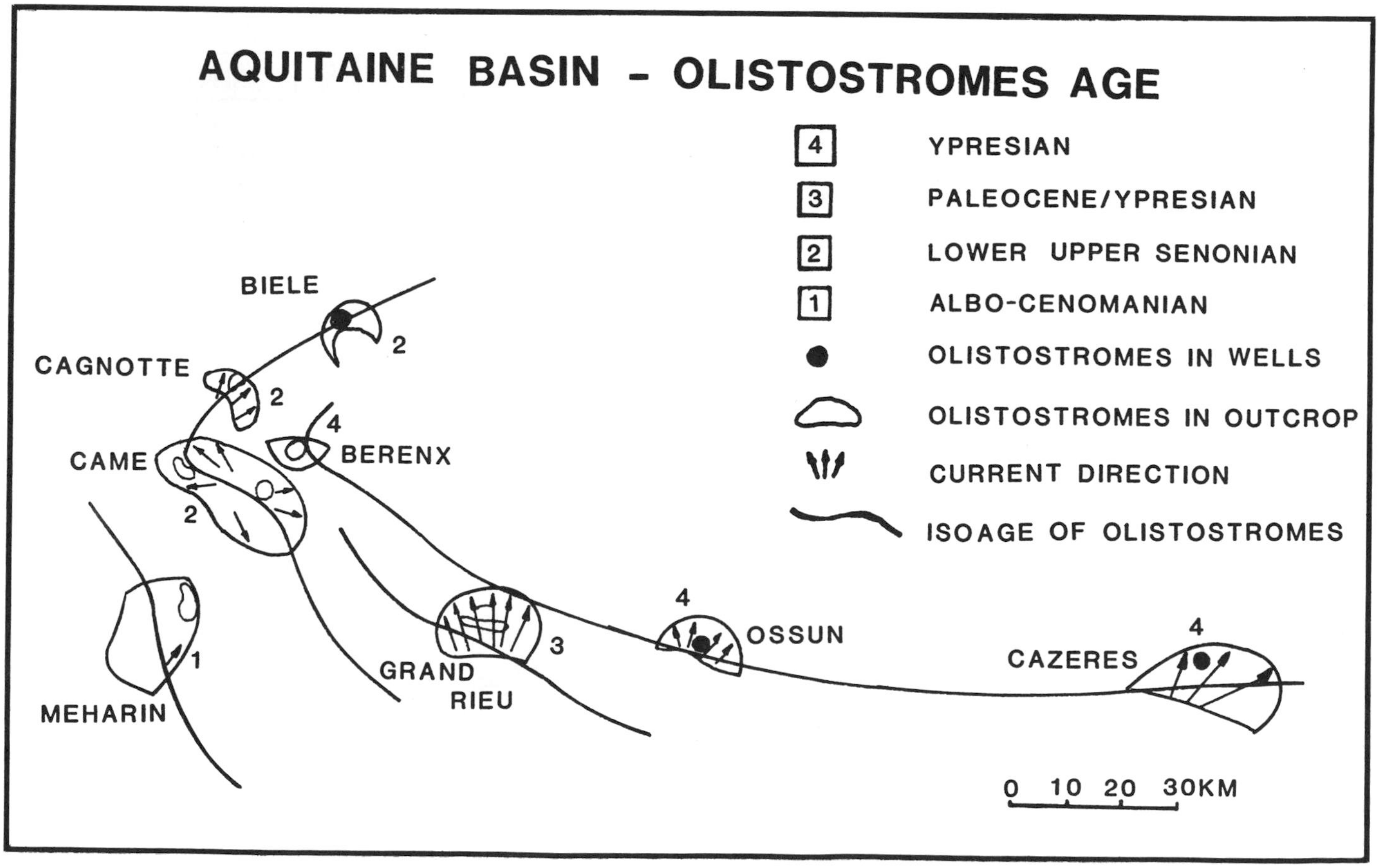

AQUITAINE BASIN - OLISTOSTROMES AGE

4 YPRESIAN
3 PALEOCENE/YPRESIAN
2 LOWER UPPER SENONIAN
1 ALBO-CENOMANIAN
OLISTOSTROMES IN WELLS
OLISTOSTROMES IN OUTCROP
CURRENT DIRECTION
ISOAGE OF OLISTOSTROMES

BIELE 2
CAGNOTTE 2
CAME 2
BERENX 4
MEHARIN 1
GRAND RIEU 3
OSSUN 4
CAZERES 4
0 10 20 30KM

Figure 17 — Synchronous periods of olistrome formation (adapted from Henry and Zolnai, 1971).

15). In each of them, the sand bypasses the platform facies through a narrow axis, which we call a canyon, and spreads in the abyssal plain, which we call a deep sea fan. This configuration proves that submarine fan deposits brought from the mainland through an erosive canyon infilled the base of the Madiela Carbonate "cliff."

THE AQUITAINE BASIN: ANOTHER POSSIBLE EXAMPLE OF SLOPE ADJUSTMENT

The Aquitaine Basin presents the characteristics of a prograding series (Figure 2); here the rate of progradation is 3 km/million years. There is an interesting publication on the Aquitaine olistostromes (Henry and Zolnai, 1971). The authors show the existence of several Cretaceous age "Triassic Salt" olistrostromes in abyssal sediments (Figure 16) and their conclusion is clear: "These Triassic sediments represent interstratified olistrostromes which have been put in place by synsedimentary flow ("ecoulement"). Important phenomena of erosion localized on the border of the basin have exposed the Triassic which later on has glided inside the basin." Another point, also very clear, is that the location in time and space of the phenomena follows the general paleogeographic evolution of the basin: "The phenomena seems to have taken place in a synchronous matter, in widely spaced area and during three distinct periods." The authors do not give an explanation of the synchronism of the phenomena, but they state that it cannot be explained by tectonic phases.

My explanation is that the synchronism of the phenomena corresponds to a period of slope adjustment. Successive growth of the basin with accompanying olistostromes is represented in Figure 17. Each lithological passage from hard to soft sediments will synchronize a period of regressive erosion and olistostrome formation.

CONCLUSION

Slope readjustment is probably a major component of sedimentation of regressive sequences. This paper is limited to three localized areas. Many illustrations could be provided from the subsurface worldwide.

It is well-known that frequently in the geological column terrigeneous clastic deposition follows a car-

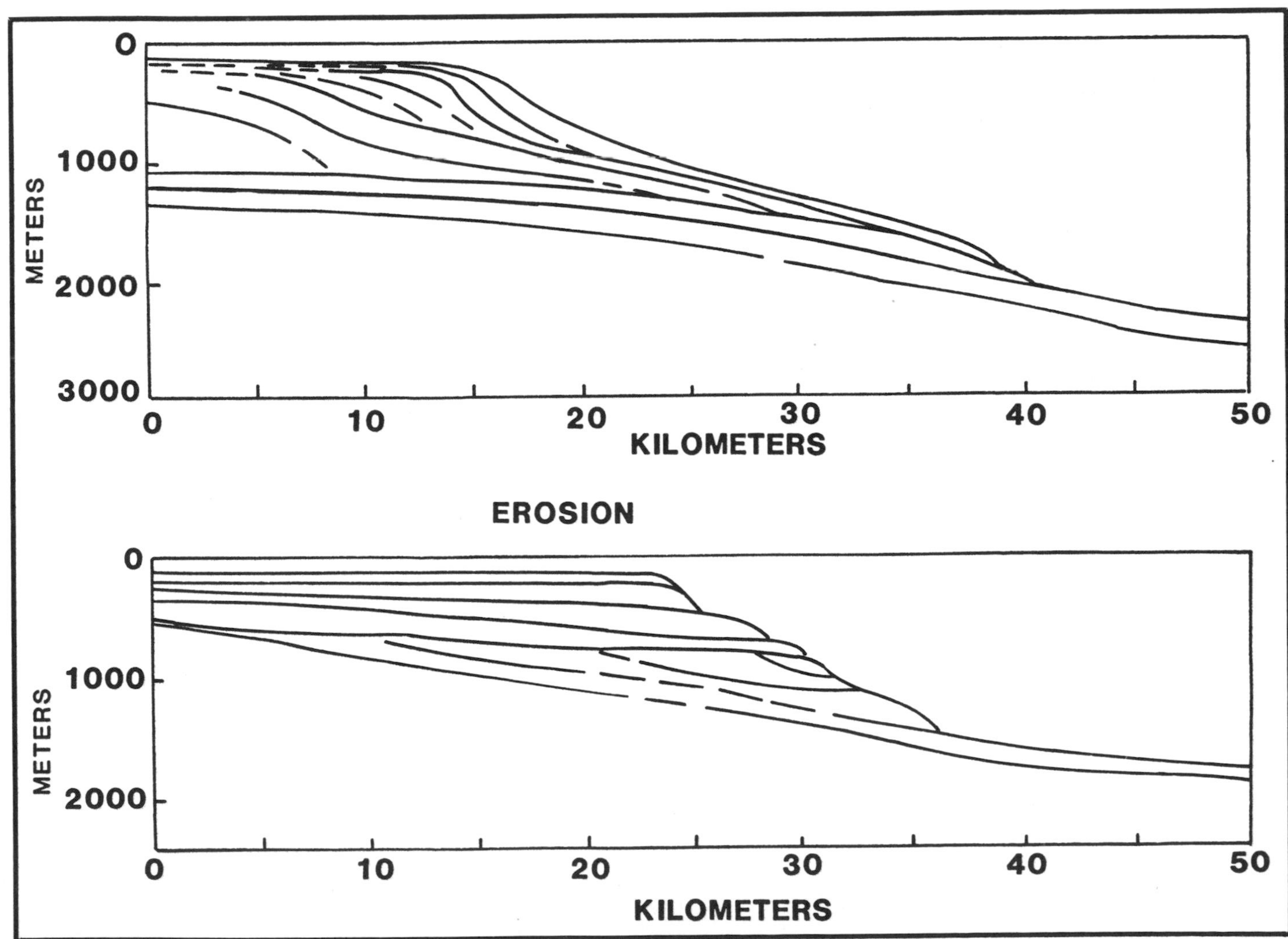

Figure 18 — Schematic classification of present continental slope (after Hedberg, 1970).

bonate shelf or platform sediment deposition. Without a tectonic phase, extensive erosion can occur and often a purely sedimentological explanation would probably be given if the theory of slope readjustment were more publicized. Even in present sedimentation, this theory seems applicable. For instance, H. D. Hedberg has classified the present continental margin in two main categories: prograded or erosional (Figure 18). If the theory of slope readjustment is admitted, the continental margins during their geological history change from prograded to erosional and so on, as a function of the nature of the sediments deposited. A period of erosion and onlap, only after a long period of time, follows deposition of sediments with steep foresets.

This paper presents the phenomenon of "slope readjustment." In nature, it is superimposed on general phenomena like tectonic phases, isostatic readjustment, eustatic variations, crust cooling, and so on. These phenomena are well described in the literature, contrary to the slope readjustment theory which has, to our knowledge, received little attention.

REFERENCES CITED

Celso Ponte, F., J. Dos Reis Fonseca, and A. V. Carozzi, 1980, Petroleum habitats in the Mesozoic-Cenozoic of the continental margin of Brazil, *in* Facts and principles of world petroleum occurrence: Canadian Society of Petroleum Geologists, Memoir 6.

Dailly G., 1975, Some remarks on regression and transgression, *in* C. J. Yorath, E. R. Parker, and D. J. Glass, eds., Deltaic sediments in Canada's continental margins and offshore petroleum: Canadian Society of Petroleum Geologists, v. 54, n. 1, p. 3-43.

Hedberg, H.D., 1970, Continental margins from the viewpoint of the petroleum geologist: AAPG Bulletin, v. 54, n. 1, p. 3-43.

Henry, J., and G. Zolnai, 1971, Resedimentation of the Triassic in the southwestern Aquitaine Basin: Bulletin Centre Recherche Pau, SNPA, n. 5, p. 389-398.

Stoakes, F. A., 1980, Nature and control of shale basin fill and its effect on reef growth and termination, *in* Upper Devonian Duvernay and Ireton formations of Alberta, Canada: Bulletin of Canadian Petroleum Geology, v. 28, n. 3, p. 345-410.

Vidal, J., R. Joyes, and J. Van Veen, 1975, L'exploration pétroliere au Gabon et au Congo: Tokyo, Proceedings, 9th World Petroleum Conference, v. 3, p. 149-165.

Neogene Palynostratigraphy of the Southern Mexico Margin, DSDP Leg 66

G. R. Fournier
Gulf Oil Exploration and Production Company
Houston Technology Center
Houston, Texas

Continuous analysis of 623.5 m of cored samples from the three holes at Site 493 resulted in the recognition of eight distinctive palynostratigraphic zones and three unconformities. The early Miocene to the Holocene assemblage succession documents vegetation changes on the mainland in the last 20 million years. Vegetational shifts are due in part to climatic fluctuations that occurred in the region. Correlation with the nannoplankton zonation and magnetostratigraphy is fair to good.

The purpose of the study was to observe the nature of the palynological assemblages, and obtain information of the climatic history from the Miocene to the Holocene in adjacent Mexico. We found that sufficient palynomorphs established a biostratigraphic zonation of the site, and recognized three hiatuses in the area. The succession of palynofloras shows enough fluctuations of the key species and relative abundances of the assemblages to recognize several events in the climatic history of the adjacent mainland.

The method of analysis used in this project is described in detail by Fournier (1981b), and it is highlighted below:

1. Representative specimens of each morphotype found in the section are picked and single mounted.
2. Counts of up to 200 palynomorphs of composite continuous samples at designated intervals are made.
3. The results of the counts are input into the computer, and three printouts record stratigraphic order of morphototypes, alphabetical order of all morphotypes, and an alphabetical order of the occurrence of all morphotypes.

We chose "sensitive species" whose distribution and abundance in the section are considered significant due to their uniqueness in distribution, intrinsic environmental value, or significance as a key species. The distribution chart of significant or "sensitive species" is the basis for zonation of the biostratigraphic section. Once the sensitive species are identified and their significance in the section recognized in terms of their presence, location, and pattern of distribution, conclusions are drawn regarding climatic changes, depositional environments, and age determinations.

Data from this study are summarized in Fournier (1981 b).

PALYNOSTRATIGRAPHIC UNITS

The criteria used for determining a palynostratigraphic unit are based on the assemblage. As long as an assemblage maintains its identity, it is an indication of stable environmental conditions. Vertical continuity of an assemblage is an index of how long a set of conditions lasted. In the horizontal or geographic sense it shows how extensive those similar conditions were. A significant gradual change in the characteristics of an assemblage through time indicates a shift from one biostratigraphic unit into another. But a geographic or horizontal change in the character of the assemblage is usually an indication of biofacies shifting. An abrupt interruption of the vertical continuity of palynological assemblage may indicate an unconformity or a fault. Reworked palynomorphs may be recognized by their erratic distribution.

Characteristics of the Assemblages

Two biological elements dominate the organic residue of all the samples from Site 493: coniferous pollen and fungal elements. Also found abundant in large parts of the section are faecal pellets, not exceeding 20 microns in diameter, derived possibly from extremely small crustacea. Although these pellets are found throughout the section, they are extremely abundant in Zones 4 and 5. The coniferous pollen must have originated in upland forest not too distant from the depositional site, and the bulk of the pollen was probably stream transported rather than carried by the wind. The bulk of pine pollen or any other pollen falls within a few miles from its source (Traverse and Ginsburg, 1966). The fungal elements probably originated in the rain forest floor or in swampy or stagnant environments sustaining a high degree of biological degradation. Marine origin is also a possibility. The large quantities of fungal elements found in all samples, except in Zone 5 where they are less abundant, suggest an endemic source, the rain forest floor plus perhaps the sea floor.

An interesting characteristic of the palynological assemblages is the scarcity of dinoflagellate cysts usually abundant in marine sediments. They are virtually absent in the section except for Zone 6, where we recorded fairly continuous occurrence of low percentages. Reworked palynomorphs, mostly pteridophyte Cretaceous spores, were found sporadically mainly in the section's lower part in Zones 7 and 8.

It is reasonable to assume that most of the palynomorphs found in Site 493 originated in the Mexican mainland and were transported to the depositional site by streams draining the mountain front and coastal areas and by wind currents.

Components of the Palynofloras

Tree pollen. The pollen assemblage from trees in this site is dominated by upland species such as pine oak (*Quercus*), and *Engelhardtia* (*Juglans*). Pollen of the family Sapotaceae, trees shrubs of tropical climates, is abundant in Zones 4 and 8. Alder pollen (*Alnus*) is abundant only in Zones 1 and 2. In the tropics, alder grows in mountain areas from 2000 to 3000 m high, as in Colombia and Venezuela (Muller, 1959). It appears for the first time in the Caribbean area in the Pleistocene (Germeraad, Hopping, and Muller, 1968). In Mexico, alder is considered an upland forest component (Sears and Clisby, 1955). The pollen of *Ilex* (holly) is sporadic in Zones 3, 4, and 5. *Ilex* is found in tropical and temperate regions and requires a wet climate. The pollen of a *Bombacaceae* (Ceiba), a tropical tree, is found only in Zones 6 and 7. *Taxodium* (cypress) pollen is found sporadically in Zones 7 and 8. *Taxodium* is edaphically controlled and grows in swampy and riverine environments.

Herbaceous pollen. Most herbaceous pollen types present in the section are from the family Compositae. The *Helianthus* type is found throughout the whole section but more abundantly in Zones 6 and 7. The *Ambrosia* and *Artemisia* types are restricted to Zones 3 and 4, with the latter showing its acme of development in Zone 3. The pollen of *Graminea,* grasses and sedges, is abundant in Zones 2 and 3 and in the upper half of Zone 4.

Ferns. Fern spores, both monolete and trilete types, are important constituents of the assemblages in this site. Ferns grow preferably in warm to cool, very moist environments. As a group they are good indicators of wet climates (Groot and Groot, 1966). They are scarce in Zone 5, become more abundant in Zone 4, and are abundant in Zones 6 and 7. In Zone 3 they are scarce again, and become abundant in Zone 2. Their abundance fluctuates with the moisture available.

PALYNOSTRATIGRAPHY

Description of the Biostratigraphic Units

The continuous analysis of the stratigraphic section of Site 493 clearly shows the succession of palynomorph assemblages through time and their response to many factors, mainly climatic fluctuations and depositional patterns. Palynostratigraphic units are described in terms of these assemblages, each one acquiring a distinctive identity for later recognition as a reference section. In a limited sense, these biostratigraphic units are time parameters. The eight zones of Site 493 are summarized in Table 1. Details of this zonation may be found in Fournier (1981a).

Table 1. Palynostratigraphy Site 493.

Zones	Depth m	Age	Mannoplankton Zone
1	21.5	Holocene	NN21
			NN20
2	88.0	Quaternary	NN13-16
3	148.5	L. Pliocene	
			NN15
4	243.5	E. Pliocene	
5	362.0	L. Miocene	NN11
6	500.0	E. Miocene	NN5
			NN4
7	585.2	E. Miocene	
			NN3
8	618.4	E. Miocene	
			NN2

Paleoclimates

Analysis of the palynofloras in Site 493 provides an opportunity to observe the changes in vegetation that took place in the Mexican mainland. The palynoflora of a depositional site reflects the vegetation shifts at the source, and is influenced by other factors such as distance from shore, depth of water, tectonic setting, drainage patterns, and marine currents.

Conditions of deposition and source areas seem to have been fairly stable from the Miocene to the Quaternary. I believe the main changes in the character of the assemblages and in the density of palynomorph population are mainly due to climatic fluctuations. I also believe that the density of land-derived palynomorphs is mainly related to transport by streams. Changes in the number of species relate mainly to fluctuations of the mean temperature. As a rule, species diversity is greater in warm climates than in cold ones.

Of the 59 sensitive species recorded, 20 are *pteridophyte* spores, most of them *filicales*. Ferns prefer warm to cool climates with abundant rainfall, the climate of tropical mountain slopes. Fluctuations in the relative abundance of fern spores are fairly good indicators of changing climatic conditions. In Groot and Groot's (1966) study of deep-sea cores off the coast of Chile they use fern spores as indicators of humid climate.

Paleoenvironments in Site 493

The main climatic events of the Site 493 cores encompass a period of about 20 million years. The eight zones described on the basis of their palynological assemblages represent a relatively stable period in the climatic history of the region.

Early Miocene basal sediments rest unconformably on basement rocks. The climate during Zone 8 was relatively dry and warm as indicated by the limited number of species and the presence of *Saportacea* pollen (15%) and scarce *Ilex*. During Zone 7 rainfall increased, as indicated by the appearance of several fern species, and the average temperature was probably higher. Tropical climate prevailed in the area until the end of Zone 6. Oak pollen is missing in Zones 8 and 7, and begins to appear sporadically in late Zone 6.

The entire middle Miocene sedimentary record is missing. Sediments of upper Miocene age rest on a hiatus at 362 m depth at the base of Zone 5.

Most of the fern spores from Zone 6 are absent in Zone 5, except for a few hardy species. Several species of oaks and abundant pine occur in Zone 5, indicating that the climate during Zone 5 time was rather dry and warm. The lowest concentration of fungal elements is in that zone, possibly another indication of drier conditions during this time.

During Zone 4, the mean temperature remained the same but an increase in the pollen concentration and in the number of species indicates an increase in precipitation and stream activity. These conditions were interrupted at the end of Zone 4 by a hiatus, expressed also by a change in lithology.

Climate became cooler and drier during Zone 3, indicated by the virtual disappearance of the tropical Sapotaceae and some species of oak, and by a notable change in floral character with a significant increase of seven fern spores, indicating increased stream activity and higher precipitation. A wetter climate is also indicated by the occurrence of high counts of oak and alder pollen in Zone 2, and a decrease in Helianthus (sunflower group) (Sears and Clisby, 1955). The climate during Zone 2 was cool and moist.

A probable hiatus separates Zone 2 from Zone 1 at 21.5 m depth. The sudden disappearance of seven species typical of Zone 2 may indicate an unconformity. Not enough information is available from samples in Zone 1. However, the limited number of species recorded in this interval indicates drier conditions and less stream activity than in Zone 2. The temperature may not have changed appreciably between these zones.

Correlation with the Nannoplankton Zonation

Fair correlation exists between the palynostratigraphic zonation established by the author in Site 493 and the nannoplankton zones determined by Stradner and Allram (1981).

Several palynological tops occur below the nannoplankton tops, indicating that the nannoplankton tops are somewhat younger: the top of Zone 2 is about 8.5 m below the tops of NN 20; the top of Zone 3 is about 30 m below the top of NN 18; the top of Zone 5 is close to 18.5 m below the top of NN 11; and the top of Zone 8 is about 19 m below the tops of NN 3. This is consistent with observations made in other parts of the world where correlations are established between zonations based on marine organisms and correlations based on detrital elements transported from the land to the marine environment.

DISCUSSION

The presence of significant numbers of dinoflagellate cysts in only one part of the section is conspicuous. Zone 6, the only interval where their numbers are significant, may represent a period of falling sea level and shallowing in the depositional areas with total emergence at the end of Zone 6. The hiatus between Zone 6 and 5 represents non-deposition or removal of all middle Miocene sediments. Dinoflagellate cysts are found again in the uppermost part of Zone 4, just below the unconformity at the top. This could be a small scale repetition of the events decribed above.

The dinoflagellate cysts, having a specific gravity more like that of water than detrital palynomorphs, may require shallow water depths and non-turbulance to sink. This may explain their absence in an area where oceanic, open sea conditions and relatively deep water have prevailed since the Miocene, except for brief periods of time.

Changes in climate and tectonic events (including volcanism) immediately affect the land floras and pollen in depositional sites. Marine forms take longer to be affected, probably because of the buffering effect of the marine environment, and the evidence is recorded later. This may explain why the tops of palynostratigraphic units appear earlier in the sedimentary column than the tops of the nannoplankton units.

ACKNOWLEDGMENTS

I am grateful to Dr. Nahum Schneidermann and to Dr. Bernard Shaffer for their valuable suggestions in the preparation of the manuscript. I am also indebted to the Gulf Oil Exploration and Production Company for giving me the opportunity and time to do this work.

REFERENCES CITED

Fournier, G., 1981a, Palynostratigraphic analysis of cores from site 493, Leg 66 *in* J. S. Watkins and J. C. Moore, eds., Initial reports of the Deep sea drilling project: Washington, D. C., U. S. Government Printing Office, v. 66, p. 661-670.

———, 1981b, Method of analysis for Tertiary basins using the computer: Lucknow, Proceedings of the 4th International Palynology Conference, p. 349-362.

Germeraad, J. H., C. A. Hopping, and J. Muller, (1968), Palynology of Tertiary sediments from tropical areas: Review of Palaeobotany and Palynology, v. 6, n. 3/4, p. 189-348.

Groot, J. J. and C. R. Groot, 1966, Pollen spectra from deep-sea sediments as indicators of climatic changes in southern South America: Marine Geology, v. 4, n. 6, p. 525-537.

Muller, J., 1959, Palynology of recent Orinoco delta and shelf sediments — reports of the Orinoco shelf expedition: Micropalentology, v. 5, p. 1-32.

Traverse, A., and R. N. Ginsburg, 1966, Palynology of the surface sediments of great Bahama Bank as related to water movement and sedimentation: Marine Geology, v. 4, n. 6, p. 417-459.

Sears, P. B., and K. H. Clisby, 1955, Palynology of southern North America — part 4: Pleistocene climate in Mexico: Geological Society of America Bulletin, n. 66, p. 521-530.

Stradner, H., and F. Allram, 1981, Nannofossil assemblages, Deep sea drilling project, Leg 66, Middle America Trench *in* J. S. Watkins and J. C. Moore, eds., Initial reports of the Deep sea drilling project: Washington, D.C., U.S. Government Printing Office, v. 66, p. 589-639.

Characteristics of Cretaceous Organic Matter in the Atlantic

Barry J. Katz
Raymond N. Pheifer
Texaco U.S.A.
Bellaire Research Laboratories
Bellaire, Texas

Drilling in the Atlantic Ocean basin has revealed large quantities of organic matter throughout Cretaceous strata. The organic-rich deposits in the North Atlantic do not appear to be genetically related to those in the South Atlantic. There are clear differences in the character of the organic matter and its possible origins.

The contrast in type of organic matter preserved in Cretaceous strata from the North and South Atlantic can be explained by continental drainage patterns and evolutionary stages of ocean basin development. During the Early Cretaceous, a less evolved, much more restricted South Atlantic exhibited conditions favorable to production and preservation of marine organic matter. This is in contrast to the more evolved, more open North Atlantic throughout the Cretaceous, and the Late Cretaceous South Atlantic. This evolutionary development of Atlantic passive margins aids in explaining the observed distribution of oil and gas discoveries.

Drilling conducted as part of the Deep Sea Drilling Project (DSDP) and Continental Offshore Stratigraphic Test (COST) program revealed widespread organic-rich horizons in Cretaceous strata of the North and South Atlantic Oceans. The quantity of organic carbon deposited in these strata has been estimated to be more than that contained in the known coal and hydrocarbon reserves of the world (Ryan and Cita, 1977; and Brumsack, 1980). The high organic carbon content characteristic of these sediments makes understanding their organic geochemistry and depositional history important to evaluating hydrocarbon resources of the deep Atlantic basin and its continental margins.

An important aspect of this evaluation relates to the nature of the organic matter contained within the sediments. Hydrogen-rich organic matter, commonly of marine origin, is necessary to generate significant quantities of liquid hydrocarbons. Geochemical studies indicate that there are clear differences in the organic geochemical character of Cretaceous organic matter from different parts of the Atlantic (Tissot, Deroo, and Herbin, 1979; Tissot et al, 1980). This paper characterizes this organic matter, determines its origins, and relates this information to the petroleum potential of the Atlantic. This work is based on data generated specifically for this study, combined with published data on numerous DSDP cores and exploratory wells in the Atlantic (Figure 1).

ORGANIC CHARACTERIZATION

Insoluble sedimentary organic matter (kerogen) has been subdivided into four categories or types (Tissot et al, 1974; Tissot, Deroo, and Herbin, 1979): type 1 kerogens, derived from algae and bacteria, are typical of oil shales and boghead coals, and are commonly associated with lacustrine depositional environments; type 2 kerogens are composed of preserved marine planktonic remains deposited and preserved under reducing conditions; type 3 kerogens are derived from partially degraded higher plants of terrestrial origin; and type 4, residual organic matter, refers to material recycled from other sedimentary deposits or severely oxidized prior to its final deposition. In a marine environment, organic matter would be expected to be either type 2, 3, or 4, or some combination of the three. Each of these kerogen types may be identified by its physical and chemical characteristics.

In general, immature organic matter with a marine origin (type 2) is amorphous, with moderatley high atomic hydrogen to carbon (H/C) and low atomic oxygen to carbon (O/C) ratios. It also has high extractable

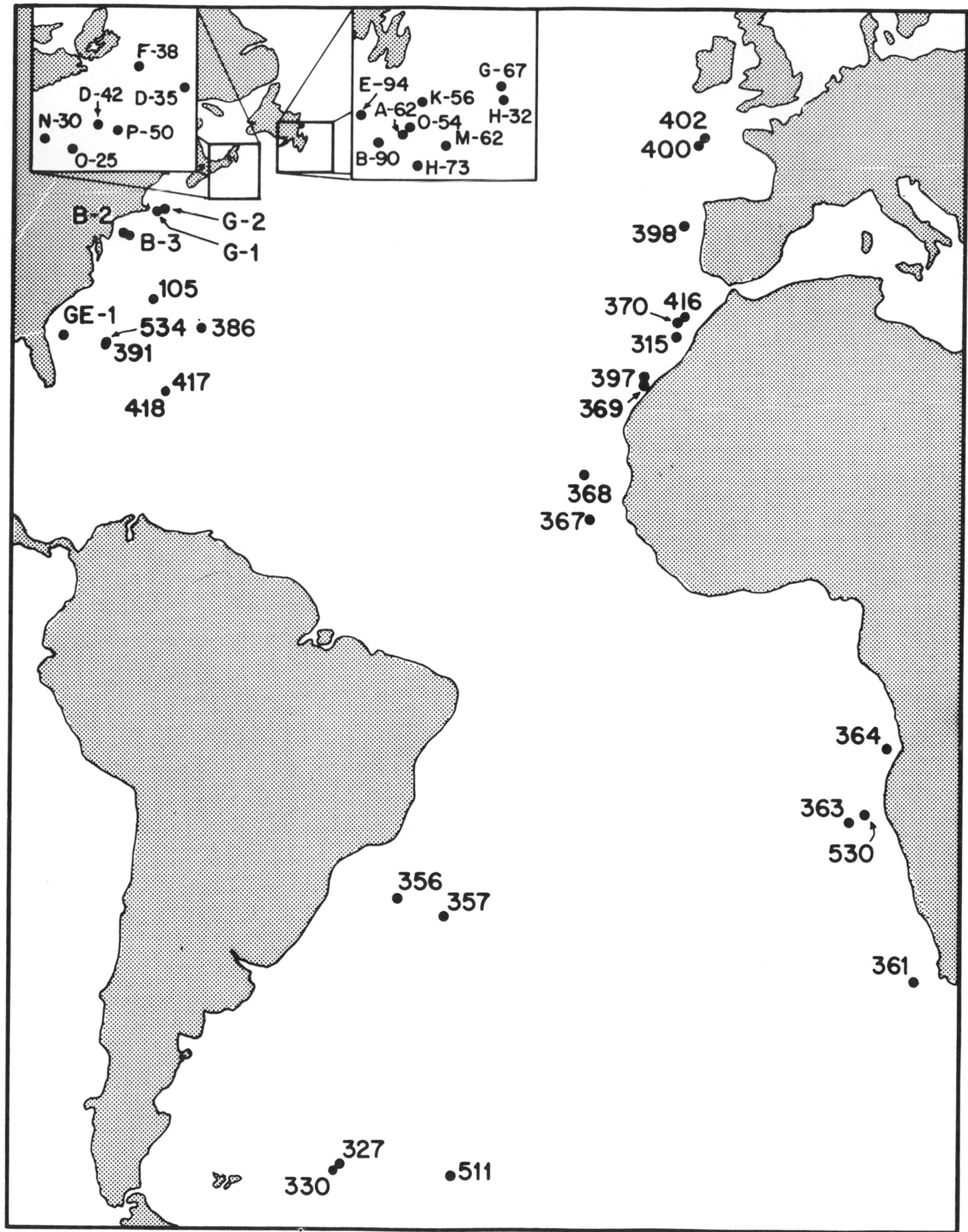

Figure 1 — Location map of DSDP, COST, and exploration wells used as part of this study. (N-30 = Naskapi N-30; 0-25 = Oneida 0-25; D-42 = Cohasset D-42; P-50 = Triumph P-50; F-38 = Argo F-38; D-35 = Dauntless D-35; E-94 = Hermine E-94; B-90 = Puffin B-90; A-62 = Petrel A-62; 0-54 = Gannet 0-54; K-56 = Coot K-56; H-73 = Heron H-73; M-62 = Bittern M-62; H-32 = Bonnition H-32; G-67 = Murre G-67)

yields relative to the organic carbon content, and kerogen δC^{13} values typically between −20 and − 24‰ relative to the PDB standard. On the other hand, organic matter with a terrestrial origin (type 3) is structured and often degraded. This results in a deficiency in hydrogen and an increased oxygen level, expressed as low H/C and moderate to high O/C values. Consequently, type 3 organic matter has lower extractable yields relative to its level of organic enrichment than does marine material. In addition, terrestrial organic matter is typically isotopically lighter than marine organic matter (i.e. its δC^{13} values are more negative). However, carbon isotope data are not completely unambiguous. Diagenesis and microbial processes may result in fractionation of carbon isotopes, resulting in an overlap of isotopic values for kerogens from different environmental origins (Waples, 1981). Residual organic matter (type 4) has even lower levels of hydrogen enrichment and higher levels of oxygen enrichment as a result of its previous thermal history and/or oxidation prior to final deposition. It commonly does not yield any extractable materials.

Four techniques characterize organic matter in this study: Rock-Eval pyrolysis, elemental analysis, visual assessment, and determination of extractable yield. Pyrolysis of crushed whole rock samples (technique described by Espitalié et al, 1977) and elemental analysis of isolated kerogen provide independent measures of the geochemical character of the kerogen. The use of pyrolysis follows that suggested by Espitalié et al (1977), in which the hydrogen index (mg hydrocarbons/gm organic carbon) and oxygen index (mg carbon dioxide/gm organic carbon) are comparable to the H/C and O/C ratios of elemental analysis. Visual assessment data obtained on isolated kerogens were used in conjunction with geochemical data to assess the origin of organic matter. Four terms are used to describe kerogen: amorphous, which includes algal material; herbaceous, which includes the softer parts of higher plants, including cuticles, spores, and pollen; woody, which represents the more lignitic structured material; and coaly, which is the inert, commonly reworked, structured material (GeoChem Laboratories, Inc., 1980). In addition to examination of kerogen, which usually accounts for 80% or more of the total organic matter of rocks, we examined the yield of soluble organic matter, or bitumen. The relative yield of organic matter is thought to be an indicator of organic matter origin, terrestrial material having lower yields than material of marine origin. Because these criteria are highly sensitive to contamination and to the thermal maturity of the sediment, caution was exercised in their use.

DATA BASE

Cuttings samples, composited over significant stratigraphic thicknesses, and core samples, which enabled analysis of discrete intervals, were used in this study. In general, data obtained on industry samples were generated from cuttings samples, whereas core samples were obtained from the DSDP.

North Atlantic

Geochemical data from the western margin of the North Atlantic appears essentially uniform, regardless of stratigraphic position within the Cretaceous. Some of the data from Sites 105, 391C, and 534A are summarized in Appendix 1. The organic matter, as observed in both Rock-Eval (Figure 2) and elemental data (Appendix 1), is characterized by low levels of hydrogen enrichment. The oxygen content is moderate to high with oxygen indices ranging upward to greater than 400 mg CO_2/gm organic carbon. Such high oxygen index values may indicate a strong "inorganic" CO_2 signal, resulting from a carbonate rock matrix (Katz, 1981). Atomic O/C ratios, which are a more reliable indicator of "organic" oxygen, are as high as 0.40 (K. K. Bissada, personal communication) but typically fall between 0.1 and 0.25. The total extractable yields are generally low, ranging from what may be considered laboratory background (200 ppm) to approximately 6800 ppm.

Visual examination shows that the organic matter is composed of predominantly herbaceous, coaly, and woody material (Figures 4 and 5). Material with an amorphous character appears to be a significant contributor to the total kerogen in the COST GE-1 well (Georgia Embayment) and the exploratory wells from the Scotian shelf and Grand Banks of Canada. Chemically, however, these kerogens appear depleted in hydrogen and typically have low extractable yields (Miller et al, 1980; and Purcell, Rashid, and Hardy, 1979) which suggests that the amorphous character is a result of degradation of previously structured material (Rashid, Purcell, and Hardy, 1980).

On the eastern North Atlantic continental margins, geochemical data from DSDP Sites 370, 415, and 416 (Moroccan Basin; Boutefeu, 1980), Sites 400 and 402 (Bay of Biscay; Deroo et al, 1979b) and Sites 397 and 398 (off northwestern Africa and Portugal; Deroo et al, 1979a and 1979c), although stratigraphically incomplete, reveal that the kerogen is hydrogen depleted similar to that observed from the North American margin (Figure 3). Recycled or residual organic matter appears to be dominant within several of the examined sections, such as DSDP Sites 400 and 402 from the Bay of Biscay (Deroo et al, 1979b). At other locations, such as DSDP Sites 397 and 416, the sedimentary organic matter is largely higher plant remains (Cornford, 1979; and Deroo et al, 1979c). Thus, the organic matter in the North Atlantic appears to be principally of a terrestrial origin.

Exceptions to the predominantly terrestrial character of North Atlantic organic matter do occur where a significant contribution of planktonic marine material is observed. DSDP Site 367 off western Africa is the most stratigraphically significant example with dominantly marine organic facies ranging from Neocomian through Coniacian age (Deroo et al, 1979; Tissot et al, 1980). Although stratigraphic sampling has been limited, there are indications that significant marine organic facies may also occur at Sites 368 and 369 (Deroo et al, 1978; and Kendrick, Hood, and Castaño, 1978a)

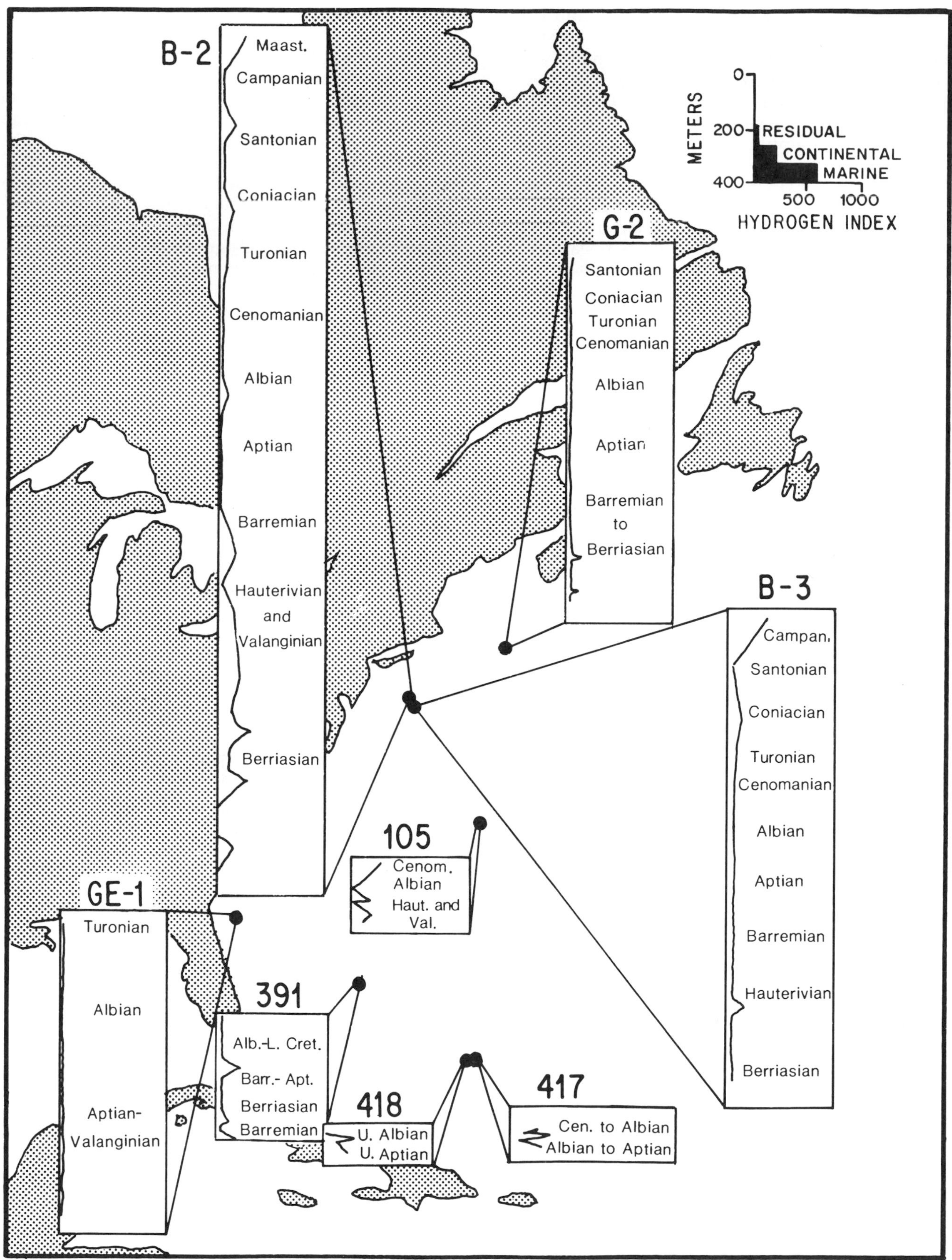

Figure 2 — Distribution of hydrogen indices as a function of stratigraphic position for the western North Atlantic. (Data sources include Deroo et al, 1980; Miller et al, 1980.)

also off northwestern Africa.

Elsewhere in the North Atlantic, stratigraphically less significant pulses of marine organic matter are present. These units appear in Cenomanian, Turonian/Coniacian, and Albian/Aptian strata of several DSDP sites, including, but not limited to, Sites 105 (Appendix 1), 370 (Boutefeu, 1980) and 386 (Tissot et al, 1980). Sampling of material from DSDP Leg 76 has also suggested that these discrete marine horizons may be present in Barremian/Hauterivian and Berriasian strata (Appendix 1). Additional sampling and analysis may reveal their presence throughout the Cretaceous. The discrete nature of these pulses, as revealed by present sampling, suggests that the duration of the depositional episodes at several sites is limited and is commonly represented by only a few centimeters of sediment.

There also have been some anomalously high H/C ratios observed in some Cretaceous coal horizons of the Baltimore Canyon Trough. These values, however, appear to have been obtained from coal seams rich in fossil resins similar to those reported in the MacKenzie delta region of Canada (Snowdon, 1980) and the Kenai Peninsula (Kelly, 1968). The coals clearly represent a terrestrial source.

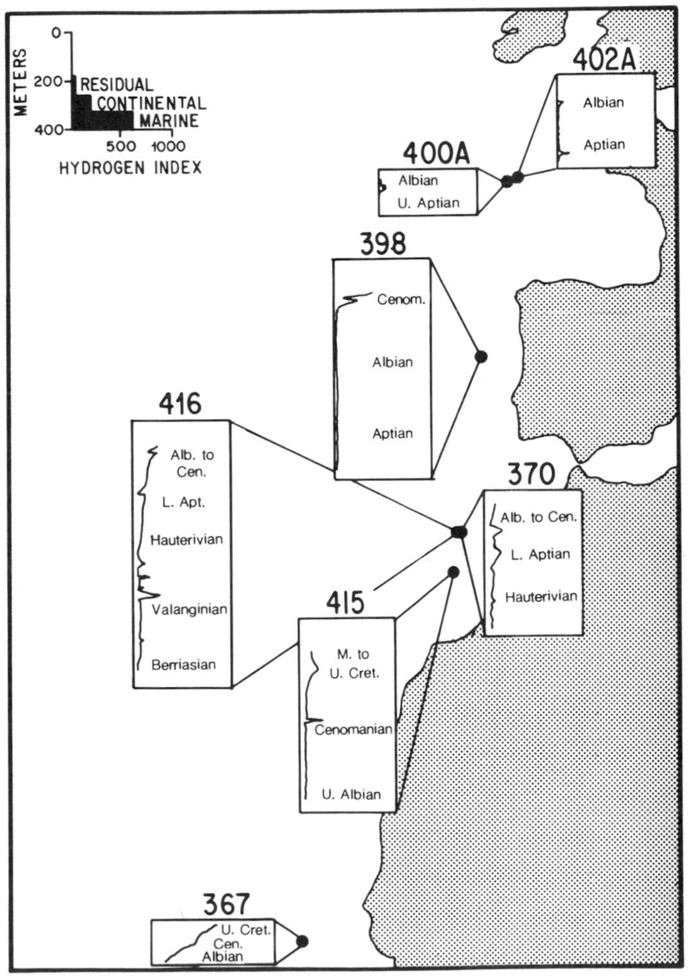

Figure 3 — Distribution of hydrogen indices as a function of stratigraphic position for the eastern North Atlantic. (Data sources include Boutefeu, 1980, Deroo et al, 1978a; Deroo et al, 1979a, 1979b,)

South Atlantic

The sedimentary environment of the South Atlantic does not appear to be as uniform as that of the North Atlantic. There appear to have been two principal stages of organic deposition and preservation (excluding the Rio Grande Rise and the Walvis Ridge). This is clearly displayed by the organic carbon content of the sediment. Strata deposited prior to Middle Albian commonly contain more than 1% organic carbon, whereas the post-Middle Albian section is characterized by less than 1% organic carbon (Herbin and Deroo, 1979). The difference in carbon content is paralleled by a marked difference in type of preserved organic matter. The pre-Middle Albian section contains primarily well-preserved marine organic matter, as indicated by elevated hydrogen indices (Figure 6), whereas the post-Middle Albian is dominated by either degraded terrestrial material or marine material that has been severely oxidized. The pre-Middle Albian section yielded hydrogen indices as high as 800 mg hydrocarbons/gm organic carbon and oxygen indices of less than 40 mg CO_2/gm organic carbon. This is in contrast to the typical sample of post-Middle Albian strata, which yielded hydrogen indices ranging from approximately 70 to 270 mg hydrocarbons/gm organic carbon and oxygen indices ranging upward to 400 mg CO_2/gm organic carbon (Robert et al, 1979).

Though visual data are stratigraphically and areally limited they also suggest a change in deposition and/or preservation of organic matter during the Cretaceous in the South Atlantic. Visual data from DSDP Sites 361 (Cape Basin) and 364 (Angola Basin), in Figure 7, show that the lower stratigraphic sections (Alb-

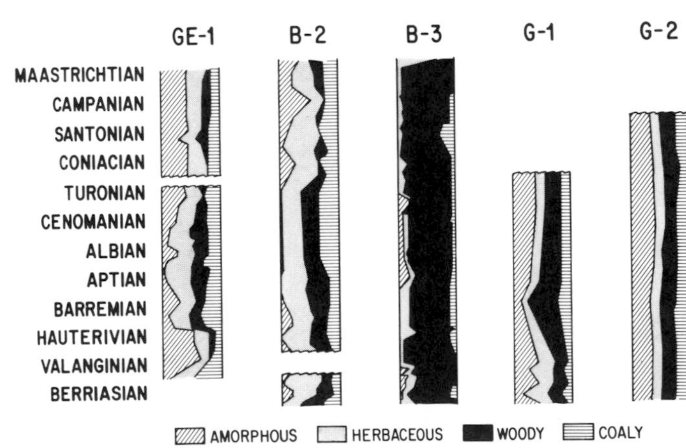

Figure 4 — Distribution of kerogen types from the eastern U.S. margin COST wells. (Data sources include Scholle, Krivoy, and Hennessy, 1978; Scholle, Krivoy, and Hennessy, 1979; Scholle, Krivoy, and Hennessy, 1980; Scholle, Schwab, and Krivoy, 1980. Miller et al, 1980.)

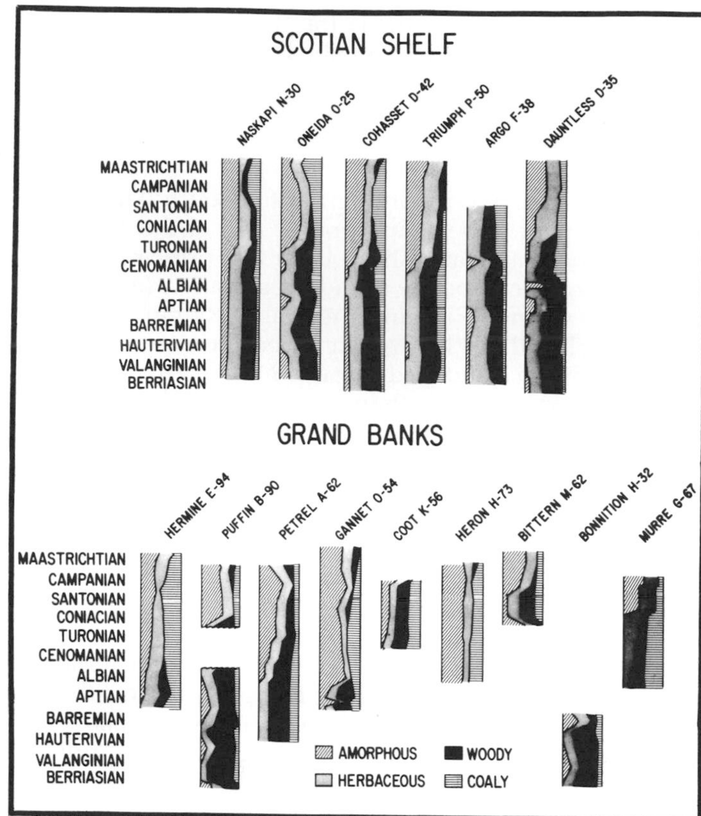

Figure 5 — Distribution of kerogen types from the Scotian shelf and Grand Banks of Canada (after Barss et al, 1980).

ian and Aptian) contain principally amorphous organic matter with only minor amounts of coaly, woody, and herbaceous material. This is in contrast to the Santonian-Coniacian samples (analyzed only at DSDP Site 364) which contain coaly material with lesser amounts of woody and herbaceous components (Kendrick, Hood, and Castaño, 1978b).

Stable carbon isotope data (-23 to $-30‰$) obtained on the total extract was not definitive but did suggest that the organic matter deposited during the Lower Cretaceous at both Sites 361 and 364 was primarily marine with pulses of terrestrial material at Site 364 (Foresman, 1978; and Erdman and Schorno, 1978). Extracted bitumens are isotopically lighter than kerogens from which they are formed. This difference increases with increasing maturity. No statisitically significant difference between carbon isotope compositions was observed between the Upper and Lower Cretaceous strata. This may imply that the low hydrocarbon yields are a result of poor preservation (oxidation) of marine organic matter rather than solely a terrestrial source.

The marine origin of the organic matter of Aptian and older strata is also supported by elemental analysis data obtained on samples from DSDP Site 327 (Comer and Littlejohn, 1977) and by pyrolysis-fluorescence results obtained on samples from Neocomian to Aptian age from Leg 71 (Ludwig et al, 1980). These results suggest petroleum source bed characteristics which usually indicate a marine origin for the organic matter.

As in the North Atlantic, there are also exceptions to general trends in the South Atlantic. The limited Lower Cretaceous strata examined on the Walvis Ridge at Site 363 appear to be organically lean (Herbin and Deroo, 1979), indicating low levels of preservation. This can be seen in the uniformly low hydrogen indices presented in Figure 6. Also within the pre-Albian section of the Cape Basin (Site 361) there are stratigraphically important intervals where terrestrial organic matter dominates. These horizons have been well-defined by Herbin and Deroo (1979) and Erdman and Schorno (1978). They apparently reflect remobilization of terrestrial organic matter associated with a deep sea fan (Dingle, 1980).

In addition, there appear to be pulses of marine planktonic material in Cenomanian to Coniacian strata of Sites 364 and 530 in the Angola Basin (Herbin and Deroo, 1979; and Dean et al, 1981). However, unlike the planktonic accumulations of the North Atlantic, these accumulations are of greater stratigraphic importance, with an overall thickness of several meters.

DISCUSSION

The distinct differences in organic matter deposited in the North and South Atlantic basins suggest that a single depositional model would be inadequate to explain the variations. Any credible model must adequately explain these differences in organic character. Several factors need to be addressed in such a model: terrestrial input, organic productivity, and organic preservation.

The amount of terrestrial input and its distribution are controlled by climate and sea level, as well as basin configuration and geometry. Warm, equable, and humid climatic conditions, capable of contributing to growth of significant quantities of terrestrial material, are thought to have existed worldwide during the Cretaceous (Barron, 1980) and were controlled principally by interaction of ocean-air systems. A relatively high sea level, resulting in reduced continentality, and a lowering of major orogenic features resulted in smaller geographic areas of arid conditions (Barron, 1980). Aridity was probably limited to areas still dominated by continental conditions. The Central Atlantic during the Early Cretaceous prior to spreading was one continental area exhibiting these arid conditions. As the South Atlantic opened, more humid climatic conditions extended into the arid region resulting in a more uniform distribution of luxuriant land floras. This resulted in a paralleling increase in the concentration of terrestrial organic matter in the Central Atlantic region as the ocean basin developed.

The distribution of terrestrial material is also controlled by sea level and basin physiography. During low sea level stands, terrestrial material may be transported farther into the basin, whereas during high sea level stands the organic matter could be effectively trapped on the basin's periphery (i.e. continental shelves) or in estuaries. This is assuming that prograding deltas, deep-sea fans, and turbidity currents

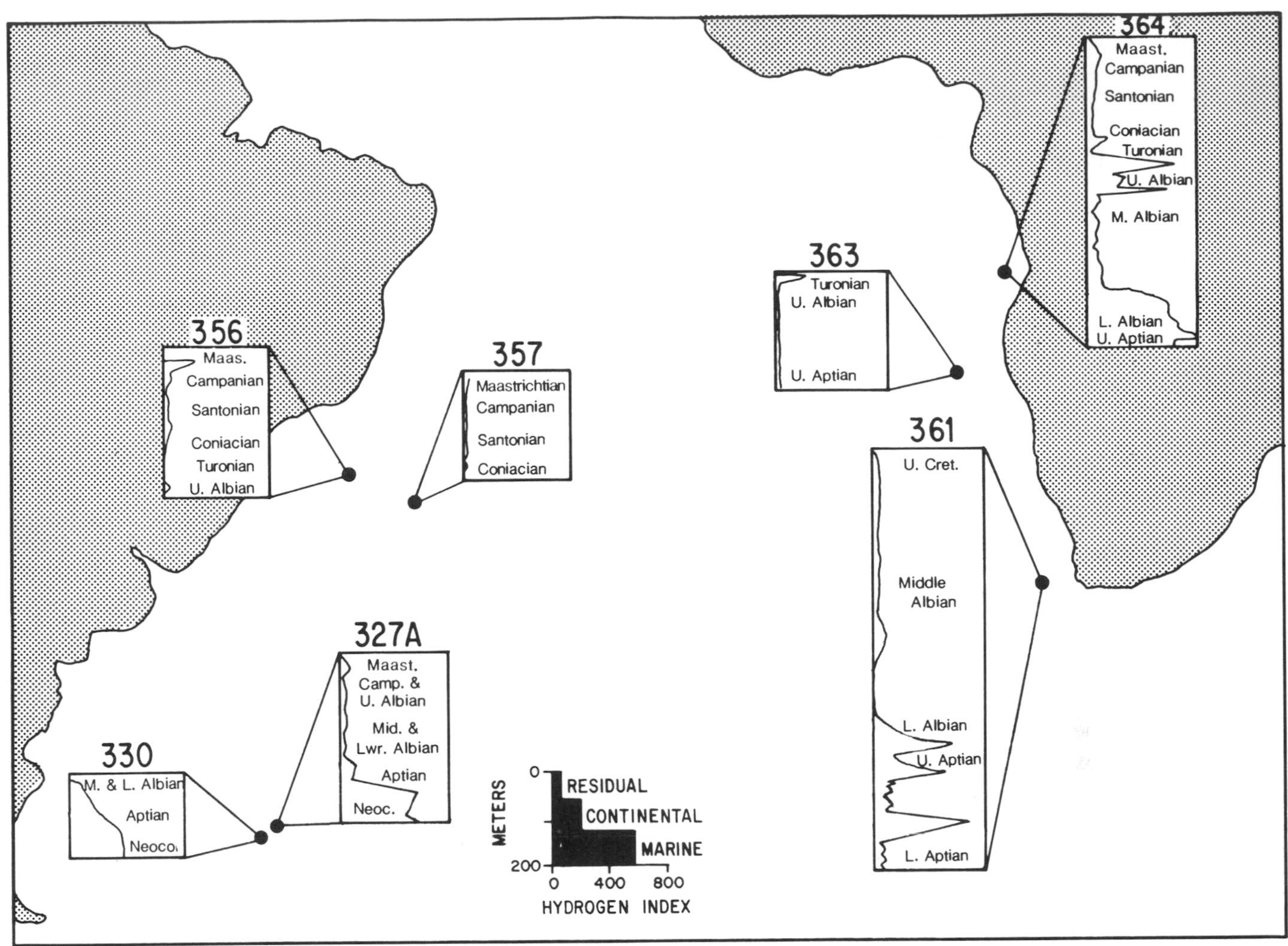

Figure 6 — Distribution of hydrogen indices as a function of stratigraphic position for the South Atlantic (after Herbin and Deroo, 1979).

do not result in large scale transport of terrestrial material to the deep ocean basin.

Similarly, the basin's configuration and geometry play a role in distribution of terrestrial material. The smaller the basin, the more universal the distribution. As the basin becomes extended, terrestrial material becomes more restricted to the continental margins.

Thus, in the North Atlantic throughout the Cretaceous, there were significant amounts of terrestrial input. Because of the generally high sea level (Vail, Mitchum and Thompson, 1978), most of this input was limited to the basin margins. However, significant amounts of terrestrial material were periodically transported by turbidity currents to the deep basin. Such deposits have been sampled at DSDP Sites 386 and 387 (Simoneit, 1979) and Site 417 (Deroo et al, 1980). Variations in relative quantities of this material are related to sampling position within the basin and the periodicity of mass transport.

In the South Atlantic, there was a background supply of terrestrial material in the Early Cretaceous which was distributed throughout the Argentine and Cape basins. This is exhibited not only in the organic matter, but also in the inorganic constituents (McCoy and Zimmerman, 1977). As in the North Atlantic, there were periods when significant quantities of terrestrial material were transported by turbidity currents deep into the marine basin. This is particularly well-defined in the Cape Basin (Dingle, 1980). As the South Atlantic evolved, two major changes occurred: (1) the supply of terrestrial matter increased toward the north, and (2) the distribution of terrestrial material in the south became more restricted to the margin's periphery as the source became further removed from the central portions of the basin.

The second source of organic matter to a basin is primary productivity within that basin. Organic productivity is closely related to nutrient supply. High productivity is commonly associated with regions of upwelling on the western margins of continental land masses. However, high organic productivity is also associated with any continental land mass where runoff and the thermally-induced overturn of waters over continental shelves supply the necessary nutrient renewal. Under these conditions, the level of productivity may be as great as that of upwelling regions (Koblenz-Mishke, Volkonsky, and Kabanova, 1970). Lower levels of primary productivity are associated

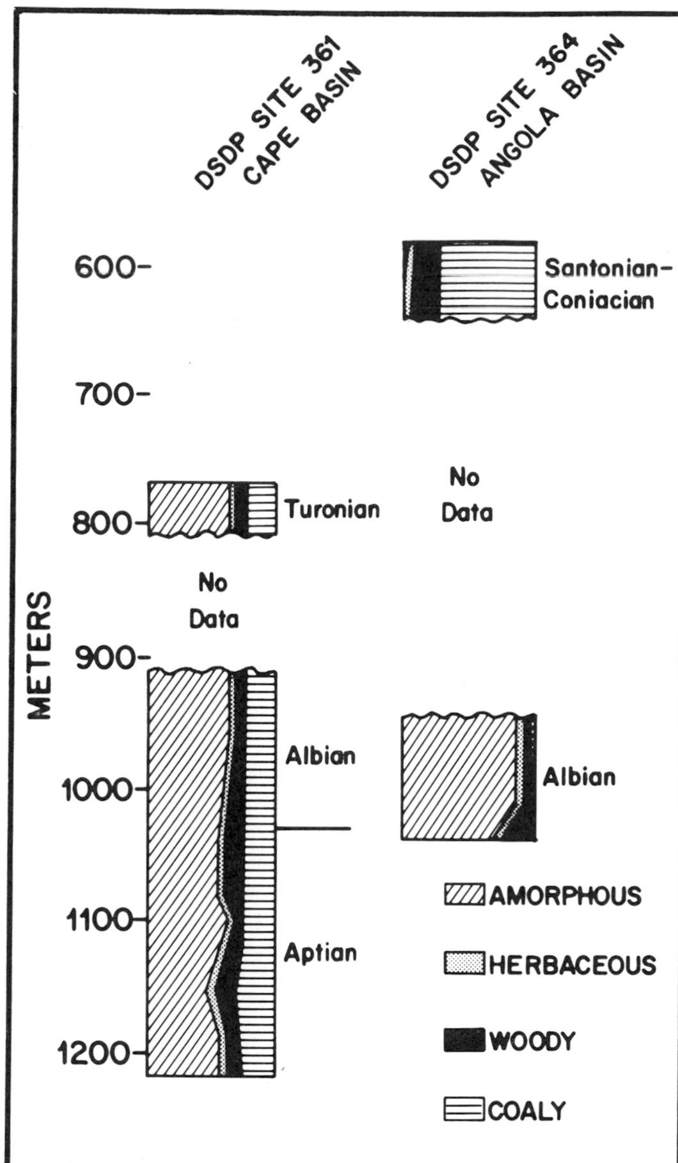

Figure 7 — Distribution of kerogen types from the southwestern African margin (after Kendrick, Hood, and Castaño, 1978b).

with central ocean gyres that are nutrient depleted.

Although estimates of paleoproductivity are controversial and difficult to make because of the criteria used, there is sufficient justification to assume that primary productivity levels during the Cretaceous were at least as great as today's, if not greater (Tappan and Loeblich, 1970, 1973). This is partially a result of the expansion of shallow seas because of higher sea level, and the expansion of the growing period because of warm, equable climates. In the North Atlantic, elevated productivities existed on the ocean margins, much as they do today, as a result of large volumes of continental runoff resupplying nutrients.

The South Atlantic during the Lower Cretaceous probably closely resembled the present day Black Sea in that it was a large, landlocked body of water. Surface productivity levels in the Black Sea basin today are comparable to elevated productivity levels associated with both upwelling and surface runoff in the major oceans (Shimkus and Trimonis, 1974). Thus, the entire South Atlantic, as a result of its relatively small size and the fact that continental runoff could supply necessary nutrients, experienced elevated productivity levels during the Early Cretaceous. It is probable that upwelling did not play a major role in nutrient renewal because of a lack of deep water circulation required to generate overturn. As the South Atlantic evolved, it began to take on characteristics which more closely resembled that of the North Atlantic. Productivity became more restricted to the periphery with the highest levels of productivity being associated with coastal upwelling and runoff. A central gyre with low levels of productivity slowly developed as the basin widened.

An understanding of input, however, is insufficient to explain the differences observed in the stratigraphic column. Critical to an understanding of the distribution pattern is an understanding of organic preservation. Organic matter is best preserved under anoxic conditions. Within the marine environment, such conditions may be associated with an open-ocean oxygen minimum layer or a silled basin (Thiede and van Andel, 1977). Alternatively, organic matter may be preserved in an oxygenated environment — though not as efficiently — by rapid burial, which removes it from this environment. Not only does rapid burial preserve available organic matter, but commonly it provides large quantities of terrestrial organic and inorganic material. Work by Heath, Moore, and Dauphin (1977) and Muller and Suess (1979) demonstrates that the preservation process may occur without terrigenous dilution, simply as a result of high organic productivity. These deposits, whether reflecting terrigenous input or marine productivity, are organically leaner than those deposited under anoxic conditions (Demaison and Moore, 1980). Organic matter deposited under oxic conditions will also typically be hydrogen depleted (gas-prone), either as a result of terrestrial input or oxidation of marine organic matter.

Deposition of organic matter throughout the Cretaceous in the North Atlantic and during the Late Cretaceous in the South Atlantic appears to have occurred under generally oxic conditions. Circulation in the North Atlantic evidently was not conducive to development of widespread long-term anoxic conditions. The North Atlantic appears to have been generally open and capable of at least intermittent oxygen renewal throughout the Cretaceous, thereby preventing establishment of a large anoxic basin. Deep-water renewal, and therefore oxygen renewal, may have occurred principally from the Pacific through a gap between the Bahamas Platform and the Guinea nose (Sclater, Hellinger, and Tapscott, 1977). Deep water circulation ceased intermittently as a result of vertical tectonic movements of a "Panamanian barrier" (Saunders et al, 1973). Flow to and from the Tethys appears to be limited to surface currents (Bergren and Hollister, 1974). Periods of reduced, deep-water flow would be more conducive to development of expanded oxygen minimum layers as suggested by Schlanger and Jenkyns (1976), Fischer and Arthur (1977), and Thiede and van Andel (1977). Oxygen minimum layers are typically associated with circulation patterns rather than high surface productivities (Demaison and Moore, 1980). Therefore, conditions were not generally favorable for extensive preservation of marine organic matter in the North Atlantic. The limited horizons in the northwestern Atlantic containing significant contributions of amorphous marine organic matter probably represent these intermittent periods when flow was reduced and the oxygen minimum layer could expand.

The Mid-Atlantic Ridge provided an effective barrier for bottom water circulating into the southeastern portion of the North Atlantic (Tissot et al, 1979). Thus, a quasi-stable anoxic environment was established and resulted in preservation of marine organic matter at Sites 367, 368, and 369. Fracture zones provided the only limited communication of bottom waters between the eastern and western basins. By Middle to Late Cretaceous time a deep water connection between the Atlantic and Tethys existed (Sclater, Hellinger, and Tapscott, 1977). This provided for water exchange and permitted oxygen renewal on the previously isolated southeastern sector of the North Atlantic, and terminated the preservation of large quantities of marine organic matter. The actual timing of this event is poorly-defined because of gaps in the rock record at these three sites.

During the Early Cretaceous in the South Atlantic, continental configuration, the presence of aseismic ridges, and extension of the Falkland Plateau beyond the coast of southern Africa provided for development of a series of silled basins (Sclater, Hellinger, and Tapscott, 1977). Circulatory exchange was largely limited to surface waters (McCoy and Zimmerman, 1977). With this limited exchange, and climatic conditions, the deep basins' waters became stratified by marked salinity variations (Arthur and Natland, 1979). This stratification and elevated level of productivity led to an anoxic environment favorable to the preservation of marine organic matter.

The vertical extent of this anoxic water mass is difficult to assess because of limited areal and stratigraphic sampling. Based on a subsidence model from Site 363 from the Walvis Ridge, it appears that oxic conditions extended to 250 m deep (Robert et al, 1979), with anoxic conditions extending possibly to the sea floor (Thiede and van Andel, 1977). About 100 m.y. ago the eastern end of the Falkland Plateau cleared the eastern margin of southern Africa (Barker et al, 1977). This permitted the influx of more oxygenated deep waters which resulted in the effective termination of high levels of organic preservation in the Cape and Argentine basins. At least intermittently stagnant conditions continued in the Angola and Brazil basins through the Coniacian (Arthur and Natland, 1979), principally as a consequence of the Walvis Ridge and the Rio Grande Rise. This explains the extension of marine horizons into younger stratigraphic units of the Angola Basin (Figure 6).

The different types of organic matter in the North and South Atlantic suggest different source rock properties. Material with a terrestrial origin tends to generate gas, whereas material with a marine origin is capable of generating oil and gas, depending on the extent of thermal evolution. If one invokes a Cretaceous source for hydrocarbons in the Atlantic, the North Atlantic and Late Cretaceous of the South Atlantic must be considered gas prone. This is supported by occurrence of gas in the Baltimore Canyon and Scotian shelf. The Lower Cretaceous strata in the South Atlantic tend to be oil prone. This contention is supported by a recent exploratory well, Ciclon X-1, in the Malvinas basin. It provided a tentative correlation of the available oil to the Lower Cretaceous section (Turic et al, 1980).

CONCLUSIONS

During the Cretaceous, there was significant terrestrial input to the North and South Atlantic. In the North Atlantic throughout the Cretaceous, and in the South Atlantic during the Late Cretaceous, marine organic matter was typically poorly-preserved because of the generally oxic environments of deposition. The organic matter was also effectively diluted by terrestrial input. Periods of intermittent anoxia during this time provided limited deposition of strata rich in marine organic matter. During the Early Cretaceous, in the South Atlantic, a more stable anoxic environment existed, permitting effective preservation of marine organic matter. This suggests that there are different hydrocarbon tendencies in the Cretaceous strata of the North and South Atlantic Oceans. The North Atlantic, throughout the Cretaceous, appears to be a dominantly gas-prone province. The available marine units appear to be so stratigraphically limited that they cannot be considered potential oil source facies. The South Atlantic includes strata that are oil-prone during the Early Cretaceous and principally gas-prone during the Late Cretaceous.

ACKNOWLEDGMENTS

The authors thank Dr. K. K. Bissada and Dr. J. E. Lacey for their critical comments, and Louise Jackson, Judith Croom, and George Mayfield for assisting in text preparation. Some of the samples used in this study were supplied by the National Science Foundation through the Deep Sea Drilling Project.

This paper, Texaco Contribution Number 2427, is published with permission of Texaco Inc.

REFERENCES CITED

Arthur, M. A., and J. H. Natland, 1979, Carbonaceous sediments in the North and South Atlantic: the role of salinity in stable stratification of Early Cretaceous basins, *in* Talwani, Hay, and Ryan, eds., Deep drilling results in the Atlantic Ocean: Continental Margins and Paleoenvironments: American Geophysical Union, Maurice

Ewing Series, n. 3, p. 375-401.
Barker, P. I., et al, 1977, The evolution of the southwestern Atlantic Ocean basin; results of leg 36, *in* Initial report of the deep sea drilling project: Washington, D.C., U.S. Government Printing Office, v. 36, p. 993-1014.
Barron, E. J., 1980, Paleogeography and climate, 180 million years to the present. Coral Gables, University of Miami, Ph.D., dissertation, 280 p.
Barss, M. S., et al, 1980, Age, stratigraphy, organic matter type and colour, and hydrocarbon occurrences in 47 wells, offshore eastern Canada: Geological Survey of Canada, Open-file Report 714.
Berggren, W. A., and C. D. Hollister, 1974, Paleogeography, paleobiogeography and the history of circulation in the Atlantic Ocean: Society of Economic Paleontologists and Mineralogists, Special Publication, 20, p. 126-186.
Boutefeu, A., 1980, Pyrolysis study of organic matter from deep sea drilling project sites 370 (leg 41), 415, and 416 (leg 50), *in* Initial reports of the deep sea drilling project: Washington, D.C., U.S. Government Printing Office, v. 50, p. 555-566.
Brumsack, H. J., 1980, Geochemistry of Cretaceous black shales from the Atlantic Ocean (DSDP legs 11, 14, 36, and 41): Chemical Geology, v. 31, p. 1-25.
Comer, J. B., and R. Littlejohn, 1977, Content, composition, and thermal history of organic matter in Mesozoic sediments, Falkland Plateau, *in* Initial reports of the deep sea drilling project: Washington, D.C., U.S. Government Printing Office, v. 36, p. 941-944.
Cornford, C., 1979, Organic deposition at a continental rise: organic geochemical interpretation and synthesis at DSDP site 397, eastern North Atlantic, *in* Initial reports of the deep sea drilling project: Washington, D.C., U.S. Government Printing Office, v. 47, part 1, p. 503-510.
Dean, W. E., et al, 1981, Cretaceous black-shale deposition within an oxidized red clay turbidite environment, southern Angola Basin, South Atlantic Ocean (Abs): AAPG Bulletin, v. 65, p. 917.
Demaison, G. J., and G. T. Moore, 1980, Anoxic environments and oil source bed genesis: AAPG Bulletin, v. 64, p. 1179-1209.
Deroo, G., et al, 1978, Organic geochemistry of some Cretaceous black shales from sites 367 and 368, leg 41, eastern North Atlantic, *in* Initial reports of the deep sea drilling project: Washington, D.C., U.S. Government Printing Office, v. 41, p. 865-873.
———, ———, 1979a, Organic geochemistry of Cretaceous shales from DSDP site 398, leg 47B, eastern North Atlantic, *in* Initial reports of the deep sea drilling project: Washington, D.C., U.S. Government Printing Office, v. 47, part 2, p. 415-422.
———, ———, 1979b, Organic geochemistry of Cretaceous mudstones and marly limestones from DSDP sites 400 and 402, leg 48, eastern North Atlantic, *in* Initial reports of the deep sea drilling project: Washington, D.C., U.S. Government Printing Office, v. 48, p. 921-930.
———, ———, 1979c, Organic geochemistry of some organic-rich shales from DSDP site 397, leg 47A, eastern North Atlantic, *in* Initial reports of the deep sea drilling project: Washington, D.C., U.S. Government Printing Office, v. 47, part 1, p. 523-529.
———, ———, 1980, Organic geochemistry of Cretaceous sediments at DSDP holes 417D (leg 51), 418A (leg 52), and 418B (leg 53) in the western North Atlantic, *in* Initial reports of the deep sea drilling project: Washington, D.C., U.S. Government Printing Office, v. 51, 52, 53, part 2, p. 737-745.
Dingle, R. V., 1980, Sedimentary basins on the continental margins of southern Africa: Originalmitteilungen, v. 33, p. 457-463.
Erdman, J. G., and K. S. Schorno, 1978, Geochemistry of carbon: Deep Sea Drilling Project Leg 40, *in* Initial reports of the deep sea drilling project: Washington, D.C., U.S. Government Printing Office, Supplement v. 38-41, p. 651-658.
Espitalié, J., et al, 1977, Source rock characterization method for petroleum exploration: Proceedings, Offshore Technology Conference, v. 3, p. 439-443.
Fischer, A. G., and M. A. Arthur, 1977, Secular variations in the pelagic realm: Society of Economic Paleontologists and Mineralogists, Special Publication 25, p. 19-50.
Foresman, J. B., 1978, Organic geochemistry DSDP leg 40, continental rise of southwest Africa, *in* Initial reports of the deep sea drilling project: Washington, D.C., U.S. Government Printing Office, v. 40, p. 557-567.
GeoChem Laboratories, Inc., 1980, Source rock evaluation reference manual: Houston, Texas, GeoChem Laboratories Inc.
Heath, G. R., T. C. Moore, Jr., and J. P. Dauphin, 1977, Organic carbon in deep-sea sediments, *in* Anderson, ed., The Fate of Fossil Fuel CO_2 in the Oceans: New York, Plenum Press, p. 627-639.
Herbin, J. P., and G. Deroo, 1979, Etude sedimentologique de la matiere organique dans les argiles noires Cretacees de l'Atlantique Sud: Lyon, Faculte des Sciences, Laboratories de Geologie, Documents, n. 7, p. 71-87.
Katz, B. J., 1981, Limitation of Rock-Eval pyrolysis for typing organic matter (Abs.): AAPG Bulletin, v. 65, p. 944.
Kelly, T. E., 1968, Gas accumulations in nonmarine strata, Cook Inlet Basin, Alaska: AAPG Memoir 9, v. 1, p. 49-64.
Kendrick, J. W., A. Hood, and J. R. Castaño, 1978a, Petroleum-potential of sediments from leg 41, Deep sea drilling project, *in* reports of the deep sea drilling project: Washington, D.C., U.S. Government Printing Office, v. 41, p. 817-819.
———, ———, ———, 1978b, Petroleum-generating potential of sediments from leg 40, Deep sea drilling project, *in* Initial reports of the deep sea drilling project: Washington, D.C., U.S. Government Printing Office, Supplement v. 38-41, p. 671-676.
Koblenz-Mishke, O. I., V. V. Volkonsky, and J. G. Kabanova, 1970, Plankton primary production of the world ocean, *in* Wooster ed., Symposium on scientific exploration of the southern Pacific: Washington, D.C., National Academy of Science, p. 183-193.
Ludwig, W. L., et al, 1980, Tertiary and Cretaceous paleoenvironments in the southwest Atlantic Ocean, *in* Preliminary results of the deep sea drilling project, leg 71: Geological Society of America Bulletin, v. 91, part I, p. 655-664.
McCoy, F. W. , and H. B. Zimmerman, 1977, A history of sediment lithofacies in the South Atlantic Ocean, *in* Initial reports of the deep sea drilling project, v. 39, p. 1047-1079.
Miller, R. E., et al, 1980, Organic geochemistry, *in* Scholle, ed., Geological studies of the COST No. B-3 Well, United States Mid-Atlantic Continental Slope Area: U.S. Geological Survey, Circular 833, p. 85-104.
Muller, P. J., and E. Suess, 1979, Productivity, sedimentation rate, and sedimentary organic matter in the oceans — I, Organic carbon preservation: Deep-Sea Research, v. 26A, p. 1347-1362.
Purcell, L. P., M. A. Rashid, and I. A. Hardy, 1979, Geochemical characteristics of sedimentary rocks in the Scotian Basin: AAPG Bulletin, v. 63, p. 87-105.
Rashid, M. A., L. P. Purcell, and I. A. Hardy, 1980, Source

rock potential for oil and gas of the east Newfoundland and Labrador Shelf areas: Canadian Society of Petroleum Geologists, Memoir 6, p. 589-608.

Robert, C., et al, 1979, L'Atlantique Sud au Crétacé d'apres l'étude des minéraux argileux et de la matiére organique (legs 39 et 40, DSDP): Oceanological Acta, v. 2, p. 209-218.

Ryan, W. B. F., and M. B. Cita, 1977, Ignorance concerning episodes of oceanwide stagnation: Marine Geology, v. 23, p. 197-215.

Saunders, J. B., et al, 1973, Cruise synthesis, *in* Initial records of the deep sea drilling project, v. 15, p. 1077-1111.

Schlanger, S. O., and H. C. Jenkyns, 1976, Cretaceous oceanic anoxic events: causes and consequences: Geologie en Mijnbouw, v. 55, p. 179-184.

Scholle, P. A., H. L. Krivoy, and J. L. Hennessy, 1978, Summary chart of geological data from the COST No. B-2 well, U.S. Mid-Atlantic outer continental shelf: U.S. Geological Survey, Oil and Gas Investigations Chart OC-79.

———, ———, ———, 1979, Summary chart of geological data from the COST No. GE-1 well, U.S. South Atlantic outer continental shelf: U.S. Geological Survey, Oil and Gas Investigations Chart OC-90.

———, ———, ———, 1980, Summary chart of geological data from the COST No. G-1 well, U.S. North Atlantic outer continental shelf: U.S. Geological Survey, Oil and Gas Investigations Chart OC-104.

———, K. A. Schwab, and H. L. Krivoy, 1980, Summary

Appendix I. Geochemical summary DSDP Sites 105, 391C, and 534A.

Site-Core Section	Age	Organic Carbon (%)	Hydrogen Index[1]	Oxygen Index[2]	H/C	O/C	Total Extract (ppm)
105-9-4	Cenomanian	7.99	256	25	1.22	.154	6808
105-11-3	Cen/Albian	2.90	187	50	1.03	.156	1199
105-15-1	Albian	1.26	38	102	.81	.168	1606
105-15-2	Albian	1.55	12	65	.70	.169	1683
105-15-3	Albian	1.56	22	76	.33	.173	1017
105-15-5	Albian	1.08	7	82	.69	.183	736
105-16-2	Aptian/Bar	.87	211	116	.84	.159	2646
105-18-1	Hauterivian	.68	208	296	1.04	.153	615
105-18-3	Hauterivian	3.38	21	59	1.05	.161	573
105-18-5	Hauterivian	.92	130	103	1.02	.161	460
105-18-6	Hauterivian	2.72	99	51	1.14	.153	597
105-19-2	Haut/Val	.60	88	195	.86	.175	624
105-19-3	Haut/Val	1.72	41	102	.83	.177	1518
105-19-4	Haut/Val	1.64	39	105	.83	.157	432
391C-8-1	L. Cret/Alb	2.15	36	63	nd	nd	1371
391C-9-1	L. Cret/Alb	1.41	44	76	.78	.159	522
391C-10CC	L. Cret/Alb	1.93	52	60	.87	.150	549
391C-11-1	L. Cret/Alb	2.12	80	130	.92	.154	658
391C-12-1	Aptian/Bar	1.78	169	114	1.17	.140	996
391C-12-3	Aptian/Bar	1.65	181	141	.96	.126	873
391C-15-2	Aptian/Bar	.51	40	369	nd	nd	232
391C-17-1	Aptian/Bar	.55	40	330	nd	nd	229
391C-24-6	Bar/Ber	.68	33	250	nd	nd	263
391C-28-2	Bar/Ber	.66	118	215	nd	nd	373
391C-31-4	Bar/Ber	.77	40	369	.88	.162	388
391C-33-2	Bar/Ber	.56	192	238	nd	nd	285
534A-28-2	Cenomanian	.84	5	98	nd	nd	312
534A-31-1	Albian	3.30	88	42	.99	.165	735
524A-33-1	Albian	.77	10	101	nd	nd	211
524A-35-2	Albian	1.63	43	47	.89	.149	281
534A-37-4	Albian	1.93	56	45	.93	.151	436
534A-39-5	Albian	3.79	125	28	.92	.153	685
534A-43-2	Aptian	3.88	133	38	.94	.151	811
534A-45-1	Barremian	1.88	71	110	.87	.152	1016
534A-48-5	Barremian	4.00	329	75	1.08	.123	2090
534A-50-1	Barremian	1.11	77	172	nd	nd	1143
534A-53-2	Hauterivian	4.62	392	49	1.25	.125	1971
534A-55-2	Hauterivian	.64	32	210	nd	nd	1985
534A-58-4	Hauterivian	.60	22	168	nd	nd	831
534A-60-4	Hauterivian	.83	23	248	nd	nd	2344
534A-64-2	Valanginian	1.10	29	87	.73	.170	1073
534A-66-2	Valanginian	.59	33	169	nd	nd	1652
534A-69-5	Valanginian	1.19	36	95	.73	.155	1795
534A-71-4	Valanginian	1.63	45	63	.87	.161	1150
534A-73-3	Valanginian	1.39	84	80	.95	.136	1224
534A-78-1	Valanginian	.65	52	120	nd	nd	2119
534A-80-2	Berriasian	1.00	192	97	1.17	.139	609
534A-83-2	Berriasian	1.20	296	97	1.22	.130	nd

[1]mg hydrocarbons /gm organic carbon
[2]mg CO_2 /gm organic carbon
nd - no data

chart of geological data from the COST No. G-2 well, U.S. North-Atlantic outer continental shelf: U.S. Geological Survey, Oil and Gas Investigations Chart OC-105.

Sclater, J. G., S. Hellinger, and C. Tapscott, 1977, The paleobathymetry of the Atlantic Ocean from the Jurassic to present: Journal of Geology, v. 85, p. 509-552.

Shimkus, K. M., and E. S. Trimonis, 1974, Modern sedimentation in the Black Sea: AAPG Memoir 20, p. 249-278.

Simoneit, B. R. T., 1979, Organic geochemistry of the shales from the northwestern proto-Atlantic, DSDP leg 43, *in* Initial reports of the deep sea drilling project, v. 43, p. 643-649.

Snowdon, L. R., 1980, Resinite — a potential petroleum source in the Upper Cretaceous/Tertiary of the Beaufort-MacKenzie Basin: Canadian Society of Petroleum Geology, Memoir 6, p. 509-521.

Tappan, H., and A. R. Loeblich, Jr., 1970, Geobiologic implication of fossil phytoplankton evolution and time-space distribution: Geological Society of America, Special Paper 127, p. 247-340.

———, ———, ———, 1973, Evolution of the oceanic plankton: Earth Science Review, v. 9, p. 207-240.

Thiede, J., and T. H. van Andel, 1977, The paleoenvironment of anaerobic sediments in the Late mesozoic South Atlantic Ocean: Earth and Planetary Science Letters, v. 33, p. 301-309.

Tissot, B., G. Deroo, and J. P. Herbin, 1979, Organic matter in Cretaceous sediments of the North Atlantic: contribution to sedimentololgy and paleogeography, *in* Talwani, Hay, and Ryan, eds., Deep drilling results in the Atlantic Ocean: Continental Margins and Paleoenvironments: American Geophysical Union, Maurice Ewing Series 3, p. 362-374.

———, et al, 1974, Influence of nature and diagenesis of organic matter in formation of petroleum: AAPG Bulletin, v. 58, p. 499-506.

———, et al, 1980, Paleoenvironment and petroleum potential of Middle Cretaceous black shales in Atlantic Basins: AAPG Bulletin, v. 64, p. 2051-2063.

Turic, M. A., et al, 1980, Malvinas Basin — offshore Argentina: Proceedings, Offshore Technology Conference, v. 4, p. 575-582.

Vail, P. R., R. M. Mitchum, Jr., and S. Thompson, 1978, Global cycles of relative changes of sea level, *in* Charles E. Payton, ed., Seismic stratigraphy-applications to hydrocarbon exploration, part 4: AAPG Memoir 26, p. 83-97.

Waples, D., 1981, Organic geochemistry for exploration geologists: Minneapolis, Minnesota, Burgess Publishing Company, 151 p.

Hydrates of Natural Gas in Continental Margins

Keith A. Kvenvolden
U.S. Geological Survey
Menlo Park, California

Leo A. Barnard
Department of Oceanography
Texas A&M University
College Station, Texas

Natural gas hydrates in continental margin sediment can be inferred from the widespread occurrence of an anomalous seismic reflector which coincides with the predicted transition boundary at the base of the gas hydrate zone. Direct evidence of gas hydrates is provided by visual observations of sediments from the landward wall of the Mid-America Trench off Mexico and Guatemala, from the Blake Outer Ridge off the southeastern United States, and from the Black Sea in the U.S.S.R. Where solid gas hydrates have been sampled, the gas is composed mainly of methane accompanied by CO_2 and low concentrations of ethane and hydrocarbons of higher molecular weight. The molecular and isotopic composition of hydrocarbons indicates that most of the methane is of biological origin. The gas was probably produced by the bacterial alteration of organic matter buried in the sediment. Organic carbon contents of the sediment containing sampled gas hydrates are higher than the average organic carbon content of marine sediments. The main economic importance of gas hydrates may reside in their ability to serve as a cap under which free gas can collect. To be producible, however, such trapped gas must occur in porous and permeable reservoirs. Although gas hydrates are common along continental margins, the degree to which they are associated with significant reservoirs remains to be investigated.

Natural gas hydrates are a special kind of clathrate in which a three-dimensional framework of water molecules includes and is stabilized by molecules of natural gas (mainly methane). In this framework, water crystallizes in the isometric system rather than in the hexagon system of normal ice. The isometric or cubic lattice contains voids, or cages, large enough to accommodate molecules of gas. Two structures of the cubic lattice are possible. In Structure I, the cages are arranged in body-centered packing and include small hydrocarbon molecules such as methane, ethane, and nonhydrocarbons such as CO_2 and H_2S. In Structure II, diamond packing is present; not only can methane and ethane be included in the voids, but, in order to stabilize the structure, propane and isobutane are also needed to occupy some of the larger voids. Apparently, gases larger than isobutane (e.g., n-butane and pentanes) cannot be included in either Structure I or II (Hand, Katz, and Verma, 1974; Hitchon, 1974).

Under an appropriate regime of pressures and temperature, natural gas, which is composed mainly of methane, can interact with water to form Structure I gas hydrates. Structure II hydrates are not expected unless the gas is rich in higher hydrocarbons such as propane (Davidson et al, 1978). The pressure-temperature stability field of natural gas hydrates (Figure 1) shows that at pressures increasingly greater than about 25 atm (depths greater than about 250 m), the maximum temperature at which gas hydrates exist is increasingly higher than the freezing point of water. The exponentially increasing pressure requirement for gas hydrate stability generally exceeds the hydrostatic pressure at temperatures determined by the geothermal gradient in deep sediments. Thus, the depths at which gas hydrates remain stable are limited. The amount of methane required for gas hydrate to form also depends on temperature and pressure. In the methane-water system, only the methane present in excess of the amount soluble in water is available for hydrate formation. The molar ratio of methane to water in a fully saturated Structure I hydrate can be nearly 1:6. Thus, one cubic meter of this ideal methane hydrate would contain about 170 cubic meters of

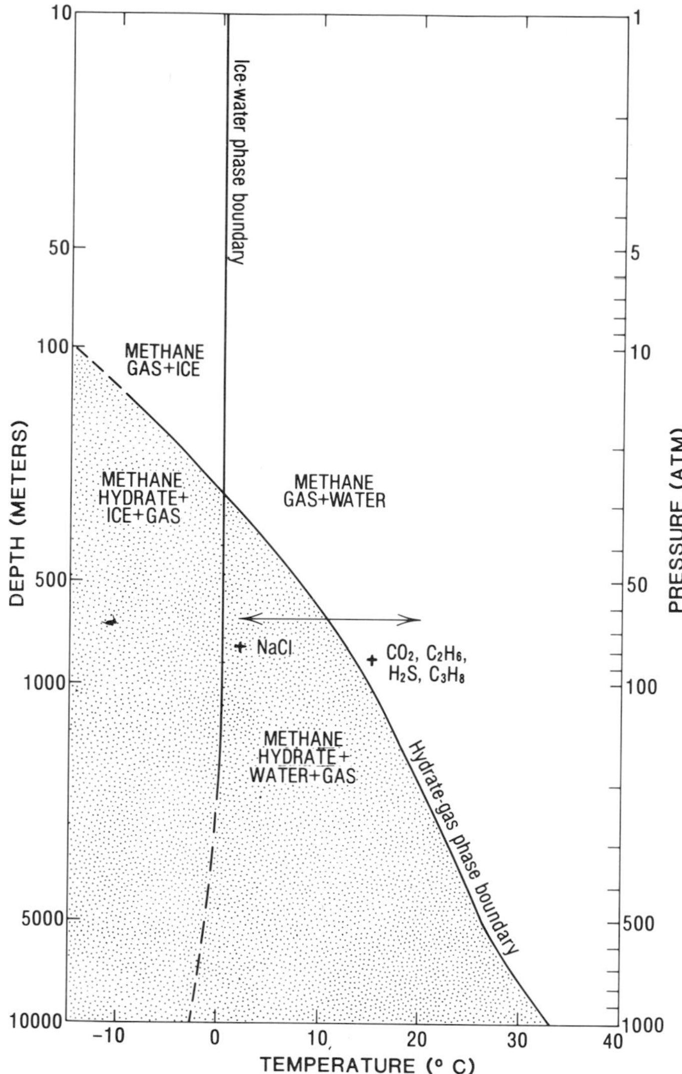

Figure 1 — Phase diagram showing boundary between free methane gas (no pattern) and methane hydrate (pattern) for a pure-water and pure-methane system. Addition of NaC1 to water shifts the curve to the left. Adding CO_2, H_2S, ethane, and propane to methane shifts the boundary to the right and thus increases the area of the hydrate stability field. Depth scale assume lithostatic and hydrostatic pressure gradients of 0.1 atm/m. Redrawn after Katz et al (1959).

methane gas. In nature, gas hydrates contain less than the ideal ratio of methane to water, probably because cages in the hydrate structure are not completely filled. Nevertheless, gas hydrates in reservoir rock can contain significant quantities of methane, which is a potential energy resource should the gas ever be recoverable. However, free gas trapped in reservoirs beneath the gas hydrate is a more promising natural resource (Hedberg, 1980).

Figure 1 shows the phase boundary for free methane gas and methane hydrate in a pure water and pure methane system. Gas extracted from continental margin sediments is not pure methane; it contains small concentrations of ethane and hydrocarbon gases of higher molecular weight, as well as CO_2 (McIver, 1974; Claypool, Presley, and Kaplan, 1973). These additional components in the gas mixture cause the phase boundary (Figure 1) to shift to the right. Pore water in normal continental margin sediments is not pure but contains salts, particularly NaCl. The presence of salts in the water shifts the phase boundary to the left (Figure 1). For natural gas in marine sediments, the shifts in position of the phase boundary are of similar magnitude but in opposite directions. Thus, the effects approximately cancel each other. The boundary for a pure water and pure methane system (Figure 1) provides a reasonable estimate of the pressure-temperature conditions under which natural gas hydrates, composed mainly of methane, will be stable on continental margins (Claypool and Kaplan, 1974).

Figure 2 shows an idealized section indicating the predicted occurrence zone of gas hydrates in continental margins with a geothermal gradient of 27.3°C/km. This section illustrates the zone of a continental margin where pressure and temperature conditions permit gas hydrate stability (Figure 1), given an adequate supply of methane. The gas hydrate zone shown in Figure 2 does not reach the continental shelf; however, where bottom water temperatures and geothermal gradients are low, gas hydrates may be found buried in continental shelf sediment. Such occurrences should be rare and confined to high latitudes. Below about 500 m of water depth (Figure 2), the potential gas hydrate zone extends from the upper continental slope through the continental rise to the abyssal basins. When interpreting seismic evidence for gas hydrates it is important to note that the thickness of the zone favorable for gas hydrate occurrence increases with the depth of water.

GEOPHYSICAL EVIDENCE FOR GAS HYDRATES

At least three lines of related geophysical evidence indicate gas hydrates present in continental margin sediments: (1) An anomalous acoustic reflector seen on seismic profiles approximately subparallels the sea floor at depths that can be predicted from information in Figures 1 and 2 (the reflector is commonly, but inaccurately, called a bottom simulating reflector or BSR); (2) Sonobuoy seismic velocity measurements show higher velocities above the anomalous reflector and lower velocities below; and (3) A consequence of this velocity inversion is that the anomalous acoustic reflector may be reversed in polarity. These three lines of geophysical evidence are well illustrated along the Blake Outer Ridge, offshore the southeastern United States. Observations in this area by Markl, Bryan and Ewing, (1970), Stoll, Ewing, and Bryan (1971), and Shipley et al (1979) showed an anomalous reflector that intersected bedding reflectors, mimicked the topography of the sea floor and appeared to deepen with increasing water depth following the pressure-temperature relationship predicted in Figure 1. A seismic profile (Figure 3) from this area shows a well-developed bottom simulating reflector. One explanation for the anomalous acoustic reflector is that it corresponds to the base of the gas hydrate zone, a transition boundary between sediment containing gas hydrate and sediment without gas hydrate (Ewing and Hollister, 1972; Tucholke, Bryan, and Ewing, 1977; Dillon, Grow, and Paull, 1980). In addition, the apparent acoustic velocity through the sediment overlying the anomalous reflector was determined to be about 2 km/sec (Lancelot and Ewing, 1972), a velocity unusually high for hemipelagic sediment. Further studies of acoustic velocities at the Blake Outer Ridge support this first observation. For example, Bryan (1974) confirmed the value of 2 km/sec by independent sonobuoy measurements. Dillon, Grow, and Paull (1980) calculated the following from multi-channel seismic velocity analyses: in the zone above the anomalous reflector, interval velocities are greater than 2.5 km/sec; in the zone below the reflector, the velocity is less than 1.5 km/sec. On the other hand, acoustic measurements by Tucholke, Bryan, and Ewing (1977) and Shipley et al (1979) show a lower range of velocities (1.6 to 1.9 km/sec) in sediments above the anomalous reflector. Nevertheless, a velocity inversion apparently occurs at the anomalous reflector resulting in a reflection polarity reversal associated with the anomalous reflector (Shipley et al, 1979). Although gas hydrates can be inferred from the presence of the anomalous acoustic reflector, gas hydrates can also exist where no anomalous reflector is identified in seismic records.

Other examples of anomalous acoustic reflectors possibly correlated with the base of the gas hydrate zone have been reviewed by Kvenvolden and McMenamin (1980) and the list of occurrences is expanded here (Table 1; Figure 4). Most of the inferred gas hydrate occurrences are on the continental margins of the North American continent, where much geophysical work has been conducted. Although more than 90% of the ocean floors are within the pressure-temperature field suitable for gas hydrate stability (Trofimuk, Cherskiy, and Tsarev, 1974), it is likely that most future discoveries of gas hydrates in

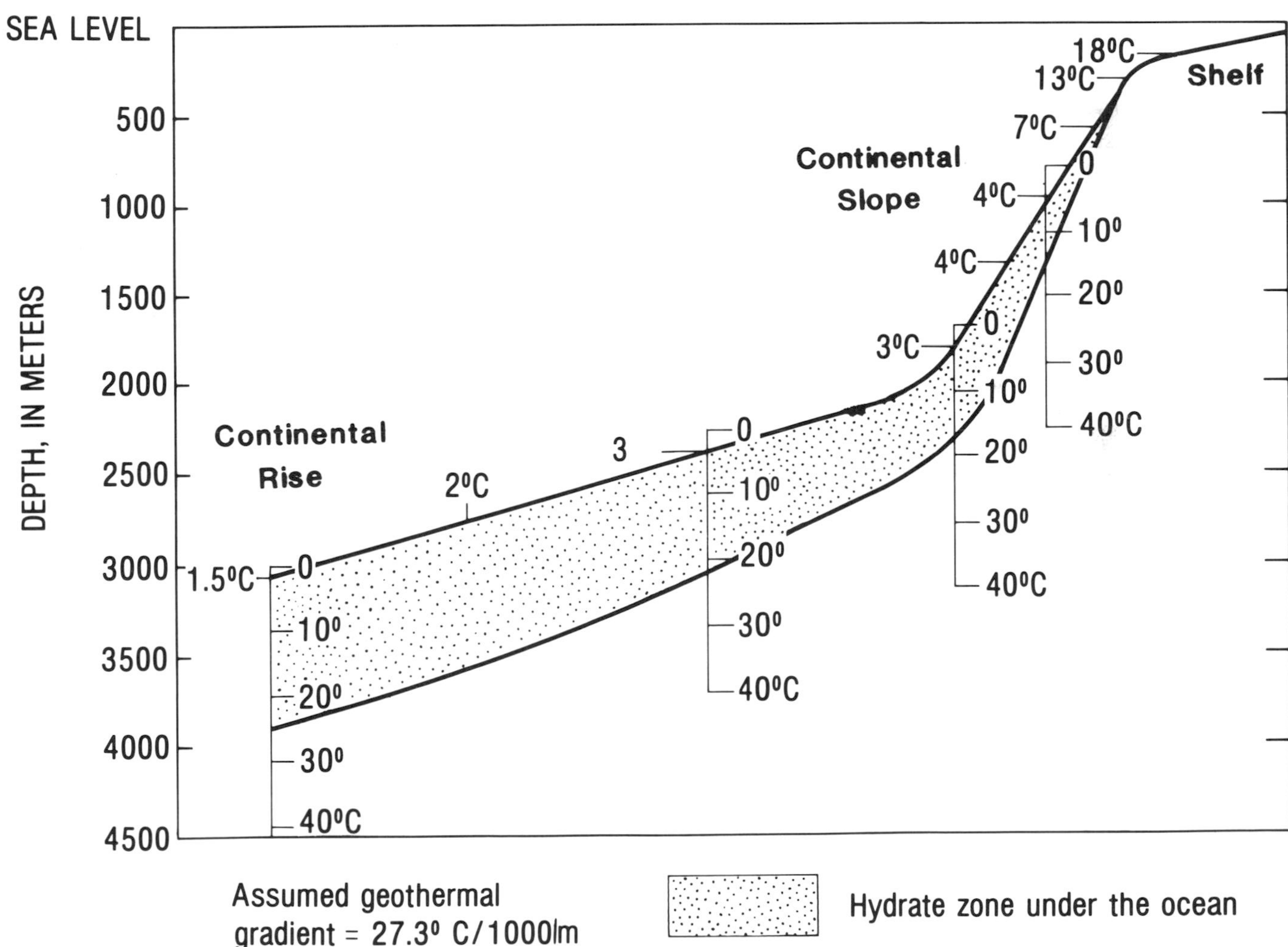

Figure 2 — Idealized section that shows the zone of gas hydrate stability for outer continental margins. Stippled area is potential region of gas hydrate formation where pressure and temperature conditions are correct for hydrate stability (Figure 1) assuming an adequate methane supply. The following assumptions apply: 1) geothermal gradient of 27.3°C/km; 2) lithostatic and hydrostatic pressure gradients of 0.1 atm/m; and, 3) bottom-water temperature range from 1.5° to 18°C depending on water depth. Redrawn from R. D. McIver, written communication, 1979.

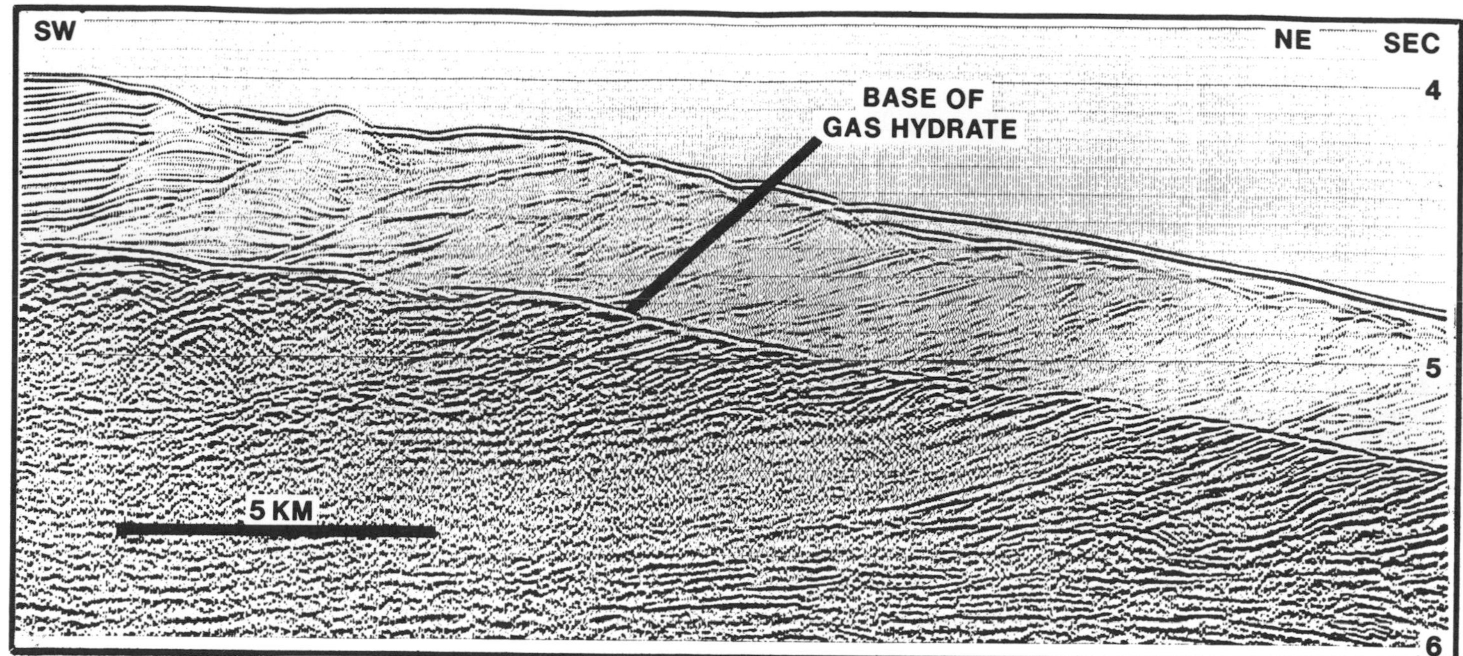

Figure 3 — A 12-fold multichannel seismic reflection profile from crest and eastern flank of Blake Outer Ridge. The reflector at the base of the gas hydrate follows the bathymetry of the sea floor and transects dipping bedding reflectors. The apparent lower applitude above the gas hydrate reflector may result from reduced acoustic impedance differences between strata caused by the presence of the gas hydrate (Shipley et al, 1979; Figure 3, p. 2206).

oceanic sediments will be confined to continental margins and oceanic basins where methane can be generated or supplied in quantities sufficient to saturate sediment pore water.

GEOCHEMICAL EVIDENCE FOR GAS HYDRATES

An objective of the Deep Sea Drilling Project (DSDP) Leg 11 (Ewing and Hollister, 1972) was to investigate the nature of the anomalous acoustic reflector observed by Markl, Bryan, and Ewing (1970) and Stoll, Ewing, and Bryan (1971) on seismic profiles along the Blake Outer Ridge (Designation 1, Table 1; Figure 4). Sediment samples recovered at three sites (102, 103, and 104) across the Ridge yielded high gas concentrations. In many cases gas expansion was sufficient to extrude sediment from core liners. Degassing occurred immediately when the core barrels were opened, and many cores continued to produce gas for several hours (Lancelot and Ewing, 1972). The dominant component in the gas was methane, ranging in concentration from about 94.2 to 99.7% of the gas phase volume with the balance being mainly CO_2. This methane had a carbon isotopic composition ranging from -70 to -88 per mil, relative to the PDB Standard (Claypool, Presley, and Kaplan, 1973). Ethane and heavier hydrocarbon gases were present only in trace amounts. Although no obvious gas hydrates were recovered at these sites, the high gas concentrations and the anomalous acoustic reflector were cited as possible evidence for the presence of gas hydrates in sediments of the Blake Outer Ridge (Ewing and Hollister, 1972; Lancelot and Ewing, 1972).

Since DSDP Leg 11, physical recovery and geochemical evidence confirmed the presence of solid gas hydrates in marine sediments at four localities: the Black Sea, the Pacific continental margin off Mexico, the Pacific continental margin off Guatemala, and the Atlantic continental margin off the southeastern United States, along the Blake Outer Ridge.

Solid gas hydrates were first observed in oceanic sediments by Yefremova and Zhizhchenko (1975) in samples from the Black Sea (Designation 23, Table 1; Figure 4). At water depths of 2000 m, microcrystalline aggregates of gas hydrates were recovered 6.5 m below the sea floor. The hydrates, which contained mainly methane and CO_2, decomposed quickly. Coring during DSDP Leg 42B in the Black Sea failed to recover solid gas hydrates, but most cores contained gases that expanded as the cores reached the surface. Some cores contained sufficient gas to blow sediment out the end of the core barrel. The gas was mainly methane with small amounts of CO_2 and other hydrocarbons. The carbon isotopic composition of the methane ranged from -63 to -72 per mil (Hunt and Whelan, 1978).

During DSDP Leg 66, in the Pacific Ocean off Mexico, gas hydrates were found in continental margin sediment on the landward wall of the Mid-America Trench (Designation 10, Table 1; Figure 4). An anomalous acoustic reflector mimics the bathymetry in this area (Figure 5) and suggests that gas hydrates may be present. Solid gas hydrate was recovered from holes at three sites in water depths ranging from about 1780 to 2870 m. At Site 490, gassy ice was recovered at a sediment depth of 137 m. Gassy, frozen sediment obtained at Sites 491 and 492 produced about 20 ml of gas per ml of pore fluid. The gas was mainly methane. The volume of methane produced was about five times the amount which is

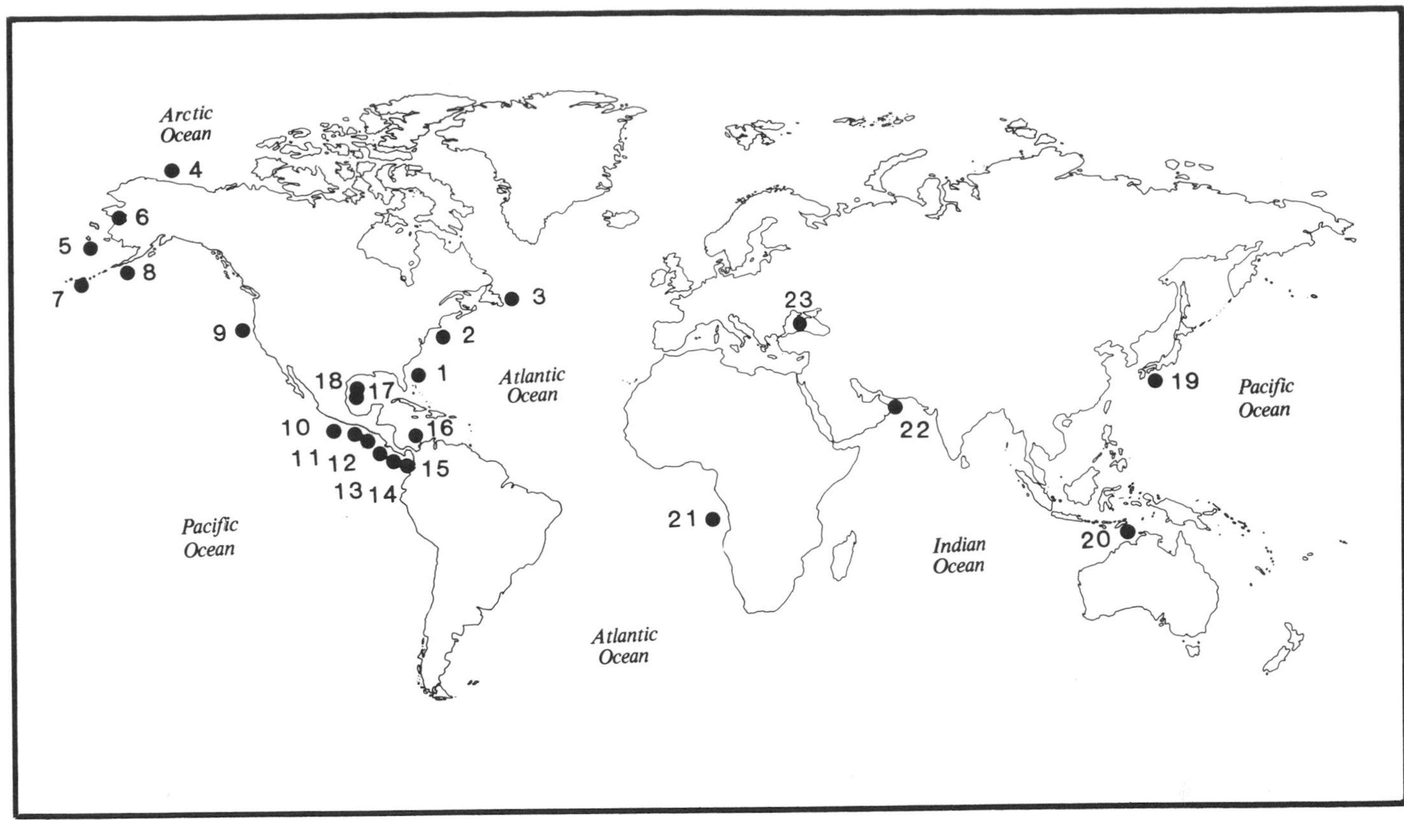

Figure 4 — Location of known or inferred gas hydrates in oceanic sediment. Inferred gas hydrates are based on the occurrence of anomalous acoustic reflectors that can be correlated with the depth of the predicted phase boundary between gas hydrate and free gas. See Table 1 for identification of the locations and pertinent references.

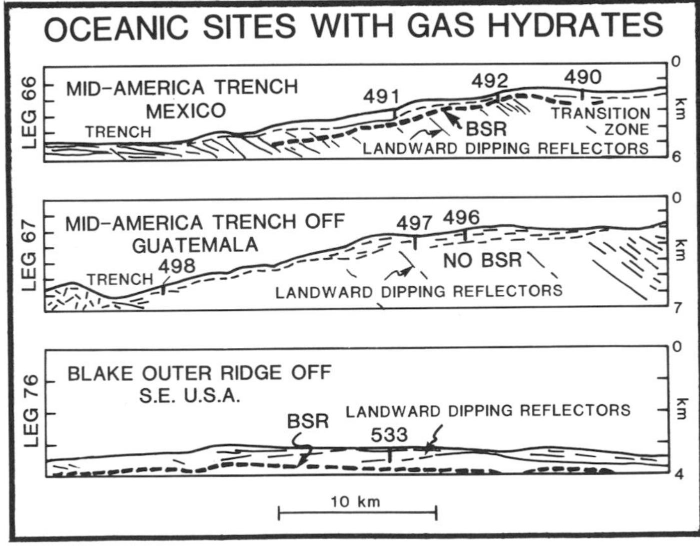

Figure 5 — Diagrammatic sections showing sites where solid gas hydrate has been recovered in cores on DSDP Legs 66, 67, and 76 (designations 10, 12, and 1 on Table 1 and Figure 4). Note that the vertical exaggeration for the Leg 76 section is twice that for Legs 66 and 67. An anomalous acoustic reflector (BSR), thought to represent the base of the gas hydrate, can be seen in sections for Legs 66 and 76. No BSR was observed on the Leg 67 transect, although detailed analyses of recently acquired seismic records suggested that a weak BSR is present (von Huene, personal communication).

soluble in seawater at equivalent conditions; such concentrations of methane indicate the presence of gas hydrates (Moore, et al, 1979).

On DSDP Leg 67 in the Pacific Ocean off Guatemala, also on the landward wall of the Mid-America Trench (Designation 12, Table 1; Figure 4), no anomalous acoustic reflector was observed (Figure 5). Solid gas hydrates were recovered in vitric sands near the bottom of the holes at Sites 497 and 498 in water depths of about 2350 and 5490 m, respectively. Decomposition of the suspected hydrates produced greater quantities of gas than are soluble in water at *in situ* pressures and temperatures, and methane/ethane ratios were large (von Huene, et al, 1980). The recovery of gas hydrates on Leg 67 amplified the observations from Leg 66 by showing that gas hydrates are widespread in sediments on the landward side of the Mid-America Trench.

Confirmation of Gas Hydrates in Sediments of the Blake Outer Ridge

The principal objectives of DSDP Leg 76, Site 533, in 3184 m of water on the Blake Outer Ridge (Figure 5; Location 1, Figure 4) were to recover samples of gas hydrate and to measure the pressure, volume, and composition of gas released during hydrate decomposition (Sheridan, et al, 1982). Expansion of cores

Table 1. Gas hydrates in oceanic sediments.

Designation on Figure 4	Location	Depth of water m = meters s = seconds (2-way)	Subbottom Depth of Inferred Bases of Gas Hydrate m = meters s = seconds (2-way)	Reference
1	Blake Outer Ridge, Western Atlantic Ocean off southwestern U.S.A.	2500-4000 m	450-600 m	Markl, Bryan, and Ewing (1970) Ewing and Hollister (1972) Tucholke, Bryan, and Ewing (1977) Shipley, Buffler, and Watkins (1978) Shipley et al (1979) Dillon, Grow, and Paull (1980) Sheridan et al (1982)
2	Continental Rise, Western Atlantic Ocean off New Jersey and Delaware	2500-3800 m	470-590 m	Tucholke, Bryan, and Ewing (1977)
3	Labrador Shelf, North-western Atlantic Ocean off Newfoundland	~ 2000	Not given	Taylor, Wehmiller, and Judge (1979)
4	Beaufort Sea, Arctic Ocean off Alaska	400-2500 m	100-800 m	Grantz, Boucher, and Whitney (1976) Grantz, USGS (personal comunication) Grantz, this volume
5	Continental Slope, Bering Sea near Navarin Basin	500-3000 m	200-500 m	Cooper, USGS (personal communication)
6	Norton Sound, Bering Sea off Alaska	Not given	Not given	R.J. Mousseau (personal communication, 1981)
7	North Pacific Ocean, south of Aleutian Islands, near DSDP Site 186	4522 m	0.5-0.7 s	Preliminary interpretation Scholl, USGS (personal communication)
8	North Pacific Ocean south of Aleutian Islands	Not available	Not available	Bruns, USGS (personal communication)
9	Pacific ocean off Northern California	800-1200 m	0.3 s	Field, USGS (personal communication)
10	Pacific Ocean off Mexico	~2.5-5.0 s	~0.5-0.7 s	Shipley et al (1979) Shipley et al (1980) Moore et al (1979)
11	Pacific Ocean off Guatemala	Not given	Not given	Shipley et al (1979)
12	Pacific Ocean off Guatemala	2000-5500 m	Not observed	von Huene et al (1980)
13	Pacific Ocean off Nicaragua	800-2400 m	~0.4-0.5 s	Shipley et al (1979)
14	Pacific Ocean off Costa Rica	~1.0-1.8 s	~0.2-0.5 s	Shipley et al (1979)
15	Pacific Ocean off southern Panama	~2.5-2.8 s	~0.4-0.5 s	Shipley et al (1979)
16	Caribbean Sea off Panama and Columbia	1500-3000 m	Not given	Shipley et al (1979) Lu et al (this volume)
17	Western Gulf of Mexico off Mexico	1200-2000 m	Not given	Shipley et al (1979)
18	Western Gulf of Mexico, location unspecified	~2.7-3.1 s	~0.5 s	Hedberg (1980)
19	Western Pacific, Nankai Trough off Japan	Not available	Not available	Aoki et al (this volume)
20	Timor Trough off northern Australia	2315 m	Not given	McKirdy and Cook (1980)
21	Eastern Atlantic Ocean off Angola	Not available	Not available	B. Tucholke (personal communication)
22	Northwest Indian Ocean, Gulf of Oman	3000 m	600-700 m	White (1979) White and Louden (this volume)
23	Black Sea, U.S.S.R.	2000 m	Not given	Yefremova and Zhizhchenko (1974)

and geochemical analyses of gases recovered through the core liner at Site 533 showed that the hemipelagic sediment in this part of the Blake Outer Ridge contains high concentrations of methane. These analyses confirm the previous observations, on Leg 11, of high gas content of sediments (Ewing and Hollister, 1972). In addition to methane, small concentrations of higher molecular weight hydrocarbons (through hexane) were found. A gradual decrease with depth of large methane/ethane ratios probably resulted from the effects of the early diagenetic alteration of organic matter. Pressure and volume measurements, visual observations, and chemical analyses of the evolved gases all provided conclusive evidence of gas hydrates at a sub-bottom depth of about 240 m. The volume of gas released upon gas hydrate decomposition was about 20 times the volume of pore fluid, a result similar to that obtained from samples of gas hydrate-containing sediment from Leg 66 (Moore, et al, 1979). The molecular distribution of the hydrocarbon gases, released by the decomposing gas hydrate, showed a preferential fractionation in which the concentrations of gases larger than isobutane abruptly decrease to trace amounts. This follows from the fact that the gas hydrate cage structures exclude hydrocarbons larger than isobutane. Gas samples taken from sediment lacking obvious gas hydrate did not show this molecular fractionation. These observations suggest that Structure I gas hydrate was present because methane and ethane were the dominant hydrocarbon gases (propane and isobutane were measured in the low parts per million range), and only traces of *n*-butane and larger molecules were observed. There may have been a few crystals of Structure II gas hydrate present to account for these observations.

At Site 533 a pressure core barrel (Larson, Robson, and Foss, 1980) was deployed five times. Three cores were recovered at *in situ* pressures of about 4500 psig. Gas was released from these three cores at intervals over about three hours. Portions of the gas were collected for analysis. The pressure release patterns of the three cores suggest the presence of gas hydrate and dissolved gas (Sheridan, et al, 1982).

Figure 6 summarizes information about the total depths (water plus sediment) and temperatures at DSDP sites of Legs 66, 67, and 76 where gas hydrates were observed. Pressure-temperature conditions for these sites fall within the gas hydrate stability field (Figure 1). Well-equilibrated, down-hole temperature measurements for the hole at Site 533 yield a temperature gradient near the bottom of the hole of about 36°C/km (Sheridan, et al, 1982). This temperature gradient, extrapolated to great depths, intersects the gas hydrate phase boundary at about 3790 m total depth, or about 600 m sub-bottom (Figure 6). This depth corresponds to the depth of the anomalous acoustic reflector, which apparently marks the transition boundary between sediment with gas hydrates and sediment without gas hydrates.

Gas hydrates recovered on Legs 66, 67, and 76 did not occur in massive, solid, thick units but rather in a few thin layers usually, but not always, associated

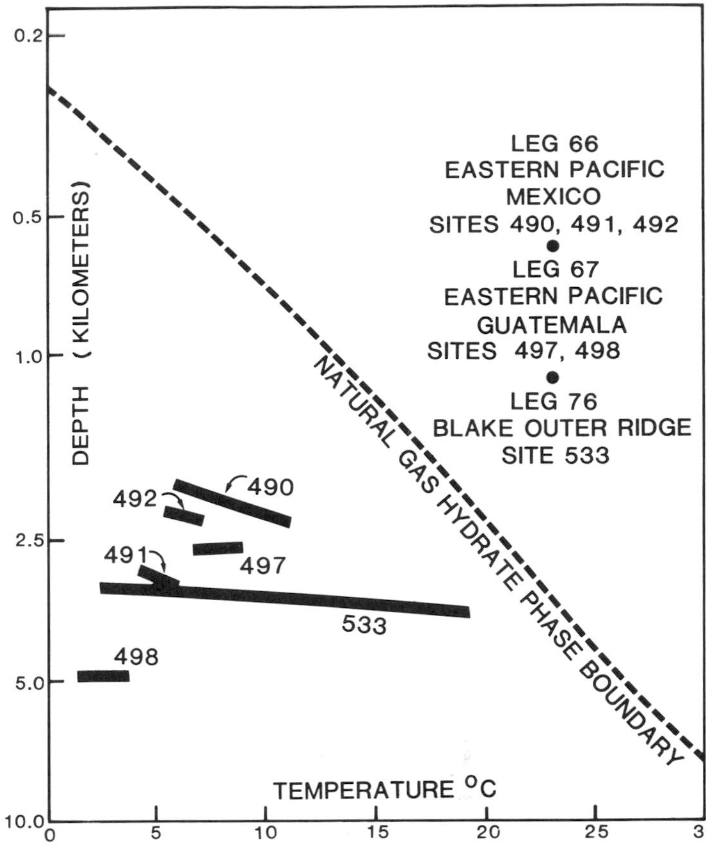

Figure 6 — Summary of preliminary depth and temperature information for DSDP sites where solid gas hydrates were recovered. Total depth (water and sediment) is indicated. All sites are within the gas hydrate stability field. The temperature gradient of about 30°C/km for Site 533 was established by three downhole temperature measurements (Sheridan, et al, 1982). Temperature gradient of 22°C/km for Sites 490, 491, and 492 are from Shipley and Didyk (in press). Pressure-temperature information for Sites 497 and 498 are from Harrison and Curiale (in preparation).

with more porous intervals. On Legs 66 and 67 the gas hydrates were associated with vitric sands. The single occurrence of solid gas hydrate noted on Leg 76 was in hemipelagic sediment not unlike the rest of the core sediment. Much gas hydrate may decompose during drilling. Such decomposition would explain why no solid gas hydrate was observed on Leg 11 and only one was noted on Leg 76.

Source of Gas and Formation of Marine Gas Hydrates

Where marine gas hydrates have been recovered from continental margin sediment (DSDP Legs 66, 67, and 76), methane, accompanied by CO_2 and small amounts of ethane and higher molecular weight hydrocarbons, is the dominant gas (Moore, et al, 1979; von Huene, et al, 1980; Sheridan, et al, 1982). In fact, methane generally constituted more than 99.9% of the hydrocarbon gases released. Similarly, the only hydrocarbon gas reportedly present in gas hydrates from sediment of the Black Sea is methane (Yefremova and Zhizhchenko, 1975). In all these locations both the molecular composition of the hydrocarbon gases and their shallow occurrence at less than 600 m subbottom suggest that microbiological processes were responsible for most, if not all, of the methane found. The carbon isotopic compositions of methane from sediments recovered from the Blake Outer Ridge (DSDP Leg 11; Claypool, Presley, and Kaplan, 1973) and from the Black Sea (DSDP Leg 42B; Hunt and Whelan, 1978) support this contention. These isotopic values range from −63 to −88 per mil (relative to the PDB Standard). These values fall within the range of −50 to −90 per mil, which is generally considered to indicate biogenic methane (Feux, 1977). Thus, both the molecular and the isotopic compositions of these hydrocarbons suggest that most of the methane in the marine gas hydrates sampled so far has resulted from bacterial alteration of organic matter buried in the sediment.

If the methane in most marine gas hydrates results from bacterial processes, then natural gas hydrates will be found only in continental margins of appropriate depth and temperature where sedimentation rates are high enough to insure adequate organic matter and suitable environmental conditions for bacterial generation of methane. Methane production requires a minimum of about 0.5% organic carbon in the form of metabolizable organic matter and pore water depleted in dissolved oxygen and sulfate (Claypool and Kaplan, 1974). At DSDP sites on Legs 66, 67, and 11 where solid gas hydrates were observed or inferred, the present organic carbon content equals or exceeds this value (Table 2; organic carbon values for Leg 76 are not available at this time). In fact, the organic carbon content of sediments from cores containing gas hydrates usually exceeds the average organic content of 0.1 to 0.6% for oceanic sediments (Hunt, 1979).

The following model describes the way in which gas hydrates may form in marine sediments. On continental margins where high sedimentation rates ensure the deposition of metabolizable organic matter, an ecological succession is established of metabolic processes. These processes vary with time and depth in the marine sedimentary column (Claypool and Kaplan, 1974). Two biochemical zones result. The upper is an aerobic zone. It is underlain by a zone in which two different anaerobic processes take place: a region of sulfate reduction overlies a region of carbonate reduction and methane production. Carbonate reduction is the favored mechanism of biogenic methane production; it probably accounts for most of the methane produced in the marine environment, although other pathways are available (Rice and Claypool, 1981).

After ecological succession is established, the biochemical zones move upward with time as new sediment is added at the sediment-water interface. If these biochemical and geological processes take place under pressure and temperature conditions suitable for gas hydrate stability (Figure 1), and if an adequate supply of metabolizable organic matter generates sufficient methane, then gas hydrates form. Methane production begins in the carbonate reduction zone at sediment depths below the sulfate reduction zone. If methane generation is sufficiently rapid to overcome the loss of methane by diffusion, then methane concentrations increase to pore water saturation levels. Additional methane forms the gas hydrate. The hydrostatic pressures, due to the overlying water column, affect methane solubility and influence the amount of methane required for forming gas hydrates. Gas hydrate formation in marine sediments may be concurrent with continued sedimentation. The zone of gas hydrate will thicken until the base of the gas hydrate zone subsides into a region where temperatures make the gas hydrate unstable. In this region free methane can occur, but it should tend to migrate back into the overlying zone of gas hydrate stability (if suitable migration pathways are available).

RESOURCE IMPLICATIONS OF MARINE GAS HYDRATES

Geochemical analyses indicate that gas associated with the marine gas hydrates thus far examined is the result of microbiological alteration of organic matter buried in the sediment. It follows that these gas hydrates probably formed *in situ* and thickened until their base reached depths where higher temperatures caused the gas hydrates to become unstable and decompose. There should be little difference between the composition of the gas in the hydrate and of the gas occurring below the hydrate. Therefore, at the base of these confirmed gas hydrates, the dissolved and possibly free gas which has not yet been collected or measured should also be biogenic.

The seal at the base of the gas hydrate can trap biogenic methane generated below the gas hydrate or released at the base of the hydrate as it moves downward into temperature regions of instability. Conceivably, the base of the gas hydrate could also form a seal for thermogenic hydrocarbons migrating toward the surface from depth. However, if significant amounts of thermogenic gas migrated without significant molecular and isotopic fractionation into the gas hydrate stability region during the gas hydrate formation process, clues to its presence should be found in the molecular and isotopic compositions of gases associated with the gas hydrates. No compelling evidence for significant amounts of thermogenic gas has been observed in any of the gas hydrates we discuss. Of course, a gas hydrate seal could trap thermogenic gas that had migrated after the gas hydrate formation. Then, there would not necessarily be clues to the thermogenic gas within the gas hydrate zone itself. Such a sequence of events is probably rare.

Gas hydrates can trap methane within the clathrate structure as well as beneath the base of the gas hydrate zone. At depths of water plus sediment totaling less than about 1700 m, gas hydrates have the potential to include within the water clathrate framework greater volumes of methane than can be trapped free in an equal volume of space. At total depths greater than about 1700 m, more methane can exist free in a

Table 2. Average organic carbon content (% OC) of sediment at DSDP sites.

	Leg 11[1]			Leg 66[2]			Leg 67[2]	
Site 102	103	104	490	491	492	496	497	498
% OC 0.6	0.5	0.7	1.4	1.1	0.4	1.8	2.2	1.6

[1]Boyce (1972).
[2]Information obtained from computer files, Deep Sea Drilling Project, Scripps Institution of Oceanography.

given volume of space than can be held within a gas hydrate of equivalent volume. Thus the volumetric advantage of gas in hydrates over free gas is limited to a specific depth range.

Even though large volumes of gas can be included within gas hydrates, current technology could not produce the hydrated gas economically even if production wells were able to reach the appropriate water and sediment depths. The main economic advantage of marine gas hydrates may reside in their ability to act as an impermeable barrier or seal under which free methane collects in accumulation traps (Hedberg, 1980). At least three kinds of structural or stratigraphic traps are possible (Dillon, Grow, and Paull, 1980): (1) Dome-shaped structures under closed topographic highs, caused by the base of the gas hydrate zone mimicking the sea floor bathymetry; (2) Dome-shaped structures at the base of the gas hydrate, created by local increases in heat flow which in turn are caused by local changes in thermal conductivity such as the occurrence of salt domes; and (3) Stratigraphic traps, where gas-containing beds dip into the sea floor and are sealed at their updip ends by the zone of gas hydrates.

To be effective traps for economically significant amounts of methane, assuming that production technology can reach the required depths of water and sediment, gas hydrates must seal porous and permeable sediments in which free gas can accumulate and be produced. Although the anomalous acoustic reflectors indicating the base of gas hydrates are common in continental margins, the indicated gas hydrates can trap significant amounts of producible methane only if adequate reservoir sediments are present. Future work should evaluate the reservoir characteristics of sediments which lie directly below the base of the gas hydrate or are intersected by it.

SUMMARY

Wide distribution of at least twelve localities, where anomalous geophysical evidence points to gas hydrate occurrence (Fig. 4 and Table 1), suggests that gas hydrates are common in continental margin sediments. This evidence includes: (1) acoustic reflectors that approximately parallel the sea floor; (2) sonobuoy seismic velocity measurements which show increased velocities within the suspected zone of gas hydrate; and (3) reversed polarity of the acoustic signal from the reflector inferred to be at the base of the gas hydrate.

The temperature-pressure stability requirements for gas hydrates are satisfied in more than 90% of the floors in the world's oceans. The limiting factor for the development of gas hydrates in marine sediment appears to be the abundance of naturally occurring gas needed to stabilize the gas hydrate structure. Where marine gas hydrates have been recovered from continental margin sediment in the Pacific Ocean off Mexico and Guatemala and in the Atlantic Ocean off the southeastern United States (DSDP Legs 66, 67, and 76), the dominant gas is methane, accompanied by CO_2 and small amounts of ethane and hydrocarbons of higher molecular weight. Gas hydrates recovered from sediment of the Black Sea also contained mainly methane and CO_2. At these locations the molecular composition of the hydrocarbon gases, dominated by methane, suggests that microbiological processes were responsible for most of the methane found. The carbon isotopic compositions of the methane support this contention.

On continental margins where gas hydrates have been identified, rates of sedimentation are high and the amount of organic matter deposited appears sufficient for bacterial generation of methane. Where pressures and temperatures are appropriate, and rates of methane generation are sufficiently rapid, gas hydrate formation probably begins below the sea floor in the zone of biological carbonate reduction and methane production. The formation process may be concurrent with continued sedimentation, so that the gas hydrate zone thickens until the zone's base has moved downward into temperature regions where the gas hydrate is unstable and methane is released.

Marine gas hydrates may form impermeable barriers or seals under which free methane can accumulate. Both structural and stratigraphic traps are possible. To be effective traps for economically significant amounts of methane, gas hydrates must seal reservoirs in which free gas can accumulate. Although gas hydrates are common in continental margins, further research is necessary to determine where these gas hydrates are accompanied by reservoirs of trapped gas.

REFERENCES CITED

Aoki, Y. T., T. Tamono, and S. Kato, 1982, Detail structure of the inner trench slopes of the Nankai Trough from migrated seismic sections: this volume.

Boyce, R. E., 1972, Carbon and carbonate analysis, leg 11, *in* C. D. Hollister et al, Initial reports of the deep sea drilling project, v. 11: Washington, D. C., U. S. Government Printing Office, p. 1059-1071.

Bryan, G. M., 1974, *In situ* indications of gas hydrates, *in* I. R. Kaplan, ed., Natural gases in marine sediments: New York, Plenum, p. 298-308.

Claypool, G. E., and I. R. Kaplan, 1974, The origin and distribution of methane in marine sediments, *in* I. R. Kaplan, ed., Natural gases in marine sediments: New York, Plenum, p. 94-129.

———, B. J., Presley, and I. R. Kaplan, 1973, Gas analysis of sediment samples from legs 10, 11, 13, 14, 15, 18 and 19, *in* J. S. Creager et al, Initial reports of the deep sea drill-

ing project, v. 19: Washington, D.C., U.S. Government Printing Office, p. 879-884.
Davidson, D. W., et al, 1978, Natural gas hydrates in northern Canada; Proceedings, 3rd International Conference on Permafrost, v. 1, p. 937-943.
Dillon, W. P., J. A. Grow, and C. K. Paull, 1980, Unconventional gas hydrate seals may trap gas off southeast U.S.: Oil and Gas Journal, v. 78, n-1, p. 124-130.
Ewing, J. I., and C. H. Hollister, 1972, Regional aspects of deep sea drilling in the western North Atlantic, *in* C. H. Hollister et al, Initial reports of the deep sea drilling project, v. 11: Washington, D.C., U.S. Government Printing Office, p. 951-973.
Fuex, A. N., 1977, The use of stable carbon isotopes in hydrocarbon exploration: Journal of Geochemical Exploration, v. 7, p. 155-188.
Grantz, A., and S. D. May, 1982, Influence of rifting geometry on the structural development of the continental margin north of Alaska: this volume.
———, G. Boucher, and O. T. Whitney, 1976, Possible solid gas hydrate and natural gas deposits beneath the continental slope of the Beaufort Sea: U.S. Geological Survey Circular 733, p. 17.
Hand, J. H., D. L. Katz, and V. K. Verma, 1974, Review of gas hydrates with implications for ocean sediments, *in* I. R. Kaplan, ed., Natural gases in marine sediments: New York, Plenum, p. 179-194.
Harrison, W. E., and J. A. Curiale, 1982, Gas hydrates in sediments of Holes 497 and 498A, Deep sea drilling project leg 67, *in* J. Aubouin et al, Initial reports of the deep sea drilling project, v. 67: Washington, D.C., U.S. Government Printing Office, p. 591-594.
Hedberg, H. D., 1980, Methane generation and petroleum migration, *in* W. H. Roberts III, and R. J. Cordell, Problems of petroleum migration: AAPG Studies in Geology, n. 10, p. 179-206.
Hitchon, B., 1974, Occurrence of natural gas hydrates in sedimentary basins, *in* I. R. Kaplan, ed., Natural gases in marine sediments: New York, Plenum, p. 195-225.
Hunt, J. M., 1979, Petroleum geochemistry and geology: San Francisco, W. H. Freeman, 617 p.
———, and J. K. Whelan, 1978, Dissolved gases in Black Sea sediments, *in* D. A. Ross, et al, Initial reports of the deep sea drilling project, v. 42, part 2: Washington, D.C., U.S. Government Printing Office, p. 661-665.
Katz, D. L., et al, 1959, Handbook of natural gas engineering: New York, McGraw-Hill, 802 p.
Kvenvolden, K. A., and M. A. McMenamin, 1980, Hydrates of natural gas: a review of their geologic occurrence: U.S. Geological Survey Circular 825, 11 p.
Lancelot, Y., and J. I. Ewing, 1972, Correlation of natural gas zonation and carbonate diagenesis in Tertiary sediments from the north-west Atlantic, *in* C. D. Hollister et al, Initial reports of the deep sea drilling project, v. 11: Washington, D.C., U. S. Government Printing Office, p. 791-799.
Larson, V. F., V. B. Robson, and G. N. Foss, 1980, Deep ocean coring — recent operational experiences of the deep sea drilling project: 55th Annual Fall Technical Conference and Exhibition, Society of Petroleum Engineers of AIME, SPE 9409, p. 1-9.
Lu, R. S., and K. J. McMillian, 1982, Multichannel seismic survey of the Columbian Basin and adjacent continental margins: this volume.
McIver, R. D., 1974, Hydrocarbon gas (methane) in canned deep sea drilling project core samples *in* I. R. Kaplan, ed., Natural gases in marine sediments: New York, Plenum, p. 65-69.
McKirdy, D. M., and P. J. Cook, 1980, Organic geochemistry of Pliocene-Pleistocene calcareous sediments, DSDP site 262, Timor Trough: AAPG Bulletin, v. 64, p. 2118-2138.
Markl, R. G., G. M. Bryan and J. I. Ewing, 1970, Structure of the Blake-Bahama Outer Ridge: Journal of Geophysical Research, v. 75, p. 4539-4555.
Moore, J. C., et al, 1979, Middle American Trench: Geotimes, v. 24, n. 2, p. 20-22.
Rice, D. D., and G. E. Claypool, 1981, Generation, accumulation, and resource potential of biogenic gas: AAPG Bulletin, v. 65, p. 5-25.
Sheridan, R. E., et al, 1982, Early history of the Atlantic Ocean and gas hydrates on the Blake Outer Ridge — results of the deep sea drilling project leg 76, Geological Society of America Bulletin, v. 93, p. 876-885.
Shipley, T. H., and B. M. Didyk, 1982, Occurrence of methane hydrates offshore southern Mexico, *in* J. S. Watkins et al, Initial reports of the deep sea drilling project, v. 66: Washington, D.C., U.S. Government Printing Office, p. 547-555.
———, R. T. Buffler, and J. S. Watkins, 1978, Seismic stratigraphy and geologic history of the Blake Plateau and adjacent western Atlantic continental margin: AAPG Bulletin, v. 62, p. 792-812.
———, et al, 1979, Seismic evidence for widespread possible gas hydrate horizons on continental slopes and rises: AAPG Bulletin, v. 63, p. 2204-2213.
———, et al, 1980, Continental margin and lower slope structures of the Middle America Trench near Acapulco, Mexico: Marine Geology, v. 35, p. 65-82.
Stoll, R. D., J. I. Ewing, and G. M. Bryan, 1971, Anomalous wave velocities in sediments containing gas hydrates: Journal of Geophysical Research, v. 76, p. 2090-2094.
Taylor, A. E., R. J. Wehmiller, and A. S. Judge, 1979, Two risks to drilling and production off the east coast of Canada — earthquakes and gas hydrates, *in* W. Denner, ed., Proceedings, Symposium on research in the Labrador coastal and offshore region: Memorial University of Newfoundland, p. 91-105.
Trofimuk, A. A., N. V. Cherskiy, and V. P. Tsarev, 1974, Accumulation of natural gases in zones of hydrate-formation in the hydrosphere: Doklady-Earth Science Section 212, p. 87-90.
Tucholke, B. E., G. M. Bryan, and J. I. Ewing, 1977, Gas-hydrate horizons detected in seismic-profiler data from the western North Atlantic: AAPG Bulletin, v. 61, p. 698-707.
von Huene, R., et al, 1980, Leg 67: The deep sea drilling project Mid-America Trench transect off Guatemala: Geological Society of America, Bulletin, part 1, v. 91, p. 421-432.
White, R. S., 1979, Gas Hydrate layers trapping free gas in Gulf of Oman: Earth and Planetary Science Letters, v. 42, p. 114-120.
———, and K. E. Louden, 1982, The Makran continental margin: structure of a thickly sedimented convergent plate boundary: this volume.
Yefremova, A. G., and B. P. Zhizhchenko, 1975, Occurrence of crystal hydrates of gases in the sediments of modern marine basins: Doklady-Earth Science Section 214, p. 219-220.

Climatic Indicators in Margin Sediments off Northwest Africa

E. Seibold
M. Sarnthein
Kiel, Federal Republic of Germany

Climatic variations, both in oceans and on continents, can be only partly recorded in marine sediments. Climatic factors more directly affect erosion and sedimentation on land, however local conditions may cause important interference of climatic changes in space and time. Seawater buffers these factors, making marine sediments more suitable to characterize regional, and sometimes global, climatic changes.

The following problems and results about climatic indicators off northwest Africa deal mainly with climatic changes on land. The present climatic zonation is given in Figure 1, and continental margin profiles are given in Figure 2.

Published information about African shelf and slope sequences is scanty because many drilling results are unavailable and sedimentation and/or recovery conditions were not ideal. For example, in Deep Sea Drilling Project Site 369, general sedimentation rates in the 500 m sequence, drilled from Aptian to Recent, are only about 5 m/million years. About 33% of the 100 million years are missing in hiatuses and, only about 33% of the sediments consist of terrigenous material. (Lancelot et al, 1978).

Further details from continental margin sequences off northwest Africa are given in other Initial Reports of the Deep Sea Drilling Project, especially in Leg 14 (Hayes et al, 1972); Leg 47A (von Rad et al, 1979); and Leg 50 (Lancelot et al, 1980). Only Sites 141 (Pliocene), 366, and 397 consist of thick pelagic and hemipelagic carbonate sediments (Figure 3), which display an almost continuous stratigraphic record with high-resolution.

Land climate is characterized today by the high aridity of the western Sahara, centered around 20 to 27°N, with transitions to more humid conditions to the north and south (Figure 1). These conditions are partly reflected by different products of weathering and soil formation and by plant particles, together

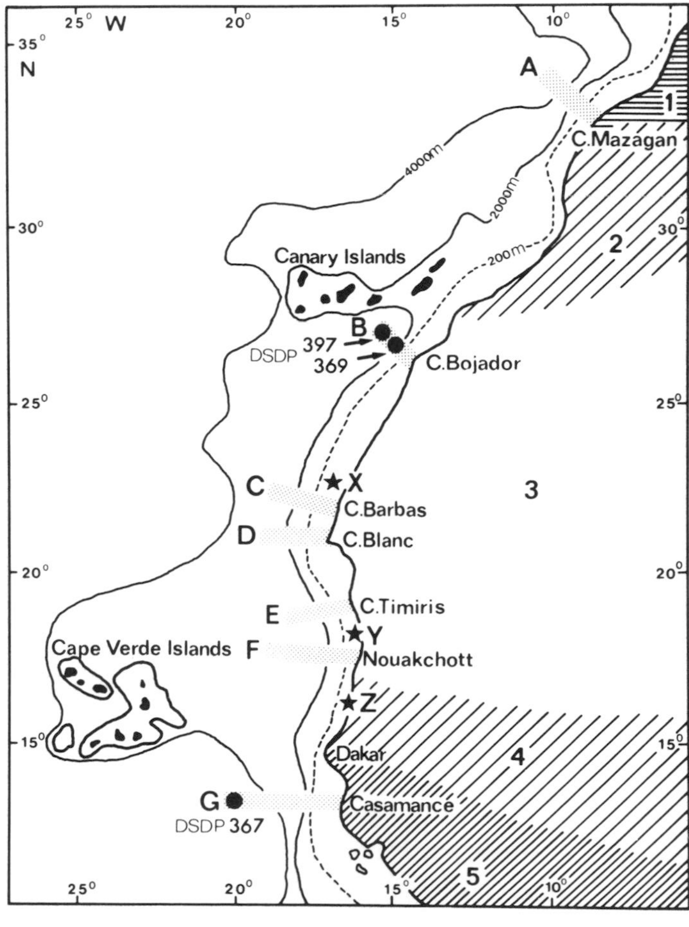

Figure 1 — General overview of northwest Africa and its continental margin with location of profiles A through G (Figure 2) and present position of climatic zones. 1—Mediterranean scrub, warm-temperate, winter rains; 2—Steppe, hot summer dry; 3—Desert, hot, dry; 4—Steppe, hot, winter dry; 5—Savannah, tropical, winter dry. (After Seibold and Fütterer, in press)

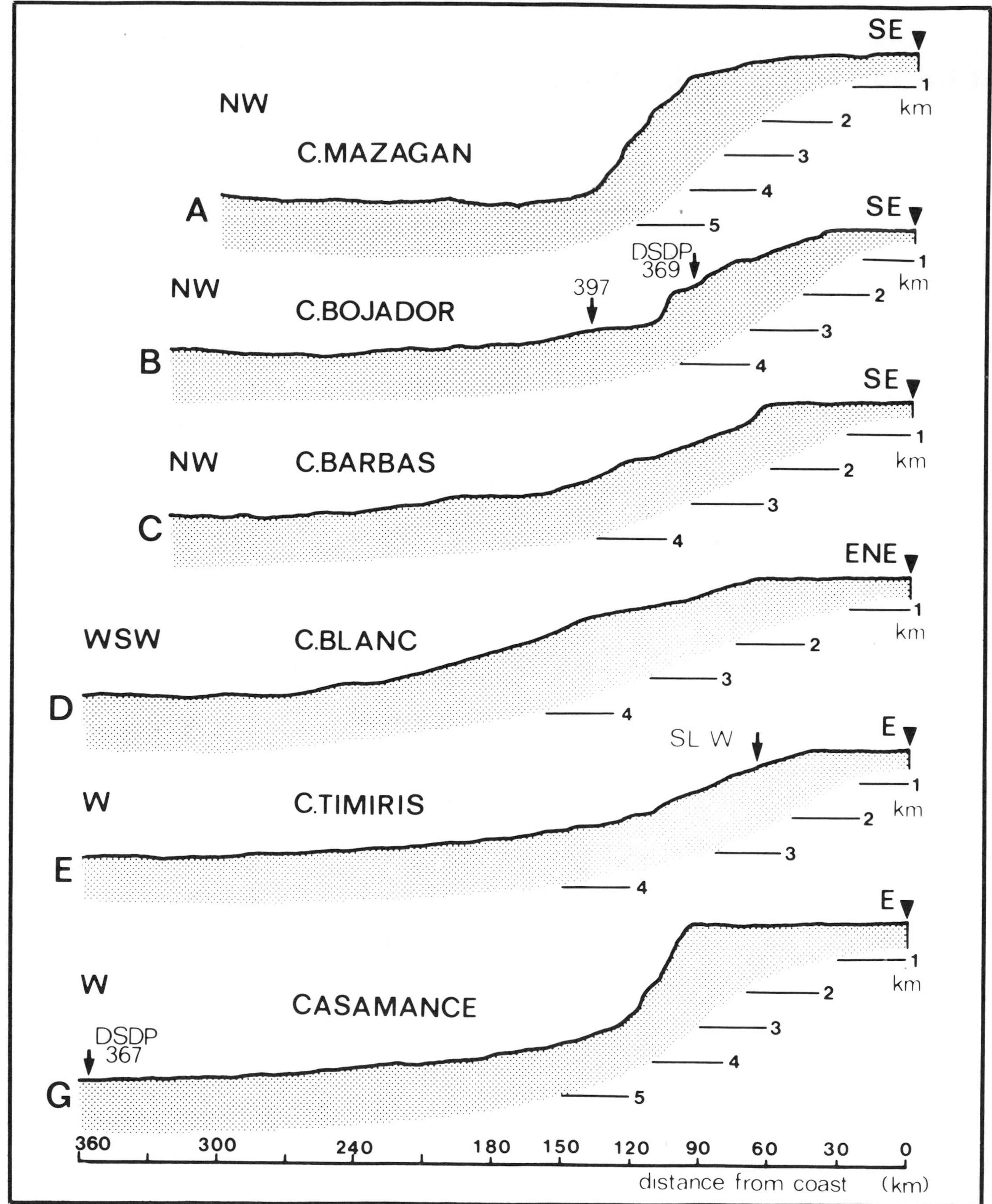

Figure 2 — Cross sections of the continental margin of northwest Africa (for location, see Figure 1). (After Seibold and Fütterer, in press)

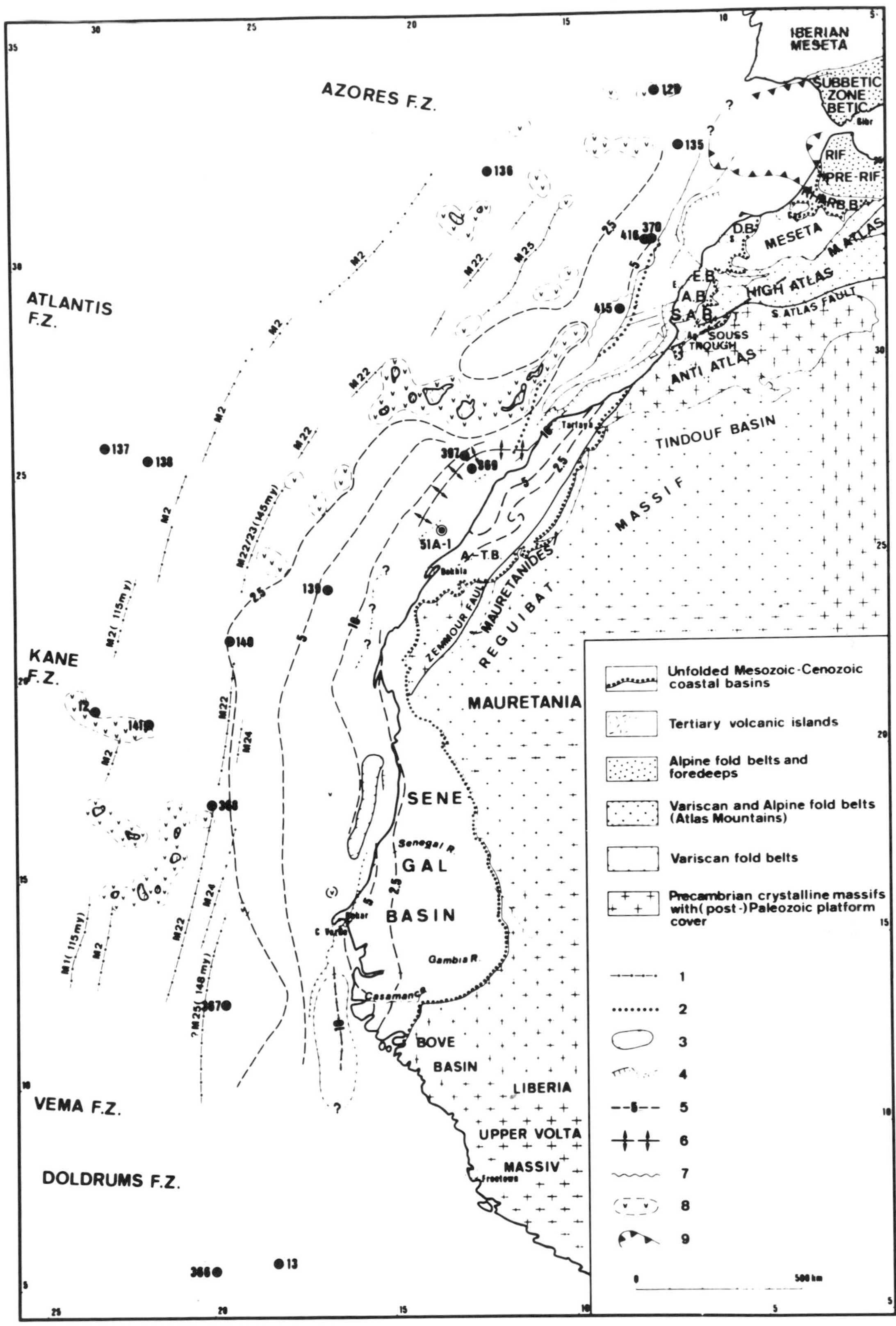

Figure 3 — Structural sketch map. • —Positions of DSDP Sites; 1,2—magnetic anomalies; 3—diapiric structures of probable evaporitic (Late Triassic/Early Jurassic) origin; 4—outer edge of Late Jurassic/Early Cretaceous carbonate platform; 5—approximate 2.5-5-10 km sediment isopachs; 6—continental slope anticline; 7—outer limit of upper Cretaceous gravity slides; 8—volcanics; 9—outer limits of Miocene nappes and olistostromes of the Rif and Betic Mountains (Seibold, 1982; from different sources, especially Bundesanstalt für Geowissenschaften und Rohstoffe, Hannover)

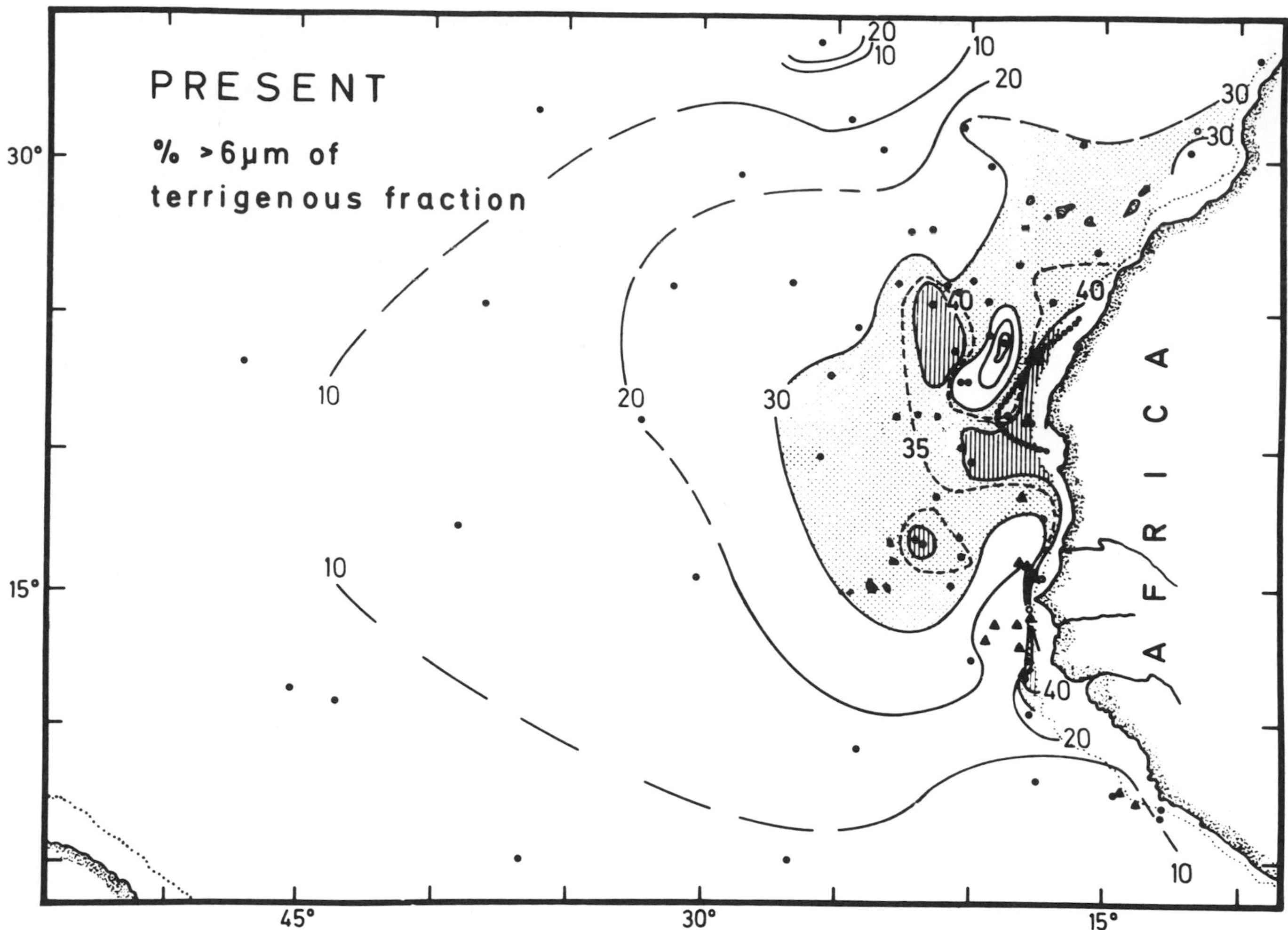

Figure 4 — Distribution map of the terrigenous sediment fractions (on a carbonate and opal free basis) in northeast Atlantic surface sediments showing the dispersal pattern of proximal Harmatton dust which originates from the southern Sahara and Sahel zones. Samples marked by triangles show a relative excess of fine fraction (<6mm) in bimodal grain size distributions and are restricted to areas seaward of river mouths. The area with dominant Trade Wind dust supply at 20 to 25°N is marked by a dotted line. (After Koopmann, 1979)

with different transport mechanisms and rates (as eolian versus river input). For example, indicators are grain size distributions where higher clay contents near the continental margin point to river input and higher silt content points to wind-borne contributions (Figures 4,5).

During the Pleistocene glacial stages, the repeated fall in sea level by approximately 100 m exposed the northwest African shelf, enlarged the Sahara region to the west, and covered it with soils and calcicrusts. In general, northwest Africa was more arid, with some ten maxima during the Brunhes magnetic epoch and a last maximum of aridity around 16,000 to 18,000 years ago (Figure 5). The wind system changed in direction and strength and eolian sediment supply was intensified. Dunes reached the present shelf edge, causing eolian-sand turbidites similar, for example, to those off Senegal. Near this edge, water dynamics and upwelling were intensified.

Neogene northwest African land climates were generally humid-subtropical with prevailing zonal wind circulation. However, since Early Miocene and a desert event in Early Oligocene, arid phases, comparable to glacial stage conditions and of different duration, are intercalated. Details are given in Sarnthein et al (1982). Glacial stage intervals with prevailing meridional winds during 24 to 20 million years, 18 to 14 million years, 13 to 9.5 million years, 5.9 to 5.3 million years, 3.2 to 1.9 million years, and since 0.73 million years to present, mark a stepwise general climatic deterioration. The Neogene northward drift of Africa resulted in a gradual southward shift of the position of wind systems and the correlated climatic belts.

Warm, humid conditions with distinct variations in humidity are indicated for the Paleogene as well. These, together with a special Eocene marine environment, are responsible for a changed silica budget and the formation of sepeolite and palygorskite (Chamley, 1979).

Information from Late Cretaceous is poor due to the widespread transgressions reducing terrigenous input

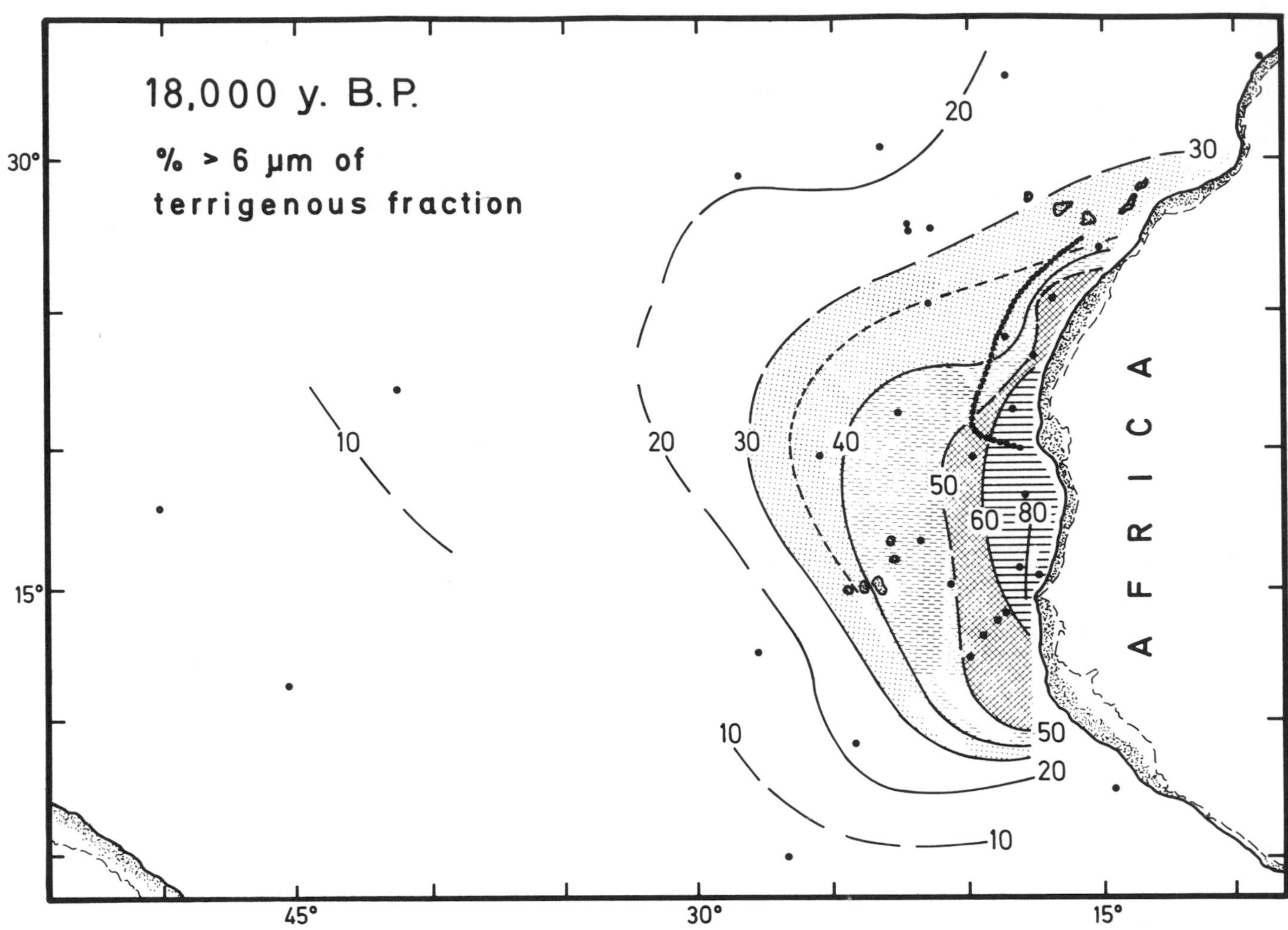

Figure 5 — Same distribution map for a time slice 18,000 years ago. Note the lowered sea level, the lack of triangle marks, and the excess of wind-borne dust. (After Koopman, 1979)

and other factors (Arthur et al, 1979). Black shale characteristics, such as abundant remains of land plants, may indicate tropical conditions with periodic variations in humidity during Middle Cretaceous (Schlanger and Jenkyns, 1976; Lancelot et al, 1978). On the other hand, Tissot, Deroo and Herbin (1979) interpret these deposits as wind-supplied from the African continent in a climate possibly comparable to that of present Central Asia (i.e., semi-desertic). Wealden-type deltaic environments during Early Cretaceous may illustrate tropical conditions (von Rad and Arthur, 1979).

These are very general trends. A detailed study of continuous sequences, accumulation rates and content of terrigenous fractions, stained quartzes, clay minerals, plant fibers and opal and oxygen isotopes resulted in interpretation of various cycles (Dean et al, 1978; Sarnthein et al, 1982). These cycles, with periods from some 10,000 to 2 million years with short pulses and different sedimentation conditions, may be attributed mainly to global oceanic circulation and/or sea level changes, but may also reflect climatic changes on land.

Information is not yet sufficient to detect possible relationships between the west African volcanism and the climate of this region.

REFERENCES CITED

Arthur, M.A., et al, 1979, Evolution and sedimentary history of the Cape Bojador continental margin, northwestern Africa, *in* R. von Rad, et al, eds., Initial reports of the deep sea drilling project, 47, part 1: Washington, D.C., U.S. Government Printing Office, p. 773-816.

Chamley, H., 1979, North Atlantic clay sedimentation and paleoenvironment since the late Jurassic, *in* M. Talwani, W. Hay, and W.B.F. Ryan, eds., Deep drilling results in the Atlantic Ocean: Washington, D.C., American Geophysical Union, p. 342-361.

Dean, W. E., et al, 1978, Cyclic sedimentation along the continental margin of northwest Africa, *in* Y. Lancelot and E. Seibold, eds., Initial Reports of the deep sea drilling project, 41: Washington, D.C., U.S. Government Printing Office, p. 965-989.

Hayes, D.E., et al, 1972, Initial reports of the deep sea drilling project, 14: Washington, D.C., U.S. Government Printing Office, p. 975.

Koopmann, B., 1979, Saharastaub in den Sedimenten des subtropisch-tropischen Nord-Atlantik während der letzten 20,000 Jahre : Ph.D. thesis, Kiel University, p. 1-107.
Lancelot, Y., et al, 1978, Initial reports of the deep sea drilling project, 41: Washington, D.C., U.S. Government Printing Office, pp. 26, 1259.
———, et al, 1980, Initial reports of the deep sea drilling project, 50: Washington, D.C., U.S. Government Printing Office, pp. 28, 868.
Sarnthein, M., et al, 1982, Atmospheric and Oceanic circulation patterns off northwest Africa during the past 25 million years, *in* U. von Rad, ed., Geology of the west African continental margin: Berlin, Springer.
Schlanger, S.O., and H.C. Jenkyns, 1976, Cretaceous oceanic anoxic events, causes and consequences: Geologieen Mijnbouw, v. 55, p. 179-184.
Seibold, E., in press, The northwest African continental margin, an introduction, *in* U. von Rad, ed., Geology of the west African continental margin: Berlin, Springer, p. 3-22.
———, and D. Fütterer, in press, Sediment dynamics on the northwest African continental margin: Wiley and Chichester, Heezen Memorial Volume.
Tissot, B., G. Deroo, and J.P. Herbin, 1979, Organic matter in Cretaceous sediments of the north Atlantic, *in* M. Talwani, et al, eds., Contribution to sedimentology and paleogeography: Maurice Ewing Series 3, American Geophysical Union, p. 362-374.
von Rad, U., and M. Arthur, 1979, Geodynamic, sedimentary and volcanic evolution of the Cape Bojador continental margin (northwest Africa), *in* M. Talwani, W. Hay, and W.B.F. Ryan, eds., Deep drilling results in the Atlantic Ocean: Washington, D.C., American Geophysical Union, p. 187-203.
———, et al, 1979, Initial reports of the deep sea drilling project, 471, part 1: Washington, D.C., U.S. Government Printing Office, pp. 25, 835.

Foraminiferal Zonation of the Cretaceous off Zaire and Cabinda, West Africa and Its Geological Significance

George A. Seiglie
Mary B. Baker
Gulf Oil Exploration and Production Company
Houston Technology Center
Houston, Texas

We describe twelve foraminiferal zones of the Cretaceous continental margin of Zaire and Cabinda (Angola). These zones correspond as follows: one to the late Albian; four to the Cenomanian (one lateral facies); two to the Turonian; one to the Coniacian; one to the Santonian; two to the Campanian; and one to the Maestrichtian. The foraminiferal fauna indicate a transgression from the late Albian to middle Campanian interrupted by three short regressions at the end of the middle Cenomanian, late Turonian, and late Santonian. A regression began from middle Campanian to late Maestrichtian. The paleobathymetric curve of Cabinda (Angola) and Zaire corresponds to sea level changes in the eastern Arabian peninsula and southern England, but differ somewhat from the curve of Vail et al (1977).

Late Cretaceous deposition offshore Zaire and Cabinda, West Africa, occurred in inner shelf to upper slope depths. The biostratigraphy of these sequences is based on twelve foraminiferal zones, most of them using planktonic foraminifers as indexes. Apart from identifying geological ages, these zones are significant because their faunal composition indicates the transgressions and regressions off West Africa, which may be related to global sea level changes. The zones range in age from late Albian, when the first marine strata occur, to late Maestrichtian.

One hundred twenty-seven foraminiferal species were used to determine the biostratigraphic zones. The samples we used consisted of well cuttings, and the study involves all the uncertainties inherent to this type of sampling. However, the reliability of the zones is enhanced by the large number of wells examined (30), most of them offshore. The study area is shown in Figure 1.

ZONATIONS AND ABSOLUTE AGES

The zonation of the Cabinda and Zaire wells was based as much as possible on established worldwide planktonic foraminiferal zones. However, index planktonic foraminifers are generally characteristic of deep waters and shallow waters covered the area for relatively long periods of time. Hence, several zones of Cabinda and Zaire are based on benthonic or benthonic and planktonic foraminifers. The table in Figure 2 compares the Cabinda and Zaire foraminiferal zonation with a worldwide zonation modified from Bolli (1957), Caron (1978), Robaszinski (1979) Sigal (1977), and the authors' experience.

In this report we use the time scale of Obradovich and Cobban (1975) from the Santonian to Maestrichtian, and of Robaszinsky, et al (1979) from the Albian to the Coniacian.

FORAMINIFERAL ZONES

The zones described in this report are the result of studying 30 wells of Zaire and Cabinda. Eleven zones were established in the late Albian to the Maestrichtian sequence. One additional zone results from lateral facies changes of two zones. Six zones are based on planktonic foraminifers, one zone on planktonic and benthonic foraminifers, and four zones on benthonic foraminifers. In three benthonic foraminiferal zones, the sporadic occurrence of planktonic foraminifers was used to determine the age (Figure 2). The foraminiferal zones described in this paper correspond to the

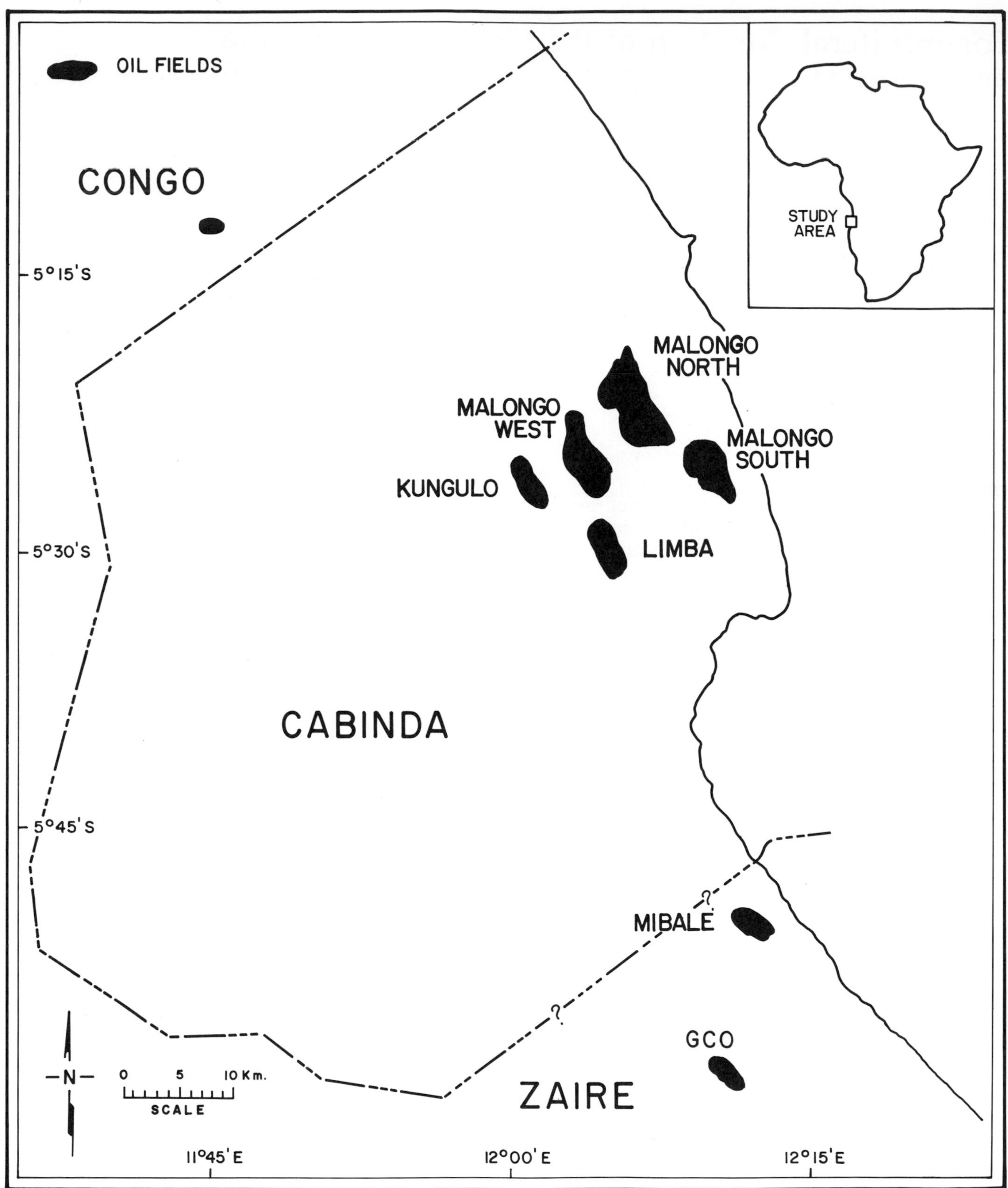

Figure 1 — Map showing the area used for this study.

stratigraphical ranges of foraminifers in West Africa. These ranges are affected in some cases by paleoecological events, such as anoxic conditions, upwelling, water depth, and turbidity.

This zonation is complete in many wells offshore Cabinda and Zaire. The type section for each zone will be published as Gulf Oil Corporation releases the information.

DESCRIPTION OF THE FORAMINIFERAL ZONES

Trocholina silvai Zone

Age: Late Albian

Description: The upper boundary of this zone is determined by the uppermost occurrence of *Trocholina silvai.* No zonal lower boundary was determined and this zone was found only with well cuttings. The paleoenvironment grades down to terrestrial sediments that are barren of foraminifers. This zone comprises the total range of *Trocholina silvai* in Cabinda and Zaire.

Paleoenvironment: Inner neritic, marginal marine. *Trocholina* occur generally in carbonate environments. The time equivalent of this zone in northern Cabinda represents mixed and terrestrial environments.

Remarks: The *Trocholina silvai* Zone corresponds to the upper part of the Pinda Limestone. The Pinda Limestone of Albian age is the only well-known formation of the studied interval. The zonal upper boundary generally coincides with the lower boundary of the *Lenticulina* cf. *ouachensis-Favusella washitensis* Zone.

Lenticulina cf. ouachensis — Favusella washitensis Concurrent Range Zone

Age: Early Cenomanian and possibly earliest middle Cenomanian

Description: The upper boundary of this zone is determined by the uppermost occurrence of *Lenticulina* cf. *ouachensis* and is associated in most wells to *Favusella washitensis. F. washitensis* is not common and occurs close to the upper boundary of the zone. The lower boundary of the zone is determined by the uppermost occurrence of *Trocholina silvai.*

Thomasinella sp. and/or *Cribratina* sp. are more common in the lower part of the zone, and *Neobulimina* sp. D is present in the upper part of this zone. *Textularia praelonga* and *Hedbergella planispira* occur throughout the *Lenticulina* cf. *ouachensis* and *Favusella washitensis* Zone. A few specimens of *Rotalipora appenninica* and *R. brotzeni* occur in some wells in this zone, and they confirm an early Cenomanian age for most of the zone. The upper part of the zone, above *Favusella washitensis,* may be middle Cenomanian.

Paleoenvironment: This zone ranges from a marginal marine paleoenvironment close to the unconformity with the Pinda Limestone, to open oceanic, neritic conditions in its middle and upper parts. The presence of inner neritic foraminifers like *Thomasinella* and *Cribratina* support this interpretation. *Favusella washitensis,* two species of *Rotalipora,* and other planktonic foraminifers indicate an open oceanic probably middle neritic paleoenvironment.

Textularia praelonga Acme Zone

Age: Late? to middle Cenomanian

Description: The upper boundary of this zone is determined by the uppermost occurrence of *Textularia praelonga,* which is abundant throughout the zone. The lower boundary is determined by the uppermost occurrence of *Lenticulina* cf. *ouachensis.* Other foraminifers occurring in the zone are *Hedbergella* cf. *planispira, Globigerinelloides bentonensis, Thomasinella* sp., and *Rotalipora* cf. *greenhornensis.* The latter species suggests a late to middle Cenomanian age.

Paleoenvironment: The presence of a few Rotalipora and other planktonic foraminifers, as well as the benthonic foraminiferal assemblage, indicate a middle to outer neritic assemblage for most of the interval. The presence in several wells of gastropods, bivalves, and ostracodes in the upper part of the zone indicates a marginal marine paleoenvironment.

Textularia praelonga — Cribratina Zone

Age: Early and middle? Cenomanian

Description: The upper boundary of this zone is determined by the uppermost occurrence of *Textularia praelonga;* the lower boundary is determined by the uppermost occurrence of *Trocholina silvai. Textularia praelonga* is associated with *Cribratina* sp. and/or *Thomasinella* sp. Mollusks, and *Ammobaculites* sp., frequently occur in the zone and small juvenile *Hedbergella* sp. are rare. The age was determined by the local chronostratigraphic range of *Textularia praelonga* as found in other wells for the *T. praelonga* Acme-Zone, and the *Lenticulina* cf. *ouachensis-Favusella washitensis* Zone.

Paleoenvironment: The assemblage of this zone indicates a shallow-water paleoenvironment. Foraminifers are rare in most of the samples, indicating a high rate of sedimentation. The time equivalent of this zone grades to mixed and terrestrial paleoenvironments in northern Cabinda.

Whiteinella brittonensis Zone

Age: Middle? and late Cenomanian and possibly earliest Turonian

Description: The upper boundary of this zone is determined by the uppermost occurrence of *Whiteinella brittonensis;* the lower boundary is determined by the uppermost occurrence of *Textularia praelonga.* A few specimens of *Rotalipora* cf. *greenhornensis* occur in here, indicating a late Cenomanian and less probably latest middle Cenomanian age. Benthonic foraminifers are rare. The most common species are *Whiteinella brittonensis, W. baltica, Hedbergella* ex gr. delrioensis and other unkeeled planktonic foraminifers.

Paleoenvironment: A middle to outer neritic environment and the possibility of upwelling waters is

CABINDA AND ZAIRE, WEST AFRICA

AGE AND MILLION YEARS (1)	PLANKTONIC FORAMINIFERAL ZONES (2)	FORAMINIFERAL ZONES CABINDA AND ZAIRE
64		
MAESTRICHTIAN	Abathomphalus mayaroesis Zone	Afrobolivina afra-Orthokarstenia clavata Zone
	Globotruncanita gansseri Zone	
	Globotruncanita stuartiformis–Globotruncana scutilla Zone	
69		?
CAMPANIAN	Globotruncanita calcarata Zone	Gabonella spp. Acme - Zone (3)
	Globotruncanita stuartiformis–Globotruncana fornicata Zone	
	Globotruncanita elevata Zone	Globotruncana fornicata Zone
82		
SANTONIAN	Dicarinella asymetrica Zone	Dicarinella asymetrica Zone
87		
CONIACIAN	Dicarinella concavata Zone	Dicarinella concavata Zone
88		
TURONIAN	Praeglobotruncana inornata Zone	Praeglobotruncana inornata — Dicarinella imbricata Zone
	Praeglobotruncana helvetica Zone	Pr. helvetica Zone
90		
CENOMANIAN	Whiteinella archaeocretacea Zone	Whiteinella brittonensis Zone
	Rotalipora cushmani Zone	Textularia praelonga Acme – Zone; Textularia praelonga-Cribratina Zone
	Rotalipora reicheli Zone	
	Rotalipora brotzeni Zone	L. cf. ouachensis-Favusella washitensis Zone
94		
ALBIAN	Rotalipora appenninica Zone	Trocholina silvai Zone
	Planomalina buxtorfi Zone	?
	Rotalipora ticinensis-R. subticinensis	
105		

(1) Absolute ages according to Robaszinski (1979) and Obradovich and Cobban (1975)

(2) World-wide zonation with planktonic foraminifers modified from Van Hinte (1976), Sigal (1977), Caron (1978), Bolli (1957) and Robaszinski (1979)

(3) Gabonella spp. acme = G. spinosa, and/or G. lata, and/or G. elongata and/or G. gigantea acme

Figure 2 — Table showing comparison of the Cabinda and Zaire foraminiferal zonation with a worldwide zonation.

suggested by: (1) the rarity of benthonic foraminifers, (2) the abundance of unkeeled planktonic foraminifers, and (3) the occurrence of Rotalipora. In northern Cabinda the time equivalent of this zone grades to mixed and terrestrial paleoenvironments.

Praeglobotruncana helvetica Partial-Range Zone

Age: Early Turonian

Description: The upper boundary of this zone is determined by the uppermost occurrence of *Praeglobotruncana helvetica*. The lower boundary is determined by the uppermost abundance of *Whiteinella brittonensis*. *Praeglobotruncana helvetica* in Cabinda and Zaire only occurs in the early Turonian. *P. helvetica* Zone is a total range zone in the world wide zonation covering the early and middle Turonian. Dalbiez (1955) used *Globotruncana helvetica* (=*Praeglobotruncana helvetica*) Zone approximately covering the middle Turonian. The zone included all of the Turonian according to Postuma (1971), but only the early and most of the middle Turonian according to Robaszinski et al, (1979). The uppermost occurrence of *P. helvetica* and the associate fauna corresponds to a decrease in the rate of sedimentation. Under normal conditions, the uppermost boundary should be in the middle Turonian. *P. helvetica* is associated in this zone with *P. praehelvetica*, *P. aumalensis*, *Dicarinella imbricata*, *Hedbergella* spp., and others. The age was determined by the presence of *Praeglobotruncana helvetica*, *P. praehelvetica*, *P. aumalensis* and *Dicarinella imbricata*.

Paleoenvironment: The abundance of keeled planktonic foraminifers and scarcity of benthonic foraminifers suggests an outer shelf to uppermost bathyal paleoenvironment. The presence of cavings does not allow a precise determination of the fauna and of the water paleodepths. This zone grades to terrestrial and mixed paleoenvironments in northern Cabinda.

Praeglobotruncana inornata — Discarinella imbricata Concurrent Range Zone

Age: Late? and middle Turonian

Description: The upper boundary of this zone is determined by the uppermost occurrence of *Praeglobotruncana inornata* and/or *Dicarinella imbricata*. This zone covers only part of the *P. inornata* and *D. imbricata* total stratigraphic range. The *Globotruncana inornata* Zone of Bolli (1957) included all of the Turonian. The associate foraminiferal species are: *Praeglobotruncana aumalensis*, *P. gibba Marginotruncana renzi*, *M. schneegansi*, *M. sigali*, *Dicarinella algeriana*, and others. The presence of *Praeglobotruncana aumalensis* close to the upper boundary of the zone in a few wells suggests that the late Turonian is represented by a thin sequence or is missing. Populations of *P. inornata* have a wide range of forms; some of them are difficult to differenciate from *P. aumalensis*. The fauna of this zone is better represented in the uppermost part of the zone, most likely due to a decrease in the rate of sedimentation. The planktonic foraminifers of this zone range in age from early to late Turonian.

Paleoenvironment: The abundance and high specific diversity of keeled planktonic foraminifers suggests an upper bathyal paleoenvironment. The rarity of benthonic foraminifers is probably caused by a high rate of sedimentation and/or low oxygen which creates difficult conditions for life on the sea bottom. The time equivalent of this zone grades to mixed and terrestrial paleoenvironments in northern Cabinda.

Dicarinella concavata Partial Range Zone

Age: Coniacian

Description: The upper boundary of this zone is determined by the uppermost occurrence of *Dicarinella concavata*, and the lower boundary is determined by the uppermost occurrence of *Praeglobotruncana inornata* and/or *Dicarinella imbricata*. This zone was originally described by Bolli (1957) covering the lower Santonian.

The occurrence of the three species in one sample interval is possible because the *D. concavata* Zone represents one million year time interval (Robaszinski et al, 1979). *Marginotruncana sigali*, *M. schneegansi*, *M. paraconcavata*, *M. angustricarinata*, and others occur in this zone.

Paleoenvironment: The keeled planktonic foraminifers suggest a bathyal paleoenvironment. The time equivalent of this zone grades to mixed and terrestrial paleoenvironments in northern Cabinda.

Dicarinella asymetrica Partial Range Zone

Age: Santonian

Description: The upper boundary of this zone is determined by the uppermost occurrence of *Dicarinella asymetrica*, and the lower boundary is indicated by the uppermost occurrence of *D. concavata*. *Globotruncana fornicata*, *G. bulloides*, *G. manaurensis*, *Marginotruncana angusticarinata*, *Hedbergella* ex gr. *delrioensis*, *Heterohelix globulosa* are common. This zone was described by Postuma (1971) covering the upper Santonian.

The time equivalent of this zone is indicated in northern Cabinda by the ostracod *Cythereis* sp. A, frequently associated with *C. reticulata*.

Paleoenvironment: The specific diversity of the keeled planktonic foraminifers suggests an upper bathyal paleoenvironment in the wells of Zaire and southern Cabinda. In central Cabinda the paleoenvironment is outer to middle shelf. The relative abundance of ostracodes in the upper part of the zone in the wells of Zaire and southern Cabinda indicates shallowing at the end of the Santonian. In northern Cabinda, the absence or presence of a few planktonic foraminifers and the abundance of ostracodes indicate middle to inner neritic paleoenvironments throughout the zone.

Globotruncana fornicata Partial Range Zone

Age: Campanian (probably early to middle)

Description: The upper boundary of this zone corresponds to the uppermost abundance of

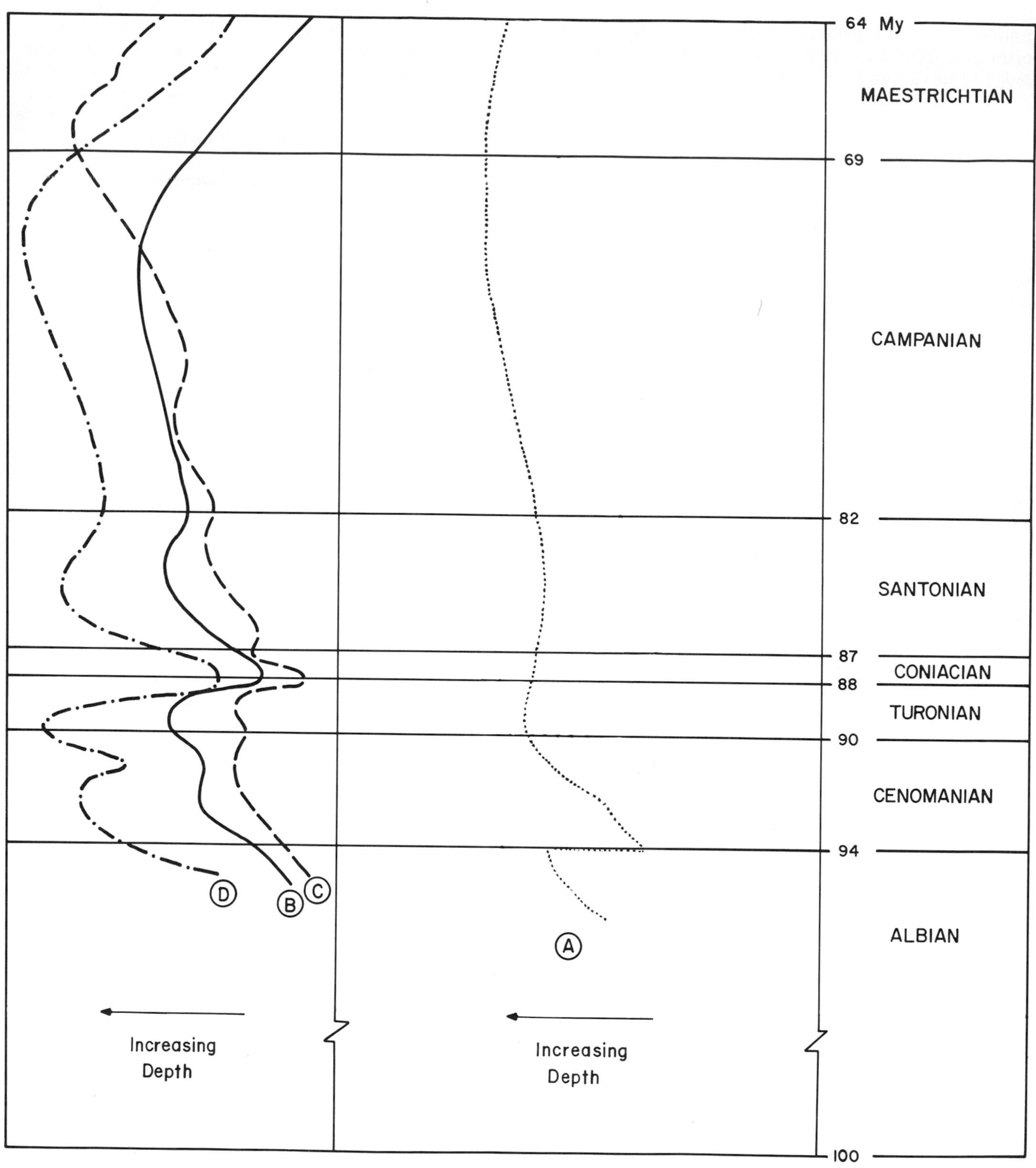

Figure 3 — Representation of the sea level changes in West Africa compared with sea level changes in other parts of the world.

Globotruncana fornicata, and the lower boundary is indicated by the uppermost occurrence of *Dicarinella asymetrica*. *Globotruncana bulloides, G. arca, G. lapparenti, Globotruncanita elevata, G. stuartiformis, Heterohelix pulchra, H. globulosa, Pseudotextularia elegans*, several species of *Gabonella* and others occur here. The *Globotruncana fornicata* — *G. stuartiformis* Zone of Dalbiez (1955) has a range approximately similar to our *G. fornicata* Zone and *Gabonella* spp. Acme Zone. This zone was described by Bolli (1957) for the upper Santonian of Trinidad as a partial range zone. We also describe it as a partial range zone, but in addition, we describe it as representing an acme zone for *G. fornicata* during the early and late Campanian of West Africa.

Paleoenvironment: This zone ranges from bathyal in the south to middle or inner neritic in northern Cabinda. The planktonic foraminifers have a high specific diversity, including the deep water species *Globotruncanita elevata* and *G. stuartiformis*, in the upper bathyal paleoenvironment. The upper part of the zone is transitional to outer or middle neritic paleoenvironments in Zaire and south and central Cabinda.

Gabonella spp. Acme — Zone

Age: Campanian and early Maestrichtian (?)

Description: The upper boundary of this zone corresponds to the uppermost abundance of *Gabonella* spp., and the lower boundary corresponds to the uppermost abundance of *Globotruncana fornicata*. A peak of abundance occurs in this zone of one or more of the following species of *Gabonella: G. spinosa, G. lata, G. elongata* and *G. "gigantea."* The uppermost occurrence of *Globotruncana fornicata* is above this zone in several wells. *Globotruncanita stuartiformis, Globotruncana bulloides, Rugoglobigerina rugosa, Heterohelix pulchra, H. globulosa*, and *Pseudotextularia elegans* occur here. This zone is Campanian but the upper part may be early Maestrichtian.

Paleoenvironment: Mostly middle neritic. The paleoenvironment may grade from outer neritic in the lower part of the zone to inner neritic in the uppermost part.

Orthokarstenia clavata — *Afrobolivina afra* Concurrent Range Zone

Age: Maestrichtian

Description: The upper boundary of this zone corresponds to the uppermost occurrence of *Afrobolivina afra* or *Orthokarstenia clavata*, and the lower boundary corresponds to the uppermost abundance of *Gabonella* spp. Occurring in this zone are *Rugoglobigerina rugosa, R. rotundata, R. macrocephala, Trinitella scotti, Pseudotextularia elegans, Heterohelix globulosa, Globotruncana aegyptiaca, G. fornicata, Platystaphyla brazoensis*, and *Racemiguembelina fructicosa*. The age of this zone ranges from early to late Maestrichtian. The presence of *Trinitella scotti* in a few wells indicates that the upper boundary of the zone is close to the top of the Maestrichtian (Van Hinte, 1976).

Paleoenvironment: The paleoenvironment in the top of this zone is indicated by only one or two foraminifers (*Afrobolivina afra* and/or *Orthokarstenia* sp. and/or *O. clavata*), indicating marine marginal depths. In some wells, a bed of ooids occurs above the beds with foraminifers. Other wells have planktonic foraminifers, mostly unkeeled, throughout the zone. The paleoenvironment ranges from middle to inner neritic.

TRANSGRESSIONS AND REGRESSIONS

The late Albian sea is characterized by a foraminiferal fauna consisting of *Trocholina silvai*, ostracodes, and small planktonic foraminifers. The sea bottom partly emerged during the transition from the Albian to the Cenomanian. A transgression occurred during the early and most of the middle Cenomanian, and was followed by a short regression at the end of middle Cenomanian. The fauna of the early and middle Cenomanian is dominated by middle and inner neritic agglutinated foraminifers, while index planktonic foraminifers occur only in the deepest part of the transgression.

The regression at the end of the middle Cenomanian is indicated by the abundance of mollusks and ostracodes. In the transgressive late Cenomanian, a strong faunal change occurred; benthic foraminifers are rare and the assemblage is dominated by nonkeeled planktonic foraminifers. This suggests anoxic conditions and abundance of nutrients at the end of the Cenomanian. This anoxic event is synchronious with other anoxic events in the world (Hart, 1980).

Relative abundance of keeled planktonic foraminifers during the early and middle Turonian indicates deeper waters, probably uppermost bathyal or outer neritic. The late Turonian is probably partly truncated by submarine and/or subaerial erosion. Also, the rate of sedimentation is extremely high during the Turonian, suggesting submarine slumping. These processes may have mixed the late and middle Turonian foraminiferal faunas, making the recognition of a late Turonian sequence difficult. The absence of a definite late Turonian foraminiferal fauna suggests the largest regression since the end of the Albian.

A new transgression occurred during the Coniacian and Santonian covering most of the offshore Cabinda. The abundance of ostracodes indicates a regression at the close of the Santonian. Another transgression in the Campanian followed, covering all the area and reaching its maximum in the middle Campanian. Part of the area was covered with the deepest water (upper bathyal) at that time. The deep water fauna is characterized by several species of single and double keeled globotruncanids. A regression took place during the late Campanian and it lasted to the end of the Maestrichtian; the onset of this is indicated by the dominance of one or two species of *Gabonella*. At the end of the Maestrichtian most of the fauna is represented by only two species, *Orthokarstenia clavata* and *Afrobolivina afra*.

WEST AFRICA AND GLOBAL CHANGES OF SEA LEVEL

The Global changes of sea level were described by Vail et al (1977). Their graphic representation of the sea level changes from the late Albian to Maestrichitian is shown, modified, in Figure 3. The absolute time scale in the figure is a combination of the scales by Obradovich and Cobban (1975), and Robaszinski et al (1979), as used above in this paper.

A curve for Cabinda-Zaire sea level changes was made on the basis of the transgressions and regressions described above (Figure 3, curve B). Curves were also included for the sea level changes in the eastern Arabian Peninsula (Harris et al, 1980) and in southern England (Hart, 1980) (Figure 3, curves C and D). Curves B, C, and D in Figure 3 are remarkably similar, while curve A of Vail et al (1977) differs in the timing of some major transgressions and regressions.

ACKNOWLEDGMENTS

Thanks to P. M. Harris, S. H. Frost, C. Kendall, B. L. Shaffer, and N. Schneidermann for the use of the curve for the Middle East and to Gulf Oil Corporation for permission to publish this paper.

REFERENCES CITED

Bolli, H. M., 1957, The genera Praeglobotruncana, Rotalipora, Globotruncana and Abathomphalus in the Upper Cretaceous of Trinidad, B. W. I., *in* A. R. Loeblich, and H. Tappan, Studies in foraminifera: U.S. National Museum, Bulletin 215, p. 51-60.

Caron, M., 1978, Cretaceous planktonic foraminifers from DSDP leg 40, southeastern Atlantic Ocean, *in* H. M. Bolli, et al, DSDP initial reports: Washington, D.C., U.S. Government Printing Office, v. 40, p. 651-678.

Dalbiez, F., 1955, The genus Globotruncana in Tunisia: Micropaleontology, v. 1, no. 2, p. 161-171.

Harris, P. M., et al, 1980, Cretaceous sea level and stratigraphy, eastern Arabian Peninsula (Abs.): AAPG Bulletin, v. 64, n. 5, p. 719.

Hart, M. B., 1980, A water depth model for the evolution of planktonic Foraminiferida: Nature, v. 286, p. 252-254.

Obradovich, J. D., and W. A. Cobban, 1975, A time-scale for the Late Cretaceous of the western interior of North America, *in* W. G. E. Caldwell, ed., Cretaceous system in western interior of North America: Geological Association of Canada, Special Paper 13, p. 31-34.

Postuma, J.A., 1971, Manual of Planktonic Foraminifera: Amsterdam, Elsevier Publishing Company, 419 p.

Robaszinski, F., et al, 1979, Atlas de foraminiferes planktoniques du Cretace Moyen (mer Boreal et Tethys): Cahiers Micropaleontologie, v. 1, 185 p., v. 2, 181 p.

Sigal, J., 1977, Essai de zonation due Cretace mediterraneen a Laide des foraminiferes planktoniques: Geologie Mediterraneene, t. 4, n. 2, p. 99-108.

Vail, P. R., et al, 1977, Seismic stratigraphy and global changes of sea level, *in* C. E. Payton, ed., Seismic Stratigraphy-applications to hydrocarbon exploration: AAPG Memoir 26, p. 49-212.

Van Hinte, J. E., 1976, A Cretaceous Time Scale: AAPG Bulletin, v. 60, n. 4, p. 269-287.

Nature and Origin of Deep-Sea Carbonate Nodules Collected from the Japan Trench

H. Wada
H. Okada
Institute of Geosciences, Shizuoka University
Shizuoka, Japan

IPOD/DSDP Leg 50 collected 18 carbonate nodules drilling the inner trench slope of the Japan Trench. They found the nodules in the sedimentary sequence deeper than 180 m where abundant methane occurs. The carbonate minerals in the nodules were comprised of magnesian calcite, dolomite, manganese calcite, and rhodochrosite. Petrographic evidence and oxygen and carbon isotopic compositions show that these carbonate nodules were authigenically formed at or near the position where they were found. Carbon isotopic compositions of many of these nodules reveal that the carbon in carbonate nodules may have been derived from carbon dioxide evolved during the sulfate reduction by microbiological activities.

Carbonate nodules in Deep Sea Drilling Project cores are not unusual in sediments deposited well below the carbonate compensation depth (CCD) recovered from inner trench slope environments of accretionary prisms (for the Japan Trench, see Okada, 1980; for the Middle America Trench, see Wada et al, 1981). These carbonate nodules resemble those frequently found in on-land eugeosynclinal sediments interpreted as accreted rocks from ancient subduction zones (Shimanto Supergroup in the Japanese Islands, Taira et al, 1982). Therefore, understanding the petrographic and chemical aspects of carbonate nodules from modern trench environments aids in understanding depositional environments of both modern trench sediments and ancient orogenic sediments.

This paper reviews petrographic, chemical, and stable isotopic characteristics of the carbonate nodules recovered in cores obtained at Site 434 of IPOD/DSDP Leg 56 in the Japan Trench transect (Figure 1). We also discuss the origin of deep sea carbonate nodules.

METHODS

Carbonate nodules were examined by microscopic observation, X-ray diffraction analysis, and carbon and oxygen isotopic analysis. Isotopic analysis of organic carbon was also carried out on the soft muds enclosing the carbonate nodules. Selected samples were examined with a scanning electron microscope (SEM).

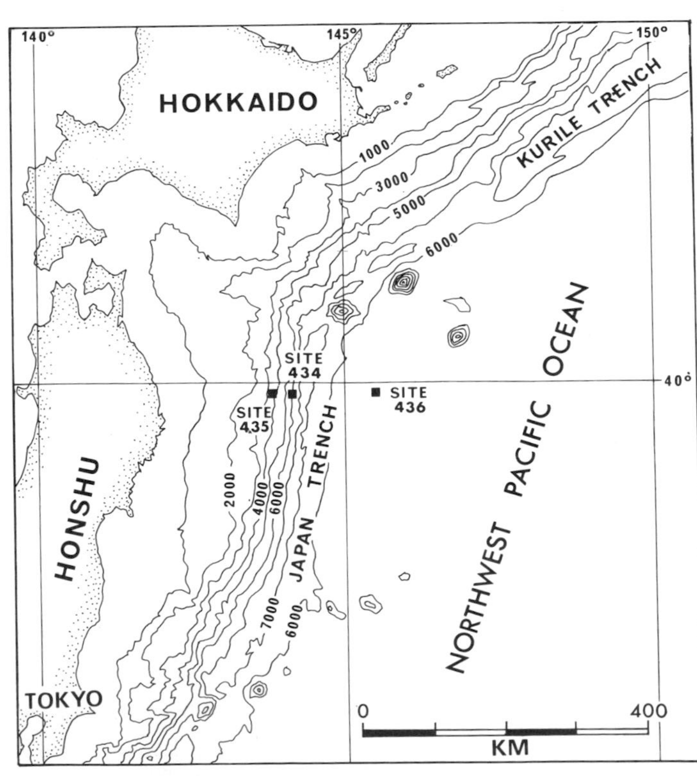

Figure 1 — Index map of the Japan Trench area showing locations of IPOD/DSDP Leg 56 drill sites. Bathymetry in meters.

The nodules were thinly sliced and rinsed with distilled water to remove NaC1. Then they were made into thin sections for microscopic observations, or dried and ground to fine powder for X-ray diffraction analysis and carbon and oxygen isotopic analyses. X-ray diffraction analysis determined mineral composition of the nodules. Chips, a few millimeters across, with fresh fracture surfaces were used for the SEM photography.

For isotopic analysis, carbonates were decomposed with 100% phosphoric acid (McCrea, 1950). To collect carbon dioxide gas, they decomposed the calcite with acid for one hour; they decomposed dolomite and rhodochrosite for 4 to 72 hours in vacuo at 25°C. Carbon dioxide was examined with a mass-spectrometer (Nier-McKinney-type Varian Mat CH-7).

Carbon and oxygen isotopic compositions are described in the conventional δ-notation in relation to the PDB standard (Craig, 1957), as shown in the following: δ (‰.) = $\{ (R_{sample} / R_{PDB}) - 1\} \times 1{,}000$. R is $^{13}C/^{12}C$ or $^{18}O/^{16}O$ in sample and PDB standard. We used a working standard NU-1 CO_2 gas, evolved from the reaction of reagent marble and phosphoric acid frequently calibrated with respect to the NBS-20 standard (Craig, 1957). According to the method of Tarutani, Clayton, and Mayeda (1957), $\delta^{18}O$ values of magnesian calcite were further subtracted 0.06‰ times of the mole % of $MgCO_3$ in calcite from the $\delta^{18}O$ values obtained by Craig's (1957) correction. Analytical errors in the carbon and oxygen isotopic measurements are less than ±0.2 ‰.

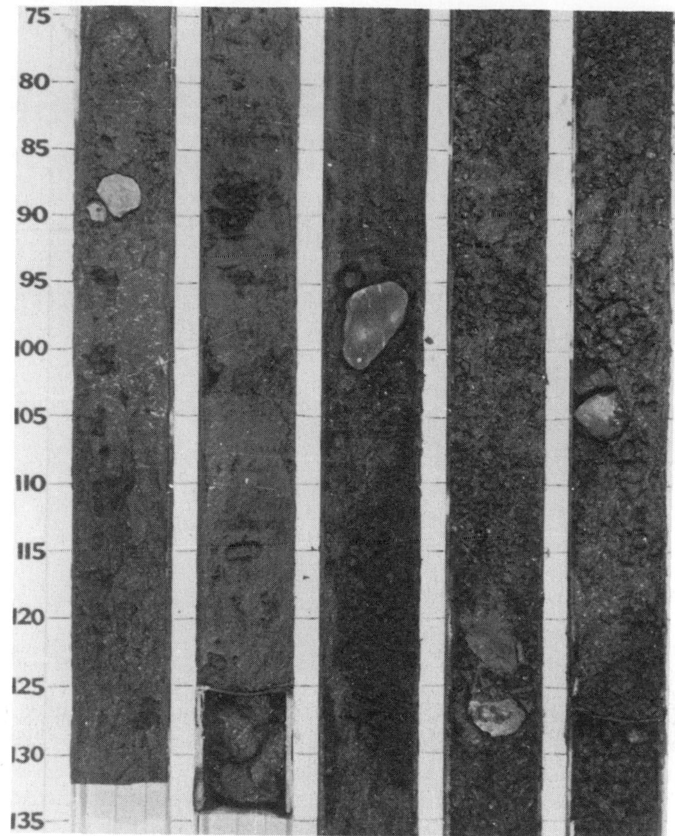

Figure 2 — Occurrence of isolated pieces of carbonate nodules in hole 434B(cf., Scientific Party, 1980). Scale in centimeters.

Table 1. Carbonate nodule lithologies.*

Sample (interval in cm)		Lithology
434-19-1,10-12	MC	Diatom-rich biomicrite with rare pumice and plagioclase
19,CC(1)	MC	Glass-shard-bearing, diatom-rich biomicrite with rare biotite, chlorite and plagioclase
19,CC(2)	RC	Diatom-pumice-chlorite-bearing, glass-shard-rich limestone with rare augite and plagioclase
23-1,140-143	MC	Diatom-rich biomicrite with rare plagioclase and glass-shards
23-3,49	MC	Plagioclase-bearing, diatom biomicrite with rare chlorite
25-2, 86-90	MC	Spicule-plagioclase-bearing, diatom-rich biomicrite with rare glass-shards
27-1, 65	MD	Glass-shard and diatom-rich micrite with rare plagioclase
28, CC	SC	Intrasparite with diatomacous micrite clasts
28-1, 104-106	MC	Diatom-augite-chlorite-bearing micrite with common glass-shards, pumice and plagioclase
29, CC, 12-14	MC	Glass-shard-rich micrite with rare biotite, chlorite, pumice, augite and hornblende and common plagioclase and diatom
434B-3-1, 30-32	MC	Biotite-chlorite-pumice-augite-bearing glass-shard-rich micrite with common diatom and plagioclase
11,CC, 2-6	MC	Diatom-bearing, glass-shard- and plagioclase-rich micrite with common biotite
19-1, 5-7	MC	Diatom- and augite-bearing micrite with rare plagioclase and common glass-shards
19-2, 33-35	MC	Biotite-chlorite-bearing sandy limestone with rare plagioclase and common glass-shards and diatom
30-1, 5-10	SC	Augite-bearing, glass-shard-rich microsparite with common plagioclase (manganoan calcite)
32,CC, 10-12	R,Rh	Diatom- and spicule-rich rhodochrosite

*M - micrite, R - recrystallized, S - sparite, C - calcite, D - dolomite, Rh - rhodochrosite.

SAMPLE DESCRIPTION

Occurrence

All samples of carbonate nodules (Table 1) are collected from soft mud or mudstone recovered at Site 434. Nodules are mixed with exotic pebbles throughout the section in sediments ranging from Late Pleistocene to Late Miocene in age. Carbonate nodules occur mostly as isolated individuals (Figure 2) or occasionally in groups (Figure 3). The groups may have been artificially concentrated by the drilling. A few of the carbonate nodules are recemented breccias (Figure 4). The original occurrence of carbonate nodules is not always clear but some pieces were definitely parts of larger rock bodies.

The upper 180 m at Site 434 are characterized by an intense odor of H_2S gas, whereas CH_4 gas is present below 180 m (Scientific Party, 1980) where carbonate nodules seem to be concentrated (Okada, 1980).

Figure 3 — Occurrence of pieces of carbonate nodules concentrated from 80 to 95 cm of section 434-25-2(cf., Scientific Party, 1980) Scale in centimeters.

Figure 4 — Brecciated carbonate nodule. Sample 434-28-CC. Scale in centimeters.

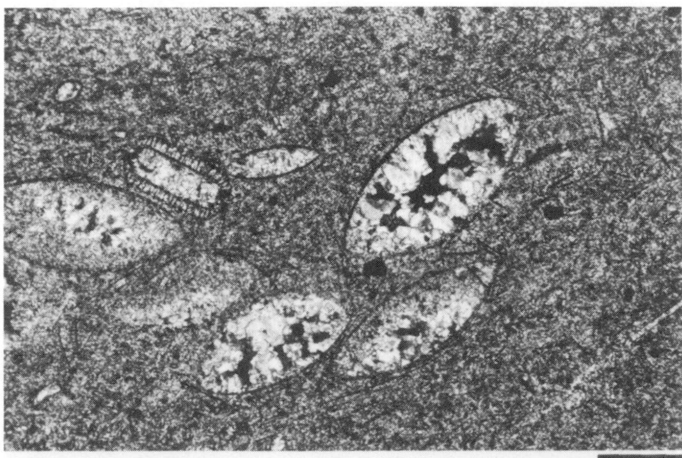

Figure 5 — Photomicrograph of diatom-rich biomicrite. Sample 434-19-1,10-12 cm. Crossed nicols. Scale bar 0.1 mm.

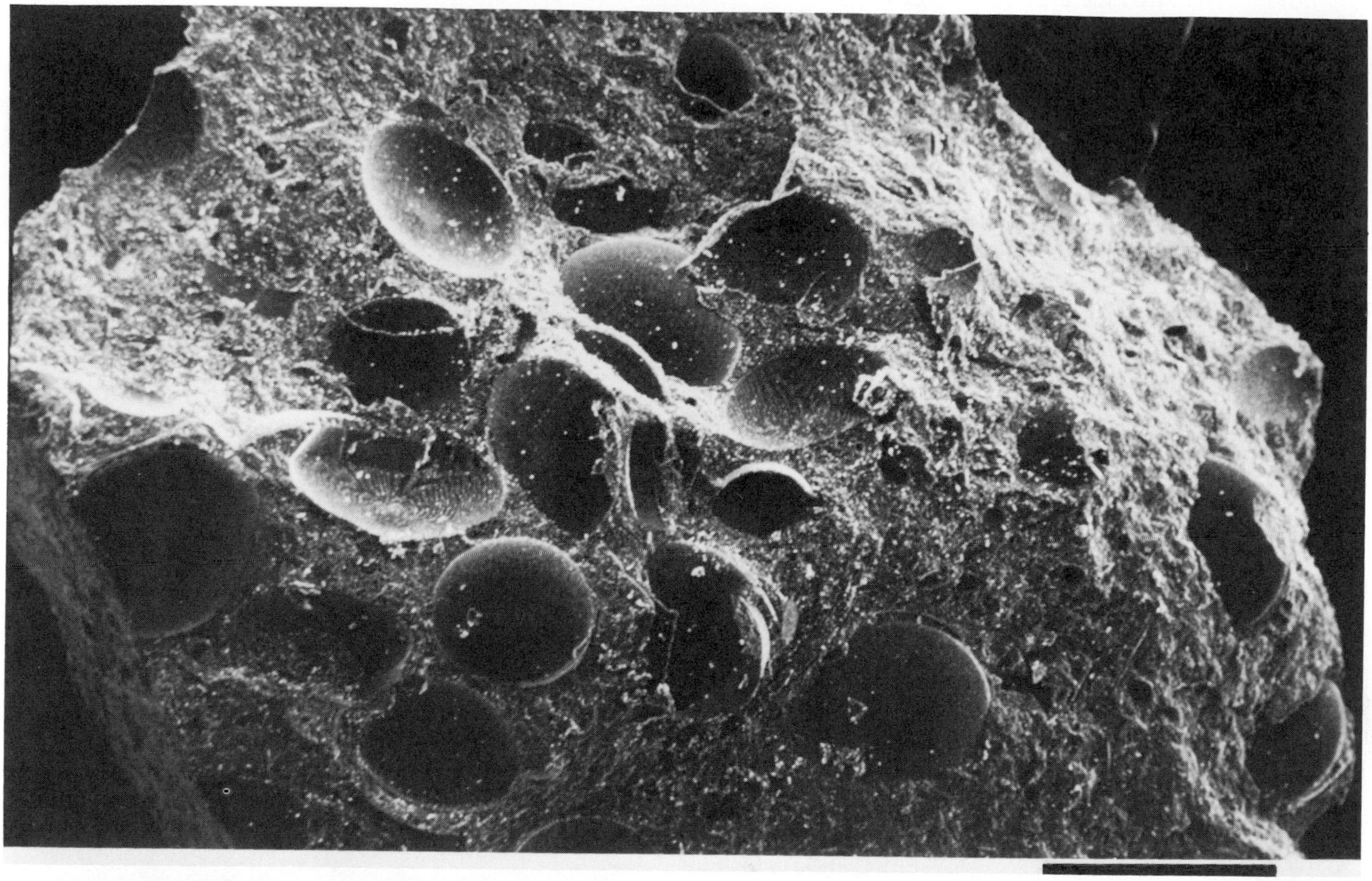

Figure 6 — SEM photograph of diatom-rich biomicrite. Sample 434-19-1,10-12 cm. Scale bar 100 μm for A and 10 μm for B.

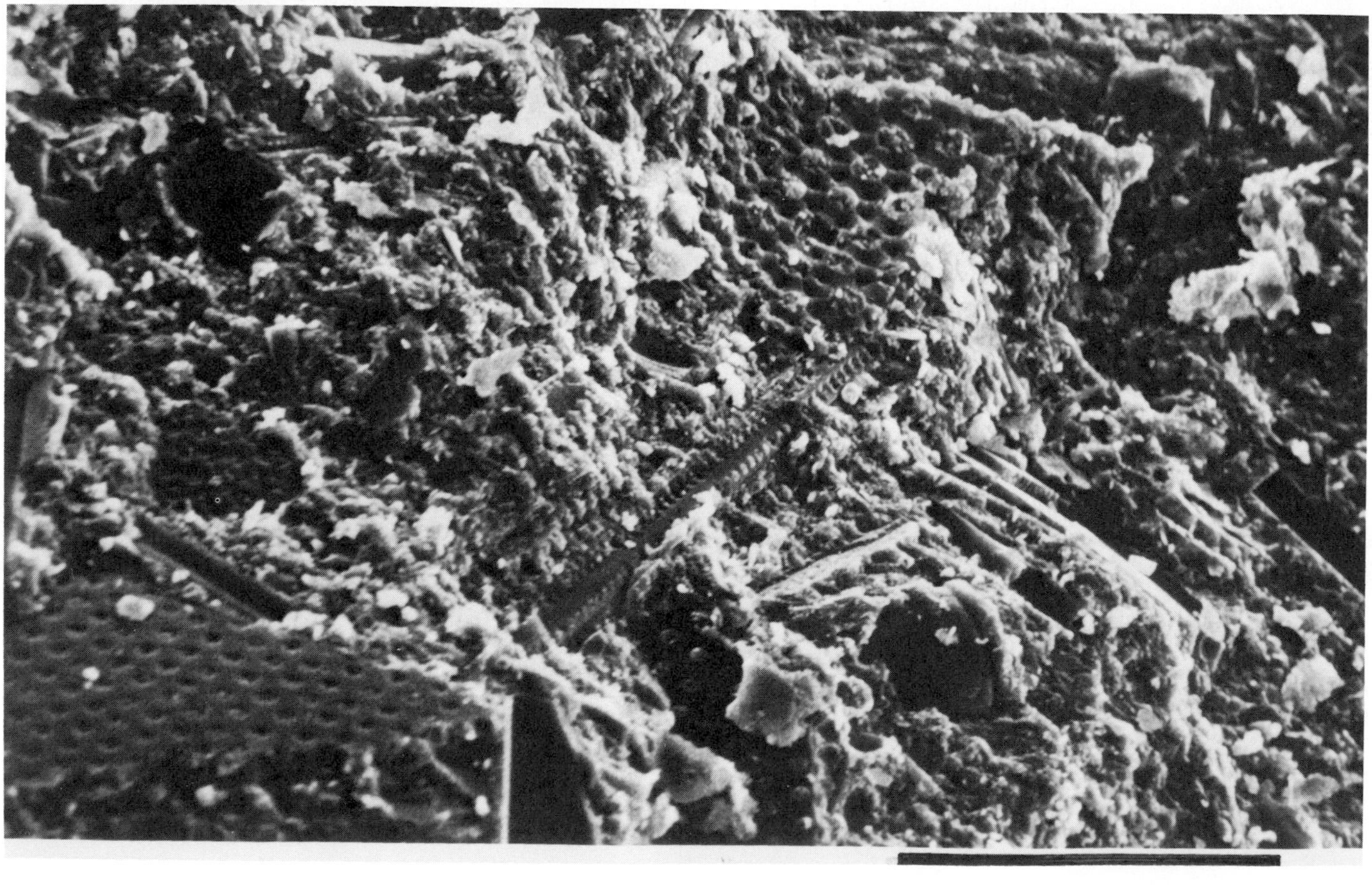

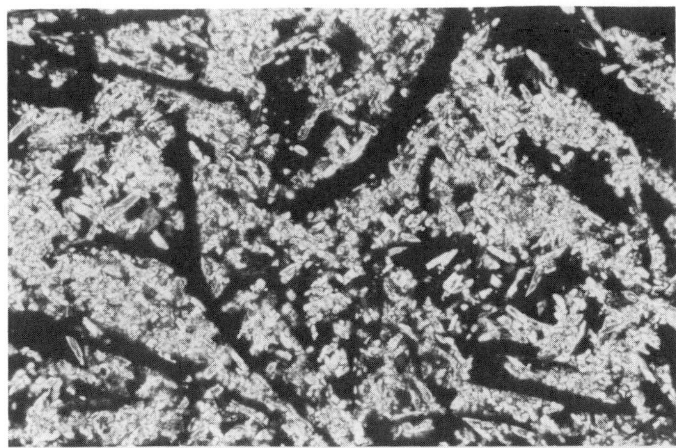

Figure 7 — Photomicrograph of glass-shard-rich microsparite. Sample 434-19,CC(2). Scale bar 0.05 mm. Crossed nicols. Note lath-shaped microspars of calcite.

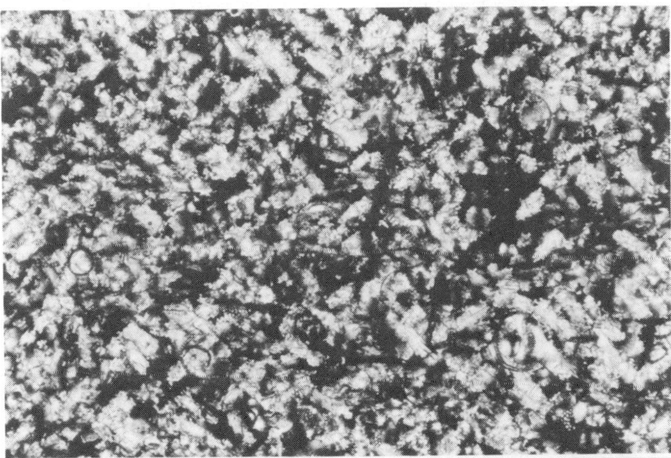

Figure 8 — Photomicrograph of rhodochrosite limestone showing recrystallized texture. Sample 434B-32,CC,10-12 cm. Scale bar 0.05 mm. Crossed nicols.

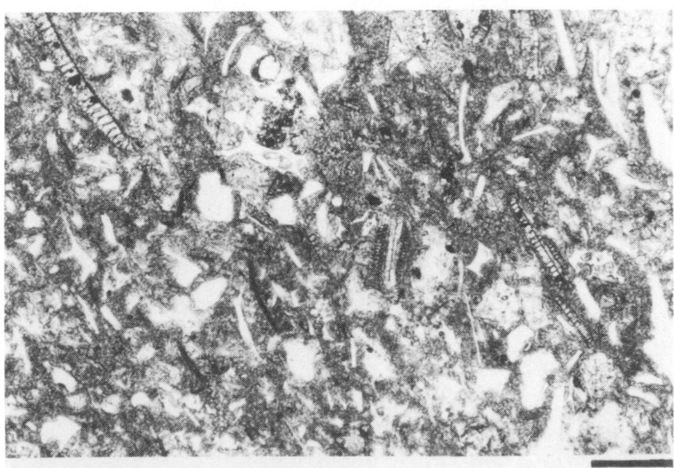

Figure 9 — Photomicrograph of glass-shard-rich micrite. Sample 434-29,CC, 12-14 cm. Scale bar 0.1 mm. Plane polarized light.

Lithologic Features

The carbonate nodules examined are grouped into three major lithologic types in terms of matrix textural features (Table 1). The first group consists of nodules with micritic texture (Figures 5, 6), the second group consists of those with sparite or microsparite texture, and the third consists of recrystallized nodules (Figures 7, 8).

Nodules with micritic texture consist mostly of biomicrite with abundant diatom-tests (Figures 5, 6), which are completely replaced by calcite (Figure 5). Glass-shard-rich micrite is also common (Figures 9, 10); it contains varying amounts of plagioclase, pumice, chlorite, augite, and biotite. The nodules are comprised of largely calcite as carbonate minerals; dolomite characterizes only one sample (434-27-1,65 cm; Figure 11). SEM micrographs show that the micrite is mostly anhedral aggregates or lamellae of carbonate minerals (Figures 6B, 10, 11).

Two samples are nodules with the sparite texture (434-28,CC and 434B-30-1,5-10 cm). One sample (434-28,CC) is a diatomaceous micrite breccia cemented by coarsely-crystalline calcite spar (Figures 12, 13) and the other (434B-30-1,5-10 cm) is a glass-shard-rich microsparite composed of manganese calcite. The brecciated nodule suggests that the carbonate rocks and enclosing sediments experienced tectonic stress (Arthur, von Huene, and Adelseck, 1980).

Recrystallized nodules include samples 434-19, CC(2) and 434B-32,CC,10-12 cm. The former is a diatom-pumice-chlorite-bearing, glass-shard-rich limestone with rare augite and plagioclase, whose matrix is wholly composed of elongate laths of calcite crystals 0.006 to 0.01 mm long (Figure 7). The latter is a diatom- and spicule-rich limestone composed of rhodochrosite with a coarsely-crystalline mosaic texture (Figure 8).

All carbonate nodules, regardless of lithologic type, contain greater or lesser amounts of volcaniclastic debris. One sample (434B-19-2,33-35 cm) contains considerable amounts of terrigenous clasts, mainly quartz.

Carbonate nodules usually contain abundant diatom remains consisting of the same assemblages found in surrounding muds and mudstones (Okada, 1980). Some nodules are rich in rhyolitic glass shards (Okada, 1980) whose refractive index is nearly the same as that of glasses in surrounding sediments (Furuta and Arai, 1980). These lines of evidence suggest that the carbonate nodules formed authigenically at or near the positions where they were recovered.

RESULTS

X-ray Diffraction Analysis

Table 2 lists interplanar spacing of the (104) plane of carbonates. Because carbonate nodules, except for rhodochrosite, contain minor amounts of quartz, the (101) reflection of quartz is used as an internal standard for correcting the carbonates d(104) value.

The d(104) value in calcite nodules ranges from

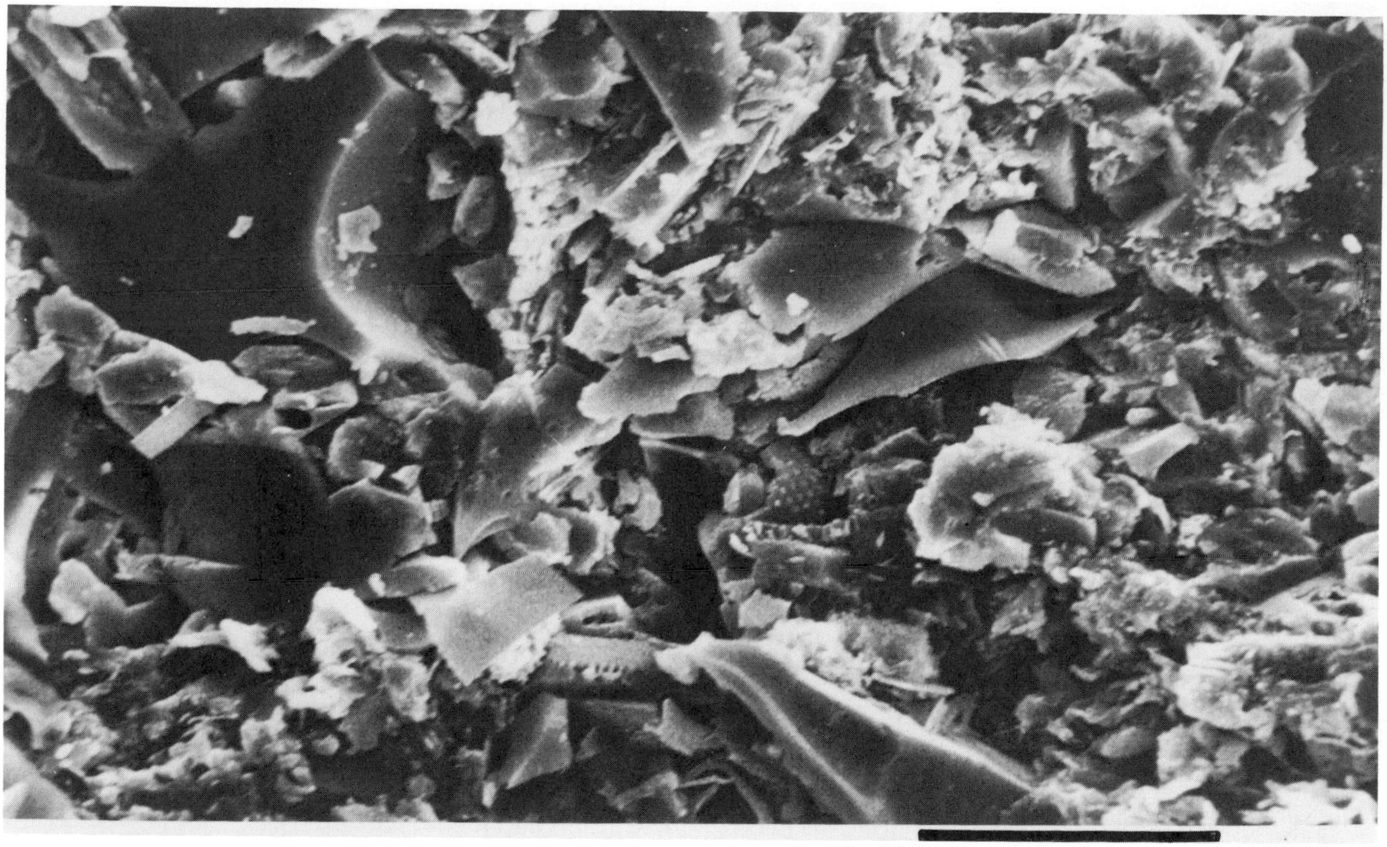

Figure 10 — SEM photograph of glass-shard-rich micrite. Sample 434-29,CC, 12-14 cm. Scale bar 5 μm.

Figure 11 — SEM photograph of glass-shard- and diatom-rich micrite. Sample 434-27-1, 65 cm. Scale bar 10 μm.

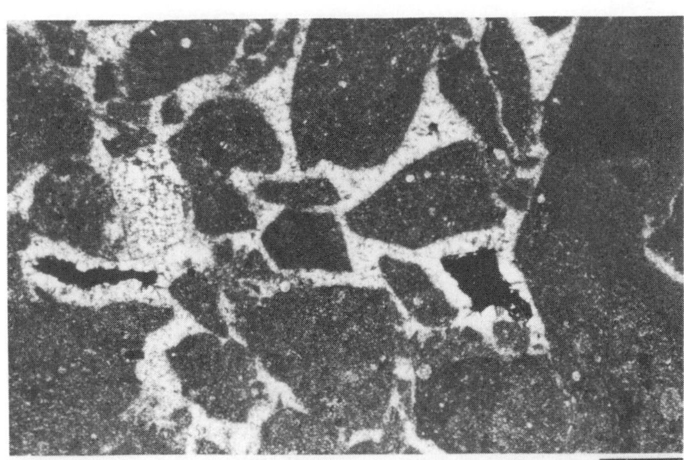

Figure 12 — Photomicrograph of micrite breccia cemented by calcite spars. Sample 434-28,CC. Scale bar 0.1 mm. Crossed nicols.

3.018 to 2.983Å. These values are smaller d(104) than those of typical pure calcite (Goldsmith, Graf, and Heard, 1961), probably due to substitution of Mg for Ca in calcite. Table 2 lists the magnesium content in calcite, calculated from X-ray (104) reflections, based on Goldsmith, Graf, and Heard (1961) data.

Figure 15 shows the relationship between the d(104) values of calcite in nodules and their stratigraphic positions in the sedimentary column. Sample 434B-32,CC, composed of rhodochrosite, is from nearly the bottom of the 434 hole. Sample 434B-30-1,5-10 cm, with the smallest d(104) value in calcite, is considered a manganese calcite (Okada, 1980). As shown in Figure 15, d(104) values generally decrease with depth. The d(104) values of calcite in nodules collected from 190 to 300 m sub-bottom show relatively large values from 3.004 to 3.018 Å. These values are similar to values reported for carbonates collected from Sites 438 and 439 on the upper inner trench slope of the Japan Trench (Matsumoto and Iijima, 1980), and Sites 490, 491, and 492 on the inner trench slope of the Middle America Trench off Mexico (Wada et al, 1981).

One dolomite specimen (434-27-1,65 cm) collected from 254m sub-bottom shows a d(104) value of 2.896 Å, which is slightly higher than that of typical dolomite (2.886 Å).

Carbon and Oxygen Isotopic Compositions

Table 2 lists carbon and oxygen isotopic compositions of carbonate minerals in nodules (Figure 15). The $\delta^{13}C$ values of calcite in nodules are highly variable, ranging from −5.9 to −35.2 ‰. Except for four calcite samples (434-28,CC; 434B-19-1,5-7 cm; 434B-19-2,33-35 cm and, 434B-30-1,5-10 cm) they are highly negative values, ranging from −23.3 to −35.2‰. At the same time, these samples give positive $\delta^{18}O$ values from +3.6 to +5.0‰. The four samples mentioned above show a large $\delta^{13}C$ range, however the $\delta^{18}O$ values show relatively constant positive values from +1.0 to +2.0‰. The $\delta^{13}C$ value of a dolomite sample (434-27-1,65 cm) is −1.8‰, and its $\delta^{18}O$ value is an unusually positive +6.1‰. The $\delta^{13}C$ and $\delta^{18}O$ values of rhodochrosite, in sample 434B-32,CC,10-12 cm, are −7.1 and +3.7‰, respectively.

DISCUSSION

Authigenic carbonates were found in IPOD/DSDP deep sea cores from inner trench walls and adjacent areas (Hein, O'Neil, and Jones, 1979; Wada et al, 1981). The samples discussed herein were collected from IPOD/DSDP Site 434 located on the floor about 6,000 m deep in the inner trench slope of the Japan Trench transect. In the following section, we discuss the origin of carbon and environmental conditions under which the nodules formed.

Origin of Carbon

As shown in Table 2 and Figure 15, carbon isotopic compositions of the examined carbonate nodules vary widely from −5.9 to −35.2‰ (excluding one dolomite specimen). Excluding the four nodules mentioned in the previous section and the dolomite and rhodochrosite nodules, the calcite nodules are characterized by highly negative $\delta^{13}C$ values and constantly positive $\delta^{18}0$ values. In lithology, all calcite nodules are classified as micrite.

Possible sources of carbon in carbonate nodules are considered as follows: 1) carbonate in ocean water. The $\delta^{13}C$ value of this carbon is nearly zero ‰; 2) biogenic calcite and/or aragonite, clasts of calcareous organisms. The $\delta^{13}C$ value of this carbon is also nearly zero ‰; 3) CO_2 derived from microbiological activity in sedimentary sequences. The $\delta^{13}C$ value of this carbon is similiar to that of organic material in the sediment, ranging from −21 to −25‰; 4) CO_2 derived from fermentation of organic matter along with methane with light carbon. The $\delta^{13}C$ values of the CO_2 are expected to be positive (for example, +18‰ reported by Nissenbaum, Presley, and Kaplan (1972); 5) CO_2 derived from oxidation of methane. The $\delta^{13}C$ value of this carbon is negative because the $\delta^{13}C$ value of methane in sedimentary column shows highly negative values, from −60 to −80‰ (Whelan and Sato, 1980).

As mentioned previously, the amount of methane seemingly increases in the sedimentary column below 100 m. The strong H_2S odor, from the surface to 180 m sub-bottom, is produced by sulfate-reducing microbiological activity; the reduction of organic carbon or carbon dioxide generally occurs below the zone of bacterial reduction of sulfate to H_2S. Carbonate concretions were only found in or below the H_2S odor zone of the sedimentary column. In the H_2S odor zone, CO_2 evolves from organic matter during the sulfate reduction by bacterial activity. The $\delta^{13}C$ value of this CO_2 may be close to that of the organic matter itself. Figure 14 shows isotopic compositions of total organic matter in soft mud enveloping the nodules. According to our data and results of Rhomankevich et

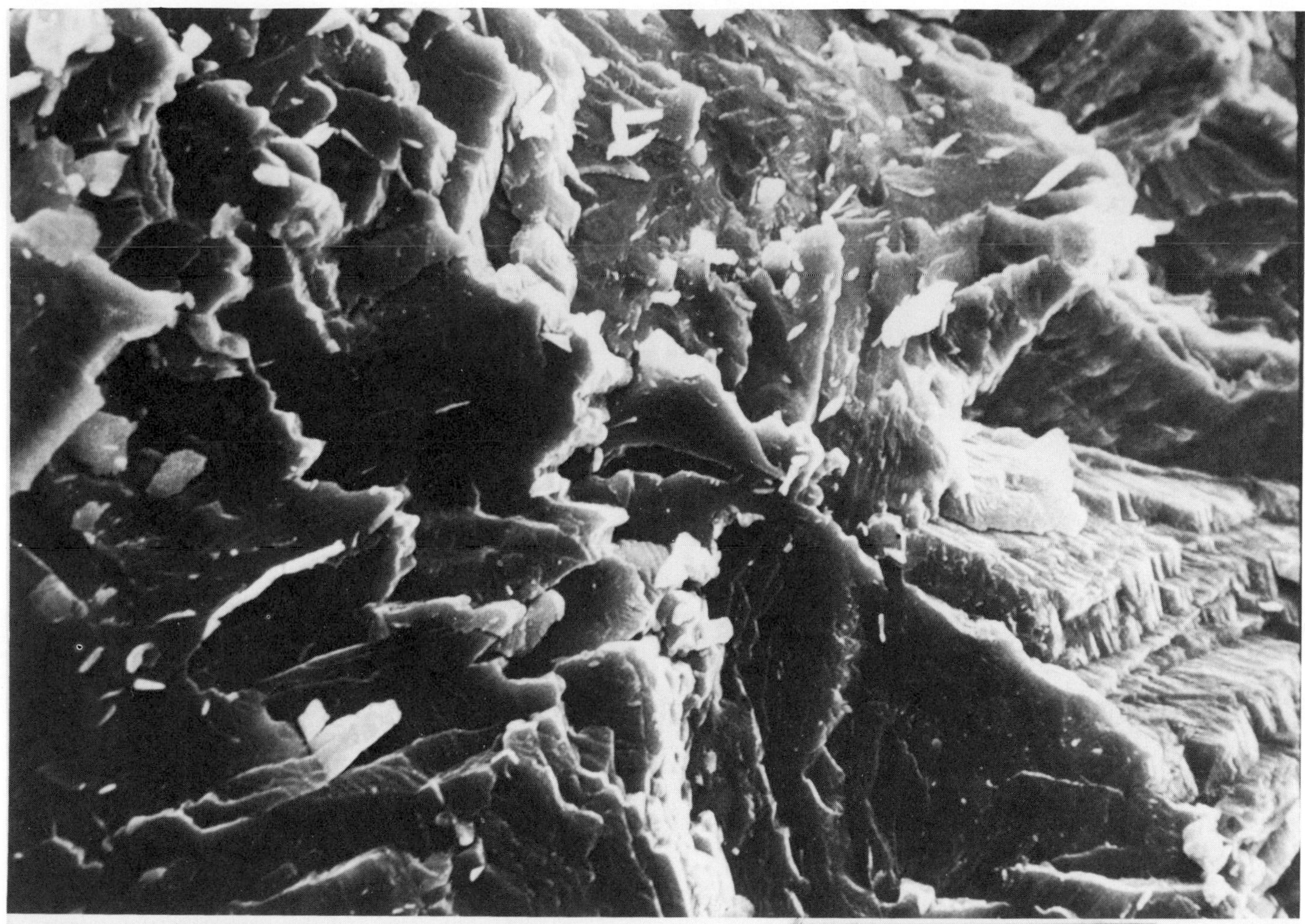

Figure 13 — SEM photograph of the calcite cement of micrite breccia. Sample 434-28,CC. Scale bar 5 μm.

al (1980), isotopic composition of total organic carbon in the IPOD/DSDP Leg 56 sample ranges from −21 to −25‰. The $\delta^{13}C$ value of calcite in nodules ranges from −23.3 to −35.2‰, averaging −26.0‰. These values are considered to be rich in ^{12}C.

Pore water in surface sediments should be undersaturated in calcium carbonate because the depth of deposition was about 6,000 m which is well below the CCD. Calcareous microfossils do not occur in these nodules nor in surrounding deposits, and we can eliminate the possibility of biogenic calcite or aragonite. Although we cannot deny the contribution of carbon dioxide from 1) and 4) above, carbon isotopic evidence suggests that these are minor sources of carbonates recovered during Leg 56. Carbon isotopic evidence further suggests that carbon in these nodules probably derives from organic carbon released during metabolic reduction of sulfates in sediments, and partly from CO_2 produced from methane oxidation. The latter case is exemplified by samples with highly negative $\delta^{13}C$ values.

Carbonate nodules found from the inner trench slope of the Middle America Trench area off Mexico (Wada et al, 1981) show negative values of $\delta^{13}C$. Some nodules with heavy carbon dolomite ($\delta^{13}C$ = +13‰) also occur along with calcite nodules with $\delta^{13}C$ values of about −30 to −42‰. Calcite nodules with highly negative $\delta^{13}C$ values are found only at drilling sites deeper than about 1,700 m. In the deep Bering Sea area, the $\delta^{13}C$ values of carbonate nodules range from −7.2 to −21.1‰ with an average of −17.2 ‰ (Hein, O'Neil, and Jones, 1979). This indicates that carbonate precipitation probably occurred in the sulfate reduction zone.

Carbon isotopic compositions of dolomite, rhodochrosite, and one calcite nodule (434-28,CC) have $\delta^{13}C$ values between those of typical marine limestone (Keith and Weber, 1964) and those of authigenic carbonate nodules. Specimen 434-28,CC is an intrasparite with diatomaceous micrite clasts. The intrasparite is characterized by recrystallized calcite, and some micrite clasts contain small amounts of glass shards, diatom remains, and plagioclase. Since the oxygen isotopic composition of these three nodules differs from others, they may be allochthonous in origin.

Condition of Carbonate Precipitation

Figure 14 shows the vertical changes in carbon and oxygen isotopic compositions and in d(104) values of

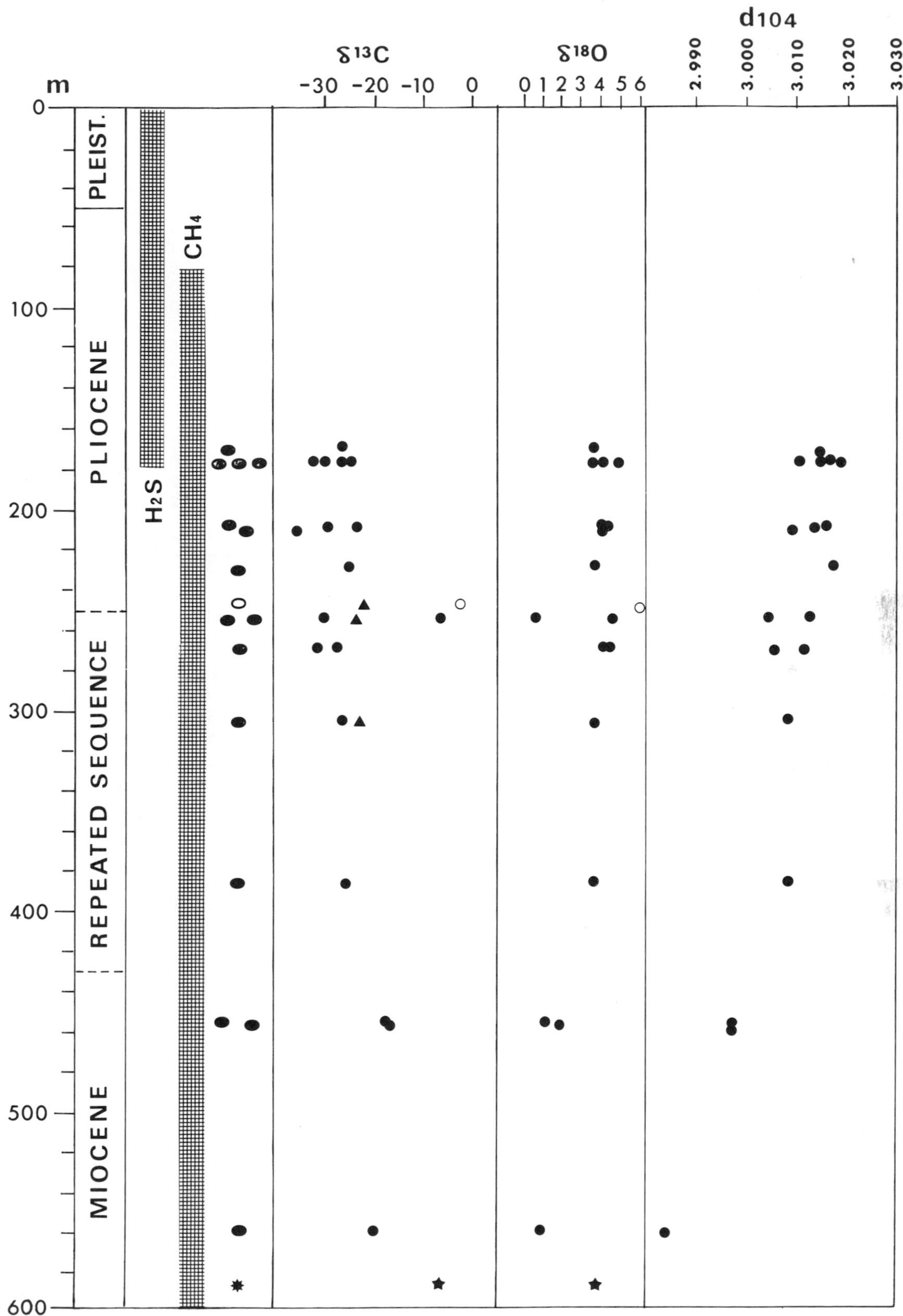

Figure 14 — Vertical change of the $\delta^{13}C$, $\delta^{18}O$, and d(104) values of carbonate minerals constituting deep sea nodules. Repeated sequence in the first column indicates repetitions of diatom zones of late Miocene and Pliocene (cf., Scientific Party, 1980, p. 359). The second column shows the presence of H_2S and CH_4 gases and occurrence of carbonate nodules. Solid circle — calcite nodules; open circle — dolomite nodule; star — rhodochrosite nodule; and, solid triangles — $\delta^{13}C$ of total organic matter in mud surrounding the nodules.

Table 2. Sample description and isotopic compositions of carbonate nodules collected from deep sea cores in IPOD/DSDP Leg 56, Japan Trench.

Hole	Core-section	Interval (cm)	Sub-bottom depth (m)	Carbonate mineralogy	d(104) (Å)	$MgCO_3$ in calcite (‰)	Carbonate $\delta^{13}C$ (‰) (PDB)	Carbonate $\delta^{18}O$ (‰) (PDB)
434	19-1	10-12	168	calcite	3.014	7.0	−26.0	+3.8
434	19CC(1)		168-178	calcite	3.018	5.6	−29.3	+4.2
434	19CC(1′)		168-178	calcite	3.014	7.0	−24.7	+4.2
434	19CC(2)		168-178	calcite	3.010	8.5	−32.0	+5.0
434	19CC(2′)		168-178	calcite	3.016	6.4	−26.9	+3.6
434	23-1	140-143A	207	calcite	3.013	7.4	−29.0	+4.1
434	23-1	140-143B	207	calcite	3.015	6.8	−23.3	+4.2
434	23-3	49	210	calcite	3.008	9.0	−35.2	+4.4
434	25-2	86-90	228	calcite	3.017	6.0	−24.6	+3.9
434	27-1	65	245	dolomite	2.896		−1.8	+6.1
434	28CC		253-263	calcite	3.004	10.8	−5.9	+1.0
434	28-1,	104-106	255	calcite	3.012	7.8	−30.1	+4.6
434	29CC	core	263-273	calcite	3.011	8.2	−27.6	+4.2
434	29CC	margin	263-273	calcite	3.005	10.4	−31.3	+4.0
434B	3-1	30-32	305	calcite	3.008	9.0	−26.1	+3.8
434B	11CC	2-6	381-391	calcite	3.008	9.0	−25.4	+3.7
434B	19-1	5-7	456	calcite	2.997	13.2	−17.0	+1.2
434B	19-2	33-35	458	calcite	2.997	13.2	−16.5	+2.0
434B	30-1	5-10	560	calcite	2.983		−19.9	+1.0
434B	32CC	10-12	580-590	rhodochrosite	2.843		−7.1	+3.7

carbonates. In carbonate nodules collected from 180 to about 300 m sub-bottom, the $\delta^{18}O$ values of calcites range from +3.5 to +5.0‰ with an average of +4.2‰, except for one specimen 434-28,CC. The $\delta^{18}O$ value of calcite depends on the temperature and the oxygen isotopic composition of the interstitial water during precipitation. If interstitial water is ordinary ocean water, we can estimate the temperature at which calcite was precipitated based on oxygen isotopic geothermometry (Craig, 1965). The temperature is estimated to be +1.6°C from the average value of +4.2 ‰. Although the temperature gradient at Site 434 was not measured, the geothermal gradient at the nearby Site 440 is estimated to be 1°C/100m (Scientific Party, 1980). If the thermal gradient has a similar value at Site 434, the equilibrium temperature may represent the formation temperature of the carbonate nodules in situ. Therefore, the $\delta^{18}O$ value seems to decrease downward. Equilibrium temperatures at 460 and 560 m sub-bottom are calculated as +10.5 and 12.3°C, respectively. These temperatures are somewhat high with respect to the thermal gradient.

As shown in Figure 14, the $\delta^{13}C$ values of carbonate nodules show an enrichment of ^{13}C with increasing sub-bottom depth. The sequence below 200 m sub-bottom is characterized by methane production (Scientific party, 1980). During the methane generation, CO_2 may also have been produced below the sulfate reduction zone. The $\delta^{13}C$ value of CO_2 evolved will be enriched in ^{13}C in comparison to original organic carbon, because the carbon isotopic composition of methane is rich in ^{12}C (Nissenbaum et al, 1972). Enrichment of ^{13}C in some calcite nodules with increasing depth (Figure 14) may be due to CO_2 evolved during the methane generation.

As shown in Figure 14, d(104) tends to decrease with increasing depth. Schlager and James (1978) cited the experiments and thermodynamic considerations led by Füchtbauer and Hardie (1976) in interpreting the origin of deep sea low magnesian calcite in Bahamas. Schlager and James (1978) attributed the low magnesian calcite to cold deep-water masses. In the present study, where oxygen isotopic evidence in carbonate nodules shows the temperature increasing with depth, the decrease in d(104) values may indicate an increase of magnesium in calcite, which may also be related to the increasing temperature.

CONCLUSION

Carbonate nodules collected from the deep inner trench slope of the Japan Trench at depths of about 6,000 m consist mainly of magnesian calcite, manganese calcite, dolomite, and rhodochrosite. These nodules occur mainly in the methane production layer which underlies the hydrogen sulfide layer and the transitional zone. The nodules fall into three categories: micrite, microsparite, and recrystallized textures. Magnesium content of the carbonate nodules seems to increase with increasing sub-bottom depths. Magnesian calcite in these nodules has highly negative $\delta^{13}C$ values of −23.3 to −35.2 ‰, averaging −26.0‰. This shows that precipitation of carbonate probably occurred in the sulfate reduction zone, or below depths where CO_2 derived from microbiological oxidation of organic matter and methane was mainly supplied for carbonate precipitation. Oxygen isotopic compositions of carbonate nodules show that the equilibrium temperature of calcite and ocean water is low and increases with sub-bottom depth. This evidence suggests that many carbonate nodules were authigenically formed at or near the positions where

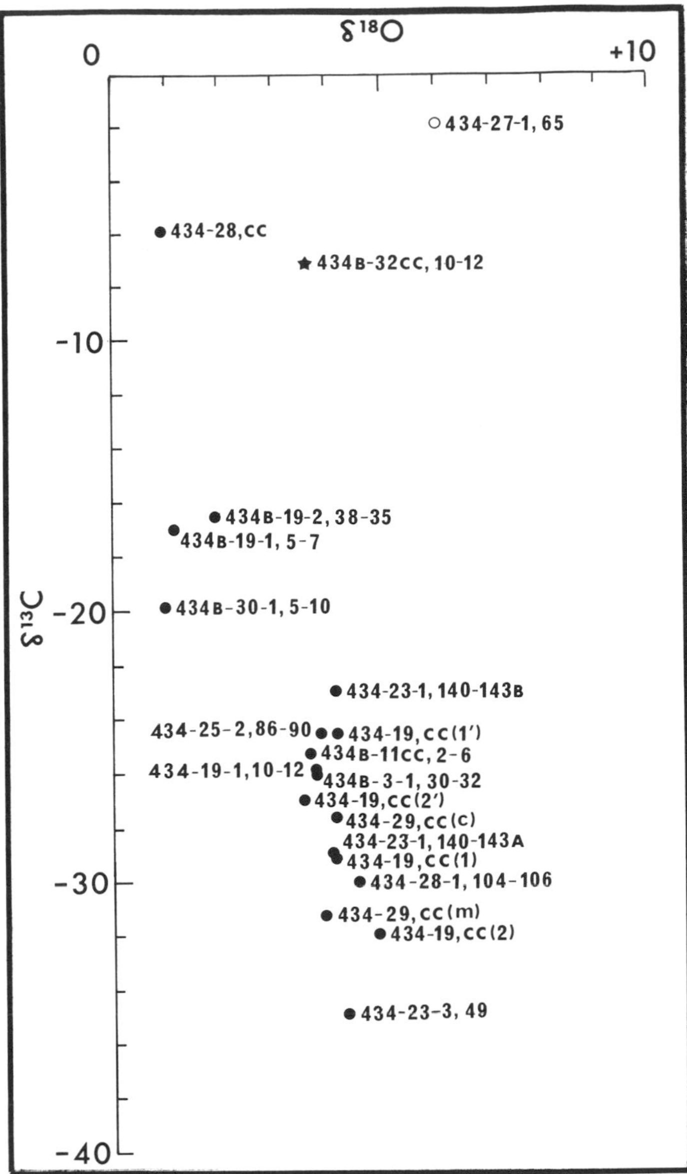

Figure 15 — Relation between the $\delta^{13}C$ and $\delta^{18}O$ values of carbonate minerals constituting deep sea nodules. Solid circle — calcite; open circle — dolomite; and, solid triangle — rhodochrosite. (c) — core position and (m) — marginal position of a nodule.

they were recovered.

As reported by Irwin, Curtis, and Coleman (1977), isotopic evidence for diagenetic carbonates in ancient sedimentary sequences corresponds to the chemical processes operating at different depths during burial diagenesis. This study shows that carbon isotopic composition of carbonate nodules in modern deep-sea sediments represents a certain depositional condition. Therefore, isotopic and microscopic studies on carbonate nodules in ancient "geosynclinal sediments" will clarify the environments of sedimentation and diagenesis.

ACKNOWLEDGMENTS

We are grateful to Joel S. Watkins for inviting us to contribute this paper and for editing the final manuscript. We thank NSF and the Japanese Scientific Advisory Board for IPOD for their support in this study. Okada thanks his IPOD/DSDP Leg 56 colleagues for their help on board D/V Glomar Challenger.

Sincere thanks also goes to Professor Nobuyuki Nakai of Nagoya University for providing us with the facilities for mass-spectrometric analysis and for his constructive discussions; Professor Noriyuki Nasu of the Ocean Research Institute, University of Tokyo, for arranging the preparation of most of the thin sections of carbonate nodules; and, Susumu Yano and Nobuhiro Kotake of Shizuoka University for their assistance in taking the SEM micrographs.

This study was partly supported by the Grant-in-Aid for Scientific Researches from the Ministry of Education, Science and Culture [Monbusho], Japan (Grant No. 434041).

REFERENCES CITED

Arthur, M. A., R. von Huene, and C. G. Adelseck, Jr., 1980, Sedimentary evolution of the Japan fore-arc region off northern Honshu, Legs 56 and 57, Deep sea drilling project, *in* Scientific party, Initial report of the deep sea drilling project, v. 56,57, part 2: Washington, D.C., U.S. Government Printing Office, p. 521-568.

Craig, H., 1957, Isotopic standards for carbon and oxygen and correction factors for mass-spectrometric analysis of carbon dioxide. Geochimica et Cosmochimica Acta, v. 12, p. 133-149.

———, 1965, The measurement of oxygen isotope paleotemperatures, *in* Proceedings, Speleto Conference on Stable Isotopes: Oceanographic studies and Paleotemperatures, n. 3.

Furuta, T., and F. Arai, 1980, Petrographic properties of tephras, Leg 56, Deep sea drilling project, *in* Scientific party, Initial reports of the deep sea drilling project, v.56,57, part 2: Washington, D.C., U.S. Government Printing Office, p. 1043-1048.

Füchtbauer, H., and L. A. Hardie, 1976, Experimentally determined homogeneous distribution coefficients for precipitated magnesian calcites. Annual Progress Meeting of the Geological Society of America (Abs.), p. 877.

Goldsmith, J. R., D. L. Graf, and H. C. Heard, 1961, Lattice constants of calcium-magnesium carbonate: American Mineralogist, v. 46, p. 453-457.

Hein, J. R., J. R. O'Neil, and M. G. Jones, 1979, Origin of authigenic carbonates in sediment from the deep Bering Sea: Sedimentology, v.26, p. 681-705.

Irwin, H., C. Curtis, and M. Coleman, 1977, Isotopic evidence for source of diagenetic carbonates formed during burial of organic rich sediments: Nature, v. 269, p. 209-213.

Keith, M. L., and J. N. Weber, 1964, Isotopic composition and environmental classification of selected limestones and fossils: Geochimica et Cosmochimica Acta, v. 28, p. 1784-1816.

Matsumoto, R., and A. Iijima, 1980, Carbonate diagenesis in cores from Sites 438 and 439 off northeast Honshu, northwest Pacific, Leg 57, Deep sea drilling project, *in* Scientific party, Initial reports of the deep sea drilling project, v. 56, 57, part 2,: Washington, D.C., U.S. Government Printing Office, p. 1117-1131.

McCrea, J. M., 1950, The isotopic chemistry of carbonates and a paleotemperature scale: Journal of Chemical Physics, v. 18, p. 849-857.

Nissenbaum, A., B. J. Presley, and I. R. Kaplan, 1972, Early diagenesis in a reducing fjord, Saanich inlet, British Columbia-I, *in* Chemical and isotopic changes in major components of interstitial water: Geochemica et Cosmochimica Acta, v. 36, p. 1007-1027.

Okada, H., 1980, Pebbles and carbonate nodules from deep sea drilling project, Leg 56 cores, *in* Scientific party, Initial reports of the deep sea drilling project, v. 56, 57, part 2: Washington, D.C., U.S. Government Printing Office, p. 1089-1106.

Romankevich, E. A., et al, 1980, Carbon and its isotopic composition in Pacific Ocean bottom sediments, Leg 56, Deep sea drilling projects, *in* Scientific party, Initial reports of the deep sea drilling project, v. 56, 57, part 2: Washington, D.C., U.S. Government Printing Office, p. 1313-1317.

Schlanger, W., and N. P. James, 1978, Low-magnesian calcite limestone forming at the deep-sea floor, Tongue of the Ocean, Bahamas: Sedimentology, v. 25, p. 675-702.

Scientific Party, 1980, Initial reports of the deep sea drilling project, v. 56, 57, part 1: Washington, D.C., U.S. Government Printing Office, p. 355-398.

Taira, A., et al, 1982, The Shimanto belt of Japan: Cretaceous-lower miocene active-margin sedimentation, *in* J. K. Leggett, ed., Trench forearc geology: Sedimentation and tectonics on modern and ancient active plate margins: Geological Society of London, p. 5-26.

Tarutani, T., R. N. Clayton, and Y. K. Mayeda, 1969, The effect of polymorphism and magnesium substitution on oxygen isotope fractionation between calcium carbonates and water: Geochemica et Cosmochimica Acta, v. 33, p. 987-996.

Wada, H., Wada, H., et al, 1981, Deep-sea 1981, Deep-sea carbonate nodules from the Middle America Trench area off Mexico, IPOD/DSDP Leg 66, *in* Initial reports of the deep sea drilling project, v. 66: Washington, D.C., U.S. Government Printing Office, p. 453-474.

Whelan, J. M., and S. Sato, 1980, C_1-C_5 hydrocarbons from core gas pockets, deep sea drilling project, Legs 56 and 57, Japan Trench transect, *in* Scientific party, Initial reports of the deep sea drilling project, v. 56, 57, part 2: Washington, D.C., U.S. Government Printing Office, p. 1335-1347.

Synthetic Seismic Sections from Biostratigraphy

Jan E. van Hinte
Vrije Universiteit
Instituut voor Aardwetenschappen
Amsterdam, The Netherlands

Linear time scales are used by exploration paleontologists to establish a numeric chronostratigraphy for marine well sections which include calculated estimates for the missing intervals at unconformities. A new kind of cross section can be constructed using isochrons, showing chronostratigraphic depositional patterns similar to the seismic record. The geochronologic resolution of isochron cross sections is normally higher than that of seismic sections, so it can be used in the calibration and interpretation of the seismic record and in its extrapolation in "bad data" areas. The depth-distance sections portray sediment accumulation rate patterns and form a useful base to plot parameters such as source and reservoir facies and geochemical and geophysical data. The numeric stratigraphic technique is particularly useful to continental margin explorationists where marine microfauna aid its application. With known vertical movements of a passive margin, the method detects sea level changes (that determined the sedimentary pattern) and local anomalies such as salt or shale movements.

Using modern time scales (Berggren, 1972; Berggren and Van Couvering, 1974; Ryan et al, 1974; Stainforth et al, 1975; Van Hinte, 1976a, 1976b; Vail, Mitchum, and Thompson, 1977; Hardenbol and Berggren, 1978; Vail and Mitchum, 1979; Van Hinte, Colin, and Lehman, 1980) exploration paleontologists can assign numeric ("absolute") geological ages to marine sedimentary successions like those of continental margins. Adding further quantitative stratigraphic elements — such as present position with respect to sea level, paleobathymetry, and unit thickness — determines the position of the geologic record of a rock-unit in space and time and makes it possible to replace much of the traditional qualitative deliberations on the geologic history of a site or an area with unambiguous geohistory analysis (Van Hinte, 1978). Viewing the geologic record in a linear time perspective facilitates evaluation of geologic events and processes because it brings them to proportion. This provides a new and rewarding approach to geology and facilitates the use of a computer. Figure 1 shows the four elements used in the construction of Cenozoic and Mesozoic time-scales.

This paper discusses a theoretical example of one application of geohistory analysis techniques originally presented at a 1979 meeting of the Petroleum Circle of the Dutch Geology and Mining Society, and at a 1980 Conference on Ancient Sea Levels at Lamont-Doherty Geological Observatory, (New York).

TIME-STRATIGRAPHY OF WELL SECTION

A conventional paleontologic report usually contains the kind of data summarized in Figure 2 and Appendix I for hypothetical well A. This information permits the geologist to make cross sections like the one of Figure 3, showing well A and two other wells B and C. Today, we can present this information in numbers, for the numeric ages at stage boundaries and of fossil markers are known. This is illustrated for well A in Figure 4 under the column MA (=Mega-annus or 1 million years). Calculating average rates of sediment accumulation draws our attention to two parts of the section where rates are anomalously low and remain low even if we take a wider average assuming the boundaries were misplaced. This suggests either that there is a section missing or that these units differ from adjacent units in facies. Figure 2 shows no changes within the units, but between them changes in lithol-

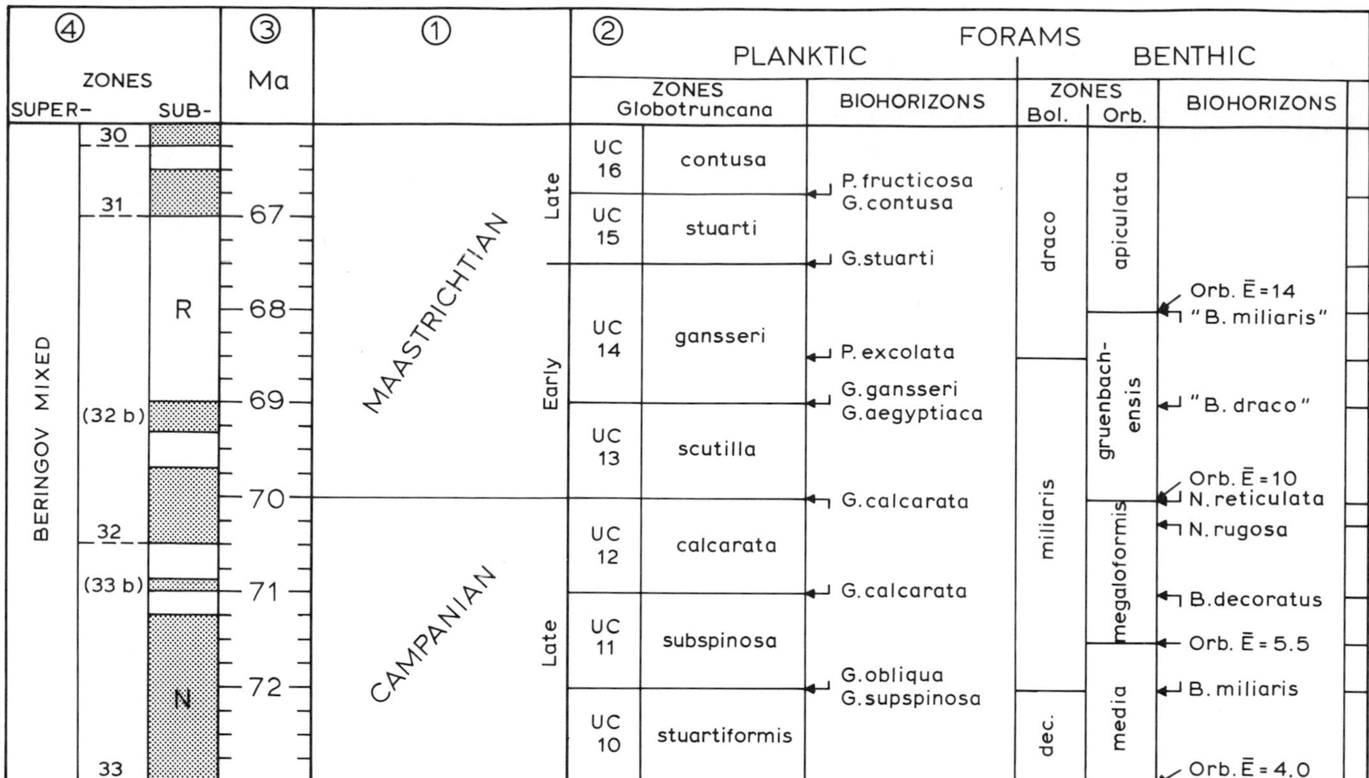

Figure 1 — Portion of Cretaceous time scale showing the four elements used during the past decade in the construction of Mesozoic and Cenozoic time scales: 1. units of "relative time scale"; 2. selected foraminiferal biostratigraphy; 3. linear numeric scale derived from first and second order radiometrically dated calibration points; 4. earth magnetic polarity reversal scale. Note that average planktic zone spans .88 Ma, that average unit of combined planktic and benthic subdivision spans .64 Ma, that only three benthic genera planktic and benthic genera (Bol.-*Bolivinoides,* Orb.,-*Orbitoides,* and N.-*Neoflabellina)* are included and that the steady increase in number of epi-auxiliary chambers (E) in *Orbitoides* makes a continuous scale possible (after Van Hinte, 1976; later work suggests changes should be made that are beyond the scope of the present paper).

DEPTH IN METER
AGE
BIOZONES AND BIOHORIZONS
LOG
ENVIRONMENT
NERITIC BATH
I M O U M
1500
2500
T.D. 3430
EOCENE
MIDDLE
EARLY
PALEOCENE
LATE
THANETIAN
EARLY
DANIAN
LATE CRETACEOUS
SENONIAN
MAASTRICHT
CAMPANIAN
"M. ARAGONENSIS"
P. PUSILLA S.S.
S. TRINIDADENSIS
BOLIVINOIDES MILIARIS
G. CALCARATA
LST.
MARL
LST. + SHALE

Figure 2 — Portion of stratigraphic log of hypothetical Well A summarizing information usually available for exploration wells: age, biostratigraphy, log patterns, gross lithology, and an environmental interpretation (Appendix 1).

ogy, log pattern, and water depth confirm our suspicion of an unconformity. After recognizing the unconformity we adjust numbers by calculating the ages at the unconformities (x and y in Figure 5); this is done by downward extrapolation of the nearest unit sedimentation rate (R1, given by markers A and B) above the unconformity, and by upward extrapolation of the nearest unit below the unconformity (R2, given by markers C and D). The age at T.D. is calculated in the same manner. Next, we determine the depth of full number ages by calculating sediment accumulation for given age and R using the same equation as in Figure 5.

Recognizing and defining unconformities in well sections can be an important contribution to the exploration effort because unconformities form the boundaries of depositional sequences that can be recognized in seismic sections. These boundaries are the primary levels for calibration between conventional stratigraphy and seismo-stratigraphy, the sequences being the basic mapping units for the exploration seismologists.

Unconformities are caused by changes in base-level, which are often changes in sea level most of which are of a global nature and referred to as eustacy. Past sea level changes have been recorded by man in history and archaeology. The rest of nature records these changes in geomorphology and in the sedimentary strata, as seen in seismostratigraphy, lithostratigraphy,

Table 1. Well Record of Paleo-Water Depth Change

	LITHOSTRATIGRAPHY	BIOSTRATIGRAPHY
Ocean floor, Lower Slope	Change in abundance or nature of TURBIDITES (GRAVITITES)	REWORKING
	TERRIGENOUS MATERIAL	DISSOLUTION FACIES
	LITHOLOGIC CHANGES: grain size, terrigenous material	REWORKING
		time: HIATUS IN STANDARD
	LITHOFACIES CHANGES: shore-line, channels	SPECIATION
Shelf,	HARDGROUND	habitat: PALEO-WATER DEPTH CHANGE
Upper Slope	RESIDUALS	CHANGE IN ORGANIC MATTER
	PEBBLES	
	DENSITY CHANGE (LOG)	
	DISSOLUTION	
	PEAT-SOIL SEQUENCES	
	REEF GROWTH	
	R-CHANGES	

and biostratigraphy. Table 1 lists some criteria for the lithologic and paleontologic recognition of sea level changes; change in rate of sediment accumulation (R, defined above) is only one of them.

TIME-STRATIGRAPHY OF CROSS SECTION

Having made calculations for wells A, B, and C (Appendix 2), we make the 1 Ma isochron cross section shown in Figure 6. At a glance, the section shows areas with high rates of sediment accumulation where isochrons are widely spaced, and areas of low rates of sediment accumulation with closely spaced isochrons. The cross section also gives an impression of the structural history and the importance of area unconformities. Such a plot can be expanded to include lithology, paleo-water depth, or other geologic information.

The magnitude of unconformities is better illustrated in the time-depth plot of Figure 6, which uses time instead of depth as an ordinate. Again, you can add lithology, paleo-water depth or other geologic information, or an interpretation of the age of the event that caused the unconformity (Figure 7). Figure 8 illustrates some applications of this approach.

IMPLICATIONS

Figure 7 resembles a figure published by Mitchum, Vail, and Thompson (1977, Figure 1b, p. 54) which illustrates the nature of the depositional sequences they recognize in seismic sections. They found that a seismic section is an image of the time-stratigraphic depositional pattern. "Primary seismic reflections are

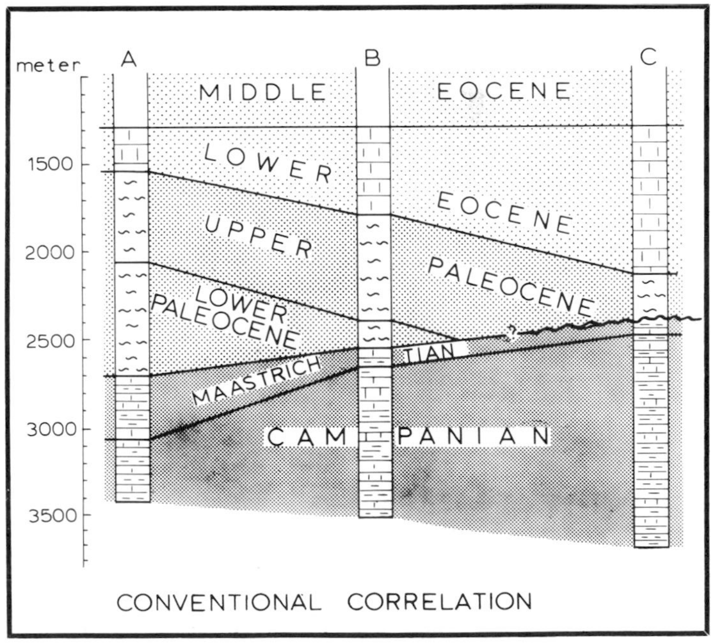

Figure 3 — Regional Cross section through Well A of Figure 1 and two other hypothetical Wells B and C showing "complete" section at A and B and an unconformity on a paleo-high at C.

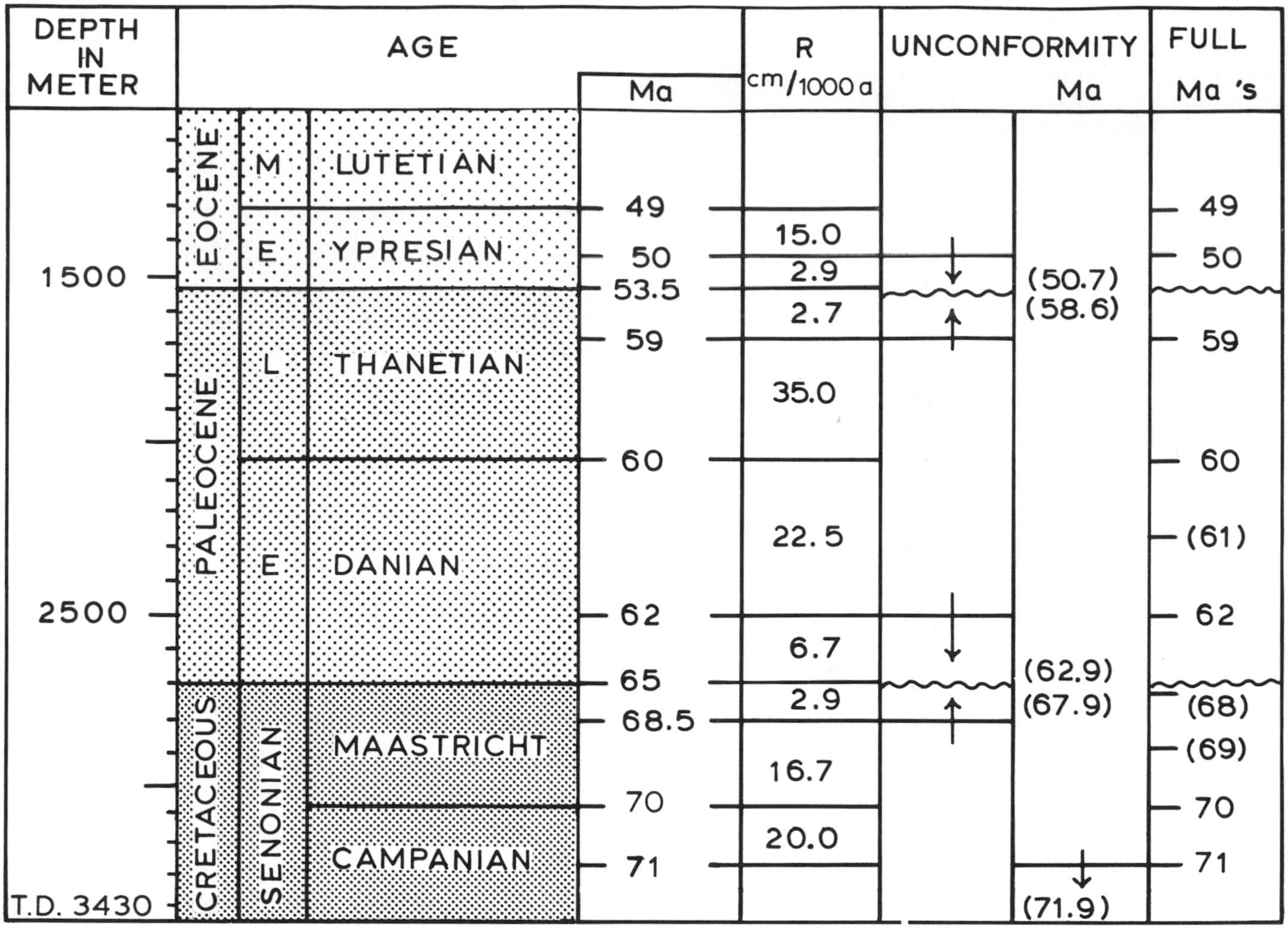

DEPTH IN METER	AGE			Ma	R cm/1000 a	UNCONFORMITY Ma	FULL Ma 's
	EOCENE	M	LUTETIAN				
				49			49
		E	YPRESIAN		15.0		
				50			50
					2.9	(50.7)	
1500				53.5			
					2.7	(58.6)	
	PALEOCENE	L	THANETIAN	59			59
					35.0		
				60			60
		E	DANIAN		22.5		(61)
2500				62			62
					6.7	(62.9)	
				65			
	CRETACEOUS	SENONIAN			2.9	(67.9)	(68)
			MAASTRICHT	68.5			(69)
					16.7		
				70			70
			CAMPANIAN		20.0		
				71			71
T.D. 3430						(71.9)	

Figure 4 — Portion of stratigraphic log of Well A. Same information as Figure 2 (Appendix 1) translated in numeric time (Ma after Hardenbol and Berggren, 1978; Van Hinte, 1976b) and used to calculate the average rate of sediment accumulation (R) of units between markers. Ages at unconformities and TD, and the depth of full number ages calculated using R = T/10A cm/ka (cf Figure 5). Calculated numbers in brackets. R = 22.5, using 61 Ma lies at 2275 m; 68 Ma and 69 Ma lie at 2717 and 2883 m respectively, using R = 16.7.

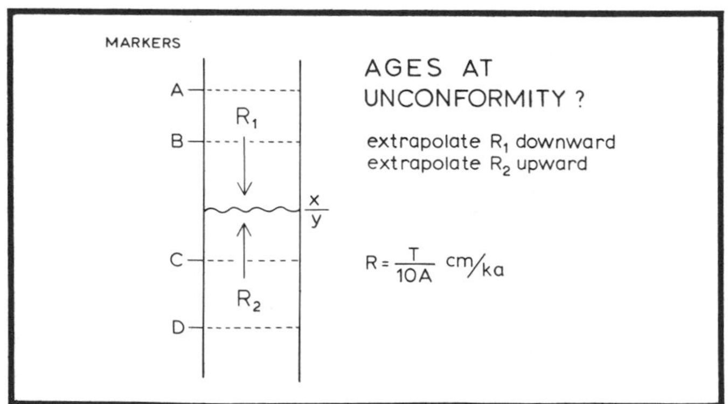

Figure 5 — Extrapolation of average rate of sediment accumulation (R1 for interval B-A, and R2 for interval D-C) to estimate ages above (x) and below (y) unconformity. Event that caused unconformity ended D-C depositional environment and new B-A depositional environment followed; calculation of R for unit C-B makes no sense. A, B, C, and D are numeric ages from markers; T = thickness of unit between markers; 10A = 10 × duration of unit in million years; ka = kilo-annus = 1000 years.

generated by physical surfaces in the rocks, consisting mainly of stratal (bedding) surfaces and unconformities with velocity-density contrasts. . . . the resulting seismic section is a record of the chronostratigraphic depositional and structural patterns and not a record of the time-transgressive lithostratigraphy." (Vail and Mitchum, 1977, p. 51)

I believe that we are looking at the same, natural time-stratigraphic units. It just so happens that stratigraphic resolutions from modern seismic data and from fossils match and are sufficient to read the sedimentary record of the effects of sea-level changes. The improving chronologic resolution of the geologic timescale has surpassed the frequency of eustatic changes and now allows for the distinction between individual eustatic events.

In most geologic settings, biochronostratigraphy may only separate the megasequences, and paleobathymetry and lithofacies interpretations are necessary to further identify individual sequences. The present technique may prove to be effective in handling such information, and achieve:

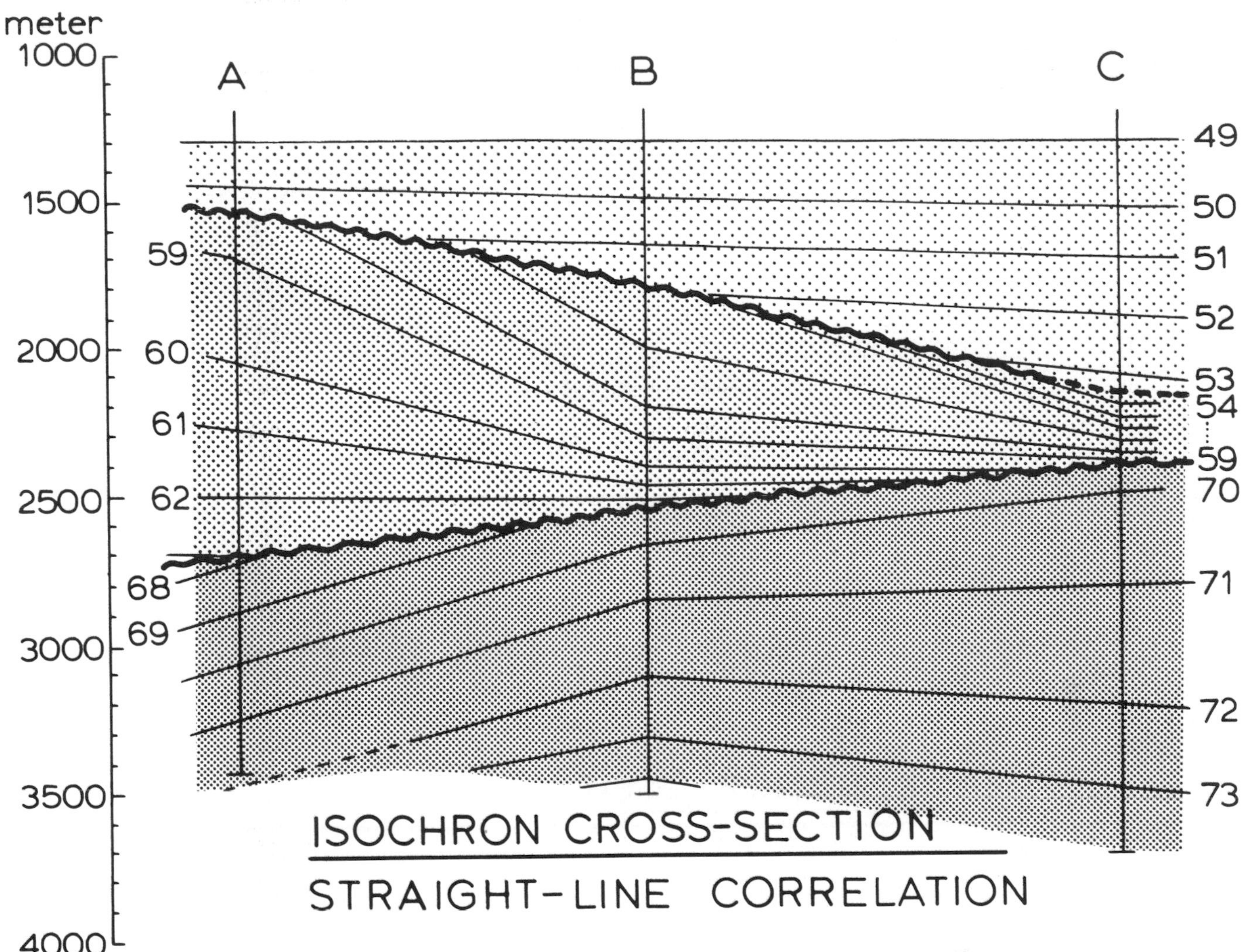

Figure 6 — Straight line isochron cross section through wells A, B and C (data in Figure 4 and Appendix 2). Interval 1 Ma (one million years). Note correlation of unconformities and difference between high and low rates of sediment accumulation. 52 Ma and 53 Ma lines between wells B and C drawn parallel with 51 Ma line, which implies geologic assumption (compare with Figure 11).

1. Better regional age calibrations for sequences recognized on seismic sections;
2. A correlation of regional seismostratigraphy to the Vail and Mitchum (1979) Global Cycle Chart (to document or to modify it) and its use for accurate time-stratigraphic prediction in undrilled areas: and,
3. A better understanding of the eustacy that caused the sequential depositional pattern.

A better correlation of the seismic sequences with the relative time-scale (points 1 and 2) is needed. This is suggested by the fact that a number of the units given on the Vail and Mitchum (1979) chart do not match the conventional stages. Yet, they should, for more than a hundred years ago d'Orbigny defined the stages in precisely the same terms, namely as "unconformity bound units." As a student of catastrophist Cuvier, d'Orbigny (1850-1852) already did relate his stage boundaries (unconformities) to global sea level changes (floods). At the turn of the century, Suess (1908) introduced the term eustacy to name this cause for Stage boundaries. Whether improvement of numeric ages is most needed for the seismic sequences or for the old Stages is a question of moving the matching problem.

Observed relative changes of sea-level may result from eustacy as well as vertical movement (Figure 9a), a fact which complicates the determination of eustatic rates and magnitudes. Figure 9b classifies possible contributions to vertical movement involving the crust or just the sediment column. For most single well data it is impossible to separate, with any degree of certainty, the eustacy signal from the mixture of local vertical movements. The present cross-section technique, however, allows distinction of relatively high frequency-low amplitude eustacy from low frequency-high amplitude regional vertical movements. Of course, this does not determine (Figure 9c) whether we are dealing with glacial, tectonic, geoidal, thermal eustacy, or the effect of an extra terrestrial body (large and hot) hitting the ocean from outer space.

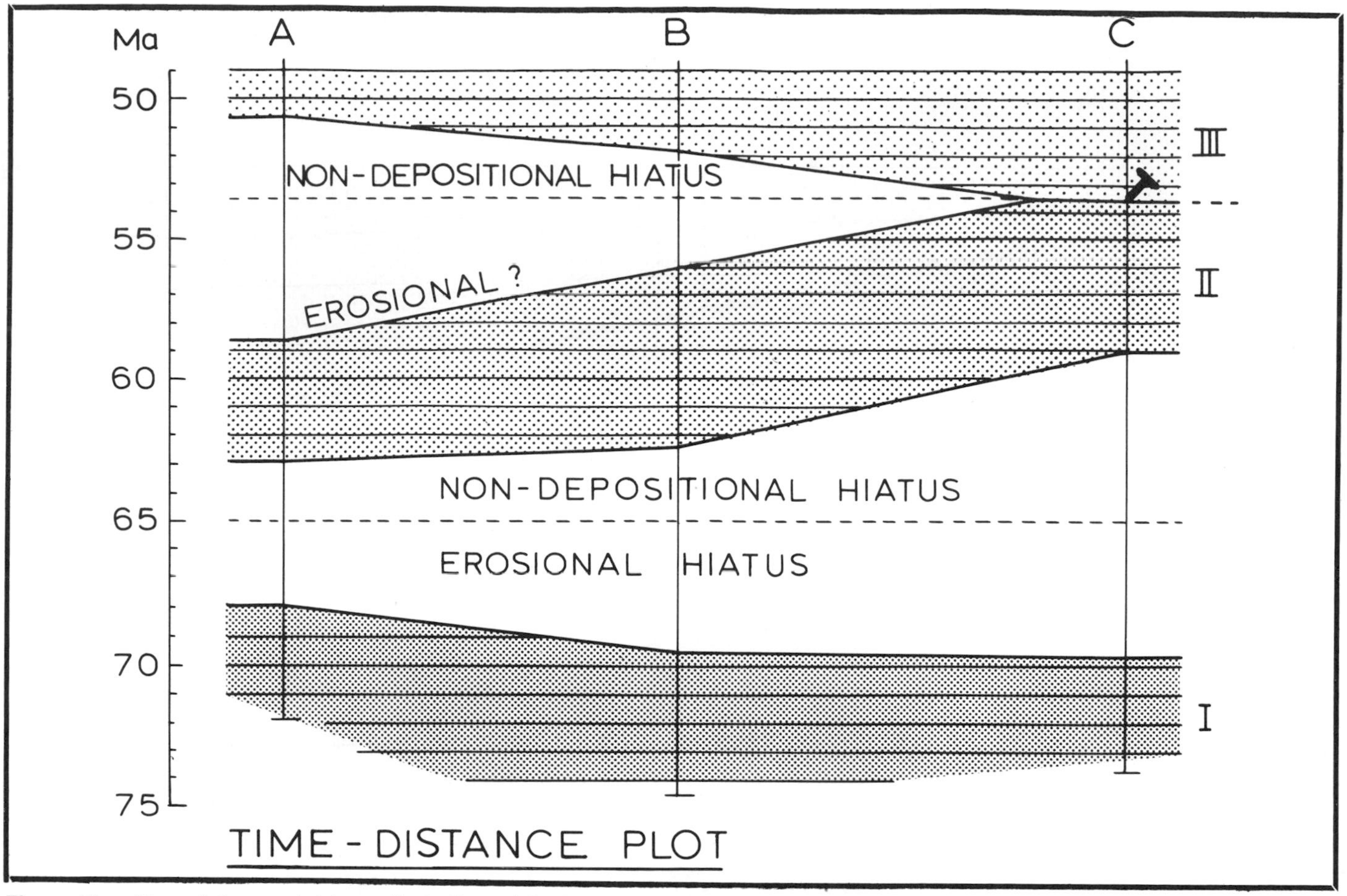

Ma
A
B
C
50
55
60
65
70
75
NON-DEPOSITIONAL HIATUS
EROSIONAL ?
NON-DEPOSITIONAL HIATUS
EROSIONAL HIATUS
III
II
I
TIME - DISTANCE PLOT

Figure 7 — Time-distance plot for wells A, B, and C of Figure 4. Note choice of 53.5 Ma and 65 Ma as ages for events that caused unconformities (compare with Figure 16 and 17 in Van Hinte, 1978). Interval I, II, and III are natural time-stratigraphic units. Nail at 53.5 Ma in C marks the only level where a boundary stratotype would not disrupt natural time-stratigraphic classification. Broken line at 65 Ma drawn at Cretaceous/Tertiary boundary event.

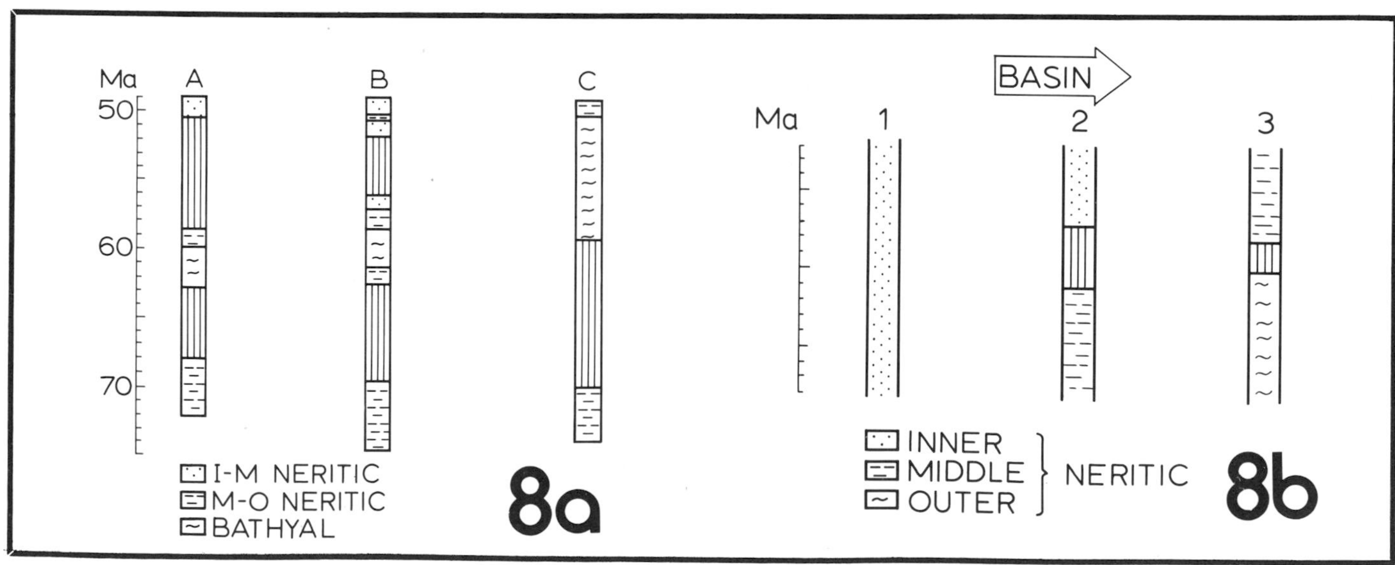

Figure 8 — Columnar time sections of wells A, B, and C of Figure 4 and of hypothetical wells 1, 2, and 3. Column shows depositional environments and presence or absence (vertical signature) of sedimentary record. Illustrates that analysis of unconformities is elementary in unraveling geologic history, and that comparison of real examples with cycle chart of Vail and Mitchum (1979) is most instructive. **A** — Upper conformity is regular effect of sea level drop, C is most basinward well. Lower unconformity suggests reversed basin position and/or C is at paleo-high in larger regional setting. Knowledge of regional geology needed for further interpretation. **B** — Completeness of Well 1 indicates that unconformity has been overlooked in shallow-water section (probably quite common in literature), or implies strong differential movement between wells 1 and 2 during time of hiatus.

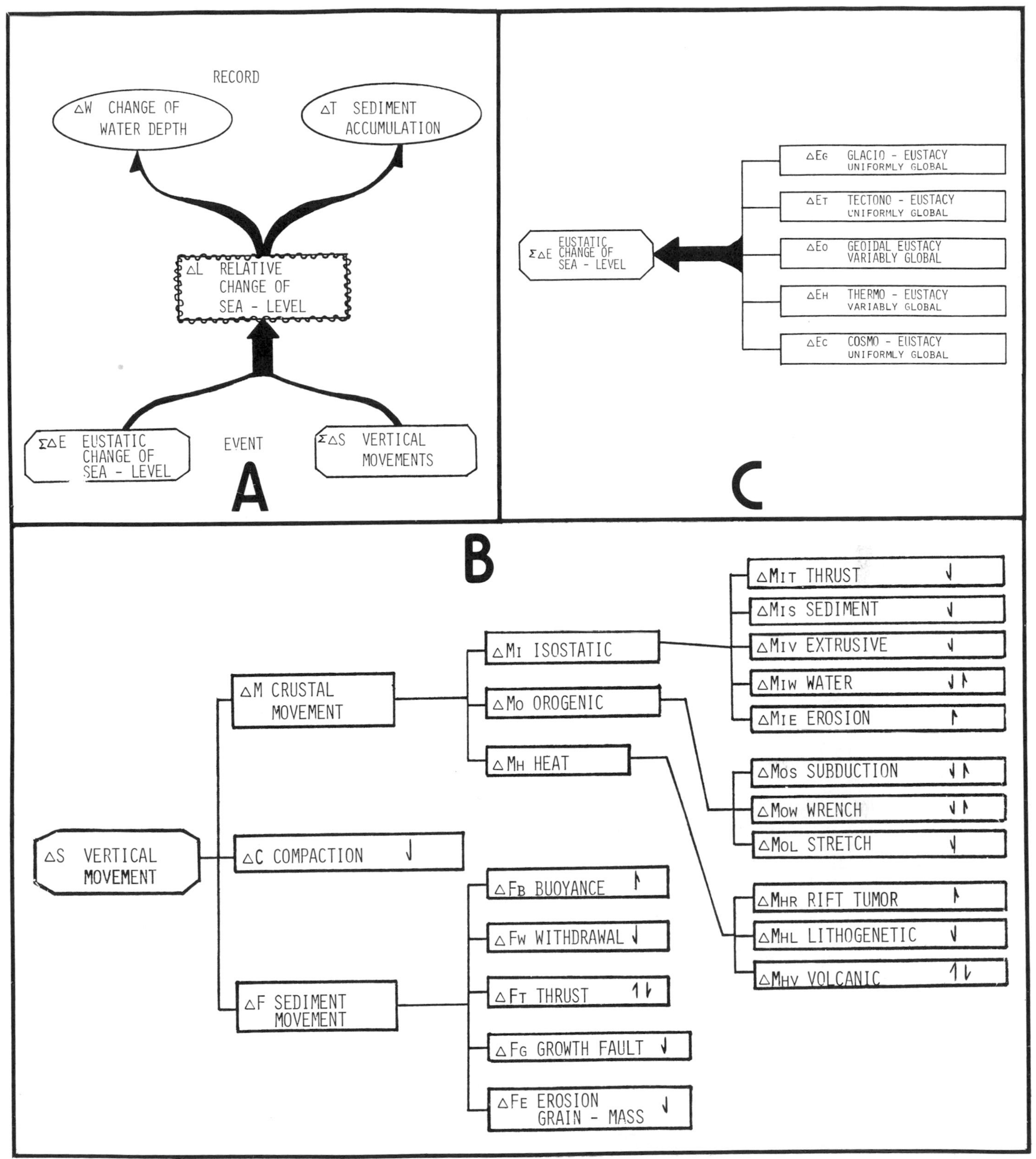

Figure 9 — Summary of possible events and processes leading to relative sea level changes, deduced from local geologic record. Chances to determine rate and magnitude of eustacy seem best at passive margins (outer shelf-upper slope) where most vertical movement elements of Figure 9b equal 0 and the remaining can be calculated from the stratigraphic record and are predictable from geophysical models. Arrows in Figure 9b indicate subsidence and uplift. Eustacy causes (Figure 9c) are complex and their magnitude is yet unknown. Geoidal eustacy, for instance, depends on changes in atmospheric circulation, earth gravity centre (ice caps, plate motions), and variations of seawater density. Cosmo-eustacy refers to effect or impact of extra-terrestrial body in ocean and is brief and probably minor also for large body.

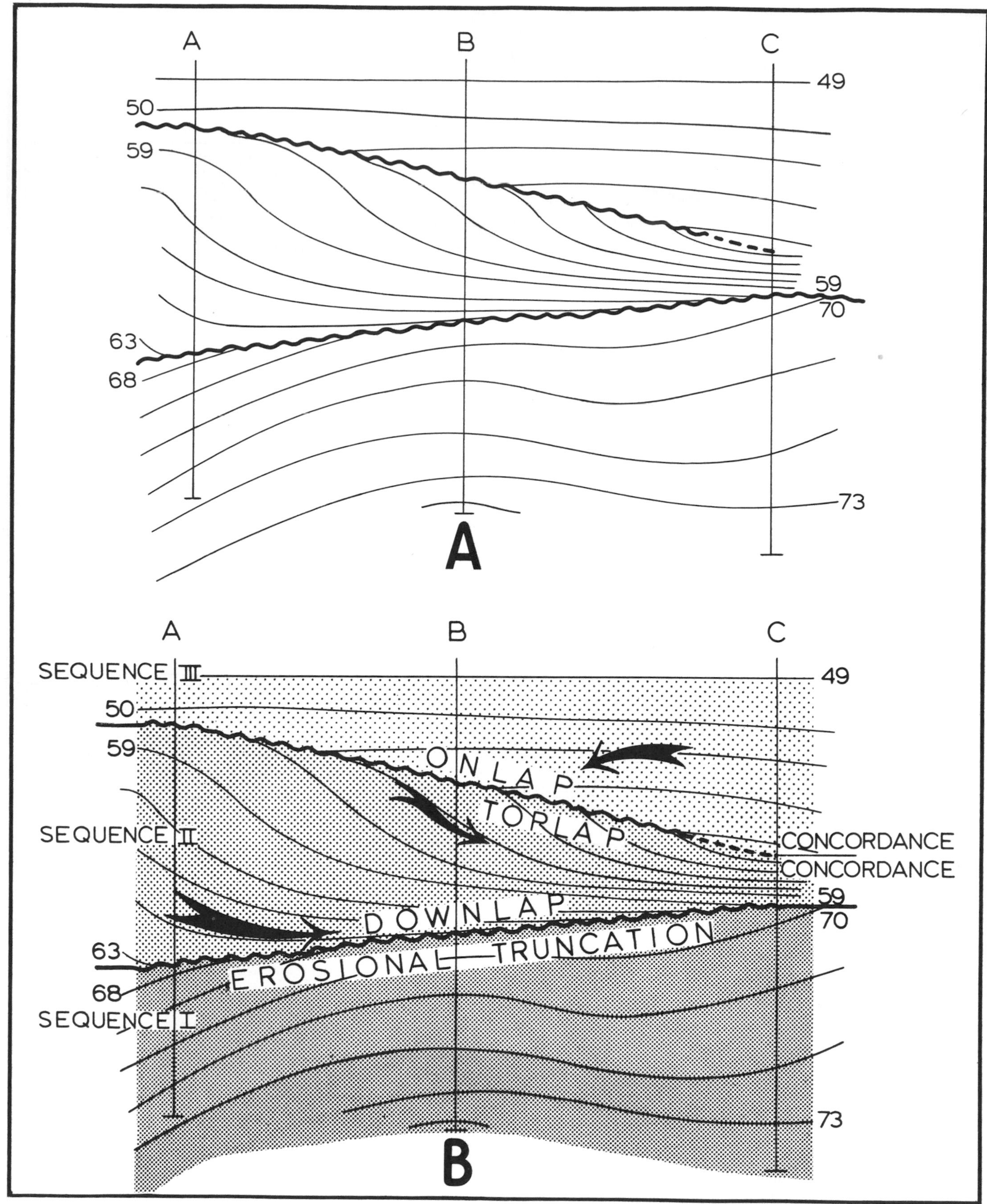

Figure 10 — Synthetic seismic section through wells A, B, and C of Figure 4. Naturally drawn isochrons (10a) make applicability of seismostratigraphic terminology obvious (10b). If A, B, and C were real wells, I through III would be supersequences rather than sequences.

SYNTHETIC SEISMIC SECTION

The isochron cross section of Figure 4 is redrawn in Figure 10 with more natural lines. With this information, we can apply seismostratigraphic terminology to the base and top of our units, including onlap, toplap, downlap, etc. This produces an image of the time-stratigraphic depositional and structural pattern similar to a seismic section that assists geophysicists in dating sequences and correlating through bad data areas. Problems with a lack of control points to handle unconformities (Figure 11) are unavoidable, however. Paleotologic data do not provide answers, and the interpreter must be guided by his ideas about local geology and the nature of the event causing the unconformity.

Our approach may also solve conflicts similar to the one shown in Figure 12. The seismic section and its interpretation are taken from Mitchum, Vail, and Sangree (1977, Figure 3, p. 120). The upper line drawing shows the seismic interpretation suggesting a substantial hiatus at the unconformity. Yet, the paleontologist who reports a "complete section" can be right, as illustrated in the hypothetical 1Ma isochron cross section of the lower line drawing. This synthetic seismic section is drawn on the same scale as the seismic section and shows lower resolution where sediment accumulation rates are high (20 cm/ka) but higher resolution where the rates are low (1.25 cm/ka). The bundle of 10 isochrons (left) corresponds to a complete section where a thick seismological pencil would draw a hiatus. This pseudo-problem arises in work on and near continental margins where low, pelagic sediment accumulation rate are not uncommon.

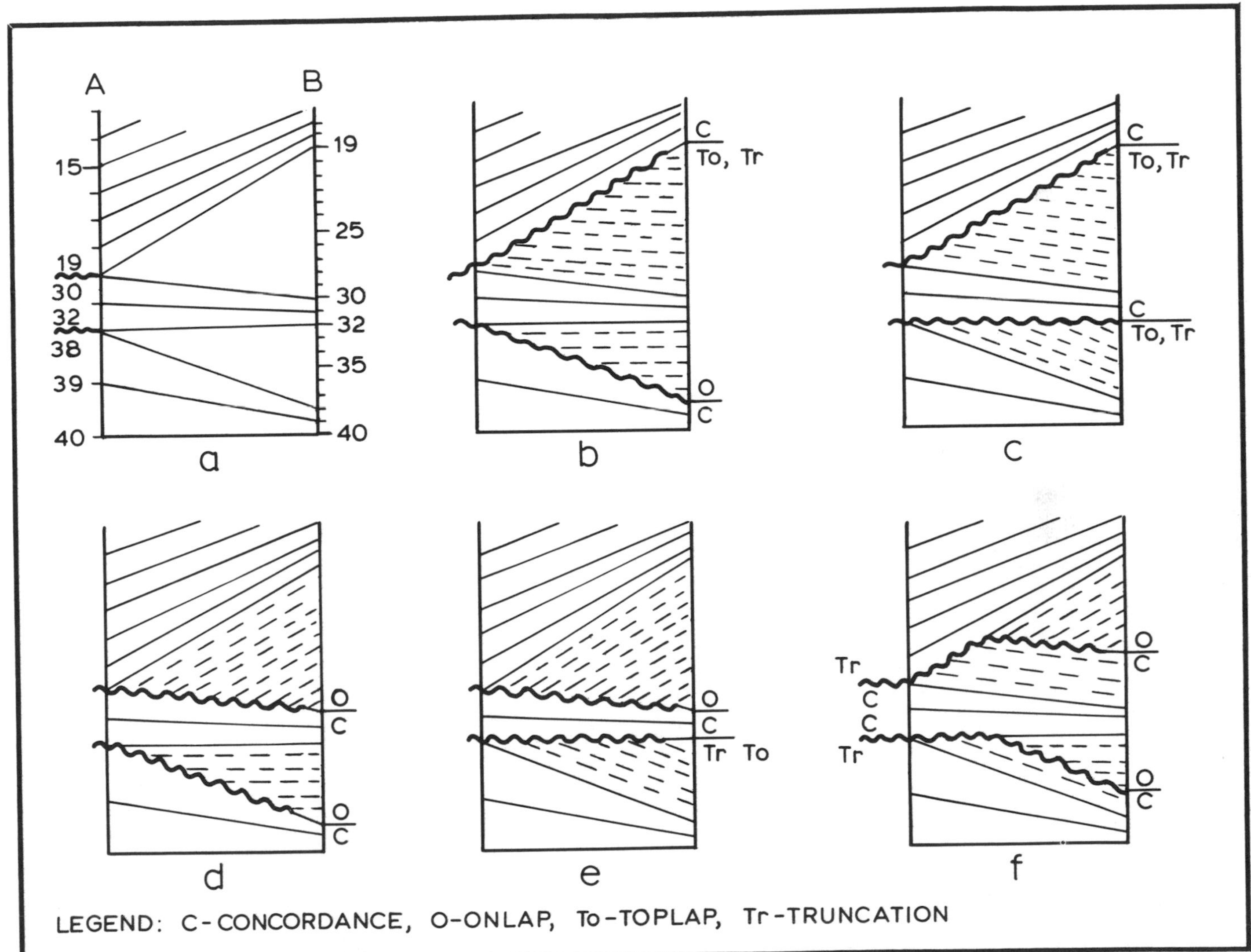

Figure 11 — Pitfalls in construction of isochron cross section on depth-distance plot (compare with Figure 6). Figure 11a shows full Ma levels found in sections A and B with straight line correlations of common ages giving isochrons at 1 Ma interval and leaving part of drawing unfinished. Stratigraphic section is complete at B. Figures b through f illustrate different ways to finish drawing, each completion having different geologic implications. Knowledge of regional geology and of depositional environment of sections, and use of Cycle Chart of Vail and Mitchum (1979), will assist in reaching an independent conclusion to be compared with seismic data.

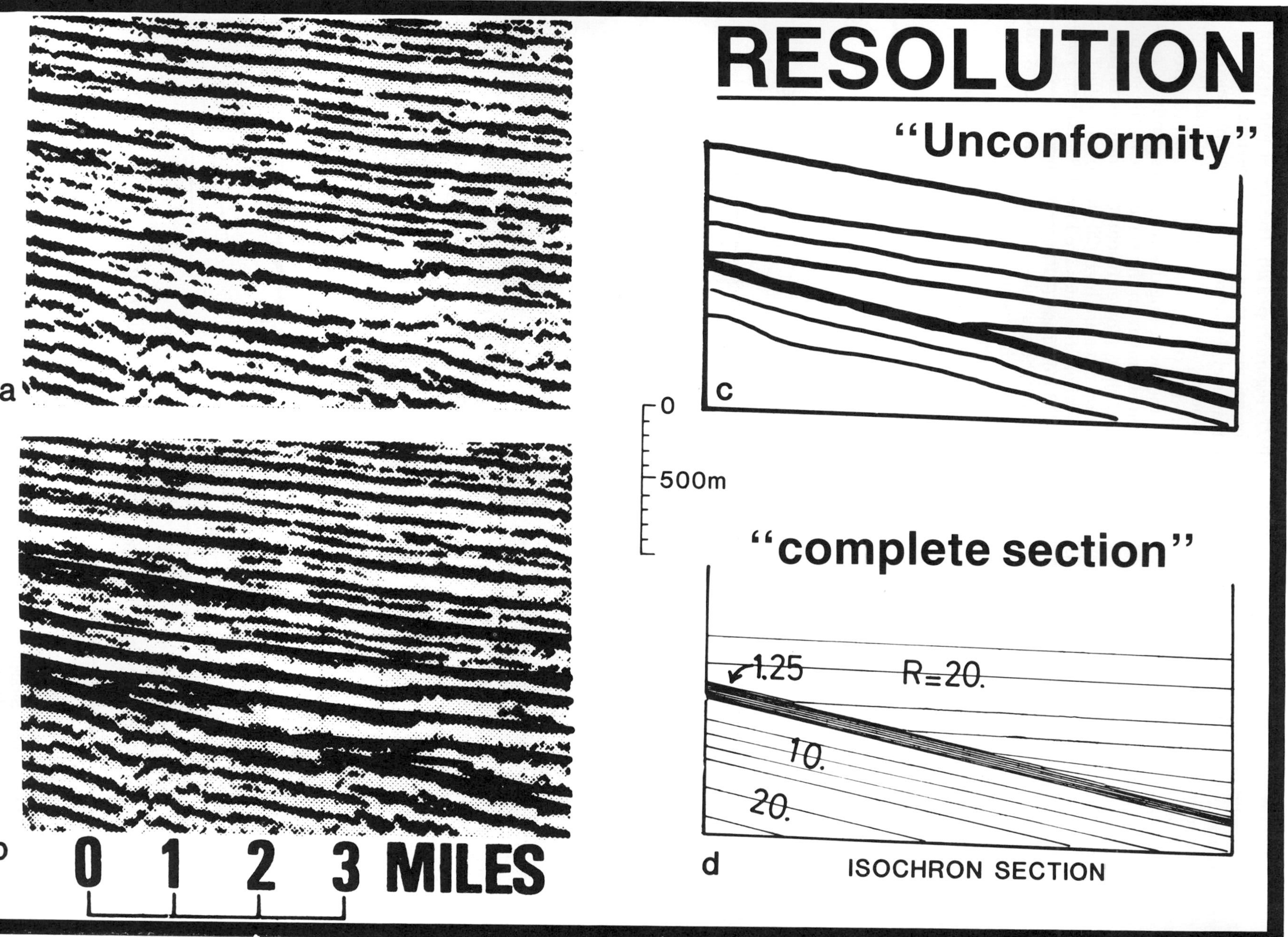

Figure 12 — Illustration of scale problem hampering communication between different explorationists. Seismologists find unconformity with missing section, whereas paleontologists find no hiatus. Seismic section (12a) and interpretation (12b,c) after Mitchum, Vail and Sangree (1977, Figure 3). Synthetic seismic section (12d), drawn to scale with isochron interval of 1 Ma. R = average rate of sediment accumulation; numbers are in cm/ka. The isochron section is hypothetical and, in fact, quite unrealistic as one would rather expect such low accumulation rates in a downlap setting.

SUMMARY

Biostratigraphic resolution is highest on continental margins. Here, biostratigraphy routinely serves exploration purposes and helps define and achieve the two research goals of calibrating seismostratigraphy and answering questions on timing, magnitude, and rate of eustacy. Biostratigraphic calibration, of the depositional sequences recognized on seismic sections, allows for correlation with the classical stages. Both kinds of units were originally defined as unconformity bound, and form the "natural" time-stratigraphic partitions of the sedimentary record. Defining such units with boundary stratotypes would disrupt a natural time-stratigraphic classification.

REFERENCES CITED

Berggren, W. A., 1972, A Cenozoic time scale; some implications for regional geology and paleobiogeography: Lethaia, v. 5, p. 195-215.

——— and J. A. van Couvering, 1974, Biostratigraphy, geochronology and paleoclimatology of the last 15 million years in marine and continental sequences: Paleogeography, Paleoclimatology, Palaeocology, v. 16, p. 1-216.

Hardenbol, J., and W. A. Berggren, 1978, A new Paleogene numerical time scale: AAPG Studies in Geology, n. 6, p. 213-234.

Mitchum, R. M., Jr., P. R. Vail, and S. Thompson, III, 1977, Seismic stratigraphy and global changes of sea level, part 2: the depositional sequence as a basic unit for stratigraphic analysis: AAPG Memoir, n. 26, p. 53-62.

———, ———, and J. B. Sangree, 1977, Seismic stratigraphy and global changes of sea level, part 6: Stratigraphic interpretation of seismic reflection patterns in depositional sequences: AAPG Memoir, n. 26, p. 117-133.

Orbigny, A.d., 1850, Cours elementaire de paleontologie et de geologie Stratigraphique: Paris, Masson, 2 volumes.

Ryan, W. B. F., et al, 1974, A paleomagnetic assignment of Neogene stage boundaries and the development of isochronous datum planes between the Mediterranean, the Pacific and Indian oceans in order to investigate the response of the world ocean to the Mediterranean salinity crisis. Rivista Italiana Paleontologia e Stratigrafia, v. 80, n. 4, p. 631-688.

Stainforth, R. M., et al, 1975, Cenozoic planktonic foraminiferal zonation and characteristics of index forms: University of Kansas Paleontological Contribution, article 62, p. 1-425.

Suess, E., 1909, Das Antlitz der Erde. Prag. Tempsky, v. 3, part 2, 789 p.

Vail, P. R., and R. M. Mitchum, Jr., 1977, Seismic stratigraphy and global changes of sea level, part 1: Overview: AAPG Memoir, n. 26, p. 51-52.

———, ———, 1979, Global cycles of relative changes of sea level from seismic stratigraphy: AAPG Memoir, n. 29, p. 469-472.

———, ———, and S. Thompson, III, 1977, Seismic stratigraphy and global changes of sea level, part 4: global cycles of relative changes of sea level: AAPG Memoir, n. 26, p. 83-97.

Van Hinte, J. E., 1976a, A Jurassic time scale: AAPG Bulletin, v. 60, p. 489-497.

———, 1976b, A Cretaceous time scale: AAPG Bulletin, v. 60, p. 498-516.

———, 1978, Geohistory analysis-application of micropaleontology in exploration geology: AAPG Bulletin, v. 62, p. 201-222.

———, J. P. Colin, and R. Lehman, 1980, Micropaleontologic record of the Messinian event at Esso Libya Inc. Well B1-NC35A on the Pelagian Platform *in* Proceedings, Second Symposium on the Geology of Libya: London, Academic Press, v. I, p. 205-244.

Appendix 1. Paleontologic report on hypothetical Well A*.

Depth	Top	Environment
X m	Middle Eocene Limestone	
1300 m	Lower Eocene Limestone	Inner-Middle Neritic
1550 m	Upper Paleocene Marl (Thanetian)	Outer Neritic - Upper Bathyal
2050 m	Lower Paleocene Marl (Danian)	Upper Bathyal
2700 m	Upper Cretaceous Limestone and Shale	
2700 m	Maastrichtian	Outer Neritic
3050 m	Campanian	

*The paleo-log further gives the top of a shallow water equivalent of the **M. aragonensis** Zone at 1450 m, a marker for the top of the **M. angulata** Zone at 1700 m and for the **S. trinidadensis** Zone at 2500 m, the top of **Bolivinoides miliaris** at 2800 m and the base of **G. calcarata** at 3250 m. No other markers have been recorded till TD at 3430 m. E-log pattern changes occur at 1550 and 2700 m.

Appendix 2. Whole number numeric age depths and the ages at unconformities of wells B and C.

Well B		Well C	
Age Ma	depth m	Age Ma	depth m
49	1300	49	1300
50	1500	50	1525
51	1650	51	1700
51.9	1800	52	1900
56.0	1800	53	2100
57	2000	54	2200
58	2200	55	2240
59	2300	56	2280
60	2400	57	2320
61	2460	58	2360
62	2500	59.0	2400
62.4	2520	69.7	2400
69.5	2520	70	2500
70	2650	71	2800
71	2850	72	3200
72	3100	73	3500
73	3300	TD 73.7	3700
74	3450		
TD 74.5	3500		

Tectonic Processes

Mechanism of Passive Margins and Inland Seas Formation

E. V. Artyushkov
S. V. Sobolev
Institute of Physics of the Earth Academy of Sciences of the USSR

Crustal subsidence on passive margins is caused by three main processes: 1) crustal thinning by horizontal extension, 2) increase in crustal density and crustal attenuation as a result of phase transitions, and 3) cooling of the crust and mantle after a continental breakup. Rapid intracontinental subsidence and formation of inland seas is mainly associated with the second process. Additional subsidence of passive margins and inland seas is produced by the load of sediments.

Continental passive margins are regions of large crustal subsidence where thinning of the crust and transformation of continental regions into deep water basins occur. For example, the present continental margin of the Atlantic ocean, about 100 to 200 km wide, originated after the breakup of Pangaea. In addition, there are some submerged continental regions, like the Rockall and Vöring plateaus, within the oceans.

Continental shelf subsidence proceeded with a somewhat lower intensity than the slope and rise. However, the crust should have thinned or become more dense.

A main problem with the continental margin formation is determining possible mechanisms for continental crust thinning. A similar process also occurs in the inland seas (Black Sea, southern part of the Caspian Sea, etc.) Thus, the mechanism of continental margin formation should be discussed in connection with that of inland seas.

POSSIBLE MECHANISMS OF THE CONTINENTAL CRUST THINNING

Crustal thinning and subsidence can be associated with both horizontal and vertical movements. Currently one of the most popular models attributes crustal thinning to lithospheric stretching (McKenzie, 1978; Sclater and Christie, 1980). According to this model the crust of continental margins is thinned as a result of extension during the rifting before the continental breakup. We discuss whether this model can be applied to all passive margins.

Rift valleys are indeed typical depressions resulting from the extension of continental crust. They form on long uplifts underlain by low-velocity and high temperature mantle. Large tensile stresses arise in the crust of such regions (Artyushkov, 1971, 1972, 1973). Heat flow from the low-velocity mantle strongly decreases the viscosity in the lower crust. Horizontal extension takes place in the latter layer under tensile stresses which results in thinning, faulting, and subsidence of the overlying upper crust. This is a mechanism for rift valley formation on the continents (Artemjev and Artyushkov, 1971; Artyushkov, 1979, 1981a).

The width of a rift valley is determined by the width, ℓ, of the region where a neck-shaped deformation is concentrated in the lower crust. From mechanical considerations, ℓ should be about the same as the thickness of the lower crust, Δ, and can only be a few times greater than Δ. The latter quantity is about 20 to 30 km, therefore the width of an extensional structure of the rift valley type cannot be greater than 100 km. All recent rift valleys are indeed very narrow with a maximum width of ~ 80 km for the case of the Baikal graben.

Continuing extension of a rift valley results in breakup of the continental crust and formation of oceanic crust between two continental blocks moving apart. The initial width of the continental slope should be about one-half the width of the rift valley which existed before the break-up, and cannot be greater than the total width of that structure. That is why the width of a passive margin formed by extension should be a few tens of kilometers and cannot exceed 100 km. In the Gulf of California and the Gulf of Aden, recently formed continental slope is not wider than a few tens of kilometers.

Crustal thinning on very narrow passive margins can be explained by an extension during the continental rifting. However, there are many passive margins, perhaps a majority of them, and submerged continental blocks more than 100 km wide. An outstanding example is the Northern Exmouth Plateau, situated on the northwestern margin of Australia. Its main part, about 400 km wide, is approximately uniformly submerged to a depth of approximately 1.5 to 2 km and is underlain with the continental crust approximately 20 km thick (Exon and Willcox, 1978). Additional examples of passive margins with a width exceeding 100 km may also be given. The crustal thinning on such broad passive margins cannot be explained by a stretching model.

There is a possibility of crustal thinning on passive margins by plastic flow under gravity from the continent toward the ocean after a breakup (Evison, 1960). This process is mechanically similar to the crust stretching and it should result in the formation of numerous deep and narrow grabens in the upper crust after breakup. These structures do exist on passive margins but, according to recent concepts, were formed before, not after, breakup of the continent (Burke, 1976). This argues against a considerable flow of passive margins toward the oceans.

Intense subsidence of continental crust, resulting in the formation of deep basins, occurs on the passive margins and within the continents in the regions of inland seas. Deep depressions of the Black Sea and of the southern part of the Caspian Sea formed as a result of recent subsidence. This is proved by the existence of thick piles of undeformed shallow-water sediments accumulated before the subsidence which created the present-day deep-water basins (Kravchenko and Mouratov, 1973; Yanshin et al, 1977). Like on passive margins, these sediments would be disturbed if deep-water basins formed by extension of continental crust or sea floor spreading.

Flow of the crust or of its lower part in the gravity field can only take place from an area of higher elevation to an area of lower elevation. Consequently, it is obvious that this process could not result in crustal thinning in seas located inside the continents. At the same time the consolidated crust is strongly thinned under the Black Sea and under the southern part of the Caspian Sea, which probably caused recent subsidence. This and other arguments (Yanshin, Artyushkov, and Schlezinger, 1977) make it clear that this phenomenon, as well as crustal thinning on broad passive margins, should be explained by a mechanism not associated with crustal stretching.

In the absence of considerable stretching, crustal thinning is produced only by the replacement of the lower crust material by the mantle. For that to happen, the lower crust should become denser than the mantle to be able to sink into the mantle. The density of mantle at depths of $\gtrsim$30 to 40 km cannot be less than 3.20 g/cu cm (Green and Liebermann, 1976; Sobolev, 1980). Among the continental crust rocks only eclogitized rocks of a mafic composition (garnet granulites and eclogites) are denser and these have seismic velocities close to the mantle's (Manghnani, Andro, and Clark, 1974).

The most probable cause for crustal thinning on broad passive margins and in inland seas is a transition of lower continental crust gabbro into garnet granulite and then into eclogite. This explanation was suggested for the formation of inland seas and intracontinental depressions with strongly attenuated crust (Artyushkov, Schlesinger, and Yanshin, 1976; Yanshin, Artyushkov, and Schlesinger, 1977); of the continental slope on passive margins (Artyushkov, 1979, 1981b); and of deeply submerged continental regions in marginal seas (Artyushkov, 1979, 1981b). The main features of this mechanism are presented here. They are given in combination with detailed analysis of the gabbro-eclogite transition and calculations of its influence on crustal movements in the inland seas and on passive margins done by S. V. Sobolev (1978, 1980).

GABBRO-ECLOGITE PHASE TRANSITION AND ITS RATE

In many continental areas the lower crust apparently consists mostly of mafic rocks.

Experimental research (Ringwood and Green, 1966; Ito and Kennedy, 1971) indicated that under pressures and temperatures characteristic for a relatively cool lower continental crust ($P \sim 5$ to 10 kilobars, $T = 500$ to 700°C) and a small water content in the fluid in mafic rocks, dense garnet granulites and eclogites are stable rather than basalt (Figure 1). The density of eclogites is about 3.45 to 3.6 g/cu cm. The density of garnet granulites of a high grade of metamorphism is >3.2 g/cu cm. Thus, if chemical equilibrium is obtained in relatively cool continental crust, the mafic rocks must be in a phase of garnet granulites and eclogites there (Ringwood and Green, 1966). Whether chemical equilibrium is really obtained or not depends on the rate of the gabbro-eclogite transition.

The gabbro-garnet granulite-eclogite transition results from a number of solid-phase reactions between minerals. When there is no water containing fluid in the rock (i.e., under dry conditions*) the rate of these reactions is limited by the rate of diffusion in the

* The term dry metamorphism is often used also to indicate metamorphism in the presence of fluid with a low content of water. We shall use this term in the above meaning only.

grains of minerals (Ahrens and Schubert, 1975).

In this connection, the gabbro-eclogite transition under non-equilibrium conditions can be described by diffusion models of the kinetic theory of solid-phase reactions (see Appendix). The time scale of the process is determined by the time, τ, necessary for the transition of gabbro into eclogite. Within the framework of diffusion kinetic models, this time can be roughly estimated from the relation (Ahrens and Schubert, 1975; Sobolev, 1978, 1980).

$$\tau(T) \sim \frac{r^2}{\Delta C_{Al^{3+}} D_{Al^{3+}}(T)}; \qquad (1)$$

$$D_{Al^{3+}}(T) = D_o e^{\frac{-E_{act}}{RT}}$$

r = the average radius of mineral grains; $\Delta C_{Al^{3+}}$ = a change in the free Al^{3+} concentration across a garnet grain or border; $D_{Al^{3+}}$; E_{act} = coefficient and activation energy of Al^{3+} diffusion in garnet; R = gas constant; and, T = absolute temperature.

Unfortunately, currently there are no experimental data about diffusion of Al^{3+} in garnet. The value of $\Delta C_{Al^{3+}}$ in (1) is also still uncertain and that is why only rough estimates of $\tau(T)$ based on indirect data are possible now.

Ahrens and Schubert (1975) estimated τ using available experimental data on Al^{3+} (and other cations) diffusion in various minerals. They obtained the values of $\tau \sim 10^7$ years at T ~ 600 to 900°C and more likely at T ~ 800 to 900°C. Our estimate of time, $\tau(T)$, from an equation like (1) is based on: 1) natural zonality of garnets, 2) transition rate under experimental conditions (Ito and Kennedy, 1971) with consideration of the difference between such conditions and natural ones, and 3) rates of reactions similar to that of eclogitization. This indicates that under dry conditions time amounts to $\tau > 5 \cdot 10^5$ or 10^7 years at T = 800°C and $E_{act} \sim$ 80 to 100 kcal/mole (Sobolev, 1976, 1978, 1980).

These estimates agree with the data on dry metamorphism. For example, in high pressure regional metamorphism a "frozen" assemblage of sapphirine+quartz is recorded (Facies of regional metamorphism of high pressure, 1974). It is only stable at a temperature of T > 800 to 900°C (Chatterjee and Schreyer, 1972; Newton, Charlu, and Kleppa, 1974), yet the cooling of these rocks must have proceeded at a very slow rate. Therefore it seems reasonable that eclogitization reactions also "freeze" at a temperature T ~ 800°C under dry conditions. Considerable changes in mineral composition of mafic rocks at temperature T < 800°C do not occur in this case.

Based on kinetic models of the gabbro-eclogite transition, it is possible to calculate the density and seismic velocities in a lower continental crust of mafic composition (see next section and Appendix). Computed values are comparable with the seismic data if the time, τ, for different kinetic models is assumed to be $\tau \sim 10^7$ to 10^9 years at T=800°C. The latter values correlate well with those mentioned above.

When rocks contain a fluid with a relatively low water content, $X_{H_2O} \sim 0.1$, the gabbro-eclogite transition proceeds much faster than under dry conditions (Ringwood and Green, 1966; Ahrens and Schubert, 1975). This is indicated by the presence of eclogites formed at T about 500 to 600°C in metamorphic complexes (Banno, 1970). Dissolution of ions in thin films of fluids between mineral grains leads to a strong acceleration in their transport and, consequently, to an intense catalysis of eclogitization reactions (Ahrens and Schubert, 1975). Other factors also may result in a catalysis of the eclogitization reactions, and one of them could be high deviatoric stresses in rocks (Ringwood and Green, 1966).

Thus the gabbro-eclogite transition can proceed in two fundamental kinetic regimes: under dry conditions, and in a regime of catalysis. Differences in transition rates in these two regimes may be several orders of magnitude.

GABBRO-ECLOGITE PHASE TRANSITION IN CONTINENTAL CRUST UNDER DRY CONDITIONS

According to many petrologists, dry conditions are normal for the lower continental crust (Facies of regional metamorphism of high pressure, 1974). Geological data also evidence that the crust of most continental areas has been subjected to strong heating several times (Hain, 1973). Assuming the lower continental crust consists of mafic rocks, consider what heating and cooling would do to them under normal dry conditions. This permits us to determine the peculiarities of development of the gabbro-eclogite transition in the crust of inland seas and broad passive margins.

Intense heating of the continental crust to T~1100 to 1200°C at its base leads to a de-eclogitization of mafic rocks. Eclogitization decreases as does the density of garnet granulites that have been present in the crust. This leads to an uplift additional to that produced by the heating of the mantle alone. Equilibrium is reached in the mafic rocks of the lower crust until they cool down to T~800°C. Under equilibrium conditions and decreasing temperature, an eclogitization of mafic rocks takes place as demonstrated in Figure 1. If this process proceeds under dry conditions, the rate of reaction becomes very low at T~800°C and the reactions become frozen. Dry conditions at such a temperature exist in the absence of a fluid influx from outside. In this case, by the time the mafic rocks cool down to T~800°C the water-containing fluid previously present either ascended into the upper crust or became entrapped into the structure of amphiboles. Further cooling of mafic rocks does not lead to their eclogitization and metastable mafic rocks remained in the relatively cool continental crust (T<800°C). The degree of eclogitization, the density, and the seismic velocities in such rocks reflect high-temperature equilibrium conditions at T>800°C that existed in the crust earlier (Sobolev, 1976, 1978).

Figure 2a shows computed frozen density (ρ) and V_p velocity versus depth (Z) relations in the lower continental crust (see Appendix). The calculations assumed that the crust copied from T = 1200°C to T = 500 at its base. The V_p values were estimated on the basis of experimental V_p (ρ) data for basalts, granulites, and eclogites (Manghnani, Andro, and Clark, 1974).

Curve 1 represents a ρ (Z), V_p (Z) relation in a cooled lower crust of mafic rocks (P, T diagram, Figure 1a) for the case of eclogitization frozen at T 800 to 850°C. Computations were made based on the kinetic model 1 (see Appendix, equations 3, 5, 6) with $\tau = 10^9$ years at T = 800°C and E_{act}=80 kcal/mole. For this model, the transformation of gabbro into garnet granulite with a density of 3.22 g/cu cm proceeds during 4×10^7 years at T=800°C under the assumed value of τ.

The densities of gabbro and garnet granulites, with low degrees of eclogitization, determined by Ito and Kennedy (1971) seem too high (Figure 1a). Reduction by 2% leads to a gabbro density of about 3.08 g/cu cm instead of 3.14 g/cu cm at P = 10 kbar. The former value of density seems more probable. The reduced frozen ρ (Z), V_p (Z) relation is shown in Figure 2a, curve 2.

Curve 3 represents the frozen $\rho_1(Z)$, $V_p(Z)$ relation for mafic rocks with P, T diagram similar to that of Figure 1b. Kinetic model 1, which corresponds to $\tau = 5 \times 10^8$ years at T=800°C and E_{act}=80 kcal/mole, was used in these calculations. Computations in a framework of kinetic model 2 (Appendix equations 4, 5, and 6) lead to similar $\rho_1(Z)$, V_p (Z) relations for $\tau = 1{\times}10^7$ to $\tau = 5{\times}10^7$ years at T=800C and E_{act}=80 kcal/mole.

Figure 2a shows that under dry conditions, the gabbro-eclogite phase transition does not lead to the formation of heavy garnet granulites of ρ>3.30 g/cu cm and of eclogites in continental crust when crustal thickness is less than 40 km. As evidenced by seismic data, relations V_p(Z) similar to those shown in Figure 2a are common for the continental crust (Seismic models of the lithosphere for the major geostructures on the territory of USSR, 1980).

Mafic rocks in a cool continental crust (T<800°C) under dry conditions are in a metastable state. After an anomalously-heated, low-velocity mantle approaches the crust, the crust becomes heated. The gabbro-eclogite transition rate increases exponentially with temperature rises. As a result, rocks in the lower crust change from a metastable into a stable state. You could expect this to be followed by an increased degree of eclogitization and density (Falvey, 1974; Sobolev, 1976, Haxby, Turcotte, and Bird, 1976; Yanshin, Artyushkov, and Schlezinger, 1977). Computations indicate that if the gabbro-eclogite transition in heated crustal rocks proceeds under dry conditions, considerable eclogitization of mafic rocks is not the rule (Sobolev, 1980). The unfreezing temperature of the phase transition during crustal heating is higher than the temperature of its freezing during the preceding stage of crustal cooling. For this reason, when rocks are heated, as soon as the reactions between the minerals become rapid they immediately result in a de-eclogitization of the rocks.

More or less intense gabbro-eclogite transition under dry conditions may occur due to crustal heating only if the lower part of a thick,> 35 km, continental crust was not earlier heated to the T>800°C earlier. The latter possibility cannot be excluded, but it seems

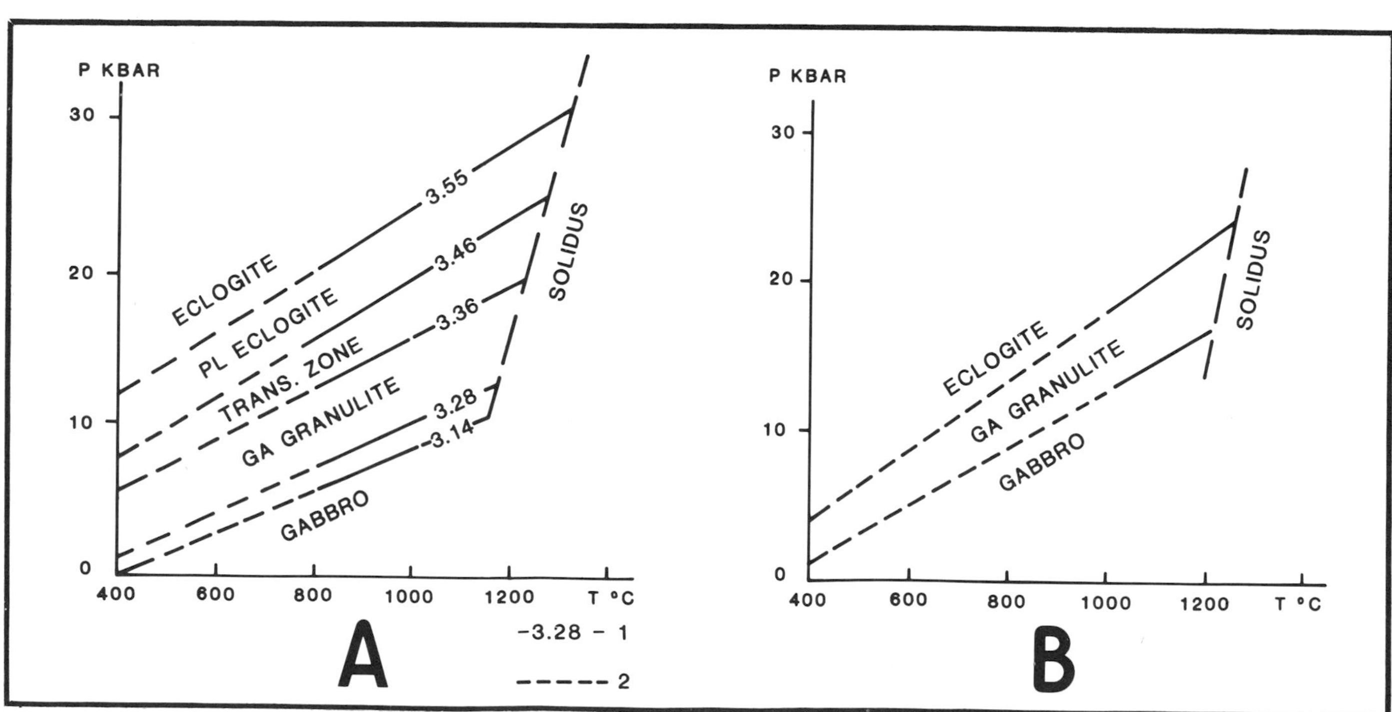

Figure 1 — Experimental P, T diagrams of basalts: **(a)** average oceanic theoleiite by Ito and Kennedy (1971); **(b)** quartz tholeiite by Ringwood and Green (1966). 1=isolines of density in g/cu cm, 2=extrapolation of phase lines.

rather unlikely. Thus, under normal dry conditions the gabbro-eclogite phase transition is not intense enough to cause a considerable crustal thinning and subsidence several kilometers below sea level. The transition occurs mainly due to the cooling of intensely heated continental crust. As a result metastable garnet granulites of relatively low density ($\rho \leq 3.20$-3.30 g/cu cm) remain in a cool lower crust.

MECHANISM OF INLAND SEAS FORMATION

Inland seas, such as the Black Sea, the Southern Caspian depression, and similar deep sea basins, formed in the main part on young platforms and their subsidence was very rapid. The Black Sea basin, 2 km deep, was formed in a few million years. The Southern Caspian depression filled with a layer of young sediments, 10 to 15 km thick, during the last 5 million years (Yanshin et al, 1977, Tectonics of oil and gas-bearing regions on the south of USSR, 1973).

Consolidated crust thickness is strongly reduced under inland seas. For example, it is only about 5 to 20 km under the Black Sea (The Earth's crust and history of development of the Black Sea Basin, 1975) and 10 to 12 km under the Southern Caspian depression. The mantle under the young inland seas and other regions of rapid intra-continental subsidence is strongly heated. This is confirmed by heat flow measurements and other data (Ashirov, Dubrovsky, and Smirnov, 1976; The Earth's crust and history of development of the Black Sea Basin, 1975; Stegena, Geczy, Horvath, 1975).

As was mentioned earlier, crustal stretching most likely did not play a significant role in inland sea formation, such as the Black Sea and the Southern Caspian depression. Their crust should have been thinned by the gabbro-eclogite phase transition in the lower crust, as suggested by Artyushkov, Schlesinger, and Yanshin, (1976), Yanshin, Artyushkov and Schlesinger, (1977) and Artyushkov, Schlesinger, and Yanshin (1979, 1980). In this model, rapid intra-continental subsidence occurs due to an approach to the crust under platform areas of anomalously-heated, low-velocity mantle. The lower continental crust becomes heated to $T \sim 700$ to 800°C and the gabbro-eclogite transition takes place there. Heavy eclogite tears off the crust and sinks into the mantle, strongly reducing crustal thickness. This mechanism explains the formation of inland seas underlain by strongly attenuated crust through vertical subsidence without horizontal extension. This hypothesis uses the kinetic effect of eclogitization of metastable gabbro due to its heating (Sobolev, 1976).

Haxby, Turcotte, and Bird (1976) suggested that the formation of the Michigan basin resulted from the gabbro-eclogite transition due to crustal heating. Sobolev (1980) showed that the kinetic effect of eclogitization due to heated rocks is, by itself, unlikely to produce a phase transition sufficiently intense to explain inland sea formation (see above). To avoid this difficulty, the hypothesis of phase transition catalysis in the lower crust of inland seas was suggested.

Presently, the hypothesis of inland sea formation is as follows. Unlike other continental areas where the gabbro-eclogite phase transition proceeds under normal dry conditions and hence it is not very intense, the inland seas represent areas where abnormal conditions arise in the lower crust. "Abnormal conditions" does not mean that these conditions are necessarily rare. Before subsidence begins, anomalously heated low-velocity mantle comes to the base of the continental lithosphere. Heat flow from the low-velocity mantle reduces the viscosity of the lithosphere. This permits convective instability to develop, replacing the normal mantle layer in the lithsophere with low-velocity mantle of decreased density. The low-velocity mantle then ascends to the base of the crust and its temperature decreases to $T \sim 800$ to 900°C at its top.

The initial temperature of a tectonically stable continental crust is low and reaches $T \sim 500°$ to 600° at its base. The low-velocity mantle heats the lower crust and creates abnormal conditions — the catalysis of the gabbro-eclogite transition. It is most likely that fluids with low water content ($X_{H_2O} \sim 0.1$), penetrating into the lower crust from the high temperature mantle, have a catalytic effect. Magmas crystallizing in the most heated regions of the low-velocity mantle as it cools can be a source of such fluids. In the absence of such fluids, the phase transition catalysis most likely does not arise.

As a result of catalysis the characteristic time of the gabbro-eclogite phase transition drops down to $\tau \sim 10^6$ years at T=600°C. This results in a rapid phase transition in relatively cool sections of the lower crust. The eclogitization of mafic rocks is accompanied by an increase in their density and as a result isostatic subsidence occurs. When the density of the lowermost crust becomes higher than that of low-velocity mantle ($\rho_{L.V.M.} \sim 3.25$ g/cu cm to 3.30 g/cu cm) blocks of heavy rocks tear away from the crust and sink into the mantle, leading to crustal thinning. When a heated mantle comes in contact with overlying mafic rocks, which are still not appreciably metamorphized, the process is repeated. Crustal thinning continues as long as 1) there are anhydrous mafic rocks in the lower crust, 2) the decreasing pressure in the lower crust is still high enough to produce a sufficient increase in the mafic rocks density, 3) the catalysis of the phase transition continues, and 4) the viscosity of the low-velocity mantle is low enough to provide rapid sinking of heavy blocks of the lower crustal material.

Sinking of crustal eclogites into the mantle is confirmed by the investigation of the carbon isotopic composition in diamonds from diamond-bearing eclogite xenoliths in kimberlite pipes. These rocks ascended from the depths > 140 km. At the same time, the isotopic composition of carbon indicates their crustal origin (Sobolev and Sobolev, 1980).

Figures 2b and 2c show the results of mathematical modelling of inland sea formation on the basis of kinetic model 1 (see Appendix). A lower crust 25 km thick was assumed to consist of mafic rocks, their P, T diagram being similar to that shown in Figure 1b. The

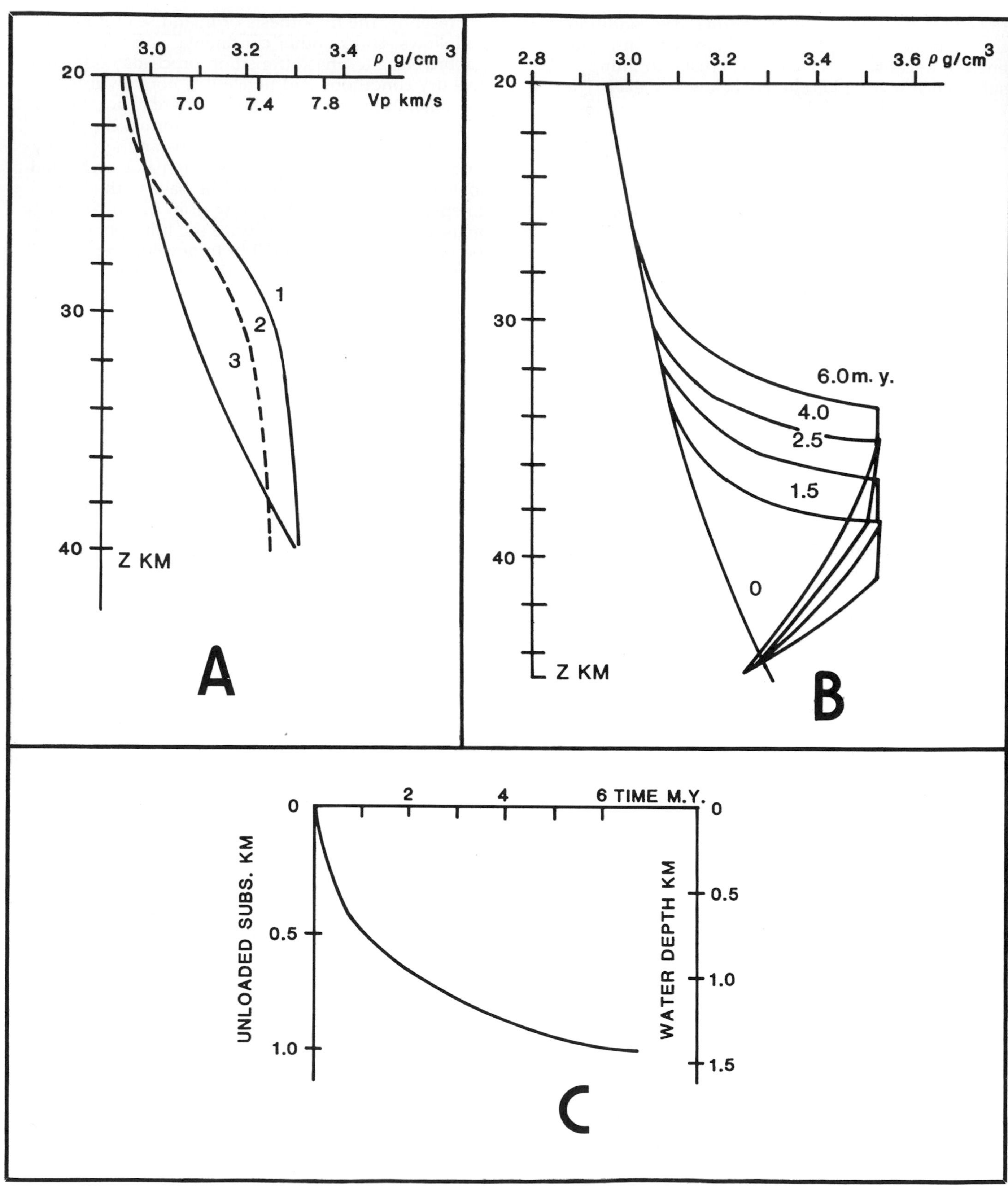

Figure 2 — Development of gabbro-eclogite transition in the continental crust under different conditions. **(a)** Computed frozen density and Vp velocity-depth relations in the mafic lower continental crust. The phase transition proceeds under dry conditions. 1=the Ito and Kennedy (1971) tholeiite, 2=the same reduced by 2%, 3=the Ringwood and Green (1966) tholeiite. **(b)** Computed density-depth relations in the mafic lower crust of an inland sea at different times moments after the commencement of catalysis. **(c)** Subsidence (km) caused by the gabbro-eclogite transition in the lower crust of an inland sea in the regime of intense catalysis.

initial density-depth relation in the lower crust was taken according to a computed frozen $\rho(Z)$ relation (Figure 2a, curve 3). It was assumed that after the freezing of the phase transition in the lower crust in the geological past, 5 km of sediments were deposited and the crust increased from 40 to 45 km thick. The temperature in the crust was assumed to be constant at the time $t<0$, reaching $T=600°C$ at the crustal base. The temperature at this latter boundary increased to $T=900°C$ at $t=0$ due to the approach of low velocity mantle, subsequently remaining constant. Catalysis of the phase transition started at $t=0$, and the time τ was equal to 1 million years at $T=600°C$, and E_{act} to 60 kcal/mole. Figure 2b shows the computed ρ (Z) relations in the lower crust at times $\tau=0$, 1.5, 2.5, 4, and 6 million years after the commencement of the catalysis.

This figure shows that a layer of heavy rocks appears in the lower crust and its thickness rapidly grows. These calculations relate to the first stage of the process preceding the sinking of high density lower crustal blocks into the mantle. The corresponding unloaded subsidence of the crust is shown in Figure 2c. If this subsidence occurs below sea level, it results in formation of a sea basin 1.5 km deep at $\tau=6$ million years. At this moment the crustal thickness may be reduced from H_C = 45 km to H_C = 26 to 30 km after the heavy lower crust sinks into the mantle. The process may repeat if the conditions mentioned above are still fulfilled.

The model fits the geological and seismic data on the central and eastern sections of the Black Sea basin (The Earth's crust and history of development of the Black Sea Basin, 1975). Similar calculations were also made to explain the origin of the Great Hungarian depression (Sobolev, in press).

In some cases, sinking of heavy lower crust into the mantle does not happen. This may be the case, for instance, if the viscosity of the mantle is too high to allow sinking. In that case, the layer of rocks with high density and high seismic velocities remain in the lower crust. According to seismic data, such layers are observed in some regions. The lower crust of the Dnepzovo-Donetskaya depression, ~10 km thick, is characterized by V_p = 7.4 to 7.6 km/sec. There are layers ~10 to 20 km thick, having V_p = 7.4 to 8.0 km/sec, in the lower crust of the Kurinskaya depression situated in Caucasian region (Kosminskaya and Pavlenkova, 1979).

MECHANISM OF PASSIVE MARGIN FORMATION

Crustal subsidence on passive margins is associated with three main processes: crustal thinning due to extension; increase in the lower crustal density caused by the phase transition followed by crustal attenuation in a broad area; and thermal contraction of the crustal and mantle matter during their cooling. An additional subsidence is produced by the sediment load.

In a narrow area, <100 km wide on each side of the ocean, crustal thinning has probably been associated with the lithosphere stretching during the continental rifting (McKenzie, 1978). However, as shown above, the subsidence of the greater part of wide passive margins and particularly of broad plateaus is most likely caused by an eclogitization of mafic rocks in the lower continental crust (Artyushkov, 1979, 1981b).

To cause appreciable crustal thinning, the gabbro-eclogite transition in the lower continental crust must be very intense and it becomes so under abnormal conditions in the lower crust when a catalysis of the eclogitization reactions occurs. That is why we believe that passive margins, like inland seas, are regions where abnormal conditions in the crust have existed.

At the stage of active rifting and breakup, continental margins are areas of intense crustal heating. However, the crust was relatively cool there ($T<800°C$) before the breakup and a few tens of millions of years after it. As shown earlier, an intense gabbro-eclogite transition can take place in such a crust under certain conditions, resulting in crustal attenuation and subsidence.

To reveal whether there has been considerable crustal thinning on a continental margin before the breakup or after it, we should determine the crustal thickness at breakup. This value can be determined with the following data: 1) Elevation of the margin outside the deep rift depression, E, just before or just after the breakup at a distance of about 50 to 100 km from the breakup line; and 2) The depth of the ocean just after its formation, h_O. Using the relation of isostatic equilibrium it is easy to show that these values are related to the crustal thickness on the passive margin H_C,

$$E = \frac{\rho_m - \rho_c}{\rho_m} \cdot H_c - \frac{\rho_m - \rho_o}{\rho_m} \cdot h_O \tag{2}$$

ρ_c, ρ_O are crustal and water densitites, ρ_m is the mean density of the mantle material at the depth Z $< H_C$ under a mid-ocean ridge. The depth of a new formed ocean *h* is the same as that of the axial zones of a newly formed mid-ocean ridge. The latter quantity is about 2.5 km for the recent mid-ocean ridges (Parsons and Sclater, 1977). The density of the low-velocity mantle is about 3.15 g/cu cm in a layer 30 km thick beneath the mid-ocean ridges (Talwani, LePichon, and Ewing, 1965). Substituting ρ_m = 3.15 g/cu cm, h_O = 2 to 3 km, ρ_O = 1.0 g/cu cm, we find that for a crustal thickness of 40 km the uplift should be E = 1.8 to 2.4 km at ρ_c = 2.85 g/cu cm, and E = 2.4 to 3.1 km at ρ_c = 2.80 g/cu cum. Similarly we find that E = 0.1 to 0.5 km at ρ_c = 2.85 g/cu cm and E = 0.2 to 0.9 km at ρ_c = 2.80 g/cu cum, when crustal thickness is 20 km.

Depending on the time of crustal attenuation (before or after breakup), two main models of passive margin evolution are suggested.

Model A

Crustal thinning on a broad passive margin takes place before continental breakup due to an intense gabbro-eclogite transition in the lower crust. The transition is caused by an approach of heated low-velocity mantle to relatively cool continental crust, T <800°C, and by a catalysis of the eclogitization reactions in mafic rocks. It probably happens at the beginning of a new rifting phase when the crust is still not strongly heated. This process is similar to the formation of inland seas and it also leads to rapid crustal subsidence.

Active rifting and continental breakup occur when large masses of intensely heated, low-velocity mantle ascend to the crust. Injection of low-velocity mantle under the crust results in its isostatic uplift. In a number of areas, where medium intensity eclogitization has taken place and the crust did not thin too much, crustal subsidence may not have resulted in deep sea basin formation. Subsidence could have happened above sea level and have been partially camouflaged by intense uplift of the territory. The Pannonian basin and the Basin and Range province are examples of areas where thinned continental crust are located above sea level and at a high altitude in some regions.

The left section of Figure 3a shows a possible type of evolution of a broad passive margin, before breakup, outside an area of strong extension. The crust thins from 40 to 20 km (curve 1) and from 40 to 15 km (curve 2) at the beginning of strong heating of the crust and mantle as in inland sea formation. Curves 3 and 4 show a supposed uplift of a non-attenuated crust due to heating and the replacement of the mantle in the lithosphere by a strongly heated low-velocity and low-density mantle. We do not know exactly how rapid this uplift could be, which makes the curves in the left section of Figure 3a more qualitative rather than quantitative.

As shown earlier, crustal thinning on passive margins more than 100 km wide cannot result from lithospheric stretching. This does not mean, however, that no faults and grabens can form on such margins during continental rifting. The main extension > 50 to 100% is concentrated in a narrow zone where the breakup then occurs. A minor extension, about several percent, may take place in the adjacent wide areas. An average extension of 5 to 10%, combined with crustal thinning due to phase transition, is enough to result in graben formation about 5 km deep bordered by faults dipping at a 45° angle and covering an area 100 to 200 km wide. This extension, however, is too small to cause appreciable crustal thinning.

It is most likely that even a relatively strongly attenuated crust ($H_C < 20$ km) will be located above sea level on passive margins during the last phase of rifting just before the breakup (see Figure 3a). The elevation of such an uplift is not very high and it is unlikely to appreciably exceed 1 km.

After the breakup, continental blocks, together with the underlying low-velocity mantle, move away from the source of new masses of strongly heated mantle matter. As a result, the crust and underlying mantle cool, leading to a slow isostatic subsidence of thinned continental crust. The latter process can be described by thermal models similar to those for the ocean floor subsidence (Parsons and Sclater, 1977). Figure 3a (right section) shows the computed time dependence of subsidence of an attenuated crust after the beginning of the crust and mantle cooling. The elevation just after breakup was calculated on the basis of equation (2) for two cases of crustal structure shown in Figure 3a. The values ρ_m = 3.15 g/cu cm and h_0 = 2.5 km were used, and the appendix shows a description of the thermal model used. An additional subsidence, with respect to that shown in Figure 3a, may be produced by the load of sediments.

The features of the A model of evolution of a broad passive margin are:

1) Subsidence of a broad passive margin before continental breakup. It may be not very pronounced because of simultaneous, intense heating of the mantle resulting in the uplift of the crust.

2) Relatively small uplift, $E \leq 1$ km, during the most intense phase of rifting just before the breakup.

3) Subsidence after the breakup which is close to that predicted by models of lithospheric cooling.

The geological history of Northern Exmouth Plateau, northwestern Australia (von Rad, personal communication), is better described by Model A than Model B.

Model B

Crustal thinning on broad passive margins occurs after continental breakup. The catalysis of gabbro-eclogite phase transition does not take place during the rifting preceding breakup.

At the stage of intense rift formation the temperature at the Moho discontinuity is very high: T~1100 to 1200°C. This is indicated by the intense basaltic volcanism typical of the most active rift zones. In the lower crust under the continents the pressure is ~ 6 to 12 kilobars at 20 to 40 km deep. As shown in Figure 1, only light garnet granulites can exist under such pressures at T~1100 to 1200°C. Thus, crustal attenuation cannot occur even if the catalysis of eclogitization takes place. The anomalously heated crust is strongly uplifted in such a case, and the uplift reaches maximum value before breakup and at that time it most likely exceeds 1.5 to 2 km. The sediments are deposited only in long narrow grabens.

After continental breakup the lithosphere of passive margins cools down. If the catalysis of the gabbro-eclogite transition is proceeding, the density of mafic rocks in the lower crust strongly increases during crustal cooling.

Figure 4 shows the $\rho(Z)$ relations in the basaltic layer of the crust at t=10, 20, 30, and 35 million years after the beginning of crustal cooling. This layer, 20 to 25 km thick, was assumed to consist of basalt with the P, T diagram of Figure lb. The density changes in the basaltic layer produced by gabbro-eclogite phase transition in the regime of intense catalysis were calculated based on kinetic model 2 (see Appendix, equations 4, 5, and 6). The values of kinetic parameters τ=1 million years, at T=600°C and E_{act}=60 kcal/mole, were used.

Figure 4 shows that the gabbro-eclogite phase transition does not appreciably develop during the first 10

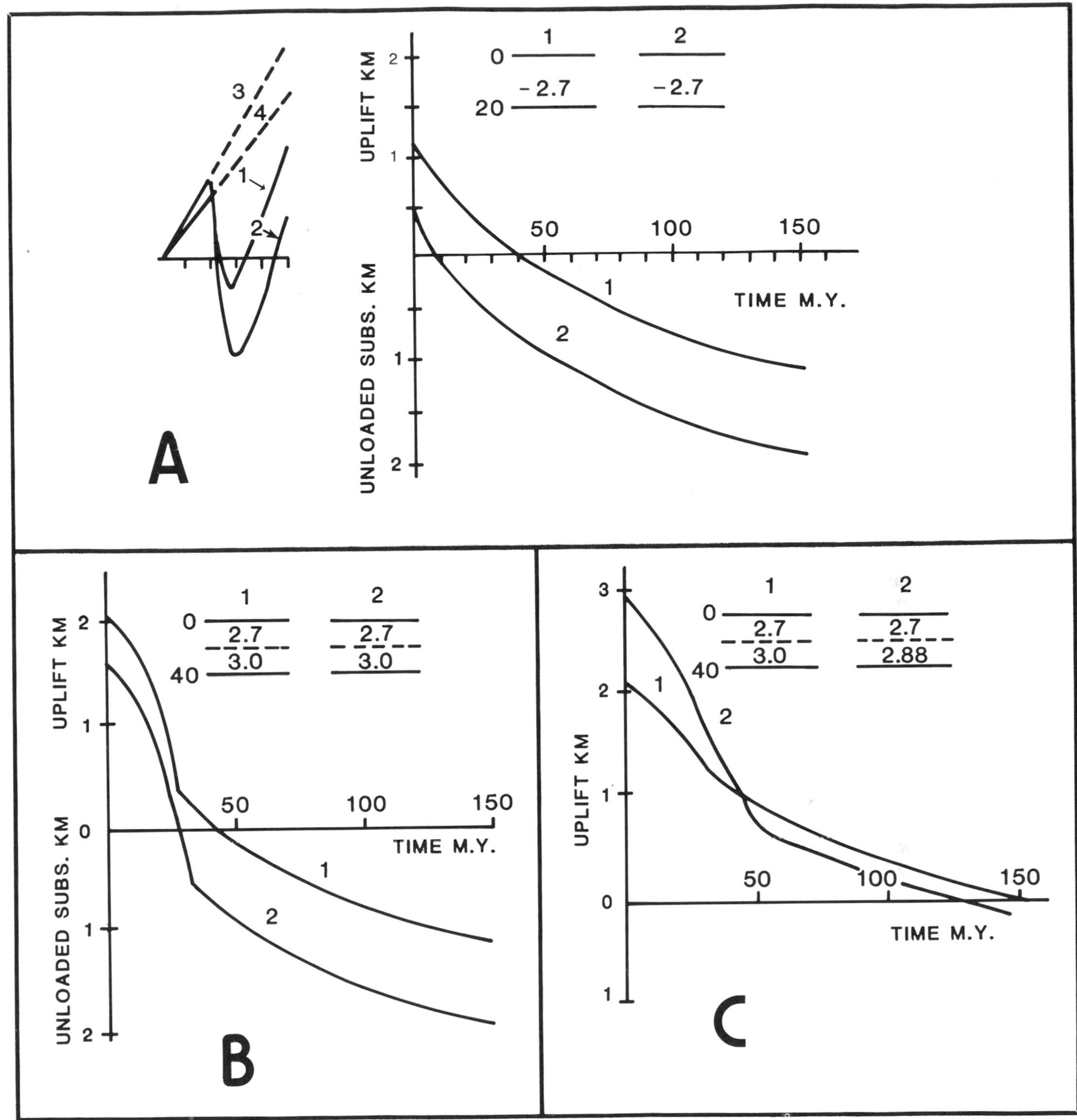

Figure 3 — Models of the evolution of broad passive margins. In the upper section of each drawing, the crustal structures assumed are shown with values of density in grams/cubic centimeter and depth in kilometers. **(a)** Possible vertical movements on passive margin before the breakup (left section). Thermal subsidence after the beginning of crust and mantle cooling (right section). 1=crustal thinning from 40 to 20 km; crustal structure at the moment of continental breakup (1). 2=crustal thinning from 40 to 15 km; crustal structure at the moment of breakup (2). 3=the supposed uplift of a non-thinned crust with the basaltic layer 20 km thick. 4=the supposed uplift of a non-thinned crust with the basaltic layer 25 km thick. **(b)** Subsidence after the breakup and beginning of the lithospheric cooling. 1=crustal thinning from 40 to 20 km; crustal structure at the moment of continental breakup (1). 2=crustal thinning from 40 to 15 km; crustal structure at the moment of continental breakup (2). **(c)** Subsidence of a broad passive margin outside the central rift valley after the breakup in a case when the gabbro-eclogite phase transition is poorly-developed. 1=the phase transition under dry conditions; initial crustal structure at the moment of continental breakup (1). 2=the intermediate composition of the lower crust; crustal structure at the moment of continental breakup (2). The phase transition proceeds in the regime of catalysis.

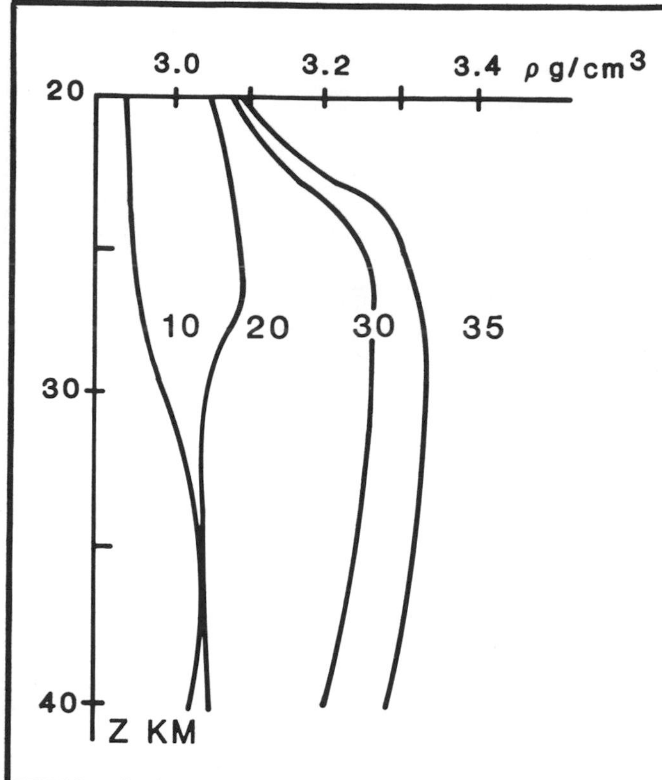

Figure 4 — Density-depth relations in the cooling crust of passive margin at different time moments. The gabbro-eclogite transition is proceeding in the regime of catalysis. Numbers near the curves indicate time intervals in million years after the beginning of cooling.

to 15 million years after the beginning of crustal cooling. At that time the temperature of the crust is still very high. The transition becomes considerable at t>20 million years, and leads to the formation of a thick layer of heavy rocks, $\rho > 3.20$-3.30 g/cu cm and crustal attenuation at t ~ 30 to 35 million years. This happens when the temperature becomes T~800 to 900°C at the base of the crust. The time necessary for such a drop in temperatures lies in the range of t ~ 20 to 50 million years for different thermal models.

Increases in the density of the mafic rocks and crustal attenuation result in subsidence (additional to thermal subsidence). Figure 3b shows computed subsidence of passive margins corresponding to two different crustal structures. The elevation of the crust just after breakup was calculated on the basis of equation (2) using the same values of ρ_m and h_o as before. The crust of structure 1 (Figure 3b) thins from 40 to 20 km (curve 1) and that of the structure 2 from 40 to 15 km (curve 2). Figure 3b shows that the total subsidence of the passive margin is much more rapid at $\tau < 40$ million years than the thermal subsidence for the model under consideration.

After crustal thinning, the subsidence of a broad passive margin is caused mainly by the thermal contraction of the underlying mantle and of the crustal material. The type of passive margin evolution is the same in A and B models at that period of time. Features of the B model of the evolution of broad passive margins are:

1) In contrast to the A model there is no subsidence of a broad area during the continental rifting. The maximum uplift is very high, E = 1.5 km. The sediments are deposited only in narrow grabens.

2) Subsidence during the time interval 10 to 20 million years < t < 30 to 60 million years after the beginning of crust and mantle cooling is much more rapid than the thermal model predicts.

The two models presented above do not describe all possible types of passive margin evolution. The real situation may be much more complicated. Crustal thinning on some passive margins may take place, for instance, both before and after the breakup, and the corresponding evolution should be between those mentioned above. Crustal thinning on broad passive margins may sometimes result from the interaction between the continental crust and the low-velocity mantle, which ascended to the crust several tens of million years after the continental breakup.

Our calculations do not consider the effect of erosion on passive margins. This process is believed to be not so important for their evolution as crustal thinning and lithospheric cooling. The 1 km erosion of the upper crust leads to an additional unloaded subsidence about 0.25 km. This process should be more important on passive margins developing according to the B model evolution than for the A model.

The width of passive margins varies along the ocean-continent boundaries, indicating that the intensity of gabbro-eclogite phase transition was different in different regions. In some cases it probably did not result in significant crustal thinning and subsidence. In such regions the continental slope is very narrow and was probably formed by lithospheric stretching during intense continental rifting.

A low-degree development of the gabbro-eclogite phase transition in the lower continental crust occurs in two main cases (Sobolev, 1980): (1) if this process develops under dry conditions, and (2) if the content of anhydrous gabbro in the lower crust is low. Figure 3c shows the unloaded subsidence of such passive margins after continental breakup. Curve 1 represents subsidence caused by the crust and mantle cooling and by eclogitization under dry conditions in a lower crust 20 km thick (P, T diagram of Figure 1b, kinetic model 2, $\tau = 1.10^7$ years at T=800°C and E_{act}=80 kcal/mole). Crustal subsidence in the case of intermediate composition of the lower crust and of the catalysis of eclogitization reactions is represented by curve 2 (kinetic model 2, τ= million year at T=600°C, E_{act} = 60 kcal/mole). The lower crustal layer, 20 km thick, was assumed to consist of gabbro (50%, P, T diagram of Figure 1b) and of acidic rock (50%, ρ_a = 2.75 g/cu cm). Crustal structure for the above cases during continental breakup is shown in Figure 3c. In both cases, there is practically no crustal thinning on passive margins and deep water basins are not formed there.

In this way, the variations of margin widths may result from variations in the composition of the lower crust and in the rate of gabbro-eclogite phase transition. The catalysis of the transition apparently takes place only in crust underlain by low-velocity mantle. Therefore, variations in the width of the low-velocity mantle belt under the continent before breakup may also result in corresponding variations of the passive margin widths.

SEISMIC VELOCITIES IN THINNED CONTINENTAL CRUST

As mentioned above, a layer of garnet granulites with $V_p \sim 7.4$ to 7.6 km/sec sometimes remains in the crust of regions of rapid intracontinental subsidence. A similar layer may exist beneath some passive margins.

The velocities $V_p \sim 7.4$ to 7.6 km/sec, intermediate between the crust and mantle, may be typical of three types of rocks: 1) relatively cool garnet granulites of rather high density (Manghani, Andro, and Clark, 1974); 2) strongly heated, low-velocity mantle, T ~ 1,100 to 1,300°C (Green and Liebermann, 1976); or 3) partially serpentinized, ultramafic rock formed by hydratation of mantle peridotite.

The last rocks are normally formed in rather small quantities at the base of the oceanic crust. They usually form a layer no more than a few kilometers thick. Passive margins are usually regions of low temperature in the crust and upper mantle; therefore, cool regions on passive margins, where a layer with P-wave velocities $V_p \sim 7.4$ to 7.6 km/sec and > 5 to 10 km thick are found, can be definitely considered as continental blocks strongly subsided by the gabbro-eclogite transition.

As a consequence of crustal attenuation by eclogitization, the consolidated crust under the inland seas and broad passive margins should mainly consist of acidic and intermediate rocks. Nevertheless P-wave velocities in the consolidated crust of inland seas are usually higher, $V_p \sim 6.4$ to 6.8 km/sec rather than the $V_p \sim 6.2$ km/sec of normal "granitic". Basaltic velocities are also typical of the consolidated crust of the continental slope of passive margins, especially of its deeper part.

The "granitic" layer of the continents generally represents a complicated mixture of rocks of various composition plus a certain amount of sediments. These rocks usually exist under rather low temperatures. After the destruction of the basaltic layer beneath the passive margins and inland seas, the granitic layer contacts a strongly heated low-velocity mantle. This can result in metamorphic reactions increasing seismic velocities and in heating rocks strongly decreasing their viscosities. Thus, most fractures can be closed under lithostatic pressure. This effect is also known to increase seismic velocities (Nur and Simmons, 1969). Note that typical P-wave velocities are considerably higher in granitic and intermediate rocks with closed microfractures (V_p no less than about 6.5 to 6.6 km/sec) than those normally observed in the granitic layer of the continents.

In this way a strongly-thinned continental crust of an acid and intermediate composition acquires velocities of elastic waves characteristic for the basaltic layer of continents or for the oceanic crust. Because of this, it can sometimes be taken for a thick oceanic crust.

In conclusion, note that crustal transformation by a gabbro-eclogite transition is fundamentally different from an oceanization. The latter process implies a change in the composition of the crust, due to an input of ultrabasic material from the mantle into the crust. As a result, the upper sialic crust supposedly becomes basic and the lower basic crust becomes ultrabasic. In contrast, the gabbro-eclogite transition is a phase transition which does not change the mean chemical composition of the rock. Heavy garnet granulites and eclogite tearing away from the crust and sinking into the mantle have the same mean chemical composition as that of basalt. Sialic rocks in the upper crust are not appreciably influenced by this transition and they are preserved under a sedimentary layer.

REFERENCES CITED

Ahrens, T. J., and G. Schubert, 1975, Gabbro-eclogite reaction rate and its geophysical significance: Reviews of Geophysics and Space Physics, v. 13, p. 383-400.

Artemjev, M. E., and E. V. Artyushkov, 1971, Structure and isostasy of the Baikal rift and the mechanism of rifting: Journal of Geophysical Research, v. 26, p. 1197-1211.

Artyushkov, E. V., 1971, Horizontal stresses in the lithosphere in the state of isostasy: Doklady Akademii, Nauk SSSR, v. 201, p. 1084-1087 (in Russian).

———, 1972, Origin of large stresses in the Earth's crust: Rizika Zemli, v. 8, p. 5-25 (in Russian).

———, 1973, Stresses in the lithosphere caused by crustal thickness inhomogeneities: Journal of Geophysical Research, v. 78, p. 7675-7708.

———, 1979, Geodynamics: Moscow, Nauka, 328 p. (in Russian).

———, 1981a, Mechanism of continental riftogenesis: Tectonophysics, v. 73, p. 9-14.

———, 1981b, Physical origin of crustal movements on passive margins, *in* Geology of continental margins: Oceanologica Acta, v. 4, p. 167-170.

———, 1981c, Mechanism of formation of active margins, *in* Geology of continental margins: Oceanologica Acta, p. 245-250.

———, A. E. Schlesinger, and A. L. Yanshin, 1976, Crustal extension and formation of sedimentary basins: Durham, Symposium on Sedimentary basins of the continental margin and Craton. Sedimentary basins of the continental margin and craton.

———, ———, and ———, 1979, Main types of mechanisms of formation of the structures on the lithospheric plates, II, *in* Marine depressions and plates and regions of deiteroorogenesis: Moscow, Moskovskoye Olschestvo Ispytateley Prirody, Byulleten', Otdel Geologicheskiy, n.3, p. 3-13 (in Russian).

———, ———, and ———, 1980, Origin of vertical crustal movements in the plate interiors, *in* Dynamics of plate interiors: Washington, D.C., American Geophysical Union, ICG WG 7 Final Report.

Ashirov, T., V. G. Dubrovsky, and Y. B. Smirnov, 1976, Geothermic and geoelectric investigations in southern Caspian depression and nature of the layer of high conductivity:

Doklady Akademii, Nauk SSSR, v. 226, n. 2, p. 401-404 (in Russian).

Banno, S., 1970, Classification of eclogites in terms of physical conditions of their origin: Physics of the Earth and Planetary Interiors, v. 3, p. 405-421.

Burke, K., 1976, Development of grabens associated with the initial rupture of the Atlantic ocean: Tectonophysics, v. 36, p. 93-112.

Chatterjee, N. D., and N. Schreyer, 1972, The reaction enstatite sillimanite = sephirine + quartz in a system Mg0 — $A1_20_3$ — $Si0_2$: Contributions to Mineralogy and Petrology, v. 36, p. 40-62.

Evison, F. F., 1960, On the growth of continents by plastic flow under gravity: Royal Astronomical Society, Geophysical Journal, v. 3, p. 155-190.

Exon, N. F., and J. B. Willcox, 1978, Geology and petroleum potential of Exmouth Plateau area of western Australia: AAPG Bulletin, v. 62, n. 1, p. 40-72.

Facies of regional metamorphism of high pressure, 1974: Moscow, Nedra, 328 p. (in Russian).

Falvey, D. A., 1974, The development of continental margins in plate tectonic theory: Australian Petroleum Exploration Association Journal, v. 14, p. 95-106.

Green, D. H., and R. C. Liebermann, 1976, Phase equilibria and elastic properties of a pyrolite model for the oceanic upper mantle: Tectonophysics, v. 32, p. 61-92.

Hain, V. E., 1973, General Geotectonics: Moscow, "Nedra," 511 p. (in Russian).

Haxby W. F., D. L. Turcotte, and J. M., Bird, 1976, Thermal and mechanical evolution of the Michigan basin: Tectonophysics, v. 36, p. 57-75.

Ito, K., and G. C. Kennedy, 1971, An experimental study of the basalt-garnet granulite-eclogite transformation, *in* the structure and physical properties of the Earth's crust: Geophysical Monograph Series, v. 14, p. 303-314.

Kononyuk, I. F., 1975, Models of solid-phase reactions in the powder mixture, *in* Heterogeneous chemical reactions and reactional ability: Minsk Science and Technology, p. 93-115 (in Russian).

Kosminskaya, I. P., and N. I. Pavlenkova, 1979, Seismic models of inner parts of the Euro-Asian continent and its margins: Tectonophysics, v. 59, p. 307-320.

Manghnani, M. H., R. R. Andro, S. P. Clark, Jr., 1974, Compressive and shear-wave velocities in granulite facies rocks and eclogites to 10 kbars: Journal of Geophysical Research, v. 79, p. 5427-5446.

McKenzie D., 1978. Some remarks on the development of sedimentary basins: Earth and Planetary Science Letters, v. 40, p. 25-32.

Newton, R. C., T. V. Charlu, O. J. Kleppa, 1974, Colorimetric investigation of the stability of anhydrous magnesium cordierite with application of granulite facies metamorphism: Contributions to Mineralogy and Petrology, v. 44, p. 295-311.

Nur, A., G. Simmons, 1969, The effect of saturation on velocity in low porosity rocks: Earth and Planetary Science Letters, v. 7, p. 183-193.

Parsons, B., and J. G. Sclater, 1977, an analysis of the variation of ocean floor bathymetry and heat flow with age: Journal of Geophysical Research, v. 82, p. 803-827.

Ringwood, A. E., and D. H. Green, 1966, An experiemental investigation of the gabbro-eclogite transformation and some geophysical implications: Tectonophysics, v. 3, p. 383-427.

Schpon, T., and H. J. Neugebauer, 1978, Metastable phase transition models and the bearing of the development of Atlantic type geosynclines: Tectonophysics, v. 50, p. 387-412.

Sclater, J. G., and P. F. Christie, 1980, Continental stretching: an explanation of the post-Mid-Cretaceous subsidence of the central North Sea basin: Journal of Geophysical Research, v. 85, p. 3711-3739.

Seismic models of the lithosphere for the major geostructures on the territory of USSR, 1980, Moscow, "Nauka", 184 p. (in Russian).

Sobolev, S. V., 1976, Seismic models of the Moho discontinuity in the case of a phase transition, *in* Materialy XIV Vsesoyusnoy studentches-koy konfezentsyi: Series Geology, Novosibirsk University publication, p. 62-69 (in Russian).

———, 1978, Models of the lower crust on continents with consideration of the gabbro-eclogite transition, *in* Petrological problems of the Earth's crust and mantle: Novosibirsk, Nauka, p. 347-355. (in Russian).

———, 1980, Physico-petrological processes in the crust and mantle resulting in vertical movements of the continental lithosphere, *in* Synopsis of candidate dissertation: Moscow, Institute of the integrated development of the interiors, Academy of Sciences of the USSR, 24 p. (in Russian).

———, in press, Physical nature of rapid intracontinental subsidence. Origin of the great Hungarian depression: Budapest, Hungary, Proceedings of the European Geophysical Society annual meeting.

———, and N. V. Sobolev, 1980, New evidence of the sinking to great depth of the ecologitized rocks of the Earth crust: Doklady Academii, Nauk SSSR, v. 250, n. 3, p. 683-685 (in Russian).

Stegena, L., B. Geczy, and F. Horvath, 1975, Late Cenozoic evolution of the Pannonian basin: Tectonophysics, v. 26, p. 71-90.

Talwani, M., X. Le Pichon, and M. Ewing, 1965, Crustal structure of the mid-ocean ridge, part 2, *in* Computed model from gravity and seismic refraction data: Journal of Geophysical Research, v. 70, p. 341-352.

Terentyev, Y. D., 1978, Solid phase reactions: Moscow, "Chemistry", 360 p. (in Russian).

Tectonics of oil and gas-bearing regions on the south of USSR, 1973, K. N. Kravchenko and M. V. Mouratov, eds: Moscow, "Nedra", 222 p. (in Russian).

The Earth's crust and history of development of the Black Sea Basin, 1975: Moscow, "Nauka", 358 p. (in Russian).

Yanshin, A. L., E. V. Artyushkov, and A. E. Schlezinger, 1977, Fundamental types of large lithosperic plate structures and possible mechanisms of their formation: Doklady Academii, Nauk SSSR, v. 234, n. 5, p. 1175-1178 (in Russian).

———, et al, 1977, Structural peculiarities of the sedimentary layer of the Black Sea depression and their bearing for understanding its formation: Moskovskoye Obschestvo Ispytate Ley Prirody Byulleten', Otdel Geologicheskiy, n. 5, p. 42-69. (in Russian).

Appendix

Kinetic models of gabbro-eclogite phase transition

Many kinetic models were suggested for the solid phase reactions (Kononyuk, 1975, Terentyev, 1978). In natural gabbro the garnet often grows around grains of pyroxene on their boundaries with grains of plagioclase. In this case the kinetic equation looks as follows (Sobolev, 1980):

$$\frac{\partial \alpha}{\partial t} = \frac{1}{\tau(T(t))}; \quad 0 < \alpha < f(P,T), \tag{3}$$

where $\alpha(t)$ is the current degree of rock eclogitization. It is defined as a ratio of current volumetric content of garnet in the rock to the garnet content in eclogite. The α value comes to 0 in gabbro and to 1 in eclogite. If chemical equilibrium is achieved, the degree of eclogitization appears to be a function of pressure and temperature only; $\alpha = f(P,T)$. The function $f(P,T)$ is an equilibrium degree of rock eclogitization, $0 \leq f(P,T) \leq 1$. It can be determined from the experimental P, T diagram of mafic rock with isolines of density or garnet content. τ $(T(t))$ is a characteristic time of the phase transition (text, equation 1). The above model does not consider the relation between the rate of phase transition and the difference between this equilibrium and current degrees of rock eclogitization (f-α). This is kinetic model 1.

The simplest kinetic model taking into account the relation between $\frac{\partial\alpha}{\partial t}$ and $(f-\alpha)$ is the following one:

$$\frac{\partial\alpha}{\partial t} = \frac{f-\alpha}{\tau(T(t))}; \quad 0 < \alpha < 1 \tag{4}$$

This is kinetic model 2.

The time $\tau(T)$ in both models is determined from the relation:

$$\tau(T) = \tau(T_o)\, e^{\frac{E_{act}}{R}\left(\frac{1}{T}-\frac{1}{T_o}\right)} \tag{5}$$

where $\tau(T_o)$ is the known value of time τ at $T = T_o$°K.

Density changes and vertical movements caused by gabbro-eclogite phase transition

The function $\alpha(Z, t)$ is found from equations 3 and 5, or 4 and 5. To do that, it is necessary to know the distribution of $\alpha(Z)$ at t=0 and the changes of pressure and temperature with time and depth at t >0. The density of mafic rocks, $\rho(Z,t)$, can be calculated from the relation

$$\rho(Z,t) = \overset{o}{\rho}_g + (\overset{o}{\rho}_e - \overset{o}{\rho}_g)\,\alpha(Z,t) + \Delta\rho_{P,T}, \tag{6}$$

where $\overset{o}{\rho}_g$, $\overset{o}{\rho}_e$ are the densities of gabbro and eclogite at T=2 P=1 bar; $\rho_{P,T}$ (a correction for rock compressibility and thermal expansion).

The density of gabbro increases due to dissolution of plagio clase in pyroxene at P > 5 kbar (Ringwood and Green, 1966). This effect can be roughly taken into account by a relation

$$\overset{o}{\rho}_g = \begin{cases} \overset{00}{\rho}_g + (\overset{01}{\rho}_g - \overset{00}{\rho}_g)\dfrac{P-5}{5} & \text{at } 5 \leq P \leq 10\text{kbar}, \\ \overset{01}{\rho}_g & \text{at } 10 \leq P \end{cases} \tag{7}$$

where $\overset{00}{\rho}_g$ is gabbro density at P=5 kbar and $\overset{01}{\rho}_g$ that at P=10 kbar.

The values of $\overset{01}{\rho}_g$ and $\overset{00}{\rho}_g$ for the basalt studied by Ito and Kennedy (1971) are known from their experiments, $\overset{01}{\rho}_g$ = 3.14g/cu cm and $\overset{00}{\rho}_g$ = 3.00 g/cu cm. The same values for the basalt studied by Ringwood and Green (1966) are probably, about $\overset{01}{\rho}_g$ = 3.05 g/cu cm at $\overset{00}{\rho}_g$ = 2.95 g/cu cm.

The density of eclogite was taken as ρ_e = 3.55 g/cm^3 for both mafic rocks according to experimental data (Ito and Kennedy 1971). The function $\Delta\rho_{P,T}$ can be approximately determined from the relation

$$\Delta\rho_{P,T} = 1.1 \cdot 10^{-4}\,(T(Z,t) - 500Z/45) \text{ g/cu cm} \tag{8}$$

where Z is a depth in km, H_C refers to crustal thickness and H_g basaltic layer thickness.

The unloaded subsidence (U(t)) caused by the gabbro-eclogite phase transition in the lower crust is given by the relation

$$U(t) = \frac{\rho_g(t) - \rho_g(O)}{\rho_g(t)} \cdot \overset{o}{H}_g \tag{9}$$

$$\rho_g(t) = \frac{1}{\overset{o}{H}_g} \int_{H_c - \overset{o}{H}_c}^{H_c} P(z,t)\, dZ,$$

where $\rho_g(t)$ is the average density of basaltic layer at the moment, and $\overset{o}{H}_g$ is the initial thickness of basaltic layer.

The temperature T(Z,t) in the lithosphere is taken from the solution of one-dimensional equation of heat transport under different boundary and initial conditions:

$$\frac{\partial T}{\partial t} = \varkappa \cdot \frac{\partial^2}{\partial Z^2}(T - T_{st}(Z)), \tag{10}$$

where T_{st} (Z) is stationary temperature in the lithosphere and $\varkappa$ is thermal diffusivity.

In the basaltic layer of the crust where the phase transition proceeds, additional terms appear in the equation of heat transport (Schpon and Neugebauer, 1978). These terms are associated with vertical crustal displacements and with the energy release (both are associated with the gabbro-eclogite phase transition). The estimates show, however, that these terms are relatively small and can be neglected (Sobolev, 1980).

Thermal subsidence of passive margins

We used a thermal model similar to that for the ocean floor subsidence (Parsons and Sclater, 1977). Equation 10 was solved under the following boundary and initial conditions:

$$T(0,t) = 0°C,\ T(H_1,t) = T_1$$

$$T(Z,0) = \begin{cases} T_1 \dfrac{Z}{H_c} & \text{if } O \leq Z \leq H_c \\ T_1 & \text{if } H_c \leq Z \leq H_1 \end{cases} \tag{11}$$

$$T_{st}(Z) = T_1 \frac{Z}{H_1} \text{ if } O \leq Z \leq H_1$$

where H_1 is the depth where temperature is constant and equal to T_1.

The unloaded thermal subsidence ($U_T(t)$) is given by the relation

$$U_T(t) = \alpha_T T_1 \left(\frac{H_1 - H_c}{2} - H_1 \cdot \sum_1^{\infty} 2(1-(-1)^n) \frac{H_1}{H_c} \frac{\sin(\pi n H_c/H_1)}{\pi^3 n^3} \right) e^{-\pi^2 n^2 \varkappa \frac{t}{H_1^2}} \tag{12}$$

where α_T is the thermal expansion coefficient.

The values of the parameters are from ranges used in a simple cooling model of the oceanic floor subsidence (Parsons and Sclater, 1977): α_T = 3.2 . 10^{-5} 1/°C, T_1 = 1350°C, $\varkappa$ = 0.0075 cm^2/sec, H_1 = 150 km.

Extension Ratio Measurements on the Galicia, Portugal, and Northern Biscay Continental Margins: Implications for Evolutionary Models of Passive Continental Margins

P. Chenet
L. Montadert
Institut Francais du Petrole
Rueil, France

H. Gairaud
Societe Nationale des Petroles d'Aquitaine
Rueil, France

D. Roberts
B.P. Exploration Company Limited
London, England

The absence of thick sediments on the margins of Galicia, Portugal, and North Biscay allows the underlying rift structure to be clearly seen on seismic reflection profiles. The rifted structure consists of a series of tilted, deformed, and rotated fault blocks bounded by listric normal faults, evidently produced by extension and subsidence. Comparison of the volume of a horizontal slice of the margin (sectional area) before and after extension suggests that the formation of the margin (rifting stage) results from two main tectonic processes acting simultaneously: extension by stretching of the lithosphere, and thinning of the crust by deep thermal processes. Simple models of margin formation dealing only with stretching processes may thus neglect the effect on heat flow and subsidence of the deep thermal processes.

The evolution of passive rifted margins is considered to take place in two main phases. The first, rifting stage, is characterized by tensional tectonics, volcanism, and both uplift and subsidence. The second, postrift or drifting stage, is contemporaneous with the formation of new ocean crust and is characterized by subsidence of the whole margin in response to thermal cooling and sediment loading. Theoretical modeling of both thermal and loading processes gives satisfactory results on subsidence and heat flow on some passive continental margins (Sleep, 1971; Watts and Steckler, 1979; Royden, Sclater, and Von Kerzen, 1980; Keen, Beaumont, and Boutilier, 1981).

On the Northern Biscay and Galicia-Portugal margins, the rift fabric is particularly well displayed and both the prerift and synrift sediments are seen on seismic profiles (Montadert et al, 1979; Sibuet and Ryan, 1978). These data clearly show tilted blocks bounded by listric normal faults and the thinning of the entire crust from continent to ocean (Figure 1). The rift fabric, thinning of the crust, and subsequent thermal subsidence have been interpreted as the consequence of an initial stretching of the lithosphere during the rifting phase; followed by cooling of the lithosphere, resulting in progressive recovery of the original thermal structure (MacKenzie, 1978). Modifications of this hypothesis incorporate variation in the rheology of the lithosphere with depth, (Watts and Steckler, 1979; Royden, Sclater, and Von Kerzen, 1980; Keen and Barett, 1981).

This paper examines the rift tectonics of Galicia-Portugal and Northern Biscay described previously by Montadert et al (1979) and Sibuet and Ryan (1978) and discusses their implications for stretching models of the lithosphere.

THE RIFTED STRUCTURE

Deep drilling and various geophysical studies (summarized in Montadert et al, 1979, and Sibuet and Ryan, 1978) show that present structure of Northern Biscay and Galicia-Portugal margins resulted from rifting of a post-Hercynian platform at Lower Cretaceous times, followed by sedimentation on the thermally subsiding rifted structure.

To obtain the best geometric description of the rift structure, geologic cross sections at scale 1/1 were constructed from migrated multi-channel seismic profiles using the seismic velocity data (Figure 2, 3a, 3b, 4).

On nearly all the sections, postrift sediments are unaffected by faulting or major subsequent deformation. The underlying structure consists of a series of

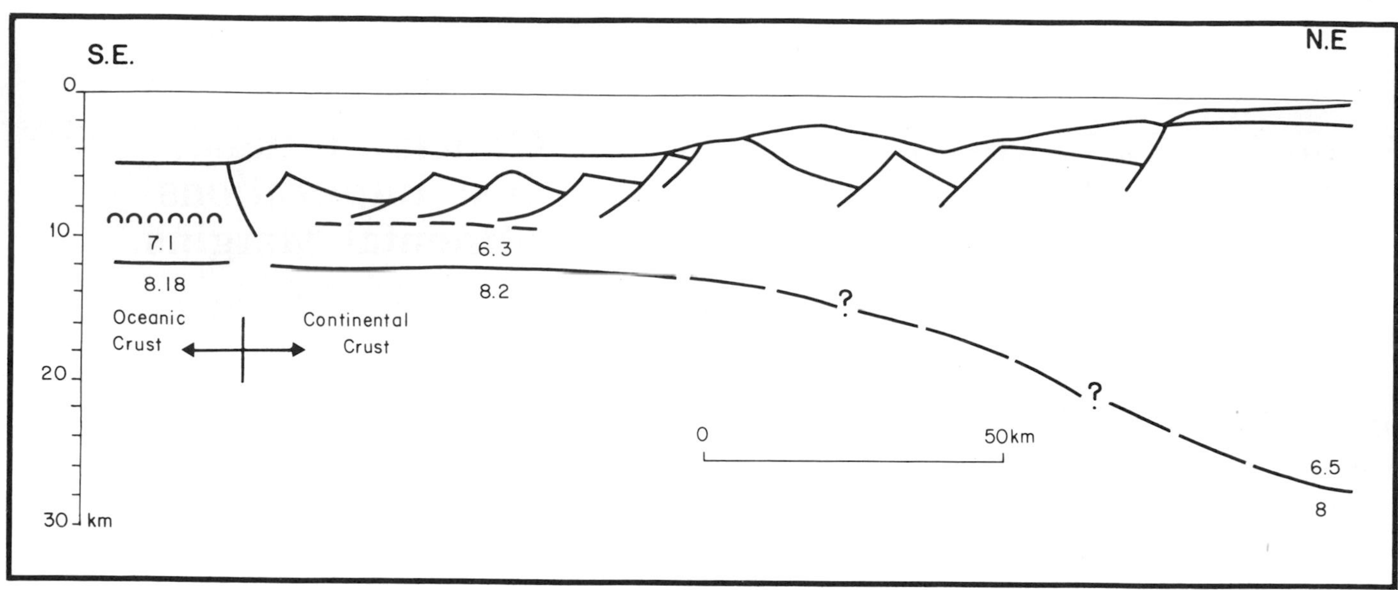

Figure 1 — Structure section across the north margin of Biscay based on seismic reflection and refraction data (from Montadert et al, 1979).

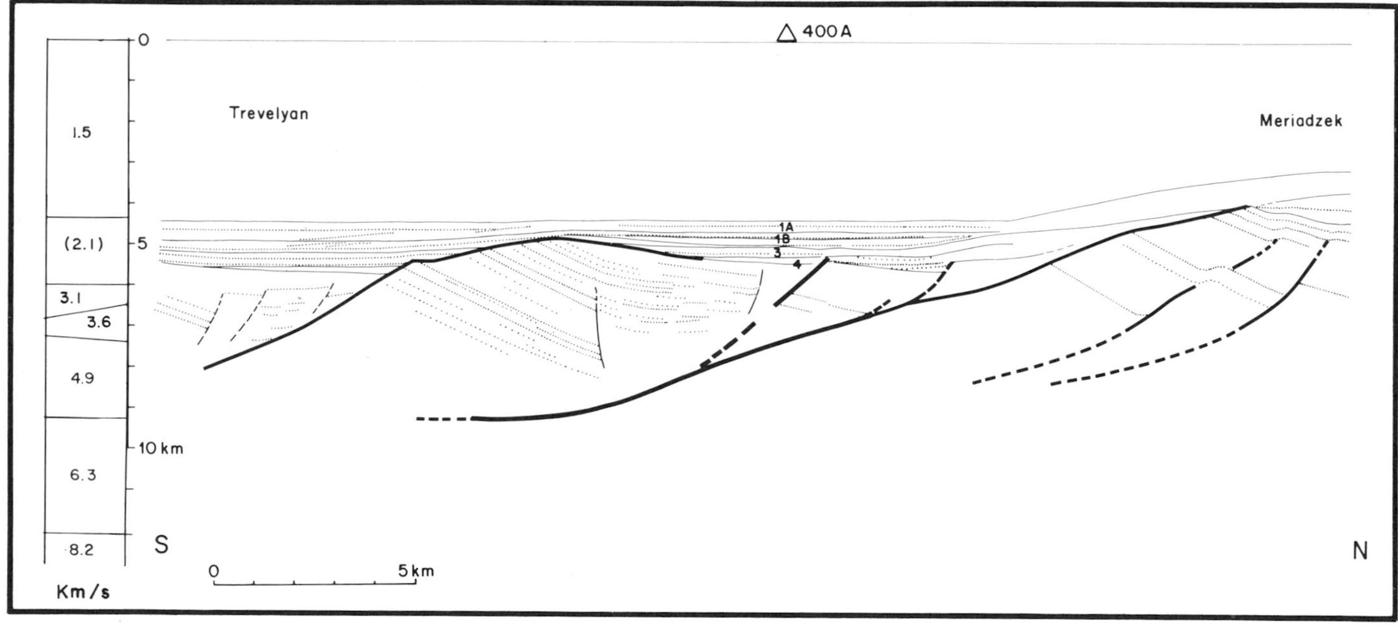

Figure 2 — Geological cross section through site 400 on Biscay margin at true scale (from Montadert et al, 1979). Note the listric faults bounding the tilted blocks.

tilted and rotated fault blocks with variable thicknesses of synrift sediments. These synrift deposits are observed only in the deep axes of the half grabens. Prerift sediments are common within the tilted blocks.

The principal feature of the rift fabric is the extension of the upper part of the crust by normal faulting. The listric faults bounding blocks have dips of 60° near the continent, but less than 20° near the oceanic crust, and become nearly horizontal at depth (Figure 4). The block tilting may reach 30° and seems to increase from the continent toward the oceanic crust (Figure 4). Deformation within the block is shown by slight folding of the prerift reflectors (Figure 2) and by the presence of numerous small normal faults whose dips vary from 45 to 90° (Figure 2). On both margins a strong reflector observed between 8 and 10 km depth is thought to correspond to a velocity discontinuity (De Charpal et al, 1978). The listric faults appear to flatten above this reflector, which does not appear faulted.

Tilting of the blocks is probably due to the block rotating as it slides down the listric fault toward the rift axis (Montadert et al, 1979). Deformation within the block is probably due to this movement along the

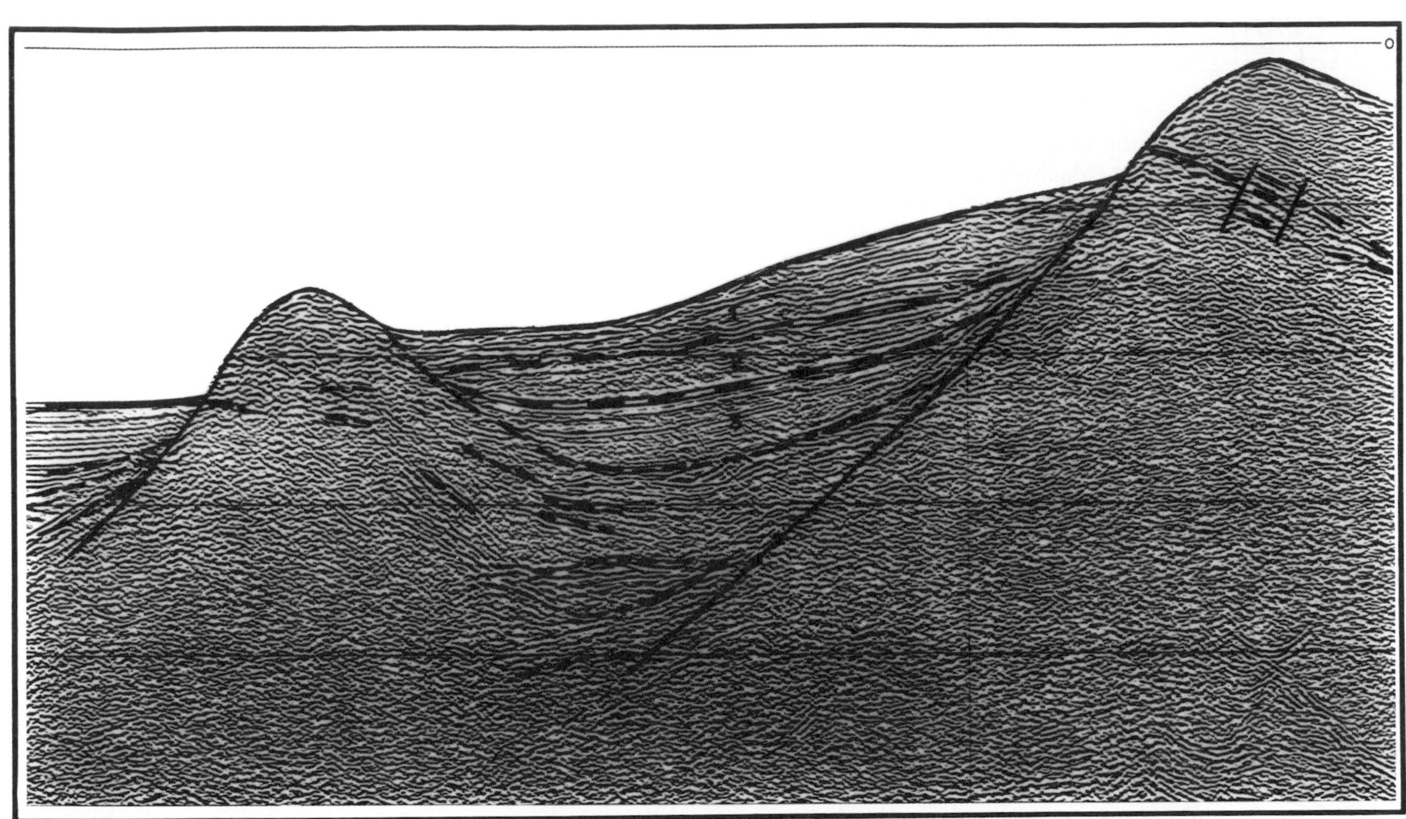

Figure 3a — Seismic reflection migrated profile south of Galicia Bank showing tilted blocks and listric faults and a reflector below (profile GP 11, processed).

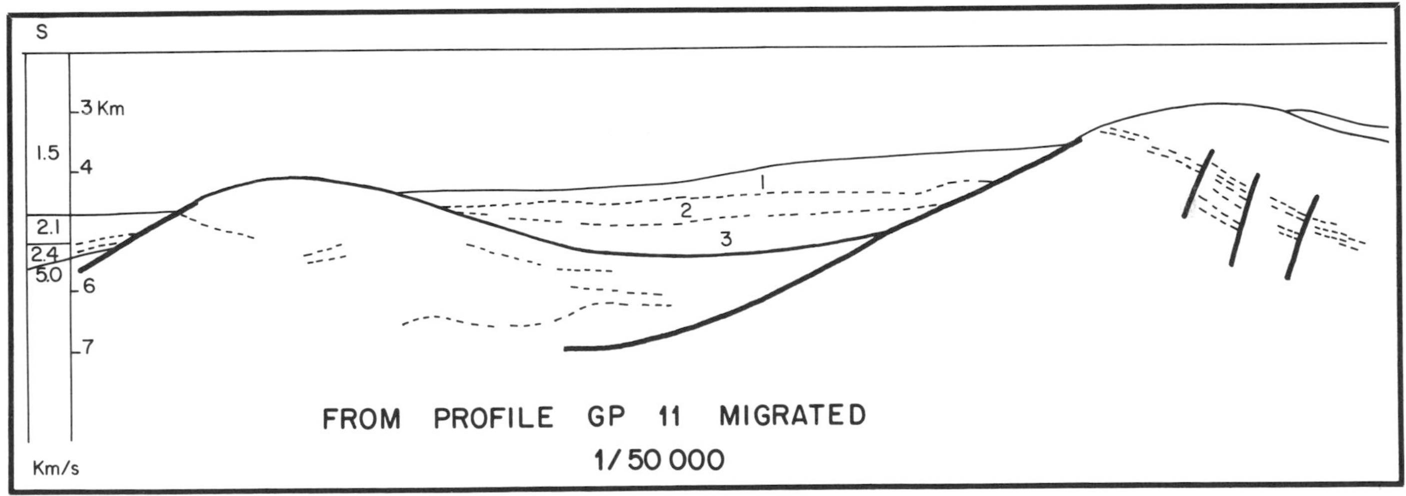

Figure 3b — Geological cross section obtained from seismic reflection profile GP 11. Note the slope of the last prerift reflector and the near horizontal listric fault at 7 to 8 km deep.

bounding listric faults (see last prerift reflector, Figure 3b). This deformation suggests the existence of lateral stresses applied across the boundary faults of the block. These stresses result from wedging of the block between the two adjacent ones. Because of the subsidence and the lateral offset due to the non-vertical dip of the fault, these local stresses are not incompatible with extensional tectonics. The dip of the listric faults decreases with depth, and this is explained partly by rheological considerations, which show that flattening of normal faults with depth occurs in response to increased confining pressure. When the maximum stress is vertical in an extensional regime, increasing confining pressure tends to lower the break angle of the system (see Mattauer, 1976).

Furthermore, the listric fault system results in lateral offset of the prerift layers. If the listric faults are not parallel, as in Figure 2, and their curvature is not constant, then lateral displacement and deformation must be invoked to keep the mass of extended layer constant.

QUANTITATIVE PARAMETERS OF THE STRETCHING AND THINNING PROCESSES

The above observations have been interpreted as a consequence of stretching of the lithosphere with mass conservation of the crust (MacKenzie, 1978). In order to examine the stretching process quantitatively, estimates of extension and crustal thinning were made from the 1/1 geological cross sections and volumes of the crust were computed before and after extension.

Thinning ratio

By definition, thinning of the crust at a given point on the margin is the ratio between the thickness of the crust on the adjacent continent and the crustal thickness at the point.

Assuming that the crust includes postrift as well as prerift material, the maximum thinning ratio for continental crust is 3, as shown by refraction data (Figure 1). New data of Avedik et al (1981; Figure 9), however, give a maximum value of 4. If postrift sediments are excluded, the thinning ratio is higher, ranging from 4 in older refraction data to 6 in new data.

During thermal subsidence after rifting, thickness of the thinned continental crust may or may not be modified. If it does not change, and the Moho is not affected by the increase of pressure due to sediment and water loading and higher temperatures at greater depths, the vertical distance between Moho and top of the prerift series represents the thickness of the crust just after rifting. In this case, the thinning ratio measured by excluding the postrift sediments is representative of the modifications in the structure of the crust.

But, if the Moho is affected by temperature and pressure changes occurring after rifting, it is difficult to reconstruct the geometry of the crust at the end of the rifting phase. We may suppose, in first approximation, that the thinning ratio of the crust just after extension is between the values obtained by exclusion or inclusion of postrift sediments thickness.

Extension ratios

The extension ratio of a column of crust of original horizontal length, *l*, and present length, L, is *L/l*. The extension ratio should be computed for various layers in the prerift section to detect variations in extension between different layers.

Here, extension ratios are computed for the last prerift isochron surface because of the ease of estimating the original horizontal length for this time. We call this ratio the superficial extension ratio, *e*. From the cross sections, superficial extension ratios are computed for blocks presently 10 to 20 km in length. Figure 5 shows the value of *e* for a single tilted block. The last prerift isochron is easily recognized and does not show evidence of contemporaneous deformation. The horizontal length is given by the prerift reflector lying just below the last prerift reflector surface. In some cases, where erosion of the crust of the tilted block occurred (Figure 27 from Montadert et al, 1979), the horizontal length (*l*) of the last prerift isochron was extrapolated; resulting in uncertainty when measuring the superficial extension ratio.

As mentioned earlier, deformation took place within the blocks (Figure 2 and 3). In cases where the prerift sediments are folded, the true extension ratio is lower than computed ratio neglecting folding (Figure 6a). On the other hand, curvature of the last prerift surface might also result from extension. The curvature can result from deformation within a graben and slight offset by undetected small faults (Figure 6b). In this case, the true extension ratio is greater than computed ratio because internal extension is excluded.

Superficial extension ratios computed for five profiles off Biscay and three off Galicia-Portugal are presented in Tables 1 and 2. The mean extension ratio for both margins lies in the range 1.10 to 1.45 because of uncertainties in the measurements. For a single block 10 km in length, the maximum extension ratio including uncertainties due to measurements does not exceed 1.70 (Figure 7). For Northern Biscay, the mean extension ratio was found for the deep part of the margin (60 km in length) where the upper surface of the tilted blocks is at a depth of about 5 ± 0.5 km (after backstripping the postrift sediment load) and the Moho depth is known. This value is in the range 1.10 to 1.40.

The extension ratio for a prerift reflector at depth was also computed for the block shown in Figure 8. The superficial extension ratio is 1.45 ± 0.15, depending on the hypothesis stated above and the un-

Table 1. Galicia Portugal Continental Margin

N° Section	Mean e Whole Margin	Max. e For a Single Block
GP 10	1.10-1.20	1.30
GP 11	1.20-1.30	1.60
GP 12	1.12-1.20	1.25
Superficial Extension Rate (e)	Whole Margin	1.10-1.30
	Maximum for a Single Block	1.60

Table 2. Northern Biscay Continental Margin

N° Section	Mean e Whole Margin	Mean e 60km Near Ocean	Max. e for a Single Block
OC 202	1.13	1.10	1.38
OC 209	1.34-1.45	1.26-1.38	1.70
OC 207		1.20-1.28	1.68
OC 205-305	1.23-1.27	1.24-1.25	1.54
Superficial Extension Rate (e)	Whole Margin		1.10-1.45
	within 60km Near Oceanic Crust		1.10-1.40
	Maximum for a Single Block		1.70

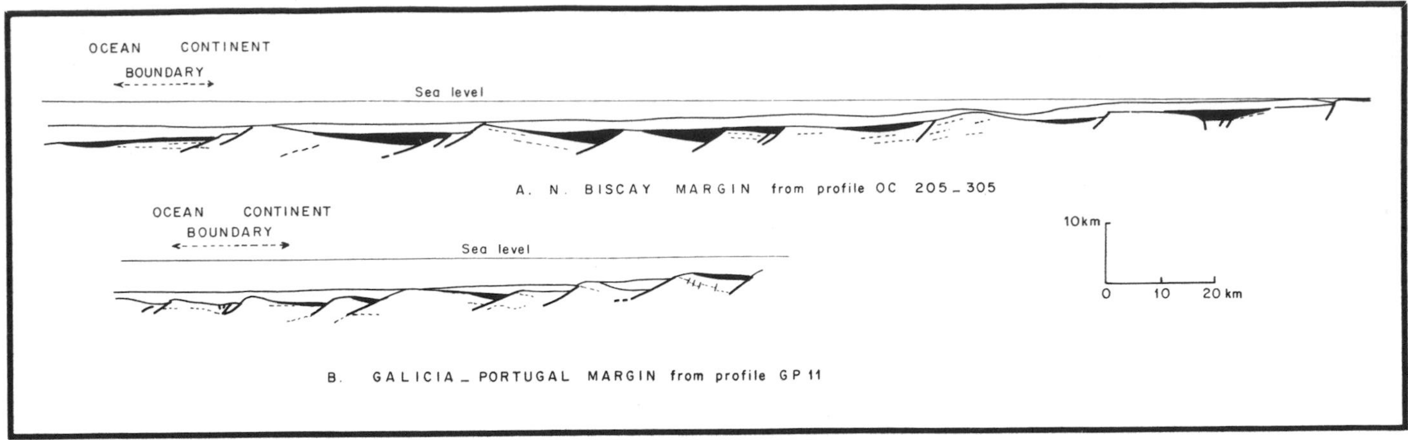

Figure 4 — Geological schematic cross sections of Northern Biscay continental margin from profile OC 205-305 and partial section across Galicia-Portugal margin from profile GP 11. Blacklayer: synrift deposits. Note strong similarities between rifted structure of both margins. Postrift deposits are thicker for Northern Biscay margin.

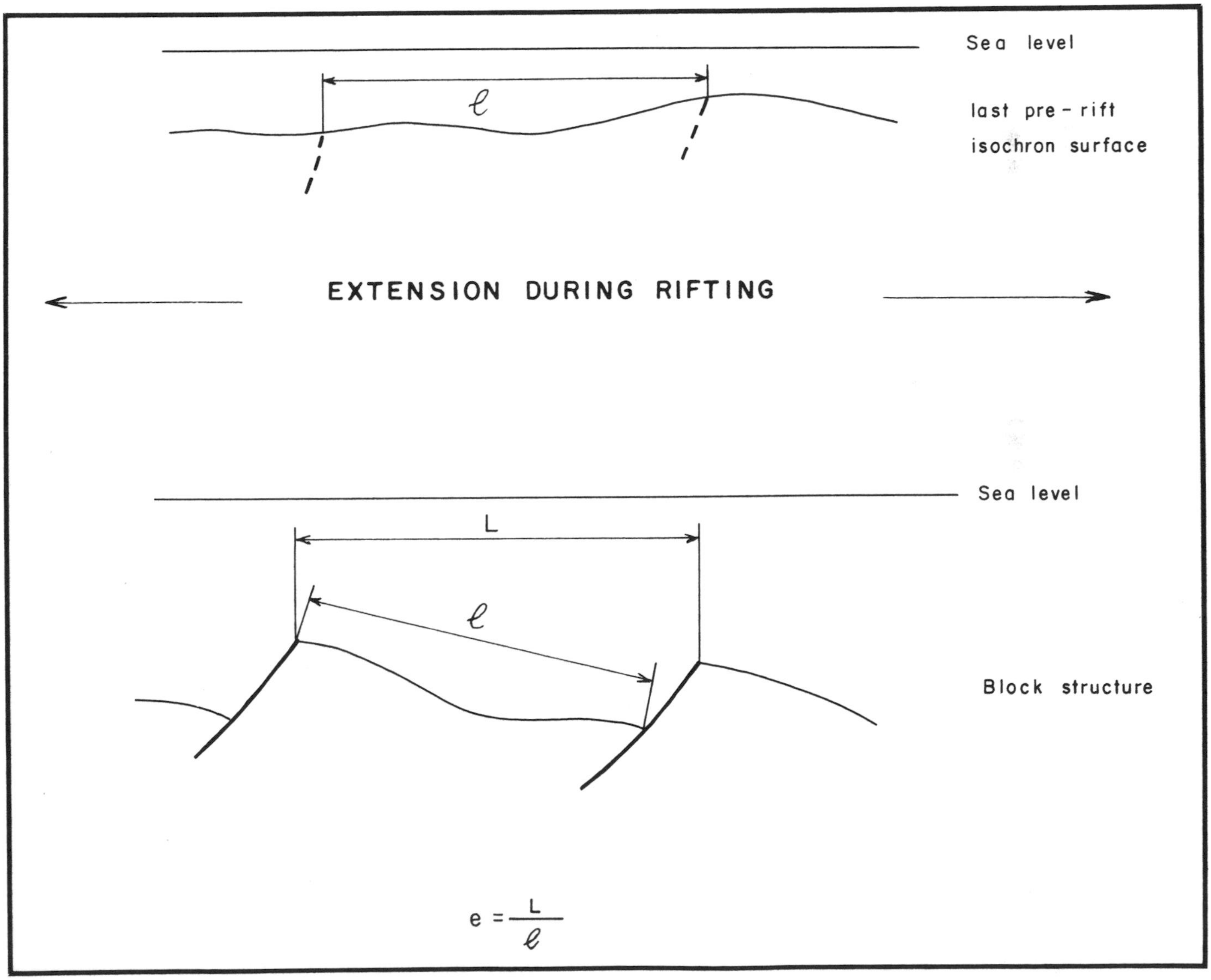

Sea level

ℓ

last pre-rift isochron surface

EXTENSION DURING RIFTING

Sea level

L

ℓ

Block structure

$$e = \frac{L}{\ell}$$

Figure 5 — An element of last prerift isochron surface of irregular topography and of horizontal length *l*, individualized as a block bounded by two faults of present horizontal length *L*. Superficial extension rate *e* will be *L*/1.

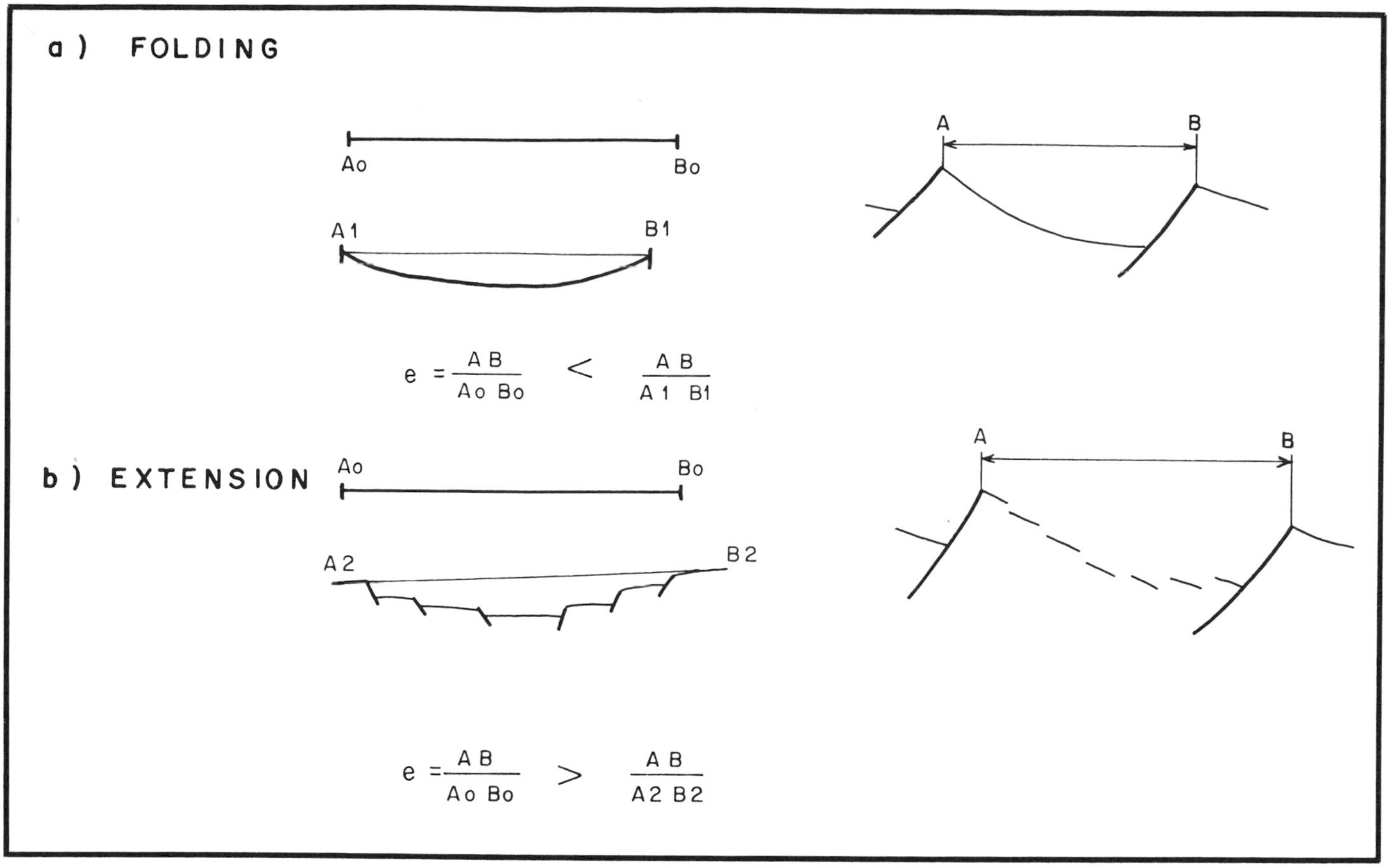

a) FOLDING

$$e = \frac{AB}{A_oB_o} < \frac{AB}{A_1B_1}$$

b) EXTENSION

$$e = \frac{AB}{A_oB_o} > \frac{AB}{A_2B_2}$$

Figure 6 — Superficial extension rate measurements for a deformed block. Hypothesis a: folding of the block. The original segment of length AoBo has been folded. Its present horizontal length is A1B1 with A1B1 < AoBo and e = AB/AoBo < AB/A1B1 Hypothesis b: extension of the block. The last prerift isochron surface is composed of little segments offset by small normal faults not visible in the seismic profile. Present horizontal length of AoBo is A2B2 and e = AB/AoBo > AB/A2B2. Estimation of AoBo can be assuming that vertical offset of the small faults is equal to the resolution interval and that dip of these faults are similar to the observed ones. See small faults within the block of Figure 2.

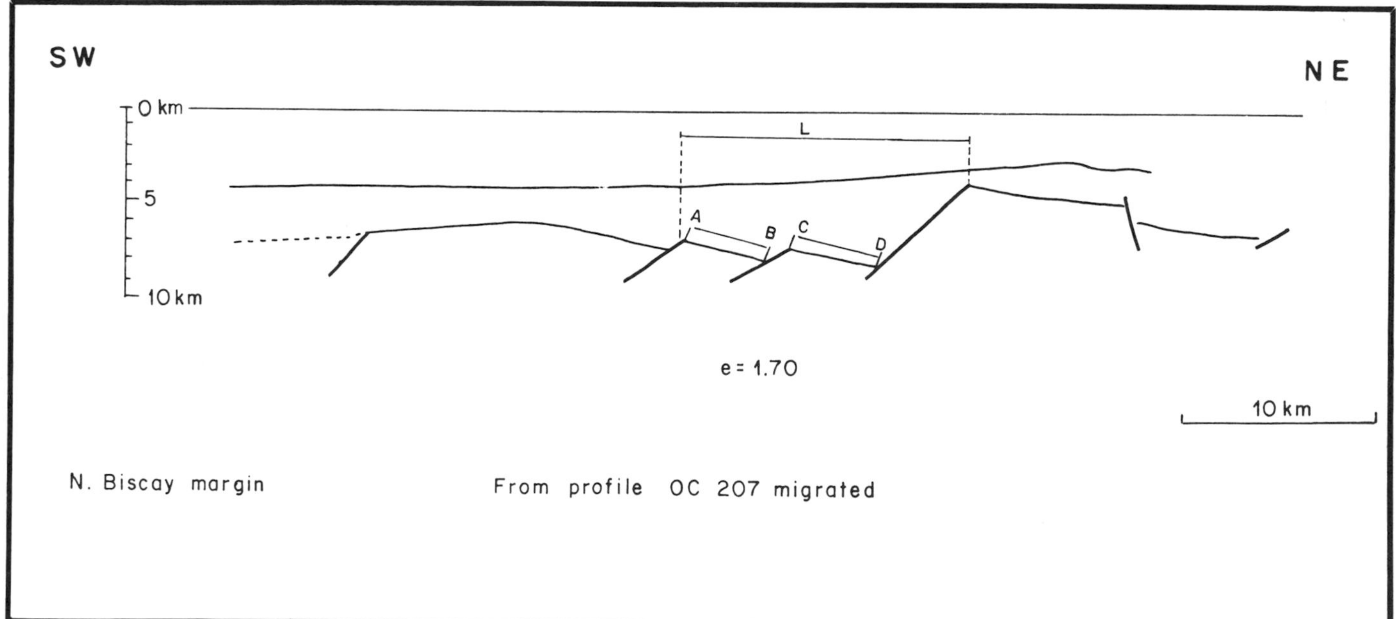

Figure 7 — Northern Biscay margin. Geological section obtained from profile OC 207. Occurrence of maximum superficial extension rate. Note this is mainly due to large vertical offset of the northeastern fault limiting the composite block.

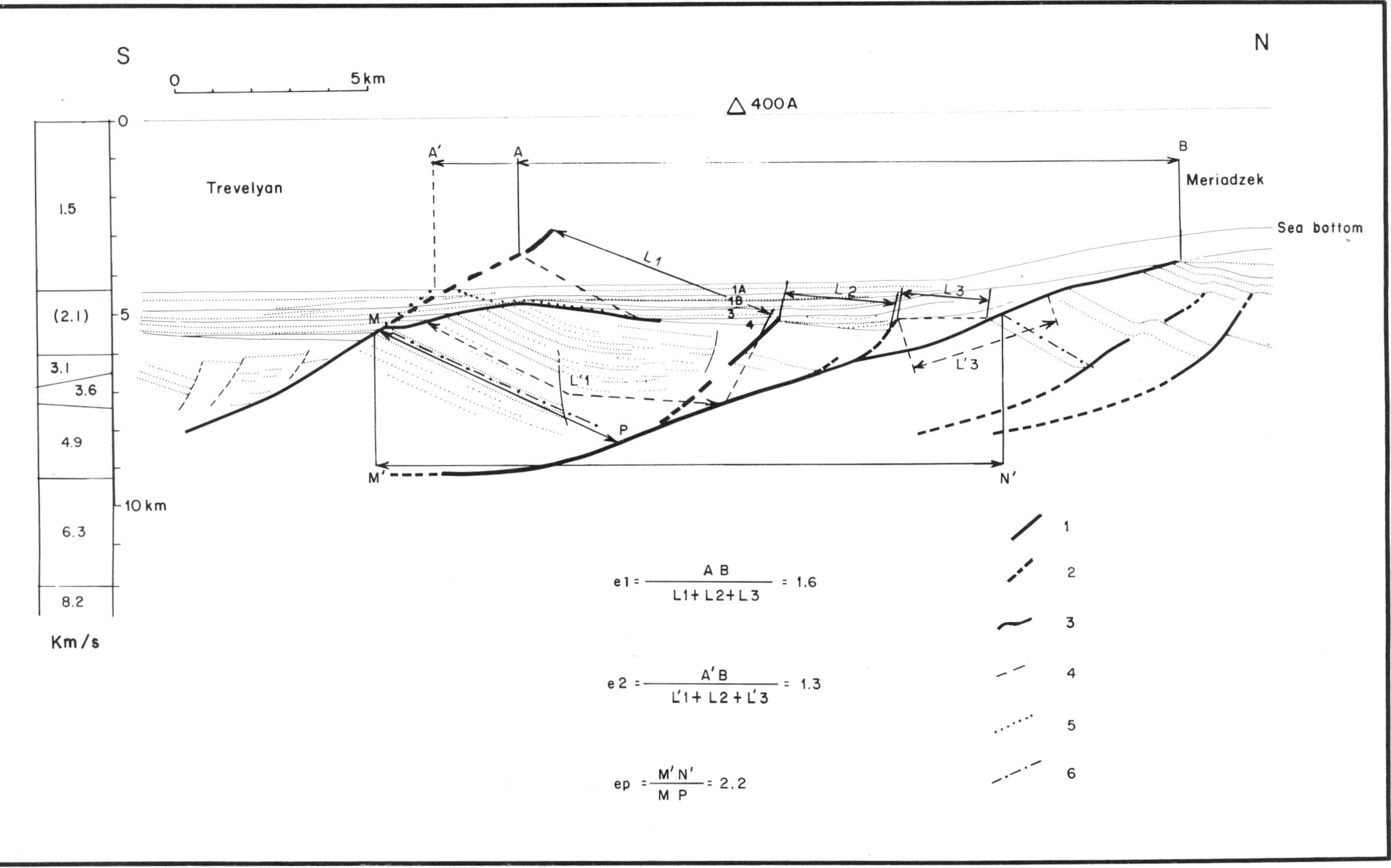

Figure 8 — Extension rate measurement for a block of Northern Biscay margin. 1: Faults; 2: Hypothetical fault; 3: upper surface of the block; 4: last prerift surface (first hypothesis); 5: last prerift surface (second hypothesis); 6: prerift isochron surface.
e 1: superficial extension rate computed in the first hypothesis; *e* 2: superficial extension rate computed in the second hypothesis; *e* p: extension rate at depth.

NE
SW
0
10
20
30
Km
S
M
CONTINENTAL
TRANSITION ZONE
OCEANIC
0
30Km
after AVEDIK et al 1981

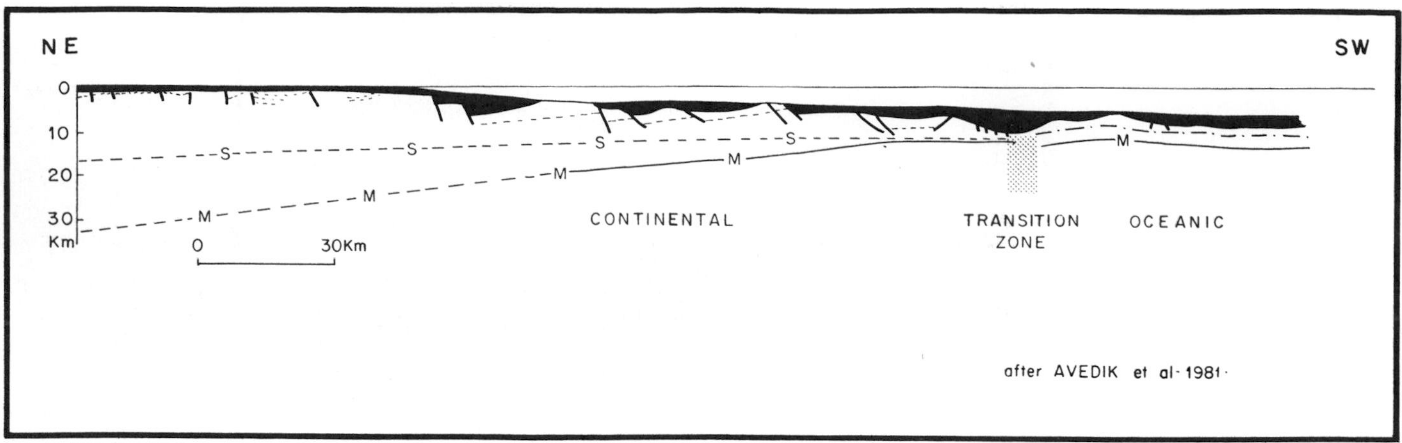

Figure 9 — Geological cross section of Northern Biscay continental margin from new refraction data of Avedik et al (1981). Vertical and horizontal scale are identical.

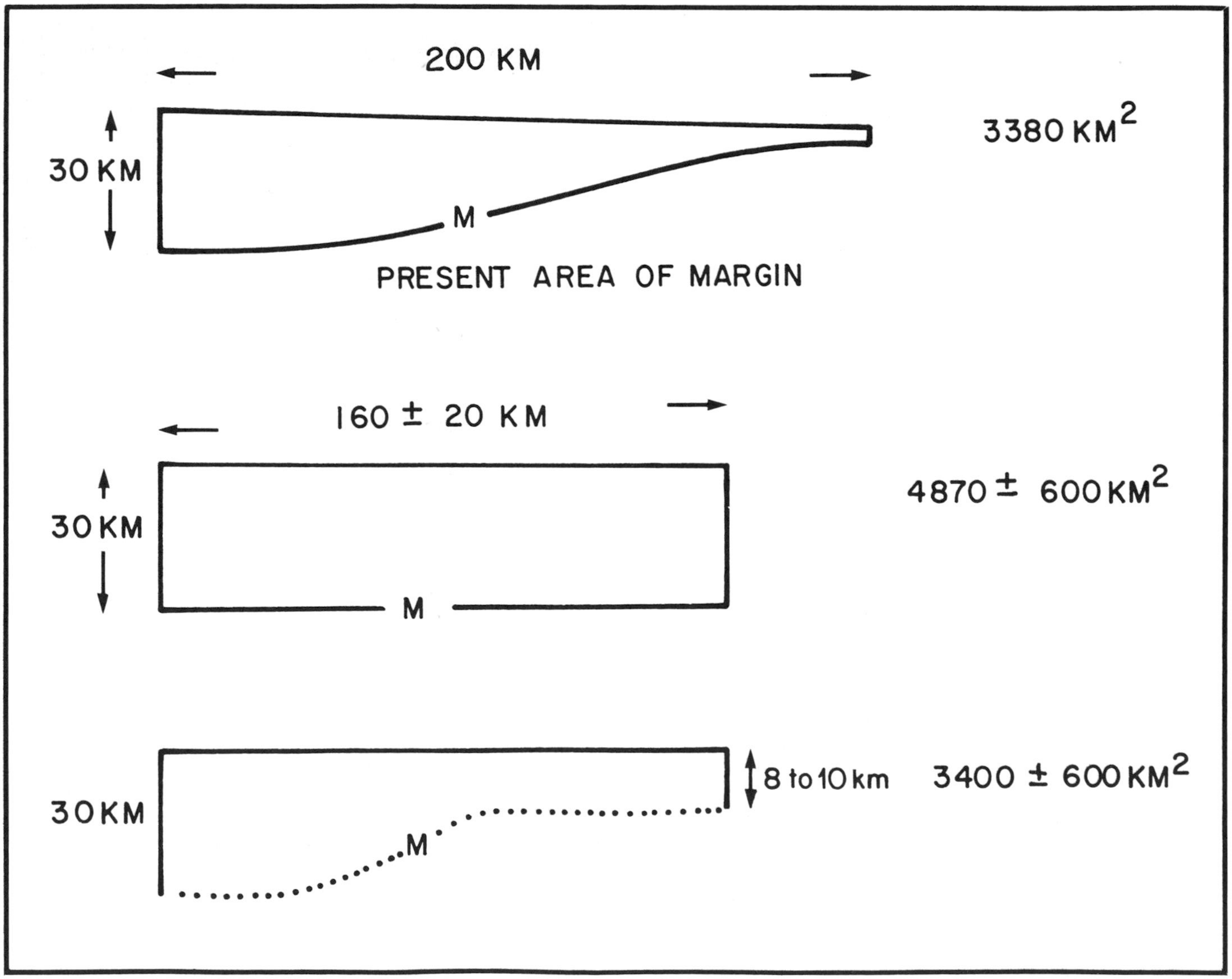

200 KM
30 KM
M
$3380 KM^2$
PRESENT AREA OF MARGIN
160 ± 20 KM
30 KM
M
$4870 \pm 600 KM^2$
30 KM
M
8 to 10 km
$3400 \pm 600 KM^2$

Figure 10 — **a.** Present cross sectional area (horizontal slice) of the entire crust of the Northern Biscay continental margin after refraction data of Avedik et al (1981; Figure 9). **b.** Sectional areas of the crust of the margin just before rifting, assuming the Moho was at 30 km depth or higher.

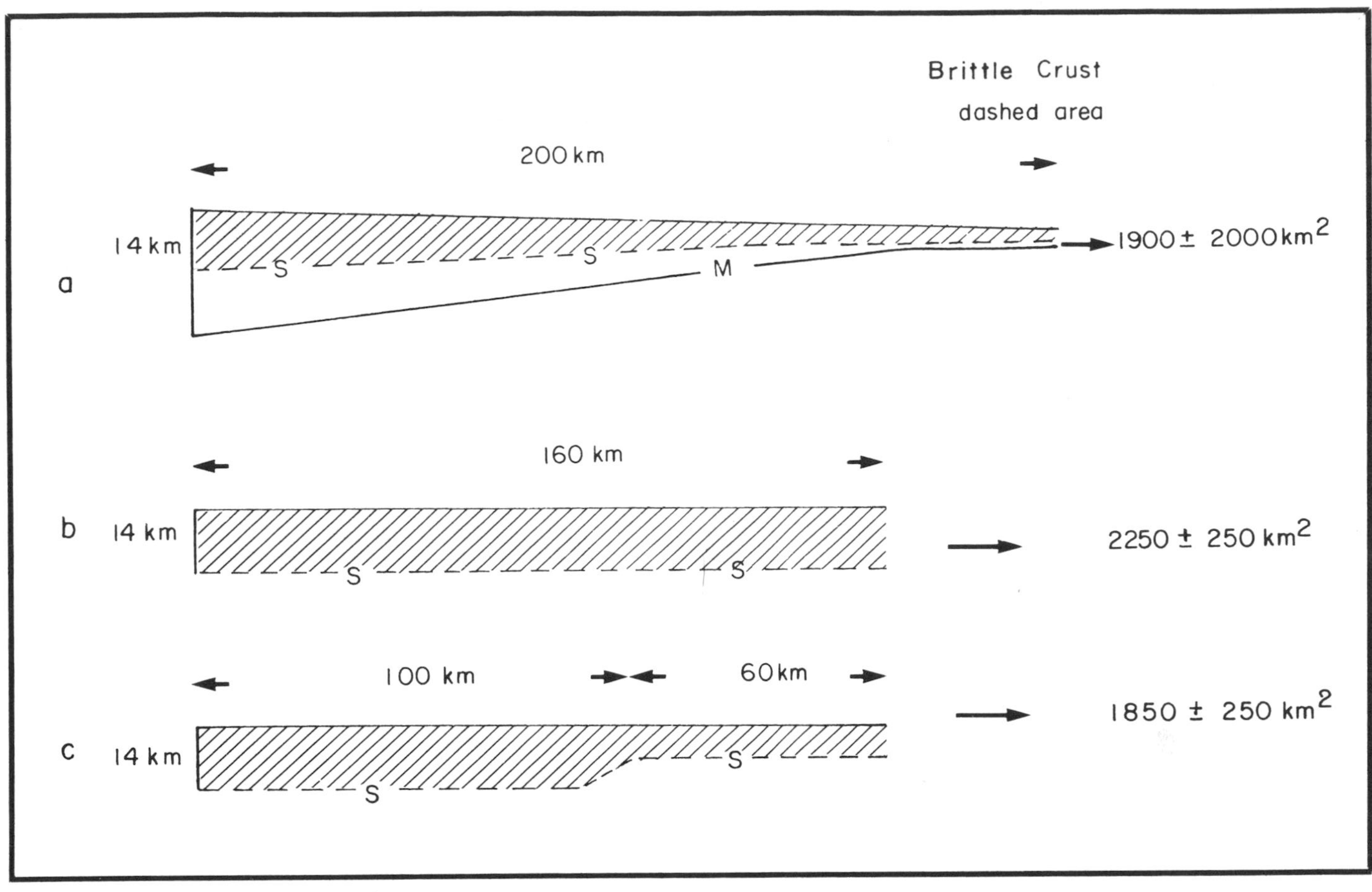

Figure 11 — a. Present sectional area (horizontal slice) of the upper brittle crust of the North Biscay margin. **b.** Original sectional area of the upper brittle crust of the margin just before rifting assuming a brittle ductile transition at 14 km depth throughout the margin. **c.** Original section area of the upper brittle crust just before rifting assuming that brittle-ductile transition is at 14 km depth near the continent and 8 km depth near the ocean.

certainty on the position of the last prerift reflector. For a prerift isochron surface, the extension ratio was about 2.2 by the same method, although position of faults bounding the block at great depth may be discussed.

The difference between the two extension ratios computed for this block is mainly due to nonparallel faults limiting the block. Le Pichon, Sibuet, and Angelier (1981) make similar measurements on the same block but interpret the difference as a strong deformation by thinning of upper layers (2 km thick) of the prerift rocks, deeper layers being less deformed. These authors state that this strong thinning cannot be observed and that upper crust was actually extended by a factor of 2. With computed values of extension ratio, this leads to an unobserved upper layer thinning of about 30%. This is unreasonable, as there are no geological observations such as low angle faults, or sedimentary reworking, supporting the hypothesis of strong thinning of the later prerift layer during rifting. Moreover, it is well known from rock mechanics principles that rocks have no resistance to traction; thus, continuous thinning in the upper layers is impossible. Deformation is more probable at much greater depth where ductile creep may reasonably occur.

PROBLEM OF MASS CONSERVATION OF THE CRUST DURING RIFTING

Whole crust (northern Biscay margin)

From the margin's present length of 200 km, an original horizontal length of 162 ± 20 km is computed from its global superficial extension. The present sectional area (horizontal slice) of the crust of the margin is 3,600 sq km (Figure 10a), obtained from refraction data of Avedik et al (1981). The original sectional area of the crust of the margin is 4,860 ± 600 sq km (Figure 10b), assuming an original crustal thickness of 30 km.

The difference between the present and the original surface is explained in two ways: (1) Some mass of the crust was lost during rifting; and 2) Moho was higher than 30 km at the beginning of rifting and mass of the crust is conserved during extension by stretching, as previously emphasized.

Upper brittle crust (Northern Biscay Margin)

The presence of the strong reflector at a depth of between 8 and 10 km, where listric faults are nearly

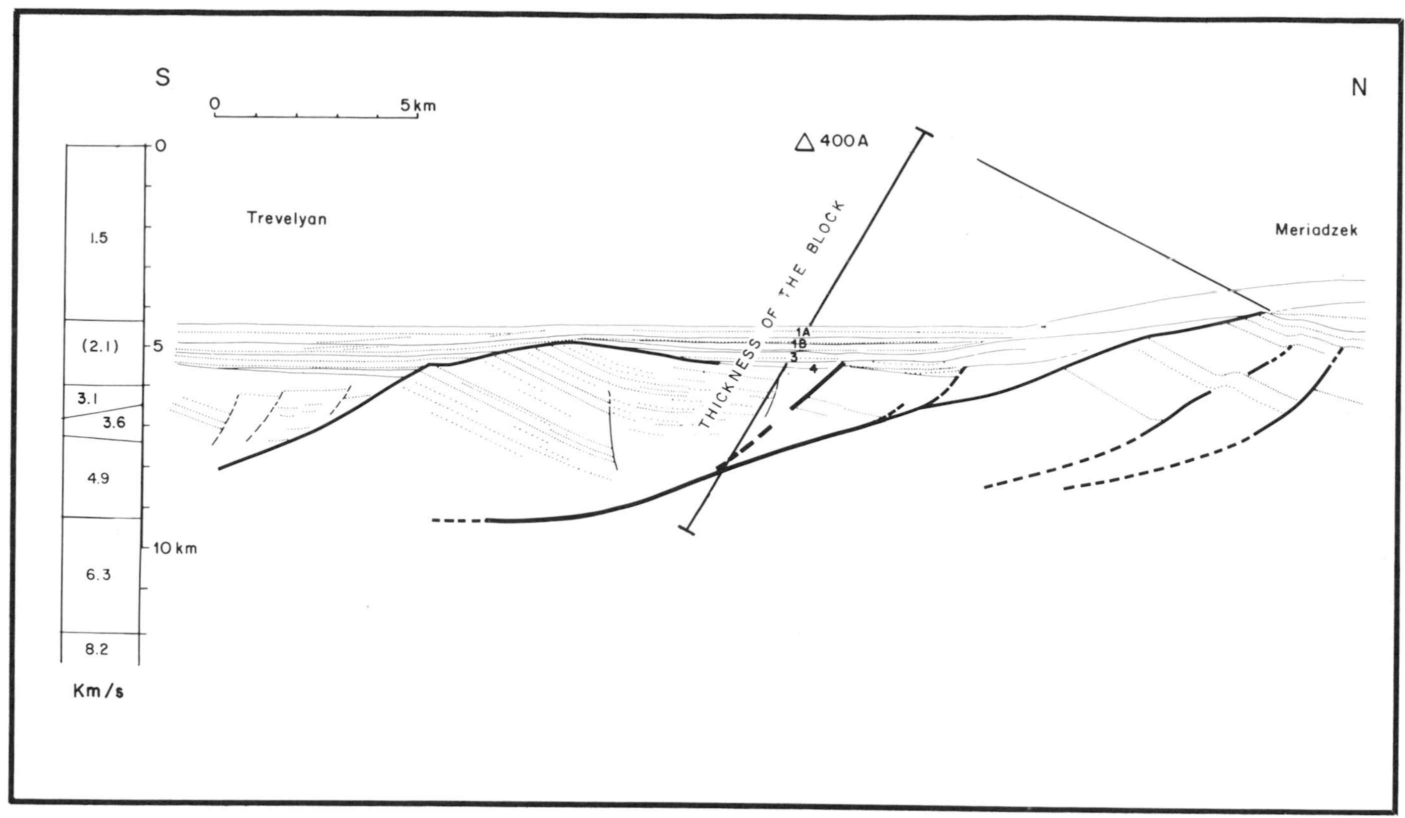

Figure 12 — Thickness of a tilted block near the ocean from profile OC 205-305 (Figure 4).

horizontal, could correspond to a surface of lateral displacement, perhaps the boundary between brittle fracture and ductile flow. Lateral decoupling in the crust at this depth may have taken place in rifting because high temperatures (350 to 450°C) would have favored ductile creep in the lower crust.

Refraction data on Northern Biscay margin (schematized on Figure 9) show the existence of a velocity discontinuity S within the crust, at about 14 km deep near the continent and around 10 km deep just above the Moho at the ocean-continent transition (Avedik et al, 1981). Assuming this S reflector corresponds to the brittle-ductile transition, we computed present sectional area of the crust above S reflector.

Present area is 1900 ± 200 sq km (Figure 11a) and present thickness of the tilted blocks near the ocean is 8 to 10 km (Figure 12). This is interpreted in two ways: (1) The original thickness of the brittle crust was 14 km throughout the margin, and lateral displacement in the brittle crust will lead to an apparent thickness of 8 to 10 km before extension near the ocean; the mass of the brittle crust is globally conserved. In this case, the original sectional area would have been 2250 ± 250 km (Figure 11b). (2) The brittle ductile transition boundary moved upwards just before extension around 8 to 10 km deep, and the original brittle crust sectional area would have been 1850 ± 250 sq km (Figure 11c).

Precision in area measurements is not sufficient to discriminate between hypotheses, but we can see that scheme (2) gives better results. The hypothesis of a shallowed brittle-ductile transition is also consistent with higher temperatures in the crust during rifting.

A hypothesis of margin formation

The above mass balance computation suggests an evolution scheme for the Northern Biscay margin. The stretching of the lithosphere may be responsible for the measured superficial extension, but cannot fully explain the shallow depth of both Moho and brittle-ductile boundaries. Thus, it is probable that apart from the stretching process the shallowing of these phase boundaries (i.e., the attenuation of the crust) is due to heating during rifting below the continental crust. Heating and stretching processes are probably linked together and act simultaneously.

IMPLICATIONS FOR PASSIVE MARGIN STRETCHING MODELS

For a given block, it was shown that deformation within the block, as well as lateral displacement of deep layers, was likely to occur because of the non-parallel listric fault pattern. Therefore, uniform stretching in the crust must not be considered at the scale of a single block. Furthermore, stretching models which assume that locally there is no lateral displacement within the crust are only an approximation of the geological pattern. Amounts of subsidence, extension, and thinning predicted by these models must be considered on a broader scale. New data on rift structure

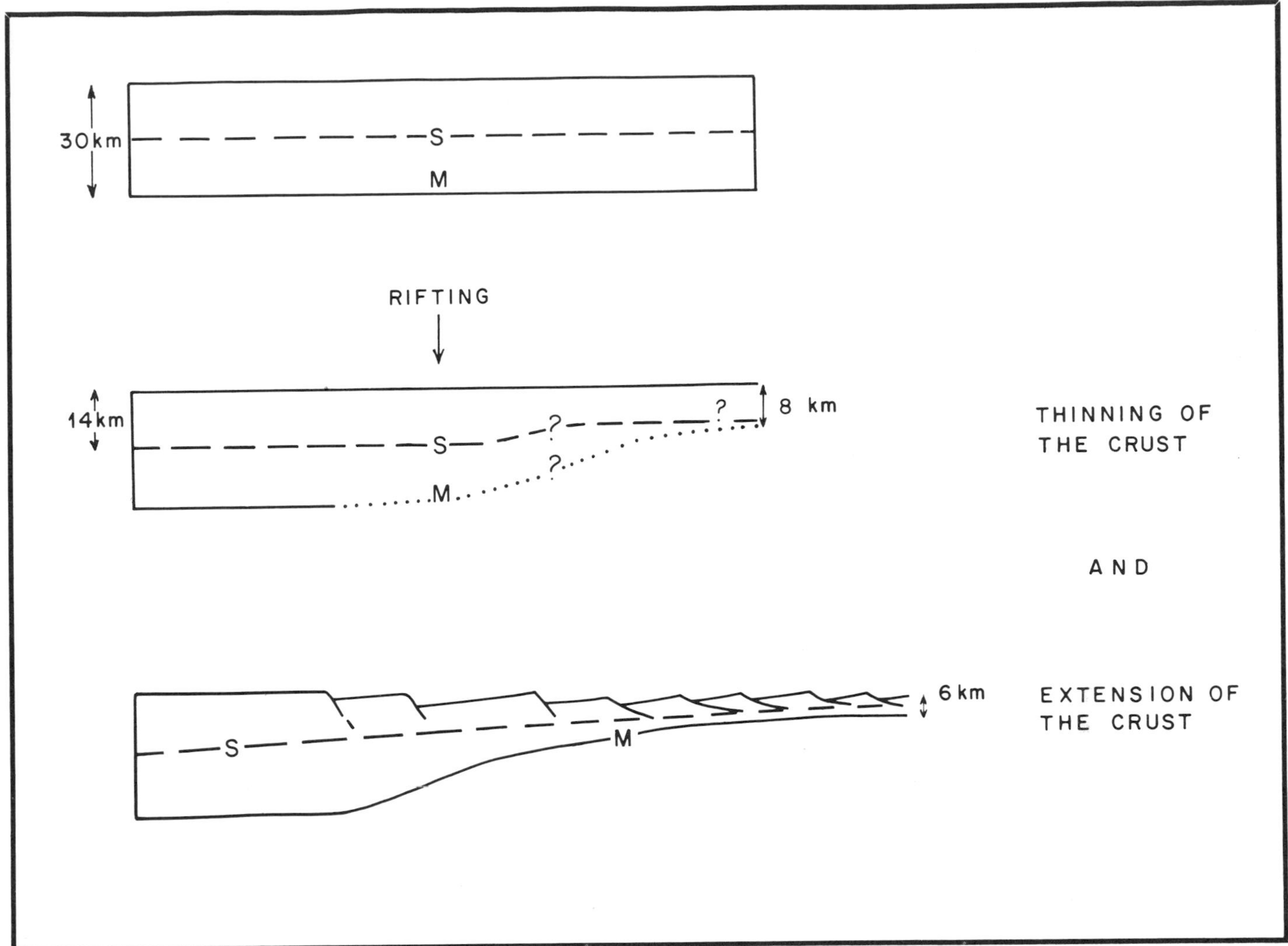

Figure 13 — Combined effects of stretching and attenuation of the crust during rifting.

at depth must be obtained to explain the structure, with special attention to the geometrical constraints implied by mass conservation hypothesis in the upper crust.

Estimation of Stretching Parameters

Stretching models of MacKenzie (1978), Royden, Sclater, and Von Kerzen (1980), and LePichon, Sibuet, and Angelier (1981) predict values for subsidence, extension ratios, and thinning ratios of the crust. Obtained values are a function of one or more stretching parameters depending on the stretching model.

For the Northern Biscay margin, the tectonic subsidence of the crust since the beginning of rifting 120 m.y. ago is not very different from the subsidence after complete cooling of the lithosphere. For stretching models depending on one parameter, LePichon, Sibuet, and Angelier (1981) showed that subsidence for an 120 m.y. old margin may be approximated by the simple formula ($Z=7.2\ (1-\frac{1}{\beta})$km; β: stretching parameter). For stretching models depending on three parameters δ, γ , and Y as described by Royden and Keen (1980) subsidence at infinite time would be: $Z_\infty = 7.83\ (1-\frac{1}{\gamma}) - 0.185\ Y\ (\frac{1}{\delta} - \frac{1}{\gamma})$ km using Royden's data on lithosphere and crust physical constants. This relationship is obtained by simple isostatic balance between original and present stretched cooled lithosphere columns. δ and γ are respectively the stretching parameters of the upper and lower layers of the lithosphere, and Y the original thickness of the upper layer (Royden and Keen, 1980). Subsidence of the Northern Biscay margin may be 10% less than subsidence at infinite time Z_∞, as shown by LePichon and Sibuet. Thus subsidence of Northern Biscay margin will be $Z = 7.2\ (1-\frac{1}{\gamma}) - 0.17\ Y\ (\frac{1}{\delta} - \frac{1}{\gamma})$ km.

Validity of stretching models is checked by comparison between stretching factors obtained by subsidence data, extension ratio, and thinning ratios.

Stretching model with uniform stretching

These models depend on a single stretching factor (see Royden, Sclater, and Von Kerzen, 1980; LePichon, Sibuet, and Angelier, 1981).

For the area 60 km wide near the oceanic floor, sub-

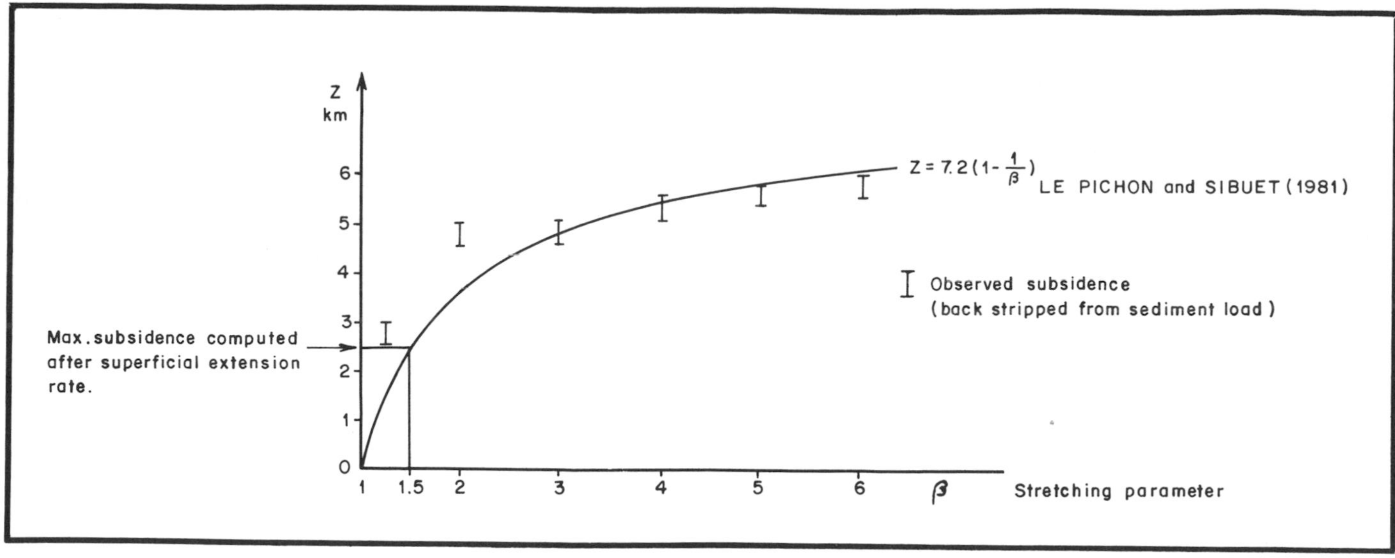

Figure 14 — Comparison between observed and predicted subsidence data on Northern Biscay margin using a uniform stretching model β parameter being the measured thinning ratio.

sidence of the crust backstripped from sediment load since rifting is around 5 ± 0.5 km, which corresponds to a stretching factor of 3.2 ± 0.5. For the same area, the mean extension ratio is 1.25 ± 0.15. The difference between the two values indicates that these models are not consistent with observations on the Northern Biscay margin.

We compared the subsidence data to the uniform extension model, using the present thinning ratio of the whole crust as β, and it appears that the model is consistent with the data. From this point of view, the uniform stretching model must be considered as a physical model where the parameter represents the thinning ratio of the crust but not the amount of superficial extension, which is much lower.

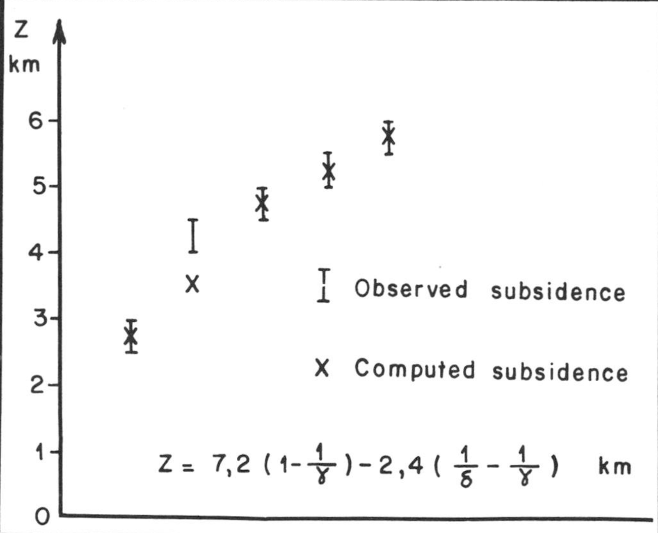

Figure 15 — Comparison between observed and predicted subsidence. The horizontal scale is schematic only, ranging from unstretched continental crust at the ordinate to the ocean-crust boundary immediately beyond the last point on the right.

Stretching models with non-uniform stretching

These models depend on three parameters: the amount of stretching, δ, of the upper layer of the lithosphere; the original thickness of the upper layer, Y ; and, the amount of stretching, γ , of the lower layer of the lithosphere.

If γ is greater than δ, the lower layer of the lithosphere must undergo additional attenuation to avoid the length problem between two differently extended layers. As noted by Royden and Keen (1980), it is more convenient to define δ as the stretching factor of the upper layer and γ as the thinning factor of the lower layer of the lithosphere.

If upper and lower layers of the lithosphere correspond respectively to upper brittle crust and lower crust plus lithosphere, the model predicts mass conservation in the brittle crust and mass loss in the lower crust. This is consistent with calculations from the Northern Biscay margin.

For the 60-km-wide zone near ocean of Northern Biscay margin, the superficial extension ratio of the upper crust is 1.25 ± 0.15 km and the thinning of the lower crust is 6 to 8 km.

Assuming that the S reflector of refraction profile in Figure 9 corresponds to the lower limit of the upper brittle crust, the original thickness of the upper crust is the same as its present thickness near the continent: 14 km. The following parameters of the non-uniform extension model correspond fairly well with observed data on subsidence (Figure 15): δ, superficial extension ratio; γ , thinning ratio of the lower crust; and, Y, thickness of upper crust near the continent.

This study of rift tectonics in Biscay and Galicia-Portugal demonstrates that rifting is characterized by extension and thinning of the crust in response to stretching of the lithosphere and attenuation of the crust. However, the stretching models based on MacKenzie's concept do not explain the superficial structure with tilted blocks bounded by listric fault because they assume no lateral displacement of the extended layer. Therefore, their usefulness is limited to large scale studies of passive margins. Simple models assuming uniform stretching are inadequate to explain the differences between stretching ratios computed from subsidence data and measured extension ratios. Extension models assuming stretching of the upper brittle crust and stretching with additional attenuation of the lower crust and the mantle are more consistent with geological observations.

In view of recent use of these stretching models in predicting heatflow, geothermal gradient, and petroleum potential, it is desirable to construct models that account for geological observations and thus yield heat-flow values that may be more realistic.

ACKNOWLEDGMENTS

We would like to thank our colleagues from Institut Francais du Petrole and Institute of Oceanographic Sciences for many helpful discussions and remarks.

REFERENCES CITED

Avedik, F., et al, 1981, A seismic refraction and reflection study of the continent-ocean transition beneath the northern margin of the Bay of Biscay: London, Proceedings, Meeting of the Royal Society.

DeCharpal, O., et al, 1978, Rifting, crustal attenuation and subsidence in the Bay of Biscay: Nature, v. 275, p. 706.

Keen, C. E., and D. L. Barett, 1981, Thinned and subsided continental crust on the rifted margin of eastern Canada; crustal structure, thermal evolution and subsidence history: Geophysical Journal of Royal Astronomical Society, p. 443-465.

———, L. Beaumont, and R. Boutilier, 1981, Preliminary results from a thermo-mechanical model for the evolution of Atlantic type continental margin: Oceanologica Acta, p. 123-128.

LePichon, X., and J. C. Sibuet, 1981, Passive margins, a model of formation: Journal of Geophysical Research, v. 86, p. 3708-3720.

———, ———, and J. Angelier, 1981, A model for passive margins of Biscay: London, Proceedings, Meeting of the Royal Society.

MacKenzie, D., 1978, Some remarks on the development of sedimentary basins: Earth Planetary Science Letters, v. 40, p. 25-32.

Mattauer, M., 1976, Les deformations de l'ecorce terrestre: Paris, Duncd Edition.

Montadert, L., et al, 1979, Rifting and subsidence of the northern continental margin of the Bay of Biscay, *in* Initial reports of deep sea drilling project: Washington, D.C., U.S. Government Printing Office, v. 48, p. 1025-1060.

Royden, L., and C. E. Keen, 1980, Rifting processes and thermal evolution of the continental margin of eastern Canada determined from subsidence curves: Earth and Planetary Science Letters, v. 51, p. 343-361.

———, J. G. Sclater, and R. P. Von Kerzen, 1980, Continental margin, subsidence and heat-flow; important parameters in formation of petroleum hydrocarbons: AAPG Bulletin, v. 64, n. 2, p. 173-187.

Sibuet, J.C., and W.B.F. Ryan, 1978, Evolution of the west Iberian passive continental margin in the framework of the early evolution of the North Atlantic Ocean, *in* Initial reports of deep sea drilling project: Washington, D.C., U.S. Government Printing Office, v. 47, p. 761-776.

Sleep, N.H., 1971, Thermal effects of the formation of Atlantic continental margins by continental break-up: Geophysical Journal of Royal Astronomical Society, p. 325-350.

Watts, A.B., and M.S. Steckler, 1979, Subsidence and Eustasy at the continental margin of eastern North America, *in* Talwani, M. W. Hay, and W.B.F. Ryan, eds., Deep drilling results of the Atlantic ocean; continental margin and paleoenvironment: Washington, D.C., American Geophysical Union, p. 218-234.

State of Stress at Passive Margins and Initiation of Subduction Zones

S. A. P. L. Cloetingh
M. J. R. Wortel
N. J. Vlaar
Vening Meinesz Laboratory
University of Utrecht
The Netherlands

We present the results of a finite element analysis of the state of stress at passive continental margins, which we performed to investigate whether these margins are potential sites for initiation of subduction. The state of stress is determined by the effect of sediment loading at the rise, and to a lesser extent by plate tectonic forces. For a model of sedimentation at the margin, coupled to the subsidence of the cooling oceanic lithosphere, stresses up to 3 kbar can be generated. Stresses of this order may cause failure of the lithosphere and initiation of subduction. In general, however, an additional cause (e.g. plate interaction and geometrical focusing effect) not included in the modelling is required to start the subduction process. If, after 100 million years of evolution, subduction has not started, continued aging of the passive margin alone does not result in conditions significantly more favorable for initiation of subduction.

Initiation of oceanic lithosphere subduction is one of the major issues in geodynamics. However, as stated by Dickinson and Seely (1979) "the mechanisms that initiate plate consumption at new subduction zones and thus generate arc-trench systems are incompletely known." These authors summarize the possible mechanisms for initiation of oceanic lithosphere subduction, and distinguish the following two classes: (1) Plate rupture, either within an oceanic plate, or at a passive margin, and (2) Reversal of the polarity of an existing subduction zone, eventually after a collision of an island arc with a passive margin (see also Speed and Sleep, 1982; Speed, this volume). Another class, although not mentioned by these authors, is initiation of subduction by inversion of transform faults into trenches (see below). We concentrate here on the mechanisms for the formation of a new plate boundary and refrain from dealing with the mechanisms of polarity reversals.

A key factor generally determining the possibility of subduction of oceanic lithosphere is its gravitational stability. The total stability of the oceanic lithosphere is the sum of the positive buoyancy of its stable petrological stratification and the negative buoyancy resulting from thermal contraction upon cooling (Oxburgh and Parmentier, 1977). Wortel (1980) shows that for a model of oceanic lithosphere consistent with observations of heat flow and topography the system is stable for ages less than 30 million years, due to the stabilizing effect of the density changes accompanying the formation of oceanic crust (according to Oxburgh and Parmentier's model). As a result of its further cooling and thermal contraction, the lithosphere becomes unstable for ages above 30 million years (Oxburgh and Parmentier, 1977). Considering the age-dependence of the lithospheric instability, it is reasonable to expect that the chances for initiation of oceanic lithosphere subduction increase with age (Vlaar and Wortel, 1976). However, to allow initiation of subduction of gravitationally unstable lithosphere, stress conditions must be favorable for the creation of a failure zone within the lithosphere. Failure of oceanic lithosphere and initiation of subduction might preferentially take place at existing weakness zones. As such, transform faults have been advocated by several authors (Uyeda and Ben-Avraham, 1972; Uyeda and Miyashiro, 1974; Dewey, 1975). Turcotte, Haxby, and Ockendon (1977) argue favorably for initiation of subduction by mantle material penetrating the lithosphere at oceanic spreading centers. Although at ridge-transform intersections high stresses might be generated (Fujita and Sleep, 1978), this young lithosphere is still gravitationally stable. Initiation of subduction of stable oceanic lithosphere is hampered by resistive forces active in trench formation (McKenzie, 1977). Apparently, spreading centers are not the most appropriate sites for initiation of subduction.

Geological evidence and speculation for margins with widely different ages, ranging from early Proterozoic to recent (Dewey, 1969; Hoffman, 1980; Williams, 1979; Karig et al, 1980), support the thesis that particular passive margins might be potential sites for the formation of plate boundaries. There are several reasons why such a process might preferentially take place at a passive margin. Important factors are the contrast in mechanical properties across the margin (in general much greater than the contrast between oceanic lithosphere adjacent to a transform fault) and pre-stressing the lithosphere by sediment loading at the margin. The above consideration led us to focus here on the possibilities for initiation of oceanic lithosphere subduction at a passive margin.

The state of stress at a passive margin is determined by its local geometric and rheologic lithospheric properties, and by the system of forces acting on the lithosphere. Of these features the thickness of the oceanic lithosphere (Parsons and McKenzie, 1978), its rheological stratification (Caldwell and Turcotte, 1979; Bodine, Steckler, and Watts, 1981), the push exerted by the elevation of the oceanic ridge (Richter and McKenzie, 1978; England and Wortel, 1980), and the forces associated with the (negative) buoyancy of the lithosphere (Vlaar and Wortel, 1976) are a function of the oceanic lithosphere's age. The sediment loading capacity of oceanic lithosphere increases with age, through continued cooling and densification of the lithosphere. One might expect a coupling between the height of the sedimentary column deposited at the passive margin and the age-dependent thermal subsidence of the underlying oceanic lithosphere (Turcotte and Ahern, 1977).

In previous studies of the state of stress at passive margins (Walcott, 1972; Bott and Dean, 1972; Turcotte, Ahern, and Bird, 1977; Neugebauer and Spohn, 1978), possible implications of age-dependent properties of the oceanic lithosphere were not considered. Present work attempts to study the interrelations between age-dependent forces, geometry, and rheology and to decipher their net effect on the state of stress at passive margins using the finite element technique for stress calculations. Models presented here deal only with gross features of passive margin evolution and no detailed modelling of a specific margin is intended. As detailed (local) stratigraphy is essential before gravity anomalies can be used as a successful constraint for flexural modeling, we refrained from incorporating gravity data as a constraint in the present work. Subsequently we discuss the implications for initiation of subduction.

MODELS FOR THE STATE OF STRESS AT A PASSIVE MARGIN

We constructed finite element models for a passive continental margin in four different stages of evolution, at 30, 60, 100 and 200 million years respectively. For all models, a half-spreading rate of 1 cm/year is taken, characteristic for oceanic lithosphere without attached downgoing slabs (Forsyth and Uyeda, 1975). Calculations were carried out with the MARC finite element package (MARC, 1980) using linear strain quadrilateral elements. For more details on the numerical aspects of the work, refer to Cloetingh and Wisse (1981). The model features are summarized in Figure 1.

Forces

We adopt a triangular sediment wedge at the continental rise as a sedimentary loading model. As our reference, we assume that the maximum sediment thickness at the continental rise corresponds with the thickness that results if the sedimentation keeps up with the subsidence of a boundary layer model of the cooling oceanic lithosphere (Turcotte and Ahern, 1977; Wortel, 1980). For the maximum height of the sedimentary wedge, this implies an increase from 4.5 km at 30 million years to 9.4 km at 200 million years, following roughly a square-root of age relation (Figure 2). The width of the sedimentary wedge is taken to be 300 km, a value typical for wedges at continental rises (Sheridan et al, 1979), and only excess densities are considered. As sediments at the rise replace water we take a $\Delta\rho = 2.4\text{-}1 = 1.4$ g/cu cm. Although related to

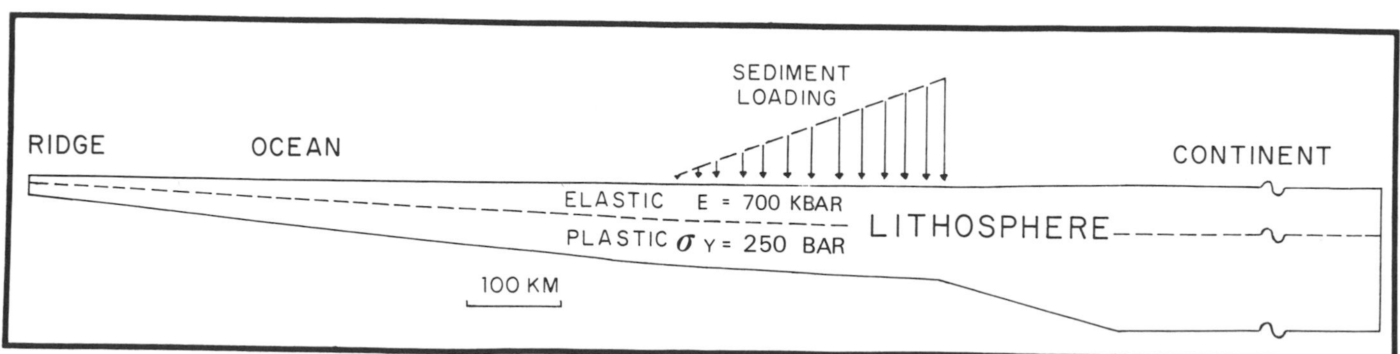

Figure 1 — Geometry and other features of the models illustrated for a passive margin 100 m.y. old. Vertical scale equals horizontal scale.

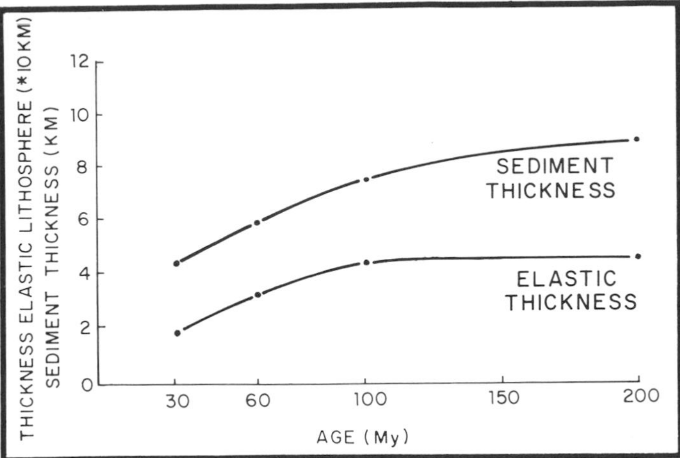

Figure 2 — Maximum height of the sedimentary prism and the thickness of the elastic part of the lithosphere as functions of age.

age in a somewhat similar way as sediment loading at the rise, sediment loading at the shelf is more difficult to generalize. The position of the shelf edge is not controlled by tectonic subsidence but probably by the local features of sedimentary processes (Watts and Steckler, 1979). On some margins (e.g. off Nova Scotia) the shelf edge has not built out far, it is located landward of the maximum sediment thickness, and the maximum sediment loading is at the rise. The amount of lithospheric thinning under the shelf varies strongly (Watts and Steckler report 10 km thinning off Nova Scotia and 20 km thinning off New York) and is therefore difficult to incorporate in our modeling. Therefore, although in more detailed modelling of specific margins careful attention should be given to sediment loading at the shelf, we refrain from incorporating this feature in our models.

Isostatic forces proportional to the deflection of the lithosphere under loading are included by modifying the stiffness matrix at the model's basal nodes. To suppress rigid body translation, zero horizontal displacements are prescribed for the right hand boundary of the model. The horizontal extent of the continental lithosphere is sufficient to ignore any artificial effects of this boundary condition in the region of interest.

Of the plate tectonic forces implemented in the models, the magnitudes of the forces associated with the ridge push and the negative buoyancy of the oceanic lithosphere are calculated on the basis of Oxburgh and Parmentier's (1977) model for the formation of oceanic crust, and Crough's (1975) model for the thermal evolution of oceanic lithosphere. Following Lister (1975) we model the ridge push not as a line force, but as a pressure gradient, excluding artificial stress concentrations at the ridge. The integrated pressure gradient per unit width along the ridge for oceanic lithosphere 100 million years old is 2.3×10^{12} N/m. For completeness we assume a resistive drag at the lithosphere's base, acting opposite the spreading direction with a magnitude per unit area of 1% of the pressure exerted by the elevation of the ridge (Solomon, Sleep, and Richardson, 1977).

Lithospheric thickness and rheology

Age-dependent lithospheric thicknesses are based on Crough's (1975) oceanic lithosphere model and are taken from Wortel (1980). A thickness of 150 km, inferred from a number of independent geophysical approaches (e.g. Pollack and Chapman, 1977), is assigned to the continental lithosphere.

Turcotte, Ahern, and Bird (1977) argue that oceanic and continental lithosphere at a passive margin are effectively decoupled due to a marginal fault system. Since depth and decoupling of such a fault zone are unclear, we model the transition between oceanic and continental lithosphere as continuous. We adopt a width of 200 km for the transition, taking into account the reported evidence for the presence of rift-stage crust at the margin (e.g. Hutchinson et al, this volume). Implications of fault systems present at the margin are briefly discussed further on.

Based on evidence from studies on seismicity (Chapple and Forsyth, 1979) and flexure (McAdoo, Caldwell, and Turcotte, 1978) of oceanic lithosphere at trenches, we model the oceanic lithosphere as a plate with an elastic upper layer (Young's modulus $E = 7 \times 10^{10}$ N/m^2, Poisson's ratio $\upsilon = .25$) and a perfectly plastic lower part. From studies on steady-state flow of rocks (Mercier, Anderson, and Carter, 1977), we adopt a yield stress of 250 bar for the lower part of the oceanic lithosphere. Based on data on the width of the outer rise seaward of oceanic trenches, Caldwell and Turcotte (1979) argue that the thickness of the elastic upper part of the oceanic lithosphere is a strong function of its age. The thickness increases approximately proportional to the square-root of age function, from a few km near the ridge to 45 km for ages round 100 million years, with negligible increase beyond 100 million years (Figure 2). These values are confirmed by estimates inferred from microrheology of olivine by Beaumont (1979). One might argue that if the base of the elastic lithosphere coincides with the depth to an isotherm, the thermal effect of sediment loading may reduce the elastic thickness of the lithosphere. However, since the reduction in elastic thickness is relatively small (at most 15%) we do not incorporate this effect in the models. As a result, the stresses inferred from our calculations are somewhat conservative.

Because of cooling, the lithosphere's elastic upper part thickens with age by a continuous transition from a plastic to an elastic constitution. Stresses induced by loading, which are higher than the yield strength, can only be accommodated once the material has undergone the plastic-elastic transition. For our reference model of sediment loading, this implies that only the part of the sedimentary wedge deposited in an interval between ages t_1 and t_2 effectively determines the state of stress at t_2 in the bottom part of the elastic lithosphere (that is created between t_1 and t_2). This follows from the fact that loading prior to t_1 only induces stresses up to the yield strength in the plastic lithosphere. Therefore, the stresses calculated with our "static" models for the bottom of the elastic lithosphere tend to be overestimated. Numerical experi-

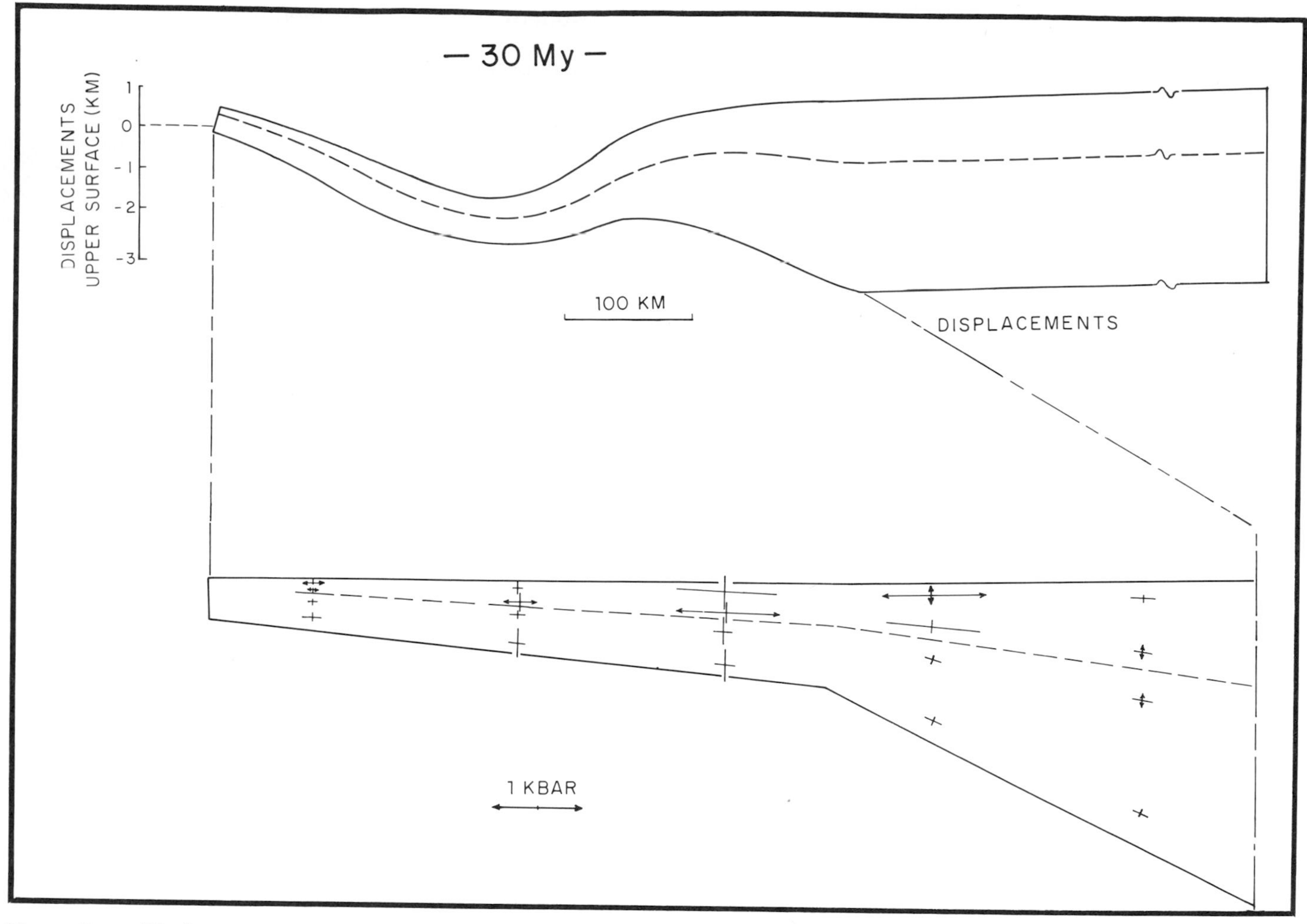

Figure 3a — Displacements and stresses calculated for a passive margin 30 m.y. old, based on the reference model of sediment loading given in Figure 2. Stresses are plotted only for the parts of the lithosphere where displacements are significant. These sections are bounded by the dashed lines. The rheological boundary between the elastic upper part and the plastic lower part of the lithosphere is marked by a broken line. Above: Displacements of the lithosphere (km). Note that the scale of displacements (vertical axis) and geometry (indicated by a horizontal bar at the bottom) are not the same. Below: Principal stresses (kbar) at the margin. Line with arrow on each end denotes tension, single line denotes compression. Scale indicated at the lower part of the figure.

ments showed an overestimation in the magnitudes of the stress maxima, for the four ages considered, varying between 15 and 22%.

A thickness of 50 km is assumed for the elastic part of the adjacent continental lithosphere, based on Haxby, Turcotte, and Bird (1976).

Results

Stress calculations for a passive margin are made in four different stages of evolution. The reference model of sediment loading and the age-dependence of the elastic thickness and other features (as given in Figures 1 and 2) are incorporated into these calculations. The results for ages of 30 and 100 million years are given in Figures 3a and 3b respectively. The deformation of the lithosphere and the resulting stress field (order of magnitude a few kbar) is dominated by sediment loading at the rise; the contribution of the plate tectonic forces to the stress field is of smaller magnitude. Differential stresses are largest at the points of maximum flexure. Tensional stresses up to 3 kbar are generated at the base of the elastic lithosphere. Note once again that the magnitudes of tensional stresses at the bottom of the elastic lithosphere are somewhat overestimated. Stresses are concentrated in the elastic part of the lithosphere due to dividing the lithosphere into an elastic and a plastic part (e.g., Kusznir and Bott, 1977). The displacements outside the bending area shown in Figure 3b express the equilibrium of the forces associated with the negative buoyancy of the (unstable) oceanic lithosphere and the opposing isostatic forces. The resulting depth profile is a well-known feature of oceanic basins (e.g. Parsons and Sclater, 1977). A comparison of Figures 3a and 3b shows that the state of stress at a passive margin depends on the age of the adjacent oceanic lithosphere. Note that as sediment loading on the shelf is related to age in a similar way as sediment loading at the rise, incorporating this feature in the models would not alter this conclusion. To summarize this dependence

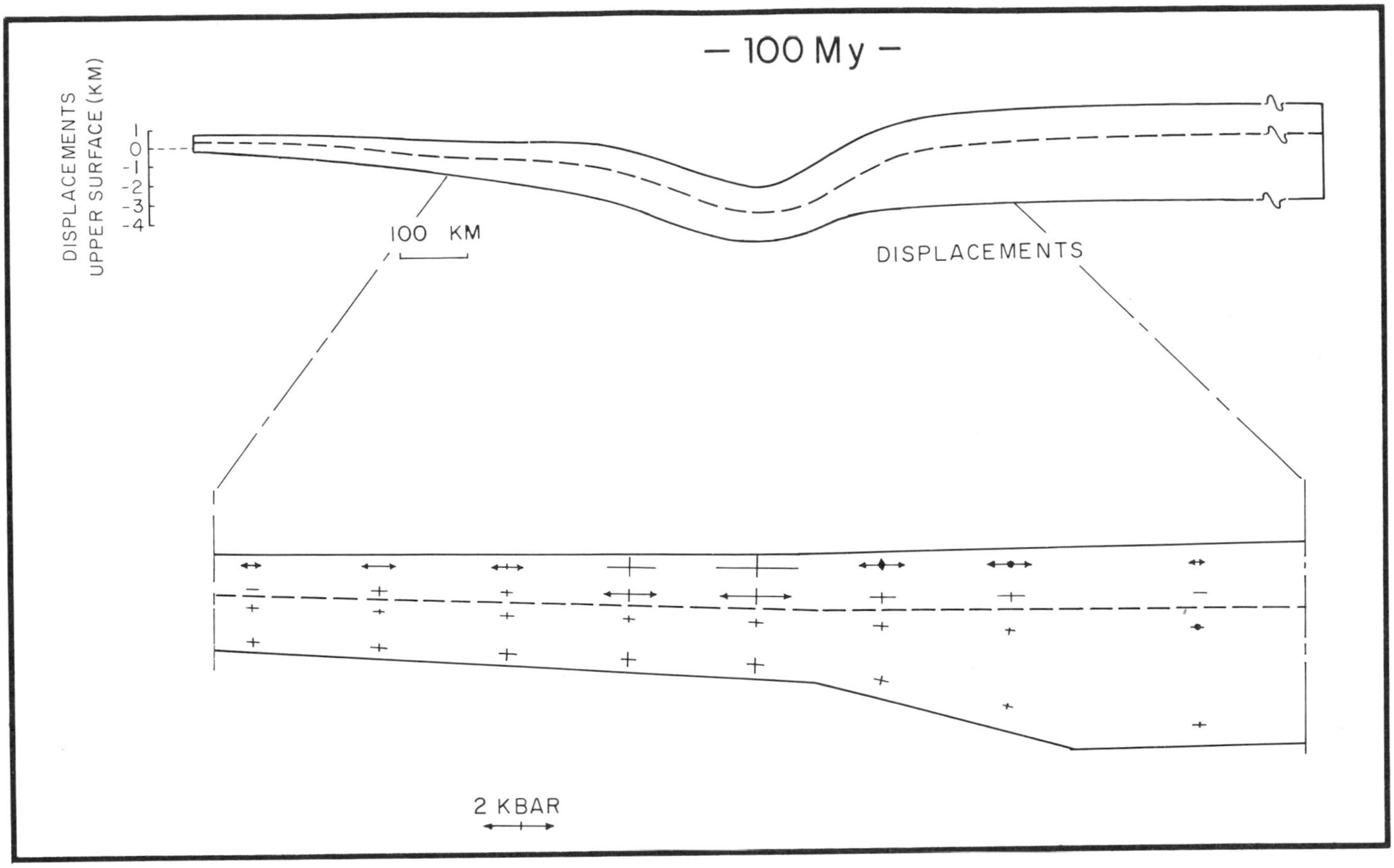

Figure 3b — Displacements and stresses calculated for a passive margin 100 m.y. old. Figure conventions as in Figure 3b.

we plotted (Figure 4) the maximum differential (tensional) stresses ($\sigma_1 - \sigma_3$) as a function of age for all four cases. The age-dependence is strongest for ages below 100 million years. From 30 to 100 million years, an interval during which both the sedimentary loading and the elastic thickness increase (Figure 2), the differential stress maxima also increase with age. From 100 to 200 million years, the elastic part of the lithosphere is at a constant thickness (Figure 2). The increase in sediment load, according to our time-dependent sedimentation model, results only in a minor increase of the stresses.

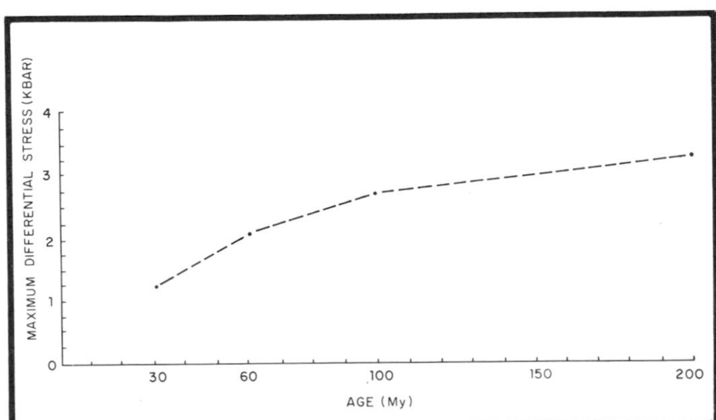

Figure 4 — Maximum differential stresses as a function of age.

DISCUSSION

Our calculations for the reference model of sediment loading show that stresses up to 3 kbar can be generated at a passive margin. However, can stresses of this magnitude create failure zones in the lithosphere? The critical point in answering this question is the strength of the elastic part of the lithosphere. Marine geophysical work on seamount loading and lithospheric bending at trenches (Watts and Cochran, 1974; Caldwell and Turcotte, 1979; Watts, Bodine, and Steckler, 1980) shows that oceanic lithosphere can support stresses of several kbars for long periods of geological time (>50 million years). Pertinent laboratory experiments on olivine (e.g., Ashby and Verrall, 1978; Evans and Goetze, 1979; Kirby, 1980) point to yield stresses of about 3 to 10 kbar. Consider the laboratory data as an upper limit on the strength of the lithosphere (Paterson, 1979); nevertheless, a gap remains between the magnitude of the stresses and the strength of the elastic part of the lithosphere. The stresses generated according to our models are for a reference model of sediment loading. Deviations from the reference model occur in thicker sediment loads (e.g. the eastern U.S. margin with up to 10 km of sediments at the continental rise) and virtually no sediment loading (e.g. the starved southwest European margin). As a result of the age-dependence of the thickness, a surplus sediment load, above the load adopted in the reference model, effectively creates higher stresses when deposited on a young margin.

Several other mechanisms, contributing to stress and not considered here, might be envisaged: thermoelastic stresses due to cooling of the plate (Turcotte, Ahern, and Bird, 1977), stresses due to crustal thickness inhomogeneities (Bott and Dean, 1972), and stresses generated by phase changes in the lithosphere under the influence of sediment loading (Neugebauer and Spohn, 1978). However, thermo-elastic stresses (order of magnitude several kbars) are probably largely relieved in the early stages of lithospheric evolution. The exact mechanism of relaxation of thermo-elastic stresses is uncertain, but fracture zones formation (Turcotte, 1974) is one possible mechanism. Stress contributions due to changes in crustal thickness at the ocean-continent boundary are of relatively minor importance. This mechanism results in shear stresses about a few hundred bars in the continental crust, while the contribution to the stresses in the oceanic lithosphere is negligible (Bott and Dean, 1972). The effect of a hypothetical phase change is even smaller, only a few tens of bars (Neugebauer and Spohn, 1978).

From the previous information you might conclude that the stresses calculated for our models do not provide upper bounds. On the other hand the strength of the lithosphere may be locally reduced by the occurrence of fossil fault systems at passive margins. So in some circumstances, stresses might be generated at passive margins quite close to, or even exceeding, lower bounds of estimates on the lithospheric strength; this allows the creation of failure zones in the lithosphere and initiates the subduction process.

In general, an additional cause must trigger the subduction process. Deviations from the two-dimensional situations represented by our models may cause geometrical focusing. At certain stages in the process of global plate reorganization, plate tectonic stresses may be concentrated locally to a level comparable with the stresses resulting from sediment loading. The short episode (latest Cretaceous-Eocene) of activation of the northern Iberian passive margin inferred by Boillot et al (1979) might have been caused by this mechanism. The northeastern Indian Ocean might provide a similar example of passive continental margin deformation. Here the onset of the deformation seems to be related to the Himalayan orogenic stage of the collision between India and Asia (Weissel, Anderson, and Geller, 1980). In this region continental rise sediments up to 12 km thick are present (Curray and Moore, 1974). This margin apparently combines numerous features, making it an interesting case for further study of the preparatory stages of subduction initiation.

In summary, our calculations show that tensional stresses up to 3 kbar can be generated at passive margins. If subduction has not begun after 100 million years of evolution, continued aging of the passive margin alone does not provide conditions significantly more favorable for initiation of the subduction process.

ACKNOWLEDGMENTS

Gerald Wisse of the Scientific Applications Group of the Delft University of Technology is gratefully acknowledged for his assistance with the finite element calculations. Discussions held with other participants at the Hollis Hedberg Research Conference, at Carleton University, and at L-DGO, are appreciated.

REFERENCES CITED

Ashby, M. F., and R. A. Verrall, 1978, Micromechanisms of flow and fracture and their relevance to the rheology of the upper mantle: Philosophical Transactions, Royal Society of London, v. A288, p. 59-95.

Beaumont, C., 1979, On rheological zonation of the lithosphere during flexure: Tectonophysics, v. 59, p. 347-365.

Bodine, J. H., M.S. Steckler, and A. B. Watts, 1981, Observations of flexure and the rheology of the oceanic lithosphere: Journal of Geophysical Research, v. 86, p. 3695-3707.

Boillot, G., et al 1979, The northwestern Iberian margin: A Cretaceous passive margin deformed during Eocene, *in* M. Talwani, et al, eds., Deep drilling results in the Atlantic ocean: Continental margins and paleoenvironment: American Geophysical Union, Maurice Ewing Series, v. 3, p. 138-153.

Bott, M. H. P., and D. S. Dean, 1972, Stress systems at young continental margins: Nature Physical Sciences, v. 235, p. 23-25.

Caldwell, J. G., and D. L. Turcotte, 1979, Dependence of the thickness of the elastic oceanic lithosphere on age: Journal of Geophysical Research, v. 84, p. 7572-7576.

Chapple, W. M. and D. W. Forsyth, 1979, Earthquakes and bending of plates at trenches: Journal of Geophysical Research, v. 84, p. 6729-6749.

Cloetingh, S., and G. Wisse, 1981, Finite element modelling in geodynamics: The state of stress at passive continental margins: Finite Element News, n. 3, p. 37-40.

Crough, S. T., 1975, Thermal model of oceanic lithosphere: Nature, v. 256, p. 388-390.

Curray, J. R., and D. G. Moore, 1974, Sedimentary and tectonic processes in the Bengal deep-sea fan and geosyncline *in* C. A. Burk, and C. L. Drake, eds., The geology of continental margins: New York, Springer Verlag, p. 617-627.

Dewey, J. F., 1969, Continental margins: A model for conversion of Atlantic-type to Andean-type: Earth and Planetary Science Letters, v. 6, p. 189-197.

———, 1975, Finite plate evolution: Some implications for the evolution of rock masses at plate margins: American Journal of Science, v. 275A, p. 260-284.

Dickinson, W. R., and D. R. Seely, 1979, Structure and stratigraphy of forearc regions: AAPG Bulletin, v. 63, p. 2-31.

England, P., and R. Wortel, 1980, Some consequences of the subduction of young slabs: Earth and Planetary Science Letters, v. 47, p. 403-415.

Evans, B., and C. Goetze, 1979, The temperature variation of hardness of olivine and its implication for polycrystalline yield stress: Journal of Geophysical Research, v. 84, p. 5505-5524.

Forsyth, D. W., and S. Uyeda, 1975, On the relative importance of driving forces of plate motion: Geophysical Journal, Royal Astronomical Society, v. 43, p. 163-200.

Fujita, K., and N. H. Sleep, 1978, Membrane stresses near mid-ocean ridge-transform intersections: Tectonophysics,

v. 50, p. 207-221.
Haxby, W. F., D. L. Turcotte, and J. M. Bird, 1976, Thermal and mechanical evolution of the Michigan basin: Tectonophysics, v. 36, p. 57-75.
Hoffman, P. F., 1980, Wopmay orogen: a Wilson cycle of early Proterozoic age in the northwest of the Canadian shield, *in* D. W. Strangway, ed., The continental crust and its mineral deposits: Geological Association of Canada Special Paper, v. 20, p. 523-549.
Hutchinson, D. R., et al, 1982, Deep structure and evolution of the Carolina trough, *in* J. S. Watkins, and C. L. Drake, eds., Continental margin processes: AAPG Memoir, v. 34.
Karig, D. E., et al, 1980, Structural framework of the fore-arc basin, N. W. Sumatra: Journal, Geological Society of London, v. 137, p. 77-91.
Kirby, S. H., 1980, Tectonic stresses in the lithosphere: constraints provided by the experimental deformation of rocks: Journal of Geophysical Research, p. 6353-6368.
Kusznir, N. J., and M. H. P. Bott, 1977, Stress concentration in the upper lithosphere caused by underlying viscoelastic creep: Tectonophysics, v. 43, p. 247-256.
Lister, C. R. B., 1975, Gravitational drive on oceanic plates caused by thermal contraction: Nature, v. 257, p. 663-665.
McAdoo, D. C., J. G. Caldwell, and D. L. Turcotte, 1978, On the elastic-perfectly plastic bending of the lithosphere under generalized loading with application to the Kuril trench: Geophysical Journal, Royal Astronomical Society, v. 54, p. 11-26.
McKenzie, D. P., 1977, The initiation of trenches: a finite amplitude instability, *in* M. Talwani, and W. Pitman III, eds., Island arcs, deep sea trenches and back-arc basins: American Geophysical Union, Maurice Ewing Series, v. 1, p. 57-62.
MARC Analysis Research Corporation, 1980, Marc general purpose finite element program: Palo Alto, user manual, v. A-E.
Mercier, J. C., D. A. Anderson, and N. L. Carter, 1977, Stress in the lithosphere: inferences from steady state flow of rocks: Pure and Applied Geophysics, v. 115, p. 199-226.
Neugebauer, H. J., and T. Spohn, 1978, Late stage development of mature Atlantic-type continental margins: Tectonophysics, v. 50, p. 275-305.
Oxburgh, E. R., and E. M. Parmentier, 1977, Compositional and density stratification in oceanic lithosphere-causes and consequences: Journal of the Geological Society of London, v. 133, p. 343-355.
Parsons, B., and J. G. Sclater, 1977, An analysis of the variation of ocean floor heat flow and bathymetry with age: Journal of Geophysical Research, v. 82, p. 803-827.
———, and D. McKenzie, 1978, Mantle convection and the thermal structure of the plates: Journal of Geophysical Research, v. 83, p. 4485-4496.
Paterson, M. S., 1979, The mechanical behaviour of rocks under crustal and mantle conditions, *in* M. N. McElhinny, ed., The earth, its origin, structure and evolution: New York, Academic Press, p. 469-489.
Pollack, H. N., and D. S. Chapman, 1977, On the regional variation of heat flow, geotherms and lithospheric thickness: Tectonophysics, v. 38, p. 279-296.
Richter, F. M., and D. McKenzie, 1978, Simple plate models of mantle convection: Journal of Geophysics, v. 44, p. 441-471.
Sheridan, R. E., et al, 1979, Seismic refraction study of the continental edge of the eastern United States, *in* C. E. Keen, ed., Crustal properties across passive margins: Amsterdam, Elsevier, p. 1-26.
Solomon, S. C., N. H. Sleep, and R. M. Richardson, 1977, Implications of absolute plate motions and intraplate stress for mantle rheology: Tectonophysics, v. 37, p. 219-231.
Speed, R. C., 1982, Passive paleozoic continental margin in the western U.S. and its modification by collisions, accretion, and truncation, *in* J. S. Watkins and C. L. Drake, eds., Continental margin processes: this volume.
———, and N. H. Sleep, in press, Antler orogeny: a model: Geological Society of America Bulletin.
Turcotte, D. L., 1974, Are transform faults thermal contraction cracks: Journal of Geophysical Research, v. 79, p. 2573-2577.
———, and J. L. Ahern, 1977, On the thermal and subsidence history of sedimentary basins: Journal of Geophysical Research, v. 82, p. 3762-3766.
———, J. L. Ahern, and J. M. Bird, 1977, The state of stress at continental margins: Tectonophysics, v. 42, p. 1-28.
———, W. F. Haxby, and J. R. Ockendon, 1977, Lithospheric instabilities, *in* M. Talwani, and W. Pitman III, eds., Island arcs, deep-sea trenches and back-arc basins: American Geophysical Union, Maurice Ewing Series, v. 1, p. 63-69.
Uyeda, S., and Z. Ben-Avraham, 1972, Origin and development of the Philippine sea: Nature, v. 240, p. 176-178.
———, and Miyashiro, A., 1974, Plate tectonics and the Japanese Islands, a synthesis: Geological Society of America Bulletin, v. 85, p. 1159-1170.
Vlaar, N. J., and M. J. R. Wortel, 1976, Lithospheric aging, instability and subduction: Tectonophysics, v. 32, p. 331-351.
Walcott, R. I., 1972, Gravity, flexure and the growth of sedimentary basins at a continental edge: Geological Society of America Bulletin, v. 83, p. 1845-1848.
Watts, A. B., and J. R. Cochran, 1974, Gravity anomalies and flexure of the lithosphere along the Hawaiian-Emperor seamount chain: Geophysical Journal, Royal Astronomical Society, v. 38, p. 119-141.
———, and M. S. Steckler, 1979, Subsidence and eustasy at the continental margin of eastern North America, *in* M. Talwani, et al, eds., Deep drilling results in the Atlantic ocean: Continental margins and paleoenvironment: American Geophysical Union, Maurice Ewing Series, v. 3, p. 218-234.
———, J. H. Bodine, and M. S. Steckler, 1980, Observations of flexure and the state of stress in the oceanic lithosphere: Journal of Geophysical Research, v. 85, p. 6369-6376.
Weissel, J. K., R. N. Anderson, and C. A. Geller, 1980, Deformation of the Indo-Australian plate: Nature, v. 287, p. 284-291.
Williams, H., 1979, Appalachian orogen in Canada: Canadian Journal of Earth Sciences, v. 16, p. 792-807.
Wortel, R., 1980, Age-dependent subduction of oceanic lithosphere: Utrecht, The Netherlands, Utrecht University, PhD dissertation, 147 p.

A Summary of Thermo-Mechanical Model Results for the Evolution of Continental Margins based on Three Rifting Processes

C. E. Keen
Geological Survey of Canada
Bedford Institute of Oceanography
Nova Scotia, Canada

C. Beaumont
R. Boutilier
Department of Oceanography
Dalhousie University
Nova Scotia, Canada

We present three models for rifting at passive continental margins which primarily involve extension, thinning, and heating of the lithosphere. These models lead to predictions of postrift evolution of the margins including consequences of conductive cooling of the lithosphere and the lithosphere's isostatic response to loading by sediments and water. The response to loading is described using an elastic plate approximating the mechanical behavior of the lithosphere. We assume that plate thickness, and therefore rigidity is a function of temperature. The lower boundary of the elastic plate corresponds to an isotherm. Thus, the rigidity of the margin increases as the margin cools, and varies with position across the margin. We compare model predictions with observed data for the Nova Scotian continental margin.

Models of the evolution of rifted, Atlantic-type continental margins are the concern of many recent studies (e.g. Watts and Steckler, 1979; Royden and Keen, 1980; Royden, Sclater, and von Herzen, 1980). This evolution appears largely controlled by the physical processes active during the rift stage, which determine thermal and subsidence histories. Subsidence also depends on the lithosphere's isostatic response to sediment load.

We describe results for a coupled thermo-mechanical model which includes rift processes and the mechanical behavior of the lithosphere. The results are for three rifting process models, all of which involved extension and thinning of the lithosphere. The models approximate the rheological properties of the lithosphere by an elastic plate with temperature dependent thickness in the same manner as described in Keen, Beaumont and Boutilier (1981). Elastic plate thickness and hence flexural rigidity are defined by the depth to an isotherm. Because the depth to this isotherm varies with time as the margin cools and with position across the margin, flexural rigidity varies both spatially and temporally. The value chosen for this isotherm is 250°C.

We apply these models to the continental margin off Nova Scotia. Sufficient observational data are available (seismic reflection, deep crustal refraction, stratigraphy from deep exploratory well, and gravity data) to constrain the model parameters and allow comparison with model predictions (Jansa and Wade, 1975; Keen and Hyndman, 1979). Beaumont, Keen, and Boutilier (1982) present a complete description of the models, methods, and results.

The Rifting Models

The three models we use to describe the rifting processes are: 1) uniform extension (Mckenzie, 1978); 2) uniform extension/melt segregation (Beaumont, Keen, and Boutilier, 1982); and 3) depth-dependent extension (Royden and Keen, 1980; Sclater et al, 1980). In all three models, extension and thinning of the lithosphere are the dominant processes during rifting (Figure 1). Uniform extension involves stretching of the crust and underlying lithosphere by a factor β, which varies with position across the margin. This model predicts immediate initial subsidence during extension followed by thermal subsidence as the lithsphere cools. The thermal subsidence is isostatically amplified by sediment and water load.

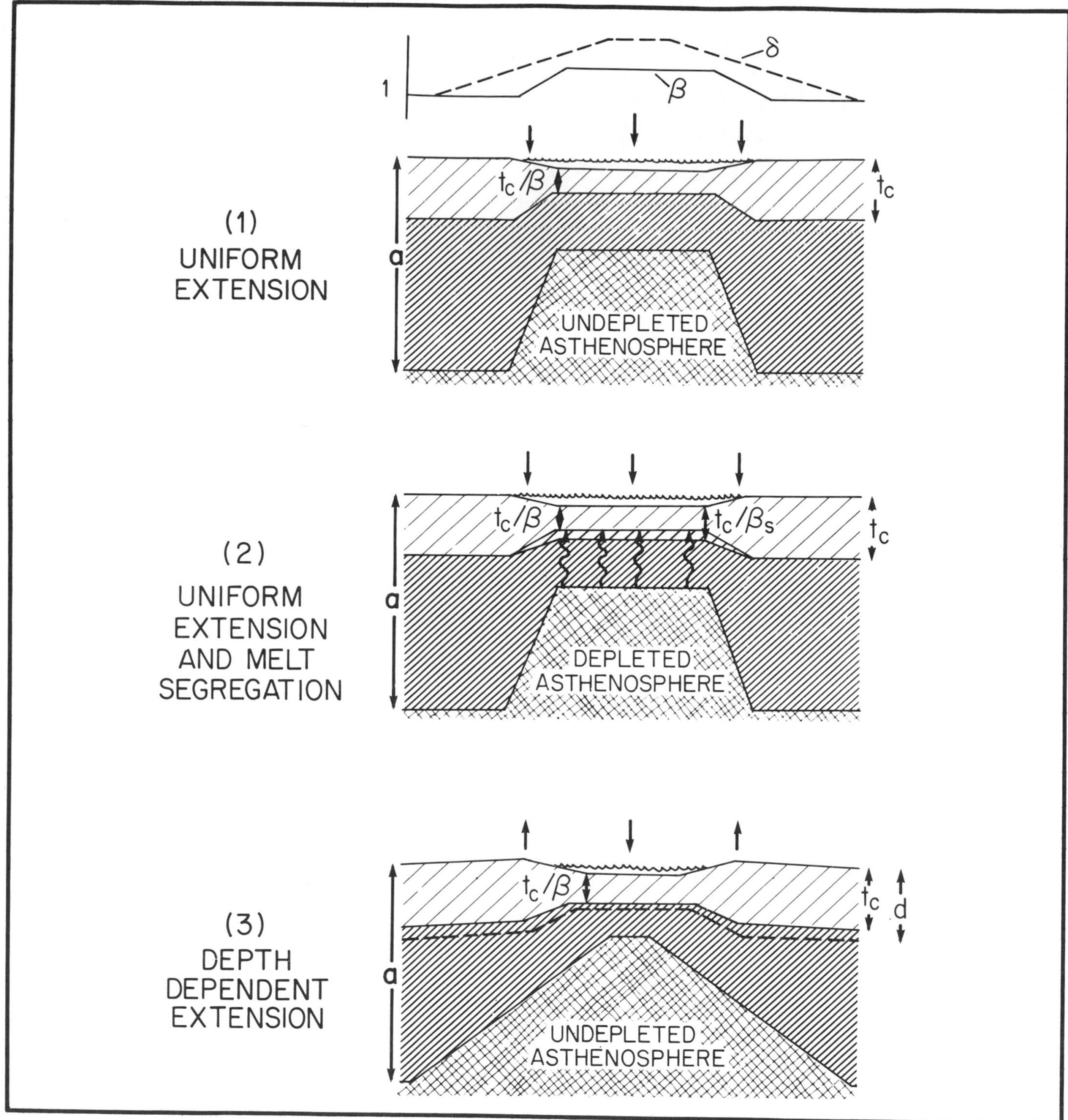

Figure 1 — Schematic representation of the three models. The arrows above each indicate whether initial uplift or subsidence is predicted. The diagonal shading is the crust; the lower crustal layer in the melt segregation model is the basaltic magma intruded from the asthenosphere. Note that in this model the crust is slightly thicker than predicted by uniform extension by β. The crustal thickness becomes t_c/β_s; where β_s is related to β through the equation

$$\beta_s = t_c/\left(\left(\frac{t_c - t_{oc}}{\beta}\right) + t_{oc}\right)$$

and t_{oc} = 6.5 km, the oceanic crustal thickness; t_c = 35 km, the unstretched continental crustal thickness; and a = 125 km, the initial thickness of the lithosphere. The variation in stretching is shown at the top of the diagram. These models depict the configuration of the margin during continental rifting.

The model involving uniform extension/melt segregation is similar to uniform extension, except that it allows realistic description of the generation of oceanic crust and assesses the relative importance of basaltic intrusion into the margin. As before, crust and lithosphere are extended by β but, in addition, fractionated basaltic magma migrates from the asthenosphere to the base of the crust. This thickens the crust from the t_c/β of the uniform extension model to t_c/β_s where

$$\beta_s = t_c/((\frac{t_c - t_{oc}}{\beta}) + t_{oc})$$

(Figure 1), a change that effectively reduces the values of β as determined from seismic measurements of crustal thickness. Melt segregation results in small changes in density, which are included in initial subsidence calculations. The thermal subsidence predicted by this model is almost identical to that predicted by uniform extension. A maximum amount of melt is generated when oceanic crust is formed, that is when $\beta = \infty$. Proportionately smaller amounts segregate as β decreases. The main difference between the uniform extension model and the uniform extension/melt segregation model is that the latter provides a smooth transition from unextended continental crust to oceanic crust (through extended continental crust with an added basaltic fraction).

The depth dependent extension model differs from the uniform extension model in that it assumes different amounts of extension in the upper and lower lithoshere, separated at a depth, d. In this paper $d = t_c$, the original thickness of continental crust. The crust is thinned by β, while the lithosphere beneath the crust is thinned by δ. This model predicts either initial uplift or initial subsidence during rifting, depending on the choice of β and δ.

Values of β are estimated from measurements of crustal thickness determined in seismic refraction experiments. Across the margin, these vary from $\beta = 1$, where no extension occurs, to values consistent with a 6.5 km thick oceanic crust. Density distribution is chosen so that the initial depth of the oceanic region in the uniform extension/melt segregation model equals the average depth of oceanic ridges with the same density values. The other two models fail to predict the correct initial subsidence in this region because they do not generate oceanic crust in a realistic manner. Nevertheless, we retain the same density distribution so that the evolution of all three models can be directly compared. Values of δ must be specified in the depth dependent extension model. Relevant data are scanty; we chose values that give initial uplift in the basin's shallow parts where observational data required an early episode of non-deposition. For the remaining regions of the basin, δ gives the minimum initial subsidence consistent with the early sedimentation pattern.

Each model predicts numerous parameters as a function of time: the depth to basement, depth to crust-mantle boundary, free-air gravity anomalies, and basin stratigraphy. These parameters are computed for a two-dimensional cross section of the Nova Scotian margin extending from the coastline out across the ocean-continental transition, and compared to observational data.

Methods

Model calculations require three sets of input variables: 1) The variation in β (and δ) as a function of position across the margin. Values of β are estimated from seismic refraction measurements of crustal thickness; 2) The value of the isotherm defining the base of the elastic plate. Here the 250°C isotherm is chosen; 3) The sediment input versus time and position. This is derived from deep exploratory well data and seismic data.

Calculations are performed by stepping the models through time, from rifting to the present. Time steps of 10 to 20 Ma are used. For each time step, the temperature distribution and thermal subsidence is computed. The depth to the 250°C isotherm and configuration of the elastic plate is then defined. Finally, the subsidence caused by the isostatic response of the plate to the sediment and water load added since the last time step is found. In this sense, the thermal and mechanical responses of the model are coupled; the thermal state of the lithosphere determines the elastic plate thickness and consequently its rigidity. The isostatic response to loading is computed using a two-dimensional finite element model. Calculations of temperature distribution and thermal subsidence are made using a one-dimensional finite difference approximation for the equation of conductive heat transport.

Keen, Beaumont, and Boutilier (1981) and Beaumont, Keen and Boutilier (1982) discuss the justification for using an elastic plate to approximate the mechanical response of the lithosphere. We believe that the mechanical behavior of the lithosphere on rifted margins is so strongly dominated by cooling effects that it is difficult to detect any properties that directly ascribe to viscous or viscoelastic relaxation. Consequently we choose the simple elastic model and assume that it is a reasonable first approximation to more complex rheological models, at least until the model predictions are proved inadequate with respect to observational data.

Results

The uniform extension and the uniform extension/melt segregation models give similar results, except in the oceanic region where the former predicts less subsidence than required and the latter shows good agreement with observational data. As noted above, the uniform extension/melt segregation model describes the generation of oceanic lithosphere more realistically than any model involving finite extension of continental lithosphere. Both uniform extension models fail to satisfy the sedimentary stratigraphy landward of the hinge line in the Nova Scotia area (Jansa and Wade, 1975; their Figure 6). The downward flexure of this region, in response to sediment loads further seaward, is insufficient to account for the total subsidence. Extension and flexure cannot explain the absence of Lower Jurassic sediments landward of the hinge line.

The depth-dependent extension model provides a good fit to the sedimentary stratigraphy and depth to basement in all parts of the margin. The value of δ chosen causes uplift landward of the hinge line, which is consistent with the absence of Early Jurassic sediments, and can satisfy the later subsidence history. It is not as successful, for reasons noted earlier, in predicting subsidence in the oceanic part of the model.

The Nova Scotian margin is old (≈185 Ma) and most of the heat generated during rifting has dissipated. Consequently, all three models predict similar depths to basement over most of the margin, and these depths agree with observations. Since all models were constrained to give the observed crustal thickness and have the same density distribution, they yielded similar free-air gravity anomalies which agree with the observed gravity anomalies. The differences the models produce in stratigraphy and gravity anomalies result from differing sedimentation patterns forced by the differences in early subsidence history.

The reduction in heat loss, due to the presence of the low-conductivity sediments which act as an insulating layer, causes the subsidence beneath the outer shelf to be about 1.8 km less than that predicted by models which do not include this important effect.

There are significant differences between these thermo-mechanical models and those which assume larger or smaller values of the isotherm that defines elastic plate thickness. Models in which the sediment and water loads are compensated by an Airy-type isostatic adjustment (0°C isotherm) cannot satisfy observations landward of the hinge line, where flexure is important, and they predict gravity anomalies unlike those observed. Beneath the outer shelf, an Airy model predicts about 1.5 km more subsidence than our 250°C thermo-mechanical models (11.0 km compared to 9.5 km). A 450°C isotherm model predicts gravity anomalies that are shaped similar to the 250°C isotherm models but larger in amplitude and wavelength than those observed.

SUMMARY

The results suggest that models in which extension is the dominant process during rifting can agree with observed sediment and crustal thicknesses, sedimentary stratigraphy, and gravity anomalies. Some margins require depth-dependent extension (such as off Nova Scotia). The oceanic region requires a model including total destruction of continental lithosphere and emplacement of basaltic magma to form oceanic crust. Airy isostatic compensation of the sediment and water load does not adequately describe lithospheric behavior; however, mechanical as well as thermal properties of the lithosphere can adequately predict the observed gravity anomalies.

REFERENCES CITED

Beaumont, C., C. E. Keen, and R. Boutilier, 1982, On the evolution of rifted continental margins: Comparison of models and observations for the Nova Scotian margin: Geophysical Journal of the Royal Astronomical Society, v. 70, p. 667-715.

Jansa, L. F., and J. A. Wade, 1975, Geology of the continental margin off Nova Scotia and Newfoundland, *in* Offshore geology of eastern Canada, v. 2: Regional Geology, Geological Survey of Canada Paper 74-30, p. 51-105.

Keen, C. E., and R. D. Hyndman, 1979, Geophysical review of the continental margins of eastern and western Canada: Canadian Journal of Earth Sciences, v. 16, p. 712-747.

———, C. Beaumont, and R. Boutilier, 1981, Preliminary results from a thermo-mechanical model for the evolution of Atlantic-type continental margins: Oceanologica Acta, Geologie des Marges Continentales, Supplement to v. 4, p. 123-128.

McKenzie, D. P., 1978, Some remarks on the development of sedimentary basins: Earth and Planetary Science Letters, v. 40, p. 25-32.

Royden, L., and C. E. Keen, 1980, Rifting processes and thermal evolution of the continental margin of eastern Canada determined from subsidence curves: Earth and Planetary Science Letters, v. 51, p. 343-361.

———, J. G. Sclater, and R. P. von Herzen, 1980, Continental margin subsidence and heat flow, important parameters in formation of petroleum hydrocarbons: AAPG Bulletin, v. 64, p. 173-187.

Sclater, J. G., et al, 1980, The formation of the intra-Carpathian basins as determined from subsidence data: Earth and Planetary Science Letters, v. 51, p. 139-162.

Watts, A. B., and M. S. Steckler, 1979, Subsidence and eustasy at the continental margin of eastern North America: Maurice Ewing Series, v. 3, American Geophysical Union, p. 218-234.

Subsidence and Stretching

Xavier Le Pichon
Jacques Angelier
Université Pierre et Marie Curie
Paris, France

Jean-Claude Sibuet
Centre Océanologique de Bretagne
Brest, France

We present a new formulation of McKenzie's simple uniform stretching model that is based on two reference levels, one near 3.6 km and the other near 7.8 km below sea level. In the absence of lithosphere, the asthenosphere would reach these levels if no oceanic crust were formed. The first level is for hot asthenosphere, the other is for asthenosphere cooled to thermal equilibrium. The instantaneous motion as well as the total vertical motion, produced by uniform stretching of the lithosphere, is expressed simply as a function of the elevation difference between the starting level and respectively the 3.6 and 7.8 km reference levels. In addition, we show that the behavior of the lithosphere under extensional strain is different above and below the 2.5 km-deep asthenosphere geoid. Below this level, oceanic accretion starts rapidly; above it, extensive thinning of the lithosphere produces subsidence until the asthenosphere geoid level is reached, enabling the asthenospheric material to break through to the surface. At low strain rate, pieces of the lower lithosphere may detach and sink in the asthenosphere. This process results in uplift and is taken into account in the formulation proposed.

This paper, following an earlier and shorter presentation made elsewhere (Le Pichon, Angelier, and Sibuet, 1982),presents and discusses a new formulation of the simple uniform stretching model of subsidence proposed by McKenzie (1978a, 1978b). The main advantage of McKenzie's model is its simplicity, which leads to a simple formulation and the use of a minimum number of parameters. We demonstrate that as the lithosphere, to a first approximation, is floating on top of the asthenosphere, the subsidence is controlled by the existence of two reference levels, one near 3.6 km and the other near 7.8 km water depth. That subsidence can be expressed as a function of the difference of elevation between the starting level and the reference levels.

To apply these models to active extensional areas, it is necessary to demonstrate that extensions as large as those predicted by the model do exist in the upper part of the crust. The model was first developed for the Aegean area by McKenzie. There, the surface extension produced by normal faulting is in fair agreement with the extension predicted by the model (Angelier, 1979; Le Pichon and Angelier, 1981; Angelier, 1981). However, the largest measured coefficient of surface extension β is only 1.4 to 1.5. The areas presumably affected by larger extension lie under water, where adequate field observations have not been made. Even in these submerged areas, the predicted β is generally smaller than 2.

The model has since been applied quite successfully to the formation of passive continental margins by Royden, Sclater, and von Herzen (1980) and Royden and Keen (1980). There, the predicted extension exceeds 3 in the deepest portions; the sedimentary thickness is too large to check whether such a large extension does indeed exist within the upper crustal layer. Le Pichon and Sibuet (1981) have independently applied this model to the Armorican continental margin. Using data of Montadert et al (1979a, 1979b), they made a quantitative analysis of the extension within the upper brittle layer and showed that it is compatible to a first approximation with the uniform stretching model. However, using the same data, Montadert et al (1981) made much lower estimates of the

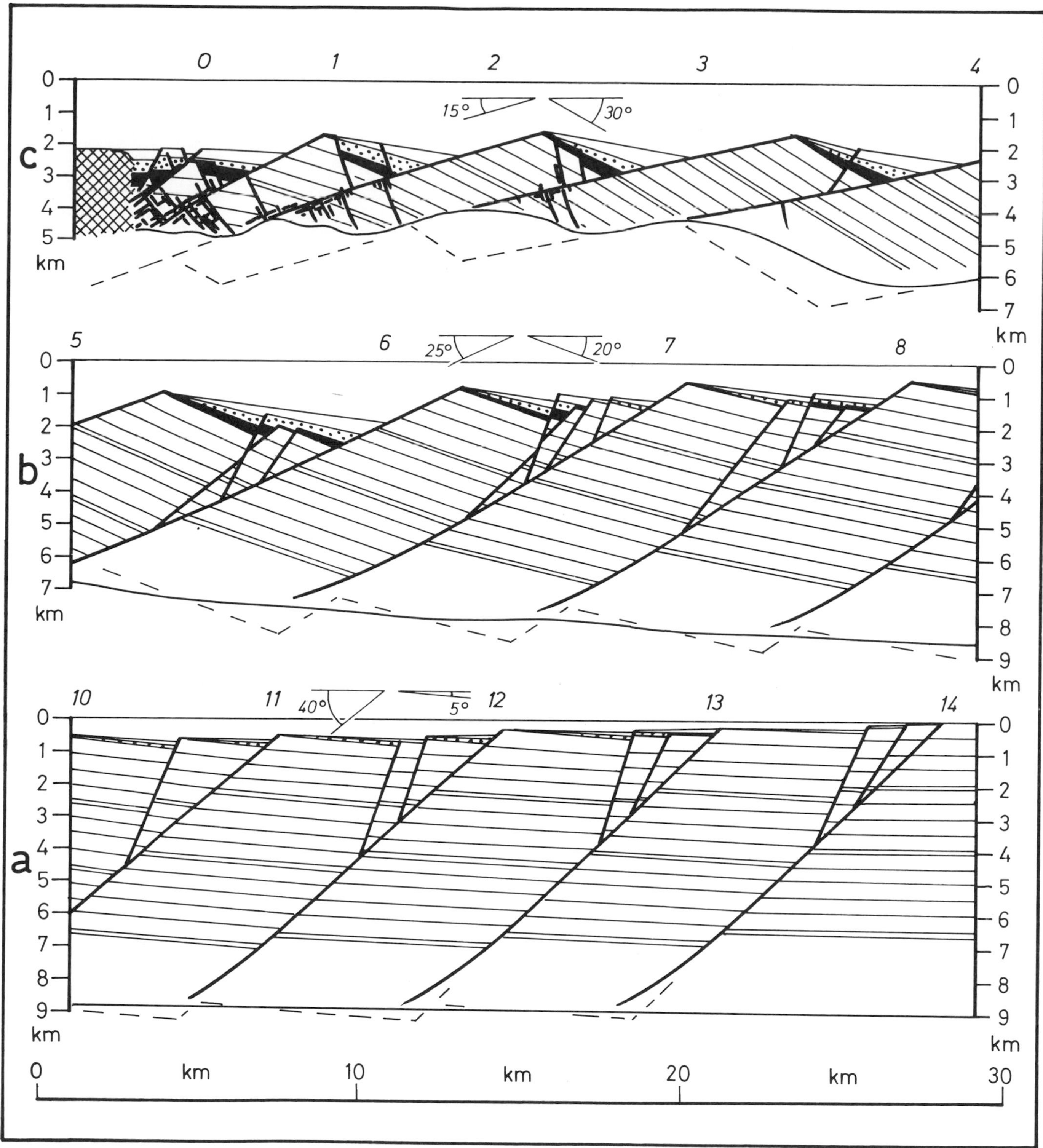

Figure 1 — Simplified geometric model of a continental margin, based on the interpretation of the Armorican margin by Le Pichon and Sibuet (1981), without vertical exaggeration. The volume remains constant during extension. Plastic deformation may occur only in the lower layers (in white), whereas upper layers (stratified) are faulted. Different parts of the margin from the continent toward the ocean are shown as *a, b, c,* with increasing extensional rates. Faults are plane, except for their lower part where plastic deformation occurs. As the tilt of blocks slightly increases from continent to ocean, the existence of steeper normal faults is geometrically indispensable. Together with the lower parts of main faults, these faults may resemble concave listric faults. The deformation is more complex in *c*. The thinning is increased by the interfingering of the two sets of normal faults with opposite dips. In addition, oceanic crust is created where β is greater than 3.3.

upper crustal extension. It is consequently necessary to discuss estimates by Le Pichon and Sibuet before going further.

EXTENSION WITHIN THE UPPER CRUSTAL LAYERS

Figure 1 is a theoretical model without vertical exaggeration proposed by Le Pichon, Angelier, and Sibuet (1982) on the basis of the interpretation of the Armorican continental margin of Le Pichon and Sibuet. This model illustrates the main features of their interpretation. The faulting pattern can be compared to a pack of cards resting at an angle on a plane, with each card (tilted block) forming a slight angle with the preceding one. The two critical factors are the original angle (α) of the fault plane with the bedding plane and the angle (θ) of tilting of the bedding plane. If the height of the two blocks is the same, then this equation is true:

$$\beta = \frac{\sin \alpha}{\sin (\alpha - \theta)}$$

In the lower portion of the northern Bay of Biscay continental margin, α and θ are remarkably constant when measured on the largest best defined blocks (Montadert et al, 1979a, 1979b).

Le Pichon and Sibuet obtain $\alpha \sim 45°$, $\theta \sim 30°$, thus $\beta \sim 2.7$; this is close to the value predicted by the simple stretching model.

The proposed model is valid if the tilted blocks are contiguous and share the same fault plane. If they are not, much lower extension factors may be measured. It becomes very difficult to explain why thc blocks are tilted in such a uniform manner if they do not share common fault planes. Clearly, then, the pack of cards model does not work and each block must be tilted independently of the others. Le Pichon and Sibuet discussed the fact that this model is substantiated by Morton and Black (1975) in the Danakil and Aisha areas near Djibouti and by Proffett (1977) in the Basin and Range province. There, contiguous fault blocks which have been tilted in the pack of cards manner result in extension factors as large as 3.

In contrast to a pack of cards, voids are not permitted so that the increase in θ from one block to the other, as the process begins, is accommodated by deformation through additional conformal normal faulting as shown in Figure 1 where an extension factor larger than 3 is obtained in this manner at the base of ther continental margin. This phenomenon reduces the apparent fault offset and tends to bias the estimate of surface extension toward lower values when they are based on actual measurements of the original length of the surface layer. In a zone of extension of this type, it is more reasonable to expect the deformation to be absorbed by extension in the upper surface layer than by compression in the lower ones. And this is what is observed in the field (Angelier, 1979). We believe that this is the origin of the discrepancy between Le Pichon and Sibuet estimates of β and those made by Montadert et al. (1979a, 1979b, 1981). Since the difference in estimates is very large we will discuss it in more detail on the single migrated section published to date by Montadert et al, which served as the type example to Le Pichon and Sibuet.

TYPE-SECTION OF THE ARMORICAN MARGIN

Figure 2 is the interpretation of the seismic section through the Armorican margin published by Montadert et al (1979a). To estimate the extension, we need to correlate the geological horizons throughout the section. We did this by correlating two levels of reflectors, shown shaded in Figure 1, on the basis of acoustic similarities suggested by the Montadert et al

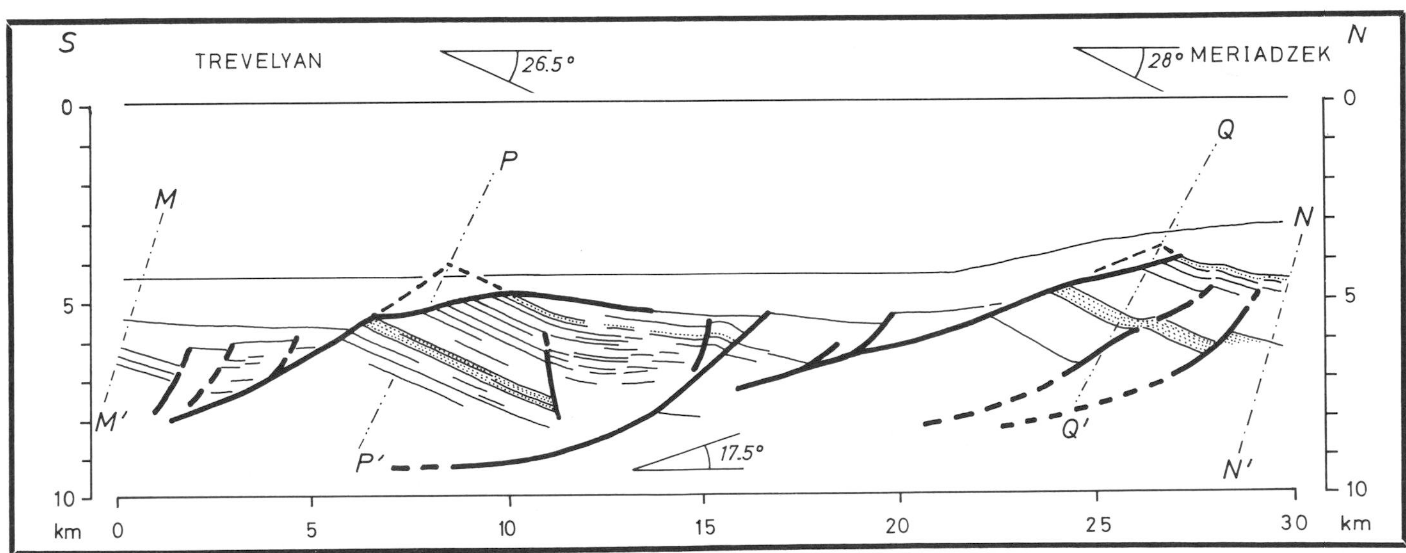

Figure 2 — Geological cross section through tilted blocks of the Armorican margin based on a migrated seismic profile. The interpretation was published by Montadert et al (1979a). Two equivalent stratigraphic levels are shaded. The angles of tilting and of the major fault plane are also indicated.

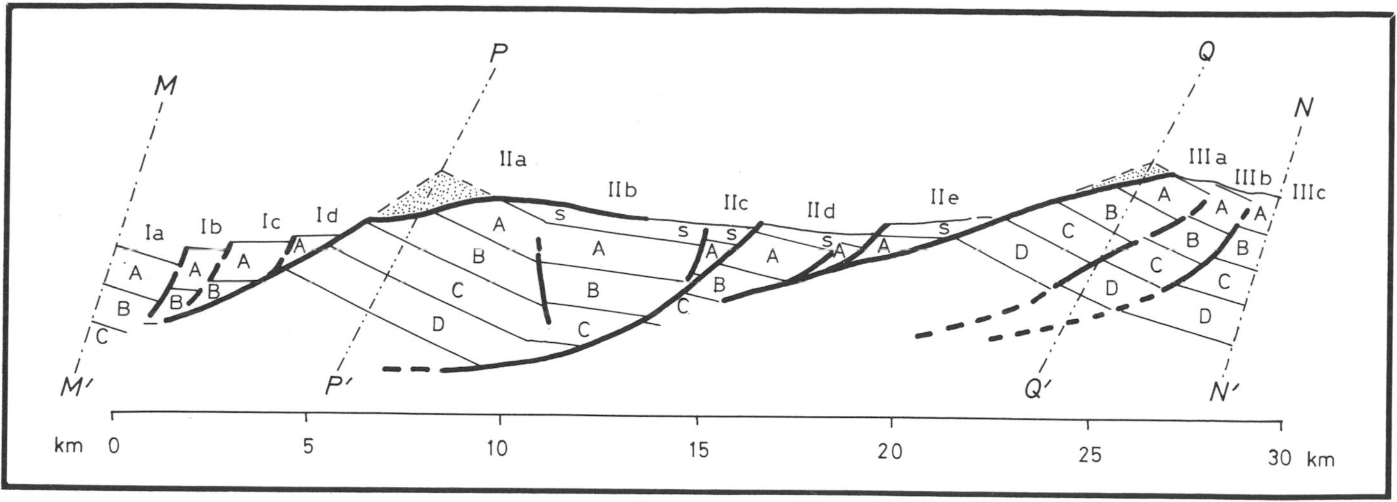

Figure 3 — The section of Figure 2 is modified here by four additional main layers A to D, one km thick each, for retrotectonic analysis purposes. The different blocks are numbered and the limits of the sections on which the extension is measured are also indicated (MM′ to NN′ and PP′ to QQ′).

(1979a) interpretation. Now we can estimate the extension, provided we can assume that no significant extension occurred perpendicular to the section and that the thicknesses of the layers did not change. The first assumption is quite reasonable because all the tilted blocks are parallel. We discuss the validity of the second assumption later.

It is then possible to divide the original section into stratigraphic levels (Figure 3). For convenience, we chose four layers, each 1 km-thick, called A to D. The two shaded reflectors correspond respectively to the top of layers A and C. These layers were presumably horizontal before the faulting and related block tilting occurred. A thin layer (Figure 3) lies on top of layer A in the central block (II), whereas it seems absent on the top of block III. The nature of blocks Ib, Ic, Id, IId, and IIe is uncertain. Since the hypothesis of their nature has noticeable effect on the computations, this problem is discussed later. Note the absence of the crests of faulted blocks (shaded areas of blocks IIa and IIIa) which may be explained by erosional processes or by tectonic deformation; this alternative is also important in the computation of β and is taken into account.

It is important to define the extremities of the section along which β is computed. Taking into account the relative homogeneity of the fault patterns and tilted blocks across the margin, these extremities must be chosen among homologous portions of the fault pattern. This is clearly the case for lines PP′ and QQ′ passing by the crests of the two adjacent main blocks. It is probably the case for lines MM′ and NN′, which cut similar portions of two main blocks. Consequently, the computations are made for both sections.

To estimate β, measure the total length of various levels on the sections of Figure 3. For example, the total length of layer A on MM′ to NN′ varies from 19 km (top) to 16 km (bottom), assuming deformation removed dotted portions in Figure 2. If erosion removed the dotted portions, the total original length is increased. Also, if we assume that layer *s* is absent on top of the small intermediate blocks (II d and e) and that the tops of these blocks are the top of layer A, we obtain the greatest possible estimate for the original length of layer A: 22 km (top) and 18 km (bottom). As the present length of section MM′ to NN′ is 29.5 km, β is 1.3 to 1.6 at the top of layer A and 1.6 to 1.8 at its bottom. The same computations made for section PP′ to QQ′ (18 km long) give β equal to 1.4 to 1.7 at the top of layer A and to 1.7 to 1.9 at its bottom. Regardless of assumptions, there is a systematic increase of computed β from top to bottom of layer A.

Let us now complete the stretching factor β for the deepest layer D. We have no problems about possible erosion and about the nature of the small intermediate blocks. The original length on MM′ to NN′ is 14 km (top) and 13 km (bottom), and 8.5 and 7 km respectively on PP′ to QQ′. The corresponding β is 2.1 at the top, in both cases, and 2.3 and 2.6 respectively at the bottom.

The discrepancy between the estimates made for layers A and D is important and suggests that the upper layers were noticeably elongated during rifting and that consequently their thickness did not remain constant. Figure 4 shows measured ratios between the present total length of the profile and the length of the different layers (described in Figure 3). The results differ slightly near the surface depending on the assumptions made (compare Figures 4a and 4b). In general, the ratio progressively increases with depth from about 1.5 at the surface to values between 2 and 2.5 below B (these ratios are slightly greater if we use PP′ to QQ′ instead of MM′ to NN′). Although the structure is not well controlled at depth, there is little doubt that the apparent stretching factor increases with depth to values close to 2.2 to 2.5.

This may take place for two reasons. First, it is likely that the uppermost layers of the tilted blocks were extensively deformed while gliding along the great normal faults with a *minimum* offset to 5 km; such deformation is commonly observed on land and

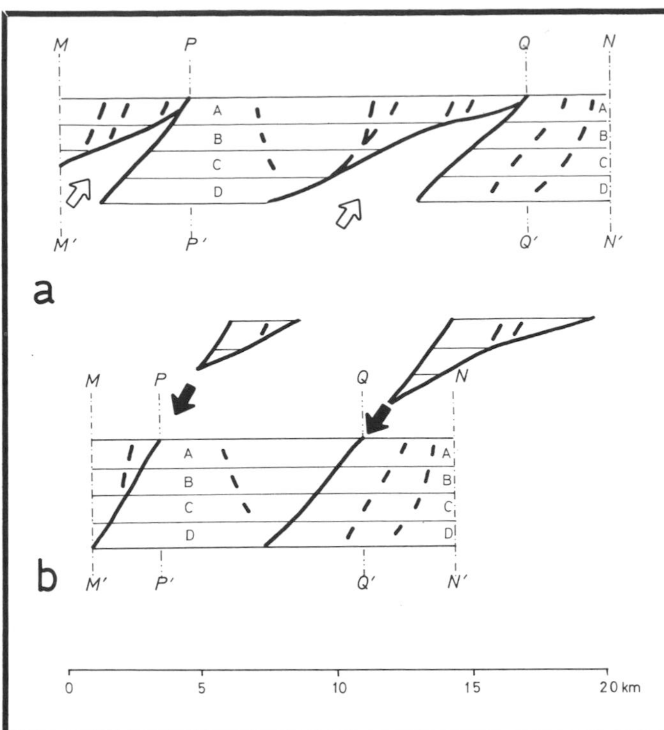

Figure 4 — Measured ratios versus depth of present length of profile MM′ to NN′ and profiles PP′ to QQ′ on original lengths of different strata as defined in Figure 3 assuming constant thickness of the layers. In *a*, erosion is assumed for the shaded areas in Figure 3; in *b*, deformation is assumed to explain this geometry.

results in strata thinning, especially in poorly consolidated sediments (corresponding small-scale normal faults and cracks as well as the continuous deformation remain undetected by seismics). Second, the identification of the acoustic stratigraphy of the small intermediate blocks is hypothetical. We chose the hypothesis leading to the minimum amount of extension. For example, parts of these blocks could correspond to synrift sedimentary deposits, as suggested by increased thickness in the lowermost areas.

Figure 5 is a simple geometric reconstruction of the initial configuration (a retrotectonic profile). It clearly illustrates the alternative: deformation by extension in the upper layers or deformation by compression in lower layers. In the first hypothesis (Figure 5a), we chose to insure the continuity of layer A and large gaps are present at depth. The corresponding extension coefficient is 1.45. The large shortening of the lower layers cannot be explained reasonably within an extensional framework. A possible explanation is that the major fault plane, with a shallow dip between blocks II and III, intersects older steeply dipping normal faults of the southern part of block III; thus, the older fault pattern is now masked by intermediate blocks II c and II d. This rather complex hypothesis, which assumes a two-phase evolution of the margin, is supported by no data and does not explain the regularity of the youngest fault pattern observed over the whole margin.

In the second hypothesis, the continuity of layer D is maintained, but larger overlaps exist for layer A and a smaller one for layer B. Earlier we pointed out that reasonable geological explanations may be found for this overlap. The corresponding values of the stretching factor are 2.2 for MM′ to NN′ and 2.4 for PP′ to QQ′. These estimations are minimum values; reconstruction assumes that the thickness of layer D remained constant and consequently ignores any possible small-scale extensional deformation. However, geological analyses of fault blocks on land suggest that small-scale faulting and continuous deformation have little effect on large blocks, except in the vicinity of large faults (e.g. Angelier, 1979). Thus, the geometrical analysis of Figures 2 to 5 leads to β values which are certainly larger than 2.2 but probably do not exceed 2.6.

This compares to the evaluation by Le Pichon and Sibuet, using a large fault block geometry with no internal deformation, which is $\beta = 2.5$. Le Pichon and Sibuet show that this stretching factor is the one predicted by the uniform stretching model to obtain the present water depth by subsidence from a sea-level pre-stretching state.

We conclude that the safest way to evaluate the extension of the upper crustal layer is to assume that it is the result of tilted blocks. Then, you need only to evaluate the angle of tilting and the angle of the fault plane with the bedding plane on the best defined blocks, thus avoiding the difficulties related to the deformation of the uppermost layer. Otherwise, use great care to make complete retrotectonic analysis of the whole thickness of the observed section and not only the uppermost layer.

Evidence now available to us indicates that the simple uniform stretching model is a good approxi-

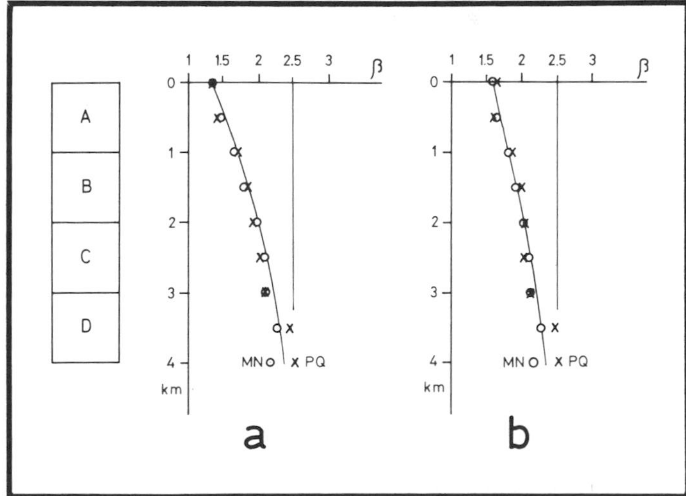

Figure 5 — Retrotectonic profiles reconstructed from Figure 3 assuming constant thickness layers. In *a*, continuity of layer A is insured. The corresponding extension is 1.45. In *b*, continuity of layer D is insured. The corresponding extension is 2.2 to 2.4 depending on the section chosen. In *a*, there are large gaps at depth (open arrows). In *b*, there is a large overlap for layer A and a smaller one for B (black arrows).

mation to the actual geological situation both in Aegea and on the Armorican continental margin. Next, we explore this uniform stretching model.

NEW FORMULATION OF THE SIMPLE STRETCHING MODEL

Consider a lithosphere of thickness h_L, composed of a crust portion of thickness h_c and of average density ρ_c and a mantle portion of average density ρ_m (Figure 6). ρ_c and ρ_m relate to the densities ρ_{co} and ρ_{mo} at 0°C through the actual distribution of temperature within the crust and mantle. We call ρ_L the average density of the lithosphere $\rho_L = f(h_c/h_L, \rho_c, \rho_m)$, as $\rho_c < \rho_m$, then $\rho_c < \rho_L < \rho_m$.

The lithosphere lies on the asthenosphere of average density ρ_a. The asthenosphere is made of the same material as the mantle portion of the lithosphere and has a constant temperature T_a, which is higher than the lithosphere average temperature so that $\rho_c < \rho_a < \rho_m$. The lithosphere floats in isostatic equilibrium on top of the asthenosphere and we define a reference level (L) as the level which the asthenosphere reaches in the absence of lithosphere. The surface elevation (E) of the lithosphere, with respect to this reference level (L), is determined by the buoyancy (B) of the lithosphere; $B = h_L (\rho_a - \rho_L)$. The sign of B depends on whether ρ_a is larger or smaller than ρ_L, which is determined primarily by the ratio between h_c and h_L. If B is positive, that is if ρ_a is larger than ρ_L, then E is positive and the lithosphere stands above the reference level and vice versa.

We can determine rather precisely the reference level depth with respect to the crest of the mid-ocean ridge, as was first proposed by Turcotte, Ahern, and Bird (1977) who call this reference level the mantle geoid. Adopting the constants of Le Pichon and Sibuet (in press), who balanced a ridge crest column with 30 km continental lithosphere column, we choose an oceanic crust 5.5 km thick with an average density of 2.765 $g\text{-}cm^{-3}$ on top of asthenospheric material at temperature, T_a. The adopted density for mantle material at 0° C, ρ_{mo}, is 3.35 $g\text{-}cm^{-3}$. Finally, we adopt the values for the thermal expansion coefficient α and the temperature of the asthenosphere T_a proposed by Parsons and Sclater (1977); $\alpha = 3.28 \times 10^{-5}$ °C^{-1}, and T_a = 1 333° C. We take 2.5 km as the water depth of the ridge crest.

By removing the oceanic crust and balancing the columns, we find that the reference level lies 3.61 km below sea level. Turcotte, Ahern, and Bird, choosing what we consider less realistic constants, obtain 3.25 km. Thus, this reference level is probably known with a precision of 300 to 400 m wherever the asthenosphere is not unusually hot (as under Iceland or Afar).

The elevation (E) of the lithosphere above this level is $E = B/(\rho_a - \rho_w)$ when it is under water, and $E = (B + 3.61\,\rho_w)/\rho_a$ when it is above water.

If uniform stretching of the whole lithosphere occurs instantaneously at time t = 0 in such a way that the thickness of the lithosphere becomes h_L/β, the average density ρ_L does not change and the buoyancy is reduced in the same ratio. The new elevation becomes $E_n = B/[\beta(\rho_a - \rho_w)]$ if E_n is below water, and $E_n = (B/\beta + 3.61\,\rho_w)/\rho_a$ if it is above water. In the simplest case where E is at or below sea level, $E_n = E/\beta$.

The instantaneous change in surface elevation is $Z_i = E - E_n$. When E is at or below sea level $Z_i = E(1 - 1/\beta)$, or $Z_i = \gamma E$ where $\gamma = 1 - 1/\beta$.

There is uplift instead of subsidence if E is negative ($\rho_L > \rho_a$), which reflects the fact that E_n is 0 when the stretching factor is infinite and the lithosphere is completely replaced by hot asthenosphere (Figure 6a).

The proportion of original thickness which is taken away by thinning is γ. Thus, we find the linear relationship in relative thinning coefficient first proposed by McKenzie (1978a). In this formulation, the amount of subsidence (Z_i) is entirely determined by the elevation (E) of the surface of the lithosphere with respect to the reference level and the thinning coefficient (γ). It does not depend explicitly on the thickness, composition, and density distributions within the lithosphere; this is important as these parameters are poorly known, whereas the starting elevation is generally much better known.

Let us take as an example a starting elevation at sea level (E = 3.61 km). We have $Z_i = 3.61\,\gamma$, which is the exact relationship found by Le Pichon and Sibuet (1981) for the Armorican continental margin. To obtain this relationship, they used the formulation proposed by McKenzie which requires defining the density distribution within the lithosphere. Thus, they chose a ρ_{co} of 2.78 $g\text{-}cm^{-3}$, an h_c of 30 km for the continental crust, an h_L of 125 km for the lithosphere, and a linear temperature distribution. The two methods of computation agree because they balanced their continental lithosphere column with the same ridge crest column. One could use different densities, thicknesses, and temperature distribution for the continental lithosphere. Provided the columns are balanced, which they must be, the formula for the subsidence is the same.

It follows that the higher the starting elevation, the larger the amount of subsidence for a given stretching factor. To reach the same level, the stretching factor must increase with the initial elevation.

The surface of the lithosphere is at the 3.6 km water depth reference level (L) when $\rho_L = \rho_a$. Adopting a linear temperature distribution and the constraints defined above, $\rho_c = \rho_L$ when $h_c/h_L = 0.13$. When h_L = 125 km, h_c = 16 km. A critical level in geodynamics is 3.6 km because it is the boundary between a lithosphere lighter than the asthenosphere, which can only be forcibly subducted, and a lithosphere denser than the asthenosphere, which is in a state of gravitational instability and should tend to subduct by itself.

However, as pointed out by Le Pichon and Sibuet, the actual hydrostatic level to which the asthenosphere rises by itself is 2.5 km and not 3.6 km. This is because as the asthenosphere rises, partial fusion occurs resulting in segregation of an oceanic crust. Consequently, the 2.5 km level is the real "asthenosphere geoid." The asthenosphere may not break to

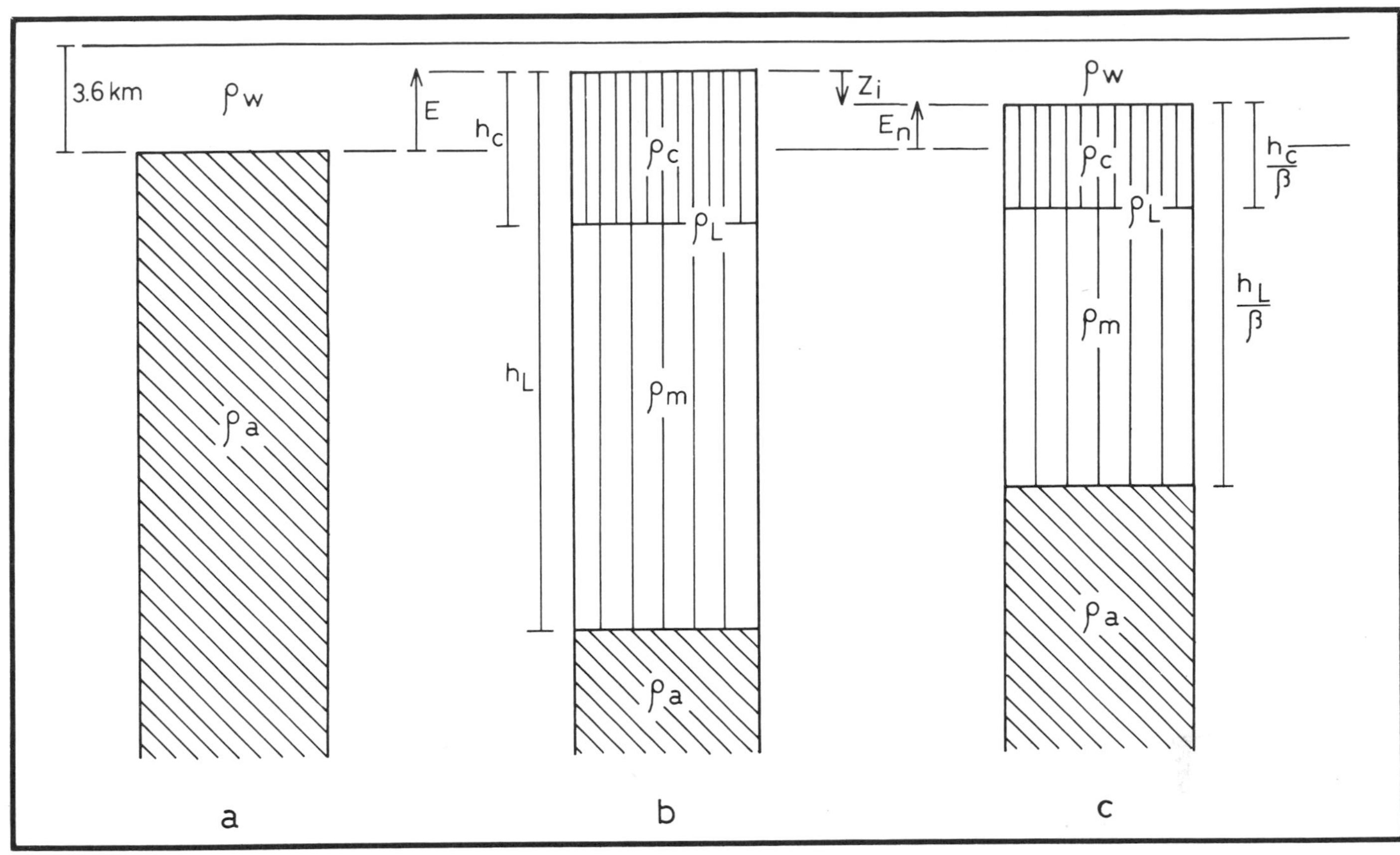

Figure 6 — Initial stretching phase and isostatic equilibrium. Oblique hatchured pattern: asthenosphere. Vertical hatchured pattern: lithosphere. Average densities are ρ_a (asthenosphere), ρ_m (mantle portion of lithosphere), ρ_c (crust), ρ_L (whole lithosphere), ρ_w (water). *a*: hypothetic column with lithosphere entirely replaced by asthenosphere (which is isostatically equivalent with $\rho_a = \rho_L$). This enables to define the "mantle geoid" and the "asthenosphere geoid". *b*: lithosphere before stretching (h_c and h_L: thicknesses of crust and lithosphere, respectively). The surface elevation with respect to *a* is E. *c*: lithosphere just after instantaneous stretching by a factor β. The new surface elevation with respect to *a* is E_n. The subsidence relatively to *b* is Z_i.

the surface while the level of the previous lithosphere remains above 2.5 km; this is the minimum depth at which new oceanic lithosphere may form. It follows that the higher the starting elevation, the larger the stretching factor must be before oceanic accretion starts and the more difficult it is to pass from continental extension to oceanic accretion.

The above considerations lead us to recognize two quite different behaviors for lithosphere under extensional strain, depending on whether the lithosphere surface is above or below the asthenosphere geoid. Below this level, the asthenospheric material tends to rise to the surface as soon as stretching proceeds. The ascension is furthered helped by the buoyancy of the melted portion produced by partial fusion. This phenomenon increases rapidly as any possible intrusion rises and consequently the crust is invaded by magma and breaks apart. Thus, no significant thinning of the lithosphere is likely to occur and the transition to accretion should be rather sharp.

On the other hand, if the surface of the lithosphere stands above the 2.5 km deep asthenosphere geoid, then the asthenospheric material cannot reach the surface level and break the continuity of the lithosphere. In addition, as crust-mantle interface is situated deeper, its density contrast is a more efficient barrier, with respect to any possible asthenospheric intrusion into which the proportion of partial fusion is much less than in the previous case. Thus stretching may proceed until subsidence reaches a water depth of 2.5 km. It is unlikely that it will greatly exceed this value as the likelihood of the asthenosphere breaking to the surface increases rapidly beyond this depth.

THE THERMAL SUBSIDENCE PHASE

As pointed out by McKenzie (1978a, 1978b), the initial stretching phase results in a crowding of the isotherms near the surface and consequently in a subsequent cooling phase which produces thermal subsidence. The asymptotic equilibrium state to which the infinitely thinned cooling lithosphere should tend is the one corresponding to a hypothetical outcropping asthenosphere cooled to a steady thermal state. Accepting an equilibrium value of 125 km for the oceanic lithosphere (Parsons and Sclater, 1977) and using a linear distribution of temperature, we find that the 3.61 km deep hot asthenosphere deepens to 7.82 km. Thus, 7.82 km is the value which should be reached asymptotically after an infinite amount of time in the zero thickness crust case; 3.61 km is the value corresponding to the same zero thickness crust case immediately after stretching. For reasons developed above, such a case is not possible and the maximum water depth is unlikely to significantly exceed the maximum depth for oceanic lithosphere, which is 6.4 km.

The role of the 7.82 km level for the steady thermal state reached after an infinite amount of time is equivalent to the role of the 3.61 km reference level for the instantaneously thinned lithosphere. The difference is that the law of subsidence at infinite time as a function of thinning cannot be determined unless the altitude of the unthinned lithosphere *at thermal equilibrium* is known. A special case exists when the thermal equilibrium prior to thinning is obtained for the same 125 km thickness of lithosphere with a linear distribution of temperature. Then the buoyancy (B′) of the initial state of unthinned lithosphere at thermal equilibrium, with respect to the 7.82 km reference level, is $B' = h_c (\rho_{mo} - \rho_{co}) (1 - 2 T_a h_c/h_L)$. Its elevation E′ above the 7.82 km level is $E' = B' / (\rho_a - \rho_w)$ under water and $E' = (B' + 7.82 \rho_w) / \rho_a$ above water.

In the simplest case where the initial pre-stretching level is at or below sea level, the final subsidence at infinite time is: $Z_\infty = (1 - 1/\beta) (E' + \varepsilon) = \gamma (E' + \varepsilon)$; where $\varepsilon = - (\alpha/2 \beta) T_a (h_c^2 / h_L) (\rho_{mo} - \rho_{co}) / (\rho_a - \rho_w)$; as $h_c < 60$ km, $\varepsilon < E'/100 \beta$; and in general $h_c \sim 30$ km, $\varepsilon \sim E'/200 \beta$. Thus, within the precision of determination of the different constants $Z_\infty = \gamma E'$. This result could be obtained directly from the equation giving B′, noting that $\alpha T_a h_c/2h_L \sim 1$ % ($h_c < 60$ km) and, in general, of the order of 0.5 % ($h_c \sim 30$ km). Neglecting this term, $B' = h_c (\rho_{mo} - \rho_{co})$ and the new buoyancy at thermal equilibrium (infinite time) after thinning, B'_n is equal to $B'_n = h_c (\rho_{mo} - \rho_{co}) / \beta = B'/\beta$.

The simple formulas developed in the first section can be applied, replacing B by B′ and E by E′. In particular, when the initial stage is under water, $Z_\infty = \gamma E'$. Putting the initial stage at sea level, we find the result obtained by Le Pichon and Sibuet for the Armorican margin, $Z_\infty = 7.82 \gamma$, which applies to any margin in which the initial pre-stretched level at equilibrium was sea level, assuming a linear distribution of temperature over a thickness of 125 km.

GENERALIZED FORMULATION

The easiest way to deal with the air-water interface discontinuity in the formulation is to convert the altitudes A above sea level to virtual water depth (D^x) by multiplying it by the isostatic factor under water: $D^x = - A \rho_a/(\rho_a - \rho_w) = - 1.47$ A. Then, the virtual elevation (E^x) above the 3.61 km level is always $E^x = 3.61 - D^x$ in km, where D^x is either the true water depth (below water) or the virtual water depth which is negative (above water). Similarly the virtual elevation (E′) above the 7.82 km level is $E^{x'} = 7.82 - D^x$. The instantaneous virtual subsidence is (1) $Z^x_i = (3.61 - D^x) \gamma$, and the total virtual subsidence is (2)$Z^x_\infty = 7.82 - D^x) \gamma$. The thermal virtual subsidence (Z^x_t) at any

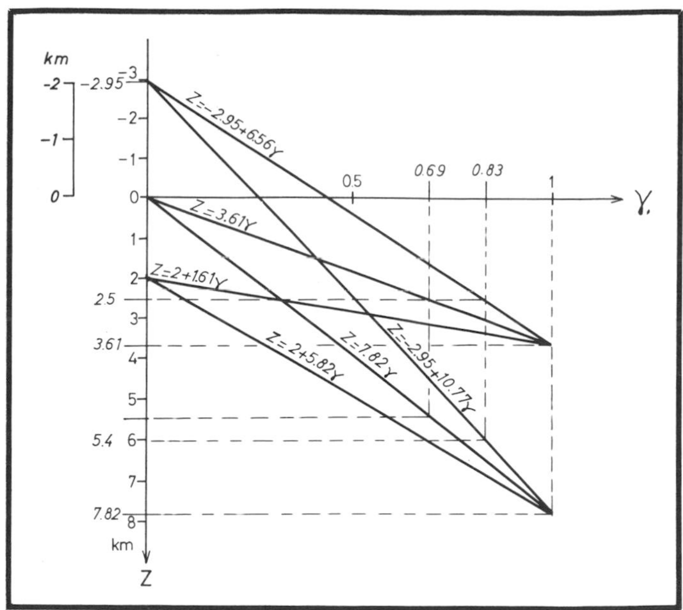

Figure 7 — Initial virtual instantaneous subsidence (Z^x_i) and virtual total subsidence (Z^x_∞) as a function of relative thinning of the lithosphere for three different starting elevations: 2 km above sea level, sea level, and 2 km below sea level. A change of scale is used above sea level to convert virtual water depth into actual altitude, by multiplying the scale by $\rho_a / (\rho_a - \rho_w)$. The 2.5, 3.61, and 7.82 km levels are identified.

given time (t) can be estimated using the 62.8 million year thermal time constant proposed by Parsons and Schlater (1977); (3) $Z^x_t \sim (Z^x_\infty - Z^x_i) (1 - e^{-t/62.8})$. Formulas (2) and (3) are only valid if the initial pre-stretching state is at thermal equilibrium and can be described in a first approximation by a linear distribution of temperature over a thickness of 125 km. Then, the virtual subsidence can be converted back into a real subsidence: $Z_i = Z^x_i$ if $Z^x_i \leqslant 3.61$; $Z_i = (Z^x_i - 3.61) (\rho_a - \rho_w)/\rho_a + 3.61$ if $Z^x_i \rangle 3.61$. Similarly: $Z_\infty = Z^x_\infty$ if $Z^x_\infty \leqslant 7.82$; $Z_\infty = (Z^x_\infty - 7.82) (\rho_a - \rho_w)/\rho_a + 7.82$ if $Z^x_\infty \rangle 7.82$.

Figure 7 illustrates this generalized formulation. The air-water discontinuity is taken care of by a change in vertical scale. Three different cases are shown with starting elevations 2 km above sea level, sea level, and 2 km below sea level. The streching factor required to reach a water depth of 2.5 km, immediately after stretching at which oceanic accretion starts is respectively 5.91, 3.25, and 1.45. Thus, the higher the initial elevation, the larger the amount of stretching necessary before oceanic accretion starts. The final water depth reached at infinite time is respectively 6.0, 5.4, and 3.8 km. Thus the maximum depth at which thinned continental crust should be expected on old continental margins, in the absence of sedimentation, is about 5.5 km if the initial steady state was close to sea level (slightly more or quite less if it was above or below sea level). Note that "old" means an age large with respect to 62.8 million years (more than 120 million years); "instantaneous" means an amount of time small with respect to 62.8 million years, (less than 20 million years Jarvis and McKenzie, 1980).

Figure 6 shows that the depth of 3.61 km at which the lithosphere is neutrally buoyant corresponds to an initial depth immediately after stretching of 1.04, 1.66, and 2.44 km and a stretching factor (β) of 2.56, 1.86, and 1.38, respectively. It is important level, as the thinned and cooled lithosphere below it may be easily subducted whereas it resists subduction above this depth. Thus, if an old marginal basin is subducted, one expects the subduction of the continental rise and slope to a depth of about 1.5 km water depth. If it is recent, only the lower portion of the continental rise is subducted.

CONTINENTAL MARGINS AND CONTINENTAL RIFTS

Since Heezen (1962), it has been explicitly assumed that the basic genetic sequence is one which evolves from a continental rift valley, similar to the African ones, to an open ocean continental margin. This leads to a major difficulty: although the measured extension in the upper brittle layer of Rift Valleys is small (a few kilometers) it is accompanied by an important thinning of the lithosphere but not apparently of the continental crust. As a result, the lower part of the lithosphere is replaced by hot asthenosphere and the continental crust is uplifted. If, following Artemjev and Artyushkov (1971) and Bott (1971), a process of necking with brittle behavior in the upper crustal layer and plastic behavior in the lower one is invoked, then it is necessary to assume a larger amount of stretching in the lower than in the upper layer, as proposed by Bott (1971) and many authors since. In any case, it is clear that the simple model of uniform stretching cannot produce continental rifts.

A significant difference between continental rifts and regions where the uniform stretching model has been applied is the strain rate. The amount of extension across continental rifts is a few kilometers over 100 km, about 10%. The extension typically has been acting for several tens of million years. Choosing 10% and 20 million years as typical values, the strain rate is $\beta_{inst} = 16 \times 10^{-17}$ sec $^{-1} \sim 10^{-16}$ sec^{-1}. In the Aegean Sea, following Le Pichon and Angelier (1979), the average surface extension is 1.4 (40%) and has been obtained over 13 million years. Thus, $\beta_{inst} = 8 \times 10^{-16}$ sec$^{-1} \sim 10^{-15}$ sec^{-1}. On the Armorican margin, the largest total extension is 3.24 according to the model of Le Pichon and Sibuet (1981) and the average one is $\beta = 2\,\beta_{max}/(1 + \beta_{max}) = 1.53$. It was obtained over a period of about 20 million years. The maximum $\beta_{inst} = 5 \times 10^{-15}$ sec^{-1} and the average one is 7×10^{-16} sec^{-1}, that is of the order of 10^{-15} sec^{-1}.

In Afar, the largest measured β is 3 (Morton and Black, 1975) over a time span of about 20 million years, and in the Basin and Range it is 2.4 (Profett, 1977) over a time span of about 7 to 13 million. The maximum strain rate is about 2×10^{-15}. In all these regions where stretching was very large, the instantaneous strain rate is, as an average, of the order of 10^{-15} sec^{-1}. It is an order of magnitude lower in the continental Rift domain.

A second significant difference is the distribution of major faults. Continental rifts have the structure of a large graben with two major faults. On the contrary, regions where a large extension was measured have numerous large faults. For example, in Crete, which is part of the Aegean extensional domain (Angelier, 1979), the largest fault delimitate blocks typically 30 km across. In the Basin and Range, Afar, and on the Armorican continental margins, tilted blocks are a few kilometers to at most 20 to 30 km across.

Another important observation is that there is no indication, in continental rifts, that the lower continental crust was thinned to a greater extent than in the upper brittle portion. For example, in the Rhine graben, which is probably the best studied continental rift, the total surface extension is about 4 to 5 km (Sittler, 1974). This gives a surface extension rate of 1.12 over the 40 km width of the central valley. The surface extension cannot be measured over the eroded uplifted shoulders of the central valley. The crust has an average thickness of about 26 km over the entire 100 to 150 km width of the rift system, instead of about 29 to 30 km on each side (Werner and Kahle, 1980). This is an average extension rate of 1.12 to 1.15, quite comparable to the measured surface extension rate. On the other hand, Werner and Kahle (1980) show that gravity requires the lithosphere to be thinned by a factor of about 1.8 from an assumed original thickness of 135 to about 75 km. Thus, everything happens as if the lower portion of lithosphere was replaced totally by hot asthenosphere whereas the upper one was only slightly extended, mostly through a set of deep master faults.

We suggest that at low strain rates, near 10^{-16} sec^{-1}, the limited extension in the upper brittle portion is mostly absorbed in the main graben structure over two large master faults, whereas it results in the lower lithosphere in the detachment and sinking of large pieces replaced by hot asthenosphere, in a manner similar to the one suggested by McKenzie (1978b). This process might be triggered by the intrusion of asthenospheric material in narrow zones immediately below the rift where the totality of the strain is being released.

Although we do not know the actual process through which the gravitationally unstable portions of lithosphere sink in the asthenosphere, our hypothesis explains rather well the present structure of the Rhine graben or African rift valley which we may schematize by a normal thickness crust (30 km) overlying a greatly thinned portion of mantle lithosphere (30 km or less). This results not in subsidence but in an uplift of about 1.0 km, making further evolution to an oceanic stage even more difficult, as pointed out earlier. At higher strain rates, the extension distributed over a much larger surface in the upper brittle portion and the entire lithosphere appears to be thinned at a uniform rate. Intermediate cases might occur at intermediate stages. In addition, a zone of low strain rate may convert later into a zone of high strain rate. Thus, extremely complex evolutions may

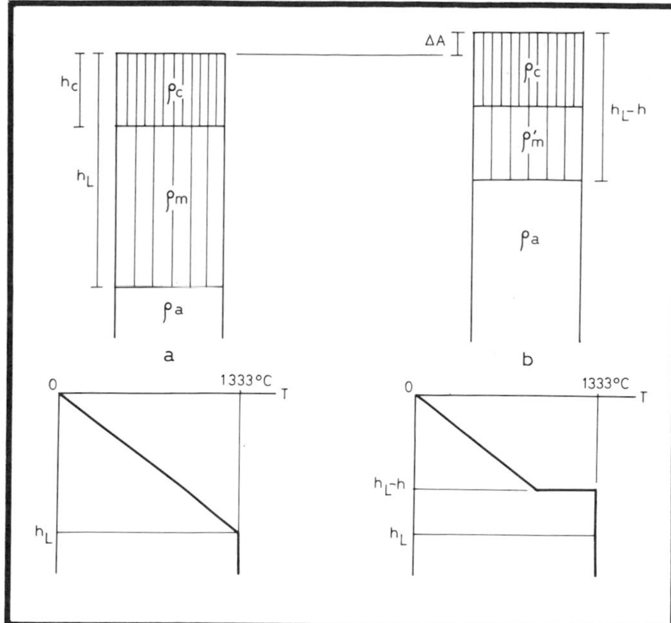

Figure 8 — Sinking of a lower portion of lithosphere of thickness h and its replacement by hot asthenosphere resulting in an uplift Δ A.

result from varying levels of stresses being applied to a given area, even if unusual asthenosphere or lithosphere conditions are absent.

Next compute the uplift due to the replacement of the lower portion of the lithosphere of thickeness h, at thermal equilibrium, by hot asthenosphere. The resulting uplift (ΔA; A is the altitude) is $\Delta A = + \alpha h^2 T_a / (2 h_L) = 1.75 h^2 \times 10^{-4}$, where h is in km. The maximum uplift is 1.58 km if the entire mantle portion (95 km) is replaced by hot asthenosphere. The change of depth under water should be multiplied by the isostatic factor $\rho_a / (\rho_a - \rho_w)$, and consequently $\Delta D^x = 2.57 h^2 \times 10^{-4}$ km where D^x is the virtual water depth and ΔD^x the part of this virtual water depth produced by the lower part of the lithosphere sinking.

Compute the subsidence resulting from thinning of this truncated lithosphere of virtual water depth D^x, situated — ΔD^x above the thermal equilibrium stage (see Figure 8), using formulas (1), (2), (3). The formula for the instantaneous subsidence does not change because the average density, and consequently the buoyancy, is not modified by thermal processes during the subsidence. The total virtual subsidence becomes (4) $Z^x_\infty = [7.82 - (D^x - \Delta D^x)] \gamma - \Delta D^x$, and the thermal virtual subsidence is (5) $Z^x_t = [\gamma (4.21 + \Delta D^x] (1 - e^{-t/62.8})$. An example is a lithosphere, in thermal equilibrium at sea level, in which the whole mantle portion sinks and is replaced by asthenosphere. Then, the uplift $D^x = \Delta D^x = -2.32$ km ($\Delta A = 1.58$ km). Thus $Z^x_i = 5.93 \gamma$. The instantaneous subsidence is larger, but the depth reached is less important than it would have been prior to the uplift for the same value of β. To reach the water depth of 2.5 km at which oceanic accretion may start, we need a β of 5.3 instead of 3.25. $Z^x_\infty = 7.82 \gamma + 2.32$. Thus, the total virtual subsidence corresponding to the 2.5 km water depth instantaneous subsidence level is: $Z^x_\infty = 8.68$ km and the actual depth reached is 6.36 km (8.68 − 2.32). Finally, $Z^x_t = [1.89 \gamma + 2.32] (1 - e^{-t/62.8})$, and for the 2.5 km instantaneous level $Z^x_t = 3.86 (1 - e^{-t/62.8})$.

This case quantitatively corresponds to a prolonged period of low strain rate, producing a continental rift stage, followed by a high strain rate period then producing continental margin stage. The results of Royden and Keen (1980) for the eastern Canada margin might be interpreted in this way. Royden and Keen use a model in which the mantle portion of the lithosphere may be stretched by a factor which is equal to, or larger than, the one applied to the continental crust. When it is larger, this is equivalent to the sinking of the lower lithosphere model although the uplift they obtain is somewhat larger. The temperature at the base of the thinned lithosphere is in their case T_a and not $T_a (1 - h/h_L)$. Royden and Keen find that uniform stretching over the entire lithosphere fits data on the subsidence of eight wells on the Nova Scotia margin. However, on the Labrador margin they need to use a mantle portion which is more highly stretched than the crust. Our hypothesis indicates that the Labrador region started as a continental rift with uplift during a low strain rate phase which was followed by a higher strain rate phase with uniform stretching. We believe that formulas (1) and (5) could fit the Royden and Keen data probably as well as they did within the accuracy of the measurements; the parameters to be fitted being D^x, ΔD^x, and γ.

CONCLUSIONS

We have confirmed the geometrical analysis of surface faulting in highly strained extensional areas, made by Le Pichon and Sibuet (1981). The geometry of faulting in the upper brittle layer, 8 to 10 km thick prior to extension, can be compared to a pack of cards resting at an angle on a plane, with each card making a slight angle with the preceding one. However, be careful to measure the extension over the entire thickness of the brittle layer, rather than on the uppermost, often disturbed layer.

We showed that the behavior of the lithosphere under extensional strain is controlled by the existence of two reference levels, one near 3.6 km and the other near 7.8 km water depth. These are the levels which would be reached by the asthenosphere in the absence of lithosphere and of formation of oceanic crust. The first one is for hot asthenosphere, the other one for asthenosphere cooled to thermal equilibrium. The instantaneous as well as total vertical motion after an infinite time produced by uniform stretching of the lithosphere is then expressed simply as a function of the difference of elevation between the starting level and respectively the 3.6 and 7.8 km reference levels.

We also showed that the behavior of the lithosphere under extensional strain is quite different above and below the 2.5 km deep asthenosphere geoid. Below this level, continuity of the old lithosphere is likely to be rapidly broken and, as a consequence, oceanic accretion begins rapidly. Above this level, the old lithosphere is thinned extensively until it reaches the level of the asthenosphere geoid, thus enabling the asthenospheric material to break through to the surface.

We suggested that the behavior of the lithosphere under extensional strain is quite different at low (10^{16} sec $^{-1}$) and high (10^{15} sec $^{-1}$) strain rates. The low strain rate situation is considered typical of the continental rifts, similar to the African rifts or the Rhine graben. The lower lithosphere is thinned by simultaneous diapiric asthenospheric instrusions and lithospheric sinking; the upper one is only slightly thinned. For high strain rates, the whole lithosphere is apparently thinned rather uniformly, as in Aegea and probably on many continental margins. Thus, the continental rift stage, as typified by the Rhine graben and African rifts, is not necessarily the early stage of evolution of a continental margin. Present day Aegea may be a better example of an early stage of passive continental margin revolution. However, composite cases may occur and we provide a simple method to quantitatively account for them.

ACKNOWLEDGMENTS

The CNRS (ATP IPOD n. 4228) and the Centre National pour l'Exploitation des Océans (contrat CNEXO No. 79-5929) supported this work. Contribution number 720 du Centre Océanologique de Bretagne.

REFERENCES CITED

Angelier, J., 1979, Néotectonique de l'Arc égéen: Société Géologique du Nord, publication n. 3, 418p.

Angelier, J., 1981, Analyse quantitative des relations entre déformation horizontale et mouvements verticaux: l'extension égéenne, la subsidence de la mer de Crète et la surrection de l'arc héllénique: Annales de Géophysique, v. 37, p. 327-346.

Artemjev, M. E., and E. V. Artyushkov, 1971, Structure and isostasy of the Baikal rift and the mechanism of rifting: Journal of Geophysical Research, v. 76, p. 1197-1212.

Bott, M. P., 1971, Evolution of young continental margins and formation of shelf basins: Tectonophysics, v. 11, p. 319-327.

Heezen, B. C., 1962, The deep-sea floor, *in* Runcorn, ed., Continental drift: New York, Academic Press, p. 235-288.

Jarvis, J. G., and D. P. McKenzie, 1980, Sedimentary formation with finite extension rates: Earth and Planetary Science Letters, v. 48, p. 42-52.

Le Pichon, X., and J. Angelier 1979, The Hellenic arc and trench system — a key to the neotectonic evolution of the eastern Mediterranean area: Tectonophysics, v. 60, p. 1-42.

———, and ———, 1981, The Aegean Sea: Proceedings, Royal Society of London, A 300, p. 357-372.

———, and J. C. Sibuet, 1981, Passive margins, a model of formation: Journal of Geophysical Research, v. 86, p. 3708-3720.

———, J. Angelier, and J. C. Sibuet, 1982, Plate boundaries and extensional tectonics: Tectonophysics, v. 81, p. 239-256.

McKenzie, D., 1978a, Some remarks on the development of sedimentary basins: Earth and Planetary Science Letters, v. 40, p. 25-32.

———, 1978b, Active tectonics of the Alpine-Himalayan belt; the Aegean Sea and surrounding regions: Geophysical Journal of the Royal Astronomical Society, v. 55, p. 217-254.

Montadert, L., et al, 1979a, Rifting and subsidence of the northern continental margin of the Bay of Biscay, *in* L. Montadert and D. G. Roberts, eds., Initial reports of the deep sea drilling project: Washington, D.C., U.S. Government Printing Office, v. 48, p. 1025-1060.

———, et al, 1979b, North-east Atlantic passive margins: rifting and subsidence processes, *in* M. Talwani, et al, eds. Deep-drilling results in the Atlantic Ocean: Washington, American Geophysical Union, Continental Margins and Paleoenvironment, series 3, p. 164-186.

———, et al, 1981, Extension rates measurements on Galicia Portugal and North Biscay continental margin, consequences for continental margins evolution models: Galveston, Texas, 1980 Hedberg Research Conference on Continental Margin Processes (Abs.).

Morton, W. H. and R. Black 1975, Crustal attenuation in Afar, *in* A. Pilger and A. Rosler, eds., Afar depression of Ethiopia: Stuttgart, Schweizerbart'sche Verlagsbuchlandlung, v. 1, p. 55-65.

Parsons, B., and J. G. Sclater, 1977, Analysis of the variation of ocean floor bathymetry and heat flow with age: Journal of Geophysical Research, v. 82, p. 803-827.

Proffett, J. M., Jr., 1977, Cenozoic geology of the Yerington district, Nevada, and implications for the nature and origin of basin and range faulting: Geological Society of America Bulletin, v. 88, p. 247-266.

Royden, L., and C. E. Keen, 1980, Rifting process and thermal evolution of two continental margins of eastern Canada determined from subsidence curves: Earth and Planetary Science Letters, v. 51, p. 343-361.

———, J. G. Sclater, and R. P. von Herzen, 1980, Continental margin subsidence and heat flow: Important parameters in formation of petroleum hydrocarbon: American Association of Petroleum Geologists Bulletin, v. 64, p. 173-187.

Sittler, C., 1974, Le fossé rhénan ou la plaine d'Alsace, *in* J. Debelmas, ed., Géologie de la France: Paris, Doin, p. 78-104.

Turcotte, D. L., J. L. Ahern, and J. M. Bird, 1977, The state of stress at continental margins: Tectonophysics, v. 42, p. 1-28.

Werner, D., and H. G. Kahle, 1980, a geophysical study of the Rhine graben, parts I and II: Geophysical. Journal of the Royal Astronomical Society, v. 62, p. 617-647.

Thermal Evolution of the Baltimore Canyon Trough and Georges Bank Basin

D. S. Sawyer*
M. N. Toksöz
J. G. Sclater
Department of Earth and Planetary Sciences
Massachusetts Institute of Technology
Cambridge, Massachusetts

B. A. Swift
U.S. Geological Survey
Woods Hole, Massachusetts

A simple, one-dimensional extensional model can explain the major features of the northeastern United States Atlantic continental margin. The extensional model allows us to predict the subsidence history of the margin and we compare that prediction with well data. Tectonic-subsidence observations indicate that the COST B-2 and B-3 wells are over highly thinned continental or oceanic crust, while the G-1 and G-2 wells are over continental crust that experienced less thinning.

A two-dimensional finite difference numerical scheme for simulating the thermal and mechanical evolution of a basin supports an extensional origin for the Baltimore Canyon trough. The simulation provides time-temperature history predictions for the sediments. From these we conclude that the trough should contain a significant volume of thermally mature sediments that are more likely to have generated natural gas than oil.

The temporal variation of ocean floor heat flow and subsidence has been modeled successfully. The age-dependent elevation of the sea floor is explained by lithospheric cooling of a plate approximately 125 km thick (Parsons and Sclater, 1977). In that model, the basement depth increases as the square root of time for the first 80 million years and thereafter as the exponential of (−time/62.8 million years). These expressions do not describe the observed basement subsidence landward of normal oceanic crust. Passive continental margins, which are transitional areas bridging continental and oceanic lithosphere, have unique origins and thermal histories.

Sleep (1971) showed that the general subsidence of the continental margin along the east coast of North America can be described in terms of exponential thermal decay. Sleep (1971) hypothesized that this decay resulted from cooling of the lithosphere following a period of heating, uplift, and erosion. This model did not provide a mechanism by which the thick sediment layers present along most of the margin could accumulate without extreme amounts of prior uplift and erosion. The Sleep (1971) model provided no continuous mechanism for the formation of the margin and adjacent ocean floor. To explain deep sedimentary basins that formed without prior major uplift, Artemyev and Artyushkov (1971) suggested crustal extension and thinning as a possible subsidence initiating mechanism.

McKenzie (1978) suggested that a one-dimensional thermal model could explain the behavior of continental crust and lithosphere during and after their extension. Extension causes thinning of the continental crust and lithosphere, which results in hot asthenosphere being brought close to the surface, thus causing the geothermal gradient to become steeper. Two types of subsidence are predicted by this model. First, the initial subsidence that accompanies extension is caused by the replacement of light continental crust and dense lower lithosphere with intermediate density hot asthenosphere. If the continental crust being thinned is less than about 18 km thick, extension is predicted to cause initial uplift rather than subsidence (McKenzie, 1978). More typically, the now

*Presently with: Institute for Geophysics, The University of Texas, Austin, Texas.

heavier rock column subsides to restore isostatic equilibrium. After active extension ceases, thermal subsidence begins. The hot asthenosphere that has been brought close to the surface cools by conduction; its contraction causes the surface to subside further. The magnitude of each of these types of subsidence increases as the amount of extension increases. Extension could continue until the continental crust had reached zero thickness. If continental crust that was originally 40 km thick reached zero thickness, the initial subsidence would be predicted to be about 2.5 km, which corresponds to the average depth of mid-ocean ridges. The predicted long run thermal subsidence is an additional 4.0 km. When this 4 km subsidence is added to the 2.5 km initial subsidence, the total is the 6.5 km of subsidence of oceanic lithosphere (Parsons and Sclater, 1977).

The United States Atlantic continental margin is a passive margin dominated by a series of northeast-trending basins filled by a sedimentary prism of Mesozoic and Tertiary age. The lateral boundaries of the margin are marked by block-faulted continental crust on the western (landward) edge (Schlee et al, 1976; Schlee, Dillon, and Grow, 1979) and by the East Coast magnetic anomaly (ECMA) on the oceanward edge (Klitgord and Behrendt, 1979). The ECMA is thought to be caused by an edge effect arising from the juxtaposition of normal oceanic crust and rift-stage crust overlain by thick sediments (Klitgord and Behrendt, 1979). Four Continental Offshore Stratigraphic Test (COST) wells have been drilled in the northern part of the margin, two in the Baltimore Canyon trough off New Jersey, and two in the Georges Bank basin off Cape Cod (Figure 1).

The Baltimore Canyon trough trends northeastward, widening and deepening along strike from Cape Hatteras to New Jersey, from 4 km to more than 15 km depth (Scholle, 1977; 1980). It is asymmetric in cross section (Figure 2); the sediment wedge increases from 5 km to 14 km in thickness. The maximum thickness is seaward of a basement hinge zone beneath the outer shelf (Grow, 1980; Schlee, 1981a). Upper Triassic and Lower Jurassic synrift sediments accumulated very rapidly and are concentrated in a northern depocenter. They are inferred to be nonmarine to marginal marine sediments, limestones interbedded

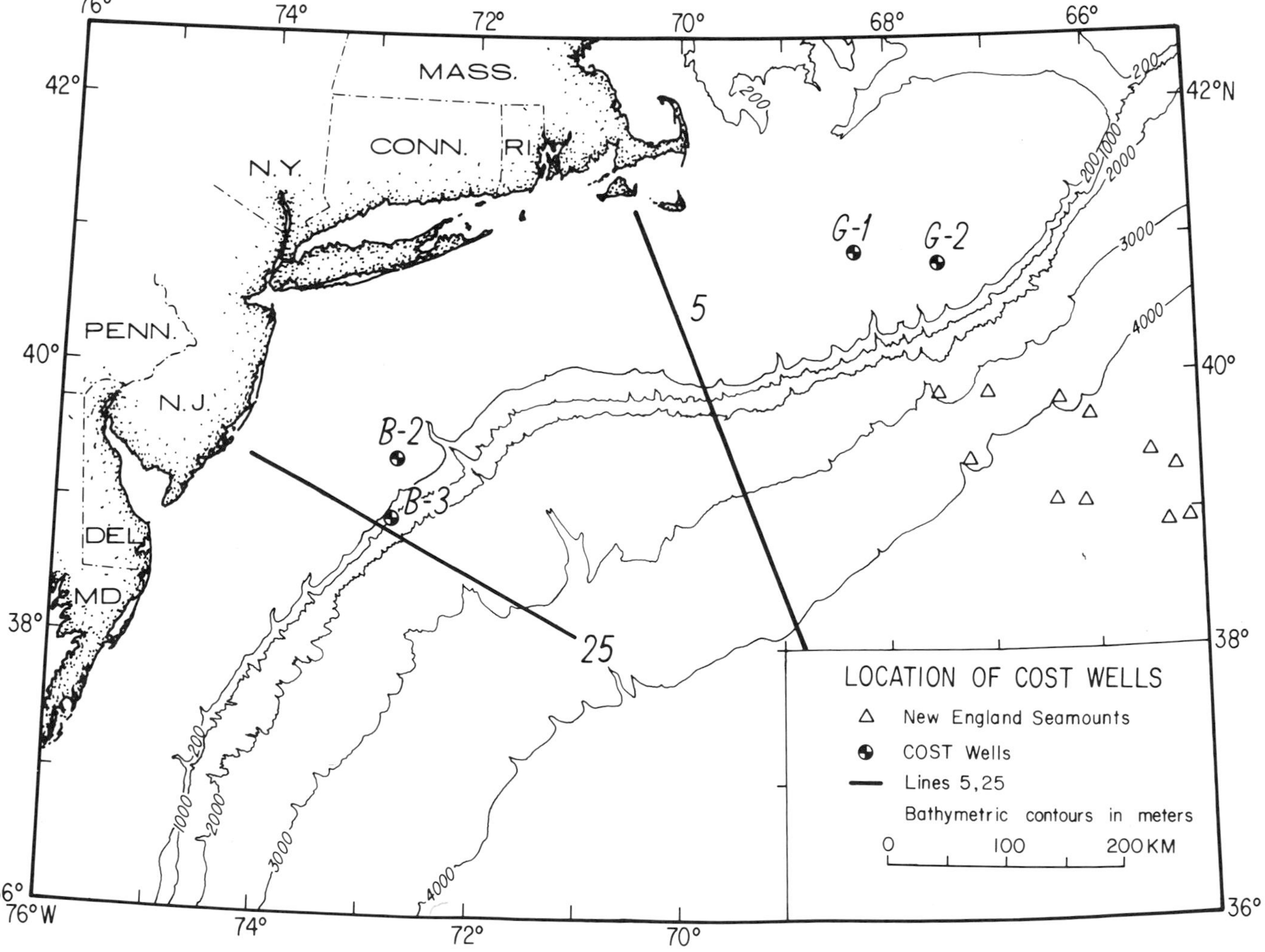

Figure 1 — Index map of the northeastern United States Atlantic continental margin showing locations of COST wells, New England seamounts, and seismic lines 5 and 25.

with evaporite deposits (Schlee, 1981a). Middle Jurassic to Cenozoic postrift sediments are separated from the earlier rift sediments by a breakup unconformity (Falvey, 1974). They are more broadly distributed and accumulated much more slowly than the synrift sediments. The Cretaceous and younger postrift sequence is dominated by clastic sediments rather than the carbonate and evaporitic rocks that are more typical of the synrift and early postrift sediments. A reflection seismic section, USGS line 25 (Figure 2), crosses the trough. The Jurassic shelf edge was about 20 km seaward of the present shelf edge and was the seaward edge of a prograding carbonate bank. The COST B-2 and B-3 wells (Figure 2) are located nearly over the boundary between oceanic and rift-stage crust. Both were drilled through the entire Cretaceous and Tertiary section and ended in Late Jurassic sediments. Modeling of gravity data indicates an unusually wide (120 km) section of rift-stage (transitional) crust landward of the ECMA (D. R. Hutchinson and J. A. Grow, unpublished data). Magnetic data for the area provide depth-to-basement estimates and suggest that the ECMA delineates the location of the boundary between oceanic crust and rift-stage crust (Klitgord and Behrendt, 1979). Together, these data indicate the thickness and nature of the crust under each part of the margin, information that we will compare to extensional-model predictions.

The Georges Bank basin is a more complex basin, which originated along an easterly trending part of the margin that formed at an oblique angle to the spreading ridge (Amato and Bebout, 1980; Amato and Simonis, 1980; Klitgord and Behrendt, 1979; Austin et al, 1980; Schlee, this volume; Klitgord, Schlee, and Hinz, 1981). The synrift sediments here are probably of similar lithology to those found in the Baltimore Canyon trough, but their deposition pattern is less regular. Georges Bank basin contains one main and several smaller depocenters that are possibly associated with grabens (Schlee, this volume). The COST G-1 well (Figure 1) drilled into phyllite, which may be continental basement. The COST G-2 well bottomed

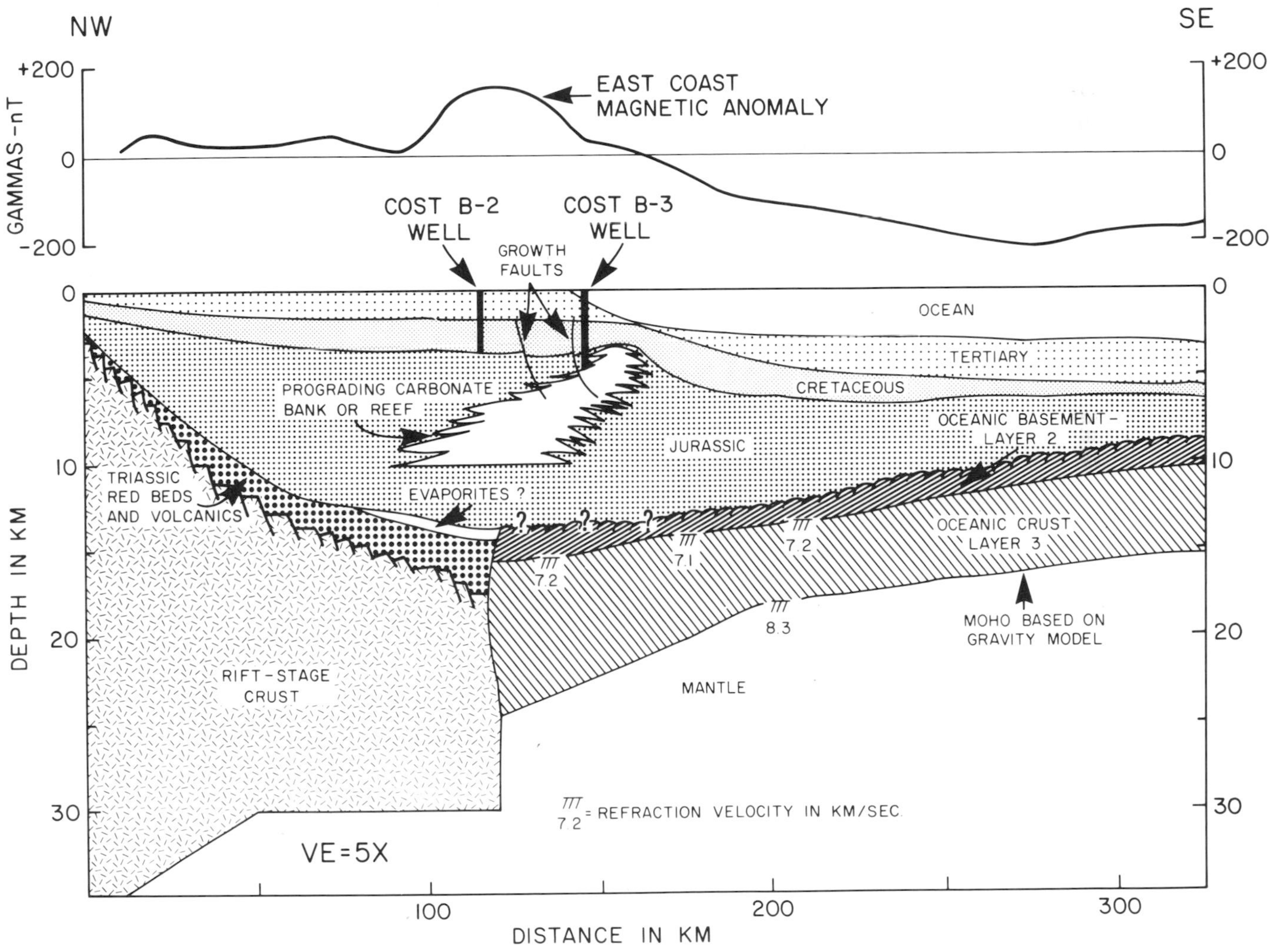

Figure 2 — Cross section of Baltimore Canyon trough along U.S. Geological Survey seismic line 25. The sediment interface information is based on seismic-reflection data (Grow, 1980). The position of the crust-mantle boundary is derived from gravity models (D. R. Hutchinson and J. A. Grow, unpublished data). In some seismically obscure areas, depth to basement has been taken from magnetic data (Klitgord and Behrendt, 1979).

in salt, whose relationship with basement is not clear. The COST wells in the Georges Bank basin (Figure 3) were drilled over fairly thick rift-stage crust, relatively landward of the wells in the Baltimore Canyon trough. The postrift sediments are much more evenly distributed over the area than are the synrift sediments. The maximum observed sediment thickness is in excess of 12 km in the large subbasin under the south-central part of the bank. USGS line 5 (Figure 3) crosses the southwestern flank of the basin. A significant part of the sediment-basement interface is not observed on seismic sections, so accurate sediment thicknesses are difficult to determine. A relatively undeformed sediment wedge thickens seaward and extends out to the Late Jurassic shelf complex 5 to 10 km seaward of the present shelf edge. The landward gradient of the ECMA is steeper here than it is under the New Jersey margin, suggesting a more abrupt transition from oceanic to rift-stage crust (Klitgord and Behrendt, 1979). Shoreward of the basin, the hinge zone appears to be broader and more complex than it is in other areas.

Sufficient data have been collected to test whether a version of the McKenzie (1978) extensional model can account for some or all of the thermal and mechanical evolution of these two continental margin basins. First, we compare the subsidence history of the two basins with the extensional model predictions; second, we model the evolution of one of them, the Baltimore Canyon trough, and derive its thermal history.

TECTONIC SUBSIDENCE OF THE MARGIN

We discuss the tectonic subsidence predicted by the extensional model and indicate how it should be distributed across a passive continental margin. The total

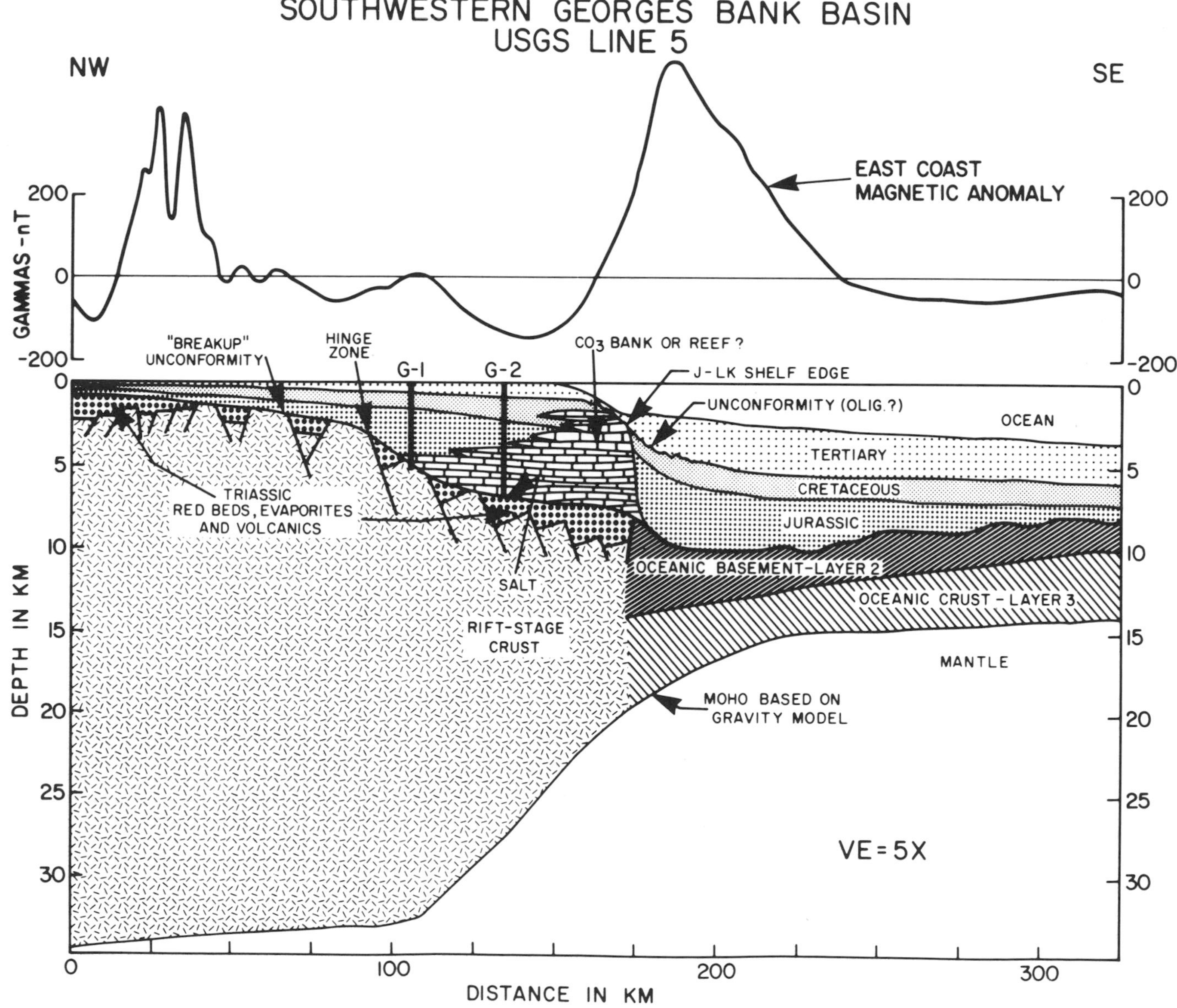

Figure 3 — Cross section of Georges Bank basin along USGS seismic line 5 taken from Grow et al (1979b).

amount of tectonic subsidence that has taken place in both the Baltimore Canyon trough and Georges Bank basin is in agreement with the model. The tectonic-subsidence histories of several parts of the margin are derived from well data. They have the character predicted by the model, and each indicates a different amount of extension. These results are supported by continental crust thickness observations.

McKenzie (1978) gave equations for the amounts of initial, S_i, and thermal, S_t, subsidence predicted by the extensional model, as a function of the amount of extension that has taken place. The amount of extension can be characterized by the variable β, where the thickness of any layer after extension is 1/β times its thickness prior to extension.

$$S_i = \frac{-t_1 \left(1 - \frac{1}{\beta}\right)\left[(\rho_m - \rho_c)(t_c/t_1)(1 - \alpha T_m t_c/2t_1) - \alpha T_m \rho_m/2\right]}{\rho_m(1 - \alpha T_m) - \rho_w} \quad (1)$$

$$S_t(t) = \frac{\alpha t_1 \rho_m T_m}{\rho_m - \rho_w} \left\{ \frac{1}{2}\left(1 - \frac{1}{\beta}\right) - \frac{4}{\pi^2} \sum_{m=0}^{\infty} \frac{1}{(2m+1)^2} \left[\frac{\sin \frac{(2m+1)\pi}{\beta}}{\frac{(2m+1)\pi}{\beta}} \right] \left[\exp\left(-(2m+1)^2 \pi^2 \frac{\kappa t}{t_1^2}\right) \right] \right\}$$

where:

$\alpha = 3.3 \times 10^{-5}\ °C^{-1}$	Thermal expansion coefficient
$\kappa = 8 \times 10^{-3}\ cm^2/sec$	Thermal diffusivity
$\rho_m = 3.33\ gm/cm^3$	Density of lower lithosphere
$\rho_c = 2.9\ gm/cm^3$	Density of continental crust
$\rho_w = 1.03\ gm/cm^3$	Density of ocean water
$t_c = 40$ km	Initial continental crust thickness
$t_1 = 120$ km	Initial depth to lithosphere asthenosphere boundary
$T_m = 1350°C$	Initial temperature of asthenosphere
t (in m.y.)	Time since rifting

The equations are used to predict subsidence-history curves for various values of β and a continental crust having an initial thickness of 40 km (Figure 4). They are based on an extensional event beginning 200 m.y. ago and continuing for 25 million years. The predicted amount of initial subsidence is distributed linearly over that 25 million year period. Initial subsidence is followed by the predicted thermal subsidence, which is assumed to begin at the end of the period of rifting, from 175 m.y. ago to the present. For each value of β, the initial subsidence accounts for about 40% of the total subsidence and takes place rapidly during the extension (Figure 4). The thermal subsidence is larger than the initial subsidence as it is 60% of the total, but it is distributed over a much longer period of time and, therefore, takes place at a slower rate than initial subsidence.

Subsidence varies laterally across a margin because the amount of extension and the time at which subsidence began varies (Figure 5). A uniform continental crust section begins to undergo extension and responds by necking. This process continues until the continental crust in the center has been thinned to zero thickness. Sea-floor spreading begins at the point of zero continental crust thickness, where β is infinite. The center of the extended region experiences initial subsidence of 2.5 km while, away from the center, the initial subsidence decreases to zero. The continental margin moves away from the newly formed spreading center as the extended continental crust and oceanic lithosphere cool and subside owing to thermal contraction. The subsidence history of rift-stage crust is characterized by the variable β and the time since extension. Oceanic crust subsidence, however, is characterized only by its time of formation. The oceanic crust formed by the spreading center follows the subsidence history appropriate for $\beta = \infty$. The total amount of subsidence decreases from the margin toward the ridge because the oceanic lithosphere is progressively younger. Therefore, the subsidence of the oceanic lithosphere is composed of the $\beta = \infty$ initial subsidence and an amount of thermal subsidence which depends on its age. That is the Parsons and Sclater (1977) result for the subsidence of the ocean floor. Landward, away from the initial point of sea-floor spreading, subsidence histories should reflect a gradually decreasing amount of extension.

The model cannot be compared directly to observed subsidence on the U.S. Atlantic continental margin because it assumes that no sediment has been deposited. Sediment thicknesses exceed 12 km in both the Baltimore Canyon trough and the Georges Bank basin. When sediment is deposited, the basement that underlies it subsides isostatically, so that the total basement subsidence observed in a sedimented area is larger than a simple model would predict. As a result, the subsidence due to this sediment loading must be separated from that due to extension or other tectonic phenomena before direct comparisons with the model can be made. This process can be quantized and its effects eliminated to at least first order. If S is the thickness of sediment, of average density $\bar{\rho}_s$, deposited at a point, the amount of subsidence there, U, due only to sediment loading, is given by:

$$U(t) = S^*(t) \times \frac{\bar{\rho}_s - \rho_w}{\rho_m - \rho_w} \quad (2)$$

where $\bar{\rho}_s$ = average density of S^* thick sediment layer

This equation is derived with the assumption that pointwise isostatic equilibrium is maintained in a basin. In many cases, a basin develops on crust that might be expected to behave more like an elastic plate floating on a liquid. We use such a model below to calculate a flexural unloading correction. For the present simple analysis, the simpler unloading correction is sufficient. Deposition of one kilometer of sediment results in about 650 meters of basement subsidence due to loading effects. In general, the density of sediments increases with depth due to compaction. As a sediment compacts, its porosity (ϕ) decreases. The depth effect on porosity has been approximated by Sclater and Christie (1980) and Ruby and Hubbert (1960):

$$\phi = \phi_0 e^{-cz} \qquad (3)$$

where z = depth below sediment surface
ϕ_0 = sediment surface porosity
c = compaction coefficient

The density of sediment may be expressed as a function of the grain density (ρ_{sg}) and porosity.

$$\rho_s = \phi\rho_w + (1-\phi)\,\rho_{sg}$$

Average sediment density in a layer between depths z_1 and z_2 is derived by integrating this density function:

$$\bar{\rho}_s = \int_{z_1}^{z_2} \frac{\rho_s}{z_2 - z_1}\,dz = \rho_{sg} + \frac{\rho_w - \rho_{sg}}{z_2 - z_1} \int_{z_1}^{z_2} \phi dz$$

$$\bar{\rho}_s = \rho_{sg} - \frac{(\rho_w - \rho_{sg})\phi_0}{(z_2 - z_1)c}\left[e^{-cz_2} - e^{-cz_1} \right]$$

The total tectonic subsidence that has taken place on the margin can be derived from these equations and observations of bathymetry and sediment thickness. These data are taken from seismic-reflection profiles. The total observed subsidence is the present depth to the sediment-basement interface. This subsidence is converted to total tectonic subsidence by subtracting U, the amount of subsidence caused by the sediment load, whose thickness and properties are known. Average sediment compaction properties are used in Equation 2 to calculate U. The grain density is assumed to be 2.65 gm/cu cm, the initial porosity is .55, and the compaction coefficient is $4.5 \times 10^{-4} m^{-1}$. The error bars represent the same calculation but

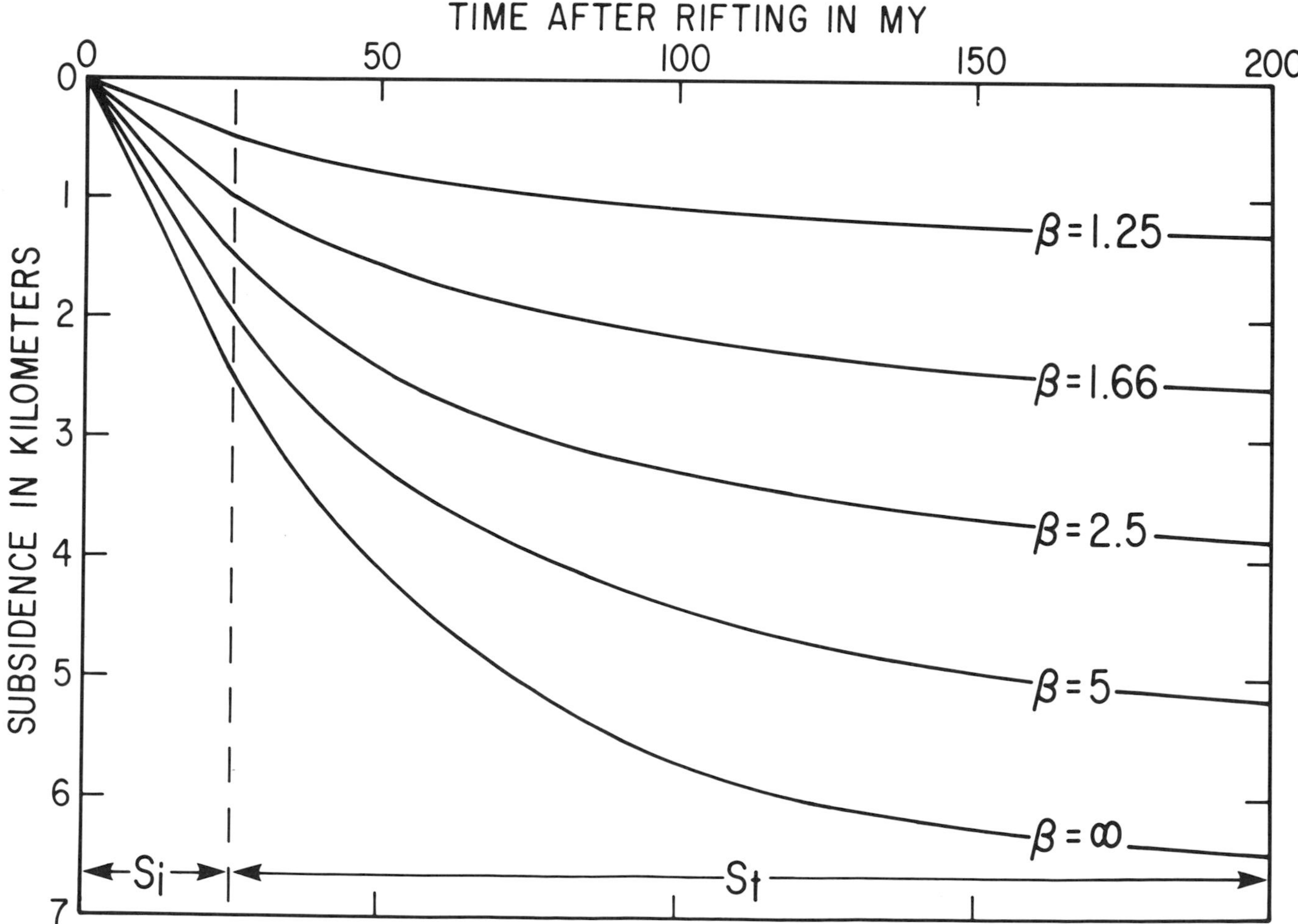

Figure 4 — Subsidence vs time predicted by the extensional model. The rift event is assumed to have begun 200 m.y. ago. The initial subsidence is distributed linearly over the first 25 million years. The thermal subsidence continues to the present.

using extreme estimates of the sediment compaction properties. The minimum subsidence is based on ρ_{sg}=2.7 gm/cu cm, ϕ_0=.75, and c=3×$10^{-4}m^{-1}$ while the maximum subsidence is based on ρ_{sg}=2.5 gm/cu cm, ϕ_0=.25, and c=6×$10^{-4}m^{-1}$. Cross sections of the Atlantic continental margin unloaded in this way (Figures 6 and 7) show subsidence comparable to the total subsidence for a mature margin shown in Figure 5. The amount of tectonic subsidence as a function of β that is expected to have taken place when a margin is 200 m.y. old is calculated by using Equation 1 and is shown in Figure 8. Using Figure 8, an amount of total tectonic subsidence can be related to the amount of crustal thinning predicted to have taken place. The distribution of extension across the two total tectonic subsidence cross sections (Figures 6 and 7) are compared and they have several features in common.

1. Normal continental crust to the west. (β=1)
2. Shallow lateral gradient of extension toward the east. (1< β <1.25)
3. Basement hinge zone, or inflection point, at which the lateral gradient of extension increases. (β=1.25)
4. Steep lateral gradient of extension toward the east. (1.25< β<15)
5. Possible basement ridge or uplifted block at the outer edge of the basin.
6. Maximum amount of subsidence on oceanward side of ridge. Total subsidence of about 6.4 km. (β=∞)

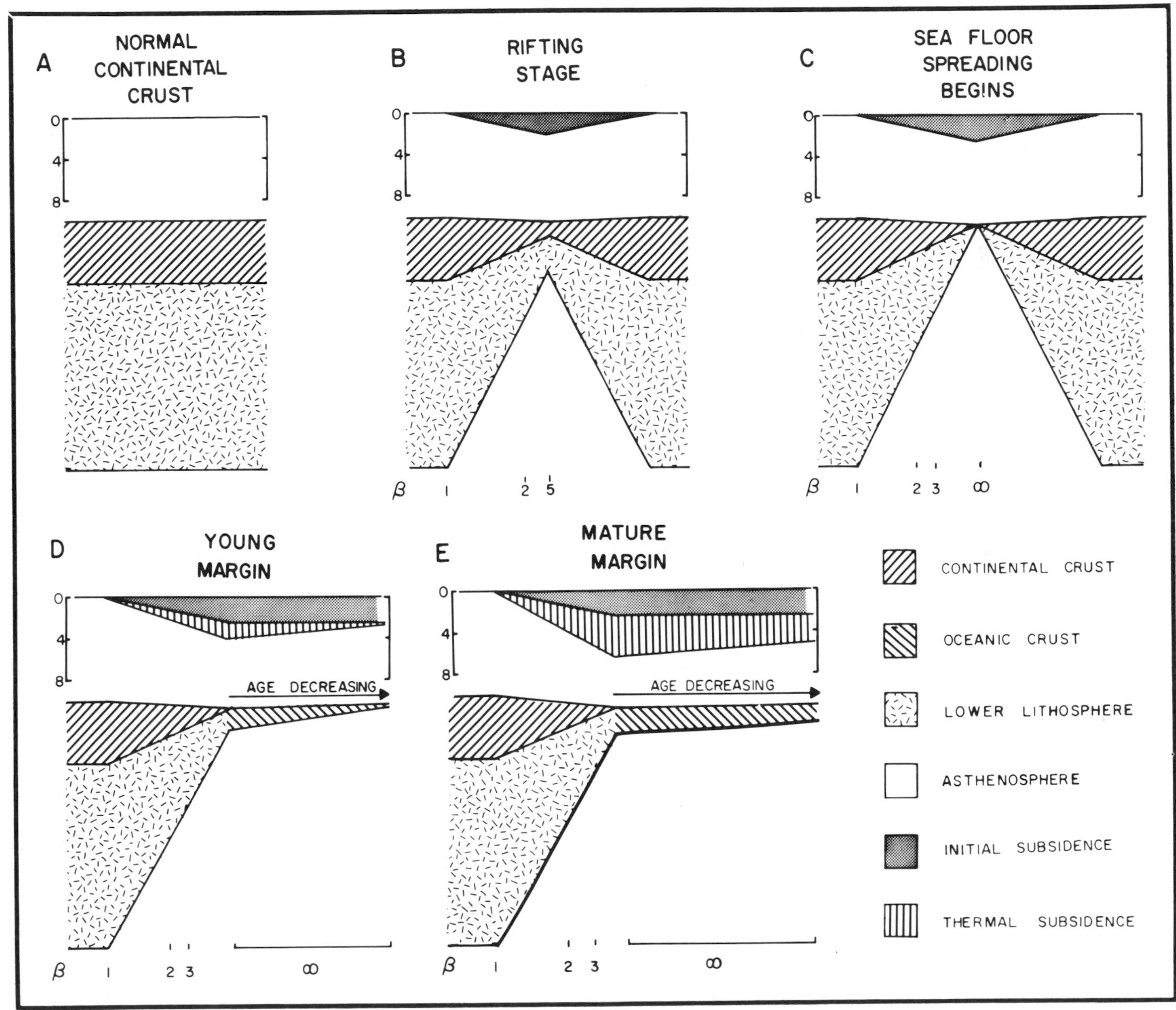

Figure 5 — Schematic diagram of the evolution of a continental margin. The thickness of the continental crust is typically 40 km and the depth to the asthenosphere is about 120 km. The upper graph in each set shows the predicted subsidence in km at an expanded vertical scale. Amount of extension is shown below the cross sections.

7. Gradually decreasing amount of subsidence to the east.

Estimates of sediment thickness are derived from reflection seismic data (USGS lines 19 and 25), and these data are not of uniformly good quality across the sections. Parts of each section shown in Figures 6 and 7 have been marked, indicating that the sediment-basement interface is not identifiable from the seismic-reflection data.

Subsidence of the basement just east of the possible basement ridge is about 6.4 km in both cross sections (Figures 6 and 7). This is the amount expected for old oceanic crust. The small decrease in subsidence toward the east is probably due to the decreasing age of the oceanic crust.

The existence of a basement ridge is not established by the subsidence data. In the line 25 section, it is in the area of poor basement depth control and is therefore based primarily on magnetic data (Klitgord and Behrendt, 1979). On line 5, the magnitude of the ridge is small and insignificant. Klitgord and Grow (1980) concluded that no major basement features parallel the margin in this area.

The region landward of the basement ridge is underlain by thinned continental (rift-stage) crust. The width of the zone of rift-stage crust is very similar on both lines. The zone is likely to be very different under other parts of the U.S. Atlantic continental margin, and its width may indicate something about the mechanical response of the crust and lithosphere to extension. The distribution of subsidence (Figures 6 and 7) is in good agreement with the qualitative model (Figure 5).

Total-subsidence cross sections show only the amount of tectonic subsidence that has taken place at one point in time, the present, because that is all that can be derived without using additional subsurface data. Where a well has been drilled, more information can be obtained about the subsidence of the margin. Paleoenvironmental, biostratigraphic, and lithologic data from the well can be used to infer the tectonic-subsidence history of the margin by means of a technique devised by Steckler and Watts (1978). They calculated the effect of progressive removal of sediment layers and restoration of the basement surface to its original depth. The result is a curve showing the tectonic subsidence as a function of time (tectonic-subsidence history).

The basement subsidence that can be directly observed, S(t), is simply the thickness of sediments in the well deposited prior to time (t). This thickness is not, however, the thickness of the sediment when it was initially deposited because sediment compacts as it is buried. As the sediment compacts, its porosity (ϕ) is reduced and is related to depth (z) by Equation 3. The values of c and ϕ_0 for shales in the COST wells

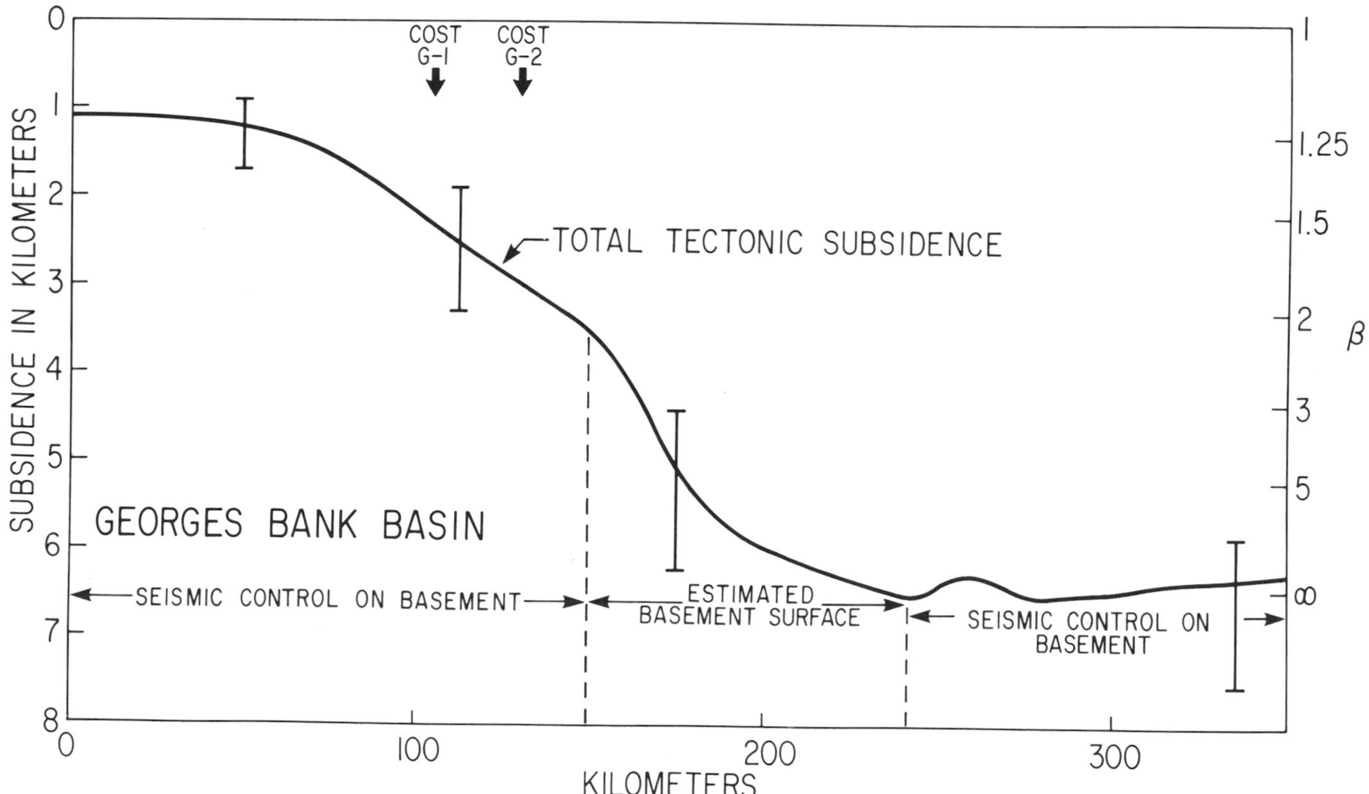

Figure 6 — Total tectonic-subsidence along U.S. Geological survey Line 5. This is the observed depth to the present sediment-basement interface minus the subsidence due to the isostatic compensation of the sediment load. The error bars are the result of considering a range of sediment compaction properties.

have been taken from Sclater and Christie (1980) and are $5.1 \times 10^{-4} m^{-1}$ and .63, respectively. The values for sandstone and limestone are derived from observed porosity-versus-depth data from the COST wells. For sandstone $c = 3.0 \times 10^{-4} m^{-1}$ and $\phi_0 = .40$. In limestone $c = 5.4 \times 10^{-4} m^{-1}$ and $\phi_0 = .45$. By use of Equation 3, the thickness of a layer at a given time can be reconstructed. The subsidence corrected for decompaction (S*) is given by the solution to:

$$S^*(t) + \frac{\phi_0}{c} e^{-cS^*(t)} = S(t) + \frac{\phi_0}{c} \left[1 + e^{-cS(0)} - e^{-cd(t)} \right]$$

where $d(t) = S(0) - S(t)$ = depth to sediment of age t

The largest correction to the observed subsidence data is for the loading effect (U) of the sediment layer. It must be calculated and removed to determine what the depth to basement would have been in the absence of sediment. The lithosphere of the margin is assumed to respond to loading as if it were a flexing elastic plate. Sawyer et al (1982) used the Airy isostatic assumption for the lithospheric response and conclude that for subsidence data on the Atlantic margin, the use of a flexing-plate assumption is desirable. Use of this assumption is warranted particularly if the wells are located near the transition from rift-stage crust to oceanic crust, as is the case with the COST B-2 and B-3 wells. The rigidity of that plate will be very low early in the margin's history and increases as the margin cools. This loading model differs from Airy isostasy in that a load at a given point causes a response (deflection) not only at that point, but also at nearby points. To calculate $U_f(t)$, the flexural unloading correction, at a well, one must know how the rigidity of the plate has varied with time and the time distribution of sediment on the margin. Watts, Bodine, and Steckler (1980) gave a range of elastic-plate thicknesses vs age for oceanic lithosphere. Since the wells are located over or near rift-stage crust that is thicker than oceanic crust, we use the maximum estimates for flexural plate thickness given by their data. The curve begins with very low values, 3 to 4 km at 5 million years, and increases rapidly to about 25 km at 30 million years. Its slope decreases dramatically with age thereafter. For our subsidence-history calculations, the plate is assumed to have had no flexural strength during the 25 million year rifting stage, and thereafter to have thickened according to the Watts, Bodine, and Steckler (1980) curve. The sedimentation history of the margin is derived from well data where they exist and from seismic data elsewhere. As this load is not a simple analytic function of time and space, it is approximated by a set of prism loads for each time interval. The loading effect of each prism on the subsidence of the basement at the well location is calculated by equations given by Hetenyi (1946) for the response to loading of a thin plate on an elastic foundation. The loading effect, w(x,t), of a prism of width 2a, thickness S*, distance from the well x, age t, and plate thickness Th, is:

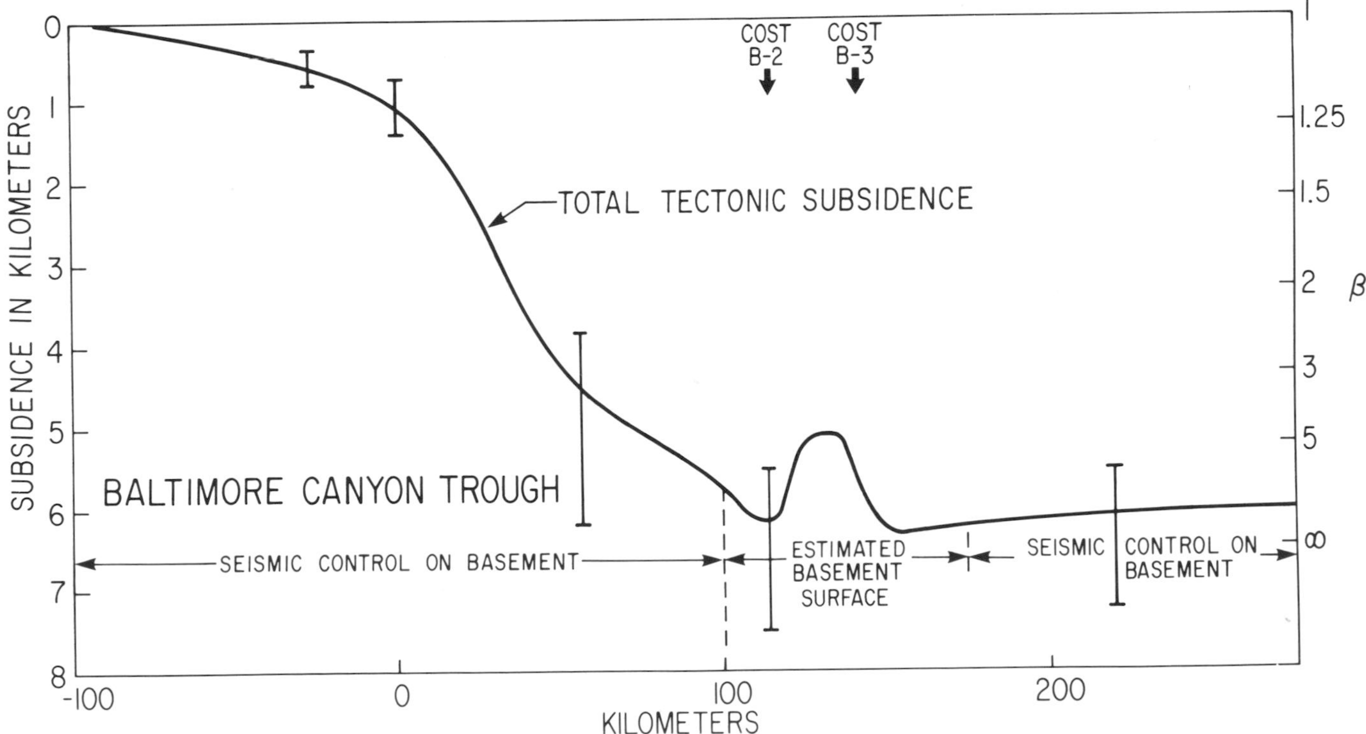

Figure 7 — Total tectonic subsidence along U.S. Geological Survey line 25 in the Baltimore Canyon trough.

$$\text{if } |x| < a \quad w(x,t) = \frac{h}{2}\,\frac{\bar{\rho}_s-\rho_w}{\rho_m-\rho_w}\left[2 - e^{-\frac{x-a}{\gamma}}\cos\left(\frac{x-a}{\gamma}\right) - e^{-\frac{x+a}{\gamma}}\cos\left(\frac{x+a}{\gamma}\right)\right]$$

$$\text{if } |x| > a \quad w(x,t) = \frac{h}{2}\,\frac{\bar{\rho}_s-\rho_w}{\rho_m-\rho_w}\left[e^{-\frac{x-a}{\gamma}}\cos\left(\frac{x-a}{\gamma}\right) - e^{-\frac{x+a}{\gamma}}\cos\left(\frac{x+a}{\gamma}\right)\right]$$

$$\text{where} \quad \gamma^4(t) = \frac{E\,Th^3(t)}{3\,(1-\sigma^2)\,(\rho_m-\rho_w)\,g}$$

$E = 6.5\times10^{11}$ dyne/cm^2	Young's modulus
$\sigma = .25$	Poisson's ratio
$g = 980.621$ cm/sec^2	Acceleration of gravity

The total unloading correction, U_f, is obtained by summing the contribution of each prism.

$$U_f(t) \approx \sum_{\substack{n=0\\ \text{time}}}^{t} \; \sum_{\substack{m=1\\ \text{sediment}\\ \text{prisms}}}^{n} w(x(m),\, n)$$

The amount of subsidence U_f, due only to sediment loading, will be subtracted from the observed and decompacted subsidence.

A correction is applied for the paleowater-depth (W_d) determined from paleoenvironmental analysis of rock samples. These data are usually given as minimum and maximum estimates of the paleowater-depth, so this uncertainty is incorporated into the subsidence-history curves. The paleo-water depth data for the COST B-2 and B-3 wells are taken from Poag (1980), for the G-1 well from Lachance, Bebout, and Bielak (1980), and for the G-2 well from Bielak and Simonis (1980).

The final correction to the observed subsidence is to account for eustatic sea level variation. Curves of eustatic sea level variation have been published by Watts and Steckler (1979) and Vail et al (1977). The Watts and Steckler (1979) estimates of eustatic sea level are used to determine the correction to the observed subsidence. Their sea level data are given by the function $\Delta SL(t)$.

The tectonic subsidence $Y(t)$ is the observed subsidence corrected for each of these disturbing effects.

$$Y(t) = S^*(t) - U_f(t) + W_d(t) - \Delta SL(t)\,\frac{\rho_m}{\rho_m-\rho_w}$$

In each step in the calculation of the tectonic-subsidence history curves, errors are likely to be made in the assumptions and observations. One source of error is in the paleowater-depth estimates for which a range of values is given. Other possible sources of error are the assumptions made about how sediment decompacts, how the lithosphere responds to sediment loading, and the variation of sea level through time. The magnitude of these errors is harder to estimate than that arising from the paleowater-depth estimates. We use a worst-case approach to estimate error bar length. The unloading correction is usually the largest correction to the observed data and therefore the subsidence is sensitive to the assumptions involved. The two critical assumptions are the sediment compaction properties and the variation of flexural rigidity. Maximum values of the unloading correction come from taking a low porosity versus depth function and high grain density. A minimum estimate is obtained by assuming a high porosity function and low grain density. The values chosen are the same as those used for the error analysis of the total tectonic subsidence. Since the maximum flexural plate thickness consistent with the data of Watts, Bodine, and Steckler (1980) is used to make the tectonic subsidence history calculations, it seems more likely that we are erring in the direction of too thick a plate. Therefore, for this error analysis the calculations are repeated using the minimum plate thicknesses allowed by the Watts, Bodine, and Steckler (1980) data. The Watts and Steckler (1979) sea level curve gives much smaller estimates for the variation of sea level through time than that of Vail et al (1977). Maximal estimates of tectonic subsidence result from using the Watts and Steckler (1979) curve, while minimal values are indicated by the Vail et al (1977) data. From among all the extreme values of each parameter, the combination that produces the maximum and minimum tectonic subsidence value for each data point are used to define the error bar length.

The B-2 and B-3 well subsidence data (Figure 9) are consistent with the extensional model for β between 5 and infinity. The subsidence in the G-1 and G-2 wells (Figure 9) is less, corresponding to extensions of 1.6 and 2.5, respectively. The differences between the observed tectonic-subsidence histories are largely explained by the different positions of the wells relative to the boundary between rift-stage crust and oceanic crust. The two Georges Bank basin wells are located over rift-stage crust (Figure 3) and have therefore subsided only about half as much as normal oceanic crust. The Baltimore Canyon trough wells are near the boundary of rift-stage crust and oceanic crust (Figure 2). The subsidence history at these locations (Figure 9) is very close to that expected of normal oceanic crust. The observed subsidence-history curves support the extensional model in two ways. First, the shape of the curves is quite like the shape predicted, and second, the indicated amounts of extension are reasonable for the well locations.

If the extensional model is valid for the margin, then the thickness of the continental crust under the wells should show the same amount of thinning as that required by the subsidence data. Gravity models (line 5: Grow, Bowin, and Hutchinson, 1979a; line 25: D. R. Hutchinson and J. A. Grow, unpublished data provide an estimate of the continental crust thickness that can be compared with the thickness of normal continental crust thinned by the amount indicated by the subsidence history curves. As stated above, the amounts of extension indicated by the subsidence data for the COST G-1 and G-2 wells are 1.6 and 2.5. If a 40 km thick continental crust had experienced extension by $\beta = 1.6$, it would have become 25 km (40 km/1.6) of rift-stage crust. The gravity model (Figure 3) indicates that the actual thickness is 27 km. For the COST G-2 well, the predicted thickness of rift-stage crust is 16 km (40 km/2.5), and the gravity model value is 19 km. The two wells in the Baltimore Canyon trough have sedimentation histories consistent with extension by $\beta=5-\infty$. The COST B-3 well is located over oceanic crust (Figure 2), and therefore the thickness of rift-stage crust is 0 km. Under the B-2 well is a thin (8 to 12 km) layer of rift-stage crust. The subsidence histories indicate that the wells should be underlain by 0 to 8 km of rift-stage crust. The deviations are very small for each well.

Schlee (1981a, 1982) mapped the depth of an unconformity that is present under most of the margin. Below the unconformity is a sequence of sediments of Latest Triassic to Early Jurassic age. The unconformity is observed on seismic-reflection profiles as a surface of landward onlap. The sediments above the unconformity are inferred to be postrift sediments; reflections in this section tend to be horizontal and continuous, suggesting slow, orderly, widespread deposition. Below the unconformity, synrift sediments appear to have been deposited in a more rapidly subsiding, block-faulted environment. The distinction between synrift and postrift sediments may correspond closely to that between sediment deposited during initial and thermal subsidence.

The coastal onlap above the breakup unconformity is probably due to the rapid increase in flexural strength of the lithosphere, which is assumed to take place at the end of the extension or rifting phase. The

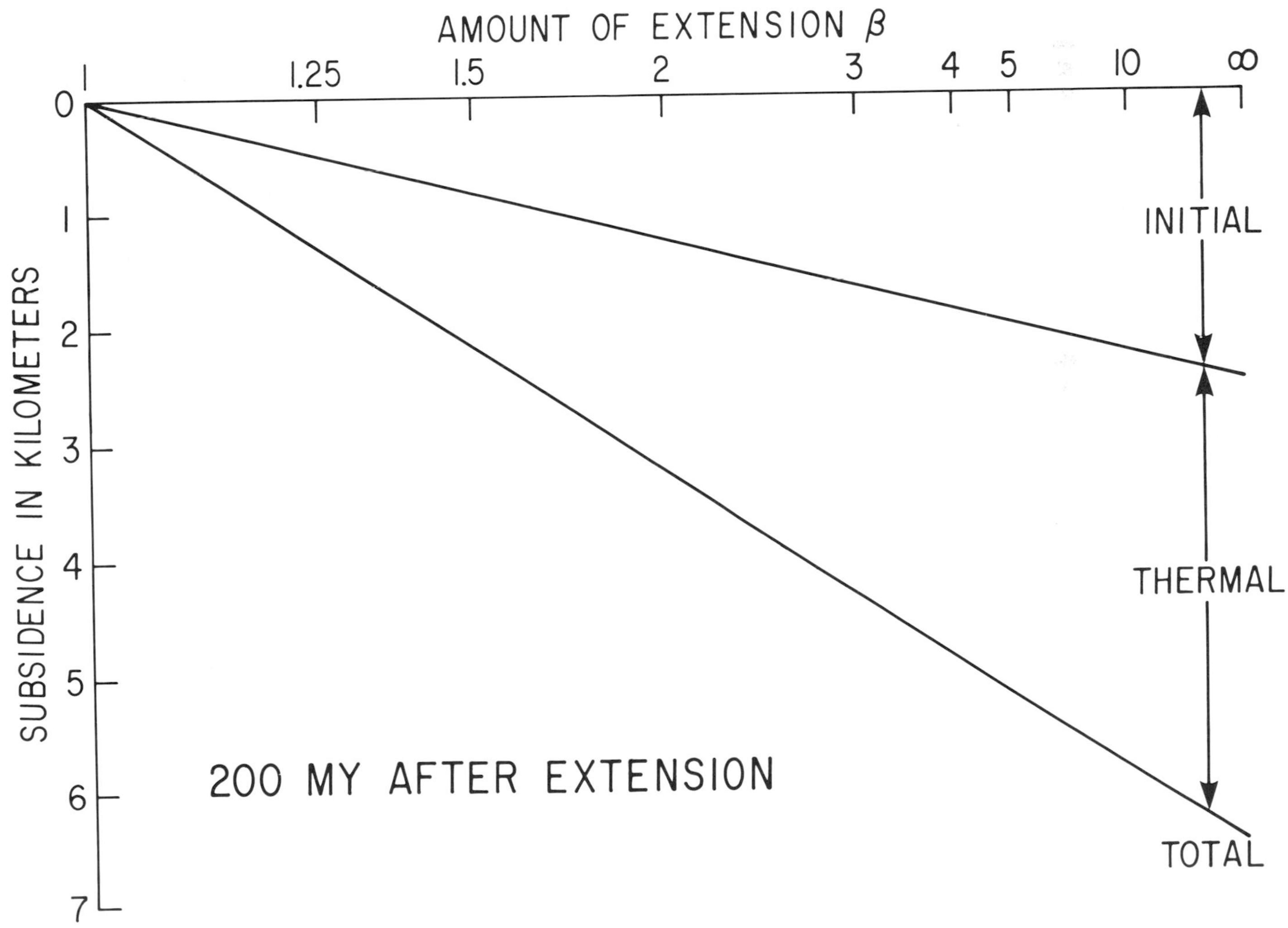

Figure 8 — Amounts of initial and thermal subsidence predicted to have taken place, 200 million years after an extension event, as a function of the amount of extension.

load of sediment would then cause downward deflection of the lithosphere gradually further from the center of deepest deposition. Tectonic tilting, down on the seaward side, would result and it would be particularly rapid at the end of the rift phase. Therefore, sediment deposition would move toward the coast, producing the surface of sediment onlap characteristic of the unconformity.

We do not suggest that this simple extensional model alone accounts for all the complexity in the evolution of this continental margin. The thermal effects of intrusive activity such as the New England seamounts (Figure 1), which were active in the Georges Bank basin about 115 m.y. ago (Vogt and Tucholke, 1979), can be seen in the subsidence-history curves for the COST G-1 and G-2 wells. The basement surface in these wells (Figure 9) was anomalously high during the last 90 million years, and we speculate this was because of the extra heat introduced into the region by the volcanic activity. The simple analytical model also assumes that the extension takes place smoothly, but the seismic data show that the surface of the crust is blocky and characterized by grabens. The extensional model does, however, account for

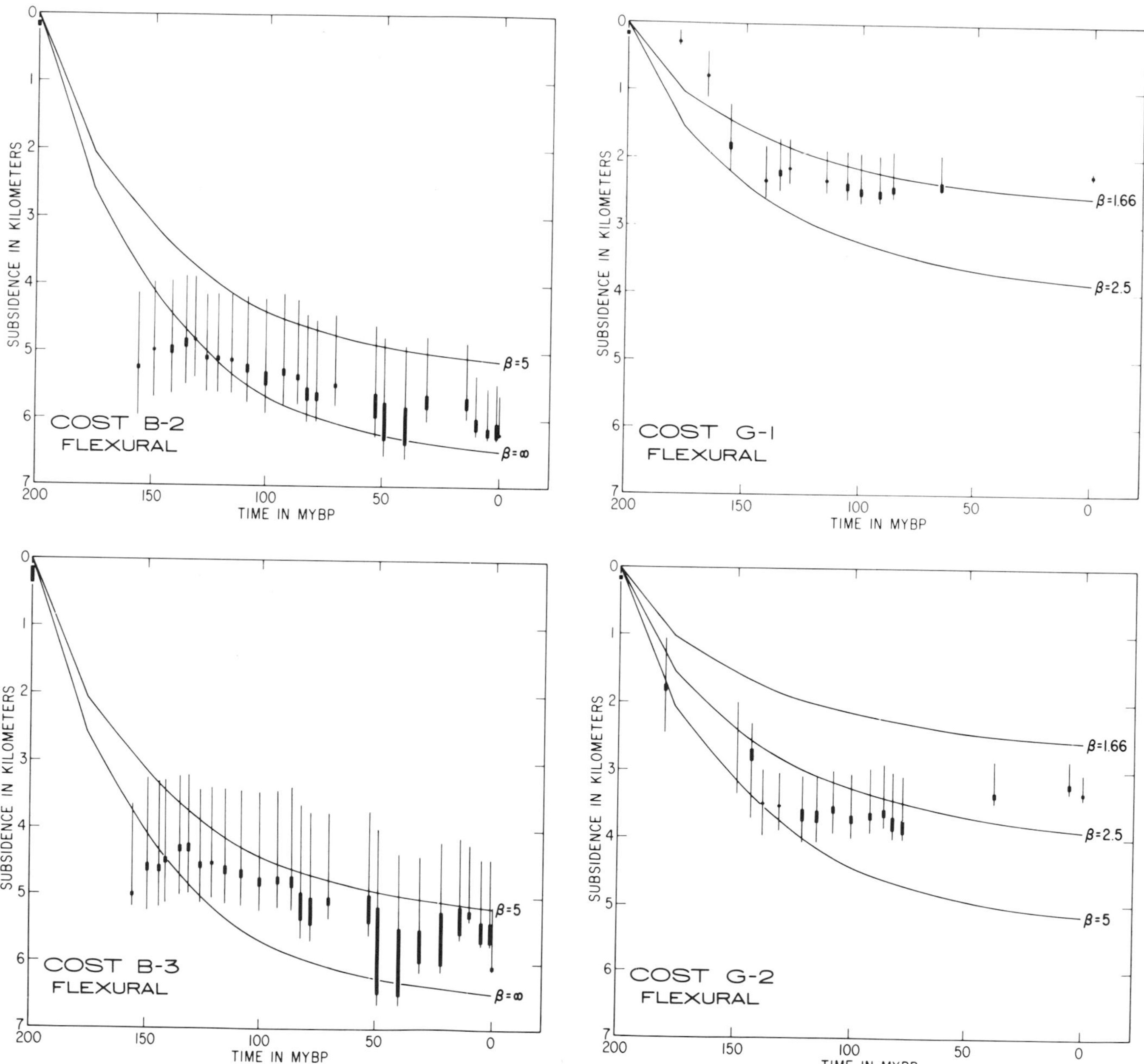

Figure 9 — Tectonic-subsidence history curves for the four COST wells, B-2, B-3, G-1, and G-2. The vertical bars represent observed values (see text for explanation of the error bars). Predicted subsidence curves for appropriate values of β are superimposed.

many large-scale features of the United States Atlantic continental margin, including crust thicknesses, sediment thicknesses, and tectonic-subsidence histories.

TWO-DIMENSIONAL BASIN EVOLUTION

All the preceding analysis of the basins use one-dimensional analytic techniques to predict subsidence resulting from the extensional model. This method ignores many effects, including those resulting from lateral heat flow and thermal blanketing by sediments. These features of basin development probably cannot be modeled by analytic techniques. A finite-difference scheme is used here to simulate the thermal and mechanical evolution of the Baltimore Canyon trough.

The simulation technique is designed to predict the temperature and subsidence of each point in the cross section of a two-dimensional basin. A two-dimensional basin or trough is one that is very long with respect to its width and depth. An initial temperature structure of the margin must be specified. For the model shown, the initial conditions were those appropriate for the start of sea-floor spreading (175 m.y. ago).

The simulation uses the full two-dimensional heat-conduction equation (Carslaw and Jaeger, 1959).

$$\rho C_p \frac{\partial T}{\partial t} = \frac{\partial}{\partial x} k \frac{\partial T}{\partial x} + \frac{\partial}{\partial z} k \frac{\partial T}{\partial z} - \rho C_p V \frac{\partial T}{\partial z} \quad (4)$$

where z = depth, x = horizontal distance, T = temperature, k = thermal conductivity, C_p = heat capacity, V = subsidence velocity

It does so by dividing the evolution of the margin into short time periods (Δt) and solving the finite-difference equivalent of the conduction equation for

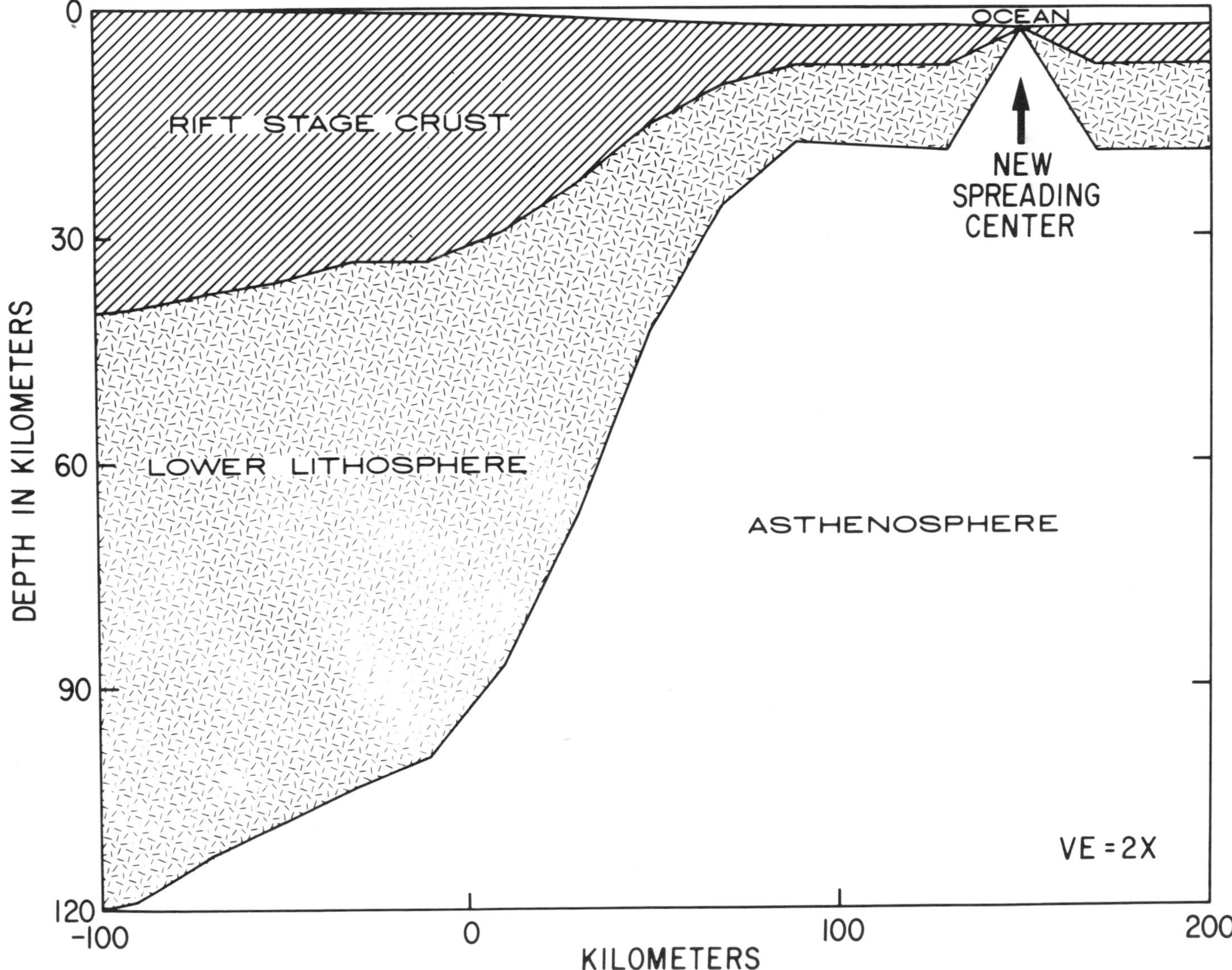

Figure 10 — The initial configuration of the Baltimore Canyon trough determined by thermal modeling. The crust-lithosphere and lithosphere-asthenosphere boundaries are shown. All isotherms would be parallel to these boundaries. The cross section is along seismic line 25.

each time step. The numerical method and choice of parameters are described in more detail in the appendix. The entire history of a basin is stepped through in this manner. At any point, the process may be stopped and the structure of the basin examined for reasonableness. For example, if sediment was being deposited on an area that the model predicts was elevated far above sea level, then the conditions that led to this model must be considered unrealistic and the model must be discarded or modified.

For the model shown here, the physical properties of all the earth materials were chosen and fixed. The distribution of extension across the margin and the original continental crust thickness were unknowns. The sedimentation rates are calculated from sediment layer thickness observations in wells and on reflection seismic sections. To be considered successful, a model must predict the present topography of the margin. It should also be consistent with the thickness of rift-stage crust under the margin predicted by gravity modeling and agree with the position of the oceanic to rift-stage crust transition given by the ECMA. Finally, the model must be geologically reasonable as described above. The model discussed below seems to be the best fit for the Baltimore Canyon trough along Line 25.

The new spreading center is predicted to have formed approximately 20 km oceanward of the B-2 well location (Figures 2 and 10) after a period of rapid extension. This new center is slightly east of the point indicated by the gravity and magnetic data to be the boundary between oceanic and rift-stage crust, but is consistent with the subsidence histories in both the B-2 and B-3 wells. The new ridge is predicted to have been at a depth of 2.5 km, in accordance with observations (Parsons and Sclater, 1977). The surface of the continental crust is smooth in the model (Figure 10). Actually, it is more likely to be block faulted in a half-graben manner. The distribution of this irregular surface should not significantly affect the long-term thermal history of the basin and therefore is neglected in the calculations.

The temperature structure and sediment wedge predicted by the model for the end of the Jurassic

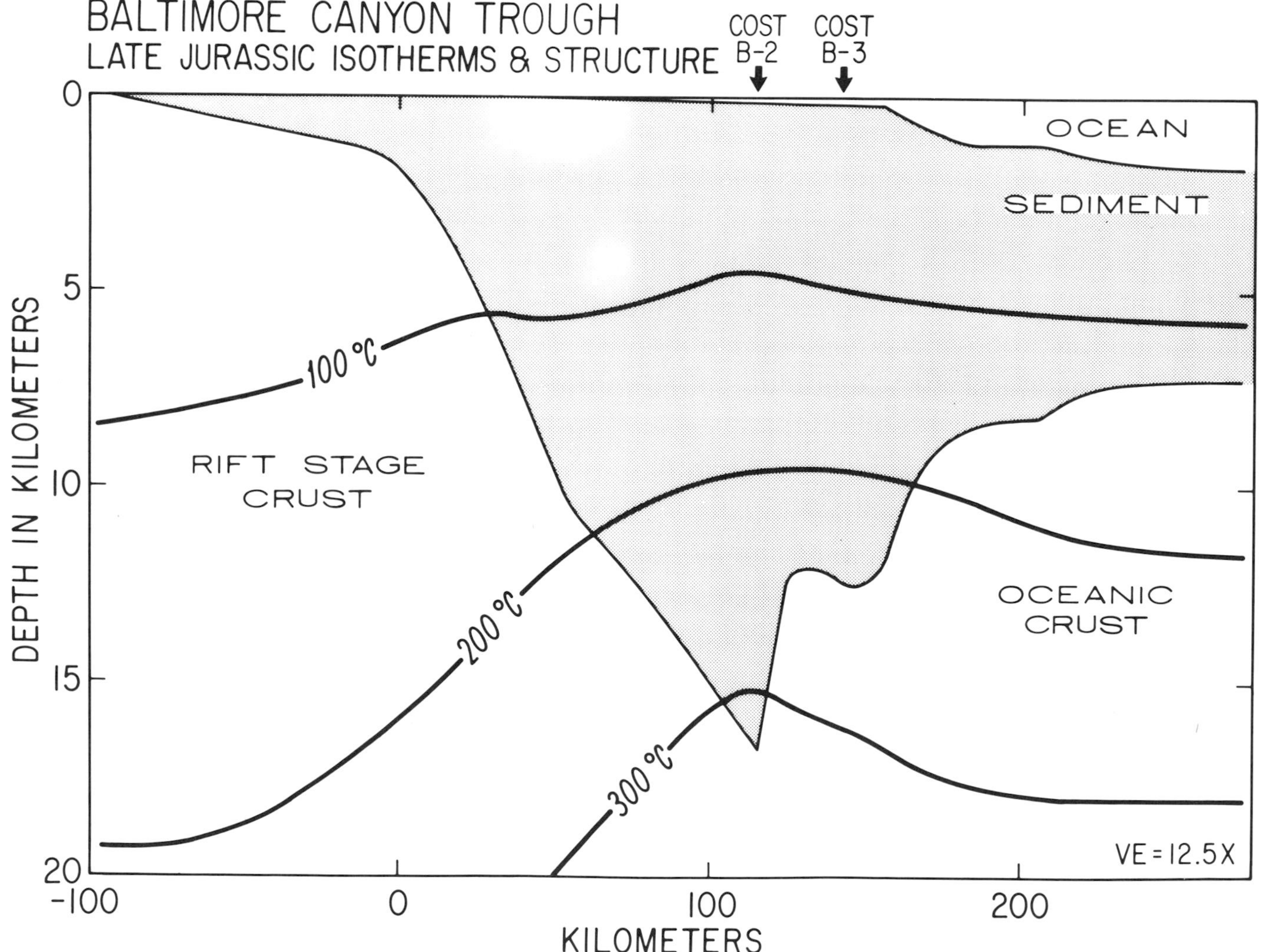

Figure 11 — Predicted temperature structure of the Baltimore Canyon trough along line 25 during the late Jurassic.

(Figure 11) show that a thick sediment pile has already accumulated. The position of the isotherms has not changed much during the later evolution of the margin (Figure 12). Seismic data indicate that the position of the shelf-slope break was oceanward of the B-3 well location during the Latest Jurassic but is presently between the wells (Figure 2). The model correctly locates the shelf edge both in the Latest Jurassic and the present (Figures 11 and 12). Most of the short-period variability in sediment thickness took place during rifting, early in the basin's history; the post-Early Jurassic sediment distribution was more gradual and even. This distribution is an expected characteristic of thermal subsidence and the extensional model. The positions of isotherms have changed only a little since the end of the Jurassic. The minor change of isotherm position indicates that the basin reached thermal equilibrium during the latter part of margin formation, as was assumed in an earlier study of passive margins (Royden and Keen, 1980).

This simulation indicates that the simple extensional model can account for many of the important features of the evolution of the Baltimore Canyon trough. The modeling was done using an Airy isostatic compensation mechanism. The amount of extension does not need to be a function of depth to account for the development of this portion of the margin, as is required further north (Royden and Keen, 1980).

The basin simulation provides time-temperature-depth information for the sediments in the basin, and these may be used to estimate hydrocarbon source rock maturity. Royden, Sclater, and Von Herzen (1980) gave an equation for a thermal alteration index (C), which is based on the empirical observation that first-order organic reaction rates double for each 10°C rise in temperature. Hydrocarbon generation is thought to be composed of many first-order reactions.

$$C = \ln \int_0^t 2^{T/10}\,dt \quad , \quad T \text{ in °C}$$

The temperature histories of representative packets of sediment are taken from the basin simulation (Figure

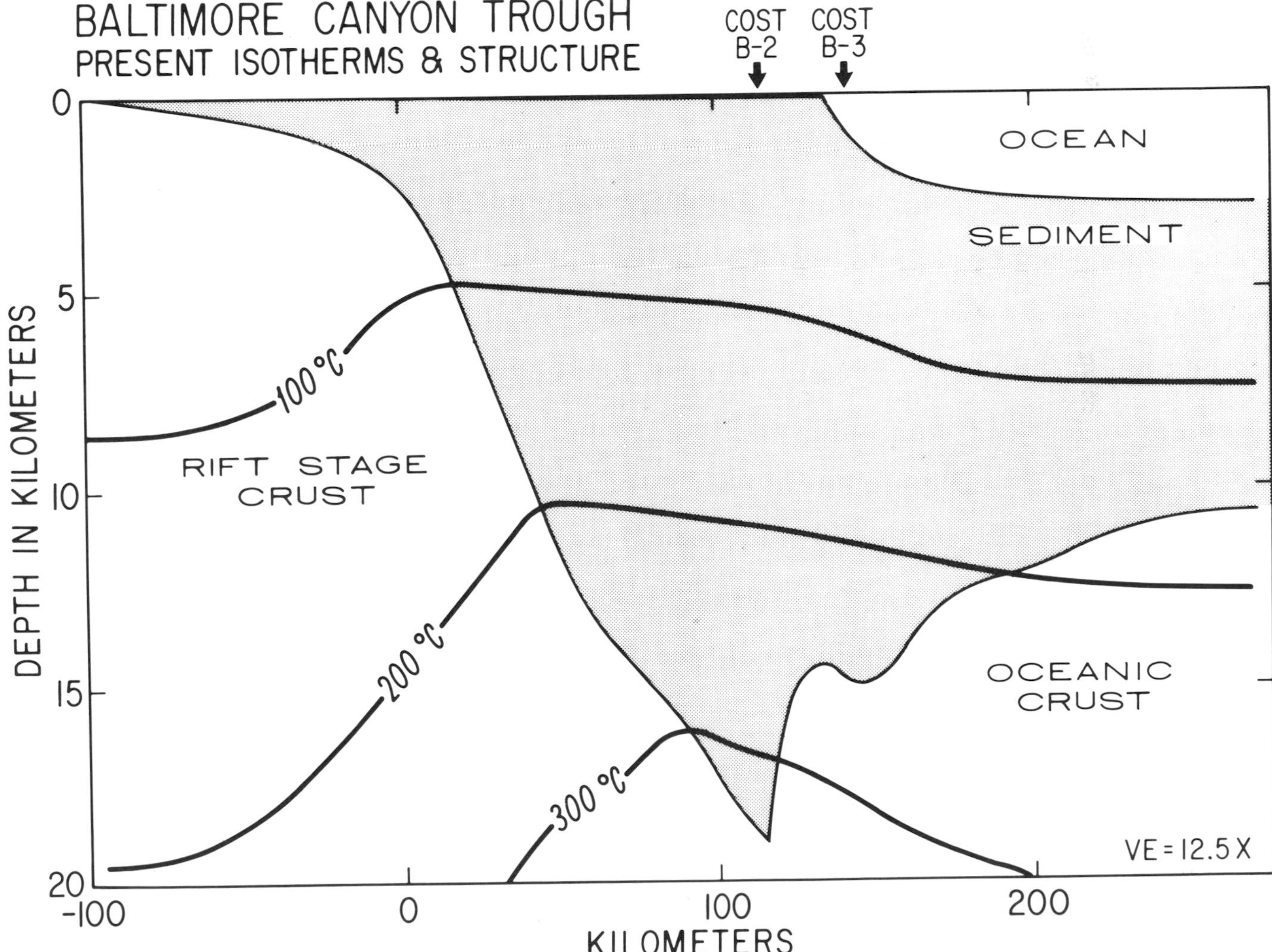

Figure 12 — The present temperature structure, predicted by the thermal model, of the Baltimore Canyon trough along line 25.

13). The C values are derived by integrating the temperature function along the time-temperature curves. The process of generating oil begins at about C=10 and continues until C=16. At values of C between 16 and 19, oil is cracked to gas, and above that point, hydrocarbons can no longer exist (Royden, Sclater, and Von Herzen, (1980). The C values predicted for one column near the COST B-3 well (Figure 13) indicate oil maturity in sediments from about Middle Jurassic to the Jurassic-Cretaceous boundary, and gas maturity in the Early Jurassic. For petroleum to have been formed in these layers, the proper chemical compounds, kerogens, must have been present. The sediments of Jurassic age in the Baltimore Canyon trough are primarily of nonmarine origin (Schlee, 1981a) and therefore are most likely to contain kerogen of type III (Miller et al, 1980). When thermally mature, these sediments are more likely to produce gas than oil (Tissot and Welte, 1978).

Vitrinite reflectance is the most widely used measure of hydrocarbon-source maturity, and it has been measured in many samples from the COST B-2 (Claypool et al, 1977) and COST B-3 (Miller et al, 1980) wells. Their values of vitrinite reflectance versus depth in each well have been converted to C values (Figure 14) by using tables prepared by Waples (1980). Values predicted by the basin simulation are overlain. The B-3 curve and data appear offset because the well is located in deeper water. The model curves are reasonably close to the observed values for each well.

A cross section of the Baltimore Canyon trough along USGS line 25 shows the predicted degree of thermal maturation of the sediments (Figure 15). The shallow regions that are predicted to be under-mature may serve as reservoirs for hydrocarbon but are unlikely to have been sources. The over-mature sediments may have been sources for hydrocarbon that later migrated upward but are not likely to be reservoirs now. The proper maturity conditions for the formation of hydrocabons would seem to have been present in the Baltimore Canyon trough, but organic kerogens and reservoir rocks are also required if significant reserves are to accumulate.

As of March, 1981, twenty-five exploratory wells have been drilled in the Baltimore Canyon trough area. Only five of these report significant hydrocarbon shows (Mattick and Ball, 1981). They are close together, about 20 km east of the COST B-2 well. Their locations project onto USGS line 25 (Figure 2) about midway between the two COST wells. Natural gas flowed at low rates from thin sandstone formations at depths ranging from 3,800 to 4,800 m in all

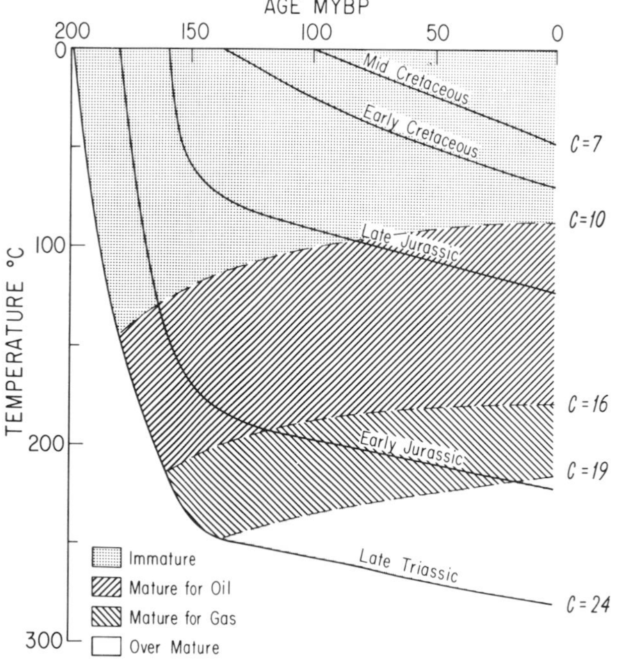

Figure 13 — The solid curves show the temperature history experienced by sediment layers deposited 200, 180, 160, 136, and 100 m.y. ago near the present site of the COST B-3 well. The shaded areas are bounded by iso-maturity (constant C) contours.

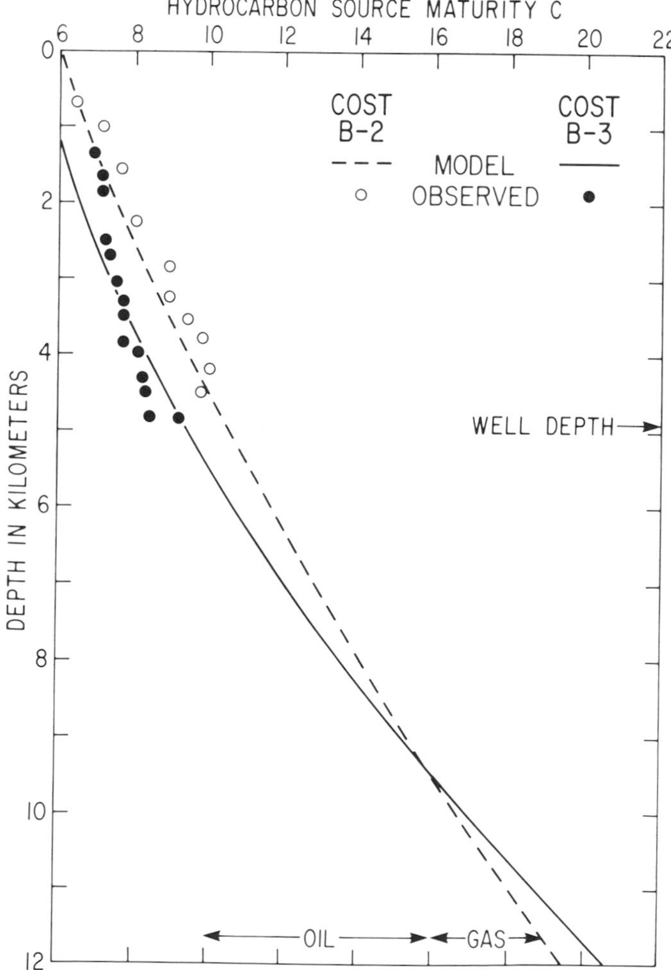

Figure 14 — Observed and predicted thermal maturity (C) versus depth for the locations of the COST B-2 and B-3 wells. Observed vitrinite-reflectance values have been converted to C values according to the tables given by Waples (1980).

five wells. One well, centrally located among the five, was tested at 2,535 m, and a small amount of oil flowed from a Lower Cretaceous sandstone bed (Mattick and Ball, 1981). All these discoveries are probably within a single structure, but it has not yet proven to be of commercial size.

The oil discovery is in Lower Cretaceous sediment that the thermal simulation of the basin indicates to consist of nearly mature oil-source rock (Figure 13). The gas-bearing horizons occur in sediments of Late Jurassic age, which should be mature for oil generation, but not for gas. It seems likely that all the oil and gas discovered so far were generated at greater depth and migrated to the sandstone reservoirs in which they are now found. Their presence supports our contention that some sediments in the Baltimore Canyon trough have experienced appropriate temperature histories for the generation of both oil and natural gas. The apparently greater amount of gas in the reservoir is consistent with the predominance of continentally derived sediment in the part of the trough that should be mature for natural gas generation.

SUMMARY

A simple, one-dimensional extensional model can explain the major features of the evolution of the northeastern United States Atlantic continental margin. Profiles of the total subsidence across the margin in the Baltimore Canyon trough and Georges Bank basin indicate that the margin can be divided into two regions. The outer part of the margin is underlain by oceanic lithosphere which, after sedimentation effects have been eliminated, behaves as predicted by the Parsons and Sclater (1977) theory. The inner part of the margin is underlain by rift-stage crust, continental crust thinned but not eliminated. The extensional model allows the prediction of the total subsidence of the margin and the subsidence history. Tectonic-subsidence histories indicate that the COST B-2 and B-3 wells are over highly-thinned continental or oceanic lithosphere, whereas the G-1 and G-2 wells are over less-thinned, rift-stage crust.

A two-dimensional thermal simulation of the evolution of the Baltimore Canyon trough confirms that the

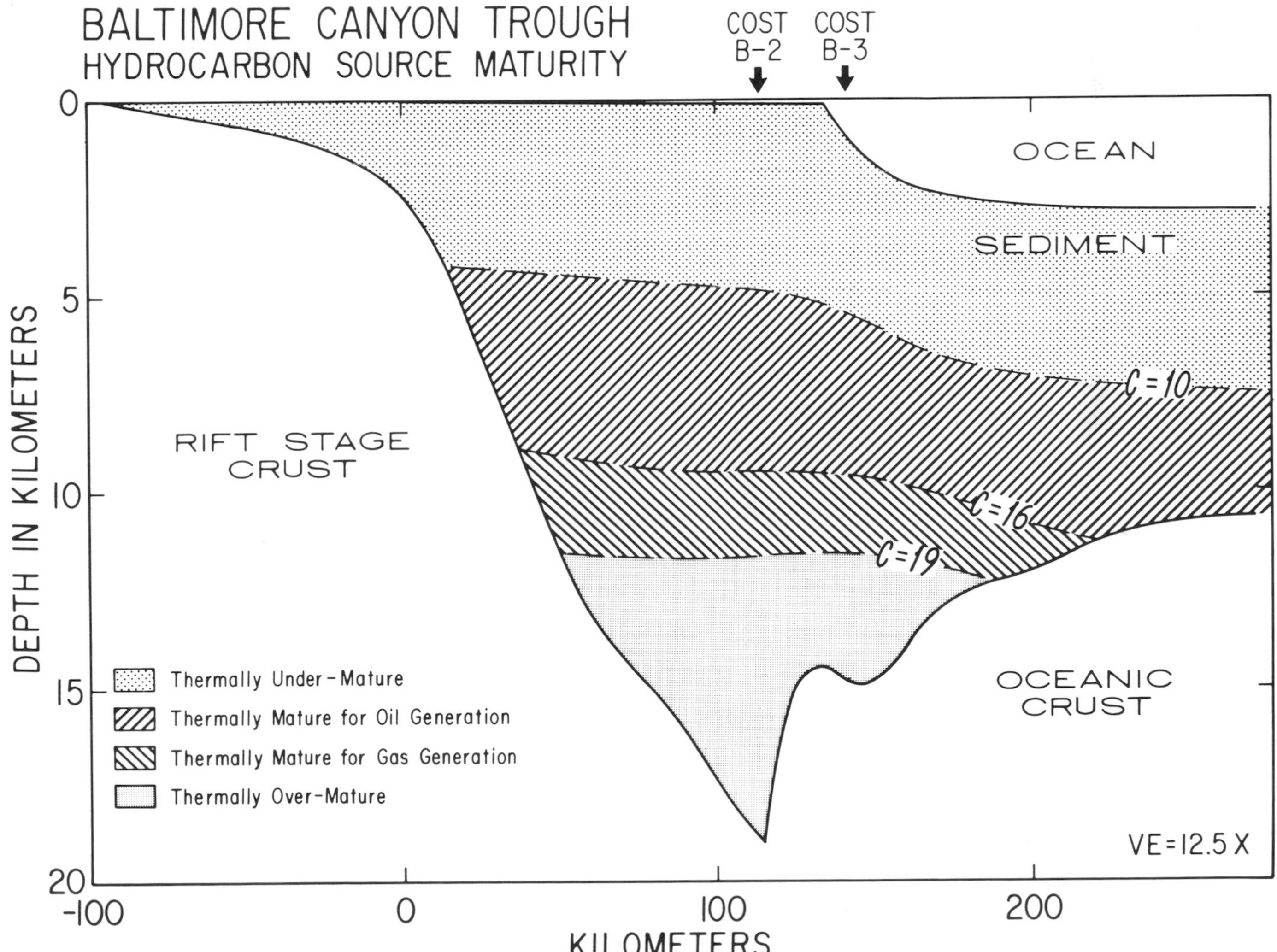

Figure 15 — Predicted contours of thermal maturity (C) in the Baltimore Canyon trough along line 25. Shaded areas indicate maturity conditions for the generation of oil and gas.

simple extensional model can explain its present and past tectonic subsidence. This model provides time-temperature histories for sediments in the basin from which their hydrocarbon-source maturity is calculated and found to agree with observed values in the COST B-2 and B-3 wells. The Baltimore Canyon trough is predicted to contain a significant volume of thermally mature sediments of probable nonmarine origin which are more likely to have produced gas than oil. Some natural gas has been found in exploratory drilling in the Baltimore Canyon trough. The wells have been drilled into sediments considered mature for oil generation but immature for gas, so the gas probably formed at greater depth and migrated upward.

ACKNOWLEDGMENTS

We appreciate helpful discussions with J. A. Grow, W. P. Dillon, and D. R. Hutchinson, and useful comments on the content of this paper by J. S. Schlee and K. D. Klitgord and reviewers Richard George and Chris Kendall. The research was supported by a grant from the United States Geological Survey, Woods Hole, Massachusetts, and by U.S. National Aeronautics and Space Administration Grant NAG 5-41.

REFERENCES CITED

Amato, R. V., and J. W. Bebout, eds., 1980, Geologic and operational summary, COST No. G-1 Well, Georges Bank area, north Atlantic OCS: U.S. Geological Survey Open-File Report 80-268, 117 p.

———, and E. K. Simonis, eds., 1980, Geologic and operational summary, COST No. G-2 Well, Georges Bank area, north Atlantic OCS: U.S. Geological Survey Open-File Report 80-269, 120 p.

Artemyev, M. E., and E. V. Artyushkov, 1971, Structure and isostasy of the Baikal rift and the mechanism of rifting: Journal of Geophysical Research, v. 76, p. 1197-1211.

Austin, J. A., Jr., et al, 1980, Geology of New England passive margin: AAPG Bulletin, v. 64, p. 501-526.

Bielak, L. E., and E. K. Simonis, 1980, Paleoenvironment analysis, *in* R. V. Amato and E. K. Simonis, eds., Geologic and operational summary, COST No. G-2 Well: U.S. Geological Survey Open-File Report 80-269, p. 29-32.

Carslaw, H. S., and J. C. Jaeger, 1959, Conduction of heat in solids, 2nd edition: London, Clarendon Press, 510 p.

Claypool, C. E., 1977, Organic Geochemistry, *in* P. A. Scholle, ed., Geological studies on the COST No. B-2 Well, U.S. Mid-Atlantic Outer Continental Shelf area: U.S. Geological Survey Circular 750, p. 46-59.

Falvey, D. A., 1974, The development of continental margins in plate tectonic theory: Australian Petroleum Exploration Association Journal, v. 14, p. 95-106.

Grow, J. A., 1980, Deep structure and evolution of the Baltimore Canyon trough in the vicinity of the COST No. B-3 well, *in* P. A. Scholle, ed., Geological studies of the COST No. B-3 Well, United States Mid-Atlantic Continental Slope area: U.S. Geological Survey Circular 833, p. 117-126.

———, C. O. Bowin, and D. R. Hutchinson, 1979a, The gravity field of the U.S. Atlantic continental margin: Tectonophysics, v. 59, p. 27-52.

———, R. E. Mattick, and J. S. Schlee, 1979b, Multichannel seismic depth sections and interval velocities over outer continental shelf and upper continental slope between Cape Hatteras and Cape Cod, *in* J. S. Watkins, L. Montadert, and P. Dickerson, eds., Geological and geophysical investigations of continental margins: AAPG Memoir 29, p. 65-83.

Hetenyi, M., 1946, Beams on elastic foundation: Ann Arbor, Michigan, The University of Michigan Press, 255 p.

Klitgord, K. D., and J. C. Behrendt, 1979, Basin structure of the U.S. Atlantic margin, *in* J. S. Watkins, L. Montadert, and P. W. Dickerson, eds., Geological and geophysical investigations of continental margins: AAPG Memoir 29, p. 85-112.

———, and J. A. Grow, 1980, Jurassic seismic stratigraphy and basement structure of western Atlantic magnetic quiet zone: AAPG Bulletin, v. 64, p. 1658-1680.

———, J. S. Schlee, and K. Hinz, 1981, Unpublished data on basement structure, sedimentation, and tectonic history of Georges Bank basin: available at U.S. Geological Survey.

Lachance, D. J., J. W. Bebout, and L. E. Bielak, 1980, Depositional environments, *in* R. V. Amato, and J. W. Bebout, eds., Geologic and operational summary, COST No. G-1 Well, Georges Bank area, north Atlantic OCS: U.S. Geological Survey Open-File Report 80-268, p. 53-58.

Mattick, R. E., and M. M. Ball, 1981, Petroleum geology, *in* J. A. Grow, ed., Summary report of the sediments, structural framework, petroleum potential and environmental condition of the United States middle and northern continental margin in area of proposed oil and gas lease sale n. 76: U.S. Geological Survey Open-File Report 81-765, p. 69-82.

McKenzie, D. P., 1978, Some remarks on the development of sedimentary basins: Earth and Planetary Science Letters, v. 40, p. 25-32.

Miller, R. E., et al, 1980, Organic geochemistry, *in* P. A. Scholle, ed., Geological studies of the COST No. B-3 Well, United States Mid-Atlantic continental slope area: U.S. Geological Survey Circular 833, p. 85-104.

Parsons, B., and J. G. Sclater, 1977, An analysis of the variation of ocean floor bathymetry and heat flow with age: Journal of Geophysical Research, v. 82, p. 803-827.

Poag, C. W., 1980, Foraminiferal stratigraphy, paleoenvironments, and depositional cycles in the outer Baltimore Canyon trough, *in* P. A. Scholle, ed., Geological studies of the COST No. B-3 Well, United States Mid-Atlantic continental slope area: U.S. Geological Survey Circular 833, p. 44-66.

Richtmyer, R. D., and K. W. Morton, 1967, Difference methods for initial value problems, 2nd edition: New York, Interscience Publishing, 405 p.

Royden, L., and C. E. Keen, 1980, Rifting process and thermal evolution of the continental margin of eastern Canada determined from subsidence curves: Earth and Planetary Science Letters, v. 51, p. 343-361.

———, J. G. Sclater, and R. P. Von Herzen, 1980, Continental margin subsidence and heat flow: Important parameters in formation of petroleum hydrocarbons: AAPG Bulletin, v. 64, p. 173-187.

Ruby, W. W., and M. K. Hubbert, 1960, Role of fluid pressure in mechanics of overthrust faulting, II overthrust belt in geosynclinal area of western Wyoming in light of fluid pressure hypothesis: Geological Society of America Bulletin, v. 60, p. 167-205.

Sawyer, D.S., et al, 1982, Extensional model for the subsidence of the northern United States Atlantic Continental margin: Geology, v. 10, p. 134-140.

Schlee, J. S., 1981a, Seismic stratigraphy of the Baltimore Canyon trough: AAPG Bulletin, v. 65, p. 26-53.

———, 1982, Seismic stratigraphy of the Georges Bank basin complex, offshore New England: this volume.
———, W. P. Dillon, and J. A. Grow, 1979, Structure of the continental slope off the eastern United States: Society of Economic Paleontologists and Mineralogists Special Publication No. 27, p. 95-117.
———, et al, 1976, Regional geologic framework off northeastern United States: AAPG Bulletin, v. 60, p. 926-951.
Scholle, P. A., ed., 1977, Geological studies on the COST No. B-2 Well, U.S. Mid-Atlantic outer continental shelf area: U.S. Geological Survey Circular 750, 132 p.
———, ed., 1980, Geological studies of the COST No. B-3 Well, United States Mid-Atlantic continental slope area: U.S. Geological Survey Circular 833, p. 1-132.
Sclater, J. G., and P.A.F. Christie, 1980, Continental stretching: An explanation of the Post-Mid-Cretaceous subsidence of the central North Sea basin: Journal of Geophysical Research, v. 85, p. 3711-3739.
Sleep, N. H., 1971, Thermal effects of the formation of Atlantic continental margins by continental breakup: Geophysical Journal of the Royal Astronomical Society, v. 24, p. 325-350.
Steckler, M. S., and A. B. Watts, 1978, Subsidence of the Atlantic-type continental margin off New York: Earth and Planetary Science Letters, v. 41, p. 1-13.
Tissot, B. P., and D. H. Welte, 1978, Petroleum formation and occurrence: New York, Springer-Verlag, 538 p.
Vail, P. R., et al, 1977, Seismic stratigraphy and global changes of sea level, *in* C. E. Payton, ed., Seismic stratigraphy — Applications to hydrocarbon exploration: AAPG Memoir 26, p. 49-212.
Vogt, P. R., and B. E. Tucholke, 1979, The New England Seamounts: Testing origins, *in* B. E. Tucholke, et al, Initial reports of the deep sea drilling project, v. 43: Washington D.C., U.S. Government Printing Office, p. 847-856.
Waples, D. W., 1980, Time and temperature in petroleum formation: Application of Lopatin's method to petroleum exploration: AAPG Bulletin, v. 64, p. 916-926.
Watts, A. B., J. H. Bodine, and M. S. Steckler, 1980, Observations of flexure and the state of stress in the oceanic lithosphere: Journal of Geophysical Research, v. 85, p. 6369-6376.
———, and M. S. Steckler, 1979, Subsidence and eustasy at the continental margin of eastern North America: American Geophysical Union Maurice Ewing Series 3, p. 218-234.

APPENDIX

An explicit, time-stepping, centered finite-difference approximation is used to model the two-dimensional heat conduction equation (Equation 4) (Richtmyer and Morton, 1967). As an example, the approximation of the first conduction term from Equation 4 is:

$$\frac{\partial}{\partial x} k \frac{\partial T}{\partial x} = \frac{k^1_{m+1/2}\,(T^1_{m+1} - T^1_m) - k^1_{m-1/2}\,(T^1_m - T^1_{m-1})}{\Delta x^2}$$

where: k^1_m is the conductivity at $x = m\Delta x$ and $z = 1\Delta z$
and
T^1_m is the temperature at $x = m\Delta x$ and $z = 1\Delta z$

The size of the time step is governed by a stability condition, which is a function of the vertical grid spacing, horizontal grid spacing, maximum thermal conductivity (k), and maximum subsidence velocity (V).

$$\Delta t < \left[2\left(\frac{k}{\rho C_p}\right)\left(\frac{1}{\Delta x^2} + \frac{1}{\Delta z^2}\right) + \frac{V}{\Delta z}\right]^{-1}$$

where: $k = 7.5\times10^{-3}$ cal/cm °C sec (3.2 W/m °C)
$C_p = 0.3$ cal/gm °C (0.07 joules/gm °C)

The time step is 5,000 years, and the grid spacings are 1 km vertically and 10 km horizontally. The depth of isostatic compensation and the bottom boundary of the grid are at 125 km. The bottom boundary is maintained at a constant temperature of 1,350°C to simulate the top of a convecting fluid asthenosphere. This temperature and the 125 km thickness are taken from Parsons and Sclater's (1977) observations of the base temperature and thickness of oceanic lithosphere. For most of the evolution of a model basin, no horizontal heat flow is allowed through the side boundaries. Early in its history, one side of the margin was bounded by a spreading center which gradually moved away. During this time, the horizontal heat flow condition for the side boundary is inappropriate. The effect of this process is modeled by a constant-temperature (1350°C) side-boundary condition, imposed on a boundary moving away at the appropriate rate. When the spreading center has moved away from the margin, the normal, zero-horizontal-heat-flow boundary condition is substituted. The sediment surface maintains a constant temperature of 0°C. The variation of bottom-water temperature from 0°C has been neglected here because its effect would be small and would introduce unlimited complexity.

The conduction equation predicts the temperature change at each grid point during a time step. Paralleling this conduction calculation, the effects of thermal contraction, sediment deposition, sediment compaction, and sea level changes are determined.

Thermal contraction of a layer is calculated using the temperature before and after the time step and a given thermal expansion coefficient (α) for the material.

$$\Delta Th = -\alpha \times \Delta T \times Th$$

Because the basin is generally cooling, thermal contraction is responsible for most of the basement's thermal subsidence. The density of the cooling material is increased in inverse proportion to its decrease in volume.

The sedimentation rates and compaction properties of the sediments are derived from seismic and well data. The thickness and time interval of deposition for each identifiable layer across the basin are converted to rates of sediment deposition. The variation in compaction with depth is taken into account, an important correction to make prior to comparing sedimentation rates through time. The sediment will be deposited as a thicker, more porous layer and then will be gradually compacted to reach its present porosity and thickness. The only sediment assumed to have been deposited on the margin is sediment presently on the margin. Thus, times of unconformities in the column are falsely reflected as periods of non-deposition.

This assumption does not lead to significant errors in the simulation for two reasons. The sedimentation rates are fairly low for these basins so great thicknesses of sediment probably have not been removed even if the time gaps are long. The time constant for thermal processes is long ($\approx$60 million years) so the effect on temperatures of deposition for 10 million years followed by erosion for 10 million years is not radically different from nondeposition for 20 million years. The appropriate amount of new sediment is added to the top of the columns of material, its initial temperature is assumed to be 0°C, and the sediment below is assumed to be compacted according to Equation 3. The sediment is assumed to have an initial porosity (ϕ_0) of 55%, porosity change coefficient (c) of $4.5x10^{-4}m^{-1}$, and grain density (ρ_{sg}) of 2.65 gm/cu cm. These values are typical of a sandy shale and are the average of the parameters from the individual lithology observations in the COST B-2 and B-3 wells. The thermal conductivity of the sediments significantly affects the subsurface temperatures throughout the evolution of a basin. It also affects the rate of heat loss from the cooling lithosphere and therefore the rate of thermal subsidence. Thermal conductivity of sediments is primarily a function of their porosity. Sclater and Christie (1980) gave data for the thermal conductivity of North Sea shale versus its porosity. This data can be approximated by a linear function of porosity given by this relation:

$$k = (-4.4 \phi + 5.35) \mu cal/cm\ s\ °C$$

When combined with Equation 3 for porosity versus depth and the sediment compaction parameters above, thermal conductivity is given as a function of depth of burial. The lithosphere is required to remain in Airy isostatic balance. After sediment is deposited, the mass of each column is integrated. If the calculated mass of the column is too great, then it subsides to lighten the column and restore the isostatic balance.

Changes in relative sea level have a small effect on the subsidence of the margin because the water adds to the load on submerged columns, but not to the load on columns elevated above sea level. In the simulation, the Watts and Steckler (1979) eustatic sea level curve is used.

A Thermal-Mechanical Model of Rifting with Implications for Outer Highs on Passive Continental Margins

R. C. Vierbuchen
R. P. George
P. R. Vail
Exxon Production Research Company
Houston, Texas

We propose a thermal-mechanical model of the rifting process that differs from previous models of lithospheric thinning (McKenzie, 1978; Sclater and Christie, 1980) by including the effect of the mechanical heterogeneity of the lithosphere. We divide the prerift continental lithosphere into ten horizontal mechanical layers that obey power-law creep. The ten layers are deformed in finite steps by extensile forces that remain constant at great distances from the rift. Power-law creep during extension causes necking of the lithosphere at mechanical instabilities beneath the rift axis. Thus, aesthenospheric upwelling is concentrated at an ever greater rate at the rift axis. We compute surface topography by assuming local isostatic compensation. At the rift axis, thinning of the crustal layers causes subsidence initially, but in some cases acceleration of aesthenospheric upwelling eventually causes thermal uplift late in the rifting process. This deformation sequence can explain the origin of one class of outer highs observed on passive continental margins (Schuepbach and Vail, 1980). Our model is consistent with data on mechanical properties of the lithosphere and with geological knowledge of rift valleys and passive margins. Furthermore, it satisfies our measurements of the scale and timing of formation of outer highs.

We suggest that many of the differences among the crustal structure and subsidence histories of passive continental margins may result from the variety of mechanical properties, geothermal gradients, and strain rates that existed during rifting, and not necessarily from fundamentally different rifting mechanisms. We show the results of a finite-element rifting model that predicts considerable variety in crustal structure and subsidence history from rifting caused by lithospheric stretching. For example, we show that outer highs on rifted margins can occur under certain conditions and will not occur under other conditions. The structure of passive continental margins is also determined in part by postrift sedimentation, such as the presence or absence of carbonate buildups. However, we concentrate on the aspects of margin structure that are related to the rifting process.

A simplified cross section of offshore eastern Canada (Figure 1) illustrates crustal thinning at rifted passive margins. Seismic reflection profiles constrain the upper crustal structure, and depth to Moho is constrained by a refraction line that crosses the section and by gravity models. Several points should be made:

1. The continental crust thinned from over 30 km to an average of 20 km and in places to 10 km.
2. The upper surface of the basement has a horst-and-graben topography, indicative of brittle behavior during extension.
3. The Moho has relatively smooth topography, suggestive of ductile behavior during extension. Although refraction and gravity modelling do not accurately resolve the relief on the crust-mantle interface, we conclude from these data that this relief is less than that on the sediment-basement interface defined by seismic reflections. Ductile behavior at depths greater than 15 km and at temperatures greater than 450°C is consistent with the extrapolation to geologic strain rates of creep experiments on crustal and mantle rock types (Carter, 1976; Kirby, 1977; Paterson, 1978; Tullis, 1979).
4. The brittle deformation in the upper crust seems to be concentrated in a zone that is at most 150 to 200 km wide; but the total width of thinned crust is over 500 km. Such a distribution of strain suggests that the total displacement on basement faults is not a direct measure of the total strain in

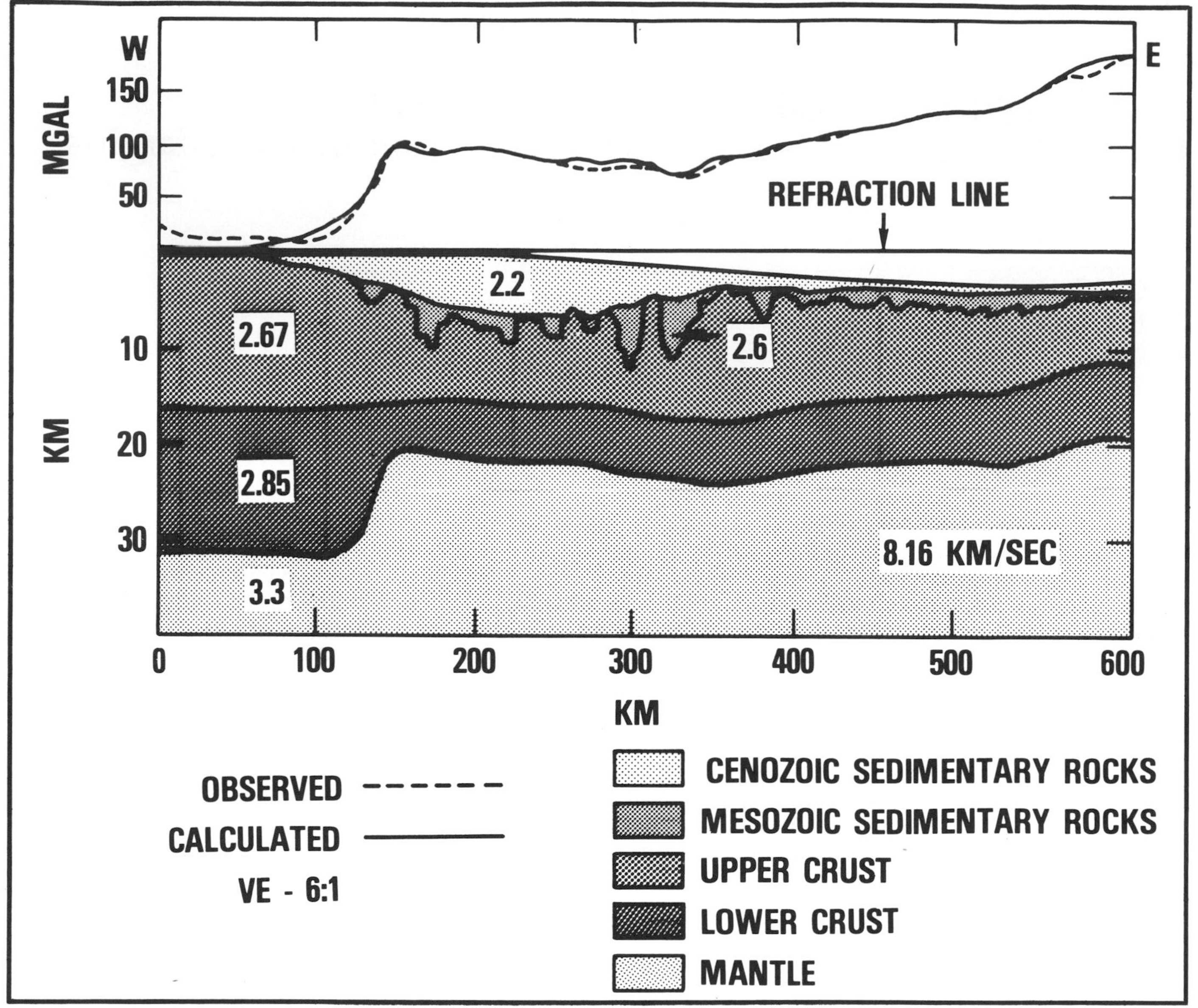

Figure 1 — East-west gravity profile (top) and geophysically-controlled cross section (bottom) across the eastern margin of Canada. Unlabeled numbers are rock densities used to calculate the Bouguer gravity model. Shallow crustal structure is controlled by seismic reflection profiles. Deep crustal structure and depth to Moho is controlled at one point by a seismic refraction profile that crosses the plane of this cross section at a high angle.

the lithosphere. This observation, that the strain in the brittle crust is concentrated into narrower zones than the zone affected by ductile deformation, is also consistent with the topography of the east African rift system where the zone of crustal doming is wider than the total width of rift valleys (Burke and Wilson, 1972; Burke and Whiteman, 1973).

5. In Figure 1, the transition from normal crust to thinned crust is abrupt. Keen and Hyndman (1979) show that in other parts of the eastern Canada margin, the transition is more gradual.

In summary, crustal thinning occurred. The upper crust is brittle and the lower crust is ductile. Distribution of strain and total amount of strain are different in the upper and lower crust. In short, continental crust is mechanically inhomogeneous during rifting.

INHOMOGENEOUS STRETCHING MODEL

Our work builds on McKenzie's (1978) concept of lithospheric stretching (Figure 2). In this paper, McKenzie made some simple assumptions: (1) instantaneous stretching; (2) homogeneous stretching of the entire lithosphere (the stretching factor, β, is constant with depth and constant with distance across the rift); (3) Local isostatic compensation; and (4) a linear geotherm reaching a maximum temperature of 1333°C at the base of the lithosphere.

His model has several notable results: (1) it predicts

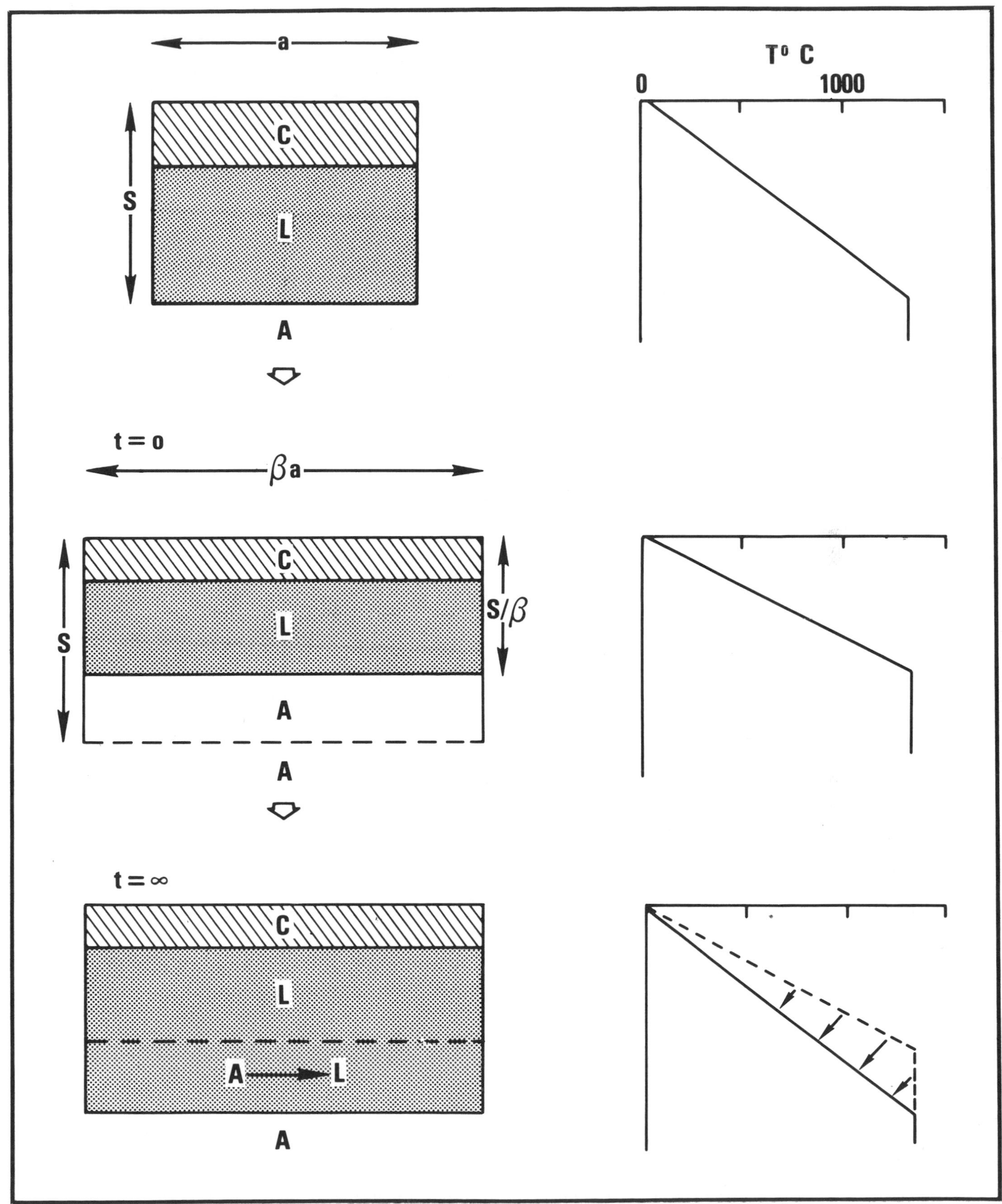

Figure 2 — Instantaneous stretching of the continental lithosphere (modified after McKenzie, 1978). C = continental crust, L = mantle portion of lithosphere, A = aesthenosphere, a = initial width, S = initial thickness, β = stretching parameter.

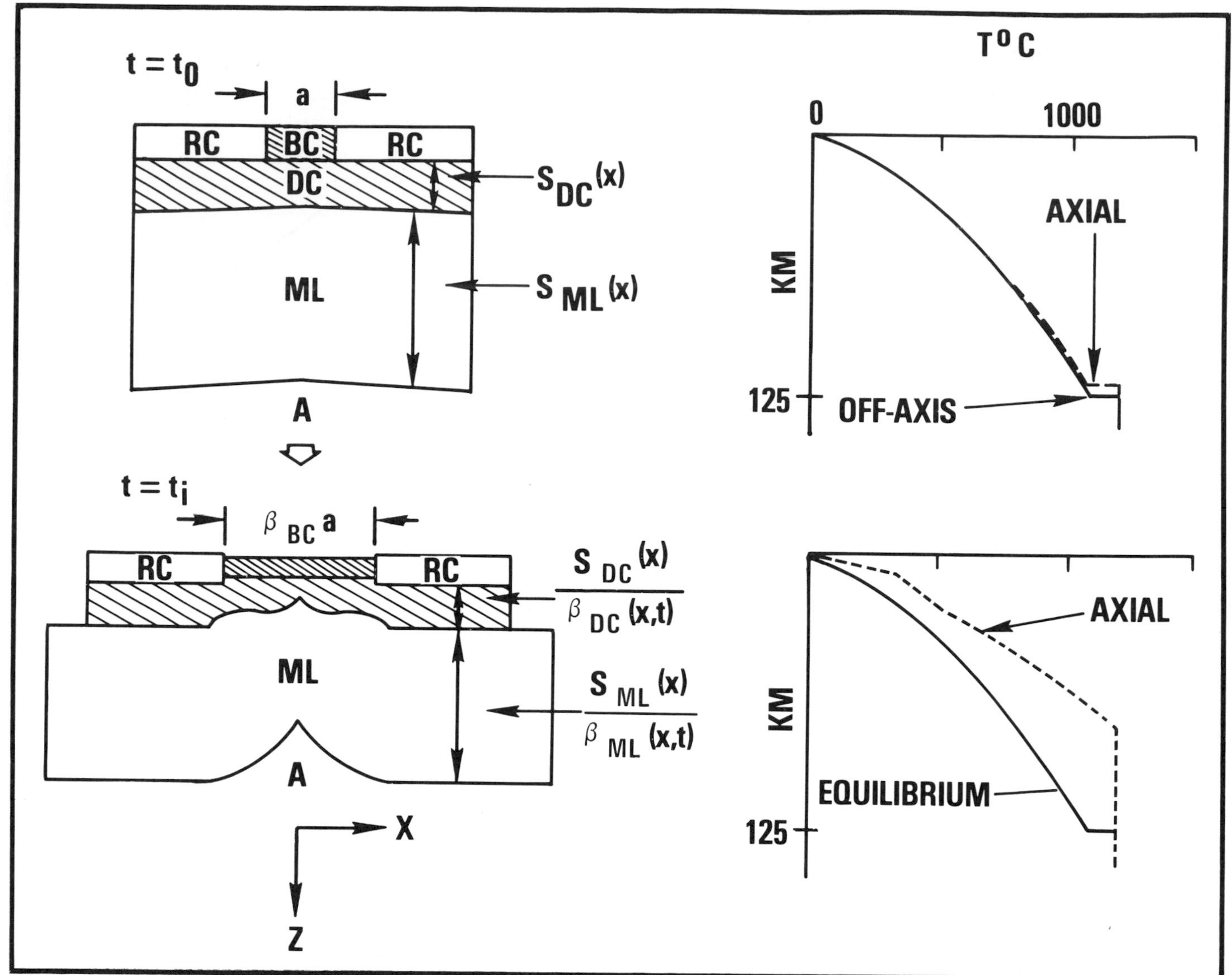

Figure 3 — Inhomogeneous stretching of the continental lithosphere. BC = brittle crust; RC = rigid crust; DC = ductile crust; ML = mantle portion of lithosphere; A = aesthenosphere, $S_{DC}(x)$ and $S_{ML}(x)$ = original thickness of ductile crust and mantle portion of lithosphere, respectively (both are functions of x, the distance from the rift axis); $\beta_{DC}(x,t)$ and $\beta_{ML}(x,t)$ = stretching parameters for ductile crust and mantle lithosphere (functions both of position and of time). Note that the curved geotherm of Mercier and Carter (top right) develops kinks at layer boundaries (bottom right), owing to differential advection of heat in different layers. Step in geotherm at lithosphere-aesthenosphere boundary reflects the assumption of constant average temperature in the aesthenosphere.

realistic amounts of postrift subsidence; (2) the heat flow and subsidence are directly controlled by the amount of stretching and loading; (3) uplift occurs only when the original crustal thickness is less than 18 km; and (4) uplift above sea level can never occur. This last conclusion implies that an outer high at the edge of a passive margin, as defined by Schuepbach and Vail (1980), cannot form if rifting occurs by lithospheric stretching as described by McKenzie.

The mechanical properties of the lithosphere change with depth because mineralogy, temperature, and pressure change. We attempted to evaluate the effects of this change on the rifting process by modelling the continental lithosphere as four major groups of mechanical facies (Figure 3). We have substituted time-dependent rheologies for McKenzie's instantaneous stretching and homogenous lithosphere. Thus, in this model, the lithospheric layers have different stretching factors (β) at different times and positions, consistent with conclusions of Keen and Barrett (in press). We assume local isostatic compensation of the lithosphere. At the strain-rates, crustal thicknesses, and temperatures considered here, we find that this assumption does not introduce large errors into the model. Finally, we have used the Mercier and Carter (1975) continental geotherm to assign the initial distribution of temperatures at layer boundaries.

From this model we see that: (1) calculated rift and postrift subsidence and subsidence rates agree with these observations; (2) the calculated subsidence is controlled by stretching, loading, mechanical properties, and strain-rate contrasts between layers; (3) uplift can occur at any original crustal thickness, even 35 km, for the right combination of variables; (4) uplift above sea level can occur.

Description of the Model

In our finite-element model, the lithosphere is divided into ten sub-horizontal layers, the lower nine of which deform independently of one another (Figure 4). The upper layer models the sedimentary accumulation, and it strains in concert with the subjacent brittle crustal layer. The brittle crust comprises three layers, each 5 km thick initially; the ductile crust comprises two layers, each 10 km thick initially; and the mantle lithosphere comprises four layers, each with different initial thicknesses (15 km for the uppermost, 20 and 25 km for the intermediate layers, and 30 km for the lowermost layer). The total initial thickness of the lithosphere is 125 km, and the boundaries of the layers are assumed to be isotherms. Although compositional boundaries such as the Moho need not parallel isotherms, by assuming that they do we can study the effect of mechanical heterogeneity in the simplest case.

Extensile forces are applied to the ends of each of the nine lithospheric layers and are held constant throughout the computer run. Deformation causes the layers to thin; initial instabilities in layer thicknesses (described below) cause the layers to neck at the rift axis, as discussed below. Only horizontal stresses are used to compute the flow rate, which introduces small errors near the ridge crest. We terminate each model when the basal (aesthenospheric) isotherm intersects the peridotite solidus, generating partial melt and thus invalidating the assumptions of the model. We presume this is geologically equivalent to changing from rifting to drifting.

The ductile layers of the crust and mantle are assigned flow-laws that are consistent with extrapolations of experimentally determined behavior of appropriate rocks to the temperatures, pressures, and strain rates of the natural deformation that we are modelling (Table 1). The flow-laws of the ductile layers are held constant throughout the run, a procedure that contains inherent assumptions and approximations as follows:

Experimentally induced ductile flow of silicates follows flow-laws of the form

$$\dot{\varepsilon} = b\,[\sinh(c\sigma)]^n\,\{\exp[-(E^* + PV^*)/RT]\},$$

where $\dot{\varepsilon}$ = strain rate (sec^{-1}), b and c = constants dependent on the material, σ = differential stress (dynes/cm^2), n = the stress exponent (a constant typically between 2 and 5), P = hydrostatic stress (dynes/cm^2), E^* = the activation energy for creep, V^* = the activation volume for creep, R = the gas constant (8.31×10^7 erg/gr mole °K), and T = temperature in °K (Carter, 1976). We assume for each layer in the mantle that the same creep mechanism operates throughout the lithospheric stretching event; that is, b, c, n, E^*, and V^* remain constant within each layer. We ignore the effect on the flow-law of depressurization of lithospheric layers as they thin and move upward toward the surface. For example, the term PV^* for the deepest lithospheric layer (the one that depressurizes the most) changes from 4.58×10^{11} erg/mole at 125 km to 1.47×10^{11} erg/mole at 40 km. The total change (3.11×10^{11} erg/mole) is, in the worst case, only 10% of the activation energy (3.97×10^{12} erg/mole). We neglect conductive migration of the isotherms that define layer boundaries, because advective transfer of heat will be more rapid than conductive transfer of heat at the strain rates that we consider in this model. The average temperature of each layer remains constant, so we treat the exponential factor within the curved brackets as a constant for each layer in the mantle. Similar arguments apply to the ductile crustal layers as well.

The flow-laws can be further simplified if we note that $\sinh(c\sigma) \cong c\sigma$ for values of $c\sigma \leq .25$. This condition is fulfilled for values of $\sigma \leq 6.25 \times 10^8$ dynes/cm^2 (625 bars). Using Kirby's (1977) values (Table 1) at a strain rate of 10^{-14}/sec and a temperature of 500°C, stresses arc of the order of 50 bars (see also Mercier, Anderson, and Carter, 1977). Thus, the expression can be further simplified to the following:

$$\dot{\varepsilon} = k\sigma^n ,$$

where k = a constant within each layer (b • exp $[(E^* + PV^*)/RT]$).

Furthermore, since the existensile stress

$$\sigma = \frac{F}{A}$$

Table 1. Data for flow-laws.

MANTLE ROCKS	$\dot{\varepsilon} = b\,[\sinh(c\sigma)]^n\{\exp[-(E^* + PV^*)/RT]\}$					
Experimental Specimens	b (1/sec)	c (cm²/dyne)	n	E*(erg/mole)	V*(cm³/mole)	Reference
Dry dunite	8.30×10^9	4×10^{-10}	3.6	3.97×10^{12}	11	Kirby (1977)

LOWER CRUSTAL ROCKS	$\dot{\varepsilon} = A\,\sigma^n\{\exp[-(E^* \times PV^*)/RT]\}$				
Experimental Specimens	A[(cm²/dyne)ⁿ/sec]	n	E*(erg/mole)	V*(cm³/mole)	Reference
Dry quartzite	2.29×10^{-27}	2.86	1.51×10^{12}	—	Koch, Christie, and George (1980)
Dry α-quartzite	$1.1(\pm.4) \times 10^{-17}$	2	1.68×10^{12}	—	Shelton and Tullis (1981)
Dry albite	$1.2(\pm.8) \times 10^{-33}$	3.9	2.35×10^{12}	—	Shelton and Tullis (1981)
Dry anorthite	—	—	2.39×10^{12}	—	Shelton and Tullis (1981)
Dry clinopyroxenite	$1.10(\pm.2) \times 10^{-17}$	2.6	3.35×10^{12}	—	Shelton and Tullis (1981)
Dry Maryland diabase	$3.5(\pm3) \times 10^{-28}$	3.4	2.60×10^{12}	—	Shelton and Tullis (1981)
Dry Maryland diabase	—	3±.3	—	—	Caristan and Goetze (1978)

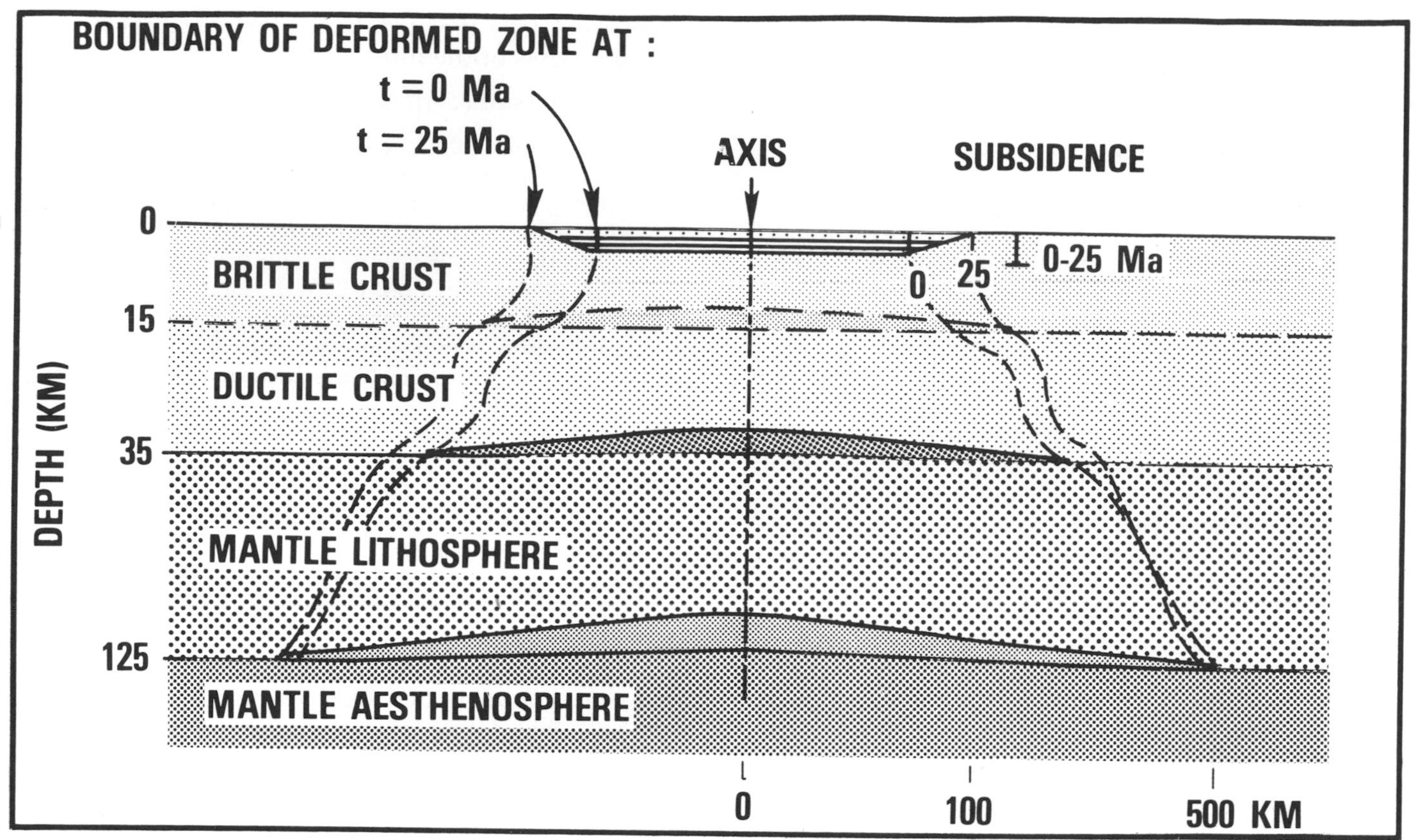

Figure 4a — Schematic diagram of evolution of rift, 0 to 25 Ma after onset of rifting. Note non-linear depth scale; horizontal scale is also non-linear. Thinning of crust dominates, causing subsidence and accumulation of sediments; zone of deformed lithosphere slowly broadens.

(where F is the extensile force, held constant through a run, and where A is the cross sectional area of a layer), it follows that

$$\dot{\varepsilon} = k'\left(\frac{A_0}{A_{j\ell}}\right)^n$$

where $k' = k(\sigma_0)^n$, A_0 equals the original cross sectional area of the layer at the *end* of the layer, and $A_{j\ell}$ equals the current cross sectional area of the layer at the jth position and the ℓth time step. The ratio $A_0/A_{j\ell}$ varies in time, because the layer becomes thinner during extension, and in position, because the initial geometry of the layers includes a taper from normal thicknesses at 1000 km from the rift axis to 90 or 95% of normal thickness at the rift axis. Note that the assumed taper is equivalent to assuming only a modest (5.3 to 11.1%) increase in geothermal gradient along the rift axis at the onset of rifting.

When an extensional force is applied to the ends of a slightly necked layer, the instability at the center acts as a stress concentrator that is inversely proportional to the reduction in cross sectional area of the layer. Because of power-law creep, as the instability grows the strain rate in the neck accelerates to the nth power of the stress. Consider a layer with an initial instability of 5%, that is, the cross sectional area of the center is 5% less than at the ends. If we apply a force at the ends of the layer, the stresses at the center will be slightly higher than at the ends, because the force is acting on a smaller cross sectional area at the center. In the time that the ends have strained 10%, power-law creep causes the center to strain 12%, because the stress exponent is 3.6. In a 10% instability, power-law creep causes an even greater disparity between strain at the center and strain at the ends. The ratio A_o/A_i is 1.11, which raised to the 3.6 power is 1.46. In other words, the center strains 1.46 times more than the ends during that time.

The simplification of neglecting thermal conduction makes the results of the numerical model independent of the absolute opening rate or strain rate. In the examples discussed below, we choose to normalize the strain rates in the ductile layers by the strain rate in the brittle crust, which is held constant at 10^{-15}/sec. This strain rate is high enough for advection to dominate conduction and also produces a geologically reasonable, half-opening rate of 3 mm/yr at a reference point 100 km from the rift axis. Upper crust that is initially more than 100 km from the rift axis is treated as rigid (Figure 3); thus a point on the brittle crust initially 1000 km from the rift axis also migrates at a rate of 3 mm/yr from the rift axis.

The extensile force applied to the lower crust is the one that produces an initial displacement rate in the lower crust equal to that of the brittle crust at 1000 km from the rift axis. This procedure prevents differential flow between crustal layers at the ends of the model and is geologically equivalent to a smooth distribution

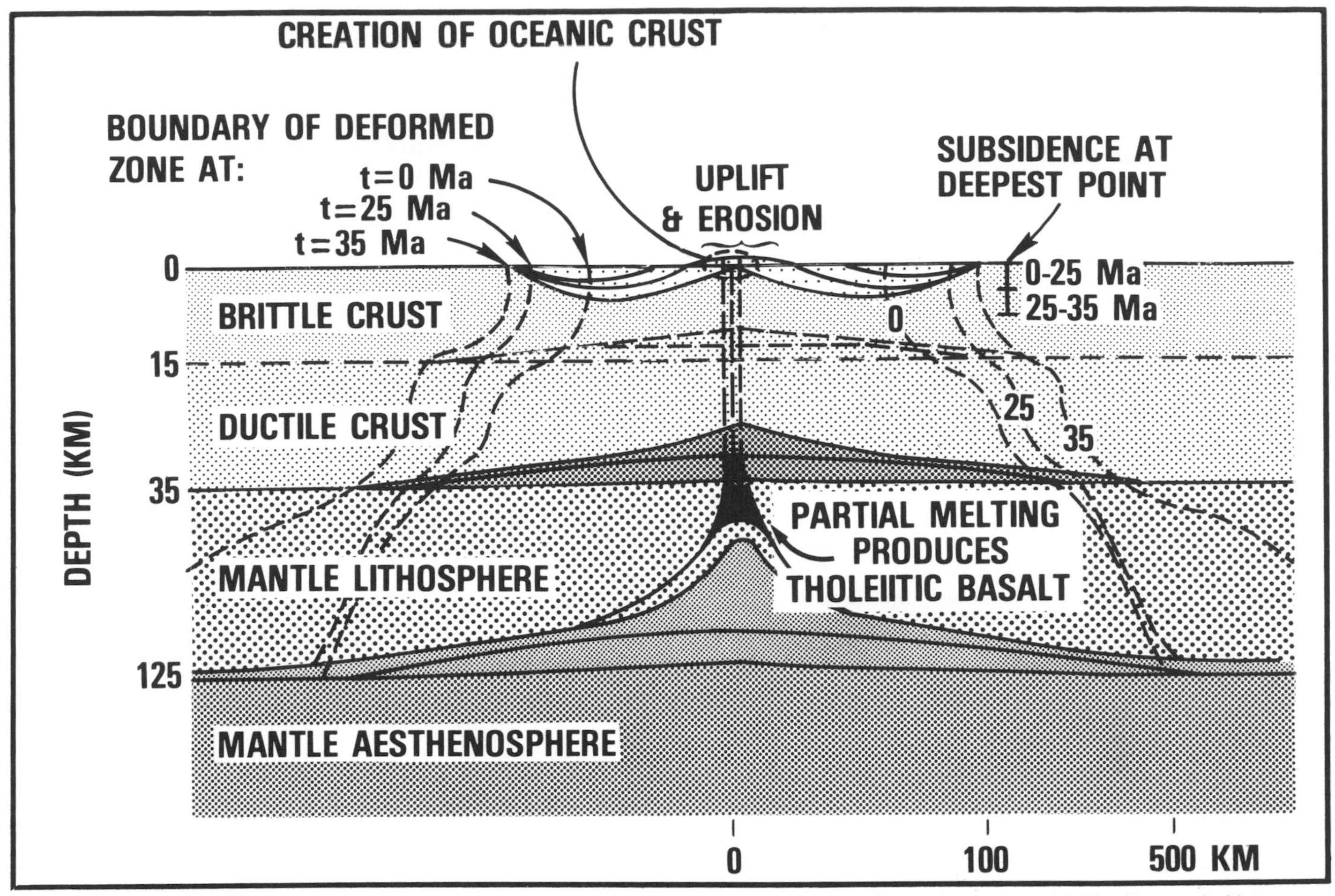

Figure 4b — Schematic diagram of evolution of rift, 25 to 35 Ma after onset of rifting (overlain on 0 to 25 Ma diagram). Note non-linear depth scale; horizontal scale is also non-linear. Thinning of crust continues, but extreme thinning of mantle dominates at the axial position (note aesthenospheric "spike" or diapir), causing local uplift and erosion. Eventually, aesthenospheric spike undergoes partial melting (black area). On flanks of axial uplift, subsidence continues.

of flow in the lower crustal rocks over a zone with a 1000-km half-width. Hence, the average initial strain rates in the ductile crust are a factor of 10 smaller than in the deforming portion (100-km half-width) of the brittle crust.

The choice of extensile force applied to the mantle layers is relatively unconstrained. The magnitudes of extensile forces that produce reasonable initial strain rates are partially dependent on the shape and scale of mechanical instabilities. In the test cases described below, the extensile forces produce initial strain rates in the mantle twice that of the ductile crust at 1000 km from the rift axis.

Isostatic subsidence is calculated with an equation that includes a term for keeping the sedimentary accommodation (the space between sea level and the sediment-water interface at the end of the previous iteration) three-quarters full of sediment:

$$SUB_{\ell+1} = \frac{\sum_{k=1}^{10} \rho_k \Delta S_k - \rho_A \sum_{k=1}^{10} \Delta S_k + \frac{3}{4}(\rho_S - \rho_w) WD_\ell}{\rho_A - \left(\frac{3\rho_S + \rho_w}{4}\right)}$$

But if uplift above sea level occurs, isostatic uplift (negative subsidence) is calculated with an equation allowing for erosion to sea level of all material uplifted above sea level:

$$SUB_{\ell+1} = \left[\sum_{k=1}^{10} \rho_k \Delta S_k - \rho_A \sum_{k=1}^{10} \Delta S_k + (\rho_S - \rho_w) WD_\ell + \sum_{k=1}^{m} \Delta S_k (\rho_m - \rho_k) \right] \Big/ \left[\rho_A - \rho_m \right]$$

(see Table 2 for description of variables).

Surface subsidence and uplift are controlled by two counteracting effects: crustal thinning, which causes subsidence, and thinning of the mantle lithosphere, which causes uplift. These effects are demonstrated by considering two simple cases: (1) Crustal thinning without concomitant mantle thinning causes subsidence of the surface ($SUB_{\ell+1}>0$). In this case, ΔS_k is negative for crustal layers and zero for mantle layers.

Table 2. Subscripts and variables.

Subscripts	
k	= layer index
ℓ	= iteration (time-step) index
A	= aesthenosphere (compensating medium)
S	= sedimentary layer
w	= water
m	= index of deepest layer touched by erosion
Variables	
ρ_k	$= \rho^\circ_k(1 - \alpha_k\bar{T}_k)$ = density of a layer *in situ;* values of ρ°_k for k = 2, 3, 4: 2.67, 2.67, 2.8; for k = 5, 6: 2.85, 2.90; for k > 6: 3.33
ρ_A	$= \rho^\circ_A(1 - \alpha_A\bar{T}_A)$ = density of the aesthenosphere *in situ;* ρ°_A = 3.31
ρ°	= S.T.P. density
α	= coefficient of thermal expansion; α_k for k = 1,10: 3.28×10^{-5}/°C; $\alpha_A = 3.29 \times 10^{-5}$/°C
$\bar{T}$	= average temperature
ρ_S	= density of sediment = 2.26
ρ_W	= density of water = 1.03
ΔS_k	$= S_{k\,\ell+1} - S_{k,\ell}$
$S_{k,}$	= thickness of layer k at start of iteration
WD	= depth below sea level of sediment-water interface at start of iteration ℓ. WD° = 2.0 m.

Therefore the quantity $\Sigma\rho_k\Delta S_k - \rho_A\Sigma\Delta S_k$ reduces to $(\rho_{crust} - \rho A)\Delta S_{crust}$. Because $\rho_A > \rho_{crust}$, SUB $_{\ell+1}$ is positive (subsidence). (2) Mantle thinning without concomitant crustal thinning causes uplift (SUB $_{\ell+1}\langle 0)$. In this case, ΔS_k is zero for crustal layers and negative for mantle layers. Therefore the quantity $\Sigma\rho_k\Delta S_k - \rho_A\Sigma\Delta S_k$ reduces to the quantity $(\rho_{\text{mantle lithosphere}} - \rho_A)\Delta S_{\text{mantle lithosphere}}$. Because the average density of the relatively cool mantle lithosphere is greater than the density of relatively hot aesthenosphere (ρ_A), SUB_{l+1} is negative (uplift). One can view uplift as resulting from thermal expansion in the mantle.

Results

In our model, subsidence and uplift are controlled indirectly by the distribution of crustal strain rates and mantle strain rates, because the strain rates control the amounts of crustal and mantle thinning. The relationships between strain rate, time, and crustal thickness and lithospheric thickness are shown in Figures 5 and 6. These graphs are used to make rough estimates of the strain rates during rifting. In Figure 5, the change in thickness of the brittle crust is shown diagrammatically for strain rates of 1×10^{-15}/sec and 5×10^{-16}/sec and for other strain rates by the contours crossing the diagram. If we know the amount of time that a rift was active, and the change in thickness of the brittle crust, then this graph is used to estimate the average strain rate in the brittle crust.

Strain rates in the ductile crust and in the mantle are not constant, but accelerate during deformation. As shown in Figure 6, an initial mantle strain rate of 1×10^{-16}/sec can double in less than 30 Ma. The rate of lithospheric thinning accelerates with the mantle strain rate. Thickness of mantle lithosphere, expressed as percent of original thickness, is shown by the contours crossing the strain rate curves in Figure 6. If the onset of rifting occurs 10 Ma to 70 Ma before initiation of sea-floor spreading, then initial mantle strain rates are between 1×10^{-16}/sec and 5×10^{-16}/sec.

Uplift occurs on the flanks of a rift basin, where mantle lithosphere is thinned but the crust is not. The magnitude of the buoyant force, and the width over which it is significant, are controlled by the density (or average temperature) assumed for the aesthenosphere and by the shape and scale of the mechanical instabilities. As deformation progresses, mantle thinning becomes focused at the axis, even though the zone of mantle thinning becomes broader.

Subsidence is dominant in the rift basin during most of the rift's evolution because crustal thinning overwhelms thermal expansion in the mantle. At the rift axis, though, uplift can occur late in the rift history, because of rapid acceleration of mantle strain rates (shown schematically in Figure 4B). Figure 6 shows that these conditions exist if the initial mantle strain rate is a factor of 5 to 10 less than the initial crustal strain rate. In this case uplift occurs late in rift history, immediately prior to the onset of sea-floor spreading. This type of rift history is similar to that suggested by Scheupbach and Vail (1980) from their study of seismic reflection profiles across passive margins. They interpreted the outer highs on some passive margins as uplifts that formed as medial highs late in rift history. Study of these features suggests uplift of 0.5-1.0 km over widths of 10 to 30 km (Vierbuchen, George, and Vail, 1980). Uplifts produced by our model, in certain ranges of parameters, closely match these dimensions (Figures 7, 8, 9). Under other conditions, however, the model predicts that uplift will not occur.

As examples of the types of subsidence histories that this model can produce, and of the relative importance of variables, we present three cases (Figures 7, 8, 9). Figures 7 and 8 illustrate the dependence of theoretical subsidence history on the temperature assumed for the aesthenosphere. In Case 1, the aesthenosphere temperature is assumed to be 1200°C. At the rift axis, the subsidence rate is uniform for the

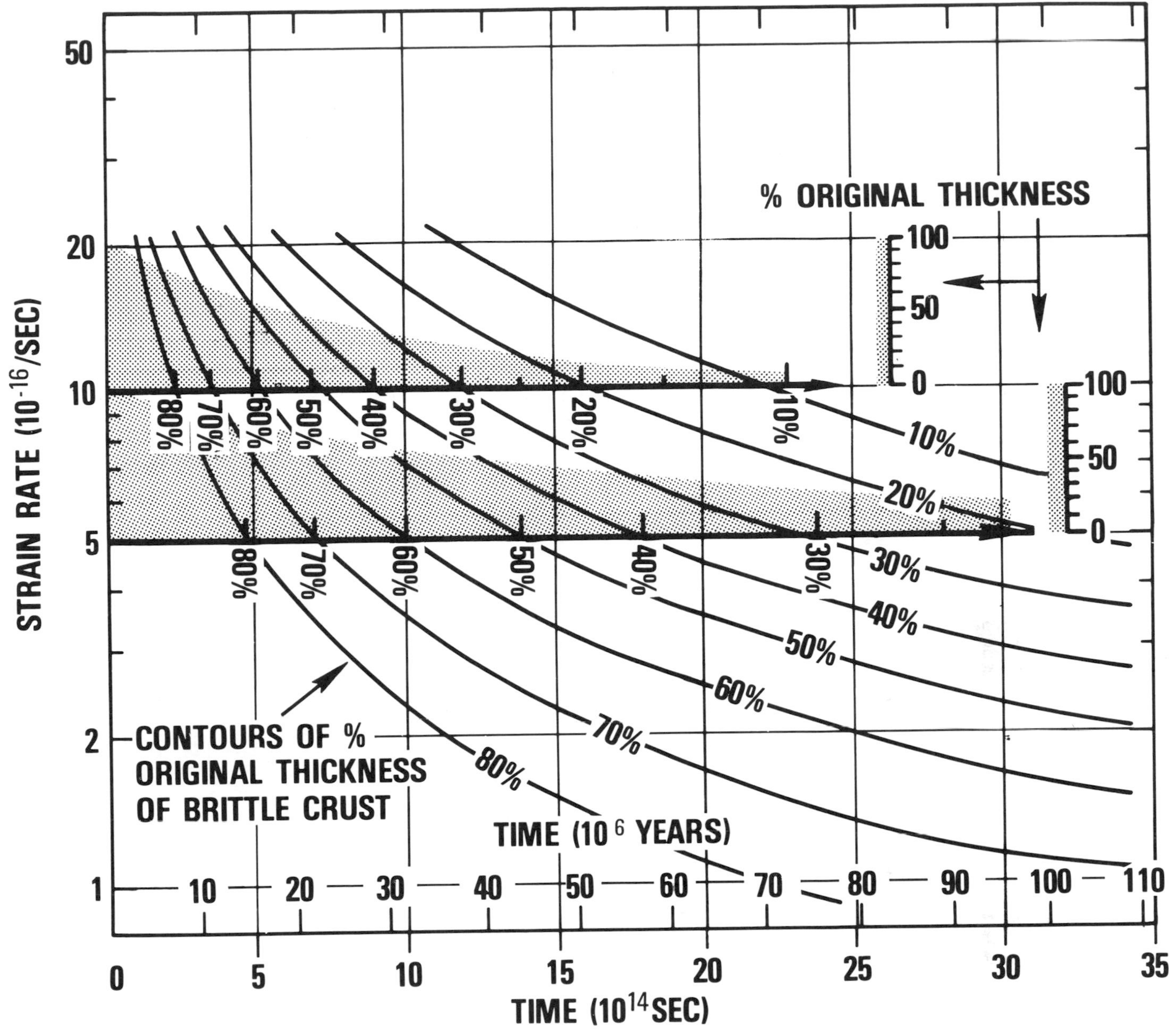

Figure 5 — Nomograph of thickness of brittle crust, deformed at *constant* strain rate, plotted in strain rate vs. time space. Upper shaded area diagramatically illustrates thickness of brittle crust deformed at 10×10^{-16}/sec; it thins to 50% of its original thickness in 22 Ma (and to 25% in 44 Ma). Lower shaded area illustrates thickness of brittle crust deformed at 5×10^{-16}/sec (half the strain rate of the upper shaded area); it thins to 50% of its original thickness in 44 Ma. These points are just two of an infinite number of points in strain rate vs. time space that define the 50% contour (curve labeled "50%" that passes through the two "50%" points). Contours for other percentages of original crustal thicknesses are similarly constructed.

first 20 Ma at which time the top of basement reaches a depth of about 3.4 km. Between 25 Ma and 35 Ma the subsidence rate slows significantly, and the maximum depth of burial attained is about 4.3 km. Uplift begins at 35 Ma. At the end of rifting and the onset of drifting, about 0.5 km of uplift has occurred. At 10 km from the rift axis the subsidence rate slows noticeably only after about 30 Ma, and the total uplift at onset of drifting is only 100 m. At 40 km, although the subsidence rate slows, uplift never occurs.

In Case 2, the aesthenospheric temperature is assumed to be 100°C higher than in Case 1. Thus the effect of thermal expansion in the mantle is seen sooner and over a broader area. At the axial position the subsidence rate begins to slow at 17 Ma, and the maximum depth of burial is only 3.9 km. Total uplift at the axis exceeds 1 km. Even at 40 km from the rift axis, uplift starts to occur at the onset of drifting. To summarize, the hotter aesthenosphere of Case 2 causes greater, broader, and earlier uplift than in Case 1.

Comparing Figures 7 and 9 illustrates the de-

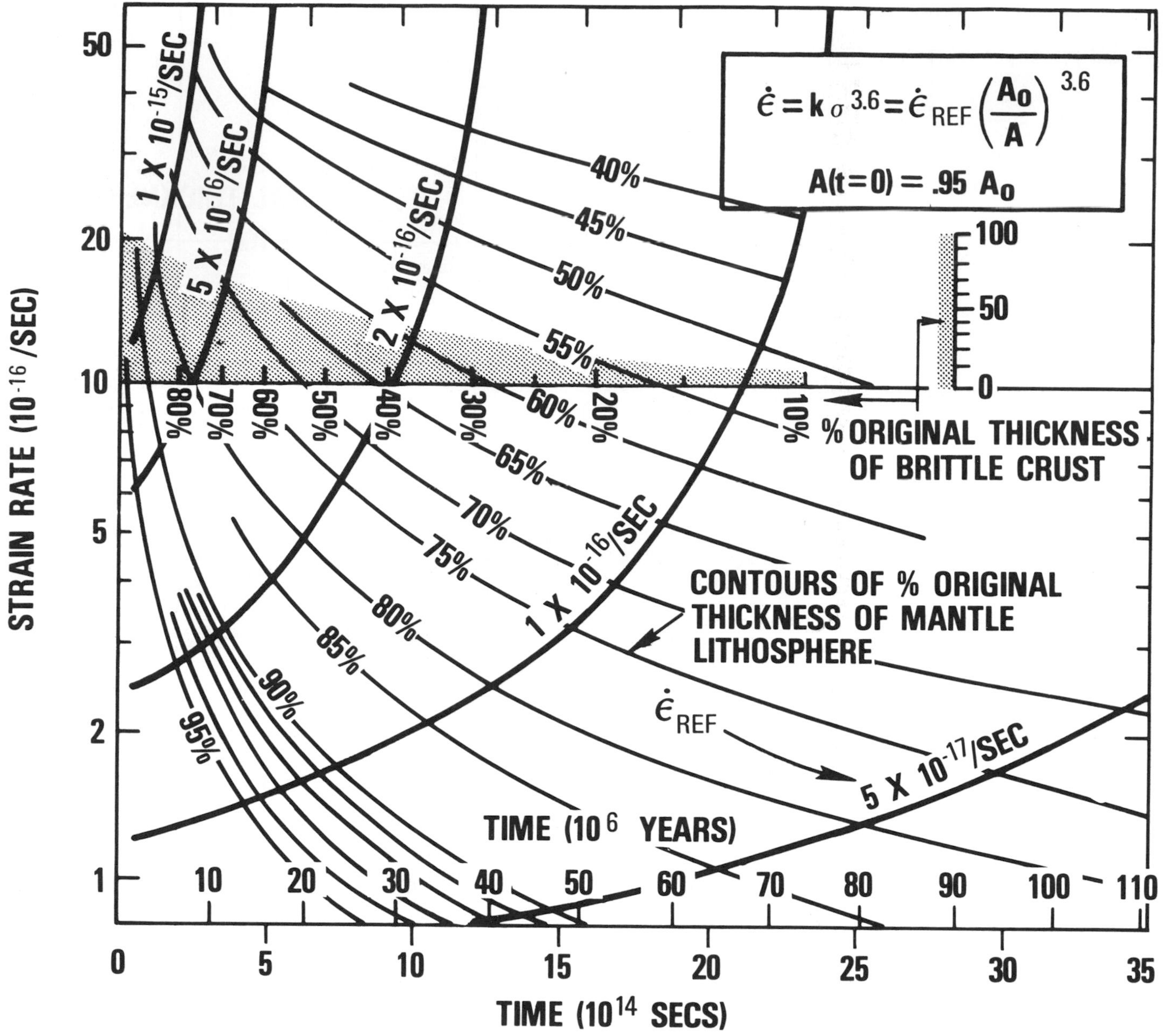

Figure 6 — Nomograph of thickness of mantle lithosphere, deformed at ever-accelerating strain rate, plotted in strain rate vs. time space. (For reference, shaded area is from Figure 5, for crust at constant strain rate of 10×10^{-16}/sec). Heavy curves labeled $\dot{\varepsilon}_{REF}$(5×10^{-17}, 1×10^{-16}, 2×10^{-16}, 5×10^{-16}, and 1×10^{-15}/sec) illustrate acceleration of strain rates at the axial position of an initially slightly necked layer ($A = .95\,A_o$) undergoing power-law creep. Note that at t = 0, $\dot{\varepsilon} = \dot{\varepsilon}_{REF}\left(\frac{A}{A_o}\right)^{3.6} = 1.2\,\dot{\varepsilon}_{REF}$.

pendence of theoretical subsidence history on the assumed size of the initial instability in the ductile layers. In Case 1 we input a relatively small instability of 5%. In Case 3 (Figure 9) we input a 10% instability. At the axis we see that thermal expansion in the mantle slows the subsidence rate in Case 3 earlier than in Case 1, and that maximum subsidence is one-half km less than in Case 1. This results from power-law creep in the mantle acting on a larger initial instability. Note that the width of the uplifted region is narrower in Case 3. The axial region is uplifted about 0.5 km, as in Case 1, but the point 10 km from the rift axis is uplifted only slightly in Case 3 (in contrast to the 100 m uplift at that same position in Case 1). This again reflects the behavior of power-law creep: the larger the instability, the more focused is its effect.

CONCLUSIONS

The lithospheric stretching model we describe incorporates rheological properties appropriate to the rocks involved, and non-instantaneous or time-

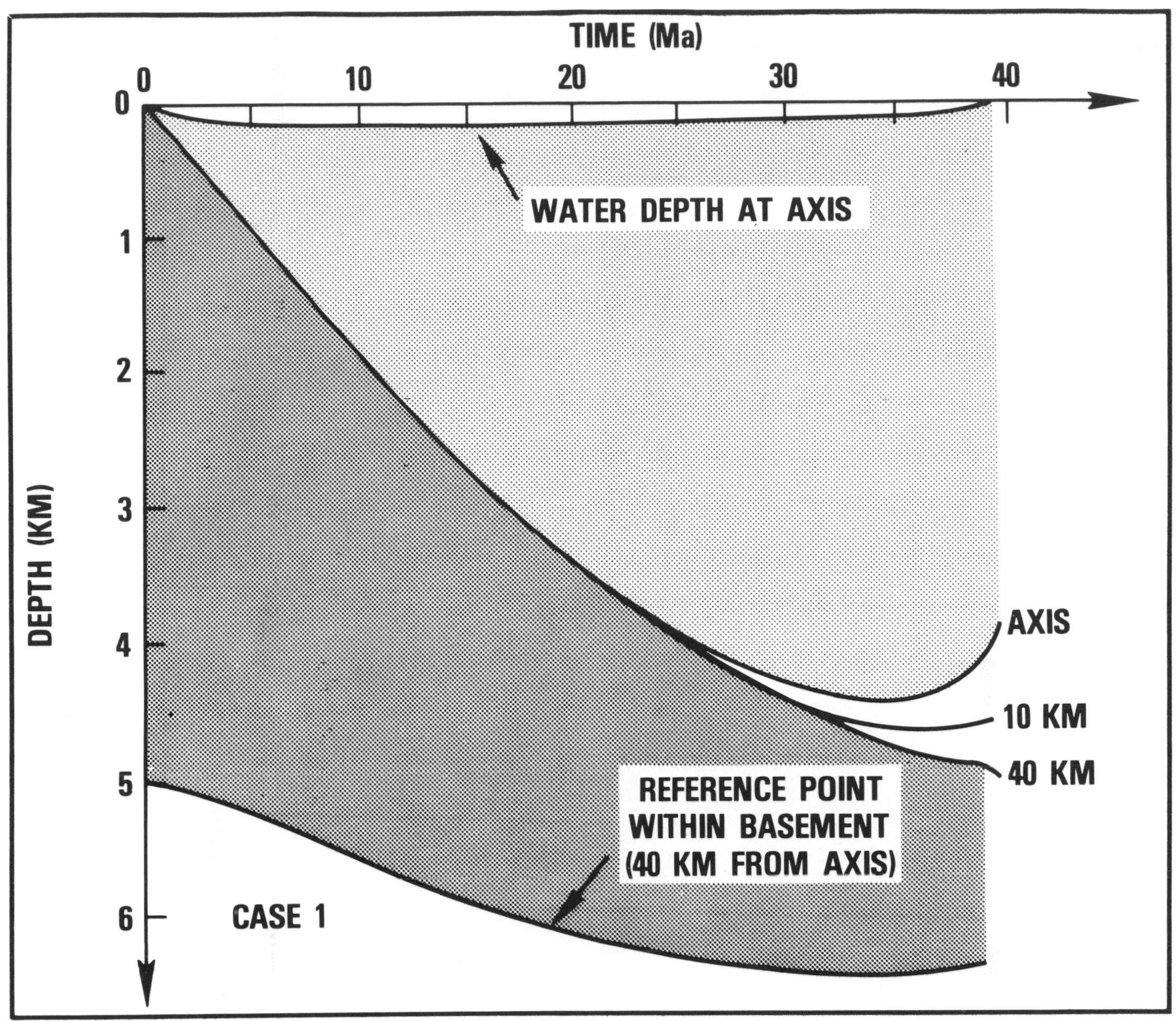

Figure 7 — History of burial and of brittle crustal thinning for Case 1. Assumed average aesthenosphere temperature = 1200°C, initial layer thickness at axis = 95% of layer thickness at ends. Light shading = sedimentary accumulation, dark area = thickness of the topmost portion (initially 5 km thick) of brittle crust at 40 km from the rift axis. The curves labeled "axis," "10 km," and "40 km" illustrate the burial history of the top of the basement at positions 0, 10, and 40 km, respectively, from the rift axis.

dependent deformation. Our purpose was not to consider all aspects and complications of the rifting process but simply to demonstrate that the particular variables we investigated are important enough to be included in models of the rifting process.

Significant results of our models are improved calculations of the variation of both the stretching parameter (β) and the initial postrift geothermal gradients across a passive margin. Our finite-element model predicts the location of isotherms throughout the rift history and can be used to predict paleo-heat-flow during rifting. The commonly used assumption of linear geothermal gradients at the end of the rift phase can produce serious errors in paleo-heat-flow predictions.

Furthermore, the subsidence histories suggested here differ greatly from those in models that ignore the rheological properties of the lithosphere. In particular, we show that necking of the lithosphere may cause uplift above sea level during the late phases of rifting. This suggestion provides a mechanism for forming the outer highs on passive margins described by Schuepbach and Vail (1980). The uplift that our finite-element model predicts at the rift axis under certain conditions has the correct magnitude to produce outer highs on continental margins (Vierbuchen, George, and Vail, 1980). However, under other conditions no uplift is predicted.

We suggest that the diversity of structures and subsidence histories of rifted passive margins may result

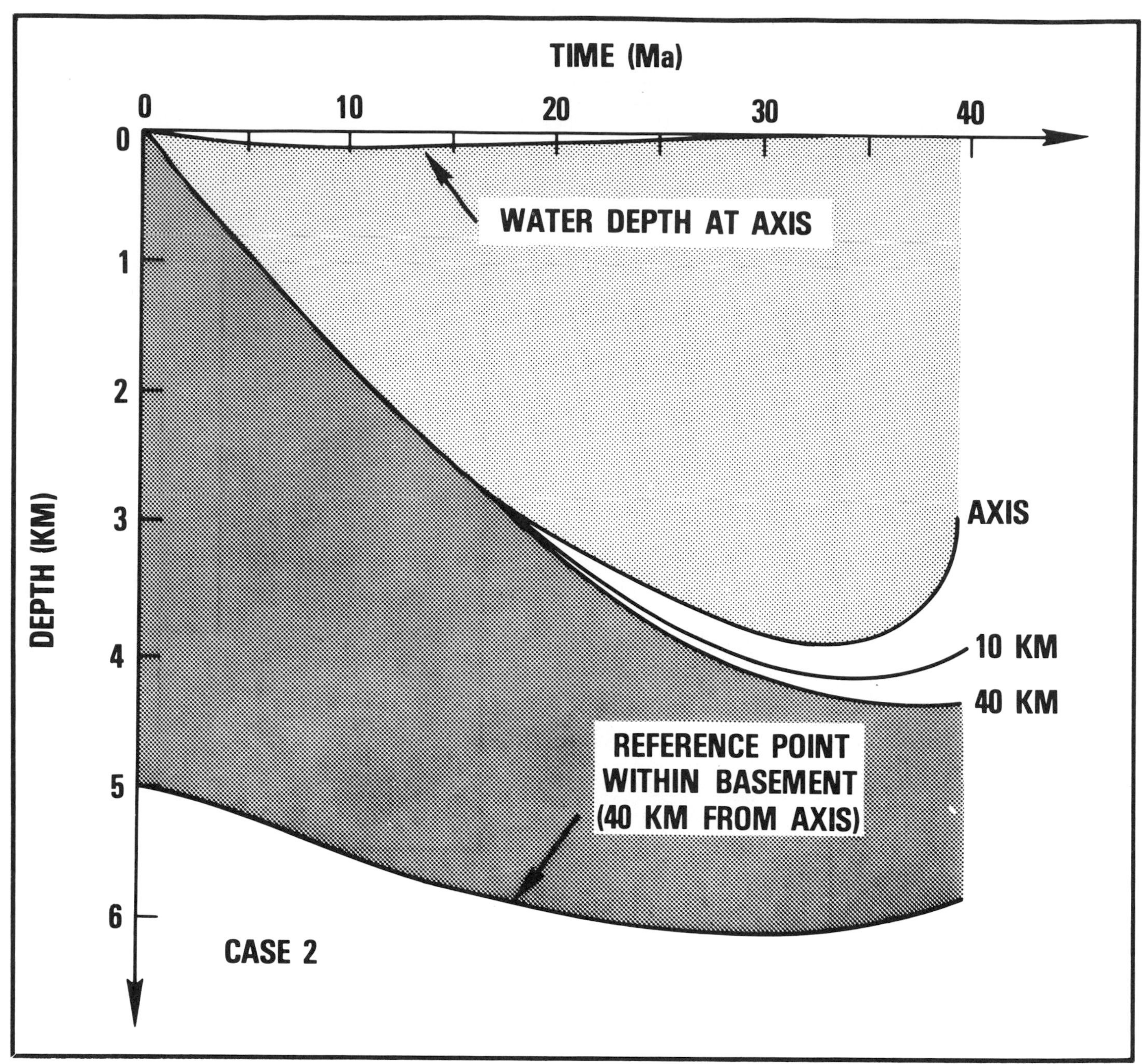

Figure 8 — History of burial and of brittle crustal thinning for Case 2. Assuming average aesthenosphere temperature = 1309°C. All other parameters and symbols as in Case 1.

from the differences in mechanical properties, temperatures, stretching and loading history, and initial strain rate contrasts of the lithosphere, not necessarily from differences in rifting mechanisms. We observe that it is possible to model rifts and passive margins of very different width, depth, crustal and lithospheric extension, lifespan, and with and without flanking or axial uplifts, simply by applying the stretching model to a range of reasonable starting conditions. Therefore, we conclude that the diversity in appearance and history of extensional rifts and rifted passive margins need not result from a diversity of rifting mechanisms.

Future attempts to model the rifting process should investigate the significance of many other complicating factors. These include the effects of partial melting, lateral heat conduction during rifting, mechanical anisotropies in the crust, phase changes in the lower crust and mantle, alternative mechanisms of isostatic compensation, depth of the brittle-ductile transition, and lateral offsets of shallow brittle deformation from the position of maximum strain in the lower lithosphere. Finally, our ability to understand these and other factors will be severely limited until more detailed observations of subsidence rates, structural style, and crustal composition and thickness at rifted margins are available.

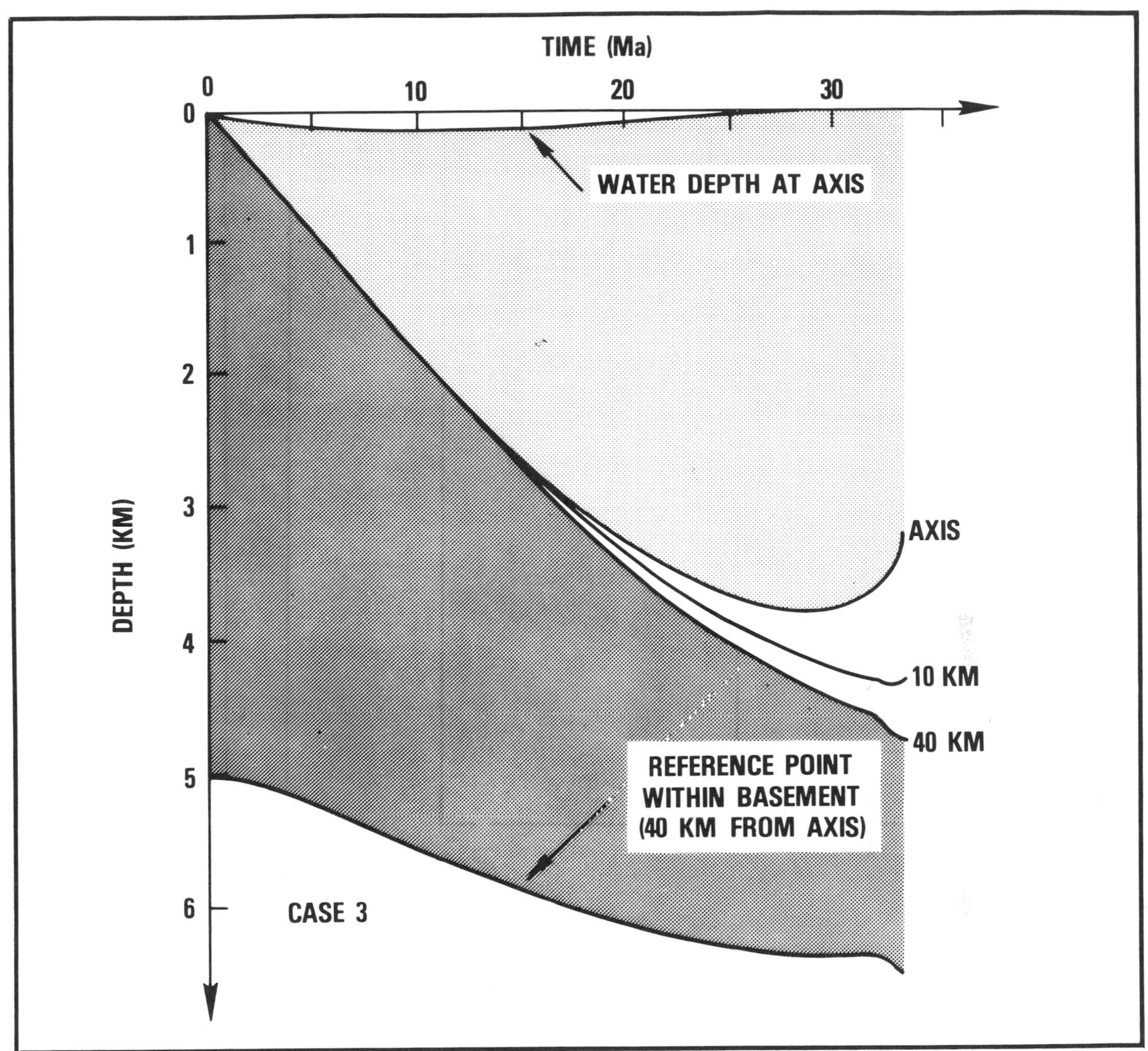

Figure 9 — History of burial and of brittle crustal thinning for Case 3: Initial layer thickness at axis = 90% of layer thickness at ends. All other parameters and symbols as in Case 1.

ACKNOWLEDGMENTS

We thank our colleagues in the Basin Systems group at Exxon Production Research for their encouragement and stimulating discussions. We are especially grateful to Ken Green for valuable insight into numerical modelling and timely assistance in debugging the computer program. We thank Pete Sherman for providing the gravity profile and model of crustal structure illustrated in Figure 1. This paper was reviewed by K. E. Green, Frank Levin, Norm Sleep, and D. J. Hall. We acknowledge their helpful comments.

REFERENCES CITED

Burke, K., and J. T. Wilson, 1972, Is the African plate stationary?: Nature, v. 239, p. 387-390.

———, and A. J. Whiteman, 1973, Uplift rifting and the breakup of Africa, *in* D. H. Tarling and S. K. Runcorn, eds., Implications of continental drift to the Earth Sciences: London, Academic Press, p. 735-755.

Caristan, Y., and C. Goetze, 1978, High temperature plasticity of Maryland diabase (Abs.): Transactions of the American Geophysical Union, v. 59, p. 375.

Carter, N. L., 1976, Steady state flow of rocks: Review of Geophysics and Space Physics, v. 14, p. 301-360.

Keen, C. E., and R. D. Hyndman, 1979, Geophysical review of the continental margins of eastern and western Canada: Canadian Journal of Earth Sciences, v. 16, p. 712-747.

———, and D. L. Barrett, in press, Thinned and subsided continental crust on the rifted margin of eastern Canada: crustal structure, thermal evolution and subsidence history.

Kirby, S. H., 1977, State of stress in the lithosphere: inferences from the flow laws of olivine: Pure and Applied Geophysics, v. 115, p. 245-258.

Koch, P. S., J. M. Christie, R. P. George, Flow law of wet quartzite in the α-quartz field (Abs.): Transactions of the American Geophysical Union, v. 61, p. 376.

McKenzie, D. P., 1978, Some remarks on the development of sedimentary basins: Earth and Planetary Science Letters, v. 40, p. 25-32.

Mercier, J.-C. C., and N. L. Carter, 1975, Pyroxene geotherms: Journal of Geophysical Research, v. 80, p. 3349-3362.

———, D. A. Anderson, and N. L. Carter, 1977, Stress in the lithosphere: inferences from steady state flow of rocks: Pure and Applied Geophysics, v. 115, p. 199-226.

Paterson, M. S., 1978, Experimental rock deformation — The brittle field: Berlin, Heidelberg, New York, Springer-Verlag, 254 p.

Schuepbach, M. A., and P. R. Vail, 1980, Evolution of outer highs on divergent continental margins, *in* Continental tectonics: Washington, D. C., National Academy of Sciences, p. 50-61.

Sclater, J. G., and P. A. F. Christie, 1980, Continental stretching: an explanation of the post-mid-Cretaceous subsidence of the central North Sea basin: Journal of Geophysical Research, v. 85, p. 3711-3739.

Shelton, G., and J. A. Tullis, 1981, Experimental flow-laws for crustal rocks (Abs.): Transactions of the American Geophysical Union, v. 62, p. 396.

Tullis, J. A., 1979, High temperature deformation of rocks and minerals: Review of Geophysics and Space Physics, v. 17, p. 1137-1154.

Vierbuchen, R. C., Jr., R. P. George, Jr., and P. R. Vail, 1980, A possible mechanism for formation of outer highs on passive continental margins (Abs.): 1980 Symposium on Rifted Margins, Geodynamics Program, Texas A & M University.

The Possible Significance of Pore Fluid Pressures in Subduction Zones

Roland von Huene
Homa Lee
U. S. Geological Survey
Menlo Park, California

The influence of pore fluids in the tectonics of modern subduction zones has been argued by only a few authors and is mentioned only casually in much of the earth science literature on subduction because of the difficulty of obtaining data on pore pressure in modern trench environments. Limiting values of physical properties can be derived along modern subduction zones where DSDP and other sampling recovered deep materials. We use two equations that bracket a conservative situation and present a simplified case where only vertical loads are considered to obtain simple estimates of pore pressure along the Aleutian, Oregon-Washington, Japan, Nankai, Middle America and Barbados subduction zones. The calculations indicate that sufficiently elevated pore pressure can develop early in the subduction process, making possible low shear strengths and low friction across faults in the front of some subduction zones. Very low friction is a convenient way to achieve the high degree of decoupling required in explanations of subduction related tectonic processes.

Explanations of tectonic mechanisms in subduction zones borrow heavily from possible tectonic analogues in mountain ranges. One enigmatic aspect of overthrusts in the Alps or Rocky Mountains is the undisturbed condition of many overthrust slices and the narrow deformed zone along fault contacts. These conditions, and the recognition that rocks in thrust sheets do not have sufficient strength to transmit the force required for thrusting, require greatly reduced fault friction. Hubbert and Rubey (1959) proposed an explanation for reduced fault friction that has long been accepted by earth scientists and has been amplified by many other authors (see Gretner 1972 and 1978). These authors postulate that thrust slices could be almost totally supported by high pore fluid pressures during fault movement. The mechanism of overpressured fluid is often assumed in geologic discussions of overthrust belts on land, especially since induced earthquakes from fluid injection have been recognized (Raleigh, 1972; and Raleigh, Healy and Bredehoeft, 1972). Similar assumptions should hold for subduction zones at ocean margins.

In most mountain belts, the thrust faults observed at the surface are generally inactive and, thus, the original elevated pore pressures cannot be measured. The proposed elevated pore pressure associated with thrust faults should certainly be found in modern subduction zones because the rates and amounts of thrusting associated with a Wadati-Benioff zone generally exceed those of Alpine or Rocky Mountain thrust belts (Bally and Snelson, 1980) and the processes are still active. Elevated pore pressures have been observed in drill holes along continental shelves and coastal areas of modern convergent margins. Structures associated with elevated pore pressure, such as mud diapirs, are common in convergent margin settings. However, elevated pore pressures are also found in non-tectonic settings and the unequivocal direct measurement of overpressure in subduction zones, or even the less convincing evidence from diapirism or piercement structure, is difficult to obtain.

Early in the Deep Sea Drilling Project (DSDP), other than normal consolidation of sediment along modern subduction zones was observed in drill cores. Emphasis was on overconsolidation, supporting convergent tectonism and the expected compressional stress. On Leg 18, overconsolidated and dewatered mudstone was sampled form the mobile core of a fold at the front of the Aleutian Trench subduction zone. Overconsolidated sediment was also sampled in the folded sequence at a similar position on the Oregon margin (von Huene and Kulm, 1973; Lee, Olsen, and von Huene, 1973). On Leg 31, overconsolidated sediment was sampled at the front of the Nankai Trough subduction zone in an area of folding and probable thrust faulting (Karig et al, 1975; Trabant, Bryant and Boumo, 1975). Underconsolidation and evidence for elevated pore pressure were reported from a study of slope deposits above the Japan Trench subduction zone (Arthur, von Huene, and Adelseck, 1980). Along the Middle America Trench, underconsolidation and overconsolidation were observed (Shepard, Bryant, and Chiou, 1982; Faas, 1982). In all DSDP drilling on convergent margins, both the slope sediment and the subducted material show other than normal consolidation and hydrostatic disequilibrium. But perhaps the most significant single observation from the DSDP drilling near active subduction zones was the measurement of near lithostatic pore pressure along a thrust fault at the front of the Barbados subduction zone, only 1.5 km from the front of the margin (Moore et al, 1982). This is probably the first direct confirmation in a modern subduction zone of the mechanism proposed by Hubbert and Rubey (1959).

Other evidence for elevated pore fluid pressure in subduction zones has been reported, but only a few authors have emphasized it. From modeling and seismic records, Seely (1977) concluded that overpressured shale and deep ocean mud diapirs were responsible for the landward vergence or seaward dipping thrusts of the Washington and Gulf of Alaska subduction zones. This same conclusion was developed from industry drilling history in the central Gulf of Alaska by Hottman, Smith, and Purcell (1979). Off Oregon, Washington, and British Columbia, shale diapirism and overpressured wells were linked with convergent tectonism by several authors (Shouldice, 1971; Rau and Grocock, 1974; Snavely, Pearl, and Lander, 1977). Carson (1977) showed that sediment from the surface of the present zone of folding at the Washington slope base was overconsolidated. Gretner (1978) and Sorokhtin and Lobkovskiy (1976) considered the role of pore pressure in the structural geology of thrust belts and subduction zones, and discussed some fundamental mechanical principles. Many earth scientists make speculations regarding elevated pore pressure in subduction zones (Karig, 1974; Moore 1975; Murauchi and Ludwig, 1980). Some propose that ancient zones of melange were formed under near fluid flow conditions (Cloos, 1982). Until recently, tectonic compression and dewatering were emphasized. However, when the diversity of tectonic mechanisms in subduction found in DSDP studies across active margins was accepted, it became clear that many subduction zones must contain areas of reduced friction. Elevated pore fluid pressures and underconsolidation can readily explain this reduced friction.

Uyeda (1981) notes the evidence for reduced friction across subduction zones in a global sense, and suggests that either changes in configuration of the Wadati-Benioff zone or motion of the landward plate are principal causes. We consider an additional explanation, reduced friction through elevated pore pressure, and through simplified computation make some quantitative assessment of the importance of fluid pressure in subduction zone tectonics. To simplify our calculations we consider only the vertical forces, realizing the importance of horizontal compression but concentrating on some simple and very conservative end-member cases that are only facilitated by added tectonic compression.

TECTONIC MECHANISMS IN SUBDUCTION ZONES

Some concepts of the tectonic processes associated with subduction crust are summarized by Scholl et al (1980) and the additional processes of underplating are discussed for specific cases by Watkins et al (1981) and Karig and Kay (1981). The simple end-member models fall into three general classes: 1) material is added to a margin (subduction accretion, underplating, allochthonous terranes); 2) material is removed from the margin (tectonic or subduction erosion); or 3) sediment on the ocean crust bypasses the front of the margin and is subducted. Most of these mechanisms require detachment, which many earth scientists attribute to the low frictional resistance produced by overpressured pore fluids. Perhaps the lack of more quantitative treatment stems from the general lack of quantitative physical properties measurements in subduction zones.

Subduction accretion is the most visible mechanism in the rock record. In its simplest form, subduction accretion consists of adding material to the front of a convergent margin by thrusting sediment and stacking it on the oceanic plate. Classical diagrams depict uplift of the youngest and most seaward imbricate thrust packet as it overrides soft trench and ocean basin sediment (Seely, Vail, and Walton, 1974). The soft sediment sampled in trenches generally has insufficient shear strength to transmit thrust forces very far and thus, a low friction on the fault plane appears necessary. The abrupt transition from undeformed trench sediment to the highly deformed accretionary complex, proposed by many authors, argues for sudden detachment. However, these are only some of the most frequently noted features and the variety of structuring in subduction complexes should rival that of onland counterparts.

Along some of the IPOD transects, the net volume of sediment accreted during a specific time is compared with the amount carried into the trench on the ocean plate. Such estimates along the Japan Trench transect indicate a net subduction of 80 to 90% of the sediment and along the Middle America Trench a 35% net subduction of sediment (Watkins et al, 1981). The lack of accreted sediment along the Mariana transect and the Middle America transect off Guatemala also indicates sediment subduction (von Huene et al, 1980a; Hussong and Uyeda, 1981; Aubouin et al, 1981). Subduction of sediment is seen in some seismic reflection records, (i.e. White and Louden, this volume), especially in those of the Nankai Trough described by Aoki et al (this volume). Evidence for the sediment subduction is now quite well-established.

The subduction of deep ocean basin sediment strata without disruption is mechanically difficult without a low friction zone above the underthrust strata.

In their analysis of the IPOD transect off Oaxaca, Watkins et al (1981) proposed accretion by underplating. They note a rapid landward tilting of units only at the foot of the margin,and not a gradual increase in tilt as shown in classical diagrams of the accretionary model. The paleobathymetry at the drill sites shows uplift of the landward tilted section without added rotation to thicken the upper plate. This motion requires adding material from below, a process these authors call "underplating." Karig and Kay (1981) cite another general body of evidence for underplating. To achieve underplating, soft sediment is first subducted beneath the front of the margin and is then detached beneath the accretionary prism. This detachment indicates a downward shift of the plane of least shear strength deep in the subduction zone.

During subduction, upper plate material may become detached and travel down the subduction zone. Subduction erosion of the Japan Trench margin off northern Honshu has been proposed by several authors (Murauchi 1971; Murauchi and Ludwig 1980; von Huene et al, 1980b). The subduction erosion concept was used to explain a major recession of the landward slope of the Japan Trench and the associated landward shift of the trench axis.

Subduction erosion is a difficult concept to explain mechanically. The disposal of material is documented by its absence, and thus direct or quantitative observations of rate or the process are not possible. It has been proposed that sea floor roughness or the horst and graben topography of the trench's seaward slope abrades the continental framework (Hilde and Sharman, 1978; Schweller and Kulm, 1978). Another mechanism suggested by Murauchi and Ludwig (1980) involves upward migration of water to soften rock along the underside of the continental framework until the rock can be carried away by the subducting slab. The absence of material along some convergent margins requires a concept other than lateral displacement to detach and then dispose of material from the front of the margin.

These are a few examples of the evidence that seems to require extensive detachment in subduction zones. Modern subduction zones are not the mass of melange they were once viewed to be; a view that has been changed by records made with multichannel seismic reflection techniques. Increasing numbers of records show broad stratal continuity at the few tens of meter scale of seismic resolution. Normally ordered landward tilted sequences of strata are common, and normal faulting is not rare in sections across forearc basins. The small amount of compressional stress that seems to be communicated across many subduction zones has been a paradoxial observation, and the concept of reduced friction from overpressures offer a likely explanation.

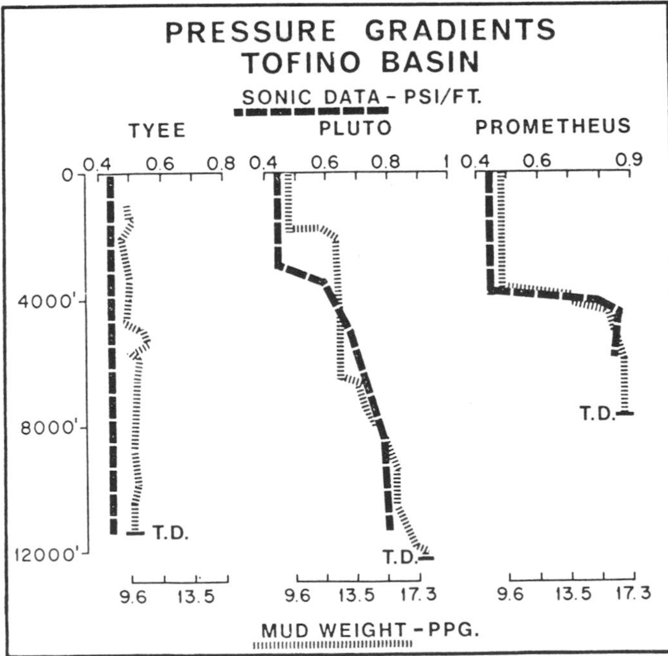

Figure 1 — Pressure gradients in selected wells, showing contrast between a hole on normal continental crust in Queen Charlotte Sound (Tyee), and two on the continental shelf (Pluto and Prometheus). Mud weights and pressures computed from sonic logs show that pressures below the shelf approach lithostatic values (Shouldice, 1971).

EVIDENCE FOR ANOMALOUS POROSITY AND OVERPRESSURE

Some drill holes along modern active margins are overpressured, as shown by the high drilling mud weights needed to control formation pressures. Off British Columbia, the mud weight needed to control formation pressure of two holes on the outer continental shelf far exceeded those used in the holes behind the Queen Charlotte Islands (Figure 1; Shouldice, 1971). Similarly high mud weights were needed in drilling in the Gulf of Alaska (Figure 2; Hottman, Smith, and Purcell, 1979). In both instances, the high pressures were encountered in forearc basins where subduction related compressional deformation was infered from seismic records. Elevated fluid pressures are indicated by logs in 3 holes on the IPOD Japan Trench transect. Here, a sudden decrease in formation density and underconsolidation is associated with the development of a highly fractured rock (von Huene et al, 1980b; Arthur, von Huene, and Adelseck, 1980; Shephard and Bryant, 1980; Carson, von Huene, and Arthur, 1982). The density decrease is attributed to fracture porosity and overpressuring.

Shale diapirism and sediment mobility are a second type of observation that suggests anomalously high formation fluid pressure along modern convergent margins. In convergent margin settings, diapirs are commonly reported: for instance, the Oregon, Washington, and British Columbia shelves (Shouldice, 1971; Rau and Grocock, 1974, Snavely, Pearl, and Lander, 1977) and the landward slope of the Aleutian Trench (von Huene, 1972).

Because diapirs and overpressures are also common

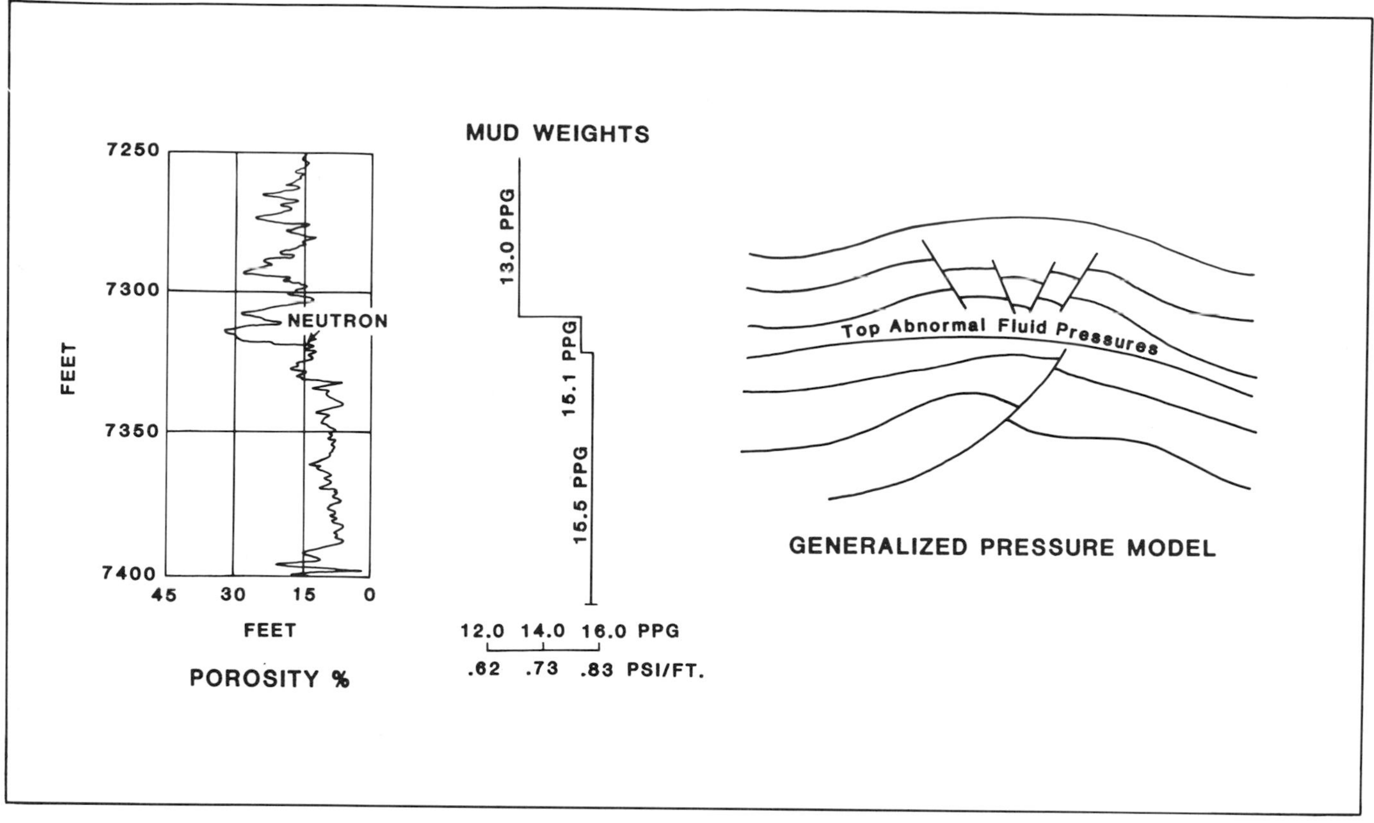

Figure 2 — Abnormal fluid pressure in wells on the continental shelf in the Gulf of Alaska required increases in mud weights during drilling up to 17.1 lb/gal (0.89) psi/ft. The first occurrence of elevated pressure was at 6,000 ft and a sharp increase at 7300 ft in shown in A. At 10,700 ft some seismic records show a change from extensional to compressional structure, B (from Hottman, Smith, and Purcell, 1979).

in non-convergent settings, especially river deltas, this type of evidence for a tectonic cause is not unequivocal. In fact, many wells along modern convergent margins are also normally pressured.

Abundant intergranular pore water has been measured in the DSDP cores recovered from three trenches. Sediment in the northeast Pacific off the Oregon and Aleutian subduction zones, sediment just seaward of the Middle America Trench, and most DSDP samples of deep-ocean sediment sequences entering subduction zones near continents are abundantly terrigeneous with porosities of about 50% or more. Considering that a 1-km-thick section of deep ocean basin and trench fill entering a subduction zone can easily contain 50% or more water, it is clear that abundant water is transported to the front of the subduction zone. Of this large quantity of water, how much remains in the subducted or accreted sediment? There are few direct observations that answer this question because only the very top of accreted material has been sampled during the IPOD program and much of the material was not suited for consolidation tests. On Leg 67, Site 500 was drilled as close to the base of the landward slope of the Middle America trench as possible; the porosities were essentially the same as those recorded from Site 499 on the seaward side of the trench, and from site 495, 20-km seaward of the trench floor (von Huene et al, 1980a). Thus, immediately in front of this subduction zone it can be inferred that the sediment has not dewatered significantly. Since the sediment sequence was dominantly beds of mud (85%), with few if any discrete layers of sand, it must have a relatively low permeability. At the high rate of present plate convergence, it seems unlikely that the water will fully drain from the sediment as accretion or subduction begins. Once this water is trapped in the subduction zone, its depth below the sea floor increases rapidly and thus high excess pore pressures are likely to develop. This is a qualitative speculation that we treat in a quantitative way with simple models and measured parameters.

MODEL STUDIES

Two simplified equations provide a quantitative estimate of the potential for overpressures to develop in sediment that is being subduced. By using several combinations of parameters associated with the geometry and sediment of specific modern trenches, the calculations show significant levels of underconsolidation which cause elevated pore pressures.

Definitions

In this section, the terminology used comes from soil mechanics terminology, which differs somewhat from common geologic terminology.

"Consolidation" is defined as the process of pore fluid expulsion resulting from the application of normal confining stresses. It is essentially analogous to the term compaction, as used in geologic literature.

"Normal consolidation" represents a situation in which a sediment is in equilibrium with its environmental loads and has never been in equilibrium with any higher loads.

"Overconsolidation" is a condition in which a sediment may be in equilibrium with its environmental loads, but has been in equilibrium with higher loads in the past.

"Underconsolidation" is a condition in which a sediment is not in equilibrium with its present environmental loads. The environmental loads, which consist of the weight of overburden and any tectonic stresses, are termed the "total stress" field. The loads which are carried by intergranular stresses are termed the "effective stress" field. The difference between the effective and the total stresses is the excess pore water pressure. In an underconsolidated sediment, the effective and total stresses are not equal and a portion of the total stress field is carried by pore water pressure. By the "effective stress principle," the mechanical behavior of a sediment, including its shearing strength or frictional resistance, is determined by the effective stress field. Therefore, an underconsolidated sediment has a strength that is proportional to its effective stress field. The shearing stresses that might cause the sediment to fail, however, are typically proportional to the total stress field. An underconsolidated sediment appears weak, relative to its environmental loads, while a normally or overconsolidated sediment appears strong. Underconsolidation and elevated pore pressure are essentially synonymous in the way they are used here.

Underconsolidation commonly results from an application of total stresses at a rate exceeding that at which water can drain from the sediment. Active river deltas are classic examples of underconsolidation in a tectonically stable environment. The more complex underconsolidation along active margins resulting from tectonism is not as well understood.

In the simplified diagram of a trench and subduction zone used to produce quantitative estimates of the degree of underconsolidation (Figure 3), the thickness of the subducted sediment layer is H and the variable vertical distance from the base of the sediment layer to the seafloor is h. The angle between the surface of the subducted layer or the surface of oceanic crust and the trench slope is α, and the variable horizontal distance from the base of the trench slope is x. The rate of convergence is identified as R. A subduction time, t, can be calculated as x/R.

Two available solutions to the basic consolidation equation (Terzaghi and Peck, 1967) are partially applicable to this situation. Gibson (1958) investigated consolidation in a sediment layer that increased in thickness linearly with time. This case effectively models normal deltaic sedimentation. The results are presented in the original paper as plots of the degree of underconsolidation (pore pressure/total overburden

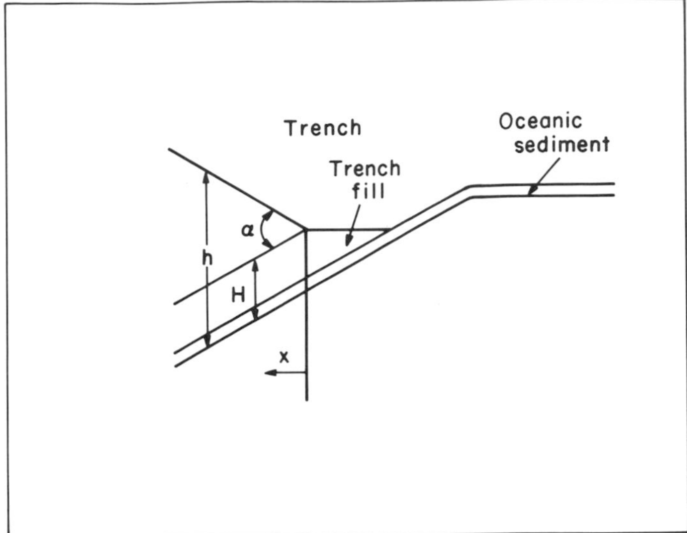

Figure 3 — Diagram of a trench and subduction zone showing notation.

stress) versus normalized depth in the sediment column. A dimensionless time parameter, T_1, defines each plot:

$$T_1 = \frac{m^2 t}{c_v} \quad (1)$$

where m = sedimentation rate; t = time since initiation of sedimentation; and, c_v = coefficient of consolidation.

The coefficient of consolidation is a measurable parameter that defines the rate at which consolidation proceeds in a given sediment. It, in turn, is related to other more basic characteristics:

$$c_v = \frac{k}{\gamma \; m_v}, \quad (2)$$

where k = permeability; γ = density of water; and, m_v = volume compressibility.

The permeability defines the amount of time needed for water to flow out of the material. The volume compressibility defines the amount of water that must be expelled to achieve equilibrium.

The second available solution is provided by Wissa et al (1971). This solution is for the case of a constrained sediment layer subjected to a constant rate of axial strain application. If a linear material is assumed, this situation becomes simlar to a constant rate of total stress application. Results can be presented as a plot of the average degree of consolidation (one minus the degree of underconsolidation) versus a time factor, T:

$$T = \frac{c_v t}{H^2} \quad (3)$$

where H = thickness of sediment layer.

Pore water drainage occurs only through the top surface.

Neither solution completely solves our problem. The distance from the base of the sediment layer to the sea floor is increasing approximately linearly with time and distance from the trench. This is similar to Gibson's problem. However, only the subducted sediment layer with thickness H is consolidating. For the sake of simplicity in modeling, the overlying material are presumably not consolidating. Therefore, consolidation in our case probably proceeds at a rate faster than Gibson's solution would predict. Wissa et als' (1971) solution represents the subducted sediment problem to the extent that it includes a single layer of sediment that retains approximately the same thickness. However, free drainage of the sediment layer is allowed at the top. In the simplified subduction model drainage must occur through all of the overlying material or along the thrust fault at the top of the subducted sediment. If the overlying material is highly fractured, the escape of pore fluid is accelerated but is less than free drainage. In general, Wissa et als' (1971) solution partially fits our model, but in their model the free drainage at the surface of the sediment layer will cause the sediment to consolidate faster than in our situation. The Wissa et al (1971) and Gibson (1958) solutions therefore approximately bracket the limits of our more complex situation.

The two solutions can be presented on a single plot through a redefinition of the dimensionsless time parameters. If the sedimentation rate, m, of Gibson's solution is taken as the rate of thickening of material over the sediment base, it may be defined as h/t.

Substituting for m in equation (1) yields:

$$T_1 = \frac{h^2}{tc_v} \qquad (4)$$

This time parameter is basically the reciprocal of Wissa et als' (1970) parameter T, except h is substituted for H. If we define a modified Gibson dimensionless time parameter, T_2, it is possible to plot both solutions together:

$$T_2 = 1/T_1 = \frac{c_v t}{h^2} \qquad (5)$$

Figure 4 shows both solutions for the base of the sediment column.

General Application

To affect the subduction process, underconsolidation must significantly decrease shear strength and must produce a deviation from the average strength profile greater than that produced by lithologic heterogeneity. The lithologic variation can be obtained from the ratio of undrained shearing strength to overburden effective stress. Lambe and Whitman (1969, p. 452) show that this ratio varies between 0.2 and 0.4 (i.e., .28 ± 30%) for marine clays. Because excess pore water pressures

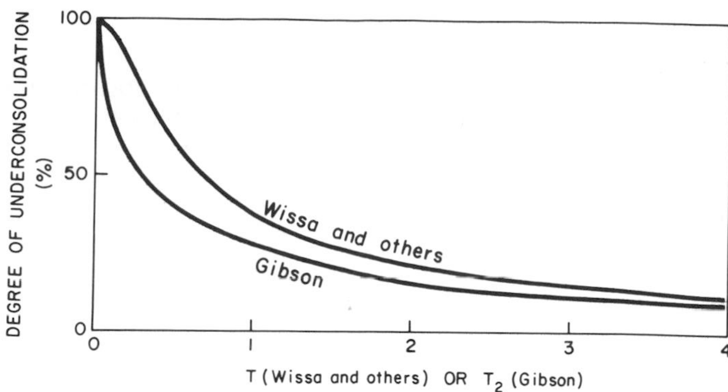

Figure 4 — Solutions of the equations of Gibson (1958) and of Wissa et al (1971) showing consolidation.

affect the overburden effective stress linearly, they effect shearing strength linearly also. Any pore water pressures in excess of about 30% of the overburden stress (a degree of underconsolidation of 30%) would have an effect on strength that would dominate over lithologic variability. Referring to Figure 4, this value corresponds to a Gibson's T_2 of .9, or Wissa et als' T of 1.3.

To use these values for our problem it is necessary to redefine the T parameters in terms of the conditions shown in Figure 3. The distance, X, may be taken as Rt where t is the time that has passed since subduction began. The depth, h, can be taken as X Tan α if H and α are relatively small. Gibson's T_2 can then be redefined as:

$$T_2 = \frac{c_v t}{h^2} = \frac{c_v}{R^2 t \tan^2\alpha} = \frac{c_v}{XR \tan^2\alpha}, \qquad (6)$$

and Wissa et als' T can be redefined as:

$$T = \frac{c_v\, t}{H^2} = \frac{c_v X}{RH^2}, \qquad (7)$$

An important difference between Equations 6 and 7 should be noted. Equation 6 indicates that T_2 decreases with time while Equation 7 shows T increasing. Equation 6 indicates an increasing degree of underconsolidation as subduction progresses while Equation 7 shows a decreasing degree. As noted previously, these solutions are approximate upper and lower bounds for our case and this seeming discrepancy may indicate that the two equations are indeed at the bounding limits. We may thus be confident of underconsolidation when it is indicated by both solutions. Overlap between the two solutions may be determined by solving both Equations 6 and 7 for X and inserting realistic values for the required parameters. If the value of X from Equation 6 is less than that from Equation 7, then the zone of overlap is specified by this range of distances from the foot of the trench slope. If the value of X from Equation 6 were greater than that from Equation 7, there would be no overlap. Underconsolidation might be occurring but of a magnitude insufficient to trigger these very conservative solutions.

Typical values for the parameters of Equation 6 and 7 are needed to specify the range of X values over which underconsolidation occurs. Measurements made on a variety of marine sediments and minerals, including values measured in subduction zone sediment, show that c_v varies between 10^{-4} and 10^{-3} cm^2/sec for most fine-grained deep sea and trench sediments. However, much higher values (as high as 10^{-2} or 10^{-1}; Terzaghi and Peck, 1967, p. 8) might occur if the sediment is sandy. Although sand has commonly been assumed in trenches, most samples that have been recovered from trenches are of fine-grained material and discrete layers of sand are rare. One exception is in the Middle America Trench off southern Mexico, where coarse to fine sand and lesser muddy sand was recovered in 3 cores during Leg 66, but from seismic data the sand appears to be associated with the mouth of a large submarine canyon emptying into the trench (Shipley, 1982). Measured ranges for the parameters c_v, R, α, and H for several modern subduction zones are given in Table 1. Calculated values of X are given for each combination of parameters. The span between the values calculated by the Gibson 1958) and Wissa et al (1971) solutions represents the range over which both solutions (upper and lower bound) predict underconsolidation. Values of X for degrees of underconsolidation of both 30 and 80% are given. The span of overlap between the solutions is shown graphically in Figure 5.

These values should be taken as a very approximate representation of the true situation. If c_v is high, as would be typical of sandy materials, underconsolidation would play a less significant role.

The results given in Table 1 show that underconsolidation at the 30% level can exist in all of the modern subduction zones considered. The length of overlap between the two solutions at this level varies between 1.3 km for the Barbados subduction zone, to as much as 173 km for the Middle America Trench. Clearly, values as high as 173 km are unrealistic because the basic simplified geometry on the model was not intended to extend this far from the trench. Such large values can mainly indicate that the potential for underconsolidation is especially high. It is notable that the subduction zone that appears least liklely to be underconsolidated (Barbados) is the one in which lithostatic pore pressures were measured only 1.5 km from the deformation front, which is equivalent to the foot of the trench slope (Moore et al, 1981; in press). This observation indicates the high level of conservation present in the model.

The subduction zones examined may be listed in order of decreasing level of expected underconsolidation as determined by the simplified model:

1. Middle America (Guatemala)
2. Japan
3. Middle America (Oaxaca)
4. Washington
5. Aleutian
6. Nankai
7. Barbados

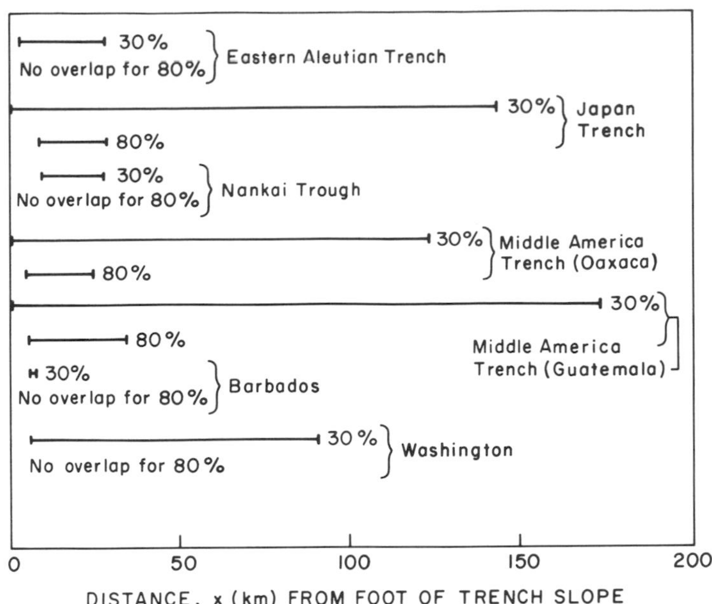

Figure 5 — Overlap span for two limiting solutions showing the conservative range over which 30% and 80% of underconsolidation may exist.

Overpressure may also build to significant levels from rapid sediment accumulation prior to the subduction of trench sediment. In the eastern Aleutian Trench, the 800 m of silt and clay ponded in the trench axis accumulated in 0.6 million years (von Huene and Kulm, 1973, Site 180). From the equation of Gibson (1958) and a c_v from the trench landward slope (Lee, Olsen, and von Huene, 1973), an overpressure of 15% of lithostatic could develop prior to any tectonism. Through a similar procedure, the degree of underconsolidation present in trench sediments before subduction was calculated for the other areas examined (Table 1). A more accurate estimate of the degree of underconsolidation present in the subducted sediment would approximate the sum of these values and the 30 to 80% used in the calculation of the X distances.

OVERPRESSURE IN EXPLANATIONS OF SUBDUCTION ZONE TECTONICS

Many tectonic mechanisms that have been proposed to explain subduction zone structure require a high degree of decoupling. For instance, decoupling is needed to explain coherent sediment strata subducted tens of kilometers beneath the foot of the Nankai Trough slope (Atoki et al, this volume). Before seismic records showing subducted sediment were available, Scholl and Marlow (1974) pointed out the sparse amount of pelagic sediment accreted, and Karig (1974) and Moore (1975) had proposed subduction of sediment in which the upper turbidite layers are accreted and the lower more pelagic sequence remains coupled to the subducting crust. Moore (1975) emphasized strong differences in physical properties between deep ocean pelagic sediment and the rapidly deposited hemipelagic and turbidite sediment. Differences in the physical properties of trench sediment are certainly important but so is decoupling. A great potential for the development of elevated pore fluid pressures, and thus decoupling, occurs in trenches where rapid sedimentation, slumping, and the rapid tectonism that superimposes one body of sediment on another, provide sudden sediment loading.

Rapid sediment accumulations, documented by drilling in the Aleutian Trench, Middle America Trench, and Nankai Trough, indicate accumulation rates sufficiently high to cause elevated pore pressure prior to subduction. Most sediment sequences passing through trenches have vertical changes in lithology to provide zones of relative low shear strength if pore fluid pressure is elevated (Figure 6). The potential for detachment can therefore begin prior to subduction from rapid sediment loading. Then, as subduction begins, the elevation of pore pressure must accelerate due to additional tectonic forces. However, in addition to any compressive forces there is also an added rate of vertical loading because as the sediment descends down the subduction zone, its depth below the ocean floor increases more rapidly than in the trench (Figure 3).

If the ponded turbidites are sandy, such as those recovered on Leg 66 off Oaxaca, they drain more readily than the fine-grained sediment, and their lower pore pressure gives them a greater shear strength. Thus, massive sand would have fewer potential detachment surfaces or zones of low shear strength and may more likely be accreted than subducted. The affect of pore pressure in selective subduction strengthens the arguments pointed out by Karig (1974) and Moore (1975).

In simple diagrams, subduction erosion is presented as a large scale abrasion of the continental framework caused by the rough topography of the ocean floor (Hilde and Sharman, 1978; Schweller and Kulm, 1978), but it might also be caused by small scale dismemberment and mobilization. Subduction erosion can be visualized as a nearly grain-by-grain detachment of material from the hanging wall of an overpressured thrust fault. Pore fluids from subducted sediment tend to flow upward toward the ocean floor and across the major thrust into the rock forming the front of the margin. Drilling in the Japan Trench and Middle America Trench recoverd a slope deposit mudstone from the front of the margin with numerous veins caused by fluid migration (Arthur, von Huene, Adelseck, 1980; Cowan, 1982). The vein spacing became increasingly close at greater depths to eventually form a highly fractured rock (von Huene et al, 1980b; Arthur, von Huene, and Adelseck, 1980; Carson, von Huene, and Arthur, 1982). This fractured rock has an apparent reduced shear strength as deduced from seismic

Table 1. Measured and calculated parameters relating to underconsolidation of subduction zones.

Area	Convergence Rate R (cm/yr)	Trench Sediment Thickness H (m)	Angle between trench slope and oceanic crust, a (°)	Coefficient of Consolidation, c_v (cm²/sec)	Distance X(km) from foot of trench slope to initiation of 30% and 80% underconsolidation (Gibson, 1958) 30%	80%	Distance X(km) from foot of trench slope to termination of 30% and 80% underconsolidation (Wissa et al, 1971) 30%	80%	Degree of underconsolidation produced by rate of trench sedimentation (Gibson, 1958) (%)	Data Source
Eastern Aleutian Trench	6	1,800	13	3×10^{-3}	2.4	132	28	5.5	15	von Huene, 1972 Unpublished USGS data
Japan Trench	10	1,000	12.7	3×10^{-4}	0.1	8	143	28.5	No terrigenous fill	Nasu et al, 1980; Shephard and Bryant, 1980
Nankai Trough	2	1,800	7.2	10^{-3}	8.9	494	28	5.5	15	Aoki, in press; Trabant, Bryant, and Bouma 1975
Middle America Trench (Oaxaca)	9	800	15.6	2×10^{-4}	0.1	4	123	24.6	28	Shipley et al, 1982; Shephard et al, 1982
Middle America Trench (Guatemala)	10	900	14	2×10^{-4}	0.1	5	173	34.6	21	Ladd et al, in press; von Huene et al, in press
Barbados	2	700	7	6×10^{-4}	5.7	316	7	1.4	<1	Moore et al, 1982
Washington	2	2,100	6.3	4×10^{-4}	5.6	320	91	18.2	25-28	Unpublished data from Bobb Carson and P.D. Snavely, 1981

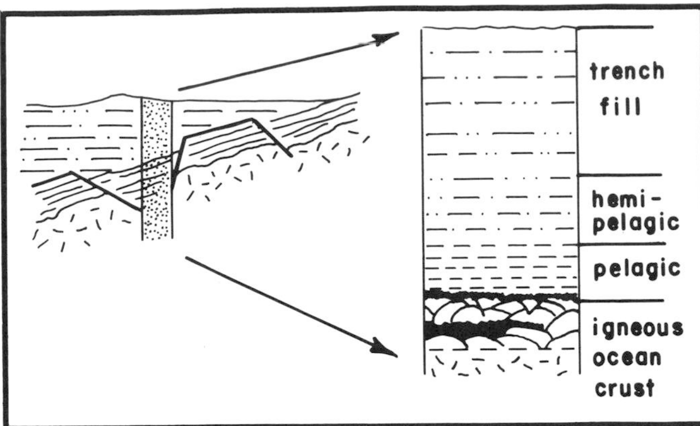

Figure 6 — Generalized sediment section in a modern trench.

parameters (Nagumo, Kasahara, and Koresawa, 1980) and was difficult to recover because it readily broke apart when disturbed by the drill. Small mudstone chips finally stopped the drilling because they formed a slurry of sand and gravel sized material that immobilized the bottom-hole assembly (Sites 434 and 441, Scientific Parties Legs 56 and 57, 1980). Off Japan, this condition occurred about 1 km above the subduction zone and in slope deposits that cover the accreted section. If the entire section immediately above the subduction zone is similarly fractured, it could readily form a slurry when disturbed by the slip along a thrust fault and then move down the subduction zone, perhaps by the type of fluid flow proposed for some Franciscan melanges (Cloos, 1982). This type of subduction erosion is similar to that proposed by Marauchi and Ludwig (1980).

Underplating requires detachment of a sediment sequence that has been carried down the subduction zone on the oceanic plate. Such detachment differs from that described above, because it occurs at the base of the section that has been dewatered to some degree, rather than at a higher level toward which water is migrating. Causes of such detachment could be the 10 to 15% of water some of the igneous oceanic basements contain (Donnelly et al, 1979), or the fluids produced during early mineral transformation. At temperatures above 82°C, smectite changes to illite and releases about 15% of its volume in water, as noted in the Gulf of Mexico (Bruce, 1973). Thus after subduction to deeper levels, a second stage of upward fluid migration might be initiated from the base of the subducting sediment section. The 82°C isotherm in the Middle America Trench off Guatemala is estimated at about 3 km deep (von Huene et al, 1980a). Thus, fluids from mineral transformation collected and formed a plane of detachment near the base of the section once it is subducted to 3 km. Interestingly, the area of underplating proposed by Watkins et al (1981), which is just below the last landward dipping reflections on seismic records (Shipley et al, 1982), is also about 3 km deep or at the level of the smectite-illite transformation (Figure 7).

SUMMARY

Extensive decoupling can explain the puzzling observation of mild deformation and sometimes extensional structure above a subduction zone where the underthrusting of hundreds of kilometers of ocean crust would seem to require large compressional structures. Low friction and low shear strength from elevated pore fluid pressure has long been assumed to explain thrust faults exposed in mountain belts. Thus, the deep water environment, rapid rates of plate convergence which cause underthrusting of fluid rich sediment, and the common fine-grained sediment, with low permeability sampled from modern trenches, are conditions favorable for developing high fluid pressures that might even rival those in ancient thrust faults of the Alps or Rockies. Petroleum exploration in forearc basins, geophysical indications of structure formed by mobile shales, and one direct measurement of lithostatic pore pressure in a subduction zone are evidence of such an overpressured condition.

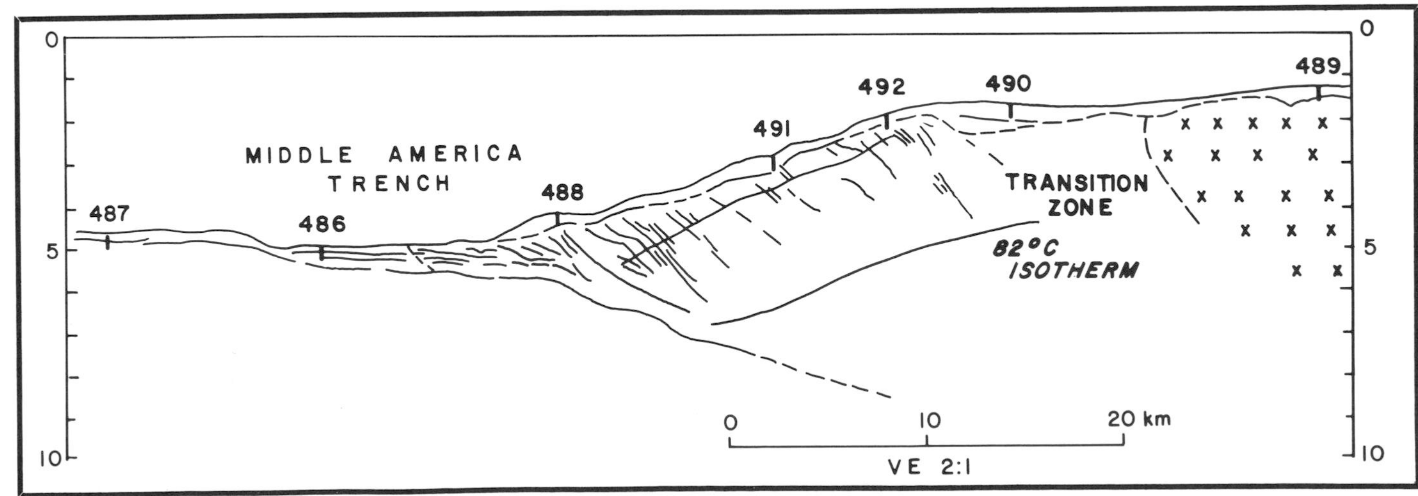

Figure 7 — Section across the Middle America Trench off Oaxaca showing a possible position of the 82°C isotherm in relation to recorded reflections. The area beneath this isotherm corresponds to the area of underplating proposed by Moore et al (1981).

Simplifying the overpressure calculation by considering only the vertical loads in a subduction zone provides conservation estimates of how far down a subduction zone overpressure may become significant. Along seven modern subduction zones, there are sufficient measurements of physical properties to make realistic estimates. We estimate that 30% of lithostatic pore pressure can develop from vertical loading at the front of some subduction zones. In other subduction zones, it develops as far as 8 km down. This estimate, when compared with the Barbados measurement, is conservative by a factor of about 3. The elevation of pore pressure early in subduction development can provide the high degree of decoupling required to explain structure or tectonic processes in many subduction zones.

REFERENCES CITED

Aoiki, Y., T. Tamano, and S. Kato, 1982, Detailed Structure of the Nankai Trough from Migrated Seismic Sections: this volume.

Arthur, M. A., R. von Huene, and C. Adelseck, 1980, Sedimentary evolution of the Japan forearc region off northern Honshu, legs 56 and 57, in Initial reports of the deep sea drilling project, v. 57: Washington, D.C., U.S. Government Printing Office, p. 521-568.

Aubouin, J., et al, 1981, Summary of Deep sea drilling project leg 67, *in* Shipboard results from the Mid-America Trench transect off Guatemala: Oceanologica Acta, Supplement, v. 4, p. 225-231.

Bally, A. W., and S. Snelson, 1980, Facts and principles of world petroleum occurrence — realms of subsidence, *in* Facts and principles of world petroleum occurrence: Canadian Society of Petroleum Geologists, p. 9-94.

Bruce, C. H., 1973, Pressured shale and related sediment formation: Mechanism for development of regional contemporaneous faults: AAPG Bulletin, v. 57, n. 5, p. 878-886.

Carson, B., 1977, Tectonically induced deformation of deep sea sediments off Washington and Northern Oregon: Mechanical consolidation: Marine Geology, v. 24, p. 289-307.

———, R. Von Huene, and M. Arthur, 1982, small-scale deformational structures and physical properties related to convergence in Japan arc-slope sediment: Tectonics, v. 1, n. 3, p. 277-302.

Cloos, M., in press, Flow melanges: numerical modeling and geologic constraints on their origin in the Franciscan subduction complex, California: Geological Society of America Bulletin, v. 93, p. 330-345.

Cowan, D. S., 1982, Origin of 'vein structures' in slope sediments on the inner slope of the Mid-America Trench off Guatemala, *in* J. Aubouin et al, Initial reports of the Deep Sea drilling project, 67: Washington, D.C., U.S. Government Printing Office, p. 645-651.

Donnelly, T. W., et al, 1979, Site 417, *in* Initial reports of the deep sea drilling project, v. 51, 52, 53: Washington, D.C., U.S. Government Printing Office, p. 23-95.

Faas, R. W., 1982, Plasticity characteristics of the Quaternary sediments of the Guatemalan continental slope, Middle America Trench, and Cocos Plate, leg 67, Deep sea drilling project: Washington, D.C., U.S. Government Printing Office, p. 639-644.

Gibson, R. E., 1958, The progress of consolidation in a clay layer increasing in thickness with time: Geotechnique, v. 8, pp. 71-182.

Gretner, P. E., 1972, Thoughts on overthrust faulting in a layered sequence: Canadian Society of Petroleum Geology, v. 20, n. 3, p. 583-607.

———, 1978, Pore Pressure: fundamentals, general ramifications, and implications for structural geology: AAPG Continuing Education Course Note Series, n. 4, 87 p.

Hilde, T. W. C., and G. F. Sharman, 1978, Fault structure of the descending plate and its influence on the subduction process: Eos, American Geophysical Union Transcripts, v. 59, p. 1182.

Hottman, C. E., J. H. Smith, and W. R. Purcell, 1979, Relationships among earth stresses, pore pressure, and drilling problems, offshore Gulf of Alaska: Journal of Petroleum Technology, v. 31, p. 1477-1484.

Hubbert, M. K., and W. W. Rubey, 1959, Role of fluid pressure in mechanics of overthrust faulting: Geological Society of America Bulletin, v. 70, p. 115-166.

Hussong, D. M., and S. Uyeda, 1981, Tectonics in the Mariana Arc: results of recent studies including DSDP leg 60: Oceanologica Acta, Supplement, v. 4, p. 203-211.

Karig, D. E., 1974, Evolution of arc systems in the western Pacific: Earth and Planetary Science Annual Review, v. 2, p. 51-76.

———, and J. C. Ingle, 1975, Cruise synthesis, *in* Initial reports of the deep sea drilling project, v. 31, part 4: Washington, D.C., U.S. Government Printing Office, p. 835-879.

———, and R. W. Kay, 1981, Fate of sediments on the descending plate at convergent margins: Philosophical Transcripts, Royal Society of London.

Ladd, J. W., et al, in press, Interpretation of seismic reflection data of the Middle America Trench offshore Guatemala, *in* J. Aubouin et al, eds., Initial reports of the deep sea drilling project: Washington, D.C., U.S. Government Printing Office, v. 67.

Lambe, T. W., and R. V. Whitman, 1969, Soil mechanics: John Wiley and Sons Inc., 553 p.

Lee, H. J., H. W. Olsen, and R. von Huene, 1973, Physical properties of deformed sediments from Site 181, *in* L. D. Kulm et al, eds., Initial reports of the deep sea drilling project, v. 18: Washington, D.C., U.S. Government Printing Office, p. 897-901.

Marauchi, S., 1971, The renewal of island arcs and the tectonics of Marginal Seas, *in* Island arc and marginal sea Tokyo, Tokai University Press, p. 39-56.

———, and W. J. Ludwig, 1980, Crustal structure of the Japan Trench: The effect of subduction of ocean crust, *in* Initial reports of the deep sea drilling project, v. 56 and 57, part 1: Washington, D.C., U.S. Government Printing Office, p. 463-470.

Moore, J. C., 1975, Selective subduction: Geology, v. 3, n. 9, p. 530-532.

———, et al, 1981, Near Barbados Ridge scraping off, subduction scrutinized: Geotimes, v. 26, n. 10, p. 21-26.

———, ———, 1982, Offscraping and underthrusting of sediment at the deformation front of the Barbados Ridge, *in* Results from leg 78A, Deep sea drilling project: Geological Society of America Bulletin, v. 93, p. 1065-1077.

Nagumo, S., J. Kasahara, and S. Koresawa, 1980, OBS Airgun seismic refraction survey near sites 441 and 434 (J-1A), 438-439 (J-12), and proposed site J-2B, *in* Scientific party, Initial reports of the deep sea drilling project, v. 56 and 57, part 1: Washington, D.C., U.S. Government Printing Office, p. 459-462.

Nasu, N., et al, 1980, Interpretation of multichannel seismic reflection data, legs 56 and 57, Japan Trench transect,

Deep sea drilling project, *in* Initial reports of the deep sea drilling project, v. 56 and 57, part 1: Washington, D.C., U.S. Government Printing Office, p. 489-504.

Raleigh, C. B., 1972, Earthquakes and fluid injection: AAPG Memoir 18, p. 273-279.

———, J. H. Healy, and J. D. Bredehoeft, 1972, Faulting and crustal stress at Rangely, Colorado: American Geophysical Union, Memoir 16, p. 275-284.

Rau, W. W., and G. R. Grocock, 1974, Piercement structure outcrips along the Washington coast: Washington Division of Mines and Geology, Information Circular, n. 51, 7 p.

Scholl, D. W., M. N. Marlow, and A. K. Cooper, 1977, Sediment subduction and offscraping at Pacific margins, *in* M. Talwani and W. C. Pittman III, eds., Island arcs, deep sea trenches and back-arc basins: American Geophysical Union, Maurice Ewing Series, n. 1, p. 199-210.

———, et al, 1980, Sedimentary masses and concepts about tectonic processes at underthrust ocean margins: Geology, v. 8, p. 564-568.

Schweller, W. J., and L. D. Kulm, 1978, Extensional rupture of oceanic crust in the Chile Trench: Marine Geology, v. 28, p. 271-291.

Seely, D. R., P. R. Vail, and G. G. Walton, 1974, Trench slope model, *in* C. A. Burk and C. L. Drake, eds., Geology of continental margins: New York, Springer-Verlag, p. 261-283.

———, 1977, The significance of landward vergence and oblique structural trends on trench inner slopes, *in* Talwani and W. D. Pitman III, eds., Island arcs, deep-sea trenches, and back-arc basins: American Geophysical Union, M. Ewing Series, n. 1, p. 187-198.

Shephard, L E., and W. R. Bryant, 1980, Consolidation characteristics of Japan Trench sediments, *in* Initial reports of the deep sea drilling project, v. 57, part 2: Washington, D.C., U.S. Government Printing Office, p. 1201-1206.

———, ———, and W. A. Chiou, 1982, Geotechnical properties of Middle America Trench sediments, leg 66, *in* Initial reports of the deep sea drilling project, v. 66: Washington, D.C., U.S. Government Printing Office, p. 475-504.

Shipley, T. H., et al, 1982, Tectonic processes along the Middle America Trench inner slope, *in* J. K. Leggett, ed., Trench forarc geology: Geological Society of London, Special Publication n. 10, p. 95-106.

Shouldice, D. H., 1971, Geology of the western Canadian continental shelf: Bulletin of Canadian Petroleum Geologists, v. 19, n. 2, p. 405-436.

Snavely, P. D., J. E. Pearl, and D. L. Lander, 1977, Interim report on petroleum resources potential and geological hazards in the outer continental shelf, Oregon and Washington Tertiary Province: U.S. Geological Survey, Open-File Report 77-282, 63 p.

Sorokhtin, O. G., and L. I. Lobkovskiy, 1976, The mechanism of subduction of oceanic sediments into a zone of underthrusting of lithospheric plates,: Earth Physics, n. 5, p. 3-10.

Trabant, P. K., W. R. Bryant, and A. H. Bouma, 1975, Consolidation characteristics of sediments from leg 31 of the deep sea drilling project, *in* Initial reports of the deep sea drilling project, v. 31: Washington, D.C., U.S. Government Printing Office, p. 569-572.

Terzaghi, K., and R. B. Peck, 1967, Soil mechanics in engineering practice: John Wiley and Sons, Inc., 729 p.

Uyeda, S., 1981, Subduction zones (introduction to comparative subductology), *in* W. T. C. Hilde, ed., Geodynamics of the western Pacific and Indonesian region: International Geodynamics Series, American Geophysical Union and Geological Society of America Publication.

von Huene, R. E., 1972, Structure of the continental margin and tectonism at the eastern Aleutian Trench: Geological Society of America Bulletin, v. 83, p. 3613-3626.

———, and L. D. Kulm, 1973, Tectonic summary of leg 18, *in* Initial reports of the deep sea drilling projects v. 18: Washington, D. C., U.S. Government Printing Office, p. 961-976.

———, et al, 1980a Leg 67, The deep sea drilling project Mid-America Trench Transect off Guatemala: Geological Society of America Bulletin, v. 91, p. 421-432.

———, ———, 1980b, Sites 438 and 439, Japan deep sea terrace leg 57, *in* Scientific part, Initial reports of the deep sea drilling project, v. 56 and 57, part 1: Washington, D.C., U.S. Government Printing Office, p. 23-192.

Watkins, J. S., et al, 1981; Accretion, underplating, subduction, and tectonic evolution, Middle America Trench, southern Mexico — results from Deep sea drilling project, leg 66: Oceanologica Acta, Supplement, v. 4, 1981.

White, R. S., and K. E. Louden, 1982, The Makran continental margin: Structure of a thickly sedimented convergent plate boundary: this volume.

Wissa, A. E. Z., et al, 1971, Consolidation at constant rate of strain: Journal of the Soil Mechanics and Foundations Division, v. 97, p. 1393-1413.

A Mechanism for Fragmentation of Oceanic Plates

Rinus Wortel
Sierd Cloetingh
Vening Meinesz Laboratory
University of Utrecht
The Netherlands

To investigate the mechanism underlying the fragmentation of oceanic plates, we study the breakup of the Farallon plate into the Cocos plate and the Nazca plate, which took place 25 to 30 m.y. ago. Using finite element methods and a reconstruction of regional plate boundaries, we calculate the stress field in the Farallon plate at approximately 30 m.y. ago (just prior to the fragmentation). The resulting stress field is dominated by high tensional stresses (maximum principal stresses of 5 to 6 kbar). North-to-south tension near present-day Panama is proposed to have caused fragmentation of the Farallon plate and inception of spreading along the new Cocos-Nazca plate boundary. The plate interaction that allowed the high level of tensional stresses in the Farallon plate is not unique for our case study, and the proposed mechanism sheds light on the problem of fragmentation of oceanic plates. The preliminary results of stress field calculations for the present-day Nazca plate show interesting stress variations along the strike of the South American trench system.

The spreading activity of oceanic ridges has been documented extensively, but the processes underlying the fragmentation of lithospheric plates and the formation of new accreting plate boundaries are not completely known. We propose a mechanism for fragmentation of oceanic plates and the subsequent inception of spreading centres. This mechanism is based on a case study of the plate tectonic evolution of the east-central Pacific.

In the last decade, marine geophysical studies have clarified many features of the apparently complex Cenozoic history of this area (Herron, 1972; Handschumacher, 1976; Hey, 1977; Mammerickx, Herron, and Dorman, 1980). These investigations show the breakup of the former Farallon plate into the Cocos plate and the Nazca plate (which took place approximately 25 to 30 m.y. ago) as a milestone in the tectonic evolution of the region. However, the amount of attention given to the mechanism causing the splitting of the Farallon plate does not reflect the importance of this event. In fact, it is limited to some tentative suggestions. For example, Hey (1977) suggested that the Cocos-Nazca spreading centre (or alternatively, the Galapagos spreading centre) "was born about 25 m.y. ago when an old Pacific-Farallon fracture zone opened in response to a new stress pattern in this area." Lonsdale and Klitgord (1978) hypothesized that the Farallon plate was subject to divergent gravitational stresses from the Middle America Trench and the South American Trench system. Furthermore, Van Andel et al (1971) postulated the splitting of an east-to-west trending ridge (the 'ancestral Carnegie Ridge') in attempting to explain the origin of the young Panama Basin. Finally, Menard (1978) discussed the fragmentation of the Farallon plate, concentrating on the relative motion of the fragmented parts rather than on the fragmentation process itself.

In the present context, it is important to realize the changes brought about by the Pacific-Farallon spreading centre approaching the western margin of the North American and South American continents. Not only did the approach imply a continuous change in geometry of the Farallon plate, but it also caused lateral and temporal variations in the age of the lithosphere at the trench systems along the eastern boundary of the plate. As is discussed below, several important forces acting on a lithospheric plate are strong functions of the age of the lithosphere in or near a subduction zone. Therefore, the stress field in the Farallon plate was subject to temporal variations as well.

We investigate the state of stress in the Farallon

plate at approximately 30 m.y. ago. (just prior to the breakup) to see whether the stress field assists in understanding the documented fragmentation. To do this, we use finite element methods and a reconstruction of regional plate boundaries and forces appropriate for 30 m.y. ago. We model the plate as elastic which allows us to derive the state of stress directly from the instantaneous forces acting on the plate.

Results obtained in previous work (Vlaar and Wortel, 1976; England and Wortel, 1980; Wortel, 1980) lead us to consider (as a new feature in numerical modelling of the state of stress in lithospheric plates) that important driving forces such as the slab pull and the ridge push depend on the age of the oceanic lithosphere. The results obtained by England and Wortel (1980), who investigated the relation between the forces acting at a convergent plate boundary and the nature of the tectonic regime in the overriding plate above the subducted slab, are of particular interest. In view of the strong dependence of the gravitational slab pull on the age of the descending lithosphere, England and Wortel (1980) suggested that below a critical age the driving forces may not be able to overcome the resistive forces acting in the subduction zone or at the plate contact. If, in a changing pattern of plate motion, lithosphere younger than this critical age would arrive at an active trench, subduction would be continued only if: 1) the young lithosphere was actively driven into the subduction zone by forces acting on adjacent segments of the plate, or 2) if the 'upper' plate, usually continental, would actively overthrust the young slab. In both cases, a compressive regime is expected in the overriding plate. If the forces promoting subduction are great enough to overcome the resistive forces, then there is no need for compression in the overriding plate and a tensional regime might be observed as a result of the retrograde motion of the subducting slab. A compilation of observations showed that a transition from tensional to compressional regimes, corresponding with a net resistive force per unit width along the trench of 8×10^{12} N/m, indeed takes place. The age at which this transition occurs varies from about 40 to 70 million years, depending on the vertical velocity of the sinking slab.

The intersections of the Pacific-Farallon spreading centre with the trench system (Figure 1) imply lateral variations in the age of the lithosphere at the trench and these findings appear relevant to modeling the forces which acted on the Farallon plate.

To test our modelling by observations, which is naturally impossible for the case of the Farallon plate, we also calculate the stress field for the present-day Nazca plate. Preliminary results of this work were reported earlier by Wortel and Cloetingh (1979).

MODEL

The plate boundaries adopted for the Farallon plate model at 30 m.y. ago and the mesh used in the calculations are shown in Figure 1. From present-day northwest Mexico to southern Chile we envisage the presence of a continuous trench system. This is justified because the subduction zones beneath Mexico and western South America have been active since at least the Late Cretaceous, whereas beneath Central America, subduction was initiated during the Eocene (Malfait and Dinkelman, 1972). The spatial discrepancy between the mesh boundary and the coastline of northwestern Mexico is not relevant in the context of our model since it is due to the opening of the Gulf of California, which took place after 30 m.y. ago. As Karig et al (1978) and Moore et al (1979) present strong evidence for truncation and accretion in some segments of the Middle America Trench, the boundary between the Farallon plate and Mexico/Central America may have been somewhat different from the one depicted in Figure 1. However, it is shown below that the basic features of the model are not affected by these uncertainties. The location of the northern ridge-trench intersection at 30 m.y. ago is taken from Atwater (1970). The location of the southern intersection is somewhat uncertain. From various lines of evidence

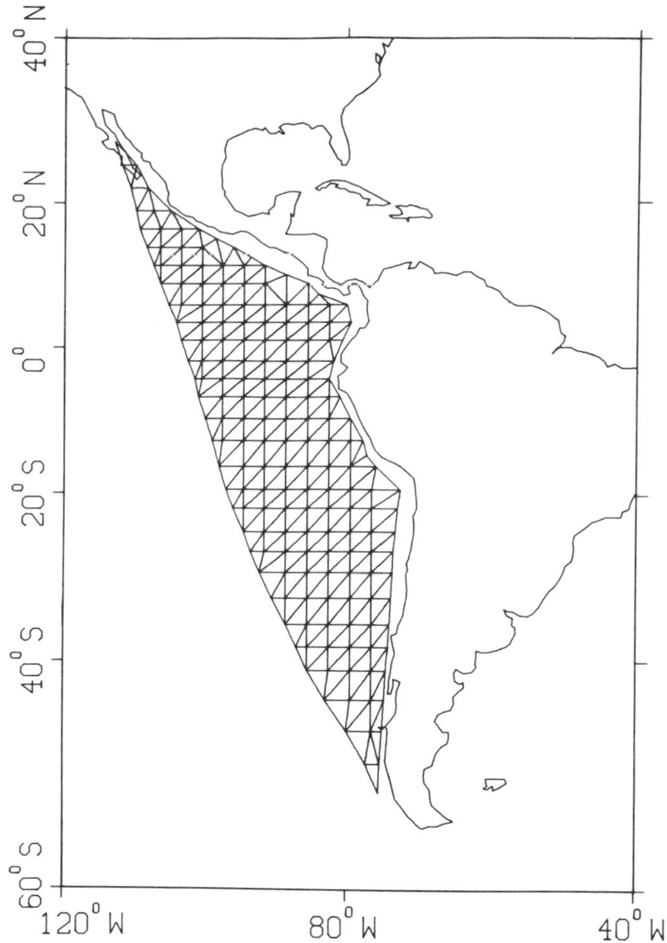

Figure 1 — Mesh used in the finite element calculations of the stress field of the Farallon plate (approximate configuration of 30 m.y. ago). The mesh consists of 311 triangular membrane elements with a quadratic displacement field (linear strain).

(e.g. the plate boundaries reconstructed for 55 m.y. ago (Jurdy and Van der Voo, 1975), the presence of a sediment filled trench off western Southern America south of 45°S), the position indicated in Figure 1 may be taken as a reasonable estimate (see Barker and Griffiths (1972) for a discussion of this topic). Because we do not model the ridge push as a line force (see below), the detailed configuration of the ridge is not very important. Therefore, the ridge is modelled without any offsets along transform faults.

The model plate is considered elastic (Young's modulus $E = 7 \times 10^{10}$ N/m^2 and Poisson's ratio $\nu = 0.25$), with a thickness of 100 km. This thickness is only a reference value. We use a plane stress approximation, but stress in an elastic plate with a thickness different from this reference value can be derived directly from those obtained for the model plate: for all thicknesses for which the plane stress approximation is valid, the product of average stress and plate thickness is constant. The spherical surface was approximated by an assembly of 311 triangular membrane elements. The stress calculations were made with the ASKA package of finite element routines (Argyris, 1979), which employs a quadratic representation of the displacement field (linear strain).

Following Richardson, Solomon and Sleep (1979), we assume that the state of stress in a lithospheric plate is determined largely by plate tectonic forces. The forces considered to act on the plate are the driving forces F_{sp} (slab pull) and F_{rp} (ridge push), and the resistive forces F_{tr} (resistance at the trench and in the subduction zone) and F_{dr} (drag at the base of the lithospheric plate). The slab pull and the ridge push were calculated according to Richter and McKenzie (1978). It can be shown that per unit width along the trench the slab pull (F_{sp}), resulting from the density contrast between the cold descending slab and the surrounding upper mantle, depends on the thickness (L) of the subducted slab according to $F_{sp} \propto L^3$. It is generally agreed upon that L depends on the square root of the lithospheric age (t) for $t < 70$ million years (e.g. Parsons and McKenzie, 1978). Hence, for these ages, $F_{sp} \propto t^{3/2}$. Owing to the reduced rate of thickening of L for $t > 70$ million years, the age-dependence of F_{sp} is somewhat weaker for old lithosphere. The possible contribution of the olivine — spinel phase change to the slab pull is neglected. As shown below, this does not seriously affect our results and conclusions. Similarly, it was found (Richter and McKenzie, 1978; England and Wortel, 1980) that for $t > 70$ million years, the ridge push F_{rp} (per unit width parallel with the ridge) depends linearly on the lithospheric age near the trench. This ridge push, acting on the oceanic lithosphere in the trench region, should be considered as the integrated (over the distance from ridge to trench) value of a horizontal pressure gradient (Lister, 1975; Hager, 1978). Accordingly, in our model calculations, the total ridge push was distributed over the area of the plate.

In view of the age-dependence of both F_{sp} and F_{rp}, we need to know the lithospheric age pattern in the Farallon plate, in particular the ages along the trenches. Since the geometry of our model is simplified (the offsets in the ridge axis are neglected), an assessment of only the gross features of the age pattern is warranted. Such an assessment was made by combining Handschumacher's (1976) Oligocene rotation pole (70°N, 145°W) for the spreading along the Pacific-Farallon ridge with the requirement that the age of the lithosphere at the trench off northern Chile had to be 75 to 80 million years. The latter value is a conservative extrapolation of the results obtained by Wortel and Vlaar (1978). These authors showed that the age of the oceanic lithosphere at the South American trench system decreased in the Late Tertiary, due to the convergence rate being higher than the half-spreading rate at which the descending lithosphere was originally created at the Pacific-Farallon spreading centre.

Furthermore, the slab pull (F_{sp}) depends on the convergence rate v_c (Richter and McKenzie, 1978) and the dip angle (ϕ) of the descending slab as $F_{sp} \propto v_c \sin\phi$; hence, F_{sp} varies linearly with v_z, the vertical velocity of the subducted slab. In a comprehensive study of the subduction process of oceanic lithosphere, Wortel (1980) showed that present-day subduction zones display the following characteristics: $2 \lesssim v_z \lesssim 3.5$ cm/yr for oceanic lithosphere younger than 65 million years and $4 \lesssim v_z < 6$ cm/yr for lithosphere older than 100 million years. In our model, we used 3 cm/yr and 5 cm/yr as representative values for these two age groups and linearly interpolated velocities for the intermediate ages, for which no observational data were available.

In other studies of stress in the lithosphere (Richardson, Solomon, and Sleep, 1979), the resistive forces acting at a convergent plate boundary appeared to be most difficult to model. Various mechanisms may contribute to the plate interaction and as yet no detailed model has been established. Therefore, we follow England and Wortel (1980) and take the total resistive force per unit width along the trench to be 8×10^{12} N/m. The buoyancy effect of the stable petrological stratification of the oceanic lithosphere, according to Oxburgh and Parmentier's (1977) model, accounts for 4×10^{12} N/m. The remaining resistance of 4×10^{12} N/m is attributed to shearing forces acting along the plate contact and along the slab's interfaces.

To ensure mechanical equilibrium, the net torque on the plate is required to vanish. The drag at the base of the lithosphere is determined from the torque balance. Without having to adjust Handschumacher's (1976) pole position we found that a constant resistive shear stress of 8 bar, acting at the base of the plate and in the direction derived from the position of the pole, balances the torques.

The relative importance of the forces F_{sp}, F_{rp}, F_{tr}, and F_{dr} is demonstrated in Table 1, in which these forces are specified (per unit width along the ridge or trench) for two cross sections from a spreading ridge to a trench system. In these examples, we assume that the ridge runs parallel with the trench; we consider the forces to be positive if they drive the oceanic plate

Table 1. Forces acting on oceanic lithosphere, calculated for two cross sections from ridge to trench. All forces are given per unit width along the trench. "Age" refers to the age of the oceanic lithosphere at the trench. $F_{sum} = F_{rp} + F_{sp} + F_{tr} + F_{dr}$ (see text for the explanation of these forces).

age (m.y.)	F_{rp} (10^{12} N/m)	F_{sp} (10^{12} N/m)	F_{tr} (10^{12} N/m)	F_{dr} (10^{12} N/m)	F_{sum} (10^{12} N/m)
30	+ 0.8	+ 3.2[(a)]	−8.0	−0.6[(b)]	−4.6
70	+ 1.7	+12.9[(a)]	−8.0	−1.4[(b)]	+5.2

(a) Values taken for the vertical velocity v_z of the subducted slabs of 30 and 70 m.y. old are 3 cm/yr and 4 cm/yr, respectively.

(b) Based on a shear stress of 8 bar and a half-spreading rate (determining the relation between age at the trench and distance from ridge to trench) appropriate for the Farallon plate.

towards the trench. In the first cross section, the oceanic lithosphere near the trench is 30 m.y. old and the sinking velocity (v_z) of the subducted slab is 3 cm/yr. The sum of the forces appears to be negative. Hence, the resistance to plate motion (F_{tr} and F_{dr}, based on a shear stress of 8 bar) is greater than are combined driving forces F_{sp} and F_{rp}. As a result, the oceanic lithosphere is in a state of compression (principal stress in the direction from ridge to trench). In the second cross section, the lithosphere at the trench is 70 m.y. old and v_z = 4 cm/yr. Owing to the strong age-dependence of the slab pull (F_{sp}), the driving forces now outweight the resistive forces, which leads to tensional stresses in the oceanic lithosphere. It can be shown that, in these examples, the transition from compression to tension takes place if the lithosphere at the trench is around 50 m.y. old. Realize that the net forces (unequal to zero) acting in cross sections of this type do not violate the required balance of forces or, on a spherical surface, the balance of torques. Such laws refer to the whole of the plate.

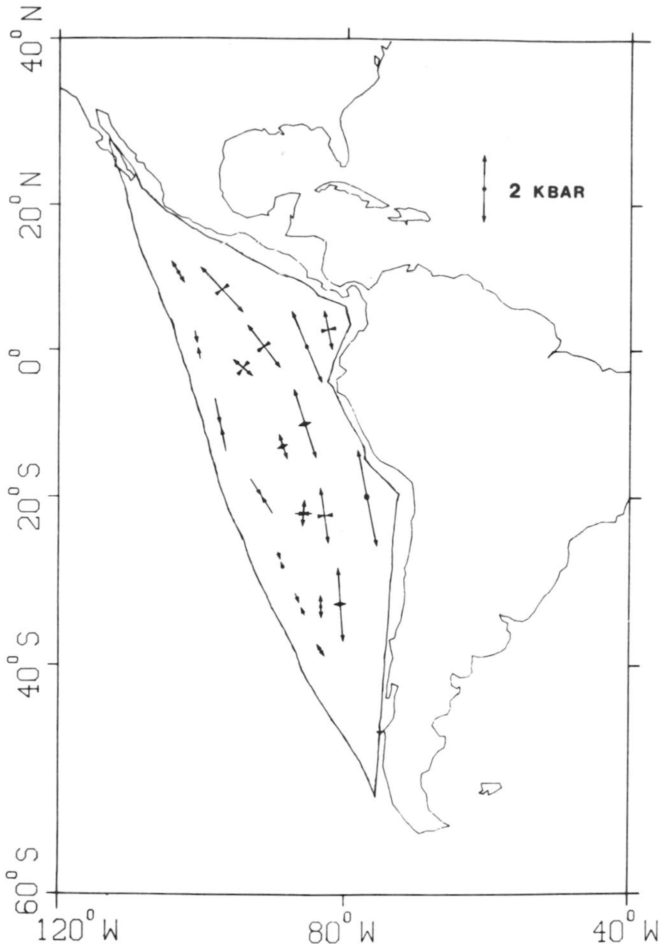

Figure 2 — Calculated stress field in the Farallon plate under the reconstructed conditions of 30 m.y. ago, prior to the breakup of the Farallon plate. The thickness of the elastic plate is 100 km (see text).

RESULTS AND ANALYSIS

Stress field in the Farallon plate

The resulting stress field in the Farallon plate under the reconstructed conditions of 30 m.y. ago is displayed in Figure 2. The accuracy of the finite element solution was checked and confirmed by convergence tests and an analysis of the internal reaction forces of the model. The high tensional stresses near the trench constitute the most conspicious feature of the stress field.

In Figure 3, two mechanisms are indicated which may account for tensional stresses in a plate (labeled plate I) attached to a subducted slab. In Figure 3a, the forces F_A, F_B, and F_C represent the downdip gravitational pull exerted by the descending lithosphere. If these slab pull forces are not totally balanced by resistive forces acting on the dipping slab (i.e. if a net pull is transferred to the horizontal part of the plate), they will cause tensional stresses in the plate. Similarly, if only, or predominantly, the driving forces in the direction of plate convergence are balanced by resistive forces (shearing forces parallel with the direction of relative motion), the components F_{A2} and F_{B2} give rise to tensional stresses as indicated in the figure. Thus, the geometry of the convergent plate boundary determines the directions of the slab pull forces F_A, F_B, and F_C. The divergence of these forces results in a tensional stress field, as was realized by Lonsdale and Klitgord (1978) for the Farallon plate and earlier, in a more general context, by Solomon and Sleep (1974). Our modelling of plate tectonic forces depending on lithospheric age implies a distribution of forces, schematically shown in Figure 3b. Here, the descending lithosphere exerts a net pull on plate I in the central part of the subduction zone, but the younger lithosphere in the upper and lower regions (near A and D, corresponding with Mexico/Central America and southern Chile), where the ridge approaches the trench system, is too buoyant to overcome resistive forces. This situation causes a stress field in plate I with tension near the plate contact and compression farther away from the trench (Figures 3b and 2).

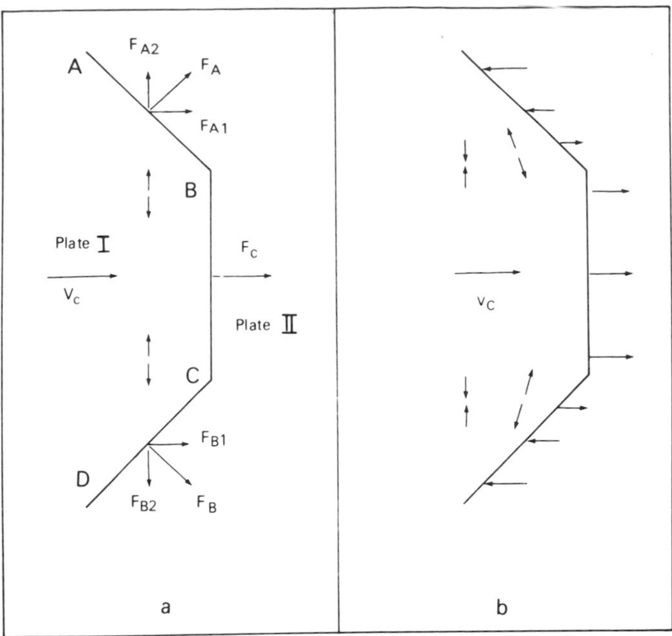

Figure 3 — Schematic representation of two mechanisms which produce tensional stresses in a plate approaching a subduction zone. Plate I underthrusts plate II. In a, the forces F_A, F_B, and F_C represent the pull exerted by the descending slab. If only, or predominantly, the components in the direction of plate convergence are balanced by resistive forces (shearing forces parallel with direction of relative motion), the components F_{A2} and F_{B2} give rise to tensional stresses as indicated in the figure. In b, the descending slab exerts a net pull on plate I in the central part of the subduction zone, but the younger lithosphere in the upper and lower regions (near A and D) is too buoyant to overcome resistive forces. This situation causes a stress field in plate I with tension near the plate contact and compression further away from the trench.

Although the curved or angular geometry of the convergent plate boundary may play a role in creating lithospheric age differences along the strike of the trench system, the age differences themselves are of more fundamental importance. In many cases they result from large offsets in a spreading ridge, from the presence of extinct spreading centres, or from other peculiarities in the spreading history of the plate, which bear no direct relation to the curvature of the plate boundary. Even along a linear trench system, descending lithosphere of widely different ages, and hence, different structures, could give rise to similar tensional stresses parallel with the trench. In short, the age-dependence of forces acting on a plate consititutes a factor which is essentially independent of the actual shape of the plate. At this stage, the pattern of forces acting on the Farallon plate (as sketched in Figure 3b) is not affected seriously by the relatively minor uncertainties regarding the configuration of the Middle America Trench in Tertiary times.

The calculated stress field represents the non-hydrostatic state of stress in an elastic plate which was initially unstressed (Richardson, 1978). As noted earlier, the thickness of 100 km for the elastic model plate is only a reference value. Caldwell and Turcotte (1979) showed that a more appropriate model for the oceanic lithosphere is a division of the lithosphere into an elastic upper part and a plastic lower part. According to their model, the thickness of the elastic upper layer, which is presumably temperature-dependent, increases linearly with the square root of lithospheric age from a few kilometers for very young lithosphere to 40 to 50 km for lithosphere of about 80 m.y. old. Similar results were obtained by Bodine, Steckler, and Watts (1981). This implies that the magnitudes of the principal stresses plotted in Figure 2 (at some distance from the ridge) should be multiplied by 2 to 3 to get the stresses in the elastic upper layer. Thus, we obtain maximum principal stresses and also maximum differential stresses of 5 to 6 kbar. Assigning a yield stress of a few hundred bars to the lower part of the lithosphere or modelling this part as visco-elastic does not alter the above conclusion significantly (Kusznir and Bott, 1977). Incorporation of the body forces associated with an elevation of the olivine-spinel phase boundary is not expected to change the nature of our calculated stress field. Subduction of the young lithosphere (< 70 million years) near the ridge-trench intersections probably does not cause such an elevation (Vlaar and Wortel, 1976), whereas the increase in gravity pull experienced by the older parts of the slab (beneath present-day Peru and northern Chile) would only lead to even greater tensional stresses in the Farallon plate.

Note that our modeling procedure is closely constrained by extensive data sets. The model for the oceanic lithosphere, used in the evaluation of the slab pull (F_{sp}) and ridge push (F_{rp}) and in the interpretation of the stress field, satisfies oceanic heat flow and bathymetry data (Wortel, 1980) and deformation of oceanic lithosphere data (Caldwell and Turcotte, 1979). The modelling of forces is constrained by observational data on the stress regime in the lithosphere above subducted slabs (England and Wortel, 1980). In our approach of calculating intraplate stress fields, we prefer to use the internally consistent parameters and results derived from these studies to making a variable parameter study.

Stress field in the Nazca plate

The maximum values of the stresses in Figure 2 are considerably higher than those calculated by Richardson, Solomon, and Sleep, (1979) for the present-day system of lithospheric plates (several hundreds of bars). For direct comparison, we calculated the stress field for the present-day Nazca plate employing the same modelling procedure as we did for the Farallon plate. Our finite element mesh for the Nazca plate and the computational results are displayed in Figures 4 and 5, respectively. More detailed work on the state of stress in the Nazca plate, in which specific local factors are incorporated, is in progress (Wortel and Cloetingh, in preparation). In the Nazca plate, the stress level appears to be significantly lower than in the Far-

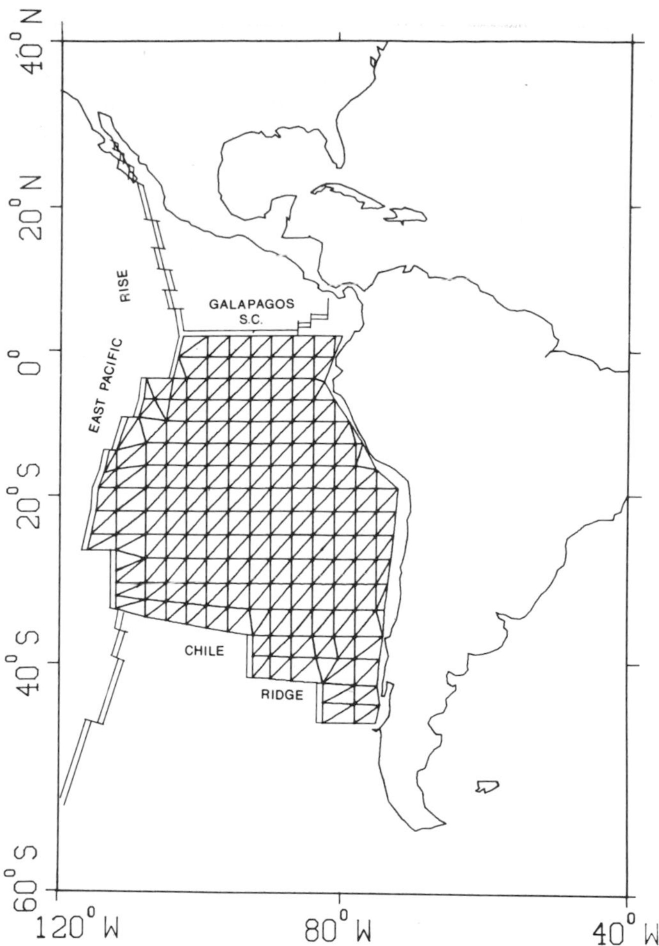

Figure 4 — Mesh used in the finite element calculation of the present-day stress field in the Nazca plate (see Figure 5).

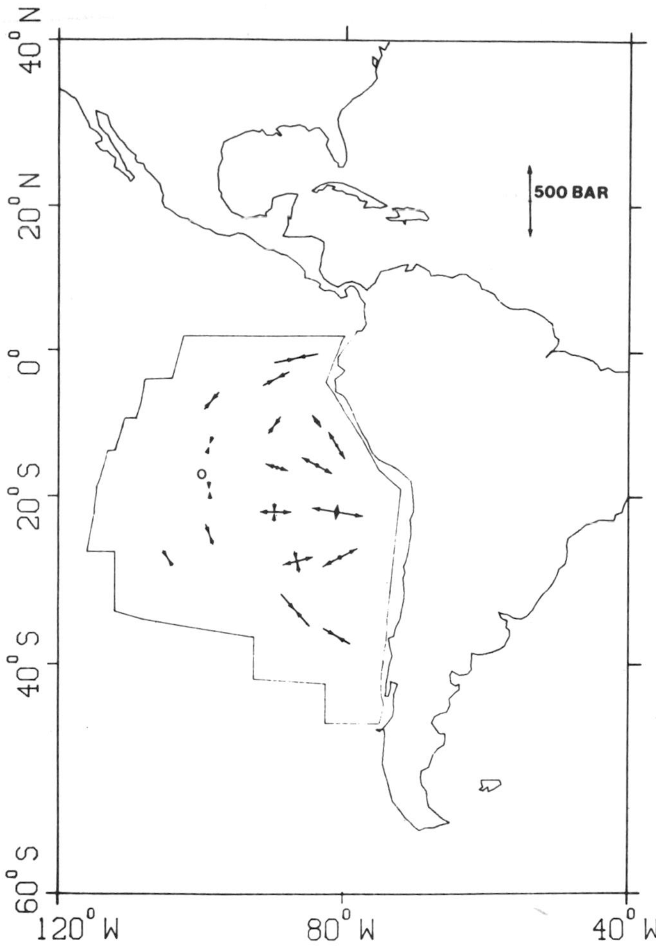

Figure 5 — Calculated stress field in the present-day Nazca plate. Note that the stresses were calculated for an elastic plate with a thickness of 100 km.

allon plate. Taking into account the difference between the thickness of the elastic model plate (100 km) and the elastic thickness of the lithosphere, we find the stresses in the eastern part of the plate to typically range from 500 to 700 bar, the maximum stress being close to 1 kbar. Thus, in comparable situations, the stresses are of the same magnitude as found by Richardson, Solomon, and Sleep (1979). Our results for the Nazca plate exhibit another noteworthy feature: a transition from approximately east-west compression near the northern segment of the South American trench system, via tension off northern Chile, to compression again along the southern part of the trench. Various data on the deformational characteristics of trench deposits and of the adjacent oceanic crust provide evidence for such a lateral variation along the South American trench system, and in particular for the presence of east-west compression (Prince and Kulm, 1975; Hussong et al, 1976; Kulm, Schweller, and Masias, 1977; Schweller and Kulm, 1978). In Richardson's (1978) modelling of the state of stress in the Nazca plate, he used the focal mechanism solution of an intraplate earthquake (east-west compresison, epicentre indicated by open circle in Figure 5) determined by Mendiguren (1971) as the most important constraint. However, the extensive marine geophysical evidence cited above might provide a much more useful constraint in numerical modelling of the state of stress in the lithosphere. From the difference in stress level between the present-day Nazca plate (and other plates; see Richardson, Solomon, and Sleep, 1979) and the Farallon plate at 30 m.y. ago, we infer that, just prior to its fragmentation, the Farallon plate attained unusually high intraplate stresses.

Fragmentation of the Farallon plate

From laboratory experiments on olivine, the yield stress for the upper part of the lithosphere is inferred to range from 3 to 10 kbar (Ashby and Verrall, 1978; Evans and Goetze, 1979; Kirby, 1980). Therefore, we propose that the Farallon plate failed in response to a stress field of the type depicted in Figure 2. This failure resulted in two smaller plates, the Cocos plate and the Nazca plate (see Figure 6) and gave birth to the Cocos-Nazca spreading centre. An interesting feature of Hey's (1977) reconstruction of the Cocos-Nazca spreading centre is that its original orientation was perpendicular with the strike of the Pacific-Farallon ridge. Later, reorganizations in the spreading pattern led to the present east-west orientation of the spreading centre.

From Figure 2 no particular preference can be inferred for failure in the region shown in Figure 6. Everywhere along the central portion of the trench system high stresses prevail. If failure occurs in one part of the plate, however, the state of stress in the plate is relaxed and the two new parts may change their rate and direction of motion. Their motion is no longer determined by the forces acting on all of the original plate, but only by those acting on each of the smaller plates separately. The young lithosphere near the trench off northwest Mexico and southern Chile will initially cause the Cocos and Nazca plates to pivot around the intersections of the ridge and the trench. In this respect, the postulated breakup (Figure 6) may have had preference over other regions as the new direction of the Cocos plate could easily be accommodated in the Central American trench system. The observed evidence for pivoting motion of the Cocos plate (see Menard, 1978) is considered to support our force modelling (Figure 3b). The pivoting motion of the Nazca plate (S. Cande, personal communication, 1981) is much less pronounced. This is probably due to the fact that the new Nazca plate did not retain the original wedge-shaped geometry shown in Figure 6. Instead, it was affected by reorganizations in the system of speading ridges in the east-central and South Pacific (Herron, 1972; Mammerickx, Herron, and Dorman, 1980) which changed the plate boundaries significantly.

Finally, we draw attention to implications for the geology of continental margins which are adjacent to a region of (oceanic) plate fragmentation. The start of spreading along the new Cocos-Nazca plate boundary has caused important changes in the subduction process near present-day Panama. Prior to breakup of the Farallon plate, the subduction zone was consuming relatively old oceanic lithosphere. After breakup, extremely young lithosphere arrived at the trench. In view of the important role of the lithospheric age in the subduction process (Wortel, 1980) and associated geological processes (De Long and Fox, 1977), we expect that accounting for this change may contribute to studies of the geology of Colombia, eastern Central America, and adjacent continental margins (e.g. Lu and McMillen, this volume).

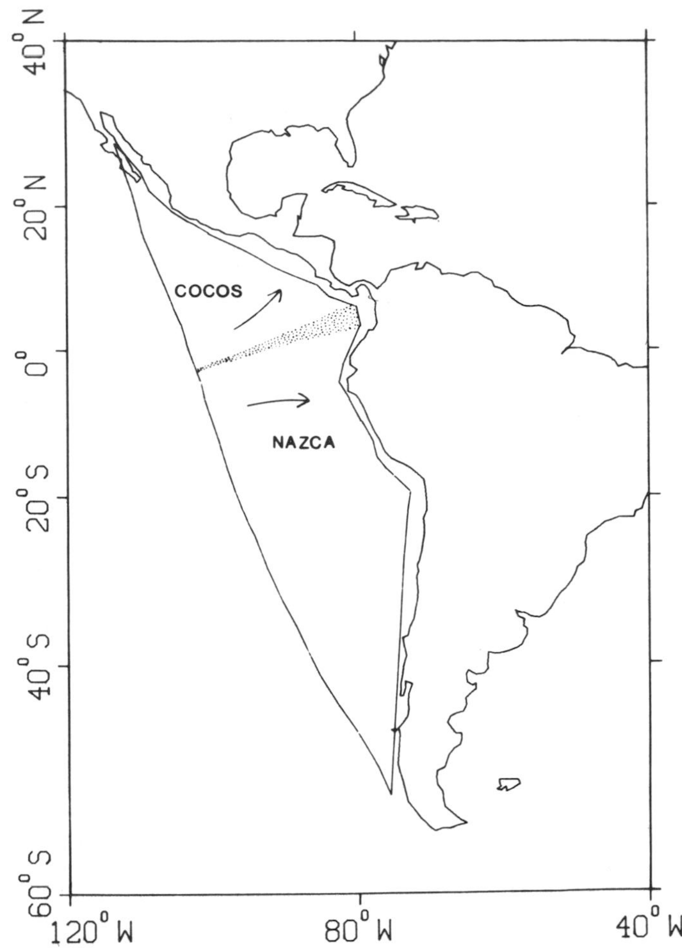

Figure 6 — Fragmentation of the Farallon plate into the Nazca plate and the Cocos plate, supposedly resulting from the state of stress depicted in Figure 2 (the strippled region indicates the zone of failure). The age-dependent forces exerted by the descending slab initially cause the Cocos plate and the Nazca plate (see text) to pivot around the points of intersection of the ridge and the trenches off California/ northwest Mexico and southernmost Chile, respectively.

CONCLUSION

The plate interaction that gave rise to the high tensional stresses in the Farallon plate is not restricted to our case study area. A curved or angular geometry of a trench system and significant lateral variations in the age of the oceanic lithosphere at the trench are encountered in several other regions. Also, in the earlier history of the Farallon plate (prior to 30 m.y. ago), when the Farallon plate extended farther north into the northeastern Pacific (Handschumacher, 1976), situations have arisen which are quite similar to the one studied in this paper, and fragmentation has repeatedly taken place (Menard, 1978; see also Riddihough, 1977). For example, the Vancouver plate and the Guadelupe plate (Menard, 1978) were broken off the northern part of the Farallon plate. We suggest that these fragmentations occurred in response to tensional stresses of the type discussed.

Intraplate stresses associated with plate tectonic forces (such as those considered in this paper) have tentatively been advanced as a cause for intraplate volcanism (e.g. Jackson and Shaw, 1975; Watts, Bodine, and Ribe, 1980). Our numerical results indicate that tensional stresses may reach a level which is sufficient to cause failure of oceanic plates. Since inception of intraplate volcanism probably requires lower tensional stresses than creation of spreading centres, we conclude that for explaining the origin of volcanic island chains, stress fields of the type calculated for the Farallon plate provide a sound alternative to hypotheses based on postulated hot spots.

We conclude that lateral variations in the age of the slab descending in a subduction zone may be the source of significant stresses in the plate to which the slab is attached. As such, they provide a possible cause for fragmentation of oceanic plates, in general, and the Farallon plate, in particular.

ACKNOWLEDGMENTS

We acknowledge Gerald Wisse of the Scientific Applications Group, Delft University of Technology, for support with the finite element calculations. Discussions held with participants at the Hollis Hedberg Research Conference, Carleton University, and at Lamont-Doherty Geological Observatory are highly appreciated.

REFERENCES CITED

Argyris, J. H., 1979, ASKA user's reference manual: Institut für Statik und Dynamik der Luft- und Raumfahrtkonstruktion, University of Stuttgart, Report n. 73.

Ashby, M. F., and R. A. Verrall, 1978, Micromechanisms of flow and fracture, and their relevance to the rheology of the upper mantle: Royal Society of London, Philosophical Transactions v. A288, p. 59-95.

Atwater, T., 1970, Implications of plate tectonics for the Cenozoic evolution of western North America: Geological Society of America Bulletin, v. 81, p.3513-3536.

Barker, P. F., and D. H. Griffiths, 1972, The evolution of the Scotia Ridge and Scotia Sea: Royal Society of London, Philosophical Transactions, v. A271, p. 151-183.

Bodine, J. H., M. S. Steckler, and A. B. Watts, 1981, Observations of flexure and the rheology of the oceanic lithosphere: Journal of Geophysical Research, v. 86, p. 3695-3707.

Caldwell, J. G., and D. L. Turcotte, 1979, Dependence of the thickness of the elastic oceanic lithosphere on age: Journal of Geophysical Research, v. 84, p. 7572-7576.

DeLong, S. E., and P. J. Fox, 1977, Geological consequences of ridge subduction *in* M. Talwani and W. C. Pitman III, eds., Island arcs, deep sea trenches and back-arc basins: Washington, D.C., American Geophysical Union, Maurice Ewing Series 1, p. 221-228.

England, P. C., and R. Wortel, 1980, Some consequences of the subduction of young slabs: Earth and Planetary Science Letters, v. 47, p. 403-415.

Evans, B., and C. Goetze, 1979, The temperature variation of hardness of olivine and its implications for polycrystalline yield stress: Journal of Geophysical Research, v. 84, p. 5505-5524.

Hager, B. H., 1978, Oceanic plate motions driven by lithospheric thickening and subduction slabs: Nature, v. 276, p. 156-159.

Handschumacher, D. W., 1976, Post-eocene plate tectonics of the Eastern Pacific, *in* G. H. Sutton, M. H. Manghnani, and R. Moberly, eds., The Geophysics of the Pacific Ocean Basin and its Margin: Washington, D.C., American Geophysical Union, Geophysical Monograph, v. 19, p. 177-202.

Herron, E. M., 1972, Sea-floor spreading and the Cenozoic history of the east-central Pacific: Geological Society of America Bulletin, v. 83, p. 1671-1692.

Hey, R., 1977, Tectonic evolution of the Cocos-Nazca spreading center: Geological Society of America Bulletin, v. 88, p. 1104-1420.

Hussong, D. M., et al, 1976, Crustal structure of the Peru-Chile Trench: 8°-12° S latitude, *in* G. H. Sutton, M. H. Manghnani, and R. Moberly, eds., The Geophysics of the Pacific ocean basin and its margin: American Geophysical Union, Geophysical Monograph, v. 19, p. 71-85.

Jackson, E. D., and H. R. Shaw, 1975, Stress fields in central portions of the Pacific plate: Delineated in time by linear volcanic chains: Journal of Geophysical Research, v. 80, p. 1861-1874.

Jurdy, D. M., and R. Van der Voo, 1975, True polar wander since the early Cretaceous: Science, v. 187, p. 1193-1196.

Karig, D. E., et al, 1978, Late Cenozoic subduction and continental margin truncation along the northern Middle America Trench: Geological Society of America Bulletin, v. 89, p. 265-276.

Kirby, S. H., 1980, Tectonic stresses in the lithosphere: Constraints provided by the experimental deformation of rocks: Journal of Geophysical Research, v. 85, p. 6353-6386.

Kulm, L. D., W. J. Schweller, and A. Masias, 1977, A preliminary analysis of the subduction processes along the Andean Continental Margin, 6° to 45° S, *in* M. Talwani and W. C. Pitman III, eds., Island arcs, deep sea trenches and back-arc basins: Washington, D. C., American Geophysical Union, Maurice Ewing Series, v. 1, p. 285-301.

Kusznir, N. J., and M. H. P. Bott, 1977, Stress concentration in the upper lithosphere caused by underlying viscoelastic creep: Tectonophysics, v. 43, p. 247-256.

Lister, C. R. B., 1975, Gravitational drive on oceanic plates caused by thermal contration: Nature, v. 257, p. 663-665.

Lonsdale, P., and K. D. Klitgord, 1978, Structure and tectonic history of the eastern Pacific Basin: Geological Society of America Bulletin, v. 89, p. 981-999.

Lu, R. S., and K. J. McMillen, 1982, Multi-channel seismic survey of the Colombian Basin and adjacent continental margins: this volume.

Malfait, B. T., and M. G. Dinkelman, 1972, Circum-Caribbean tectonic and igneous activity and the evolution of the Caribbean plate: Geological Society of America Bulletin, v. 83, p. 251-272.

Mammerickx, J., E. Herron, and L. Dorman, 1980, Evidence for two fossil spreading ridges in the southeast Pacific: Geological Society of America Bulletin, part I, v. 91, p. 263-271.

Menard, H. W., 1978, Fragmentation of the Farallon plate by pivoting subduction: Journal of Geology, v. 86, p. 99-110.

Mendiguren, J. A., 1971, Focal mechanism of a shock in the middle of the Nazca-plate: Journal of Geophysical Research, v. 76, p. 3861-3879.

Moore, J. C., et al, 1979, Progressive accretion in the Middle America Trench, southern Mexico: Nature, v. 281, p. 638-642.

Oxburgh, E. R., and E. M. Parmentier, 1977, Compositional and density stratification in oceanic lithosphere: Geological Society of London, Journal, v. 133, p. 343-355.

Parsons, B., and D. P. McKenzie, 1978, Mantle convection and the thermal structure of plates: Journal of Geophysical Research, v. 83, p. 4485-4496.

Prince, R. A., and L. D. Kulm, 1975, Crustal rupture and the initiation of imbricate thrusting in the Peru-Chile Trench: Geological Society of America Bulletin, v. 86, p. 1639-1653.

Richardson, R. M., 1978, Finite element modelling of stress in the Nazca-plate: Driving forces and plate boundary earthquakes: Tectonophysics, v. 50, p. 223-248.

———, S. C. Solomon, and N. H. Sleep, 1979, Tectonic stress in plates: Review of Geophysics and Space Physics,

v. 17, p. 981-1019.

Richter, F. M., and D. P. McKenzie, 1978, Simple plate models of mantle convection: Journal of Geophysics, v. 44, p. 441-471.

Riddihough, R. P., 1977, A model for recent plate interactions off Canada's west coast: Canadian Journal of Earth Science, v. 14, p. 384-396.

Schweller, W. J., and L. D. Kulm, 1978, Extensional rupture of oceanic crust in the Chile trench: Marine Geology, v. 28, p. 271-291.

Solomon, S. C., and N. H. Sleep, 1974, Some simple physical models for absolute plate motions, Journal of Geophysical Research, v. 79, 2557-2567.

Van Andel, Tj. H., et al, 1971, Tectonics of the Panama Basin, eastern equatorial Pacific: Geological Society of America Bulletin, v. 82, p. 1489-1508.

Vlaar, N. J., and M. J. R. Wortel, 1976, Lithospheric aging, instability and subduction: Tectonophysics, v. 32, p. 331-351.

Watts, A. B., J. H. Bodine, and N. M. Ribe, 1980, Observations of flexure and the geological evolution of the Pacific Ocean basin: Nature, v. 283, p. 532-537.

Wortel, R., 1980, Age-dependent subduction of oceanic lithosphere: Utrecht, University of Utrecht, Ph.D. dissertation, p. 147.

———, and S. Cloetingh, 1979, Changing plate boundaries and stress patterns in oceanic plates: Washington, D.C., EOS, American Geophysical Union, Transactions, v. 60, p. 586.

Wortel, M. J. R., and N. J. Vlaar, 1978, Age-dependent subduction of oceanic lithosphere beneath western South America: Physics of Earth and Planetary Interiors, v. 17, p. 201-208.

Index

Index

Index

Index